LIBRARY
ZOOLOGICAL MUSEUM
UNIVERSITY OF WISCONSIN-MADISON

FLORA OF THE GREAT PLAINS

Publication made possible by subventions from

Chadron State College Research Institute
Kansas State University
North Dakota State University
South Dakota State University
University of Kansas
University of Nebraska–Lincoln
University of Nebraska at Omaha
University of South Dakota
University of Wyoming

FLORA OF THE GREAT PLAINS

by the
GREAT PLAINS FLORA ASSOCIATION

Ronald L. McGregor, *Coordinator*
T. M. Barkley, *Editor*
Ralph E. Brooks, *Associate Editor*
Eileen K. Schofield, *Associate Editor*

MEMBERS OF THE GREAT PLAINS FLORA ASSOCIATION

William T. Barker, *North Dakota State University*
T. M. Barkley, *Kansas State University*
Margaret Bolick, *University of Nebraska–Lincoln*
Ralph E. Brooks, *University of Kansas*
Steven P. Churchill, *New York Botanical Garden*
Ronald L. Hartman, *University of Wyoming*
Robert B. Kaul, *University of Nebraska–Lincoln*
Ole A. Kolstad, *Kearney State College*
Gary E. Larson, *South Dakota State University*
Ronald L. McGregor, *University of Kansas*
David M. Sutherland, *University of Nebraska at Omaha*
Theodore Van Bruggen, *University of South Dakota*
Ronald R. Weedon, *Chadron State College*
Dieter H. Wilken, *Colorado State University*

UNIVERSITY PRESS OF KANSAS

© 1986 by the University Press of Kansas
All rights reserved

Published by the University Press of Kansas (Lawrence, Kansas 66045), which was organized by the Kansas Board of Regents and is operated and funded by Emporia State University, Fort Hays State University, Kansas State University, Pittsburg State University, the University of Kansas, and Wichita State University.

Library of Congress Cataloging in Publication Data

Flora of the Great Plains.
Includes index.
121. Botany—Great Plains—Classification.
I. McGregor, Ronald L. II. Barkley, T. M. (Theodore Mitchell), 1934–   . III. Great Plains Flora Association (U.S.)
QK135.F55   1986   582.0978   86-23
ISBN 0-7006-0295-X

Printed in the United States of America

Partially supported by the National Science Foundation, Grants DEB 76-08508 with supplements and DEB 82-06545, to Kansas State University; T. M. Barkley, Principal Investigator.

Floristic research at the University of Kansas Herbarium has been supported by grants from the Joseph S. Bridwell Foundation of Wichita Falls, Texas, to R. L. McGregor, Principal Investigator.

Contribution No. 84-135-B, Division of Biology, Kansas Agricultural Experiment Station, Kansas State University
Contribution No. 1254, Department of Botany, North Dakota Agricultural Experiment Station, North Dakota State University

Contents

Acknowledgments	vii
Introduction	1
Sources of Data 3	
Sequence of Taxa 3	
Species Concepts 3	
Taxonomy and Nomenclature 4	
Colloquial Names 4	
Systematic Descriptions 4	
Distribution 4	
Synonymy 5	
Authorship and Citation of the *Flora* 5	
Principal References 5	
Physical and Floristic Characteristics of the Great Plains	7
List of Families in Sequence	11
Abbreviations	13
The Keys	15
The Systematic Descriptions	39
Abbreviations for Nomenclatural Authorities	1285
Glossary	1317
Index	1329

Acknowledgments

This book is based on the researches and experiences of many people, but the chief instigator of the project has been Ronald L. McGregor, coordinator of the GPFA and for many years a faculty member of the University of Kansas. Dr. McGregor saw the need for a modern floristic treatment for the region in the early 1940s when he began basic studies on the flora of Kansas. It was early evident to him that a flora covering the Great Plains would be a more significant contribution than a state flora for Kansas, although its preparation would be a much more time-consuming undertaking. For the next forty years, Dr. McGregor, with his students and colleagues at the University of Kansas, made some 200,000 collections and built a sizeable herbarium in anticipation of this project. The *Flora* could have been written by Dr. McGregor alone, for his fund of information and experience exceed that of any of us; but administrative duties, and the realization that a collaborative effort would produce a better product, led him to approach me in 1972 with the notion of a group project. Our discussions led in turn to the formation of the GPFA, the publication of the *Atlas*, and eventually to this *Flora*.

H. A. "Steve" Stephens was employed for a number of years by the Herbarium of the University of Kansas, and he made more than 90,000 collections throughout the Great Plains. These specimens were used extensively by those preparing the manuscript.

Eileen K. Schofield, an associate editor, contributed her superb editorial skills and many long, tedious hours in checking, polishing, and collating the manuscript. She also compiled the glossary and index.

Ralph Brooks, an associate editor, has participated in the project since its inception and has provided much careful editorial expertise to the preparation of the final manuscript. He is also the contributor of numerous treatments, which are so attributed in the text.

JoAnn Luehring served as secretary to the GPFA in the early years of the project and contributed to the organizational efforts.

Dr. Janice Coffey acted as shepherdess to our application for support from the National Science Foundation during her tenure with that agency. In addition, she contributed the treatment of *Luzula* (Juncaceae).

Dr. R. Glenn Bellah participated in the early stages of the preparation of the treatments for several families. His participation was supported by the National Science Foundation.

Dr. Arthur Cronquist and Dr. Noel Holmgren, both of the New York Botanical Garden, provided much advice and generously shared their floristic knowledge.

To all of the above-named people, and to others too numerous to list who offered assistance to the project, we are particularly grateful. In addition, we thank the National Science Foundation for partial support for the project and the administrations of our respective home institutions for support during the preparation of this book.

T. M. Barkley
Editor, GPFA

Introduction

by T. M. Barkley

Flora of the Great Plains treats all of the vascular plants known to occur spontaneously in the Great Plains of central North America. Included are the native members of the flora plus those introductions that appear to be permanently established outside of intentional cultivation.

The Great Plains is a natural and floristically coherent region, extending from the base of the Rocky Mountains east to the beginnings of potentially continuous forest and from the Canadian border south to the Texas panhandle region (Figure 1). Thus included are the states of Kansas, Nebraska, South Dakota, North Dakota, Wyoming, and Colorado, plus eastern Montana and northeastern New Mexico. Also included are the western two to three tiers of counties in Minnesota, Iowa, and Missouri, plus northwest Oklahoma and the adjacent portion of Texas. For convenience the region is called the Great Plains, although this name has sometimes been used with a narrower sense.

The eastern edge of the range passes into the area treated by the principal manuals for the flora of the northeastern quarter of the U.S.A., i.e., *Gray's Manual of Botany*, 8th ed., by M. L. Fernald (1950), *The New Britton & Brown Illustrated Flora* by H. A. Gleason (1952), and its derivative *Manual of Vascular Plants* by H. A. Gleason and A. Cronquist (1963). The northwestern boundary abuts the eastern edge of the treatment in *Vascular Plants of the Pacific Northwest* by C. L. Hitchcock et al. (1955–1969), and its derivative *Flora of the Pacific Northwest* by C. L. Hitchcock and A. Cronquist (1973). The area covered by *Flora of the Great Plains* thus extends somewhat farther to the northwest than the area treated in *Atlas of the Flora of the Great Plains* (1977).

The Great Plains comprises about a fifth of the land area of the conterminous United States, and it is characterized as principally flat or gently rolling, with only the Black Hills rising to what can be called mountains. The flora is generally of great uniformity to the casual observer, and it did not attract the attention of aggressive botanical explorers in the nineteenth century as did the mountainous regions farther to the west. In this century the Great Plains has become an exceptionally productive agricultural region, and a knowledge of the flora is of more than academic interest. It is of primary concern to those who farm or ranch in the area, for they either remove the native vegetation to cultivate their land or manipulate the flora to provide fodder for their livestock.

In the latter decades of the nineteenth century the various states in the Great Plains founded land-grant colleges and state universities, and numerous botanists at these institutions engaged in pioneer floristic studies. It was evident that the flora was similar throughout the region, but there was scant cooperation among the states. The only previous floristic treatment for the Great Plains was published at the New York Botanical Garden in 1932

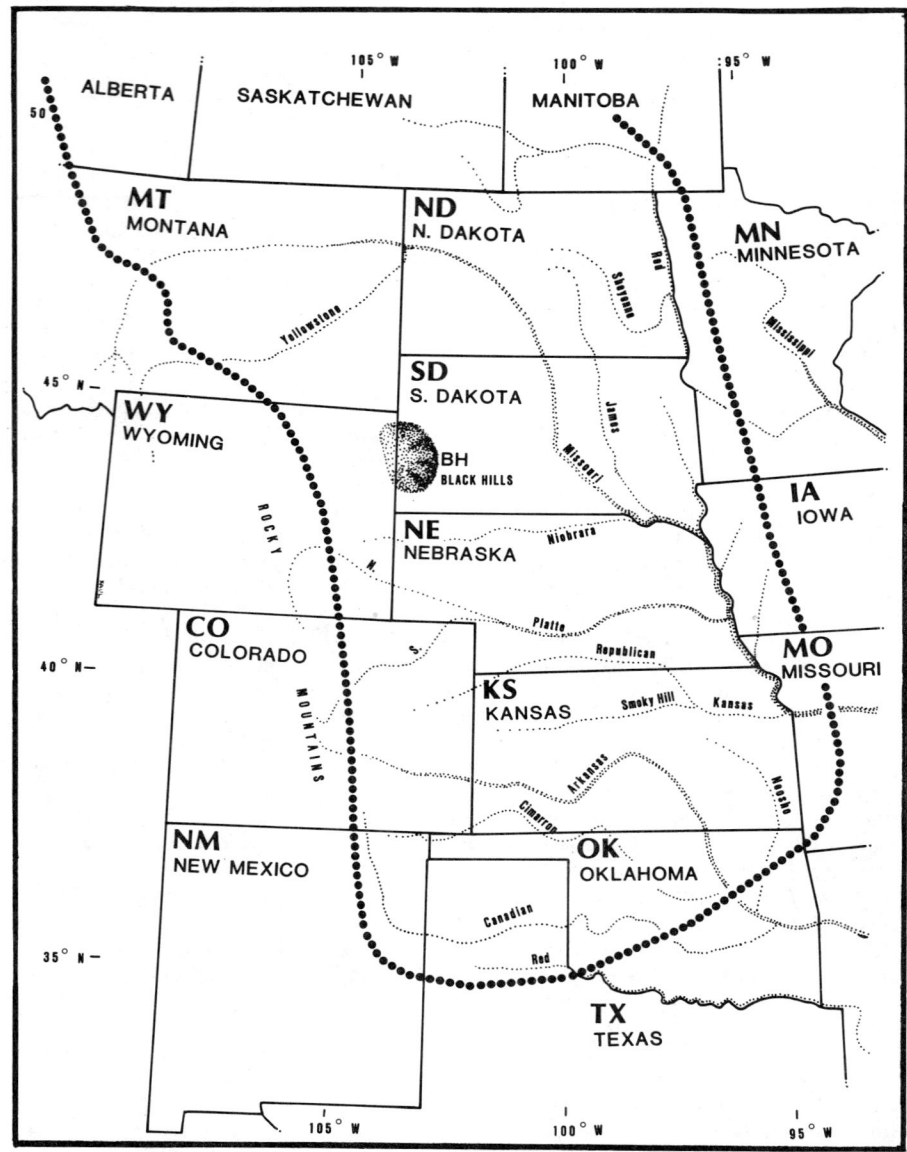

Figure 1. Geographical area covered in *Flora of the Great Plains*.

by P. A. Rydberg. He compiled his book from limited herbarium materials, mostly those of the New York Botanical Garden, aided by some field experience in the Great Plains. His work is the classic treatment, but it was effectively out-of-date soon after its publication. Furthermore, he held to a concept of species that is considerably narrower than that of more recent botanists. Subsequent studies have shown Rydberg's work to be admirable but, at best, a pioneer effort.

Floristic studies after Rydberg were confined largely to state floras and county checklists. These state floras and other works were based on better and more abundant collections than those available to Rydberg, but the works fell short of nomenclatural uniformity and taxonomic accuracy. Recent state floras are listed among the principal references.

Flora of the Great Plains has been written by the Great Plains Flora Association (GPFA), with collaborating specialists. The GPFA was formed in 1973, and it now includes the botanists listed on the title page. The initial collaboration of the group resulted in a book entitled *Atlas of the Flora of the Great Plains* (1977). A collection of distribution maps for the vascular plants of the Great Plains, it led directly to the present work. The GPFA is organized through a Memorandum of Cooperation, a quasi-legal agreement among the participants in the project. The Memorandum made it convenient for us to seek and obtain partial support from the National Science Foundation, through a single grant. Furthermore, the Memorandum signified to our administrations that we were engaged in a project beyond the scope or resources of any one of our home institutions.

SOURCES OF DATA

The *Flora* is based on herbarium and field studies plus the pertinent literature. The combined herbaria of the participating institutions in the GPFA number more than a million specimens, nearly all of which are from the Great Plains and adjacent regions. In addition, specimens have been freely borrowed from other herbaria to ensure the accuracy of the taxonomic conceptions and descriptions, and several GPFA members have visited major herbaria in the course of this project. We have tried to be current with the published literature through 1982.

SEQUENCE OF TAXA

The circumscription and sequence of the families of the flowering plants follow Cronquist (1981). Genera are presented alphabetically within the family, and species are presented alphabetically within the genus. When there is but a single species within a genus, the generic and specific descriptions are combined; likewise there is but a single extended description for unispecific families. Infraspecific taxa are keyed and described when three or more are recognized in our region, otherwise they are treated in discussion.

SPECIES CONCEPTS

This book employs a traditional and fundamentally conservative taxonomic concept. Species and infraspecific taxa are conceived as entities with at least some degree of populational integrity. Color forms and incidental morphotypes that occur "with the species" are accorded no formal recognition. Each contributing author is responsible for the botanical concepts of his/her contribution. Therefore, a certain amount of inconsistency is inevitable, and we acknowledge this.

TAXONOMY AND NOMENCLATURE

Taxonomic interpretations and nomenclature are generally in accord with the recent fully developed monographs for the various groups, and these are cited in the text. We have chosen to maintain traditional taxonomic concepts and usages whenever possible to do so, while admitting the necessity of change when such change clearly reflects a better understanding. No nomenclatural innovations are intentionally published in the *Flora*.

Citation of a monograph or revision implies a consideration of that work during the preparation of this book. However, the treatment presented here sometimes deviates significantly from the cited monograph, particularly in matters of distribution. The GPFA has better distributional data for our region than those available to monographers, at least until recently.

COLLOQUIAL NAMES

Colloquial names are employed as appropriate, derived from our *Atlas of the Flora of the Great Plains* and other published sources, as well as our own experience. No attempt has been made to compile all of the colloquial names from the region, for that would be an extensive project in folklore, and no consistent attempt has been made to provide a colloquial name for each taxon. For a plant lacking a colloquial name, the generic name may serve. This is a practice well entrenched in the English language (e.g., petunia, chrysanthemum, and philodendron are generic names also used colloquially).

SYSTEMATIC DESCRIPTIONS

Descriptions are written with the presumption that the plant is green and photosynthetic, that a woody plant is perennial, stems are erect, roots are fibrous, leaves of perennials are deciduous, leaves are simple and petiolate, flowers are perfect, regular, and pedicellate, perianth parts are separate, ovary is superior, carpels are fused, and the floral parts are inserted hypogynously, unless otherwise stated. Descriptions are for the plants as they occur in our range, unless stated to the contrary.

The chromosome numbers given in the species descriptions have been taken indiscriminately from many sources. They are provided merely as supplementary information, and the numbers do not represent an exhaustive, summary statement.

Ecological notes and flowering times are generalities for the plants as they occur in our range. Some reasonable freedom must be employed in using the flowering times, for there is considerable variation from year-to-year in the Great Plains, as well as from south to north for widespread species.

DISTRIBUTION

Distribution for taxa that occur throughout the Great Plains is cited as GP (cf. List of Abbreviations). Widespread distribution through a portion of the region is cited by the abbreviated states' names and is intended to provide a general and accurate statement. Taxa of limited distribution are cited by the abbreviation for the state's name followed by the counties in which the entity is known (e.g., KS: Cherokee, Labette). Distribution in the Black Hills is cited as BH. Taxa that are apparently rare, threatened, or perhaps endangered are so indicated.

Distributions outside the Great Plains are given in parentheses following the Great Plains distributions. These statements are generalities, based on monographs, other floristic studies, and herbarium data. Every attempt has been made for accuracy, but without exhaustive research.

SYNONYMY

Synonyms are cited for names that appear in other floristic works that are relevant to the Great Plains. When an entity is known by another name in another work, the other name (synonym) is cited, followed by the surname of the author in whose work the name is used: *Drymocallis fissa* (Nutt.) Rydb.—Rydberg. Names relegated to synonymy in this book may occur in more than one regional floristic work, but citing more than one work is impractical; therefore Rydberg is cited if he used the name as a valid entity. If the name is not employed in Rydberg, but is in other floristic works, the citation is for Fernald, then Gleason & Cronquist, and finally Gleason, in that order of preference. If the name is not employed in one of these, the appropriate local, state, or other regional flora is cited.

Every attempt has been made to account for the appropriate synonymy. However, it should be noted that not all of the names used by Rydberg are accounted for here. Rydberg's range extended farther to the east than does that of the *Flora of the Great Plains*, and therefore it included some taxa that are properly excluded here. Some names employed by Rydberg are improperly applied and some are simply problematic to us; solution of these matters is beyond the scope of the present work.

AUTHORSHIP AND CITATION

The various contributors to this project are identified throughout the text. Family treatments are presented with the authorship attributed with a by-line under the family name. Authorship of a smaller group (e.g., of a genus) is attributed with a by-line in the appropriate place. Thus, some large families have a general author, plus several cited authors for occasional generic treatments.

Future researchers will undoubtedly need to cite the *Flora* as a reference. Citation of multiple authorships is cumbersome, so we prefer citation as a book with corporate authorship by the Great Plains Flora Association:

> Great Plains Flora Association. 1986. *Flora of the Great Plains.* University Press of Kansas. Lawrence, Kansas.

When citing a particular treatment, it may be cited as a contribution to the *Flora:*

> Kaul, R. B. 1986. Annonaceae, *in* Great Plains Flora Association, *Flora of the Great Plains.* University Press of Kansas. Lawrence, Kansas.

PRINCIPAL REFERENCES

Listed here are the references cited in the Introduction and those cited frequently in the descriptions, particularly in the synonymies. References cited in the text but not listed here are given expanded bibliographic notation wherever cited.

Barkley, T. M. 1968. *Manual of the Flowering Plants of Kansas.* Kansas State University Endowment Association, Manhattan.

Boivin, B. 1967–present. *Flora of the Prairie Provinces.* Published as a series of papers in *Phytologia;* reprinted in *Provancheria.*
Budd, A. C., and K. F. Best. 1964. *Wild Plants of the Canadian Prairies.* Canada Department of Agriculture, Ottawa.
Correll, D., and M. C. Johnston. 1970. *Manual of the Vascular Plants of Texas.* Texas Research Foundation, Renner.
Cronquist, A. 1981. *An Integrated System of Classification of Flowering Plants.* Columbia University Press, New York.
Dorn, R. D. 1977. *Manual of the Vascular Plants of Wyoming.* Garland Publishing Company, New York.
Fernald, M. L. 1950. *Gray's Manual of Botany,* 8th ed. American Book Company, New York.
Flora Europaea. See Tutin, T. G., et al.
Gates, F. C. 1940. *Flora of Kansas.* Kansas Agricultural Experiment Station, Manhattan.
Gleason, H. A., 1952. *The New Britton and Brown Illustrated Flora.* New York Botanical Garden, New York.
Gleason, H. A., and A. Cronquist. 1963. *Manual of the Vascular Plants of the Northeastern United States and Adjacent Canada.* D. Van Nostrand Company, Princeton.
Great Plains Flora Association. 1977. *Atlas of the Flora of the Great Plains.* Iowa State University Press, Ames.
Harrington, H. D. 1954. *Manual of the Plants of Colorado.* Sage Books, Denver.
Hitchcock, C. L., A. Cronquist, M. Ownbey, and J. Thompson. 1955–1969. *Vascular Plants of the Pacific Northwest.* 5 vols. University of Washington Press, Seattle.
Hitchcock, C. L., and A. Cronquist. 1973. *Flora of the Pacific Northwest.* University of Washington Press, Seattle.
Kartesz, J. T., and R. Kartesz. 1980. *A Synonymized Checklist of the Vascular Flora of the United States, Canada and Greenland.* University of North Carolina Press, Chapel Hill.
Martin, W. C., and C. R. Hutchins. 1980. *A Flora of New Mexico.* J. Cramer, Vaduz.
Munz, P. A., and D. D. Keck. 1965. [1st ed., 1959]. *A California Flora.* University of California Press, Berkeley.
Rydberg, P. A. 1932. *Flora of the Prairies and Plains of Central North America.* New York Botanical Garden, New York.
Scoggan, H. J. 1957. *Flora of Manitoba.* National Museum of Canada, Ottawa.
Scoggan, H. J. 1978. *Flora of Canada.* National Museum of Natural Sciences, Ottawa.
Stevens, O. A. 1963. *Handbook of North Dakota Plants.* North Dakota Institute for Regional Studies, Fargo.
Steyermark, J. A. 1963. *Flora of Missouri.* Iowa State University Press, Ames.
Tutin, T. G., et al., eds. 1964–1980. *Flora Europaea.* 5 vols. Cambridge University Press, Cambridge, U. K.
Van Bruggen, T. 1976. *Vascular Plants of South Dakota.* Iowa State University Press, Ames.
Waterfall, U. T. 1969. *Keys to the Flora of Oklahoma,* 4th ed. Student Union Bookstore, Oklahoma State University, Stillwater.
Weber, W. A. 1976. *Rocky Mountain Flora.* University of Colorado Press, Boulder.
Winter, J. M. 1936. *An Analysis of the Flowering Plants of Nebraska.* Nebraska Department of Conservation, Lincoln.

Physical and Floristic Characteristics of the Great Plains

by Robert B. Kaul

The area covered in this book (see Figure 1) extends from the Rocky Mountains eastward to the forest edge in Minnesota and Missouri, and it cuts arbitrarily across the prairies of southwestern Minnesota and western Iowa. On the east it includes the western part of the vegetation province often known as the true prairie or tall-grass prairie, rather than as the Great Plains, a term more often restricted to our central and western grasslands.

We have included all native and introduced plants growing without cultivation in our area, whether or not they are grassland species. The many forest extensions from west and east into our grasslands bring numerous nonprairie species, and other areas, such as the Black Hills, add to our flora.

Nearly all species found in the southern parts of the prairie provinces of Canada can be identified with this book.

Figure 2 shows the major zones of vegetation in the Great Plains and adjacent areas. The grasslands are bordered on the west by coniferous, evergreen forests of the Rocky Mountains and their outliers. Elements of these mountain forests extend halfway across our area on hills and escarpments. Along much of our eastern border are the deciduous hardwood forests of eastern North America, and the transition from prairie to forest is often abrupt. At our border, these forests are mostly dominated by oaks, hickories, sugar maples, and basswoods. The deciduous forests extend west along our rivers well out into the grasslands, where they become improverished in species.

The coniferous forest extending east from the Rocky Mountains meets the eastern deciduous forest extending west only in the valley of the Niobrara River in north-central Nebraska. Both forest types are found also in the Black Hills of South Dakota as disjunct stands. Both these areas are the home of many relict northern species persisting from early postglacial times, when the climate was cooler.

Throughout our area the flood plains of the rivers are often at least partially covered by trees such as the willow, cottonwood, elm, ash, box elder, and, here and there, bur oak, walnut, and sycamore.

Our grasslands are often considered to consist of three parallel north-south zones: short grasses in the west, tall grasses in the east, and mixed short and tall grasses in the middle. These zones are presumed to be largely influenced by combinations of rainfall, evaporation, and soils. Within these zones are numerous subzones, some of which have been studied

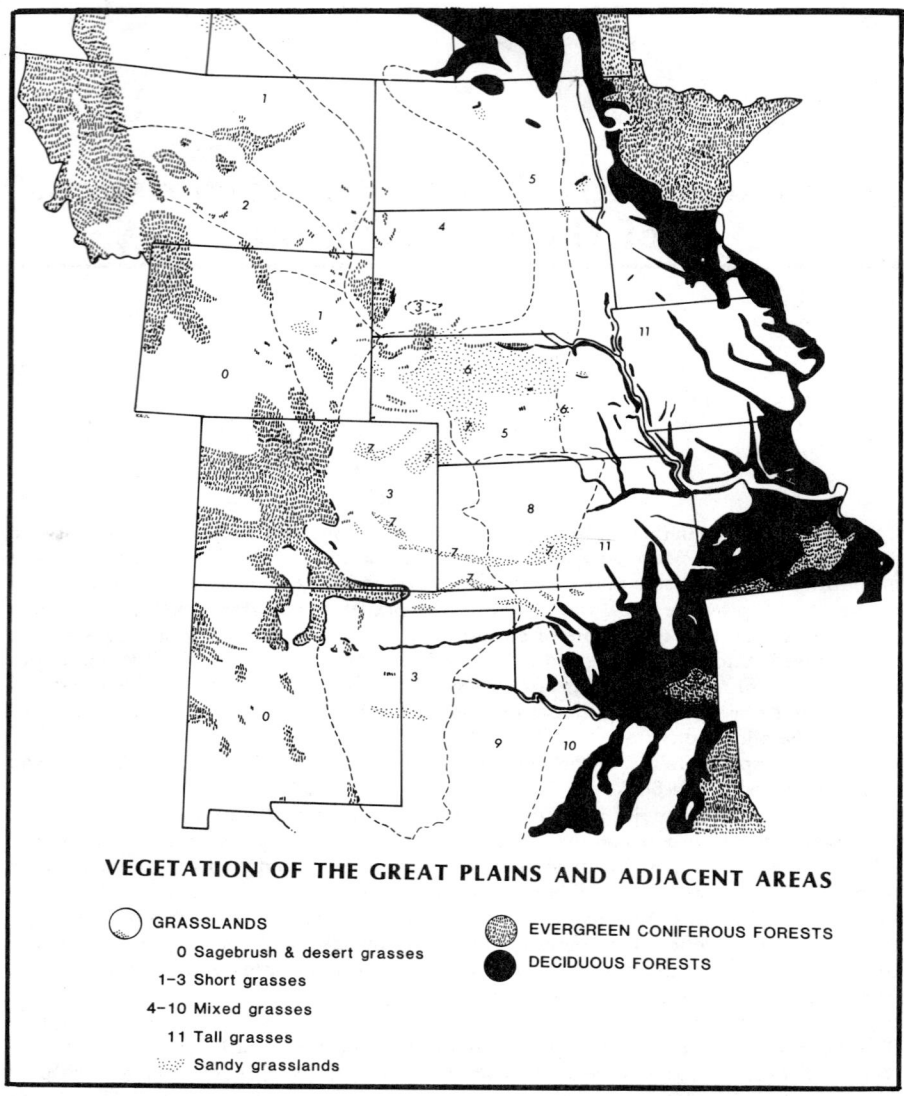

Figure 2. Vegetation of the Great Plains and adjacent areas. GRASSLANDS. 0: Sagebrush (*Artemisia*) and desert grasses in open to dense stands, including various grasses in different areas and also shrubs. 1–3: SHORT-GRASS PRAIRIES. (1) grama-, needle-, and wheat-grasses (*Bouteloua, Stipa, Agropyron*) in sparse to dense stands. (2) Wheat-, fescue-, and needle-grasses (*Agropyron, Festuca, Stipa*). (3) Grama- and buffalo-grasses (*Bouteloua, Buchloë*) with *Agropyron, Aristida, Opuntia, Yucca*, and other forbs; rather dense. 4–10: MIXED SHORT AND TALL GRASSES. (4) Wheat- and needle-grasses (*Agropyron, Stipa*) mixed with other grasses and some forbs, mostly Asteraceae and Fabaceae.

in detail, while others await delineation. Areas of sandy soil in particular have unique floristic combinations that set them sharply apart from adjacent grasslands. Such sandy soils occupy thousands of square miles in the Great Plains.

Our area may be considered a vast piedmont, with elevations of as much as 5,000 feet (1,600 m) at the foot of the mountains on our west, sloping eastward at about 10 feet per mile (2m/km) to less than 800 feet (290 m) at some places along our eastern border. Most of our major rivers arise in or near the mountains and flow east, eventually into the Mississippi River, but the Red River of the north and its tributaries flow north to Hudson Bay.

Our climate is wetter in the east than west and is wettest of all in the southeast. The heaviest precipitation occurs in the growing season, but winter rain and snow are common. Our moisture varies widely from year to year everywhere in our area, and short- and long-term cycles are thought to exist. It is not unusual for abnormally wet periods to be followed by abnormally dry ones, even within a year. Droughts of several years' duration sometimes occur over large parts of our area. Major droughts, such as those of the 1930s, have had profound effects on the flora. These droughts must be considered normal parts of our environment and may be useful in explaining plant distributions. In some places, recovery of the flora after droughts is thought to be rapid, while elsewhere recovery is much slower. Natural seed banks and prolonged dormancy of underground organs allow many grassland species to survive years of unfavorable rainfall and then make a quick recovery when adequate rains return.

Our summers are hot to very hot, even in our far north, and winters are cold and rather dry, but snowfall is erratically heavy and rain is not unusual. The growing season is, of course, shorter in the north and at higher elevations in our west. Thus a spring-flowering species blooming in March in Oklahoma may not bloom until May or June in North Dakota, and a fall-flowering species may bloom in August in North Dakota but in October in Oklahoma. This northward compression of the growing season means that some species that flower sequentially in our south do so synchronously in our north. There is also a westward compression of the growing season in our area, but it is less pronounced.

The soils of our eastern grasslands are among the most fertile in the world and consequently are heavily used for agriculture. This has resulted in the elimination of native vegetation in vast areas, approaching a hundred percent in some of our eastern counties and making some once-common species rare. The drier, less fertile soils in the central and western Great Plains are more often used for grazing, and the flora there is less disturbed. Everywhere in our area there is need to preserve good samples of the native vegetation before agricultural pursuits and the current almost unrestricted use of herbicides extirpate many of our species over even greater areas.

The vegetational history of our area is poorly known in detail. While the most recent continental glaciers were at their maxima (covering areas north and east of the Missouri River, as well as the eastern one-fifth of Nebraska and the northeastern one-sixth of Kan-

---

(5) Wheat-, bluestem-, and needle-grasses (*Agropyron, Andropogon, Stipa*) mixed with other grasses and forbs, mostly Asteraceae and Fabaceae; dense. (6) Bluestem-, sandreed-, grama-, and needle-grasses (*Andropogon, Calamovilfa, Bouteloua, Stipa*) on sand with abundant forbs, mostly Asteraceae and Fabaceae; *Yucca* and *Opuntia* common; dense to open and in some places without vegetation. (7) Sandsage, bluestem- and grama-grasses (*Artemisia, Andropogon, Bouteloua*) with other grasses and *Yucca*; moderately dense. (8) Bluestem- and grama-grasses (*Angropogon, Stipa*) with other grasses; dense. (9) Buffalo-grass and mesquite (*Buchloë* and *Prosopis*) with other grasses and scattered shrubs. (10) Bluestem- and needle-grasses (*Andropogon, Stipa*) with other grasses; dense. 11: TALL-GRASS PRAIRIES. Bluestem-, switch-, and Indian-grasses (*Andropogon, Panicum, Sorghastrum*) with numerous other grasses and forbs, mostly Asteraceae and Fabaceae; dense; traversed by oak forests in the river valleys. FORESTS. Evergreen coniferous forests (*Pinus* at lower elevations). Deciduous forests (*Quercus, Carya, Tilia, Acer*) with numerous understory herbs. (Modified after Küchler, 1964: *Potential Natural Vegetation of the Conterminous United States.* Amer. Geog. Soc. Spec. Pub. 36).

sas), much of our area was covered by cool-temperate forests. As the glaciers melted and the climate warmed, these forests gave way directly to grasslands. Some of the forest and wetland plants of these earlier times persist today in favored sites in our area; these species are included in this book.

Our flora is thus thought to be recent and adventive in origin. Nearly all our species have extensive ranges beyond our borders and appear to have colonized our area from elsewhere, especially from the southwest and southeast. Very many of our species are at the edge of their range with us, and they often show attenuated clinal variation over vast distances. These variations have barely been studied and should provide rich sources of information about the fates of adventive floras. Many western species extend east to about 100° west longitude—about the middle of our area—and many eastern species reach west to that meridian, especially in our uplands. In our river valleys, eastern and western species extend even farther to the west and east, respectively. Our flora, then, is a combination of elements from adjacent floras in the earliest stages of differentiation into a distinctive flora.

Fires and grazing have undoubtedly played a part in the maintenance of our original grasslands. Today, fires are much more limited and are far less influential than in earlier times. Grazing continues, although with a different set of grazers than in preagricultural times. In the absence of fire, woody plants can rapidly invade our grasslands from nearby forests to the west and east of us and from the wooded valleys. In many parts of our area, and especially in the eastern grasslands, there are today far more woody plants than existed before agrarian development.

Many influences continue to affect our grasslands, some with unforeseeable consequences. Rapidly increasing irrigation of cropland is causing water tables to rise in some areas and fall in others. It is also raising the humidity in areas once noted for their dry air. Heavy grazing, mowing, and herbicide spraying are certain to favor some species at the expense of others, either by elimination of some or by suppression of their sexual reproductive cycles. Exotic forage grasses, such as smooth brome, and a host of exotic weeds that have found homes in our disturbed soils are providing competition with native species. Recent widespread use of insecticides has almost certainly altered the populations of desirable pollinating insects. Restriction of some of our species to a few widely separated relict areas will undoubtedly have long-term effects on the gene pools of those species.

Almost every plant species in our area is in need of intensive, modern biological study. Our treatments in this book are based on the best available data that are consistent with our field and herbarium observations. Such phenomena as clinal variation, incipient speciation, and endemism, and reproductive isolation in species with extensive ranges, all need attention here. Basic floristic information is still lacking for large parts of our area, most notably the grasslands of Montana, Wyoming, Colorado, and the Dakotas. The westward impoverishment of the eastern deciduous forest is well known but undocumented, and the similar fate of Rocky Mountain species in our western forests needs delineation. The persistence of boreal and cool-temperate species from glacial times needs careful analysis and explanation. The role of short-term fluctuations in rainfall was documented for the great drought of the 1930s by J. E. Weaver and his associates, but further work is needed. It is not clear whether the Great Plains are today being invaded from adjacent areas, whether those possibly invading species have reached their limits, or whether they may once have been even more widespread.

In recent years numerous small tracts of virgin grassland have been designated as preserves, and several much larger tracts also have been set aside. All these are collectively of great importance in providing refuges for grassland plants and animals and in facilitating long-term studies.

List of Families in Sequence

Colloquial names are in parentheses, followed by page numbers for each family. The Family sequence for Magnoliophyta is that of Cronquist (1981).

VASCULAR CRYPTOGAMS (Pteridophytes)
1. Lycopodiaceae (Clubmoss), 39
2. Selaginellaceae (Spikemoss), 39–41
3. Isoetaceae (Quillwort), 41
4. Equisetaceae (Horsetail), 42–45
5. Ophioglossaceae (Adder's-tongue), 45–48
6. Osmundaceae (Royal Fern), 48–49
7. Polypodiaceae (True Fern), 49–69
8. Marsileaceae (Pepperwort), 70
9. Salviniaceae (Water Fern), 71

Division PINOPHYTA (Gymnosperms)
10. Cupressaceae (Cypress), 71–74
11. Pinaceae (Pine), 74–76
12. Ephedraceae (Ephedra), 76–77

Division MAGNOLIOPHYTA (Flowering Plants)
Class Magnoliopsida (Dicots)
13. Annonaceae (Custard Apple), 77–78
14. Lauraceae (Laurel), 78–79
15. Saururaceae (Lizard's Tail), 79–80
16. Aristolochiaceae (Birthwort), 80–81
17. Nelumbonaceae (Lotus), 81
18. Nymphaeaceae (Waterlily), 81–83
19. Cabombaceae (Water Shield), 83–84
20. Ceratophyllaceae (Hornwort), 84
21. Ranunculaceae (Buttercup), 84–107
22. Berberidaceae (Barberry), 107–109
23. Menispermaceae (Moonseed), 109–110
24. Papaveraceae (Poppy), 110–114
25. Fumariaceae (Fumitory), 114–118
26. Platanaceae (Sycamore), 119
27. Ulmaceae (Elm), 119–123
28. Cannabaceae (Hemp), 123–125
29. Moraceae (Mulberry), 126–127
30. Urticaceae (Nettle), 127–131
31. Juglandaceae (Walnut), 131–134
32. Fagaceae (Oak), 134–141
33. Betulaceae (Birch), 141–144
34. Phytolaccaceae (Pokeweed), 144–145
35. Nyctaginaceae (Four-o'clock), 145–152
36. Aizoaceae (Fig-marigold), 152–153
37. Cactaceae (Cactus), 153–160
38. Chenopodiaceae (Goosefoot), 160–179
39. Amaranthaceae (Pigweed), 179–187
40. Portulacaceae (Purslane), 187–190
41. Molluginaceae (Carpetweed), 191
42. Caryophyllaceae (Pink), 192–214
43. Polygonaceae (Buckwheat), 214–235
44. Plumbaginaceae (Leadwort), 235
45. Elatinaceae (Waterwort), 236
46. Clusiaceae (St. John's-Wort), 236–239
47. Tiliaceae (Linden), 239–240
48. Malvaceae (Mallow), 240–252
49. Droseraceae (Sundew), 252–253
50. Cistaceae (Rockrose), 253–255
51. Violaceae (Violet), 255–264
52. Tamaricaceae (Tamarix), 265–266
53. Passifloraceae (Passion-flower), 266–267
54. Cucurbitaceae (Cucumber), 267–269
55. Loasaceae (Stickleaf), 269–273
56. Salicaceae (Willow), 273–291
57. Capparaceae (Caper), 291–293
58. Brassicaceae (Mustard), 293–333
59. Resedaceae (Mignonette), 333
60. Ericaceae (Heath), 334–336
61. Pyrolaceae (Wintergreen), 336–340
62. Monotropaceae (Indian Pipe), 340–341
63. Sapotaceae (Sapodilla), 341–342
64. Ebenaceae (Ebony), 342
65. Primulaceae (Primrose), 342–351
66. Hydrangeaceae (Hydrangea), 351
67. Grossulariaceae (Currant), 352–356
68. Crassulaceae (Stonecrop), 356–358
69. Saxifragaceae (Saxifrage), 358–364
70. Rosaceae (Rose), 364–406
71. Crossosomataceae (Crossosoma), 406–407
72. Mimosaceae (Mimosa), 407–411

LIST OF FAMILIES IN SEQUENCE

73. Caesalpiniaceae (Caesalpinia), 411–416
74. Fabaceae (Bean), 416–490
75. Elaeagnaceae (Oleaster), 490–492
76. Haloragaceae (Water Milfoil), 492–494
77. Lythraceae (Loosestrife), 494–498
78. Thymelaeaceae (Mezereum), 498
79. Onagraceae (Evening Primrose), 498–526
80. Melastomataceae (Melastome), 526–527
81. Cornaceae (Dogwood), 527–530
82. Garryaceae (Silk Tassel), 530–531
83. Santalaceae (Sandalwood), 531–532
84. Viscaceae (Christmas Mistletoe), 532–533
85. Rafflesiaceae (Rafflesia), 533–534
86. Celastraceae (Staff Tree), 534–535
87. Aquifoliaceae (Holly), 535
88. Euphorbiaceae (Spurge), 535–553
89. Rhamnaceae (Buckthorn), 554–557
90. Vitaceae (Grape), 557–561
91. Linaceae (Flax), 561–564
92. Polygalaceae (Milkwort), 564–566
93. Krameriaceae (Ratany), 566
94. Staphyleaceae (Bladdernut), 567
95. Sapindaceae (Soapberry), 567–568
96. Hippocastanaceae (Buckeye), 568
97. Aceraceae (Maple), 569–570
98. Anacardiaceae (Cashew), 571–575
99. Simaroubaceae (Quassia), 575
100. Rutaceae (Citrus), 575–577
101. Zygophyllaceae (Caltrop), 577–578
102. Oxalidaceae (Wood Sorrel), 578–579
103. Geraniaceae (Geranium), 580–582
104. Balsaminaceae (Touch-me-not), 582–583
105. Araliaceae (Ginseng), 583–584
106. Apiaceae (Parsley), 584–604
107. Loganiaceae (Logania), 604
108. Gentianaceae (Gentian), 604–609
109. Apocynaceae (Dogbane), 610–613
110. Asclepiadaceae (Milkweed), 614–637
111. Solanaceae (Nightshade), 637–651
112. Convolvulaceae (Morning Glory), 652–661
113. Cuscutaceae (Dodder), 661–666
114. Menyanthaceae (Buckbean), 666
115. Polemoniaceae (Polemonium), 666–677
116. Hydrophyllaceae (Waterleaf), 678–683
117. Boraginaceae (Borage), 683–701
118. Verbenaceae (Vervain), 701–708
119. Lamiaceae (Mint), 708–740
120. Hippuridaceae (Mare's Tail), 740
121. Callitrichaceae (Water Starwort), 741–742
122. Plantaginaceae (Plantain), 742–747
123. Oleaceae (Olive), 747–750
124. Scrophulariaceae (Figwort), 751–797
125. Orobanchaceae (Broomrape), 798–799
126. Acanthaceae (Acanthus), 800–802
127. Pedaliaceae (Unicorn-plant), 803
128. Bignoniaceae (Bignonia), 803–805
129. Lentibulariaceae (Bladderwort), 806–807
130. Campanulaceae (Bellflower), 808–816
131. Rubiaceae (Madder), 816–823
132. Caprifoliaceae (Honeysuckle), 823–832
133. Adoxaceae (Moschatel), 832–833
134. Valerianaceae (Valerian), 833–836
135. Dipsacaceae (Teasel), 836–837
136. Asteraceae (Sunflower), 838–1021

Class Liliopsida (Monocots)

137. Butomaceae (Flowering Rush), 1021–1022
138. Alismataceae (Water Plantain), 1022–1028
139. Hydrocharitaceae (Frog's-bit), 1028–1031
140. Scheuchzeriaceae (Scheuchzeria), 1031
141. Juncaginaceae (Arrowgrass), 1031–1032
142. Potamogetonaceae (Pondweed), 1032–1039
143. Ruppiaceae (Ditchgrass), 1040
144. Najadaceae (Naiad), 1040–1041
145. Zannichelliaceae (Horned Pondweed), 1041–1042
146. Araceae (Arum), 1042–1043
147. Lemnaceae (Duckweed), 1043–1046
148. Commelinaceae (Spiderwort), 1046–1049
149. Juncaceae (Rush), 1049–1059
150. Cyperaceae (Sedge), 1059–1113
151. Poaceae (Grass), 1113–1235
152. Sparganiaceae (Bur-reed), 1235–1237
153. Typhaceae (Cat-tail), 1237–1239
154. Pontederiaceae (Pickerel-weed), 1239–1241
155. Liliaceae (Lily), 1241–1258
156. Iridaceae (Iris), 1258–1263
157. Agavaceae (Agave), 1263–1265
158. Smilacaceae (Catbrier), 1266–1267
159. Dioscoreaceae (Yam), 1267–1268
160. Orchidaceae (Orchid), 1268–1284

Abbreviations

Numerous abbreviations are used throughout the book, according to these definitions:

adj.	adjacent	ID	Idaho
AK	Alaska	IL	Illinois
AL	Alabama	IN	Indiana
Alta.	Alberta (Can.)	Is.	Island
AR	Arkansas	km	kilometer
AZ	Arizona	KS	Kansas
Baja Calif.	Baja California (Mex.)	KY	Kentucky
B.C.	British Columbia (Can.)	LA	Louisiana
BH	Black Hills distribution	Lab.	Labrador
ca	circa (about)	lat.	latitude
C. Amer.	Central America	long.	longitude
CA	California	m	meter
Can.	Canada	Man.	Manitoba (Can.)
cen	central	MA	Massachusetts
cf.	compare	MD	Maryland
Chih.	Chihuahua (Mex.	ME	Maine
cm	centimeter	Medit.	Mediterranean
CO	Colorado	Mex.	Mexico
Coah.	Coahuila (Mex.)	MI	Michigan
CT	Connecticut	mm	millimeter
cult.	cultivated	μm	micrometer
d.b.h.	diameter breast high	MN	Minnesota
DC	District of Columbia	MO	Missouri
DE	Delaware	MS	Mississippi
diam	diameter	mt., mts.	mountain, mountains
dm	decimeter	MT	Montana
Dur.	Durango (Mex.)	n	north, northern
e	east, eastern	n =	number of chromosomes
elev.	elevation	N. Amer.	North America
et seq.	et sequentes (and the following)	N.B.	nota bene (take careful note)
		N.B.	New Brunswick (Can.)
f.	forma (form)	NC	North Carolina
FL	Florida	ND	North Dakota
GA	Georgia	NE	Nebraska
GP	Great Plains distribution	Newf.	Newfoundland (Can.)
Greenl.	Greenland	NJ	New Jersey
Gto.	Guanajuato (Mex.)	N.L.	Nuevo Leon (Mex.)
Guat.	Guatemala	NM	New Mexico
HI	Hawaii	no.	number
IA	Iowa	N.S.	Nova Scotia (Can.)

N.T.	Northwest Territory (Can.)	sensu	in the sense of
NV	Nevada	s.n.	sine numero (without number)
NY	New York	Son.	Sonora (Mex.)
OH	Ohio	sp.	species (singular)
OK	Oklahoma	spp.	species (plural)
Ont.	Ontario (Can.)	subsp.	subspecies (singular)
op. cit.	opere citato (in the work cited)	subspp.	subspecies (plural)
		Tam.	Tamaulipas (Mex.)
OR	Oregon	temp.	temperate
PA	Pennsylvania	TN	Tennessee
P.E.I.	Prince Edward Island (Can.)	trop.	tropical
pers. comm.	personal communication	TX	Texas
±	more or less	UT	Utah
p.p.	pro parte (as a part)	VA	Virginia
Que.	Quebec (Can.)	var.	variety (singular)
q.v.	quo vide (which see)	vars.	varieties (plural)
R.	River	vs.	versus
s	south, southern	VT	Vermont
S. Amer.	South America	w	west, western
Sask.	Saskatchewan (Can.)	WA	Washington
SC	South Carolina	WI	Wisconsin
SD	South Dakota	W.I.	West Indies
sect.	section	WV	West Virginia
sens. lat.	sensu lato (in the broad sense)	WY	Wyoming
sens. str.	sensu stricto (in the narrow sense)	×	times (multiplied by)
		Zac.	Zacatecas (Mex.)

The Keys

by T. M. Barkley

The purpose of these keys is to aid in the identification of the plants in our flora. The keys are based on characters that are readily observable, and they are not designed to show natural relationships among the groups or to show the correct dispositions among orders, classes, or divisions. They lead to families and exceptional genera, and the large, diverse families may be reached in several places.

The keys begin with a general key, through which a plant will be referred to a subsequent key where its family will be identified. Some exceptional groups are reached directly in the general key. Keys to genera and species are incorporated within the family treatments in the text.

Keys are seldom as precise as one would wish them to be, and effective keying requires considerable practice. The beginner should recognize that plants are variable, and that families, genera, and species do not always have exact boundaries.

The keys are derived from the research and experience of many people and they reflect conventional wisdom in botany. However, the organization is ultimately based on that of Gleason in *The New Britton & Brown Illustrated Flora* (1952).

GENERAL KEY

1 Plants reproducing by microscopic spores that are borne in sporangia, usually on leaves or leaflike structures; flowers and seeds absent (Vascular Cryptogams, i.e., ferns and fern allies) .... Key A.
1 Plants reproducing by seeds.
    2 Cone-bearing plants, with seeds borne on the surface of an open scale (becoming fleshy in *Juniperus*); leaves needlelike or appressed to the stem, or reduced to short, scarious scales in whorls on jointed stems in *Ephedra* (Division Pinophyta, Gymnosperms) .. Key B.
    2 Flowering plants with seeds borne in closed ovaries that give rise to fruit; leaves rarely needlelike or scalelike (Division Magnoliophyta, flowering plants).
        3 Plants parasitic or saprophytic, often lacking chlorophyll and then whitish, ochroleucous or yellow-brown, or if green then clearly growing upon and absorbing nutrients from a host plant; plants sometimes greatly reduced .... Key C.
        3 Plants green and chlorophyllous; rooted in the soil or sometimes aquatic.
            4 Plants obligate aquatics, growing completely submersed or floating, or at most with flowers or a few leaves emersed .... Key D.
            4 Plants terrestrial, or at most merely rooted in shallow water; inflorescence always aerial.

5 Plants succulent, leafless and variously covered with spines that arise from distinctive areoles .... Family 37. Cactaceae
5 Plants not at once succulent, leafless and spiny.
  6 Basic unit of the inflorescence a flowerlike cyathium, consisting of a cuplike involucre with 1–5 marginal glands, sometimes alternating with petallike appendages; within the involucre are several to numerous stamens and 1 prominent stipitate ovary, with 3 fused carpels and 3 or more style branches; juice milky (*Euphorbia*) .... Family 88. Euphorbiaceae
  6 Flowers and inflorescences various, but not of cyathia.
    7 Flowers individually small but aggregated into a common receptacle and subtended by an involucre that ± resembles a calyx, forming a head (capitulum) that often resembles a single, large flower; corolla gamopetalous and either regular or expanded to one side, rarely absent; stamens united by the anthers into a sleeve around the style, or separate in a few genera; ovary inferior; calyx replaced by a whorl of hairs, bristles, or scales, (pappus) or sometimes absent. [N.B. *Dipsacus* (Dipsacaceae, q.v.) has an elongate, involucrate head, but 4 free stamens and a short calyx] .... Family 136. Asteraceae
    7 Flowers small to large, but mostly individually conspicuous, or if aggregated into heads, without the above combination of characters.
      8 Stamens and pistil greatly modified from the normal structures, so as to be scarcely recognizable, attached to each other and forming a specialized structure in the center of the flower.
        9 Flowers irregular, the lower petal differing markedly from the others; stamens and style fused to form a distinctive *column;* ovary inferior .... Family 160. Orchidaceae
        9 Flowers regular, with petals all alike; stamens, style and stigma modified into a *gynostegium;* ovary superior .... Family 110. Asclepiadaceae
      8 Stamens and pistil (or pistils) normal and readily recognizable.
        10 Trees, shrubs, or woody vines .... Key E.
        10 Herbs, the aboveground portions dying at the end of each growing season, although the basal and underground portions may be woody and perennial.
          11 Leaves usually parallel veined, often sheathing at the base; flower parts commonly in 3s (or 6s); seed with a single cotyledon (Monocots) .... Key M.
          11 Leaves usually net-veined; flower parts usually in 4s or 5s; seeds with 2 cotyledons (Dicots).
            12 Flowers all unisexual; no perfect flowers present .... Key F.
            12 Flowers normally perfect, but some unisexual flowers may be present among the perfect flowers.
              13 Flowers with 1 perianth whorl, or perianth absent (some plants may have 2 perianth whorls, but 1 whorl is early deciduous; such plants are reached on both legs of this key) .... Key G.
              13 Flowers with 2 perianth whorls, i.e., calyx and corolla.
                14 Ovaries 2 or more and separate .... Key H.
                14 Ovary 1, although often compound and having 2 or more styles and/or stigmas.
                  15 Ovary inferior .... Key I.
                  15 Ovary superior.
                    16 Stamens more numerous than the petals or corolla lobes .... Key J.
                    16 Stamens the same number as the petals or corolla lobes, or fewer.
                      17 Corolla of separate petals .... Key K.
                      17 Corolla of fused petals (i.e., gamopetalous) .... Key L.

Key A. Vascular Cryptograms (Ferns and Fern Allies)

1 Plants small floating aquatics .... 9. Salviniaceae
1 Plants terrestrial, rooted in the soil (mud) or on rocks or tree trunks.
  2 Stem conspicuously jointed and longitudinally ribbed, simple or with branches in whorls at the nodes; leaves reduced to small scales at the nodes; spores borne in a terminal strobilus .... 4. Equisetaceae
  2 Stems not conspicuously jointed, often a rhizome; leaves various but at least some of them green.
    3 Leaves long linear-filiform, terete or angled, ± grasslike; plants often rooted in mud.
      4 Leaves expanded at or just below the ground level and clustered on a cormlike stock, resembling an onion; sporangia axillary .... 3. Isoetaceae
      4 Leaves separate, scattered along a creeping rhizome; sporangia in stalked sporocarps *(Pilularia)* .... 8. Marsileaceae
    3 Leaves variously expanded, compound or simple, but not elongate-filiform.
      5 Leaves regularly 4-foliolate, resembling a 4-leaved clover *(Marsilea)* .... 8. Marsileaceae
      5 Leaves compound or simple but not regularly 4-foliolate.
        6 Stems ± elongate and ropy, well covered with numerous short, overlapping or divergent, linear to lanceolate or scalelike leaves (microphylls).
          7 Sterile leaves with a basal, transverse ligule; strobili ± 4-angled .... 2. Selaginellaceae
          7 Sterile leaves eligulate; strobili cylindrical .... 1. Lycopodiaceae
        6 Stem an underground rhizome or short crown or caudex, relatively inconspicuous compared to the relatively few large leaves (fronds).
          8 Fertile (sporangia-bearing) and sterile fronds, or fertile and sterile portions of the same frond, basically similar in structure and appearance .... 7. Polypodiaceae
          8 Fertile fronds, or fertile portions of the fronds, conspicuously different from the sterile fronds, or sterile portions of the fronds.
            9 Fronds with stipules at the very base of the petiole .... 6. Osmundaceae
            9 Fronds without stipules.
              10 Fertile and sterile fronds obviously dimorphic .. 9. Polypodiaceae
              10 Fertile and sterile segments present on the same frond .... 5. Ophioglossaceae

Key B. Division Pinophyta (Gymnosperms)

1 Leaves scalelike and forming membranaceous sheaths at the nodes; stems green and jointed .... 12. Ephedraceae
1 Leaves free or appressed but not forming membranaceous sheaths; stems neither green nor jointed.
  2 Leaves linear or needlelike, alternate or in fascicles; ovulate cones woody at maturity .... 11. Pinaceae
  2 Leaves scalelike and opposite or needlelike and whorled; ovulate cones fleshy at maturity .... 10. Cupressaceae

Key C. Epiphytes, Parasites, or Saprophytes

1 Monocots with zygomorphic flowers; anthers modified and united with the style to form a distinctive column; erect saprophytes. *(Corallorhiza, Hexalectris)* .... 160. Orchidaceae
1 Dicots with flowers regular or zygomorphic; stamens and pistils distinct from each other.
  2 Thalloid parasites on certain legumes *(Dalea)*, with vegetative parts imbedded in the host and with only small reddish-brown flowers and a few scalelike leaves evident on the surface of the host .... 85. Rafflesiaceae
  2 Plants without this combination of characters.
    3 Green-leaved, weakly woody parasites on the branches of tress .... 84. Viscaceae
    3 Achlorophyllous saprophytes or parasites.
      4 Corolla bilabiate; stamens 4, didynamous; root-parasites .... 125. Orobanchaceae

4 Corolla regular or nearly so; stamens 5 or more, or if 4 then all of the same length.
   5 Plants clumped or matted, saprophytic or parasitic on decaying soil-matter or roots; fruit a 1-celled capsule, with numerous seeds .... 62. Monotropaceae
   5 Scrambling or twining parasites, growing over the stems of the host plants; fruit a 2-celled capsule, usually with 4 seeds .... 113. Cuscutaceae

Key D. Aquatics (Growing Submersed or Floating, or at Most with the Inflorescence and a Few Leaves Emersed)

1 Plants floating on the surface or just beneath, occurring as single plants or as small chains, the entire plant body rarely exceeding 2 cm.
  2 Plants with branching stems and numerous, minute leaves .... 9. Salviniaceae
  2 Plant body thallose, without definite stems or leaves .... 147. Lemnaceae
1 Plants free-floating or rooted, submersed or emersed, much more than 3 cm long.
  3 Reproduction by spores produced in basal sporangia/sporocarps; leaves either narrow-filiform or 4-foliate and suggesting a 4-leaved clover (Ferns and allies).
    4 Leaves linear but basally expanded, with sporangia in the axils of the leaves; plant tufted and onionlike .... 3. Isoetaceae
    4 Leaves 4-foliate or filiform, arising from a creeping rhizome; sporangia produced in lateral sporocarps .... 8. Marsileaceae
  3 Reproduction by flowers, fruits, and seeds; leaves various.
    5 Leaves all or mostly deeply divided into narrow segments.
      6 Plants mostly free-floating; roots absent or rarely produced.
        7 Leaves alternate, with numerous small bladders borne on the leaf segments .... 129. Lentibulariaceae
        7 Leaves mostly whorled; bladders absent .... 20. Ceratophyllaceae
      6 Plants rooted.
        8 Primary leaf divisions pinnate .... 76. Haloragaceae
        8 Primary leaf divisions palmate.
          9 Leaves alternate *(Ranunculus)* .... 21. Ranunculaceae
          9 Leaves opposite or whorled *(Cabomba)* .... 19. Cabombaceae
    5 Leaves simple, margins entire, toothed or shallow-lobed.
     10 Leaves peltate or with the petiole attached in a deep, basal sinus.
       11 Leaves peltate.
         12 Leaf blade oval, less than 10 cm across *(Brasenia)* .... 19. Cabombaceae
         12 Leaf blade ± round, 30–70 cm across .... 17. Nelumbonaceae
       11 Leaves with petiole attached in a deep, basal sinus.
         13 Blade mostly 3–5 cm across; margins ± crenate *(Hydrocotyle)* .... 106. Apiaceae
         13 Blade 10 cm or more across; margins entire .... 18. Nymphaeaceae
     10 Leaves various but neither peltate nor with a deep, basal sinus.
      14 Leaves in whorls, giving the stem a "bottle brush" appearance.
       15 Whorls with 3(4–5) leaves, all submersed, blade 5–15 mm long .... 139. Hydrocharitaceae
       15 Whorls with 5(6–13) leaves; submersed leaves sometimes replaced by short scales; emersed leaves 10–30(40) mm long .... 120. Hippuridaceae
      14 Leaves clearly either alternate or opposite.
       16 Leaves, or at least some of them, floating or emersed.
        17 Leaves opposite.
         18 Floating leaves ± broadly spatulate; flowers unisexual .... 121. Callitrichaceae
         18 Emersed leaves narrowly elliptic or lanceolate; flowers perfect *(Didiplis)* .... 77. Lythraceae
        17 Leaves alternate.
         19 Leaves with pinnate venation .... 43. Polygonaceae
         19 Leaves with parallel venation.
          20 Midvein not at all evident .... 154. Pontederiaceae

```
          20 Midvein distinct and prominent.
             21 Flowers perfect, in axillary or terminal
                spikes ........................................... 142. Potamogetonaceae
             21 Flowers in unisexual globose heads, dispersed laterally along a
                zigzag rachis ..................................... 152. Sparganiaceae
       16 Leaves submersed only.
          22 Leaves with distinct petioles .......................... 142. Potamogetonaceae
          22 Leaves essentially sessile.
             23 Leaf margins finely dentate.
                24 Leaves very narrow, less than 3 mm wide ......... 144. Najadaceae
                24 Leaves more than 5 mm wide ................ 142. Potamogetonaceae
             23 Leaf margins entire.
                25 Leaves filiform but with an expanded, tubelike stipule; flowers in
                   subumbellate clusters .................................. 143. Ruppiaceae
                25 Leaves broader than filiform, or if very narrow then without an
                   expanded tubelike stipule; flowers solitary or in capitate or
                   spikelike inflorescences.
                   26 Leaves 2–6 mm wide, without evident
                      midvein ........................................... 154. Pontederiaceae
                   26 Leaves less than 2 mm wide, or if wider, with a prominent
                      midvein.
                      27 Leaves alternate ....................... 142. Potamogetonaceae
                      27 Leaves opposite or mostly so.
                         28 Leaf apex retuse (notched) .......... 121. Callitrichaceae
                         28 Leaf apex rounded or acute.
                            29 Leaves linear-lanceolate to narrowly elliptic; fruit
                               globose, 1.5–2.5 mm across
                               (Didiplis) ................................. 77. Lythraceae
                            29 Leaves linear-filiform; fruit ellipsoid, 2–3 mm long;
                               with a prominent beak to 1.5 mm
                               long ............................... 145. Zannichelliaceae
```

Key E. Trees, Shrubs, and Woody Vines

```
1 Leaves and leaf scars opposite or whorled.
   2 Flowers present before leaves present or fully expanded.
      3 Perianth of both calyx and corolla.
         4 Corolla of separate petals, flowers mostly unisexual, stamens ± 8 ... 97. Aceraceae
         4 Corolla gamopetalous, flowers perfect, stamens 5 (Lonicera) .... 132. Caprifoliaceae
      3 Perianth of a single whorl or absent.
         5 Flowers staminate or perfect.
            6 Stamens 2 or 4 ............................................................. 123. Oleaceae
            6 Stamens (5)8(10).
               7 Perianth lobes 4, yellowish; shrub (Shepherdia) ............. 75. Elaeagnaceae
               7 Perianth lobes ±5, often red; trees ................................. 97. Aceraceae
         5 Flowers pistillate.
            8 Perianth parts borne at the edge of an expanded hypanthium
               (Shepherdia) ......................................................... 75. Elaeagnaceae
            8 Perianth borne at the base of the ovary.
               9 Ovary prominently 2-lobed, producing 2 samaras ............... 97. Aceraceae
               9 Ovary not obviously lobed, producing a single samara ......... 123. Oleaceae
   2 Flowers present after leaves fully developed.
      10 Leaves compound.
         11 Plants climbing vines.
            12 Corolla none, sepals 4, petaloid; stamens numerous
               (Clematis) ............................................................ 21. Ranunculaceae
            12 Corolla present, 5-lobed; stamens 4 (Campsis) .................. 128. Bignoniaceae
         11 Plants erect trees or shrubs.
            13 Perianth of 1 whorl, or greatly reduced.
```

                14 Stamens 2–4, ovary not lobed .... 123. Oleaceae
                14 Stamens ± 8, ovary 2-lobed .... 97. Aceraceae
            13 Perianth of 2 whorls, petals prominent.
                15 Corolla gamopetalous; shrubs.
                    16 Leaves pinnately compound; stamens 5 *(Sambucus)* . 132. Caprifoliaceae
                    16 Leaves palmately compound; stamens 4 *(Vitex)* .... 118. Verbenaceae
                15 Corolla of separate petals.
                    17 Leaflets 3; stamens 5; flowers in pendulous racemes .. 94. Staphyleaceae
                    17 Leaflets 5–7; stamens 6–8; inflorescence erect ..... 96. Hippocastanaceae
        10 Leaves simple.
            18 Dioecious evergreen shrubs; inflorescences pendulous, catkinlike .... 82. Garryaceae
            18 Without this combination of characters.
                19 Perianth a single whorl, or calyx and corolla not differentiated.
                    20 Leaves palmately lobed .... 97. Aceraceae
                    20 Leaves entire or nearly so.
                        21 Flowers unisexual, variously perigynous; stamens 8 .... 75. Elaeagnaceae
                        21 Flowers perfect, hypogynous, often cleistogamous; stamens 5
                            (Selinocarpus) .... 35. Nyctaginaceae
                19 Perianth of both calyx and corolla, but the calyx may be reduced.
                    22 Stamens more numerous than the petals.
                        23 Stamens 10 or fewer; leaves smooth and not at all punctate.
                            24 Leaves palmately lobed .... 97. Aceraceae
                            24 Leave sserrate to pinnately lobed .... 66. Hydrangeaceae
                        23 Stamens more than 10; leaves punctate with translucent
                            dots .... 46. Clusiaceae
                    22 Stamens as many as petals or corolla lobes, or fewer.
                        25 Corolla of separate petals.
                            26 Flowers in terminal clusters or cymes .... 81. Cornaceae
                            26 Flowers in axillary inflorescences.
                                27 Flowers in axillary cymes *(Euonymus)* .... 86. Celastraceae
                                27 Flowers in subcorymbiform axillary clusters
                                    (Rhamnus) .... 89. Rhamnaceae
                        25 Corolla gamopetalous.
                            28 Ovary inferior.
                                29 Flowers numerous, in dense, globose heads; leaves entire
                                    (Cephalanthus) .... 131. Rubiaceae
                                29 Flowers variously disposed but not in globose heads; leaves lobed,
                                    toothed, or entire .... 132. Caprifoliaceae
                            28 Ovary superior.
                                30 Corolla zygomorphic; trees with broadly cordate leaves
                                    (Catalpa) .... 128. Bignoniaceae
                                30 Corolla regular or nearly so.
                                    31 Stem twining or trailing; ovaries 2
                                        (Periploca) .... 110. Asclepiadaceae
                                    31 Stem erect; ovary 1.
                                        32 Stamens 2; corolla lobes 4 .... 123. Oleaceae
                                        32 Stamens and corolla lobes both 4
                                            (Callicarpa) .... 118. Verbenaceae
1 Leaves and leaf scars alternate.
    33 Plants of the sunflower family (Asteraceae), with numerous small florets clustered into
        distinctive heads subtended by involucral bracts (cf. discussion of that
        family) .... 136. Asteraceae
    33 Plants with flowers variously arranged but not in distinctive involucrate heads.
        34 Plants dioecious, i.e., an individual bearing either staminate or pistillate flowers, but
            not both.
            35 Plants climbing vines, or at least leaning.
                36 Plants producing tendrils, or if tendrils absent the plant weakly if at all
                    twining.

THE KEYS

37 Tendrils arising from stipules (or tendrils absent in *Smilax ecirrhata*); perianth of 6 parts; stems mostly spiny .... 158. Smilacaceae
37 Tendrils arising on the stem opposite the leaves; petals 4 or 5 and early deciduous; stems without spines .... 90. Vitaceae
36 Plants twining but without tendrils.
 38 Stamens 5, pistil 1; leaves pinnate-veined .... 86. Celastraceae
 38 Stamens 6, pistils 3(+); leaves palmately veined .... 23. Menispermaceae
35 Plants erect trees or shrubs, or if weakly prostrate then not at all climbing.
 39 Flowers grouped in catkins or tight globose to elongate clusters; perianth a single whorl of inconspicuous sepals, or lacking.
  40 Juice milky, calyx present but minute .... 29. Moraceae
  40 Juice watery, calyx absent .... 56. Salicaceae
 39 Flowers variously disposed but not in catkins or tight clusters; perianth nearly always present, often of both calyx and corolla; individual flowers usually conspicuous.
  41 Leaves compound and present at flowering.
   42 Flowers in terminal or lateral racemes. .... 73. Caesalpiniaceae
   42 Flowers in various clusters, panicles, or cymes but not in racemes.
    43 Flowers in small, lateral clusters .... 100. Rutaceae
    43 Flowers in terminal panicles or cymes.
     44 Stems thorny .... 100. Rutaceae
     44 Stems thornless.
      45 Leaflets typically 3; stamens 5 .... 98. Anacardiaceae
      45 Leaflets more than 3; stamens more than 5.
       46 Stamens 8; ovary 1 .... 95. Sapindaceae
       46 Stamens 10.
        47 Ovary lobed, producing samaralike mericarps; leaves 1-pinnate .... 99. Simaroubaceae
        47 Ovary single, producing a legume; leaves bipinnately compound *(Gymnocladus)* .... 73. Caesalpiniaceae
  41 Leaves simple, or absent (immature) at flowering.
   48 Flowers pistillate.
    49 Perianth undifferentiated into calyx and corolla, or perianth absent.
     50 Perianth lobes 6; wood and crushed herbage fragrant .... 14. Lauraceae
     50 Perianth lobes (1)3–5, or perianth absent.
      51 Leaves inequalateral at base *(Celtis)* .... 27. Ulmaceae
      51 Leaves essentially equilateral.
       52 Fruit a drupe; small trees or shrubs .... 89. Rhamnaceae
       52 Fruit a utricle or achene; shrubs or subshrubs .. 38. Chenopodiaceae
    49 Perianth clearly differentiated into calyx and corolla, although the calyx may be reduced.
     53 Flowers in loose or dense terminal panicles .... 98. Anacardiaceae
     53 Flowers axillary or in axillary clusters.
      54 Corolla 1.5–4 cm across, gamopetalous; styles (2)4(6) ..... 64. Ebenaceae
      54 Corolla less than 1 cm across; petals separate; style 1.
       55 Style very short, stigma 1 .... 87. Aquifoliaceae
       55 Style divided, stigmas 2–4 .... 89. Rhamnaceae
   48 Flowers staminate.
    56 Stamens more numerous than the petals or sepals.
     57 Perianth of 6 yellow lobes; stamens 9; wood and crushed herbage fragrant .... 14. Lauraceae
     57 Perianth of a 4-parted calyx and gamopetalous corolla; stamens ± 16 .... 64. Ebenaceae
    56 Stamens as many as the calyx lobes or the petals, or petals sometimes absent.
     58 Flowers in loose or dense terminal panicles .... 98. Anacardiaceae
     58 Flowers, or at least some of them, axillary or in axillary clusters.
      59 Stamens alternate with the sepals and opposite (i.e., directly in front of) the petals, or petals sometimes absent .... 89. Rhamnaceae

> 59 Stamens opposite the sepals and alternate with the petals, or petals sometimes absent.
>> 60 Leaves inequilateral at the base *(Celtis)* .... 27. Ulmaceae
>> 60 Leaves equal at the base.
>>> 61 Perianth of 4 sepals and 4 petals; small trees .... 87. Aquifoliaceae
>>> 61 Perianth of a single whorl, mostly 5-parted; shrubs *(Atriplex)* .... 38. Chenopodiaceae

34 Plants not dioecious; flowers perfect or unisexual, but if unisexual then both staminate and pistillate flowers on the same plant, although sometimes in dissimilar inflorescences.
> 62 Flowers, or at least some of them (especially the staminate), in catkins, aments, or globose to elongate clusters, always unisexual and individually small and inconspicuous.
>> 63 Staminate and pistillate flowers both in dense, globose heads .... 26. Platanaceae
>> 63 Staminate flowers in elongate, cylindric, or ellipsoidal catkins or aments.
>>> 64 Pistillate flowers solitary or in small clusters.
>>>> 65 Leaves pinnately compound .... 31. Juglandaceae
>>>> 65 Leaves simple, at most lobed .... 32. Fagaceae
>>> 64 Pistillate flowers in catkins, heads, or conelike structures.
>>>> 66 Juice milky, calyx present .... 29. Moraceae
>>>> 66 Juice watery, calyx absent .... 33. Betulaceae
> 62 Flowers variously disposed but not in catkins, aments, or dense globose heads, usually perfect and often individually large and conspicuous.
>> 67 Perianth of a single series, or not obviously differentiated into calyx and corolla, or perianth absent.
>>> 68 Leaves compound; trees .... 99. Simaroubaceae
>>> 68 Leaves simple, or absent at flowering; trees, shrubs, or woody vines.
>>>> 69 Stamens more numerous than the lobes of the perianth.
>>>>> 70 Climbing vines .... 16. Aristolochiaceae
>>>>> 70 Erect trees and shrubs.
>>>>>> 71 Perianth lobes 4, stamens 8 .... 75. Elaeagnaceae
>>>>>> 71 Perianth lobes 6, stamens typically 9 .... 14. Lauraceae
>>>> 69 Stamens as many as the lobes of the perianth.
>>>>> 72 Styles clearly 2 or 3.
>>>>>> 73 Leaves inequilateral at the base; fruit a drupe or samara; trees .... 27. Ulmaceae
>>>>>> 73 Leaves equal at the base or very narrow; fruit a utricle or achene; shrubs .... 38. Chenopodiaceae
>>>>> 72 Style 1, with 1 stigma or branched only toward the tip.
>>>>>> 74 Flowers in a branching terminal cyme .... 81. Cornaceae
>>>>>> 74 Flowers in lateral or axillary clusters.
>>>>>>> 75 Tendril-bearing vines .... 90. Vitaceae
>>>>>>> 75 Erect shrubs or small trees.
>>>>>>>> 76 Style branched at the apex into 2–4 stigmas .... 89. Rhamnaceae
>>>>>>>> 76 Style unbranched .... 75. Elaeagnaceae
>> 67 Perianth present and clearly differentiated into calyx and corolla.
>>> 77 Pistils 3 or more, separate or nearly so.
>>>> 78 Stamens 10 or fewer, leaves pinnately compound .... 99. Simaroubaceae
>>>> 78 Stamens more than 10.
>>>>> 79 Sepals 3, petals 6, leaves simple .... 13. Annonaceae
>>>>> 79 Sepals 5, petals 5(+), leaves simple or compound .... 70. Rosaceae
>>> 77 Pistil 1, but styles may be more than 1.
>>>> 80 Corolla obviously irregular, or reduced to 1 petal.
>>>>> 81 Petal 1, blue; flowers in spikelike terminal inflorescences *(Amorpha)* .... 74. Fabaceae
>>>>> 81 Petals typically 5.
>>>>>> 82 Posterior petal in front of the lateral petals, i.e., the upper median petal is innermost in the bud; leaves simple, broadly cordate-rotund *(Cercis)* .... 73. Caesalpiniaceae

  82 Posterior petal behind the lateral ones, i.e., the upper, median petal is outermost in the bud; leaves compound .... 74. Fabaceae
 80 Corolla regular or essentially so.
  83 Corolla gamopetalous.
   84 Styles or style branches 4 or 5 .... 64. Ebenaceae
   84 Style and stigma 1.
    85 Stamens attached basally and apparently free from the corolla, as many or 2× as many as the corolla lobes.
     86 Style very short, the stigma nearly sessile .... 87. Aquifoliaceae
     86 Style prominent and well developed .... 60. Ericaceae
    85 Stamens inserted on the corolla tube, as many as the corolla lobes.
     87 Stamens directly in front of the corolla lobes, i.e., opposite the corolla lobes; petaloid staminodia present .... 63. Sapotaceae
     87 Stamens alternate with the corolla lobes; staminodia absent.
      88 Stamens 4 .... 118. Verbenaceae
      88 Stamens 5 .... 111. Solanaceae
  83 Corolla of separate petals.
   89 Ovary inferior, or at least appearing so.
    90 Stamens 2× as many as petals, or more.
     91 Style 1 *(Vaccinium)* .... 60. Ericaceae
     91 Styles 2–5 .... 70. Rosaceae
    90 Stamens the same number as the petals.
     92 Petals 4, white; flowers in terminal cymes .... 81. Cornaceae
     92 Petals 5; flowers mostly axillary or in axillary clusters .... 67. Grossulariaceae
   89 Ovary superior.
    93 Mature leaves scalelike and appressed to the stem .... 52. Tamaricaceae
    93 Mature leaves flat, expanded and typically petiolate.
     94 Flowers typically present in early spring, before the leaves.
      95 Styles 3 .... 98. Anacardiaceae
      95 Style 1 *(Prunus)* .... 70. Rosaceae
     94 Flowers and leaves present together.
      96 Stamens more than 2× as many as the petals.
       97 Leaves compound .... 72. Mimosaceae
       97 Leaves simple.
        98 Flowers yellow, style 1 .... 50. Cistaceae
        98 Flowers white or pink.
         99 Inflorescence arising from a woody or herbaceous branch *(Prunus)* .... 70. Rosaceae
         99 Inflorescence pedunculate and apparently arising from the middle of a narrowly oblong foliaceous bract .... 47. Tiliaceae
      96 Stamens 2× as many as the petals, or fewer.
       100 Leaves compound.
        101 Leaves even-pinnate or even-bipinnate; trees.
         102 Flowers in small, lateral spikelike racemes; stems often thorny *(Gleditsia)* .... 73. Caesalpiniaceae
         102 Flowers in a terminal raceme or panicle; stems thornless.
          103 Leaves bipinnate, or at least partly so *(Gymnocladus)* .... 73. Caesalpiniaceae
          103 Leaves once-pinnate .... 95. Sapindaceae
        101 Leaves odd-pinnate, odd-bipinnate, trifoliate, or palmate.
         104 Inflorescences terminal.
          105 Flowers in loose, open cymes; leaves with small, translucent dots *(Ptelea)* .... 100. Rutaceae
          105 Flowers in dense, spikelike clusters or panicles .... 98. Anacardiaceae

104 Inflorescences lateral or axillary.
   106 Stamens directly in front of the petals, i.e., opposite the petals; mostly woody vines .... 90. Vitaceae
   106 Stamens alternate with the petals; mostly erect shrubs or small trees, but if climbing then the leaves trifoliate and the terminal leaflet on a longer stalk than the lateral leaflets .... 98. Anacardiaceae
100 Leaves simple.
   107 Stamens more numerous than the petals.
      108 Flowers yellow; low, gray-tomentose subshrubs .... 50. Cistaceae
      108 Flowers white to pinkish; low green subshrubs .... 61. Pyrolaceae
   107 Stamens the same number or fewer than the petals, or if more, plants clearly erect and not low subshrubs.
      109 Flowers yellow, 6-merous; stems spiny .... 22. Berberidaceae
      109 Flowers white, or faintly greenish.
         110 Flowers in slender, elongate racemes; leaves evergreen, white-tomentose beneath, margins revolute *(Ledum)* .... 60. Ericaceae
         110 Without this combination of characters.
            111 Petals 5, stamens typically 8, unequal .... 71. Crossosomataceae
            111 Petals and stamens equal in number, stamens all alike.
               112 Stamens directly in front of the petals, i.e., opposite the petals; styles 2-3 lobed .... 89. Rhamnaceae
               112 Stamens alternate with the petals; style very short, stigma nearly sessile .... 87. Aquifoliaceae

Key F. Dicot Herbs with Unisexual Flowers

1 Stems fleshy jointed; leaves reduced to small scales; plants of saline sites *(Salicornia)* .... 38. Chenopodiaceae
1 Stems not succulent; leaves various.
  2 Leaves compound.
    3 Leaves 3-foliate or palmately compound.
      4 Flowers on a fleshy spadix or in a dense head, variously subtended or enfolded by the spathe, a modified leaf (Monocots with net-veined leaves, keyed here for convenience) .... 146. Araceae
      4 Flowers in umbels, spikes, or panicles.
        5 Flowers in umbels or compound umbels.
          6 Leaves alternate or basal *(Sanicula)* .... 106. Apiaceae
          6 Leaves in a single, cauline whorl *(Panax)* .... 105. Araliaceae
        5 Flowers in spikes or panicles.
          7 Perianth conspicuous; stamens numerous; pistils few to numerous *(Clematis)* .... 21. Ranunculaceae
          7 Perianth minute; stamens 5; pistil 1 .... 28. Cannabaceae
    3 Leaves pinnately compound, or pinnately to ternately decompound.
      8 Flowers in globose heads or short, subglobose spikes.
        9 Leaves once-pinnately compound *(Sanguisorba)* .... 70. Rosaceae
        9 Leaves twice-pinnately compound .... 72. Mimosaceae
      8 Flowers in umbels or panicles, or solitary.
        10 Flowers in 1 to many umbels *(Aralia)* .... 105. Araliaceae

THE KEYS

        10 Flowers solitary or in panicles.
           11 Stem leaves opposite.
              12 Stamens numerous; pistils few to numerous
                  (*Clematis*) .... 21. Ranunculaceae
              12 Stamens 3; pistil 1 with 1 style .... 134. Valerianaceae
           11 Stem leaves alternate.
              13 Plant twining-climbing (*Cardiospermum*) .... 95. Sapindaceae
              13 Plants erect or at most leaning (*Thalictrum*) .... 21. Ranunculaceae
2 Leaves simple.
    14 Leaves all basal.
        15 Flowers in short or elongate spikes .... 122. Plantaginaceae
        15 Flowers in small, open panicles (*Rumex*) .... 43. Polygonaceae
    14 Leaves all or chiefly cauline.
        16 Leaves opposite or whorled.
           17 Foliage densely stellate-pubescent or lepidote (*Croton,
              Crotonopsis*) .... 88. Euphorbiaceae
           17 Foliage glabrous to variously pubescent, but not as above.
              18 Flowers solitary; plants of muddy ground, especially recently submersed sites, or on moist, wooded slopes.
                  19 Leaves in whorls of 6–12 .... 120. Hippuridaceae
                  19 Leaves merely opposite .... 121. Callitrichaceae
              18 Flowers in axillary or terminal clusters.
                20 Inflorescences axillary.
                  21 Leaves obviously serrate .... 30. Utricaceae
                  21 Leaves entire or essentially so.
                    22 Flowers in pedunculate spikes .... 122. Plantaginaceae
                    22 Flowers in sessile or subsessile heads
                      (*Alternanthera*) .... 39. Amaranthaceae
                20 Inflorescences terminal.
                  23 Petals absent (*Iresine*) .... 39. Amaranthaceae
                  23 Petals present.
                    24 Stamens 10; styles 3(5) (*Lychnis*) .... 42. Caryophyllaceae
                    24 Stamens 3; styles 1 .... 134. Valerianaceae
        16 Leaves alternate.
           25 Foliage densely stellate-pubsecent or lepidote (*Croton,
              Crotonopsis*) .... 88. Euphorbiaceae
           25 Foliage glabrous to variously pubescent, but not as above.
              26 Perianth of 2 whorls; petals usually white or colored.
                27 Stamens 3; pistil 1; trailing or plants climbing by
                  tendrils .... 54. Cucurbitaceae
                27 Stamens 6 or more; pistils 3 or more.
                  28 Sepals, petals, and stamens each 6; pistils 3 or 6; plants twining
                  (*Cocculus*) .... 23. Menispermaceae
                  28 Stamens more than 6; pistils 4 or more, separate or ± fused.
                    29 Leaves sessile or nearly so, fleshy; stamens 8 or 10; pistils 4 or 5
                    (*Sedum*) .... 68. Crassulaceae
                    29 Leaves petioled; pistils and stamens numerous
                    (*Rubus*) .... 70. Rosaceae
              26 Perianth of 1 whorl or absent; perianth segments rarely petallike.
              30 Flowers in small axillary clusters; individual flowers always small.
                31 Plants with pistillate flowers.
                  32 Style 1, unbranched (*Parietaria*) .... 30. Urticaceae
                  32 Styles 2 or 3.
                    33 Styles 3, each of them 2-lobed or branched (*Acalypha,
                    Phyllanthus*) .... 88. Euphorbiaceae
                    33 Styles 2 or 3, unbranched.
                      34 Sepals and bracts acute, scarious .... 39. Amaranthaceae
                      34 Sepals (often absent) and bracts herbaceous
                      (*Atriplex*) .... 38. Chenopodiaceae

               31 Plants with staminate flowers.
                  35 Flowers and flower clusters subtended and often exceeded by
                     bracts.
                     36 Bracts foliaceous and lobed or cleft
                        (Acalypha) .... 88. Euphorbiaceae
                     36 Bracts entire.
                        37 Sepals and bracts acute and scarious . 39. Amaranthaceae
                        37 Sepals and bracts herbaceous *(Parietaria)* .. 30. Urticaceae
                  35 Flowers and flower clusters without associated bracts.
                     38 Leaves conspicuously stipulate
                        (Phyllanthus) .... 88. Euphorbiaceae
                     38 Leaves without stipules *(Atriplex)* .... 38. Chenopodiaceae
            30 Flowers in spikes, racemes, or panicles; inflorescence usually terminal,
               but sometimes opposite the leaves.
               39 Perianth ± petallike in size and color.
                  40 Leaves entire; stem glabrous .... 34. Phytolaccaceae
                  40 Leaves lobed; stem ± spiny *(Cnidoscolus)* ..... 88. Euphorbiaceae
               39 Perianth segments minute or none, or green and pale, but not
                  petallike.
                  41 Perianth segments 6, obscurely in 2 whorls
                     (Rumex) .... 43. Polygonaceae
                  41 Perianth segments 5 or fewer, or absent.
                     42 Perianth segments acute, scarious, intermixed with or
                        subtended by similar scarious bracts ...... 39. Amaranthaceae
                     42 Perianth segments neither acute nor scarious, nor subtended
                        by scarious bracts.
                        43 Plants with pistillate flowers or with fruit.
                           44 Ovary with 1 locule; fruit a 1-seeded
                              utricle .... 38. Chenopodiaceae
                           44 Ovary with 3 locules; fruit a
                              capsule .... 88. Euphorbiaceae
                        43 Plants with staminate flowers.
                           45 Perianth parts all separate .... 38. Chenopodiaceae
                           45 Perianth parts ± united into a calyx with 2–5
                              lobes .... 88. Euphorbiaceae

               Key G. Dicot Herbs with Perfect Flowers;
               Perianth Whorl 1, or Perianth Absent

1 Perianth absent.
   2 Leaves deeply lobed or compound .... 21. Ranunculaceae
   2 Leaves entire, or merely toothed.
      3 Leaves alternate, cordate, entire .... 15. Saururaceae
      3 Leaves whorled, linear to linear-lanceolate .... 120. Hippuridaceae
1 Perianth present, in 1 whorl.
   4 Ovary inferior.
      5 Stamens more numerous than the lobes or sections of the perianth.
         6 Perianth lobes 1 or 3; stamens 6 or 12 .... 16. Aristolochiaceae
         6 Perianth lobes 4; stamens 4 *(Sanguisorba)* .... 70. Rosaceae
      5 Stamens the same number as the sections of the perianth, or fewer.
         7 Leaves opposite or whorled.
            8 Flowers in umbels or dense terminal heads.
               9 Flowers in umbels; leaves in a single whorl *(Panax)* .... 105. Araliaceae
               9 Flowers in terminal heads.
                  10 Heads subtended by large white bracts .... 81. Cornaceae
                  10 Heads subtended by green bracts, or bracts absent ..... 135. Dipsacaceae
            8 Flowers neither in umbels nor heads.
               11 Leaves whorled *(Asperula, Galium)* .... 131. Rubiaceae

THE KEYS

    11 Leaves opposite *(Ludwigia)* .... 79. Onagraceae
  7 Leaves alternate or basal.
    12 Sections of the perianth 3 or 4; stamens 4 or fewer.
      13 Leaves with large, conspicuous stipules *(Alchemilla,
        Sanguisorba)* .... 70. Rosaceae
      13 Leaves without stipules *(Ludwigia)* .... 79. Onagraceae
    12 Sections of the perianth 5; stamens 5.
      14 Flowers in terminal or subterminal cymules; leaves entire or nearly
        so .... 83. Santalaceae
      14 Flowers in heads or umbels; leaves mostly compound or deeply lobed.
        15 Styles 2 .... 106. Apiaceae
        15 Styles 5 *(Aralia)* .... 105. Araliaceae
4 Ovary superior.
  16 Pistils more than 1.
    17 Pistils united at the base or nearly to their middle.
      18 Leaves serrate; flowers in cymes *(Penthorum)* .... 68. Crassulaceae
      18 Leaves entire; flowers in racemes .... 34. Phytolaccaceae
    17 Pistils clearly distinct from each other.
      19 Leaves with conspicuous stipules *(Alchemilla, Sanguisorba).*
      19 Leaves without stipules .... 21. Ranunculaceae
  16 Pistil 1; styles and stigmas 1-several.
    20 Stamens more than 2× as many as the perianth segments.
      21 Perianth small, inconspicuous, pale or greenish.
        22 Leaves compound, alternate; flowers in terminal racemes
          (Actaea) .... 21. Ranunculaceae
        22 Leaves simple, opposite; flowers in axillary clusters
          (Sesuvium) .... 36. Aizoaceae
      21 Perianth well developed, colored.
        23 Leaves simple, entire *(Talinum)* .... 40. Portulacaceae
        23 Leaves compound, dissected, lobed or toothed.
          24 Perianth segments 4 or 8; juice milky or colored .... 24. Papaveraceae
          24 Perianth segments 5; leaves bipinnately compound; juice watery
            (Schrankia) .... 72. Mimosaceae
    20 Stamens 2× as many as the perianth segments or fewer.
      25 Styles 2 or more.
        26 Leaves reduced to small scales; stems succulent and jointed; plants of saline
          sites. .... 38. *Salicornia*
        26 Without the above combination of characters.
          27 Leaves chiefly basal or crowded near the base.
            28 Leaves broadly lobed, palmately veined
              (Heuchera) .... 69. Saxifragaceae
            28 Leaves narrow, succulent, terete *(Talinum)* .... 40. Portulacaceae
          27 Leaves all or mostly cauline.
            29 Leaves opposite or whorled.
              30 Ovary and capsule with 1 locule .... 42. Caryophyllaceae
              30 Ovary and capsule with 3–5 locules *(Mollugo)* ..... 41. Molluginaceae
            29 Leaves alternate.
              31 Stamens 10, styles 5 or 10.
                32 Leaves entire; flowers in racemes .... 34. Phytolaccaceae
                32 Leaves serrate; flowers in cymes *(Pen-
                  thorum)* .... 68. Crassulaceae
              31 Stamens fewer than 10; styles 2–5.
                33 Stipules papery and sheathing the stem at the
                  nodes .... 43. Polygonaceae
                33 Stipules none.
                  34 Stamens as many as the perianth segments or
                    fewer .... 38. Chenopodiaceae

            34 Stamens more than the number of perianth segments but fewer than 2× as
               many *(Eriogonum)* .... 43. Polygonaceae
25 Style 1 or none (i.e., the stigma sessile); 2 or more stigmas may be discernible.
   35 Stamens more numerous than the perianth segments.
      36 Perianth segments 5.
         37 Leaves compound.
            38 Leaves 3-foliate (cleistogamous flowers of *Amphicarpaea,*
               Lespedeza) .... 74. Fabaceae
            38 Leaves bipinnately compound *(Schrankia)* .... 72. Mimosaceae
         37 Leaves simple.
            39 Perianth segments unequal, 2 of them smaller than the other 3 .. 50. Cistaceae
            39 Perianth segments all alike or essentially so.
               40 Flowers axillary and sessile or subsessile *(Glinus)* .... 41. Molluginaceae
               40 Flowers in terminal, pedunculate clusters *(Talinum)* .... 40. Portulacaceae
      36 Perianth segments 4.
         41 Flowers irregular; leaves variously compound to dissected .... 25. Fumariaceae
         41 Flowers regular; perianth greenish and sepallike.
            42 Leaves opposite *(Ammannia)* .... 77. Lythraceae
            42 Leaves alternate.
               43 Stamens 8, in 2 whorls .... 78. Thymelaeaceae
               43 Stamens 6 (apetalous phases of several genera) .... 58. Brassicaceae
   35 Stamens as many as the perianth segments or fewer.
      44 Leaves alternate or basal.
         45 Perianth segments 6 or 8; stamens 6 or 8 .... 22. Berberidaceae
         45 Perianth segments 4 or 5; stamens 5 or fewer.
            46 Leaves lobed or compound; flowers in terminal spikes or in headlike clusters
               (Alchemilla, Sanguisorba) .... 70. Rosaceae
            46 Leaves simple, entire or nearly so, of if deeply lobed then the flowers chiefly
               basal.
               47 Stamens 1–3, fewer than the perianth segments (cleistogamous flowers of
                  Viola) .... 51. Violaceae
               47 Stamens 4–5, as many as the perianth segments.
                  48 Flowers pedicellate, solitary or in loose cymules .... 83. Santalaceae
                  48 Flowers sessile, in small axillary clusters *(Parietaria)* .... 30. Urticaceae
      44 Leaves opposite, rarely whorled.
         49 Flowers or flower clusters axillary and sessile, or nearly so.
            50 Perianth and subtending bracts scarious; flowers numerous in heads or clusters
               (Alternanthera, Tidestroemia) .... 39. Amaranthaceae
            50 Perianth herbaceous and not subtended by conspicuous scarious bracts; flowers
               solitary or few in the axils.
               51 Perianth segments 4 .... 77. Lythraceae
               51 Perianth segments 5.
                  52 Leaves sessile *(Glaux)* .... 65. Primulaceae
                  52 Leaves petiolate .... 36. Aizoaceae
         49 Flowers or inflorescences terminal.
            53 Flowers in open cymes or panicles.
               54 Leaves linear or narrowly elliptic, sessile and often
                  stipulate .... 42. Caryophyllaceae
               54 Leaves broad, petiolate and without stipules
                  (Boerhaavia) .... 35. Nyctaginaceae
            53 Flowers variously clustered in heads, spikes, or umbels.
               55 Flowers in umbellate or capitate clusters, subtended by a 5-parted or lobed
                  involucre .... 35. Nyctaginaceae
               55 Flowers in dense spikes or heads; perianth parts and subtending bracts
                  scarious *(Froelichia)* .... 39. Amaranthaceae

Key H. Dicot Herbs with Perfect Flowers, Perianth Whorls 2; Pistils 2 or More

1 Style apparently single, although sometimes divided above into separate stigmas.
  2 Ovaries 2; corolla gamopetalous; stamens 5 .... 109. Apocynaceae
  2 Ovaries 4–5.
    3 Ovaries 5 +; stamens numerous, monadelphous; petals separate or but weakly united at the base .... 48. Malvaceae
    3 Ovaries 4; stamens separate; corolla gamopetalous.
      4 Leaves opposite; stamens 2 or 4; corolla usually irregular .... 119. Lamiaceae
      4 Leaves alternate; stamens 5; corolla usually ± regular .... 117. Boraginaceae
1 Styles as many as the ovaries, sometimes very short, with the stigma sessile or nearly so.
  5 Flowers irregular.
    6 Stamens 5; leaves variously lobed *(Heuchera)* .... 69. Saxifragaceae
    6 Stamens numerous; leaves deeply divided or compound *(Aconitum, Delphinium)* .... 21. Ranunculaceae
  5 Flowers regular.
    7 Petals united, at least toward the base, into a tubular or funnelform corolla .... 109. Apocynaceae
    7 Petals separate.
      8 Cauline leaves opposite or whorled.
        9 Ovaries 2; petals deeply dissected *(Mitella)* .... 69. Saxifragaceae
        9 Ovaries 3–5; petals ± entire *(Sedum)* .... 68. Crassulaceae
      8 Cauline leaves alternate, or leaves all basal.
        10 Hypanthium none; sepals separate to the base .... 21. Ranunculaceae
        10 Hypanthium apparent as a calyx-tube or an expanded disk bearing the perianth at the margin.
          11 Pistils the same number as the petals, or more.
            12 Leaves compound, or at least distinctly lobed .... 70. Rosaceae
            12 Leaves simple, entire or serrate .... 68. Crassulaceae
          11 Pistils fewer than the petals.
            13 Leaves compound *(Agrimonia)* .... 70. Rosaceae
            13 Leaves simple, or but shallowly lobed .... 69. Saxifragaceae

KEY I. Dicot Herbs with Perfect Flowers; Perianth Whorls 2; Pistil 1; Ovary Inferior (Family 136. Asteraceae, the sunflower family, is reached in the General Key.)

1 Stamens more numerous than the petals.
  2 Stamens more than 2× as many as the petals, often 20 or more .... 55. Loasaceae
  2 Stamens 2× as many as the petals, commonly 8 or 10.
    3 Style clearly 1 (the stigma may be divided) .... 79. Onagraceae
    3 Styles 2 or more.
      4 Styles 2 .... 69. Saxifragaceae
      4 Styles 3 or more.
        5 Leaves ternately compound .... 133. Adoxaceae
        5 Leaves simple .... 40. Portulacaceae
1 Stamens as many as the petals or fewer.
  6 Petals separate.
    7 Petals 2; stamens 2 *(Circaea)* .... 79. Onagraceae
    7 Petals and stamens each more than 2.
      8 Petals 4; stamens 4.
        9 Flowers clustered in heads, subtended by large white bracts ...... 81. Cornaceae
        9 Flowers separate and not subtended by white bracts *(Ludwigia)* .... 79. Onagraceae
      8 Petals 5.
        10 Flowers in panicles or cymes, or solitary on scapelike peduncles; leaves simple .... 69. Saxifragaceae

                10 Flowers in racemes, heads, umbels, or solitary at the base of the petioles.
                    11 Flowers in slender terminal racemes; leaves pinnately compound
                        (Agrimonia) .... 70. Rosaceae
                    11 Flowers in heads, umbels, or solitary.
                        12 Fruit dry, separating at maturity into 2 mericarps; styles
                            2 .... 106. Apiaceae
                        12 Fruit berry or drupelike, styles 2–5, ± fused or
                            separate .... 105. Araliaceae
        6 Petals fused, forming a gamopetalous corolla.
            13 Cauline leaves alternate.
                14 Corolla irregular, split along the upper side; anthers fused into a sleeve or collar
                    around the style *(Lobelia)* .... 130. Campanulaceae
                14 Corolla regular.
                    15 Corolla 2–3 mm across; flowers in terminal racemes
                        (Samolus) .... 65. Primulaceae
                    15 Corolla clearly more than 3 mm across, or flowers ±
                        clustered .... 130. Campanulaceae
            13 Cauline leaves opposite or whorled, or leaves all basal.
                16 Leaves in whorls of 4–8 .... 131. Rubiaceae
                16 Leaves opposite or basal.
                    17 Stipules present .... 131. Rubiaceae
                    17 Stipules absent.
                        18 Stamens 3 .... 134. Valerianaceae
                        18 Stamens 4 or 5.
                            19 Flowers in dense, elongate terminal heads .... 135. Dipsacaceae
                            19 Flowers in pairs, terminating a short scape *(Linnaea)*, or axillary and
                                sessile *(Triosteum)* .... 132. Caprifoliaceae

                    Key J. Dicot Herbs with Perfect Flowers; Perianth Whorls 2; Pistil 1;
                        Ovary Superior; Stamens More Numerous Than the Petals
1 Flowers clearly irregular.
    2 Sepals, or at least some of them, petallike or prolonged into a spur.
        3 Spur absent; stamens 6 or 8; leaves ± entire .... 92. Polygalaceae
        3 One of the sepals prolonged into a spur or sac.
            4 Stamens 5; petals 3, 2 of them bi-lobed; leaves crenate to
                serrate .... 104. Balsaminaceae
            4 Stamens numerous; leaves deeply parted *(Delphinium)* .... 21. Ranunculaceae
    2 Sepals not at all petallike, normally green.
        5 Sepals 2, separate, usually early-deciduous; leaves dissected or
            compound .... 25. Fumariaceae
        5 Sepals 4 or more, often ± united.
            6 Lower 2 petals fused or closely appressed, enclosing the 10(9) stamens; leaves sim-
                ple or compound .... 74. Fabaceae
            6 Petals separate, or sometimes only 1, not enclosing the stamens.
                7 Flowers in dense terminal spikes, each with 1 petal, 5 anther-bearing stamens
                    and 5 petallike staminodia (some spp. of *Dalea*) .... 74. Fabaceae
                7 Inflorescences various, but never of dense spikes.
                    8 Leaves compound.
                        9 Petals and sepals each 4.
                            10 Plant climbing by tendrils; leaflets of the principal leaves more than
                                3 *(Cardiospermum)* .... 95. Sapindaceae
                            10 Plants erect, leaflets 3 .... 57. Capparaceae
                        9 Petals and sepals each 5.
                            11 Leaves palmately compound *(Geranium)* .... 103. Geraniaceae
                            11 Leaves pinnately or bipinnately compound *(Cassia,
                                Hoffmanseggia)* .... 73. Caesalpiniaceae
                    8 Leaves simple, entire to variously lobed or parted.

                12 Upper petals larger than the lower .... 59. Resedaceae
                12 Lower petals larger than the upper.
                    13 Styles 2 *(Saxifraga)* .... 69. Saxifragaceae
                    13 Styles 1, but sometimes branching at the apex.
                        14 Leaves palmately parted *(Geranium)* .... 103. Geraniaceae
                        14 Leaves entire or nearly so .... 77. Lythraceae
1 Flowers regular or nearly so.
    15 Sepals 2.
        16 Leaves entire, ± thick and succulent .... 40. Portulacaceae
        16 Leaves variously serrate, lobed or compound.
            17 Flowers regular; petals separate and spreading; juice milky or
                colored .... 24. Papaveraceae
            17 Flowers closed and flattened; petals ± united; juice watery .... 25. Fumariaceae
    15 Sepals 3 or more.
        18 Stamens more than 2× the number of petals.
            19 Leaves simple.
                20 Style 1 .... 50. Cistaceae
                20 Styles 2 or more.
                    21 Leaves alternate; flowers of various colors .... 48. Malvaceae
                    21 Leaves opposite; flowers yellow .... 46. Clusiaceae
            19 Leaves compound.
                22 Leaves once-palmately compound *(Polanisia)* .... 57. Capparaceae
                22 Leaves 2–3× compound.
                    23 Flowers in slender, open racemes *(Actaea)* .... 21. Ranunculaceae
                    23 Flowers in dense heads or spikelike racemes *(Acacia,
                        Schrankia)* .... 72. Mimosaceae
        18 Stamens 2× as many as the petals or fewer.
            24 Stamens exactly 2× the number of petals.
                25 Petals 3; flowers large, solitary, and terminal; leaves in a single whorl of 3
                    (*Trillium*, a monocot with net-veined leaves, keyed here for
                    convenience) .... 155. Liliaceae
                25 Petals 4 or more.
                    26 Sepals and petals 6 or more.
                        27 Leaves a single, opposite pair, peltate, with a large solitary flower be-
                            tween them *(Podophyllum)* .... 22. Berberidaceae
                        27 Leaves more than 2; flowers normally several–many .... 77. Lythraceae
                    26 Sepals and petals 4 or 5.
                        28 Leaves compound or divided nearly to the base.
                            29 Leaves opposite.
                                30 Flowers yellow; leaves pinnately compound ... 101. Zygophyllaceae
                                30 Flowers pink to purple or whitish; leaves palmately dissected
                                    (Geranium) .... 103. Geraniaceae
                            29 Leaves alternate.
                                31 Styles 5; leaves 3-foliate .... 102. Oxalidaceae
                                31 Style 1.
                                    32 Leaves palmately compound .... 57. Capparaceae
                                    32 Leaves pinnately or bipinnately compound.
                                        33 Flowers in dense axillary heads
                                            (Schrankia) .... 72. Mimosaceae
                                        33 Flowers in axillary racemes *(Cassia, Hoffman-
                                            seggia)* .... 73. Caesalpiniaceae
                        28 Leaves simple, entire or dentate to shallowly lobed.
                            34 Style 1.
                                35 Hypanthium present as a disk-shaped, expanded receptacle, bear-
                                    ing the petals and sepals on or near the margin.
                                    36 Anthers opening longitudinally .... 77. Lythraceae
                                    36 Anthers opening by terminal pores ...... 80. Melastomataceae
                                35 Hypanthium none.

           37 Sepals obviously not alike in size and shape, or 2 pairs of
              them partly united .... 50. Cistaceae
           37 Sepals essentially equal in size and shape .... 61. Pyrolaceae
        34 Styles 2 or more.
           38 Ovary lobed; each lobe with a separate style; leaves alternate.
              39 Styles 2; leaves chiefly basal, the cauline ones few and
                 reduced upwards *(Mitella, Saxifraga)* .... 69. Saxifragaceae
              39 Styles 4 or 5; leaves chiefly cauline .... 68. Crassulaceae
           38 Ovary not lobed, styles arising together from the summit; leaves
              opposite.
              40 Leaves serrulate; flowers minute, sessile or nearly so in the ax-
                 ils *(Bergia)* .... 45. Elatinaceae
              40 Leaves entire or but obscurely serrulate.
                 41 Flowers yellow .... 46. Clusiaceae
                 41 Flowers white to pink or purplish ..... 42. Caryophyllaceae
  24 Stamens more than the number of petals, but fewer than 2 × the number.
     42 Style 1; leaves alternate or opposite.
        43 Sepals and petals each 6; stamens 11 *(Cuphea)* .... 77. Lythraceae
        43 Sepals and petals 5 or fewer.
           44 Sepals or calyx lobes 5; petals 5 or 3.
              45 Flowers in dense heads; leaves bipinnately compound; stems spiny
                 (Schrankia) .... 72. Mimosaceae
              45 Flowers in open inflorescences; leaves once-pinnately compound;
                 stems unarmed *(Cassia)* .... 73. Caesalpiniaceae
           44 Sepals 4, petals 4.
              46 Stamens 6, 4 long and 2 short .... 58. Brassicaceae
              46 Stamens 6 or more, but not of 2 distinct lengths ... 57. Capparaceae
     42 Styles 2–5; leaves opposite or whorled.
        47 Flowers yellow .... 46. Clusiaceae
        47 Flowers other than yellow.
           48 Stamens 9, in 3 fascicles of 3 stamens, alternating with 3 glands
              (Triadenum) .... 46. Clusiaceae
           48 Stamens separate and not in fascicles .... 42. Caryophyllaceae

Key K. Dicot Herbs with Perfect Flowers; Perianth Whorls 2; Pistil 1; Ovary Superior; Stamens as Many as the Petals or Fewer; Petals Separate

1 Leaves compound or deeply dissected.
  2 Flowers borne on leafless scapes .... 51. Violaceae
  2 Flowers borne on leafy or bracteate stems, not on scapes.
     3 Flowers sessile or nearly so in the leaf axils *(Cassia)* .... 73. Caesalpiniaceae
     3 Flowers variously arranged in pedunculate clusters, spikes, or heads.
        4 Flowers in dense pedunculate axillary heads; leaves bipinnately
           compound .... 72. Mimosaceae
        4 Flowers in terminal spikes or open clusters; leaves other than bipinnately
           compound.
           5 Flowers in dense terminal spikes; leaves once-pinnately compound
              (Dalea) .... 74. Fabaceae
           5 Flowers in loose open clusters.
              6 Petals and stamens each 6; leaves ternately compound
                 (Caulophyllum) .... 22. Berberidaceae
              6 Petals and stamens each 5; leaves deeply dissected .... 103. Geraniaceae
1 Leaves entire, or at most lobed.
  7 Leaves opposite.
     8 Sepals 2 or 3; petals 2–5.
        9 Sepals 2; petals 3 or 5 .... 40. Portulacaceae
        9 Sepals and petals each 2 or 3 .... 45. Elatinaceae
     8 Sepals and petals each 4–6, rarely more.

                10 Leaves deeply palmately lobed *(Geranium)* .... 103. Geraniaceae
                10 Leaves entire or dentate-serrate.
                    11 Style 1.
                        12 Hypanthium well developed, ± dish-shaped or short-tubular, with the
                            petals and sepals borne on or near the margin .... 77. Lythraceae
                        12 Hypanthium none.
                            13 Stamens alternate with the petals .... 108. Gentianaceae
                            13 Stamens opposite the petals, i.e., directly in front of the
                                petals .... 65. Primulaceae
                    11 Styles 2–5.
                        14 Leaves glandular-serrate *(Bergia)* .... 45. Elatinaceae
                        14 Leaves entire.
                            15 Ovary and capsule with 4–5 locules .... 91. Linaceae
                            15 Ovary and capsule with 1 locule.
                                16 Flowers yellow .... 46. Clusiaceae
                                16 Flowers pink to purple, or white.
                                    17 Petals clearly all separate .... 42. Caryophyllaceae
                                    17 Petals ± united at the base *(Sabatia)* .... 108. Gentianaceae
    7 Leaves alternate or all basal.
        18 Leaves shallowly to deeply palmately lobed.
            19 Flowers irregular, with 1 petal prolonged into a basal spur .... 51. Violaceae
            19 Flowers regular.
                20 Flowers with a conspicuous fringed corona; vines .... 53. Passifloraceae
                20 Flowers without a corona; erect herbs *(Heuchera)* .... 69. Saxifragaceae
        18 Leaves entire to variously dentate or serrate, or pinnately lobed.
            21 Styles 2 or more.
                22 Leaves cauline. .... 91. Linaceae
                22 Leaves basal or mostly so.
                    23 Leaves densely stipitate-glandular .... 49. Droseraceae
                    23 Leaves glabrous or nearly so .... 44. Plumbaginaceae
            21 Style 1, sometimes very short and the stigmatic area essentially sessile.
                24 Hypanthium well developed, ± tubular, with the sepals and petals borne at or
                    near the margin .... 77. Lythraceae
                24 Hypanthium absent, or very short and inconspicuous.
                    25 Flowers irregular.
                        26 Lowermost petal prolonged into a spur .... 51. Violaceae
                        26 Lowermost 2 petals small, thickish and glandlike, not at all
                            spurlike .... 93. Krameriaceae
                    25 Flowers regular.
                        27 Petals and sepals each 4 *(Lepidium)* .... 58. Brassicaceae
                        27 Petals and sepals each 5.
                            28 Leaves pinnately lobed *(Erodium)* .... 103. Geraniaceae
                            28 Leaves entire or ± dentate.
                                29 Flowers solitary, terminal *(Parnassia)* .... 69. Saxifragaceae
                                29 Flowers axillary or, if terminal, in an umbel or raceme.
                                    30 Flowers axillary *(Cubelium)* .... 51. Violaceae
                                    30 Flowers in a terminal umbel or raceme .... 65. Primulaceae

Key L. Dicot Herbs with Perfect Flowers; Perianth Whorls 2; Pistil 1; Ovary Superior; Stamens as Many as the Petals (Corolla Lobes) or Fewer; Petals Fused

1 Corolla regular *and* the stamens as many as the corolla lobes (petals).
    2 Leaves all basal and covered with stipitate-glandular hairs .... 49. Droseraceae
    2 Leaves otherwise.
        3 Leaves bipinnately compound; flowers in dense, pedunculate axillary heads; stamens
            exserted beyond the corolla *(Desmanthus)* .... 72. Mimosaceae
        3 Without this combination of characters.

- 4 Stamens opposite the lobes of the corolla, i.e., directly in front of each petal.
  - 5 Style 1 .... 65. Primulaceae
  - 5 Styles 5 .... 44. Plumbaginaceae
- 4 Stamens alternate with the lobes of the corolla.
  - 6 Ovary deeply lobed, superficially resembling 2 or 4 separate ovaries, separating into 4 fruits at maturity.
    - 7 Leaves alternate .... 117. Boraginaceae
    - 7 Leaves opposite .... 119. Lamiaceae
  - 6 Ovary not conspicuously lobed.
    - 8 Ovary with 1 locule; fruit a capsule.
      - 9 Styles 2 or, if 1, then deeply parted .... 116. Hydrophyllaceae
      - 9 Style 1, sometimes very short and the stigma nearly sessile.
        - 10 Leaves simple, opposite or whorled .... 108. Gentianaceae
        - 10 Leaves trifoliate, basal .... 114. Menyanthaceae
    - 8 Ovary with 2, 3 or 4 locules.
      - 11 Ovary with 3 locules.
        - 12 Plants twining or scrambling; leaves alternate ... 112. Convolvulaceae
        - 12 Plants erect or diffuse; or, if prostrate, then not twining or scrambling .... 115. Polemoniaceae
      - 11 Ovary with 2 or 4 locules.
        - 13 Styles 2, or if 1, then deeply 2-parted, or styles 2 and each 2-parted.
          - 14 Corolla limb entire or but weakly 5-lobed ... 112. Convolvulaceae
          - 14 Corolla limb distinctly lobed or parted .... 115. Polemoniaceae
        - 13 Style 1, or 1 with 2 stigmas and at most obscurely parted at the apex.
          - 15 Leaves opposite or basal.
            - 16 Corolla scarious, 4-lobed .... 122. Plantaginaceae
            - 16 Corolla petaloid and herbaceous, 4- or 5-lobed.
              - 17 Leaves connected at the base by prominent triangular stipules; flowers in branching helicoid cymes .... 107. Loganiaceae
              - 17 Stipules absent; flowers in axillary heads or spikes *(Phyla)* .... 118. Verbenaceae
          - 15 Leaves alternate.
            - 18 Fruit a many-seeded berry .... 111. Solanaceae
            - 18 Fruit dry at maturity.
              - 19 Fruit separating into distinct nutlets *(Heliotropium)* .... 117. Boraginaceae
              - 19 Fruit a capsule.
                - 20 Capsule with 4 seeds .... 112. Convolvulaceae
                - 20 Capsule many-seeded.
                  - 21 Corolla rotate or saucer-shaped, not at all prolonged into a tube; at least some of the stamen-filaments villous *(Verbascum)* .... 124. Scrophulariaceae
                  - 21 Corolla variously campanulate, funnelform, tubular or salverform .... 111. Solanaceae
- 1 Corolla irregular *or* the stamens fewer than the corolla lobes.
  - 22 Anther-bearing stamens 5.
    - 23 Ovary deeply 4-lobed around the base of the central style *(Echium, Lycopsis)* .... 117. Boraginaceae
    - 23 Ovary ± evenly expanded and not at all 4-lobed.
      - 24 Corolla rotate or saucer-shaped, and not at all tubular *(Verbascum)* .... 124. Scrophulariaceae
      - 24 Corolla narrowly campanulate to funnelform or salverform .... 111. Solanaceae
  - 22 Anther bearing stamens 2 or 4 (rarely 3); 1 or more sterile stamens (staminodia) may also be present.
    - 25 Corolla prolonged backwards into a spur or sac.

THE KEYS 35

26 Calyx 2-parted and apparently of 2 sepals; rootless mud-plants and aquatics .... 129. Lentibulariaceae
26 Calyx 5-parted; normally rooted terrestrial plants .... 124. Scrophulariaceae
25 Corolla without a basal spur or sac.
27 Leaves alternate or basal.
28 Stamens 2.
29 Flowers borne on leafy stems or bracteate scapes *(Veronica, Besseya)* .... 124. Scrophulariaceae
29 Flowers borne in spikes on naked scapes .... 122. Plantaginaceae
28 Stamens 4.
30 Flowers solitary on scapes *(Limosella)* .... 124. Scrophulariaceae
30 Flowers in terminal racemes or spikes, often conspicuously bracteate.
31 Calyx split to the base along the lower side; corolla 3–5 cm long and nearly as wide .... 127. Pedaliaceae
31 Calyx 4- or 5-parted or bilaterally symmetrical but not split to the base; corolla less than 3–5 cm long and wide .... 124. Scrophulariaceae
27 Leaves opposite or whorled.
32 Ovary deeply 4-lobed, appearing as 4 separate ovaries around the base of a central style, separating into 4 nutlets at maturity; plants often with square stems and aromatic .... 119. Lamiaceae
32 Ovary not prominently 4-lobed.
33 Stamens 2.
34 Corolla scarious .... 122. Plantaginaceae
34 Corolla herbaceous and petaloid.
35 Flowers in terminal racemes or spikes, or solitary or in pairs in the leaf axils .... 124. Scrophulariaceae
35 Flowers in axillary racemes or spikes.
36 Corolla ± regular; corolla limb longer than the tube .... 124. Scrophulariaceae
36 Corolla distinctly bilabiate *(Justicia, Dicliptera)* .... 126. Acanthaceae
33 Stamens 4.
37 Ovary with 1 locule, but becoming falsely several-loculed at maturity by the intrusion of the parietal placentae; fruit an elongate, ± woody capsule with 2 prominent recurved beaks; corolla weakly bilabiate .... 127. Pedaliaceae
37 Ovary with 1 or 2–4 locules, but not as above.
38 Ovary with 1 locule and 1 ovule; fruit an achene; inflorescence a slender, terminal spike with the mature calyx reflexed and closely appressed to the axis *(Phryma)* .... 118. Verbenaceae
38 Ovary with 2–4 locules.
39 Ovules 2–4; maturing into 1-seeded nutlets.
40 Corolla conspicuously bilabiate; flowers not in axillary heads or spikes .... 119. Lamiaceae
40 Corolla weakly bilabiate, or flowers in axillary heads or spikes .... 118. Verbenaceae
39 Ovules 1–many in each locule; fruit a capsule.
41 Capsule longitudinally dehiscent to the base; seeds borne on distinctive, hooked projections from the placentae, seeds usually 2–12 .... 126. Acanthaceae
41 Capsule variously dehiscent but seldom splitting below the middle; seeds not borne on hooks, usually numerous .... 124. Scrophulariaceae

Key M. Monocot Herbs (Obligate aquatics are reached in Key D; Family 160. Orchidaceae is reached in the General Key.)

1 Perianth chaffy, scalelike or of bristles, never petallike in texture or color, or perianth absent.

2 Flowers borne in the axils of chaffy scales and concealed by them, at least when young; styles and stamens protruding at anthesis; perianth none or represented by bristles or small scales.
   3 Leaves usually 2-ranked, with sheaths normally split lengthwise on the side opposite the blade; stems typically round or flat in cross-section but never triangular, typically jointed, with hollow internodes .... 151. Poaceae
   3 Leaves usually 3-ranked, with sheaths continuous around the stem or splitting only in age, or leaves reduced to sheaths only; stems often triangular, typically not jointed, solid .... 150. Cyperaceae
2 Flowers not subtended by chaffy bracts, or if bracteate, the flower parts clearly exceeding the bracts and not at all concealed.
   4 Flowers minute, clustered in elongate or short spikes, surrounded or subtended by a single large white or colored modified leaf (spathe); leaves broad and often net-veined, never grasslike .... 146. Araceae
   4 Flowers or flower clusters not subtended by a spathe; leaves mostly linear or nearly so, sometimes grasslike.
      5 Inflorescence an elongate spike.
         6 Spike apparently lateral on an erect stem *(Acorus)* .... 146. Araceae
         6 Spike terminal.
            7 Spike differentiated into a large, lower portion of pistillate flowers, and a smaller, upper portion of staminate flowers .... 153. Typhaceae
            7 Spike uniform of perfect flowers .... 141. Juncaginaceae
      5 Inflorescence of racemes, loose or open clusters or of subglobose heads.
         8 Flowers unisexual, in globose heads; lower heads pistillate, upper ones staminate .... 152. Sparganiaceae
         8 Flowers perfect, not in heads.
            9 Pistils 3 or 6, ovaries ± fused but separating at maturity .... 140. Scheuchzeriaceae
            9 Pistil 1; fruit a capsule .... 149. Juncaceae
1 Perianth with all or some of the segments petallike in size, texture or color.
  10 Flowers unisexual; plants either monoecious or dioecious.
     11 Perianth clearly differentiated into 3 green sepals and 3 white or pinkish petals; stamens more than 6 .... 138. Alismataceae
     11 Perianth segments essentially alike.
        12 Ovary inferior; stamens 3 or 6; stems twining .... 159. Dioscoreaceae
        12 Ovary superior; stamens 3 or 6.
           13 Leaves narrow, parallel veined or all basal; plants erect .... 155. Liliaceae
           13 Leaves broad and net-veined; stems climbing, leaning or erect .... 158. Smilacaceae
  10 Flowers perfect (perfect flowers rare in Dioscoreaceae).
     14 Ovary inferior.
        15 Stems twining; stamens 3 or 6; perfect flowers rare .... 159. Dioscoreaceae
        15 Stems erect; stamens 3 .... 156. Iridaceae
     14 Ovary superior.
        16 Pistils 3 or more.
           17 Pistils 3; leaves both cauline and basal .... 140. Scheuchzeriaceae
           17 Pistils more than 3, often numerous; leaves all basal.
              18 Perianth of 3 green sepals and 3 white or pinkish petals; fruit an achene; flowers in panicles or umbels .... 138. Alismataceae
              18 Perianth of 6 pink segments, the inner smaller and more deeply colored; fruit a follicle; flowers in umbels .... 137. Butomaceae
        16 Pistil 1; leaves cauline or basal.
           19 Flowers irregular.
              20 Sepals green; petals separate, colored, the lower one much larger than the upper 2; fertile stamens 2 or 3 .... 148. Commelinaceae
              20 Sepals and petals both colored, united at the base; stamens 6 .... 154. Pontederiaceae
           19 Flowers regular.

21 Perianth clearly differentiated into calyx and corolla; stamens 6.
   22 Leaves alternate or basal; flowers in umbels .... 148. Commelinaceae
   22 Leaves in a single whorl of 3; flowers solitary, terminal
      (*Trillium*) .... 155. Liliaceae
21 Perianth not differentiated into distinct calyx and corolla, the sections similar in size, texture, and color.
   23 Stamens 3; perianth segments 6, ± united at the
      base .... 154. Pontederiaceae
   23 Stamens 6; perianth segments 6, separate.
      24 Leaves perennial, elongate, ± leathery .... 157. Agavaceae
      24 Leaves dying each year, herbaceous .... 155. Liliaceae

The Systematic Descriptions

VASCULAR CRYPTOGAMS (Pteridophytes)

1. LYCOPODIACEAE Reichb., the Clubmoss Family

by Ralph E. Brooks

1. LYCOPODIUM L., Clubmoss

1. *Lycopodium dendroideum* Michx., ground pine. Evergreen herb; rhizomes deeply subterranean, widely creeping, and branched; aerial stems scattered, simple below, then upwardly bushy forking with spreading branches. Leaves on the lower portions of the aerial stems strongly divergent; leaves of the lateral branchlets in 2 dorsal, 2 ventral, and 2 lateral ranks and all equally divergent; leaves linear-lanceolate, 3–6 mm long, 0.6–1.2 mm wide, glabrous, apex acuminate, margins entire, base decurrent. Cones 0–10 at the tips of branches, erect, cylindric, 1–6.5 cm long, 4–6 mm diam, sessile; sporophylls yellowish, tightly appressed and imbricate, broadly ovate, apex acuminate, margins inconspicuously erose; sporangia 1-celled, reniform, 1.5–2 mm wide, solitary in the axils of the sporophylls; spores uniform, tetrahedral. (2n = 68) Jul–Nov. Wooded areas along streams or moist ravines; SD: Lawrence; (circumboreal; s in N. Amer. to WV, MI, IL, e MN, sw SD, w MT, & OR). *L. obscurum* of reports — Rydberg.

Reference: Hickey, R. J. 1977. The *Lycopodium obscurum* complex in North America. Amer. Fern J. 67: 45–48.

Ground pine has been collected three times in the BH of SD: by Rydberg in 1892, Over sometime around 1920, and Hammon in 1946. Since then the species has not been collected in our region and it may be extirpated.

Lycopodium annotinum was reported for NE in the Atlas GP but it was discovered that the locality data for the specimen was in error. Dorn (Amer. Fern J. 73: 2. 1983) recently reported *L. annotinum* L. and *L. complanatum* L. from WY: Crook, in the BH.

2. SELAGINELLACEAE Mett., the Spikemoss Family

by Ralph E. Brooks

1. SELAGINELLA Beauv., Spikemoss

Evergreen plants of spreading habit, mosslike; stems prostrate or erect, branching dichotomously or monopodially, leafy throughout; roots adventitious, often borne on descending rhizophores. Leaves simple, scalelike, or less than 4 mm long, sessile, 1-nerved, imbricate and spirally arranged in several ranks. Sporophylls slightly differentiated from the vegetative leaves, green, arranged in terminal cones, the cones compact, sessile, and usually 4-angled; sporangia of two kinds, 1-celled, solitary, axillary; microsporangia containing numerous microspores, usually borne on the upper sporophylls of the cone; megasporangia with 1–4 megaspores, usually borne on the lower sporophylls.

Reference: Tryon, R. M. 1955. *Selaginella rupestris* and its allies. Ann. Missouri Bot. Gard. 42: 1–99.

1 Leaves on erect branches mostly adnate at the base; prostrate stems strongly dorsiventral, the leaves turned sharply upward .... 2. *S. peruviana*
1 Leaves on erect branches decurrent for about 1/5 of the length; prostrate stems radially symmetrical or only slightly dorsiventral.
  2 Upper and under leaves on the same portion of stem essentially equal in length; stems forming open, spreading mats.
    3 Sporophylls 2× wider than the leaves; leaves eciliate or with strongly ascending cilia .... 4. *S. underwoodii*
    3 Sporophylls 4× wider than the leaves; leaves with lax or slightly ascending cilia .... 3. *S. rupestris*
  2 Upper and under leaves on the same portion of stem unequal; stems compactly branched, forming dense mats .... 1. *S. densa*

1. Selaginella densa Rydb. Stems prostrate with short, erect branches, compactly branched, forming dense mats, slightly dorsiventral. Leaves appressed, subulate to linear-lanceolate, 1.7–2.5 mm long, 0.2–0.3 mm wide, under leaves longer than the upper on the same portion of stem but may be nearly equal near the branch ends, glabrous, apex narrowly acute to abruptly setaceous, setae whitish, 1/4× to 3/4× as long as the blade and forming conspicuous tufts at the branch tips, margins eciliate or with cilia ascending or spreading, bases decurrent. Cones to 3.5 cm long, 1–2 mm wide, sharply 4-angled; sporophylls triangular-ovate, keeled, setae short. May–Sep. Dry, sandy or rocky soil in upland prairies or subalpine meadows; nw GP to w OK to w NM; (Alta. to w TX, w to B.C., n CA, UT, & w AZ). *S. engelmannii* Hieron. — Petrik-Ott, Rhodora 77:505. 1975.

2. Selaginella peruviana (Milde) Hieron., Peruvian spikemoss. Stems prostrate with short, erect branches, prostrate portions dorsiventral, the leaves strongly turned upward. Leaves loosely appressed, subulate to lanceolate, 2–3.3 mm long, 2.5–4 mm wide, the under leaves slightly longer than the upper, setae about 1 mm long, margin ciliate, bases adnate. Cones to 2 cm long, 1–2.5 mm wide, sharply acuminate, short ciliate to serrate. Apr–Sep. Exposed bluffs, rocky slopes or ledges of igneous rocks or sandstone; e NM & sw OK; (OK & NM, s to S. Amer.). *S. sheldonii* Maxon — Dittmer, et al. Univ. New Mexico Publ. Biol. 6: 121. 1954.

3. Selaginella rupestris (L.) Spring, rock spikemoss. Stems widely creeping, forming loose, open mats, radially symmetrical, erect branches to 5 cm tall. Leaves appressed, subulate, 1.5–2 mm long, 0.2–0.3 mm wide, under and upper leaves of equal length, glabrous, setae 1/2× to 3/4× as long as the leaves, margins with lax or slightly ascending cilia, bases decurrent. Cones to 5 cm long, 1–1.7 mm wide; sporophylls ovate, about 4× wider than the leaves, keeled setae short, margins ciliate. (n = 6) Apr–Oct. Shaded or exposed sandy soils, sandstone, or granite outcrops; scattered e GP, also w SD, ne WY, n NE, & w OK; (Que. w to Alta., s to cen & e U.S.).

4. Selaginella underwoodii Hieron., Underwood's spikemoss. Stems prostrate with short, erect branches, slightly dorsiventral. Leaves loosely appressed, subulate to linear-lanceolate,

1.5–2.5 mm long, 2-3 mm wide, under and upper leaves of equal length, glabrous, setae 1/4× to 3/4× as long as the blade, margins eciliate or with strongly ascending cilia, bases decurrent. Cones to 2 cm long, 1–1.5 mm wide; sporophylls narrowly ovate, about 2× wider than the leaves, keeled, setae short, margins eciliate or finely dentate. Jun–Sep. Exposed sandstone & granite outcrops; WY: Laramie; CO: Baca; NM: Union; OK: Cimarron; (se WY, s to OK, TX, w to AZ).

3. ISOETACEAE Reichb., the Quillwort Family

by Ralph E. Brooks

1. ISOETES L., Quillwort

Perennial herbs; stock cormlike, distinctly or obscurely 2-lobed in our species, short, bearing numerous branched roots. Leaves arising in a rosette, elongate, cylindric with a broadened base, containing 4 tranversely septate, longitudinal air-channels around the central vascular bundle. Sporangia oblong, sessile, solitary and adaxial on the leaf base; microsporangia borne on the inner leaves; megasporangia borne on the outer leaves; microspores minute, megaspores much larger, spherical with variously sculptured walls. Aquatic, amphibious, or terrestrial in moist places.

Reference: Taylor, W. C., R. H. Mohlenbrock, & J. A. Murray. 1975. The spores and taxonomy of *Isoetes butleri* and *I. melanopoda*. Amer. Fern J. 65: 33–38.

1 Megaspores mostly 480–700 μm in diam .... 1. *I. butleri*
1 Megaspores mostly 240–440 μm in diam .... 2. *I. melanopoda*

1. *Isoetes butleri* Engelm., Butler's quillwort. Leaves arch-spreading 8–20 cm long, 0.4–1.2 mm wide. Sporangia 5–15 mm long, usually with brown lines; megaspores mostly 480–700 μm in diam, tuberculate and densely arachnoid; microspores ca 30 μm long. Mar–Jun. Moist, sandy or calcareous, often shallow soil in prairie swales or around surfacing rocks; e 1/4 KS, s MO, OK: Comanche; (TN w to se KS & e OK).

While Taylor et al. (op. cit.) state that *I. butleri* is found on calcareous soils, ours sometimes grow in sandy soil. The plants are most easily found in the early spring since, by late spring when the spores mature, the sporophylls are concealed by surrounding vegetation or more often have withered away.

2. *Isoetes melanopoda* Gay & Durieu, midland quillwort. Leaves erect, 10–50 cm long, 0.5–3.5 mm wide. Sporangia 5–30 mm long, usually brown spotted; megaspores mostly 240–440 μm in diam, echinate or echinate-tuberculate; microspores ca 25 μm long. Jun–Aug. Submerged in shallow water or swales or ponds, or terrestrial in wet soil; scattered, MN & SD, s to MO & OK; (NJ s to GA, w to MN, SD, & OK).

The dubious f. *pallida* (Engelm.) Fern. occasionally is found mixed in populations of typical plants.

4. EQUISETACEAE Rich., the Horsetail Family

by Ralph E. Brooks

1. EQUISETUM L., Horsetail, Scouring Rush

Rushlike perennial herbs with widely creeping, branching rhizomes; aerial stems mostly erect, articulated, simple or with branches in whorls at the nodes, surfaces with silica deposits in the form of tubercles, transverse ridges, rosettelike aggregations, or uniformly distributed. Leaves reduced to small scales more or less coalesced into a cylindrical sheath at each node. Cones (strobili) terminal on the main stem (rarely on branches), solitary, ellipsoid, of peltate crowded sporangiophores, each bearing several oblong to subglobose, thin-walled sporangia on the undersurface near the margins of the shield; spores numerous, uniform, spherical, minute, each with 4 spirally wound bands (elaters) with enlarged tips.

References: Hauke, R. L. 1963. A taxonomic monograph of the genus *Equisetum* subgenus Hippochaete. Beih. Nova Hedwigia 8: 1–123; Hauke, R. L. 1966. A systematic study of *Equisetum arvense*. Nova Hedwigia 8: 81–109.

1 Aerial stems mostly achlorophyllous, fleshy, unbranched, and fertile.
  2 Teeth of sheaths coherent in 3 or 4 broad lobes .... 9. *E. sylvaticum*
  2 Teeth of sheaths distinct and separate, not coherent in lobes.
    3 Teeth of sheaths firm, dark with a narrow hyaline margin; stems withering after shedding spores .... 1. *E. arvense*
    3 Teeth of sheaths thin, dark in the center with a broad hyaline margin; stems becoming green and branched in age .... 7. *E. pratense*
1 Aerial stems green, firm, simple or branched, and fertile or sterile.
  4 Aerial stems regularly branched in whorls.
    5 Teeth of sheaths coherent in 3 or 4 broad lobes .... 9. *E. sylvaticum*
    5 Teeth of sheaths distinct and separate, not coherent in lobes.
      6 Central cavity of the main stem at least 4/5 the diam of the stem .... 3. *E. fluviatile*
      6 Central cavity of the main stem less than 3/4 the diam of the stem.
        7 Sheaths of branches with 5 or 6 teeth .... 6. *E. palustre*
        7 Sheaths of branches with 3 or 4 teeth.
          8 Teeth of the sheath on the main stem firm and dark; ridges on the main stem smooth or with small silica tubercles .... 1. *E. arvense*
          8 Teeth of the sheaths on the main stem thin and dark in the center with a broad hyaline margin; ridges on the main stem with spicules of silica .... 7. *E. pratense*
  4 Aerial stems unbranched, or sometimes irregularly branched.
    9 Stems flexuous, prostrate or laxly ascending; central cavity of stem lacking; teeth 3 per sheath .... 8. *E. scirpoides*
    9 Stems erect; central cavity usually present; teeth more than 3 per sheath.
      10 Wall of the stem papery thin, easily crushed; central cavity of stem at least 4/5 diam of stem .... 3. *E. fluviatile*
      10 Wall of the stem firm; central cavity of stem 3/4 or less diam of stem.
        11 Teeth of sheaths not articulated, always retained; ridges of stem each with 2 rows of silica tubercles; sheaths green .... 10. *E. variegatum*
        11 Teeth of sheaths articulated, deciduous or not; ridges of stem smooth, with silica tubercles in 1 or 2 rows, or in transverse bars; sheaths green to blackish.
          12 All of the sheaths on the main stem bearing a black band, sheaths about as long as broad; spores normal .... 4. *E. hyemale*
          12 None or only part of the sheaths on the main stem bearing a black band, sheaths mostly longer than broad; spores normal or abortive.

13 Spores normal, well formed; sheaths of the main stem all green or the basal 1 or 2 irregularly banded or dark colored; stems smooth .... 5. *E. laevigatum*
13 Spores abortive; sheaths on at least the lower half of the main stem black banded; stems mostly rough with silica tubercles on the ridges .... 2. *E. x ferrissii*

1. **Equisetum arvense** L., field horsetail. Aerial stems annual, solitary or several in a clump, dimorphic; sterile stems erect or decumbent, branching in whorls, 0.5-6(8) dm tall, ridges smooth or with silica tubercles, sheaths appressed or slightly flared, green to blackish, 2-7 mm long, teeth persistent, brown to blackish (rarely pale green), firm, deltoid to lanceolate, 1-3 mm long, acute to acuminate, branches simple or ramified arch-spreading to sharply ascending, to 20 cm long, 0.6-1.5 mm diam, 3 or 4 angles (often on the same plant); fertile stems erect, precocious and soon withering, unbranched, fleshy, and achlorophyllus, to 20 cm tall, sheaths flaring or inflated, pale or brownish 4-10 mm long, teeth dark with a narrow hyaline margin, lanceolate, 3-9 mm long. Cones 1-3.5 cm long, long peduncled, blunt. (n = 108) Apr-Jun. Moist soils along lakeshores & streams or riverbanks, in low pastures, meadows, woodland thickets, & disturbed areas; common in n GP, scattered in se & rare in sw; (cosmopolitan). *E. arvense* f. *boreale* (Bong.) Klinge-Scoggan; *E. arvense* f. *diffusum* (A.A. Eat.) Clute, f. *ramulosum* (Rupr.) Klinge-Steyermark; *E. arvense* f. *varium* (Milde) Klinge-Fernald.

Many of the named infraspecific taxa of *E. arvense* can be found in the GP, however Hauke (1966) appears to be correct in concluding the forms are mere superficial modifications not deserving of taxonomic recognition.

In the n GP this species sometimes becomes weedy. It is frequently reported to cause the poisoning of horses when mixed in hay.

2. **Equisetum x ferrissii** Clute, intermediate scouring rush. Aerial stems semievergreen, dying back from the tip in winter with the bases persisting, scattered or in groups, unbranched or irregularly branched with age or if injured, to 1.7 dm tall, 4-16 mm diam, ridges numerous and usually with crossbands of silica; cross section with central cavity about 4/5 diam of stem; sheaths appressed or slightly flared, those lowest on the stem with a dark band at the middle, the midsection with a white or tan band, and the uppermost sheaths uniformly green, 6-14 mm long, about 1.5 × longer than wide, teeth numerous, articulated and deciduous or persistant, brownish with a hyaline margin, firm to thin, lanceolate, 2-3.8 mm long, acuminate; branches, if present, similar to the main stem but smaller, sheaths retaining the teeth. Cones 0.6-2.7 cm long, subsessile or short peduncled, apiculate or obtuse, shriveling and drying without shedding spores; spores abortive. Jun-Aug. Wet banks, ditches, pastures, & lakeshores; GP, less frequent in w; (scattered in N. Amer.). *E. intermedium* (A. A. Eat.) Rydb.—Rydberg; *E. hyemale* var. *intermedium* A. A. Eat.—Fernald.

This is a frequently encountered hybrid between *E. hyemale* and *E. laevigatum*. Although the plants are sterile, they have become widely distributed through vegetative propagation and are not always found in the presence of the parental species.

3. **Equisetum fluviatile** L., water horsetail. Aerial stems annual, mostly solitary, unbranched or branching in whorls, to 1 m or more tall, 2-9 mm diam, 12-30 ridged, ridges smooth; cross section with central cavity at least 4/5 diam of stem; sheaths appressed, green, 6-10 mm long, teeth 15-20, persistent, dark brown with a thin hyaline margin, lanceolate, 2-3.5 mm long, acuminate; branches spreading, to 20 cm long, 4-6 angled, ridges smooth or with a row of silica tubercles, 4-6 teeth per sheath. Cones 1-3.5 cm long, peduncled, blunt. (n = 108) May-Jul. In shallow water or on wet ground along lakeshores, backwater areas, & marshes; common in n ND & scattered to n-cen & e NE & IA; (circumboreal, s in N. Amer. to PA, MN, IA, ND, & OR). *E. fluviatile* f. *linnaenum* (Doll) Broun—Scoggan.

4. Equisetum hyemale L., common scouring rush. Aerial stems evergreen, clustered, often in dense stands, unbranched or irregularly branching with age or if injured, to 2 m tall, 2-19 mm diam, ridges numerous, each with 2 rows of silica tubercles, the tubercles often cross-joined to form a series of transverse bars; cross section with central cavity about 3/4 diam of stem; sheaths appressed or flaring at the top, green when young but soon developing a dark brown or blackish ring at the middle and the portion above the ring becoming tan, ashy-gray, or white, 4-18 mm long, length about equaling the stem diam, teeth numerous, articulated and deciduous or persistent, usually coherent for about 1/2 their length, blackish with a wide hyaline margin, firm and stiff to thin, papery, and twisted or wrinkled, lanceolate, 1.8-4.3 mm long, acuminate; branches smaller than the stem but similar, frequently bearing cones. Cones 0.5-2.4 cm long, sessile or short peduncled, apiculate, often overwintering; spores well formed. (n = 108) Apr-Oct. Shaded or exposed, wet or moist areas along rivers, streams, lakes, ditches, railroad embankments, or in woodlands; GP, common in e, less frequent in w; (circumboreal, s in N. Amer. to FL, TX, & CA). *E. affine* Engelm., *E. robustum* A. Br.—Rydberg; *E. praealtum* Raf.—Gates; *E. hyemale* var. *elatum* (Engelm.) Morton—Steyermark; *E. hyemale* var. *pseudohyemale* (Farw.) Morton—Gleason & Cronquist.

Ours is the American var. *affine* (Englem.) A. A. Eat. It readily hybridizes with *E. laevigatum*, producing the sterile hybrid *E.* × *ferrissii*.

5. Equisetum laevigatum A. Br., smooth scouring rush. Aerial stems annual, scattered or in colonies, unbranched except in old stems, injured stems or those in harsh environments, to 1.4 dm tall, 2-8 mm diam, ridges numerous, usually smooth; cross section with central cavity 2/3 to 3/4 diam of stem; sheaths gradually flared from the base, green or the lower sheaths sometimes black banded or totally black, 4-15 mm long, 1.5-2 × longer than wide, teeth numerous, articulated, and promptly deciduous leaving a ring of persistent dark tooth bases, blackish with a hyaline margin, thin, lanceolate, 1-2.5 mm long, acuminate; branches similar to the main stem but smaller, often bearing cones. Cones 0.5-2 cm long, sessile or short peduncled, apiculate or obtuse; spores well formed. May-Aug. Sandy riverbanks or streams, lakeshores, meadows, pastures, upland prairies, & disturbed sites; GP, common; (scattered in N. Amer.). *E. kansanum* J. H. Schaffn.—Rydberg; *E. kansanum* f. *ramosum* (A. A. Eat.)Broun—Fernald; *E. laevigatum* subsp. *funstonii* (A. A. Eat.) E. Hartman—Hartman, Trans. Kansas Acad. Sci. 61: 144. 1958.

This species readily hybridizes with *E. hyemale*, producing the sterile hybrid *E.* × ferrissii.

6. Equisetum palustre L., marsh horsetail. Aerial stems annual, mostly solitary, branching in whorls, to 1 m or more tall, 3-8 mm diam, 8-10 ridged, ridges smooth or with scattered silica tubercles; cross section with central cavity about 1/4 diam of stem; sheaths slightly flared, green, 9-14 mm long, teeth 8-10, persistent, dark brown on lower sheaths to green with dark tips on the upper sheaths, margin hyaline, lanceolate, 1.5-3 mm long, acuminate; branches simple or sometimes ramified, to 35 cm long, 5-6 angled, ridges with a row of silica tubercles, 5-6 teeth per sheath. Cones 1-2.5 cm long, peduncled, blunt. (n = 108) Jul-Aug. Shaded boggy areas around lakes or along rivers; infrequent in extreme ne GP; MN: Kittsen; ND: Ransom, Richland; (circumboreal, s in N. Amer. to PA, ND, & CA).

7. Equisetum pratense Ehrh., meadow horsetail. Aerial stems annual, solitary or several in a clump, dimorphic; sterile stems branching in whorls, 1.5-4 dm tall, ridges with spicules of silica, sheaths slightly flared, green, 2.5-4 mm long, teeth persistent, hyaline with a thin brown or blackish center, membranaceous, lanceolate, 1.5-3.2 mm long, acuminate; branches slightly ascending or spreading at nearly right angles, to 12 cm long, 0.6-1 mm diam, 3(4)-angled; fertile stems precocious, at first unbranched, fleshy, and achlorophyllous, sheaths larger and flaring, teeth longer but similar to those of the sterile stems, with age stems branching, becoming green, the cone deciduous in the process or shortly afterwards. Cones 1–2.5 cm long, long peduncled, blunt. (n = 108) May–Jun. Moist or wet soil in woods or thickets, shaded stream banks; infrequent in extreme ne GP & BH; (circumboreal, s in N. Amer. to NJ, IA, SD, & B.C.).

8. *Equisetum scirpoides* Michx., dwarf scouring rush. Aerial stems evergreen, caespitose and prostrate or laxly ascending, unbranched 5–16 cm long, 0.5 mm diam, flexuous and firm, 6-ridged, each ridge with a row of silica tubercles; cross section with central cavity lacking; sheaths short, flaring, green at the base and becoming black above, teeth 3, persistent hyaline with a blackish central portion, deltoid, 1.5–2 mm long, apex apiculate and deciduous, about 1 mm long. Cones 2–3 mm long, subsessile or short peduncled, apiculate, often overwintering. (n = 108) May–Aug. Low, wet rich areas in woods; rare in BH; SD: Lawrence; WY: Crook; (circumboreal, s in N. Amer. to NY, MI, MN, SD, & WA).

9. *Equisetum sylvaticum* L., wood horsetail. Aerial stems annual, mostly solitary, dimorphic; sterile stems branching in whorls, 2–8 dm tall, sheaths inflated and flaring upward, green at the base and tan above, 0.7–2.3 cm long, teeth numerous and coherent in 3 or 4 broad lobes, persistent, castaneous, membranaceous; branches numerous, simple or ramified, spreading and recurving, to 15 cm long, 3 teeth per sheath; fertile stems precocious, at first unbranched, fleshy and achlorophyllous, later green, developing branches and persisting after the cone has fallen, slightly shorter than the sterile stems. Cones 0.8–2.3 cm long, long peduncled, blunt, deciduous. (n = 108) May–Jul. Wet shaded stream banks & boggy woods; locally common in extreme ne GP & BH; (circumboreal, s in N. Amer. to NC, KY, IA, SD, MT, & WA).

10. *Equisetum variegatum* Schleich., variegated scouring rush. Aerial stems evergreen, clustered, unbranched, 6–40 cm tall, 0.5–3 mm diam, 6–12 ridged, each ridge with 2 rows of silica tubercles; cross section with central cavity well developed to absent in narrow stems; sheaths slightly flared, green with a black rim, teeth 5–10, persistent, hyaline with a brown central portion, rounded, 1–2 mm long, apex apiculate and deciduous, about 1 mm long. Cones 3–5 mm long, short peduncled, usually apiculate, often overwintering. (n = 108) Jun–Sep. Wet lakeshores, on stream banks, swampy ground; scattered in n MN; (circumboreal, s in N. Amer. to PA, IL, MN, CO, UT, & OR).

Reports of this species have previously been made for KS, NE, ND, OK, & SD, but all were based on depauperate plants of *E. laevigatum*.

5. OPHIOGLOSSACEAE Presl, the Adder's-tongue Family

by Ralph E. Brooks

Perennial, terrestrial, soft fleshy herbs without sclerenchymatous tissue; stem short, upright, and subterranean; roots stout or narrow, fleshy, mycorrhizal. Leaves 1 or occasionally several, divided into a sterile leaflike segment (sterile blade) and a lateral fertile segment (fertile spike) joined on a common stalk, leaf segments erect or reflexed in vernation, not circinate; common stalk enlarged at the base and partially or wholly enclosing the bud; fertile spike solitary and stalked with the sporangia arranged in a terminal spike or panicle. Sporangia large, 0.5–1 mm in diam, borne in 2 rows on the spike or fertile branches, bivalvate, lacking an annulus, thick walled, not indusiate; spores numerous in each sporangium, yellowish. Gametophyte subterranean, mycorrhizal.

Reference: Clausen, R. T. 1938. A monograph of the Ophioglossaceae. Mem. Torrey Bot. Club 19: 1–176; Wagner, W. H., Jr., & L. P. Lord, 1956. The morphological and cytological distinctness of *Botrychium minganense* and *B. lunaria* in Michigan. Bull. Torrey Bot. Club 83: 261–280. W. H. Wagner, Jr., kindly reviewed the manuscript for this treatment.

1 Sterile blade lobed or variously dissected, veins free and forking .... 1. *Botrychium*
1 Sterile blade simple, entire, veins reticulate .... 2. *Ophioglossum*

1. BOTRYCHIUM Sw., Grape-fern

Leaves solitary; sterile blade pinnatifid to several times pinnate, veins free and forking; fertile spike stalked, paniculate. Sporangia borne laterally on branches of the fertile spike, separate, globose, sessile or short stalked.

Botrychium lanceolatum (Gmel.) Angstr. has been reported from both SD and ND but neither report is supported by specimen documentation.

1 Sterile blade 2–4 × pinnately compound.
  2 Common stalk usually more than 10 cm long, equal to or longer than the sterile blade, glabrous .... 7. *B. virginianum*
  2 Common stalk less than 8 cm long, much shorter than the sterile blades, pilose, often sparsely so.
    3 Ultimate segments of the sterile blade ovate or elliptic with entire or rarely crenulate margins .... 5. *B. multifidum*
    3 Ultimate segments of the sterile blade lanceolate to narrowly ovate, margins serrate or lacerate into linear often forked teeth .... 1. *B. dissectum*
1 Sterile blade simple to pinnate-pinnatifid.
  4 Fertile spike attached at or near the base of the plant .... 6. *B. simplex*
  4 Fertile spike attached nearer the sterile blade than to the base of the plant.
    5 Sterile blade once-pinnate, pinnae usually imbricate, lunate, decidedly wider than long .... 2. *B. lunaria*
    5 Plants without the above combination of characters.
      6 Middle pinnae obovate or widely so, often incised .... 4. *B. minganense*
      6 Middle pinnae oblong to elliptic, toothed to pinnatified ... 3. *B. matricariifolium*

1. ***Botrychium dissectum*** Spreng., cut-leaved grape-fern. Plants 1.5–6 dm tall, pilose to glabrate; common stalk 1–5(7) cm long, bud hairy and entirely enclosed; sterile blade semievergreen, often persisting into the next season, bright green at first and usually becoming bronze after frost, deltoid, 5–23 cm long, 5–26 cm wide, pinnately or ternately 2–3× compound, long stalked (to 12 cm) and attached at or near ground level; primary divisions deltoid, secondary segments narrowly deltoid to ovate, ultimate segments lanceolate to narrowly ovate, margins serrulate to lacerate into linear, often forked, teeth; fertile stalk 8–25 cm long; fertile panicle 3–15 cm long, 1–3× pinnate. Sporangia sessile, crowded. (n = 45) Aug–Nov. Rich wooded valleys or occasionally rocky ridges; MO & e KS; (N.S. w to MN, s to FL & TX; also AZ).

Two varieties may be recognized in the GP, although W. H. Wagner, University of Michigan (pers. comm.), has suggested these may best be treated as extreme forms.

1a. var. *dissectum*. Ultimate segments lanceolate to narrowly ovate, lacerate into linear, often forked, teeth. Collected once in a rocky, wooded creek valley; KS: Cherokee.

1b. var. *obliquum* (Muhl.) Clute. Ultimate segments lanceolate to narrowly ovate, margins serrulate. Scattered in rich, low woodlands or occasionally on wooded ridges; MO & e KS. *B. obliquum* Muhl.—Rydberg; *B. dissectum* f. *obliquum* (Muhl.) Fern.—Fernald.

2. ***Botrychium lunaria*** (L.) Sw., moonwort. Plants 5–22 cm tall, glabrous; common stalk 2–10 cm long, bud entirely enclosed; sterile blade reflexed in vernation, oblong, 2–5 cm

long, 1–2 cm wide, pinnate, sessile or nearly so, attached near the summit of the plant; pinnae imbricate, lunate, 2–8 mm long, decidedly wider than long, outer margins entire to erose; fertile stalk 1.5–8 cm long; fertile panicle 1–6 cm long, branches short and often distant, erect in vernation. Sporangia sessile or nearly so, crowded. (n = 45) Jun–Aug. Moist, gravelly places, usually disturbed in the past; SD: Lawrence; (circumboreal, s in N. Amer. to NY, MN, w SD, AZ, & NV).

Moonwort has been connected to numerous superstitions in the past. It is said that locks would open from mere contact with the plant, and that it would unshoe horses that stepped on the plants. It also has been regarded as a cure for lunacy.

3. *Botrychium matricariifolium* A. Br., matricary grape-fern. Plants 5–20 cm tall, glabrous; common stalk 3–18 cm long, entirely enclosing the bud; sterile blade reflexed in vernation, oblong to narrowly triangular, 1–3.5 cm long, 0.5–1.5 cm wide, pinnate, the apical segment bifurcate at the tip, pinnae oblong to elliptic, toothed to pinnatifid, sessile or on short stalk to 7 mm long and attached near the summit of the plant; fertile stalk 1–3 cm long; fertile panicle 1–4 cm long, branching compact or loose, erect in vernation. Sporangia usually short stalked. Jun–Aug. Moist woods; MN: Lac Qui Parle; NE: Brown; BH; (circumboreal, in N. Amer. from Newf. w to Alta., s to NC, OH, MN, SD, & CO). *B. neglectum* Wood—Rydberg.

4. *Botrychium minganense* Vict. Plants 5–20 cm tall, glabrous, reflexed in vernation; common stalk 3–11 cm long, entirely enclosing the bud; sterile blade lanceolate to linear-lanceolate, 2–9 cm long, 1–2 cm wide, short stalked and attached near the summit of the plant; upper pinnae widely obovate, entire to shallowly lobed on the outer margin, middle pinnae obovate or widely so, shallowly to deeply incised, lower pinnae roundish to ovate, pinnatifid with 1 or 2 pairs of lobes to pinnate; fertile stalk 2–9 cm long; fertile panicle 1–7 cm long, branching, loose. Sporangia short stalked to sessile. (n = 90) May–Jun. Low sandy prairies; IA: Woodbury; ND: Burke, McHenry; e WY(?); (Lab. w to AK, s to MI, IA, ND, CO, NV). *B. lunaria* f. *minganense* (Vict.) Clute—Fernald; *B. lunaria* var. *minganense* (Vict.) Dole—Britton & Brown.

According to W. H. Wagner (pers. comm.), our plants are not morphologically consistent with those found outside the GP. Ours tend to be somewhere between *B. minganense* and *B. matricariifolium* but closer to the first. Wagner suggests that these plants may, in fact, be a distinct prairie species that has not been recognized.

5. *Botrychium multifidum* (S. G. Gmel.) Trevis, leathery grape-fern. Plants 8–30 cm tall, pilose, often glabrate, reflexed in vernation; common stalk 2–4 cm long, partially enclosing the bud; sterile blade semievergreen, often persisting into a second season, deltoid, 2–8 cm long, 3–10 cm wide, usually wider than long, pinnately or ternately 2–3 × compound, long stalked (up to 8 cm) and attached near ground level; ultimate segments ovate to elliptic, margins entire to rarely crenulate; fertile stalk 6–20 cm long; fertile panicle 1–5 cm long, 1–3 × pinnate. Sporangia crowded, sessile. (n = 45) Aug–Nov. Moist meadows & rich woods; BH & MN; (circumboreal, s in N. Amer. to NC, IA, SD, WY, & CA). *B. multifidum* (S. G. Gmel.) Rupr.—Fernald; *B. silaifolium* Presl—Rydberg.

6. *Botrychium simplex* E. Hitchc., little grape-fern. Plants 3–16 cm tall, glabrous, erect in vernation or the sterile blade sometimes reflexed; common stalk 0.5–2 cm long, entirely enclosing the bud; sterile blade ovate to oblong, 1–4 cm long, 0.5–3 cm wide, pinnatifid with subequal ovate or triangular pinnae, the apical segment undivided at the tip, conspicuously stalked, attached at or near ground level; fertile stalk 2–8 cm long; fertile panicle sparingly branched, 1–3 cm long. (n = 45) Jun–Jul. Moist, open woodlands; SD: Lawrence; (circumboreal, s in N. Amer. to NC, IN, NM, & CA).

7. *Botrychium virginianum* (L.) Sw., rattlesnake fern. Plants 1-8 dm tall, reflexed in vernation; common stalk 7-40 cm long, bud hairy and partially enclosed; sterile blade broadly deltoid, 3-25 cm long, 4-40 cm wide, pinnately 3-5 × compound, glabrous, sessile and attached near the summit of the plant; primary divisions narrowly triangular, lanceolate, elliptic or rhombic, secondary divisions oblong to lanceolate to ovate, ultimate divisions elliptic, incised; fertile stalk 3-25 cm long; fertile panicle 1-20 cm long, 1-3 × pinnate, diffuse. Sporangia crowded, short stalked. (n = 92) Apr-Jun (Jul). Moist, rich woods; MN w to ND, s to MO, e to OK, & nw NE; (Lab. w to B.C., s to FL, & CA; Mex.; Eurasia).

2. OPHIOGLOSSUM L., Adder's-tongue

Leaves 1 or 2(3), glabrous, erect in vernation; buds borne at the apex of the rhizome, exposed; common stalk slender, about as long as the sterile blade; sterile blade simple, entire (in our species), veins reticulate; fertile stalk long and slender, arising from the base of the sterile blade; fertile segment a spike with the sporangia arranged in 2 coherent rows. Sporangia subglobose, sessile or embedded.

1 Sterile blades with veins forming principal areoles surrounding secondary areoles, apex acute or apiculate .... 1. *O. engelmannii*
1 Sterile blades with veins forming principal areoles only, apex usually rounded .... 2. *O. vulgatum*

1. *Ophioglossum engelmannii* Prantl, limestone adder's-tongue. Leaves 1 or 2(3) per plant, 5-25 cm tall, basal sheath usually persistent, brown, 3-7 mm long; sterile blade ovate to narrowly elliptic, 2.5-10 cm long, 1.5-3.5 cm wide; veins reticulate with primary and secondary areoles, apex acute or apiculate, base attenuate to sometimes rounded; fertile stalk 3.5-15 cm long, usually exceeding the blade in length; fertile spike 1-3.5 cm long. Apr-Jun, occasionally Sep-Oct. Shallow soil over limestone or rarely sandstone in prairies, pastures, or open woods; e KS, MO, OK; TX: Bailey; (WV w to OH, KS, & AR, s to FL & Mex.).

2. *Ophioglossum vulgatum* L. Leaves 1-2 per plant, 10-25 cm tall; basal sheath membranaceous, ephemeral; sterile blade elliptic ovate, 2.5-9 cm long, 1-3 cm wide, veins reticulate and forming only principal areoles with free veins within, apex rounded, base attenuate or sometimes rounded; fertile stalk 7-17 cm long, usually exceeding the blade in length; fertile spike 1.5-4 cm long. (n = 240-260) Jun-Sep.

Two varieties occur in the GP.

2a. var. *pseudopodum* (Blake) Farw., northern adder's-tongue. Sterile blade usually elliptic, widest near the middle, gradually tapering at the base. Low, usually sandy meadows, marshes; ND: Richland; NE: Cherry; (e Can., w to s AK, s to VA, IN, IL, NE, & CA).

2b. var. *pycnostichum* Fern., southern adder's-tongue. Sterile blade usually ovate, widest in the lower half, abruptly tapering at the base. Wooded lowlands; KS: Crawford; (GA w to e TX, n to NJ, s MI, IL, MO, & se KS).

6. OSMUNDACEAE R. Br., the Royal Fern Family

By Ralph E. Brooks

1. OSMUNDA L., Royal Fern

1. *Osmunda regalis* L. var. ***spectablilis*** (Willd.) A. Gray, royal fern. Perennial herb; rhizomes stout, short-creeping, thickly beset with old petiole bases and roots. Leaves

2-pinnate, 5–13 dm tall, numerous, in a dense crown; blade broadly ovate to lance-ovate, wholly sterile or divided into sterile and fertile parts; sterile pinnae in 4–7 pairs, subopposite, the lowermost slightly reduced; pinnules alternate 2–10 per side, oblong to lance-oblong, 2.5–7 cm long, 0.7–1.5 cm wide, glabrous or a few lanceolate blackish scales at the base of the midvein, apex obtuse to acute, margins subentire to serrulate, base oblique or sometimes auricled, subsessile; petiole about equaling blade in length, glabrous, winged (stipulate) at the base. Fertile pinnae borne at the tip of some blades, 3–5 pairs, pinnules strongly contracted to 2 cm long, 2–3 mm wide, densely crowded with sporangia, green at first and turning brown later; sporangia naked, globose, mostly pediceled, the annulus poorly developed, opening into 2 valves by a longitudinal slit. (n = 22) Apr–Jul. Rich wooded ravines & stream banks; extreme se GP; (Newf. s to FL, w to Sask. & TX).

Rydberg (1932) reported *O. regalis* for NE, however this and later reports for the state are erroneous, being based on inaccurate label data or cultivated specimens, which are not uncommon in our region.

7. POLYPODIACEAE S. F. Gray, the True Fern Family

by Ralph E. Brooks

Rhizomes creeping or erect, branched or not, often clothed with scales. Leaves circinate in vernation, erect or pendent, fertile and sterile leaves uniform or dimorphic, exstipulate; blade simple or several times compound, veins numerous, pinnate and free or sometimes anastomosing. Sporangia grouped in sori, long stalked, provided with an incomplete vertical annulus and opening transversely, elastically dehiscent through the action of the hygroscopic cells of the annulus; sori rotund or elongate and borne upon the veins on the lower surface of the blades or borne in lines or clusters near the margins, occasionally borne on wholly fertile leaves or segments of sterile leaves; sori naked or covered by an indusium that develops from the vein or modified leaf margin; gametophytes green.

Rydberg (1932) reported *Cryptogramma acrostichoides* R. Br. (*C. crispa* R. Br. subsp. *acrostichoides* (R. Br.) Hulten) from NE but the record is apparently based on erroneous data.

1 Sterile and fertile leaves distinctly dimorphic; sporgania borne in berrylike or cylindrical modified pinnae.
  2 Rhizomes widely creeping; sterile blade pinnatifid .... 10. *Onoclea*
  2 Rhizomes nearly erect (stolons may be present); sterile blade pinnate-pinnatifid .... 8. *Matteuccia*
1 Leaves uniform or nearly so.
  3 Blade simple or pinnatifid.
    4 Blade simple or divided into several linear segments, to 2.5 cm long, barely wider than the petiole .... 2. *Asplenium*
    4 Blade more than 2.5 cm long, many times wider than the petiole.
      5 Blade simple .... 2. *Asplenium*
      5 Blade pinnatifid.
        6 Veins anastomosing .... 10. *Onoclea*
        6 Veins free .... 12. *Polypodium*
  3 Blade at least 1-pinnate.
    7 Pinnae whitish or yellowish waxy on the lower side .... 9. *Notholaena*
    7 Pinnae glabrous, scaly or hairy on the lower surface.
      8 Blade broadly triangular.
        9 Sori continuous along the margins of the ultimate segments ..... 14 *Pteridium*

       9 Sori discrete, round, median to submarginal.
        10 Rachis distinctly winged, pilose .... 15. *Thelypteris*
        10 Rachis never winged, glabrous .... 7. *Gymnocarpium*
 8 Blade linear, lanceolate, or otherwise but never broadly triangular.
    11 Blade 1-pinnate.
       12 Pinnae about as wide as they are long, densely scaly at least on the lower
           surface .... 9. *Notholaena*
       12 Pinnae mostly several times longer than wide, if shorter then glabrous on both
           surfaces.
          13 Sori continuous along the reflexed or revolute margin of the pinnae 11.*Pellaea*
          13 Sori discrete, median on the pinnae.
             14 Sori round .... 13. *Polystichum*
             14 Sori elongate.
                15 Petiole dark and shining, at least at the base; sori
                    straight .... 2. *Asplenium*
                15 Petiole stramineous, not shining; sori curved or hooked
                    distally .... 3. *Athyrium*
    11 Blade pinnate-pinnatifid or 2-pinnate.
       16 Petiole divided at its tip into 2 arch-spreading rachises .... 1. *Adiantum*
       16 Petiole unbranched.
          17 Blade widest above its middle .... 8. *Matteuccia*
          17 Blade widest at or below its middle.
             18 Sori continuous along the margins of the segments.
                19 Pinnules jointed at the base .... 11. *Pellaea*
                19 Pinnules and segments not jointed at the base .... 4. *Cheilanthes*
             18 Sori discrete.
                20 Blade densely hairy or scaly, or both, on one or both surfaces.
                   21 Petiole jointed .... 16. *Woodsia*
                   21 Petiole continuous .... 4. *Cheilanthes*
                20 Blade lacking indument or ramentum of any kind or if present only sparse-
                    ly so and mostly limited to the veins.
                   22 Indusium reniform, attached at the sinus.
                      23 Rhizome compact or only short creeping; rachis
                          glabrous .... 6. *Dryopteris*
                      23 Rhizome slender and extensively creeping; rachis usually at least
                          sparsely pilose .... 15. *Thelypteris*
                   22 Indusium various.
                      24 Indusium attached on the basiscopic side only, elsewhere free, ovate-
                          lanceolate to broadly cup-shaped .... 5. *Cystopteris*
                      24 Indusium attached beneath the sorus, early opening and splitting in-
                          to spreading segments .... 16. *Woodsia*

1. ADIANTUM L., Maidenhair Fern

Medium-sized mesophytic, perennial herbs; rhizomes short to long creeping, heavily beset with tan or darker subulate to triangular scales. Leaves erect or lax and drooping; blades 2 or 3-pinnate, glabrous; veins free; petiole slender, purplish, lustrous, glabrous or with a few scattered scales near the base. Sori borne on the reflexed margins of the lobes on the ultimate segments, true indusia wanting; spores tetrahedral.

1 Blade ovate-lanceolate, much longer than wide .... 1. *A. capillus-veneris*
1 Blade reniform-orbicular, about as long as wide .... 2. *A. pedatum*

1. *Adiantum capillus-veneris* L., Venus'-hair fern. Rhizomes short to long creeping; scales brown, lanceolate, 2-4 mm long. Leaves scattered, lax and usually drooping, 1-6 dm long; blade ovate-lanceolate, 2-3 pinnate, 7-45 cm long, 3-15 cm wide, the rachis continuous with the petiole and bearing pinnae on both sides; pinnae alternate, at least the lower ones pinnate or rarely 2-pinnate; ultimate segments widely obovate to flabellate, 7-28 mm long, 7-30 mm wide, membranaceous, irregularly round lobed and dentate on the ends, base attenuate, short stalked; petiole 3-20 cm long. (n = 30) Jul–Sep. Wet stream banks or rocky ledges, usually around springs; SD: Fall River; se OK & TX; (trop. & temp. areas in the New World & Old World, n in N. Amer. to VA, MO, SD, CO, & B.C.). *A. modestum* Underw. — Rydberg.

2. Adiantum pedatum L., maidenhair fern. Rhizomes short to long creeping; scales tan to dark brown, lanceolate to triangular, 3-8 mm long. Leaves scattered, erect, 1.5-7 dm tall; blades reniform-orbicular, 2 or rarely 3-pinnate, 5-30 cm long, 8-45 cm wide, pinnae produced from the upper sides of the forked, arch-recurving branches of the rachis; pinnae oblong to linear-oblong, 3-30 cm long, 1.5-5 cm wide; ultimate segments variable, obliquely oblong to transversely rhombic, the terminal flabellate, 8-25 mm long, 5-12 mm wide, outer edge variusly lobed and dentate, base cuneate, short stalked; petiole 8-45 cm long. (n = 29) Jun–Sep. Banks & hillsides in rich woods, moist rocky ledges; SD: Lawrence; IA, e NE, MO, e KS, & e OK; (N.S. w to AK, s to GA, e OK, w SD, NV, & CA).

2. ASPLENIUM L., Spleenwort

Small to medium-sized, mostly evergreen, perennial plants; rhizomes erect or short creeping, covered with lanceolate to linear-lanceolate dark scales. Leaves clustered, erect or spreading; blade simple or 1-pinnate; veins free, simple or forked, not reaching the margin; petioles slender, wiry, green to black, glabrous. Sori linear, oblique, the indusia attached lengthwise to the upper side of the fertile veinlet; spore bilateral, with perispore. *Camptosorus* Link — Atlas GP.

1 Blade simple.
  2 Blade 1-3 mm wide .... 4. *A. septentrionale*
  2 Blade more than 5 mm wide .... 3. *A. rhizophyllum*
1 Blade 1-pinnate, more than 5 mm wide, not forked at the apex.
  3 Rachis green throughout .... 6. *A. viride*
  3 Rachis brown, purple, or black at least on the lower half.
    4 Pinnae roundish, broadest above the base, without basal
      auricles .... 5. *A. trichomanes*
    4 Pinnae elongate, broadest at the base, with 1 or 2 basal auricles.
      5 Pinnae alternate, sessile, their bases usually overlapping the
        rachis .... 1. *A. platyneuron*
      5 Pinnae nearly opposite, subsessile or short stalked, their bases not overlapping the
        rachis .... 2. *A. resiliens*

1. *Asplenium platyneuron* (L.) Oakes ex D.C. Eat., ebony spleenwort. Plants 10-50 cm tall; rhizomes short, ascending; scales linear-lanceolate, dark castaneous, 4-5 mm long. Leaves slightly dimorphic, the larger fertile leaves erect from a rosette of shorter, spreading sterile leaves; blade linear-lanceolate to narrowly elliptic-lanceolate, 1-pinnate, 9-45 cm long, 1-5 cm wide, apex subobtuse to acuminate, base gradually tapering; pinnae alternate, oblong to lanceolate, 4-2.5 mm long, 3-6 mm wide, apex obtuse to acute, margins crenate to deeply serrate, base auricled at least on the upper margin, sessile and slightly overlapping the rachis; the lower pinnae greatly reduced and obliquely triangular; veins simple or once forked; petiole reddish-brown to atropurpureous, lustrous, 1-10 cm long, rachis similar. Sori elliptic to linear oblong or confluent with age, nearer the midvein than the margin. (n = 36) May–Sep. Shaded, mosit rocky ledges, hillsides, or occasionally stripmine tailings; MO, e KS & OK; NM: Union(?);

(s Que. & ME, w to WI, IA, & KS, s to FL & TX). *A. platyneuron* f. *serratum* (E.S. Miller) Hoffm.—Steyermark.

2. *Asplenium resiliens* Kunze, black-stem spleenwort. Plants 7–30 cm tall; rhizomes short creeping; rhizome scales blackish, narrow-lanceolate, 1–2 mm long, acuminate, sinuate. Leaves clustered; blade linear-oblong to narrowly linear-oblanceolate, 1-pinnate, 6–25 cm long, 0.8–2.3 cm wide, apex acutish; gradually tapering base; pinnae nearly opposite, linear-oblong, 4–11 mm long, 1.5–5 mm wide, apex rounded or obtuse, margins subentire to crenulate or serrate, base oblique to cuneate, auricled at least on the upper margin, subsessile or short stalked; lower pinnae gradually reduced in size, nearly triangular; veins mostly once forked; petiole blackish, lustrous, 1–6 cm long, rachis similar. Sori elliptic, straight or slightly curved, borne midway between the midvein and margin, confluent in age. (n = 108; 2n = 108) May–Sep. Moist, shaded rocky areas; MO, se KS, OK; CO: Baca; (PA w to se CO & AZ, s to FL & Mex.).

3. *Asplenium rhizophyllum* (L.) Link, walking fern. Evergreen fern with short, erect, and scaly rhizomes. Leaves tufted, arching or nearly prostrate, to 45 cm long; blade lanceolate, 4–30 cm long, 0.7–3 cm wide, subcoriaceous, glabrous, apex long attenuate and often rooting and vegetatively producing a new plant at the tip, margins entire to sinuate, base cordate with rounded auricles or rarely truncate; veins anastomosing near the midvein and free near the margin; petiole green or stramineous with a castaneous and scaly base, flattened, 1–15 cm long. Sori elongate, oblong or linear, parallel or oblique to the midvein, sori whose indusia open towards each other often become confluent, thus appearing as double sori; indusium inconspicuous, barely arching over the sporangia. (n = 36; 2n = 72) May–Sep. Rocky wooded hillsides or ledges, especially limestone; IA, MO, e KS, e OK; (Que. s to GA, w to Ont., e MN, sw IA, e KS, e OK, & MS). *Camptosorus rhizophyllus* (L.) Link—Atlas GP.

4. *Asplenium septentrionale* (L.) Huds., forked spleenwort. Rhizomes short creeping, branched; rhizome scales blackish, linear-lanceolate, 3–5 mm long. Leaves tufted, to 15 cm tall; blade simple or forking into several linear segments, the segments 1–2.5 cm long, 1–2(3) mm wide, apex acuminate or bidentate, margins entire to irregularly spinose; petiole castaneous at the base, green above, 3–13 cm long. Each segment with a single compound sorus about as long as the segment, indusium more or less continuous just within the margin on each side. (2n = 144) Jul–Oct. Shaded rock crevices in canyons or hills; BH; CO: Baca; NM: Union; OK: Cimarron; (w SD, w to OR, s to w OK, NM, & Baja Calif.).

5. *Asplenium trichomanes* L., maidenhair spleenwort. Rhizome short, mostly erect, paleaceous at the apex; scales dark brown with a black median stripe, linear, 1–3 mm long. Leaves clustered, wide spreading, to 20 cm tall; blades narrowly linear-oblong, 1-pinnate, arcuate, 5–18 cm long, 0.6–1.5 mm wide; pinnae subopposite to alternate, orbicular, broadly elliptic or obovate, margins crenulate, base cuneate, subsessile; petiole brown to blackish, 1–5 cm long, glabrous, rachis similar. Sori medial, elliptic to linear-oblong, 2–6 pairs on each pinna. (n = 36, 72, 2n = 72, 144) May–Sep. Moist rock crevices; SD: Pennington; se KS, OK; (circumboreal, s in N. Amer. to GA, TX, AZ, & OR).

6. *Asplenium viride* Huds., green spleenwort. Rhizome short creeping; scales blackish, lanceolate, 1–3 mm long. Leaves semievergreen, clustered, spreading, to 15 cm tall; blade linear-lanceolate or linear-oblong, 1-pinnate, arcuate, 1.5–12 cm long, 0.5–1.3 cm wide; pinnae subopposite to alternate, widely rhombic to obovate, 2–6 mm long, margins coarsely toothed, base cuneate, subsessile or short stalked; petiole castaneous at the base, becoming green above, 0.3–4 cm long, glabrous or with a few brown, glandular hairs; rachis

green. Sori median, oblong or elliptic, 2–6 pairs, confluent with age. (2n = 72) Jun–Sep. Moist, shaded crevices of limestone rocks; SD: Lawrence; (circumboreal, s in N. Amer. to NY, WI, w SD, CO, UT, NV, & WA).

3. ATHYRIUM Roth, Lady-fern

Medium-sized mesophytic perennial herbs; rhizomes erect and stout to slender and short creeping, scaly. Leaves usually clustered; blades elliptic-lanceolate to lanceolate, 1- to 4-pinnate, mostly glabrous; veins free or rarely anastomosing, simple or forked; petioles slender. Sori elongate along one or both sides of the veins; indusium crescentiform or linear, opening toward the midvein, glabrous to pubescent or glandular, entire to fimbriate; spore bilateral.

Athyrium thelypteroides (Michx.) Desv. has been reported for the GP, but those reports were based on cultivated specimens and misdeterminations.

1 Blade pinnate-pinnatifid to 3-pinnate .... 1. *A. filix-femina*
1 Blade 1-pinnate .... 2. *A. pycnocarpon*

1. ***Athyrium filix-femina*** (L.) Roth, lady-fern. Rhizome suberect to horizontal short creeping, beset with numerous old petiole bases, sparsely scaly; scales brown, lanceolate, 4–6 mm long. Leaves 1.5–10 dm tall; blades ovate-lanceolate to elliptic-lanceolate, pinnate-pinnatifid to 3-pinnate, 1–7 dm long, 8–30 cm wide, apex acute to acuminate, base tapering or not, essentially glabrous; pinnae alternate, lanceolate, the larger pinnae 4–15 cm long, 0.8–4 cm wide, short stalked; ultimate segments ovate-lanceolate or elliptic, 4–20 mm long, 2–7 mm wide, apex obtuse to acuminate, margins serrate, incised, or lobulate, the lobes often serrate, bases mostly decurrent on the rachis or subsessile; stramineous or reddened, 5–45 cm long, with large, brownish, lanceolate-ovate scales at least near the base. Sori 3–9 pairs per segment, elongate, about 2 mm long; indusium crescentiform or appearing hooked distally, glabrous or sparsely pubescent, glandular or not, margins entire to fimbriate, often deciduous. (n = 40; 2n = 80) Jun–Sep. Moist thickets, bogs., open woods, & stream banks; cen & n GP; (circumboreal, in most of N. Amer.).

Athyrium filix-femina is a highly variable species in North America. While most of the GP material is assignable to one of three varieties, intermediates between the western var. *cyclosorum* Rupr. and the northeastern var. *angustum* (Willd.) Moore are present in the Black Hills and northeast plains.

1 Indusia margins mostly ciliate.
  2 Indusia eglandular; spores yellow .... 1a. var. *angustum*
  2 Indusia at least sparsely glandular; spores black .... 1b. var. *asplenioides*
1 Indusia margins mostly entire .... 1c. var. *cyclosorum*

1a. var. *angustum* (Willd.) Moore, northern lady-fern. Blade somewhat tapering toward the base. Indusium eglandular, margins short-ciliate; spores yellow. MN, e ND, BH, & e SD., IA, & MO. *A. angustum* (Willd.) Presl—Rydberg; *A. filix-femina* var. *michauxii* (Spreng.) Farw. f. *rubellum* (Giebert) Farw.—Fernald.

1b. var. *asplenioides* (Michx.) Farw., southern lady-fern. Blade usually broadest near the base. Indusium glandular, margins ciliate; spores black. MO & se KS.

1c. var. *cyclosorum* Rupr., western lady-fern. Blade usually tapering toward the base. Indusium eglandular, margins entire; spores yellow. BH.

2. ***Athyrium pycnocarpon*** (Spreng.) Tidest., glade fern. Rhizome short creeping, usually with several coiled leaves at the apex and just back of that a cluster of leaves, scales few or absent. Leaves 2–10 dm tall, slightly dimorphic, the sterile shorter and spreading more than the larger fertile leaves; blade lanceolate to elliptic-lanceolate, 1-pinnate, 15–70 cm long, 10–20 cm wide, apex acuminate, base slightly tapering, glabrous; pinnae alternate, lanceolate to linear-lanceolate, 5–10 cm long, 0.5–1.5 cm wide, apex acuminate, margins crenulate, base broadly cuneate, rarely auricled on the upper side; veins mostly 1-forked, free or rarely anastomosing; petiole stramineous, 5–30 cm long, glabrous except for a tuft of long brownish hairs at the very base, scales scattered, brownish, lanceolate-ovate. Sori elongate, straight; indusium entire, glabrous. (n = 40) Aug–Sep. Mosit, rich wooded ravines; KS: Leavenworth, Wyandotte; MO; (Que. w to Ont. & MN, s to GA, LA, & e KS). *Asplenium pycnocarpon* Spreng.—Rydberg.

4. CHEILANTHES Sw. Lip Fern

Our species small, evergreen, lithophytic, and perennial, mostly xerophytes; rhizomes short to widely creeping, multicipital, scaly. Leaves mostly hygroscopic; blades lanceolate to triangular, pinnate-pinnatifid to 4-pinnate, coriaceous; veins free, thickened at the tips; ultimate segments small, glabrous or pubescent, scaly or not, margins reflexed or revolute; petioles slender, wiry. Sori borne at the vein tips, marginal, roundish, and separate or narrowly confluent, indusium formed by the membranaceous reflexed edge of the ultimate segment; spores tetrahedral.

1 Pinnae glabrous or with a few scattered hairs on the upper surface .... 1. *C. alabamensis*
1 Pinnae pubescent or scaly, or both on at least one surface.
  2 Pinnae pubescent on one or both surfaces, scales absent or rare on the rachises.
    3 Indument of noticeably articulate hairs at least on the stipe and rachis.
      4 Blade pinnate-pinnatifid to 2-pinnate, hirsute below; rhizome scales at least sparingly dentate .... 5. *C. lanosa*
      4 Blade 2-pinnate-pinnatifid to 3-pinnate, tomentose below; rhizome scales entire .... 3. *C. feei*
    3 Pubescence not articulate .... 7. *C. tomentosa*
  2 Pinnae noticeably scaly on at least one surface, pubescent or not.
    5 Pinnae glabrous above, scaly below, pubescence lacking.
      6 Pinnae scales entire .... 4. *C. fendleri*
      6 Pinnae scales long-ciliate .... 8. *C. wootonii*
    5 Pinnae pubescent above, pubescent and scaly below.
      7 Rhizome thick, short creeping; scales on the blade entire .... 2. *C. eatonii*
      7 Rhizome slender, long creeping; scales on the blade long-ciliate . 6. *C. lindheimeri*

1. **Cheilanthes alabamensis** (Buckl.) Kunze, Alabama lip fern. Rhizome short creeping, 1–5 cm long; scales concolorous, ferruginous, linear-lanceolate, 1.5–4 mm long, the tips often entangled, margins entire. Leaves more or less clustered, 8–30 cm tall; blade narrowly elliptic to lanceolate, 2-pinnate or rarely 2-pinnate-pinnatifid, 5–20 cm long, 2–5 cm wide, apex acuminate, glabrous, or rarely with sparse pubescence above, glabrous below, pinnae subopposite to mostly alternate, triangular to lanceolate, 10–35 mm long, 4–15 mm wide, apex acuminate to sometimes obtuse; short stalked or sessile; ultimate segments elliptic to broadly lanceolate, apex acute to obtuse, margins entire to shallowly lobed, reflexed; petiole blackish, 2.5–14 cm long, smooth or with scattered long hairs or scales at the base. (n = 87; 2n = 58,87) Jun–Sep. Dry limestone & sandstone bluffs; KS: Cherokee; OK: Ottawa; (VA w to MO & se KS, s to GA & Mex.).

2. **Cheilanthes eatonii** Baker, Eaton's lip fern. Rhizome thick, short, 1–3 cm long; scales ferruginous with a dark median stripe, linear-lanceolate to narrowly triangular, 3–5 mm long, entire. Leaves clustered, 12–35 cm tall; blade narrowly elliptic to elliptic-lanceolate or lanceolate, 2-pinnate-pinnatifid to 3-pinnate, 6–20 cm long, 2–4 cm wide, apex acute to acuminate, usually whitish tomentose above or becoming glabrescent with age, rusty

tomentose below mixed with lanceolate to narrowly triangular, membranaceous, entire scales especially on the rachis and rachises; pinnae lanceolate to narrowly ovate, 10–18 mm long, 6–12 mm wide, apex acute to obtuse, short stalked; ultimate segments broadly ovate or elliptic, margins entire, revolute; petiole castaneous, 7–17 cm long, with brownish, linear-lanceolate scales. (2n = 87) Jun–Sep. Dry, exposed rocky ledges & hillsides; se CO, OK, & NM; (OK & s CO, w to UT, s to NM & Mex.). *C. eatonii* f. *castanea* (Maxon) Correll — Correll & Johnston.

Forma *castanea* is described as a phase of *C. eatonii* that has blades with the upper surface green and only sparsely villous. Field observations, however, suggest that the glabrescent condition is a function of leaf age. Mixed populations are not uncommon in OK & CO; both tomentose and glabrescent leaves can be found on the same plant.

3. *Cheilanthes feei* Moore, slender lip fern. Rhizomes short, 1–3 cm long; scales brown with a dark midstripe, linear-lanceolate, 2.5–3.5 mm long, entire. Leaves clustered, 4–25 cm tall; blade deltoid to linear-lanceolate, 2-pinnate-pinnatifid to 3-pinnate, 2.5–15 cm long, 1–4 cm wide, apex acute, green and sparsely villose above, brownish or whitish tomentose below; pinnae triangular to ovate, 5–20 mm long, 5–12 mm wide, short stalked; ultimate segment oblong to nearly round, sessile or subsessile, petioles castaneous or nearly black, 1–10 cm long, bearing loosely spreading, articulate hairs, scaly at the base, rachis similar. (n = 87; 2n = 87) Jun–Sep. Dry, rocky slopes or crevices of boulders & cliffs, limestone or sandstone; e SD, WY, w NE, MO, KS, CO, OK, TX, NM; (NY w to WA & B.C., s to GA, CA, & Mex.).

4. *Cheilanthes fendleri* Hook., Fendler's lip fern. Similar to *C. wootonii*, but the scales on the lower side of the pinnae entire, not long-ciliate. (n = 30) Jun–Sep. Dry, exposed rocky ledges & hillsides; OK: Comanche; NM: Union; (sw OK, NM, & AZ, s to Mex.).

5. *Cheilanthes lanosa* (Michx.) D.C. Eat., hairy lip fern. Rhizomes slender, short creeping, to 8 cm long; scales brownish with a darker median stripe, lanceolate, 2–3.5 mm long, margins at least sparingly dentate. Leaves clustered at the apex of the rhizome, 7–35 cm tall; blade oblong-elliptic to oblong-lanceloate, pinnate-pinnatifid to 2-pinnate, 5–20 cm long, 1.5–4 cm wide, apex gradually attenuate, greenish and sparingly villous above, brownish or rarely white hirsute below; pinnae opposite or alternate, narrowly triangular to ovate, 7–27 mm long, 4–11 mm wide, short stalked; ultimate segments oblong to ovate, apex obtuse or rounded, margins entire to shallowly 2- or 3-lobed, sessile; petiole dark brown or purplish, 2–16 cm long, pilose with whitish or brownish articulate hairs; rachis similar to the petiole. (n = 30) Jun–Sep. Exposed sandstone or limestone ledges or rocky hillsides; MO, se KS, e & sw OK; (NY w to MN, s to FL & TX; also NM & AZ). *C. vestita* (Spreng.) Sw. — Fernald.

6. *Cheilanthes lindheimeri* (Sm.) Hook., Lindheimer's lip fern. Rhizomes slender, widely creeping, to 15 cm or more long; scales brownish, concolorous, broadly ovate to triangular, 1.5–3 mm long, entire. Leaves borne somewhat distant along the rhizome, to 35 cm tall; blades ovate-lanceolate to elliptic-lanceolate, 3- to 4-pinnate, 8–20 cm long, 3–5 cm wide, apex acute, white tomentose above, brownish tomentose mixed with scales below, the scales rusty brown, lanceolate to triangular, 1.5–2.5 mm long, attenuate at the apex and entire to long ciliate; pinnae triangular, lanceolate, 8–27 mm long, 8–15 mm wide, apex acuminate or acute, sessile; pinnules crowded, linear-oblong, pinnatifid or divided again into 3–7 ultimate segments; ultimate segments crowded, roundish, 0.5–1 mm diam; petiole dark brown, 4–25 cm long, wooly and scaly, the scales appressed, linear-lanceolate; rachis similar to the stipe. Jun–Sep. Dry, rocky ledges, often among rubble; OK: Comanche; (sw OK w to AZ, s to TX & Mex.).

7. Cheilanthes tomentosa Link, woolly lip fern. Rhizomes short, 1-3 cm long; scales brown with a dark median stripe or sometimes concolorous, linear-lanceolate, 3-5 mm long. Leaves clustered, 20-40 cm tall; blade linear-lanceolate to oblong-lanceolate, 2-pinnate-pinnatifid to 3-pinnate, 14-25 cm long, 3-6 cm wide, apex acuminate, upper surface green and sparsely villous, lower surface brownish tomentose or rarely mixed with linear-lanceolate scales; pinnae alternate, triangular to ovate, 1.5-3.5 cm long, 1-2 cm wide, apex acute to obtuse, subsessile to short stalked; ultimate segments mostly elliptic, apex broadly acute to rounded, margins entire or lobed, sessile; petiole dark brown, 6-18 cm long, pilose with simple brownish hairs; rachis similar to the petiole. (n = 90; 2n = 90) Jun-Sep. Crevices of dry limestone or granite rocks on cliffs or hillsides; KS: Cherokee; e & sw OK; (VA w to se KS, OK, & TX, s to GA & Mex.).

8. Cheilanthes wootonii Maxon, beaded lip fern. Rhizomes slender, widely creeping, to 15 cm or more long; scales rusty brown, concolorous, ovate to lanceolate, 1.5-3 mm long, entire to irregularly toothed. Leaves somewhat distant on the rhizome, 10-25 cm tall; blade narrowly oblong to oblong-lanceolate, 3-pinnate to 3-pinnate-pinnatifid, 4-13 cm long, 2-3.5 cm wide, apex acuminate, green and glabrous above, the lower surface usually completely covered with lanceolate, acuminate and long-ciliate scales; pinnae mostly alternate, lanceolate to narrowly triangular, 10-18 mm long, 5-9 mm wide, apex acuminate, base short stalked or subsessile; ultimate segments roundish or obovate, about 1 mm diam, sessile; petiole castaneous or dark brown, 5-10 cm long, sparsely hirsute, scaly, the scales numerous, linear-lanceolate and attenuate; rachis similar to the petiole. Jun-Sep. Dry exposed ledges & rocky hillsides; CO: Baca; NM: Union; OK: Cimarron, Comanche, Kiowa; (w OK & se CO, w to AZ, s to TX & Mex.).

5. CYSTOPTERIS Bernh., Fragile Fern

Small to medium-sized perennial herbs; rhizomes compact to widely creeping, scaly. Leaves tufted or scattered on the rhizome; blade lanceolate to deltoid-attenuate, 2- to 3-pinnate, membranaceous; veins free; petiole stramineous to darkened at the base, slender. Sori supramedial, round; indusium attached to the receptacle on the basiscopic side only, elsewhere free, thin, ovate-lanceolate to broadly cup-shaped; spores bilateral.

Reference: Blasdell, R. F. 1963. A monograph of the fern genus *Cystopteris*. Mem. Torrey Bot. Club 21: 1-102.

1 Blade broadest at the base, apex acuminate to attenuate; bulblet sometimes produced on the rachis and rachises.
  2 Blade apex long-attenuate; bulblets 1.6-6 mm diam .... 1. *C. bulbifera*
  2 Blade apex acuminate; bulblets 0.5-1.5 mm diam .... 4. *C. tennesseensis*
1 Blade broadest above the base, apex acuminate; bulblets never produced.
  3 Rhizome long creeping, internodes distinct, growing tip extended 2-5 cm beyond the leaves of the season; basal pinnae petiolate .... 3. *C. protrusa*
  3 Rhizome short creeping, internodes short and not distinct, growing tip not much extended beyond leaves of the season; basal pinnae sessile or nearly so .... 2. *C. fragilis*

1. Cystopteris bulbifera (L.) Bernh., bulblet bladder fern. Rhizome short creeping, to 15 cm long, sometimes erect near the tip, internodes short, glabrous, scaly near the tip; scales brownish, lanceolate, 2-3 mm long, entire. Leaves mostly tufted, 2.5-7 dm tall; blade narrowly deltoid, 2- or 3-pinnate, 12-55 cm long, 6-15 cm wide, glabrous, apex usually long attenuate; bulblets frequently produced in or near the axils of the pinnae on the underside of the blade, globose, 1.6-6 mm diam, whitish or brownish scaly when young, glandular or not, soon falling and growing into young plants with favorable conditions; pinnae subopposite to alternate, narrowly triangular, ovate-lanceolate, or oblong, 3–7.5 cm long, 1–3.5 cm wide, apex acute to acuminate, short petiolate; ultimate segments ovate, oval, or oblong, veins directed into the emarginations of the segment, apex obtuse to rounded, margins double serrulate, the teeth often rounded, sessile; petiole 6–20 cm long, smooth. Sori discrete; indusium cup-shaped, glandular or not, apex truncate, margins entire or erose; spores echinate. (n = 42) May–Sep. Moist, shaded ledges or rich soil on partially exposed seepy slopes; SD: Roberts; NE: Richardson; (Newf. w to Man., s to GA, TX, NM, & AZ). *Filix bulbifera* (L.) Underw.—Rydberg.

2. *Cystopteris fragilis* (L.) Bernh., fragile fern. Rhizome short to long creeping, usually less than 15 cm long with the old petiole bases crowded, rarely to more than 25 cm long with the internodes to about 5 mm long, glabrous, apex densely scaly, scales broadly lanceolate, 2.5–4 mm long, apex long acuminate, margins entire. Leaves usually borne at the apex of the rhizome (rarely up to 3 cm behind the apex in those plants with extensive rhizomes and rugose-verrucose spores), 1–6 dm tall; blade linear-lanceolate to ovate, pinnate-pinnatifid to 2-pinnate-pinnatifid, 5–30 cm long, 2–11 cm wide, widest above the base, glabrous or rarely glandular, in the axils of the pinnae, apex acute to acuminate; bulblets never present; pinnae subopposite to alternate, narrowly triangular to narrowly ovate or lanceolate, 1–8 cm long, 0.5–3 cm wide, apex acute or less often acuminate, base short petiolate to subsessile; ultimate segments broadly ovate to oblong or elliptic, veins directed into the teeth of the segment, apex obtuse to acute (especially in n GP), margins mostly serrate, the teeth sometimes slightly rounded, sessile; petiole 4–27 cm long, smooth or rarely with a few scattered scales toward the base. Sori discrete; indusium fragile, ovate, glabrous, apex acute to acuminate, frequently bifurcate, margins entire or irregularly toothed; spores rugose and (40)50–60 µm long or echinate and 35–45(50) µm long, rarely intermediate. (n = 84, 126) Jun–Sep. In soil on banks, hillsides, & ledges or on rocks; common in n GP, scattered s; (widespread in Northern Hemisphere, scattered in Southern Hemisphere). *Filix fragilis* (L.) Gilib.—Rydberg.

Cystopteris fragilis is a poorly understood taxon throughout its distribution. In the GP two races are evidently present. Typical *C. fragilis* var. *fragilis* with echinate spores occurs along the eastern fringe of the GP, nearly always on rocks, but is rare elsewhere in our region. The common race in the GP has rugose spores, occurs mostly in soil and is widespread in the central and western GP from central Kansas northward. Most of the plants in the Black Hills had rugose spores, but a small percentage were found to be *C. fragilis* var. *fragilis*, and two specimens appeared to be intermediate. One intermediate also was discovered from NE: Knox, where both races occur. Our rugose-spored pants have been referred to as *C. fragilis* var. *dickieana* (Sim) Lindberg by some authors. Var. *dickieana*, however, usually grows on rocks and has close-set or often imbricate pinnae, neither of which are the case with the GP plants. In view of this, referring our plants to var *dickieana* seems premature until further studies of this complex are made.

3. *Cystopteris protrusa* (Weath.) Blasd., lowland fragile fern. Rhizomes extensively creeping, to 3 dm or more long, internodes mostly 4–7 mm long, glabrous, scaly at the apex, scales lanceolate to ovate-lanceolate, 3–5 mm long. Leaves borne 2–5 cm behind the growing tip especially late in the season, 2.5–5 dm tall; blade lanceolate or rarely ovate, pinnate-pinnatifid to 2-pinnate-pinnatifid, 10–25 cm long, 3–13 cm wide, widest above the base, glabrous or occasionally with a few whitish hairs, apex acute to acuminate; bulblets never present; pinnae subopposite to alternate, lanceolate to narrowly ovate, 1.5–8 cm long, 0.8–3.5 cm wide, apex acute or acuminate, short petiolate; ultimate segments elliptic to oval or obovate, veins ending in the teeth or rarely the emarginations of the segment, apex mostly obtuse, less often rounded or acute, margins serrulate or double serrulate, sessile; petiole 10–27 cm long, smooth. Sori discrete; indusium fugacious, broadly ovate or less often nearly oval, glabrous, apex acute to rounded or sometimes notched, margins entire; spores echinate. (n = 42) Jun–Sep. Low woodlands in rich soil; IA, e NE, MO, e KS, e OK

(NY w to MN, s to GA & e TX). *C. fragilis* (L.) Bernh. var. *protrusa* Weath.—Fernald.

4. Cystopteris tennesseensis Shaver, Tennessee bladder fern. Rhizome short creeping, to 15 cm long, internodes short, old petiole bases congested, glabrous, few scales near the tip; scales brownish, linear-lanceolate, 2–3 mm long, entire. Leaves mostly borne at the apex of the rhizome, 1–4(5) dm tall; blade narrowly deltoid nearly to lanceolate, pinnate-pinnatifid to 2-pinnate-pinnatifid, 7–25(35) cm long, 3–14 cm wide, widest at the base, glabrous, apex acuminate; bulblets infrequently produced in or near the axils of the pinnae on the underside of the blade, clavate to nearly globose, mostly less than 1.5 mm diam, covered with a few brownish, lanceolate scales, sparsely glandular, falling to the ground and producing young plants; pinnae mostly subopposite, deltoid to lanceolate, 1.5–7.5 cm long, 1–3.5 cm wide, apex acute to acuminate, short petiolate; ultimate segments ovate, oval, or oblong, veins directed into the teeth and emarginations of the segment, apex obtuse or rarely acute, margins irregularly serrulate to double serrulate, sessile; petiole 3–25 cm long, smooth. Sori discrete; indusium ovate to ovate-lanceolate, sparsely glandular or not, apex acute or acuminate and often bifurcate, margins entire or with several shallow teeth; spores echinate. (n = 84) May–Sep. Shaded limestone or sandstone, rock crevices or ledges; MO, cen & e KS, & e OK; (VA & OH w to cen KS, s to NC, AL, AR, & e TX; NM). *C. fragilis* (L.) Bernh. f. *simulans* Weath.—Fernald; *C. fragilis* (L.) Bernh. var. *simulans* (Weath.) R. L. McGreg., var. *tennesseensis* (Shaver) R. L. McGreg.—McGregor, Amer. Fern J. 40: 202–203. 1950.)

This species is a proposed hybrid derived from a cross between *C. protrusa* and *C. bulbifera* based only on morphological characters. In spore germination studies, *C. tennesseensis* spores germinate normally and produce sporophytes indicating this is an allotetraploid capable of reproducing itself sexually as a good species.

6. DRYOPTERIS Adans., Wood Fern

Our species medium- to large-sized, mesophytic plants; rhizomes stout, short creeping or erect, covered with old petiole bases, scaly. Leaves clustered, forming a crown at the apex of the rhizome; blades pinnate-pinnatifid, subcoriaceous to coriaceous, glabrous; veins free, simple or forked; petiole stramineous, usually scaly, with 3–7 free bundles. Sori inframedial to submarginal, roundish; indusia reniform and open at the outer edges, persistent, glabrous; spores bilateral.

```
1   Blade at least 2-pinnate ............................................................. 5. D. spinulosa
1   Blade pinnate-pinnatifid.
    2   Sori submarginal ............................................................... 4. D. marginalis
    2   Sori median.
        3   Larger pinnae with 4–10 pairs of lobes .......................................... 1. D. cristata
        3   Larger pinnae with 12 or more pairs of lobes.
            4   Petiole scales denticulate; blade widest above the base, basal pinnae
                reduced ............................................................... 2. D. filix-mas
            4   Petiole scales entire; basal pinnae not noticeably reduced ............. 3. D. goldiana
```

1. Dryopteris cristata (L.) A. Gray, crested shield fern. Rhizomes suberect to short creeping, to 7 cm long; scales few and mostly near the apex of the rhizome, ovate, 3–6 mm long, membranaceous, acuminate, entire. Leaves slightly dimorphic, 3–6 dm tall, the sterile leaves evergreen, spreading, and shorter than the erect, herbaceous fertile leaves; blades lanceolate to elliptic-lanceolate, pinnate-pinnatifid, 20–40 cm long, 5–12 cm wide, subcoriaceous, glabrous, apex acuminate, base gradually tapering; pinnae subopposite, narrowly triangular to lanceolate in the upper portion of the blade and becoming gradually shorter and wider toward the base, pinnatifid with 4–10 pairs of lobes, to 7 cm long, 1.3–3 cm wide, a few linear-lanceolate scales sometimes scattered on the lower side, apex acute to acuminate, short petiolate to subsessile; pinnae lobes mostly oblong, apex rounded or broadly acute, margins serrulate; spinulose; petiole stramineous to reddish-brown, 4–20 cm long, glabrous, scaly, the scales brownish, ovate, 4–9 mm long, larger near the base, subentire; the rachis similar. Sori median on the pinnae lobes, round, about 1.5 mm diam; indusia glabrous, irregularly toothed to entire. (n = 82) Jun–Aug. Boggy woods & wet meadows; MN, e ND, & cen NE; (Newf. w to Sask., s to NC, TN, AR, NE, MT, & ID; Europe).

D. cristata × *spinulosa* was collected several times in Pembina Co., ND. The hybrid is morphologically intermediate between the parental species.

2. *Dryopteris filix-mas* (L.) Schott, male fern. Rhizomes short and erect or horizontal; scales mostly towards the tip of the rhizome, brownish, lanceolate to elliptic lanceolate, to 1 cm long, membranaceous, margins entire to denticulate. Leaves deciduous, to 2 m tall; blade broadly lanceolate, pinnate-pinnatifid, 2–16 dm long, 6–35 cm wide, glabrous, apex gradually acuminate; pinnae alternate, lanceolate to narrowly so, pinnatifid with 12–30 pairs of lobes, to 17 cm long, 1–4 cm wide, apex acuminate, base subsessile to short-petiolate; pinnae lobes oblong, apex obtuse, margins serrulate; petiole stramineous with a reddish-brown or darker base, 6–45 cm long, glabrous, scaly, the scales numerous, reddish-brown, ovate and to 1 cm long at the base but becoming subulate and much shorter above, margins denticulate, the rachis similar. Sori usually closer to the midvein than the margin of the pinnae lobe, roundish, about 1 mm diam, glabrous, margins entire or irregularly toothed. (n = 82) Jun–Sep. Rich, moist soil in canyons or crevices of rocks in protected sites; w SD, WY; NM: Union; OK: Cimarron; (circumboreal, s in N. Amer. to VT, MI, w SD, NM, AZ, & OR; s CA).

3. *Dryopteris goldiana* (Hook.) A. Gray, Goldie's fern. Rhizomes short creeping to nearly erect, scaly at the apex; scales brown, lanceolate, to 14 mm long. Leaves deciduous, 4–10 dm tall; blades broadly ovate, pinnate-pinnatifid, 3–5 dm long, 18–35 cm wide, glabrous, the lower surface with a few subulate, brownish scales, apex abruptly acuminate; pinnae alternate, mostly broadly lanceolate, rarely broadly oblanceolate, pinnatifid with 14–20 or more pairs of lobes, to 18 cm long, 2.5–4.5 cm wide, apex acuminate, base short petiolate; pinnae lobes oblong, apex obtuse to acute, margins serrulate to crenulate; petiole stramineous, darkened toward the base, 15–45 cm long, thickly beset with dark brown lanceolate scales to 1.5 cm long near the base, scales on the upper petiole and rachis tan, narrowly lanceolate to subulate and to 9 mm long, margins mostly entire. Sori borne along the midvein of the pinnae lobes, roundish, about 1 mm diam; indusia glabrous, margins subentire. (2n = 82) Jun–Aug. Moist, rich soil in steep-sided, wooded ravines; MO: Jackson; (N.B. w to MN, s to VA, NC, TN, & MO).

A specimen labeled KS: Leavenworth, collected in 1871, is believed to be from a cultivated plant.

4. *Dryopteris marginalis* (L.) A. Gray, marginal shield fern. Rhizomes erect, densely scaly; scales brownish, ovate to lanceolate, to 1 cm long, apex long acuminate, margins entire. Leaves semievergreen, to 10 dm tall; blades broadly lanceolate to ovate, pinnate-pinnatifid, 1–7.5 dm long, 6–30 cm wide, glabrous, small subulate scales occasionally present on the lower surface, apex acuminate; pinnae subopposite, narrowly triangular to lanceolate, 3–15 cm long, 1–7 cm wide, apex acuminate, short petiolate to subsessile; pinnae lobes oblong, apex obtuse or rounded, margins subentire to serrulate; petiole stramineous, darkened at the base, 8–35 cm long, glabrous, densely scaly at the base and less so above into the rachis, scales ovate to lanceolate, to 1 cm long at the base and smaller above, margins entire. Sori submarginal on the pinnae lobes, roundish, about 1 mm diam; indusia glabrous, margins entire. (n = 41) May–Sep. Rocky woodlands or shaded ledges & cliffs; e KS, MO, s & e OK; (N.S. w to MN, s to GA, AL, AR, & OK).

5. *Dryopteris spinulosa* (O. F. Muell.) Watt, spinulose wood fern. Rhizomes erect to mostly creeping, to 10 cm long; scales few, similar to those on the petiole. Leaves deciduous or rarely semievergreen, to 8 dm tall; blades broadly lanceolate, narrowly triangular, or ovate, 2-pinnate to 2-pinnate-pinnatifid, 2.5–4.5 dm long, 10–20 cm wide, glabrous, apex acuminate; pinnae alternate or subopposite, lanceolate to narrowly triangular, 5–13 cm long, 2–6 cm wide, apex long acuminate, base oblique, the lower pinnule longer than the upper pinnule, short petiolate; larger pinnules narrowly triangular to lanceolate, pinnatifid, apex mostly acute, sessile or subsessile; smaller pinnae and ultimate segments oblong or elliptic, apex obtuse to rounded, margins serrate, the teeth spinulose; petiole stramineous above and reddish-brown toward the base, 10–35 cm long, moderately scaly, the scales broadly ovate to lanceolate, to 1.5 cm long, smaller above and into the rachis, margins entire. Sori median, roundish, about 0.5 mm diam; indusia glabrous, margins entire or nearly so. (n = 82) Jun–Aug. Low, rich woodlands & stream ravines; MN, e ND, n-cen & e NE; (N.S. w to AK, s to WV, KY, MO, ND, MT, & OR; Eurasia). *D. austriaca* (Jacq.) Woynar var. *spinulosa* (Muell.) Fiori—Britton & Brown; *D. carthusiana* (Vill.) H. P. Fuchs—Atlas GP Flora; *D. spinulosa* (Muell.) O. Ktze.—Rydberg.

This species hybridizes with *D. cristata*. See discussion under no. 1.

7. GYMNOCARPIUM Newm., Oak Fern

1. *Gymnocarpium dryopteris* (L.) Newm., oak fern. Perennial herb; rhizomes slender and widely creeping, bearing scattered lanceolate to ovate, membranaceous scales, especially near the nodes. Leaves scattered and borne singly along the rhizome, to 40 cm long; blade broadly deltoid-pentagonal, 2- or 3-pinnate-pinnatifid, 5–17 cm long, 6–20 cm wide, glabrous, pinnae opposite, the lowermost pair asymmetrically deltoid, pinnate-pinnatifid, long petiolulate, and articulate, each about equaling the rest of the blade in size, the pinnae above oblong to lanceolate, pinnatifid, apex obtuse, base sessile or subsessile; ultimate segments oblong, to 1.5 cm long, apex obtuse to acute, margins crenate to slightly pinnatifid; veins free; petiole stramineous and darkened at the base, slender, to 24 cm long, glabrous, scaly near the base, the scales similar to those on the rhizome. Sori medial to submarginal, round, exindusiate. (n = 80) Jul–Sep. Damp, shaded granite rock ledges & crevices & rocky, wooded slopes; MN: Mahnomen; SD: Custer, Lawrence; (circumboreal, s in N. Amer. to VA, WV, e IA, sw SD, AZ, & OR). *Phegopteris dryopteris* (L.) Fee—Rydberg; *Dryopteris disjuncta* (Ledeb.) Morton—Fernald; *Carpogymnia dryopteris* (L.) Love & Love—Petrik, Rhodora 77: 478–511. 1975.

8. MATTEUCCIA Todaro, Ostrich Fern

1. *Matteuccia struthiopteris* (L.) Todaro, ostrich fern. Medium-sized perennial plant; rhizomes erect, stout, covered with persistent petiole bases, black stolons arising from the base and forming new plants. Leaves dimorphic, arising in a spiral from the apex of the rhizome, glabrous; sterile leaves herbaceous, to 1.5 mm tall; blade obtrullate, pinnate-pinnatifid, 2–12 dm long, 1–4 dm wide, widest above the middle, apex pinnatifid, abruptly short-acuminate, base long tapering with the pinnae gradually diminutive and clasping; pinnae alternate, spreading-ascending, linear-acuminate, deeply pinnatifid into oblong,

blunt segments; veins free; petiole stramineous, short, 5–30 cm long, grooved on the upper surface, dark brown at the flattened base with thin brownish scales, glabrous above; fertile leaves persistent, infrequently produced, 25–35(50) cm tall; blade olivaceous, reduced, oblanceolate, pinnate-pinnatifid, 15–25 cm long, 3–8 cm wide; pinnae linear-moniliform, formed of the tightly revolute margins which enclose the sori; petiole like that of the sterile leaf. (n = 40) Jul–Oct. Boggy meadows or woods & stream banks; MN, cen & e ND, BH, IA, MO; (Newf. w to AK, s to VA, IN, MO, SD, & B.C.). *Pteretis nodulosa* (Michx.) Nieuw.—Rydberg; *P. pennsylvanica* (Willd.) Fern.—Stevens.

9. NOTHOLAENA R. Br., Cloak-fern

Our species small, evergreen, lithophytic, perennial xerophytes; rhizome short, branching, forming a multicipital caudex heavily beset with long, linear-lanceolate, brown scales. Blades linear to pentagonal, 1- to 4-pinnate, thick-membranaceous to coriaceous, veins free; ultimate segments scaly or waxy at least below, margins usually flat and unmodified, sometimes revolute, forming a marginal indusium. Sori mostly submarginal, borne at the tips of the veins, oblong or oval; indusia wanting although the leaf margins sometimes revolute and protecting the sporangia which are hidden by the scaly or waxy indument.

Reference: Tryon, R. M. 1956. A revision of the American species of *Notholaena*. Contr. Gray Herb. 179: 1–106.

The difficulties in defining *Notholaena* are well known, wherein what appear to many as unrelated species and closely related species are assembled into a single genus. Obviously, it merges with *Pellaea* and *Cheilanthes*, but those limits are uncertain and will not be easily discovered.

```
1  Blade linear; petiole scaly .................................................................. 3. N. sinuata
1  Blade pentagonal or ovate to deltoid; petiole smooth.
   2  Blade pentagonal, underside yellow waxy ........................................ 4. N. standleyi
   2  Blade ovate to deltoid, underside white waxy.
      3  Rachis and pinna-rachises markedly flexuous, the branching of the pinnae appearing
         dichotomous ...................................................................... 2. N. fendleri
      3  Rachis and pinna-rachises straight, or essentially so, the branching
         pinnate ............................................................................. 1. N. dealbata
```

1. ***Notholaena dealbata*** (Pursh) Kunze, false cloak-fern. Rhizomes slender, short; scales castaneous, concolorous, linear, 1–4 mm long, thin entire. Leaves clustered, numerous, 4–17 cm tall; blade narrowly to broadly deltoid, 3- to 5-pinnate, with 3–5 pairs of pinnae, 2–8 cm long, upper surface glabrous, green to glaucous, lower surface white waxy; rachis and pinna-rachises straight with pinnate branching; pinnae deltoid, long petiolate, ultimate segments ovate to oblong or obovate, 1–3 mm long, the terminal segments usually slightly larger, apex obtuse to rounded, margins entire, flat or slightly revolute, base cuneate, sessile or nearly so; petiole shining, castaneous, about equal to or longer than the blade, glabrous. Sori submarginal, borne back of the vein-tip; spores 64 per sporangium. (n = 27) May–Sep. Shaded or exposed calcareous outcrops; se NE, MO, e KS, & OK; (MO & se NE, s to cen TX).

2. ***Notholaena fendleri*** Kunze, Fendler cloak-fern. Rhizomes stout, often horizontal; scales castaneous, concolorous, lanceolate to linear-lanceolate, 2–6 mm long, thin entire. Leaves clustered, numerous, 5–25 cm tall; blade broadly to narrowly deltoid, 3- to 6-pinnate, 2.5–14 cm long, upper surface glabrous, green to glaucous, lower surface white waxy; rachis and pinna-rachises flexuous with the branching of the pinna appearing dichotomous; pinnae 5–11, alternate, deltoid, long-petiolate; ultimate segments ovate to oval, 2–4 mm long, obtuse, margins entire to crenulate, flat or sometimes revolute, base cuneate to attenuate,

sessile or nearly so; petiole shining, castaneous, about equal to the blade in length, glabrous. Sori submarginal, borne back of the vein-tip; spores 64 per sporangium. (n = 27) May–Sep. Dry, rocky bluffs & cliffs; WY: Laramie; NM: Union; (se WY, s to NM). *Cheilanthes cancellata* Mickel—Mickel, Phytologia 41: 433. 1979; *Pellaea fendleri* (Kunze) Prantl — Martin & Hutchins.

3. ***Notholaena sinuata*** (Sw.) Kaulf., long cloak-fern. Rhizomes short, horizontal; scales castaneous to tan, linear, 4–8 mm long, pectinate-ciliate to entire. Leaves clustered, numerous, arch-spreading, to 30 cm tall, blade linear, 1-pinnate, 5–25 cm long, 0.7–2 cm wide, coriaceous, upper surface at least sparsely beset with whitish, pectinate or often stellate scales, lower surface densely covered with castaneous to light brown, imbricate, lanceolate, short-fimbriate scales (rachis similarly scaly); pinnae numerous, broadly ovate to quadrate rhombic, 3–9 mm long, 2–5 mm wide, apex obtuse, margin flat, entire or shallowly 2- or 3-lobed, short petiolate; petiole 1.5–4 cm long, densely clothed with lanceolate, pectinate scales. Sporangia borne at the vein tips forming a continuous but hidden marginial band, each sporangium containing 32 spores. May–Sep. Exposed rocky hillsides, cliffs, and ravines.

Notholaena sinuata is a complex of three closely related races (two occurring in the GP) that have been variously ranked in modern works. While most plants encountered are easily assigned to one of the races, some populations show intergradation of morphological characters between var. *cochisensis* and var. *integerrima*. Because of this, it is felt the races would best be recognized at the variety level until further studies are made.

3a. var. *cochisensis* (Goodd.) Weath. Rhizome scales castaneous, linear and entire. Pinnae broadly ovate and usually with 2 or 3 pairs of shallow lobes. NM: Quay; TX: Motley; (n TX w to CA, s to Mex.). *Cheilanthes cochisensis* (Goodd.) Mickel—Mickel, Phytologia 41: 433. 1979.

3b. var. *integerrima* Hook. Rhizome scales usually tan, linear and pectinate-ciliate. Pinnae quadrate-rhombic, usually entire. (2n = ca 87). OK: Murray; NM: Union; TX: Motley; (s OK w to AZ, s to Mex.). *Cheilanthes integerrima* (Hook.) Mickel—Mickel, Phytologia 41: 434. 1979.

4. ***Notholaena standleyi*** Maxon, star cloak-fern. Rhizomes short horizontal; scales with a shining blackish central band and lighter margin, thin, linear-lanceolate to lanceolate, 4–7 mm long, entire. Leaves clustered, numerous, 5–30 cm tall; blade pentagonal, pinnate-pinnatifid, 2–7 cm long and wide, coriaceous, upper surface green and glabrous, lower surface yellowish waxy; basal pinnae pinnate-pinnatifid and larger than the upper portion of the blade, unequilateral, the basal pinnules on the lower side much elongated and pinnatifid; ultimate segments linear-oblong to elliptic, margins entire to undulate, flat; petiole shining, castaneous, 3–23 cm long, glabrous, sometimes with a few cordate, membranaceous scales near the base. Sori appearing marginal, borne at the vein tips; sporangium with 16–32 spores. (2n = 60) May–Sep. On dry cliffs & rocky slopes, often with southern exposures; CO: Baca, Las Animas; OK: Cimarron, Grady, Greer; NM: Union; (se CO w to NV, s to sw OK, AZ, & Mex.). *Cheilanthes standleyi* (Maxon) Mickel—Mickel, Phytologia 41: 436. 1979.

10. ONOCLEA L., Sensitive Fern

1. ***Onoclea sensibilis*** L., sensitive fern. Perennial plants with cylindrical, long creeping and branching rhizomes, 2–10 diam, glabrous. Leaves dimorphic, scattered on the rhizomes, glabrous; sterile leaves herbaceous, to 1 m tall; blade broadly deltoid to trullate, pinnate with a broadly winged rachis, 5–35 cm long, 5–40 cm wide, thin membranaceous, veins reticulate; pinnae subopposite, spreading ascending, lanceolate or lanceolate-oblong, entire to undulate or the lower and middle pinnae sinuately lobed (rarely the pinnae deeply

pinnatifid or pinnate and sometimes partially fertile), apex acuminate to obtuse; petiole terete to convex on the upper side, 5–70 cm long; fertile leaves persistent, 15–60 cm tall; blade reduced, lance-oblong, 2-pinnate, 2–18 cm long, 1.5–3 cm wide, pinnules rolled into globular divisions, dark brown or black at maturity. Sori enclosed by the inrolled pinnules; spores bilateral, green. (n = 32, 37, 40; 2n = 64, 74) Jul–Oct. Exposed or shaded areas in boggy woods, marshes, stream ravines, or ditches; MN w to ND, s to MO & KS; (Newf. w to Sask., s to FL & TX). *O. sensibilis* var. *obtusilobata* (Schkuhr) Gilbert — Fernald.

11. PELLAEA Link, Cliff-brake

Ours small, evergreen, perennial ferns; rhizomes moderately stout, decumbent, multicipital, heavily beset with appressed scales. Blade 1- to 4-pinnate, usually coriaceous, veins free; ultimate segments jointed and deciduous; petiole and rachis slender, wiry. Sori intramarginal, borne at the vein tips, forming a marginal line protected by the reflexed margins of the fertile segments; sporangia mostly long stalked; spores tan to brown.

References: Knobloch, I. W., & D. M. Britton. 1963. The chromosome number and possible ancestry of *Pellaea wrightiana.* Amer. J. Bot. 50: 52–55; Tryon, A. F. 1958. A revision of the fern genus *Pellaea* section *Pellaea.* Ann. Missouri Bot. Gard. 44: 125–193.

1 Rhizome scales bicolorous with a dark center stripe; petiole convex on the upper surface .... 3. *P. wrightiana*
1 Rhizome scales concolorous; petiole terete.
   2 Petiole and rachis appressed pubescent, scurfy .... 1. *P. atropurpurea*
   2 Petiole and rachis glabrous or with a few scattered hairs .... 2. *P. glabella*

1. ***Pellaea atropurpurea*** (L.) Link, purple-stemmed cliff-brake. Rhizome compact; scales tan to rust colored, concolorous, linear-subulate, the tips entangled, to 8 mm long. Leaves clustered, numerous, 7–45(50) cm tall, dimorphic; blades lanceolate to narrowly triangular, 1- or 2-pinnate, 4–30 cm long, 2–12 cm wide; lower pinnae simple to pinnate and at least short petiolate, half again as long as the upper pinnae, the upper pinnae simple, entire and sessile or subsessile; segments dark green above and lighter below, lanceolate to linear-oblong, 1–7 cm long, 0.4–1 cm wide, glabrous above, sparsely pubescent with articulate hairs on the main vein below, those of the sterile blade flat with acute to rounded, mucronate apices, margins entire to crenulate and often whitened, and bases rounded to truncate, sessile or subsessile, the fertile segments with mostly acute, mucronate apices, margins entire, revolute, and bases truncate to oblique; petiole and rachis terete, atropurpureous, appressed pubescent. (n = 87, 2n = 87) May–Oct. Exposed or shaded rocky slopes, ledges, or cliffs, calcareous or sandy soils; sw SD, nw & se NE, MO, e KS, se CO, OK, TX, & NM; (VT w to w Can., s to FL, AZ; Mex.).

2. ***Pellaea glabella*** Mett. ex Kuhn. Rhizome compact; scales rust colored, concolorous, linear-subulate, the tips free or entangled, to 6 mm long. Leaves clustered, numerous, 3–24 cm tall, monomorphic; blades narrowly oblong to lanceolate or ovate, 1- to 2-pinnate, 1–15 cm long, 1–7 cm wide; lower pinnae simple to pinnate, sessile or short petiolate, the upper pinnae simple, entire and usually sessile; segments narrowly oblong to elliptic, 5–25 mm long, 2–8 mm wide, glabrous or with a few scattered hairs below, apex rounded to acute, often mucronulate, margins revolute, entire or crenulate, bases cuneate, sessile; petiole and rachis terete, castaneous or sometimes blackish, glabrous or rarely sparsely hairy. May–Oct. Cliffs and ledges, usually calcareous rocks.

Two morphologically distinct and geographically separated varieties occur in the GP.

2a. var. *glabella,* smooth cliff-brake. Leaves usually arch spreading, 1-pinnate, 3–24 cm tall; pinnae deeply lobed or with 3–7 segments. Spores 32 per sporangium. (n = 116). SD: Union; MO, KS, e OK;

(VT w to e MN, e SD, s to TN, AR, OK, & nw TX). *P. atropurpurea* var. *bushii* Mack. ex Mack. & Bush—Gates.

2b. var. *occidentalis* (E. Nels.) Butt., dwarf cliff-brake. Leaves usually erect, 1- or 2-pinnate, 3–15 cm tall; pinnae entire or with 2 or 3 lobes. Spores 64 per sporangium. (n = 29). w ND, MT, w SD, & WY; (Alta. & w ND, s to w SD & CO). *P. pumila* Rydb.—Rydberg.

3. *Pellaea wrightiana* Hook., Wright's cliff-brake. Rhizomes somewhat elongate; scales bicolorous, tan or castaneous with a dark center stripe, subulate, to 8 mm long, margins dentate. Leaves clustered, 5–45 cm tall, monomorphic; blades lanceolate to narrowly triangular, bipinnate, 3–30 cm long, 0.5–6 cm wide; pinnae all of nearly the same length, sessile or subsessile, the uppermost simple and the lower pinnate; segments lanceolate to narrowly oblong, 0.5–4 cm long, 0.5–1 cm wide, glabrous, apex mostly acute, mucronulate, margins entire to crenulate, revolute, base sessile to short stalked; petiole and rachis castaneous to atropurpureous, convex on the upper surface, glabrous. (n = 58) Apr–Oct. Exposed or partially shaded, igneous or limestone rocks on cliffs or hillsides; CO: Baca; OK: Cimarron, Comanche, Greer; TX: Hardeman; (OK w to Baja Calif., s to TX & n Mex.). *P. mucronata* D. C. Eat.—Rydberg; *P. ternifolia* (Cav.) Link var. *wrightiana* (Hook.) A. Tryon—Atlas GP.

12. POLYPODIUM L., Polypody

Small evergreen, mesophytic perennials; rhizomes widely creeping and branched, scaly. Leaves scattered and jointed to knoblike protuberances on the rhizome, erect to pendent; blades in our species pinnatifid with toothed or entire segments; veins free and forking 1–3 times; petiole slender, wiry, with 3 vascular bundles near the base. Sori orbicular or slightly elliptic, laminar, borne depressed and dorsal at or near vein forks in a distinct row on each side of the costa; indusium wanting; spores bilateral, lacking perispore.

References: Brooks, R. E. 1975. Notes on the *Polypodium vulgare* complex in South Dakota. Amer. Fern J. 65: 11–12; Lang, F. A. 1971. The *Polypodium vulgare* complex in the Pacific Northwest. Madrono 21: 235–254.

1 Lower surface of blade densely scaly .... 2. *P. polypodioides*
1 Lower surface of blade glabrous.
    2 Paraphyses present among sporangia; rhizome scales usually with a dark median stripe .... 3. *P. virginianum*
    2 Paraphyses not present among sporangia; rhizome scales uniformly colored .... 1. *P. hesperium*

1. *Polypodium hesperium* Maxon, western polypody. Rhizomes to more than 2 dm long, 1.5–3 mm diam excluding scales; rhizome scales tan to castaneous, uniformly colored, ovate, 2–5 mm long, somewhat crenate. Leaves 4–25 cm long, glabrous; blades oblong-linear to lanceolate, deeply pinnatifid, 3–18 cm long, 1.5–4 cm wide, coriaceous; segments in 5–18 alternate pairs, oblong, 8–22 mm long, 2–9 mm wide, apex rounded to obtuse, margins entire to crenulate or rarely serrulate; petiole 1.5–8 cm long. Sori mostly orbicular or oval, paraphyses absent or very rare. (n = 74) Jun–Aug. Shaded crevices in granite rocks; SD: Custer, Pennington, Shannon; (B.C. & AK s to sw SD & NM, w to CA).

Our plants may be more correctly referred to *P. amorphum* Suksd.

2. *Polypodium polypodioides* (L.) Watt var. *michauxianum* Weath., resurrection fern. Rhizomes to more than 2 dm long, 1–2 mm diam and covered with linear to linear-lanceolate, subulate scales, these dark brown or black at the center with a narrow, tawny margin. Leaves hygroscopic, 3–15 cm long; blades oblong to triangular-oblong, deeply pin-

natifid, 2–11 cm long, 1.5–4 cm wide, dark green and smooth above, densely covered with peltate circular and subentire scales below, coriaceous; segments alternate to subopposite, linear to linear-spatulate, 5–20 mm long, 2–4 mm wide, apex obtuse, margins entire; petiole 1.5–7 cm long, appressed scaly. Sori orbicular, borne near the margin of the leaf. (n = 37) Jul–Sep. In our area growing on shaded sandstone ledges & boulders; KS: Chautauqua; (NJ w to IA, KS, s to FL & C. Amer.).

3. **Polypodium virginianum** L., common polypody. Rhizomes to more than 2 dm long, 1.5–3 mm diam excluding scales; rhizome scales tan to castaneous with a dark median stripe, ovate, 2–4.5 mm long. Leaves 5–20 cm long, glabrous; blades oblong-linear to lanceolate, deeply pinnatifid, 3–20 cm long, 1.5–4 cm wide, coriaceous; segments in 4–20 alternate pairs, oblong, 7–20 mm long, 2–8 mm wide, apex obtuse to acute, margins entire to denticulate; petiole 1.5–9 cm long. Sori mostly orbicular, paraphyses mixed with sporangia. (n = 37, 74) Jul–Aug. Shaded crevices of quartzite boulders; SD: Minnihaha; MN; (Circumboreal, s in N. Amer. to GA, AL, AR, e SD, Alta., & B.C.). *P. vulgare* L. var. *virginianum* (L.) Eat.—Gleason & Cronquist.

13. POLYSTICHUM Roth, Holly Fern, Christmas Fern

Medium-size, coarse perennials; rhizomes stout and erect to short creeping, copiously scaly. Leaves tufted, those toward the center usually semievergreen erect and fertile, the outer leaves evergreen, often nearly decumbent and sterile; blades in our species pinnate, scaly especially on the rachis and costae, otherwise glabrous, pinnae subsessile, asymmetrical at the base, the basipetal side cuneate and the acropetal side auricled; veins free, mostly 1-forked. Sori round, laminar, borne dorsal on veins along one or both sides of the costa in a definite row; indusium round, centrally peltate, persistent, rarely wanting; spores bilateral, with perispore.

1 Blade distinctly narrowed toward the base, lower pinnae about as long as they are wide .... 2. *P. lonchites*
1 Blade not much reduced toward the base, lower pinnae at least 2× longer than wide.
   2 Pinnae linear-oblong, acute to obtuse; fertile upper pinnae abruptly reduced in size .... 1. *P. acrostichoides*
   2 Pinnae linear-lanceolate, acuminate to attenuate; fertile pinnae not reduced .... 3. *P. munitum*

1. **Polystichum acrostichoides** (Michx.) Schott, Christmas fern. Leaves 3–8 dm long; blades linear-lanceolate, acuminate, subcoriaceous; pinnae mostly alternate, 17–40 pairs, linear-oblong, 3–8 cm long, 5–15 mm wide above the basal tooth, apex acute to obtuse, margins spinulose toothed or deeply incised in rare forms, fertile upper pinnae abruptly reduced in size; petiole 1/4–1/3 the length of the leaf, scaly. Sori borne in one to two rows on either side of the costa. (n = 41) Jul–Oct. Moist wooded hillsides or shaded sandstone or limestone cliffs; extreme se GP; (N.S. s to FL, w to WI, IA, KS, OK, & TX).

Plants are occasionally encountered with deeply incised pinnae and fertile pinnae bearing sori at their tips. These have been referred to the dubious forma *incisum* (A. Gray) Gilbert.

2. **Polystichum lonchites** (L.) Roth, holly fern. Leaves 3–5 dm long; blades linear-oblanceolate, distinctly narrowed toward the base, acuminate, subcoriaceous; pinnae alternate, 25–45 pairs, the basal pinnae deltoid and equilateral, the middle and upper pinnae oblong-lanceolate and falcate, the larger pinnae 1–4.5 cm long, 3–13 mm wide above the basal tooth, margins unevenly serrate with the teeth spreading spinulose; petiole less than 1/6 the length of the leaf, scaly. Sori in one or two rows nearly midway between the costa

and margin and mostly confined to the upper portion of the leaf. (n = 41) Jul–Sep. Damp rocky ravines; SD: Lawrence; WY: Crook; (circumboreal, s in N. Amer. to Newf., Que., w SD, WY, NM, AZ, & CA).

3. Polystichum munitum (Kaulf.) Presl, sword fern. Leaves 3–10 dm long; blades linear-lanceolate, acuminate, subcoriaceous; pinnae alternate, 35–70 pairs, lanceolate to linear-attenuate, subfalcate, 2–10 cm long, 8–15 mm wide above the basal tooth, margins closely serrate or sometimes double serrate, the teeth incurved and spinulose; petiole about 1/4 the length of the leaf, scaly. Sori in one or several rows nearly midway between the costa and margin and confined to the upper portion of the leaf. (n = 41) Jun–Sep. Rich soil of moist shaded banks; SD: Pennington; (AK s to CA, ID, & w MT; SD).

The Black Hills station in SD represents a disjunction of some 550 miles from the metropolis of the species which lies to the northwest.

14. PTERIDIUM Gleditsch ex. Scop., Bracken

1. *Pteridium aquilinum* (L.) Kuhn, bracken, brake. Large perennial herb; rhizomes deeply subterranean, extensively creeping and branched, blackish, to 1 cm diam, reddish-brown pilose, often glabrate. Leaves distant along the rhizome, to 1.5 m tall; blade broadly triangular, sometimes appearing ternate, 2-pinnate-pinnatifid (less often 2- or 3-pinnate), 2–8 dm long, 2–9 dm wide, coriaceous, glabrous or pubescent; pinnae opposite, the lowest pair triangular to ovate, often each nearly as large as the upper portion of the blade, long petiolate, upper pinnae smaller, narrowly triangular to lanceolate, petiolate or sessile; pinnules lanceolate to elliptic-lanceolate, apex acuminate, short petiolate or sessile; ultimate segments or lobes oblong, apex obtuse, margins membranaceous, reflexed and entire; veins free except for a marginal strand; petiole stramineous, usually darker at the base, 2–7 dm long, glabrous or with a few scattered dark hairs. Sori marginal, mostly continuous, borne on the connecting vein; indusium double, the outer false indusium formed by the reflexed margin of the segment, the inner true indusium developed or obsolescent; spores tetrahedral-globose. (n = 52) Jul–Sep. Poor soil in open woods, cut-over areas, & pastures, less often in moist, rich woodlands.

Reference: Tryon, Jr., R. M. 1941. A revision of the genus *Pteridium*. Rhodora 43: 1–31, 37–67.
Three varieties occur in the GP. While intermediate plants are occasionally encountered outside the GP, our plants appear sufficiently distinct to warrant the recognition of the three varieties.

1 Ultimate segments decidedly pubescent between the midvein and margin; indusia pubescent .... 1c. var. *pubescens*
1 Ultimate segments glabrous or sometimes with scattered hairs between the midvein and margin; indusia glabrous.
   2 Margin of the ultimate segments pubescent; terminal segments 5–8 mm wide .... 1a. var. *latiusculum*
   2 Margin of the ultimate segments glabrous; terminal segments 2–4.5 mm wide .... 1b. var. *pseudocaudatum*

1a. var. *latiusculum* (Desv.) Underw. ex Heller. Blade often ternate; pinnules usually at an oblique angle to the rachises; terminal segment of pinnules not much longer than the longer lateral segments and 5–8 mm wide, segments pubescent along the margins, otherwise glabrous or only sparsely pubescent. MN, ne ND, BH, nw WY, & MO; (Newf. w to ne ND, s to TN & OK; isolated in MS, WY, SD, & CO; Eurasia).

1b. var. *pseudocaudatum* (Clute) Heller. Blade usually pinnate, rarely ternate; pinnules at oblique angle to the rachises; terminal segment of pinnules usually much longer than the lateral segments and 2–4.5 mm wide, segments usually glabrous. MO, se KS, & e OK; (MA w to OH, MO, & se KS, s to FL & TX).

1c. var. *pubescens* Underw. Blade pinnate but not appearing ternate; pinnules usually at about right angles to the costae; terminal segments of pinnules not much longer than the lateral segments and 5–8 mm wide, segments glabrous to pubescent above, pubescent below between the midvein and margins. SD: Lawrence; (Que. w to AK, s to MI, w SD, w TX, CA, & Mex.).

15. THELYPTERIS Schmidel, Beech Fern

Ours medium-sized perennial herbs; rhizomes slender, extensively creeping. Leaves scattered along the rhizome; blade bipinnatifid or pinnate-pinnatifid at least sparsely pubescent on the rachis; veins few, simple or 1-forked and reaching the margin; petiole stramineous, darkened at the base. Sori median to nearly marginal, roundish; indusium wanting or small, reniform and ciliate; spores bilateral. *Phegopteris* Fee—Rydberg.

1 Blade broadly triangular, bipinnatifid .... 1. *T. hexagonoptera*
1 Blade lanceolate, pinnate-pinnatifid .... 2. *T. palustris*

1. *Thelypteris hexagonoptera* (Michx.) Weath., broad beech fern. Rhizomes slender, wide-creeping, to more than 3 dm long, densely scaly towards the growing tip to smooth on older segments, scales mixed with scattered long hairs; scales pale brown, lanceolate, 2–5 mm long, apex long acuminate, margins entire. Leaves uniform, 2–6.5 dm tall; blade broadly triangular, bipinnatifid with the rachis broadly and continuously winged, 10–30 cm long, 12–27 cm wide, often wider than long, pilose on the rachis, costae, and margins, sparsely pubescent on other surfaces; primary segments subopposite, the upper and middle lanceolate, apex acuminate, their lobes oblong and entire, the lower primary segments basiscopic, narrowly ovate to narrowly rhombic, deeply lobed, 6–15 cm long, 2.5–5 cm wide, widest near the middle, apex long acuminate, margins crenate to entire, base acuminate, veins mostly simple; petiole 10–35 cm long, with scales near the base similar to those on the rhizome, smooth otherwise. Sori nearly marginal, roundish; indusium lacking. (n = 30) Jun–Sep. Moist wooded, rocky hillsides & ravines; KS: Cherokee; sw MO & e OK; (Que. w to MN, s to FL & TX). *Dryopteris hexagonoptera* (Michx.) C. Chr.—Fernald; *Phegopteris hexagonoptera* (Michx.) Fee—Rydberg.

2. *Thelypteris palustris* Schott, marsh fern. Rhizome slender, extensively creeping and branched, to more than 3 dm long, scales few or lacking. Leaves slightly dimorphic; sterile leaves 1 m tall, blade membranaceous, lanceolate or less often narrowly ovate, pinnate-pinnatifid, 2.5–5 dm long, 6–20 cm wide, pilose especially on the lower surface and costae to nearly glabrous, minute brownish, ovate scales sometimes present of the lower surface, apex acute; pinnae subopposite to alternate, lanceolate to rarely narrowly ovate, 3–10 cm long, 0.7–2 cm wide, apex acuminate; pinnae segments ovate to elliptic, apex rounded to acute, margins flat, entire or rarely serrate, base sometimes auriculate, petiole stramineous, blackish at the base, 2–6 dm long, usually smooth; fertile leaves similar to the sterile except coriaceous, to 1.5 m tall; blade lanceolate, slightly longer and narrower; pinnae segments with reflexed or revolute margins partially hiding the sori. Sori median, roundish, often confluent with age, forming a line around the pinnae margins; indusia persistent, reniform, ciliate, often glandular. (n = 35) Jun–Oct. Usually sandy soil in wet ditches & prairie ravines, marshes, low woods, or thickets; MN, ND, sw SD, NE, MO, e KS; (Newf. w to Man. & cen ND, s to FL & TX; Eurasia). *Dryopteris thelypteris* (L.) A. Gray var. *pubescens* (Lawson) Nakai—Fernald.

16. WOODSIA R. Br., Woodsia

Small to medium-sized perennial herbs; rhizomes short creeping, covered with old petiole bases, scaly. Leaves tufted; blade membranaceous to subcoriaceous, ovate to linear-lanceolate, 1- or 2-pinnate; veins free; petiole slender, usually shorter than the blade. Sori medial, discrete or confluent with age, roundish; indusia basal, thin, globose, early opening at the top and splitting into spreading segments, segments filiform or wider; spores bilateral.

Reference: Brown, D. F. 1964. A monograph of the fern genus *Woodsia*. Beih. Nova Hedwigia 16: 1–154.

1 Petiole jointed 1.5 cm above the base .... 1. *W. ilvensis*
1 Petiole continuous, not jointed.
  2 Lower surface of the blade pilose and glandular, especially on the veins .... 5. *W. scopulina*
  2 Lower surface of the blade glabrous or sparsely glandular but never pilose.
    3 Indusium segments broad, platelike; rachis usually scaly; petiole stramineous .... 3. *W. obtusa*
    3 Indusium segments filamentous; rachis usually not scaly; petiole castaneous at the base and becoming stramineous above, rarely wholly stramineous.
      4 Segments of the indusium 10 or more, much longer than the sporangia, flaccid, intertwined .... 2. *W. mexicana*
      4 Segments of the indusium fewer than 10, mostly shorter than the sporangia and concealed at maturity .... 4. *W. oregana*

1. ***Woodsia ilvensis*** (L.) R. Br., rusty woodsia. Rhizome to 6 cm long, 3–7 mm diam; scales brownish with a dark median stripe, ovate lanceolate, 4–5(7) mm long, entire. Leaves 5–30 cm tall; blade subcoriaceous, lanceolate to oblong-lanceolate, pinnate-pinnatifid to 2-pinnate, 4–10(15) cm long, 1.5–5 cm wide, sparsely strigose or pilose above, pilose and chaffy with brownish, linear-lanceolate scales below, apex acuminate; pinnae subopposite, ovate to lanceolate, pinnatifid or pinnate, 0.7–2.5 cm long, 3–12 mm wide, apex acute to obtuse, lower pinnae sessile or petiolate, upper pinnae sessile; pinnae lobes or pinnules elliptic, apex obtuse, margins subentire to repand, often reflexed, petiole castaneous into the rachis or less often castaneous at the base and stramineous above, 2–10 cm long, jointed 1–5 cm above the base, pilose, densely scaly at the base with brownish, linear-lanceolate scales, less scaly above. Sori often becoming confluent; indusium of 10–20 filamentous segments, segments mostly longer than the sporangia but often appearing interwoven among them. (n = 41) Jun–Sep. Dry, rocky ledges & cliffs; MN; IA: Lyon; (circumboreal, s in N. Amer. to NC, IA, & B.C.).

2. ***Woodsia mexicana*** Fee. Rhizome to 4 cm long, 3–5 mm diam; scales brownish with a dark median stripe, lanceolate, 3–5 mm long, entire. Leaves 7–25 cm tall, blade membranaceous, oblong-lanceolate, 1-pinnate to pinnate-pinnatifid, 4–15 cm long, 1.5–4.8 cm wide, sparsely to densely glandular, apex acuminate; pinnae subopposite or sometimes alternate, simple or sparingly lobed pinnae oblong to narrowly ovate, pinnatifid pinnae ovate to lanceolate, 0.7–2.4 cm long, 3–11 mm wide, apex obtuse to rounded, rarely acute, margins erose, flat, base short petiolate on the lower portion of the blade and becoming sessile above; pinnae lobes oval to oblong, apex obtuse or rounded; petiole stramineous, may be castaneous at the base, 1.5–7(10) cm long, continuous, sparsely to densely glandular, usually chaffy with pale brown, linear-lanceolate scales. Sori median to submarginal, often confluent; indusium of 7–15 filamentous to lanceolate lobes, glandular-ciliate, segments longer than the sporangia. (n = 82) Jun–Sep. Dry, often exposed, rocky hillsides & ledges; CO: Baca, Las Animas; OK: Cimarron; NM: Union; TX: Armstrong; (w OK & se CO, w to AZ, s to w TX & Mex.).

The KS record reported in the Atlas GP is *W. obtusa* × *W. oregana*.

3. **Woodsia obtusa** (Spreng.) Torr., blunt-lobed woodsia. Rhizomes to 4 cm long, 2–5 mm diam; scales brownish with a dark median stripe, lanceolate, 2–5 mm long, entire. Leaves (1)2–6 dm tall; blade membranaceous to subcoriaceous, ovate to oblong-lanceolate, pinnate-pinnatifid to 2-pinnate, 7–45 cm long, 3.5–11 cm wide, glabrous to glandular, glands stalked or sessile, apex acute to acuminate; pinnae subopposite, ovate to ovate-lanceolate, pinnatifid or pinnate, 1.8–6 cm long, 1–3 cm wide, apex obtuse to acute, lower pinnae short petiolate, the upper mostly sessile; pinnules or pinnae lobes oval, elliptic, or oblong, apex obtuse or rounded, margins flat, serrate; petiole stramineous, rarely darkened at the base, 3–17 cm long, continuous, glandular or not, scaly, the scales light brown, linear-lanceolate, to 6 mm long, entire. Sori discrete or sometimes nearly confluent on small leaves; indusium of 4 or 5 broad segments, segments about as long as the sporangia, nearly entire or with several filamentous lobes at or near the apex. (n = 76) Jun–Sep. Moist shaded banks, wooded hillsides, rocky ledges, & cliffs; sw MN, IA, e NE, MO, e KS, OK; (Que. w to MN, s to FL & TX).

The only known report of a hybrid between *W. obtusa* and *W. oregana* var. *oregana* (*W.* × *kansana* R. E. Brooks) is from McPherson and Ellsworth counties, Kansas (Brooks, Amer. Fern J. 72: 79–84. 1982.). The hybrid exhibits morphological characteristics intermediate with those of the parental species and its spores are abortive. Among the numerous parental plants growing at the station only a few hybrids were discovered.

4. **Woodsia oregana** D. C. Eat., Oregon woodsia. Rhizomes to 5 cm long, 3–5 mm diam; scales few, brown with a dark median stripe, lanceolate, 2.5–4 mm long, entire. Leaves 5–15(30) cm tall; blade coriaceous to subcoriaceous, oblong-lanceolate to linear-lanceolate, pinnate-pinnatifid to 2-pinnate, 4–20 cm long, 1–5 cm wide, glabrous or sparsely glandular, apex acute to acuminate; pinnae subopposite to alternate, deltoid to lanceolate, pinnatifid or pinnate (at least the lower pinnae), 0.5–2.5 cm long, 4–10(15) mm wide, apex obtuse to acute, short petiolate or sessile; pinnae lobes or pinnules oval, obovate or oblong, apex obtuse to rounded, margins flat or sometimes reflexed, serrate; petiole dark red-brown on about the lower third and becoming stramineous above, 1–10 cm long, continuous, glabrous or sparsely glandular, scales few or none except at the base, scales brownish, lanceolate, entire. Sori discrete to confluent, medial; indusium of 5–9 filamentous segments, the segments 1 or 2 cells wide, mostly shorter than the sporangia. (n = 38) Jun–Sep. Rocky prairie hillsides, shaded banks, & rocky ledges; se MN w to MT, s to cen KS, w OK, & NM; (Que. w to B.C., s to VT, MN, w OK, NM, AZ, & CA).

This species hybridizes with *W. obtusa;* see no. 3.

5. **Woodsia scopulina** D. C. Eat., Rocky Mountain woodsia. Rhizomes to 5 cm long, 4–6 mm diam; scales few, brown with a dark median stripe, lanceolate, 4–5 mm long, entire. Leaves 0.5–4 dm tall; blade membranaceous, oblong-lanceolate, pinnate-pinnatifid to 2-pinnate, 4–25 cm long, 2–6 cm wide, glabrous or pilose and glandular above, whitish pilose and glandular below and on the rachis, apex acuminate; pinnae subopposite to alternate, lanceolate or rarely ovate, 1–3 cm long, 4–10(16) mm wide, apex acute to acuminate, mostly subsessile; pinnae lobes or pinnules oval or elliptic, apex obtuse, margins serrate, flat; petiole reddish-brown, often becoming stramineous at or above the middle, 1–15 cm long, continuous, whitish pilose and glandular or glabrate, scaly only near the base, scales brownish, ovate-lanceolate, entire. Sori mostly discrete, median; indusium of 5–9 straplike segments, segments about as long as the sporangia, 3–5 cells wide, often forked at the apex, glandular. (n = 38) Jun–Sep. Rock crevices & ledges; w SD & WY; (Que. w to AK, s to MN, w SD, NM, AZ, & CA).

8. MARSILEACEAE R. Br., the Pepperwort Family

by Ralph E. Brooks

Small aquatic, perennial herbs; rhizomes slender and wide creeping. Leaves alternate, erect, and borne distant along the rhizome. Sori borne within a hard short-pediceled sporocarp which represents a modified pinna with connate margins, each a sorus bearing microsporangia and megasporangia; microsporangia with numerous minute microspores, megasporangia with a single larger megaspore.

1 Leaves petioled and with 4-foliate blades .... 1. *Marsilea*
1 Leaves filiform, without an expanded blade .... 2. *Pilularia*

1. MARSILEA L., Pepperwort; Water Clover

1. ***Marsilea vestita*** Hook. & Grev., western water clover. Rhizomes less than 1 mm diam, branching. Leaves long petiolate with cruciform (4-foliolate) blades, to 35 cm tall; blades emergent and often folded, floating or submersed; pinnae obdeltoid, 5–25 mm long and nearly as wide, glabrous to densely pubescent with short to long appressed hairs, margins entire, sessile. Sporocarps solitary on short peduncles attached to the rhizomes, ellipsoid, 5–8 mm long, 1.5–3 mm thick, usually with a pair of projections near the base, densely pubescent, with short appressed hairs or eventually glabrate, splitting into 2 valves at maturity and discharging 10–20 sori on a gelatinous receptacle, each sorus with 5–20 megasporangia and numerous microsporangia. (n = 16) May–Oct. Shallow water or stranded in roadside ditches, temporary pools, ponds, irrigated fields, & rarely shallow creeks; GP, common in the w 1/2 & scattered e; (Sask. & Alta. s to TX, AZ, & Mex.). *M. mucronata* A. Br. — Gleason & Cronquist.

Aquatic forms are generally glabrous with relatively long leaf petioles and usually produce few sporocarps, while strand plants tend to be densely hairy with short leaf petioles and produce an abundance of sporocarps.

Marsilea quadrifolia L. is a native of Europe and southeast Asia that has been collected once in KS: Cherokee. It can be recognized by its glabrous sporocarps that are borne on peduncles attached to the leaf petiole above its base. Rhizomes of the species do not produce fascicled branches.

Marsilea tenuifolia Kunze is an endemic of central TX that has been previously reported from OK: Osage numerous times but the specimens are a narrow-leaf form of *M. vestita*.

Marsilea uncinata A. Br. is known from TX: Potter and could be expected in other neighboring areas. It is distinguished by rhizomes which produce fascicled branches that are paleaceous at their tips and pubescent sporocarps borne on peduncles attached to the leaf petiole above its base.

2. PILULARIA L., Pillwort

1. ***Pilularia americana*** A. Br., American pillwort. Inconspicuous mat-forming plants; rhizomes filiform. Leaves filiform, 1.5–11 cm long, glabrous. Sporocarps solitary on short (less than 2 mm long), deflexed peduncles at the nodes, brown or blackish, spherical, 2–3 mm diam, appressed woolly, 2- to 4-celled, each cell with a single sorus; sorus on an elongate receptacle with lateral microsporangia and apical megasporangia; spores tetrahedral. Jul–Oct. Submerged in shallow water of small lakes or temporary pools in sandhill regions, sometimes stranded; NE: Cherry; KS: Harvey, Reno; OK: Comanche; (GA & AR; NE s to TX; OR & CA).

9. SALVINIACEAE Dum., the Water Fern Family

by Ralph E. Brooks

1. AZOLLA Lam., Water Fern, Mosquito Fern

1. *Azolla mexicana* Presl. Delicate annual, free floating; stems compactly and dichotomously branched, forming mats to 3 cm diam; roots unbranched. Leaves sessile, 2-lobed, crowded; upper lobe emersed, rhombic-ovate to obovate, 0.7–1.3 mm long, green to dark red; lower lobe floating, broadly ovate, larger than the upper lobe, mostly achlorophyllus. Sporocarps of two kinds, borne in like or occasionally mixed pairs in axils of lower leaf lobes; microsporocarps about 1.3 mm long, containing numerous microsporangia in which the microspores are aggregated into several massulae, each massula bearing septate glochidia; megasporocarps about 0.4 mm long, containing a single megasporangium and megaspore. Jul–Nov. Backwater areas, marshes, ponds, & ditches; occasionally stranded; NE, KS, OK, IA, MO; (WI s to TX, Mex., & S. Amer., w to B.C. & CA).

Azolla shares a symbiotic relation in the leaf with a blue-green alga, *Anabaena azolla*.
Our plants have often been misidentified as *A. caroliniana* Willd., a more eastern species with nonseptate glochidia.

Division PINOPHYTA (Gymnosperms)

10. CUPRESSACEAE Bartl., the Cypress Family

by Ralph E. Brooks

1. JUNIPERUS L., Juniper

Aromatic, evergreen, dioecious or sometimes monoecious, prostrate, decumbent, or erect shrubs or trees; stems with thin shredded or thick checkered or furrowed bark and soft close-grained white or reddish wood; most with deep taproots. Leaves persistent for several years, scalelike to linear, opposite or in whorls of 3, decurrent or sometimes jointed at the base, dimorphic in some species, i.e., the juvenile leaves subulate or linear and the older leaves scalelike. Cones axillary and short stalked or terminal, solitary or sometimes clustered; staminate cones catkinlike, spherical to ovoid, the microsporophylls opposite or in whorls of 3, peltate, and with 3–6 anther-locules attached to the abaxial side; pistillate cones of 2–6, coalescent scales, the lower sterile and the terminal sometimes fertile, the scales at maturity becoming fleshy and forming a bluish to reddish-brown, glaucous, berrylike fruit; seeds 1–6, ovoid, wingless; cotyledons usually 2. *Sabina* Haller—Rydberg.

References: Adams, R. P. 1972. Chemosystematics and numerical studies of natural populations of *Juniperus pinchotii* Sudw. Taxon 21: 407–427; Adams, R. P., & T. A. Zanoni. 1979. The distribution, synonymy, and taxonomy of three junipers of southwestern United States and northern Mexico. Southw. Naturalist 24: 323–329; Flake, R. H., L. Urbatsch, & B. L. Turner. 1978. Chemical documentation of allopatric introgression in *Juniperus*. Syst. Bot. 3: 129–144.

1. Leaves all needlelike, spreading, jointed at the base, and not decurrent on the stem; staminate cones axillary .... 1. *J. communis*
1. Leaves scalelike (the juvenile leaves may be needlelike), not jointed at the base and not decurrent on the stem; staminate cones terminal.
   2. Prostrate, mat-forming shrubs; peduncles recurved. .... 2. *J. horizontalis*
   2. Upright shrubs or small trees; peduncles straight.
      3. Leaf margins minutely cellular-serrulate or cellular-denticulate.
         4. Ovulate cones bluish-brown to dark blue, glaucous; glands on the leaves rarely leaving a white crystalline exudate when crushed and dried ...... 3. *J. monosperma*
         4. Ovulate cones copper to reddish-brown, not glaucous; glands on the leaves leaving a white crystalline exudate when crushed and dried .... 4. *J. pinchotii*
      3. Leaf margins entire and smooth, sometimes hyaline.
         5. Leaves distinctly overlapping; dorsal gland on leaves usually shorter than the distance between the gland and leaf apex. .... 6. *J. virginiana*
         5. Leaves barely, if at all, overlapping; dorsal gland on leaves usually longer than the distance between the gland and leaf apex. .... 5. *J. scopulorum*

1. *Juniperus communis* L., common or dwarf juniper. Low, often decumbent dioecious or rarely monoecious shrubs to 1.5 m tall, often forming clumps; young twigs yellowish and 3-angled, older stems grayish or reddish-brown with the bark papery and shredding. Leaves in whorls of 3, spreading at nearly right angles to the stem and curved sharply upward just above the base, linear-lanceolate, acerose, flat or slightly conduplicate with ventral sides facing each other, (7)10–18 mm long, 0.9–1.5 mm wide, apex of upper surface with a broad glaucescent or whitish bank of stomata, apex sharp pointed, margins entire, base jointed but not decurrent. Staminate cones mostly solitary, axillary, 3–5 mm long, 1–2 mm wide, sessile; ovulate cones solitary, dark blue with a glaucous bloom, succulent, globose, 5–10 mm diam, short stalked, maturing the second season and persisting for a year or more, sessile; seeds 1–3 per cone, light brown, broadly ovoid, 4–5 mm long, irregularly tuberculate, apex apiculate or not; cotyledons 2. (2n = 22) May–Jun. Usually on rocky or sandy wooded hillsides, less often in exposed sites; w 1/2 ND to MT, s to nw NE & WY; also ND: Bottineau; SD: Spink; n MN; (circumboreal, s in N. Amer. to SC, IN, IL, MN, nw NE, MT, NM, & CA). *J. communis* var. *depressa* Pursh—Fernald.

2. *Juniperus horizontalis* Moench, creeping juniper. Dioecious, glabrous shrubs, prostrate and mat forming, the main stems often trailing 3–5 m or more; young twigs reddish-brown to yellowish, older stems dark reddish-brown or grayish with shredding bark. Leaves dark green or yellow-green with a recessed dorsal gland, imbricate, margins entire; leaves of young stems frequently glaucous, usually slightly spreading, acerose, 3–7 mm long, 0.6–1.3 mm wide, base decurrent, dorsal gland often obscure; older stems with leaves tightly appressed, scalelike, mostly ovate to triangular, 0.7–2.5 mm long, 0.4–1 mm wide, apex acute to acuminate, dorsal gland round or in the nw GP often elliptic. Staminate cones solitary at the tips of branchlets, yellowish at first and becoming purplish, ellipsoid, 3.5–4.5 mm long, 1–2 mm diam, sessile; ovulate cones maturing in the second season, solitary on usually recurved tips of branchlets, dark blue and glaucous, succulent, globose 6–8(10) mm diam, sessile; seeds 2–4(6) per cone, red-brown, ovoid and usually flat on one side, 3–5 mm long, 2–4 mm diam, muricate. (2n = 22) May–Jun. Usually dry, rocky, open hillsides or wash areas, less often in open woods; n MN w to MT, s to nw & n-cen NE & WY; CO: Baca, Las Animas; (N.S. w to AK, s to NY, IL, IA, MN, NE, MT, CO, & cen B.C.). *Sabina horizontalis* (Moench) Rydb.—Rydberg.

    Hybrids with *J. scopulorum* have been reported at the nw periphery of our region (Van Haverbeck, Univ. Nebraska Studies, n. ser. 38. 1968).

3. *Juniperus monosperma* (Engelm.) Sarg., one-seeded juniper. Small to medium dioecious or rarely monoecious shrubs or small trees to about 6 m tall, usually with several branches from the base or just above the base, crown rounded, bark light ashy-gray, ridged and shredding; branches grayish to reddish-brown, usually roughened by numerous old leaf bases and shredding bark, glabrous. Leaves usually yellow-green with a round to broadly elliptic, recessed gland on the dorsal side, glands if ruptured rarely with white crystalline exudate, tightly appressed and imbricate, glabrous, margins cellular serrulate; leaves or juvenile stems acerose and somewhat spreading, 2–5 mm long, 0.6–1 mm wide, decurrent at the base; leaves of mature branches ovate, 1–3 mm long, 0.5–1.2 mm wide, much thicker than the juvenile leaves, apex acute or sometimes acuminate. Staminate cones brownish or yellowish, solitary at the tips of branchlets, ellipsoid, 3–4 mm long, about 2 mm diam, sesslie; ovulate cones maturing in the first season, solitary at the tips of branchlets, bluish-brown to dark blue, glaucous, somewhat fleshy, globose, 5–7 mm diam, sessile or subsessile; seeds 1 per cone, shiny reddish-brown, ovoid to nearly globose, 3.5–4.5 mm long, smooth. Mar–Apr. Gravelly, rocky, or sandy soils (infrequently gypsiferous) on open flats, hillsides, or in canyons; se CO; OK: Cimarron; n 1/2 TX panhandle; NM; (se CO, w OK, n 1/2 TX panhandle w to AZ & s through NM to w TX).

    Previous reports of this species from Oklahoma were based on specimens of *J. pinchotii*. *Juniperus monosperma* and *J. pinchotii* are very similar morphologically, with fruit color the only apparent character that can be used consistently to separate them. Hybridization between *J. monosperma* and *J. pinchotii* has been detected in the Palo Duro Canyon area of Texas (Adams, op. cit.).

4. *Juniperus pinchotii* Sudw., Pinchot juniper. Similar to *J. monosperma* except: glands on leaves if ruptured usually leaving a white crystalline exudate; ovulate cones copper to reddish-brown, not glaucous. Sep–Oct. Exposed hillsides, canyons, caprock regions, often in gypsiferous soil; s 1/2 TX panhandle & w OK (except panhandle); (w OK, sw 1/2 TX panhandle, se NM, s to n Mex.).

    See discussion on no. 3.

5. *Juniperus scopulorum* Sarg., Rocky Mountain juniper. Medium-sized dioecious or rarely monoecious trees to 10 m tall, scraggly with rounded crowns to pyramidal, rarely rigid columnar; bark dark reddish-brown to grayish, thin, fibrous and usually shredding in age, furrowed; branches grayish to reddish-brown, mostly smooth. Leaves green or blue-green, usually barely overlapping and closely appressed, 0.8–4.3 mm long, 0.7–1.5 mm wide, recessed gland on dorsal side elliptic to occasionally oval and mostly longer than the distance between the gland and leaf apex, apex obtuse to weakly acute, margins entire; leaves of juvenile branches spreading to appressed, acerose, to 7 mm long. Staminate cones yellowish-brown, solitary at the tips of branchlets, mostly ellipsoid, 2–4 mm long, 1–2 mm diam; sessile; ovulate cones maturing in the second season, solitary at the tips of branchlets, blue to purplish, glaucous, somewhat fleshy, with a resinous pulp, globose to subglobose, 5–7 mm diam, sessile; seeds (1)2(3) per cone, yellowish to tan, ovoid 3–4 mm long, smooth or weakly ridged toward the base. (n = 11) Apr–May. Dry, rocky, or sandy hillsides, canyons, & wash areas; w 1/2 ND, MT, w 2/3 SD, WY, w 1/3 NE; CO: El Paso, Elbert, Las Animas; OK: Cimarron; ne NM; TX: Armstrong, Briscoe, Randall; (w ND w to WA, s to TX, NM, AZ, & NV; n Mex.). *J. scopulorum* var. *columnaris* Fassett—Fassett, Bull. Torrey Bot. Club 72: 482. 1945; *Sabina scopulorum* (Sarg.) Rydb.—Rydberg.

6. *Juniperus virginiana* L., red cedar. Medium-sized dioecious or rarely monoecious trees to 10(20) m tall, pyramidal to occasionally columnar; bark reddish-brown to grayish, thin, fibrous, and shredding; branches usually reddish-brown. Leaves green or blue-green, closely appressed and overlapping the leaf above, 1.5–3.5 mm long, 0.8–1.5 mm wide, dorsal gland oval to elliptic and mostly shorter than the distance between the gland and leaf apex, apex acute, to acuminate, margins entire; leaves of juvenile branches widely spreading, acerose, to 11 mm long. Staminate cones yellowish-brown, solitary at the tips of branchlets, ovoid to ellipsoid, 2.5–4 mm long, 1–2 mm diam, sessile; ovulate cones maturing the first season, solitary at the tips of branchlets, dark blue or bluish-purple, glaucous, globose to subglobose or ovoid, with a thin resinous pulp, 4–7 mm diam, sessile; seeds (1)2(3) per cone, yellowish-brown, ovoid, 2–4 mm diam, ridged near the base and sometimes shallowly pitted. (2n = 22) Apr–May. Prairie hillsides, fields, pastures, & occasionally in woodlands, rocky, sandy, or clay soils; MN, e edge ND, IA, e 2/3 SD, NE, MO, KS, ne CO, OK, except panhandle; (MA w to e ND & SD, CO, s to FL & e TX). *Sabina virginiana* (L.) Antoine — Rydberg.

Where unchecked by appropriate range management procedures, red cedar is becoming an increasing menace in pasture and rangeland areas. Densely covered hillsides are a frequent sight in the se GP, where the plant has invaded poorly managed or unmanaged areas.

11. PINACEAE Lindl., the Pine Familiy

by Ralph E. Brooks

Ours evergreen, monoecious trees with resinous wood and foliage; mostly with deep taproots. Leaves persistent for several years, spirally arranged or fascicled, linear or needlelike. Staminate cones consisting of numerous spirally arranged microsporophylls each with 2 pollen sacs on the abaxial side, soon deciduous after anthesis. Ovulate cones consisting of many spirally arranged scales (secondary scales) each with 2 naked ovules on the adaxial side, each scale subtended by a distinct bract (primary scale); fruit a woody cone much larger than the staminate cones and maturing in the first or second season; seeds winged or not; embryo axile in the endosperm; cotyledons several.

Martin & Hutchins (1980) report *Abies concolor* (Gord. & Glend.) Lindl. ex Hildebr. from NM: Union. It can be recognized by its erect cones with deciduous scales and leaves that are flattened in cross section.

1 Leaves spirally arranged and borne on more or less elongated branches .... 1. *Picea*
1 Leaves borne in fascicles of 2–5 surrounded at the base by a papery sheath .... 2. *Pinus*

1. PICEA A. Dietr., Spruce

1. *Picea glauca* (Moench) Voss, white or Black Hills spruce. Evergreen pyramidal trees to 25 m tall; trunk gradually tapering, bark thin and scaly; young twigs yellowish but turning gray with age, glabrous, roughened with persistent leaf bases. Leaves spirally arranged and borne on elongate branches, appearing upcurved and distributed more on the upper side of the branches, dark green, linear, 4-sided, 7–18 mm long, 1 mm wide, stomatiferous on all sides, glabrous, deciduous on dried specimens. Cones borne on the twigs of the previous year and mostly in the upper portion of the tree. Staminate cones 1–4 in axillary clusters, surrounded at first by pale brown membranaceous bud scales that are soon deciduous; mature cones yellowish, ovoid to cylindric, 10–25 mm long, short stalked; pollen with winglike extensions. Ovulate cones covered by membranaceous, soon deciduous scales in bud, young cones reddish, ovoid, ca 10 mm long, sessile; at maturity (first season) cones deciduous as a whole, pendulous, light brown, oblong-cylindrical, 3–5 cm long, 2–2.5 cm wide; secondary scales thin, woody, persistent, obovate to broadly so, entire or rarely emarginate at the apex, primary bracts much smaller, thin, erose. Seeds ca 2 mm long, the wing about 3 × as long. (2n = 24) May–Jun. Moist, rich soils in coniferous woods, often on n-facing

hillsides or in moist boggy areas; BH & nw MN; (Lab. w to AK, s to n NY, MI, MN, w SD, & WY).

Picea pungens Englm. (blue spruce) is reported to occur in the Sierra Grande area in NM: Union (pers. comm., J. P. Hubbard, NM Fish & Game, Santa Fe). It can be distinguished from *P. glauca* by its glaucous leaves and large ovulate cones (6–10 cm long).

2. PINUS L., Pine

Evergreen trees with scaly or furrowed bark; branches mostly whorled. Leaves dimorphic; the conspicous green, needlelike leaves borne in fascicles of 2–5 (ours) on deciduous spurs in the axils of scalelike primary leaves and in bud enclosed by scales that elongate and form a membranaceous sheath at the base of each fascicle. Staminate cones in axillary clusters. Ovulate cones maturing at the end of the second or third season and opening to release seeds or remaining closed for years; each scale of the cone with a thickened exposed part (apophysis) terminated by a brown protuberance (umbo); seeds winged or not. *Apinus* Necker—Rydberg.

Early literature for the region occasionally reported the presence of *Pinus echinata* P. Mill. in se KS; while the species occurs 25 miles to the s in OK, it does not occur in KS today, nor do specimens exist to document past reports. *Pinus banksiana* Lamb., jack pine, is occasionally planted in the ne GP where it has been reported to escape in the sandhills region of ND.

1 Leaves 5 to a fascicle .... 3. *P. flexilis*
1 Leaves 2 or 3 to a fascicle.
  2 Leaves mostly 3 to a fascicle, 8–20 cm long .... 4. *P. ponderosa*
  2 Leaves mostly 2 to a fascicle, 2–7 cm long.
    3 Leaf sheaths mostly deciduous; dorsal umbo of cone scales unarmed, seeds not winged .... 2. *P. edulis*
    3 Leaf sheaths persistent; dorsal umbo of cone scales armed with a sharp prickle; seeds prominently winged .... 1. *P. contorta* var. *latifolia*

1. ***Pinus contorta*** Dougl. ex Loud. var. ***latifolia*** Engelm. ex Wats., lodgepole pine. Trees to 25 m tall, mostly with few lower branches and a pyramidal crown; trunk with gray or orange-brown, thin bark. Leaves in fascicles of 2, 3–5.5 cm long, margins serrulate, the membranaceous sheaths persistent. Staminate cones in clusters of 25–50, yellowish, cylindric 8–11 mm long; ovulate cones borne mostly on upper branches, some opening at the end of the second season and others remaining closed for several years, ovoid and slightly asymmetrical, 3–6 cm long, 3–3.5 cm wide, subsessile, scales brown with a lighter apex, dorsal umbo armed with a sharp prickle. Seeds 4–4.5 mm long with a prominent wing about $3 \times$ as long as the seed. ($2n = 24$) Jun. Well-drained soils on mt. slopes; SD: Lawrence; (w Alta. to s Yukon, s to w SD, s CO, ne UT, ID, & OR). *P. murrayana* Balf.—Rydberg.

Lodgepole pine occurs in NE: Thomas, in the Nebraska National Forest. These specimens, however, were planted and are not native to that area.

2. ***Pinus edulis*** Engelm., pinyon pine. In our area small pyramidal trees to 10 m tall with branches often borne near ground level; trunk frequently crooked and twisted; bark irregularly furrowed and with small scales. Leaves in fascicles of 2(3), 2.5–4 cm long, often curved upward on the branches, margins entire, membranaceous sheaths mostly deciduous. Staminate cones in clusters of 20–40, purplish-red to yellow, ovoid, 4–6 mm long; ovulate cones maturing the first season, ovoid, 2–5 cm long and about as wide, short peduncled, scales light brown and thickened toward the apex, dorsal umbo unarmed. Seeds thick shelled and edible, brown, ovate, 10–16 mm long, wingless. ($n = 12$) May–Jun. Rocky soil of mesas & dry slopes; se CO, ne NM, & OK: Cimarron; (s WY w to UT, s to w OK, w TX, e AZ, & Mex.).

3. **Pinus flexilis** James, limber pine. Ours broadly pyramidal trees to 12 m tall with crooked trunks; bark smooth at first and becoming gray-black with deep furrows and thick scales; branches pubescent when young and becoming glabrous, usually light gray. Leaves 5 to a fascicle, 4–7 cm long, usually curved slightly, margins entire, membranaceous sheaths deciduous. Staminate cones in clusters of 30–60, pale yellow, ovoid to cylindrical, 7–10 mm long; ovulate cones maturing during the second season, green into the second season and then becoming light brown, ovoid to narrowly so, often curved, 9–11 cm long, 5–6 cm wide, scales thick especially near the apex, dorsal umbo rounded, unarmed; seeds dark brown or mottled, ovoid, 10–11 mm long, wingless. (2n = 24) May–Jun. Exposed ridges or foothills; sw NE, sw WY, & ne CO; ND: Slope; SD: Custer; (Rocky Mts. from Alta. w to B.C., w GP, s to NM & AZ; s CA). *Apinus flexilis* (James) Rydb.—Rydberg.

4. **Pinus ponderosa** Laws., ponderosa pine. Large trees with pyramidal crowns, crown often becoming open with age, to 35 m tall; trunk straight, in young trees bark gray-brown with deep furrows, with age bark becomes gray mixed with orange-brown and scaly; branches gray-black, glabrous. Leaves (2)3 to a fascicle clustered near the tips of the branches, (5)8–20 cm long, margins serrulate, membranaceous sheath deciduous, orange-brown, 1.5–2.5 cm long. Staminate cones 10–20 per cluster, yellowish-orange, cylindrical, 1.5–3 cm long; ovulate cones maturing during the second season, brown, broadly ovoid, 6–12 cm long, 6–8 cm wide, subsessile, scales with a slender prickle. Seeds 6–7 mm long with a prominent papery wing 3–4 × as long as the seed. (2n = 24) May–Jun. Rocky hills & low mts.; nw GP, scattered from w-cen ND, w to MT, s to w NE & WY; infrequent e to s-cen SD & cen NE; OK: Cimarron; CO: Elbert, El Paso; (w ND w to s B.C., s to w OK, w TX, s CA; n Mex.). *P. scopulorum* (Engelm.) Lemm.—Rydberg.

12. EPHEDRACEAE Dum., the Ephedra Family

by Ralph E. Brooks

1. EPHEDRA L., Mormon Tea, Mexican Tea

Medium-sized, erect or profusely branched, dioecious (rarely monoecious) shrubs; stems with jointed, opposite or whorled branches, the branches grooved, greenish, and photosynthetic; usually with deep taproots. Leaves opposite at first and later opposite or whorled, reduced and scalelike, forming a membranaceous sheath. Staminate cones borne (1)2 or 3 at the nodes of young branches, each cone with 2–8 stamens free or united into a column and each with 1–8 terminal bilocular pollen sacs. Ovulate cones borne 1–3 at the nodes of young branches, sessile or pedunculate, cones with 3–12 whorled, membranaceous or fleshy bracts surrounding 1–3 central ovules, the ovules enclosed in an urn-shaped envelope formed by the union of two lower bracts to form an outer integument and two upper bracts to form a tubular inner integument which extends through the opening of the outer integument forming a pollen chamber at the tip of the ovule; embryo with 2 cotyledons.

Reference: Cutler, H. C. 1939. Monograph of the North American species of the genus *Ephedra*. Ann. Missouri Bot. Gard. 26: 373–423.

Ephedra coryi E. L. Reed was erroneously reported in the Atlas GP from TX: Briscoe based on a specimen of *E. antisyphilitica* Berl. ex C. A. Mey. *Ephedra coryi* is endemic to w-cen TX immediately south of our area and may be distinguished by its distinctly peduncled (more than 5 mm long) ovulate cones and opposite leaves that are connate for less than 1/2 their length, and upon falling leave a dark brown persistent base.

1 Leaves opposite .... 1. *E. antisyphilitica*
1 Leaves in whorls of 3 .... 2. *E. torreyana*

1. ***Ephedra antisphyilitica*** Berl. ex C. A. Mey., clapweed. Erect or spreading shrub to 1 m tall, branches green, gray-green, or yellowish-green, rigid and hard; bark on older stems cinereus or grayish, cracked and irregularly fissured, often becoming shredded; plants from deep taproots. Leaves usually soon deciduous, opposite (rare plants have been seen with several branches having leaves in whorls of 5 or 6, apparently altered by insect activity), connate from 1/2 to nearly their total length, pale green or yellowish, 1–4 (11, in aberrant forms) mm long, apex acute, margins entire, base not thickened or only slightly so (often appearing thickened because of the swollen node) and with a narrow tan to yellowish-orange band encircling the stem. Staminate cones 1 or 2 per node, ellipsoid to ovoid, 4–8 mm long, 3–5 mm wide, subsessile; bracts in 5–8 opposite pairs, greenish to slightly reddened, obovate 2–3 mm long, 2–3 mm wide, slightly thickened with hyaline margins, lower bracts empty; perianth slightly exceeding the bract; staminal column 3–4.5 mm long, about 1/2 exserted. Ovulate cones 1 or 2 per node, caducous, ellipsoid, 6–11 mm long, 5–8 mm wide, subsessile; bracts in 4–6 pairs, the inner pairs becoming brilliant red and fleshy at maturity, ovate, 2.5–6 mm long, 2–4 mm wide, base connate. Seeds 1(2) per cone, chestnut or sometimes light brown, lanceoloid, 3(4)-sided, 6–9 mm long, 2–3.5 mm wide, partially exserted from the cone. Apr–early May. Sandy flats, gravelly hills, or canyons; extreme s GP; s TX panhandle & sw OK; s GP s to e-cen Mex.).

2. ***Ephedra torreyana*** S. Wats., Torrey ephedra. Erect shrub to 2 m tall; branches green to yellowish-green, rigid and hard; bark of older stems grayish, cracked, and shredded; plants from deep taproots. Leaves mostly persistent, in whorls of 3, connate 1/2 to 1/3 their length, at first hyaline with a greenish median stripe, obovate, apex mucronate, and the margins erose, soon after the leaves grayish with a dark brown median stripe, triangular, 1.5–3 mm long, apex narrowly acuminate, margins irregular and shredded, base somewhat thickened. Staminate cones 1–4 per node, nearly spherical to slightly ovoid, 3–7 mm long, 3–5 mm wide, sessile; bracts in 3s, in 6–9 whorls, yellowish-brown, ovate and slightly clawed, 1.5–2.3 mm long, 1–1.7 mm wide, membranaceous, the lowermost whorls empty; perianth exceeding the subtending bract; staminal column 1.5–2.5 mm long, partially to wholly exserted. Ovulate cones 1–3 per node, caducous, ovoid, 6–11 mm long, 4–8 mm wide, sessile or nearly so; bracts in 3s in 5 or 6 whorls, brownish or reddish-brown with broad hyaline margins, membranaceous, 4–7 mm long, 3–6 mm wide, apex rounded and sometimes mucronulate, margins erose. Seeds 1 or 2(3) per cone, light brown, lanceoloid, trigonous, 7–11 mm long, usually slightly exceeding the bracts. Apr–early May. Deep sandy soil of mesquite flats, gravelly breaks, canyons; extreme sw GP; TX panhandle, e-cen NM; (w TX nw to UT & NV).

Division MAGNOLIOPHYTA (Flowering Plants)
Class Magnoliopsida (Dicots)

13. ANNONACEAE Juss., the Custard Apple Family

by Robert B. Kaul

1. ASIMINA Adans., Pawpaw

1. *Asimina triloba* (L.) Dun. Shrub or small tree, often gregarious. Leaves alternate, simple, entire, oblong-obovate, to 30 cm long, acuminate, tapering at the base to short petioles.

Buds dark brown and invested with golden hairs. Flowers appearing with the leaves and below themon stems of the previous year, perfect, regular, hypogynous, 3-4 cm wide, pediceled; sepals 3, soon falling; petals 6 in 2 series, lurid purplish, orbicular to ovate, the outer ones somewhat recurved, the inner ones erect, shorter, the veins prominent; stamens numerous, free, the filaments short; pistils 3-15, not all maturing. Fruits pendent, fleshy, bananalike, to 15 cm long and 4 cm thick, greenish, brownish, sometimes purplish-black, edible; seeds several to many, flattened. (2n = 18) Apr–May, fruits Aug–Sep. Rich deciduous forests & along streams, usually shaded by taller trees; se NE & sw IA s through e KS & w MO to ne OK; (NY to MI & se NE, s to ne TX & FL).

14. LAURACEAE Juss., the Laurel Family

by Robert B. Kaul

Ours aromatic shrubs and small trees of the se GP. Leaves alternate, simple, sometimes lobed. Flowers small, clustered, appearing before or with the leaves, perfect or imperfect, hypogynous; sepals 4-6, free, petals 0; stamens 0 or 9, or staminodial, opening by apical valves; pistil 1, uniovulate, style 1. Drupe blue or red, 1-seeded.

1 Leaves often lobed; flowers produced only at the branch tips, appearing with the leaves .... 2. *Sassafras*
1 Leaves never lobed; flowers produced along the branches and at the branch tips, appearing before the leaves .... 1. *Lindera*

1. LINDERA Thunb., Spice Bush

1. *Lindera benzoin* (L.) Bl. Much-branched shrub or small tree to 5 m tall. Leaves 1-15 cm long, 1.5-6 cm wide (the early leaves of the season small), ovate to mostly obovate or elliptic, acute to acuminate or obtuse, the base cuneate or tapering, green above, glaucous below, the petioles slender, short. Flowers yellow, in subsessile clusters at the nodes, the clusters numerous. Sepals 6, stamens of staminate flowers 9, staminodia of pistillate flowers 12 or more; pistil globular. Drupe red, ovoid to subglobose, 6-10 mm long, 3-5 mm wide, on pedicels 2-5 mm long. (2n = 24) Mar–Apr. Moist deciduous forests, often on rocky soil; se KS, sw MO, ne OK; (VA to MO & se KS, s to TX & FL). *Benzoin aestivale* (L.) Nees— Rydberg.

2. SASSAFRAS Trew, Sassafras

1. *Sassafras albidum* (Nutt.) Nees. Dioecious colonial shrub or tree to 40 m tall (much less than that in our area), the bark aromatic, the twigs pale green. Petioles 1-3 cm long, pubescent; blades variable on the plant, elliptic, ovate, or obovate, obtuse to acute, entire to deeply 3-lobed, the lobes variable; blades 3-12 cm long, 1.5-9 cm wide, the larger ones lobed, green above, glaucous below, more or less puberulent on the veins, especially below. Flowers greenish-yellow, clustered in corymbose racemes, appearing with the leaves and borne at the stem tips. Sepals 6, 3-4 mm long, ca 1 mm wide, linear, obtuse to acute;

stamens of staminate flowers 9, staminodia of pistillate flowers 6. Inflorescences much enlarged in fruit, the pedicels red, elongate and clavate; fruit blue-black, ovoid, 7–10 mm long, 5–6 mm wide. (2n = 48) Mar–Apr. Forest edges, roadsides, often in rocky soil; se KS, sw MO, ne OK; (ME to se KS, s to TX & FL). *S. variifolium* (Salisb.) O. Ktze.—Rydberg.

15. SAURURACEAE E. Mey., the Lizard's Tail Family

by Robert B. Kaul

Perennial aromatic herbs of damp meadows and marshes. Stems erect or ascending from creeping rhizomes. Leaves alternate, simple, entire, mostly petioled, the stipules adnate to the petioles. Inflorescences spikes or spiciform racemes, sometimes with white involucral bracts at the base of the floriferous part. Flowers perfect, hypogynous to apparently epigynous; perianth absent; stamens 4–8; pistils 3–5, basally or completely fused, 1–many ovulate. Fruit a dry compound capsule or fleshy capsulelike berry.

1 Leaves cauline, the tips acuminate, the bases cordate; spikes elongate, slender, without an involucre; se GP .... 2. *Saururus*
1 Leaves mostly basal, the tips rounded, the bases truncate; spikes stout, with a white perianthlike involucre; sw GP .... 1. *Anemopsis*

1. ANEMOPSIS H. & A., Yerba Mansa

1. *Anemopsis californica* (Nutt.) H. & A. Plant colonial, from stout aromatic rhizomes, the flowering stems erect. Leaves mostly basal, few, to 30 cm long, long petioled, the largest blades to 12 cm long, 6 cm wide; basal leaf blades oblong-elliptic, rounded, truncate at the base; cauline leaves smaller, their tips and bases acute. Flowering stems to 50 cm tall, usually with 1 or 2 remote sessile foliaceous bracts 1–4 cm long. Spike stout, 1–3 cm long, held above the leaves, pseudanthial and resembling an anemone flower, subtended by 4–8 unequal white petallike bracts, these 1–2 cm long, to 1 cm wide, rounded; each flower subtended by a small bracteole. Stamens usually 6; pistils usually 3, fused, the ovary embedded in the rachis, ovules many on 3 projecting parietal placentae. Fruit a compound conelike capsule, dehiscent, ferruginous. (2n = 22) Jun–Jul. Saline & alkaline wet meadows & stream sides; cen KS, n-cen OK, TX panhandle, CO at the foot of the Front Range; (OK to CO & CA, s to AZ, TX; n Mex.).

2. SAURURUS L., Lizard's Tail

1. *Saururus cernuus* L. Plant erect from slender creeping rhizomes, to 1 m tall. Stems simple or few-branched, leafy above, often naked below; leaves, including petiole, to 15 cm long and 9 cm wide, the blade usually longer than the petiole; blade ovate, the tip acuminate, the base cordate to reniform. Spikes/spiciform racemes terminal and/or arising at the upper nodes opposite a leaf, erect, scorpioid or apically nodding, 7–15 cm long, not subtended by a showy involucre; flowers to 300, each subtended by a small bract. Stamens 4–8; pistils 3–5, scarcely fused below; ovules 1 or 2, placentation lateral. Fruits somewhat fleshy, wrinkled, probably indehiscent. (2n = 22) Jun–Jul. In shallow water of

swamps & ditches, along lakeshores; se KS, sw MO, ne OK; (s Que. to WI, KS, s to e TX & FL).

16. ARISTOLOCHIACEAE Juss., the Birthwort Family

by Robert B. Kaul

Ours herbs and vines of deciduous forests in the e and se GP. Leaves 2, basal and paired, or numerous, cauline, and alternate, usually cordate at the base. Flowers perfect, regular or irregular, 3-merous; sepals fused below into a tube, the tube straight or curved, the lobes flaring, luridly colored; ovary inferior, usually 6-celled; stamens 6–12, the anthers connivent or adnate to the style. Fruit a capsule, the seeds numerous.

1 Low creeping herbs, the leaves paired; flowers regular; e GP .... 2. *Asarum*
1 Erect herbs and twining vines, the leaves alternate, numerous; flowers irregular, curved; se GP .... 1. *Aristolochia*

1. ARISTOLOCHIA L.

Ours perennial herbs and vines. Leaves alternate, petioled, palmately veined. Flowers solitary or clustered, axillary, perfect, irregular, 3-merous, epigynous, the calyx tubular and conspicuously bent, corollalike; petals 0; stamens 6, the anthers sessile and adnate to the style; ovary (4)6-locular, the ovules numerous, the placentation axile. Fruit a septicidal capsule; seeds numerous, horizontal.

1 Erect herbs to 6 dm tall, not strongly pubescent; flowers borne on short branches at the lower nodes .... 1. *A. serpentaria*
1 High-climbing pubescent woody twining vines; flowers borne at the nodes opposite a leaf .... 2. *A. tomentosa*

1. *Aristolochia serpentaria* L., Virginia snakeroot. Erect rhizomatous herb to 6 dm tall. Leaves slender-petioled, the lower blades reduced, the blades to 14 cm long and 8 cm wide, acuminate, the base cordate to sagittate. Flowers solitary on short scaly branches arising somewhat below the leaves. Calyx tube U-shaped, to 2 cm long, purple to brownish, pubescent, flaring and red-purple above, 3-lobed, the lobes obtuse to rounded. Capsules ellipsoid to almost globose, 1–1.5 cm wide. (2n = 28) May–Jul. Woods, often on rocky soil, barely entering our area from the e; KS: Cherokee; MO: Cass, Jackson, Jasper; (CT to KS, sw to TX, s to FL).

2. *Aristolochia tomentosa* Sims, woolly pipevine. Vine woody, twining, high-climbing, pubescent throughout, especially when young. Petioles of largest leaves shorter than the blades; blades suborbicular to ovate, obtuse to rounded, the base truncate to usually cordate, 4–15 cm long, 3–12 cm wide, becoming glabrate with age. Flowers solitary (paired) at the nodes, arising opposite a leaf, pediceled. Calyx to 7 cm long, J-shaped, flaring and 3-lobed above, the lobes rugose, greenish-yellow, the throat purple. Capsules cylindric, 4–6(8) cm long, 2.5–3 cm in diam, with prominent ribs. (2n = 28) May–Jun. Moist thickets & bottomland forests; sw MO, se KS, cen & ne OK; (NC to KS, s to e TX & FL).

A somewhat similar species, *A. durior* Hill, (*A. macrophylla* Lam.—Rydberg) the Dutchman's pipevine, native well e of our area, is often grown as an ornamental and may be expected to persist or escape here and there in the e GP; it is glabrous throughout.

2. ASARUM L., Wild Ginger

1. *Asarum canadense* L. Creeping perennial, colonial herb from shallow aromatic rhizomes. Leaves basal, paired, erect, long petioled, to 2 dm tall, soft-pubescent; petiole 5-20 cm long; blades cordate-reniform, entire, obtuse to rounded, 4-15 cm wide, becoming glabrate above. Flower solitary on the ground between the petiole bases, hidden by the leaves; pedicel 1-3 cm long, pubescent; flower regular, 1-2 cm wide; sepals 3, fused below to the ovary, the lobes spreading to reflexed, brownish-purple, acute (KS, MO, OK) to long acuminate (ND), to 1.5 cm long; stamens 12, connivent with the styles; ovary 6-locular. Fruit a dehiscent many-seeded capsule, the sepals persisting. (2n = 26) Apr-Jun. Moist deciduous forests in extreme e GP; e ND & n MN, s to sw MO & ne OK; (Que. to e ND, s to e OK, AR, & NC). *A. acuminatum* (Ashe) Bickn., *A. reflexum* Bickn. — Rydberg.

The varieties and species erected on the basis of variations in sepal lobes are yet to be validated. Western and European species, as well as this one, are sometimes cultivated as ground covers in shady places.

17. NELUMBONACEAE Dum., the Lotus Family

by Robert B. Kaul

1. NELUMBO Adans., Lotus

1. *Nelumbo lutea* (Willd.) Pers., lotus, water chinkapin. Plant robust, from bananalike rhizomes. Leaves long petioled, the blades floating or held well above the water, orbicular, without a sinus, centrally peltate, with a single large radial cicatrix, to 6 dm wide, centrally depressed, often glaucous. Flowers solitary, held well above the water, showy, sometimes 25 cm wide, pale yellow, hypogynous; sepals and petals numerous, similar, ovate, obtuse to rounded; stamens numerous, introrse; receptacle obconic, the numerous pistils embedded in its flat end; pistils uniovulate, style very short. Fruits indehiscent, nutlike, ca 1 cm in diam, embedded in the dry receptacle, each with a single seed; receptacle much expanded in fruit, dry, brown, nodding. (2n = 16) Jul-Sep. Quiet water of lakes & ponds; sw IA, se NE, e KS, MO, e & cen OK; (MA to se MN, w to NE, s to TX & FL).

Although this plant grows vigorously and is sometimes a pest, it seems to be disappearing from our area because of extermination by humans and habitat loss. The similar, but pink-flowered, Asiatic *N. nucifera* Gaertn. has been collected from an established population in MO: Platte.

18. NYMPHAEACEAE Salisb., the Waterlily Family

by Robert B. Kaul

Perennial aquatic rhizomatous herbs. Leaves mostly floating, occasionally held just above the water, to ca 30 cm long, orbicular to ovate or subsagittate, entire, with a single deep sinus extending to the more or less centrally attached petiole; submersed leaves occasionally present, thin, sagittate. Flowers borne singly on long peduncles, floating or held above the water, white, yellow or reddish, 3.5-20 cm wide, sometimes fragrant, hypogynous to nearly epigynous; sepals or tepals 4-6, free, green on the outside, green, yellow, or reddish on the inside; petals 0 (in yellow or reddish flowers) to numerous (in white flowers); stamens numerous, the filaments sometimes broadened; pistil compound; ovules numerous, the placentation laminar; stigmas raylike. Fruit fleshy, dehiscing under water.

This family, Cabombaceae, and Nelumbonaceae are often treated as a single family, Nymphaeaceae *sensu lato*.

1 Flowers white; sepals 4; floating leaves orbicular .... 2. *Nymphaea*
1 Flowers yellow to reddish; tepals 5 or 6; leaves suborbicular to mostly ovate, sometimes subsagittate .... 1. *Nuphar*

1. NUPHAR Small, Cowlily, Spatterdock

1. *Nuphar luteum* (L.) Sibth. & Small. Plant robust, from stout yellow rhizomes. Leaf blades floating or held more or less erect above the water surface, less than twice as long as wide in ours, to 30 cm long and 20 cm wide, suborbicular to ovate, with a deep sinus extending to the more or less centrally attached petiole, glabrous to pubescent below, sometimes a few small, thin submersed leaves present; petiole terete or flattened above, winged or wingless. Flowers solitary at or just above the surface, sometimes fragrant, to 3.5 cm wide, hypogynous; sepals 6, ca 2 cm long, rounded, truncate, or emarginate, persistent, yellow or red within; stamens numerous, the outer ones staminodial ("petals") but not petaloid, yellow or reddish, thick, oblong, emarginate, the anthers recurving, introrse, less than 1 cm long and not exceeding the stigmatic disc; pistil compound, ovules numerous, laminar, styles fused, stigmas raylike, embedded on the flat or depressed disk. Fruit fleshy, maturing underwater; dehiscence irregularly circumcissile near the base; seeds ovoid, smooth. (2n = 34) Jun–Oct. Quiet, shallow or deeper waters, sometimes stranded.

Two subspecies in our area:

1a. subsp. *macrophyllum* (Small) E. O. Beal. Petioles terete to flattened above, not winged; blades floating or erect. Sepals yellow within, seldom reddish; stigmatic rays mostly not extending to the disk margin. sw MO, se KS, e OK; (ME to WI, KS, s to Mex. & Cuba). *N. advenum* Ait.—Rydberg.

1b. subsp. *variegatum* (Engelm.) E. O. Beal. Petioles flattened above, winged; blades floating. Sepals reddish within; stigmatic rays often extending to the disk margin. ND & SD e of the Missouri R., NE (sandhills), MN, IA; (Yukon to Newf., s to w MT, NE, IA, e to DE). *N. variegatum* Engelm.—Rydberg.

2. NYMPHAEA L., Waterlily

Our plants robust, stoutly rhizomatous, sometimes these bearing lateral tubers. Blades submersed, floating, or held just above the water, orbicular or nearly so, peltate, with a deep sinus extending to the petiole. Flowers borne singly at the surface or held above it, showy, opening in the morning and closing in late afternoon, remaining open all day when cloudy. Sepals 4, essentially free, spreading; petals numerous, white, grading to the stamens, the outer ones about equaling the sepals; stamens numerous, borne at the sides of the ovary, the filaments broad; pistil 1, compound, almost inferior, the concave top with a single globular protuberance in the center from which the stigmas radiate, these overarched by numerous fingerlike projections; ovules numerous, laminar. Fruit fleshy, globose, invested by persistent perianth bases; seeds arillate. Quiet, clear shallow waters.

1 Leaves usually reddish below; petals elliptic, tapering to a subacute point; rhizome not bearing tubers .... 1. *N. odorata*

1 Leaves green below; petals oblanceolate to spatulate, rounded; rhizome bearing lateral tubers .... 2. *N. tuberosa*

1. ***Nymphaea odorata*** Ait., fragrant white waterlily. Rhizome elongate, thick, not bearing lateral tubers. Blades to 20 cm or more wide, green above, tinged reddish below. Flowers pleasantly and noticeably fragrant at least on some days of opening, to 12 cm wide; sepals often purplish on the back, to 8 cm long; petals elliptic, almost acute. (2n = 84) May–Sep. Uncommon to rare in NE sandhills, e NE, e KS, MO, cen OK; (Newf. to Man., GP s to LA & FL).

This species is probably far less common now than in the past because of grazing by cattle.

2. ***Nymphaea tuberosa*** Paine, white waterlily. Rhizome bearing stout lateral tubers. Blades to 30 cm wide, green above and usually green below. Flowers to 20 cm or more wide, odorless or fragrant; sepals green on the back; petals spatulate or oblanceolate, rounded. (2n = 84) Jun–Oct. Uncommon to rare in extreme e GP, & w in NE to 100° long; (Que. to n Ont., s to MN, NE, OK, & MD).

19. CABOMBACEAE A. Rich., the Water Shield Family

by Robert B. Kaul

Perennial rhizomatous aquatic herbs. Floating leaves alternate, entire to sagittate, peltate, usually distinctly mucilaginous below; submersed leaves, if any, opposite or whorled, deeply dissected. Flowers single in the axils, perfect, regular, hypogynous; sepals (2)3(4), free, green; petals (2)3(4), white with yellow auricles or reddish-purple; stamens 6–36; pistils 2–18, free; ovules 1–4. Fruits 1–several in a cluster, indehiscent.

1 Leaves all floating, peltate, entire, 4–12 cm long; flowers reddish-purple .... 1. *Brasenia*
1 Most leaves submersed, deeply dissected; floating leaves appearing with the flowers, peltate, entire to sagittate, to 1.5 cm long; flowers white .... 2. *Cabomba*

1. BRASENIA Schreb., Water Shield

1. ***Brasenia schreberi*** J. F. Gmel. Stems rising to the surface from slender stolons, gelatinous. Leaf blades floating, long petioled, centrally peltate, oval to elliptic, the ends rounded, 4–12 cm long, 3–8 cm wide, the underside and the petiole thickly gelatinous. Flowers held just above the water, to 2 cm wide, hypogynous; sepals and petals (2)3(4) each, similar, linear, 1–1.5 cm long, reddish-purple; stamens 12–36, filaments slender, anthers latrorse but apparently extrorse, pistils free, 4–18, style short, ovules 1–4, pendulous, parietal. Fruit maturing under water, indehiscent, beaked, ca 6–8 mm long. (2n = 80) May–Sep. Clear ponds & quiet streams, on mud; rare in our area; KS: Johnson; MN: Ottertail, but common eastward; MO: Jasper, Barton; OK: Mayes, Ottawa; (across most of N. Amer. except GP & the Southwest; sporadic on all continents except Europe).

2. CABOMBA Aubl., Fanwort

1. ***Cabomba caroliniana*** A. Gray. Plant submersed except a few floating leaves at flower-time. Stems slender, to 1 m or more long. Leaves opposite or whorled, the petiole to 3 cm long, the submersed blades rounded in outline, di- or trichotomously dissected into narrow linear lobes; floating blades produced at flowering time, peltate, elongate-oval or sagittate, to ca 1.5 cm long. Flowers borne singly on elongated pedicels in axils of the upper leaves, 7–12 mm long, hypogynous. Sepals 3, green; petals 3, white, clawed, the limb rounded at the tip and basally auriculate, the auricles incurved, yellow; stamens 6; pistils 2–4, free, 1–3 ovulate. Fruit indehiscent. (2n = 24, 104) May–Sep. Ponds & quiet streams, rare & possibly only introduced in our area; KS: Johnson; (MA to MI, KS, s to cen TX & FL, probably naturalized in the n part of the range).

This species is commonly sold as an ornamental for aquaria, and it may be expected occasionally here and there in our area as an escape.

20. CERATOPHYLLACEAE S. F. Gray, the Hornwort Family

by Robert B. Kaul

1. CERATOPHYLLUM L., Hornwort, Coontail

1. *Ceratophyllum demersum* L. Submersed, monoecious, rootless branching herbs to 2 m or more long, often lime-encrusted and brittle. Leaves in crowded whorls of 3–10, once or twice dichotomously branched into linear-filiform segments, the segments irregularly serrate, variable in length, to 2 cm. Flowers axillary, sessile or nearly so; perianth/involucre of 8–12 elliptic-lanceolate segments; staminate flowers with 12–16 stamens, the filaments short or lacking, the anthers large and apically 2- or 3-pointed; pistillate flowers with a simple 1-celled pistil containing one suspended ovule; style and stigma 1. Achene ellipsoid, smooth, ca 5 mm long, with 2 basal spines and the style persistent as a slender beak. (2n = 24) May–Sep. Quiet, often brackish or calcareous waters; GP; (Que. to B.C., s to FL & Mex., also Old World).

C. echinatum A. Gray has been found once in our range, (KS: Lyon). It resembles *C. demersum* but the leaves are branched 2–4 times, the segments essentially entire, the achenes with 3–5 spines on their sides. Its range approaches ours, and it is to be sought along our eastern borders.

21. RANUNCULACEAE Juss., the Buttercup Family

by David Sutherland

Herbaceous or somewhat shrubby perennials, rarely annuals. Stems erect to decumbent or climbing. Leaves cauline or basal, sessile or petiolate, alternate, or less commonly, opposite or whorled, entire and simple to variously toothed, lobed, or compound; stipules usually absent, but the base of the petiole sometimes expanded into stipulelike projections. Flowers solitary or in racemes or panicles, bisexual or, rarely, unisexual, actinomorphic or, less commonly, zygomorphic; sepals (3)4–8, inconspicuous and green to petaloid and colorful, spurless or one spurred or saccate; petals variable in number, usually 5, but not uncommonly absent, and then the sepals generally petaloid, often nectariferous at the base, sometimes 2 or 5 of them bearing elongated spurs; stamens numerous (sometimes as few as 9 or 10), the outer ones sometimes modified as staminodia; carpels (1)3–numerous, all separate, ovules 1–many. Fruit consisting of (1)3–many achenes or follicles or, rarely, a solitary berry.

1 Flowers imperfect; plants dioecious herbs .... 12. *Thalictrum*
1 Flowers perfect or, if imperfect, then plants vining and somewhat woody.
  2 Ovule solitary in each carpel; fruit and achene.
    3 Achenes borne on elongated, slender receptacles; leaves linear; sepals spurred .... 10. *Myosurus*
    3 Achenes borne on a relatively short receptacle or plants not otherwise as above.
      4 Petals lacking or very inconspicuous; sepals petaloid.
        5 Sepals valvate in bud, usually 4 .... 7. *Clematis*
        5 Sepals overlapping in bud, 5 or more.
          6 Leaves compound, the leaflets distinct .... 4. *Anemonella*
          6 Leaves lobed or dissected but not truly compound .... 3. *Anemone*
      4 Petals present, often showy; sepals not petaloid .... 11. *Ranunculus*
  2 Ovules 3-numerous in each carpel; fruit a follicle or berry.
    7 Sepals small and deciduous; fruit a berry .... 2. *Actaea*
    7 Sepals petaloid; fruit a follicle.
      8 Flowers strongly zygomorphic.
        9 Uppermost sepal and 2 petals spurred .... 8. *Delphinium*
        9 Uppermost sepal saccate or hoodlike, not spurred; petals not spurred .... 1. *Aconitum*
      8 Flowers actinomorphic.
        10 Petals present, each one strongly spurred .... 5. *Aquilegia*
        10 Petals lacking.
          11 Leaves compound .... 9. *Isopyrum*
          11 Leaves simple .... 6. *Caltha*

1. ACONITUM L., Monkshood

1. ***Aconitum columbianum*** Nutt. Slender, weak stemmed to rather coarse, stiffly erect perennial herbs 2–12(21) dm tall. Stems fistulose, glabrous to slightly crisp-puberulent below, usually pubescent above with spreading, often glandular, hairs. Roots fibrous from 1–several tuberous thickenings at the crown. Leaves alternate, mostly cauline, the blades strongly palmately (3)5(7)-lobed, the lobes toothed to deeply incised, the larger leaves with blades mostly 3–10 cm long from petiole to tip, 5–15 cm wide; petioles gradually reduced upward, the lowest mostly 5–20 cm long, the uppermost short or nearly absent. Inflorescence a simple raceme 2–20 cm long, varying to highly branched in larger plants with lateral racemes arising in the axils of upper leaves and the overall inflorescence up to 70 cm long; pedicels spreading and arcuate to abruptly ascending. Flowers zygomorphic; sepals 5, usually deep purplish-blue but varying to pale blue (rarely white), usually pubescent externally, the uppermost one hoodlike, (10)12–20(33) mm high, the lateral 2 reniform or oval, 8–17 mm long, the lowest 2 narrow, 7–12 mm long; petals 2, whitish, ascending beneath the hoodlike upper sepal, each bearing a pendent blade and a coiled spur; stamens numerous, the filaments broadened; carpels 3. Fruit consisting of 3 follicles, these glabrous to pubescent, erect to somewhat divergent, 13–19 mm long at maturity; seeds brownish or blackish with a translucent longitudinal wing on one side and several delicate ruffled transverse wings on the other side. (n = 8) Jun–Aug. Meadows & open woods, usually in moist soil; se SD to WY & NM; (B.C. to AK, s to SD, NM, & CA). *A. porrectum* Rydb., *A. ramosum* A. Nels., *A. tenue* Rydb.—Rydberg.

2. ACTAEA L., Baneberry

1. *Actaea rubra* (Ait.) Willd. Perennial herbs 5-9 dm tall. Stems usually unbranched, glabrous below, usually crisp-puberulent above. Leaves (1)2(3) in number, cauline, large, the largest one with the petiole 1-12(16) cm long and the compound blade 15-35 cm long, ternate-pinnate, biternate-pinnate, or triternate-pinnate, the ultimate leaflets broad, irregularly toothed. Inflorescence a terminal raceme about 1-3 cm long in flower and 2-10 cm long in fruit. Flowers small, whitish; sepals 3 or 4(5), caducous, mostly 2.4-3.7 mm long; petals 3-5(10), caducous, spatulate, mostly 1.2-3.2 mm long; stamens numerous, the filaments widened toward the summit; pistil 1, its stigma sessile, narrower than the broadest part of the ovary. Fruit a fleshy, elliptical, red or white, 9- to 16-seeded berry 7.5-13 mm long. (2n = 16) May-Jun. In moist soil in woods; MN, ND, MT, IA, se & w SD, e & nw NE, WY, ne KS, CO, NM; (Lab. to B.C., s to NJ, KS, NM, & CA). *A. arguta* Nutt.—Rydberg, *A. rubra* f. *neglecta* (Gillman) Robins.—Fernald, *A. alba* (L.) P. Mill.—Rydberg, misapplied.

Actaea alba (L.) P. Mill. (*A. pachypoda* Ell. of many authors, *A. brachypoda* Ell.—Rydberg) has been collected at the edge of our range in habitats similar to that of *A. rubra* (IA: Adair; KS: Doniphan; MO: Atchison, Holt, Worth; NE: Richardson). It may be recognized in fruit by its prominently thickened pedicels or, in flower, by its stigma, which is broader than the broadest part of the pistil. Fruit color is not especially helpful as a diagnostic feature, since *A. rubra* sometimes has white berries and *A. alba* red ones.

3. ANEMONE L., Wind Flower

Perennial herbs with erect stems arising from a caudex or from along slender or thickened rhizomes. Leaves of the stem lobed, dissected, or nearly compound, 3 or more, sessile or petiolate, borne in a single whorl as an involucre below the peduncle or peduncles; basal leaves absent to numerous, lobed to deeply dissected, long petiolate. Flowers solitary or borne in cymes or umbels, the peduncles of all but the central flower sometimes bearing secondary involucres. Flowers regular, conspicuous; sepals 4-20, petaloid; petals absent or present as glandlike staminodia; stamens numerous; pistils numerous in a cylindrical to spherical cluster, becoming strigose to white-woolly achenes. *Pulsatilla* Adans.—Rydberg.

Reference: Keener, C. S. 1975. Studies in the Ranunculaceae of the southeastern United States. I. *Anemone* L. Castanea 40: 36-44.

```
1 Style long, 2-4 cm at maturity, becoming plumose; glandlike staminodia usually
    present ............................................................................................................... 6. A. patens
1 Styles much shorter, not becoming plumose; staminodia absent.
    2 Achenes concealed by long, cottony hairs.
        3 Involucral leaves clearly petioled.
            4 Styles 1-2 mm long; achenes borne in an ovoid cluster .............. 8. A. virginiana
            4 Styles shorter; achenes borne in a cylindrical cluster .................. 4. A. cylindrica
        3 Involucral leaves sessile or nearly so.
            5 Sepals 4-9; plants from a stout caudex ..................................... 5. A. multifida
            5 Sepals 10-35; plants from a tuber or tuberous rhizome.
                6 Plants rhizomatous; achene bodies mostly less than 2.5 mm
                    long .............................................................................. 3. A. caroliniana
                6 Plants nonrhizomatous; achene bodies mostly more than 2.5 mm
                    long .............................................................................. 1. A. berlandieri
    2 Achenes nearly glabrous to strigose or short-hirsute, not concealed by long, cottony hairs.
        7 Involucral leaves petiolate; basal leaves absent to few in
            number ...................................................................................... 7. A. quinquefolia
        7 Involucral leaves sessile; basal leaves conspicuous and often
            numerous ................................................................................... 2. A. canadensis
```

1. **Anemone berlandieri** Pritz., tenpetal anemone. Nonrhizomatous perennials 1–3(4.5) dm tall from a tuberous root. Stems loosely spreading-pubescent below, becoming densely appressed-pubescent above. Leaves rather thinly hairy, those of the involucre sessile and deeply palmately dissected, 1.8–4.5(6) cm long, the basal ones petiolate and highly variable in size and shape, even on the same plant, usually ternate, with segments varying from elliptic and serrate to laciniate, the largest blades 3–5.5 cm wide, the longest petioles 5–16 cm long. Flowers solitary, about 2–4.3 cm across, the sepals mostly 10–20(33) in number, usually violet on the outside and whitish internally, 11–21 mm long. Fruiting head cylindrical, 11–28 mm long, 5.5–11 mm wide; achenes nearly covered by white, woolly hairs, 2.7–3.5 mm long, the styles slender, 1.3–2.3 mm long, often largely concealed by the pubescence. ($2n = 16$) Mar–May. Open grasslands & rocky hillsides, often on limestone; s-cen KS, OK, & TX; (NC to VA, & in MS, AL, AR, KS, OK, & TX). *A. decapetala* Ard.—Atlas GP, misapplied, *A. heterophylla* Nutt.—Correll & Johnston.

Although this species has sometimes been confused with *A. caroliniana*, the distinctions were made clear by Joseph and Heimburger (Canad. J. Bot. 44: 899–928. 1966). The species has been commonly known as *A. decapetala* or *A. heterophylla*, but Keener (op. cit.) concludes that *A. berlandieri* is the correct name.

2. **Anemone canadensis** L., meadow anemone. Strongly rhizomatous perennials (1.5)2–6 dm tall. Stems pubescent with spreading to ascending hairs, usually most densely so above. Leaves appressed-pubescent, especially beneath, those of the involucre sessile, 3–10 cm long, deeply 3-lobed, the segments irregularly and sharply toothed, the basal ones petiolate, the blades 3- to 5-lobed, the lobes irregularly toothed, (5)8–15(20) cm wide, the petioles 8–22(37) cm long. Flowers solitary or, more commonly, in cymes, about 1.5–4.6 cm across, the peduncles of all but the central flower normally bearing secondary involucres; sepals mostly (4)5(6) in number, white, 8–19(22) mm long. Fruiting head often broader than long, 9–16 mm long, (9)12–19 mm wide; achenes merely strigose, (2.5)3–4 mm long, their styles 2.5–4 mm long. ($2n = 14$) May–Jul. In wet prairies, low ditches, wet woodlands, & a variety of other moist habitats; MN, ND, MT, SD, IA, NE, WY, KS (mostly ne), & CO; (e Que. to B.C., s to MD & NM).

3. **Anemone caroliniana** Walt., Carolina anemone. Rhizomatous perennials 0.5–2(3) dm tall, the rhizomes slender, but the stems and leaves arising from tuberous thickenings. Stems glabrous below, becoming loosely spreading-pubescent to densely appressed-pubescent above. Leaves thinly pubescent to glabrous, those of the involucre sessile and deeply dissected, 1–3 cm long, the basal ones petiolate and similarly laciniate or the lowest ones sometimes merely ternate with serrate leaflets, the largest blades 2–5 cm wide, the longest petioles 2–6(10) cm long. Flowers solitary, about 1.5–4 cm across, the sepals mostly (11)14–16(27) in number, purple, violet, or white, (6)11–15(19) mm long. Fruiting head more or less elliptic, (8)13–19(22) mm long, 7–10 mm wide; achenes nearly covered by white-woolly pubescence, 1.5–2(2.5) mm long, the styles 1.3–2 mm long, often clearly projecting beyond the hairs. ($2n = 16$) Apr–May. Prairies, pastures & meadows; MN, SD, IA, NE, MO, KS, OK, TX; (SC to SD, s to GA & TX).

4. **Anemone cylindrica** A. Gray, candle anemone. Nonrhizomatous (short-rhizomatous) perennials (2.5)3–5.5(7) dm tall, from a stout caudex. Stems more or less densely pubescent throughout with spreading to ascending hairs. Leaves appressed-pubescent, especially beneath, those of the involucre petiolate, 3- to 7-lobed, the lobes themselves jaggedly toothed and lobed, usually 3 in number but often apparently more because secondary involucres of lateral peduncles arise at the same point, the largest involucral leaves with blades 2.5–6.5 cm long and petioles 1–4.5 cm long, the basal leaves similar in shape, the largest blades 4.5–11(14) cm wide, the longest petioles 9–21 cm long. Flowers solitary (rarely)

or, more commonly, 2–7 in number, the peduncles mostly apparently naked, the secondary involucres of lateral ones being incorporated into the main stem whorl, but sometimes one or more of them bearing a secondary involucre above the base; flowers relatively inconspicuous, 1.4–2.0 cm across; sepals mostly 4–6, whitish, 7–12 mm long. Fruiting head cylindrical, 17–34 mm long, 7–11 mm wide; achenes nearly covered by white, woolly pubescence, 1.8–3 mm long, the styles often largely concealed by the pubescence, 0.3–1.0 mm long. (2n = 16) Jun–Jul. Mostly in open prairies & pastures; MN, ND, MT, SD, IA, NE, & ne KS; (ME to Alta., s to PA, KS, & AZ).

5. *Anemone multifida* Poir. Nonrhizomatous perennials (1.5)3–5(6) dm tall from a stout caudex. Stems more or less spreading-pubescent, the hairs often becoming appressed above. Leaves appressed- to spreading-pubescent, those of the involucre sessile to subsessile, 3- to 5-lobed and further divided into narrow segments, mostly about 3–6 cm long, the basal ones similar but petiolate, the largest blades 2–10 cm wide, the longest petioles 8–20 cm long. Flowers 1–3, the lateral peduncles often bearing secondary involucres; flowers not showy, 1.0–2.0 cm across; sepals mostly 4–6(9) in number, greenish, whitish, or (usually) purplish, (5)6–10(15) mm long. Fruiting head more or less globose, about 8–14 mm long; achenes densely covered with long, woolly pubescence, 1.5–3.2 mm long, the styles 0.7–1.5 mm long. (2n = 32) May–Jul. Meadows, prairies, & open woods, often in rocky soil; MN, ND, MT, w SD, w NE, WY, CO; (Newf. to AK, s to NY, NE, NM, & CA). *A. hudsoniana* Richards., *A. globosa* Nutt.—Rydberg.

6. *Anemone patens* L., pasque flower. Perennials (0.5)1–3(4.5) dm tall from a stout branched or unbranched caudex. Stems and leaves commonly villous throughout. Leaves with the blades 3- to 7-lobed and the lobes further dissected into narrowly lanceolate or linear segments, the involucral leaves sessile, 2–5 cm long, the basal leaves few to numerous, expanding late, the largest ones with blades (3)5–10 cm wide and petioles 5–10(13) cm long at maturity. Flowers solitary, very showy, about 4–8 cm across; sepals 5–8 in number, blue to lavender (white), 1.8–4.0 cm long; glandlike staminodia usually present with stamens, but not conspicuous. Fruiting head large, about 3–6 cm long and 4–8 cm wide; achenes about 3–6 mm long, bearing plumose styles 2–3.6 cm long. (2n = 16) Apr–Jun. Open prairie, often in rocky soil; MN, ND, MT, SD, IA, NE, WY, CO; (WI to AK, s to IL, MO, NM, & WA; also in Eurasia). *Pulsatilla ludoviciana* (Nutt.) Heller—Rydberg.

7. *Anemone quinquefolia* L., wood anemone. Perennials 0.5–3 dm tall, arising from slender rhizomes. Stems sparsely villous to nearly glabrous. Leaves glabrous to sparsely appressed-pubescent, those of the involucre usually 3, petiolate, with 3–5 principal lobes, the margins of the lobes irregularly serrate, the blades 1.5–4(5.5) cm long, the petioles 0.5–3 cm long, basal leaves usually arising after flowering or a solitary one sometimes present at the base of the flowering stem, the blade similar to the blade of the involucral leaves, the petiole longer. Flower solitary, 1–3 cm across, the sepals (4)5(6) in number, whitish, 7–15 mm long. Fruiting head subglobose, about 1 cm in diam; achenes short-hirsute, about 3–4.5 mm long, the styles mostly 1–2 mm long. (2n = 32) Apr–Jun. Uncommon, in moist ground in woods; MN, ND, ne SD, e NE; (Que. to e Man., s to KY, IL, & NE). *A. quinquefolia* var. *interior* Fern.—Fernald.

8. *Anemone virginiana* L., tall anemone. Nonrhizomatous perennials (3)4.5–9(11) dm tall from a stout caudex. Stems more or less densely pubescent throughout with spreading or ascending hairs. Leaves appressed-pubescent, especially beneath, those of the involucre petiolate, usually 3 in number, the blades with 3–5 main lobes, the lobes themselves somewhat lobed and serrate, but the ultimate segments relatively broad, blades 5–14 cm long, petioles 2.5–7.5 cm long; basal leaves similar in appearance, the largest blades 6–15(21)

cm wide, the longest petioles 12–24(37) cm long. Flowers solitary or in cymes, 1.3–2.5 cm across, the peduncles of all but the first-blooming flower normally bearing secondary involucres; sepals normally 5, whitish or greenish, 6.5–13 mm long. Fruiting head elliptic, 12–30 mm long, 9–18 mm wide; achenes nearly covered by white, woolly pubescence, 2.1–3.7 mm long, the styles sturdy and strongly protruding, 1.3–2 mm long. (2n = 16) Jun–Jul. In moist or dry woods; MN, ND, SD, IA, e & n NE, MO, e KS; (Newf. to B.C., s to GA, OK, & ND). *A. riparia* Fern.—Fernald.

4. ANEMONELLA Spach

1. *Anemonella thalictroides* (L.) Spach, rue anemone. Slender glabrous perennials 0.7–3 dm tall, arising from a cluster of tuberous roots. Leaves of the flowering stem 2 or 3, opposite or whorled, sessile, forming an involucre, ternately compound, the leaflets elliptic to round, usually trilobed toward the tip, the largest leaflets 0.5–3.5 cm long; basal leaves biternate (triternate), petiolate, petioles 3–14 cm long, the leaflets similar to those of the involucral leaves in size and shape. Flowers (1)–several in an umbel, more or less regular, 8–22(25) mm across; sepals 5–8, petaloid, white or pink, 4–12 mm long; petals absent; stamens numerous; carpels 4–15. Fruit a cluster of achenes 8–11 mm wide, the individual achenes strongly 8–10 ribbed, fusiform, 3.4–5 mm long. (2n = 14, 42) Mar–May. Wooded hillsides; IA, MO, e KS; e OK; (ME to MN, s to FL, AR, & OK). *Syndesmon thalictroides* (L.) Hoffmsg.—Rydberg; *Thalictrum thalictroides* (L.) Eames & Boivin—Boivin, Bull. Soc. Roy. Bot. Belgique 89: 315–318. 1957.

5. AQUILEGIA L., Columbine

1. *Aquilegia canadensis* L., wild columbine. Perennial herbs 3–8(10) dm tall from a stout caudex. Stems fistulose, glabrous throughout to glandular-villous, especially on the upper parts. Leaves alternate and basal, glabrous to finely villous, especially beneath, compound, the basal ones (ternate) biternate (triternate) with the ultimate segments cuneate and shallowly or deeply lobed; stem leaves similar but gradually reduced and with shorter petioles upward. Flowers nodding, showy, regular, about 2–5 cm long from the tip of the stamens to the ends of the petal spurs, 1.7–4.3 cm wide; sepals 5, dull rose colored or reddish, 0.9–2 cm long; petals 5, reddish or rose colored basally, yellowish on the upper parts, alternating with the sepals, the lower part prolonged into a narrow spur which is enlarged slightly at the tip, petals 2–3.6 cm long from the outside edge to the tip of the spur; stamens numerous, prominently exserted; carpels 5. Fruit consisting of 5 erect follicles 1.5–3 cm long, each bearing a stylar beak 0.9–1.8 cm long. (2n = 14) Apr–Jun. In woods, often in moist soil; MN, ND, SD, IA, e & n NE, MO, e KS, e OK; (Newf. to Man., s to GA, TN, & OK). *A. canadensis* var. *latiuscula* (Greene) Munz and var. *hybrida* Hook.—Fernald, *A. latiuscula* Greene—Rydberg.

Another species, *Aquilegia brevistylus* Hook., is known from MN and from the BH of SD (MN: Clay; SD: Custer, Lawrence, and Pennington). It may be distinguished by its blue sepals and spurs and by its short stylar beaks, which are only 2–5 mm long.

6. CALTHA L., Marsh Marigold

1. *Caltha palustris* L. Glabrous, fleshy, herbaceous perennials 2–6(8) dm tall, from rather fleshy, fibrous roots. Stems hollow. Leaves basal and alternate, reduced upward, the blades

reniform to nearly circular and cordate at the base, dentate to crenate, the largest blades 3–10(12) cm long and 4–15(20) cm wide, the longest petioles 8–30(57) cm long; petioles with a sheathing, stipulelike base. Flowers regular, showy, 2–5 cm wide, axillary or terminal; sepals petaloid, yellow, mostly 5–6(9) in number, 1–2.3 cm long; petals absent; stamens numerous; carpels 5–10(12). Fruits recurved divergent follicles 0.8–1.7 cm long; seeds 1.7–2.4 mm long. (2n = 28–70, 32 and 56 the most frequent counts) Apr–May. Wet woods, marshes, & bogs, often rooted in shallow water; MN, e ND, e SD, IA, ne NE (rare); (Newf. to AK, s to VA & NE).

7. CLEMATIS L., Virgin's Bower

Dioecious to perfect-flowered perennial herbs to somewhat woody vines. Leaves compound to simple, petiolate to sessile, opposite. Flowers solitary to paniculate, relatively inconspicuous to very showy, regular; sepals petaloid, usually 4, separate to connivent at the base, spreading to cupped forward, leathery to thin, valvate in the bud; petals lacking; stamens numerous, the outer ones very rarely modified as staminodia, or all stamens modified as staminodia in the pistillate flowers of the dioecious species; pistils numerous, maturing into achenes with long, often plumose, persistent styles. *Atragene* L., *Viorna* Reichb.— Rydberg.

1 Sepals thin, spreading or directed forward but the calyx not urn-shaped.
  2 Flowers numerous, paniculate, white or cream, not showy.
    3 Plants perfect-flowered; leaflets entire to somewhat lobed .... 5. *C. terniflora*
    3 Plants dioecious; leaflets coarsely toothed to serrate.
      4 Leaflets 3 on most leaves, ovate .... 6. *C. virginiana*
      4 Leaflets 5 more more on most leaves, lanceolate to
        lance-ovate .... 2. *C. ligusticifolia*
  2 Flowers solitary, blue to reddish-purple (white), very showy .... 4. *C. tenuiloba*
1 Sepals thick, directed forward, the calyx urn-shaped; flowers solitary or in small clusters.
  5 Plants viny; leaves compound, petiolate .... 3. *C. pitcheri*
  5 Plants not viny; leaves simple, sessile .... 1. *C. fremontii*

1. *Clematis fremontii* S. Wats., Fremont's clematis. Perennial herbs 2–4 dm tall from a stout caudex. Stems erect, not at all vining, firm, villous more or less throughout. Leaves simple, sessile, leathery, reticulate-veined, ovate, entire to sparingly toothed, the largest ones 7–11 cm long and 4–7.5 cm wide. Flowers solitary, terminating main stems or branches, somewhat urn-shaped, mostly not over 3.5 cm wide; sepals leathery, slightly connivent at the base, purplish, pubescent on the margins, recurved at the tips, 1.8–4 cm long; anthers narrow, tapering. Achene head globular, about 1.6–3 cm across exclusive of the styles; achenes pubescent, (4)5–6.5(7.5) mm long, the styles glabrous or silky near the base, variously curved or contorted, 1.8–3.2 cm long, often breaking off at maturity. (2n = 16) Apr–early Jun. Rocky prairie hillsides, usually on limestone soil; in our range restricted to n-cen KS & adj. NE; (s-cen NE, n-cen KS, disjunct in s & e MO). *Viorna fremontii* (S.Wats.) Heller— Rydberg.

It has been customary to recognize the MO populations as the distinct var. *riehlii* Erickson, but Keener (J. Elisha Mitchell Soc. 83: 38–39. 1967) does not maintain that distinction. The related *C. hirsutissima* Pursh (*Viorna scottii* (Porter) Rydb.— Rydberg) has also been collected a few times within our range (NE: Sheridan; SD: Fall River; WY: Niobrara). It may be readily distinguished by its pinnately compound leaves and plumose styles.

2. *Clematis ligusticifolia* Nutt., western clematis. Dioecious, somewhat woody vines often many meters in length. Stems glabrous to pubescent. Leaves petiolate, pinnately compound (rarely twice compound) mostly with 5–7 leaflets, the leaflets coarsely toothed, variable in shape but usually lance-ovate in general outline, the largest leaflets 3.5–8.5 cm long and 1–5 cm wide. Flowers paniculate, usually numerous, about 1–2.5 cm across; sepals cream, spreading, 0.5–1.3 cm long; pistillate flowers with normal-sized, but sterile, stamens; staminate flowers without pistils. Achene head about 0.7–1.2 cm across, excluding the styles, 3–8 cm across, including the styles; achenes pubescent, 2–4.5 mm long, the styles plumose, persistent, 1.5–5.5 cm long. (2n = 16) Jul–Aug. Climbing on banks, bushes, & trees in a variety of habitats; w ND, MT, w SD, WY, w NE, CO, NM; (ND to B.C., s to NM & CA).

    The related species, *C. drummondii* T. & G., which has been collected at the s edge of our range (OK: Harmon; TX: Briscoe), should also key here. It may be distinguished by its longer styles (8 cm or more long). In addition, it has more sparsely flowered panicles and flowers which are, on the average, somewhat larger. See note under *C. virginiana*.

3. *Clematis pitcheri* T. & G., Pitcher's clematis. Somewhat woody vines extending to many meters in length. Stems pubescent to nearly glabrous. Leaves, except those of the peduncles, compound, usually with (3)5–9(11) leaflets, these extremely variable in shape, not serrate, but the lowest and largest sometimes bi- or trilobed, occasionally very deeply so, leathery, reticulate veined, the largest often 3–9 cm long and 2.5–5 cm wide (wider when lobed), the petiolules of the smaller leaflets often tendrillike. Flowers solitary on peduncles which normally bear a pair of simple bracts near the base, somewhat urn-shaped, mostly not over 4 cm wide. Sepals leathery, slightly connivent at the base, purplish, recurved at the tip, pubescent, especially on the margins, 1–3.5 cm long. Achene head globular, about 1.6–2.8 cm across, exclusive of the styles. Achenes pubescent, 8–13 mm long, the styles glabrous, silky near the base, variously curved or contorted, mostly 1.5–2.5 cm long. May–Aug. Climbing on other vegetation or scrambling along the ground, often near streams or creeks; s IA, se NE, MO, e KS, e OK, TX; (IN to NE, s to MO & TX). *Viorna pitcheri* (T. & G.) Britt.—Rydberg.

    The nonviny *Clematis hirsutissima* Pursh, which occurs at the w edge of our range, may key here. See the discussion following *C. fremontii*.

4. *Clematis tenuiloba* (A. Gray) C. L. Hitchc. Mat-forming suffrutescent herbs 1–3.5 dm tall. Stems thinly pubescent to nearly glabrous, the woody basal portion decumbent but not scandent. Leaves petiolate, generally triternately compound or nearly so, the blades of the largest ones 3–10 cm long and about as broad. Flowers solitary on long peduncles, nodding, very showy, mostly 2.5–7.5 cm across; sepals blue to purplish (white), long tapering, thin, not recurved nor connivent but directed forward, 3–5.5 cm long; staminodia often present outside the normal stamens. Achene head, excluding the styles, about 1.2–2.5 cm across, including the styles, about 5.5–9 cm across; achenes short-pubescent, (2)3–4.2 mm long, the styles plumose, variously contorted, about 3.5–5 cm long. Mostly Jun. Open woods & prairies, usually in rocky (often limestone) soil; relatively common in the BH of SD; (SD, n WY, MT). *Atragene tenuiloba* (A. Gray) Britt.—Rydberg.

    This showy species is similar to *C. pseudoalpina* (O. Ktze.) Nels., which also occurs at the margin of our range (ND: Dunn; SD: Lawrence and Pennington) and should also key here. It differs in being decidedly vining and in having biternately compound leaves.

5. *Clematis terniflora* DC. Somewhat woody vines often many meters in length. Stems usually glabrous. Leaves mostly pinnately compound with 5 leaflets, the leaflets thin to leathery, variable in shape, the margins entire or undulate, the tips obtuse to acute, some of the larger ones sometimes lobed, the largest mostly 2–7 cm long and 1.5–5 cm wide (wider if lobed). Flowers perfect, paniculate, numerous, about 2–4 cm across; sepals cream, spreading, 0.8–2 cm long. Achenes fewer in number than in the similar species *C. ligusticifolia* and *C. virginiana* and achene head therefore not as showy, achene body glabrous to pubescent, fusiform, up to about 8 mm long, the mature style plumose, variously contorted, up to about 4 cm long. (2n = 16, 48; n = 32) Aug–Oct. Climbing on banks, bushes, & trees along roads & in woods; scattered locations in KS, MO, NE, & OK; (widely introduced in e N. Amer.; e Asia). *Naturalized. C. dioscoreifolia* Levl. & Van.—Fernald, *C. maximowicziana* Franch. & Sav.—Keener, Sida 6:33–47. 1975.

For an explanation of the nomenclatural problems surrounding this species see Keener, Sida 7: 397. 1978.

6. **Clematis virginiana** L., virgin's bower. Dioecious, somewhat woody vines often many meters in length. Stems pubescent to nearly glabrous. Leaves petiolate, usually ternately compound, the 3 leaflets ovate, coarsely toothed to serrate, the largest ones 4–11 cm long and 3–9 cm wide. Flowers paniculate, usually numerous, 1–2.5 cm across; sepals cream, spreading, 0.5–1.3 cm long; pistillate flowers with normal-sized, but sterile, stamens; staminate flowers without pistils. Achene head about 0.8–1.3 cm across, excluding the styles, 3–6 cm across, including the styles; achenes silky, 2.7–4.5 mm long, the styles plumose, persistent, variously contorted, 1.5–4 cm long. (2n = 16) Jul–Aug. Climbing on road banks & in bushes in a variety of habitats; MN, e ND, e SD, IA, e NE, e KS; (Que. to Man., s to GA & e KS). *C. missouriensis* Rydb.—Rydberg.

This species is sometimes distinguished with difficulty from the closely related *C. ligusticifolia* Nutt. and *C. drummondii* T. & G.

8. DELPHINIUM L., Larkspur

Perennial (rarely annual) herbs with fibrous or tuberlike roots (rarely taprooted). Stems solid to fistulose, simple to branched, single or multiple. Leaves alternate, often reniform, the lower ones petiolate, with the blades palmately divided into 3–7 main lobes, each lobe often further compounded, the ultimate segments linear to broadly rounded or cuneate, usually white-apiculate. Flowers zygomorphic, few to numerous in a simple or compound raceme. Sepals 5, blue to white or pink, the uppermost one bearing a long tapered spur; petals 4, dimorphic, usually separate, the upper pair glabrous, cartilaginous, spurred, concave, forming a nectar tunnel, the spurs extending into the sepal spur and each bearing a nectar cup, the lower pair attached laterally, clawed, the blades elliptic to orbicular, often bifid, bearded, at least partly concealing the stamens and carpels (in one introduced species all petals united into a single, spurred, beardless structure); stamens numerous, the filaments dilated below; carpels 3(1). Fruits many-seeded follicles; seeds often irregularly winged, sometimes scaly.

Reference: Warnock, M. J. 1981. Biosystematics of the *Delphinium carolinianum* complex (Ranunculaceae). Syst. Bot. 6: 38–54.

```
1  Plants annual; carpel 1; petals united ..................................................... 1. D. ajacis
1  Plants perennial; carpels 3; petals distinct.
   2  Plants low-growing species entering our range in the nw part; raceme broad, its lowest
      flower on a spreading pedicel, with its spur not normally overlapping the raceme axis;
      flowers few, often 3–15; seeds usually bearing irregular light-colored wings formed by the
      loose outer coats, but not scaly.
      3  Flowers showy; lower petals large, skirtlike, mostly 6–10 mm wide, covering the
         stamens completely, the sinus absent or shallow or at least not gaping ... 2. D. bicolor
      3  Flowers less showy; lower petals narrower, mostly 3–6 mm wide, only partly concealing
         the stamens, the sinus usually deep and gaping ........................ 3. D. nuttallianum
   2  Plants often taller, general throughout our range or limited to the se part; raceme narrower, the lowest flower generally on a short or ascending pedicel, its spur overlapping
      the raceme axis; seeds either bearing transverse scales or dark and irregularly wrinkled
      but not with light-colored wings.
```

4 Flowers usually deep blue; plants often of woodland habitats; stems succulent; roots tuberous; follicles divergent; seeds dark, not scaly .... 4. *D. tricorne*
4 Flowers white to pale blue in most of our material; plants often of open areas; stems slender to rather stout, not succulent; roots fibrous to tuberous; follicles erect; seeds scaly. .... 5. *D. virescens*

1. ***Delphinium ajacis*** L., rocket larkspur. Erect, branched annuals 1–7 dm tall. Stems crisp-puberulent, sometimes also glandular, at least in the upper portion. Leaves numerous, mostly very short-petiolate, dissected into linear segments, usually pubescent. Flowers blue to various shades of pink or white, 3–19 in number in the principal raceme; lower sepals 0.8–1.5 cm long, spur 1–2 cm long; petals united into a single, 4-lobed, beardless structure which arches over the stamens and carpels and extends into the sepal spur; carpel 1. Follicle pubescent, usually 1–2 cm long; seeds bearing numerous transverse, ruffled scales. (2n = 16) Jul–Aug. A garden annual, often escaping & sometimes persisting for a time; scattered in GP, most common in the se part; (native of Europe, widely introduced elsewhere). *Naturalized. Consolida ambigua* (L.) Ball & Heywood — Keener, Castanea 41: 12–20. 1976.

2. ***Delphinium bicolor*** Nutt., little larkspur. Erect to weak-stemmed perennials (1)1.5–4.5(9) dm tall from a fibrous or, rarely, a tuberlike root system. Stems seldom fistulose, glabrous throughout or variously pubescent above, usually unbranched. Main stem leaves 3–6, often mostly near the base, but sometimes more evenly spaced, blades with the ultimate segments linear to lanceolate or oblanceolate, lowest petioles long, gradually to abruptly reduced above. Inflorescence a short raceme of (2)3–15(21) flowers, the pedicels spreading, usually the lowest one(s) exceeding the calyx spurs. Flowers showy; sepals blue to blue-violet, rarely white or pink, the lower pair (9)15–21(26) mm long, the spurs mostly (9)13–20 mm long; upper petals white, usually blue-tinged; lower petals skirtlike, concealing the stamens, bearded, the blade elliptical to orbicular, (5.5)7.5–10.5(13) mm long, (4.5)6–9.5(11) mm wide, the sinus absent, shallow or sometimes relatively deep but not gaping. Follicles usually divergent, up to 20(25) mm long, seeds irregularly winged. May–Jul. In openings in woods & on gravelly banks, road cuts, etc., extreme w ND, w SD, MT; (Alta., ND, & MT, w to ID, s to w NE & WY).

This species is closely related to *D. nuttallianum*, with which it intergrades in some parts of its range. It is possible that both taxa should be regarded as subspecies of the more western *D. menziesii* DC. Although *D. bicolor* is not recorded for SD in the Atlas GP, it is more common in the BH than *D. nuttallianum*.

3. ***Delphinium nuttallianum*** Pritz., blue larkspur. Erect to weak-stemmed perennials (0.5)1.5–5(9) dm tall from a tuberlike or, rarely, a fibrous root system. Stems with a narrow to stout connection to the root system, simple or sometimes branched, glabrous throughout or variously pubescent, especially above. Main stem leaves (2)3–5(9), the lower usually withering by anthesis, blades with the ultimate segments linear to lanceolate or oblanceolate, the lowest petioles long, gradually reduced upward. Inflorescence a short, broad raceme of (1)4–15(27) flowers, the pedicels spreading, at least the lowest one(s) exceeding the calyx spurs; sepals blue or violet, sometimes white or pink, the lower pair (10)11–17(20) mm long, the spurs mostly (12)13–18(24) mm long; upper petals white, often tinged and penciled with blue; lower petals usually blue or violet (white), the blades seldom completely covering the stamens, elliptical, bearded, (3.5)4.5–8(13) mm long and (2.5)4–6.2(10) mm wide, the sinus gaping widely hence the surface area of the blade considerably smaller than in *D. bicolor.* Follicles usually divergent; seeds irregularly winged. (n = 8) Apr–Jul. In a variety of mesic to xeric habitats; w SD, MT, WY, w NE; (Alta. to B.C., s to NE, WY, NV, UT, & CA). *D. nelsonii* Greene — Rydberg.

This species is separable into several indistinct varieties in the Pacific Northwest. Our material belongs to the typical variety. *D. nuttallianum* is closely related to *D. bicolor,* which see.

4. ***Delphinium tricorne*** Michx., dwarf larkspur. Erect to weak-stemmed perennials (1)3–7 dm tall from a tuberlike root. Stems slender to, usually, rather stout, fistulose, somewhat succulent, unbranched to sparingly branched, usually glabrous below and more or less crisp-puberulent in the inflorescence. Main stem leaves 4–9(12), the blades with the ultimate segments usually lanceolate or oblanceolate (linear), the petioles of the lower leaves long, those of the upper leaves gradually reduced. Inflorescence relatively slender, but not spiciform, with (6)8–25(30) flowers, the spurs of even the lowest flowers normally intersecting the axis of the raceme; sepals blue or violet, rarely pink or white, the lower ones mostly 10–21 mm long, the spurs mostly 10–20 mm long; upper petals white, often tinged with blue; lower petals usually blue (white), the blades elliptical, bearded, 5–8 mm long and 3–6 mm wide, slightly to prominently bifid. Follicles divergent, up to about 20 mm long; seeds dark, irregularly wrinkled. Apr–Jun. In rich woods or, rarely, in moist prairie; IA, se NE, MO, e KS, e OK; (PA to s MN, s to NC, AR, & OK).

5. ***Delphinium virescens*** Nutt., prairie larkspur. Erect perennials mostly (2.5)5–12 dm tall from a fibrous, or sometimes tuberlike, root system. Stems relatively sturdy, branched or unbranched, pubescent, some of the hairs often glandular. Leaves usually both basal and stem-borne, 6 to very numerous, the lowest often rather crowded and persistent, the blades with the ultimate segments lanceolate (linear); petioles of the lowest leaves long, gradually reduced upward. Inflorescence a rather spikelike raceme of 5–30(50) flowers; sepals white or whitish, the lower pair 7–16 mm long, the spur 11–20 mm long, often curved upward; upper petals whitish, lower petals whitish, bearded, the blades elliptical, 4.3–7.5 mm long, 3–5.8 mm wide, bifid. Follicles erect; seeds scaly. (n = 8) May–Jun. Prairies & pastures; GP, except nw part; (Alta. to MN & ND, s to OK & TX).

Delphinium carolinianum Walter may be distinguished from *D. virescens* because it has shorter pedicels, the upper ones under 1 cm long, linear leaf segments, mostly blue or blue-tinged flowers, and few basal leaves, these withering at maturity. Warnock has identified *D. carolinianum* from the following KS counties: Marshall, Wyandotte, Johnson, Franklin, Harper, Neosho, and Cherokee. Most of these collections show some degree of intermediacy to *D. virescens*. They usually have pale blue or bluish-white flowers and crisp, eglandular pubescence. Such collections were recognized by Steyermark in the Flora of Missouri as *D. carolinianum* var. *crispum* Perry. Since *D. virescens* is intergradient with *D. carolinianum*, Warnock (op. cit.) proposes that it be considered a subspecies of that taxon. Such an approach may be biologically sound, although inconsistent with the traditional treatment in this genus, in which many of the "species" are somewhat intergradient.

According to R. McGregor (pers. comm.), among members of the *D. virescens* group there is a general tendency for plants from e WY, ND, SD, NE, and KS to be glandular on the upper stem and inflorescence axis and to have the spurs spreading or ascending, but in the high plains of KS, sw NE, CO, and s into the TX panhandle, the plants tend to be glandular only below the inflorescence, with flowers slightly larger and the spurs erect. These western plants also tend to have larger, darker seeds. The former variant probably should be considered typical *D. virescens*, and the latter one may, if desired, be called *D. penardii* Huth or *D. virescens* subsp. *penardii* (Huth) Ewan. In addition, specimens from e NM, se CO, and the TX panhandle may show intermediacy to *D. wootonii* Rydb. (= *D. virescens* subsp. *wootonii* (Rydb.) Ewan), a species with heavy fibrous roots, pubescence consisting entirely of crisp, eglandular hairs, mostly basal leaves, narrow leaf segments, and dark brown seeds. Warnock has annotated some material from the edge of our range as *D. wootonii* (NM: Roosevelt and Quay; CO: Baca and Bent; TX: Potter); but at least some of this material seems clearly transitional to *D. virescens* and fits comfortably within the broad concept adopted here.

It is apparent that more study should be undertaken to work out relationships within this complex and to assess the relationship of this group to more western species such as *D. geyeri* Greene.

9. ISOPYRUM L.

1. ***Isopyrum biternatum*** (Raf.) T. & G., false rue anemone. Slender, glabrous, weak-stemmed perennials 1.5–3.5(4) dm tall, arising from a fibrous root system. Leaves alter-

nate and basal, the basal ones biternate (triternate), petiolate, petioles mostly 4–14 cm long, the leaflets usually trilobed and 1–2.8 cm long, the stem leaves similar, but short-petiolate to sessile, the uppermost usually merely ternate. Flowers terminal and axillary, regular, 1–2.5 cm across; sepals (4)5, petaloid, white, 6–11 mm long; petals absent; stamens numerous, carpels 4–7. Fruit consisting of a cluster of divergent follicles, these ovate, compressed, 3–5 mm long, bearing (straight) curved styles 1.2–2 mm long, each 2- to 3-seeded; seeds smooth. (2n = 14) Mar–May. Moist to rocky soil in woods; MN, IA, MO, e KS, e OK, also reported from the se & BH regions of SD; (s Ont. to SD, s to FL & TX). *Enemion biternatum* Raf.—Keener, Sida 7: 1–12. 1977.

10. MYOSURUS L., Mouse-tail

1. *Myosurus minimus* L. Acaulescent, glabrous annuals 0.2–1.5 dm tall. Leaves linear, 1–10 cm long, 0.3–1.7(2.2) mm wide at the widest part, gradually narrowing toward the base. Flowers borne singly above the leaves on leafless scapes, inconspicuous, actinomorphic; sepals usually 5, greenish, (1.3)2–3.5(5.5) mm long, bearing membranaceous spurs (0.7)1–3 mm long; petals often 5, linear, whitish, 2–3.5 mm long, often quickly deciduous, sometimes lacking; stamens about 10; carpels numerous, usually more than 100, on a slender receptacle which becomes greatly elongated in fruit. Achenes rhombic in face view with a central keel which projects as a short stylar beak, rectangular or trapezoidal in side view; the spicate achene cluster (1.5)2–5 cm long, 1.5–2.5 mm wide, the individual achenes 0.9–2 mm long with the beaks not usually more than 0.5 mm long. (2n = 16) Mostly Mar–May (Feb–Jun). Moist areas, often near open water; known from scattered locations throughout the GP, probably often overlooked; (Ont. to B.C., s to VA, TX, s CA).

Myosurus aristatus Benth., a species with fewer carpels (normally 20–50) and more prominent stylar beaks, is known from within our range in ND: Slope, Ward, & Williams; and in MT: Daniels.

11. RANUNCULUS L, Buttercup, Crowfoot

Contributed by G. E. Larson

Perennial and annual herbs of aquatic, palustrine, and terrestrial habitats; roots usually fibrous, sometimes fleshy-thickened. Stems erect to procumbent, flexuous and floating in aquatic plants, often fistulose, rooting at the nodes in some spp., branching or simple. Leaves simple or more often ternately compound or divided, finely dissected into filiform segments in some, frequently variable on the same individual, with basal leaves usually longer petioled and less dissected than the alternate cauline leaves, the leaves mostly or entirely basal in some spp.; petioles often dilated at the base with stipulelike wings. Flowers terminal, sometimes appearing axillary, perfect, regular, hypogynous, inconspicuous to showy, borne above the water surface in aquatic spp.; sepals 5, greenish to yellowish, or purplish-tinged, usually deciduous during or shortly after flowering, rarely persistent; petals 5, seldom more (individuals of some spp. occasionally double-flowered by conversion of some stamens to additional petals), yellow to pale yellow or white, rarely pinkish, often fading from yellow to whitish with age, bearing a minute nectary pit near the base on the upper surface, the pit typically covered or surrounded by a petaloid scale, rarely naked or absent; stamens (5)10–many; pistils (5)10–many, usually numerous, rapidly developing into a globose to cylindrical head of achenes; receptacle globose to ovoid or cylindrical, often much expanded in fruit, glabrous or pubescent. Achenes turgid and only slightly to not at all keeled, or flattened with a sharp, differentiated, winglike border (turgid and 3-chambered with an empty chamber on each side of the seed in *R. testiculatus*), the coat smooth, ridged, striate, papillate or spiny, glabrous or pubescent, the style usually persistent as a terminal or lateral beak. *Batrachium* S. F. Gray, *Coptidium* Beurl., *Halerpestes* Green — Rydberg.

References: Benson, L. 1948. A treatise on the North American ranunculi. Amer. Midl. Naturalist 40: 1–264; Cook, C. D. K. 1966. A monographic study of *Ranunculus* subgenus *Batrachium* (DC.) A. Gray. Mitt. Bot. Staatssamml. München 6: 47–237; Drew, W. B. 1936. The North American representatives of *Ranunculus* sect. *Batrachium*. Rhodora 38: 1–47; Duncan, T. 1980. A taxonomic study of the *Ranunculus hispidus* Michx. complex in the Western Hemisphere. Univ. California Publ. Bot. 77: 1–125; Keener, C. S. 1976. Studies in the Ranunculaceae of the southeastern United States. V. *Ranunculus* L. Sida 6: 266–283.

1 Petals white or only the claw sometimes yellowish; achenes transversely ridged; plants aquatic, with leaves all finely divided into filiform segments.
  2 Achenes (7)15–25; achene bodies averaging 1.5 mm long, with a prominent beak 0.7–1.1 mm long (often shorter when dried) .... 15. *R. longirostris*
  2 Achenes 30–45(80); achene bodies averaging 1.25 mm or less long, with a beak 0.2–0.5 mm long (often nearly lacking when dried) .... 23. *R. subrigidus*
1 Petals entirely yellow (white or pinkish in *R. testiculatus* which is not aquatic); achenes not transversely ridged (except sometimes in *R. sceleratus*); plants aquatic or terrestrial, with leaves simple to compound; finely divided into flat segments in some.
  3 Sepals persistent in fruit; achenes tomentose, conspicuously 3-chambered with an empty pouchlike chamber on each side of the seed, the beak lanceolate, 2.5–4 mm long .... 24. *R. testiculatus*
  3 Sepals deciduous during or soon after flowering; achenes not tomentose, only 1-chambered, the beak not both lanceolate and 2.5–4 mm long.
    4 Leaves all simple and entire or crenate to crenate-lobed.
      5 Leaves ovate to round or reniform, crenate to crenate-lobed; achenes striate with longitudinal ribs .... 6. *R. cymbalaria*
      5 Leaves elliptic to lanceolate or linear, entire to denticulate; achenes not striate.
        6 Plants perennial, 4–15 cm tall; stems decumbent to prostrate, freely rooting at the nodes; achenes 1.2–1.7 mm long .... 9. *R. flammula*
        6 Plants annual, 1.5–8 dm tall; stems erect to reclining, rooting only at the lowest nodes if at all; achenes 0.6–1.1 mm long .... 14. *R. laxicaulis*
    4 All, or at least the cauline leaves deeply lobed, divided or compound.
      7 Basal and cauline leaves distinctly different in shape, the basal leaves mostly entire or crenate, not lobed or divided, the cauline leaves deeply divided.
        8 Petals 1.5–3.5 mm long, shorter and narrower than the sepals.
          9 Achenes (40)50–100, often finely pubescent; achene beak 0.5–0.9 mm long; stem simple or sparingly branched above, 1- to several-flowered .... 13. *R. inamoenus*
          9 Achenes 10–40(50), glabrous; achene beak minute, to 0.3(0.4) mm long; stem eventually freely branched above, several- to many-flowered.
            10 Plants glabrous throughout or sparsely pubescent mainly above; basal leaves mostly cordate .... 1. *R. abortivus*
            10 Plants villous below; basal leaves proximally truncate, rounded or tapered to the petiole .... 17. *R. micranthus*
        8 Petals 3.5–15 mm long, longer and wider than the sepals.
          11 Oldest basal leaves entire or at most shallowly 3-lobed .... 10. *R. glaberrimus*
          11 Basal leaves all crenate or some variously lobed, none entire.
            12 Petals 8–15 mm long; nectary scale and often the adjacent petal surface ciliate, with hairs mostly 0.5–1 mm long .... 5. *R. cardiophyllus*
            12 Petals 3.5–8 mm long; nectary scale and adjacent petal surface glabrous.
              13 Achenes glabrous, tapered to a winged-stipitate base, the beak to 0.3 mm long .... 20. *R. rhomboideus*
              13 Achenes glabrous or often pubescent, not winged or stipitate at the base, the beak 0.5–0.9 mm long .... 13. *R. inamoenus*

7 Basal and cauline leaves essentially similar in shape and form, all variously lobed, deeply divided or compound, none merely crenate.
  14 Achenes turgid, without a sharp border.
    15 Petals 4–14 mm long; achenes 1.2–2.5 mm long, prominently beaked; aquatic or amphibious perennials.
      16 Achene margin thickened and corky below the middle; petals 6–15 mm long .... 8. *R. flabellaris*
      16 Achene margin rounded but not thickened; petals 3–8 mm long ... 11. *R. gmelinii*
    15 Petals 2–5 mm long; achenes 0.8–1.2 mm long, nearly beakless; weedy annual, usually emersed .... 22. *R. sceleratus*
  14 Achenes flattened, with a sharp or winglike border.
    17 Achenes papillate (rarely smooth) or spiny; introduced annuals.
      18 Achenes 2–3 mm long, merely papillate (rarely smooth), the beak 0.2–0.5 mm long .... 21. *R. sardous*
      18 Achenes 4–5 mm long, strongly spiny, the beak 1.5–3 mm long ...... 3. *R. arvensis*
    17 Achenes smooth or minutely pitted; native and introduced perennials.
      19 Petals 2–6 mm long; anthers 0.7–1 mm long (or slightly larger in *R. recurvatus*).
        20 Achene beaks strongly recurved; terminal lobe of the principal leaves sessile, connected to the lateral lobes by leaf tissue .... 19. *R. recurvatus*
        20 Achene beaks straight to slightly curved; terminal lobe of the principal leaves stalked.
          21 Petals distinctly shorter than the sepals; heads of achenes ovoid-cylindric to cylindric .... 18. *R. pensylvanicus*
          21 Petals equaling or longer than the sepals; heads of achenes ovoid to globose .... 16. *R. macounii*
      19 Petals 7–17 mm long; anthers 1.2 mm or more long (or slightly shorter in *R. acris*).
        22 Principal leaves deeply dissected but not compound, the lobes connected basally to one another by leaf tissue .... 2. *R. acris*
        22 Principal leaves ternately compound, at least the terminal segment stalked.
          23 Plants with cormlike thickening at the base; styles (or beaks) recurved, stigmatic along the upper side; introduced species .... 4. *R. bulbosus*
          23 Plants lacking cormlike thickening at the base (roots sometimes tuberous thickened); styles straight or nearly so, stigmatic apically; native species.
            24 Most of the ultimate leaf divisions narrow, linear to oblong, mostly 2–8 mm wide, entire or remotely and shallowly lobed, obtuse to rounded; roots of 2 types, some short and tuberous thickened, others long-filiform .... 7. *R. fascicularis*
            24 Main lobes of the principal leaves broad, ovate to rhombic, mostly sharply toothed and acute-tipped; roots all elongate, none short and thickened .... 12. *R. hispidus*

1. Ranunculus abortivus L., early wood buttercup. Erect to spreading biennial or short-lived perennial, (1)2–7 dm tall, glabrous throughout or sparsely pubescent mainly above; roots filiform, sometimes enlarged at their bases. Stems usually single (to several), eventually freely branched above, fistulose, several- to many-flowered. Basal leaves with petioles 2–15 cm long, the blades simple, reniform, mostly cordate, some occasionally lobed, 1–6(8) cm across, crenate; some basal leaves transitional to the sessile to subsessile cauline leaves, the latter 3- to 5-cleft, the segments obovate and irregularly toothed or lobed to linear-lanceolate and entire, reduced to bracts above. Sepals spreading, greenish-yellow with purplish or whitish, membranaceous margins, elliptic, 2–3.5 mm long, glabrous to hispidulous dorsally; petals 5, pale yellow, fading white, elliptic, 1–3 mm long, shorter and narrower than the sepals; nectary scale pocketlike, the lateral margins free for no more than 1/3 of their length, glabrous; stamens 15–30. Achenes 10–40(50) in a globose-ovoid to short-cylindric head 3–8 mm long, 3–5 mm thick; achene body obovate, turgid, with a narrow inconspicuous margin, (1)1.2–1.8 mm long, about 1/2 as thick as broad, smooth; beak minute, 0.1–0.2 mm long; receptacle conical-fusiform, 2.5–8 mm long in fruit, sparsely hairy; pedicels to 10 cm long in fruit. (2n = 16) Apr–Jun. Moist woods & flood plains; GP, except sw portion, more common e; (Lab. to Sask., s to FL, TX, CO, & WA; W.I.). *R. abortivus* var. *acrolasius* Fern.—Fernald; *R. abortivus* f. *giganteus* F. C. Gates—Gates.

2. **Ranunculus acris** L., tall buttercup. Hirsute to appressed-hairy perennial, 4–10 dm tall; roots fibrous, not thickened. Stems 1–several, erect, fistulose, branched above. Basal leaves with petioles 5–25 cm long, the blades oval to reniform and broadly cordate in outline, 3–6 cm long, 3–9 cm across, pedately 3-cleft, with the lateral lobes often deeply divided nearly to the base to give the blade a 5-lobed appearance, the divisions of the lateral lobes more shallowly incised (2 or 3 times) and toothed, the segments acute-tipped, the central lobe ternately incised and toothed like the lateral lobes; cauline leaves reduced, transitional to the 3- to 5-lobed bracts. Sepals spreading, green, ovate, 4–7 mm long, hirsute dorsally; petals usually 5, occasionally more in double-flowered specimens, yellow, broadly obovate, (7)8–14 mm long; nectary scale truncate, the lateral margins free for about 2/3 of their length, glabrous; stamens 30–70, anthers (1)1.2–1.5 mm long. Achenes 15–40 in a globose head 5–6 mm in diam; achene body obovate, flattened, narrowly wing-margined, 2–2.5 mm long, about 1/3 as thick as broad, smooth; beak flattened at the base and curved, 0.3–0.7 mm long; receptacle pyriform, 2–4 mm long in fruit, glabrous; pedicels to 12 cm long in fruit. (2n = 14, 28) May–Jul. Meadows, woods, & roadsides; scattered locations mainly in e GP & BH; (throughout much of N. Amer.; Europe). *Adventive.*

3. **Ranunculus arvensis** L. Erect, sparingly hirsute annual, 1.5–5 dm tall; roots fibrous, rather stout. Stem single, freely branched above. Basal leaves with blades varying from simple, obovate and coarsely toothed apically to deeply 3-parted or ternate, mostly 1.5–5 cm long and about as wide, when ternate, the segments obovate, oblanceolate or linear and themselves usually ternately divided; petioles 3–6 cm long; cauline leaves reduced upward, ternately or biternately divided into linear to oblanceolate segments which are 1.5–5 mm wide, short petioled to sessile upward. Sepals greenish, narrowly elliptic, 3.5–6 mm long, membranaceous, hirsute dorsally; petals 5, yellow, obovate, 4–8 mm long; nectary scale rounded, about as wide as the petal, free except at the base, glabrous; stamens 10–15. Achenes 5–8(10), loosely spreading in a hemispheric to globose head 12–15 mm in diam; achene body obovate, strongly flattened, wing-margined and stipitate, 4–5 mm long, strongly spiny on the faces, the largest spines adjacent to the margins, also finely papillate; beak stout, spinelike, straight to slightly curved, 1.5–3 mm long, stigmatic laterally; receptacle hemispheric, 1–2 mm long in fruit, sparsely hairy; pedicels 3–6 cm long in fruit. (2n = 32) May–Jun. Waste places & disturbed areas; KS: Douglas, locally introduced and persisting for a few years on the University of Kansas campus; (widely introduced in N. Amer.; Europe). *Adventive.*

4. **Ranunculus bulbosus** L., bulbous buttercup. Hirsute to appressed-hairy perennial, 1–5(7) dm tall, from a cormose-thickened base; roots fibrous, rather stout. Stems 1–few, erect, simple or usually branched, 1- to several-flowered. Basal leaves with petioles 3–15(20) cm long, the stipular-winged bases overlapping around the cormlike stem base; blades ternate, with at least the terminal leaflet stalked, broadly oval to ovate and usually cordate in outline, 2–5 cm long, 1.5–4 cm wide, the leaflets themselves trifid and each lobe less deeply 1- or 2-toothed or lobed; cauline leaves short petioled to sessile upward, more deeply dissected into narrower lobes, reduced to bracts upward. Sepals greenish, usually reflexed, ovate, 5–7 mm long, hirsute dorsally; petals 5, yellow, broadly obovate, 7–14 mm long; nectary scale broad and rounded, the margins free for about 2/3 of their length, glabrous; stamens 40–80, anthers mostly 2–3 mm long. Achenes 15–40 in a globose to

globose-ovoid head 5–8 mm long, 5–7 mm thick; achene body obovate, 2.5–3 mm long, about 1/4 as thick as broad, smooth; beak flattened, recurved, 0.3–0.7 mm long, stigmatic on the upper side; receptacle ellipsoid, to 6 mm long and stipitate in fruit, pubescent; pedicels 3–6 cm long in fruit. (2n = 16) May–Jun. Meadows, waste places, & roadsides; KS: Cherokee, Logan, Wyandotte; NE: Dawes, Franklin; (introduced & widely established over much of N. Amer.; Europe). *Adventive.*

5. Ranunculus cardiophyllus Hook. Erect, pilose to glabrate perennial from fibrous roots, 1.5–4 dm tall. Stem single, simple or branched above, 1- to several-flowered. Basal leaves with petioles 2–12 cm long, the previous year's petioles often persistent at the base of the stem; blades simple, reniform to mostly ovate-cordate, 2–6 cm long, about as wide, crenate, 1 or 2 sometimes shallowly to deeply cleft; cauline leaves few, subsessile to sessile, deeply parted into several linear lobes. Sepals yellowish, often tinged with purple, spreading, obovate, 6–10 mm long, pilose dorsally, petaloid on the margins; petals 5 or sometimes a few more, yellow, obovate, 8–15 mm long; nectary scale adnate to the petal along the margins, forming a pocket, truncate, conspicuously ciliate with hairs mostly 0.5–1 mm long, the adjacent petal surface often hairy also; stamens 35–80. Achenes 20–100 in an oblong-cylindrical head 8–15 mm long, 7–9 mm thick; achene body obovate, turgid with an inconspicuous margin, 1.5–2.5 mm long, about 1/2 as thick as broad, puberulent; beak straight or recurved at the tip, 0.5–1 mm long; receptacle oblong-ovoid, 4–14 mm long in fruit, hairy; pedicels erect, 5–14 cm long in fruit. (2n = 32, 64) Jun–Jul. Mt. meadows & seepage areas; BH; ND: McKenzie; (Sask. to N.T. & B.C., s to SD, NM, AZ, & WA).

6. Ranunculus cymbalaria Pursh, shore buttercup. Low, extensively stoloniferous, scapose perennial from fibrous roots, often forming dense mats, 3–15(25) cm tall, glabrous or villous mainly on the scapes and petioles. Scapes surpassing the leaves, simple or sparingly branched with 1–several small, yellow flowers. Leaves all basal, with petioles mostly 2–5 cm long, the blades simple, ovate to round or reniform, cordate or truncate at the base, 5–30(40) mm long, 4–20(35) mm wide, crenate to shallowly crenate-lobed, often with 3 prominent lobes at the tip. Sepals greenish-yellow, sometimes with purplish margins, spreading, elliptic, 3–5 mm long; petals usually 5(12), pale yellow, fading white, obovate, 3–8 mm long; nectary scale pocketlike, with the margins completely adnate to the petal, glabrous; stamens 10–30. Achenes 40–200 in an ovoid to cylindrical head 3–15 mm long, 3–6 mm thick; achene body cuneate-oblong, ca 1.5 mm long, striate, with 2–4 longitudinal nerves on each face; beak triangular, straight, 0.2–0.3 mm long. (2n = 16) Apr–Sep. Marshes, muddy shores, stream banks, ditches, & seepage areas; GP, except se KS & cen OK; (throughout most of N. & S. Amer.; Eurasia). *Halerpestes cymbalaria* (Pursh) Greene—Rydberg; *R. cymbalaria* f. *hebecaulis* Fern.—Fernald; *R. cymbalaria* var. *saximontanus* Fern.—Waterfall.

7. Ranunculus fascicularis Muhl. ex Bigel., early buttercup. Hirsute or usually appressed-hairy, tufted perennial (0.5)1–3 dm tall; roots of 2 types, some short (to 5 cm long) and tuberous thickened, others long-filiform. Stems erect to ascending, scapose or with 1–few reduced leaves, simple or little branched, 1- to 4-flowered, canescent below. Basal leaves with petioles mostly 2–10 cm long, the blades ovate in outline, 1.5–5.5 cm long, (1)1.5–4 cm wide, pinnately 3- to 5-parted or some earlier leaves merely 3-lobed; leaf segments linear to oblong and entire or shallowly toothed, or more often the segments themselves trifid, the ultimate segments narrow, linear to oblong, mostly 2–8 mm wide, entire or remotely and shallowly lobed, obtuse to rounded; cauline leaves (when present) reduced, of 1–3 narrow segments. Sepals greenish-yellow, spreading, ovate, 6–9 mm long, appressed-hairy; petals 5(10), yellow, elliptic to oblong, 8–16 mm long; nectary scale truncate, not pocketlike, free for nearly its whole length, glabrous; stamens usually ca 40–50, anthers 1.3–1.8 mm long. Achenes 10–35 in a globose-ovoid head 5–11 mm long, 5–9 mm thick; achene body rotund, sharp margined, 1.3–3.5 mm long, about 1/3 as thick as wide, smooth; beak slender, straight or nearly so, 1–2 mm long. (2n = 32) Apr–May. Prairie & dry woods; MN & e NE, s to e OK; (MA to s Ont., s to PA, WV, TN, LA, & TX). *R. fascicularis* var. *apricus* (Greene) Fern.—Fernald.

8. ***Ranunculus flabellaris*** Raf., threadleaf buttercup. Aquatic or amphibious perennial, usually submersed but occasionally stranded on mud, glabrous or commonly hirsute when emersed. Stems floating and inflated or, when stranded, erect from a decumbent base, rooting at the lower nodes, branching, 3–7 dm long or shorter when emersed, 1- to several-flowered. Leaves all cauline, the lower ones with petioles to several times longer than the blades, the upper subsessile, with broad stipular bases 3–10 mm long; blades semicircular to reniform in outline, 1.5–10 cm long, 2–12 cm wide, finely ternately dissected into flat, narrow segments mostly 0.2–2 mm wide, averaging smaller in size and not as finely dissected on emersed plants. Sepals greenish-yellow, spreading, ovate, 5–8 mm long, early deciduous; petals 5, occasionally more, yellow, obovate, 6–15 mm long; nectary scale ovate, glabrous, adnate to the petal for most of its length but the margins and tip free, forming a pocket around the gland which is borne on the exposed surface of the scale; stamens (30)50–80. Achenes (30)40–75 in an ovoid to globose head 6–10 mm long, 5–8 mm thick; achene body turgid, obovate, 1.5–2 mm long, corky thickened around the margins and especially below the middle, pustulose roughened on the faces; beak broad and flat, straight or nearly so, 1–1.7 mm long; receptacle ovoid-cylindrical, 5–7 mm long in fruit, sparsely hairy. Apr–Jul, occasionally Aug–Sep. Shallow water or mud of marshes, ponds, ditches, & sluggish streams; GP, mainly in the e & cen parts; (ME to B.C., s to NJ, VA, OH, IN, IL, LA, UT, NV, & CA). *R. delphinifolius* Torr.—Rydberg; *R. flabellaris* f. *riparius* Fern.—Fernald; *R. purshii* Richards. var. *schizanthus* Lunell, var. *polymorphus* Lunell—Lunell, Bull. Leeds Herb. 2: 6. 1908.

The latter two synonyms were applied to some ambiguous collections of Lunell from the area of Leeds, ND. Benson (op. cit., p. 213) admits that these plants may represent variations of *R. gmelinii* rather than *R. flabellaris*.

9. ***Ranunculus flammula*** L., spearwort. Low, stoloniferous perennial, often appressed-hairy. Stems decumbent to prostrate, rooting at the nodes, simple to sparingly branched, the upright tips 4–15 cm tall, 1- to several-flowered. Leaves clustered at rooting nodes, reduced and shorter petioled on upper portions of the stem; blades simple, elliptic to lanceolate or linear, 1–5 cm long, 1.5–7(20) mm wide, entire, or very slightly toothed, tapered to slender petioles mostly 5–15 mm long. Sepals yellowish-green, ovate, 1.5–3(5) mm long, strigose; petals 5, yellow, obovate, 3–5 mm long; nectary scale forming a shallow pocket, adnate along the lateral margins, truncate to obtuse, glabrous; stamens 25–50. Achenes 10–25 in a subglobose head 2.5–3.5 mm long, 3–4.5 mm thick; achene body turgid, obovate, 1.3–1.7 mm long, about 1/2 as thick as broad, inconspicuously margined, smooth, the beak 0.2–0.6 mm long; receptacle obovoid, ca 1 mm long in fruit, glabrous. (2n = 32) Jun–Aug. Marshes & muddy shores; ND: Burke; MT: Sheridan; (circumboreal, in N. Amer. s to MA, PA, MI, MN, ND, NM, AZ, & CA). *R. reptans* L.—Rydberg; *R. reptans* var. *ovalis* (Bigel.) Torr. & Gray—Fernald.

Our plants as described above are *R. flammula* var. *ovalis* (Bigel.) L. Benson.

10. ***Ranunculus glaberrimus*** Hook. Low, tufted perennial from rather fleshy, tomentose roots 2–3 mm thick. Stems usually few to several, erect to decumbent, 0.5–2 dm long, glabrous or rarely sparsely hirsute. Leaves mostly basal, with petioles 1–7 cm long, the stipular-winged bases often long-hirsute; blades simple, elliptic to ovate or obovate, 1–4 cm long, 0.8–2.5 cm wide, those of the oldest leaves entire or shallowly 3-lobed, not crenate, later basal leaves occasionally with more deeply lobed blades; cauline leaves 1-several,

sessile or nearly so, shallowly to deeply 3-parted, the lobes entire or sparingly lobed or toothed. Sepals usually purplish-tinged, elliptic, 4-8 mm long, glabrous to hirsute; petals 5(8), yellow, fading white with age, obovate, 6-15 mm long; nectary scale truncate or retuse, forming a deep pocket, with the lateral margins entirely adnate or free only near the tip, usually ciliate; stamens 40-80. Achenes 30-150 in a subglobose to globose head 7-20 mm in diam; achene body obliquely obovoid, slightly winged ventrally, winged-stipitate at the base, 2-2.5 mm long, about 1/2 as thick as broad, usually finely puberulent; beak straight or curved, flattened at the base, 0.5-1 mm long; receptacle globose, 7-17 mm long in fruit, glabrous or sparsely puberulent; pedicels 1-10 cm long in fruit. Apr-Jun. Ponderosa pine woodland, prairie, & pastures; cen & w ND & MT, s to nw NE & CO; (Alta. to B.C., s to NM & n CA). *R. waldronii* Lunell, —Lunell, Amer. Midl. Naturalist 3: 12. 1913.

The majority of our plants are var. *ellipticus* Greene, with the blades of the basal leaves usually entire and elliptic to oblanceolate. A few collections approach var. *glaberrimus*, which has basal leaves with blades ovate to obovate and usually 3-lobed.

11. *Ranunculus gmelinii* DC., small yellow buttercup. Similar in habit to no. 8 but usually emersed, glabrous to hirsute. Stems usually procumbent and rooting at the nodes, floating when submersed, sparingly branched, 1-5 dm long, 1- to several-flowered. Leaves all cauline or a few clustered at rooted nodes, the lower ones with petioles to 5 cm long, the upper leaves commonly floating on submersed plants; stipular bases broad, 3-6 mm long; blades deeply 3-lobed or dissected, the primary divisions dichotomously forked 2 or 3 times or sometimes dissected into flat, linear segments when submersed, not triternately dissected, pentagonal in outline, usually 0.8-2 cm long, 1.5-2.5 cm wide, or submersed leaves often larger, to 6(9) cm across. Sepals greenish-yellow, spreading, ovate to broadly ovate, 2.5-6 mm long, glabrous or pubescent; petals 5-8, rarely more, obovate to orbiculate (occasionally notched), 3-8 mm long; nectary scale entire to deeply notched, free nearly to the base, glabrous, the gland borne on the upper surface; stamens 20-45. Achenes 40-80 in an ovoid to globose head 7-10 mm long, 5-8 mm thick; achene body turgid, obovate, 1-1.5 mm long, the margin rounded or inconspicuously keeled, not corky thickened although the basal and ventral portions whitish and callous thickened; beak flat, narrowly triangular, (0.4)0.6-0.8 mm long; receptacle ovoid to obovoid, 3-6 mm long in fruit, short-hairy. ($2n = 16$, 32, 64) Late May-Jul, occasionally Aug-Sep. Shallow water or mud of ponds, marshes, streams, & ditches; n, cen, & e ND; NE: Douglas; (boreal, s to ME, MI, IA, NE, CO, NV, & OR; Asia). *R. purshii* Richards., *R. limosus* Nutt.—Rydberg; *R. gmelinii* var. *terrestris* (Ledeb.) L. Benson, var. *prolificus* (Fern.) Hara—Gleason; *R. purshii* var. *dissectus* Lunell, var. *geranioides* Lunell, var. *radicans* Lunell—Lunell, Bull. Leeds Herb. 2: 6. 1908.

Two intergrading varieties are found in our range. Of the two, the first described below is more frequent; the second seems most common in n ND.

11a. var. *hookeri* (D. Don) L. Benson. Plants usually glabrous. Leaf blades mostly 1-3 cm long, deeply 3-parted or divided, the primary divisions dichotomously forked 2 or 3 times or dissected into linear segments. Sepals 4-6 mm long; petals 4-8 mm long.

11b. var. *limosus* (Nutt.) Hara. Plants hirsute. Leaf blades mostly 0.8-1.5 cm long, rather shallowly lobed with the lobes themselves shallowly lobed and rounded. Sepals 2.5-4 mm long; petals mostly 4-5 mm long.

12. *Ranunculus hispidus* Michx., bristly buttercup, marsh buttercup. Nearly glabrous to strongly hirsute or hispid perennial 1.5-7 dm tall, the pubescence appressed to spreading or retrorse on the stems; roots mostly filiform, some often thickened to 3 mm but these elongate (over 5 cm long). Stems erect, repent or stolonlike, mostly 1- to 8-flowered, the stolonlike stems (when present) arched back to the ground and rooting at an upper node. Basal leaves with petioles 0.5-3 dm long, prominently stipular winged; blades ternate or those of the earliest leaves merely 3-lobed, broadly ovate-cordate in outline, (2)3–14 cm long, (2)3–20 cm wide, the segments broad, ovate to rhombic, 2- or 3-lobed or cleft and irregularly toothed; cauline leaves shorter petioled upward, the blades ternate, reduced upward. Sepals yellowish-green, spreading or reflexed, ovate or oblong-ovate, mostly 5–8 mm long, glabrous to appressed-hairy or pilose dorsally; petals 5 (rarely to 10), yellow, obovate to elliptic, 7–16 mm long; nectary scale flabellate, free nearly its whole length, glabrous; stamens mostly 40–70. Achenes (7)15–60 in an ovoid to globose head 6–12 mm long, 6–12 mm thick; achene body obovate, narrowly to broadly wing margined, 2–4.5 mm long, about 1/3 as thick as broad; beak flat at the base, straight or nearly so, 1–3 mm long; receptacle cylindrical to clavate, 4–8.5 mm long in fruit, sparsely hispidulous; pedicels 3.5–15(20) cm long in fruit. (2n = 32, 64) Apr–early Jul.

The *R. hispidus* complex has been recently re-worked by Duncan (1980, op. cit.) who merges *R. septentrionalis* and *R. carolinianus* with *R. hispidus*. Given the frequent difficulty in distinguishing between these three as species, Duncan's broad interpretation of *R. hispidus* seems a logical approach. Accordingly, *R. hispidus* is divisible as follows into three varieties that occur in the GP.

1 Achenes narrowly winged with a margin to 0.3 mm wide; stems erect or repent at time of fruiting.
   2 Plants of wet habitats, some stems stolonlike, arching back to the ground and rooting at the nodes; aerial shoots 5–8(9) dm long at time of fruiting .... 12a. var. *caricetorum*
   2 Plants of drier soil, with stems stout, all erect, not becoming repent and never rooting at the nodes; aerial shoots 1.5–4.5(6) dm long at time of fruiting .... 12b. var. *hispidus*
1 Achenes broadly winged with margin 0.4–1 mm wide; stems partly repent and usually stolonlike at the time of fruiting .... 12c. var. *nitidus*

12a. var *caricetorum* (Greene) T. Duncan. Some stems stolonlike, arching back to the ground and rooting at the nodes; aerial shoots 5–8(9) dm long at fruiting time. Achenes narrowly winged, with margin 0.3 mm wide. Wet meadows, springs, swamps, shores, & stream banks; MN to cen ND, s to w MO & possibly e KS; (Lab. to s Man., s to VA, OH, & MO). *R. caricetorum* Greene — Rydberg; *R. septentrionalis* Poir. var. *caricetorum* (Greene) Fern. — Fernald.

12b. var. *hispidus*. Stems all erect, not becoming repent and never rooting at nodes; aerial shoots 1.5–4.5(6) dm long at fruiting time. Achenes narrowly winged, with margin 0.3 mm wide. Moist or dry woods, often where rocky; MO, e KS, & e OK; (MA, NY, & s Ont. to IL, MO, & KS, s to GA, AR, & OK). *R. hispidus* var. *falsus* Fern. — Fernald; *R. hispidus* var. *marylandicus* (Poir.) L. Benson — Gleason & Cronquist.

12c. var. *nitidus* (Ell.) T. Duncan. Stems partly repent, usually stolonlike at fruiting time. Achenes broadly winged, with sharply demarcated wing 0.4–1 mm wide. Same habitats as 12a; sw MN & e SD, s to MO, e KS, & e OK; (NY to MN, s to FL & TX). *R. septentrionalis* Poir. — Rydberg; *R. septentrionalis* var. *pterocarpus* L. Benson — Gleason & Cronquist.

13. *Ranunculus inamoenus* Greene. Slender, sparingly hirsute to glabrous perennial 0.9–2(3) dm tall; roots fibrous, filiform. Stems single or occasionally several, simple or sparingly branched, fistulose, 1- to several-flowered. Basal leaves with petioles 2–10 cm long, the blades simple, ovate to obovate or orbiculate, attenuate to rounded or subcordate at the base, 1–3.5(4) cm long, about as wide, crenate to crenate lobed; cauline leaves short petioled to sessile, deeply 3- to 5(7)-cleft into linear or oblanceolate segments. Sepals greenish-yellow, spreading, narrowly obovate, 3–6 mm long, pilose dorsally; petals 5, yellow, narrowly elliptic to obovate, 2–8 mm long; nectary scale pocketlike, with the margins adnate to the petal, glabrous; stamens 25–50. Achenes (40)50–100 in an elongate-cylindrical head 6–17 mm long, 5–8 mm thick; achene body obovate, turgid with an inconspicuous margin, 1.5–2 mm long, 1/3 to 1/2 as thick as broad, finely pubescent or glabrous; beak 0.5–0.9 mm long, often recurved at the tip; receptacle elongate, 6–12 mm long in fruit, usually hairy; pedicels 1–7 cm long in fruit. (2n = 32, 48) Jun–Jul. Moist meadows & stream banks, usually in mts.; NE: Cheyenne; SD: Pennington; (B.C. to BH, s to NM & AZ).

14. *Ranunculus laxicaulis* (T. & G.) Darby, water plantain spearwort. Slender, glabrous, fibrous-rooted annual 1.5–8 dm tall, terrestrial or emergent in shallow water. Stems erect to reclining, sometimes rooting at the lowest nodes, freely branching especially above, fistulose at the base. Leaves all simple; basal leaves with petioles 1–7 cm long, or sometimes longer on floating leaves, the blades ovate to oblong, 1–4.5 cm long, 6–18 mm wide, rounded to truncate at the base, obtuse to truncate at the tip, shallowly dentate or entire; cauline leaves sessile, narrower than the basal, linear-elliptic to lanceolate, acute, denticulate. Sepals greenish-yellow, spreading, ovate, 1.5–3 mm long, glabrous or sparsely hairy; petals 5 or rarely to 10, yellow, oblong, 3–9 mm long; nectary scale forming a pocket, adnate along the lateral margins, truncate or prolonged on the margins, glabrous; stamens 10–30. Achenes 15–50 in a hemispheric to ovoid head 2–4 mm long, 2–2.5 mm thick; achene body turgid, obovate to subrotund, 0.6–1.1 mm long, about 1/2 as thick as broad, smooth, the margin inconspicuous, the style deciduous, leaving a beak 0.1–0.2 mm long; receptacle pyriform or spherical, 1.5–3 mm long in fruit, glabrous; pedicels 1–6 cm long in fruit. Apr–Jun. Marshes, ditches, & shores; KS: Miami; MO: Barton, Jasper; (CT to e KS, s to FL & TX). *R. obtusiusculus* Raf.—Rydberg, misapplied; *R. texensis* Engelm.—Gleason & Cronquist. *Possibly extinct* in the GP.

15. *Ranunculus longirostris* Godr., white water crowfoot. Aquatic, mostly glabrous perennial. Stems floating, elongate, flexuous, mostly 3–8 dm long, simple or sparingly branched, rooting from the lower nodes. Leaves all cauline, the blades rather rigid or flaccid, finely divided into filiform segments, once or twice trichotomous, then dichotomous, globular in outline, reniform when flattened, 1–3(4) cm long, 1.5–4(5) cm wide; petioles consisting mainly of the inflated stipular bases, 2–4 mm long, glabrous or pubescent. Flowers appearing axillary on the upper portion of the stem; sepals 5(6), purplish-green, spreading, narrowly elliptic, 2–4 mm long, deciduous shortly before the petals; petals 5, white, suffused with yellow at the base, obovate, 4–10 mm long; nectary scale pyriform to lunate or absent, glabrous; stamens 10–20. Achenes (7)15–25 in a hemispheric to globose head 3–5 mm long, 3–6 mm thick; achene body obovoid, 1.3–1.7 mm long (averaging 1.5 mm), transversely ridged, glabrous or slightly hispid; beak prominent, slender, straight or curved, 0.7–1.1 mm long (often shorter when dried); receptacle ovoid, 1–2 mm long in fruit, pubescent; pedicels 1–9 cm long and usually recurved in fruit. (2n = 48) Apr–Jul, occasionally Aug–Sep. Shallow water of streams, ponds, marshes, & ditches; GP; (Que., to Sask., s to DE, TN, AL, TX, NM, & NV). *Batrachium longirostre* (Godr.) Schultz—Rydberg; *Ranunculus circinatus* Sibth.—Gleason, misapplied.

The names *Ranunculus aquatilis* L. and *R. trichophyllus* Chaix have been mistakenly applied to our white-flowered crowfoots (*R. longirostris* and *R. subrigidus*) in previous works. Neither of these species has been encountered in the GP.

16. *Ranunculus macounii* Britt., Macoun's buttercup. Sparsely to densely hirsute annual or short-lived perennial 2–7 dm tall; roots rather thick and fleshy. Stems erect or decumbent, dichotomously branched 1–few times above, fistulose, the branches few- to several-flowered. Basal and cauline leaves similar, the basal usually larger and longer petioled than the cauline, with petioles to 3 dm long, prominently stipular winged; blades deltoid or broadly so in outline, 4–14 cm long, 6–16 cm wide, ternate, the terminal segment 3-lobed, the lateral 2- or 3-lobed, the segments coarsely and irregularly toothed as well, subsessile to prominently stalked. Sepals yellowish-green, reflexed, ovate, 3–5(7) mm long, glabrous or pilose dorsally; petals 5, yellow, obovate, 3–6(8) mm long, equaling or surpassing the sepals; nectary scale oval to truncate, free for about 1/2 its length, glabrous; stamens 15–35, anthers 0.7–1 mm long. Achenes (20)30–60 in an ovoid to globose head 7–12 mm long, 8–12 mm thick; achene body flattened, obovate, 2–3(3.5) mm long, smooth or shallowly and finely pitted, narrowly keel margined; beak stout, straight or slightly curved, 0.7–1.2

mm long; receptacle ovoid to ellipsoid, 4–7 mm long in fruit, hispidulous; pedicels 3–10 cm long in fruit. (2n = 32, 48) Jun–Jul. Wet meadows, shores, stream banks, & other wet places; MN, ND, MT, n IA, SD, cen & w NE, WY, & CO; (Lab. to AK, s to Que., n MI, IA, NE, NM, AZ, & CA). *R. rivularis* Rydb.—Rydberg.

17. *Ranunculus micranthus* Nutt. Much like no. 1, differing mainly as follows: Some of the roots often thickened. Lower parts of the stem and petioles of basal leaves villous. Basal leaves with blades orbiculate to oblong or obovate in outline, some often deeply 3- to 5-lobed or parted, 1–5 cm long, 1–5 cm wide, attenuate to truncate at the base, rarely subcordate, crenate. Achene beaks 0.2–0.3 mm long. (2n = 16) Apr–May. Dry or moist woods; e KS, e OK; (MA to e KS, s to NC, WV, OH, TN, & OK).

Reports of this species from the Black Hills are likely based on *R. abortivus,* and in particular, on plants with short pubescence in the upper parts. This form of *R. abortivus* has been distinguished as var. *acrolasius* Fern. Fernald (1950) attributed this variety to the BH, but not *R. micranthus.*

18. *Ranunculus pensylvanicus* L. f., bristly crowfoot. Strongly hirsute annual or short-lived perennial, (1)2–10 dm tall, resembling no. 16; roots rather fleshy. Stems usually single, erect, fistulose. Basal and cauline leaves similar, the basal leaves often withering early, usually larger than the cauline, with petioles to 2.5 dm long; blades ternate, 4–12 cm long, 4–20 cm wide, the terminal segment 3-lobed, the lateral 2- to 3-lobed, the segments coarsely toothed, distinctly stalked. Sepals yellowish-green, reflexed, narrowly elliptic, (3)4–5 mm long, sparsely hirsute dorsally; petals 5, yellow, obovate, (1.5)2–3 mm long, distinctly shorter than the sepals; nectary scale usually truncate, with the margins free for at least 2/3 of their length, glabrous; stamens 15–25, anthers 0.7–1 mm long. Achenes 50–80 in an ovoid-cylindric to cylindrical head 10–15 mm long, 6–9 mm thick; achene body flattened, obovate, (1.5)2–2.5 mm long, smooth, narrowly keel margined; beak stout, deltoid, 0.6–0.9 mm long; receptacle elongate-cylindrical, 5–13 mm long in fruit, short-hairy; pedicels 1–6 cm long. (2n = 16) Jun–Aug. Wet meadows, shores, stream banks, & other wet places, often where wooded; MN & ND, s to IA, n NE & ne WY; (Newf. to AK, s to NJ, OH, n IL, NE, CO, AZ, & OR; also e Asia).

19. *Ranunculus recurvatus* Poir., hooked buttercup. Hirsute perennial 1.5–5 dm tall; roots fibrous, filiform. Stems 1–few from a cormlike, thickened base, branched above. Basal and cauline leaves similar, the basal leaves with petioles to 2 dm long; blades ovate to reniform in outline, cordate, 2–9 cm long, 2–12 cm wide, 3-lobed, the terminal lobe shallowly 3-lobed, the lateral lobes shallowly 2-lobed, margins bluntly toothed; cauline leaves shorter petioled and more deeply 3-parted than the basal. Sepals green, reflexed, narrowly ovate to lanceolate, 3–6 mm long, hirsute; petals 5, pale yellow, lanceolate to oblanceolate, 2–5 mm long, usually shorter than the sepals; nectary scale forming a pocket, truncate, with the lateral margins adnate to the petal for all or most of their length, glabrous; stamens 10–25, anthers 0.7–1(+) mm long. Achenes 15–50 in an ovoid-globose head 5–8 mm long, 5–7 mm thick; achene body flattened, obovate, 1.7–2.3 mm long, minutely pitted, narrowly keel margined; beak slender, strongly recurved, 0.7–1 mm long; receptacle ellipsoid to subglobose, 2–3 mm long in fruit, hispid; pedicels 1–6 cm long in fruit. (2n = 28, 32) Apr–May. Moist or dry woods; ND: Richland; se NE & e KS; (Que. & ME to n MN, ND, s to FL, LA, & e TX).

20. *Ranunculus rhomboideus* Goldie, prairie buttercup. Solitary or tufted, pilose perennial, 0.5–2.5 dm tall; roots fibrous, some rather thickened, not tomentose. Stems usually erect, 1- to several-flowered. Basal leaves with hairy petioles 2–10 cm long; blades elliptic-ovate to obovate, some occasionally irregularly lobed, otherwise crenate, 1–4.5 cm long, 0.7–3.5 cm wide, cuneate to rounded or seldom truncate at the base, cauline leaves shorter petioled to sessile upward, deeply 3- to 5-lobed or parted into linear segments. Sepals yellowish-green, sometimes purplish, spreading to reflexed, ovate, 4–6 mm long, densely pilose dorsally; petals 5(6), yellow, obovate, 5–7 mm long; nectary scale truncate and pocketlike, with the margins entirely adnate, glabrous; stamens 25–50. Achenes 25–50(120) in a globose head 6–10 mm in diam; achene body obovoid, tapered to a winged-stipitate base, 1.8–2.5 mm long, about 1/2 as thick as broad, glabrous; beak minute to 0.3 mm long; receptacle ovoid to globose, 2–6 mm long in fruit, hirsute; pedicels 2–6 cm long in fruit. (2n = 16) Apr–Jun. Prairie & dry open woods; MN to MT, s to cen IA & NE; (w Ont. to Sask., MI s to IL, IA, NE, & MT). *R. ovalis* Raf.—Rydberg.

21. **Ranunculus sardous** Crantz. Sparingly to copiously hirsute annual 0.5–5 dm tall; roots fibrous, not thickened. Stem single or branched from the base, erect to decumbent. Basal leaves sometimes withering by fruiting time, with petioles 5–15 cm long; blades ternate, the leaflets shallowly to deeply 3-lobed, 5–15 mm long, the ultimate segments crenate to laciniate lobed; cauline leaves often larger than the basal, shorter petioled to sessile, ternate to 3-lobed, reduced upward to mostly 3-parted bracts with linear segments. Sepals yellowish-green, reflexed, ovate-attenuate, 3–6 mm long, sparsely hirsute dorsally; petals 5, yellow, obovate, 4–10 mm long; nectary scale truncate to obovate, with the margins free for about 3/4 of their length, glabrous; stamens 25–50. Achenes 10–40 in a globose-ovoid head 5–8 mm long, 5–7 mm thick; achene body flattened, rotund-obovate, 2–3 mm long, narrowly thin winged, papillose to seldom smooth on the faces; beak flattened, triangular, 0.2–0.5 mm long; receptacle pyriform, ca 2 mm long in fruit, hirsute; pedicels slender, 2–8 cm long in fruit. (2n = 16, 32, 48) May–Jun. Disturbed places & roadsides; KS: Neosho; (sporadic in many parts of N. Amer., especially the Atlantic Coast; Europe). *Adventive.*

22. **Ranunculus sceleratus** L., cursed crowfoot. Glabrous or rarely hirsute annual, emersed or seldom aquatic, 1–5 dm tall, or the stem to 10 dm long when submersed; roots fibrous, somewhat fleshy. Stems 1–several, erect, fistulose, usually strongly inflated, sparsely to profusely branched, especially above. Basal leaves with petioles 2–15 cm long or much longer when submersed; blades shallowly to usually deeply 3(or apparently 5)-parted, reniform in outline, 1–7 cm long, 3–9 cm wide, the terminal lobe more shallowly 3-lobed, the lateral 2- or 3-lobed, the lobes otherwise with more shallow, obtuse to rounded lobes or teeth; cauline leaves shorter petioled to sessile upward, the blades more deeply parted with narrower segments, the uppermost bracts simple or divided into oblanceolate or linear segments. Sepals yellowish-green, spreading to reflexed, ovate, 2–3 mm long, usually glabrous; petals 5, pale yellow, fading white, obovate, 2–5 mm long; nectary scale glabrous, the margins mostly adnate to the petal and prolonged to form a shallow pocket around the gland; stamens usually 10–25. Achenes 40–300 in an ovoid-cylindric to rarely globose head 4–11 mm long, 3.5–7 mm thick; achene body obovoid, 0.8–1.2 mm long, about 1/2 as thick as broad, glabrous, corky thickened around the margins at maturity; beak minute, blunt; receptacle obovoid to cylindric, 3–12 mm long in fruit, sparingly hirsute or glabrous; pedicels 1–3(4) cm long in fruit. (2n = 32, 64) May–Sept. Margins of ponds, wet ground; GP; (circumboreal, in N. Amer. s to FL, AR, LA, TX, NM, & CA). *Native* and *introduced.*

Two phases of *R. sceleratus* occur in the GP. The widespread var. *sceleratus* may be partly naturalized in our region, whereas var. *multifidus* Nutt. is distinctly American. The two varieties are distinguished as follows:

22a. var. *multifidus* Nutt. Basal leaves with broad sinuses, the 3 main lobes often with deep secondary lobing. Achenes with smooth central areas on each face surrounded with minute pinprick depressions. Mostly n & w GP.

22b. var. *sceleratus.* Basal leaves deeply divided into 3 segments, the sinuses between the lobes narrow, the secondary lobing shallow. Achenes cross-corrugated or reticulate on the central area of each face. Mostly e GP, occasional w.

23. Ranunculus subrigidus Drew, white water crowfoot. Very similar to no. 15, occasionally stranded on mud and assuming a semiterrestrial growth habit. Stems 2–6(10) dm long, shorter when emersed, simple or sparingly branched, rooting at the lower nodes. Leaves all cauline or some basal on semiterrestrial plants, the blades finely divided as in no. 15, roughly globular in outline, usually 1–3 cm long and about as wide, often smaller, the filamentous leaf segments flattened on emersed plants; petioles consisting of the stipular leaf bases or extending slightly beyond the dilated base, 2–5 mm long, glabrous or pubescent. Flowers as in no. 15, but the nectary scale a small round pit or absent; stamens about 10(5–12). Achenes 30–45(80) in a globose head 3–5 mm in diam; achene body obovoid, 1–1.5 mm long (averaging 1.25 mm or less), transversely ridged, hispidulous on the back; beak 0.2–0.5 mm long (often nearly beakless when dried); receptacle ovoid, 1–2 mm long in fruit, pubescent; pedicels usually strongly recurved in fruit, (1)3–10 cm long. ($2n = 16$, 32) Late May–Aug. Shallow water of streams, ponds, marshes, & ditches, often where brackish; MN to MT, s to n IA, SD, & CO; (Que. to N.T. & B.C., s to MA, MI, IA, CA; n Mex.). *Batrachium divaricatum* (Schrank) Wimm.—Rydberg; *R. circinatus* Sibth. var. *subrigidus* (Drew) L. Benson—Gleason & Cronquist.

See comments under no. 15.

24. Ranunculus testiculatus Crantz. Small, scapose, shallowly rooted annual 2–12 cm tall, thinly to densely gray-tomentose. Scapes 2–10 cm long. Leaves all basal, the blades ternate to biternate, 1–4 cm long, 0.5–2 cm wide, the segments flat, linear, usually 0.5–3 mm wide; petioles mostly 1–2 cm long, winged above by the decurrent base of the leaf blade. Sepals green, often purplish on the membranaceous margins, persistent and reflexed in fruit, ovate, 2–4 mm long, tomentose dorsally; petals 5, white or pinkish (perhaps yellowish when fresh), oblanceolate or narrowly so, 3.5–6 mm long; nectary scale oblong-elliptic, free along the margins, glabrous; stamens 10–15. Achenes (25)35–70, spreading in a subcylindric to cylindric head 1–2.5 cm long, 1–1.5 cm thick; achene body conspicuously 3-chambered, with 2 empty pouchlike chambers lateral to the central, seed-containing chamber, sparingly to densely tomentose; beak prominent, stiff, straight, lanceolate, 2.5–4 mm long. Apr–Jun. A weed of lawns, campgrounds, roadsides, & disturbed places; sporadically introduced in w GP & likely spreading; (throughout most of w & nw U.S. & adj. Can.; Eurasia). *Adventive*.

In *Flora Europaea* this species and another, *R. falcatus* L., are segregated from *Ranunculus* as species of *Ceratocephalus* Pers. by token of the empty chamber on each side of the seed and the prominent achene beak. Under this interpretation, *R. testiculatus* becomes *Ceratocephalus testiculatus* (L.) Pers. Of the genera proposed as segregates from *Ranunculus sensu lato*, this segregation is perhaps most justifiable.

12. THALICTRUM L., Meadow Rue

Dioecious, rhizomatous to nonrhizomatous perennial herbs. Stems fistulose. Lowest leaves petiolate, ternately decompound, the leaflets variously lobed or toothed, the petioles expanded and sheathing at the base; stem leaves similar but progressively less compounded, the uppermost sometimes short-petiolate or sessile or all of them subsessile or all petiolate. Flowers borne in panicles, not especially showy, although the staminate ones may be very conspicuous from the stamens at anthesis; sepals 4 or 5, greenish or purplish or whitish, caducous; petals none; stamens numerous, yellowish or purplish, anthers slender, apiculate, filaments slender, becoming elongated and, often, tangled together; carpels 4–17. Fruits 1-seeded, ribbed achenes.

1 Middle and upper leaves long-petioled; plants spring blooming; achenes mostly 2–3.5 mm long .... 2. *T. dioicum*

1 Middle and upper leaves short-petiolate to sessile; plants usually summer blooming; achenes often (3)3.5–5.5 mm long.
   2 Most leaflets longer than wide and entire or 3-lobed; plants arising from a stout caudex .... 1. *T. dasycarpum*
   2 Most leaflets wider than long and bearing 4 or more lobes or teeth; plants rhizomatous .... 3. *T. venulosum*

1. **Thalictrum dasycarpum** Fisch. & Ave-Lall., purple meadow rue. Sturdy dioecious perennials 5–14(20) dm tall from a stout caudex. Stems fistulose, glabrous to inconspicuously pubescent. Leaves glabrous or pubescent beneath, the lowest petiolate, the middle and upper ones subsessile to sessile, the leaflets entire to 3-lobed, sometimes some of the leaves with some leaflets bearing additional lobes or teeth, leaflets usually longer than wide, the longest ones 1.5–4.5(5.5) cm long, the margins often clearly revolute. Panicles open; sepals 4 or 5, purplish to whitish, 2–4 mm long, often caducous; stamens numerous, the anthers mostly 1.3–2.5 mm long; carpels 5–14. Fruits dark, ribbed achenes (2.5)3.8–5.5 mm long with styles 2–4 mm long. (2n = 28, 42, ca 100) (May)Jun–Jul. In a variety of moist, shaded to open habitats; MN, ND, MT, SD, WY, IA, NE, CO, MO, e KS, e OK; (Ont. to Alta., s to OH, OK, NM, & AZ). *T. hypoglaucum* Rydb.—Rydberg, *T. dasycarpum* var. *hypoglaucum* (Rydb.) Boivin—Fernald.

2. **Thalictrum dioicum** L., quicksilver-weed. Slender, glabrous or subglabrous perennials 2–7(9) dm tall. Stems fistulose. All leaves long-petiolate, even the uppermost relatively large and well developed, leaflets with more than 3 lobes or often with 3 main lobes across the tip, but these further lobed or crenate, most leaflets wider than long, the largest ones 1.7–2.8 cm long. Panicles relatively open; sepals usually 4, purplish to greenish or whitish, 1.5–3.5 mm long, often caducous; stamens numerous, the anthers mostly 2.5–4.2 mm long; carpels 7–13. Fruits ribbed achenes (2.2)2.5–3.5 mm long with styles 1–2 mm long. (2n = 42) Mostly Apr–May. In wooded areas on the e edge of our range; MN, e ND, e SD, IA, MO, ne KS, & reportedly also in the BH of SD; (Que. to s ND, s to GA, MO, & ND).

3. **Thalictrum venulosum** Trel., early meadow rue. Slender, rhizomatous perennials 3.5–9.5 dm tall. Stem fistulose, glabrous to sparsely pubescent. Leaves glabrous or somewhat pubescent beneath, the lowest long-petiolate, but the petioles gradually reduced upward and the uppermost leaves sessile or short-petiolate, leaflets with more than 3 lobes, or with 3 lobes but these further lobed or crenate, most leaflets wider than long, the largest ones mostly 0.7–3.0 cm long. Panicles open to relatively narrow; sepals usually 4, purplish to whitish, 1.4–4.5 mm long, often caducous; stamens numerous, the anthers mostly 2.2–3.8 mm long; carpels 5–17. Fruits ribbed achenes 3–5 mm long with styles 1.5–2.5 mm long. Jun–Jul. Moist soil in woods or in the open; MN, ND, MT, ne & sw SD, w NE, CO; (sw Ungava Distr. to B. C., s to WI, MI, NE, & NM). *T. lunellii* Greene—Rydberg; *T. nigromontanum* Boivin—Boivin, Rhodora 46:454. 1944.

    Rydberg listed *T. megacarpum* Torr. (a synonym of *T. occidentale* Gray) for SD. This record is evidently an error and may have been based on material of *T. venulosum*.

22. BERBERIDACEAE Juss., the Barberry Family

by Robert B. Kaul

    Ours perennial herbs and evergreen shrubs. Leaves alternate, lobed or compound, stipulate. Flowers hypogynous; sepals and petals 3–9 each, free; stamens (3)6–18, opening by apical valves or lateral slits; pistil 1, style short to absent, with 1–many ovules. Fruit a berry, or withering and replaced by a drupelike seed (*Caulophyllum*).

1 Woody shrubs; leaves compound, coriaceous, spiny; sw and nw GP .... 1. *Berberis*
1 Herbaceous perennials; leaves simple or compound, not coriaceous or spiny; e GP.
    2 Leaves triternately compound; flowers several in a panicle just above the leaves .... 2. *Caulophyllum*
    2 Leaves simple, peltate, deeply lobed; flower solitary below the leaves ...... 3. *Podophyllum*

1. BERBERIS L., Barberry

Ours evergreen shrubs of the w GP, the stems without spines. Leaves compound, trifoliolate or odd-pinnate, the leaflets hollylike, coarsely serrate, the serrations spinulose-tipped. Racemes and cymes axillary, erect or drooping. Sepals and petals 6 each, the petals with a pair of basal glands; stamens 6, opening by apical valves; placentation basal, the ovules 1–few. Fruits berries. *Mahonia* Nutt.—Rydberg.

Our native barberries have compound, evergreen, spiny leaves. Several oriental species and their hybrids are commonly cultivated in our area; their leaves are mostly deciduous and simple, but their stems are spiny.

1 Leaflets 3, often sessile; TX panhandle .... 3. *B. trifoliolata*
1 Leaflets 3–7, the leaf pinnately compound; TX panhandle and nw GP.
    2 Inflorescence bracts foliaceous; berries white or red; TX panhandle .... 2. *B. swaseyi*
    2 Inflorescence bracts small; berries blue-black; nw GP .... 1. *B. repens*

1. *Berberis repens* Lindl., Oregon grape. Low or prostrate shrub. Leaves odd-pinnate, borne mostly near the branch tips, the leaflets 3–7, 6–20 cm long, sessile, ovate to orbicular, acute to rounded, the base oblique, serrate, the serrations spinulose-tipped, green or glaucous. Racemes several among the upper leaves, 3–6 cm long, the bracts ovate, acute to apiculate, 2–4 mm long, the pedicels slender. Sepals 2, 1–2 mm long, yellow; petals pale to bright yellow, acute, ca 6 mm long; pistil ca 3 mm long, ovoid, the stigma capitate. Berries borne in grapelike clusters, globose, to 1 cm in diam, blue-black, glaucous. ($2n=28$) Jun. Sandy, chalky, or granitic soil in coniferous forests; e MT, e WY, w SD (BH), nw NE; (B.C. to w SD, s to CA & w TX). *Mahonia aquifolium* (Pursh) Nutt.—Rydberg, misapplied.

2. *Berberis swaseyi* Buckl., Texas barberry. Small shrub to 1 m tall. Leaves borne on short shoots, odd-pinnate, the leaflets 3–5(9), to 20 mm long and 6 mm wide, the terminal leaflet the largest, pale greenish above, paler below, coarsely serrate or dentate, each lobe spinulose-tipped. Raceme bracts foliaceous. Berries globose, ca 1 cm in diam, white or red. ($2n=28$) Apr, fruits Jun. Calcareous canyons & ridges; TX: Bailey; (endemic to TX).

3. *Berberis trifoliolata* Moric. Erect shrub to 2 m tall, bark gray, often exfoliating. Leaves borne on lateral short-shoots and terminal on the stems. Leaves mostly trifoliolate, this not always immediately apparent when the petiole is short or absent; leaflets 2–5 cm long, to 3 cm wide, more or less elliptic to linear, coarsely serrate, the 3–7 teeth with a short sharp spine at the tip, pale green to glaucous. Racemes corymbose, few-flowered, axillary at the upper axils and also terminating the short shoots, much shorter than the leaves. Flowers yellowish. Fruit globose, 8–12 mm in diam, black. ($2n=28$) Mar–Apr; fruits Jun. Open woods & pastures, usually on rocky soil; TX panhandle; (TX w of 100° long., w to AZ & Mex.).

2. CAULOPHYLLUM Michx., Blue Cohosh

1. *Caulophyllum thalictroides* (L.) Michx. Perennial herbs from rhizomes. Leaf 1, very large, to 5 dm wide, usually triternately compound, the erect stem to 7 dm tall. Leaflets obovate, 2- or 3-lobed. Panicle arising just below the leaf as the leaf matures, to 6 cm long. Flowers yellowish-green; sepals and petals 6 each; petals thick, hooded, glandlike, smaller than the sepals; stamens 6, opening by 2 apical values. Pistil gibbous, with 1 or 2 basal ovules, soon bursting when the ripening seed enlarges, the seed then becoming drupelike, globose, blue, the pedicels becoming subclavate. (2n = 16) Apr–Jun. Damp deciduous forests; extreme e parts of ND, SD, NE, KS, & OK; MN, IA, MO; (N.B. to Man., s to OK & SC).

3. PODOPHYLLUM L., May-apple

1. *Podophyllum peltatum* L. Colonial, erect perennial herbs from creeping rhizomes. Stem to 5 dm tall, bearing 1 or 2 large terminal leaves, the 2-leaved stems with a single nodding flower between the petiole bases. Blades peltate, to 3.5 dm wide, more or less orbicular, deeply 3- to 9-lobed. Flower short pediceled; sepals 6, soon falling; petals 6 or 9, obovate; stamens twice as many as the petals, opening laterally; pistil ovoid, the stigma sessile; ovules numerous. Berry yellowish-green to occasionally purplish, 4–5 cm in diam. (n = 6) Apr–Jun. Rich damp woods; e-cen NE & sw IA s through e KS & w MO to ne OK; (Que. to s MN & e NE, s to e TX & FL).

23. MENISPERMACEAE Juss., the Moonseed Family

by Robert B. Kaul

Dioecious twining vines of forests and thickets of the e and s GP. Stems herbaceous or suffrutescent to mostly woody, without tendrils or spines. Leaves alternate, simple, petiolate, estipulate; blades palmately veined, entire or deeply lobed. Inflorescences single in or just above the leaf axils, mostly shorter than the leaves, racemose to mostly narrowly paniculate, the bracts small or lacking. Flowers unisexual, small, greenish or whitish, hypogynous, the parts free and often trimerous; sepals and petals similar, in 2–4 whorls, the outer whorl(s) exceeding the inner. Stamens usually 6, 12, or 24; pistils 2–6, free, uniovulate, usually only one maturing to fruit. Fruit a 1-seeded black or red drupe, the stone variously flattened or concave, rough and coiled.

1 Petioles attached a short distance inside the edge of the blade, the leaf thus peltate; drupes spherical and black; throughout GP e of 100° long. .... 3. *Menispermum*
1 Petioles attached at the edge of the blade; drupes ellipsoidal and black or spherical and red; se GP.
  2 Leaves unlobed or shallowly lobed; drupe red, the stone flattened on both sides; petals present .... 2. *Cocculus*
  2 Leaves shallowly lobed to deeply 3–5 lobed; drupe black, the stone hollowed on one side; petals lacking .... 1. *Calycocarpum*

1. CALYCOCARPUM Nutt.

1. *Calycocarpum lyonii* (Pursh) Nutt., cupseed. Vine woody, high climbing, often to the treetops. Leaves to 25 cm long, the blades 5–10(24) cm wide, the petioles 3–10 cm long,

slender; blades variable on the vine, subrhombic to shallowly 1- to 3-lobed but mostly deeply 3- to 5-lobed, the lobes acuminate, the sinuses rounded; margins entire to irregularly shallow-serrate on larger leaves; base truncate to mostly cordate, the petiole attached at the edge of the blade. Inflorescences narrow-paniculate, about 2/3 as long as the subtending leaf. Flowers ca 5 mm wide, sepals 6, petals 0; stamens 12; pistils 3, the stigma cleft and dilated. Drupe ellipsoidal, 13–25 mm long, black when ripe, the stone hollowed on one side and the margin rough. May–Jun, fruits Jul–Sep. Open or dense woods & along wooded stream banks; se KS, sw MO, e OK; (e TX to FL, n to KY, MO, e KS).

This is the only species in the genus.

2. COCCULUS DC.

1. *Cocculus carolinus* (L.) DC., Carolina moonseed, snailseed. Vine herbaceous to woody, climbing 3–4 m. Leaves to 15 cm long and wide, variously shaped, the blades ovate to deltoid to hastately 3-lobed, the lobes shallow, acute to retuse and mucronate; base truncate to cordate; petiole attached at the edge of the blade and equal to or shorter than the blade. Inflorescences narrowly paniculate, shorter than subtending leaves except near the tip of the stem, the staminate panicles longer than the pistillate. Flowers 3–4 mm wide; sepals, petals, and stamens 6 each, in alternating whorls; pistils 3 or 6. Drupes borne in grapelike clusters, spherical, 4–8 mm in diam, red; stone flattened to shallowly concave on both sides, once-coiled. ($2n = 78$) Jul–Sep, fruits Jul–Oct. Forests & forest edges & along wooded streams; se KS, sw MO, cen & w OK, n TX; (se KS to VA, s to FL & cen TX).

3. MENISPERMUM L.

1. *Menispermum canadense* L., moonseed. Vine woody, wiry, climbing to 4 m. Leaves 8–17 cm long and nearly as wide, the blades ovate to suborbicular, entire to shallowly 2- to 6-lobed, the slender petiole usually attached a short distance inside the edge of the blade and the leaf thus eccentrically peltate. Inflorescences narrowly paniculate. Sepals and petals 4–8 each; stamens 12–24; pistils 2–4(5). Drupes borne in small clusters, blackish, about spherical, the stigmatic scar not terminal; stone crescent-shaped, flattened on both sides, the margin roughened and grooved. ($2n = 52$) May–Jun, fruits Jul–Oct. Damp woods & woodland edges, thickets; GP e of 100° long.; (Man. to Que., s to GA & OK).

24. PAPAVERACEAE Juss., the Poppy Family

by Ralph E. Brooks and Ronald L. McGregor

Ours annual, biennial, or perennial herbs, stems, leaves, and roots with secretory canals which produce yellow, milky, or watery juice. Leaves alternate, simple, palmately lobed, or pinnately compound. Flowers conspicuous, solitary, terminal or axillary, perfect, regular, hypogynous; sepals 2 or in *Argemone* usually 3, distinct, enclosing the bud before anthesis, caducous; petals fugacious, usually twice as many as the sepals but sometimes 6, 8, or 12, distinct, imbricate in bud; stamens numerous, distinct, anthers dithecal, longitudinally dehiscing, and basifixed or subbasifixed; gynoecium of 2–many fused carpels, a superior, unilocular ovary, stigmas as many as the carpels and forming a discoid or lobed structure,

POPPY FAMILY

ovules numerous on parietal placentae, anatropous. Fruit a capsule opening by valves or pores; seeds with a small embryo and oily endosperm.

Chelidonium majus L., celandine, is a native of Europe that has been collected in MO: Jackson. It has pinnate leaves, white flowers about 1 cm across borne in several flowered umbels, and linear capsules 3-5 cm long.

1 Foliage prickly .... 1. *Argemone*
1 Foliage not prickly.
   2 Leaves entire or palmately lobed; flowers white .... 4. *Sanguinaria*
   2 Leaves pinnatifid or pinnate; flowers usually salmon, orange, or red.
      3 Peduncles mostly over 10 cm long; capsules subglobose to obconic, opening by pores .... 3. *Papaver*
      3 Peduncles less than 5 cm long; capsules linear, opening by valves .... 2. *Glaucium*

1. ARGEMONE L., Prickly Poppy

Ours annual, biennial, or perennial herbs; foliage prickly; latex golden yellow; stems solitary or several, simple or cymosely branched. Leaves deeply pinnately lobed to merely shallowly lobed, margins spinose toothed. Flowers white, yellow, or pale lavender, large and conspicuous; sepals 3(2-6) imbricate, abruptly cuspidate spinose (sepal horn); petals 6 (rarely to 12), in two whorls; anthers becoming coiled after anthesis; carpels 3-5, stigmas 3- to 5-lobed, style short. Capsules persistent, elliptical, variously prickly, somewhat fluted on the sutures, the sutures dehiscing from the apex toward the base for about 1/3 their length, with the exposed vascular elements forming a cagelike structure.

Reference: Ownbey, G. B. 1958. Monograph of the genus *Argemone* for North America and the West Indies. Mem. Torrey Bot. Club 21: 1-159.

The yellow-flowered *A. mexicana* L., a native of the West Indies, has been cultivated in our region and has been reported from KS: Douglas.

1 Stems and leaves hispid and prickly.
   2 Capsules armed with spine-tipped tubercles, the tubercles bearing smaller prickles at the base .... 3. *A. squarrosa*
   2 Capsules armed with spine-tipped tubercles but their bases lacking smaller prickles .... 1. *A. hispida*
1 Stems and leaves only prickly and usually sparsely so .... 2. *A. polyanthemos*

1. Argemone hispida A. Gray, hairy prickly poppy. Similar to no. 3 except as noted: Stems 3-6 dm tall. Capsules hispid and prickly, spine-tipped tubercles present but not bearing smaller prickles at the base. Jun-Aug. Exposed gravelly hillsides, wash areas, & stream banks; CO: Baca, Las Animas; NM: Union; (foothills of the Rocky Mts. from se WY, s to NM, e on the plains, to se CO & ne NM). *A. platyceras* Link & Otto var. *hispida* (A. Gray) Prain—Harrington.

Relatively mature capsules are essential to distinguishing *A. hispida* from *A. squarrosa*. In the canyon regions of ne NM and se CO the two species may grow in close proximity.

2. Argemone polyanthemos (Fedde) G. Ownbey, prickly poppy. Annual, winter annual, or biennial (?); stems solitary or few from a deep taproot, unbranched or usually cymosely branched, 4-15 cm tall, 3-20 mm in diam, glaucous, usually sparingly prickly, prickles yellowish, spreading or recurved, 2-7 mm long. Leaves glaucous, succulent, upper surface smooth or sometimes with scattered prickles on the veins, lower surface prickly on the veins; lower leaves oblanceolate, deeply pinnately lobed, 7-20(25) cm long, 3-10 cm wide, lobes oblong, elliptic or obovate with undulate, somewhat sinuate and irregularly spinose-toothed margins, base long attenuate to the winged petiole; middle and upper leaves elliptic, oblong, or ovate, upwards gradually reduced and only shallowly lobed, margins as the lower leaves, base usually sessile and clasping. Flowers (3)5-10 cm in diam, subtended by a pair of foliar bracts usually shorter than the sepals; peduncles 1-4 cm long; sepals widely elliptic and abruptly cuspidate spinose, body 10-25 mm long, 10-15 mm wide, glaucous, prickly, margins entire, sepal horns yellowish, terete, (4)6-10 mm long, often recurved; petals white (drying yellowish), rarely lavender, widely obovate, the outer ones sometimes suborbicular, (1.5)2.5-5 cm long, 2.5-5.5 cm wide, glabrous; stamens yellow (drying brownish), 6-12 cm long, anthers 1-1.5 cm long. Capsules narrowly to widely elliptic, 2.5-4 cm long, 1-1.5 cm wide (excluding armament), beset with stout yellow spines 3-8 mm long, the largest mostly restricted to the ridges; seeds dark brown, shiny, globose with a 2-horned crest along one side, 2-2.5 mm long, alveolate-reticulate. (2n = 28) Late May-Aug (Sep). Usually sandy soil in prairies, flood plains, & along roadsides; MT; ND: McKenzie; sw & s-cen SD, NE, WY, KS, CO, w 1/2 OK, TX, NM; (w ND & e MT, s to TX & e NM). *A. intermedia* Sweet — Rydberg.

Steyermark (1963) reports *A. albiflora* Hornem. for MO: Jackson. This taxon is similar to *A. polyanthemos* except the sepal horns average much shorter (3-6 mm long). The western MO record is surely an introduction, since this species is normally found much farther east and south of our region.

3. *Argemone squarrosa* Greene, hedgehog prickly poppy. Perennial; stems solitary or several from a stout caudex, unbranched or usually cymosely branched, 3-12 dm tall, 3-20 dm in diam, glaucous, hispid and densely covered with yellowish, uneven-sized spreading prickles 1-8 mm long. Leaves glaucous, succulent, hispid, sparsely prickly above and moderately so below, especially on the veins; lower leaves oblanceolate, deeply pinnately lobed, 6-12 cm long, 3-7 cm wide, lobes usually widely oblong or elliptic with undulate, irregularly toothed margins, base short attenuate to a winged petiole; middle and upper leaves similar to the lower ones but gradually reduced, the uppermost often ovate and shallowly lobed, base mostly sessile. Flowers 4-10 cm in diam, subtended by 1 or 2 foliar bracts shorter than the sepals; peduncles 1-3 cm long; sepals widely elliptic and abruptly cuspidate spinose, body 12-20 mm long, 15-20 mm wide, glaucous, hispid, prickly, margins entire, sepal horns yellowish, angular in cross section, 3-8 mm long, erect or slightly recurved; petals white (drying yellowish), widely obovate, 2-5 cm long, 3-5 cm wide, glabrous; stamens yellow (drying brown), 5-8 mm long, anthers 1-1.5 mm long; capsules elliptic or narrowly so, 3-4 cm long, 1-1.5 cm wide (excluding armament), hispid, prickly, the ridges beset with stout, spine-tipped tubercles to 15 mm long, the bases of which bear smaller prickles making the entire structure superficially appear as a compound spine at maturity; seeds dark brown, shiny, globose with a 2-horned crest along one side, 2-2.5 mm long, alveolate-reticulate. (2n = ca 112) Jun-Jul. Usually sandy prairies, pastures, & roadsides; sw 1/4 KS, se 1/4 CO, OK & TX panhandles, e NM; (sw KS & se CO, s to w OK, n TX, & e NM). *A. hispida* of KS reports — Rydberg.

2. GLAUCIUM P. Mill., Horned Poppy

1. *Glaucium corniculatum* (L.) J. H. Rudolph, red horned poppy. Winter annual or biennial herb, foliage sparsely to densely covered with whitish, coarse, multicellular hairs, latex yellowish; stems several from a deep, stout taproot, sparingly branched, 3-6 dm tall. Rosulate at first, the leaves oblanceolate and deeply pinnatifid with 4-8 pairs of lobes, 4-18 cm long, 1.5-7 cm wide, the lobes elliptic to ovate, margins irregularly lobed or toothed, apices acute to obtuse, petiole narrowly winged; rosette leaves often withered by flowering time; lower cauline leaves similar to rosette leaves, to 20 cm long, 8 cm wide; middle and upper cauline leaves mostly broadly elliptic to ovate and pinnatifid, 4-12 cm long, 3-8 cm wide,

and sessile, otherwise similar to the lower leaves. Flowers solitary, terminal and axillary; peduncles 1-3(4) cm long; sepals 2, lanceolate-ovate, 2-2.8 cm long, 4-7 mm wide, entire, acuminate; petals 4, red (drying blackish) with a dark blackish spot at the base or infrequently pale orange with a dark basal spot, broadly obovate, 2-3.5 cm long, 1.5-2.5 cm wide; stamens 7-13 mm long, anthers blackish, 1.5-2 mm long; stigma divergently 2-lobed, subsessile. Capsule 2-celled, dehiscing from the base upward, linear, 15-20 cm long, 4-6 mm in diam, moderately pubescent, the hairs antrorsely appressed, brownish, and glandular at the base; seeds black, lustrous, ellipsoid to subglobose, 12-15 mm long, alveolate-reticulate. Late Apr-May. Exposed sites in upland fields & pastures; KS: Clark; (KS, MT; Europe). *Adventive.*

Glaucium corniculatum was first discovered in KS: Clark in 1979, in pasture and cropland and represents the first known station for the plant in North America. It has increased in abundance locally each year and may well become an established weed in the future.

3. PAPAVER L., Poppy

Ours annual herbs with slender taproots, latex white; stems erect or ascending, solitary and simple or sparingly branched above. Leaves pinnatifid or pinnate-pinnatifid, usually petiolate. Flowers solitary, terminal or axillary on long peduncles, perfect; sepals 2, caducous; petals fugacious, 4(6) in 2 whorls, membranaceous; stamens numerous; carpels 4-many, stigmas as many as the carpels and radiating in a sessile discoid structure terminating the ovary. Capsule opening by valves just beneath the persistent stigmatic disk; seeds minute.

Papaver somniferum L., opium poppy, is a native of Europe that Rydberg (1932) reported for ND and Steyermark (1963) for MO: Jackson. The robust hispid perennial with large red or orange flowers that is frequently cultivated in our region is *P. orientale* L. (oriental poppy).

1 Capsules 2× or more longer than wide; pubescence on the upper portion of the peduncle appressed .... 1. *P. dubium*
1 Capsules about as long as wide; pubescence on the upper portion of the peduncle spreading .... 2. *P. rhoeas*

1. *Papaver dubium* L., longhead poppy, blind eyes. Annual, 2.5-6 dm tall, stems erect, typically unbranched or sparingly branched, hispid. Basal leaves with blades pinnate or pinnate-pinnatifid, the rachis at least narrowly winged, oblanceolate to elliptic-oblanceolate, 3-8 cm long, 1-4 cm wide, hispid, especially on the veins below, leaflets in 4-8 pairs, obovate, elliptic, or ovate, margins entire or with a few coarse teeth, acute, petiole 2-10 cm long, often longer than the blade; cauline leaves similar to the basal ones but slightly reduced and the petioles becoming much shorter. Peduncles 12-30 cm long, appressed hispid at least in the upper portion, may be spreading below; sepals broadly ovate, 10-15 mm long, 6-10 mm wide, hispid apex acutish, margins entire; petals salmon, pale scarlet, or orange, widely obovate, 1.5-2.5 cm long, nearly as wide, glabrous, apically minutely erose; stamens black or dark purplish, 3-5 mm long, anthers ca 1 mm long. Capsules narrowly obovoid or ellipsoid, 10-15 mm long, 4-6 mm in diam, at least twice as long as wide, glabrous and occasionally glaucous at first; seeds dark brown, kidney-shaped, ca 0.5 mm long, alveolate-reticulate. (2n = 42) May (Jun). Usually gravelly or sandy soil in exposed sites including farm lots, roadsides, along railroads, or other waste sites; KS: Labette, Leavenworth; MO: Jackson; (naturalized in scattered areas of the e U.S., Europe). *Adventive.*

2. *Papaver rhoeas* L., field poppy. Similar to no. 1 except as noted: Stems usually branched. Upper cauline leaves mostly pinnatifid, margins irregularly serrate or erose, sessile. Petals scarlet or purple, 2-4 cm long, about as wide. Capsules widely obovoid to subglobose, 7-15 mm long, nearly as wide, glabrous, glaucous. (2n = 14) May-Jul. Open or more often shaded sites in sandy or gravelly soil, usually roadsides or waste areas; e KS, MO: Jasper; reported, but specimens not seen from NE & ND; (introduced in scattered areas of the e U.S.; Europe). *Adventive.*

4. SANGUINARIA L., Bloodroot

1. *Sanguinaria canadensis* L., bloodroot. Perennial, glabrous, red-juiced herbs with shallow, extensively branched rhizomes, each growing tip usually bearing a scapose flower and single leaf; rhizome 6–15 mm in diam. Leaf at anthesis usually shorter than the scape but the blades soon expanding and the petiole lengthening; blades green above, somewhat glaucous below, reniform to suborbicular, palmately 3- to 7-lobed, infrequently entire, 6–15(20) cm long, 8–20 cm wide, margins subentire to sinuate, the primary lobes often shallowly 3-lobed and the sinuses entire, basal sinus wide and open to narrow or closed by the overlapping lobes; petiole 1–3.5 dm long, longer than the blade. Flower conspicuous; peduncles 5–12 cm long; sepals 2, elliptic-ovate, 8–12 mm long, 5–8 mm wide, membranaceous, apex obtuse, margins entire; petals 8, infrequently 12, 14, or 16, white, oblanceolate to elliptic, 1–3 cm long, 5–12 mm wide, 4 of the petals usually larger than the others; stamens 5–10 mm long, anthers yellow, 2–2.5 mm long; stigmas 2-lobed, style stout, about 1 mm long; ovary with 2 placentae. Capsule dehiscing longitudinally, glaucous, fusiform and crowned by the persistent style, 3–5 cm long, 7–11 mm in diam; peduncle 8–18 cm long; seeds reddish-brown, lustrous, ovoid to subglobose, 3–3.8 mm long, smooth with a prominent funicular crest. (n = 9) Late Mar–Apr. Rich, often rocky soil on woodland slopes; MN, se ND, e 1/6 SD, SD: Lawrence; IA, e 1/6 NE, MO, e 1/3 KS, ne OK; (N.S. w to Man. & e ND, s to FL & e OK).

25. FUMARIACEAE DC., the Fumitory Family

by Ralph E. Brooks

Annual or perennial herbs with watery sap. Leaves rosulate or alternate, infrequently subopposite, usually divided, exstipulate. Inflorescences cymose or racemose, terminal or axillary, usually few-flowered. Flowers perfect, hypogynous, strongly irregular; sepals 2, bractlike, not enclosing the developing bud; petals 4, in 2 series, one or both outer petals with a prominent basal spur or pouch, inner petals more or less connate over the stigmas at the tip; stamens 6, diadelphous, the 2 groups opposite the outer petals, anthers dimorphic, the middle one of each group with 4 microsporangia and 2 pollen sacs, the lateral ones with 2 microsporangia and 1 pollen sac; carpels 2, united into a compound, 1-locular ovary, stigmas 2 or 1 and several-lobed, style 1, ovules 2–many. Fruit a capsule, 2-valved, rarely indehiscent.

Adlumia fungosa (Ait.) Greene has been reported for Kansas, but specimen evidence is lacking and it is doubtful that the plant ever grew there other than in a garden.

1 Both outer petals of the flower spurred at the base .... 2. *Dicentra*
1 Only 1 of the outer petals of the flower spurred at the base.
   2 Ovary elongate; fruit an elongate dehiscent capsule .... 1. *Corydalis*
   2 Ovary subglobose; fruit a subglobose indehiscent capsule .... 3. *Fumaria*

1. CORYDALIS Vent., Corydalis, Fumewort

Ours annual or winter annual herbs from a taproot. Leaves pinnate, the primary segments once or twice divided and incised. Inflorescence a terminal raceme, congested at first and soon elongating, bracteate; chasmogamous-flowered racemes usually present (especially in *C. micrantha*). Flowers bilaterally symmetrical; sepals 2, scarious, fugacious; petals 4, free or somewhat coherent at the base, outer petals dissimilar, one spurred, the other not, both more or less keeled or hooded at the apex, inner petals similar, connate at the apices, clawed; stigma persistent, flattened, obscurely 2-lobed, style distinct, slender. Capsule many seeded.

Reference: Ownbey, G. B. 1947. Monograph of the North American species of *Corydalis*. Ann. Missouri Bot. Gard. 34: 187–259.

1 Spurred petal less than 10 mm long, the spur incurved and about 2 mm long; fruits pendent; pedicels (6)10–18 mm long .... 4. *C. flavula*
1 Spurred petal more than 10 mm long, the spur straight or only slightly incurved and usually 4–8 mm long; fruits erect or pendent; pedicels 1–4(6) mm long.
  2 Fruits densely beset with transparent, clavate pustules; spurred petal 16–22 mm long .... 2. *C. crystallina*
  2 Fruits glabrous; spurred petal 11–18 mm long.
    3 Seeds 1.4–1.6 mm long; plants often bearing racemes of cleistogamous flowers; spurred petal of chasmogamous flowers 10–15 mm long .... 5. *C. micrantha*
    3 Seeds 1.8–2.1 mm long; plants seldom producing cleistogamous flowers; spurred petal of chasmogamous flowers 14–18 mm long.
      4 Capsules usually pendent; spur of spurred petal 4–5 mm long; racemes usually not surpassing the leaves .... 1a. *C. aurea* subsp. *aurea*
      4 Capsules usually erect; spur of spurred petal 5–9 mm long; racemes usually surpassing the leaves.
        5 Seeds smooth or obscurely muricate; bracts mostly 4–10 mm long .... 1b. *C. aurea* subsp. *occidentalis*
        5 Seeds distinctly muricate; bracts mostly 10–17 mm long .... 3. *C. curvisiliqua* subsp. *grandibracteata*

1. **Corydalis aurea** Willd., golden corydalis. Green or glaucous, glabrous winter annual; stems several, simple or sparingly branched, 1–5 dm tall, usually becoming prostrate with age. Leaves with petioles to 6 cm long, the basal and lower ones long petioled, the upper short petioled or sessile; blades 2–10 cm long, oblong to ovate, pinnate with (5)7–11(13) primary segments, these pinnatifid and again divided, the ultimate segments widely to narrowly elliptic. Bracts of the inflorescence elliptic to linear, 4–10 mm long, 1–2 mm wide, reduced upwards; pedicels 2–4 mm long. Sepals fugacious, ovate, 1–3 mm long, acuminate to attenuate, usually irregularly toothed; petals pale to bright yellow; spurred petal 13–18 mm long, the hood crested or not, the crest, if present, low and incised, the wing margin moderately to well developed, the spur straight or slightly curved, 4–9 mm long, subglobose at the tip; spurless outer petal 8–13 mm long, the hood and crest as in the spurred petal; inner petals 8–11 mm long, the claw about 1/2 the total length; stamen spur 2–6 mm long; stigma 2× as wide as high. Capsules erect to pendent, straight or incurved, 15–30 mm long, 2–2.5 mm in diam; seeds shiny black, circular or subcircular, slightly compressed, 1.8–2.1 mm in diam.

Two subspecies can be recognized in the GP:

1a. subsp. *aurea*. Racemes usually surpassed by the leaves, 10- to 30-flowered. Spurred petals with the hood sometimes crested, the spur 4–5 mm long. Capsules usually pendent at maturity, straight to arcuate, 18–25(30) mm long; seeds smooth or obscurely muricate. (May) Jun–Jul. Usually gravelly soil, open woodlands, moist prairies; common in n GP, s to n NE, infrequent in TX & sw OK; (Que. w to AK, s to PA, IL, NE, sw OK, w TX, AZ, & CA). *C. aurea* var. *aurea*—Atlas GP.

1b. subsp. *occidentalis* (Engelm.) Ownbey. Racemes usually surpassing the leaves, 5- to 12(20)-flowered. Spurred petal with the hood not crested, the spur 5–9 mm long. Capsules erect, arcuate, 15–18(20) mm long; seeds smooth or obscurely muricate. May–Jun. Usually open, sandy sites, prairies, roadsides, bottomlands; sw SD & s WY, s to w OK, n TX, & NM, ND(?); (sw SD s to cen TX, w to UT, s NV, & AZ). *C. aurea* var. *occidentalis* Engelm.—Atlas GP; *C. montanum* Engelm.—Rydberg.

Seeds provide the most desirable means of separating this variety from the similar *C. curvisiliqua* subsp. *grandibracteata*.

2. **Corydalis crystallina** Engelm., mealy corydalis. Green or glaucous winter annual, foliage glabrous; stems several, simple or sparingly branched, 1–3(4) dm tall. Leaves with petioles to 6 cm long, the basal and lower ones long petioled, the upper short petioled or sessile; blades (1.5)2–8 cm long, elliptic-ovate to ovate, pinnate with (5)7–9 primary segments, these pinnatifid and again divided, the ultimate lobes widely lanceolate to linear-lanceolate. Racemes usually surpassing the leaves, 8- to 20-flowered (secondary racemes with fewer flowers); bracts ovate to narrowly so, 5–12 mm long, 3–6 mm wide, reduced upwards; pedicels stout, erect, 1–2 mm long. Sepals fugacious, widely ovate to cordate, 1.5–2 mm long, attenuate, sometimes incised; petals bright yellow; spurred petal 16–22 mm long, the hood always crested, the crest high, undulate or toothed, the wing margin wide and reflexed upon the hood, spur 6–8 mm long, the blunt tip distinctly globose; spurless outer petal 12–14 mm long, the wing margin wide, not reflexed upon the hood, enclosing the margins of the spurred petal in the bud, the crest as in the spurred petal; inner petals oblanceolate, 9–11 mm long, the narrow claw 4–5 mm long, the blade 2× as wide at the apex than at the base, basal lobes small; stamen spur 3.5–5.5 mm long, clavate, curved or bent near the apex; stigma about 2× as wide as long. Capsules erect; straight or slightly incurved, 10–20 mm long, 2–2.5 mm diam, densely beset with transparent, clavate pustules; seeds shiny black, circular or subcircular, slightly compressed, 2–2.3 mm in diam, concentrically finely muricate. Apr–Jun. Usually exposed sites, old fields, pastures, prairies, disturbed areas; se KS, MO, e OK; (w MO & se KS, s to AR & e TX).

3. **Corydalis curvisiliqua** Engelm. subsp. ***grandibracteata*** (Fedde) G. Ownbey, large-bracted corydalis. Green or glaucous, glabrous winter annual; stems several, simple or sparingly branched, 1–5 dm tall, sometimes becoming prostrate with age. Leaves with petioles to 7 cm long, the basal ones long petioled and the cauline progressively shorter petioled to sessile; blades 3–9 cm long, oblong to oblong-ovate, pinnate with (5)7–11 primary segments, these pinnatifid and again incised, ultimate segments ellipic to obovate. Racemes surpassing the leaves, mostly 8- to 20-flowered; bracts conspicuous, ovate to narrowly so, mostly 10–17 mm long, 4–6 mm wide, reduced upwards, acuminate, entire; pedicels erect; 2–3(4) mm long. Sepals ovate, 0.8–1.2 mm long, variously toothed; petals bright yellow; spurred petal 15–18 mm long, the wing margin well developed, the hood crested, the crest conspicuous, regular or undulate, the spur 7–9 mm long, subglobose at the tip; spurless outer petal 12–15 mm long, geniculate, the basal portion clawlike, the crest similar to that of the spurred petal; inner petals oblanceolate, 9–11 mm long, the claw slender, 4–5 mm long; stamen spur about 2/3 as long as the petal spur; stigma 2× wider than long. Capsules erect, incurved, 20–25(30) mm long, 2–3 mm in diam; seeds shiny black, circular to subcircular, slightly compressed, 1.8–1.9 mm in diam, concentrically finely muricate. Apr–Jun. Usually sandy soil, open ground, prairies, hillsides, alluvial plains or occasionally disturbed areas; s-cen & sw KS, w OK, TX; (s-cen & sw KS, s to n TX).

This taxon is often difficult to distinguish from the closely related *C. aurea* subsp. *occidentalis* where their distributions overlap in sw KS and w OK. One wonders if their frequent lack of distinction and apparent close relation (Ownbey, op. cit.) might not be grounds for including the *curvisiliqua* complex as an infraspecific taxon of *aurea*.

4. **Corydalis flavula** (Raf.) DC., yellow harlequin, pale corydalis. Green or glaucous, essentially glabrous winter annual; stems 1–several, simple or sparingly branched, 1–2(3) dm tall, erect or ascending, sometimes prostrate with age. Basal and lower cauline leaves with petioles 4–8 cm long, middle and upper cauline leaves short-petiolate or sessile; blades

1-4 cm long, ovate, pinnate with 5-7 primary segments that are again pinnatifid into about 5 lobes, these again divided, ultimate lobes narrowly to widely elliptical. Racemes equaling or barely exceeding the leaves, mostly 6- to 12-flowered; cleistogamous-flowered racemes, if present, 1- to 5-flowered and inconspicuous; bracts widely to narrowly elliptic, 6-12 mm long, 3-7 mm wide, the lowermost often foliaceous; pedicels slender, erect or ascending at anthesis, reflexed in fruit, (6)10-18 mm long. Sepals fugacious, lanceolate, 0.8-1.3 mm long; petals pale yellow; spurred petal 7-9 mm long, the hood crested, the crest undulate or toothed, the wing margin well developed and undulate or toothed, the incurved spur 1.5-2.5 mm long; spurless outer petal 6-8 mm long, the crest and wing margin as in the spurred petal; clawed inner petals 5-7 mm long, the claw 2-3 mm long, the blade about 2× wider near the apex than at the lobed base; stamen spur less than 1 mm long; stigma wider than long. Capsule straight, 14-25 mm long, ca 2 mm in diam; seeds shiny black, circular or subcircular, slightly compressed, 1.9-2.1 mm in diam, concentrically finely muricate. Mar-Apr (May). Shaded or occasionally open sites, including flood plains, stream banks, moist wooded hillsides; s IA, se NE, e 1/4 KS, MO, e OK; (CT w to MI & se NE, s to NC, AL, LA, & ne TX).

5. *Corydalis micrantha* (Engelm.) A. Gray, slender fumewort. Green or glaucous, glabrous annual or winter annual; stems several, simple or sparingly branched, 1-3.5 dm tall, often becoming prostrate with age. Leaves with petioles to 6 cm long, the basal and lower ones long petioled, the upper ones short petioled or sessile; blades oblong to ovate, with 5-7(9) cm long, primary segments, these pinnatifid and again divided, the ultimate segments oblong-elliptic to obovate. Chasmogamous-flowered racemes usually present, slightly to much exceeding the leaves, 6- to 20-flowered; bracts elliptical to ovate, 5-8 mm long, 2-4 mm wide, reduced upwards, acute to acuminate, entire or sparingly toothed; cleistogamous-flowered racemes, if present, inconspicuous, 1- to 6-flowered, the bracts reduced; pedicels erect, 2-6 mm long. Sepals fugacious, ovate, 1-1.5 mm long, toothed or undulate; petals pale yellow; spurred petal 11-15 mm long, the hood crested, the crest low, regular or undulate, the wing margin well developed, the spur 4-6 mm long; spurless outer petal 9-11 mm long, geniculate, the crest low; inner petals 7-10 mm long, oblanceolate, the claw 3-4 mm long, the blade 2× as wide at the apex as at the obscurely lobed base; stamen spur 2.5-4 mm long, straight or curved, sometimes clavate; stigma 2-lobed, rectangular, 2× as wide as high. Capsules erect, 10-30 mm long, straight or slightly incurved; seeds shiny black, circular or subcircular, slightly flattened, 1.4-1.6 mm in diam, obscurely concentrically muricate.

Two subspecies occur in the GP:

5a. subsp. *australis* (Chapm.) G. Ownbey. Racemes greatly exceeding the leaves; spur of the spurred petal blunt, not distinctly globose; capsule 15-30 mm long. (Mar) Apr-May. Disturbed, usually sandy soil, fields, roadsides, open woods; s MO, se KS; (s MO & se KS, s to e TX, e on the Coastal Plain to NC). *C. campestris* (Britt.) Rydb.—Rydberg.

5b. subsp. *micrantha*. Racemes barely exceeding the leaves; spur of the spurred petal distinctly globose; capsules 10-15 mm long. (Mar) Apr-May. Sandy or calcareous soils of open woods, roadsides, fields, & other disturbed sites; s MN & se SD, s to MO & e OK; (WI w to e SD, s to AR & e TX).

2. DICENTRA L., Bleeding Hearts

1. *Dicentra cucullaria* L., Dutchman's breeches. Glabrous perennial, leaves and scapes from a dense cluster of whitish, ovate, flattened tubers. Leaves erect to ascending, to 2.5(3) dm long; blades depressed ovate to widely triangular, 3- or 4-pinnate-pinnatifid and again incised, 4-15 cm long, nearly as wide or wider, ultimate segments narrowly elliptic or oblong to oblanceolate, 1-3(5) mm wide, margins entire, apex acute or obtuse; petiole 8-15 cm

long, usually longer than the blade. Scapes 1-3 dm long, usually surpassing the leaves, terminated by a 3- to 15-flowered raceme; bracts whitish, ovate-lanceolate, ca 1.5 mm long, membranaceous; bracteoles similar but slightly smaller than the bracts, borne in pairs usually several mm below the flower; pedicels at first ascending and later pendent, 3-10 mm long. Sepals whitish, sometimes streaked with purple, widely ovate to triangular, 2-3.5 mm long, 1.5-2.5 mm wide, membranaceous; petals white tipped with cream or yellow, rarely purplish; outer petals spurred, 10-15 mm long, the hood crested, reflexed at anthesis, the spurs pouchlike and inflated, divergent, 7-12 mm long, inner petals coherent at the apex, oblanceolate to spatulate, 7-12 mm long, the hood slightly crested and undulate, winged; stigma 2-lobed, style 5-9 mm long, slender; stamens 6-10 mm long, anthers ca 1 mm long. Capsules fusiform, glaucous, 10-15 mm long, 3-6 mm in diam, base stipelike, terminated by the persistent style; seeds shiny black, depressed ovate, compressed 1.5-2.3 mm long, finely muriculate. (2n = 32) Mar-May. Rich, usually moist woods; MN & se ND, s to e OK & MO; (e Que., w to se ND, s to NC, AL, AR, & e OK). *D. cucullaria* f. *purpuritineta* Eames—Fernald.

Dicentra canadensis (Goldie) Walp. has flowers with rounded nondivergent spurs, borne from a horizontal rhizome bearing roundish, pealike tubers, but is otherwise similar to *D. cucullaria*. It was erroneously reported by Rydberg (1932) from NE but is known to occur in MO: Jackson and areas adjacent to the se GP.

3. FUMARIA L., Fumitory

1. **Fumaria vaillentii** Lois. Glabrous annual; stems weak, erect or ascending, 0.5-3(5) dm tall, sparingly- to much-branched. Leaves all cauline, alternate; blades mostly ovate, 2- or 3-pinnate-pinnatifid and again incised, 2-5 cm long, 1-3 cm wide, reduced upwards, ultimate segments linear or linear-lanceolate, usually 1-2 mm wide, apex acute; petioles shorter than or about equaling the blade in length. Racemes terminal or axillary, usually surpassed by the leaves, (6)8-14(18)-flowered, dense at first and later elongated slightly; bracts lanceolate, 1/2-3/4 × the length of pedicels, membranaceous; pedicels 1.5-2.5(3) mm long, ascending to erect. Sepals fugacious, whitish, ovate to deltoid, 0.5-1 mm long, apex acuminate, margins irregularly toothed; petals lavender with an obviously darker apex; spurred petal 4.5-6 mm long, the hood slightly crested and undulate, spur 1-1.5 mm long, subglobose at the tip, slightly incurved; spurless outer petal panduriform, 3.5-4.5 mm long, basal portion lanceolate, crest similar to that of spurred petal; inner petals oblanceolate, 3-4.5 mm long; stigma wider than long, 2-lobed. Capsules erect, subglobose, 1.7-2.1 mm in diam, obscurely keeled, weakly rugose, apex rounded; seeds reddish-brown, shiny, oblong, ca 1.5 mm long, slightly compressed, smooth to slightly wrinkled. (2n = 32) May-Jun. Exposed, disturbed sites, including fields, roadsides, & railroad rights-of-way; ND, se SD; (ND & SD to MN(?) & IA(?); Europe). *Adventive. F. officinalis* of GP reports—Atlas GP.

Our n GP *Fumaria* have been referred erroneously to *F. officinalis*, thus the occurrence of *F. vaillentii* has gone undetected in N. Amer. until now. This species should be expected in w MN and nw IA.

F. officinalis L. is native to Europe, but now is established in scattered areas of e N. Amer. In the GP it is known only from KS: Atchison, where it was collected in 1961. The species is recognized by the following characters: stems reclining, decumbent or weakly ascending, to 10 dm long; bracts 1.5-2.2 mm long; spurred petal 7-8(9) mm long; fruit obovoid, apex truncate or retuse.

26. PLATANACEAE Dum., the Sycamore Family

by William T. Barker

1. PLATANUS L., Sycamore, Plane-tree

1. *Platanus occidentalis* L. Large monoecious tree to 50 m tall, with a trunk to 3 m diam, the bark of young trees greenish-gray and white mottled due to exfoliation, becoming shallowly fissured with brownish flakes on trunks of older trees; twigs tan to brown, often tomentose when young but soon glabrous. Leaves quite large, simple, alternate, the blades broadly ovate to reniform in outline, with 3 or 5 broad lobes, mostly 1-2 dm across, glabrous or stellate-pubescent mainly on the principal veins, palmately 3- or 5-veined, the lobes broadly triangular, acuminate, entire or usually with a few remote pointed teeth; petioles ca 1/2 the length of the blade or less, dilated and hollow at the base to completely enclose the axillary bud, sparingly to densely stellate-pubescent; stipules adnate to form a membranaceous cylinder around the twig, foliaceous and spreading at the summit. Flowers imperfect, borne in dense, globose, unisexual heads; perianth insignificant or none; staminate heads greenish-yellow, 8-10 mm diam, short peduncled, containing numerous subsessile anthers ca 2 mm long, the anthers laterally adnate to the central connective which has a disklike apex; staminodes and abortive stamens often numerous; pistillate heads reddish, 10-12 mm diam in flower, long peduncled, containing numerous simple pistils intermixed with scattered linear bracts; stigma unilateral on the linear-clavate style, ovary slender, 1-celled; 1-ovuled. Pistillate head 2-3 cm diam in fruit, pendulous on a slender peduncle 8-15 cm long, persistent into winter; fruit a linear-clavate achene 7-8 cm long, subtended by numerous tawny bristles nearly equaling the achene body. (2n = 42) Apr–May, fruits Sep–Oct. Moist woods & flood plains, s IA, e NE, MO e KS, e & cen OK, frequently planted as a shade tree farther n & w of its natural range in the GP; (ME to s MI & e NE, s to n FL, LA, & TX). *P. occidentalis* var. *glabrata* (Fern.) Sarg.—Fernald; *P. occidentalis* f. *attenuata* Sarg.—Gates.

27. ULMACEAE Mirb., the Elm Family

by William T. Barker

Trees or shrubs with watery sap. Leaves with conduplicate vernation, alternate (rarely opposite), simple, the blade oblique at the base, entire to variously serrate, petiolate; stipules paired, caducous. Flowers perfect or imperfect with both sexes on the same plant, actinomorphic to slightly zygomorphic, solitary, cymose, or fasciculate, arising from branchlets of the previous season (*Ulmus*) or of the current season (*Celtis*); sepals (2)5(9), distinct or connate, imbricate, persistent; petals none; stamens erect in bud, hypogynous, usually the same number as and opposite the calyx lobes, filaments curved or sigmoid and distinct, the anthers bilocular with longitudinal dehiscence; gynoecium of 2(3) fused carpels forming a 1-loculate (sometimes 2-loculate in *Ulmus*) ovary with 1 ovule. Fruit a samara, dry or thinly fleshy, or a drupe.

Reference: Elias, Thomas S. 1970. The genera of Ulmaceae in the Southeastern United States. J. Arnold Arbor. 51: 18–40.

1 Flowers usually perfect; fruit a flat samara; leaves usually biserrate .... 2. *Ulmus*
1 Flowers usually imperfect; fruit drupaceous; leaves entire or serrate .... 1. *Celtis*

1. CELTIS L., Hackberry

Trees or rarely shrubs, bark usually gray, smooth or often fissured and conspicuously warty; branches unarmed or spinose; buds scaly or naked. Leaves distichous, serrate or entire, often oblique at base, pinnately 3(5)-veined, petiolate, membranaceous to coriaceous, deciduous (in GP); stipules lateral, free, usually scarious, caducous. Plants polygamomonoecious, the flowers vernal, small, pedicellate on branches of the current year, staminate flowers cymose or fascicled in axils of lower leaves, pistillate flowers solitary or in few-flowered fascicles in the axils of upper leaves. Calyx imbricate, slightly to deeply 4(5)-lobed; petals absent; stamens as many as calyx lobes, inserted on a pilose receptacle, filaments in staminate flowers incurved in bud, exserted after anthesis, in pistillate flowers, filaments usually shorter, included and often nonfunctional, rarely wanting, anthers ovate, face to face in bud and extrorse; gynoecium in staminate flowers minute and rudimentary, in pistillate flowers style short, sessile, divided into 2 divergent, elongate, reflexed lobes, lobes entire or bifid, papillate-stigmatic on the inner face, ovary ovate, sessile, 1-loculate. Fruit a fleshy drupe, ovoid or globose, the outer mesocarp thick and firm, the inner thin and fleshy; the stone thick-walled, bony, smooth or rugose, ovule filling the locule, ripening in autumn and persisting after the leaves fall.

1 Leaves typically elliptic-lanceolate to ovate-lanceolate and apex mostly sharply acute to acuminate .... 1. *C. laevigata*
1 Leaves typically broadly to narrowly ovate and apex obtuse to abruptly long-acuminate.
  2 Leaves typically 45 mm long or less, the margins usually entire .... 3. *C. reticulata*
  2 Leaves mostly 50 mm long or more, the margins coarsely serrate.
    3 Leaves evidently serrate to well below the middle .... 2. *C. occidentalis*
    3 Leaves mostly entire with a few scattered teeth above the middle ...... 4. *C. tenuifolia*

1. ***Celtis laevigata*** Willd., sugarberry. Tree to 30 m high, with spreading, often pendulous branches forming a broad crown, bark light gray, smooth or covered with corky warts; young branches pubescent at first, then glabrous. Leaves with petioles 6–10 mm long, lanceolate, 4–8 cm, long-acuminate, margin entire or rarely with a few long teeth, broadly cuneate to rounded at the base, glabrous or nearly so except for ciliate margins, thin and membranaceous to coriaceous. Drupes subspherical, 5–8 mm in diam, beakless, orange to brown to red when ripe, on pedicels 6–15 mm long; stone 4.5–7 mm long, 5–6 mm broad. (2n = 20) May–Oct. Rich bottomlands, stream banks, flood plains, & rocky hillsides near streams; e TX, n to OK, MO, & se KS; (se VA to FL & TX, n to MN & OK, MO, KS, s IN). *C. mississippiensis* Bosc.—Rydberg; *C. laevigata* Willd. var. *texana* (Scheele) Sarg.—Gates; *C. laevigata* Willd. var. *smallii* (Beadle) Sarg.—Fernald; *C. smallii* Beadle—Gleason.

Three varieties of this species have been described for our area, but they are often not recognizable. Var. *texana* Sarg. has been distinguished by its leaves, coriaceous, entire, scabrous above, and its pubescent petioles. Var. *laevigata* and var. *smallii* (Beadle) Sarg. are distinguished by having thin membranaceous or submembranaceous leaves with upper surface smooth or only sparingly hirtellous and glabrous petioles. Var. *laevigata* has leaf blades without teeth, whereas var. *smallii* (Beadle) Sarg. has leaf blades with many teeth. Further taxonomic work with the genus *Celtis* in the GP may establish these varieties.

2. **Celtis occidentalis** L., hackberry. Large or small tree or shrub, varying greatly in response to habitat, older bark gray, deeply furrowed, checkered and warty, young branches mostly pubescent. Leaves alternate, lance-ovate to broadly ovate or deltoid, coriaceous, scabrous, those on the fertile branches 5–12 cm long, 3–6(9) cm wide, margin conspicuously serrate with 10–40 teeth, tip acuminate, base oblique or obliquely subcordate. Drupe dark orange to red when ripe, on pedicels to 15 mm long, spherical, 8–11 mm in diam, commonly with a thick beak; stone 7–9 mm long and 5–8 mm thick, cream colored and reticulate. (2n = 20, 28) Apr–May. Rich, moist soil along stream banks, on flood plains, & on rocky hillsides in open woodlands; throughout e 3/4 GP; (s Ont., NH, MA to NY, cen MI, s Man., MN, cen ND s to e WY, NE, ne CO, KS, OK, e to AR, AL, & GA). *C. canina* Raf., *C. crassifolia* Lam.—Rydberg; *C. occidentalis* L. var. *canina* (Raf.) Sarg., var. *crassifolia* (Lam.) A. Gray—Gates; *C. occidentalis* L. var. *pumila* (Pursh) A. Gray—Fernald.

Three varieties of this species have been described for our area. The variation encountered within the species makes the recognition of these varieties difficult. Var. *occidentalis* has coriaceous, scabrous leaves and spherical, orange-red to fuscous drupes. Var. *pumila* (Pursh) A. Gray is distinguished by thin, smooth leaves whose base is uneven and brown to deep purple fruits. Var. *canina* (Raf.) Sarg. has thin, smooth, narrow leaves with symmetrical leaf bases and brown to purplish fruits.

3. **Celtis reticulata** Torr., netleaf hackberry. Small tree or shrub, 7(16) m high, trunk rarely 6 dm in diam, branches usually crooked, bark gray with corky ridges, young branches villous. Leaves with stout petioles 3–8 mm long, grooved above, pubescent, ovate, 3–4.5(7) cm long, 1.5–4 cm wide, tip obtuse to acute or subacuminate, base cordate or occasionally oblique, margin entire or somewhat serrate above middle, upper surface gray-green and scabrous, lower surface yellow-green, thick and rigid. Drupe reddish or reddish-black when ripe, spherical, 8–10 mm in diam, beaked, borne on pedicels 10–14 mm long. Apr–Sep. On dry limestone hills, ravine banks, & occasionally in sandy soils; TX n to OK, KS, CO; (TX, OK, KS, CO, w to ID, WA, CA; n Mex.). *C. rugulosa* Rydb., *C. rugosa* Rydb. not Willd.—Rydberg; *C. laevigata* Willd. var. *reticulata* (Torr.) Benson—Correll and Johnston.

4. **Celtis tenuifolia** Nutt., dwarf hackberry. Small tree or shrub to 8 m high, bark light gray, furrowed, warty. Leaves entire or very sparingly toothed toward tip (leaves on shoots mostly serrate), ovate to occasionally ovate-elliptic, 2–8 cm long, (1)3–4 cm wide, tip blunt, acute or short acuminate, base unequal, one side rounded, upper surface dark gray-green, scabrous, lower surface gray-green, pubescent; petioles 6–10 mm long. Drupes orange to brown or cherry red, glaucous, spherical, 5–8 mm in diam, beakless, with pedicels 3–13 mm long; stones 5–7 mm long and 5–6 mm wide, cream-colored and reticulate. Apr–May. On hardwood slopes & along streams in open woodlands; TX n to e OK, se KS, MO; (NJ to IN, MO, se KS, OK, s to FL & TX). *C. pumila* Pursh var. *georgiana* Small—Gates; *C. pumila* of most Amer. authors, not Pursh, *C. tenuifolia* Nutt. var. *georgiana* (Small) Fern. & Schub., *C. georgiana* Small—Fernald.

2. ULMUS L., Elm

Trees (infrequently shrubs), bark usually deeply furrowed, branches unarmed, slender, terete, often with corky wings; buds conspicuous, axillary, covered with numerous ovate to rounded chestnut-brown, glabrous to pubescent, closely imbricated scales. Leaves distichous, petiolate, usually oblique at base, margin simply or doubly (most) serrate, pinnately veined, deciduous; stipules lateral, linear-lanceolate to obovate, entire, free or connate at the base, scarious, enclosing the leaf in bud, caducous. Flowers perfect, vernal (in ours), minute, produced in axillary subsessile or pedicellate cymes or racemes. Calyx campanulate, slightly to deeply (4)5(9)-lobed and membranaceous; petals absent; stamens equal in number to calyx lobes, filaments filiform to more or less flattened, exserted after anthesis, anthers oblong and extrorse; style deeply 2-lobed, ovary sessile or stipitate, compressed, glabrous or hirsute, usually 1-locular by abortion. Fruit a flattened samara ripening a few weeks after flowering.

1 Flowers with slender pedicels, drooping; samaras elliptical, ovate to oblong, densely ciliolate-fringed on each margin.
 2 Flowers in fascicles, the axis only slightly elongating; samaras glabrous over the seed; branches not corky thickened .... 2. *U. americana*
 2 Flowers in racemes; samaras usually pubescent on the sides; branches often corky thickened.
  3 Petioles 1–3 mm (rarely 5 mm) long; samaras lance-ovate, 3–5 mm broad; young branches often corky .... 1. *U. alata*
  3 Petioles 3–10 mm long; samaras broadly elliptic, 9–15 mm broad; only the older branches corky .... 5. *U. thomasi*
1 Flowers nearly sessile or short pedicelled, not drooping; samaras suborbicular to broadly elliptic, their margins not ciliate.
 4 Leaf buds covered with long rusty hairs, obtuse; leaves usually 10 cm or more long, scabrous with stiff hairs on the upper surface, doubly serrate; fruit noticeably pubescent in the middle .... 4. *U. rubra*
 4 Leaf buds pubescent or glabrous; leaves 7.5 cm or less long, smooth on the upper surface, simply serrate; fruits glabrous. .... 3. *U. pumila*

1. *Ulmus alata* Michx., winged elm. Small tree with spreading branches, forming a round-topped oblong crown, to 20 m high, bark thin, light brown, irregularly glabrous, usually developing 2 broad opposite wings beginning the first or second season of growth; buds acute, glabrous or finely pubescent. Leaves subsessile with petioles 1–3 mm long, ovate-oblong and oblong-lanceolate or elliptic (3)5–7(10) cm long, 2–3(4) cm wide, tip acute to acuminate, rounded to subcordate at the oblique base, margin biserrate to triserrate, mostly shiny dark green and glabrous above, soft pubescent and paler green below. Racemes short, few-flowered; stamens usually 5. Samaras ovate-elliptic to oblong, flat, 7–8 mm long, 3–3.5 mm wide, narrow winged with slender incurved beaks at the apex, villous. Late Feb–May. Along streams & in woodlands on rocky hillsides; KS: Cherokee; OK: Creek, Ottawa, Tulsa; (VA to KY, s IN, s IL, MO, se KS s to cen OK, e & se TX, e to cen FL).

2. *Ulmus americana* L., American elm. Tall tree with branches gradually curving outward to form a wide-spreading crown with pendulous branches, to 30 m high, bark light gray, scaly and deeply fissured with broad forking scaly ridges; young branches pubescent or sometimes nearly glabrous, not corky thickened; buds ovoid, obtuse to acute, glabrous or only slightly whitish short-pubescent. Leaves ovate to elliptic, 4–12 cm long, 2–6 cm wide, tip acuminate, unequal at the base, margin doubly serrate, dark green, glabrous and more or less scabrous above, paler green, pubescent or nearly glabrous below; petiole 2–6 mm long, glabrous or pubescent; stipules lance-linear, 5–7 mm long, pubescent, caducous. Flowers with elongated, unequal pedicels 1–2 cm long; calyx with 5–9 lobes; stamens 5–9 and exserted; stigmas white. Samaras elliptic, flat, glabrous with a hairy margin, 1–1.5 cm long, deeply notched at the apex, the incision reaching the seed. Feb–May. In rich alluvial soil along streams and in open woodlands; commonly planted as a shade tree, currently threatened by Dutch elm disease; throughout GP; (Newf. to Man., s to FL & TX).

3. *Ulmus pumila* L., Siberian elm. Small tree to 15 m high with a broad rounded crown, bark gray, rough, with shallow furrows and long flat ridges, young branches slender, glabrous or glabrate; leaf buds glabrous, ovoid, 2–3 mm long, scales dark brown, ciliate, flower buds globose, 3.5–4 mm in diam, dark red-brown, lustrous, long ciliate. Leaves narrowly

narrowly elliptic to lanceolate, (2.5)4–7(7.5) cm long, (1.5)2–3(3.5) cm wide, short pointed, essentially symmetrical, margin serrate, thick, smooth, and dark green, glabrous with some pubescence in the vein axils; stipules broad based, lanceolate, 2–4 mm long, caducous. Flowers small, greenish, in clusters, pedicels 0.5 mm long; calyx campanulate, 4 or 5 rounded calyx lobes, reddish-brown, pubescent; petals none; stamens 4–5(8). Samaras light brown, flattened, round-obovate, glabrous, 11–14 mm in diam, the wing all around, notched at the apex with overlapping tips. (2n = 28) (Feb) Mar (Apr). Widely cultivated tree, often escapes & persists in disturbed wooded sites GP; (cen U.S.; Asia).

4. *Ulmus rubra* Muhl., slippery elm. Tree with spreading branches forming a broad open crown, to 20 m high; bark thick, gray-brown, with shallow fissures and long, flat, often loose plates, inner bark mucilaginous; young branches stiff, pubescent and scabrous, red-brown to orange in color; leaf buds narrowly ovoid, obtuse; flower buds broadly ovoid to obovoid, 4–5 mm long, fulvous-pubescent with long hairs. Leaves with petioles 4–8 mm long, obovate to ovate or elliptic, 1–2(3) dm long, 5–7(7.5) cm wide, tip long-acuminate, oblique base, margin doubly serrate, dark green and very scabrous above, paler green and densely soft-pubescent beneath. Flowers clustered, subsessile, green, pubescent; calyx lobes 5–9, green, 0.5–0.7 mm long, rounded, fringed with red-brown hairs; petals none; stamens 5–8(9), filaments 4–5 mm long, white, glabrous; stigmas pinkish. Samaras suborbicular to obovate or broadly elliptic, flat, 1–2 cm long, slightly notched, rufous-pubescent in the center. (2n = 28) Mar–May. Rich, moist soil along stream banks & in bottomlands; e 1/2 TX, OK, KS, w MO, IA, MN, e 1/3 NE, SD & ND; (FL to TX, n to Que., & w to SD & ND). *U. fulva* Michx.—Rydberg.

5. *Ulmus thomasi* Sarg., rock elm. Tree with spreading crown to 20 m high; bark dark gray, with shallow furrows and flat-topped ridges with vertical sides, young branches thinly pubescent, often becoming irregularly winged with 2 or more plates of cork after their second year; buds narrowly ovoid, 5–7 mm long, 2.3–2.8 mm in diam, slightly flattened, acute, red-brown in color, pubescence appressed, outer scales without cilia, inner scales ciliate. Leaves broadly elliptic or oval, thick and stiff, usually 8–14 cm long, distinctly cordate on one side at the base, upper surface dark green, pubescent especially on the veins; petiole 3–10 mm long; stipules lanceolate, 8–10 mm long, deciduous. Flowers in slender racemes to 1–3(4) cm long, 3–9 flowers; pedicels 2–4 mm long, with brown ciliate, obovate bract at base; calyx obconic with 7–9 lobes; petals none; stamens 7–9. Samaras elliptic, (1.4)1.5–1.8(2) cm long, flat, densely pubescent. (2n = 28) Apr–May. Rich woodlands on hillsides & along flood plains; e ND, e SD, e NE, ne KS, MO, IA, MN; (Que. to s MI & ND, SD, s to NJ, w VA, TN, & MO). *U. racemosa* Thomas—Fernald.

28. CANNABACEAE Endl., the Hemp Family

by William T. Barker and Ralph E. Brooks

Erect or twining herbs with watery juice, dioecious. Leaves simple and unlobed, palmately 3- to 7-lobed, or palmately compound, opposite (at least below); stipules free, persistent. Staminate flowers few to many in loose compound cymes or panicles; sepals 5; petals none; stamens 5, opposite the sepals, filaments much shorter than the anthers. Pistillate flowers borne in somewhat compact, few-flowered cymose inflorescences, short spikes, or drooping amentlike structures; calyx tube short, membranaceous, and enclosing the ovary; carpels

2; stigmas 2; style much shorter than the stigmas; ovary unilocular with a single pendulous, anatropous ovule. Fruit an achene.

1 Stems erect; leaves palmately compound .... 1. *Cannabis*
1 Stems twining; leaves simple, unlobed or 3- to 7-lobed .... 2. *Humulus*

1. CANNABIS L., Hemp, Marijuana

1. *Cannabis sativa* L., hemp, marijuana. Large, erect annual herb, dioecious (rarely monoecious), odoriferous; stems solitary, well branched, 0.5–4 m tall, antrorsely appressed hirsute. Leaves opposite at least below, the upper ones mostly alternate, palmately compound; leaflets 5–9(11), linear to linear-lanceolate, 4–15 cm long, 3–20 mm wide, the middle leaflet the longest and the outer ones progressively shorter, strigose above, sparsely to densely pubescent below, apex acuminate, margins serrate; petiole 2–11 cm long; stipules lanceolate to linear-lanceolate, 5–20 mm long. Inflorescences numerous small cymose clusters borne on short leafy branches in the upper axils. Staminate flowers on pedicels 0.5–2 mm long; sepals ovate to lanceolate, 2.5–4 mm long, puberulent, margins usually hyaline; stamens slightly shorter than the sepals, filaments 0.5–1 mm long. Pistillate flowers sessile, partially surrounded to nearly enclosed by a closely subtending bract, the bract abruptly acuminate and stipitate glandular; calyx usually short, or if longer, hyaline and becoming adnate to the pericarp. Achenes buff or greenish, variously mottled or marbled with purple, ellipsoid to ovoid, 2.7–4.2 mm long, 2–2.8 mm wide, smooth. (2n = 20) Jul–Oct. Usually exposed sites including pastures, flood plains, roadsides, & waste areas; MN, e 1/6 ND, IA, e 1/2 SD, NE, MO, KS, e OK; (Que. to B.C. & s in U.S.; Asia). *Naturalized*.

Reference: Small, E., & A. Cronquist. 1976. A practical and natural taxonomy for *Cannabis*. Taxon 25: 405–435.

The *Cannabis sativa* complex contains a northern element, subsp. *sativa*, used for fiber (hemp) production, and a southern element, subsp. *indica* (Lam.) E. Small & Cronq., harvested for hallucinogenic drugs (marijuana and hashish). Plants growing in the GP today are nearly all descendents of the hemp introduced here during the latter half of the 19th century as a field crop and may be referred to as subsp. *sativa* var. *sativa*. These plants appear to be regressing towards a similar wild-adapted strain, subsp. *sativa* var. *spontanea* Vavilov. Var. *sativa* has mostly unmottled achenes that are not constricted basally while var. *spontanea* has heavily mottled achenes that are constricted at the base. To date, the illegal subsp. *indica*, with achenes averaging more than 3.8 mm long, has not been found growing outside of cultivation in the GP.

2. HUMULUS L., Hops

Twining, harshly scabrous perennial herbs, dioecious. Leaves simple, unlobed or 3- to 7-lobed, opposite throughout; stipules narrowly triangular to lanceolate, membranaceous. Staminate flowers numerous in loose axillary panicles; anthers nearly as long as the sepals; filaments minute. Pistillate flowers sessile, paired under large imbricate bracts in short spikes or drooping amentlike structures or hops; calyx entire and surrounding the ovary. Achene enclosed by the persistent calyx.

Reference: Small, E. 1978. A numerical and nomenclatural analysis of morph-geographic taxa of *Humulus*. Syst. Bot. 3: 37–76.

1 Veins on the lower leaf surface armed with rigid, spinulose hairs .... 1. *H. japonicus*
1 Veins on the lower leaf surface lacking rigid, spinulose hairs, lax weak hairs may be present .... 2. *H. lupulus*

1. **Humulus japonicus** Sieb. & Zucc., Japanese hops. Vines to 10 m or more long; stems 1-5 mm in diam, roughly scabrous on the angle, pubescent to glabrous on the faces. Leaf blades widely ovate, 5- to 9-lobed, 5-15(20) cm long, 7-18(23) cm wide, sinuses of the upper lobes narrow or closed, sparsely hirsute-scabrous above, below the veins with rigid spinose hairs and otherwise sparsely pubescent to glabrous, apices acute to short acuminate, margins serrate or sometimes double-serrate, base cordate; petiole to 25 cm long, nearly always longer than the blade. Staminate panicles 1-3 dm long, 5-10 cm wide; pedicels 0.5-3 mm long; sepals ovate-lanceolate, 2-3 mm long, pubescent. Pistillate spikes 1-2.5 cm long, the mid-portion usually bearing only 1 flower in the axils of the stipular bracts; bracts ovate to widely so, 8-12 mm long, 2.5-7 mm wide, at least sparsely hirsute, apex attenuate to abruptly so, margins hirsute-ciliate. Achenes buff or mottled with dark brown, widely ovoid to subspherical, 4-5 mm long, 4-5 mm wide, smooth, apex abruptly blunt tipped. Aug-Oct. Open or wooded, usually moist sites, especially roadsides, streams, & riverbanks; IA: Page; NE: Douglas, Nance; MO & ne KS; (se Can. to SC, w to NE & KS; e Asia). *Naturalized.*

2. **Humulus lupulus** L., common hops. Vines to 10 m or more long; stems 1-5 mm in diam, scabrous on the angles, puberulent to glabrous on the faces. Leaf blades narrowly to widely ovate, unlobed or 3-5 lobes, the middle and lowermost primary lobes sometimes shallow, 5-15 cm long, 4-16 cm wide, sinuses of the upper lobes mostly broad and open, scabrous or weakly so above, below the veins glabrous or variously hirsute but not armed with rigid spinose hairs, pubescent to glabrous between the veins, margins serrate or infrequently double-serrate, base cordate or occasionally rounded; petiole 2.5-12 cm long, barely equaling to mostly shorter than the blade. Staminate panicles 7-15 cm long, 1.5-5 cm wide; pedicels 0.5-3.5 mm long; sepals ovate to lanceolate or narrowly oblong, 1.5-3 mm long, pubescent to glabrous. Amentlike pistillate inflorescences at anthesis 5-10 mm long with the stigmas mostly exceeding the bracts, at maturity 2-5 cm long, 2-3 cm in diam; mature bracts ovate or widely so to elliptic, 7-20 mm long, 5-10 mm wide, membranaceous, glabrous to puberulent, apex obtuse, acute or abruptly acuminate, margins entire and ciliate or not. Achenes brownish, widely ovoid, 2-2.7 mm long, 2-2.5 mm wide, smooth, apex abruptly blunt tipped. (2n = 20, 40) Jul-Sep. Shaded or exposed, usually moist sites including stream banks, ditches, low woodlands, ravines, or in thickets.

Humulus lupulus var. *lupulus,* a native of Asia, has long been cultivated for the resinous material of the pistillate inflorescences which is used in beer. While the plant has been found immediately to the east and north of our region, we have yet to document its occurrence within the GP. Three other native varieties do occur in the GP and may be recognized as follows:

1 Largest leaves 5-lobed or if 3-lobed then 5 primary veins visible from the base .... 2b var. *neomexicanus*
1 Largest leaves unlobed or 3-lobed with only 3 primary veins visible.
  2 Leaves glabrous to sparsely pubescent between the veins on the lower surface .... 2a. var. *lupuloides*
  2 Leaves conspicuously pubescent between the veins on the lower surface .... 2c. var. *pubescens*

2a. var. *lupuloides* E. Small. Leaves 3-lobed, glabrous to sparsely pubescent on the lower surface. MN, ND, MT, SD, IA, n NE, e WY; KS: Atchison; (N.S. w to Alta., s to NC, n IL, IA, ne KS, & e WY). *H. americanus* Nutt.—Rydberg; *H. lupulus* subsp. *americanus* (Nutt.) Löve & Löve—Löve, A. and D. Löve. Taxon 31: 121. 1982.

2b. var. *neomexicanus* A. Nels. & Cockll. Leaves 5-lobed or, if 3-lobed, then 5 primary veins visible from the base, glabrous to densely pubescent below. ND, MT, w SD, WY, CO, NM; KS: Kearny; (Man. & Sask. s to KS, w SD, NM, w TX, & n Mex., w to MT & CA).

2c. var. *pubescens* E. Small. Leaves unlobed or 3-lobed, conspicuously pubescent between the veins below. IA, e NE, MO, e KS; (NY w to s MN & e NE, s to NC & e KS).

29. MORACEAE Link, the Mulberry Family

by William T. Barker

Monoecious or dioecious trees and shrubs with milky sap. Leaves alternate (or some opposite in *Broussonetia*), petiolate, stipulate, the blades simple and undivided or lobed. Flowers small, imperfect, in globose to oblong clusters or cylindrical catkins, the pistillate flowers often confluent by the receptacles, especially in fruit; calyx of 4 sepals united near the base; corolla none; staminate flowers with 4 stamens opposite the sepals; pistillate flowers with a somewhat reduced but accrescent calyx; pistil 1, 2-carpellate or 1 carpel abortive, style simple or 2-branched, stigmatic for most of the length, ovary superior to inferior, 1-celled, 1-ovuled. Fruit an achene, usually enclosed by the ultimately fleshy calyx, the fruits of an entire pistillate inflorescence confluent by the receptacles to form a corky or juicy multiple fruit.

1 Leaves entire; stems often thorny .... 2. *Maclura*
1 Leaves serrate and often lobed; stems unarmed.
    2 Twigs pubescent; multiple fruits globose; styles simple; bark smooth .... 1. *Broussonetia*
    2 Twigs glabrous or nearly so; multiple fruits cylindric; styles 2-branches; bark scaly or furrowed .... 3. *Morus*

1. BROUSSONETIA L'Her. Paper Mulberry

1. *Broussonetia papyrifera* (L.) Vent. Dioecious shrub or small tree to 15 m tall, with smooth bark and pubescent twigs. Leaves alternate or occasionally some opposite, long petioled, the blades broadly ovate in outline, undivided or irregularly lobed, 6–20 cm long, scabrous above, densely pubescent beneath, acuminate, serrate, broadly rounded to cordate at the base. Flowers greenish; staminate flowers in slender pendulous catkins 4–8 cm long; pistillate flowers in dense, globose, tomentose clusters 1–2 cm thick; style simple, long and exserted. Multiple fruits orange-red, globose, 2–3 cm diam, the red achenes protruding from the accrescent fleshy calyces. (2n = 26) May–Jun. An ornamental, occasionally escaping to roadsides, clearings, & waste places; se GP; NE: Richardson; KS: Leavenworth, Neosho, Wilson; OK: Grady; (NY to se GP, s to FL & TX; e Asia). *Introduced* and *naturalized*.

2. MACLURA Nutt., Osage Orange

1. *Maclura pomifera* (Raf.) Schneid. Small to medium-sized tree to 12(20) m tall, with deeply furrowed bark and thorny branches, sometimes forming thickets; thorns stout, to 2.5 cm long. Leaves alternate or clustered on lateral spurs of the branches, the blades ovate to elliptic-lanceolate, 4–12 cm long, 2–6 cm wide, subcoriaceous, usually pubescent on the midrib and veins, acuminate at the tip, entire, broadly cuneate to rounded or subcordate at the base; petioles usually pubescent, 2–5 cm long; stipules minute, caducous. Staminate flowers numerous, pedicellate, in globose to oblong clusters 1.5–3.5 cm long, the clusters axillary, peduncled, 1–several on short, lateral spurs of young branches. Pistillate flowers sessile and confluent in globose heads 1.5–2.5 cm diam, single and short peduncled from the axils; styles conspicuous, simple, filiform and elongate. Multiple fruits large, yellowish-green, globose, 6–15 cm diam, with a wrinkled corky rind comprised of the fleshy accrescent calyces; achenes deeply embedded in the multiple fruit, oval to oblong, 8–12 mm long. May, Fruit Aug–Sep. Woods, fence rows, ravines, field margins, waste places, often

planted in hedgerows & shelter belts along roads & fields; IA, se SD, e & s NE, s to TX; (native to OK, TX, & AR, adventive elsewhere in the GP & in e & to a lesser extent w U.S.). *Toxylon pomiferum* Raf.—Winter.

3. MORUS L., Mulberry

Monoecious to dioecious, large shrubs and trees with scaly or furrowed bark; twigs glabrous or only sparsely pubescent, the branchlets glabrous to sparingly pubescent. Leaves alternate, the blades generally ovate to rotund-ovate or elliptic-ovate, unlobed or irregularly lobed with 1-4 sinuses, serrate; stipules lanceolate to linear-lanceolate, deciduous. Staminate and pistillate flowers rarely mixed in the same cluster; staminate flowers small and numerous in axillary pendulous catkins; pistillate flowers in short-cylindrical to elongate-cylindrical, axillary clusters; styles 2-branched, persistent in fruit. Multiple fruits comprised of the fleshy calyces, sweet, juicy, edible, resembling a blackberry, white, pink, purple, or nearly black when ripe; achenes somewhat flattened, ovate, ca 2 mm long, enclosed by the fleshy calyces, the persistent styles protruding.

1 Lower surface of the leaves glabrous or pubescent on the veins only .... 1. *M. alba*
1 Lower surface of the leaves entirely pubescent .... 2. *M. rubra*

1. *Morus alba* L., white mulberry. Large shrub or tree mostly 3-15 m tall. Leaf blades ovate to elliptic-ovate or rotund-ovate in outline, unlobed or irregularly 2- to 5-lobed and coarsely serrate, often mitten-shaped, 4-10 cm long, 3-6 cm wide, often larger on vigorous shoots, glabrous on both surfaces or usually spreading-pubescent on the veins of the lower surface, acute to acuminate, rounded or truncate to cordate at the base; petioles 2-5 cm long, pubescent. Staminate catkins drooping, 1-4 cm long; pistillate clusters 5-20 mm long. Multiple fruits white, pink, purple, or nearly black, 5-25 mm long. (2n=28) Apr-May, fruits late May-Jun. Flood plains, roadsides, fence rows, wooded areas, often planted in yards & shelter belts; s MN, se SD, NE, MO, KS, OK, & TX; (established in much of e & cen U.S. & some places in the w; e Asia). *Naturalized*, especially in the s GP. *M. nigra* L.—Rydberg, misapplied.

2. *Morus rubra* L., red mulberry. Tree to 20 m tall. Leaf blades similar in shape to no. 1 though less often lobed, 6-18 cm long, 4-12 cm wide, glabrous or commonly scabrous on the upper surface, soft pubescent over the entire lower surface, cuspidate to caudate at the tip, rounded to cordate at the base; petioles 2-3 cm long, pubescent. Staminate catkins drooping, 2-5 cm long; pistillate clusters 8-20 mm long. Multiple fruits mostly dark purple, 10-25 mm long. (2n=28) Apr-May; fruits Jun-Jul. Rich woods, flood plains, & stream banks; IA, e & cen NE, s to MO, OK, & TX; (VT to MI & NE, s to FL & TX).

30. URTICACEAE Juss., the Nettle Family

by William T. Barker

Monoecious or dioecious, annual or perennial herbs (ours) with watery juice, sometimes beset with stinging hairs, the leaves alternate or opposite, simple, petiolate, usually stipulate. Flowers greenish, small and inconspicuous, borne in simple or branched axillary clusters. Flowers usually imperfect (some perfect in *Parietaria*); perianth consisting of a 2- to 5-parted or toothed calyx, that of the staminate flowers usually more deeply parted; staminate flowers with stamens equal in number and arranged opposite to the calyx lobes, a vestigial pistil sometimes present; pistillate flowers containing a unicarpellate pistil and sometimes with scalelike rudiments of stamens, style 1, ovary superior, 1-celled, ovule 1. Fruit an achene, often enclosed by an accrescent calyx.

1 Leaves entire .... 3. *Parietaria*
1 Leaves serrate.
  2 Flowers in simple, axillary, spikelike clusters (1 spike per axil) which often become leafy at the tips .... 1. *Boehmeria*
  2 Flowers in branched clusters in the leaf axils.
    3 Plants glabrous or nearly so, not armed with stinging hairs .... 4. *Pilea*
    3 Plants armed with stiff stinging hairs.
      4 Leaves opposite .... 5. *Urtica*
      4 Leaves alternate .... 2. *Laportea*

1. BOEHMERIA Jacq.

1. *Boehmeria cylindrica* (L.) Sw., false nettle. Dioecious or monoecious, erect perennial, 4–10 dm tall, usually unbranched, glabrate to sparingly strigose, not armed with stinging hairs. Leaves opposite, long petioled, the blades ovate to ovate-lanceolate, 4–15 cm long, 2–8 cm wide, puncticulate above, conspicuously reticulate beneath, 3-nerved from the base, acuminate to caudate at the tip, coarsely serrate, obtuse to broadly rounded or subcordate at the base; stipules brownish, lanceolate, deciduous. Flowers minute, imperfect, hispidulous, glomerulate in continuous or interrupted, simple, axillary, spikelike clusters, these borne singly in the leaf axils, often leafy at the tips; staminate flowers consisting of a 4-parted calyx and 4 stamens; pistillate flowers with a tubular, flattened to ovoid, minutely 2- to 4-toothed calyx enclosing the ovary, style exserted, brown, papillose along one side. Achene rotund-ovate, beaked, ca 1 mm diam. (2n = 28) Jul–Oct. Shores, stream banks, marsh borders, & moist or wet woods; s MN, se SD, NE, MO, KS, OK, & n TX; (Que. to SD, s to FL & TX). *B. drummondiana* Wedd.—Rydberg; *B. cylindrica* var. *drummondiana* Wedd.—Fernald.

2. LAPORTEA Gaud.

1. *Laportea canadensis* (L.) Wedd., wood nettle. Monoecious, unbranched perennial 4–10 dm tall, sparsely to densely armed with stiff stinging hairs on the stems, petioles, inflorescence branches and sometimes on the midrib of the leaves. Leaves alternate, the blades ovate to elliptic or broadly so, 8–20 cm long, 5–13 cm wide, nearly glabrous to hispidulous, puncticulate, acuminate or caudate at the tip, coarsely serrate, subcordate to rounded or obtuse at the base; petioles mostly 3–11 cm long; stipules brownish, chartaceous, lanceolate, soon deciduous. Flowers minute, in cymose clusters in the upper leaf axils, the lower clusters staminate, seldom surpassing the subtending petiole, the uppermost clusters pistillate, ultimately divaricate-spreading and rather open. Staminate flowers with a 5-parted calyx and 5 stamens. Pistillate flowers with a (2 or 3)4-parted calyx, the outer pair of sepals much smaller than the inner pair, or one or both of the outer sepals absent by abortion, style filiform, 3–4 mm long. Achene loosely enclosed toward the base by the accrescent pair of inner sepals, tawny to dark brown, sometimes mottled, flat, D-shaped, 2–3 mm long, about as wide. (2n = 26) Jul–Sep. Rich, moist woods; MN, e ND, s to MO, e KS, & OK; (N.S. to Man., s to GA & OK). *Urticastrum divaricatum* (L.) O. Ktze.—Winter.

3. PARIETARIA L., Pellitory

1. *Parietaria pensylvanica* Muhl., Pennsylvania pellitory. Monoecious or polygamous, puberulent to pubescent annual 0.5–4.5 dm tall, sprawling-ascending to erect, simple or branched, lacking stinging hairs. Leaves alternate, the blades ovate-lanceolate to lanceolate, 1.5–7 cm long, 0.8–2 cm wide, punctate with cystoliths, mostly glabrous except for the pubescent, entire margin, gradually tapered to a rounded or acutish tip, cuneate at the base; petioles 5–25 mm long, stipules none. Flowers in short, few-flowered clusters in the middle and upper axils, the flowers subtended and exceeded by green, linear to linear-lanceolate bracts, the staminate and pistillate flowers mixed or a few flowers often perfect; calyx deeply 4-parted in the staminate flowers, 4-parted to about the middle in the pistillate flowers, brownish, 1–2 mm long, puberulent; stamens 4; stigma tufted, subsessile. Achene loosely enclosed by the calyx, light green to brown, lustrous, ovoid, flattened, 0.9–1.2 mm long. (2n = 16) May–Sep. Woods, banks, & brushy or rocky places, usually where partially to heavily shaded; GP; (ME to B.C., s to VA, AL, TX, n Mex., & CA).

4. PILEA Lindl., Clearweed

Monoecious or usually dioecious, glabrous annuals; stems erect to decumbent, simple or branched, brittle and watery, translucent; cystoliths appearing as numerous minute, whitish or dark lines on foliage of dried specimens. Leaves opposite, the blades thin and translucent, ovate, with 3 major veins from the base, broadly cuneate to rounded at the base, prominently serrate, the teeth obtuse to rounded, the terminal tooth short to elongate; petioles subtending the inconspicuous, connate, membranaceous stipules. Flowers clustered in axillary cymes, imperfect, the staminate and pistillate flowers usually mixed; staminate flowers with a deeply 4-parted calyx and 4 stamens; pistillate flowers with a deeply 3-parted calyx, the segments often unequal; staminodes present, minute and scalelike; stigma sessile. Achene flattened, ovate, subtended by the persistent calyx.

Reference: Fernald, M. L. 1938. *Pilea* in eastern North America. Rhodora 38: 169–170.

1 Achenes dark olivaceous to nearly black with a narrow pale margin .... 1. *P. fontana*
1 Achenes green, often marked with purple .... 2. *P. pumila*

1. *Pilea fontana* (Lunell) Rydb. Plants 1–4 dm tall, often decumbent. Leaf blades 1.5–6 cm long, 1–4 cm wide; petioles mostly 0.5–5 cm long. Flower clusters spreading 0.5–5 cm from the stem, the male flowers usually innermost in the clusters when mixed with female flowers. Achenes dark olivaceous to nearly black with a narrow pale margin, 1.3–2 mm long, mostly 3/4 or more as wide, the persistent calyx shorter than to slightly exceeding the achene. Jul–Sep. Springs, fens, swamps, & wet shores; MN: Clay; ND: Barnes, Richland, Stutsman; SD: Roberts; NE: Cuming; (P.E.I. to ND, s to VA, IN, & NE; also FL). *P. opaca* (Lunell) Rydb.—Rydberg; *Adicea fontana* Lunell, *A. opaca* Lunell—Lunell, Amer. Midl. Naturalist 3: 7–9. 1913.

2. *Pilea pumila* (L.) A. Gray. Very similar to no. 1, differing mainly as follows: Plants sometimes larger, to 7 dm tall. Leaf blades to 15 cm long, 10 cm wide, thinner and more translucent than no. 1; petioles to 8 cm long. Achenes green, often marked with purple, 1.3–2 mm long, mostly 1/2 to 3/4 as wide. (2n = 26) Jul–Sep. Moist or wet shaded places in rich soil; MN & e ND, s to MO, cen & e NE, KS, & OK; (Que. to e ND, s to FL, LA, & e TX) *P. pumila* var. *deamii* (Lunell) Fern.—Fernald.

5. URTICA L., Nettle

Dioecious or monoecious, annuals and perennials with erect, fibrous stems and opposite, stipulate leaves, with few to many stinging hairs mainly on the stems, petioles, and main nerves of the leaves, otherwise glabrous to pubescent. Flower imperfect, clustered in panicles or branched spikes in the middle and upper axils; staminate flowers with a 4-parted calyx, the segments subequal; stamens 4, surrounding a minute, rudimentary pistil; calyx of the pistillate flowers unequally divided into 4 segments, the outer 2 small and inconspicuous, the inner 2 larger, accrescent; stigma sessile, tufted. Achenes ovate, lenticular, enclosed by the 2 inner sepals.

References: Fernald, M. L. 1926. *Urtica gracilis* and some related North American species. Rhodora 28: 191–199; Hermann, F. J. 1946. The perennial species of *Urtica* in the United States east of the Rocky Mountains. Amer. Midl. Naturalist 35: 773–778; Woodland, D. W., I. J. Bassett, & C. W. Crompton. 1976. The annual species of stinging nettle (*Hesperocnide* and *Urtica*) in North America. Canad. J. Bot. 54: 374–383.

1 Plants annual; stipules 1–4 mm long, spreading or deflexed; flower clusters shorter than the petioles .... 1. *U. chamaedryoides*
1 Plants perennial, strongly rhizomatous; stipules 5–15 mm long, erect; flower clusters usually surpassing the petioles .... 2. *U. dioica*

1. **Urtica chamaedryoides** Pursh, weak nettle. Monoecious, slender erect annual 2–10 dm tall, sometimes leaning on surrounding vegetation, sparingly armed with stinging hairs, otherwise glabrous; stems simple or branched from the base, sometimes with short opposite branches above. Leaves sometimes purple on the underside, the blades broadly ovate to subrotund on the lower portion of the stem, ovate to lanceolate and smaller upward, 1–6 cm long, 1–4 cm wide, marked by numerous, minute, short-cylindric to spherical cystoliths when dried, crenate-serrate, cordate to truncate or cuneate at the base; petioles about equal to or shorter than the blade; stipules linear-lanceolate, 1–4 mm long, spreading or deflexed. Flower clusters usually 2(1–3) per axil, androgynous, globose to short-spicate, 3–6 mm long, shorter than the subtending petioles. Achenes tan to brown, ovate-elliptic, 1–1.5 mm long, ca 1 mm wide, enclosed by the inner pair of sepals which are 1–1.5 mm long, the outer pair less than 1/2 as long. Mar–Jun. Moist, rich woods, thickets, & flood plains; s MO, se KS, & e OK; (s OH to se KS, s to FL, TX, & into Mex.).

2. **Urtica dioica** L., stinging nettle. Monoecious or dioecious, stout perennial 8–25 dm tall, often forming dense patches, sparsely to moderately clothed with stinging hairs, otherwise glabrous or sparsely to densely puberulent, the stems usually simple. Leaf blades ovate to lanceolate, sometimes conduplicate, mostly 5–15 cm long, 2–8 cm wide, puncticulate with cystoliths when dried, acute to acuminate, coarsely serrate, cordate to truncate or rounded at the base; petioles mostly 1–6 cm long, ca 1/10 to 1/2 the length of the blade; stipules linear-lanceolate, 5–15 mm long. Flower clusters branched and spreading, usually surpassing the subtending petioles, the clusters unisexual or some staminate and some pistillate, the pistillate clusters usually above the staminate when both are present. Achenes tan, ovate, 1–1.2(1.5) mm long, ca 1/2 as wide, enclosed by the inner pair of sepals which are 1–1.5 mm long, the outer pair ca 1/2 as long. ($2n = 26$) Jun–Sep. Moist woods, thickets, ditches, shores, stream banks, & disturbed areas; GP, except sw KS, cen & w OK; (Lab. to AK, s through most of the U.S. & Mex.; also S. Amer. & Eurasia). *U. procera* Muhl., *U. viridis* Rydb., *U. gracilenta* Greene (misapplied) — Rydberg; *U. dioica* var. *procera* (Muhl.) Wedd. — Barkley; *U. gracilis* Ait. — Winter; *U. dioica* subsp. *gracilis* (Ait.) Seland. var. *gracilis*, var. *procera* (Muhl.) Wedd. — Hitchcock et al.

American *U. dioica* ($2n = 26$) is designated subsp. *gracilis* (Ait.) Seland. The European *U. dioica* subsp. *dioica* differs in chromosome number ($2n = 48, 52$) and is typically dioecious, whereas subsp.

gracilis is predominately monoecious. Although subsp. *dioica* has been naturalized in e U.S., it has not been detected in our range.

31. JUGLANDACEAE A. Rich. ex Kunth, the Walnut Family

by Robert B. Kaul

Ours monoecious trees with large odd- or even-pinnately compound alternate leaves without stipules. Male flowers small, numerous, in elongate pendulous catkins, the few-many stamens attached to a 2- to 6-lobed calyx/bracteole, the filaments very short; female flowers terminal on the branches, the flowers solitary to several in clusters, each in a cup-shaped perianthlike green involucre, the perianth tiny; pistil 1, ovary inferior, apparently unilocular, the single erect ovule apparently basal, styles 2 or at least 2-lobed. Fruit a drupelike nut enveloped by a fleshy or woody husk (the ripened involucre and perianth), this dehiscent (*Carya*) or indehiscent (*Juglans*).

1 Pith solid, not chambered; husk dehiscent; nut smooth or angled; terminal 3 leaflets the largest or at least equaling the lateral leaflets, the leaves odd-pinnate .... 1. *Carya*
1 Pith chambered; husk indehiscent; nut furrowed, rugose; median lateral leaflets the largest, the leaves odd- or even-pinnate .... 2. *Juglans*

1. CARYA Nutt., Hickory, Pecan

Trees of the se GP, the bark sometimes shaggy, the twigs often pubescent at least when young, their pith solid. Leaflets mostly 5–13, the 3 terminal ones usually the largest or at least equaling the laterals; leaflets sessile or short petiolate, opposite, lanceolate to elliptic or obovate, acute to acuminate, serrate, glabrous above, sometimes pubescent below. Male catkins borne in 3s, the calyx 2- or 3-lobed and shorter than the adherent bracteole, the stamens 4(10); female flowers solitary or 2–10 in clusters, the involucre 4-lobed, the calyx apparently 1-lobed. Fruits borne singly or a few in a cluster, the dehiscent husklike matured involucre enveloping the single brown nut; husk 4-angled or grooved and splitting half to all the way to the base. Nut brown, globose to cylindric, smooth or angled, 2–6 cm long. Flowers appearing in spring as the leaves expand; fruits autumnal.

There are six species in our southeastern woodlands. A seventh, *C. glabra* (P. Mill.) Sweet has been reported from se KS but we have no records of it. It grows in nearby sw MO, however, at the edge of our area.

1 Bark conspicuously shaggy and exfoliating in large vertical strips; terminal leaflets the largest; leaflets with hairs on or between the teeth.
   2 Leaves with 7–9 leaflets; twigs orangish; leaflets with hairs all along the teeth .... 3. *C. laciniosa*
   2 Leaves with 5(3–7) leaflets; twigs grayish; teeth of leaflets with tufts of hairs at each side .... 4. *C. ovata*
1 Bark grooved and ridged but not shaggy or exfoliating; terminal leaflets the largest or equaling the laterals; leaflets with or without hairs along the margins.
   3 Husks with vertical ridges nearly to the base; nuts smooth or nearly so; leaflets glabrous to pubescent below, without hairs on the margins.
      4 Husks dehiscent to about halfway to the base; fruits and nuts more or less globose; winter buds yellow; leaflets mostly 7–9, not falcate, the terminal 3 the largest .... 1. *C. cordiformis*

4   Husks dehiscent to the base, fruits and nuts ellipsoidal or cylindric, pointed at both ends; buds not yellow; leaflets 11 or more and often falcate, about equal .... 2. *C. illinoensis*
3 Husks with vertical grooves, or at least partly grooved; nuts angled; leaflets usually pubescent below, with marginal hairs.
    5   Leaflets rusty-pubescent below, at least when young, the hairs not grouped; husk grooved its entire length .... 5. *C. texana*
    5   Leaflets pale-pubescent below, the hairs grouped; husks at least partially grooved .... 6. *C. tomentosa*

1. Carya cordiformis (Wang.) K. Koch, bitternut hickory. Bark merely scaly and shallow-furrowed, rather smooth for a hickory. Winter buds yellow-orange, scurfy, blunt, to 15 mm long. Leaves to 30 cm long, the leaflets 7–9, once-serrate, without marginal hairs, the terminal 3 the largest and obovate, the laterals lanceolate, sometimes narrowly so. Fruits solitary or paired, subglobose to somewhat obovate, 2.5–3.5 cm in diam, with 4 vertical ridges, the husk thin and splitting about halfway to the base; nut globose to obovate, 1.5–3 cm long, somewhat laterally compressed, obscurely angled, thin-shelled, the flesh bitter. (2n = 32). May. Upland & bottomland forests; nw IA & e NE s through e KS to e & cen OK; (Que. to e MN, sw to se NE & e TX, e to n FL).

   This species is reported to hybridize with *C. illinoensis*, the pecan, to form the "hickon," *C.* × *brownii* Sarg. (KS), and with *C. ovata*, the shagbark hickory, to produce *C.* × *laneyi* Sarg.
   In some parts of our area this tree is severely afflicted with fungal blight.

2. Carya illinoensis (Wang.) K. Koch, pecan. Bark shallow-furrowed, with long vertical flat-topped ridges, but not shaggy. Terminal buds pointed, laterally compressed, pubescent, to 10 mm long. Leaves 30–50 cm long, the leaflets (9)11 or more and about equal, often falcate, lanceolate, mostly doubly-serrate, without hairs on the margins. Fruit in groups of 3–10, oblong to ellipsoid, with 4 vertical ridges, splitting to the base to reveal the large ellipsoid or cylindric nut, to 4 cm long and pointed at both ends, marked irregularly with black, the shell thin, the flesh sweet. (2n = 32). Apr–May. Bottomland forests & along streams; w-cen MO & se KS to sw OK & n-cen TX; (sw OH to n IL, sw to se KS & sw OK, s to cen TX, e to AL). *C. pecan* (Marsh.) Schneid.—Rydberg.

   This species is reported to hybridize with no. 1 (q.v.) and no. 3. It is occasionally planted in our area n to cen IA, e NE, & w to w-cen KS.

3. Carya laciniosa (Michx. f.) Loud., kingnut or big shellbark hickory. Bark shaggy, exfoliating in long vertical strips, the twigs glabrous, orangish when young. Terminal buds in winter ovoid, 1–2 cm long, blunt, brownish. Leaves 30–50 cm long, the rachis glabrous, with (5)7–9 leaflets, the terminal ones the largest, these evenly pubescent along the margin, pubescent below. Fruit solitary or paired, ellipsoid to globose, to 7 cm long, apically depressed, not prominently ridged, splitting to the base, the husk thick; nut compressed, more or less 4- to 6-angled, to 6 cm long, thick-shelled, the flesh sweet. (2n = 32) Apr–May. Bottomland forests; w MO, e KS, & cen KS in the Arkansas R. valley, ne OK; (NY to MI & e KS, s to ne OK, AR, & GA).

   This species reportedly hybridizes with *C. illinoensis*, pecan, q.v.

4. Carya ovata (P. Mill.) K. Koch, shagbark or shellbark hickory. Resembling no. 3, the bark very shaggy, the exfoliating strips curling outward below; twigs reddish-brown when young, the terminal bud in winter ovoid, 1–2 cm long, brown, blunt. Leaves 20–35 cm long, with 5(3–7) leaflets, the terminal ones the largest, a tuft of hairs at each side of the serrations, glabrous or not strongly pubescent below when mature. Fruits solitary or 2 or 3 in a cluster, to 6 cm long, with 4 vertical shallow grooves, the husk rather thick and splitting to the base to reveal the 4-angled nut, compressed-globose, thin-shelled, the flesh sweet.

(2n = 32) Apr–May. Moist upland forests, occasionally in bottomlands; sw IA, w MO, e KS, se NE, ne OK; (ME to se MN & se NE, s to e TX, e to GA & NC).

This species is rather variable in leaf pubescence and nut shape. It reportedly hybridizes with *C. cordiformis*, q.v.

5. ***Carya texana*** Buckl., black hickory. Small trees in our area, the bark of older trunks dark, coarsely ridged and checkered; twigs reddish-gray, tomentose, becoming glabrous with age, the mature terminal bud ovoid, acute to blunt, to 8 mm long, rusty-tomentose. Leaves 20–28 cm long, with 5–7 leaflets, the terminal ones the largest, rusty-tomentose below, especially when young, the marginal hairs in groups. Fruit solitary or paired, globose to obovoid, to 5 cm long, with 4 shallow vertical grooves, splitting to the base, the husk valves of 2 different sizes, not very thick; nut globose, 4-angled above, the shell thickish, the flesh sweet. (2n = 64) Apr–May. Upland forests on well-drained slopes; cen & sw MO, e-cen & se KS, cen & ne OK; (se IN to se KS, s to cen TX, e to LA & AR). *C. buckleyi* Durand—Gleason & Cronquist.

6. ***Carya tomentosa*** Nutt., mockernut hickory. Small to medium-sized trees in our area, the bark furrowed, with short ridges, becoming dark with age. Twigs and petioles usually tomentose, the mature terminal buds rusty-tomentose, subglobose, to 1.5 cm long. Leaves 20–35 cm long, the rachis pubescent, leaflets mostly 5–9, the terminal ones the largest, pale-pubescent below, the hairs in groups on the margins. Fruit solitary or a few in a group, globose to obovoid, to 5 cm long the husk thick, barely 4-ridged at one end, becoming grooved at the other, splitting to the middle or base; nut globose to obovoid, more or less 4-angled, the shell thick, the flesh sweet. (2n = 32, 64) Apr–May. Well-drained slopes in upland forests; e KS, w MO, ne OK, & possibly also in extreme se NE & sw IA; (NH to IA, KS, sw to e TX, e to FL).

2. JUGLANS L., Walnut

Ours trees and shrubs, the bark fragrant, the pith chambered (separating in plates). Leaves with 7–23 leaflets, these usually odd-pinnate (even-pinnate leaves sometimes present), to 50 cm long, some of the lateral leaflets the largest; leaflets mostly opposite but sometimes subopposite or alternate, oblong-lanceolate or oblong-ovate, acuminate, often falcate, serrate. Male catkins pendulous, just below the maturing leaves on wood of the previous year, green, sepals (3)4(6), adhering to the bracteole, stamens 8–40, pendulous, filaments short, free, anthers green. Female flowers 1–several in a cluster terminating the new growth, each with 3 bracteoles, pistil 1, ovary inferior, tomentose, surmounted by 4 small green sepals, style very short, stigmas 2(3), large, reflexed, yellowish or reddish. Fruits solitary or in groups of 2 or 3, indehiscent but 2-valved, ellipsoid or globose, pungent, the exocarp leathery, the endocarp stony, rough furrowed, and rugose.

There are three species in our area. At least two Old World species are occasionally planted with us: *J. regia* L., English walnut, with the leaflets entire, of questionable hardiness in GP, is the source of commercial walnuts elsewhere; *J. sieboldiana* Maxim. is hardier, with very large leaves.

1 Medium-sized or large trees mostly of GP e of 100° long.
   2 Fruits more or less oblong, pointed; mature leaves glandular-pubescent below, rare in extreme e GP .... 1. *J. cinerea*
   2 Fruits globose, not pointed (but often the style beak persistent); mature leaves glabrate; common .... 3. *J. nigra*
1 Graceful shrubs and small clumped trees of sw GP (from w tier of s-cen KS counties, sw across OK) .... 2. *J. microcarpa*

1. *Juglans cinerea* L., butternut. Small to medium-sized tree, the bark grayish, the ridges smooth, the pith dark brown. Leaves to 50 cm long; leaflets (7)11–17, oblong-lanceolate, glandular-pubescent, the hairs beneath more or less clustered. Male catkins to 8 cm long, subsessile, the receptacles elongated. Female flowers ca 1 cm long, stigmas red, ca 6 mm long. Fruits oblong-ovoid, pointed, obscurely ridged, 4–7 cm long, rather clammy. Nut more or less ovoid. (2n = 32) Apr–May. Rich bottomland forests & lower forest slopes; sw IA(?), w MO just at the edge of our range; (N.B. to e MN, s to cen AR, n GA, & NC).

2. *Juglans microcarpa* Berl., Texas or little walnut. Rather graceful large shrub or small, clumped, low-branched tree to 6 m tall. Bark rather smooth, becoming rough with age. Leaves to 30 cm long, the leaflets (11)17–23, narrowly lanceolate, glabrous when mature. Stigmas greenish-red. Fruits globose, to 2 cm in diam, brownish pubescent, becoming glabrous with age; nut globose to ovoid. (2n = 32) Mar–Apr. Rocky stream beds, s tier of sw KS counties, sw through OK; (sw KS & w OK to cen NM, e to cen TX, s to Mex.). *J. rupestris* Engelm.—Waterfall.

3. *Juglans nigra* L., black walnut. Large tree, the bark brown, rough, the pith light brown. Young twigs glandular-pubescent. Leaves to 50 cm long, with 11–17(23) leaflets, glabrate at maturity. Male catkins to 12 cm long, the flowers pedicelled. Female flowers 1–1.5 cm long, the stigmas yellow-green. Fruits globose, roughish, pale yellow-green turning brown, to 5 cm in diam. Nut ovoid, oblong, or globose, to 4 cm in diam. (2n = 32) Apr–May. Upland & bottomland forests; often planted in windbreaks and fence rows; e ND & w MN (Red R. valley), se SD, NE (Niobrara, Missouri, & Republican R. valleys), IA, MO, e & cen KS, e & cen OK; rare in TX panhandle; (MA to se ND, s to e TX & n FL).

32. FAGACEAE Dum., the Oak Family

by Robert B. Kaul

1. QUERCUS L., Oak

Monoecious trees and shrubs, the buds with deciduous stipules. Leaves alternate, petiolate, simple, entire to deeply pinnately lobed. Male flowers in pendulous catkins; calyx 5-lobed; stamens 5–10. Female flowers solitary or several in axils of leaves on new growth, epigynous; sepals 6; ovary inferior, 3-carpellate and 3-locular, each locule with 2 ovules; styles 3, free. Fruit a sessile or pedunculate 1-seeded nut (acorn) partially enveloped by a scaly involucral cup. Flowers appearing with the leaves in spring; fruits maturing in autumn of the same or the next year.

References: Elias, T. S. 1980. Trees of North America. Van Nostrand Reinhold, New York; Muller, C. H. 1951. The oaks of Texas. Contr. Texas Res. Found. Bot. Stud. 1: 21–311; Rosendahl, C. O. 1955. Trees and Shrubs of the Upper Midwest. Univ. Minnesota Press, Minneapolis; Stephens, H. A. 1973. Woody Plants of the North Central Great Plains. Univ. Press of Kansas, Lawrence; Steyermark, J. A. 1963. Flora of Missouri. Iowa State Univ. Press, Ames.

As treated here, there are 18 native species in the GP; others just reach our borders and various non-GP species are planted as ornamentals in our area. Other Fagaceae are also cultivated with us: *Castanea* (chestnut), especially *C. dentata* Borkh., American c., and *C. mollissima* Bl., Chinese c., trees with coarsely serrate leaves and large, cream-colored fragrant semierect spikes bearing both male and female flowers, and several nuts enclosed in a spiny involucre. *Castanea ozarkensis* Ashe, a shrubby species, approaches our border in sw MO and ne OK. *Fagus*, the beech, is sometimes planted in se GP; it bears several 3-angled nuts in a 4-parted involucre.

Mature leaves and mature fruits are needed to identify some of our oaks. Leaf shapes are inconstant even on a single plant, and care should be taken to sample leaves from several branches. Leaves of sucker shoots, stump sprouts, and shoots in deep shade are sometimes very different from typical leaves and will not key here. The cup in immature fruits completely encloses the nut; the key is based upon mature fruits, when the nut has emerged from the cup. Hybrids that show characteristics intermediate between the parents can be found in our area. Hybrids form in each of the subgenera where the parents are sympatric. Not all possible hybrids have been reported from GP, but they can be expected where parental ranges overlap. Hybrids and varieties often occupy habitats different from those of the parent species.

1 Oaks of the GP, but excluding those of sw GP (i.e., excluding se CO, ne NM, TX & OK panhandles, w 1/3 of OK).
  2 Leaves serrate to shallowly or deeply lobed, the lobes rounded or truncate, the serrations rounded to acute and not distinctly bristle-tipped (but sometimes with a mucro); fruits maturing the first year on the current season's branches, the interior of the acorn shell glabrous (The white oaks).
    3 Leaves coarsely serrate to shallowly lobed.
      4 Leaves serrate, the teeth sharp to blunt; fruits sessile or nearly so; trees and shrubs of uplands.
        5 Tree; leaves lanceolate to elliptic or obovate, to 20 cm long, the teeth acute .... 11. *Q. muehlenbergii*
        5 Coarse shrub to small tree; leaves obovate and mostly less than 10 cm long, the teeth acute to obtuse or blunt .... 13. *Q. prinoides*
      4 Leaves with rounded teeth, obovate; fruits long peduncled; tree of lowlands and swamps .... 2. *Q. bicolor*
    3 Leaves distinctly lobed, some or all sinuses deep.
      6 Cup distinctly fringed; leaves often with a large rounded terminal lobe, or panduriform, lyrate-pinnatifid, cruciform, or digitately lobed, these often mixed on the same tree; median lateral sinuses often the deepest; over much of GP .... 8. *Q. macrocarpa*
      6 Cup not distinctly fringed; leaf lobes fingerlike, rounded, or the leaves rather cruciform and the distal lobes much larger and expanded and truncate; se GP.
        7 Leaves more or less cruciform; distal lateral lobes much larger and somewhat constricted at the base, their tips retuse or truncate (rounded in var. *margaretta*) .... 15. *Q. stellata*
        7 Leaves not at all cruciform; lobes usually fingerlike, those above not much wider than those below; median lateral lobes often the longest .... 1. *Q. alba*
  2 Leaves entire to deeply lobed, the lobes mucronulate or bristle-tipped, or at least acuminate; fruits maturing in autumn of the second year, the trees then often with older and younger fruits on the previous and current seasons' twigs respectively; interior of the acorn shell tomentose (The black oaks).
    8 Leaves entire.
      9 Leaves more or less oblong .... 7. *Q. imbricaria*
      9 Leaves triangular-obovate, obdeltate, or clavate .... 9. *Q. marilandica*
    8 Leaves pinnately lobed or with 2-3 broad terminal lobes separated by shallow sinuses.
      10 Leaves with 3(2-5) broad terminal mucronulate lobes (occasionally with a few small lobes below), the sinuses shallow; blade distinctly widened distally, clavate to obdeltate in outline .... 9. *Q. marilandica*
      10 Leaves pinnately lobed, the teeth bristle-tipped, the sinuses mostly extending halfway or more to the midrib.
        11 Sinuses mostly halfway to the midrib; lobes broadest at the base and usually tapering distally.
          12 Cup shallow, saucer-shaped, covering ca 1/5 of the nut, to 3 cm wide; se GP .... 3b. *Q. borealis* var. *maxima*
          12 Cup deeper, cup-shaped, covering ca 1/3 of the nut, to 2 cm wide; ne GP .... 3a. *Q. borealis* var. *borealis*

11 Sinuses mostly reaching more than halfway to the midrib; lobes often narrowest at the base and distally widened.
  13 Cup shallow, saucer-shaped, covering 1/3 or less of the nut; mature buds glabrous; mesic lowlands.
    14 Mature buds reddish, 1–5 mm long; nut subglobose, reddish, to 14 mm long; lower branches often notably declined .... 12. *Q. palustris*
    14 Mature buds grayish, 3–7 mm long; nut ovoid, brown, 20–30 mm long; lower branches not declined .... 14b. *Q. shumardii* var. *shumardii*
  13 Cup deeper, cup-shaped to turbinate, covering 1/3 or more of the nut; mature buds glabrous or tomentose; dry uplands.
    15 Mature buds tomentose, angular; upper scales of cup loose, almost fringed; inner bark yellow-orange; leaves rather leathery, tomentose beneath .... 17. *Q. velutina*
    15 Mature buds essentially glabrous, ovoid; cup scales tight; inner bark reddish-gray or yellow; leaves essentially glabrous beneath.
      16 Cup less than 15 mm wide; leaves often 5-lobed; inner bark yellow; dry, sandy soils, ne GP .... 4. *Q. ellipsoidalis*
      16 Cup 15–25 mm wide; leaves mostly 7-lobed; inner bark reddish-gray; dry, often rocky soils, se GP .... 14a. *Q. shumardii* var. *schneckii*
1 Oaks of the sw GP (se CO, ne NM, TX & OK panhandles, w 1/3 of OK).
  17 Leaves entire to deeply lobed, the lobes rounded, or if toothed the teeth not bristle-tipped (but often with a mucro); fruits maturing the first year on the current season's branches, the interior of the acorn shell glabrous (The white oaks).
    18 Leaves lobed, at least some of the sinuses extending more than halfway to the midrib.
      19 Cup distinctly fringed on the margin; leaves often with a large rounded terminal lobe or panduriform, lyrate-pinnatifid, or cruciform, sometimes merely shallow-lobed; w OK, but not OK panhandle .... 8. *Q. macrocarpa*
      19 Cup not fringed; leaves not as above, the lobes usually digitate, the median lateral lobes often the largest; NM & CO, OK panhandle .... 5. *Q. gambelii*
    18 Leaves entire, serrate, or shallowly lobed or toothed.
      20 Leaves not leathery, the margins serrate, not revolute.
        21 Tree; leaves lanceolate to elliptic or obovate, to 20 cm long, the teeth mostly acute .... 11. *Q. muehlenbergii*
        21 Coarse shrub or small tree; leaves mostly obovate and less than 10 cm long, the teeth acute to obtuse or blunt .... 13. *Q. prinoides*
      20 Leaves leathery, the margins entire to variously serrate or shallow-lobed, sometimes revolute.
        22 Leaves bluish-green and sparingly pubescent above; cut to 12 mm wide, covering ca 1/3 of the nut; se CO, ne NM, OK panhandle .... 16. *Q. undulata*
        22 Leaves green and glabrous above; cup covering more than 1/3 of the nut, to 25 mm wide.
          23 Leaves tomentose beneath; fruits subsessile to short-pedunculate.
            24 Leaves uniformly tomentose beneath, the veins beneath not especially prominent; cups to 1.8 cm wide; tree or shrub on limestone soils .... 10. *Q. mohriana*
            24 Leaves not uniformly tomentose beneath, the veins beneath rather prominent; cups mostly more than 1.5 cm wide; low shrub of sandy soils .... 6. *Q. havardii*
          23 Leaves canescent beneath, the hairs stellate, appressed; fruits long-pedunculate; sw OK .... 18. *Q. virginiana*
  17 Leaves entire to deeply lobed, the lobes and tips sharp and bristle-tipped; fruits maturing in autumn of the second year, the trees often then with older and younger fruits on the previous and current seasons' growth respectively, the interior of the acorn shell tomentose (The black oaks).
    25 Leaves entire to apically 3-lobed, the lobes separated by very shallow sinuses; leaves much wider near the tip than at the base (obovate-triangular, obdeltate, or clavate), seldom slightly pinnatifid; petioles mostly less than 1 cm long .... 9. *Q. marilandica*
    25 Leaves clearly pinnatifid, with numerous teeth; petioles often more than 1 cm long.

26 Buds 6–12 mm long, angular, tomentose or strigose; cup hemispheric, deep; inner bark yellow-orange .... 17. *Q. velutina*
26 Buds less than 5 mm long, sparingly pubescent or glabrous; cup shallow to turbinate; inner bark reddish-gray.
   27 Cup shallow, enclosing ca 1/4 of the nut; mesic woods .... 14b. *Q. shumardii* var. *shumardii*
   27 Cup deep, rounded to turbinate, enclosing up to 2/5 of the nut; dry uplands .... 14a. *Q. shumardii* var. *schneckii*

1. ***Quercus alba*** L., white oak. Large tree with ashy gray bark in flat blocky ridges. Leaves elliptic to obovate in outline, with 5–9 oblong fingerlike rounded lobes; terminal lobe resembling the laterals and barely or not at all expanded, the median lateral lobes usually the longest; sinuses mostly extending at least 1/2 way to the midrib and often much more; glabrous and dull green above, essentially glabrous and paler beneath. Fruits maturing the first year; cup gray, warty, pubescent within, to 22 mm wide, enclosing up to 1/3 of the nut; nut ellipsoid to ovoid, 1.5–3 cm long, 1.5–2 cm wide. (2n = 24) May. Heavy, well-drained noncalcareous soils; sw IA & extreme se NE, e KS, w MO, ne OK; (ME to se MN, NE s to e TX & n FL).

2. ***Quercus bicolor*** Willd., swamp white oak. Tree with dark brownish flat-ridged bark. Leaves to 18 cm long, obovate, unlobed, the teeth rounded, coarse, the tip acute to obtuse, the base tapering; dark green and moderately lustrous above, silver-white below. Fruits maturing the first year, solitary to mostly 2–4 on a peduncle 3–7 cm long; cup enclosing ca 1/4 of the nut, sometimes fringed at the rim; nut ellipsoid to subovoid, light brown, to 13 mm long. (2n = 24) Flood plains & wet meadows, swamps; sometimes planted as an ornamental in parks & along streets; w-cen MO (ME to MN, s to MO, TN, & NC).

Reports of this species from near Topeka, KS, are based on cultivated trees. This species is readily distinguished from our other native oaks by its long-peduncled acorns. A European species, *Q. robur* L., English oak, is sometimes cultivated as a shade tree in se GP and might key here. It too has long-peduncled acorns but can be distinguished by the smaller leaves that are auricled at the base and by the nut which is up to 3 cm long and is less than 1/4 enclosed by the unfringed cup.

3. ***Quercus borealis*** Michx. f., red oak. Tree with gray to brown furrowed (not blocky) bark. Buds reddish. Leaves ovate to obovate, pinnatifid; lobes 5–11, acute, bristle-tipped, somewhat tapering or at least not broadened toward the tip, each with a few large bristle-tipped teeth; sinuses rounded and barely reaching 1/2 way to the midrib; blade dark green and lustrous above, paler with a few hairs below. Fruit maturing in autumn of the second year; cup shallow and enclosing only ca 1/5 of the nut, to deeper and enclosing ca 1/3 of the nut, reddish; nut ovoid, reddish, to 2.5 cm long. Apr–May.

Two varieties occur in our area:

3a. var. *borealis*. Cup deep, to 2 cm wide, enclosing ca 1/3 of the nut. Upland woods & hillsides; nw MN; (Que. to nw MN, s to n IA & PA). Probably *Q. coccinea* Wang., misapplied—Rydberg.

3b. var. *maxima* (Marsh.) Ashe. Cup shallow, to 3 cm wide, enclosing ca 1/5 of the nut. Moist, rich, calcareous soils; MN, IA, e NE, e KS, MO, e OK; (N.B. to MN, s through e NE to e OK & GA). *Q. maxima* (Marsh.) Ashe—Rydberg; *Q. rubra* Du Roi—Little, Atlas U.S. Trees, Vol. 1. 1971.

4. ***Quercus ellipsoidalis*** E. J. Hill, Hill's oak, northern pin oak. Tree or shrub, the bark smooth at first, shallow-furrowed with large plates later, the inner bark yellow. Leaves thin, elliptic to obovate in outline, pinnatifid; lobes usually 5, bristle-tipped, each with a few coarse bristle-tipped lobes or teeth; sinuses rounded; bright green and somewhat lustrous above, paler and essentially glabrous below. Fruits maturing in fall of the second year; cup turbinate, barely pubescent, to 1.4 cm wide, enclosing 1/3–1/2 of the nut; nut subglobose to ellipsoid, small, to 2 cm long, glabrous or barely pubescent, light brown, often with darker stripes. May. Dry, acid, sandy soil, sometimes forming copses; nw MN, extreme se ND, nw & sw IA; (MI to MN, e ND, s to n MO & nw OH).

This species can best be distinguished from *Q. velutina* by its tight cup scales and thinner and less lustrous leaves; the ranges of the two species overlap only in sw IA in our area.

5. ***Quercus gambelii*** Nutt., Gambel's oak. Shrub or small tree with thick gray bark. Leaves very variable, to 10(15) cm long, ovate, obovate, oblong, or elliptic in outline, shallowly to usually deeply 5- to 9-lobed, the middle lateral lobes usually the largest; lobes oblong, entire or with a few secondary lobes, rounded to subacute, or with 2 or 3 coarse teeth; sinuses rather narrow; blades green, glossy, essentially glabrous above, densely tomentose to nearly glabrous-glaucous beneath. Fruits maturing the first year; cup deep, to 15 mm wide, tomentulose, covering 1/3–1/2 of the nut; nut ovoid to ellipsoid, to 15 mm long. Mar–Apr. Valleys & canyon sides; NM & CO near the mts. at the edge of the plains, extending e along the CO-NM border to OK: Cimarron; (s-cen WY & n UT, s to s NV, AZ, NM, w TX).

This species shows evidence of past sympatry and introgression with *Q. macrocarpa* in ne NM, BH & ne WY, but is not sympatric there now.

6. ***Quercus havardii*** Rydb., shinnery oak. Low rhizomatous shrub to 1 m tall. Leaves leathery, variable on the plant, to 10 cm long and 3.5 cm wide but usually less; blade oblong to elliptic, or lanceolate to oblanceolate, or ovate to obovate, acute to rounded, entire, undulate, toothed, or shallowly lobed, sometimes revolute, shiny above, tomentose beneath, the veins beneath prominent, visible through the tomentum. Fruits maturing the first year, variable in size and shape; cup to 2.5 cm wide, enclosing 1/3–1/2 of the nut; nut large, to 2.5 cm long, variable, often broadly ovoid and truncate. Apr. Sandy plains & sand dunes; w OK, & OK & TX panhandles, ne NM; (w OK to ne AZ & se UT, s to se NM & TX panhandle).

7. ***Quercus imbricaria*** Michx., shingle oak. Tree, the bark gray-brown, shallowly furrowed, the ridges long. Leaves elliptic to lanceolate, entire, acute, the base rounded to acute, to 15 cm long and 5 cm wide, lustrous above, paler and sparsely tomentose beneath. Fruits maturing in the fall of the second year; cup somewhat turbinate, to 1.5 cm wide, reddish, enclosing 1/3–1/2 of the nut; nut subglobose. ($2n = 24$) Apr–May. Upland forests, occasionally in bottomlands; sw IA, MO, e-cen KS, & along the Arkansas R. in s-cen KS; (PA to sw IA, s to e KS, nw AR, & scattered locations from LA to DE).

8. ***Quercus macrocarpa*** Michx., bur oak. Coarse shrub to large tree, the bark dark, deeply furrowed. Leaves rather variable on the plant, sometimes very large (to 25 cm long), mostly obovate in outline, usually deeply lobed, often panduriform, lyrate-pinnatifid, cruciform, or digitately-lobed, occasionally merely shallow-lobed to subentire, especially on stump sprouts or branches from deep shade; terminal lobe, when present, often much the largest and rounded; median lateral sinuses often deepest, sometimes approaching the midrib; lustrous dark green above, silvery stellate-pubescent below. Fruits maturing the first year; cup deep, the margin and often the sides distinctly fringed, to 5 cm wide, enclosing at least 1/2 and usually 2/3 or more of the nut; nut ovoid, to 4 cm long and 4 cm broad. ($2n = 24$) Apr–May. Upland woods & valley floors, often forming savannas; MN, ND, extreme se MT & ne WY, SD, NE, e & cen KS, e & cen OK; (N.B. to se Sask., s to se MT & NE, then e of 100° long. to e TX, AR, & MD).

This species shows evidence of past introgression with *Q. gambelii* in ne NM, BH, & ne WY.

This is our most wide-ranging oak, locally abundant along our rivers (mostly on the upland slopes in our e, but often in bottom lands in our w), draws, and escarpments farther west than our other eastern oaks. It is much less robust in the west of our area, where it is often a very small tree or shrub.

These have smaller and less fringed cups and notably corky twigs, and have been called *Q. mandanensis* Rydb.—Rydberg, and *Q. macrocarpa*. var. *depressa* (Nutt.) Engelm. A northern form with thinner and smaller (to 2 cm wide) cups, the fringe sparse, is *Q. macrocarpa* var. *olivaeformis* (Michx. f.) A. Gray (IA, MN, ND, SD).

9. ***Quercus marilandica*** Muench., blackjack oak. Small tree or coarse shrub, the bark thick, furrowed, inner bark whitish or pinkish. Leaves very short petioled, thick, stiff, broadly obovate-clavate to obdeltate in outline, usually only shallowly 3-lobed at the tip; lobes very broad, bristle-tipped, occasionally a few narrow lobes below, these not exceeding the upper lobes, or sometimes the blade entire, unlobed; blade dark yellow-green and lustrous above, brown pubescent below. Fruit ripening in autumn of the second year; cup deep, to 18 mm wide, reddish, sericeous, the scales relatively few, large, enclosing 1/3-1/2 of the nut; nut ovoid, yellowish-brown, to 17 mm long. (2n = 24) Apr. Dry, sterile clay & sometimes sandy soils; nw MO & extreme se NE, s through e KS to w OK; (NJ to se NE, s to w OK, cen TX, & FL).

Specimens from the sw part of our range have smaller leaves.

10. ***Quercus mohriana*** Buckl., shin oak. Shrub or small tree, often thicket-forming. Leaves evergreen or deciduous, coriaceous, to 8 cm long, ovate to elliptic, the margins sometimes revolute or crisped, entire or with a few coarse teeth near the tip; dark green and essentially glabrous above, gray-white tomentose beneath, the veins beneath hidden, not prominent. Fruits maturing the first year; cup turbinate or shallowly to deeply cup-shaped, to 18 mm wide, tomentose toward the base, enclosing 1/2-2/3 of the nut; nut ovoid to oblong, rounded, 5-15 mm long. Apr. Dry limestone hills; w OK, OK & TX panhandles, ne NM; (w OK to ne NM, s to w & cen TX & Mex.).

11. ***Quercus muehlenbergii*** Engelm., yellow chestnut oak, chinkapin oak. Tree with flaky gray bark. Leaves to 20 cm long, lanceolate to elliptic or obovate, acute to acuminate, the margin regularly and coarsely serrate, the teeth (6)8-13 on each side, pointed; veins entering the teeth from the midrib unusually straight and parallel; lustrous yellow-green to dark green above, white pubescent below when young, becoming glabrous to glaucous below with age. Fruits ripening the first of year, sessile or nearly so; cup enclosing 1/3-1/2 of the nut, to 1.5 cm wide, the scales rather thick, tomentose; nut ovoid, light brown, to 22 mm long. (2n = 24) Apr-May. Calcareous bluffs & upland woods, often near the forest edge, or sometimes in valleys & then usually near limestone; sw IA, se NE, e KS, e, cen & w OK, MO; (VT to se MN, se NE, w OK, e NM, Mex., e to FL & the Appalachians). *Q. prinoides* var. *acuminata* (Michx.) Gl.—Steyermark.

Individuals intermediate with *Q. prinoides* can be found, and the two species are sometimes considered to be varieties of a single species. When grown together in GP arboretums for many years, each retains its distinctiveness.

12. ***Quercus palustris*** Muench., pin oak. Tree, the bark smooth or shallow-furrowed, the ridges flat-tipped; buds reddish, 1-5 mm long. Trees in the open or at the forest edge usually conical until quite old, their lower branches declined, their upper branches inclined. Leaves often rather small, more or less ovate in outline, pinnatifid; lobes 5-7, slender, tapering or the sides parallel, sometimes expanding toward the tip, acute, bristle-tipped, each with 0-3 coarse bristle-tipped teeth; sinuses deep (often approaching the midrib), wide, rounded to squarish. Fruits maturing in the fall of the second year; cup shallow, to 2 cm wide, ca 1/4 of the nut; nut subglobose to 14 mm long, reddish. (2n = 24) Apr-May. Bottomland forests. MO & e KS, s to e OK; (MA to se IA & e KS, s to e OK, n AR, & NC).

This species is commonly planted in GP well beyond its natural range. It is prone to chlorosis on calcareous soils, the leaves then very pale green to yellow, the trees stunted in extreme cases.

13. **Quercus prinoides** Willd., dwarf chinkapin oak. Coarse shrub or small tree. Leaves mostly less than 10 cm long, obovate, acute, serrate to almost undulate, the teeth mostly 3–8 on each side and often rather blunt; yellow-green to dark green and rather dull above, pale greenish pubescent below at least when young. Fruits ripening the first year, nearly sessile; cup enclosing ca 1/3 of the nut, to 1.4 cm wide, the scales appressed, floccose; nut ovoid, light brown, to ca 20 mm long. Apr–May. Dry exposed soils in woods, on bluffs, & along roadsides & forest margins; sw IA, se NE, e KS, e, cen, & w OK, w 1/2 MO; (NY to se MN, se NE & cen TX, e to n FL & the Appalachians).

Specimens intermediate with *Q. muehlenbergii* are sometimes found, and the two species are sometimes considered to be varieties of a single species.

14. **Quercus shumardii** Buckl. Tree with gray bark, the furrows broken, the inner bark reddish-gray; buds gray, glabrous when mature. Leaves pinnatifid, the lobes mostly 7, bristle-tipped, the sinuses extending 1/2 way or more to the midrib, dark green and lustrous above, paler and essentially glabrous beneath. Fruits maturing in fall of the second year, the cup shallow or deeper to turbinate, enclosing ca 1/4 or up to 2/5 of the nut; nut ovoid, grayish-brown, 2–3 cm long. Apr.

Two edaphically separated varieties occur in our area:

14a. var. *schneckii* (Britt.) Sarg., Schneck oak. Cup deep to turbinate, to 2 cm wide and enclosing up to 2/5 of the nut, the nut 2 cm long. Dry rocky uplands & bluffs; se KS, w-cen to sw MO, e, cen & sw OK; (OH to KS, s to OK & TN).
This var. may intergrade with *Q. texana* Buckl. in sw OK.

14b. var. *shumardii*, Shumard's oak. Cup shallow, enclosing ca 1/4 of the nut, the nut 2–3 cm long. Bottomland forests, occasionally in moist upland woods; extreme se NE, e KS, MO, w, cen, & e OK; (PA to se NE, s to cen TX & FL).

15. **Quercus stellata** Wang. post-oak. Tree or shrub with thick gray rough blocky bark, the twigs pubescent. Leaves broadly obovate in outline, rather thick, variably lyrate-pinnatifid, or cruciform, usually the 3 main lobes the largest and the lateral 2 of these truncate and somewhat constricted at the base, each sometimes with a few large teeth; the apical and lower lobes rounded to subacute; blade glossy and sparsely pubescent (becoming glabrate) above, golden brown tomentose below, becoming less so with age. Fruit maturing the first year; cup hemispheric to pyriform, covering ca 1/2 of the nut, to 1.5 cm wide; nut ovoid, dome-shaped, 1–1.5 cm long. Apr. Dry upland sandy clays & gravel; MO, e KS, OK (except panhandle); (MA to se IA & e KS, s to cen TX & FL).

The sand post-oak, *Q. stellata* var. *margaretta* (Ashe) Sarg. (*Q. margaretta* Ashe), a shrub of low woodlands on sandy soils in OK and TX, has the cruciform leaves rather small, ovate to narrowly obovate in outline, the lobes rounded and not constricted at the base.

16. **Quercus undulata** Torr., wavy-leaf oak. Shrub or small tree with rough gray bark. Leaves deciduous or evergreen, leathery, usually crowded near the stem tips, rather small, 2–6 cm long, 1–3 cm wide, elliptic to oblong, the tip usually acute to obtuse; margin coarsely serrate and the teeth pointed, to undulate or shallowly lobed, the lobes rounded or occasionally entire; leaves dark bluish-green and sparingly pubescent above, densely pubescent and dull green below, especially when young, the veins beneath prominent. Fruits maturing the first year; cup to 12 mm wide, tomentose, enclosing ca 1/3 of the nut, the scale tips often reddish; nut ovoid to oblong, to 1(1.5) cm long. Apr. Sandstone soils & dry rocky slopes; se CO: Baca; ne NM; OK: Cimarron; (OK panhandle w to se CO, UT, AZ, NM, Mex.).

17. ***Quercus velutina*** Lam., black oak. Tree, the bark dark, deeply furrowed, blocky, the inner bark yellow-orange; mature buds tomentose, or strigose, angular, 6–12 mm long. Leaves usually leathery, elliptic to obovate in outline, pinnatifid, with 5–7 bristle-tipped lobes; largest lobes widened toward the tip, each with a few coarse bristle-tipped teeth; sinuses shallow to rather deep, rounded; dark green and lustrous above, paler and more or less brown tomentose beneath. Fruits ripening in the fall of the second year; cup deep, to 2.5 cm wide, enclosing 1/3–3/4 of the nut, reddish, tomentose, the upper scales rather loose and almost short-fringed; nut ovoid to depressed-globose, to 2 cm long. Apr–May. Rocky hillsides & dry woods on noncalcareous soils; sw IA, se NE, e KS, MO, e & sw OK; (ME to se MN & se NE, s to e TX & n FL).

This species is distinguished by its yellow-orange inner bark and loose, almost fringed, upper cup scales.

18. ***Quercus virginiana*** P. Mill. var. ***fusiformis*** (Small) Sarg., live oak. Shrub with evergreen, leathery leaves. Blades linear-oblong, elliptic or oblanceolate, acute to obtuse or rounded, entire or with a few remote shallow teeth, to 6 cm long and 2.5 cm wide, bright green and glabrous above, paler and minutely pubescent beneath, the hairs appressed-stellate. Fruits maturing the first year, long-pedunculate; cup turbinate to 2 cm wide, about 1/2 enclosing the nut; nut oblong-ovoid, to 2.5 cm long. Mar–Apr. Forming thickets in rocky woods & on brushy hillsides; OK: Comanche, Greer, Kiowa; (sw OK, cen TX; Mex.).

33. BETULACEAE S. F. Gray, the Birch Family

by Robert B. Kaul

Monoecious trees and shrubs. Leaves alternate, petioled, simple, serrate. Flowers borne in unisexual catkins, tiny, crowded, subtended by bracteoles, the perinath minute or absent; stamens 2–many, the filaments short, free or connate, the anther sacs connate or separate; female flowers grouped in 2s or 3s on the catkins, the catkin sometimes reduced to 1–few flowers; pistil 1, inferior or apparently superior, ovules 1–4, styles 2 or apparently so. Fruits nuts, nutlets, or winged samaras, 1-seeded, indehiscent, sometimes enveloped by the expanded involucre.

There are four genera in our area; a fifth entity, *Carpinus caroliniana* Walt., the blue beech, is reported from the Kansas City area. It is a small tree with smooth gray bark, the winged nutlets attached in loose leafy-bracteate drooping spikes, the bracts flaring, lobed. *Carpinus betulus* L., European hornbeam, is sometimes planted as an ornamental in our area, especially its fastigiate form.

1 Fruits nuts 1–1.5 cm wide, partly or completely enclosed by a conspicuous involucre, solitary or a few in a cluster; shrubs .... 3. *Corylus*
1 Fruits nutlets or winged samaras, borne in catkins, conelike catkins, or papery cones; trees or shrubs.
  2 Fruits nutlets borne in papery cones to 5 cm long .... 4. *Ostrya*
  2 Fruits tiny samaras borne in catkins or small conelike catkins.
    3 Bracteoles of fruiting catkins 3-lobed, papery and finally deciduous in fruit; bark sometimes exfoliating .... 2. *Betula*
    3 Bracteoles of fruiting catkins not lobed, becoming indurate and persistent in the small conelike catkin; bark not exfoliating .... 1. *Alnus*

1. ALNUS Ehrh., Alder

Coarse shrubs and small trees in wet places, the bark smooth. Leaves broad, ovate to obovate, serrate to dentate. Male catkins slender, pendulous, the flowers in 3s, each group subtended by a bracteole, each flower of 4 stamens attached to 4 sepals; female catkins pendulous to nearly erect, stout, much shorter than the males, the flowers paired, the pistils apparently naked but subtended by several bracteoles, these becoming woody and the catkin conelike in fruit. Fruit a thin-winged nutlet.

1 Blades doubly-serrate, the teeth unequal; e ND, w MN .... 1. *A. incana* subsp. *rugosa*
1 Blades once-serrate, the teeth fine; se KS, sw MO .... 2. *A. serrulata*

1. *Alnus incana* (L.) Moench subsp. ***rugosa*** (Du Roi) Clausen, speckled alder. Coarse shrub to small tree, the twigs with white lenticels. Blades elliptic to ovate, usually widest below the middle, to 10 cm long, acuminate to obtuse, the base obtuse to rounded, doubly serrate, the teeth unequal, paler green to glaucous below, the veins below hairy. Fruits 2–3.5 mm wide, narrow winged. (2n = 28) (Apr)–Jun. Wet woods; e ND, nw MN; (Newf. to B.C., s to ND, MN, IN, & MD). *A. rugosa* (Du Roi) Spreng., *A. tenuifolia* Nutt., *A. incana* (L.) Moench–Rydberg.

2. *Alnus serrulata* (Ait.) Willd., smooth alder. Shrub or small tree. Blades elliptic to obovate, usually widest above the middle, obtuse to rounded, the base obtuse or cuneate, finely serrate, the serrations equal, green and glabrous below. Fruit narrow winged. Mar–Apr. Wet woods, rare with us; KS: Cherokee; MO: Vernon; ne OK just beyond our range; (ME to se KS & e OK, s to e TX & n FL).

2. BETULA L., Birch

Trees and shrubs of mesic woods, flood plains, and bogs. Bark smooth or shaggy, sometimes resinous. Immature male catkins usually present in winter; male catkins pendulous in spring anthesis, the flowers in 3s, each triad subtended by a bracteole; calyx minute; stamens 2, the anther sacs separate. Female catkins emerging in spring, ovoid or cylindric, more or less erect in anthesis and often drooping in fruit, the flowers in 3s, each triad subtended by a trifid bracteole; perianth absent, the flower consisting of a single apparently superior pistil; fruiting catkins shedding their bracteoles and fruits in the current or following growing season; samara 2-winged, the styles usually persistent.

1 Trees with white, tan, or reddish bark, the bark often peeling.
  2 Bark white; blade serrate for most of its length; fruits ripe in autumn and often persistent in winter; nutlet much narrower than the wings; n GP .... 4. *B. papyrifera*
  2 Bark tan to reddish-salmon; blade usually entire toward the base and distally serrate; fruits ripe and dehiscing in spring of the current year, not persistent in winter; nutlet wider than the wings; se GP .... 2. *B. nigra*
1 Shrubs or small gregarious trees with brown or bronzed bark, the bark not shaggy or peeling.
  3 Blade orbicular to obovate, rounded, leathery; lateral lobes of fruiting bracteoles somewhat ascending; nutlet wider than the wings; bark dull; ne GP & BH .... 1. *B. glandulosa* var. *glandulifera*
  3 Blade ovate, acute to acuminate, thin; lateral lobes of fruiting bracteoles divergent; nutlet narrower than the wings; bark lustrous; nw GP & BH .... 3. *B. occidentalis*

1. *Betula glandulosa* Michx. var. ***glandulifera*** (Regel) Gl., bog birch. Erect or prostrate shrub to 2(3) m tall, the branches ascending, the twigs resinous-glandular. Bark brownish, dull, not peeling. Blade rather leathery, orbicular to obovate, rounded, the base rounded,

2–4 cm long, to 2 cm wide, crenate to serrate, often resinous, glandular dotted below, dark green and lustrous above. Staminate catkins in winter less than 1 cm long. Bracteoles of fruiting catkins with divergent lateral lobes; nutlet wider than the wings. (2n = 28, 56) May—Jun(Aug), fruits Aug. Bogs & stream sides, often with grasses & sedges, sometimes in shallow water; e ND, BH, nw MN; (NY to Yukon, s to OR, CO, MN, IN). *B. glandulifera* (Regel) Butler—Rydberg.

2. ***Betula nigra*** L., river birch. Tree with drooping branches, sometimes in clumps. Bark of the leaf-bearing trunk and major branches reddish, tan, and salmon, prominently peeling in shaggy curling layers; bark of older trunks grayish, scaly, fissured. Blade ovate to almost deltoid or rhomboid, 4–8 cm long, acute, doubly serrate at least above the middle, usually entire below the middle. Fruiting catkins peduncled, to 4 cm long, pubescent; nutlet as wide as or wider than the wings. (2n = 28, 56) Apr, fruits ripening and falling May-Jun of the same year and not present in winter. Flood plains & swamps; sw IA, sw MO, extreme se KS, ne OK; often planted in parks & yards well beyond its range; (NH to se MN & se KS, s to e TX, e to FL).

Betula lutea Michx., yellow birch, resembles the river birch but has yellowish shaggy bark, the inner bark wintergreen-flavored, the fruiting catkins sessile or nearly so; it reaches our ne border in nw and w-cen MN and can be expected there just within our area.

3. ***Betula occidentalis*** Hook., mountain birch, water birch. Large shrub or small tree, often gregarious. Bark smooth, bronzy, lustrous, not peeling, the lenticels horizontal, large, pale; twigs resinous-glandular. Blade thin, firm, ovate, acute (rounded), the base truncate, rounded, or cuneate, mostly 3–4.5 cm long, usually sharply double-serrate. Staminate catkins in winter 1–2 cm long. Lateral lobes of fruiting bracteoles rather acute, more or less ascending; nutlet narrower than the wings. (2n = 28, 88) May–Jun; fruits Aug. Ravines, stream sides, bogs; w ND, MT, WY, w SD, extreme nw NE, CO at the edge of our area; (Man. to AK, s to CA, AZ, CO & nw NE). *B. fontinalis* Sarg.—Rydberg.

4. ***Betula papyrifera*** Marsh., paper birch, canoe birch. Tree with drooping branches, sometimes in clumps. Bark white (grayish), smooth or peeling in strips, the lenticels horizontal, large; bark of young trees and twigs reddish. Blade elliptic to ovate, acute to acuminate, the base more or less rounded (cordate), 4–7(10) cm long, serrate or double-serrate for most of its length. Fruiting catkins to 5 cm long, samaras very broad winged, retuse, the nutlet much narrower than the wings. (2n = 56–84) Apr–Jun; fruits ripe in autumn and often persisting in winter. Forest slopes, cool canyon walls, shady or open bottomlands; nw MN, e & w ND & SD, se MT, ne WY, & in the Niobrara R. valley of n NE; often planted in parks & yards over much of GP; (Lab. to AK, s to OR, MT, CO, n NE, & NJ).

Specimens with cordate leaves and spreading teeth are var. *cordifolia* Fern. (n ND). Hybrids with *B. glandulosa* are reported from ne GP. Our species is sometimes treated as a variety of the European *B. alba* L., which is similar but with smaller leaves and catkins. Weeping birch, *B. pendula* Roth, from Europe, is commonly planted in GP; it has fine pendulous twigs and, in some forms, deeply lacerate leaves; the bark is creamy-white. The Japanese birch, *B. platyphylla* Suk., has the leaves lustrous, long-acuminate, with the base truncate, and the bark is white. It is increasingly cultivated in our area, and might key here.

3. CORYLUS L., Hazelnut

Monoecious shrubs to ca 3 m tall, flowering in spring before the leaves appear. Male catkins present in winter, cylindric, much larger than the females, the numerous crowded flowers borne singly in the bracteoles, stamens 4, the anthers divided at the top; female catkins few-flowered, ovoid, the flowers borne singly in the bracteoles. Nuts solitary or clustered, each more or less enveloped by the conspicuously expanded foliaceous involucre.

1 Young twigs and petioles usually glandular-pubescent; nuts subtended by a campanulate to somewhat flaring laciniate involucre; male catkins peduncled .... 1. *C. americana*
1 Young twigs and petioles glabrous to sparingly villous; nuts enveloped by a bristly, beaked involucre; male catkins sessile or subsessile .... 2. *C. cornuta*

There are two species in our area; a third, *C. avellana* L., from Europe, is occasionally cultivated as an ornamental, especially its contorted-twig cultivar, "Harry Lauder's Walking Stick"; the species is a large shrub, the nuts barely enclosed by the deeply dissected involucre.

1. **Corylus americana** Walt., hazelnut. Young twigs and petioles glandular-pubescent (nongladular). Blade ovate, abruptly acuminate, rounded to cordate at the base, 1–12 cm long, doubly serrate, pubescent below. Male catkins pedunculate. Nuts solitary or clustered, each partially enveloped by a laciniate pubescent involucre, the 2 bracts essentially free and campanulate at first, opening with age, or united on one side and only partly open. Nut brown, oblate, 1–1.5 cm in diam. (2n = 22, 28) Mar–May, fruits Sep. Upland forests & thickets; ND, MN, e SD, e NE, IA, e KS, ne OK; (ME to Sask. s to OK & GA).

Two subspecific taxa are found in sw MO and se KS: var. *indehiscens* Palm. & Steyerm., with the bracts united on one side and not opening widely in fruit; f. *missouriensis* (A. DC.) Fern., with the young twigs and petioles pubescent, but not glandular-pubescent.

2. **Corylus cornuta** Marsh., beaked hazelnut. Young twigs and petioles glabrous to sparsely villous, but not conspicuously glandular. Blades ovate to obovate, acuminate, the base rounded to almost cordate, 4–10 cm long, coarsely doubly-serrate, variously pubescent below. Male catkins sessile or subsessile. Nuts each enveloped by a bristly involucre, the bracts partially connate and forming a long beak, nuts 1–1.5 cm in diam. (2n = 22) Apr–May, fruits Sep. Upland forests & thickets; ND, MN, ne SD, BH; (Newf. to B.C., s to OR, SD, & GA). *C. rostrata* Ait.—Rydberg.

4. OSTRYA Scop., Ironwood, Hop-hornbeam

1. **Ostrya virginiana** (P. Mill.) K. Koch. Small understory trees, the older bark scaly, rough or shaggy. Petioles short, pubescent. Blades oblong to ovate, acuminate, the base tapering or rounded, 7–12 cm long, pubescent below, often doubly-serrate, the teeth fine, sharp. Male catkins clustered, pendulous in anthesis, dense, the pointed bracteoles subtending a group of apically divided stamens; female catkins slender, ca 1 cm long, loose, the bracteoles each subtending 2 flowers in a bract/pouch, this inflated in fruit and collectively forming a papery cone to 5 cm long. Nutlets ovoid, flattened, 5–8 mm long. (2n = 16) Apr–May; fruits Jun–Jul. Upland forests, often abundant on slopes; e ND, MN s through e NE & e KS to ne OK, & w across n NE & s SD to BH; (N.S. to Man., s to e TX & FL).

The immature male catkins are visible among the persistent dry leaves in winter. Specimens with the twigs densely villous are var. *lasia* Fern.

34. PHYTOLACCACEAE, the Pokeweed Family

by William T. Barker

1. **Phytolacca americana** L., pokeweed, pokeberry. Glabrous perennial herb to 3 m tall, odoriferous; stems solitary or several, branched above, purplish, to 3 cm in diam. Leaves

alternate, simple, blade lance-oblong to ovate, 10–30 cm long, 4–10 cm wide, apex acute to acuminate, margins entire, base attenuate; petioles 1–5 cm long. Flowers borne in racemes 1–2 dm long, bisexual or unisexual, regular, hypogynous; sepals 5, greenish-white to pink and becoming darker in fruit, ovate, 2–3 mm long; petals absent; stamens 10; stigmas and styles usually 10, forming a compound pistil; pedicels stout, 5–10 mm long. Fruit a dark purple, juicy berry, subglobose, 6–9 mm in diam; seeds black, shiny, oval, slightly flattened, 2.5–3.5 mm long, smooth. (2n = 36) Jun–Oct. Rich soil in waste places, farms, roadsides, fields, low wooded areas; IA, se NE, KS, MO, OK, & TX; (s Ont. & s Que. s to FL & TX). *P. decandra* L.—Rydberg. POISONOUS

All parts of pokeberry are poisonous; however, the *young* shoots are often prepared as a cooked vegetable after several boilings.

35. NYCTAGINACEAE Juss., the Four-O'Clock Family

by Ronald L. McGregor and Ralph E. Brooks

Ours annual or perennial herbs, one species a low shrubs; stems erect, procumbent or decumbent, often dichotomously or trichotomously branched, glabrous or pubescent, often swollen at the nodes. Leaves opposite, simple, sessile or petiolate; exstipulate. Inflorescences terminal or lateral, usually in paniculate or corymbose cymes, heads racemose or solitary, bracteate to involucrate; involucre enclosing 1–many flowers, often accrescent in age. Flowers perfect, hypogynous; sepals 5, united, constricted above the ovary, limb ephemeral, tube long or short, usually indurate at base; stamens 1–7, included or exserted, exserted filaments often united at base, often unequal; anthers opening by lateral slits; ovary 1-celled; ovule 1, basal. Fruit (anthocarp) accessory, the accrescent calyx base usually hard and leathery, enclosing the achene.

Reference: Bogle, A. Linn. 1974. The genera of Nyctaginaceae in the Southeastern United States. J. Arnold Arbor. 55: 1–37; Standley, Paul C. 1918. Allioniaceae. N. Amer. Fl. 21: 171–254.

1 Involucre of distinct bracts or each flower subtended by 1–3 bracts.
  2 Fruits without wings; inflorescence cymose and much branched .... 3. *Boerhavia*
  2 Fruits winged; flowers axillary or in heads.
    3 Fruits with obvious thin scarious wings.
      4 Flowers axillary, sessile or short pedicellate, often all
        cleistogamous .... 5. *Selinocarpus*
      4 Flowers in heads; peduncles 1–2.5 cm long, shorter than leaves, none
        cleistogamous .... 6. *Tripterocalyx*
    3 Fruits with thick, opaque wings which are often interrupted; peduncles usually much
      exceeding leaves; flowers in heads .... 1. *Abronia*
1 Bracts of involucre united at least at base.
  5 Fruit compressed, the often dentate margins strongly inflexed over outer (seemingly inner) face; outer face with 2 rows of stipitate glands .... 2. *Allionia*
  5 Fruit not compressed, usually angled, rarely terete, without inflexed
    margins .... 4. *Mirabilis*

1. ABRONIA Juss., Sand Verbena

1. *Abronia fragans* Nutt. ex Hook., sweet sand verbena. Perennial herbs with deep elongate, woody taproot, sometimes with subterranean branching caudex, some branches lateral and appearing as creeping; stems few to numerous, 3–7(10) dm tall or long, from nearly erect to usually decumbent or procumbent, branched below and above, sometimes whitish, finely

glandular-puberulent and usually also more or less hirsute, rarely glabrate. Leaves opposite, sparsely viscid-pubescent to glabrous, paler on lower surface, blades variable in shape, ovate to triangular or lanceolate, mostly ovate-oblong, narrowed rather abruptly to petiole or base truncate to rounded or rarely subcordate, (1)2–9(11) cm long, apex rounded to acute; petioles 1–4(7) cm long, viscid-puberulent to rarely glabrate. Flowers perfect, fragrant, borne in pedunculate heads, several to many flowered, peduncles axillary, usually considerably exceeding leaves; involucral bracts distinct, usually 5, whitish, membranaceous, sometimes pinkish or purplish, ovate to broadly lanceolate, (8)10–20 mm long, (2)4–12 mm wide; calyx funnelform to salverform, the tube 16–20 mm long, above the ovary, puberulent to pubescent, limb 5-lobate, 7–10 mm wide, white or pinkish tinged, withering and often persistent after anthesis; corolla lacking; stamens (3)5, not exserted; ovary ovoid, style filiform, stigmas fusiform, included. Fruit accessory, the accrescent calyx base hard and leathery, enclosing the achene, (5)8–10 mm long, pubescent, often coarsely reticulate, tapered at both ends or apically truncate, narrowly 3- to 5-winged; achenes obovate, 2.5–3 mm long, lustrous, black or dark brown. Mar–Aug. Sandy prairies, dunes, stream valleys, roadsides, waste places; w ND, e MT, sw SD, WY, w NE, CO, w KS, s to TX, NM; (ND, ID, s to TX, NM, AZ; n Mex.).

2. ALLIONIA L., Trailing-four-o'clock

1. *Allionia incarnata* L., trailing-four-o'clock. Perennial herbs with woody taproot and short branching caudex. Stems prostrate, sprawling or decumbent, rarely somewhat erect, 2–10 dm long, rather densely viscid-villous or glandular-puberulent, rarely glabrate, often reddish. Leaves opposite, often unequal; blades broadly deltoid-orbicular to oval, ovate, or oblong, 1–4(6) cm long, usually rounded or subcordate at the oblique base, green above, paler below, apex rounded to acute, margins entire or sinuate, surfaces viscid-villous or glandular-puberulent, becoming glabrate; petioles 0.5–10 mm long, viscid-villous. Peduncles axillary, short or to 3–5 cm long; lobes of involucre 4–6(9) mm long, obovate-orbicular, villous-viscid; calyx 5–10(15)mm long, short-funnelform, 4- to 5-lobed, purplish-red or pink, rarely white; corolla absent; stamens 4–7; stigmas capitate. Fruit flattened, 3–4.5 mm long, inner (apparent outer) side 3-nerved, the incurved margins either dentate or entire, strongly inflexed over the outer face, the latter with 2 distinct rows of stipitate glands, achenes yellowish, embryo curved. Apr–Sep. Dry prairie hills, flats, & valleys, roadsides, waste places, on sandy, gravelly or rarely gypsiferous soils; OK: Cimarron; TX panhandle, NM; (OK, s CO, TX, NM, AZ, CA; Mex., S. Amer.).

3. BOERHAAVIA L., Spiderlings

1. *Boerhaavia erecta* L., spiderling. Annual herbs with taproots; stems erect or decumbent 2–10 dm tall or long, usually branching at base, branches spreading or decumbent, puberulent below to nearly glabrous above, middle internodes sometimes with viscid bands. Leaves opposite, blades ovate-rhombic, deltoid-ovate or ovate-lanceolate to linear above, 2–7 cm long, 1.5–5 cm wide, acute or obtuse to rounded and apiculate at apex, truncate to rounded or slightly cuneate at base, margins entire or undulate, green above, pale below and usually with small brown spots on both surfaces, glabrescent; petioles 4–40 mm long. Inflorescences axillary and terminal, usually much branched; flowers without an involucre, 2–6 in a cluster, pedicels 1–5 mm long, irregularly cymose or subracemose; bracts linear or lanceolate, reddish, 0.6–1.2 mm long. Calyx white or tinged with pink, 1–1.5 mm long, tube glabrous, limb campanulate; corolla absent; stamens 2 or 3, exserted; stigma peltate. Fruit green or yellowish-green, obpyramidal, truncate, glabrous, 3–4 mm long, 5-angled,

angles obtuse. Sandy or dry fields & waste places; OK: Cleveland; (a pantropical weed, north in N. Amer. to SC, MO, AR, cen OK, TX, NM, AZ).

Boerhaavia spicata Choisy was reported for our area in the Atlas GP but is just south of our area. It has racemose flowers and fruits rounded at apex.

4. MIRABILIS L., Four-o'clock

Ours perennial herbs, with fleshy tuberous, thickened or woody roots, caudex with short erect branches; stems erect, ascending, or sprawling, simple or branched from the base or above, glabrous to pubescent, often glaucous, the branches dichotomous, nodes usually swollen. Leaves opposite, usually somewhat fleshy, various in shape, glabrous to pubescent, often glaucous, sessile or petiolate. Inflorescences terminal or axillary, or both, of loose or congested thyrsiform or paniculate cymes of involucrate flowers, or involucres solitary in the axils; involucres 1- to 10-flowered, calyxlike, 5-lobed, lobes equal or unequal, green or margins tinged with purple, somewhat campanulate, sometimes accrescent, usually becoming rotate and veiny in fruit. Flowers perfect, calyx tubular, corollalike, constricted above the ovary, limb campanulate to rotate or funnelform, 5-lobed, the perianth falling after anthesis but the base persistent around ovary and thickening in fruit; corolla absent; stamens 3–5, filaments unequal, anthers exserted; ovary superior, of 1 carpel, style filiform, stigma capitate, ovule 1. Fruit (anthocarp) accessory, the persistent calyx tube thickened, hard, 5-angled or ribbed, the angles and surfaces smooth, rugose, varrucose, pubescent, or glabrous, ovary wall membranaceous; seed filling the pericarp to which seed coats adhere. *Allionia* Loefl. — Rydberg.

Mirabilis jalapa L., the common garden four-o'clock, sometimes self-seeds and has been found as nonpersisting waifs. It is here excluded from consideration. *M. pauciflora* (Buckl.) Standl., a species of the Edwards Plateau in Texas, was listed in the Atlas GP from OK: Greer, Kiowa. Since voucher specimens have not been located the species is excluded in the following treatment. In collecting species of *Mirabilis* every effort should be made to secure mature fruits.

1 Principal leaves with petioles 1–5 cm long.
  2 Fruits glabrous, 2.5–3 mm long; involucres 5–6 mm long, slightly or not enlarged in fruit, not membranaceous, deeply 5-cleft .... 9. *M. oxybaphoides*
  2 Fruits pilose, 4–6 mm long; involucres membranous, 10–15 mm long in fruit, 5-lobed .... 8. *M. nyctaginea*
1 Principal leaves sessile or with petioles usually less than 10 mm long.
  3 Stems usually densely short-pilose, long-villous, or hirsute, at least at base.
    4 Fruits glabrous .... 2. *M. carletonii*
    4 Fruits pubescent.
      5 Leaf blades linear to lance-linear, 1.5–5 mm wide .... 4. *M. gausapoides*
      5 Leaf blades mostly ovate-oblong, sometimes lanceolate to broadly ovate, usually well over 5 mm wide .... 6. *M. hirsuta*
  3 Stems glabrous or only puberulent to glabrate.
    6 Fruits glabrous.
      7 Principal leaf blades 2–7 mm wide .... 5. *M. glabra*
      7 Principal leaf blades 10–20 mm wide .... 3. *M. exaltata*
    6 Fruits pubescent.
      8 Angles of fruit not tuberculate, sides tranversely rugose; leaves 1–5 mm wide .... 7. *M. linearis*
      8 Angles of fruit tuberculate, sides tuberculate; leaves 3–25 mm wide .... 1. *M. albida*

1. *Mirabilis albida* (Walt.) Heimerl., white four-o'clock. Perennial herbs with often woody taproots crowned by a short, branching caudex; stems 1–few, erect or ascending, 2–8(12) dm tall, simple or branched below and above, each internode below the inflorescence with

2 longitudinal strips of incurved hairs usually less than 0.5 mm long, or these strips indistinct to absent and the stem nearly glabrous below the inflorescence, often whitish. Leaves opposite; blades variable in shape, from thin to thick and succuluent, from oblong, oblong-lanceolate, linear-lanceolate, elliptic-lanceolate, to linear or ovate, (3)4–10(12) cm long, 3–20(25) mm wide, apex blunt or rounded to acute or short-acuminate, base attenuate or narrow-cuneate, margin ciliate or eciliate, entire to subsinuate, surfaces glabrous or somewhat pilose, green on both sides or usually bright green above and glaucous or whitened below; sessile or petioles to 10 mm long. Involucres terminating solitary axillary peduncles or usually in much branched cymose-paniculate inflorescences; branches of inflorescence or peduncles viscid-pubescent to short-pubescent with incurved hairs, or these mixed; involucres 1- to 5-flowered, 3.8–4 mm long at anthesis, (8)10–12(15) mm long in fruit, from nearly glabrous or with short hairs to often densely or sparsely viscid pilose, lobes acute or rounded. Calyx pink, rose or whitish, (6)8–10 mm long, lobes shorter than tube, glabrous to sparsely pilose; stamens 3–5, exserted. Fruits obovoid, 5–6 mm long, angles wide, with flattened or cylindrical tubercles, each tubercle terminated with a tuft of short, silvery hairs, the sides also with similar tubercles; seed yellowish-brown, obovoid, 3.3–3.7 mm long. May–Oct. Infrequent to common on prairies, pastures, rocky bluffs, open wooded hillsides, roadsides; ND to w IA, e NE, e 1/2 & s-cen KS, OK, TX; (SC, TN, to Man., ND, s to LA, TX). *Allionia bracteata* Rydb., *A. decumbens* Nutt.—Rydberg.

This is a highly variable species in our area, and we are uncertain of its true distribution. It has often been confused with *M. glabra* and *M. linearis* with which it often integrades in foliage characters. The mature fruits, however, apparently readily separate the 3 taxa. In e NE & e KS from May–June there are plants in which involucres terminate solitary axillary peduncles, leaves are relatively thin and green on both sides, the internodes always with 2 longitudinal strips of incurved hairs on both sides, and plants mature by July. Such plants have been named *Mirabilis decumbens* (Nutt.) Daniels or *M. albida* var. *uniflora* Heimerl. Some authors consider *M. decumbens* to be young plants of *M. albida* which later develop cymose-paniculate inflorescence, strips of longitudinal hairs less obvious or absent, and leaves become thick, dark green above, white or glaucous below. A few of our specimens, of late Jun–Jul, have the foliage characters of *decumbens* and develop the branched inflorescences of *albida*, but we have not observed the earlier maturing form to develop into the usual *albida* type. The complex merits critical field and experimental studies. Our treatment is a tentative approach.

2. **Mirabilis carletonii** (Standl.) Standl., Carleton four-o'clock. Perennial herbs with deep, often woody, taproots crowned by branching caudex; stems erect, 1–few, 4–8(12) dm tall, simple or sometimes branching above, densely viscid-pilose to often glabrate below, whitish or yellowish-green. Leaves opposite; blades (2)4–8(10) cm long, 1–4 cm wide, ovate, ovate-oblong, to lanceolate or lance-oblong, often inequilateral, cuneate, rounded, subcordate or truncate at base, rounded, obtuse, acute, or short-acuminate at apex, margin entire or slightly undulate, short-pilose to puberulent on both surfaces, rarely glabrate, usually thickish, dark green above and whitened below; petioles absent or to 5 mm long. Inflorescence cymose-paniculate, open viscid-pubescent; involucres 4–6 mm long at anthesis, enlarging to 10–12 mm long in fruit, viscid-pilose, densely ciliate. Flowers usually 3 per involucre; calyx pink or whitish, pubescent, (4)5–6 mm long, lobes shorter than tube; stamens usually 3, exserted. Fruit obovoid, (4)5–6 mm long, brownish, glabrous, 5-angled, angles obtuse, smooth, sides obscurely tuberculate, rugulose or smooth, conspicuously raphidulous; seed obovoid, (2.5)3–3.5 mm long, brownish. May–Sep. Infrequent to locally common on sandy prairies, stream valleys, sand dunes, often more common on Permian Red Beds or Rolling Red Plains; cen & s-cen KS, w OK, se CO, nw TX & panhandle, cen NM; (endemic in area). *Allionia carletonii* Standl.—Rydberg.

3. **Mirabilis exaltata** (Standl.) Standl., tall four-o'clock. Perennial herbs with deep, stout, woody taproot crowned by branching caudex; stems 1–few, erect, 6–15 dm tall, simple or sparingly branched above, glabrous, glaucous below. Leaves opposite; blades lanceolate,

3-7 cm long, 8-18 mm wide, usually 3-4 × longer than wide, glabrous, acute at apex, attenuate at base, margins entire or slightly undulate; petioles absent or 1-3 mm long. Inflorescence paniculate, open; involucres 3-4 mm long at anthesis, 10-12 mm long in fruit, glabrous or with sparse, very short hairs; flowers usually 3 per involucre. Fruit obovoid, 4.5-4mm long, brownish, glabrous, 5-angled, angles acute, narrow, smooth, sides transverse-rugulose. May-Sep. Rather rare on Permian Red Bed prairies & hillsides; sandy prairies; w OK, s in TX to Comanche & Ward; (endemic in area).

Specimens of this species were too rare to enable preparation of a definitive description for plants in our area.

4. **Mirabilis gausapoides** (Standl.) Standl. Perennial herbs with deep, woody taproot and subterranean caudex; stems erect to ascending or decumbent, 3-8 dm tall, rarley branched, densely hirsute below to short-pilose or viscid-puberulent above, glaucous. Leaves opposite, remote to crowded; blades linear to linear-lanceolate (3)5-10 cm long, 1.5-5(6) mm wide, entire, attenuate at both ends, rather densely hirsute or short-pilose, rarely glabrate. Inflorescence terminal, cymose-paniculate, sparsely branched, branches densely villous or viscid-puberulent; involucres comparatively few, solitary or in clusters of several, 3.5-4.8 mm long at anthesis, 8-12 mm long in fruit, densely to sparsely villous-hirsute or viscid-villous, lobes rounded. Flowers usually 3 per involucre; calyx pink, 10 mm long, limb deeply lobed, lobes retuse, sparsely short-pilose; stamens 5, exserted. Fruit obovoid, 4-5 mm long, olivaceous, short silvery hirtellous, angles wide, smooth, sides coarsely transverse-rugose; seed obovoid, 3 mm long, yellowish-brown. May-Sep. Infrequent to common on prairies, rocky hillsides, roadsides; OK: Cimarron; TX panhandle, sw CO, NM; (OK, CO, w TX, NM; Mex.).

This species is often referred to *Mirabilis linearis* var. *subhispida* Heimerl. and there is evidence of intergradation between the two species. Careful field and experimental studies are necessary before a taxonomic decision can be made.

5. **Mirabilis glabra** (S. Wats.) Standl., smooth four-o'clock. Perennial herbs with deep, usually woody taproots, crowned by subterranean, branched caudex; stems erect, stout, 1-few, (6)8-10(15) dm tall, rarely branched below, glabrous, glaucous. Leaves opposite, remote or crowded near base, sessile; blades linear, lanceolate, or oblanceolate, (4)6-10(13) cm long, 2-7(10) mm wide, entire, thickish, glabrous, glaucous, attenuate at base, apex obtuse or acute. Inflorescence terminal, cymose-paniculate, often much branched, branches glabrous or short-pilose, often with glandular hairs; involucres 3.2-3.7 mm long at anthesis, becoming 10-12 mm long in fruit, ciliate, glabrous, or viscid-pilose, lobes shallow, rounded. Flowers 1 or 2(3) per involucre, most usually cleistogamous; calyx 7-10 mm long, pink, rose, or white, glabrous, stamens usually 5, exserted. Fruit obovate, 4-5 mm long, olivaceous, glabrous, angles narrow, smooth, acute, sides with obscure rounded tubercles or slightly rugose, seed obovoid, yellowish-brown, 2.8-3.3 mm long. Jun-Sep. Locally common in sandy prairies, sand dunes, rocky prairies, stream valleys, & roadsides; NE: Garden, Grant; w 1/2 KS, e CO, OK, TX, NM; (NE, KS, UT, s to TX; Mex.). *Allionia glabra* (S. Wats.) O. Ktze.—Rydberg.

Some specimens are difficult to distinguish from *M. exaltata* and we suspect intergradation.

6. **Mirabilis hirsuta** (Pursh) MacM., hairy four-o'clock. Perennial herbs with stout, usually woody, taproots, crowned by branching caudex; stems erect, 1-few, ascending or rarely decumbent, 2-8(12) dm tall, simple or rarely branched below the inflorescence, densely long-pilose or hirsute below with hairs 1.5-2.5(5) mm long, upper internodes sometimes only puberulent but pilose at nodes. Leaves opposite; blades (2)4-10(12) cm long, rather variable in shape, lanceolate, ovate, ovate-oblong, oblong-lanceolate, or somewhat rhombic, cuneate to subcordate at base, apex acute, obtuse, sometimes rounded, densely hir-

sute to viscid-puberulent, rarely glabrate; petioles absent or to 5 mm long. Inflorescences terminal and axillary, sometimes involucre solitary on axillary peduncles, usually cymose-paniculate, involucres often in clusters or inflorescence more open; involucres 4-5 mm long at anthesis, 8-10 mm long in fruit, viscid-pilose. Flowers usually 3 per involucre; calyx 8-10(12) mm long, pink, rose, or purplish-red, usually sparsely pilose; stamens 3-5, long, exserted. Fruits obovoid, 4-5 mm long, olivaceous to brownish, densely to rather sparsely pilose, 5-angled, angles wide, rather smooth, sides rugose to short-tuberculate; seed obovoid, 2.9-3.2 mm long, yellowish-brown. (2n = 58) May-Oct. Infrequent to common in prairies, pastures, rocky hillsides, stream valleys, open woodlands, roadsides, on a variety of soils; w MN, ND, MT, s to MO, KS, CO, TX panhandle, NM; (WI to Man., MT, s to MO, LA, TX, NM, rarely adventive elsewhere). *Allionia hirsuta* Pursh—Rydberg.

This species becomes rare in KS, MO, and has not been reported from OK. The few Kansas specimens are usually much less pubescent than those of the n and w GP.

7. **Mirabilis linearis** (Pursh) Heimerl., narrowleaf four-o'clock. Perennial herb with deep, elongate, woody taproot, crowned with branching caudex; stems erect to ascending or procumbent, 2-10 dm tall, simple or branched below, often branched above, usually very whitish and glaucous, glabrous or rarely puberulent below, glandular-pubescent and sometimes short-villous above especially in the inflorescence. Leaves opposite, few to many, often crowded, 3-10 cm long, 1-5 mm wide, linear to linear-lanceolate, thick, gray-green, usually glaucous, glabrous, or rarely glandular puberulent, sessile or long-cuneate at base to short petiole, narrowed to acute or obtuse apex, margins entire to rarely undulate or sparsely dentate. Inflorescence usually a rather open, freely branched panicle of involucres or cymose-paniculate; involucres 3.8-4.2 mm long at anthesis, rotate-campanulate, glandular-pubescent, becoming (6)8-10(12) mm long in fruit, usually 3-flowered. Calyx 8-12 mm long, pilose, purple-red to pink or rarely white, limb deeply lobed, lobes emarginate; stamens exserted. Fruits 4-5 mm long, obovoid, 5-angled, angles obtuse, brownish-olivaceous, densely pubescent to sparsely strigose, sides transversely rugose; achenes 2.8-3.2 mm long, yellowish-brown. (n = 26) May-Sep. Infrequent to common on prairie plains & hillsides, stream valleys, roadsides, pastures, sandy or rocky soils; GP, but rare in ne & e 1/4, common on high plains; (MN to MT, s to MO, TX, AZ; Mex.; sometimes adventive elsewhere). *Allionia linearis* Pursh, *A. diffusa* Heller—Rydberg; *M. diffusa* (Heller) Reed—Correll & Johnston.

In the western 1/4 of the GP the calyx is usually pink to nearly white and in some plants the stems are much branched from the base and tend to be prostrate to procumbent or ascending. Such plants have been referred to *Mirabilis diffusa* (Heller) Reed but they are always found with erect forms and all degrees of transition exist even in the same colony. Thus, we consider *M. diffusa* as lacking taxonomic significance.

8. **Mirabilis nyctaginea** (Michx.) MacM., wild four-o'clock. Perennial herbs with thick fleshy taproots crowned by a somewhat branching caudex; stems erect or ascending, swollen at nodes, 3-8(12) dm tall, branching below or above, branches usually forking, usually glabrous or very sparsely pubescent, sometimes slightly glaucous. Leaves opposite, entire, glabrous or nearly so, blades ovate, ovate-lanceolate, to ovate-oblong, (3)5-12(15) cm long, base obtuse, truncate, rounded, or cordate, apex obtuse or short-acuminate; petioles 1-5(8) cm long, the upper leaves often sessile. Involucres borne on pubescent peduncles that are grouped in umbels in forked terminal clusters, often appearing paniculate; involucre 5-6 mm long at anthesis, united at least 2/3 its length, lobes 5, enlarging to (8)10-15(17) mm long in fruit, persistent, pilose on margins and sometimes at the base and sparsely so on outer surfaces, often reddish. Flowers perfect, 3-5 per involucre; calyx pink to purple or reddish, rarely white, campanulate, tube 2 mm long at anthesis, limb (8)10-15(18) mm wide; stamens 3-5, exserted. Fruit 4-6 mm long, hard, cylindric-obovoid, or narrowly elliptic, pilose, rugose, or warty, 5-ribbed and angled, grayish-brown to nearly black; achenes

3–3.5 mm long, obovoid, yellowish or brown. (2n = 58) May–Oct. Often a weedy species of pastures, prairies, fields, stream valleys, roadsides, waste places; GP but infrequent to absent along w 1/4; (WI & Man., w to MT, s to TN, TX, NM; Mex.; adventive to e coast & in CA). *Allionia nyctaginea* Michx., *A. ovata* Pursh — Rydberg.

9. ***Mirabilis oxybaphoides*** (A. Gray) A. Gray, spreading four-o'clock. Perennial herb from a thick fleshy root crowned with short caudex; stems ascending to decumbent, 3–6 dm tall or long, usually much branched, often forming clumps 4–12 dm in diam, viscid-pubescent to glabrate. Leaves opposite; blades deltoid or ovate, usually cordate at base or sometimes truncate, acute or short-acuminate at apex, entire or undulate, viscid-pubescent to glabrate; petioles 1–4 cm long. Inflorescence cymose or axillary; peduncles of involucres solitary; involucres 5–6 mm long at anthesis, subrotate, deeply 5-cleft, pubescent, ciliate, not or slightly enlarged in fruit. Calyx (5)7–9 mm long, white to pink or purple, sparsely hairy to glabrate. Fruits 2.4–3 mm long, obscurely ridged or smooth, sometimes obscurely black-spotted or transversely ridged. Jun–Sep. Dry hillsides, canyons, stream valleys; OK: Cimarron; NM: Union; s CO; (OK, s CO, UT, NV, s to TX, AZ; Mex.).

This species is known from very few specimens in the GP and the description is based on specimens largely from outside our area.

5. SELINOCARPUS A. Gray, Moonpod

1. ***Selinocarpus diffusus*** A. Gray, spreading moonpod. Low shrub (?) or suffrutescent plant with a branching, woody caudex; stems dichotomously much branched from the base and above, erect or decumbent; 1–3 dm tall or long, slender, glandular-puberulent, the younger portions usually covered by whitish, inflated hairs. Leaves opposite, blade ovate, ovate-lanceolate, or elliptic, 1–2.5 cm long, 5–15 mm wide, sparsely strigillose to glabrate above, below glaucous, strigillose, and sometimes glandular-puberulent, the veins usually strigose below, apex acute to obtuse, margins denticulate-ciliolate; base cuneate or infrequently rounded; petioles 3–25 mm long, covered with whitish, inflated hairs and usually strigose. Flowers perfect, sometimes nearly all cleistogamous, solitary or in 2s or 3s, axillary, sessile or subsessile; bracts 2, subulate or linear, 2–5 mm long, 0.5–1 mm wide. Chasmagamous flowers conspicuous; calyx tubular-funnelform, 35–45 mm long, densely glandular-hirtellous without, the tube constricted slightly above the ovary, the limb greenish-yellow, shallowly 5-lobed, 10–15 mm wide, corolla lacking; stamens 5, slightly exserted, ovary oblong, stigma peltate and exserted, style filiform. Cleistogamous flowers minute and inconspicuous; calyx tubular, 2–4 mm long, obscurely 5-lobed, usually persistent in fruit. Fruit widely obovate in outline, usually 5-winged, 6–7 mm long, 5–6.5 mm wide, the body narrowly obovate and strongly 5-ribbed, 2.5–3.5 mm in diam, the walls membranaceous, strigillose, the wings membranaceous and not evidently veined; seeds light brown, oblanceolate in outline, 4–4.5 mm long, smooth. (May) Jun–Aug. Dry sandy or gypsiferous soil; OK: Harmon; TX: Oldham; (sw OK, n-cen & w TX, NM, s UT, s NV, & s CA).

6. TRIPTEROCALYX Hook. ex Standl., Sand Puffs

1. ***Tripterocalyx micranthus*** (Torr.) Hook., sand puffs. Annual herbs; stems few to many, 0.5–4 dm tall or long, erect to mostly decumbent or procumbent, sparingly branched beyond the base, glandular-pubescent and usually sparsely puberulent. Leaves opposite; blade narrowly ovate to elliptic or oblong, 2.5–6.0 cm long, 0.7–3.5 cm wide, glabrous above, scabridulous and glaucous below, apex usually obtuse, margins viscid-ciliate, base cuneate,

infrequently attenuate or rounded; petiole 1.5–3.5 cm long, glandular-pubescent, sometimes becoming glabrate. Flowers perfect, borne in pedunculate, several to many-flowered heads; peduncles axillary, 1–2.5 cm long, exceeded by the leaves; involucral bracts distinct, usually 5, ovate to ovate-lanceolate, 4–7 mm long, 1.5–2.5 mm wide, vestiture as the leaves. Calyx funnelform or slightly salverform, the tube 6–10 mm long and constricted above the ovary, sparsely scabridulous and viscid-puberulent, limb whitish, 5-lobed, 3–5.5 mm wide; corolla lacking; stamens 5, included; ovary ellipsoid, stigmas fusiform and included, style filiform. Fruit elliptic or widely so in outline, 3(rarely 5)-winged, 1.5–2.5(3) cm long, 1–2 cm wide, the body fusiform, 4–6 mm in diam, coriaceous, coarsely rugose to nearly smooth, usually scabridulous, wings at first pinkish, later blackish, membranaceous, reticulate-veined, sparsely scabridulous, ciliolate; seeds yellow-brown, dull, narrowly obovoid, 6–7 mm long, smooth. May–Jun (Aug). Usually sandy soil in flood plains, prairie pastures, or on hillsides, occasionally on roadsides; w 1/3 ND, MT, WY, CO; KS: Hamilton; NE: Cheyenne; SD: Harding; (s Alta., Sask. (?), w ND & e MT, s to sw KS, NM, AZ, & NV). *Abronia micrantha* Torr. — Gates.

36. AIZOACEAE Rudolphi, the Fig-marigold Family

by William T. Barker

Slightly fleshy to succulent annual or perennial (ours perennial) herbs with glabrous, papillose, or pubescent, prostrate to erect and usually much-branched stems spreading from a small taproot, often forming mats. Leaves more or less succulent, simple, entire, cauline, alternate or opposite; stipulate or exstipulate, sessile or petiolate; petioles often dilated, with membranaceous margins sometimes decurrent, or connate in a sheath about the node. Flowers inconscpicuous, regular, perigynous to semiepigynous, perfect; axillary in a cymose or spikelike cluster. Calyx united below, lobes (3)5; petals absent; stamens 5–10 or numerous, alternate with or opposite the calyx lobes, perigynous; carpels 1 or 2, usually 3–5, united, styles as many as the carpels, free or united below, ovary hypogynous to perigynous, (1)2(5) loculate. Fruit a loculicidal or circumscissile capsule; seeds numerous, small, reniform to pyriform, smooth to wrinkled or tuberculate, brown to black. Tetragoniaceae — Rydberg.

Reference: Bogle, A. L. 1970. The genera of Molluginaceae and Aizoaceae in the Southeastern United States. J. Arnold Arbor. 51:431–462.

1 Stipules absent; ovary 3- to 5-celled .... 1. *Sesuvium*
1 Stipules present; ovary 1- to 2-celled .... 2. *Trianthema*

1. SESUVIUM L., Sea Purslane

1. *Sesuvium verrucosum* Raf., sea purslane. Perennial herb, freely branched, prostrate, succulent, glabrous; stems to 9 cm long, smooth and usually more or less verrucose with crystalline globules; roots fibrous. Leaves opposite, equal, entire, oblanceolate to oblong ovate or some linear-oblong, apex rounded to somewhat subacute, tapering to an expanded scarious, clasping base, to 3 cm long and 1 cm wide above the middle; exstipulate. Flowers perfect, sessile to short, stout-pedicillate. Calyx lobes 5, broadly ovate-elliptic to ovate-lanceolate, 4.5–7 mm long, hooded with a subapical dorsal prolonged appendage; stamens numerous; ovary 3–5 loculate, with numerous ovules. Capsule membranaceous, enclosed

by a persistent perianth, circumscissile near the middle, conical, about 5 mm long and 3 mm in diam; seeds black, arillate, smooth and lustrous, about 1 mm in diam. Apr–Aug. In saline and alkaline soils about lakes, in creek bottoms, on mud flats & clay dunes; TX, w OK & cen KS; (MO, KS, AR to OK, w to NM, AZ, CA; n Mex.).

2. TRIANTHEMA L., Horse Purslane

2. *Trianthema portulacastrum* L., horse purslane. Perennial succulent herb or suffrutescent, diffusely branched from the base, decumbent or ascending, to 1 m tall. Leaves opposite, members of the pair unequal, round-obovate 1–4 cm long, 3 cm wide, abruptly narrowed to a long slender petiole scariously winged at the base; stipulate. Flowers solitary or few in axils, sessile or short-pedicellate, 4 mm wide, pink or purple inside, green outside. Calyx lobes 5, ovate-lanceolate to lanceolate, concave, about 2.5 mm long; stamens 5–10, perigynous in two cycles, alternate and opposite calyx lobes, shorter than and falling with the calyx; ovary superior, 1- or incompletely 2-celled. Fruit a capsule about 4 mm long, cylindrical, somewhat curved, with winged appendages at the apex, circumscissile near the middle; seeds (1) several, round-reniform, arillate, wrinkled, reddish-brown to black, about 2 mm in diam. (2n = 26, 36) May–Oct. Waste places; s & w TX, OK, w MO; (FL, TX, s CA, n to NJ, OK, & MO; Africa, Australia, weed in New World tropics). Introduced.

37. CACTACEAE Juss., the Cactus Family

by Robert B. Kaul

Perennial succulent herbs, shrubs, and small trees, ours all apparently leafless, the leaves small and soon falling. Stems simple or branched, often segmented (the segments sometimes called joints), cylindric to globose or flattened, smooth, ribbed or tuberculate, naked or woolly, bearing spiniferous areoles at the tips of the tubercles, on the ribs, or scattered regularly over the surface of smooth stems; spines usually definable as central spines that project from the stem axis and radial spines that are more or less parallel to the stem surface and, in *Opuntia*, tiny barbed spines (glochids) among the central and radial spines. Flowers borne singly at or near the areoles or, in some tuberculate species, formed at the base of the tubercles. Flowers small to showy, often brilliantly colored, epigynous, the ovary naked, spiny or woolly; perianth parts numerous, intergrading, the outer sepaloid parts green to partially colored, the inner petaloid parts larger and green to vividly colored; perianth parts borne on a short or long floral tube or cup (hypanthium) to which the numerous stamens are attached; style 1, stigmas more than 1; ovules numerous, parietal. Fruit fleshy or dry, dehiscent or indehiscent, naked or spiny, often with glochidiate areoles.

References: Benson, L. 1982. The Cacti of the United States and Canada. Stanford Univ. Press, Stanford, CA; Weniger, D. 1970. Cacti of the Southwest. Univ. of Texas Press, Austin, TX.

In the absence of extensive cactus collections from the GP, the treatment here follows Benson (1982) rather closely. Many of our cacti show clinal variation and hybridization and are in need of field and laboratory study.

1 Stems smooth or tuberculate, but not ribbed.
   2 Stems articulated into segments, these cylindric or bilaterally flattened; areoles bearing spines and glochids .... 5. *Opuntia*

    2   Stem not articulated into segments, prominently tuberculate; areoles bearing spines only.
        3   Most tubercles with a longitudinal groove on the upper side from the areole to near
            the base; flowers borne at the base of the grooves .... 1. *Coryphantha*
        3   Tubercles without a groove on the upper side; flowers borne on the apical areoles or
            between the tubercles.
            4   Flowers produced at the base of the tubercles and not attached to them; central
                spine 1(0–4); stem tip usually flattened to concave .... 4. *Mammillaria*
            4   Flowers produced beside the areoles at the tip of the tubercles; central spines 5–9;
                stem tip convex .... 6. *Pediocactus*
1   Stems prominently ribbed, not tuberculate or the ribs barely tuberculate.
    5   Spines slender, smooth; fruit spiny .... 3. *Echinocereus*
    5   Spines coarse, annulate; fruit scaly .... 2. *Echinocactus*

1. CORYPHANTHA Engelm.

Stems 1–several, not jointed above ground, sometimes forming low mounds, subglobose to cylindroid, mostly less than 10 cm high, covered with spirally arranged tubercles, these mostly with a longitudinal groove on the upper side. Areoles apical on the tubercles, the central spines 0–4, the radial spines 12–40. Flowers and fruits on new growth of the current season, at the base of the upper side of a tubercle and not on the spiniferous areole, the tubercle groove connecting the flower with the areole. Flowers 2–6 cm long, greenish to pale yellow or pinkish. Fruit fleshy, green to red, naked, 1–2.5 cm long; seeds brown to black, punctate. *Mammillaria* Haw.—Fernald; *Neomammillaria* Britt. & Rose—Rydberg.

1   Central spines 0(1–4), radial spines 10–20; fruit red, subglobose; flowers greenish to
    yellowish .... 1. *C. missouriensis*
1   Central spines 3 or 4, radial spines 12–40; fruit green, oblong to clavate; flowers pink to
    purplish .... 2. *C. vivipara*

1. *Coryphantha missouriensis* (Sweet) Britt. & Rose. Stems 1–several in a cluster, to 7(10) cm tall. Tubercles mostly with a groove on the upper side; areoles with 0(1–4) central spines and 10–20 radial spines. Flowers pale yellow to greenish, often tinged pinkish to reddish below, the perianth fimbriate. Fruit red or reddish when mature, nearly globose, to 1(2) cm diam; seeds black, punctate, 1–2.5 mm long. (2n = 44) Jun–Jul. Dry and rocky prairies, often on lime soils.

There are two varieties in our area, according to Benson (op. cit.):

1a. var. *caespitosa* (Engelm.) L. Benson. Radial spines 12–15 per areole. Flowers 5–6 cm long and wide. Fruit 1.5–2 cm long; seeds 2–2.5 mm long. e 1/3 KS; OK except panhandle; (ne KS s to s-cen TX, e to LA). *Neomammillaria similis* (Engelm.) Britt. & Rose—Rydberg.

1b. var. *missouriensis*. Radial spines 13–20 per areole. Flowers ca 2.5 cm long and wide. Fruits ca 1 cm diam; seeds 1 mm long. MT, ND, SD, w 1/2 NE, KS, & to be expected sporadically in w MN, w IA, w MO; (ID to ND, s to s KS, w to AZ). *Neomammillaria missouriensis* (Sweet) Britt. & Rose—Rydberg; *Mammillaria missouriensis* Sweet—Harrington; *Neobesseya missouriensis* (Sweet) Britt. & Rose—Barkley.

2. *Coryphantha vivipara* (Nutt.) Britt. & Rose, pincushion cactus. Stems 1–several, more or less globose to cylindroid, turbinate at the base, to 7 cm tall. Most tubercles with a groove on the upper side; areoles with 3–4(12) central spines, one of them turned downward, and ca 12–40 smaller radial spines. Flowers 2.5–4 cm long, pink to reddish purple. Mature fruit green, oblong to clavate, to 2.5 cm long; seeds brown, reticulo-punctate, 1.5–2 mm wide. (n = 11) May–Aug. Dry sandy or rocky prairies; GP except e NE, e KS, e OK, & w IA; (e OR to Alta. & w MN, s to cen TX & CA). *Mammillaria vivipara* (Nutt.) Haw.—Fernald.

Benson (op. cit.) distinguishes two varieties in our area:

2a. var. *radiosa* (Engelm.) Backeb. Central spines not prominent among the radials; radial spines 20–40, 12–19 mm long. Fruit 19–25 mm long. w OK, TX panhandle; (w OK to NV & CA, s to Mex.). *Neomammillaria radiosa* (Engelm.) Rydb.—Rydberg.

2b. var. *vivipara*. Central spines prominent among the radials; radial spines 12–20 per areole, 9–12 mm long. Fruit 12–20 mm long. GP except e NE, e KS, w IA, & nw MO; (e OR to Alta. & w MN, s to w OK, n NM, UT).

2. ECHINOCACTUS Link & Otto

1. *Echinocactus texensis* Höpffer., barrel cactus, devil's head. Stem usually single, hemispheroidal, not jointed, to 2 dm tall and to 3 dm diam, prominently ribbed, the ribs 13–27, acute; spines reddish, whitened basally, somewhat flattened, heavy, ringed with ridges; central spine 1, stout, curving downward and longer than the radial spines, sometimes recurved or hooked; radial spines ca 6. Flower ca 6 cm wide, showy, the colored members of the perianth white to lavender or salmon, a reddish line extending from the red base to the mucronate apex, the petals fringed, the floral tube woolly. Fruits globose to ovoid, to 4 cm long, fleshy, red; seeds black, reniform. Apr–Jul. Dry to mesic prairies, often among grasses; sw OK: Harmon; TX s panhandle: Bailey, Cottle; (TX w of 100° long., sw OK, se NM to Mex.).

3. ECHINOCEREUS Englem., Hedgehog Cactus, Lace Cactus

Stems 1–several, mostly unbranched, to ca 20 cm tall, ovoid to cylindroid, ribbed, the ribs somewhat tuberculate. Areoles elongate, borne on the ribs, with 0–5 erect or inclined central spines and few to many radial spines, these usually appressed to the stem and touching or interlocking with those of adjacent ribs and often more or less obscuring the stem. Flowers borne on the ribs adjacent to the areoles, on old growth, in ours green or greenish to pink or purplish, the ovary and fruit spiny and sometimes woolly. Fruit globose to ellipsoid, green (reddish), the spines falling; seeds dark, tuberculate.

1 Flowers green or greenish, to 2.5 cm long, not especially showy; ovary and fruit not woolly; stems mostly less than 5 cm tall .... 2. *E. viridiflorus*
1 Flowers pink or purplish, 5–15 cm long, showy; ovary and fruit woolly; stems to 20 cm tall .... 1. *E. reichenbachii*

1. *Echinocereus reichenbachii* (Terscheck) Haage, lace cactus. Stems 1–many, cylindroid, to 20 or more cm tall, to 5 cm diam. Areoles woolly when young; central spines 0–1(3), much shorter than the radial spines; radial spines 12–32, to 6(25) mm long, often spreading, slender, pectinately arranged and imparting a lacy appearance to the stem, variously colored. Flowers showy, 5–15 cm long, bright pink to purple, the petaloids more or less erose, obtuse to acuminate. Fruit globose to ovoid, dry, green, woolly, spiny. (2n = 22) May–Jun. Dry rocky hillsides, mostly in limestone areas but sometimes on granite; (s CO s to Mex., e to OK & e TX). *E. caespitosus* Engelm. & Gray—Rydberg.

Some of the variants of this species have been given varietal or specific names, especially those of sw OK, and Benson (op. cit.) recognizes 3 varieties in our area:

1 Radial spines 22–32 per areole; central spines 0 .... 1c. var. *reichenbachii*
1 Radial spines mostly 12–16 per areole; central spines 0–3.
  2 Central spines 1–3, to 3 mm long .... 1a. var. *albispinus*
  2 Central spine 1, ca 1 mm long .... 1b. var. *perbellus*

1a. var. *albispinus* (Lahman) L. Benson. Central spines 1-3, ca 3 mm long; radial spines 12-14 per areole, to 12 mm long, usually dark but sometimes white, yellow, or brown. Granitic rocks and soils; sw OK: Comanche, Greer, Kiowa; TX panhandle: Childress; (s OK, TX panhandle). *E. baileyi* Rose — Atlas GP.

1b var. *perbellus* (Britt. & Rose) L. Benson. Central spine 1, ca 1 mm long; radial spines 12-16(20) per areole, to 6 mm long, straw colored to pinkish. Limestone soils; se CO; ne NM: Union; TX panhandle: Randall, Armstrong, in the Palo Duro Canyon; (s CO to cen TX, w to se NM).

1c. var. *reichenbachii*. Central spines absent; radial spines 22-32 per areole, to 6 mm long, straw colored to gray at the base, pink at the tip. Limestone sands & gravels, perhaps occasionally on granitic soils; KS: Morton; s & w OK, TX panhandle; (TX panhandle to s-cen OK, s. to Mex.).

2. Echinocereus viridiflorus Engelm., hedgehog cactus. Stem 1(4), to 5 cm tall, rarely to 12 cm tall, less than 4 cm diam, cylindroid-ovoid, domed, prominently ribbed, the ribs with more or less obvious spiniferous tubercles. Areoles distinctly elliptic; central spine 0-1, ca 1 cm long, longer and thicker than the radial spines; radial spines 8-16 or more, pectinately disposed on the areole, ca 5 mm long, appressed to the ribs or sometimes recurved into the grooves, the upper mostly shorter than the lower in the same areole, whitish to purplish and often the upper and lower similarly colored but different from the middle spines, the entire plant then color-patterned. Flower ca 2.5 cm long, green or greenish, the petaloids entire, acute to rounded. Fruit spiny, globose to ovoid, green, ca 1 cm long; seeds black, densely tuberculate. May–Jun. Rocky, gravelly, & sandy prairies; se WY & sw SD, s through w NE, w 1/6 KS, e CO, OK & TX panhandles, & w to the mountains (GP w to mts.).

4. MAMMILLARIA Haw., Nipple Cactus

1. Mammillaria heyderi Muehlenpfordt. Stem 1, simple, oblate to hemispheroidal, the tip flattened to concave, to 5 cm high and 12 cm diam, the base not prolonged into a taproot, the juice milky. Tubercles prominent, more or less conical from a pyramidal base, without a longitudinal groove on the upper side, the areole apical on the tubercle. Central spine 1(0-4), reddish, shorter than the radial spines; radial spines 6-13 or more, to 1 cm long, the lower the longest, reddish to white, slender, weak, evenly spreading. Flowers and fruits borne on growth of previous years and between the tubercles, not connected by a groove to the areoles. Flowers 2-3 cm long and wide, pinkish or whitish. Fruits clavate, fleshy, naked, red; seeds brownish, reticulate. May–Jun. Dry to mesic prairies, among grasses & shrubs, often on limestone soils; sw OK, TX panhandle, ne NM, & reported from se CO; (sw OK to s TX, w to AZ & Mex.).

5. OPUNTIA P. Mill., Prickly Pear, Cholla

Coarse shrubs and prostrate clump-forming perennials, the stems jointed, the segments cylindrical (chollas) or bilaterally flattened (prickly pears), ribless, with or without tubercles. Leaves cylindrical to subulate, fleshy, green, soon falling. Spines present on the areoles, barbed or not; glochids (tiny barbed spines) usually abundant among the larger spines. Flowers produced on marginal areoles of old stem segments; floral tube short and the stamens borne just above the ovary. Fruit fleshy or dry, naked or spiny.

The prickly pears are extremely variable in many characteristics and they sometimes hybridize, so it is not always possible to identify a given specimen with certainty.

1 Stem segments somewhat to distinctly bilaterally flattened; spines without a sheath; plant low to prostrate, less than 15 cm high.

2   Spines weakly to strongly barbed, fruit dry at maturity, spiny.
    3   Spines strongly barbed, the longest spines often longer than the stem is wide; stem segments readily detaching, less than 5 cm long .... 1. *O. fragilis*
    3   Spines weakly barbed, not nearly as long as the stem is wide (except some very long and hairlike in var. *trichophora*); stem segments not readily detaching, to 12 cm long .... 8. *O. polyacantha*
2   Spines without barbs; fruit fleshy at maturity, not spiny (but glochid-bearing areoles usually present).
    4   Spines round in section, not flattened at the base, present only on the uppermost areoles (rarely absent); plant prostrate, most or all segments on the ground.
        5   Plant bluish-green; roots sometimes with large tubercles; seed margin ca 1 mm broad, corky, irregular .... 6. *O. macrorhiza*
        5   Plant green; roots without tubercles; seed margin much less than 1 mm broad, firm, even .... 2. *O. humifusa*
    4   At least some spines basally flattened or at least elliptic in section, present on at least the upper half of the stem segment; plant erect to sprawling to nearly prostrate.
        6   Spines yellow; stem segments green; NM .... 5. *O. lindheimeri*
        6   Spines brown; stem segments often bluish-green; sw 1/4 GP .... 7. *O. phaeacantha*
1   Stem segments cylindrical, smooth to prominently tuberculate; spines with a sheath; plant erect, shrubby, or matted.
    7   Stem segments smooth or nearly so, less than 1 cm thick; spine 1(0–3) per areole .... 4. *O. leptocaulis*
    7   Stem segments distinctly tuberculate, 1–3(5) cm thick; spines 2–30 per areole.
        8   Spine sheaths barely wider than the spines, tan, dull; spines to 3 cm long; flowers pink to purplish; shrub or small tree .... 3. *O. imbricata*
        8   Spine sheaths distinctly wider than the spines, golden, prominent; spines to 5 cm long; flowers yellow to greenish; plant bushy or matted ...... 9. *O. tunicata* var. *davisii*

1. *Opuntia fragilis* (Nutt.) Haw., little prickly pear. Plant low, sometimes clumped, to 12 cm tall. Stems with 1 or a few segments, these ovoid to obovoid, at least the largest ones somewhat flattened, mostly less than 4 cm long but in some of our northern specimens to 5 cm long, readily detaching and clinging to fur and clothing by barbed spines. Spines 1–10 per areole, on all areoles, to 2.5 cm long, barbed, spreading, the uppermost areoles with the longest spines; glochids ca 2 mm long. Flowers to 4 cm wide, yellow to greenish, seldom observed. Fruits ovoid, green, spiny, becoming tan on drying, to 3 cm long; seeds whitish to gray, with a conscpicuous margin, to ca 6 mm long. (2n = 66) Jun–Jul. Sandy to rocky prairies & hillsides; MT to sw MN, s across cen NE, cen KS to TX panhandle, w to the mts.; (n B.C. to MI, s to IL, KS, n TX, w to n CA).

This inconspicuous little cactus makes its presence known by its easily detached stem segments that cling tenaciously to clothing, including shoes. Sometimes the spines are longer than the stem is wide.

2. *Opuntia humifusa* (Raf.) Raf., eastern prickly pear. Plant small, creeping, less than 10 cm high, the roots without prominent tubercles. Stem segments green and sometimes glaucous, flattened, orbicular to obovoid, 5–7.5 cm long, 4–6 cm wide (occasionally much larger or smaller); areoles not strongly raised. Spines present only on upper areoles, often only on the marginal areoles, mostly 1 per areole, 2–3.5 or more cm long, straight, spreading, one of them sometimes deflexed; glochids 3 mm long, yellow to brown. Flowers 4–6 cm long and wide, yellow, sometimes pinkish. Fruit obovoid to clavate, 2–4 cm long, red to purple at maturity, bearing scattered glochid-bearing areoles; seeds tan to gray, flattened, orbicular, with a hard narrow rim, ca 4.5 cm in diam, smooth. May–Jun. Prairies, open woodlands, rock ledges, on various soils in the moister parts of GP; e 1/3 KS, e 1/3 OK, w MO; (e KS & e IA n to WI, MI, MA, s to FL & Mex., with widely scattered stations beyond the Rocky Mts.). *O. compressa* (Salisb.) Macbr. of authors, name invalid according to Benson (1982).

Specimens with characters intermediate with *O. macrorhiza* are common in much of GP.

3. Opuntia imbricata (Haw.) DC., tree cholla. Branching shrub to small low-branched tree to 2.5 m tall. Stem segments dark green, cylindrical, to 35 cm long, ca 3 cm thick; tubercles prominent. Spines 10–30 per areole, to 3 cm long, red to pink, barbed; central spines 1–8, to 3 cm long, slender to strong, loosely sheathed, the sheath dull tan; radial spines ca 1 cm long, tightly sheathed, glochids inconspicuous, ca 1 mm long. Flowers to 5 cm long, pink to purplish. Fruits globose to obovoid, prominently tuberculate, spineless, yellowish, to 4 cm long; seeds tan. (2n = 22) May–Jul. Sandy to gravelly soils; se 1/4 CO; sw & s-cen KS: Comanche, Kiowa; s to TX panhandle; OK: Cimarron, Woods; & ne NM; (CO & w KS s to s TX & Mex., w to AZ).

4. Opuntia leptocaulis DC., pencil cholla. Erect shrub to 1.5 m tall, the stems much branched, the branches rather crowded. Stem segments cylindrical, to 30 cm long, less than 1 cm thick, smooth or faintly tuberculate. Spine 1(0–3) per areole, straight, often somewhat downturned, to 5 cm long, the sheath tan, loose, deciduous; glochids few, tiny. Flower to ca 2 cm long and wide, opening widely, green to yellow, the petaloids 6–9, rather broad. Fruit red, fleshy, spineless but glochidiate, smooth, ca 1 cm long; seeds tan, irregular. (n = 22) May–Jun. Heavy soils of the plains; ne NM, TX s panhandle, s OK; (AZ to OK s to Mex.).

5. Opuntia lindheimeri Engelm., Texas prickly pear. Large, erect or sprawling shrub to 1 m or more high. Stem segments flattened, orbicular to obovate, to 2.5 dm or more long, green. Spines 1–6 per areole, on all but the lowest areoles (rarely absent from most or all areoles), straight, not barbed, some of them narrowly elliptic in section, yellowish, usually less than 4 cm long; glochids yellow to brown, 3–6 mm long. Flowers 5–7.5 cm long and wide, greenish to yellow (red). Fruit fleshy, purple when mature, obovoid, 3–7 cm long, with small areoles and glochids; seeds tan, 3–4 mm long. May–Jun. Sandy, gravelly, or loamy soils in the prairies; e-cen NM; OK: Murray; (NM to cen OK s to Mex.).

6. Opuntia macrorhiza Engelm., plains prickly pear. Plant low, clumped, less than 12 cm high, the roots often with prominent large tubercles. Stem segments flattened, often glaucous, blue-green, orbicular to obovate, to 10 cm long, mostly less than 6 cm wide. Spines mostly on the uppermost areoles and often on only the marginal areoles, 1–6 per areole, mostly straight, occasionally twisted, not barbed, to 5 cm or more long. Glochids yellow to brown, ca 3 mm long. Flower 5–6 cm long, yellow to reddish, the center sometimes red. Fruit fleshy, purple or red at maturity, with glochidiate areoles scattered over it; seeds irregularly discoid, ca 4.5 mm in diam, the rim rather broad. May–Jun. Sandy, gravelly, or rocky soils in the grasslands, sometimes persisting for years under invading trees. *O. tortispina* Engelm.—Rydberg, especially for specimens with twisted spines.

This species probably hydridizes with *O. phaeacantha* where sympatric. Many of our specimens show some characteristics of *O. humifusa*.

According to Benson (op. cit.) we have two varieties:

6a. var. *macrorhiza*. Plant moderately glaucous; stem segments 7–10 cm long. Spines slender, ca 0.5 mm in diam. Flowers yellow, the center sometimes red. GP except ND & e MT; (ID to MI s to LA & Mex.).

6b. var. *pottsii* (Salm-Dyck) L. Benson. Plant distinctly glaucous; stem segments 5–6 cm long. Spines very slender, ca 0.25 mm diam. Flowers reddish. TX: Randall; e NM: Curry, Quay; (se AZ to TX panhandle, s to Mex.).

7. Opuntia phaeacantha Engelm., prickly pear. Plant spreading to nearly prostrate, to 30 cm tall, forming large clumps. Stem segments large, flattened, to 22 cm long, ovate to clavate, with or without a constricted base, often bluish-green; areoles not conspicuously raised, more than 1 cm apart. Spines present on at least the upper 1/3 of the areoles and often on all areoles, 1–8 per areole, brownish, to 5(7) cm long, not barbed, some of them basally flattened; glochids absent to numerous, stramineous to brownish. Flowers to 9 cm long, yellow to orange, the center red to maroon. Fruit ovoid to obovoid, to 7 cm long, bright to dull red, fleshy, spineless, smooth but with scattered glochidiate areoles; seeds tan, discoid. (n = 33) May–Jun. Dry prairies & open woodlands, often on sandy soil; se CO & w 1/2 KS s through cen OK to cen TX, w to the mts.; (s CA to cen KS & cen OK s to Mex.).

Hybrids with *O. macrorhiza*, and *O. polyacantha*, can be expected where the ranges overlap. Benson (op. cit.) recognizes three somewhat intergradient varieties in our area:

1 Plant not low-growing; stem segments to 25 cm long, to 20 cm wide; spines 1–3 per areole .... 7b. var. *major*
1 Plant low-growing; stem segments to 17 cm long, to 14 cm wide; spines 3–8 per areole.
   2 Spines present on most areoles, 5–8 on upper areoles .... 7a. var. *camanchica*
   2 Spines present on upper 3/4 or more areoles, 3–5 per areole .... 7c. var. *phaeacantha*

7a. var. *camanchica* (Engelm. & Bigel.) L. Benson. Plant low-growing. Stem segments 14–17 cm long, 11–14 cm wide. Spines present on most areoles, 5–8 on upper areoles. w 1/2 OK, OK & TX panhandles, e-cen NM; (cen NM to cen OK, s to se NM).

7b. var. *major* Engelm. Plant large, not low-growing. Stem segments 12–25 cm long, 10–20 cm wide. Spines 1–3 per areole, the spiniferous areoles mostly on the upper half of the stem segment. se Co, cen & w KS, w 1/2 OK, TX & OK panhandles, ne NM; also in SD: Fall River; (s CA to cen KS, s to Mex., & in SD).

7c. var. *phaeacantha*. Plant low-growing. Stem segments 10–15 cm long, 7–10 cm wide. Spines present on upper 3/4 or more areoles, 3–5(9) on upper areoles, 1 or 2 on lower areoles. cen & se CO, ne NM; (s UT to n-cen CO s to Mex.).

8. ***Opuntia polyacantha*** Haw., plains prickly pear. Plant prostrate, rising only the height of one stem segment, clumped or mat-forming. Stem segments bluish-green to gray, to 12 cm long, flattened, orbicular to ovate; areoles rather crowded, mostly less than 1 cm apart. Spines 1–10 or more per areole, present on all areoles, the upper areoles with more spines than the lower. Spines on upper areoles 2–4(5) cm long, much less to very much more on lower areoles; spines rigid, deflexed, weakly barbed or, in one variety, long and hairlike on the lower areoles; glochids yellow (pink). Flowers 4–7 cm wide, occasionally larger, yellow to pink or red. Fruit globose to obovoid, 2–4 cm long, dry, spiny all over; seeds discoid, with a prominent, irregular margin, tan to white. (2n = 44, 66) May–Jun (Sep). Dry prairies, pastures, & roadsides, favoring sandy soils. *O. rutila* Nutt.—Rydberg, for red-flowered forms.

There are two varieties in our area:

8a. var. *polyacantha*. Spines of the lower areoles rigid, less than 4 cm long. GP w of 100° w long., and eastward in SD: Brule, Douglas; NE: Cuming; KS: Geary; MO: Jasper; (B.C. to Sask. & SD, s to NM, w to NV).

8b. var. *trichophora* (Engelm. & Bigel.) Coult. Spines of the lower areoles flexuous, hairlike, to 12 cm long. CO: El Paso, Kit Carson, Las Animas; NM: Curry, Quay, Union; OK: Cimarron; TX panhandle: Armstrong, Hutchinson, Potter; (NV to e CO s to Mex.; also LA).

A similar entity, *O. erinacea* Engelm. & Bigel. var. *utahensis* (Engelm.) L. Benson is known in NM. It is taller, has some spines to 10 cm long and flexuous, their base somewhat flattened; the fruits are spiny only on the upper part.

9. ***Opuntia tunicata*** (Lehm.) Link & Otto var. ***davisii*** (Engelm. & Bigel.) L. Benson. Plant bushy to matted, the stems crowded, cylindric, 3–6 dm high, much branched, the segments

not detaching readily. Stem segments to 15 cm long and 1–2 cm wide, woody, strongly tuberculate, the tubercles 1–1.5 cm long. Spines 6–10 per areole, to 5 cm long, straight, barbed, brownish, their conspicuous sheaths papery, golden, obviously wider than the spines; glochids yellow, ca 1 mm long. Flowers ca 5 cm long, 3 cm wide, yellow to yellow-green, sometimes tinged red. Fruit ca 3 cm long, 2.5 cm diam, yellow, tuberculate; seeds tan, to 2.5 mm long. May–Jun. Sandy soils among grasses or by themselves; ne NM, TX panhandle, extreme sw OK: Harmon, Greer; (NM to sw OK s to cen & s TX). *O. davisii* Engelm. & Bigel.—Atlas GP.

The conspicuous golden sheaths impart a golden hue to this uncommon cactus.

6. PEDIOCACTUS Britt. & Rose, Nipple Cactus

1. *Pediocactus simpsonii* (Engelm.) Britt. & Rose. Stem usually solitary, 2–12 cm long, 2–10 cm diam, globose to oblate or ovoid, the tip convex; tubercles spirally arranged, pyramidal, not fused, to 1 cm long, the areole terminal. Central spines 5–9, to 12 mm long, straight, spreading, brownish; radial spines 12–30, mostly ca 1 cm long and obscuring the stem, straight, slender, whitish. Flowers borne on new growth of the current season near the stem tip, produced very near the areoles, 18–25 mm wide, opening wide, pink to white or yellowish. Fruit green, becoming dry and dehiscent, naked except for a few scales, cylindroid clavate, to ca 1 cm long; seeds black, tuberculate, the hilum apparently lateral. May–Jun (Jul). Fine soils on mesas & escarpments; BH, se WY; (e OR, cen ID, & s WY & BH s to cen NM & n AZ). *Echinocactus simpsonii* Engelm.—Harrington.

Rydberg reported this species from w KS but we have no specimens from there.

38. CHENOPODIACEAE Vent., the Goosefoot Family

by Ronald L. McGregor

Ours annual herbs or perennials woody at base or throughout, stems sometimes nodally jointed, sometimes succulent. Leaves simple, opposite at lowest nodes and alternate above or alternate throughout, sometimes scalelike, often succulent, glabrous, pubescent, sometimes stellate-pubescent, or characteristically farinose with small to much inflated hairs, which collapse with age or drying and appear scalelike. Flowers small, usually green or greenish, 1–many and glomerate in axils or in spikes or panicles, regular, perfect or unisexual, the plants then dioecious or monoecious, apetalous, perianth (2–4) or 5(6)-lobed or parted, or reduced to a single scale, often lacking in pistillate flowers, stamens as many as perianth parts, pistil 1, ovary 1-celled, 1-ovuled, superior, sometimes with perianth adherent. Fruit an indehiscent or irregularly rupturing utricle; seed erect or suspended, embryo encircling the endosperm or spiral and endosperm reduced or lacking.

1 Plants perennial.
  2 Herbage stellate-pubescent .... 4. *Ceratoides*
  2 Herbage not stellate-pubescent.
    3 Blades of leaves flat .... 1. *Atriplex*
    3 Blades terete or subterete.
      4 Branches woody, spinose; flowers ebracteate .... 12. *Sarcobatus*
      4 Branches not spinose; bracteoles scalelike .... 13. *Suaeda*
1 Plants annual.
  5 Stems jointed; leaves reduced to small scales .... 10. *Salicornia*

5   Stems not jointed; leaves not scalelike.
    6   Fruiting perianth segments with dorsal, curved, or hooked spines .... 3. *Bassia*
    6   Fruiting perianth without spines or perianth absent.
        7   Flowers all unisexual.
            8   Pistillate flowers without perianth.
                9   Bracts of pistillate flowers conduplicate, strongly compressed in fruit, carinate dorsally .... 14. *Suckleya*
                9   Bracts of pistillate flowers not strongly compressed or carinate .... 1. *Atriplex*
            8   Pistillate flowers with 3- or 4-parted perianth .... 2. *Axyris*
        7   Flowers perfect or some unisexual.
            10   Flowers subtended by bracts that may be very small.
                11   Bracts small, scalelike, shorter than the perianth .... 13. *Suaeda*
                11   Bracts equaling or longer than the perianth .... 11. *Salsola*
            10   Flowers ebracteate.
                12   Fruiting perianth with dorsal, horizontal, usually hyaline wings.
                    13   Leaves entire; wings interrupted .... 8. *Kochia*
                    13   Leaves sinuate-dentate; wings continuous .... 7. *Cycloloma*
                12   Fruiting perianth without dorsal, horizontal wings.
                    14   Perianth of 1(3) sepals.
                        15   Larger leaves hastate; flowers glomerate in axils .... 9. *Monolepsis*
                        15   Larger leaves linear, flowers mostly solitary in the axils .... 6. *Corispermum*
                    14   Perianth 5(4)-parted .... 5. *Chenopodium*

1. ATRIPLEX L., Saltbush, Orache

Annual, monoecious herbs or perennial, dioecious, rather woody, sometimes spinescent shrubs, glabrous to usually farinose or scurfy with short, inflated hairs that collapse with age or drying. Leaves often opposite at lowest nodes, alternate above or alternate throughout, sessile or petioled, entire, dentate or irregularly lobed, sometimes hastate. Flowers unisexual, solitary to usually clustered in axils or terminal spikes that sometimes appear paniculate; staminate flowers ebracteate with (4)5-parted perianth, stamens (3)5; pistillate flowers subtended by 2 bracteoles (rarely without bracteoles) that enclose the fruit, perianth none or of 1-5 minute squamellae; ovary ovoid or depressed-globose, stigmas 2(3). Pericarp membranous, free from the erect or inverted seed; seed with radicle pointing away from base of ovary (superior) or toward it (inferior) or intermediate in orientation (lateral or ascending).

The genus *Atriplex* is taxonomically difficult, and there is little consensus on delimiting species. Care should be taken to gather mature fruiting specimens, as much of our herbarium material is too young for careful study. In the past, *A. patula* L. and *A. patula* var. *hastata* have been catch-all taxa to which many of our specimens have been referred. I have been unable, following recent treatments, to detect these taxa in the GP but careful field studies may reveal their presence. The following treatment is purely provisional until more adequate studies can be made.

1   Plants perennial, dioecious.
    2   Lateral branches rigid, sharply spinose; principal leaves evidently petiolate; pistillate bracteoles neither winged nor appendiculate .... 3. *A. confertiflora*
    2   Lateral branches scarcely spinose; principal leaves sessile or subsessile; pistillate bracteoles strongly winged lengthwise or appendiculate.
        3   Plants woody nearly throughout; fruiting bracteoles prominently winged lengthwise .... 2. *A. canescens*
        3   Plants woody only at base; fruiting bracteoles not winged lengthwise .... 7. *A. nuttallii*
1   Plants annual, monoecious or rarely dioecious.
    4   Mature fruiting bracteoles free to base or nearly so (or absent in some flowers of *A. hortensis*).

5 Fruiting bracteoles triangular and tuberculate .... 11. *A. subspicata*
 5 Fruiting bracteoles rounded, ovoid-rhombic or triangular ovate, without tuberculate appendages.
   6 Lower leaf surfaces abundantly mealy at maturity; fruiting bracteoles ovoid-rhombic .... 8. *A. oblongifolia*
   6 Upper and lower leaf surfaces of mature leaves nearly equally green.
     7 Veins of bracteoles meeting at base; all pistillate flowers with bracteoles .... 5. *A. heterosperma*
     7 Veins of bracteoles merging above base; some pistillate flowers without bracteoles .... 6. *A. hortensis*
4 Mature fruiting bracteoles evidently united above the base or to the summit.
 8 All parts of the plant glabrous .... 4. *A. dioica*
 8 Herbage and inflorescence densely to sparsely scurfy.
   9 Radicle of the seed superior.
     10 Fruiting bracteoles united to the summit .... 9. *A. powellii*
     10 Fruiting bracteoles united to above the middle, tips free .... 1. *A. argentea*
   9 Radicle of the seed inferior.
     11 Bracteoles becoming firmly indurate around the fruit; rare adventive species .... 10. *A. rosea*
     11 Bracteoles remaining relatively thin, easily separated from the fruit; common native species .... 11. *A. subspicata*

1. Atriplex argentea Nutt., silver-scale saltbush. Erect monoecious annual herbs, stems 1.5–6 dm tall, branched from the base and usually forming a globular shape, branches angled, scurfy when young. Leaves opposite below, alternate above, blades 2–5 cm long, triangular-ovate to rounded-ovate, margins smooth, surfaces gray-scurfy but sometimes glabrate, upper surface usually somewhat greener; leaves subsessile or with petioles to 2 cm long. Flowers unisexual; staminate flowers in upper axils or in short spikes, or mixed with pistillate flowers in axillary clusters, usually near middle of plant; staminate flowers with 5-parted perianth, stamens 5; pistillate flowers subtended by 2 bracteoles, these cuneate rotund, 4–7 mm long and wide, united to above the middle, irregularly toothed across the rounded apex, central part indurate, faces smooth or with a few tubercles to 2 mm long. Seed brown, 1.5–2 mm long, radicle superior. (2n = 36) Jul–Sep. Plains & valleys, usually on alkaline soils; GP but absent to rare in e ND s to cen OK; (w MN to Sask., OR, s to TX, NM, CA, introduced in e U.S.).

As described above, our plants are subsp. *argentea*. In the TX panhandle and westward there are plants with upper leaves sessile, the lowest alternate, and the margins irregularly dentate or entire. These are subsp. *expansa* (Wats.) Hall & Clem., which is more common in w U.S.

2. Atriplex canescens (Pursh) Nutt., four-wing saltbush. Erect, dioecious or monoecious perennials 4–10(15) dm tall, usually woody throughout, loosely to densely branched, lateral branches seldom spinose, gray-scurfy. Leaves alternate, simple, sessile or nearly so, linear-spatulate to narrowly oblong, obtuse at apex, cuneate at base, entire, thickish, both surfaces gray-scurfy, becoming glabrous. Staminate flowers in rather dense glomerules, arranged in dense spikes of terminal panicles, often leafy below, each with 5 perianth parts united at base, stamens 5; pistillate flowers in short axillary spikes, sometimes appearing paniculate, subtended by 2 bracteoles, these united nearly to summit, conspicuously 4-winged from sides and backs of bracteoles, faces smooth or with small appendages, whole bracteoles (4)8–10(15) mm long. Seed brown, 1.5–2.5 mm in diam, radicle superior. (2n = 36) May–Aug. Dry barren flats, slopes, bluffs, usually on alkaline soils; w 1/2 GP; (ND w to Alta., WA, s to TX, CA; Mex.).

Most of our plants are dioecious, but usually a few monoecious ones will be found in large colonies.

3. *Atriplex confertiflora* (Torr. & Frem.) S. Wats., spiny saltbush. Dioecious, perennial shrubs, 4–7 dm tall, much branched, compact and rounded, branches rigid, sharply spinose. Leaves alternate, simple, ovate to obovate or elliptic or triangular, commonly early deciduous, 1–2(4) cm long, 8–12(20) mm wide, entire, apex obtuse or rounded, abruptly cuneate at base, surfaces dull, scurfy, light gray-green, axillary bud leaves ovate, obovate to elliptic, 1–12 mm long, 1–7 mm wide, persistent; petioles 1–7 mm long. Staminate flowers sessile in dense axillary glomerules or short, leafy terminal spikes, perianth segments (4)5, distinct nearly to base, stamens (4)5; pistillate flowers in axillary glomerules or subpaniculate spikes, bracteoles sessile, ovate or elliptic, 5–13 mm long, 5–10 mm wide, margins united 1/3 length, free and divergent above, indurate around fruit, margins herbaceous, denticulate, sides sparsely tuberculate or somewhat crested near base, rarely smooth. Seed reddish-brown, circular, flat, 1.5–2 mm wide, radicle superior. Jul–Aug. Dry, rocky, valleys, hillsides, bluffs, usually on alkaline soils; w ND, MT, s to TX panhandle, NM; (ND to e OR, s to TX, AZ, Mex.).

4. *Atriplex dioica* (Nutt.) Macbr., sillscale. Annual, monoecious herbs, stems erect or spreading, usually branched from the base, forming plants 1–3 dm tall and equally as wide, glabrous. Leaves alternate, pale green, glabrous, sessile, blades lanceolate, variable in size on some plants to 3 cm long, on others to 1 cm or less, tapering to base, apex acute to obtuse. Staminate flowers usually in short spikes at tips of stems, perianth 4- to 5-parted, lobes with fleshy crest, stamens often reddish; pistillate flowers few to solitary in axils, very small, subtended by 2 sessile, ovate, bracteoles united to summit, 2 mm long, 1 mm wide, without appendages, scurfy, membranaceous. Seed brown, 1.2 mm wide, radicle superior. (2n = 18) Jun–Aug. Prairies, pastures, hillsides, stream valleys, & waste places; w 1/2 ND, SD; NE: Thomas; MT, WY; (Alta. to Sask. s to ND, NE, WY). *Endolepis dioica* (Nutt.) Standl., *E. suckeyi* Torr. — Rydberg.

5. *Atriplex heterosperma* Bunge. Erect, monoecious, annual herbs, stems (3)5–10(15) dm tall, usually branched. Leaves opposite near base, alternate above, blades triangular-ovate or triangular-oblong, with hastate or truncate to rarely slightly cuneate base, apex acuminate, usually as wide as long, 2–8(12) cm long, margins entire to variously toothed or lobed, young leaves whitish-mealy on lower surface, older leaves usually uniformly green on both surfaces; petioles 2–3 cm long. Flowers in a loosely branched terminal or axillary spike or paniculate, unisexual; staminate flowers with 5-parted perianth, stamens usually 5; pistillate flowers without perianth but with 2 bracteoles, free to base, rounded, smooth, entire, without appendages, veins meeting at base, bracteoles of 2 sizes, larger 5–6 mm long, 5 mm wide, enclosing a flat, yellowish seed 2–3 mm wide, smaller ones 1.8–2.1 mm long and as wide, enclosing a shiny, biconvex black seed 1.3–1.5 mm wide. Seed vertically attached, with easily removed membranaceous pericarp, radicle inferior. (2n = 36) Aug–Sep. Waste places, stream banks, prairie ravines, roadsides; scattered in ND, SD; NE: Lancaster; MT, WY, CO; (Ont., Man., Alta., B.C., s to NE, CO, WA; Eurasia and Chinese Turkestan).

6. *Atriplex hortensis* L., orache. Erect or decumbent monoecious annual herbs, stems 6–15(20) dm tall, usually branched. Leaves opposite near base, alternate above, blades triangular or ovate-triangular with hastate base, principal well-developed ones 8(15) cm long, 7(9) cm wide, with hastate base, acuminate, entire or irregularly toothed, mealy at first but soon glabrous on both surfaces. Flowers unisexual, arranged in terminal or axillary spikes that are paniculate; staminate flowers with 5-parted perianth, stamens 5; pistillate flowers of two kinds, some ebracteolate and with a 5-parted perianth containing a horizontal, biconvex, black, shiny seed 2 mm in diam; more commonly without perianth but with 2 bracteoles, bracteoles orbicular, veins merging above the base, of 2 sizes, the smaller ones containing black, vertical seeds, larger bracteoles with a flat, dull, yellowish-brown vertical

seed, 3.5–4 mm in diam, radicle inferior. (2n = 18) Jul–Sep. Cult. as an herb & found as casual escapes in GP; (widespread in N. Amer.; Asia). *A. nitens* Sch.—Rydberg.

7. *Atriplex nuttallii* S. Wats., moundscale. Dioecious suffruticose or shrubby perennials, woody at base, 1–5 dm tall, stems terete, herbaceous, and more or less ascending, densely grayish-scurfy. Leaves alternate, numerous, linear-spatulate to oblong or narrowly obovate, 2–4 cm long, 2–12 mm wide, entire, rounded at apex, cuneate at base, rather thick and green or grayish-green, closely scurfy. Staminate flowers in dense glomerules forming dense, stout, naked or sparsely leafy single or paniculate spikes, the perianth 5-lobed nearly to base; pistillate flowers glomerate in short axillary spikes and in terminal leafless spikes or panicles, bracteoles 2, lanceolate, ovate to ovate-orbicular, 3–5 mm long, united to above the middle, indurate, irregularly dentate along the margins, sides smooth to prominently tuberculate or with linear appendages. Seed reddish-brown, 2 mm in diam, radicle superior. Jun–Aug. Usually on saline soils of plains, valleys, badlands; ND, w 1/2 SD, nw NE, MT, WY, CO; (Man., Sask., s to NE, CO, NM, AZ). *A. oblanceolata* Rydb.—Rydberg.

It is possible that the correct name for this highly variable species is *A. tridentata* O. Ktze.; or if *A. gardneri* (Moq.) Standl., a plant with nearly smooth bracteoles found in ne WY, is considered to be the same species, the latter name would have priority over *A. nuttallii*.

8. *Atriplex oblongifolia* Waldst. & Kit. Erect, monoecious annual herbs, stems 1.5–8(12) dm tall, simple or branched, whitish farinose above. Leaves opposite near base, alternate above, blades mostly narrow, lance-oblong, 3–7 cm long, 2–3 cm wide, entire or with a few teeth, acuminate, cuneate at base, lower surface mealy, this persisting on upper leaves; petioles 7–10 mm long. Flowers unisexual, in spiciform axillary or terminal, paniculate-appearing inflorescences; staminate flowers with 5-parted perianth, stamens 5; pistillate flowers without a perianth, but with 2 bracteoles that are free to base; bracteoles ovate or ovate-rhombic to triangular-ovate, thin, entire, without appendages, nearly smooth on back; bracteoles of 2 sizes, larger 9–13 mm long and enclosing a yellowish-brown seed 2.8–3 mm long and wide, smaller ones about 1/2 size of larger and enclosing a black, convex, shiny seed 1.5–2 mm in diam, radicle inferior. Aug–Sep. SD: Lawrence; (B.C. & SD; cen Asia & Eurasia).

This species could well be more common in our area than records indicate. Its usually farinose leaves and large bracteoles distinguish it from *A. heterosperma*.

9. *Atriplex powellii* S. Wats., Powell's saltbush. Erect, monoecious annual herbs, stems 1–10 dm tall, strict or branched from the base, whitish-scurfy throughout. Leaves alternate or lowest opposite, blades 1–3(4) cm long, ovate or rhombic-ovate, rounded or abruptly cuneate at base, apex acute, margins entire, 3-nerved from base, finely gray-scurfy below, lighter above; upper leaves sessile, lower with petioles to as long as blade. Flowers in small axillary clusters, staminate and pistillate mixed, or staminate ones above; staminate flowers with 5-parted perianth, stamens 5; pistillate flowers subtended by 2 bracteoles, 3–4 mm long and as wide, broadly spatulate or oblong, united to apex, irregularly toothed, faces with short, thick, appendages. Seed brown, flat, 1.5 mm long, radicle superior. (2n = 18) Jul–Aug. Alkaline plains, hillsides, badlands; w SD, MT, WY; (SD to Alta., s to NM, AZ, UT).

10. *Atriplex rosea* L., red scale, red orache. Erect, annual monoecious herbs, stems 3–10 dm tall, divaricately branched from base, farinose at least above. Leaves alternate, blades lanceolate to rhombic-ovate, 2–7 cm long, coarsely sinuate-dentate, thinly gray-scurfy, entire and cuneate at base, obtuse at apex, thinly gray-scurfy at least below, uppermost reduced and entire, short-petiolate. Staminate flowers in upper axils, clustered, and in terminal spikes to 1 cm long, with 5-parted perianth, stamens 5; pistillate flowers in axillary glomerules of 5–10, bracteoles 2, these broadly ovate-triangular, ovate or rhombic, 4–10 mm long and wide, united in lower half, free and toothed above, evidently 3- to 5-nerved, central parts indurate about the seed, tubercles none to several. Seed dark brown, round, shiny, 1.7–2 mm diam, radicle inferior. (n = 9) Jul–Sep. A weedy plant in alkaline waste areas; scattered in n 2/3 GP; (scattered in w 1/2 U.S. & occasionally in e 1/2; Eurasia). *Adventive.*

11. *Atriplex subspicata* (Nutt.) Rydb., spearscale. Monoecious erect herbs, stems 3–10(15) dm tall, angular, with light green to green or sometimes reddish stripes, somewhat woody at base, branches 1–many, lower 2–5 opposite, upper alternate. Leaves opposite below, alternate above, somewhat succulent, usually with small scurfy scales, green to grayish-green, blades lanceolate to lance-linear to ovate or oblong, 3–12 cm long, 2–6 cm wide, often with a pair of obtuse lobes near base, margins entire or irregularly toothed; petioles 1–3 cm long. Flowers somewhat loosely arranged in interrupted short glomerules on short to long stalks in axils of upper leaves, inflorescences leafless except at base; staminate flowers with 5-parted perianth, stamens 5; pistillate flowers subtended by 2 bracteoles, these sessile, thick, green, becoming blackish in age, margins entire or sometimes with short teeth, broadly triangular to ovate-triangular, 3–10 mm long, longer than wide, dorsal surface with 1 or more tubercles, inner surface with spongy layer. Seeds of 2 types: brown, dull, 1.5–3 mm wide, flattened at base; black, shiny 1–2 mm wide, wider than long, radicle inferior to lateral or ascending. (2n = 36, 54) Jun–Aug. Usually on saline or alkaline soils in prairies, stream valleys, along shores, waste places; GP; (Newf. to B.C., s to NC, TX, UT). *A. hastata* L., *A. lapathifolia* Rydb.—Rydberg; *A. patula* L.—Atlas GP.

2. AXYRIS L.

1. *Axyris amaranthoides* L., Russian pigweed. Erect, monoecious annual herbs; stems (1.5)4–8(12) dm tall, simple below, usually with divergently ascending filiform branches above, these naked below, leafy or floriferous above, stems and branches sparsely to densely stellate pubescent. Leaves simple, alternate, lower early deciduous; blades elliptic, narrowly lanceolate, or ovate lanceolate, entire, 2–7(10) cm long, stellate pubescent below, glabrate above; petioles 2–15 mm long. Inflorescence paniculate, flowers unisexual; staminate spike slender 2–15(20) mm long, remotely glomerate below, at ends of stem and branches, calyx segments mostly 3, oblong to oblong-oval, obtuse, membranaceous, 0.4–0.7 mm long, stellate-hairy on back, stamens 3; pistillate flowers in axils of bractlike leaves below staminate inflorescence, each 3-bracteate, calyx segments 3(4), whitish-scarious, oblong, to 3 mm long, covered with stellate hairs interspersed with long simple hairs, styles 2. Utricles enclosed by accrescent sepals, obovate or oval, 2–3 mm long, compressed laterally, with 2-lobed wing at apex or unappendaged, brown to blackish, sometimes silvery. Jul–Sep. A weed in cult. fields, waste places, lakeshores, along streams; MN, ND, MO: Jackson; (across Can.; GP, CO, & ME; Siberia). *Naturalized.*

3. BASSIA All.

1. *Bassia hyssopifolia* (Pall.) O. Ktze., five-hook bassia. Erect annual herbs, stems simple to freely branched, 1–10(20) dm tall, hirsute or villous, sometimes becoming glabrate, often reddish-tinged. Leaves alternate, linear to linear-lanceolate, flat, attenuate at base, acute, pubescent, 1–4 cm long, 1–2(4) mm wide, the floral ones much reduced and oblong. Flowers mostly glomerate in terminal spikes, in lateral spikes and in axillary clusters, perfect, pistillate and sterile flowers commonly mixed; calyx-segments 5, villous-lanate, enlarging in fruit, each segment with a dorsal spine up to 2 mm long, spines subulate, curved, hooked,

glabrous; stamens 5, stigmas 2; fruiting calyx membranaceous, enclosing the fruit. Fruit flattened, plano-convex, brown to black, obovate, 1.5-2 mm long, pericarp membranaceous, free of seeds. Jul-Sep. A weed of fields, waste places, roadsides; MT: Phillips; SD: Jones; WY: Albany; CO: Bent, Boulder; (MA, NY, VA, w 1/2 U.S.; Eurasia). *Introduced.*

Bassia hirsuta (L.) Asch. may be expected in our area. It differs from *B. hyssopifolia* chiefly in having leaves semicylindrical and fleshy, and having only 3 conical tubercules per fruit.

4. CERATOIDES Gagnebin

1. ***Ceratoides lanata*** (Pursh) Howell, white sage, winter fat. Perennial, monoecious or dioecious spreading shrublet, with a gray-brown exfoliating bark, the woody base 0.5-1(2) dm tall, with many erect annual stems to 5 dm long, these densely stellate-pubescent and villous-tomentose, early grayish but becoming rufescent. Leaves alternate, linear or narrowly lanceolate, 1-4 cm long, 1.7-2.5 mm wide, entire, revolute, with dense stellate and simple hairs, the primary ones with reduced secondary branches fascicled in their axils. Annual branches floriferous 1/3-2/3 their length; staminate flowers uppermost on the stems, few to many in short capitate-spicate axillary clusters, the lower clusters often with 1-3 pistillate flowers at base, calyx lobes 4, obovate, 1.5-2 mm long, stamens 4, exserted; pistillate flowers 2-4 in the axils, naked, each enclosed by 2 ovate, connate, densely hirsute bracteoles that become 4-6 mm long, each with a hornlike tip; stigmas 2, exserted. Fruit oval, flat, 1.8-2.2 mm long, white pubescent. May-Jul. Plains & foothills, dry clay or chalky soils, often in saline or alkaline soil; w 1/2 ND, SD, NE, nw KS, MT s to NM; (e WA, s CA, e to ND, KS, NM). *Eurotia lanata* (Pursh) Moq.—Rydberg.

5. CHENOPODIUM L., Goosefoot, Lamb's Quarter

Contributed by Daniel J. Crawford and Hugh D. Wilson

Annual (rarely biennial or perennial) herbs; usually farinose, rarely glandular, pubescent, or glabrous; stem erect to spreading, solitary to branched from the base, or branched only above. Leaves alternate, blades linear to deltoid, entire to variously toothed or lobed. Inflorescence of glomerules (rarely large heads of dichotomous cymes), these variously arranged and terminal or axillary. Flowers small, inconspicuous, with 5 or rarely fewer sepals, petals lacking, stamens mostly 5, sometimes fewer, pistils with 2 or rarely up to 5 stigmas, ovaries with 1 basal ovule. Fruit a utricle, pericarp separable or attached; seeds with smooth to roughened or "honeycombed" surface, embryo curved around the central perisperm.

References: Standley, P. C. 1916. *Chenopodium, in* N. Am. Flora 21(1): 9-31; Aellen, P., & T. Just. 1943. Key and synopsis of the American species of the genus *Chenopodium.* Amer. Midl. Naturalist 30: 47-76; Wahl, H. A. 1952-53 (issued 1954). A preliminary study of the genus *Chenopodium* in North America. Bartonia 27: 1-46; Gleason, H. A., & A. Cronquist. 1963. *Chenopodium, in* Manual of Vascular Plants of Northeastern United States and Adjacent Canada, pp. 273-275, D. Van Nostrand Co., Princeton, NJ; Van Bruggen, T. 1976. *Chenopodium, in* The Vascular Plants of South Dakota, pp. 195-200, Iowa State Univ. Press, Ames.

This is a difficult genus in our area. Environmental conditions such as moisture, day length, and shading can cause the same species to exhibit marked differences in growth form. Leaf blade morphology, also phenotypically and developmentally plastic, is often of taxonomic value, especially size, shape, and margins of lower cauline leaves. Fruits and seeds are much less affected by the environment and provide useful features for distinguishing species. In fact, in many instances identifications are difficult, if not impossible, without mature fruits. Material possessing both mature fruits and lower, cauline leaves is ideal for determination but not always available.

1 Plants glandular and/or variously pubescent.
  2 Leaves mostly 3 cm or less long; flowers in dichotomous cymes .... 5. *C. botrys*
  2 Leaves mostly 6 cm or more long; flowers in small glomerules, rarely solitary, these disposed in spikes .... 2. *C. ambrosioides*
1 Plants farinose to glabrous, never pubescent or glandular.
  3 Flowers in large dense globose heads, often forming terminal spikes; fruits vertical .... 7. *C. capitatum*
  3 Flowers in glomerules, these separate, congested into spikes that may be disposed in panicles, axillary or terminal; fruits vertical and horizontal, or horizontal.
    4 Sepals usually 3; fruits vertical and horizontal.
      5 Leaves densely farinose beneath; perianth glabrous .... 12. *C. glaucum*
      5 Leaves glabrous beneath; perianth farinose to nearly glabrous .... 18. *C. rubrum*
    4 Sepals 5; fruits horizontal.
      6 Leaves with 1 vein from base.
        7 Pericarp readily separable .... 21. *C. subglabrum*
        7 Pericarp attached.
          8 Fruits 0.9–1.2 mm in diam .... 14. *C. leptophyllum*
          8 Fruits 1.3–1.6 mm in diam.
            9 Sepals enlarging slightly in fruit, fused for over half their length, with an undulate collar from the sinuses .... 8. *C. cycloides*
            9 Sepals not enlarging in fruit, fused for half their length or less, lacking an undulate collar from the sinuses .... 16. *C. pallescens*
      6 Leaves with 3 or more veins from base.
        10 Pericarp distinctly alveolate, fruits appearing "honeycombed" on surface.
          11 Pericarp white at maturity; fruits ± 1.0 mm in diam .... 22. *C. watsonii*
          11 Pericarp transparent, fruits always dark, 1.0–2.3 mm in diam.
            12 Fruits 1.0–1.5 mm in diam; style bases persistent and conspicuous on fruits; sepals with a winged keel equaling half the width of the sepal .... 4. *C. berlandieri*
            12 Fruits 1.5–2.3 mm in diam, style bases not prominent or sometimes lacking on fruits; sepals with a winged keel less than half the width of the sepal .... 6. *C. bushianum*
        10 Pericarp smooth to variously roughened, fruits appearing smooth to slightly roughened, never alveolate.
          13 Lower leaves toothed above the base, never either entire or with only basal lobes or teeth.
            14 Fruits 1.5–2.5 mm in diam; leaves with 1–4 large, divaricate, prominent teeth .... 11. *C. gigantospermum*
            14 Fruits usually 1.6 mm or less in diam; leaves without prominent teeth.
              15 Pericarp readily separable .... 19. *C. standleyanum*
              15 Pericarp usually attached, sometimes separable.
                16 Sepals exposing fruits at maturity .... 20. *C. strictum*
                16 Sepals mostly covering fruits at maturity.
                  17 Fruits about 1.0 mm in diam; blades of lower leaves less than 1 1/2 × longer than wide .... 15. *C. missouriense*
                  17 Fruits 1.1–1.5 mm in diam; blades of lower leaves 1 1/2 × or more longer than wide .... 1. *C. album*
          13 Lower leaves nearly always entire above the base, sometimes with 1 or 2 lobes or teeth at base.
            18 Lower leaves triangular to sometimes rhombic.
              19 Fruits 0.9–1.1 mm in diam; sepals covering fruits at maturity .... 13. *C. incanum*
              19 Fruits (1.1)1.2–1.4 mm in diam; sepals exposing fruits at maturity .... 10. *C. fremontii*
            18 Lower leaves ovate or narrower, never triangular.
              20 Fruits 1.3–1.6 mm in diam .... 16. *C. standleyanum*
              20 Fruits 0.9–1.4 mm in diam.

21 Lower leaves oval to oblong, entire, thick to nearly fleshy, moderately to densely farinose above; stem branched from base, usually spreading .... 9. *C. desiccatum*
21 Lower leaves broadly ovate to oblong, lanceolate or rarely nearly linear, entire or with basal lobes or teeth, usually thin, never fleshy, sparsely farinose to nearly glabrous above; stem solitary or sometimes branched from base, usually upright.
   22 Lower leaves mostly broadly ovate to oblong, 1.5–3 × longer than wide .. 3. *C. atrovirens*
   22 Lower leaves mostly lanceolate to rarely nearly linear, 3–5 × longer than wide .... 17. *C. pratericola*

1. **Chenopodium album** L., lamb's quarters. Erect annual to 1(1.5) m tall, stem solitary with well developed, often ascending and compacted lateral branches. Blades variable, often narrowly trullate to lanceolate, more than 1 1/2 × longer than wide, 3–5(6) cm long, 2–3(4) cm wide, moderately to heavily farinose, acute, margin irregularly sinuate-dentate to entire, cuneate. Inflorescence of glomerules, typically clustered into dense, paniculate spikes, often ascending at maturity, occasionally spreading. Sepals 5, moderately to densely farinose, median keel either absent or not strongly developed, usually enclosing the fruit at maturity; stamens 5; stigmas 2. Fruits horizontal, 1.1–1.5 mm in diam, pericarp lightly roughened, nonalveolate, usually attached to the seed. (2n = 54) Jun–Sep. Disturbed soil of open habitats; GP, scattered, but most abundant in n & e; (a cosmopolitan weed, probably European origin).

    This is probably one of the most widespread and variable angiosperm species, although its presumed abundance in our flora is based on the frequent inclusion of *C. berlandieri*, *C. bushianum*, and *C. missouriense* in *C. album*, and the tendency to place immature material which is usually impossible to identify under *C. album*. Eliminating these inclusions, we still find *C. album* to be present and sporadically distributed in disturbed, open habitats of the GP. It is often confused with *C. berlandieri* and *C. missouriense*. However, it can be readily distinguished from *C. berlandieri* by the lack of regular, minute honeycomb depressions on the pericarp. It differs from *C. missouriense* by its larger fruit, stricter habit, wider variation in flowering time, and lack of the dark nodal and infructescence pigmentation that is typical of many *C. missouriense* populations.

2. **Chenopodium ambrosoides** L., Mexican tea. Annual, biennial, or perennial herb to 1 m tall, with unpleasant odor; stem with several ascending branches. Blades of lower leaves oblong, ovate, or lanceolate, 2–14 cm long, 1–6 cm wide, densely yellow-glandular to glabrous, margin serrate, dentate or sinuate to shallowly sinuate-pinnatifid, cuneate, upper leaves progressively smaller, narrower, and becoming sessile and entire. Inflorescence of glomerules (flowers rarely solitary) disposed in bracted or bractless spikes. Sepals 5, sparsely glandular to glabrous; stamens 5; stigmas 2. Fruits horizontal or vertical, 0.7–1.0 mm in diam, pericarp readily separable from seed; seeds dark brown to black. (2n = 16, 32, 36, 48?, 64 have been reported) Aug–Oct. Weed of roadsides, gardens, pastures, waste ground; extreme se SD, e NE, e KS, OK; (naturalized throughout much of U.S.; trop. Amer.).

    The species is widespread and variable. Aellen and Just (op. cit.) discussed the geographic distribution of morphological variants. Whether morphological forms are associated with different chromosome numbers remains to be determined.

3. **Chenopodium atrovirens** Rydb. Annual, to 1 m tall, stem usually solitary or sometimes branched from the base, often branched above. Blades broadly ovate to oblong, those of lower leaves 1.5–3.5 cm long, 3–12 mm wide, sparsely to moderately farinose below, thin, entire or with basal lobes or teeth. Inflorescence of glomerules usually widely spaced, often disposed in open spreading panicles. Sepals 5, sparsely to moderately farinose, exposing fruit at maturity; stamens 5; stigmas 2. Fruits horizontal, 1.1–1.3 mm in diam, pericarp readily separable from seed or rarely lightly attached to seed. (2n = 18) Jul–Sep. Moist, open or disturbed sites; primarily a western montane species, barely reaching w GP; e WY, e CO, & e NM; (e CO to CA, MO to NM).

4. Chenopodium berlandieri Moq., pitseed goosefoot. Erect annual to 1.5 m tall, mostly less than 1 m. Blades variable, populations of n and w GP tend to show rhombic-ovate blades with greatest width toward the middle, (1)2–4(6) cm long, (0.5)1–3(4) cm wide, thick, tending toward fleshy, moderately to copiously farinose, often obtuse and mucronulate with irregularly sinuate margins, broadly cuneate or rounded at the base, occasionally ill-scented; scattered throughout, but more frequent in s GP, are populations producing rhombic-lanceolate, acute blades with irregular sinuate-serrate or sinuate-dentate margins. Inflorescence of glomerules, typically clustered into dense, paniculate, erect heads, although narrow-leaved forms tend to show a more diffuse configuration. Sepals 5, moderately to densely farinose, sharply keeled, enclosing the fruit at maturity; stamens 5; stigmas 2. Fruits horizontal, 1.2–1.5 mm in diam, pericarp alveolate, attached to the seed, although partial separation often occurs in fully mature fruits at the style-base, producing a distinctive yellow-white spot. ($2n = 36$) Jul–Sep. Disturbed, open ground; GP; (common throughout most of N. Amer.). *C. dacoticum* Standl., *C. petiolare* H.B.K., *C. ferulatum* (in part)—Rydberg; *C. album* L.—Gleason.

This is the most commonly encountered chenopod of our flora. Our plants are assignable to var. *zschackei* (Murr) Murr (sensu Wahl, op. cit.). This includes variants mentioned above, plus large-leaved, copiously farinose populations of sw SD (*C. dacoticum* Standl.). *C. berlandieri* is closely related to *C. bushianum* Aellen. Typical populations of each taxon are differentiated easily by fruit and leaf size, although intergrading populations, which occur in the eastern GP (especially e ND), may reflect introgressive hybridization (Wilson, Syst. Bot. 5: 253–263. 1980). Often confused with *C. album* L., *C. berlandieri* can be differentiated readily by the presence of strongly keeled sepals and the alveolate pericarp. Both characters show strongest expression in fully mature plants.

5. Chenopodium botrys L., Jerusalem oak. Annual, mostly 2–6 dm tall, stem solitary to much branched, glandular-pubescent. Blades of lower leaves oblong, oval or ovate, 1–8 cm long, 0.5–4 cm wide, sparsely to densely glandular-pubescent, margin shallowly sinuate to sinuate-pinnatifid, upper leaves progressively smaller, narrower, becoming nearly sessile and entire. Inflorescence of dense cymes, these disposed in axillary panicles which collectively may form a large terminal inflorescence. Sepals 5, conspicuously glandular-pubescent; stamens 5; stigmas 2. Fruits mostly horizontal, 0.6–0.8 mm in diam, pericarp attached to seed. ($2n = 18$) Jul–Sep. Weed of dry, sandy or gravelly areas; occurring very sporadically in ND, SD, NE, KS; (throughout much of U.S.; Europe). *Naturalized.*

6. Chenopodium bushianum Aellen. Erect annual to 2 m tall, often more than 1 m, stem solitary, lateral branches well developed, usually weakly ascending or spreading. Blades typically trullate to widely trullate (4)5–10(12) cm long, (2)3–8(9) cm wide, thin, lightly farinose, typically acute with irregular dentate or dentate-sinuate margins. Inflorescence of glomerules, typically clustered into dense, paniculate spikes, often pendulous and lead gray at maturity. Sepals 5, moderately farinose, sharply keeled and enclosing the fruit at maturity; stamens 5; stigmas 2. Fruits horizontal, 1.5–2.3 mm in diam, pericarp alveolate, attached to the seed. ($2n = 36$) Aug–Sep. Disturbed, open ground, often in alluvial soil of cult. fields; e GP; (se Can., s to VA, MO). *C. album* L.—Gleason, *C. paganum* Reich.—Rydberg.

Often confused with *C. album* L., this species can be distinguished readily by the alveolate pericarp, larger fruits, and strongly keeled sepals. It is closely related to *C. berlandieri* Moq., but typical *C. bushianum* is distinguished by large fruit with no obvious style-base spot, wider sepals, and larger, thinner leaves. In addition, *C. bushianum* plants are usually more robust, tend to occupy only alluvial soil and tend to flower only in late Aug or early Sep. Both species are cross-compatible and sympatric in the eastern GP. Hybridization may be responsible for atypical populations present in our flora, especially in eastern ND.

7. Chenopodium capitatum (L.) Asch., strawberry blite. Annual, stem solitary to much branched from base, to 6 dm tall. Blades triangular to triangular-hastate, 2–10 cm long, nearly as wide, glabrous, margin entire or sparsely sinuate or dentate. Inflorescence of large (up to 1 cm), dense heads, these in axils of upper leaves or forming terminal leafless spikes. Sepals 3(4–5), sometimes red and becoming deliquescent at maturity; stamens (1–2)3–4(5); stigmas 2. Fruits vertical, 0.8–1.5 mm long, pericarp attached to seed. ($2n = 16$, 18) Jun–Aug. Often on recently disturbed sites in woodlands; rare in GP, known from several collections in w SD & w MN; (sporadic in much of U.S.; Eurasia).

Wahl (op. cit.) recognized plants with weakly toothed or entire leaves, smaller heads (6 mm or less in diameter) and shorter, more squarrose stigmas as *C. overi* Heller. He suggested that these plants are "apparently native" in the western U.S. Since *C. capitatum* is naturalized in the U.S., Wahl's statement implies that the two species originated in different hemispheres. Critical study of the situation is needed.

8. ***Chenopodium cycloides*** A. Nels., sandhill goosefoot. Annual, to 1 m tall; stem usually much branched above. Blades mostly linear, with 1 vein from base, 1–2 cm long, 1–1.5 mm wide, glabrous above, slightly farinose below. Inflorescence of glomerules disposed in congested or interrupted spikes which may be narrow panicles. Sepals 5, sparsely farinose, fused for over half their length, with an undulate collar from the sinuses, enlarging in fruit and spreading to expose the fruit, stamens 5, stigmas 2. Fruits horizontal, 1.3–1.6 mm in diam, pericarp firmly attached to seed. Aug–Oct. Sandy soil; extreme sw KS; (s NM through w TX & n TX, to KS).

This is a rare and little known, though distinctive, species. No collections made within the last 25 years have been seen.

9. ***Chenopodium desiccatum*** A. Nels. Annual, to 4 dm tall, stem usually branched from the base and branches mostly somewhat spreading. Blades oval to oblong, those of lower leaves 1–2.5 cm long, 3–9 mm wide, moderately to densely farinose below, thick to almost fleshly, entire. Inflorescence of glomerules mostly crowded into dense spikes, these often disposed in panicles. Sepals 5, densely to sparsely farinose, mostly covering fruit at maturity; stamens 5; stigmas 2. Fruits horizontal, 0.9–1.1 in diam, pericarp readily separable from seed. ($2n = 18$) Jul–Sep. Dry soil & disturbed areas; in high plains & mts.; e CO & WY; (sporadic in sw U.S. to WY).

10. ***Chenopodium fremontii*** S. Wats., Fremont goosefoot. Annual, stem solitary or branched from base, usually branched above, up to 6 dm tall. Blades of lower leaves deltoid, with one or more lobes or teeth at base, usually entire above, to 4 cm long, about as wide, varying in thickness, sparsely farinose to glabrous above, sparsely to moderately farinose below. Inflorescence of glomerules or short spikes, often disposed in open terminal or axillary panicles. Sepals 5, sparsely to moderately farinose, opening to expose fruit at maturity; stamens 5; stigmas 2. Fruits horizontal, 1.1–1.4 mm in diam, pericarp readily separable from seed. ($2n = 18$) Jul–Sep. In shaded areas; GP, infrequent throughout; (widely distributed in U.S. but much more common in sw).

This is a widespread and variable species, although it is almost always recognizable by the large, thin, deltoid leaves. Much of the variation may be attributable to environmental factors.

11. ***Chenopodium gigantospermum*** Aellen, maple-leaved goosefoot. Erect annual to 1.5(2) m tall, stem solitary, usually with well-developed, spreading, lateral branches. Blades broadly ovate, 7–20 cm long, 5–15 cm wide, lightly farinose to glabrous, thin, acute to acuminate, margin deeply sinuate-dentate with 1–4 large teeth separated by rounded sinuses, rounded to subcordate at the base. Inflorescence a terminal panicle of short glomerate spikes, becoming diffuse at maturity. Sepals 5, sparsely farinose or glabrous, not enclosing the fruit at maturity; stamens 5; stigmas 2. Fruit horizontal, fusiform-lenticular, 1.5–2.4(2.7) mm in diam, pericarp cellular-reticulate, readily separable or attached. ($2n = 36$) Jul–Sep. Disturbed ground, occasionally in shaded, woodland situations; GP; (circumboreal). *C. hybridum* L.—Atlas GP; *C. hybridum* var. *gigantospermum* (Aellen) Rouleau—Fernald.

These distinctive plants are not confused readily with other species of the genus. Our populations appear to be native to North America. They are morphologically (Baranov, Rhodora 66: 168–171. 1964) and cytologically (Uotila and Suominem, Ann. Bot. Fennici 13: 1–25. 1976) distinct from the *C. hybridum* complex of Eurasia.

12. *Chenopodium glaucum* L., oak-leaved goosefoot. Annual, stem solitary to usually much branched from base, branches prostrate to ascending, 2–7 dm long. Blades oblong, ovate or lanceolate, 0.7–4 cm long, 0.4–2 cm wide, very densely farinose beneath, margin entire to undulate or sparsely toothed, cuneate at base. Inflorescence of small glomerules, disposed in axillary spikes or rarely in a terminal panicle. Sepals 3(4-5), glabrous; stamens (1)3(4-5); stigmas 2, short. Fruits horizontal or vertical, 0.8–1.0 mm in diam, pericarp moderately to readily separable from seed. (2n = 18) Jul–Sep. Weed of primarily alkaline habitats; GP, more common in n; (widely naturalized in U.S.; Europe). *C. salinum* Standl. — Rydberg.

This species is variable in features of the leaf margins and the inflorescence. The forms with sharper serrations on the blades and more leafy (bracteate) inflorescences have been recognized as distinct at the specific or varietal level. The question of whether such forms are deserving of taxonomic recognition can be answered only after additional study.

13. *Chenopodium incanum* (S. Wats.) Heller. Annual, stem usually much branched from the base, branches spreading to sometimes ascending, mostly less than 25 cm long. Blades of lower leaves narrowly triangular to rhombic, 7–15 mm long, about as wide, moderately to sparsely farinose above, densely to moderately farinose below, very thick, with one or two lobes or teeth at base of each side, entire above. Inflorescence of glomerules congested into spikes, these disposed in terminal or axillary panicles. Sepals 5, densely to moderately farinose, covering fruit at maturity; stamens 5; stigmas 2. Fruits horizontal 0.9–1.1 mm in diam, pericarp readily separable from seed. (2n = 18) Jul–Aug. Dry plains, often on disturbed soil; GP, more common in s; (widely distributed in w U.S., but more common in sw).

This widespread species consists of a Great Plains element, a form that occurs in the Great Basin and Mohave Desert, and a form found in w TX and s NM and AZ. These have been recognized as distinct taxonomically at the varietal level (Crawford, D. J. Brittonia 29: 291–296. 1976). Our GP plants, which are low and usually profusely branched, constitute var. *incanum*.

14. *Chenopodium leptophyllum* Nutt. ex Moq. Annual, to 9.0 dm tall, stem solitary or branched from the base and branches upright. Blades linear, with single vein from base, those of lower leaves 2–4 cm long, mostly less than 1.5 cm wide, margin entire. Inflorescence of glomerules mostly spaced or sometimes crowded. Sepals 5, moderately to densely farinose, exposing or covering fruit at maturity; stamens 5; stigmas 2. Fruits horizontal, 0.9–1(1.1) mm in diam, pericarp attached to seed. (2n = 18) Jul–Sep. Open sites; e NM; (widely distributed in w U.S., but much more common in sw U.S., sporadically introduced in e U.S.).

15. *Chenopodium missouriense* Aellen. Erect annual to 1.5 m tall, often over 1 m, stem solitary, lateral branches usually well developed, spreading, and flexuous at maturity. Blades variable, often trullate to rhombic-ovate, not more than 1 1/2 × longer than wide, (2)3–4(5) cm long, (1)2–3(4) cm wide, lightly to moderately farinose, rounded to acute, margin irregularly and coarsely sinuate-dentate, often with large teeth near the broadly cuneate base. Infloresence of relatively small, delicate glomerules, typically diffuse, often highly pigmented and pendulous at maturity. Sepals 5, median keel not well developed, lightly farinose, enclosing the fruit at maturity; stamens 5; stigmas 2. Fruits horizontal, 0.9–1.2 mm in diam, pericarp lightly roughened, not alveolate, usually attached to the seed. (2n = 54) Sep. Disturbed, open areas; GP, more frequent in the e & s; (e U.S. to GP). *C. album* L. — Rydberg, Gleason, Fernald.

This species is distinguished readily from *C. berlandieri* by its small, nonalveolate fruit. It is separated from *C. album* by its smaller fruit and less compacted, more or less delicate, flexuous inflorescence which is produced only in Sep. Pigmentation at the node, which may be present, absent, or segregating in populations of *C. berlandieri*, *C. bushianum* and *C. album*, is usually present and strongly expressed in *C. missouriense*.

16. **Chenopodium pallescens** Standl. Annual, to 5 dm tall, stem much branched, ascending. Blades linear, with one vein from base, 1–4 cm long, 1–6 mm wide, sparsely farinose to glabrous. Inflorescence of glomerules arranged in spikes which are disposed in cymes or panicles. Sepals 5, sparsely farinose, enclosing fruit at maturity; stamens 5; stigmas 2. Fruits horizontal, 1.3–1.6 mm in diam, pericarp firmly attached to seed. (2n = 18) Jun–Aug. Primarily in rocky or sandy habitats; w MO & e KS to cen WY; (NM, OK, GP; also known from one collection in IN).

This species is rare, with only a few collections made during the past 60 years. Several reports of the species from w KS are based upon misidentifications.

17. **Chenopodium pratericola** Rydb. Annual, up to 1 m tall, stem usually solitary from base and branched or unbranched above, branches strongly ascending. Blades lanceolate to rarely nearly linear, those of lower leaves 1.5–4(4.5) cm long, 5–15 mm wide, thin, often with 1 or 2 basal lobes or teeth. Inflorescence of glomerules often crowded in a compact panicle. Sepals 5, moderately to densely farinose, exposing fruit at maturity; stamens 5; stigmas 2. Fruits horizontal, 1.0–1.3 mm in diam, pericarp readily separable from seed. (2n = 18) Jul–Sep. Dry soil, open & disturbed areas; GP; (widely distributed over w 1/2 of U.S., occurring sporadically in e U.S. where introduced). *C. albescens* Small — Rydberg. *C. dessicatum,* misapplied in reports from GP.

18. **Chenopodium rubrum** L., alkali blite. Annual, to 1 m tall, stem solitary or branched from the base, prostrate to erect. Blades rhombic-ovate or lanceolate, 2–12 cm long, to about as wide, glabrous, 1–several teeth on each side, cuneate below the teeth, upper leaves becoming smaller, often entire and nearly sessile. Inflorescence of glomerules aggregated either into axillary spikes, or terminal spikes disposed in panicles. Sepals 3, lightly farinose; stamens often 3; stigmas 2. Fruits vertical, 0.6–1.0 mm in diam, pericarp readily separable from seeds. (2n = 36) Jun–Sep. Occurs primarily in alkaline or saline situations; ND, SD, e MN, NE, cen KS; (widely distributed in U.S.). *C. humile* Hook.— Rydberg.

Plants with larger seeds (0.8–1.0 mm), entire to few-toothed blades, and spreading stems have been segregated as *C. humile* or treated as a variety of *C. rubrum*, by Wahl (op. cit.).

19. **Chenopodium standleyanum** Aellen. Annual, to 2 m tall (mostly less than 1 m), stem often branched above. Blades lanceolate to sometimes ovate or rhombic, 2–10 cm long, 0.5–5 cm wide, sparsely farinose to mostly glabrous above, moderately to sparsely farinose or nearly glabrous below, very thin, entire or the larger ones often with several teeth. Inflorescence of glomerules disposed in spikes that form open, sometimes nodding panicles. Sepals 5, moderately to sparsely farinose, exposing fruit at maturity; stamens 5; stigmas 2. Fruit horizontal, 1.3–1.6 mm in diam, pericarp readily separable from seed. (2n = 18) Jul–Sep. Most common in wooded areas; e GP; (widely distributed in e U.S. to GP). *C. boscianum* Moq.— Rydberg.

20. **Chenopodium strictum** Roth. Erect annual to 1 m tall, stem solitary, often with strongly developed basal branches that project horizontally from the main stem and ascend sharply. Blades distinctive, showing strong developmental polymorphism, basal blades oblong-ovate, about 2 × longer than wide, 3–5 cm long and 1–2 cm wide, obtuse, margin finely serrate, broadly cuneate, upper blades tending toward lanceolate and entire, plants lightly

farinose. Inflorescence of small glomerules, often ascending and distinctively spicate. Sepals 5, lightly farinose, median keel weakly developed, reflexing to expose the fruit at maturity; stamens 5; stigmas 2. Fruits horizontal, rarely exceeding 1 mm in diam, pericarp smooth to lightly roughened, nonalveolate, usually attached to the seed. ($2n = 36$) Aug–Sep. Disturbed soil in open habitats; GP, more frequent toward n; (circumboreal).

Our material has been assigned to subsp. *glaucophyllum* (Aellen) Aellen & Just. Wahl (op. cit.) treated this at the varietal rank indicating that northern populations are naturalized and southern populations (IL, MO, AR) may represent another taxon. As is the case with several other wide-ranging and polymorphic species complexes of the genus, additional systematic work is needed.

21. Chenopodium subglabrum (S. Wats.) A. Nels. Annual, stem solitary, or branched from base, sometimes branched above, up to 8 dm tall. Blades linear, entire, to 3 cm long, with single vein from base, glabrous. Inflorescence of glomerules disposed in terminal and axillary panicles. Sepals 5, glabrous, exposing fruit at maturity; stamens 5; stigmas 2. Fruits horizontal, 1.2–1.6 mm in diam, pericarp readily separable from seed. Jul–Sep. Sandy areas; GP, very infrequent; (very sporadic & infrequent, MT, WY, CO, UT).

This species is easily recognizable on the basis of its linear, 1-veined glabrous leaves (more glabrous than any other species in our area), and large fruits with readily separable pericarps. However, it is quite rare, and has been collected very infrequently during the last several decades.

22. Chenopodium watsonii A. Nels. Annual, very ill-scented, stem usually much branched from base, sometimes solitary or weakly branched above, branches ascending to spreading, up to 8 dm long. Blades of lower leaves ovate, rhombic to nearly orbicular, 1–4 cm long, about as wide, moderately farinose to nearly glabrous above, densely to moderately farinose below, thick to fleshy, margin with 1 or 2 teeth or lobes on each side near base and entire above, or these absent and blades entire. Inflorescence of dense glomerules aggregated into loose or dense spikes, disposed in panicles. Sepals 5, densely to moderately farinose, covering fruit at maturity; stamens 5; stigmas 2. Fruits horizontal, 0.9–1.1 mm in diam, pericarp alveolate, white at maturity, firmly attached to seeds. ($2n = 18$) Jul–Sep. Dry prairies & plains, or disturbed soil at higher elevations in mts.; most common in w-cen GP; (widely distributed in sw U.S., very rare n of CO & UT).

6. CORISPERMUM L.

Contributed by Ralph E. Brooks

Annual branching herbs. Leaves alternate, simple, linear or lanceolate. Flowers small, perfect, solitary in the axils of bracts, together forming congested to loose spiciform inflorescences; perianth wanting or consisting of 1–5 scarious segments, hypogynous; stamens 1–3(5); anthers oval to subglobose; stigmas 2, style obsolete. Fruits oval to obovate, strongly flattened, dorsally convex, ventrally concave, the margins winged or not, pericarp adherent to the vertical seed, embryo annular.

Reference: Maihle, N. J., & W. H. Blackwell. 1978. A synopsis of North American *Corispermun* (Chenopodiaceae). Sida 7(4): 382–391.

1 Spiciform inflorescences at maturity dense, the bracts concealing the stem .... 1. *C. hyssopifolium*
1 Spiciform inflorescences at maturity loose and usually somewhat lax, the bracts not concealing the stem .... 2. *C. nitidum*

1. Corispermum hyssopifolium L., hyssopleaf tickseed. Annual, sparsely to densely stellate pubescent; stems solitary, freely branching, 1.5–6 dm tall, frequently reddish with age.

Leaves linear to linear-lanceolate, 2–5 cm long, 1–4 mm wide, apex acute, margins entire. Flowers borne in dense, erect, spiciform inflorescences 1–5(10) cm long; lowest bracts leaflike, upper bracts imbricate and hiding the stem, widely ovate, (4)5–7(10) mm long, 3–4 mm wide, the body wider than the fruit, apex acuminate to caudate, straight or slightly recurved, margin entire, hyaline. Perianth segments 1(–3), membranaceous, translucent, oval to oblong, ca 1 mm long, apex rounded to truncate, margins erose. Fruit obovate to oval, 2.5–3.5 mm long, smooth, the edge with a distinct lighter pellucid rim or narrow wing (to 0.2 mm wide; best observed from the ventral side). Aug–Sep. Sandy soil in railroad yards, on river bars, or waste areas; MN: Clay; ND: Cass; SD: Fall River; KS: Harvey, Meade; (Que. to AK, s to NJ, OH, TX, & CA; Mex.; Eurasia). *Adventive. C. marginata* Rydb., *C. simplicissimum* Lunell — Rydberg.

Closely related to *C. hyssopifolium* is *C. orientale* Lam. (= *C. emarginatum* Rydb., *C. villosum* Rydb. — Rydberg), which has been reported for the n GP. While similar in habit, *C. orientale* is described as having wingless fruits. This at first seems clear enough until one examines a series of fruits from "orientale-hyssopifolium" type plants. On a single plant the fruits may vary from wingless to distinctly winged. Because of the small number of specimens at hand, the adventive nature of the species, and the variability of the fruit character, I believe further taxonomic disposition of our plants (or N. Amer. plants, as done by Maihle and Blackwell, op. cit.) is presumptuous without consideration of comparative materials from Eurasia.

Corispermum sibericum Iljin has been collected once in ND: Richland. It is similar to *C. hyssopifolium* in habit but the wings on the fruit are quite wide, about 1/3 as wide as the kernel. This appears to represent the first N. Amer. report for the species.

2. ***Corispermum nitidum*** Kit., bugseed. Annual, sparsely hirsute or stellate, pubescent to glabrate; stems solitary, freely branching, 2–5 dm tall, often reddish with age. Leaves linear, 1–4 cm long, 1(3) mm wide, apex acute, margins entire. Flowers borne in loose, somewhat lax long-spiciform inflorescences to 15 cm long; lower bracts leaflike and remote, upper bracts partially overlapping but the stem usually exposed, ovate 3.5–6(8) mm long, 1.5–2.5 mm wide, the body usually narrower than the fruit, margin hyaline, entire, apex acuminate, straight to slightly recurved. Perianth segments 1(3), membranaceous, translucent, oblong to obovate, ca 1 mm long, apex rounded, margin erose. Fruit oval, 2.5–3.5 mm long, smooth, winged, the wings 0.2–0.5 mm wide. Aug–Sep. Deep sandy soil in prairies, dunes, or disturbed sites; widely scattered in n & w GP, perhaps more common than records indicate; (widely scattered in N. Amer., especially the cen & w; Eurasia). *Naturalized.*

7. **CYCLOLOMA** Moq.

1. ***Cycloloma atriplicifolium*** (Spreng.) Coult., tumble ringwing, winged pigweed. Freely branched, bushy, annual herbs 1–8 dm tall and of equal diam, of tumbleweed habit, stems erect or spreading, divaricately branched, striate, branches slender, angled, loosely and finely woolly, becoming glabrate in age. Leaves alternate, principal stem leaves to 6–7 cm long, 1–1.5 cm wide (these usually absent at fruiting time), blades lanceolate to ovate, irregularly sinuately dentate, acute at apex, cuneate at base; petioles 0–15 mm long. Flowers in terminal, interrupted spikes, appearing paniculate, perfect or a few pistillate; sepals 5, keeled, the calyx developing below its lobes a continuous, horizontal, irregularly lobed and toothed membranaceous wing; stamens 5; ovary densely tomentulose; styles 2 or 3, partially united. Fruit depressed-globose, enclosed in calyx, pericarp free from the seed; seed flat, 1.3–1.7 mm wide, black, horizontal, bearing scattered, white, silky hairs. May–Oct. Weedy, most often found in sandy places; GP; (Man. to IN, AR, TX, w to WY, UT, NM).

8. KOCHIA Roth

1. *Kochia scoparia* (L.) Schrad., kochia, fire-weed, summer or mock cypress, Belvedere, Mexican fire-bush. Annual herbs, stems erect, 3–20(40) dm tall, usually branched from base, branches erect or spreading, stems and branches yellowish-green, green or streaked with red, sometimes purplish-red in autumn, short-villous to pilose with silvery or rusty hairs, sometimes nearly glabrous, usually glabrous or glabrescent below. Leaves alternate, 2–7(10) cm long, 0.5–8(12) mm wide, with 1 or 3(5) veins, lower leaves linear to lanceolate, often oblanceolate to narrowly obovate, acute or obtuse to rounded at apex, narrowed to a distinct petiole, those upward becoming linear, elliptical, narrowly lanceolate or oblanceolate, acute or acuminate at apex, narrowing slightly at sessile base, all entire, flat, ciliate, glabrate especially above, to usually villous or pilose with hairs to 6 mm long. Inflorescences remotely long-spiciform, densely long-spiciform, to short, compact cylindric or oblong-claviform, some plants floriferous for much of their length. Flowers perfect, or functionally pistillate or staminate, or some strictly pistillate on same plant, usually paired in axils of leaflike but reduced bracts 3–18 mm long, sometimes solitary or rarely 3–5, each enveloped by tufts of short to long hairs arising below each flower; calyx campanulate to urceolate, 0.3–0.6 mm long at anthesis, glabrous except for the 5 ciliate lobes, at maturity 2.3–3 mm wide, with sepals incurved over the fruit, and bearing 5 highly variable horizontal dorsal lobes or wings, varying from short and tuberclelike to usually flat, oblong-rotund or rotund, semimembranaceous, cellular-reticulate, often striate structures, to 2 mm long, which may be variously lobed of bifid; stamens 5, exserted to included; styles 2(3), distinct or usually united for 0.3 mm, stigmatic nearly their full length. Fruit a depressed-globose utricle with membranaceous persistent pericarp free from seed; seed horizontal, obovate, (1.5)2–3 mm long, faces concave, surfaces brown to black, dull, smooth to granular. (n = 9) Jul–Oct. A weed of roadsides, pastures, fields, waste areas; GP but infrequent in extreme se; (cosmopolitan weed; Asia & Europe). *Naturalized.*

The common plants in our area vary much in degree of hairiness, size and shape, form of the inflorescence, and coloration, particularly late in the season. They differ markedly from the cultivated kochia, which usually has a dense conical or nearly globular shape, leaves usually linear or linear-lanceolate, all parts becoming purple-red in autumn, and flowers along lateral branches usually pistillate while those near the ends of principal branches are perfect. This cultivar has commonly been called *K. scoparia* var. *culta* Farw. or forma *trichophila* (A. Voss) Stapf ex Schinz & Thell. It is quite possible, however, that the cultivar is actually the *scoparia* of Linnaeus while our common plant should be referred to *K. siversiana* (Pall.) C. A. Mey.

While our plants usually are considered as objectionable weeds, they are readily grazed by livestock. It has been reported that the protein digestibility of kochia is nearly equal to that of alfalfa. Recently seed has become available commercially to grow the plant for grazing, hay, and ensilage. However, kochia may cause a photosensitization syndrome in cattle.

9. MONOLEPIS Schrad.

1. *Monolepsis nuttalliana* (R. & S.) Greene, poverty weed. Annual herbs, freely branched from base, stems 1–3 dm long, prostrate to ascending, slightly succulent, sparsely mealy to glabrate. Leaves alternate, fleshy, 1–6 cm long, gradually reduced upward, lanceolate to ovate or triangular, with a pair of divergent lobes at middle or below, sometimes with a few teeth above, otherwise entire, cuneate to the base. Flowers in dense, sessile, axillary clusters, often reddish, perfect or a few pistillate; calyx of 1, entire, persistent, herbaceous and bractlike, oblanceolate, acute sepal, 1.5–2.5 mm long; stamen 1 or lacking in pistillate flowers, styles 2, short; ovary compressed, ovoid. Pericarp of fruit finely pitted at maturity, adherent to seed when dry; seed vertical, dark brown to black, 0.8–1.4 mm wide. (2n = 18)

Apr–Sep. Dry to moist soils of fields, prairie ravines, roadsides, & waste places, often on saline or alkaline soils; GP; (Man., to B.C., s to MO, TX, w to Pacific, adventive eastward).

10. SALICORNIA L.

1. **Salicornia rubra** A. Nels., saltwort. Annual herbs; stems erect or ascending, 1–2(3) dm tall, usually freely branched, branches opposite, articulate, joints dilated at apex into a short sheath, stems and branches succulent, glabrous, often reddish at maturity. Leaves scalelike, opposite. Flowers perfect or some imperfect, borne in groups of 3(1–7), sessile and sunken in the joints of fleshy spikes, the flowering joints forming cylindric terminal spikes, flowers usually connate and adnate to joints, perianth of lateral flowers pyramidal, that of the central flower longer than others, nearly obpyramidal, nearly closed above but with a slit through which essential organs barely protrude, mature calyx carinate; stamens 2(1), only anthers exserted; style branches 2. Fruit compressed laterally, enclosed in the spongy perianth; pericarp thin and papery; seed vertical, puberulent, yellowish-brown, 1–1.5 mm long. (2n = 18) Jul–Sep. Moist saline or alkaline soils; w MN, ND, SD, NE; KS: Stafford; MN s to NM; (w MN to s B.C., s to cen KS, NV, NM).

11. SALSOLA L.

Contributed by Ralph D. Brooks

Annual herbs (ours), shrubs, or trees. Leaves alternate, or rarely opposite, simple. Flowers small, perfect, solitary in the leaf axils and together forming short- to long-spiciform inflorescences; perianth (4)5-merous, segments distinct (rarely united at the very base), thickened and closely enveloping the ovary below, membranaceous and erect above, and with a median horizontal protuberance that in fruit develops into a winglike projection in some species; stamens (3)5, hypogynous, inserted at the edge of a minute lobed disk, exserted; stigmas 2(3). Fruit enveloped by the persistent calyx, obovoid, depressed at the apex, usually horizontally ridged or winged; seeds horizontal with a spiral embryo.

Reference: Beadle, J.C. 1973. Russian-thistle *(Salsola)* species in western United States. J. Range Manag. 26: 225–226.

The vast majority of *Salsola* specimens examined in various herbaria were found to be poorly collected, i.e., too young or fragmentary. Every effort should be made to procure specimens late in the season after the inflorescences have developed fully and preferably when fruiting has started. Many members of this genus are by nature weedy and the introduction of several species to the west of the GP make these prime candidates to infiltrate the region in years to come.

1 Inflorescences long-spiciform, (7)15–40 cm long; bracts at maturity appressed or only the apical portion slightly recurved; fruits lacking prominent wings .... 1. *S. collina*
1 Inflorescences short-spiciform, 1–7(10) cm long; bracts at maturity spreading and usually recurved; fruits prominently winged .... 2. *S. iberica*

1. **Salsola collina** Pall., tumbleweed. Glabrous or lightly hirsute annual herb to 1 m tall; stems green or sometimes red-streaked at maturity, freely branching but the lateral branches much reduced (monopodial), forming rounded bushes that late in the season break off and roll in the wind. Leaves alternate, filiform, 2–6 cm long, 1 mm wide, reduced upwards, apex spinose, margins entire or crenulate, base broad, coriaceous and hyaline margined. Inflorescences long-spiciform, (7)15–40 cm long, frequently nodding toward the tip; bracts ovate, 4–8 mm long, apex straight or slightly recurved, weakly spinose, margins entire to crenulate, base appressed; bracteoles somewhat shorter and narrower. Perianth 2.5–3.5

mm long, segments ovate to oval below the somewhat inconspicuous protuberance, above the protuberance ovate to narrowly deltoid, apex acute, margins entire to crenulate; stamens included at first, the later ones often exserted, anthers ca 0.6 mm long; stigmas filiform, ca 1.5 mm long, style much shorter. Fruit obovoid and horizontally ridged at the apex, 1.5–2.5 mm in diam; seeds blackish, shiny, cochleate, ca 1.5 mm in diam, smooth. (n = 9) Aug–Oct. Cultivated fields, roadsides, along railroads, & other open disturbed sites; w GP & scattered e; (scattered areas of N. Amer., especially cen & w U.S.; w Asia). *Naturalized.*

While collections of *S. collina* were made as early as 1922 in the U.S., the species went undetected until 1958 when it was reported for CO, IA, & MN (Schapaugh, W. 1958. Proc. Iowa Acad. Sci. 65: 118–121). Today in areas such as w KS and e CO the species is equally as common as *S. iberica*. The two species frequently grow together with what appear to be sterile hybrids sometimes present, however, this has not been formally documented.

2. Salsola iberica Senn. & Pau, Russian-thistle, tumbleweed. Annual herb to 1 m tall; stems usually red-streaked, freely branching without a main axis but with many, more or less equal, lateral branches, forming rounded bushes that late in the season break off and roll in the wind, sparsely and coarsely hirsute to glabrous. Leaves alternate, filiform, 2–8 cm long, 1 mm wide, reduced upwards, apex spinose, margins entire to denticulate, base broad, coriaceous, and hyaline margined. Inflorescence short-spiciform, 1–7(10) cm long, rigid; bracts spreading and often recurved, ovate to narrowly deltoid, 3–8 mm long, glabrous or pubescent at the base, apex strongly spinose, margins entire to mostly denticulate-crenulate; bracteoles similar but small. Perianth 2.5–3.5 mm long, segments ovate to oval or oblong below the median protuberance, above the protuberance narrowly deltoid, apex acute, margins entire to crenulate; stamens at first included and later exserted, anthers ca 1 mm long; stigmas filiform, ca 1.5 mm long, style much shorter. Fruit obovoid with a prominently membranaceous winged apex, body 1.5–2.5 mm in diam, including wings 3–5(6) mm across; seeds blackish, shiny, cochleate, ca 1.5 mm in diam, smooth. (n = 18) Aug–Oct. Cult. fields, roadsides, along railroads, & other open disturbed sites; w GP & scattered eastward; (well established in w & cen N. Amer. & scattered e; Eurasia). *Naturalized.* *S. kali* L. var. *tenuifolia* Tausch — Fernald; *S. pestifer* A. Nels. — Rydberg.

This species was first introduced with flaxseed in SD around 1873, from Russia, where it was then a serious weed.

12. SARCOBATUS Nees

1. Sarcobatus vermiculatus (Hook.) Torr., greasewood. Perennial monoecious shrubs, stems erect, 0.3–3 m tall, much branched, branches rigidly stout, divaricate, younger ones yellowish-white, becoming grayish, glabrous or with short, white, branched hairs, many branches spine-tipped. Leaves opposite below, alternate above, deciduous, (1)1.5–4 cm long, fleshy, entire, linear and subterete, obtuse or acute at apex, narrowed to base, rounded on upper surface, glabrous or with branched hairs. Flowers imperfect, in numerous axillary spikes, 1–3 cm long; staminate flowers uppermost in spike, more numerous than pistillate flowers, perianth lacking, each flower subtended by a peltate, stipitate bract, stamens 2 or 3, nearly concealed by bract; pistillate flowers sessile, 1 or 2 together in axils of leaves, pistil surrounded by a cuplike, shallowly lobed or nearly entire perianth that in fruit develops into wide membranaceous wings, 6–12 mm wide; styles 2, short. Fruit turbinate, winged above the middle, body 4–5 mm long, to 1 cm wide (including wing margin); seed erect, orbicular, 1.8–2.2 mm wide. May–Aug. Usually at base of eroded hills, flood plains, often in barren saline or alkaline soils; w ND, SD, nw NE, MT, s to NM; (Sask., ND, in w NE, WY, TX, w to e WA, s CA).

13. SUAEDA Forsch. ex. Scop., Seepweed, Sea Blite

Annual herbs or shrubs, more or less fleshy; leaves alternate, linear, terete or dorsiventrally flattened, commonly glaucous. Flowers perfect or imperfect, sessile, solitary to clustered, in upper axils of a single bract, in short to elongate spikes, each flower usually with 2(4) very small membranaceous bracteoles; calyx of 5 sepals, fleshy, united for 1/2 their length or more, deeply lobed, erect or ascending, usually keeled or narrowly winged at maturity, persistent in fruit; stamens usually 5; styles often 2(3–5). Fruit compressed, surrounded by the calyx; seed horizontal or vertical, smooth, tuberculate or reticulate, often glossy.

1 Plants annual; calyx lobes unequal .... 1. *S. depressa*
1 Plants perennial; calyx lobes equal.
   2 Herbage glabrous below the inflorescence .... 2. *S. moquinii*
   2 Herbage evidently tomentose .... 3. *S. suffrutescens*

1. Suaeda depressa (Pursh) S. Wats., sea blite. Annual herbs, stems erect to decumbent, often low and spreading, simple or freely branched from base, glabrous, glaucous, 1–6(10) dm tall. Leaves alternate, linear, semiterete, 1–4 cm long, glabrous, glaucous, often crowded, much reduced in the inflorescence. Spikes slender, flowers congested, 3–7 in each axil, bracts 2–3 mm long, ovate-lanceolate; calyx parted to the middle, lobes distinctly unequal, 1.5 mm long, 1.5–2 mm wide at maturity, all cucullate but upper 1 or 3 sepals much more strongly corrugate-corniculate transversely; stamens 5; stigmas 2–5. Seed horizontal, slightly reticulate, 1 mm in diam, black. ($2n = 54$, $n = 27$) Jul–Oct. On usually saline or alkaline soils; GP but less common or absent in e NE & e KS; (MN, Sask., B.C., s to KS, MO, TX, NV, CA, & occasionally introduced eastward to IL, MI). *S. erecta* (S. Wats.) Nels.—Rydberg; *S. calceoliformis* (Hook.) Moq.—McNeill, Basset & Crompton, Rhodora 79: 133–138. 1977.

2. Suaeda moquinii (Torr.) Greene, seepweed. Erect, ascending, or spreading, perennial with woody base, stems 2–7 dm tall, glabrous to puberulent above, branches numerous. Leaves alternate, numerous, linear, flat or subterete, mostly 1–2 cm long, fleshy, somewhat shorter above. Flowers 1–3(5) in upper axils, not crowded, perfect; calyx lobes 5, with 3 or 4 small bracteoles at base, all alike, erect, slightly cucullate but not carinate on the back, obtuse or rounded, rarely shortly acute; stamens 5, exserted; styles 2- to 3-parted. Seeds horizontal or rarely vertical, black, 1.5–2 mm wide; pericarp adherent to seed, seed coat minutely reticulate. ($n = 9$) Apr–Oct. Usually on moist saline or alkaline soils; w ND, nw SD, MT; TX: Bailey; (Alta., ND w to OR, s to w TX, s CA). *S. intermedia* Wats., *S. torreyana* Wats.—Atlas GP.

3. Suaeda suffrutescens S. Wats., desert seepweed. Perennial shrub; stems woody, or suffrutescent below, erect, much branched, 5–9 dm tall, paniculately branched, often forming mounds 12–15 dm wide, herbage tomentose. Leaves numerous, alternate, pubescent or glabrous, terete, linear, 1–2 cm long, acute, much reduced upwards, ascending or spreading. Flowers in clusters of 3–9 in the axils, crowded in spikes; calyx divided to below the middle, lobes rounded on dorsal side, densely pubescent, bracteoles acuminate to attenuate; stamens exserted. Seed usually vertical, 0.7–1 mm wide, black, minutely tuberculate. Apr–Sep. Plains & valleys, usually on saline or alkaline soils; OK: Major, Tillman; TX panhandle; (w OK, nw TX, to AZ, s CA, s to Mex.). *S. nigrescens* var. *glabra* I. M. Johnst.—Atlas GP.

As described above, our plants are the var. *suffrutescens*. Just s of our area is the var. *detonsa* I. M. Johnst. with stems and leaves glabrous or nearly so. It extends north to TX: Bailey and may be expected in our area.

14. SUCKLEYA A. Gray

1. *Suckleya suckleyana* (Torr.) Rydb., poison suckleya. Annual monoecious herbs, stems succulent, stout, prostrate or ascending, freely branched, 1–4(5) dm long, glabrate or sparingly mealy, often reddish. Leaves alternate; blades rhombic-ovate to orbicular, rounded at apex, 1–3 cm long, repand-dentate with short triangular acute to obtuse teeth, scurfy with scales when young, soon glabrate; petioles equal to or exceeding blades. Flowers unisexual, in dense clusters in axils of leaves; staminate flowers in upper axils, bractless, calyx nearly globose, 3- to 4-parted, 2 segments longer than others, stamens 3 or 4; pistillate flowers with 2 bracts, these conduplicate, ovate-rhombic, slightly hastate, carinate, united below middle, in fruit with narrow, dorsal, crenulate wings, stigmas 2, short. Fruit enclosed in bracts, compressed, pericarp membranaceous, free; seed ovate, compressed, 2.8–3.1 mm long, reddish-brown, smooth. Jun–Sep. Dried lakeshores, stream valleys, roadsides, waste places; w ND, MT, s to cen KS, TX, NM; (s Sask., s Alta., s to w ND, e MT, TX, NM).

It is reported that this species causes cyanide poisoning in cattle.

39. AMARANTHACEAE Juss., the Pigweed Family

by Ronald L. McGregor

Ours monoecious, dioecious, or perfect flowered annuals or perennials; stems erect or prostrate to ascending. Leaves alternate or opposite, simple, petiolate or sessile; stipules absent. Inflorescence of solitary flowers or usually flowers glomerulate, racemose, spicate, or capitate. Each flower subtended by a bract and 2 bracteoles; calyx of usually 5 distinct to united sepals, or sepals rarely wanting; calyx scarious or indurate at maturity; corolla absent; stamens 2–5, filaments distinct or united, anthers 2- or 4-celled; ovary short, compressed, 1-celled; styles 1 or 2, or nearly absent, stigmas elongate or capitate; ovule 1. Fruit a dehiscent or indehiscent utricle; seeds smooth, usually shiny.

Excluded from this treatment are *Celosia argentea* L. and *C. cristata* L., the cultivated cockscombs, which have been reported as nonpersisting escapes. *Amaranthus caudatus* L., the tassel flower, has also been found as a temporary escape and is omitted from consideration.

```
1  Leaves alternate ........................................................................ 2. Amaranthus
1  Leaves opposite.
   2  Stems with branched or stellate hairs .................................... 6. Tidestromia
   2  Stems without branched or stellate hairs.
      3  Plants dioecious rhizomatous perennials ............................. 5. Iresine
      3  Plants with perfect flowers; annual or perennial.
         4  Erect or ascending herbs; inflorescences mostly terminal, racemose-
            spicate ................................................................. 3. Froelichia
         4  Usually prostrate herbs, inflorescences axillary.
            5  Sepals with long hairs having glochidiate tips ............ 1. Alternanthera
            5  Sepals densely woolly, the surfaces hidden by the hairs ........... 4. Guilleminea
```

1. ALTERNANTHERA Forsk., Chaff-flower

1. *Alternanthera caracasana* H.B.K., mat chaff-flower. Perennial herb from an elongate tuberous root; stems prostrate, branched at base and above, 1–5 dm long, hirsute with

ascending or spreading hairs to glabrate. Leaves opposite, appearing clustered, spatulate to suborbicular, 8–20 mm long, entire, sparsely pilose to glabrate, apex obtuse or abruptly pointed, blade narrowed into a petiole 1–5 mm long. Inflorescence an axillary ovoid to short-cylindric head. Flowers perfect, subtended by whitish ovate bracts; sepals 5, distinct, unequal, 3–5 mm long, lanceolate to ovate, 1-nerved, awn-tipped, with glochidiate tipped hairs on back and margins; petals absent; stamens 5(3), alternating with staminodia, filaments partially united into a cuplike tube; ovary 1-celled, style short, stigma capitate, ovule solitary. Fruit a flattened, indehiscent, membranaceous utricle; seed 1.2–1.6 mm long, reddish-brown, minutely pitted, lustrous. Jul–Sep. Infrequent along roadsides, edge of fields, & waste places; s OK, s TX panhandle; (SC to FL, w to OK, TX, CA).

2. AMARANTHUS L., Pigweed

Annual monoecious or dioecious herbs with taproots; stems prostrate to usually erect, mostly much branched. Leaves alternate, simple, entire or sinuate, petiolate. Inflorescences of dense terminal or axillary spikes or clusters. Each flower subtended by a bract and 2 bracteoles, these often colored and concealing perianth in some species; sepals 1–5, scarious or membranaceous, distinct to base, often aristate, or absent from pistillate flowers in *A. tuberculatus;* stamens 5(1–3), distinct, anthers 4-celled but appearing 2-celled after dehiscence; ovary 1-celled, short, compressed, ovule 1; style short or obsolete, stigmas 2–3. Fruit a 1-seeded circumscissile, irregularly splitting or indehiscent utricle; seed flattened or lenticular, erect, smooth, usually lustrous. *Acnida* L. — Rydberg.

Reference: Sauer, Jonathan. 1955. Revision of the dioecious Amaranths. Madroño 13: 5–46.

1 Plants monoecious with staminate and pistillate flowers intermingled or in nearly separate inflorescences.
  2 Stems with a pair of rigid sharp spines at most nodes .... 9. *A. spinosus*
  2 Stems without spines at nodes.
    3 Flowers in axillary clusters, or glomerules.
      4 Stems prostrate; bracts of pistillate flowers about as long as calyx and fruit; seeds 1.4–1.7 mm wide .... 3. *A. graecizans*
      4 Stems erect or ascending; bracts of pistillate flowers 2–3× longer than the calyx or fruit; seed 0.7–1 mm wide .... 1. *A. albus*
    3 Flowers in dense terminal or axillary erect spikes which are often paniculate, sometimes also with smaller axillary clusters of flowers.
      5 Sepals all rounded to truncate at apex, usually emarginate, often mucronate .... 7. *A. retroflexus*
      5 Outer sepals acute or acuminate.
        6 Inflorescence lax, usually with many branches; bracts 3–4 mm long; sepals and stamens 5; e 1/2 GP, scattered elsewhere .... 4. *A. hybridus*
        6 Inflorescence stiff, with few branches; bracts about 5 mm long; sepals and stamens 3–5; rare in w GP .... 6. *A. powellii*
1 Plants dioecious.
  7 Plants pistillate.
    8 Sepals lacking or very rudimentary .... 10. *A. tuberculatus*
    8 Sepals present and well developed.
      9 Sepals 1 or 2 with one very rudimentary .... 8. *A. rudis*
      9 Sepals 5.
        10 All sepals obtuse or retuse, midveins slightly if at all excurrent; bract and outer sepals scarcely longer than inner sepals .... 2. *A. arenicola*
        10 Outer sepal acute or acuminate with midvein excurrent into a rigid spine; bract and outer sepals much longer than inner sepals .... 5. *A. palmeri*
  7 Plants staminate.
    11 Bracts with midveins scarcely excurrent .... 2. *A. arenicola*
    11 Bracts with midveins clearly excurrent.

12 Bracts 4–6 mm long .... 5. *A. palmeri*
12 Bracts 1–2.5 mm long.
   13 Outer sepals with midveins not excurrent .... 10. *A. tuberculatus*
   13 Outer sepals with conspicuous excurrent midveins .... 8. *A. rudis*

1. **Amaranthus albus** L., tumbleweed. Monoecious annual herb with taproots; stems erect, 2–8(12) dm tall, bushy branched with divaricate or ascending branches; stems glabrous to villous, whitish or pale green. Leaves alternate; principal stem leaves with blades oblanceolate, obovate, rhombic-ovate to elliptic or lanceolate, 1.5–6 cm long, obtuse to rounded, often emarginate, cuneate at base, green or purplish below, petioles as long as blade, usually disappearing by flowering time; branch leaves with blades elliptic to oblong or obovate, 5–30 mm long, pale green, obtuse or rounded, attenuate to a petiole 1/4 to as long as blade. Flowers in short axillary clusters, often floriferous to base; male and female flowers on same plant, occasional flowers perfect, the staminate ones few. Bracts green, rigid, 2–4 mm long, oblong-lanceolate, sharp-pointed, spreading; sepals 3, staminate ones oblong, cuspidate, scarious; pistillate sepals oblong to linear, acute, 1-nerved, often reddish; stamens 3; style branches 3. Utricle lenticular, 1.2–1.8 mm long, circumscissile at the middle, rugulose at maturity; seed lenticular, 0.6–0.8(1) mm in diam, margin ridged, black, lustrous. (n = 16) Jun–Oct. Dry prairies, fields, stream valleys, roadsides, waste places, on a variety of soils; GP; (widely distributed in N. Amer., adventive in Europe, Asia, S. Amer., Africa). *A. graecizans* L.—Rydberg.

    Plants which are densely viscid-pubescent have been called *A. albus* var. *pubescens* (Uline & Bray) Fern. and while they are more common in the sw GP, the distinction appears unwarranted as this pubescent form is to be found throughout the GP.

2. **Amaranthus arenicola** I. M. Johnst., sandhills pigweed. Dioecious annual herb with taproots; stems erect, (0.5)1–2(3) m tall, simple or branched at base, branched above, glabrous to minutely pubescent, sometimes scabrous, whitish and striate. Leaves alternate, blades oval-oblong to oblong-linear, 1.5–8 cm long, rounded to acute at apex, obtuse to attenuate at base, yellowish-green, glabrous, often disappearing by fruiting; branch leaves similar but smaller. Flowers in slender to thick, dense or interrupted spikes or thyrses 1–4(6) dm long, either all terminal on leafy branches, or on leafless branches subtended by leaves, rarely some in lower axillary glomerules. Bracts 1.5–2.5 mm long, lanceolate, acuminate, with midrib barely if at all excurrent; male flowers with 5 stamens and 5 nearly equal sepals 3–5 mm long, inner sepals emarginate or obtuse, outer obtuse or acute, all apiculate with dark midveins not excurrent; female flowers with 5 recurved spatulate sepals, each with a branched midvein, inner sepals 1.5–2 mm long, emarginate or obtuse, outer sepals 2–2.5 mm long, obtuse-apiculate, style branches usually 3. Utricle 1.4–1.7 mm long, thin, circumscissile, rather smooth; seed round, lenticular, 1–1.3 mm in diam, dark reddish-brown to black, smooth, lustrous. (n = 16) Jul–Oct. Common on sand dunes, sandy prairies, stream valleys, fields, roadsides, waste places, less common on hard soils; sw SD & s in w 2/3 GP; (IA, sw SD, e WY, s to OK, TX, CO, NM, AZ, introduced elsewhere). *A. torreyi* (A. Gray) Benth.—Rydberg.

    In the GP this species is known to hybridize with *A. palmeri* and *A. retroflexus* and is suspected to hybridize with *A. albus* and *A. rudis*. Some hybrids with *A. palmeri* grow to 4 m tall and are completely sterile. Those hybrids with *retroflexus* form large sterile bushy plants to 2 m tall with the aspect of *retroflexus*. Both of these hybrids cause problems with mechanical harvest machines.

3. **Amaranthus graecizans** L., prostrate pigweed. Monoecious annual herb with taproots; stems stout, prostrate or rarely ascending, branched at base and often above, 1–6(10) dm long, forming mats, glabrous or sparsely pubescent, whitish or green, sometimes tinged with red. Leaves alternate, numerous, usually crowded; blades obovate, oval, elliptic to

spatulate, 8-40 mm long, rounded to acute at apex, cuneate to attenuate at base, pale green, glabrous, usually white-margined; petioles 2-20 mm long. Flowers in dense axillary clusters which are usually shorter than petiole, often floriferous to base. Bracts oblong to lanceolate, erect, attenuate at apex to a short spinose tip, 2.5-3 mm long; sepals of staminate flowers (4)5, scarious, oblong, acute, 2.5-3 mm long; sepals of pistillate flowers (4)5, ovate to oblong, 2.5-3 mm long, acuminate, 1-nerved, green, white margined; stamens 3; style branches usually 3. Utricle subglobose, 2.5-3.2 mm long, smooth, circumscissile at the middle, sometimes tinged with red; seed circular, 1.3-1.7 mm in diam, black, dull to lustrous, margin with a narrow ridge. (n = 16) Jul-Oct. Infrequent to locally common in dry prairies, pastures, fields, roadsides, stream valleys, waste places; GP; (w 1/2 N. Amer., adventive elsewhere). *A. blitoides* S. Wats.—Rydberg.

4. ***Amaranthus hybridus*** L., slender pigweed, green pigweed. Monoecious weedy annual herb with taproots; stems rather stout, erect, usually freely branched, 0.5-1.5(2.5) m tall, glabrous to rough-pubescent, often villous in inflorescence. Leaves alternate, ovate to lanceolate or rhombic-ovate to 15 cm long, acute or rounded at apex, cuneate or rounded at base, pubescent or glabrous, darker green above; petioles shorter than to as long or longer than blade. Flowers in many cylindric spikes about 1 cm thick, lateral spikes 1-7 cm long, the terminal to 20 cm long, aggregated into a terminal panicle, often with smaller panicles or single spikes in upper axils. Bracts lance-attenuate, 3-4 mm long, spinulose-tipped; sepals of staminate flowers narrowly oblong to ovate, acute, midvein excurrent, 1.7-2 mm long; sepals of pistillate flowers 1.5-2 mm long, oblong or linear-oblong, acute, nerve usually excurrent; stamens 5; style 3-branched. Utricle subglobose, 1.3-2 mm long, circumscissile at the middle, usually smooth; seed round, 1-1.3 mm in diam, black, lustrous. (n = 16) Jun-Oct. Infrequent to locally common or abundant in river valleys, fields, roadsides, waste places; GP but rare in n 1/2 & w 1/2; (now a weed found in much of N. Amer. & elsewhere). Said to be a native of trop. Amer.

This species hybridizes with *A. rudis* and swarms are frequently encountered. The name *Amaranthus tamariscinus* Nutt. was apparently based on one of these hybrids.

5. ***Amaranthus palmeri*** S. Wats., Palmer's pigweed. Dioecious weedy herb with long stout taproots; stems erect, 1-2(3) m tall, simple to branched at base and usually much branched above with ascending branches, glabrous to rarely villous-pubescent. Leaves alternate; blades rhombic-ovate to rhombic-lanceolate, 1-6(10) cm long, acute to often abruptly acuminate at apex, cuneate or rounded at base; petioles usually nearly as long to longer than blades. Inflorescence of dense to interrupted, often lax or drooping, simple to racemose spikes, 2-7(12) dm long, terminating the stem or branches or, if on leafless branches, these subtended by a leaf, lateral spikes shorter, some flowers often in lower axillary clusters; the mature inflorescence stiff, spiny. Bracts 4-6 mm long, 2 × as long as sepals, midrib excurrent into a spine; male flowers with 5 sepals, the inner ones 2.5-3 mm long, obtuse or emarginate, the outer sepals 3.5-4 mm long, acuminate, with conspicuous long-excurrent midveins; stamens 5; female flowers with 5 recurved sepals each with a branched midvein, inner sepals 2-3 mm long, spatulate, emarginate and often denticulate, outer sepals 3-4.5 mm long, acute, with midvein excurrent into a rigid spine, style branches 2(3). Utricle 1.5-2.2 mm long, circumscissile near the middle, somewhat rugose; seeds obovate, lenticular, dark reddish-brown to black, 1-1.3 mm long, smooth, lustrous. (n = 16) Jul-Oct. Infrequent to locally abundant in sandy fields, pastures, stream valleys, roadsides, waste places, less common on hard soils; s 1/2 GP; (s NE, KS; CO: Cheyenne & Prowers; s to OK, TX, NM, CA, Mex.; introduced IL, MO, AR, LA, NY, PA, WV, SC, FL, & elsewhere). *A. torreyi* S. Wats.—Rydberg.

This is one of our more aggressive weedy pigweeds and has become more common in the s 1/2 of the GP particularly in irrigated fields. It hybridizes with *A. rudis, A. arenicola,* and *A. retroflexus* and perhaps other species in our area.

6. *Amaranthus powellii* Wats., Powell's pigweed. Monoecious annual herb with taproots; stems rather stout, 0.6–2 m tall, erect, striate and often reddish, simple to freely branched, glabrous to rough-pubescent and sometimes villous in inflorescence. Leaves alternate, blades rhombic-ovate to lanceolate or deltoid-elliptic, 2–10(12) cm long, usually glabrous to sparsely pubescent, acute or rounded at apex, cuneate or rounded at the base; petioles often nearly as long as blade. Inflorescence paniculate-spicate or of simple spikes and sometimes axillary clusters; terminal panicle of a few stiff spikes, the central one often to 10–20 cm long or longer, the lateral ones 4–10 cm long. Bracts linear-lanceolate and spinose, 2.5–5 mm long or about 2 × as long as sepals; sepals of staminate flowers scarious, narrowly oblong to ovate, acute, 1.2–3 mm long, midvein excurrent, stamens 3(4–5); sepals of pistillate flowers 1.5–2 mm long, oblong or linear oblong, scarious, acute, the midvein usually excurrent; style 3-branched. Utricle subglobose, circumscissile at about the middle, usually rugulose; seeds round-lenticular, 1–1.3 mm in diam, reddish-brown to usually black, lustrous. Jul–Sep. In fields, waste places, along streams, roadsides; WY: Goshen, Platte; KS: Saline; CO: Lincoln; TX, NM; (probably w 1/2 of U.S.; Mex., & introduced elsewhere).

This species is very rare in the GP and at least the KS record was probably introduced. The plant is said to hybridize with *A. hybridus* and *A. retroflexus*.

7. *Amaranthus retroflexus* L., rough pigweed. Monoecious annual herb with often reddish taproots; stems stout, erect, 0.3–3 m tall, simple to more commonly freely branched, usually obviously roughish villous-puberulent, white or reddish striate. Leaves alternate; blades lanceolate or ovate-lanceolate to obovate-oblanceolate, 2–8(10) cm long, usually hairy on lower surface especially along the veins; petioles as long as to somewhat shorter than blades. Inflorescence of terminal or axillary, usually paniculate, densely crowded, erect spikes 5–20 cm long and 10–20 mm thick, spikes often lobulate, often dense clusters of flowers also in axils of upper leaves. Bracts lanceolate to ovate, tapering into a short green tip or acicular to spinose, 1-nerved, sparsely villous, 3.5–5 mm long, sepals of staminate flowers ovate-oblong to lanceolate, 3 mm long, acute or nearly so, nerve shortly excurrent, stamens 5; sepals of pistillate flowers linear-oblong, 2.5–3.2 mm long, rounded to truncate at apex, often emarginate and mucronate, scarious; styles 3-branched. Utricle subglobose, circumscissile near middle, somewhat rugose; seeds round, 0.9–1.2 mm in diam, black, lustrous. (n = 16) Jul–Oct. A common plant in cult. fields, fallow land, stream valleys, prairie ravines, roadsides, & waste places; GP; (a semicosmopolitan weed).

This species hybridizes commonly with *A. hybridus*, *A. palmeri*, and apparently with *A. arenicola* and *A. rudis*.

8. *Amaranthus rudis* Sauer, water-hemp. Dioecious annual herb with a taproot; stems usually stout, erect or ascending, 0.5–2 m tall, simple or usually with ascending branches, glabrous or nearly so. Leaves alternate; blades usually oblong to lance-oblong, sometimes lanceolate or rhombic-oblong, 2–10 cm long, rounded or obtuse at apex; sometimes notched, attenuate at base; upper leaves often much reduced and narrowly oblong; petioles to 5 cm long. Inflorescence spicate to usually paniculate spicate, usually with axillary clusters of flowers. Bracts 1.5–2 mm long, with moderate to heavy excurrent midribs; male flowers with 5 sepals, the inner 2.2–2.5 mm long, obtuse or emarginate, outer sepals 2.6–3 mm long, acuminate, midribs excurrent, stamens 5; female flowers with 1 or 2 sepals, one rudimentary or less than 1 mm long, the longer sepal 2 mm long, narrowly lanceolate, acuminate, midrib sometimes branched. Utricle 1.2–1.6 mm long, cirumscissile at middle or slightly above, thin, usually rugose, often with faint ridges of tubercles above; style branches 3 or 4; seed round, lenticular, 0.9–1.1 mm in diam, reddish-brown to usually black, smooth, lustrous. (n = 16) Jun–Oct. Common along stream banks, lakeshores, flood plains, & often abundant in fields, waste places; GP but rare to absent in n & w; (WI, IN, ND, s to LA, TX, adventive elsewhere). *Acnida tamariscina* (Nutt.) Wood — Rydberg; *Amaranthus tamariscinus* (Nutt.) Wood — Gleason & Cronquist.

This is a highly variable and very weedy species in se GP. It regularly hybridizes with several other species of *Amaranthus*, and intermediates are all too common.

9. *Amaranthus spinosus* L., spiny pigweed. Monoecious annual herb with a long taproot; stems erect, usually much branched, 0.3–1.2 m tall, stout and somewhat succulent, usually reddish, glabrous or sparsely puberulent, bearing at most nodes a pair of divergent spines 5–10 mm long. Leaves alternate, blades ovate-lanceolate to ovate, or rhombic-ovate, 3–10 cm long, glabrous to sparsely pubescent, narrowed to an obtuse and mucronate tip, broadly cuneate to long petioled. Inflorescence of terminal and axillary spikes 5–15 cm long, 6–10 mm thick, the terminal spike often wholly or mainly staminate, the basal portion and axillary clusters of flowers usually pistillate. Bracts lanceolate or subulate, 0.5–1 mm long; sepals of staminate flowers lance-oblong, acute or slightly acuminate, 1–1.6 mm long, stamens 5; sepals of pistillate flowers 5, oblong or acutish, 1–1.5 mm long. Utricle 1.5–2 mm long, indehiscent or bursting irregularly, roughened above; seed nearly circular, 0.7–1 mm in diam, black, lustrous. (n = 17) Jun–Oct. Locally common in over-grazed pastures, feedlots, fields, roadsides, waste places; se GP but sporadic elsewhere; (a pantrop. weed extending n to NY, PA, ME, IN, MO, NE).

10. *Amaranthus tuberculatus* (Moq.) Sauer, tall water-hemp. Dioecious annual herb with taproots; stems prostrate, ascending to usually erect, simple to usually much branched, 0.2–1(2) m tall, glabrous to rarely sparsely puberulent above. Leaves alternate, highly variable, the shorter ones 1–4 cm long with blades oblong or spatulate, longer leaf blades ovate or lanceolate, 4–10 cm long, glabrous; petioles short or as long as blades. Inflorescence highly variable, of simple terminal and lateral spikes, and some axillary clusters, to paniculate-spicate. Bracts 1–1.5 mm long, those of male flowers slender and midrib excurrent, those of pistillate flowers with midrib excurrent well beyond lamina; staminate flowers with 5 sepals, nearly equal, 2.5–3 mm long, the inner obtuse or emarginate, outer acuminate, midveins not excurrent, stamens 5; pistillate flowers usually without sepals, rarely with 1 or 2 rudimentary ones; style branches 3 or 4. Utricle 1.5–2 mm long, indehiscent or bursting irregularly, smooth or usually irregularly tuberculate, sometimes with 3 or 4 faint ridges above; seeds 0.8–1 mm in diam, round or somewhat obovoid, lenticular, dark reddish-brown to usually black, lustrous. (n = 16) Jul–Oct. Margins of lakes, riverbanks, sand bars, marshes, fields, & waste places; scattered in MN, ND, IA, SD, NE, w MO; (VT, ND, s to NJ, OH, KY, TN, ne AR, NE, MO). *Acnida altissima* Ridd., *A. subnuda* (S. Wats.) Standl. — Rydberg; *A. altissima* Ridd. var. *subnuda* (S. Wats.) Fern., var. *prostrata* (Uline & Bray) Fern. — Fernald.

In the Atlas GP all records for Kansas should be eliminated and a few in ND, SD, NE, and IA appear atypical and perhaps are hybrids, or hybrid segregates, with *A. rudis* or one of the monoecious species.

3. **FROELICHIA** Moench, Snake-Cotton

Ours annual herbs with taproots; stems erect, simple or branched, lowest branches procumbent to ascending, white or brownish-pubescent, sometimes viscid above. Leaves opposite, entire, sericeous or tomentose below, canescent or silky above, rarely glabrate; sessile or petiolate. Inflorescences usually racemose-spicate. Flowers perfect, sessile, each subtended by a scarious bract and 2 bractlets; calyx tubular, densely woolly with 5 glabrate lobes, becoming conic or flask-shaped and indurate at maturity, bearing 2 lateral rows of distinct spines or 2 deeply dentate to nearly entire crests or wings, the faces bearing one or more tubercles or spines near base; petals absent; stamens 5, filaments united to form a tube as long as calyx tube, bearing 5 1-locular anthers and 5 ligulate appendages; ovary ovoid, ovule 1; style elongate, stigma capitate. Fruit a membranaceous indehiscent utricle included in the tube of filaments and the whole surrounded by indurate calyx; seed lenticular in outline, yellowish-brown or brown, often lustrous.

1 Stem usually divergently several branched from base; lateral spikes sessile; mature calyx-tube with lateral rows of distinct and rather sharp spines .... 2. *F. gracilis*
1 Stem simple or with few erect branches; some of lateral spikes usually peduncled; mature calyx-tube with lateral deeply dentate crests or wings .... 1. *F. floridana*

1. *Froelichia floridana* (Nutt.) Moq., field snake-cotton. Annual herb with taproots; stems rather stout, erect, wandlike, (3)4–10(14) dm tall, usually simple at base, often with sparse branches above, canescent or tomentose, branches sericeous-tomentose with white or brownish hairs and slightly viscid above. Leaves opposite, entire, principal ones narrowly oblong, elliptic, oblanceolate to commonly spatulate, widest at or below the middle, (2)3–10(14) cm long, (0.5)1–2(3) cm wide, obtuse or acute at apex, narrowed to base, canescent or silky above, sericeous-tomentose below; sessile or with petioles to 3 cm long. Inflorescences terminal, racemose-spicate, with some of lateral spikes often peduncled; spikes 1–7(10) cm long, 1–1.5 cm thick, whitish. Bractlets scarious, rotund, white to stramineous or blackish, shorter than calyx; mature calyx flask-shaped, 5–6 mm long, densely woolly, with 2 lateral deeply dentate crests or wings, the faces often with tubercles near base; calyx lobes oblong to narrowly rhombic, 1–2 mm long, obtuse. Seed reddish-brown, 1.5–2 mm long, often lustrous. (n = ca 39) Jun–Sep. Locally common on sand dunes, sandy prairies, stream valleys, roadsides, less common in sandy or rocky open woodlands; s 2/3 GP; (WI, IL, IN, MN, SD, s to AR, CO, OK, TX, NM). *F. campestris* Small—Rydberg.

Our plants normally have leaves widest above the middle and with hairs on peduncle commonly at least 2 mm long. These are commonly referred to the var. *campestris* (Small) Fern. Var. *floridana* of the coastal plain has leaves widest below the middle and hairs of peduncle usually less than 0.5 mm long. Some of our specimens, however, are intermediate in these characters.

2. *Froelichia gracilis* (Hook.) Moq., slender snake-cotton. Annual herb with taproot; stem slender, (1)2–5(7) dm tall, simple or usually much branched at the base, the branches ascending or somewhat procumbent, densely or sparsely villous-tomentose and sometimes viscid above. Leaves opposite, entire, often clustered toward base, linear, linear-lanceolate, narrowly oblanceolate, to elliptic-lanceolate, 3–7(12) cm long, 2–7(12) mm wide, acute or acuminate at apex, cuneate at base, sericeous or tomentose to silky on both surfaces, sessile or short petioled. Inflorescence racemose-spicate; spikes 0.7–3 cm long, (5)7–8(10) mm in diam, lateral spikes sessile. Bractlets rotund, sometimes acuminate, scarious, shorter than calyx, white, yellowish, brownish, or blackish; mature calyx conic to flask-shaped, 3.5–5 mm long, densely woolly, with 2 lateral rows of distinct spines, the faces with 1–3 blunt tubercles, or these spinelike; calyx lobes oblong-linear, obtuse or acute, (0.5)1.2–2 mm long, glabrate. Seed yellowish or reddish-brown, 1.2–1.6 mm long, usually lustrous. Jun–Oct. Locally common on sand dunes, sandy prairies, & pastures, stream valleys, less common in sandy or rocky open woodlands; s 2/3 GP; (IN, IA, s SD, CO, s to AR, TX, AZ & Mex., adventive e to NY, NJ, MD, VA, SC, GA, AL).

This species is usually found with *F. floridana* and nearly always putative hybrids between the two are all too frequent. Such plants usually have more of the habit of *floridana* and the fruiting calyx characters of *gracilis*. The two taxa merit careful field and experimental studies.

4. GUILLEMINEA H.B.K., Cottonflower

Ours perennial herbs with subterranean or superficial often-branching caudex; stems procumbent to ascending, usually branched, lanate to nearly glabrous. Basal leaves in rosettes, linear to oblanceolate or spatulate; cauline leaves opposite, lanceolate to oval or rounded-obovate, surfaces pilose-sericeous to glabrate; sessile of short petiolate. Inflorescence a small axillary head or spike, these often densely aggregated at the nodes and subtended by small leaves. Flowers perfect, with 2 bracts and a bractlet; sepals 5, 2 more concave than other 3, covered with silky hairs; stamens 5, filaments united into a tube which is free or adnate to calyx tube, anthers 2-celled but appearing unilocular after dehiscence, staminodia absent; ovary unilocular, ovule 1; style 1; stigmas capitate and bilobed. Fruit a membranaceous, indehiscent utricle; seed 0.6–1 mm long, brownish, lustrous. *Brayulinea* Small and *Gossypianthus* Hook.—Waterfall.

1 Basal and stem leaves similar, blades oval; filament tube adnate to calyx .... 1. *G. densa*
1 Basal leaves linear to oblong-obovate, much longer than stem leaves; filament tube free from calyx .... 2. *G. lanuginosa*

1. Guilleminea densa (Willd.) Moq., dense cottonflower. Perennial herb with subterranean to superficial branching caudex, stems several, much branched, prostrate, forming mats, 0.5–3(6) dm long, usually lanate. Basal leaves rosulate, 10–25 mm long, 5–15 mm wide, lanceolate to spatulate acute, soon disappearing; stem leaves opposite, of unequal size in the pair, lanceolate to lance-ovate, 3–30 mm long, 1–12 mm wide, densely lanate below, often glabrate above; petiole short, broadly winged. Inflorescence of up to 10 perfect flowers in axillary small heads of cylindrical spikes. Flowers 1.2–1.5 mm long or 1.8–2.2 mm long; sepals 5, united into tube with free lobes, tube densely lanate; lobes with an incomplete median nerve, lance-ovate, glabrous, scarious; stamens 5, filament tube adnate to calyx tube; bracts ovate, scarious, glabrous, white, acute, bractlets obtuse. Utricle glabrous; seed ovoid, somewhat flattened, 0.5–0.7 mm long, brown, lustrous. Jun–Sep. Dry sandy, rocky prairies, pastures, roadsides, & waste places; OK: Cimarron; TX panhandle, e NM; (OK, TX, NM, AZ, to s Mex., also S. Amer.). *Brayulinea densa* (H. & B.) Small—Waterfall.

The common plant with leaves not aggregated at the node and mature flowers less than 1.5 mm long is the var. *densa*. The less common form with leaves densely aggregated, mature flowers 1.8–2.2 mm long, and plant more robust has been referred to the var. *aggregata* Uline & Bray, but these distinctions often appear arbitrary.

2. Guilleminea lanuginosa (Poir.) Hook. f., cottonflower. Perennial herb with superficial caudex; stems forming above ground, procumbent or ascending, 1–3 dm tall, sparsely to densely pilose with long silky hairs. Rosulate leaves 1–9 cm long, linear, lanceolate, oblanceolate or spatulate, acute or obtuse, densely pilose to nearly glabrous; stem leaves opposite, linear to ovate, (5)8–15(20) mm long, (3)5–10(13) mm wide, pubescent to nearly glabrous; petiole scarious winged. Inflorescence a spike of 6–12 axillary flowers. Flowers with 2 bracts and a bractlet, perfect; calyx 2.5–3 mm long, densely woolly; sepals 3-nerved; stamens 5, filaments united into a tube, free from calyx tube; styles to 0.4 mm long. Seed 1 mm long, brown, shiny. Jun–Sep. Dry sandy or rocky prairies, stream valleys, open woodlands, waste places; w 2/3 OK, TX; (AR, OK, TX, Mex.). *Gossypianthus lanuginosus* (Poir.) Moq., *G. tenuiflorus* Hook.—Waterfall; *Guilleminea lanuginosa* var. *sheldonii* (Uline & Bray) Mears, var. *tenuiflora* (Hook.) Mears—Atlas GP.

This is a highly variable species in which several varieties have been recognized on the basis of leaf size and shape, degree of pubescence, stem thickness, and similar characters. Additional population studies are necessary in our area before conclusions can be made on the status of the complex.

5. IRESINE P. Br., Bloodleaf

1. *Iresine rhizomatosa* Standl., bloodleaf. Dioecious perennial herb spreading by slender rhizomes; stems erect, 0.5-1(1.5) m tall, usually solitary and simple to inflorescence, glabrous or sparsely pubescent, sometimes short-pilose at the somewhat swollen nodes. Leaves opposite, thin, ovate to ovate-lanceolate, entire, glabrous or sparsely pubescent, acute to acuminate, abruptly narrowed at base and decurrent into slender petioles. Panicles terminal and from upper axils, lax or somewhat pyramidal, 1-3 dm long, with numerous flowers; staminate panicle usually laxly branched, the flowers in short spikelets; bracts and bractlets ovate, silvery-white, shorter than sepals; sepals ovate-lanceolate, 1.2-1.5 mm long, obscurely 1-nerved; stamens 5; pistillate pannicle: usually with erect or ascending branches, appearing pyramidal, spikelets densely flowered, 5-20 mm long. Bracts and bractlets ovate, shorter than sepals; sepals lance-ovate, 1-nerved, silvery-white, 1.2-1.5 mm long, flowers subtended by hairs to 3-5 mm long; ovary compressed, ovule 1, styles 2, short. Utricle compressed, membranaceous, indehiscent, 2-2.5 mm long; seed 0.5 mm long, suborbicular, reddish-black, smooth, lustrous. Aug-Oct. Infrequent to locally common in low wet woods & thickets near streams; s 1/2 MO, se KS; (MD to s IL, s MO, se KS, s to E VA, GA, FL, AL, e 1/2 TX). *I. celosia* L.—Rydberg.

6. TIDESTROMIA Standl.

1. *Tidestromia lanuginosa* (Nutt.) Standl. Annual herb with taproots; stems prostrate to rarely decumbent or ascending, much branched, radiating from the root, 1-6(10) dm long, densely stellate-pubescent, becoming glabrate in age. Leaves opposite, gray-green, both surfaces densely stellate-pubescent but becoming glabrate in age, blades obovate to rhombic-ovate, entire, 5-20(30) mm long; petioles equaling or shorter than blades; stipules absent. Flowers perfect, in small axillary clusters, subtended by 3 hyaline and pubescent bracts; sepals 5, unequal, 1-3 mm long, about 3 × longer than bracts, the outer 3 wider than inner 2, 1-nerved, ovate-lanceolate, pubescent; stamens 5, filaments united at base to form a short cup; ovary globose, with 1 ovule; style short, stigma capitate or obscurely 2-lobed. Utricle subglobose, indehiscent, glabrous; seed 1.2-1.6 mm long, yellowish-brown, lustrous. Jun-Oct. Locally common on sand dunes, open dry rocky prairie, stream valleys, waste places; SD: Washabaugh; w 1/3 KS, se CO, w OK, NM, TX; (SD, KS, UT, s to AZ, n Mex.).

This species is reported as one of the host plants of the beet leaf-hopper.

40. PORTULACACEAE Juss., the Purslane Family

by William T. Barker

Annual or perennial herbs or rarely shrubs or half shrubs, often somewhat succulent, glabrous or rarely pilose at the nodes. Leaves opposite, alternate or in basal rosettes, entire, often fleshy; stipules scarious, or modified into tufts of hair, or wanting. Flowers solitary, racemose, paniculate or cymose, terminal or axillary, perfect, regular or nearly so, hypogynous to semi-epigynous. Sepals usually 2, persistent or deciduous, scarious or herbaceous; petals usually 4 or 5, often fugacious or dehiscent; stamens 4-many, inserted with petals, sometimes adnate at the base, filaments filiform, anthers 2-celled, dehiscent longitudinally; ovary 1-celled, superior to partly or wholly inferior, styles 2–7, more or less united; ovules 2–many, on central or basal placenta. Fruit a loculicidal or circumscissile capsule, valves as many as the styles; seeds 3–many or 1 or 2, commonly round-reniform, lenticular, often smooth and shining, but sometimes tuberculate or otherwise roughened and often strophiolate.

1 Ovary partly or wholly inferior; capsule circumscissile .... 2. *Portulaca*
1 Ovary superior; capsule opening lengthwise.
   2 Calyx persistent in fruit; leaves opposite .... 1. *Claytonia*
   2 Calyx deciduous; basal rosette of slender leaves .... 3. *Talinum*

1. CLAYTONIA L., Spring Beauty

1. *Claytonia virginica* L., Virginia spring beauty. Perennial herb with globose corm, 1–3 dm high, glabrous, succulent, erect or ascending. Basal leaves petioled, 6–20 cm long, blade as long as to 2× as long as the petiole, succulent, linear-oblanceolate, 5–15 mm wide, indistinctly 3-ribbed, acute at both ends; stem leaves short petioled, opposite, 9–15 cm long. Raceme (5)6- to 15(19)-flowered, 4–16 cm long, with a single small oval bract below the lowest pedicel; pedicel 1.5–2.4 cm long, in fruit recurved. Flowers regular, sepals 2, rounded-ovate or oval, usually obtuse or rounded at the apex, herbaceous, persistent, 5–7 mm long; petals 5, oval, 9–14 mm long, rounded to obtuse or rarely retuse at the apex, white or rose-colored with pink or purple veins; stamens 5, opposite the petals and adnate to them at the base; ovary 1-celled, 3-valved, 6-ovuled, styles 3, united to near the apex. Fruit a capsule about 4 mm long, rounded-ovoid, membranaceous, the edges of the valves inrolling at dehiscence; seeds blackish-brown, orbicular, shining, 2 mm across. (2n = 12 – ca 191) Feb–Jul. Rich woods, thickets, clearings; TX, OK, e 1/2 KS, MO, e NE & w IA; (N.S. to MN, s to GA, LA, & TX). *C. robusta* (Somes) Rydb. – Rydberg.

2. PORTULACA L., Purslane

Prostrate, ascending to erect annual (ours) or perennial succulent herbs. Leaves alternate or opposite, flat or terete, often the uppermost crowded and forming an involucre; stipules scarious, reduced to hairy tufts or lacking. Flowers perfect, solitary or crowded at top of stem and branches; sepals 2; petals 4–6, usually 5; stamens 8–many, inserted at the base of the petals; ovary partly or wholly inferior; styles 3–9; ovules numerous. Capsules 1-celled, membranaceous, circumscissle, many-seeded; seeds reniform, with a smooth or minutely tuberculate or sometimes echinate surface.

Portulaca grandiflora Hook., a native of Argentina, commonly called moss rose, has been introduced throughout the U.S. as an ornamental and occasionally escapes cultivation.

1 Lower valve of capsule with an extended ciruclar membranaceous wing just below its rim .... 5. *P. umbraticola*
1 Lower valve of capsule without a subtending wing.
   2 Leaf axils glabrous; leaves flat, cuneate to obovate-cuneate.
      3 Leaves broadly obtuse to truncate, rarely retuse; calyx lobes pointed in bud by the projecting keel; seeds rounded-tuberculate .... 2. *P. oleracea*
      3 Leaves often retuse; calyx lobes obtuse in bud; seeds echinate-tuberculate .... 4. *P. retusa*
   2 Leaf axils conspicuously villous with long white hairs; leaves terete.
      4 Petals yellow to bronze, less than 3 mm long; capsules 2 mm or less in diam .... 3. *P. parvula*
      4 Petals red to purple, 3 mm or more long; capsules more than 2 mm in diam .... 1. *P. mundula*

1. ***Portulaca mundula*** I. M. Johnst. Stems 3-6, emerging annually from an often thickened taproot, prostrate, laxly decumbent or laxly ascending, spreading branches 5-15 cm long, upward-directed branches 1-5 cm long, succulent, with decidedly shortened internodes. Leaves alternate, fleshy, linear to oblanceolate-linear, 5-15 mm long, 0.5-1.5 mm wide, terete or nearly so; hairs of leaf axils conspicuously kinky, woolly, often white, to 5-7 mm long. Flowers terminal, subsessile, 2-8 clustered in villous heads; leaves of the involucre 6-10, linear, 5-12 mm long, succulent. Calyx persisting on upper part of capsule 4(6) mm long, lobes triangular to triangular-oblanceolate; petals purple, obovate, 6(7.5) mm long, 3-4.5 mm wide, retuse at the apex; stamens 10-15(30); styles 3-5. Capsules ovate-globose, circumscissile below the middle; seeds black, stellate-tuberculate, 0.3-0.5 mm in diam. May-Oct. Gravelly or sandy soil; w MO, KS, OK, NM, & w TX; (MO to KS, OK, TX, w to CA; n Mex.).

2. ***Portulaca oleracea*** L., common purslane. Stem prostrate, fleshy, usually purplish-red, glabrous, repeatedly branched, forming large mats. Leaves alternate, succulent but flat, spatulate to obovate-cuneate, (0.6)1-3 cm long, 0.2-13 mm wide, rounded or nearly truncate at apex. Flower buds flattened, acute; flowers yellow, sessile, 5-10 mm wide, solitary or in small terminal glomerules; sepals broadly ovate to orbicular, 2.8-4.5 mm long, 2.8-3.8 mm wide, keeled; petals 3-4.6 mm long, 1.8-3 mm wide; stamens 6-10; styles lobes 4-6. Capsules 5-9 mm tall, circumscissile at or about the middle; seeds black and granulate. (2n = 54) May-Nov. Cult. & waste ground throughout the GP; (cosmopolitan weed; probably native to w Asia). *P. neglecta* Mack. & Bush—Rydberg, Gates.

3. ***Portulaca parvula*** A. Gray, slenderleaf purslane. Prostrate or ascending from a slender annual root; stem slender to 15 cm long, 1-2 mm thick, loosely branched, copiously hairy in the axils. Leaves linear-subulate, nearly terete, 6-12(13) mm long and 2 mm wide, succulent, somewhat compressed. Inflorescence terminal, a capitate cluster of 2-10 flowers, involucral bracts 3-8 mm long; sepals becoming reddish, about 2.5 mm long; petals yellow, orange or bronze, 2-2.5 mm long. Capsules 1.5-2 mm wide, basal portion saucer-shaped, with a stipe 1-1.5 mm long; seeds about 0.5 mm in diam, brown but becoming black when mature, covered with minute flattened stellate roughenings. Mar-Nov. Sandy or gravelly, open or brushy areas; TX & OK; (w MO to CO, s to OK, TX, NM, AZ; Mex.).

4. ***Portulaca retusa*** Engelm. Glabrous, rather stout, annual with prostrate or ascending branches, forming mats similar to *P. oleracea* but rather more slender and open. Leaves cuneate to cuneate-obovate, mostly retuse or emarginate at the apex, some rounded or nearly truncate, 2.5(3.5) cm long, 1(2) cm wide above the middle. Calyx lobes obtuse in the bud, carinate-winged; petals yellow, usually 2.5-4 mm long; stamens 7-19; style lobes 3 or 4. Capsules 5-6 mm tall; seeds iridescent black, ca 1 mm in diam, conspicuously echinate, tuberculate or sharply granulate. Jun-Sep. Gravelly soils, on clay mounds & ledges; w MO, KS, OK, & w TX; (w MO to UT, s to AR, TX, & AZ).

5. ***Portulaca umbraticola*** H.B.K. Glabrous, prostrate to erect fleshy annual, with angled stems. Leaves rather few, blades flat, sessile, the lower spatulate or obovate and obtuse to rounded, the upper ones oblanceolate to oblong and often acute, 1-3 cm long, 2-11 mm wide. Flowers clustered at ends of branches; sepals ovate, obscurely carinate; petals yellow or orange and partly red, spatulate or obovate, acute or cuspidate; stamens 7 to many; styles 3-6. Capsules circumscissile at the middle or above, the rim of the lower part crowned by a wing; seeds gray, tuberculate. Mar-Nov. Sandy prairie soils, mesquite thickets, salt marsh areas; TX; OK: Comanche; (GA w to TX, AZ, s CA; also Cuba, Jamaica).

3. TALINUM Adans., Fameflower

Herbs (ours) or shrubby plants, often with fleshy tuberous roots; stems sometimes very short or elongate, glabrous, succulent. Leaves fleshy, alternate or nearly opposite, entire, flat or terete. Flowers often showy, borne in long or short peduncled cymes or sometimes 1–several in the axils of leaves; sepals 2, distinct and free, deciduous; petals 5 (rarely more), ephemeral; stamens 5–many; ovary superior, style 3-lobed at the apex. Mature capsule 1-celled, 3-valved, with many seeds; seeds flattened, round-reniform, smooth or minutely roughened.

1 Flowers axillary; leaves flat or nearly so .... 1. *T. aurantiacum*
1 Flowers in terminal cymes, leaves terete.
   2 Petals 10–15(16) mm long, deep rose pink or rose red; stamens more than 30; capsules 6–8 mm tall .... 2. *T. calycinum*
   2 Petals 4–8 mm long, pale or dull pink; stamens less than 30; capsules 3–5 mm tall.
      3 Stamens 4–8; capsule ellipsoidal .... 3. *T. parviflorum*
      3 Stamens 12–25; capsule globose .... 4. *T. rugospermum*

1. *Talinum aurantiacum* Engelm. Erect or ascending herb, 15–35(40) cm tall, with stout leafy stems and tuberous roots. Leaves linear or linear-lanceolate, 5.5(6) cm long, 1.5–3.5 mm wide, fleshy. Flowers solitary in axils of leaves; pedicels bracted below the middle, reflexed in fruit; sepals ovate, 6–9 mm long, 3.2–4.5 mm wide, cuspidate; petals orange or reddish, obovate, 9–13 mm long, 5–6 mm wide; stamens 20 or more; stigmas linear. Capsule ovoid, 5–7 mm long, 4.5–5.2 mm in diam; seeds black, with several concentric subcircular ridges on the side. (2n = 48) Jun–Oct. Dry plains & rocky slopes; TX; (TX to AZ; n Mex.).

2. *Talinum calycinum* Engelm., rockpink, fameflower. Erect herb with a thick reddish-brown fleshy rootstock, stems 3–10 cm tall. Leaves terete or nearly so, 2.5–7 cm long, 1–2 mm thick, acute. Flowers in a terminal bracteate cyme; peduncle slender, 6–23 cm long; pedicels slender; sepals ovate to ovate-orbicular, (4)4.5–6(8) mm long, 5–6 mm wide, persistent in fruit; petals 8–10, pink to red, broadly obovate, 10–15(16) mm long, 5–9 mm wide; stamens 30 or more; stigma capitate. Capsule globose-ovoid, 6–7(8) mm long, 4.5 mm in diam; seeds black, smooth. May–Jun. Sandy & rocky soil; TX, OK, MO, KS, CO, & NE; (IL to CO, s to AR & TX).

3. *Talinum parviflorum* Nutt., prairie fameflower. Herb 5–19(20) cm tall, with fleshy roots, short stemmed or subacaulescent. Leaves terete or nearly so, linear, 1.5–5 cm long, 0.8–2.5 mm thick, broadened at the base. Flowers in a terminal bracteate cyme; peduncles slender, 3–15 cm long; pedicels slender; sepals ovate or oval-ovate (2.7)3–4 mm long, 1.5–2 mm wide, deciduous; petals pink to purplish, (4)5–7 mm long, 2.2–2.6 mm wide; stamens 4–8; style longer than the stamens; stigma capitate. Capsule ellipsoidal, 3.5–4.5(5) mm long, 2.8–3 mm in diam; seeds smooth. (2n = 48) Apr–Aug. Sandy acidic soil, overlying rocks; throughout GP; (AZ, TX, & AR, n to ND & MN).

4. *Talinum rugospermum* Holz. Erect or ascending herb, to about 20 cm tall, with 1–few short or elongate caudices from a slender taproot. Leaves linear, terete, 2–5 cm long, with a curved mucronate tip. Flowers in terminal bracteate cyme, often many bracts subtending aborted flowers; sepals ovate, deciduous, 4 mm long; petals rose colored, about 8 mm long; stamens 12–25, anthers spherical; style cleft 1/3 its length. Capsule 4 mm in diam; seeds minutely roughened and strongly wrinkled. May–Aug. Dry sandy prairie; KS: Harvey; (nw IN, IL, WI, & MN, s to IA & KS).

    This species is often confused with the eastern *T. teretifolium* Pursh, which differs in having a nearly capitate stigma, seeds that are only minutely roughened, and a thicker, more fleshy taproot.

41. MOLLUGINACEAE Hutchins., the Carpetweed Family

by William T. Barker

Annual herbs, scarcely or nonsucculent, prostrate or ascending, glabrous or stellately pubescent. Leaves simple, narrow, glabrous or pubescent, alternate to whorled; exstipulate. Flowers small, borne on pedicels or essentially sessile in the axils of leaves, regular, perfect, hypogynous. Perianth uniseriate, inconspicuous; sepals 5, persistent, distinct; petals absent; stamens 3–10, filaments distinct or more or less connate at the base, anthers 2-loculate, dehiscing by longitudinal slits; ovary of 3–5 free or united carpels, when united the ovary hypogynous, with usually distinct styles (style solitary in *Glinus*); ovules 1 per carpel and basal or several to many on axillary placentae. Fruit an achene or 3- to 5-valved loculicidal capsule; seeds reniform to roundish, sometimes arillate. Tetragoniaceae—Rydberg.

Reference: Bogle, A. L. 1970. The genera of Molluginaceae and Aizoaceae in the Southeastern United States. J. Arnold Arbor. 51: 431–462.

The Molluginaceae were formerly included in the Aizoaceae, from which they differ in being mostly nonsucculent and in having hypogynous flowers with distinct sepals. The few members of the Molluginaceae that have been tested have anthocyanins and lack betalins, the reverse of the condition existing in tested members of the Aizoaceae.

1 Plants glabrous; flowers with filiform pedicels; sepals distinct to the base .... 2. *Mollugo*
1 Plants thinly to densely stellate-pubescent; flowers sessile or with stout pedicels; calyx cleft only to the middle .... 1. *Glinus*

1. GLINUS L., Glinus

1. *Glinus lotoides* L., glinus. Prostrate or ascending, annual herb, cinereous-tomentose with stellate pubescence. Leaves alternate to whorled, narrowly to broadly obovate, rounded or abruptly acute at the apex, narrowed to a slender petiole, to 25 mm long and 15 mm broad. Flowers stoutly pedicellate or essentially sessile, in axillary glomerules. Sepals 5, free, lanceolate, stellate tomentose, to 7 mm long and 3 mm broad; stamens 5–10, alternate with and shorter than the sepals, hypogynous, filaments filiform, anthers versatile, 2-loculate at anthesis, dehiscence dorsilateral by longitudinal slits; carpels 3(5), united, styles wanting, stigmas 3(5) and sessile; ovules numerous, axillary placentation. Fruit an ovoid, loculicidal capsule, to 4.5 mm long; seeds numerous, with a short funiculus enclosed within a bladderlike caruncle and a long filiform appendage coiled around the seed; seed coat tuberculate to nearly or quite smooth and shiny, reddish- to blackish-brown. ($2n = 36$) Jul–Sep. Muddy ground along lakes & streams; KS: Wilson; MO: Jasper, Vernon; (LA to SC, n to w MO, OK, & e KS; s Europe). *Introduced* locally.

2. MOLLUGO L., Carpetweed

1. *Mollugo verticillata* L., carpetweed. Prostrate or ascending annual, glabrous throughout; stems dichotomously branched, radiating from a short slender, tapering taproot, slender, wiry, with slightly swollen nodes. Leaves in whorls of 3–8, simple, unequal, narrowly to broadly oblanceolate, 1–3 cm long and 1 cm wide, tapering to a short petiolar base; exstipulate. Flowers 2–5, inconspicuous, short, filiform pedicels 5–15 mm long. Sepals 5, pale green to white, oblong to elliptic, to 2.5 mm long and 1 mm broad; stamens 3 or 4, alternate with the carpels and alternate with the sepals, united basally by a ring of filament tissue. Capsule ovoid to ellipsoid, slightly exceeding the sepals; seeds many, minute, reniform, dark reddish-brown, smooth and shining. ($2n = 54$) Jun–Sep. Waste places & cult. ground; GP; (a common weed throughout temp. N. Amer., apparently native of trop. Amer.).

42. CARYOPHYLLACEAE Juss., the Pink Family

by G. E. Larson

Annual (often weedy) or perennial herbs (ours), seldom biennial, sometimes from a woody-thickened base when perennial, commonly with swollen nodes. Leaves simple, entire, opposite (partly in whorls of 4 in *Silene stellata*), mostly sessile, usually rather narrow, often connected at the base by a transverse line, exstipulate or (in *Loeflingia, Paronychia, Spergula,* and *Spergularia*) with scarious or hyaline stipules. Flowers commonly in dichasial cymes (umbellate in *Holosteum*) or sometimes axillary or terminal and solitary, usually perfect, regular, hypogynous or rarely perigynous (*Scleranthus*), (4)5-merous. Sepals distinct or nearly so, or united to form a campanulate, cylindric or often inflated calyx tube; petals distinct, sometimes much reduced or absent, often 2-lobed at the tip, sometimes lobed laterally as well, sometimes long clawed and then often with a pair of appendages positioned ventrally at the juncture between the claw and blade; stamens commonly 5 or 10, seldom 1–4, filaments free and distinct, or basally adnate to the petals to form a short or elongate tube that is free or adnate to the carpophore, or sometimes inserted on a nectary disk surrounding the ovary, arising from the summit of a cupulate hypanthium in *Scleranthus*; pistil 2–5 carpellate, styles 1–5, distinct or more or less united, ovary often borne on a carpophore, unilocular, sometimes 3–5 partitioned basally or nearly to the tip, ovules 1–many. Fruit usually a few- to many-seeded capsule splitting longitudinally by valves or apically by teeth, the valves or teeth numbering the same or twice as many as the styles, the fruit less often a 1-seeded utricle; seeds flattened to plump, smooth to tuberculate, sometimes wing margined; placentation usually free-central or basal, axile when the ovary is partitioned. Alsinaceae Bartling, Corrigiolaceae Reichb.—Rydberg.

1 Leaves with membranaceous or hyaline stipules.
  2 Outer 2 or 3 sepals leaflike, with a setaceous, hyaline tooth on each margin .... 7. *Loeflingia*
  2 Sepals not leaflike, entire.
    3 Fruit a 1-seeded utricle; sepals cucullate, awned or apiculate behind the hooded tip; petals absent or apparently so .... 9. *Paronychia*
    3 Fruit a several- to many-seeded capsule; sepals not cucullate, awned or apiculate; petals present, about equaling the sepals.
      4 Leaves apparently whorled; styles and valves of the fruit usually 5. .... 14. *Spergula*
      4 Leaves opposite; styles and valves of the fruit usually 3 .... 5. *Spergularia*
1 Leaves exstipulate.
  5 Sepals free or united only at the base (arising from a cupulate hypanthium in *Scleranthus*); petals present or absent, not clawed.
    6 Fruit a 1-seeded utricle; flowers perigynous, with a cupulate hypanthium resembling a calyx tube; petals absent .... 12. *Scleranthus*
    6 Fruit a few- to many-seeded capsule; flowers hypogynous; petals present or absent.
      7 Petals entire, emarginate or fimbriate at the tip (shallowly notched in *Arenaria drummondii*), sometimes absent.
        8 Styles 4 or 5, as many as the sepals .... 10. *Sagina*
        8 Styles 3, fewer than the sepals.
          9 Inflorescence cymose or capitate, or of solitary flowers; petals entire or emarginate (shallowly notched in *A. drummondii*) .... 2. *Arenaria*
          9 Inflorescence umbelliform; petals erose at the tip .... 6. *Holosteum*
      7 Petals 2-lobed, seldom absent.
        10 Capsules laterally dehiscent to about the middle by 6 valves (5 in *S. aquatica*) with each valve notched or shallowly bifid; styles mostly 3 (5 in *S. aquatica*) .... 16. *Stellaria*
        10 Capsules terminally dehiscent to produce 10 short teeth; styles mostly 5 .... 3. *Cerastium*

5 Sepals united to form a campanulate, cylindric or often inflated calyx tube; petals present, clawed, or at least weakly so.
  11 Styles 3–5(6).
    12 Styles mostly 5 (occasionally 4 in some spp., absent in male plants of *Silene pratensis*).
      13 Calyx lobes much longer than the tube at anthesis; petals not auricled or appendaged .... 1. *Agrostemma*
      13 Calyx lobes much shorter than the tube; petals auricled and ventrally appendaged at the juncture between the claw and expanded blade.
        14 Calyx cylindric, with multicellular eglandular hairs .... 8. *Lychnis*
        14 Calyx cylindric or often inflated, glandular-pubescent .... 13. *Silene*
    12 Styles mostly 3 (occasionally 4 in some spp.) .... 13. *Silene*
  11 Styles 2.
    15 Calyx subtended by 1–3 pairs of bracts .... 4. *Dianthus*
    15 Calyx ebracteate.
      16 Flowers 1 cm or less long .... 5. *Gypsophila*
      16 Flowers 2 cm or more long.
        17 Calyx cylindric, 20-nerved; petals ventrally appendaged at the juncture between the claw and blade .... 11. *Saponaria*
        17 Calyx inflated, 5-angled and ribbed; petals not appendaged .... 17. *Vaccaria*

1. AGROSTEMMA L., Corn Cockle

1. *Agrostemma githago* L. Erect, taprooted, softly whitish-hirsute annual to 10 dm tall; stem simple to freely branched, often prominently 4-angled. Leaves sessile, exstipulate, linear to linear-lanceolate, 4–12 cm long, 3–7 mm wide. Flowers usually several, showy, borne on slender pedicels to 2 dm long, perfect, gamosepalous, the calyx tube 12–16 mm long, strongly 10-ribbed, becoming inflated and hardened in fruit, the 5 calyx lobes linear-lanceolate, 1.5–2 × as long as the tube (at anthesis) or longer; petals 5, red or reddish-purple, oblanceolate, shallowly 2-lobed, lacking auricles or appendages, about 2 × as long as the calyx tube; stamens 10; styles (4)5, alternate with the sepals, ovary sessile, 1-celled. Fruit a capsule, dehiscent by (4)5 broad, apical teeth, as long as or somewhat longer than the calyx; seeds black, tuberculate, ca 3 mm long. (2n = 24, 48) May–Jul. A weed of grain fields, roadsides, & waste places; occurring sporadically in the n & e GP; (widely established throughout much of N. Amer., especially the n part; Eurasia). *Naturalized.*

2. ARENARIA L., Sandwort

Low annual and perennial herbs, when perennial sometimes forming mats or cushions from a woody caudex, often glandular-pubescent. Leaves sessile, exstipulate, the blades narrow and linear to acicular, or broad and flattened. Flowers usually few to many in open and diffuse to contracted or capitate cymes, occasionally solitary, perfect. Sepals 5, free or connate only at the base; petals 5, white or yellowish, entire or emarginate (shallowly notched in *A. drumondii*); stamens usually 10, inserted with the petals at the edge of a weakly- to well-developed perigynous disk; styles 3. Fruit a 1-celled, few- to many-seeded capsule, dehiscent by 3 valves, or in some, each valve 2-toothed so that the capsule is 6-toothed or valved; seeds plump to flattened, round to more or less reniform, smooth (and strophiolate in *A. lateriflora*) to tuberculate. *Moehringia* L., *Sabulina* Reichb.—Rydberg; *Minuartia* L.—Weber; *Eremongone* Fenzl—Weber et al. Brittonia 33: 326. 1981; *Alsinopsis* Small—Small.

    References: Maguire, B. 1947. Studies in the Caryophyllaceae—III. A synopsis of the North American species of *Arenaria*, Sect. *Eremogone* Fenzl. Bull. Torrey Bot. Club 74: 38–56; Maguire, B., 1951. Studies in the Caryophyllaceae—V. *Arenaria* in America north of Mexico. A conspectus. Amer. Midl. Naturalist 46: 493–511; Maguire, B. 1958. *Arenaria rossii* and some of its relatives in America. Rhodora

60: 44–53; McNeill, J. 1980. The delimitation of *Arenaria* (Caryophyllaceae) and related genera in North America, with 11 new combinations in *Minuartia*. Rhodora 82: 495–502; Shinners, L. H. 1949. *Arenaria drummondii* Shinners, nom. nov. Field and Lab. 17: 89; Shinners, L. H. 1962. New names in *Arenaria* (Caryophyllaceae). Sida 1: 49–52; Weber, W. A., B. C. Johnston, & R. Wittman. 1981. Additions to the flora of Colorado—VII. Brittonia 33: 325–331; Wofford, B. E. 1981. External seed morphology of *Arenaria* (Caryophyllaceae) of the southeastern United States. Syst. Bot. 6: 126–135.

Many new name combinations are being created by some American authors who would follow European taxonomists in elevating sections of *Arenaria sensu lato* to generic rank, e.g., *Minuartia* (McNeill, 1980) and *Eremogone* (Weber et al. 1981). As Wofford (1981) contends, this diverse assemblage of species probably needs more detailed study on a worldwide basis before genera can be delimited with confidence.

1 Principal leaves ovate, elliptic or linear, 2 mm or more wide.
  2 Plants annual, with a slender taproot; leaves 2–7 mm wide.
    3 Leaves ovate, 2–6 mm long .... 7. *A. serpyllifolia*
    3 Leaves linear to linear-lanceolate or linear-oblong, mostly 10–30 mm long .... 1. *A. drummondii*
  2 Plants perennial from slender rhizomes; leaves 5–10 mm wide .... 4. *A. lateriflora*
1 Principal leaves linear to acicular, 1.5 mm or less wide.
  4 Sepals strongly ribbed, with 3–5 evenly distributed nerves; capsules 3-valved.
    5 Primary leaves subtending conspicuous fascicles of small secondary leaves; seeds 0.8–1 mm long .... 8. *A. stricta* subsp. *texana*
    5 Primary leaves only present (or fascicles often weakly developed and inconspicuous in *A. rubella*, mainly on sterile branches); seeds 0.4–0.7 mm long.
      6 Plants annual, not tufted or cushion-forming; stems glabrous or with few scattered, glandular hairs; sepals 3–5 ribbed .... 5. *A. patula*
      6 Plants usually perennial, loosely to densely tufted, often cushion-forming; stems usually glandular-puberulent above, seldom glabrous; sepals 3-ribbed .... 6. *A. rubella*
  4 Sepals weakly nerved, with a green midstripe and white-hyaline margins; capsules 6-toothed or valved.
    7 Flowers in open cymes; basal leaves grasslike, (2)4–10 cm long .... 2. *A. fendleri*
    7 Flowers in congested, headlike cymes; basal leaves 0.4–2(3) cm long ..... 3. *A. hookeri*

1. Arenaria drummondii Shinners. Annual with a slender taproot, 0.5–2 dm tall; stem single or sparingly branched from the base, erect to ascending, often forking above, glandular-pubescent into the inforescence. Leaves few to several pairs per stem, linear to linear-oblong or rarely linear-oblanceolate, mostly 10–35 mm long, 2–7 mm wide, somewhat succulent, glabrous, obtuse to acutish, the upper leaves much reduced. Flowers in open cymes, the pedicels widely spreading in flower, usually reflexed in fruit, 10–25 mm long. Sepals ovate to elliptic, 5–7 mm long, faintly nerved, glandular-pubescent, obtuse, narrowly scarious-margined; petals white, sometimes greenish at the base, obcordate, with a shallow notch at the apex, 9–15 mm long, usually much surpassing the calyx. Capsule 3-valved, ovoid, 4–5 mm long; seeds brownish, ca 1 mm long, muriculate. Apr–May. Moist sandy prairies & open woods; OK: Craig, Oklahoma, Tulsa; (AR, OK, LA, & TX). *Alsinopsis nuttallii* (T. & G.) Small—Small; *Minuartia drummondii* (Shinners) McNeill—J. McNeill *in* Kartesz, J. T., & R. Kartesz. 1980. A synonymized checklist of the vascular flora of the United States, Canada and Greenland. Univ. of North Carolina Press, Chapel Hill. p. 151.

2. Arenaria fendleri A. Gray. Rather pale and glaucous, grasslike perennial, mostly 1–3 dm tall, finely glandular-pubescent above, the foliage rather pungent; stems often numerous and cowded in densely leafy tufts, arising from a thick, branched, woody rootstock. Basal leaves setaceous, gramineous, (2)4–10 cm long, mostly less than 1 mm wide; cauline leaves usually of 3–5 pairs, reduced upward, connate at the base. Inflorescence a few- to many-flowered, dichotomously branched and spreading cyme, glandular-pubescent; bracts subulate, scarious-margined; pedicels mostly to 1(1.5) cm long, glandular-pubescent. Sepals lanceolate to linear-lanceolate, 4–6 mm long, glandular-pubescent, attenuate, broadly hyaline-margined with a green midstripe; petals white or pale yellowish, oblong with an emarginate tip, equaling to usually exceeding the sepals. Capsule 6-toothed, ellipsoid, shorter than the sepals; seeds blackish, 1–1.2 mm long, finely tuberculate. (2n = 40, 44) Jun–Aug. Exposed rocky hillsides & rock crevices in foothills & higher elevations of the Rocky Mts.; barely entering our range in e CO & se WY; (WY, CO, to UT, s to TX, NM, & AZ).

3. *Arenaria hookeri* Nutt. ex T. & G. Tufted perennial from a branched caudex, forming dense cushions to 15 cm across; flowering stems (1)2–10(15) cm tall, puberulent and more or less glandular. Leaves linear, rigid and acicular, 4–20(30) mm long, 1–1.5 mm or less wide, sharp-tipped, those of the flowering stems 1–4 pairs, often somewhat longer than the crowded basal leaves. Flowers few to many in congested, headlike cymes; pedicels seldom over 1–2 mm long. Sepals lanceolate, (5)6–9(10) mm long, glabrous or glandular-puberulent, acute to acuminate, hyaline-margined with a green midstripe; petals white, oblanceolate, slightly shorter than to 1 1/2 × the length of the sepals, marcescent-persistent. Capsule 6-toothed or valved nearly to the middle, ovoid, shorter than the sepals; seeds blackish, ca 1.5 mm long, finely papillate. (2n = 44, 66) Jun–Aug. Sandy, gravelly or rocky hillsides & ledges; sw SD & e WY, s to extreme w KS, OK, & adj. CO; (sw SD to cen MT & ne UT, s to TX).

Two varieties occur in our range and are differentiated as follows:

3a. var. *hookeri*. Plants pulvinate; principal leaves strict or recurved, 0.5–1.5 cm long; sepals 5–7(9) mm long, glabrous. e WY, w NE, extreme w KS & OK, & e CO.

3b. var. *pinetorum* (A. Nels.) Maguire. Plants loosely pulvinate or merely caespitose; principal leaves strict or flexuous, 1.5–3(4) cm long; sepals (5)7–10 mm long, glabrous or often puberulent. sw SD, se WY, & nw NE, also supposedly as far s as CO: El Paso(?).

4. *Arenaria lateriflora* L., grove sandwort. Low perennial from slender rhizomes; stems slender, ascending or decumbent, often branched, 5–20 cm long, retrorsely puberulent. Leaves elliptic to oblong-elliptic, 10–30 mm long, 5–10 mm wide, minutely pustulate on both surfaces, puberulent mainly on the midvein and margins, obtuse-tipped; lower leaves much reduced, oval to obovate. Flowers usually paired (1–5) in terminal or lateral, minutely bracteate cymes; peduncle slender, 1–3 cm long; pedicels 5–15 mm long. Sepals ovate or obovate, 2–3 mm long, faintly 3- to 5-nerved, obtuse or acutish, scarious-margined; petals white, obovate, 3–6 mm long, entire. Capsule 6-valved nearly to the base, subglobose, 3–5 mm long; seeds few, strophiolate, black, smooth and shiny, round and lenticular, ca 1 mm in diam. (2n = 48) May–Aug. Moist or dry woods & thickets, occasionally open areas; MN to MT, s to IA, n NE, & WY; (circumboreal, s in Amer. to PA, ne MO, NE, NM, & CA). *Moehringia lateriflora* (L.) Fenzl — Rydberg.

5. *Arenaria patula* Michx. Taprooted annual 5–20 cm tall, few- to many-branched from the base; stems slender, erect to decumbent, sparingly to freely branched and spreading upward, glabrous or occasionally with few, scattered glandular hairs. Leaves mostly cauline, several pairs per stem, evanescent from the base, linear, 5–20(25) mm long, 0.5–1 mm wide, acute. Flowers usually numerous, in open, spreading, bracteate cymes, often extending to below the middle; pedicels 5–50 mm long. Sepals lanceolate, (3)4–6(7) mm long, 3- to 5-ribbed with narrow scarious margins, acute; petals white, spatulate, 5–8(9) mm long, irregularly emarginate. Capsule 3-valved to about the middle, oblong, shorter than to about equaling the sepals; seeds dark brown, nearly round, 0.5–0.7 mm long, low tuberculate. Apr–Jun. Limestone & sandy barrens in pastures & grasslands; e KS, s & e OK; (VA to IN & MN, s to AL & TX). *Sabulina patula* (Michx.) Small — Rydberg; *A. patula* f. *pitcheri* (Nutt.) Steyerm., *A. patula* f. *meadia* Steyerm. — Steyermark.

Our plants are var. *patula*, having predominantly 5-nerved sepals and seeds 0.5–0.7 mm long. Var. *robusta* (Steyerm.) Maguire, with 3-nerved sepals and seeds 0.7–0.9 mm long, occurs east of our range.

6. *Arenaria rubella* (Wahl.) J. E. Sm. Compact or loosely tufted, delicate perennial, possibly sometimes flowering the first year, 1–10(25) cm tall; often forming cushions to

10 cm across, taprooted with a branched caudex; stems usually numerous, decumbent or mostly prostrate, slender, simple (and 1-flowered) or dichotomously branched, glandular-puberulent mainly above or seldom glabrous, some stems usually sterile. Leaves mostly basal and overlapping, only a few pairs upward on flowering stems, much shorter than the internodes, linear to linear-lanceolate, 3–10(20) mm long, 0.2–1 mm wide, 3-nerved, blunt or somewhat acicular; basal leaves and the primary leaves of sterile shoots often subtending weakly developed fascicles of secondary leaves. Flowers solitary on stems or in narrow, bracteate, 2- to 5(7)-flowered cymes; pedicels slender, 5–15(35) mm long, glandular-puberulent or seldom glabrous. Sepals lanceolate, 2.5–4 mm long, 3-ribbed with narrow scarious margins, glabrous or glandular-puberulent, acute; petals white, elliptic to obovate, usually shorter than the sepals. Capsule 3-valved to below the middle, narrowly ovoid, about equaling to considerably exceeding the sepals; seeds dark brown, 0.4–0.7 mm long, very finely tuberculate. (n = 12, 2n = 24) Jun–Aug. Subalpine rocky or gravelly slopes & hilltops, often in limestone soils; BH; (circumboreal, in N. Amer. s to Que., n VT, Ont., NM, NV, & CA). *Sabulina propinqua* (Richard.) Rydb.—Rydberg; *Minuartia rubella* (Wahl.) Graebn.—Weber.

7. *Arenaria serpyllifolia* L., thyme-leaved sandwort. Erect to spreading annual with a slender taproot, 5–25 cm tall; stems 1–many, erect to decumbent, leafy, retrorsely puberulent, often glandular above. Leaves ovate, much shorter than the internodes, 2–6(8) mm long, 1–4 mm wide, often minutely pustulate, glabrous or puberulent, acute or broadly so, the lower leaves short-petiolate and slightly connate. Flowers several to many per stem, in open, leafy-bracteate cymes; pedicels 3–7(12) mm long. Sepals lanceolate to ovate-lanceolate, 2–4 mm long, faintly 3- or 5-nerved, glabrous or scabrid-puberulent, scarious-margined; petals white, oblong to narrowly spatulate, 1/2–3/4 as long as the sepals. Capsule 6-toothed, pyriform, about equaling to exceeding the sepals; seeds plump, blackish-brown, low-tuberculate in concentric rows. (n = 14, 28; 2n = 10, 20, 28, 40, 44) Apr–Jun. Lawns, roadsides, alluvium, & disturbed places, especially where sandy or rocky; MO & e NE, e & cen KS & OK; (widely established throughout much of temp. N. Amer.; Eurasia). *Naturalized. A. serpyllifolia* var. *tenuior* Mert. & Koch—Gleason & Cronquist.

8. *Arenaria stricta* Michx. subsp. *texana* (Robins.) Maguire, rock sandwort. Loosely tufted, diffuse perennial (ours) with a stout taproot and branched caudex, glabrous, (0.7)1–2 dm tall; stems usually numerous, decumbent to erect, branching from a marcescent-leafy base, leafy to about the middle or below, with short, leafy, sterile axillary shoots, these appearing as uniformly developed fascicles of secondary leaves. Primary leaves subulate-setaceous, 4–10(15) mm long, ca 0.5 mm wide, 3-nerved, somewhat involute. Flowers (1)–several to many in an open, minutely bracteate cyme, often extending to the middle of the stem; pedicels slender, 3–15(25) mm long. Sepals broadly lanceolate, 3–5.5 mm long, firm-textured, 3- to 5-ribbed, acute, scarious-margined; petals spatulate to oblong or obovate, 5–8 mm long, entire. Capsule 3-valved to the middle or beyond, ovoid, shorter than to about equaling the sepals; seeds brownish-black, plump, 0.8–1 mm long. (n = 11-15) May–Jun. Open gravelly or rocky places, often in limestone areas; s-cen & sw NE, cen & w KS, s to cen TX; (OH to MO & NE, s to AR & TX). *Sabulina texana* (Robins.) Rydb.—Rydberg; *Arenaria stricta* Michx. var. *texana* Robins.—Fernald; *Minuartia stricta* (Sw.) Hiern—Weber; *Minuartia michauxii* (Fern.) Farw. var. *texana* (Robins.) Mattf.—J. McNeill *in* Kartesz, J. T. & R. Kartesz. A synonymized checklist of the vascular flora of the United States, Canada, and Greenland. Univ. of North Carolina Press, Chapel Hill. p. 151.

Rydberg also attributes *A. stricta* subsp. *dawsonensis* (Britt.) Maguire to SD, under the name of *Sabulina dawsonensis* (Britt.) Rydb., but the range of this subsp. supposedly lies far to the n of the GP. It differs from subsp. *texana* in having the petals equaling or mostly shorter than the calyx and the capsule exserted beyond the calyx.

3. CERASTIUM L., Mouse-ear Chickweed

Low, erect to spreading annuals and perennials, usually with short and often glandular-viscid pubescence. Leaves usually rather small, sessile or essentially so, sometimes narrowed to a flat, subpetiolate base, exstipulate, often short-hairy on both surfaces. Flowers usually few to many in compact to open, short-bracteate, terminal cymes, or seldom solitary, perfect. Sepals 5, separate; petals 5, sometimes absent, white, shorter than to exceeding the sepals, 2-lobed, weakly clawed; stamens (5)10; styles mostly 5. Fruit a 1-celled, many-seeded, membranaceous, cylindrical capsule, terminally dehiscent to produce 10 small teeth, straight or upcurved, usually exceeding the calyx; seeds angular-obovoid, dorsally grooved, papillate-tuberculate.

References: Fernald, M. L., & K. M. Wiegand. 1920. Studies of some boreal American cerastiums of the section *Orthodon*. Rhodora 22: 169–179; Shinners, L. H. 1966. *Cerastium glutinosum* Fries (Caryophyllaceae) in Mississippi: new to North America. Sida 2: 392–393.

1 Plant a perennial; stems usually trailing and matted, conspicuously marcescent-leafy at the base, with dried leafy fascicles or short lateral branches present at lower nodes; petals 1.5–2 × as long as the sepals .... 1. *C. arvense*
1 Plant either a shallowly taprooted annual, or a weedy perennial with more or less fibrous roots; stems without leafy fascicles or sterile branches, or these mostly herbaceous; petals seldom to 2 × as long as the sepals, usually much shorter, frequently absent.
  2 Sepals with some long, eglandular hairs on the back projecting well beyond the tip.
    3 Flowers in compact clusters, the pedicels shorter than the calyx, rarely to 5 mm long; filaments of the stamens and claws of the petals glabrous .... 4. *C. glomeratum*
    3 Flowers in open dichasia, at least some of the pedicels longer than the calyx; filaments of the stamens and claws of the petals ciliate .... 2. *C. brachypetalum*
  2 Sepals with hairs becoming shorter toward the tip and not projecting beyond it.
    4 Uppermost bracts of the inflorescence scarious-margined; plant perennial (but often flowering the first year), more or less fibrous-rooted; stems often matted and tending to root at the lower nodes .... 6. *C. vulgatum*
    4 Uppermost bracts of the inflorescence completely herbaceous; plant a shallowly taprooted annual; stems mostly erect to ascending, not matted and not rooting at the nodes.
      5 Pedicels 0.5–1.25 × the length of the calyx in flower, to 3 × the calyx length and straight or only slightly curved in age .... 3. *C. brachypodum*
      5 Pedicels 1–3 × the length of the calyx in flower, to 5 × the calyx length and sharply curved below the calyx in age .... 5. *C. nutans*

1. **Cerastium arvense** L., prairie chickweed. Clumped or mat-forming perennial 0.5–3 dm tall, short-pubescent, usually glandular-pubescent at least in the inflorescence; flowering stems erect above the trailing base, sparsely to densely retrorse-strigose below, more spreading-hairy and usually glandular upward, conspicuously marcescent-leafy at the base, with leafy fascicles or short lateral branches produced from lower nodes, these eventually drying. Leaves mostly linear to lanceolate or narrowly oblong, 7–30 mm long, 1–5 mm wide, 1-nerved, acute, those of trailing stem bases and spreading lateral branches generally smaller, oblong or oblong-ovate. Flowers usually few to several in an open to somewhat compact cyme, rarely single; uppermost bracts scarious-margined; pedicels 5–30 mm long, usually curved below the calyx in age. Sepals ovate-lanceolate to lanceolate, 4–6 mm long, glandular-puberulent to glandular-pilose, acute to subacute, scarious-margined; petals 1.5–2 × as long as the sepals, deeply bifid. Capsule maturing slowly, globose to ovoid when immature, cylindric and 6–10 mm long when ripe; seeds reddish-brown, 0.7–0.9 mm long. (n = 18, 36; 2n = 36, 38, 72, 90). Flowers May–Jun; fruits Jun–Aug. Prairies, pastures, meadows, subalpine slopes, often weedy; MN to MT, s to n IA, n & w NE, e WY; (circumboreal, in Amer. s to GA, MN, NE, NM, & CA, also possibly in S. Amer.). *C. campestre* Greene, *C. oreophilum* Greene, *C. strictum* L., *C. velutinum* Raf.—Rydberg; *C. arvense* var. *villosum* (Muhl.) Hollick & Britt., *C. arvense* var. *viscidulum* Gremli—Fernald.

C. arvense is a highly variable species, perhaps more so elsewhere than in the GP. Attempts to segregate species and varieties within the complex have been thwarted by instability of characters.

2. Cerastium brachypetalum Pers. Annual 0.5–3.5 dm tall, finely hirsute and becoming glandular-pubescent upward, single stemmed or usually branched from the base; stems erect to decumbent but not rooting at the nodes, often with spreading or ascending lateral branches. Leaves mostly elliptic to elliptic-ovate, 5–20 mm long, 3–7(9) mm wide, acute to obtuse, the basal and lower leaves obovate to spatulate, obtuse to rounded. Flowers in open dichasia; bracts herbaceous, long-hairy beyond the tip, mostly eglandular; pedicels 1–3 × as long as the calyx, hirsute and glandular-pubescent, mostly curved in age. Sepals lanceolate, 4–5 mm long, with some long eglandular hairs projecting well beyond the tip, usually with some short-glandular hairs as well, acute, scarious-margined; petals shorter than to subequal to the sepals, ciliate on the claw; stamen filaments long-ciliate. Capsule 6–8(10) mm long, straight or upcurved; seeds brown, 0.4–0.5 mm long. (n = 45, 52, ca 39; 2n = 90, ca 78). Apr–May. A weed of pastures, roadsides, & disturbed open areas, usually in sandy or rocky soil, recently extending into the s GP; KS: Cherokee, Cowley, Montgomery; OK: Pittsburg; (VA, NC, & SC, w to KS & OK; Eurasia). *Naturalized* and likely spreading in the s GP.

3. Cerastium brachypodum (Engelm. ex A. Gray) Robins. Shallowly taprooted annual 0.5–3.5 dm tall, short glandular-pubescent especially above, single stemmed or branched from the base, erect to ascending. Leaves mostly lanceolate to narrowly elliptic or oblanceolate, lower ones sometimes ovate to obovate or spatulate, 5–30 mm long, 2–8 mm wide, acute to obtuse. Flowers (rarely single) few to many in a rather compact to open dichasial cyme, the flowers more clustered at the tips of dichotomous branches when numerous; bracts herbaceous, glandular-puberulent; pedicels 0.5–1.25 × the length of the calyx in flower, to 3 × the calyx length in fruit, straight or only slightly curved in age. Sepals lanceolate, 3–4.5 mm long, glandular-puberulent or sometimes glabrate, acute or subacute, herbaceous or narrowly scarious-margined; petals shorter than to 2 × as long as the sepals, or frequently absent (especially in plants of n GP). Capsules 6–12 mm long, straight or upcurved; seeds golden-brown, 0.4–0.7 mm long. (n = 17) Apr–Jul. Grasslands, fields, meadows, open woods, roadsides, & waste places, in wet or dry soil, often where rocky or sandy; GP; (Que. to N.T., s to GA, TX, AZ, OR; Mex. & C. Amer.). *C. nutans* Raf. var. *brachypodum* Engelm.—Gleason.

C. brachypodum is not distinguished from *C. nutans* by some authors, or it is considered merely as a variety of the latter. The two entities seem morphologically distinct in our region and are thus treated here as separate species.

4. Cerastium glomeratum Thuill. Annual 0.5–3 dm tall, finely hirsute throughout and glandular-pubescent mainly above, single stemmed or branched from the base, erect or decumbent. Leaves obovate to spatulate below, becoming oval to ovate or elliptic-ovate upward, 5–20(30) mm long, 3–9(15) mm wide, obtuse to rounded. Flowers in dense clusters at the tips of dichotomous branches; bracts herbaceous, hirsute and glandular-pubescent; pedicels shorter than the calyx, rarely to 5 mm long. Sepals usually purple-spotted at the tip, lanceolate, 3–5 mm long, hirsute and glandular-pubescent, with some eglandular hairs projecting beyond the tip, acute, scarious-margined; petals shorter than to slightly exceeding the sepals, sometimes absent, the claws of the petals and the filaments of the stamens glabrous. Capsule 5–8 mm long, straight or upcurved; seeds light brown, 0.4–0.5 mm long. (n = 36; 2n = 72). Apr. A weed of fields, roadsides, pastures, open woods, & disturbed places, usually in rocky or sandy soil; w MO, se KS, e & cen OK, possibly of sporadic occurrence northward in e GP; (throughout much of e & cen N. Amer. & along the Pacific Coast; Eurasia). *Naturalized*. *C. viscosum* L.—Gleason & Cronquist; *C. vulgatum* L., in part—Gleason.

5. **Cerastium nutans** Raf., nodding chickweed. Slender annual 1-5 dm tall, finely glandular-pubescent, single-stemmed or branched at or near the base, erect or decumbent. Leaves oblanceolate to spatulate below, becoming lanceolate to linear-lanceolate upward, occasionally elliptic, 10-60 mm long, 3-15 mm wide, mostly acute. Flowers in a rather open and ultimately widely branched cyme; bracts herbaceous, glandular-puberulent; pedicels 1-3 × the length of the calyx in flower, to 5 × the calyx length and sharply curved below the calyx in age. Sepals lanceolate, 3-6 mm long, glandular-puberulent or sparingly so, acute to subacute, scarious-margined; petals shorter than to 2 × as long as the sepals, sometimes absent. Capsule 8-13 mm long, mostly upcurved; seeds golden-brown, 0.5-0.8 mm long. (n = 18; 2n = 36). Apr-Jun. Moist woods, stream banks, meadows, shores, & boggy places; MN to nw & cen ND, s to MO, e & cen KS & e OK; (Que. to B.C., s to FL, AL, AR, TX, NM, & AZ). *C. longipedunculatum* Muhl.—Winter.

Records of *C. nutans* from sw ND, w SD, & w NE likely are based on *C. brachypodum*.

6. **Cerastium vulgatum** L., common mouse-ear chickweed. Weedy perennial (often flowering the first year) 0.5-3.5 dm tall, hirsute, also glandular-pubescent mainly above or throughout, single stemmed or usually few- to many-branched from the base; stems often purplish, the flowering stems erect to decumbent, with lateral branches or short leafy fascicles developing from lower nodes, these normally remaining green, the prostrate portions of flowering stems and lateral branches often becoming matted and then tending to root at the nodes. Lower leaves and those of lateral branches obovate to spatulate or elliptic, obtuse to rounded; main cauline leaves larger, ovate to elliptic-ovate or oblong, 5-30 mm long, 2-10 mm wide, obtuse to acute. Flowers in ultimately spreading dichasia; bracts often purplish-tipped, the uppermost slightly to prominently scarious-margined and tipped; pedicels 1-3 × as long as the calyx, straight or curved below the calyx. Sepals usually purplish-tipped, lanceolate, 4-6 mm long, hirsute, often eglandular; petals shorter than to slightly exceeding the sepals. Capsule 6-10 mm long, straight or upcurved; seeds reddish-brown, 0.5-0.7 mm long. (2n = 70, 72, 126, 144, 180). Apr-Jun, sometimes flowering again Sep-Oct. Roadsides, pastures, meadows, prairies, & open woods, often weedy in lawns; GP, especially e part; (established over much of subarctic & temp. N. Amer.; Eurasia). *Naturalized. C. vulgatum* var. *hirsuta* Fries—Harrington; *C. vulgatum* f. *glandulosum* (Boenn.) Druce—Steyermark; *C. holosteoides* Fries—Shinners, L. H., Sida 2: 392. 1966; *C. fontanum* Baumg. subsp. *triviale* (Link) Jalas—J. McNeill *in* Kartesz & Kartesz, op. cit.

4. DIANTHUS L., Pink

1. **Dianthus armeria** L., Deptford pink. Slender, taprooted annual or biennial 2-7 dm tall, with 1-several stiffly erect stems, dichotomously branched above, glabrous or crisp-pubescent mainly at the nodes and in the inflorescence. Cauline leaves sessile, connate for 2-4 mm, linear, 3-10 cm long, 1-5 mm wide, strigillose, attenuate or blunt tipped, the basal leaves sometimes oblanceolate, to 8 mm wide. Inflorescence of terminal, headlike cymes containing 2-several flowers, some flowers occasionally solitary, the flowers sessile or nearly so, each subtended by 1-3 pairs of linear-attenuate bracts, these somewhat shorter than to surpassing the calyx. Calyx gamosepalous, the tube cylindrical, 10-15 mm long, 20- to 25-nerved, strigillose, the teeth 3-6 mm long; petals pink or rose, dotted with white, drying purple, long clawed, without auricles or appendages, the blade rhombic-obovate, 4-5 mm long, dentate; stamens 10; pistil borne on a carpophore ca 1 mm long, styles 2. Capsule dehiscent by 4 teeth, about equaling the calyx; seeds dark brown, concavo-convex, 1-1.3 mm long, minutely tuberculate. (2n = 30) Late May-early Jul. A garden ornamental now established as a weed in pastures, fields, roadsides, & disturbed areas; w MO, e KS & e OK, sporadically n & w to SD; (Que. to B.C., s to GA, AR, OK, MT, & WA; Europe). *Naturalized.*

5. GYPSOPHILA L.

Introduced annual or perennial herbs, often diffusely branched. Leaves sessile, exstipulate, linear to lanceolate or oblong, connected at the swollen nodes by a transverse ridge. Inflorescence of numerous small flowers (less than 1 cm long) in dichasial cymes, often diffuse and paniculate; bracts scarious or scarious-margined; pedicels capillary. Calyx gamosepalous, campanulate to turbinate, prominently 5-nerved, white-scarious between the nerves and around the margins of the lobes; petals white to pink, truncate to emarginate at the tip, without auricles or appendages, weakly clawed; stamens 10; styles 2, ovary globose. Capsule globose to oblong, terminally dehiscent by 4 broad teeth; seeds subglobose to subreniform, flattened on the sides, tuberculate in rows.

Reference: Barkoudah, Y. I. 1962. A revision of *Gypsophila, Bolanthus, Ankyropetalum* and *Phryna*. Wentia 9: 1–203.

1 Plants annual, with a slender taproot; pedicels several to many times as long as the calyx .... 1. *G. muralis*
1 Plants perennial, with a thick, woody taproot; pedicels 2–3 × as long as the calyx .... 2. *G. paniculata*

1. Gypsophila muralis L. Glaucous annual 0.5–1.5(2) dm tall, branched throughout; stems glabrous above, puberulent or sparsely so below. Leaves linear or nearly so, 5–15(25) mm long, 0.5–3 mm wide, acute. Inflorescence lax and open, diffuse, comprising most of the height of the plant, with flowers arising from the axils of all but the lowest leaves; bracts reduced but foliaceous; pedicels spreading to ascending, 1–2 cm long. Calyx green along the nerves and white-scarious between, long-campanulate, 2.5–4 mm long, incised for about 1/4–1/3 of its length, the lobes broadly rounded and erose; petals pink, oblong, 4–8 mm long, emarginate. Capsule oblong, 3–4 mm long; seeds obovoid, 0.5–0.6 mm in diam. (2n = 34) Jun–Aug. Cult. & occasionally escaping & persisting as a weed in disturbed places & along roadsides; mainly in the BH; MN: Ottertail; SD: Custer, Lawrence, Pennington; (parts of n U.S. & s Can.; Eurasia). *Escape*.

Gypsophila elegans Bieb. is another annual (or biennial) species from se Europe that occasionally escapes from garden plantings in e N. Amer. It is larger than the above, 1.5–6 dm tall, with leaves 2–4 cm long, 2–10(15) mm wide, a broadly campanulate calyx 3–5 mm long, incised to about the middle, a more globose capsule and larger seeds ca 1.2 mm in diam. We have only one record of *G. elegans* from the ne part of our range; MN: Kittson.

2. Gypsophila paniculata L., baby's breath. Diffusely branched, glaucous, globose-shaped perennial 4–10 dm tall, from a thick, woody taproot, glabrous throughout or glandular-puberulent on the main stems toward the base. Leaves linear to lanceolate, 1.5–6 cm long, 2–8 mm wide, reduced above. Inflorescence cymose-paniculate, with numerous flowers; bracts scarious-margined, with a green or purplish midvein; pedicels relatively short, 3–6 mm long. Calyx purple or purplish-green on the nerves, white-scarious between the nerves, campanulate to turbinate, 1.5–2 mm long, incised to about the middle or slightly below, the lobes broadly rounded and white-scarious around the margin; petals usually white (pinkish), 2.5–3.5 mm long, truncate to emarginate at the tip. Capsule globose, about equaling to exceeding the calyx; seeds subreniform, 1.2–1.5 mm long. (2n = 28, 34) Jul–Sep. Introduced and frequently escaping from yards & cemeteries, persisting in pastures, waste places, & along roadsides, often where sandy or rocky; mainly n GP, also KS: Graham; (locally distributed over much of n U.S. & s Can.; Eurasia). *Escape*.

The distinctive inflorescences of *G. paniculata*, particularly double-flowered cultivars, often are used as fillers in fresh and dried floral arrangements.

6. HOLOSTEUM L., Jagged Chickweed

1. Holosteum umbellatum L. Tufted annual (0.5)1–3 dm tall; stems 1–several, simple, erect to decumbent, glandular-pubescent near the middle, generally glabrous above and toward the base. Leaves mostly tufted at the base, these spatulate to oblong-lanceolate, mostly 8–20 mm long, 3–5 mm wide; cauline leaves of 2 or 3 pairs, remote from the inflorescence, sessile, elliptic-ovate to oblong, mostly 5–25 mm long, 3–7 mm wide, blunt-tipped, the leaves mostly glabrous or especially the cauline ones glandular-stipitate along the margins. Inflorescence a minutely bracteate, umbellate cyme of 3–16 flowers; pedicels erect to spreading or some deflexed, mostly 10–25 mm long in fruit, glabrous or rarely with some glandular hairs. Sepals 5, distinct, green, often purple-tinged, elliptic, 3–5 mm long, finely nerved; petals 5, white; oblanceolate, a little longer than the sepals, erose-fimbriate at the tip; stamens 3–5; styles 3. Capsule terminally dehiscent by 6 recurved teeth, short-cylindric, 5–7 mm long, membranaceous; seeds 0.8–1 mm in diam, low-papillate. (2n = 20) Apr–May. A weed in sandy or rocky, often calcareous soil of roadsides & waste places; MO, s NE, to e & cen OK, perhaps occasional as a waif farther n; (throughout much of the U.S. & parts of s Can.; n Africa & Eurasia). *Introduced*.

Reference: Shinners, L. H. 1965. *Holosteum umbellatum* (Caryophyllaceae) in the United States: population explosion and fractionated suicide. Sida 2: 119–128.

7. LOEFLINGIA L.

1. **Loeflingia squarrosa** Nutt. Small globose-shaped annual to 15 cm tall; stems freely branched, glandular-puberulent. Leaves mostly crowded in fascicles at stem tips and on short lateral branches, subulate, tapering to a subspinose tip, 4–10 mm long, straight or recurved-spreading; stipules hyaline, setaceous, broadened and adnate to the leaf at the base. Flowers inconspicuous, axillary, sessile, solitary or few together in fascicles, somewhat camouflaged among the crowded leaves. Sepals 5, distinct, lanceolate, awn-tipped, the outer 2 or 3 leaflike, with a setaceous (stipulelike), hyaline tooth on each margin, the inner ones somewhat shorter; petals 3–5 and minute or sometimes lacking; stamens 3–5; style short or lacking, stigmas 3. Capsules slender, 3-valved, shorter than or equaling the sepals; seeds several, obovate, 0.3–0.4 mm long. Apr–May. Dry soil or barren sand in grasslands; NE: Dawes; cen OK & s into TX; (sw WY to se OR, s to AZ & CA). *L. texana* Hook.—Rydberg.

Reference: Barneby, R. C., & E. C. Twisselmann. 1970. Notes on *Loeflingia* (Caryophyllaceae). Madroño 20: 398–407.

8. LYCHNIS L.

1. **Lychnis chalcedonica** L., scarlet lychnis. Stout-based perennial 5–10 dm tall; stem simple, hirsute. Basal leaves spatulate or oblanceolate to lanceolate, usually withered by flowering time; cauline leaves of 10–20 pairs, ovate to ovate-lanceolate, 5–12 cm long, 1.5–5 cm wide, sparsely pubescent at least on the lower surface, margin serrulate-ciliate, cordate-clasping at the base, often with smaller leaves in the axils. Inflorescence a dense, terminal, capitate or subcapitate cyme, mostly 10- to 60-flowered. Flowers showy, rose to scarlet or seldom white; calyx tubular, with 5 short teeth, 14–17 mm long, the tube 10-nerved, with eglandular, multicellular hairs on the nerves; petals 5, long clawed with a deeply bilobed blade, auricled and ventrally appendaged at the juncture between the claw and expanded blade, 14–18 mm long, the claw ciliate, the appendages tubular, 2–3 mm long, the blade 7–9 mm long; stamens 10; styles 5 or occasionally 4 or 6 in some individuals, ovary 1-celled, borne on a carpophore 4–6 mm long. Capsule dehiscing by (4)5(6) terminal teeth (teeth numbering the same as styles), ca 1 cm long; seeds grayish-brown, round-reniform, flattened on the sides, ca 1 mm in diam, tuberculate. (2n = 18, 24, 48) Jun–Aug. An introduced garden ornamental occasionally escaping in the n GP; ND: Benson, Rolette, Sheridan; (scattered from e Can. to N. Eng. w to GP; Asia). *Escape*.

Most species formerly included in *Lychnis* are now placed in *Silene*.

9. PARONYCHIA P. Mill., Whitlow-wort, Nailwort

Small annual and perennial herbs, the perennials tufted from a branched caudex and often forming low mats. Leaves small, stipulate, linear and often spinulose-tipped or oblong to elliptic with an acute to obtuse tip; stipules small to conspicuous, hyaline. Flowers clustered in terminal, bracteate cymes (single or paired in *P. sessiliflora*) or in the forks of the stem, perfect, hypogynous or nearly so. Sepals 5, united toward the base to form a short tube, 1- to 3-nerved, cucullate, awned or apiculate behind the hooded tip; petals none or apparently so; stamens usually 5, opposite the sepals and included in the calyx, attached to a short perigynous disk lining the basal, tubular portion of the calyx, sometimes alternating with tiny, filamentlike staminodia (vestigial petals?); style 2-parted, included in the calyx, ovary 1-celled, with 1 basal ovule, maturing as a membranaceous, 1-seeded utricle enclosed by the calyx. *Anychia* Michx.—Rydberg.

References: Core, E. L. 1941. The North American species of *Paronychia*. Amer. Midl. Naturalist 26: 396–397; Hartman, R. L. 1974. Rocky Mountain species of *Paronychia*, a morphological and chemical study. Brittonia 26: 256–263.

1 Plants annual.
  2 Stems glabrous; sepals 1-nerved .... 1. *P. canadensis*
  2 Stems puberulent; sepals mostly (1)3-nerved .... 3. *P. fastigiata*
1 Plants perennial.
  3 Flowers single or paired at stem tips; ovary and fruit pubescent on the upper 1/2; leaves 3–6 mm long, little if any exceeding the bracts, the stipules nearly as long .... 5. *P. sessiliflora*
  3 Flowers more or less clustered; ovary and fruit glabrous; leaves 5–25 mm long, longer than the bracts and stipules.
    4 Stems mostly prostrate and crowded with ascending or erect tips, mat-forming, the entire plant not more than 10 cm high above the ground; leaves crowded, the internodes 3–10(14) mm long .... 2. *P. depressa*
    4 Stems erect or ascending in tufts, not mat-forming, the plant (7)10–25(30) cm tall; leaves not crowded, the internodes 6–30 mm long .... 4. *P. jamesii*

1. Paronychia canadensis (L.) Wood, forked chickweed. Glabrous annual 10–30 cm tall, simple below, dichotomously branched and spreading above. Leaves green above, slightly glaucous beneath, usually white-punctate, sometimes reddish-brown mottled when young, elliptic to oval, 5–20 mm long, 3–10 mm wide, thin-textured, obtuse to rounded at the tip, narrowed to a short-petiolate base; stipules triangular, 0.5–2 mm long. Inflorescence open and spreading, leafy-bracteate, the flowers on short, curved pedicels in the axils and at the tips of the slender branches. Calyx 0.8–1.2 mm long, the tube short, sepals oblong-ovate, 1-nerved, with a short-hooded tip, low apiculate behind the hood, scarious-margined; style branches minute, free nearly to the base, recurved. Utricle globose, exceeding the calyx, granular at the tip; seed reddish-black and shiny, round, ca 1 mm in diam. Jun–Sep. Open woods & moist prairie, in rocky or sandy, often calcareous soil; cen IA to se NE, s to AR, e KS & OK; (NH to MN, NE, s to VA, GA, AL, & OK). *Anychia canadensis* (L.) B.S.P.—Rydberg.

2. Paronychia depressa Nutt. ex T. & G. Tufted, mat-forming perennial 3–10 cm tall, tinged with brown, taprooted, with a much-branched, rather woody caudex; stems mostly prostrate and crowded with ascending or erect tips, scaberulous-puberulent, internodes short, 3–10(14) mm long. Leaves crowded, erect or some recurved, linear-subulate, 5–15(20) mm long, less than 1 mm wide, scaberulous, 3-nerved, cuspidate to bristle-tipped; stipules silvery-white, lanceolate-attenuate, 3–8 mm long. Flowers clustered at branch tips; bracts shorter than to slightly exceeding the flowers, much shorter than the leaves. Calyx 2.5–3.5(4) mm long (including awns), strigose below mainly on the campanulate tube, sepals oblong, 3-nerved and corrugated, usually scaberulous on the nerves, cucullate, with an awn 0.5–1 mm long; style divided to near the middle or above. Utricle ovoid, glabrous; seed yellow, ovoid, ca 1 mm long. (2n = 32) Jun–Aug. Sandy or rocky hills in prairie; sw SD & e WY, s to w NE, extreme w KS & e CO; (SD & WY s to w KS & CO). *P. diffusa* A. Nels.—Rydberg.

3. ***Paronychia fastigiata*** (Raf.) Fern., forked chickweed. Green or more commonly reddish-brown annual 5–30 cm tall, simple below, dichotomously branched above, the branches erect or spreading; stem puberulent, with spreading-decurved hairs. Leaves green or brown-tinged, often white-punctate, elliptic-lanceolate to oblanceolate, 5–12(20) mm long, 1–4 mm wide, mostly acute, usually ciliate-serrulate; stipules narrowly triangular, mostly 1–3.5 mm long. Flowers clustered in repeatedly forked cymes, subtended by hyaline-stipulate, foliar bracts. Calyx 1–1.5 mm long, the tube very short, sepals linear-lanceolate, mostly (1)3-nerved and usually corrugated, glabrous or short-pubescent on the nerves, apiculate behind the hooded tip; style divided to about the middle, the branches recurved. Utricle globose, included or slightly exserted, minutely granular at the tip; seed reddish-black and shiny, round, 0.8–1 mm in diam. Jul–Oct. Rocky or sandy soil in open woods, pastures, & prairies; MO, e KS, & e OK; (MA to WI, s to FL & TX). *Anychia polygonoides* Raf.—Rydberg; *P. fastigiata* var. *paleacea* Fern.—Fernald.

4. ***Paronychia jamesii*** T. & G., James' nailwort. Brownish-tinged, tufted perennial (7)10–25(30) cm tall, taprooted, with a branched, woody caudex; stems erect to ascending in tufts, not forming mats, forked above, scaberulous to scaberulous-puberulent at least above, internodes 6–30 mm long. Leaves erect or some recurved, linear, 10–25 mm long, to 1 mm wide, scaberulous, 3-nerved, blunt to mucronate or cuspidate; stipules silvery-white, linear-lanceolate, 6–15 mm long. Flowers clustered in terminal cymes, the uppermost flowers commonly in 3s; bracts much shorter than the leaves, the upper ones shorter than to slightly surpassing the flowers. Calyx 2–3.2 mm long (including awns), strigose mainly on the campanulate tube, the sepals oblong, 3-nerved and corrugated, strigillose to scaberulous, cucullate, with an awn 0.3–0.8 mm long; style branched well above the middle or near the tip. Utricle obovoid, glabrous; seed yellowish-brown, ovoid, ca 1 mm long. Jun–Sep. Rocky or sandy hills in prairie & pastures, often where calcareous; s-cen & w NE & se WY, s through cen & w KS, OK, n TX, e CO, & NM; (NE & WY, s to TX, AZ, & into Mex.). *P. wardii* Rydb.—Rydberg.

5. ***Paronychia sessiliflora*** Nutt. Densely tufted, dark yellow-green perennial, taprooted with a greatly branched caudex, forming cushiony mats 2–6(10) cm high and to 2 dm across; stems numerous, short-pubescent, internodes to 5 mm long, older branches marcescent-leafy. Leaves very crowded, the lower ones appressed and imbricate, the upper ones (and the bracts) recurved-spreading, linear-subulate, 3–6 mm long, less than 1 mm wide, 3-nerved, acute to mucronate, scaberulous on the margins; stipules white, subulate or oblong-linear, slightly shorter than the leaves, usually cleft. Flowers single or sometimes paired at stem tips; bracts much like the leaves, little if any shorter. Calyx turning yellowish-brown, 3–4 mm long (including awns), pubescent at the base, mainly on the campanulate tube, sepals oblong, 3-nerved and corrugated, glabrous or scaberulous on the nerves, cucullate, with an awn ca 1 mm long; style divided in the upper 1/4, the branches erect or ascending. Utricle globose, pubescent on the upper 1/2; seed yellow to brown, roughly oblong-ovoid, to ca 1 mm long. (2n = 64) Jun–Sep. Barren, rocky outcrops, prairie hills, & slopes; w & cen ND & e MT, s to w NE, e CO, w OK, n TX, & NM; (Sask. & Alta., s to TX, NM, & UT).

10. SAGINA L., Pearlwort

1. ***Sagina decumbens*** (Ell.) T. & G., trailing pearlwort. Low annual 2–10 cm tall, branched from the base and often tufted, glabrous or sometimes with sparse glandular hairs above; stems erect to decumbent and spreading, green or often purple-tinged. Leaves linear, 3–10

mm long, to 1 mm wide, reduced upward, apiculate, hyaline-margined and connate at the base to form a short tube around the stem. Flowers usually numerous and both terminal and axillary, seldom solitary on the stems, perfect, hypogynous, 5- or 4-merous, sometimes both on the same plant; pedicels erect to ascending, filiform, 3–15 mm long. Sepals 5 or 4, green and often with purple at the tip, oval to ovate, 1.4–2.2 mm long, obscurely nerved, obtuse, narrowly hyaline-margined; petals numbering the same as the sepals, white, ovate to obovate or elliptic, to about equaling the sepals or slightly longer, entire; stamens twice as many as the sepals, included; styles numbering the same as the sepals. Fruit a many-seeded capsule with free-central placentation, 5- or 4-valved from the tip, ovoid, ca 1.5 × as long as the sepals, membranaceous; seeds light tan, obovoid-triangular, 0.25–0.3 mm long, minutely tuberculate. ($2n = 36$) Apr–May. Moist or dry sandy soil of open woods, roadsides, & disturbed places; sw MO, se KS, e & cen OK; (MA & VT to IL & KS, s to n FL & e TX, with disjuncts in N.B., Alta. & AZ).

References: Crow, G. E. 1978. A taxonomic revision of *Sagina* (Caryophyllaceae) in North America. Rhodora 80: 1–91; Crow, G. E. 1979. The systematic significance of seed morphology in *Sagina* (Caryophyllaceae) under scanning electron microscopy. Brittonia 31: 52–63.

Our plants are typical *S. decumbens*. The subsp. *occidentalis* (S. Wats.) Crow, with light brown, more or less smooth seeds, occurs along the West Coast from s B.C. to s CA. The apetalous form of *S. decumbens*, called var. *smithii* (A. Gray) S. Wats. by some authors, has not been found in our range.

11. SAPONARIA L., Soapwort, Bouncing Bet

1. *Saponaria officinalis* L. Glabrous or subglabrous perennial 3–8 dm tall, colonial from rhizomes; stem simple, erect, leafy. Leaves usually 10–20 pairs, the lower ones withering and usually deciduous by flowering time, elliptic to elliptic-ovate or elliptic-lanceolate, 4–10 cm long, 1–4 cm wide, prominently 3-nerved, acute, narrowed to a short-petiolate or sessile base, connected at the base by a transverse ridge. Flowers fragrant and showy, often double, perfect, hypogynous, usually many in a condensed, subcapitate, terminal cyme, some subterminal cymes (or even single flowers) often produced from the axils of the upper leaves, the bracts leafy but much reduced. Calyx green or sometimes purple-tinged, tubular, cylindric (not inflated), 15–20 mm long in flower, to 25 mm long in fruit, membranaceous, 20-nerved, with 5 triangular-attenuate teeth, deeply cleft in 1 or more of the sinuses; petals 5 (or more when double-flowered), white or pink, long clawed, with a pair of subulate appendages at the juncture between claw and blade, lacking auricles, the blades spreading, 8–15 mm long, retuse to emarginate; stamens 10, exserted, basally coherent with the petals around a short carpophore; styles 2 (rarely 3). Fruit a membranaceous, many-seeded, 1-celled capsule, included in the calyx, dehiscent by 4(6) teeth; seeds dark brown, rotund-reniform, 1.5–1.8 mm in diam, finely reticulate-papillose in concentric rows. ($2n = 28$) Jun–Sep. Introduced and cult. as a garden flower, frequently escaped to roadsides, shelter belts, & waste places, rather weedy; GP, but uncommon in the drier w parts; (established over much of temp. N. Amer.; Europe). *Escape*.

12. SCLERANTHUS L., Knawel

1. *Scleranthus annuus* L., annual knawel. Low, glabrous or usually crisp-puberulent annual (biennial), branched from the base and prostrate to ascending, or occasionally simple from the base and erect, the stems 3–15 cm long. Leaves exstipulate, spreading, linear, the principal ones 5–25 mm long, less than 1 mm wide, subulate-pointed, broadened and connate at the base with membranaceous, often puberulent-ciliate margins. Flowers usually numerous, sessile or subsessile in compact, terminal and axillary cymes, small, perfect, perigynous, 2.5–4 mm long, the hypanthium comprising ca 1/3–1/2 the flower length. Sepals 5, persistent atop the cupulate hypanthium, narrowly lanceolate, 1-nerved; petals none; stamens 5–10 (rarely fewer), the filaments broadened at the base and attached to

a membranaceous disk lining the rim of the hypanthium; styles 2, ovary superior; hypanthium 10-ribbed, becoming indurate in fruit and enclosing the membranous, 1-seeded utricle; seed pale yellow, obovoid to globose and beaked at the micropylar end, 0.7–1 mm long. (2n–22, 44) May–Oct. An introduced weed of gardens, lawns, fields, roadsides, & waste places; KS: Cherokee, Crawford, Wilson; OK: Grady; (established from Que. to WI, s to SC, MO, & OK; also B.C. to ID & CA; Europe). *Naturalized.*

13. SILENE L., Catchfly, Campion

Annual, biennial and perennial herbs, many introduced, often weedy, glabrous or pubescent, frequently glandular. Leaves opposite (middle cauline leaves in whorls of 4 in *S. stellata*), sessile or especially the lower ones often short-petiolate, exstipulate, basally connected by a transverse ridge. Flowers (1) few to many in leafy-bracteate, open to contracted, dichasial cymes, sometimes appearing in 1-sided racemes due to suppression of one branch at each node of the dichasium, often showy, perfect or imperfect (in the dioecious *S. pratensis* and *S. menziesii* and in occasional flowers of *S. noctiflora* and *S. dichotoma*), hypogynous. Calyx tubular, campanulate to cylindric, sometimes much inflated, 5-toothed, 10- or 20-nerved (ours), often accrescent in fruit; petals 5 (sometimes absent in *S. antirrhina*), white to pink or purple, rarely red, exserted from or sometimes included in the calyx, narrowly clawed, often with a pair of auricles at the juncture between blade and claw, usually with a pair of appendages on the ventral surface at the juncture, the blade 2-lobed, or less often, fimbriate-lobed or toothed, seldom entire; stamens 10, united with the petals near the base to form a short tube around the carpophore; styles 3 or 5 (rarely 4 or 6), ovary borne on a short to prominent carpophore. Fruit a firm-textured capsule, 1-locular or essentially 3-locular by the formation of partitions, usually about equaling the calyx and enclosed by it, opening by twice as many teeth as styles, or less often, by as many teeth as styles; seeds reniform to globose, usually papillate in rows. *Melandrium* Roehl, *Wahlbergella* Fries — Rydberg; *Lychnis* L., in large part — Fernald.

References: Hitchcock, C. L., & B. Maguire. 1947. A revision of the North American species of *Silene*. Univ. Wash. Publ. 13: 1–73; Maguire, B. 1950. Studies in the Caryophyllaceae — IV. A synopsis of the subfamily Silenoideae. Rhodora 52: 233–245; McNeill, J., & H. C. Prentice. 1981. *Silene pratensis* (Raf.) Godron & Gren., the correct name for white campion or white cockle (*Silene alba* (Miller) E. H. L. Krause, *nom. illeg.*). Taxon 30: 27–32; Williams, F. N. 1896. A revision of the genus *Silene*. J. Linn. Soc. 32: 1–196.

```
1  Plants strictly male, the flowers all staminate, with no evidence of styles, ovary or
   fruit ........................................................................................................ 8. S. pratensis
1  Plants monoecious or female, or if male, the styles and ovary reduced but evident.
   2  Styles mostly 5 (rarely 4 or 6 in some flowers), the capsule opening by 5 or 10 teeth.
      3  Flowers strictly female, with no evidence of stamens; calyx much inflated, at least in
         age; petals prominently exserted, the blade 7–10 mm long ............... 8. S. pratensis
      3  Flowers perfect; calyx not inflated; petals usually included in the calyx, the blade less
         than 5 mm long ................................................................... 4. S. drummondii
   2  Styles mostly 3 (rarely 4 in some flowers), the capsule opening by 6 teeth.
      4  Petals bright red, entire to irregularly dentate ..................................... 9. S. regia
      4  Petals white to pink or purplish, never red, 2-lobed or fimbriate-lobed, sometimes in-
         cluded in the calyx or absent.
         5  Leaves mostly in whorls of 4 ................................................. 10. S. stellata
         5  Leaves all opposite.
            6  Calyx 20-nerved at least below the middle.
               7  Calyx campanulate, not constricted at the mouth, becoming much inflated,
                  12–15(20) mm long in fruit, papery-textured, with 20 equal nerves and con-
                  spicuously reticulate-veiny between the nerves, wrinkled when dried; in-
                  florescence open, spreading ......................................... 11. S. vulgaris
               7  Calyx cylindric-ellipsoid, constricted at the mouth, not becoming much in-
                  flated, 9–12 mm long in fruit, firm-textured, with 10 long nerves and 10
```

        shorter nerves, with few, if any, weak veins between the nerves, remaining smooth when
        dried; inflorescence narrow, elongate .... 2. *S. cserei*
6 Calyx 10-nerved for its entire length, the nerves sometimes faint.
    8 Plants perennial from rhizomes.
        9 Calyx 5-8 mm long; petals 6-10 mm long .... 5. *S. menziesii*
        9 Calyx 14-18 mm long; petals 13-18 mm long .... 6. *S. nivea*
    8 Plants annual, with a taproot.
        10 Flowers appearing in 1-sided racemes (actually cymes with 1 branch suppressed at
            each node) .... 3. *S. dichotoma*
        10 Flowers in open to contracted, nonracemose cymes.
            11 Calyx 4-10 mm long, glabrous .... 1. *S. antirrhina*
            11 Calyx 15-30 mm long, glandular-pubescent .... 7. *S. noctiflora*

1. **Silene antirrhina** L., sleepy catchfly. Slender, erect, taprooted annual (0.5)1.5-8 dm tall, simple or few-branched from the base, usually forked above, mostly glabrous, the stems usually retrorse-puberulent near the base, the internodes, or at least the upper ones, often with a broad, dark, glutinous band. Basal and lower cauline leaves spatulate to oblanceolate, mostly 3-6 cm long, 2-15(20) mm wide; cauline leaves of 5-10 pairs, linear to oblanceolate, 2-5 cm long, 1-15 mm wide, acute, ciliate near the base. Flowers (1) few to many in a compact to open, forked cyme, the pedicels stiffly erect to ascending, 4-25 mm long. Calyx tubular, often constricted at the mouth, 4-10 mm long, 10-nerved, glabrous, the teeth short-triangular, purple-tipped; petals 5 (occasionally none), white to pink or purplish, shorter than to exceeding the calyx, 2-lobed, the appendages minute or absent; carpophore ca 1 mm long; stamens included; styles 3, included. Capsule 3-locular except near the tip, 6-toothed, about equaling the calyx; seeds brownish or grayish-black, rotund-reniform, 0.5-0.8 mm long, papillate. (n = 12) May-Aug. Fields, roadsides, waste places, & sandy areas; GP, except for the drier w part; (throughout the U.S. & s Can.). *S. antirrhina* f. *deaneana* Fern., f. *bicolor* Farw. and f. *apetala* Farw.—Fernald.

2. **Silene cserei** Baumg., smooth catchfly. Erect or ascending, taprooted biennial 3-8 dm tall, with 1-few stems from the base, often branched above, strongly glaucous, glabrous. Leaves thick-textured, the basal ones oblanceolate to spatulate, mostly withered by flowering time, the cauline leaves of 6-9 pairs, broadly elliptic to ovate, 2-8.5 cm long, 5-30(40) mm wide, mostly acute, more or less connate-clasping at the base. Inflorescence a narrow, usually elongate, paniculate cyme of several to many flowers, forked at the base, the flowers loosely clustered along the elongate branches; bracts lanceolate, much reduced; pedicels 5-20(40) mm long. Calyx glaucous or often purple-tinged, cylindric-ellipsoid, constricted at the mouth, 8-11 mm long in flower, 9-12 mm long and not much inflated in fruit, 20-nerved, with 10 long nerves and 10 shorter nerves, the latter reaching about the middle of the calyx or more before fading, with a few, inconspicuous anastomosing veins between the nerves, firm-textured and remaining smooth when dried, glabrous or the low, obtuse teeth often fringed with a short tomentum; petals white, long-clawed, broadened or slightly auriculate at the juncture between claw and blade, appendages none, the blade exserted 4-7 mm beyond the calyx, deeply cleft; carpophore 2-3 mm long; stamens strongly exserted, the filaments pale to purple; styles 3, slightly surpassing the filaments. Capsule 3-locular, 6-toothed, slightly exserted, closely enveloped by the calyx; seeds brown, rotund-reniform, 1-1.2 mm long, papillate. (2n = 24) Jun-Aug (Sep). An introduced weed of roadsides, waste places, often spreading to prairie hillsides, especially common on railroad embankments; w MN, ND, & MT, s to n MO, & NE, likely on the increase in the GP; (established in parts of n U.S. & s Can.; Europe). *Naturalized. S. fabaria* (L.) Sibth. & Sm.—Rydberg.

   This species is often mistaken for *Silene vulgaris*. Although similar, the two are fairly easy to distinguish on the basis of the criteria given in the key. The broader leaves and biennial nature of *S. cserei* are also helpful in differentiating it from the perennial *S. vulgaris*. *S. cserei* is by far the more common of the two in the n part of our range.

3. **Silene dichotoma** Ehrh., forked catchfly. Hirsute taprooted annual 2.5-10 dm tall; stems 1-few, erect, frequently branched above the base, retrorsely hirsute or sometimes sparing-

ly so, with multicellular hairs. Leaves hirsute on both surfaces, 3-nerved, the basal ones ciliate-petiolate, oblanceolate to lanceolate, 3–8(12) cm long, 5–25(45) mm wide; cauline leaves of 5–8 pairs, little reduced, lanceolate to elliptic-lanceolate or oblanceolate, mostly acute, short-petiolate to subsessile and ciliate at the base. Inflorescence of several to many flowers, once or twice forked, the branches racemose and 1-sided (actually cymose with a branch suppressed at each node); bracts 2 per flower, lanceolate to linear-lanceolate, 5–25 mm long; pedicels 1–3 mm long. Flowers mostly perfect (some occasionally female); calyx narrowly tubular and 10–12 mm long in flower, to ellipsoid and 12–15 mm long in fruit, not inflated, constricted at the mouth, strongly 10-nerved, hirsute, with multicellular hairs on the nerves, membranaceous between the nerves, the teeth triangular, 2–3 mm long; petals white or seldom pink, long clawed, without auricles, appendages broadly truncate, 0.2–0.4 mm long, the blade 5–9 mm long, deeply cleft; carpophore 2–4 mm long; stamens exserted (or vestigial); styles usually 3, strongly exserted. Capsule 3-locular nearly to the top, 6-toothed, about equaling the calyx; seeds dark grayish-brown, rotund-reniform, 1–1.3 mm long, papillate. (2n = 24) Jun–Aug. A weed sporadically introduced into fields (especially with forage crops) & waste places, apparently not long persisting in our range; w MN, ND, & NE; Cass; (established as a weed in much of e U.S. & se Can., sporadic elsewhere; Eurasia). *Introduced.*

4. Silene drummondii Hook. Erect perennial 2–6 dm tall from an erect rootstock, simple or with a few stems from the base, retrorsely puberulent below, becoming densely glandular-pubescent and viscid above. Basal leaves petiolate, lanceolate or elliptic to oblanceolate, (1.5)3–10 cm long, 4–12 mm wide; cauline leaves of 2–5 remote pairs, linear-lanceolate to linear, 3–9 cm long, 2–7 mm wide. Inflorescence a narrow, contracted cyme of few to several flowers (rarely 1-flowered); bracts much reduced; pedicels erect-appressed, 5–40 mm long. Calyx oblong-cylindric, slightly expanded but not inflated in fruit, 10–13 mm long in flower, 12–15 mm long in fruit, 10-nerved, glandular-puberulent, membranous between the nerves, the teeth triangular, 1–2 mm long; petals white or pinkish, included in the calyx or seldom slightly exserted, the claw broadened above and wider than the blade, often auriculate, appendages minute or none, the blade retuse or shallowly lobed; carpophore ca 1 mm long; stamens included; styles (4)5. Capsule 1-locular, (4)5-toothed, equaling the calyx; seeds dark brown, subreniform, 0.7–0.9 mm long, prominently papillate. (n = 24, 2n = 48) Jun–Aug. Dry prairie & pine forest; w MN: Clay, ND, n & w SD, w NE & WY; (Sask. & w MN to B.C., s to NE & AZ). *Wahlbergella drummondii* (Hook.) Rydb.— Rydberg; *Lychnis drummondii* (Hook.) Wats.—Fernald; *Melandrium drummondii* (Hook.) Hulten—Weber.

5. Silene menziesii Hook. Functionally dioecious perennial from slender rhizomes; stems decumbent to ascending, often somewhat matted, simple to freely branched, 5–30(70) cm long, retrorsely pubescent below, glandular-puberulent above, especially in the inflorescence. Leaves mainly cauline, of usually 3–7 pairs below the inflorescence, ovate or elliptic to obovate or oblanceolate, mostly 2–6 cm long, 5–25 mm wide, glabrous or puberulent, acute to acuminate, cuneate at the sessile to subsessile base. Inflorescence of (1)few to several flowers in a leafy, open cyme; pedicels 5–30 mm long, glandular-puberulent. Flowers with either the stamens or the pistil reduced and nonfunctional; calyx green, tubular-campanulate, 5–8 mm long, rather obscurely 10-nerved, glandular-puberulent, the teeth triangular, 1–3 mm long; petals white, 6–10 mm long, long clawed, the blade 1.5–3 mm long, lobed to about the middle or more, with or without a small tooth on each margin just below the level of the sinus, auricles none, appendages 0.1–0.4 mm long or fused to the blade; carpophore 1–1.5 mm long; stamens exserted in functionally male flowers, vestigial or absent in female flowers; styles 3(4), exceeding the petals in female flowers, much shorter and included in male flowers. Capsule 1-locular, 6(8)-toothed; seeds ca 1 mm long, smooth or lightly reticulate. (n = 12, 24; 2n = 24, 48) Jun–Aug. Occurring in a variety of habitats throughout its range, but apparently confined to woods in our region; NE: Cherry, Sioux; also possibly BH; (MT & nw NE to B.C., s to NM & CA).

Our plants are typical *S. menziesii*, with the lower portion of the stems covered with nonglandular, septate hairs. The var. *viscosa* (Greene) Hitchc. & Maguire, with glandular hairs on the lower portion of the stem, occurs mainly w of our range.

6. ***Silene nivea*** (Nutt.) Otth, snowy campion. Rhizomatous perennial 2–7 dm tall; stems leafy, erect to ascending, usually decumbent at the base, simple or sparingly branched, glabrous or seldom puberulent, especially above. Leaves mainly or wholly cauline, lanceolate to oblong-lanceolate, 4–10 cm long, 8–35 mm wide, long-acuminate, sessile or short-petiolate, glabrous or puberulent. Flowers 1–few, when more than 1, mostly appearing axillary in an elongate and open, leafy cyme; pedicels 10–50 mm long. Calyx cylindric to tubular-campanulate, 14–18 mm long, glabrous or hirsutulous, faintly 10-nerved, with some anastomosing veins, the teeth triangular-ovate, 2–3 mm long; petals white, 13–18 mm long, the claw broadened and rounded above, not auricled, appendages 1–1.6 mm long, erose or entire, the blade 6–7 mm long, 2-lobed to ca 1/4 of its length, with or without a small tooth on each margin; carpophore 6–8 mm long; stamens exserted; styles 3, exserted. Capsule 1-locular, 6-toothed; seeds black, ca 1 mm long, rugose-papillate. (2n = 48) Jun–Aug. Moist woods & flood plains; NE: Dodge; (PA & MD to MN & NE, s to TN & n MO). *S. alba* Muhl.—Winter.

Our single record for *Silene nivea* is based on an 1895 collection from near Fremont, NE. The plant may now be extinct in our range. Other reports of the species in GP were based upon misidentified specimens of *Silene vulgaris*.

7. ***Silene noctiflora*** L., night-flowering catchfly, sticky cockle. Coarse, taprooted annual (1)2.5–6(8) dm tall, simple or few-branched from the base, often branched above, erect to decumbent or spreading; stems hirsute below, with multicellular hairs ca 2 mm long, these long hairs becoming intermixed with short, glandular-viscid hairs above, especially in the inflorescence. Leaves hirsute to short-pubescent on both surfaces, the basal and often the lower cauline leaves spatulate to oblanceolate or obovate, usually withering by flowering time; cauline leaves of 3–9 pairs, ovate to elliptic-oblanceolate or lanceolate, 3–12 cm long, 5–50 mm wide, acute. Inflorescence of (1)few to several flowers in a rather open, loosely branched cyme; bracts lanceolate, reduced; pedicels erect to ascending, 3–40 mm long. Flowers perfect or mostly so; calyx tubular and 15–25 mm long in flower, becoming much inflated (ovoid-campanulate) and 20–30 mm long in fruit, with 10 strong, green nerves, these interconnected by a few anastomosing veins, the calyx otherwise membranaceous between the nerves, hirsute and glandular-pubescent on the nerves, the teeth linear-lanceolate to subulate, 5–12 mm long; petals white to pinkish, the claws about equaling the calyx, with auricles 1–1.5 mm long, the appendages broad, 0.5–1.5 mm long, entire or erose, the blade inrolled during the day, opening at night, 7–10 mm long, 2-lobed to about the middle; carpophore 1–3 mm long; stamens about equalling the claws of the petals, styles 3. Capsule 3-locular, 6-toothed; seeds grayish-brown, rotund-reniform, 1–1.3 mm long, tuberculate and reticulate. (2n = 24) Jun–Sep. An introduced weed of fields, roadsides, disturbed areas, & waste places; scattered over the GP, but especially common from MN & ND, s to NE, sparse in w MO, n & cen KS; (established throughout most of the U.S. & s Can.; Europe). *Naturalized*.

See the discussion under no. 8 regarding the distinctions between *S. noctiflora* and *S. pratensis*.

8. ***Silene pratensis*** (Rafn) Godr. & Gren., white campion, white cockle. Dioecious, short-lived perennial or biennial 3–8(12) dm tall, simple or branched from the base, erect to decumbent, with a stout, often laterally branched taproot; stems finely hirsute below, with multicellular hairs ca 1 mm long or less, becoming glandular-puberulent and hirsute above, especially in the inflorescence. Leaves hirsutulous on both surfaces, the basal leaves petiolate, oblong-lanceolate to elliptic, often withering by flowering time, the cauline leaves of 5–10 pairs, lanceolate to broadly elliptic, 3–12(15) cm long, 6–30(40) mm wide, acute, sessile to short-petiolate. Inflorescence of (1)several to many flowers in a usually freely branched cyme; bracts ovate to ovate-lanceolate, much reduced, the flowers subsessile to short pedi-

celled in male plants or mostly on pedicels 10–50 mm long in female plants. Flowers odorous, opening in the evening; calyx in male flowers tubular, 15–20 mm long, 10-nerved, in female flowers becoming much inflated in fruit, 20–30 mm long, 20-nerved, with a few anastomosing veins, the calyces of both male and female flowers hirsute and short glandular-pubescent, especially on the nerves, the teeth triangular to linear, 3–6 mm long; petals white, the claw exserted beyond the calyx, auriculate, appendages 1–1.5 mm long, erose, the blade 7–10 mm long, 2-lobed to about the middle; carpophore 1–2 mm long in female flowers; stamens about equaling the claws of the petals; styles 5(4 or 6), exserted. Capsule 1-locular, with 5 main teeth that are themselves bifid, the capsule thus 10-toothed; seeds grayish-black, 0.8–1.2 mm long, bluntly tuberculate. (n = 12, 2n = 24) Jun–Sep. An introduced weed of fields, roadsides, disturbed areas, & waste places, also sometimes encountered in prairie & woodland; MN to MT, s to w MO & KS, most common in the n part; (established in much of n & cen U.S. & s Can.; Europe). *Naturalized. Melandrium album* (P. Mill.) Garcke — Rydberg; *Lychnis alba* P. Mill. — Fernald, *Silene alba* (P. Mill.) E.H.L. Krause — Kruckeberg *in* Kartesz & Kartesz, op. cit.

Silene pratensis and *S. noctiflora* are superficially similar and commonly confused in herbarium collections. In flowering condition, the dioecious nature of *S. pratensis* plainly distinguishes it from the mostly perfect-flowered *S. noctiflora*. In fruit, the female plants of *S. pratensis* have a 20-nerved calyx and a 10-toothed capsule, in contrast to the 10-nerved calyx and 6-toothed capsule of *S. noctiflora*.

9. *Silene regia* Sims, royal catchfly. Erect, taprooted perennial 5–16 dm tall; stem usually single, simple, glabrous or puberulent below to glandular-puberulent above, especially in the inflorescence. Leaves mainly cauline, 10–20 pairs, broadly ovate to lanceolate, 3.5–12 cm long, 1.5–7 cm wide, glabrous to densely puberulent on both surfaces, bluntly acute to acuminate, sessile and rounded to cordate-clasping at the base. Inflorescence a narrow, compound cyme of usually numerous showy flowers; bracts ovate, much reduced; pedicels 5–20 mm long, densely glandular-pubescent. Calyx tubular, somewhat expanded in fruit, 17–27 mm long, prominently 10-nerved, membranous between the nerves, glandular-pubescent, the teeth triangular, 2–4 mm long; petals bright red, the claws about equaling the calyx, auricles rounded, 1–2 mm long, entire or erose, sometimes lacking, appendages tubular, 2–4 mm long, the blade (9)12–20 mm long, entire to irregularly toothed; carpophore 3–5 mm long; stamens exserted; styles 3(2–5). Capsule 1-locular, 6-toothed, slightly exserted; seeds brown and shiny, subreniform, 1.8–2.2 mm long, low-tuberculate. (n = 24) Jul–Aug. Prairie, open woods, & thickets; KS: Cherokee; MO: Jasper; (OH to MO & se KS, s to GA, AL, & OK).

10. *Silene stellata* (L.) Ait. f., starry campion. Erect perennial 3–12 dm tall; stems several, from a thickened taproot, simple below the inflorescence, often purple below, sparsely to densely crisp-puberulent. Leaves all cauline, the lower ones paired, the middle and upper in whorls of 4, lanceolate to ovate-lanceolate, 4–11 cm long, 10–45 mm wide, usually puberulent to hirsutulous on both surfaces, white-punctate, acuminate, sessile to subsessile. Inflorescence of several to many flowers, loosely paniculate, oblong to pyramidal; bracts lanceolate to linear, much reduced; pedicels mostly 3–8(15) mm long, puberulent. Calyx campanulate to funnelform, 5–13 mm long, 10-nerved, with freely anastomosing veins between the nerves, finely puberulent to seldom glabrous, the teeth broadly triangular, 2–5 mm long; petals white, 8–18 mm long, the claw gradually broadened above and not clearly differentiated from the blade, auricles and appendages none, the blade fimbriately 8- to 12-lobed; carpophore 1–3 mm long; stamens exserted; styles 3, exserted. Capsule 1-locular, 6-toothed, subglobose and included in the calyx at maturity; seeds dark purplish-brown, ca 1 mm long, papillate. (n = 24, 2n = 48) Jun–Aug. Woods, thickets, & moist, open areas; sw MN, e SD, IA, n-cen & e NE, MO, e KS, e & cen OK; (CT & NY to s MN & SD, s to GA, LA, & TX). *S. stellata* var. *scabrella* (Niewl.) Palm. & Steyerm. — Gleason.

11. *Silene vulgaris* (Moench) Garcke, bladder campion. Perennial 2–10 dm tall, glabrous and glaucous, seldom sparingly hirsutulous; stems usually several to many, erect to decumbent, simple or sparingly branched below the inflorescence, arising from a thick, often woody taproot. Leaves mainly or wholly cauline, the uppermost often remote from the inflorescence, linear-oblong or oblanceolate, seldom ovate, 2–8 cm long, 5–30 mm wide, rather thin-textured, acute to acuminate, often somewhat clasping. Inflorescence an open, spreading, dichasial cyme of several to many flowers; bracts very much reduced, lanceolate to ovate; pedicels 5–30 mm long. Calyx pale green or purplish, campanulate, not constricted at the mouth, 9–12 mm long in flower, 12–15(20) mm long and much inflated in fruit, loosely enclosing the capsule, papery-textured, with 20 more or less equal nerves and conspicuously cross-veined between the nerves, wrinkled when dried, glabrous or with a tomentum on the margins of the broadly triangular teeth which are 2–3 mm long; petals white, the claws about equaling the calyx, broadened and rounded above, lacking true auricles, appendages minute or none, the blade 3.5–6 mm long, deeply 2-lobed, carpophore 2–3 mm long; stamens exserted; styles 3(4), exserted. Capsule 3-locular, 6-toothed, ovoid-globose, included in the calyx; seeds dark grayish-brown, globose-reniform, 1–1.5 mm long, low-tuberculate. (n = 12; 2n = 24, 48) Jun–Aug. An occasional introduced weed of roadsides & waste places; widely scattered locations, mainly in the e & cen GP as far s as KS, also BH; (widely established as a weed throughout most of the temperate U.S. & s Can.; Europe). *Naturalized. S. cucubalus* Wibel—Fernald; *S. latifolia* (P. Mill.) Britt. & Rendle—Winter.

14. SPERGULA L., Spurry

1. *Spergula arvensis* L., corn spurry. Erect to spreading, rather succulent, taprooted annual 1–4 dm tall, simple or branched from the base, sparingly puberulent to glandular-puberulent. Leaves appearing whorled, with 2 opposite sets of 6–8 clustered at each node, narrowly linear or subulate, mostly 1–5 cm long, 1 mm or less wide; stipules small, 1 mm or less long, scarious. Flowers inconspicuous, perfect, hypogynous or slightly perigynous; usually numerous, in diffuse, divaricately branched terminal cymes; bracts ovate, 1 mm or less long, scarious; pedicels erect to spreading or mostly reflexed, 4–25(40) mm long, glandular-puberulent. Sepals 5, ovate, 2.5–4.5 mm long, glandular-puberulent, scarious-margined; petals 5, white, obovate, shorter to longer than the sepals; stamens 10 or 5; styles usually 5, minute. Capsule broadly ovoid, opening by 5 valves from the tip, 3.5–5 mm long; seeds subrotund and thickly biconvex, black with a white to tawny winged margin, 1–1.5 mm in diam, smooth or covered with short, clavate papillae. (2n = 18, 36) May–Sep. A sparingly introduced weed of fields & waste places, apparently not long persisting in our range; KS: Riley; MO: Jackson, Jasper; ND: Cass; SD: Brookings; (N.S. to AK, s to FL & CA; Europe). *Introduced. S. arvensis* var. *sativa* (Boenn.) Mert. & Koch—Fernald.

15. SPERGULARIA (Pers.) J. & K. Presl, Sand Spurry

1. *Spergularia marina* (L.) Griseb., salt-marsh sand spurry. Low taprooted annual 5–20 cm tall, simple and erect to usually branched from the base and decumbent or spreading, usually glandular-puberulent throughout. Leaves opposite, linear, 8–25(40) mm long, to 1.5 mm wide, fleshy, blunt or mucronulate; stipules broadly deltate, 1.5–4 mm long, with a short, acuminate tip and entire or lacerate margins, often connate. Flowers usually numerous, small, perfect, hypogynous, in divaricate, leafy-bracteate cymes, the inflorescence comprising most or nearly all of the plant; pedicels 1–10 mm long in age, often reflexed. Sepals 5, ovate, 2.5–4(5) mm long, obtuse, scarious-margined; petals 5, pink (white), ovate to oblong, 1/2 to nearly as long as the sepals; stamens 2–5; styles 3. Capsule ovoid, 3–6 mm long, 3-valved from the tip, membranaceous; seeds brown, obliquely obovate, 0.6–0.8 mm long, smooth or glandular-papillate, wingless or occasionally with an erose wing to 0.4 mm wide. (n = 18, 2n = 18, 36) Jul–Sep. Apparently introduced, along brackish or saline shores & wet alkali flats; scattered locations in ND, probably to be expected farther s; (along

coasts from Que. to FL & w to TX, & B.C., s to CA, sporadically inland; Eurasia). *Naturalized. S. salina* J. & K. Presl—Rydberg; *S. marina* var. *leiosperma* (Kindb.) Gurke—Fernald.

Earlier reports of *Spergularia rubra* (L.) J. & K. Presl for ND were based upon specimens of *S. marina*.

Reference: Rossbach, R. P. 1940. *Spergularia* in North America and South America. Rhodora 42: 57-83, 105-143, 158-193, 203-213.

16. STELLARIA L., Chickweed, Stitchwort

Low, slender, perennial and annual herbs, often with 4-angled stems. Leaves rather small, opposite, exstipulate. Flowers small, in terminal or axillary, bracteate cymes, or solitary and terminal, axillary or in forks of the stem, borne on slender pedicels, perfect, hypogynous. Sepals 5, separate; petals 5, sometimes much reduced or absent, white, deeply to shallowly 2-lobed, often so deeply cleft that they appear superficially as 10 petals; stamens usually 10, often 5-10 or sometimes fewer; styles mostly 3(4) but 5 in *S. aquatica*. Fruit a 1-celled, several- to many-seeded capsule, dehiscent to about the middle or beyond by 6 valves, but in *S. aquatica* with 5 notched or shortly bifid valves; seeds brown, nearly smooth to rugose-tuberculate. *Alsine* (Tourn.) L.—Winter; *Myosoton* Moench—Fernald.

References: Chinnappa, C. C., & J. K. Morton. 1976. Studies on the *Stellaria longipes* Goldie complex—variation in wild populations. Rhodora 78: 488-502; Fernald, M. L. 1940. *Stellaria calycantha*. Rhodora 42: 254-259; Hulten, E. 1943. *Stellaria longipes* Goldie and its allies. Bot. Notiser 1943: 251-270; Porsild, A. E. 1963. *Stellaria longipes* and its allies in North America. Nat. Mus. Canada Bull. 186: 1-35.

1 Styles 5; leaves ovate or ovate-lanceolate, 2-8 cm long; flowers in open, leafy cymes .... 1. *S. aquatica*
1 Styles 3(4); leaves and inflorescence various.
  2 Leaves broadly elliptic to ovate or rotund-ovate, the middle and lower ones petiolate; stems with pubescence in longitudinal lines .... 7. *S. media*
  2 Leaves linear to lanceolate, sessile or only the basal ones short-petiolate; stems glabrous or hairy, but the pubescence not in lines.
    3 Flowers solitary, appearing terminal or axillary, not in cymes.
      4 Petals shorter than the sepals or lacking; seeds smooth or nearly so .... 2. *S. calycantha*
      4 Petals about equaling or surpassing the sepals; seeds obviously rugose, with concentric ridges .... 3. *S. crassifolia*
    3 Flowers few to many (sometimes single and terminal in *S. longipes*), in terminal, bracteate, usually dichotomous cymes.
      5 Cymes leafy-bracteate; petals shorter than the sepals or lacking ... 2. *S. calycantha*
      5 Cymes membranaceous-bracteate; petals equaling or surpassing the sepals.
        6 Inflorescence rather narrow, the pedicels erect to ascending; leaves strongly ascending .... 6. *S. longipes*
        6 Inflorescence widely spreading, the pedicels spreading or reflexed; leaves spreading to ascending.
          7 Flowers mostly few; sepals 2-4(5) mm long; weakly 3-nerved, glabrous; leaf margins and angles of the stem minutely tuberculate-scaberulous (visible under 20-30× magnification) .... 5. *S. longifolia*
          7 Flowers mostly many; sepals 3-6 mm long, the margins usually ciliolate; leaf margins and angles of the stem smooth .... 4. *S. graminea*

1. **Stellaria aquatica** (L.) Scop., giant chickweed. Weak-stemmed, decumbent perennial, rhizomatous and usually rooting at the lower nodes; stems 4-angled, glabrous below, becoming hirsute (with multicellular hairs) upward, also glandular-pubescent above. Leaves sessile or short-petiolate, ovate to ovate-lanceolate, mostly 2-8 cm long, 1-4 cm wide, glabrous

or the upper ones pubescent, whitish-punctate, acute to acuminate, rounded to cordate at the base. Flowers in open, leafy-bracteate, terminal cymes; pedicels 0.8-3 cm long, eventually reflexed. Sepals ovate-lanceolate, 5-7(9.5) mm long, glandular-pubescent, acute to obtuse; petals exceeding the sepals, cleft nearly to the base; styles 5. Capsule ovoid, 5-valved, each valve notched or shallowly bifid at the tip; seeds subrotund, 0.8-1 mm in diam, papillate in concentric rows. (n = 10, 2n = 28) May-Oct. An introduced weed of moist woods, thickets, flood plains, & waste places; KS: Doniphan; MN: Lac Qui Parle; (Que. & Ont. to MN, s to NC, LA, & KS, also B.C.; Europe). *Naturalized. Myosoton aquaticum* (L.) Moench — Fernald.

2. *Stellaria calycantha* (Ledeb.) Bong., northern stitchwort. Rhizomatous perennial, often matted; stems prostrate to ascending or erect, usually freely branched, sharply angled, (0.5)1-5 dm long, mostly more than 2.5 dm long, glabrous to scabrous or hairy. Leaves linear-lanceolate to ovate-lanceolate or elliptic-lanceolate, 1-5 cm long, 2-8 mm wide, often ciliate or scabrous on the margins, especially toward the base, usually narrowed at the base. Flowers solitary and both terminal and axillary, or in few-flowered, leafy-bracteate, terminal cymes; pedicels 1-4 cm long, usually reflexed with maturity. Sepals lanceolate, 2-3(4) mm long, faintly 3-nerved, scarious-margined; petals shorter than the sepals or more often lacking. Capsule often dark, ovoid, rather firm, surpassing the sepals; seeds obovoid, 0.7-0.9 mm long, smooth or only slightly roughened. (2n = 44-52) Moist, usually shaded places; barely entering our range in w MN: Becker; (circumboreal, in Amer. s to NY, WV, MI, MN, WY, UT, & CA). *S. borealis* Bigel. — Rydberg; *S. calycantha* var. *floribunda* Fern., var. *isophylla* Fern. — Fernald.

3. *Stellaria crassifolia* Ehrh., fleshy stitchwort. Glabrous, often matted, somewhat fleshy perennial from filiform rhizomes; stems many, very slender and angled, decumbent to sprawling, often supported by surrounding vegetation, freely branched, (5)10-25(30) cm long. Leaves sessile, elliptic to oblong-lanceolate or linear-oblong, or the lowest ones often oblanceolate or obovate, 5-24 mm long, 1.5-4 mm wide, thin and soft, acute or obtusish. Flowers solitary, axillary or in forks of the stem, usually nodding; pedicels 0.5-4(6) cm long, filiform, spreading or reflexed. Sepals ovate-lanceolate, 2-3.5 mm long, acute, narrowly scarious-margined; petals equaling the sepals or usually surpassing them by 1-2 mm, deeply 2-lobed. Capsule several-seeded, oblong-ovoid, surpassing the sepals; seeds subrotund, 0.8-1 mm in diam, rugose, with low, concentric ridges. (2n = 26) Jun-Aug. Fresh marshes, swamps, springs, fens, & stream banks; w MN to cen & nw ND, n-cen & ne SD; (circumboreal, in Amer. s to the Gaspé Peninsula, MN, SD, CO, ID & CA).

4. *Stellaria graminea* L., common stitchwort. Ascending to decumbent, spreading perennial; stems weak, strongly 4-angled, usually freely branched, 3-7(10) dm long, glabrous. Leaves sessile, linear to lanceolate, 1.5-4(5) cm long, 1.5-7 mm wide, glabrous or longciliate at the base, acute. Flowers many, in terminal, membranaceous-bracteate, diffusely spreading cymes; bracts ovate to ovate-lanceolate, 2-5 mm long, ciliolate; pedicels slender, spreading to reflexed, 1-6 mm long. Sepals lanceolate, 3-6 mm long, strongly 3-nerved, acute or acuminate, scarious-margined and usually ciliolate, at least basally; petals conspicuous, usually considerably longer than the sepals, deeply cleft. Capsule greenish-yellow, ellipsoid, about equaling or slightly surpassing the sepals; seeds oblong-subreniform to subrotund, 0.8-1.2 mm long, prominently rugose-tuberculate in concentric rows. (n = 26; 2n = 26, 39, 52) Jun-Jul. An introduced weed of lawns, fields, & roadsides, rare in our range, with only one record; ND: Cass; (established from Newf. & Que. to MN, s to NC, also coastal & sporadically inland in the Pacific Northwest; Eurasia). *Introduced.*

5. **Stellaria longifolia** Muhl. ex Willd., long-leaved stitchwort. Ascending to decumbent perennial, usually branched and spreading; stems weak, 1-4.5 dm long, minutely scaberulous on the 4 angles, otherwise glabrous. Leaves spreading to ascending, linear to narrowly lanceolate, or especially the lower ones elliptic to oblanceolate, (1)1.5-5 cm long, (0.8)1.5-4(7) mm wide, minutely tuberculate-scaberulous on the margins (visible with 20-30 × magnification), sometimes ciliate at the base. Flowers few to several, in an open, loosely spreading, dichotomous cyme; bracts ovate to lanceolate, 2-5 mm long, membranaceous, sometimes ciliolate; pedicels slender, 7-25 mm long, spreading or reflexed. Sepals lanceolate, 2-4(5) mm long, faintly 3-nerved, acutish, scarious-margined, glabrous; petals about equaling or surpassing the sepals, deeply 2-lobed. Capsule greenish-yellow or dark brown, ovoid or oblong-ovoid, somewhat to much surpassing the sepals; seeds oblong, 0.7-1 mm long, obscurely pitted. (2n = 26) May-Jul. Wet meadows, shores, stream banks, fens, & damp woods; MN to nw & cen ND, s to e KS: Douglas & n NE; (circumboreal, in Amer. s to SC, LA, KS, NM, & CA). *Alsine longifolia* (Muhl.) Britt.—Winter.

6. **Stellaria longipes** Goldie, long-stalked stitchwort. Low, often matted perennial from filiform rhizomes; stems many, erect to decumbent, (0.5)1-2.5 dm tall, 4-angled, glabrous. Leaves strongly ascending, rather stiff and shiny, linear to linear-lanceolate, 1-3(4) cm long, 0.8-3(5) mm wide, acute to attenuate. Flowers few to several in a rather narrow, dichotomous cyme, or occasionally, some stems with only a single terminal flower; bracts ovate to lanceolate, 2.5-5 mm long, membranaceous; pedicels slender, erect to ascending, 0.8-3(7) cm long. Sepals lanceolate or ovate-lanceolate, 3-5 mm long, obtuse or acutish, scarious-margined; petals slightly shorter to slightly longer than the sepals, 2-lobed to about the middle. Capsule yellow-green to usually dark purple and shiny, equaling to somewhat surpassing the sepals; seeds oblong to ovoid, 0.7-1 mm long, lightly reticulate. (2n = 51-107, most often 52, 78, 104) Late May-Aug. Moist thickets, stream banks, & meadows; MN, n & e ND & the BH; (circumpolar, in Amer. s to N.S., NY, IN, MN, BH, NM, AZ, & CA). *S. laeta* Richard.—Rydberg.

Our plants are typical *S. longipes*, which in itself is highly variable. Several other varieties and closely related species occur n & w of our range.

7. **Stellaria media** (L.) Cyr., common chickweed. Low, often matted annual, possibly overwintering in moist, protected places, usually much branched from the base; stems decumbent and rooting at the nodes, 0.7-5 dm long, glabrous below, becoming finely hirsute in 1 or 2 longitudinal lines upward, the hairs multicellular. Leaves sessile or the middle and lower ones usually with winged-ciliate petioles to as long as the blade, elliptic to ovate or rotund-ovate, 1-3 cm long, 3-15 mm wide, acute to short-acuminate. Flowers axillary or also in terminal, leafy-bracteate cymes of 3-7; pedicels slender, 3-30 mm long, ascending to eventually reflexed. Sepals oblong-lanceolate, 3.5-6 mm long, acute to obtuse, scarious-margined, hirsute and more or less glandular; petals shorter than the sepals, 2-lobed nearly to the base; stamens 3-5(10). Capsule ovoid, surpassing the sepals by 1-2 mm; seeds subrotund, 1-1.2 mm in diam, tuberculate. (n = 20, 22; 2n = 28, 40, 42, 44) (Jan) Mar-Oct (Dec). An introduced weed of lawns, gardens, & waste places, especially where shaded; GP, especially in urban areas; (widely established in N. Amer.; Eurasia). *Naturalized. Alsine media* L.—Winter.

17. VACCARIA Medic.

1. **Vaccaria pyramidata** Medic., cowherb, cow-cockle. Glaucous and glabrous taprooted annual 2-8 dm tall, branched above. Leaves mainly cauline, the lowest often short-petiolate,

otherwise sessile or the lower ones slightly connate, lanceolate to ovate-lanceolate, 4–9(12) cm long, 0.5–4(5) cm wide, acute, cordate or auriculate-clasping. Flowers showy, perfect, hypogynous; usually numerous, in a diffuse, open, often flat-topped paniculate cyme, the lowest bracts leafy, much reduced upward. Calyx tubular in flower, much inflated and broadly ovoid or flask-shaped in fruit, often purplish, 11–17 mm long, strongly 5-ribbed and wing-angled, the 5 teeth triangular, 2–3 mm long; petals pink, 18–22 mm long, lacking auricles and appendages, the claw usually exceeding the calyx, the blade obovate, 5–8 mm long, retuse; stamens 10, exserted; styles 2(3), exserted. Fruit a firm, 1-celled capsule, dehiscent by 4(6) ascending teeth; seeds reddish-brown to black, globose, 2–2.6 mm in diam, minutely tuberculate. (n = 15, 2n = 30, 60) Jun–Aug. An introduced weed of fields, roadsides, & railways; MN to MT, s to MO & cen KS; (widely established throughout most of temp. N. Amer.; Europe). *Naturalized*. *V. vulgaris* Host — Rydberg; *Saponaria vaccaria* L. — Fernald; *V. vaccaria* (L.) Britt. — Winter; *V. segetalis* (Neck.) Garcke — Gleason & Cronquist.

43. POLYGONACEAE Juss., the Buckwheat Family

by Robert B. Kaul

Ours annual, biennial, or perennial herbs, suffrutescent herbs, and herbaceous vines. Stems erect to sprawling or twining, the nodes sometimes swollen; leaves cauline and/or basal, alternate, simple, entire, petiolate to sessile; stipules absent to prominent, sometimes sheathing the stem (ocreae). Flowers subsessile to pedicellate, the pedicels often jointed, single or in groups from sheathing bracts (ocreolae), these axillary or grouped in spiciform or racemose condensed panicles, or in open panicles or capitate clusters. Flowers perfect, imperfect, or sometimes functionally unisexual, hypogynous; perianth undifferentiated, the 4–6 members variously connate but the lobes free, greenish and small to red and becoming showy in fruit; stamens 6–9, free; pistil 1, unilocular, the single ovule basal. Fruit an achene that is included or exserted from the persistent perianth.

1 Leaves without stipules; inflorescences with involucres of partially united bracts; dry places in w GP .... 1. *Eriogonum*
1 Leaves with stipules or stipulelike structures, these often large and sheathing the stem; inflorescences without involucres.
   2 Leaves mostly basal and huge, the blades to 50 cm long; achenes winged .... 5. *Rheum*
   2 Leaves mostly less than 20 cm long and usually much less, cauline or basal; achenes not winged (but the persistent perianth around the achene sometimes winged).
      3 Erect, prostrate, or vining herbs; flowers 1–several in the bracts, the pedicels erect or reflexed; leaves not needlelike.
         4 Basal leaves usually much larger than the cauline leaves and crowded; perianth parts 6, 3 of them forming winglike valves enclosing the achene, the valves often with a dorsal tubercle .... 6. *Rumex*
         4 Cauline leaves about the same size as the basal leaves, or basal or cauline leaves absent; perianth parts fewer than 6, winged or wingless in fruit; achene included or exserted from the persistent perianth.
            5 Achene included to somewhat exserted; blade with or without basal lobes (if lobed, the plant a vine or barbed herb) .... 4. *Polygonum*
            5 Achene strongly exserted; blade sagittate, cordate, or deltoid; erect to sprawling herb without barbs .... 2. *Fagopyrum*
      3 Erect, suffrutescent, much-branched perennial; flowers single in the bracts, the pedicel reflexed; leaves linear, needlelike; TX panhandle .... 3. *Polygonella*

1. ERIOGONUM Michx., Wild Buckwheat

Annual, biennial, or perennial herbs and subshrubs, often cespitose or matted, from slender taproots or thick caudices and stout woody taproots, the flowering stems erect, the plants glabrous or variously tomentose. Leaves cauline and/or basal, mostly alternate and entire, sessile or petiolate, without stipules. Inflorescences scapose, usually much branched open and paniculate to capitate, basically cymose, often long-pedunculate and held well above the leaves; flowers 1–several in a membranaceous, tubular, turbinate or campanulate calyxlike toothed involucre; involucres solitary or variously aggregated, sessile or pedunculate. Flowers perfect or imperfect; pedicels slender, jointed at the base of the perianth; perianth rather persistent, white, whitish, or yellow, sometimes tinged pink or purple, often changing color after anthesis; perianth members 6, free to the base or partially connate, the base slender and pedicellike (stipitate) or merely narrowed, glabrous, tomentose, or pilose, the outer 3 members similar to the inner 3, ovate to obovate or oblong, cordate or cuneate at the base; stamens 9, longer or shorter than the perianth, pilose at the base; pistil 3-carpellate, styles 3, stigmas capitate, ovary 1-celled, uniovulate. Achene 3-angular or lenticular, sometimes winged on the angles, glabrous or tomentose. Plants of dry places, often on sandy, gravelly, or rocky mesas and hills and in gypsum flats throughout the western half of our area.

References: Reveal, J. L. 1967a. Notes on *Eriogonum*. II. On the status of *Eriogonum pauciflorum* Pursh. Great Basin Naturalist 27: 102–117; Reveal, J. L. 1967b. Notes on *Eriogonum*. V. A revision of the *Eriogonum corymbosum* complex. Great Basin Naturalist 27: 183–229; Reveal, J. L. 1968. Notes on the Texas *Eriogonums* (Polygonaceae). Sida 3: 195–205; Reveal, J. L. 1971. Notes on *Eriogonum*. VI. A revision of the *Eriogonum microthecum* complex (Polygonaceae). Brigham Young Univ. Sci. Bull. Biol. Ser. 13: 1–45; Stokes, S. G. 1936. The genus *Eriogonum*. J. H. Neblett Press, San Francisco.

1 Plant annual, without old leaves or leaf bases, from a slender taproot, usually with a single flowering stem.
  2 Leaves cauline, the blades oblanceolate, short-petiolate .... 2. *E. annuum*
  2 Leaves basal, the blades ovate to orbicular to reniform, mostly long-petiolate.
    3 Involucral peduncles strongly reflexed .... 4. *E. cernuum*
    3 Involucres sessile or on erect or inclined peduncles.
      4 Flowers white to pinkish in anthesis; bracts of the inflorescence tiny, scalelike .... 8. *E. gordonii*
      4 Flowers yellow in anthesis; bracts of the inflorescence foliaceous, elliptic .... 14. *E. visheri*
1 Plant perennial, usually with old leaves and leaf bases, from thick, woody caudices, the branches sometimes cespitose or mat-forming.
  5 Flowering stems glabrous.
    6 Leaves basal .... 13. *E. tenellum*
    6 At least a few leaves cauline.
      7 Perianth yellow in anthesis; achenes 3-winged .... 1. *E. alatum*
      7 Perianth white or pinkish in anthesis; achenes not winged .... 6. *E. effusum*
  5 Flowering stems not glabrous.
    8 Perianth basally narrowed to a pedicellike stipe.
      9 Leaves cauline; achenes densely tomentose throughout .... 11. *E. longifolium* var. *lindheimeri*
      9 Leaves basal; achenes pubescent only at the top.
        10 Perianth whitish or greenish in anthesis .... 9. *E. jamesii*
        10 Perianth yellow in anthesis.
          11 Leaves crowded and mat-forming; n GP .... 7. *E. flavum*
          11 Leaves not distinctly mat-forming; TX & OK panhandles ..... 5. *E. correllii*
    8 Perianth basally narrowed but not to a pedicellike stipe.
      12 Perianth glabrous on the outside.
        13 Perianth white or pinkish in anthesis .... 6. *E. effusum*
        13 Perianth yellow or yellowish in anthesis.

     14 Achenes large (4–9 mm long), 3-winged, reddish; lowest floral bracts
        foliaceous .... 1. *E. alatum*
     14 Achenes smaller, not winged; floral bracts all tiny, scalelike .... 3. *E. brevicaule*
  12 Perianth pubescent or pilose on the outside.
     15 Leaves cauline; perianth white or pinkish in anthesis (yellow in var. *canum*, MT: Treasure
        only) .... 12. *E. pauciflorum*
     15 Leaves basal or essentially so; perianth yellow in anthesis.
        16 Achenes tomentose throughout; s GP .... 10. *E. lachnogynum*
        16 Achenes pubescent only at the top; nw GP .... 7. *E. flavum*

1. **Eriogonum alatum** Torr., winged eriogonum. Plant perennial, erect, stout, from an unbranched or few-branched woody caudex with a deep taproot. Stem l(few), to 1 m or more tall, alternate-branching above, strigose to long-tomentose at least below. Leaves mostly basal, a few sometimes cauline and smaller, linear-oblanceolate, acute or rounded, to 15 cm long and 1 cm wide, gradually tapering to a petiole that is shorter than the blade, the margins and midrib with long flexible hairs. Inflorescence weakly to distinctly paniculate, the terminal branch systems cymose; bracts small and foliaceous below, acuminate scales above; involucre campanulate, several-flowered, strigose. Perianth obconic, yellowish, the members lanceolate, glabrous on the outside. Achenes 4–6(9) mm long, reddish, glabrous, 3-angular, the angles winged for their entire length. (n = 20) Jun–Oct. Prairies & rocky hillsides; se WY, w NE s to ne NM, TX panhandle, w OK; (UT to NE, WY s to w TX & AZ).

   The var. *glabriusculum* Torr. can be distinguished by its glabrous inflorescences and involucres. It enters our range in TX panhandle, w OK.

2. **Eriogonum anuum** Nutt., annual eriogonum. Herbaceous annual (biennial) with 1–few erect simple or few-branched stems 1–10 dm tall, silver-gray tomentose throughout. Basal leaves few, oblanceolate, 2–5 cm long, often withering before anthesis; cauline leaves similar to the basal leaves, numerous below but few–none above. Inflorescence terminal, sometimes a few smaller ones on lateral branches, open-cymose, di- or trichotomously branched, the branches inclined or ascending and almost helicoid, the bracts tiny; involucres erect, turbinate, 2.5–3 mm long, sessile and pedunculate, shallow-toothed. Perianth white or whitish, sometimes pink-tinged, the inside densely hairy, the outside glabrous, the members half-connate from the base, the outer members broader than the inner. Achenes glabrous. Jul–Sep. Dry open grasslands, sandy & rocky slopes; e MT, e WY, w ND, w SD, w & cen NE, w & cen KS, OK, TX, NM; (e MT to w ND s to TX & Mex.).

3. **Eriogonum brevicaule** Nutt., shortstem eriogonum. Perennial mat-forming subshrub, the stems woody and branching below. Leaves numerous, somewhat crowded at the base, linear to narrowly oblanceoalte, sometimes revolute, the blade 2–7 cm long, 5–8 mm wide, short- or long-petiolate, tomentose below, greenish but floccose above. Flowering stems to 20 cm tall, 2–5 times di- or trichotomously branched, the branches ascending and often flat-topped, the first branching often subtended by a small leaf, the bracts above small to minute scales; involucres mostly pedunculate, 2–3 mm long, usually glabrous, campanulate, the lobes rounded, short. Perianth yellow in anthesis, 3 mm long, glabrous on the outside, not narrowed to a pedicellike stipe, campanulate, the members similar, basally connate. Achenes papillate. Jul–Aug. Open sandy grasslands & dry hillsides; e WY, w NE, (ID to w NE s to CO & UT). *E. campanulatum* Nutt.—Rydberg.

4. **Eriogonum cernuum** Nutt., nodding wild buckwheat. Annual herb 10–45 cm tall, delicate to robust, from a slender taproot. Leaves basal, long-petiolate, the blade ovate to mostly orbicular, its base rounded to subcordate, 0.5–2 cm long, densely tomentose below, barely floccose above. Flowering stem dichotomous or trichotomous above, glabrous, glaucescent, the inflorescence paniculate; involucres mostly solitary at the nodes, 3–10 mm long,

on slender and strongly reflexed peduncles, glabrous, turbinate, to 2 mm long, several-flowered. Perianth 1.5–2 mm long, glabrous on the outside, narrowed toward the base, white or pink-tinged in anthesis, the members basally connate, the outer members more or less quadrangular, the apices crisped, retuse, or truncate with their bases nearly truncate. Jun–Aug. Open sandy grasslands & hillsides; e WY, w SD, w NE; (s OR to w SD s to NM & CA).

5. *Eriogonum correllii* Reveal, Correll's eriogonum. Perennial herb from a stout woody caudex. Leaves basal but not distinctly tufted or matted, 3–15 cm long, the largest 1.5–2 cm wide, long-petiolate, the blade lanceolate to oblong, glabrous or sparingly tomentose to floccose above, white-tomentose below. Flowering stems 20–40 cm tall, tomentose; inflorescence flat-topped or somewhat rounded, dichotomously branched 3–6 times, the nodes with 3 or 4 foliaceous short-petiolate bracts, the bracts tomentose, 0.5–1.5 cm long, to 0.5 cm wide, the upper ones oblong to ovate and often obtuse; involucres turbinate or campanulate, 3–5 mm long, with shallow lobes, tomentose. Perianth bright yellow to mustard-greenish in anthesis, 3–7 mm long, with a short narrow pedicellike base, pubescent on the outside, the members of the inner whorl long and wider than those of the outer whorl, this pronounced in fruit. Achenes to 6 mm long, barely pubescent at the top. (n = 20) Jul–Oct. Rocky or clay slopes, caprock ledges, gypsum flats; TX & OK panhandles. *Endemic.* *E. jamesii* Benth. in part — Rydberg.

6. *Eriogonum effusum* Nutt., spreading wild buckwheat. Plant perennial, from a branching woody or suffrutescent base, the bark exfoliating. Leaves cauline, overlapping, linear to oblong or narrowly obovate, 1–3(4) cm long, sometimes revolute, the base narrowed to a short petiole, sparingly tomentose above, gray-tomentose below. Flowering stems 5–20 cm long, leafy about half their length, glabrous, floccose, tomentose, or lanate; inflorescence 2- to 4-chotomously branched, rather flat-topped, the bracts scalelike; involucres sessile and short-pedunuculate, turbinate, 2–2.5 mm long, glabrous or somewhat tomentose on the outside, the teeth rounded or acute, sometimes ciliate. Perianth turbinate, 2–2.5 mm long, glabrous on the outside, white or pink in anthesis in ours, the outer members wider than the inner, ovate to obovate, their bases rounded to almost cuneate, basally connate for about 1/3 of their length. Achenes glabrous. (Jul) Aug–Sep. Sandy sagelands & grasslands. *E. helichrysoides* (Gand.) Rydb., *E. microthecum* Nutt. — Rydberg.

This extremely variable species has two varieties with us:

6a. var. *effusum*. Leaves 2–4 cm long, not revolute. Inflorescence 6–20 cm long. e WY, e CO, w NE, sw SD.

6b var. *rosmarinoides* Benth. in DC. Leaves less than 2 cm long, usually revolute. Inflorescence to 5 cm long. e MT, e WY, e CO, w NE, w KS.

A third entity, subsp. *fendlerianum* S. Stokes, recognized by Reveal (1967) as *E. fendlerianum* (Benth. in DC.). Small, barely enters our range in se CO & ne NM; it has longer and revolute petioles, larger involucres, and the perianth members more alike.

7. *Eriogonum flavum* Nutt., yellow wild buckwheat. Perennial herb from a woody taproot, silvery-gray tomentose throughout, the caudex woody, branched, the branches thick and invested with old leaf bases, mat-forming. Leaves crowded at the base, 3–8 cm long, 3–14 mm wide, rather thick, the blades linear-oblong, the petioles slender, shorter or longer than the blades, gray tomentose or greenish above, silvery-gray tomentose below. Flowering stem leafless, 4–25 cm tall; inflorescence umbellate, the rays to 3 cm long, subtended by a few foliaceous or reduced bracts; involucres tomentose to pilose, obconic to campanulate, ca 4–7 mm long, the lobes shallow or absent. Perianth 4–6 mm long, yellow in anthesis, sometimes tinged pinkish, pilose on the outside, attentuate to narrowed and short pedicellike

at the base. Achenes pubescent at the top. (2n = 36, 76–80) May–Sep. Dry plains & ridges; GP only; cen & e MT, e WY, cen CO, w ND, w SD, w NE.

Most of our specimens are the narrow-leaved *(crassifolium)* phase.

8. ***Eriogonum gordonii*** Benth. in DC., Gordon's eriogonum. Annual herb from a slender taproot, glabrous or sparingly hirsute. Leaves basal, 1–3 cm long, the blades mostly ovate to orbicular to reniform, the petioles about equaling the blades. Flowering stem to 35 cm tall, slender, much branched, cymose, the bracts tiny; involucres all pedunculate, campanulate, more or less glabrous, shallow-toothed, the teeth rounded. Perianth white or pinkish in anthesis, 1–2.5 mm long, the members nearly alike. Achenes glabrous. Jun–Sep. Dry hills & prairies; nw NE: Sioux; (UT, CO, WY, to nw NE).

9. ***Eriogonum jamesii*** Benth., James' wild buckwheat. Perennial herb from a branching, spreading caudex, sometimes mat-forming. Leaves mostly basal, 1–4(7) cm long, mostly less than 1 cm wide, the blades elliptic to obovate, rarely orbicular, the petioles mostly as long as or longer than the blades, green and glabrate or sparingly tomentose above, densely whitish-tomentose below. Flowering stems 10–25 cm tall, tomentose, the inflorescence cymose, dichotomously 2- to 4-branched, occasionally unbranched, open or becoming flat-topped, the terminal units often subcapitate and the heads to 1.5 cm wide; the bracts foliaceous, 3–5 at each node, smaller than but otherwise like the basal leaves; involucres turbinate, 3–7 mm long, tomentose, the lobes shallow. Perianth whitish to greenish-white (yellow) in anthesis, changing to pinkish in fruit, the members alike, 3–6 mm long, more or less obovate, pubescent on the outside. Achenes sparingly pubescent at the top. (n = 20) Jun–Oct. Dry & rocky slopes & shale mesas; se CO, w KS, OK & TX panhandles, ne NM; (se WY to w KS s to AZ, n Mex.).

10. ***Eriogonum lachnogynum*** Torr. Perennial herb from a branching woody caudex. Leaves basal, tufted or mat-forming, the blades elliptic, lanceolate, or oblanceolate, sometimes revolute, 1–4.5(6) cm long, 3(5) mm wide, somewhat long-tomentose above, densely white short-tomentose below, the petioles equal to or shorter than the blades. Flowering stems 10–20 cm tall, gray-tomentose, the smaller ones unbranched, the larger ones 1- or 2-branched, the terminal units subcapitate; involucres campanulate, 3–5 mm long, sessile and pedunculate, tomentose on the outside. Perianth yellow, pubescent on the outside, the members alike, lanceolate. Achenes densely tomentose throughout, 3–4 mm long. Jun–Oct. Clay banks, rocky slopes, shale mesas; se CO, sw KS, OK & TX panhandles, ne NM; (se CO, KS, s to TX panhandle & sw to AZ).

11. ***Eriogonum longifolium*** Nutt. var. ***lindheimeri*** Gand., Lindheimer's longleaf eriogonum. Perennial herb, erect, tall, leafy, the stems pubescent, to 1.75 m tall. Basal leaves petiolate, the blades oblong to lanceolate to oblanceolate, 1–20 cm long, the largest about 1.5 cm wide, glabrous to tomentose above, densely tomentose below, the petioles somewhat winged toward the base; cauline leaves smaller, to scalelike above, sometimes revolute. Inflorescences cymose, the branching irregularly dichotomous, the branches few; involucres sessile and pedunculate, more or less campanulate, 4–6 mm long, tomentose. Perianth narrow and pedicellike at the base, atop a pedicel to 8 mm long, densely white pubescent on the outside, yellow inside at anthesis, ca 5 mm long, the members alike, lanceolate. Achenes densely white-tomentose, 4–6 mm long. Jun–Oct. Dry sandy & clay soils; GP only; sw KS, se CO, OK & TX panhandles, w & cen OK, ne NM.

12. ***Eriogonum pauciflorum*** Pursh. Mat-forming perennial, the stem to 20 cm tall, leafy below, tomentose. Leaves linear to oblanceolate or narrowly spatulate, their bases partly sheathing the stem, 2–5(8) cm long, 2–10 mm wide, densely gray-tomentose below, variously

tomentose above. Inflorescences mostly solitary, of capitate clusters, or umbellate with unbranched rays to 3 cm long, the rays then terminating in capitate clusters, or several-branched, the first node with 2-6 linear-lanceolate rather leafy, tomentose bracts to 2 cm long; involucres narrowly turbinate, 4-5 mm long, tomentose, the teeth acute. Perianth whitish, yellow, pink, or pinkish or brownish tinged, the lower part densely pubescent (glabrous), 2-2.5 mm long, the members alike, oblong, connate at the base for about 1/3 of their length, the bases and tips rounded. Achenes glabrous, brownish, 3-angular above, globose below, ca 2 mm long. Jun-Aug. Rocky slopes & dry prairies; w GP, MT to CO; (s Sask., GP w to ID).

This is an extremely variable species with 4 varieties recognized by Reveal, op. cit., in our area. Intermediate forms can be found.

1 Inflorescences capitate.
  2 Heads ca 1 cm wide, more or less obconic or hemispherical; plants loosely matted .... 12d. var. *pauciflorum*
  2 Heads stout, to 2.5 cm wide, subspherical; plants densely matted .. 12b. var. *gnaphalodes*
1 Inflorescences cymose, the branching compound.
  3 Flowers white; extr. w NE, ne CO .... 12c. var. *nebraskense*
  3 Flowers yellow; cen MT .... 12a. var. *canum*

12a var. *canum* (S. Stokes) Reveal. Inflorescences several-branched, cymose, flowers yellow. MT: Treasure.

12b. var. *gnaphalodes* (Benth.) Reveal. Plant densely matted; inflorescence capitate, the heads stout, subspherical, to 2.5 cm wide. e MT, e WY, w NE, ne CO.

12c. var. *nebraskense* (Rydb.) Reveal. Inflorescences several-branched, cymose; flowers white. Extreme w NE, ne CO.

12d. var. *pauciflorum*. Plant loosely matted; inflorescence capitate, the heads more or less obconic or hemispherical, ca 1 cm wide, e MT, w ND, w SD, ne WY, extreme w NE, n-cen CO.

13. *Erigonum tenellum* Torr., matted wild buckwheat. Perennial herb, cespitose, from woody branched caudices invested with old leaves. Leaves small, basal, crowded or matted, blades elliptic to orbicular, to 1.5 cm long, mostly less than 6 mm wide, densely white- or gray-tomentose on both sides, petioles of all but the smallest leaves longer than the blades. Flowering stems 15-50 cm tall, glabrous, glaucous, delicate to slender, branching dichotomous, the branches inclined or erect, and the inflorescence open and not flat-topped, the bracts minute; involucres long-pedunculate, the peduncles slender; involucres obconic, glabrous, glaucous, the teeth acute, ca 3 mm long. Perianth white or pinkish, glabrous, the whorls unlike, the members of the outer whorl ovate to obovate, 2 mm wide, the inner members shorter, narrowed, oblong. Achenes glabrous. Jun-Oct. Rocky slopes, shale mesas, & prairies; GP only; se CO, ne NM, OK & TX panhandles.

14. *Eriogonum visheri* A. Nels., Visher's eriogonum. Annual herb, from a slender taproot, glabrous, the single stem much branched, 15-25 cm tall. Leaves in a basal rosette, the blades ovate to reniform, 15-20 mm long, the petioles sometimes longer than the blades. Bracts of the inflorescence elliptic; involucres turbinate, sessile and slender-pedunculate, erect. Flowers 1-few/involucre, yellow; perianth members narrowly oblong, 1-1.5 mm long. Achene ovoid-acuminate. Jul-Sep. Dry open prairies & slopes (in disturbed areas?); GP only; s-cen ND, w SD. *E. trichopes* Torr., misapplied — Stevens.

2. FAGOPYRUM P. Mill., Buckwheat

1. *Fagopyrum esculentum* Moench, buckwheat. Annual erect herb from a taproot, to 5 dm or more tall. Stem glabrous to pubescent at the nodes. Leaves long-petiolate below,

becoming sessile above, acuminate, the base broadly sagittate, cordate, or truncate; stipules glabrous to puberulent, acute, sheathing but open on one side. Racemes axillary, single or in corymbiform to paniculate groups, these 1 per leaf axil, long-pedunculate and usually exceeding the subtending leaf; bracts glabrous, ovate, acute; pedicels not joined. Perianth white to creamy, streaked tan, 5-lobed, connate below, 3–4 mm long; stamens 8; styles 3, free. Achene strongly exserted, 3-angular, shiny, dark brown. (n = 8, 2n = 16) Jun–Sep (Oct). Cult. fields, disturbed places, waste ground, escaped from cultivation but seldom persisting; scattered in GP, mostly in e ½; (escaped in many parts of U.S. & the world; Asia). *Introduced*.

3. POLYGONELLA Michx.

1. *Polygonella americana* (Fisch. & Mey.) Small, jointweed. Suffrutescent, much branched, erect or prostrate perennial to nearly 1 m tall, from a taproot, the branches usually somewhat supra-axillary. Leaves linear, needlelike, the lower ones to 1.5 cm long, ca 1 mm wide, the tips often erose; stipules acute, entire. Racemes spiciform, numerous in a panicle, erect, mostly at the top of the plant, the bracts overlapping, acuminate, their margins hyaline; pedicels with a joint just above the base, solitary in the bracts. Perianth white, 2.5–4 mm long in fruit, the outer members 2, ovate, becoming reflexed in fruit, the inner members 3, accrescent, orbicular, retuse, enlarging in fruit and turning pinkish; stamens 8, filaments dilated. Achenes included, reddish-brown, 3-angular. Jun—Sep (Oct). Dry sandy places, open sandy forests, rocky depressions; TX (s panhandle), OK just beyond our range; (SC to OK s to GA, TX, & NM).

4. POLYGONUM L., Knotweed, Smartweed

Annual and perennial herbs and vines from taproots, fibrous roots, or rhizomes, sometimes weedy. Stems glabrous to pubescent, sometimes with retrorse barbs; nodes often swollen. Leaves cauline or basal, alternate, entire, petiolate or sessile; stipules 2-lobed or tubular (ocreae), often rupturing. Inflorescences of inconspicuous axillary fascicles or of axillary and terminal spiciform or racemiform panicles; bracts of the inflorescence (ocreolae) usually subtending several flowers. Flowers pedicellate, perfect (sometimes functionally unisexual); perianth undifferentiated, petaloid, 4- to 5-lobed, the lobes connate below, the outer lobes often enclosing the inner in fruit and sometimes becoming winged then; perianth green, white, pink, or red, or several of these colors, often collectively showy; stamens 3–9, often arising near a basal nectariferous disk; styles 2 or 3, free or variously connate; pistil unilocular, the single basal ovule orthotropous. Achene lenticular or angular, included or somewhat exserted from the persistent perianth. *Tovara* Adans., *Persicaria* (Bauh.) P. Mill.), *Bistorta* (Bauh.) P. Mill., *Tracaulon* Raf., *Bilderdykia* Dum.—Rydberg; *Tiniaria* Webb. & Moq.—Winter.

References: Hitchcock, C. L. 1964. *Polygonum* L. *In* C. L. Hitchcock et al., Vascular Plants of the Pacific Northwest, Part 2. Univ. Wash. Press, Seattle. pp. 139–168; Mertens, T. R., & P. H. Raven. 1965. Taxonomy of *Polygonum* section *Polygonum (Avicularia)* in North America. Madroño 18: 85–92; Mitchell, R. S. 1968. Variation in the *Polygonum amphibium* complex and its taxonomic significance. Univ. Calif. Publ. Bot. 45: 1–65; Mitchell, R. S. 1971. Comparative leaf structure of aquatic *Polygonum* species. Amer. J. Bot. 58: 342–360; Mitchell, R. S. 1976. Submergence experiments on nine species of semiaquatic *Polygonum*. Amer. J. Bot. 63: 1158–1165; Mitchell, R. S., & J. K. Dean. 1978. Contributions to a flora of New York State, I. Polygonaceae (Buckwheat Family) of New York State. New York State Mus. Bull. 431. 81 pp.

1 Largest leaves less than 4 cm long, short-petiolate to sessile, leaves with a joint near the base of the blade; flowers axillary or in racemelike inflorescences, inconspicuous; the inflorescences not showy. .... *Group I*, p. 221
1 Largest leaf blades longer than 4 cm, long-petiolate to sessile, leaves without a joint near the base of the blade; flowers in racemelike or paniculate inflorescences, these often conspicuous or showy.
   2 Leaves mostly basal; inflorescences with a few bulblets in the axils of the lower bracts; erect herbs; BH *(P. viviparum)* .... *Group III*, p. 228
   2 Leaves cauline; inflorescences without bulblets; erect or sprawling herbs and vines.
      3 Twining or sprawling vines, or scandent to reclining barbed herbs; leaves sagittate, cordate, or truncate .... *Group III*, p. 228
      3 Erect, sprawling, or floating herbs, never barbed; leaves never sagittate or truncate or strongly cordate .... *Group II*, p. 224

POLYGONUM—Group I

Erect to prostrate, often much-branched annuals from a slender taproot. Leaves mostly less than 4 cm long, a joint near the base of the blade; often whitish when infected with fungi, especially in late season. Flowers solitary or in few-flowered clusters, axillary or in spiciform inflorescences, not showy.

1 Perianth parts connate to just above to just below the middle, but not at the base only; upper leaves equaling or smaller than the lower leaves; flowers axillary (in spiciform racemes in *P. sawatchense*).
   2 Leaves acute to rounded, more than 1.5 mm wide, not linear; perianth parts cucullate or not.
      3 Flowers axillary; leaves overlapping to densely crowded; plant erect to sprawling.
         4 Perianth parts connate to above the middle and there constricted; leaves ovate to obovate, the tip rounded; plant, or at least the ultimate branches, erect .... 1. *P. achoreum*
         4 Perianth parts connate to below the middle or to near the middle, not constricted above; leaf shapes various, the tip acute to rounded; plant erect to prostrate.
            5 Perianth with a protruding pouch on one side near the base .. 4. *P. buxiforme*
            5 Perianth without a pouch.
               6 Plant erect; perianth connate below the middle, the outer lobes cucullate .... 6. *P. erectum*
               6 Plant erect to prostrate; perianth connate to about the middle, the lobes sometimes cucullate; common lawn weed .... 2. *P. arenastrum*
      3 Flowers in elongate spiciform racemes; leaves not at all crowded; plant erect; BH .... 9. *P. sawatchense*
   2 Leaves linear to linear-lanceolate, acute, less than 1.5 mm wide; perianth parts not cucullate .... 7. *P. neglectum*
1 Perianth parts connate only at the base and not to near the middle; upper leaves sometimes gradually reduced, the inflorescence then becoming almost spiciform above.
   7 Plant prostrate or sprawling, the ultimate branches often erect, or occasionally the entire plant erect; flowers axillary .... 3. *P. aviculare*
   7 Plant erect, the branches inclined to ascending; flowers axillary or those above in spiciform racemes.
      8 Leaves with 2 distinct pleats parallel to the midrib; pedicels erect in fruit .... 10. *P. tenue*
      8 Leaves without pleats parallel to the midrib; pedicels erect or reflexed in fruit.
         9 Pedicels obviously reflexed in fruit; leaves above reduced; nw GP ... 5. *P. douglasii*
         9 Pedicels erect or not strongly reflexed in fruit; leaves above reduced or not; GP .... 8. *P. ramosissimum*

1. **Polygonum achoreum** Blake, knotweed. Plant annual, from a taproot, erect, becoming basally decumbent with age, the stems stout, to 40 cm or more tall. Leaves numerous, overlapping, short-petiolate, little reduced above, inclined or erect, bluish-green, 0.7–3 cm long, about half as wide, ovate to obovate, the tops rounded. Inflorescences axillary, 1- to 3-flowered, the pedicels erect. Perianth 5-lobed, yellowish-green (pinkish) connate to above the middle, constricted above, urceolate below, the outer 3 members longer and cucullate, the tips obtuse. Achene included (exserted), yellowish-brown, 3-angular, the faces concave. (2n = 20) Jul–Sep. Dry, weedy places, often in packed soil; GP n of n KS; (OR to Que. s to NY & CO). Probably *introduced*, origin unknown.

This species is barely distinguishable from *P. erectum*, but the perianth is connate to above the middle and there constricted.

2. **Polygonum arenastrum** Jord. ex Bor., knotweed. Erect to prostrate mat-forming annual from a taproot, the ultimate branches often erect, or occasionally the entire plant inclined or erect. Leaves numerous, bluish-green (whitish when infected with mildew fungi), 5–20(30) mm long, to 4 mm wide, little to somewhat reduced above, ovate to elliptic, the tips acute to rounded. Inflorescences axillary, 1- to 3-flowered, the pedicels erect. Perianth members alike, connate to about the middle, the tips obtuse, the outer members often cucullate, green with white or pink margins. Achene included or partly exserted, dull brown, less than 2.3 mm long, 3-angular, the narrow side concave, the wide sides convex, or two sides concave, or all 3 sides flat. (2n = 40) Jun–Oct. Weedy places, lawns, often in packed soil or pavement cracks; GP; (cosmopolitan weed; Eurasia). *P. aviculare* L. *sens. lat.*

Linear-leaved forms, sometimes referred to as *P. neglectum* Bess. or *P. aviculare* L. var. *angustissimum* Meisn., are possibly variants of this or the next species.

3. **Polygonum aviculare** L., *sens. str.*, knotweed. Annual mat-forming (erect) herb to 1 m or more long, from a taproot, the ultimate branches often erect. Leaves numerous, bluish-green (gray-green when infected with mildew fungi), reduced above, the upper leaves 1/2 the size of the lower; lower leaves elliptic, to narrowly ovate 1.5–3 cm long, the tip acute or obtuse; upper leaves linear or lanceolate. Inflorescences axillary, 3- to 6-flowered, the pedicels erect, short, and the flowers barely exserted from the sheaths. Perianth members 5, green with white or pink margins, similar, connate only at the very base, barely keeled, the outer members sometimes cucullate. Achene included, 2.5–3.5 mm long, dull brown, 3-angular, 2 or 3 sides concave. (2n = 60) Jun–Oct. Weedy places, cult. fields; GP? (cosmopolitan; Eurasia).

This variable species resembles *P. arenastrum*.

4. **Polygonum buxiforme** Small, knotweed. Resembling *P. arenastrum*, but always prostrate, the leaves not reduced above. Basal portion of the perianth with a protruding pouch on one side, the lobes cucullate. (2n = 20, 60) Jun–Nov. Sandy soils, often near marshes; GP; (B.C. to Ont. s to TX & NV).

5. **Polygonum douglasii** Greene, knotweed. Erect, slender annual, 0.5–4 dm tall, the branches few, ascending and sometimes crowded, from a taproot. Leaves not crowded, inclined, the lower ones to 4 cm long, 8 mm wide, linear to lanceolate, very short-petiolate, reduced above. Inflorescences terminal on the branches, of single open spiciform racemes, or the racemes in a small panicle, the flowers 2–4 at the nodes. Flowers pedicellate, the pedicels becoming distinctly reflexed in fruit; perianth 5-lobed, connate only at the base, the tips obtuse and cucullate, green with white (pinkish) margins. Achene included, black, 3-angular, the sides concave, sometimes 1 or 2 sides flat. (2n = 40) Late Jul–Sep. Rocky soils in the open; e MT, w ND, w SD, e WY, cen CO; (B.C. to NY s to SD, NM, & CA).

6. ***Polygonum erectum*** L., erect knotweed. Leafy annual herb from a taproot, much branched, erect to somewhat spreading, to 5 dm tall. Leaves numerous, overlapping, very short-petiolate, green to yellow-green, the largest 2-5 cm long, 1-2.5 cm wide, ovate, obovate, or broadly elliptic, the apex rounded or obtuse, the base tapering, the ultimate branches with leaves much smaller. Inflorescences axillary, 2- to 3-flowered, mostly on short axillary branches; pedicels erect. Perianth 5-lobed, connate below the middle, the outer lobes cucullate and barely keeled, green with yellowish margins. Achene included to partially exserted, brown, 3-angular, the sides concave. (2n = 40) Jul-Sep. Dry open ground & waste places; GP; (U.S. & s Can.). Not *P. aviculare* L. var. *erectum* (Roth) Koch.

This species closely resembles *P. achoreum*, but the perianth is connate below the middle.

7. ***Polygonum neglectum*** Bess., knotweed. Annual, from a taproot. Stems slender, flexuous, wiry, erect, inclined, or prostrate, to 35 cm long. Leaves subsessile, linear to linear-lanceolate, 5-25 mm long, less than 1.5 cm long, acute. Inflorescences axillary, few-flowered, the pedicels erect. Perianth to 3 mm long, 5-lobed, not cucullate, the lobes equal and connate below the middle, green or roseate, not overlapping in fruit. Achene usually partly exserted, one side convex. (2n = 40) May-Nov. Disturbed places; GP; (widespread in N. Amer. e of the Rocky Mts.; Europe). *Introduced.*

8. ***Polygonum ramosissimum*** Michx., knotweed. Erect, branched annual to 8 dm tall, from a taproot. Leaves bluish-green or sometimes yellowish, more or less alike or sometimes reduced above, often falling early, 0.5-3 cm long, 1-5 mm wide, linear-lanceolate, the tip acute to rounded. Inflorescences axillary, few-flowered, sometimes hidden by the sheaths, sometimes becoming spiciform above, the pedicels erect in anthesis and usually also in fruit. Perianth 5- to 6-lobed, 2-3 mm long, basally connate, the inner lobes almost as long as the outer, the outer cucullate. Achenes dimorphic: included, 3-angular, brown, or exserted and lens-shaped and yellowish, both types present in late season. (2n = 20, 60) Jul-Sep. Damp or dry, often brackish soils & shores; GP; (B.C. to Newf. s to VA & w to CA). *P. exsertum* Small, *P. latum* Small, *P. leptocarpum* Robins., *P. prolificum* (Small) Robins., *P. ramosissimum* Michx. — Rydberg.

This is a highly variable species. Much-branched forms with the leaves alike, their tips obtuse, and the pedicels less than 2 mm long have been designated as *P. ramosissimum* var. *prolificum* Small and as *P. prolificum* (Small) Robins. Specimens with the upper leaves reduced, the leaf tips acute to acuminate, and the pedicels longer than 2 mm have been called var. *ramosissimum*, while specimens of that variety with the achenes at least partially exserted have been called *P. exsertum* Small. The complex needs critical study to determine the role of environment, season, and genetics in such phenotypic plasticity.

9. ***Polygonum sawatchense*** Small, knotweed. Erect annual to 4 dm tall, from a taproot, the stem simple or few-branched. Leaves not at all crowded, sometimes revolute, the lower ones oblong to oblanceolate, 1.5-2.5 cm long, acute, reduced to bracts above. Inflorescences axillary, the flowers 1-4 in the axils of most leaves, becoming spiciform above; pedicels erect. Perianth 5-lobed, ca 2.5 mm long, the lobes connate below the middle, cucullate, green with lighter margins. Achene included, black, 3-angular, 2 sides concave. Jul-Sep. Open slopes; BH; NE: Banner, Sioux; ne NM; (WA to ND s to NM & CA).

10. ***Polygonum tenue*** Michx., knotweed. Erect, branching slender annual, the stems to 3 dm or more tall, ascending, wiry, more or less 4-angled. Leaves not crowded, ascending, 0.5-2(3) cm long, 0.5-3(5) mm wide, linear to lanceolate, the apex acute to acuminate, the leaves above reduced; blade with 2 longitudinal pleats parallel to the midrib. Inflorescences axillary, becoming erect spiciform racemes above, slender, elongate, rather open, the flowers single or in groups of 2 or 3. Pedicels short, erect, the flowers subsessile,

erect. Perianth 5-lobed, basally connate, green with white (pinkish) margins, the outer lobes cucullate, the inner lobes shorter, the tips obtuse. Achene included, black or brown, 3-angular, the sides concave. (2n = 20, 30, 32) Jun–Oct. Dry open places, usually on acid soils; e GP except ND, & w in NE to WY & BH; (WY to ME s to GA, OK, & TX).

POLYGONUM—Group II

Erect, sprawling, or floating annuals and perennials to 2 m tall. Nodes usually swollen. Leaves not jointed near the base of the blade, sessile to long-petiolate; stipules (ocreae) tubular, sheathing, becoming lacerate with age; leaf blades to 15 cm long. Flowers loose or crowded and fascicled in spiciform racemes that are terminal on the branches or grouped in panicles, often showy; bracts (ocreolae) subtending the groups of flowers. Perianth members variously connate, the tips free; stamens and styles included or sometimes exserted from the perianth.

1 Racemes 1 or 2, terminal on the branches; plant perennial, terrestrial or aquatic, sometimes floating; flowers pink to red, heterostylous, the stamens or styles exserted .. 11. *P. amphibium*
1 Racemes several to numerous, terminal and axillary; plant annual or perennial, terrestrial or emersed in shallow water; flowers greenish, white, pink, or red, the stamens and styles included in the perianth or exserted.
  2 Stipules entire or merely lacerate, not fringed with cilia or bristles; bracts of the inflorescence oblique, tubular only at the base, the ovate limb acute.
    3 Racemes nodding or drooping; outer perianth members with the midvein divided at the top into two short recurved veins; flowers green, white, or pinkish .... 16. *P. lapathifolium*
    3 Racemes erect or ascending or arching; outer perianth members without such a vein; flowers white, pink, or roseate.
      4 Stamens and styles unequal, either the stamens or the styles distinctly exserted from the perianth, but not both .... 12. *P. bicorne*
      4 Stamens and styles equal or nearly so, not exserted from the perianth .... 18. *P. pensylvanicum*
  2 Stipules fringed with cilia or bristles; bracts of the inflorescence truncate or oblique at the top.
    5 Stipules pubescent but not strigose.
      6 Plant rooted in or near water; leaves short-petiolate to sessile, linear to lanceolate; racemes erect to arching; flowers greenish to pink .... 15. *P. hydropiperoides*
      6 Upland plant; lower leaves long-petiolate, the blades ovate; racemes often nodding, especially in fruit; flowers pink to rose .... 17. *P. orientale*
    5 Stipules glabrous, or if strigose the hairs adnate at the base.
      7 Perianth glandular-punctate, the glands tiny and depressed.
        8 Glandular punctae abundant throughout the perianth and especially below.
          9 Stipules cylindric; racemes irregularly and distinctly interrupted, erect to arching; perennial .... 20. *P. punctatum*
          9 Stipules often gibbous; racemes loose but not markedly interrupted, often drooping; annual .... 14. *P. hydropiper*
        8 Glandular punctae only on the inner perianth members .... 15. *P. hydropiperoides*
      7 Perianth not glandular-punctate; glands, if any, few, superficial, resinous, not punctate.
        10 Bracts of the inflorescence not crowded, barely if at all overlapping, the pedicels exserted from them; racemes slender, loose ...... 15. *P. hydropiperoides*
        10 Bracts of the inflorescence crowded, overlapping, the pedicels not exserted from them; raceme slender to stout.
          11 Cilia of the bract margins as long as the bracts and often exceeding the flowers; leaves never with a purple spot; e NE .... 13. *P. cespitosum*

11 Cilia of the bract margins, when present, less than 1 mm long and much shorter than the bracts; leaves sometimes with a prominent purple spot near the middle; GP .... 19. *P. persicaria*

11. *Polygonum amphibium* L., *sens. lat.*, water smartweed. Perennial, rhizomatous, floating, prostrate or erect herbs of quiet waters, shores, and uplands. Leaves and stipules variable. Inflorescences 1 or 2, erect, terminal on the stems. Perianth 4–5 mm long, pink or rose; stamens 8, included or exserted, and always of different length from the styles; styles 2, included or exserted (heterostylous), united to about the middle. Achene lenticular, 2.5 mm long, dark, glossy. (2n = 66, 96) Jun–Sep. Cosmopolitan.

For convenience, 2 varieties may be recognized in our area; intermediates are readily found in the northern GP, where both varieties occur. Cf. Mitchell, op. cit.

1   Stems and leaves floating, glabrous. .... 11a. var. *stipulaceum* Colem., natant form
1   Stems and leaves mostly not floating; glabrous, strigose, or hirsute.
  2   Plant sprawling on the shore, pubescent to hirsute, the stipules often with a ciliate collar .... 11a. var. *stipulaceum* Colem., land form
  2   Plant erect, glabrous to strigose, stipules entire; in swamps, meadows, or dry uplands .... 11b. var. *emersum* Michx.

11a. var. *stipulaceum* Colem. NATANT FORM: Stems and leaves floating, the stems rooting at the nodes, to 1 m long. Leaves long-petiolate, glabrous, elliptic, obtuse to rounded (acute in forms intermediate with *P. a.* var. *emersum*), the bases acute to rounded (nearly cordate in intermediate forms), 2–15 cm long, 1–5 cm wide; stipules oblique, glabrous, entire. Racemes spiciform, held above the water, stout, thick, cylindrical in full anthesis and fruit, 1.5–3 cm long, 0.5–1.5 cm thick, the bracts pink, ovate, entire, glabrous; peduncles shorter to longer than the inflorescence, glabrous. Mid Jun–Sep. Quiet shallow water; GP mostly n of the Platte R. LAND FORM: Plant sprawling on mud or sand, to 1 m long. Stems and leaves pubescent to hirsute. Leaves short-petiolate to sessile, elliptic, acute, the bases tapering, cuneate, or acute, occasionally narrowly cordate, never truncate, to 10 cm long, 3 cm wide; stipules hirsute, often with flared collar, the margin ciliate. Inflorescences as in the natant form, but apparently appearing less often. Mid Jun–Sep. Sandy or muddy shores; GP mostly n of the Platte R. *Persicaria mesochora* Greene, *P. nebraskensis* Greene, *P. psycrophila* Greene — Rydberg; *Polygonum natans* Eat., *P. hartwrightii* Gray, *P. fluitans* Eat. — Fassett, A Manual of Aquatic Plants. 1940.

11b. var. *emersum* Michx., swamp smartweed. Erect or basally decumbent perennial of marshes and upland habitats, to 1 m tall. Stems and leaves glabrous (especially when rooted in water) to strigose (especially in drier situations), becoming ferruginous with age. Leaves ovate-lanceolate to oblong-lanceolate, usually not reduced above, often undulate-margined, acuminate, the bases acute, truncate, or cordate, to 25 cm long, to 6 cm wide; petioles 1–3(7) cm long; stipules glabrous to strigose, entire, almost truncate, without a flaring collar. Inflorescences 1 or 2 at the branch tips, erect, 2–15 cm long, often tapering, the peduncles glandular-pubescent, stout to rather slender; bracts overlapping, ovate, strigose to ciliate, often long-ciliate on the margin. In shallow water, shoreline marshes, roadside ditches, and common along railroad embankments and in dry upland pastures, where seldom flowering. Jul–Sep. GP; (N. Amer.). *Persicaria coccinea* (Muhl.) Greene, *P. iowensis* Rydb., *P. mesochora* Greene, *P. pratincola* Greene, *P. rigidula* (Sheld.) Greene, *P. vestita* Greene — Rydberg.

This variety has long been known as *Polygonum coccineum* Muhl. Mitchell (op. cit.) has shown experimentally that it is an extreme form of the *P. amphibium* complex and that there is a continuous series of forms between it and var. *stipulaceum* Coleman. This variation is manifested in our area mostly north of the Platte R. while south of the Platte R. only var. *emersum* Michx. is abundant and the var. *stipulaceum*, in either of its forms, is rarely found.

12. *Polygonum bicorne* Raf., pink smartweed. Similar to *P. pensylvanicum* but the perianth always pink, the stamens included or exserted, the styles then exserted or included, respectively (heterostylous); these protruding parts giving the inflorescence a fringed appearance. Achene with a hump in the center of the concavity. Jul–Oct. Disturbed soil in marshes, ditches, & fields; e CO to sw MN & southward; (CO to IL s to LA, CA, & Mex.). *Persicaria longistyla* Small — Rydberg, *Polygonum longistylum* Small — Gleason & Cronquist.

13. Polygonum cespitosum Bl. var. *longisetum* (De Bruyn) Stewart, smartweed. Erect or sprawling, rather delicate annual, rooting at the lower nodes, to ca 5 cm tall. Leaves 2-8 cm long, 1-2 cm wide, ovate-lanceolate to almost rhombic, acute, the base tapering to a short, narrowly winged petiole. Stipules long-ciliate on the margin, the cilia often equaling the stipules. Inflorescences slender, the spiciform racemes in an open panicle; bracts rather crowded, roseate, long-ciliate on the margins, the cilia as long as the bracts and often equaling the flowers in anthesis. Perianth rose to purple, the margins paler; stamens and styles not exserted. Achene included, 3-angular, 1.5-2.5 mm long. Aug-Sep. Disturbed moist ground; established as a weed in Omaha (NE: Douglas) & IA: Mills; undoubtedly elsewhere along our border; (e U.S.; se Asia). *Introduced.*

14. Polygonum hydropiper L., water pepper. Erect or spreading annual (perennial) from fibrous roots, to 1 m tall. Leaves short-petiolate, lanceolate to ovate-lanceolate, acuminate, 3-10(15) cm long, to 2 cm wide, gradually reduced above; stipules tubular to gibbous, truncate to slightly oblique, glabrous or scabrous, the margins slender-bristly. Inflorescences slender, often arching or drooping, the flowers not crowded but mostly adjacent and the inflorescence not markedly interrupted; bracts turbinate, glabrous, entire or slightly ciliate. Perianth greenish with white or pinkish margins, glandular-dotted; 2.5-4 mm long; styles (2)3. Achene included to partially exserted, 3-angular or lenticular. (2n = 20, 22) Aug-Oct. Damp disturbed areas, marshes; GP, mostly e of 100° long.; (most of N. Amer.; Eurasia). *Introduced. Persicaria hydropiper* (L.) Opiz — Rydberg.

This species resembles *P. punctatum,* which is perennial, and whose inflorescence is interrupted.

15. Polygonum hydropiperoides Michx., mild water pepper. Erect perennial to 1 m tall, from a rhizome, the stems often basally decumbent and there rooting at the nodes, glabrous to somewhat pubescent. Leaves short-petiolate to sessile, linear to mostly lanceolate, 3-20 cm long, 0.5-5 cm wide. Stipules pubescent to long-strigose, the hairs partially adnate at the base, the margins ciliate, the cilia 3-5(8) mm long. Inflorescences of robust plants numerous at the ends of the stems, the spiciform racemes on slender to delicate peduncles, in strict panicles, erect to arching, slender, elongate, often interrupted near the base but otherwise the flowers somewhat crowded, the bracts barely or not at all overlapping in anthesis, nearly truncate, pinkish, glabrous, the margin with bristles to 2 mm long but not longer than the bract. Perianth pinkish or greenish, the inner lobes sometimes punctate, 2.5-3 mm long; styles 3, partly connate. Achene included, 3-angular. (2n = 40) Jul-Oct. Rooted in or near the water; SD & southward; (most of N. Amer.). *Persicaria hydropiperoides* (Michx.) Small, *P. setacea* (Baldw.) Small, *P. opelousana* (Ridd.) Small of reports — Rydberg.

Numerous varieties have been recognized based upon vestiture. Ours are closest to var. *hydropiperoides.* Specimens with the stipules long-ciliate have been found here and there in KS and OK; these have been called *Polygonum persicarioides* H.B.K. (*P. hydropiperoides* var. *persicarioides* Stanford).

16. Polgonum lapathifolium L., pale smartweed. Erect annual from a taproot, to 8 dm or more tall. Leaves ovate-lanceolate to elliptic-lanceolate, occasionally subrhombic, the tips acuminate, the bases cuneate or tapering, mostly 5-20 cm long, to 3.5 cm wide, glabrous (tomentose below in aquatic forms), glandular-punctate below; stipules tubular, obliquetruncate, the margin entire or ciliate. Racemes numerous, nodding or drooping, dense, in loose panicles at the ends of the stems, to 7 cm long, less than 1 cm wide, the flowers rather crowded, the peduncles sometimes stipitate-glandular; bracts entire, glabrous. Perianth white, green, or roseate, 2-3(4) mm long, often constricted above the achene, often glandular but not punctate; sepals distinctly 3-nerved, most of the nerves divided at the top and the branch-nerves recurved and anchor-shaped; styles 2, free almost to the base. Achene included, lenticular, flat or biconcave. (2n = 22) Late Jul-Oct. Damp soils in disturbed places, often in ditches, fields, wet meadows; GP; (N. Amer.; Eurasia). *Persicaria lapathifolia* (L.) S.F. Gray, *P. tomentosa* (Schrank) Bickn.—Rydberg.

This is a common and rather variable species that is best distinguished by the anchor-shaped veins in the outer members of the perianth.

17. ***Polygonum orientale*** L., kiss-me-over-the-garden-gate. Erect annual to 2 m or more tall, from a stout rhizome. Stem simple to branched above, strigose to tomentose, becoming reddish, the nodes rather swollen. Lower leaves long-petiolate, the petioles to 3/4 the length of the blade and winged above, the stipules tubular, sheathing, truncate, the largest ca 18 mm long, strigose, the margin bristly-ciliate; blades of the largest leaves ovate, to 20 cm long, 14 cm wide, sparsely to densely tomentose, the tips acuminate to sometimes acute, the bases truncate, rounded, or cordate; upper leaves reduced, short-petiolate to subsessile, ovate, 2–6 cm long, 0.5–5 cm wide, elliptic, or lanceolate, the tips acuminate, the bases acute to truncate. Inflorescences terminal, of pedunculate spiciform racemes grouped in an open panicle; fertile portion erect to nodding, 4–7 cm long in anthesis, elongating in fruit, and then nodding or drooping; bracts crowded, pink or red, ovate, the tips acute, the margins bristly-ciliate; flowers several in the bracts, the pedicels equaling or longer than the perianth. Perianth members 5, pink or red, 4–5 mm long, connate near the base, cucullate. Achene included, compressed-spheroid, both (1) sides concave in the middle, 2–3 mm long and wide. (2n = 22, 24) Aug–Sep. Moist ground, usually near habitations, where escaping from cultivation; IA: Adair; NE: Butler, Douglas; ne KS, w MO; (widespread in northern latitudes; India). *Introduced. Persicaria orientalis* (L.) Spach.—Rydberg.

18. ***Polygonum pensylvanicum*** L., Pennsylvania smartweed. Erect, much-branched annual from a taproot, to 1.5 m tall. Leaves lanceolate, acute to acuminate, the bases acuminate, to 15 cm long and to 4 cm wide, the lower ones with petioles to 3 cm long; stipules tubular, truncate, entire, glabrous to scarcely pubescent. Inflorescences numerous, erect, throughout much of the plant, cylindrical, dense, blunt, rather thick, 0.5–5 cm long, 0.5–1.5 cm thick; bracts sheathing, overlapping, glabrous to sparingly glandular, entire to minutely ciliate; peduncles glabrous or glandular. Perianth 3–5 mm long, rose or pink (white), sometimes glandular, the glands resinous, superficial, this variable on the plant; stamens and styles equal, not exserted from the perianth; a glandular disk at the base of the ovary; styles 2 or 3, partially connate. Achene lenticular (3-angular), one or both sides concave, the concavity without a hump in the middle. Jul–Oct. Disturbed soil in fields & along roadsides, often a troublesome weed; GP; (N. Amer.). *Persicaria pensylvanica* (L.) Small, *P. omissa* Greene—Rydberg.

19. ***Polygonum persicaria*** L., lady's thumb. Erect annual from a taproot, usually branched, to 1 m tall. Leaves short-petiolate, lanceolate to linear-lanceolate, acuminate, often with a prominent purplish spot or sagittate crescent near the middle, this not always obvious on dry specimens, 3–14 cm long, 4–15 mm wide; stipules strigose, the margin short-ciliate. Racemes erect, cylindric, dense, less than 3 cm long, numerous in leafy panicles, the lower nodes of the stem with small groups of flowers in the stipules; bracts overlapping, strigose, nearly truncate, the margins entire or short-bristly. Perianth dusky pink to greenish-pink, sometimes with a few superficial resinous glands, 2–3.5 mm long; styles 2(3); stamens and styles not exserted. Achene included, lenticular or 3-angular, 2–3 mm long. (n = 22, 2n = 44) Jul–Oct. Damp disturbed soil, often a weed in cult. fields; GP; (N. Amer.; Europe) *Introduced. Persicaria maculosa* (Schrank) Bickn.—Rydberg.

20. ***Polygonum punctatum*** Ell., water smartweed. Erect or ascending perennial, the bases of the lower branches often decumbent and rooting at the nodes, rhizomatous-stoloniferous,

usually less than 1 m tall. Leaves lanceolate, elliptic, or becoming subrhombic, acuminate, to 10(15) cm long, to 2 cm wide; stipules obconic, glabrous or scarcely strigose, the margin ciliate. Inflorescences single or few, erect or arching, slender, to 10 cm long, 5 mm wide, interrupted, the flowers not crowded, irregularly spaced, especially below; bracts not overlapping, glabrous, the margins ciliate. Perianth greenish, the lobes whitish, glandular-dotted; styles 2 or 3; stamens and styles not exserted. Achene dark, 3-angular or lenticular. Jul–Oct. Marshes, wet ditches, lakeshores; e ND, w MN, e SD, w IA across NE to WY & CO, s to TX; (Western Hemisphere). *Persicaria punctata* (Ell.) Small—Rydberg.

POLYGONUM—Group III

Vines and scandent or reclining herbs with sagittate to cordate leaves, and erect herbs with basal or cauline leaves with mostly nonsagittate, noncordate bases; leaves without a joint at the base of the blade.

1 Twining or sprawling vines, or scandent or reclining barbed herbs, the leaves sagittate to cordate.
  2 Stems and petioles with numerous retrorse barbs painful to the touch; plant scandent or reclining .... 23. *P. sagittatum*
  2 Stems and petioles glabrous to minutely scabrous; twining or sprawling vines.
    3 Perianth becoming winged on the angles in fruit .... 24. *P. scandens*
    3 Perianth becoming angled but not winged in fruit .... 21. *P. convolvulus*
1 Erect, often robust herbs; leaves not sagittate or cordate.
  4 Leaves mostly basal, linear to oblong; raceme spiciform, thick, the lower few bracts bearing bulblets; BH .... 26. *P. viviparum*
  4 Leaves cauline, at least the lower ones ovate to obovate; racemes spiciform and slender or paniculate.
    5 Racemes spiciform, slender, the flowers sometimes remote .... 25. *P. virginianum*
    5 Racemes in panicles in the axils of at least the upper leaves .... 22. *P. cuspidatum*

21. ***Polygonum convolvulus*** L., climbing or wild buckwheat. Twining or trailing annual vine to 1 m or more long, the stems and leaves often minutely scabrous. Leaves slender-petiolate, the petioles of the lower leaves about equaling the blades, the upper leaves gradually reduced; blades sagittate to cordate, the largest 6 cm long, 5 cm wide, acuminate or acute; stipules small, sheathing, glabrous, oblique, entire. Racemes terminal and axillary in the upper leaves, the racemes loose, most flowers in small clusters, sometimes with a small leaf below the lowest flowers; pedicels shorter than the perianths, jointed above the middle, erect or reflexed. Perianth greenish, sometimes purple-spotted, angular but not winged in fruit, 1–2 mm long, 5-lobed above, the lobes equal; styles 3, united. Achene included, 3-angled, black, dull. (2n = 20, 40) (May)Jun–Sep. Cult. fields, roadsides, waste places, where often a troublesome weed; abundant; GP; (widespread in the Northern Hemisphere; Europe). *Bilderdykia convolvulus* (L.) Dum.—Rydberg; *Tiniaria convolvulus* (L.) Webb. & Moq.—Winter.

22. ***Polygonum cuspidatum*** Sieb. & Zucc., Japanese bamboo, Mexican bamboo. Strong erect perennial to 2 m tall, the stems often rather woody below, the flowering branches becoming horizontal and zigzag. Leaves little reduced above, to 20 cm long, 10 cm wide, ovate, acuminate to attenuate, the bases of the larger leaves truncate; petioles 1–4 cm long, ridged; stipules dark, sheathing, glabrous, not persisting. Inflorescences numerous, the racemes in panicles to 18 cm long in the axils of most or only the upper leaves; bracts sheathing, oblique; pedicels jointed at the middle. Perianth 3–6 mm long, greenish-white, the outer 3 members cucullate and expanding in fruit and becoming winged; styles 3. Achenes 3-angular, included. (2n = 88) Aug–Sep. Roadsides, old gardens, & banks, usu-

ally near habitations; escaping from cultivation here & there in our range; (established throughout temp. N. Amer.; Europe; Asia).

A similar robust species, *P. sachalinense* F. Schmidt, with the leaves cordate, has escaped from cultivation elsewhere and can be expected in our area. *Polygonum cuspidatum* var. *compactum* (Hook. f.) Bailey (*P. compactum* Hook. f.; *P. reynoutria* Hort., not Makino), a horticultural dwarf form, has recently been extensively planted as a ground cover and bank-holder in our area. It is much smaller than the species (to 7 dm tall) and is floriferous throughout; flowers are sterile, the outer perianth members are white and turn red and showy in fruit.

23. *Polygonum sagittatum* L., tear-thumb. Erect, scandent, or reclining annual to 2 m long, the stem 4-angled, the angles with numerous retrorse yellow barbs painful to the touch. Leaves petiolate below, often sessile above, the petioles seldom more than 1/2 the length of the blade; blades narrowly sagittate, 2–12 cm long, to 2.5 cm wide, the lobes short, sometimes incurved-cuspidate, the tips mostly acute, sometimes obtuse, the petioles and midribs with retrorse barbs, the margins scabrous below; stipules sheathing, oblique, entire. Inflorescences terminating the branches, long-pedunculate, erect, capitate, the peduncle with small barbs or none; bracts lanceolate, their margins hyaline. Flowers pedicellate, erect, white or green with pinkish or reddish tinges, the 5 lobes connate below the middle; styles 3, connate below. Achene included, 3-angular, dark. ($2n=40$) Jun–Sep. Marshes, damp meadows; e GP & across NE to w SD; (Sask. to Newf. s to GA & TX). *Tracaulon sagittatum* (L.) Small—Rydberg.

24. *Polygonum scandens* L., false buckwheat. Twining or trailing perennial vine to 5 m long; the stems somewhat angular. Lower leaves long-petiolate, the petioles sometimes equaling the blades; blades mostly cordate, sometimes truncate, acuminate, to 12 cm long, to 8 cm wide, reduced above; stipules tubular, 1–6 mm long. Racemes loose, axillary, 3–20 cm long, with a few small leaves below, these reduced to bracts above; pedicels delicate, jointed at the middle. Flowers divergent or drooping, greenish to whitish, the wings paler, 5–11 mm long, to 8 mm wide in fruit, 5-lobed, 3 of them undulate winged and connate below. Achene included, 3-angular, black or dark brown. ($2n=20, 44$) Jun–Oct. Cult. fields, brushy & waste places; GP; (ND to Que. s to FL & TX). *Bilderdykia scandens* (L.) Greene, *B. dumetorum* (L.) Dum.—Rydberg. *Tiniaria scandens* (L.) Sm.—Winter.

Several varieties have been recognized, based mostly upon perianth and achene sizes. Two of these may be found in our area, although distinctions between them are not always clear.

25a. var. *dumetorum* (L.) Gl. Wings of the perianth in fruit flat, perianth in fruit 7–10 mm long. Introduced from Europe.

25b. var. *scandens.* Wings of the perianth in fruit flat or crisped, perianth in fruit 10–15 mm long. Native.

25. *Polygonum virginianum* L. Erect, rhizomatous perennials to 1.3 m tall, the stem often naked below. Leaves short-petiolate, the petioles 1–2 cm long, becoming subsessile above; blades ovate to obovate, reduced above and there lanceolate, acuminate, the base acute; blades of largest leaves to 16 cm long, 9 cm wide, sparingly tomentose above, somewhat strigose below; stipules strigose to long-tomentose, tubular, sheathing, truncate, the margins ciliate-bristly. Racemes slender, spiciform, terminal and axillary in the upper leaves, the axis strigose, the flowers remote below and sometimes throughout; bracts truncate, glabrous, the margins ciliate; pedicel jointed just below the flower. Perianth greenish (pinkish), ca 3 mm long, the 4 members connate below the middle, the inner members slightly longer than the outer; styles 2, free, exserted. Achene included or the tip exserted, ovoid, shiny brown, the styles persisting and terminally reflexed. ($2n=44$) Aug–Sep. Woodlands & shores; w IA, se NE, e KS, w MO, cen OK; (NE & MN to NH s to FL & TX). *Tovara virginiana* (L.) Adans.—Rydberg.

26. Polygonum viviparum L., bistort. Erect perennial from a thick erect rhizome, the stem simple, to 3 dm tall. Leaves mostly basal, long-petiolate, the longest petioles about equaling the blades, petioles delicate; blades of the basal leaves linear to oblong-lanceolate, 1.5–8 cm long, 5–15 mm wide, obtuse to rounded; cauline leaves narrower, becoming sessile above; blade margins exhibiting raised "stitched" veins in dry specimens; stipules brown, sheathing, 1–5 cm long, oblique, entire. Raceme spiciform, solitary, terminal, rather thick, 2–6 cm long, the lower few bracts remote and bearing bulblets. Flowers functionally imperfect, some without fully developed stamens, the developed stamens exserted; perianth white to pink, the 5 members connate toward the base, the members equal; styles 3, free, long. Achene 3-angled, dark brown. (2n = 88–132) Jun–Aug. Moist alkaline soil; BH; (AK to Newf. s to ME, MN, SD, & in the mts. to NM & OR; Eurasia). *Bistorta vivipara* (L.) S. F. Gray—Rydberg.

5. RHEUM L., Rhubarb

1. Rheum rhabarbarum L., rhubarb. Robust perennial to 1.5 m tall, from thick roots. Leaves very large, basal, the blades to 50 cm long, suborbicular to broadly ovate, the petioles flat above; stipules sheathing. Panicle erect, the branches strict, ebracteate, the flowers in fascicles at the nodes, the pedicels with an obscure joint near the middle. Perianth white, the 6 members not enlarging in fruit; stamens (6)9, styles 3. Pedicels elongating and reflexing in fruit. Achene completely exserted, brown, prominently 3-winged, ca 1 cm long, 0.75 cm wide. (2n = 44) Jun–Jul. Old gardens, abandoned farms; cult. & persisting but apparently not spreading; occasional in GP, especially in n 1/2; (widely cultivated in U.S. & Can., especially in e & n; e Asia). *R. rhaponticum* L.—Gleason & Cronquist, misapplied.

6. RUMEX L., Dock, Sorrel

Annual and perennial erect leafy herbs from fibrous roots, taproots, or rhizomes, mostly glabrous, often reddish-tinged. Plants monoecious, dioecious, or the flowers perfect. Leaves basal and/or cauline, alternate, mostly petiolate, with prominent stipular membranes sheathing the stem and usually disintegrating by fruiting time; blades flat or crisped, entire to very irregularly and shallowly toothed, the apices acute to obtuse or rounded, the bases acute, truncate, or cordate. Inflorescences terminal, of verticillate racemes in a panicle, conspicuous, greenish to pinkish in anthesis, pinkish or golden to mostly reddish-brown and open to mostly dense in fruit, the branches inclined or ascending, the bracts foliaceous to small or absent; flowers numerous in verticels, the pedicels slender, once-jointed. Flowers perfect or imperfect; perianth parts (sepals) 6 in two whorls of 3, basally connate, more or less alike in anthesis, persistent, the outer whorl remaining small in fruit, the inner whorl usually greatly enlarging and then called valves, the valves membranaceous, reticulate, accrescent, often one or more of them developing a single median tubercle, entire or with marginal teeth, reddish or pinkish; stamens 6; pistil 1, superior, styles and stigmas 3, ovary 3-angled, unilocular, the single ovule basal and orthotropous. Fruit a 3-angled achene enveloped by the valves.

References: Hitchcock, C. L. 1964. *Rumex, in* Hitchcock, C. L., et al., Vascular Plants of the Pacific Northwest, Part 2, pp. 168–182. Univ. Wash. Press, Seattle; Mitchell, R. S., & J. K. Dean. 1978. Contributions to a Flora of New York State. I. Polygonaceae (Buckwheat Family) of New York State. New York State Mus. Bull. 431, 81 pp.; Rechinger, K. H., Jr. 1937. The North American species of *Rumex*. Field Mus. Nat. Hist. Bot. Ser. 17: 1–151.

This is a genus mostly of the temperate regions. Many of our species are introduced and some are troublesome weeds. Hybrids are probably frequent and, thus, positive identification to the species given here is not always possible. For most of our species accurate identification requires mature fruits.

The shoots of some species are used as potherbs or in folk-medicine, but their astringent properties dictate careful use.

1. At least some of the basal leaves hastate; valves without tubercles.
    2. Valves expanded in fruit and exceeding the achene, 2–3 mm long; joint of the pedicel below the middle .... 5. *R. hastatulus*
    2. Valves not expanded in fruit, equaling the achene, ca 1 mm long; pedicel jointed just below the flower .... 1. *R. acetosella*
1. Leaves never hastate; valves with or without tubercles.
    3. Valves large, 2–5 cm long, without tubercles .... 14. *R. venosus*
    3. Valves smaller, 2–10 mm long, with or without tubercles.
        4. Pedicels several times longer than the valves and jointed very near the base and drooping in fruit, the fruiting verticels distinct .... 15. *R. verticillatus*
        4. Pedicels equaling or longer than the valves, but not several times longer, the joint not very near the base; inflorescences mostly dense in fruit.
            5. Valves with a few coarse or spinulose teeth on the margins.
                6. Valve teeth spinulose; leaves cauline .... 7. *R. maritimus*
                6. Valves coarse-toothed; leaves basal and cauline.
                    7. All valves with a tubercle; leaf bases not cordate .... 13. *R. stenophyllus*
                    7. One (2) valves with a tubercle; bases of the largest leaves cordate .... 9. *R. obtusifolius*
            5. Valves entire or merely fine toothed or crenulate.
                8. Valves without tubercles.
                    9. Leaves flat .... 2. *R. altissimus*
                    9. Leaves crisped.
                        10. Outer sepals reflexed in fruit .... 6. *R. hymenosepalus*
                        10. Outer sepals not reflexed in fruit.
                            11. Valves reniform to rotund; pedicel joint distinct, below the middle .... 4. *R. domesticus*
                            11. Valves deltoid to ovate; pedicel joint at about the middle, obscure .... 10. *R. occidentalis*
                8. One, 2, or 3 valves with a tubercle.
                    12. Outer sepals reflexed in fruit; only 1 valve with a tubercle .... 12. *R. patientia*
                    12. Outer sepals not reflexed in fruit; 1, 2, or 3 valves with a tubercle.
                        13. Leaves distinctly crisped; largest tubercle 1/2 or less the length of the valve .... 3. *R. crispus*
                        13. Leaves flat or undulate but not crisped; largest tubercle 1/2 or more the length of the valve.
                            14. Tubercle base above the valve base; tubercle about 1/2 the length of the valve; valves orbicular .... 11. *R. orbiculatus*
                            14. Tubercle base even with the valve base; largest tubercle greater than 1/2 the length of the valve.
                                15. Valves 4–5 mm long, the tubercles 1–3, often unequal; leaf blades not narrow and willowlike .... 2. *R. altissimus*
                                15. Valves ca 2 mm long, the tubercles usually 3, equal; leaf blades relatively narrow, willowlike .... 8. *R. mexicanus*

1. **Rumex acetosella** L., sheep sorrel. Dioecious or polygamous, slender, colonial perennial from long, slender rhizomes. Stems 10–30 or more cm tall, mostly simple below the inflorescence. Leaves variable on the plant, rather fleshy, the basal ones mostly long-petiolate, the blades hastate or sagittate, the terminal lobe much longer than the basal lobes, sometimes the cauline and some basal leaves with 1 lobe or unlobed; the largest leaves to 12 cm long, the blades to 5 cm long but usually smaller; cauline leaves few, smaller. Inflorescence 5–15 cm long, variously yellowish to reddish, the branches slender, ascending, the bracts small or absent; fruiting verticels not crowded; pedicels slender, shorter than or equal to the valves, the joint just below the flower. Flowers imperfect; outer sepals not reflexed in fruit;

valves ca 1 mm long, barely exceeding the achene, entire, without tubercles or other ornamentation. Achene 1–1.5 mm long, yellowish or brownish red. (n = 21; 2n = 14, 28, 42) Apr–Aug. Waste places & open woodlands; mostly in sandy acid soil; throughout GP, more abundant n & e; (N. Amer., Europe). *Introduced.*

2. Rumex altissimus Wood., pale dock. Perennial erect herb to 1 m tall, the stem ribbed and often branched below the inflorescence. Basal leaves long-petiolate, the blades flat, lanceolate to nearly ovate, the largest to 20 cm long, 5 cm wide, the margins entire to irregularly crenulate, acute to acuminate, the base acute to rounded or subcordate; cauline leaves numerous, smaller. Inflorescence to 30 cm long, the branches ascending, spiciform and dense in fruit; pedicel joint below the middle. Flowers imperfect; outer sepals not reflexed in fruit; valves greenish to reddish-brown, broad, 4–5(6) mm long and wide, the tip obtuse, the base truncate, the margin entire, without tubercles or 1 or more valves with a more or less ovoid tubercle, the largest more than 1/2 as long as the valve. Achene dark brown. (2n = 20) Apr–Jul. Open wet places, sometimes in shallow water; throughout GP except MT & ND; (SD, MN, to NH s to GA, TX, & AZ).

3. Rumex crispus L., curly dock. Coarse erect perennial, to 1 m or more tall. Stem simple below the inflorescence. Leaves obviously and irregularly crisped, the basal ones long-petiolate, to 30 cm long and 5 cm wide, the blade oblong-lanceolate and acute, rounded or subcordate at the base; the cauline leaves similar but smaller. Inflorescence large, to 30 cm or more long, the branches ascending, dense in fruit, a few foliaceous bracts below; pedicels slender and flexuous, longer than the valves in fruit, the joint below the middle. Flowers perfect; outer sepals not reflexed in fruit; valves 4–5 mm long, deltoid to ovate, obtuse, the base truncate, entire or denticulate, mostly each (1) with a small tubercle 1/2 or less than 1/2 the length of the valve. Achene reddish-brown. (n = 30; 2n = 60) Apr–Jul. Disturbed areas, in the open; throughout GP; (naturalized in much of the world; Eurasia). *Introduced.*

Hybrids with *R. obtusifolius, R. patientia,* and *R. stenophyllus* are known in and beyond our range.

4. Rumex domesticus Hartm., yard dock. Perennial herb to 1 m or more tall, the stem slender. Leaves flat to mostly undulate or crisped; basal leaves long-petiolate, to 7 dm long, the blade narrowly oblong, widest near the middle, acute to subobtuse, the base acute; cauline leaves smaller, their bases cuneate to almost truncate. Inflorescence few branched, the branches ascending, rather dense in fruit; the bracts foliaceous and gradually reduced upward; pedicels longer than the valves, slender, the joint below the middle. Outer sepals not reflexed in fruit; valves 4–6 mm long, somewhat wider, reniform to rotund, sometimes cordate, the margins entire or nearly so, without tubercles. Achene dark brown. (2n = 40, 60, 80) Jun–Aug. Disturbed places in the open; ND, MN; (introduced & naturalized in the cooler parts of the Northern Hemisphere; Eurasia). *Introduced.*

5. Rumex hastatulus Baldw., heartwing sorrel. Erect perennial from a taproot, rather slender, to 1 m tall but usually much less, branched or unbranched below the inflorescence. Basal leaves crowded, long-petiolate, to 10 cm long, the blade linear to oblanceolate to mostly hastate, the terminal lobe linear to oblanceolate and much longer than the basal lobes; basal lobes 1 or 2, often unequal, divergent; base of blade tapering; cauline leaves similar but shorter-petiolate or narrowly lanceolate and sessile. Inflorescence 10–30 cm long, narrow, with a few small bracts, the branches inclined or ascending, not dense in fruit; pedicels delicate, about equaling the valves in fruit, the joint below the middle obscure. Flowers imperfect; outer sepals reflexed in fruit; valves 2–3 mm long and about as wide, orbicular, without tubercles, the base cordate, yellow or pinkish. Achene brownish, 1 mm long. (2n = 9, 2n = 12 + XX:female, 2n = 12 + XYY:male) Apr–Jul. Sandy soil, in the open;

se KS, w MO, cen OK (on the Coastal Plain from MA to FL & TX, n in the Mississippi valley to OK, KS, & IL).

6. **Rumex hymenosepalus** Torr., wild rhubarb. Stout perennial to 1 m tall, from clustered, tuberous roots, the stem simple below the inflorescence. Basal leaves to 35 cm long, 10 cm wide, the petioles shorter than the blades; blade fleshy, crisped, oblong to narrowly obovate, the midrib thick at the base and strongly tapering to the tip; cauline leaves smaller, lanceolate. Inflorescence to 30 cm long, the branches ascending, the bracts small; pedicels about equal to the valves, jointed near or below the middle. Outer sepals reflexed in fruit; valves 8–14(17) mm long, 7–8(12) mm wide, ovate, entire, without tubercles, the base cordate, pinkish and showy. Achene 4–6(9) mm long, brown. (2n = 40, 100) Mar–May. Dry sandy places in the open; ne NM, TX panhandle, sw OK; (CA to CO s to OK, TX, & Mex.).

7. **Rumex maritimus** L., golden dock. Erect or basally decumbent annual to 65 cm tall, the stems hollow and sometimes rooting at the nodes of the decumbent base, often shorthairy. Leaves cauline, rather thick, short-petiolate, to 15 cm long, smaller above and below, the blade linear-oblong to lanceolate, the apex acute to obtuse, the base acute to truncate, the margins undulate to irregularly crisped, entire to unevenly crenulate. Inflorescence leafy-bracteate, large, many branched, the verticels crowded above in fruit and golden or reddish at maturity; pedicels longer than the valves, the joint below the middle. Flowers perfect; outer sepals often somewhat reflexed in fruit; valves 2–3 mm long, greenish to golden, ovate to long-triangular, with 4–6 long bristlelike teeth on the margins, a prominent tubercle on each valve, the tubercle light-colored, fusiform to ellipsoid and more than 1/2 as long as the valve. Achene golden, sharp-angular. (n = 20; 2 = 40) Jun–Aug. Sandy shores, often near brackish water & sometimes in the water; throughout GP except OK & TX; (AK to Que. s to NC, OH, AR, KS, NM, & CA(?); S. Amer., Eurasia). *R. persicarioides* L. misapplied—Rydberg.

Our material is sometimes recognized as *R. maritimus* var. *fueginus* (Phil.) Dusen.

8. **Rumex mexicanus** Meisn., willow-leaved dock. Perennial to 1 m or more tall, the stem sometimes much branched. Leaves cauline, thin, flat or undulate, sometimes glaucous, the blades to 15 cm long and to 4 cm wide, the petioles of lower leaves long, those of the upper leaves much shorter than the blades; blades below ovate-lanceolate, blades above lanceolate, apex acute to acuminate or obtuse, the base cuneate to acuminate, the margins entire. Inflorescence small or large, the branches ascending, leafy-bracteate below, dense in fruit; pedicels longer than the valves, the joint near the base obscure. Flowers imperfect; outer sepals not reflexed in fruit; valves ca 3 mm long, deltoid, the margins entire or nearly so, the bases truncate to subcordate, the valves mostly all with an elongate tubercle, the tubercles equal, acute-tipped, plump, more than 1/2 as long as the valve. Achene dark brown, sharply 3-angular. (2n = 20) Jun–Aug. Moist, often brackish and sandy soils, in the open; GP except TX, OK; (B.C. to Que. s to NY, MO, & Mex.).

This species is a member of a highly variable and probably often-hybridizing complex that produces variations in tubercle and valve morphology.

9. **Rumex obtusifolius** L., bitter dock. Robust perennial from a taproot, to 1 m or more tall, the stem simple below the inflorescence, reddish, ribbed. Basal leaves large, longpetiolate, to 30 cm long and 15 cm wide, the blade oblong to ovate-oblong, obtuse to rounded, the base distinctly cordate (truncate), the margin usually slightly crisped; cauline leaves similar but smaller, shorter-petiolate, and sometimes acute. Inflorescence branches ascending, subtended by foliaceous bracts, reduced upward, rather dense in fruit; pedicels longer than the valves in fruit, the joint below the middle. Flowers perfect; outer sepals not reflexed in fruit; valves 3–5 mm long, triangular-ovate, the margins with 4–8 promi-

nent teeth, 1(2 or 3) valve with a prominent ovoid tubercle, the largest more than 1/2 as long as the valve. Achene golden to brown, the angles acute. (2n = 20, 40, 50) May–Jul. Marshes & damp waste places in the open; NE s to TX (native to Europe, *naturalized* AK to Que. s to FL, AZ; Mex., S. Amer.; Eurasia; Africa). *R. brittanica* L., misapplied — Rydberg.

10. ***Rumex occidentalis*** S. Wats., western dock. Stout perennial 0.5–1.5 m tall or more from a taproot, glabrous to sparsely hairy. Stem simple below the inflorescence, reddish. Basal leaves long-petiolate, the blade 10–25 cm long, oblong-ovate to oblong-lanceolate, the apex acute to obtuse, the base cordate, the margins somewhat crisped; cauline leaves few, wide-spaced, long-petiolate, the blades narrow. Inflorescence narrow, the branches erect, with foliaceous bracts below; pedicels slender, longer than the valves, the joint near the middle obscure. Outer sepals not reflexed in fruit; valves deltoid to reniform, cordate or truncate at the base, entire or nearly so, without tubercles. Achene 3–4 mm long. (2n = ca 140–160) May–Jul. Infrequent in wet places in the open; e MT, ND, w MN, SD, WY; (AK to Que. s to SD, NM, & CA).

11. ***Rumex orbiculatus*** A. Gray, great water dock. Stout erect perennial to 2 m tall, the stem simple below the inflorescence. Basal leaves to 40 cm long, petiolate, the blade flat, irregularly undulate or crisped, oblong-lanceolate, the base rounded; cauline leaves similar but reduced above. Inflorescence branches erect, subtended by foliaceous bracts; pedicels slender, long, the joint below the middle obscure. Flowers imperfect; outer sepals not reflexed in fruit; valves orbicular, to 7 mm long and wide, the base truncate, the margin mostly entire, each valve with a narrow tubercle about 1/2 as long as the valve, the tubercle base above the valve base. Achene brown. (2n = 160) May–Jul. Wet meadows & shallow water; ND, w MN, SD, IA, NE; (ND to Newf. s to NJ, OH, & NE).

12. ***Rumex patientia*** L., patience dock. Coarse erect perennial to 2 m or more, from a taproot, the stem simple below the inflorescence. Basal leaves long-petiolate, the blade flat or undulate, ovate to oblong-lanceolate, to 30 cm long and about 1/2 as wide, the apex acute, the base acute to subcordate or truncate; cauline leaves gradually reduced above, short-petiolate. Inflorescence stout, dense, the branches ascending, the lower ones subtended by foliaceous bracts; pedicels slender, about equaling the valves, the joint below the middle prominent. Flowers perfect; outer sepals reflexed in fruit; valves 4–9 or more mm long and wide, ovate to rounded, the base cordate, the margin entire or denticulate, without tubercles or one valve with a small tubercle less than 1/2 the length of the valve. Achene brown. (n = 30; 2n = 60) May–Jun. GP; throughout much of northern N. Amer.; s Europe & Asia). *Naturalized.*

Hybrids with *R. crispus* have been found in our area.
A similar entity, *R. cristatus* DC., adventive from Europe, has been collected in KS: Leavenworth, and can be expected elsewhere in our area. It differs in having veins in the middle of the leaf at an angle of 45–60° with the midrib (60–90° in *R. patientia*) and the perianth segments with frequent irregular, acute teeth to 1 mm long (segments entire to denticulate in *R. patientia*).

13. ***Rumex stenophyllus*** Ledeb. Erect perennial from a taproot, to 7 dm or more tall, simple or few branched below the inflorescence. Basal leaves long-petiolate, the petioles above equaling the blades; blade flat or undulate, to 25 cm long, acute to acuminate; cauline leaves somewhat reduced. Inflorescence dense in fruit, the branches ascending,

the bracts foliaceous; pedicel jointed below the middle. Outer sepals not reflexed in fruit; valves ca 5 mm long and wide, triangular with cordate or truncate bases, the margins coarse-toothed, each valve with an ovoid tubercle about 1/2 as long as the valve. Achene light reddish-brown, 2 mm long. ($2n = 60$) May–Jul. Damp meadows & ditches, & along streams; throughout GP; (Man., ND, & MN s; Eurasia). *Introduced. R. alluvius* Gates & R. McGreg.—Gates and McGregor, Trans. Kansas Acad. Sci. 53: 186. 1950.

14. *Rumex venosus* Pursh, wild begonia. Perennial from a wide-spreading rhizome. Flowering stems erect, often branching and reddish tinged. Leaves cauline, the petioles shorter than the blades; blades flat, leathery, variable, the lowest ones reduced, the upper ones ovate-lanceolate to oblong-elliptic, the apex and base acute. Inflorescences small in flower, becoming showy in fruit, unbranched or with a few branches, the flowers not especially crowded; pedicels in fruit slender, to 2 cm long, the joint near the middle. Flowers perfect; outer sepals reflexed in fruit; valves very large, 2–4.5 cm wide, not quite as long, orbicular, the apex retuse to obcordate, the base cordate, and the lobes often overlapping, the margin entire, without tubercles, reddish and showy. Achene 5–7 mm long, light brown. ($2n = 40$) Apr–Jul. Sandy dunes & sandy riverbanks; GP w of the Red R. & Missouri R., more common westward; (s B.C. to Sask. s to NM & n CA).

15. *Rumex verticillatus* L., water dock. Erect or basally decumbent perennial to 1 m or more tall, sometimes with a few short branches below the inflorescence; stem often purplish. Basal leaves short-petiolate, the blades to 30 cm long and 4 cm wide, flat, linear to broadly lanceolate or oblong-ovate, entire, the cauline leaves reduced above. Inflorescence rather open, the branches ascending and each subtended by a foliaceous bract, the verticels remaining rather distinct in fruit; pedicels longer than the valves and conspicuously drooping, jointed very near to the base. Flowers imperfect; outer sepals not reflexed in fruit; valves 3–5 mm long, deltoid-ovate, the apex rounded, the base truncate, the margins entire, each valve with a narrow tapering tubercle about 2/3 the length of the valve. Achene dark brown, sharply 3-angular. ($2n = 24, 48, 60$) May–Jul. Wet meadows & watersides; cen & e NE, w IA, w MO, e KS; (NE to Que. s to FL & e TX).

44. PLUMBAGINACEAE Juss., the Leadwort Family

by Robert B. Kaul

LIMONIUM P. Mill., Marsh Rosemary

1. *Limonium limbatum* Small. Perennial herb from woody rootstocks, to 6 dm tall. Leaves basal, entire, simple, bluish-green, long petioled, to 3 dm long, the blade thick, leathery, to 6.5 cm wide, obovate to elliptic, rounded to mucronulate, tapering below. Panicle large, the scape naked, the flowers distichously paired and the pairs grouped in spikes on the ultimate branches. Flowers perfect, regular, hypogynous; sepals 5, fused, the lobes flaring, ca 4 mm long, the lobes acute to obtuse; petals 5, clawed, bright blue; stamens 5; pistil 1, uniovulate, styles 5. Fruit indehiscent, included in the persistent calyx. Jun–Sep. Saline flats; TX panhandle, w OK; (w OK & TX panhandle sw to trans-Pecos).

45. ELATINACEAE Dum., the Waterwort Family

by Robert B. Kaul

Plants small, herbaceous, creeping or ascending annuals of marshes and other wet places. Leaves cauline, opposite, entire to crenulate, smooth or glandular-pubescent, stipulate. Flowers axillary, hypogynous, the perianth parts free; stamens 1 or 2 times as many as the petals; pistil 1, of 2-4 fused carpels; placentation axile. Fruits capsular; seeds few to many, pitted.

1 Plants erect to ascending, glandular-pubescent throughout; flower parts 5-merous .. 1. *Bergia*
1 Plants creeping, or some branches ascending, glabrous; flower parts mostly 3-merous .... 2. *Elatine*

1. BERGIA L.

1. ***Bergia texana*** (Hook.) Walp., Texas bergia. Plant annual, from a taproot, branching from the base, the stems ascending and often reddish, 1-2.5(4) dm tall, spreading to 3 dm, glandular-pubescent. Leaves elliptic to oblong, acute, tapering almost to the base of the petiole, serrulate, glandular-pubescent, the pairs sometimes crowded above and there apparently whorled, to 3 cm long and 1.5 cm wide, the stipules serrate, scarious, ca 1 mm long. Flowers solitary in the axils, short pedicelled; sepals 5, acuminate, 3-4 mm long, the midrib green, the margins white, scarious; petals 5, white, oblong, not exceeding the sepals; stamens 5 or 10. Fruit globose or nearly so, to 3 mm wide. (n = 6) Jul-Sep. Swamps & other wet places, dry playa lakes on sandy soil; s SD, ne NE, KS, w MO, OK, TX; (s IL to WA s to s CA & s TX).

2. ELATINE L., Waterwort

1. ***Elatine triandra*** Schkuhr. Plant creeping, the branches sometimes erect, to ca 1 dm tall and spreading to ca 2 dm. Stems rather succulent, rooting at the nodes, glabrous. Leaves rather succulent, glabrous, linear to spatulate, entire, the tips truncate or usually emarginate, 3-6(12) mm long, 1.5-3 mm wide, punctate above, papillate below. Flowers solitary in the axils, sessile, 1.5-2 mm wide, 3-merous. Capsules globose, ca 3 mm wide; seeds numerous. Jun-Aug. Muddy shores & in shallow water; occasional throughout GP; (WI to WA s to CA, TX, Mex.). *E. americana* (Pursh) Arn.—Rydberg.

E. brachysperma A. Gray, more diminutive, the leaves not emarginate, is known in our area from OK: Comanche.

46. CLUSIACEAE Lindl., the St. John's-Wort Family

by Robert B. Kaul

Annual and perennial glabrous herbs and shrubs, the stems simple or branched, the bark sometimes exfoliating. Leaves opposite, sessile, tapering to clasping at the base, pellucid- or black-punctate. Flowers solitary or cymose, hypogynous. Sepals 2, 4, or 5, equal or unequal, free, persistent; petals 4 or 5, free, yellow or pink. Stamens few to numerous, sometimes grouped in 3–5 fascicles; pistil 1, 1- to 5-locular, the placentation parietal or axile; styles 2–5, free or basally united. Fruits capsular, dehiscence septicidal; seeds few–numerous. Hypericaceae — Rydberg.

1 Sepals 2 or 4, petals 4; styles 2; sprawling heathlike shrubs .... 1. *Ascyrum*
1 Sepals and petals 5 each; styles 3–5; plants erect.
   2 Petals yellow, convoluted in the bud; stamens free or fascicled, numerous .. 2. *Hypericum*
   2 Petals flesh-pink to purplish to greenish, imbricated in the bud; stamens ca 9, fascicled .... 3. *Triadenum*

1. ASCYRUM L., St. Andrew's Cross

1. *Ascyrum hypericoides* L. var. *multicaule* (Michx.) Fern. Slender, more or less straggling much-branched heathlike glabrous shrub from a thickish rootstock, to 3 dm tall, the reddish bark exfoliating. Leaves opposite (but apparently whorled where short axillary branches are present), evergreen, punctate above and below, sessile, entire, linear-oblanceolate, to 2 cm long and 3–8 mm wide, glaucescent beneath, obtuse or rounded, the base tapering. Flowers scattered in axils of the upper leaves, pedicelled, subtended by 2 narrow bracteoles, ca 1–2 cm broad; sepals 2 or 4, the outer 2 ovate and sometimes subcordate, ca 7 mm long and 4 mm wide, obtuse, the inner 2 much smaller or obsolete; petals 4, pale yellow, narrowly elliptic, barely exceeding the outer sepals; stamens numerous, free; styles 2. Capsule ovoid. (n = 9) Jul–Aug. Dry, acid, sandy & rocky soil; extreme se KS, sw MO, e OK; (MA to se KS s to e TX & GA).

2. HYPERICUM L., St. John's-wort

Ours erect glabrous herbs and subshrubs, annual or perennial. Leaves opposite, entire, pellucid- or black-punctate, sessile. Flowers solitary or cymose. Sepals 5, equal, persistent; petals 5, yellow, convoluted in the bud; stamens few (less than 20) to numerous (more than 20), free or basally united into fascicles; pistil 1, compound, of 3–5 carpels, unilocular and the placentae parietal or protruding or 3- to 5-locular and the placentation axile; ovules few to many; styles 3–5(6), free or connate below, persistent (or at least persisting as a single beak), the stigmas small. Capsules globose, ovoid, or conical, the seeds numerous, cylindrical, with a sticky caruncle.

There are eight species in our area; several others are in cultivation here for their showy flowers, and the semievergreen shrub *H. kalmianum* L., with very numerous free stamens and 5 styles, is increasingly used as a landscape subject in the se GP. *Sarothra* L.—Rydberg.

1 Petals (at least their margins) and sometimes the leaves and stems conspicuously black-punctate; perennials with leafy-bracteate inflorescences; styles 3.
   2 Petal margins black-punctate; stamens united below into 3 groups .... 4. *H. perforatum*
   2 Petals, undersides of the leaves, and often the stems black-punctate; stamens free .... 5. *H. punctatum*
1 Plants without black punctae (pale punctae sometimes present); annuals and perennials; inflorescences leafy-bracteate or not, or the flowers solitary; styles 3 or 5.
   3 Flowers solitary in the upper leaf axils; stamens fascicled or free.
      4 Flowers to 5 cm wide; styles 5; stamens in 5 fascicles .... 6. *H. pyramidatum*
      4 Flowers to 1.2 cm wide; styles 3; stamens free .... 1. *H. drummondii*
   3 Flowers in cymes; stamens free; styles 3.
      5 Leaves tapering to the base; capsules globose or nearly so; stamens numerous .... 7. *H. sphaerocarpum*
      5 Leaves rounded at the base to clasping, capsules ovoid to ellipsoid; stamens few to numerous.

6 Flowers very small (3–4 mm wide); cymes many flowered; leaves often ovate .. 3. *H. mutilum*
6 Flowers ca 1 cm wide; cymes few- to several-flowered; leaves lanceolate .... 2. *H. majus*

1. **Hypericum drummondii** (Grev. & Hook.) T. & G., nits-and-lice. Erect annual to 35(80) cm tall; stems slender, simple or mostly branched above, the branches ascending, the bark below reddish, exfoliating. Leaves linear to oblanceolate, acute, erect or ascending, to 1.5 cm long and 2 mm wide, not much reduced above. Flowers solitary in the axils of the upper leaves, not obviously grouped in cymes. Sepals and often the petals persistent in fruit; sepals to 6 mm long, exceeding the petals; stamens to 12, free; styles 3. Capsule ovoid, 3–5 mm long, exceeding the persistent calyx. Jul–Sep. Open oak woods, forest edges, prairies, & weedy pastures; se KS, sw MO, s-cen, cen, & ne OK; (MD to se KS & e TX, e to FL). *Sarothra drummondii* Grev. & Hook.—Rydberg.

2. **Hypericum majus** (A. Gray) Britt., greater St. John's-wort. Erect rhizomatous perennial to 5(7) dm tall; stems simple or branched above. Leaves ascending, linear to lanceolate, acute to obtuse, glaucous below, the largest ca 4 cm long and 1 cm wide, gradually reduced below. Cymes not leafy, collectively rather flat-topped. Flowers to 1 cm wide; sepals equal, lanceolate, to ca 6 mm long; stamens to 20, free. Capsule ovoid, purplish, 3–7 mm long, included or barely exceeding the sepals. (2n = 16) Aug–Sep. Wet sandy meadows & shores; e ND & w MN s to cen KS, & w to w NE; (N.B. to B.C. s to WA, CO, KS, & PA).

This species is reported to hybridize with no. 3. A similar species, *H. canadense* L., was reported erroneously in GP (Atlas GP).

3. **Hypericum mutilum** L., dwarf St. John's-wort. Erect annual-perennial to 3 dm tall, the stem few branched above, 2- to 4-angled, the ultimate branches slender, inclined, the bark smooth. Leaves not strongly inclined or erect, more or less glaucous below, to 3 cm long and 1 cm wide, clasping, the larger ones elliptic to ovate and obtuse to rounded, rounded below, the smaller ones lanceolate and acute to obtuse. Cymes open, rather leafy, in the axils of the upper leaves. Flowers 3–4(5) mm wide; sepals linear-lanceolate; stamens 12, free. Capsules ovoid or ellipsoid, 2–4 mm long, shiny, included or exserted. (n = 8) Jul–Aug. Creek banks, lakeshores, wet soil of open woods; e KS, w MO, ne, cen & sw OK; (Newf. to Man. s to e TX & FL).

This species is reported to hybridize with no. 2.

4. **Hypericum perforatum** L., common St. John's-wort. Erect rhizomatous perennial to 7(15) dm tall, the stems branched above and often at the base. Leaves elliptic to linear, more or less acute, pellucid-perforate, reduced above, mostly 1–1.5 cm long, 2–5 mm wide. Flowers numerous in leafy-bracteate cymes at the branch tips, the cymes collectively rounded to somewhat flat-topped. Flowers to 2 cm wide; sepals to 6 mm long; petals 8–12 mm long, with marginal black dots; stamens numerous, in 3 fascicles. Capsule exceeding the sepals. (2n = 32) Jun–Aug. Roadsides, pastures, & prairies, often on sandy soil; occasional in GP; (N.B. to B.C. s to CA, TX, & NC; Europe). *Introduced.*

This introduced species has become a troublesome weed in the west, where it is known as Klamath-weed. It is less of a problem in GP.

5. **Hypericum punctatum** Lam., spotted St. John's-wort. Erect perennial to 8(10) dm tall, from a woody rhizome; stem usually simple below the inflorescence, smooth; stem, undersides of leaves, and perianth conspicuously black-punctate. Leaves elliptic to oblanceolate, acute to rounded, 2–4 cm long, 4–15 mm wide, those above reduced. Cymes terminal, leafy-bracteate, crowded, compact, collectively rounded to flat-topped. Flowers to 1.5 cm wide; petals 6–7 mm long; stamens numerous, free. Capsule ovoid, 6–8 mm long, usually much exceeding the calyx. (n = 8) Jun–Aug. Thickets, open woods, damp fields; sw IA

& se NE, e KS, MO, cen & ne OK; (Que. to se MN sw to e TX, e to FL). *H. punctatum* Lam., *H. subpetiolatum* Bickn., *H. pseudomaculatum* Bush — Rydberg.

6. *Hypericum pyramidatum* Ait., great St. John's-wort. Erect perennial herb to 15 dm tall, the base becoming suffrutescent and the bark there exfoliating, the stem few branched above. Leaves clasping at the base, 3–8 cm long, 1–2 cm wide, not reduced above, lanceolate to elliptic, acute to obtuse. Flowers few, solitary in the upper leaf axils, showy, 3–5 cm wide; sepals ca 1 cm long, 3–5 mm wide, acute; petals 1.5–2 cm long, 1–1.5 cm wide, rounded; stamens numerous, in 5 fascicles; styles 5, united below. Capsule ovoid, 1–3 cm long, apparently 5-locular; seeds with a broad longitudinal wing. (n = 9) Jun–Aug. Wet meadows & rich woods; sw IA, se NE, ne KS, nw MO; (Que. to Man. s to ne KS & MD). *H. ascyron* L. — Rydberg.

7. *Hypericum sphaerocarpum* Michx., roundfruit St. John's-wort. Erect perennial from a slender woody rhizome; stem slender, suffrutescent, simple or few branched above, to 6 dm tall, the branches ascending, the bark exfoliating below. Leaves linear-oblong to narrowly ovate, acute to obtuse, to 4(7) cm long and 1(1.5) cm wide. Cymes rather crowded at the top of the plant, not leafy. Sepals ovate, to 5 mm long; petals 5–9 mm long; stamens numerous, free. Capsule globose or nearly so, to 5 mm in diam and much exceeding the persistent sepals. (n = 9) Jun–Sep. Dry rocky soil, rocky woodlands, roadsides; se NE, sw IA, e KS, w MO, ne OK; (OH to se NE, s to e TX to AL). *H. cistifolium* Lam. — Rydberg.

3. TRIADENUM Raf., Marsh St. John's-wort

1. *Triadenum fraseri* (Spach) Gl. Erect perennial herb to 6 dm tall, from creeping rhizomes; stem simple up to the floriferous portion. Leaves sessile to clasping, ovate to oblong, to 6 cm long and 3 cm wide, rounded to emarginate, glaucous and black-punctate beneath. Flowers pedicelled, axillary in the upper leaves and terminal, 1–2 cm wide. Sepals 5, ovate-lanceolate, acute; petals 5, 5–10 mm long, imbricated in the bud, exceeding the sepals, pink to greenish to purplish; stamens 9(12), united below into 3 fascicles, these alternate with orange glands; styles 3, free. Capsule ovoid to oblong, to 12 mm long. (n = 18; 2n = 38) Jul–Aug. Marshes & shores; nw MN, n-cen NE: sandhills; (Newf. to MN, w to n-cen NE, e to NY). *T. virginicum* (L.) Raf. of NE reports — Rydberg; *Hypericum virginicum* L. var. *fraseri* (Spach) Fern. — Fernald.

47. TILIACEAE Juss., the Linden Family

by Robert B. Kaul

1. TILIA L., Basswood, Linden

1. *Tilia americana* L. Tree to 40 m tall, twigs reddish. Leaves alternate, simple, petiolate, the blade ovate to nearly orbicular, to 20 cm long, acute to abruptly acuminate, the base more or less obliquely cordate or truncate, coarsely serrate, glabrous to sparingly stellate-pubescent, green above, light green below. Cymes solitary, pendulous at the nodes, the peduncle adnate for ca half its length to a linear, foliaceous bract, the bract acute to obtuse, tapering to rounded at the base. Flowers perfect, hypogynous, regular, to 1.5 cm wide;

sepals and petals 5 each, free, yellowish; stamens numerous, free, not exceeding the petals, staminodia present; pistil 1, 5-celled, with several ovules. Fruit nutlike, spherical to ovoid, tomentose, indehiscent, with 1–2 seeds. (2n = 82) Late Jun–Jul. Rich upland woods, often on n-facing slopes; from e ND to e OK & eastward, & w in the Niobrara R. valley of NE to w NE & BH; (Que. to nw NE, BH, s to ne TX, AL, & VA).

This tree is rather variable throughout its full range, the variants sometimes given at least varietal status. The species and its cultivars are commonly planted as ornamentals, as are various European and Asiatic species and their hybrids, which often have pubescent leaves and the stamens fused in groups.

48. MALVACEAE Juss., the Mallow Family

by Joan Colette Thompson and William T. Barker

Ours herbaceous or suffrutescent, the sap often mucilaginous, usually with stellate or lepidote pubescence. Leaves alternate, entire or variously lobed, usually palmately nerved; stipulate. Flowers often solitary or clustered in axils of leaves, sometimes in a terminal raceme, panicle, corymb, thyrse or spike. Flowers perfect (in ours), rarely dioecious or polygamous, sepals 3–5, ours 5, valvate, usually united below, persistent, sometimes subtended by an involucel of bracts (epicalyx); petals 5, free from each other but often attached at the base of the stamen-column; stamens numerous, monadelphous, anthers reniform, 1-celled, longitudinally dehiscent, pollen grains large, spiny; pistil typically 1, 2–many locules and 2–many carpels, with locules usually in a ring, placentation axile, stigmas 1–2 × as many as carpels, style 1 and apically branched (in ours), rarely clavate. Fruit a loculicidal capsule or schizocarp (in ours), rarely a berry or samara, seeds reniform, embryo straight or curved, endosperm usually present.

This family is found throughout the world except in very cold regions.

1 Fruit a loculicidal capsule; carpels 5, united into a compound ovary; stamen-column bearing anthers along the sides, the summit 5-toothed or truncate .... 5. *Hibiscus*
1 Fruit a schizocarp; carpels 5–many, loosely united in a ring around a central axis and separating at maturity; stamen-column bearing anthers at the summit, sometimes also along the sides.
  2 Style branches filiform or clavate, stigmas decurrent along the inner side; carpels 1-ovulate.
    3 Involucel of 6–9 bracts, united basally .... 2. *Althaea*
    3 Involucel of 3 free bracts or lacking.
      4 Petals truncate; stamen-column bearing anthers below as well as at the summit .... 4. *Callirhoe*
      4 Petals obcordate; stamen-column bearing anthers only at the summit .... 6. *Malva*
  2 Style branches terminating in capitate to discoid or truncate stigmas; carpels 1–9 ovulate.
    5 Carpels sharply differentiated into 2 parts, the upper part seedless, smooth, dehiscent, the lower part seed-containing, reticulate, indehiscent .... 10. *Sphaeralcea*
    5 Carpels not differentiated into upper and lower parts.
      6 Ovules 2 or more in each carpel .... 1. *Abutilon*
      6 Ovules 1 in each carpel.
        7 Involucel none; silvery-lepidote tomentum never present.
          8 Calyx explanate under fruit; lateral carpel walls becoming obliterated at the breaking up of the fruit at maturity .... 3. *Anoda*
          8 Calyx enclosing fruit; lateral carpel walls firm, enduring .... 9. *Sida*
        7 Involucel of 1–3 bracts or foliage silvery-lepidote (sometimes both).

9 Calyx at maturity inflated; leaves symmetrical, linear to narrow-oblong or lanceolate .... 7. *Malvastrum*
9 Calyx at maturity not inflated; leaves asymmetrical, reniform or hastate .... 8. *Malvella*

1. ABUTILON P. Mill., Indian Mallow

Annual or perennial herbs. Leaf blades entire, toothed or lobed, cordate. Flowers mostly solitary in the axils or in leafy panicles; involucel none. Sepals united below; petals distinct, usually orange or yellow to pink or red, oblique, entire; stamen-column bearing anthers at the summit; carpels 5–many, united in a ring around a central axis and separating at maturity, rounded or beaked at the summit, dehiscent nearly to the base, stigmas terminal, capitate, style-branches filiform or clavate, ovules 2 or more in each carpel. Fruit a schizocarp, truncate-cylindric or truncate-ovoid.

1 Carpels 10–15, 10–18 mm long, carpel beaks divergent, 2–5 mm long .... 3. *A. theophrasti*
1 Carpels fewer than 10, shorter than 10 mm; carpels muticous or acuminate to cuspidate.
  2 Stems erect or nearly so; flowers orange to yellow; leaves thick .... 1. *A. incanum*
  2 Stems slender and spreading, trailing or ascending; flowers pink or red, occasionally orange or yellow; leaves thin .... 2. *A. parvulum*

1. Abutilon incanum (Link) Sweet. Perennial, with soft stellate pubescence; stems erect or nearly so, branched, to ca 6 dm tall; rootstalks large. Leaves (2)4–10(16) cm long, leaf blades ovate to triangular, (1)2.5–5(9.5) cm wide, rather thick, usually tomentose above and below, acute to acuminate, irregularly dentate to crenate-dentate; stipules linear, 2–3 mm long, sometimes deciduous. Flowers solitary in the axils; peduncles 1–3 cm long, usually articulated at or above the middle. Calyx tube 2–5 mm long, the lobes ca 3 mm long, acute to acuminate, sepals usually reflexed in fruit; petals orange to yellow, 5–10 mm long; carpels 5–9, 6–9 mm long, minutely pubescent at maturity, acute to rounded, muticous. Seeds black, flattened, rounded-triangular, ca 2 mm long, pubescent. (n = 7) May–Nov. Dry, usually rocky soil on cliffs, slopes, prairies, & in open woods & chaparral, occasionally in moist areas; OK: Arbuckle Mts., to CO, s to TX & NM; (OK to CO, s to TX & AZ; Mex.).

2. Abutilon parvulum A. Gray. Perennial, more or less cinereous-tomentose with stellate trichomes, branches pilose; stems slender and spreading, trailing or ascending; root woody. Leaves seldom over 3(5) cm long including petiole, leaf blades ovate-lanceolate to ovate, thin, large leaves usually obtuse to rounded, smaller leaves acute to acuminate, dentate; stipules linear, 1–3 mm long, sometimes deciduous. Flowers solitary in the axils; peduncles filiform, mostly longer than the leaves, 1–2 cm long, usually articulated at or above the middle. Calyx 4 mm long, reflexed in fruit, the lobes ovate, 2 mm long, acuminate, corolla ca 1 cm wide, petals pink or red, occasionally orange or yellowish, 4–6 mm long, exceeding the calyx; carpels 5, to 8(9) mm long, acuminate or cuspidate. Fruit 7–8 mm long. (n = 7) May–Nov. Limestone hills, dry ledges, & mesas; CO s to TX & NM; (CO to CA, s to TX & Mex.).

3. Abutilon theophrasti Medic., velvet-leaf. Annual, with soft stellate pubescence throughout; stem rather thick, sparingly branched, 2–20 dm tall. Leaf blades ovate to nearly orbicular, (3)5–15(17.5) cm long, to 16(21) cm wide, velvety pubescent, acuminate, dentate to crenate, often only slightly so, petioles long, stipules deciduous. Flowers scattered, mostly in the upper axils; peduncles 1.5–5 cm long, shorter than petioles, often with an articulation at or above the middle, bracts subtending peduncles to ca 1 cm long. Calyx accrescent, 5-parted nearly to the base, 8–14 mm long, calyx lobes ovate, acuminate to cuspidate; corolla yellow, 15–25 mm wide, petals 6–15 mm long, truncate to retuse; carpels usually 10–15, 10–17 mm long, villous, truncate or retuse with divergent beaks 2–5 mm long. Fruit 1–2 cm wide; seeds black, flattened, rounded-triangular, 3–4 mm long. (2n = 42) Late Jun–Oct. Weed of fields, pastures, waste ground, roadsides; MN to MT & SD, s to OK, TX, & NM, expected in CO; (P.E.I. to N.S., s Que., to s Sask., sw B.C., ME to MT & SD, s to FL, TX, & CA; Eurasia). *Introduced* and now well established in the n and w. *A. abutilon* (L.) Rusby—Winter.

2. ALTHAEA L., Hollyhock

Biennial or perennial, pubescent herbs. Flowers conspicuous in upper leaf axils; bracts of involucel 6–9, united basally. Corolla white, pink to purple; stamen-column bearing anthers at the summit; carpels numerous, usually 15 or more, united in a ring around a central axis, indehiscent, ours beakless, style branches elongate, slender, stigmatic along the inner side, ovule 1. Fruit a schizocarp.

1 Pubescence velvety; bracts of involucel narrowly lanceolate .... 1. *A. officinalis*
1 Pubescence coarse; bracts of involucel broadly triangular .... 2. *A. rosea*

1. *Althaea officinalis* L., marshmallow. Perennial with soft velvety-pubescence of stellate trichomes; stem 6–12 dm tall, branched; roots thick, mucilaginous. Leaves ovate, commonly shallowly 3-lobed, 5–14 cm long, crenate or irregularly serrate, base rounded to slightly cordate. Flowers several in a peduncled cluster; bracts of involucel narrowly lanceolate. Corolla pink to pale rose, ca 3 cm wide. (2n = 42) Jul–Oct. Cult. & locally escaping to borders of fresh to saline marshes; scattered in GP; (N.B. to s Man., s to VA & AR, locally westward; Eurasia). *Introduced.*

The roots were formerly used in confectionery.

2. *Althaea rosea* Cav., hollyhock. Biennial or perennial with coarse stellate pubescence; stems strict, mostly unbranched, 1–3 m tall. Leaves suborbicular, slightly 5- to 7-lobed or sinuate, 8–30 cm wide, rugose, crenate to dentate, base cordate; petiole long. Flowers in a terminal spikelike raceme; peduncles short; bracts of involucel broadly triangular. Corolla variably colored, purple to white, ca 10 cm wide, petals ca 5 cm long. (2n = 42) May–Sep. Cult. & escaped around homesteads & into waste places; GP; (N.S. to s Man., s B.C., ne U.S. to MT, s to MO & TX; Eurasia). *Introduced.*

3. ANODA Cav.

1. *Anoda cristata* (L.) Schlecht. Annual herb, sparsely hirsute to hirsute with mostly simple hairs; stem branched from near base, to 1 m tall. Leaf blades deltoid to triangular-ovate, triangular-lanceolate or hastate, exceedingly variable even on the same individual, some or all leaves with shallow to deep lobes or divisions, 4.5–10 cm long, acute to acuminate, irregularly dentate, crenate or entire, truncate to broadly cuneate; stipules linear. Flowers solitary, axillary; peduncles long; involucel none. Calyx often purplish-red, explanate under the fruit, in fruit 2–3 cm wide, calyx lobes narrowly ovate to triangular-lanceolate, acuminate; corolla pale blue or lavender to violet, petals commonly cuneate, 1–2.5 cm long, retuse; stamen-column bearing anthers at the summit; carpels 8–20, united in a ring around a central axis, dark green, the backs separated by pale bands, conspicuously beaked with an elongate dorsal spur. Fruit a schizocarp, flattened and disklike, lateral walls separating carpels obliterated in the breaking up of the fruit at maturity, the firmer dorsal part embracing the seed; seeds pubescent. Jun–Nov. Waste places, roadsides, gravelly banks, & open woods along streams; IA, cen KS, & sw MO, s to TX & NM; (PA, IA, & sw MO s to TX & AZ; S. Amer.). Native in the sw U.S., adventive n & e.

4. CALLIRHOE Nutt., Poppy Mallow

Perennial herbs from a thick root or annual from a slender taproot, pubescent with both simple and branched hairs or glabrous. Leaf blades ovate or triangular to orbicular, crenate to 3- to 7-divided palmately or pedately, the divisions entire or variously toothed, lobed, parted, or incised; stipules sometimes deciduous. Flowers solitary in the upper leaf axils or sometimes in a terminal raceme or corymb; involucel of 3 bracts or lacking. Calyx deeply 5-parted, each lobe usually with 3–5 conspicuous nerves; petals red to purple, occasionally white or pink, cuneate, truncate, erose; stamen-column in ours bearing anthers along the sides as well as at the summit; carpels 10–20(25), rounded or with a short inflexed beak, often rugose or pubescent, indehiscent or partly 2-valved, style branches filiform, longitudinally stigmatic, ovule 1 in each unilocular carpel. Fruit a schizocarp, made up of a ring of carpels united around a central axis and separating at maturity, depressed.

References: Waterfall, U. T. 1951. The genus *Callirhoe* (Malavaceae) in Texas. Field & Lab. 19: 107–119; Waterfall, U. T. 1959. *Callirhoe bushii* (Malvaceae), a variety of *C. papaver.* Southw. Naturalist 3: 215–216.

1 Involucel of 3 bracts.
  2 Involucel bracts separated from the calyx by 1–3 mm .... 5. *C. papaver* var. *bushii*
  2 Involucel bracts not noticeably separated from the calyx .... 3. *C. involucrata*
1 Involucel lacking.
  3 Carpels strigose-pubescent, at least on the beaks .... 1. *C. alcaeoides*
  3 Carpels glabrous.
    4 Carpel beaks not or only slightly elevated above the body of the mature carpel, hardly visible above the body of the compound fruit when viewed from the side, back of mature carpel body not or only slightly prolonged over the base of the beak; plant perennial .... 2. *C. digitata*
    4 Carpel beaks usually protruding above the body of the mature carpels to form ca 1/3 of the upper part of the fruit, back of carpel body prolonged ca 1 mm into a conspicuous whitish chartaceous "collar" covering the base of the beak; plant annual .... 4. *C. leiocarpa*

1. *Callirhoe alcaeoides* (Michx.) A. Gray, pink poppy mallow. Perennial from a thickened to napiform root; stems several, branched from the base, to 5.5 dm tall, slender, more or less strigose with 4-rayed hairs, sometimes glabrate. Leaf blades mostly triangular to ovate, incised or 5- to 7-cleft palmately to divided, the basal blades often merely crenate, 2–8(10) cm long, 2–6(10.5) cm wide, the leaf segments mostly laciniate-cleft, with the ultimate leaf divisions linear to oblong; petioles of cauline leaves usually shorter than the blades to essentially lacking in the uppermost, petioles of basal leaves usually longer than the blades; stipules persistent, usually ovate or lanceolate, to ca 1 cm long. Flowers in a raceme, corymb or solitary; peduncles to ca 11 cm long, usually with an articulation above the middle; involucel none. Calyx ca 1 cm long, the lower 1/3–1/2 united, calyx lobes triangular to lanceolate, 3-nerved, acute to attenuate; corolla rose, pink or white, 1.5–4 cm wide, petals 10–25 mm long; carpels 10–14(17), rugose, at least the beaks strigose-pubescent, indehiscent. Fruit 3–5 mm tall, the upper 1/4–1/3 formed by the protruding carpel beaks, 6–8 mm wide; seeds dark, ca 2 mm long, smooth. Mar–Aug. Usually in dry or sandy soil, commonly in prairies and plains; IA & s SD, s to OK & n-cen TX; (KY, IL, IA, & s SD, s to TN & AL, MO, OK, & TX, introduced in ID).

2. **Callirhoe digitatea** Nutt., finger poppy mallow. Erect to reclining perennial from a thick root; stems 3–10(13) dm tall, glabrous to pilose-strigose. Leaf blades ovate to suborbicular, palmately or pedately 3- to 7-parted, basal blades sometimes merely crenate, to 11 cm long, leaf segments sometimes again parted, the ultimate divisions linear or oblong; petioles of cauline leaves short, petioles of basal leaves long; stipules persistent or early deciduous and commonly absent at anthesis, linear-lanceolate to ovate-lanceolate, 5–10(12) mm long. Flowers mostly solitary or subracemose; peduncles 2–15 cm long, sometimes with an articulation; involucel none. Calyx lobes lanceolate or triangular; corolla red to purple or sometimes white, 3–4.5 cm wide, petals 1–2 cm long; carpels 12–14(15), rugose, glabrous, indehiscent, the carpel beaks not or only slightly elevated above the body of the mature carpel, hardly visible above the body of the compound fruit when viewed from the side, back of the mature carpel body not or only slightly prolonged over the base of the beak. Fruit (3)4–5 mm tall, 6–10 mm wide. Mar–Jul. Usually in dry or sandy soil; plains, prairies, open woods, & barrens; MO to the se 1/4 of KS, s to OK & TX; (s Ont., MO, & se KS, s to AR & TX, adventive ne to IN & IL). *C. digitata* Nutt. var. *stipulata* Waterfall — Waterfall.

3. **Callirhoe involucrata** (T. & G.) A. Gray, purple poppy mallow. Perennial from an elongate to napiform root; stems mostly procumbent or decumbent, to 7(8) dm long, more or less strigose or hirsute in addition to 4-rayed hairs. Leaf blades rounded, 5- to 7-cleft palmately or pedately to divided, the segments variously toothed, incised, lobed or parted, the ultimate divisions usually linear, lanceolate or oblong, blades of principal cauline leaves 2.5–5(7) cm long, 3–6(9.5) cm wide, petioles equal to or longer than the blades; stipules persistent, mostly ovate, large, 5–15(20) mm long, 4–8(14) mm wide. Flowers solitary; peduncles 3–10(21) cm long, usually surpassing the leaves, occasionally with an articulation to 1 cm below the persistent involucel; involucel of 3 bracts, not noticeably separated from the calyx, bracts linear to lanceolate or oblanceolate, 6–15 mm long. Calyx divided to near base, calyx lobes lanceolate, 7–13 mm long, 3- to 5-nerved; corolla rose to purple, usually drying or fading to purple, (2)3–6 cm wide, petals 1.5–3 cm long; carpels (14)15–20(25), rugose, strigose, sometimes glabrous toward the summit, indehiscent, carpel beaks prominent, 1–1.5 mm tall. Fruit 3–5 mm tall, 8–10 mm wide. Feb–Aug. Usually dry & often sandy soil in prairies, plains, & open woods; MN to ND & WY, s to MO, TX, & NM; (MN to ND, WY, & UT, s to MO, TX, & NM, s to n Mex.; adventive in waste places e to OH). *C. involucrata* (Nutt. ex Torr.) A. Gray var. *lineariloba* (T. & G.) A. Gray — Waterfall.

4. **Callirhoe leiocarpa** Martin. Stems several from a slender annual taproot, 3–8.5 dm tall, glabrous or slightly pubescent with small 4-rayed hairs. Leaf blades reniform-cordate to ovate, crenate to 3- to 6-parted palmately or pedately to divided, 1–6 cm long, 1–8 cm wide, leaf segments cuneate to oblong, lanceolate, falcate to linear, entire to lobed, the upper leaves usually more deeply parted and with narrower lobes than the lower leaves; petioles shorter than to 3 × as long as the blades. Flowers solitary; peduncles many, 3–10(18) cm long, sometimes with an articulation, in fruit diverging from the upper 1/3–2/3 of the stem; involucel none. Fruiting calyx (6)9–15 mm long, 5–8 mm wide, the lower 1/4–1/3 united, calyx lobes linear-lanceolate to lanceolate; petals light-pink to red-purple, to ca 23 mm long; carpels 10–12, smooth to slightly wrinkled, glabrous, beak large, hollow, usually protruding above the body of the mature carpel to form about 1/3 of the upper part of the fruit, dehiscent, back of the carpel body prolonged ca 1 mm into a conspicuous whitish, chartaceous "collar" covering the base of the beak. Fruit 3–4 mm tall, 4–5 mm wide. (n = 14, 15) Mar–Aug. Prairies, plains, woods, & mesquite groves; s-cen KS s to OK & cen TX; (KS, OK, to s TX).

5. **Callirhoe papaver** (Cav.) A. Gray var. **bushii** (Fern.) Waterfall. Perennial with varying amounts of pilose spreading hairs in addition to small appressed 4-rayed stellate hairs, sometimes nearly glabrous; stems usually erect or ascending, sometimes more or less decumbent, 30–60 cm tall; root long (to 20 cm), narrow (to 2.5 cm), woody. Leaf blades hastate, cordate, triangular or ovate, 3–8 cm long, 4–9 cm wide, 3- to 5-lobed deeply palmately or pedately, the upper blades often less deeply lobed than the lower ones, blades of cauline leaves usually wider than long, usually deeply lobed into ovate-lanceolate, lance-falcate, oblong or linear-oblong divisions which are often entire but may be sparsely toothed; upper petioles mostly equaling to somewhat longer than the blades, basal petioles many times longer than the blades to about equaling the blades; stipules ovate or rhombic-ovate to oblong, 4–12 mm long and 3–7 mm wide. Peduncles long and slender, commonly 2 or 3 × as long as the subtending leaves; involucel bracts 3, usually narrowly linear and ca 1/2 as long as the calyx, 1 or 2 of the bracts 1–3 mm removed from the calyx. Calyx 15–25 mm long, the lower 1/5–1/4 united, calyx-lobes lanceolate, somewhat attenuate, basal part of the calyx varying from slightly to rather densely hispid with simple hairs 1.5–3 mm long; petals red, 2.5–3.5 cm long; carpels ca 20, ca 4 mm tall, ca 3.5 mm wide, the sides reticulate, backs glabrous, beaks slightly strigose. Fruit ca 4 mm tall, ca 1 cm in diam. May–Aug. Open woods, wooded valleys, shaded creek banks, wet prairies, ravine bottoms; sw MO, adj. KS & e OK; (MO & KS s to adj. AR & OK).

5. HIBISCUS L., Rose Mallow

Ours herbs, annual or perennial. Leaf blades mostly large, lobed, pedately cleft or merely crenate or dentate. Flowers solitary, axillary, involucel bracts usually ca 12, distinct or united, linear. Sepals more or less united; petals usually 2 cm long or more, showy; stamen-column bearing anthers along the sides, the summit 5-toothed or truncate; carpels 5, permanently united, essentially glabrous to long-hairy, stigmas 5, capitate or peltate, style bearing 5 short branches, ovules several in each locule. Fruit a loculicidal capsule, subglobose to ovoid or prismatic, subtended or enclosed by the persistent accrescent calyx; seeds without lint.

1 Leaves deeply incised or 3- to 7-cleft to divided; calyx bladdery-inflated, soon becoming scarious, loosely enclosing the capsule .... 4. *H. trionum*
1 Leaves merely toothed or lobed, not deeply incised or parted; calyx not inflated, herbaceous, closely applied to or filled by the capsule.
  2 Stem, calyx, and lower leaf blade surface glabrous or very nearly so; median and upper leaf blades often hastate at base .... 1. *H. laevis*
  2 Stem, calyx, and lower leaf blade surface densely covered with fine stellate hairs; leaf blades rounded to cordate at base.
    3 Upper leaf blade surface glabrous, glabrate, or soon glabrescent, lower leaf blade surface gray-pannose; capsules glabrous, conic-ovoid, tapering to an erect beak .... 3. *H. moscheutos*
    3 Leaf blades permanently pubescent on both surfaces, upper leaf blade surface closely pubescent, lower leaf blade surface loosely and coarsely stellate-tomentose; capsules densely villous-hirsute, short-cylindric, subtruncate or rounded at the apex .... 2. *H. lasiocarpos*

1. **Hibiscus laevis** All., halberd-leaved rose mallow. Perennial; often tinged with red, essentially glabrous; stems to 25 dm tall. Leaf blades triangular, ovate or broadly lanceolate, 5–15 cm long, often hastately 3- to 5-lobed, glabrous, acute or acuminate, serrate, truncate, base rounded or cordate; petioles to ca 1 dm long or more, slender; stipules deciduous. Involucel bracts 1.5–3 cm long, tapering to a filiform point. Calyx herbaceous, not inflated but somewhat accrescent, in fruit becoming oblong-campanulate and at length ovoid, glabrous or nearly so, calyx lobes ovate; petals obovate, pink or white with a dark purple

base, (4)5–8 cm long; stamen-column 5-toothed. Capsule ovoid, 1.5–2.5(3) cm long, puberulent to glabrous, pointed; seeds pubescent. May–Nov. In wet soil or in water along streams, riverbanks, marshes, & in wooded swamps; MN & e NE, s to MO & TX; (s PA & WV to IL, MN & e NE, s to FL & TX). *H. militaris* Cav.—Rydberg.

2. *Hibiscus lasiocarpos* Cav., rose mallow. Perennial; stems to ca 2 m tall, pubescent. Leaf blades broadly to narrowly ovate, occasionally some angulate or obscurely lobed, 1–2 dm long, more or less softly pubescent on both surfaces, upper surface closely pubescent with many simple or subsimple hairs, lower surface loosely and coarsely stellate-tomentose, acute to acuminate, crenately dentate, base cordate to subcordate; petioles to 1 dm long. Calyx at maturity herbaceous, not inflated, closely applied to or filled by the capsule, prominently 5- or 7-nerved; corolla 10–20 cm wide, petals white to rose with crimson blotch at the base, 7.5–10 cm long. Capsule short-cylindric, densely villous-hirsute, subtruncate or rounded. Jun–Oct. Marshes, flood plains, ditches, streams, & rivers, in wet soil or in water; MO s to OK & nw TX; (KY, IN, IL, & MO s to FL & TX).

3. *Hibiscus moscheutos* L., rose mallow. Perennial, stems 1–2.5 m tall, minutely stellate-pubescent to glabrescent. Leaf blades lanceolate to ovate, upper surface dark green, to ca 22 cm long, 3–9 cm wide, averaging 2.5 × as long as wide, unlobed or the middle and lower leaf blades tricuspidate, upper surface glabrous, glabrate or soon glabrescent, lower surface gray-pannose, acuminate, coarsely incised-dentate to crenate, base broadly cuneate to rounded or subcordate; petioles to ca 5 cm long, slender; stipules deciduous. Peduncles jointed near or above middle, usually fused for as much as 3/4 their length to the subtending petiole; involucel bracts and calyx canescent but not hairy. Calyx herbaceous, not inflated, closely applied to the capsule; corolla 10–20 cm wide, petals white or creamy-yellow with red or purple base 5–10 cm long; stamen-column averaging 2 cm thick; style averaging 6 cm long, style-branches glabrous or remotely hispid. Capsule conic-ovoid 2.5–3 cm long, glabrous except the sutures, tapering to an erect beak. Jun–Oct. Marshes, swamps, & wet meadows; escapes or introductions in se KS; (MD to OH & IN, KS s to FL & TX).

4. *Hibiscus trionum* L., flower-of-an-hour, Venice mallow. Low annual, more or less hirsute with simple and stellate hairs; stem widely branched, spreading at the base, 1–6 dm tall. Leaf blades ovate to orbicular, (1)2–6 cm long, palmately 3- to 7-cleft to divided, lobes lanceolate, oblong, ovate or obovate, the middle segment the largest, incised or lobed, serrate to crenate; petiole usually about as long as the blade; stipules linear, to 6 mm long. Peduncle elongating in fruit, articulated above the middle; involucel bracts 10–15, linear, 2–10 mm long, hirsute, persistent. Calyx hirsute, soon becoming scarious, bladdery-inflated in fruit, loosely enclosing the capsule, 5-winged, nerves prominent, dark colored, numerous, hispid, calyx lobes triangular; corolla 2–7.5 cm wide, expanded for only a few hours, petals yellow or whitish with the base and one edge dark purple, 1.5–4 cm long; stamen-column bearing anthers along the upper 1/2, 5-toothed at the summit; stigmas capitate. Fruit ca 1–3 cm tall, globose-ovoid, hirsute; seeds black, 2 mm long, rough. (2n = 56) Mid Jun–Sep. Often a troublesome weed of cult. & waste ground, pastures, along roadsides & railroads; GP; (N.S. to Sask., MT, & SD, s to FL & TX, also UT & OR; Europe). *Introduced.*

6. MALVA L., Mallow

Annual, biennial or perennial herbs, often ascending, procumbent or decumbent, glabrate or sparsely pubescent often with stellate hairs. Leaf blades reniform or orbicular, ours lobed, crenate, dentate or serrate. Flowers small or large, solitary or fascicled in the leaf axils, rarely in terminal racemes or spikes; involucel bracts 3, linear to ovate or obovate, persistent. Sepals partly united; petals whitish, pink, or purple, truncate to obcordate in ours; stamen-column bearing anthers only at the summit; carpels 10–20, round-reniform, beakless, the sides incompletely covering the seed, smooth to rugose, glabrous to pubescent, 1-seeded, 1-celled, indehiscent, stigmas decurrent along the inner sides of the filiform style branches. Fruit a schizocarp, disklike, depressed, the carpels united in a ring around a central axis and separating at maturity.

1 Petals red-purple or rose-purple, (2)3–5 × the length of the calyx; involucel bracts oblong to ovate or obovate .... 4. *M. sylvestris*
1 Petals white to pale pink, purple or blue, to 2 × the length of the calyx; involucel bracts linear to linear-lanceolate or oblong.
  2 Leaves distinctly but shallowly lobed to ca 1/3 the blade length, margins often crisp; flowers fascicled in only the upper axils, most nearly sessile .... 5. *M. verticillata*
  2 Leaves shallowly to obscurely lobed; floriferous nearly to the base, peduncles of some of the flowers well developed.
    3 Mature carpels smooth or faintly reticulate; fruit with a crenate outline; flowers usually 2 × the length of the calyx .... 1. *M. neglecta*
    3 Mature carpels rugose-reticulate; fruit with a circular outline; flowers usually less than 2 × the length of the calyx.
      4 Carpel angles somewhat wavy-winged and denticulate; calyx lobes broadly ovate, mucronate .... 2. *M. parviflora*
      4 Carpels sharply angled but not winged; calyx lobes narrowly triangular to triangular .... 3. *M. rotundifolia*

1. Malva neglecta Wallr., common mallow. Annual or long-lived; stem procumbent to ascending, to 1 m long, slender, usually branched from the base, central stem often erect, more or less pubescent, often with both simple and stellate hairs; root deep and thickened. Leaf blades reniform to orbicular, 1–4.5 cm long, 1.5–6(7.5) cm wide, lobes 5–7(9), rounded, shallow or obscure, crenate to dentate, base cordate or subcordate; petiole long, to ca 18 cm, upper petioles of the main axis usually 3–5 × the blade length; stipules narrowly lanceolate to ovate, ca 3–6 mm long. Flowers fascicled in the axils; peduncles to 3(5) cm long, reflexed in fruit; involucel bracts linear to lanceolate or oblong, ca 3–5 mm long. Calyx united to about the middle, 4–7 mm long, not much accrescent, not reticulate, calyx lobes ovate to triangular, in fruit curving over the carpel ring; corolla to 12(20) mm wide, petals white to slightly pink, purple or blue, usually 2 × the calyx length, to ca 14 mm long, claw pubescent; carpels (10)12–15, usually 13, at maturity smooth or faintly reticulate, usually puberulent, distal surface rounded, lateral sides not radially veined. Fruit with a crenate outline, 1–2 mm tall, 4–7 mm in diam, the central depressed area making up ca 1/3 the diam. (2n = 42) Apr–Oct. A weed of waste places & cult. soils; GP; (throughout temp. N. Amer.; Eurasia, N. Africa). *Introduced. M. rotundifolia* L.—Rydberg.

2. Malva parviflora L., small-fruited mallow. Annual, erect or ascending, glabrous or sparsely pubescent with stellate and sometimes also simple hairs; stem branched, to ca 2 m tall. Leaf blades reniform to orbicular, to 6 cm long, 2–10 cm wide, usually wider than long, usually 5- to 7-lobed, the lobes shallow, rounded-angulate to obscure, crenate to dentate, base cordate or subcordate; petioles of the main axis usually less than 3 × the blade length; stipules linear to narrowly triangular. Flowers fascicled in the axils; peduncles short, 2–15 mm long; involucel bracts linear to linear-lanceolate. Calyx widely spreading or forming a loose cup around the carpels, veiny-reticulate, calyx-lobes broadly ovate, mucronate; corolla not quite equaling to slightly exceeding the calyx, (3)4–6 mm long, (4)8–12 mm wide, petals whitish to pink, lilac or pale blue, claw glabrous; carpels rugose-reticulate, glabrous or puberulent, angles of the carpels somewhat wavy-winged and denticulate. Fruit with a circular outline. (n = 21) Feb–Oct. In thickets, along roadsides, in wasteground & cult. soils; GP; (s Que. to s B.C., s to NJ, MO, TX, & NM; Mex., Europe). *Introduced.*

3. **Malva rotundifolia** L., common mallow. Very similar to no. 1. Stems glabrous or sparsely pubescent. Leaf blades (1)1.5–5(6.5) cm long, 2–9 cm wide; petioles to 16(21.5) cm long. Peduncles to 3(4.5) cm long; involucel bracts usually linear. Calyx 3–4(5) mm long, calyx lobes usually narrowly triangular to triangular; corolla usually scarcely exceeding the calyx, 2–13 mm long, 3–12 mm wide; lilac to whitish; carpels 8–13, commonly 10, mature carpels rugose-reticulate, more or less pubescent, distal surface flattened, lateral surface radially veined, margins sharply angled but not winged. Fruit with a circular outline, usually 1–2 mm tall, 5–7 mm in diam, the central depressed area making up ca 1/5 the diam. May–Nov. Established as a weed in waste places & cultivated soils; GP; (P.E.I. to s B.C., locally in ne U.S., MI, & IN to WA & OR, s to GA, TX, & CA; Europe). *Introduced. Malva pusilla* Sm. — Rydberg.

4. **Malva sylvestris** L., high mallow. Biennial, 2–10 dm tall, glabrous to hirtellous or hirsute, often with both simple and stellate hairs. Leaf blades reniform to orbicular, conspicuously shallowly 3- to 7-lobed, lobes broadly rounded, triangular or ovate, crenate to dentate; petioles long, usually hirsute mostly in a single line on the upper side; stipules lanceolate to ovate. Flowers fascicled in the upper axils or in a terminal spikelike raceme; peduncles slender, to 5 cm long; involucel bracts oblong to ovate or obovate. Calyx 5(6) mm long, calyx lobes triangular to broadly triangular; corolla (2)3–5 × the length of the calyx, (1.5)2.5 cm wide, petals red-purple or rose-purple with darker veins, 1.5–3(3.5) cm long, claw pubescent; carpels rugose-reticulate, glabrous or sparsely pubescent. Fruit ca 2 mm tall, (6)7–10 mm in diam. (2n = 42) May–Oct. Escape from cultivation, roadsides, waste places; GP; (throughout much of N. Amer.; Eurasia). *M. sylvestris* L. var. *mauritania* (L.) Boiss. — Fernald.

5. **Malva verticillata** L., clustered mallow. Annual, more or less glabrous; stem to 2 m tall, usually more than 1 cm thick. Leaf blades rounded to reniform, (4.5)5–20 cm wide, distinctly but shallowly lobed to ca 1/3 the blade length, lobes rounded to triangular, double crenate, margins sometimes very crisp, base cordate or subcordate; petioles often shorter than the blades, upper petioles of the main axis progressively shorter, usually with pubescence mostly in a line on the upper side. Flowers fascicled in only the upper axils, most nearly sessile; peduncles less than 2 × the calyx length; involucel bracts linear to linear-lanceolate. Calyx accrescent, tending to close in fruit, finally thin and veiny, calyx-lobes triangular; corolla ca 2 × the length of the calyx, ca 1 cm wide, petals white or purplish, claw glabrous; carpels smoothish or with weak transverse simple ridges. (2n = 84) Jul–Sep. Escaped or persisting after cultivation, waste places, & around dwellings; MN to MT, s to MO & CO; (s Que. to MT, s to NJ & DE, IL & KS; Eurasia). *M. crispa* L. — Rydberg; *M. verticillata* L. var. *crispa* L. — Fernald.

7. MALVASTRUM A. Gray

1. **Malvastrum hispidum** (Pursh) Hochr. Annual herbs (1.0)1.5–4.5(7.5) dm tall; simple or branched, the branches ascending, often arising from near the base, progressively shorter distally, rarely exceeding the leader, stems and branches slender, pubescent with mostly 4(6)-armed hairs. Leaf blades symmetrical, linear to narrow-oblong or lanceolate, up to 5.5 cm long by 1.3 cm wide, acute or obscurely mucronate, serrate or rarely subentire, base cuneate to narrow-obtuse; petiole generally less than 1/3 the blade length; stipules spreading, linear, 2–7 mm long, drying and turning brownish early. Flowers solitary in the axils but sometimes glomerate by reduction in internode lengths; peduncles erect at anthesis, mostly less than 10 mm long to 15(17) mm and finally spreading in fruit; involucel bracts 3, free, (2.6)3–5(6.8) mm long, linear, in fruit becoming brownish, reflexed. Calyx

at maturity inflated, 8–12.5 mm long, copiously hispid, calyx lobes in bud through fruit plicate-winged at the margins, forming a 5-winged calyx, overhanging the whitish tube, 8–12 mm wide in fruit, (narrowly to) broadly cordate-ovate, abruptly acuminate to subcuspidate, drying brown and scarious; corolla within or slightly exceeding the calyx, petals yellow, 2.8–4.5 mm long, obovate, obliquely shallow-emarginate; stamen-column shorter than the petals, 2–3 mm long, bearing anthers at the summit; carpels 5(6) in a flat whorl, apical and basal portions not differentiated, reniform, at maturity conforming to the shape of the single seed and not much larger than it, beakless, thin-walled, at length loculicidally dehiscent into 2 free valves, stigmas capitate, papillate, style 1 with as many branches as carpels. Fruit a starlike whorl with broad, deep sinuses between the adjacent carpels, 3–3.5 mm tall, 5.6–7.6 mm wide, all free surfaces copiously and minutely pubescent. (n = 18, 2n = 36) Jun–Sep. Dry soil in prairies & rocky and gravelly barrens, usually near limestone outcrops, occasionally in open alluvial ground in valleys & along gravel bars, exceedingly local throughout the range; IA & NE s to MO & OK; (IL to IA & NE, e to KY, TN, & AL, s to AR & OK). *Sphaeralcea* St. Hil.—Steyermark; *S. angusta* (A. Gray) Fern.—Fernald; *Sidopsis hispida* (Ell.) Rydb.—Rydberg; *S. hispida* (Pursh) Rydb., emend. Kearney—Waterfall; *Malvastrum angustum* A. Gray—Gleason & Cronquist.

Reference: Bates, D. M. 1967. A reconsideration of *Sidopsis* Rydberg and notes on *Malvastrum* A. Gray (Malvaceae). Rhodora 69: 9–28.

8. MALVELLA Jaub. & Spach

Perennial, suffrutescent or (in ours) herbaceous, decumbent or prostrate, stellate-pubescent or silvery-lepidote or both. Leaf blades often asymmetrical, reniform or hastate, pedately nerved, margins costate. Flowers solitary or clustered in the axils; either with 1–3 involucel bracts or foliage silvery-lepidote (sometimes both). Calyx at maturity not inflated, sepals united into a usually angular base; petals stellate-pubescent without, where exposed in bud; stamen-column bearing anthers at the summit; carpels 5–many, united in a single whorl around a central axis and separating at maturity, 1-celled, muticous, obtuse or short acuminate, indehiscent, stigmas capitate, style 1, style-branches filiform, ovule 1, seeds pendulous, 3-angled. *Disella* Greene—Rydberg; *Sida* L.—Harrington.

References: Clement, I. D. 1957. Studies in *Sida* (Malvaceae). Contr. Gray Herb. 180: 50–54; Fryxell, P. A. 1974. The North American *Malvellas* (Malvaceae). Southw. Naturalist 19(1): 97–103.

1 Leaf blades reniform or suborbicular to triangular-ovate (wider than long), stellate pubescence predominating; involucel usually present; calyx-lobes ovate .... 1. *M. leprosa*
1 Leaf blades narrowly triangular or lanceolate to almost linear, 3–5 (rarely 2–6) × longer than wide; indument silvery-lepidote, involucel absent; calyx-lobes cordate .... 2. *M. sagittifolia*

1. ***Malvella leprosa*** (Ortega) Krapov. Low perennial, silvery, densely scurfy-canescent with free-rayed stellate hairs, becoming partially or completely lepidote; stem branched, decumbent or prostrate, the tips often ascending, 5–40(43) cm long; cespitose caudex or rootstock. Leaf blades asymmetrical, reniform or suborbicular to triangular-ovate, to 40(43) mm long, 50 mm wide, stellate pubescence predominating, variably and irregularly crenate to acute-serrate. Flowers solitary, axillary; peduncles usually slightly longer than the petioles, 10–50 mm long, not noticeably articulated; involucel bracts usually present, (0)1–3, linear, 3(4) mm long. Calyx hardly angulate, 4–7 mm long, more or less lepidote, deeply divided into ovate lobes, 2–6 mm long, acute or acuminate; corolla white, cream colored or rose, often drying to pale brown or pink, usually 2–2.5 cm wide, petals 10–16 mm long, obtuse, rounded; carpels 6–10, 2.7–3.2 mm tall, tomentulose to glabrate. (n = 11) Mar–Oct. Rocky or silty soils & moist saline soils, low banks & alkali flats; sw KS to se CO, s to OK, TX, & NM; (KS to WA, s to TX & CA; Mex., S. Amer.). *Disella hederacea* (Dougl.) Greene—Rydberg;

Sida hederacea (Dougl.) Torr.—Gates; *S. hederacea* (Hook.) A. Gray—Correll & Johnston; *S. leprosa* (Ort.) K. Schum. var. *hederacea* (Dougl.) K. Schum.—Waterfall.

2. Malvella sagittifolia (Gray) Fryxell. Low, silvery-lepidote throughout; stems commonly prostrate, to ca 4 dm long; cespitose. Leaf blades asymmetrical, narrowly triangular or lanceolate to almost linear, 3-5 (rarely 2-6) × longer than wide, mostly 2-3(5.4) cm long, 2-6(10) mm wide, especially lepidote when young, acute, entire except for a few hastate teeth at the base, base strongly oblique; petioles slender. Flowers solitary, axillary; peduncles usually slightly longer than the petioles, 10-50 mm long, not noticeably articulated; involucel bracts absent. Calyx 3-8 mm long, somewhat enlarged and angulate in age, deeply divided into cordate lobes; corolla purple to yellow or white with a purple tinge, (9)12-15(17) mm long; carpels 8 or 9, glabrous, thin-walled. Mar-Oct. Igneous soils, on clay flats & rocky areas; s CO, s to w TX & e NM; (s CO, s to TX, w to AZ, s to Dur. & Son.). *Sida lepidota* A. Gray var. *sagittaefolia* A. Gray—Harrington; *S. leprosa* (Ort.) K. Schum. var. *sagittaefolia* (A. Gray) I. Clem.—Correll & Johnston.

9. SIDA L.

Annual or perennial herbs, more or less pubescent with stellate hairs, silvery-lepidote tomentum never present; stems sometimes decumbent or prostrate, but usually erect. Leaf blades variously shaped, usually symmetrical, usually narrow, palmately (rarely pinnately) nerved, not marginally costate. Flowers mostly small, solitary or fascicled in the axils or forming a terminal panicle; involucel bracts lacking. Calyx enclosing the fruit; petals of various colors, glabrous (except sometimes ciliate on the claw), usually oblique, entire; stamen-column bearing anthers at the summit; carpels 5-15, united in a single whorl around a central axis and separating at maturity, glabrous to minutely pubescent, generally with 2 apical spines, usually dehiscent above, walls at maturity thin or thick, firm, enduring, stigmas capitate, style 1, style-branches filiform, ovule 1, pendulous. Fruit ovoid to disklike.

Reference: Fryxell, P. A. 1974. The North American *Malvellas* (Malvaceae). Southw. Naturalist 19(1): 97-103.

1 Calyx slightly accrescent, ca 5 mm wide; small spinelike tubercle at base of well developed leaves .... 2. *S. spinosa*
1 Calyx much accrescent, ca 10-15 mm wide; no tubercle at base of leaves .... 1. *S. physocalyx*

1. Sida physocalyx A. Gray. Herbaceous above ground, loosely and coarsely stellate-hirsute, partly glabrate; stems spreading or decumbent, to ca 4 dm long; rootstock perennial, fleshy-ligneous. Leaf blades rather succulent, suborbicular to oblong, to 6 cm long and 5 cm wide, obtuse to broadly rounded, crenate to serrate, base cordate; petioles to 3 cm long. Peduncles solitary, axillary, soon recurved. Calyx lobes cordate in appearance, with a small apical ligule, at anthesis ca 8 mm long, later much accrescent, 10-12 mm long, ca 10-15 mm wide, membranaceous and veiny, connivent and forming a vescicular, globular, and wing-angled loose covering over the fruit; petals yellowish or buff, scarcely exceeding the calyx; carpels 10-14, blackish when mature, ovate with a short beaklike apex, very thin-membranaceous and reticulate-veiny. Fruit disklike. May-Oct. Sandy, gravelly, or rocky soils on prairies, in washes & waste ground; sw OK s to TX & NM; (sw OK s to TX, w to AZ, s to n Mex., S. Amer.).

2. Sida spinosa L., prickly sida. Low annual, minutely and softly pubescent; stem branching, to ca 7 dm tall. Leaf blades ovate to lanceolate or oblong, to ca 5.5 cm long and 3 cm wide, obtuse to acute, crenate to serrate, base rounded or truncate to cordate; petiole to 3 cm long, longer than the flowers, petiole of well-developed leaves commonly with a small spinelike tubercle at the base dorsally; stipules linear. Flowers axillary, often subsessile; peduncles 2-12(13) mm long, longer ones with an articulation near or above the middle. Calyx 5-7 mm long, slightly accrescent, in fruit ca 5 mm wide, calyx lobes triangular, acute or short acuminate; corolla pale yellow, 1-1.5 cm wide, slightly exceeding the calyx, petals 4-6 mm long; carpels 5, 4 mm long, dehiscent at the apex into 2 prominent, erect, hispidulous beaks. Fruit ovoid. Jun-Oct. Fields, waste places, & roadsides, rarely in open woods; IA & NE, s to OK & TX; (MA to MI & NE, s to FL, TX, & trop. Amer.). *Naturalized.*

10. SPHAERALCEA St. Hil., False Mallow, Globe Mallow

Mostly perennial, herbaceous or suffrutescent, stellate-pubescent. Leaf blades divided to merely toothed. Inflorescence a raceme or thyrse, terminal; involucel bracts 0-3, subulate, usually caducous. Sepals united below; petals usually red and emarginate; stamen-column 1/2 to nearly as long as the petals, anther-bearing at the summit, filaments separate above the column; carpels 10-15, united in a ring around a central axis, often remaining attached to the axis after maturity by a threadlike extension of the dorsal nerve, densely stellate-pubescent, sharply differentiated into 2 parts, the upper part seedless, smooth, dehiscent, the lower part seed-containing, the sides reticulate, indehiscent, forming 1/2 or more of the carpel length, stigmas capitate or truncate, style 1, style branches as many as carpels, ovules and seeds 1-4 in each carpel. Fruit hemispheric to truncate-conic.

1 Leaves divided; indehiscent and reticulate part of carpel conspicuously wider than the dehiscent part, forming 2/3 or more of the carpel .... 2. *S. coccinea*
1 Leaves toothed or lobed, if lobed, these not extending 1/2 way to the midrib; indehiscent and reticulate part of the carpel not conspicuously wider than the dehiscent part, forming less than 2/3 of the carpel.
   2 Leaves oblong-lanceolate to linear-lanceolate, at most merely angulate or toothed near the base, not more than 1/3 as wide as long .... 1. *S. angustifolia* var. *cuspidata*
   2 Leaves ovate-oblong to very widely ovate, more or less distinctly lobed, more than 1/3 as wide as long .... 3. *S. fendleri*

1. ***Sphaeralcea augustifolia*** (Cav.) D. Don var. ***cuspidata*** A Gray, narrowleaf globe mallow. Perennial, rather densely stellate-canescent throughout ; stems to ca 18 dm tall. Leaf blades usually thickish, oblong-lanceolate to linear-lanceolate, to ca 15 cm long, usually much smaller, not more than 1/3 as wide as long, not lobed, usually crenulate, crenate or crenate-dentate with subhastate teeth usually less than 1/10 as long as the midlobe, base cuneate; petioles rarely longer than 1/4 the blade length. Inflorescence a many-flowered thyrse, conspicuously leafy nearly to the apex, the lower branches not more than 4 cm long; involucel bracts short, narrow and inconspicuous; pedicels stout, usually much shorter than the calyx. Calyx 5-9 mm long at anthesis, calyx lobes lanceolate to deltoid-ovate, acute to acuminate; petals usually bright pinkish-red, often drying violet, 6-20 mm long; carpels 10-15, 3.5-6.5 mm tall, 1/2 to 3/5 as wide as long, walls chartaceous, usually strongly connate at maturity, upper part erect, usually acutish, mucronate or cuspidate, lower part 1/10-2/5 the entire length, the back very finely reticulate, the sides reticulate. Fruit ellipsoid to ovoid. Apr-Aug. Sandy or rocky soils, mostly limestone & gypsum, waste places & along roadsides; sw KS & se CO, s to TX & NM; (KS to CO, s to TX, CA, & n Mex.). *S. cuspidata* Torr.—Rydberg; *S. angustifolia* (Cav.) Don—Barkley.

2. ***Sphaeralcea coccinea*** (Pursh) Rydb., red false mallow. Low perennial herb, usually densely pubescent; stems simple to clustered from a woody caudex, usually decumbent or ascending, to 5 dm long; root deep-seated, running. Leaf blades often gray-pubescent, deltate or suborbicular to ovate, 1-6 cm long, usually wider than long, 3- to 5-divided, the divi-

sions usually irregularly lobed to divided, the final segments oblong, obovate or oblanceolate to spatulate, acute to rounded, base cuneate; petioles of the lower leaves equal to or longer than the blade; stipules deciduous. Inflorescence a raceme (or terminal cluster), 2–6(11.5) cm long; involucel bracts usually lacking; pedicels shorter than the calyx and usually rather stout, lower ones sometimes more elongate and slender. Calyx at anthesis 3–10 mm long, not angulate, not inflated, shorter than the mature fruit, calyx lobes narrowly triangular to ovate, usually 1–2 × the length of the tube, villous, acuminate; corolla deep orange or brick red to pinkish, much exceeding the calyx, petals (0.8)1–2 cm long, emarginate; carpels 10 or more, reniform, 3–3.5 mm tall and wide, at maturity densely stellate pubescent, the walls coriaceous, upper part muticous, with an internal, dorsal palatelike fold near the base, tardily and incompletely dehiscent, lower part ca 3/4 the length of the carpel, much wider than the upper part, sides prominently reticulate, back tuberculate, stigmas capitate. (n = 5) Apr–Aug. Dry prairies & plains, hills; ND & MT, s to w IA, OK, TX, & NM; (s Man. to s B.C., s to w IA, TX, AZ, & OR). *Malvastrum coccineum* (Pursh) A. Gray—Winter.

3. *Sphaeralcea fendleri* A. Gray. Perennial with several erect or spreading stems from a woody crown, sparsely to densely stellate-canescent with usually very short hairs; stems to 15 dm. Leaves thin or thickish, usually green above, lighter below, ovate-oblong to very widely ovate, 2–6 cm long, shallowly to deeply 3-lobed below the middle, the lateral lobes relatively narrow-triangular or nearly rectangular, usually ascending, ca 1/4 as long as the midlobe, somewhat acute, crenate, crenate-dentate or cleft; petiole usually less than 1/2 as long as the blade, slender. Inflorescence an interrupted many-flowered thyrse, the lower branches not exceeding 3 cm long and usually much shorter; involucel bracts small, narrow, thin; pedicels usually equal to or longer than the calyx, slender but very tough and persistent. Calyx 4–6 mm long at anthesis, calyx lobes deltate to ovate-lanceolate, about as long as the tube, acute to short acuminate; petals bright pinkish-red, drying violet, 8–13 mm long; carpels 11–15, 4–5 mm tall, ca 3/5 as wide, walls chartaceous, often somewhat connate at maturity, upper part usually cuspidate, lower part 1/5–1/3 of the whole carpel, usually faintly and very finely reticulate. Apr–Nov. Gravelly soils, banks, & slopes, often in forest openings, among boulders in mts.; s KS & s CO, s to TX & NM; (KS, CO s to TX, NM, AZ; n. Mex.).

49. DROSERACEAE Salisb., the Sundew Family

by G. E. Larson

1. DROSERA L., Sundew

Small, scapose, perennial or biennial (rarely annual), insectivorous herbs of wet or damp soil. Leaves all basal and rosulate, circinate in bud, green or usually red, covered with long, glandular-viscid hairs, each hair tipped with a shiny droplet of clear, glutinous fluid; stipules scarious or rudimentary. Inflorescence (rarely a single flower) a 2–many flowered, racemelike cyme borne on a naked scape, simple or forked, mostly 1-sided, nodding at the developing apex. Flowers perfect, regular, usually 5(4–8)-merous, the sepals, petals, and stamens withering-persistent; sepals connate at the base, imbricate; petals white to rose, distinct or slightly united at the base; stamens opposite the sepals, anthers versatile; pistil of 3(5) united carpels, styles 3(5), deeply 2-parted, ovary superior, 1-celled, with 3(5) parietal placentae. Fruit a many-seeded, 3(5)-valved, loculicidal capsule.

1 Scape glandular-hairy; leaf blades cuneate, usually longer than wide .... 1. *D. brevifolia*
1 Scape glabrous; leaf blades rotund to depressed obovate, about as wide or wider than long .... 2. *D. rotundifolia*

1. *Drosera brevifolia* Pursh. Leaves spatulate, the blades oblong to suborbicular, 4–10 mm long, usually longer than wide and usually longer than the dilated glandular petiole, cuneate; stipules absent or setaceous. Scape 1- to 8-flowered, 4–12 cm tall, glandular-hairy. Flowers ca 1.5 cm wide; sepals ovate, 2.5–4 mm long, glandular-pubescent, subacute; petals rose to white, obovate, 4–9 mm long. Capsule about equaling the calyx; seeds black, obovate-oblong, 0.3–0.4 mm long, cratered, with pits in 10–12 rows. (n = 10, 2n = 20) May–Jun. Damp sand in pine or mixed forests & in open bogs; KS: Cherokee; (s VA & TN to se KS & e OK, s to FL, AL, LA, & TX). *D. annua* E. L. Reed—Correll & Johnston.

2. *Drosera rotundifolia* L., round-leaved sundew. Leaves with blades broadly obovate to rotund or depressed obovate, 2–10 mm long and about as wide or wider, much shorter than the petiole and abruptly tapered to it; petioles 1.5–5(9) cm long, glandular-pilose; stipules adnate to the petiole, 4–6 mm long, fimbriate. Scape 2- to 15(25)-flowered, 7–25(35) cm tall, glabrous. Flowers 4–7 mm wide; sepals narrowly oblong, 4–5 mm long, obtuse and erose at the tip; petals white to pink, oblong, slightly to considerably longer than the sepals. Capsule about equaling the calyx; seeds light brown, finely striate and shiny, sigmoid-fusiform, 1–1.5 mm long. (n = 10, 2n = 20) Jul–Aug. Bogs & swamps; MN: Becker, Pope; ND: Bottineau; (circumboreal, in Amer. s to SC, GA, TN, IL, MN, n ND, MT, ID, NV, & CA).

50. CISTACEAE Juss., the Rockrose Family

by William T. Barker

Herbs or low shrubs. Leaves simple, alternate, opposite or whorled, stipulate or exstipulate, entire. Flowers hypogynous, regular, perfect, variously arranged; sepals 5 with the 2 outer smaller than the inner 3; petals 3, 5, or absent in cleistogamous flowers; stamens few to numerous, irregular in number, filaments distinct; ovary superior, 1-celled, of (2)3(5–10) united carpels. Fruit a capsule; seeds few to many.

1 Petals in the petaliferous flowers 5, yellow, conspicuous.
  2 Leaves lanceolate or oblong; style very short .... 1. *Helianthemum*
  2 Leaves narrowly linear or scalelike; style slender, elongate .... 2. *Hudsonia*
1 Petals 3, dark red, minute; style none .... 3. *Lechea*

1. HELIANTHEMUM P. Mill., Frostweed

1. *Helianthemum bicknellii* Fern., frostweed. Stems clustered at the tips of multicipital caudices, without creeping rootstocks, mostly 2–6 dm tall, hoary-pubescent, erect or arching. Leaves linear-oblong to oblanceolate or narrowly elliptic, commonly 2–3 cm long on the main axis, much smaller on the branches. Flowers of two sorts: the primary or earlier ones with large yellow petals and indefinitely numerous stamens, opening on a sunny day and losing petals by the next day, borne 5–12 in a loose terminal raceme, 15–25 mm wide, sepals densely stellate pubescent, the outer ones nearly as long as or equal to the inner; the secondary or later flowers much smaller and cleistogamous, numerous and crowded on short axillary branches, with small petals and 3–10 stamens. Capsules of petaliferous flowers 4–5 mm in diam, with many seeds; capsules of cleistogamous flowers about 2 mm in diam, strongly triquetrous, with 3–few seeds. Jun–Jul. Dry sandy soil of open woods & plains; e KS, NE, MO, IA, MN, SD, & ND; (ME w to MN, ND, SD, CO, s to MD, NC, OH, IN, MO, & KS). *Crocanthemum bicknellii* (Fern.) Britt.—Rydberg.

2. HUDSONIA L., Poverty Grass

1. *Hudsonia tomentosa* Nutt., beach heather. Prostrate or bushy, much branched, heathlike plant, forming dense mats or bushes 2–6 dm across. Leaves small, scalelike or linear, 1–4 mm long. Numerous yellow flowers at the stem tips, each solitary at the end of a short, leafy, lateral branch; on pedicels 1–5 mm long. Flowers 6–10 mm wide; outer sepals each connate with one of the inner, which appears as a subulate lobe above the middle; petals 5, much larger than sepals, fugacious, narrowly elliptic; stamens 10–30, anthers ovoid; ovary 1-celled with 3 narrowly intruded placentae. Capsule strictly 1-locular with 1 or 2 seeds attached to each placenta. May–Jul. Dunes & sandy soil; MN: Polk; ND: Ransom; (Que. to NC w to Alta., IN, IL, MN, & ND).

3. LECHEA L., Pinweed

Perennial herbs, with solitary or few erect stems. Leaves small, alternate, occasionally falsely opposite or whorled, entire, sessile to short petioled, 1-nerved. Inflorescence a leafy panicle of very numerous minute, red flowers. Sepals 5, the outer linear to lanceolate, the inner 3 broadly ovate to obovate, concave in conformity to the capsule; petals 3, smaller than the sepals, marcescent; stamens 5–15; anthers minute, broadly ovate; ovary 1-celled, style none, stigmas 3, plumose; ovules 2 on each parietal placenta. Capsule 3-valved, maturing 1–6 seeds, enclosed wholly or largely by the persistent calyx.

1 Outer sepals about as long as inner or distinctly longer.
  2 Pubescence of the stem spreading; outer sepals about equaling the inner; midcauline leaves wider than 3 mm .... 2. *L. mucronata*
  2 Pubescence of the stem appressed or strongly ascending; outer sepals usually distinctly exceeding the inner; midcauline leaves less than 3 mm wide .... 4. *L. tenuifolia*
1 Outer sepals notably shorter than the inner, extending usually to slightly beyond the middle of the calyx.
  3 Herbage canescent with copious appressed pubescence; lower surface of the leaf more or less pubescent, as well as the midrib and the margin .... 3. *L. stricta*
  3 Herbage green; leaves pubescent benath only on the midrib and margin 1. *L. intermedia*

1. *Lechea intermedia* Leggett, pinweed. Stem 2–6 dm, thinly pubescent with appressed hairs. Leaves of basal shoots oblong-lanceolate to narrowly elliptic, 3–7 mm long, sparsely pilose beneath on the midrib and margin, or glabrous; cauline leaves linear-oblong, with a few hairs on the midrib beneath, or glabrous. Panicle 1/3 to 1/2 the plant, narrowly cylindric, the lateral branches seldom over 5 cm; pedicels equaling or surpassing the sepals. Calyx subglobose, with obtuse to abruptly rounded base. Fruit subglobose to slightly ovoid to obovoid, barely exceeding the sepals; seeds shaped like sections of an orange, pale brown, partly covered with a gray membrane. Jul–Sep. Dry sandy soil; SD: Pennington; (N.S. to Ont. & MN, s to PA, VA, OH, IL, NE, & SD).

2. *Lechea mucronata* Raf., hairy pinweed. Stems (2)3–8(9) dm, spreading villous. Leaves villous beneath on the margins and midvein, otherwise glabrous or nearly so, often some

in whorls, those on basal shoots ovate-elliptic, to 15 mm long, cauline leaves lanceolate to oblanceolate to elliptic, 1-3 cm long. Flowers densely aggregated on short lateral branches; pedicels mostly 0.5-1(1.5) mm long. Calyx subglobose, inner sepals deeply concave, scariously margined, 1.3-2.2 mm long, outer sepals slightly shorter to somewhat longer than the inner, 1.9-2.2 mm long. Capsule subglobose, 1.4-2.1 mm long, (2)3(4)-seeded. Jun-Nov. Dry sandy soil of open woods & prairies; e TX, OK, KS, MO, IA, & NE; (NH to MI, IL, MO, IA, KS, & OK, s to FL & TX). *L. villosa* Ell.—Rydberg.

3. ***Lechea stricta*** Leggett, pinweed. Plants canescent, gray green. Stems commonly procumbent for 1-5 cm at base, thence abruptly erect, 1-4 dm tall. Leaves of basal shoots crowded but not conspicuously whorled, 3-5 mm long, canescent beneath with hairs on the surface as well as the midrib and margin; cauline leaves similarly pubescent, narrowly lanceolate to oblanceolate, up to 15 mm long. Panicle branching from above the middle of the stem, branches erect or strongly ascending, strongly pilose. Calyx subglobose, strongly pilose with spreading pubescence, 1.6-1.8 mm long, 1.2-1.3 mm wide, inner sepals ovate to obovate, obtuse, outer sepals linear and rarely 2/3 as long as inner ones. Capsule subglobose; seeds 3 or 4, smooth. May-Jul. Dry sandy woods & prairies; w MN, ND: Richland, Bowman; NE: Holt, Brown; (MN, WI, s to IN, IL, IA, & ne NE, e to NY & Ont.).

4. ***Lechea tenuifolia*** Michx., pinweed. Stems 1-3 dm tall, sparsely pubescent with erect or ascending hairs. Leaves of basal shoots linear, crowded, 3-6 mm long; cauline leaves linear, to 2 cm long, usually 1(1.5) mm wide, sparsely hairy only beneath, soon deciduous. Panicle occupying 1/2 the plant or more, its numerous branches often racemiform and secund. Calyx subglobose, completely covering capsule, 1.6-1.9 mm long, inner sepals spreading-pilose, dull green, subacute, with a hard keel, outer sepals 2-3 mm long, equaling to much exceeding inner sepals. Capsule ovoid, 1.3-1.5 mm wide; seeds 2-5, opaque, yellow to reddish-brown, broad and greatly thickened toward the base. Jun-Nov. Dry soil, upland woods, & barrens; n TX, OK, e 1/2 KS, MO, & e NE; (s ME to SC, MN to IN, s & sw to MS & TX).

51. VIOLACEAE Batsch, the Violet Family

by Ralph E. Brooks and Ronald L. McGregor

Ours small annual or usually perennial acaulescent or caulescent herbs with entire or lobed stipulate leaves which are alternate or rarely whorled. Flowers axillary, hypogynous, perfect, strongly irregular, usually nodding; sepals 5, persistent, distinct, with or without basal auricles; petals 5, distinct, the lowermost one spurred or gibbous; stamens 5, filaments short, distinct or connate, with adnate introrse anthers, usually the 2 lower ones with a nectary at base, connective prolonged into a membranaceous appendage; ovary superior, 1-celled, 3-carpellate, with 3 parietal placentae; style solitary, often distally enlarged or modified, sometimes hollow; stigma simple, often oblique or hooked. During the summer and autumn most species produce very fruitful reduced cleistogamous flowers with 5 closed sepals; (0)1-5 rudimentary petals; (1)2 stamens; a 3-carpellate ovary, the style shortened, and the tip reflexed. Fruit a loculicidal 3-valved capsule, each valve with a several-seeded placenta on its middle; after opening each valve, as it dries, folds firmly together lengthwise, projecting the seeds from a few to several dm; seeds usually arillate.

We acknowledge with much appreciation the assistance of Arthur Cronquist, who provided us the opportunity of studying the draft of his manuscript on Violaceae prepared for the revision of the Gleason and Cronquist Manual of the Vascular Plants of Northeastern United States and Adjacent Canada. While we have made use of the draft manuscript, the material here presented is the responsibility of the authors, in particular the recognition of some taxa, which does not always follow the philosophy of Cronquist.

1 Sepals without auricles; lower petals merely gibbous, not spurred; stamens united, lower 2 not spurred .... 1. *Hybanthus*
1 Sepals with auricles; lower petal spurred or prominently gibbous; stamens distinct, the lower 2 spurred .... 2. *Viola*

1. HYBANTHUS Jacq., Green Violet

Perennial, caulescent herbs with leafy stems. Leaves alternate or the lower sometimes opposite or subopposite, venation pinnate; stipulate, stipules sometimes foliaceous and mistaken for leaves. Flowers solitary (rarely 2), axillary on short peduncles, sepals not auricled, subequal; petals subequal, the lowest longer than the upper 4; anthers united into a sheath and enclosing the ovary at anthesis. Fruit a 3-valved capsule. *Calceolaria* Loefl., *Cubelium* Raf.—Rydberg.

1 Leaves at least 1.5 cm wide .... 1. *H. concolor*
1 Leaves less than 1 cm wide .... 2. *H. verticillatus*

1. *Hybanthus concolor* (T. F. Forst.) Spreng., green violet. Plants 4–10 dm tall; stems solitary or clustered, simple, hirsute to nearly glabrous on the upper portion and less so below; rhizomes to 1 dm long, 3–5 mm in diam and heavily beset with fibrous roots. Leaves alternate, blade elliptic to obovate, 5–17 cm long, 1.5–5.5 cm wide, sparingly hirsute but sometimes only on the veins below, apex acuminate to attenuate, margins entire or serrate near the apex, ciliate, base attenuate to cuneate; petiole narrowly winged, 0.5–1.5 cm long; stipules linear-lanceolate to subulate, entire, 4–15 mm long, 0.5–2 mm wide, vestiture as the leaf blades. Flowers 1(2) in the leaf axils, 4–4.5 mm long and about as wide; sepals lanceolate to linear-lanceolate, 3.5–5 mm long, 0.5–1 mm wide, glabrous, apex attenuate, margins entire; petals deciduous, greenish-white, glabrous, emarginate at the apex and recurved; upper petals (4) oblong, 3.5–4.5 mm long, lowest petal slightly clawed and spurred at the base, 4–5.5 mm long; stamens 2–2.5 mm long; anthers with a bilobed gland at their base, the glands of the 2 lowest filaments connate into a single gland about 1.5 mm long, anther appendages colorless or whitish; style more or less s-shaped, about 1.5–2 mm long; the stigma directed downward; peduncles jointed above the middle, 5–13 mm long, recurved at anthesis and becoming erect to 2.3 cm long in fruit, vestiture usually as on the upper stem; cleistogamous flowers reported but not seen in GP material. Capsules shallowly lobed, olbong-ellipsoid, 1.5–2 cm long, 5–8 mm diam, glabrous; seeds white or cream-colored, globose, 3–4 mm diam, smooth. Apr–May. Rich, usually rocky (calcareous) soil in woodlands; MO; e edge KS; (Ont. w to MI & e KS, s to GA, AL, & e OK). *Cubelium concolor* (T. F. Forst.) Raf.—Rydberg; *H. concolor* f. *subglaber* (Eames) Zenkert—Steyermark.

2. *Hybanthus vertcillatus* (Ort.) Baill., nodding green violet. Plants 1–2.5(3.5) dm tall with a much-branched subterranean caudex; stems mostly clustered and forming small mounds, simple or branching from the lower nodes, 1–2 mm in diam, retrosely scabrous, remotely so at times; roots often bearing adventitious buds. Leaves opposite or subopposite below to mostly alternate above, sometimes smaller ones fascicled in their axils or appearing fascicled where stipules are foliaceous, lanceolate to linear-lanceolate, 1–4 cm long, 1–5 mm wide, lower leaves often oblanceolate and larger, to 6 cm long and 11 mm wide, glabrous, apex acute to acuminate, margins entire or remotely serrate on the largest leaves, cilate, base attenuate, sessile; stipules subulate and minute to foliaceous and frequently mistaken for the leaf. Flowers solitary in the leaf axils, nodding, 2.5–6 mm long; sepals ovate to lanceolate, subequal, 1.5–2.3 mm long, 0.3–0.7 mm wide, glabrous or scabrous on the midvein, apex acute, margins entire and usually ciliate; petals often persistent, greenish-white or cream-colored with purplish tips; upper petals (4) oblong, 2–2.5 mm long, glabrous; lower petal panduriform, 2.5–5 mm long, bearded within, minutely scabrous on the midvein without, slightly gibbous at the base; stamens 1.5–2 mm long; anthers united at anthesis but soon separating, with a bilobed gland at their base, about 1–1.5 mm long, anther appendages reddish-brown; style clavate with the stigma directed upward; peduncles jointed above the middle, mostly ascending, 3–5 mm long, in age to 13 mm long and sometimes recurved. Capsules ovoid to subglobose, 4–6 mm long, glabrous; seeds shiny black, subglobose, slightly flattened, 1.8–2.3 mm long. Late Apr–Jun. Usually gravelly or sandy soils in prairies; w 4/5 KS plus Wyandotte, se CO, ne NM, w OK, TX; (KS & e CO, s to TX, AZ; Mex.). *Calceolaria verticillata* (Ort.) O. Ktze.—Rydberg; *H. linearis* (Torr.) Shinners—Barkley.

2. VIOLA L., Violet

Ours annual caulescent herbs with taproots, or usually acaulescent or caulescent herbs with simple or branched erect, oblique, to prostrate caudex, some species with slender rhizomes or stolons. Some with subepidermal mucilage cells on lower leaf surface stipules or stems that appear as brown dots or splotches on dried specimens. Caulescent taxa with stems erect, ascending, or procumbent; leaves alternate or lower sometimes opposite, blades simple; stipules usually conspicuous, foliaceous or membranaceous, entire, lobed, toothed, fimbriate. Acaulescent taxa with usually long-petiolate leaves, blades simple or variously lobed. Flowers solitary on axillary or basal peduncles which bear 2 small bracts variously located on axis; sepals 5, distinct or nearly so, all or at least the lower 2 with basal auricles; persistent petals 5, unequal, the lower one with a basal spur or gibbous, glabrous or bearded in throat, the lateral 2 petals bearded in throat or glabrous, upper 2 petals usually glabrous; ovary unilocular, with 3 parietal placentae; style 1, usually clavate, with a capitate or somewhat flattened summit, sometimes hollow with a variously located aperture which leads to the stigma; stamens 5, closely surrounding the ovary, often slightly cohering with each other by the broad filaments which continue beyond anther-locules as conspicuous appendages, 2 lower ones bearing spurs which project into spur of corolla. Fruit a 3-valved 1-celled capsule; seeds numerous, arillate.

Reference: Russell, N. H. 1965. Violets *(Viola)* of central and eastern United States: An introductory survey. Sida 2(1): 1–113.

In addition to the conspicuous, usually vernal, chasmogamous flowers, all species, except for *V. pedata* and *V. arvensis*, produce cleistogamous flowers either on peduncles or stolons; these are found with and after the petaliferous ones and develop abundant seeds.

Several species reported for the GP are excluded because specimen evidence has not been located. These are *V. cucullata* Ait.—Fernald; *V. lanceolata* L.—Rydberg; *V. sarmentosa* Dougl.—Stevens; *V. septentrionalis* Greene—Fernald; and *V. striata* Ait.—Barkley. In addition, *V. villosa* Walt., the woolly violet, with leaves flat on the ground and tending to be evergreen, the plant downy throughout, is excluded though it is known in central OK and might occur in our area.

The taxonomy of *Viola* is made difficult in the GP by environmentally induced morphological variation in several species; by hybridization between related species though this appears less than the literature indicates; by changes in leaf morphology, in several species, as the season progresses; and by the extensive variation in the more isolated colonies of a species. However, with a little experience the basic taxa are readily recognized and the hybrids and variants appropriately named.

The following key is designed to aid in the identification of normal flowering, usually vernal, specimens in our region. Care should be given to noting presence or absence of pubescence on petals before pressing specimens.

1 Plants caulescent, may appear acaulescent when young.
    2 Plants winter annual or annual .... 12. *V. rafinesquii*
    2 Plants perennial.
        3 Flowers white, lavender, or violet.
            4 Petals white inside, usually purplish on back .... 2. *V. canadensis* var. *rugulosa*
            4 Petals lavender to violet, except for albinos .... 1. *V. adunca*
        3 Flowers yellow with brown-purple veins near base or purplish-brown tinged on back.
            5 Leaf blades broadly cordate; petals brown-purple veined near base; plants of woodlands .... 11. *V. pubescens*
            5 Leaf blades lanceolate, lance-elliptic, or ovate, truncate or subcordate at base; petals usually brown-purplish on back; prairie plants .... 5. *V. nuttallii*
1 Plants acaulescent.
    6 Style thick-clavate; stamens prominently protruding; all petals beardless; cleistogamous flowers absent .... 8. *V. pedata*
    6 Style not thick-clavate; stamens not protruding; lateral petals bearded in blue-flowered spp. or beardless in white flowered spp.; cleistogamous flowers usually present.
        7 Caudex usually short and rather thick, erect, oblique, often branched; stolons absent; lateral petals always bearded.
            8 Principal leaf blades 1.5–4× longer than wide, ovate to narrowly ovate or oblong, subtruncate to subcordate at base, crenate-serrate and with prominent teeth or lobulate near base; plants of se GP .... 14. *V. sagittata*
            8 Principal leaf blades mostly less than 1.5× longer than wide, subtruncate or cordate at base, varying from toothed to lobed or dissected.
                9 Spurred petal beardless.
                    10 Petioles and leaf surfaces glabrous .... 10. *V. pratincola*
                    10 Petioles and leaf surfaces of other than first leaves pubescent . 16. *V. sororia*
                9 Spurred petal bearded at throat.
                    11 Foliage pubescent .... 6. *V. palmata*
                    11 Foliage glabrous.
                        12 Leaf blades unlobed .... 4. *V. nephrophylla*
                        12 Leaf blades deeply cleft or lobed .... 9. *V. pedatifida*
        7 Plants with slender and elongate rhizomes, with or without stolons; plants of Black Hills or ne ND.
            13 Petals pale lilac or lavender, rarely nearly white.
                14 Stolons usually present; spurred petal with spur 1.5–2.3 mm long .... 7. *V. palustris*
                14 Stolons absent; spurred petal with spur 5–7 mm long .... 15. *V. selkirkii*
            13 Petals white, the lower with purplish lines near base.
                15 Leaf blades reniform, usually wider than long; stolons absent .. 13. *V. renifolia*
                15 Leaf blades cordate-ovate to cordate-orbicular, as wide as long; stolons present at least late in season .... 3. *V. macloskeyi*

1. Viola adunca, J. E. Sm., hook-spurred violet. Perennial caulescent herbs with short to usually elongate, slender rhizomes; early in season plants appear as erect tufts 2–6(8) cm tall, and acaulescent, later the stems become 4–12(18) cm long, leafy, glabrous to puberulent, and ascending to prostrate. Leaves alternate, blades ovate to suborbicular or oblong, (1)2–3(4) cm long, 1–2.5(3) cm wide, crenulate, subtruncate to subcordate at base, rounded to obtuse at apex, puberulent or rarely glabrous in our area; petioles shorter to longer than blades, puberulent or glabrous; stipules lance-linear or elliptic, green, 5–20 mm long, entire or usually with a few lobes or fimbriate incised teeth, these near base, above the middle or sometimes throughout, sometimes pectinate. Peduncles shorter to longer than leaves; sepals (4)5–6(8) mm long, lanceolate or lance-attenuate, auricles 0.5–2 mm long, glabrous or sparsely ciliate; petals 8–15 mm long, light to deep violet or lavender-violet, the lower three white at base and veined with dark violet, lateral 2 petals bearded, spur 4–6 mm long, straight and blunt, or sometimes tapering to a short, incurved or hooked point; style slender, summit rounded, aperture at end of upturned beak. Cleistogamous flowers produced from late anthesis till autumn. Capsules 4–5 mm long; seeds, including caruncle, 1.5–2 mm long, dark brown. (2n = 20, 40) May–Aug. Locally common in meadows, woodlands, stream banks; MN, ND, MT, w SD, WY, CO, NM; (Greenl. & Lab. to AK, s to MA, NY, MI, MN, ND, w SD, CO, NM, CA). *V. subvestita* Greene — Rydberg.

Most of our specimens are variously puberulent and have been referred to var. *adunca*. Rare specimens from ND and the Black Hills are entirely glabrous and are the var. *minor* (Hook.) Fern.

Viola conspersa Reichb. the American dog-violet, a related species of the e U.S. and Can., has been reported for e ND. It differs from *adunca* in being somewhat larger, glabrous throughout, stipules ovate-lanceolate, and flowers usually surpassing the leaves.

2. *Viola canadensis* L. var. *rugulosa* (Greene) C. L. Hitchc., tall white violet. Caulescent perennial herbs spreading by long stolons or superficial rhizomes, foliage glabrate or usually variously pubescent; stems (1)2–4(4) dm tall, 1–several from the base. Basal leaves usually 3–5, cordate-ovate to reniform-ovate, crenate-serrate, to 10 cm wide, often wider than long, abruptly acuminate or gradually narrowed to apex, pubescent to glabrate, petioles long; cauline leaves similar but more broadly ovate and more often abruptly tipped; stipules lance-acuminate or ovate-lanceolate, more or less scarious, 8–15(20) mm long. Flowers from axils of upper leaves, 8–15 mm long; sepals lanceolate or lance-acuminate, 4–6(9) mm long, auricles short, rounded, often ciliate; petals white inside, with a yellow base, the 3 lower with purplish lines toward base, the lateral petals bearded, all purplish tinged on outside and sometimes less so on inside, spur rounded and short; style bearded at or near apex. Closed flowers borne for a short time in axils of upper leaves following open ones. Capsules (5)6–8(10) mm long, glabrous or puberulent; seeds brown to purplish-black, 1.9–2.2 mm long. (2n = 24) May–Aug. Frequent to common in woodlands; n 1/2 GP, s along e side Rocky Mts. to NM; (w Ont. to B.C., s to WI, IA, NE, NM, AZ, UT, disjunct in sw VA, nw NC, & e TN). *V. rugulosa* Greene — Rydberg.

The var. *canadensis* differs in having a short, stout rhizome; foliage glabrous or less pubescent; leaves often longer than wide, and seeds often smaller. All our adequately collected specimens are var. *rugulosa* but it is possible var. *canadensis* will be found from ne NE to e ND.

3. *Viola macloskeyi* Lloyd, wild white violet. Perennial acaulescent herbs spreading by slender rhizomes and long stolons, especially by midseason. Leaves several from the base; cordate-ovate to reniform, (1)2–3 cm wide, later to 5 cm wide, rounded above or rarely short pointed, glabrous or puberulent below, margins nearly entire to inconspicuously crenate; petioles 0.7–5 cm long. Peduncles 1.5–7(12) cm long; sepals lanceolate, 3–4 mm long, auricle short; petals white, 7–11 mm long, the lower 3 with purple veins, the lateral 2 petals glabrous to slightly bearded, spur short; style nearly truncate. Cleistogamous flowers from axils of leaves and on stolons. Capsules 3–5 mm long, greenish-yellow; seeds 1–1.3 mm long, dark brown to black. (2n = 48) May–Jun. Rare along streams & wet places; BH of SD; (Alta. to B.C., s to BH, CO, CA). *V. incognita* Brainerd, *V. pallens* (Banks) Brainerd — Rydberg.

As described our plants are the var. *macloskeyi*.

4. *Viola nephrophylla* Greene, northern bog violet. Perennial acaulescent herbs from rather thick, erect, oblique or branching caudex, without spreading rhizomes or stolons; foliage usually glabrous. Leaf blades ovate-cordate to cordate-triangular, sometimes nearly reniform, 1.5–5(7) cm wide, margins crenate-serrate, early leaves sometimes with minute stiff hairs on upper surface, more rarely so on veins below and petiole; petioles 5–20 cm long; stipules linear-lanceolate, entire. Peduncles equaling to usually longer than leaves; sepals lanceolate, to oblong-lanceolate, 5–7 mm long, glabrous, auricles short; petals blue, 10–15(20) mm long; the lower 3 often whitish at base, conspicuously bearded, the upper 2 bearded or glabrous; spur short; tip of style nearly truncate, glabrous. Cleistogamous flowers on erect

to prostrate peduncles. Capsules greenish-yellow, 6–8 mm long, glabrous; seeds 2–2.3 mm long, brown. (2n = 54) May–Jun. Infrequent to common in moist prairies, along streams, & around lakes; MN, ND, SD, n NE, MT, s to NM; (Man., B.C., s to SD, CO, NM, CA). *V. retusa* Greene — Rydberg.

As here interpreted, *V. nephrophylla* is largely geographically separated from the closely related *V. pratincola* which has the spurred petal without a beard. The reports in the Atlas GP of *V. nephrophylla* in KS and NE we interpret as being hybrid segregates of *V. pedatifida* x *V. pratincola* in which the entire leaved plants retain the character of beard on the spur petal. Perhaps *nephrophylla* should be considered as a subsp. of *V. pratincola* or both as subspp. of *V. sororia*.

5. **Viola nuttallii** Pursh, Nuttall's violet or yellow prairie violet. Perennial caulescent herbs with short, stout rootstocks which bear several stout lateral roots; stems often largely subterranean but aerial portions 2–12(20) cm long, erect or decumbent, glabrous or sometimes puberulent. Leaves alternate, blades elliptic-lanceolate to lanceolate or ovate, cuneate to rounded or subcordate to truncate at base, 2–10 cm long, entire to sinuate or remotely crenate-dentate, glabrous to puberulent, eciliate or ciliate; petioles 2–10(15) mm long, somewhat wing margined, glabrous or puberulent; lower stipules adnate much of their length, scarious, free portion entire to few-toothed or lobed, upper stipules less adnate to nearly free, sometimes foliaceous. Flowers solitary in axils, peduncles equaling to usually shorter than leaves; sepals lanceolate, 4–7 mm long, spur 0.5–1.5 mm long; petals 5–15 mm long, spur short; petals yellow, the upper 2 usually brownish-purple on the back, the lower 3 veined with dark brown to brownish-purple, the lateral 2 bearded; style head bearded. Cleistogamous flowers borne on short peduncles in upper axils. Capsules elliptic-ovoid, 6–9 mm long, glabrous or rarely minutely puberulent; seeds 2–2.5 mm long, buff colored. (2n = 24, 54) Apr–Jul. Locally common on dry prairies, bluffs, stream valleys, thickets; ND, MT, s to nw 1/4 KS, WY, CO, NM; (MN to B.C., s to w KS, NM, CA). *V. vallicola* A. Nels. — Rydberg.

Some of our plants have a few to all leaves usually less than 3× as long as wide, bases truncate to subcordate, and have been called *V. nuttallii* var. *vallicola* (A. Nels.) St. John. Such variations are to be found in most colonies and the taxonomic distinction is without value in our area.

6. **Viola palmata** L., wood-violet. Perennial acaulescent herb with a stout horizontal, erect, oblique, sometimes branched caudex; rhizomes and stolons absent. Early leaves with blades entire and merely toothed and later ones both simple and lobed or dissected, or all leaves lobed or dissected from the first; lobed leaves 3-lobed, with central lobe broadly oblong, obovate, oblanceolate or broadly elliptic, entire or toothed, sometimes with 1–4 narrow lobes in lower 1/2, the 2 basal lobes entire or shallowly to deeply 1- to 4-lobed, lower inner margins clearly recurving, blades usually pubescent; petioles 0.5–2 dm long, usually pubescent. Peduncles shorter to equaling or surpassing early leaves; sepals 6–10 mm long, glabrous or usually ciliate and puberulent, auricles 0.5–2 mm long; petals 1.5–2.5(3) cm long, deep to pale violet, lower 3 petals bearded or spurred petal only sparsely so. Cleistogamous flowers with peduncles erect or ascending, sometimes prostrate or subterranean. Capsules purplish-green, (5)8–10(15) mm long; seeds 1.8–2.3 mm long, buff colored to light brown and sometimes mottled. (2n = 54) Apr–Jun. Rare in rich woods, stream valleys, bluffs; KS: Cherokee; MO, OK; (ME to MN s to FL, se KS, e OK, e TX). *V. lovelliana* Brainerd, *V. triloba* Schwein. var. *dilatata* (Ell.) Brainerd — Atlas GP.

Our vernal plants, as here described, are var. *triloba* (Schwein.) Ging. ex DC. in contrast to the more eastern var. *palmata* with leaves reportedly palmately 5- to 11-lobed. In late summer our plants are frequently without lobed leaves and are impossible to separate from *V. sororia*.

7. **Viola palustris** L., northern marsh-violet. Small perennial, glabrous, acaulescent herbs with elongate slender rhizomes and stolons. Leaves erect or ascending, thin, 2.5–3.5 cm

wide, orbicular to reniform, crenulate, cordate to subcordate at base, basal sinus rather wide or distinct. Peduncles equaling or longer than leaves; sepals ovate to lanceolate, 2-3 mm long, glabrous or sparsely ciliate, auricles 1-1.5 mm long; petals 9-13 mm long, lilac to pale lilac or nearly white, with purple veins, lateral 2 sparsely bearded, lower petal sometimes larger than others, spur 1.5-2.3 mm long; tip of style with a beak on lower side. Cleistogamous flowers axillary and on stolons, usually common. Capsules 4-5 mm long, greenish-yellow; seeds brown, often mottled, 1.6-1.8 mm long. (2n = 48) Jun-Jul. Along cool streams & in moist places; of SD(?); WY: Albany, Johnson, Laramie, Sheridan; CO; (Lab. to AK, s to ME, NH, MT, CO, CA).

8. *Viola pedata* L., bird's-foot violet. Perennial acaulescent herbs with a stout, erect, caudex, without rhizomes or stolons. Early leaves merely trilobed to usually deeply parted; principal leaves 3-parted, 2-5(7) cm long, lateral segments again 3- to 5(7)-cleft into linear, elliptic, lanceolate, or narrowly oblanceolate divisions, these often with 2-4 teeth or lobes near the apex, surfaces glabrous, margins sometimes ciliate; petioles to 15 cm long; stipules usually adnate to petiole 2/3 their length, free portions toothed or fimbriate. Peduncles equaling to often longer than leaves; sepals 8-12 cm long, lanceolate, glabrous or sparsely ciliate, auricles 3-4 mm long; petals 12-18(22) mm long, all lilac purple except in albinos or rarely the upper 2 petals dark violet, lower 3 petals veined with dark purple and lower petal white at base, all petals beardless or lower one rarely sparsely bearded; orange tips of stamens conspicuously exerted; style thick-clavate, orifice on one side near the summit, stigma near center of the cavity. Cleistogamous flowers none. Capsules yellowish-brown, 6-9 mm long, glabrous; seeds reddish, tan or coppery, 1.4-1.7 mm long. (2n = 56) Apr-Jun; Sep-Oct. Infrequent to locally common in rocky open woodlands, rocky or sandy prairies, roadside banks, usually in acid soils; w-cen IA, MO, e 1/4 KS, e OK; (ME to MN s to FL, e KS, TX). *V. pedata* var. *lineariloba* DC.—Fernald.

The species has been subdivided into several varieties and forms based on flower color and leaf shape. The most striking variants in our area are rare plants with the upper two petals dark volet and the lower three lilac-purple.

9. *Viola pedatifida* G. Don, prairie violet, larkspur-violet. Perennial acaulescent herbs with slender to usually stout erect, ascending, or sometimes branched caudex, without rhizomes or stolons. Early leaves strongly dissected, these usually 3-parted with each division again cleft into linear lobes, and these usually again cut into 2-4 segments, margins ciliate, lower surface glabrous or obscurely puberulent especially on the veins; summer leaves usually deeply incised; petioles to 15 cm long, glabrous or pubescent; stipules lance-attenuate, adnate to petiole below, rarely toothed or fimbriate. Peduncles nearly equaling to usually longer than leaves, glabrous or rarely glabrate; sepals lanceolate to ovate-lanceolate, obtuse or acute, (5)6-8(10) mm long, usually ciliate; petals violet or reddish-violet, 10-18(20) mm long, lower ones white at base and veined with dark violet, lateral 2 less strongly dark veined, lower 3 petals densely bearded; style truncate at summit, orifice at end of a short lateral beak. Cleistogamous flowers on short erect or ascending peduncles. Capsules yellow-green, 8-12 mm long; glabrous; seeds light brown, 1.7-2 mm long. (2n = 54) Apr-Jun. Infrequent to common in prairies & open woodlands; MN, ND, SD, e 1/2 NE, & KS; (s Ont. to Alta. s to MI, IN, AR, MT, ne WY, NE, KS, OK).

In our area *V. pedatifida* hybridizes to some extent with *V. pratincola* and *V. sororia*. Some of these hybrids and segregates have been called *V. viarum* Pollard and a few simulate *V. palmata*. The late spring and summer leaves of *pedatifida* are often only shallowly lobed or incised and have frequently been considered as hybrids of various species.

10. *Viola pratincola* Greene, blue prairie violet. Perennial acaulescent herb, with stout erect, oblique, or commonly branched caudex; rhizomes and stolons absent. Foliage glabrous; leaves entire, highly variable, blades at anthesis ovate to deltoid-ovate, or sometimes nearly reniform or nearly orbicular, base usually cordate with wide sinus, or subtruncate, tapering or rounded to blunt or acute apex, or abruptly wide acuminate, margin from nearly entire to crenate, or crenate-serrate, sometimes sinuate-dentate; later blades often much wider than long and more abruptly acuminate, under surface often becoming purplish late in the fall and early winter. Peduncles at first much longer than leaves, becoming much over-topped by leaves; sepals lanceolate or oblong-lanceolate, to ovate lanceolate or lance-attenuate, eciliate, 8–12 mm long, auricles 1–2(3) mm long; petals dark violet, (8)12–18(22) mm long, violet to nearly white (white in albinos), white to greenish-yellow at base, the 3 lower veined with darker violet, only lateral 2 petals bearded, spur 2–3 mm long; style truncate at summit, orifice at end of short lateral beak; stamens not erect. Cleistogamous flowers with erect, ascending or prostrate peduncles which are rarely subterranean. Capsules greenish-yellow or brown, (8)10–12(14) mm long, glabrous; seeds light buff colored to dark brown or purplish, often mottled, 1.8–2.2 mm long. (2n = 54) Mar–Jun and sporadic till frost. Infrequent to common in open woodlands, stream valleys, prairie hillsides, & canyons, roadsides, pastures, waste places; GP but becoming rare or absent in w 1/2; (MN, ND, MT, s to IL, MO, OK, CO, TX). *V. affinis* Le Conte — Barkley; *V. missouriensis* Greene, *V. papilionacea* Pursh — Rydberg.

This is a rather stable species over much of our area and most variation found can be attributed to habitat modifications and hybridization with *V. nephrophylla*, *V. pedatifida*, and *V. sororia*. The correct name for our plant is open to question and it would be reasonable to consider our plants as a subsp. of *V. sororia*. We have excluded *V. affinis*, *V. missouriensis*, and *V. papilionacea*, though they have priority, because all three were originally described as having spurred petal bearded, a character found in our region only in putative hybrid segregates of *V. pratincola* with *V. nephrophylla* or *V. pedatifida*. The rare glabrate specimens of our eastern woodlands are interpreted as hybrids with *V. sororia*.

11. **Viola pubescens** Ait., downy or smooth yellow violet. Ours caulescent perennial herbs with short erect, oblique or horizontal rhizomes; stems 1–several, erect or decumbent, 1–4 dm tall, densely to sparsely pubescent or glabrous, with prominent leafless stipules on lower 1/2. Basal leaves 0–several, long petioled, blades reniform-cordate to broadly ovate and short pointed, crenate-dentate, pubescent to glabrous; cauline leaves 2–4, near the top, orbicular-ovate to broadly ovate, short pointed, crenate-dentate, 4–10 cm long, to 3 cm wider than long, pubescent to glabrate or glabrous; petioles shorter to longer than blade, pubescent to glabrous; stipules broadly ovate to lanceolate, 9–15 mm long, pubescent to glabrous, sometimes ciliate. Flowers solitary in the axils; peduncles shorter to longer than leaves, pubescent or glabrous; sepals 6–10 mm long, lanceolate to oblong-lanceolate, acute to short-acuminate or obtuse, with a narrow white margin, pubescent to glabrous, sometimes ciliate, spurs short; petals 8–12 mm long, yellow, lower 3-veined with brownish-purple, lateral 2 bearded, the lower one gibbous. Cleistogamous flowers from upper axils. Capsules oblong-ovoid, 8–12(20) mm long, dark green to yellowish-green, conspicuously woolly to sparsely pubescent or glabrous; seeds buff colored to brown, 1.8–2.4 mm long. (2n = 12) Mar–Jun. Locally common in woodlands, stream banks, thickets, rarely in the open; e 1/2 GP but w across n NE & SD to the BH; (N.S. to Man., s to GA, TX). *V. eriocarpa* Schwein. — Rydberg; *V. pensylvanica* Michx., *V. pensylvanica* var. *leiocarpa* (Fern. & Wieg.) Fern., *V. pubescens* Ait. var. *peckii* House — Fernald; *V. eriocarpa* Schwein. var. *leiocarpa* Fern. & Wieg. — Gates.

From ne KS n to the Dakotas are found occasional colonies in which plants are erect, with basal leaf 1 or lacking, and with stems and leaves densely pubescent. These represent the variety *pubescens*.

Most of our plants are sparsely pubescent to glabrous, have 2–several basal leaves, stems often decumbent, and are the var. *eriocarpa* (Schwein.) Russell. In KS and NE plants with glabrous fruits and with pubescent fruits occur together.

12. **Viola rafinesquii** Greene, Johnny-jump-up, wild pansy. Ours slender winter annual or annual herbs; stems (4)7–15(25) cm tall, simple and erect or often branched from the base and above; lower branches decumbent to ascending; glabrous or often with short reflexed hairs on the angles and sometimes on the surface; internodes usually longer than leaves. Cotyledons usually persistent; leaves alternate, simple, the lower blades orbicular to ovate, 4–12 mm long, nearly entire but with remote notches, base abruptly attenuate to truncate or subcordate, apex rounded; upper leaves spatulate to broadly elliptic or lanceolate, nearly entire; petioles 3–15(20) mm long; stipules foliaceous, 10–20 mm long, pectinate-palmately divided, middle lobe oblanceolate to elongate, spatulate, margins ciliate or ecilate. Flowers solitary in axils; peduncles longer than leaves; chasmogamous flowers variable, the complete ones with sepals 5, oblong-lanceolate, 4–6 mm long, with rounded basal spur 0.5–2 mm long, acute at apex, margins ciliate or eciliate; petals 5, bluish-white or blue, 2 upper oblanceolate, 8–10 mm long, lateral 2 obovate, 8–10 mm long, bearded, spurred petal broadly obovate, 8–10 mm long, spur 1–1.5 mm long; ovary 3 mm long, style 1–1.5 mm long, stigma globose and hollow; stamens 5. Chasmogamous flowers (gradually) changing to true cleistogamous ones which have 5 closed sepals, petals absent or very abortive, stamen 1, style twisted with stigma under anther. Capsules oblong, yellowish, 4–7 mm long, shorter than mature sepals; seeds 1–1.5 mm long, yellowish, smooth. (n = 17) Chasmogamous flowers Mar–May; cleistogamous flowers Mar–Jul. Locally common in prairies, open woodlands, fields, pastures, roadsides, waste places; SD: Clay, SW IA, s 1/2 NE, n-cen CO, MO, KS, OK, TX; (NY, MI, NE, s to SC, GA, TX). *V. kitaibeliana* R. & S. var. *rafinesquii* (Greene) Fern.—Fernald.

The winter annual plants of this species produce chasmogamous flowers in early spring. As later flowers develop they exhibit a progressive reduction in petal size until they become obsolete, stamens are reduced to one, and the style becomes twisted so that the stigma orifice is directly below the anther. Flowers of late spring and summer are all fully cleistogamous and are responsible for most seed produced. It appears that all flowers on annual plants are cleistogamous.

Viola arvensis Murr., a native of Europe, has been reported from KS: Marshall, Saline; NE: Webster. It is similar to *V. rafinesquii* but is more robust; leaves definitely crenate; stipules pinnate; all flowers chasmogamous; petals light yellow or ivory and sepals as long as petals.

13. **Viola renifolia** A. Gray, kidney-leaved violet. Perennial acaulescent herbs with relatively slender erect, oblique, creeping, often branched caudex; stolons absent. Leaves with blades reniform to reniform-cordate, 1.5–3(4) cm wide, margins crenate-serrate, rounded above or more rarely with a short, blunt tip, basal sinus shallow, wider than long, minutely puberulent or glabrous; stipules ovate-lanceolate, scarious, 8–12 mm long, with slender teeth or narrow lobes. Peduncles shorter to equaling or surpassing leaves; sepals ovate-lanceolate to lanceolate 3–4(5) mm long, auricles 0.5–1 mm long; petals white, beardless, 6–8 mm long, the lower 3 with prominent purple veins, spur short; style truncate at apex with orifice at end of short lateral beak. Cleistogamous flowers on short erect to prostrate peduncles. Capsules yellowish-green to brown, 3–5(6) mm long; seeds brown, 2–2.2 mm long. (2n = 24) May–June. Rare in cool, moist woodlands; BH; (Newf. to B.C., s to MA, CT, NY, MI, MN, BH, CO, w WY). *V. renifolia* var. *brainerdii* (Greene) Fern.—Fernald.

The above description was based largely on specimens outside our region as only a few very poor specimens from the Black Hills were available. It is possible the BH specimens may actually be *V. macloskeyi*.

14. Viola sagittata Ait., arrowhead violet. Perennial acaulescent herbs with usually stout erect, oblique, often short-branching caudex. Early leaves narrlowly ovate, later ones narrowly ovate, triangular-ovate, to nearly oblong, generally 1.5–4× longer than wide, subtruncate to normally cordate at base, margins crenate-serrate to remotely so above, usually with short basal lobes or prominent large serrate or dentate teeth, apex gradually narrowed to obtuse tip, sometimes somewhat attenuate, rarely acute, surfaces glabrous, glabrate, or rarely pubescent; petioles long, glabrous; stipules lance-attenuate, (1)1.5–2(3) cm long, scarious, with several narrow teeth. Peduncles shorter to longer than leaves, usually glabrous; sepals (9)12–15 mm long, lanceolate, glabrous, auricles 0.5–2 mm long to 7 mm long in fruit; petals 12–18 mm long, violet-purple, the lower 3 white at base, veined with purple and bearded, spur 1–3 mm long; style truncate at apex with orifice at tip of a short lateral beak. Cleistogamous flowers on erect or ascending to prostrate peduncles. Capsules yellowish, glabrous, 9–13(18) mm long; seeds dark brown, 1.3–1.6 mm long. (2n = 54) Apr–Jun. Infrequent to common in moist prairies, open woodlands, stream banks, roadsides; se KS, MO, OK; (ME to MN, s to FL, se KS, OK, LA, e TX). *V. emarginata* (Nutt.) Le Conte — Fernald.

Our plants are reasonably distinct, and we have not found evidence of possible hybridization with other species.

15. Viola selkirkii Pursh, great spurred violet. Perennial small, acaulescent herbs with long, slender rhizomes, stolons absent. Leaf blades at anthesis 1.5–3.5 cm wide, broadly cordate-ovate, crenate, apex obtuse to acute, minutely puberulent above, glabrous below; later leaves larger with narrow sinus and converging or overlapping lobes; petioles glabrous; stipules lance-attenuate, greenish-scarious, 8–14 mm long, entire or with a few narrow teeth. Peduncles equaling or longer than leaves; sepals 6–7 mm long, lanceolate, glabrous; petals 7–10 mm long, pale violet, beardless, the lower 3 veined with dark violet, spur 5–7 mm long, blunt; style enlarged upwards, truncate, orifice at tip of a lateral beak. Cleistogamous flowers on erect peduncles. Capsules 4–6 mm long, yellowish-green, glabrous; seeds 1.5–1.7 mm long, grayish-brown. (2n = 24) May–Jun. Rare in shaded ravines; BH; (circumboreal; Lab. to AK, s to PA, n OH, WI, MN, SD, B.C.).

16. Viola sororia Willd., downy blue violet. Perennial acaulescent herbs with stout, erect, oblique, horizontal, and frequently branching caudex, without stolons. Leaf blades at anthesis crenate, ovate, orbicular, reniform or ovate-triangular, obtuse to acute at tip, cordate at base, (1)2–3(5) cm wide, first developed ones glabrous to glabrate or evidently pubescent on lower surface and petiole, later ones evidently spreading pubescent, 6–10(12) cm wide and usually wider than long. First peduncles often longer than leaves, soon shorter to equaling leaves; sepals (5)6–8(9) mm long, lanceolate to ovate-lanceolate, glabrous or sparsely ciliate, auricles 0.5–2 mm long; petals deep violet to lavender, except albinos, 10–15 mm long (8)10–15(18) mm long, the 3 lower white at base and veined with purple, the 2 lateral with dense beard, spurred petal usually glabrous, spur 0.5–2 mm long, blunt; style nearly truncate at apex, orifice at tip of lateral beak. Cleistogamous flowers erect, to ascending, or prostrate and sometimes subterranean. Capsules 6–10(12) mm long, yellowish, glabrous; seeds buff colored, dark brown or purplish, 1.8–2.3 mm long. (2n = 54) Apr–Jun, sporadic till frost. Common to infrequent in woodlands, stream valleys, sometimes in woodland-prairie borders, rarely in prairie ravines; MN, e SD, e 1/4 NE, & KS, MO, OK; (Que., ME, MN, s to FL, e SD, KS, TX).

Plants in early anthesis, with glabrous to glabrate leaves, are frequently identified as *V. pratincola*. The hydrid, *V. sororia* × *V. pratincola*, is generally variously gabrate throughout and is rarely found. The hybrid with *V. pedatifida* is uncommon and often simulates *V. palmata*.

52. TAMARICACEAE Link, the Tamarix Family

by William T. Barker

1. TAMARIX L., Salt Cedar

Shrubs or small trees, with slender branches, often evergreen, mostly halophytic or xerophytic; monoecious or dioecious. Leaves alternate, small, commonly scalelike or subulate, mostly sessile, with salt-excreting, multicellular, external glands. Flowers small, solitary or more often in slender, scaly-bracteate racemes, spikes or panicles, hypogynous, without bracteoles. Sepals 4–5(6), distinct or less often connate below, imbricate, persistent; petals as many as and alternate with the sepals, white, pink or red, seated with stamens on a fleshy nectary-disk; stamens as many or often 2 × as many as petals, anthers with longitudinal dehiscence; ovary 1-celled, with 3–5 basal placentae; stigmas 2–5, distinct; ovules 2–many on each placenta. Fruit a loculicidal capsule; seeds erect, terminating in a single tuft of hairs.

References: Baum, Bernard R. 1967. Introduced and naturalized tamarisks in the United States and Canada (Tamaricaceae). Baileya 15(1):19-25; Baum, Bernard R. 1978. The Genus *Tamarix*. Israel Acad. Sci. & Hum., Jerusalem, Israel. 209 pp.; McClintock, E. 1951. Studies in California ornamental plants, 3. The tamarisks. J. California Hort. Soc. 12: 76–83; Shinners, L. H. 1957. Salt cedars (*Tamarix*, Tamaricaceae) of the Soviet Union. Appendix: Notes on *Tamarix* in southwestern United States. Southw. Naturalist 2: 48–73.

Tamarix gallica L. was considered the most common species in the U.S. by Shinners (op. cit.) and by almost all the local floras. However, Baum (op. cit.), based on studies of herbarium material for his monographic work on the genus, considers this species to be rare in the U.S. He has identified specimens of *T. gallica* from Texas, a very few from California, South Carolina and Georgia, but none from the GP area. *T. ramosissima* Ledeb. and *T. chinensis* Lour. have been identified as *T. pentandra* Pall. by many American authors.

1 Flowers 4-merous, occasionally with more than 4 stamens .... 2. *T. parviflora*
1 Flowers 5-merous, occasionally with more than 5 stamens.
   2 Sepals more or less entire; petals ovate to elliptic; flowers of racemes occurring on green branches with 1 or 2 of the filaments inserted between lobes of disk .... 1. *T. chinensis*
   2 Sepals denticulate; petals obovate; all filaments of all flowers inserted below disk near margins .... 3. *T. ramosissima*

1. ***Tamarix chinensis*** Lour. Tree with brown to black bark, entirely glabrous. Leaves sessile with a narrow base, 1.5–3 mm long. Vernal inflorescences pyramidal, of many dense racemes, aestival inflorescences loose, of slender racemes; racemes 2–6 cm long, 5–7 mm wide; pedicels about as long as calyx; bracts equaling pedicels to slightly longer, linear to linear-oblong. Flowers pentamerous; sepals 0.75–1.25 mm long, subentire, trullate-ovate to narrowly so, acute, somewhat connate at base in aestival inflorescences, the outer 2 keeled; filaments inserted between lobes of disk, but from its lower part near margin; in aestival flowers 1 or 2 filaments inserted in the sinuses between the lobes and the other 3 or 4 under the disk near the margin. May–Sep. Escapes from cultivation to waste places & river flood plains; OK: Blaine, Greer, Oklahoma; (widespread in U.S. & Can.; Mongolia, China, Japan). *Introduced.*

2. ***Tamarix parviflora*** DC., salt cedar. Shrub 2–3 m high, with brown to deep purple bark, entirely glabrous. Leaves sessile with narrow base, 2–2.5 mm long. Vernal inflorescences simple, aestival inflorescences rare; racemes 1.5–4 cm long, 3–5 mm wide, densely flowered, most often on last year's branches; bracts triangular-acuminate, blunt, boat-shaped, almost completely transparent, longer than pedicels; pedicels much shorter than calyx. Flowers

tetramerous; sepals adnate at base, erose-denticulate, 1.25–1.5 mm long, the outer 2 trullate-ovate, acute and keeled, the inner ovate, obtuse; petals subpersistent, ± ovate, 2 mm long, subentire or faintly erose; staminal filaments emerging gradually from the disk lobes. (2n = 24) May–Jul. Escapes from cultivation to waste places & along river flood plains; widely scattered in TX, OK, & KS; (widespread in U.S. & Can.; Medit. region). *Introduced.*

3. *Tamarix ramosissima* Ledeb., salt cedar. Shrub 1–5(6) m high, with reddish-brown bark, entirely glabrous. Leaves sessile with narrow base, 1.5–3.5 mm long. Aestival inflorescences densely composed of racemes, the vernal ones usually simple, loose and not as common; racemes 1.5–7 cm long, 3–4 mm wide; bracts longer than pedicels, triangular-trullate to narrowly trullate, acuminate, margins more or less denticulate; pedicels shorter than calyx. Flowers pentamerous; sepals narrowly trullate, acute, denticulate to erose, 0.5–1 mm long, not connate at the base; petals 1–1.75 mm long, obovate to broadly elliptic-obovate; filaments inserted under the disk near the margin between the usually emarginate lobes. May–Oct. River flood plains, salt marshes, & roadsides; TX, OK, KS, w NE, w SD, & w ND; (widespread in s U.S. to GP; Eurasia). *Introduced.*

53. PASSIFLORACEAE Juss., the Passion-flower Family

by William T. Barker

1. PASSIFLORA L., Passion-flower

Perennial vines that climb by axillary, simple tendrils. Leaves simple, alternate, stipulate, usually palmately lobed. Flowers regular, polypetalous, usually perfect, perigynous with saucer-shaped to tubular hypanthium; sepals 3–5, alternating with the petals and attached to the edge of hypanthium which also bears the corona; stamens 3–10, inserted in the center of the hypanthium; styles 3, ovary on a gynophore, 1-celled, with 3–5 parietal placentae and numerous ovules. Fruit a berry; seeds arillate.

1 Lobes of leaves blunt or rounded, without teeth; flowers greenish-yellow, small, 1.5–2.5 cm wide; fruit 1.0–1.2 cm in diam .... 2. *P. lutea* var. *glabriflora*
1 Lobes of leaves pointed, serrulate; flowers white and purple to pink, large, 4–8 cm wide; fruit 3–5 cm in diam .... 1. *P. incarnata*

1. *Passiflora incarnata* L., May-pop. Perennial climbing or trailing vine to 8 mm long, glabrous or usually finely pilose pubescent, glandular near or at the summit. Leaves 6–15 cm long along midnerve, 5–12 cm long along lateral nerves, 7–15 cm wide between apices of the lateral lobes, 3-lobed from 3/4 to 4/5 their length, finely serrate, 3-nerved, membranaceous, dark green above, glaucescent beneath; petioles to 8 cm long, biglandular at apex, the sessile glands suborbicular; stipules setaceous, 2–3 mm long, very early deciduous. Flowers 4–6(7) cm wide; peduncles to 1 dm long, stout; bracts spatulate to oblong, 4–7 mm long, 2.5–4 mm wide, obtuse to acute, minutely glandular-serrulate toward apex, conspicuously biglandular at the base. Petals and sepals white or pale lavender; corona filaments in several series, purple or pink, rarely pure white, those of outer 2 series filiform, 1.5–2 cm long, those of the next 3 series capillary, about 2 mm long, radiate or suberect, the innermost series membranaceous at base, filamentose, with the filaments about 4 mm long; ovary ovoid, densely brownish or whitish-velutinous-tomentose. Fruit ovoid to subglobose, to 5 cm long, orange-yellow when ripe, edible; seeds obovate to nearly obcor-

date, 4–5 mm long, 3–4 mm wide, truncate at apex, reticulate. (2n = 18) Jun–Aug. Fields, roadsides, thickets, & open woods; TX, OK, MO, se KS; (VA to s OH, s IL, & OK, s to FL & TX).

2. **Passiflora lutea** L. var. **glabriflora** Fern., passion-flower. Perennial climbing or trailing vine to 3 mm long, glabrous to sparingly pilose. Leaves usually wider than long, 3–7 cm long, 4–10 cm wide, 3-lobed, lobes blunt or rounded, without teeth; petioles to 5 cm long, glandless; stipules setaceous, 3–5 mm long, deciduous. Flowers greenish-yellow, (1)2(2.5) cm wide, exterior corona yellow; ebracteate; peduncles solitary or in pairs, 1.5–4 cm long, very slender. Calyx tube patelliform, sepals linear-oblong, 5–10 mm long, 2–3 mm wide, obtuse, pale green; petals linear, 3–5 mm long, about 1 mm wide, acutish, white; corona filaments in 2 series, outer ones about 30, narrowly linear or almost filiform, 5–10 mm long, the inner ones narrowly liguliform, 1.5–2.5 mm long, slightly thickened toward apex, white above, pink-tinged at the base. Fruit globose-ovoid, glabrous, about 15 mm long and 1 cm wide, purple; seeds broadly obcordate to suborbicular, 4.5–5.5 mm long, 3 mm wide, transversely sulcate with 6–7 grooves, the ridges strongly rugulose. (2n = 84) May–Aug. In the shade of low moist woods; TX, OK, MO, & se KS; (OH to MO and OK, s to AL & TX).

54. **CUCURBITACEAE** Juss., the Cucumber Family

by William T. Barker

Herbs or sometimes softly woody, mostly climbing and trailing, with watery juice, commonly with spirally coiled, often branched tendrils; monoecious or dioecious. Leaves alternate, usually palmately veined and often lobed, or sometimes palmately compound, frequently with extrafloral nectaries. Flowers mostly in axillary inflorescences or solitary in axils, unisexual, regular or irregular, epigynous or rarely semiepigynous, hypanthium shortly or strongly prolonged beyond the ovary. Calyx tubular, lobes (3)5(6), imbricate or open; petals (3)5(6), distinct or more often connate, usually yellow or white, imbricate or induplicate-valvate; corolla sometimes unlike in staminate and pistillate flowers; stamens free or variously united, (1–4)5, attached to the hypanthium or the summit of ovary, filaments distinct or connate, anthers distinct or connate, extrorse and opening by longitudinal slits; gynoecium mostly of (2)3(5) carpels united to form a compound ovary, style solitary with 1–3(5) usually bilobed stigma lobes; ovules (1) to numerous. Fruit usually a berry or a pepo, less often a dry or fleshy capsule or rarely samaroid.

```
1  Seeds less than 10 in each fruit; fruit with spines or prickles.
    2  Fruit filled with a solitary seed, indehiscent ............................................... 6. Sicyos
    2  Fruit with several seeds, variously dehiscent.
        3  Fruit oblique, gibbous, elastically rupturing; leaves compound .......... 2. Cyclanthera
        3  Fruit not gibbous, rupturing irregularly at the summit; leaves lobed ... 3. Echinocystis
1  Seeds numerous in each fruit; fruit smooth.
        4  Corolla campanulate, 5-lobed to about the middle ........................ 1. Cucurbita
        4  Corolla rotate to campanulate, 5-lobed to near or at the base.
            5  Sinuses between leaf lobes extending 1/2 way to the petiole or less; fruits green
               when mature ................................................................. 5. Melothria
            5  Sinuses between leaf lobes extending 3/4 way to petiole or farther; fruits reddish or orange at maturity ................................................. 4. Ibervillea
```

1. CUCURBITA L., Gourd

1. *Cucurbita foetidissima* H.B.K., buffalo-gourd. Plant rank-growing, stems rough, trailing from a thick perennial root, often to several m. Leaves coarse and thick, triangular-ovate, broadly rounded to cordate at the base, acute to acuminate at the tip, 1–2(3) dm long, irregularly and finely toothed and often angularly lobed, grayish-green in color, ill-smelling especially when bruised. Monoecious, flowers solitary in axils, the staminate flowers long peduncled. Rather large corolla to 1 dm long, yellow, 5-lobed to about or above the middle; anthers united; ovary 1-celled with 3–5 placentas, the 3–5 stigmas bilobed. Fruit subglobose, 5–10 cm across, smooth, greenish-orange; seeds cream colored, obovate, flattened, smooth, 6–9 mm long. (2n = 40) Jun–Aug. In sandy & gravelly soils of waste places; TX, OK, KS, & s NE; (s NE, MO to TX & w to CA; Mex.; introduced in waste places farther e). *Pepo foetidissimus* (H.B.K.) Britt., *C. perennis* A. Gray—Rydberg.

2. CYCLANTHERA Schrad., Cyclanthera

1. *Cyclanthera dissecta* (T. & G.) Arn., cyclanthera. Annual, glabrous, with slender climbing stems to 3 m long or more, climbing over trees and shrubs; tendrils simple to trifid. Leaves 3- to 7-foliolate, the elliptic-lanceolate leaflets somewhat lobed or toothed, leaflets to 6 cm long; leaves and leaflets stalked. Both staminate and pistillate flowers from the same axils, white, staminate flowers in racemes or panicles, pistillate flowers solitary; corolla rotate, deeply 5-parted; stamens united into a central column and with annular anthers; ovary 1- to 3-celled, with few erect or ascending ovules. Fruit 2–3 cm long, on peduncles 1–2 cm long, narrowly ovoid, somewhat asymmetric, rostrate, with long, slender, smooth spines, bursting irregularly; seeds brownish, obovoid, flattened, granular and irregularly tuberculate, 6–8 mm long. May–Oct. In open woodland in rocky soil; cen KS, w OK, & TX; (s NE, KS to LA, TX, s AZ; Mex.).

3. ECHINOCYSTIS T. & G., Wild Cucumber

1. *Echinocystis lobata* (Michx.) T. & G., wild cucumber. Climbing annual, essentially glabrous throughout; tendrils forked. Leaves suborbicular-ovate in outline, with (3)5(7) sharp triangular lobes. Monoecious, flowers 6-merous, greenish or white, the staminate flowers in racemes or panicles, pistillate flowers short peduncled, solitary or in small clusters in the same axils as staminate flowers. Calyx small; corolla rotate, 0.8–1 cm wide, with lanceolate lobes; stamens 3, filaments united into a column, anthers connivent; ovary 2-celled with 2 erect ovules in each cell; style with broad-lobed stigma. Fruit ovoid, 3–5 cm long, to 2.5 cm in diam, green, bladdery-inflated, beaked, weakly prickly, bursting irregularly at the summit; seeds brown with darker mottling, elliptic, stipitate at base, flattened, irregularly pitted, 12–20 mm long. (2n = 32) Jun–Oct. Moist rich soil of alluvial woods & open woodlands, cultivated for arbors & freely escaping; throughout GP; (N.B. to Sask., s to FL & TX). *Micrampelis lobata* (Michx.) Greene—Rydberg.

4. IBERVILLEA Greene, Globe-berry

1. *Ibervillea lindheimeri* (A. Gray) Greene, globe-berry. Climbing glabrous herb from a perennial rootstock, stems slender and branched. Leaves broadly ovate in outline, most deeply 3(5)-lobed, the divisions 1 cm or more wide, cuneate to flabellate or rhombic-ovate, toothed to lobulate. Dioecious, pistillate flowers solitary, staminate flowers racemose to fasci-

cled or sometimes solitary. Flowers greenish to yellow, tubular, glandular-puberulent, 6–8 mm long and 2mm wide, calyx tube with 5 short lobes; petals 5, oblong or linear; stamens 3, inserted in tube; stigma 3-lobed, ovary with 2 or 3 placenta. Fruit globose, 2.5–3.5 cm in diam, orange to bright red when ripe; seeds to 6 mm long. (2n = 22, 24) Apr–Jul. Open dry woodlands or thickets, among brush, in fence rows; TX: Armstrong, Hardeman; OK: Comanche, Greer, Harmon; (cen TX n to s OK). *Sicydium lindheimeri* A. Gray, *Maximowiczia lindheimeri* (A. Gray) Cogn.—Correll & Johnston.

5. MELOTHRIA L., Creeping Cucumber

1. *Melothria pendula* L., creeping cucumber. Slender, glabrous climbing vine from a perennial root, with mostly undivided tendrils. Leaves orbicular in outline, 3–7 cm long, shallowly or deeply 3(5)-lobed, with a cordate base, scabrous. Monoecious, staminate flowers racemose or corymbose, pistillate flowers solitary or clustered. Flowers small, yellow or greenish, calyx campanulate, 5-dentate; corolla rotate to campanulate, deeply 5-parted; stamens 3, with short, distinct filaments, straight, oblong anthers free or barely connivent; hypanthium greatly constricted above 3-celled ovary, ovules numerous; style short, with 3 slender stigmas. Fruit ovoid, pulpy, green, 1–2 cm in diam; seeds white, obovate, flattened, smooth, 5–7 mm long. Jun–Sep. Stream banks, roadbanks, other exposed sites; TX, se KS, se OK, & s MO; (s VA, s IN & se KS to FL; n Mex.). *M. pendula* L. var. *chlorocarpa* (Engelm.) Cogn., *M. chlorocarpa* Engelm.—Correll & Johnston.

6. SICYOS L., Bur Cucumber

1. *Sicyos angulatus* L., bur cucumber. Climbing annual vine with forked tendrils, viscid-hairy, hairs weak and distinctly articulated. Leaves petiolate, orbicular in outline, cordate at base, to 2 dm long and wide, shallowly 3- to 5-lobed, usually with a deep basal sinus, the lobes denticulate, acuminate. Monoecious, staminate flowers in racemes and corymbs, pistillate flowers usually from the same axils, in a long-peduncled capitate cluster. Flowers 5-merous, calyx small, 5-toothed; corolla rotate, 5-lobed; stamens and anthers united into a column; ovary 1-celled, with 1 suspended ovule. Fruit ovoid, dry, indehiscent, adorned with prickly bristles or rarely smooth and variously pubescent; seed olivaceous, oval flattened, smooth, 7–10 mm long. (2n = 24) May–Sep. Damp soil along rivers & streams, brushy waste areas; e 1/2 GP; (s ME & w Que. to ND, s to AZ & FL).

55. LOASACEAE Dum., the Stickleaf Family

by Robert B. Kaul

Ours annual, biennial, and perennial herbs and suffrutescents of dry sandy and rocky places mostly in the w GP. Stems simple or branched, smooth, puberulent, scabrous, prickly, or exfoliating. Leaves mostly cauline, alternate, sessile to short-petiolate, without stipules entire, sinuate, lobed, or pinnatifid and variable on the plant, rough-scabrous and adhering to fur and clothing, or spiny on the margins. Flowers borne singly or in leafy cymes or panicles. Flowers perfect, epigynous, often with an involucre; sepals 5, usually persisting in fruit; petals 5, or apparently more in those with petaloid staminodia, attached with the sepals on the hypanthium, white to orange; stamens 5, with prominent appendages (*Cevallia*) or 10–300, without appendages (*Mentzelia*); ovary unilocular, placentae 1–5, ovules 1–many. Capsules indehiscent (*Cevallia*) or dehiscent by apical valves (*Mentzelia*); seeds pendulous or horizontal, in 1 or 2 rows on the placentae, rhomboidal to prismatic, wingless, or flattened and winged.

1. Plant rough-scabrous but not invested with prickly hairs; stamens 10 or more, without appendages; flowers single or in loose but not capitate clusters; capsules apically dehiscent; w and s GP .... 2. *Mentzelia*
1. Plant invested with sharp prickly hairs; stamens 5, with prominent appendages; flowers in villous capitate clusters; capsules indehiscent; sw GP .... 1. *Cevallia*

1. CEVALLIA Lag.

1. Cevallia sinuata Lag. Suffrutescent to almost shrubby branching perennial to 6 dm tall. All vegetative parts with sharp prickly hairs. Leaves cauline, numerous, the largest 5 cm long and 2.5 cm wide, sessile, sinuate to coarsely dentate, greenish and scabrous above, whitish tomentose below, the midrib below with long sharp hairs, the margins with irregularly disposed sharp bristles. Flowers in capitate clusters; involucre and calyx densely villous. Sepals 5, yellow, similar to the 5 darker yellow petals, 5–8 mm long; stamens 5, the appendage spatulate, inflated and extending beyond the anther, filament short and basally adnate to the perianth; style short, stigma large, conical and obscurely 3-lobed. Capsule indehiscent, 1-seeded. (n = 7, 13) Jun–Aug (Oct). Gravelly, rocky, or clay soils, caprock ledges; sw OK, TX panhandle, ne NM; (sw OK to se AZ s to Mex.).

2. MENTZELIA L., Blazing Star, Sand Lily, Stickleaf

Annual, biennial, and perennial mostly suffrutescent herbs. Stems simple or branched, often strict, smooth, puberulent, scabrous, or exfoliating. Leaves sessile to short-petiolate, sometimes basally clasping, often somewhat appressed to the stems, entire to sinuate, dentate, lobed, or pinnatifid, usually variable on the plant, scabrous above and below and adhering to fur and clothing, the 3 kinds of hairs spinelike, glochidiate, and harpoon-topped, these mixed. Flowers solitary or usually in cymose groups, diurnal or crepuscular, subtended by leafy green bracts that are sometimes fused to the ovary. Sepals usually persistent, apiculate; petals 5 or apparently 8 or more in those with petaloid staminodia, white to deep yellow or orangish, borne at the top of a short hypanthium; stamens 10–very numerous, inserted on the hypanthium, sometimes exceeding the petals; filaments of the outer stamens sometimes broadened; petaloid staminodia, when present, much narrower than to as wide and long as the petals; disk concave; style elongate; placentae 1–5; ovules pendulous or horizontal, in 1 or 2 rows on each placenta. Capsules dehiscent by apical valves, cylindrical to somewhat clavate; seeds pendulous and prismatic and wingless, or horizontal and flattened and winged. *Acrolasia* Presl, *Nuttallia* Raf.—Rydberg.

In addition to the species given, others more typical of the regions to our west are reported from time to time along our extreme western border. These can be identified using appropriate manuals for the state or region in which they are found.

1. Petals 5, yellow to orange, 2–10 mm long, the flowers not showy; filaments narrow, staminodia lacking; seeds pendulous, wingless, not flattened.
    2. Annuals with smooth, puberulent, or merely scabrid stems; petals 2–4(6) mm long; seeds 10–40, rhomboidal or prismatic, angular, often grooved on the angles.
        3. Seeds apparently smooth; flowers rather crowded; bracts and at least the upper leaves often ovate, entire, acute .... 4. *M. dispersa*

  3 Seeds tuberculate; flowers not especially crowded; bracts and leaves often lanceolate, lobed .... 2. *M. albicaulis*
 2 Perennial with rough-scabrous stems; petals 6–10 mm long; seeds 1–4, oblong or elliptical, not sharply angular .... 7. *M. oligosperma*
1 Petals (including petaloid staminodia) 8 or more, white to yellow, 5–70 mm long, the flowers somewhat to very showy; filaments of outer fertile stamens sometimes broadened, staminodia sometimes present; seeds horizontal, barely to strongly winged, flattened.
 4 Petals white to cream, 2–7 cm long.
  5 Petals 1.5–4 cm long, 0.3–1 cm wide, not overlapping in anthesis, usually white; bracts not fused to ovary; mostly on sandy soil .... 6. *M. nuda*
  5 Petals (2.5)5–7 cm long, 1–2 cm wide, touching or overlapping in anthesis, usually cream; bracts partly fused to ovary; not often on sandy soil .... 3. *M. decapetala*
 4 Petals pale yellow to golden, less than 3 cm long.
  6 Petals 6–7(10) mm long, pale yellow to whitish; stamens fewer than 40 .... 1. *M. albescens*
  6 Petals (5)10–30 mm long (white) yellowish to golden; stamens more than 40.
   7 Some outer fertile stamens with broadened filaments; capsules rather urceolate .... 5. *M. multiflora*
   7 All fertile stamens with narrow filaments; capsules not urceolate .... 8. *M. reverchonii*

1. *Mentzelia albescens* (Gill. & Arn.) Griseb. Biennial-perennial from a taproot. Stems to 6 dm tall, usually simple, puberulent, somewhat whitish. Leaves 3–9 cm long, 1–2.5 cm wide, linear to ovate-lanceolate, coarsely sinuate-dentate, the teeth and tip acute, obtuse, or rounded. Floral bracts linear to lanceolate, entire to somewhat lobed. Flowers opening in late afternoon; sepals 3–6 mm long; petals apparently 10, 6–7(10) mm long, 1–2 mm wide, lanceolate to spatulate, acute, pale yellowish to whitish; stamens to 40, 5 of the outer ones petaloid and the flower thus appearing to have 10 petals. Capsule 1.5–3 cm long; seeds numerous, flattened, winged. (n = 11) May–Aug. Dry, often granitic hills; s GP; sw OK, TX panhandle; (cen & w TX to Mex.). Reported as a waif on mine wastes in sw MO: Jasper; & se KS: Cherokee, but possibly not persisting there; reported from CO.

2. *Mentzelia albicaulis* (Hook.) T. & G. Annual, 1.5–4 dm tall and often branching from the base. Stems white, glabrous below, glabrous to scabrid above. Leaves basal and cauline, sessile, entire to nearly pectinate, variable on the plant, or some plants with all leaves entire, others with all leaves lobed, 1–7 cm long, 2–12 mm wide, scaberulous. Flowers solitary or few in loose corymbs, opening in the morning; bracts linear, entire to somewhat dentate; sepals 2–3 mm long; petals 5, (2)3–5(7) mm long, 3–4 mm wide, obovate, rounded, overlapping in anthesis, yellow; stamens numerous, all fertile, shorter than the petals; placentae 3, parietal. Seeds 10–40, pendulous, angular, tuberculate. (n = 36) May–Jul. Dry gravelly & sandy soil; WY, CO, sw SD, NE panhandle; (B.C. to Baja Calif., e to w NE).

3. *Mentzelia decapetala* (Pursh) Urban & Gilg, ten-petal mentzelia. Resembling *M. nuda* but coarser, with larger flowers and the petals usually overlapping in anthesis, and not favoring sandy soil. Course erect biennial or perennial from a taproot. Stems 1–several, to 1 m tall, strict, branched above, the bark whitish and exfoliating below. Leaves below short-petiolate and more or less lanceolate, leaves above sessile; leaves 5–15 cm long, 1.5–4 cm wide, not especially reduced above, the margins regularly or irregularly sinuate to serrate, the serrations acute to acuminate but not dentate, scabrous, acuminate. Flowers showy near the branch tips, 8–15 cm wide, opening in late afternoon and closing about midnight; bracts more or less pectinate, some of them partially fused to the ovary; sepals 1–5 cm long, acute to acuminate; petals apparently about 10, white to cream, 5–7 cm long, 1–2 cm wide, oblanceolate to spatulate and often touching or overlapping in anthesis, acute to acuminate or cuspidate; stamens numerous, shorter than the petals, nonpetaloid staminoida absent. Capsules cylindric, (1.5)3–5 cm long, (1)1.5–2 cm wide; seeds numerous on 5–7 parietal placentae, flattened, the wing rudimentary, in rows on each placenta. (n = 11) Jul–Sep. Locally common on roadsides & other disturbed places, mostly not on sandy soil; GP mostly w of 98° long & s along the Missouri R. to IA: Pottawatamie; (extreme w IA to Alta. & s, e of the Rocky Mts. and w of 98° long, to Mex.). *Nuttallia decapetala* (Pursh) Greene — Rydberg.

A somewhat similar species, *M. laevicaulis* (Dougl.) T. & G., with the petaloid staminodia much narrower than the petals and the bracts not fused to the ovary, is sympatric in s MT and ne WY and barely enters our range.

4. *Mentzelia dispersa* S. Wats. Slender annual 1–3 dm tall, stem simple or branched above and/or below, whitish, puberulent. Leaves basal and cauline, 2–7 cm long, 4–10(15) mm wide, sessile, often variable on the plant, lanceolate, linear, or ovate, often ovate and entire at least above, often linear to lanceolate and entire to serrate or dentate below. Flowers rather crowded near the tips, small, numerous; sepals acuminate, 1.5–2.5 mm long; petals 5, yellow, 3–5 mm long, 2–3 mm wide, obovate to spatulate or suborbicular, rounded; stamens fewer than 40, all fertile. Capsules 1–2 cm long, 1–3 mm wide, linear; seeds few, uniseriate on the placentae, pendulous, wingless, angular and usually grooved on the angles, appearing smooth under ordinary magnification. (2n = 36) Jun–Sep. Disturbed & usually sandy places; MT, w ND, w SD, WY; (WA to MT & WY s to CO & CA). *Acrolasia compacta* (A. Nels.) Rydb. — Rydberg.

5. *Mentzelia multiflora* (Nutt.) A. Gray. Rather coarse perennial from a taproot; stem usually branched above, to 9 dm tall, whitish, glabrous or pubescent, the bark exfoliating below. Leaves 2–6(15) cm long, 0.5–1.5 cm wide, scabrous, linear to lanceolate, sinuate to coarsely dentate, pinnatifid, or even pectinate, the teeth obtuse to rounded. Flowers several at the branch tips, opening in late afternoon; flowers bractless or sometimes with small linear bracts; sepals 6–9(12) mm long, acuminate; petals apparently 10 or so, (whitish) yellow; 1–2.5 cm long, to 5 mm wide, obovate, mostly obtuse to rounded; stamens numerous, mostly shorter than the petals, outer fertile stamens with broader (but not petaloid) filaments. Capsules cylindric-urceolate, 1–2 cm long, less than 1 cm wide, the persistent sepals flaring in fruit; seeds biseriate on the 3 placentae, horizontal, winged, flattened. (2n = 18) Jul–Aug. Dry, sandy soils, occasionally in hard clays & rocky soil; se WY, w NE, e CO, OK & TX panhandles, ne NM; (CA to WY s to TX & Mex.).

This rather variable entity is sometimes regarded as a species-complex that includes *M. pumila* (Nutt.) T. & G. and other yellow-flowered species that occur along our western border.

6. *Mentzelia nuda* (Pursh) T. & G. Resembling *M. decapetala* but less coarse, the flowers smaller and the petals not touching in anthesis, and favoring sandy soil. Erect biennial or perennial to 1 m tall, from a taproot; stems 1–few, strict, usually branched above, yellowish to whitish, pubescent at least above. Leaves below short-petiolate and oblanceolate, leaves above sessile; leaves 4–10 cm long, 1.5–2 cm wide, usually somewhat reduced above, coarsely serrate or dentate, acute to obtuse, scabrous. Flowers near the branch tips, opening in mid to late afternoon and closing around sunset; bracts not fused to the ovary, serrate to laciniate but not pectinate. Flowers 4–9 cm wide; sepals lanceolate, acuminate, 10–25 mm long; petals apparently about 10, white to pale cream, oblanceolate to almost spatulate, 2–5 cm long, 3–10 mm wide, and mostly not touching in anthesis; stamens very numerous, the longest about equaling the petals; narrow staminodia nearly as long as the petals. Capsules more or less cylindric, (1)2–3 cm long, 8–10 mm wide; seeds numerous, biseriate on 3(4) parietal placentae, horizontal, broad-winged, flattened. (n = 10) Jul–Sep. Common on sandy & gravelly hillsides, pastures, roadsides, gypsum banks; e MT & WY, w SD, w

& cen NE, KS & OK, e CO, TX panhandle; (from the Rocky Mts. e to w SD, southward w of 98° long. to TX). *Nuttallia nuda* (Pursh) Greene, *N. stricta* (Osterh.) Greene — Rydberg.

7. ***Mentzelia oligosperma*** Nutt., stickleaf. Perennial from a thick woody caudex; stem 2–6(10) dm tall, often much branched, scabrous, whitish, the bark usually exfoliating below. Leaves sessile, lanceolate-ovate to rhombic, 1–6 cm long, 0.5–3 cm wide, coarsely dentate or serrate to basally lobed, scabrous. Flowers few to many, opening in the morning; sepals lanceolate, attenuate to acicular, often persisting in fruit, ca 5 mm long; petals 5, (4)7–10 mm long, 3–4 mm wide, pale yellow to orangish, obovate, sometimes mucronate, barely if at all overlapping in anthesis; stamens 15–40, all fertile, about equaling the petals. Capsules cylindrical to clavate, 7–12 mm long, 2–3 mm wide; seeds 1–3(4), pendulous, wingless, not sharply angular. (n = 10) Jun–Aug. Dry rocky prairies in sand or clay, gypsum & limestone bluffs & ledges; e WY, sw SD, NE panhandle, e CO, across KS into w MO, OK, TX panhandle; (WY to AR s to TX & NM).

8. ***Mentzelia reverchonii*** (Urban & Gilg) Thomps. & Zavortink. Perennial to 1 m tall; stems strict, whitish, scabrid-pubescent, branched above. Basal and lower cauline leaves 3–8 cm long, linear to lanceolate, leaves above gradually reduced and somewhat clasping basally; leaves shallowly lobed to mostly regularly fine- or coarse-toothed. Flowers rather showy, borne at the branch tips; sepals ca 10 mm long; petals apparently more than 10, yellow, spatulate, 10–30 mm long; stamens numerous, with narrow or only barely expanded filaments, staminodia narrow. Capsules cylindrical, 1.5–3 cm long; seeds horizontal, broad-winged, flattened. May–Sep. Gravelly & limestone soils; se CO, sw OK, TX panhandle, e NM; (GP to Mex.).

56. SALICACEAE Mirb., the Willow Family

by G. E. Larson

Dioecious spring-flowering trees and shrubs with simple, alternate, usually stipulate leaves, the stipules often deciduous. Flowers in erect or pendulous, bracteate catkins (aments), the bracts small and scalelike, often deciduous, each flower provided with either 1 or 2 enlarged basal glands (*Salix*) or an oblique, cup-shaped disk (*Populus*) positioned just inside the bract; perianth none; staminate flowers each consisting of 2–many stamens; pistillate flowers comprised of a single pistil; carpels 2–4; stigmas equaling the number of carpels, or sometimes bifid, sessile or with a common style, ovary 1-celled, ovules many. Fruit a many-seeded capsule, dehiscent by 2–4 valves; seeds minute, lacking endosperm, tufted with long, white, silky hairs.

1 Trees; leaf buds covered by several bud scales; flowers subtended by an obliquely cup-shaped disk; catkin bracts laciniate or fimbriate; stamens (4)7–80; catkins pendulous ...... 1. *Populus*
1 Shrubs and trees; leaf buds covered by a single bud scale; flowers subtended by 1 or 2 enlarged glands; catkin bracts entire; stamens commonly 2 or 3–8(12); catkins erect to pendulous .... 2. *Salix*

1. POPULUS L., Aspen, Cottonwood, Poplar

Contributed by James E. Eckenwalder

Trees, among the largest in the GP, some soboliferous. Bark light, smooth at first, often becoming deeply furrowed with age. Winter buds of several scales, often resinous and/or pubescent; vegetative and inflorescence buds separate. Leaf blades 3-nerved from base, lanceolate to orbicular or palmately 5-lobed, glandular-serrate or nearly entire, sometimes with 1–5 basilaminar glands; petioles usually 1/3 or more the length of the blade; stipules minute, caducous. Shoot system heterophyllous. Leaves of spring flush (early leaves) separated by internodes generally less than 1 cm long, with fewer, less conspicuously glandular teeth than leaves produced during late spring and summer on long shoots (late leaves), the latter separated by internodes more than 2 cm long. Inflorescences pendulous, leafless catkins, appearing before the leaves; bracts caducous, entire, fimbriate, or laciniate, membranaceous. Floral disk cup-shaped, nonnectariferous; stamens (4)7–80, filaments free; stigmas 2–4, bifurcated, linear or dilated. Fruits variously ovoid, 2- to 4-valved, thin walled. Moist soils from sea level to 4000 m; chiefly riparian. 20–30 spp. of Northern Hemisphere (1 sp. in E. Africa).

Populus trichocarpa T. & G. (Sect. *Tacamahaca*) has been recorded erroneously for our area (Stevens; Little, E. L., Jr. 1971. Atlas of United States Trees. Vol. 1. U.S.D.A. Misc. Publ. 1146; Atlas GP). The specimens on which these reports are based are *P. x jackii* (no. 10), a hybrid between *P. balsamifera* (no. 4) and *P. deltoides* (no. 8). This hybrid, and others recorded here, sometimes occur outside of the present range of either or both parents. These distributional anomalies, as well as those of species not involved in hybridization, are probably a consequence of Late Pleistocene and Recent climatic changes. Hybrids constitute a minute, but conspicuous and awkward fraction of the total number of individual poplars in the GP. Certain Eurasian balsam poplars and their hybrids, such as *P. x berolinensis* Dipp., *P. maximowiczii* A. Henry, and *P. simonii* Carr., may be found in shelter-belt plantings but appear not to have become established here as they have in Canada. Twig colors are useful for separating species but are variable, and I have cited only the most common colors below.

1 Winter buds, first year twigs, and lower surface of leaf blades tomentose.
  2 Tomentum dense and persistent; blades of late (long shoot) leaves palmately 5-lobed .... 2. *P. alba*
  2 Tomentum thin and usually transient, especially on leaves; blades of late leaves ovate to angular-dentate.
    3 Blades of early (short shoot) leaves with bluntly rounded teeth; blades of late leaves angular-dentate; floral bracts shallowly toothed .... 7. *P. x canescens*
    3 Blades of early leaves with sharply pointed teeth; blades of late leaves crenate-serrate; floral bracts laciniate .... 9. *P. grandidentata*
1 Winter buds, first year twigs, and lower surface of leaf blades glabrous or variously pubescent, but not tomentose.
  4 Blades of most leaves lanceolate to lance-ovate, at least 1.5 × as long as wide; mature terminal winter buds usually 6–10 mm long, reddish-brown, very resinous.
    5 Leaf blades serrate in middle portion with teeth up to 1.5 mm deep; capsules 2- or 3-valved; stamens 25–40 .... 1. *P. x acuminata*
    5 Leaf blades serrate along whole margin with teeth less than 0.6 mm deep; capsules 2-valved; stamens 10–25(30).
      6 Petioles usually less than 1 cm long; third-year twigs light tan ... 3. *P. angustifolia*
      6 Petioles usually more than 1 cm long; third-year twigs light gray.
        7 Leaf blades pale green to white but not strongly glaucous beneath; base usually cuneate; first-year twigs orange-brown; capsules orbiculoid, with 4–12 seeds per placenta .... 5. *P. x brayshawii*
        7 Leaf blades white and strongly glaucous beneath; base usually rounded; first year twigs reddish-brown; capsules ovoid, with 15–22 seeds per placenta .... 4. *P. balsamifera*
  4 Blades of most leaves ovate to deltoid ovate or orbicular, no more than 2× as long as wide; mature terminal winter buds usually longer than 10 mm or, if shorter, sparingly resinous.
    8 Blades of early (short) shoot leaves acuminulate, either ovate to orbicular and subentire to minutely crenate-serrate, or ovate and coarsely sharp-serrate, often with round or cup-shaped basilaminar glands; petiole strongly transversely flattened, flexuous; mature terminal winter buds shorter than 10 mm, sparingly resinous; capsules lance-ovoid; stamens 6–12.

9 Blades of early leaves subentire to finely crenate-serrate with (12)18–31(42) teeth on each side, the largest of these 0.1–1 mm deep; bud scales and lower surface of blade of late (long shoot) leaves glabrous .... 12. *P. tremuloides*
9 Blades of early leaves coarsely sharp-serrate with (1)5–12(16) teeth on each side, the largest of these 1.5–6 mm deep; bud scales and lower surface of blade of late leaves pubescent .... 9. *P. grandidentata*
8 Blades of early leaves acute or long-acuminate, subentire to coarsely crenate-serrate, with or without basilaminar glands, these never cup-shaped; petiole flattened or not; mature terminal winter buds usually longer than 10 mm, very resinous; capsules ovoid or elliptic-ovoid; stamens 15–80.
  10 Blades of early leaves ovate (rarely deltoid-ovate), acute (rarely acuminulate), subentire to evenly crenate-serrate along the whole margin with teeth no more than 0.5 mm deep; petioles terete; winter buds fragrant; capsules 2-valved, with 15–22 seeds per placenta .... 4. *P. balsamifera*
  10 Blades of early leaves usually deltoid-ovate to deltoid, acuminate, unevenly finely to coarsely crenate-serrate in the middle portion of the margin with the largest teeth 1–6 mm deep, usually entire at base and apex; petioles weakly to strongly transversely flattened at junction with blades; winter buds not fragrant; capsules 2- to 4-valved, or if exclusively 2-valved, with 4–8 seeds per placenta.
    11 Blades of early leaves concolorous, light green on both surfaces, very coarsely crenate-serrate, the largest teeth (1.5)2–6 mm deep, often with 2(5) prominent, round (or tubular) basilaminar glands; mature winter buds brown or tan; stamens (30)40–80; capsules 3- to 4-valved .... 8. *P. deltoides* subsp. *monilifera*
    11 Blades of early leaves darker green above, paler beneath, coarsely crenate-serrate, the largest teeth 1–2 mm deep, sometimes with 1 or 2(3) usually weak, round or marginal basilaminar glands; mature winter buds reddish; stamens 15–30(40); capsules 2- or 3-valved.
      12 Petioles weakly transversely flattened and slightly channeled above at junction with blade; base of early leaf blades usually truncate to subcordate, not shouldered, often with 2(3) basilaminar glands; petioles and shoots glabrous or pubescent .... 10. *P. x jackii*
      12 Petioles strongly transversely flattened and not channeled above at junction with blade; base of early leaf blades usually cuneate to rounded, shouldered if truncate, rarely with 1 or 2 basilaminar glands; petioles and shoots glabrous.
        13 Blades of early leaves generally rounded at base, shouldered, the largest teeth 1.5–2 mm deep; blades of late (long shoot) leaves deltoid-ovate, about as broad as long; crown sometimes narrow but not columnar; capsules 2- to 3-valved .... 6. *P. x canadensis*
        13 Blades of early leaves cuneate at base, not shouldered, the largest teeth ca 1 mm deep; blades of late leaves broadly deltoid, usually broader than long; crown often narrowly columnar; capsules exclusively 2-valved .... 11. *P. nigra*

1. ***Populus x acuminata*** Rydb. (Sect. *Aigeiros x Tacamahaca*), lanceleaf cottonwood. Single-stemmed tree 10–20(25) m tall, 2–5(8) dm d.b.h.; branches horizontal to shallowly ascending, forming a narrowly spreading flat-topped crown; first-year twigs orange-tan, turning grayish-tan by the third year, glabrous; bark tan, furrowed on trunk; winter buds reddish, lanceoloid, very resinous, sparsely setose; terminal buds (5)9–12 mm long; flowering buds (6)11–14 mm long. Leaf blades concolorous, light green, angular lanceolate, (1)5–9(13.5) cm long, (0.6)3–6(7.5) cm wide, minutely to coarsely crenate-serrate along middle 2/3 of margin, apex acute, base cuneate, sometimes with 1 or 2 weak, round basilaminar glands, petiole slightly compressed at junction with blade, slightly channeled above, (1)2–4.5 cm long; blades of early leaves with (6)15–25(30) teeth on each side, these 0.1–1.2 mm deep; blades of late leaves with (25)30–45(51) teeth on each side, these 0.2–1.3 mm deep. Catkins loose, 6.5–9 cm long (–16.5 cm in fruit), rachis glabrous, pedicels 1–2 mm long (–4 mm in fruit). Floral disks 1.5–2.5 mm wide; staminate flowers with 25–40 stamens;

pistillate flowers with broad, compact stigmas. Capsules 2- or 3-valved, broadly ovoid, 5–7 mm long, with (6)8–10 seeds per placenta. (n = 19) Apr–May, fruits May–Jul. Riparian in westernmost GP; e CO, sw ND, nw NE, ne NM, sw SD, e WY; (w ND, MT, & Alta., s to w NE, TX, sw NM, ne AZ, & se UT).

Lanceleaf cottonwood, as early suspected and recently demonstrated, is a hybrid between *P. angustifolia* (no. 3) and *P. deltoides* (no. 8). It occurs primarily along the foot of the Rocky Mts. Front Range where these species grow together, but also extends out onto the plains. Today, there are no trees of *P. angustifolia* in the ND localities of this hybrid, nor in some other outlying localities.

2. *Populus alba* L. (sect. *Populus*), silver poplar. Soboliferous tree 5–20(43) m tall, 2–10(20) dm d.b.h.; trunk sometimes dividing into large branches near base, branches ascending, forming a globose crown; first-year twigs olive-brown, turning orange-brown by the third year, densely white-tomentose at first, becoming glabrate, bark light gray and furrowed on lower trunk, white and smooth on branches and upper portion, with rhomboid gray rough areas surrounding former branch attachments; winter buds reddish, ovoid, not resinous, tomentose; terminal buds (3)5–8 mm long; flowering buds 6–10 mm long. Leaf blades dark green above, densely white-tomentose beneath; blades of early leaves ovate, (1)3–5(7) cm long, (0.7)2–4(6) cm wide, coarsely sinuate-dentate with 3–8 doubly convex teeth on each side, these 0.5–2 mm deep, apex obtuse, base rounded to subcordate, often with 2 round basilaminar glands, petiole terete, (0.5)1–2.5 cm long, tomentose; blades of late leaves palmately 5-lobed. Catkins dense, 2–8 cm long (–9 cm in fruit), rachis tomentose, bracts shallowly toothed, pedicels 0.5–1 mm long (–1.2 mm in fruit). Floral disks 0.7–1.1 mm wide; staminate flowers with 6–10 stamens; pistillate flowers with filiform, erect stigmas. Capsules narrowly ovoid, (2)3–5 mm long, with (1)2(3) seeds per placenta, these usually not maturing here. (n = 19, 2n = 38, 57) Mar–Apr, fruits Apr–Jun. Hedgerows, margins of woodlots, & other disturbed areas; GP; (throughout U.S. & Can., Eurasia, N. Africa). *Cultivated, persisting and adventive alien. P. bolleana* Mast.—Van Bruggen.

Most seemingly spontaneous GP silver poplars are members of a single pistillate clone. However, fastigiate individuals of a second pistillate clone are occasionally cultivated. The latter are *P. alba* "Bolleana." The silver poplar hybridizes naturally with the native aspens, *P. grandidentata* (no. 9) and *P. tremuloides* (no. 12), e of the GP, but there is no present evidence of such hybrids within the region. It is also a parent (along with the Eurasian aspen, *P. tremula* L.) of the introduced hybrid, *P. x canescens* (no. 7).

3. *Populus angustifolia* James (sect. *Tacamahaca*), narrowleaf cottonwood. Single-stemmed tree 5–8(20) m tall, 2–5(7) dm d.b.h.; branches ascending in upper half of trunk, forming a narrowly spreading crown; first-year twigs orange-brown, turning light tan by the third year, glabrous; bark light brown, shallowly furrowed on lower portion of trunk, smooth above; winter buds reddish, ovoid to narrowly ovoid, resinous, glabrous; terminal buds (3)6–9(13) mm long; flowering buds 8–12(18) mm long. Leaf blades dark green above, light green beneath, lanceolate to narrowly ovate (1.5)4–9(13.5) cm. long, 0.8–2.5(4) cm wide, glabrous, minutely crenate-serrate along the whole margin, apex acute, base acute to rounded, without basilaminar glands, petiole terete, slightly channeled above, 0.2–0.8(1.7) cm long; blades of early leaves with (14)23–35(65) teeth on each side, these 0.1–0.3 mm deep; blades of late leaves with 35–65(80) teeth on each side, these 0.1–0.6 mm deep. Catkins loose, 3–8 cm long (–9 cm in fruit), rachis thinly pubescent or glabrous, pedicels 0.5–1.5 mm long (–3 mm in fruit). Floral disk 1–2 mm wide; staminate flowers with 10–20 stamens; pistillate flowers with broad, compact stigmas. Capsules 2-valved, orbiculoid, 3–5 mm long, with 4–8 seeds per placenta. (2n = 38) Apr–May, fruits Jun. Riparian in canyons in conifer zone of w edge of GP; n-cen CO, se MT, nw NE, ne NM, sw SD, se WY; (Sask. to B.C., s through SD & ID to TX; Mex).

Narrowleaf cottonwood, although uncommon in the GP, is abundant in the Mts. westward. It hybridizes with *P. balsamifera* (no. 4) to form *P. x brayshawii* (no. 5) and with *P. deltoides* (no. 8) to form *P. x acuminata* (no. 1).

4. *Populus balsamifera* L. (sect. *Tacamahaca*), balsam poplar. Single-stemmed tree 10–30 m tall, 3–10(21) dm d.b.h.; branches ascending in upper 2/3 of trunk, forming a narrow obovoid crown; first-year twigs reddish-brown, turning grayish-brown by the third year, glabrous to densely pubescent; bark reddish-gray, furrowed on trunk; winter buds reddish, lance-ovoid, very resinous, fragrant, glabrous; terminal buds (8)12–18(20) mm long; flowering buds 15–19 mm long. Leaf blades dark green above, white and glaucous beneath, with prominent veins and often marked by orange streaks of resin, ovate to broadly ovate, (2.5)5–8.5(15) cm long, (0.7)3–5.5(10) cm wide, glabrous or pubescent, subentire to finely crenate-serrate along the whole margin, apex generally obtuse, base rounded to subcordate, sometimes with 2(5) prominent, marginal or round basilaminar glands, petioles terete, channeled above, (0.2)1.5–5 cm long; blades of early leaves with (9)20–35(45) teeth on each side, these 0.1–0.4 mm deep; blades of late leaves with (20)30–50(60) teeth on each side, these 0.2–0.6 mm deep. Catkins loose, 7.5–15 cm long (–18 cm in fruit), rachis pubescent, pedicels 0.5–2 mm long (–3.5 mm in fruit). Floral disks 2–3(4) mm wide; staminate flowers with 20–30 stamens; pistillate flowers with platelike, reflexed stigmas. Capsules 2-valved, ovoid, (3)5–8 mm long, with 15–22 seeds per placenta. (n = 19, 2n = 38) Mar–Jun, fruits May–Jul. Riparian in boreal & montane conifer forests; n-cen CO, nw IA, n MN, NE, nw NE, ne & sw SD, WY; (Newf. w to AK, s to WV, IA, NE, & CO). *P. balsamifera* var. *subcordata* Hyl.—Fernald; *P. tacamahaca* P. Mill.—Rydberg.

Balsam poplar is more common n and e of the n GP where it forms extensive flood-plain forests, and it also extends s sparingly in the Rocky Mts. w of the GP to CO. It hybridizes with *P. angustifolia* (no. 3) to form *P. x brayshawii* (no. 5) and with *P. deltoides* (no. 8) to form *P. x jackii* (no. 10).

5. *Populus x brayshawii* Boivin (sect. *Tacamahaca*), hybrid balsam poplar. Single-stemmed tree 8–20 m tall, 3–8 dm d.b.h.; branches ascending in upper half of trunk, forming a narrow obovoid crown; first year twigs orange-brown, turning light gray by the third year, glabrous to thinly pubescent; bark light gray-brown, shallowly furrowed on trunk; winter buds reddish, lance-ovoid, resinous, glabrous; terminal buds 7–15 mm long; flowering buds 15–18 mm long. Leaf blades dark green above, white but not glaucous beneath, lanceolate to narrowly ovate, (2.5)5–9(11.5) cm long, (0.7)1.5–3.5(5) cm wide, glabrous, minutely crenate-serrate along the whole margin, apex acute, base cuneate to rounded, occasionally with 1 or 2 weak marginal basilaminar glands, petiole terete, (0.3)1–2.5(4.2) cm long; blades of early leaves with (10)15–35(45) teeth on each side, these 0.1–0.3 mm deep; blades of late leaves with (30)50–60(70) teeth on each side, these 0.2–0.5 mm deep. Catkins loose, 2.5–9 cm long (–10 cm in fruit), rachis pubescent, pedicels 0.5–2 mm long (–4 mm in fruit). Floral disks 1–2 mm wide; staminate flowers with 15–25 stamens; pistillate flowers with broad, compact stigmas. Capsules 2-valved, broadly ovoid to orbiculoid, 4–5.5 mm long, with (4)7–9(12) seeds per placenta. Apr–Jun, fruits May–Jul. Riparian in canyons of conifer zone of BH; SD: Custer, Lawrence, WY: Crook; (Alta. se to w SD & WY, s to CO, UT).

This hybrid is intermediate in many respects between its closely related parents, *P. angustifolia* (no. 3) and *P. balsamifera* (no. 4), and complicates their recognition. It is most similar to narrowleaf cottonwood but differs in petiole length and twig color, characters in which it approaches balsam poplar. Some trees of balsam poplar from ND: Bottineau and Divide, also show influence of narrowleaf cottonwood, although far from the range of the latter species.

6. *Populus x canadensis* Moench (sect. *Aigeiros*), Carolina poplar. Single-stemmed tree 15–30(45) m tall, 3–6(24) dm d.b.h.; branches horizontal to ascending in upper 2/3 of trunk, forming an obovoid crown; first-year twigs olive-brown, turning grayish-orange by the third year, glabrous to sparsely pubescent; bark tan, furrowed on trunk; winter buds reddish, ovoid, very resinous, glabrous; terminal buds (2)7–15(22) mm long; flowering buds (8)13–20(24) mm long. Leaf blades dark green above, light green beneath, broadly ovate, (2)5–9(12) cm long, (2)4–8(10.5) cm wide, glabrous, crenate-serrate along middle 2/3 of margin, apex acuminate, base broadly cuneate to subcordate, shouldered, sometimes with 1 or 2 weak round basilaminar glands, petiole laterally compressed at junction with lamina, (0.7)3–7(9.5) cm long; blades of early leaves with 13–25(30) teeth on each side, these 0.3–1.3(2) mm deep; blades of late leaves with (25)30–45 teeth on each side, these 0.4–1(2) mm deep. Catkins loose, (2.5)7–10 cm long (–19.5 cm in fruit), rachis glabrous, pedicels 1.5–3 mm long (–4 mm in fruit). Floral disks 2–3.5 mm wide; staminate flowers with 15–30 stamens; pistillate flowers with broad, compact stigmas. Capsules 2- or 3-valved, ovoid, 6–9 mm long, with 6–11 seeds per placenta. (n = 19) Mar–May, fruits May–Jun. Former homesites & other disturbed areas; GP; (throughout U.S., & Can.). *Cult. and persisting alien*. *P. canadensis* var. *eugenei* (Simon-Louis) Schelle — Gates; *P. x eugenei* Simon-Louis — Steyermark.

Most GP Carolina poplars are members of a single staminate clone, *P. x canadensis* 'Eugenei', which shares something of the narrow-crowned habit of its staminate parent, the lombardy poplar (no. 11). Other clones of *P. x canadensis* which may be cultivated in the GP are also the result of spontaneous and artificial hybridization between different lines of *P. deltoides* (no. 8) and *P. nigra* (no. 11).

7. *Populus x canescens* (Ait.) Sm. (sect. *Populus*), gray poplar. Soboliferous tree 8–25(37) m tall, 2–10(19) dm d.b.h.; branches ascending in upper half of trunk, forming a globose crown; first-year twigs reddish, turning grayish-brown by the third year, thinly tomentose at first, becoming glabrate; bark gray and furrowed on lower trunk, light gray and smooth on branches and upper portion, with scattered rhomboid dark rough areas; winter buds reddish, ovoid, not resinous, thinly tomentose; terminal buds (2)3–5(9) mm long; flowering buds 6–9(11) mm long. Leaf blades dark green above, light green and thinly tomentose beneath, becoming glabrate, broadly ovate, (2)3–6.5(8) cm long, (1.5)3–6(7) cm wide, apex obtuse, base broadly cuneate to subcordate, usually with 2(4) prominent round basilaminar glands, petiole slightly compressed at junction with blade, (0.5)2–4.5(5) cm long, tomentose to glabrate; blades of early leaves coarsely sinuate-dentate with 4–11 rounded teeth on each side, these 0.5–3.3 mm deep; blades of late leaves irregularly angular-dentate with (5)10–25(30) teeth on each side, these 0.5–7 mm deep. Catkins dense, 1.5–6.5 cm long (–10 cm in fruit), rachis tomentose, bracts shallowly toothed, pedicels 0.5–1 mm long (–1.5 mm in fruit). Floral disks 0.7–1.2 mm wide; staminate flowers with 8–15 stamens; pistillate flowers with 4-lobed, slender stigmas. Capsules 2-valved, narrowly ovoid, 3–5 mm long, with 3 or 4 seeds per placenta. (n = 19, 2n = 57) Apr, fruits May. Former homesites & other disturbed areas; GP; (Que. to Man., s to GA, AR & NE). *Cult. and persisting alien.*

The gray poplar is a natural hybrid between *P. alba* (no. 2) and the Eurasian aspen (*P. tremula* L.). It is much less commonly cultivated in the GP than its parent, the silver poplar, and seemingly not yet naturalized here. Nor has it hybridized with either of the native aspens here, as it has east of the GP.

8. *Populus deltoides* Marsh. subsp. **monilifera** (Ait.) Eckenw. (sect. *Aigeiros*), cottonwood. Single-stemmed tree 20–30(40) m tall, 2–12(20) dm d.b.h.; trunk dividing into large branches near base, ascending at a moderate angle to form a very broad, rounded crown; first year twigs olive-brown to orange-tan, turning grayish-tan by the third year, glabrous; bark tan, deeply furrowed on the trunk and major branches; winter buds tan, ovoid, very resinous, setose; terminal buds (6)8–14(21) mm long; flowering buds (8)10–18(28) mm long. Leaf blades concolorous, light green, deltoid-ovate, (1)4–9(14) cm long, (1.5)4–10(15) cm wide, glabrous, finely to coarsely crenate-serrate in middle portion, apex caudate-acuminate, base shallowly cuneate to cordate, usually with 2(6) prominent round basilaminar glands, petioles laterally compressed at junction with bade, (1)3–8(13) cm long; blades of early leaves with (3)7–15(21) teeth on each side, these (0.4)0.7–4.5(7) mm deep; blades of late leaves with (19)25–35(55) teeth on each side, these 0.5–1.3(2.2) mm deep. Catkins loose, (0.7)5–13 cm long (–20.5 cm in fruit), rachis glabrous, pedicels 1–6 mm long (–13 mm in fruit). Floral disks 1.5–4 mm wide; staminate flowers with (30)40–80 stamens; pistillate flowers with platelike, spreading stigmas. Capsules 3- or 4-valved, elliptic-ovoid, (4)8–10(16) mm long, with (3)7–10(18) seeds per placenta. (n = 19) Mar–Jun, fruits May–Jul. Riparian along rivers, streams, & lakes, & generally in moist soils; GP; (Que. to Alta. & WA, s to PA, MO, TX, NM). *P. besseyana* Dode, *P. sargentii* Dode, *P. virginiana* Foug. — Rydberg; *P. sargentii* var. *texana* (Sarg.) Correll — Correll & Johnston.

Cottonwood is the commonest poplar in the GP and one of the largest trees in the region. It hybridizes naturally with *P. angustifolia* (no. 3) to form *P. x acuminata* (no. 1) and with *P. balsamifera* (no. 4) to form *P. x jackii* (no. 10). It is also a parent of the introduced hybrid, *P. x canadensis* (no. 6). It intergrades with *P. deltoides* subsp. *deltoides* along the se margin of the GP, where there are trees with longer pedicels and 3–5 elongate basilaminar glands, and also intergrades with *P. deltoides* subsp. *wislizenii* (S. Wats.) Eckenw. along the sw margin of the region, where there are plants with longer pedicels, and early leaves with fewer teeth and without basilaminar glands.

9. *Populus grandidentata* Michx. (sect. *Populus*), bigtooth aspen. Soboliferous tree 8–15(30) m tall, 2–6(10) dm d.b.h.; branches horizontal below, ascending above to form a narrow cylindrical crown; first year twigs reddish-brown turning reddish-gray by the third year, nearly glabrous; bark brown and furrowed on lower trunk, light gray and smooth on branches and upper portion; winter buds reddish, ovoid, slightly resinous, pubescent; terminal buds 2.5–7(10) mm long; flowering buds 6–9(13) mm long. Leaf blades dark green above, white and glaucous beneath, thinly tomentose beneath at first, becoming glabrate, ovate, apex acute, often with (1)2(4) prominent cup-shaped basilaminar glands, petiole laterally compressed at junction with blade, 1.5–6(11.5) cm long, glabrate; blades of early leaves (2)3–9(11.5) cm long, (2)2.5–7(10.5) cm wide, coarsely serrate with (1)5–12(16) teeth on each side, these 0.3–4.5(6) mm deep, base shallowly cuneate; blades of late leaves (3)5–10.5(17.5) cm long, (2.5)3.5–8.5(18) cm wide, finely crenate-serrate with (5)15–50(65) teeth on each side, these 0.8–1.5(2.5) mm deep, base subcordate. Catkins dense, 4–10 cm long (–14 cm in fruit), rachis densely pubescent, bracts laciniate, pedicels 0.5–1.5 mm long (–2 mm in fruit). Floral disks 1–2 mm wide; staminate flowers with 6–12 stamens; pistillate flowers with slender, upright stigmas. Capsules 2-valved, 2–5(6) mm long, with (3)5–8 seeds per placenta. (n = 19) Mar–May, fruits May–Jun. Moist upland woodlands; nw IA, nw MN; (N.S. w to Man., s to NC, TN, & MO).

Bigtooth aspen is a species of the Great Lakes region that barely enters the ne corner of the GP. It hybridizes naturally with the introduced *P. alba* (no. 2) and the native *P. tremuloides* (no. 12) e of the GP. Aspens in some groves of the Niobrara R. valley, more than 350 km w of the nearest present station of *P. grandidentata*, are a hybrid between this species and *P. tremuloides*.

10. *Populus x jackii* Sarg. (sect. *Aigeiros x Tacamahaca*), balm-of-gilead. Single-stemmed tree 10–30 m tall, 2–10 dm d.b.h.; branches shallowly ascending to form an obovoid crown; first-year twigs reddish-brown, turning orange-tan by the third year, glabrous or pubescent; bark grayish-brown, furrowed on trunk; winter buds reddish, ovoid, very resinous, slightly setose; terminal buds (7)9–13(20) mm long; flowering buds (7)9–15(20) mm long. Leaf blades dark green above, pale green and slightly glaucous beneath, deltoid-ovate, (2.5)4.5–8(17) cm long, (2)4–7(15) cm wide, glabrous or pubescent on the veins beneath, finely crenate-serrate along most of the margin, apex acuminate, base broadly rounded to subcordate, often with 2(3) prominent round or submarginal basilaminar glands, petiole slightly compressed at junction with blade, (1)2.5–5(7.5) cm long, glabrous or pubescent; early leaves with (14)17–30(45) teeth on each side, these 0.1–1.4 mm deep; late leaves with

25–50(80) teeth on each side, these 0.3–0.6(1.2) mm deep. Catkins loose, (3)5–15 cm long (–20.5 cm in fruit), rachis pubescent, pedicels 1–3 mm long (–5 mm in fruit). Floral disks 2.5–4 mm wide; staminate flowers with 25–40 stamens; pistillate flowers with platelike, reflexed stigmas. Capsules 2- or 3-valved, broadly ovoid, 5–9 mm long, with 7–14(20) seeds per placenta. (n = 19, 2n = 38) Apr–May, fruits May–Jul. Riparian along rivers & streams; n-cen IA, nw MN, ND; (Que. w to Alta. s to WV, IN, NE, & CO). *Native and introduced.* *P. x andrewsii* Sarg.—Harrington; *P. candicans* auct. non Ait.—Rydberg; *P. x gileadensis* Rouleau—Fernald.

Wild balm-of-gilead trees are uncommon hybrids between *P. balsamifera* (no. 4) and *P. deltoides* (no. 8). They were the source of the report by Stevens of *P. trichocarpa* T. & G. in ND. The cultivated balm-of-gilead is a single pistillate clone of this parentage which is often cultivated e of the GP, and which may appear here sparingly.

11. *Populus nigra* L. (sect. *Aigeiros*), black poplar, lombardy poplar. Single-stemmed or sparingly soboliferous tree 20–30(40) m tall, 2–10(20) dm d.b.h.; branches shallowly ascending to patent, forming a broad or columnar crown; first-year twigs orange-tan, turning reddish-gray by the third year, glabrous or pubescent; bark light brown, deeply furrowed on trunk; winter buds reddish-brown, lance-ovoid, resinous, glabrous; terminal buds (4)10–15 mm long; flowering buds (7)10–19 mm long. Leaf blades bright green above, paler beneath, glabrous, (1.5)3–7(10) cm long, finely crenate-serrate along most of the margin, apex acuminate, rarely with 1 or 2 minute round basilaminar glands, petiole laterally flattened at junction with blade, (1)2–4.5(6.5) cm long; blades of early leaves rhomboid-ovate, (1.2)2.5–6(6.5) cm wide, with 10–22 teeth on each side, these 0.3–1.2 mm deep, base cuneate; blades of late leaves broadly deltoid-ovate, (2.5)4.5–7(11) cm wide, with (15)20–25(37) teeth on each side, these 0.2–0.6(1.5) mm deep, base truncate. Catkins loose, (1.5)3–6 cm long (–10 cm in fruit), rachis glabrous or pubescent, pedicels 0.5–1.5 mm long (–2 mm in fruit). Floral disks 1–2 mm wide; staminate flowers with 20–30 stamens; pistillate flowers with broad, reflexed stigmas. Capsules 2-valved, ovoid, 5–9 mm long, with 4–8 seeds per placenta. (n = 19) Apr–May, fruits May–Jun. Shelter belts, former homesites, & other disturbed areas; GP; (throughout U.S. & Can., Eurasia). *Cult. and persisting alien.* *P. nigra* var. *italica* Moench—Gleason & Cronquist.

Most black poplars planted in the GP are *P. nigra* 'Italica', a single staminate clone. These lombardy poplars are very popular ornamentals because of their tall, fastigiate habit. The clone is one of the parents of the Carolina poplar (no. 6). A few black poplars of ordinary habit and either sex may be encountered in the GP.

12. *Populus tremuloides* Michx. (sect. *Populus*), quaking aspen. Soboliferous tree 6–15(30) m tall, 1–5(9) dm d.b.h.; branches horizontal to shallowly ascending in upper half of trunk, forming a narrow, globose crown; first-year twigs reddish-brown, turning grayish-brown or grayish-orange by the third year, glabrous; bark dark gray and shallowly furrowed at base of trunk, pale greenish or yellowish white, or light gray and smooth above; winter buds reddish-brown, slightly resinous, glabrous; terminal buds lance-ovoid, acute, (2.5)4–6(9) mm long; flowering buds globose, obtuse, (4.5)6–10(11) mm long. Leaf blades dark green above, light green and slightly glaucous beneath, glabrous, apex acuminate, base shallowly cuneate to subcordate, shouldered, sometimes with 1 or 2 prominent round basilaminar glands, petioles laterally compressed at junction with blade, (0.7)1–6 cm long, glabrous; blades of early leaves ovate to suborbicular, (1)2.5–5(7) cm long, (0.6)2–5.5(7) cm wide, subentire to finely crenate-serrate with (12)18–30(42) teeth on each side, these 0.1–1 mm deep; blades of late leaves broadly ovate, (2)4–7.5(12) cm long, (1.2)3.5–7(10.5) cm wide, finely crenate-serrate with (20)25–40(50) teeth on each side, these 0.1–1.3 mm deep. Catkins dense, (1.7)4–7 cm long (–11 cm in fruit), rachis thinly pubescent, bracts laciniate, pedicels 0.5–1.5 mm long (–2 mm in fruit). Floral disks 1.3–1.8 mm wide; staminate flowers with

6–12 stamens; pistillate flowers with 2- or 3-lobed, slender stigmas. Capsules 2-valved, lance-ovoid, 2.5–4.5(7) mm long, with 3–6 seeds per placenta. (n = 19, 38; 2n = 38, 57) Apr–Jun, fruits May–Jul. Moist upland woodlands, mt. slopes, & stream sides; CO: Elbert, Las Animas; IA, MN, MT, ND, n NE, NM: Union; SD, WY; (Newf. to AK, s to VA, MO, NE, TX, Baja Calif., & Mex.). *P. tremuloides* var. *aurea* (Tidestr.) Daniels—Harrington.

Quaking aspen is common in the mts. just w of the GP, far s of its n GP occurrences. It hybridizes with *P. alba* (no. 2) and with *P. grandidentata* (no. 9) e of the GP, and remnants of the latter hybridization occur in Niobrara R. Valley (see discussion under *P. grandidentata*).

2. SALIX L., Willow

Trees and shrubs of typically wet or moist habitats. Leaves variable in shape, from ovate to lanceolate or linear-lanceolate, or obovate to oblanceolate, the margin serrate, crenate-serrate or entire; petioles much shorter than the blades, sometimes glandular or glandular-viscid at the summit; stipules persistent or caducous, occasionally lacking. Catkins sessile or on leafy branchlets, erect to pendulous, often precocious; bracts entire, usually pubescent, often apparently ciliate on the margins. Flowers each subtended by 1 or 2 enlarged basal glands; staminate flowers of commonly 2 or 3–8(12) stamens, the filaments sometimes connate; pistillate flowers each comprised of a bicarpellate pistil, stigmas 2- or 4-lobed, styles well developed to none. Capsules 2-valved, sessile or stipitate.

References: Argus, G. W. 1980. The typification and identity of *Salix eriocephala* Michx. (Salicaceae). Brittonia 32: 170–177; Dorn, R. D. 1975. A systematic study of *Salix* section *Cordatae* in North America. Can. J. Bot. 53: 1491–1522; Dorn, R. D. 1977. Willows of the Rocky Mountain states. Rhodora 79: 390–429; Froiland, S. G. 1962. The genus *Salix* in the Black Hills of South Dakota. U.S.D.A. Forest Serv. Tech. Bull. 1269. 75 pp.; Raup, H. M. 1959. The willows of boreal western America. Contr. Gray Herb. 185: 1–95; Stephens, H. A. 1973. Woody plants of the North Central Plains, Univ. Press of Kansas, Lawrence. 530 pp. George Argus provided much valuable assistance, and his contributions are gratefully acknowledged.

Willows are taxonomically one of our most perplexing groups. In our region most problems of identification are caused by the relatively small differences between species, coupled with considerable variability displayed within species. Determinations are actually fairly reliable if based upon fertile specimens, especially pistillate ones with mature capsules. Willow catkins, however, are deciduous structures which, in precocious species, appear in spring and drop off even before the leaves are fully expanded. Vegetative characters must usually be employed to identify summer or autumn material. The following key provides for the identification of pistillate catkin-bearing specimens as well as vegetative ones. Whether specimens are fertile or vegetative, correct determinations are more easily achieved if the collector records pertinent field data, including the plant size and growth habit (indicate whether a tree or shrub and estimate height), bark color and fresh leaf and twig color. In addition, obtaining twig samples from different parts of the tree or shrub is helpful for recognizing variability within individuals and species, e.g., the foliage of vigorous sucker shoots often differs considerably from that of normal vegetative branchlets. Collecting from the same individual at regular intervals during the growing season also provides a record of seasonable variation and permits relating of reproductive and vegetative structures.

Excluded taxa:

Salix babylonica L., the weeping willow, is widely cultivated as a shade and ornamental tree throughout the GP but shows no inclination to escape.

S. drummondiana Barr. ex Hook., *S. geyeriana* Anderss. and *S. glauca* L. are three Rocky Mt. willows included in the *Flora of the Black Hills* (Dorn, R. D., & J. Dorn. 1977. Published by the authors, pp. 284–287.). These species have not been detected among the numerous BH collections studied for this treatment.

S. lasiandra Benth. var. *caudata* (Nutt.) Sudw. is attributed to the BH of SD in *Vascular plants of the Pacific Northwest* (Hitchcock, C. L., A. Cronquist, M. Ownbey, & J. W. Thompson. 1964. Univ. of Wash. Press, vol. 2, p. 57.), but no specimens were seen among BH material.

1 Plants with pistillate catkins.
  2 Capsules glabrous (ovary pubescent in no. 8, normally glabrous at maturity).
    3 Petioles bearing conspicuous, irregularly lobate glands at or near the attachment to the blade.
      4 Leaves ovate-lanceolate, green on both surfaces, paler beneath but not white-glaucous; capsules 4–6.5 mm long at maturity.
        5 Leaves acute to short-acuminate, dark green and glossy above, thick and rather leathery .... 16. *S. pentandra*
        5 Leaves mostly long-acuminate, bright green and semiglossy above, not especially thick .... 11. *S. lucida*
      4 Leaves elliptic-lanceolate, white-glaucous beneath; capsules 7–10 mm long at maturity .... 21. *S. serissima*
    3 Petioles lacking glands or sometimes with minute vestiges of glands, or the petioles only glandular-viscid when young (often persistently glandular-viscid in *S. fragilis*).
      6 Leaves conspicuously or inconspicuously toothed.
        7 Leaves linear-lanceolate, mostly 8–20 × longer than wide; colonial shrub often forming dense thickets .... 8. *S. exigua*
        7 Leaves usually less than 10 × longer than wide (except often proportionately longer in *S. nigra* which is a tree); trees and noncolonial shrubs.
          8 Shrubs or small trees to 5(10) m tall; catkins emerging before or with the leaves, sessile or on short branchlets with a few small leaves; bracts brown to black, persistent after capsule maturity.
            9 Catkins sessile; leaves ovate to obovate .... 19. *S. pseudomonticola*
            9 Catkins sessile or often on short branchlets bearing a few small leaves; leaves lanceolate to somewhat oblanceolate.
              10 Twigs gray-brown to dark brown, closely gray-pubescent the first year and often into the second .... 7. *S. eriocephala*
              10 Twigs yellow or yellowish-gray to yellowish-brown, glabrous .... 12. *S. lutea*
          8 Trees to 10–20 m tall; catkins emerging after the leaves, on leafy branchlets; bracts yellowish-green or pale yellow, deciduous before capsule maturity.
            11 Capsules on stipes 1 mm or less long; introduced trees frequently escaping.
              12 Twigs olive to brown; petioles glandular-viscid near the summit on vigorous shoots .... 9. *S. fragilis*
              12 Twigs golden-yellow to orange; petioles lacking glands or with minute traces of glands only .... 1. *S. alba* var. *vitellina*
            11 Capsules on stipes 1–2(3) mm long; native trees.
              13 Leaves green on both sides .... 14. *S. nigra*
              13 Leaves noticeably pale to white-glaucous beneath.
                14 Flowering branchlet commonly branched from the axil of the uppermost leaf so that the catkin (although terminal) appears lateral on the branchlet; twigs and branchlets often gray pubescent; leaves dark green and shiny above, with petioles mostly 3–8 mm long, straight .... 5. *S. caroliniana*
                14 Flowering branchlet simple, the catkin not surpassed by a branch from the axil of the uppermost leaf; twigs and branchlets glabrous; leaves pale to yellow-green and dull above, with petioles mostly 5–20 mm long, commonly recurved ... 2. *S. amygdaloides*
      6 Leaves entire .... 15. *S. pedicellaris*
  2 Capsules pubescent.
    15 Catkins emerging and maturing ahead of the leaves.
      16 Capsules nearly sessile, on stipes 0.5 mm or less long .... 18. *S. planifolia*
      16 Capsules on stipes 0.5–4 mm long.
        17 Shrubs of rather dry upland sites, often where sandy; twigs of the previous year gray-pubescent or mostly so; leaves persistently gray-pubescent beneath (rarely glabrate); catkins 1–3(4) cm long .... 10. *S. humilis*
        17 Shrubs of low, wet sites; twigs of the previous year glabrous (rarely pubescent in

S. *discolor*); mature leaves glabrous, or if pubescent, then some of the hairs reddish-brown; catkins 2-6(9) cm long.
- 18 Catkins sessile; stipes of the capsules 1.5-4 mm long; leaves elliptic to narrowly ovate or narrowly obovate, acute at the apex, glabrous . 6. *S. discolor*
- 18 Catkins sessile or on very short, bracteate peduncles; stipes of the capsules 0.5-1.5 mm long; leaves oblanceolate or obovate, usually rounded or obtuse at the apex, with some reddish-brown hairs on one or both surfaces .... 20. *S. scouleriana*
- 15 Catkins emerging and maturing with the leaves.
  - 19 Leaves entire or merely crenate-serrate, the shallow teeth unevenly distributed around the margin.
    - 20 Leaves persistently white-tomentose beneath, linear-oblong to oblong or narrowly lanceolate; stipes ca 1 mm long .... 4. *S. candida*
    - 20 Leaves grayish-pubescent to glabrate beneath, elliptic to narrowly ovate or narrowly obovate; stipes 2-5 mm long .... 3. *S. bebbiana*
  - 19 Leaves evenly serrate or mostly so.
    - 21 Capsules 8-10 mm long, gray-tomentose; leaves somewhat paler green beneath but not glaucous, mostly 2-3.5 × longer than wide .... 13. *S. maccalliana*
    - 21 Capsules 5-7 mm long, closely pubescent mostly toward the base; leaves white-glaucous beneath, mostly 3-6 × longer than wide .... 17. *S. petiolaris*
1 Plants vegetative, or with male catkins, but with fully expanded leaves.
  - 22 Petioles bearing lobate glands at or near the attachment to the blade; leaves finely glandular-serrate.
    - 23 Leaves ovate-lanceolate, green on both surfaces, paler beneath but not white-glaucous.
      - 24 Leaves acute to short-acuminate, glossy above, thick and rather leathery .... 16. *S. pentandra*
      - 24 Leaves mostly long-acuminate, semiglossy above, not especially thick .... 11. *S. lucida*
    - 23 Leaves elliptic-lanceolate, white-glaucous beneath .... 21. *S. serissima*
  - 22 Petioles lacking glands or sometimes with minute vestiges of glands, or the petioles only glandular-viscid, in which case the leaves are narrowly lanceolate to lanceolate; leaves serrate or entire, occasionally glandular-serrate.
    - 25 Leaves linear-lanceolate, mostly 8-20 × longer than wide, entire to remotely serrulate; colonial shrub often forming dense thickets .... 8. *S. exigua*
    - 25 Leaves mostly broader in proportion to their length (often up to 10-12 × longer than wide in *S. nigra* which is a large tree).
      - 26 Leaves acuminate, gradually or abruptly tapered to a long, slender tip.
        - 27 Leaves the same shade of green on both surfaces, linear-lanceolate to lanceolate, often up to 10-12 × longer than wide .... 14. *S. nigra*
        - 27 Leaves pale to white-glaucous beneath, narrowly lanceolate, lanceolate, ovate-lanceolate or somewhat oblanceolate, less than 10 × longer than wide.
          - 28 Leaves dark green and shiny above; twigs brittle and easily snapping off at the base.
            - 29 Leaves coarsely serrate, with 4-6 glandular teeth per cm of leaf margin; petioles glandular-viscid at the summit; twigs olive to brown .... 9. *S. fragilis*
            - 29 Leaves more finely serrate, with 7-10 or more teeth per cm of leaf margin; petioles not glandular-viscid at the summit, or only with minute vestiges of glands.
              - 30 Twigs golden-yellow to orange, glabrous; large tree to 20 m tall .... 1. *S. alba* var. *vitellina*
              - 30 Twigs reddish-brown to grayish-brown, glabrous or often gray-pubescent; small or medium-sized tree to 10 m tall .... 5. *S. caroliniana*
          - 28 Leaves yellowish-green to dark green and dull above; twigs flexible, not easily snapping off at the base.
            - 31 Leaves ovate-lanceolate to lanceolate, mostly long-acuminate with taillike tips; petioles commonly recurved; branchlets flexuous, somewhat

                    drooping .... 2. *S. amygdaloides*
            31 Leaves lanceolate or somewhat oblanceolate, acuminate; petioles straight; branchlets erect to spreading, not drooping.
                32 Twigs gray-brown to dark brown, closely gray-pubescent the first year and often into the second .... 7. *S. eriocephala*
                32 Twigs yellow or yellowish-gray to yellowish-brown, glabrous .... 12. *S. lutea*
26 Leaves acute, obtuse, rounded or only short-acuminate at the tip.
    33 Leaves persistently pubescent, especially on the lower surface (rarely glabrate in age in *S. humilis*).
        34 Leaves elliptic, narrowly ovate or narrowly obovate, sparsely to densely pubescent beneath; leaf margins flat.
            35 Pubescence including some reddish-brown hairs intermixed with silvery ones on one or both of the leaf surfaces; leaves obtuse to rounded at the tip, commonly arranged so that they appear fanlike on the branchlets .... 20. *S. scouleriana*
            35 Pubescence all gray; leaves acute to short-acuminate at the tip, more or less regularly alternate on the branchlets .... 3. *S. bebbiana*
        34 Leaves linear-oblong to narrowly lanceolate or oblanceolate to narrowly obovate, densely pubescent (rarely glabrate in *S. humilis*) or white-tomentose beneath; leaf margins usually revolute.
            36 Shrubs of cold springs or fens; leaves linear-oblong to oblong or narrowly lanceolate, white-tomentose beneath; leaf margins revolute .... 4. *S. candida*
            36 Shrubs of upland habitats; leaves oblanceolate to narrowly obovate, densely pubescent (rarely glabrate) and greenish beneath; leaf margins flat to slightly revolute .... 10. *S. humilis*
    33 Leaves glabrous or glabrate with age.
        37 Leaves entire or nearly so, or with a few scattered inconspicuous teeth, sometimes to crenate-serrate with the teeth distributed unevenly around the margins.
            38 Small bog shrub 4–10 dm tall; leaves elliptic-lanceolate to oblanceolate, acute to rounded and often apiculate at the tip, 2–4.5 cm long .... 15. *S. pedicellaris*
            38 Larger shrubs and small trees of various habitats, mostly 2–7 m tall; leaves of various shapes, never apiculate at the tip, mostly 3–10 cm long.
                39 Mature leaves with some minute reddish-brown hairs persistent on one or both surfaces; leaves arranged in a fanlike manner on the branchlets, mostly obtuse to rounded at the tip .... 20. *S. scouleriana*
                39 Mature leaves glabrous or glabrate, lacking reddish-brown hairs; leaves rather regularly alternate on the branchlets, acute to short-acuminate.
                    40 Leaves dull grayish-green above, the lower surface usually rugose, with the veins raised prominently on the lower surface (except in var. *perrostrata*, with the lower leaf surface smooth, without raised veins) .... 3. *S. bebbiana*
                    40 Leaves bright to dark green above, smooth beneath, only the primary veins, if any, raised on the lower surface.
                        41 Leaves entire or nearly so, 3–6 cm long; twigs reddish-brown to nearly black, shiny .... 18. *S. planifolia*
                        41 Leaves, or at least the larger ones, crenate-serrate, 4–10 cm long; twigs yellowish-brown to dark brown, dull .... 6. *S. discolor*
        37 Leaves mostly serrate or finely serrate, the teeth evenly distributed around the margins.
            42 Stipules persistent, often prominent; leaves ovate to obovate, rounded to cordate at the base .... 19. *S. pseudomonticola*
            42 Stipules lacking or caducous; leaves generally lanceolate, acute to obtuse at the base.
                43 Leaves paler below than above, but not glaucous, elliptic-lanceolate to oblanceolate, mostly 2–3.5× longer than wide .... 13. *S. maccalliana*
                43 Leaves white-glaucous beneath, narrowly lanceolate or narrowly oblanceolate, mostly 3–6× longer than wide .... 17. *S. petiolaris*

1. **Salix alba** L. var. *vitellina* (L.) Stokes, yellowstem white willow. Large tree to 20 m tall; twigs (in ours) golden-yellow to orange, brittle and easily snapping off at the base; branchlets spreading, golden yellow to dark brown, glabrous with age. Leaves dark green and shiny above, white-glaucous beneath, glabrous to sparsely sericeous beneath, lanceolate to narrowly lanceolate, acuminate and often asymmetric at the tip, cuneate at the base, mostly 4–10 cm long, 1–2.5 cm wide, serrate, mostly with 7–10 glandular teeth per cm of margin; petioles glandless or with minute vestiges of glands at the summit, 0.5–1.5 cm long; stipules caducous, lanceolate, entire, 2–4 mm long, sericeous. Catkins appearing with the leaves; pistillate catkins 3–6 cm long, on leafy branchlets 1–3(5) cm long; bracts caducous, pale yellow, pubescent, ciliate at the tip; stamens 2. Capsules ovoid-conic, 3.5–5 mm long, glabrous, nearly sessile or on stipes to 1 mm long. (2n = 76) May, fruits early Jun. Introduced & frequently escaping to wet areas from shelter belts & ornamental plantings; mostly n GP; (scattered localities U.S., Eurasia). *Naturalized.*

Typical *S. alba*, which also has been introduced in the GP, has flexible, gray or brown twigs and persistently white-sericeous leaves, but this is apparently planted less often and escapes rarely, if at all.

2. **Salix amygdaloides** Anderss., peachleaf willow. Small to medium-sized tree with 1–several trunks, to 12 m tall; twigs gray to light yellow, shiny, flexible; branchlets spreading to drooping, yellow to dark brown, glabrous. Leaves yellowish-green above, pale to white-glaucous beneath, glabrous, lanceolate to ovate-lanceolate, mostly 3–8 cm long, 1–3 cm wide, occasionally much larger on vigorous shoots, short to mostly long-acuminate with taillike tips, acute to nearly rounded at the base, finely serrate; petioles glandless or rarely with vestiges of glands on vigorous shoots, often recurved, 5–20(30) mm long; stipules minute and caducous, occasionally well developed and persistent on vigorous shoots, reniform, 3–12 mm long, serrate. Catkins emerging with the leaves; pistillate catkins 3–8 cm long, on leafy branchlets 1–4 cm long; bracts deciduous, pale yellow, villous on the inside; stamens 4–7. Capsules ovoid, 3–5 mm long, glabrous, uncrowded on the axis giving the catkin a loose, open appearance; stipes 1–2 mm long. (2n = 38) May, fruits Jun. Flood plains, stream banks, lake & pond borders, moist ravines, ditches, & other wet or damp places; GP; (Que. & NY to se B.C. & WA, s to PA, KY, MO, n TX, NM, & AZ).

3. **Salix bebbiana** Sarg., beaked willow. Shrub to 4 m tall; twigs grayish-brown, closely pubescent to eventually glabrous, gnarled and rough in appearance owing to jutting leaf scars, irregular growth and dieback; branchlets spreading, yellowish-brown to dark brown, tomentulose, occasionally glabrate toward the base. Leaves dull grayish-green and glabrate to pubescent above, pale to gray-pubescent and usually rugose beneath, with the veins raised prominently on the lower surface, elliptic to narrowly ovate or narrowly obovate, acute to short-acuminate, cuneate at the base, mostly 3–6 cm long, 1–3 cm wide, entire to shallowly toothed; petioles glandless, 5–10(15) mm long; stipules deciduous or persistent on vigorous shoots, ovate to reniform, 2–6 mm long, 1–3 mm wide, shallowly dentate. Catkins emerging and maturing with the leaves; pistillate catkins persistent for some time after capsule dehiscence, 2–5 cm long, on short leafy branchlets 0.5–2 cm long, with 2–4 small leaves; bracts persistent, pale with a reddish or darkened tip when young, yellowish to brown with age, villous; stamens 2. Capsules ovoid-conic, 5–8 mm long, finely pubescent; stipes 2–5 mm long. (2n = 38) Late Apr–May, fruits late May–Jun. Wet meadows, stream banks, moist wooded ravines & hillsides, marsh borders, & seepage areas; MN, n IA, ND, SD, nw NE, MT, WY, CO, & NM; (Newf. to AK, s to MD, OH, IL, NM, AZ, & CA). *S. perrostrata* Rydb.—Rydberg.

Some plants in our range are var. *perrostrata* (Rydb.) Schneid., differing from the typical in having leaves thinner, more often glabrous, more entire-margined, and smooth, not rugose, beneath. This variety is more characteristic of drier woodland habitats.

4. **Salix candida** Fluegge, hoary willow. Low shrub to 1.5 m tall; twigs yellow to reddish-brown or brown, usually with patches of white tomentum; branchlets strongly ascending, yellow to brown, mostly white-tomentose. Leaves dark green and glabrate or thinly white-tomentose above, densely white-tomentose beneath, linear-oblong to oblong or narrowly lanceolate, acute at the tip, cuneate at the base, mostly 3–9(11) cm long, 0.5–1.5(2) cm wide, the margin revolute; petioles glandless, 3–10 mm long; stipules persistent, obliquely ovate to lanceolate, 2–10 mm long, tomentose, entire or serrulate. Catkins emerging with the leaves; pistillate catkins 1.5–4.5 cm long, on leafy branchlets 4–15 mm long, with 2 or 3 small leaves; bracts persistent, yellow to brown, villous; stamens 2. Capsules narrowly ovoid, 4–8 mm long, white-tomentose; stipes 1 mm long. ($2n = 38$) May, fruits May–Jun. Fen areas associated with marshes, streams, & springs; MN, n IA, ND, SD, MT, WY, CO; (Lab. to AK, s to NJ, PA, OH, IL, ID, & s B.C.).

5. **Salix caroliniana** Michx., Carolina willow. Small to medium-sized tree with 1–few trunks, to 10 m tall; twigs reddish-brown to grayish-brown, glabrous or often closely gray-pubescent, brittle; branchlets spreading, reddish-brown to brown, glabrous or gray-pubescent. Leaves dark green and shiny above, strongly white-glaucous beneath, glabrous or pubescent mainly on the veins, lanceolate, acuminate, cuneate to rounded or auriculate at the base, mostly 5–17 cm long, 7–20(25) mm wide, finely serrate; petioles glandless, 3–8(20) mm long; stipules persistent on vigorous branchlets, reniform or broadly ovate, 2–15 mm long, up to twice as wide, minutely serrate. Catkins emerging with the leaves; pistillate catkins 5–15 cm long, on leafy branchlets 2–4 cm long, the branchlet commonly branched from the axil of the uppermost leaf, the branch eventually surpassing the catkin so that the catkin, although terminal, appears lateral on the branchlet; bracts caducous, yellowish, villous; stamens 4–8. Capsules ovoid, 4.5–5.5 mm long, glabrous; stipes 1–2 mm long. Apr–May, fruits Jun. Shores, stream banks, sand & gravel bars; MO, e KS & e OK; (PA & s IN to ne KS, s to FL, Tex; & Cuba). *S. wardii* Bebb — Rydberg; *S. longipes* Shuttlew. var. *wardii* (Bebb) Schneid. — Gates.

6. **Salix discolor** Muhl., pussy willow. Shrub or small tree to 5 m tall; twigs reddish-brown to dark brown, dull, glabrous to slightly pubescent, rarely densely pubescent; branchlets spreading, yellowish-brown to nearly black, tomentulose, often glabrous with age. Leaves bright to dark green above, pale to white-glaucous beneath, glabrous, not rugose beneath, only the primary veins, if any, raised on the lower surface, elliptic to narrowly ovate or narrowly obovate, acute to short-acuminate, cuneate to narrowly rounded at the base, mostly 3–10 cm long, 1–3 cm wide, subentire to more often shallowly and irregularly crenate-serrate, more or less regularly alternate on the branchlets; petioles glandless, 5–20 mm long; stipules deciduous, often persistent on vigorous shoots, obliquely ovate to flabellate, 3–10 mm long, about as wide, glabrous, sometimes deeply lobed. Catkins emerging and maturing before the leaves; pistillate catkins sessile, sometimes with 2 or 3 minute, bractlike leaves at the base, soon deciduous after capsule dehiscence, 2–6(9) cm long; bracts persistent, black or very dark brown, villous; stamens 2. Capsules ovoid with a long neck, 5–10 mm long, finely pubescent; stipes 1.5–4 mm long. ($2n = 76, 114$) Mid Apr–early May, fruits mid May–early Jun. Swamps, fens, stream banks, slough borders, & ditches; MN, n IA, ND, SD, MT, & WY; (Newf. to B.C., s to DE, n GA, KY, IL, n MO, & ID). *S. prinoides* Pursh — Rydberg; *S. discolor* var. *latifolia* Anderss., *S. discolor* var. *overi* C. R. Ball — Fernald.

7. **Salix eriocephala** Michx., diamond willow, Missouri willow. Shrub or small tree to 7 m tall; twigs gray-brown to dark brown, closely gray-pubescent, the pubescence often patchy; branchlets reddish-brown, gray-pubescent. Leaves dark green to yellowish-green above, pale to weakly glaucous beneath, glabrous on both sides or pubescent beneath, lanceolate to somewhat oblanceolate, acuminate at the tip, cuneate, rounded or cordate at the base,

3–8(12) cm long, 1–3(4) cm wide, finely serrate; petioles glandless, 3–15 cm long; stipules persistent on vigorous shoots, semicordate, ovate or reniform, to 12 mm long, glabrous, serrate. Catkins emerging with or prior to the leaves; pistillate catkins 2–8 cm long, on short leafy or bracteate branchlets to 1.5 cm long; bracts persistent, brown to nearly black, pubescent; stamens 2. Capsules ovoid with a long neck, 4–7 mm long, glabrous; stipes 1–2 mm long. (2n = 38 Apr–early May, fruits May–early Jun. Shores, stream banks, flood plains, ditches, & wet meadows, especially along major river courses; MN, IA, MO, ND, SD, NE, KS, & MT; (N.S. & s Que. to Sask. & MT, s to VA, MO, & KS, also GA & AR). *S. missouriensis* Bebb — Rydberg; *S. rigida* Muhl. var. *vestita* Anderss. — Gleason & Cronquist; *S. rigida* var. *rigida* — Hitchcock et al.

8. *Salix exigua* Nutt., sandbar willow, coyote willow. Colonial, rhizomatous shrub to 4 m tall, often forming dense thickets; twigs light yellow to orange, glabrous; branchlets erect, yellow to orange, glabrous. Leaves yellowish-green above, the same or paler beneath, initially pubescent and soon glabrous (rarely persistently silvery-pubescent) or persistently gray-pubescent, linear-lanceolate, slowly tapered to an acute tip, acuminate at the base, 4–10 cm long, 2–10 mm wide, remotely and irregularly dentate; petioles glandless, 1–5 mm long; stipules minute or absent. Catkins emerging after the leaves, borne on leafy branchlets 0.5–10 cm long, these often branched; pistillate catkins 1.5–8 cm long; bracts deciduous, yellowish; stamens 2. Capsules narrowly ovoid, 4–8 mm long, glabrous (although pubescent when immature); stipes 0.5–1 mm long. (2n = 38) May–early Jun, fruits Jun–early Jul. Shores, stream banks, alluvial bars, ditches, & other wet places.

Two phases of *S. exigua* occur within our range:

8a. subsp. *exigua*, coyote willow. Leaves persistently gray-pubescent, at least beneath; pistillate catkins mostly dense and short; capsules 3–5(6) mm long, sessile or nearly so. The western phase; w SD, MT, WY, w NE, CO, w KS, w & cen OK, n TX, & NM; (MT to B.C., s to NM, AZ, w TX; n Mex.).

8b. subsp. *interior* (Rowlee) Cronq., sandbar willow. Leaves usually glabrous at maturity, although rarely silvery-pubescent; pistillate catkins rather loose and elongate; capsules 5–8 mm long, distinctly stipitate. Very common throughout GP; (N.B. and Que. to AK & B.C., s to VA, TN, LA, TX, CO, & MT). *S. interior* Rowlee, *S. linearifolia* Rydb., *S. wheeleri* (Rowlee) Rydb. — Rydberg; *S. interior* f. *wheeleri* (Rowlee) Rouleau — Fernald; *S. interior* var. *pedicellata* (Anderss.) C. R. Ball, *S. interior* var. *wheeleri* Rowlee — Gleason & Cronquist.

9. *Salix fragilis* L., crack willow. Large tree to 20 m tall; twigs olive to yellowish-brown, brittle, easily snapping off at the base; branchlets spreading, green to reddish-brown, eventually glabrous. Leaves dark to yellowish-green and shiny above, pale to white-glaucous beneath, glabrous, lanceolate to narrowly lanceolate, acuminate, often asymmetric at the tip, acute at the base, mostly 7–13 cm long, 1.5–3 cm wide, coarsely serrate, mostly with 4–6 glandular teeth per cm of margin; petioles 0.5–1.5 cm long, glandular-viscid at the summit, the glands often stipitate; stipules caducous, narrowly lanceolate, 2–3 mm long when well developed, pubescent, entire. Catkins appearing with the leaves; pistillate catkins 3–6 cm long, on leafy branchlets 1–3(5) cm long; bracts caducous, yellowish, pubescent, ciliate at the tip; stamens 2. Capsules narrowly conic, 4–5.5 mm long, glabrous, subsessile or on stipes to 1 mm long. (2n = 38, 76, 114) May, fruits early Jun. Often planted as a shade tree, rarely escaping to wet places; MN, IA, ND, SD, & NE; (Newf. to Ont. s to VA, KY, w to GP, Rocky Mts.; Eurasia). *Naturalized.*

10. *Salix humilis* Marsh., prairie willow. Shrub to 3 m tall; twigs yellowish-brown to dark brown, gray-pubescent or mostly so; branchlets strongly ascending, brown, gray-pubescent. Leaves dark green and usually glabrous above, glaucous and densely short-pubescent (rarely glabrate) beneath, with the golden-yellow veins raised prominently on the lower surface, oblanceolate to narrowly obovate, acute, cuneate at the base, mostly (1.5)4–8 cm long,

7-25 mm wide, the margins coarsely and irregularly serrate to subentire, flat to slightly revolute; petioles glandless, 3-10 mm long; stipules commonly persistent on vigorous branchlets, lanceolate to ovate, 3-7 mm long, pubescent, sparsely serrate. Catkins emerging and maturing before the leaves; pistillate catkins sessile, 1-3(4) cm long; bracts persistent, dark brown or purplish, villous on the back; stamens 2. Capsules ovoid-conic, 4-6(8) mm long, pubescent; stipes 0.5-1.5 mm long. (n = 38) Apr-May, fruits May-early Jun. Upland sites in prairies & sparse woods, especially in sandy soil; MN, IA, MO, & e & cen ND, SD, NE, KS, & OK; (Newf. & s Que. to ND, s to FL & TX). *S. humilis* var. *hyporhysa* Fern.—Fernald; *S. humilis* var. *rigidiuscula* (Anderss.) Robins. & Fern.—Gleason & Cronquist.

Var. *microphylla* (Anderss.) Fern. occurs sparingly in our region, differing from the typical as follows: shorter in stature, only 0.5-1 m tall; leaves narrowly oblanceolate, 2-5 cm long, 7-12 mm wide, mostly entire; stipules absent or very small. *S. tristis* Ait.—Rydberg.

11. Salix lucida Muhl., shining willow. Shrub or small tree to 4 m tall; twigs gray to yellowish-brown; branchlets ascending, yellowish-brown to dark brown, glabrous. Leaves yellowish-green to green and semiglossy above, pale beneath, initially reddish-pubescent, soon glabrous, lanceolate to ovate-lanceolate, acuminate to long-acuminate and falcate at the tip, broadly cuneate to nearly rounded at the base, 4-8(12) cm long, 1.2-2.5(4) cm wide, finely glandular-serrate; petioles glandular above, usually with few to several lobate glands, 0.5-1.5(2) cm long; stipules often persistent for some time toward the tips of branchlets, flabellate, well-developed ones 2-3 mm long, 3-4 mm wide, strongly glandular. Catkins produced with the leaves; pistillate catkins 1-3 cm long, on leafy branchlets 1-3 cm long; bracts caducous, yellowish, pubescent; stamens 3-5 or more. Capsules ovoid with a long neck, 4-6.5 mm long, glabrous; stipes 0.5-1.5 mm long. (2n = 76) May, fruits Jun. Swamps, shores, & wet meadows; MN, n IA, n & e ND, & BH; (Lab. & Newf. to Sask., s to DE, OH, IA, & SD).

12. S. lutea Nutt., yellow willow. Very similar to no. 7 and perhaps better treated as a variety of it. More often shrubby, to 5 m tall; twigs yellow or yellowish-gray to yellowish-brown, glabrous or nearly so. Shores, stream banks, flood plains, ditches, & wet meadows; more widespread & common than no. 7, MN w to MT, s to MO, KS, & CO; (n Que. to Alta., s to w IA, NE, NM, AZ, & CA). *S. cordata* Muhl., not Michx., *S. rigida* Muhl. var. *angustata* (Pursh) Fern.—Fernald; *S. lutea* var. *platyphylla* C. R. Ball, *S. lutea* var. *famelica* C. R. Ball—Harrington; *S. rigida* Muhl. var. *watsonii* (Bebb) Cronq.—Hitchcock et al.

13. Salix maccalliana Rowlee. Upright shrub, mostly 1-2 m tall; twigs reddish-brown to purplish-brown, glabrous; branchlets spreading, yellowish to purplish-brown, glabrous. Leaves rather firm and leathery, dark green and glossy above, somewhat paler (but not glaucous) and conspicuously reticulate beneath, glabrous, elliptic-lanceolate to oblanceolate, acute to short-acuminate at the tip, acute to obtuse at the base, mostly 4-8 cm long, 12-25 mm wide, the margin glandular-serrate; petioles glandless, 4-10 mm long; stipules lacking. Catkins emerging with the leaves; pistillate catkins on short, leafy branchlets; bracts persistent, dark brown to yellowish, pubescent on the outside, especially toward the base; stamens 2. Capsules elongate-conic, 8-10 mm long, gray-tomentose; stipes 1-2 mm long. (2n = ca 190, ca 224) May, fruits Jun. Swamps & bogs; ND: Bottineau; (Que. to Alta., s to n ND & s B.C.).

14. Salix nigra Marsh., black willow. Tree to 20 m tall, usually with a single trunk; twigs light reddish-brown to darker grayish-brown, brittle at the base, glabrous or pubescent; branchlets spreading, yellow to reddish-brown, glabrous or pubescent. Leaves the same shade of green on both sides, glabrous, linear-lanceolate to lanceolate, long-acuminate, cuneate at the base, 4–15 cm long, 7–20 mm wide, often up to 10–12 × longer than wide, glandular-serrate; petioles glandless or sometimes with only minute vestiges of glands, 5–10 mm long; stipules caducous or sometimes persistent on vigorous shoots, ovate to lanceolate, 1–4 mm long, glandular on the margin. Catkins emerging with the leaves; pistillate catkins 4–10 cm long, on leafy branchlets 1–4 cm long, the branchlet occasionally branched from the axil of the uppermost leaf, the branch eventually surpassing the catkin so that the catkin, although terminal, appears lateral on the branchlet; bracts caducous, yellowish, pubescent; stamens 3–7. Capsules ovoid to ovoid-conic, 3–5 mm long, glabrous; stipes 1–2(3) mm long. Apr–May, fruits Jun. Flood plains, stream banks, meadows, shores, ditches, & other wet places; s MN, IA, MO, se NE, all but nw KS, OK, TX, & NM; (s N.B. to s MN, s to FL & TX). *S. nigra* var. *lindheimeri* Schneid.—Gates; *S. gooddingii* C. R. Ball—Correll & Johnston.

15. *Salix pedicellaris* Pursh, bog willow. Slender shrub 4–10 dm tall; twigs grayish-brown, glabrous; branchlets erect to spreading, dark brown, glabrous. Leaves green above, white-glaucous beneath, glabrous, narrowly oblanceolate, oblanceolate or obovate, acute to obtuse and often apiculate at the tip, acute to obtuse at the base, 2–4(6) cm long, 1–1.5(2) cm wide, the margin entire, often slightly revolute; petioles glandless, 2–8 mm long; stipules absent. Catkins emerging with the leaves; pistillate catkins 1–3 cm long, on leafy branchlets 1–3 cm long; bracts persistent, yellow to brown, glabrous or pubescent only at the tip; stamens 2. Capsules narrowly conic, 5–8 mm long, glabrous; stipes 2–4 mm long. ($2n = 38, 57, 76$) Late May–early Jun, fruits Jun–early Jul. Sphagnum bogs & swamps; ND: Bottineau, McHenry, Ransom; (Newf. to N.T. & B.C., s to NJ, n IA, n ID, & OR).

16. *Salix pentandra* L., laurel-leaved willow. Small tree or shrub 2–6 m tall; twigs yellowish-green, shiny; branchlets spreading, dark brown and shiny. Leaves dark green and glossy above, light green and dull beneath, glabrous, thick and leathery, ovate to ovate-lanceolate, acute to short-acuminate at the tip, obtuse to rounded at the base, mostly 4–10 cm long, 2–3 cm wide, finely glandular-serrate; petioles strongly glandular at the summit, 5–10 mm long; stipules deciduous or persistent for a short time on vigorous shoots, reniform, ca 1 mm long, 2 mm wide, glandular-dentate. Catkins produced after the leaves; pistillate catkins 3–5 cm long, on leafy branchlets 2–3 cm long; bracts caducous, yellowish, pubescent; stamens (4)5(12). Capsules ovoid-conic, the 2 halves bulged at the base, 4–5 mm long, glabrous, subsessile, the stipes 0.5–1 mm long. ($2n = 57, 76$) May–early Jun, fruits Jun–Jul. Introduced from Europe as an ornamental, occasionally escaping to marsh borders, ditches, stream banks, ravines, & other moist places, especially n GP; MN, IA, ND, SD, NE, KS (rare), MT, WY; (N.S. to Ont., s to PA, MD, w to GP; Eurasia). *Naturalized.*

17. *Salix petiolaris* J. E. Sm., meadow willow. Clumped or few-stemmed shrub to 3 m tall; twigs reddish-brown to dark brown or almost black, glabrous; branchlets spreading to erect, yellowish-green to dark brown, tomentulose, often glabrous with age. Leaves dark green above, white-glaucous beneath, pubescent when young, becoming glabrous with age, narrowly lanceolate to narrowly oblanceolate, acute to abruptly short-acuminate at the tip, acute to slightly rounded at the base, 2.5–8 cm long, 4–15 mm wide, entire to closely serrate; petioles glandless, 3–10 mm long; stipules absent. Catkins emerging with the leaves; pistillate catkins 1–3(5) cm long, sessile or on short branchlets to 1.5 cm long, these naked or with 2 or 3 small leaves; bracts persistent, brown, villous; stamens 2. Capsules narrowly conic, 5–7 mm long, closely pubescent mostly toward the base; stipes 1–3 mm long. ($2n = 38$) May, fruits Jun. Wet meadows, stream banks, shores, ditches, & other wet places; MN, n IA, n & e ND, ne & s SD, BH, n-cen & nw NE, n MT, & CO; (N.B. to Alta. s to NJ, OH, IL, & n MT). *S. gracilis* Anderss.—Rydberg; *S. gracilis* var. *textoris* Fern.—Fernald; *S. petiolaris* J. E. Sm. var. *angustifolia* Anderss.—Gleason & Cronquist.

18. Salix planifolia Pursh, planeleaf willow. Shrub or shrubby tree with clustered trunks, to 3 m tall; twigs dark reddish-brown to nearly black, shiny, glabrous; branchlets spreading to ascending, brown, glabrous. Leaves green above, paler to glaucous beneath, initially short-pubescent but soon glabrous, elliptic to oblanceolate, acute or occasionally obtuse at the tip, rounded to acute at the base, 3–6 cm long, 12–20 mm wide, entire or only sparsely crenulate; petioles glandless, 3–6 mm long; stipules minute, deciduous. Catkins emerging slightly before or with the leaves; pistillate catkins 2–4 cm long, sessile or on short branchlets with 1–3 bractlike leaves; bracts persistent, black, villous; stamens 2. Capsules ovoid with a long neck, 5–8 mm long, pubescent, nearly sessile, on stipes 0.5 mm or less long. (2n = 57, 76, 114) May, fruits Jun. Stream banks, meadows, & moist hillsides at higher elevations in the BH; SD: Custer, Lawrence, Pennington; (Newf. & Lab. to AK, s to ME, NH, VT, n MN, w SD, NM, & n CA). *S. nelsonii* C. R. Ball—Rydberg; *S. planifolia* var. *monica* (Bebb) Schneid., *S. planifolia* var. *nelsonii* (C. R. Ball) C. R. Ball—Harrington; *S. phylicifolia* L. subsp. *planifolia* (Pursh) Hiitonen—Hitchcock et al.

19. Salix pseudomonticola C. R. Ball, serviceberry willow. Shrub to 3(5) m tall, twigs light brown to dark grayish-brown, dull, glabrous; branchlets spreading, brown, glabrous. Leaves dull green above, paler to glaucous beneath, glabrous, ovate to obovate, acute at the tip, rounded to cordate at the base, 3–8 cm long, 12–35 mm wide, serrate; petioles glandless, 3–15 mm long; stipules persistent, often prominent, broadly ovate, cordate at the base, 5–15 mm long and about as wide, serrate. Catkins emerging before the leaves; pistillate catkins 3–7 cm long, sessile or very short peduncled with 1–few leafy bracts at the base; bracts persistent, dark brown to black, long-villous on the back; stamens 2. Capsules ovoid with a narrow neck, 6–8 mm long, glabrous; stipes 1–1.5 mm long. (2n = 38, 114) May, fruits Jun. Open meadows & stream banks at high elevations in the BH; SD: Lawrence, Pennington; (Lab. & Que. to AK, s to w SD, NM, & cen ID). *S. monticola* Bebb ex Coult.—Van Bruggen; *S. pseudomonticola* var. *padophylla* (Rydb.) C. R. Ball—Harrington; *S. padophylla* Rydb.—Froiland, U.S.D.A. Forest Serv. Tech. Bull. 1269: 40. 1962.

20. Salix scouleriana Barr., western pussy willow. Shrub or occasionally a small tree to 4 m tall; twigs reddish-brown to brown, puberulent; branchlets spreading, yellowish to dark brown, pubescent. Leaves dull green above, glaucous beneath, only the primary veins, if any, raised on the lower surface, pubescent on one or both surfaces, with some reddish-brown hairs mixed with silvery hairs, oblanceolate or obovate, usually rounded or obtuse (sometimes abruptly acute) at the tip, cuneate at the base, 2–6(8) cm long, 10–25(32) mm wide, entire to shallowly crenate-serrate; petioles glandless, 2–10 mm long; stipules persistent on vigorous branchlets, ovate to reniform, to 10 mm long, sparsely serrulate. Catkins emerging before the leaves; pistillate catkins to 5 cm long; bracts persistent, dark brown to black, villous toward the tip on both sides; stamens 2. Capsules ovoid with a long neck, 6–8 mm long, pubescent; stipes 0.5–1.5 mm long. (2n = 76, ca 114). May, fruits Jun. Moist slopes, often where shaded, at higher elevations in the BH; SD: Custer, Lawrence, Meade, Pennington; (Man. to AK, s to w SD, NM, AZ, & CA).

21. Salix serissima (Bailey) Fern., autumn willow. Shrub to 3 m tall; twigs gray to yellowish-brown, shiny, glabrous; branchlets erect to spreading, yellow to dark brown, glabrous. Leaves yellowish-green to green and semiglossy above, white-glaucous beneath, glabrous, elliptic-lanceolate, acute to short-acuminate at the tip, cuneate to narrowly rounded at the base, 4–8 cm long, 1–2.5 cm wide, finely glandular-serrate; petioles glandular at the summit, 0.5–1 cm long; stipules rarely present, flabellate, 1.5 mm long, 2 mm wide, glandular-serrate. Catkins emerging after the leaves; pistillate catkins 2–4 cm long, on leafy branchlets 1.5–3 cm long; bracts deciduous, light yellow, pubescent; stamens 3–5 or more. Capsules ovoid with a long neck, 7–10 mm long, glabrous; stipes 0.5–2 mm long. Jun–early Jul, fruits late Jun–Aug. Swamps, fens, & bogs; MN, n & e ND; rare in the BH, also highly localized in MT & CO; (Newf. to Alta., s to PA, n OH, n IL, MN, MT, & CO).

57. CAPPARACEAE Juss., the Caper Family

by William T. Barker

Herbs (ours), shrubs or trees, often with rank odor. Leaves alternate, petiolate, palmately (1)3–9(11) foliolate; stipules minute or absent. Flowers single and axillary or ours usually in terminal, bracteate or seldom ebracteate many-flowered racemes. Flowers perfect (ours), ± regular or zygomorphic, hypogynous, 4-merous; sepals free or partially fused; petals free, subequal, ± ovate to spatulate, often slenderly clawed; stamens 6 or more, as long or longer than the petals, nectariferous or adaxial gland frequently present between the corolla and stamens; pistil 1. Fruit a 2-valved, unilocular capsule, usually borne on a slender gynophore, or subsessile; seeds many, reniform. *Capparidaceae*—Rydberg.

The assistance of Hugh H. Iltis is gratefully acknowledged.

1 Fruits rhomboid-obdeltoid capsules, about as long as broad or broader, less than 1 cm long; flowers yellow .... 2. *Cleomella*
1 Fruits elongate capsules, much longer than broad, 2–8 cm long; flowers white, pink to purple or yellow.
   2 Plants ± glabrous; stamens 6, of equal length, covered in the bud by the overlapping pink to purple (rarely white) or yellow petals; receptacular gland a flattened adaxial projection, or obsolete; fruit pendent or deflexed, borne on an elongate gynophore; the valves deciduous and releasing the seeds .... 1. *Cleome*
   2 Plants viscid-pubescent; stamens 6–many, of unequal lengths; petals white to rose-purple in the bud stage, very short and exposing the stamens; receptacular gland well defined, solid or cylindric, truncate; fruit sessile or short-stipitate, erect; the valves persistent, retaining the seeds .... 3. *Polanisia*

1. CLEOME L., Bee Plant, Spider Flower

Ours erect annual herbs, glabrous or nearly so. Leaves palmately (1)3–5(11) foliolate; leaflets entire or serrulate, flat or folded; stipules none or minute. Racemes terminal, elongate in fruit, bracteate, or flowers single in axils of cauline leaves; sepals 4, free or fused at the base; petals 4, free, pink to purple (rarely white) or yellow; stamens (ours) 6, or equal length; anthers elongate, longitudinally dehiscent. Capsules elongate, pendent or deflexed, borne on a slender gynophore; seeds many.

Cleome hassleriana Chod. (*C. spinosa* L., non Jacq.), spider flower or pink queen, is a robust herb native to tropical and subtropical southeastern South America that is widely cultivated as an ornamental in North America, including the GP. It has showy, bicolored inflorescences, the petals a rich pink or purple (rarely white) the first day, fading to very pale pink the second day. It is spiny and has 5–7 leaflets per leaf. The plant sometimes escapes from cultivation but does not persist. (2n = 20).

1 Petals yellow; leaflets 5 .... 1. *C. lutea*
1 Petals pink to purplish, rarely white; leaflets 3 .... 2. *C. serrulata*

1. *Cleome lutea* Hook., yellow cleome. Simple to freely branched annual, 5–10 dm tall, glaucous, glabrous to sparsely pilose. Leaflets 5, or 3 in upper leaves, lanceolate to oblanceolate, entire, 3-5 cm long. Racemes elongating in fruit; bracts simple, or the lower 3-foliolate; pedicels slender, ascending, 10-20 mm long. Petals yellow, 5-8 mm long, not clawed; stamens 6, longer than the petals. Capsules pendent, linear, nearly terete, 15-35 mm long; gynophore slender, 10-15 mm long. (2n = 34) May-Jul. Sandy soil in river bottoms & along stream banks; NE: Lancaster; (cen WA to CA, e to WY, CO, & w NM). *Peritoma luteum* (Hook.) Raf.—Rydberg.

2. *Cleome serrulata* Pursh, Rocky Mountain bee plant. Erect, branched annual, 2-15 dm tall, glabrous, glaucous, unarmed. Leaflets 3, narrowly lanceolate, 2-6 cm long, 5-15 mm wide, acuminate or cuspidate, entire. Flowers in dense elongated, many-flowered racemes; bracts narrow, simple; pedicels 14-20 mm long. Sepals united for 1/2-2/3 their length, ± persistent in fruit; petals bright pink to purplish, rarely white, 8-12 mm long, short clawed; stamens 13-20 mm long; disk adaxially prolonged into an elaborate, flattened scale. Capsule linear-cylindric to fusiform, sharply pointed at each end, 2-8 cm long, 3-9 mm wide; gynophore 11-23 mm long; seeds several to many, ovoid, pointed, 3-4 mm long, black-brownish mottled. (2n = 34, 60). Jun-Aug. Prairies & open woodlands, especially wash areas or disturbed sites, locally very common; GP except se; (MN & Man. to MO & OK, w to WA & AZ, rarely adventive to the e). *Peritoma serrulatum* (Pursh) DC.— Rydberg.

2. CLEOMELLA DC., Cleomella

1. *Cleomella angustifolia* Torr., eastern cleomella. Glabrous, erect, often bushy annual, 6-26 dm tall. Leaves alternate, petiolate, palmately 3-foliolate, leaflets entire, linear-elliptic, acute, 25-60 mm long, 2-8 mm wide; stipules minute. Racemes terminal, bracteate, pedicels 7-12 mm long. Flowers regular, small; sepals 4, distinct, thin, imbricate, deciduous; petals yellow, 4-6 mm long; stamens 6, equal. Mature capsules rhomboid-obdeltoid, 5-10 mm long, 5-9 mm wide, borne on a gynophore 4-7 mm long; seeds 4-6 per capsules, dark brown mottled. Jun-Oct. Sandy prairie soils, river bottoms, & gravel bars, roadsides; w NE, e CO, s to TX; (w NE to TX).

This species is similar in flower to *Cleome lutea,* but it has 3 leaflets and very short fruit.

3. POLANISIA Raf., Clammy-weed

Herbaceous viscid-pubescent annuals with strong, rank odor. Leaves petiolate, palmately 3-foliolate. Flowers in bracteate racemes; sepals free nearly to the base; corolla open in the bud and not covering the stamens, zygomorphic; petals obtuse or emarginate to laciniate, white to purplish tinged, the adaxial pair longer; gland prominent between corolla and stamens, strictly adaxial, solid or tubular, usually orange; stamens 6-20, of unequal length and with staggered maturation. Capsule elongate, erect, sessile or short-stipitate; seeds many.

1 Petals emarginate or obtuse; gland solid with concave apex, wider than long; style slender, purple, withering in fruit. .... 1. *P. dodecandra* subsp. *trachysperma*
1 Petals laciniate; gland a hollow tube; style short, setaceous, persistent in fruit. ... 2. *P. jamesii*

1. *Polanisia dodecandra* (L.) DC. subsp. *trachysperma* (T. & G.) Iltis, clammy-weed. Viscid-pubescent, branched annual, 2-8 dm tall. Leaves petiolate, palmately 3-foliolate; leaflets oblanceolate, 2-4 cm long, 4-20 mm wide. Flowers in bracteate racemes. Petals white or purplish tinged, especially the claw, ± equal, obovate to spatulate, usually retuse (emarginate), (8)10-16 mm long; stamens 10-20, exserted, 9-30 mm long, pink to purple;

style 5–12 mm long, purple, irregularly twisting and breaking off in fruit. Capsule 2–7 cm long, 5–10 mm wide, somewhat inflated, the valves persistent, releasing the seeds by shaking; seeds many, reddish to dark brown, minutely wrinkled or tuberculate, dull. (2n = 20) May–Oct. Sandy, gravelly soils of prairies & plains; GP; (s Sask., MN, AR, to n Mex., w to Rocky Mts.; introduced elsewhere). *P. trachysperma* T. & G. — Rydberg.

Subsp. *dodecandra*, an eastern entity with smaller flowers and fruits (petals 5–8 mm long, stamens 8–16, style only 1–4 mm long and ± persistent in fruit) occurs sporadically along the e edge of our range, from MN to e KS, e to Que. & MD. *P. graveolens* Raf. — Rydberg.

2. Polanisia jamesii (T. & G.) Iltis, cristatella. Viscid and puberulent annual, 1–4 dm tall, branching. Leaves 3-foliolate, short petioled; leaflets linear or linear-lanceolate, 2–4 cm long. Racemes short and few-flowered, flowers appearing axillary; petals 4, unequal, 2–4 mm long, laciniate, cuneate-flabelliform, white or ochroleucous, clawed; stamens 6–14, purplish; receptacle with an orange, tubular gland 2–3 mm long on the upper side of the inclined ovary. Capsule stipitate, ascending, linear-cylindric, 2–3 cm long, glandular-pubescent; seeds numerous. Jun–Jul. In sandy or gravelly soils, or in pure sand; s 1/2 GP from s-cen SD to OK; (w WI & w IL, w to GP & w TX). *Cristatella jamesii* T. & G. — Rydberg.

58. BRASSICACEAE Burnett, the Mustard Family

by William T. Barker

Annual, biennial or perennial herbs or infrequently suffrutescent. Leaves alternate or rarely subopposite, simple to often pinnately ± dissected, but only seldom with distinct, articulated leaflets; stipules absent. Flowers bisexual, usually tetradynamous, hypogynous, mostly regular and ebracteate in terminal racemes, infrequently solitary; sepals 4, in 2 opposite, decussate pairs, deciduous, erect and appressed to the corolla or spreading at anthesis; petals 4, diagonal to the sepals, forming a cross, commonly with an elongate claw and abruptly spreading blade, or seldom absent; stamens usually 6, the outer 2 shorter than the inner 4, the outer almost always and the inner usually subtended by minute glands; ovary normally with 2 locules, separated by a thin septum, with 1–many parietal ovules per locule, sometimes ovary with a single locule and 1 ovule. Fruit dry, usually dehiscent, generally a silique (elongate) or a silicle (short) with the valves falling away leaving the septum (replum) persistent on the pedicel, varying to indehiscent, or uniloculate, or transversely septate; seeds with scanty endosperm and large embryos; embryo curved with the cotyledons lying against the radicle either edgewise (accumbent) or flatwise (incumbent), or the embryo rarley straight. Cruciferae — Fernald.

Reference: Rollins, R. C. 1981. Weeds of the Cruciferae (Brassicaceae) in North America. J. Arnold Arbor. 62: 517–540.

The taxonomy of the Brassicaceae relies heavily on characters of the fruit, and the following key to genera requires mature fruit.

```
1  Mature fruit borne on a conspicuous stipe, usually more than 1 cm long ......... 33. Stanleya
1  Mature fruit sessile, or borne on a stipe less than 3 mm long.
   2  Fruit transversely divided into 2 unlike segments, one of which is indehiscent.
      3  Upper segment elongate and distinctly thicker than the lower segment . 29. Raphanus
      3  Upper segment a flat or angled beak, usually shorter than the lower segment.
         4  Valves clearly 3-nerved; beak 3-nerved on each side; seeds in 1 row in each
            locule ................................................................ 8. Brassica
```

           4 Valves 1-nerved; beak 1-nerved on each side; seeds in 2 rows in each locule . 19. *Eruca*
  2 Fruit not tranversely segmented into 2 unlike segments, but sometimes the style persistent as
    a conspicuous beak.
     5 Fruit a silicle, i.e., less than 3 × longer than wide.
        6 Silicle flattened at right septum (replum), the septum therefore much narrower than
          the width of the silicle, and its presence indicated by a nerve or suture on the median
          of both faces of the silicle.
           7 Seeds 3 or more in each silicle.
              8 Silicle oval or circular, winged on the margin, rounded or obtuse at the
                base. .... 36. *Thlaspi*
              8 Silicle obcordate or obtriangular, wingless or weakly winged only at the sum-
                mit, acute at the base .... 10. *Capsella*
           7 Seeds 1 or 2 in each silicle.
              9 Silicle deeply notched at both summit and base .... 16. *Dimorphocarpa*
              9 Silicle at most shallowly notched at the apex.
                 10 Silicles dehiscent, ± winged on the margins, often weakly notched at the
                    apex. .... 24. *Lepidium*
                 10 Silicles indehiscent, firm walled, acute-acuminate at the
                    apex .... 12. *Cardaria*
        6 Silicle inflated to terete, or flattened parallel to the septum; the septum therefore as
          wide as the silicle.
           11 Silice very thin and flat, 1–2 cm long and 6–8 mm wide; flowers
              yellow .... 31. *Selenia*
           11 Silicle inflated to terete, or if flattened then smaller than above; flower color
              various.
              12 Fruit indehiscent, ± woody or spongy, often irregular and wrinkled; seeds few,
                 1–4.
                 13 Flowers yellow; cauline leaves clasping. .... 27. *Neslia*
                 13 Flower white to pink or purple.
                    14 Cauline leaves auriculate-clasping .... 12. *Cardaria*
                    14 Cauline leaves not at all clasping .... 29. *Raphanus*
              12 Fruit dehiscent, the walls variously thin and membranaceous or herbaceous;
                 seeds usually more numerous.
                 15 Silicles inflated, didymous (i.e., of 2 equal halves, side-by-side), cordate at
                    the base and deeply notched at the apex .... 28. *Physaria*
                 15 Silicles variously globose or short cylindric, terete or flattened but not in-
                    flated as above.
                    16 Petals yellow.
                       17 Cauline leaves usually bipinnate; fruits ± narrow-clavate,
                          elongate .... 15. *Descurainia*
                       17 Cauline leaves entire to pinnatifid.
                          18 Silicles distinctly flattened.
                             19 Silicles broadly oblong to orbicular, less than 2 × longer than
                                wide .... 2. *Alyssum*
                             19 Silicles oblong, 2–3 × longer than wide .... 18. *Draba*
                          18 Silicles globose, ovoid, short-cylindric to pyriform.
                             20 Pubescence of simple hairs or absent .... 30. *Rorippa*
                             20 Pubescence of stellate hairs.
                                21 Cauline leaves sessile, auriculate at the base, margins en-
                                   tire or nearly so .... 9. *Camelina*
                                21 Cauline leaves tapering to the base, not auriculate, or if
                                   auriculate then the margins ±
                                   toothed. .... 25. *Lesquerella*
                    16 Petals white.
                       22 Petals distinctly bifid (notched) at the apex; plants with stems
                          leafy .... 7. *Berteroa*
                       22 Petals rounded or emarginate at apex, plants scapose or leafy-
                          stemmed.

23 Silicle ovoid to obovoid, nearly as thick as wide.
   24 Foliage hairy, cauline leaves auriculate-clasping ...... 12. *Cardaria*
   24 Foliage glabrous, cauline leaves not auriculate ...... 5. *Armoracia*
23 Silicle orbicular to elliptic oblong, distinctly flattened.
   25 Silicle orbicular or nearly so, notched at the apex, with a thin, winglike margin .... 2. *Alyssum*
   25 Silicle longer than wide, obtuse to acute .... 18. *Draba*
5 Fruit a silique, i.e., 3–50(100)× longer than wide.
  26 Flowers yellow or light orange.
    27 Plants ± pubescent with 2-pronged, stellate or otherwise branched hairs.
      28 Leaves variously pinnate, bipinnate to tripinnatifid .... 15. *Descurainia*
      28 Leaves entire or merely dentate.
        29 Silique not more than 5× longer than wide, flattened .... 18. *Draba*
        29 Silique 7× or more longer than wide, subterete or 4-angled ...... 21. *Erysimum*
    27 Plants pubescent with simple hairs or glabrous.
      30 Silique indehiscent, with no external sutures, tipped with a prominent beak 1–3 cm long .... 29. *Raphanus*
      30 Silique dehiscent, with evident sutures.
        31 Leaves entire, cordate-clasping at the base .... 14. *Conringia*
        31 Leaves dentate to pinnatifid.
          32 Seeds in 2 rows in each locule of the silique.
            33 Beak 5–12 mm long, flat, as wide as the body of the silique .. 19. *Eruca*
            33 Beak 4 mm or less, slender.
              34 Flowers 2–5 mm across .... 30. *Rorippa*
              34 Flowers 10 mm or more across .... 17. *Diplotaxis*
          32 Seeds in 1 row in each locule.
            35 Cauline leaves clearly clasping at the base.
              36 Beak of the silique 8–15 mm long .... 8. *Brassica*
              36 Beak of the silique 1–3 mm long .... 6. *Barbarea*
            35 Cauline leaves not at all clasping.
              37 Fruiting pedicels less than 5 mm long; siliques closely appressed.
                38 Flowers ca 3 mm across; siliques subulate .... 32. *Sisymbrium*
                38 Flowers ca 1 cm across; siliques linear .... 8. *Brassica*
              37 Fruiting pedicels 5+ mm long; siliques divergent.
                39 Beak of the silique more than 5 mm long .... 8. *Brassica*
                39 Beak of the silique not more than 3 mm long.
                  40 Each valve of the silique with 3 nerves, the midvein the most prominent .... 32. *Sisymbrium*
                  40 Each valve of the silique with but 1 nerve .... 20. *Erucastrum*
  26 Flowers white, pink, blue or purple.
    41 Principal leaves, or at least some of them, distinctly and deeply lobed or parted or compound.
      42 Silique short-stipitate; sepals spreading at anthesis .... 35. *Thelypodium*
      42 Silique sessile or very nearly so and/or sepals erect at anthesis.
        43 Leaves palmately parted into 3 or more divisions .... 11. *Cardamine*
        43 Leaves pinnately lobed, parted or compound.
          44 Beak of the silique 5–20 mm long, rather wide or thick.
            45 Beak conic, hardly distinct from the spongy indehiscent silique .... 29. *Raphanus*
            45 Beak flat, about as wide as the body of the silique .... 19. *Eruca*
          44 Beak of the silique 5 mm long or less, slender.
            46 Seeds in 2 rows in each locule; plants of muddy sites or in shallow water .... 26. *Nasturtium*
            46 Seeds in 1 row in each locule; plants terrestrial.
              47 Lobing of the leaves restricted to a few laciniate divisions below the middle, the terminal segment the largest, acuminate, sharply serrate; flowers usually pale violet .... 23. *Iodanthus*
              47 Lobing of the leaves variable, the latter and terminal segments often

        about equal in size, or the terminal segment broad and not sharply
        serrate.
      48 Cauline leaves entire or with only a few small
        teeth .... 4. *Arabis*
      48 Cauline leaves pinnate or pinnatifid.
        49 Seeds narrowly winged .... 4. *Arabis*
        49 Seeds clearly wingless .... 11. *Cardamine*
 41 Principal leaves entire or merely toothed.
    50 Beak persistent and conspicuous, 4-15 mm long; silique breaking into 2-seeded segments
      at maturity .... 13. *Chorispora*
    50 Beak small or absent, siliques dehiscing by valves.
      51 Siliques distinctly flattened.
        52 Siliques less than 1.5 cm long, or if longer then 1.5-2 mm wide and clearly
          beakless .... 18. *Draba*
        52 Siliques 2 cm long or longer, of if shorter than ca 1 mm wide or with a beak 1
          mm or more in length.
          53 Cauline leaves both petiolate and auriculate-
            clasping .... 11. *Cardamine*
          53 Cauline leaves not both petiolate and auriculate-clasping.
            54 Cauline leaves sessile and auriculate at the base .... 4. *Arabis*
            54 Cauline leaves sessile or petiolate but not at all auriculate.
              55 Siliques less than 3 cm long .... 11. *Cardamine*
              55 Siliques 5-11 cm long.
                56 Basal leaves present, often in a rosette; anthers ± straight 4. *Arabis*
                56 Basal leaves absent; anthers twisted and
                  sagittate .... 34. *Streptanthus*
      51 Siliques terete or 4-angled.
        57 Cauline leaves sessile and auriculate-clasping; silique 5-12 cm long.
          58 Siliques appressed; foliage ± pubescent .... 4. *Arabis*
          58 Siliques spreading; foliage glabrous or nearly so.
            59 Leaves entire, obtuse .... 14. *Conringia*
            59 Leaves sharply dentate, acute or acuminate .... 23. *Iodanthus*
        57 Cauline leaves petiolate or sessile, not auriculate-clasping.
          60 At least some of the pubescence of branched hairs.
            61 Petals 2-2.5 cm long, usually bluish-purple .... 22. *Hesperis*
            61 Petals 2-4 mm long, white .... 3. *Arabidopsis*
          60 Pubescence of simple hairs only, or none.
            62 Cauline leaves deltoid; herbage with the odor of onions .... 1. *Alliaria*
            62 Cauline leaves rotund or ovate to oblong or narrower; without an odor of
              onions.
              63 Cauline leaves with many upward-pointing teeth .... 23. *Iodanthus*
              63 Cauline leaves entire to sinuate with at most 1-6 teeth on each side.
                64 Silique sessile .... 11. *Cardamine*
                64 Silique tapering to a definite short stipe, ca 1 mm
                  long .... 35. *Thelypodium*

1. ALLIARIA Heist. ex Fabr., Garlic Mustard

1. ***Alliaria petiolata*** (Bieb.) Cavara & Grande, garlic mustard. Erect biennial to 1 m tall, simple or little branched, glabrous or with a few simple hairs. Lower leaves reniform, the other deltoid, 3-6 cm long and wide, acute, coarsely toothed. Mature pedicels stout, 5 mm long; petals white, spatulate, gradually narrowed to the claw; short stamens surrounded at the base by an annular gland, each pair of long stamens separated by a trigonous gland, filaments flattened, anthers oval. Fruit a silique, linear, 4-angled, widely divergent, 4-6 cm long; seeds nearly cylindric, black, 3 mm long, in one row in each locule. (2n = 36)

Apr–Jun. Roadsides, fields, & open woods; e 1/3 KS; ND: Cass; MN: Clay; (Que., Ont., MN, & ND, s to VA, KY, & e KS; Europe). *Introduced and naturalized. A. officinalis* Andrz. — Rydberg.

2. ALYSSUM L., Alyssum

Annual, biennial or perennial, stellate pubescent herbs. Leaves simple, entire (ours) to serrate. Flowers in elongate, bractless racemes; sepals persistent to quickly deciduous; petals yellowish or cream-colored fading to white; stamens 6, the shorter 2 flanked by linear glands about 1 mm long, anthers oval; style very short. Silicle elliptic to orbicular, flattened parallel to the septum, each cell with 1 or 2 seeds.

Most species are Eurasian but several are introduced in N. America as weeds or ornamentals.

References: Dudley, T. R. 1964. Studies in *Alyssum:* Near Eastern representatives and their allies, I. J. Arnold Arbor. 45: 57–100; Dudley, T. R. 1968. *Alyssum* (Cruciferae) introduced in North America. Rhodora 70: 298–300.

1. Silicles glabrous; styles at least 0.5 mm long; sepals deciduous soon after anthesis .... 2. *A. desertorum*
1. Silicles with stellate pubescence; style scarcely 0.5 mm; sepals persistent until the fruit is nearly mature .... 1. *A. alyssoides*

1. *Alyssum alyssoides* L., pale alyssum. Annual with stems simple or usually several from the base, grayish, with appressed stellate pubescence, 1–3 dm tall. Leaves linear-spatulate. Flowers pale yellow or nearly white, 2 mm wide; petals narrowly oblong, little exceeding the persistent calyx, styles 0.2–0.4 mm long. Fruit on widely divergent pedicels, orbicular, 3–4 mm long, flat at margin, convex toward the center, stellate pubescent; seeds 2 per cell. (2n = 32) May–Jun. Dry soil along roadsides & in waste places; scattered in MT, WY, & w SD, sporadic s to ne KS & MO: Cass; (scattered in N. Amer., Europe). *Naturalized.*

Alyssum minus (L.) Rothmaler var. *micranthus* (C. A. Mey.) Dudley has been collected in CO: Arapahoe, Douglas, and Huerfano; and KS: Logan. It is easily confused with *A. alyssoides* but has styles 1–2 mm long while those of *A. alyssoides* are 0.2–0.4 mm long.

2. *Alyssum desertorum* Stapf., alyssum. Annual with stems simple or usually several from the base, 1–3 dm tall. Leaves linear-spatulate. Flowers yellow or nearly white, 1–1.5 mm wide; sepals deciduous soon after anthesis. Fruit on widely divergent pedicels, orbicular, 3–4 mm long, flat at margin, convex toward the center, glabrous; seeds 2 per cell. (2n = 32) May–Jun. Pastures, weedy prairies, waste areas; nw NE, CO, MT, & w ND; (introduced in w N. Amer.; Europe). *Naturalized.*

3. ARABIDOPSIS Heynh., Mouse-ear Cress

1. *Arabidopsis thaliana* (L.) Heynh., mouse-ear cress. Annual herbs, pubescent with branched hairs, 1–4 dm tall. Leaves chiefly in a basal rosette, oblong to spatulate, 1–5 cm long, stellate-hairy; cauline leaves smaller, linear to narrowly oblong. Mature racemes very open, the widely divergent pedicels 5–10 mm long. Sepals oblong, obtuse, pilose; petals white, spatulate, gradually narrowed to the base, 2–4 mm long; filaments capillary, anthers ovate, short stamens subtended by a minute circular gland, each pair of long stamens separated by a minute gland; ovary cylindric; ovules numerous. Silique divergent or ascending, 1–2 cm long, 1 mm wide, glabrous; seeds numerous, uniseriate. (2n = 10) May–Jun. Moist, wooded ravines; SD: Lawrence, Pennington; MO: Jackson; (introduced in scattered areas of the U.S.; Europe). *Adventive.*

4. ARABIS L., Rock Cress

Biennial or perennial herbs from herbaceous taproots or well-developed woody caudices, stems usually erect, stiff, simple or branched, glabrous to pubescent. Basal leaves either in dense rosettes or are few to many not aggregated in thick clusters, petiolate; cauline leaves petiolate or sessile, nonclasping or with an amplexicaul base. Inflorescence racemose, ebracteate. Sepals erect or spreading, outermost sometimes saccate at base, oblong; petals spatulate to oblong, white, cream colored, lavender or pink; stamens 6, tetradynamous, anthers oblong; ovary cylindric; ovules numerous; style short and scarcely differentiated; stigma truncate. Silique linear, elongate, flat or subterete, straight to curved, the valves often with a midvein or reticulately veined, erect to pendulous; styles evident, entire; seeds flattened, orbicular to oblong, often winged; cotyledons accumbent.

References: Hopkins, M. 1937. *Arabis* in eastern and central North America. Rhodora 39: 63–98, 106–148, 155–186; Rollins, R. C. 1941. A monographic study of *Arabis* in western North America. Rhodora 43: 289–325, 348–411, 425–481.

```
1  Plants perennial, stems 1 to several from a simple woody caudex ................. 4. A. fendleri
1  Plants biennial or somewhat perennial, stems herbaceous from a taproot.
   2  Pedicels and fruits at maturity erect or nearly so.
      3  Fruits terete or nearly so, 0.8–1.3 mm wide ....................................... 5. A. glabra
      3  Fruits flat.
         4  Seeds uniseriate in each locule; siliques 0.7–1.0 mm wide, stem hairy
            below ......................................................................... 6. A. hirsuta var. pycnocarpa
         4  Seeds biseriate in each locule; siliques 1.5–2.3 mm wide; stem commonly
            glabrous ...................................................................... 3. A. drummondii
   2  Pedicels and fruits at maturity ascending, spreading or reflexed.
      5  Cauline leaves not auricled or sagittate at base.
         6  Siliques less than 5 cm long, ascending .................................... 10. A. virginica
         6  Siliques more than 5 cm, widely spreading, recurved or reflexed.
            7  Cauline leaves hairy; siliques 2–4 mm wide ........................ 1. A. canadensis
            7  Cauline leaves glabrous; siliques 1.2–2 mm wide ................... 8. A. laevigata
      5  Cauline leaves auricled or sagittate at base.
         8  Siliques and pedicels ascending, spreading or somewhat decurved.
            9  Cauline leaves (except possibly the very lowest) glabrous on both sides.
               10  Basal leaves glabrous or sparsely pubescent with simple hairs
                   only ..................................................................... 8. A. laevigata
               10  Basal leaves with stellate pubescence on both sides ......... 2. A. divaricarpa
            9  Cauline leaves distinctly pubescent ......................................... 9. A. shortii
         8  Siliques pendent; mature pedicels abruptly reflexed ................... 7. A. holboellii
```

1. *Arabis canadensis* L., sicklepod. Biennial from a thick taproot; stem erect, 3–9 dm tall, simple or rarely sparingly branched above, sparsely hirsute at the base usually with simple (rarely bifurcate) hairs, passing to glabrous above. Basal leaves short-petiolate, obovate to lanceolate, 2.5–13 cm long, 1.5–4 cm wide, entire to serrate-dentate, hirsute on both surfaces, especially along the midrib, with simple and bifurcate hairs or more rarely entirely glabrous, soon disappearing; cauline leaves imbricate to subremote, oblong-lanceolate to elliptic, 2.5–12 cm long, 0.5–2.5 cm wide, attenuate to a sessile or subsessile base or the lowermost short petioled, acuminate, denticulate or more rarely subentire, lowermost villous hirsute, uppermost hirsutulous with simple and forked hairs, to entirely glabrous. Flowers small, lowermost often pendulous in very long loose racemes; pedicels 7–10(12) mm at anthesis, glabrous or often hirsutulous with simple hairs, first erect but becoming pendulous at anthesis; sepals 2–4 mm long, 1–1.25 mm wide, membranaceous, acute or obtuse, yellowish or purplish, hirsutulous with simple and bifurcate hairs, only slightly shorter than the petals; petals white to cream, narrowly oblanceolate to oblong, 3–5 mm long. Siliques falcate to arcuate, never straight, pendulous or recurved 7–10 cm long, 2–4 mm wide, glabrous, distinctly 1-nerved to the top or slightly below the top, prominently reticulate-veined; fruiting pedicels slender, at first divaricate or ascending, deflexed and subgeniculate at maturity, hirsutulous to glabrous, 8–12(15) mm long at maturity; seeds uniseriate, narrowly winged. (2n = 14) Apr–Sep. Woodlands; eastern 1/3 GP; (ME to s Ont., MN, ND s to GA, AL, OK, & TX).

2. *Arabis divaricarpa* A. Nels. Biennial, stem erect, 2–9 dm tall, branched at base or above or simple, finely and sparingly hirsute at extreme base with appressed simple or forked hairs or glabrous throughout. Basal leaves in rosettes, oblanceolate-spatulate to narrowly oblanceolate, 2–6 cm long, 4–10 mm wide, acute, dentate to denticulate or very rarely subentire, finely and evenly pubescent on both surfaces with minute stellate hairs, petiolate, petioles very narrowly winged and stellate-pubescent; cauline leaves narrowly oblong to linear-lanceolate, imbricate to subremote, erect or strongly ascending, 1.5–6 cm long, 3–10 mm wide, sessile with an auriculate or sagittate base, acute, lowermost subentire to entire, uppermost entire, glabrous on both surfaces or rarely the lowermost sparingly stellate pubescent. Flowers in loose racemes; pedicels ascending when young, becoming wide-spreading or somewhat reflexed at anthesis, glabrous or more rarely slightly stellate pubescent, 6–7 mm long at anthesis; sepals 2–4 mm long, 1–1.25 mm wide, 1/2 the length of petals, linear to narrowly oblong, herbaceous, glabrous or rarely with a few scattered stellate hairs, green with a whitish or hyaline margin; petals pinkish to pale purple, rarely white, oblanceolate-spatulate, (5)5.5–8 mm long, 0.5–1.5 mm wide, at apex. Siliques straight or subarcuate, the uppermost and youngest suberect, the lowermost and older suberect to widespreading or subarcuate or subreflexed, glabrous, 2.5–9 cm long, 1.25–2.5 mm wide, prominently 1-nerved 2/3 of their length or often to the tip; fruiting pedicels ascending or divaricately spreading or more rarely subdeflexed, glabrous, 5–12(14) mm long at maturity; young seeds in 2 rows, when mature becoming somewhat uniseriate, 1–1.5 mm in diam, narrowly winged. (2n = 14, 20, 21, 28) Jun–Jul. Sandy & rocky soil; CO, WY, w NE, MT, SD, ND, MN, & IA; (Que. to Man., nw to Yukon, s to N.B., NH, VT, NY, OH, s MI, WI, n IA, NE, CO, & CA). *A. brachycarpa* (T. & G.) Britt., *A. confinis* S. Wats.—Rydberg.

3. *Arabis drummondii* A. Gray. Biennial, stem erect, 2–9 dm tall, simple or branching at base and above, glabrous throughout to somewhat glaucous or rarely scantily appressed-pubescent at the extreme base. Basal leaves spatulate to oblanceolate, 3–9 cm long, 5–20 mm wide, dentate to serrate to subentire, acute to subacuminate, tapering at base to a slender winged petiole, glabrous or rarely sparingly ciliate on the petioles with simple or two-forked hairs; cauline leaves linear-lanceolate to lanceolate to oblong, imbricate to subremote, 2–9 cm long, 4–15 mm wide, sessile with a sagittate or very auriculate base, acute to acuminate, sparingly dentate to entire, glabrous on both sides. Flowers in loose racemes; pedicels glabrous, erect, 7–10 mm long at anthesis; sepals linear-oblong, 1/2 as long as the petals, 3–4 mm long, glabrous, acute to subacute, herbaceous; petals pink to purple (often white when dried), 5–10 mm long, 0.5–2 mm wide at apex. Siliques straight or rarely slightly curved, flattish, erect or ascending, often subappressed, 4–10 cm long, 1.5–2.3 mm wide, obtuse or rarely subacute, glabrous, 1-nerved at least beyond the middle and frequently to the top; fruiting pedicels strictly erect, appressed to subappressed, glabrous, 9–15 mm long at maturity; seeds in 2 rows, broadly elliptical to orbicular, averaging 1 mm in diam, narrowly winged all around. (2n = 14, 20) May–Aug. Limestone & calcareous ledges, gravelly soil of open woods, thickets, & meadows; MN, ND, w SD, MT, WY, CO, w NE, & NM; (Lab. to B.C. s to DE, IA, NE, NM, & CA). *A. connexa* Greene—Rydberg.

4. *Arabis fendleri* (S. Wats.) Greene. Perennial, stems 1 to several from a simple, woody caudex, simple or branched above, hirsute near the base with simple spreading trichomes, glabrous above, 2.5–6 dm tall. Basal leaves oblanceolate to linear-oblanceolate, entire to

dentate, sparsely to densely hirsute with coarse, simple or forked hairs or surfaces nearly glabrous, margins ciliate, 2-6 cm long, (2)3-15 mm wide; cauline leaves sessile, oblong to lanceolate, auriculate, lower ones pubescent and usually imbricate, upper ones glabrate, entire or rarely dentate, 1-4 cm long, 2-8 mm wide. Flowers racemose; pedicels slender, ascending at anthesis, arched downward in fruit, glabrous, 1-2 cm long; sepals glabrous or with a few trichomes, oblong, 3-5 mm long, 1.5-2 mm wide; petals spatulate, white to pink, 5-8 mm long, 2-3 mm wide. Siliques glabrous, pendulous, nerved to the middle or slightly above, obtuse, 3-6 cm long, 1.5-2.5 mm wide; seeds orbicular to slightly oblong, narrowly winged or rarely almost wingless, 1-1.5 mm wide, biseriate. ($2n = 14$) Apr-Jun. Open rocky hillsides and in open, rocky pine woods.

Two varieties of this species occur at the western edge of the GP.

4a. var. *fendleri*. Basal leaves dentate, oblanceolate, obtuse; petals pink. CO, NM, & TX; (CO to NV, s to TX, NM, AZ, & Mex.).

4b. var. *spatifolia* (Rydb.) Rollins. Basal leaves entire, linear-oblanceolate, acute; petals white. WY, s to CO, NM; (s WY to NM & e UT).

5. *Arabis glabra* (L.) Bernh., tower mustard. Stout biennial, from a stout taproot, stem erect, usually simple below, rarely branching at base, 6-12 dm tall, hirsute at base with simple or bifurcate, spreading to subappressed hairs, glabrous and glaucous above or rarely glabrous throughout. Basal leaves spatulate to oblanceolate, rarely lyrate-pinnatifid, entire or irregularly dentate, petiolate, acutish, 5-12 cm long, 1-3 cm wide, those of the first year stellately pubescent on both surfaces, those of the second year less so, or often pubescent only along the midrib or glabrous; cauline leaves lanceolate to elliptic-oblong, sessile with an amplexicaul-sagittate or auriculate base, imbricate, passing upwards to subimbricate or rarely subremote, entire or slightly denticulate, acutish, quite variable in size, 2-12 cm long, 1-3.5 cm wide, usually glabrous on both surfaces or the lowermost sometimes slightly hirsute or stellate-pubescent along the midrib. Flowers small, in close or loose racemes; pedicels glabrous, 0.5-1 cm long at anthesis, slender, erect or ascending, appressed to subappressed; sepals membranaceous, 2-5 mm long, glabrous, obtuse to subacuminate, oblong, greenish or frequently purple; petals cream to yellowish, rarely pale lavender, 2.5-6 mm long, narrowly oblanceolate to linear. Siliques (4)5-9.5 cm long, 0.8-1.3 mm wide, terete, narrow, straight or slightly curved, appressed close to stem, distinctly erect and ascending, 1-nerved beyond the middle and usually to the tip; fruiting pedicels erect and appressed to subappressed, glabrous, 7-18 mm long at maturity; mature and fertile seeds irregular in outline, mostly elliptical to oblong, sparingly winged all around, averaging 1 mm long and 0.15 mm wide, in either 1 or 2 rows. ($2n = 12, 16, 32$) May-Jun. Dry soil of ledges, cliffs, thickets, woods, & fields, often a weed; MT, ND, MN, WY, SD, NE, & ne KS; (circumboreal, in N. Amer. s to NC, AR, NM, & CA).

6. *Arabis hirsuta* (L.) Scop. var. ***pycnocarpa*** (Hopkins) Rollins, rock cress. Biennial or perennial, stems erect, one to several from a simple or branching caudex, simple or branched above, hirsute with coarse, spreading, simple or forked hairs, often glabrous above, 2-7 dm tall. Basal leaves oblong to oblanceolate or broadly spatulate, short petioled, obtuse to rarely acutish, entire, dentate or repand, hirsute on both surfaces with coarse simple or forked hairs or rarely glabrous, 2-8 cm long, 1-3 cm wide; cauline leaves lanceolate to oblanceolate or nearly spatulate, acute or obtuse, sessile, auriculate, entire to coarsely dentate, hirsute on both surfaces or the upper glabrous, 1-5(7) cm long, 0.5-2.5 cm wide. Racemes slender, becoming lax; pedicels erect to divaricately ascending, glabrous or rarely sparsely hirsute, 0.5-1.5 cm long; sepals herbaceous, oblong, glabrous or rarely with a few trichomes, 2.5-4.5 mm long, about 1 mm wide; petals white to rarely pinkish, oblong to spatulate, 3-5 mm long. Siliques erect, strict to divaricately ascending, glabrous, nerved nearly the entire length, 3-6 cm long, 0.7-1 mm wide; seeds brown to blackish, suborbicular to nearly rectangular, prominently winged on the distal end to narrowly winged or wingless, 1–1.5 mm long, about 1 mm wide, uniseriate. (2n = 8, 16, 32) May–Jul. Woods, banks, & ledges, chiefly in calcareous soils; MT, ND, MN, s to n KS, CO, & MO; (Que. to AK, s to PA, n GA, IN, AR, KS, & AZ).

7. *Arabis holboellii* Hornem., rock cress. Biennial or perennial, stems erect, 1 to several from a simple or branching caudex, simple or branched above, pubescent throughout with appressed or spreading trichomes to glabrous above, 1–9 dm tall. Basal leaves linear-oblanceolate to broadly spatulate, entire to somewhat dentate, densely pubescent with fine to coarse stellate trichomes, acute to obtuse, 1–5 cm long, 1.5–6(8) mm wide; cauline leaves auriculate and clasping to nonauriculate, with a narrowed base, entire, oblong to lanceolate, 1–4 cm long, 1.5–6 mm wide, lower surface densely pubescent, upper surface pubescent to glabrous. Flowers loosely racemose; sepals oblong, scarious-margined, pubescent or glabrous, 2–4(5) mm long, 1–2 mm wide; petals spatulate with a narrow claw, purplish-pink to whitish, (5)6–10 mm long, 2–3.5 mm wide. Siliques glabrous, straight to slightly curved, strictly reflexed to loosely pendulous, nerved below or to slightly above middle, obtuse to acute, 3–7 cm long, 1–2.5 mm wide; fruiting pedicels straight to somewhat curved, often geniculate, strictly reflexed to loosely descending, pubescent or glabrous, slender, 6–16 mm long; seeds orbicular, narrowly winged all around, about 1 mm wide, uniseriate or imperfectly biseriate. (2n = 14, 21, 28, 35, 42) May–Jul. Dry, sandy or rocky or gravelly, calcareous soil.

Two varieties of this species are found in the GP:

7a. var. *collinsii* (Fern.) Rollins. Pedicels geniculate near base, usually straight or at least not uniformly curved; siliques strictly reflexed to somewhat spreading, but not loosely pendulous, straight or nearly so; pubescence of basal leaves fine. MT, ND, SD, WY, & NE; (Que., Man. to Alta., s to NE & WY).

7b. var. *pinetorum* (Tidest.) Rollins. Pedicels gently curved downward; siliques pendulous, usually somewhat curved inward; pubescence of basal leaves coarse. MT, ND, MN, s to SD, NE, WY, & CO; (Sask. to B.C., s to NE, CO, NM, & CA).

8. *Arabis laevigata* (Muhl.) Poir., smooth rock cress. Biennial from a branched taproot, stems 3–10 dm tall, branched at the base and above, or simple, glabrous and glaucous throughout. Basal leaves rosulate, soon disappearing, spatulate-obovate to narrowly oblanceolate, first-year leaves sparingly pilose with short simple hairs, second-year leaves glabrous, dentate to serrate, 3–11 cm long, 0.5–2.2(3) cm wide, acute to subacuminate, petiolate; cauline leaves oblong-lanceolate to linear, spreading to subappressed, imbricate, 3–20 cm long, 3–15 mm wide, sessile with a saggitate or sometimes auriculate base, glabrous, serrate-dentate to entire, acute to obtuse or somewhat acuminate. Flowers small, in long, loose racemes; flowering pedicels ascending, often divergent, glabrous, 5–9 mm long at anthesis; sepals membranaceous, greenish, 2.5–4.5 mm long, nearly the length of the petals, glabrous, spatulate to oblong; petals white, 3–5 mm long, spatulate to oblanceolate. Siliques irregularly downward-curved to subarcuate or more rarely straightish, ascending when young, recurved-spreading at maturity, compressed, attenuate, glabrous, faintly 1-nerved below the middle or only toward the base, 5–10 cm long, 1.2–2 mm wide; fruiting pedicels ascending, divergent, glabrous, 7–14 mm long at maturity; seeds in 1 row, quadrate to oblong, averaging 1 mm long, 0.5 mm wide, winged all around. (2n = 14) May–Jun. Woods & hillsides; KS: Cherokee, Crawford, Douglas; MO: Buchanan, Jasper, Vernon; OK: Commanche; (sw Que. to SD, s to GA, AL, AR, & OK).

9. *Arabis shortii* (Fern.) Gl., rock cress. Biennial from a simple taproot, stem branching at base or rarely from the top or simple, often decumbent, 2–5(6) dm tall, thinly pubescent throughout with appressed to subappressed simple or forked hairs. Basal leaves spatulate or obovate to oblanceolate, 4–15 cm long, 1–4.5(6) cm wide, acutish, petiolate, irregularly

dentate to sinuate or rarely lyrate-pinnatifid, finely and evenly stellate-pubsecent on the lower surface, strigose to strigillose on the upper surface; cauline leaves oblanceolate to lanceolate or narrowly obovate, 1–6 cm long, 0.5–2.5 cm wide, imbricate to subimbricate, sessile with an amplexicaul base, irregularly dentate or more rarely sinuate, acutish, finely and evenly stellate-pubescent on the lower surface, strigillose to glabrous on the upper surface. Flowers small, in close racemes; flowering pedicels erect or ascending, 0.8–2 mm long at anthesis, strongly hirsute with simple and forked trichomes; sepals membranaceous, 1.5–2.5 mm long, 1/2 the length of the petals, greenish, finely stellate-pubescent; petals white to cream colored, 2–3 mm long, narrowly oblanceolate to broadly linear. Siliques 1.5–4 cm long, 0.75–1.25 mm wide, nearly straight or only very slightly curved, more or less finely stellate-pubescent on both surfaces, divaricately spreading or slightly ascending, faintly 1-nerved at the base or more often nerveless; fruiting pedicels divaricately spreading or slightly ascending, coarsely pubescent with simple and forked trichomes, 2–3.5 mm long at maturity; seeds oblong to subelliptical, uniseriate, wingless. Rich moist woods; MN, IA, MO, e SD, NE, KS, & OK; (cen NY to MN, SD, s to OK, AR, TN, KY, & VA). *A. dentata* T. & G.—Rydberg.

10. *Arabis virginica* (L.) Poir., rock cress. biennial from a long taproot, stem spreading from the base, decumbent or ascending, 1–2.5 dm tall, usually hirsute at the base with short simple or bifurcate trichomes, passing to glabrous above or hirsute throughout. Basal leaves narrowly oblong, 3–8 cm long, 7–12 mm wide, lyrate-pinnatifid with nearly even oblong to linear segments, segments nearly all 1-toothed or entire, petioled, hirsute or glabrous; cauline leaves similar to basal ones but smaller, or the uppermost lanceolate and subentire, 3–7 cm long, 7–10 mm wide, either short petioled or sessile, nearly always glabrous, rarely sparingly hirsute with simple hairs. Flowers small, inconspicuous, in close racemes; flowering pedicels short, 2–3 mm long at anthesis, glabrous; sepals membranaceous, 1–2 mm long, 1/2–2/3 × the length of the petals, glabrous or rarely with a few scattered simple hairs, often tinged purplish or pinkish; petals 1.5–3 mm long, white to faintly pinkish, oblanceolate to narrowly oblong and rarely almost linear. Siliques 2–2.5 cm long, (1)1.25–1.72(2) mm wide, nearly straight or very rarely slightly curved, erect or ascending or more rarely somewhat spreading, glabrous, faintly 1-nerved at the extreme base or more rarely entirely nerveless; fruiting pedicel stoutish, erect or ascending, glabrous, 3–7 mm long at maturity; seeds orbicular to suborbicular or more rarely subelliptic, uniseriate, evenly winged all around, averaging 1.5 mm long, 1.25 mm wide. Mar–May. Woods, fields, & roadsides; TX, OK, KS, MO, & s NE; (FL to TX, n to VA, OH, IN, IL, MO, & NE; also CA).

5. ARMORACIA Gaertn., B. Mey. & Scherb., Horseradish

1. *Armoracia rusticana* Gaertn., B. Mey & Scherb., horseradish. Erect, coarse perennial, 0.6–1.3 m tall, from thick vertical roots. Basal leaves oblong to oblong-ovate, 1–3(4) dm long, base cordate, long petioled; upper leaves smaller, lanceolate, short petioled to sessile. Racemes several, terminal and from the upper axils. Petals 6–8 mm long; mature pedicels spreading-ascending, 8–12 mm long. Silicle subglobose to globose-obovoid, 2-celled, to 6 mm diam; style 0.3 mm long, with persistent broad stigma; seeds apparently never perfected. May–Jul. Moist ground, spread from cultivation; sporadic in the GP; (established in scattered areas of N. Amer.; se Europe & w Asia). *Adventive. A. lapathifolia* Gilib.—Fernald; *Rorippa armoracia* (L.) Hitchc.—Stevens.

6. BARBAREA R. Br., Winter Cress

Biennial or perennial herbs with angled stems, glabrous or with a few simple hairs. Basal leaves mostly pinnatifid with a large terminal lobe and 2–several smaller lateral ones; cauline leaves smaller, entire to pinnatifid. Flowers in racemes or panicles; sepals oblong, ascending; petals yellow, spatulate to obovate; short stamens partly surrounded at base by a semicircular gland, long stamens separated by a short erect gland, anthers oblong; ovary cylindric, narrowed to a slender style; ovules several per locule; stigma truncate. Silique linear, terete, or somewhat 4-angled, tipped by the persistent short style, valves keeled by a midnerve.

1 Beak of the fruit slender, 2–3 mm long; petals 6–8 mm long .... 2. *B. vulgaris*
1 Beak of the fruit thickish, 0.5–1 (rarely 2) mm long; petals 2.5–5 mm long . 1. *B. orthoceras*

1. ***Barbarea orthoceras*** Ledeb., northern winter cress. Stem simple or branched above, 3–8 dm tall. Basal leaves simple, oblong to elliptic, or with 2 or 4 small basal lobes; middle and upper leaves deeply toothed or lyrately lobed, often with 3 or 4 pairs of leaflets. Racemes dense in anthesis; petals pale yellow, 2.5–5 mm long; mature pedicels clavate-thickened, to 1 mm long. Siliques subterete or compressed, not angled, 2–3.5 mm long, somewhat crowded, appressed or strongly ascending; the beak 0.5–1.5(2) mm long; seeds 0.8–1 mm wide, brown, minutely rugulose. (2n = 16) Jun–Jul. Swamps & wet woods; SD: Clay, Custer, Lawrence, Pennington; (circumboreal, s in N. Amer. to Ont., ME, NH, SD, & CA). *B. americana* Rydb.—Rydberg.

2. ***Barbarea vulgaris*** R. Br., winter cress. Erect, branched above, 2–8 dm tall. Basal leaves petiolate, lyrate with 1–4 pairs of small, elliptic to ovate lateral lobes and a large, ovate to rotund terminal one; cauline leaves progressively reduced, the upper sessile and generally lobed or the uppermost entire or merely toothed. Flowers crowded at anthesis; petals 6–8 mm long; mature pedicels 3–6 mm long, 0.5 mm in diam. Siliques erect and appressed to ascending or spreading, the beak (1.5)2–3 mm long; seeds grayish and rugulose. (2n = 16) Apr–Jun. Weed in wet meadows, fields, roadsides, & gardens; OK, MO, KS, IA, NE, SD, & ND; (scattered across U.S.; Eurasia). *Naturalized*.

7. BERTEROA DC., Hoary False Alyssum

1. ***Berteroa incana*** (L.) DC. Stellate-canescent annual, stiffly erect, usually branched above, to 7 dm tall. Leaves alternate, oblanceolate to occasionally elliptic, (1)2–5 cm long, 0.5–1 cm wide, apex obtuse to acute, entire. Flowers racemose; sepals equal, ascending; petals white, deeply bifid, about 3 mm long, more than 2 × longer than the sepals; short filaments flanked on each side by a short semicircular gland, anthers oblong; ovary with 2–6 ovules per locule; style elongate, persistent. Silicle elliptical, slightly flattened parallel to the septum (sides convex), 5–8 mm long, 3–4 mm wide, stellate-pubescent, often slightly so; seeds 3–6 per locule, brown, roundish, 1.5–2 mm long, narrowly winged. (2n = 16) May–Sep. Disturbed & waste places, roadsides, along railroads; scattered in the GP, especially in the n; (established as a weed in N. Amer. from N.S. to WA, s to NJ, WV, OH, IN, IL, MO, OK, ID, & CA; Europe).

Berteroa mutabilis (Vent.) DC. has been reported for Kansas (Fernald, 1950) but specimens cannot be found. It differs from *B. incana* in having silicles that are distinctly flattened, 4.5–6 mm wide, and usually glabrous.

8. BRASSICA L., Mustard

Coarse, weedy annuals and biennials. Caulescent, with alternate leaves, at least the lower leaves pinnatifid. Flowers usually conspicuous; sepals erect or spreading, saccate at the base; petals yellow (ours), obovate, clawed; staminal glands 4, rounded, evident; anthers oblong; ovary subcylindric, scarcely narrowed to the short style; ovules 2–many in each locule; stigma capitate. Fruit subterete or angled, more or less elongate, often torulose, terminated by a conspicuous beak sometimes containing a basal seed; valves conspicuously nerved; cotyledons conduplicate; seeds large, subglobose, in one row. *Sinapsis* L.—Rydberg.

1. Beak of silique large, flat or conspicuously angled, usually containing 1 seed in an indehiscent locule; valves of the silique with 3 parallel nerves about equally strong.
  2. Leaves all pinnatifid; fruit bristly, 4 mm in diam; beak 1–2× longer than the body of the fruit; seeds 4–8 .... 2. *B. hirta*
  2. Middle and upper leaves rhombic to oblong, merely toothed; fruit glabrous or sparsely bristly, 2 mm in diam; beak 1/3–1× as long as the body of the fruit; seeds 7–13 .... 4. *B. kaber*
1. Beak of silique terete or slenderly conical, empty; valves of the silique with a prominent midnerve, the other nerves much weaker, scarcely parallel, often anastomosing.
  3. Upper leaves not clasping at the base.
    4. Plants glabrous or nearly so, glaucous, pedicels elongate, slender, spreading; silique terete, loosely ascending, 3–6 cm long .... 3. *B. juncea*
    4. Plant hirsute, green; pedicels short and thick, appressed; siliques more or less 4 angled, closely appressed, 1–2 cm long .... 5. *B. nigra*
  3. Upper leaves clasping and sessile .... 1. *B. campestris*

1. **Brassica campestris** L., wild turnip. Annual, winter annual, or biennial with erect stems, 3–8(10) dm tall, glabrous or nearly so and commonly glaucous or very slightly pubescent below. Basal leaves petioled, lyrate-pinnatifid, upper leaves oblong to lanceolate, entire to shallowly dentate, sessile and clasping. Flowers to 1 cm wide; pedicels divaricately ascending, 1–3 cm long; petals yellow, 6–10 mm long. Siliques (3)4–7 cm long, about 3 mm in diam, terete or nearly so, the beak (8)10–15(20) mm long; seeds round or slightly compressed laterally, reddish-gray, 1.5–2 mm in diam, minutely roughened. ($2n = 20$) Apr–Oct. A weed of fields & waste ground; throughout the GP; (throughout the much of N. Amer., Europe). *Naturalized.*

Brassica rapa L. (turnip) and *B. napus* L. (rape) occasionally escape from gardens or persist after cultivation in our region.

2. **Brassica hirta** Moench, white mustard. Rough-hairy annual, 3–7 dm tall. Leaves obovate in outline, deeply lyrate-pinnatifid, the lobes sinuate-dentate, petiolate. Flowers 1.5 cm wide; mature pedicels divergent, 1 cm long. Siliques divergent or ascending, commonly bristly at least when young, 1.5–3.5 cm long, the valves prominently 3-nerved, beak 1–2 cm, flat, often curved; seeds pale, smooth, 2 mm diam. ($2n = 24$) Apr–Aug. Waste places & roadsides; scattered locations in KS, MO, NE, SD, & ND; (scattered throughout N. Amer.; Eurasia). *Introduced. Sinapsis alba* L.—Rydberg.

This species is sometimes cultivated as a source of mustard and salad greens.

3. **Brassica juncea** (L.) Czern., Indian mustard. Glabrous, often glaucous annual, 3–10 dm tall. Lower leaves to 2 dm long, lyrate-pinnatifid and dentate, petiolate; upper leaves oblong, entire or dentate, short petioled or sessile. Flowers 12–15 mm wide; mature pedicels slender and ascending. Siliques ascending, subterete, 3–6 cm long, valves strongly 1-nerved; seeds 2 mm in diam, conspicuously and evenly reticulate. ($2n = 36$) Apr–Oct. Fields & waste places; GP; (throughout temp. N. Amer., Eurasia). *Naturalized.*

4. ***Brassica kaber*** (DC.) Wheeler, charlock. More or less hispid, especially at the base, to glabrous annual, 2-8 dm tall. Lower leaves on hispid petioles, obovate, lyrate-pinnatifid; middle and upper leaves nearly or quite sessile, oblong to ovate or rhombic, acute, dentate, sparsely pilose. Flowers 1.5 cm wide; mature pedicels ascending, 5(7) mm long. Siliques ascending, linear, subterete, the body 1-2 cm long, 1.5-2.5 mm in diam, smooth or rarely bristly; beak flattened-quadrangular, commonly 1/2 as long as the body; seeds 7-13, smooth, 1-1.5 mm in diam. (2n = 18) May-Jul. A weed of fields, gardens, & waste ground; GP; (throughout N. Amer.; Eurasia). *Naturalized. Sinapsis arvensis* L.—Rydberg; *B. arvensis* (L.) Rabenh.—Stevens.

5. ***Brassica nigra*** (L.) Koch, black mustard. Annual, to 15(20) dm tall, rough-hairy below, glabrous and glaucous above. Leaves all petioled, ovate to obovate, the lower pinnatifid, with a terminal lobe and a few small lateral ones. Flowers 1 cm wide; mature pedicels erect, 3-4 mm long. Siliques erect, quadrangular because of the stout midnerves of the valves, 1-2 cm long, smooth; beak slender, 2.5-4 mm long; seeds black, 1.5-2 mm in diam, minutely rough-reticulate. (2n = 16) May-Aug. A weed in fields & waste ground; GP; (throughout N. Amer., Eurasia). *Naturalized.*

9. CAMELINA Crantz, False Flax

Annual, caulescent herbs, glabrous or with branching hairs. Leaves alternate, entire or toothed, often clasping. Flowers in elongate racemes; sepals erect, obtuse, the outer slightly saccate at base; petals yellow, spatulate; filaments linear; anthers ovate; short stamens flanked at base by a pair of semicircular glands. Silicle obovoid or pyriform, pointed, margined, with filiform style; partition broad; valves 1-nerved; ovules 4-12 per cell; cotyledons incumbent.

1 Herbage rough-hairy, the short stellate hairs exceeded by simple hairs 1-2 mm long .... 1. *C. microcarpa*
1 Herbage sparsely hairy or glabrate, the few simple hairs not longer than the stellate .... 2. *C. sativa*

1. ***Camelina microcarpa*** Andrz. ex DC., small-seeded false flax. Erect annual, 3-7 dm tall, rough-hairy, 5.5-8 mm long, 3-4.5 mm wide, obscurely rugulose, strongly margined; seeds 1.0-1.3 mm long, brown. (2n = 40) Apr-Jun. Fields & waste places; GP; (throughout U.S. & s. Can., Europe). *Naturalized.*

2. ***Camelina sativa*** (L.) Crantz, gold-of-pleasure. Annual, 3-9 dm high, erect, usually with ascending branches; stem glabrous or with minute, closely appressed stellate hairs, the simple hairs not exceeding the stellate. Silicles 7-10 mm long, 5-7 mm wide, inconspicuously veined; seeds 1-2 mm long, pale yellowish-brown. (2n = 40) Apr-Aug. Fields, waste ground, & along roadsides; ne KS, NE, IA, CO, SD, MN, MT, & ND; (casual weed in n U.S. & adj. Can.; Europe).

This species is less common than *C. microcarpa.*

10. CAPSELLA Medic., Shepherd's Purse

1. ***Capsella bursa-pastoris*** (L.) Medic., shepherd's purse. Sparingly branched annual or biennial, 1-6 dm tall. Basal leaves oblong, 5-10 cm long, pinnately lobate; cauline leaves much smaller, lanceolate to linear, entire to denticulate, auriculate. Racemes becoming much elongated; pedicels divaricate at maturity, 1-2 cm long; sepals 2 mm long; petals

1.5–4 mm long, white; short stamens flanked by a pair of minute glands; ovary flat, obovate; ovules 6–many per locule; style very short. Silicles strongly flattened at right angles to the septum, triangular-obcordate, truncate to broadly notched at the summit, narrowly winged distally. (2n = 16, 32) Mar–Nov. Roadsides, fields, & waste ground; GP; (a cosmopolitan & ubiquitous weed; s Europe).

11. CARDAMINE L., Bitter Cress

Annual, biennial or perennial herbs, glabrous to sparsely hirsute with simple trichomes. Leaves alternate, entire to variously lobed or compound, petiolate. Flowers in racemes, corymbs or panicles; petals white or purple, obovate to spatulate; short stamens subtended by a semicircular gland; anthers oblong; ovary cylindric, gradually tapering to the short, slender style; ovules several per locule; stigma truncate. Siliques linear, straight, slightly compressed parallel to the septum; valves opening elastically from the base; replum margin extending partially over the valves; seeds uniseriate, marginless, plump, longer than broad; cotyledons accumbent.

```
1  Leaves deeply palmate-parted or compound; rhizome fleshy and jointed. .. 2. C. concatenata
1  Leaves simple or pinnate-dissected.
    2  Principal cauline leaves entire or nearly so; rhizome short, fleshy but
       unjointed ................................................................................ 1. C. bulbosa
    2  Principal cauline leaves deeply lobed to ± compound.
        3  Cauline leaves mostly 2–4 cm long; leaflets narrow and not decurrent on the
           rachis ........................................................... 3. C. parviflora var. arenicola
        3  Cauline leaves mostly 4–8 cm long; leaflets broad, oval to broadly oblong, decurrent
           on the rachis ............................................................... 4. C. pensylvanica
```

1. Cardamine bulbosa (Schreb.) B.S.P., spring cress. Stems erect, borne singly or few together from a short, stout rhizome, 2–6 dm tall, simple or branched above, with fine pubescence on the lower part. Leaves all simple, basal leaves long petioled and obovate to cordate-ovate, lower cauline leaves progressively shorter petioled; the upper leaves sessile or nearly so, entire to remotely dentate, ovate or rounded to lanceolate. Sepals greenish, with white margins, turning yellow after anthesis; petals white, 7–16 mm long; pedicels at maturity ascending to divergent, the lower to 4 cm long. Siliques slender, 1.5–2.5 cm long, often abortive, beak 2–4.5 mm long; seeds oval. (2n = 64, 80, 86) Feb–May. Moist or wet woods, creek bottoms, & marshy meadows; throughout e edge of GP; (Que. to MN, ND, & SD, s to FL & TX).

2. Cardamine concatenata (Michx.) O. Schwarz, toothcup. Erect, subscapose perennial; rhizome thick and fleshy but constricted at intervals. Leaves aplmately divided or compound, 3- to 5-parted, the basal ones long petioled; cauline leaves only 2 or 3, subopposite or alternate, laciniate. Inflorescence a corymb or short raceme; sepals ascending, usually saccate at the base, 5–8 mm long; petals white or pink, 12–19 mm long; ovary cylindric, gradually tapering to a slender style. Siliques 2–4(5) cm long; seeds in 1 row in each locule, somewhat flattened, not margined. (2n = 128, 256) Apr–May. Rich, moist woods; e edge GP from MN to e OK; (s Que. to MN, s to FL & OK). *Dentaria laciniata* Muhl. — Rydberg.

3. Cardamine parviflora L. var. **arenicola** (Britt.) Schulz, smallflower bitter cress. Glabrous annual or biennial, 1–2(4) dm tall, erect, simple to much-branched stem from fibrous roots. Basal leaves with about 5 pairs of oblong, mostly entire distinct leaflets; cauline leaves with 5–8 pairs of distinct, nondecurrent, linear, entire leaflets, the terminal leaflet scarcely broader than others. Flowers small, crowded; petals white, spatulate, 2–3.5 mm long; pedicels slender, ascending 5–8 mm long. Siliques erect, 2–3 cm long, less than 1 mm in diam, beakless or with style up to 0.7 mm long; seeds 0.7–0.9 mm long. (2n = 16) Mar–Jul. Dry woods, shaded or exposed ledges, & sandy soils; IA, MO, KS, OK, & TX; (n FL to TX, n to se Can.; OR to B.C.).

4. **Cardamine pensylvanica** Muhl., bitter cress. Erect or spreading annual or biennial, 0.5–6(7.5) dm tall, stem hispid at the base, otherwise mostly glabrous, glabrous throughout if wholly or partly submersed, simple or much branched. Basal leaves forming a rosette, these leaves with 1–6 pairs of elliptic, obovate or rounded glabrous leaflets with bases decurrent along or confluent with the rachis, dentate or undulate, the terminal leaflet largest, 1–2 cm wide; cauline leaves commonly (3)4–8 cm long with membranaceous leaflets linear-oblanceolate to obovate, usually decurrent on the rachis, entire or often with a few teeth or shallow lobes, the terminal leaflet cuneate-obovate. Petals white, 1.5–4 mm long; mature pedicels ascending. Siliques narrowly linear, 1–3 cm long, beaked by a tapering style 0.5–2 mm long; seeds 1–1.5 mm long. (2n = 32, 74) Apr–Jun. Wet woods, along springs & streams; e 1/3 of the GP; (Newf. & Que. to B.C., s to FL, GP, & CA).

12. CARDARIA Desv., Hoary Cress

Erect perennial herbs, stems arising from horizontal rootstocks, usually at least sparsely pubescent with short, simple hairs. Basal and lower cauline leaves petiolate, sinuate or lyrate when young but almost entire at flowering; middle and upper cauline leaves sessile, elliptic to lanceolate, usually sinuate-toothed, auriculate and cordate or sagittate at the base. Racemes corymbose; pedicels to 1 cm long, spreading; sepals elliptical, blunt; petals white to cream colored, about 2× longer than sepals, clawed; nectar glands large and well developed, completely surrounding bases of individual stamens; filaments free. Silicle ovoid, subglobose or cordate, more or less inflated, indehiscent or only tardily dehiscent, style persistent, slender; seeds 2–4, pendulous, wingless; cotyledons incumbent.

References: Mulligan, G. A., & C. Frankton. 1962. Taxonomy of the genus *Cardaria* with particular reference to the species introduced into North America. Canad. J. Bot. 40: 1411–1425; Rollins, R. C. 1940. On two weedy crucifers. Rhodora 42: 302–306.

1 Sepals and silicles short-hairy, the silicle evidently inflated, not compressed .. 3. *C. pubescens*
1 Sepals and silicles glabrous or nearly so, the silicles usually somewhat compressed.
   2 Silicles cordate, usually constricted at the septum especially towards the base .... 2. *C. draba*
   2 Silicles subreniform to widely obovate .... 1. *C. chalepensis*

1. **Cardaria chalepensis** (L.) Handel-Mazzetti, lens-padded hoary cress. Plants similar to the common *C. draba* (no. 2) except as follows: silicles inflated, subreniform to widely obovate, 2.5–6(8) mm long, 4–6(7) mm wide, glabrous or nearly so. (2n = 80) Apr–Aug. Roadsides, pastures, waste areas; scattered in GP from KS, CO, & MO northward; (MI w to B.C., s to MO, KS, CO, NV, & CA; Eurasia). *Naturalized. C. draba* var. *repens* (Schrenk) Schulz—Stevens.

2. **Cardaria draba** (L.) Desv., hoary cress. Erect perennial to 6 dm tall, spreading by rootstalks, stems short-hairy below, less so or glabrate upwards. Leaves oblong, mostly dentate, the basal 4–10 cm long and petioled; the cauline sessile and auriculate. Racemes in terminal corymbs; mature pedicels 10–15 mm long; sepals 2–2.5 mm long; petals 3–4 mm long. Silicles cordate, 2.5–3 mm long, 3–5 mm wide, somewhat inflated but usually constricted at the narrow partition; seeds red-brown, oval, compressed, about 1.5 mm long and 2 mm wide. (2n = 64) Apr–Aug. Fields, roadsides, & waste places; scattered in GP; (N.S. w to B.C., s to VA, MO, OK, NM, & CA; Eurasia).

3. ***Cardaria pubescens*** (C. A. Mey.) Jarmolenko, whitetop. Erect perennial to 4 dm tall, spreading by rootstalks, minutely pubescent from stem bases extending to the sepals and silicles. Leaves sagittate or cordate clasping. Racemes 4–10 cm long; pedicels slender, spreading-ascending; sepals short-hairy. Silicles evidently inflated, obovoid to subglobose, 3–4.5 mm long, 2.5–4.5(5) mm wide, short hairy, nearly equaled by the slender persistent style. (2n = 16) May–Aug. Fields, roadsides, & waste places; KS: Wichita; NE: Lancaster; ND: Cass, Mackenzie; WY; (established in w N. Amer. & locally e to PA; Eurasia). *C. pubescens* var. *elongata* Rollins—Fernald.

13. CHORISPORA DC., Blue Mustard

1. ***Chorispora tenella*** (Pall.) DC., blue mustard. Sparsely glandular-hirtellous annual, mostly branched, 2–5 dm tall. Lower leaves runcinate; middle and upper lanceolate to oblong, petiolate, undulate-dentate. Sepals erect, saccate at base, 3.5–6 mm long; petals blue-purple, rarely white, with claws evidently longer than blades; ovary linear, gradually tapering into the slender style, 3–4 cm long. Siliques slightly constricted at short intervals, transversely septate at the constrictions, at maturity breaking transversely into 2-seeded segments; ovules numerous. (2n = 14) May–Jul. Fields, roadsides, & waste areas; GP but most common in the west; (established locally as a weed in U.S.; cen Asia).

Malcolmia africana R. Br., a native of Europe, is superficially similar but eglandular and without cross-partitions in the fruit. It is established in the w U.S. and has been collected in KS: Ellsworth.

14. CONRINGIA Adans., Hare's-ear Mustard

1. ***Conringia orientalis*** (L.) Dum., hare's-ear mustard. Erect annual, winter annual, or biennial, glabrous herb, often glaucous, to 8 dm tall, somewhat succulent. Leaves pale, the lower narrowed to the base, the upper oval-elliptic, or oblong, broadly rounded above, cordate-clasping, entire. Flowers in elongate racemes; sepals erect, saccate at base; petals long clawed, ochroleucous, narrowly obovate, 10–12 mm long; short stamens subtended by a U-shaped gland; ovary cylindric, gradually tapering to the short style; ovules many; pedicels widely divergent. Siliques 8–12 cm long, 2–3 mm in diam, 4-angled; seeds in 1 row in each locule, oblong, granular-roughened. (2n = 14) May–Aug. Weed of fields, roadsides, & waste places; scattered n GP, sporadic s; (throughout U.S., more common in nw; Eurasia). *Introduced.*

15. DESCURAINIA Webb & Berth., Tansy Mustard

Herbaceous annuals or biennials, pubescent with branched trichomes, these intermixed with simple trichomes and/or stalked glands, the latter especially frequent on the axes of the racemes. Leaves pinnate to tripinnate, the basal leaves forming a rosette, these withering as the plant matures, usually absent by flowering; cauline leaves becoming reduced in size and less compound on the upper portion of the stem. Inflorescence racemose, elongating on maturing; flowers small, not over 3 mm long; sepals ovate, acute, green or yellow, frequently rose-tinged; petals yellow or whitish, clawed, the blades obovate, obtuse; stamens yellow, included or slightly exserted. Siliques narrow, linear or clavate, terete or subterete, the valves opening from below upward, these more or less prominently 1-nerved; style short or obsolete; stigma truncate or somewhat capitate, entire; seeds in 1 or 2 series in the locule, elliptical, yellowish to reddish-brown. *Sophia* Adans.—Rydberg.

References: Detling, Leroy E. 1939. A revision of the North American species of *Descurainia*. Amer. Midl. Naturalist 22: 481–520; Shinners, Lloyd H. 1949. Nomenclature of Texas varieties of *Descurainia pinnata* (Cruciferae). Field & Lab. 17: 145.

1 Siliques clavate or subclavate, 1–2 mm wide, 3.5–8 × longer than wide, spreading from the axis of the raceme; inflorescence usually glandular; upper leaves simply pinnate, lower leaves 1- to 2-pinnate .... 1. *D. pinnata*
1 Siliques linear, 0.5–1.3 mm wide, 6–30 × longer than wide; inflorescence not glandular.
  2 Leaves 2- to 3-pinnate; siliques 10–30 mm long, 20- to 40-seeded .... 3. *D. sophia*
  2 Leaves 1-pinnate, the leaflets often deeply incised; siliques 5–10(12) mm long, 4- to 30-seeded .... 2. *D. richardsonii*

1. *Descurainia pinnata* (Walt.) Britt., tansy mustard. Annual, sparsely pubescent to densely canescent with usually branched trichomes, sometimes glandular on the stems, inflorescences, and leaves, stems to 8 dm high, several from the base and branched or simple but then often branched above. Basal leaves to 1 dm long, usually bipinnate, the segments often deeply incised, the upper leaves gradually reduced in size and usually simply pinnate, the lobes narrow and linear to broadly obovate, amount of pubescence on the leaves varying from sparse to densely canescent. Sepals more or less oblong to ovate, sometimes with rose or magenta pigmentation; petals spatulate to obovate, clawed, whitish to bright-yellow; fruiting pedicels to 25 mm long, ascending to horizontal or slightly reflexed. Infructescences elongated and rather loose; siliques 4–20 mm long, clavate, sometimes broadly so, slightly latisept, usually straight; style very short or obsolete, seeds to 1 mm long, more or less ellipsoid, flattened, biseriate or rarely crowded so as to appear uniseriate. ($2n = 14$, 28, 42) Mar–Aug. Dry, open prairie or sparsely wooded areas, roadsides, fields, & waste places.

Shinners, op. cit., considers the subspecies of Detling, op. cit., to be varieties. Detling's subspecies are used in this treatment.

1 Fruiting pedicels spreading about 75°; herbage mostly canescent .... 1b. subsp. *halictorum*
1 Fruiting pedicels spreading about 45°; herbage not canescent.
  2 Stems glandular and moderately pubescent .... 1a. subsp. *brachycarpa*
  2 Stems not glandular, or if glandular then otherwise only sparsely pubescent or glabrous .... 1c. subsp. *intermedia*

1a. subsp. *brachycarpa* (Richard.) Detling. Plants erect with simple or branched stems, glandular pubescent. Sepals 1.5–2.5 mm long; petals about 0.5 mm longer than the sepals. Siliques 5–12 mm long; pedicels spreading at about 45°. GP; (Que. to Alta., s to n NH, VT, WV, TN, AR, CO, NM, & TX). *D. pinnata* var. *brachycarpa* (Richard.) Fern.—Fernald; *D. magna* (Rydb.) F. C. Gates—Gates; *Sophia brachycarpa* (Richard.) Rydb., *S. magna* Rydb., *S. pinnata* (Walt.) Howell—Rydberg.

1b. subsp. *halictorum* (Cockll.) Detling. Plants much branched, the lower stem and foliage more or less canescent, subglabrous to canescent and usually glandular in the inflorescences. Flowers small, 1–2 mm long, whitish or pale yellow. Fruiting pedicels 8–12 mm long, spreading at about 75°. W KS, OK, CO, NM, & cen TX; (OR e to AR, KS, & cen TX, s to AZ, CA, & Mex.). *D. pinnata* var. *osmiarum* (Cockll.) Shinners—Correll & Johnston.

1c. subsp. *intermedia* (Rydb.) Detling. Plants sparingly pubescent and sparsely glandular. Sepals 1.5–2.5 mm long; petals yellow, about 0.5 mm longer than the sepals. Siliques clavate or rarely broadly linear, 8–12 mm long, standing more erect than the pedicels; pedicels spreading at 45–70°; seeds biseriate, 5–13 per locule, about 1 mm long. MT: McCone, Sheridan; SD: Fall River, Lawrence; WY: Laramie, Niobrara; (s B.C. & Alta., s to CO, WY, ne CA, & w NV). *D. intermedia* (Rydb.) Daniels—Gates; *D. pinnata* var. *intermedia* (Rydb.) C. L. Hitchc.—Atlas GP; *Sophia intermedia* Rydb.—Rydberg.

This subspecies intergrades with subsp. *brachycarpa*, making separation of the two sometimes difficult.

2. *Descurainia richardsonii* (Sweet) O. E. Schulz. Biennial, stems simple below, usually branching above, 3–12 dm tall. Leaves ovate to broadly lanceolate or oblanceolate in outline,

1.5–10 cm long, basal leaves simple pinnate to bipinnate, the ultimate segments mostly broad and obtuse, the upper leaves simply pinnate with entire or lobed margins. Mature racemes dense; flowers 3 mm wide, yellow; calyx 1–1.5 mm long, equaled or exceeded by the petals. Siliques narrowly linear, 5–10(12) mm long, 0.5–1.5 mm wide, tapering rather abruptly at either end, usually more or less arcuate but sometimes straight, apex of the valves acutish, style short and blunt; fruiting pedicels strongly ascending, 3–6 mm long; seeds strictly uniseriate, 4–14 in each locule, reddish-brown, oblong to elliptical. (2n = 14, 28, 41) May–Jul. Prairies, calcareous gravel areas, & roadsides; MT, ND, w MN, SD, WY, & n CO; (Que. to Yukon, s to n Mex., e to n CO & MN). *Sophia richardsoniana* (Sweet) Rydb.—Rydberg.

3. *Descurainia sophia* (L.) Webb, flixweed. Annual or biennial, (2.5)3–7.5(8) dm tall, short branched above. Leaves broadly ovate or obovate to oblanceolate in outline, 1–9 cm long, 2- to 3-pinnate, the ultimate lobes commonly linear, sometimes narrowly oblanceolate or even obovate. Leaves, stems and axis of the racemes and occasionally the pedicels from sparingly to quite densely stellate pubescent, eglandular. Mature racemes loose, pedicels 8–14 mm long, widely ascending; flowers 3 mm wide, petals yellow. Siliques linear, somewhat torulose, 1–3 cm long, about 1 mm wide, arcuate or less commonly straight, style short and blunt; seeds uniseriate, 10–20 in each locule, oblong-elliptic, 0.75–1.5 mm long; septum of silique 2- to 3-nerved. (2n = 28) May–Jul. Weed in fields & waste places, GP except extreme se; (throughout U.S. & Can., Eurasia). Introduced. *Sophia multifida* Gilib.—Rydberg.

16. DIMORPHOCARPA Rollins, Spectacle Pod

1. *Dimorphocarpa palmeri* (Pays.) Rollins, spectacle pod. Densely pubescent annual or biennial, hirsute with dichotomously or irregularly branched trichomes; one to several erect stems, sometimes branches near the base, to 1 m tall. Basal leaves to 14 cm long, 4 cm wide, sessile on a cuneate base or short petiolate, oblong to elliptic, rounded to truncate, repand-sinuate, coarsely dentate or pinnatifid, densely pubescent; cauline leaves 2–12 cm long, 8–30 mm wide, sessile on a cuneate base or short petiolate, lanceolate, entire to repand or coarsely dentate, densely pubescent, often numerous and overlapping on the stem. Flowers white to pale yellow, borne in elongated terminal racemes; sepals oblong to narrowly ovate, the outer usually slightly saccate, the inner slightly cucullate, densely pubescent; petals with blades obovate, the claw expanded at the base and occasionally toothed, white or with purplish streaks. Infructescences loose; pedicels 1–2 cm long, usually horizontally spreading, sometimes recurved, straight or ascending, densely pubescent; silicles 7–10 mm long, 10–16 mm wide, oblong, densley pubescent, ovules 1 per locule, seeds 2.5–3.5 mm long, 2–3 mm wide, broadly oblong to nearly orbicular, much flattened with a narrow wing, cotyledons accumbent. (2n = 18) Apr–Aug. Sandy soil, especially dunes; nw TX, OK; KS: Clark, Comanche; (NV, UT, NM, & s KS s to Mex.). *Dithyrea wislizenii* Engelm. var. *palmeri* Pays.—Atlas GP.

Reference: Rollins, R.C. 1979. *Dithyrea* and a related genus (Cruciferae). Bussey Inst. Harvard Univ. (Misc. Publ): 3–32.

17. DIPLOTAXIS DC., Sand Rocket

1. *Diplotaxis muralis* DC., sand rocket. Perennial herb, stems usually several from the base, erect or decumbent, 2–4 dm tall, branched, sparsely hispid to glabrous. Leaves mostly basal, sinuate, dentate to pinnatifid or lyrately lobed, rarely spatulate and nearly entire;

cauline leaves few or none, cuneate at base when present. Flowers few in elongated racemes; sepals erect or ascending, not saccate, petals yellow to lavender, gradually narrowed to the claw, 5–7 mm long; short stamens subtended by a hemispheric or reniform gland; each pair of long stamens subtended by a short, prismatic gland; ovary linear, with many ovules; style scarcely differentiated; stigma capitate. Siliques ascending, 2–3 cm long, 2 mm in diam, not stipitate; lower pedicels becoming remote, 1–1.5 cm long. (2n = 44) May–Sep. Fields, roadsides, & waste places; KS: Geary; NE: Douglas; SD: Clay; (N.S. to NJ, WI to MN, GP & w U.S.; Europe). *Naturalized.*

18. DRABA L., Whitlow grass

Annual, biennial or perennial herbs, stems leafy or scapose, usually pubescent with simple branched trichomes. Leaves entire or dentate. Racemes short to elongate; sepals ascending or erect, blunt; petals white or yellow, rounded, emarginate, or bifid, narrowed belowt to a claw, or in some reduced or wanting; anthers short oval or oblong; ovary ovoid, with 2–many ovules per locule; style short or none; stigma capitate; glands various. Fruit a silicle (approaching a silique), elliptic to linear, 2–5× longer than wide, latisept, flat or sometimes twisted; seeds numerous, biseriate to irregularly seriate; cotyledons accumbent.

1 Plants annual or winter-annual with bractless racemes; flowering stems leafy or at least with 1 pair of leaves above the base.
  2 Silicles small, only 2.5–4 mm long, 0.8–1.4 mm wide, 6- to 16-seeded . 2. *D. brachycarpa*
  2 Silicles larger, mostly 5–20 mm long, 1.5–2.5 mm wide.
    3 Rachis and pedicels (and usually silicles) glabrous.
      4 Pedicels short, less than 2–3× as long as the silicles; petals white .... 6. *D. reptans*
      4 Pedicels as long as or generally longer than the silicles; petals yellow .... 5. *D. nemorosa*
    3 Rachis, pedicels and silicles pubescent .... 3. *D. cuneifolia*
1 Plants perennial, often with a multicipital caudex.
  5 Petals white .... 4. *D. lanceolata*
  5 Petals yellow.
    6 Style present and distinguishable; petals 4–6 mm long .... 1. *D. aurea*
    6 Style very short or lacking; petals 2–4 mm long .... 7. *D. stenoloba*

1. *Draba aurea* Vahl. Perennial, stems erect to decumbent, the caudex simple or branched, 10–50 cm tall, densely hirsute with simple or branched trichomes. Basal leaves 1–5 cm long, petiolate, spatulate to oblanceolate, usually entire; cauline leaves oblanceolate to ovate, densely pubescent. Racemes often many-flowered; sepals 2–3.5 mm long, pilose or stellate; petals 4–6 mm long, pale to bright yellow; stipes about 1 mm long. Silicles longer than the pedicels, 8–16 mm long, 2–4 mm wide, flat or contorted, pubescent; pedicels 3–20 mm long, ascending to erect in fruit. (2n = 64) Jun–Jul. Wooded slopes & meadows; BH; (AK s to BH, NM, & AZ). *D. aureiformis* Rydb., *D. surculifera* A. Nels.—Rydberg.

2. *Draba brachycarpa* Nutt., shortpod draba. Annual or winter annual, pubescent with sessile branched trichomes; stems to 2 dm tall, erect, or several from the base, usually branched above. Basal leaves rosulate, to 15 mm long, obovate to ovate, entire or remotely toothed, tapering into a short petiole, cauline leaves sessile, sometimes slightly auriculate, lanceolate or narrowly ovate, usually entire. Sepals 1–1.5 mm long, ovate to linear, often with some purple pigmentation; petals white, 2–3 mm long, blade obovate and submarginate or very short and linear or absent. Silicles 2–6 mm long, narrowly elliptic to oblanceolate, tapered and acute at both ends, glabrous; style very short or obsolete; racemes lax in fruit; pedicels 2–3 mm long, spreading-ascending; seeds 0.5–1.5 mm long, ellipsoid, biseriate. (2n = 24) Mar–Jun. Roadsides, open woodlands, & pastures; ne TX,

OK, MO, KS; (FL to TX, n to VA, IN, s IL, MO, & KS).

3. **Draba cuneifolia** Nutt., wedgeleaf draba. Annual or winter annual, 1–2.5 dm tall, simple or branched at the base, hispid from base to summit with mixed simple, bifurcate and minute stellate trichomes. Basal leaves oblanceolate to narrowly obovate, 1–3 cm long, coarsely dentate, rough hairy, cauline leaves few and only near the base. Racemes dense and congested, 5–10 cm long maturity; sepals 1–2.5 mm long, oblong-ovate to linear; petals white, usually 3.5–4(5) mm long, spatulate and emarginate, less often 1–2 mm long and linear or absent; pedicels 2–7 mm long, usually spreading. Silicles 5–15 mm long, elliptic to oblong, hispid with simple trichomes or glabrous; styles nearly obsolete; seeds about 0.7 mm long, biseriate to several-rowed. (2n = 32) Mar–Apr (May). Dry, rocky & sandy soil; TX, OK, KS, MO, CO, se NE, e SD; (FL to TX, n to KY, PA, s OH, SD, se NE, KS w & s to CA; n Mex.).

4. **Draba lanceolata** Royle. Perennial herb, stems to 3 dm tall, simple or branched, grayish, with mostly soft stellate hairs. Basal rosettes very dense, the numerous leaves oblanceolate or spatulate, densely stellate-pubescent, commonly less than 3 cm long, cauline leaves lanceolate to ovate, usually 5–20 mm long. Racemes often leafy-bracted at base, in maturity 1/3–4/5 full height of plant; sepals pilose; petals white, 3–4(5) mm long. Silicles oblong-lanceolate or oblong, 7–12(14) mm long, 1.5–2(2.5) mm wide, twisted or flat, stellate pubescent; style up to 0.75 mm long; pedicels strongly ascending in fruit; seeds 20–48. (2n = 32) May–Jul. Rocky & gravelly soils in open woodlands; BH; (circumboreal; in N. Amer.; AK & n Que. s to CO, UT, NV, NM, e WI, & NY). *D. cana* Rydb.—Rydberg.

5. **Draba nemorosa** L., yellow whitlowort. Annual or winter-annual, stem leafy, up to 3 dm tall, branched below. Basal leaves ovate-lanceolate to oval, elliptic, or obovate, 1–2.5 cm long, cauline leaves, similar, few to several, all below the middle of the stem, often dentate, pubescent with both simple and branched hairs. Inflorescence racemose; petals yellow, turning white with age, 2 mm long. Silicles ascending to erect, glabrous or hirsutulous, linear-oblong, 5–10 mm long; raceme elongate in fruit, to 2 dm long; pedicels glabrous, widely spreading and ascending, to 3 cm long. (2n = 16) Apr–Jun. Dry soil, prairies & hillsides; CO, WY, w NE, SD, MT, ND, MN; (w Can. e to w Ont.; s to CO, NE, & MI; Europe). *D. lutea* Gilib.—Rydberg.

6. **Draba reptans** (Lam.) Fern., white whitlowort. Annual or winter-annual, stems simple or branched at the base, 5–15(20) cm tall, pubescent with stipitate, stellate trichomes, glabrous above. Basal leaves rosulate, oblanceolate to spatulate or obovate, 1–3 cm long, blunt, entire, pubescent with mostly simple hairs above, stellate below, cauline leaves few and near the base of the stem, obovate, sessile. Racemes at maturity congested, the glabrous axis 5–20 mm long or rarely longer; sepals 1–2 mm long, linear to oblong, often with purple pigmentation; petals white, 3–4 mm long, clawed, the blade obovate and subemarginate or much reduced and linear or none. Silicles 10–15 mm long, linear but slightly narrowed at the base, slightly curved, glabrous to strigulose; style lacking; pedicels glabrous, ascending, 1/3 or 1/2 as long as the fruit; seeds 0.2–0.5 mm long, elliptic-oblong, biseriate. (2n = 32) Mar–May. Dry sterile sandy soil; GP; (MA, RI to s Ont., MN, & WA, s to NC, GA, GP, & CA). *D. caroliniana* Walt., *D. coloradensis* Rydb., *D. micrantha* Nutt.—Rydberg.

Our plants are usually treated as var. *reptans,* however the materials with strigulose to hispid silicles have sometimes have recognized as var. *micrantha* (Nutt.) Fern.

7. **Draba stenoloba** Ledeb. Annual, biennial or short-lived perennial, stem 5–30(40) cm tall, simple or branched, glabrous or hirsute at base, glabrous above. Basal leaves rosulate, obovate to oblanceolate, 0.5–5 cm long, 2–15 mm wide, denticulate, cauline leaves becoming

lanceolate or ovate. Racemes 10- to 20-flowered; sepals about 1.5–2 mm long, pilose; petals 2–4.5 mm long, yellow to white; style lacking or not over 0.1 mm long. Silicles 8–22 mm long, 1.5–2.5 mm wide, linear to oblong, usually glabrous; pedicels exceeding the fruit in length, glabrous. (2n = 24) Jun–Jul. Meadows, hills, & slopes; WY; SD: Lawrence, Pennington; (Alta. to B.C., s to SD, CO, & CA).

19. ERUCA P. Mill., Rocket-salad

1. *Eruca sativa* P. Mill., rocket-salad. Coarse, erect, annual herb with branched stems, 3–6 dm tall, hirsute below with simple spreading or retrorse trichomes, often glabrous above. Lower leaves pinnatifid or lobed, the upper merely dentate, lower leaves petiolate, the upper sessile. Flowers in racemes; sepals erect, the inner somewhat saccate at base, caducous, the inner not cucullate; petals white to ochroleucous, with a few conspicuous purple veins, 1.5–2 cm long, the claw surpassing the calyx; anthers obtuse, staminal glands minute. Siliques 12–25 mm long, 3–6 mm in diam, erect on short, stout pedicels, glabrous, slightly flattened, tipped with a prominent flattened beak 5–12 mm long; seeds numerous in 2 rows, nearly globose and wingless. (2n = 22) May–Jun. A weed of fields, roadsides, & waste places; IA: Mills, Pottawattamie; KS: Riley; MO: Jackson; ND: Cass, Pierce, Rolette; NE: Garfield; (e Ont. to ND, s to NJ, PA, IL, & KS; Europe). *Naturalized. E. versicaria* (L.) Cav. subsp. *sativa* (P. Mill.) Thell.—Rollins, J. Arnold Arbor. 62: 523. 1981

In Europe this species is cultivated for salad greens.

20. ERUCASTRUM Presl, Dog Mustard

1. *Erucastrum gallicum* (Willd.) O. E. Schulz., dog mustard. Annual or winter annual, stem erect or ascending, branched from the lower nodes, 3–6(8) dm tall. Basal and lower leaves oblanceolate, to 15 cm long, sparsely pubescent, deeply pinnatifid, segments crenately or angularly dentate, terminal segment the largest; cauline leaves progressively reduced, the uppermost 1–2 cm long. Flowers in racemes, bracteate below; sepals erect to spreading, somewhat cucullate at the tip; petals pale yellow or whitish, spatulate, 4–8 mm long; each pair of long stamens subtended by a short, pyramidal gland; a short gland between the ovary and each short stamen; ovary cylindric; ovules numerous. Siliques usually upcurved, the body 2–4 cm long, 4-angled, beak 3 mm long; seeds in 1 row in each locule. (2n = 30) Apr–Sep. Fields, roadsides, & waste places; n KS, MO to SD, ND, & MN; (much of e U.S. to GP & s Can.; Europe). *Naturalized. E. pollichii* Spenner—Rydberg.

21. ERYSIMUM L., Wallflower

Plants annual, biennial or perennial, rather coarse, with appressed branched trichomes. Leaves simple, narrow, entire to dentate, more or less pubescent with appressed branched trichomes. Inflorescence racemose; sepals erect, outer pair saccate at base; petals mostly large, clawed, yellow to orange; obovate or spatulate; short stamens subtended by a semicircular or annular gland; each pair of long stamens also subtended by a gland; anthers linear-oblong; ovary linear-cylindric, hairy; ovules numerous; style very short; stigma capitate, 2-lobed. Siliques linear, more or less 4-angled, hairy, the valves with a prominent midnerve; seeds in 1 row, numerous; cotyledons (ours) incumbent. *Cheirinia* Link.—Rydberg.

Reference: Rossbach, G. B. 1958. The genus *Erysimum* in North American north of Mexico. Madroño 14: 261–267.

1 Petals 15–25 mm long; mature siliques 5–12 cm long.
  2 Fruits erect or ascending .... 2. *E. capitatum*

```
    2  Fruits stiffly spreading ................................................................. 1. E. asperum
1  Petals 10 mm long or less.
    3  Sepals 2-3.5 mm long; petals 3.5-5.5 mm long; siliques 1.2-3 cm
       long ...................................................................................... 3. E. cheiranthoides
    3  Sepals 4.5-7 mm long; petals 6-10 mm long.
       4  Siliques widely spreading, 5-12 cm long; anthers 1 mm long ........... 5. E. repandum
       4  Siliques erect or nearly so, 1.5-5 cm long; anthers 2-2.5 mm
          long ................................................................................. 4. E. inconspicuum
```

1. *Erysimum asperum* (Nutt.) DC., western wallflower. Biennial or perennial, stems erect, simple or branched above, 2-10 dm tall, strigose from the presence of a dense covering of appressed bifurcate trichomes. Leaves numerous, crowded, linear to lanceolate, cuneate at the base, remotely dentate to entire. Flowers showy in dense racemes; sepals about 10 mm long; petals yellow to orange-yellow, 15-25 mm long, the blade about 1/2 as long as the very slender exserted claw; racemes at maturity greatly elongate, pedicels stout, divergent, 7-15 mm long. Siliques widely spreading, 4-angled, 8-12 cm long, densely pubescent on 4 sides, keels evident as longitudinal lines with a less dense covering of trichomes; seeds plump, oblong, often angular, with a short distal wing, 1.7-2 mm long. ($2n = 36$) Apr-Jun. Prairies, sandhills, & open woods; GP; (MT to NM, e Man., MN, NE, KS, OK, TX). *Cheirinia argillosa* (Greene) Rydb., *C. aspera* (Nutt.) Rydb., *C. asperrima* (Greene) Rydb., *C. elata* (Nutt.) Rydb.—Rydberg.

2. *Erysimum capitatum* (Dougl.) Greene, western wallflower. Coarse biennial, densely pubescent with appressed bifurcate or trifurcate trichomes, stems erect, usually single from the base, branched above, 4-10 dm tall. Leaves numerous, remotely dentate to entire, basal leaves lanceolate, 4-15 cm long, 4-10 mm wide, petiolate, upper leaves gradually becoming cuneate. Inflorescences dense, elongating in fruit; sepals 8-12 mm long; petals yellow to orange or maroon, 12-20 mm long; pedicels spreading to divaricately ascending, 6-9 mm long. Siliques erect or less frequently divaricately ascending, 5-8 cm long; styles 1-2 mm long, persistent stigma large and deeply 2-lobed; seeds oblong, plump, 1.5-2 mm long, wingless or with a short distal wing. Apr-Jul. Open slopes, rocky outcrops, road cuts, dry stream beds, & open wooded hillsides; NM: Union, Curry; n & w TX; (Sask. to WA, s to TX, NM, AZ, & CA).

3. *Erysimum cheiranthoides* L., wormseed wallflower. Annual or winter annual, stem erect, simple or sparingly branched, 2-10 dm tall. Leaves linear to oblanceolate, entire to barely sinuate, thinly pubescent but bright green, tapering to the base. Flowers small, on divergent to spreading, ascending filiform pedicels; sepals 2-3.5 mm long; petals yellow, 3.5-5.5 mm long; mature racemes elongate, the rachis straight, the pedicels very slender, widely divergent, (6)8-12(14) mm long. Siliques ascending to erect, subterete, slender, (1.2)1.5-2.5(3) cm long, green and glabrous or only sparsely pubescent; seeds oblong, brown, about 1 mm long. ($2n = 16$) Jun-Aug. Waste places, cult. fields, & rich meadows; common in WY, MT, ND, MN, SD, less so in CO, NE, IA, MO; (circumboreal, s in N. Amer. to NC, AR, CO, & OR). *Cheirinia cheiranthoides* (L.) Link—Rydb.

4. *Erysimum inconspicuum* (S. Wats.) MacM., smallflower wallflower. Erect perennial, 3-6(8) dm tall, cinereous and scabrous, stems simple or with ascending branches. Basal leaves crowded, sometimes repand to dentate, or entire, linear; cauline leaves linear to oblanceolate, entire or remotely sinuate dentate, rarely over 5 mm wide. Inflorescence racemose; pedicels thick, with dilated tip; calyx densely stellate, 6-7(8) mm long, with linear to oblong sepals; petals pale yellow, 6-10 mm long; anthers 2-2.5 mm long. Mature racemes elongate, 2-3 dm long; fruiting pedicels stout, ascending, 3-9 mm long; siliques erect or nearly so, 1.5-4 cm long, 4-angled; seeds 1-1.3 mm long. ($2n = 54, 162$) May-Aug.

Dry soil of prairies, plains, & upland woods; common in n GP & sporadic in s GP, except OK & TX; (N.S. to B.C. s to NY, MO, KS, NM, & AZ). *Cheirinia inconspicua* (S. Wats.) Rydb., *C. syrticola* (Sheldon) Rydb.—Rydberg.

5. ***Erysimum repandum*** L., bushy wallflower. Annual, stems simple or more often divergently branched, sparsely to densely pubescent with appressed bifurcate trichomes, 1–4(6) dm tall. Leaves linear to linear-oblanceolate, mostly repand-dentate, sometimes entire, the lower leaves short petiolate, the upper leaves cuneate at the base. Flower racemose; pedicels thick, not enlarged at the summit, spreading at right angles to rachis, 2–4 mm long; sepals densely stellate, 4.5–5.5 mm long; petals pale yellow, 6–10 mm long; anthers 1 mm long. Siliques widely spreading, nearly terete, 5–12 cm long, about 1 mm in diam; seeds oblong, compressed, 1–1.4 mm long. (2n = 16) May–Jul. Weed of prairies, fields, roadsides, & waste places; common in s GP, sporadic n of KS, & absent in ND, MN, & IA; (MA to OR, s to AL, AR, TX, NM, CA; Europe). *Adventive. Cheirinia repanda* (L.) Link—Rydberg.

22. HESPERIS L., Dame's Rocket

1. ***Hesperis matronalis*** L., dame's rocket. Biennial or perennial herb, stems erect, often branched above, 5–10 dm tall, pubescence of both simple and branched hairs. Leaves lanceolate or deltoid-lanceolate to oblong-lanceolate, short-petioled to sessile, remotely and sharply denticulate, pubescent above with simple hairs, below chiefly with branched hairs. Flowers racemose, fragrant; sepals erect, outer narrow, crested near the summit, inner ones broad, saccate at base; petals bluish-purple, rarely white, the blade obovate, the claw exceeding the calyx, 2–2.5 cm long; short stamens subtended by a U-shaped gland; long stamens glandless; anthers linear-oblong; ovary cylindric; style scarcely narrower; stigma 2-lobed, decurrent along the style; ovules numerous. Siliques 5–14 cm long, widely spreading on stout pedicels, somewhat constricted between the seeds; seeds large, 3–4 mm long, angularly fusiform. (2n = 16, 26, 29, 32) May–Aug. Roadsides, thickets, & open woods; KS, CO, MO, NE, IA, SD, WY, ND, & MN; (s Can. & U.S.; Europe). Introduced as an ornamental and now widely *naturalized*.

23. IODANTHUS T. & G., Purple Rocket

1. ***Iodanthus pinnatifidus*** (Michx.) Steud., purple rocket. Perennial, glabrous or nearly so, to 1 m tall, stem simple below the inflorescence. Leaves thin, lanceolate to elliptic or oblong, acute or acuminate, commonly sharply dentate, often laciniate, rarely double-dentate or merely crenate, tapering to a petiolelike base, frequently auriculate, the larger or lower leaves often pinnatifid at base with 1–4 pairs of small segments. Racemes elongating before anthesis; sepals erect, the inner somewhat saccate at base, obtuse, 3–5 mm long; petals pale violet to nearly white, 10–13 mm long; the blade triangular-obovate, gradually narrowed to the exserted claw; short stamens surrounded at base by an annular gland, anthers linear-oblong; pedicels 6–10 mm long. Siliques slender, widely divergent, 2–4 cm long, the valves covered with minute transparent papillae, beaked, slightly constricted between the seeds; seeds numerous, in 1 row. May–Jun. Moist or wet alluvial woods; MN, IA, KS, MO, OK, & TX; (w PA & WV to IL, MN, KS, AR, OK, TX, & AL).

Reference: Rollins, R. C. 1942. A systematic study of *Iodanthus*. Contrb. Dudley Herb. 3: 209–215.

24. LEPIDIUM L., Peppergrass

Annual, biennial or perennial herbs, glabrous to hispid, pubescence, if present, of simple hairs. Leaves entire to finely divided, sometimes amplexicaul. Flowers in racemes or panicles; sepals 4, equal, oblong to round; petals small, white, yellow, or greenish, linear to spatulate, or often lacking; stamens 6, or by abortion 4 or 2; anthers small, oval; ovary flat; ovules 1 per cell, suspended; style short or none; stigma capitate. Fruit a silicle, flattened contrary to the narrow septum, thin or somewhat distended over the seeds, ovate to orbicular or obovate, often winged, commonly retuse, tipped by the persistent style or stigma; cotyledons usually incumbent, rarely accumbent or oblique.

References: Hitchcock, C. L. 1936. The genus *Lepidium* in the United States. Madroño 3: 265–320; Mulligan, G. A. 1961. The genus *Lepidium* in Canada. Madroño 16: 77–90.

1 Middle and upper leaves suborbicular, deeply cordate-clasping with a closed sinus and slightly overlapping lobes, thus appearing perfoliate .... 6. *L. perfoliatum*
1 Middle and upper leaves narrower, linear to broadly lanceolate, if clasping, not appearing as if perfoliate.
  2 Silicles 5–6 mm long.
    3 Middle and upper leaves clasping the stem; silicles on spreading pedicels .... 2. *L. campestre*
    3 Middle and upper leaves not clasping; silicles on strongly ascending to appressed pedicels .... 8. *L. sativum*
  2 Silicles 2–3.5 mm long.
    4 Glaucous perennial, 5–13 dm tall, with rhizomes; leaves thickish and rugose, lanceolate to broadly lanceolate .... 4. *L. latifolium*
    4 Annual or biennial, 5–40 cm tall; leaves not thickish and rugose, linear to lanceolate.
      5 Sepals persistent until fruits are nearly mature; pedicels slightly flattened and wing margined .... 5. *L. oblongum*
      5 Sepals deciduous with the petals and stamens or soon after; pedicels often decidedly flattened, but scarcely wing margined.
        6 Plant hispid; basal leaves entire or but toothed; fruits hirsute, the hairs somewhat appressed .... 1. *L. austrinum*
        6 Plant not hispid; basal leaves usually pinnatifid or at least cleft.
          7 Silicles puberulent, at least on the margin.
            8 Silicles nearly elliptic, widest near the middle, narrowed into acute apical teeth .... 7. *L. ramosissimum*
            8 Silicles round-obcordate to short oblong-obovate, widest above the middle, rounded to abruptly curved into obtuse apical teeth .... 3. *L. densiflorum*
          7 Silicles glabrous.
            9 Silicles oval, orbicular to rotund; petals conspicuous, as long as or slightly longer than sepals .... 9. *L. virginicum*
            9 Silicles round obcordate to short-obovate, rounded to abruptly curved into obtuse apical teeth; petals shorter than sepals or lacking .... 3. *L. densiflorum*

1. Lepidium austrinum Small, peppergrass. Annual or biennial, the entire plant more or less hispid, stems usually single but branched, erect, often sulcate, to 6 dm tall. Basal leaves to 9 cm long, lyrate-pinnatisect with the segments dentate to serrate, tapering to a slender and often winged petiole; lower and middle cauline leaves to 6 cm long, more or less lanceolate, rarely entire, usually dentate to serrate or somewhat pinnatifid; upper cauline leaves to about 2 cm long, linear-lanceolate, remotely dentate to serrate. Sepals narrowly elliptic, with white margins; petals narrowly spatulate and about 2× longer than the sepals or rudimentary and much shorter than the sepals, white; stamens 2. Infructescences densely fruited and elongate; pedicels to 5 mm long, slender and terete or slightly flattened, spreading to slightly recurved. Silicles 2.5–3.5 mm long, 2–2.8 mm wide, elliptic to ovate, compressed, winged on the distal half, the wings forming a shallow notch, sparsely pubescent with short appressed trichomes, often ciliate on the margins; seeds to about 1.6 mm long and 1 mm wide, ovoid but compressed, usually with a narrow wing; cotyledons incumbent. Feb–May. In sandy or loamy soils; KS: Allen; OK: Blaine, Cleveland, Osage; n-cen TX; (se KS, OK, s to TX, & Mex.).

2. **Lepidium campestre** (L.) R. Br., field peppergrass. Densely short-villous annual or biennial, simple to freely branching, 1.5–6(7) dm tall. Basal leaves petiolate, elongate, oblanceolate, entire to pinnatifid; cauline leaves numerous, overlapping, sagittate-clasping, suberect or ascending, lanceolate to narrowly oblong, 2–4 cm long, entire or denticulate, sessile. Racemes dense, to 15 cm long; sepals 1.5–2 mm long; petals white, 2–2.5 mm long; the mature pedicels divergent, 4–8 mm long. Silicles oblong-ovate, 5–6 mm long, broadly winged above, the short style barely exserted. (2n = 16) May–Sep. Weed of fields, roadsides, & waste places; s IA, MO, e KS, & e OK; widely established in U.S.; Europe). *Introduced.*

3. **Lepidium densiflorum** Schrad., peppergrass. Thinly short-hairy annual or biennial, 2–5 dm tall. Basal leaves 4–7 cm long, oblanceolate, coarsely toothed to pinnatifid; cauline leaves shorter, linear or narrowly oblanceolate, mostly entire, sharply acute. Racemes numerous, erect, 5–10 cm long; petals none, or if present, white and shorter than sepals. Silicles broadly oval to obovate, 2–3.3 mm long, nearly or fully as wide, narrowly winged above; stigma included in the notch; cotyledons incumbent. (2n = 32) Feb–Jun. Dry or moist soil of waste grounds, roadsides, & pastures; GP; (throughout U.S., more common e of Rocky Mts.). *L. bourgeauanum* Thell., *L. Fletcheri* Rydb., *L. neglectum* Thell.—Rydberg.

4. **Lepidium latifolium** L. Perennial plants from widely spreading root systems, stems to 1.3 m tall, glabrous or nearly so. Leaves entire to dentate; basal leaves oblong, 10–30 cm long, 5–8 cm wide, petiolate; cauline leaves 1–4 cm wide, reduced in size but numerous, upper ones nearly sessile. Racemes many flowered; sepals shorter than the petals, less than 1 mm long, oval, somewhat pilose on back; petals spatulate, white, about 1.5 mm long; stamens 6. Silicles 2 mm long, ovate-rotund, sparsely pilose, not emarginate, tipped by stigma and almost obsolete style; pedicels terete, longer than the fruit, 3–4 mm long. (2n = 24) Jun–Aug. Fields & waste places; CO: Sedgwick; NE: Hooker; (occasional weed in many parts of U.S.; Europe). *Introduced.*

5. **Lepidium oblongum** Small. Annual or biennial, stems diffusely branched, 5–20 cm tall, hirtellous to villous. Basal leaves to 3(5) cm long, bipinnatifid to pinnatifid, segments usually dentate, often pubescent only on the vein; cauline leaves laciniate. Racemes numerous; pedicels somewhat flattened; sepals ovate to deltate, often magenta with white margins, about 1 mm long, not persistent; petals absent or minute; stamens 2. Infructescences dense, with elongated pedicels to 5 mm long. Silicles glabrous, 2.5–3.5 mm long, 2–3 mm wide, usually elliptic to obovate or nearly orbicular, narrowly winged at the distal end, wings forming a notch; style obsolete or nearly so. Mar–May. Roadsides & waste places; s NE, KS, OK, & TX; (AR, KS, OK, TX, AZ; Mex. & Guat.).

6. **Lepidium perfoliatum** L., clasping peppergrass. Erect, usually glabrous to sparsely hairy annual, 2–5 dm tall, branched above. Basal and lower cauline leaves pinnately dissected into linear segments; midcauline leaves entire, auriculate, broadly ovate to subrotund; upper leaves appearing perfoliate. Pedicels spreading, terete; sepals firm, pilose, broadly oval, 1–1.3 mm long; petals narrowly spatulate, yellow, slightly longer than the sepals, 1–1.5 mm long; stamens mostly 6. Silicles rhombic-elliptic, about 4 mm long, minutely notched

at the apex, style about equal to the notch; seeds nearly 2 mm long, narrowly winged. (2n = 16) Mar–Jun. Weed of waste & disturbed places; scattered, GP; (w U.S., e to MI & OH; Europe). *Introduced.*

7. Lepidium ramosissimum A. Nels., bushy peppergrass. Profusely branched biennial or possible annual, stems 1.5–5 dm tall, densely but finely puberulent. Basal leaves usually pinnatifid, lobes often toothed; upper cauline leaves entire and linear. Inflorescences with many short corymbiform racemes in the leaf axils, as well as terminal longer racemes, 2–4 dm long; sepals 1 mm long; petals shorter than sepals, white, linear; stamens 2; style lacking or very short. Silicles 2.5–3.5 mm long, often ciliate or pubescent, elliptic, shallowly notched and winged at apex; pedicels about equaling the silicles, slightly flattened to somewhat wing margined. Apr–Oct. Dry plains; MT, ND, WY, CO, & NM; (Man. to Alta. s to ND, NM).

8. Lepidium sativum L., garden cress. Glabrous, more or less glaucous annual, 2–4 dm tall. Leaves all dissected, the lower bipinnatifid, segments linear, oblong or oblanceolate. Racemes strict, elongate; petals 2 mm long, white; stamens 6. Silicles oblong to ovate, 5–6(7) mm long, broadly winged, deeply notched at apex; style half as long as the notch; mature pedicels erect or closely ascending, 2–4 mm long. (2n = 16, 32) Jul–Sep. Escapes from cultivation to roadsides & waste places; NE: Antelope, Valley; (scattered in N. Amer.; w Asia). *Adventive.*

9. Lepidium virginicum L., peppergrass. Erect annual or biennial, glabrous or minutely pubescent, 1–5 dm tall. Basal leaves oblanceolate, sharply toothed to pinnatifid or even bipinnatifid; upper cauline leaves smaller, oblanceolate to linear, dentate to entire, acute, narrowed to the base. Racemes numerous, to 1 dm long; petals equaling to 2× longer than the sepals, white; stamens 2, rarely 4. Silicles nearly orbicular, emarginate at the apex, 2.5–4 mm long; style included in the notch. (2n = 32) Jun–Nov. Dry or moist soil of fields, gardens, roadsides, & waste places; GP; (Newf. to FL, w to GP, Pacific Coast).

25. LESQUERELLA S. Wats., Bladderpod

Herbaceous annuals, biennials or perennials, more or less densely pubescent with branched, stellate or sometimes simple trichomes. Basal leaves entire to pinnatifid, linear to oblanceolate, less frequently suborbicular; cauline leaves cuneate with a short petiole or less frequently sessile and auriculate at base, entire to dentate, mostly oblanceolate but often oblong to narrowly ovate in the auriculate species. Sepals erect to spreading, obtuse; petals yellow or white, entire, broadly obovate to narrowly spatulate; filaments linear, dilated at base only in a few species. Infructescences usually elongated, congested and nearly subumbellate in a few species. Silicles most commonly globose to obovoid or flattened in some species, sessile or stipitate, glabrous or stellate-pubescent; ovules 1–15 per locule; styles slender, persistent or rarely deciduous; seeds biseriate, usually somewhat flattened; cotyledons accumbent.

Reference: Rollins, Reed C., & Elizabeth A. Shaw. 1973. The genus *Lesquerella* (Cruciferae) in North America. Harvard Univ. Press, Cambridge, MA. 288 pp.

1 Some or all cauline leaves auriculate .... 3. *L. auriculata*
1 Cauline leaves petiolate or cuneate at the base, never auriculate.
  2 Silicles glabrous on the exterior.
    3 Trichome rays free and distinct or slightly fused, sometimes very small, the rays then contiguous, but not fused.
      4 Trichomes small, less than 0.13 mm in diam, the rays contiguous and forked or bifurcate .... 6. *L. gracilis* subsp. *nuttalii*

              4 Trichomes large, 0.3–0.6 mm in diam, the rays distinct to their bases or slightly fused.
                 5 Pedicels straight or only slightly curved, ascending to divaricately
                    spreading .... 9. *L. ovalifolia*
                 5 Pedicels sigmoid or recurved; infructescences loose and often
                    secund .... 5. *L. gordonii*
           3 Trichome rays conspicuously fused for 1/2 their lengths or more .... 4. *L. fendleri*
      2 Silicles pubescent on the exterior.
           6 Pedicels uniformly recurved downward, not straight or sigmoid; fruit globose or nearly so.
                 7 Plants perennial; inner basal leaves involute and usually entire; infructescences not se-
                    cund; valves often pubescent on the interior .... 7. *L. ludoviciana*
                 7 Plants annual or short-lived perennials; inner basal leaves flat and generally dentate
                    or at least angular; infructescence usually secund; valves usually glabrous on the in-
                    terior .... 2. *L. arenosa*
           6 Pedicels sigmoid or straight; fruit ovate, obovate to oblong.
                 8 Basal leaves linear to narrowly oblanceolate .... 1. *L. alpina*
                 8 Basal leaves oblanceolate to obovate .... 8. *L. montana*

1. ***Lesquerella alpina*** (Nutt.) S. Wats. Caespitose perennial, densley pubescent with appressed or spreading stellate trichomes, trichomes sessile or on a short stalk, smooth or glandular, caudex much branched, stems 0.1–2.5 dm long, erect and usually simple, buried among the lower leaves or more or less exserted beyond them, often very sparsely leaved. Basal or outer leaves, when present, 0.5–7 cm long, 1–10 mm wide, linear-oblanceolate and gradually narrowed to the petiole, or with a rhombic, hastate, or ovate blade narrowed abruptly to the petiole; cauline leaves 0.5–5 cm long, 1–5 mm wide, oblanceolate to linear, often tufted at the bases of the stems. Inflorescences usually dense, buds ovoid to ellipsoid; pedicels 4–15 mm long, straight or slightly curved to sharply sigmoid; sepals about (3.5)4–5(7) mm long, stellate pubescent; petals yellow, 4.5–8(10) mm long, 1.5–3.5 mm wide, ligulate to spatulate; filaments linear, usually more or less abruptly dilated at the base, paired stamens 3–7 mm long, single stamens 2.5–6 mm long; glandular tissue pentagonal to hexagonal around the single stamens and subtending the paired stamens. Silicles 2.5–8 mm long, sessile or nearly so, ovate, more or less compressed and flattened at the acute apex, the valves densely pubescent on the exterior, glabrous on the interior or rarely sparsely pubescent; septum smooth, often perforate, the funicles attached about 1/3 their lengths; styles 1.5–6 mm long, slender, sometimes sparsely pubescent, the stigmas expanded or not; ovules 2–6 per locule; seeds 2–2.7 mm long, oblong and plump, orange-brown to reddish-brown, lacking margins or wings. (2n = 10) May–Jul. Dry plains & on open slopes; MT, w ND, w SD, w NE, WY, & CO; (s Alta. & sw Sask., s to ID, MT, ND, n UT, CO, & w NE).

2. ***Lesquerella arenosa*** (Richardson) Rydb. Plants annual to short-lived perennials; more or less densely stellate pubescent, trichomes sessile or on a short stalk, roughly granular, stems (0.5)1–2(3) dm long, several, prostrate and straggling to erect, usually unbranched, often with purplish pigmentation, arising from a several-branched caudex. Basal leaves 1.5–5(7) cm long, 2–10 mm wide, entire to shallowly dentate, oblanceolate and narrowing to a slender petiole, up to 2 × longer than the blade, or the blades obdeltate or narrowly rhombic, subacute to acute and flat; cauline leaves (0.5)1–2.5(3) cm long, 1–4 mm wide, elliptic to linear, usually entire. Inflorescences dense, buds ovoid to ellipsoid; sepals 4–6(7) mm long, elliptic to oblong, the lateral ones barely saccate, the median thickened at the apex and cucullate; petals 6–8.5(9.5) mm long, 1.5–3(3.5) mm wide, yellow or with varying areas of reddish or lavender pigmentation when dry, spatulate with little distinction between blade and claw or with blades obovate and narrowing gradually to a broad claw; filaments little dilated at the base, paired stamens 4.5–7(8) mm long, single stamens 3.5–6(7) mm long; glandular tissue roughly pentagonal around the single stamens and subtending the paired, absent between the latter. Infructescences elongated, loose and usually secund;

pedicels 5–15(20) mm long, stout and usually sharply recurved, but occasionally only divaricately spreading or nearly horizontal. Silicles (3.5)4–5.5(6.5) mm long, sessile or nearly so, subglobose, obovoid or broadly ellipsoid, the valve exterior densely pubescent with spreading or closely appressed trichomes, the valve interior glabrous, or rarely very sparsely pubescent; septum entire and smooth or slightly wrinkled longitudinally, the funicles attached about 1/2 their lengths; styles 3–5.5(6.5) mm long and slender, the stigmas expanded; seeds 2–2.7 mm long, flattened and suborbicular or slightly wider than long, red-brown, lacking margins or wings. (2n = 10, 18, 26, 30) May–Jun. Open,dry, sandy or gravelly soils.

Two varieties of this species occur in the GP:

2a. var. *arenosa*. Stellate trichomes on the silicles with the rays spreading and not closely appressed. MT, ND, SD, WY, & w NE; (s Man., Sask., & Alta., s to ND, MT, w SD, s WY, & nw NE).

2b. var. *argillosa* Rollins & Shaw. Stellate trichomes on the silicles with the rays closely appressed. Restricted to e WY, nw NE.

3. *Lesquerella auriculata* (Engelm. & Gray) S. Wats. Annual, hirsute with simple trichomes and pubescent with an understory of branched trichomes, stems several to numerous, erect or decumbent, 0.5–2 dm tall. Basal leaves 2–5 cm long, 8–15 mm wide, short petiolate, dentate to lyrate, sometimes nearly entire, hirsute and/or pubescent with simple or branched trichomes; cauline leaves 1–4 cm long, 3–10 mm wide, sessile and auriculate, oblong to sagittate, entire to dentate, usually overlapping on the stem. Inflorescences dense; pedicels 7–15 mm long, ascending to spreading at about 45°, densely hirsute; sepals narrowly oblong, the outer pair slightly saccate; petals 7–11 mm long, 4–5 mm wide, yellow, obovate, entire or slightly emarginate. Silicles 4–6(8) mm long, subsessile, globose to slightly longer than wide, glabrous on the exterior and interior septum usually entire; ovules 4–8(10) per locule; seeds about 2 mm long, suborbicular, flattened and margined. (2n = 16) Mar–May. Dry prairies & disturbed soils; cen TX, cen & ne OK; s cen KS (cen TX, cen & ne OK; s cen KS).

4. *Lesquerella fendleri* (A. Gray) S. Wats. Perennial, usually densely stellate pubescent and silvery, trichomes sessile or on a short stalk, the rays numerous and unforked, fused for 1/2 or more their lengths, but never fully fused, stems .5–2.5(4) dm long, several, usually unbranched, erect or the lateral ones decumbent, arising from a usually divided caudex. Basal leaves 1–4(8) cm long, 1–6(10) mm wide, the blades roughly elliptic, entire to coarsely dentate, narrowing gradually to the slender petiole; cauline leaves 0.5–2.5 cm long, 1–5(15) mm wide, linear to narrowly elliptic, entire and sometimes involute or the blades elliptic to rhombic, entire or remotely dentate, narrowing gradually to a slender petiole. Inflorescences loose, buds ellipsoid; sepalş 5–8 mm long, elliptic to oblong, not saccate, but the median often thickened at the apex and more or less cucullate; petals (6)8–12(15) mm long, (3)4.5–7.5 mm wide, yellow, sometimes with orange guide lines, obdeltate to obovate or with an obdeltate or obovate blade narrowed to a short claw; filaments slender and not or only slightly expanded at the base, paired stamens 6–9.5 mm long, single stamens 4.5–7 mm long; glandular tissue fleshy, roughly V-shaped around the single stamens and subtending but absent between the paired stamens. Infructescence dense and usually elongated, but sometimes not extending beyond the leaves; pedicels 8–20(40) mm long, straight or slightly curved, occasionally sigmoid, divaricately spreading to erect. Silicles 4.5–8 mm long, sessile or nearly so, globose or broadly ellipsoid or ovoid, then usually quite acute, the valves glabrous on the exterior and interior, often reddish at maturity; septum entire and smooth; the funicles attached for 1/2 or more their lengths; styles (2)3–6 mm long, slender, stigmas expanded; ovules (6)10–16(20) per locule; seeds 1.3–2.0 mm long, flattened, suborbicular to broadly ovate, yellow to orange-brown, lacking margins or wings. (2n = 12) Mar–May. In sandy & rocky calcareous soil; CO, sw KS, w OK, NM, & cen TX; (sw KS to UT, s to TX & n Mex.).

5. **Lesquerella gordonii** (A. Gray) S. Wats. Annual, usually densely stellate pubescent, trichomes sessile or short stipitate, smooth or finely granular, rays 4–7, distinct and forked or bifurcate, usually 0.5–0.6 mm in diam; stems 1–3.5(4.5) dm long, several, erect, the outer decumbent, or sometimes prostrate, sometimes branched and quite densely leaved. Basal leaves 1.5–5(8) cm long, 4–15 mm wide, elliptic or obovate, lyrate-pinnatifid to entire, terminal lobe often rhombic; cauline leaves 1–4(7) cm long, 1–10 mm wide, linear to oblanceolate, often falcate, entire to repand or shallowly dentate, the upper sessile, the lower gradually narrowing to the slender petiole. Inflorescences dense, buds ellipsoid; sepals 3.0–6.5 mm long, elliptic or oblong, the lateral slightly saccate, the median thickened at the apex and cucullate; petals widely spreading at anthesis, (4)5–8(10) mm long, (1.5)2–4(5.5) mm wide, yellow to orange, the claw sometimes whitish, cuneate or with an obdeltate or obovate blade tapering to the broad claw, this often widened at the base; filaments slender and slightly expanded at the base or those of the lateral stamens sometimes spatulate, paired stamens (3)4–8 mm long, single stamens (2.5)3.5–6.5 mm long; glandular tissue more or less surrounding the single stamens, and subtending but absent between the paired stamens. Infructescences elongated but of varying density, more or less secund; pedicels 5–15(25) mm long, normally sigmoid, sometimes straight and spreading. Silicles (3.5)4–8 mm long, on a short gynophore 0.5–1.5 mm long, subglobose or a little longer than wide, glabrous on the exterior and interior; septum entire and smooth or a little wrinkled, the funicles attached about 1/2 their lengths; styles (1.5)2–4(5) mm long, slender, stigmas expanded and sometimes bilobed; ovules (4)6–10(13) per locule; seeds 1.5–2.5 mm long, orbicular or slightly broader than long, orange to reddish-brown, flattened, lacking margins or wings. (2n = 12) Mar–Jun. In sandy and gravelly soils, especially when they are disturbed; s-cen KS, w OK, & w TX; (cen KS, s to w OK, NM, AZ, cen & sw TX).

6. **Lesquerella gracilis** (Hook.) S. Wats. subsp. **nuttallii** (T. & G.) Rollins & Shaw, spreading bladderpod. Annual or biennial, stellate pubescent, trichomes sessile or nearly so, small and 4- to 7-rayed, rays distinct and usually forked near their bases, occasionally bifurcate, granular, those on the stems several-rayed, the rays usually bilaterally oriented; stems 1–3 dm tall,1 single to several, the outer stems decumbent, the inner erect, simple or branched near the top, these branches slender and flexuous. Basal leaves 1.5–8(11.5) cm long, 2–16 mm wide, the blades oblanceolate to elliptic, lyrate-pinnatifid to dentate or merely repand, tapering gradually to the petiole, sparsely pubescent on the adaxial surface, more densely below; cauline leaves 1.7 cm long, 2–20 mm wide, obovate or elliptic, dentate to repand or entire, upper ones sessile or nearly so, lower narrowing gradually to the short slender petiole. Inflorescences rather loose, buds broadly ellipsoid; sepals 3–6(8) mm long, elliptic or broadly ovate, median ones cucullate and slightly thickened at the apex, lateral ones saccate; petals (4)6–11 mm long, 3–7 mm wide, yellow to orange, blades broadly obovate and narrowing gradually to the short claw; filaments slender, sometimes slightly dilated at the base, paired stamens 4–8.5 mm long, single stamens (2)3–7 mm long; glandular tissue U- or V-shaped around the single stamens and subtending the paired stamens. Infructescences elongated and loose; pedicels (7)10–20(25) mm long, straight or slightly curved, usually divaricately spreading, sometimes horizontal or shallowly recurved. Silicles 3–10 mm long, obpyriform and truncate at the base, on a slender gynophore 0.5–1.5(2) mm long, glabrous on the exterior, glabrous or sparsely pubescent on the interior; septum entire and thick, dirty-white, smooth and slightly wrinkled, the funicles attached about 1/2 their lengths; styles 2–4.5 mm long, slender and glabrous, stigmas slightly expanded; ovules 4–10(14) per locule; seeds about 2 mm long, more or less oblong, slightly flattened, red-brown, lacking margins or wings. (2n = 12) Mar–May. On sandy, clay, or heavy black soils in prairies, heavily grazed pastures & along roadsides; e NE, e KS, w MO, e OK, & e TX; (e NE, e KS, OK, e & e cen TX, sw AR). *L. repanda* (Nutt.) S. Wats.—Rydberg.

7. Lesquerella ludoviciana (Nutt.) S. Wats., bladderpod. Perennial, densely stellate pubescent, trichomes sessile or on a stalk, rough-glandular, the rays 4–7, these distinct or slightly fused at their bases, forked and often bifurcate; stems 1–3.5(5) dm long, usually few from the simple to elaborately divided caudex, the outer stems usually decumbent. Basal leaves (1)2–6(9) cm long, 4–10 mm wide, the outer leaves more or less oblanceolate and obtuse, usually flat, the inner leaves erect, narrowly elliptic to linear and entire or shallowly dentate, frequently involute, usually with simple trichomes at the leaf bases; cauline leaves (1)2–4(8) mm long, 1–6 mm wide, narrowly elliptic to oblanceolate, flat or more or less involute. Inflorescences compact and densely flowered, buds ovoid to ellipsoid; sepals 4–7(8) mm long, oblong to broadly elliptic, the lateral ones slightly saccate, the median ones cucullate; petals yellow (5)6.5–9.5(11) mm long, 1.5–3 mm wide, oblanceolate with no distinction between blade and claw or with a broadly obovate blade gradually narrowed to a broad claw; filaments stout but not dilated at their bases, paired stamens (4)5–7.5(9) mm long, single stamens 3–5.5(7) mm long; glandular tissue pentagonal to hexagonal around the single stamens and subtending but apparently not developed between the paired. Infructescences elongated and loose, pedicels (5)10–20(25) mm long, after anthesis usually becoming recurved. Silicles (3)4–5.5(6) mm long, usually subglobose or shortly obovoid, sometimes slightly compressed, the valves densely pubescent on the exterior with spreading trichomes, usually pubescent on the interior with sessile appressed trichomes; styles 3–4.5(6.5) mm long, slender; stigmas not or only slightly expanded; septum entire and little wrinkled, the funicles attached about 1/2 their lengths; ovules (2)4–6(8) per locule; seeds 2–2.5 mm long, slightly flattened, suborbicular or broadly oblong, red-brown, neither margined or winged. (2n = 10) Apr–Aug. Sandy and gravelly soils; GP except TX, NM, MO, & IA; (ND to MT, s to IL, OK, & AZ). *L. argentea* (Pursh) MacM.—Stevens.

8. Lesquerella montana (A. Gray) S. Wats. Perennial, usually densely stellate-pubescent, trichomes sessile or on a short stalk, smooth or finely granular, the rays 4–7, distinct or slightly fused at their bases, forked or bifurcate; stems 0.5–2(3.5) dm long, prostrate or erect, arising from a sometimes much enlarged and divided woody caudex. Basal leaves (1)2–5(7) cm long, 5–15 mm wide, blades suborbicular or deltate and abruptly narrowed to the petiole or obovate to elliptic and gradually narrowed to the petiole, entire to sinuate or shallowly dentate; cauline leaves 1–2.5(4) cm long, 2–6 mm wide, often secund, linear to obovate or rhombic, entire or shallowly dentate, lower ones short-petiolate, upper ones sessile. Inflorescences dense and compact, buds ellipsoid; sepals 5–8.5 mm long, elliptic, median ones thickened at the tip and cucullate, lateral ones boat-shaped, saccate; petals (6)7.5–12 mm long, 2–3.5 mm wide, yellow to orange and sometimes fading purplish, narrowly spatulate with no distinction between blade and claw or the blade obovate and gradually narrowed to the claw; filaments not dilated at the base, paired stamens (5)6.5–9.5 mm long, single stamens 4–8 mm long; glandular tissue roughly pentagonal around the single stamens and subtending and sometimes developed between the paired stamens. Infructescences usually elongated; pedicels 5–15(20) mm long, stout and sharply sigmoid and the siliques erect, occasionally nearly straight and divaricately spreading. Silicles (6)7–12 mm long, sessile or subsessile, ellipsoid or ovoid, usually slightly obcompressed, the valves densely pubescent on the exterior with spreading trichomes, glabrous or sparsely pubescent on the interior; septum entire, little wrinkled, the funicles attached 1/3 their lengths or less; styles 3–7 mm long and slender, sometimes pubescent, the stigmas expanded; ovules (4)6–10(12) per locule; seeds about 2 mm long, oblong to ovate, flattened, dark red-brown, lacking margins or wings. (2n = 10) May–Jun. Dry, open places of western plains & foothills; NE: Kimball, Sioux; NM: Union; SD: Fall River; WY: Laramie, Platte; (sw SD, NE, se WY, s to e CO & ne NM).

9. Lesquerella ovalifolia Rydb., oval-leaf bladderpod. Perennial, densely stellate-pubescent and scabrous, trichomes sessile or short stipitate, often umbonate, glandular or smooth, rays numerous, often forked, usually fused at their bases, generally symmetrically disposed but occasionally forming a U-shaped notch on one side of the trichomes, stems 0.5–3 dm high, few to several from each division of the caudex, the caudex sometimes simple but frequently elaborately branched producing mounded "pin-cushion" plants. Basal leaves 0.5–2(6.5) cm long, 0.5–1(2) cm wide, blades suborbicular to elliptic or ovate or deltate, entire to sinuate to dentate, narrowing abruptly or gradually to the petiole, the oldest leaves often grayish; cauline leaves (0.5)1–2.5(4) cm long, 2–4(6) mm wide, narrowly elliptic or obovate, the lower ones short petiolate, upper ones generally sessile. Inflorescences dense and compact, buds ellipsoid; sepals 4.5–7.0(8.5) mm long, more or less elliptic, lateral ones saccate, median ones thickened at the tip; petals (7)8.5–15 mm long, yellow or white, blades suborbicular to obovate or obdeltate, sometimes emarginate, narrowing to a rather broad claw; filaments of the paired stamens slightly expanded at the base, those of the single stamens sometimes expanded, paired stamens 6–11 mm long, single stamens 4–6.5(8) mm long; glandular tissue V-shaped to pentagonal around the single stamens and subtending but usually absent between the paired. Infructescences subumbellate or elongated to nearly 10 cm; pedicels 5–15(20) mm long, stout, usually spreading at about 45°, but sometimes horizontal or nearly erect. Silicles (3.5)4.5–7.5(9) mm long, sessile or on a short stipe, subglobose to broadly ellipsoid, the valves glabrous on the exterior and interior; septum smooth and entire, the funicles attached 1/2 or more of their lengths; styles 4–8(9) mm long, stigmas expanded; ovules 4–8 per locule; seeds about 2 mm long, flattened, suborbicular, dull red-brown, lacking margins or wings. Apr–Jun. On rocky knolls, limestone, or gypsiferous outcrops & in rock crevices.

Two subspecies occur in the GP:

9a. subsp. *alba* (Goodman) Rollins & Shaw. White petals (9)11–15 mm long, often 2× or more longer than sepals; caudex sometimes enlarged, but rarely branched; infructescences elongated; blades of outer basal leaves often broadly elliptic, entire to rarely sinuate and narrowing gradually to petiole or blades small and deltate and then abruptly narrowing to the petiole. (n = 6, 12) Restricted to s & s-cen OK; KS: Osborne.

9b. subsp. *ovalifolia*. Yellow petals 6.5–12(14) mm long, usually only 1.5× longer than or less than the length of sepals; caudex usually elaborately branched; infructescences usually subumbellate; blades of outer basal leaves suborbicular or ovate to elliptic and entire, narrowing abruptly to petiole. (n = 6, 12, 18, 24, 36) Restricted to sw NE, w KS, e CO, e NM, cen & w OK & w TX.

26. NASTURTIUM R. Br., Watercress

1. Nasturtium officinale R. Br., watercress. Aquatic perennial, with creeping or floating stems, rooting at the nodes. Leaves pinnately compound, of 3–9(11) obtuse, entire, somewhat fleshy segments, the later ovate to rotund, the terminal one larger, usually rotund. Flowers in racemes, 5 mm wide; sepals oblong, ascending, one pair saccate at base, about 2 mm long; petals white, spatulate, about 2× as long as the sepals; short stamens flanked by a pair of reniform glands; anthers ovate; ovary cylindric, style stout, scarcely differentiated; stigma capitate, 2-lobed. Siliques linear, subterete, somewhat falcate, tipped with a short persistent style, 10–15(20) mm long, valves with inconspicuous midnerve; fruiting pedicels divergent; seeds in 2 rows, flattened, suborbicular, areolate, areolae about 35 per side. (2n = 32) Apr–Oct. In water & mud along springs, streams, & lakes; throughout GP, except in MT, ND, & MN; (widely distributed throughout N. Amer., Eurasia). *Rorippa nasturtium-aquaticum* (L.) Hayek.—Green.

Reference: Green, P. S. 1962. Watercress in the New World. Rhodora 64: 32–43.

A similar species, *N. microphyllum* (Boenn.) Reichb., has been collected in Sioux Co., NE. It is distinguished by its slender, longer fruits (17–26 mm), seeds that have about 150 tiny areolae per side, and a chromosome number of 2n = 64. Since most watercress specimens are collected without fruit it is impossible to determine how common this species might be in the GP. *Nasturtium microphyllum* is native to Europe and is now established in scattered areas of the ne & nw U.S. & s Can.

27. NESLIA Desv., Ball Mustard

1. *Neslia paniculata* (L.) Desv., ball mustard. Slender annual or biennial, stellate pubescent, usually simple up to the inflorescence then much branched, to 8 dm tall. Cauline leaves oblong-lanceolate, sagittate-clasping, 3–6 cm long, acute, entire or nearly so. Flowers minute, 1.5 mm wide; sepals oblong, obtuse; petals yellow, spatulate, gradually tapering into a claw; each short stamen flanked by a pair of fleshy U-shaped glands; ovules 2 per cell. Silicles globose, thick walled, indehiscent, slightly compressed, reticulate and pitted, 2–3 mm in diam; mature pedicels slender, divergent, to 1 cm long. (2n = 14) Jun–Sep. A weed in fields & waste places; scattered places in ND, MN, SD, & IA; (Newf. to B.C., s to N.S., NJ, PA, OH, IN, IL, & SD; Europe). *Introduced.*

28. PHYSARIA (Nutt.) A. Gray, Double Bladderpod

1. *Physaria brassicoides* Rydb., double bladderpod. Heavy rooted perennial, often with a branched caudex, silvery-stellate pubescent, stems usually many, somewhat decumbent-based, 2–17 cm long. Basal leaves numerous, arranged in a rosettelike pattern, marcescent, 2–8 cm long, the blades obovate or broadly oblanceolate to somewhat rhombic or even ovate, leaf surfaces whitened with dense, stellate hairs. Racemes somewhat closely flowered; pedicels slender to rather stout 7–18 mm long, ascending; sepals not saccate at base; petals yellow, broadly spatulate-obovate, 9–12 mm long. Silicles much inflated, 1–2 cm long and at least as wide, didymous, the base slightly cordate, the apical sinus narrow, nearly closed, 2–4 mm deep; replum obovate or oblanceolate, more nearly obtuse than acute at the apex, 3–6 mm long, 2–3 mm wide; style 6–9 mm long; seeds 2–3 per cell. (2n = 8, 16, 24) May–Jun. Dry & sandy soil; ND: Billings, McKenzie, Slope; SD: Haakon, Harding, Jackson, Sheridan; (ne WA & ID e to w WY, SD, ND, MT, & sw Alta.). *P. didymocarpa* (Hook.) A. Gray—Van Bruggen.

Reference: Rollins, R. C. 1939. The cruciferous genus *Physaria*. Rhodora 41: 392–415.

29. RAPHANUS L., Radish

1. *Raphanus sativus* L., radish. Annual, stem branched, 4–5 dm tall, glabrous or nearly so, with a thickened taproot. Lower leaves lyrate-pinnatifid with rounded, crenate divisions, more or less hairy, the uppermost lanceolate. Flowers in racemes; sepals erect, the lateral ones somewhat saccate at the base; petals large, white or light purple, 15–20 mm long, with long claws. Siliques terete, 4–5 cm long, 6–10 mm in diam, tapering into a weakly distinct beak, 1-loculed, few-seeded, filled with a spongy tissue between the seeds; seeds spherical, cotyledons conduplicate. (2n = 18) Apr–Sep. Occasionally escaping from cultivation; sporadic in GP; (widely escaped in N. Amer.; Europe).

30. RORIPPA Scop., Yellow Cress

Prepared with the assistance of Monica Rhode-Fulton

Annuals or occasionally biennials from a vertical taproot, or perennials from spreading rhizomes, stems sometimes from a basal rosette, erect, decumbent, or prostrate, glabrous or hirsute or with vescicular trichomes. Leaves alternate, sessile to short-petiolate, oblong, obovate, oblanceolate, to spatulate in outline, gradually reduced upwards, the margins entire to pinnatifid, either or both surfaces glabrous to hirsute or with vescicular trichomes. Racemes terminal and axillary, developing during or after stem elongation, the siliques nearly equal in age at corresponding points on all the racemes, or the oldest siliques on the lower portion of the terminal raceme, or racemes lateral, developing during stem elongation in the lower leaf axils and progressing upward, without the formation of a true terminal raceme, the oldest siliques on the lower axillary racemes; sepals strongly saccate to flat, caducous or rarely persistent in fruit; petals absent or present, pale to bright sulfur yellow, the blade gradually narrowed into the claw, shorter, equal to, or longer than the sepals, erect during full anthesis, caducous in fruit; stamens 6, slightly tetradynamous, rarely 3–5, anthers introrse, crowded against the stigma at anthesis, retained within the flower; ovary cylindrical, elongating in fruit, style very short, elongating and persistent in fruit, stigma capitate. Siliques (approaching a silicle) plump, globose to cylindrical, sometimes slightly constricted at the middle, the long ones straight or slightly to strong falcate, as long as to several times longer than wide, shorter than, equal to, or longer than the subtending pedicels, valves 2, rarely 3 or 4, thin, usually readily dehiscent, nerveless or obscurely nerved, glabrous, rough with minute papillae or vescicular trichomes, or densely strigose; fruiting pedicels slightly to strongly recurved, divergent, or ascending; seeds regularly to angularly cordiform, wingless, numerous and crowded into 2 irregular rows filling the locules, or sometimes no mature seeds formed, surface reddish or yellowish-brown, colliculose or foveolate; cotyledons accumbent. *Radicula* Hill—Winter.

Reference: Stucky, R. L., 1972. Taxonomy and distribution of the genus *Rorippa* (Cruciferae) in North America. Sida 4: 279–430.

1 Petals absent; stamens 3–6 .... 5. *R. sessiliflora*
1 Petals present; stamens always 6.
  2 Petals longer than the sepals.
    3 Stems erect; mature siliques without seeds; leaves and stems glabrous or sparingly pubescent, but lacking trichomes.
      4 Leaf margins dentate or serrate to nearly entire; siliques globose to subglobose, less than 2 mm long; pedicels 4–8× longer than the siliques .... 1. *R. austriaca*
      4 Leaf margins pinnatisect, the lobes toothed; siliques linear-cylindrical, 3 mm or more long; pedicels 0.9–2× longer than the siliques .... 7. *R. sylvestris*
    3 Stems decumbent to prostrate; mature siliques containing seeds; leaves and stems with trichomes.
      5 Siliques short- to elongate-cylindrical, more than 4 mm long, usually more than 3× longer than wide; trichomes hemispherical, vesicular .... 6. *R. sinuata*
      5 Siliques globose to subglobose, less than 3.5 mm long, up to 2× longer than wide; trichomes elongate, pointed .... 2. *R. calycina*
  2 Petals shorter than or equal to the sepals.
    6 Oldest siliques on lower portion of the terminal racemes; stems sparingly to densely hirsute below .... 4. *R. palustris*
    6 Oldest siliques on lower axillary racemes, or siliques nearly equal in age at corresponding points on the terminal and axillary racemes; stems glabrous throughout.
      7 Pedicels strongly recurved, or sometimes divergent at right angles to the raceme axis; stems mostly taller than 3 dm .... 3. *R. curvipes*
      7 Pedicels divergent to ascending; stems mostly shorter than 3 dm.

8   Style in fruit straight, abruptly attached to the obtuse to truncate silique apex; siliques glabrous, constricted at the center, 3.5–5.5 mm long, usually 2–4 × longer than wide .... 9. *R. truncata*
8   Style in fruit tapering toward the apex, gradually merging into the pointed silique apex; siliques rough with minute papillae, not constricted at the center, 4–6(9) mm long, more than 4 × longer than wide .... 8. *R. tenerrima*

1. **Rorippa austriaca** (Crantz) Bess., Austrian field cress. Perennial from a thick, fleshy rhizome, stems single, erect, 5–9 dm tall, glabrous or sparingly pubescent below. Middle cauline leaves sessile, sagittate and clasping, narrowly oblong to oblanceolate, 3–10 cm long, 1–2 cm wide, sharply and irregularly dentate or serrate or nearly entire, glabrous or rarely sparingly pubescent above and below, the apex mostly obtuse. Racemes terminal and axillary, 0.5–1.2 dm long; sepals 2–2.5 mm long, saccate, caducous in fruit; petals 2.5–3.5 mm long. Siliques globose to subglobose, 1.1–1.6 mm long, 1–1.2 mm wide, 1–1.3 × longer than wide, the valves smooth and glabrous, style in fruit straight, 0.8–1.2 mm long, abruptly attached to the rounded silique apex; fruiting pedicels 6–10 mm long, 4–8 × longer than the siliques, mostly divergent, mature seeds rarely formed. (2n = 16) May–Aug. Roadsides & cult. fields; MN: Lincoln; ND: Cass, Pembina, Ransom, Richland; NE: Lancaster; (NY to WI, MN, IA, NE, & ND; Europe). *Adventive. R. austriaca* (Jacq.) Crantz — Stevens.

2. **Rorippa calycina** (Engelm.) Rydb. Perennial from slender rhizomes, forming large clones, stems few to numerous, decumbent to prostrate, 1–4 dm long, moderately to densely hirsute with slender trichomes expanded at the base and pointed at the apex. Middle cauline leaves sessile, auriculate and clasping, oblong to oblanceolate, 2.5–5 cm long, 0.5–1 cm wide, shallow to strongly sinuate, hirsute above and below, especially on the midrib, the apex acute to obtuse. Racemes terminal and axillary, 0.5–1.5 dm long, or lateral; sepals 1.2–3.1 mm long, saccate, persistent in fruit; petals (2)2.5–3.7 mm long. Siliques globose to subglobose, 2.3–3.4 mm long, 1.1–2.3 mm wide, 1.3–2 times longer than wide, the valves densely strigose with very short trichomes expanded at the base and pointed at the apex; style in fruit straight, 1.2–2 mm long, abruptly attached or somewhat gradually merging into the rounded silique apex; fruiting pedicels 3.5–6.5 mm long, glabrous to sparingly hirsute, 1.5–2 × longer than the siliques, strongly recurved, often in the same direction and giving the siliques the appearance of being borne unilaterally; seeds 0.7–0.8 mm long, about 20 per silique, the surface prominently colliculose. May–Jul. Riverbanks, moist sandy soil; MT: Custer; ND: McKenzie. *Endemic. Radicula columbiae* (Suksd.) Howell — Rydberg.

3. **Rorippa curvipes** Greene. Annual, biennial, or short-lived perennial from a vertical taproot, stems 1 to several, prostrate, decumbent, or erect, 2–5 dm long, glabrous. Basal and lower cauline leaves sessile, somewhat auriculate to nonauriculate, nonclasping, oblong, obovate, or spatulate, (2)2.5–7(9) cm long, 0.5–1.5(2) cm wide, pinnatifid, with toothed lobes, or occasionally irregularly serrate, crenate, or repand, glabrous or sparingly hirsute above, glabrous below, the apex obtuse to acute. Racemes terminal and axillary, 0.5–1.1 dm long; sepals 1–1.2(1.6) mm long, flat to slightly saccate, caducous in fruit; petals 0.7–1.2 mm long. Siliques short to elongate-cylindrical, slightly to strongly curved upward and inward toward the raceme axis, tapering to the apex, somewhat constricted in the center, 2.5–5 mm long, 0.8–1.4(2) mm wide, 2–4.5 × longer than wide, 0.7–1.7 × longer than the pedicels, the valves smooth; style in the fruit straight, 0.5–1 mm long, abruptly attached or gradually merging into the obtuse to acute silique apex; fruiting pedicels 2.5–5 mm long, divergent at right angles to the raceme axis, or strongly recurved, often in the same direction and giving the siliques the appearance of being borne unilaterally; seeds 0.5–0.7 mm long, 20–50 per silique, the surface finely colliculose. (2n = 16) Jul–Sep. Wet meadows, muddy shores of drying ponds & streams, occasionally along roadsides; CO: Weld;

KS: Ford; ND: Benson; SD: Brown, Faulk; WY: Platte; (s Alta., w MT, & ID, s to NM, sparingly w to CA) *R. obtusa* (Nutt.) Britt.—Rydberg.

4. Rorippa palustris (L.) Bess., bog yellow cress. Annual or occasionally biennial or perennial from a vertical taproot, stems 1 to several, erect, rarely decumbent or prostrate, simple or branched, 3–9 dm tall, glabrous to hirsute. Basal and lower cauline leaves sessile or short-petiolate, auriculate and partly clasping to nonauriculate and nonclasping, oblong to oblanceolate, (3)5–15(18) cm long, 1–5 cm wide, pinnatifid, the lobes toothed, glabrous to hirsute above and below, the apex acute to obtuse. Racemes terminal and axillary, 0.3–1.3(1.8) dm long; sepals 1.4–2 mm long, flat to slightly saccate, caducous in fruit; petals 1–1.8 mm long. Siliques globose to cylindrical, straight to slightly curved upward and inward toward the raceme axis, not at all to slightly tapering toward the apex, not at all to somewhat constricted in the center, 2–6.5 mm long, 1–3 mm wide, 1–4× longer than wide, 0.5–1.4× longer than the pedicels, the valves smooth and glabrous; style in fruit straight, 0.2–1 mm long, abruptly attached to the obtuse silique apex; fruiting pedicels 2.7–7.5 mm long, slightly recurved, divergent, or ascending; seeds 0.5–0.9 mm long, 20–80 per silique, the surface finely colliculose. (2n = 32) May–Oct. Edges of rivers, streams, lakes, & ponds, moist roadsides.

This species is extremely variable, especially in leaf lobing and texture, fruit size and shape, pedicel length, and position and distribution of trichomes. In recent North American treatments some authors recognize this complex as one species under the name of *R. palustris* or *R. islandica*, while others have recognized two species, a glabrous one, *R. palustris* or *R. islandica*, and a hirsute species, *R. hispida*. Stuckey (op. cit.) considers *R. palustris* as one highly polymorphic species and uses both subspecific and varietal ranks to portray levels of morphological complexity and geographical segregation. He has shown *R. islandica* to be another species, which has not been found on the North American continent.

Stuckey (op. cit.) lists fours infraspecific taxa that occur in the GP, distinguishing them as follows:

1 Leaves glabrous on the lower surface; stems glabrous or sparingly hirsute.
  2 Siliques elongate-cylindrical; replum margin straight or convex, remaining flat with age and upon drying .... 4b. subsp. *glabra* var. *glabrata*
  2 Siliques mostly short-cylindrical, constricted at the center; replum margin concave, becoming twisted with age and upon drying .... 4a. subsp. *glabra* var. *fernaldiana*
1 Leaves hirsute on lower surface; stems hirsute usually up to the terminal raceme.
  3 Siliques elongate-cylindrical, (3)5.2–7.8(8.5) mm long, (1.4)2.1–3.1 mm wide, ca 2× or more longer than wide; replum oblong in outline .... 4c. subsp. *hispida* var. *elongata*
  3 Siliques globose to subglobose, (2.2)2.6–5(6.7) mm long, 1.3–2.6(3.1) mm wide, ca 1–2× longer than wide; replum circular to elliptic in outline .... 4d. subsp. *hispida* var. *hispida*

4a. subsp. *glabra* (Schulz) Stuckey var. *fernaldiana* (Butt. & Abbe) Stuckey. Pedicels divergent to slightly recurved. Stigma expanded in fruit. GP; (ME to Alta., s to FL, TX, CO, & MT; Puerto Rico, C. Amer., Colombia). *R. islandica* (Oeder) Borbas var. *fernaldiana* Butt. & Abbe—Fernald.

4b. subsp. *glabra* (Schulz) Stuckey var. *glabrata* (Lunell) Stuckey. Pedicels divergent. Stigma unexpanded in fruit. ND: Benson; (s Can., n MI w to ID & ne OR).

4c. subsp. *hispida* (Desv.) Jonsell var. *elongata* Stuckey. Stem usually sparingly hirsute below. Pedicels (3.8)4.2–7.8(12) mm long, usually divergent or recurved. Petals 1.5–2(2.8) mm long. ND: Benson, Bottineau, Emmons; NE: Morrill, Wheeler; SD: Minnehaha, Stanley; (ND to AK, s to w NE, CO, & n CA).

4d. subsp. *hispida* (Desv.) Jonsell var. *hispida*. Stem usually densely hirsute below. Pedicels 2.7–5.5(8.6) mm long, usually divergent to ascending. Petals 1–1.5(2) mm long. MN w to MT, s to IA, NE, & CO; (N.B. w to AK, s to VA, IA, NE, sw NM, n UT, & n CA). *R. hispida* (DC.) Britt.—Rydberg; *R. islandica* (Oeder) Borbas var. *hispida* (Desv.) Butt. & Abbe.—Fernald.

5. Rorippa sessiliflora (Nutt.) Hitchc., yellow cress. Annual or biennials from a vertical taproot, stems 1 to many, erect, 2–5 dm tall, glabrous. Basal and lower cauline leaves short-petiolate to cuneate and sessile, slightly auriculate to nonauriculate, nonclasping, obovate, oblanceolate, or spatulate, 1.5–6(10) cm long, (0.4)1–2 cm wide, crenate, irregularly serrate, or repand, the apex broadly obtuse to somewhat acute. Racemes terminal and axillary, 0.7–1.2(1.6) dm long; sepals 1.3–2 mm long, strongly saccate, caducous in fruit; petals absent; stamens 3–6, the number varying among flowers on the same plant, the filaments sometimes fused. Siliques elongate-cylindrical, straight to slightly curved inward toward the raceme axis, becoming slightly wider toward the apex, 5.5–8.5(10) mm long, 1.5–2.2 mm wide, 3–5× longer than wide, 3–7(8)× longer than the pedicels, the valves smooth to rough with minute hyaline ridges; style in fruit strongly expanded below the stigma, absent to 0.7 mm long, gradually merging into the obtuse silique apex; fruiting pedicels 0.8–2 mm long, rather thick, divergent to ascending; seeds 0.4–0.5 mm long, 150–200 per silique, the surface deeply foveolate. (2n = 16) May–Oct. Muddy or sandy places along rivers, streams, creeks, & ponds, occasionally on roadsides; s IA & se NE, s to MO & OK; (e VA, IN to NE, s to FL & TX). *Radicula sessiliflora* (Nutt.) Greene—Winter.

6. Rorippa sinuata (Nutt.) Hitchc., spreading yellow cress. Perennial from slender rhizomes, large clones, stems few to numerous, decumbent to prostrate, 1–4.5 dm long, sparsely to densely pubescent with hemispherical vescicular trichomes. Middle cauline leaves sessile, auriculate and partly clasping to nonauriculate and nonclasping, oblong to oblanceolate, 3–8 cm long, 0.5–1.5(2) cm wide, deeply sinuate, pinnatifid to subpinnatifid, the lobes entire to minutely toothed, glabrous or rarely with sparse vescicular trichomes above, sparsely to densely covered with vesicular trichomes below, mostly on the midrib, the apex acute. Racemes terminal and axillary, 0.5–1.5 dm long; sepals 2.5–4.5 mm long, strongly saccate, caducous in fruit; petals (3.5)4–5.5(6) mm long. Siliques short to elongate-cylindrical, slightly to strongly falcate toward the raceme axis, (4)5.5–8(11) mm long, 1–2 mm wide, (3)4–7.5× longer than wide, 0.6–1.5× as long as the pedicels, the valves glabrous or rough with vescicular trichomes over the entire surface or only on the edges; style in fruit tapering to the apex, 1–2 mm long, gradually merging into the pointed silique apex; fruiting pedicels 5–10(12) mm long, glabrous or sparingly to densely pubescent with vescicular trichomes, strongly recurved to ascending or occasionally divergent at right angles to the raceme axis and becoming recurved or ascending at the distal end; seeds 0.6–0.9 mm long, 20–60 per silique, the surface finely colliculose. Apr–Aug. Dry or more often wet (or at least damp) roadsides & railroad ditches & along rivers, streams, & lakes; throughout GP; (w Ont. to WA, s to AR & CA). *Radicula sinuata* (Nutt.) Greene—Winter.

7. Rorippa sylvestris (L.) Bess., creeping yellow cress. Perennial from slender rhizomes, forming clones, stems 1 to several, erect, 3.5–5 dm tall, glabrous or sometimes sparingly hirsute below. Middle cauline leaves short-petiolate, auriculate to nonauriculate, nonclasping, oblong to obovate, 2.5–11 cm long, 1–4 cm wide, pinnatisect, the lobes toothed, the apex acute. Racemes terminal and axillary, 0.8–1.8 dm long; sepals 2–2.8 mm long, flat, caducous in fruit; petals 2.5–4 mm long. Siliques linear-cylindrical, straight or somewhat curved inward toward the raceme axis, (3)5–8 mm long, 0.5–1 mm wide, 3.5–7× longer than wide, the valves smooth; style in fruit straight, 0.5–1 mm long, gradually merging into the pointed silique apex; fruiting pedicels 5–8(10) mm long, 0.9–2× longer than the siliques, divergent; mature seeds rarely formed. (2n = 32, 40, 48) Jun–Oct. Along streams, rivers, & ditches, about ponds & dumps, & in gardens; ND: Cass; IA: Page; NE: Dakota; KS: Doniphan, Douglas, Franklin, Lyon; (Que. & Newf. to ND, s to NC & LA; Europe). Introduced.

8. Rorippa tenerrima Greene. Glabrous annual or possible biennial from a vertical taproot, stems many, decumbent to prostrate, 1–2 dm long. Basal and cauline leaves short-petiolate, slightly auriculate to nonauriculate, nonclasping, oblanceolate to spatulate, 2–6 cm long,

0.5–1.5 cm wide, lyrate-divided nearly to the midrib, the apex obtuse. Racemes terminal and axillary, 0.6–1.3 dm long, or lateral; sepals 0.8–1.4 mm long, flat, caducous in fruit; petals 0.7–0.9 mm long. Siliques short to elongate-cylindrical, slightly curved inward, occasionally curved outward, tapering to the apex, 4–6(9) mm long, 1–1.5 mm wide, 4–5.5(7) × longer than wide, 1.5–2 × longer than the pedicels, the valves rough with minute papillae; style in fruit tapering to the apex, 0.3–0.9 mm long, gradually merging into the pointed silique apex; fruiting pedicels 2–4(5) mm long, filiform, slightly divergent to ascending, glabrous to rough with minute papillae; seeds 0.6–0.8 mm long, 20–40 per silique, the surface finely colliculose. Jun–Sep. Riverbanks; CO: Weld; MO: Clay, Jackson; ND: McLean, Oliver; NE: Deuel; SD: Fall River; (ND to WA, s to MO, NM, & s CA; n Mex.). *R. obtusa* (Nutt.) Britt.; *R. lyrata* (Nutt.) Rydb.—Rydberg.

9. *Rorippa truncata* (Jeps.) Stuckey. Glabrous annual or possibly biennial from a slender taproot, stems several to many, decumbent to prostrate, 1–3.5 dm long. Basal and lower cauline leaves short-petiolate, slightly auriculate to nonauriculate, nonclasping, narrowly oblanceolate, 4–10 cm long, 0.7–1.8 cm wide, pinnately divided, sometimes nearly lyrate, the lobes sinuate or angularly toothed, glabrous or occasionally sparingly hirsute above, smooth below or with hyaline ridges on the midrib, the apex mostly obtuse. Racemes lateral, 0.4–1 dm long; sepals 0.9–1.5 mm long, slightly saccate, caducous in fruit; petals 0.9–1.2 mm long. Siliques short-cylindrical, rarely elongate-cylindrical, straight, not at all or slightly tapering toward the apex, constricted in the middle, 3.5–5.5 mm long, 1.2–1.8 mm wide, 2–4(5) × longer than wide, 1.8–2.5 × longer than the pedicels, the valves smooth to slightly rough with minute hyaline ridges; style in fruit straight, nearly absent to 0.7 mm long, abruptly attached to the obtuse to truncate silique apex; fruiting pedicels 1.5–3 mm long, divergent; seeds 0.5–0.6 mm long, 30–80 per silique, the surface finely colliculose. Jul–Oct. Edges of rivers, streams, lakes, & ponds, dry lake beds; ND s to NE & ne KS; (ND to WA, s to MO & e KS, NM w to s CA). *R. obtusa* (Nutt.) Britt.—Rydberg.

31. SELENIA Nutt., Golden Selenia

1. *Selenia aurea* Nutt., golden selenia. Glabrous, annual herb, stems 5–30 cm tall, simple or branched from the base. Leaves 2–6 cm long, deeply pinnately dissected, the basal short-petioled, the cauline sessile, the several lateral lobes divergent at right angles, linear or narrowly oblong, entire or with a small basal tooth. Flowers about 8 mm wide, in leafy-bracted racemes, mature racemes up to 1 dm long; pedicels widely ascending, 1–2 cm long; outer 2 sepals narrower than the inner 2, without appendages; petals yellow, oblanceolate or narrowly obovate, gradually tapering to the base; projecting receptacular glands present between the petals and the stamens; stamens short, each with a horseshoe-shaped gland on the inner side at base; ovary slender; ovules 6–8 in each cell; style elongate. Silicle erect, very flat, 1–2(2.5) cm long, 6–8(10) cm wide, subacute at both ends; the stout style 5 mm long; seeds widely margined. (2n = 46) Apr–May. Open barrens, bluffs, & sandy soil; restricted to ne TX, e OK, sw MO, AR, & se KS.

32. SISYMBRIUM L., Hedge Mustard

Annual or biennial herbs, glabrous or with simple trichomes; stems usually branched above the base. Leaves monomorphic or dimorphic, green or glaucous, at least the lower leaves deeply pinnatifid. Inflorescences dense, elongating in fruit; sepals nonsaccate, obtuse, ascending; petals small, yellow (ours), white or purple, obovate to spatulate, gradually narrowed to the claw; staminal glands of the short stamens usually annular; filaments slender; anthers oblong; ovary cylindric; style short, scarcely differentiated; stigma capitate; ovules numerous. Siliques elongate, linear or subulate, terete or slightly quadrangular, tipped with the minute persistent style; valves 3-nerved, with conspicuous midnerve and thinner lateral nerves; seeds in 1 row, oblong, smooth or nearly so, marginless; cotyledons incumbent.

1 Siliques subulate, 1-2 cm long, on short erect pedicels 2-3 mm long, closely appressed to the rachis .... 4. *S. officinale*
1 Siliques linear, elongate, 2-10 cm long, on divergent pedicels 5-10 mm long, widely spreading or loosely ascending.
   2 Pedicels at maturity nearly or quite as thick as the siliques; siliques 5-10 cm long .... 1. *S. altissimum*
   2 Pedicels at maturity obviously thinner than the siliques.
      3 Petals 5-8 mm long; sepals 3-4 m long; siliques 2-3.5 cm long .... 3. *S. loeselii*
      3 Petals 3-4 mm long; sepals 2-3 mm long; siliques 4-6 cm long .... 2. *S. irio*

1. Sisymbrium altissimum L., tumbling mustard. Annual, stems loosely branched above, to 1.5 m tall, sparsely spreading hirsute at the base with simple trichomes. Lower leaves petiolate, hirsute, pinnately lobed, the lobes oblong and dentate, gradually changing upward on the plant to leaves with linear-filiform entire segments. Flowers loosely racemose; petals pale yellow, 6-9 mm long, longer than the sepals. Siliques terete, straight, long-linear, widely spreading and extending at the same angle as the pedicle, 5-10 cm long, glabrous, style 1-2 mm long; pedicels straight, widely spreading, about as thick as the siliques; seeds wingless, plump, oblong, about 1 mm long. (2n = 14) May-Aug. Weed of fields, roadsides, & waste places; GP; (U.S. & adj. Can.; Europe). *Introduced.*

2. Sisymbrium irio L., London rocket. Annual, stems erect, branched near base and above, up to 6 dm tall, sparsely to densely hirsute below with spreading trichomes. Lower leaves deeply pinnatifid with oblong to ovate, entire to dentate or angularly lobed segments, the upper leaves with fewer smaller lobes, glabrous to sparsely hirsute with simple trichomes on petioles and lobe margins, terminal lobe larger than the laterals. Flowers small; pedicels slender, divaricately ascending, 5-10 mm long, petals pale yellow, 3-4 mm long, oblanceolate, barely exceeding the sepals. Young siliques elongating rapidly and projecting beyond the flowers; mature siliques slender, terete, glabrous, straight, divaricately ascending, 3-5 cm long, about 1 mm wide; seeds oblong, wingless, less than 1 mm long. (2n = 14, 16, 28, 42, 56) Apr-Jun. Fields, roadsides, & waste places; CO: Logan, TX: Swisher; (occasionally *adventive* in U.S. & becoming a weed in the Pacific states; Europe).

3. Sisymbrium loeselii L., tall hedge mustard. Stiffly branched annual, stems 5-10 dm tall, hirsute toward the base with reflexed hairs. Lower leaves petiolate, lyrate-pinnatifid and dentate, usually hirsute, the lateral segments triangular to ovate, acute, spreading or deflexed, the terminal larger, triangular. Flowers loosely racemose; sepals 3-4 mm long; petals yellow, 5-8 mm long. Siliques linear, 2-3.5 cm long, about 0.7 mm wide; pedicels at maturity divergent, 5-10 mm long; seeds scarcely 1 mm long. (2n = 14) May-Aug. Fields, roadsides, & waste places; MT, ND, MN, SD, WY, IA, NE, CO; (occasionally *adventive* in e U.S. & becoming a weed in w U.S.; se Europe & w Asia).

4. Sisymbrium officinale (L.) Scop., hedge mustard. Widely branching annual, stems to 1 m tall, hirsute at the base. Lower leaves petiolate, deeply pinnatifid, segments oblong to ovate or the terminal segment sometimes rotund, angularly toothed, the upper leaves sessile or nearly so, few-lobed or 3-lobed or entire, the lateral lobes widely divergent. Flowers small, in stiffly erect racemes; racemes simple or with straight, widely divergent branches, greatly elongating in fruit; petals pale yellow, about 3 mm long. Siliques subulate, closely appressed to rachis, 1-2 cm long, pubescent with simple trichomes to glabrous; style 1-2 mm long; pedicels at maturity 2-3 mm long, closely appressed, distally thickened and as wide as the silique at the summit; seeds plump and wingless. (2n = 14) Mar–Sep. Weed of fields, roadsides, & waste places; throughout e 1/2 of GP & occasionally elsewhere; (throughout most of U.S.; Europe). *Introduced.*

Two varieties of this species have been recognized:

4a. var. *leiocarpum* DC. Racemes, pedicels, and fruits glabrous. More common in GP.

4b. var. *officinale.* Racemes, pedicels, and fruits softly pubescent.

33. STANLEYA Nutt., Prince's Plume

1. ***Stanleya pinnata*** (Pursh) Britt., prince's plume. Subshrubby, mostly glabrous perennial, stems several to many from a woody base, usually branched above, to 1.5 m tall. Basal leaves absent, cauline leaves petioled, glabrous or sparsely pubescent, glaucous, thick, ovate to linear-oblanceolate, usually entire. Inflorescence dense, racemose, 1–3.5 dm long, elongating (to 5 dm) in fruit; buds clavate; sepals yellow at anthesis, spreading, linear-oblong; petals yellow, with a brownish claw, pilose on inner face, 10–15 mm long; stamens nearly equal, exserted; anthers long and narrow, coiled after pollen discharge; pedicels widely spreading, about 1 cm long; siliques linear, nearly terete, 2–8 cm long, supported on a slender stipe (gynophore) 1–3 cm long; stigma entire, nearly sessile; seeds oblong, wingless; cotyledons incumbent. Apr–Aug. Dry hills, plains, & valleys; considered to be indicative of selenium in the soil.

Reference: Rollins, R. C. 1939. The cruciferous genus *Stanleya.* Lloydia 2(2): 109–127.

Two varieties occur in the GP:

1a. var. *integrifolia* (James) Rollins. Upper stem leaves broadly ovate, entire, lower stem leaves entire or somewhat divided, (2n = 24, 48). nw KS; (KS, CO, WY, UT). *S. integrifolia* James – Rydberg.

1b. var. *pinnata.* Upper stem leaves oblanceolate to narrow, entire or divided, lower leaves pinnate or rarely bipinnate. (2n = 24). KS: Hamilton; nw NE, w SD, w ND, CO, WY, MT; (w KS to ND, w to MT & UT). *S. bipinnata* Greene, *S. glauca* Rydb.—Rydberg.

34. STREPTANTHUS Nutt., Twist-Flower

1. ***Streptanthus hyacinthoides*** Hook. Annual herb to 10 dm tall; stems simple or rarely branched above, often purplish, glabrous. Basal leaves absent; cauline leaves alternate, spreading, linear-lanceolate, 2–10(15) cm long, 2–10(15) mm wide, glabrous, apex acute, margins entire or remotely denticulate, base narrowly cuneate; sessile or short-petioled. Inflorescence a 10- to 30-flowered racemes; pedicels 2–5 mm long. Flowers spreading to pendent; calyx somewhat urn-shaped sepals dark purple (at least toward the apex), ovate to lance-ovate, 6–10 mm long, 2–3 mm wide, glabrous, acuminate, entire; petals lavender to dark purple, panduriform, 12–20 mm long, about 2× longer than the sepals, entire to emarginate; stamens 6, the 2 outer included and their filaments distinct, the 4 inner exserted and their filaments connate below in pairs, anthers basifixed, sagittate at the base, longitudinally dehiscing, pubescent. Siliques divergently ascending, flattened parallel to the septum, dehiscing longitudinally, 6–10 cm long, 1.5–2.5 mm wide; seeds brownish, circular, flattened, 1.2–1.5 mm long, smooth, with a membranaceous wing. Apr–May (Jun). Sandy & sometimes gravelly soil on prairie hillsides or in pastures; s-cen KS, cen OK; (s-cen KS s to LA & TX).

Streptanthus cordatus Nutt., a species of the western U.S., occurs in NM: Union. It is a biennial that can be recognized by its basal rosette of spatulate leaves, ovate and usually clasping cauline leaves, and siliques 5–6 mm wide.

35. THELYPODIUM Endl.

Erect, glabrous biennials, stems to 1 m or more tall, branched. Leaves simple, entire to pinnatifid, basal leaves petiolate, oblanceolate or spatulate, cauline leaves petiolate to sessile. Flowers borne in racemes; sepals ascending, thin, more or less petaloid; petals white to purple (ours) or yellow, clawed, sometimes crisped; filaments subulate, white, with conspicuous glands at their bases, anthers linear, sagittate at the base, curved; siliques sessile or very short-stipitate, elongate, slender, terete or nearly so, erect to reflexed; seeds somewhat flattened, not winged. *Pleurophragma* Rydb.—Rydberg.

1 Sepals spreading at anthesis .... 2. *T. wrightii*
1 Sepals erect to ascending at anthesis .... 1. *T. integrifolium*

1. Thelypodium integrifolium (Nutt.) Endl. Glabrous biennial, arising from a taproot, stems erect, 60–100 cm tall. Basal leaves petiolate, oblong to oblanceolate, cauline leaves linear-lanceolate to lanceolate, sessile, the upper leaves becoming linear. Flowers pale rose, purple, or white, crowded; the pedicels divaricate; petals 5–8 mm long. Siliques slender, terete, torulose, tapering at the base to a short stipe about 1 mm long, curving upward, or nearly erect, 20–30 mm long. (2n = 13) Apr–Aug. Damp soil; NM, CO, WY, w NE, w SD, & w ND; (NE to WA, s to NM, AZ, & CA). *Pleurophragma lilacinum* (Greene) Rydb.—Rydberg; *T. lilacinum* Greene—Stevens.

2. Thelypodium wrightii A. Gray. Paniculately branching biennial, to 2 m tall, branches slender. Basal leaves 10–15 cm long, lyrate-pinnatifid, cauline leaves becoming smaller upward, pinnatifid to dentate or entire, linear-lanceolate. Flowers in corymbose clusters, terminating the branches, the pedicels widely spreading to somewhat reflexed; sepals about 5 mm long; petals exceeding the sepals, white or pale purple, oblanceolate, short clawed, sepals and petals spreading at anthesis; stamens nearly equal, exserted, spreading at anthesis; anthers coiled at maturity, sagittate at the base. Siliques stipitate, narrowly linear, torulose, widely spreading, 4–7 cm long, less than 1 mm wide; styles about 1 mm long; seeds oblong, wingless. Aug–Oct. Cliffs, canyon walls, & lower montane slopes; NM: Union; OK: Cimarron; (OK & TX to NV & AZ).

36. THLASPI L., Pennycress

Glabrous annual or perennial herbs; stems solitary or several from the base, simple or branched. Leaves simple, alternate; rosette leaves petiolate; cauline leaves sessile, usually auriculate. Inflorescence racemose, congested in flower and elongating in fruit, ebracteate. Flowers numerous, small, pediceled; sepals erect, entire; petals white, spatulate, entire; stamens 6, subequal, anthers basifixed and obscurely sagittate, longitudinally dehiscing, filaments distinct; style obsolete to slender and conspicuous. Fruit a silicle, much compressed contrary to the septum, obovate, oval, or obcordate, usually winged on the margin; seeds 2–several per cell, wingless.

References: Payson, E. B. 1926. The genus *Thlaspi* in North America. Univ. Wyoming Publ. Bot. 1: 145–186.

1 Foliage green; cauline leaves oblong to lanceolate; fruist 10–18 mm long .... 1. *T. arvense*
1 Foliage glaucous; cauline leaves ovate to ovate-oblong; fruits 4–6 mm long .... 2. *T. perfoliatum*

1. Thlaspi arvense L., field pennycress. Glabrous annual to 7 dm tall, foliage green; stems solitary or several, branching above. Basal leaves evanescent, spatulate to oblanceolate, 3–9 cm long, 7–15 mm wide, apex rounded or obtuse, margins entire or irregularly toothed;

basal leaves passing abruptly or gradually into the cauline leaves; cauline leaves oblong to lanceolate; 1.5–7 cm long, 3–12(15) mm wide, gradually reduced upwards, apex obtuse to acute, margins sinuate to coarsely dentate, base auriculate. Flowers divergent to ascending, pedicels 2–3 mm long, elongating in fruit; sepals greenish-white, ovate to slightly obovate, 1.5–2.5 mm long, occasionally lightly pubescent, obtuse to acutish; petals white, 2–4 mm long; style obscure. Silicle oval to obcordate, apically cleft, 10–18 mm long; pedicels 7–15 mm long; seeds dark purple or blackish, oval, slightly flattened, 1.5–2 mm long, concentrically striate-corrugate. ($2n = 14$) Apr–Jun, sporadically later and in the fall. Pastures, roadside, & other open waste areas; GP; (widely established in N. Amer.; Europe). *Naturalized.*

2. **Thlaspi perfoliatum** L. perfoliate pennycress. Glarous annual to 3 (4) dm tall, foliage glaucous; stems solitary or several, branching above. Basal leaves evanescent, blade obovate, elliptic, or oval, 1–3.5 cm long, 8–17 mm wide, rounded to acute, irregularly serrate, cuneate to attenuate, petiole 0.5–2.5 cm long, cauline leaves ovate to ovate-oblong, gradually reduced upwards, apex acute, margins entire to weakly serrate, base auriculate. Flowers divergent, pedicels 1–3 mm long, soon elongating; sepals greenish-white, sometimes streaked with purple, ovate to elliptic, 1–2 mm long, glabrous, obtuse to acutish; petals white, 2–3 mm long; style obscure. Silicles broadly obcordate, apically cleft, 4–6 mm long; pedicels 3–7 mm long; seeds yellow, obovoid, slightly flattened, 1–1.5 mm long, smooth. ($2n = 42$) Mar–May. Fields, roadsides, & other open disturbed sites; IA, se NE, MO, e 1/3 KS, e OK; (established in much of ne N. Amer., to GP; Europe). *Naturalized.*

Thlaspi fendleri A. Gray of western N. Amer. occurs in NM: Union. It will key here to *T. perfoliatum* but is distinguished by being perennial and having a distinct style (2–3 mm long in fruit).

59. RESEDACEAE S. F. Gray, the Mignonette Family

by William T. Barker and Ralph E. Brooks

1. RESEDA L.

1. *Reseda lutea* L., wild mignonette, reseda. Biennial or short-lived perennial herb with a deep taproot, mostly glabrous and usually somewhat glaucous; stems erect or ascending, diffusely branched, 2–8 dm tall. Basal leaves rosulate, narrowly oblanceolate, simple or irregularly pinnatifid, 4–10 cm long, 1–2 cm wide; cauline leaves numerous, alternate, oblanceolate to obovate, deeply pinnatifid with mostly 1–3 pairs of narrowly oblong, entire to pinnatifid lobes, 2.5–8 cm long, 1.5–5 cm wide, margins undulate or not and serrulate; stipules small, modified into glands. Inflorescence a spikelike raceme, 5–20 cm long; pedicels 2–7 mm long. Flowers hypogynous, perfect, irregular; sepals 6, linear, subequal, 1–3 mm long; petals 6, greenish-yellow, 1.5–3 mm long, with rounded claws, the upper 2 with trifid linear limbs, the lateral 2 with bi- to trifid linear limbs, and the 2 lower entire; stamens 15–22, often recurved; filaments short; anthers dithecal, opening by longitudinal slits; carpels 3, united below, each with an apical stigma-bearing lobe, ovary superior and open between the stigma lobes, 1-celled. Fruit a 1-celled capsule opening by the spreading apical lobes, ellipsoid to clavate, 7–11 mm long, 3–4 mm in diam; seeds black, shining, 1.5–1.9 mm long, smooth. ($2n = 48$) May–Jun. Pasture hillsides, fields, & waste places; IA: Harrison; NE; Dixon; (scattered in the ne U.S. w to e NE: Eurasia & N. Africa). *Introduced.*

60. ERICACEAE Juss., the Heath Family

by T. Van Bruggen

Trailing or upright shrubs, the stems with several to many branches. Leaves simple, alternate, entire or toothed, evergreen or deciduous, exstipulate, thin-herbaceous to leathery. Inflorescence of terminal racemes, pendulous fascicles or flowers solitary in leaf axils. Flowers perfect, (4)5-merous, radially symmetrical, the sepals distinct or fused below; corolla campanulate to urceolate, 4- to 5-lobed, white to pink or green; stamens 8–10, or twice the number of calyx lobes, free, included in the corolla or exserted, anthers 2-celled, upright, each opening by a pore; ovary superior or inferior, 4- to 5-loculed. Fruit a drupe or many-seeded berry.

A specimen of *Ledum groenlandicum* Oeder is filed in the herbarium at Black Hills State College, Spearfish, SD. It was collected by F. L. Bennett, 1931, on the canyon floor of the gulch s of Spearfish. Habitat data lists a mixed woodland-talus slope, an unlikely place for *Ledum*. There is no other historical record of this species having been collected in the BH. If this is an authentic collection, I suspect the plant may now be extinct.

1. Ovary superior; leaves evergreen; stems trailing; plants forming dense mats .... 1. *Arctostaphylos*
1. Ovary inferior; leaves deciduous; stems ascending or upright; shrubby .... 2. *Vaccinium*

1. ARCTOSTAPHYLOS Adans., Bearberry

1. *Arctostaphylos uva-ursi* (L.) Spreng. Perennial with depressed or trailing flexible stems forming mostly prostrate mats 1–2 m across; bark reddish on younger stems, light colored and exfoliating on older stems. Leaves alternate, short petioled, the blades entire, spatulate to obovate with rounded summits, evergreen, 1–3 cm long, the upper surface leathery and lustrous. Flowers in dense terminal racemes or panicles, bracteate, often pendulous; sepals 5, 1.0–1.5 mm long, distinct, persisting in fruit, pink or white; corolla white to pink, ovoid to urceolate, 4–8 mm long, with 5 short, reflexed lobes; stamens (8)10, shorter than the corolla, the filaments pubescent, enlarged below, anthers subglobose, with 2 terminal pores and 2 short deflexed terminal awns; ovary superior, 5-celled, ovoid-conic, on a 10-lobed disk, with a short, columnar style. Fruit a mealy or fleshy drupe, red, 4–10 mm across, with (4)5 bony nutlets, each 1-seeded. May–Jul. Rocky or open woods to dry open or sandy hillsides; n MN, ND, w SD, NE: Custer; WY; (Newf., NJ to AK, s to GP, CA).

Two varieties have been recognized by Fernald and Macbride (Rhodora 16: 212. 1914). Both occur scattered throughout our range.

1a. var. *adenotricha* Fern. & Macbr. Plants having branchlets with long spreading hairs, viscid glandular, intermixed with black stipitate glands.

1b. var. *coactilis* Fern. & Macbr. Plants with branchlets persisting, densely tomentose, not viscid or glandular.

2. VACCINIUM L., Blueberry

Perennial shrubs with trailing to upright stems or small trees, ours mostly with deciduous leaves. Leaves alternate, thin to coriaceous, mostly ovate with acute to obtuse tips, the margins entire or serrate. Flowers solitary in axils, in racemes or panicles. Calyx (4)5-merous, fused below, with small, persistent lobes; corolla open-campanulate to ovoid or urceolate, variously connate, (4)5-lobed above, white to rose-colored; stamens 8–10, twice the calyx lobes, filaments distinct, glabrous or hairy, the anthers awned or awnless, opening by slender tubes or terminal pores; ovary inferior, with a distinct hypanthium, (4)5-celled. Fruit a variously colored berry, 4- to 5-loculed (or incompletely 10-celled by false partitions) with many seeds.

    References: Camp, W. H. 1945. The North American blueberries with notes on other groups of Vacciniaceae. Brittonia 5: 203–275; Vander Kloet, S. P. 1978. The taxonomic status of *Vaccinium pallidum*, the hillside blueberries, including *Vaccinium vacillans*. Canad. J. Bot. 56: 1559–1574.

1. Mature leaves less than 1.5 cm long, many on green, angled branches; small shrub 1–3 dm tall; berries red .... 4. *V. scoparium*
1. Mature leaves exceeding 2 cm long; branches not green; shrubs 0.3–9 m tall; berries green, blue, purple, or black.
   2. Flowers and fruits on filiform pedicels, in bracted racemes or panicles, corolla open-campanulate; berries green to black.
      3. Stamens included in the corolla; small trees to 9 m tall; leaves coriaceous, margins frequently revolute; berries black .... 1. *V. arboreum*
      3. Stamens exserted from the corolla; shrubs 1–4(5) m tall; leaves thin, margins not revolute; berries greenish to purple .... 5. *V. stamineum*
   2. Flowers and fruits on short, few-flowered racemes or single in the axils of leaves; corolla urceolate, with short lobes; berries black or blue.
      4. Anthers awned; flowers and fruits single in the axils on pedicels 5–10 mm long; branchlets angled; berry blue to black, not glaucous .... 2. *V. membranaceum*
      4. Anthers not awned; flowers and fruits on short, terminal or lateral racemes, branchlets terete to slightly angled; berry blue, glaucous .... 3. *V. pallidum*

1. *Vaccinium arboreum* Marsh., sparkleberry. Shrub or small tree to 9 m tall, much branched, twigs terete, glabrate or puberulent, the bark becoming gray. Leaves alternate, ovate to obovate, deciduous but evergreen southward, coriaceous, lustrous above, margins revolute, entire to denticulate, 2–7 cm long, 1–4 cm wide. Flowers in leafy-bracted racemes or panicles, 2–7 cm long, usually on second-year wood, pedicels filiform, jointed. Sepals (4)5, fused below, the lobes acute, persistent in fruit; corolla open, campanulate, 4–6 mm long, fused most of its length, the lobes 0.5–1 mm long, recurved; stamens (8)10, included in the corolla, the filaments distinct, pubescent; anthers with 2 awns 1 mm long, about ½ length of the tubular openings; ovary (4)5-merous, the style exserted. Fruit a black berry 4–6 mm in diam, becoming dry and hard with several seeds, not edible. May–Jun, fruits Aug–Sep. Rocky or sandy open woods & along wooded streams; se GP; KS, MO, OK, TX; (VA to KS, s to FL & TX). *Batodendron arboreum* (Marsh.) Nutt.—Rydberg.

2. *Vaccinium membranaceum* Dougl., mountain huckleberry. Shrub 0.5–1.5 m tall, the branches spreading, young twigs angled or ridged, puberulent to glabrous, the older bark gray, exfoliating. Leaves thin, ovate, 2–7 cm long, apex acutely pointed, the margins finely serrulate with incurved teeth, pale on the lower surface. Flowers 4–6.5 mm long, single in the axils, pedicels 5–8 mm long; calyx short, saucer-shaped, only slightly lobed; corolla yellow to pink, ovoid to urceolate, 4–6 mm long, obscurely lobed above; stamens included, filaments glabrous, the anthers opening by 2 extended tubes, awned on the back. Ovary and fruit 4- to 5-celled, bluish or more commonly black, 7–9 mm in diam, edible, not glaucous, several-seeded. Jun–Jul, fruits Aug–Sep. Dry rocky woods & openings; MN, BH, WY; (Ont. to B.C., s to CA).

3. *Vaccinium pallidum* Ait., hillside blueberry. Stoloniferous shrub (0.8)2–6(8) dm tall, in small to extensive colonies, twigs green to yellow, glaucous, terete to more commonly angled. Leaves variable, ovate to elliptic, 2–4 cm long, 1.2–3.5 cm wide, pale green, usually glabrous above, glaucous to pubescent beneath, entire to serrate, apex acuminate-apiculate.

Flowers in 4–11 short, terminal or lateral racemes, at first dense, but the raceme elongating with the season's growth, bracts caducous; calyx and pedicel glaucous, the lobes acute, persisting in fruit; corolla 5–8 mm long, greenish-white with pink striping, urceolate to ovoid, slightly constricted at the throat, the lobes becoming reflexed; stamens included in the corolla, filaments glabrous to ciliate on the margins, anther sacs not awned, the 2 pores extending as tubes 2 × the length of the sacs; ovary incompletely 10-celled by false partitions. Fruit a globose berry, blue, glaucous, 5–7(8) mm in diam, 8- to 14-seeded, the seeds dimorphic, viable ones brown, larger than the pale, smaller, imperfect ones. ($n = 12$, rarely tetraploid) Apr–Jun, fruits Jul–Sep. Rocky woods & dry chert rock or gravelly places on hillsides; KS: Cherokee; MO, OK; (MN to ME; s to GA, AL, AR, & KS). *Cyanococcus vacillans* (Kalm) Rydb. — Rydberg; *V. vacillans* Torr. — Fernald.

4. Vaccinium scoparium Leib., grouseberry. Shrub 1–3 dm tall, much branched from the base, branches green, sharply angled, the grooves glabrous. Leaves 0.7–1.5 cm long, light green, thin, the blades lance-ovate to oblong, glabrous to minutely puberulent, apex acute, margins finely serrulate, deciduous. Flowers solitary in the axils of upper leaves of current year's wood. Calyx reduced, 1 mm high, shallowly lobed; corolla white or pink, ovoid to urceolate, 2–4 mm long, the 5 short lobes recurved; stamens (8)10, included in the corolla, the filaments glabrous, anthers with pore-bearing tubes 1 mm long, equaling the sac, awned on the back, the awns 0.5 mm long; style slightly exceeding the throat of the corolla. Fruit bright red, 3–5 mm in diam, not glaucous, several seeded, edible. May–Jul, fruits Aug–Sep. Upland loamy or rocky woods; BH; (Alta. to sw Can., BH, CO to s CA).

5. Vaccinium stamineum L., deerberry. Shrub 1–2(4) m, much branched, the branches glabrous, terete, younger twigs pubescent. Leaves petioled, deciduous, pubescent when young, becoming glabrous, ovate to oblong, 3–8 cm long, 1.5–4.5 cm wide, the margin entire to remotely denticulate or pubescent, blade thin and flat, acute at the apex, cuneate or slightly cordate at the base, glaucous beneath. Flowers in bracteal racemes on second year wood, 1–5 cm long, the foliaceous bracts 0.5–2 cm long, usually smaller than the vegetative leaves. Calyx 5-lobed, persistent, the lobes acute; corolla open, campanulate, 4–7 mm long, white to green or purple, the lobes spreading, obtusely rounded, stamens (8)10, filaments broad, pubescent, the anthers 2-awned, extended into 2 long tubes that exceed the corolla by 3–5 mm. Fruit green to purple, 0.6–1.5 cm in diam, juicy, tart, several-seeded. Apr–Jun, fruits Jul–Sep. Dry, rocky, open woods; KS: Cherokee; MO, OK; (ME to Ont., s to FL & TX). *Polycodium stamineum* (L.) Greene (pubescent), *P. neglectum* Small (glabrous) — Rydberg; *V. neglectum* (Small) Fern. — Fernald.

This is a polymorphic species divided into varieties based on pubescence of leaves and branchlets, or lack thereof, and color of fruit.

61. PYROLACEAE Dum., the Wintergreen Family

by T. Van Bruggen

Small perennial herbs 0.4–3 dm tall, from slender, stoloniferous rootstocks, scapose to short stemmed; stems terete, glabrous. Leaves persistent, mostly evergreen, simple, alternate to opposite, subverticillate to whorled or crowded basally on the stem, coriaceous to leathery bright green, entire or toothed, oblanceolate to orbicular, sessile to petioled. Flowers sometimes slightly irregular, on peduncled corymbs or on scapes, solitary or in bracted racemes; sepals (4)5, basally fused, spreading, persistent; petals (4)5, distinct, or slightly united at the base, waxy, white to pink or greenish; stamens (8)10, twice the number of petals, the filaments narrow and subulate to dilated below, usually included, anthers incurved, subapically attached, 2-celled, opening by pores; ovary (4)5-celled, the styles united, straight or curved, persistent; stigma 5-lobed, thicker than the style; hypogynous disk present or lacking. Fruit a depressed globose capsule, loculicidally dehiscent from the summit or the base; seeds many.

1 Flowers in corymbs; stems leafy; leaves at least 3× longer than wide; staminal filaments dilated at bases .... 1. *Chimaphila*
1 Flowers in racemes or solitary; stems scapose or nearly so, the leaves mostly basal, less than 3× longer than wide; staminal filaments slender .... 2. *Pyrola*

1. CHIMAPHILA Pursh, Prince's Pine

1. ***Chimaphila umbellata*** (L.) Bart. Perennial plants, evergreen, suffruticose, stems spreading-ascending, the fertile branches erect, from creeping rootstocks, 1–3 dm tall. Leaves simple, subverticillate or whorled, 3–6 at a node, 3–9 cm long, 1 cm wide, oblanceolate, cuneate, the margins sharply serrate and slightly revolute; upper surface glabrous, bright leathery green, pale green below; petioles 3–6 mm long, exstipulate. Flowers 1–2 mm in diam, 4–8 in terminal corymbs, the peduncles erect, 4–7 cm long. Perianth 5-merous; sepals 5, pink-purple, basally fused, the lobes 1–2 mm, ovate, erose-ciliate, persistent; petals 5, 4–5 mm long, distinct, white to roseate, waxy, concave, broadly ovate with ciliate margins; stamens 10, the filaments dilated below, ciliolate, abruptly narrowed to a short, glabrous upper portion 2 mm long; anthers plump, 2-celled, incurved, the sacs horn-like, opening by 2 wide, basal pores, appearing apical by inversion; ovary 5-celled, the style short; stigma 2 mm wide, peltate, 5-lobed, persistent. Fruit 5–6 mm in diam, loculicidally dehiscent from the summit down; seeds numerous. Jun–Aug. Rich, rocky or sandy woods; MN: Clay, Polk; SD: Lawrence; (N.S. to B.C. & CA, Que. & Ont. s to GA). *C. occidentalis* Rydb.— Rydberg.

2. PYROLA L., Wintergreen

Scapose, mostly simple stemmed herbs, 0.5–3 dm tall, glabrous. Leaves basal, persistent, simple, coriaceous, 6–12, alternate or opposite, clustered at the upper part of the stem; blades broadly ovate to orbicular, entire to toothed, glabrous, the petioles ½ as long to much exceeding the blade. Flowers 2–several in bracted scapose racemes or solitary and terminal, the pedicels 8–12 mm long, spreading or recurved. Calyx (4)5-lobed, persistent; corolla broadly bowl-shaped, the petals spreading or concave, quite distinct, white to pink or greenish; stamens (8)10, the filaments narrow or basally dilated, erect or declined; anthers 2-celled, opening by 2 pores or irregular slits; ovary 5-merous, the style straight or declined and curved upward at the end, with 5 radiating stigmatic lobes, persisting in fruit.

1 Flower solitary at the end of a singly bracted, naked recurved scape; petals spreading, plants usually less than 1 dm tall .... 6. *P. uniflora*
1 Flowers 2 or more in racemes, scapes bracted; plants commonly 1–3 dm tall.
  2 Styles straight; racemes 1-sided; corolla longer than wide .... 5. *P. secunda*
  2 Styles declined or otherwise curved; racemes spiraled; corolla wider than long.
    3 Leaves mottled on the upper surface and with pale streaks associated with the main veins; blades broadly ovate, the apex acutely pointed .... 3 *P. picta*
    3 Leaves uniformly green above, blades elliptic to orbicular.

4 Principal leaf blades less than 3 cm long or wide; racemes with 8 flowers or less .... 7. *P. virens*
4 Principal leaf blades exceeding 3 cm long or wide; racemes usually with more than 8 flowers.
   5 Petals rose-pink to purple; leaves orbicular in outline, usually as wide or wider than long, with a slightly cordate base .... 1. *P. asarifolia*
   5 Petals white to greenish; leaves broadly oval or suborbicular in outline, slightly longer than wide.
      6 Sepals lanceolate, 3–3.5 mm long, distinctly longer than wide; leaves suborbicular; petals white .... 4. *P. rotundifolia*
      6 Sepals acute, 1.5–2 mm long, as wide as long, leaves oval to elliptical, petals greenish-white .... 2. *P. elliptica*

1. **Pyrola asarifolia** Michx., round-leaved wintergreen. Plant with extensively creeping scaly rhizomes; scape 1.5–3 dm tall, with 2 or 3 scale leaves widely spaced. Leaves 3–7, simple, basal, persistent, the blades 3–6 cm wide, reniform-orbicular with a rounded apex, entire to slightly crenulate; bright green above, lustrous-coriaceous, pale below; petioles mostly longer than the blade. Racemes elongate, (8)10- to 20-flowered, pedicels 4–8 mm long, slightly exceeding the lanceolate bractlets. Flowers 1–1.5 cm wide, pink to lavender; calyx similar in color to the petals, the lobes 2–3 mm long, lance-deltoid; corolla campanulate, the petals broadly ovate, 6–8 mm long, veiny; stamens included, the filaments glabrous, declined, anthers 2 mm long, contracted to an apiculate tip at the lower end; style 5–7 mm long, sharply curved downward immediately above the ovary, tapered to a collar which is broader than and just below the stigma. Fruit 4–8 mm in diam, opening from below, many-seeded. Jul–Aug. Moist, shady woods; MN, ND, SD, WY; (Newf. to P.E.I. & AK, Yukon to GP, NM). *P. uliginosa* Torr.—Rydberg.

2. **Pyrola elliptica** Nutt., wild lily of the valley. Plant from slender rhizomes; scape 1–2.5 dm tall, usually solitary, with 1 or 2 lance-subulate bracts. Leaves 3–7, basal, rosulate, the blades thin, bright green, 2–7 cm long, oval to elliptical, acute at the base, rounded to retuse at apex, exceeding the petioles. raceme cylindric, (6)8- to 13-flowered, the pedicels 3–7 mm long, arched, as long as or slightly exceeding the lanceolate bractlets. Flowers 1–1.2 cm in diam; calyx green, the lobes acute to acuminate, 1.5–2 mm long, almost as wide as long, reflexed, persistent; petals 6–8 mm long, fragrant, white to greenish white, spreading obovate, forming a campanulate corolla; stamens ascending, incurved, the filaments 5 mm long, declined, anthers 2 mm long, scarcely narrowed above the pores; style 6–9 mm at maturity, arched and declined, forming a collar immediately below the stigma, persistent. Fruit depressed-globose, 5–7 mm in diam, opening from below, with many seeds. Jun–Aug. Dry to moist, rich woods; MN, ND, SD, NE; (N.S. & Newf. to B.C., s to GP, NM).

3. **Pyrola picta** Sm. Plant from slender, scaly rhizomes, the stems reddish-brown; flowering scape solitary or 2, 1–2.5 dm tall, with 1–3 lanceolate bracts. Leaves 2–several on sterile stems, 0–4 on fertile stems, basal, the blades 2–7 cm long, coriaceous, dull green with mottled gray areas along the main veins, lavender on the lower surface; ovate to elliptic, acute at both ends, the margins thick, entire to denticulate, the petiole shorter than the blade. Raceme cylindric, 8- to 20-flowered, the pedicels 4–7 mm long, spreading, with lanceolate bractlets almost as long. Flowers about 1 cm in diam; sepals reddish-green, the lobes triangular-acute, 1–1.5 mm long; petals 6–7 mm long, ovate, creamy-yellow to purple or greenish; stamens included, curved inward, the anthers contracted below the pores; style 6–7 mm long, deflexed at the base and arched upward, forming a collar immediately below the stigma, which is smaller, persistent. Fruit 5–7 mm in diam, globose, opening from below; seeds numerous. Jul–Aug. Moist woods; SD: Lawrence; (SD to B.C., s to AZ).

4. **Pyrola rotundifolia** L., round-leaved wintergreen. Stems solitary, rarely 2, from slender rhizomes, the scape 1-3 dm tall, 1-3 lanceolate bracts 6-8 mm long, scattered. Leaves 3-7, 2.5-6 cm long, mostly basal, rosulate, lustrous green, leathery, the blades rounded ovate to suborbicular, margins entire to remotely crenate, revolute, apex rounded to retuse, base short-cuneate, tapering decurrently to the petiole; petioles shorter than the blade. Racemes cylindric, (6)8- to 13-flowered, pedicels spreading to arching, 6-11 mm long, the lanceolate bracts scarious, almost equaling the pedicel. Flowers 1-1.5 cm in diam, calyx spreading, the lobes oblong-lanceolate, 3-3.5 mm long, twice as long as wide, acutely tapered, persistent; petals 5-7 mm long, creamy white, fragrant, leathery, rounded obovate; stamens included, 4-6 mm long, filaments flat, anthers 2-3 mm long, contracted below the pores into short necks; style exserted, deflexed at the base and arching upward, at maturity 6-8 mm long, tapered to distinct collar below the 5-lobed stigma, persisting. Fruit 5-7 mm in diam, globose, depressed, opening from below; seeds numerous. Jul-Aug. Moist woods; SD: Custer; (Greenl. & Newf. to MN, SD s to NC). *P. americana* Sweet — Rydberg.

5. **Pyrola secunda** L., one-sided wintergreen. Stems solitary, rarely 3, from slender, creeping rhizomes, the scape 0.6-1.5 dm tall, with 1-3 lanceolate bracts; basal bracts below and among the mostly basal leaves. Leaves 2-10, somewhat scattered with conspicuous internodes on the ascending stem, blade thin, light green, ovate to elliptic, 1-5 cm long, crenulate-serrulate, rounded to slightly tapering at the base, acute to rounded at the apex, longer than the petiole. Racemes secund, 2- to 12-flowered, pedicels slender, 3-7 mm long, at first horizontal, then drooping, bractlets ciliate, light green, ascending, lanceolate, 2-4 mm long. Flowers 4-6 mm wide, campanulate; sepals spreading, persistent, the lobes ciliate-triangular, 0.5-1.2 mm long; petals white to greenish-yellow, oblong, longer than wide, 3.5-4 mm long, each with 2 small tubercles basally on the inner face; stamens included, about as long as the petals, filaments slender, not declined, the anthers 1.5 mm long, not apiculate, the pores terminal; style straight, slender, 3-5 mm long, in fruit to 9 mm long, exserted, lacking a collar but the stigma peltate, 5-lobed; ovary with a 10-lobed, hypogynous disk. Fruit 4-5 mm in diam, globose, depressed, opening from below; seeds numerous. Jun-Aug. Moist woods; MN, ND, SD, WY, CO; (Greenl. & Lab. to AK, s to GP, VA; Mex.). *Orthilia secunda* (L.) House — Rydberg.

6. **Pyrola uniflora** L., one-flowered wintergreen. Stems solitary from slender rootstocks, ascending to a slender scape (peduncle) 3-10(12) cm tall, recurved above and bearing a single flower, 1 mm in diam, 1(2) ovate scales 2-3 mm long at or below the recurved portion. Leaves 2-10, opposite or in clusters of 3, crowded, 1-2.5 cm long, ovate-elliptic to obovate, crenate to finely serrate or subentire, retuse or rounded at the apex, tapered decurrently to a petiole 5-10 mm long, as long as to shorter than the blade. Flower solitary, nodding, 1.5-2.5 cm in diam, fragrant, waxy-white or tinged with pink; sepals (4)5, 2 mm long, greenish-white to yellowish, the lobes ovate, reflexed, erose-ciliate, the apex rounded, persistent; petals (4)5, 8-12 mm long, spreading, orbicular or broadly ovate; stamens (3)1, filaments 5-6 mm long, subulate, dilated below, incurved; anthers 2-3.5 mm long, attached subapically, each theca narrowed above the filament to a tubular opening; style prominent, 5-6 mm long, straight, persistent, enlarged above to a peltate, erect to later radiating, (4)5-lobed stigma. Fruit (4)5-celled, 5-7 mm in diam, depressed-globose, opening from the summit; seeds numerous. Jul-Aug. Boggy or rich, cool, & moist woods; SD: Lawrence, Pennington; CO, MN; (Lab. & Newf. to AK, s to WV, CO, UT). *Moneses uniflora* (L.) A. Gray — Rydberg.

7. **Pyrola virens** Schweigg. Stem solitary or occasionally 2 from widely spreading, slender rhizomes, scape 0.6-2.5 dm tall, with 1 or 2 lance-shaped bracts. Leaves 4-11, densely clustered near the soil level, reniform to suborbicular, thick, coriaceous, pale green, 1-3

cm long and wide, obscurely crenate, apex rounded, base rounded or broadly cuneate, not decurrent on the petiole; petiole 1–2 × longer than the blade. Raceme cylindric, 3- to 8-flowered, rarely more, pedicels spreading, 4–7 mm long, bractlets almost as long. Flowers 8–12 mm across, greenish but becoming pale yellow-white at anthesis; calyx lobes about 1 mm long, rounded, broader than long, persisting; corolla open-campanulate; petals 4–6 mm long, spreading-ascending, oval, obtuse; stamens mostly included, 4 mm long; filaments subulate, dilated below; anthers 2–3 mm long, constructed below the terminal pores to form well developed tubes; style deflexed and then arched upward towards the apex with a collar below the stigma, persisting. Fruit 5–8 mm in diam, depressed-globose, opening from below. (n = 23) Jun–Jul. Moist to dry, upland woods; NE: Cherry, Sioux; SD: Lawrence, Pennington; (Lab. & Newf. to AK, s to MD, NE, AZ). *P. chlorantha* Sw.—Rydberg.

62. MONOTROPACEAE Nutt., the Indian Pipe Family

by T. Van Bruggen

Achlorophyllous herbs, the roots fibrous, clumped or matted, saprophytic or parasitic on fungi or roots of other vegetation, stems glabrous to pubescent, unbranched, white to crimson or brownish-purple, 0.5–15 dm tall, scaly, the leaflike scales alternate, appressed or ascending, upper ones bracted to involucrate, also without chlorophyll and nonfunctional. Flowers in racemes on the stem or solitary, erect to reflexed. Sepals 2–6, sometimes wanting, distinct, deciduous, not unlike the upper bracts; petals 3–5(6), rarely wanting, distinct to united; stamens 6–12, anthers 2-celled, free or awned, opening by variously dehiscent valves or pores; ovary 1- to 6-celled, the styles united to form a broad capitate or funnelform stigma, ovules many. Fruit a membranaceous capsule, erect or drooping, 1-celled with loculicidal dehiscence; seeds minute, numerous.

1 Petals separate; anthers awnless; fruits erect; plants usually less than 4 dm tall .... 1. *Monotropa*
1 Petals fused, forming globose corolla; anthers awned; fruits on reflexed pedicels; plants commonly over 4 dm tall .... 2. *Pterospora*

1. MONOTROPA L., Indian Pipe

Fleshy herbs, leafless, the roots fibrous, poorly developed, matted in decaying vegetation or on fungal mycelia; stems white to crimson, the entire plant turning black upon drying, unbranched, 1–3 dm tall, with alternate, leaflike scales instead of leaves, the upper scales becoming bracteate. Flowers solitary or several in a dense, bracteate raceme, on short pedicels. Sepals 2–5, 0.7–2 cm long, distinct, scalelike, deciduous, or lacking; petals 3–5(6), 0.6–2 cm long, distinct, with saccate base, forming an urceolate or tubular corolla; stamens 8–10, the filaments pubescent, subulate; anthers opening by transverse or irregular lines, becoming 1-celled at anthesis; ovary 3- to 5-celled, with a 10- or 12-toothed hypogynous disk, style 1–5 mm long, with a capitate stigma. Fruit a capsule, 4- to 5-celled, slightly grooved, erect, opening with 4–5 valves on the locules; seeds many.

1 Flower solitary, terminating the stem apex, 5-merous; stems glabrous, white to pink .... 2. *M. uniflora*
1 Flowers few to several in terminal racemes, the lateral ones below the terminal one, 3- or 4-merous; stems pubescent, red to yellow-brown .... 1. *M. hypopithys*

1. ***Monotropa hypopithys*** L., pinesap. Plants pink-red to yellow or tawny brown, commonly aromatic, 1–3 dm tall, the stems pubescent, becoming darkened upon drying; stems

simple, sometimes few-clustered from the mass of roots; scales sessile, partially clasping the stem, 0.7–1.2 cm long, appressed, oblanceolate, with erose margins. Flowers few to several in bracted, terminal racemes, at first drooping but becoming erect (occasionally with a solitary terminal flower), the terminal one 5-merous, lower lateral ones 4-merous. Sepals strap-shaped, bractlike, 6–12 mm long, as many as the petals; petals 4 on lateral flowers, 5 on the terminal one, 6–15 mm long, with saccate bases and rounded apices, pubescent on the inner face; stamens 8–10, twice the number of petals, almost equaling the petals, filaments slender, the anthers opening with a line into 2 unequal valves; style as long as or longer than the ovary, the stigma capitate, hollowed out, ciliate. Fruit a membranaceous capsule, erect, 5–8 mm long, the style persistent; seeds many. Jun–Oct. Rich, upland woods; e KS, MO, NE; (circumboreal, in N. Amer. s to FL, e KS; Mex.). *Hypopithys lanuginosa* (Michx.) Nutt.—Rydberg.

2. ***Monotropa uniflora*** L., Indian pipe. Plants off-white to pink, without odor, 0.7–3 dm tall, stems glabrous, becoming blackened upon drying; stems simple or rarely paired from the mass of roots; scales sessile, 0.6–1.2 cm long, lanceolate, ascending, with erose margins. Flower solitary, terminal, bracted by upper scales, at first nodding but erect in fruit, 1.4–3 cm long, 5-merous; sepals often lacking, if present, sepals 2–4, strap-shaped, glabrous, similar to bracts below; petals 0.8–2 cm long, oblong, slightly gibbous at the base, abruptly widened above, pubescent only on the inner face; stamens 10, included in the corolla, the filaments pubescent-strigillose; anther opening by apical, transverse slits; ovary 5-celled, the style short and thick, topped by a broad, capitate stigma. Fruit a membranaceous capsule 8–10 mm long, topped by the persistent style, erect; many-seeded. Jul–Sep. Rich woods; e GP; (Newf. to AK, s to GP; Mex.).

2. **PTEROSPORA** Nutt., Pine Drops

1. ***Pterospora andromedea*** Nutt. Herbs with simple, unbranched, purple-brown stems, glandular-hairy, 2–9 dm tall, at first fleshy but becoming fibrous and persisting as dried stalks for many months, with scattered leaflike scales 1–3 cm long on the lower ½. Flowers many in a long raceme, ½ or more of the upper part of the stem, pendulous from axils of small linear bracts, the pedicels recurved, 5–12 mm long, glandular-hairy; sepals 5, glandular-hairy, fused below, the lobes oblong, 3–5 mm long; corolla globose, sometimes urceolate, connate nearly to the top, 4–7 mm long, the short lobes 1–2 mm long, spreading, white to pink, glabrous, persisting in fruit; stamens 10, included in the corolla, the filaments flattened, linear, 2–3 mm long, anthers ovoid, dehiscent along the exterior face, each with a pair of reflexed awns, equaling the length of the anther; ovary 5-lobed, globose, 6–10 mm in diam, 5-celled, the style short and columnar with a flared, capitate, shallowly 5-lobed stigma. Fruit a depressed globose capsule about 1 cm in diam with loculicidal dehiscence; seeds many, 0.2–0.3 mm long, each with a reticulate wing at one end. Jul–Sep. Rich humus in coniferous woods, especially under ponderosa pine; SD, WY, CO: El Paso, Las Animas; (P.E.I. & NY to B.C., s to GP, Mex.).

63. **SAPOTACEAE** Juss., the Sapodilla Family

by William T. Barker

1. ***Bumelia lanuginosa*** (Michx.) Pers. var. ***oblongifolia*** (Nutt.) Clark, woolly buckthorn, chittimwood, gum-elastic. Shrub or small tree to 15 m or more tall, more or less thorny. Leaves alternate, sometimes fascicled, petiolate, simple, entire, exstipulate, oblanceolate

to sometimes obovate or elliptic, broadly rounded to sometimes acute at the apex, 4–10 cm long, 1.5–3.5 cm wide, upper surface dark green, glossy with some pubescence on the midrib, lower surface densely woolly with rusty or nearly white hairs, reticulate-veiny on both sides. Flowers numerous, small, perfect, usually in axillary clusters, borne on hairy to subglabrous pedicels 6–15 cm long; calyx free, persistent, 5 lobed, distinct, 1.5–3.2 mm long, strongly hairy or nearly glabrous; corolla 3–4.7 mm long, petals 5, united, tube 1.3–2 mm long; staminodes 5, white, 1.9–2.7 mm long, nearly equaling the corolla lobes, alternate with them; stamens 5, anthers yellow, 1–1.5 mm long; ovary globose, pilose, 1 mm long; style 1–1.5 mm long. Fruits single or clustered, obovoid to broadly ellipsoid, commonly purplish-black at maturity, 7–13 mm long, 1-seeded. Jul–Oct. Wooded uplands & hillsides & along streams; se KS, MO, OK, TX; (MO & se KS s to FL & TX; n Mex.). *B. lanuginosa* (Michx.) Pers. var. *albicans* Sarg.—Steyermark.

64. EBENACEAE Gurke, the Ebony Family

by William T. Barker

1. DIOSPYROS L., Persimmon

1. *Diospyros virginiana* L. Dioecious or occasionally polygamo-dioecious tree to 20 m tall, usually with a rather open crown, sometimes colonial; twigs dark gray or brown, pubescent or glabrous. Leaves simple, alternate, exstipulate, the blades rather thick, dark green and glabrate above, pale and glabrous to pubescent below, oval to oblong, 7–15 cm long, 3.5–8 cm wide, acuminate to short-acuminate, minutely ciliate on the margin, acute to rounded or obliquely truncate to subcordate at the base; petioles glabrous or pubescent, mostly 1–2 cm long. Flowers yellowish-green, axillary on short peduncles, unisexual or an occasional flower perfect, 4-merous; calyx deeply lobed; corolla urceolate-campanulate, the lobes spreading to recurved; staminate flowers solitary or in clusters of 2 or 3, 5–8 mm long; stamens usually 16, in 2 rows attached to the base of the corolla tube, about equaling the corolla tube, surrounding a vestigial pistil; pistillate flowers solitary, larger than the staminate, 1–1.5 cm long, with usually 6–10 sterile stamens; styles 4, connate ca ½ their length, bilobed at the stigmatic tip, ovary superior, 4- or 8-celled. Fruit a yellowish-brown, globose berry, 2–6 cm diam, sweet and edible when ripe, usually after frost; seeds few to several. (2n = 60, 90) May–Jun, fruits Sep–Oct. Dry, open woods, clearings, & old fields; MO, e & cen KS, OK; (CT & NY to KS, s to FL, OK, TX). *D. virginiana* var. *pubescens* (Pursh) Dipp., var. *platycarpa* Sarg.—Fernald.

65. PRIMULACEAE Vent., the Primrose Family

by Gerald Seiler and William T. Barker

Annual or perennial herbs, sometimes scapose, subtended by small or leafy bracts. Leaves simple, exstipulate, mostly entire, opposite or mostly so, sometimes appearing whorled due to leaf fascicles in the axils. Flowers single, or in axillary racemes, sometimes on leafless scapes, perfect, regular. Calyx (3)5(9)-parted, free from or partially adherent to the ovary, merely toothed or divided nearly to base; corolla (absent in *Glaux*) gamopetalous, (3)5(9)-lobed, rotate and often very deeply lobed to tubular or salverform and shallowly lobed; stamens typically 5, epipetalous (free in *Glaux*) opposite the petals, occasionally alternating with staminodia, included; stigma capitate at the tip of the single slender style, ovary superior (except *Samolus* partially inferior), 1-celled with free-central placentation. Fruit a 5-valved capsule or circumscissile; seeds few to many.

Trientalis borealis Raf. has been reported from the northern and eastern edge of the region. While this species does make its appearance in the transitional area of the prairie and the Eastern Deciduous Forest, it is not a characteristic species of the Great Plains and is excluded.

1. Ovary partially inferior, its base definitely adnate to the calyx tube; plant caulescent; inflorescence an elongate naked raceme, the pedicels bracteate or ebracteate .... 8. *Samolus*
1. Ovary superior, wholly free calyx tube; plants caulescent or acaulescent; inflorescence of axillary flowers, umbels or leafy racemes.
   2. Plants scapose; the leaves radial or clustered near base of plant in rosettes; scapes terminated by an umbellike inflorescence rarely reduced to 1 flower.
      3. Corolla lobes reflexed, several times as long as the tube; stamens exserted, protruding their full length, connivent in a cone .... 4. *Dodecatheon*
      3. Corolla lobes erect or merely spreading, less than 2 × longer than the tube; stamens usually included, distinct.
         4. Corolla-tube longer than calyx; flowers numerous in open inflorescence, usually well over 5 mm long; style slender .... 7. *Primula*
         4. Corolla-tube equaling or shorter than calyx; flowers usually few in a rather close inflorescence, less than 5 mm long; style minute .... 2. *Androsace*
   2. Plants caulescent; leaves scattered along stem; flowers axillary, solitary in leaf axis or in raceme.
      5. Flowers sessile in leaf axils.
         6. Corolla absent; ascending perennial with mostly opposite leaves; capsule valvate .... 5. *Glaux*
         6. Corolla present; minute prostrate annual with alternate leaves; capsules circumscissile .... 3. *Centunculus*
      5. Flowers pedicellate in leaf axils or terminal.
         7. Plants perennial with rootstocks; leaves not noticeably clasping, well over 2 cm long; capsule valvate .... 6. *Lysimachia*
         7. Plants annual, prostrate to ascending; leaves somewhat clasping, less than 2 cm long; capsule circumscissile .... 1. *Anagallis*

1. ANAGALLIS L., Poorman's Weatherglass, Pimpernel

1. *Anagallis arvensis* L., poorman's weatherglass, pimpernel. Annual or rarely perennial herb, low spreading or procumbent, stems 4-angled, much branched, forming loose prostrate mats. Leaves opposite, elliptic to ovate, to 2 cm long and 1 cm wide, entire, sessile or somewhat clasping the stem. Flowers solitary in the axils, variable in size and color, from scarlet to salmon color and sometimes almost white or blue; pedicels slender, ascending at anthesis, recurved in fruit, usually exceeding the leaves. Calyx 5-parted, lobes lanceolate, 3–4 mm long, persistent; corolla rotate, petals deeply 5-parted, the lobes convolute in bud, equaling or somewhat surpassing the calyx, obovate to cuneate-obovate, somewhat fringed with minute teeth and stalked glands at the obtuse to rounded apex; stamens 5, inserted on slender filaments near base of corolla tube, filament sometimes pubescent; stigma capitate, style slender, filiform, ovary superior, globose, many-ovuled. Capsule globose, about 4 mm in diam, circumscissile, membranaceous, many-seeded. (2n = 40, n = 20) Jun–Aug. Usually moist places, wet depressions, waste places, sandy soil; KS; OK: Cleveland; (Newf. to B.C., s to FL, OK, CA; Mex.; Eurasia). *Naturalized*.

Two forms of *Anagallis* have been recognized in the area: the more common forma *arvensis* with flower color red, scarlet, or rarely white and forma *caerulea* (Schreb.) Baumg, with flowers blue, known from KS: Neosho.

2. ANDROSACE L., Rock Jasmine

Small erect annual herb; scapes erect, 1 to several, leafless, bearing terminal involucrate umbels; fibrous root system. Leaves in basal rosettes, linear-lanceolate to oblanceolate or ovate-obovate, slightly or indistinctly petiolate. Flowers very small, in terminal involucrate umbels, subcompact or diffuse and few- to many-flowered, subtended by bracts. Calyx tube hemispherical to subglobose-obconic, equaling or exceeding the triangular to deltoid lobes, calyx 5-cleft, lobes narrowly subulate to deltoid, (1.5)3.9(5.5) mm long, 5-ribbed or smooth; corolla 5-parted, salverform or funnelform, inflated around the ovary, constricted at the throat, shorter than calyx, white to pink; stamens 5, attached opposite corolla lobes at or below the middle of the tube, anthers oval or oblong, filaments very short, not exceeding the ovary; style solitary, short and included within corolla, bearing a small capitate stigma, ovary superior, 1-celled, with few to many ovules with free central placentation. Capsule subglobose or globose, shorter than, equal to, or exceeding the persistent calyx-tube, dehiscing longitudinally from apex to base by 5 valves, few- to many-seeded; seeds ovoid to elliptical, dark or pale brown, minutely pitted.

Reference: Robbins, G. T. 1944. North American species of *Androsace*. Amer. Midl. Naturalist. 32: 137–163.

1 Involucral bracts lanceolate-ovate to ovate-obovate; calyx lobes inear-elliptic to triangular-lanceolate, equal to or slightly longer than calyx-tube .... 1. *A. occidentalis*
1 Involucral bracts narrowly triangular to lanceolate, attenuate, or subulate to acerose acute; calyx lobes deltoid to acerose, much shorter than the calyx-tube .... 2. *A. septentrionalis*

1. ***Androsace occidentalis*** Pursh, western rock jasmine. Small, rather inconspicuous annual; scapes numerous, 2–6(10) cm tall, scabrous with stellate hairs, often with erect central scape and several arched-ascending lateral ones, sometimes flexuous. Leaves basal in a rosette, ovate-lanceolate to nearly linear, obtuse, to 2 cm long and 6 mm wide, entire or sparingly denticulate above middle, upper surface and margin covered with short, stiff, usually simple white hairs, glabrous beneath, slightly petiolate. Flowers small, in terminal involucrate umbels, pedicels 5–9, 2- to 10-flowered, unequal, erect or ascending up to 3 cm long, mostly less, stellate-pubescent; involucral bracts ovate-lanceolate to ovate, (2)4(6.6) mm long, (0.7)1.7(5) mm wide, acute, more or less puberulent, often reddish-tipped calyx-tube (1.5)4(5.5) mm long, subglobose-campanulate, whitish or greenish-white noticeably contrasting with the lobes, broadly ridged, nearly glabrous; calyx lobes linear, elliptic to narrowly triangular lanceolate, (1)2(4) mm long, bright green, 1/3–1/2 length of tube, more or less puberulent, corolla white to pink, small, obtuse, 2.5 mm wide, shorter than or equal to calyx. Capsule rotund, included within calyx-tube, nearly or quite equaling the calyx-tube; seeds small, many, blackish. (2n = 20) Mar–Jun. Dry or sandy soil, open prairies, sometimes rocky, open woods & ravines & hillsides; GP; (Ont. to B.C. s to TX, NM, & AR). *A. simplex* Rydb.—Rydberg.

2. ***Androsace septentrionalis*** L., northern rock jasmine. Erect annual herb; plant to 25 cm tall, scapes several, erect, central one well developed with many weak supplementary ones, glabrous to more or less puberulent with reddish glandular or nonglandular hairs. Leaves in basal rosette, linear to lanceolate-oblanceolate, 1–3(4) cm long and 6 mm wide above middle, upper surfaces puberulent, appearing scaberulous, often with stellate pubescence, and denticulate or sparingly so at apex and margins. Flowers small, compact to diffuse in terminal involucral umbel, involucral bracts subulate to narrowly lanceolate, (1.5)2.6(6.6) mm long, (0.3)0.6(1) mm wide, pedicels 6–20, strictly ascending or erect, to at least 55 mm long, soon glabrate; calyx tube narrowly to broadly campanulate, greenish, (2.5)3.3(4.2) mm long, prominently 5-ridged, nearly glabrous, calyx lobes subulate to narrowly lanceolate or broad and foliaceous in fruit, (0.9)1.3(2) mm long, ridged or rounded, usually greenish, 1/3–1/2 as long as the tube, often puberulent; corolla white, fading to pink, small, 3–4 mm wide, shorter than or equaling the calyx. Capsule globose, included within and nearly or quite equaling the calyx-tube; seeds small, 5–15. (2n = 20, n = 10) May–Jul. Upland prairies, hillsides, roadsides, open woods, rocky soil of ravines, occasionally moist areas about ponds & seepage areas; MN, ND, MT, SD, WY, CO; (Sask., Alta., & B.C., GP, cen & s Rocky Mts., NM, AZ). *A. puberulenta* Rydb.—Rydberg.

The plants from the BH tend toward var. *subulifera* A. Gray, which is glabrous, taller, and has a shorter calyx lobe. Other GP plants have characters more in common with var. *puberulenta* and are best assignable to that variety.

3. CENTUNCULUS L., Chaffweed

1. ***Centunculus minimus*** L., chaffweed. Low, glabrous annual herb; stems decumbent to erect and rooting at the nodes, often forming mats or small clumps 2–12 cm across. Leaves chiefly alternate or lower ones opposite, subsessile, entire, obovate to oblanceolate or oblong-spatulate, 5–10 mm long, mm wide. Flowers solitary, minute, sessile-subsessile in axils of most leaves, ephemeral, pedicels about 1 mm long; calyx 4- or occasionally 5-merous, lobed nearly to the base, linear-lanceolate, minutely serrulate, 2–3 mm long; corolla tube short, about 2 × the length of the lobes, withering on the apex of the capsule; corolla pinkish, rotate, ovate-lanceolate lobes about 1 mm wide, when expanded and petals are erect about 1.5 mm long; stamens 4 or 5, borne at the throat of the corolla, not exserted, the filaments beardless. Capsule globose-subglobose, about 1.5–2 mm in diam, circumscissile; seeds brown, pitted, numerous. (2n = 22) Jun–Aug. Moist ground, seepage areas along streams; ND; SD: Custer; NE: Holt, Nemaha; e KS; OK: Comanche, Tulsa, Payne; (irregularly cosmopolitan, N.S. to OH, MN, ND, s to OK).

4. DODECATHEON L., Shooting Star

Scapose herbaceous perennial from slender to thick rhizomes or very short caudices; stem erect, scapose, not branched, glabrous to conspicuously glandular-pubescent; roots fibrous-fleshy. Leaves simple, entire to dentate, petioled, in a basal rosette originating from the upper portion of the caudex. Flowers showy, 4- or 5-merous, borne in terminal involucrate umbels or sometimes solitary, on slender pedicels, erect in bud, nodding in flower and erect in fruit. Calyx tubular, very short, the lobes lanceolate, 5-parted (sometimes 4-parted); corolla showy, short tubular, white to purple, lobes erect in bud and strongly reflexed in flower; stamens usually 5, inserted on the corolla tube opposite the lobes, connivent around the style, protruding their full length, free or connected by a membrane, anthers long and slender, basally attached, dehiscent on the inner surface, connective prominent, highly colored, smooth to transversely rugose, pollen sac yellow or dark maroon, filaments short, free or connected by a membrane; style exserted, slightly exceeding the stamens, stigma single, capitate, sometimes rather conspicuously enlarged, ovary superior, free central placentation. Capsule 1-celled, valvate at tip, or the tip operculate with the style and the walls valvate below; seeds many, 2–3 mm long, spherical or ovoid with reticulate coat.

References: Thompson, H. J. 1953. Biosystematics of *Dodecatheon*. Contr. Dudley Herb. 4(5): 73-154; Thompson, H. J. 1960. A new combination in *Dodecatheon* (Primulaceae). Leafl. W. Bot. 9(6): 91.

1 Capsule wall thick, ligneous, not flexible .... 1. *D. meadia*
1 Capsule wall thin, sometimes almost membranaceous, flexible under the slightest pressure .... 2. *D. pulchellum*

1. Dodecatheon meadia L., shooting star. Perennial herb, glabrous or the inflorescence glandular-pubescent; stem erect, scapose to 55 cm tall, not branched; roots fibrous, white, without bulblets. Leaves simple, in a basal rosette, ovate to spatulate, narrowly elliptic-oblong to oblanceolate, 10-20(30) cm long (including petiole), 2.5-4(8) cm wide above middle, obtuse to rounded at apex, entire or with callous points along the margin, blade gradually tapered into petiole. Flowers showy, few to many (4-125) in umbel, subtended by an involucre of small bracts to 1 cm long, deltoid to lanceolate, acute; pedicels slender, erect, 3-7 cm long in flower, recurving in anthesis and longer in fruit; calyx tube short, 2-4 mm long, lobes 5, subulate to lanceolate, usually about 5(8) mm long at anthesis, slightly longer and persistent in fruit; corolla tube very short with thickened throat, maroon, yellow above, lobes 5, magenta, lavender to white, all intermediates sometimes present, oblong-elliptic, to 12-20(25) mm long and 1 cm wide, strongly reflexed; anthers linear-lanceolate, acute, 7-10 mm long, pollen sacs yellow, connivent to form a slender cone, filaments 1-3 mm long, free or united into a tube, yellow, connective dark maroon, purple, or black, smooth (often longitudinally wrinkled upon drying); stigma not enlarged. Capsule ovoid to cylindric, 7.5-8 mm long, 4-6 mm wide, dark reddish-brown, with firm ligneous walls, opening by 5 short terminal valves; seeds many. (2n = ca 88) Apr-May. Moist or dry woods & prairies, sometimes rocky limestone soil; MO, KS, OK; (sw VA, w to WI, s to GA & TX).

Most of our plants belong to the typical variety *meadia* var. *brachycarpum* (Small) Fassett, with a more southern distribution, occurs in MO: Jasper; OK: Caddo, Creek, Tulsa. It differs in being shorter (to 40 cm) and in having pedicels 2-6 cm long; calyx tube 2-3 mm long, calyx lobes 3-5 mm long; anthers 4-7 mm long; capsules 7-10 mm long and 4-5 mm wide.

2. Dodecatheon pulchellum (Raf.) Merrill, shooting star. Perennial herb, stem erect, scapose, to 50 cm tall; roots white, fibrous, without bulblets. Leaves simple in a basal rosette, oblanceolate, ovate or spatulate, acute, obtuse or rounded, blade gradually tapered into petiole, 4-25 cm long (including petiole), 1-6 cm wide, usually entire, occasionally sinuate or crenate. Flowers few to many (3-25) in umbel, showy, subtended by an involucre of small bracts, deltoid to lanceolate or spatulate, acute to obtuse, to 1.5 cm long; pedicels 1-5 cm long in flower, longer in fruit; calyx tube 2-3.5 mm long, the lobes 2.5-6 mm long, lanceolate to subulate, acute to acuminate; corolla tube maroon, yellow above, the lobes 9-20 mm long, magenta to lavender; filaments 0.5-3.5 mm long, united into a tube or nearly free, yellow, smooth or rugulose, anthers lanceolate, obtuse or acute, (3.2)5(7.6) mm long, pollen sacs yellow, sometimes red or maroon, connective dark maroon to black, rarely yellow, smooth (often longitudinally wrinkled upon drying); stigma not enlarged. Capsule 7-17 mm long, 4-7 mm wide, cylindric to ovoid, thin walled, valvate; seeds many. (2n = ca 44) May-Jun. Low prairie meadows, moist hillsides, & open woods; w ND, w SD, w NE, MT, WY, CO; (e U.S., WI to AK s to AR, NE). *D. amethystinum* Fassett—Fernald; *D. media* L.—Stevens; *D. pauciflorum* (Durand) Greene, *D. radicatum* Greene, *D. salinum* A. Nels, *D. thornense* Lunell—Rydberg.

The morphological variation in this species is not well understood. This has resulted in several different taxonomic treatments and several species names. The ratio of anther length/filament length which has been used to separate the various species was examined for the plants of our area. This ratio showed no clear separation into groups. There does appear to be a north-south clinal variation for this character (increasing from north to south) except for the South Dakota plants which appear more like the western Cordilleran plants. This species is treated as a wide ranging and highly variable one.

5. GLAUX L., Sea Milkwort

1. *Glaux maritima* L., sea milkwort. Low, glabrous and glaucous perennial herb, 3–25 cm tall, from shallow horizontal slender rhizomes; stems leafy to tip, simple, slender, erect to branched and spreading and sometimes prostrate. Lower leaves opposite or mostly so, the upper ones usually becoming alternate, sessile, rather succulent, elliptic to oblong to oblanceolate, 3–20 mm long, 1–5 mm wide, obtuse to subacute. Flowers small, sessile or subsessile in leaf axils; calyx campanulate, 3–4 mm long, the 5 lobes rounded, extending to about the middle, more or less petaloid, white to pinkish; corolla absent; stamens 5, free of the calyx, inserted at base of calyx tube and alternate with its lobes, anthers hairy, the filaments as long or longer than the calyx lobes. Capsule ovoid to globose, 2.5–3 mm long, beaked with persistent style, valvate the full length; seeds few, black, roughly elliptic and flattened, 1–1.5 mm long, coherent to the placenta and shed with it as a unit. (2n = 30) Jun–Aug. Moist alkaline or saline soils, seepage areas, stream margins; MN, MT, ND, NE; (circumboreal in N. Amer., s to VA, NE, NM, & OR).

6. LYSIMACHIA L., Loosestrife

Typically erect, rhizomatous perennial; stems leafy, simple or branched, usually glabrous, but punctate. Cauline leaves opposite (occasionally appearing whorled in *L. quadriflora* due to leaf fascicles in the axils), leaf blades entire, varying in shape from ovate, ovate-lanceolate, lanceolate to linear, sessile to long petiolate, petioles or leaf bases frequently fringed with ciliate hairs. Flowers yellow, solitary and pedicellate from leaf axils, terminal or in axillary racemes; calyx green, deeply (3)5(9)-parted, nearly to the base, imbricate or valvate, generally herbaceous, persistent sepals subulate to deltoid; corolla yellow to pale yellow, rotate, deeply (3)5(9)-parted, nearly to base, tube very short, the lobes convolute, sometimes each division convolute around a stamen, petals ovate to ovate-lanceolate, densely glandular basally on internal surface, sometimes with basal red-purple coloration; stamens adnate to corolla near the base; staminodia lanceolate to narrowly membranaceous, deltoid and alternate with the stamens; pistil superior, compound, unilocular, 5-carpellate, placentation free central. Capsule globose to ovoid, dehiscent usually by 5 valves, rarely indehiscent; seeds few to many, oblong to angular.

References: Coffey, V. J., & S. B. Jones, Jr. 1980. Biosystematics of *Lysimachia* section *Seleucia* (Primulaceae). Brittonia 32: 309–322; Ray, J. D. 1956. The genus *Lysimachia* in the New World. Illinois Biol. Monogr. 24: 1–160.

Several species of *Lysimachia* have been excluded from the present treatment. *L. radicans* Hook. has been reported twice as occurring in the area. One collection was an error in location recording and the other an error in identification. *L. lanceolata* Walt. has been reported from areas next to the GP but no documented specimens have been located. It may occur along the extreme southeastern part of the area. *L. punctata* L. has been reported from one location in Kansas, but the specimen could not be found. The species is a weed in temperate regions and one that probably occurs as a waif in many areas.

1 Flowers clustered in dense axillary racemes; foliage punctate with dark glands .... 5. *L. thyrsiflora*
1 Flowers single in the axils, borne on slender pedicels; foliage not punctate.
   2 Leaves nearly round, heart-shaped at base, 1–3.5 cm long; stems creeping and trailing .... 3. *L. nummularia*

2 Leaves of other shapes, linear, lanceolate, elliptic or ovate, mostly 3-15 cm long; stems upright, spreading, or reclining.
   3 Pubescence of petioles of median cauline leaves restricted to basal portion; median cauline leaves subsessile, linear, 1-6 mm wide with prominent midveins, and lateral veins obscure .... 4. *L. quadriflora*
   3 Pubescence of petioles of median cauline leaves variable, basal ½ of entire petiole pubescent; median cauline leaves lanceolate or ovate with prominent midveins and lateral veins.
     4 Median cauline leaf blade ovate to ovate-lanceolate, 1.8-5.7 cm wide; distinctly petioled, densely fringed with cilia from node to leaf blade base .... 1. *L. ciliata*
     4 Median cauline leaf blade ovate-lanceolate to linear-lanceolate, 0.6-2.2 cm wide; petioles of upper cauline leaves short or essentially sessile, petioles of median cauline leaves ciliate at least on basal ½ .... 2. *L. hybrida*

1. Lysimachia ciliata L., fringed loosestrife. Erect, perennial herb; stem simple or sometimes sparingly branched above, 3-10 cm tall, apparently glabrous but upper part, especially nodal regions and pedicels, glandular-puberulent; rhizomes few, long, slender. Median leaves opposite, distinctly petiolate, the blades dark green to bright green above, lighter beneath; cauline leaves ovate to ovate-lanceolate (rarely lanceolate), (3.5)4-13(15) cm long, apex acute to acuminate, margins minutely ciliate apically, base obtuse to cuneate or somewhat attenuate; petioles (0.5)0.8-5(7.3) cm long, densely ciliate from the node to the base of leaf blade. Flowers solitary in apical leaf axils and inserted in leaf axils of flowering branches; pedicels (0.5)2-6(8) cm long; calyx tube short, about 1 mm long, lobes lanceolate, entire, (2)4-8(9) mm long, (1)1.5-3(7) mm wide, 3(6) reddish-brown veins parallelling the long axis; corolla tube short, flat, 1-2 mm long, sometimes with red blotches at base or lobes, lobes rotund to obovate, 5-12 mm long, 3-9 mm wide, erose near apex, apiculate, densely yellow glandular within; stamens with anthers 2-3.5 mm long; staminodes present, 1-2 mm long, triangular to subulate, glandular, almost hyaline; ovary globose to ovoid, minutely glandular-puberulent at style base, style 3.5-5 mm long, persistent, tip slightly swollen, stigmatic surface with minute hyaline trichomes, ovules numerous. Capsule subglobose to ovoid, glabrous, (1.5)4-6.5(8) mm in diam, containing 25-40 trigonous seeds, 1.9-2.2 mm long, rufescent, finely reticulate. (2n = 34, 72, 92, 96, 100, 108, 112) Jun-Aug. Swamps, marshes, wet meadows, stream banks, ditches, flood plains, moist woods, thickets, & shaded stream banks; GP; (Que. to AK, s to FL, TX, NM, UT, & OR). *Steironema ciliatum* (L.) Raf., *S. pumilum* Greene — Rydberg.

2. Lysimachia hybrida Michx., loosestrife. Erect or rarely reclining perennial herb; stem ascending, simple or often branched from below the middle, (1.5)2.5-7(8) dm tall; rhizomes ascending, up to 5 cm long. Basal leaves ovate to lanceolate, rounded or obtuse at base, seldom persistent as a rosette, cauline leaves ovate to lanceolate, green above and beneath, (4.5)5.8(7.3 cm long, (1)1.2(2.4 cm wide, apex acute, margins minutely papillate, base rounded to cuneate; petioles of basal leaves (0.6)0.9(1.1) cm long, length decreasing from basal to apical portion of the plant, becoming subsessile to subverticillate, pubescent along basal ½, sometimes over the entire length, but more sparingly toward the blade. Flowers solitary in the axils, appearing clustered in apical leaf axils because of the close spacing of nodes, and inserted in leaf axils of flowering branches; pedicels slender, 0.8-4 cm long, somewhat glandular-puberulent; calyx lobes lanceolate, (4.2)4.9(5.4) mm long, (1.6)2(2.3) mm wide, apex acute to attenute, venation conspicuous, 3(5)-nerved; corolla lobes obovate to suborbicular, 5-10.5 mm long, 4-10 mm wide, weakly erose, apiculate; stamens with anthers 1.5-2 mm long; staminodes present, 1-2 mm long, triangular to subulate glandular, almost hyaline. Capsules 3.5-5 mm in diam, subequal to or shorter than the calyx, containing 20-35 trigonous seeds, 1.2-1.8 mm long. (2n = 34) Jul-Aug. Wet meadows & prairies, margin of marshes, wet ditches, shores, & swamps, often in shallow water; MN, ND, SD, NE, IA,

MO, KS; (ME to s Sask., s to n FL, IN, AR, KS, also NM). *Steironema hybridum* (Michx.) Ref., *S. verticillatum* Greene — Rydberg.

3. ***Lysimachia nummularia*** L., moneywort. Prostrate perennial herb with creeping stems, rooting at the nodes; stem glabrous or nearly so, finely punctate with tiny red-blackish dots, repent or becoming so. Leaves opposite, dark green, with red glandular punctae, suborbicular to oblong-oval, (1)1.5–2(3.5) cm long, nearly as wide, petioles short, 1.5–4 mm long, narrow wings weakly decurrent. Flowers solitary from medial axils; pedicels 1–5 cm long; calyx lobes cordate-lanceolate, 6–9 mm long, 2.5–5 mm wide, weakly keeled, apex acuminate, margins entire, with cordate base; corolla rotate, yellow, lobes obovate to oblanceolate, 1–1.5 cm long, 5–8 mm wide, rounded to acute, somewhat erose, glandular-ciliolate and glandular-pubescent near the base on the upper surfaces; staminal tube about 1 mm long, densely glandular; sinuses rounded, often with a small dentation, stamens much exceeding corolla, anthers linear, about 1.5 mm long, filament 4–5 mm long, unequal, glandular-pubescent, shortly connate at the base; staminodes none. Capsule included in calyx. (2n = 30, 32, 34, 36, 43) May–Jul. Moist soil along lakes, rivers & ponds, wooded areas along rivers; e KS; (sparingly naturalized throughout cen & e U.S.; Europe).

4. ***Lysimachia quadriflora*** Sims, whorled loosestrife. Erect perennial herb, stem (1.2)2–7 dm tall, simple or branched; rhizomes few, slender, forming subsessile offshoots. Basal leaves when persistent as a rosette, elliptical to obovate, 2–3 cm long, 0.5–1 cm wide, tapering at base, petioles longer than blade, cauline leaves opposite, linear, narrowly lanceolate, 3–7(9) cm long, (1)3–6 mm wide, shiny above, dull below, apex acute, base cuneate to tapering, margins revolute, midrib prominent, lateral veins not in evidence or obscure, petiole 1–5 mm long, pubescent only at base. Flower pendent, solitary in apical leaf axils of flowering branches; pedicels 0.5–3.5 cm long; calyx tube short, lobes lanceolate, attentuate, entire, 3.5–6 mm long, 1–2 mm wide; corolla yellow, rotate, lobes oval to obovate, 7–12 mm long, 5–9 mm wide, erose or entire, apiculate, with glandular tipped hairs within; stamens with anthers linear, notched below, about 2 mm long, filaments glandular, about 2 mm long; staminodia triangular to subulate, weakly joined by a membranaceous line at base of the corolla tube; ovary subglobose, glabrous or with a few glandular tipped hairs at summit, style 4.5–5 mm long, ovules numerous. Capsule short, ovate to subglobose, 3.5–6 mm in diam, containing 30–44 trigonous seeds, about 1.2 mm long, outer surface flat, somewhat angular in outline, adjacent surface concave, shiny rufescent coat with thin scarious covering. (2n = 34) Jul–Aug. Moist or wet soils, especially prairie meadows & ponds; MN, ND; (ME w to WI, IL, ND, s to SC, KY, & AL). *L. longifolia* (Pursh) A. Gray—Stevens; *Steironema quadriflorum* (Sims) Hitchc.—Rydberg.

5. ***Lysimachia thyrsiflora*** L., tufted loosestrife. Stout, erect perennial herb 2–8 dm tall; rhizomes creeping, usually smooth and rather thick, finely dark purplish or blackish-maculate almost throughout, somewhat pithy; stem simple or occasionally branched from lower nodes, glandular-punctate throughout, glabrous below, puberulent-brownish-villous in patches above. Lower leaves opposite, sessile, much reduced, scalelike and scarious, the upper cauline leaves opposite to verticillate, lanceolate, elliptic or oblanceolate, (4)5–13(16) cm long, (0.5)1.2(6) cm wide, glabrous above, glabrous or sparingly villous with fine septate hairs beneath along the midrib and base, subsessile or petiolate, apex acute to attenuate, margins entire to weakly sinuate, the base tapering, dark glandular-punctate. Flowers small in dense, capitate or spikelike pedunculate racemes, 1–3 cm long from medial or upper axils, flower 1–2 cm wide, peduncules 2–5.5 cm long, raceme bracteate, bracts linear-subulate, 3–5 mm long, dark glandular-punctate; pedicels 0.4–4 mm long, glabrous to sparingly villous. Calyx tube very short, dark streaked or dotted, lobes deeply (3)5- to 7(9)-parted, lance-attenuate, 1.5–3 mm long, 0.5–1 mm wide; corolla lobes usually the

same number as the calyx lobes, linear attenuate, 3–5 mm long, 1–2 mm wide, obtuse or acute at the apex, somewhat clawed at the base; stamens usually 5(7), much exceeding the corolla, anthers oblong about 0.7 mm long, on long slender filaments 4–5 mm long, unequal; staminodes none; ovary dark glandular verruculose, sparingly puberulent with slender segmented trichomes, style 4.5–6 mm long. Capsules 2–4 mm in diam, dark glandular punctate; seeds few, trigonous, dark brown, 1.2–1.5 mm long, outer surface usually convex, adjacent surfaces concave. (2n = ca 40, 42, 54, n = 21) May–Aug. Fens, bogs, springs, marshes, wet meadows, & shores, usually growing in fresh shallow water; MN, ND, SD, NE, IA, KS, MO; (circumboreal, s in N. Amer. to NJ, OH, IL, IN, KS). *Naumburgia thyrsiflora* (L.) Reich.—Rydberg.

7. PRIMULA L., Primrose

1. ***Primula incana*** M. E. Jones., primrose. Perennail, scapose herb, farinose, erect scapes 6–40 cm tall. Leaves oblanceolate to spatulate, 1.5–8 cm long, 0.5–2 cm wide, shallowly denticulate to subentire, farinose below. Flowers 3–12, bracts several, linear-lanceolate, 6–10 mm long, flat, gibbous at base; calyx 6–8 mm long, farinose to some degree, nearly equal to slightly shorter than the corolla tube, lobes oblong, obtuse or somewhat acute; corolla salverform, lilac, 8–11 mm long, lobes 2–3 mm long, obcordate, deeply bifid; stamens attached in the upper 1/3 of the corolla tube, included, filaments very short; style usually included, ovary superior. Capsule valvate, about equaling or slightly exceeding the calyx. (2n = 72) Jun. Meadows, bogs, damp places, sometimes alkaline soils; ND: Burke; MT: Daniels; (n Can., Rocky Mts., w CO, MT, w ND, UT).

8. SAMOLUS L., Water Pimpernel, Brookweed

Somewhat succulent, glabrous perennial herb; stems caulescent. Leaves in basal rosettes, usually alternate on stem, entire. Flowers in terminal racemes or panicles, on wiry pedicels, panicles branched or naked; calyx campanulate, herbaceous, 5-lobed, persistent, its broadly triangular lobes shorter than tube; corolla white or pink, tube short campanulate, the 5 rounded lobes imbricate, bearing a minute scalelike staminodia in each sinus; stamens 5, included, about equaling the corolla tube, adnate near the base, sometimes alternating with the staminodia in the sinuses, anther oval, cordate, erect, filament short; ovary 1/2–2/3 inferior, 1-celled, stigma capitellate, style very short or obsolete, ovules numerous. Capsule globose or ovoid, dehiscent by 5 valves as far as the base of the calyx lobes; seeds numerous.

Reference: Henrickson, J. 1983. A revision of *Samolus ebracteatus* sensu lato (Primulaceae). Southw. Naturalist. 28: 303–314.

1 Pedicels bractless; racemes long peduncled; corolla lobes usually much shorter than tube; the staminodia wanting .... 1. *S. ebracteatus* var. *cuneatus*
1 Pedicels with small bracts; racemes sessile or nearly so; corolla lobes longer than tube; staminodia in the sinuses between the lobes .... 2. *S. parviflorus*

1. ***Samolus ebracteatus*** H.B.K. var. ***cuneatus*** (Small) Henrickson., water pimpernel. Deep green perennial fleshy herb, to 6 cm tall; stem solitary or tufted, simple to usually sparingly branched, ascending or reclining. Leaves opposite or mainly so, obovate to oblanceolate or broadly spatulate, to 15 cm long, and 6 cm wide, mostly glaucous, truncate or coarsely mucronate at the apex, bases decurrent as broad wings. Flowers in racemes, 1–3 cm long, stoutish straight peduncles longer than stem, together with the racemes more or less glandular-pilose, pedicels slender, stipitate-glandular, spreading or ascending, 1–3 cm long; calyx campanulate, lobes deltate to oblong-ovate, stipitate glands at base, acute, longer than or about as long as tube, to shorter than tube at maturity, and often purple-tinged;

corolla white, 4–6 mm wide, corolla tube (1.5)1.8–2.7(3.5) mm long, 1.4–2.1 mm wide at base, corolla lobes obovate-cuneate to obovate, flattish or truncate at the apex, toothed as long as the tube; stamens included, anthers 0.5–0.9 mm long, filaments 0.3–0.5 mm long; styles 0.5–1.5(2) mm long. Capsules depressed globose, 3–4 mm in diam; seeds many, 0.4 mm in diam. May–Oct. Marsh & seepage areas, moist soil along streams & rivers, sometimes wet limestone; KS: Barber; OK, TX: (GP, also NM). *S. cuneatus* Small — Correll & Johnston.

2. ***Samolus parviflorus*** Raf., water pimpernel. Light green, glabrous perennial herb; stem to 6 dm tall, simple or diffusely branched in the upper 1/2 and usually also from the base. Leaves alternate and clustered at its base, spatulate ovate to oblanceolate, sessile or narrowed into a winged petiole, upper stem leaves subsessile and ovate, to 15 cm long and 4 cm wide, rounded to obtuse at apex. Flowers white, racemes sessile or nearly so, 3–15 cm long, slender rachis, straight or flexuous; pedicels filiform, widely spreading or ascending, the lower usually 10–15 mm long; calyx tube about 1.5 mm long, about as long as, or shorter than the lobes, calyx lobes ovate to triangular ovate; corolla white, 2–3 mm wide, lobes oblong rounded or emarginate at apex and longer than the tube; staminodia 5, at the sinuses of the corolla. Capsule 2–3 mm in diam, 5-valved at the apex; seeds many. (2n = 26, n = 13) May–Oct. Moist soils, along streams, marshes, seepage areas, lakes, wet rocks, sandy soil; KS, OK, MO, TX; (FL to CA, n to e Can., MI, IL, KS, & B.C., also Mex., trop. Amer.). *S. floribundus* H.B.K. — Rydberg.

66. HYDRANGEACEAE Dum., the Hydrangea Family

by T. Van Bruggen

1. HYDRANGEA L.

1. ***Hydrangea arborescens*** L., wild hydrangea. Shrubs 1–3 m tall, stems sometimes straggly, young branches lightly pubescent, older branches with pale, exfoliating bark. Leaves opposite, ovate to suborbicular, with a broadly rounded to cordate base, the apex abruptly acuminate, margins with 3 or 4 serrations per cm, blade 10–18 cm long, nearly as wide, upper surface green, glabrous, lower surface lighter green, mostly glabrous except for strigillose veins; petioles 3–10 cm long, exstipulate. Inflorescence of compound cymes, terminal, flattened to convex, 5–15 cm across, all flowers small and perfect except for several (less than 10 percent) at the margin enlarged and sterile. Marginal sterile flowers with (3)4(5) radially spreading sepals, 3–8 mm long, white or cream colored, membranaceous, broadly ovate, persisting; perfect flowers 1–3 mm wide at anthesis; calyx fused below and to the ovary, the tube hemispherical, 0.5–2.0 mm long, 5 lobes, 0.2–0.5 mm long, persistent; petals 5, white to cream colored, 1–2 mm long, valvate in bud, soon deciduous; stamens 8–10, exserted, spreading, 1–2.5 mm long; pistil of usually 2 carpels, occasionally 3 or 4, ovary inferior, styles short, diverging. Fruit capsular, about 2 mm long and fully as wide, strongly 8- to 12-ribbed vertically with a connecting horizontal rib at the top, 2-loculed below, opening by a pore at the top between the 2 styles; seeds many, brown, longitudinally ribbed, ovoid, 0.2–0.5 mm long. Jun–Jul. Woods & rocky, wooded slopes; OK: Ottawa; KS: Cherokee; (NY to KS & OK, s to GA).

Reference: McClintock, E. 1957. A monograph of the genus *Hydrangea*. Proc. Calif. Acad. Sci. 29: 147–256.

67. GROSSULARIACEAE DC., the Currant Family

by T. Van Bruggen

1. RIBES L., Currant, Gooseberry

Perennial shrubs, the main stems 0.4–2.0 m tall, erect to spreading or reclining, branches glabrous to glandular-pubescent, unarmed or with 1–several nodal spines and few to many sharp, internodal bristles. Leaves alternate, often fascicled on short, lateral branches, palmately veined and palmately 3- to 5(7)-lobed and toothed, generally rotund to broadly angular in outline, glabrous to densely pubescent or glandular-pubescent, petioles shorter to longer than the blades, stipules lacking or adnate to the petioles. Flowers about 1 cm long, solitary, in small clusters, or in bracteate racemes on short, axillary branches, perfect, epigynous. Calyx of 5 erect, spreading or sharply reflexed lobes, hypanthium tubular to saucer-shaped, variously colored; petals 5, shorter than the sepals, erect to spreading, not sharply clawed, inserted at or near the top of the hypanthium; stamens 5, inserted on the hypanthium, alternate with the petals, included to long exserted from the calyx lobes, the filaments flattened to terete, anthers oval; carpels 2, the ovary mostly inferior, 1-loculed, with 2 parietal placentae, styles 2, distinct from below the middle to connate almost to the capitate stigmas. Fruit a berry, 0.5–1.5 cm long, globose or subglobose, many-seeded, yellow or red to blue or black, often glaucous, glabrous to pubescent or glandular or spiny, usually with the persistent floral remains at the apex; seeds angular, red or lavender, 2–3 mm long.

Reference: Berger, Alwin. 1924. A taxonomic review of currants and gooseberries. New York Agric. Exp. Sta. Bull. 109: 1–118.

1 Pedicels jointed below the ovary; plants without spines or prickles, except for *R. lacustre;* inflorescence racemose, 3- to 6(8)-flowered.
  2 Racemes, pedicels, or fruits with purple glandular hairs; free part of the hypanthium shallow, saucer-shaped, purple or purple-tinged.
    3 Stems spiny, upright or spreading; leaves 3- to 5-lobed, deeply cleft ..... 5. *R. lacustre*
    3 Stems without spines, straggly or decumbent; leaves 6–10 cm wide, principally 3-lobed, not deeply cleft .... 10. *R. triste*
  2 Racemes, pedicels, and fruits without purple glandular hairs; free part of the hypanthium campanulate to cylindrical.
    4 Lower leaf surfaces with yellow, resinous dots; calyx campanulate; flowers greenish-white to cream colored .... 1. *R. americanum*
    4 Lower leaf surfaces not resinous-dotted; calyx cylindrical.
      5 Flowers bright yellow, glabrous; leaves principally 3-lobed, with two prominent indentations .... 7. *R. odoratum*
      5 Flowers white to tinged with pink or red, glandular pubescent; leaves round-reniform, crenately lobed .... 2. *R. cereum*
1 Pedicels not jointed below the ovary; older stems with spines or prickles (younger branches often may be unarmed); inflorescence in corymbiform clusters of 2 or 3 flowers or few-flowered racemes.
  6 Ovary and fruit with prickles or bristles; calyx lobes shorter than the hypanthium .... 3. *R. cynosbati*
  6 Ovary and fruit glabrous or sparsely pubescent; calyx lobes equaling or longer than the hypanthium.
    7 Stamens at full anthesis long exserted, at least 3× longer than the petals; peduncle longer than the pedicles, pendulous .... 6. *R. missouriense*
    7 Stamens at full anthesis included to slightly exserted, not more than twice the length of the petals; peduncle not pendulous.
      8 Flowers 5–8 mm long; hypanthium obconic to narrowly campanulate, spreading upwards.

9   Stamens almost twice the length of the petals, as long as or slightly exceeding the sepals; leaf bracts ciliate-hairy. .... 4. *R. hirtellum*
9   Stamens about as long as the petals but shorter than the sepals; leaf bracts often glandular-ciliate .... 8. *R. oxyacanthoides*
8   Flowers 10–12 mm long; hypanthium narrowly cylindric; stamens as long as the petals .... 9. *R. setosum*

1. **Ribes americanum** P. Mill., wild black currant. Shrub 1–1.5 m tall, branches unarmed, erect to spreading, the younger ones pubescent and dotted with sessile, yellow, crystalline glands, older ones becoming smooth, gray or blackened. Leaves simple, suborbicular in outline, the blades 3–8 cm long and fully as wide, broadly truncate to shallowly cordate, 3-lobed almost ½ the length, or with 3 principal lobes and 2 much smaller basal lobes, the margin coarsely serrate to crenate, glabrate above, pubescent and glandular-dotted below, petioles slender, shorter to longer than the blades. Inflorescence racemose, drooping, 6- to 15-flowered, pubescent, as long as the leaves, the pedicels jointed, 1–2 mm long, their bracts linear, 6–10 mm long, pubescent. Calyx greenish-white above, becoming tawny brown below, 8–10 mm long, fused below, the hypanthium campanulate, 3–4.5 mm long, pubescent, sepal lobes 5, erect to spreading or reflexed, 4–5 mm long, spatulate to oblong, pubescent on the outer surface; petals 5, 2–3 mm long, creamy white, inserted at the top of the hypanthium, oblong, blunt at the apex; stamens 5, inserted at the level of the petals on the hypanthium opposite the sepals, subequal to the petals, the filaments with a broad base, narrowed above, anthers oval, 1 mm long; pistil 2-carpeled, smooth, fused to the hypanthium on the lower part, the styles connate nearly to the stigma, equaling the stamens or slightly exceeding them. Fruit an ovoid berry 6–10 mm long, black, crowned with the persistent floral remnants, not very palatable; seeds several, smooth; May–Jun. Mesic thickets & edges of woods, moist ravines, & stream banks. n GP, s to NE; (ne U.S. to Alta., s to GP, NM).

2. **Ribes cereum** Dougl., western red currant. Shrub 0.5–1.5(2) m tall, branches numerous, unarmed, scraggling or spreading, the younger ones densely pubescent and with short, stipitate glands; older branches becoming smooth, the bark grayish brown. Leaves simple, reniform to rounded in outline, 1–2.5(3) cm wide, broadly truncate to cordate at the base, shallowly 3- to 5-lobed, almost equally crenate-dentate along the entire margin, glabrate to densely pubescent or stipitate glandular on both surfaces, petioles almost as long as the blades, vestiture as on the blades. Inflorescence of short, 2- to 8-flowered racemes terminating young branches, pendulous, densely pubescent to stipitate glandular, the pedicels jointed immediately below the ovary, bracts ovate to obovate in ours, longer than the pedicels. Calyx narrowly cylindrical, white to more often tinged with pink, pubescent and stipitate glandular, the hypanthium 6–8(10) mm long, lobes 5, ovate, spreading or reflexed, 1.5–3 mm long; petals 5, 1–2 mm long, about ½ as long as the sepals, mostly erect, white to tinged with pink; stamens 5, inserted at the top of the hypanthium opposite the sepals, equaling to shorter than the petals; filaments slender, anthers 0.5–1 mm long, tipped with a small, cup-shaped gland; pistil 2-carpeled, mostly inferior, fused to the hypanthium for most of its length, the style free, tapered, connate almost to the stigmas, slightly exserted from the hypanthium tube, glabrate to stipitate glandular. Fruit a spherical to ovoid berry 6–8 mm long, smooth to glandular, dull to bright red, crowned with the persistent floral remnants, bitter; seeds few, angular, about 1 mm long. (n = 8) Jun–Jul. Dry to moist rocky or sandy soil in woods or openings in woods; w GP; (N.B. to ID, MT to NM). *R. inebrians* Lindl. — Rydberg.

C. L. Hitchcock et al., *in* Vascular plants of the Pacific Northwest, Part 3. Univ. Wash. Press, 1961, recognize 3 varieties. Our plants can be referred to var. *inebrians* (Lindl.) C. Hitchc., having leaves mostly 2 cm wide and pubescent, calyx pubescent and glandular and pedicel bracts ovate and pointed.

3. **Ribes cynosbati** L., dogberry. Shrub 0.5–1.5 m tall, stems branched, arching, usually with 1–3 subulate spines 0.5–1 m long at the nodes, internodes glabrous to sometimes with dense prickles; bark gray, exfoliating on older branches, becoming dark brown or black. Leaves simple, orbicular in outline, 2–5 cm long and 3–6 cm wide, deeply 3-lobed with 2 smaller basal lobes, the lobes crenate-dentate, the base truncate to cordate, pubescent on both surfaces, petioles slender, shorter than to much exceeding the blades, densely pubescent and stipitate-glandular. Flowers solitary or 2 or 3 in clusters laterally on the stem, the pedicels slender, 6–15 mm long, not jointed, peduncles exceeding the pedicels, pubescent and glandular, Calyx campanulate, the hypanthium 3–4 mm long, greenish-white or cream colored, sepals 5, 1.5–2.5 mm long. broadly oblong, soon reflexed; petals 5, 1–2 mm long, white, erect, obovate; stamens 5, inserted opposite the sepals at the top of the hypanthium, 1–2 mm long, equaling to shorter than the petals, included, filaments slender, anthers 0.5–1 mm long; pistil 2-carpeled, the ovary distinctly inferior, subulate spiny; style equaling the sepals, tapered, hairy below, connate to the stigma. Fruit a spherical to ovoid berry 0.7–2 mm long, covered with subulate spines up to 2 cm long, red, the floral remnants persisting; seeds many, 1–2 mm long, angular. (n = 8) May–Jun. Rocky to loamy rich wooded hillsides; e ND, e SD, MN, IA, MO; (N.B. to Man. s to MO, GA). *Grossularia cynosbati* (L.) P. Mill. — Rydberg.

4. **Ribes hirtellum** Michx. Shrub 0.5–1.5 m tall, branches erect or ascending with 1–3 nodal spines 5–10 mm long or often wanting, internodal prickls few when present, bark gray, exfoliating, the older branches dark brown. Leaves simple, usually clustered on short, lateral shoots, the leaf bracts narrowly ovate, tapered to acute tips, with ciliate to villous margins, the blades suborbicular in outline, 1.5–3.5 cm long, 2–4 cm wide, cuneate to truncate, deeply 3-lobed to ½ their length, with 2 smaller basal lobes, margins crenately lobed, glabrous to softly hairy on both surfaces, petioles slender, densely pubescent, as long as to exceeding the blades by ½. Flowers solitary or in small corymbose clusters along the stem, peduncles shorter than the pedicels, the pedicels slender, 5–8 mm long, not jointed below the flower, densely pubescent. Calyx obconic to narrowly campanulate, flared above, the tubular portion of the hypanthium and sepals 5–8 mm long, dull purple to yellowish, lightly villous to glabrous; sepals 5, light green, 3–4 mm long, erect to spreading or reflexed; petals 5, 1–2 mm long, erect, obovate; stamens 5, erect, 3–5 mm long, distinctly longer than the petals, equaling or surpassing the sepals, inserted on the top of the hypanthium opposite the sepals, the filaments broad and flat at the base, narrowed above; pistil 2-carpeled, with long, villous hairs on the lower ½, ovary inferior, the 2 styles gradually tapered to the stigmas, equaling the stamens. Fruit an ovoid or spherical berry 8–10 mm long, smooth, blue to black, topped with the persisting floral remnants; seeds several, about 1 mm long, angular. May–Jun. Rocky woods, ravines, & thickets; n GP; (N.B. to Man., s to IN). *Grossularia hirtella* (Michx.) Spach — Rydberg.

5. **Ribes lacustre** (Pers.) Poir., swamp currant. Shrub 0.5–1.5 m tall, branches ascending to arched or declining, spiny on the internodes, nodal spines 3–8 mm long, in clusters of 2–5, bark light brown or gray, becoming deep brown or black. Leaves simple, pentagonal in outline, the blades 1.5–3 cm long and fully as wide, (3)5-lobed, the 3 central lobes cleft nearly ¾ the length of the blade, segments rhombic, coarsely toothed, shallowly cordate, nearly glabrous, veins and margins sparsely ciliate, petioles slender, equaling the blade, with glandular hairs. Inflorescence racemose, 3- to 7(9)-flowered, spreading or drooping, 2–6 cm long, all parts beset with purple glandular hairs or bristles; pedicels jointed just below the flower, with 1 or 2 minute leafy bractlets just below the node. Calyx flattened, the hypanthium above the ovary saucer-shaped and spreading, 1–1.5 mm long, light green and tinged with lavender; the sepals 5, obovate, spreading, 2–2.5 mm long, wider than long; petals 5, reniform and clawed, erect, 1–2 mm long, shorter than the sepals, equaling to slightly longer than the stamens; stamens 5, inserted on the hypanthium tube just above the ovary, opposite the sepal lobes, the filaments slender, anthers 0.5 mm long; pistil 2-carpeled, the ovary inferior, covered with glandular bristles where fused to the hypan-

thium, styles separate above the middle, equaling the petals. Fruit an ovoid or spherical berry 6–8 mm long, bristly-glandular, dark purple to black; seeds several, oval, about 2 mm long, smooth. May–Jun. Swampy or wet woods; BH; (Newf. to Alta., s to TN, MN, BH, & CO). *Limnobotrya lacustris* (Pers.) Rydb.—Rydberg.

6. Ribes missouriense Nutt., Missouri gooseberry. Shrub 1–2 m tall, the branches arching, armed at the nodes with 1–4 stiff, red to brown spines 8–16 mm long, internodes mostly glabrous, rarely with prickles, bark gray, exfoliating on older branches. Leaves simple, alternate, often fascicled on short, lateral branches, blades rotund in general outline, 2–6 cm long and almost as wide, 3-lobed with 2 smaller basal lobes, the 2 principal sinuses reaching almost to the middle, the lobes crenate, broadly wedged to truncate at the base, glabrate above, softly pubescent beneath, petioles shorter than the blades, densely pubescent with branched hairs. Flowers green to white, solitary or in clusters of 2–4, the peduncles as long as or exceeding the pedicels, the latter with ciliate leafy bracts, glabrous, not jointed below the ovary. Calyx cylindric, the hypanthium tube 2–2.5 mm long, sepals 5, their lobes 4–6 mm long or 2–3× longer than the tube, linear-oblong, erect early but later sharply reflexed; petals white or cream, erect, obovate to oblong, erose, 2–3.5 mm long or usually less than ½ the length of the sepals; stamens 5, inserted near the top of the hypanthium opposite the sepals, the filaments long-exserted, spreading, 8–12 mm long, at least 3× longer than the petals, anthers oval, about 1 mm long; pistil 2-carpeled, the ovary distinctly inferior, styles equaling the stamens, fused below from ½ to their entire length, pubescent on the lower ½. Fruit a spherical berry about 1 cm in diam, purple to brown at maturity, smooth, crowned with the floral remnants; seeds numerous, ovoid, 2–2.5 mm long, smooth. Apr–May. Dry to moist open woods, thickets, & fence rows; e GP; (CT to SD, s to KS & TN). *Grossularia missouriensis* (Nutt.) Cov. & Britt.—Rydberg.

7. Ribes odoratum Wendl., buffalo currant. Shrub 1–2 m tall, unarmed, the branches erect to arching, younger ones light brown, older ones becoming dark brown. Leaves simple, alternate on branches of the year, fascicled on side branches of older wood, rotund in outline, the blades 2–5 cm long, fully as wide, glabrous, 3-lobed, the 2 sinues to below the middle, often with 2 smaller basal lobes, entire to coarsely dentate, cuneate to truncate at the base, petioles equaling the blades, glabrous on both surfaces, exstipulate. Inflorescence racemose, (2) 3- to 8-flowered, the peduncles 3–6 mm long, pubescent, pedicels pubescent, shorter than their green, leafy bracts, jointed below the ovary. Calyx bright yellow, the tube 10–14 mm long, slightly oblique, tubular, the 5 lobes 3–4.5 mm long, obovate, spreading or reflexed; petals 5, 2–3.5 mm long, yellow to bright red, erect, erose; stamens 5, inserted on the top of the hypanthium opposite the seals, erect, equaling the petals, filaments 1–2 mm long, anthers oval, about 1 mm long; pistil 2-carpeled, the ovary inferior, smooth, styles connate to the stigma, smooth, as long as the petals and stamens. Fruit a globose berry 7–9 mm in diam, at first yellowish but turning black; seeds several, 2–3 mm long, ovoid with a longitudinal rib, smooth. Apr–May. Dry, open wooded hillsides, edges of thickets, & stream banks; GP; (MN to CO, s to AR & TX). *Chrysobotrya odorata* (Wendl.) Rydb.—Rydberg.

8. Ribes oxyacanthoides L. Shrub 0.5–1.5 m tall, irregularly branched, stems ascending to sprawling, when young light yellow or brown with dense prickles on the internodes, the nodes with 1–4 stout, pale brown spines 5–14 mm long, older branches becoming blackened and unarmed. Leaves alternate, often fascicled at the ends of lateral shoots, the blades broadly ovate to subrotund in outline, 1–3 cm long and about as wide, 3-lobed with 2 smaller basal lobes, crenate-dentate with irregular teeth, subtruncate to cordate at the base, upper surface dark green, mostly pubescent, lower surface pubescent or glandular-pubescent, petioles much shorter to equaling the blades, glandular-pubescent; leaf bracts often glandular-ciliate. Inflorescence of 1–3(4) flowers in corymbose clusters, rarely racemelike, the peduncles shorter than the pedicels, glandular-ciliate, pedicels bracted, not jointed below the ovary. Calyx green to white, glabrous, the hypanthium tube obconic to campanulate, 2-3 mm long, sepals 5, flared or spreading, 3-4.5 mm long, narrowly oblong to ovate; petals 5, obovate, erose, 2-2.5 mm long, mostly erect and distinctly shorter than the sepals; stamens 5, inserted at the top of the hypanthium opposite the sepals, 2-3 mm long, mostly equaling the petals, filaments stout, glabrous, anthers about 1 mm long; pistil 2-carpeled, the ovary inferior, styles tapered, connate for about ½ their length, pilose below the middle, as long as or slightly shorter than the sepals. Fruit a spherical berry 10-12 mm in diam, smooth, deep blue or purple; seeds several to many, angular, 1-2 mm long. (n = 8) May-Jun. Dry to moist woods, thickets, & rocky places; MN to MT, s to NE; (Newf. to B.C., s to CO & NE). *Grossularia oxyacanthoides* (L.) P. Mill.—Rydberg.

9. ***Ribes setosum*** Lindl., bristly gooseberry. Shrub 0.5-2 m tall, the branches slender and spreading, the internodes from sparsely to densely covered with prickles, nodes with 1-4(5) stout spines 0.5-1.5 cm long, bark light brown when young, exfoliating, the older branches dark brown. Leaves simple, the blades orbicular in outline, 1-4 cm long and wide, (3)5-lobed ½ their length, the lobes with 4-7 crenate-dentations, glabrate to pubescent on the upper surface, petioles slender, from shorter to much longer than the blades; flowers solitary or in clusters of 2 or 3, peduncles 0.5-1 cm long, shorter than the petioles, densely pubescent, pedicels 1-2 cm long, glandular-pubescent, bracteate, not jointed below the ovary. Calyx greenish-white, glabrous, the hypanthium narrowly cylindric, 4-5 mm long, sepals 5, erect to spreading, oblong to obovate, 2-3.5 mm long, erect, obovate, erose, ½ to ⅔ as long as the sepals; stamens 5, inserted on the hypanthium tube opposite the sepals at the level of the petals and equal to them, the filaments slender, anthers 1-1.2 mm long; pistil 2-carpeled, the ovary inferior, smooth or infrequently villous, styles shorter to longer than the sepals, pilose on the lower ½, connate to midlength or above, sometimes to the stigmas. Fruit a red to black globose berry 8-12 mm in diam, mostly glabrous, crowned with the floral remnants, palatable; seeds numerous, angular. May-June. Rocky hillsides, edges of woods, & thickets in ravines; nw GP, BH; (MI to Ont. & Alta., s to CO). *Grossularia setosa* (Lindl.) Cov. & Britt.—Rydberg.

10. ***Ribes triste*** Pall., swamp currant. Shrub 0.4-0.8(1) m tall, the branches weak, declining or decumbent, the spreading branches rooting at the lower nodes, unarmed, young growth pubescent and with stipitate glands, yellowish-green, older branches red-brown, the bark exfoliating. Leaves alternate, the blades thin, 6-10 cm wide and almost as long, broadly cordate, with 3 prominent shallow lobes and 2 smaller basal lobes, the segments broadly triangular with coarse dentate-serrations, glabrate above, softly pubescent below, especially along the veins, petioles as long as the blades, flattened and winged at the base, the margins long-ciliate. Inflorescence a 6- to 12-flowered drooping raceme 4-9 cm long, produced at the base of new growth of the year, the rachis with glandular hairs, pedicels 1-4 mm long, jointed immediately below the ovary with purple, cordate bracts at the juncture with the rachis. Calyx purple or purple-tinged, glabrous, sepals 5, 1-2 mm long, obtuse, as wide as long, spreading, the hypanthium tube shallow, saucer-shaped, fused to the ovary its entire length, with a prominent 5-lobed disk covering the top of the ovary; petals 5, erect, red to purple, about 1 mm long, cuneate; stamens 5, inserted just outside the disk opposite the sepals, the filaments not longer than the petals, anthers retuse, wider than long; pistil 2-carpeled, ovary inferior, glabrous, styles scarcely 1 mm long, separate ⅓ to ½ their length, glabrous. Fruit red, ovoid, smooth, 8-10 mm long, many-seeded, sour. May-Jun. Bogs, wet or marshy soils in woods; ND, MN; (Lab. to AK, s to MI, WI, OR).

68. CRASSULACEAE DC., the Stonecrop Family

by T. Van Bruggen

Annual or perennial herbs, some succulent. Leaves alternate, toothed or entire, without stipules. Flowers in terminal cymes with secund, spreading branches; calyx persistent; sepals

5(4–6), only slightly connate at the base; petals same number as sepals, inconspicuous or absent; stamens twice the number of sepals; pistils 4 or 5, free for most of their length or united to the middle. Fruit a follicle or several united below to form a capsule.

1 Petals inconspicuous or absent; stems and leaves not succulent; pistils united to the middle; fruit a capsule that dehisces in a circumscissile fashion .... 1. *Penthorum*
1 Petals 5; stems and leaves succulent; pistils separate or only slightly connivent at their bases, becoming follicles in fruit .... 2. *Sedum*

1. PENTHORUM L., Ditch Stonecrop

1. *Penthorum sedoides* L. Erect to ascending perennial herb, 2–6(8) dm tall, glabrous-shiny below but stipitate-glandular on the rachises of the inflorescence. Stem simple or sparingly branched above from branched, stoloniferous roots. Leaves petiolate, the blades lanceolate to narrowly elliptical, acuminate at base and apex, sharply serrate, 5–10(13) cm long and 2–4 cm wide. Inflorescence a 2- to 6-branched cyme, the greenish flowers secund on the upper sides of the branches. Sepals 5, rarely 6 or 7, green, erect to slightly spreading; petals absent or rarely present and inconspicuous; pistils 5, rarely 7, 3–4 mm long, united from the middle downward to form a ringlike, 5-angled capsule, stamens 10. Fruit a 5-horned, 5-loculed capsule circumscissile below the diverging beaks. Seeds many, pink, ellipsoid, 0.7 mm long. Jul–Sep. Ditches, marshes, stream banks, & lakeshores; e NE, e SD, KS, OK; (N.B. to MN, s to FL & TX).

2. SEDUM L., Stonecrop

Annual or perennial, the stems ascending from decumbent bases or erect, simple to bushy-branched. Leaves alternate, simple, mostly entire, tending to be succulent, the lower ones deciduous at the time of anthesis. Flowers in tight to open terminal, paniculate cymes; sepals 4 or 5, rarely 6; petals the same number as sepals, narrow; stamens 8–10, twice the number of sepals; pistils 4 or 5, rarely 6, distinct or slightly united at their very bases. Fruit a dry follicle with tapered beak, 4 or 5(6) arranged in a circle with erect to diverging tips, many-seeded.

1 Plants perennial, with basal offshoots and short, sterile rosettes of telescoped succulent leaves; petals yellow; follicles erect in fruit, only the beaks spreading .... 1. *S. lanceolatum*
1 Plants annual, lacking offshoots or sterile rosettes; follicles in fruit divergent or ascending.
  2 Petals yellow; follicles widely divergent in fruit; leaves usually not exceeding 9 mm .... 2. *S. nuttallianum*
  2 Petals off-white to pink or lavender; follicles ascending in fruit; leaves commonly exceeding 9 mm .... 3. *S. pulchellum*

1. *Sedum lanceolatum* Torr. Perennial, stems glabrous, 6–15(18) cm tall, tufted from slender rootstocks that have sterile shoots with basal rosettes of crowded, succulent leaves. Cauline leaves alternate, sessile, linear and terete, minutely papillate at their distal ends, 6–15 mm long. Flowers in compact, several branched cymes; mostly pentamerous; sepals lance-acuminate, persistent in fruit, 2–5 mm long; petals yellow to tinged with pink, lanceolate with acuminate tips, separate, 7–10 mm long, mostly exceeding stamens by 1–2 mm; pistils 5, basally connate, the follicles erect in fruit, becoming striate and tawny brown, the beaks 1.2 mm long, divergent. Seeds 0.7 mm long, 5–25 produced in each follicle. Jun–Aug. Open, exposed, & rocky places or in thin soils overlaying rock; w SD, w NE, e CO, e WY; (SD & NE to AK, s to CA & CO). *S. stenopetalum* Pursh — Rydberg.

2. *Sedum nuttallianum* Raf. Annual, with thin, glabrous 1–several stems 4–13 cm tall that fork at or above the middle into several leafy branches, each bearing 6–15 flowers; roots slender and superficial. Leaves alternate, succulent-fleshy, subterete with obtuse apices, 6–19(23) mm long. Flowers remote on the inflorescence branches, alternating with reduced

leaves. Sepals fleshy, ovate with obtuse tips, 2-3 mm long; petals yellow, slightly exceeding the sepals, lanceolate with acute tips; pistils 4 or 5, 2-3 mm long, their beaks 0.3 mm long. Follicles becoming widely divergent in fruit. May-Jul. Dry, rocky soils of limestone or sandstone origin in open, exposed areas or in open woods; e KS, w MO, OK, TX; (MO & AR to KS, s to TX).

3. *Sedum pulchellum* Michx. Annual herb with glabrous stems single or several, ascending, 1-3 dm tall. Lower leaves cuneate, soon deciduous, the upper leaves 0.7-2.5(3.5) cm long, crowded and numerous, linear terete. Inflorescence 2- to 7-forked, widely diverging with closely flowered, secund to recurving branches 2-6 cm long. Flowers white to pink or lavender; sepals lanceolate, 2-3 mm long; petals 4-6 mm long, linear to lanceolate with acuminate tips; pistils 4 or 5. Follicles 4-7 mm long, spreading-ascending at maturity, with prominent tapered beaks. May-Jul. Rocky or sandy soils of open places or open woods; KS, MO OK, TX; (VA to KS, s to GA, TX).

69. SAXIFRAGACEAE Juss., the Saxifrage Family

by T. Van Bruggen

Perennial herbs, sometimes succulent, stems sometimes scapelike. Leaves simple, often lobed, alternate, sometimes all basal, often palmate-lobed and veined, stipules absent or when present, adnate to the petiole. Inflorescence basically cymose but sometimes racemose, or flowers solitary. Flowers perfect, regular or somewhat irregular, with a well-developed hypanthium, free or often adnate to the ovary; calyx lobes (3)4 or 5(6); petals 4 or 5(6) or rarely absent; stamens as many to twice as many as sepals, often inserted on the rim of the hypanthium; pistil 1, compound, usually 2-carpeled or carpels 3 or 4, separate or connate towards the base to form a more or less deeply lobed ovary, ovary superior to more commonly ½ inferior or inferior, often tapered above into beaklike styles, each carpel usually with many ovules; placentation axile or parietal. Fruit dry, capsular, dehiscent or, if the carpels are distinct, follicles, many-seeded.

1 Staminodia present; flowers solitary on scapelike peduncles; leaves entire .... 4. *Parnassia*
1 Staminodia absent; flowers in cymes, panicles or racemes; leaves variously toothed.
  2 Stamens 5; hypanthium oblique .... 1. *Heuchera*
  2 Stamens 10; hypanthium tubular or flattened, not oblique.
    3 Petals digitately or pectinately cleft; placentae parietal.
      4 Stem essentially naked except for a reduced leaf at the base; petals pinnately cleft .... 3. *Mitella*
      4 Stem leafy; petals digitately 3- or 5-cleft .... 2. *Lithophragma*
    3 Petals entire; placentae axial.
      5 Styles distinct; petals clawless; plants with cormlike tubers or with slender rootstocks .... 5. *Saxifraga*
      5 Styles partially connate; petals clawed; plants from thick rootstocks ... 6. *Telesonix*

1. HEUCHERA L., Alumroot

Perennial herbs, stems erect, 1-6(7) dm tall, scapose, usually solitary, rarely with small bracts, from branched crowns on thick, scaly rootstocks. Leaves basal, densely clustered from the crown, the blades cordate-rounded, palmately veined, with 5-9 crenate to truncate lobes, the stipule united to the petiole at its base, petiole 1-6 × the blade length. Inflorescence at first spikelike, congested, becoming a loose to open raceme or a narrow panicle, the main axis and branches with glandular hairs. Sepals 5(6), green to yellowish, fused most of their length to a flared or campanulate, usually oblique hypanthial tube, the lobes unequal, petaloid; petals 5(6), green to yellow, rarely absent, spatulate to lanceolate, clawed, entire; stamens 5(6), opposite the sepal lobes, inserted at the level of the petals near the top of the hypanthium; ovary 2-carpeled, ½ inferior, fused below to slightly above the line of adnation to the hypanthium, the 2 upper portions free, beaked, 1-celled with 2 parietal placentae. Fruit a capsule, dehiscing between the carpel beaks; seeds numerous, spiny in ours.

Reference: Rosendahl, Carl Otto, Frederick K. Butters, & Olga Lakela. 1936. A monograph on the genus *Heuchera*. Minnesota Stud. Pl., Sci. 2: 1–180.

1 Calyx 4–6 mm long, moderately oblique; stamens strongly exserted from the calyx .... 1. *H. hirsuticaulis*
1 Calyx 6–10 mm long, decidedly oblique; stamens usually enclosed or slightly exserted from the calyx .... 2. *H. richardsonii*

1. *Heuchera hirsuticaulis* (Wheelock) Rydb. Stems 2–6(7) dm tall, robust, from a branched caudex, especially white hispid with hairs to 3 mm; cauline bracts, when present, erect or spreading, green, divided into 4–6 tapered segments 3–8 mm long. Leaves basal, 4–many, the blades cordate-rounded with 5–9 crenate lobes, 5–9 cm long and 4–7 cm wide, green and becoming glabrate on the upper surface, lighter and more hispid-hairy on the lower surface, petioles as long as to 8× the blade length, hairs as on the stem. Inflorescence racemose to narrowly paniculate, 1–2.5 dm long at anthesis, panicle branches and pedicels densely covered with glandular hairs. Calyx green to rose-purple, glandular-puberulent in early anthesis, campanulate and only slightly oblique, 4–6 mm long, the 5 lobes 2–3 mm long, oblong, erect; petals 5, alternate to the sepal lobes and nearly equaling them, green to purple, obovate to spatulate; stamens 5, opposite the sepals, at anthesis exserted 2–4 mm beyond the sepal lobes; ovary ⅓–½ inferior, the 2 upper free portions tapered to beaklike styles 1–2.5 mm long; free portion of the hypanthium shorter than to longer than the lower adnate portion. Seeds numerous, ovoid, 0.6 mm long, echinate. May–Jul. Dry, rocky woodlands, prairie openings in woods, & on rocky ledges & cliffs; e GP; NE, KS, IA, MO, OK; (MI to NE, OK, s to TN & AR). *H. americana* L., in part, *H. hispida* Pursh in part, Rydberg.

Gleason (1952) believes this species to be a hybrid swarm derived from *H. americana* L., which is widespread in the Mississippi valley, and *H. richardsonii* R. Br., widely distributed in the northern plains and south to MO. The characters intergrade between these two entities to such a degree that it is difficult to separate them with certainty. Others, such as Steyermark (1963), prefer to follow the treatment of Rosendahl, Butters, & Lakela, op. cit., referring this taxon to var. *hirsuticaulis* under *H. americana* L.

2. *Heuchera richardsonii* R. Br. Stems 1.5–7.0 dm tall, rarely more than one from a stout, branched caudex, usually softly hispid below, glabrate to glandular-puberulent above and into the inflorescence. Leaves basal, the blades 3–8 cm wide, cordate-rounded to reniform, with 5–7(9) prominent crenate to dentate lobes, green-glabrate or glabrous above, lighter in color and hirsute-hispid below; petioles variable in length, up to 6× or more longer than the width of the blade, hispid to nearly glabrous. Inflorescence racemose or narrowly paniculate, 0.5–2 dm long at anthesis, all parts glandular-puberulent, the cymules ascending, 2- to 6-flowered. Calyx green, the lobes petaloid, purple or yellowish, 6–10 mm long; calyx tube cylindrical to campanulate, obliquely irregular, adnate to the ovary, the hypanthium gibbous, 4–5 mm long on the longer side, to less than ½ as long on the other, the 5 lobes 2–4 mm long, erect, oblong to spatulate, slightly clawed, nearly equaling the calyx lobes or longer; stamens 5, opposite the sepals, included to slightly exserted in ours; ovary 2-carpeled, ⅓–½ inferior, the 2 upper portions free, tapered to styles 1.5–2.5 mm long, included or exserted, diverging at maturity. Seeds numerous, 0.5–0.7 mm long, brown to black, ovoid to ellipsoidal, densely echinate. Jun–Jul. Prairies, hillsides, rocky woods, & openings in woods; n GP, s to NE & IA; (MI to WI to Alta. & CO, s to MO & NE).

Rosendahl, Butters, & Lakela, op. cit., list several varieties of this species for the eastern part of the GP. Their key to the three varieties in our area follows:

1   Capsules included; stamens barely exserted; petal glandular but not papillose.
     2   Flowering scapes and petioles moderately hispid with short (1.5 mm or less) glandular hairs; leaves 3–6 cm wide .... 2c. var. *richardsonii*
     2   Flowering scapes and petioles densely hispid with long (2–3.5 mm) glandular hairs; leaves 4–8 cm wide .... 2b. var. *hispidior*
1   Capsules more or less exserted; stamens obviously exserted; petals both glandular and papillose; flowers 6–10 mm long; hypanthium strongly oblique .... 2a. var. *grayana*

2a. var. *grayana* Rosend., Butt. & Lak. Flowers 6–10 mm long, hypanthium strongly oblique. Fruiting capsule exserted from the hypanthium, sometimes 2–3 × the length of the sepal lobes. Jun–Jul. Rocky slopes, woods, & prairie openings in se GP; IA, NE, MO, KS.

The above authors consider this variety a hybrid form linking this species complex to *H. americana* L., ranging s and e of the GP.

2b. var. *hispidior* Rosend., Butt. & Lak. Basal leaves 4–8 cm wide. Flowering scapes and petioles densely hispid with glandular hairs 2–3.5 mm long. Petals glandular; stamens slightly exserted from the hypanthium. Capsules included in the hypanthium. Jun–Jul. Throughout most of the species range in prairies, rocky slopes, & woods; ND, SD, MN, IA, NE, MO.

Most of our specimens can be referred to this variety.

2c. var. *richardsonii*. Basal leaves 3–6 cm wide. Flowering scapes and petioles moderately hispid, the glandular hairs 1.5 mm long or less. Petals glandular but lacking papillose projections; stamens slightly exserted from the hypanthium. Capsules at maturity not extending from the persistent hypanthium. Jun–Jul. Prairies, open woods, & hillsides; ND, SD, NE.

2. LITHOPHRAGMA Nutt.

1. **Lithophragma parviflora** (Hook.) Nutt., prairie star. Herbs, stems 1–several, sparingly leafy, 1–4 dm tall, densely glandular-pubescent, almost canescent, from slender rootstocks with white, ricelike bulbs. Basal leaves 4–8, clustered, alternate, the blades 1–3 cm across, orbicular-reniform with (3)5 divisions that are biternately or ternately lobed, sometimes cleft nearly to the base, copiously pubescent on both surfaces; petioles slender, 1–5 cm long, with dilated, membranous stipuled bases; cauline leaves 1–3, mostly short petioled or subsessile, the main divisions palmate to laciniately 3(5)-lobed, smaller than the basal leaves. Inflorescence at first a congested, terminal raceme, (3)5- to 10-flowered, in fruit elongating to 7–12 cm, the pedicels ascending to erect, 2–5 mm long. Flowers white to pink, showy; calyx obconic, tapering to the pedicel, forming a tube that is adnate to the ovary most of its length, at anthesis 4–6 mm long, gradually elongating to 6–10 mm in fruit, the 5 lobes triangular-ovate, persistent, 1–2 mm long, white-canescent on their outer surfaces; petals 5, slightly unequal, digitately 3-cleft, middle segment longest, abruptly narrowed to a claw equaling the calyx lobes, 4–8 mm long; stamens 10, inserted at the edge of the ovary, short, mostly included within the hypanthium; ovary 2- to 3-carpeled, ⅔ or more inferior, 1-celled with 2 or 3 parietal placentae, styles 2 or 3, short-divergent. Fruit capsular, 2- or 3-cleft at the apex; seeds many, 0.5 mm long, brown, smooth, ovoid. May–Jun. Prairie, grassland openings, or open woods; NE: Dawes, Sioux; BH; WY: Weston, Niobrara; (SD to B.C., s to CO & CA).

The species *L. bulbifera* Rydb. has been reported from the BH by Rydberg. It is found in the Bighorns of Wyoming and the Rocky Mts. west. It differs from the above species by having petals 5-cleft, muricate bulbils in the axils of cauline leaves, basal leaves glabrous to sparsely pubescent, and muricate seeds. I have not seen a voucher from the GP, but it may occur in the BH.

3. MITELLA L.

1. **Mitella nuda** L., Bishop's cap. Herbs, 1–several stems 3–18 cm tall, essentially naked except for occasional reduced, sessile leaf near the base, pubescent-glandular on the lower portion, becoming glandular above, from rhizomes that are filiform, brown, scaly; perennation partly by creeping stolons produced from middle to late in the growing season. Leaves

several–10, alternate, the slender petioles 1.5–3.0 × the blade length, arising basally at the scape-stolon junction, blades reniform to cordate or suborbicular, 1- to 3-lobed, obscurely thin, light green above, paler to whitish beneath, sparingly hirsute above, glabrate below, the petioles densely glandular-pubescent when young, later becoming glabrate. Inflorescence a raceme 3–12 cm long, remotely 2- to 10-flowered, the pedicels 1–5 mm long. Flowers greenish-yellow, the calyx broadly saucer-shaped, fused below to ½ its length, the 5 lobes ovate with obtuse tips, 1.5 mm long; upper portion of the hypanthium almost free, translucent, flared outwardly into an irregular crenately margined disk reaching the bases of the sepal lobes; petals 5, 3–5 mm long modified into slender pectinately pinnatifid segments with 4 or 5 filiform divisions 1–2 mm long on each side; stamens 10, opposite and alternate with the sepal lobes, spreading radially from the margin of the hypanthium, shorter than the sepal lobes; ovary 2-carpeled, 1-celled with 2 parietal placentae, ⅓–½ inferior, 1–2 mm across, subspherical, the surface papillose, styles 0.5–1.0 mm long, widely divergent. Fruit capsular, dehiscing along the ventral sutures, each side opening widely to form the "miter" or bishop's cap-shaped structure. Seeds 5–20, about 1 mm long, black and shining, ellipsoidal. May–Jun. Bogs, swamps, damp woods, & along stream banks; MN, ND; (across Can, s to n U.S.).

4. PARNASSIA L.

Perennial herbs, stems scapelike, 1–3.5 dm tall, solitary to several from short rhizomes, glabrous. Leaves mostly basal, glabrous, entire, petioled, the blades elliptical to ovate-rounded. Cauline leaves, when present, reduced, bractlike, at midstem or below, sometimes clasping the stem. Flowers solitary, erect, terminal on scapelike peduncles. Sepals 5, slightly united at their bases, the calyx lobes ascending, persistent; petals 5, showy, white or cream colored with green or yellowish veins; stamens 5, inserted on the calyx, alternate with the petals, persistent; staminodia in 5 clusters opposite the petals, each with 3–15 or more slender divisions, each division ending in a glandular knoblike tip; pistil 4-carpeled, the ovary superior to partly inferior, 1-celled, with 4 parietal placentae; styles 4, very short, outcurved, the stigmas capitate. Fruit an ovoid, membranaceous capsule 0.7–1.0 cm long, apically dehiscent; seeds many, anatropous, 1 mm long.

References: Hitchcock, C. L., A. Cronquist, M. Ownbey, & J. W. Thompson. 1961. Vascular plants of the Pacific Northwest. Part 3. Univ. Wash. Press, Seattle. pp. 27–31; Scoggan, H. J., 1978. The Flora of Canada. Part 3. National Mus. Natural Sci., Ottawa. pp. 870–871.

1 Staminodia with 3 segments, their bases not dilated; petals 3 or more × as long as the sepals; leaves coriaceous; scape naked or with a small, bractlike leaf near the base .... 1. *P. glauca*
1 Staminodia with 5–13(15) segments, their bases dilated; petals as long as to 1.5 × longer than the sepals; leaves membranaceous; scape with a bractlike leaf near or below the middle.
  2 Basal and cauline leaves cordate or rounded at their bases; the cauline leaf often clasping the stem; staminodia with 9–13(15) segments; petals 7- to 13-veined .... 2. *P. palustris*
  2 Basal and cauline leaves tapered at their bases; the cauline leaf not at all clasping the stem, erect; staminodia with (5)7(9) segments; petals usually 5-veined ..... 3. *P. parviflora*

1. *Parnassia glauca* Raf., grass-of-Parnassus. Herbs, the stems (scapes) glaucous, 1.5–4.0 dm tall when in flower, solitary or several from fibrous, stringy roots. Principal leaves basal, coriaceous, the blades broadly ovate to oblong or rounded, usually tapered but sometimes subcordate at their bases, decurrent on the upper part of the petiole, (5)7- to 9-veined, 1.5–5.0 cm long; cauline leaf, when present, much below the middle of the stem, bractlike, sessile, the blade not unlike those of the basal leaves. Sepals imbricate, coriaceous, with hyaline margins, the apex rounded, oblong to ovate, 3- to 5(7)-nerved, spreading in flower, reflexed in fruit, persistent, 3–5 mm long; petals 1.0–1.8 cm long, creamy white to greenish, broadly ovate, spreading, conspicuously 9-nerved, the center ones unbranched; staminodia of 3 lance-shaped segments or prongs 3–7 mm long, with rounded tips, separate nearly to their bases, shorter than the stamens; stamens 4–9 mm long, the anthers 1.0–2.0 mm

long; ovary 4-9 mm long, ovoid to conical at maturity, sessile to slightly inferior. Seeds many, angular, about 1 mm long. Jul-Oct. Moist calcareous soils, swamps, bogs & wet meadows; MN, ND; (Newf. & N.B. to Sask., s to GP, VA). *P. americana* Muhl.—Rydberg.

2. ***Parnassia palustris*** L., northern grass-of-Parnassus. Herbs, the stems about 1 mm in diam, 0.6-3.5 dm tall, solitary or more commonly clustered from fibrous roots, a sessile, ovate leaf borne on the lower 1/3, clasping the stem. Principal leaves basal, thin, ovate to deltoid ovate, their bases cordate to reniform, (0.5)1-2(3) cm long, the petioles 0.5-4.0(5) cm long. Calyx lobes ascending in flower and spreading widely in fruit, withering persistent, lanceolate to oblong, 4-11 mm long, 5- to 7-veined; petals ovate to elliptic-obovate, slightly clawed, 7-11(12) mm long, 7- to 13-veined, persistent; staminodia with clawlike bases, dilated upward to 2.0-2.5 mm wide, bearing 9-13(15) filamentlike segments with capitate knobs, 3-7(8) mm, from shorter to longer than the stamens; anthers 1-2 mm long, ovoid; ovary becoming ovoid in fruit, about 1 cm long, only slightly inferior. Seeds numerous, angular, about 1 mm long. Jul-Aug. Moist to wet calcareous soils; MN, ND; (Newf. & Lab. to AK, s to CA & CO).

Three varieties of this wide-ranging northern species are recognized by Scoggan, op. cit. Ours is referred to var. *neogaea* Fern., differing from the other two varieties by petals 7- to 11-veined, withering persistent, and staminodia with clawlike bases.

3. ***Parnassia parviflora*** DC., small-flowered grass-of-Parnassus. Herbs, the slender, scaposelike stems several from a fibrous rootstock, 0.8-3.0 dm tall, bearing a sessile leaf 7-15 mm long near or below the middle, not clasping the stem. Leaves thin, the blades lance-ovate, 1-2(2.5) cm long, 0.7-2.0 cm wide, tapered at both ends, petioles slender, from shorter than to much longer than the blades. Calyx lobes lanceolate to narrowly oblong, 3-6 mm long, herbaceous, ascending in flower and fruit; petals ovate to oblong, surpassing the calyx lobes, 4-8 mm long, usually 5(7)-veined; staminodia of (5)7(9) segments, the common base narrowly clawed, dilated upward to an obovate scale that is longer than the longest central filament; stamens 4-7 mm long, surpassing the staminodes; ovary ovoid, at maturity 7-10 mm long. Jul-Aug. Along streams & in moist rock crevices in the BH; SD: Lawrence, Pennington; (across Can., n U.S., especially mountainous regions, s to NM).

5. SAXIFRAGA L.

Perennial herbs, scapose to leafy-stemmed, with basal leaves or a basal rosette, from bulbous to rhizomatous rootstocks; stems in ours 5-25 cm tall, erect to weakly ascending, glabrous to glandular pilose. Leaves ovate-oblong to reniform, petioled, alternate or basal. Inflorescence terminal, cymose-paniculate, cymules contracted into tight heads or of a solitary flower. Calyx fused below, forming a turbinate to campanulate hypanthium, free or adnate to the base of the pistil; sepals 5, erect to reflexed; petals white to pink or purple tinged; stamens 10, inserted with the petals on the calyx around the pistil; ovary free to ½ inferior, 2- to 4(5)-carpeled, tapered upward into slender, often recurving beaklike styles that are distinct. Fruit follicular to distinctly capsular, dehiscent down or between the ventral sutures of the stylar beaks; placentation axile, seeds numerous, fusiform, variously wrinkled. *Micranthes* Haw.—Rydberg.

References: Correll, D. K., & Mc C. Johnston. 1970. Manual of the vascular plants of Texas. Texas Res. Foundation, Renner. pp. 718-719; Hitchcock, C. L., A. Cronquist, M. Owenbey, & J. W. Thompson. 1961. Vascular plants of the Pacific Northwest. Part 3. Univ. Wash. Press, Seattle. pp. 31-56; Scoggan, H. J. 1978. The flora of Canada. Part 3. National Mus. Natural Sci., Ottawa. pp. 880-894.

1 Flowering stems leafy; principal basal leaves petioled, reniform to orbicular reniform, 5- to 7-lobed; bulbils replacing flowers in all or some axils of the upper cauline leaves; inflorescence falsely racemose, with a single, terminal flower .... 1. *S. cernua*
1 Flowering stems scapose; principal leaves in basal rosettes, lanceolate to obovate, with undulate to crenate-serrated margins; bulbils absent; flowers many in terminal contracted to open paniculate cymes.

2 Pistil 2-carpeled; leaves 4–10 cm long, cymes opening at maturity, the panicle branches 1–4 cm long; stems 1–2.5 dm tall .... 2. *S. occidentalis*
2 Pistil 3- to 4-carpeled; leaves usually less than 4 cm long; cymes remaining aggregated in relatively tight heads at maturity; stems 0.5–1.5 dm tall .... 3. *S. texana*

1. Saxifraga cernua L., nodding saxifrage. Herbs, with 1–several, somewhat cespitose, simple, weakly ascending leafy stems, 0.8–2.0 dm tall; glandular-pubescent above to pilose or sparingly rusty-pilose below. Basal leaves several, reniform, cordately based, 5- to 7-lobed palmately, blades 0.8–2.0 cm across, petioles slender, 2–4 cm long; often bearing several ricelike bulblets in the axils at soil level; cauline leaves several, reduced, few-lobed to entire, the petioles shorter, exstipulate. Inflorescence falsely racemose, on most only the terminal flower develops, the lower ones, or all, replaced by 1–several reddish purple bulbils. Calyx campanulate, 3–5 mm long, erect; petals white to pink, 3-nerved, 7–10 mm long, obovate, retuse, not clawed, deciduous; stamens slightly exceeding the calyx lobes, the filaments subulate; ovary 2-carpeled, ¼–⅓ inferior, the styles free above the middle, slightly spreading. Fruit capsular; seeds 0.5–1 mm long, fusiform, numerous. (n = 25) Jun–Jul. Moist, rocky woods & ledges at higher elevation of BH; SD: Lawrence, Pennington; (across Can., s to SD, CO, UT, MT). *S. simulata* Small — Rydberg.

2. Saxifraga occidentalis S. Wats. Herbs, from short, horizontal rhizomes, forming clumps of 3–6 plants, each with 1 or rarely 2 leafless flowering stems 1–2.5 dm tall, reddish glandular-pubescent throughout. Leaves 4–8, basal, 4–8(10) cm long, the blade lance-ovate to elliptic, gradually tapering to the slightly winged petiole, crenate to coarsely serrate, glabrous above, reddish tomentose below. Inflorescence of 3–6 paniculate cymes, contracted to openly divaricate, at anthesis compact, less than 5 cm long but the branches elongating in fruit. Calyx divided 1/10–1/5 up from the base into diverging ovate to lanceolate purplish lobes, 1.9–2.5 mm long; hypanthium shallow; petals white to pink, ovate to oblong, 1.5–3.5 mm long, nearly clawless, deciduous; stamens inserted around the pistil at the level of adnation to the calyx, equaling of slightly exceeding the petals, the filaments reddish, slightly clavate, persistent; pistil reddish, 2-carpeled, the carpels nearly distinct, the ovary adnate to the calyx at its base, tapered upward to the diverging styles, 2.5–5 mm long; stigmas capitate, slightly decurrent. Fruit more follicular than capsular, each carpel dehiscing longitudinally along the inner surface; seeds numerous, brown, 0.5–0.7 mm long, wrinkled. Jun–Jul. Moist wooded hillsides & banks; BH; (nw WY & MT, BH to B.C. s to NV).

3. Saxifraga texana Buckl. Herb, from a cormlike base with many fibrous roots immediately beneath soil level, plants 1–several in clumps, flowering scapes solitary, 5–15 cm tall, glabrate to coarsely white pubescent below. Leaves simple, 4–8(10), 1–4 cm long, in a spreading, basal rosette, fleshy, broadly ovate to oblong, the apex obtuse, margins undulate, tapering to narrow, petiolate bases, glabrous. Inflorescence terminal, of 3–6(8) cymules aggregated into compact heads 1–2 cm across, the branches of the cymules 3–5 mm long, less than twice that length in fruit. Flowers rose or pinkish at anthesis; calyx lobes oblong to ovate, 1.5–2 mm long, glabrous, erect to slightly diverging; petals white, nerves pink near the base, 2–3 mm long, elliptic to obovate, somewhat clawed; stamens inserted at the same level with the petals, equal to or exceeding them in length; filaments subulate, persistent; anthers 1 mm long, oval; ovary ½ inferior, 3 mm long, 3- or 4(5)-carpeled, carpels adnate to the calyx and partially united to each other, the folliclelike carpel tips narrowed upward into the spreading styles, stigmas flat-capitate. Fruit dehiscing along the inner face of each carpel; seeds many, 0.7–1 mm long, fusiform. Mar–Apr. Moist, acid sandstone outcrops or open woods; s GP; KS, MO, OK, TX; (MO & AR to TX). *Micranthes texana* (Buckl.) Sm. — Rydberg.

6. TELESONIX Raf.

1. Telesonix jamesii (Torr.) Raf., James' saxifrage. Herbs, 1–several stems 5–15 cm tall, mostly unbranched from short, thick rootstocks; older plants with a scaly caudex; stems glandular-pubescent, the hairs longer in the upper parts; inflorescence and upper stem dark red to purple. Leaves alternate, simple, mostly basal, the stem leaves 1 or 2, considerably reduced, subsessile; blades rounded to reniform, doubly crenate to shallowly lobed and crenate-dentate, 2–6 cm across at the widest, light green and hirsute on both surfaces; petioles slender, 2–6(7) cm long; stipules membranaceous, persisting for some time. Inflorescence a terminal, leafy bracted panicle, contracted, somewhat secund, 5- to 20-flowered; pedicels 0.5–1.2 cm long, with glandular pubescent hairs to 1 mm; bracts 1–3 on each pedicel, the lower one leafy, upper ones linear, 5–10 mm long. Flowers showy, perfect, radially symmetrical; calyx 9–12 mm long, campanulate, fused below and adnate to the lower part of the ovary, upwards forming a free tubular hypanthium; calyx lobes 5, 5–7 mm long, lanceolate, erect, nearly ½ the length of the calyx; petals 5, red-purple, inserted at the top of the hypanthium, alternate with the calyx lobes and equaling them, ovate to spatulate, the blades about 3 mm long, but with claws as long as or longer than the blade; stamens 10, equaling the petals or longer, inserted at the top of the hypanthium opposite the petals and sepal lobes; filaments subulate, the anthers 1–1.5 mm long; pistil 2-carpeled and 2-celled, adnate to the hypanthium below, 6–8 mm long at anthesis but surpassing the calyx lobes in fruit; styles partially connate above; ovary ½ inferior, gradually tapering upward to the beaklike styles; placentation axile. Fruit a capsule, dehiscing between the 2 beaklike parts; seeds many, oblong, 1–1.5 mm long, brown and shiny. (n = 7) Jun–Aug. Moist rock crevices & outcrops at higher elevations; SD: Meade; CO, MT; (SD to Alta; s to CO & UT). *T. heucheraeformis* Rydb. — Rydberg.

Reference: Hitchcock, C. L., A. Cronquist, M. Ownbey, & J. W. Thompson. 1961. Vascular plants of the Pacific Northwest. Part 3. Univ. Wash. Press, Seattle. pp. 58–59.

Our material has been referred to var. *heucheriformis* (Rydb.) Bacig., Hitchcock et al., op. cit.

70. ROSACEAE Juss., the Rose Family

by Ronald L. McGregor

Annual to perennial herbs, shrubs, or trees, sometimes armed. Leaves alternate or basal (rarely opposite), simple to ternately, pinnately, or palmately compound, usually stipulate, the stipules from free to nearly completely adnate to petiole, sometimes caducous to obsolete or absent. Inflorescences terminal or axillary cymes, corymbs, umbels, racemes, spikes, or panicles, infrequently reduced to a single flower. Flowers perfect, rarely imperfect, mostly regular, perigynous to epigynous, seldom nearly hypogenous; calyx usually (4)5-merous, frequently with bracteoles alternate with lobes, appearing to arise from rim of a floral cup (hypanthium), persistent or the lobes (sometimes the hypanthium) deciduous; hypanthium flat, cup-shaped, cylindric, campanulate, turbinate, or urceolate, free from or adnate to the carpels, sometimes enlarging in fruit; petals same number as calyx-lobes (sometimes appearing numerous by doubling), occasionally much reduced or lacking, usually deciduous, white, yellow, pink, purple, or orange; stamens commonly from 15 to numerous and usually in indistinct series of 5 per whorl, frequently 10, rarely fewer, inserted near edge of hypanthium, often persistent; pistils one to many, distinct or united and sometimes adnate to floral cup or hypanthium; ovules 1 to several per carpel. Fruits very diverse; follicles (sometimes dehiscing along both sutures), achenes exposed or enclosed within floral cup, pomes, drupes, or aggregate or accessory with achenes or drupelets.

Reference: Robertson, Kenneth R. 1974. The Genera of Rosaceae on the Southeastern United States. J. Arnold Arbor. 55(2): 303–332, (3): 344–401, (4): 611–662.

The Rosaceae is a rather heterogeneous assemblage of genera held together largely by general common resemblances. Rydberg divided the group into three families: Amygdalaceae *(Prunus)*; Malaceae *(Amelanchier, Crataegus, Malus, Pyrus, Sorbus);* and Rosaceae (the remaining genera treated by him).

Others usually have divided the group into at least seven tribes. In our area, the family can be divided into four subfamilies, based mainly on fruit characters, as follows:

1 Pistil of 1 carpel; fruit a drupe .... Amygdaloideae
1 Pistil of 2–many carpels, if 1 then not maturing into a drupe; fruit a follicle, pome, achene, aggregate, or accessory.
   2 Ovaries inferior; fruit a pome .... Maloideae
   2 Ovaries superior.
      3 Fruit a follicle .... Spiraeoideae
      3 Fruit indehiscent, a single achene or aggregate or accessory with several to many drupelets or achenes .... Rosoideae

For the GP the Amygdaloideae includes *Prunus;* the Maloideae consists of *Amelanchier, Crataegus, Pyrus, Sorbus;* Spiraeoideae includes *Petrophytum, Physocarpus, Porteranthus, Spiraea;* the remaining genera are in the Rosoideae.

Duchesnea indica (Andr.) Focke, Indian strawberry, is infrequently found in lawns and gardens in se GP. It differs from *Fragaria* in that the bracts subtending the calyx are clearly 3-toothed at the summit (instead of entire), the petals are yellow and the red, mature receptacles do not become juicy and are inedible. The plant is a native of e Asia and is now adventive across e N. Amer.

The following artificial key is provided to facilitate identification of specimens to genus.

1 Principal leaves simple.
   2 Leaves palmately veined with 3–5 veins at base, usually palmately lobed.
      3 Pistils 2–5; fruit follicular; inflorescence without glandular hairs ...... 10. *Physocarpus*
      3 Pistils many; fruit an aggregate cluster of drupelets; inflorescence with glandular hairs .... 16. *Rubus*
   2 Leaves pinnately veined.
      4 Leaves deeply cleft or pinnatifid; extreme sw GP .... 6. *Fallugia*
      4 Leaves not deeply cleft or pinnatifid.
         5 Pistils 2–many; fruit of follicles.
            6 Plant an erect shrub .... 19. *Spiraea*
            6 Plant a prostrate, matted shrub .... 9. *Petrophytum*
         5 Pistil 1; fruit not follicular.
            7 Ovary superior; style 1; fruit a drupe or achene.
               8 Petals present; fruit a fleshy drupe .... 13. *Prunus*
               8 Petals absent; fruit a terete elongated pilose-villous achene with a plumose style .... 3. *Cercocarpus*
            7 Ovary inferior; styles 2–5; fruit a pome.
               9 Flowers in racemes; ovary and fruit 10-celled by intrusion of false septae .... 2. *Amelanchier*
               9 Flowers not racemose; ovary and fruit 5-celled.
                  10 Styles distinct; ovules 1 per cell; actual mature carpels within the pericarp hard and bony, seedlike; plants usually with thorns .... 5. *Crataegus*
                  10 Styles united at base (distinct in *P. communis*); ovules 2 per cell; actual mature carpels within the pericarp leathery or papery, easily opened to expose the seeds; often with spinescent branchlets .... 14. *Pyrus*
1 Principal leaves compound.
   11 Leaves 2–4× ternately dissected .... 4. *Chamaerhodos*
   11 Leaves not 2–4× ternately dissected.
      12 Leaves pinnately once-compound.
         13 Plants trees, shrubs, or woody vines.
            14 Fruit a pome; carpels 2–4; plants of the Black Hills .... 18. *Sorbus*
            14 Fruit of achenes or drupelets; carpels many.
               15 Hypanthium globose to urceolate, with a constricted orifice, concealing the achenes .... 15. *Rosa*
               15 Hypanthium flat or hemispheric, ovules and drupelets exposed .. 16. *Rubus*
         13 Plants herbs.
            16 Sepals 4; petals 0; flowers in dense subglobose to cylindric heads; annual .... 17. *Sanguisorba*
            16 Sepals and petals 5; flowers not in dense heads; perennials.

17 Hypanthium with uncinate prickles at least around the apex; enclosing 1 achene .... 1. *Agrimonia*
17 Hypanthium without uncinate prickles.
  18 Styles inconspicuous, not elongating after anthesis, usually deciduous .... 12. *Potentilla*
  18 Styles elongating after anthesis, jointed or plumose; persistent .... 8. *Geum*
12 Leaves trifoliolate or palmately once-compound.
  19 Pistils 5; fruit of follicles; se GP .... 11. *Porteranthus*
  19 Pistils numerous; fruit not follicular.
    20 Styles persistent and elongating after anthesis, plumose or jointed .... 8. *Geum*
    20 Styles not elongate after anthesis, mostly deciduous.
      21 Bractlets between sepals absent; plants woody, usually with prickles.
        22 Hypanthium globose to urceolate, orifice constricted, concealing ovaries and achenes .... 15. *Rosa*
        22 Hypanthium flat to hemispheric, ovaries and drupelets exposed ...... 16. *Rubus*
      21 Bractlets between sepals present; plants herbaceous (woody in *Potentilla fruticosa* which lacks prickles).
        23 Receptacle enlarged, hemispheric, fleshy, edible; petals white .... 7. *Fragaria*
        23 Receptacle not enlarged and fleshy; petals yellow or white .... 12. *Potentilla*

1. AGRIMONIA L., Agrimony

Herbaceous perennials from short, stout knobby horizontal rhizomes; roots fibrous or tuberous or fusiform-thickened; stems erect, often glandular, variously pubescent to glabrous or glabrate. Leaves alternate, pinnate; leaflets of stem leaves dimorphic with small leaflets interspersed with larger primary ones; stipules foliaceous, those of basal leaves (rarely collected) elongate, adnate to either side of petiole and forming a wing, the tips free and acuminate, those of stem leaves larger, only basally adnate, toothed. Inflorescences terminal, spicate-racemose; pedicels ascending at anthesis, subtended by a 3-cleft bract, later spreading or deflexed. Hypanthium turbinate or hemispheric, with a conical nectar-ring nearly closing orifice, armed with uncinate bristles above, becoming indurate and 10-grooved; sepals 5, spreading at anthesis, later incurved and forming a beak on the fruit; petals 5, yellow, 2–5 mm long; stamens 5–15; pistils 2; ovaries concealed by hypanthium; styles and stigmas terminal. Fruit accessory, of 1 achene enclosed within hardened hypanthium.

Some manuals state that the hardened hypanthium or fruit encloses 2 achenes. In the GP at least 1 ovary regularly aborts and the fruit encloses but 1 mature achene, which is sometimes referred to as a nutlet.

1 Hypanthium minutely to conspicuously strigose in the furrows; axis of inflorescence eglandular or glands obscured by pubescence.
  2 Leaves conspicuously glandular and glabrous to sparsely pubescent below .... 5. *A. striata*
  2 Leaves velvety-pubescent and eglandular or nearly so below .... 3. *A. pubescens*
1 Hypanthium not strigose in furrows; axis of inflorescence usually glandular.
  3 Larger leaves with 11–23 principal leaflets, axis of inflorescence glandular, finely pubescent with ascending hairs, sometimes with long spreading hairs .... 2. *A. parviflora*
  3 Larger leaves with 3–9 principal leaflets; axis of inflorescence usually glandular, with or without sparse short and long hairs.
    4 Mature hypanthium 3–5 mm long; axis of inflorescence glandular and with sparse hairs longer than diam of axis .... 1. *A. gryposepala*
    4 Mature hypanthium 2–2.5 mm long; axis of inflorescence glandular or nearly eglandular, sometimes with short hairs and scattered long ones .... 4. *A. rostellata*

1. ***Agrimonia gryposepala*** Wallr., hooked agrimony. Perennial herbs with short rhizomes and long fibrous roots; stems usually in small clumps, 3–15 dm tall, glandular and sparsely to densely spreading-hirsute. Principal leaflets or larger leaves 5–9, lance-ovate, elliptic,

obovate, or rhombic, coarsely toothed, glabrous or glabrate above, glandular-dotted below and sparsely hirsute mostly on veins; stipules 1–2 cm wide, foliaceous, semicordate, entire or coarsely toothed. Axis of inflorescence minutely glandular and usually with sparse divergent hairs longer than diam of axis intermixed. Hypanthium turbinate to hemispheric, glandular only or with short stiff hairs below, 3–5 mm long at maturity, the uncinate bristles in several rows, outer ones often reflexed; petals 3–5.5 mm long; stamens mostly 15; achene globose, 2.8–3.3 mm in diam. (2n = 56) Jul–Aug. Infrequent in moist open woodlands, prairie ravines, stream valleys; n 1/2 & extreme e GP; (ME, Ont., ND, s to NC, IN, n KS, MO, also NM, CA).

2. **Agrimonia parviflora** Ait., many-flowered agrimony. Perennial herb from short, stout rhizomes, with fibrous and sometimes fusiform-thickened roots; stems usually in clumps, 3–15 dm tall, stout, hirsute below, pubescent and glandular above. Principal leaflets of larger leaves 11–19(25), elliptic, lanceolate to oblanceolate, often attenuate, sharply serrate, glabrous above, conspicuously glandular-dotted and pubescent on the veins beneath. Axis of inflorescence glandular, finely canescent and often with long spreading hairs; pedicels erect at anthesis, becoming abruptly deflexed at maturity. Hypanthium turbinate, ribbed, 2–3 mm long, glandular, sometimes slightly canescent and with stiff hairs below, outer uncinate bristles much shorter than inner; petals 5, oblong, 2–6 mm long; stamens 5–10. Fruit 4–5 mm long; achene globose, 2–2.5 mm in diam. (2n = 28) Jul–Sep. Locally common around marshes, moist prairie ravines, lakeshores, stream valleys; SD: Todd; se NE, IA, e ⅔ KS, MO, OK, TX: Hemphill; (ME, s Ont., NY, s MI, IL, se NE, s to FL, TX).

3. **Agrimonia pubescens** Wallr., downy agrimony. Perennial herbs with short rhizomes and tuberous-thickened as well as fibrous roots; stems usually solitary, 3–10 dm tall, densely short-hairy and ascending, spreading, or retrorsely hirsute. Principal leaflets of larger leaves 5–9(13), lanceolate to elliptic or obovate, coarsely serrate, glabrous or sparsely pubescent above, soft-pubescent below. Axis of the inflorescence densely short-pubescent and usually with some long hairs, not glandular; pedicels erect at anthesis, becoming deflexed. Hypanthium turbinate or hemispheric at maturity, 2.5–5 mm long, strigose in the furrows, often with stiff hairs below, eglandular; petals 3–4 mm long stamens (8)10(12). Achene globose, 2–2.5 mm in diam. Jul–Aug. Frequent but never abundant on dry, open or rocky woodland hilltops, slopes, & valleys; e GP from se SD southward; (ME, MI, se NE, s to VA, KY, GA, e OK).

A few specimens from se KS have only 3–5(7) principal leaflets, fruits 1.5–3 mm long with usually all bristles ascending. These appear close to the more southeastern *A. microcarpa* Wallr. but intergrade completely with *A. pubescens* in our area, and are not here recognized as distinct.

4. **Agrimonia rostellata** Wallr., woodland agrimony. Perennial herbs from short, knobby rhizomes; roots fibrous or some fusiform-thickened; stems erect, usually slender 3–6(10) dm tall, usually solitary or 2 or 3, sparsely appressed-pubescent to glabrous, rarely with remote stiff spreading hairs. Principal leaflets of larger leaves 3–7(9), oblong-obovate, coarsely serrate, glabrous to glabrate above, glandular-dotted below to nearly eglandular, glabrous or with veins short-hairy, stipules lanceolate, entire to toothed, 1 cm or less long. Axis of inflorescence from conspicuously glandular to nearly eglandular, otherwise glabrous or sparsely pubescent. Hypanthium hemispheric, ribbed to nearly smooth, 2–2.5(3)mm long, glandular, otherwise usually glabrous; petals (2)3–4 mm long; stamens 10–15. Achene 2–2.5 mm in diam. Jul–Aug. Infrequent & never common on dry, rocky open woodland hilltops & slopes; extreme se GP; (MA, NY, OH, IN, MO, e KS, s to SC, LA, ne TX).

5. **Agrimonia striata** Michx., striate agrimony. Stout herbaceous perennial with well-developed rhizomes and fibrous roots; stems erect, 5–10(15)dm tall, papillate-hirsute below to hirsute, puberulent, and sometimes glandular above; principal leaflets of larger leaves 5–13, the upper to 6 cm long, often long-acuminate, coarsely serrate-dentate, glabrous or sparsely strigose above, usually conspicuously glandular-dotted and pubescent especial-

ly on the veins below; stipules lanceolate to ovate, 1–2 cm long. Axis of the inflorescence densely pubescent and usually with some long crooked hairs, eglandular. Hypanthium turbinate, to 5 mm long at maturity, deflexed, minutely strigose in furrows, eglandular or glandular, uncinate bristles ascending; petals yellow or ochroleucous, rarely white, (2)3–4 mm long; stamens 10–15. Achene globose, 2.5–3.2 mm in diam. Jun–Aug. Infrequent to locally common on rocky open wooded hillsides & stream valleys; GP from s-cen NE northward; (Newf., to e B.C., s to RI, NJ, PA, WV, OH, WI, NE, NM, AZ).

2. AMELANCHIER Medic., June-berry, Shad-bush, Service-berry, Sugar-plum

Ours unarmed shrubs or small trees. Leaves alternate, simple, deciduous, serrate to subentire; petiolate; stipules linear-attenuate, quickly caducous. Flowers in short terminal racemes at end of branches, regular, perfect, usually appearing with or a little before the leaves. Hypanthium campanulate to urceolate; sepals 5, spreading to recurved, persistent; petals 5, white to pinkish; stamens usually 20, filaments free, persistent, unequal in length; pistil compound, carpels 5, ovary inferior; styles 5, distinct or partly united; ovules 2 in each cell, soon separated by a partition that grows inward from the outer carpel wall. Fruit a pome, in section appearing 10-celled and 10-seeded due to the false septa.

References: Jones, George Neville. 1946. American species of *Amelanchier*. Ill. Biol. Monogr. 20: 1–126; Nielsen, Etlar L. 1939. A taxonomic study of the genus *Amelanchier* in Minnesota. Amer. Midl. Naturalist 22: 160–206.

1 Top of ovary glabrous or sparsely pubescent; petals (12)14–20(24) mm long; mature leaves finely serrate with 5–9 teeth per cm at midleaf; young folded leaves densely tomentose on both surfaces; plant a small tree or fastigiate shrub, not stoloniferous; plants of se NE and e KS .... 2. *A. arborea*
1 Top of ovary strongly hairy; petals usually less than 12 mm long; mature leaves coarsely serrate to nearly entire or serrate in upper half, teeth usually 2–5 per cm; young folded leaves glabrous to sparsely pubescent above; plants usually stoloniferous; plants of n 1/2 GP.
  2 Leaves mostly truncate or broadly rounded at apex, typically orbicular or quadrate; plants of open habitats .... 1. *A. alnifolia*
  2 Leaves mostly acute or obtuse at apex, oblong, sometimes ovate; plants of more wooded areas .... 3. *A. humilis*

1. ***Amelanchier alnifolia*** Nutt., Saskatoon service-berry. Stoloniferous erect shrubs 1–3(5) m tall; young branches silky-pubescent, becoming glabrous or glabrate and reddish-brown to grayish. Leaves alternate, longitudinally folded at anthesis, glabrous to sparsely pubescent above, yellowish-tomentose below, soon expanded and glabrous or glabrate, mature blades broadly elliptic to quadrangular, (2)2.5–5 cm long, 2.5–3.5(5) cm wide, apically broadly rounded to truncate, rarely subacute, basally rounded to less commonly subcordate, margin serrate mostly above the middle with 2–5(8) teeth per cm, sometimes serrate to near base, more rarely nearly entire, lateral veins or their 2 or 3 distal branches entering the teeth, lateral veins sometimes distally obscure on entire blades. Racemes erect, 3- to 20-flowered, 1.5–4 cm long at anthesis, axis silky-pubescent, becoming glabrate to glabrous; pedicels 5–11 mm long. Sepals 5, (1)2.5–3 mm long, triangular, the tips recurving; petals 5, (5)6–8(12) mm long, obovate to spatulate; hypanthium just after anthesis (3)3.5–4.2 mm in diam, shallowly cup-shaped; summit of ovary tomentose; styles usually 5, from nearly distinct to united nearly full length; stamens (10)15–20. Pome usually globose, 10–14 mm long, 8–11 mm in diam, usually dark purple at maturity, sweet and juicy, often with slight bloom; seed 3–4 mm long, reddish-brown, rarely white, often slightly hooked at one end, faintly striate. Apr–May. Frequent on open hilltops & ledges, open prairie ravines, lakeshores, open wooded or brushy areas; n 1/2 GP; (s Man., B.C., Yukon, s to w MN, nw IA, n NE, CO, UT). *A. carrii* Rydb. and *A. macrocarpa* Lunell—Rydberg; *A. alnifolia* var. *dakotensis* Nielsen—Nielsen op. cit.; *A. leptodendron* Lunell, Amer. Midl. Naturalist. 5: 237. 1918.

The plants are highly variable in pubescence and leaf morphology and many specimens appear to merge with *A. humilis*.

2. **Amelanchier arborea** (Michx. f.) Fern., June-berry, service-berry, shad-berry. Small trees or fastigiate shrubs 2-8 m tall, usually with one trunk; twigs red-brown to grayish, essentially glabrous. Leaves alternate, young folded leaves tomentose on both surfaces, becoming green above, sparsely pubescent below; mature blades ovate, broadly elliptical, more rarely slightly obovate, often acuminate at apex, cordate or rounded basally, 4-10 cm long, 3-5 cm wide, finely to coarsely serrate, with 5-9 teeth per cm, sometimes doubly serrate, primary lateral veins anastomosing distally and becoming somewhat indistinct. Racemes somewhat nodding, 3-5 cm long, 3- to 15-flowered, pedicels 8-15 mm long. Hypanthium 2.5-3.5 mm in diam; sepals 5, oblong-triangular, obtuse or abruptly pointed, strongly reflexed, tomentose; petals 5, white to roseate, linear or oblong, (10)14-20(25) mm long; summit of ovary glabrous or nearly so, styles 5, partly united; stamens 15-20. Pome globose, 6-12 mm in diam, glabrous to sparsely pubescent, reddish-purple, dry, insipid; seed 3-4 mm long, dark purple, smooth to slightly granular, obliquely elliptic. (n = 34) Mar-Apr. Infrequent to locally common in open to rocky woods, slopes, & bluffs; se NE, e KS; (ME, sw Que., s Ont., n MI, w to NE, s to n FL, LA, e OK). *A. canadensis* (L.) Medic., *A. laevis* Wieg.—Rydberg.

3. **Amelanchier humilis** Wieg., low service-berry. An erect stoloniferous or strongly surculose shrub 2-3(5) m tall with reddish-orange to grayish twigs. Very similar to *A. alnifolia* from which it differs in having leaves mostly acute or obtuse at apex, oblong to elliptic or ovate; fruits black, glaucous; and a preference for a more wooded habitat. Apr-May. Infrequent to locally common in wooded hillsides, bluffs, brushy prairie ravines, stream valleys; e ND, MT, SD, sporadic elsewhere & perhaps escaped from cultivation; (sw Que., w Ont., s to VT, PA, SC, OH, MI, WI, MN, SC). *A. sanguinea* (Pursh) DC.—Stephens; *A. humilis* var. *campestris* Nielsen, *A. humilis* var. *compacta* Nielsen, and *A. humilis* var. *exserrata* Nielsen—Fernald.

This very variable service-berry has been reduced to synonymy with *A. sanguinea* (Pursh) DC., which apparently is nonstoloniferous, or essentially so, and is east of our area. In the Dakotas *A. humilis* is often nearly impossible to separate from *A. alnifolia*. Some colonies appear like hybrid swarms and the two could be considered as variants of one species. Much more careful collecting of these forms is necessary for the Dakotas.

3. CERCOCARPUS H.B.K., Mountain Mahogany

1. **Cercocarpus montanus** Raf. Shrubs or rarely small treelike plants 1-3(6) m tall, with smooth, grayish to brown bark and short spur branches. Leaves simple, alternate, lanceolate to oblanceolate or elliptic, sometimes ovate to obovate or subrotund, apically acute to obtuse or rounded, usually cuneate basally, entire on lower 1/3-1/2, serrate or dentate above, usually 1-3(5) cm long, 1-2(4) cm wide, lateral veins 3-10, usually prominent, densely fine tomentose below, strigose to pilose or glabrate above at maturity; petioles 1-8 mm long; stipules 2-4 mm long, soon deciduous. Flowers solitary or in fascicles of 2 or 3 in axils of short spur shoots, fascicles sometimes crowded on spur branchlets. Hypanthium salverformlike, tube 3-9 mm long, 8-14 mm long in fruit, limb shallow, 4-6 mm wide, the 5 lobes recurved, appressed silky-villous or tomentose, rarely glabrate, the limb deciduous after anthesis; petals absent; stamens 20-40, born at several levels on limb, anthers hairy, pistil single, 1-carpellary; ovary free from hypanthium, style terminal, exserted. Fruit a terete, pilose-villous achene, the style greatly elongate and plumose, to 8 cm long; seed slender, 8-10(12) mm long. (n = 9) Mar-Jun. Infrequent but locally common on rocky hillsides, cliffs, canyon breaks, & open woodlands; SD: Custer, Fall River; extreme w NE, WY; s to OK: Cimarron; TX panhandle, NM; (throughout most of w 1/2 of U.S. & n Mex).

Most of our plants have mature leaves densely pubescent and can be assigned to var. *argenteus* (Rydb.) F. L. Martin, but some in NE, WY, and SD have mature leaves glabrous or nearly so and approach the more western var. *glaber* (S. Wats.) F. L. Martin. The species appears to be highly variable, and it is doubtful that recognition of varieties in our area is meaningful.

4. CHAMAERHODOS Bunge

1. *Chamaerhodos erecta* (L.) Bunge, little ground rose. Small short-lived perennial herbs from a taproot, usually reddish-tinged, glandular-pubescent and hirsute throughout; stems usually one, 1–3 dm tall, branched above. Basal leaves numerous, rosulate, marcescent, slender petiolate, the blades 2–4 × ternately dissected into linear segments, the cauline ones alternate, 2–3 × ternately dissected, sessile above. Flowers in dichotomous, much-branched to congested bracteate cymes, complete and perigynous. Hypanthium turbinate, 1.5–2.5 mm in diam, lobes erect, 1.5–2 mm long; petals 5, white to purplish-tinged, subequal to calyx-lobes; stamens 5, opposite the petals; pistils 5–10 or more, free from hypanthium, 1-carpellate; ovaries with 1 ovule. Fruit an ovoid-pyriform achene, olivaceous to blackish, 1.5 mm long. Jun–Jul. Infrequent to common on gravelly prairie hillsides, ravines, open woodlands, waste places; w MN, ND e MT; SD: Harding; WY: Albany, Campbell, Laramie; (Man., Sask., Alta., Yukon, s to w MN, ND, WY, CO, UT; Asia). *Chamaerhodos nuttallii* Pick.—Rydberg.

Our plants are the variety *parviflora* (Nutt.) C. L. Hitchc. and differ from the Asiatic var. *erecta* primarily in having flowers a little smaller. Many of our specimens appear to be biennial, and this should be carefully checked in the field.

5. CRATAEGUS L., Hawthorn

Ours deciduous small trees or shrubs with usually flexuous banches and armed with auxillary spines. Leaves alternate, petiolate, simple; blades usually with gland-tipped teeth and lobed; those of vegetative shoots usually larger, more deeply lobed and differently shaped than those of flowering branches; stipules small and deciduous on flowering branches, large and more persistent on vegetative branches. Inflorescences terminating short lateral branches, of few to many flowers in compound or simple cymes. Flowers perfect; hypanthium campanulate or obconic; sepals 5, bractlets between them absent; petals 5, white or rarely pinkish; stamens 5–20, in 1–3 series, anthers white, yellow, or red; ovary inferior or free at tip, the 5 carpels with persistent free styles; each ovary with 2 ovules. Fruit a small variously colored and shaped pome with 1–5 bony nutlets, usually 1-seeded.

An understanding of the genus *Crataegus* is complicated by the presumed occurrence of hybridization, polyploidy, and apomixis. Apparently, fertile apomict triploids are not uncommon. Crataegiologists have named over 1,000 taxa, which more conservative taxonomists have reduced to a number of large complex species groups. While the number of taxa that have been ascribed to our area is relatively small, the evolutionary noise and segregates created, following disturbance of the pristine environment, are nevertheless all too evident. The following treatment for the GP is merely an effort to assist those in need of handling *Crataegus* material for our area. My conclusions were influenced considerably by field trips in sw MO with the late E. J. Palmer. Unless otherwise indicated, all description of leaf morphology refers to flowering or fruiting branchlets.

I have been unable to find voucher records in support of records for *C. douglasii* Lindl. in our area. Reports of escapes of *C. monogyna* Jacq. and the report of *C. munsoniana* for KS—Rydberg (attributed to E. J. Palmer) have not been substantiated.

1 Plants in flowering condition.
  2 Stamens 10.
    3 Corymbs glabrous.
      4 Styles usually 2; common in our area .... 3. *C. crus-galli*
      4 Styles usually 3; rare plant in se KS .... 6. *C. palmeri*
    3 Corymbs pubescent.
      5 Leaves with blades widest above the middle; glands of leaf teeth small and inconspicuous .... 10. *C. succulenta*
      5 Blades widest below middle; glands on teeth dark and distinct .... 9. *C. rotundifolia*

2 Stamens (15)20.
    6 Corymbs glabrous.
      7 Flowers about 2.5 cm wide, 3–7 in corymb .... 2. *C. coccinioides*
      7 Flowers 1.2–2 cm wide.
        8 Flowers more than 10 per corymb .... 11. *C. viridis*
        8 Flowers 5–10 per corymb .... 7. *C. pruinosa*
    6 Corymbs pubescent.
      9 Floral leaves mostly obovate .... 8. *C. punctata*
      9 Floral leaves broad at rounded or truncate base.
        10 Anthesis in late May or early June .... 1. *C. calpodendron*
        10 Anthesis late April to mid-May.
          11 Thorns simple, few to nearly absent; plants common .... 5. *C. mollis*
          11 Thorns compound, numerous; plants very rare .... 4. *C. lanuginosa*
1 Plants in fruiting condition.
  12 Leaves conspicuously to thinly hairy beneath at maturity.
    13 Nutlets with a concave cavity on inner face.
      14 Leaves dull yellow-green .... 1. *C. calpodendron*
      14 Leaves dark green above and much paler beneath (var.
        pertomentosa) .... 10. *C. succulenta*
    13 Nutlets with inner face flat.
      15 Leaves yellow-green and slightly pubescent beneath; common
        species .... 5. *C. mollis*
      15 Leaves blue-green and velvety beneath; rare species .... 4. *C. lanuginosa*
  12 Leaves glabrous below or nearly so at maturity.
    16 At least some of the leaves crisped .... 2. *C. coccinioides*
    16 None of the leaves crisped.
      17 Nutlets with a concave cavity on inner face .... 10. *C. succulenta*
      17 Nutlets with inner face flat.
        18 Leaf teeth tipped by a conspicuous dark gland .... 9. *C. rotundifolia*
        18 Teeth without conspicuous dark glands at tip.
          19 Leaves obtuse, rounded or cordate at base .... 7. *C. pruinosa*
          19 Leaves narrowed or attenuate at base.
            20 Leaves ovate or rhombic .... 11. *C. viridis*
            20 Leaves obovate or oblong-obovate.
              21 Leaves dull green, veins distinctly impressed
                above .... 8. *C. punctata*
              21 Leaves usually glossy above, veins inconspicuous or rarely slightly impressed above.
                22 Nutlets usually 3–5; pomes 6–8 mm in diam; nutlets 3–4 mm
                  long; leaves thin and slightly lustrous above .... 6. *C. palmeri*
                22 Nutlets usually 1 or 2; pomes 8–11 mm in diam; nutlets 6–8 mm
                  long; leaves coriaceous, glossy above .... 3. *C. crus-galli*

1. **Crataegus calpodendron** (Ehrh.) Medic., urn-tree hawthorn. Tree 4–6 m tall or often a shrub; branchlets slender, sparsely thorny to nearly thornless, tomentose while young, becoming glabrous and brownish-gray. Leaves alternate, ovate, oblong-elliptic or rhombic, 5–9 cm long, 4–8 cm wide, coarsely serrate in upper 2/3, often with 3–5 pairs of shallow lateral lobes, villous above when young, and persistently pubescent beneath; petioles 1–3 cm long, pubescent; stipules linear, 4–5 mm long, pubescent. Corymbs densely villous. Hypanthium 4–5 mm long, villous; sepals 5, triangular, 3–4 mm long, villous, often pectinate and glandular; petals 5, white, obovate, 8–10 mm long, abruptly contracted to a very short claw; stamens 20, in 2 series; anthers pink to white; ovary inferior; carpels 2 or 3. Pomes pubescent when young, 7–9 mm in diam, oblong, obovoid, to nearly globose, red or orange-red, with sweet succulent flesh; nutlets 2 or 3, yellowish, 4–5 mm long, deeply pitted on inner faces. May–Jun. Very rare in rocky open woodlands & on stream banks; e KS; (s Ont., NY, MN, s to GA, AL, MO, KS, TX). *C. globosa* Sarg.—Rydberg; *C.*

calpodendron var. *globosa* (Sarg.) E. J. Palm. and var. *hispidula* (Sarg.) E. J. Palm.—Fernald; *C. calpodendron* var. *obesa* (Ashe) E. J. Palm.—Gates.

The rarity of this species in our area precludes careful study, and it might better be considered a variety of *C. succulenta*.

2. ***Crataegus coccinioides*** Ashe, hawthorn. A shrub or small tree 5–7 m tall with slender and usually very thorny branches. Leaves alternate, blades broadly ovate or deltoid, truncate or subcordate at base, sharply and deeply serrate and with 4 or 5 pairs of triangular lateral lobes, glabrous except for sparse appressed hairs above while young and a few persistent hairs at base of midrib on under side, edges often crisped at maturity. Corymbs glabrous, with 3–7 flowers. Hypanthium 3–4 mm long, glabrous; sepals 5, triangular-lanceolate, 4–5 mm long, laciniately glandular-serrate; petals 5, obovate, 12–15 mm long; stamens 20, anthers pinkish; ovary inferior; carpels 5. Pomes subglobose, 1.3–1.7 cm in diam, red, flesh firm, juicy; nutlets 5, yellowish, somewhat glossy, 4 mm long. Early May. Very rare on rocky open wooded hillsides; KS: Cherokee; (IL, MO, se KS, nw AR).

3. ***Crataegus crus-galli*** L., cockspur hawthorn. A small tree or shrub 4–6(8) m tall with rounded or depressed crown and wide-spreading branches, branchlets with thorns 3–8 cm long. Leaves alternate, not lobed; those of flowering branchlets usually obovate, serrate above, 2–6 cm long, 1–3.5 cm wide, glabrous, at maturity thick and shiny above; leaves of shoots to 2× larger and sometimes lobed; petioles 0.3–3 cm long, glabrate, sparsely glandular; stipules on shoots narrowly oblanceolate, sparsely glandular, early deciduous. Inflorescences terminal on spur branches, of nearly glabrous corymbs. Hypanthium turbinate, 2–2.5 mm long, glabrous; sepals 5, lanceolate to lance-liner, 3.5–5 mm long, entire to sparsely glandular-serrate; petals 5, white, orbicular to obovate, 7–8 mm long; stamens 10. Pome short-oblong to slightly obovoid, 8–11 mm in diam, greenish to dull red, with thin dry flesh; nutlets 1 or 2, very rarely 3–5, yellowish-brown, 6–8 mm long, ventral face flat. Apr–May. Locally common to infrequent on rocky open woodlands, rocky prairie hillsides, pastures, with preference for calcareous soils; se GP; (Que., MN, s to FL, IA, e KS, TX). *C. berberifolia* T. & G.—Gleason & Cronquist; *C. crus-galli* var. *barrettiana* (Sarg.) E. J. Palm. and var. *pyracanthifolia* Ait., *C. engelmanni* Sarg., *C. hannibalensis* E. J. Palm., *C. munita* Sarg., *C. regalis* var. *paradoxa* (Sarg.) E. J. Palm., *C. reverchoni* var. *discolor* (Sarg.)E. J. Palm., *C. tantula* Sarg., *C. vallicola* Sarg.—Steyermark; *C. stevensiana* Sarg., *C. discolor* Sarg.—Gates.

This is a highly variable composite species from which a number of taxa have been segregated, based on minor variations in pubescence, leaf morphology, comparison of leaves between flowering and vegetative branches, number of nutlets per fruit, and fruit shape. In studying this complex in the field and in the herbarium, some specimens can be assigned to various of the taxa placed in synonymy. However, there are many which, in various ways, are intermediate between two or more taxa.

4. ***Crataegus lanuginosa*** Sarg., woolly hawthorn. Tree 5–9 m tall or sometimes a shrub; branches stout, flexuous, very thorny, villous when young; larger branches with many compound thorns. Leaves alternate, ovate to suborbicular, blades 4–8 cm long, 3–6 cm wide, sharply serrate, obscurely lobed, those of shoots often incised below, densely short-pilose above, and densely tomentose beneath while young, becoming coriaceous, scabrous above and velutinous below at maturity; petioles 1–2 cm long, villous; stipules linear, pubescent, with a basal glandular lobe, early deciduous. Corymbs dense, tomentose. Hypanthium 4–5 mm long, tomentose; sepals 5, triangular-acuminate, 4–5 mm long, glandular-toothed; petals 5, white, obovate, 8–10 mm long; stamens 20, anthers pinkish-red; ovary inferior. Pomes subglobose, 1–1.5 cm in diam, dark red, pubescent, with thick hardened flesh; nutlets usually 5, yellowish, 6–7 mm long, inner faces flat. Apr–May. Very rare on calcareous rocky, open wooded hillsides & ravines; KS: Cherokee; (sw MO, se KS, e OK, cen AR). *C. dasyphylla* Sarg.—Rydberg.

This appears to be a rather distinctive species, but it is too rare and infrequently collected to allow for careful study. It may be only an extremely pubescent form of *C. mollis*.

5. *Crataegus mollis* (T. & G.) Scheele, summer haw, downy hawthorn. A small tree 3–6(10) m tall with rounded crown and flexuous branches; branches with scattered stout thorns or nearly thornless; young twigs villous, becoming glabrous. Leaves alternate, variable in shape, mostly ovate or deltoid in outline, (3)5–7(10) cm long, 3–8 cm wide, sharply or coarsely serrate, usually with 3–5 pairs of lateral lobes, sometimes laciniate on vegetative shoots, short appressed-pubescent above and tomentose especially along veins beneath while young, becoming glabrous above and slightly pubescent beneath at maturity; petioles villous; stipules 10–15 mm long, pubescent, glandular. Inflorescences of tomentose corymbs terminating spur branches. Hypanthium 4–5 mm long, villous; sepals 5, lanceolate, 4–6 mm long, coarsely glandular-toothed; petals 5, orbicular, 9–12 mm long, white; stamens usually 20, in 2 series; anthers yellowish; ovary inferior. Pome subglobose to rarely oblong or obovoid, 1.3–1.8 cm in diam, red, fleshy at maturity, often pubescent near the ends; nutlets usually 5, yellowish, 7–7.5 mm long, inner faces flat or slightly concave. Apr–May. Infrequent to locally common on open wooded hillsides, stream valleys, pastures, waste places, usually on calcareous soils; e 1/4 GP, but rare in the Dakotas; (s Ont., MI, MN, s to MS, e KS, e OK, TX). *C. lasiantha* Sarg.—Palmer, J. Arnold Arbor. 6: 44. 1925.

6. *Crataegus palmeri* Sarg., Palmer's hawthorn. Tree 6–8(10) m tall with thornless or sparingly thorny branches. Leaves alternate, broadly obovate or elliptic, serrate except on cuneate base, those on vegetative shoots ovate to nearly orbicular and some usually lobed near base, glabrous or sparingly pubescent when young; petioles 5–12 mm long, glabrous; stipules linear, 4–5 mm long, glabrous, promptly deciduous. Corymbs glabrous. Sepals 5, triangular-lanceolate, 3–4 mm long, glabrous; petals 5, white, 7–8 mm long, obovate to suborbicular; stamens 10, anthers pale yellow; ovary inferior; carpels 3–5. Pomes subglobose, 6–8 mm in diam, dull red, with thin firm flesh at maturity; nutlets 3–5, yellowish, 3–4 mm long. Late Apr–May. Rare on rocky wooded hillsides & along small rocky streams; KS: Cherokee; (sw MO, se KS, ne OK, nw AR).

This plant is included here largely in deference to the late E. J. Palmer who "introduced" me to it in the field. It is a rather rare plant and could well be a segregate from a hybrid between *C. crus-galli* and *C. viridis* or another extreme variant of *C. crus-galli*.

7. *Crataegus pruinosa* (Wendl.) K. Koch, frosty hawthorn. Shrubs or small trees 4–8 m tall with somewhat tangled thorny branches, branchlets glabrous while young. Leaves alternate, blades mostly ovate or deltoid, glabrous, 3–5 cm long, 2.5–4 cm wide, sharply and deeply serrate, with 3 or 4 pairs of shallow lobes, bluish-green at maturity. Corymbs nearly simple, glabrous except for glandular bractlets. Hypanthium 3–4 mm long, glabrous; sepals 5, triangular at base, narrow or linear, entire or rarely with a few glands; petals 5, white, obovate, 7–9 mm long; stamens 20, anthers pink or whitish-yellow; ovary inferior; carpels 3–5. Pomes dull red, pruinose, flesh dry, 0.8–1.5 cm in diam; nutlets 3–5, yellowish-brown, 6–7 mm long. Apr–May. Infrequent to rare on rocky wooded hillsides & in stream valleys; se KS; (Newf., WI, s to NC, se IA, MO, se KS, AR, OK). *C. mackenzii* Sarg. var. *bracteata* (Sarg.) E. J. Palm.—Fernald; *C. disjuncta* Sarg.—Steyermark.

8. *Crataegus punctata* Jacq., hillside hawthorn. Tree 4–8 m tall or a shrub; branches stout, thorny, appressed-pubescent to villous when young, becoming glabrous and grayish. Leaves mostly obovate, 3–4 cm long, 2–3 cm wide, finely serrate except near base, unlobed or those of shoots slightly lobed, appressed-pubescent above, densely pubescent along veins below while young, soon glabrous or glabrate at maturity; petioles 3–9 mm long; stipules narrowly linear, 3–5 mm long, very early deciduous. Corymbs pubescent to nearly glabrous. Hypanthium pubescent to nearly glabrous, 3–4 mm long; sepals 5, lanceolate to lance-linear, entire or glandular-serrate, 3–4 mm long; petals 5, white, obovate or suborbicular, 6–8 mm long; stamens 15 or 20 (often variable on same plant); ovary inferior; carpels 3–5. Pomes short-oblong or subglobose, 8–12 mm in diam, dull red; nutlets 3–5, yellowish, 5–6 mm long, inner faces flat. Apr–May. Rare on calcareous wooded hillsides & along small streams; KS: Cherokee; (Newf., WI, MO, s to NC, OK). *C. collina* Chapm. var. *collicola* (Ashe) E. J. Palm. and var. *secta* (Sarg.) E. J. Palm.—Steyermark.

9. **Crataegus rotundifolia** Moench, northern hawthorn. Shrub or small tree 2–4(7) m tall with rounded top; branches stout, armed with spines 2–7 cm long. Leaves alternate, blades oval, elliptic or suborbicular to rhombic-ovate or obovate in outline, 3.2–6.8 cm long, 2.6–5.5 cm wide, usually widest above the middle, with compound teeth in upper half and simple toothed or entire below, larger teeth conspicuously glandular-tipped, glabrous to strigose or short-villous on one or both sides while young, glabrous or glabrate at maturity; petioles 1–4 cm long, sparsely pubescent or glabrous, often glandular; stipules 5–6 mm long, linear, glandular, early deciduous. Corymbs loose villous. Hypanthium turbinate, 3–4 mm long, loose villous; sepals 5, triangular-acuminate, 4 mm long, with sessile glands; petals 5, white, orbicular, 6–8 mm long, abruptly narrowed to short claw; stamens 10, pale yellow; ovary inferior; carpels 3 or 4(5). Pomes 8–15 mm in diam, red to yellowish; nutlets usually 3 or 4, yellowish, 6–8 mm long, inner faces slightly rough or occasionally with shallow pits. Late May–Jun. Locally common on stream banks, open wooded hillsides, prairie ravines; ND, SD; (Newf., Que., Man., s to NY, PA, w to MT). *C. chrysocarpa* Ashe, *C. columbiana* Howell—Rydberg.

The more shrublike habit and teeth of leaves conspicuously glandular-tipped usually serve to distinguish this species from *C. succulenta*, but frequently specimens can only arbitrarily be assigned to one or the other species.

10. **Crataegus succulenta** Link, succulent hawthorn. Tree 3–5(8) m tall or sometimes a shrub; branches slender, flexuous, glabrous or pubescent when young, armed with thorns 3–5 cm long. Leaves alternate, elliptic, oval, rhombic to ovate or oblong-obovate, 3–6 cm long, 2–5 cm wide, blades widest below the middle, slightly or obscurely lobed, coarsely serrate, with short appressed hairs above and pubescent below when young, glabrous at maturity or velvety-pubescent below; petioles 1–3 cm long, villous when young, usually eglandular; stipules linear, 6–9 mm long, early deciduous. Corymbs villous to tomentose. Hypanthium turbinate, 3–4 mm long, villous; sepals 5, lanceolate, 4–5 mm long, pubescent; petals 5, white, nearly orbicular, 6–8 mm long, sometimes with a short claw; stamens 10, anthers yellow or pinkish; ovary inferior; carpels (2)3 or 4. Pome subglobose or short-oblong, 7–9(12) mm in diam, usually pubescent at both ends, red; nutlets 2 or 3(4), yellowish-brown, 5.5–6.5 mm long, inner faces with conspicuous pits or cavities.

Two varieties are found in our area:

10a. var. *occidentalis* (Britt.) E. J. Palm. Leaves elliptic, rhombic or oblong-obovate, usually acute at both ends, slightly paler below, firm to subcoriaceous, glabrous or nearly so at maturity. May–Jun. Infrequent to locally common in rocky open woodlands, stream banks, prairie ravines; n 1/2 GP; (MI, Man., MT, s to NE, CO). *C. macrantha* Lodd. var. *occidentalis* (Britt.) Eggl. and var. *colorado* (Ashe) Kruschke—Kruschke, Milwaukee Public Mus. Publ. Bot. 3: 198. 1964.

10b. var. *pertomentosa* (Ashe) E. J. Palm. Leaves oval, rhombic or rarely ovate, slightly or obscurely lobed, at maturity coriaceous, dark green above, much paler and velvety-pubescent below. Late May–Jun. Rare in rocky open woodlands & along streams; w MN, IA, e KS; (IL, MN, IA, MO, e KS). *C. macrantha* var. *pertomentosa* (Ashe) Kruschke—Kruschke, Milwaukee Public Mus. Publ. Bot. 3: 199. 1964.

These two varieties differ from var. *succulenta* in having 10 rather than 20 stamens and in other minor characters. Var. *pertomentosa* is similar to *C. calpodendron*, but the latter has 20 stamens. In the Dakotas, var. *occidentalis* is sometimes difficult to distinguish from *C. rotundifolia*.

11. **Crataegus viridis** L., green haw. A small tree 3–6(10) m tall with slender unarmed to sometimes thorny branches. Leaves alternate, highly variable and usually asymmetrical, thin, glabrous at maturity except for tufts of white hairs in axils of veins below, those of flowering branchlets rhombic to oblong-elliptic, finely serrate above the middle or nearly to base, those on shoots ovate and sharply serrate and often lobed toward base; petioles 1–5 cm long, glabrous or sparsely pubescent while young; stipules 5–6 mm long, falcate, with a stalk 1–1.5 mm long. Corymbs terminating spur branches, glabrous. Hypanthium turbinate, 1.5–2.2 mm long, glabrous; sepals 5, triangular, obtuse, 1.5–2 mm long, glabrous, sometimes gland-tipped; stamens 20, anthers yellow or reddish; ovary inferior. Pomes subglobose, 5–8 mm in diam, red to orange-red, with thin juicy flesh; nutlets usually 5, yellowish-brown, 5–6 mm long, ventral face flat. Apr–May. Rare, but locally common in low woods, stream valleys, pastures; extreme se GP; (VA, IL, MO, se KS, s to FL, e TX). *C. viridis* var. *lanceolata* (Sarg.) E. J. Palm. var. *lutensis* (Sarg.) E. J. Palm. and var. *ovata* (Sarg.) E. J. Palm.—Fernald.

6. FALLUGIA Endl.

1. **Fallugia paradoxa** (D. Don) Endl., Apache-plume. Slender erect shrubs 0.5–2 m tall, much branched; branchlets grayish-white tomentose, bark exfoliating, and branchlets then reddish to gray. Leaves somewhat evergreen, simple, alternate, often fascicled, cuneate, with 3–7 narrow lobes which are decurrent into the petiole, 7–15 mm long, revolute, white or yellowish-tomentose below; stipules lanceolate, deciduous. Flowers terminal 1–3(4), conspicuous; hypanthium hemispheric, with 5 narrow bractlets alternating with the sepals; sepals 5, ovate, imbricate, acute or cuspidate, densely pubescent; petals 5, white, suborbicular to broadly elliptical, 1–1.6 cm long; stamens numerous; pistils numerous, simple, free from hypanthium, on a short conical axis, pubescent; style slender and elongating in fruit. Achenes 3–5 mm long, persistent style purplish, plumose, 2–5 cm long. May–Sep. Dry, rocky hillsides, bluffs, open woodlands, stream valleys; OK: Cimarron; s CO, NM; (CO to CA, s to TX, Mex.).

This shrub is browsed by livestock and appears to decrease under grazing.

7. FRAGARIA L., Strawberry

Perennial rosulate herbs with rhizomes terminating in simple or branched crowns, spreading and forming colonies by stolons that root and produce plantlets at the nodes. Leaves basal, trifoliolate or simple in the inflorescence, long-petiolate; leaflets serrate at least above, usually short-petiolulate; stipules adnate to base of petioles, persistent and appearing as scales on the crown. Inflorescences usually 1–3 per crown, with scapelike peduncles, few- to several-flowered cymes, sometimes becoming racemiform, bracts conspicuous and often foliaceous. Flowers perfect or to varying degrees imperfect; hypanthium saucer-shaped; sepals 5, alternating with foliaceous bracts of nearly equal size; petals 5, white, obovate, early deciduous; stamens usually 20–40, in 3 whorls, sometimes a few to all reduced to staminodia; pistils many, simple, on a hemispheric to conical receptacle that enlarges in fruit, styles inserted near base, persistent; ovaries with 1 ovule. Fruits accessory, red or rarely whitish, fleshy, subtended by calyx; achenes yellowish-brown, numerous, on surface or in pits, persistent.

1 Terminal tooth of leaflets usually a little longer then the adjacent lateral ones; achenes superficial on the mature receptacle; inflorescence eventually racemiform or paniculiform .... 1. *F. vesca*

1   Terminal tooth usually shorter than the adjacent lateral ones; achenes in pits on the mature receptacle; inflorescence forming a corymbiform cluster .... 2. *F. virginiana*

1. *Fragaria vesca* L., woodland strawberry. Perennial rosulate herbs with unbranched to more rarely branching rhizomes and strongly stoloniferous. Leaflets sessile or on petiolules to 3 mm long, broadly elliptic, somewhat rhombic to obovate-oblong, thin, yellowish-green, usually sparsely hairy above and pilose-silky below, the serrate teeth sharp, a few usually doubly serrate, the terminal tooth usually 1/2 or more as wide as adjacent lateral ones and longer; petioles sparsely to densely pubescent with appressed, ascending or spreading hairs. Peduncles usually shorter than leaves at anthesis, later becoming longer than leaves, pubescence similar to petioles; inflorescence eventually racemiform or paniculiform; pedicels usually unequal. Sepals 4–5 mm long, acuminate, pubescent, spreading to erect in fruit, bractlets 4 mm long; petals white, 5–8(11) mm long. Achenes superficial on mature receptacle. (n = 7) Apr–May. Frequent in thickets & woodlands, stream banks, prairie ravines; n 1/2 GP; (Newf., Man., s to VA, NC, IN, SD, NE). *F. americana* (Porter) Britt.—Rydberg.

Our plants as described above are usually recognized as var. *americana* Porter but some specimens are only weakly separated from the chiefly European var. *vesca* and other varieties to the north and west of our area. Perhaps some of our plants, with widely spreading to retrorse hairs on peduncles and petioles, represent introduced var. *vesca*, though the pubescence character is most variable. It is reported that pollen of *Potentilla anserina* will stimulate the production of normal plants in *F. vesca*.

2. *Fragaria virginiana* Duchn., wild strawberry. Perennial rosulate herbs with thicker rhizomes and less stoloniferous then *F. vesca*; leaflets usually short-petiolulate, glabrous to sericeous below, sparsely pubescent to glabrate above, broadly elliptic, rhombic or cuneate-obovate, serrate with rather blunt teeth, the terminal tooth usually less than 1/2 as wide as adjacent lateral ones and shorter; petioles glabrous to more often appressed-strigose or with spreading to ascending hairs. Peduncles usually shorter than leaves even at maturity, pedicels nearly equal in length, pubescence similar to petioles; flowers in a corymbiform cluster. Flowers perfect or unisexual (at least functionally so), the pistillate flowers much smaller than staminate ones; sepals 4–10 mm long; petals white, 6–14 mm long. Achenes in pits on mature receptacle. (n = 28) Mar–Jun. Prairies, open woodlands, stream valleys, roadsides; e & n 1/2 GP; (throughout much of U.S.). *F. glauca* (S. Wats) Rydb., *F. grayana* E. Vilm., *F. pauciflora* Rydb., *F. pumila* Rydb.—Rydberg; *F. virginiana* var. *illinoensis* (Prince) A. Gray—Fernald; *F. virginiana* var. *glauca* S. Wats.—Atlas GP.

A number of taxa have been segregated from this variable species but intergrade rather completely. They are not at all clearly delimitable in our area.

The cultivated strawberry, *Fragaria x ananassa* Duchn., is a hybrid between *F. virginiana* and *F. chiloensis* (L.) Duchn. It has persisted or escaped from cultivation in a few places in the se GP.

8. GEUM L., Avens

Perennial herbs, usually with basal rosettes of leaves terminating vertical caudices or horizontal rhizomes. Basal leaves long-petiolate, simple, ternate, or pinnate; cauline leaves alternate, rarely opposite, ternate or odd-pinnate, becoming less compound to often simple above; stipules of basal leaves adnate entire length to petioles, those of cauline leaves free. Inflorescences terminating stems arising from rosettes; flowers in open cymes or solitary. Hypanthium saucer-shaped to campanulate or turbinate, free from carpels, with a nectar-ring surrounding mouth or at base of carpels; sepals 5, erect, spreading, or reflexed at anthesis, usually reflexed in fruit, bractlets alternating with sepals or absent; petals 5, white or yellow, spreading or erect at anthesis, soon deciduous; stamens 20 to many, in several series; pistils several to many, free, on a hemispheric to globose or cylindric receptacle which may be stipitate, receptacle glabrous or pubescent; styles entire and persistent, or jointed and geniculate near or above the middle with the apical part deciduous; each ovary with one ovule. Fruit an agregation of achenes. *Sieversia* Willd.—Rydberg.

Our species of *Geum* are rather easily recognized in the field even though most are quite variable. The construction of a key to species, however, is difficult because the possible combinations of definitive characters is limited. It is essential that flowering and mature fruiting specimens be made.

1   Leaves of flowering stem 2(4), opposite; style not obviously jointed, 3–5 cm long at maturity .... 6. *G. triflorum*
1   Leaves of flowering stem alternate; style obviously jointed at or above middle, persistent portion hooked at tip, usually less than 3 cm long.
   2   Bractlets between sepals absent; head of achenes on a stipe raised well above calyx .... 7. *G. vernum*
   2   Bractlets between sepals usually present; head of achenes sessile or on a stipe shorter than calyx in *G. rivale*.
      3   Sepals purple or crimson, spreading or ascending at anthesis; flowers more or less nodding; head of achenes on a stipe shorter than calyx .... 5. *G. rivale*
      3   Sepals green or greenish, reflexed at anthesis; flowers erect; head of achenes sessile.
         4   Plants with flowers.
            5   Petals golden yellow.
               6   Terminal segment of basal leaves somewhat larger than principal lateral lobes, both terminal and lateral lobes cuneate-based; persistent style segment eglandular .... 1. *G. aleppicum*
               6   Terminal segment of basal leaves much larger than principal lateral lobes and usually rounded or to subcordate basally; persistent segment of style usually glandular-pubescent .... 4. *G. macrophyllum*
            5   Petals white, often fading yellowish in pressed specimens.
               7   Pedicels minutely velvety-puberulent, with or without longer hairs; receptacle densely hirsute among achenes and ovaries .... 2. *G. canadense*
               7   Pedicels hirsute; receptacle glabrous or inconspicuously short-hairy .... 3. *G. laciniatum*
         4   Plants with fruits.
            8   The lower persistent portion of the style sparsely to conspicuously glandular-pubescent .... 4. *G. macrophyllum*
            8   The lower portion of style eglandular.
               9   Receptacle hirsute with hairs 2/3 as long as achenes .... 2. *G. canadense*
               9   Receptacle glabrous or with short bristles.
                  10   Receptacle with rather dense short pubescence; common in n 1/2 GP .... 1. *G. aleppicum*
                  10   Receptacle glabrous or essentially so; very rare in extreme e-cen GP .... 3. *G. laciniatum*

1. ***Geum aleppicum*** Jacq., yellow avens. Perennial herbs with a short caudex or horizontal rhizome; stems 1–several, simple, 3–1.2 dm tall, strongly hirsute to puberulent, especially below, rarely glandular-pubescent above. Leaves highly variable, the basal ones lyrate-pinnatifid, main segments 5–9, cuneate-obovate, strongly cleft and toothed, the terminal segment usually lobed over 1/2 the length; lower cauline leaves pinnatifid and with large leafletlike stipules, the upper leaves becoming only 3-lobed; petioles usually hirsute. Flowers several in a leafy-bracteate, unsymmetrical cyme. Hypanthium saucer-shaped, 3–4 mm long; sepals 5, reflexed, 5–8 mm long; petals 5, golden yellow spreading, subequal with sepals; stamens numerous. Mature fruiting heads globose-ovoid, 1.4–2.5 cm wide; achenes flattened, elliptic, 3–4 mm long, divaricately hispid toward apex; persistent segment of style glabrous or slightly hirsute near the base, 4–5 mm long, uncinate; deciduous segment of style 1.5 mm long, hirsute. (n = 21) Jun–Aug. Frequent in woodlands, thickets, moist ravines; n 1/2 GP; (Newf., Que., N.T., s to NJ, IN, IL, IA, NE, NM, CA; Eurasia). *G. strictum* Soland.—Rydberg; *G. aleppicum* var. *strictum* (Ait.) Fern.—Fernald.

The American plants of this highly variable species have been segregated from the Eurasian forms in having less hairy achenes, but this distinction appears unwarranted.

2. **Geum canadense** Jacq., white avens. Perennial herbs with short caudices or horizontal rhizomes; stems 1–several, simple, 3–10 dm tall, glabrous or sparingly hirsute below, becoming densely velvety-puberulent above and on pedicels, with or without scattered long hairs, rarely glandular-pubescent on upper stem and pedicels. Basal leaves simple and undivided to usually with 3–5 larger leaflets, serrate, long petioled; cauline leaves similar, but with shorter petioles and becoming ternately cleft or simple above, serrate, acute, stipules 1–2 cm long, ovate-oblong, entire or cleft. Flowers solitary, few or several in a leafy-bracteate unsymmetrical cyme. Pedicels velvety-pubescent and with very sparse to somewhat dense long hairs, rarely glandular; hypanthium 2.5–3 mm long, sparsely to densely pubescent; sepals 5, lanceolate to lance-ovate, acuminate 4–8(10) mm long; petals 5, white, but fading yellowish, oblong, 5–9 mm long, equaling to longer than sepals; stamens numerous. Mature fruiting heads spherical, 1–2 cm in diam, with numerous achenes; receptacle densely bristly with long hairs; body of achene 2.5–3.5 mm long, usually hairy above; persistent portion of style 4–7 mm long, glabrous or sparsely hirsute below; deciduous segment of style 1–2 mm long, sparsely bearded at base. (n = 21) May–Oct. Rather common in woodlands, ravines thickets, stream valleys; GP but scattered to absent in w 1/2; (N.S. to ND, ne WY, s to GA, AL, e TX). *G. camporum* Rydb.—Rydberg; *C. canadense* var. *camporum* (Rydb.) Fern. & Weath.—Fernald.

3. **Geum laciniatum** Murray, rough avens. Very similar to *G. canadense* from which it differs in having pedicels hirsute; receptacle glabrous or inconspicuously short-hairy; petals 3–5 mm long, usually shorter than sepals. (n = 21) May–Jul. Rare in moist woodlands, alluvial soils, marshes; scattered in w IA, MO: Jackson; (N.S., s Ont., MN, s to NC, AL, IL, MO). *G. virginianum* L.—Rydberg; *G. laciniatum* var. *trichocarpum* Fern.—Fernald.

Previous reports of this species for KS and NE were based on specimens of *G. canadense*, in which pedicels have rather frequent long hairs, but are otherwise velvety-pubescent and receptacles are densely bristly.

4. **Geum macrophyllum** Willd., large-leaved avens. Perennial herbs with a short caudex and horizontal rhizome; stems 1–3, erect, 0.3–1 dm tall, densely to sparsely hirsute below, becoming sparsely hirsute above and puberulent, often with very short glandular hairs; basal leaves rosulate, blades interruptedly lyrate-pinnatifid, to 3 dm long, terminal segment much larger than principal lateral lobes and normally rounded at base, sparsely hirsute; cauline leaves 2–5(7), deeply 3-lobed or trifoliolate. Inflorescence an asymmetrical cyme; pedicels conspicuously glandular to eglandular. Hypanthium saucer-shaped, 3–4 mm long; sepals 5, reflexed, 4–5 mm long, often glandular or finely pubescent and with a few long hairs; petals 5, golden yellow, 4–7 mm long; stamens numerous. Fruiting heads globose, 1–2 cm in diam; receptacle glabrous or short-hispid; mature achenes somewhat flattened, elliptic, 2.8–3.2 mm long; persistent segment of style sparsely to conspicuously glandular-puberulent to eglandular; the deciduous segment 1–1.5 mm long, minutely pubescent to nearly glabrous. (n = 21) Jun–Aug. Rather rare in woodlands, stream banks, brushy areas; e ND, e SD & BH; (Newf. to AK, s to ME, VT, MI, MN, SD, WY, CO, CA; Mex.; e Asia). *G. oregonense* (Scheutz) Rydb., *G. perincisum* Rydb.—Rydberg.

Our plants with the terminal segment of basal leaves lobed to 1/2 the length and coarsely 1- or 2-toothed or cleft and the peduncles and pedicels usually conspicuously glandular are usually segregated as var. *perincisum* (Rydb.) Raup. A few specimens have the terminal segment of basal leaves rather shallowly rounded-lobed and short-serrate and peduncles and pedicels very sparsely glandular to eglandular. These may be referred to var. *macrophyllum*. More and better collections are much needed from our area.

5. Geum rivale L., water or purple avens. Perennial herbs with short to usually well-developed rhizomes; stems 5–6(10) dm tall, sparsely hirsute throughout and puberulent above with a few glandular hairs. Basal leaves somewhat rosulate, to 3 dm long, lyrate-pinnatifid, leaflets (5)7–15, crenate-serrate, with 1–3 larger than others, terminal one cuneate-obovate, sparsely pubescent; cauline leaves 2–5, alternate, much smaller than basal onces, blades pinnatifid below to 3-lobed above, stipules foliaceous. Inflorescence 3- to 9-flowered, cymose but alternately branched, the flowers nodding in bud but becoming erect; peduncle and pedicels densely hirsute, tomentulose and glandular. Hypanthium campanulate, 3.5–4.5 mm long, purplish, pubescent and glandular; sepals ascending, purplish, 7–10 mm long, lanceolate and acute to acuminate, pubescent and often glandular; petals 5, erect, contracted to claw, retuse, yellowish to pinkish and with purplish veins, normally a little shorter than calyx; achenes elliptic, 3–4 mm long, very hirsute; persistent segment of style 6–8 mm long, plumose below; deciduous segment 3–4 mm long, sparsely hirsute. (n = 21) May–Jul. Rare in marshes & moist meadows; MN: Otter Tail; ND: Pembina; reported from SD: Day, Meade; (Newf., Que., Alta., s to NJ, PA, IN, SD, NM, WA; Eurasia).

6. Geum triflorum Pursh, torch flower, maidenhair. Perennial herb with thick rhizomes and often forming clumps 2–4 or more dm wide; flowering stems 2–4 dm tall, usually purplish at least in upper 1/2, soft-hairy and with scattered longer hairs, with a pair of opposite and much reduced laciniate leaves about midlength. Basal leaves with blades 5–15(20) cm long, unequally pinnate to pinnatifid or lyrate; leaflets or lobes 7–19, to 5 cm long, becoming larger toward apex, irregularly laciniate or lobed, terminal leaflet not much larger than upper laterals. Inflorescence mostly cymose; peduncles to 1 dm long. Flowers (1)3–4(9), usually nodding; hypanthium nearly hemispheric, 4–5 mm long; sepals 5, purplish, triangular to ovate-lanceolate, 8–12 mm long; bractlets between sepals, linear to narrowly elliptic, simple to bifid or trifid, from a little shorter to conspicuously longer than sepals; petals 5, yellowish to pinkish or purplish, elliptic or elliptic-obovate, longer or shorter than calyx bractlets; stamens numerous. Achenes pyriform, 2.8–3.3 mm long; lower part of style 2.5–5 cm long, purplish, plumose, the terminal segment glabrous, 3–6 mm long, sometimes a little geniculate at point of juncture with lower segment, persistent or tardily deciduous. (n = 21) Apr–Jun. Common on prairies & open woodlands; MN, ND, MT, SD, WY; (Newf., B.C., s NY, n IL, IA, SD, WY, NM, UT, NV, CA). *Sieversia ciliata* (Pursh) D. Don, and *S. triflora* (Pursh) R. Br.—Rydberg.

Our plants are the var. *triflorum*. A few specimens from the western GP approach the more western var. *ciliatum* (Pursh) Fassett, which has larger leaflets of basal leaves more deeply and repeatedly cleft or divided into ultimate linear or narrowly oblong segments; terminal segment of style usually deciduous and the persistent segment normally to 3.5 cm long.

7. Geum vernum (Raf.) T. & G., heartleaf avens. Short-lived perennial herbs with short caudex terminating a horizontal rhizome; stems 3–6 dm tall, usually branched from base, erect or arched-ascending, glabrous to somewhat villous. Basal leaves somewhat rosulate, some simple, roundish-cordate, often 3- to 5-lobed, crenate-serrate, long petioled, others pinnate to lyrate-pinnatifid and smaller; stem leaves alternate, pinnate to trifoliate or simple above; stipules foliaceous. Inflorescence terminal, subumbellately arranged. Hypanthium campanulate to turbinate, 1.5–2 mm long at anthesis; sepals 5, reflexed at anthesis, triangular, 2–3 mm long, without bractlets between them; petals 5, yellow, obovate, 1.5–2 mm long; stamens 20–30. Fruiting heads 8–14 mm in diam, on a stipe (1)2–3(4) mm long and longer than hypanthium at maturity; achenes 2–3 mm long, minutely pubescent, sometimes with glandular hairs at apex; persistent segment of style 1.5–3 mm long, uncinate; deciduous segment 0.7 mm long, glabrous. (n = 21) Apr–Jun. Infrequent to locally common in moist woods, stream valleys, roadside ditches, & thickets; extreme se GP; (NY, s Ont., MI, IA, s to MD, NC, TN, MO, e KS, e OK).

9. PETROPHYTUM Rydb., Rock-spiraea

1. *Petrophytum caespitosum* (Nutt.) Rydb., rock-spiraea. A densely cespitose undershrub with short stout branches, forming cushions or mats 3–10 dm in diam; season's shoots 1–3 cm tall. Leaves persistent, simple, evergreen, crowded on short branches, spatulate to oblanceolate, 5–12(14) mm long, (1)1.5–3 mm wide, sericeous and grayish-green on both surfaces, 1-nerved, obtuse or mucronate. Racemes dense, spikelike, 1–5 cm long, usually simple; peduncles (1)2–5(8) cm tall with several bractlike leaves; pedicels 0.5–2 mm long. Flowers perfect; hypanthium hemispheric, 1 mm long, densely silky; sepals 5, triangular-ovate, acute, 1.5 mm long; petals 5, white, spatulate-oblanceolate to linear, 1.5–2.5 mm long; stamens 20, 2× as long as petals, glabrous; disk with a prominent entire margin; pistils (3)5; styles 2–3 mm long, pubescent below; ovules 2–4. Follicles 3–5, 2 mm long, dehiscent along both sutures; seeds fusiform, 1–1.5 mm long, irregularly striate. Jun–Aug. Rare on limestone ledges & boulders; SD: Lawrence; (MT, SD, ne OR, CA, s to NM, TX, AZ).

10. PHYSOCARPUS (Camb.) Raf., Ninebark

Deciduous shrubs with erect, spreading, or decumbent principal branches and short lateral branches, the bark in several layers and peeling off in strips. Leaves simple, alternate, palmately 3-lobed, distinctly petiolate; stipules linear, early deciduous, pubescence of stellate hairs. Inflorescences many flowered, bracteate, of umbellike corymbs terminating lateral branches of the season. Hypanthium shallowly cupulate or hemispheric; sepals 5, triangular, ebracteolate; petals 5, white; stamens 20–40, exserted, arising from nectar-ring that surrounds mouth of hypanthium, some persistent; pistils (1)2–5, united at base or 1/2 way, free from hypanthium; ovaries with 2–4 ovules. Follicles somewhat inflated, often dehiscent along both sutures; seeds pyriform, shiny. *Physocarpa* Raf.—Rydberg.

1 Plant usually less than 1 m tall, most often branching near base and with some decumbent stems; follicles (1)2–3(4–5), united to above middle; leaves 2–4 cm long ...... 1. *P. monogynus*
1 Plant to 3 m tall, usually branching above middle; follicles 3–5, weakly united below, leaves 4–12 cm long .... 2. *P. opulifolius*

1. *Physocarpus monogynus* (Torr.) Coult., mountain ninebark. Shrub usually less than 1 m tall, mostly with decumbent stems and exfoliating bark; branches brownish or red-brown, glabrous to sparsely stellate-pubescent. Leaves alternate, ovate to reniform, glabrous to sparsely pubescent, normally palmately 3- to 5-lobed, margins biserrate or incised, 2–4 cm long, rounded or subcordate basally, acute to rounded apically; petioles 3–15(20) mm long; stipules lanceolate to oblong, 2–4(6) mm long, deciduous. Inflorescences of terminal corymbs; bracts lanceolate, 2–4 mm long, soon deciduous; pedicels 1–1.5 cm long. Hypanthium hemispheric, 2.5–3 mm long and about as wide, stellate-pubescent; sepals 5, persistent, ovate-lanceolate to elliptic, 2.5–3.2 mm long, stellate-pubescent on both sides; petals 5, orbicular, white, 2.5–4 mm long, spreading; stamens 20–40, on a disk covering mouth of the hypanthium, filaments 3–4 mm long. Follicles (1)2–3(4–5), united to above the middle, stellate-pubescent, 4–6 mm long, with spreading beaks, dehiscent along both sutures; seeds 1–3, obliquely pyriform, 2–2.5 mm long, yellowish-white to brownish, smooth, glossy, coat bony. Jun–Jul. Infrequent on ledges, rocky open wooded slopes; SD: Custer, Lawrence, Pennington; WY: Albany, Goshen, Platte, Weston; OK: Cimarron; (BH, WY, CO, NM, TX). *Physocarpa monogyna* (Torr.) Coult.—Rydberg.

2. *Physocarpus opulifolius* (L.) Raf., ninebark. Shrub 1–3 m tall, usually branching above the middle, the bark exfoliating in narrow strips; branches yellow-brown, glabrous. Leaves

alternate, ovate to obovate, glabrous above, sparsely stellate-pubescent below, larger ones palmately 3- to 5-lobed, irregularly serrate, 4–12 cm long, basally cuneate to truncate or rounded, apically acute; petioles 1–3 cm long, glabrous; stipules green, 5–7 mm long, ovate and usually acuminate, semipersistent. Inflorescences of terminal corymbs; bracts lanceolate, 2–4 mm long, soon deciduous; pedicels 1–2 cm long, glabrous. Hypanthium shallowly cupulate, glabrous; sepals 5, persistent, 2–2.5 mm long, pubescent on both sides; petals 5, orbicular, white, 3.8–4.2 mm long, spreading to reflexed; stamens 20–40, on a disk covering mouth of hypanthium, filaments 4–5.5 mm long. Follicles (2)3–5, stellate-pubescent, weakly united basally, 8–12 mm long, with spreading beaks, dehiscent along both sutures; seeds 2–2.5 mm long, obliquely pyriform, ivory-colored to brownish, smooth, glossy. (n = 9) May–Jun. Infrequent on rocky banks & bluffs of streams & woodlands; se SD, BH, n-cen NE, se KS; (Que., NY, MN, SD, CO, s to NC, TN, AR, OK). *Physocarpa intermedia* (Rydb.) Schneid.—Rydberg.

Our plants with the fruit permanently stellate-pubescent are var. *intermedius* (Rydb.) Robins., as compared to the more eastern form with mature fruits glabrous.

11. PORTERANTHUS Britt., Indian-physic

1. ***Porteranthus stipulatus*** (Muhl. ex Willd.) Britt., Indian-physic, American ipecac. Erect perennial herbs from woody caudex, rhizomatose and also propagating by adventitious root shoots; stems 4–10 dm tall, often reddish, glabrous to sparsely hairy, often branched above. Leaves alternate, trifoliolate; lower leaves with pinnatifid leaflets, upper leaves with serrate leaflets; leaflets conspicuously glandular-hairy and usually with sparse simple hairs; petioles 0–10 mm long; stipules foliaceous, 1–3 cm long, ovate to orbicular, serrate to laciniate, persistent. Flowers terminal or subterminal, long pediceled, solitary or in a few-flowered panicle. Hypanthium-tube cylindric or narrowly campanulate, 4–5 mm long, 10-nearved, glabrous; sepals 5, lanceolate, 1–1.5 mm long; petals 5, white, spreading, 10–13 mm long, linear to oblanceolate, clawed, deciduous, convolute in bud; stamens usually 20, the outer 10 longer than lower 2 whorls of 5 each; pistils 5, simple, basally connate but becoming distinct in fruit; ovaries with 2–4 ovules. Follicles 2- to 4-seeded, glabrous, 6–8 mm long, dehiscing completely on one side and partially on other side; seeds 3–3.5 mm long, reddish, longitudinally rugulose. May–Jul. Locally common on dry, rocky, open wooded hillsides, stream valleys, prairie thickets, with preference for acid soils; extreme se GP; (NY, OH, IL, MO, se KS, s to GA, TX, OK). *Gillenia stipulata* (Muhl.) Baill.— Fernald.

12. POTENTILLA L., Cinquefoil

Annual, biennial, or perennial herbs, rarely shrubs *(P. fruticosa)*, or woody based *(P. tridentata)*, often with a caudex covered with persistent leaf bases, the plants rosulate, cespitose, or with creeping stems rooting at nodes, or with true stolons rooting at tips. Leaves basal and/or cauline, alternate, digitate, trifoliolate, or odd-pinnate, lower ones long-petiolate; leaflets toothed to dissected, sessile or short-petiolulate; basal leaves with stipules adnate for most of their length to petioles, those of cauline leaves mostly free, entire, or deeply parted. Flowers solitary on naked peduncles from nodes of a usually creeping stem or few to many in cymose inflorescences often with foliaceous bracts. Hypanthium saucer-shaped to hemispheric, a nectar-ring often prominent; sepals 5 in ours, alternating with 5 somewhat shorter and narrower foliaceous bractlets (epicalyx); petals 5, spreading, deciduous, usually imbricate, variously shaped, often emarginate, bases rounded, ours yellow or white; stamens 5–many, often 20, inserted at edge of nectar-ring; carpels 10–many, inserted on prolongation of receptacle, ovary short, ovule 1, style terminal, lateral to nearly basal, basally articulated, deciduous, filaments sometimes basally thickened. Fruit a head of achenes, often enclosed by accrescent calyx, surface of achenes smooth or variously textured, glabrous or pubescent, sometimes appendaged. *Argentina* Lam., *Comarum* L., *Dasiphora* Raf., *Drymocallis* Fourr., *Sibbaldiopsis* Rydb.—Rydberg.

Potentilla is a taxonomically difficult genus because of the frequent occurrence of polyploidy, hybridization, apomixis, and biological species about which little is known. Much more field work and experimental study are necessary before more than a provisional treatment can be presented.

1 Carpels and achenes pubescent; plant a shrub or suffruticose at base.
  2 Plant definitely a shrub; leaves pinnate; flowers yellow .... 8. *P. fruticosa*
  2 Plant woody at base; leaves 3-foliolate; flowers white .... 19. *P. tridentata*
1 Carpels and achenes glabrous, stems herbaceous.
  3 Flowers solitary on naked peduncles arising from nodes of usually creeping stems.
    4 Leaves pinnately compound, tomentose below .... 1. *P. anserina*
    4 Leaves palmately compound, greenish to rarely silvery-silky below .... 18. *P. simplex*
  3 Flowers few to numerous, usually cymose.
    5 Plants annual or biennial, rarely short-lived perennials, without well-developed rootstocks.
      6 Mature achenes with a corky thickening on the inner face nearly as large as rest of achene .... 13. *P. paradoxa*
      6 Mature achenes without an obvious corky thickening.
        7 Calyx mealy-glandular; rare plants of w SD .... 4. *P. biennis*
        7 Calyx eglandular; plants generally distributed.
          8 Stems stiffly hirsute below; petals and sepals subequal; achenes usually ridged .... 12. *P. norvegica*
          8 Stems soft pubescent below; petals much shorter than sepals; achenes smooth .... 17. *P. rivalis*
    5 Plants perennial with well-developed rootstocks.
      9 Styles nearly basal or attached somewhat below middle of ovary, slenderly fusiform; leaves pinnate.
        10 Flowers in open cymes, sometimes glomerate, the lateral branches not tightly appressed .... 9. *P. glandulosa*
        10 Flowers in strict and narrow cymes, the lateral branches nearly erect.
          11 Leaflets regularly decreasing in size from apex to base; lateral leaflets rarely over 2 cm long .... 7. *P. fissa*
          11 Leaflets usually irregularly decreasing in size from apex to base; lateral leaflets usually over 2 cm long .... 3. *P. arguta*
      9 Styles terminal or nearly so, usually tapered from the base or filiform; leaves pinnate or palmate.
        12 Most of the basal leaves odd-pinnate rarely a few digitate or ternate.
          13 Styles shorter than to as long as mature achene, thickened or roughened at base, tapered to tip; stipules deeply cleft .... 14. *P. pensylvanica*
          13 Styles much longer than mature achenes, slender, only slightly tapered, not roughened; stipules entire or only shallowly lobed or toothed.
            14 Leaflets 2–5 cm long, grayish-tomentose beneath, rarely toothed over 1/2 way to midvein.
              15 Leaves usually clearly pinnate .... 11. *P. hippiana*
              15 Leavse subdigitate .... 10. *P. gracilis*
            14 Leaflets usually 2 cm long or shorter, greenish or dissected much more than 1/2 way to midvein and then white-tomentose below.
              16 Leaflets mostly 7–17; greenish .... 15. *P. plattensis*
              16 Leaflets mostly 5–7; greenish or white-tomentose below.
                17 Leaflets usually greenish on both surfaces .... 6. *P. diversifolia*
                17 Leaflets white-tomentose below .... 5. *P. concinna*
        12 Most of basal leaves digitate or ternate.
          18 Leaves whitish-tomentose or whitish-lanate beneath.
            19 Leaves greenish above .... 2. *P. argentea*

19 Leaves never greenish above .... 5. *P. concinna*
18 Leaves not whitish-tomentose or lanate beneath.
20 Leaves mainly cauline; stamens 25 or 30; mature achenes obviously reticulate .... 16. *P. recta*
20 Leaves mainly basal; stamens usually 20; mature achenes smooth or obscurely reticulate.
21 Anthers 0.4–0.6 mm long; leaflets usually greenish or almost equally grayish-sericeus on both sides .... 6. *P. diversifolia*
21 Anthers 0.8–1.3 mm long; leaflets often much paler or tomentose on lower surface .... 10. *P. gracilis*

1. *Potentilla anserina* L., silverweed. Perennial, at first acaulescent or very short caulescent, soon developing long prostrate stolons which root and produce clusters of leaves at the nodes, the plant usually grayish, silky-tomentose. Leaves oblanceolate, (0.5)1–3(3.5) dm long, odd-pinnate, blades whitish-silky-lanate on both sides to greenish above; leaflets (7)15–29(31), obovate to oblong, sharply and coarsely serrate, 0.5–3.5 cm long, interspersed with much smaller leaflets which are entire; stipules of basal leaves 1–4 cm long, membranaceous, hairy, those of the stolons connate-sheathing and deeply linear-lobed, more foliaceous. Flowers solitary at the nodes of the stolons on peduncles 3–15 cm long or from original plant; hypanthium shallowly bowl-shaped, pubescent; sepals ovate, often somewhat abruptly acuminate, (3)4–6 mm long, spreading to reflexed at anthesis, but erect and to 1 cm long in fruit; bractlets elliptic, often toothed; petals yellow, oblong-obovate to oval, 6–12 mm long, rounded at apex, abruptly short clawed at base; stamens 20–25, filaments broadened at base, anthers 1–1.2 mm long; carpels numerous, style slender, smooth, midlaterally attached to ovary. Achenes yellowish-brown, plump obliquely ovoid, 2–2.3 mm long, dorsal surface with a sulcus, base somewhat corky, surface rugulose or obscurely ribbed and areolate. (n = 14, 21) May–Aug. Locally common on low, especially saline ground, stream banks, beaches, roadsides; MN, ND, e SD, LA, MT, WY, sw NE, CO; (circumboreal; Newf. to AK, s to NY, IN, IA, NM, CA). *Argentina anserina* (L.) Rydb.—Rydberg.

2. *Potentilla argentea* L., silvery cinquefoil. Herbaceous perennial with a slender to heavy woody caudex; stems numerous, depressed or ascending, 1–3(5) dm tall, often paniculately branched at summit, with a grayish tomentum. Leaves digitate, the larger with 5 leaflets; leaflets linear-oblanceolate to narrowly obovate, 1.5–3(5) cm long, lobate-serrate 1/2–3/4 way to midvein into 2–4(9) lanceolate or oblong teeth with revolute margins upper surface green and with a mixture of light tomentum and silky to stiff hairs, lower surface with a dense gray tomentum; lower leaves petioled, becoming nearly sessile above; stipules of cauline leaves lanceolate, 4–9 mm long, entire. Cymes compound, many flowered, with leafy bracts particularly at lower nodes. Hypanthium cupuliform, with short and long silky hairs; sepals ovate-lanceolate, 2–3 mm long, somewhat silky-pubescent; bractlets elliptic-lanceolate, 2–3 mm long, petals yellow, obovate, rounded to slightly emarginate at apex, basally cuneate, equaling or slightly longer than sepals; stamens 20, anthers 0.5–0.75 mm long; carpels numerous, styles thickened and somewhat warty at base, subterminal. Achenes 0.5–0.8 mm long, surface with obscure to obvious often branching ribs. (n = 7,14,21,28) Jun–Aug. Very infrequent on fields, roadsides, & waste places with preference for sandy soils; MN, ND, SD, IA; (scattered localities of s Can., e U.S., GP, WA, ID; Europe).

3. *Potentilla arguta* Pursh, tall cinquefoil. Herbaceous perennials from a simple to more often branched stout caudex. Stems erect, (2)3–10 dm tall, simple below the inflorescence, brownish viscid-pilose. Basal leaves long-petiolate, weakly rosulate, pinnate; leaflets usually (5)7–9(11), increasing in size distally and to 8 cm long, usually alternating with much smaller leaflets; cauline leaves with fewer and smaller leaflets, short petioled to nearly sessile, ovate to obovate, oblong, or elliptic, deeply serrate-dentate to shallowly incised, heavily short-hirsute and glandular-puberulent, infrequently sparsely hairy to nearly glabrate;

stipules of cauline leaves foliaceous, entire or toothed, 1-3 cm long. Inflorescence cymose, usually narrow and crowded, lateral branches nearly strictly erect, the whole sometimes nearly flat-topped, or the flowers semiglomerate. Hypanthium saucer-shaped, glandular, pubescent; sepals oblong-lanceolate, cuspidate, (5)6-8(10)mm long, to 12-16 mm long and erect in fruit; bractlets elliptic to lanceolate, shorter than sepals; petals ochroleucous or white, oblong-obovate to obovate, shorter or 2-3 mm longer than sepals; stamens 25(30), anthers 0.75-1 mm long; carpels numerous, styles slenderly fusiform, glandular-roughened near middle, subbasal. Achenes 0.9-1.3 mm long, slightly beaked, surface brown and faintly ribbed or rugulose. (n = 7) Jun-Aug. Common on prairies, open wooded areas, roadsides; n 1/2 GP, e 1/3 KS, in w GP s to NM; (e Que. to AK, s to IN, MO, OK, OR, AZ, NM). *Drymocallis agrimonioides* Pursh—Rydberg.

Some plants from the western edge of our area have the stems only slightly anthocyanus or villous, leaves sparingly hairy, and cymes condensed. These have been called var. *convallaria* (Rydb.) T. Wolf. In the n and w GP, however, the separation of var. *convallaria* and var. *arguta* becomes impossible.

4. *Potentilla biennis* Greene. Annual or biennial herb with a taproot and a simple to slightly branched caudex; stems 1-several from the base, ascending to erect, (1)3-5(6) dm tall, with ascending branches, finely and sometimes densely pubescent, often reddish tinged, glandular. Leaves mostly cauline, all ternate, reduced upwards; leaflets broadly obovate to oblanceolate, coarsely crenate-serrate, 1-4 cm long, pubescent and glandular-puberulent; stipules entire, 7-12 mm long. Inflorescences terminating branches, of leafy-bracteate, many flowered, rather open cymes. Hypanthium shallowly cup-shaped, 4-7 mm wide at anthesis, glandular-puberulent, often appressed-hirsute; sepals erect, ovate, 3.5-4 mm long, longer than hypanthium; bractlets ovate-lanceolate or elliptic-oblong, a little shorter than sepals; petals yellow, cuneate-obovate, 1.5-2 mm long; stamens 10 or 15; carpels numerous, style basally thickened, terminal. Achenes yellow or whitish, 0.7-0.9 mm long, smooth. Rare in moist, open woodlands & streams in BH, SD; (Sask. to B.C., s to SD, WY, CO, CA, AZ).

5. *Potentilla concinna* Richards. Perennial with a stout taproot and sometimes branched, thick caudex; stems spreading to ascending, 4-15 cm long, hirsute-strigose, grayish to sparsely strigose and greenish. Leaves varying from usually digitate to more rarely pinnate; basal leaves numerous, mostly pinnately 5- or 7-foliolate; leaflets from crowded and nearly digitate to scattered, oblong to oblanceolate, or cuneate-oblanceolate, 1-3 cm long, shallowly toothed or deeply cleft; cauline leaves 1 or 2(3), little reduced; lower surface of leaflets white-tomentose and strigose-hirsute, upper surface strigose or hirsute and grayish to light green, or tomentose and sericeous; stipules lanceolate, 3-11 mm long, entire or cleft. Cymes 2- to 7-flowered, inconspicuously bracteate. Hypanthium cupuliform, 8-14 mm wide, sericeous-villous; sepals lanceolate, 3-6 mm long; bractlets linear-oblong, slightly shorter than sepals; petals yellow, obovate, slightly emarginate, 5-9 mm long; stamens 20, anthers 0.6-0.7 mm long; carpels numerous, style slender, subapical. Achenes about 1.5 mm long, yellowish, smooth. Apr-Jul. Infrequent to common in sandy prairies, rocky hillsides, open woodlands; ND, w SD, MT, WY, CO; (Sask. to Alta., s to SD, NM, UT, ID, AZ).

Most of our specimens seem to be var. *concinna* with leaves more digitate than pinnate and leaflets toothed about 1/2 way to midrib. A few specimens along the w GP have leaflets dissected over 1/2 way to midrib and could be referred to the dubious var. *divisa* Rydb. Near the mountains some plants have leaves more clearly pinnate and are referred to var. *macounii* (Rydb.) C. L. Hitchc. More careful collecting in the GP is desirable.

6. *Potentilla diversifolia* Lehm. Perennial herbs with short thick rootstocks and usually branching caudex; stems several, spreading to erect, 1-3 dm tall, glabrous to sparsely strigose, often tinged with red. Basal leaves digitate or sometimes nearly pinnate or truly pinnate,

usually with (5)7 leaflets; leaflets oblanceolate, cuneate, or sometimes obovate, 1–3(5) cm long, shallowly toothed or dissected to near midvein into linear-oblong segments; cauline leaves 1 or 2(3), lower surface of leaves from sparsely hirsute-strigose to grayish-sericeous; stipules usually entire, ovate-lanceolate, 1–2 cm long. Cymes open, rather few-flowered, bracteate. Hypanthium saucer-shaped, villous-sericeous, 4–7 mm wide; sepals 3–6 mm long, lanceolate; bractlets narrower than sepals, about as long; petals obcordate or obovate, emarginate, (4)6–7(9) mm long; stamens 20, anthers about 0.7 mm long; carpels numerous, style subapical. Achenes 1.4–1.6 mm long, light yellowish to brown, obscurely reticulate. (2n = 90,91,101) May–Jul. Rare in meadows and rocky slopes; ND: Billings, Slope; higher elevations of BH of SD: Custer, Lawrence, Pennington; (Sask. to B.C, s to SD, WY, NM, WA, CA, NV). *P. glaucophylla* Lehm.—Rydberg.

Most of our BH specimens have greenish and somwhat glabrate leaves and are referred to var. *perdissecta* (Rydb.) C. L. Hitchc. The few plants of the GP appear very close to *P. plattensis* Nutt., and if the lower leaf surfaces were tomentose they would be similar to forms of *P. concinna* Richards. More carefully prepared collections are needed for our area.

7. ***Potentilla fissa*** Nutt. Perennial herbs with taproots, developing a usually branched caudex and lateral rhizomes; stems erect, 2–3(4) dm tall, usually pilose or villous with multicellular brownish or white hairs, often glandular. Basal leaves odd-pinnate; leaflets 9–13, 1–4 cm long, increasing in size from base to apex, with rudimentary leaflets sometimes interspersed, nearly orbicular, except for terminal leaflet which is often rhombic, incised, hairy or glabrate above, short-hirsute especially on veins beneath, usually glandular; cauline leaves similar but reduced; stipules of cauline leaves ovate-lanceolate, 6–12 mm long, hirsute and glandular. Cymes narrow to somewhat open and diffuse. Hypanthium saucer-shaped, 4–6 mm wide; sepals triangular-lanceolate, acuminate, 5–7(10) mm long; bractlets lanceolate-oblong, shorter than sepals; petals yellow, obovate, with a short claw, 7–12(14) mm long; stamens about 30, anthers 1 mm long; carpels numerous, style slenderly fusiform, somewhat glandular-roughened near middle, subbasal. Achenes 1.2–1.5 mm long, slightly beaked, brownish-yellow, obscurely and irregularly striate. Jun–Aug. Locally common on rocky open wooded hillsides; SD: Custer, Lawrence, Pennington; e WY, CO; (SD, WY, CO, NM). *Drymocallis fissa* (Nutt.) Rydb.—Rydberg.

8. ***Potentilla fruticosa*** L., shrubby cinquefoil. Bushy branched spreading to erect shrub, (2)4–7(10) dm tall, young branches silky-pilose but soon becoming glabrate and with reddish-brown shredding bark. Leaves alternate, odd-pinnate, (3)5(7) foliolate; leaflets crowded, (5)10–20 mm long, linear to elliptic-oblong or narrowly obovate, entire, often revolute, appressed or somewhat spreading silky pubescence and grayish beneath, less pubescent above; petioles to 12 mm long, pubescent; stipules brownish-scarious, 6–12 mm long, sheathing about 1/2 length, with ovate to lanceolate tips, persistent. Flowers single in leaf axils, or 2–5(9) clustered in small terminal cymes. Hypanthium saucer-shaped, 3.5–5 mm wide; sepals 4–6 mm long, ovate-triangular, somewhat acuminate, spreading; bractlets narrow-lanceolate, shorter to longer than sepals; petals yellow, spreading, rotund to ovate-oblong, (6)8–10(13) mm long; stamens (20)25(30), anthers 1 mm long; carpels numerous, styles clavate, midlaterally attached; ovary hirsute. Achenes ovoid, light brown, 1.5–1.8 mm long, densely whitish hirsute. (n = 7,14) Jun–Aug. Infrequent to common on dry or moist rocky hillsides, canyons, meadows; nw MN, w ND, MT, w SD, WY, CO, NM; (circumboreal, s to NJ, n IL, SD, CO, NM, AZ). *Dasiphora fruticosa* (L.) Rydb.—Rydberg.

A number of horticultural variants of this species have been developed for cultivation.

9. ***Potentilla glandulosa*** Lindl. Ours perennial herbs with a branched caudex and short to well-developed rhizomes; stems 1–several, erect, simple up to inflorescence, 3–4 dm tall, somewhat densely pilose-glandular and glandular-puberulent with white to rarely brownish

hairs, some hairs often multicellular. Leaves odd-pinnate; leaflets of basal leaves 5–9, flabellate-cuneate to rhombic, obovate, 1–3 cm long, sharply serrate or biserrate, sparingly and finely pubescent or glabrate; only slightly decreasing in size from apex to base; cauline leaves few, reduced upward; stipules ovate-lanceolate, 4–7 mm long, entire or toothed. Inflorescence an open cyme, few-flowered, bracteate. Hypanthium shallowly bowl-shaped; sepals lanceolate to ovate-oblong, 4–8 mm long to 12 mm long and accrescent in fruit; bractlets as long as or slightly shorter than sepals; petals broadly ovate, yellow to whitish, from shorter to 1/2 again as long as sepals; stamens usually 25, anthers 1–1.3 mm long; carpels numerous, styles thickened and glandular below the middle, tapering to each end, attached well below the middle of ovary. Achenes 0.8–1.3 mm long, yellowish-brown, smooth or very obscurely striate, somewhat laterally attached. (n = 7) Jun–Aug. Common on hillsides & rocky open woods; BH of SD & WY; (Alta. to B.C., s to MT, SD, WY, CO, UT, AZ, & CA). *Drymocallis glandulosa* (Lindl.) Rydb., *D. pseudorupestris* Rydb.—Rydberg.

Our plants with definitely yellow petals to 1.5 mm longer than sepals and stems with many nonglandular hairs could be referred to var. *intermedia* (Rydb.) C. L. Hitchc. Those with light yellow to whitish petals could be referred to var. *pseudorupestris* (Rydb.) Breitung. A few specimens having petals shorter than to only slightly longer than sepals would be var. *glandulosa*. Unfortunately, these characters intergrade nearly completely and recognition of infraspecific categories, in our area, must await careful and experimental studies. The frequency of glands should be determined with fresh specimens. Specimens may easily be confused with *P. arguta* and *P. fissa*, which have been treated by some authors as variants of *P. glandulosa*.

10. Potentilla gracilis Dougl. ex Hook., cinquefoil. Ours perennial herbs with a stout branching caudex; stems several, ascending to more commonly erect, (3)4–8 dm tall, sparsely appressed to spreading-hirsute, often reddish. Leaves digitate; basal leaves numerous, digitate or semipinnate; leaflets (5)7–9, usually cuneate-oblanceolate or oblong-elliptic, 3–8 cm long, crenate or variously dissected, eglandular, sparsely pubescent and greenish below or more densely pubescent and grayish, usually greenish above; petioles 0.5–10 dm long; stipules lanceolate, 1–2.5 cm long, entire or toothed; cauline leaves 1–3, smaller. Cymes open, many flowered, conspicuously bracteate, somewhat flat-topped. Hypanthium cupuliform, 6–10 mm wide, strigose or hirsute; sepals 5–10 mm long, lanceolate to ovate-lanceolate and mostly acuminate; bractlets lanceolate, shorter than sepals; petals yellow, obovate to obcordate, 5–15 mm long; stamens 20, anthers 1 mm long; carpels numerous, styles subapical, slightly enlarged at base. Achenes greenish-yellow, 1–1.3 mm long, faintly areolate or veined. (2n = 52–109) May–Aug. Infrequent to locally common in meadows, open wooded hillsides; w 1/2 ND, BH, MT, WY, CO; (Sask. to AK, s to SD, WY, CO, NM, AZ, WA, n CA). *P. camporum* Rydb., *P. flabelliformis* Lehm., *P. nuttallii* Lehm., *P. viridescens* Rydb.—Rydberg.

This species is one of our most variable, reported to be largely apomictic, and is most difficult to treat taxonomically. Of the several proposed segregates, the var. *glabrata* (Lehm.) C. L. Hitchc., with lower surface of leaves variously pubescent but greenish, seems the more common form in our area.

11. Potentilla hippiana Lehm., cinquefoil. Perennial herb with a stout branched caudex; stems erect or ascending, 1.5–5 dm tall, often freely branched; plants more or less grayish-hirsute and tomentose throughout, but leaves usually green on upper surface, eglandular. Leaves odd-pinnate; leaflets rather crowded, (5)7–11(13), oblong to oblong-lanceolate, 2–5 cm long, grayish on both sides or more often greenish above, toothed 1/2 way or less to midvein; cauline leaves 2–5; stipules ovate to lanceolate, 1–3 cm long, entire. Inflorescence cymose but freely branched with ascending to erect branches, prominently bracteate. Hypanthium shallow saucer-shaped, hirsute and grayish-tomentose, 4–7 mm wide; sepals triangular-lanceolate, 4–6 mm long, hirsute-strigose; bractlets linear and 1/2 length of sepals, to nearly as wide and as long as sepals; petals yellow, 4–8 mm long, obovate, rounded or retuse; stamens 20(25), anthers 0.8–1.1 mm long, carpels numerous, style slender, subapical. Achenes 1.4–1.6 mm long, yellowish-brown, smooth. Jun–Aug. Frequent in prairies & open wooded hillsides; ND, w SD, MT, WY, CO, NM; (Sask. to Alta., s to SD, NM, AZ). *P. argyrea* Rydb., *P. effusa* Dougl., *P. propinqua* Rydb.—Rydberg.

The variation in shape and size of the epicalyx segments is striking, and *P. effusa* has been recognized for its smaller bractlets. However, every degree of intergradation is to be found among our plants.

12. **Potentilla norvegica** L., Norwegian cinquefoil. Annual, biennial, or perhaps short-lived perennial herbs with taproots; stems erect or ascending, simple to usually much branched, hirsute below, subtomentose above, hairs slightly pustulose, eglandular. Cauline leaves 3(5)-foliolate; leaflets elliptic to broadly ovate to obovate 3–6(8) cm long, crenate-serrate, spreading to appressed-hirsute; stipules prominent, ovate, entire or commonly toothed. Cymes leafy-bracteate and rather long pedunculate, but rather compact, often floriferous for much of stem length. Hypanthium saucer-shaped, 4–5 mm wide, strigose to hirsute, sometimes glandular-puberulent; sepals erect, 4–6 mm long, broadly lanceolate; bractlets oblong-elliptic to lanceolate, as long as or longer than sepals; calyx strongly accrescent in fruit; petals yellow, obovate, often retuse, 3/4 as long to nearly as long as sepals; stamens usually 20, anthers 0.3–0.5 mm long; carpels numerous, style terminal, basally thickened. Achenes light brown, 1–1.3 mm long, surface with prominent, longitudinal, branching ribs and obscurely areolate, rarely smooth. (n = 35) May–Sep. Locally common in moist prairies, stream banks, lakeshores, low fields, waste places; GP except sw 1/3; (circumboreal, s to NC, TX, CA). *P. monspeliensis* L.—Rydberg.

13. **Potentilla paradoxa** Nutt., bushy cinquefoil. Annual, biennial, or short-lived perennial with taproots and simple to branched caudex; stems spreading to ascending, 2–4 dm long, often much-branched, glabrous below to hirsute above. Principal leaves odd-pinnate, lower ones with 2–5 pairs of elliptic to oblong, crenate-serrate leaflets 1–3 cm long, strigose-hirsute below, less so above, upper leaves sometimes serrate; petioles 1–6 cm long; stipules well developed, variously toothed or lobed. Inflorescence diffusely cymose, leafy bracteate, stem floriferous for over 1/2 the length. Hypanthium shallowly saucer-shaped, 3–4 mm wide, hirsute; sepals 3–4 mm long, ovate-triangular, erect, hirsute; bractlets similar to sepals, subequal; petals yellow, obovate, 3–4 mm long; stamens 10, 15, 20, anthers 0.3–0.5 mm long, many often abortive; carpels numerous, style terminal usually somewhat thickened basally. Achenes 1.1–1.3 mm long, smooth or obscurely longitudinally ridged, bearing a corky wedge-shaped appendage on adaxial side, this often nearly as large as rest of fruit. May–Oct. Locally common on sandy lakeshores, riverbanks, low moist areas; GP but absent in se 1/4 KS; (Ont. to B.C., s to NY, PA, IL, IA, MO, KS, WA, NM; also in e Asia). *P. nicolletii* (S. Wats.) Sheld.—Rydberg.

This species will flower the first season but, where protected, persists as a perennial.

14. **Potentilla pensylvanica** L., cinquefoil. Perennial herbs with slender to stout, often branched, ascending caudex, covered with remains of leaves and stipules. Stems 1–several, (1)2–5(7) dm tall, erect, spreading or decumbent, simple or with erect or spreading branches, tomentose mixed with spreading longer hairs, occasionally glabrate or sparingly puberulent. Basal leaves with 5–19 leaflets, in remote or approximate pairs, incised 1/2 way or nearly to midrib, glabrous to sericeous above, subglabrous to more commonly white tomentose beneath; cauline leaves progressively reduced; petioles 2–15 cm long, glabrate to densely pubescent; stipules ovate, 1–2 cm long, green or scarious, entire or pectinately divided. Inflorescence usually a compact cyme with erect or ascending branches. Hypanthium grayish-strigose or tomentose, often glandular, 4–10 mm wide at maturity; sepals triangular-lanceolate to ovate, 4–6 mm long; bractlets shorter than to as long as or longer than sepals; petals yellow, 2–6 mm long; stamens usually 20; carpels numerous; style subapical, thickened and glandular-roughened at base. Achenes 1 mm long, obliquely ovate,

surface somewhat irregularly ribbed. (2n = 14, 28, 56) Jun–Aug. Infrequent to common on dry sandy or rocky prairies, open woodlands, sandy ridges; MN, ND, MT, s to nw IA, w NE, WY, CO, NM; (Greenl. to AK, s to NH, MN, NE, CO, NM, NV). *P. atrovirens* Rydb., *P. bipinnatifida* Dougl., *P. glabella* Rydb., *P. lasiodonta* Rydb., *P. strigosa* Pall.— Rydberg; *P. finitima* Kohli & Parker, Canad. J. Bot. 54: 706–719. 1976; *P. pensylvanica* var. *glabrata* (Hook.) S. Wats., var. *bipinnatifida* (Dougl). T. & G.—Fernald.

This polymorphic species has often been divided into several varieties or segregate species of wide or local distributions. In our area plants with generally 5–7 leaflets arranged in approximate pairs and appearing almost digitate (2n = 56) have been called *P. bipinnatifida* Dougl. ex Hook. Those with usually 7–9 or more leaflets arranged in remote pairs have been separated into two species. Plants with upper leaflet surface conspicuously reticulate-veined, bractlets of epicalyx generally much longer than sepals and petals 3–5 mm long (2n = 14) have been called *P. finitima* Kohli & Parker. Plants with upper surface of leaflets often obscurely veined, bractlets of calyx about as long as sepals and petals 4–6 mm long (2n = 28) have been called *P. pensylvanica* L.

While the extremes seem quite clear-cut, and rather easily distinguished, too many specimens exhibit all degrees of intergradation between the taxa, making it most difficult to arrive at satisfactory determinations unless chromosome numbers are known.

15. *Potentilla plattensis* Rydb., cinquefoil. Perennial herbs with taproots and stout branching caudex; stems 1–several, erect, ascending or prostrate, 1–2(3) dm tall, often branched, sparsely strigose or glabrate, often reddish. Basal leaves many, odd-pinnate, with 7–17 leaflets; leaflets obovate-oblong in outline, incised to near midrib into linear or narrowly oblong lobes 5–8(9) mm long, light green on both sides, strigose or glabrate; petioles of basal leaves 1–7 cm long, strigose, upper leaves sessile; stipules conspicuous, foliaceous, 5–12 mm long, lanceolate to elliptic-lanceolate, entire to rarely lobed, strigose. Flowers in rather open cymes; the upper 1/2 of plant floriferous. Hypanthium shallowly saucer-shaped, 3–4 mm wide at anthesis; sepals 3–4 mm long, triangular-lanceolate or lanceolate, strigose; bractlets oblong, 1/2 as long to nearly as long as sepals; petals yellow, obovate, slightly retuse, 4–6 mm long; stamens 20, anthers 0.6–0.8 mm long; carpels numerous, style filiform, subterminal. Achenes 1.5 mm long, oblivaceous to dark brown, surface smooth to obscurely pitted. Jun–Aug. In moist to dry prairie meadows & hillsides; ND: Burke, Divide, Mountrail, Williams; NE: Cheyenne; WY: Albany, Laramie; n-cen CO; (ND to Sask., s to NM, AZ).

This species was reported for SD by Rydberg, but specimens have not been located.

16. *Potentilla recta* L., sulphur cinquefoil. Perennial herbs with simple to branched stout caudex; stems erect, 1–several, (2)3–5(8) dm tall, simple up to inflorescence, sparsely hirsute below, becoming more hirsute above and with shorter, spreading, often glandular hairs above, longer hairs with slight pustulose base. Principal leaves palmately compound, 5- to 7-foliolate; leaflets narrowly oblanceolate, 3–8 cm long, serrate-lobate to about 1/2 way to midrib, teeth divergent-antrorse, upper surface glabrate, or short-appressed pubescent, lower surface sparsely to densely hirsute; lower leaves with long petioles, the upper with shorter or no petiole and 3-foliolate; stipules conspicuous, lanceolate to ovate, 1–3 cm long, laciniate above with slender lobes often as long as body. Cymes large, many flowered, branches ascending, leafy-bracteate at lower nodes, the whole nearly flat-topped. Hypanthium cupuliform, to 12 mm wide, hirsute; sepals 5–9 mm long, lanceolate to ovate-lanceolate, usually acuminate; bractlets narrower than sepals, about same length; calyx accrescent in fruit; petals yellow, obovate, emarginate, 7–11 mm long; stamens 25 or 30, anthers 1–1.3 mm long; carpels numerous, style subapical, thickened and warty near base. Achenes slightly obliquely ovate, 1–1.3 mm long, margin slightly winged, surface with distinct, curved, branching ribs, obviously reticulate. (2n = 28, 42) May–Jul. Infrequent to locally common along roadsides, pastures, old fields, waste ground, & less frequent in prairies; MN, e SD, IA, e 1/2 NE & KS, MO, ne OK; (Newf., Que., Ont., s to MN, GA, AL, KS, TX, CO, WA, MT; Europe). *P. sulphurea* Lam.—Rydberg.

This species is spreading westward in our region and has become somewhat noxious in some areas.

17. **Potentilla rivalis** Nutt., brook cinquefoil. Annual or biennial (winter annual?) herbs with a taproot and simple or rarely branched caudex; stems freely branched, ascending to erect, (2)4–8 dm tall, glabrate to strongly pubescent with fine spreading or appressed eglandular hairs. Basal or lower leaves pinnate with 2 or 3 approximate pairs of leaflets or a single pair with the terminal leaflet 3-parted, or basal leaves digitate with 3 leaflets; leaflets narrowly obovate-oblong to broadly obovate or oval, 1–5 cm long, crenate-serrate; upper leaves always with 3 leaflets; petioles reduced upwards to nearly absent; stipules of cauline leaves ovate-lanceolate, acuminate, 6–12 mm long, entire or toothed. Inflorescence diffusely branched, leafy-bracteate, many flowered, usually long pedunculate. Hypanthium cup-shaped, 4–5 mm wide at anthesis; sepals ovate-triangular, erect, 2–3 mm long at anthesis, becoming accrescent; bractlets elliptic, a little shorter to as long as sepals; petals pale yellow, cuneate-obovate to oblanceolate, usually about 1/2 as long as sepals; stamens usually 10, rarely 15, anthers 0.2 mm long; carpels numerous, style apical, somewhat thickened at base. Achenes yellowish, 0.7–0.8 mm long, smooth or obscurely ridged. May–Oct. Infrequent to locally common on stream banks & around lakes, less common in moist pastures, roadside ditches, & waste places; GP but less frequent w; (Sask. to B.C, s to MN, IL, MO, TX, OR, CA, NM, Mex.). *P. millegrana* Engelm., *P. pentandra* Engelm.—Rydberg; *P. rivalis* var. *millegrana* (Engelm.) S. Wats. and *P. rivalis* var. *pentandra* (Engelm.) S. Wats.—Gates.

Most of our plants have no more than 3 leaflets on all leaves and have sometimes been referred to the species or var. *millegrana*. The more rare and scattered plants with basal or lower leaves pinnately compound have been referred to var. *rivalis*. Plants with basal and lower leaves digitately compound and leaflets 5 or only 3, with the lower leaflets deeply divided, have been referred to the species or var. *pentandra*. A careful search of nearly all large colonies of this species will reveal a few plants with each of these characters, and taxonomic distinctions seem without value.

18. **Potentilla simplex** Michx., old-field cinquefoil. Perennial herb with a short rhizome 2–8 cm long; stems at first erect or ascending and 2–5 dm tall, then widely spreading to 1–1.2 m long, arching, branching, rooting at tip to produce small tubers which form the rhizome of following seasons; stems usually hirsute or villous-hirsute with spreading, rarely appressed, hairs, leaves palmately compound, principal leaves 5-foliolate; leaflets oblanceolate to elliptic, 2–7 cm long, usually less than 1/2 as wide, toothed in distal 1/2–3/4, strigose-pubescent and sometimes whitened beneath; petioles spreading or rarely appressed-pubescent; stipules of stem leaves 0.5–3 cm long. Flowers solitary on slender peduncles from the nodes, the first flower from end of second well-developed internode; hypanthium saucer-shaped to shallowly cupuliform, 3–5 mm wide at anthesis, pubescent; sepals triangular-lanceolate, 4–6 mm long; bractlets narrower, sometimes apically lobed, subequal with sepals; petals yellow, obovate, rounded at summit or retuse, 4–7 mm long; stamens 20, anthers 0.7–1.0 mm long; carpels numerous, style lateral. Achenes 0.9–1.2 mm long, yellowish-brown, surface with obscure ribs. Apr–Jun. Locally common in dry open woods, prairie hillsides, roadsides, old fields, & waste places; s MN, IA, e 1/3 KS, MO, e OK; (Newf., Que., MN, s to NC, TN, MO, KS, OK, ne TX). *P. simplex* var. *argyrisma* Fern., *P. simplex* var. *calvescens* Fern.—Steyermark.

In se KS some specimens with the lower surface of leaflets densely sericeous have been called var. *argyrisma* Fern., but these grade completely into the more common form with lower surface green or grayish-green. Scattered plants with hairs on stems and petioles appressed have been named var. *calvescens* Fern., but these are always associated with the more common plants with spreading hairs. These distinctions seem taxonomically unwarranted.

19. **Potentilla tridentata** Ait., three-toothed cinquefoil. Perennial with spreading subterranean stems to 10 cm long and erect branches 0.5–3 dm tall which are woody at base

and herbaceous above, terminating in an inflorescence; flowering stems strigose. Leaves palmate, with 3 firm, oblong-oblanceolate leaflets, 1–2.5 cm long, entire below, 3-toothed at truncate apex, glabrous above, sparsely strigose below, bright green on both surfaces, evergreen; petioles of lower leaves 1–2(5) cm long, strigose; stipules 8–15 mm long, free tips 3–5 mm long. Inflorescence in an open, flattened, rather stiff cyme. Hypanthium saucer-shaped, 3–4 mm wide at anthesis, hirsute; sepals triangular-lanceolate, 4–5 mm long; bractlets narrowly oblong or lanceolate, usually shorter than sepals; petals white, obovate, 6.5–7.5 mm long; stamens 20; carpels numerous, ovary villous, style subbasal. Achenes 0.9–1.2 mm long, obliquely ovate, surface with white hairs at least at base, otherwise smooth. (2n = 28) Jun–Aug. Rare in GP on dry, shale outcrop of a prairie hillside; ND: Cavalier; (Greenl. to N.T., s to CT, MI, GA, IA, ND). *Sibbaldiopsis tridentata* (Soland.) Rydb.—Rydberg.

13. PRUNUS L., Cherry, Peach, Plum

Deciduous trees or shrubs, sometimes producing root sprouts. Leaves alternate, simple, usually serrate, often glossy above, convolute or conduplicate in bud; petioles often with large glands; stipules paired, free from petiole, small, soon deciduous. Flowers 5-merous, mostly perfect, in racemes terminating shoots of the season, in racemes from axils of previous season leaves, or in corymbs or umbels from dwarf branchlets and appearing before or with leaves, rarely solitary. Calyx ebracteate, lobes triangular; hypanthium cup-shaped, obconic, or urceolate, disk thin; sepals spreading or reflexed; petals white to pink or red, elliptic to obovate, spreading, quickly deciduous; stamens 15–20, in 2 or more whorls of 5, inserted at edge of hypanthium; pistil 1, simple, 2-ovuled, inserted at bottom of receptacle. Fruit a 1-seeded drupe, exocarp fleshy or juicy, endocarp hard, indehiscent, nearly globose or compressed.

Excluded from this treatment are *Prunus armeniaca* L., apricot; *P. avium* L., sweet cherry; *P. cerasus* L., sour cherry; *P. domestica* L., common plum; and *P. insititia* L., damson plum. These are cultivated species which sometimes are found around old homesites but none have become naturalized. Also excluded is *P. nigra* Ait., Canada-plum, which has been reported for SD (cf. Gleason & Cronquist), but specimen evidence has not been located. It is known from nw MN, & a ne ND specimen may belong to this species.

```
1   Plants bearing flowers.
    2   Petals pink or rose colored; ovary velvety hairy ..................................... 9. P. persica
    2   Petals white, ovary glabrous.
        3   Flowers terminating shoots of the season.
            4   Flowers 15–30, in enlongated racemes.
                5   Sepals definitely glandular-erose, soon deciduous; leaf teeth
                    ascending .............................................................................. 13. P. virginiana
                5   Sepals entire or inconspicuously glandular-erose, persistent; leaf teeth appressed
                    or incurved ............................................................................... 12. P. serotina
            4   Flowers 4–12, in a corymblike raceme on short leafy branches ....... 5. P. mahaleb
        3   Flowers from buds of previous year.
            6   Plants with decumbent or ascending stems
                1–4(7) dm tall ..................................................... 10. P. pumila var. besseyi
            6   Plants erect, usually well over 7 dm tall.
                7   Glands absent or esentially so on calyx lobes.
                    8   Petals 3–7 mm long.
                        9   Calyx lobes ciliate ............................................... 2. P. angustifolia
                        9   Calyx lobes not ciliate ......................................... 8. P. pensylvanica
                    8   Petals 6–15 mm long.
                        10  Pedicels and hypanthium more or less hairy; plants usually
                            solitary ........................................................... 6. P. mexicana
```

              10 Pedicels and hypanthium glabrous; plants usually forming
                  thickets .... 1. *P. americana*
       7 Glands definitely present on calyx lobes.
              11 Pedicels densely short pubescent .... 3. *P. gracilis*
              11 Pedicels glabrous or nearly so.
                  12 Plants usually solitary and not thicket forming .... 4. *P. hortulana*
                  12 Plants usually forming thickets.
                      13 Calyx-lobes as long as tube .... 7. *P. munsoniana*
                      13 Calyx-lobes shorter than tube .... 11. *P. rivularis*
1 Plants with mature leaves and fruits.
   14 Stone somewhat compressed, longer than wide (plums, peach).
       15 Fruit velvety or tomentose; peach .... 9. *P. persica*
       15 Fruit not velvety or tomentose; plums.
           16 Teeth of leaves without glands.
              17 Branchlets finely hairy; plants small trees occurring singly or
                  solitary .... 6. *P. mexicana*
              17 Branchlets glabrous; plants occurring as thickets .... 1. *P. americana*
           16 Teeth of leaves usually with glands (at maturity often with a scar marking presence
              of a gland).
              18 Leaves mostly more than 7 cm long.
                  19 Glands of leaf teeth adjacent to sinus, teeth low and nearly
                      pointless .... 7. *P. munsoniana*
                  19 Glands of leaf teeth terminal; teeth ascending .... 4. *P. hortulana*
              18 Leaves mostly less than 7 cm long.
                  20 Lower leaf surface reticulate and densely pubescent .... 3. *P. gracilis*
                  20 Lower leaf surface not reticulate; glabrous or pubescent only on veins.
                      21 Leaf teeth capped with a large permanent gland; common
                          plant .... 2. *P. angustifolia*
                      21 Leaf teeth with or without a small caducous gland; rare in
                          s GP .... 11. *P. rivularis*
   14 Stone globose or subglobose (cherries).
       22 Plants with decumbent or ascending stems
           1–4(7) dm tall .... 10. *P. pumila* var. *besseyi*
       22 Plants erect, usually well over 1 m tall.
           23 Leaves broadly ovate to nearly round, about as wide as long .... 5. *P. mahaleb*
           23 Leaves usually longer than wide.
              24 Teeth of leaves short, appressed or incurved .... 12. *P. serotina*
              24 Teeth of leaves not appressed or incurved.
                  25 Leaves sharply serrate with ascending teeth; pedicels
                      3–6 mm long .... 13. *P. virginiana*
                  25 Leaves finely and unevenly serrate; pedicels 1–1.5 cm
                      long .... 8. *P. pensylvanica*

1. *Prunus americana* Marsh., wild plum. A shrub or small tree 3–8 m tall, usually forming thickets or colonies from root-suckers and shoots; branchlets often spinescent, glabrous or glabrate. Leaves obovate to obovate-oblong or lanceolate-ovate, 6–10 cm long, abruptly acuminate, acute to rounded at base, sharply and often doubly serrate, green and glabrous above, slightly pubescent below especially on veins or glabrate; petioles 8–20 mm long, slightly pubescent on upper side, sometimes with glands at summit; stipules linear-aristate, 5–14 mm long, usually with 1–3 linear teeth, pubescent, caducous. Flowers in fascicles of 2–5 at tip of spur-branchlets or from axillary buds; usually appearing before leaves; pedicels 7–20 mm long, glabrous. Hypanthium obconic, 2.5–3 mm long, glabrous; sepals 5, oblong to lance-attenuate, 3–4 mm long, promptly reflexed, pubescent on upper side, ciliate, sometimes distally toothed; petals 5, white, oblong-ovate, clawed, 8–12 mm long; stamens 20–30; ovary glabrous, style 12–15 mm long; stigma capitate. Fruit 2–2.7 cm long, 2–2.5 cm in diam, reddish-purplish or yellowish, with a slight bloom; stone compressed,

1.5–1.8 mm long, 1–1.2 cm wide, 1/2 margin ridged, the other sulcate, surface veined or slightly rugulose. (n = 8) Apr–May. Woodlands, thickets, prairie ravines, stream banks, pastures, roadsides; GP but rare westward & absent in TX panhandle; (MA to Man., MT, s to FL, OK).

A number of horticultural varieties have been derived from this species.

2. **Prunus angustifolia** Marsh., Chickasaw plum, sandhill plum. A much-branched shrub 1–3(5) m tall, often forming extensive thickets from root-suckers and shoots; branchlets usually zigzag, sometimes spinulose, glabrous, reddish. Leaves lanceolate to oblong-lanceolate, strongly trough-shaped or conduplicate, 2–6(8) cm long, 1–2 cm wide, acute or short-acuminate, cuneate or rounded at base, lustrous and glabrous above, paler beneath and glabrous or somewhat pubescent along lower midrib, margins with appressed gland-tipped teeth; petioles with 1 or 2 glands at summit or glandless. Flowers in clusters of 2–4, usually expanding with leaves; pedicels 3–6(10) mm long, glabrous. Hypanthium obconic, 1–2 mm long, glabrous, lobes ovate, 1–1.5 mm long, ciliate, glandless, a few hairs at base; petals 3.5–6 mm long, white to creamy-white, obovate, short clawed; stamens 20; ovary glabrous, style 4–6 mm long, stigma capitate. Fruit subglobose to ellipsoid, 2–2.5 cm long, 1–2 cm in diam, red or yellowish, with a slight bloom; stone ovoid, plump, 1–1.5 cm long, obtuse at both ends or acute at apex, surface roughened or somewhat pitted. (n = 8) Apr. Sand dunes, prairie ravines, stream valleys, pastures, old fields, roadsides; s 1/2 GP; (s NJ, WV, IN, s NE, s to FL, TX). *P. rugosa* Rydb., *P. watsonii* Sarg.—Rydberg.

In the w 1/2 of its distribution in the GP plants usually have leaves 2–5 cm long, elliptic to elliptic-oblong, less conspicuously serrate, flowers smaller, and fruit with a thicker skin. These have been called var. *watsonii* (Sarg.) Waugh. Plants of the sw GP are often more robust, have leaves larger, pedicels longer, and stones pointed at apex. These have been named var. *varians* Wight & Hedr. While these variants are readily recognizable, intergradation seems too frequent even in the same area.

Presumed hybrids of *P. angustifolia* and *P. americana* have been named *P. x othosepala* Koehne while *P. x slavinii* E. J. Palm. is a presumed hybrid between *P. angustifolia* var. *varians* and *P. gracilis* Engelm. & Gray. Horticultural varieties have been developed from this species complex.

3. **Prunus gracilis** Engelm. & Gray, Oklahoma plum. Somewhat straggling shrub 0.5–1.5 m tall, usually forming thickets or rarely as a single plant; young branchlets pubescent, becoming reddish-brown and glaucous. Leaves alternate, blades broadly elliptic to ovate-elliptic, 2–5 cm long, 1–2.5 cm wide, obtuse to acute or short-acuminate, cuneate or rounded at base, margin finely serrate with obtuse or acute teeth which are usually gland-tipped when young, lightly pubescent above, dark yellow-green, lower surface reticulate, paler, densely pubescent; petioles 4–16 mm long, pubescent, glandless; stipules narrowly lanceolate, 2.5–5 mm long, glandular serrate, sometimes lobed, caducous. Flowers before or with young leaves in axillary clusters of 2–4(8); pedicels 6–12 mm long, pubescent. Hypanthium obconic, 2.5–3 mm long, pubescent; calyx lobes triangular, 1.2–1.7 mm long, obtuse or acute, pubescent, slightly glandular-toothed; petals white, obovate, 5–6.5 mm long, 4–5 mm wide, short clawed, spreading; stamens usually 20; ovary ovoid, 2 mm long. Fruit subglobose to ellipsoid, 15–18 mm long, 12–15 mm in diam, yellow-red, with a slight bloom; stone oval, compressed, 10–12 mm long, 8–9 mm wide somewhat obtuse at both ends, slightly ridged on one margin, surface rough. (n = 8) Mar–Apr. Infrequent to common on sandy or dryish soils on open hillsides, stream valleys, roadsides, open woodlands; restricted to KS, OK, TX, e NM.

This species is reported to hybridize with *P. angustifolia*.

4. **Prunus hortulana** Bailey, wild goose plum. Small tree 4–6(10) m tall, rarely forming thickets by root-suckers but usually as a dense colony of individual plants; twigs glabrous,

reddish-brown, somewhat glaucous. Leaves with lanceolate or oblong-lanceolate blades 5–10 cm long, 3–5 cm wide, gradually acuminate, obtuse to rounded at base, pubescent below at least on veins, very finely serrate, teeth gland-tipped; petioles 1–2 cm long, slightly pubescent above, normally with 2 glands at apex; stipules 5–8 mm long, linear-aristate, glandular, sometimes with linear-lobes, caducous. Flowers opening when leaves are ca. 1/2 developed, usually in clusters of (2)4 or 5 from short spurs; pedicels 10–18 mm long, glabrous. Hypanthium obconic, 4–5 mm long, glabrous; calyx lobes 3–3.7 mm long, ovate-lanceolate, glandular-serrate, pubescent at base of inner surface; petals ovate, white, 5–7(9) mm long, with a short claw, rounded or notched at apex; stamens (20)30; ovary 1–1.5 mm long, glabrous. Fruit red or yellowish-red, yellow-dotted, bloom slight, globose, 2–3 cm in diam; stone 1 cm long and nearly as wide, more or less 2-sided, shallowly pitted and with irregular ridges on one margin, and grooved on other margin. (n = 8) Mar–May. Rather infrequent in open woodlands, edge of woods, stream valleys, thickets; se NE, e 1/3 KS, MO, e OK; (IN, IA, e KS, s to AL, TN, AR, OK).

Reputed hybrids with *P. mexicana* are known as *P. x palmeri* Sarg. The species has given rise to a number of horticultural varieties, which are sometimes found near old homesites.

5. ***Prunus mahaleb*** L., mahaleb or perfumed cherry. A shrub or small tree with short trunk and spreading branches forming a loose crown; young branches tomentulose, becoming glabrous and pruinose. Leaf blades broadly ovate to orbicular, 3–5(7) cm long, more than half as wide, abruptly short-acuminate, rounded to subcordate at base, glabrous above, hairy on veins below, finely serrate with gland near the sinus; petiole 6–12 mm long, glabrous, sometimes with 1 or 2 glands at apex; stipules ovate-lanceolate, glandular, caducous. Flowers terminating new shoots, appearing when leaves are 1/2 or more grown, in corymblike racemes of 4–12, branches of inflorescence with leafy bracts; pedicels 10–14 mm long, glabrous, bract at base glabrous, glandular-margined. Hypanthium obconic, 2.2–3 mm long, glabrous; calyx lobes ovate, 1.6–2 mm long, entire, glabrous, becoming sharply reflexed; petals white, 5–8 mm long, obovate, narrowed to base, rounded above; stamens 20; ovary 1 mm long, glabrous. Fruit dark red to black, ovoid, 6 mm in diam, glossy; stone smooth, subglobose, 6.5–7 mm long, tip acute, dorsal margin ridged, ventral margin with a small groove. (n = 8) Apr–May. Infrequent around old farmsteads & as escapes in open woodlands, stream banks, thickets; e 1/3 KS, MO, OK; (MA to Ont., s to DE, IN, MO, KS, AR, OK; Europe). Introduced.

The species has been important in horticulture as stock for grafting cherries and for hedges. Its aromatic bark has been used in the manufacture of pipestems.

6. ***Prunus mexicana*** S. Wats., big-tree plum, Mexican plum. Usually a solitary tree 3–6(12) m tall but known to form thickets; branchlets pubescent to glabrous, light red-brown, becoming gray. Leaf blades obovate to oblong-obovate, 5–12 cm long, 3–6 cm wide, abruptly acuminate, base rounded or subcordate, sharply and often doubly serrate, short-pubescent above and rugose, lower surface pubescent and somewhat reticulate, thickish at maturity; petioles 1–2 cm long, pubescent all around, usually glandular at apex; stipules lanceolate, 3–6 mm long, pubescent, often toothed or lobed, caducous. Flowers with young leaves, in clusters of 2–4(6), from previous year's buds; pedicels glabrous or pubescent. Hypanthium obconic, 3–3.5 mm long, finely pubescent; calyx lobes oblong, 3.5 mm long, rounded or dentate at apex, slightly glandular, pubescent within and often on outside, reflexed after anthesis; petals white, obovate, 6–7.5 mm long; stamens usually 30; ovary 0.5–1 mm long, puberulous or glabrous. Fruit globose or ellipsoid, 2–3 cm long, purplish-red, with a gray-glaucous bloom; stone obovoid to subglobose, 1.2–1.8 cm long, 1–1.5 cm wide, smooth, tip rounded, one margin with a ridge, the other grooved. (n = 8) Apr–May. Locally common on prairie hillsides, open woodlands, stream valleys, thickets; se SD, NE, KS, IA, MO, OK; (IN, IA, se SD, s to TN, AR, TX, Mex.). *P. lanata* Mack. & Bush—Rydberg; *P. americana* Marsh. var. *lanata* Sudw.—Fernald.

Several horticultural varieties have been developed from this species.

7. Prunus munsoniana Wight & Hedr., wild goose plum. Shrub or small tree 3–5(8) m tall, with suckering habit and forming dense thickets; branchlets red-brown, glabrous, becoming grayish. Leaf blades oblong-lanceolate to lanceolate or somewhat broadly elliptic, 6–10 cm long, acute or acuminate, rounded to cuneate at base, finely glandular-serrate, bright green above paler below and densely villous when young, becoming glabrate and sparingly pubescent on veins, commonly conduplicate, rather thin at maturity; petiole 1.5–2 cm long, pubescent above, usually biglandular at apex; stipules 4–7 mm long, linear, often lobed, glandular, caducous. Flowers opening before leaves expand, on short lateral spurs, in clusters of 2–4(5); pedicels 10–15 mm long, glabrous, often with a small bract at base. Hypanthium obconic, 2–4 mm long, glabrous; calyx lobes ovate-oblong, 2–4 mm long, glandular-ciliate, glabrous or sparsely pubescent outside, pubescent inside near base; petals white, obovate to oblong-obovate, 4–7 mm long; stamens ca 30; ovary 1 mm long, ovoid, glabrous. Fruit globose to short-ellipsoid, 1.5–3 cm long, 2 cm in diam, red or reddish-yellow, with a slight bloom; stone oval, obiquely truncate at base, apex acute, ridged on dorsal margin, a narrow groove on the ventral. (n = 8) Mar–May. Locally common on rocky, open wooded hillsides, border of woods, prairie ravines, stream valleys, roadsides; se 1/4 KS, MO, OK; (OH, KY, KS, s to LA, OK, TX).

Several horticultural varieties have been developed from this species.

8. Prunus pensylvanica L. f., pin-cherry or bird cherry. A small tree 5–10 m tall with widely spreading branches; young branches red-brown, glabrous, usually glaucous. Leaf blades lanceolate or oblong-lanceolate to elliptic, oblong or obovate, 6–12 cm long, usually less than 1/2 as wide, gradually or abruptly long-acuminate, obtuse to rounded at base, finely and irregularly serrate, with a gland near the sinus; petiole 1–2 cm long, glabrous, usually glandular at summit; stipules narrowly lanceolate, 2–4 mm long, glandular-serrate, caducous. Flowers expanding with the leaves, in umbellike clusters of (1)2–5; pedicels 1–1.5 cm long, glabrous. Hypanthium campanulate, 1.5–1.8 mm long, glabrous; calyx lobes triangular, obtuse, glabrous, spreading; petals white, obovate, 5–7 mm long, with a short claw; stamens 10 or 20; ovary ovoid, 1.5–2 mm long, glabrous. Fruit red, globose, 5–7 mm in diam; stone subglobose, 6 mm long, 4.5 mm wide, base rounded, tip acute, with a ridge on ventral side. (n = 8) May–Jun. Dry or moist woods, ravines; MN, ND, e MT, BH, WY; (Newf., Que., to e B.C., s to NJ, WV, n IN, IA, SD, CO).

9. Prunus persica (L.) Batsch, peach. Small tree 3–10 m tall with glabrous branches and pubescent buds. Leaf blades elliptic-lanceolate, lance-oblong, to oblong-lanceolate, 7–15 cm long, usually widest about or slightly above the middle, long-acuminate and attenuate at apex, broadly cuneate at base, serrate or serrulate, glabrous; petiole 1–2 cm long, glabrous, glandular at summit or these on lower leaf margin; stipules lanceolate, 6–12 mm long, glabrous, caducous. Flowers usually solitary, sessile or subsessile, appearing before the leaves, 2.5–3.5 cm across; sepals pubescent on outer surface; petals pink to red. Fruit subglobose, 5–8 cm in diam, velvety, fleshy; stone very hard, deeply pitted and furrowed. (n = 8) Mar–May. Infrequent to locally common as an escape along roadsides, fence rows, edge of woods, stream valleys, & waste places; se GP; (s Ont. s through e U.S., w to GP; China). *Seminaturalized.*

10. Prunus pumila L. var. **besseyi** (Bailey) Gl., sand cherry, dwarf cherry. Low, much-branched, decumbent or ascending, rarely prostrate shrubs, 1–4(7) dm tall; branches red, glabrous, becoming grayish. Leaf blades oblanceolate, obovate, elliptical, 4–6.5 cm long, 1–2.6 cm wide, finely and remotely serrate to nearly entire especially below, acute to obtuse at apex, base cuneate, upper surface dark green, glabrous, lower surface paler, glaucous, margin somewhat cartilaginous; petioles 5–14 mm long, glabrous, with or without glands at summit; stipules narrowly lanceolate, 4–10 mm long, often with narrow lobes, glandular, glabrous. Flowers in clusters of 2–4, opening with 1/2 grown leaves; pedicels 4–12 mm long, glabrous. Hypanthium campanulate, 2.5–3.5 mm long, glabrous; calyx-lobes oblong, 1.2–1.7 mm long, toothed at apex, often glandular, glabrous; petals white, elliptic to oval, 6.5–7.6 mm long, abruptly short clawed; stamens 25–30; ovary ovoid, 1.2 mm long, glabrous. Fruit subglobose, 13–15 mm long, dark purple; stone oval, compressed, 7.6–8 mm long, 6.5–7.2 mm wide, apex acute, base rounded, surface slightly rugulose, margin with a ridge and a sulcus. (n = 8) Apr–May. Locally common on sandy or rocky prairies; w MN, ND, MT, WY, NE, IA, n-cen KS, ne CO; (restricted to GP). *P. besseyi* Bailey — Rydberg.

11. *Prunus rivularis* Scheele, creek plum, hog plum. A shrub 1–2(3) m tall, forming dense thickets from root-suckers; young branchlets glabrous or rarely pubescent, reddish-brown to grayish, somewhat glaucous. Leaf blades lanceolate to ovate-lanceolate or elliptic, 5–6(7) cm long, 2–4 cm wide, acuminate, cuneate or rounded at base, crenately glandular-serrate, normally conduplicate or trough-shaped, rarely flat, green and glabrous above, paler and pubescent along veins below; petioles (7)10–12 mm long, somewhat pubescent above, with 2(4) glands at summit or glands absent; stipules narrowly lanceolate, 5–9 mm long, glandular, puberulent or glabrate. Flowers opening with unfolding leaves, in clusters of 2–4(8); pedicels 6–15 mm long, glabrous to sparsely pubescent. Hypanthium obconic, 2–2.7 mm long, glabrous; calyx lobes oblong, 1.5–2 mm long, glandular, pubescent near base within, pubescent to glabrate outside; petals white, obovate or oblong-obovate, 5–6.3 mm long, abruptly short clawed; stamens 20; ovary ovoid, 1–1.5 mm long, glabrous. Fruit globose to subglobose, 1.3–2 cm long, yellowish-orange with a reddish cheek, or rarely red; stone oblong, 1.3–1.6 cm long, pointed at both ends, smooth or obscurely reticulate, margin with a ridge and a sulcus. Apr–May. Infrequent along creeks, edge of woods, prairie hillsides, roadsides; KS: Chautauqua, Cowley, Neosho; OK: Beckman, Caddo, Harper, Hughes, Lincoln; n-cen TX; (s KS, OK, cen TX, reported for w AR).

12. *Prunus serotina* Ehrh., wild black cherry. A tree 10–15 m tall in our area but often somewhat shrubby in fence rows and along roadsides; young branches reddish-brown, glabrous or puberulent. Leaf blades firm, lanceolate to oblong or oblanceolate, rarely ovate, 6–12 cm long, 2–5 cm wide, gradually or abruptly acuminate, acute or obtuse at base, finely incurved-serrate, with callous-tipped teeth, upper surface dark green, lustrous, glabrous, lower surface paler, glabrous, petioles 1–1.5 cm long on fertile branches to much longer on foliage branches, glabrous, usually with 2 glands at summit; stipules lanceolate, 4–7(10) mm long, glandular-toothed, caducous. Flowers in racemes (5)8–15 cm long, terminating leafy twigs of season; pedicels 3–6 mm long, glabrous. Hypanthium campanulate, 1.2–1.7 mm long, glabrous; calyx lobes oblong or triangular, 0.6–1.2 mm long, sometimes glandular-erose, glabrous, persistent; petals 2.5–4 mm long, blade subrotund, tapering into a claw 0.5–1.5 mm long; stamens (15)20; ovary ovoid, 1.6 mm long, glabrous. Fruit globose, 8–12 mm in diam, dark purple, glossy; stone subglobose, 5–7(9) mm long, surface smooth or rugulose, margin ridged on one side, sulcate on the other. (n = 16) Apr–May. Low or upland woods, along streams, fence rows, roadsides; se NE, e 1/3 KS, IA, MO; (N.S., Que., Ont., MN, s to FL, TX). *P. virginiana* L. — Rydberg.

Few trees of any size remain in our area and most stands are somewhat of a weed-tree of fence rows, roadsides, and woodland borders. Cattle are poisoned from eating leaves, particularly wilted ones that contain hydrocyanic or prussic acid.

13. *Prunus virginiana* L., choke cherry. A shrub or small tree 2–6(10) m tall, often forming thickets; branchlets red-brown to dark brown, glabrous. Leaf blades ovate to broadly elliptic or obovate, 4–12 cm long, 3–6 cm wide, abruptly or gradually acuminate at apex, broady cuneate to rounded at base, margin serrate with ascending teeth, dark green and somewhat lustrous above, glabrous, glaucescent or grayish-green beneath, glabrous or pubescent on veins and with axillary tufts of hairs, rarely puberulent on surface; petioles 1–3 cm long, glabrous, glandular at summit; stipules linear-lanceolate, 2–4 mm long, with slender teeth, caducous. Flowers in rather dense racemes, 4–9(15) cm long, terminating leafy twigs of the season; pedicels 3–6 mm long, pubescent. Hypanthium somewhat campanulate, 1.2–1.6 mm long, glabrous; calyx lobes broadly triangular, to nearly circular, 1–1.5 mm long; glandular-erose, deciduous soon after anthesis; petals white, 3–4 mm long, blade subrotund; stamens 20–30; ovary ovoid, 1.5–1.7 mm long, glabrous. Fruit globose, 8–11 mm in diam, crimson to deep red or bluish-purple to early black; stone globose, 6.8–7.4 mm in diam, smooth or slightly rugulose, margin ridged on one side, sulcate on the other. (n = 16) Apr–May. Locally common in open woodlands, prairie hillsides, rocky bluffs, canyons, fence rows, roadsides; GP except for se KS; (Newf. to Sask., s to NC, TN, MO, KS, OK, TX, NM). *P. melanocarpa* (A. Nels.) Rydb., *P. demissa* (Nutt.) Walp., *P. nana* Du Roi—Rydberg; *P. virginiana* var. *melanocarpa* (A. Nels.) Sarg.—Gates.

In ne KS, se NE, and scattered elsewhere in our area, some plants grow to 10 m tall, have thinner leaves with principal lateral veins relatively prominent, and mature fruit crimson to deep red. These have been referred to var. *virginiana* while those with lower leaf surface somewhat pubescent to forma *deamii* G. N. Jones. In most of our area plants are 2–6 m tall, with thicker leaves, principal lateral veins more obscure, and mature fruit bluish-purple or nearly black. These have been referred to var. *melanocarpa* (A. Nels.) Sarg. While the extremes are rather easily recognizable, intergradation is too common over our area to make subspecific distinctions meaningful. Perhaps careful field studies through the season would provide useful information.

14. PYRUS L., Crabapple, Pear, Apple

1. **Pyrus ioensis** (Wood) Carruth, Iowa or Bechtel crab. A small tree or shrub of irregular habit, sometimes forming thickets from sucker shoots, to 8 m tall; branches tomentose at first, becoming glabrous and reddish-brown; spur branches sometimes thorny. Leaves alternate, simple, ovate-oblong to broadly elliptic, 6–10 cm long, obtuse to short-acuminate, serrate and usually shallowly lobed; young leaves folded in bud, densely tomentose on both sides, upper surface becoming dark green and glabrate, the lower persistently pubescent; petioles 1–2.5(4) cm long; stipules (3)4–6(8) mm long, linear-attenuate, early deciduous. Flowers in terminal corymbs or umbellike cymes on spur branches, perfect, fragrant; pedicels 1–3.5 cm long, tomentose. Hypanthium turbinate, 4–5 mm long, densely tomentose; sepals 5, lanceolate, tomentose on both sides; petals 5, rose to whitish, obovate, clawed, 12–18 mm long, promptly deciduous; stamens many, anthers pink or salmon colored; ovary inferior, 5-celled; styles 5, densely pubescent below, basally connate. Pome subglobose, green or greenish-yellow, 2–3(4) cm in diam; seeds 6.5–8 mm long, dark brown, smooth. (2n = 34) Apr–May. Infrequent to locally common in open woods, thickets, pastures, along streams, with a preference for calcareous soils; se SD, s along e GP; (IN, MN, se SD, s to KY, LA, TX). *Malus ioensis* (Wood) Britt.—Rydberg.

Reports of *P. angustifolia* Ait. and *P. coronaria* L. (as species of *Malus*) by Rydberg and Gates were based on variants of *P. ioensis*. Various cultivars of *P. malus* L., apple, and *P. communis* L., pear, have commonly been planted and often persist around homesites but appear to never escape and become a part of the flora in our area.

15. ROSA L., Rose

Ours woody, suffruticose or subherbaceous shrubs with erect, arching, climbing or rarely decumbent to trailing stems; stems strict or branched; stem petioles, rachis, and inflorescence variously armed with prickles or bristles; sessile or stalked glands often present on various structures. Leaves alternate, petiolate, odd-pinnate or trifoliolate; leaflets toothed; stipules paired, conspicuous, usually persistent, adnate to petiole and forming wings, entire or rarely pinnatifid, glabrous to pubescent, sometimes glandular. Flowers solitary, or in bracteate corymbose or paniculate inflorescences terminating primary or lateral stems, the flowers mostly on short lateral shoots from previous years growth. Hypanthium urceolate to globose, the opening constricted; disk forming a ring around the opening; sepals 5, all entire or 2 outer and upper half of middle one toothed or with lateral appendages, apices acute to long attenuate or dilated; petals 5 (double-forms rare), various shades of rose, white, (yellow only in cultivated escapes), usually obovate, apices usually emarginate, inserted at outer edge of disk; stamens in several whorls, 50–200; ovaries numerous, free, inserted on the bottom or also on the sides of the hypanthium; styles distinct, barely exserted or long exserted from mouth of hypanthium or united into a column above; each ovary with one ovule. Fruit accessory (hip), of numerous achenes enclosed by the enlarged, fleshy hypanthium; achene bony, often pubescent; seed with thin coat.

Each of our wild rose species exhibits a considerable amount of variation and barriers to interspecific hybridization are weak. A number of hybrids and morphological variants have been named for our area and often several of these are to be found in a given colony. The following treatment focuses on the more clear-cut taxa which, with experience, can usually be recognized in the field. Casually collected specimens often will be most difficult to name, and more careful population studies should be made.

In addition to the taxa treated below a few introduced species have been found in and around abandoned farmyards. Such plants have not truly escaped nor become even a minor part of our flora, and they are increasingly more difficult to locate. The more common of these have been *Rosa canina* L., the dog rose, in se KS; *Rosa eglanteria* L. (*R. rubiginosa* L.—Rydberg), the sweetbriar, in e NE and KS; and *Rosa spinosissima* L. (*R. pimpinellifolia* L.—Rydberg), the Scotch rose, in KS.

1 Styles united into a column which is exserted from the orifice of the hypanthium.
  2 Leaflets usually 3(5); petals usually pink, 2–3 cm long; stipules stipitate-glandular; native plants .... 7. *R. setigera*
  2 Leaflets usually 7–9; petals usually white, 1–2 cm long; stipules pectinate-serrate and glandular-serrate; introduced plants .... 6. *R. multiflora*
1 Styles distinct, slightly or not at all exserted, the stigmas often closing mouth of hypanthium.
  3 Pedicels and hypanthium usually with stipitate glands which are more or less persistent; sepals usually spreading or reflexed after anthesis and eventually deciduous.
    4 Stems borne singly from stolons; flowers rose colored; 8–14 mm wide .... 4. *R. carolina*
    4 Stems not from stolons; flowers white or rose; leaflets 3–7 mm wide .... 5. *R. foliolosa*
  3 Pedicels and hypanthium glabrous; sepals usually somewhat erect, long persistent.
    5 Stems usually with well-defined infrastipular prickles .... 8. *R. woodsii*
    5 Stems without infrastipular prickles or these not differentiated from those of internodes.
      6 Leaflets usually 5–7; lowers borne on lateral branches from stems of previous year.
        7 Stems usually without prickles except at lower nodes .... 3. *R. blanda*
        7 Stems with most internodes prickly .... 1. *R. acicularis*
      6 Leaflets usually (7)9–11; flowers borne at summit of stem and from lateral branches of stems of previous year .... 2. *R. arkansana*

1. **Rosa acicularis** Lindl., prickly wild rose. Shrub 0.2–12 dm tall; stems usually more or less densely bristly with straight, slender, unequal prickles, the infrastipular prickles, when present, not clearly differentiated from those of internodes. Leaves alternate, odd-pinnate; rachis usually pubescent and glandular to subglabrous; leaflets 5–7(9), oblong-elliptic to ovate or obovate, 1.5–5(8) cm long, 1–3(5) cm wide, serrate or biserrate, teeth often gland-tipped, glabrous or glabrate above, lower surface paler, villous, an often glandular, sometimes glabrate; petioles with or without bristles, glandular to eglandular, usually puberulent, rarely glabrous; stipules with few to many glands. Flowers usually 1(2 or 3) on lateral branches of the season; pedicels without bristles, eglandular to rarely glandular.

Hypanthium glabrous, often with a distinct neck; sepals 5, 1.5–3(4) cm long, 2–4 mm wide at base, tapering into a slender portion which is often dilated above, glandular or more rarely eglandular on back, pubescent or glabrous, persistent and erect in fruit; petals 5, pink to rose, (1.5)2–3.2 cm long; stamens numerous. Achenes usually 15–30, 3.8–4.5 mm long, yellowish, with stiff hairs along one side or toward tip. ($2n = 42$) May–Jul. Infrequent to locally common on wooded hillsides, stream banks, & rocky bluffs; n ND, BH, & scattered in SD; (across Can., s to MN, NY, WV, MI, WI, SD, NM, ID). *R. bourgeauiana* Crep., *R. engelmannii* S. Wats.—Rydberg.

Our plants are the subsp. *sayi* (Schwein.) Lewis. Subsp. *acicularis* is Eurasian, extending into AK, and has pedicels conspicuously glandular ($2n = 56$). In addition to showing a considerable amount of minor variations, subsp. *sayi* hybridizes with *R. arkansana*, *R. blanda*, and *R. woodsii*.

2. Rosa arkansana Porter, prairie wild rose. Plants woody, suffruticose, or less commonly subherbaceous shrubs from stout horizontal roots; stems 1–5 dm tall and dying back partly or completely to the ground each year, or becoming truly shrubby to 1 m tall, usually densely to sparsely beset with slender, unequal prickles, the infrastipular ones, when present, not clearly differentiated from those of internodes. Leaves alternate, odd-pinnate; rachis glabrous or villous-puberulent, sometimes glandular; leaflets (7)9–11, obovate or obovate-oblong to elliptic, 1–4(6) cm long, 1–2.5(3.5) cm wide, serrate in upper 2/3, teeth rarely glandular-tipped, glabrous or more commonly villous-puberulent beneath; acute or cuneate at base, rounded, obtuse or acute at apex; petioles glabrous or pubescent, rarely glandular; stipules pubescent to glabrous, sometimes glandular-toothed. Flowers generally 3 or more, corymbose, terminating herbaceous stems of the season or short lateral branches from older stems. Hypanthium and pedicel glabrous; sepals 5, 1.5–2(3) cm long, 3–5 mm wide at base, stipitate-glandular on back, sometimes sparsely so, persistent and erect or spreading in fruit; petals 5, pink to white, rarely deep rose, 1.5–2.5(3) cm long; stamens numerous. Achenes usually 15–30, 3.5–5 mm long, plump, stiffly long hairy along one side. ($2n = 28$) May–Aug. Relatively common in prairies, bluffs, open woodlands, thickets, roadsides; GP but less common sw; (NY, Alta., s to IN, WI, MO, KS, OK, CO, NM, TX). *R. alcea* Greene, *R. conjuncta* Rydb., *R. lunellii* Greene, *R. polyanthema* Lunell, *R. subglauca* Rydb., *R. suffulta* Greene—Rydberg.

This is our most common and variable wild rose. Most of our specimens have the underside of the leaflet blades villous-puberulent and have been recognized as *R. suffulta* Greene or as *R. arkansana* var. *suffulta* (Greene) Cockll. A few plants in the w GP have leaflets glabrous beneath and would be referred to the Rocky Mountain var. *arkansana*, but all degrees of transition exist. In eastern Kansas, *R. arkansana* grows with the less common *R. carolina* and fertile hybrid swarms frequently exist. These hybrids have been referred to *Rosa* x *rudiuscula* Greene. Suspected hybrids between *R. arkansana* and *R. acicularis* appear commonly, while hybrids with *R. blanda* and *R. woodsii* are less frequent.

3. Rosa blanda Ait., smooth wild rose. An erect, rather stout, colony forming shrub, 0.5–1.5 m tall; stems branched above, unarmed or with few to many prickles toward the base, these rarely found sparsely on flowering branches. Leaves alternate, odd-pinnate; leaflets usually 5–7, elliptic-oblong to obovate, 1.5–4(5) cm long, 1–3 cm wide, coarsely serrate especially in upper 2/3, acute or cuneate basally, acute or obtuse to rounded apically, glabrous and yellowish-green above, pale and tomentulose to rarely glabrate below; stipules usually tomentulose. Flowers solitary or corymbose on lateral branches from stems of previous year. Pedicels and hypanthium glabrous to rarely sparsely hairy; sepals 5, 1.5–2 cm long, pubescent and glandular on back, erect after anthesis and persistent; petals 5, obovate, 2–3.2 cm long, pink and sometimes streaked with red, rarely white; stamens numerous. Achenes 20–40, ellipsoid-ovoid, 4–5 mm long, long-hairy along one side. ($2n = 14$) May–Jun. Rather rare to locally common on dry rocky hillsides, open woodlands, roadsides; ND, SD, e 1/2 NE & ne 1/4 KS; (Que. to Man., s to NY, PA, IN, n MO, ND, ne KS). *R. subblanda* Rydb.—Rydberg.

This species appears to hybridize easily with *R. woodsii* and less successfully with *R. arkansana* and *R. acicularis*. It is highly variable, and a number of trivial morphological variants have been named.

4. ***Rosa carolina*** L., pasture rose. Slender shrub 2–7(10) dm tall; stems usually borne singly from stolons, erect to ascending often semiherbaceous, simple or little branched above, usually armed with variable-sized terete prickles, infrastipular prickles, when evident, scarcely differentiated; stipules entire to glandular-dentate. Leaves alternate, odd-pinnate; leaflets (3)5–7(9), elliptic, obovate or obovate-oblong, rarely ovate-lanceolate, acute or obtuse apically, acute basally; blades 1.5–4 cm long, usually widest at middle, serrate in upper 2/3 and sometimes with marginal glands, dark green and glabrous above, lower surface paler, softhairy to glabrous; rachis glabrous or stipitate-glandular; stipules entire or glandular-dentate near tip, pubescent. Flowers usually borne singly at tip of stems of the season or rarely in corymbs. Pedicels and hypanthium conspicuously stipitate-glandular, rarely smooth; sepals 5, 2.6–2.2 cm long, 3–4 mm wide at base, attenuate into a linear tip, sometimes with an apical foliaceous appendage, glabrous or stipitate-glandular; petals 5, pink, obovate, 2–2.5(3.2) cm long; stamens numerous. Achenes 15–35, semiobovoid (straight on one side), yellowish to reddish-brown, 4–5 mm long, long-hairy on one side. (2n = 28) May–Jun. Locally common in open woodlands, prairies, thickets, roadsides, pastures; e 1/4 KS, MO, e 1/2 OK: (e 1/2 U.S., w to MN, MO, e KS, e TX). *R. lyoni* Pursh, *R. serrulata* Raf. — Rydberg; *R. carolina* var. *grandiflora* (Baker) Rehd., var. *villosa* (Best) Rehd., *R. carolina* var. *carolina* f. *glandulosa* (Crep.) Fern. — Steyermark.

This is a highly variable species, and in our area it is nearly always found with *R. arkansana* with which it apparently hybridizes freely.

5. ***Rosa foliolosa*** Nutt. ex T. & G., leafy rose, white prairie rose. Low shrubs 2–6 dm tall; stems from rhizomes or running roots, erect or ascending, slender, usually with slender infrastipular and internodal prickles or nearly unarmed, rarely bristly. Leaves alternate, odd-pinnate; leaflets 7–9(11), blades elliptic, lanceolate or narrowly oblong, 1–3 cm long, (3)4–6(8) mm wide, finely serrate, teeth with few or no glands, glabrous above, glabrous or pubescent on veins below. Petioles with few or many bristles, glandular, glabrous or somewhat pubescent; stipules narrow, slightly glandular on margin and back or eglandular. Flowers usually solitary or 2–5, terminating shoots or branches of the season. Pedicels and hypanthium glandular-hispid or hypanthium smooth; sepals 5, glandular-hispid, 12–16 mm long, 3–4 mm wide at base, narrowly attenuate; petals 5, white or light pink, obovate, 15–20 mm long; stamens numerous. Achenes 10–20, yellowish to reddish-brown, 3–3.5 mm long, plump, straight on one side, with long white hairs on one side. (n = 7) May–Jun! Rare to locally common in sandy or rocky open woodlands & prairies, thickets, roadsides, & stream valleys; KS: Cherokee; extreme se GP; (se KS, w AR, e 1/2 OK, e & ne TX).

This is one of our most distinctive wild rose species and exhibits little morphological variation. It has been confused with small-leaved forms of *R. carolina* and rare sterile specimens may be hybrids between *R. carolina* and *R. foliolosa*.

6. ***Rosa multiflora*** Thunb., multiflora rose, Japanese rose. Rather stout shrub 1–5 dm tall; stems few to many from base, much branched, erect and arching to more or less trailing or sprawling, armed with stout, recurved prickles. Leaves alternate, odd-pinnate; leaflets (5)7–9(11), rather membranaceous, ovate, obovate, or oblong, 2–4 cm long, obtuse to acute or acuminate, glabrous or nearly so above, usually soft-pubescent below; petioles pubescent; stipules fimbriate-pectinate and usually stipitate-glandular. Flowers few to abundant in rounded or pyramidal inflorescences; pedicels with stipitate glands or eglandular. Hypanthium glabrous to usually loosely pubescent; sepals 5, 7–10 mm long, glabrous or stipitateglandular, often with attenuate tip; petals 5, white, rarely pink, obovate, 7–10 mm long; stamens numerous; styles connate above and exserted from orifice of hypanthium. Achenes

yellowish, 4-4.5 mm long, somewhat flattened, with hairs above on one side. (n = 7) May-Jun. Locally common along roadsides, in pastures, open woodlands, stream valleys, & waste places; se NE, e 1/2 KS; (e 1/2 U.S.; introduced from Asia).

This species was formerly recommended and much planted as a living fence and wild-life cover. Unfortunately, the plant readily escapes and becomes weedy. A number of minor variants are found, of which the more common are double-flowered, pink-flowered, and large-leaved forms.

7. *Rosa setigera* Michx., climbing rose, prairie rose. Climbing, sprawling or trailing shrubs or in the open an erect shrub with arching to drooping stems; stems 2-5(10) m long, armed with remote rather stout prickles, occasionally hispid, rarely unarmed. Leaves alternate, 3(5)-foliolate; blades of leaflets ovate or ovate-oblong, 3.5-5(8) cm long, 1.5-4 cm wide, sharply serrate, acute or acuminate, rounded at base, usually green, lustrous, and glabrous above, tomentose below to pubescent only on veins or glabrous, eglandular or slightly glandular; petioles pubescent and glandular or glabrous or eglandular; stipules with a few stipitate glands. Flowers 5-15 in corymbs. Pedicels and hypanthium glandular-hispid; sepals 5, 1-1.6 cm long, ovate to lanceolate, attenuate, glandular-hispid and pubescent on outside, tomentose within, deciduous; petals 5, roseate and fading to whitish, obcordate, 3-3.5 cm long; styles united into a column and exserted from orifice of hypanthium. Achenes yellowish, rounded on one side, somewhat angular, 3.8-4.2 mm long, with long white hairs on one side. (n = 7) May-Jul. Locally common in open woodlands, prairie thickets, pastures, stream valleys, roadsides; se GP; (MA, NY, MI, IA, s to ne FL.). *R. rubifolia* R. Br.— Rydberg; *R. setigera* var. *tomentosa* T. & G.—Fernald.

The plants of our area normally have the undersurface of the leaflets variously pubescent and have been referred to var. *tomentosa*. A few specimens from se KS have leaflets glabrous below and have been referred to var. *setigera*, but they are infrequent and mixed with the common pubescent form.

8. *Rosa woodsii* Lindl., western wild rose. A much-branched shrub 2-15 dm tall and usually forming thickets; stems commonly with infrastipular prickles clearly different from those of internodes, with reddish-brown bark on branchlets. Leaves alternate, odd-pinnate; leaflets 5-9, elliptic to oval or elliptic-obovate, 2-2.5 cm long, coarsely serrate, teeth often gland-tipped at least on one side, puberulent or glandular below to glabrous; petioles 1.5 cm long, pubescent; stipules glabrous or pubescent, glandular to eglandular on the margins. Flowers in corymbiform cymes terminating growth of the season, sometimes solitary; pedicels glabrous or stipitate-glandular. Hypanthium glabrous; sepals 5, 1-2 cm long, 1.5-3.5 mm wide at base, attenuate tip often dilated, pubescent, rarely glandular, persistent; petals 5, obovate, 2-2.5 cm long, pink to deep rose. Achenes 15-35, 3-4 mm long, hairy along one side. (2n = 14) May-Jul. Common in rocky prairie ravines, open woodlands, roadsides, stream valleys; n 2/3 GP, w KS, nw OK; (Ont., MN, B.C., s to IA, w KS, TX, Mex.). *R. fendleri* Crep., *R. macounii* Greene, *R. pyrifera* Rydb., *R. terrens* Lunell—Rydberg.

This quite variable species is known to hybridize with *R. acicularis*, *R. arkansana*, and *R. blanda*.

16. RUBUS L., Blackberries, Dewberries, Raspberries

Armed or unarmed erect, trailing, or scrambling plants with stems from perennial rootstocks or creeping stems, semiwoody or rarely herbaceous; shoots mostly biennial, those of first year (primocanes) from buds at or below ground level, bearing leaves in whose axils the buds produce lateral branches with leaves and flowers the second year, the cane usually dying after fruit matures (the whole second-year cane called a floricane); reproducing vegetatively by root or stem suckers or rooting stem tips; armature of canes, inflorescence, petioles, and calyx lobes of prickles or bristles or absent; indumentum of simple and/or sessile or stalked glandular trichomes, or these nearly absent. Leaves deciduous or rarely persistent through winter; those of primocanes compound; those of the floricane often partly simple, usually smaller, and often of a different shape, or principal foliage-leaves simple and lobed; stipules linear, small to conspicuous, free to adnate to base of petiole, persistent or caducous. Flowers solitary or in simple or compound cymes, racemes, or panicles, usually showy, 5-merous, perfect (rarely plants dioecious). Hypanthium small, flat to hemispheric, nectar-ring often prominent; calyx of 5-valvate, ascending to reflexed sepals, persistent, commonly ending in a caudate appendage, ebracteate; petals as many as the sepals, erect or spreading; stamens numerous; pistils simple, many, inserted on a convex to conic receptacle which often elongates in fruit; ovules 2, only one maturing; style filiform or clavate. Fruit a cluster of red to black 1-seeded drupelets on a dry or spongy, often elongated receptacle, the drupelets falling individually or coalescent and either falling from the receptacle as a unit or with it; stones hard, variously textured; seeds filling the stones. *Rubacer* Rydb.—Rydberg.

The taxonomy of *Rubus* is complicated by hybridization, polyploidy, and apomixis, with the blackberries being particularly difficult. In addition, collectors have not gathered both primocanes and floricanes when preparing specimens, though both are essential to identifying most species. Much more careful collecting with notes on habit is necessary in our region. The following treatment is conservative and no doubt will be much modified when our plants are better studied.

1 Stems without prickles or stiff bristles.
  2 Leaves compound .... 9. *R. pubescens*
  2 Leaves simple, usually lobed.
    3 Flowers solitary; principal leaves 3-6 cm wide .... 2. *R. deliciosus*
    3 Flowers in clusters of 2-9; principal leaves 11-18 cm wide .... 7. *R. parviflorus*
1 Stems armed at least with some prickles or stiff bristles.
  4 Primocanes normally trailing or lying on ground or low-arching, rooting at the tip.
    5 Stems with glandular bristles particularly near cane tips .... 10. *R. trivialis*
    5 Stems without glandular bristles .... 3. *R. flagellaris*
  4 Primocanes erect to ascending, often with arching tips which may root.
    6 Lower surface of leaflets distinctly white or whitened; fruit separating from the receptacle which is persistent on pedicel.
      7 Pedicels with bristles and glands; 5-foliolate leaves of primocanes, when present, pinnate; stems usually not glaucous; fruit red .... 4. *R. idaeus* subsp. *sachalinensis*
      7 Pedicels eglandular; 5-foliolate leaves of primocanes, when present, digitate; stems often very glaucous; fruit black .... 5. *R. occidentalis*
    6 Lower surface of leaves green or grayish-green, not whitened; fruit falling with the receptacle.
      8 Pedicels with numerous glandular hairs and these often on other parts of inflorescence and young stem tips of primocanes .... 1. *R. allegheniensis*
      8 Pedicels and young growth eglandular or glands inconspicuous.
        9 Terminal leaflet of primocane leaves widest near the middle; main inflorescence exserted above subtending foliage, of a short raceme type ..... 6. *R. ostryifolius*
        9 Terminal leaflet of primocane leaves widest well below the middle; main inflorescence usually hidden in subtending foliage, of a corymbiform type with lowest pedicels elongated .... 8. *R. pensilvanicus*

1. ***Rubus allegheniensis*** Porter, common blackberry. Stems erect or nearly so, 0.5-3 m long, young primocanes usually sparsely glandular, canes with nearly straight prickles which spread at right angles or are barely reflexed and strongly flattened at base; prickles of petioles, pedicels, and midveins similar but clearly hooked. Primocane leaves soft pubescent below; terminal leaflet ovate-oblong to ovate, 1-2 dm long, widest near or just below middle, acuminate, finely and sharply serrate, rounded, truncate or subcordate at base; lateral leaflets usually 4, smaller. Inflorescence racemose, elongate, somewhat cylindric, many flowered, lower 1-3 flowers subtended by leaves, the remainder by stipules; pedicels stipitate-glandular and tomentulose. Lobes of calyx ovate to ovate-lanceolate, commonly short-caudate or rarely acute; petals 1-2 cm long, 4-12 mm wide, white, cuneate and distinct

at base. Fruit globose to elongate, purplish-black, 1–2 cm long; stones yellowish, 2.5–3 mm long, with coarse reticulations. (n = 7) May–Jun. Relatively rare in dry pastures, prairies, margins of woodlands, roadsides, rocky open wooded hillsides, stream valleys; e 1/3 KS; (N.S., Que. to MN, s to NC, TN, MO, e KS). *R. nigrobaccus* Bailey—Rydberg.

2. ***Rubus deliciosus*** Torr., thimbleberry, boulder raspberry. Shrub to 1.5 m tall, stems unarmed, erect or ascending, branches woody, older bark exfoliating, young branches pubescent and stipitate-glandular. Leaves simple, 3–6 cm wide, usually 3- to 5-lobed, orbicular-reniform, dentate, cordate at base, sparsely pubescent and becoming glabrate above, puberulent to glabrous below; petioles pubescent and stipitate-glandular; stipules lanceolate to ovate-lanceolate, 7–12 mm long, attenuate, pubescent, sparsely glandular, semipersistent. Inflorescence a single terminal flower on a short leafy branchlet. Hypanthium flat, nectar-ring prominent; sepals 9–18 mm long, pubescent, often glandular, ovate to ovate-lanceolate, acuminate or with foliaceous tips; petals white, 1.5–3 cm long, obovate. Fruit hemispheric, 8–12 mm in diam, hardly edible; drupelets dryish; stones 3–3.5 mm long, areolate to reticulately ridged. May–Jun. Infrequent to common on rocky canyon slopes of the plains, foothills, & montane zones; WY: Platte, se CO, OK: Cimarron, ne NM; (se WY, CO, OK, NM).

3. ***Rubus flagellaris*** L., northern dewberry. Primocanes prostrate or low-arched, 0.5–5 m long, usually rooting at tips, armed with few to many stout, curved, somewhat recurved prickles with expanded bases, never bristly; floricanes trailing, flowering branches erect, armature similar to that of primocanes. Leaves of primocanes with 5(3) leaflets, those of floricanes with 3 leaflets; leaflets ovate to lance-elliptic, 2–7 cm long, 0.5–6 cm wide, finely pubescent to glabrous or rarely densely pubescent beneath, serrate, acute to acuminate, base rounded to cuneate; terminal leaflet of 3-foliolate floricane-leaves ovate, broadly rounded to subcordate at base, varying to oblong, oblanceolate or obovate and narrowly to broadly cuneate below. Inflorescence a flat-topped or elongate cyme, or flowers solitary near tip of floricane; pedicels and rachis pubescent and often stipitate-glandular. Sepals lance-ovate, 5–8 mm long, densely pubescent; petals white or rarely pinkish, obovate, 1.5–3 cm long. Fruit 1–2.5 cm long and nearly as wide, black, juicy and sweet; stones 3–3.4 mm long, yellowish, strongly areolate or reticulate. Apr–Jun. Locally common in rocky open woods, thickets, prairies, pastures, roadsides, & railroad embankments; MN, IA, se NE, e 1/2 KS, OK; (e Can. to MN, s to GA, AR, e KS, e TX). *R. aboriginum* Rydb., *R. baileyanus* Britt., *R. plicatifolius* Blanch.—Rydberg; *R. occidualis* Bailey—Fernald; *R. oppositus* Bailey—Bailey, Gentes Herb. 7: 510. 1949; *R. hancinianus* Bailey—Bailey, Gentes Herb. 5: 289. 1943.

This is a highly variable species in our area, and the named taxa do not acount for all the variants to be found. Plants with the terminal leaflet of 3-foliolate floricane-leaves cuneate below and with sides below the middle nearly straight could be named *R. enslenii* Tratt.; plants with floricane leaves soft or velvety-hairy have been named *R. occidualis* Bailey; central KS plants with primocanes at first erect but recurving and becoming prostrate have been named *R. hancinianus* Bailey; a similar plant of central KS but with many prickles has been named *R. oppositus* Bailey. All of these intergrade so completely as to make distinctions meaningless.

4. ***Rubus idaeus*** L. subsp. ***sachalinensis*** (Levl.) Focke var. ***sachalinensis***, red raspberry. Stems erect or arching, 0.5–2(3) m tall, sparsely to copiously bristly and prickly, otherwise glabrous to sparsely pubescent, bark exfoliating; floricanes and inflorescence eglandular or stipitate-glandular. Leaves 3- to 5-foliolate; leaflets ovate to ovate-lanceolate, 4–10 cm long, usually acuminate, base rounded to cuneate, margins serrate or biserrate, upper surface dark green, sparsely pubescent, lower surface permanently grayish-lanate; stipules linear-subulate, 4–10 mm long, usually caducous. Inflorescence leafy, racemose to thyrsoid; flowers several, usually 1–4 per axil. Calyx more or less lanate, eglandular or stipitate-

glandular, lobes reflexed, lanceolate to lanceolate-caudate, 4–7 mm long; petals white, erect or ascending, narrowly oblong to spatulate, 4–6 mm long. Fruit hemispheric, 12–18 mm in diam; drupelets easily separating, reddish, finely tomentulose; stones 2.4–2.6 mm long, whitish, reticulate. May–Jul. Open wooded hillsides, ravines, stream banks, often in rocky places; MN, ND, e MT, WY, s to e SD, nw MO; (Newf. to AK, s to IA, TN, NC, & in w to NM, AZ, WA, OR). *R. melanolasius* Focke, *R. strigosus* Michx.—Rydberg.

The mainly Eurasian subsp. *idaeus* is an eglandular plant often cultivated in N. America and sometimes escaping. The sporadic eglandular native N. American plants are considered to be mutants.

5. ***Rubus occidentalis*** L., black raspberry. Stems erect or ascending, sometimes arching and short-trailing, then rooting at tip, 0.5–2.5 mm long, eglandular; primocanes distinctly glaucous, with sparse, stout, straight, or hooked prickles with enlarged bases, as are petioles and pedicels. Primocane-leaves with 3–5 leaflets, green and glabrous above, densely white tomentose below; leaflets ovate to ovate-lanceolate, 5–9 cm long, 3–7 cm wide, base rounded or subcordate, doubly serrate, sometimes with small lobes; floricane-leaves usually with 3 leaflets or uppermost simple, otherwise similar to those of primocanes. Inflorescence of 3–7 flowers in an umbelliform cluster. Sepals lanceolate, tomentose, 6–8 mm long, elongating in fruit; petals white, obovate, 3–4.5 mm long, usually less than 1/2 as long as sepals. Fruit hemispherical, 12–15 mm in diam; drupelets black, glaucous, separating as a unit from receptacle; stones 2.5 mm long, yellowish, areolate-reticulate. Apr–Jun. Locally common in open woods, thickets, along hillsides, pastures, roadsides, often in dry rocky sites; MN, se ND, e & s SD, e 2/3 NE, e 1/2 KS, IA, MO, OK; (Que. to ND, s to GA, AR, OK, KS).

A number of horticultural varieties have been developed from this species and are frequently cultivated.

6. ***Rubus ostryifolius*** Rydb., high-bush blackberry. Stems 1–2(2.5) m tall, erect or ascending, often arched, glabrous; primocanes usually with sparse straight, slender prickles with bases slightly expanded, often some prickles stout and with widened bases. Leaves of primocanes with (3)5 leaflets, terminal leaflet usually oblong-ovate or oblong-ovate to narrowly ovate, 6–10 cm long, 3–5 cm wide, normally widest near or above the middle, short-acuminate, base obtuse to truncate, softly pubescent below, midrib with a few hooked prickles; floricane-leaves with 3 leaflets, similar to those of primocane but smaller; petioles 3–6 mm long, pubescent, with slender recurved prickles. Inflorescence terminal, usually a short raceme surpassing leaves, with 1–7 flowers, lower flowers subtended by leaflike bracts. Calyx lobes 5–6 mm long, acute, pubescent on both sides; petals white, obovate, 1.7–2 cm long. Fruit cylindric or globose, 1.5–2.5 cm long; drupelets black at maturity, falling with receptacle; stones 2.6–3 mm long, yellowish, rather uniformly areolate-reticulate. Apr–Jun. Open woodlands, prairie ravines, thickets, pastures, roadsides, railroad embankments; se NE, IA, MO, e 1/3 KS; (ME, MI, e MN, s to VA, KY, AR, KS, OK). *R. argutus* Link—Rydberg(?); *R. laudatus* Berger—Gates.

This species is very similar to *R. pensilvanicus* Poir. which has terminal leaflet widest near the base and main inflorescence shorter and rather hidden in subtending foliage. Many specimens can only be arbitrarily assigned to either of these two species.

7. ***Rubus parviflorus*** Nutt., thimbleberry. An erect unarmed shrub 0.5–1.5(3) m tall; stems stipitate-glandular, puberulent, becoming glabrate, with gray exfoliating bark. Leaves simple, palmately (3)5(7) lobed, 6–15(20) cm long, 11–18 cm wide, doubly serrate, glabrous to somewhat hairy, sometimes with a few stipitate glands; petioles 6–15 cm long, puberulent and with stipitate glands or glabrous; stipules membranaceous, lanceolate or ovate-lanceolate, 5–12 mm long, pubescent to glabrate, stipitate-glandular. Inflorescence a terminal corymb with (2)3–7(10) flowers or a flat-topped panicle. Calyx pubescent to villous, often stipitate-glandular, the lobes spreading, 10–16 mm long, with a caudate appendage

about 1/2 the length; petals white or rarely pinkish, oblong-ovate to obovate, (1)2-2.5 cm long. Fruit a thimblelike aggregate; drupelets reddish, pubescent, falling as a unit from persistent receptacle; stones 1.8-2.2 mm long, areolate-reticulate. May-Jul. Open wooded hillsides, stream banks, canyons; BH, WY, MT; (Ont. to n MN, AK, s to s CA, WY, CO, NM, n Mex.). *Rubacer parviflorum* (Nutt.) Rydb.—Rydberg.

Rubus odoratus L., the flowering raspberry, similar to this species but with rose-purple flowers, was reported by Gates as sometimes escaping in KS, but we have found no evidence for this statement.

8. ***Rubus pensilvanicus*** Poir., high-bush blackberry. Stems stout, 1-3 m tall, nearly erect, floricanes becoming arched; prickles of primocanes straight or slightly reflexed, with expanded bases. Primocane leaves (3)5-foliolate, softly pubescent below, terminal leaflet ovate, 6-12 cm long, 6-8 cm wide, widest well below middle, acuminate, rounded or subcordate at base, coarsely serrate to doubly serrate; floricane leaves usually with 3 leaflets or some of them simple, blades elliptic to rhombic or obovate, coarsely toothed. Inflorescence usually a short compact raceme with 3-9 flowers, usually shorter than leaves, pedicels 2-4 cm long, each or nearly all flowers subtended by a large bract. Calyx lobes ovate, apiculate, tomentose inside and on margins, 6-10 mm long; petals white, obovate, 15-20 mm long, 8-12 mm wide. Fruit 1-1.5 cm long, 1 cm in diam, falling with receptacle; drupelets black at maturity, juicy; stones yellowish, 3 mm long, areolate-reticulate. Apr-Jun. Open woodlands, prairie ravines, thickets, pastures, roadsides; se NE, e 1/3 KS; (Newf. to Ont., MN, s to VA, AL, TN, KS, AR, OK). *R. frondosus* Bigel.—Rydberg; *R. mollior* Bailey, *R. orarius* Blanch.—Steyermark; *R. alumus* Bailey—Fernald.

Some rare plants with elongated racemes have been referred to *R. mollior* Bailey, and a few plants with numerous stipitate-glands on the pedicels have been named *R. orarius* Blanch. Only dedicated study and field work can determine if these distinctions are meaningful in our area.

9. ***Rubus pubescens*** Raf., creeping blackberry, dwarf blackberry. Horizontal stems slender, long-creeping at or near surface of soil, often rooting at nodes or at the tip, sometimes loosely ascending, unarmed, bearing erect herbaceous flowering branches 1-4(6) dm tall, slightly pubescent above, rarely with a few bristles. Leaves 2-5 on herbaceous shoots, 3(5)-foliolate; leaflets rhombic-ovate to obovate or lanceolate, 2-6 cm long, coarsely and doubly serrate, often entire along cuneate lower portion, glabrous or nearly so, rarely closely pilose, stipules oblanceolate, to 1 cm long. Inflorescence terminal, flowers usually 1-2(7), axis often stipitate-glandular. Calyx pilose, often glandular, the lobes lanceolate, 3-6 mm long, reflexed; petals white, rarely pinkish, or greenish-white, oblong-lanceolate to spatulate, often erect, 4-8 mm long. Fruit dark red, hemispheric, 5-12 mm in diam, drupelets tardily separating from receptacle; stones white, 2.6-2.8 mm long, smooth or irregularly reticulate with low ridges. May-Jul. Moist woods, stream banks, open woodlands, bog areas; MN, e ND, ne SD, BH; WY: Crook; NE: Cherry; CO; (Lab. to Yukon, B.C., s to NJ, WV, IN, NE, CO, WA). *R. pubescens* Raf. var. *pilosifolius* A. F. Hill—Fernald.

10. ***Rubus trivialis*** Michx., southern dewberry. Trailing or low-arching, with primocanes rooting at the tip, armed with short, stout, somewhat recurved prickles with expanded bases, and with sparse to usually numerous prominent glandular bristles particularly near apex. Primocane leaves 5-foliolate, somewhat coriaceous, semievergreen and turning reddish in the winter; terminal leaflet with a prickly or hispid petiolule to 1/3 length of blade, blade elliptic or oblong, to 1 dm long, coarsely toothed; lateral leaflets sessile; floricane leaves 3-foliolate, smaller. Flowers solitary or in 3-flowered cymes. Sepals 5-7 mm long, pubescent to glabrescent, or glandular; petals obovate, white, 1.5-2.5 cm long. Fruit subglobose to elongate, black, 1-3 cm long, very juicy; stones oblong, 3 mm long, with irregular ridges. Apr-Jun. Stream banks, roadsides, thickets, old fields; s 1/2 MO, KS: Cherokee, Labette; e OK; (se VA, MO, KS, s to NC, GA, FL, TX).

Rubus hispidus L. was reported for KS by Rydberg, but we have not found specimen evidence. It differs from *R. trivialis* in having eglandular bristles on the stem and a racemiform inflorescence.

17. SANGUISORBA L., Burnet

1. ***Sanguisorba annua*** (Nutt. ex Hook.) T. & G., prairie burnet. Glabrous winter annuals from a taproot; stems erect, freely branched, 1.5-4 dm tall. Leaves alternate, pinnately compound, 2-7 cm long; leaflets (7)11-15(17), obovate, 7-15 mm long, deeply pectinately divided into narrow lobes; stipules adnate to the petiole, pectinately lobed and resembling leaflets. Flowers in dense subglobose to cylindric heads 1-3 cm long, on naked peduncles terminating the stems. Hypanthium turbinate, constricted at throat, small; sepals 4, greenish, 2-3 mm long, scarious-margined; petals 0; stamens 4; pistils 1(3), free from hypanthium; ovary with 1 ovule. Fruit accessory, of 1(2) achenes enclosed within the 4-winged, dry hypanthium tube; achene globose, 1.5-2 mm in diam. (2n = 14) Mar-Jun. Infrequent to common on sandy or gravelly soils in prairies, pastures, open woodlands, roadsides, & waste places; KS: Morton; s GP; (AR, KS, OK, TX, adventive in SC, MA). *Poteridium annuum* (Nutt.) Spach — Rydberg.

18. SORBUS L., Mountain Ash

1. ***Sorbus scopulina*** Greene, mountain ash. Erect, several-stemmed shrub 1-4(5) m tall; young twigs pubescent becoming glabrous and yellowish to red-brown or grayish-red; winter buds glutinous, glabrous to sparsely pubescent. Leaves alternate, odd-pinnate; leaflets 9-13, nearly sessile, oblong-eliptic to oblong-lanceolate, 3-7 cm long, acute to shortly acuminate at apex, cuneate to acute basally, finely serrate, glabrous or nearly so above, paler and glabrous to sparsely pubescent below; petioles 2.5-4 cm long; stipules ovate-lanceolate to linear, 6-9 mm long, deciduous before end of anthesis. Flowers in terminal or axillary compound nearly flat-topped corymbs, complete, regular. Hypanthium turbinate-obconic, adnate nearly to top of ovary, free portion disk-lined; the 5 sepals triangular, 1.5-2 mm long, persistent; petals 5, white or yellowish, orbicular, 5-7 mm long, pubescent near base on upper side; stamens 15-20; carpels 3-4. Pome subglobose, orange-brown, glossy, 9-12 mm in diam; seeds 4-4.5 mm long, brown, smooth. May-Jun. Infrequent on moist rocky wooded hillsides & stream banks; SD: Lawrence; WY: Crook, Converse, Sheridan; (w Alta., AK, s to NM, TX, UT, WA, OR).

The European *Sorbus aucuparia* L., a small tree with some leaves with over 13 leaflets, is planted in our area, but has not truly escaped. The genus *Sorbus* is often merged, and perhaps correctly so, with *Pyrus*.

19. SPIRAEA L., Meadow-sweet

Ours erect, sometimes rhizomatous shrubs. Leaves alternate, simple, usually serrate, petiolate and without stipules. Inflorescences corymbose or paniculate; flowers perfect; hypanthium cupulate or turbinate; sepals 5, persistent; petals 5, white or pinkish; stamens numerous, persistent; pistils 5, styles terminal; ovules 2-several; follicles 2- to several-seeded, ventrally dehiscent; seeds fusiform.

Reports by Rydberg of *Spiraea densiflora* Nutt. in SD, *S. salicifolia* L. for KS, *S. latifolia* (Ait.) Borkh. for NE, and *S. tomentosa* L. for KS are without specimen evidence in our herbaria.

Spiraea prunifolia Sieb. & Zucc., *S. salicifolia* L., *S. thunbergii* Sieb., and more commonly *S. vanhouttei* (Briot.) Zabel have been cultivated and are sometimes found persisting near old homesites but not as spreading escapes.

1 Leaves lance-oblong, usually 2× or more longer than wide; inflorescence elongate and paniculate; plants of ND and e SD .... 1. *S. alba*
1 Leaves ovate-oblong or obovate, usually about 1/2 as wide as long; inflorescence a nearly flat-topped corymb; plants of the BH .... 2. *S. betulifolia*

1. **Spiraea alba** Du Roi, meadow-sweet. Erect shrubs 0.5–1.2 m tall, usually forming colonies; branches puberulous while young, reddish-brown, striate and somewhat angular. Leaves alternate, elliptic to oblanceolate, 3–6 cm long, 10–20 mm wide, finely serrate, glabrous or sparsely puberulous. Flowers in terminal thyrsoid panicles 5–20 cm long, branches puberulent and subtended by lance-linear bracts; pedicels 2–5 mm long, with a linear bractlet usually midway on axis. Hypanthium cupulate, 1–1.8 mm long, 1.5–2 mm wide, glabrous or puberulent, 10-nerved; sepals 5, triangular, obtuse, 1–1.5 mm long, puberulent; petals 5, suborbicular, white, 2.7–3.5 mm long; stamens 25–50, inserted at edge of hypanthium between perianth and nectar-ring; pistils simple, usually 5 distinct; ovaries with 2-several ovules. Follicles glabrous, 3–3.5 mm long, dehiscent completely along one suture and apically on the other; seeds irregularly fusiform, 2–2.3 mm long, faintly areolate. (2n = 36) Jun–Aug. Locally common in moist meadow, marshes, moist open low woodlands; often in sandy soils; ND except for sw 1/6, extreme e SD, reported for BH; (w Que. to Alta., s to VA, NC, IN, IL, MO, ND, SD).

2. **Spiraea betulifolia** Pall., wild spiraea. Shrubs 3–5(10) dm tall, usually rhizomatose, simple or sparsely branched, glabrous or nearly so; branches smooth, reddish-brown. Leaves alternate, ovate-oblong or obovate, 2–5(7) cm long and usually 1/2 as wide, rounded to subcordate basally, obtuse or rounded apically, irregularly serrate and sometimes slightly lobed, green above, pale on lower surface; petioles 4–10 mm long; stipules absent. Inflorescences of terminal corymbs 3–8 cm wide, nearly flat-topped; branches glabrous, subtended by lance-ovate bracts; pedicels 3–8 mm long, glabrous, with a linear bractlet along axis. Hypanthium cupulate, 1–1.5 mm wide, glabrous outside, sometimes slightly pubescent within; sepals 5, triangular, 0.5–1 mm long, sparsely hairy at tip and on inner surface; petals 5, white or with a pinkish tinge, obovate to oblong, 1.8–2.2 mm long, often notched at tip; stamens 15–25, inserted at edge of hypanthium between perianth and nectar-ring; pistils 5, simple, distinct; ovaries with 2–several ovules. Follicles glabrous or ciliate along sutures, 2.8–3.5 mm long, dehiscent completely along one suture and apically on the other; seeds irregularly fusiform, 1.7–2 mm long, yellowish to reddish-brown, somewhat striate. (2n = 36) Jul–Aug. Infrequent to locally common on open rocky wooded hillsides, bluffs, & stream banks; SD: Brule, Custer, Lawrence, Pennington; WY: Crook, Johnson, Sheridan, Weston; MT: Rosebud; (Sask. to B.C., s to SD, WY, MT, n-cen OR). *S. lucida* Dougl.—Rydberg

Our plants are the var. *lucida* (Dougl. ex Greene) C. L. Hitchc. and differ from the Asiatic var. *betulifolia*, which has pubescent fruits and less toothed leaves. Our plants differ from var. *corymbosa* (Raf.) S. Wats., of the eastern U.S. in being glabrous rather than somewhat pubescent.

71. CROSSOSOMATACEAE A. Gray, the Crossosoma Family

by Ronald L. McGregor

1. **Glossopetalon planitierum** (Ensign) St. John, grease-bush. Much-branched shrubs 4–12 dm tall; branches greenish, angled, with decurrent lines from the nodes, usually glabrous,

becoming spinescent. Leaves alternate, simple, entire, narrowly elliptic to oblanceolate, acute, blades (4)6–12(14) mm long, 2–4.5 mm wide, minutely pubescent with short straight hairs or glabrous, tapering to petioles 0.5–1.5 mm long; petiole usually with a swollen reddish base; stipules subulate, reddish-black, to 0.5 mm long, persistent. Flowers axillary, 1(3) in axils; pedicels usually 1–3 mm long, rarely to 14 mm long; sepals 5, united below, lobes ovate, 1.6–2.2 mm long, usually hyaline-margined; petals 5, distinct, white, oblanceolate, clawed, 4–6 mm long, 1–2 mm wide; stamens usually 8, unequal; disk crenate-lobed; carpels 1(2), distinct, ovate, sessile on disk; ovary superior, 1-celled, ovules 1 or 2. Fruit an asymmetrical ovoid, somewhat coriaceous, striated follicle, 5–7 mm long, opening along ventral suture; seeds 1 or 2, 4 mm long, reddish-yellow, with a small white aril. Mar–May. Rocky prairie hillsides, bluffs, & canyons; OK: Cimarron; CO: Las Animas; NM: Quay; TX panhandle s to Lubbock; (*endemic* to area). *Forsellesia planitierum* Ensign — Atlas GP.

72. MIMOSACEAE R. Br., the Mimosa Family

by Ronald L. McGregor

Ours trees, shrubs, or herbaceous perennials; stems armed or unarmed. Leaves alternate, usually even-bipinnately compound; petiole and rachis usually bearing 1–several depressed or stalked glands somewhere along the axis; pinnae and leaflets few to many pairs; stipules evident or inconspicuous. Inflorescences capitate or spicate, usually pedunculate, from leaf axils or spurs. Flowers regular, hypogynous, usually perfect; sepals 5, united nearly their full length; petals 5, distinct or united; stamens 5, 10, or many, distinct or united basally, usually long exserted and giving the inflorescences their characteristic color and appearance; ovary of 1 unilocular carpel, ovules usually several. Legume dehiscent or indehiscent, sometimes stipitate, occasionally transversely septate between seeds.

Reference: Isely, Duane. 1973. Leguminosae of the United States: I. Subfamily Mimosoideae. Mem. New York Bot. Gard. 25 (1): 1–152.

The so-called mimosa or silk tree, *Albizia julibrissin* Durazz., a native of Asia, is often grown in the se GP for its fragrant clusters of pink flowers, lacy foliage, and flat-topped spreading branches. While it sometimes reproduces by seed in gardens and waste places in towns, it has not become naturalized in the GP and is not treated here.

1 Stems herbaceous (rarely partly persistent below in *Acacia*).
  2 Lowest pair of pinnae with a gland between them .... 2. *Desmanthus*
  2 Lowest pair of pinnae without a gland between them.
    3 Plants armed with prickles .... 6. *Schrankia*
    3 Plants unarmed.
      4 Pinnae 4–12 pairs; secondary venation of leaflets rarely evident; stipules inconspicuous or absent, not striate .... 1. *Acacia*
      4 Pinnae (2)4 or 5(9) pairs; secondary venation of leaflets evident; stipules evident, striate .... 4. *Neptunia*
1 Shrubs or trees.
  5 Flowers in heads; leaflets 2–6 mm long; prickles 3–8 mm long, flattened basally, recurved apically .... 3. *Mimosa*
  5 Flowers in spikes; leaflets 15–62 mm long; thorns, when present, usually 2–4 cm long, terete, straight .... 5. *Prosopis*

1. ACACIA P. Mill.

Acacia angustissima (P. Mill.) O. Ktze. var. *hirta* (Nutt.) Robins., prairie acacia. Ours herbaceous perennials 4-6 dm tall, from woody caudex, often forming localized colonies by long rhizomes; stems unarmed, sparingly or not branched, wandlike, ridged, glabrate or with spreading hairs. Leaves alternate, bipinnate, 5-12 cm long; petiole 1-4 cm long, appressed-pubescent or pilose; stipules linear-lanceolate, 3-4 mm long; pinnae (7)10-14 or more pairs, each with 10-35 pairs of leaflets; leaflets sessile or nearly so, crowded, linear or oblong. Inflorescence axillary, pedunculate; heads ca 1 cm diam, whitish. Calyx mostly campanulate, 4- to 5-lobed; petals (4)5, distinct or united below, stamens numerous; ovary often stipitate. Fruit brownish, flat, 4-7 cm long, 6-8 mm wide, stipitate, promptly dehiscent; seeds flat, orbicular, brown, 4.5-5 mm long. Jun-Jul. Dry rocky prairie hillsides, open woodlands, thickets, & roadsides; KS: Cowley, Chautauqua, Montgomery; OK, TX; (MO, KS, s to TX, disjunct e to FL; Mex. & C. Amer.). *Acaciella hirta* (Nutt.) Britt. & Rose — Rydberg.

Cattle selectively graze this species, and no doubt it was formerly more abundant. In some colonies plants are found with stems and petioles spreading-hirsute, hairs 0.5-1.5 mm long while others are glabrate to glabrous.

2. DESMANTHUS Willd.

Herbaceous unarmed perennial herbs from turnip-shaped roots, caudex somewhat woody; stems several from crown, 2-10 dm long, erect to spreading; leaves bipinnate, 2-10 cm long; leaflets small and numerous; stipules subulate, small. Peduncles axillary, each bearing a head of few to several very small flowers. Flowers white; sepals 5, essentially distinct as are the 5 petals; stamens 5(10 in *D. cooleyi*), some flowers in each head with sterile stamens. Fruit a dry, flattened, 1- to several-seeded, dehiscent legume.

1 Pods falcate, 2-5× longer than wide; stamens 5, stems erect .... 2. *D. illinoensis*
1 Pods linear, at least 7× longer than wide; stamens 5 or 10; stems decumbent or ascending.
  2 Stipules less than 2 mm long or wanting; stamens 10; seeds quadrate-rhombic .... 1. *D. cooleyi*
  2 Stipules 4-6 mm long; stamens 5; seeds narrowly obovate .... 3. *D. leptolobus*

1. *Desmanthus cooleyi* (Eat.) Trel. Herbaceous perennials from a woody caudex; stems decumbent or ascending, 2-7 dm long, angles glabrous or minutely strigose. Leaves bipinnate; pinnae 2-7 pairs, with an orbicular gland 0.4-2 mm in diam between lower pair; stipules less than 2 mm long; leaflets 6-15 pairs per pinna, oblong or linear-oblong, 3-4 mm long, ciliate and sparsely pubescent when young. Peduncles 2 cm long or less; stamens 10. Pod nearly straight, linear, 3-7 cm long, at least 7× longer than wide; seeds quadrate-rhombic, 4-4.5 mm long, dark reddish-brown. ($2n = 28$) Jun-Sep. Infrequent to locally common on dry, rocky or sandy prairie hillsides, ravines, roadsides, & waste places; KS: Morton; CO: Baca; OK: Cimarron, Texas; w TX, NM; (KS, w TX, w to AZ, adjacent Mex.).

2. *Desmanthus illinoensis* (Michx.) MacM., Illinois bundleflower, false sensitive plant. Herbaceous perennials from a somewhat woody caudex; stems erect, 3-20 dm tall, minutely strigose to short-pilose on angles, or glabrate. Leaves bipinnate, 5-10 cm long; pinnae 6-12(16) pairs, with a small gland between them or between lower pair only; leaflets 15-30 pairs, 2-5 mm long, glabrous or ciliate; stipules filiform, 6-10 mm long. Peduncles ascending, 2-7.5 cm long in fruit. Flowers whitish; calyx short-cylindric to campanulate, 1-1.5 mm long, lobes shorter than tube; petals 5, appearing united to middle, becoming separate, ca 2 mm long; stamens 5, distinct. Pods broadly oblong and flat, 3-4× longer than wide, usually strongly falcate, brown or blackish; seeds 3–5 mm long, nearly as wide, brown. (2n = 28) Jun–Aug. Rocky, open wooded slopes, prairies, ravines, stream banks, roadsides, waste places; GP but infrequent to absent n & w; (OH, MN, ND, CO, s to FL, NM).

This species is readily eaten by all classes of livestock and is rated by some authorities as the most important native legume. It is frequently used in range revegetation programs.

3. *Desmanthus leptolobus* T. & G., slender-lobed bundleflower. Herbaceous perennials; stems several from the base, spreading or ascending, 6–12 dm long, angles sparsely hairy to glabrate. Leaves bipinnate; stipules setiform, 4–6 mm long; pinnae 4–12 pairs, with a small gland between lowest pair or gland absent; leaflets 8–26 pairs, linear-lanceolate, acute, sparsely ciliate or glabrous. Peduncles 5–30 mm long; flowers whitish, calyx cylindric, 1.5–2 mm long; petals 2–3 mm long; stamens 5. Pod straight or nearly so, linear, at least 7 × longer than wide, glabrous; seeds ca 5 mm long, narrowly obovate, yellowish, with long axis oriented parallel to the long axis of the pod. (2n = 28) May–Sep. Infrequent to locally common in rocky prairies, sandy areas, roadsides, stream banks, fields, with preference for calcareous soils; s 1/3 GP but infrequent e & w; (KS to s-cen TX, local in e-cen MO).

3. MIMOSA L., Mimosa, Catclaw

Ours shrubs about 1 m tall, much branched, armed with stout and usually apically recurved prickles, branches slightly zigzag. Leaves bipinnate; pinnae (1)2–8(10) pairs; leaflets 3–12 pairs per pinna; stipules small and subulate. Flowers in globose heads, axillary; sepals 5, united; petals 5, united or distinct; stamens 10. Legumes flattened, margins separate from valves at maturity, the valves often breaking transversely into 1-seeded segments.

1 Young flower buds pubescent; flowers sessile or nearly so; petals united; legumes 3–4 mm wide; young branches, rachis, and usually the leaflets pubescent .... 1. *M. biuncifera*
1 Young flower buds glabrous; flowers with pedicels 0.5–1 mm long; petals distinct, clawed; legumes 4–8 mm wide; young branches, rachis, and usually the leaflets glabrous .... 2. *M. borealis*

1. *Mimosa biuncifera* Benth., cat's claw mimosa. Rounded shrub to 1 m tall; numerous branches often slightly pubescent at least when young, zigzag, prickles flattened basally and apically distinctly recurved. Leaves bipinnate, alternate; pinna (2)4–8(10) pairs; leaflets 5–12 pairs per pinna, 2–3 mm long, linear-oblong, usually pubescent. Inflorescence axillary, globose, peduncles 0.5–2 cm long. Flowers essentially sessile; calyx short-tubular or cup-shaped, the 5 lobes small, pubescent; petals united 1/2 length or more, pubescent; stamens 10, filaments whitish or pinkish; ovary pubescent. Legumes linear, curved to nearly straight, 2–4 cm long, 3–4 mm wide, the valves separating at maturity from margin but not breaking up into joints, pubescent or glabrate, margins with or without recurved prickles; seeds flattened, 3–3.5 mm long, 2–2.5 mm wide, dark brown. (2n = 52) Apr–Sep. Infrequent to locally common on rocky or sandy prairie hillsides, flats, ravines, roadsides; s TX panhandle, e NM; (TX, NM, AZ, se to Son., Chih. & Coah.).

2. *Mimosa borealis* A. Gray, pink mimosa. Shrubs 1(2) m tall, much branched, branches often slightly zigzag, glabrous, prickles basally flattened, recurved apically. Leaves bipinnate, alternate; pinnae (1)2 or 3 pairs per leaf; leaflets 3–8 pairs per pinna, oblong to ovate, 2–6 mm long, glabrous; petioles 5–15 mm long; stipules 1.5 mm long. Inflorescence of pinkish globose heads, axillary, peduncles 1.5–2 cm long. Calyx glabrous, cuplike, lobes short; petals distinct, short clawed, limb obovate, glabrous, pinkish; stamens 10, filaments pink; ovary glabrous. Legumes linear-oblong, 2.5–6 cm long, 6–8 mm wide, glabrous, brown, more or less constricted between seeds, margins with or without prickles, valves separating

from margins and breaking into 1-seeded joints; seeds 5–6 mm long, 4–5 mm wide, smooth, light brown. May–Jul. Infrequent to locally common on rocky or gravelly hillsides, canyons, & brushy areas; KS: Barber, Clark, Meade; se CO, w OK, TX panhandle, NM; (sw KS, se CO, w OK, TX, NM, n Mex.).

4. NEPTUNIA Lour., Neptunia

1. *Neptunia lutea* (Leavenw.) Benth., yellow-puff. Herbaceous perennials, from usually orange taproots; stems prostrate, 1 m or more long, usually hirsutulous with hairs to 1 mm long, or glabrate. Leaves alternate, bipinnate; pinnae (2)4 or 5(9) pairs, without a gland between lowest pair; leaflets 8–18 pairs per pinna, 3–4 mm long, oblong, ciliate; stipules thin, lanceolate, to 4 mm long, petiolate, Flowering peduncles axillary, to 2 cm long; flowers in ovoid to short-cylindric, bright yellow heads, 2 cm long, 1.5 cm wide; calyx campanulate, ca 1 mm long; petals 5, distinct nearly to base, 2–3 mm long; stamens 10, all functional but pistil lacking in lower flowers. Legume with a stipe exceeding the calyx, body broadly oblong, flat, 2.5–5 cm long, 10–15 mm wide; seed 6–7 mm long, 4 mm wide, brown, glossy. ($2n = 28$) Apr–Sep. Open woodlands, creek valleys, dry prairies, & roadsides; extreme se GP; (OK, e TX, AR, LA, w AL).

5. PROSOPIS L., Mesquite

1. *Prosopis glandulosa* Torr., honey mesquite. Shrubs or small trees 4–6 m tall, often armed with 1 or 2 stout spines at the nodes. Leaves alternate, bipinnate, petiolate; pinnae usually 1 pair per leaf; leaflets 6–15(18) pairs per pinna, linear or narrowly elliptic, 2–6 cm long, entire, usually glabrous. Flowers in axillary spikes, 7–9 cm long, yellow; peduncles 1–3 cm long. Flowers sessile or with pedicel to 0.5 mm long; calyx shallow cuplike, ca 1 mm long, with 5 short lobes finely pubescent at tips; petals 5, elliptic to obovate, ca 3 mm long, distinct, pubescent above; stamens 10; ovary densely white woolly. Legumes essentially straight, 7–20 cm long, indehiscent, constricted between seeds; seeds 6–6.5 mm long, brownish. ($2n = 56$) May–Jul. Infrequent to abundant on a variety of soils but most common on dry ranges; KS: Barber, Clark, Comanche, Meade, Morton; w OK, TX, NM: (sw KS, w OK, e NM, s to Tam., N.L. & Coah.). *Neltuma glandulosa* (Torr.) Britt. & Rose — Rydberg; *P. chilensis glandulosa* (Torr.) Standl. — Gates.

The pods furnish important fodder for various classes of livestock but the plant has increased in abundance on disturbed grassland and has become noxious.

6. SCHRANKIA Willd., Sensitive Brier

Ours perennial herbs from napiform taproots; caudex somewhat woody; stems strongly ribbed, 6–12 dm long, sprawling, armed with numerous recurved prickles. Leaves bipinnate with 3–15 pairs of pinnae, rachis and rachillas usually with recurved prickles; flowers in heads, pink, perfect or unisexual, regular; calyx minute, 5-toothed; petals united into a funnelform 5-lobed corolla; stamens 8–10, usually united below. Legume linear, with numerous recurved prickles, dehiscent into 2 valves, or each valve later splitting into 2 merivalves; seeds few to many. *Leptoglottis* DC. — Rydberg.

1 Mature leaflets with only midrib evident; seeds obovate, 6–7 mm long; young stems, peduncles, rachis, and legumes usually (90%) puberulent .... 2. *S. occidentalis*
1 Mature leaflets with midrib prominent, at least upper half with raised arcing or anchor-shaped lateral veins; seeds quadrate, 4 mm long; herbage glabrous .... 1. *S. nuttallii*

1. **Schrankia nuttallii** (DC.) Standl., sensitive brier, catclaw sensitive brier. Sprawling herbaceous perennials; stems armed with recurved prickles, 0.5-2 m long, glabrous; leaves bipinnate; pinnae 4-8 pairs; leaflets 8-15 pairs, 3-9 mm long, midrib prominent, at least upper portion with raised, arcing or anchor-shaped lateral veins, glabrous; stipules 4-6 mm long. Peduncles axillary, 3-9 cm long; heads many-flowered. Flowers pink or lavender, sessile; calyx minute; petals 5, united, ca 4 mm long, lobes 1 mm long; stamens 8-12. Legumes 3-12 cm long, linear, strongly ribbed, densely or sparsely prickly, glabrous; seeds ca 4 mm long, quadrate-rhombic, smooth. ($2n=26$) Apr-Sep. Rocky or sandy open woodlands, prairies, ravines, roadsides; GP from cen SD s, but infrequent to absent w; (IL, SD, s to AL, TX). *Leptoglottis nuttallii* DC.—Rydberg; *S. uncinata* Willd.—Gates

2. **Schrankia occidentalis** (Woot. & Standl.) Standl., western sensitive brier. Sprawling herbaceous perennials; stems armed with recurved prickles, 0.5-1 m long, puberulent; leaves bipinnate; pinnae 3-4(8) pairs; leaflets usually 12-16 pairs per pinna, smooth beneath or only midvein prominent, rachis and petiole usually puberulent; stipules subulate, 3-6 mm long. Peduncles axillary, 2-4 cm long; heads many-flowered. Flowers pinkish, sessile; calyx minute; petals 5, united, 4 mm long, lobes ca 1 mm long; stamens 8-12. Legumes 6-8(12) cm long, tetragonal, with numerous recurved prickles; seeds 6-7 mm long, obovate, smooth. ($2n=26$) May-Jul. rocky or sandy prairies, ravines, dry stream banks, roadsides; KS: Stevens; OK: Beaver, Cimarron; TX panhandle, e NM; (sw KS, OK & TX panhandle, e NM).

The leaves of this and the preceding species respond immediately to any contact by folding their leaflets and lowering the petiole. The young plants are readily eaten by all classes of livestock, and the seeds have been used as an ingredient in some laxative medicines.

73. CAESALPINIACEAE R. Br., the Caesalpinia Family

by Ronald L. McGregor

Ours trees, shrubs, perennial or annual herbs; stems unarmed or armed in *Gleditsia*. Leaves alternate, pinnate or bipinnate, or simple in *Cercis*; petiole and rachis eglandular except in *Cassia*; stipules persistent or early deciduous. Inflorescences of axillary racemes or these in terminal racemes or panicles, sometimes of apparent axillary fascicles or of fascicled spikes in *Gleditsia*. Flowers regular or irregular, perfect or unisexual in *Gleditsia*, hypanthium usually well developed, often irregular; sepals commonly 5, distinct or basally united; petals 5, distinct, clawed or not, subequal to definitely unequal; stamens 10 or less, often of unequal sets or of different sizes and forms; ovary sessile or stipitate, inserted in the bottom or on side of hypanthium, unicarpellate, 1-celled, ovules 1-many; legume usually elongate and several seeded, dehiscent or indehiscent.

Reference: Isely, Duane. 1975. Leguminosae of the United States II. Subfamily Caesalpinioidae. Mem. New York Bot. Gard. 25 (2): 1-228.

1 Plants herbaceous annuals or perennials.
  2 Leaves once-pinnate .... 2. *Cassia*
  2 Leaves bipinnate.
    3 Lower leaf surface with conspicuous orange glands, these becoming black on dried specimens .... 1. *Caesalpinia*
    3 Lower leaf surface without orange glands, dried specimens not black-dotted .... 6. *Hoffmanseggia*

1   Plants trees or tall shrubs.
   4   Leaves simple, preceded by pink flowers .... 3. *Cercis*
   4   Leaves compound; flowers white or greenish, appearing with or after the leaves.
      5   Leaflets 1 cm wide or less; stems usually thorny; flowers in axillary spikelike racemes .... 4. *Gleditsia*
      5   Leaflets 2-3 cm wide; stems unarmed; flowers in terminal panicles or racemes .... 5. *Gymnocladus*

1. CAESALPINIA L.

1. Caesalpinia jamesii (T. & G.) Fisher, James rush-pea. Herbaceous perennial from a somewhat woody and often branching caudex, root 1-5 dm long, spindle-shaped; stems erect, 1-4 dm tall, usually clustered, often branched above, ridged, retrorsely appressed-pubescent, glandular. Leaves alternate, bipinnate; pinnae 5-7; leaflets 5-10 pairs, ovate or oblong, 4-5 mm long, pubescent or glabrate above, pubescent and glandular below, glands orange, drying black; petioles 1-3 cm long, pubescent, stipules lanceolate or subulate, 3-6 mm long, pubescent, deciduous. Inflorescence of axillary racemes, 5-10 cm long, with 3-15 flowers, often overtopped by branches. Calyx tube 1.5-2 mm long, lobes lanceolate, 5-8 mm long, acute, pubescent, glandular; petals 5, yellow, glandular, 6-8 mm long, clawed, upper two dissimilar; stamens 10, shorter than petals, distinct, filaments pubescent. Legume lunate, widest above middle, 2-2.5 cm long, 7-10 mm wide, with hairy emergences on margins, glandular; seeds 2 or 3, 6-6.5 mm long, nearly as wide, glossy, obovate. May-Sep. Infrequent to locally common in dry rocky, gravelly, or sandy prairies, stream valleys, roadsides; NE: Dundy; sw 1/4 GP; (sw NE, w 1/4 KS, CO, s to TX, CA). *Larrea jamesii* (T. & G.) Britt.—Rydberg; *Hoffmanseggia jamesii* T. & G.—Gates.

2. CASSIA L., Senna

Ours annual or perennial herbs; leaves spiral or distichous, once-compound with 2 to numerous leaflets; the petiole bearing a gland between or below leaflets; stipules caducous or persistent. Flowers in supra-axillary racemes or clusters or in terminal panicles. Sepals 5, united near base, subequal; petals 5, yellow, subequal or 1 larger; stamens 10, trimorphic (the uppermost 3 reduced, the lower 3 with longer anthers and/or filaments or stamens 10 or 5, equal); ovary pubescent. Pods erect or pendent, elastically dehiscent or latently but not elastically dehiscent. *Chamaecrista* Moench—Rydberg.

1   Plants annual; leaflets 6-20 mm long.
   2   Petals 10-20 mm long; pedicels 10-25 mm long; stamens 10 .... 1. *C. chamaecrista*
   2   Petals 3-8 mm long; pedicels 1-3 mm long; stamens 5 .... 3. *C. nictitans*
1   Plants perennial; leaflets 20-80 mm long.
   3   Leaflets 6-10 pairs per leaf .... 2. *C. marilandica*
   3   Leaflets 1 pair per leaf .... 4. *C. roemeriana*

1. Cassia chamaecrista L., showy partridge pea. Erect annual herbs 1-12 dm tall; stems slender and few branched to diffusely branched, with short appressed incurved hairs, or with spreading hairs 1-2 mm long, or with a mixture of these. Leaves 1-pinnate, distichous; leaflets usually 8-15 pairs, linear-oblong, mucronate, 1-2 cm long, 2-4 mm wide, glabrous, ciliolate; petiole with a single sessile or stipitate gland, 0.5-1.5 mm across, below lowest pair of leaflets, appressed or spreading-pubescent; stipules attenuate, ciliolate, 5-10 mm long, persistent. Flowers in short 2- to 7-flowered supra-axillary racemes; pedicels 1-2.5 cm long, appressed or spreading-pubescent. Sepals lanceolate, acuminate, 10-15 mm long; petals obovate, yellow, upper 4 with reddish spot at base, the lower larger, incurved 10-18 mm long; stamens 10, yellow, or purple, or 4 yellow and others purple, subequal. Legumes erect, linear-oblong, 2.5–7 cm long, 4–6 mm wide, glabrous, short appressed-pubescent, or rarely with spreading hairs 1–2 mm long, elastically dehiscent with spirally coiling valves; seeds 4–20, flattened oval or somewhat triangular, 3.5–5 mm long, smooth, lustrous, faintly marked with pits in longitudinal rows. (n = 8) Jun–Oct. Infrequent to often abundant on rocky or more commonly sandy soils of open woodlands, prairies, fields, roadsides; GP from s MN, se SD sw to e CO & s; (most of e 1/2 U.S., w to GP). *Chamaecrista fasciculata* (Michx.) Greene—Rydberg; *Cassia fasciculata* Michx.—Fernald.

For many years the showy partridge pea has been known as *C. fasciculata* Michx., but Irwin and Barneby (Brittonia 28: 381–389. 1976) have provided convincing evidence that *C. chamaecrista* L. is the correct name. In the GP most of the plants have stems, petioles, pedicels, and fruits with short appressed incurved hairs; petiolar gland saucer-shaped, sessile or short-stipitate, 0.5–1.5 mm wide. In se KS some specimens have stems, petioles, pedicels, and fruits (or just pedicels and fruits) with white or yellowish spreading hairs 0.5–1.5 mm long in addition to the short appressed hairs. Such plants have been referred to variety *robusta* Pollard, which I feel does not merit taxonomic recognition. In sw KS and s, specimens are found which are short, much branched, leaflets 10 or less per leaf, and petiolar gland clearly stipitate and 0.5 mm wide. Similar short-branched plants with fewed leaflets are found on dry habitats throughout the GP, and as far as se SD some have clearly stipitate glands 0.5–1.0 mm wide. The short plants of the sw have been referred to *C. fasciculata* var. *rostrata* (Woot. & Standl.) B. L. Turner, but the combination with *C. chamaecrista* has not been made. Because of the complete intergradation with the characteristic plant of the GP, I do not feel var. *rostrata* merits taxonomic recognition.

The showy partridge pea is occasionally grown as an ornamental, and because the roots bear nodules it has been used in soil-improving cover crops. It is readily eaten by livestock, but one report considers the fresh or dried plant and its seed to be toxic to animals.

2. *Cassia marilandica* L., Maryland senna. Perennial herb from woody caudex; stems erect, 0.5–2 m tall, mostly simple, glabrous or very sparingly pubescent, sometimes glaucous. Leaves alternate, spirally arranged, once-pinnate; leaflets usually 6–10 pairs, oblong to elliptic, mucronate, glabrous, glaucous beneath, 3–6 cm long, 1–2.5 cm wide, subequal; petiolar gland ovoid-conic, sessile or short-stipitate, at or just above pulvinus; stipules linear-lanceolate, caducous. Flowers in short several-flowered racemes in upper axils or in terminal panicles. Sepals ovate, obtuse, light yellow, 4–6 mm long, glabrous, ciliolate; petals obovate or elliptic, yellow, 9–12 mm long, glabrous; stamens 10, the 3 lowest with larger anthers and filaments, the 3 uppermost much reduced; ovary pubescent. Legumes linear, usually curved downward, flat, 7–11 cm long, 8–11 mm wide, sparsely puberulent, black at maturity, septate between seeds; seeds ovate, 4–5 mm long, 2–2.5 mm wide, black lustrous. (n = 14) Jul–Sep. Infrequent to common in moist prairie ravines, creek banks, & alluvial thickets, open woodlands, base of slopes & bluffs; se 1/4 GP; (PA, IA, se NE, s to FL, TX). *C. medsgeri* Shafer—Rydberg.

Two similar species have been reported from the GP:

C. obtusifolia L., with petiolar gland between or just above lowest pair of leaflets, leaflets 4–6, widest above the middle, was collected as a waif in KS: Wyandotte in 1896 and in Elk Co. in 1940. It is found in se U.S. and s through the American tropics. *C. tora* L.—Rydberg.

C. occidentalis L., with leaflets conspicuously acuminate and flowers 3 or fewer per axil, was found as a waif in KS: Wyandotte in 1896. It is a pantropic weed found in se US.

3. *Cassia nictitans* L., sensitive partridge pea. annual herb 1–4 dm tall, usually branched and somewhat spreading; stems slender, puberulent with short, appressed, incurved hairs. Leaves distichous, usually touch sensitive, alternate, pinnate; leaflets 8–22 pairs, oblong, mucronate, 6–16 mm long, glabrous, ciliate; petiolar gland stipitate, less than 0.75 mm wide; stipules linear-lanceolate, acuminate, ciliate, 4–8 mm long, persistent. Flowers inconspicuous, 1–3 in supra-axillary clusters, on puberulent pedicels 1–3 mm long. Sepals

lanceolate, puberulent; petals unequal, 3–8 mm long, yellow; stamens 5; ovary pubescent. Legume erect, flat, linear-oblong, 2–5 cm long, 3–5 mm wide, appressed-pubescent or glabrate, elastically dehiscent with spirally coiled valves; seeds quadrate, 3–3.5 mm long, black, lustrous, faintly marked by pits in rows. (n = 8) May–Sep. Infrequent to common in rocky or sandy soils of open woods, hillsides, prairies, thickets, & roadsides; extreme se GP; (MA, NY, OH, IL se KS, s to FL & TX). *Chamaecrista nictitans* (L.) Moench — Rydberg.

The leaflets of this species and of *C. chamaecrista* are sensitive to the touch and slowly fold against one another.

4. Cassia roemeriana Scheele, two-leaved senna. Perennial herbs; stems erect or ascending, 3–6 dm tall, gray-puberulent. Leaflets 1 pair, spirally arranged, lanceolate or lance-linear, 3–6 cm long, 6–12 mm wide, base inequilateral, sparsely appressed-puberulent on both sides; petiole shorter than leaflets, pubescent; stipules setaceous, 6–9 mm long. Flowers 2–5 on axillary peduncles, or in terminal panicles. Sepals ovate, thin, 6–7 mm long, not persisting to fruit; petals 7–9 mm long, yellow; stamens 10, upper 3 much reduced; ovary pubescent. Legume turgid, compressed, 2–3 cm long, 5–6 mm wide, sparsely striose, dehiscent only part way down each suture; seeds ca 3 mm long. glossy, slightly rugulose. (n = 14) Apr–May, again in Sep. Rocky soils of open woodlands, pastures, fields, roadsides; extreme s GP; (sw OK, NM & s to N.L. & Coah.).

3. CERCIS L., Redbud

1. Cercis canadensis L., redbud. Deciduous unarmed tree to 8(12) m tall. Leaves alternate, simple, cordate, to somewhat reniform, 8–14 cm long, 5–12 cm wide, entire, apex abruptly short-acuminate or acute, upper surface dark green, glabrous, lower surface pale, glabrous or more or less pubescent; petiole 4–7 cm long, glabrous; stipules minute. Flowers appearing before leaves or rarely as leaves expand, rose or pink-purplish, in umbellike clusters along branches of last or preceding year's growth. Pedicels 10–18 mm long; calyx purplish, basally oblique, irregularly campanulte, enlarged on lower side, the 5 minute lobes broadly triangular, ciliate at tip; corolla imperfectly papilionaceous, standard smaller than wings, keel petals larger and not united; petals 5; stamens 10, distinct. Legume slightly stipitate, oblong; seeds ca 5 mm long, 4–4.5 mm wide, light to dark brown, smooth, somewhat glossy. (2n = 14) Mar–Apr. Infrequent to common in rocky open woodlands, borders of woods, stream banks, thickets, roadsides; se 1/4 GP; (e 1/2 U.S. to GP).

This is the state tree of Oklahoma and is often cultivated even in areas beyond its native distribution. Rare plants with mature leaves glabrous below have been named forma *glabrifolia* Fern. — Fernald.

4. GLEDITSIA L., Honey Locust

1. Gleditsia triacanthos L., honey locust. Polygamous tree to 15 m tall; trunk and large branches armed (rarely unarmed in the wild) with thorns to 2 dm long, these simple to much branched, terete but flattened at base, shiny brown, often dense. Leaves pinnate and bipinnate; pinnate leaves 3–6, fascicled from spurs on older wood, with 10–14 pairs of leaflets 1.5–4 cm long, lanceolate to ovate-oblong; bipinnate leaves from new growth, with 2–8 pairs of pinnae, leaflets 5–10 pairs, usually elliptic-oblong, 1.5–2.5 cm long, puberulent when young; stipules obsolete. Staminate inflorescences predominate, 1–7 from spurs, racemose (rarely branched apically), 2–7 cm long, flowers numerous, crowded, sessile to short pedicelled, densely yellow-pubescent, sepals (3)5, stamens (4)5–7. Fertile inflorescences usually solitary, axillary, from spurs, perfect or pistillate; flowers pedicellate, hypanthium short-campanulate; sepals 3(5); petals 3(5), slightly larger than sepals. Legumes 1–3 per peduncle, essentially indehiscent, stipitate, oblong, laterally compressd but plump, straight or curved, 2–4 dm long, 2.5–4 cm wide, in age becoming dry and twisted or contorted; seeds several, 9–11 mm long, 5–7 mm wide, dull, smooth. (2n = 28) Apr–Jun. Infrequent to common in moist woodlands of river & creek valleys & adjacent slopes, but also on upland slopes, pastures, roadsides & fence rows; GP but rare to absent n & w (e 1/2 U.S., to GP, introduced elsewhere).

The honey locust was native along the streams of the eastern GP from NE s. Its natural range is obscured because it was much planted by the early settlers and escaped to become somewhat of a weed. Plants without thorns are the forma *inermis* Scheider and are rarely found wild. A number of cultivars have been developed from forma *inermis* and are now much planted in the GP region. Some of these do not produce fruit and others have younger foliage yellow or yellowish instead of green.

5. GYMNOCLADUS Lam., Kentucky Coffee-tree

1. *Gymnocladus dioica* (L.) K. Koch., Kentucky coffee-tree. Polygamous unarmed trees to 23 m tall, with rough bark. Leaves alternate, bipinnate, 3–9 dm long, 3–6 dm wide; pinnae 3–6 pairs, each more than 1 dm long and with 4–7 pairs of leaflets; leaflets 4–7 cm long, 2–3 cm wide; stipules small, deciduous. Flowers greenish-white, in terminal panicles or racemes; racemes 0.5–3 dm long. Hypanthium tubular-obconic, 6–10 mm; sepals 5, equal or subequal, oblong; petals 3–5, subequal, 4–5 mm long, narrow, yellowish, stamens 10, those opposite sepals ca 3 mm long, those opposite petals ca 4 mm long. Legume indehiscent, oblong, straight or slightly curved, 5–15 cm long, 3–5 cm wide; seeds 15–20 mm long, embedded in glutinous material, dark olive-brown. (2n = 28) May–Jun. Low or rich woods, base of bluffs, along streams & rocky open wooded hillsides; se SD to w-cen OK in GP; (MA, NY, WS, se SD, s to n NC, AL, AR, OK).

This species was formerly much planted around farm homes. The plant sometimes occurs in small colonies resulting from root-suckers.

6. HOFFMANSEGGIA Cav., Rush-pea

Herbaceous perennials from spreading roots or a woody caudex. Leaves odd-bipinnate, not glandular-punctate; petiolate; stipules persistent but usually inconspicuous. Inflorescence a terminal raceme. Flowers pedicellate; calyx tube short, lobes 5, valvate; petals 5, yellow, uppermost dissimilar; stamens 10, distinct, filaments subequal to petals, often glandular at base. Legume flat, straight or falcate, often glandular, tardily dehiscent; seeds 2–8, ovate.

1 Stems and inflorescence not stipitate-glandular .... 1. *H. drepanocarpa*
1 Upper stem and inflorescences stipitate-glandular .... 2. *H. glauca*

1. *Hoffmanseggia drepanocarpa* A. Gray, sicklepod rush-pea. Subscapose to caulescent herbaceous perennials; stems 6–20 cm tall from branched woody caudex and taproot, glandless, pubescent; pinnae 2–5 pairs; leaflets crowded, 4–8 pairs, obovate or elliptic-oblong, 1.5–6 mm long, pubescent or glabrate; stipules ovate or deltoid, very small; petioles 4–10 cm long. Flowers 2–10 in short or exserted terminal, eventually decumbent racemes, pedicels 2–5 mm long; calyx lobes lanceolate, 3–5 mm long, acute; petals yellowish to orange-yellow, 5–6 mm long, subequal with short claws; stamens 10, filaments shorter than petals. Legume indehiscent, slightly to strongly falcate, 2–4 cm long, 5–8 mm wide, valves papery, eventually impressed between seeds; seeds 6–11, ca 3 mm long, brown. (2n = 24) Apr–Jul.

Infrequent to common on rocky or sandy prairie hillsides, ravines, stream valleys, roadsides; KS: Seward; se CO, s TX, NM; (sw KS, se CO, s to TX, AZ, & Mex.).

2. *Hoffmanseggia glauca* (Ort.) Eifert, Indian rush-pea, pignut. Subscapose to caulescent herbaceous perennials; stems erect or spreading, simple or branched, 0.5–4 dm tall, from subterranean caudices developed from a spreading root system which bears roundish tubers, pubescent to glabrate, with stipitate glands above. Leaves subbasal or partly cauline, odd-bipinnate; pinnae 2–6 pairs plus 1; leaflets 6–11 pairs, nearly sessile, elliptic or oblong, 2.5–6 mm long, 2–3 mm wide, blades minutely pubescent or glabrate; stipules ovate-deltoid, ciliate, persistent; petioles as long or longer than rachis, pubescent or glabrate. Inflorescence a terminal raceme, glandular, pubescent, 1–2 dm long, 5- to 15-flowered. Pedicels 2–5 mm long, pubescent, glandular; calyx pubescent and glandular, lobes 6–7 mm long; petals orange-red, 10–13 mm long, with glandular claws; stamens 10, shorter than petals, filaments glandular and pubescent, red. Legume indehiscent, persistent, flat, 2–4 cm long, lustrous. (n–12) May–Sep. Infrequent to common on rocky or sandy prairies, stream valleys, fields, roadsides; sw GP; (sw KS, s to TX, w to CA, Mex., S. Amer.). *Larrea densiflora* (Benth.) Britt.—Rydberg; *Hoffmanseggia densiflora* Benth.—Gates.

This is an aggressive species and a noxious weed by the laws of Kansas and other states. The underground tubers are eaten by hogs, and they also were roasted and eaten by the Indians.

74. FABACEAE Lindl., the Bean Family

by Ronald L. McGregor

Ours herbs, vines, shrubs, or trees of various habit. Leaves alternate, variously compound or simple in *Crotalaria*, usually stipulate; leaflets sometimes with stipels. Flowers somewhat perigynous but floral cup often difficult to distinguish and flowers appearing hypogynous, usually strongly zygomorphic. Calyx prolonged into a tube, often irregular, lobes 5, equal or unequal, in some genera only 4 lobes apparent through fusion of 2 in lower lobe; corolla typically papilionaceous, the upper median petal, the standard (banner), exterior and usually larger than the others, the 2 lateral petals, termed the wings, exterior to the 2 lowest ones, the keel petals, which are often partly coherent to enclose stamens and style (exceptions found in *Amorpha* and *Dalea*); stamens typically 10, rarely less, their filaments monadelphous, diadelphous, or distinct; pistil 1, simple, with 1 style and stigma. Fruit commonly a 1-celled legume (pod) dehiscent along both sutures, several-seeded, but in some genera indehiscent and 1-seeded or transversely divided into 1-seeded joints.

In addition to the taxa treated in the text a number of cultivated or introduced species have been reported for the GP. Since they have not been found to persist, the following have been excluded: *Anthyllis vulneria* L., lady's fingers; *Arachis hypogaea* L., peanut; *Cicer arietinum* L., chick-pea; *Glycine max* (L.) Merr., soy-bean; *Lathyrus odoratus* L., sweet-pea; *Lens esculenta* Moench., lentil; *Onobrychis viciaefolia* Scop., sainfoin; *Phaseolus coccinea* L., scarlet runner; *Phaseolus vulgaris* L., common beans; and *Pisum sativum* L., garden pea. In addition, reports of *Phaseolus polystachios* (L.) B.S.P. for our area were based on specimens of *Amphicarpaea bracteata* (L.) Fern.

```
1  Leaves all simple (see also Astragalus spatulatus and A. ceramicus) ................ 9. Crotalaria
1  Leaves variously compound.
   2  Plants trees or shrubs.
      3  Leaves even-pinnate; flowers yellow; introduced sp ........................... 6. Caragana
      3  Leaves odd-pinnate; flowers white, pinkish or purplish.
         4  Plant a tree; flowers white ...................................................... 25. Robinia
```

```
        4   Plant a shrub; flowers usually rose or purple.
            5   Flowers with only 1 petal; flowers purple ................................. 1. Amorpha
            5   Flowers with 5 petals; flowers rose, purplish, white or yellow ........... 10. Dalea
    2   Plants not trees or shrubs.
        6   Stamens 10, distinct.
            7   Leaves pinnate ...................................................................... 27. Sophora
            7   Leaves 3-foliolate.
                8   Pods plump, inflated; flowers white, ochroleucous, or purple ......... 5. Baptisia
                8   Pods elongate, straight or recurved to annular, flattened; flowers
                    yellow ................................................................................ 32. Thermopsis
        6   Stamens monadelphous or diadelphous.
            9   Leaves with rachis terminating in a simple or branched tendril.
                10  Style filiform with a dense ring of hairs just below the stigma .......... 34. Vicia
                10  Style flattened, with a line of hairs down the inner surface ......... 16. Lathyrus
            9   Leaves without tendrils.
                11  Leaves even-pinnately compound; se GP ................................. 26. Sesbania
                11  Leaves odd-pinnately compound or digitate.                                Key 1
                    12  Leaves odd-pinnately compound.
                    12  Leaves digitately-pinnate or palmately compound.                      Key 2
```

                                Key 1
1   Ovaries and fruits with uncinate prickles .... 13. Glycyrrhiza
1   Ovaries and fruits without uncinate prickles.
    2   Plant a climbing or scrambling vine; flowers purple-brown .... 3. Apios
    2   Plant not a vine.
        3   Pods constricted between seeds.
            4   Flowers in umbels or headlike inflorescences; pod segments
                subcylindric .... 8. Coronilla
            4   Flowers in racemes; pod segments nearly oval .... 14. Hedysarum
        3   Pods not constricted between seeds.
            5   Pods 1-seeded; leaves usually gland-dotted .... 10. Dalea
            5   Pods 2- to several-seeded; leaves usually not gland-dotted.
                6   Flowers in terminal spikes, heads, or racemes.
                    7   Flowers 10-14(20) mm long, brick red to pinkish, not
                        bicolored .... 15. Indigofera
                    7   Flowers 14-21 mm long, bicolored with yellow banner and rose wings and
                        keel .... 31. Tephrosia
                6   Flowers on axillary peduncles or these opposite leaves or flowers subsessile in
                    leaf axils.
                    8   Stipules glandlike; lowest pair or leaflets simulating stipules .... 18. Lotus
                    8   Lowest pair of leaflets not simulating stipules.
                        9   Keel into a point or curved beak 0.5-2.5 mm long; plants
                            acaulescent .... 22. Oxytropis
                        9   Keel blunt or rounded; plants acaulescent or caulescent.
                            10 Flowers brick-red; pod membranaceous and conspicuously inflated;
                                stems 4-10 dm tall; rare introduction .... 28. Sphaerophysa
                            10 Flowers not brick-red and pods conspicuously inflated; stems usually
                                less than 5 dm tall .... 4. Astragalus

                                Key 2
1   Plants vines, usually climbing or twining.
    2   Principal leaflets usually 10-20 cm long; stems to 20 m long; rare introduction
        in GP .... 24. Pueraria
    2   Principal leaflets usually 2-9 cm long.
        3   Flowers 4-5 cm long .... 7. Clitoria
        3   Flowers about 1 cm or less long.
            4   Inflorescence umbellate or headlike .... 29. Strophostyles
            4   Inflorescence racemose.

     5 Calyx nearly equally toothed, not subtended by a pair of bracteoles; flowers 10–12 mm long .... 2. *Amphicarpaea*
     5 Calyx unequally toothed, subtended by a pair of bracteoles; flowers 6–8 mm long .... 12. *Galactia*
1 Plants not vines, stems may be procumbent.
  6 Pod a loment; stems and leaves usually uncinulately pubescent .... 11. *Desmodium*
  6 Pod not a loment; uncinate pubescence absent.
     7 Leaflets 5 or more.
        8 Foliage usually glandular-dotted; pods 1-seeded, indehiscent .... 23. *Psoralea*
        8 Foliage not glandular-dotted; pods several-seeded, dehiscent .... 19. *Lupinus*
     7 Leaflets usually 3.
        9 Leaflets denticulate or serrulate at least on upper 1/2
          10 Flowers in elongated racemes 5–15 cm long .... 21. *Melilotus*
          10 Flowers in dense short racemes or umbellike heads.
             11 Corolla withering and persistent; pod straight and usually included in calyx .... 33. *Trifolium*
             11 Corolla soon deciduous; pod usually curved or coiled, prickly in some species .... 20. *Medicago*
        9 Leaflets entire.
          12 Plants acaulescent or short stem covered with scarious stipules .... 4. *Astragalus*
          12 Plants definitely caulescent.
             13 Stipules adnate to petiole and connate into a tube around stem, often setose at base .... 30. *Stylosanthes*
             13 Stipules not forming a tube around stem, not basally setose.
                14 Flowers 1–3 on axillary peduncles .... 18. *Lotus*
                14 Flowers in several-flowered inflorescences.
                   15 Herbage, calyces, or fruits not gland-dotted .... 17. *Lespedeza*
                   15 Herbage, calyces, or fruits with glandular dots.
                      16 Leaflets with all petiolules of equal length .... 23. *Psoralea*
                      16 Leaflets with petiolule of middle leaflet longer than lateral ones .... 10. *Dalea*

1. AMORPHA L.

Erect suffrutescent shrubs or undershrubs, sometimes rhizomatose. Leaves odd-pinnate, alternate, sessile or petiolate; stipules setaceous, caducous; leaflets entire to rarely crenulate, glandular-punctate or epunctate, petiolulate, stipellate on upper side. Inflorescence a spikelike terminal raceme or racemes clustered and appearing paniculate. Flowers pedicellate from axil of a caducous bract; calyx obconic, persistent, 5-lobed; corolla reduced to a single petal which is erect, clawed, obovate, purple, and wrapped around stamens and style; stamens 10, monadelphous at very base, distinct above, exserted. Fruit a 1-seeded, indehiscent pod, gland-dotted.

Reference: Wilbur, Robert L. 1975. A revision of the North American genus *Amorpha* (Leguminosae-Psoraleae). Rhodora 77: 337–409.

1 Plants usually less than 1 m tall; petioles 0.5–5 mm long; leaflets 0.5–1.5 cm long.
  2 Foliage and calyces conspicuously pubescent and often canescent; racemes usually several in axils of upper leaves, forming a compound cluster .... 1. *A. canescens*
  2 Foliage and calyces glabrous or nearly so, never canescent; racemes usually solitary at tips of stems and branches .... 3. *A. nana*
1 Plants usually 1–3.5 m tall; petioles 2–5 cm long; leaflets 2–5 cm long .... 2. *A. fruticosa*

1. *Amorpha canescens* Pursh, lead plant. Erect or ascendant shrub, often rhizomatose, 3–8(12) dm tall; stems 1–several, often branched, usually tomentose but becoming glabrate. Leaves alternate, odd-pinnate; leaflets (5)13–20(24) pairs plus 1; leaflet blades ovate-oblong,

oblong-elliptic, elliptic, or rarely ovate, usually canescent, marginally entire or slightly revolute, apex obtuse to broadly rounded or emarginate with a slender brownish mucro, basally obtuse or rounded; petioles 0.5–3(5) mm long, densely pubescent; stipules inconspicuous, caducous, 1–3.5 mm long, pustular glands absent; petiolules 0.5–1 mm long, stipels acicular. Inflorescence of several racemes in axils of upper leaves and forming a compound cluster, mostly (1)3–30(30) or more in number; racemes densely flowered, (2)7–15(23) cm long. Pedicels 0.5–1.5 mm long, subtended by a caducous bract; calyx tube turbinate, ca 2 mm long, densely pubescent, hairs somewhat obscuring resinous glands, lobes 5, triangular lanceolate; petal 1, bright violet, broadly obcordate, with a slender claw, incurved and enclosing stamens and pistil; stamens 10, lower 2 mm of filaments united in to a tube; ovary ca 1 m long, densely pilose, style 4–6 mm long, densely antrorse pubescent. Legume (3)3.5–4(5) mm long, ca 2 mm wide, with a stipelike base, villous-canescent, punctate-glandular, apex terminated by 0.5–1.5 mm long persistent base of style; seeds 2–2.8 mm long, 1–1.4 mm wide, olive-brown, smooth. ($2n = 20$) May–Aug. Infrequent to common in prairies, hillsides, open woodlands, roadsides; GP but infrequent westward; (n IN, MI, WI, MN, ND, s Man., nw WY, s to IL, AR, TX, NM). *A. canescens* f. *glabrata* (A. Gray) Fassett—Fernald.

Lead plant is one of our most palatable range plants, decreases with grazing, and is rarely found in overgrazed habitats. In pastures and mowed meadows it appears as an herbaceous perennial, but in undisturbed places it often reaches a height of 1–2 m with stems 1–2 cm in diameter. We have not found the glabrous form in the GP, and somewhat glabrate individuals are always in colonies of more characteristic plants.

2. *Amorpha fruticosa* L., false indigo. Erect shrubs 1–3.5 m tall with 1–several stems from the base, often branching and bushy-topped, sometimes proliferating from lateral root sprouts, herbage of current season densely pilosulose, usually 1–3 dm long; petioles 2–4 cm long; stipules narrowly linear, caducous, 2–4 mm long; leaflets 4–10(15) pairs plus 1, oblong, elliptic-oblong or elliptic (1)2–4(5) cm long, (0.4)1–2(3) cm wide, opposite or subopposite, entire, basally acute or rounded, rounded to acute or emarginate apically, often with short mucro, lower surface variously hairy or glabrate, eglandular or rarely conspicuously glandular-punctate; upper surface puberulent or glabrous; petiolules 2–4 mm long; stipels setaceous, 2–4 mm long. Racemes erect, solitary or more typically with 3–several clustered together, densely flowered, (0.5)1–2 dm long. Pedicels 1–2 mm long; calyx tube usually obconic, 2–3 mm long, glabrous or variously hairy, lobes broadly rounded or triangular, 0.2–0.5 mm long, ciliate; petal 1, 5–6 mm long, obovate, claw indistinct, folded to enclose stamens and pistil, reddish-purple; stamens 10, 6–8 mm long, lower 1–3 mm of filaments united; ovary usually glabrous. Legume 5–7 mm long, 2–3 mm wide, curved, glabrous or short-pubescent, conspicuously glandular or eglandular; seeds 3.5–4.5 mm long, 1.5 mm wide, smooth, reddish-brown. ($2n = 40$) May–Jun. Infrequent to locally common in moist ground along streams, rocky banks, open wet woods, pond shores, ravines, & roadsides; GP but infrequent w; (e 1/2 U.S., WY, CO, NM, w to s CA). *A. fragrans* Sweet—Rydberg; *A. fruticosa* var. *angustifolia* Pursh, var. *tennesseensis* (Shuttlew.) E. J. Palm.—Fernald.

This species is highly variable as regards leaf shape and pubescence. The named varieties in our area are to be found in any sizeable colony and do not account for all the variability. Often young specimens of a plant can be assigned to one variety but older ones to another.

3. *Amorpha nana* Nutt., dwarf wild indigo. Erect or ascendant shrubs (1)3–6(10) dm tall, branched above, sometimes rhizomatose; stems moderately strigulose but becoming glabrate. Leaves alternate, odd-pinnate, (1.5)3–7(10) cm long; petioles (2)4–8(10) cm long; stipules setaceous, 3–5 mm long; leaflets usually (3)6–13(15) pairs plus 1, (2)6–13(16) mm long, 3–6(7) mm wide, crowded; blades narrowly to broadly oblong, or somewhat elliptical, round-

ed or cuneate at base, apex broadly rounded or truncate and usually emarginate with a mucro to 1.5 mm long; marginally entire to inconspicuously crenate and somewhat revolute; puberulent to glabrate; punctate glands conspicuous. Racemes solitary at tips of season's growth, densely flowered, 3–9 cm long. Pedicels ca 2–3 mm long; calyx tube turbinate, ca 2 mm long, glabrous, with punctate glands, lobes triangular-lanceolate, 1–2 mm long, glabrous, punctate-glandular; petal 1, 4.5–6 mm long, obcordate, with a slender claw, enveloping stamens and pistil, dark purple; stamens 10, filaments united below; ovary glabrous. Legume 4.5–5.5 mm long, 2–3 mm wide, glabrous, conspicuously punctate-glandular in upper 2/3, terminated by persistent base of style, ca 0.5 mm long; seed 2.5–3 mm long, 1.5 mm wide, olive-brown. May–Jul. Dry prairies & rocky or sandy hillsides; MN, ND, SD, nw IA, NE: Boyd, Cedar, Keya Paha; KS: Clark, Geary, Riley, Rooks, Saline, Wabaunsee; (s Man. & Sask., MN, ND, s CO; NM, KS).

2. AMPHICARPAEA Ell. ex Nutt.

Reference: Turner, B. L., & O. S. Fearing. 1964. A taxonomic study of the genus *Amphicarpaea* (Leguminosae). Southw. Naturalist 9 (4): 207–218.

1. *Amphicarpaea bracteata* (L.) Fern., hog peanut. Taprooted annual; stems 3–20 dm long, twining on herbs and undershrubs or rarely sprawling on moist banks, glabrous, retrorsely appressed-pubescent or densely villous-hirsute. Leaves alternate, pinnately trifoliate; petioles 2–10 cm long; stipules membranaceous, persistent, 3–8 mm long, lanceolate to ovate; leaflets thin, broadly lanceolate to ovate or rhombic-ovate, appressed-pubescent or glabrate, acute at apex, 2–10 cm long, 1.8–7 cm wide; terminal petiolules 5–40 mm long, lateral ones 1–2.5 mm. Flowers of two sorts; chasmogamous flowers in axillary racemes 1.5–9(13) cm long; peduncle 1–6 cm long; pedicels 2–5 mm long, each subtended by 2 bracts; bracts 2–4 mm long, widest above the middle, obtuse or truncate at apex; sepals 5, united ca 2/3 length, 2 upper completely so, 4–5 mm long, tube somewhat gibbous at base; lobes 4, lanceolate or deltoid, 0.5–2 mm long; corolla papilionaceous, lilac colored or white, blades of keel petals longer than claws; stamens diadelphous; legumes 1.5–4 cm long, flattened, often falcate; seeds 4–6 mm long, reniform, brown. Cleistogamous flowers near base of stem or subterranean, usually produced on creeping branches near ground and lacking well-developed petals; legumes fleshy, 6–12 mm in diam, indehiscent, usually with 1 seed. (2n = 20) Aug–Oct. Infrequent to locally common in dry or moist woodlands & thickets, roadside banks, & bushy prairie ravines; e 1/2 GP but infrequent to absent w; (e 1/2 U.S.). *Amphicarpa comosa* (L.) Nieuw. & Lunell, *A. pitcheri* T. & G.— Rydberg.

The hog peanut varies considerably in type of pubescence and size of leaflets, and the extreme forms have been recognized as species or varieties. This variation is present throughout the GP, often in the same woodland population, and taxonomic recognition is without meaning. The subterranean fleshy fruits are edible when cooked by those with the patience required for gathering them in the fall.

3. APIOS Fabr., Groundnut

1. *Apios americana* Medic., American potato bean, groundnut. Herbaceous perennials from slender rhizomes with tuberous thickenings to 6 cm in diam; stems twining and high climbing or sprawling over shrubs and herbs, 1–5 m long, glabrous or sparsely to densely pubescent. Leaves alternate, once-pinnately 5- to 7-foliolate; petioles 1.5–8 cm long, glabrous or pubescent; stipules setaceous, 4–7 mm long, soon deciduous; leaflets usually ovate to lance-ovate, acute or acuminate, rounded at base, 2–10 cm long, glabrate or pubescent;

stipels 1-2 mm long, deciduous. Flowers in axillary peduncled racemes, peduncles 2-5 cm long, flowers loosely or compactly arranged on pedicels 2-5 mm long. Calyx tube broadly hemispheric, 3-5 mm long, the 4 upper lobes very short or obsolete, the lowest one lance-triangular, 1/4-1/2 as long as tube; corolla papilionaceous, 10-14 mm long; banner rotund, soon reflexed, whitish dorsally, brown-red ventrally; wings down-curved, brown-purple; keel sickle-shaped, slender, short clawed, brownish-red; stamens 10, diadelphous. Legume straight or slightly curved, linear, 5-10 mm long, 4-6 mm wide, valves coiling after dehiscence; seeds 4-5 mm long, oblong or quadrate, dark brown, surface wrinkled. (2n = 40) Jul-Sep. Infrequent to locally common in moist woods, low thickets, stream & pond banks, moist prairie ravines; GP but infrequent or absent n & w; (Que. to MN, ND, s to n-cen CO, FL, TX). *Apios tuberosa* Moench—Rydberg; *A. americana* f. *pilosa* Steyerm., var. *turrigera* Fern.—Fernald.

This species varies considerably in amount and type of pubescence, density of flowers in raceme, and size of leaflets. Though more northern plants have a tendency toward flowers loosely arranged, we find that no named variants account for the variations in any of the larger colonies. Occasional plants have cleistogamous flowers and such plants have been referred to forma *cleistogama* Fern. Livestock, especially horses, readily graze the foliage, and the underground tubers were either roasted or fried and eaten by Indians and early explorers.

4. ASTRAGALUS L., Milk Vetch

Perennial or annual herbs, caulescent or not, with a taproot, and after first season an aerial, cespitose, or subterranean caudex which is sometimes stoloniform or rhizomatose and adventitiously rooting; pubescent with basifixed or dolabriform hairs or nearly glabrous. Leaves alternate, usually odd-pinnate, more rarely palmately 3-foliate, or in *A. spatulatus* essentially simple, or in *A. ceramicus* most leaves reduced to filiform rachis; petioles absent or present; stipules connate of distinct. Flowers usually several in pedunculate axillary racemes, papilionaceous, generally white or yellowish to reddish or purple; calyx campanulate to tubular; banner usually reflexed from wings; wings generally shorter than banner but usually longer than keel; stamens 10, diadelphous. Legumes sessile to conspicuously stipitate, very variable in size, shape, texture, compression, and dehiscence; seeds 1 to many. *Atelophragma* Rydb., *Batidophaca* Rydb., *Cnemidophacos* Rydb., *Diholcos* Rydb., *Geoprumnon* Rydb., *Hamosa* Medic., *Holcophacos* Rydb., *Homalobus* Nutt., *Kentrophyta* Nutt., *Microphacos* Rydb., *Orophaca* (T. & G.) Britt., *Phaca* L., *Pisophaca* Rydb., *Tium* Medic., and *Xylophacos* Rydb.—Rydberg.

Reference: Barneby, R. C. 1964. Atlas of North American *Astragulus*. Mem. New York Bot. Gard. 13: 1-1188.

1 Leaves palmately trifoliate.
  2 Stipules densely long pilose dorsally .... 32. *A. sericoleucus*
  2 Stipules not long pilose, glabrous or glabrescent dorsally.
    3 Flowers pedunculate; calyx tube 2.8-5 mm long .... 6. *A. barrii*
    3 Flowers sessile; calyx tube 6-16 mm long.
      4 Banner tapering from tip to base; petals glabrous or banner rarely puberulent .... 15. *A. gilvilflorus*
      4 Banner fiddle-shaped; all petals dorsally villous .... 17. *A. hyalinus*
1 Leaves not palmately trifoliate.
  5 Plants annuals or winter annuals.
    6 Flowers with banner 4.5-7.5 mm long .... 25. *A. nuttallianus*
    6 Flowers with banner 12-18 mm long .... 19. *A. lindheimeri*
  5 Plants perennial, rarely flowering first year.
    7 Stipules at the lower leafless nodes united on the side corresponding with the suppressed petiole into an entire blade; plants stoloniferous, with nodding flowers, and pendulous, stipitate, inflated, unilocular pods of papery texture ..... 5. *A. americanus*

7 Stipules at lower nodes free or connate, if united then connate opposite petiole and sheath emarginate or bidentate; other characters different.
   8 Leaves principally unifoliate or appearing so; occasionally a few compound leaves present.
      9 Plants acaulescent; leaves spatulate; pod laterally compressed .... 34. *A. spatulatus*
      9 Plants caulescent; all or at least upper leaves reduced to a naked, linear-filiform rachis; pod bladdery-inflated .... 10. *A. ceramicus* var. *filifolius*
   8 Leaves principally odd-pinnate.
      10 Leaflets 3–7, linear-elliptic, continuous with the rachis and mucronate or spinulose at apex, usually becoming stiff and prickly in age .... 18. *A. kentrophyta*
      10 Leaflets either more numerous, or jointed to the rachis, commonly both, never spinulose or prickly.
         11 Pubescence of the leaves dolabriform or mostly so.
            12 Plants definitely caulescent; stipules, at least those at lower nodes, connate opposite the petiole.
               13 Fruits unilocular; rare in Black Hills .... 21. *A. miser* var. *hylophilus*
               13 Fruits bilocular.
                  14 Flowers declined and retrorsely imbricated, usually greenish-white or yellowish; stems from oblique or creeping rhizomes .... 9. *A. canadensis*
                  14 Flowers erect or ascending, usually purplish; stems from superficial root-crown or shortly branched caudex .... 2. *A. adsurgens* var. *robustior*
            12 Plants subacaulescent or stems no longer than inflorescences and leaves; stipules distinct, not connate opposite leaves.
               15 Banner 16–22 mm long; calyx tube 6–9 mm long ...... 22. *A. missouriensis*
               15 Banner not over 14 mm long; calyx tube 2–4.5 mm long .. 20. *A. lotiflorus*
         11 Pubescence of leaves basifixed.
            16 Leaves and stems gray-hirsute with widely spreading, filiform but minutely bulbous-based (use 30×) hairs to 1.1–1.8 mm long .... 13. *A. drummondii*
            16 Leaves and stems variably pubescent but hairs shorter and not thickened at base.
               17 Stipules all distinct.
                  18 Wing-petals bidentate at apex; fruit pendulous, stipitate, laterally compressed .... 1. *A. aboriginum*
                  18 Wing-petals entire or obscurely emarginate, fruit sessile or nearly so, if at all stipitate then dorsiventrally compressed.
                     19 Low, tufted, subacaulescent or shortly caulescent plants with stems usually no longer than inflorescences and longer leaves.
                       20 Pubescence villous-tomentulose, of shorter, curly hairs, together with longer, ascending, twisted hairs, turning rusty on drying .... 23. *A. mollissimus*
                       20 Pubescence strigulose, or if tomentose then softly cottony with hairs all very fine and sinuous, or all hairs alike, not changing color on drying.
                          21 Fruit with valves densely strigulose .... 33. *A. shortianus*
                          21 Fruit very densely villous or hirsute-tomentose sinuous hairs to 5 mm long .... 30. *A. purshii*
                  19 Caulescent plants, inflorescence shorter than stems.
                     22 Stems erect; flowers white or ochroleucus.
                       23 Pods greatly inflated, valves papery; plant not ill-scented; extreme ne GP .... 24. *A. neglectus*
                       23 Pods fleshy, becoming leathery; plants strongly ill-scented; extreme sw GP .... 28. *A. praelongus* var. *ellisiae*
                     22 Stems decumbent, prostrate or ascending; flowers usually purplish.
                       24 Ovary and fruit pubescent .... 27. *A. plattensis*
                       24 Ovary and fruit glabrous.
                          25 Stems essentially glabrous; flowers 9–15 mm long; fruit dry; plants of extreme se GP .... 12. *A. distortus*
                          25 Stems pubescent at least above; flowers 15–25 mm long; fruit fleshy; plants widespread in GP .. 11. *A. crassicarpus*

17 Stipules at least those of lowest 1–3 nodes connate into a bidentate sheath.
  26 Flowers small, calyx tube 1.4–4.3 mm long, banner 4.3–12.6 mm long.
    27 Wing-petals bidentate at apex, pod pendulous, stipitate, laterally compressed .... 1. *A. aboriginum*
    27 Wing-petals entire or obscurely emarginate; pods various.
      28 Pod strongly compressed laterally, 2-sided, bicarinate.
        29 Pod stipitate, stipe 0.6–7 mm long; petals white with purple keel tip .... 35. *A. tenellus*
        29 Pod sessile or nearly so; petals bright pink-purple .... 36. *A. vexilliflexus*
      28 Pod either terete or more or less dorsiventrally compressed, never bicarinate.
        30 Pod pendulous, stipitate.
          31 Stems arising from buried points of renewal on subterranean rhizomelike caudex branches; BH in GP .... 4. *A. alpinus*
          31 Stems arising together from a determinate, superficial root-crown or caudex .... 7. *A. bisulcatus*
        30 Pods sessile.
          32 Stipules all connate-amplexicaul .... 8. *A. bodini*
          32 Upper stipules nearly distinct.
            33 Pods 12–24 mm long .... 14. *A. flexuosus*
            33 Pods 4–9 m long .... 16. *A. gracilis*
  26 Flowers larger, calyx tube at least 4.5 mm long, banner 13 mm long or more.
    34 Leaflets without petiolules; pods sessile; petals white or ochroleucous .... 26. *A. pectinatus*
    34 Leaflets with petiolules; petals purple, if white then pod stipitate.
      35 Flowers crowded into ovoid or short oblong heads; pods bilocular .... 3. *A. agrestis*
      35 Flowers loosely racemose; pods unilocular.
        36 Stems erect, from a determinate root crown; pod stipitate.
          37 Body of pod dorsiventrally compressed; doubly grooved lengthwise .... 7. *A. bisulcatus*
          37 Body of pod trigonously compressed, angles all narrow and acute, the 3 faces of equal width or nearly so, all flat or slightly concave .... 31. *A. racemosus*
        36 Stems diffuse or prostrate, from subterranean caudex branches; pod sessile.
          38 Pod broadly and plumply ovoid, 1–1.3 cm in diam, valves fleshy, green .... 27. *A. plattensis*
          38 Pod obliquely oblong or ovoid-ellipsoid, 5–9.5 mm in diam, not fleshy, usually red-mottled .... 9. *A. puniceus*

1. *Astragalus aboriginum* Richards., Indian milk-vetch. Perennial, with a woody taproot and branching caudex; stems, several, 0.5–3 dm tall, erect and ascending, or decumbent, rarely branched, sparsely to densely silky-strigose, crisp-strigulose, or villous, all hairs basifixed. Leaves alternate, odd-pinnate, sessile or lower with petioles to 3 cm; lower stipules ovate, amplexicaul, 4–7 mm long, upper lanceolate, ca 7 mm long; leaflets 7–15, elliptic or linear, glabrous or pubescent above, pubescent beneath. Peduncles axillary, 2–10 cm long; racemes 5- to 30-flowered. Pedicels 1–2 mm long; flowers whitish but with purple keel to overall purplish; calyx 4–8 mm long, black-hairy, teeth subulate, 1–3 mm long; banner obovate or oblanceolate, 8–12 mm long; wings 7–12 mm long, clawed, apex lobed; keel 6–9 mm long, claw ca 1/2 the length, apex rounded. Legumes pendulous, somewhat falcate, stipitate, stipe 3–8 mm long, the body glabrous or pubescent, 2–3 cm long, 4–5 mm wide; seeds 2.3–3 mm long, smooth, dull. (n = 8) Jun–Jul. Infrequent to locally common on open wooded hillsides, prairie hills, river bluffs; w ND & SD, MT, WY; (se Man. to ne B.C., AK s to NM, UT, NV). *Atelophragma aboriginum* (Richards.) Rydb., *A. glabriusculum* (Hook.) Rydb., *A. forwoodii* (S. Wats.) Rydb., *A. herriotii* Rydb.—Rydberg.

2. *Astragalus adsurgens* Pall. var. *robustior* Hook., standing milk-vetch. Cespitose perennials from woody taproot and branching woody caudex; stems 1–4 dm tall, herbage grayish-strigulose with dolabriform hairs. Leaves 4-15 cm long, alternate, odd-pinnate, lower petioled, upper sessile; leaflets 9-25, narrowly oblong to oblong-obovate, 1-2(3) cm long, 3-8 mm wide, strigose; stipules 5-15 mm long, connate 1/3 or more their length. Peduncles axillary, 4-14 cm long; racemes oblong, congested, somewhat spikelike, 15- to 50-flowered. Pedicels ca 1 mm long; flowers dark purple, dull blue, reddish-lilac, or whitish; calyx 5-9 mm long, teeth ca 1/3 as long to as long as tube, with mixed black and white strigulose hairs; banner 13-19 mm long, 4-8 mm wide; wings 10-17 mm long, claws 5-8 mm long; keel 9-15 mm long, claws 5-8 mm long; stamens 10, diadelphous. Legume nearly sessile, 7-12 mm long, 3-4 mm wide, strigulose with white basifixed hairs; seeds 2-2.5 mm long, smooth, brown. (n = 16) Jun-Sep. Common on dry, rocky prairie or open wooded hillsides, pastures, ravines, roadsides; GP n of KS; (Man., MN to AK, s nw IA, NE, NM, ID, WA). *A. striatus* Nutt., *A. chandonetti* Greene — Rydberg.

    This is a highly variable species especially in flower color, length of calyx teeth, amount of pubescence, and shape of raceme.

3. *Astragalus agrestis* Dougl. ex G. Don, field milk-vetch. Low perennials with a taproot ending in a crown or subterranean branching caudex; stems decumbent to weakly erect, 1-3 dm tall, thinly strigulose with basifixed hairs or glabrate. Leaves alternate, pinnate, petiolate or subsessile; stipules linear to ovate, usually connate basally, 2-10 mm long; leaflets 11-21, lanceolate, oblong-elliptic, usually retuse, 4-20 mm long. Peduncles longer or shorter than leaves; racemes densely 5- to 15-flowered, these crowded into an ovoid, oblong, or subglobose head. Pedicels to 1.5 mm long; calyx tube cylindric or campanulate, 5-8 mm long, teeth 3-5 mm long, with black and white hairs, corolla purplish, bluish, whitish, or yellowish; banner oblanceolate or elliptic, 17-22 mm long, shallowly or deeply notched; wings 15-18 mm long, claws 6-8 mm long; keel 11-14 mm long, claws 6-8 m long, apex acute; stamens 10, diadelphous. Legume nearly sessile, erect, straight, 8-10 mm long, densely pubescent with white hairs; seeds 1.5-2 mm long, smooth, brown. (n = 8) May-Aug. Infrequent to locally common in moist prairies, open wooded hillsides, roadsides; n 1/2 GP; (Man., Yukon, s to nw IA, n NE, NM, UT, NV, n CA). *A. goniatus* Nutt.—Rydberg.

    This species is sometimes confused with small forms of *A. adsurgens* which has dolabriform pubescence rather than basifixed hairs.

4. *Astragalus alpinus* L., alpine milk-vetch. Low, weak-stemmed caulescent or acaulescent perennials with horizontal or ascending, branching subterranean caudex-branches, stems decumbent or ascending, 5-20 cm long, often forming patches, greenish or sparsely to densely gray strigulose. Leaves alternate, pinnate, 5-15 cm long; leaflets 13-25, ovate to oblong-elliptic, often retuse, 2-20 mm long, to 10 mm wide; petioles 8-20 mm long; stipules 2-8 mm long, connate to 1/2 their length. Peduncles axillary, longer or shorter than leaf; racemes 10- to 20-flowered, elongate, lax. Pedicels 0.5-1.5 mm long; flowers pale lilac to purplish, or bicolored; calyx tube 2.5-4 mm long, teeth subulate triangular 1-3 mm long, both usually black strigulose; banner 7-13 mm long; wings 6-11 mm long, clawed; keel usually darkest in color, a little shorter than banner. Legumes somewhat pendulous, stipe 1.4-3 mm long, body ellipsoid, 8-12 mm long, with strigulose black hairs; seeds 2-3 mm long, brownish or yellowish, smooth. (n = ca 28) Jun-Aug. Moist, rocky, open wooded hillsides; known only from BH in GP; (across n Can., AK, s in mts. to n NM, BH). *Antelophragma alpinum* (L.) Rydb.—Rydberg.

5. *Astragalus americanus* (Hook.) M.E. Jones, American milk-vetch. Perennial herb with a woody caudex and often rhizomatose; stems erect, 3-10 dm tall, striate, usually hollow, glabrous or with sparse hairs. Leaves alternate, odd-pinnate, petiolate below to sessile above; stipules 1-3 cm long, deflexed, not connate or lowest united; leaflets 9-17, ovate to oblong or elliptic-oblanceolate, 2-5 cm long, 7-15 mm wide, lighter green beneath, glabrous or sparsely strigulose at least beneath. Peduncles axillary, usually shorter than leaves; racemes loosely 10- to 25-flowered. Pedicels 3-10 mm long; flowers usually reflexed, whitish or ochroleucous, 10-15 mm long; calyx tube campanulate 4-5 m long, membranaceous, nearly glabrous, teeth less than 1 mm long; banner 11-14 mm long,m spatulate; wings 10-13 mm long, blade shorter than claw; keels 10-12 mm long, claws 6-7 mm long; stamens 10, diadelphous. Legumes pendulous; stipe longer than calyx, body 2-3 cm long, inflated, usually glabrous; seeds olive-brown, smooth, 2.5-3 mm long. (n = 8) Jun–Jul. Stream banks & moist meadows; known only from BH in GP; (AK & s in mts. through B.C., MT, BH, n CO, e to Ont., Que.). *Phaca americana* (Hook.) Rydb.—Rydberg.

6. *Astragalus barrii* Barneby, Barr orphaca. Perennial subacaulescent densely tufted and mounded herbs from branching caudex; stems reduced to leafy crowns. Leaves 1-4 cm long, palmately trifoliate, silvery with dolabriform hairs; leaflets linear-lanceolate, oblanceolate, acute, usually involute, 3-12 mm long; petioles slender, 1-3 cm long; stipules 4-8 mm long, hyaline, connate-sheathing. Peduncles 7-24 mm long, erect, or spreading; racemes 2- to 5-flowered. Pedicels 0.5-1.5 mm long; calyx tube 3.5-5 mm long, short-cylindric, teeth linear, 2-5 mm long; corolla pink-purple; banner 10-16 mm long, clawed; wings 9-13 mm long, clawed; keels 7.5-10 mm long, clawed; stamens 10, diadelphous. Legume sessile, 4.5-8 mm long, beaked, silvery-strigulose; seeds 1-3, ca 2 mm long. Apr–Jun. Local on dry, rocky prairie knolls, hillsides, or barren areas; MT: Carter, Powder River; SD: Fall River, Shannon; WY: Johnson, endemic in area.

After several seasons of field work in the region, additional localities for the Barr orophaca have not been found, though many likely sites have yet to be investigated.

7. *Astragalus bisulcatus* (Hook.) A. Gray, two-grooved vetch. Stout perennial herbs from a woody branching caudex; stems few to many, 2-7 dm tall, erect and ascending or decumbent, forming clumps, sparingly strigulose or glabrate. Leaves alternate, odd-pinnate, 4-12 cm long, lower short petioled, sessile above; stipules 3-10 mm long, scarious, connate; leaflets 15-35, ovate-oblong, oblong-elliptic, lance-elliptic or oblanceolate, acute, obtuse and often mucronulate, 5-35 mm long. Peduncles 3-12 cm long; racemes many-flowered, ovoid to cylindric, flowers reflexed. Pedicels 1-3 mm long; calyx tube obliquely campanulate, 3-6 mm long, strigulose, teeth variable; corolla whitish to purple; banner 10-17 mm long, oblanceolate to obovate; wings 8-15 mm long; keel 6-13 mm long; stamens 10, diadelphous. Legumes pendulous, stipitate, 7-20 mm long, deeply 2-grooved, strigulose or glabrous; seeds 3-3.5 mm long, brown, smooth. (n = 12) Jun–Aug. Locally common on prairies, stream valleys, hillsides, roadsides; ND, SD, nw NE, MT, WY, CO, NM, w KS; (Man., Sask., s to ND, sw SD, NM, CO). *Diholcos bisulcatus* (Hook.) Rydb.—Rydberg.

This handsome floriferous plant is considered to be one of the most troublesome of obligate selenium species but, fortunately, animals rarely eat it.

8. *Astragalus bodini* Sheld., Bodin milk-vetch. Slender perennials with branching caudex; stems 1-4 dm long, decumbent, branched, glabrous or sparsely strigulose, often forming mats. Leaves alternate, odd-pinnate, petioled below, subsessile above; stipules 1-7 mm long, deltoid to ovate, connate; leaflets 7-17, lance-oblong, oblanceolate, retuse, 7-18 mm long, glabrous or sparingly strigose beneath. Peduncles axillary, longer than leaves; racemes 3- to 15-flowered. Pedicels 1-2 mm long; calyx tube 3 mm long, strigose with black hairs, teeth subulate, 2 mm long; corolla purple; banner 8-11 mm long; wings 7-10 mm long; keel 5-8 mm long. Legume sessile, body ellipsoid 6-10 mm long, with beak to 2 mm long, strigulose with black and white hairs; seeds ca 2 mm long, brown, smooth. Jul–Aug. Sandy or gravelly stream banks & meadows. NE: Scotts Bluff; (s WY, cen CO, s-cen UT, NE). *Phaca bodinii* (Sheld.) Rydb.—Rydberg.

This species is included for the plains on the basis of an old western NE specimen, which may have come from plants briefly established along the North Platte R. The species is known to the west in WY.

9. *Astragalus canadensis* L., Canada milk-vetch. Caulescent perennial from rhizomes, often forming patches; stems solid or hollow, 3–12 dm tall, with appressed dolabriform hairs, often branched. Leaves alternate, odd-pinnate, 5–35 cm long, short petioled to sessile, stipules 3–18 mm long; leaflets 15–35, lanceolate, oblong, ovate, or elliptic, 1–4 cm long. Peduncles 4–10 cm long; racemes dense, usually 4–20 cm long. Pedicels short, to 2.5 mm in fruit; calyx tube 4–7 mm long, teeth linear, 1–4 mm long, strigose or glabrate; corolla greenish-white or ochroleucous; banner 11–16 mm long; wings 10–14 mm long, clawed; keel 9–13 mm long, clawed. Legume 10–15 mm long, terete or nearly so, cusp 2–4 mm long, glabrous or rarely puberulent; seeds ca 2 mm long, smooth, brown. (2n = 16) May–Aug. Infrequent to locally common in moist prairies, riverbanks, open wooded hillsides, & dry bluffs; GP but absent in sw; (e 1/2 U.S. to GP).

10. *Astragalus ceramicus* Sheld. var. *filifolius* (A. Gray) Herm., painted milk-vetch. Slender herbaceous perennials from buried caudex and spreading rhizomatose caudex-branches, with appearance of being leafless and broomlike; stems weak, few, decumbent or ascending, 3–40 cm long, simple or few-branched, silvery-canescent, with basifixed hairs. Leaves alternate, 2–17 cm long, odd-pinnate or usually reduced to rachis, some lower leaves with 3–5(7) leaflets. Peduncles axillary, 1.5–7 cm long; racemes 2- to 7-flowered. Flowers whitish or purplish; calyx tube 2–3 mm long, white or black-strigose, teeth 1–1.5 mm long; banner 6–9 mm long, clawed, wings 6–8 mm long, clawed; keel 6–8 mm long, clawed. Legume 3–5 cm long, bladder-inflated, ellipsoid, ovoid, or subglobose, stipe 1–3 mm long, valves red or purple-mottled, glabrous; seeds 2–3 mm long, brown, smooth. Jun–Jul. Infrequent to locally common on sand dunes or sandy soils in prairies & along streams; w ND, e MT, NE sandhills, e WY, CO, sw KS, OK: Cimarron, Texas; NM: Union; TX: Briscoe; (w ND, e MT, s to TX & NM). *Phaca longifolia* (Pursh) Nutt.—Rydberg; *A. longifolius* (Pursh) Rydb.—Gates.

This variety of the painted milk-vetch is somewhat common in the Nebraska sandhill region but rare elsewhere in the GP.

11. *Astragalus crassicarpus* Nutt., ground-plum. Perennial with a thick woody taproot and branched caudex; stems several, usually decumbent, ascending, or suberect, 1–6 dm long, with appressed or somewhat ascending hairs. Leaves alternate, odd-pinnate, 4–13 cm long, mostly subsessile, stipules 3–9 mm long; leaflets 15–27, oblanceolate, elliptic, oblong-elliptic, or suborbicular, 3–17 mm long, 3–6 mm wide, strigose beneath, sparsely hairy or glabrate above. Peduncles axillary 2–10 cm long; racemes 5- to 25-flowered. Calyx tube usually 6–9 mm long, black-strigose, teeth 1.5–4 mm long; corolla purple, light blue, pinkish, rarely white or ochroleucous, or white and keel tip purplish; banner 16–24 mm long, notched; wings 16–18 mm long; keel 11–15 mm long, clawed. Legumes globose or plumply obovoid-oblong, 1.5–2.5 cm long, nearly as thick, glabrous; seeds 2–3 mm long, black. (2n = 22)

There are three vars. in the GP:

1 Stems decumbent; calyx strigulose; petals purple, bluish, pinkish, or rarely white.
  2 Petals all purple or bluish .... 11a. var. *crassicarpus*
  2 Petals white except for purplish keel tip .... 11b. var. *paysoni*
1 Stems ascending to erect; calyx usually tomentulose; petals greenish-white or yellowish .... 11c. var. *trichocalyx*

11a. var. *crassicarpus*. Petals all purple or bluish. Mar–Jun. Commonly encountered in a variety of soils in prairies, rocky open wooded hillsides, stream valleys, & roadsides; GP; (Man., Alta., s to MN, w MO, TX). *Geoprumnon crassicarpum* (Nutt.) Rydb. and *G. succulentum* (Richards.) Rydb.—Rydberg; *A. caryocarpus* Ker—Fernald. In the high plains from se SD to extreme sw KS many plants have flowers much lighter in color and foliage light green to yellowish-green. These plants intergrade with the next variety westward, and often the two cannot be separated satisfactorily.

11b var. *paysoni* (Kelso) Barneby. Petals white except for the purple keel tip, in drying, however, the white petals often become a light blue similar to the color of petals in var. *crassicarpus*. May–Jun. Prairie hillsides & knolls; extreme w edge of high plains, se SD, WY, s to sw KS; OK: Cimarron; CO, NM; (s Alta., s to s-cen CO).

11c var. *trichocalyx* (Nutt.) Barneby. Ascending stems, tomentulose calyx, yellowish petals. Mar–May. Rocky open wooded hillsides, rocky prairie banks, & bluffs; KS: Cherokee; (s IL, s MO, e OK, w AR, e TX). *Geoprumnon trichocalyx* (Nutt.) Rydb.—Rydberg; *A. caryocarpus* var. *trichocalyx* (Nutt.) Fern.

12. *Astragalus distortus* T. & G., Ozark milk-vetch. Low, diffuse, perennials with a taproot and short branching caudex; stems several, decumbent or prostrate, usually 1–3 dm long, often forming mats, glabrate below, sparsely strigulose above. Leaves alternate, odd-pinnate, 4–12 cm long; stipules 2–6 mm long, becoming membranous; petioles 0.5–2 cm long below, subsessile above; leaflets 11–27, oval, obovate, elliptic-oblanceolate, 3–12 mm long, retuse or truncate, sparsely strigulose. Peduncles axillary, 3–15 cm long; racemes 5- to 20-flowered. Calyx tube campanulate, 3–4 mm long, strigose with white or black hairs, teeth 1–2 mm long; corolla reddish-purple, lilac, pale blue, or white; banner 11–15 mm long, deeply notched; wings 9–13 mm long; keel 6–9 mm long. Legumes sessile, 1.5–2.5 cm long, lunate; seeds 2–2.5 mm long, brown. Apr–Jun. Infrequent to locally common on rocky, open wooded hillsides, rocky or sandy prairie banks, pastures, roadsides; se GP; (e IA, w IL, s to AR, e OK, e KS). *Holcophacos distortus* (T. & G.) Rydb.—Rydberg.

Plants of the Ozark milk-vetch vary in color of corolla from one colony to another and within colonies. The pure albino form is conspicuous and not unusual.

13. *Astragalus drummondii* Doug. ex Hook., Drummond milk-vetch. Caulescent perennials, with short-branched caudex; stems several, erect or ascending in clumps, 3–5 dm tall, hollow, ribbed, simple, villous-hirsute. Leaves alternate, odd-pinnate, 4–12 cm long, villous-hirsute; all but lowest subsessile; stipules 3–12 mm long, distinct or connate; leaflets 12–33, oblong-oblanceolate, oval-oblong, or obovate, 4–27 mm long, obtuse, truncate, or emarginate, flat or somewhat folded. Peduncles axillary, 5–12 cm long; racemes 14- to 35-flowered. Calyx tube campanulate, 5–8 mm long, teeth 2–5 mm long, with black, white, or mixed hairs; corolla white or pale yellowish, keel tip purplish; banner 18–26 mm long; wings 15–21 mm long, claws 7–9 mm long; keel 12–15 mm long, claws 6–8 mm long. Legumes pendulous, stipitate, stipe 5–11 mm long, body linear-oblong, 2–3 cm long, straight or curved, bilocular or nearly so; seeds 2–3 mm long, brown. May–Jul. Prairie plains & hills, open wooded or brushy hillsides & ravines, in a variety of soils; ND: Williams; e MT, sw SD, w NE, e WY & CO; NM: Union; (Sask., Alta., s to w ND, CO, NM, UT, ID). *Tium drummondii* (Dougl.) Rydb.—Rydberg.

14. *Astragalus flexuosus* (Hook.) G. Don, pliant milk-vetch. Caulescent perennials, with subterreanean branching caudex, rarely short rhizomatose; stems 1.5–6 dm long, slender, decumbent with ascending tips; herbage greenish or silky-canescent, with basifixed hairs. Leaves alternate, odd-pinnate; petiolate below to subsessile above; stipules 1.5–7 mm long, connate below, distinct above; leaflets 11–25, linear or narrowly oblong-lanceolate, obtuse, truncate, or retuse. Peduncles axillary, 2.5–6 cm long; racemes 12–26 mm, flowered. Calyx tube ca 2.5 mm long, teeth ca 1 mm long; corolla pale purple; banner 7–9 mm long; wings 7–8 mm long; keels 5–5.5 mm long, claws ca 2.4 mm long. Legume subsessile, oblong-ellipsoid, slightly compressed, strigulose; seeds 2.–2.5 mm long, pale brown, smooth. (n = 11) Jun–Jul. Often common on prairies, bluffs, open wooded hillsides, stream & canyon valleys, roadsides; ND, e SD & w SD, w edge of GP; (Man., Sask., Alta., s to SD, MT, WY, NM). *Pisophaca elongata* (Hook.) Rydb., *P. flexuosa* (Dougl.) Rydb.—Rydberg.

The report of this species for KS; Harper, by Gates, was based on a young specimen of *Indigofera miniata* Ort.

15. ***Astragalus gilviflorus*** Sheld., plains orophaca. Perennial, acaulescent, tufted or mounded herbs, with branching cespitose caudex; caudex with dense persistent leaf-stalks; herbage densely silvery-strigose with dolabriform hairs. Leaves trifoliate, 2–13 cm long; those expanding with the flowers having shorter petioles than later ones; leaflets obovate-cuneate to rhombic-obovate or narrowly oblanceolate, 3–35 mm long, acute or acuminate, rarely obtuse; stipules ovate or oblong-obovate, 6–13 mm long, often transversely corrugated. Peduncle obsolete; racemes capitately (1)2(3)-flowered. Pedicels 0–1.5 mm long; calyx tube cylindric, 6.5–15 mm long, teeth 1.5–4 mm long; corolla white, becoming yellowish on drying, keel tip purplish or rarely all bluish; banner 16–28 mm long; wings 12–24 mm long; keel 10–21 mm long, claws 7–15 mm long. Legume erect, ovoid-ellipsoid, 6–10 mm long, body strigose-hirsutulous; seeds 1.5–2 mm long, yellowish to blackish. (n = 12) Apr–May. Infrequent to locally common on rocky prairie hilltops, slopes, or barren flats; in a variety of soils; e-cen ND, sw to sw NE; CO: Kit Carson; KS: Greeley, Logan, Sherman, Wallace, Wichita; OK: Beaver; (se Man., Alta., s to ND, nw KS, ne CO, WY, MT). *Orophaca caespitosa* (Nutt.) Britt.—Rydberg: *Astragalus triphyllus* Pursh—Stevens.

16. ***Astragalus gracilis*** Nutt., slender milk-vetch. Caulescent perennials with subterranean caudex; caudex a knobby crown or often proliferating radially underground (infrequently collected) 2–20 cm and producing aboveground stems at intervals; herbage strigulose or glabrous. Stems few to several, 1.5–8 dm long, branched, or rarely simple, erect, ascending, or decumbent. Leaves (2)3–5(7) cm long, alternate, odd-pinnate; petioles 0–3 cm long; stipules 1–4 mm long, the lowest connate and amplexicaul, progressively less united upward to distinct or united by a stipular line; leaflets 7–21, linear, linear-oblong, oblanceolate, or cuneate-oblong, apically obtuse, truncate, or retuse, (3)5–18(27) mm long. Peduncles axillary or terminal on stem and branches, 4–17(23) cm long; racemes (3)12- to 40(50)-flowered. Pedicels ca 1 mm long; calyx tube campanulate, 1.5–2.6 mm long, teeth 0.5–1 mm long; corolla pale to dark purple, or whitish and keel tip purplish, often drying yellowish. Banner 6–8(9) mm long, ovate-cuneate; wings 5–8 mm long, claws 1.5–2.5 mm long; keel 3.5–6 mm long, claws 1.5–3 mm long. Legumes somewhat reflexed, 4–8 mm long, plumply ovate or ovate-elliptic, abruptly cuspidate to apex; seeds 2.5–3.5 mm long, brownish, smooth. May–Oct. Infrequent to common on prairie hilltops, slopes, ravines, stream valleys, open wooded or brushy hillsides, roadsides; w 1/2 of GP but uncommon in n; (sw ND, MT, s to w OK, TX, ne NM, e 1/2 CO). *Microphacos gracilis* (Nutt.) Rydb., *M. parviflora* (Pursh) Rydb.—Rydberg; *Astragalus gracilis parviflorus* (Pursh) F. C. Gates—Gates.

The slender milk-vetch, as described above, is a highly variable species in habit; number, size, and shape of leaflets; amount of pubescence; size of flowers and legume characters. Some authorities have recognized either two species or two varieties in the complex but these are always sympatric. The extremes of the two variants are quite distinct, and if they had an allopatric distribution they would without doubt be recognized as "good" species or varieties. The complex merits careful field and experimental studies.

17. ***Astragalus hyalinus*** M. E. Jones, summer orophaca. Perennial low, acaulescent cespitose plants, eventually forming cushions or rounded mounds 1–4 dm wide; taproot thickened; caudex branching, woody, with thatch of persistent petioles and stipules; stems reduced to crowns or to 2 cm long, with a cluster of rosulate leaves, and imbricated stipules; pubescence of herbage villous-strigose throughout and silvery with dolabriform hairs to 2

cm long. Leaves trifoliate, 0.5–4 cm long; stipules hyaline, 6–10 mm long, connate, sheath transversely wrinkled, ciliate; leaflets oblanceolate or obovate, 2.5–14 mm long, acute or obtuse. Peduncles obsolete or to 3.5 mm long, hidden by stipules; racemes 1- to 2(3)-flowered, flowers erect. Pedicels obsolete; calyx tube cylindric 5.5–7 mm long, teeth 1.5–3.5 mm long; corolla whitish, drying yellowish, tips of keels and wings often light purple, petals all villous dorsally; banner erect, 12–18 mm long, narrowly fiddle-shaped, claw oblanceolate, 6–8 mm long, blades 5.5–9 mm long, linear-oblong; wings 10–17 mm long, claws 5–7 mm long; keel 10–13 mm long, blades ovate-elliptic, 4–5 mm long, claws 6.5–9 mm long. Legumes erect, concealed by imbricated stipules, ovoid-ellipsoid, 5–6 mm long, rounded at base, with short cusp, densely strigose-hirsutulous; seeds 2.5–3 mm long, brownish, smooth, often 1 or 2 per fruit. Jun–Jul. Rare to very locally common on shallow soil above escarpments, rocky prairie knolls, & open brushy hillsides; Mt: Powder River; SD: Pennington; e WY, w NE; KS: Cheyenne; CO: Yuma; (se MT, sw SD, s to ne KS, ne CO, e 1/2 WY). *Orophaca argophylla* (Nutt.) Rydb.—Rydberg.

At anthesis the flowers of this species are whitish but soon wither to yellowish. These persist for 2–3 weeks and account for labels recording the flower color as yellow.

18. *Astragalus kentrophyta* A. Gray, Nuttall's kentrophyta. Perennial decumbent plants (some flowering first season) forming loose mats to 3 dm in diam; stems freely branching, canescent or loosely strigulose with basifixed or obscurely dolabriform hairs to 1 mm long. Leaves 7–25 mm long, (3)5- to 7-pinnate; petiole 0–4 mm long; stipules 1.5–6 mm long, lowest connate into a short sheath, upper longest, becoming acerose; leaflets linear-lanceolate, 7–12 mm long, with a terminal spine to 1.2 mm long, becoming rigid and prickly, persistent. Peduncles very short, usually concealed by stipules; racemes 1- to 3-flowered. Pedicels to 1.7 mm long; calyx tube 2–2.5 mm long, teeth 1.5–2 mm long; corolla whitish; banner 4–5 mm long; wings 3.5–5 mm long; keel 3–4 mm long, claws 1.2–2 m long, often purplish at tip. Legumes 3.5–7 mm long, laterally compressed, short cuspidate, strigulose; seeds 2–3 mm long, brownish to blackish, smooth, often minutely pitted, 1–3 per fruit. Jun–Jul. Rare but locally common on rocky, gravelly, or sandy prairie hilltops, slopes, & badlands; ND: Billings, Slope; MT: Custer; SD: Fall River, Mellette, Washabaugh; w NE, se WY; (se Sask., sw Alta., se to w ND, w NE, se WY). *Kentrophyta montana* Nutt., *K. viridis* Nutt.—Rydberg; *A. viridis* (Nutt.) Sheld.—Stevens.

Nuttall's kentrophyta is a rare plant in the GP, but where found it is often locally common to abundant. In most colonies a few to many plants will flower the first season and then perish, but these are apparently rarely collected. Others persist for 2–4 years and some form mats to 3 dm in diameter.

19. *Astragalus lindheimeri* Engelm. ex A. Gray, Lindheimer milk-vetch. Annual herbs with a taproot and sparse appressed basifixed hairs 0.5 mm long or less. Stems 1–7(20), ascending or prostrate, erect when single, 0.5–4 dm long. Leaves alternate, odd-pinnate, 2–7 cm long, petioled below to subsessile above; stipules 1.5–5 mm long; leaflets 11–21, 2–14 mm long, broadly to narrowly cuneate, cuneate-oblong, or oblanceolate, apex truncate-emarginate or deeply retuse. Peduncles axillary, 1.5–7 cm long; racemes 2- to 8-flowered. Pedicels 1–2.5 mm long; calyx tube 2.4–3.3 mm long, teeth 2.3–5 mm long; corolla bicolored with banner 12–18 mm long, purple-margined around white purple-striate eye; wings white, 11–17 mm long, but purplish below, keels white and purple-spotted, 9.5–13 mm long. Legume stipitate, stipe 1–3 mm long, body linear-oblong or oblanceolate, somewhat lunate, 1.5–3 cm long, compressed triquetrous, 3.5–6 mm in diam, glabrous, brownish to blackish; seeds 3–4 mm long, brownish and often purple-dotted. Apr–Jun. Often common on rocky or sandy prairie hills, bluffs, & open disturbed areas, roadsides; s GP; (sw OK, n-cen TX, with isolated stations in TX: Ochiltree; OK: Payne).

20. Astragalus lotiflorus Hook., lotus milk-vetch. Perennial (sometimes flowering first year) low, loosely or densely tufted plants with a taproot and eventually a short branching caudex which may be subterranean for 1-6 cm; pubescence dolabriform, strigulose, pilosulous, or hirsute; stems 1-several, 0-8(12) cm long, outer decumbent or prostrate (erect in first-year flowering plants), central ones erect or nearly so. Leaves alternate, odd-pinnate, 2-9(14) cm long; petioles 0-2 cm long; stipules 2-8 mm long, distinct; leaflets (3)7-17, narrowly to broadly elliptic, oblanceolate, oval, or obovate, 4-15 mm long, variously pubescent. Inflorescences chasmogamous or cleistogamous, these usually on separate plants or the later produced later in season: chasmogamous racemes on peduncles 3-9(12) cm long, flowers 3-17, usually congested into an ovoid or subcapitate head with pedicels 0.5-2.5 mm long, calyx tube campanulate 3-4.5 mm long, teeth 2-5 mm long, corollas whitish, yellowish, sometimes purplish-veined or tipped with purple, banner 8-14 mm long, wings 7.5-12 mm long, claws 3-4.5 mm long, keel 6.5-9(10) mm long, claws 3.5-4.5 mm long. Cleistogamous on short racemes on peduncles 0-2 cm long, flowers 1-3(5), calyx tube 2.5-3.5 mm long, teeth to 4 mm long, accrescent, corolla whitish, not expanding, banner 4-7 mm long, wings 4-6 mm long, keel 4-6 mm long. Legumes similar on both inflorescence types, ascending to prostrate; 1.2-4 cm long, 5-9 mm in diam, straight or curved, ovoid-acuminate, or oblong-ellipsoid, rounded or cuneate basally, apically abruptly contracted or gradually tapering into a beak, body strigulose or villosulous; seeds 1.5-2.5 mm long, brownish and often with purplish spots. Mar-Jun, and sporadic later. Infrequent to common on a variety of soils in prairies, bluffs, open wooded or bushy hillsides & canyons, stream valleys, ponds, lake shores, roadsides, & waste places; GP but absent se & rarer in n; (s Man., s Alta., s to e MN, nw MO, cen & w TX, w MT, cen NM). *Batidophaca cretacea* (Buckl.) Rydb., *B. lotiflora* (Hook.) Rydb., *B. nebraskensis* (Bates) Rydb.—Rydberg, *A. lotiflorus cretaceous* (Buckl.) F. C. Gates, *A. lotiflorus nebraskensis* Bates—Gates.

The lotus milk-vetch is a highly variable species and none of the described taxa account for the variation to be found in populations, particularly in the w-cen GP. The perplexing variability is well described in Barney (op. cit.), but the species merits much more careful field and experimental study.

21. Astragalus miser Dougl. ex Hook. var. **hylophilus** (Rydb.) Barneby, woodland weedy milk-vetch. Perennial, low, tufted plants with a branching caudex; pubescence of herbage of dolabriform hairs, or mostly so; stems 1-12 cm long, several. Leaves alternate, odd-pinnate, 4-19 cm long; leaflets 11-21, elliptic or oval, 5-26 mm long, glabrous or glabrate. Racemes 6- to 16-flowered, 2-7 cm long. Calyx tube 2.5-3.5 mm long, teeth 1-2.2 mm long; corolla whitish, but keel tip purplish; banner 6.5-13 mm long, wings 7-9.5 mm long, claws 2.5-3.5 mm long; keels 8-10 mm long, claws 2.5-3.5 mm long. Legumes 1.5-2.5 cm long, linear or linear-lanceolate, pendulous, glabrous; seeds 2-3 mm long, brownish, often with purple dots. Jun. Wooded hilltops; SD: Pennington; (w MT, w WY, e ID, SD). *Homalobus hylophilus* Rydb.—Rydberg.

This species is included for the GP on the basis of one or two collections from the BH of SD, but it apparently has not been recently collected.

22. Astragalus missouriensis Nutt., Missouri milk-vetch. Low perennials, loosely tufted or prostrate, shortly caulescent or sometimes subacaulescent, with a taproot and branched caudex, densely strigulose or strigose throughout with dolabriform hairs; herbage silvery-white to greenish-gray. Stems 1-4(15) cm long, usually prostrate and radiating; leaves alternate, odd-pinnate, (2)4-10(14) cm long; stipules 2-9 mm long, lanceolate, distinct; leaflets (7)9-17)21), elliptic to narrowly obovate, acute, obtuse, or mucronulate, (4)7-13(17) mm long. Peduncles equaling or exceeding the leaves, becoming prostrate in fruit; racemes 3- to 9(15)-flowered. Pedicels to 2 mm long; flowers 3-15, rose-purple; calyx tube 6.5-9.5 mm long, grayish and usually blackish—strigulose, teeth 1.5-3.5(5) mm long; banner 14-24 mm long; wings 13-19 mm long, claws 7-10 mm long; keel 12-18 mm long, claws 7-10 mm long. Legumes 1.5-3 cm long, 6-8 mm wide, sessile, oblong-elliptic, straight or nearly so, subcylindric but becoming laterally compressed, abruptly contracted into a beak to 4 mm long, body cross-corrugated and strigulose; seeds 2-3 mm long, brown, usually wrinkled and lustrous. (n = 11) Mar-Jul. Infrequent to common on a variety of soils on prairies, rocky bluffs, brushy ravines, stream valleys, roadsides; GP but absent in se; (s Man., Alta., s to e MN, sw OK, w TX, MT, e WY, CO, NM). *Xylophacos missouriensis* (Nutt.) Rydb. — Rydberg.

The peduncles and blackened legumes often persist well into the second season. The flowers of southern GP plants are of the larger size and become progressively smaller northward.

23. *Astragalus mollissimus* Torr., woolly locoweed. Low, somewhat robust, densely or loosely tufted perennial with woody taproot and sometimes a shortly branching caudex; the stems and leaves villous-tomentose throughout the basifixed hairs; stems 1-several, 1.5-14 cm long, usually simple, outer ones prostrate and inner ascending, often reduced to crowns and covered with imbricated stipules; stipules (5)7-17 mm long. Leaves 7-25 cm long, odd-pinnate; leaflets 15-27(33), 5-22 mm long, oval, ovate, obovate, or rhombic-elliptic. Peduncles 6-21 cm long; racemes oblong, dense, 10- to 40-flowered, axis elongating and 4-17 cm long in fruit. Calyx tube 6.5-9.5 mm long, 3.4-4.5 in diam, teeth 3-5 mm long; corolla yellowish-purplish, reddish-purple, rarely yellow; banner 17-22 mm long; wings 15-21 mm long, claws 7.3-11 mm long; keel 14-18 mm long, claws 8-11 mm long. Legumes oblong-ellipsoid to lance-ellipsoid, 1.5-2.5 cm long, 4-7 mm in diam, usually lunate, with a short beak, glabrous or rarely puberulent, humistrate; seeds 2-3 mm long, brownish, smooth. (2n = 24) Apr-Jun. Infrequent to locally common on a variety of soils in prairies, plains, hillsides, stream valleys, pastures, roadsides; SD: Shannon; s in w 1/2 GP; (sw SD, se WY, s to w OK, w TX, e 1/2 NM).

Some plants in the TX panhandle and e NM have a calyx tube 3 mm or less in diam, shorter flower, and often slightly pubescent legume. These are intermediates with var. *earlei* (Greene ex Rydb.) Tidest., which is found to the south and into Mex. Rare plants in KS and NE have strictly yellow flowers, but occur as individual plants in normal colonies.

The woolly locoweed constitutes one of the more serious sources of livestock loss in the plains. All classes of livestock may develop a craving for the plant and contract loco poisoning.

24. *Astragalus neglectus* (T. & G.) Sheld., Cooper milk-vetch. Tall, caulescent perennials with a taproot and knobby caudex; herbage with basifixed appressed or ascending hairs; stems erect, hollow, 3-9 dm tall, strict or branched. Leaves alternate, odd-pinnate, 4-12 cm long, short petioled or upper sessile; stipules distinct, 2-6 mm long; leaflets 11-25, oblong-elliptic, oblong, or oblong-obovate, 7-30 mm long. Peduncles axillary, 3-7 cm long; racemes loosely 10- to 20-flowered, flowers nodding. Pedicels to 2 mm long; calyx tube campanulate, 3.7-5 mm long, teeth 2-3 mm long; corolla whitish or pale yellow; banner 11.6-14 mm long; wings 10-13 mm long; keel 10-12.5 mm long. Legumes erect, obliquely ovoid or ovoid-ellipsoid, inflated, 1.5-3 cm long, with a short beak; seeds 2-3 mm long, yellowish, smooth. Jun-Jul. Rare along stream banks, lakeshores, & moist areas; ND: Pembina; MN: Marshall, Otter Tail, Polk; SD? (se Man., ne ND, e-cen SD, e Ont., MI, NY). *Phaca neglecta* T. & G. — Rydberg.

Apparently, this is a very rare plant in the GP that has not been collected in recent years.

25. *Astragalus nuttallianus* DC., small-flowered milk-vetch. Small slender annuals with a taproot; herbage green, and nearly glabrous to gray or silvery-strigose. Stems 0.3-3 dm tall; leaves alternate, odd-pinnate, 1-7 cm long, petioled or subsessile above; stipules 1.5-6 mm long; leaflets 7-23, linear-oblong, oblong-cuneate, ovate-cuneate, or narrowly elliptic, retuse, truncate-emarginate, or obtuse to acute at apex. Peduncles axillary, 0.2-10 cm long; racemes mostly 1- to 7-flowered capitately. Calyx tube 1.3-2.5 mm long, teeth 1.9-3.2

mm long; corolla whitish, pale purple or reddish-purple; banner 4-7.5 mm long; wings 4.7-6.7 mm long, keel 4-5.8 mm long. Legumes sessile, 1.6-2.6 cm long, glabrous or pubescent; seeds 2-3 mm long, yellowish, purple-spotted, somewhat wrinkled, shiny.

Two varieties are found in the GP as follows:

25a. var. *austrinus* (Small) Barneby. Leaflets narrowed elliptic, never retuse or truncate-emarginate, gray or silvery-strigulose. Legume valves glabrous or strigulose. Apr-Jun. Habitat similar to that of var. *nuttallianus* but drier; KS: Harper, Kiowa, w OK, TX; (s KS, w OK, w TX, NM, AZ, Mex.).

25b. var. *nuttallianus*. Leaflets of all leaves retuse or truncate-emarginate, dark green glabrous above, linear-oblong, oblong, oblong-cuneate, or ovate-cuneate. Legume valves glabrous. Apr-Jun. Rocky or sandy prairies, pastures, roadsides, open wooded areas; KS: Harper, Kingman, Sedgwick, Sumner; cen & s OK, TX; (s-cen KS, s to s TX). *Hamosa leptocarpa* (T. & G.) Rydb.—Rydberg.

While the two varieties overlap somewhat in distribution, they have not been found in mixed colonies in the GP.

26. *Astragalus pectinatus* Doug. ex G. Don, tine-leaved milk-vetch, narrow-leaved poison-vetch. Coarse, caulescent perennials with a woody taproot, and subterranean branching caudex; some underground caudex-branches horizontal and to 1 dm long, without adventitious roots; plants strigulose nearly throughout. Stems stout, 1-7 dm long, decumbent or prostrate with ascending tips, diffusely branched at or below middle, forming bushy mats; stipules 1.5-10 mm long, lower ones amplexicaul and connate into a bidentate sheath, becoming connate at base only and distinct above. Leaves alternate, pinnate, sessile or nearly so, 4-11 cm long, 4-10 pairs of opposite or scattered, linear, filiform, stiff, and somewhat falcate incurved leaflets (0.7)1.5-7 cm long, the terminal leaflet continuous with the rachis. Peduncles axillary, stout, often incurved-ascending 2-11 cm long; racemes 7- to 30-flowered, these ascending but eventually declined. Pedicels 1-3.5 mm long; calyx tube cylindric, 6-9.5 mm long, teeth subulate or lanceolate, 1.5-3.5 mm long, both black, white or mixed strigulose; corolla ochroleucous, drying yellowish; banner 21-24 mm long, deeply notched; wings 17-20 mm long, claws 7-9 mm long; keel 14-16 mm long, claws 8-9 mm long. Legume declined or deflexed, sessile, plumply ellipsoid, straight or slightly decurved, 1.5-2.5 cm long, 5-8 mm in diam, long-cuspidate at apex; seeds 3-3.8 mm long, pale brownish, smooth, shiny. May-Jun. Infrequent to locally common on shale, chalk, or gravelly hillsides & flats; w 1/2 GP, s to s KS & se CO; (s Alta., sw Man., s to s KS, se CO, MT, WY). *Cnemidophacos pectinatus* (Hook.) Rydb.—Rydberg.

The plants from sw NE to MT show a strong tendency to have calyx a little larger and densely black-strigose, while plants of w KS and e CO have calyx smaller and usually with only white hairs. The tine-leaved milk-vetch is one of the more troublesome obligate selenium plants involved in selenium poisoning of livestock, though animals seldom graze it under range conditions.

27. *Astragalus plattensis* Nutt. ex T. & G., Platte River milk-vetch. Low, diffuse or decumbent perennials with a deeply buried taproot giving rise to slender, branched, and often creeping subterranean caudex-branches, these often with adventitious rootlets at nodes, the stems emerging and forming loosely entangled mats or more widely spaced, the whole plant usually thinly to densely pilose with ascending or spreading shiny hairs, these mixed with shorter, curled ones, the herbage green or grayish-green. Stems of the year subterranean for 1-20 cm, emerging to branch below, simple above, above ground axis 0.5-4 dm long; stipules at leafless nodes, connate, papery, median ones green, wider than stem, upper narrow and mostly distinct except for base. Leaves 2.5-11.5 cm long, short petioled to subsessile above; leaflets (11)15-27, broadly to narrowly elliptic, oval, or oblong, obtuse or acute, rarely some or all oblong-obovate and truncate-emarginate, 3-17 mm long. Peduncles axillary, ascending, humistrate in fruit, to 7.5 cm long, usually shorter than leaves; racemes (3)6- to 15-flowered. Pedicels to 3 mm long; calyx tube 5.4-7.8 mm long, teeth 2-6 mm long; corolla reddish-purple, drying whitish-blue; banner usually deeply notched, 14–20 mm long; wings 13–18 mm long, claws 6–9 mm long; keel 12–16 mm long, claws 6–8.5 mm long. Legumes ascending or prostrate, sessile, fleshy, 1–1.7 cm long, 1–1.3 cm in diam, with beak 1–3.5 mm long, valves strigulose or with ascending hairs, becoming leathery in age; seeds 2.5 mm long, purplish-black, smooth. Mar–Jul. Infrequent to locally common on rocky or gravelly prairie plains, hillsides, ravines, open wooded or brushy areas, roadsides; GP but infrequent to absent n & e; (ND s to e NE, cen OK, TX, e WY, e CO). *Geoprumnon plattense* (Nutt.) Rydb. — Rydberg.

The Platte R. milk-vetch is readily eaten by all classes of livestock and decreases rapidly under grazing.

28. *Astragalus praelongus* Sheld. var. *ellisiae* (Rydb.) Barneby, stinking milk-vetch. Rather coarse, tall, ill-scented perennial, from superficial root-crown, usually nearly glabrous or upper stems, lower leaf surfaces, and inflorescence strigulose with basifixed hairs. Stems several, erect, in clumps, 2–6 dm tall; stipules 2.5–7 mm long, distinct. Leaves 4–22 cm long, petioled or uppermost sessile; leaflets 7–27, obovate or oblong-obovate to lanceolate or oblanceolate, obtuse or retuse, or some upper elliptic and acute, (3)8–35(50) mm long. Peduncles axillary; racemes 10- to 25(33)-flowered; pedicels ascending in fruit. Flowers ochroleucous; calyx 5.8–14 mm long, glabrous or strigulose with black or white hairs; banner 15–23.5 mm long; wings 15–22 mm long, claws 5.3–8.4 mm long; keel 11–17 mm long, usually purplish at tip. Legume with stipe 1–2.5 mm long, body ellipsoid, oblong-ellipsoid, or narrowly clavate-ellipsoid, turgid or inflated body 1.8–3.4 cm long, 5–11 mm in diam, strigulose or puberulent along sutures or glabrous; seeds 3–3.8 mm long, brown, smooth. (2n = 24) Apr–Jun. Rocky or sandy prairie plains, hillsides, ravines, pastures, roadsides; extreme sw GP; (w TX, e 1/2 NM; Colorado R. basin of UT & CO).

This plant, which barely enters our area, is said to be very toxic to sheep.

29. *Astragalus puniceus* Osterh., Trinidad milk-vetch. Diffuse perennials from woody taproot and subterranean branching caudex, stems and herbage villosulous, becoming glabrate. Stems several to many, 1.5–5 dm long, decumbent and spreading to form depressed clumps, often matlike, buried for 3–17 cm, branching aboveground, zigzag above; stipules 2–6 mm long, lowest connate and membranaceous, upper green, connate 1/2 length or distinct. Leaves 2–11 cm long, lower shortly petioled, upper subsessile, with (7)17–27 broadly to narrowly elliptic, oblong-oblanceolate, obovate-cuneate, or linear-oblong, usually truncate-emarginate, sometimes obtuse leaflets (3)5–16 mm long. Peduncles 2–11.5 cm long; racemes 5- to 19(27)-flowered. Pedicels 1–2 mm long; corolla pink-purple; calyx tube cylindro-campanulate, 4.9–8 mm long, teeth 1.1–2.8 mm long; banner 16–21 mm long; wings 13.5–18 mm long, claws 6.2–7.5 mm long; keels 10.4–14 mm long, claws 5.8–8 mm long. Legume sessile, oblong-ellipsoid, 1.5–2.4 cm long, (4.5)5–8 mm in diam, turgid or somewhat inflated, with a short cuspidate beak, valves red-mottled; seed 2.5–3 mm long, brown, pitted. May–Jun. Dry rocky or sandy bluffs & mesas, stabilized dunes, roadsides; OK: Cimarron; CO: Las Animas; NM: Union, Colfax; (*endemic* to area).

Our plant is the var. *puniceus*. The plant in the Black Mesa of OK: Cimarron differs from the CO and NM specimens in having shorter calyx and smaller petals. Another variety, var. *gertrudis* (Greene) Barneby, is endemic in southern NM: Taos. This little-known complex merits additional field study.

30. *Astragalus purshii* Dougl. ex Hook., Pursh milk-vetch. Low cespitose perennials from a taproot and branched caudex, densely tufted or matted. Stems 0–10 cm long, usually prostrate, densely villose. Leaves 1–12 cm long, odd-pinnate, slender petioled, white-tomentose; stipules membranaceous, 3–15 mm long, usually exceeding internodes; leaflets 5–17, elliptic, elliptic-oblanceolate, or rhombic-elliptic, usually acute. Peduncles 0–10 cm long; racemes 1- to 6-flowered, subcapitate at anthesis. Corolla whitish or ochroleucous with keel usually purple-tipped; calyx tube 9–13 mm long, lanate, teeth 2–7 mm long;

banner 19–26 mm long; wings 17–22 mm long, claws 10–13.5 m long; keel 15–21 mm long, claws 10–14 mm long. Legume obliquely ovoid or broadly lance-elliptic, 1.3–2.7 mm long, 5–10 mm in diam, densley hirsute-tomentose; seeds 2–3 mm long, brownish and often purple-spotted, smooth. (n = 11) May–Jun. Rocky or gravelly prairie plains, slopes, hilltops, & open brushy areas; sw ND, w SD, MT, WY; (sw Sask., se Alta., sw ND, s to n CO, UT, NV, WA, OR, CA). *Xylophacos purshii* (Dougl.) Rydb.—Rydberg.

31. *Astragalus racemosus* Pursh, alkali milk-vetch, creamy poison-vetch. Perennial rather coarse caulescent herbs with a woody taproot and a branched caudex at or below ground level, thinly to densely appressed-pubescent with basifixed hairs. Stems few to many, erect or ascending in clumps, 1.5–7 dm tall, usually branched above. Leaves 4–15 cm long, alternate, odd-pinnate, shortly petioled below, subsessile above; stipules 3–12 mm long, lowest connate becoming less united to distinct above; leaflets 11–31, 1–4 cm long, lance-elliptic to ovate, obtuse and mucronulate, or upper ones linear-lanceolate or narrowly elliptic and acute to acuminate. Peduncles axillary, 3–11 cm long; racemes 15- to 70-flowered, flowers dense, nodding, pedicels 2–3.5 mm long. Calyx tube campanulate, 4.7–9 mm long, truncate or gibbous at base, the subulate or subulate-setaceous teeth 1.5–10 mm long; corolla white or whitish and keel tip purplish, or banner whitish with purple streaks, wings light purple or purple-tipped, and keel purplish; banner 16–21 mm long; wings 12–19 mm long, claws 5–8 mm long; keel 10.5–15.5 mm long, claws 4.7–8 mm long. Legumes pendulous, stipe 3.5–7 mm long, body linear-oblong to narrowly oblong-ellipsoid, 1.5–3 cm long, 3–5.7 mm in diam, triquetrously compressed with acute ventral and narrow lateral angles, the 3 faces usually flat, glabrous or strigulose; seeds 2.6–3.5 mm long, dark brown, often with purple spots, smooth. (2n = 24) Mar–Jul. Infrequent to locally common on clay, shale, gypsum, or chalk soils of prairie plains, hillsides, knolls, stream valleys, open wooded or brushy areas, roadsides; GP but infrequent to absent in e 1/3; (s Sask., ND, SD, s to w OK, TX panhandle, e MT, e WY, e 1/2 CO, e NM). *Tium racemosum* (Pursh) Rydb.—Rydberg.

Plants with calyx tubes produced into a gibbous sac behind the insertion of the pedicel, its teeth (4.5)5–10 mm long, and at least lower leaflets broadly elliptic to ovate-oblong have been recognized as var. *longisetus* M. E. Jones. Such plants are found from sw SD, nw NE, and in eastern foothills of CO and n NM. The variety *racemosus* has calyx tube truncate at base, with pedicel inserted at the ventral corner, its teeth 1.5–3.5 mm long, leaflets mostly narrow, elliptic or linear-elliptic, and is the common plant in the GP. I find, however, that many plants in the high plains of KS, CO, and NE have characters variously intermediate between the two taxa, and they often can be only arbitrarily distinguished. Numerous populations in KS and NE also have flowers more purplish than white. The complex deserves more critical field study with attention given to flower color of living plants.

The alkali milk-vetch is another obligate selenium plant which livestock rarely eat in range conditions but which can be a problem.

32. *Astragalus sericoleucus* A. Gray, silky orophaca. Perennial caulescent, prostrate and matted herbs with woody taproot and branching caudices; pubescence silvery, dense throughout, with dolabriform hairs. Stems freely branching from caudex-branches and forming mats 1.5–12 dm in diam; those of the year short knobs or to 5 cm long; stipules 2–8 mm long, connate sheathing. Leaves palmately trifoliate, 1–4 cm long, the leaflets narrowly to broadly oblanceolate, rarely obovate-cuneate, leaflets 3–13 mm long. Peduncles 0.5–2.5 cm long, exserted or included in stipular sheath; racemes 2- to 5-flowered. Pedicels 1–1.5 mm long; corolla pink-purple or rarely whitish, drying yellowish; calyx tube campanulate, 2–2.5 mm long, teeth 0.5–1.5 mm long; banner 5–6 mm long, ca 4 mm wide; wings 5–5.8 mm long, claws 2–2.4 mm long; keel 4–4.5 mm long, claws 1.9–2.3 mm long. Legumes 4.5–7 mm long, body ovoid-ellipsoid, densely silky-strigose, 3–4.5 m long, mostly included in calyx, tapering into a slender beak; seeds olive-brown, 1.3–1.7 mm long, smooth, usually 1 or 2 maturing per fruit. May–Jun, rarely again Sep–Oct. Infrequent to locally common on shallow soil above escarpments, rocky open ridges, hilltops, slopes,

& prairie barrens; NE panhandle, se WY, nw KS, ne CO (*endemic* in area). *Orophaca sericea* (Nutt.) Britt. — Rydberg.

33. **Astragalus shortianus** Nutt. ex T. & G., Short's milk-vetch. Perennial low, tufted, essentially acaulescent plants with knobby or shortly forking caudex; herbage densely silky-strigose with basifixed hairs. Stems 0–2.5 cm, concealed by stipules; stipules 5–12 mm long, amplexicaul but distinct. Leaves odd-pinnate, (4)6–15(21) cm long, petiolate; leaflets 7–19, obovate, rhombic-obovate, or elliptic-ovate, obtuse to subacute, 5–20 mm long. Peduncles 2–15 cm long; racemes with 5–16 flowers. Pedicels 1.5–3 mm long; calyx tube cylindric, villous, 6–8(10) mm long, teeth 3–5.5 mm long; corolla pink-purple; banner 16–22 mm long, 7.5–11 mm wide; wings 15–20 mm long, claws 8–10 mm long; keel 13–17 mm long, claws 9–10 mm long. Legumes ascending or prostrate, obliquely ovoid or ellipsoid, 2.5–4.5 cm long, 9–18 mm in diam, laterally compressed, rounded or cuneate at base, apically with a rigid beak, valves leathery, brown, strigulose; seed 2.5–3 mm long, brown, often wrinkled. May–Jul. Dry rocky prairie plains, hilltops, ridges, gravelly ravines, & roadsides; NE: Banner, Kimball; se WY, ne CO; (nw & se WY, extreme sw NE, s to cen NM). *Xylophacos shortianus* (Nutt.) Rydb. — Rydberg.

This is a rare plant in the GP but is common just to the west of our area.

34. **Astragalus spatulatus** Sheld., draba milk-vetch. Perennial low, tufted, matted plants with a taproot and repeatedly branching caudex; herbage densely silvery-strigose with dolabriform hairs. Stems short knobs to 1.5 cm long, covered by imbricated stipules; stipules 2–7 mm long, papery, amplexicaul and connate. Leaves simple or a few with 1 or 2 pairs of leaflets, lower leaves 0.3–1.5 cm long, oblanceolate, blade as long or longer than petiole, later ones similar but more erect, 1.5–6 cm long. Peduncles erect or outer prostrate in fruit, 0.4–7 cm long; racemes 1- to 11-flowered, loosely or subcapitately arranged. Pedicels to 1.7 mm long in flower, to 3 mm in fruit; calyx tube campanulate, 2–3.5 mm long, teeth 0.5–2.5 mm long; corolla pink-purple with white wing tips, rarely light yellowish-purplish or white; banner 6–9.5 mm long, 4.5–6.5 mm wide; wings 6–8 mm long, claws 2–3 mm long; keel 4.5–6 mm long, claws 2–3 mm long. Legumes 4–13 mm long, 1.5–3 mm wide, nearly straight, laterally compressed, valves strigulose, becoming papery; seeds 2–2.5 cm long, purplish-brown, wrinkled or rugulose. (n = 12) May–Jul. Locally common on open rocky hilltops, knolls, & prairie hillsides, usually on shallow soil over surfacing outcrops; w ND, SD, NE; KS: Cheyenne; MT s to n CO; (se Alta., sw Sask., w ND, s to nw KS, n CO, WY). *Homolobus caespitosus* Nutt. — Rydberg.

The draba milk-vetch varies considerably in stature and size of parts, even within the same population. The length of leaves and peduncles is often longer in wet years, and in dry years the plants appear more matted.

35. **Astragalus tenellus** Pursh, pulse milk-vetch. Erect or diffuse caulescent perennials with a woody taproot, knobby crown, and usually short branching caudex; strigulose throughout with basifixed hairs. Stems several to many in bushy clumps, 1–5 dm tall, usually branched; stipules 1.5–8 mm long, lower blackish, amplexicaul and connate, upper ones progressively less connate. Leaves alternate, odd-pinnate, 2–9 cm long, lower petiolate, upper sessile; leaflets (7)11–21(25), linear-oblong, linear-elliptic, linear, oblanceolate, or oblong-cuneate, apex acute, obtuse or mucronulate, 3–25 mm long. Peduncles, axillary, 0.2–5 cm long; racemes sometimes 2 per node, loosely 4- to 20-flowered or shorter ones 1- to 3-flowered. Pedicels 0.5–2.5 mm long in flower; calyx tube campanulate, 2–3 mm long, teeth 0.7–3 mm long; corolla white or pale yellowish, the keel tip often purplish, banner and wings sometimes pink-purple, or rarely all purplish; banner (5.5)6–8.8(10) mm long, wings 5.5–7.5(8) mm long, claws 2–3 mm long; keel 3–5.7(6.5) mm long, claws 1.8–3.2 mm long. Legumes pendulous, stipitate, or sessile, stipe 0.6–7 mm long, body narrowly

oblong-elliptic or elliptic lanceolate, straight, 7–17 mm long, 2.5–4.5 mm in diam, strongly compressed, glabrate, often reddish-spotted, blackish at maturity; seeds 2.2–3 mm long, brown, often purple-spotted, and somewhat shiny. (n = 12) May–Jul. Frequent & often locally abundant on open rocky or gravelly prairie plains, hillsides, bluffs, badlands, open wooded hillsides, stream banks, & lakeshores; ND, w-cen MN to nw NE, MT, WY, CO foothills; (ne Man., Yukon, s to e-cen MI, nw NE, n NM, UT, NV, ID). *Homalobus dispar* Nutt., *H. stipitatus* Rydb., *H. tenellus* (Pursh) Britt.—Rydberg.

36. *Astragalus vexilliflexus* Sheld., bent-flowered milk-vetch. Perennial, low, tufted, or matted bushy-branched plants with woody taproot and somewhat woody branching caudex; strigulose throughout with basifixed hairs. Stems several, incurved-ascending, decumbent, or prostrate, 3–30 cm long, much branched; stipules 2–5 mm long, amplexicaul and connate below to nearly distinct at base above. Leaves alternate, odd-pinnate, 1–5.5 cm long, short petioled to subsessile above; leaflets 5–17, elliptic or linear-elliptic, rarely lanceolate, 3–15 mm long, often glabrate above. Peduncles axillary, 0.5–4 cm long; racemes loosely 5- to 11-flowered. Pedicels 1–2.5 mm long; calyx tube campanulate, 1.5–2 mm long, teeth 1–3 mm long; corolla pink-purple; banner 5–9 mm long; wings 4–7 mm long, claws 1–2 mm long; keel 3–5 mm long, claws 1–2 mm long. Legume declined, essentially sessile, 5–12 mm long, body elliptic, obovate-elliptic or oblong-elliptic, strongly compressed, valves strigulose; seeds 2–2.5 mm long, olivaceous to blackish, smooth. May–Jul. Rare but locally common in rocky prairie knolls, ridges, badlands, & open wooded hillsides; ND: Dunn, Slope, Stark; SD: Harding, Pennington; MT, WY; (s Sask., w Alta, s to w ND, w SD, n WY, sw MT). *Homalobus vexilliflexus* (Sheld.) Rydb.—Rydberg.

5. BAPTISIA Vent., False Indigo

Ours herbaceous perennials from tough knobby caudex, often rhizomatose with adventitious roots; stems erect, usually much branched above; lowest leafless nodes with conspicuous stipules, these connate at base or to 3/4 length with entire or bidentate blade, stipules progressively less connate to distinct above but often amplexicaul. Leaves alternate, usually trifoliate, but lower sometimes simple, bilobed, or lateral ones decurrent on short petiole and united to form a semitrilobed leaf. Flowers in terminal racemes, pedicellate, white, purple, or yellowish, subtended by fugacious or persistent bracts; papilionaceous; calyx bilabiate, upper lip entire to 2-lobed, the lower clearly 3-lobed; banner not longer than wings, its sides reflexed; wings straight; keel petals nearly separate, straight; stamens 10, distinct. Pod stipitate, rounded or subcylindric, terminating in a curved beak.

1 Leaves, ovaries, and pods glabrous.
  2 Flowers blue .... 1. *B. australis* var. *minor*
  2 Flowers white .... 3. *B. lactea*
1 Leaves, ovaries, and pods pubescent .... 2. *B. bracteata* var. *glabresans*

1. *Baptisia australis* (L.) R. Br. var. *minor* (Lehm) S. Wats., blue false indigo. Plants 4–12 dm tall, glabrous throughout; stems divaricately branched, glaucous. Leaves trifoliate, with petioles 5–18 mm long or subsessile; stipules lanceolate, lance-acuminate, rarely ovate, 6–20 mm long, fugacious or persistent; leaflets oblanceolate, oblong-cuneate, obovate, sometimes elliptic, 2–4 cm long. Racemes terminal, erect, 1–5 dm long. Pedicels 5–30 cm long at anthesis, bracts lanceolate, or ovate-lanceolate, 6–15 mm long, usually deciduous before or during anthesis; calyx tube 5–9 mm long, glabrous, upper lip entire to slightly 2-lobed, teeth of lower lip 2–5 mm long, ciliate, finely pubescent within; petals purple or dull violet-blue; banner 2–2.7 mm long, its sides reflexed, emarginate, claw 2–5 mm long; wings and keel 2.5–3 cm long, claws 1 cm long. Mature pod oblong, 3–6 cm long,

1.5–2.5 cm in diam, brownish-black, inflated, tough, tapering into a short beak, with stipe 2× length of calyx; seed 4–4.5 mm long, 2–2.5 mm wide, brown, verrucose. (2n = 18) Apr–Jun. Infrequent to locally common on rocky prairie plains & hillsides, rocky open woodlands, occasional on sandy or gravelly stream valleys, roadsides; se NE, e 2/3 KS, w MO, OK, e TX panhandle; (se NE, MO, KS, e AR, OK, TX). *B. vespertina* Small — Rydberg; *B. minor* Lehm. — Gates.

A presumed hybrid between this species and *B. bracteata* var. *glabrescens* is *B. x bicolor* Greenm. & Larisey, found rarely in KS, MO, and OK. It has the banner blue-purple, wings and keel yellowish, bracts of spike persistent, and herbage pilosulous.

2. *Baptisia bracteata* Muhl. ex Ell. var. ***glabrescens*** (Larisey) Isely, long bracted wild indigo, plains wild indigo. Plants 2–4(8) dm tall, villous-pilosulous throughout; stem with divergent branches. Leaves subsessile or petioles to 3 mm long, usually 3-foliate, rarely some lower simple or bilobed to some with lateral leaflets subpalmately lobed; leaflets oblanceolate, spatulate, or elliptic, 3–10 cm long, lateral ones often decurrent on petioles; stipules ovate to lanceolate, often acuminate, 2–4 cm long, 1/3 as large as leaflets, persistent. Racemes terminal, usually solitary, declined, 1–2 dm long, secund. Bracts persistent, lanceolate or oblong, 1–3 cm long pedicels 1–4 cm long; calyx tube 4–6 mm long, strigose, upper lip entire to shortly 2-lobed, teeth of lower lip deltoid, 3–4 mm long; petals yellowish or whitish; banner 1.5–2.2 cm long, wings and keel 2–2.8 cm long. Pods ellipsoid or ovoid, 3–5 cm long, pubescent, tapering to a short thick stipe not exceeding calyx, apically into a slender beak; seed 4–4.5 mm long, olivaceous to brown, verrucose. (n = 9) Apr–Jun. IA, se NE, e 1/2 KS, MO, OK, n TX; (MI, MN, se NE, s to KY, AR, TX). *B. leucophaea* Nutt. — Rydberg.

The var. *bracteata* is the plant of the southeastern coastal plain. Unfortunately, our pubescent plant is an element in the central states which must take its name from a less pubescent form, including the type, just to the east of our area.

3. *Baptisia lactea* (Raf.) Thieret, white wild indigo. Plants 0.5–2 m tall, glabrous throughout, glaucous; stem solitary with ascending branches. Leaves 3-foliolate; petioles 5–25 mm long; stipules 5–30 mm long, lanceolate to ovate, usually deciduous before or during anthesis; leaflets narrowly obovate, oblanceolate, cuneate-obovate, or elliptic, 2.5–8 cm long. Racemes 1–several, stout, central one 2–6 dm long. Pedicels 3–12 mm long; bracts ovate-lanceolate or lanceolate, 6–14 mm long, usually deciduous before or during anthesis; calyx tube 7–9 mm long, upper lip entire or notched, teeth of lower lip deltoid, 3–4 mm long, densely white-pubescent within; petals white; banner 1–1.6 cm long; wings and keel 2–2.5 cm long. Mature pods black, drooping, ellipsoid-oblong, 2.5–4 cm long, stipe 2–3 × as long as calyx, apex abruptly narrowed to short beak; seed 4.5–5.2 mm long, 3 mm wide, oliv aceous to yellowish-brown, verrucose. (2n = 18) May–Jul. Infrequent to locally common on rocky prairie hillsides & ravines, low prairies, stream valleys, roadsides; IA, se NE, e 1/4 KS, MO, e 1/2 OK; (OH, MI, MN, se NE, s to MS, TX). *B. leucantha* T. & G. — Rydberg.

6. CARAGANA Lam., Pea-shrub

1. *Caragana arborescens* Lam., Siberian pea-shrub. Shrub or small tree to 6 m tall, young branches pubescent. Leaves alternate or fascicled, even-pinnate, petiolate; stipules spiny, 5–9 mm long; leaflets 8–12, obovate to elliptic-oblong, 1–3 cm long, rounded or truncate at apex and mucronate, pubescent when young, later glabrescent. Flowers 1–4, fascicled on short spur branches, yellow, pedicels 1–4 cm long; calyx tube 5–5.5 mm long, minutely pubescent, membranaceous, with teeth 1 mm or less long; banner 20 mm long, 18 mm wide, notched apically, narrowed abruptly to claw 4–5 mm long; wings 20 mm long; keel 18 mm long. Pod 3.5–5 cm long, 5 mm wide, sessile, compressed, narrowed to slender beak; seed 4.5–5 mm long, 3 mm wide, dark brown, smooth. May–Jun. Commonly planted in the outside rows of multiple-row shelter belts in the Dakotas & escaping into grazed pastures & waste places particularly in nw ND; n GP; (n U.S. & Can.; Siberia).

7. CLITORIA L., Butterfly Pea

1. *Clitoria mariana* L., butterfly pea, pigeon wings. Perennial herbs from long taproots and shortly branching caudex; stems slender, 2–4(10) dm long, suberect to trailing or climbing, glabrous to sparsely pubescent. Leaves alternate, pinnately trifoliate, 5–10 cm long; stipules subulate, 3–10 mm long; petioles (1)3–7 cm long; leaflets entire, mostly ovate to lanceolate to somewhat elliptic, 2–7 cm long, glabrous or sparingly pubescent below, stipellate. Peduncles axillary, 0.5–4 cm long, 1- to 3-flowered. Pedicels 2–8 mm long; calyx tube cylindric with flaring tube, 10–14 mm long, usually glabrous, the 5 teeth 4–7 mm long; corolla papilionaceous, bluish or lavender; banner 4–6 cm long, 3–4 cm wide, much exceeding other petals; keel strongly incurved, shorter than wings and coherent with them to middle; stamens 10, diadelphous (9 + 1). Pod flattened, stipitate, oblong-linear, 3–6 cm long, valves longitudinally twisting upon dehiscence; seeds sticky, 4–4.5 mm long, brown, smooth and shiny. May–Sep. Low or upland rocky open wooded hillsides, ravines, & stream valleys; KS: Cherokee, where not collected since 1949; s MO, e OK, TX; (NY, OH, IL, MO, s to FL, TX).

It is frequently stated that the stamens are monadelphous below, but our plants are diadelphous with the 10th stamen free. Late in the season cleistogamous flowers often are produced.

8. CORONILLA L., Crown Vetch

1. *Coronilla varia* L., crown vetch. Perennial herbs with taproots and shortly branching caudex; stems trailing to ascendent, glabrate, branched, 3–5(10) dm long. Leaves alternate, odd-pinnate, usually 5–10 cm long; petiolate below to sessile above; stipules persistent, 2–3 mm long, narowly lanceolate; leaflets 9–25, mostly (5)10–15(25) mm long, obovate, to oblong. Peduncles axillary, 5–15 cm long; umbels with 5–15(20) flowers. Pedicels 3–7 mm long, each subtended by a linear bract to 1 mm long; calyx tube campanulate to hemispheric, 1–1.5 mm long, bilabiate, the 2 upper lobes nearly united, the 3 lower distinct but very short; petals pinkish (drying lavender) or white; banner 10–13 mm long, 7–8 mm wide, claw 1–2 mm long; wings 11–15 mm long, claw 2–4 mm long; keel 10–12 mm long, claw 3 mm long, body beaked and usually purple-tipped; stamens 10, diadelphous (9 + 1). Loments usually 2–5 cm long with stipe 3–10 mm long, linear, 4-angled, with 3–12 segments; seed 3.5–4 mm long, cylindrical, dull brown, smooth. (n = 12) May–Aug. Infrequent to locally abundant in waste ground along roads & open areas; e 1/2 of NE, KS, & scattered elsewhere; (ME, SD s to WV, KY, KS; Europe, sw Asia, n Africa).

Crown vetch is frequently planted as a bank cover along highways and to some extent as an ornamental. It seldom spreads from plantings and cannot be considered as truly naturalized in the GP. The plant is highly variable as regards pubescence, flower color, and leaflet size and shape. The seeds are reported to be poisonous.

9. CROTALARIA L., Rattlepod

1. *Crotalaria sagittalis* L., rattlebox. Ours annual herbs from a taproot; stems erect or ascending, often bushy-branched, 1–4 dm tall, with spreading hairs to 2 mm long and

short strigose hairs intermixed. Leaves alternate, simple, lanceolate, or linear to elliptic, 1.5-8 cm long, 3-15 mm wide, pubescent, sessile or shortly petiolate; stipules usually present, decurrent inversely sagittate. Peduncles terminal and axillary, 1-4 cm long; racemes (1)2- to 4-flowered. Calyx gamosepalus, tube obliquely campanulate, floral cup 2 m long, the unequal lobes lanceolate to linear, to 9 mm long, villous or hirsute; petals yellow, fading whitish, about as long as calyx; banner 8 mm long, longer than wings and keel; stamens 10, monadelphous, 5 filaments short and with linear anthers, the longer 5 with subglobose anthers. Pod oblong, glabrous, 20-35 mm long, much inflated; seed 2.5-3 mm wide, brown, shiny. (n = 16) May-Sep. Infrequent to locally common in rocky prairies, open wooded slopes, sandy open areas, dunes, fields; e 1/3 of GP s of se SD, sporadic elsewhere; (e 1/2 U.S., Mex., S. & C. Amer.).

South of our area this species also is found as a short-lived perennial. Cases of fatal poisoning have been reported for horses that have eaten fresh plants or dried plants contained in hay.

10. DALEA Lucanus

Annual or perennial herbs or low shrubs, unarmed; leaves alternate, odd-pinnate, or 3-foliate, gland-dotted; stipules herbaceous or subglandular. Flowers in terminal spikes or racemes, of various colors; calyx subtended by a deciduous or persistent bract, usually gland-dotted, tube 10-ribbed, with 5 variously shaped teeth; corolla papilionaceous or very obscurely so; banner (upper petal) inserted on the hypanthium rim; wings and keel attached at various levels to sockets on the androecial column, all free or keel blades narrowly imbricate and adherent; stamens monadelphous, 5-10, or 5. Pods dry, indehiscent, included in the persistent calyx, or somewhat exserted, 1-seeded. *Petalostemon* Michx.—Fernald.

Reference: Barneby, R. C. 1977. Daleae Imagines. Mem. New York Bot. Gard. 27: 1-891.

Rydberg listed *Petalostemon tenuis* (Coult.) Heller and *P. standsfieldii* Small for Kansas. These are *Dalea tenuis* (Coult.) Shinners which is found se of our area. He also listed *Petalostemon pulcherrimum* Heller for Kansas which is *Dalea compacta* Spreng. var. *pubescens* (A. Gray) Barneby and is also to the se of our area.

1 Plants shrubs 2-10 dm tall.
  2 Calyx lobes as long as the tube or longer; leaves 2.5-11(14) mm long .... 5. *D. formosa*
  2 Calyx lobes much shorter than tube; leaves 10-20 mm long .... 6. *D. frutescens*
1 Plants annuals or herbaceous perennials.
  3 Plants annual .... 9. *D. leporina*
  3 Plants herbaceous perennials.
    4 Stems prostrate .... 8. *D. lanata*
    4 Stems erect, suberect, or decumbent.
      5 Leaves essentially 3-foliate .... 7. *D. jamesii*
      5 Leaves 5 or more foliate.
        6 Plants silky-pilose, villous-tomentose throughout or pilosulous to softly villous-tomentose.
          7 Calyx teeth 2.3-5 mm long; aristate; flowers yellow at anthesis.
            8 Stems usually diffuse and branched, 0.5-3(3.5) dm long; petals fading pink or brown; sw GP .... 11. *D. nana*
            8 Stems erect, 2-7.5 dm long; petals permanently yellow; widespread .... 1. *D. aurea*
          7 Calyx teeth 0.8-2.1 mm long; not aristate; petals red, pinkish, or whitish.
            9 Leaflets of midstem leaves 5-8(12) pairs; foliage villous-pilose throughout; plants of dunes or loose sandy areas .... 14. *D. villosa*
            9 Leaflets of midstem leaves 1-4(5) pairs; foliage rarely villous-pilose throughout; usually not in loose sand.
              10 Spikes relatively loose, especially after flowering, not conelike, axis revealed in pressed specimens; calyx tube and teeth with dense spreading usually tawny pubescence .... 13. *D. tenuifolia*

             10 Spikes permanently dense and conelike, axis concealed; calyx tube and teeth with
                 short white appressed hairs .... 12. *D. purpurea*
  6 Plants essentially glabrous to spikes.
         11 Spike remotely flowered, its axis always clearly visible; bract subtending calyx with a
             broad conspicuous pallid thin margin; stamens 9 .... 4. *D. enneandra*
         11 Spike densely flowered, its axis not clearly visible; bract subtending calyx without con-
             spicuous pallid margin.
             12 Petals rose or purplish.
                 13 Spikes relatively loose especially often flowering, not conelike, axis visible in press-
                     ed specimens; calyx tube and teeth with dense spreading usually tawny
                     pubescence .... 13. *D. tenuifolia*
                 13 Spikes permanently dense and conelike, axis concealed; calyx tube and teeth with
                     short white appressed hairs .... 12. *D. purpurea*
             12 Petals white or yellowish-white
                 14 Spikes headlike, subglobose, 4–15 mm long .... 10. *D. multiflora*
                 14 Spikes elongate, cylindric, over 15 mm long.
                     15 Spikes 6–16(24) cm long, axis densely pilose; calyx tube densely pilose with
                         hairs to 1.6–2.4 mm long; petals white or yellowish-white .... 3. *D. cylindriceps*
                     15 Spikes 1.5–5(7.5) cm long; axis minutely pilosulous; calyx tube glabrous or with
                         hairs to 0.3 mm long .... 2. *D. candida*

1. **Dalea aurea** Nutt. ex Pursh, golden prairie-clover, silktop dalea. Herbaceous perennials with yellow taproot and short caudex; stems 1–several, erect or virgately ascending, 2–7.5 dm tall, simple or branched above; plants silky-pilose or silky-canescent almost throughout. Leaves alternate, odd-pinnate, 1–4 cm long, scattered and reduced in size above; leaflets (3)5–7, obovate, oblanceoalte, obtuse, or obtuse and mucronulate, rarely acute, 4–16(20) mm long; stipules subulate; petiole 3–14 mm long. Spikes densely many-flowered, conelike, becoming oblong-cylindroid, 1.5–7 cm long, 12–21 mm in diam (less petals), axis densely pilosulous. Lowest bracts ovate-cuneate, inner ones lance-elliptic, acuminate, slender tip reddish; calyx tube turbinate, 2.2–3 mm long, densely silky-pilose, teeth 3.5–5 mm long, aristiform and plumose; petals yellow; banner 6.5–8.6 mm long, claws 3.5–5 mm long; wings 5–6 mm long, claws ca 1 mm long; keel 5.7–8.5 mm long, claws 1.4–2 mm long; stamens 10. Pod 3–4 mm long, silky-villous; seed 2–2.6 mm long, yellowish or dark brown, smooth and lustrous. (2n = 14) Jun–Sep. Infrequent to locally common on prairie plains & hillsides, open wooded or brushy hillsides & ravines, stream valleys; sw & s SD, s in w 4/5 GP; (SE, e WY, s to TX Gulf Coastal Plain, e CO, NM, disjunct e-cen AZ, Coah., Chih.). *Parosela aurea* (Nutt.) Britt.—Rydberg.

   In our area this is usually a short-lived plant that is browsed by livestock, and therefore has become less common. A few plants have been found flowering late in the first season.

2. **Dalea candida** Michx. ex Willd., white prairie-clover. Herbaceous perennials from an ultimately thick taproot and superficial or subterranean caudex. Stems 1–several, 3–10 dm tall, erect or diffuse, simple or branched above, ribbed, minutely punctate or glandless, glabrous. Leaves alternate, odd-pinnate, short petioled or subsessile, 1.5–6 cm long; leaflets (2)3–5(6) pairs, 0.5–3.5 cm long, obovate to elliptic-oblanceolate or linear-oblong, acute, short-acuminate, or emarginate, minutely punctate. Spikes lax or densely flowered, ovoid to cylindroid (rarely subglobose), 1–5.5(7.5) cm long, axis glabrous or minutely pilosulous. Calyx tube 1.9–2.7 mm long, glabrous or pubescent, glandular, teeth 0.6–1.8 mm long; petals white; banner 4–5.7 mm long, claws 2–3.8 m long; wings and keels 3.2–5.3 mm long, claws 1–2.3 mm long; stamens 5. Pod 2.6–4.5 mm long, usually exserted, glandular; seed 1.5–2.5 mm long.

   There are two varieties in the GP as follows:

2a. var. *candida*. Stems usually virgate, 5–10 dm long. Largest leaflets of primary stem leaves to (12)15–40 mm long. Spikes dense and conelike during and after anthesis, the axis permanently concealed. Calyx tube glabrous outside. (2n = 14) May–Aug. Prairies, open woodlands, roadsides, waste places; e 1/2 GP, infrequent w; (se Sask., s Man., s Ont. s to WS, IL, TN, w AL, e 1/2 KS, TX). *Petalostemon candidus* (Willd.) Michx. — Rydberg.

2b. var. *oligophylla* (Torr.) Shinners. Stems diffuse, virgately ascending or decumbent, 2.5–7 dm long. Largest leaflets of primary stem leaves to (6)9–20(24) mm long. Spikes relatively lax or becoming so by full anthesis, axis partly visible in pressed specimens. Calyx tube pubescent outside or often glabrous. (2n = 14) May–Sep. Rocky prairies, stream valleys, roadsides; w 1/2 GP, infrequent se; (s Man., Sask., Alta., s to w MN, w IA, w TX, UT, AZ, ne Son., nw Dur.). *Petalostemon oligophyllus* (Torr.) Rydb., *P. occidentale* (A. Gray) Fern. — Fernald.

The character of spikes being lax or permanently dense is the only one that works reasonably well in separating these two varieties, as other characters are quite variable. Where the two grow in the same area, var. *oligophylla* is found in drier sites. White prairie-clover is browsed by all classes of livestock and soon disappears with overgrazing.

3. ***Dalea cylindriceps*** Barneby, massive spike prairie-clover. Herbaceous perennials from an orange taproot with knobby crown or shortly branching superificial subterranean caudex. Stems 1–several, erect, simple or few branched, 2–6(8) dm tall, glabrous to inflorescence. Leaves alternate, odd-pinnate, primary stem ones 3–8 cm long; leaflets 3–4(5) pairs, 1–2.5 cm long, oblanceolate, or oblong-elliptic, or narrowly elliptic, acute or acuminate, rarely obtuse and mucronulate, glabrous, glandular. Spikes dense, conelike, (2)3–16)23) cm long, 8–12 mm in diam. Calyx tube campanulate, densely pilose on outside with hairs to 2.4 mm long, teeth 1.5–2.5 mm long; petals white or light pink, fading light yellow; banner 4.7–6.2 mm long, claws 2.8–3.8 mm long; wings and keel 2–4.7 mm long; stamens 5. Pod 2.5–3 mm long, not exsert, valves in lower 2/3 hyaline, pilose; seed 1.7–2.3 mm long, olivaceous or yellowish, smooth, sublustrous. May–Sep. Rare to locally common on sand dunes, sandy prairies, & stream valleys; SD: Fall River; w NE, se WY; KS: Grant, Morton, Stevens; e CO; OK: Cimarron; w TX, ne NM; (sw SD, s to w TX, ne NM, e CO). *Petalostemon compactus* (Spreng.) Swezey — Rydberg.

On dunes and loose sand areas this species sometimes flowers the first year and frequently expires at end of the second or third season. On more stable sandy areas it is often a longer-lived perennial. It has been collected infrequently and needs more careful study.

4. ***Dalea enneandra*** Nutt., nine-anther prairie clover. Perennial herbs with a yellow taproot and a knobby or shortly branching caudex. Stems usually 1–3, erect, (1.5–4)5–10(12) dm tall, usually simple in lower 1/2–2/3, much branched above, glabrous up to silky calyces. Leaves subsessile, primary stem ones 1.3–2.6 cm long, usually deciduous by anthesis, branches retaining leaves; leaflets (2)3–6 pairs, narrowly oblanceolate or elliptic, obtuse or acute, glaucous, glandular, 4–12 mm long. Spikes loosely or remotely (2)5- to 35-flowered, appearing 2-ranked and clearly revealing glabrous axis. Bracts subtending and folded around calyx prominent, broadly ovate, obovate, or obovate-truncate, 3–4 mm long, with prominent pale or white membranaceous margin, body glabrous, glandular; calyx tube 3–3.7 mm long, silky-pilosulous, teeth 3.3–4.6 mm long, becoming divergent, plumose; petals white; banner 5.7–7 mm long, claw 2.5–3.5 mm long; wings 2.8–4.1 mm long, claw to 1.2 mm; keel 5.5–7 mm long, claws 1–2 mm long, stamens 9. Pod 3–3.7 mm long, hyaline in lower 1/2; seed 2.5–3 mm long, yellowish, smooth. (2n = 14) Jun–Sep. Infrequent to locally abundant on a variety of soils (but with preference for calcareous types) in prairie plains, hillsides, stream valleys, roadsides; ND: Grant, Morton, Sioux; w 1/2 SD, & GP except se; (s-cen ND, to nw MO, w 2/3 of KS & OK, e WY & CO, ne NM, TX). *Parosela enneandra* (Nutt.) Britt. — Rydberg. *D. laxiflora* Pursh — Shinners, Field & Lab. 17: 86–87, 1949.

In sandy areas, plants occasionally have the caudex buried several centimeters and the stems above ground only 1.5–2 dm long. Such dwarf plants have been called var. *pumila* (Shinners) B. L. Turner. In our area this variety does not merit recognition.

5. **Dalea formosa** Torr., feather plume. Shrub 1.5–10 dm tall, divaricately branched, branches glabrous, sparingly gland-dotted on young twigs. Leaves alternate, odd-pinnate, 2.5–11(14) mm long, glabrous, gland-dotted; leaflets (2)3–6(7) pairs, obovate-cuneate, oblanceolate, or narrowly obovate, (0.5)1–7 m long, mostly folded. Spikes subcapitate, 2- to 9-flowered, terminal on branches and spurs. Calyx tube 3–5 mm long, pilose with hairs 1–2.5 mm long, teeth 4.5–8.5 mm long, filiform, plumose; banner yellowish but soon rose-purple, 7–9 mm long, claw 4–5 mm long; wings and keel rose-purple; wings 8–10 mm long, claws 3–3.8 mm long; keel 8–11 mm long, claws 4–5 mm long; stamens 10. Pod 3–3.5 mm long, pilose above, and glandular; seed 2.5–3 mm long, yellowish, smooth. (2n = 14) Jun–Sep. Infrequent to common on dry rocky hillsides, shrubby barren places, prairie knolls; sw GP; (extreme w OK, se CO, w TX, NM, se AZ, nw Son., Coah.).

This species is browsed by livestock and is one of the host species of the parasitic *Pilostyles thurberi* A. Gray.

6. **Dalea frutescens** A. Gray, black dalea. Small rounded or spreading shrub 2–10 dm tall; stems erect and ascending, or decumbent at base, rooting, and forming small thickets. Leaves alternate, odd-pinnate, 10–20 mm long, short petioled; leaflets 4–9(10) pairs, obovate to broadly oblanceolate, obtuse or emarginate, 1.5–3.5(5) mm long, glabrous, glaucous, minutely gland-dotted; stipules 2 mm long. Spikes subcapitate, 3- to 25-flowered. Calyx tube 2.6–3.5 mm long, glabrous, teeth deltoid, much shorter than tube; petals purple; banner 5–6 mm long, claw 2.5–3 mm long; wings 5.5–6 mm long, claws 1–1.6 mm long; keel 6.5–8 mm long, claws 1.5–2.2 mm long; stamens 10. Pod 2.8–3.5 mm long, glabrous, glandular; seed yellowish, 1.7–2.5 mm long, smooth. (n = 7) Jul–Oct. Dry, rocky hillsides; OK: Comanche; (s OK, TX, NM, Chih., Coah., N.L.).

This is reported to be an excellent browse plant that has been largely eliminated in ranges.

7. **Dalea jamesii** (Torr.) T. & G., James' dalea. Low, loosely tufted herbaceous perennials, with thick woody roots and woody, branching superficial or subterranean caudex; stems 1–7(15), 1–12 cm long, shorter ones erect, longer ones decumbent, essentially simple, silky-pilose. Leaves alternate, 3-foliolate; petiole 3–15(18) mm long; stipules 2–6 mm long; leaflets obovate to broadly oblanceolate, obtuse, (3)5–18 mm long, densely silky-canescent on both sides. Spikes sessile or nearly so, dense, 1.5–6 cm long, 15–18 mm in diam. Calyx tube 2.5–3.5 mm long, pilose with spreading hairs to 2.5 mm long, teeth deltoid aristate, plumose, 5–10 mm long; corolla yellow, fading purplish or brown; banner 5–6.5 mm long, claw 2.5–3.5 mm long; wings 5.5–6.8 mm long, claw 1.5–2 long; keel 6.5–8.5 mm long, claws 1.5–2.8 mm long; stamens 10. Pod 3.5–4 mm long; seed 2.5–3 mm long, yellowish, smooth. (n = 7) Apr–Aug. Infrequent to locally common on rocky prairie plains & hillsides; KS: Morton, Stanton; OK: Cimarron, Texas; w TX panhandle, e NM; (sw KS, se CO, s to w TX, NM, se AZ, Chih.). *Parosela jamesii* (T. & G.) Vail — Rydberg.

8. **Dalea lanata** Spreng., woolly dalea. Prostrate herbaceous perennial with yellow taproot and buried crown or shortly branching caudex; stems 1–several, 2.5–10 dm long, divaricately branching upward from near the base, sometimes forming mats 2 m in diam, densely and shortly villous-tomentose. Leaves alternate, odd-pinnate, 1–3 cm long; short petioled or subsessile; stipules 1–2.5 lmm long; leaflets 4–7 pairs, obovate, obovate-cuneate, or broadly oblanceolate, usually emarginate, 3–10 mm long, densely short-villous. Peduncles terminal and axillary, 4–40 mm long; spikes narrow, loose, axis 1.5–9 cm long. Bracts subtending flowers persistent, ovate, acuminate; glandular; calyx tube 2–2.5 mm long, villosulous; teeth triangular to lanceolate, 1.5–2.3 mm long; petals red-violet; banner 3–4.3 mm long, claw 1–3 mm long; wing and keel petals 2–4 mm long, claw to 0.8 mm long; stamens 8, 9, or 10. Pod 2.5–3 mm long, valves hyaline at base, pilosulous; seed greenish-yellow or brown, smooth, somewhat shiny, 2–2.5 mm long. (2n = 14) Jun–Oct. Infrequent to locally

common on sand dunes, sandy river valleys, roadsides; s GP s of Arkansas R.; (s KS, s-cen CO, s to w AR, n & w TX, e NM). *Parosela lanata* (Spreng.) Britt.—Rydberg.

9. Dalea leporina (Ait.) Bullock, foxtail dalea, hare's-foot dalea. Erect annual herbs with slender taproot; stems 1.5–10 dm tall, branching from near base to more often at midstem, striate, glandular above, glabrous. Leaves alternate, odd-pinnate, 2–10 cm long; petiole 0–8 mm long; stipules 1–3 mm long; main stem leaves with 8–24 pairs of leaflets; leaflets oblanceolate, oblong-lanceolate, or obovate, emarginate or retuse, 3–12 mm long, gland-dotted beneath, glabrous, sometimes with red margins. Peduncles at end of branches and opposite leaves, 3–10 cm long; spikes dense, ovoid or cylindric, 2–8 cm long, 8–10 mm in diam, axis pilosulous. Calyx tube 2–2.5 mm long, pilose, with ascending hairs, teeth 1–3 mm long, ovate-triangular and apiculate to lance-acuminate; corolla white or tinged with blue; banner 4–6 mm long, claw 1.5–3 mm long; wings 2–3 mm long, claw to 1 mm long; keel 2–3 mm long, claws to 1 mm long; stamens 9 or 10. Pod 2.5–3 mm long, lower 3/4 of valves hyaline, papery; seed 2–2.5 mm long, brown, shiny, smooth. (2n = 14) Jul–Sep. Infrequent to locally common on disturbed soils along roads, edges of fields, stream banks, open wooded areas, rarely in prairie ravines & pastures; e 1/2 GP to s KS infrequent n & w; (WS, IL, MN, ND, s to MO, KS, CO, NM, trans-Pecos TX, Mex., S. Amer.). *Parosela alopecurioides* (Willd.) Rydb.—Rydberg; *D. alopecurioides* Willd.—Fernald.

This is a weedy species that varies in stature with the habitat. In some years it is relatively common in some areas, but cannot be found in other years. Livestock have been observed to browse the plant.

10. Dalea multiflora (Nutt.) Shinners, round-headed prairie clover. Perennial herbs from an ultimately thick, blackish root system, terminated with a branching, woody, subterranean caudex; stems several, usually erect and ascending in clumps, 3–8 dm long, much branched, glabrous, punctate, drying striate; stipules lance-subulate, 1–2 mm long, usually fugacious. Leaves alternate, odd-pinnate, main stem ones 2–4 cm long; leaflets (5)7–13, narrowly oblong to elliptic-oblanceolate or linear-oblong, 6–14 mm long, usually obtuse or mucronate, glabrous, gland-dotted below. Spikes usually numerous, permanently dense, headlike, subglobose to oblong, to 15 mm long, axis glabrous. Bracts subtending calyx shorter than calyx tube at anthesis, 2–2.4 mm long, early deciduous,; calyx tube white, glabrous, 2.2–2.6 mm long, teeth deltoid, 1.0–1.4 mm long, terminating in a gland; petals white; banner 5–5.5 mm long, claw 3 mm long; wings and keel 3.5–5 mm long, claw 1.3–1.6 mm long; stamens 5. Pod obliquely obovoid, 3.5–5 mm long, valves hyaline in lower 1/2, included to 1/3 exserted; seed 2–2.5 mm long, olivaceous or brown, smooth, sublustrous. (2n = 14) Mid May–early Aug. Infrequent to locally common on dry rocky prairies & brushy or open wooded hillsides, roadsides; NE: Pawnee, Richardson; MO: Jackson; e 1/2 KS, OK, TX; (se NE, e KS, through cen OK, cen & e-cen TX). *Petalostemon multiflorus* Nutt.—Rydberg.

This species is selectively grazed by all classes of livestock and is soon eliminated from ranges. It often can be found only in relic or roadside undisturbed prairie remnants. It frequently grows side by side with *D. candida* var. *candida*, and occasional intermediate forms are found. Westward it is found to a limited extent in the same area as *D. candida* var. *oligophylla*, but on more moist sites. Although var. *oligophylla* often has somewhat subglobose spikes, we have found no evidence of putative intermediates.

11. Dalea nana Torr. ex A. Gray, dwarf dalea. Slender herbaceous perennials from slender taproot and superficial to usually subterranean branching caudex; stems always diffuse or prostrate, simple to usually branched, silky-canescent, 1–2(3) dm long; stipules subulate-setaceous 0.5–2 mm long. Leaves 1–3 cm long, alternate, usually odd-pinnate, mostly 5–7 foliate, but some trifoliate; petiole 2–8 mm long; leaflets obovate, oblong-obovate, elliptic, or oblanceolate, obtuse or emarginate, 3–15 mm long, sericeous on both sides but less so above. Peduncles at ends of branches and opposite leaves, 5–20 mm long; spikes 1–5 cm long, relatively loose or becoming so. Calyx tube campanulate, 2–2.7 mm long, sericeous teeth filiform, plumose, 2–4.2 mm long; petals yellow at anthesis, fading brownish, often purplish on pressed specimens; banner 4.5–5.5 mm long, claw 2.6–3.5 mm long; wings 2.5–4.3 mm long, claw less than 1 mm long; keel 3.8–6 mm long, claws 0.8–2.2 mm long, brown, smooth. (2n = 14) May–Sep. Sandy prairie plains, dunes, stream valleys, roadsides; sw KS, se CO; OK: Cimarron, Texas; TX panhandle, e NM; (sw KS, se CO s to NM, se AZ, Tam., N.L., Chih.). *Parosela nana* (Torr.) Heller—Rydberg.

In recent years this species has become more common in sw KS in dune areas and in the Cimmaron R. valley.

12. *Dalea purpurea* Vent., purple prairie clover. Perennial herbs from a tough root system, eventually developing a subterranean caudex 2–12 cm tall, sometimes with lateral subsurface caudex-branches 3–7 cm long and producing aboveground shoots at intervals. Stems 2–9 dm tall, erect and virgately ascending, or bushy, or prostrate, either glabrous or thinly pilosulous, to densely and softly villous-tomentose, striate-ribbed when dry. Leaves alternate, odd-pinnate, main stem ones 1.5–4.5 cm long, with (3)5(7) linear-oblanceolate, linear, or linear-elliptic leaflets (7)10–24(28) mm long, acute, usually involute; glabrous, sparingly pilosulous, to villous-tomentose. Peduncles 0–9(15) cm long; spikes dense, conelike ovoid becoming oblong-cylindrical, the axis 1–7 cm long, densely villosulus, 7–14 mm in diam (less petals). Bracts subtending calyx 2.5–5.8 mm long, 1–2 mm wide, abruptly contracted into a subulate, erect or recurved tail, densely pilosulous and ciliolate; calyx tube campanulate, densely pilosulous with ascending hairs to 0.7 mm long, 2–2.9 mm long, teeth 1–2 mm long, shorter than tube; petals rose or dark purple, pinkish (white in rare albino forms); banner 4.5–7 mm long, claw 2.5–4.5 mm long; wing and keel 3–5 mm long, claw 0.5–1.5 mm long; stamens 5. Pod obliquely obovoid, 2.1–2.5 mm long, valves glabrous, hyaline in lower 1/2; seed brown, smooth, 1.5–2 mm long.

This species is rather uniform in the e 2/3 of the GP but varies considerably in stature, pubescence, length of peduncle, and diam of spike, particularly in sw GP. Two varieties are recognized as follows:

12a. var. *arenicola* (Wemple) Barneby. Plants usually less than 3.5 dm tall; herbage glabrous or glabrate. Diam of spikes at anthesis 7–9 mm (less petals); peduncles usually 3–15 cm long, sometimes exceeding 1/4 length of stem. (2n = 14) Jun–Aug. Local populations in dunes or loose sand of stream valleys & hillsides; w NE, w KS, e CO, w OK, e NM, TX panhandle & n plains; (*endemic* in area). *Petalostemon arenicola* Wemple—Atlas GP.

12b. var. *purpurea*. Plants 3–9 dm tall; herbage glabrous, pilosulous, to villous-tomentose. Diam of spikes at anthesis 9.5–12 mm (less petals); peduncles short, less than 1/4 length of stems, or stems leafy to spikes. (2n = 14) May–Aug. Infrequent to common on rocky prairie plains & hillsides, open wooded areas, stream valleys, roadsides; GP but infrequent w; (s Man., s Alta., s to MN, IL, AL, LA, MT, WY, CO, NM, n 1/2 TX). *Petalostemon purpureus* (Vent.) Rydb., *P. mollis* Rydb.—Rydberg.

These two varieties at first appear reasonably distinct, but we find only the diameter of spikes to be useful in separation. Even then many specimens are borderline cases, particularly in areas of transition from loose sand to hard soils.

Purple prairie clover is readily eaten by all livestock and therefore decreases when ranges are overgrazed.

13. *Dalea tenuifolia* (A. Gray) Shinners, slimleaf prairie clover. Perennial herbs from tough roots and subterranean branched caudex, these sometimes short-rhizomatose but without adventitious roots; stems (1)2–4(5) dm tall, decumbent or diffusely ascending, densely pilosulous below, pilosulous to glabrate above. Leaves alternate, odd-pinnate, short petioled, pilosulous or glabrate, punctate, those of main stem 2–4 cm long; stipules subulate, to 5 mm long; 3–5 leaflets linear, usually involute, 1–2 cm long. Peduncles 1–12 cm long; spikes globose in bud, becoming columnar and loose during and after anthesis, up to 10 cm long, 7–10 mm in diam, axis densely pilosulous. Bracts subtending calyx 2.5–5 mm long, obovate to broadly oblanceolate, 1–2.5 mm wide, pilosulous, contracted into subulate or subulate-attenuate tail 1–3 mm long, deciduous by or during anthesis; calyx tube campanulate, 1.5–2.5 mm long; densely pilose with spreading-ascending hairs 1–1.5 mm long, teeth 1.5–2 mm long; petals rose-purple; banner 5.5–6.5 mm long, claw 3–4 mm long; wings and keel 4–5 mm long, claw 1–1.5 mm long; stamens 5. Pods 2.8–3.5 mm long; valves hyaline in lower 1/3; seed 1.8–2.2 mm long, brown, smooth. (2n = 14) May–Jul. Locally common on dry rocky hilltops, hillsides, badlands; w KS, se CO plus Larimer, ne NM, panhandles of OK, TX; (*endemic* in area). *Petalostemon tenuifolius* A. Gray, *P. porterianus* Small — Rydberg.

Specimens in the northern parts of the range in the GP often have the lower portion of the stem more sparsely pilose and the whole plant more glabrate than plants of sw KS and s.

14. **Dalea villosa** (Nutt.) Spreng., silky prairie clover. Perennial herbs from red-orange roots and branching caudex; stems 1–several, ascending or diffuse, usually branched above, densely villous or villous-tomentose, 2–3.5(5) dm tall. Leaves alternate, odd-pinnate, usually crowded, densely villous or villous-tomentose, 2–4 cm long; shortly petiolate; stipules subulate, 5–7 mm long; leaflets 11–21, elliptic to elliptic-oblanceolate, obtuse, 5–11 mm long. Peduncles terminating stem and branches, 0–2.5 cm long; spikes moderately dense, becoming lax and oblong-cylindric, 3–12 cm long, 7–10 mm in diam. Bracts subtending calyx linear-lanceolate, 1.5–5 mm long, deciduous by full anthesis; calyx tube 2–2.7 mm long, densely spreading-pilosulous, teeth to 1.4 mm long; petals rose-purple, pinkish, or nearly white; banner 4.5–5.5 mm long, claw 2–3 mm long; wings and keel 2.5–4.5 mm long, claw ca 0.5 mm long; stamens 5. Pod 2.5–3 mm long, densely villosulous; seed 2–2.5 mm long, brown, smooth. (2n = 14) Jun–Aug. Locally common in loose sand of dunes, stream valleys, very sandy prairies, sandy open woodlands; GP but infrequent to absent e; (s Sas., s Man., s to w WI, MN, w 3/4 KS, e-cen OK, w-cen TX, e MN, se CO, ne NM). *Petalostemon villosus* Nutt. — Rydberg.

Sometimes when roots are exposed by shifting sand, adventitious stems are formed and the plant appears stoloniferous. Plants from the more northern GP average shorter than those of southern areas. Flower color will vary in the same population from bright rose-purple, to light pink, and more rarely to white or bluish-white.

11. DESMODIUM Desv., Tickclover, Tick-trefoil

Ours herbaceous perennials; stems erect, trailing or prostrate, usually with hooked hairs. Leaves alternate or subverticillate, pinnately trifoliolate, stipels obsolescent or persistent; stipules deciduous or persistent. Inflorescence racemose or paniculate, axis pubescence including or consisting entirely of hooked hairs. Calyx more or less 2-lipped, the upper 2 lobes connate for most their length, the lower 3 separate; petals pink, red-purple, lavender, whitish, often drying blue-green, bluish, or yellowish; banner oblong to suborbicular, narrowed at base; wings oblong; keel nearly straight; stamens monadelphous or diadelphous. Loments stipitate, indehiscent, seed-bearing portion flattened and constricted into 2–8 1-seeded segments, sutures between segments usually deeper on ventral margin than dorsal; segments generally separating into 1-seeded, mostly uncinulate-pubescent parts. *Meibomia* Heister — Rydberg.

Reference: Isley, Duane. 1955. Desmodium, *in* The Leguminosae of the North-Central United States II: Hedysareae. Iowa State Coll. J. Sci. 30 (1): 38–73.

1   Calyx lobes less than 1/2 as long as tube; stamens monadelphous; stipe of loment longer than remains of stamens and 3× as long as calyx; dorsal margin of loment glabrous.

2   Flowering stem arising at ground level from base of plant, usually
    leafless .... 8. *D. nudiflorum*
2   Flowering branches from leafy stems.
    3   Terminal leaflet as wide as long or nearly so; loment stipe glabrous; leaves mostly
        crowded toward summit; flowers pinkish or purple .... 5. *D. glutinosum*
    3   Terminal leaflet longer than wide; loment stipe usually uncinulate-puberulent; leaves
        scattered, flowers white .... 11. *D. pauciflorum*
1   Calyx lobes more than 1/2 as long as tube; stamens diadelphous; loment stipe shorter than
    remains of stamens and less than 2× longer than calyx; dorsal margin of loment pubescent.
    4   Stems prostrate .... 12. *D. rotundifolium*
    4   Stems erect.
        5   Leaves without petioles or these to 3 mm or less long .... 13. *D. sessilifolium*
        5   Leaves at midstem with petioles 3 mm or more long.
            6   Leaflets with uncinate hairs beneath
                7   Axis of inflorescence with spreading hairs longer than diam of axis; stem usual-
                    ly much branched, with several inflorescences; loment segments semirhombic;
                    leaflets not strongly reticulate-veined beneath .... 2. *D. canescens*
                7   Axis of inflorescence with short-uncinate or glandular hairs; stem usually un-
                    branched and with one inflorescence; loment segments oval to orbicular;
                    leaflets strongly reticulate-veined beneath .... 6. *D. illinoense*
            6   Leaflets without uncinate hairs beneath or only with a few along veins.
                8   Lower margin of loment segments abruptly curved near the middle, the
                    segments nearly triangular or semirhombic.
                    9   Stipules 8–17 mm long, usually persistent; loment segments 7–11 mm
                        long .... 4. *D. cuspidatum*
                    9   Stipules 2–8 mm long, usually deciduous; loment segments 4.5–9 mm
                        long .... 10. *D. paniculatum*
                8   Lower margin of loment segments gradually curved, segments nearly
                    semicircular.
                    10 Flowers 8–13 mm long; stipe of loment 2–2.5 mm long ..... 1. *D. canadense*
                    10 Flowers 4–6 mm long; stipe of loment (1)1.5–2(3) mm long.
                        11 Lateral leaflets about as long as petioles; stems and leaves essentially
                            glabrous or sparsely uncinulate-puberulent .... 7. *D. marilandicum*
                        11 Lateral leaflets distinctly longer than petioles; stems and leaves pubes-
                            cent with pilose and/or dense uncinulate puberulence.
                            12 Stems and leaves sparsely to densely pilose; terminal leaflets 1.5–2.5
                                cm long .... 3. *D. ciliare*
                            12 Stems and leaves without pilose hairs or these rare; terminal leaflets
                                3–7 cm long .... 9. *D. obtusum*

1. *Desmodium canadense* (L.) DC., Canada tickclover. Perennial herbs with woody taproot and knobby to shortly branching caudex. Stems erect, usually clustered, 0.5–2 m tall, normally simple below inflorescence, uncinulate-puberulent and spreading-pilose at least above. Leaves alternate, pinnately trifoliolate; petioles to 2–3 cm long below, to only 2 mm long above; stipules lanceolate to lance-attenuate, 4.5–10 mm long, striate, puberulent and often pilose on lower surface; stipels lanceolate, 2–4 mm long, persistent; leaflets of midstem leaves ovate to ovate-lanceolate, obtuse to short-acute, uncinulate-puberulent and appressed-pilose on upper surface, pilose below, ciliate; blades 4.5–10.5 cm long. Inflorescence terminal, racemose-paniculate; racemes densely flowered, with lance to ovate-attenuate bracts 4–10 mm long, these conspicuous at anthesis but soon deciduous. Calyx tube 1.5–2 mm long, upper bifid lobe 4.5–5 mm long, middle tooth of lower lobe 5–7 mm long and lateral 4.5–6 m long; petals red-violet, fading to dark blue, rarely white; 10–12 mm long; stamens diadelphous. Loment with stipe 2–2.5 mm long, (1)3- to 5-jointed, segments rounded above, obtusely angled below, densely uncinulate-pubescent, often with long hairs, 5–7 mm long, 4–5 mm wide; seed 3.5–4 mm long, brown, smooth. (n = 11) July–Sep. Infrequent to common in rocky or sandy prairies, ravines, stream valleys, thickets, roadsides, & waste places;

e 1/2 GP, infrequent w; (N.S. to s Sask, s to N.E., MD, WV, OH, IN, ND, MO, OK). *Meibomia canadensis* (L.) O. Ktze. — Rydberg.

Young plants of Canada tickclover are browsed by livestock and the species disappears with heavy grazing.

2. ***Desmodium canescens*** (L.) DC., hoary tickclover. Perennial herb with taproot and knobby or shortly branching caudex; stems erect, 0.5-1.5 m tall, usually branched, 1-8 per clump, spreading-pilose and uncinulate-puberulent. Leaves alternate, pinnately trifoliolate; petioles 1-11 cm long with pubescence like the stem; stipules 5-12 mm long, deltoid to ovate-acuminate and often attenuate, 3-5(7) mm wide at base and slightly clasping the stem, becoming spreading or reflexed; stipels linear to lance-subulate, 1-5 mm long; leaflets of midstem leaves, ovate or lance-ovate, rounded at base, acute or acuminate at apex, usually thick and scabrous, finely pubescent beneath with minute hooked hairs, blades of terminal leaflet 3-9(13) cm long, lateral blades 2-7(12) cm long. Inflorescence racemose-paniculate, becoming diffuse, raceme bracts ovate-attenuate, 3-6 mm long, early deciduous. Calyx-tube campanulate, 1.5-2 mm long, puberulent and pilose, upper lobe notched, 3-5 mm long, middle tooth of lower lobe 4.5-6 mm long, lateral teeth 3-4.5 mm long; petals pink or whitish, drying blue-green or purplish, rarely white, 10-12 mm long; stamens diadelphous. Loment with stipe 2-6 mm long; segments (1)3-6, 6-9(11) mm long, 4-5 mm wide, uncinulate-pubescent, convex above, angled below; seed 4-4.5 mm long, 2 mm wide, smooth, brown. (n = 11) Jul-Sep. Dry open woods, thickets, stream valleys, roadsides; se 1/4 GP; (MA, Ont., WS, IA, NE, s to FL, AL, TX). *Meibomia canescens* (L.) O. Ktze.— Rydberg, *D. canescens hirsutum* (Hook.) Robins.—Gates.

Very hairy plants have been referred to var. *hirsutum*, but the degree of pubescence is too variable for taxonomic consideration. Many specimens in the GP have leaf measurements in the lower range of those given above, and the lower leaf surface is definitely reticulate and approaching that of *D. illinoense*.

3. ***Desmodium ciliare*** (Muhl. ex Willd.) DC., slender tickclover. Slender perennial herbs with a stout taproot and shortly branching caudex; stems erect or ascending, clustered, often branched from near base, with somewhat densely spreading-pilose hairs and short uncinulate-puberulence intermixed, or the pilose hairs sparse to absent, becoming glabrate in age. Leaves alternate, pinnately trifoliolate; petioles 1-10(15) mm long, pubescent; stipules slenderly subulate, 2-5 mm long; stipels subulate, 0.5-2 mm long; leaflets ovate-oblong, elliptic or nearly rhombic, obtuse, mucronate, pilose and puberulent above, less so below, ciliate; terminal leaflet 1-3 cm long, 5-17 mm wide; lateral ones nearly as large. Inflorescence racemose-paniculate, bracts ovate-acuminate, 1-3 mm long, early deciduous, usually subtending 1-4 pedicels. Calyx tube 1 mm long, puberulent, upper bifid lobe 1.5-2 mm long, middle tooth of lower lobe 2-3 mm long, teeth ciliate; petals 3-4 mm long, pink or whitish; stamens diadelphous. Loment with stipe 1-2 mm long; segments (1)2(3), 3.5-5.5 mm long, 2.7-4 mm long, somewhat convex above, gradually rounded below, somewhat sparsely uncinulate-puberulent on faces, margins more densely so; seed 3 mm long, light brown, smooth. (n = 11) Aug-Oct. Local on dry, usually acid soils of rocky or sandy open woodlands & clearings; extreme se GP; (MA, NY, MI, OH, IL, MO, KS, s to FL, TX, Mex.).

Some specimens of this species with petioles to 15 mm long and leaflets nearly glabrous closely resemble *D. marilandicum*, and separation is sometimes arbitrary. Such specimens may be hybrids and the complex merits critical study.

4. ***Desmodium cuspidatum*** (Muhl. ex Willd.) Loud., long-leaf tickclover. Perennial herbs from stout taproots and knobby or shortly branched caudex; stems erect, stout, 0.5-2 m tall, simple or branched, grooved when dry, glabrous or sparsely uncinulate-puberulent, rarely sparsely pilose. Leaves alternate, pinnately trifoliolate; petioles 4-8(10) cm long, nearly

glabrous; stipules conspicuous, lance-attenuate, striate, 1-2 cm long, 1-2 mm wide at base, essentially glabrous, persistent; stipels linear, 4-6(9) mm long; leaflets ovate-acuminate, rounded to cuneate at base, essentially glabrous, glaucous below; terminal leaflet 6-14 cm long, 4-7 cm wide, acuminate; lateral leaflets similar but smaller. Inflorescence racemose-paniculate; rachis densely uncinate-pubescent; bracts subtending pedicels ovate-attenuate, (6)8-10(12) mm long, 4-5 mm wide, conspicuous before anthesis, soon deciduous. Calyx tube campanulate, 1-2 mm long, puberulent, upper bifid lobe 4-4.5 mm long; middle tooth of lower lobe 4.5-6 mm long, lateral teeth 4-5 mm long; petals pink, 8-10 m long; stamens diadelphous. Loment on stipe 2-3 mm long; segments (3)4, triangular or rhomboidal, 7-11 mm long, 4-5 mm wide, margins densely uncinulate-puberulent, surfaces less so; seed 4-4.5 mm long, brown, smooth. (n = 11) Jul-Sep. Locally common in rich woodlands, less common in thickets, roadsides; se GP from se NE s; (NH, VT to MI, WS, IA, se NE, s to FL, AR, TX). *Meibomia bracteosa* (Michx.) O. Ktze, *M. longifolia* (T. & G.) Vail — Rydberg, *D. bracteosum* (Michx.) DC., *D. bracteosum longifolium* (T. & G.) Robins. — Gates.

The more hairy forms of this species have been named var. *longifolium* (T. & G.) Schub. and supposedly are the common phase in our area. In eastern Kansas, however, glabrous and pubescent plants are usually found together, with glabrous plants more common.

5. ***Desmodium glutinosum*** (Muhl. ex Willd.) Wood, large-flowered tickclover. Perennial herbs with slender root and knobby caudex; stems erect, simple below inflorescence, 4-10 dm tall, striate when dry, sparsely spreading-pilose and minutely uncinulate-puberulent or glabrous. Leaves trifoliolate, usually nearly whorled at base of inflorescence, rarely somewhat scattered; petioles 6-15 cm long, spreading-pilose and uncinulate-puberulent; stipules linear, 8-12 mm long, semipersistent, pilose, ciliate; stipels obsolescent; terminal leaflet broadly ovate-acuminate, 7-15 cm long, 6-13 mm wide, apex conspicuously acuminate; lateral leaflets smaller; surfaces of leaflets sparsely pubescent. Inflorescence terminating the stem, simple racemose to branched; rachis densely uncinulate-puberulent; bracts early deciduous, 5-9 mm long. Calyx shallowly 2-lobed, 3-3.5 mm long, teeth shorter than tube, puberulent; corolla pink, purple, or rarely white, 6-8 mm long; stamens monadelphous. Loment with stipe 4-10 mm long, segments (1)2-3(4), triangular or semiovate, 8-12 mm long, 4.5-6 mm wide, margins essentially glabrous, surfaces uncinulate-puberulent; seed 6-7 mm long, 4 mm wide, flat, brown. (n = 11) Jun-Aug. Rather common in rocky open or rich woods, wooded stream valleys, & thickets; e 1/3 GP but less common n; (ME, Que., MN, ND, s to FL, TX, Mex.). *Meibomia acuminata* (Michx.) Blake — Rydberg; *D. acuminatum* (Michx.) DC. — Gates.

This plant is often browsed by deer, and the seeds are eaten by a variety of birds.

6. ***Desmodium illinoense*** A. Gray, Illinois tickclover. Perennial herbs with rather slender taproot and knobby caudex; stems erect, simple below inflorescence, with uncinate and glandular pubescence, and rarely sprasely spreading-pilose. Leaves alternate, pinnately trifoliolate; petioles nearly as long as leaflets; stipules linear or ovate-acuminate, 10-15 mm long, semipersistent, semiclasping at base; leaflets narrowly ovate to ovate-lanceolate, uncinate hairs on both sides, and glandular ones below, strongly reticulate below; terminal leaflet 6-10 cm long, lateral ones somewhat smaller. Inflorescence terminal, simple or slightly branched, axis with glandular hairs; bracts lanceolate, 10 mm long, early deciduous. Pedicels 10-25 mm long, calyx tube 1-2 mm long, upper lobe bifid (2)4-4.5 mm long, middle tooth of lower lobe (3)5.5-6(7) mm long, and lateral teeth (2)3.7-4.2 mm long, sparsely pubescent; petals white or pinkish, fading purplish; stamens diadelphous. Loments with (2)3-6(7) segments elliptic or rounded, 4-8 mm long, and 3.5-5 mm wide, elliptic or rounded margins, surface, and margins densely uncinulate-pubescent; seed 3.5 mm long, 2 mm wide, brown, smooth. (n = 11) Jun-Sep. Locally common in prairie ravines, hillsides,

rarely open woodlands, roadsides; SD: Clay, Union; e 1/2 NE & KS, cen OK; TX: Hemphill, Wheeler; (n OH, s MI, to se SD, s to n AR, OK, e TX panhandle). *Meibomia illinoense* (A. Gray) O. Ktze.—Rydberg.

Some plants from s-cen KS and s in OK have teeth of calyx lobes only 2–4 mm long, loment segments 6–8 mm long, and seem close to *D. tweedyi* Britt. of n-cen TX. Our plants, however, lack the large pale blotches on either side of the midrib as described for *D. tweedyi*. The complex merits additional field and experimental studies.

The Illinois tickclover, in young stage of growth, is browsed by livestock and soon disappears from ranges. Thus, it is most frequently encountered in prairie remnants along roads and in other relic areas.

7. ***Desmodium marilandicum*** (L.) DC., Maryland tickclover. Herbaceous perennials with stout taproot and knobby or shortly branching caudex; stems erect, slender, several from base, simple or branching below, 4–10(14) dm tall, striate, glabrous to moderately uncinulate-puberulent. Leaves alternate, pinnately trifoliolate; (1.2)1.5–3 cm long; stipules lance-attenuate, 2–4.5 mm long, early or tardily deciduous; stipels 1–2 mm long; leaflets glabrous above, glabrate or sparsely pubescent beneath; blade of terminal leaflet ovate to oval, 1.9–2.7 cm long, 1–1.5 cm wide, obtuse and mucronulate, rounded to subtruncate basally; lateral leaflet blades similar but more elliptic to deltoid-ovate, 1.5–2.5 cm long. Inflorescence usually terminal, somewhat diffusely racemose-paniculate; rachis uncinulate-puberulent; bracts ovate, acuminate, 2 mm long, early deciduous. Pedicels 1–2 cm long; calyx tube campanulate, 1 mm long; upper lobe notched, 1.5–2 mm long; middle tooth of lower lobe 2–3.5 mm long and lateral teeth 1.5–2.5 mm long, calyx minutely puberulent throughout and scattered pilose; corolla reddish, fading purplish or bluish-green, 5–7 mm long; stamens diadelphous. Loments with (1)2(3–4) obovate segments, 4.5–5 mm long, 3–4 mm wide, convex above, obtusely rounded below, surface and margins uncinulate-pubescent; seeds 2–2.5 mm long, olivaceous to brown, smooth. (n = 11) Jul–Sep. Local colonies in dry acid soils of rocky or sandy open woodlands, hillsides, ridges, less often in open sandy prairies; extreme se GP; (MA to Ont., MI, IL, MO, se KS, s to GA, TX). *Meibomia marilandica* (L.) DC.—Rydberg.

See *D. ciliare* for note on relationship with that species.

8. ***Desmodium nudiflorum*** (L.) DC., scapose tickclover. Perennial herbs with taproots but eventually fibrous root system and knobby caudex; stems erect or ascending, usually branched at base; central stem leafy and flowerless; lateral 1–4 branches usually flowering scapes, these rarely with 1–3 leaves; flowering scapes slightly to 3 × the length of leafy stems; leafy stems usually 1–3 dm tall, glabrate or sparsely pilose, bearing a whorl or scattered cluster of leaves at apex. Leaves pinnately trifoliolate, 4–9; petioles 4–13 cm long, glabrous or sparsely pilose; stipules early deciduous; stipels obsolescent; terminal leaflet blade ovate, elliptic or rhombic-ovate, sometimes abruptly acuminate, 4–12 cm long, 3–8 cm wide, sparsely puberulent and pilose above, glaucous and sparsely pilose below; lateral blades inequilateral basally, semiovate, a little smaller. Inflorescence leafless except in occasional plants, racemose or racemose-paniculate, axis sparsely pilose below and uncinulate-pubescent above or glabrate; bracts linear-lanceolate, 5 mm long, early-deciduous. Pedicels 10–25 mm long; calyx tube 1.7–2.5 mm long, glabrous or puberulent, teeth barely evident; petals pink to purple, rarely white, 5–7(9) mm long; stamens monadelphous. Loment stipe 9–12 mm long, segments 2–4, semiovate, 6–8(12) mm long, 4–5 mm wide, margins glabrous, upper nearly straight, surfaces uncinulate-puberulent; seed 5–6 mm long, flat, fragile, smooth, yellowish. (n = 11) Jun–Aug. Rare in our area in rocky acid soils of open oak-hickory wooded hilltops & slopes; KS: Cherokee; (NY, OH, IL, MO, se KS, s to FL, AL, TX). *Meibomia nudiflora* (L.) O. Ktze.—Rydberg.

9. *Desmodium obtusum* (Muhl. ex Willd.) DC. Perennial herbs with taproots and superficial or subterranean knobby or shortly branching caudex; stems erect, 1–several from base, 0.5–1.2 m tall, usually simple to inflorescence, somewhat densely uncinulate-puberulent with spreading tapering hairs only near nodes. Leaves alternate, palmately trifoliolate; leaflets narrowly to broadly ovate to lanceolate, blades 4.5–7.5 cm long, 1.8–3.5 cm wide, sparsely to densely short-pubescent, acute to obtuse and mucronate at apex, rather thick and conspicuously reticulate-veined below; stipels lance-attenuate, soon deciduous; petioles 0.5–2 cm long, densely uncinulate-puberulent; stipules ovate-attenuate, 3–6 mm long, early deciduous. Inflorescence usually racemose-paniculate, densely uncinulate-puberulent, sometimes with long gland-tipped hairs. Pedicels 4–20 mm long; bracts ovate-attenuate, usually subtending 3–5 flowers; calyx tube campanulate, 1–1.5 mm long, middle tooth of lower lobe 2–3 mm long, lateral teeth 1.5–2.5 mm long; petals purplish, blue, or white, 5–6(8) mm long. Loment with stipe 1.5–3.5 mm long, segments 1 or 2(3–4), upper margin convex, lower rounded-obtuse, 4–5 mm long, 2.5–4 mm wide, margins and surfaces uncinulate-puberulent; seed reniform, 3–3.5 mm long, brown, shiny. (n = 11) Aug–Sep. Very rare in dry, rocky open woodlands & rocky prairies; KS: Linn; MO: s to Missouri R.; e OK; (MA, NY, IL, se KS, s to SC, FL, TX). *Meibomia rigida* (Ell.) O. Ktze.—Rydberg; *D. rigidum* (Ell.) DC.—Fernald.

10. *Desmodium paniculatum* (L.) DC. panicled tickclover. Perennial herbs from rather thick taproot and knobby to shortly branching caudex; stems erect or ascending, 0.5–1.5 m tall, 1–several from base, usually with spreading branches, glabrous, uncinulate-puberulent or moderately spreading-pilose. Leaves alternate, pinnately trifoliolate; petioles 1.5–6 cm long; stipules 2–6 mm long, subulate to lance-attenuate, early deciduous or semipersistent. Leaflets quite variable; terminal leaflet 3–8 × as long as wide, narrowly lanceolate, elliptic, or oblong, undersurface glabrate or with appressed hairs, or terminal leaflets 1.5–2.5 × as long as wide, lower surface mostly spreading or loose pilose, apically acute, obtuse, or shortly acuminate; lateral leaflets similar but smaller. Inflorescence axillary and terminal, racemose-paniculate and often diffuse, bracts small and early deciduous. Pedicels 4–11 mm long, ascending, uncinulate-puberulent; calyx tube 1.5–2 mm long, minutely puberulent, teeth pilose, upper lobe entire or notched, 1.5–3.5 mm long, middle tooth of lower lobe 2.5–5.5 mm long, lateral teeth 2–4 mm long; flowers 5–8 mm long; petals lavender, red-purple, rarely white, fading blue- or greenish-purple. Loments with stipes 2–3.5 mm long, segments 3–6, 5.5–8 mm long, triangular to rhombic or subrhombic, usually slightly curved above, angled or obtusely rounded below, margins and surface uncinulate-puberulent; seed 4–4.5 mm long, light brown, smooth. (n = 11) Jul–Sep. Infrequent to locally common in rich or rocky upland woods, stream valleys, prairie ravine thickets, pond shores, roadsides; se NE, e 1/2 KS, OK, e TX panhandle; (ME, Ont., MI, WS, MN, e NE, s to FL, TX). *Meibomia paniculata* (L.) O. Ktze., *M. dillenii* (Darl.) O. Ktze., *M. pubens* (T. & G.) Rydg.—Rydberg; *D. dillenii* Darl.—Gates; *D. paniculatum* var. *dillenii* (Darl.) Isely, *D. glabellum* (Michx.) DC., *D. perplexum* Schub.—Fernald.

Our plants are extremely variable as regards leaflet shape and pubescence. The complex has been recognized as one polymorphic species, as 4 species, or a species consisting of 2 varieties. In our area the extreme phenotypes appear distinct and may be recognized as follows:

10a. var. *dillenii* (Darl.) Isley. Terminal leaflets mostly 1.5–2.5(3) × longer than wide; pubescence of lower surface spreading-pilose. Stems with a mixture of uncinulate puberulence and pilose hairs.

10b. var. *paniculatum*. Terminal leaflets 3–8 × longer than wide; pubescence of lower surface sparse, appressed. Stems glabrate or uncinulate-puberulent.

In most habitats these distinctive elements are obscured by phenotypes blending these characters in all possible combinations, and populations appear as hybrid swarms. In general, var. *dillenii* is found in more open and drier sites and var. *paniculatum* on more moist and shaded places. The group merits more critical field and experimental study.

11. ***Desmodium pauciflorum*** (Nutt.) DC., few-flowered tickclover. Perennial herb with taproots and branching root system and slender shortly branching caudex, lateral roots with round nodules 1-3 mm in diam, larger ones with fusiform thickenings 0.5-1.5 cm long and 3-7 mm wide; stems erect to decumbent, 2-6 dm long, often branched, glabrous to sparsely puberulent or spreading-pilose. Leaves alternate, pinnately trifoliolate, scattered from near base upward; petioles 5-7 cm long; stipules linear-lanceolate, 3-7 mm long, very early deciduous; stipels obsolescent; terminal leaflet rhombic-ovate, 5-9 cm long, 4-6 cm wide, apically acute or abruptly acuminate, cuneate basally, whitened below, sparsely pubescent; lateral leaflets ovate, inequilateral at base, 3.5-6 cm long, 3-6 cm wide. Inflorescences terminal or lateral, racemose or slightly branched; bracts linear-lanceolate 1-4 mm long, early deciduous. Pedicels 4-6(12) mm long, erect or recurved; calyx tube 1.5-2 mm long, puberulent, teeth very short; flowers 4-6 mm long; petals white; stamens monadelphous. Loment stipe 4-6 mm long, segments 1 or 2(3), more or less triangular, 9-12 mm long, 5-8 mm wide, margins glabrous or scattered pilose, surface uncinulate-puberulent; seed 6-6.5 mm long, 4 mm wide, flat, dark brown, smooth (n = 11) Jun-Sep. Rare but locally common in rich or rocky open woodlands & stream valleys; KS: Cherokee; (NY, OH, IL, s MO, se KS, s to FL, AL, LA, TX). *Meibomia pauciflora* (Nutt.) O. Ktze.—Rydberg.

In se KS this species flowers sporadically during the season and in some years does not flower at all. Nodules on the roots are always present and often abundant.

12. ***Desmodium rotundifolium*** DC., dollarleaf. Perennial herbs with taproot and slender, shortly branching caudex; stems prostrate, trailing, branched from base, 0.5-1 m long, forming mats 2 m wide, spreading-pilose. Leaves alternate, pinnately trifoliolate; petioles 2-5 cm long, pilose; stipules ovate-acuminate, 5-10 mm long, ciliate, amplexicaul at base, persistent; stipels linear, 1-3 mm long; terminal leaflet orbicular to widely rhombic or obovate, 3-7 cm long, appressed to spreading-pilose; lateral leaflets somewhat inequilateral, truncate to subcordate basally, 2.5-4 cm long. Inflorescences primarily axillary, racemose to racemose-paniculate, pilose; bracts ovate-acuminate, 3-7 mm long, persistent. Pedicels 6-13 mm long, uncinulate-puberulent; calyx tube 1.5-2 mm long, puberulent to sparsely pilose, upper notched lobe 3-5 mm long, middle tooth of lower lobe 3-5 mm long, lateral teeth 3.5-4.5 mm long; petals 8-10 m long, purple. Loment stipe 3-6 mm long, segments 3-6, 5-7 mm long, 4-5 mm wide, elliptic or subrhomboidal, margins and surface uncinulate-puberulent; seed 2.5-3 mm long, brown, smooth. (n = 11) Jul-Sep. Acid soils of dry open woodland hilltops & slopes; KS: Cherokee; (MA, s VT, NY, s Ont., MI, IL, MO, s FL, TX). *Meibomia michauxii* Vail—Rydberg.

13. ***Desmodium sessilifolium*** (Torr.) T. & G., sessile-leaved tickclover. Perennial herbs from stout taproot and knobby caudex; stems erect or ascending, 1-several from base, striate, densely uncinulate-puberulent, simple to inflorescence, rarely branched below. Leaves alternate, trifoliolate; petioles 0-3 mm long; stipules lance-attenuate, 4-10 mm long, rather persistent; terminal leaflets 3.5-9 cm long, 7-18 mm wide, narrowly elliptic to oblong-lanceolate, rounded at base and apex, mucronate, rather prominently reticulate and short-pilose below, uncinulate-puberulent above, lateral leaflets 3-6 cm long. Inflorescence racemose-paniculate with virgate branches; bracts ovate-attenuate, 2.5-3.5 mm long. Pedicels 1-4.5 mm long; calyx tube 1-1.5 mm long, puberulent, upper lobe entire or notched, 2.5 mm long, middle tooth of lower lobe 2.5-3.5 mm long, lateral teeth 2-3 mm long; petals pinkish, pale lavender, or white, fading yellowish-white, 5-6 mm long. Loment stipe 1.5-3 mm long, segments (1)2 or 3(4), 4-6 mm long, 3-4 mm wide, rounded above and below, margins and surface uncinulate-puberulent; seed 3-3.5 mm long, olivaceous to brown, smooth. (n = 11) Jul-Sep. Locally common in dry rocky open woodlands, prairie hillsides & ravines, dunes, & sandy stream valleys, roadsides; e 1/2 KS, OK; TX:

Potter; (MA, MI, IL, MO, KS, s to SC, GA, AL, TX). *Meibomia sessilifolia* (Torr.) O. Ktze. — Rydberg.

In its young growth stages this plant is readily eaten by livestock, and it soon disappears from grazed areas. Some colonies have white flowers at anthesis, others have pale lavender or pinkish, but flower color is not correlated with other characters.

12. GALACTIA P. Br., Milk-pea

1. ***Galactia regularis*** (L.) B.S.P., downy milk-pea. Perennial prostrate, trailing, twining, or climbing herbaceous vines from taproots and slender, shortly branching caudex; stems 5–10 dm long, pilose with spreading to reflexed hairs, more rarely with retrorse canescent hairs. Leaves alternate, pinnately 3-foliolate; leaflets oblong, oblong-ovate or elliptic (1)2–4(6) cm long, stipulate, usually strigose-pilose above to glabrate, strigose, or short-pilose beneath; petioles 0.5–4 cm long, pilose; stipules 1–3 mm long, early deciduous. Peduncles axillary, 1 or 2 per node, 1–40 mm long, spreading-pilose; racemes 3–10(15) cm long, spreading-pilose, flowers not congested, pedicels 1–4 mm long. Calyx greenish-yellow, tube campanulate, 2–2.5 mm long, spreading short-pubescent, lobes 2–3.5 mm long, upper 2 lobes connate; petals pink or roseate; banner 6–9 mm long; keels and wings 6–8 mm long. Pod 2–5.5 cm long, 4–5 mm wide, compressed, nearly straight, with scattered divergent to antrorse-appressed hairs, coiling during dehiscence; seed 3–4 mm long, 1.5–2 mm wide, yellow to dark brown, often purple-mottled, smooth. (n = 10) Jul–Sep. Infrequent to locally common on rocky open wooded slopes; extreme se GP; KS: Chautauqua, Cherokee, Crawford; (NY, NJ, PA, IN, MO, se KS, s FL, AL, TX). *G. mississippiensis* (Vail) Rydb. — Rydberg, *G. volubilis* (L.) Britt. var. *volubilis* and var. *mississippiensis* Vail — Fernald.

Duncan (Sida 8(2):170–180. 1979) has shown that the name of our plant should be as here indicated.

13. GLYCYRRHIZA L., Licorice

1. ***Glycyrrhiza lepidota*** Pursh, wild licorice. Herbaceous perennial. Stems 3–10(12) dm tall, erect, striate when dry, minutely pubescent to glabrous, glandular, arising from deep, extensive woody rhizomes. Leaves alternate, odd-pinnate; leaflets 7–21, oblong, lanceolate or oblong-lanceolate, 2–5(7) cm long, mucronate pointed, sprinkled with small scales when young and with corresponding dots later, glandular-viscid; petioles 0.5–5 cm long; stipules 3–7 mm long, slender, deciduous. Peduncles axillary, 1–7 cm long; racemes spiciform, erect, flowers numerous. Calyx tube tubular-campanulate, 2–2.5 mm long, lobes 2.5–3 mm long, upper 2 teeth connate 1/2 length or more, tube and teeth with stalked glands; petals ochroleucous or whitish; banner 10–14 mm long; wings and keel 8–12 mm long; stamens 10, diadelphous. Pods 1–2 cm long, densely beset with hooked prickles, tardily dehiscent; seeds 3.2–4 mm long, olivaceous or brown, smooth. (n = 8) May–Aug. Infrequent to locally abundant in prairie ravines, stream valleys, lakeshores, moist areas, roadsides; GP, less frequent e; (MN, Alta., WA, s to AR, TX, NM, AZ, CA).

The root of this plant is sweet and somewhat licorice-flavored and was eaten raw or baked by the Indians. In MT, WY, and the BH some plants have stalked glands throughout the inflorescence and on petioles and upper main stem. These are the var. *glutinosa* (Nutt.) S. Wats., which otherwise is to the west of our area.

14. HEDYSARUM L., Hedysarum, Sweet-broom

Herbaceous perennials with a woody taproot and shortly branching caudex; stems usually several, erect to decumbent, usually branched above, more or less longitudinally grooved, usually appressed-pubescent. Leaves alternate, odd-pinnate; leaflets nearly sessile, puncticulate above; petiolate; stipules connate to distinct. Inflorescence axillary, peduncled, racemose. Flowers erect to reflexed, pink to purple or rarley white; calyx campanulate, 5-toothed; banner and wings shorter than keel, wings with basal auricles; stamens 10, diadelphous. Fruit a loment, segments elliptical to suborbicular, valves areolate.

Reference: Rollins, Reed C. 1940. Studies in the genus *Hedysarum* in North America. Rhodora 42: 217–239.

1 Calyx teeth clearly unequal, deltoid, shorter than tube; segments of loment wing-margined, reticulations on surface about as wide as long; leaflets with rather distinct secondary veins. .... 1. *H. alpinum*
1 Calyx teeth nearly equal, linear-subulate, longer than tube; segments of loment wingless, reticulations laterally elongated; leaflets with secondary veins very obscure ...... 2. *H. boreale*

1. Hedysarum alpinum L., sweet-broom. Perennial herbs with taproots and shortly branching caudex; stems clustered; erect or ascending, often branched above, 2–8 tall, striate, finely strigose or glabrate. Leaves alternate, odd-pinnate; petiolate below to subsessile above; stipules connate, 5–18 mm long, lance-acuminate, brown, canescent, notched to deeply lobed apically; leaflets (9)13–21, broadly lanceolate to oblong, obtuse to rarely acute, apiculate, 10–35 mm long, 5–10 mm wide, glabrous and obscurely brown-puncticulate above, sparsely canescent below. Peduncles axillary, 5–15(20) cm long; racemes densely flowered, often somewhat secund, flowers usually deflexed. Calyx tube campanulate, 2–2.5 mm long, teeth unequal 1–2 mm long, upper shorter than lower, pubescent; petals pink to reddish-purple, rarely white; banner 10–15 mm long, obovate, emarginate; wings 10–13 mm long, 2 mm wide, claw 2–3 mm long, wing-auricles linear, united beneath standard, equaling claw; keel 13–18 mm long; stamens diadelphous. Loment stipitate, pendulous, segments (1)2–5, 5–7 mm long, 3.5–6 mm wide, glabrous to appressed-pubescent, with wing margin to 0.5 mm, reticulations about as wide as long; seed 3.5–4 mm long, 2–2.5 mm wide, brown, smooth (n = 7). Jun–Aug. Infrequent to locally common on rocky open wooded hillsides, creek banks, moist meadows; SD: Custer, Lawrence, Pennington; WY: Albany, Weston; (Newf. to n ME & VT, Man. to B.C., s to SD, WY; Europe). *H. americanum* (Michx.) Britt., *H. boreale*—Rydberg; *H. alpinum* var. *philoscia* (A. Nels.) Rollins—Rollins, Rhodora 42: 224. 1940.

Our plants have been named var. *philoscia* by Rollins and separated from other American forms by having loments densely pubescent on both surfaces. Most of our specimens, however, have loments glabrous; some have margins pubescent, and a few have surfaces somewhat densely strigose. These variants are to be found in the same populations and thus our plants would appear better referred to var. *americanum* Michx. ex Pursh.

Reports of *H. occidentale* Greene for the Black Hills were based on specimens of *H. alpinum* with larger loment segments, and it is a plant to the west of our area.

2. Hedysarum boreale Nutt., sweet-broom. Perennial herbs with taproots and shortly branching caudex; stems several to many, erect or ascending, 2–4(6) dm tall, striate, often branched above, glabrate to silvery-canescent. Leaves alternate, odd-pinnate; petiolate below to subsessile above; stipules connate below to nearly distinct above, triangular with a subulate tip, 3–10 mm long; leaflets 7–13(17), linear-oblong, elliptic, or rarely obovate, 1–3 cm long, 3–7 mm wide, obtuse, glabrous or pubescent above to densely pubescent on both surfaces, always brown-puncticulate above. Peduncles axillary, often appearing terminal, 0.5–1.5 dm long; racemes 5- to many-flowered, compact to elongate, flowers erect, spreading, or lower tardily reflexed. Calyx tube campanulate, 2–2.5 mm long, pubescent, teeth nearly equal, subulate, 3–5 mm long; petals reddish-purple, rarely white; banner obovate to cuneate-emarginate, 12–18 mm long, 7–12 mm wide; wings 10–14 mm long, 2.5–4 mm wide, claw wide, 2–3 mm long, auricle 1 mm long, distinct; keel 13–20 mm long; stamens 10, diadelphous. Loments pendulous to somewhat divaricate, segments 2–5, 6–8 mm long, 5–7 mm wide, orbicular, appressed-pubescent, not wing-margined, reticulations laterally elongated, short tubercles often present on segment face; seed 3-3.5 mm long, dark brown, often shiny. (2n = 16) Jun-Aug. Locally common on dry, rocky prairie plains & hillsides, sandy stream valleys, open wooded hillsides, roadsides, & waste places; w 1/2 ND, MT, e WY, CO; OK: Cimarron, Ellis, Roger Mills; TX: Hutchinson; NM; (Man. to Sask., s to ND, WY, CO, OK, TX, NM, AZ, UT, OR). *H. cinerascens* Rydb., *H. mackenzii* Richards.—Rydberg.

Many specimens in ND with leaflets and stems silvery-pubescent and with short tubercles on the loment segments could be referred to var. *cinerascens* (Rydb.) Rollins, but all degrees of intergradation to var. *boreale* with glabrous to sparsely pubescent upper leaflet surfaces and lack of tubercles are to be found, even in the same population.

15. INDIGOFERA L., Indigo

1. *Indigofera miniata* Ort. var. *leptosepala* (Nutt.) B. L. Turner, scarlet pea. Perennial herbs with woody taproot and superficial or subterranean, shortly branching caudex; stems procumbent or prostrate to somewhat ascending, 1-8(12) dm long, gray-pubescent with appressed dolabriform hairs. Leaves alternate, odd-pinnate; leaflets 5-9, alternate on rachis, oblanceolate to obovate, 1-2(3) cm long, acute or apiculate, gray-pubescent with appressed hairs; petioles short; stipules setaceous, 6-9 mm long. Peduncles axillary, 3-8 cm long, usually surpassing subtending leaf; racemes loosely flowered. Calyx tube 1.5-2 mm long, teeth 4.5-6 mm long; petals brick-red to pinkish; banner 10-14(20) mm long; wings and keel 8-12 mm long; stamens 10, diadelphous. Pod 1-3 cm long, 2-3 mm in diam; seed 3-3.5 mm long, more or less prismatic, often purple-dotted. (n = 16) May-Aug. Infrequent to abundant on rocky or sandy prairies, dunes, stream valleys, roadsides; s-cen KS, OK, TX; (s-cen KS, OK, e 2/3 TX, rare w to e panhandle). *I. leptosepala* Nutt.—Rydberg.

16. LATHYRUS L., Vetchling

Ours annual or perennial herbs with taproots, short caudex, or rhizomes; stems usually trailing or climbing vines, or erect. Leaves alternate, pinnate with rachis extending into a tendril or short spine; leaflets 2-18, paired or scattered; stipules persistent, usually semisagittate, lobes entire or toothed. Flowers few to several in axillary pendunculate racemes; calyx with campanulate tube, lobes and teeth often unequal; corolla papilionaceous, purplish, pink, or ochroleucous; banner usually with blade and claw, blade reflexed; wings free from keel; stamens 10, diadelphous; style flattened and hairy on upper surface for 1/2 length, persistent. Fruit a linear to oblong 2-valved pod, thin-walled or coriaceous, usually somewhat flattened, promptly or tardily dehiscent *(L. venosus).*

Reference: Hitchcock, C. Leo. 1952. A revision of the North American species of *Lathyrus.* Univ. Wash. Publ. Biol. 15: 1-104.

1 Leaflets of mature leaves 2.
  2 Plants annual; petioles wingless; flowers less than 6-12 mm long; pods 2-5 cm long .... 5. *L. pusillus*
  2 Plants perennial; petioles winged; flowers 15-25 mm long; pods 6-10 cm long .... 1. *L. latifolius*
1 Leaflets of mature leaves 4 or more.
  3 Tendrils bristlelike, simple, not prehensile .... 4. *L. polymorphus*
  3 Tendrils branched, at least some prehensile.
    4 Stipules with basal lobe broadly rounded; flowers white or ochroleucous .... 2. *L. ochroleucous*
    4 Stipules with sharp or narrowed basal lobe; flowers red-purple.

5   Leaflets usually (4)6(8); racemes with 2–6(9) flowers .... 3. *L. palustris*
5   Leaflets usually (8)10–14; racemes with (5)10–30 flowers .... 6. *L. venosus* var. *intonsus*

1. **Lathyrus latifolius** L., perennial sweetpea, everlasting pea. Herbaceous perennial herbs from rhizomes; stems climbing or trailing, 1–2 m tall, glabrous, slightly glaucous, conspicuously 2-winged. Leaves pinnate, the rachis terminating in a stout much-branched tendril; leaflets 2, lanceolate, elliptic or oblong, usually 4–9(15) cm long, 1–5 cm wide, glabrous; stipules (1)3–5(8) cm long, lanceolate to ovate, upper lobe 2–3 × as long as lower; petioles 2-winged. Peduncles axillary, often exceeding the leaves; racemes 5- to 15-flowered. Flowers odorless; pedicels 1–2.5 cm long; calyx tube campanulate, 4–6 mm long, teeth of upper lobe triangular, 3–3.5 mm long, middle tooth of lower lobe lance-attenuate, 6.5–7.5 mm long, lateral teeth 4.5–5.5 mm long; petals 15–25 mm long, usually purple, or red, pink, white, sometimes striped; banner nearly as wide as long, short clawed like wings and keel. Pod 6–10 cm long, 7–10 mm wide, flattened, glabrous; seed oblong, 5 mm long, surface dark brown and rugose. (n = 7) May–Sep. Infrequent along roadsides, fence rows, & waste places, usually near homesites; scattered in e 1/2 GP; (escaped from cultivation over much of U.S., Europe).

Most colonies of this species persist for several years but eventually disappear in our area.

2. **Lathyrus ochroleucous** Hook., yellow vetchling. Perennial herbs from slender rhizomes; stems climbing or sprawling, slender, wingless, 3–8(10) dm tall, glabrous. Leaves alternate, pinnate, rachis terminating in a branched tendril; leaflets (4)6(8), ovate to lance-ovate, or obovate, 2–5(7) cm long, glabrous, glaucous below, 1–4 cm wide; petioles 1–3 cm long; stipules semicordate, ovate to ovate-lanceolate, 1.5–3 cm long, rounded basal portion entire to dentate, not constricted into 2 lobes. Peduncles axillary, shorter than leaves; racemes mostly with 5–10(15) flowers. Pedicels 1–6 mm long; calyx tube campanulate, 4–5 mm long, teeth of upper lobes triangular, 1–2 mm long, middle tooth of lower lobe lanceolate, 4–5 mm long, lateral teeth ovate-lanceolate, 3–4 mm long; petals white to ochroleucous; banner 14–18 mm long, claw 5–7 mm long; wings and keel shorter than banner. Pods 4–5(7) cm long, 4–6 mm wide, glabrous; seed 3–3.5 mm long, olivaceous to brown, smooth. (n = 7) Jun–Aug. Infrequent to locally common in dry rocky woodlands, brushy ravines, stream valleys, roadsides; MI, ND; SD: Custer, Lawrence, Marshall, Meade, Pennington; WY: Crook, Weston; NE: Box Butte, Dawes, Sheridan, Sioux; (VT, Man., nw Can., s to PA, IA, NE, ID, WA).

3. **Lathyrus palustris** L., marsh vetchling. Perennial herbs from rhizomes; stems climbing, 3–10 dm tall, angled or narrowly winged, glabrous or sparsely puberulent. Leaves alternate, pinnate, terminating in a branched tendril; leaflets 4–6(8), linear to elliptic, 2–8 cm long, 3–20 mm wide, glabrous or pubescent; petioles 1–3 cm long; stipules 1–3 cm long, with upper and lower lobes entire or dentate. Peduncles axillary, shorter than or exceeding leaves; racemes 2- to 5(9)-flowered. Pedicels 2–6 mm long; calyx tube campanulate, 3–4 mm long, glabrous or pubescent, teeth of upper lobe triangular, 1–2 mm long, middle tooth of lower lobes lanceolate, 4–6 mm long, lateral teeth 3–5 mm long; petals reddish-purple or rarley whitish; banner 12–20 mm long, wings and keel nearly as long. Pods 4–6 cm long, 4–5 mm wide, with short red glandular hairs; seeds 3–3.5 mm long, olivaceous, mottled. (n = 7) Jun–Jul. Infrequent to locally common in low prairies, stream valleys, lakeshores; n & e ND, MN, e SD, NE, IA, nw MO; (circumboreal, s to NJ, PA, OH, MO, NE, CO, CA). *L. macranthus* (White) Rydb.—Rydberg.

Occasional plants have calyx and leaves somewhat pubescent and have been recognized as var. *pilosus* (Cham.) Ledeb. but this distinction does not seem significant.

4. **Lathyrus polymorphus** Nutt., hoary vetchling. Perennial herbs from rhizomes and subterreanean shortly branching caudex; stems erect or ascending, usually branched, 1–3(5)

dm tall, glabrous to densely hirsute-pilose or villous, angled but not winged. Leaves alternate, pinnate, rachis prolonged as a straight or curved bristle, without tendrils; leaflets 4–8(12), paired to scattered, 1.5–3(5) cm long, 1–3(6) mm wide, linear-elliptic, linear-lanceolate, or oblong-elliptic, usually prominently veined, glabrous or pubescent; petioles 0.5–1.5 cm long; stipules 7–20 mm long, upper lobe lanceolate, lower shorter and triangular or lanceolate. Peduncles axillary, 6–7 cm long, usually surpassing leaves; racemes 2- to 5(8)-flowered. Flowers fragrant; calyx tube 4–6 mm long, teeth of upper lobe triangular, 2–3 mm long, middle tooth of lower lobe lanceolate, 4–6 mm long, lateral teeth 3–4 mm long; petals purple, or banner purple and wings bluish or white, or rarely corolla ochroleucous; banner 20–27 mm long, obcordate, blade reflexed, claw not differentiated; wings and keel 17–23 mm long, claws ca as long as blade. Pods 2–6 cm long, 5–10 mm wide, glabrous, coriaceous; seeds nearly spherical, 5–6 mm in diam, brown, smooth. (n = 7) May–Jun. Locally common on dry, sandy prairie hillsides & plains, rocky open wooded areas, sand dunes, stream valleys; s 2/3 GP but infrequent in e 1/4; (sw SD, NE, e WY, w 2/3 KS, CO, n TX). *L. hapemanii* A. Nels. *L. incanus* (Sm. & Rydb.) Rydb., *L. decaphyllus* Pursh, *L. stipulaceous* (Pursh) Butt. & St. John—Rydberg.

Our glabrous plants are recognized as subsp. *polymorphus* and are the common plants in the eastern 1/2 of the GP. The pubescent plants of more western distribution are subsp. *incanus* (Sm. & Rydb.) C. L. Hitchc. It is not unusual, however, to find mixed populations with intermediate plants common. The variations in corolla color do not merit recognition.

Seeds of this species are reported to be poisonous to livestock.

5. Lathyrus pusillus Ell., singletary vetchling. Annual herb, with slender taproot; stems prostrate to clambering or erect, 2–4(7) dm long, 2-winged, glabrate to sparsely pubescent. Leaves alternate, pinnate with rachis extended into a branched tendril; leaflets 2, narrowly lanceolate to narrowly elliptic, 2–5(7) cm long, glabrate; petioles 1–2 cm long, wingless, or with ridges; stipules lanceolate to lance-ovate, 1–3 cm long, upper lobe 2–3 × as long as lower. Peduncles axillary, 1–5 cm long; racemes usually 2-flowered. Pedicels 1–3 mm long; calyx tube glabrous, 2–2.5 mm long; lobes nearly equal, linear-lanceolate, 3–5 mm long; petals purple, banner obcordate, 6–10(12) mm long, claw hardly evident; wings and keel narrow, claws shorter than blades. Pods linear, 2–5 cm long, 3–5 cm wide, glabrous, valves twisting in dehiscence; seed nearly spherical, 2 mm in diam, olivaceous to dark brown, surface rugulose. (n = 11) May–Jul. Infrequent to locally common in low moist pastures, roadsides; KS: Cherokee, Labette, Montgomery, Neosho; (NC, se KS, s to FL, TX, n Mex.).

It is possible that this species was introduced into se KS, as it has been grown as a cover crop in the Gulf States. It has been reported to be responsible for livestock poisoning.

6. Lathyrus venosus Muhl. ex Willd. var *intonsus* Butt. & St. John, bushy vetchling. Rather coarse perennial herbs from rhizomes; stems sprawling to erect, 0.4–1(2) m tall, angled but not winged, sparsely to densely hairy, branched. Leaves alternate, pinnate, rachis terminating in a branched tendril; leaflets 8–14, usually scattered, more rarely paired, 3–6 cm long, 1–3 cm wide, narrowly to broadly elliptic or ovate-elliptic, paler on lower surface, densely to sparsely pubescent; petioles 0.5–3 cm long; stipules 0.5–3.5 cm long, linear-lanceolate to lanceolate. Peduncles axillary, 1/2 to as long as leaves; racemes with 5–20(30) rather densely clustered flowers. Calyx tube 3.5–4.5 mm long, pubescent to glabrate, teeth of upper lobe 1–1.5 mm long, middle tooth of lower lobe 3–4.5 mm long, lateral ones 3–3.5 mm long; petals purplish; banner 12–22 mm long, obcordate, claw almost as long as blade; wings a little shorter than banner; keel shorter than wings. Pods 4–6 cm long, 5–8 mm wide; seed 4–5 mm long, 2–3 mm wide, dark brown, smooth. (n = 7) May–Jul. Infrequent to locally common on rocky open woodlands, brushy prairie ravines, stream valleys, lakeshores, roadsides; ND, e SD, MN, IA; (Que., Sask., s to GA, AL, MO, TX).

Our plants differ from the more eastern var. *venosus* in having calyx and herbage densely to sparesely hairy. Some of our more glabrate plants, however, make this distinction somewhat questionable.

17. LESPEDEZA Michx., Bush Clover, Lespedeza

Ours introduced annual herbs and native perennial herbs with woody rhizomes (often flowering first year); stems erect, procumbent or ascending, often branched; glabrous or with appressed or spreading pubescence. Leaves alternate, pinnately trifoliolate; without stipels; leaflets entire, about equal in shape, often mucro-tipped; stipules persistent, ovate, linear, setaceous or subulate. Flowers in axillary spicate or capitate racemes, or in axillary fascicles, rarely paniculate, borne in pairs. Usually both chasmogamous and cleistogamous flowers present; calyx campanulate, lobes subequal, persistent in fruit; petals purple, pinkish, ochroleucous or white; banner suborbicular to oblong, clawed; wings about equaling keel; stamens diadelphous; style long, filiform or strongly recurved in cleistogamous flowers which also have shorter calyx and much-reduced corolla. Pods 1-seeded.

Reference: Clewell, Andre F. 1966. Native North American species of *Lespedeza* (Leguminosae). Rhodora 68: 359–405.

Identification is sometimes complicated by the presence of hybrids between any two of our native species. Presumed hybrids between *L. capitata* and *L. virginica* are found where the two grow together. Less frequent are the hybrids *L. capitata* x *violacea*, *L. violacea* x *virginica*, and *L. procumbens* x *virginica*.

Most manuals list bush clover as the common name of the genus, but the general public, in our area, refers to it as lespedeza.

1 Plants annual; stipules ovate to ovate-lanceolate.
  2 Stems retrorsely appressed-pubescent; petioles of main stem leaves 1–3 mm long; leaflets not conspicuously ciliate .... 8. *L. striata*
  2 Stems antrorsely appressed-pubescent; petioles of main stem leaves 4–10 mm long; leaflets conspicuously ciliate .... 7. *L. stipulacea*
1 Plants perennial; stipules narrowly subulate to setaceous.
  3 Stems procumbent to weakly ascending.
    4 Stems spreading-pubescent .... 5. *L. procumbens*
    4 Stems appressed-pubescent.
      5 Plants trailing or procumbent; stipules mostly less than 4 mm long; keel included within wings .... 6. *L. repens*
      5 Plants ascending; some stipules 4 mm or more long; keel extending beyond wings .... 10. *L. violacea*
  3 Stems erect or strongly ascending.
    6 Flowers white or ochroleucous, with or without purple spot; calyx equaling or exceeding mature pod; racemes with 10–45 flowers.
      7 Leaflets cuneate; flowers 1–4 in axillary clusters; wings and keel equal in length .... 2. *L. cuneata*
      7 Leaflets not cuneate; flowers in spikes or heads; wings larger than keel.
        8 Larger terminal leaflets more than 1/2 as wide as long .... 3. *L. hirta*
        8 Larger terminal leaflets less than 1/2 as wide as long.
          9 Inflorescence of short-ovoid heads; stalk of terminal leaflet longer than petiole .... 1. *L. capitata*
          9 Inflorescence open and spike interrupted; stalk of terminal leaflet shorter than petiole .... 4. *L. leptostachya*
    6 Flowers violet or purple; calyx 1/2 as long as mature pod or shorter; racemes with 4–14 flowers.
      10 Leaflets linear to linear-oblong, more than 3× longer than wide .. 11. *L. virginica*
      10 Leaflets oblong, elliptic or ovate, less than 3× longer than wide.
        11 Stem and upper surface of leaflets glabrous or glabrate; keel 1–2 mm longer than wing petals .... 10. *L. violacea*
        11 Stem and upper surface of leaflets conspicuously appressed-pubescent or pilose; keel included within wing petals .... 9. *L. stuevei*

1. *Lespedeza capitata* Michx., round-head lespedeza. Perennial herbs with stout subterranean knobby or shortly branching caudex; stems erect or ascending, 0.5–2 m tall; simple or branched above; densely villous or appressed-pubescent and pilose. Leaves alternate, trifoliolate; leaflets elliptic, oblong or narrowly rhombic, 2-5 cm long, glabrous, glabrate, appressed-pubescent or sericeous above, densely pubescent below; stalk of terminal leaflet usually longer than petiole; petioles 1-5 mm long; stipules filiform to narrowly subulate, 3-8 mm long, often persisting after leaf fall. Racemes capitate or short-ovoid, several to numerous, often forming a thyrsoid inflorescence. Pedicels 1-3 mm long; calyx tube 0.5-1 mm long, teeth 7-13 mm long; corolla white or ochroleucous and with purple throat, 7-12 mm long; banner longer than keel, and longer than to equaling wings; pods elliptic to oblong-ovate, usually somewhat asymmetrical, 5-7 mm long, pubescent, conspicuously shorter than calyx; cleistogamous flowers hidden among the others. Pods 4-5 mm long; seed 2.5-3 mm long, olivaceous, brown to black, smooth. (n = 10) Jun–Aug. Frequent on upland prairies, sand dunes, open wooded hillsides, old fields, & roadsides; sw MN, se SD, e 3/4 NE & KS, OK, e TX panhandle; (ME, Que., Ont., MI, WI, MN, NE, s to SC, AL, LA, TX). *L. capitata* var. *vulgaris* T. & G. — Fernald.

   Round-head lespedeza is browsed by livestock and is considered an excellent native legume in ranges. *Lespedeza capitata* hydridizes with *L. hirta* producing *L. x longifolia* DC. — Rydberg; with *L. violacea* producing *L. x manniana* Mack. & Bush — Rydberg; with *L. virginica* producing *L. x simulata* Mack. & Bush — Rydberg; and with *L. intermedia* producing *L. x nuttallii* Darl. — Rydberg.

2. ***Lespedeza cuneata*** (Dumont) G. Don, sericea lespedeza, Chinese bush clover. Perennial but short-lived herb from knobby caudex; stems erect with ascending branches, 0.5-2 m tall, puberulent on ridges, grooves glabrate. Leaves alternate, crowded; leaflets narrowly cuneate, 1-2 cm long, gray-green or silvery-silky beneath, apices of larger leaflets retuse or truncate, apiculate; petioles 0-5 mm long; stipules setaceous, 3-11 mm long. Flowers 1-4 in short axillary clusters. Calyx tube 0.5-1 mm long, teeth 3-5 mm long; petals white or ochroleucous, marked with purple along veins of banner; banner 6-9 mm long, 1-2 mm longer than keel and wings; cleistogamous flowers common, calyx teeth 2-3 mm long. Pods 2.5-3 mm long, glabrate or appressed-pubescent; seed 1.5-2 mm long, olivaceous to brown, often mottled, smooth. (n = 19) Jul–Oct. Local in low open woolands, thickets, stream valleys, around lakes & ponds, waste places & roadsides; se 1/4 GP; (NY, PA, OH, MI, IL, MO, s to FL, TX; e Asia). *Introduced.*

   Beginning in the late 1930s this plant was used for erosion control, wildlife cover and food, and to lesser extent in our area, for forage and hay. It has escaped and apparently naturalized. In some places it is an undesirable weed.

3. ***Lespedeza hirta*** (L.) Hornem., hairy lespedeza. Perennial herbs from knobby and eventually stout branched caudex; stems erect or ascending, 0.5-1.8 m tall, branched above, rather densely spreading-villous or pilose. Leaves alternate; leaflets oval or elliptic to somewhat ovate or obovate, 1.5-4 cm long, 7-30 mm wide, glabrous, appressed-pubescent or pilose above, appressed-pubescent or pilose beneath; principal petioles 5-20 mm long; stipules filiform, 3-6 mm long. Racemes open, short-cylindric, 1-4 × longer than subtending leaves, bearing 16-44 flowers. Pedicels 1-2 mm long; calyx tube 1-2 mm long, teeth 5-8 mm long, eventually equaling or exceeding pod; corolla ochroleucous or white with a purple throat, equaling or exceeding calyx teeth; banner 6-8 mm long. Pod oblong-ovate to elliptic, 4-7 mm long, densely spreading short-pubescent; seed 3 mm long, brown or purplish-black, smooth, shiny. (n = 10) Jul–Oct. Rare in our area on acid soils of rocky open woods & stream valleys; KS: Cherokee; (ME, Ont., MI, IL, s 1/2 MO, se KS, s to FL, AL, ne TX, OK).

4. ***Lespedeza leptostachya*** Engelm., slender spike lespedeza. Perennial herbs from knobby caudex; stems erect or ascending, simple or branched above, 0.5-1 m tall, appressed-pubescent. Leaves alternate; leaflets narrowly oblong, 2-4 cm long, 3-7 mm wide, obtuse and mucronate, sparsely appressed-pubescent above, sericeus below, stalk of terminal leaflet

shorter than petiole; petioles 3–6 mm long. Peduncles 1–2 cm long; spikes or racemes open, slender, interrupted, 2–3 cm long. Calyx 4.5–5 mm long, sepals distinct nearly to base; corolla white, ochroleucous, or light purple; banner and wings about same length, keel somewhat shorter, all about as long as calyx; cleistogamous flowers common. Pods orbicular, 3–4 mm long, densely pubescent, exceeded by calyx; seeds 2.5–3 mm long, olivaceous to light brown, smooth. (n = 10) Jul–Aug. Very rare on gravelly prairies; nw IA, sw MN; (cen MN, s to ne IL, & nw IA). *Endangered.*

This is a very rare plant at the edge of our area. I have only seen the plant in IA: Dickinson at the site reported by Clewell (Rhodora 68: 382.1966).

5. *Lespedeza procumbens* Michx., trailing lespedeza. Perennial herbs with taproot and superficial or subterranean, shortly branching caudex; stems procumbent or trailing, several radiating from base, 3–12 dm long, short-pilose to nearly villous, often dense. Leaves alternate; leaflets ovate, elliptic, or oblong, 1.2–2.5 cm long, 6–13 mm wide, sparsely to densely soft-pubescent; petioles to 2.5 cm long, pilose; stipules filiform, 2–4 mm long, persistent. Peduncles and racemes 2–7 × longer than subtending leaves; flowers usually 8–12 per raceme. Pedicels 0.5–2 mm long; calyx tube 0.8–1.2 mm long, teeth 2–3 mm long, appressed-pubescent or pilose; corolla purple; banner to 8 mm long; banner a little longer than wings and keel; pods elliptic or orbicular, 4–5 mm long, appressed-pubescent to glabrate; cleistogamous flowers in axillary clusters and at apex of racemes, calyx 1–2 mm long. Seeds 2–3 mm long, olivaceous to light brown, smooth, shiny. (n = 10) Aug–Oct. Infrequent to locally common on acid soils of dry woodland hilltops & slopes; extreme se GP; (MA, NH, NY, PA, IL, MO, se KS, s to nw FL, AL, LA, TX).

6. *Lespedeza repens* (L.) Bart., creeping lespedeza. Very similar to *L. procumbens* and differing chiefly in having stems and peduncles sparsely short appressed-pubescent; flowers usually 4–8 per raceme, and cleistogamous pods usually pedunculate. (n = 11) May–Oct. Infrequent to locally common in acid soils of dry rocky woodland hilltops & slopes; e 1/4 KS; (CT, NY, PA, OH, IL, sw WI, e KS, s to FL, AL, LA, TX).

This species often grows mixed with *L. procumbens,* but is usually more common. Some mature herbarium specimens of *L. violacea* are difficult to separate from *L. repens.* Normally the cleistogamous pods of *repens* have subtending calyx extending at least 1/4 of the way up the pod, while the calyx of *violacea* is usually much less. In the field the erect to ascending stem of *violacea* is a good diagnostic character and should be recorded by collectors.

7. *Lespedeza stipulacea* Maxim., Korean lespedeza. Annual herbs with a taproot; stems erect or ascending, bushy-branched, 1–6 dm tall, with sparse to dense antrorsely appressed pubescence, sometimes with longer ascending hairs. Leaves alternate, often appearing subpalmate; leaflets spatulate to obovate, 0.8–2.5 cm long, 2/3 as wide, glabrous or glabrate, conspicuously ciliate; petioles of main leaves 4–10 mm long, glabrate to sparsely antrorse appressed-pubescent; stipules ovate to ovate-lanceolate, those of main leaves 4–8 mm long, striate, glabrous to glabrate, persistent. Flowers 1–3 in leaf axils; chasmogamous and cleistogamous ones intermingled. Calyx tube 1 mm long, teeth ovate, 1 mm long, those of upper lobe nearly completely connate; corollas pink to purple; petals 6–7 mm long. Pod 3 mm long, oval or obovate, apically rounded, strongly reticulate, minutely appressed-pubescent, 1/2 covered by calyx; seed 1.5–2 mm long, brown or black, smooth, shiny. (n = 11) Jul–Oct. Along roadsides, open woods, fields, pastures; se GP; (scattered across e 1/2 of N. Amer., e Asia). *Introduced.*

This species was introduced into the U.S. in 1919 and since the 1930s has been planted as a pasture, hay, and cover crop. Improved strains have been developed. It has spread rather quickly, and the newer variants have progressively become adapted northward. The flowers appear to be self-fertilized and produce abundant seed in both chasmogamous and cleistogamous types.

8. *Lespedeza striata* (Thunb.) H. & A., Japanese or common lespedeza. Annual herbs with taproots; stems erect or nearly prostrate in shade, 1–4 dm tall, bushy-branched, with sparse to dense retrorsely appressed pubescence. Leaves alternate, often appearing subpalmate; leaflets obovate, narrowly elliptic or oblong, 1–2 cm long, 1/3 as wide, glabrous, inconspicuously appressed-ciliate; petioles of main leaves 1–2(5) mm long, retrorsely appressed-pubescent; stipules of main leaves narrowly ovate-lanceolate, 3–6 mm long, striate, glabrous, persistent. Flowers (1)2 or 3(5) in short axillary racemes; pedicels to 1 mm long; chasmogamous and cleistogamous ones intermingled. Calyx tube 1 mm long, pubescent, teeth ovate, ciliate, 1 mm long, teeth of upper lobe connate 1/2–2/3 length; corollas pink to purple; petals 4.5–6 mm long. Pod 3–4 mm long, acuminate, inconspicuously reticulate, 1/2–3/4 covered by calyx; seed 2 mm long, brown or black, smooth, often mottled. (n = 11) Jul–Oct. Infrequent to locally abundant in dry open woods, rocky open areas, gravelly stream banks, roadsides, & waste places; se GP; (NJ, PA, OH, KS s to FL, TX; e Asia).

    This species was introduced into the U.S. in 1846 and has been planted as a hay, pasture, and cover crop. In some of our open woodlands with acid soil, it has become well established and locally abundant.

9. *Lespedeza stuevei* Nutt., tall bush lespedeza. Perennial herb from stout, woody, superficial or subterranean caudex; stems erect or ascending, 0.5–1.5 m tall, densely pilose or appressed-pubescent, ridged. Leaves alternate, pinnately trifoliolate; leaflets usually oblong-elliptic to oblong but less often ovate or obovate, (0.5)1–3 cm long, 1.5–3 × longer than wide, usually spreading-pilosulose above but sometimes sparse and appressed, densley spreading-pilose beneath; petioles 0.5–30 mm long; stipules linear-subulate to setaceous, 3–5 mm long. Chasmogamous flowers in dense racemes in upper axils, on peduncles shorter than or 1.5 × as long as leaves, 6- to 14-flowered; calyx 3–5 mm long, lobes 1.7–3.5 mm long, the upper 2 lobes connate less than 1/2 length; corollas purple, 5–8 mm long; banner and wings about equal, both longer than keel; pods ovate or elliptic, 5–6 mm long, sparse to densely appressed-pilose. Cleistogamous flowers in dense axillary clusters forming most of mature pods on plant; pods 4–7 mm long. Seed 2–3 mm long, olivaceous to dark brown, smooth, shiny. (n = 10) Jul–Sep. Local colonies in dry rocky open woodlands, sandy prairies, & dunes; se GP; (MA, PA, VA, s IN, s 1/2 MO, e KS, s to FL, AL, LA, OK, TX).

    L. intermedia (L.) Britt., has been reported from the GP but is just to the e of our area. It differs in having stems appressed-pubescent, leaflets glabrous above, and pods glabrate. In part, our records of *intermedia* are based on specimens of *L. stuevei* and in part on *L. virginica* specimens with wider than usual terminal leaflets.

10. *Lespedeza violacea* (L.) Pers., prairie lespedeza. Perennial herbs from knobby or shortly branching caudex; stems several from base, erect to arching-ascendent, 2–7 dm tall, branched above, sparsely appressed-pubescent, or with ascending hairs. Leaves alternate, pinnately trifoliolate; leaflets elliptic to broadly oblong, usually 2–5 cm long, about 1/2 as wide, thin, light green below, glabrous above or nearly so, appressed-hairy below; sometimes successive nodes with leaves of different size; petioles of main leaves 1–4 cm long; stipules filiform, 2–6 mm long. Chasmogamous flowers in racemes usually longer than leaves, loosely few-flowered; pedicels 2–6 mm long; calyx 3–6 mm long, teeth 1.5–3 mm long, upper 2 connate above the middle; corolla purple; banner and wings usually distinctly shorter than keel; pods 5–7 mm long, glabrate. Cleistogamous flowers and pods usually present, in axillary clusters; pods 3–6 mm long. Seed 3 mm long, olivaceous to purple-brown, smooth. (n = 10) Jul–Sep. locally common in rocky, open upland woods, thickets, rocky prairie, roadsides; se GP; (MA, NY, MI, sw WI, e 1/2 KS, s to n GA, n MS, ne TX, & OK). *L. prairea* (Mack. & Bush) Britt., *L. frutescens* (L.) Britt.—Rydberg.

    Some plants found in full sun have leaflets about 1/2 as long as usual and are correspondingly narrower and have been described as *L. prairea*. Rare hybrids between *L. violacea* and *L. capitata* are known as *L. x manniana* Mack. & Bush—Rydberg.

11. **Lespedeza virginica** (L.) Britt., slender bush lespedeza. Perennial herbs from stout knobby or shortly branching caudex; stems erect or ascending, usually branched above, 0.3–1 m tall, appressed-pubescent to rarely ascending-pilose. Leaves alternate, pinnately trifoliolate, rather crowded, usually somewhat ascending; leaflets linear to narrowly oblong, 0.6–2.5(4) cm long, 3–7 × longer than wide, glabrate above to appressed-pubescent on both sides; petioles 3–25 mm long; stipules filiform, mostly 3–6 mm long, persistent. Chasmogamous flowers in racemes shorter than or 1.5 × longer than leaves, from upper axils, crowded; calyx lobes 1.7–3 mm long, the 2 upper connate to middle or above; corolla purple, 6–8 mm long, keel usually a little longer than banner and wings; pod 5–7 mm long, appressed-pubescent. Cleistogamous flowers usually in small axillary clusters along midstem' pods 4–5 mm long. Seed 2.5–3 mm long, olivaceous to brown, smooth. (n = 10) May–Sep. Infrequent to common on a variety of soils in rocky or dry open woodlands, sandy stream valleys, thickets, prairies, roadsides; se GP; (NH, MA, s NY, PA, MI, sw WI, se IA, e KS, s to FL, AL, LA, e OK, TX).

Hybrids with *L. capitata* are nearly always present where the two species grow together. Slender bush lespedeza is readily browsed by livestock, and it soon disappears from grazed areas.

18. LOTUS L., Trefoil

Ours annual or perennial herbs; stems erect or decumbent. Leaves alternate, 3- or 5-foliolate; petiolate or usually subsessile; stipules small or glandular. Flowers in pedunculate axillary clusters or solitary, yellow to orange-red or yellowish-white; calyx campanulate, teeth unequal; corolla papilionaceous, petals clawed; banner obovate; wings obovate or oblong and adhering to incurved keel; stamens 10, diadelphous, the filaments dilated just below anthers. Pod terete, oblong to linear, dehiscent. *Acmispon* Raf.—Rydberg.

1 Plants perennial; leaflets 5-foliolate, flowers 3–8, in umbellate clusters.
  2 Leaflets about 1/2 as wide as long or wider, obovate to lanceolate ...... 1. *L. corniculatus*
  2 Leaflets less than 1/2 as wide as long, linear to narrowly lanceolate .... 3. *L. tenuis*
1 Plants annual; leaves 3-foliolate; flowers usually solitary .... 2. *L. purshianus*

1. **Lotus corniculatus** L., bird's-foot trefoil. Perennial herbs with woody taproot; stems several from base, decumbent or erect, to 6 dm tall, glabrous or pubescent. Leaves alternate; pinnately 5-foliolate, the lower pair of leaflets basally on rachis (often interpreted as stipules), the remaining 3 apical; leaflets glabrous to sparsely pubescent, obovate to somewhat broadly lanceolate; stipules small glandlike structures; leaves subsessile. Peduncles axillary, exceeding the leaves; umbels 3- to 8-flowered. Pedicels obsolete or to 1 mm long; calyx tube campanulate or obconic, 2–3 mm long, teeth subequal, 1–2 mm long; petals yellow to orange-red, sometimes fading greenish in drying; banner 12–16 mm long, ovate or roundish; wings oblong 10–14 mm long, claws 3–4 mm long; keel 12–14 mm long, incurved, claw 2–4 mm long; stamens 10, diadelphous. Pods 2.5–4 cm long, straight, terete, twisting in dehiscence; seeds 1.5 mm long, as wide as long, olivaceous to dark brown, often mottled. (2n = 24) May–Sep. Roadsides, fields, waste places; local in GP; (e 1/2 U.S., Pacific Coast; Europe). *Introduced.*

This quite variable species has been planted as a pasture legume and is found as an escape along roads and waste places. It remains to be determined whether it is truly naturalized in much of the GP.

2. **Lotus purshianus** Clem. & Clem., prairie trefoil, deer vetch. Annual herb; stems erect, 2–8(12) dm tall, bushy-branched, glabrous to densely pubescent. Leaves alternate, subsessile, trifoliolate; leaflets lance-elliptic to ovate-lanceolate, mostly 1–2(3) cm long, pubescent, the terminal one stalked; stipules glandlike. Peduncles solitary in the upper axils, 1–2 cm long, with a leaflike unifoliate bract subtending 1(2) flower, bract as long as flower. Calyx tube tubular-campanulate, 1.5–2 mm long, pilose, teeth narrowly lanceolate, 4–5 mm long;

petals white with pink veins, or pinkish with darker veins, rarely yellowish-white; banner 5–7(8) mm long; wings and keel 4–6 mm long. Pod narrowly oblong, 2–4 cm long, glabrous; seed 2.7–3.2 mm long, olivaceous to light brown, often mottled, somewhat shiny. (n = 7) May–Oct. Prairie plains & rocky hillsides, dunes, & sandy stream valleys, open rocky wooded hillsides, roadsides, & waste places; GP but absent in w 1/4 KS & TX panhandle; (MN, ND, Man., B.C., s to AR, TX, NM, AZ, CA, nw Mex.). *Acmispon americanus* (Nutt.) Rydb.—Rydberg; *L. americanus* (Nutt.) Bisch.—Fernald.

Quail have been observed to eat seed of this species, and livestock browse the plant during early stages of growth.

3. ***Lotus tenuis*** Waldst. & Kit. ex Willd., narrow-leaved trefoil. Much like *L. corniculatus* but differing in having leaflets linear to lanceolate, less than 1/2 as wide as long. Umbels 1- to 4(6)-flowered. (2n = 12) May–Sep. Roadsides & waste places; KS: Atchison, Brown, Doniphan, Jackson, Jefferson, Saline; (locally escaped in scattered areas of U.S.; Europe). Introduced.

In recent years this species has appeared on road right-of-ways that have been planted with brome grass. Local escapes appear to persist but are not actively spreading.

19. LUPINUS L., Lupine

Annual or perennial herbs with taproots or developing shortly branched superficial or subterranean caudex; stems erect or ascending, simple or much branched; herbage glabrous, appressed-pubescent or spreading-hairy. Leaves alternate, palmately compound, petiolate; stipules partially adnate to petioles or nearly distinct; leaflets (3)5–9(10). Flowers in terminal racemes, verticillate to scattered, papilionaceous, white, blue, bluish-white, or purple; calyx bilabiate, tube short and asymmetrical, upper lobe slightly bulged to saccate or short spurred at base, bidentate at apex, lower lobe of 3 united teeth, sometimes slightly lobed apically; banner usually with median groove, the sides reflexed; wings usually connivent distally, and usually enclosing the keel; keel falcate, glabrous or ciliate on upper margins; stamens 10, monadelphous, anthers alternately of different size and shape. Pods somewhat flattened, hairy, (1)2- to 12-seeded, commonly somewhat constricted between seeds.

In the BH and adjacent areas, particularly to the w & sw, our perennial lupines are highly variable in stature, leaf size and shape, pubescence, flower size and color, and the shape of the lower portion of the upper calyx lobe. Efforts to sort specimens into meaningful taxa are frustrating but result in recognition of the forms treated below. Many specimens, however, are intermediate between various of the taxa and frequent hybridization is presumed.

1 Plant an annual; seeds and ovules usually 2 .... 4. *L. pusillus*
1 Plant a perennial; seeds and ovules 3 or more.
    2 Either calyx spurred or banner conspicuously hairy over much of the back (adaxial surface).
        3 Calyx spurred; pubescence of banner not extending to upper 1/3 of surface .... 2. *L. caudatus*
        3 Calyx not spurred; banner hairy over at least 2/3 of the back .... 5. *L. sericeus*
    2 Calyx not spurred (may be saccate); banner not conspicuously hairy.
        4 Plant rhizomatose .... 3. *L. plattensis*
        4 Plant with a superficial or subterranean knobby or shortly branching caudex .... 1. *L. argenteus*

1. ***Lupinus argenteus*** Pursh, silvery lupine. Perennial herbs with knobby superficial or branching subterranean caudex; stems 1–several, simple to branched, erect or ascending,

1–6 dm tall, sparsely to densely strigulose to subsilky, usually grayish but sometimes reddish below midstem. Leaves alternate, nearly all cauline at anthesis and later; petioles from shorter to 2 × as long as blades; stipules united 1/2 or more of length to petioles, to nearly free above; leaflets (5)6–9(10), narrowly lanceolate to broadly oblanceolate, usually 2–5 cm long, upper surface glabrous, or strigulose, lower surface strigulose to sericeus, apically acute to acuminate or rounded to obtuse. Racemes 1–2 dm long, terminal, loose to congested; flowers separated or subverticillate, (6)7–9(10) or 9–11(12) mm long. Calyx bilabiate, tube 1.5–2 mm long, sericeous, base of upper side bulged or saccate, upper lobe shortly bidentate, lower entire; corolla white, pinkish-white, to usually light or dark blue; banner reflexed, whitish-centered, glabrous or with a median line of hairs concealed by calyx-lobe; wings glabrous; keel usually ciliate, sometimes glabrous, usually included within wings or slightly exserted. Pods 1.5–3 cm long, very hairy; seeds 3.7–4.5 mm long, gray to light brown, smooth. (n = 24).

Two somewhat distinctive but imperfectly separable varieties are in our area as follows:

1a. var. *argenteus*. Leaflets thicker and usually drying folded, elliptic or narrowly oblanceolate, 2–4 cm long, 3–6(10) mm wide, glabrous to strigulose above, moderately strigulose below, apically acute or acuminate, rarely rounded. Flowers 9–12 mm long. Jun–Aug. Infrequent to common on rocky prairie hillsides, stream valleys, roadsides; w 1/4 ND, w SD, nw NE, MT, WY; (much of w 1/2 U.S.). *L. aduncus* Greene — Rydberg.

This is a highly variable taxon with some plants having a well-developed saccate calyx, making them difficult to distinguish from *L. caudatus*. Some of the more densely pubescent plants appear to merge with *L. sericeus*.

1b. var. *parviflorus* (Nutt.) C. L. Hitchc. Leaflets thin and usually drying flat, mostly oblanceolate, 2.5–4 cm long, 6–10 mm wide, glabrous above, rather thinly strigulose below, apex mostly rounded and apiculate. Flowers (6)7–8(9) mm long. Jun–Aug. Scattered to common on open wooded hillsides, particularly at higher elevations in the BH; SD: Custer, Lawrence, Pennington; WY: Weston; (SD, MT, WY, ID s to NM, UT). *L. parviflorus* Nutt., *L. floribundus* Greene — Rydberg.

A few specimens of this variety with leaves narrowly lanceolate or elliptic and acute or acuminate could be referred to either var. *stenophyllus* (Rydb.) R.J. Davis or var. *tenellus* (Dougl. ex G. Don) Dunn, but they appear more as intermediates with smaller-flowered forms of var. *argenteus*.

The seeds of *L. argenteus* are reported to be poisonous to sheep and may cause severe losses.

2. **Lupinus caudatus** Kell., tailcup lupine. Perennial herbs with shortly branching caudex; stems 1–several, erect or ascending, 2–4(6) dm tall, simple to branched, strigose. Leaves alternate, mainly cauline but basal ones often present; petioles 2 × longer than blades or upper shorter than blades; leaflets 7–9, elliptic to narrowly oblanceolate, 2–5 cm long, acute or rarely rounded, equally strigose to silky on both surfaces. Racemes terminal, 6–20 cm long. Flowers (9)10–12 mm long; calyx 3–4 mm long, base spurred or strongly saccate, upper lobe bidentate, lower lobe entire; corolla light to deep blue or purple; banner lighter blue or whitish in the middle, reflexed, normally hairy in the middle of lower 1/2; wings glabrous or hairy near base; keel ciliate. Pods 2–2.5 cm long, silky; seeds 4–4.5 mm long, brownish, smooth. (n = 24) Jun–Aug. Rare in rocky prairie & gravelly stream valleys; sw SD, nw NE, WY; (SD, MT, OR, s to CO, NM, CA).

Our few specimens have a distinct spur 0.5–1 mm long but otherwise seem all too close to *L. argenteus*.

3. **Lupinus plattensis** S. Wats., Platte lupine. Perennial herbs with deep root system and branching subterranean rhizomatous caudex without adventitious roots; stems erect or ascending, (0.5)1.5–4(6) dm long, simple or branched, sparsely to usually densely short-appressed pubescent and with ascending hairs to 2 mm long. Leaves alternate, some basal ones usually present at and after anthesis; leaflets (5)7–9(11), spatulate to usually broadly or narrowly oblanceolate, (1.5)2.5–4(5) cm long, glabrous above and sericeous below, ciliate,

rounded, retuse, acute or acuminate, often apiculate; petioles of principal leaves 2–4(7) cm long. Racemes 0.5–1.5(2.5) dm long, flowers closely approximate to often separated. Flowers (11)12–13(14) mm long, concolorous, or usually banner with a dark purple median large spot, wings and keel white or suffused with blue, keel often maculate; calyx tube asymmetrical, 2–3 mm long, sericeous, often saccate, upper lobe 6–7 mm long, entire or shortly bifid, lower lobe entire, 7–9 mm long; banner strongly reflexed, glabrous or slightly hairy in lower 1/4 of median line; wings glabrous, enclosing keel or keel slightly exsert; keel ciliate. Pods 2–5 cm long, densely hairy, 3- to 8-seeded; seed 6–7 mm long, 4 mm wide, flat, yellowish, smooth. (n = 24) May–Aug. Infrequent to locally abundant on sandy prairies, dunes, stream valleys, more rarely on rocky open wooded or sandy hillsides; w NE, w WY, s in CO; OK: Cimarron, Texas; TX: Hartley; ne NM; (*endemic* in area). *L. perennis* L. subsp. *plattensis* (S. Wats.) Phillips—Phillips, Res. Stud. State Coll. Washington 23:161–201. 1955.

This is a rather distinct species with usually bicolored flowers and is suspected of being poisonous to livestock. In the Atlas GP this species was omitted and inadvertently mapped with the distribution of *L. argenteus* var. *argenteus*. Thus the dots s of w cen NE represent *L. plattensis*.

4. *Lupinus pusillus* Pursh, small or rusty lupine. Plant annual with a taproot which often has fusiform or close clusters of warty thickenings; stems simple or diffusely branched, 0.5–2 dm tall, densely pilose to hirsute with hairs 2–4 mm long, cotyledons sessile, connate-perfoliate, basal portions semipersistent. Leaves alternate; leaflets (5)6–8(9), oblanceolate to elliptic, principal ones 1.5–4 cm long, 3–8 mm wide, obtuse or acute apically and often very shortly apiculate, glabrous above, somewhat sparsely pilose below, ciliate; petioles 2–5 cm long; lower stipules 6–7 mm long, mostly united with petiole. Peduncles 1–3 cm long, racemes 3–7 cm long, usually equal to or exceeding leaves, pedicels 1–2 mm long at anthesis. Calyx bilabiate, tube 2 mm long, thinly pilose, upper lobe 1.5–2 mm long, bidentate, lower lobe entire, 5–6 mm long; corolla glabrous, white, ochroleucous, or usually pale blue, purplish or pink; banner reflexed, 8.5–10 mm long; wings 8–10 mm long, glabrous; keel glabrous, often purple-spotted at tip. Pods 1–2.5 cm long, 5–7 mm wide, hirsute or pilose; seeds usually 2, circular, 4 mm long, 1.5 mm thick, with raised marginal rims, sides convex, slightly rugulose, whitish or light green. (n = 24) May–Aug. Relatively common in sandy prairies, dunes, stream valleys, badlands, & roadsides; w 1/2 GP s to OK panhandle; (s Sask. & Alta., s to e 1/2 SD, nw OK, MT, CO, NM).

In dry seasons this plant often is only 0.5 dm tall, unbranched with very short peduncles. Otherwise, it is a rather uniform species. The plant is reported to be poisonous to livestock.

5. *Lupinus sericeus* Pursh, silky lupine. Perennial herbs with branching caudex; stems 1–several, 2–5 dm tall, simple or sparsely branched, with usually spreading white or rust-colored hairs or appressed pubescence. Leaves alternate, mostly cauline at anthesis; leaflets 7–9, oblanceolate, 3–7 cm long, 3–7 mm wide, sericeous on both sides; lower petioles to 3 × length of blades, becoming subequal above. Peduncles terminal, 3–5 cm long; racemes 5–15 cm long; pedicels 4–7 mm long at anthesis; flowers scattered to semiverticillate. Flowers 10–12 mm long, bluish-white to purplish; calyx tube 2–2.5 mm long, silky, often saccate, upper lobe 4.5–5 mm long, bifid, lower lobe 6–8 mm long, entire; banner reflexed, silky-hairy over much of the back, with a whitish center; wings glabrous or sparsely hairy near base; keel ciliate. Pods 2–3 cm long, 8–10 mm wide, silky; seeds 4 mm long, yellowish-brown, smooth. (n = 24) Jun–Aug. Rare in our area on rocky open wooded hillsides & rocky prairie hillsides; SD: Custer, Lawrence, Pennington; NE: Sioux; WY: Campbell, Crook, Johnson, Niobrara, Weston; (sw Alta., s B.C., s to NE, CO, NM, AZ, CA). *L. leucopsis* Agardh—Rydberg.

This species is known in our area from comparatively few collections, but these vary considerably in stem pubescence. The plant is suspected to be involved in the crooked calf disease of cattle.

20. MEDICAGO L., Medick, Alfalfa

Annual or perennial herbs with taproots; stems prostrate, ascending or erect, glabrous or pubescent. Leaves alternate, pinnately trifoliolate; leaflets with distal margin variously serrulate; stipules partially united with petioles. Flowers papilionaceous, solitary or more commonly in axillary clusters or in crowded spikelike racemes in upper axils; calyx campanulate, lobes almost equal; corolla yellow, violet-blue, rarely white, petals not adnate to stamen column; banner obovate to oblong; wings oblong; keel obtuse; stamens 10, diadelphous (9 + 1), anthers all alike. Pods not enclosed in calyx, straight or coiled, usually indehiscent. None native to U.S.

1 Pods without uncinate spines.
  2 Plant usually annual; flowers 2–3 mm long; pods reniform, black at maturity, 1-seeded .... 1. *M. lupulina*
  2 Plant perennial; flowers (5)7–12 mm long; pods straight, falcate, or coiled, with 2 or more seeds .... 3. *M. sativa*
1 Pods with unicinate spines 1–2.5 mm long .... 2. *M. minima*

1. Medicago lupulina L., black medick. Annual or short-lived perennials with taproots; stems prostrate or ascending, often much branched at base, (0.5)3–6(8) dm long, glabrate, finely strigulose or villous. Leaves alternate, pinnately trifoliolate; petioles 0–3 cm long; stipules lanceolate to ovate, entire or serrate, united to petiole for 1/4–1/2 of length; leaflets elliptic, obovate, rhombic or oblong-cuneate, 5–20 mm long, rounded to emarginate and usually apiculate, minutely toothed in upper half, sparsely to densely hairy. Peduncles axillary, 1–4 × longer than leaves; racemes closely 10- to 50-flowered, ovoid, becoming short-cylindrical in fruit. Flowers 2–3 mm long, yellow or ochroleucous. Pods 1.5–3 mm long, reniform, reticulate, black at maturity; seed 1, olivaceous, brown, or black, 2 mm long. (2n = 16) Apr–Nov. Rather common in lawns, pastures, fields, stream valleys, prairie ravines, roadsides, waste places; GP but less common in w 1/2; (nearly all temperate regions of the world; Eurasia). *Naturalized. M. lupulina* var. *glandulosa* Neilr.—Fernald.

Many of our specimens have glandular hairs on pods, pedicels, stems, and leaves and have been recognized as var. *glandulosa*, but this character varies from none to dense. Most of our plants are annual but some are perennial for at least 4 years.

2. Medicago minima (L.) Bartal., prickly medick, small bur-clover. Annual herb with taproot; stems decumbent, 1–4 dm long, villous, much branched from the base. Leaves alternate, pinnately trifoliolate; petioles 5–20 mm long; stipules lanceolate to ovate-lanceolate, entire or short toothed, 4–7 mm long, adnate to petiole less than 1/2 length; leaflets obovate to cuneate-oblong, villous, 5–10(16) mm long, rounded to emarginate, minutely toothed above and rarely near base. Peduncles axillary, 1–2.5 cm long; racemes capitate, (1)3- to 6(8)-flowered. Flowers 3–5 mm long, yellow. Pods 3–5 mm in diam, coiled with 3–5 turns, with many uncinate prickles 1–2.5 mm long, often also sparsely villous and glandular; seed reniform, 2 mm long, yellowish to dark brown, smooth. (2n = 16) Apr–Aug. Rare, in local colonies on lawns, parks, roadsides, & waste places; e 1/2 KS, cen OK, n-cen TX; (e 1/2 KS, OK, AR, TX, VA, NC; Eurasia). *Naturalized.*

This species was first collected in our area in 1974 and is found principally in se and n-cen KS. It is increasing in abundance and no doubt will become more common, though it appears somewhat intolerant of dry summer periods.

3. Medicago sativa L., alfalfa. Perennial herb with deep stout taproot and knobby or shortly branching superficial caudex; stems erect or decumbent, few to many from the base, 3–6(10) dm tall, glabrous or finely hairy. Leaves alternate, pinnately trifoliolate; petioles of principal leaves 1–5 cm long; stipules lanceolate, slightly toothed, 5–20 mm long, basally united with petiole. Peduncles axillary, 1-3 cm long, in upper axils; racemes subglobose to short-cylindric, 5- to 40-flowered. Flowers 5-11 mm long, purplish, yellowish-green or brownish, yellow, rarely white. Pods several-seeded, glabrous or sparsely hairy, usually in spiral of 1-3 turns, or nearly straight or falcate. May–Sep. Roadsides, old fields, waste areas; (cult. in temp. N. Amer.; Eurasia).

This highly variable complex of introduced and improved cultivated strains has two principal variants in our area.

3a. subsp. *falcata* (L.) Arcang. Leaflets oblong and often nearly linear to narrowly oblanceolate. Flowers 5-9 mm long, yellow or yellowish-green. Pod straight, falcately curved or slightly twisted. (2n = 16, 32). Scattered in n & w 1/2 GP.

3b. subsp. *sativa*. Leaflets usually obovate to oblanceolate. Flowers 7-11 mm long, various hues of blue and purple (rarely white). Pod spirally twisted. (2n = 32). Common throughout GP.

A number of specimens referred to subsp. *falcata* have pods curved into 1-1½ coils, petals blue-tipped in bud and flowers in the larger size range. These often are found as scattered individuals associated with plantings of subsp. *sativa* and are presumed to be hybrid segregates. Both taxa are known to hybridize freely and have been used in plant improvement programs in North America.

M. sativa often is reported as naturalized in the U.S., but my experience indicates that plants found along roadsides, waste places, etc., arise from repeated seeding and casual escapes and do not form reproducing and spreading populations.

21. MELILOTUS P. Mill., Sweet Clover

Ours annual, biennial, or short-lived perennial herbs with taproots; stems erect or ascending and freely branched; glabrous or sparsely pubescent. Leaves alternate; pinnately trifoliolate; leaflets mostly oblanceolate to obovate, serrulate on distal margin; stipules lanceolate or more commonly subulate, partially united with petiole. Peduncles axillary, usually several cm long; flowers in elongate spikelike racemes, white or yellow, papilionaceus; calyx short, campanulate, with subequal lobes; corolla deciduous after anthesis, free from staminal tube, wing and keel petals more or less coherent; stamens 10, diadelphous. Pod compressed, ovoid, usually 1-seeded, indehiscent or nearly so.

1 Flowers white, usually 4-5 mm long; pod reticulate-veined .... 1. *M. alba*
1 Flowers yellow, usually 4.5-7 mm long; pod cross-ribbed or irregularly rugose .... 2. *M. officinalis*

1. *Melilotus alba* Medic., white sweet clover. Annual or biennial herb; stems erect, branched, 0.5-2 m tall, glabrous to sparsely pubescent above. Leaflets oblanceolate or obovate to lanceolate or oblong, 1-2.5(4) cm long, glabrous above, appressed-pubescent below, short toothed apically and along sides; stipules 6-10 mm long. Racemes 4-15 cm long, calyx 1.5-2 mm long, tube gradually tapering to base, teeth deltoid to subulate, 0.5-1 mm long; corolla 4-5 mm long, white. Pod obovoid or ovoid, 2.5-5 mm long, glabrous, reticulate-veined, very short-stipitate. (n = 8, 12, 16) May–Oct. Common along roadsides, fields, & waste places; GP; (much of N. Amer.; Eurasia). *Introduced.*

When first introduced into our area, the plant was considered as a weed and often is still so considered. Its use as a crop began in the 1920s, and it has now invaded disturbed areas.

2. *Melilotus officinalis* (L.) Pall., yellow sweet clover. Very similar to no. 1, but differing in having yellow flowers 4.5-7 mm long; calyx slightly saccate or rounded below. Pod cross-veined or irregular rugose or veined. (n = 8) May–Oct. Weed along roadsides & in waste places; GP; (much of N. Amer.; Eurasia). *Introduced.*

As a crop plant this species generally is considered to be superior to *M. alba*.

Melilotus indica (L.) All. is similar to *M. officinalis* but is strictly an annual with flowers only 2–3 mm long. It was reported for the GP by Rydberg, but has not been found in the area.

22. OXYTROPIS DC., Locoweed

Ours cespitose perennials with taproots and knobby or branching superficial or subterranean caudex; acaulescent or with short leafy stems in *O. deflexa*. Leaves odd-pinnate, often dimorphic; leaflets scattered, opposite, subopposite or with some leaflets fasciculate and appearing verticillate in *O. spendens;* base of leaflets inequilateral though often nearly imperceptibly so; petiolate; stipules adnate to petiole and amplexicaul or connate opposite it into a close sheath in early unfolding. Flowers in pedunculate spikes or racemes; papilionaceous; calyx usually cylindro-campanulate or campanulate, becoming inflated and enclosing pod in *O. multiceps* and *O. nana;* petals white, ochroleucous, reddish-purple, white with purple keel-tips, or a variety of colors in hybrid populations; banner usually erect; wings usually longer than keel; keel with a terminal cusp consisting of a knob, point, or tooth, or a straight or curved beak; stamens 10, diadelphous. Pod sessile or stipitate, membranaceous to leathery or very hardened, sometimes inflated; several-seeded.

Reference: Barneby, R. C. 1952. A revision of the North American species of *Oxytropis*. Proc. Calif. Acad. Sci. IV, 27: 177–309.

1 Plant usually at least shortly caulescent; stipules adnate to petiole 1–3 mm of their length; pods pendulous .... 3. *O. deflexa* var. *sericea*
1 Plant acaulescent; stipules adnate for 4 mm or more of their length; pods erect or less commonly spreading.
  2 At least some leaflets fascicled at some point on one side of rachis, often appearing verticillate .... 9. *O. splendens*
  2 All leaflets subopposite or scattered.
    3 Pubescence of herbage with dolabriform hairs .... 5. *O. lambertii*
    3 Pubescence of basifixed hairs.
      4 Racemes 1- to 5-flowered; calyx conspicuously inflated in fruit.
        5 Mature pod papery, not rigid; valves villous .... 6. *O. multiceps*
        5 Mature pod leathery, rigid; valves strigose .... 7. *O. nana*
      4 Racemes 6- to many-flowered; calyx not conspicuously inflated in fruit.
        6 Corolla purple.
          7 Calyx shaggy-villous or hispid-hirsute with spreading hairs.
            8 Hairs of calyx shaggy-villous nearly concealing surface of tube at anthesis; pubescence of herbage silky-villous; bracts lanceolate, membranous .... 4. *O. lagopus* var. *atropurpurea*
            8 Hairs of calyx hispid-hirsute at anthesis, not concealing surface of tube; herbage nearly appressed-pilose; bracts rhombic-lanceolate .... 1. *O. besseyi*
          7 Hairs of calyx appressed silky-pilose, not shaggy or hispid.
            9 Pods leathery, rigid; leaflets 11–19; keel 15 mm or more long; plants not in ND .... 8. *O. sericea*
            9 Pods thin, hardly rigid; leaflets 19–25; keel usually about 13 mm long; plants of ND .... 2. *O. campestris* var. *gracilis*
        6 Corolla white, ochroleucous, or pale yellow, the keel sometimes purple-tipped.
          10 Pods leathery at maturity, very rigid and hard, fleshy when young; flowers usually over 15 mm long .... 8. *O. sericea*
          10 Pods somewhat papery at maturity, not rigid nor fleshy when young; flowers often less than 15 mm long .... 2. *O. campestris* var. *gracilis*

1. ***Oxytropis besseyi*** (Rydb.) Blank., Bessey's locoweed, red loco. Perennial cespitose herbs with stout taproots and subterranean much-branched caudex; ultimate caudex-branches

often weakly ascending at ground level and producing foliage and scapes of the season. Leaves odd-pinnate, 3–12 cm long; leaflets 9–19, scattered, elliptic, lanceolate, or lance-ovate, 4–20 mm long, 1–4 mm wide, pilose on both sides; petioles 1–6 cm long, with appressed and some ascending hairs; stipules thin, 6–14 mm long, glabrous above, sparsely pilose dorsally, prominently ciliate. Scapes (10)14–19 cm long, racemes 2–8 cm long at complete anthesis, flowers 7–20. Calyx tube 6–7 mm long, hispid-hirsute, teeth 3–6 mm long; petals pink or reddish-purple, fading bluish; banner 18–25 mm long enlarged above broad claw into an ovate-oblong blade 6–9 mm wide; wings with blades 9–15 mm long, 5–7 mm wide near apex, emarginate; keel 13–18 mm long, with straight or curved appendage to 2 mm long. Pod covered by calyx, short-stipitate, body 10–14 mm long, villous, with a beak 4–5 mm long; seeds brown and often with purple spots, 2–2.5 mm in diam. May–Jul. Infrequent to locally common on gravelly or sandy hilltops, ravines, & badlands; NE: Kimball; e WY, e MT; (MT, n & e WY, sw NE).

2. **Oxytropis campestris** (L.) DC. var. *gracilis* (A. Nels.) Barneby, slender locoweed. Perennial herbs with stout taproots and knobby or much-branched subterranean caudex, these sometimes weakly ascending at ground level. Leaves odd-pinnate, weakly to strongly dimorphic, 5–18(25) cm long, the primaries short and with ovate leaflets, the secondary leaves with (13)17–25(33) elliptic, linear-oblong, or oblong to obovate leaflets; herbage sparingly or finely pilose to silky-pilose; stipules glabrate or pilose dorsally, sparingly to densely ciliate, 4–16 mm long. Scapes erect (5)10–19(30) cm long; racemes 8- to 20(30)-flowered. Calyx tube (4.5)6–6.5 mm long, teeth 1.5–2.5 mm long, the tube white-pilose and with brown or black short appressed hairs; petals whitish, ochroleucous or pink, blue, white, or purple; banner 12–19 mm long, 6–9 mm wide; wings 10–16 mm long, 2.5–6 mm wide near emarginate apex; keel 10–14 mm long, the distal appendage 0.5–1 mm long. Pod nearly sessile, oblong to obovoid-ellipsoid, 8–16 mm long with a beak to 5 mm long, short-pilose, membranaceous to somewhat coriaceous; seed 2–2.4 mm long, dark brown, smooth. (n = 16) May–Jul. Rather common in open woodlands, prairies, brushy ravines; w 3/4 ND, BH, ne WY, MT; (sw Man. to B.C., s to ND, SD, ne CO, MT, WA). *O. dispar* (A. Nels.) K. Schum, *O. gracilis* (A. Nels.) K. Schum, *O. macounii* (Greene) Rydb. (?), *O. villosa* (Rydb.) K. Schum.—Rydberg.

Plants of the BH tend to be sparingly pilose, with leaflets wider and thinner, and scapes longer than plants of ND which are commonly silky-pilose, have narrower and thicker leaflets, and shorter scapes. Some ND colonies have flowers of several shades from white to purple, and the ND populations have been recognized as the var. *dispar* (A. Nels.) Barneby. This appears to be a rather weak variety but merits careful study in the field. The possibility of its hybridizing with *O. lambertii*, with which it grows, should be determined.

3. **Oxytropis deflexa** (Pall.) DC. var. *sericea* T. & G., pendulous-pod loco. Ours perennial caulescent herbs with taproots and shortly branching caudex; stems erect or usually ascending, somewhat flexuous, 5–20 cm long, sparingly branched; stem and leaves villous-pilose with spreading or retrorse hairs. Leaves odd-pinnate, 5–20 cm long; leaflets 21–41, opposite or subopposite, ovate, lanceolate to lance-oblong, 3–20 mm long. Peduncles 15–25 cm long; racemes 10- to 15(25)-flowered. Calyx campanulate, tube 2–3.5 mm long, teeth usually 2.4 mm long, tube and teeth with appressed black hairs and longer spreading white hairs; petals pale blue or purplish; banner 4.5–9 mm long, 2.5–3.5 mm wide; wings 5–8 mm long, 0.7–1.2 mm wide; keel 4.5–8 mm long, distal appendage to 0.5 mm long. Pods pendulous, linear-oblong or oblong-ellipsoid, 10–16 mm long, 3–4 mm wide, with mixed white and black hairs; seeds olivaceous to brown, 1.5 mm long, smooth. (n = 16) Lake shores & stream banks, moist prairie swales, mountain open woodlands & meadows; ND: Bottineau, Cavalier, Pembina; (Man. to AK, s to ND, n NM, WA, OR, CA).

4. **Oxytropis lagopus** Nutt. var. *atropurpurea* (Rydb.) Barneby, hare's locoweed. Low cespitose perennial with taproot and clustered shortly branching caudex, silky or loosely-

pilose. Leaves odd-pinnate, dimorphic, 2-7(11) cm long; leaflets 9-17, lanceolate to narrowly elliptic, 3-13 mm long; petioles 1-5 cm long, villous or appressed-pubescent; stipules somewhat membranaceous, 7-17 mm long, silky-villous on back, ciliate. Peduncles erect to curved ascending, 1-9 cm long, villous, sometimes thinly so or appressed-pubescent, often with mixture of black hairs above; racemes subcapitate to oblong in full flower, 5- to 18-flowered, axis at maturity 1-3(4) cm long, with mixture of white and black hairs, bracts lanceolate, 3-13 mm long, somewhat involute below, with white and black hairs dorsally. Calyx tube 5-7 mm long, teeth 2-4.5 mm long, usually densely silky-hirsute, usually concealing surface of tube, little or not at all inflated at maturity and ruptured by enlarging pod; petals reddish-purple, drying bluish; banner 14-19 mm long, blade 7-10 mm long, emarginate; wings 13-16 mm long, 2-6 mm wide distally; keel 11-14 mm long, blades 6-7 mm long, distal appendage 0.5-1.2 mm long. Pod with body 8-15 mm long, oblong or ovoid, 4.5-6.5 mm in diam, with beak 4-5 mm long, valves usually somewhat membranaceous; seeds brownish, 1.5-2 mm long, smooth. May-Jun. Sandy or usually gravelly prairie hilltops & hillsides; WY: Albany, Campbell, Johnson, Platte, Sheridan; (WY, s-cen MT).

5. **Oxytropis lambertii** Pursh, purple locoweed. Perennial cespitose herbs with stout taproot and a knobby or branching caudex; pubescence of stem and leaves with dolabriform hairs, these with one branch usually very short; hairs subappressed-pilose to hirsute or villous, sparse to dense. Leaves odd-pinnate, usually strongly dimorphic; principal leaves (3)10-17(21) cm long; leaflets 7-19, linear, lance-linear, elliptic, sometimes falcate, 0.5-4 cm long, 1-4(6) mm wide, acute or short-acuminate. Scapes usually erect, 5-30 cm long; racemes (6)10- to 20(25)-flowered, flowers spreading to erect. Calyx cylindric, tube (4.5)6-7(8) mm long, teeth (1.2)1.5-3(4) mm long, tube silky-strigose, pilose, and sometimes with blackish hairs intermixed; petals usually reddish-purple (albino forms not infrequent) or various shades of rose, blue, and purple; banner (12)15-25 mm long, blade (6)8-12 mm wide; wings 12-20 mm long, 4.5-8 mm wide near truncate or emarginate apex; keel (11)13-19 mm long, the appendage 0.5-2.5 mm long, usually straight to arched. Pod sessile or shortly-stipitate, 8-15(25) mm long, body ovoid to lance-acuminate, with a straight or divergent beak 3-7 mm long, body strigose-silky or strigose; seeds brown, 2 mm long, smooth. (n = 24) May-Aug. Commonly encountered on prairies, plains, & hillsides; river bluffs, badlands, open wooded hillsides, roadsides; GP but absent in se; (s Man. & Sask., s to MN, w IA, e TX, e MT, e WY, UT, AZ, NM). *O. hookeriana* Nutt., *O. involuta* (A. Nels.) K. Schum, *O. plattensis* Nutt. — Rydberg.

In n-cen TX, s 1/2 of TX panhandle, w 1/2 OK, and into sw KS there are plants with calyx tube 4-5× longer than teeth, corolla with banner 18-25 mm long, and pod about equaling the calyx or a little exserted. These populations are recognized as var. *articulata* (Greene) Barneby, while plants to the n in the GP are the var. *lambertii* with calyx tube about 3× longer than teeth, banner usually 18 mm long or less and body of pod about 2× longer than calyx. In w 1/2 of OK, n TX panhandle, and sw KS many populations exhibit all degrees of intergradation and the larger banner is found well to the north in the GP.

Along the western edge of our area and into the mountains is found the var. *bigelovii* A. Gray with pods often short-stipitate, leaves wider or 3-5× longer than wide, and pods thinner. This is a rather distinctive geographical variant, but there is complete intergradation between the plains and montane elements.

Throughout out area there appear to be many races and forms of *O. lambertii*, but no combination of characters has been found to indicate geographical trends. In w NE and e WY *O. lambertii* mingles with *O. sericea*, and large hybrid swarms are apparent in which many shades of flower color are found. These populations merit careful experimental study.

Purple locoweed has long been known to be responsible for loco poisoning of livestock, particularly horses.

6. **Oxytropis multiceps** T. & G., dwarf locoweed. Low cespitose perennial herbs with taproots and much-branched caudex; herbage usually silky-pilose; plants forming con-

spicuous silvery mounds 0.5-2 dm in diam. Leaves odd-pinnate, 1-5 cm long; petioles 0.5-3 cm long; leaflets 5-9, lanceolate, elliptic, oblanceolate to ovoid-oblong, (3)5-12 mm long, rather congested with rachis about as long as lower leaflet. Scapes spreading, 1.5-3 cm long, becoming prostrate; racemes with (1)2 or 3(4) flowers. Calyx turgid at anthesis, membranaceous, reddish, villous-hirsute, tube 6-10 mm long, teeth unequal, 2-3 mm long; calyx becoming inflated in fruit, 8-18 mm long, 5-9 mm in diam, constricted above; petals red-purple, drying whitish-blue; banner 17-24 mm long, oblong-obovate, 7-9 mm wide, emarginate; wings widened distally, 4.5-5.5 mm wide near emarginate apex; keel 13-18 mm long, distal appendage straight or curved, 0.5-1.4 mm long. Pod enclosed by inflated calyx, stipe to 1.5 mm long, ovoid body 6-10 mm long, contracted into a beak, valves thin, short-villous; seeds 1.5-2 mm long, reddish-brown, often with purple mottling. May-Jun. Local colonies on gravelly prairie hilltops, ridges, & rocky open wooded hillsides, infrequently collected; NE: Banner, Chase, Dundy, Kimball; WY: Laramie, Platte, Niobrara; (sw NE, se 1/4 WY, cen CO, ne UT).

7. *Oxytropis nana* Nutt., dwarf locoweed. Short densely cespitose perennial herbs with taproot and branching caudex, silky-pilose throughout. Leaves odd-pinnate, 2-9 cm long; petioles 1-4 mm long; leaflets 5-11, lance-oblong, lanceolate or elliptic, 5-25 mm long, acute. Scapes 3-9 cm long, erect to curved-ascending; racemes 5- to 10(15)-flowered, often subcapitate. Calyx densely shaggy-hirsute; tube tubular-campanulate but becoming inflated or urceolate, 9-11 mm long, teeth 1.5-3 mm long; petals purple or white with keel purple-spotted distally; banner 18-22 mm long, oblong-obovate, 8-15 mm wide; wings 15-19 mm long, widened distally, 5-9 wide, truncate-emarginate; keel 15-17 mm long, distal appendage straight or curved, 0.5-1.5 mm long. Pod included in calyx, body 7-10 mm long and 4-5 mm in diam, with beak 3-5 mm long, valves coriaceous, strigose; seeds 2 mm long, olivaceous to brown, rarely purple-mottled, smooth. May-Jul. Local colonies on dry rocky prairie hilltops & ridges; WY: Converse, Niobrara; (apparently *endemic* in e-cen & c WY).

This species barely enters our area and has not been collected often. Our few specimens have been frequently incorrectly determined as *O. multiceps, O. besseyi,* and *O. lagopus* Nutt.

8. *Oxytropis sericea* Nutt., white locoweed. Cespitose, somewhat robust perennial herbs with stout taproot and much-branched caudex, silky-pilose throughout with basifixed hairs. Leaves usually dimorphic, 4-30 cm long; leaflets (7)11-19(25), opposite or scattered, elliptic or narrowly lanceolate to ovate-oblong, acute or obtuse, (0.5)1-3(4) cm long, 2-10 mm wide; petiole 1-15 cm long; stipules 7-28 mm long. Scapes erect or ascending, 5-30 cm long; racemes with (6)10-30 flowers. Calyx tube tubular-campanulate 8-12 mm long, teeth unequal 2-5 mm long, tube and teeth with white and black hairs, strigose, appressed-pilose, or with spreading hairs; petals white, ochroleucous, fading yellowish, the keel often with purple tip; banner 15-20(26) mm long, claw broad, blade oblong-obovate, 8-10 mm wide, deeply lobed or emarginate, wings 15-20 mm long, blade 9-13 mm long, widening distally and 5-8 mm wide near emarginate apex; keel 12-17 mm long, with distal straight or curved appendage 1-2 mm long. Pods erect, sessile, body oblong or ovoid-oblong, 1-2.5 cm long, 4-7 mm in diam, with a short beak, valves rigid and coriaceous to woody, silky-strigose or somewhat short-pilose; seeds 2-2.5 mm long, brown, smooth. Apr-Jun. Infrequent to common in rocky prairie plains & hillsides, gravelly banks & stream valleys, open wooded hillsides; w 3/4 MT, sw SD, w NE, WY, w KS, nw OK, CO, NM; (sw SD to w MT, s to OK, cen NM, ne UT, s ID). *O. pinetorum* (Heller) K. Schum.—Rydberg.

Our plant is the var. *sericea,* quite variable in stature, flower color, and pubescence, and it appears to hybridize with *O. lambertii* and *O. campestris.* White-flowered specimens of *O. lambertii* have frequently been determined as *O. sericea* but can be separated on the basis of dolabriform hairs present in *lambertii.* It is usually nearly impossible to separate some specimens of *O. sericiea* and *campestris* where they intermingle. The group merits careful field and experimental studies. *O. sericea* has long been known to be responsible for loco poisoning of livestock.

9. **Oxytropis splendens** Dougl. ex Hook., showy locoweed. Cespitose perennial herbs with taproot and much-branched caudex, silky-villous throughout or thinly so on leaflets. Leaves odd-pinnate, 7–23 cm long, often dimorphic; leaflets in verticels of 3–6, some often solitary, opposite or scattered, lanceolate to elliptic, 3–20 mm long; petioles 1–9 cm long; stipules thin, 10–15 mm long, adnate to petiole for 2/3 of their length and tubular-connate. Scapes erect, 10–20(30) cm long; racemes at first cylindric spikelike, densely 20- to 35(80)-flowered, 5–10 cm long but elongating in fruit. Flowers reddish-purple; calyx tube 5–7 mm long, silky-villous, teeth 2–4 mm long; banner oblanceolate to obovate, 12–16 mm long, 4.5–6 mm wide; wings 10–12 mm long, 2.5–3 mm wide near truncate-emarginate apex; keel 10–12 mm long with distal appendage to 1 mm long. Pod 10–15 mm long, 3–4 mm wide, with a beak 3–4 mm long; seeds 1.5–2 mm long, brownish, smooth. (n = 8) Jun–Sep. Moist gravelly or sandy prairies, stream banks, wooded hillsides, mountain meadows; n & ne ND; WY: Laramie, Niobrara; (n Ont., Sask., AK, s to e-cen MN, n ND, MT, CO, n NM). *O. richardsonii* (Hook.) K. Schum. — Rydberg.

23. PSORALEA L., Scurf-pea

Ours caulescent or acaulescent perennial herbs with globose, fusiform-tuberous, or elongate-thickened taproots; caudex subterranean, erect and often branched, or laterally produced and branched; some species propagating by shoots from spreading roots; stems erect, ascending, or procumbent, usually branched; herbage and calyx glandular-punctate and variously pubescent, or eglandular. Leaves alternate, pinnately or usually palmately foliolate; leaflets entire; petioles well developed; stipules of lower leafless nodes connate, amplexicaul, deeply bilobed to nearly entire upward, usually persistent. Inflorescences of axillary or terminal spikelike racemes, flowers dense or loose, usually 2 or 3 flowers subtended by a conspicuous bract. Calyx campanulate, often gibbous, sometimes oblique, 2-lipped, middle tooth of lower lobe longer and wider than others or less commonly the calyx nearly regular, often enlarging in fruit; corolla papilionaceous, whitish, bluish, purplish, fading brownish, not associated with androecium; banner tapering into a short claw, blade obovate; wings short clawed, blades oblong or oblanceolate with a distinct basal lobe, often partially connivent; keel-petals rounded and distally united, each attached to the base of adjacent wing-petal; stamens 10, diadelphous. Pod 1-seeded, enclosed within or exserted beyond calyx, indehiscent or rupturing irregularly or tardily circumscissile; seeds ovoid to ellipsoid. *Orbexilum* Raf., *Pediomelum* Rydb., *Psoralidium* Rydb. — Rydberg.

1 Leaves pinnately 3-foliolate, the petiolule of terminal leaflet conspicuously longer than the other two .... 8. *P. psoralioides* var. *eglandulosa*
1 Leaves palmately 3- to 7-foliolate, all leaflets sessile or on petiolules of about equal length.
  2 Inflorescence a dense spikelike raceme; bracts of raceme 5–15 mm or more long; calyx including longest tooth 8–17 mm or more long at anthesis.
    3 Pubescence of stem and petioles spreading .... 4. *P. esculenta*
    3 Pubescence appressed.
      4 Plant acaulescent; seeds with conspicuous ridges .... 5. *P. hypogaea*
      4 Plant with stems 3–8 dm long, procumbent to ascending; seeds smooth .... 2. *P. cuspidata*
  2 Inflorescence a slender loose raceme or narrow interrupted spike; bracts of raceme to 4 mm long; calyx including longest tooth 2–7 mm long at anthesis.
    5 Flowers at anthesis in well-separated whorls, not pedicellate.
      6 Bracts obovate to spatulate; calyx inflated in fruit .... 3. *P. digitata*

                6   Bracts ovate-lanceolate, acuminate; calyx not inflated in fruit .... 1. *P. argophylla*
         5   Flowers at anthesis in loose or dense racemes, distinctly pedicelled.
                7   Pods longer than wide, flowers bluish to purple.
                        8   Leaflets 7–15 × longer than wide; pods gradually tapering into a
                            beak .... 7. *P. linearifolia*
                        8   Leaflets 2–6 × longer than wide; pods abruptly short beaked .... 9. *P. tenuiflora*
                7   Pods subglobose; flowers whitish except for purple keel tip .... 6. *P. lanceolata*

1. Psoralea argophylla Pursh, silver-leaf scurf pea. Perennial herbs with taproots and subterranean erect shortly branching caudex; forming colonies by shoots from spreading roots (most herbarium specimens from such shoots); stems erect or ascending, (2)4–5(8) dm tall, divaricately branched, sometimes flexuous, silvery-sericeus to silky-villous, or glabrate below. Leaves alternate, those of main stem usually 4- or 5-foliolate, branch leaves usually 3-foliolate; leaflets elliptic, oblanceolate, obovate, or oblong, apically acute, obtuse, apiculate, or somewhat truncate-retuse, usually densely silvery-sericeus below, rarely glabrate and greenish, yellowish-green and less hairy to glabrous but glandular-punctate above; petioles 1–5 cm long, usually sericeous or silky-villous; stipules of lower leafless nodes 1–2 cm long, becoming shorter above. Peduncles axillary, 2–8 cm long; spikes of 2–5(8) well-separated whorls of (2)3–6(8) flowers each; bracts ovate-lanceolate, pointed or acuminate, semipersistent. Calyx tube campanulate, 2-lipped, 2–3 mm long at anthesis, upper 4 teeth 2.5–3 mm long, lower tooth 7–8(10) mm long, elongating in fruit, the whole usually densely sericeus; corolla purple fading bluish, yellowish or brown; banner obovate, spur 1–2 mm long, 5–7 mm long, blade 3–5 mm wide, with 2 basal processes; wings 4–6 mm long; keel 4–5 mm long. Pod 7–9 mm long, ovoid, with a short straight beak, body largely enclosed by the adherent calyx, densely strigose, pericarp rigid; seed 4–4.5 mm long, reniform, olivaceous to dark brown, smooth. (n = 11) Jun–Sep. Generally common on prairies, rocky hillsides, open woodlands, sand dunes, stream valleys; GP but infrequent in sw KS & not known in TX panhandle; (s Man., s Alta., MN, WI, s to MO, OK, NM). *Psoralidium collinum* Rydb., *P. argophyllum* (Pursh) Rydb.—Rydberg; *Psoralea argophylla robustior* Bates—Bates, Amer. Bot. 20: 16. 1914.

   This species varies considerably in several characters but its usually silvery coloration makes it conspicuous and rather easily recognized. Plants with lower surface of leaves thinly strigose, lower calyx lobe not much elongating, and flowers about 6 mm long have been recognized as *P. collina* Rydb. These are found somewhat frequently in the w 1/2 GP, but intergrade completely with the more usual form and do not have a distinct geographic distribution.

   Although this species is often reported to be rhizomatose, I find the plant to propagate from adventitious buds formed on roots.

2. Psoralea cuspidata Pursh, tall-bread scurf-pea. Perennial herbs with elongate or fusiform-thickened taproot and shortly branching caudex; stems procumbent to ascending or rarely erect, (2)3–6(9) dm long, branched above, sparsely appressed-pubescent. Leaves alternate, palmately (3)5-foliolate; leaflets broadly to narrowly elliptic, obovate, or somewhat rhombic, acute to obtuse or rounded, short apiculate, (0.5)1–2.5(3) cm wide, 2.5–6 cm long, upper surface glabrate and conspicuously glandular-punctate, lower surface appressed-pubescent and sparsely glandular; petioles 1–6 cm long; stipules of lower leafless nodes 1–2 cm long, ovate, amplexicaul, becoming narrowly lanceolate above and 1 cm long, semipersistent. Peduncles stout, axillary, longer than petioles; racemes dense, 4–9 cm long; bracts ovate, lanceolate, or elliptic, long acuminate or cuspidate, elongating in age, conspicuously glandular. Calyx tube 4–7 mm long at anthesis, gibbous at base, upper 4 lobes 3–7 mm long, lower lobe 7–12(15) mm long, tube strigose and glandular, becoming inflated in fruit; corolla blue, purple, or blue-lavender, rarely white; blade of banner 12–15 mm long, with short basal lobes, claw 4–6 mm long, curved; wings with blades 1 cm long, claws 5–7 mm long; free portion of keel 4–6 mm long. Pods enclosed in enlarged calyx,

6–8 cm long, with the curved beak 2 mm long, pericarp thin; seed 4–5 mm long, olivaceous to dark brown, smooth. (n = 11) May–Jul. Infrequent to locally common on dry prairies, gravelly hilltops & slopes, stream valleys, chalk bluffs, badlands; cen GP from cen SD, s to TX; (MT, SD s to TX). *Pediomelum cuspidatum* (Pursh) Rydb.—Rydberg.

Except for considerable variation in the characters of the inflated calyx, this species is relatively consistent in our area. Though cattle rarely browse the plant, it decreases under grazing.

3. *Psoralea digitata* Nutt., palm-leaved scurf-pea. Perennial herbs with taproots and erect knobby or shortly branching caudex; stems erect, 3–8(11) dm tall, branched above, appressed-canescent throughout. Leaves alternate, usually palmately 5-foliolate, or some on branches 3-foliolate; leaflets linear, linear-oblong, or linear-oblanceolate, 2–7 cm long, 2–8 mm wide, glabrous and glandular above except for strigose midvein area, densely strigose or silky below; petioles 2–7 cm long; stipules lanceolate, 5–10 mm long, strigose. Peduncles axillary, (5)10–18 cm long; spikes 3–9 cm long, interrupted, flowers in whorls of 3–7(11), sessile or subsessile; bracts obovate to spatulate. Calyx tube 2.5–3.2 mm long at anthesis, villous, upper 4 lobes 3–4 mm long, lower lobe 5–6 mm long, calyx conspicuously enlarging in fruit; corolla bluish, purple, or rarely white, fading brownish; banner obovate, blade 7–9 mm long, spur 2–3 mm long; curved; wings with blades 6–7 mm long, spur 2–3 mm long; free portion of keel 4 mm long. Pod 7–8 mm long, with a flat straight beak, entirely enclosed in calyx, pericarp thin; seed ellipsoid, olivaceous to dark brown, 4–5 mm long. (n = 11) May–Jul. Infrequent to locally common in sandy prairies, gravelly hillsides, sand dunes, sandy open wooded hillsides, stream valleys; s 2/3 GP but absent in e & infrequent w; (SD, NE, e CO, s to n-cen TX). *Psoralidium digitatum* (Nutt.) Rydb.—Rydberg.

Some specimens from s KS and OK have principal stem leaves with leaflets 2–4 mm wide and could be the var. *parvifolia* Shinners, which is reported to be endemic in e TX. However, they are always found with the more common wider leaflet form and hardly merit recognition in our area.

4. *Psoralea esculenta* Pursh, breadroot scurf-pea, prairie-turnip. Perennial herb with a deep taproot, this fusiform-thickened 4–10 cm below ground level, thickened portion 3–8 cm in diam and 4–15 cm long; from top of thickening 1–3 vertical caudices extend to soil surface, apically with conspicuous ovate stipules; stems 1–3, simple or rarely branched, 5–15(20) cm long, flexuous, densely villous-hirsute. Leaves alternate, often appearing clustered, palmately 5-foliolate; leaflets elliptic, oblanceolate to obovate, apically acute, obtuse, rounded to rarely truncate, upper surface glabrous or glabrate, lower surface loosely appressed-pubescent, without glands; petioles (2)5–10(15) cm long, villous-hirsute; stipules oblong, ovate-oblong or lanceolate, 1–1.5(2) cm long, sometimes falcate. Peduncles axillary, shorter or longer than leaves; spikes dense, 2–2.5 cm wide; bracts ovate, acuminate, 1–1.5 cm long. Corolla blue at anthesis, fading yellowish; calyx tube at anthesis gibbous, hirsute, without glands, 5–6.5 mm long, upper middle 2 of 4 teeth partially united, 5–7.5 mm long, lower tooth 6–8 mm long, conspicuously enlarging in fruit; banner blade oblong, 9–13 mm long, with 2 basal processes, claw 4–6 mm long; wing blade 9–11 mm long, with a basal lobe, claw 5–7 mm long; free portion of keel 4–5 mm long. Pod with ovoid body 5–7 mm long and enclosed in calyx, glabrous and sparsely glandular, abruptly tapering into a beak 1–2 cm long, sparsely hirsute, usually subequal with elongated calyx-teeth, pericarp thick; seed 4–5 mm long, oblong, plump, oblivaceous and often purple-spotted or dark brown, smooth. (n = 11) May–Jul. Infrequent to common on prairie plains & hillsides, bluffs, stream valleys, open woodlands, roadsides; GP but rare to absent sw; (s Man., Sask., Alta., s to WI, MN, MO, AR, OK, ne CO, MT). *Pediomelum esculentum* (Pursh) Rydb.—Rydberg.

This species varies in the length and branching of the stem and the amount of stem growth following flowering, perhaps reflecting growing conditions. In dry years or in very dry habitats the plants are often subscapose. The plant increases during early grazing of prairies but decreases rather quickly with continued use.

A number of common names have been used for this species such as: Indian bread root, Indian turnip, pomme de prairie, prairie potato, ground apple, and prairie turnip. These reflect the widespread use of the plant for food by the Indians of the GP, early explorers and settlers. Wedel, Nebraska Hist. 59:1–25. 1978, presents a comprehensive account of this subject.

5. *Psoralea hypogaea* Nutt., little breadroot scurf-pea. Perennial acaulescent herbs or rarely with stems 1–2 cm long and internodes very short; taproot apically thickened into a globose or fusiform storage organ 3–8 cm below soil surface; caudex a knobby usually conical structure at apex of storage organ, producing 1(2–4) vertical slender stipule-bearing branches which enlarge at surface to produce crown of the season, this congested or elongating 1–2 cm. Leaves palmately (3)5(7)-foliolate, all parts with densely appressed white pubescence or upper surface of leaflets becoming glabrate; leaflets linear-elliptic, linear-lanceolate to rarely narrowly obovate, (1.5)2.5–5 cm long, 4–9 mm wide; petioles (2)4–9(12) cm long; stipules lanceolate to ovate, 0.5–2 cm long, scarious. Peduncles very short to rarely as long as leaves in fruit; spikes dense; bracts, ovate-acuminate or lanceolate, scarious, 5–8 mm long, glabrous above, strigose below, ciliate. Calyx tube 4–5 mm long, hirsute, 4 upper teeth lanceolate, 5–7 mm long, lower tooth 9–11 mm long and 3–4 mm wide, 3-veined; corolla blue or violet, fading yellowish; banner ovate-oblong, blade 10–13 mm long, claw 2–3 mm long; wing blades narrowly oblong, 9–10 mm long, claw 2–3 mm long; free portion of keel 3–4 mm long. Pod ovoid, 5–6 mm long, hirsute, tapering into a beak 8–13 mm long, pericarp thin in lower 1/3, leathery above, irregularly circumscissile; seed 4–5 mm long, 2.5–3 mm wide, 2 mm in diam, faces with conspicuous irregular rounded ridges, olivaceous or pinkish-gray. (n = 11) May–Jun. Rare on rocky or sandy prairies, bluffs, stream valleys; NE: Deuel, Morrill, Sioux; WY: Laramie, Platte; w 1/2 KS, e CO, w OK, TX panhandle, ne NM; (*endemic* in area). *Pediomelum hypogaeum* (Nutt.) Rydb.—Rydberg.

Our plant is the var. *hypogaea*. In the sw part of our area a few specimens with globose roots and peduncles about the same length as the petioles are close to var. *scaposa* Gray of n-cen TX. Globose roots, however, are found to w NE, and the length of peduncles varies too much to make distinction meaningful. The seed with its prominent irregular ridges on the faces is unlike any other species of the genus in our area. Like *P. esculenta*, this species was used as food by the Indians.

6. *Psoralea lanceolata* Pursh, lemon scurf-pea. Perennial herbs with taproots and developing extensive root system, producing colonies from adventitious buds on roots; stems erect or ascending, single or clustered, much branched, 1–4(8) dm tall, sparsely pubescent, punctate-glandular throughout. Leaves alternate, usually palmately 3-foliolate; leaflets narrowly linear, oblanceolate, or obovate to elliptic, 1–4 cm long, sparsely appressed-pubescent or glabrate, punctate with glands somewhat unequal in size; petioles 1–2 cm long; stipules linear-lanceolate, 3–10 mm long. Peduncles axillary, 2–5 cm long; racemes slender, loose, 10–25(30) mm long; Pedicels 0.2–3 mm long; bracts scarious, 1–1.5 mm long, early deciduous. Calyx tube campanulate, strigose, 2–2.5 mm long, teeth about equal, but 2 united and shortly bifid, 1–1.5 mm long, not enlarging in fruit; corolla white with keel purple-tipped; banner with nearly orbicular blade 5–6 mm long, claw 1 mm long; wings obliquely oblong-oblanceolate, blade 3–4 mm long, claw 1 mm long; free portion of keel; 2–2.5 mm long. Pod globose, 4–6 mm long, with short slender beak, sparingly to densely strigose, conspicuously glandular-warty, pericarp rigid; seed 4–5 mm long, 3–3.5 mm wide, reddish-brown, smooth. (n = 11) Jun–Aug. Locally common to abundant on sandy prairies, sand dunes, stream valleys, waste places, roadsides; w 2/3 GP & along river valleys eastward; (s Sask., s Alta., s to w IA, KS, OK, TX, WA, AZ, NM). *Psoralidium lanceolatum* (Pursh) Rydb., *P. micranthum* (A. Gray) Rydb.—Rydberg.

Plants in the sw portion of our area with leaflets narrowly linear have been recognized as *P. micrantha* A. Gray ex Torr. but such specimens are found over much of our area and usually with plants having leaflets lance-oblong to linear. It is not unusual to have both leaflet shapes on a single plant. The plant has been reported to be rhizomatose, but all that I have dug produce shoots from the roots.

Such shoots often root and form a knobby or shortly branching secondary caudex resulting in a complex colony. The bruised foliage sometimes has a lemon odor.

7. Psoralea linearifolia T. & G., slimleaf scurf-pea. Perennial herbs with stout taproot and superficial or subterranean knobby or shortly branching caudex, often producing shoots and secondary caudices from adventitious buds on lateral roots; stems erect, 3–8(12) dm tall, sparingly strigose and glandular-punctate or glabrate, striate, with long slender branches above the middle. Leaves usually palmately 3(4–5)-foliolate, those of ultimate branches often 1- to 2-foliolate; leaflets linear to lanceolate or narrowly oblanceolate, 2–6 cm long, 1–4(5) mm wide, with many small glands, glabrous above, sparingly strigose below, acute at both ends; petioles 1–5 mm long; stipules lanceolate or subulate, 3–7(9) mm long. Peduncles (2)4–8 cm long; racemes loose, 1–4 flowers per node, 3–6 cm long; bracts ovate to lanceolate, acuminate, 1–4 mm long; pedicels 4–8(10) mm long. Calyx tube conspicuously glandular, strigose, at anthesis 3–4 mm long, upper 3 teeth 1.5–2 mm long, lower 2 nearly united, bifid, 2–2.5 mm long, calyx slightly enlarging in fruit; corolla bluish to purple; banner rounded-obovate, blade 6–8 mm long, claw 1–2 mm long; wings with blades 5–6 mm long, claws 2–3 mm long; free portion of spur 3–4 mm long. Pod 6–8 mm long, body ovoid-elliptic, glandular, flattened, tapering into a beak 2 mm long, pericarp rather thin; seed 5–5.5 mm long, reniform, dark brown or blackish, smooth. May–Aug. Infrequent to common on calcareous rocky prairies hillsides, ravines, bluffs, more rarely on sandy or gravelly slopes & stream valleys; SD: Mellette; NE: Deuel; WY: Laramie; e CO, w KS, OK, TX; (SD, WY, NE s to TX). *Psoralidium linearifolium* (T. & G.) Rydb. — Rydberg.

This species is much more common in the sw GP than has been recognized, even though it is one of our taller and more easily identified species.

8. Psoralea psoralioides (Walt.) Cory var. **eglandulosa** (Ell.) F. L. Freeman, Samson's snakeroot. Perennial herbs with fusiform-tuberous taproots, these apically producing 1(2–4) erect, somewhat slender rhizomes with stipules and adventitious roots, at soil surface each with knobby or shortly branching caudex; stems 1–several from base, erect or ascending, 3–8 dm tall, sparingly branched, glabrous or sparingly strigose. Leaves alternate, pinnately trifoliolate; leaflets 4–7 cm long, 1–2 cm wide, elliptic-lanceolate, 4–7 × as long as wide, glandless or sparingly dotted with minute glands, sparingly strigose; petioles 1–6 cm long petioles 1–6 cm long to nearly absent above; stipules (0.5)4–5(10) mm long. Peduncles 8–15 cm long, exceeding leaves; racemes 4–10 cm long, dense, spikelike, loosening in fruit; bracts ovate-acuminate to lanceolate, 4–6 mm long, strigose, early deciduous. Calyx tube campanulate, strigose, not enlarging in fruit, 1.5–2 mm long, teeth of upper lobe 1.5–2 mm long, lower lobe 3–3.5 mm long; corolla lilac or lavender, keel purple-tipped; banner obovate, 5–7 mm long, scarcely clawed; wings 4–6 mm long; keel 2.5–3 mm long. Pod obliquely orbicular, flattish, 4–5 mm long, transversely ridged, with a short-incurved beak from the corner, pericarp rather rigid; seed ovoid, 3–3.5 mm long, maroon or dark brown. (n = 11) May–Jul. Locally common on acid soils of rocky open woods, bluffs, & prairies; extreme se GP; (OH, IL, MO, se KS, s to GA, TX). *Orbexilum pedunculatum* (Mill.) Rydb. — Rydberg.

The leaves of this species are somewhat dimorphic in that the early primary ones are essentially palmately 3-foliolate with leaflets about as wide as long, but these are rarely collected. The adventitious roots of the vertical rhizome usually are nodule-bearing. This is our only species with pods conspicuously transversely ridged.

9. Psoralea tenuiflora Pursh, wild alfalfa, scurfy pea. Perennial herbs with deep taproots terminated by superficial knobby caudex or caudex subterranean with erect or ascending branches; stems erect or ascending, much branched, 1–several from base, (2)4–6(12) dm tall, striate, sparsely to densely short-strigose and sometimes with spreading or ascending hairs, sparsely glandular or glands absent. Leaves alternate, usually palmately 5-foliolate below, becoming 3(4)-foliolate above and 3-foliolate on branches; leaflets elliptic to oblanceolate, rarely linear or obovate, acute, to usually obtuse or rounded and mucronate, 1–5 cm long, 5–12 mm wide, glabrate above, strigose below, conspicuously glandular above, less so below; petioles 3–15(20) mm long, appressed or spreading-pubescent; stipules deltoid 6–9 mm long and deltoid below to lanceolate and 2–3 mm long above. Peduncles axillary, longer than leaves, appressed or spreading-pubescent; racemes loose or dense, in clusters of 2 or 3(7) or some single; pedicels 1–2(4) mm long, appressed or spreading-pubescent; bracts ovate-deltoid, ovate or lanceolate, acuminate, alternate or paired, 2–6 mm long, semipersistent. Calyx tube campanulate, (1)1.5–2(3) mm long, conspicuously glandular, appressed or spreading-pubescent, not enlarging in fruit, two middle teeth of upper lobe united, bidentate, upper teeth 1–1.5 mm long, lower tooth 1.8–2.5 mm long; corolla light blue, purple, rarely white with keel purple-tipped; banner rounded-obovate, nearly orbicular, blade 4–8 mm long, claw 0.5–2 mm long; wings obliquely obovate or oblong, blade 3–6 mm long, claw 1–2 mm long; free portion of keel 1.5–3 mm long. Pods 7–8(9) mm long with an abruptly tapering short beak, body slightly compressed, often asymmetric, elliptic to ovoid, glabrous, conspicuously glandular; seed 3.8–4.2 mm long, plump, reniform, olivaceous to dark brown, smooth, often shiny.

Somewhat fruitless attempts have been made to recognize two taxa based on density of racemes, number of flowers per whorl, and flower size. However, two geographic varieties can be recognized in our area as follows:

9. var. *floribunda* (Nutt.) Rydb. Plants with calyx tube and often the pedicels and branchlets loosely villous-hirsute; stems and branchlets sparsely glandular. Racemes usually with crowded flowers; flowers usually 6–8 mm long. (n = 11) May–Jul (Sep). Infrequent to locally abundant on tall-grass prairies, open woodlands, roadsides; s-cen MN, IA, se 1/4 NE, e 1/2 KS, MO, e 1/2 OK; (e of our area to IL & reportedly introduced in IN). *Psoralidium batesii* Rydb., *P. floribundum* (Nutt.) Rydb.—Rydberg; *Psoralea floribunda* Nutt.—Gates.

9b. var. *tenuiflora*. Plants with calyx tube, pedicels, and branchlets strigose, rarely a few hairs divergently ascending; stems and branches more densely glandular. Racemes rather loose; flowers usually 5–6 mm long. (n = 11) May–Jul (Sep). Infrequent to locally common in mid- and short-grass prairies, bluffs, stream valleys, roadsides; se MT, ND: Bowman; sw 1/2 SD, e WY, w 1/2 NE, w 1/2 KS, CO, w 1/2 OK, TX panhandle, NM; (GP, s to n Mex.). *Psoralidium tenuiflorum* (Pursh) Rydb.—Rydberg.

Our varieties of wild alfalfa are browsed by livestock in early stages of growth but are not relished later. In range lands it increases at first but thins out or disappears when grazing pressure forces its use. When cured in hay it is readily consumed

24. PUERARIA DC., Kudzu-vine

1. ***Pueraria lobata*** (Willd.) Ohwi, kudzu-vine. Perennial vine from farinaceous tuberous roots and woody crowns; stems trailing or climbing, semiwoody, 5–20 m long, sparsely to densely fuscous-villous. Leaves alternate, pinnately 3-foliolate, stipellate; leaflets ovate-rhombic or ovate to more or less rotund, entire or often 2- to 3-lobed, abruptly tapering to an acuminate tip, pubescent beneath, 0.5–2 dm long; petioles 10–20 mm long, often as long as rest of leaf; stipules herbaceous, lanceolate or ovate-lanceolate, medifixed, 15–20 mm long. Flowers in axillary racemes 5–20 cm long; pedicels 2–8 mm long, subtended by bracts 2–3 mm long, rachis and pedicels appressed-pubescent. Calyx campanulate, densely appressed-pubescent, tube 2.5–3.5 mm long, upper lobes united and 5–7 mm long, middle tooth of lower lobe 6–12 mm long; petals reddish; banner obovate, 1.5–2.5 cm long; wings united to, and a little longer than, the keel petals; stamens monadelphous, the uppermost one coherent with others only near the middle. Pod linear-oblong, 4–5 cm long, flattish, several-seeded, dehiscent, reddish-brown villous. (n = 11,12) Rarely flowering in our area. Rarely planted as a ground cover on embankments in e & se KS; (se U.S. to KS; e Asia). *Introduced.*

Introduced for its edible starchy root and fiber, this species was later planted as a ground cover, for green manure, hay and forage. It has escaped over much of the se U.S. and has become a serious noxious weed. Though the plant has not been found as a certain escape in our area, I include the species because of the possibility of its later becoming established. In the GP the above-ground parts normally winterkill. This description is based largely upon plants from beyond our area.

25. ROBINIA L., Locust

1. ***Robinia pseudo-acacia*** L., black locust. Deciduous tree up to 15 m tall, forming a rounded or oblong crown of irregular, more or less flexuous branches; usually forming colonies by means of root sprouts. Leaves alternate, odd-pinnate, 1–2.5 dm long; leaflets 7–21(27), elliptic to oblong-ovate or ovate, opposite, subopposite, or alternate, 2–5 cm long, 1–2(3) cm wide, at first densely puberulent, becoming glabrate, entire; stipels early deciduous; petioles 0.5–5 cm long; stipules linear-subulate, at first membranaceous but often developing into woody, persistent spines 3–25 mm long. Racemes drooping, 0.5–2 dm long, with 10–35 fragrant flowers. Pedicels 5–10 mm long; calyx tube campanulate, bilabiate, finely pubescent, 4–5 mm long, lower 3 teeth 1.5–2 mm long, upper 2 teeth connate, 2–3 mm long; petals white except for yellowish patch on banner; banner blade obcordate or suborbicular 15–25 mm long, more or less reflexed, claw 4–6 mm long; wings and keel nearly as long, claw 1/3–1/4 length of blades; stamens diadelphous. Pods 5–10 cm long, 1–1.5 cm wide, straight, flat, glabrous, short-stipitate, with dry thin valves, often persistent through winter; seeds (3)4–8(12), reniform, 5–5.5 mm long, 3 mm wide, dark brown, often mottled with purple-brown spots. (n = 10) May–Jun. Dry or moist open woodlands, stream valleys, pastures, thickets, roadsides; GP but infrequent n & w; (PA, s IN, s MO, GP, s to GA, LA, OK).

This species was often planted by early settlers in our area, and our records are based on continued plantings and escapes. It is native to the se of the GP.

Robinia hispida L., bristly locust, has occasionally been planted in our area but escapes and persistent colonies are rare. It differs in having pink or purple flowers, hispid-setose twigs, densely hispid pods, and is a shrub.

26. SESBANIA Scop.

1. ***Sesbania macrocarpa*** Muhl., bequilla. Robust annual herb; stems 7–20 dm or more tall, erect or ascending, rarely branched, glabrous; leaves alternate, even-pinnate, 1–2(3) dm long; leaflets numerous (up to 70), usually 1–3 cm long, 2–6 mm wide, narrowly oblong, Peduncles axillary, 2–4 cm long; racemes few-flowered. Pedicels 0.5–1 cm long; calyx tube campanulate, glabrous, 3–4 mm long, lobes 1–1.5 mm long, nearly equal; petals yellow, often streaked or spotted with red; banner reflexed, blade suborbicular, 12–15 mm long, claw short; wings oblanceolate to oblong, claws 1/4–1/2 as long as blades; keel with claw about as long as blade, Pods linear, glabrous, 1–2 dm long, 3–4 mm wide, short stipitate, glabrous, 30- to 40-seeded. (n = 6) Jul–Oct. Border of ox-bow lakes, stream valleys, roadside ditches; scattered in se GP; (VA, IL, MO, se KS, OK, s to GA, FL, TX). *Sesbania exaltata* (Raf.) Rydb.—Barkley.

27. SOPHORA L.

1. ***Sophora nuttalliana*** B. L. Turner, white loco. Perennial herbs with taproots but soon forming extensive colonies by means of shoots developing from adventitious buds on thick lateral roots; stems erect or ascending, 1–4(7) dm tall, single or much branched from base,

usually branching above, stems and herbage silky-canescent. Leaves alternate, odd-pinnate; leaflets 15–23(31), elliptic, narrowly oblong or obovate, 4–10(17) mm long, acute to obtuse or rarely lower retuse; petioles short; stipules bristlelike or obsolete. Peduncles axillary, 1–4 cm long; racemes loosely to densely flowered, 4–10 cm long, pedicels 1–3 mm long, bracts narrowly lanceolate, 4–10 mm long, semipersistent. Calyx tube campanulate, gibbous, 5–7 mm long, strigose, becoming membranaceous, 2 teeth of upper lobe connivent 1/2 length and 1.5–2 mm long, lower 3 teeth 2–2.5 mm long; corolla papilionaceous, petals white to ochroleucous; banner with blade oblanceolate, widened above middle, 9–12 mm long, tapering into claws 5–7 mm long; wing blades asymmetric, oblong, 8–11 mm long, claws 5–6 mm long, blades of keel 7–8 mm long, connivent in distal 1/2, claws 5–6 mm long, with two apical beaks 1–3 mm long; stamens 10, distinct to base. Pods 3–5(7) cm long, short-stipitate, constricted between seeds, beaked, terete, indehiscent or tardily so, with 1–5(7) seeds; seeds 4.5–5 mm long, oblong, 2.5–3 mm in diam, olivaceous to brown, smooth. May–Jul. Common on dry prairie hills & plains, stream valleys, badlands, roadsides; w 2/3 GP from cen SD s; (SD, e WY, to KS, OK, TX, NM, AZ). *S. sericea* Nutt.—Rydberg.

This plant has been reported to be toxic to horses, but feeding experiments failed to produce symptoms.

28. SPHAEROPHYSA DC.

1. ***Spaerophysa salsula*** (Pall.) DC. Perennial herb forming extensive colonies by means of extensive woody rootstocks; stems erect or ascending, 1–several from base, glabrous or sparsely strigose above, 4–9 dm tall, often with ascending branches above. Leaves alternate, odd-pinnate, 3–10 cm long; leaflets (9)15–25, oblong-obovate to narrowly oblong, or narrowly elliptic above, 3–18 mm long, glabrous above, strigulose below; petioles 3–20 mm long; stipules lanceolate or linear, 2–6 mm long, basally united to petiole. Peduncles axillary; racemes loosely 5- to 12-flowered, about equaling leaves; stem continuing growth in length above inflorescences after anthesis. Calyx tube campanulate, 4–5 mm long, teeth 1.2–2 mm long, the calyx becoming papery; corolla papilionaceous, brick-red, fading purplish or brownish; banner recurved, 12–15 mm long, 11–14 mm wide; wings with blades 7.5–8.5 mm long, claws 3–3.4 mm long; keel blades 6.5–7. mm long, claws 4.5–5 mm long; stamens 10, diadelphous. Pod inflated, globose and bladderlike with papery walls, body 12–24 mm long, 10–20 mm in diam, stipe 6–12 mm long seeds oval, 2.5–3 mm in diam, olivaceous to brown, smooth. Sandy or alkaline soil in Cimarron valley; KS: Clark, Meade; (well established in nw U.S., MT, WY, CO, to KS, TX; n Asia). *Introduced.*

Where established, this species is considered to be a serious weed and bears checking in our area.

29. STROPHOSTYLES Ell., Wild Bean

Ours annual, vinelike herbs with taproots; stem 1, early erect but becoming trailing or twining and alternately branching above, or more often the lowest 1–4(6) nodes with opposite branches, each alternate branched above. Leaves dimorphic; those of lower 1–4 nodes, simple, opposite, entire, reniform, blades 8–12 mm long, 12–15 mm wide, rarely collected; rest of leaves alternate, pinnately trifoliolate; stipels present, semipersistent; petioles 0.5–4(7) cm long; stipules 2–5 mm long, persistent. Peduncles axillary, 4–15 cm long; racemes subcapitate, few-flowered; each flower subtended by 2 lanceolate bracts. Calyx tube campanulate, bilabiate, upper lobe with 2 nearly united teeth, middle tooth of lower lobe the longest; corolla papilionaceous, pink to purplish or ochroleucous, often fading greenish or yellowish; banner 5–15 mm long, its sides folded over other petals; wings shorter than keel; keel widest near middle, abruptly contracted and curved upward into a beak which points back into flower or rarely contorted with a 1/2–1 spiral; stamens 10, diadelphous. Pods sessile, linear, subterete, several-seeded, valves coiling in dehiscence; seeds smooth, scurfy, or pubescent.

1 Flowers 8–14 mm long; seeds 5–10 mm long, scurfy or woolly; leaflets usually less than 2 × longer than wide .... 1. *S. helvola*
1 Flowers 5–8 mm long; seeds 3–4 mm long, glabrous; leaflets usually 2 × or more longer than wide .... 2. *S. leiosperma*

1. *Strophostyles helvola* (L.) Ell. Annual herbs with taproots; stem 1 or several from branching base, 3–12(20) dm long, trailing or twining, retrorsely spreading-pilose or glabrate, often branching above. Leaves of lowest 1–4 nodes usually opposite, simple, entire, reniform, 8–12 mm long, rarely collected; the principal leaves alternate, pinnately trifoliolate; leaflets ovate or rhombic-ovate to ovate-oblong, often 3-lobed or with a contraction on one or both sides, or entire, 2–6.5 cm long, glabrous to sparsely strigose on both sides; petioles 1–8 cm long; stipules lanceolate, 4–6 mm long, persistent. Peduncles axillary, 6–15 cm long, usually longer than leaves; racemes subcapitate, few-flowered; bracts subtending flowers lanceolate, shorter to longer than calyx-tube. Calyx tube campanulate, glabrous to sparsely hirsute, 1.7–3 mm long, upper lobe with 2 united teeth, 1.7–2.3 mm long, middle tooth of lower lobe 4–6 mm long, laterals 2.5–3 mm long; petals roseate to purplish, often greenish or fading yellowish; banner 10–14 mm long, 10–12 mm wide, claw 1–1.5 mm long; wings with blades 6–8 mm long, claws 1.5–2 mm long; keel blade 12–14 mm long, widest near middle, abruptly contracted and curved upward into a beak which points back into flower, claw 2–3 cm long. Pods (3)5–10 mm long, 5–8 mm wide, subterete, sparsely appressed-pubescent; seeds 6–10 mm long, permanently scurfy or woolly. (n = 11) Jun–Oct. Infrequent to locally common in rocky woodlands & thickets, stream banks, sandy lakeshores, moist prairie ravines, roadsides; e 1/2 GP from e SD southward; (e MA, sw Que., s Ont., MI, WI, e SD, s to FL, TX). *S. missouriensis* (S. Wats.) Small — Rydberg.

High climbing plants with entire leaflets and seeds to 12 mm long have been recognized as var. *missourienses* (S. Wats.) Small — Fernald, but such distinction appears to have no taxonomic merit.

2. *Strophostyles leiosperma* (T. & G.) Piper, slick-seed bean. Annual herbs with taproots; stem 1 or several from branching base, (2)4–10(18) dm long, trailing or climbing, retrorsely spreading-pilose, sometimes glabrate, often branching above. Leaves of lowest 1–4 nodes usually opposite, simple, entire, reniform, 8–10 mm long, rarely collected; the principal leaves alternate, pinnately trifoliolate; leaflets narrowly elliptic, lanceolate, lance-oblong, to rarely ovate or rhombic-ovate, rarely with a contraction on one side or sinuately shallowly lobed, somewhat retrorsely pilose or hirsute or with appressed pubescence. Peduncles axillary 3–10 cm long; racemes subcapitate, few-flowered; bracts subtending flowers, lanceolate, subequal to calyx tube. Calyx tube campanulate, usually densely hirsute or strigose, 1.5–2 mm long, upper lobe with 2 teeth united or slightly dentate, 1.5–2 mm long, lower lobe with middle tooth 2–4 mm long, laterals 1.7–2.4 mm long; petals light rose to purplish, fading yellowish or rarely greenish; banner blade 7–8 mm long, nearly as wide, claw 1 mm long; wing blades 5–6 mm long, claw 1.5–2 mm long; keel blade 7–8 mm long, abruptly contracted and curved upward into a beak which points back into flower or rarely with a 1/2–1 spiral. Pods (1.5)3–4(5) cm long, subterete, appressed or ascending-pubescent, rarely glabrate; seeds 3–4(4.5) mm long, early scurfy but becoming smooth and shiny, grayish or brownish and with black or purplish markings. (n = 11) May–Oct. Infrequent to locally abundant on sandy prairies, sand dunes, stream valleys, fields, roadsides; less frequent in rocky open woodlands, often weedy; GP s from s 1/2 ND; (OH, IN, WI, ND, s to MS, CO, NM, TX).

Some colonies of this species have leaflets ovate or rhombic-ovate, and without flowers and fruits are difficult to separate from *S. helvola*. In some of these the flowers are also intermediate in size and the seeds tardily lose the scrufiness. Such plants look like hybrids, but often are found much beyond the range of *S. helvola*. Rarely in our area, some plants have taproots much thickened and may be perennials. Some have bracts subtending flowers shorter than calyx tube and rounded apically. These are close to *S. umbellata* (Muhl. ex Willd.) Britt., but our plants with these characters are 200–300 mi w of known *umbellata* sites. Our few plants with partially spiral and contorted keel beaks make the separation of *Strophostyles* from *Phaseolus* somewhat questionable. These problems merit careful field and experimental studies.

30. STYLOSANTHES Sw., Pencil-flower

1. ***Stylosanthes biflora*** (L.) B.S.P., pencil-flower. Perennial herbs with taproots and superficial or subterranean, shortly branching caudex; stems usually several, erect, ascending or rarely sprawling, sometimes bushy-branched, 1–4(6) dm tall, glabrous to finely puberulent or sparsely to densely hispid with yellowish bristles. Leaves alternate, pinnately trifoliolate, petiolule of terminal leaflet 1–3 mm long; leaflets elliptic, oblanceolate to rarely obovate, 1.5–4 cm long, acute to obtuse apically, glabrous on both sides, entire, often with spinulose teeth on upper margins, prominently veined below, without stipels; petioles 1–3 mm long; stipules adnate nearly length of petiole, free apical portions 3–8 mm long, often hispid. Flowers solitary in axils of distal leaves, floriferous nodes 1–8, usually crowded into a subcapitate or spikelike raceme. Calyx tube campanulate above slender hypanthium, upper lobe of 2 united teeth 1.2–1.8 mm long, middle tooth of lower lobe 2–3 mm long, lateral 2 teeth 1–1.5 mm long; petals orange-yellow to whitish, often fading pinkish; banner 5–9 mm long, to 5–6 mm wide, orbicular; wings 3.5–4.5 mm long, short clawed; keel curved upwards, about as long as wings; stamens monadelphous, anthers alternately oblong and subglobose. Fruit a 2-segment loment, the lowest segment generally sterile and stipelike, sometimes fertile, the fertile segment obliquely ovate, 3–4(5) mm long, 1-seeded, reticulate, puberulent to glabrate, with curved beak 0.5–1 mm long; seed 2–2.5 mm long, with irregular rounded raised ridges, yellowish to brown. (n = 10) May–Sept. Locally common in acid soils of rocky or sandy open woodlands & prairies or in sand-dune prairies; se GP; (PA, NJ, IL, MO, KS, s to FL, AL, TX). *S. biflora* var. *hispidissima* (Michx.) Pollard & Ball, *S. riparia* Kearn.—Fernald.

31. TEPHROSIA Pers., Hoary Pea

1. ***Tephrosia virginiana*** (L.) Pers., goat's rue, catgut. Perennial herbs with deep woody roots and superficial or subterranean knobby or branching caudex; stems erect, one to several from base, 2–7 dm tall, unbranched or weakly branched above, sparsely to densely strigose to villous. Leaves alternate, odd-pinnate, 5–14 cm long; leaflets (7)13–25(31), those of principal leaves elliptic to linear-oblong, 10–30 mm long, 2–8 mm wide, acute to rounded-truncate, apiculate, densely short-pubescent to glabrous above, appressed to spreading-pubescent or villous below; petioles usually shorter than lowermost leaflets; stipules 8–11 mm long. Inflorescence terminal on main axis or somewhat rarely on axillary branches; peduncles short, racemes 3–10 cm long; pedicels 4–15(20) mm long; bracts subtending pedicels 0.5–2 cm long. Calyx tube campanulate, 3–5 mm long, densely strigose, pilose or villous, upper lobe with subulate to deltoid teeth 3–6 mm long, these variously acuminate, middle tooth of lower lobe 4–7 mm long, lateral teeth 3–6 mm long; corolla usually bicolored, banner lemon-yellow outside, white within, wings and keel rose, all fading brownish on drying; banner orbicular or reniform, 14–19 mm long and nearly as wide, claw 2–3 mm long, finely pubescent on back; wings 15–20 mm long, claw 2–3 mm long;

keel 14–15 mm long, nearly oval, claw 2–3 mm long; stamens 10, monadelphous but 10th stamen free below and united with others near middle; base of ovary surrounded by a collarlike disk. Pod 6- to 11-seeded, linear, straight, or downward falcate, 2.5–5.5 cm long, 3.5–5.5 mm wide, flattened, sparsely strigulose to densely villous; seeds reniform, 3–4.5 mm long, brown, mottled with black. (n = 11) May–Jul. Infrequent to locally common in sandy soils of open woodlands, prairies, sand dunes, roadsides; w IA, e 1/2 KS, OK, TX; (NH, WI, KS, s to FL, Tx). *Cracca virginiana* L., *C. leucosericea* Rydb.—Rydberg; *T. leucosericea* (Rydb.) Cory—Gates.

Under grazing the plant soon disappears. Reportedly, the Indians used the plant as a fish poison and the roots as a vermifuge. The xylem contains a small amount of rotenone, but this is absent in the bark.

32. THERMOPSIS R. Br., Buck Bean

1. ***Thermopsis rhombifolia*** Nutt. ex Richards., prairie buck bean, yellow pea. Perennial rhizomatose herbs; stems erect or ascending one to several from base, 1.5–4(6) dm tall, glabrous or appressed-pubescent, often branched above. Leaves alternate, palmately trifoliolate; leaflets broadly elliptic, oblanceolate to obovate, apically acute, or rounded and apiculate, 1.5–3(5) cm long, 1–2(3) cm wide, sparsely to densely appressed-pubescent. Flowers in subterminal axillary racemes to 1 dm long, flowers 10–30, dense; pedicels 4–10 mm long, each subtended by oblanceolate or obovate bracts 5–10 mm long. Calyx tube campanulate, bilabiate, 4–5 mm long, strigulose, somewhat gibbous, upper lobe with 2 united teeth, 3–4 mm long, teeth of lower lobe 2–3 mm long; petals yellow, banner often with median purplish dots; banner obovate, 17–20 mm long, claw 2–3 mm long; wings oblong, auriculate, 15–18 mm long; keel oblong, auriculate, 14–16 mm long; stamens 10, distinct. Pods coriaceous, recurved to annular, sparsely to densely pubescent, becoming glabrate, constricted between seeds; seeds 4.5–5.5 mm long, reniform, yellowish to dark brown, smooth, shiny. (n = 9) Apr–Jun. Infrequent to common on prairie plains & hillsides, rocky open woodlands, badlands, roadsides; w 1/2 GP, s to NE: Lincoln, foothills in CO; (ND to Alta., s to CO, NE). *T. arenosa* A. Nels.—Rydberg.

Our plants are the variety *rhombifolia*. Along the western limits of our area from se WY, s to NM, plants with pods nearly straight and ascending to divaricate are the var. *divaricata* (A. Nels.) Isely.

This species is known as buck bean, false lupine, yellow bean, and golden banner. It has been suspected of causing death in cattle and horses and the seeds have been reputed to be poisonous to children.

33. TRIFOLIUM L., Clover

Annual, biennial, or perennial herbs with taproots; stems weak, erect, ascending, decumbent or stoloniferous; leaves alternate, palmately trifoliolate or pinnately trifoliolate in 2 species, rarely with 4 or 5 leaflets; leaflets elliptic to oblong or obovate to ovate, serrulate or denticulate, glabrous or pubescent; petioles well developed; stipules conspicuous, persistent, foliaceous or membranaceous, usually at least partly fused to base of petiole. Inflorescence racemose, spicate or capitate, axillary or terminal. Flowers sessile or pedicellate, papilionaceous; calyx persistent, tube campanulate or cylindrical, usually bilabiate, teeth setaceous to triangular, often unequal; petals white to reddish or yellow, distinct in yellow-flowered species, claws united below with stamen tube in other species; petals usually withering and persistent after anthesis; stamens diadelphous. Pod often enclosed within the calyx tube, usually membranaceous, 1- to 4-seeded, dehiscent or indehiscent.

1 Flowers pale to bright yellow; terminal leaflet distinctly stalked.
  2 Flowers 3–4 mm long, usually 20–40 per head; banner striate; petioles mostly longer than leaflets .... 3. *T. campestre*
  2 Flowers 2.5–3.5 mm long, usually 5–15 per head, banner not striate; petioles mostly shorter than leaflets .... 5. *T. dubium*
1 Flowers white, pink, red, not yellow; terminal leaflet not stalked or stalks of equal length in all leaflets.
  3 Flowers with pedicels 2 mm or more long, usually recurved in fruit.
    4 Stems creeping and rooting at nodes or stoloniferous.
      5 Stipules pale and thin, usually less than 1 cm long; calyx teeth 1–1.5× longer than tube; plants common .... 11. *T. repens*
      5 Stipules foliaceous, 1–2 cm long; calyx teeth 2–4× longer than tube; plants very rare if in our area .... 13. *T. stoloniferum*
    4 Stems erect or ascending, not creeping and rooting at nodes nor stoloniferous.
      6 Heads 1.5–4.5 cm in diam; flowers 8–14 mm long; calyx barely bilabiate; banner without prolonged subulate tip.
        7 Calyx tube 5-nerved, teeth 1–1.5× longer than tube; plants common .... 7. *T. hybridum*
        7 Calyx tube 10-nerved, teeth 2.5–4× longer than tube; plants rare .... 10. *T. reflexum*
      6 Heads 1–1.5 cm in diam; flowers 4.5–7 mm long; calyx distinctly bilabiate; banner with a distinct subulate tip .... 4. *T. carolinianum*
  3 Flowers sessile or subsessile with pedicels 1 mm or less long, not recurved in fruit or only the lower ones recurved.
    8 Plants annual.
      9 Calyx tube densely pubescent on one side (upper lobe), glabrous or nearly so on other side, greatly inflated in fruit .... 12. *T. resupinatum*
      9 Calyx tube glabrous or uniformly pubescent on both sides, not greatly inflated in fruit.
        10 Corolla 4–6 mm long, white to pinkish, shorter than calyx teeth; leaflets 3× or more longer than wide .... 1. *T. arvense*
        10 Corolla 8–12 mm long, usually bright red, longer than calyx teeth; leaflets as wide as long or to 2× longer than wide .... 8. *T. incarnatum*
    8 Plants perennial.
      11 Plant glabrous throughout; rare in e-cen SD .... 2. *T. beckwithii*
      11 Plant variously pubescent.
        12 Heads subtended by a pair of leaves, short-pedunculate; plants not rhizomatose, common .... 9. *T. pratense*
        12 Heads not subtended by a pair of leaves, long-pedunculate; plants usually rhizomatose; w NE, WY, CO .... 6. *T. fragiferum*

1. *Trifolium arvense* L., rabbit-foot clover. Annual herbs with taproots, stems erect, 1–4 dm tall, freely branched, usually villous-pubescent. Leaves alternate, palmately trifoliolate; leaflets narrowly oblong to oblanceolate or linear to elliptic, 8–25 mm long, acute, rounded or truncate apically, soft-pubescent; petiolate below to subsessile above; stipules united to petioles, with free tips 5–10 mm long. Inflorescence on axillary and terminal peduncles, densely flowered, at first subcapitate, elongating into cylindrical spikes 5–25 mm long, 10–15 mm thick, grayish-pubescent. Flowers sessile, (3)4–5(6) mm long; calyx tube 1.6–2 mm long, 10-nerved, silky-pilose, teeth setaceous, 3–5 mm long, plumose, surpassing corolla; petals pale rose, pinkish, or white, fading brownish, marcescent. Pod ovoid, enclosed within calyx tube, 1(2)-seeded; seeds ovoid, yellowish, 1 mm long. (n=7) May–Oct. Rarely found on roadsides, pastures, newly planted lawns, waste ground; ND: Cass; KS: Jefferson, Lyon; MO: Buchanan, Jasper; (e 1/2 U.S., Pacific Coast; Europe).

    This species is found rarely on newly planted lawns, roadsides, and tame-grass pastures. It apparently does not persist and is not naturalized in our area.

2. **Trifolium beckwithii** Brew. ex S. Wats., Beckwith's clover. Perennial herbs with taproots and superficial knobby or shortly branching caudex, glabrous throughout; stems several from base, ascending or erect, simple, 1–4 dm tall. Leaves palmately trifoliolate, alternate; leaflets elliptic to ovate-lanceolate or oblong-ovate, 2–5 cm long, strongly nerved, finely serrulate; stipules 1–2.5 cm long, membranaceous, ovate-lanceolate. Peduncles from axils of upper 2–several stipules, heads globose 2–3 cm in diam. Calyx tube 4–6 mm long, 5-veined, teeth about same length as tube; corolla 10–15 mm long, reddish or light purplish. Pod 1- to 3-seeded. Jun–Jul. SD: Coddington, Deuel, Brookings; (MT, to se OR, s to CA; e SD).

This clover, known from se OR to w-cen MT and s to the middle Sierra Nevada of CA, is remarkable for its disjunct occurrence in e SD.

3. **Trifolium campestre** Schreb., low hop-clover. Annual herbs with taproots; stems decumbent, ascending or erect, usually much branched, 1–4 dm tall, finely appressed-pubescent to glabrate. Leaves alternate, pinnately trifoliolate; leaflets oblong to obovate or oblanceolate, 6–15 mm long, glabrous to sparsely pubescent, usually denticulate above middle; petioles to 3 cm long below, becoming 1–2 mm long above; stipules ovate to ovate-lanceolate, usually 5–8 mm long. Peduncles axillary, often longer than leaves; heads globose to short-cylindric, 8–15 mm long, flowers 20–40 per head, 3–4 mm long. Calyx tube 0.7–1 mm long, membranaceous, 5-nerved, upper 2 teeth 0.1–0.2 mm long, the lower 3 teeth 0.8–1.5 mm long; petals yellow, becoming brown and marcescent, the banner conspicuously striate. Pod with stipe 0.8–1 mm long, 1-seeded; seed 0.8–1.2 mm long, yellowish, smooth, shiny. (n = 7) May–Sep. Infrequent to locally common in rocky or sandy open pastures, open woodlands, lawns, roadsides; ND: Benson, Richland; MN: Rock, Douglas; NE: Knox, Lancaster, Richardson; e 1/2 KS; (introduced over much of N. Amer.; Europe). Probably *naturalized* in se KS, not persistent in n GP. *T. procumbens* L.—Rydberg.

4. **Trifolium carolinianum** Michx., Carolina clover. Ours an annual or winter annual herb with taproot; stems erect or ascending, often branched from base, 1–3 dm tall, glabrous or sparsely pubescent. Leaves alternate, palmately trifoliolate; leaflets obovate to obcordate, 4–10 mm long, 3–10 mm wide, denticulate, glabrous or sparsely pubescent; petioles 1–5 cm long; stipules ovate to ovate-lanceolate, adnate to petioles 1/3 their length. Peduncles terminal and axillary, 3–10 cm long; heads globose, 10–15 mm in diam at anthesis; pedicels 1–4 mm long, to 4 mm in fruit. Calyx tube 10-nerved, 0.7–1.2 mm long, pubescent, teeth 1–3 mm long, unequal; petals yellowish-white or purplish, fading brown, 5–7 mm long, with longitudinal lines. Pods oblong, short-stipitate, 2.5–3.5 mm long, 2- to 4-seeded; seeds 1–1.2 mm long, dark olivaceous to blackish, smooth, dull. May–Jun. Rare in rocky open woods & sandy prairies; KS: Cherokee; MO: Jasper, Newton; (s VA to MO, se KS, s to FL & e TX).

This species was last collected in Kansas in 1896 from open places in Cherokee County. Repeated field studies have failed to locate the species. The above description is based on collections from AR and e OK.

5. **Trifolium dubium** Sibth., small hop-clover. Annual herbs with taproots; stems erect or decumbent, 0.5–3.5 dm tall, appressed-pubescent to glabrate, much branched. Leaves alternate, pinnately trifoliolate; leaflets obovate, 5–12 mm long, nearly glabrous, serrulate to denticulate in upper 1/2; petioles 2–10 mm long; stipules ovate, 4–8 mm long. Peduncles axillary, 5–15(20) mm long; heads globose, 5–8 mm in diam, usually 5- to 15-flowered. Calyx tube 0.8–1.2 mm long, 5-nerved, glabrous, 2 upper teeth 0.2–0.5 mm long, the lower 3 teeth 1–1.2 mm long; petals yellow, fading brown, marcescent; banner 2.5–3 mm long, narrowly oblong, not conspicuously veined, not striate. Pods oblong, 2.5–3 mm long, including 1 mm long stipe, 1-seeded; seed 1–1.3 mm long, yellowish, smooth, shiny. (n = 7, 14)

May–Sep. Rare to locally common in rocky open woodlands, rocky or gravelly banks, pastures, roadsides, & waste places; extreme se GP; (introduced over se & nw N. Amer.; Europe).

This plant sometimes appears in newly seeded lawns to the n of its naturalized range in our area but does not persist.

6. **Trifolium fragiferum** L., strawberry clover. Perennial herbs with taproots; stems several from the base, (3)10–30(40) cm tall, decumbent to creeping and rooting at nodes, rarely cespitose with short stems and not rooting, sparsely to somewhat densely pubescent. Leaves alternate, palmately trifoliolate; leaflets obovate to broadly elliptical, 8–25 mm long, serrulate, retuse; petioles 3–15 cm long or longer; stipules lanceolate-subulate, 13–20 mm long, free portion acuminate. Peduncles 5–10(20) cm long, often pubescent, exceeding leaves; heads 10–14 mm wide at anthesis becoming 12–20(30) mm wide and globose, ellipsoid or nearly cylindrical in fruit; heads subtended by whorled bracts 3–5 mm long, the lowest united and forming a dissected involucre. Calyx 3.5–4.5 mm long at anthesis, bilabiate, teeth as long as or longer than tube in fruit, upper lip much inflated in fruit and reticulate veined; corolla pale pink, 6–7 mm long, persistent, mostly concealed by calyx or exserted by 2–3 mm. Pods 1- to 2(3)-seeded; seeds 2–2.3 mm long, olivaceous to dark brown, smooth, dull. (n = 8) May–Sep. Infrequent but becoming locally common in moist sandy pastures of river valleys; MT: Treasure; WY: Goshen, Johnson, Platte; NE: Hall, Garden, Grant, Morrill; CO: Yuma; (becoming established in WA, OR, ID, GP, & e U.S.; Europe).

Some of our plants with calyx 3.5–4 mm at anthesis, teeth not longer than tube, and corolla exserted 2–3 mm beyond calyx are subsp. *bonannii* (Presl) Sojak. The more common plant with calyx 4–4.5 mm at anthesis, teeth longer than tube, and flowers not exserted are subsp. *fragiferum*. These subspecies are often found together in a colony, and intermediates are frequent. The value of the distinction appears questionable.

7. **Trifolium hybridum** L., Alsike clover. Perennial herbs with taproots, sometimes appearing stolonous but without adventitious roots; stems ascending to erect, 3–8 dm tall, glabrous to sparsely pubescent, often fistulose. Leaves alternate, palmately trifoliolate, glabrous; leaflets oval to usually broadly elliptic or obovate, rounded to retuse apically, 1–6 cm long, serrulate to denticulate; petioles well developed; stipules clasping, 1–3 cm long, ovate-lanceolate, acuminate, somewhat foliaceous. Peduncles from upper leaf axils, usually longer than leaves; heads globose, pedicels to 4–5 mm long, deflexed after anthesis. Calyx tube 5-nerved, 1.5–2.5 mm long, glabrous to sparsely puberulent, teeth somewhat unequal, slightly longer than tube; petals white or pinkish, brownish in age, 6–12 mm long, banner 2–3 mm longer than wings and keel. Pod 2- or 3(4)-seeded, subreniform, olivaceous to blackish, smooth. (n = 8) May–Oct. Found in fields, pastures, stream valleys, roadsides, & waste places; e 1/2 & n 1/2 GP; (scattered over much of U.S., more common northward; Europe). *Introduced.*

Alsike clover has been planted as a substitute for red clover in moist soils. Subsp. *hybridum*, with stems sparingly branched, erect, fistulose, and heads ca 25 mm in diam, has been planted in Canada and our northern states but is apparently rare in our area. Except for a few plants in ND and SD our plants are the subsp. *elegans* (Savi) Asch. & Graebn. with stems usually much branched, ascending, hardly fistulose and heads 15–19(22) mm in diam.

8. **Trifolium incarnatum** L., crimson clover. Annual herbs with taproots; stems erect, simple or branched at base, 2–4(8) dm tall, appressed-pubescent to villous. Leaves alternate, palmately trifoliolate, pubescent; leaflets broadly obovate to almost orbicular, 1–4 cm long, denticulate in upper 1/2; petioles well developed; stipules 1–2 cm long, mostly adnate to petiole, free portion often reddish-margined. Peduncles terminal, 4–10 cm long; heads ovoid to cylindrical, 2–7 cm long; 1–2.5 cm in diam, flowers sessile. Calyx tube densely

villous, 10-nerved, 2.7–3.2 mm long, teeth subequal and 5–7 mm long; petals scarlet to deep red, rarely white, 8–12 mm long, banner longer than wings and keel. Pod ovoid, 1-seeded; seed ovoid, yellow or reddish-brown, 2–2.5 mm long, smooth. (n = 7) May–Jul. Fields & nearby waste places & roadsides; ND: Cass; SD: Custer; NE: Lancaster; (established in se & nw U.S.; Medit. Europe).

Our records are based on infrequent escapes from plantings. The plant is not hardy in our area and does not persist.

9. *Trifolium pratense* L., red clover. Cespitose short-lived perennial herbs with taproots; stems several from the base, 3–10 dm tall, often fistulose, glabrous to glabrate but usually pilose or appressed-hairy to villous, branched. Leaves alternate, palmately trifoliolate; leaflets elliptic to obovate, 2–6 cm long, pubescent on both sides or glabrate above, often with a reddish or darkened spot; stipules 1–3 cm long, fused to petiole for more than 1/2 length, abruptly contracted into a setaceous bristle, glabrous or variously pubescent; petioles well developed, becoming short to absent above. Heads sessile or on peduncles to 2 cm long, globose, subtended by reduced leaves, flowers 25–80 per head, sessile, not reflexed after anthesis. Calyx tube 3–4.5 mm long, glabrous to sparsely pilose, 10-veined, middle tooth of lower lobe 4–8 mm long, other 4 teeth 2–5 mm long; corolla pink, reddish-purple, rarely ochroleucus or white, 12–20 mm long; banner longer than long-clawed wings and keel. Pods ovate, thickened above, 1- to 2-seeded, irregularly circumscissile; seed ovoid, with a distinct lateral lobe, olivaceous to yellowish-brown, sometimes purple-mottled, 1.5–2 mm long. (n = 7, 14) May–Sep. Found in fields, pastures, roadsides, waste places; GP except sw 1/4; (introduced over much of U.S.; Europe). *T. medium* L.—Rydberg.

This is a highly variable species that has been divided into several subspecific taxa, but I fail to find any combination of characters that can be used for meaningful taxonomic distinctions. The plant is always found in cultivated fields or waste areas, is weakly persistent, and doubtfully truly naturalized in our area.

10. *Trifolium reflexum* L., buffalo clover. Annual or possibly biennial herbs with taproots; stems erect or ascending, 1–several from base, 2–3(5) dm tall, branched, pilose to nearly glabrous. Leaves alternate, palmately trifoliolate, pubescent to glabrate; leaflets 1–4 cm long, ovate, elliptic, oblong to obovate, serrulate; petioles 0.5–15 cm long; stipules foliaceous, 1–3 cm long, lanceolate to ovate, acuminate, united to petiole for 1/4 their length. Peduncles terminal and axillary, 1–10 cm long; heads globose, 2–2.5 cm in diam at anthesis, mature ones 2.5–4 cm in diam; pedicels 2–5 mm long at anthesis, elongating to 10–12 mm long in fruit and recurving. Calyx tube 1–1.6 mm long, 10-nerved, glabrous or short-pubescent, teeth slightly unequal, 4–7 mm long, teeth sometimes 3-nerved, petals 7–12 mm long, banner red or pinkish, longer than white to pinkish wings and keel, all becoming brownish. Pod with stipe 1–1.5 mm long, body 3–5 mm long, oblong, 2- to 4-seeded; seed 1.2–1.5 mm long, yellowish-brown, with minute irregular ridges or obscurely verruculose. (n = 8) May–Aug. Rare in rocky open woods & prairies, gravelly stream valleys, roadsides; se NE, s IA, e 1/3 KS: (w NY, s Ont., IA, se NE, s to FL, TX). *T. reflexum* var. *glabrum* Loj.—Fernald.

Most of our specimens have nearly glabrous stems and branches and have been referred to var. *glabrum* Lojacono, but the pubescence varies to dense in the extreme se GP where both glabrous and pubescent plants are found in the same colony.

11. *Trifolium repens* L., white clover, ladino clover. Perennial herbs with taproots; stems creeping, rooting at nodes, often mat-forming, 1–4 dm long, glabrous to sparsely pubescent. Leaves palmately trifoliolate, glabrous or nearly so; leaflets elliptic-obovate to obcordate, mostly 1–3(4) cm long, denticulate to serrulate; petioles well developed, 5–20 cm long; stipules united into a membranaceous sheath, 3–10 mm long, abruptly tapering into a setaceous tip. Peduncles axillary, 5–15(20) cm long; heads nearly globose, 1–3 cm in diam, 40- to 85-flowered; pedicels 1–2 mm long, elongating to 5–6 mm in fruit and recurving. Calyx tube cylindrical, 10-nerved, 2–3 mm long, teeth 2–3 mm long; petals white or pinkish, turning brown, 6–12 mm long; banner longer than wings and keel. Pod oblong-linear, 4–5 mm long, (2)3(4)-seeded; seed yellowish, 1–1.3 mm long. (n = 16, 24, 32) May–Oct. Common in lawns, pastures, fields, stream valleys, roadsides, waste places; GP but less common westward; (introduced nearly throughout N. Amer.; Europe).

A number of cultivated strains of this species have been developed and the plant is highly variable in size of heads, length of peduncles, pubescence, and other characters. Attempts to delimit subspecific taxa have been without taxonomic merit.

12. *Trifolium resupinatum* L., Persian clover. Annual herbs with taproots; stems decumbent or ascending, glabrous, 1–4 dm tall, branched. Leaves alternate, palmately trifoliolate, glabrous or nearly so; leaflets obovate to oblanceolate, 1–1.5(2) cm long, denticulate to serrulate in upper 2/3; petioles 0.3–8 cm long; stipules lanceolate to ovate-lanceolate, acuminate, united 2/3 length with petiole. Peduncles axillary, 1–6 cm long; heads nearly globose, 5–10 mm in diam at anthesis, 15–20 mm wide in fruit, with 6–18 subsessile flowers. Calyx tube 1.5–2 mm long, bilabiate, lower lip glabrous or nearly so, upper lip densely pubescent and becoming much inflated and papery in fruit; petals pinkish to purplish, 4–6 mm long, banner longer than wings and keel; flower basally twisted and banner apearing as lowest petal. Pod oblong to ovoid, enclosed in bladdery calyx, 1- to 2-seeded; seed olivaceous to blackish, 1.5 mm long shiny on margins. (n = 8) May–Sep. Rare in newly seeded lawns & pastures; ND: Cass; KS: Riley, Saline; (introduced in e 1/2 U.S. where infrequent as an escape; Europe).

Our few records, including a report for SD, are based on infrequent escapes from lawns, pastures, and experimental plantings. Apparently, the plant is not hardy or persistent in our area.

13. *Trifolium stoloniferum* Muhl. ex Eaton, running buffalo clover. Perennial herbs with taproots; stems stoloniferous, rooting at nodes, glabrous or slightly hairy; flowering stems ascending, 1–3 dm tall, scapose below, with 2 large leaves near apex. Leaves alternate, palmately trifoliolate, glabrous or sparsely pubescent; leaflets broadly obovate, 2–5 cm long, retuse, minutely denticulate; petioles, except those subtending peduncles, 4–12 or more cm long; stipules foliaceous, oblong or ovate on scape leaves, pointed or attenuate. Scapes longer than leaves, wider than stolons, with a pair of short-petioled opposite leaves at apex which subtend 1 or 2 peduncles 2–8 cm long; heads nearly globose, 2–3.5 cm in diam. Calyx tube 1.5–2 mm long, 10-nerved, minutely pubescent, lobes subequal, 3–5 mm long; corolla white, tinged with purple, becoming brownish; flowers 8–12 mm long; banner obovate, much exceeding wings and keel. Pod stipitate, 3–6 mm long, 2- to 4-seeded; seed 1.5 mm long, yellowish-brown, minutely and irregularly ridged. May–Aug. Rare in rocky open woodlands & prairies; KS: Miami; MO: Jasper; (WV & KY, w to e KS & n AR).

All records of this species mapped in the Atlas GP, except those mentioned above were based on incorrect determinations. The species apparently is very rare throughout its range.

34. VICIA L., Vetch

Annuals, winter annuals, biennial or short-lived perennial vinelike herbs; stems erect or more commonly decumbent or climbing. Leaves pinnate with the terminal leaflet modified into simple or branched tendrils; stipels absent; leaflets mostly (2)6–24, entire, opposite, subopposite to commonly alternate on rachis; petioles short or obsolete; stipules herbaceous, often 1/2 sagittate, persistent. Flowers in few-flowered axillary clusters or in pedunculate axillary spikelike racemes; racemes loose to compact; pedicels 0–3 mm long,

Calyx tube campanulate to obconic, often gibbous, almost regular to bilabiate, the 5 teeth equal to usually unequal with the middle tooth of lower lobe the longest; corolla papilionaceous, white, blue, or reddish-purple; banner obovate with a wide claw overlapping wings; wings adhering to middle of keel and longer than keel; stamens 10, diadelphous; style filiform, with a tuft or ring of hairs at the tip. Pods 2- to several-seeded, laterally compressed, 2-valved.

References: Gunn, Charles R. 1968. The *Vicia americana* complex (Leguminosae). Iowa State Coll. J. Sci. 42(3): 171–214; Herman, F. T. 1960. Vetches of the United States—Native, Naturalized and Cultivated. U.S.D.A. Handbook 168: 1–84.

Our vetches are all quite variable, and often the taxa cannot easily be separated. Care should be taken to collect well-grown specimens in full-flower and fruit. Reports of *V. caroliniana* Walt., in the GP are without specimen evidence.

1 Peduncles obsolete, or much shorter than leaves .... 6. *V. sativa*
1 Peduncles well developed.
   2 Flowers 12–25 mm long or 8–12 mm long and racemes 25- to 50-flowered.
      3 Calyx conspicuously gibbous at base, the pedicel appearing ventral .... 7. *V. villosa*
      3 Calyx slightly gibbous, the pedicel appearing basal or from one side at base.
         4 Flowers 8–12 mm long, recurved; ovules 4–8; stipules entire; rare escapes .... 2. *V. cracca*
         4 Flowers 12–25 mm long, not recurved; ovules 8–14; stipules mostly serrate; native plants .... 1. *V. americana*
   2 Flowers 4.5–9 mm long, racemes (1)3- to 12-flowered.
      5 Peduncles (2)5- to 12-flowered, at anthesis more than 1/2 length of leaves; flowers broad and showy .... 5. *V. ludoviciana*
      5 Peduncles 1- to 4(5)-flowered, at anthesis 1/2 length of leaves or less; flowers narrow, rather inconspicuous.
         6 Corolla 4.5–7 mm long; upper calyx teeth not appreciably shorter or wider at base than the lower; peduncles (1)2- to 5-flowered .... 4. *V. leavenworthii*
         6 Corolla 6–9 mm long; upper calyx teeth much shorter and wider at base than the lower; peduncles 1- or 2-flowered .... 3. *V. exigua*

1. ***Vicia americana*** Muhl. ex Willd., American vetch. Sprawling to climbing perennial vines; stems 2–10 dm long, glabrous to sparsely pilose, often flexuous above. Leaves pinnate, terminating in a simple or branched tendril; leaflets 4–14, entire or rarely toothed, thin to coriaceous, oval to linear, apically rounded, acute or truncate, mucronulate; stipules usually sharply serrate. Peduncles axillary; racemes (2)3- to 10-flowered, usually shorter than subtending leaves. Calyx tube 3.5–5.5 mm long, slightly gibbous at base, oblique at apex, teeth variable, variously unequal, the lower usually lance-attenuate, 1.2–4 mm long, the upper short and broad; corollas bluish-purple to rarely white; banner 12–25 mm long, longer than wings and keel. Pods 2.5–3.5 cm long, glabrous, with 8–14 seeds.

Our plants can be separated into 2 varieties as follows:

1a. var. *americana*. Plant vining, usually 4–10 dm in length. Leaflets more than 4 mm wide or over 30 mm long if narrower, thin, veins branched; tendril with several branches. Racemes 5- to 9-flowered (n = 7) May–Jun. Infrequent to locally common in woodlands, thickets, stream banks, tall-grass prairies; GP but rare to absent in NE, KS & sw; (Que. to AK, s to VA, IN, MO, GP, CA, Mex.). *V. oregana* Nutt.—Rydberg.

1b. var. *minor* Hook. Plant usually not vining; stems rarely to 3.5 dm in length. Leaflets usually less than 30 mm long and 4 mm or less wide, coriaceous or stiff, with conspicuous nonbranched veins; tendrils usually not branched. Raceme mostly 3- to 4(5)-flowered. (n = 7) May–Aug. Infrequent to locally common on dry prairies, bluffs, badlands, roadsides; GP but absent in se; (Man. to B.C. s to MN, IA, nw MO, OK, TX panhandle, NM, AZ, e NV). *V. dissitifolia* (Nutt.) Rydb., *V. sparsifolia* Nutt., *V. trifida* Dietr.—Rydberg.

The American vetch is grazed by livestock, and the plant soon disappears from ranges.

2. **Vicia cracca** L., bird vetch. Perennial herb; stems 0.5–1 mm tall, glabrous or appressed-pubescent, trailing or climbing. Leaves 6–12 cm long; leaflets 12–30, 5–30 mm long, 1–6 mm wide, linear to ovate-oblong; trendrils well developed; stipules entire. Racemes long peduncled, dense, 10- to 30-flowered, equaling or exceeding subtending leaf. Flowers 9–12 mm long, pendulous, bluish-purple; calyx tube oblique, gibbous at base, 2–3 mm long, upper teeth 0.3–0.7 mm long, triangular, lower teeth 1.3–2.5 mm long; banner with blade about as long as claw. Pod 2–3 cm long, 5–6 mm wide, stipe shorter than calyx, 4- to 8-seeded; seed globose, 2.5–3 mm in diam, brownish. (n = 6, 7, 14, 15) Jun–Aug. Fields, roadsides, & waste places; rare at scattered localities in n 1/2 GP; (Newf. to B.C., s to VA, IL, NE, WA; possibly native in n N. amer. & Eurasia).

Our few scattered records are old, and the species apparently has not been collected recently in the GP.

3. **Vicia exigua** T. & G., little vetch. Slender winter annual with a taproot; stems decumbent, ascending or erect, often branching at base, 2–7 dm tall, stem and foliage glabrous or sparsely strigose. Leaves 2–6 cm long, terminating in a usually branched tendril, leaflets 4–12, narrowly linear to oblong, 5–25 mm long, from rounded to acute or rarely emarginate and mucronate at apex; stipules narrow, semisagittate, entire or incisely serrate. Peduncles 2–4 cm long, shorter than the leaves, 1- to 2-flowered, aristate. Flowers yellowish-white to purplish, (5)6–8(9) mm long; calyx sparingly hirsute-strigose, tube 1–1.5 mm long, upper teeth 0.5–1 mm long, triangular, lower teeth 0.8–1.5 mm long. Pods oblong, 2–3 cm long, 4–6 mm wide, glabrous, 4- to 7-seeded; seeds 2.5–3 mm in diam, purplish-black. May–Jun. Infrequent to locally common on dry rocky prairies & ravines; KS: Barber; TX panhandle, NM; (n CO, sw OR, s to TX, NM, AZ, CA).

This species, along with *V. leavenworthii* and *V. ludoviciana*, form a complex in which specimens often are difficult to separate since intermediate forms are very frequent. Perhaps *exigua* and *leavenworthii* should be treated as varieties of *V. ludoviciana*.

4. **Vicia leavenworthii** T. & G., Leavenworth's vetch. Slender winter annual herbs with taproots; stems often branched at base, spreading, reclining or climbing, 3–6 dm long, glabrescent to sparsely pilose or puberulent. Leaves 3–6 cm long, terminating in forked or branching tendril, glabrous or pubescent, usually aristate; leaflets 8–14, narrowly elliptic-lanceolate to oblong or elliptic to oblanceolate, 5–12 mm long, acute or rounded and mucronulate or shallowly emarginate at apex; stipules lanceolate, occasionally with 1 basal lobe. Peduncles from shorter to longer than subtending leaves, (1)2- to 5-flowered. Flowers bluish-lavender, 4.5–7 mm long; calyx tube 1–1.5 mm long, pubescent, teeth equal to slightly longer than tube. Pods oblong, 2–2.5 cm long, 5–7 mm wide, glabrous, 4- to 6-seeded; seeds 2–2.5(3) mm in diam, purplish-brown. Apr–Jun. Infrequent to locally common on dry rocky prairies, open wooded areas, roadsides; se GP; (se KS, sw MO, OK, sw AR, e 1/2 TX).

See note under *V. exigua*.

5. **Vicia ludoviciana** Nutt., deer pea vetch. Rather stout to slender glabrous or pubescent winter annuals with taproots; stems decumbent or climbing, 3–10 dm long, branched at base. Leaves 3–9 cm long, rachis terminating in a forked or branching tendril; leaflets 6–12, linear-oblong or elliptic to broadly elliptic or oval, 6–25 mm long, apex rounded or emarginate; stipuls lanceolate, with 1 or 2 basal lobes. Peduncles 2–4 cm long, shorter than or exceeding leaves; flowers (1)2–13 in a lax or dense raceme. Flowers blue to purplish, 5–8 mm long, the folded banner 1.5–4 mm long; calyx tube 1–2 mm long, sometimes pilose,

upper teeth 0.5–1 mm long, lower teeth 1.5–3 mm long. Pods 2–3 cm long, 4- to 8-seeded, oblong; seeds 2–2.5 mm in diam, brownish-white, often mottled with purple. (n = 7) Apr–Jun. Rocky or sandy open woodlands, prairie hillsides, & ravines, roadsides; se GP; (sw MO, OK, NM, s to AL, LA, TX).

Two varieties have been recognized in our area, though many specimens are intermediate and some are difficult to distinguish from *V. exigua* and *V. leavenworthii.*

5a. var. *laxiflora* Shinners. Plants usually grayish-pubescent. Racemes loosely 3- to 13-flowered; flowers 5.5–8 mm long, folded banner 1.5–2.5 mm long.

5b. var. *ludoviciana.* Plants glabrous. Racemes with 2–12 flowers, rather compactly arranged; folded banner 2.5–4 mm long.

6. *Vicia sativa* L., common vetch. Annual herbs with taproots; stems 0.3–1 m long, simple or branched, erect-ascending or climbing, glabrate to strigulose-villous. Leaves with 8–16 leaflets, terminating in a branched tendril; leaflets oblong to elliptic or linear, obovate, or oblanceolate, 1.5–3(5) cm long, emarginate to rounded or truncate, apiculate, glabrate or sparsely strigose; stipules semisagittate, often sharply lobed or serrate, lower surface with or without a purplish glandular spot beneath. Flowers 1–3 in upper axils, subsessile or very short pedunculate; calyx tube 4–7 mm long, teeth 3–7 mm long, the lower longer than upper; corolla blue, violet, purple, or whitish, wings either bluish or reddish; flowers 1–1.8 cm long or 1.8–3 cm long. Pods pale brown or almost black at maturity, 4- to 12-seeded; seeds 3–5 mm in diam, olivaceous, yellowish or brownish-black. (n = 6, 7) May–Aug. Rarely found in fields, roadsides, & waste places; very scattered in GP; (introduced over much of N. Amer.; Europe).

Two rather distinct but intergrading varieties have been recognized as follows:

6a. var. *angustifolia* (L.) Wahl. Leaflets variable but commonly linear; nectaries on stipules absent or inconspicuous. Flowers 1–1.8 cm long, wings usually not red. Pods almost black at maturity. Apparently the more common var. in GP. *V. angustifolia* (L.) Reich.—Rydberg.

6b. var. *sativa.* Leaflets usually oblanceolate; nectaries on stipules usually conspicuous. Flowers 1.8–3 cm long, wings often red. Pods pale brown at maturity.

7. *Vicia villosa* Roth, hairy vetch, woollypod vetch. Annual, possibly biennial, or perennial herbs with taproots; stems 0.5–2 m long, clambering or climbing, spreading-villous, especially above, with hairs 1–2 mm long or with sparse appressed or incurved hairs to nearly glabrous. Leaves 6–15 cm long, terminating in a branched tendril; leaflets 10–20, narrowly oblong to linear-lanceolate, obtuse and mucronate to acute, 1–2.5 cm long; stipules lanceolate, semisagittate. Peduncles elongate; racemes dense, usually secured, about as long as peduncles, 5- to 40(60)-flowered, rachis spreading-villous or of appressed or incurved hairs. Calyx tube 2.3–4.2 mm long, gibbous at base below upper lobe; pedicel appearing ventral, lower teeth 1–5 mm long, glabrous or long-villous, the upper 0.5–1.5 mm long; corolla violet and white to rose or rarely white; blade of banner less than 1/2 as long as claw. Pods 2–3 cm long, 7–10 mm wide, glabrous or pubescent; seeds globose 3.5–5 mm in diam, brownish to black.

Two reasonably distinct but intergrading varieties are found in our area as follows:

7a. var. *glabrescens* Koch. Raceme with appressed or incurved hairs, sometimes ± spreading above, or nearly glabrous. Lowest calyx tooth short-pubescent to glabrescent, 1–2.5 mm long. (n = 7) May–Jul. Infrequent but locally common in fields, roadsides, waste places, often on sandy soil; se 1/3 GP; (much of N. Amer.; Europe). *Introduced. V. dasycarpa* Ten.—Fernald.

This plant is less common than var. *villosa* in our area, but may have been overlooked. In Flora Europaea it is recognized as subsp. *varia* (Host) Corb.

7b. var. *villosa*. Rachis of raceme spreading-villous. Lowest calyx tooth long-villous, 2–5 mm long. (n = 7) Apr–Aug. Infrequent to locally abundant in fields, roadsides, stream valleys, waste places, often on sandy soil; GP but less frequent n & w; (much of N. Amer.; Europe). *Introduced.*
This is our most common introduced vetch.

75. ELAEAGNACEAE Juss., the Oleaster Family

by T. Van Bruggen

Erect shrubs or small trees, the young branches and vegetative parts densely covered with a silvery or golden-brown–colored lepidote indumentum or stellate hairs; branches sometimes becoming spiny; bark on older branches scaly, gray-brown. Leaves alternate or opposite, entire, petioled, estipulate, densely covered on one or both surfaces with the silvery or golden-brown indumentum. Flowers perfect or plants dioecious or polygamus; solitary, racemose or in axillary clusters; sepals 4, green or colored, the fused base cylindric-campanulate, constricted upward as a hypanthium around the ovary in pistillate or perfect flowers, saucer-shaped in staminate flowers, the four lobes valvate, deciduous; petals none; stamens 4 or 8, inserted on a shallowly lobed perigynous disk (hypanthium) within the calyx tube or at the base of the saucer in staminate flowers, if 4, alternating with the calyx lobes, if 8, alternate and opposite the lobes, filaments free or adnate upward 4–5 mm on the calyx tube, the anthers 2-celled, opening lengthwise; pistil 1, superior, but appearing perigynous, the ovary 1-carpeled and 1-celled, the ovule solitary, basal. Fruit a drupelike achene, enveloped by the persistent, fleshy base of the hypanthium.

1 Leaves alternate; flowers perfect or some plants polygamous; stamens 4 .... 1. *Elaeagnus*
1 Leaves opposite; flowers unisexual, the plants mostly dioecious; stamens 8. ...... 2. *Shepherdia*

1. ELAEAGNUS L., Oleaster

Erect, usually much-branched shrubs or trees to 5 m, the young branchlets covered with a silvery gray or golden-brown indumentum; older bark gray-brown, scaly; branches with spines or unarmed. Leaves alternate, petioled, densely covered with a silvery, scaly or stellate pubescence. Flowers solitary or in small lateral or axillary clusters on twigs of the current year, perfect or unisexual (plants polygamous); sepals 4, spreading, yellow on the inner surface, fragrant, deciduous; stamens 4, the style base inserted at the summit of the hypanthium which is constricted into a tube above the ovary, the lower 4–5 mm of the filaments adnate to the calyx tube, only the upper 1–2 mm free. Fruit a mealy, drupelike achene, round to ovoid, densely covered with the silvery-gray indumentum.

1 Leaves lanceolate, 3–8× longer than wide; branchlets and petioles silvery-gray; small branches often ending in spines .... 1. *E. angustifolia*
1 Leaves oblong, 2–3× longer than wide; branchlets and petioles with golden-brown scales; branches unarmed .... 2. *E. commutata*

1. ***Elaeagnus angustifolia*** L., Russian olive. Shrub or small tree to 5 m, profusely and irregularly branched, sometimes the smaller branches ending in sharp, simple spines; current-year branchlets, petioles and leaves with a silvery-gray indumentum, older twigs becoming tawny brown, glabrous. Leaves light green, deciduous but often may persist through the winter; the blades lanceolate, simple, 3–8 cm long, 0.5–2 cm wide, entire,

obtusely tapered at apex, narrowly tapered to the petiole. Flowers on pedicels 3-6 mm, solitary or 1-3 in axils of leaves or laterally at the base of twigs of the current season; sepals 4, fused below and enclosing the pistil, 6-8 mm long, the lobes 2-3 mm long, spreading to erect; stamens 4, nearly sessile, alternate with the lobes, the anthers 2-2.5 mm long, not exserted; pistil enclosed by the persistent calyx base; style 5-7 mm long, surrounded at its base by a tubular disk 1-2 mm above the constriction of the hypanthium. Fruit yellow, oval, 1-1.5 cm long, dry and mealy but sweet, edible. May-Jun, fruits Aug-Oct. Dry to moist soils of all types, specially on sandy flood plains; GP; (Man. to B.C., s to MN, IA, & KS, w to Rocky Mts.; Eurasia). *Naturalized*. Planted for windbreaks and ornament, now firmly established.

2. *Elaeagnus commutata* Bernh., silverberry. Shrub or rarely treelike, to 5 m, compactly branched, unarmed, the internodes tending to be shorter than in the preceding species; young twigs with brown-scurfy scales, older wood dark grey. Leaves light green, densely silvery scrufy on both sides, ovate to oblong, 2-8(10) cm long and 1/2 or more as wide, entire, the scurfy-brown scales extending from the petiole on the underside of the leaf blade along the midvein to 1/2 the distance to the apex; petiole 2-6 mm. Flowers 1-4 in lateral and axillary clusters near the base of twigs of the current season; fragrant; sepals 4, fused below and enclosing the pistil, 6-10 mm long, the lobes 2-4 mm long, incurved early but spreading at maturity, yellow on the inside, the outside with a dense silvery-gray indumentum; stamens 4, alternate with the lobes, the filaments adnate to the calyx tube except for the upper 1-2 mm which is free, the anthers 1-1.5 mm long, enclosed; pistil closely surrounded by the persistent calyx base; style 7-11 mm long, lacking a tubular disk at its base. Fruit ovate to ellipsoid, 0.8-1.5 cm long, the covering dry and mealy, the achene 8-striate. Jun-Jul, fruits Aug-Oct. Dry to moist, sandy soils, along water courses, on banks & hillsides, MN, ND, SD, MT; (Que. to AK, Yukon, s to MN, SD, UT).

2. SHEPHERDIA Nutt., Buffaloberry

Shrubs or small trees 0.3-6 m tall, much branched, with brown or silvery, scaly or stellate pubescence on young twigs, older branches gray-brown, scaly. Leaves opposite, entire, petioled, glabrate to densely beset with the silvery or brown indumentum. Flowers 1-4 in clusters at the nodes of 1-year-old twigs, unisexual, the plants dioecious, pistillate flowers few; sepals 4, valvate, fused at the base in the pistillate flowers, forming an urn-shaped hypanthium; in the staminate, open and saucer-shaped, both with an 8-lobed disk at the mouth of the hypanthium; stamens 8, alternate with the lobes on the disk, the filaments free. Fruit succulent, drupelike, not scaly.

1 Leaves silvery-scurfy on both sides; blades oblanceolate; branches often spiny .... 1. *S. argentea*
1 Leaves green-glabrate above, silvery-brown scurfy below; blades lanceovate; branches unarmed .... 2. *S. canadensis*

1. *Shepherdia argentea* (Pursh) Nutt., buffaloberry. Erect shrub or small tree 1-6 m, the branches irregular, 2- or 3-year-old twigs often ending in spines; young twigs with a scaly or stellate pubescence, the older branches dark gray. Leaves opposite, gray-green above and below, entire, oblanceolate to oblong, rounded at the apex, acutely narrowed at the base, 2-5 cm long, 7-12 mm wide, petiole 3-6 mm. Flowers in small clusters on 1-year-old twigs, brown, unisexual; sepals 4, the lobes of staminate flowers 2-3 mm long, spreading, rounded at the apex; sepal lobes of pistillate flowes 1-1.5 mm long, erect to incurved, their bases fused around the pistil; hypanthial disk 8-lobed; stamens 8, the filaments free, erect to spreading, anthers 1-1.2 mm long; pistillate flowers with hypanthium mouth nearly

closed with a dense tomentum; style 1-2 mm long, stigma 1-sided. Fruit red, 5-7 mm long, ovoid, succulent, edible. May–Jun, fruits Jul–Sep. Hillsides, banks, ravines of rocky, sandy, or clayey soils; MN, ND, MT, SD, NB, WY, n GP; (MN to B.C., s to N.B., NV, & NM).

2. **Shepherdia canadensis** (L.) Nutt., rabbitberry. Erect shrub 1-3 m tall, branches unarmed, golden-brown scurfy when young; older branches gray, scaly. Leaves opposite, dark green above, only sparingly glabrate, underside with gray-green, dense, stellate pubescence with scattered golden-brown scurfy scales; blades ovate to lance-ovate, 2-5 cm long, apex obtusely rounded, the base rounded to subcordate; petioles 3-5 mm. Flowers 1-4 in small clusters on 1-year-old twigs, densely covered with brown, scurfy scales, unisexual; sepals 4, on staminate flowers the lobes 1-2 mm long, spreading, lightly veined, yellow-green on the inner surface; pistillate flowers with sepal lobes 0.8-1.5 mm long, spreading, with 3 prominent veins on the yellowish inner surface, the throat of the hypanthium with a dense stellate pubescence; stamens 8, the filaments erect, shorter than the sepal lobes, arising alternate with the 8-lobed hypanthium disk; anthers 0.5-1 mm; pistil enclosed by the fused calyx, the style erect, slightly exserted from the sepal lobes, persistent in fruit, the stigma 1-sided. Fruit yellow to red, ovoid, 5-7 mm long, fleshy, insipid, and bitter. May–Jun, fruits July–Sep. Moist slopes, wooded, rocky hillsides, & openings in pine forest; n GP; ND, SD, MT, WY, CO, MN; (Newf. to AK, s to MI, CO, & NM).

76. HALORAGACEAE R. Br., the Water Milfoil Family

by T. Van Bruggen

1. MYRIOPHYLLUM L., Water Milfoil

Perennial aquatic herbs with slender, elongate, floating stems, simple or branching, to 2 m or more long, leafy throughout, the upper portion usually emersed. Leaves whorled, alternate, subverticillate to scattered, the lower submersed, pinnately divided, with 3-13 opposite to subopposite, flaccid, capillary divisions; upper emersed leaves entire to pectinately lobed. Flowers inconspicuous, symmetrical, in whorls of 4-6, in the axils of emersed leaves, perfect, or more commonly unisexual, the staminate above and pistillate below; sepals 4-parted; petals 4 or absent; stamens 4-8, the pistils (2)4, fused most of their length; stigmas 4, recurved and minutely plumose. Fruit subglobose, splitting into 2 or 4 nutlets, each 1-seeded, the mericarps with or without ridges.

1 Emersed leaves reduced, bractlike, the upper ones mostly entire and shorter than the flowers they subtend, the lower ones serrate; leaf whorls on the lower to middle part of stem mostly 1 cm or more apart; stems becoming white on drying .... 1. *M. exalbescens*
1 Emersed leaves pinnatifid or serrate, as long as to much longer than the flowers they subtend; leaf whorls spaced mostly less than 1 cm apart; stems not becoming white on drying.
   2 Fruiting mericarps with 2 sharply tuberculate ridges; emersed leaves thin and narrow, irregularly pinnatifid with unequal lobes or clefts; stamens 4 .... 2. *M. pinnatum*
   2 Fruiting mericarps usually smooth or with keeled ridges, not prominently tuberulate; emersed leaves lanceolate to oblanceolate-elliptic, their margins serrate or pectinate.
      3 Emersed leaves with serrate margins, much longer than the flowers; anthers 4; fruiting mericarps usually less than 2 mm long .... 3. *M. heterophyllum*
      3 Emersed leaves with pectinate margins, as long as to slightly longer than the flowers; anthers 8; fruiting mericarps 2-3 mm long .... 4. *M. verticillatum*

1. ***Myriophyllum exalbescens*** Fern., American milfoil. Perennial from spreading rhizomes; stems branching, to 2 m or more long, leafy throughout, becoming off-white upon drying; leaves mostly whorled, spaced 1 cm or more on the lower to middle section of the stem. Submersed leaves flaccid, pinnate, 1–4 cm long, with 7–13 pairs of capillary divisions; transition upward to emersed leaves abrupt; emersed leaves reduced, bractlike and shorter than the flowers they subtend, 2–2.5 mm long, the upper ones entire, the lower ones becoming serrate. Turions, specialized overwintering buds subtended by many telescoped fleshy turion leaves, formed apically and laterally on summer plants during late summer and fall, remaining attached to senescent parent plants and germinating the next spring. Flowers 4–6 in axils of emersed leaves, 4-merous, petals caducous or absent; anthers 8, 2–2.5 mm long; pistils 4-parted. Fruit subglobose, 4-lobed, 2.5–3.5 mm long, the mericarps rounded on the dorsal side, smooth to slightly tuberculate. May–Sep. Lakes, ponds, marshes, sloughs, or potholes; GP, less frequent s; (N.S. w to CA, AK s to AZ). *M. spicatum* L. var. *exalbescens* (Fern.) Jesps.—Gleason & Cronquist.

Aiken and Walz (Aquatic Bot., 6: 357–363. 1980.) state that *M. exalbescens* Fern. can be distinguished from the Eurasian species *M. spicatum* L., with which it is often confused, and that the latter does not form turions. *M. spicatum* is widely naturalized as a weed of shallow lakes in eastern and western North America.

2. ***Myriophyllum pinnatum*** (Walt.) B.S.P., green parrot's feather. Perennial, from rhizomes, the branches simple to spreading branched, up to 1 m, leafy. Leaves whorled in groups of 3–6 or scattered, the submersed ones flaccid, 0.7–3.0 cm long, pinnately divided into 3–6 pairs of capillary segments; transition to emersed leaves upward on the stem gradual; upper emersed leaves rigid, linear and narrow, irregularly pinnatifid with 2–4 clefts of branches on each side, 0.5–1.7 cm long, 1–4 mm wide. Flowers 4–6 in axils of emersed leaves, 4-merous; petals 1.5–2.0 mm long, purplish; anthers 4, 2 mm long. Fruit subglobose, 1–2 mm long, 4-lobed, each mericarp beaked, at maturity with 2 prominent tuberculate ridges on the dorsal side, 1-seeded. Jun–Sep. Lakes, ponds, marshes, sloughs, shallow water, or muddy shores; GP; (RI to ND, MA to FL, TX; Mex. & C. Amer.).

3. ***Myriophyllum heterophyllum*** Michx., water milfoil. Perennial, from spreading rhizomes; stems simple or branching, to 1 m or more long, leafy throughout. Leaves mostly whorled or alternate, usually spaced on the stem less than 1 cm apart; submersed leaves pinnately divided, 1.5–5.0 cm long, 5–10 pairs of capillary divisions on each side; transition to emersed leaves abrupt; emersed leaves lanceolate to elliptic or oblanceolate, rigid, 3–12 mm long, mostly serrate (dentate), the lower more deeply toothed. Flowers clustered in groups of 3–6 in axils of emersed leaves, 4-merous, the petals 1–3 mm long in staminate flowers; anthers 4, about 2 mm long. Fruit subglobose, 1–2 mm long, rounded or 2-keeled on the dorsal side, beaked. Jun–Aug. Ponds, ditches, & slow-moving streams; KS, OK, to MN, e & s; (Que. to FL, KS, TX).

4. ***Myriophyllum verticillatum*** L. Perennial, stems simple or branched, 1–2.5 m long, from rhizomes; producing turions in late fall, abscissing from the parent plant and sinking to the bottom. Submersed leaves in 4s or 5s, 2.5–5.0 cm long, mostly whorled, flaccid, pinnate, usually spaced less than 1 cm apart on the stem, with 8–13 opposite or subopposite pairs of capillary divisions 1–3 cm long; transition to emersed leaves abrupt; emersed leaves 0.7–1.8 cm long, about as long as the flowers or exceeding them lanceolate to elliptic, with 8–15 pectinate teeth on each margin. Flowers in clusters of 3–6 in the axils of emersed leaves, subtended by a pair of 5- to 7-lobed bracteoles; petals 2.0–2.5 mm long in staminate flowers, rudimentary in pistillate flowers; anthers 8, about 2 mm long. Fruit 2–2.5 mm long, subglobose, the mericarps mostly rounded on the back. Jun–Sep. Shallow lakes, ponds, ditches, & potholes; ND, SD, NE; (Newf., MA to B.C., s to tx).

Myriophyllum brasiliense Camb., parrot's feather, a native of S. Amer., was collected once in KS: Saline in 1935. The leaves are all similar in appearance, in whorls of 4s, 5s and 6s, featherlike, each one with 10–18 pectinate divisions 3–5 mm long. Flowers are produced in clusters in the axils of the submerged foliage leaves. May–Sep. Escaped from cultivation in outdoor fish tanks, ponds, & lily pools, it persists in ponds, ditches, & streams in s U.S.; (NY, NJ, to MO, OK, & TX). *Naturalized*.

77. LYTHRACEAE J. St.-Hil., the Loosestrife Family

by Shirley A. Graham

Ours annual or perennial herbs or subshrubs of wet habitats, often with 4-angled stems. Leaves entire, opposite, seldom alternate, glabrous, exstipulate. Flowers axillary, in terminal spikes, or in lateral racemes or cymes, perfect, regular or irregular, homomorphic, or heteromorphic with styles and stamens of 2 or 3 lengths, 4- to 6-merous, with perianth and stamens perigynous; bracteoles 2, opposite on the pedicels. Floral tubes campanulate to tubular, persistent, often conspicuously nerved; calyx lobes 4–6, alternating with appendages or appendages none; petals 0–6, deep purple to pale pink, crumpled, deciduous, inserted on the inner surface of the floral tube between the calyx lobes; stamens generally as many as or twice as many as the petals, often alternately unequal, inserted on the inner surface of the floral tube below the petals, anthers versatile; pistil sometimes subtended at the base by a gland, stigma capitate, style filiform, rarely reduced or wanting, ovary superior, free in the floral tube, 2- to 4-locular. Fruit membranaceous capsules, enclosed by the persistent floral tubes, septicidally, septifragally, or loculicidally dehiscent, or indehiscent, splitting irregularly. Seeds in ours 7 to many, less than 1–3 mm long, pyramidal, ovoid, or discoid, endosperm none; placentation axile.

Reference: Koehne, E. 1903. Lythraceae *in* Engler, A., Das Pflanzenreich. Heft 17. IV. 216. H. R. Engelmann, Weinhein.

1 Floral tubes campanulate to globose, in fruit about as long as wide.
  2 Petals 4, purple to pale pink; appendages alternating with the calyx lobes; flowers 1–many in the axils of leaves.
    3 Leaves above midstem attenuate at base; flowers solitary in the axils of leaves; capsules finely and densely transversely striate on outer wall .... 5. *Rotala*
    3 Leaves above midstem cordate to auriculate at base; flowers (1)3–many in the axils of leaves; capsules smooth on outer wall .... 1. *Ammannia*
  2 Petals 0; appendages wanting; flowers solitary in the axils of leaves .... 3. *Didiplis*
1 Floral tubes cylindrical, in fruit about 2× longer than wide.
  4 Flowers regular; floral tubes entire in fruit; capsules dehiscing septicidally from the apex; seeds numerous, more than 10 .... 4. *Lythrum*
  4 Flowers irregular; floral tubes and capsules splitting longitudinally along the upper side in fruit; seeds 7–10 .... 2. *Cuphea*

1. AMMANNIA L., Toothcup

Annual glabrous herbs, 1–10 dm tall. Leaves decussate, linear to lanceolate or oblanceolate, sessile, bases above midstem cordate to auriculate. Flowers in sessile or pedunculate axillary cymes, (1)3–15 flowers per cyme, 4-merous, homomorphic. Floral tubes campanulate to urceolate, globose in fruit, 1.5–6 mm long; calyx lobes 4, alternating with appendages shorter than to equaling the lobes; petals 4, early deciduous; stamens 4(8); style, in ours, filiform, at least as long as the ovary; ovary incompletely 2- to 4-locular, gland at base of the ovary absent. Fruits irregularly dehiscent capsules, the outer walls smooth, not striate. Seeds many, 1 mm long, ovoid.

1 Slender plants with many-flowered, simple or compound, long-pedunculate cymes, peduncles filiform, 3–9 mm long; flowers mostly at least 7 per axil, petals deep rose-purple; capsules mostly 2.5 mm in diam or less .... 1. *A. auriculata*
1 Robust plants with short to long-pedunculate, 1–many flowered cymes, peduncles stout, 0–4(9) mm long; flowers mostly 1–5 per axil, petals deep rose-purple, rose with deep purple midvein, or pale lavender; capsules mostly 3.5 mm in diam or more.
   2 Inflorescence sessile; flowers mostly 1–3 per axil, petals pale lavender, anthers yellow; capsules 4–6 mm in diam .... 3. *A. robusta*
   2 Inflorescence a short to long-pedunculate cyme, rarely sessile; flowers mostly 3–5 per axil, petals deep rose-purple or rose with deep purple midvein; anthers deep yellow; capsules mostly 3.5–5 mm in diam .... 2. *A. coccinea*

1. Ammannia auriculata Willd. Slender herbs to 8 dm tall, unbranched to pyramidally multi-branched, the branches ascending. Leaves narrowly linear-lanceolate to linear-oblong, the largest 17–64 mm long, 2–10 mm wide, the bases auriculate, clasping. Flowers in axillary, long-pedunculate, simple or compound cymes, peduncles filiform, 3–9 mm long, (1)3–12(15) flowers per cyme, commonly 7, pedicellate; pedicels 1–3(6) mm long. Floral tubes 1–3 mm long; calyx lobes alternating with minute, thickened appendages <1 mm long or appendages absent; petals 4, deep rose-purple, 1.5 mm long, 1.5 mm wide; stamens 4(8), well exserted, anthers deep yellow, Capsules at maturity 1.5–3 mm in diam, equal to or exceeding the calyx lobes. (n = 15, 16) Jul–Oct. Grassy swales, pond margins, similar wet habitats; TX, OK, KS, NE, SD: Spink; (GP, s U.S., s to Argentina, world-wide).

Although the most widely distributed species in the genus, *A. auriculata* is sporadic in appearance and most common, in this hemisphere, in warmer regions to the south. It is infrequent in the GP and here is sometimes confused with the closely related *A. coccinea*. *A. auriculata* must be distinguished from that species by the entire suite of characters separating them, rather than only by the traditionally used character of elongate peduncles. The American plants of *A. auriculata* have been referred to var. *arenaria* forma *brasiliensis* Koehne, but these are weakly delimited entities generally no longer formally recognized.

2. Ammannia coccinea Rottb. Robust herbs to 10 dm tall, branching freely above the base, infrequently branching from the base, basal branches semidecumbent. Leaves linear-lanceolate to linear-oblong, rarely elliptic to spatulate, the largest 20–80 mm long, 2–15 mm wide, the bases auriculate to cordate, cuneate on lowermost leaves. Flowers in nearly sessile to long-pedunculate cymes, peduncles sturdy, 0–9 mm long, 3–5(14) flowers per cyme; pedicels 2 mm long or less. Floral tubes urceolate to slightly campanulate, (2.5)3–5 mm long; calyx lobes alternating with thickened appendages, the appendages about equaling lobes in length; petals 4(5), deep rose-purple, sometimes with deeper purple midvein at base, mostly 2 mm long, 2 mm wide; stamens 4(7), exserted, anthers deep yellow. Capsules at maturity 3.5–5 mm in diam, equal to or exceeding the calyx lobes. (n = 33) Jul–Oct. Wet habitats throughout GP in SD, NE, KS, IA, MO, OK, TX; (e & w U.S., Coastal States, s in Latin Amer. & Caribbean).

This exceedingly variable species is the most common *Ammannia* in the GP. It is believed to have originated as an amphidiploid derivative of *A. auriculata* and *A. robusta*. Some specimens may closely approach either parent in morphology and then are distinguished with difficulty. *A. coccinea* and *A. robusta* frequently occur together in mixed colonies; less often either may occur with *A. auriculata* (Graham, Taxon 28: 169–178. 1979.).

3. Ammannia robusta Heer & Regel. Robust herbs to 10 dm tall, unbranched or branched, the lowest branches decumbent from the base, often equaling the height of the main stem, the upper branches few, short. Leaves linear-lanceolate, rarely elliptic to spatulate, the largest 15–80 mm long, 4–15 mm wide, the bases auriculate-cordate, clasping, occasionally cuneate on lowermost leaves. Flowers in axillary, sessile cymes, 1–3(5) flowers per cyme. Floral tubes urceolate, often prominently 4-ridged, mostly 3.5 mm long, 2 mm wide; calyx lobes alternating with thickened appendages equaling lobes in length; petals 4(8), pale lavender, sometimes with deep rose midvein at base, 2.5 mm long, 3 mm wide; stamens 4(5–12), exserted, anthers pale yellow to yellow. Capsules at maturity 4–6 mm in diam, enclosed in or equaling calyx lobes. (n = 17) Jul–Oct. Wet habitats as in no. 1; throughout GP from s ND, excepting WY & NM; MT: Garfield, Phillips(?), & CO: Boulder, Denver; (cen midwest, scattered localities in w states, widespread in CA, s in Mex.).

Until recently, *A. robusta* was not recognized in the North American flora. Differences between *A. robusta* and *A. coccinea* were attributed to morphological variation in *A. coccinea*, cf. Graham, Taxon 28: 169–178. 1979.

2. CUPHEA P. Br.

1. ***Cuphea viscosissima*** Jacq., blue waxweed. Erect, much-branched annuals, 1–6 dm tall, the viscid upper stems often purplish red. Leaves ovate to narrowly lanceolate; blades 20–55 mm long, 6–20 mm wide; petioles 3–15 mm long. Flowers in leafy racemes, 1–3 per node, one always interaxillary, 6-merous, homomorphic. Floral tubes irregular, gibbous at the base, 12-nerved, deep purple above, lighter below, especially in fruit, covered with purple, dense, viscid-glandular hairs, 8–10 mm long, the upper calyx lobe larger than the others; petals 6, purple, the upper two largest, 3–5.5 mm long; stamens 11, alternately unequal; ovary subtended by a curved gland. Capsules and persistent floral tubes splitting longitudinally on upper side at maturity, placenta projecting outwards; seeds 7–10, discoid. (n = 6) Jul–Oct. Fields, pastures, roadsides; e NE, IA, KS, MO, OK; (e & s U.S., to GP). *C. petiolata* (L.) Koehne — Rydberg.

3. DIDIPLIS Raf., Water Purslane

1. ***Didiplis diandra*** (DC.) Wood. Amphibious or aquatic, weak-stemmed herbs, 2–40 cm long. Leaves thin, opposite to alternate, numerous, 5–30 mm long, 0.5–4 mm wide, the submersed ones linear with truncate base, the emersed ones narrowly elliptic, shorter, with tapering base. Flowers solitary in the axils, 2–3 mm long and wide, greenish, 4-merous; calyx lobes 4, appendages absent; petals 0; stamens 2–4, included; style short or none, ovary 2-locular. Capsules globose, 3 mm wide, indehiscent, irregularly splitting. Seeds numerous, <1 mm long, spatulate. (n = 16) May–Sep. Shallow waters & margins of lakes, ponds, & temporal pools, rare or at least infrequently collected; KS: Cherokee, Jackson, Saline; MO: Barton, Bates, Cass, Clay, Jackson, Jasper; NE: Hamilton; (e U.S., w to GP, & UT). *D. diandra* (Nutt.) Wood — Rydberg; *Peplis diandra* Nutt. — Fernald.

4. LYTHRUM L., Loosestrife

Ours perennial herbs, woody-based with basal offshoots, 4-angled stems, 1–12 dm tall. Leaves opposite, alternate, or whorled, ovate to linear, sessile or shortly petiolate, reduced in the inflorescence. Flowers solitary or paired in the axils or clustered in cymules and forming terminal spikes, regular or nearly so, 6-merous, homomorphic or heteromorphic. Floral tubes cylindrical, greenish, 8- to 12-nerved, 3–7 mm long; calyx lobes alternating with appendages, the appendages longer than the lobes; petals 6, ours purple or rose-purple;

stamens 6 or 12; ovary 2-locular, with a thickened ring at the base or the ring absent. Capsules septicidal or septifragal. Seeds many, ovoid, 1 mm long or less.

1 Flowers solitary or paired in the axils; stamens 6(8).
   2 Leaves ovate to linear-lanceolate, green, membranaceous .... 1. *L. alatum*
   2 Leaves linear-oblong to linear-lanceolate, gray-green, glaucous, often somewhat fleshy .... 2. *L. californicum*
1 Flowers numerous in showy terminal spikes; stamens 12 .... 3. *L. salicaria*

1. **Lythrum alatum** Pursh, winged loosestrife. Perennial herbs to 12 dm tall, virgately branched. Leaves dark green above, green to gray-green below, membranaceous, sessile, ovate to oblong with rounded to subcordate bases or lanceolate with tapering bases, the lower stem leaves opposite to subopposite, 1.5-6 cm long, 7-15 mm wide, the upper leaves subopposite to mostly alternate, reduced in size. Flowers solitary or paired in the axils, dimorphic. Floral tubes 4-6 mm long; petals purple, 3-6 mm long; stamens 6; thickened ring at the base of the ovary narrowed on one side. (n = 10) Jun-Sep. Wet soils, ditches, roadsides, pastures, flood plains; GP; (widespread in e U.S., w to GP).

Two varieties may be recognized in the GP.

1a. var. *alatum*. Slender stems to 8 dm tall. Leaves ovate to oblong with rounded to subcordate bases, 15-40 mm long, 7-15 mm wide. (n = 10) Jun-Sep. GP from se ND & w MN, s & w in SD, NE, KS, MO, to ne OK; WY: Laramie; CO: Denver, Boulder; (widespread in cen & ne U.S.). *L. dacotanum* Nieuw. —Van Bruggen.

1b. var. *lanceolatum* (Ell.) Rothr. Robust stems to 12 dm tall. Leaves lanceolate with tapering bases, 20-60 mm long, 7-10 mm wide. (n = 10) Jun-Sep. OK; (TX, widespread in se U.S.). *L. lanceolatum* Ell. —Fernald.

2. **Lythrum californicum** T. & G. Perennials with slender stems, mostly 2-6 dm tall, occasionally to 10 dm tall, stems erect to lax, virgately branched. Leaves pale, gray-green, glaucous, firm to somewhat fleshy, linear-oblong to linear-lanceolate, the bases mostly rounded, the lower stem leaves opposite to subopposite, 1.0-2.5 cm long, 3-8 mm wide, the upper stem leaves subopposite to mostly alternate, reduced in size, linear. Flowers solitary or paired in the axils, dimorphic. Floral tubes 4-7 mm long; petals purple, 4-6 mm long; stamens 6; thickened ring at the base of the ovary narrowed on one side. (n = 10) Jun-Sep. Wet soils, ditches, roadsides, pastures, lake margins; KS, OK, TX; (widespread GP w to CA & n Mex.).

Differences between *L. californicum* and *L. alatum* are few and primarily qualitative. Where the species are sympatric in KS and OK, hybridization probably occurs and then identification, especially of dry specimens, is difficult. *L. californicum* in our area differs from *L. alatum* generally in its smaller stature, more slender, numerous stems, and paler, linear leaves which tend to be less crowded and overlapping in the inflorescence. The taxa may ultimately be viewed as part of a single wide-ranging species, consisting of several geographical races.

3. **Lythrum salicaria** L., purple loosestrife. Erect, robust perennials, often pubescent, to 12 dm tall. Leaves opposite or whorled, lanceolate, 2-10 cm long, 5-15 mm wide, the bases rounded to cordate. Flowers in showy terminal spikes, trimorphic. Floral tubes 4-6 mm long; petals rose-purple; stamens mostly 12; ring at the base of the ovary absent. (2n = 30, 50, 60; n = 15, 30) Jul-Sep. Cult. & escaping from gardens to become established in wet meadows, marshes, ditches; sporadic & infrequent in GP; ND: Barnes, Cass; SD: Union; NE: Fillmore; MN: Kittson, Polk, Mahnomen; KS: Lyon; IA: Buena Vista; (common in ne U.S. & adj. Can., aggressively spreading w in recent years, sometimes forming large marshland colonies that eliminate native species; also sporadic in w U.S. as escapes from cultivation; Europe, Africa, Asia). *Naturalized*.

5. ROTALA L.

1. **Rotala ramosior** (L.) Koehne, toothcup. Ours glabrous herbs to 4 dm tall. Leaves opposite, linear to oblanceolate, 10–50 mm long, 2–12 mm wide, the bases attenuate. Flowers solitary, axillary, sessile, 4(6)-merous, ours homomorphic. Floral tubes 2–5 mm long, calyx lobes alternating with appendages equaling the length of the lobes; petals 4(6), white to pink, scarcely exceeding the calyx lobes; stamens 4(6). Capsules 2- to 4-locular, finely and densely, transversely striate on outer wall; placentation becoming free central at maturity. Seeds many, 1 mm long or less, ovoid. (n = 16) Jul–Oct. Muddy or sandy shores or damp depressions; s SD, NE, KS, MO, OK; (e U.S. especially se GP, w U.S. Coastal States, s to S. Amer. & Caribbean).

Inland plants with slightly larger leaves, bracteoles, and capsules than those of coastal specimens have been called var. *interior* Fern. & Grisc. The morphological distinctions are inadequate to justify varietal status.

78. THYMELAEACEAE Juss., the Mezereum Family

by Ronald L. McGregor

1. THYMELAEA P. Mill., Spurge-flax

1. **Thymelaea passerina** (L.) Coss. & Germ. Slender annual herb (10)20–40(7) cm tall. Stem simple or commonly branched above, glabrous or remotely pubescent. Leaves sessile, alternate, linear-lanceolate, (6)8–14(20) mm long, 1–2 mm wide, progressively reduced upward, glabrous, acute. Flowers axillary, greenish, solitary or in clusters of 2 or 3, arising from a tuft of silky hairs and subtended by 2 lanceolate bracts 2–3 mm long. Hypanthium urceolate or tubular, greenish, 2–3 mm long, strigose; sepals 4, ovate, 1 mm long; petals absent; stamens 8, inserted in 2 whorls in hypanthium, filaments obsolete; ovary superior, at base of hypanthium but free from it, ovule 1. Fruit dry, indehiscent, strigose, enclosed in persistent hypanthium. Jul–Oct. Dry pastures, bluffs, & flood plains; IA: Harrison, Monona, Plymouth, Pottawatomie, Woodbury; NE: Cedar, Merrick, Morrill; KS: Cloud; (restricted to GP; Europe). *Introduced*. *Passerina annua* Wikst.—Kartesz & Kartesz.

79. ONAGRACEAE Juss., the Evening Primrose Family

by Ronald L. McGregor

Annual, biennial or perennial, sometimes rhizomatous herbs, sometimes woody near the base. Leaves alternate or opposite, sometimes rosulate, simple, entire, toothed or pinnatifid; petioles present or absent; stipules minute or absent. Flowers regular to slightly irregular, perfect, (2 or 3) usually 4(5)-merous, borne in axils of usually reduced leaves, often in more or less distinct inflorescences; floral tube prolonged beyond summit of ovary or not; sepals free; petals free, sometimes absent or minute, yellow, white, pinkish to rose-purple; stamens as many as sepals or commonly 2 × number of sepals; ovary inferior, 2-

to 4 or 5-locular, with 1-many ovules, style 1, stigma capitate or deeply lobed, rarely short-lobed and subcapitate. Fruit a dehiscent capsule, or an indehiscent woody nutlike capsule; seeds without a coma except in *Epilobium*.

References: Munz, Philip A. 1965. Onagraceae. North Amer. Fl., ser. 2, 5: 1-278; Raven, P. H. 1964. The generic subdivision of Onagraceae, tribe Onagreae. Brittonia 16: 276-288. Peter Raven, David E. Boufford, Howard F. Towner, and Warren L. Wagner read my draft manuscript and kindly provided much constructive criticism and information, which I acknowledge with sincere appreciation.

1 Flowers 2-merous; ovary with hooked hairs .... 3. *Circaea*
1 Flowers (3)4(5)-merous; ovary without hooked hairs.
  2 Stigmas deeply 4-lobed.
    3 Flowers yellow at anthesis .... 8. *Oenothera*
    3 Flowers white, pinkish, or rose to purple at anthesis.
      4 Petals rounded or narrowed to base .... 8. *Oenothera*
      4 Petals more or less abruptly long or short clawed.
        5 Floral tube essentially lacking; fruit a dehiscent capsule; seed with a coma .... 4. *Epilobium angustifolium*
        5 Floral tube obvious; fruit an indehiscent nutlike capsule; seed ecomose.
          6 Floral tube funnelform; each staminal filament with a small scale at base (except *G. parviflora*) .... 5. *Gaura*
          6 Floral tube filiform; filaments unappendaged .... 9. *Stenosiphon*

  2 Stigmas entire; capitate, peltate, clavate or subcapitate with 4 very short lobes in *Boisduvalia*.
    7 Petals yellow, or petals absent or minute in some *Ludwigia* sp.
      8 Floral tube not prolonged beyond apex of ovary; sepals persistent in fruit .... 7. *Ludwigia*
      8 Floral tube obviously well prolonged beyond ovary; sepals deciduous in fruit .... 2. *Calylophus*
    7 Petals white or pinkish to reddish-purple.
      9 Ovary 2-celled; leaves usually less than 3 mm wide .... 6. *Gayophytum*
      9 Ovary 4-celled; leaves usually over 3 mm wide.
        10 Plants annual, of seasonally moist habitats which are dry at flowering time; seed ecomose .... 1. *Boisduvalia*
        10 Plants perennial, of permanently moist habitats; seeds comose ... 4. *Epilobium*

1. BOISDUVALIA Spach

1. *Boisduvalia glabella* (Nutt.) Walp. Annual herbs; stems suberect or decumbent, or matted and rooting at nodes, usually branched from base, 1-3 dm tall, glabrous below, with fine white strigulose or villous hairs above. Lowest leaves opposite, glabrous, narrowly connate at base, the rest alternate, subsessile, somewhat crowded, lanceolate or elliptic-lanceolate, 8-20(30) mm long, 3-6(9) mm wide, sparsely denticulate, loosely villous or strigulose to glabrate, often ciliate. Inflorescence erect, each flower borne in axil of a bract which appears as slightly reduced upper foliage leaves. Floral tube 0.3-1 mm long, short-funnelform; sepals 4, erect, 0.7-1.8 mm long; petals 4, pale pink, 1-3 mm long, bilobed about 1/3 length; stamens 8, the 4 episepalous ones inserted at mouth of floral tube, the 4 epipetalous ones with shorter filaments and inserted below its summit; anthers 0.5-0.8 mm long; ovary 4-loculate; stigma 4-lobed, subcapitate. Capsule about 7 mm long, nearly straight, pointed, usually hidden by subtending bract, dehiscing in upper 1/3; seeds 1-1.3 mm long, brownish, 6-14 per locule. (n = 15) Jun-Jul. Along small streams & edge of vernal pools which dry by flowering time; rare in GP; ND: Billings, Hettinger; MT: Wibaux; SD: Butte; (s Sask., Alta., B.C., s to ND, SD, MT, NV, CA).

Reference: Raven, Peter H. & David M. Moore. 1965. A revision of *Boisduvalia* (Onagraceae). Brittonia 17: 238–254.

2. CALYLOPHUS Spach

Herbaceous to suffrutescent perennials from a woody, often branching caudex, rarely flowering in first year in GP. Stems nearly prostrate or decumbent to erect, sometimes with exfoliating epidermis. Leaves cauline, alternate, entire to serrate; petioles absent or nearly so; stipules absent. Flowers from axils of upper leaves, regular, 4-merous. Floral tube well developed and prolonged beyond ovary, deciduous after anthesis; sepals 4, greenish-yellow, reflexed separately, often with purple or red spots or streaks; petals 4, reflexed in anthesis, yellow but often fading to reddish, orange or purplish; stamens 8, subequal or biseriate and of 2 lengths; anthers versatile. Capsule sessile, cylindrical, with 4 locules; seeds in 2 rows in each locule. *Galpinsia* Britt., and *Meriolix* Raf.—Rydberg.

Reference: Towner, Howard F. 1977. The biosystematics of *Calylophus* (Onagraceae). Ann. Missouri Bot. Gard. 64: 48–120.

1. Sepals with a conspicuous keeled midrib; stamens biseriate with epipetalous filaments about 2× longer than epipetalous ones.
  2. Stigma well exserted, usually to end of episepalous stamens or beyond; petals 9–25 mm long; pollen fertility 85–100% .... 1. *C. berlandieri*
  2. Stigma located near or slightly beyond apex of floral tube or within circle of anthers; petals 5–20 mm long; pollen 30–80% aborted .... 4. *C. serrulatus*
1. Sepals plane, lacking a keeled midrib; stamens subequal.
  3. Plants 0.4–2(3) dm tall, frequently cespitose; densely grayish-strigulose; sepal tips 0.3–3 mm long .... 3. *C. lavandulifolius*
  3. Plants 0.4–4(5) dm tall, not cespitose, variously pubescent or glabrous, if strigulose then sepal tips 2–6 mm long .... 2. *C. hartwegii*

1. *Calylophus berlandieri* Spach, Berlandier evening primrose. Herbaceous to suffrutescent perennial from a subterranean or superficial woody, often branched, caudex. Stems 1–several, simple to branched, subdecumbent to erect, 1–5(8) dm tall, glabrous to strigulose or strigulose-canescent, especially above, at least epidermis of lower stem exfoliating, sometimes flowering first year. Basal leaves absent; lowest leaves narrowly lanceolate to oblanceolate or spatulate; cauline leaves alternate, sometimes crowded, sessile or slightly petiolate, spreading to ascending, linear or more usually narrowly lanceolate or oblanceolate, often conduplicate, (1)3–7(9) cm long, 1–7(10) mm wide, surfaces glabrous to strigulose or strigulose-canescent, margin subentire to spinulose-serrate; cauline leaves often with fascicles of small leaves or much-reduced branchlets in axils. Inflorescence of flowers borne in axils of upper leaves, usually somewhat compact. Buds squarish in cross-section; floral tube funnelform, tubular in lower 1/3–1/2, 5–20 mm long, subglabrous to strigulose-canescent, especially on midribs, pale yellow; sepals 4, free, 5–12 mm long, 2–7 mm wide, free tips 0–4 mm long, with raised or keeled midribs, glabrate to strigulose-canescent, yellow-green; petals 4, free, suborbicular to obovate-truncate or obcordate (6)9–20(25) mm long, 7–30 mm wide, sometimes fading orangish to purplish; stamens 8, biseriate, the epipetalous filaments 2–8 mm long, epipetalous ones 1–4 mm long; anthers (2)4–7 mm long; pollen fertility 85–100%; style 9–30 mm long; stigmas discoid to squarish, 1–3 mm wide, exserted beyond throat to ends of anthers or beyond; ovary 6–20 mm long, minutely strigulose to strigulose-canescent. Capsule 1–3.5 cm long, 1–2 mm wide, thick-walled, completely and often tardily dehiscent; seeds 1–2 mm long, brown, angled, obliquely truncate at apex to truncate and sometimes pointed. (n = 7)

Two subspecies in our area:

1a. subsp. *berlandieri*. Stems several to many, subdecumbent to ascending, 1-4 dm tall. Leaves 1-4 cm long; plants somewhat bushy. Apr-Aug. Infrequent to locally common on calcareous or arenaceous hillsides, canyons, stream valleys, roadsides; sw KS, se CO, w OK, e NM, TX panhandle; (sw KS, se CO, e NM, s to n Mex.). *C. drummondianus* Spach subsp. *berlandieri* (Spach) Towner & Raven — Atlas GP.

1b. subsp. *pinifolius* (Engelm. ex A. Gray) Towner. Stems 1-several, suberect to erect, 3-8 dm tall. Leaves 2.5-9 cm long; plants usually not bushy. Apr-Jul. Infrequent to common on rocky prairies, rocky or sandy prairies in open woodlands, roadsides; cen OK; (cen OK, s to cen TX, w LA). *Meriolix melanoglottis* Rydb.—Rydberg; *C. drummondianus* Spach subsp. *drummondianus*—Atlas GP.

These subspecies intergrade. In the tall-grass prairies of the east half of KS and ne OK a tall, strict form of *C. serrulatus* is most difficult to separate from subsp. *pinifolius*. It has the habit and large flowers of *berlandieri*, but stigma only exserted to within the circle of anthers, and pollen 30-60% aborted. Unfortunately, I have had to resort to pollen fertility counts in naming many of these specimens.

2. Calyophus hartwegii (Benth.) Raven, Hartweg evening primrose. Herbaceous to suffrutescent perennials with a woody, sometimes branching caudex. Stems 1-several, usually branched, decumbent and spreading to somewhat ascending or erect 0.5-4(5) dm tall, strigulose, glandular-pubescent, with spreading hairs or nearly glabrous. Leaves alternate, simple, spreading to ascending, linear to ovate or oblanceolate to lanceolate or narrowly elliptic, 3-50 mm long, variously pubescent or glabrate, margin entire or serrate; petioles absent or indistinct; cauline leaves sometimes with fascicles of small leaves in axils. Inflorescence sparse to dense. Buds terete; floral tube tubular in lower 1/2 or more, gradually expanded to apex, 1.5-6 cm long, variously pubescent or glabrous; sepals 4, free, 7-25 mm long, free tips 0.5-6 mm long, without raised vein or midrib, pale yellow, sometimes fading pinkish or orange; petals 4, suborbicular to rhomboidal, 10-35 mm long, yellow, often fading pinkish or purplish; stamens 8, subequal, filaments glabrous, 4-13 mm long; anthers 5-13 mm long; style 2.5-7 cm long, mostly larger than stamens; stigma squarish, flat, 1.5-5 mm wide. Capsules 6-40 mm long, 2-4 mm wide, dehiscent; seeds 1-2.5 mm long.

Although some degree of intergradation is found between subspecies of this self-incompatible species, three subspecies may be recognized in our area as follows:

1 Spreading trichomes present on ovary and often on upper stem and leaf margins; cauline leaves abruptly narrowed to truncate or somewhat clasping at base ..... 2c. subsp. *pubescens*
1 Plant without spreading trichomes; cauline leaves gradually narrowed to base or narrow throughout.
    2 Plant glabrous or nearly so throughout. .... 2a. subsp. *fendleri*
    2 Plant pubescent on ovary and upper stem but without spreading trichomes .... 2b. subsp. *filifolus*

2a. subsp. *fendleri* (A. Gray) Towner & Raven. Plant glabrous or nearly so throughout, infrequently minutely and sparingly glandular-pubescent. Leaves more or less ascending, linear to oblanceolate or lanceolate, 1-50 mm long, 1.5-10 mm wide, apex acute, base acute-attenuate to obtuse, rarely nearly clasping. (n = 7) Apr-Aug. Infrequent to locally common on rocky, gravelly, clay soils of prairie hillsides, valleys, roadsides, & open wooded areas; sw KS, se CO, w OK, TX; (sw KS, se CO, s to w TX, NM, AZ; Chih.). *Galpinsia fendleri* (A. Gray) Heller—Rydberg; *Oenothera hartwegii fendleri* A. Gray—Gates.

2b. subsp. *filifolius* (Eastw.) Towner & Raven. Plant minutely glandular pubescent throughout, more densely so on ovary and inflorescence, rarely sparsely strigose on ovary and leaves. Leaves spreading to ascending, filiform to narrowly lanceolate, 3-45 mm long, 0.5-4 mm wide, apex acute, base acute-attenuate. (n = 7) Apr-Aug. Locally common on gypsum or sandy prairie hillsides, stream valleys, & roadsides; s TX panhandle; (w TX, sw NM; Chih., Coah., n Zac.).

2c. subsp. *pubescens* (A. Gray) Towner & Raven. Plant usually covered throughout with long spreading trichomes but these more dense on ovary, inflorescence and upper stem. Leaves spreading to reflexed, narrowly elliptic or narrowly lanceolate to ovate, 5–40 mm long, 1.5–12 mm wide, apex acute, base acute to truncate or subcordate and clasping. (n = 7, 14) Apr–Jul. Locally common on rocky, sandy, or gravelly prairie hillsides & plains; sw KS, se CO, w OK, NM, TX; (sw KS, se CO, w OK, cen TX, NM, AZ; cen Coah., ne Dur.) *Galpinsia interior* Small — Rydberg; *Oenothera greggii* A. Gray — Gates.

3. **Calylophus lavandulifolius** (T. & G.) Raven, lavender leaf primrose. Low, often cespitose, perennial herbaceous to usually suffrutescent plants with a stout, often branching, subterranean or superficial caudex. Stems few to many, often branched, spreading-decumbent to more or less ascending, 0.5–2(3) dm tall, densely gray-strigulose. Leaves linear to narrowly lanceolate or narrowly oblanceolate, 5–40(50) mm long, 1–6 mm wide, usually somewhat crowded and ascending, gray-strigulose, margin entire, base acute-attenuate; petioles absent; small axillary leaves usually present. Inflorescence lax, flowers from axils of upper leaves. Floral tube 2.5–4(6) cm long, to 15 mm wide at throat, strigulose or glandular-pubescent outside, sometimes with longitudinal purplish lines; sepals 4, free, 8–20 mm long, 3–8 mm wide, usually with purple marginal stripes, free tips 0.3–3 mm long; petals 4, 1.2–3 cm long, about as wide, yellow but usually fading pinkish to purplish; stamens 8, filaments 6–12 mm long, anthers 5–11 mm long; ovary 4–16 mm long, 1–2 mm wide, densely gray-strigulose, style 3–7.5 cm long, stigma 2–5 mm wide. Capsule 6–25 mm long, 1–3 mm wide, gray-strigulose; seeds 1.5–2.5 mm long. (n = 7) May–Aug. Infrequent to locally common on rocky ledges & prairie hillsides, stream valleys, open wooded hillsides, roadsides; sw SD, w NE, se WY, w 1/2 KS, nw OK, e CO, ne NM, TX panhandle; (sw SD, se WY s to w TX, CO, UT, NV, NM, AZ, N.L.). *C. hartwegii* (Benth.) Raven subsp. *lavandulifolius* (T. & G.) Towner & Raven — Atlas GP; *Galpinsia lavandulaefolia* (T. & G.) Small — Rydberg; *Oenothera lavandulaefolia* T. & G. — Gates.

4. **Calylophus serrulatus** (Nutt.) Raven, plains yellow primrose. Herbaceous to suffrutescent perennials with a stout, often much branched, woody caudex. Stems few to many, simple to branched, decumbent to erect, 0.5–8 dm tall, glabrous to strigulose or strigulose-canescent, especially above. Leaves alternate, spreading to more or less ascending, linear to narrowly lanceolate or narrowly oblanceolate, often conduplicate, 1–10 cm long, 0.1–12 mm wide, glabrate to strigulose above, sparsely to densely strigulose below, margin entire to variously serrulate, acute or tapering to base, apex acute or rounded; sessile or semipetiolate; with or without fascicles of small leaves in the axils. Flowers from axils of upper leaves; floral tube 2–16 mm long, 3–12 mm wide at throat, greenish-yellow, glabrous to sparsely strigulose especially on ribs; sepals 4, 1.5–9 mm long, 2–6 mm wide, distinctly keeled, strigulose below, free tips 0–3(4) mm long; petals 4, 5–14(20) mm long, about as wide, yellow, fading to light yellow or pinkish; stamens 8, biseriate; epipetalous filaments 0.5–3 mm long, episepalous filaments 1–5(6) mm long; anthers 1.5–4(6) mm long, pollen 30–80% aborted; style 2–15(18) mm long; stigma 1–2 mm wide, positioned at apex of floral tube mouth or exserted to level of anthers but not beyond, ovary 4–13 mm long, strigulose. Capsule 10–30 mm long, 1–3 mm wide, thick-walled, often tardily dehiscent; seeds 1–2 mm long, obliquely truncate or rarely pointed at apex. (n = 7) May–Sep. Commonly found on rocky, gravelly, or sandy prairie plains, & hillsides, stream valleys, open woodlands, roadside; GP but absent to rare in se; (Man., Sask., to Alta., s to MN, IA, IL, MO, e TX, MT, NM, AZ, Chih.). *Oenothera serrulata* Nutt. — Fernald; *O. serrulata oblanceolata* (Rydb.) Gates — Gates; *Meriolix intermedia* Rydb., *M. oblanceolata* Rydb., *M. serrulata* (Nutt.) Walp. — Rydberg.

This is the only self-compatible species in the genus and exhibits considerable variation in habit, degree of woodiness at base, flower size, and leaf characters. Plants of the e 1/3 of the GP tend to be taller with fewer stems and much less cespitose, have larger flowers, and flowering period of May–Jun in contrast to the lower, more bushy, smaller-flowered plants of the rest of the GP. This distinction is not clear-cut enough to allow for recognition of geographic variants. The range of variation of this species overlaps that of *Calylophus berlandieri*, and the relative position of the stigma and pollen fertility are the only reliable diagnostic characters to separate the two species.

3. CIRCAEA L., Enchanter's Nightshade

Delicate to coarse perennial herbs with slender rhizomes. Stems erect or decumbent at base and rooting at the nodes, glabrous. Leaves cauline, petiolate, opposite and decussate, the lowest usually deciduous by flowering time; stipules setaceous or glandlike, caducous or rarely persistent. Inflorescence of simple or branched racemes, terminal on main stem and often at tips of upper axillary branches; pedicels erect in bud, clustered at apex of racemes, the racemes elongating and pedicels becoming distantly spaced, divergent, or perpendicular to the raceme axis prior to or just after anthesis, reflexed or perpendicular to raceme axis in fruit. Flowers 2-merous, bilaterally symmetrical; floral tube subcylindric to funnelform, soon deciduous; sepals 2, spreading or reflexed in flower; petals 2, notched at apex; stamens 2; style filiform, stigma bilobed; ovary 1- or 2-celled, 1 ovule per cell. Fruit an indehiscent capsule, covered with soft to firm uncinate hairs.

1 Rhizomes terminated by tubers; fruits unilocular; flowers opening before elongation of raceme axis, borne on erect or ascending pedicels .... 1. *C. alpina*
1 Rhizomes not terminated by tubers; fruits bilocular; flowers opening after elongation of raceme axis, borne or spreading pedicels .... 2. *C. lutetiana* subsp. *canadensis*

1. *C. alpina* L. Herbaceous perennials with slender rhizomes producing tubers at the tips. Stems erect or decumbent at base and rooting at the nodes, 0.3–3 dm tall, glabrous. Leaf blades ovate to broadly ovate, short acuminate to acute, cordate to subcordate, less often truncate to rounded at base; dentate, 1.5–7 cm long, 1.5–5 cm wide; petioles 0.3–4 cm long, semiterete (flattened and appearing winged in pressed specimens). Racemes 1–17 cm long, glabrous or with stipitate glands, flowering pedicels ascending or erect, flowers opening before racemes elongate. Flowers with floral tube funnelform, minute to 0.5 mm long; sepals 2, 1–2.2 mm long, 0.6–1.2 mm wide, white or pink; petals 2, 0.6–1.9 mm long, 0.8–1.7 mm wide, white, apical notch 0.3–0.7 mm deep; anthers 0.2–0.4 mm long; style 1.1–2.1 mm long. Mature fruit 2–2.5 mm long, clavate, unilocular; fruiting pedicels perpendicular to raceme axis or slightly reflexed. (n = 11) Jul–Aug. Rare to locally common in moist woodlands, often on moss covered rocks & logs; MN, e ND, BH; (circumboreal, s in N. Amer. to NC, TN, IA, SD, CO, NM, CA).

Our plants are subsp. *alpina*. *C. x intermedia* Ehrh., a hybrid between *C. alpina* and *C. lutetiana* subsp. *canadensis*, is known from SD: Pennington, and MN. It has flowers opening after elongation of the raceme axis, borne on spreading pedicels; all ovaries aborting shortly after anthesis; and pollen fertility of less than 5%.

2. *Circaea lutetiana* L. subsp. *canadensis* (L.) Asch. & Mag. Perennial herbs with slender rhizomes lacking tubers at the tips. Stems erect (1.2(3–7(9) dm tall, glabrous. Leaf blades narrowly to broadly ovate to oblong-ovate, short to long acuminate to obtuse apex, rounded, truncate or rarely subcordate at base, denticulate, 5–12(16) cm long, 2.5–8.5 cm wide; petioles terete or subterete, not flattened in pressing, 1.5–5.5 cm long. Racemes (2)5–20(30) cm long, stipitate glandular; flowering pedicels perpendicular to raceme axis, flowers opening after elongation of raceme axis. Flowers with floral tube funnelform, 0.4–.2 mm long; sepals 2, 2–3.8 mm long, 1.2–2.3 mm wide, green or purple; petals 2, 1.3–2.9 mm long, 1.5–4 mm wide, white or roseate, apical notch 0.4–1.7 mm deep; anthers 0.6–0.8 mm long; style 2.5–5.5 mm long. Mature fruit 2.8–4.5 mm long, pyriform to subglobose, with prominent corky ribs, bilocular; fruiting pedicels reflexed or recurved. (n = 11) Jun–Sep. Rare to locally common in moist woodlands, often along streams; MN, e ND, e SD & BH, IA, e NE & scattered nw, e 1/4 KS, MO, OK; (Que., Ont., s to NC, TN, OK, KS; Europe & Asia). *C. lutetiana* L.—Rydberg; *C. quadrisulcata* (Maxim.) French & Sav. var. *canadensis* (L.) Hara, *C. canadensis* Hill var. *virginiana* Fern.—Fernald.

4. EPILOBIUM L., Willow-herb, Fireweed

Contributed by Peter C. Hoch

Perennial or rarely annual herbs, the perennials spreading by rosettes, sobols, bulblike offsets (turions), or filiform stolons tipped with turions, or by horizontal roots giving rise to new shoots. Stems erect to rarely decumbent, pubescent to glabrous, the hairs usually in decurrent lines below the inflorescence. Leaves opposite at least below, or alternate, simple, entire or toothed. Flowers solitary in the leaf axils or in terminal racemes or panicles; floral tube short or absent; sepals 4; petals 4, purplish, pink or white, usually notched; stamens 8, usually unequal, the episepalous ones longer; stigma clavate, capitate, or 4-cleft. Capsule linear, elongate, 4-valved, dehiscent; seeds many, with a terminal tuft of silky hairs, the coma. *Chamaenerion* (Gesn.) Ludw.—Rydberg.

The following species were listed for SD by Rydberg but are here omitted for lack of specimen evidence: *E. alpinum* L. (=*E. anagalidifolium* Lam.) and *E. latifolium* L. (*Chamaenerion latifolium* (L.) Sweet—Rydberg). *E. strictum* Muhl. was lsted for KS by Rydberg, but the westernmost station for this species is Hubbard Co., MN, just east of our region.

1 Stigma deeply 4-cleft; petals entire, 1 cm or more long; leaves spirally arranged .... 1. *E. angustifolium* subsp. *circumvagum*
1 Stigma entire or nearly so; petals notched, usually less than 1 cm long; leaves opposite, at least below.
  2 Plant an annual with a taproot; epidermis on lower stem usually exfoliating .... 8. *E. paniculatum*
  2 Plant perennial with spreading roots and no taproot, epidermis not exfoliating.
    3 Plants with basal threadlike stolons, terminating in compact fleshy turions; inflorescence white-canescent, sometimes mixed with glandular hairs; cauline leaves ±linear, ±entire, often revolute.
      4 Leaves linear to lanceolate or oblong, ±glabrous to sparsely strigillose above; inflorescence often nodding in bud, eglandular .... 7. *E. palustre*
      4 Leaves linear, rarely wider, densely strigillose above; inflorescence erect or nearly so, often with a mixture of strigillose and glandular hairs. .... 6. *E. leptophyllum*
    3 Plants lacking stolons; inflorescence predominantly glandular-pubescent; cauline leaves not linear and rarely entire, never revolute.
      5 Plants clumped or cespitose, forming short, leafy, epigeous shoots; stem base decumbent to ascending; leaves petiolate .... 5. *E. hornemannii*
      5 Plants not clumped, or if so loosely, forming sessile, leafy rosettes or fleshy turions; stem base erect; leaves sessile to occasionally petiolate.
        6 Leaves lanceolate-linear, the margins irregularly serrate with 4–9 teeth per cm; seed coma cinnamon or reddish; inflorescence ±paniculate to corymbiform .... 3. *E. coloratum*
        6 Leaves lanceolate to ovate, the margins sparsely denticulate with 1–5 teeth per cm; seed coma whitish or dingy; inflorescence racemose to paniculate.
          7 Plants 0.3–19.0 dm high, forming rosettes or large, hypogeous turions; seeds 0.8–1.6(1.9) mm long, longitudinally striate with hyaline crests or ridges but lacking distinct papillae .... 2. *E. ciliatum*
          7 Plants 0.2–6 dm high, lacking rosettes, forming only compact hypogeous turions; seeds 1.1–1.6(1.8) mm long, distinctly papillose, the pappillae often in longitudinal rows.

8 Leaves sessile, clasping, mostly narrow-ovate, denticulate; turions fleshy, elongate; capsules subsessile, appressed; seed collar conspicuous .... 9. *E. saximontanum*
8 Leaves petiolate or subsessile, not clasping, lanceolate or narrower, subentire or denticulate; turions compact, round; capsules on pedicels 0.8–3.8 cm long; seed collar inconspicuous .... 4. *E. halleanum*

1. ***Epilobium angustifolium*** L. subsp. ***circumvagum*** Mosquin. Willow-herb. Stout erect perennial herb, forming extensive colonies by vigorous growth from a thick rootstock and by sprouting from spreading lateral roots. Stems simple or branched, (13)30–250(300) cm tall, glabrous below or throughout or densely strigillose on the upper stem. Leaves spirally arranged, narrowly lanceolate to lanceolate, acute, (1.5)2.5–20 cm long, 0.4–3.5(5) cm wide, glabrous or sometimes with strigillose hairs on the abaxial midrib, subentire or obscurely denticulate, strong midrib with 10–20 obscure secondary veins, anastomosing toward margins, sometimes with marked submarginal vein, subsessile or attenuate to petiole 1–4 mm long. Inflorescence a simple elongate lax raceme with bracts resembling leaves, smaller or lacking. Flowers many, drooping in bud, ±zygomorphic, strongly protandrous; floral tube absent; sepals 4, spreading, narrowly lanceolate, 7–16 mm long, 1.6–2.5 mm wide, acute, canescent, green or purplish; petals 4, spreading, obovate, tapering to short claw, apex slightly emarginate, 10–16(20) mm long, 6–11 mm wide, deep pink to magenta or rarely white; stamens 8, subequal, initially spreading, flaccid later; anthers 1.5–2.2 mm long, 0.8–1 mm wide, purplish; pollen bluish grey, mostly triporate, released as monads with adhering viscid threads; filaments glabrous, white, 4–12 mm long, bases flattened and dilated, forming an almost enclosed chamber above nectar-secreting disk; ovaries densely white-canescent, 8–18 mm long, on pedicels 4–12 mm long; style 10–20 mm long, white to purplish, with a ring of dense incurved hairs near the base, initially sharply deflexed, becoming erect after anthers dehisce and reflex; stigma deeply 4-lobed, the receptive surfaces exposed as the lobes recurve, exserted beyond anthers. Capsules 4–10 cm long, canescent, on pedicels 0.7–1.5(2.5) cm long; seeds many, fusiform, 1–1.3 mm long, 0.32–0.4 mm wide, irregularly reticulate, the chalazal collar short and inconspicuous; coma 9–14 mm long, white, not markedly deciduous. (n = 36) Jun–Sep. Usually found in dry, open, disturbed areas, especially after fires; MN, ND, MT, w SD, nw NE, WY, CO; (s Can. from Newf. to B.C., s to NC, OH, IA, SD, NM, CA; n. Mex.; Asia).

2. ***Epilobium ciliatum*** Ref., willow-herb. Variable erect perennial herbs, stems mostly 0.5–12 dm tall, simple or branched, arising from rosettes, turions or unmodified basal shoots. Lower stem with decurrent lines of strigillose hairs, becoming densely pubescent above with recurved strigillose and long glandular hairs, rarely villous or sericeous throughout. Leaves mostly 3–12 cm long, 0.5–4.5 cm wide, the 4–8 lateral veins conspicuous, the upper narrowly lanceolate to ovate, denticulate, petioles 0–4 mm long, lower leaves often obovate to broadly elliptic with longer petioles. Inflorescence erect, rarely nodding. Floral tube 0.5–2.6 mm deep, 0.9–3.5 mm wide; sepals 1.5–5 mm long, 0.7–2.5 mm wide, often strigillose and glandular; anthers cream to white, 0.4–1.8 mm long, filaments unequal, white to purple, 0.3–4.5 mm long; style 1.1–8.2 mm long, stigma clavate to subcapitate. Capsule slender, 4–10 cm long, pubescent, on a pedicel 2–40 mm long; seeds mostly 0.8–1.6 mm long, 0.3–0.64 mm wide, distinctly ridged, very rarely reticulate or papillose; coma 2–8 mm long, white, readily detaching.

Two subspecies are found in our area:

2a. subsp. *ciliatum*. Usually with epigeous, leafy rosettes, cauline leaves narrowly lanceolate to narrowly ovate; inflorescence nonleafy, branched; petals white to pink, 1.5–5(8) mm long; seeds (0.6)0.8–1.2(1.5) mm long. (n = 18) Jul–Sep. Along roadsides, sandy stream banks, mesic, often disturbed sites; IA, NE, w OK, nw TX; (Newf. to AK, s to NC, TN, OH, IA, w OK, TX; Mex., Guat.; rare in Korea & Japan). *E. americanum* Hausskn.—Rydberg.

2b. subsp. *glandulosum* (Lehm.) Hoch & Raven. Usually with hypogeous, fleshy, turions; cauline leaves wider; inflorescence mostly leafy, unbranched; petals pink to rose-purple, rarely white, 3.5–10(12) mm long; seeds 1.1–1.6(1.9) mm long. (n = 18) Jul–Sep. Moist, often disturbed habitats; only BH, SD, & nw MN; (Newf. to AK s to NY, MI, IA, MN, SD, NM, AZ, CA; Japan & Korea; introduced to n Europe). *E. occidentale* (Trel.) Rydb — Rydberg.

3. **Epilobium coloratum** Biehler, purple-leaved willow-herb. Perennial herbs with fibrous roots, spreading by very short rhizomes that produce a basal rosette of leaves by autumn; turions absent. Stems erect (2)4–8(12) dm tall, simple below and freely branched above, glabrous below, minutely pubescent above with short incurved hairs, pubescence tending to be in lines, eglandular. Leaves opposite, principal leaves narrowly lanceolate or oblong-lanceolate, 3–8(12) cm long, 0.5–3 cm wide, usually long-acuminate, sometimes rugose-veiny below, glabrous or minutely pubescent on veins, margin sharply and irregularly serrate with 4–8 teeth per cm, or somewhat remotely serrulate, often minutely ciliate; often distinctly short petioled. Inflorescence more or less paniculate to corymbiform, with appressed or incurved hairs. Flowers usually numerous; pedicels to 1 cm long, canescent; floral tube 0.3–0.6 mm long; sepals 4, 1.3–3 mm long, canescent, tips free in bud; petals 4, pink or whitish, 3–5 mm long; stamens 8, longer ones about 1/2 length of petals; stigma subclavate. Capsules slender, 3–4(6) cm long, canescent; seeds 1.2–1.7 mm long, abruptly rounded and beakless at apex, surface evenly pappillose; coma cinnamon reddish. (n = 18) Aug–Oct. Infrequent to locally common in marshes, around springs, streams, & lake shores, ditches; MN, e SD, NE, IA, MO, KS, e OK; (ME, Que., MN, s to NC, AL, AR, NE, OK).

This is a rather distinctive species except where it grows with *E. ciliatum* and some hybridization (*E. x wisconsinensis* Ugent) is documented. Records in the Atlas GP from the BH, WY, CO, NM, and several from w NE and KS should be deleted, as they are *E. ciliatum*.

4. **Epilobium halleanum** Hausskn. Slender perennial herbs, sometimes delicate and ascending with nodding inflorescence, sometimes more robust and erect, perennating by small compact fleshy turions, usually 3–6 cm underground. Stems rarely branched, 2–60 cm tall, subglabrous below the inflorescence except for raised lines of strigillose hairs decurrent from the margins of the petioles, with a mixture of strigillose and glandular hairs on the inflorescence, or subglabrous. Leaves opposite below the inflorescence, 0.5–4.7 cm long, 0.24–1.4 cm wide, mostly about as long as internodes, becoming much reduced in size in upper pairs, mostly glabrous with strigillose margins, ovate and obtuse basally to lanceolate or narrowly elliptic, often subacute in upper pairs, subentire especially on basal leaves, or denticulate with 8–20 teeth on each side, lateral veins inconspicuous, 3–6 on each side of the midrib, subsessile or with petioles 1–1.5 mm long on lower leaves. Inflorescence nodding in bud, erect later, often crowded with bracts quite reduced in size. Flowers erect; tube 0.5–1.7 mm deep, 0.8–1.6 mm across, subglabrous or with scattered glandular hairs outside, a ring of strigillose hairs inside at the mouth; sepals 1.2–2.8 mm long, 0.5–1 mm wide, green; petals 1.6–5.5 mm long, 1.2–3 mm wide, obcordate with an apical notch 0.3–1.2 mm deep, white, often fading pink, rarely pink throughout; anthers 0.25–0.9 mm long, cream, filaments white or cream, unequal, 0.4–2.5 mm long, at least the longer stamens usually shedding pollen directly into stigma; ovaries 10–14 mm long, on pedicels 3–8 mm long, densely glandular- and strigillose-pubescent, or sometimes subglabrous, style 0.8–5 mm long, rarely exserted beyond the anthers. Capsules 2.4–6 cm long, slender, subglabrous to quite pubescent, on pdicels 0.8–3.8 cm long; seeds 1.1–1.6 mm long, attenuate, narrowly obovoid, light brown, papillose, the chalazal collar not well pronounced, 0.04–0.12 mm long, 0.12–0.22 mm wide; coma 3–6 mm long, white, easily detached. (n = 18) Jul–Sep. Seasonally wet sites, vernal pools; SD: Custer, Pennington; (w Sask. to cen B.C., s to BH, CO, AZ, s CA).

5. ***Epilobium hornemannii*** Reichenb. Ascending perennial herb, 10–45 cm tall, often densely clumped, simple or branched from the base in larger plants, spreading by short, leafy epigeous soboles. Stems subglabrous below the inflorescence except for raised more or less densely strigillose lines decurrent from the margins of the petioles, in the inflorescence covered all around with a mixture of strigillose and glandular hairs. Leaves glabrous on both surfaces, occasionally with strigillose hairs along the margins, usually alternate only in inflorescence, mostly longer than the internodes they subtend, and quite reduced into the inflorescence, broadly elliptic or ovate to narrowly ovate and obtuse below, lanceolate to narrowly ovate and acute above, 1.5–5.5 cm long, 0.7–2.9 cm wide, denticulate with 6–14 teeth per side, the base attenuate to 5–8 mm petiole below, subsessile above, the lateral veins inconspicuous, 3–5 on each side of the midrib. Inflorescence suberect and extended. Flowers erect; floral tube 1–2.2 mm deep, 1.3–2.8 mm across, the inner surface glabrous; sepals sometimes red-tipped, 2–4.5 mm long, 1–2.2 mm wide, sparsely glandular-pubescent, sometimes with an admixture of strigillose hairs; petals rose-purple to light pink or rarely white, 3–9 mm long, 2–5.5 mm wide, the notch 0.7–2.4 mm deep; anthers light yellow to cream, 0.4–1.2 mm long, filaments white to cream to light pink, unequal, 0.6–2.6 mm long, the anthers of both sets of stamens shedding pollen directly onto the stigma; ovary covered with mixture of strigillose and glandular hairs, or more often glandular-pubescent, 15–25 mm long, on a pedicel 2–5 mm long, style white or yellow, 1.7–4.2(8.3) mm long, stigma white or cream, clavate or cylindrical, rarely exserted beyond anthers. Capsule slender, green, with scattered glandular and strigillose hairs, 4–6.5 cm long, on a glandular-pubescent pedicel 5–15 mm long; seeds blond to light brown, 0.9–1.2 mm long, 0.3–0.45 mm thick, very narrowly obovoid, distinctly papillose, chalazal neck short, 0.05–0.1 mm long, coma white, 6–11 mm long, detaching readily. (n = 18) Jul–Sep. Wet sites on creek banks; SD: Lawrence; (circumboreal, Greenl., Que., N.T., Yukon, AK s to SD, NM, AZ, CA).

6. ***Epilobium leptophyllum*** Ref., narrow-leaved willow-herb. Erect perennial herb, often robust and rank, spreading by threadlike, nearly leafless epigeous stolons that terminate in compact fleshy turions. Stems erect, terete, 15–95 cm tall, simple to well branched in larger plants, covered throughout with dense incurved strigillose hairs, sometimes mixed with glandular ones on the inflorescence, lacking decurrent lines from the leaf bases. Leaves linear to very narrowly elliptic, 2–7.5 cm long, 0.15–0.7 cm wide, not much reduced in size on inflorescence, usually longer than internodes, densely strigillose-pubescent both sides, increasing up stem, sometimes revulote, subacute especially in upper pairs, subentire with inconspicuous lateral veins, subsessile, occasionally fascicled. Inflorescence erect, crowded, leafy. Flowers erect; floral tube 0.8–1.5 mm long, 1.2–1.8 mm wide, densely strigillose outside, with a ring of spreading hairs inside at the mouth; sepals 2.5–4.5 mm long, 0.9–1.3 mm wide, green, strigillose; petals 3.5–7 mm long, 1.6–4 mm wide, white to light pink, obcordate with apical notch 1–1.8 mm deep; anthers 0.60–0.85 mm wide, cream, filaments white or cream, unequal, 0.8–3.5 mm long, at least the longer stamens usually shedding pollen directly onto stigma; ovaries 12–18 mm long, densely white canescent, rarely with an admixture of gland-tipped hairs, on pedicels 5–12 mm long, style 2–3.8 mm long, deep cream, stigma cream, narrowly clavate, rarely exserted beyond stamens, Capsules 3.5–8 mm long, slender, canescent, on pedicels 1.5–3.5 cm long; seeds 1.5–2.2 mm long, 0.5–0.65 mm wide, attenuate, narrowly obovoid, densely papillose, chalazal collar sometimes quite pronounced, 0.08–0.2 mm long; coma 6–8 mm long, dingy white, persistent. (n = 18) Jul–Sep. Infrequent to locally common in marshes, seepage areas, & mesic disturbed stites; GP, but rare in MO, KS; (e N. Amer. s to NC, WV, n OH, n IL, GP, scattered from Sask, B.C., to s CO, CA).

7. ***Epilobium palustre*** L. Erect variable perennial herb, spreading by filiform epigeous stolons with widely spaced small leaves, terminating in condensed, dark, small turions.

Stems 5–80 cm tall, terete, simple or well branched in larger plants, subglabrous on lower internodes except for strigillose lines decurrent from the margins of the petioles, densely strigillose on upper internodes. Leaves opposite below the inflorescence, lanceolate to linear, 1.5–7 mm long, 0.2–1.9 cm wide, subglabrous except for strigillose margins and veins, or sparsely strigillose, especially on inflorescence, entire to obscurely denticulate, apex acute or acuminate, base cuneate, subsessile. Inflorescence nodding in bud, densely strigillose, usually eglandular. Flowers erect, few to many; floral tube 0.6–1.8 mm deep, 1.3–2.2 mm across, tomentose outside with a ring of spreading hairs inside at the mouth; sepals 1.4–4.5 mm long, 0.8–1.5 mm wide, strigillose; petals 2–9 mm long, 1.8–5 mm wide, obcordate with an apical notch 0.6–1.6 mm deep, white or rarely pink; anthers 0.4–1 mm long, densely strigillose-tomentose, style 1–4.5 mm long, cream; stigma clavate to cylindrical, rarely exserted beyond anthers. capsules 3–9 mm long, strigillose, slender, on pedicels 1.5–3.5(6) cm long; seeds 1.4–2.2 mm long, 0.4–0.54 mm wide, elliptic-attenuate to narrow-fusiform, finely papillose, chalazal collar 0.08–0.24 mm long, quite conspicuous; coma 5–7 mm long, white, persistent. (n = 18) Jul–Sep. Low, boggy, wet places & shores, swamps & mossy meadows; MN: Pope; SD: Custer, Pennington; (circumboreal, s in N. Amer. to PA, MI, MN, BH, WY, CO, NV, OR, cen CA). *E. wyomingense* A. Nels.—Rydberg.

8. ***Epilobium paniculatum*** Nutt. ex T. & G. Erect annual herb with simple taproot and exfoliating epidermis below. Stem subsimple, with paniculate branching above in larger plants, 0.2–2 m tall, glabrous at least below, often glandular pubescent on the inflorescence. Leaves opposite below, alternate or sometimes fascicled above, very narrow elliptic to linear, often folded along the midrib, 1.5–5 cm long, 0.2–0.8 cm wide, very reduced in the inflorescence, apex acuminate, base attenuate, subsessile or with petiole 1–4 mm long, serrulate to remotely so, with an early deciduous cluster of apical oil cells, glabrous or rarely with sparse strigillose pubescence along the margins, in the inflorescence sometimes glandular pubescent on both sides; reduced subulate bracts on pedicel from base to 15 mm up. Flowers in lax racemes on filiform branches, or on larger plants in open panicles. Floral tube funnelform, 1–16 mm long, 0.8–2.9 mm wide, with ring of strigillose hairs near mouth, mouth in larger flowers 0.2–1.5 mm wide; sepals 2–8 mm long, 0.8–2 mm wide, green to reddish, glandular-pubescent to glabrous; petals white to deep rose-purple, 1–15(20) mm long, 1–9.5(12) mm wide, with notch 1–8 mm deep; anthers, 0.5–4 mm long, cream, with yellow pollen shed singly or rarely in tetrads, filaments white or cream, rarely pinkish, unequal, 0.5–9.5 mm long, the longer stamens shedding pollen directly onto stigma; ovaries 4–16 mm long, covered with gland-tipped hairs or glabrous, on pedicels 2–12 mm long, glandular to glabrous, style 1.3–25 mm long, cream, stigma white or cream, clavate to deeply 4-lobed, lobes 0.8–1.2 mm long and often exserted beyond the anthers. Capsules 1.5–3.2 cm long, 2–3 mm wide, fusiform, beaked, glabrous to glandular-pubescent, on pedicels 0.3–2 cm long; seeds brown or gray, often flecked with darker spots, 1.5–2.7 mm long, 0.8–1.25 mm wide, broadly obovoid with a constriction 0.5–0.7 mm from the micropylar end, low papillose, chalazal neck very inconspicuous; coma white, 7–8 mm long and readily deciduous. (n = 12) Jul–Sep. Dry, open disturbed ground, often along roads or railroad beds, clearings of rocky slopes; w MN, ND, SD; (w MN to B.C., s to SD, cen NM, s CA). *E. adenocladon* (Hausskn.) Rydb.—Rydberg.

9. ***Epilobium saximontanum*** Hausskn. Erect perennial herb 4–55 cm tall, simple, or well branched particularly late in the season, perennating from fleshy underground turions or

thick elongated shoots formed late in the season. Stems terete, often thick, light colored, usually covered on the apical half of the stem with short glandular and strigillose hairs, occasionally subglabrous, with raised lines of strigillose pubescence decurrent from the margins of the petioles. Leaves subglabrous with strigillose margins, mostly opposite below the inflorescence, reduced and alternate above, lanceolate or narrowly elliptic to ovate, the lower leaves often subobovate, the base rounded or obtuse, the apex subacute, 1–5.5(6.6) cm long, 0.4–2.0(2.4) cm wide, about as long as the internodes they subtend, scarcely denticulate, with 9–30 teeth per side, the lateral veins more or less conspicuous, 3–6 on each side of the midrib; subsessile or rarely petiole 1–3 mm long, often clasping and appressed. Inflorescence erect, or sometimes nodding in bud, 2–30 cm long. Flowers erect; floral tube 0.8–1.4 mm deep, 0.8–1.9 mm wide, with a sparse ring of strigillose hairs inside; sepals rarely reddish, 1.2–3.5 mm long, 0.6–1.4 mm wide, covered with strigillose and sometimes glandular hairs; petals white, infrequently pink to rose-purple, 2.2–5(7) mm long, 1.7–3.2 mm wide, the notch 0.4–1.5 mm deep; anthers cream to light yellow, 0.3–0.78 mm long, filaments cream or light yellow to rarely pink, unequal, 0.5–3 mm long, the longer stamens shedding pollen directly on stigma; ovaries 9–30 mm long, covered with a dense mixture of strigillose and glandular hairs, pedicel 0.4 mm long, style cream or yellow, 1.6–2.8 mm long, stigma deep cream or yellow, narrowly to broadly clavate or rarely subcapitate. Capsules 2–5.5(7) cm long, 1.5–3 mm wide, covered with a mixture of strigillose and glandular hairs, subsessile or rarely on pedicels 1–5 mm long; seeds light brown or gray, (0.9)1.1–1.6(1.8) mm long, (0.3)0.4–0.5(0.6) mm wide, very narrowly obovoid, rugose to papillose, chalazal collar 0.04–0.18 mm long, the micropylar end acute; coma white, 3–7 mm long, usually readily detached. (n = 18) Jul–Sep. Meadows & stream banks; SD: Lawrence; (s Alta., e OR, SD, s to n NM, AZ, NV, CA). *E. drummondii* Hausskn.— Rydberg.

5. GAURA L., Butterfly Weed

Annual, biennial, or perennial herbs with taproots, or with woody branching caudex, sometimes rhizomatous. Stems 1–several from base, simple to inflorescence, or much branched. Leaves decreasing in size upwards; basal or rosette leaves the largest, usually lyrate, gradually narrowed to winged petioles; cauline leaves subsessile, usually apically acuminate. Inflorescence a spicate raceme, not leafy, more or less clearly pedunculate, flowers bracteate, usually somewhat zygomorphic, (3)4-merous; floral tube prolonged above ovary, tubular; sepals reflexed; petals white, becoming reddish, usually abruptly clawed; stamens subequal, each filament with a scale 0.2–0.5 mm long at base (except in *G. parviflora*); stigma lobed: Fruit an indehiscent, nutlike capsule with hard walls, (3)4-locular but this not evident at maturity; seeds (1)2–4(5) per capsule.

Reference: Raven, Peter H., & David P. Gregory. 1972. A revision of the genus *Gaura* (Onagraceae). Mem. Torrey Bot. Club 23(1): 1–96.

Excluded from this treatment are 2 species listed for our region by Rydberg. They are *G. biennis*

L., which is east of our area, and *G. filipes* Spach (*G. michauxii* Spach—Rydberg) of the se U.S.

1 Flowers usually 3-merous; fruits 3-angled; plants of cen OK to
  n-cen TX .... 7. *G. triangulata*
1 Flowers 4-merous; fruits usually 4-angled.
  2 Sepals 2–3.5 mm long; plants annual .... 4. *G. parviflora*
  2 Sepals 4.5 mm long or longer.
    3 Fruits with a slender stipe 2–8 mm long; plants perennial.
      4 Plants subglabrous or sparsely pubescent .... 5. *G. sinuata*
      4 Plants densely villous with hairs 2–3 mm long .... 8. *G. villosa*
    3 Fruits subsessile or with a thick cylindrical stipe about 1/2 length of body; plants annual, biennial or perennial.
      5 Plant biennial; endemic in WY and CO ... 3. *G. neomexicana* subsp. *coloradensis*
      5 Plants annual or perennial.
        6 Plants annual with taproots; fruits not abruptly constricted to a stipe.
          7 Inflorescence glabrous or with a few long hairs on bracts; plants of cen OK and TX .... 6. *G. suffulta*
          7 Inflorescence strigulose or glandular pubescent; sepals strigulose; plants of se GP .... 2. *G. longiflora*
        6 Plant perennial; fruits abruptly constricted to a thick cylindrical stipe; plants of all but se GP .... 1. *G. coccinea*

1. Gaura coccinea Pursh, scarlet gaura. Perennial herb with a deep thick taproot and a subterranean to superficial branching caudex; spreading by horizontal underground stems which give rise to new plants and forming large colonies; entire plant densely strigulose and with long spreading hairs at base or plant subglabrous with strigulose pubescence on floral tube ovary and sepals. Stems usually several from base, 2–5(10) dm tall, branched below to strict and branched above. Leaves linear to narrowly elliptic, 0.5–4(6) cm long, 1–7(15) mm wide, entire or remotely serrate-denticulate. Inflorescence spicate, 5–40(60) cm long, peduncle 1–6 cm long, sometimes branched. Flowers sessile, each subtended by a subulate bract, 2–5 mm long, 0.5–1 mm wide; floral tube 4–10(12) mm long; sepals 4, linear-lanceolate, 5–10 mm long; petals 4, white fading to orange-red or dark reddish-brown, rarely yellowish, 3–7(9) mm long, blade 2–4 mm wide, distinctly clawed; stamens 8, filaments 3–7 mm long; anthers yellow or red, 2.5–5 mm long; style 10–20 mm long; stigma deeply 4-lobed. Capsules 4–8(9) mm long, more or less abruptly constricted to a thick cylindrical stipe about 1/2 length of body, 1–3 mm thick at widest point; seeds (1)3–4, 1.5–2.5 mm long, reddish brown. (n = 7, 14, 21, 28) Apr–Sep. Locally common on dry prairies, open wooded hillsides, stream valleys, roadsides, on a variety of soils; GP but absent in se 1/4 KS; (Man. to se B.C., s to w MN, nw MO, w 1/2 OK, TX, MT, WY, CO, s UT, se NV, AZ, s CA, to s Mex.) *G. glabra* Lehm., *G. parviflora* Torr.—Rydberg; *G. coccinea parviflora* (Torr.) Rickett—Gates.

This is a highly variable species in habit, pubescence, and flower color and size. The two most striking forms are the densely pubescent phase often referred to as var. *coccinea* and the nearly glabrous phase, which has been called var. *glabra* (Lehm.) T. & G. or var. *parviflora* (Torr.) Rickett. These variants are found together and differ in no other characters.

2. Gaura longiflora Spach, large-flowered gaura. Annual or winter annual herbs from a fleshy taproot; stems erect, 0.5–2 m tall, usually branched above, epidermis of lower stem often exfoliating, stems densely strigulose with some having a mixture of glandular or villous pubescence. Principal stem leaves elliptic or lance-elliptic, 2–13 cm long, 0.2–2.5(3) cm wide, often acuminate, entire or shallowly undulate-denticulate, surface minutely strigulose to glabrate, veins often villous; leaves usually with fasicles of small leaves in axils which remain after principal leaves have dropped; winter rosette leaves 5–15(30) cm long, remotely undulate-dentate, withering by late spring. Inflorescence usually branched and paniculately-spicate, densely strigulose, hirtellous or glandular. Flowers sessile, subtended by lanceolate to ovate bracts, 1-6 mm long; floral tube 4-13 mm long; sepals 4, 7-15 mm long, 1-2 mm wide, strigulose or glandular; petals 6-15 mm long, 2-7 mm wide, clawed, white but fading pinkish or rose; stamens 8, filaments 5-12 mm long, anthers 1.5-5 mm long, yellowish or reddish; style longer than petals, stigma 4-lobed. Capsule 4.5-7 mm long, 1.5-2.5 mm in diam, angled but not winged; seeds 2-4, 1.2-3 mm long, yellowish or reddish brown. (n = 7) Jul-Oct. Infrequent to common on rocky prairie hillsides, open woodlands, roadsides, waste places, often somewhat weedy; sw IA, se NE, w 1/2 KS, MO, ne OK; (IL, IA, se NE, s to AL, e TX, introduced elsewhere). *G. pitcheri* (T. & G.) Small — Rydberg.

In most plants of our area the sepals are glandular pubescent but scattered plants have sepals strigulose. The strigulose type becomes more common southward in s KS and is about equally as common as the glandular or strigose type in OK.

3. Gaura neomexicana Woot. subsp *coloradensis* (Rydb.) Raven & Gregory. Biennial herb from a fleshy taproot; stems several from the base, erect, 5-8 dm tall, strigulose, strict to branched. Leaves elliptic, 2-13 cm long, 1-4 cm wide, subglabrous or strigulose, subentire to repand-denticulate. Inflorescence spicate or paniculately spicate. Flowers sessile, subtended by ovate-lanceolate bracts 5-8 mm long; floral tube 6-12 mm long, glandular-pubescent; sepals 9-13 mm long, 1.5-2 mm wide; petals shorter than sepals; stamens 8, filaments 7-8 mm long; anthers 3-4 mm long; style 19-25 mm long, stigma 4-lobed. Capsule 6-8 mm long, pubescent; seeds 2-4. (n = 7) Jul-Sep. Prairie or open wooded hillsides; WY: Laramie; CO: Larimer, Weld; (Subsp. endemic in area).

4. Gaura parviflora Dougl., velvety gaura. Annual or winter annual herbs with stout taproots; stem erect, (0.3)0.5-2(3) m tall, usually simple to inflorescence, densely glandular with a few longer spreading hairs. Lower stem leaves usually deciduous by flowering time; principal leaves narrowly to broadly elliptic, lance-elliptic to ovate, usually acuminate, 2-13 cm long, 0.5-4.5 cm wide, margins somewhat sinuate-dentate, surface strigulose, main veins and margins glandular pubescent with a few spreading longer hairs; rosette leaves broadly oblanceolate, 4-15 cm long, 2-4 cm wide, blade tapering into a winged petiole. Inflorescence a terminal dense spike or spikes also terminating branches; spikes 0.5-5 dm long, usually densely glandular below and with long spreading hairs to glabrous above, rarely pubescent throughout; bracts lanceolate to linear, 1-6 mm long, with long spreading hairs. Floral tube 1.5-3(5) mm long, glabrous to short pubescent; sepals 4, 2-3.5 mm long, lance-oblong, reflexed, glabrous to puberulent; petals 4, pink to rose, spatulate, 1.5-3 mm long; stamens 8, slightly shorter than sepals, basal scales reduced to obscure papillae; anthers 0.5-1 mm long, yellow or reddish; style 3-9 mm long; stigma with 4 short lobes. Capsule 5-11 mm long, somewhat fusiform, narrowed to base, glabrous or rarely short hairy; seeds 3 or 4, reddish-brown, 2-3 mm long. (n = 7) May-Oct. Common on rocky prairie hillsides, open woodlands, pastures, old fields, roadsides, & waste areas, often weedy; GP but absent in most of ND, MN, & ne SD; (IN, IA, SD, MT, to se WA, s to AR, AL, TX, NV, AZ, n Mex., introduced locally elsewhere & in Argentina, China, Australia). *G. panviflora* var. *lachnocarpa* Weath.—Munz.

Rare specimens in our area have ovary and capsule short-pubescent to glabrate and have been names *G. panviflora* var. *lachnocarpa* Weath. Most of our specimens have floral tube, sepals, and ovary glabrous and have been recognized as forma *glabra* Munz. These slight differences in pubescence can be found in the same colony and have no taxonomic validity.

5. Gaura sinuata Nutt. ex Ser., sinuate-leaved gaura. Perennial herbs with taproots, spreading by rhizomes and often forming mats; stems 1-several from base, simple or more often well branched, erect to ascending, 2-6 dm tall, subglabrous or strigulose and with long spreading hairs particularly below. Basal leaves oblanceolate to oblong-lanceolate, 3–9 cm long, 1–2 cm wide, entire or usually sinuate-dentate, with short-winged petioles, glabrous to puberulent; principal stem leaves linear to narrowly oblanceolate, 1–10 cm long, remotely sinuate-dentate, rarely entire, sometimes undulate, glabrate to densely strigulose. Inflorescence simple or branched, 1–5(10) dm long, with naked peduncles 1–2 dm long; bracts lanceolate, 1–5 mm long, subglabrous or ciliate. Floral tube 2.5–3(5) mm long, subglabrous or sparsely strigulose; sepals 7–14 mm long, strigulose, reflexed; petals white, fading red, 7–12(14) mm long, blade 3–7 mm wide, abruptly long clawed; stamens 8, filaments lanate at base, 5–12 mm long; anthers reddish, (3)4–5 mm long; style 12–18 mm long, pubescent at base; stigma lobes 0.4–0.6 mm long. Capsule with a slender stipe (2)3–8 mm long, body 8–15 mm long and 1.5–3.5 mm thick, glabrous or rarely sparsely strigulose; seeds (1)2–4, reddish brown, 2–3 mm long. (n = 14) Apr–Aug. Often too common in sandy prairies & open woodlands, stream valleys, roadsides, waste places; OK, n TX; (AR, OK, e 1/2 TX; introduced in GA, FL, AL, MO, CA; Italy & S. Africa).

This is an aggressive weed, which is a problem in some areas.

6. **Gaura suffulta** Engelm. ex A. Gray, gaura. Annual herbs with slender to stout taproots; stems erect, usually well branched, 2–10(12) dm tall, villous to the inflorescence with hairs 1.5–2.5 mm long. Rosette leaves 7–10 cm long, 1.5–2 cm wide; principal stem leaves narrowly lanceolate to lanceolate, 1–9 cm long, 1–20 mm wide, sinuate-dentate to nearly entire, subglabrous or villous on veins and near margins. Inflorescence simple to sparsely branched, 0.5–8 dm long, glabrous or glandular-pubescent; bracts lanceolate to lance-ovate, 2–6 mm long. Floral tube, 7–14 mm long, usually glabrous; sepals 4, 8–10 mm long, glabrous; petals 4, 10–15 mm long, 4–6 mm wide, clawed, white, fading reddish or yellowish-brown; stamens 8, filaments 6–9 mm long, anthers 3–4 mm long, reddish; style 1.5–3 cm long; stigma 4-lobed. Capsule subsessile, 4-angled above, not abruptly constricted to a stipe, 4.5–8 mm long, 3–5 mm in diam; seeds (1)2–4, light brown, 2–2.5 mm long. (n = 7) Apr–Aug. Infrequent to locally common on sandy prairies, open woodlands, roadsides; OK, TX; (endemic).

Our plants are subsp. *suffulta*.

7. **Gaura triangulata** Buckl. Annual herbs with taproots; stems erect, 2–6 dm tall, simple or usually branched from base, villous to inflorescence with hairs 1.5–2.5 mm long. Leaves narrowly elliptic to oblanceolate, 1.5–8 cm long, 2–15 mm wide, subsessile, entire or somewhat sinuate-dentate, subglabrous or sparsely villous on main veins and near margins. Inflorescence paniculately spicate, 1–3.5 dm long, glabrous or ovary, floral tube and sepals strigulose; bracts lanceolate or lance-ovate, 2–4 mm long, 1–1.5 mm wide. Floral tube 4–5 mm long, glabrous or strigose; sepals 3(4), 4.5–8 mm long, linear, usually glabrous; petals 3(4), white, fading reddish, 4–6 mm long; stamens 6(8), filaments 2–6 mm long, anthers red, 3–4 mm long; style 8–10 mm long, stigma 3(4)-lobed. Capsule sessile, ovoid, 6–9 mm long, 3–5 mm in diam, glabrous; seeds 2–5, light brown, 1.5–3.5 mm long. (n = 7) Apr–Jul. Infrequent in sandy prairies & open woodlands; ne OK to n-cen TX (endemic to area).

8. **Gaura villosa** Torr., hairy gaura. Perennial herbs from a subterranean woody and often branched caudex; stems 1–several from base, erect or ascending, 0.5–1.5 m tall, often branching near the base, branching again where inflorescences originate to form a cluster of branches; stems densely or sparsely villous below the inflorescence with hairs 2–3 mm long and with an understory of strigulose or glandular hairs. Lower leaves rather crowded, spatulate to ovate or lanceolate, 3–7 cm long, sinuate-serrate or denticulate, acuminate to obtuse, canescent and appressed villous; upper leaves narrowly elliptic to linear or lance-ovate, 0.5–8 cm long, subentire or sinuate-dentate, often undulate, acuminate, sessile, loosely to densely strigulose or villous, often glandular. Inflorescence usually branched, 2–6(10)

dm long, hardly pedunculate, subglabrous to more rarely strigulose; bracts lanceolate to lance-ovate, 1-9 mm long. Floral tube funnelform, 1.5-3(5) mm long, canescent within and outside; sepals 4, lance-linear, 6-10(14) mm long; petals 4, white, fading red, broadly elliptic, with an abrupt slender claw, blade 7-10(13) mm long; stamens 8, subequal, filaments 5-10 mm long; anthers reddish, 2.5-4.5 mm long; style 10-18 mm long; stigma lobes 0.2-0.4 mm long. Capsule with stipe 2-8 mm long, body 10-18 mm long; seeds (1)2-4, reddish-brown, 0.75-1.2 mm long. (n = 7) May-Aug. Infrequent to locally common on sand dunes, sandy prairies, stream valleys, roadsides; sw KS, se CO, w 1/2 OK, nw TX, e NM; (range as given). *G. villosa* Torr. var. *arenicola* Munz — Munz.

Plants with glandular hairs on ovary and sepals not strigulose have been recognized as var. *arenicola*, but these differences are not consistent. Our plants are subsp. *villosa*.

6. GAYOPHYTUM Juss.,

1. ***Gayophytum diffusum*** T. & G. subsp. ***parviflorum*** Lewis & Szweykowski. Annual herbs with taproots; stems 1.5-3(6) dm tall, branched or unbranched at base, dichotomously branched above, glabrous or rarely pubescent, epidermis usually exfoliating near base. Leaves alternate, narrowly lanceolate, 1-6 cm long, 1-5 mm wide, entire, usually glabrous, somewhat reduced above; sessile or with petioles to 10 mm long. Flowers axillary, subsessile to pedicellate. Floral tube essentially lacking; sepals 4, divided nearly to ovary, 1-1.5 mm long; petals 4, 1.2-3 mm long, white to pinkish, short clawed; stamens 8, unequal, the longer 1-2 mm long; anthers adhering to stigma; style 0.7-1.5 mm long; stigma subglobose, 0.5 mm wide; ovary 2-celled. Capsules 3-15 mm long, 1-1.5 mm wide, constricted between seeds, usually pubescent with short hairs, dehiscing into 4 valves; seeds 3-18 per capsule, brown or mottled, 1-1.5 mm long, about 1/2 as wide. (n = 14) Jun-Jul. Open sandy or rocky areas of prairies & open woodlands; BH, e WY; (SD, w MT, ID, B.C., s to n-cen NM, AZ, CA, Baja Calif.).

Reference: Lewis, H., & J. Szweykowski. 1964. The genus *Gayophytum* (Onagraceae). Brittonia 16: 343-391.

Gayophytum racemosum T. & G. is known from a collection in SD: Custer but is otherwise w of our area. It differs from *G. diffusum* subsp. *parviflorum* in having leaves only to 25 mm long and leaves not reduced above. Rydberg reported *G. humile* Juss., (as *G. nuttalii* T. & G.), which approaches our area in se WY and differs from *G. diffusum* subsp. *parviflorum* in having flowers with petals only 0.7-1.2 mm long and in being diploid. The report of this species from WY: Weston in Atlas GP was based on a specimen of *G. diffusum* subsp. *parviflorum*.

7. LUDWIGIA L., False Loosestrife, Seedbox

Ours perennial herbs of wet places with fibrous or thickened roots, sometimes stoloniferous; stems prostrate, creeping, or floating to ascending or erect. Leaves alternate or opposite, simple, essentially entire; sessile or petiolate; stipules minute or deciduous. Flowers borne in axils of upper leaves, regular, 4- or 5-merous; floral tube not prolonged beyond apex of ovary; ovary cylindric to obconic, often 4-angled or winged, usually with 2 bracteoles at base; sepals green, persistent in fruit; petals yellow or absent, caducous; stamens as many or twice as many as sepals; stigma capitate or globose, not divided. Capsules 4-celled, dehiscent longitudinally or by a terminal pore, sometimes tardily and irregularly dehiscent; seeds many, lacking a coma. *Isnardia* L., *Jussiaea* L. — Rydberg.

Ludwigia decurrens Walt., with alternate leaves and stems conspicuously 4-winged, has been reported for Kansas but is excluded for lack of specimen evidence. It is a plant of the se U.S.

1 Leaves opposite; stems prostrate or floating
  2 Capsules with broad green bands at each corner; petals absent .... 3. *L. palustris*
  2 Capsules lacking green bands; petals present .... 6. *L. repens*
1 Leaves alternate; stems erect or ascending.
  3 Flowers 5-merous; stems prostrate or floating .... 4. *L. peploides*
  3 Flowers 4-merous; stems erect.
    4 Flowers definitely pedicelled; petals yellow, 8–10 mm long .... 1. *L. alternifolia*
    4 Flowes sessile or on obscure pedicels to 1 mm long; petals none or minute and greenish.
      5 Capsules nearly as wide as long at maturity; bracteoles inserted well above base of fruit, about as long as floral tube .... 5. *L. polycarpa*
      5 Capsules about 2× or more longer than wide; bracteoles insertered at or just above base of fruit, much shorter than floral tube .... 2. *L. glandulosa*

1. **Ludwigia alternifolia** L., bushy seedbox. Perennial herbs with somewhat fascicled and often fusiform roots; stems erect, branched, 5–10 dm tall, often reddish, more or less angled above, glabrate to strigulose or with minute spreading hairs. Leaves alternate, entire, lanceolate, 5–10 cm long, acute or acuminate, blade gradually tapering to sessile base or with obscure petiole 3–10 mm long, paler below; leaves gradually reduced in size up the stem. Flowers solitary in upper axils, from few to many; pedicels 2–5 mm long, with a pair of lanceolate braceoles at upper end; floral tube not prolonged beyond ovary; sepals 4, ovate to lance-ovate, spreading, abruptly acuminate, 7–10 mm long, 4–6 mm wide, often reddish, 5-veined; petals 4, yellow, oblong-obovate, about as long as sepals, easily shed; ovary 2.5–3 mm long and nearly as wide at anthesis; stamens 4, 2–3 mm long, anthers 1.3–1.6 mm long; style 2–3 mm long, from a hairy disk. Capsule 4–5 mm long and about as wide, squarish above, angled or narrowly winged, rounded below, opening by a terminal pore; seed 0.6–0.8(1) mm long, oblong, yellowish-brown, lustrous. (n = 8) Jul–Sep. Infrequent to locally common in moist prairie ravines, swamps, lakeshores, stream banks, roadside ditches; MO, se NE, e 1/2 KS, e OK; (MA, Ont., IA s to FL, se NE, e TX). *L. alternifolia* var. *pubescens* Palm. and Steyerm. — Fernald.

In the Atlas GP all of our specimens were referred to var. *pubescens* which has stems, leaves and pedicels often pubescent with some spreading hairs. Many of our specimens, however, are glabrate and all degrees of variation are found to plants which are rather densely strigulose or have short spreading hairs. Thus, recognizing infraspecific taxa is arbitrary.

2. **Ludwigia glandulosa** Walt., cylindric-fruited seedbox. Perennial herbs with fibrous roots; stems at first prostrate and rooting at nodes, then erect, freely branched, 3–9 dm tall, glabrous or minutely strigulose especially above. Leaves alternate; primary stem leaves usually lanceolate or elliptical, 2.5–9 cm long, 6–20 mm wide, entire, glabrous; petioles winged; 2–10 mm long; leaves gradually reduced upward. Flowers solitary in axils above, few to numerous, sessile; floral tube not prolonged beyond ovary; ovary at anthesis nearly cylindric, 2.7–3 mm long; bracteoles at base of ovary linear, to 0.9 mm long; sepals 4, spreading, deltoid, 1–2 mm long, subacuminate; petals none; stamens 4, 0.7–1 mm long, disk elevated, glabrous; anthers 0.3–0.5 mm long, Capsule nearly cylindric, 4–8 mm long, 1.5–2 mm in diam, sessile; seeds yellowish-brown, 0.6–0.7 mm long, lustrous, slightly curved-oblong, surface minutely punctate. (n = 16) Jun–Aug. Rare in GP in swamps; KS: Cherokee, Cowley; (se VA, s IN, s IL, se MO, se KS, s to FL, e TX).

3. **Ludwigia palustris** (L.) Ell., marsh seedbox. Perennial herbs (but sometimes flowering in first season); stems creeping and rooting at nodes, sometimes floating, 1–5 dm long, glabrous, simple to much branched. Leaves opposite; blades broadly elliptical or elliptic ovate, entire, shiny green or reddish, acute to acuminate at apex, 3–25(40) mm long, 4–20 mm wide, tapeing to base; petiole 3–25 mm long. Flowers sessile, solitary in the axils; floral

tube not extended beyond ovary; sepals 4, deltoid, 1.5-2 mm long; petals absent; stamens 4, from a 4-lobed, glabrous elevated disk; ovary 1.5-2 mm long, with 4 longitudinal green bands; bracteoles at base of ovary minute or to 1 mm long. Capsule elongate-globose, somewhat 4-sided, rounded at base, 2-5 mm long, 2-3 mm in diam, sessile; seeds 0.5-0.9 mm long, straight along one side, whitish or yellowish-brown, somewhat lustrous. (n = 8) Jun-Oct. Locally common in shallow water & margins of ponds, lakes, slow streams, moist depressions, ditches, & around springs; e 1/2 NE, KS, IA, MO; (N. Amer. except area of Rocky Mts.; w Eurasia, Africa). *L. palustris* var. *americana* (DC.) Fern. & Grisc.—Fernald; *Isnardia palustris* L.—Rydberg.

4. **Ludwigia peploides** (H.B.K.) Raven, floating evening primrose. Perennial herb; stems creeping or floating, 2-6 dm long, rooting at nodes, freely branched especially in land forms, often with ascending flowering branches, glabrous to sparsely or moderately pubescent. Leaves alternate, oblong to oblong-spatulate, 1-10 cm long, 5-40 mm wide, entire, glabrous or sparsely hairy, obtuse to acute, narrowed to base into a flattened or winged petiole 0.5-3(4) cm long; stipules to 1.5 mm long, scalelike. Flowers solitary in upper axils; pedicels 1-6 cm long; floral tube not prolonged beyond summit of ovary; bracteoles deltoid, 0.5-1 mm long; sepals 5, linear-lanceolate, 4-12 mm long, glabrous or hairy; petals 5, yellow, obovate, emarginate, 7-24 mm long; stamens 10, 3-6 mm long, anthers 0.8-1.2 mm long; disk flat, hairy; style 3-5 mm long; stigma flattened, shallowly 5-lobed. Capsules 1-4 cm long, 3-4 mm in diam, cylindrical, glabrous or sparsely hairy, tardily dehiscent; seeds included in endocarp, 1-1.5 mm long. (n = 8) May-Oct. Locally abundant in shallow water or margins of ponds, lakes, slow streams, ditches; MO, se NE, e 2/3 KS, e OK; (s IN, MO, KS, s to NC, LA, TX, introduced elsewhere). *Jussiaea diffusa* Forsk.—Rydberg; *J. repens* L. var. *glabrescens* O. Ktze.—Fernald; *L. peploides* (H.B.K.) Raven subsp. *glabrescens* (O. Ktze.) Raven—Atlas GP.

5. **Ludwigia polycarpa** Short & Peter, manyseed seedbox. Perennial herbs with stolons; stems erect or ascending, simple or much branched, 1-9 dm tall, glabrous, usually 4-angled. Leaves alternate, entire; principal stem leaves lanceolate to oblanceolate, 3-12 cm long, 5-15 mm wide, entire, tapering to both ends; petiole winged, 2-8 mm long. Flowers sessile, in axils, often borne to base of stem; floral tube not prolonged beyond ovary; bracteoles well above base of ovary, linear-lanceolate, 2-5 mm long; sepals 4, lance-deltoid, 2.5-4 mm long, slightly acuminate; petals 4 and minute or usually absent; stamens 4, filaments 1-2 mm long, anthers 1 mm long; style 1 mm long, stigma 1 mm thick. Capsules short-cylindric or usually somewhat turbinate, 4-7 mm long, 3-5 mm in diam, the bractlets about as long; seeds light yellow, 0.6-0.9 mm long, slightly curved, minutely punctate. (n = 8) Jun-Sep. Infrequent in low woodlands, marshes, margins of ponds & lakes, wet depressions in prairies, along streams; IA, e 1/3 NE, e KS, MO; (MA, s Ont., MN, s to TN, NE, KS).

6. **Ludwigia repens** T.F. Forst., water primrose. Perennial herbs; stems creeping or floating, rooting at nodes, 1-5 dm long, glabrous or essentially so. Leaves opposite, narrowly elliptic to subrotund, 1-4 cm long, 2-20 mm wide, entire, obtuse or abruptly acute at apex, narrowed to base; petioles 3-20 mm long, flowers in axils, sessile or very short pedicelled; floral tube not prolonged beyond ovary; sepals 4, deltoid, 2.5-4 mm long, acute to acuminate; petals 4, yellow, 4-5 mm long, promptly deciduous; stamens 4; disk elevated, 4-lobed, glabrous; style and stigma 1 mm long. Capsule 3-7 mm long, short-cylindrical, tapering at base, pedicel to 1.5 mm long, tardily and irregularly dehiscent; seeds 0.6-0.8 mm long, yellowish-brown, straight along one edge, lustrous. (n = 24) Jul-Sep. Rare in GP around ponds, lakes, & along streams; KS: Cowley; OK, TX; (NC, FL, OK, TX, Mex.). *L. natans stipitata* Fern.—Gates.

This species is often grown as an aquarium plant and may possibly be found as an escape in s GP.

8. OENOTHERA L., Evening Primrose

Annual, biennial, or more commonly perennial herbs; caulescent, acaulescent or subacaulescent. Leaves alternate or basal, entire to pinnatifid; stipules absent. Flowers regular, 4-merous, borne in axils of upper leaves, on in a more or less distinct inflorescence; with a well-developed floral tube prolonged beyond the ovary 0.5–2.5 cm, deciduous after anthesis; sepals 4, reflexed at anthesis, not persistent in fruit; petals 4, yellow or white, rarely rose-purple; stamens 8; stigma deeply 4-lobed. Fruit a loculicidally dehiscent or nutlike indehiscent capsule; seeds in 1 or 2 rarely 3 definite rows in each locule or rarely clustered and not in rows; seed without a coma. *Anogra* Spach, *Gaurella* Spach, *Hartmannia* Spach, *Kneiffia* Spach, *Lavauxia* Spach, *Megapterium* Spach, *Pachylophus* Spach, and *Peniophyllum* Penn.—Rydberg.

Excluded from this treatment are the following species listed by Rydberg for our area with the currently accepted name followed by Rydberg's name in parenthesis. *Oenothera perennis* L. (*Kneiffia perennis* (L.) Penn.) found east of GP; *P. pilosella* Raf. (*K. pratensis* Small) found each of GP; *O. spachiana* T. & G. (*K. spachiana* Small) found se of our area and to cen OK.

```
1   Plants annual.
    2   Petals white.
        3   Stem densely and uniformly villous with an understory of appressed hairs; petals 1–2.5
            cm long; seeds in 1 row in each locule ........................................ 7. O. engelmanii
        3   Stem strigulose or sparsely villous; petals 1.5–4 cm long; seeds in 2 rows in each
            locule ........................................................................................ 1. O. albicaulis
    2   Petals yellow.
        4   Leaves linear, usually less than 1 mm wide, unlobed ................... 14. O. linifolia
        4   Leaves not linear, over 5 mm wide, usually lobed.
            5   Sepals 5–12 mm long; petals 5–18 mm long ........................ 12. O. laciniata
            5   Sepals 20–30 mm long; petals 20–35 mm long .................... 9. O. grandis
1   Plants biennial or perennial.
    6   Flowers in terminal braceate spikes; petals yellow; plants usually 6 dm or more tall.
        7   Floral tube 6–10 cm long .................................................. 11. O. jamesii
        7   Floral tube less than 6 cm long.
            8   Petals rhombic-obovate .......................................... 17. O. rhombipetala
            8   Petals obovate.
                9   Petals 2–4 cm long; plants of sw GP ............... 6. O. elata subsp. hirsutissima
                9   Petals 1–2 cm long.
                    10  Plants green, sparsely pubescent with long spreading hairs admixed with
                        minute appressed and glandless hairs; inflorescence densely
                        flowered ................................................................ 2. O. biennis
                    10  Plants usually grayish, densely appressed pubescent or if less pubescent and
                        with glandular and spreading hairs, then the inflorescence few flowered and
                        open .................................................................... 20. O. villosa
    6   Flowers in axils of stem or basal leaves; petals yellow or white.
        11  Plants rhizomatous; capsules without tubercles on angles or distinct wings.
            12  Stems white; epidermis usually exfoliating.
                13  Floral tube glandular-pubescent, otherwise plant nearly
                    glabrous ......................................................... 16. O. nuttallii
                13  Floral tube not glandular-pubescent, plants densely strigose ..... 13. O. latifolia
            12  Stems greenish; epidermis usually not exfoliating.
                14  Petals 2.5–4 cm long ........................................ 18. O. speciosa
                14  Petals less than 1.5 cm long.
                    15  Capsule ovoid-pyramidal, sharply 4-angled, indehiscent; leaves denticulate
                        to entire ...................................................... 4. O. canescens
```

            15 Capsule oblong-fusiform, not sharply angled, dehiscent at maturity; leaves sinuate-serrate to pectinate-pinnatifid .... 5. *O. coronopifolia*
11 Plants not rhizomatous, sometimes producing new shoots from lateral roots; capsules tuberculate on angles or with distinct wings.
      16 Petals white at anthesis, fading rose purple; angle of capsules tuberculate or with sinuate ridges .... 3. *O. caespitosa*
      16 Petals yellow or pale yellow at anthesis, fading reddish, purplish or whitish; capsules winged.
            17 Capsules winged in upper 2/3, gradually tapering to base; seeds with a narrow entire wing on one adaxial margin, reddish-brown to black, surface beaded.
                  18 Petals light yellow, fading pale orange, drying pinkish; wings of capsules 5-10 mm wide, often terminating in a hooked tooth at widest part; plants of s 1/3 GP .... 19. *O. triloba*
                  18 Petals bright yellow, fading orange, drying purplish; wings of capsules 2-5 mm wide, never with a hooked tooth; plants of nw GP .... 8. *O. flava*
            17 Capsules distinctly winged their entire length; seeds with an erose wing distally or sometimes extending part way along both adaxial margins, surface rugose.
                  19 Plants acaulescent; leaves subentire to lobed; rare plants of extreme w KS, se OK .... 10. *O. howardii*
                  19 Plants usually distinctly caulescent; leaves entire or remotely sinuate-dentate; absent in extreme w KS .... 15. *O. macrocarpa*

1. ***Oenothera albicaulis*** Pursh, prairie or pale evening primrose. Annual herbs with slender taproots; stems erect, simple or branched above, and usually with several decumbent branches from base, 1-3(5) dm tall, white epidermis not exfoliating, strigulose and sometimes with sparse, stiff, spreading hairs. Basal leaves (often absent at anthesis) appearing rosulate, oblanceolate, obovate, or spatulate, 2-10(14) cm long, entire, to sinuate-pinnatifid, with petioles often longer than blades; stem leaves alternate, lanceolate, elliptic to oblanceolate, 1-6(10) cm long, 4-20 mm wide, sinuate pinnatified, to pectinate-pinnatifid, strigulose and somewhat villous; petioles 0-8 mm long. Flowers solitary in axils, often with unpleasant odor, buds nodding; floral tube slender, slightly enlarged at summit, 1.5-3 cm long, cinereous and with sparse spreading hairs; sepals linear-lanceolate, without free tips in bud, distinct or separating in pairs at anthesis, 12-25 mm long, margins with a broad pale reddish-purple stripe, cinereous and sparsely pilose; petals white, fading pink, obcordate, 1.5-4 cm long, nearly as wide, with a terminal notch to 5 mm deep; filaments glabrous, to 1/2 as long as petals; anthers 4-10 mm long; style equaling or slightly surpassing stamens; stigma lobes 5-7 mm long. Capsule sessile, cylindric, ribbed, ascending, straight or slightly curved, abruptly tapering to apex, 2-4 cm long, 2.5-3.5 mm in diam, pubescent, dehiscent full length of capsule at maturity; seeds in 2 rows in each locule, ellipsoid, 0.9-1.1 mm long, nearly as thick, yellowish-brown, blunt tipped, surface areolate with areolae in rows, self-incompatible. (n = 7) Apr-Aug. Locally common in sandy prairies, stream valleys; roadsides, waste places, often weedy; w 1/2 GP; (ND, MT, s to TX, NM, AZ, s UT, Mex.). *Anogra albicaulis* Britt., *A. braduriana* Rydb., *A. perplexa* Rydb.—Rydberg.

2. ***Oenothera biennis*** L., common evening primrose. Biennial herbs with taproots; stems 0.5-2 m tall, erect, usually openly branched, greenish but with short appressed hairs and usually somewhat spreading hairs some of which are reddish pustulate. Rosette leaves somewhat sinuate-lobed, denticulate or nearly entire, often red-spotted, 0.6-3 dm long, 1-7 cm wide with long petioles; principal stem leaves alternate, narrowly to broadly lanceolate, acuminate, 5-15 cm long, 1.5-4 cm wide, often denticulate, sometimes with wavy margins, surfaces sparsely to moderately pubescent, passing upward gradually into bracts of inflorescence. Inflorescence a terminal bracted spike, often with short congested branches, sparsely to moderately strigose, often with longer hairs of which some are gland-tipped; bracts spreading, semipersistent, lanceolate, oblong, or ovate, 1-3 cm long. Mature

buds erect; flowers opening near sunset; floral tube 2-5 cm long, slender, greenish-yellow, with sparse, spreading, eglandular or some glandular hairs; sepals 1-2.5 cm long, linear-lanceolate, the subulate free tips 1-4 mm long; petals yellow, fading pale reddish yellow 1-2.5 cm long, obovate, slightly notched at apex; filaments 9-12 mm long; stamens 6-7 mm long; style and stigmas shorter than to as long as stamens; stigmas 5-6 mm long. Capsules cylindric, 1.4-3.5 cm long, 3.5-6 mm thick near base, tapering toward apex, strigose to subglabrous; seeds in 2 rows in each locule, ascending, 1.3-1.6 mm long, rather sharply angled, reddish-brown; self-pollinating. (n = 7, ring of 14 chromosomes at meiotic metaphase 1) Jul-Oct. Infrequent to locally common along streams, open woodlands, lakeshores, roadsides, & waste places; e 1/4 GP, a little more westward to the n; (MI, WI, Alta. s to LA, ne TX). *O. biennis* L. subsp. *centralis* Munz—Atlas GP; *O. muricata* L.— Rydberg.

It does not appear that subdivision of this variable species is warranted. Much of the distribution as dotted in the Atlas GP included plants I refer to *O. villosa* Thunb.

3. Oenothera caespitosa Nutt., gumbo evening primrose or gumbo lily. Ours perennial, cespitose, acaulescent herbs with thick taproots and simple to branching subterranean to superficial caudex, lateral roots often giving rise to new shoots. Leaves oblanceolate; blades 3-16(21) cm long, (0.3)1-3(5) cm wide, coarsely and irregularly serrate or dentate, sometimes pinnately lobed or subentire, glabrous or pubescent with appressed hairs on veins and margins; petioles winged, often as long as blades. Flowers solitary in axils, usually with sweet, somewhat rubbery fragrance, opening near sunset; buds erect; floral tube 3-6(8) cm long, often reddish, sparsely pubescent, slender but flared above; sepals 2.5-3.5 cm long, linear-lanceolate to lanceolate and acuminate, without free tips in bud; petals white fading rose-purple, obcordate, (1.6)2.5-4(4.8) cm long; stamens subequal, 2/3 as long as petals, glabrous; anthers 9-14 mm long; style nearly as long as petals; stigmas (3)4-6 mm long. Capsules lance-ovoid, falcate or sigmoid especially when young, hard, (1)2-4(5) cm long, attenuate into a stout beak, 4-angled, angles with distinct tubercles; seeds in 2 rows in each locule, 2.5-3.9 mm long, narrowly obovoid, reddish-brown, surface minutely papillose; adaxial side with outgrowth forming a hollow cavity which constitutes the upper 1/2 of seed volume, a longitudinal membrane sealing cavity along the raphe; self-incompatible. (n = 7) May-Aug. Infrequent to locally common on dry rocky or gravelly prairie hillsides or open woodlands; w 1/2 of ND & SD, nw NE, MT, WY, CO, NM (over much of w 1/2 U.S.). *O. caespitosa* subsp. *purpurea* (S. Wats.) Munz—Munz. op. cit. *Pacylophus canescens* Piper, *P. caespitosus* (Nutt.) Raim. *P. montanus* (Nutt.) A. Nels.— Rydberg.

Our plants, as here described, are all subsp. *caespitosa*. In the w-cen GP and WY: Laramie in deep canyons or in the mountains on granite are found plants without tubercles on the angles or ridges, flowers that fade pink and with at least some glandular hairs. These are *O. caespitosa* subsp. *macroglottis* (Rydb.) Wagner, Stockhouse & Klein. In CO: Otero a similar but caulescent hirtellous plant, with large tubercles on the angles of the fruit; is found: *O. harringtonii* Wagner, Stockhouse & Klein (*Pachylophus eximus* (A. Gray) Woot. & Standl.—Rydberg; *Oenothera caespitosa* subsp. *eximia* (A. Gray) Munz—Munz op. cit.).

4. Oenothera canescens Torr. & Frem., spotted evening primrose. Perennial colonial herbs with subterranean to superficial caudex, spreading by adventitious shoots from lateral roots; stems diffusely branched from base, decumbent or ascending, 1-2 dm tall, densely strigulose to canescent. Leaves lanceolate, sinuate-denticulate to almost entire, 5-15 mm long, 2-7 mm wide, strigulose, with small fascicles in axils, subsessile. Flowers axillary, sessile; opening near sunset; buds erect; floral tube 5-15 mm long, strigulose; sepals lanceolate, canescent, 8-9(12) mm long, without free tips; petals pinkish, rarely white, obovate, 8-12 mm long, red-spotted or streaked all over; stamens 2/3 as long as petals; anthers 4-5 mm long;

styles as long as stamens or as long as petals; stigmas 2-3 mm long. Capsules ovoid-pyramidal, sharply 4-angled, beaked, 7-8 mm long, canescent; seeds in several rows in each locule; seeds obovid, 0.9-1.1 mm long, brown; self-compatible but outcrossing. (n = 7) May-Jul. Infrequent to locally common in prairie depressions, dried ponds, ditches; sw NE, se WY, w 1/4 KS, e CO, OK: Beaver; (sw NE, se WY s to cen TX, e NM). *Gaurella canescens* (Torr. & Frem.) A. Nels.—Rydberg.

This species is frequently found as dense colonies in "buffalo wallows," around dried ponds and other places with temporary water, and is rare on other sites.

5. ***Oenothera coronopifolia*** T. & G., combleaf evening primrose. Perennial herbs with slender caudex, spreading by adventitious shoots from lateral roots; stems often clustered, simple or branched from base or above, 0.5-2.5 dm tall, strigulose and usually with some spreading hairs to 2 mm long. Leaves alternate, often crowded, with fascicles of smaller ones; basal leaves oblong-lanceolate to oblanceolate, sinuate-dentate, 1.5-2.5 cm long, early deciduous; principal stem leaves oblong-lanceolate or oblanceolate, deeply and regularly sinuate-pinnatifid or pectinate-pinnatifid, sessile or subsessile. Flowers solitary in upper axils, opening near sunset; mature buds nodding; floral tube slender, 1.5-3 cm long, strigulose and with scattered spreading hairs, abruptly expanded at very summit, throat with conspicuous white or tawny hairs; sepals separate and reflexed at anthesis, lanceolate, often reddish, 8-15 mm long, free tips 0.5-1 mm long; petals white, fading pink or rose, obcordate, 7-12 mm long, with a terminal notch 1-1.5 mm deep; stamens 2/3 as long as petals; anthers 5-8 mm long; style as long as petals, stigmas 2.5-4 mm long. Capsules oblong-fusiform, 8-20 mm long, somewhat 4-sided, short beaked, strigulose and with long hairs, seeds in 2 rows in each locule; seeds ellipsoid, brown, 2 mm long, minutely pitted; both self-incompatible and self-compatible populations known. (n = 7, 14) May-Jul. Infrequent to locally common in sandy or rocky pairies, open woodlands; sw SD, e WY, w NE, w KS, CO, ne NM; (SD to ID, s to KS, NM, n AZ). *Anogra coronopifolia* (T. & G.) Britt.—Rydberg.

6. ***Oenothera elata*** H.B.K. subsp. ***hirsutissima*** (A. Gray ex S. Wats.) Dietrich. Biennial herbs with taproots; stems erect, usually branched from the base, 0.5-2 m tall, densely to sparsely strigulose, usually with few to many spreading hairs 2-3 mm long, gland-tipped hairs intermixed. Rosulate leaves, oblanceolate, entire or sinuate-dentate, acute at apex, blades 5-15 cm long, 1.5-4 cm wide, densely strigose on both surfaces; base narrowed into petioles; petioles to as long as blades, strigose and with spreading hairs; primary stem leaves elliptic-lanceolate, 5-15 cm long, 5-25 mm wide, acute to short-acuminate, subentire to serrulate, densely pubescent on both surfaces, leaves gradually reduced upward. Inflorescence a terminal bracted spike, rarely branched, 2-5 dm long, strigose, sparsely hirsute, eglandular; bracts lanceolate or elliptic, the upper often twisted, 1-4 cm long. Buds erect; floral tube 2-4(5) cm long, often reddish, pilose and pubescent; sepals remaining coherent at anthesis and reflexed to one side, 2-4 cm long, with free tips 2-4 mm long, lance-acuminate, glabrous on upper surface, appressed to spreading hairy on outer surface; petals yellow, fading reddish, 2-4 cm long, obovate, with or without an apical notch; stamens 2/3 as long as petals; anthers 9-12 mm long; style as long as petals; stigmas 4-5 mm long. Capsules 2.5-4 cm long, 4.5-5 mm thick, cylindric, strigose and hirsute; seeds in 2 rows in each locule, rather sharply angled, 1.2-1.5 mm long, reddish-brown; self-compatible but out-crossing. (n = 7) Jul-Oct. Infrequent to locally common along sandy stream banks & low marshy areas; sw 1/4 KS, se CO, w OK, TX, NM; (sw KS, s CO, UT, s to TX, AZ, Mex.). *O. hookeri* T. & G.—Rydberg; *O. hookeri* T. & G. subsp. *hirsutissima* (A. Gray) Munz—Atlas GP.

7. ***Oenothera engelmannii*** (Small) Munz, Engelmann's evening primrose. Annual or winter annual villous herbs with taproots; stems simple, erect, or sometimes with ascending branches

from the base, 2–5(8) dm tall, epidermis white, exfoliating, conspicuously spreading hairy. Basal leaves somewhat rosulate, blades oblanceolate, entire, sinuate-dentate, or pinnatifid in lower portion, 2–8(12) cm long, glabrate or strigulose, petioles as long as to much longer than blade, usually disappearing by anthesis; stem leaves alternate, lanceolate to oblong-lanceolate, 2–8 cm long, 1–1.5(2.5) cm wide, coarsely sinuate-dentate to nearly pinnatifid, sparsely to densely strigulose and with scattered long hairs above, more densely pubescent below. Flowers solitary, axillary, opening near sunset; mature buds nodding; floral tube slender, 2–3 cm long, pilose and strigulose; sepals separate and reflexed at anthesis, (1)1.5–2 cm long, lance-attenuate, lacking free tips, strigulose and pilose; petals white, fading pink, obovate, rounded or slightly emarginate, 1.3–2.2 cm long; filaments 1/2–2.3 as long as petals, anthers 6–8 mm long; style as long as to longer than petals, stigmas 4–5 mm long. Capsules spreading, sessile; tapering from base to apex, somewhat 4-angled, 2.5–5 cm long, often hard at least near base, strigulose and with spreading long hairs; seeds in 1 row in each locule, 1–1.6 mm long, linear-obovoid, light reddish-brown, smooth or obscurely longitudinally striate; self-incompatible. (n = 7) May–Sep. Infrequent to locally common in sand dunes, sandy prairies, stream valleys, roadsides, waste places; KS: Morton; CO: Baca; w OK, TX panhandle, e NM; (endemic in area).

8. **Oenothera flava** (A. Nels.) Garrett. Ours perennial, cespitose, acaulescent plants, with thick fleshy taproot often crowned with fruits of previous years. Leaves in tufts or rosettes, sparsely strigulose or glabrate, oblanceolate or oblong-liner (3.4)5–25(34) cm long, deeply runcinate or runcinate-pinnatifid on lower 1/3 of blade, terminal lobe entire to sinuate dentate, rarely subentire; petioles shorter than blades. Flowers sessile among the leaves, opening near sunset; floral tube slender, 2–8(12) cm long; sepals remaining coherent and reflexed to one side at anthesis, lanceolate, 1–1.8 cm long, green, often drying purplish, sometimes glandular-pubescent and strigulose, free tips 1–5 mm long; petals bright yellow or pale yellow, fading orange, drying purplish, obovate, (0.8) 1.5–2(2.2) cm long; stamens as long as petals, anthers 4–8 mm long; style as long as petals; stigmas 3–4 mm long. Capsules narrowly ovoid to ellipsoid, (1)1.5–3.5(4) cm long, 4-winged, the wings 2–5 mm wide especially widest toward apex; seeds in 2 rows in each locule, seeds asymmetrically cuneiform, dark reddish-brown to nearly black, 1.8–2.4(2.6) mm long, with a narrow wing on 1 adaxial margin, this extending also along the apex; self-pollinating, rarely cleistogamous. (n = 7) May–Aug. Rare to infrequent or locally common in hard soil of prairie swales or vernal pools in prairies & open wooded areas, sometimes in stream valleys; w ND, MT, nw SD, WY; (ND, Sask., s to nw SD, WY, w 2/3 CO, ID, CA, Mex.) *Lavauxia flava* A. Nels.— Rydberg.

9. **Oenothera grandis** (Birtt.) Smyth, large-flowered cut-leaved evening primrose. Annual herbs with slender taproots; stems erect, simple, or usually much branched from base and above, lower branches usually decumbent, 2–6(8) dm tall, strigulose and with sparse to dense long spreading hairs. Basal leaves appearing rosulate, blades oblanceolate or broadly elliptic, entire to sinuate-dentate, often pinnitifid below, 1.5–6 cm long, glabrate or strigulose; petioles often longer than blades; stem leaves alternate, blades lanceolate to oblanceolate or elliptic, 2–6 cm long, 5–17 mm wide, denticulate, sinuate-dentate to sinuate-pinnatifid, rarely entire, sometimes deeply lobed, glabrate to strigulose or villous-hirsute; petioles 0–6 mm long, usually very short. Flowers solitary in upper axils, opening near sunset; mature buds erect but stem tip often nodding; floral tube 2.5–5 cm long, slender, sparsely to densely villous; sepals remaining coherent at anthesis and reflexed to one side, (1.5)2–3 cm long, lance-attenuate, with free tips 2–5 mm long; petals yellow, often fading pinkish, (2)2.5–3.5(4) cm long. obovate rounded apically; stamens 2/3 as long as petals, anthers 7–9 mm long; styles longer than stamens, stigmas 7–8 mm long. Capsules cylindric (1)2–3.5(4) cm long, strigulose and with spreading hairs; seeds in 2 rows in each locule,

0.9–1.2 mm long, ovate, yellowish-brown, surfaces areolate or irregularly rugose; self-incompatible. (n = 7) Apr–Oct. Infrequent to common, often abundant in sand dunes, sandy prairies, stream valleys, fields, lawns, roadsides, & waste places, often weedy; cen & sw KS, w 2/3 OK, TX; (KS, OK, TX, Mex., introduced elsewhere). *O. laciniata* Hill. var. *grandiflora* (S. Wats.) Robins.—Fernald.

In some years this plant is a winter annual, an quite often late summer seedlings flower and mature in late Sep and Oct. Though often considered a var. of *O. lacinata*, with which it often grows, we have not located intermediates. Artificial hybrids between the two species are reported to have a much-reduced fertility.

10. Oenothera howardii (A. Nels.) W. L. Wagner. Acaulescent to subacaulescent with short congested stems 1–4 cm long, perennials with stout, simple or branched caudex; stems usually strigulose or with a few long spreading hairs; leaves linear to oblanceolate, thick in texture, usually tufted, (5)8–19(25) cm long, 1–3 cm wide, entire to sinuate-dentate, often irregularly lobed, glabrous to strigulose and sometimes with a few minute glandular hairs; petioles 1–6 cm long. Flowers in leaf axils, opening near sunset, with a strong sweet fragrance; mature buds erect; floral tubes (4.3)6–11(12.5) cm long, gradually enlarged toward apex, strigulose; sepals lance-linear, 3–6(8) cm long, strigulose and sometimes with scattered long spreading hairs, with free tips 1–3 mm long; petals brilliant yellow, fading deep red, obovate, (1.8)4–6(7) cm long; stamens 2/3 as long as petals; anthers 10–17 mm long; style as long as stamens or as long as petals; stigmas 5–10 mm long. Capsule body narrowly ovoid, (2)2.5–5(8) cm long, with 4 wings, 4–7(11) mm wide, along entire length, somewhat narrowed at either end; seeds in 1 or 2 rows toward base per locule; seeds 3–8 mm long, brown, corky, the surface coarsely rugose, summit with a erose wing; self-incompatible. (n = 7) Apr–Jul. Dry, rocky, prairie hillsides, & bluffs; KS: Hamilton; CO: Boulder, Jefferson, Larimer, Otero; (sw KS, CO, UT, WY). *Megapterium brachycarpum* (A. Gray) Rydb.—Rydberg.

This species was last collected in Kansas in 1892.

11. Oenothera jamesii T. & G. Biennial herbs with stout taproots; stems stout, simple and erect to usually branched from base and above, 1–2 m tall, often reddish, strigulose and with sparse to moderate ascending longer hairs. Rosette leaves elliptic-lanceolate, denticulate to sinuate-pinnatifid, blades 6–20 cm long, 2–4 cm wide, appressed pubescent; petioles 2–6 cm long; stem leaves alternate, blades lanceolate or broadly elliptical, remotely denticulate, 5–12 cm long, 2–3 cm wide, appressed-pubescent, rather abruptly reduced upward to bracts of inflorescence; petioles 1–2 cm long, to leaves nearly sessile above. Inflorescence terminal, simple or few branched, spicate; bracts lance-ovate to lanceolate, 1–5 cm long, rather persistent; flowers opening near sunset; mature buds erect; floral tube 6–10 cm long, strigose and with spreading hairs; sepals lanceolate, 4–6 cm long, with free tips 3–6 mm long; petals yellow, fading reddish, obovate, 3.5–5 cm long; stamens about 2/3 as long as petals; anthers 15–20 mm long; style about as long as petals; stigmas 5–8 long. Capsules cylindric, thick, 2–5 cm long, strigose and with sparse ascending hairs; seeds in 2 rows in each locule, 1.5–2 mm long, reddish-brown, rather sharply angled; self-compatible but out-crossing. (n = 7) Jul–Oct. Stream banks, marshy areas, ditches; w 1/2 OK, TX; (OK, TX to Coah.).

This species is very similar to *O. elata* but differs in floral tube 5–10 cm long rather than 2–4.5(5) cm long.

12. Oenothera laciniata Hill, cut-leaved evening primrose. Annual or in s GP a winter annual, with slender to stout taproots; stems simple and erect to usually much-branched at base and sometimes above, (0.5)1–3(8) dm tall, glabrate to densely strigulose and sometimes villous. Basal leaves appearing rosulate, blades oblanceolate, entire to usually sinuate-dentate or pinnatifid, (1)3–7(9) cm long, glabrate to strigulose or spreading hairy; petioles often as long as blades soon disappearing; stem leaves alternate, highly variable, blades oblanceolate to oblong-lanceolate, 2–8(12) cm long, (0.5)1–2(3) cm wide, sinuate-dentate to sinuate-pinnatifid, sometimes nearly entire, glabrate to strigulose and sometimes villous; petioles 0–3 mm long. Flowers solitary in axils, opening near sunset; stem tip and buds erect to usually nodding; floral tube 1.5–2.5(3.5) cm long, slender sparsely pubescent; sepals remaining coherent at anthesis and reflexed to one side, lance-attenuate, (5)8–12 mm long, free tips obscure or usually to 2–3 mm long; petals yellow, fading pinkish, obovate, 5–18 mm long; stamens subequal, nearly as long as petals, anthers 5–7 mm long; style as long as or longer than stamens, stigmas 2–4 mm long. Capsules cylindric, usually spreading, (1)2–4 cm long, strigulose and with sparse long, spreading, hairs; seeds narrowly ovate, 0.9–1.1 mm long, brownish, surface areolate; self-pollinating. (n = 7, ring of 14 chromosomes at meiosis) Apr–Oct. A common weed of fields, pastures, stream valleys, open woodlands, prairies ravines, roadsides, & waste places; sw ND, w 2/3 SD; WY: Crook; NE & s in GP, becoming rare westward; (generally e 1/2 U.S. to GP).

13. *Oenothera latifolia* (Rydb.) Munz, pale evening primrose. Perennial with subterranean to superficial caudex and spreading by adventitious shoots from lateral roots appearing rhizomatous; stems 1–5 dm tall, usually erect with branches from base and above, or with spreading or ascending stems, subglabrous or sparsely to densely strigose-canescent, sometimes with sparse long hairs above, white, epidermis usually exfoliating. Leaves alternate, blades ovate to lanceolate, lance-linear or oblong-lanceolate, subentire to remotely denticulate or deeply sinuate-dentate to sinuate-pinnatifid, subglabrous to cinereous, sometimes with sparse ascending longer hairs. Flowers solitary in the axils, fragrant, opening near sunset; mature buds nodding; bracts of reduced leaves; floral tubes slender, often reddish, 1.5–4 cm long, glabrous to strigulose and sometimes with long spreading hairs; sepals separate and reflexed at anthesis, lance-attenuate, 1–3 cm long, glabrous, strigose, sometimes with dense long spreading hairs, free tips obsolete or 0.5–3 mm long; petals obovate, white fading pink or reddish, 1–4 cm long; stamens with filaments 2/3 as long as petals, anthers 5–10 mm long; style about as long as petals, stigmas 3–8 mm long. Capsules cylindric, straight or curved, sometimes contorted, 1.5–5 cm long; seeds in 1 row per locule, 1.5–2 mm long, dark brown to blackish, with purplish dots or mottling, minutely pitted. (n = 7) Jun–Sep. Infrequent to locally common on sand dunes, sandy or rocky prairie, stream valleys, roadsides, often in disturbed places; w 2/3 NE, e WY, w 1/2 KS, CO, nw KS, CO, nw OK, TX, NM; (species as a whole from w NE, WY, ID, WA, s to TX, UT, AZ, Mex.). *Anogra cinerea* Rydb., *A. latifolia* Rydb.—Rydberg; *O. pallida* Lindl. subsp. *latifolia* (Rydb.) Munz—Atlas GP.

14. *Oenothera linifolia* Nutt., narrow-leaved evening primrose. Annual herbs with taproots; stems erect, 1–3(5) dm tall, simple or branched from the base or above with ascending branches, sparsely hirsute below while young, glabrous or glabrate to inflorescence. Basal leaves ovate to obovate or narrowly elliptic, 1–2(4) cm long, 2–6 mm wide, subentire to remotely dentate, glabrous, sparsely strigulose, sometimes with glandular hairs, narrowed to a winged petiole, usually disappearing by flowering time; stem leaves usually crowded, linear, alternate, subsessile, 1–4 cm long, less than 1 mm wide, ascending, glabrous to sparsely pubescent, sometimes with axillary fascicles or abortive branches. Inflorescence of terminal unbranched spikes (1)3–6(12) cm long, strigulose to glandular pubescent. Flowers opening near sunset; mature buds erect; bracts ovate to deltoid-ovate, 0.5–2 mm long; floral tube 1.5–2.2 mm long, sparsely pubescent; sepals 1.5–2 mm long, 0.2–0.5 mm wide, ovate-lanceolate, lacking free tips; petals yellow, obcordate, often conspicuously notched, 3–5 mm long; stamens with filaments 1–2 mm long, anthers 0.5–1 mm long. Capsules sessile or with short stipe, 4–6(8) mm long, 1.5–3 mm in diam, ellipsoid-rhomboid, 4-ridged,

strigulose and often glandular-puberulent, seeds in several indistinct rows in each locule; seeds reddish-brown, 0.9–1.4 mm long, surface smooth or minutely tuberculate; self-pollinating, often cleistogamous. (n = 7) May–Jul. Common in sandy prairies, rocky open woodlands, pastures, fallow fields, & waste places; se KS, sw MO, ne OK; (s NC, TN, s IL, s 1/2 MO, se KS, s to n FL, e TX). *Peniophyllum linifolium* (Nutt.) Penn.—Rydberg.

15. Oenothera macrocarpa Nutt. Perennial caulescent or subacaulescent herbs with subterranean to superficial branching caudex, sometimes tufted, glabrous, strigose-canescent, to silky-strigose or hoary pubescent throughout; stems to 6 dm long, decumbent to erect, usually simple, often reddish. Leaves elliptic-lanceolate, linear-lanceolate, obovate, or suborbicular, acuminate to obtuse or rounded, entire to denticulate or serrulate; blades 2.5–14 cm long, 0.5–4 cm wide, gradually narrowed into slightly winged petioles to 1/2 as long as blade. Flowers few per stem, solitary in axils, opening near sunset; mature buds erect; floral tube 2.4–13(15) cm long, somewhat enlarged at summit; sepals remaining coherent at anthesis and reflexed to one side, 2–5.6 cm long, lance-linear or lanceolate, often purple spotted, with free tips 1–8 mm long, sometimes subterminal and spreading; petals yellow, fading reddish, obovate, 1.5–6(6.8) cm long and nearly as wide, emarginate with a small tooth; stamens subequal, filaments 2/3 to nearly as long as petals, somewhat flattened below, anthers 6–22 mm long; style about equal or slightly exceeding petals; stigma lobes 3–13 mm long. Capsules 1.5–8(10.5) cm long, with 4 wings 3–32 mm wide; seeds in 1 row in each locule, 2–4 mm long, usually obscurely to definitely corky, coarsely rugose, with lateral and terminal wing obscure or often to 1 mm wide, wing toothed or nearly entire; self-incompatible. (n = 7). Four intergrading subsp. occur in our region as follows:

1 Plants glabrous throughout; leaves usually conspicuously remotely denticulate to serrate .... 15d. subsp. *oklahomensis*
1 Plants with appressed pubescence on stems, buds, leaves, or only younger leaves pubescent, rarely glabrous; leaves entire to remotely denticulate,
  2 Petals 1.5–2.2 cm long; floral tube 2.4–6.5 cm long; wings of fruits 3–7 mm wide .... 15a. subsp. *fremontii*
  2 Petals 2–5 cm long; floral tube 5–15.2 cm long; wings of fruits 7–30 mm wide.
    3 Leaves lance-elliptic to oblanceolate; plant greenish, only young leaves, stems and buds grayish appressed pubescent .... 15c. subsp. *macrocarpa*
    3 Leaves broadly elliptic to suborbicular; plant permanently invested with dense silvery strigose or hoary pubescence, rarely glabrous .... 15b. subsp. *incana*

15a. subsp. fremontii (S. Wats.) W. L. Wagner, Fremont's evening primrose. Plants tufted or short-caulescent, strigose-canescent throughout. Leaves elliptic-lanceolate to oblanceolate or nearly linear. Capsules 1.3–3 cm long, sometimes twisted, wings 3–7 mm wide; seeds 2–3 mm long, wings obscure. May–Jul, sporadic till frost. Locally common in chalk badlands, rocky prairie hillsides, & bluffs; NE: Antelope, Cedar, Franklin, Webster; w-cen KS; (endemic in area). *Megapterium fremontii* (S. Wats.) Britt.—Rydberg; *O. fremontii* S. Wats.—Atlas GP.

15b. subsp. incana (A. Gray) W. L. Wagner, hoary evening primrose. Plant densely strigose or hoary pubescent throughout or rarely glabrous; stems to 2 dm long. Leaves usually broadly elliptic to suborbicular, sometimes oblanceolate. Capsules 3–4(5) cm long; seeds 3–4 mm long, wing prominent. May–Jul. Rather abundant but local on limestone escarpments & dry prairie hillsides; KS: Clark, Comanche, Kiowa, Marion, Meade; w OK, TX panhandle, ne NM? (endemic). *Megapterium argophyllum* R. R. Gates—Rydberg.

15c. subsp. macrocarpa, Missouri evening primrose. Stems to 6 dm long, strigose-canescent. Leaves lance-elliptic to oblanceolate, sometimes lanceolate, green and strigose at least when young. Capsules 4–8(10.5) cm long, wings 15–30 mm wide; seeds 4–5 mm long, wing usually prominent. May–Jul, sporadic till frost. Locally common on limestone glades, limestone prairie hillsides, & rocky prairies; se NE, e 1/2 KS, s 1/2 MO, n AR, OK, TX; (locally introduced elsewhere). *Megapterium missouriense* (Sims) Spach—Rydberg; *O. missourienses* Sims—Fernald.

15d. subsp. oklahomensis (Nort.) W. L. Wagner, Oklahoma evening primrose. Stems to 4 dm long, plant glabrous, throughout. Leaves semi-succulent, often undulate, elliptic-lanceolate, to oblanceolate or nearly linear, conspicuously denticulate, Capsules 2.7–5.2 cm long, wing (8)12–16 mm wide; seeds 3.5–4.2 mm long, wing usually prominent. May–Jul, sporadic till frost. Locally common on gypsiferous or red clay prairie hillsides, knolls, & escarpments; KS: Barber, Comanche, Clark, Meade; w-cen OK; TX: Cooke, Knox (endemic to area). *Megapterium oklahomense* Norton—Rydberg; *O. missouriensis* Sims var. *oklahomensis* (Nort.) Munz—Munz op cit.; *O. macrocarpa* Nutt. var. *oklahomensis* (Nort.) Reveal—Atlas GP.

16. ***Oenothera nuttallii*** Sweet, white-stemmed evening primrose. Perennial herbs with a caudex and spreading by adventitious shoots from lateral roots, appearing rhizomatous; stems erect, 1–several from base, 3–6(10) dm tall, simple to usually branched, glabrous, white, epidermis usually exfoliating. Leaves alternate, pale green, linear to lanceolate or oblong-linear to oblong lanceolate, 2–6(10) cm long, 3–7(10) mm wide, entire to remotely denticulate or repand-denticulate, acute, glabrous above, strigose below gradually reduced upward to floral bracts; sessile or with petioles to 2 cm long. Flowers solitary, axillary, with an unpleasant odor; mature buds nodding; bracts linear lanceolate to elliptic-lanceolate, glandular-puberulent, shorter to as long as floral tube, tube 1.5–2.7 cm long; sepals 2–3 cm long, lanceolate, glandular-pubescent, and usually reddish, with free tips 1–2.5 mm long, adhering in anthesis and reflexed to one side; petals obovate, white, fading pinkish, 1.5–2.5 cm long; stamens 2/3 length of petals; anthers 8–10 mm long; style longer than stamens and often longer than petals; stigmas linear, (4)5–6 mm long. Capsules erect, straight, tapered slightly above, glandular-pubescent, 2–3 cm long, 2.5–3 mm in diam; seeds in 1 row in each locule, narrow-obovoid 1.9–2.2 mm long, reddish-brown, surface with minute purplish dots and obscurely pitted. (n = 7, 14) Jun–Sep. Infrequent to common on dry sandy or rocky prairies, open wooded hillsides, stream valleys, & roadsides; e MN, ND, MT, s to e 1/2 NE, WY, CO; (MI, to Sask., Alta., s to e MN, SD, WY, CO, ID). *Onagra nuttallii* (Sweet) A. Nels.—Rydberg.

17. ***Oenothera rhombipetala*** Nutt. ex T. & G., fourpoint evening primrose. Biennial or winter annual herbs with taproots; stems erect 3–8(12) dm tall, simple or branched near base and above, strigose, sometimes with ascending to spreading hairs. Basal leaves rosulate, narrowly oblanceolate, sinuate-dentate to sinuate-pinnatifid, rarely subentire, 3–8 cm long, 0.3–1.5 cm wide, strigose, petioles often as long as blade; principal stem leaves narrowly oblong-lanceolate to lance-ovate below, 2–8 cm long, 3–20 mm wide, usually crowded, subentire to remotely dentate, strigose, becoming lanceolate above to lanceolate bracts of inflorescence; petioles absent or very short. Inflorescence a dense terminal spike 1–3 dm long. Mature buds erect; flowers opening near sunset; bracts lanceolate, sparsely strigose, 1–2 cm long, reduced above; floral tube slender, strigose, 2–4 cm long; sepals lance-attenuate, (1)1.5–2.2 cm long, strigose, free tips 0–4 mm long; petals yellow, rhombic-ovate, (1.2)1.5–2(2.4) cm long; stamens shorter than petals, anthers 4–8 mm long; style as long or longer than petals, stigmas 2–3 mm long. Capsules cylindric, often curved, tapering toward apex, (1)1.2–1.6(2) cm long, strigose; seeds in 2 rows in each locule, 1–1.5 mm long, reddish-brown, minutely areolate. (n–7) Jun–Oct. Common in sand dunes, sandy prairies, stream valleys, roadsides, waste places; ND: Richland; SD: Shannon, Tripp; NE, KS, OK, TX; (IN, WI, MN, s to TX, introduced elsewhere).

18. ***Oenothera speciosa*** Nutt., showy white evening primrose. Perennial herbs with taproots and superficial caudex, spreading by adventitious shoots from lateral roots, often flowering first year. Stems 1–several from base, erect to ascending, simple or variously branched above, 1–5 dm tall, sparsley to densely strigose. Basal leaves oblanceolate to obovate, 2–9 cm long, with lanceolate or ovate lateral lobes, need a larger terminal lobe, soon disappearing; stem leaves oblong-lanceolate, oblanceolate or elliptic lanceolate, 2–10 cm long, subentire, sinuate-dentate to variously deeply pinnatifid, strigose, reduced upwards to lance-linear bracts of inflorescence. Flowers solitary in upper axils; inflorescence sharply nodding;

flowers opening in evening or in morning; bracts lance-linear to elliptic, 1–2 cm long; floral tube 1–2 cm long, strigose; sepals remaining coherent and reflexed to one side at anthesis, lanceolate, 1.5–3 cm long, strigose, free tips 1–4(5) mm long; petals white fading reddish or rose-purple, obcordate, 2.5–4 cm long; stamens 2/3 as long as to nearly as long as petals, anthers 10–12 mm long, style as long as petals, stigmas 3–6 mm long. Capsules narrowly obovoid, tough, 1–1.5 cm long, sessile, strigose terminal portion ribbed, attenuated basally; seeds in several indistinct rows in each locule, fusiform, 0.9–1.2 mm long, brown; self-incompatible. (n = 7,14,21) May–Jul. Locally common on rocky prairies, open woodlands, roadsides, waste places; se NE, e 2/3 KS, MO, OK, TX; (MO, NE s to LA, TX). *Hartmannia speciosa* (Nutt.) Small — Rydberg.

Most of our plants are diploid, with white flowers that open in the evening. In se KS rare roadside colonies are tetraploid, with rose-puple flowers that open in the morning. This variant, more common southward, is often cultivated in our area and may have escaped.

19. *Oenothera triloba* Nutt., stemless evening primrose. Annual or winter annuals, rarely biennial, acaulescent to subacaulescent herbs with taproots and rarely a stem 0.5–1(2) dm tall. Leaves in rosettes, blades elliptic to oblanceolate or oblong-oblanceolate, (2.5)6–25(31) cm long, (0.6)1.5–4(5) cm wide, irregularly and deeply runcinate pinnatifid to subentire, lobes dentate to serrate, sometimes recurved, usually with secondary lobes in between, terminal lobe larger than lateral one, glabrous to strigulose and often finely glandular pubescent; petioles (0.5)1–8 cm long, usually shorter than blades. Flowers axillary, the earliest sessile, the later ones on stout pedicels to 1–2 cm long; flowers opening near sunset; mature buds erect; floral tube slender (2.5)3.5–12(15) cm long, gradually expanded above; sepals lanceolate to lance-linear, united in anthesis, 10–15(18) mm long, glabrous or finely strigulose and sometimes glandular, free tips 3–7 mm long; petals obovate, pale yellow fading pale orange or drying pinkish (0.6)1.5–2.5(3) cm long, often with a small toothlike lobe in terminal sinus and appearing shallowly 3-lobed; stamens alternately unequal, about 2/3 as long as petals; anthers 5–9 mm long; style about as long as to longer than stamens, stigmas 3–4 mm long. Capsules very hard, long persistent, obpyramidal, obovoid, 1–2.5(3) cm long, with 4 wings, 5–10 mm wide, near summit, each wing often with a lateral tooth or angle at or just above middle; seeds in 2 rows in each locule, seeds asymmetrically cuneiform, 3–4 mm long, reddish-brown to nearly black, often mottled with light tan, adaxial side concave and with a carinate ridge, abaxial side convex, summit with a wing margin; commonly self-pollinating. (n = 7) May–Jul. Infrequent to locally common on barren hillsides, prairies, open woodlands, fields, roadsides, waste places; w MO, e 2/3 KS, OK, TX; (IN, KS, s to TN, n AL, TX). *Lauvauxia triloba* (Nutt.) Spach, *L. watsonii* (Britt.) Small — Rydberg; *O. triloba watsonii* (Britt.) F. C. Gates — Gates.

The capsules of this species are persistent, and in the winter it is not unusual to find pine-conelike clusters of a few–120 capsules terminating the taproot.

20. *Oenothera villosa* Thunb., common evening primrose. Biennial herbs with taproots; stems erect, (0.3)0.6–1.5(2) m tall, simple or much branched, strigose and hirsute, often reddish, longer hairs with or without reddish pustulate bases. Rosette leaves 1–2 cm long, 2–5 cm wide, oblanceolate, shallowly sinuate-denticulate, strigose, principal stem leaves lanceolate or elliptic-lanceolate, 7–15 cm long, 1–4 cm wide, strigose and somewhat hirsute especially on veins, sinuate-dentate to subentire, sessile or with short petioles. Flowers solitary, axillary, in terminal spikes, spikes simple or branched; flowers opening in evening; mature buds erect; bracts linear-lanceolate to lanceolate, 1–5 cm long, semipersistent, strigose, hirsute, sometimes glandular; floral tube 2–3.5 cm long, strigulose, often villous, sometimes with glandular hairs; sepals lance-attenuate, 1–2 cm long, glabrate to strigulose and villous, glandular or eglandular, free tips 1–3 mm long; petals yellow, often fading orange, obovate, 8–15(20) mm long; stamens nearly as long as petals, anthers 4–8 mm long; style as long as to longer than stamens; stigmas 4–7 mm long. Capsules cylindric, 2–4.5 cm long, tapering slightly toward apex, strigose, usually with ascending or spreading longer hairs, sometimes glandular-pubescent, seeds in 2 rows in each locule; seeds 1.5–2 mm long, dark brown to blackish, mostly obliquely obovate, margins with narrow wings, surfaces rugulose to obscurely areolate, often purplish mottled. (n = 7, ring of 14 at meiosis) Jul–Oct. Common on prairie hillsides, open woodlands, stream valleys, lakeshores, roadsides, fields, wasteland, often weedy; GP but rare to absent in sw; (MI, MN, Man., B.C., s to AR, OK, AZ, CA). *O. strigosa* subsp. *strigosa*, subsp. *canovirens* (Steele) Munz—Atlas GP; *O. biennis* L. var. *canescens* T. & G.—Gleason & Cronquist.

Most of our plants are densely strigose, long hairs rarely red pustular-based, the inflorescence eglandular, and are the subsp. *villosa*. Along the w GP, s to CO, are plants referred to subsp. *strigosa* (Rydb.) Dietrich & Raven, which have some gland-tipped hairs in the inflorescence, and usually the stem with obvious red pustulate-based hairs. All too many speciments, however, appear intermediate, particularly in the n & w GP. In general *O. villosa* differs from *O. biennis* in being more densely hairy but often separation of the two is quite arbitrary, particularly in the e GP.

9. STENOSIPHON Spach, Stenosiphon

1. ***Stenosiphon linifolius*** (Nutt.) Heynh. Plant biennial with a stout taproot and stout fusiform lateral branches; stems usually 1, rarely 2–3, erect, 0.3–3 m tall, rarely branched below, usually strict to inflorescence, woody near base, cortex exfoliating, glaucous, glabrous to inflorescence. Rosette leaves developing by fall, entire, oblong to oblong-lanceolate, 3–7 cm long, glabrous; lower stem leaves alternate, oblong to oblong-lanceolate, 2–4 cm long, entire, glabrous, sessile, usually somewhat auriculate; principal stem leaves alternate, sessile, entire, lanceolate, 3–8(10) cm long, 4–18 mm wide, glabrous, with a narrow whitish cartilaginous margin, acute or somewhat acuminate at apex, the upper leaves gradually reduced and becoming linear-subulate in inflorescence. Inflorescence a long wandlike spike or often an open paniculate structure terminating in spikes, axis sparsely to densely glandular-pubescent especially above; flowers solitary, sessile, subtended by a linear-subulate bract 4–8(12) mm long, this ciliate or glandular-pubescent. Floral tube filiform, 6–13 mm long, whitish, with short spreading hairs; sepals 4, free at anthesis, lance-oblong, whitish, 4–6 mm long, sparsely pubescent, often ciliate; petals 4, white, blade somewhat rhombic, 4–6 mm long, abruptly clawed, claw 1–3 mm long; stamens 8, somewhat unequal, filaments 5–8 mm long, anthers 1.5–2 mm long; ovary 3–5 mm long at anthesis, densely pubescent; style as long as or longer than stamens; stigma shortly 4-lobed. Fruit an indehiscent obovoid nutlike 1-seeded capsule, 3–4 mm long, 1.5–2.3 mm in diam, somewhat square in cross section and with prominent ribs on the angles, a less prominent rib on each face, faces cross-ridged, the whole pubescent; seed whitish-yellow, oblanceolate, 2.4–2.6 mm long, obscurely longitudinally striate; self-incompatible. (n = 7) May–Oct. Locally common on rocky prairie hillsides & cliffs, stream valleys, roadsides, with a preference for limestone; s-cen NE, KS, OK, TX; (s MO, KS, s NE, s to AR, TX; NM: DeBaca, Quay). *S. virgatus* Spach—Waterfall.

This plant is commonly described as being perennial, but at least in our area it is a true biennial.

80. MELASTOMATACEAE Juss., the Melastome Family

by T. Van Bruggen

1. RHEXIA L., Meadow Beauty

1. ***Rhexia mariana*** L. var. ***interior*** (Penn.) Kral & Bostick. Ours weedy, perennial herbs forming extensive clones from stout, horizontal rhizomes 2–3 mm in diam, 2–10 dm tall,

simple to freely branching from the midportion upwards, obscurely quadrangular to 4-angled, the 4 stem faces at midstem approximately equal, almost flat, the angles sharp or narrowly winged; stem surfaces sparingly to moderately hirsute, the hairs 2–3 mm long, glandular-tipped, usually more dense near the node. Leaves opposite, elliptical or ovate to ovate-lanceolate, those at midstem mostly 2–4(5) cm long, 2–2.5 cm wide, widest near the middle, acutely tapered at both ends, the upper leaves becoming shorter and smaller, those at the base of the inflorescence ovate to linear and bractlike, sparsely hirsute on both faces, 3-nerved, sessile to short petioled, exstipulate. Inflorescence few- to many-flowered, open to congested cyme, terminal, with deciduous, pedicellate bracts that are consistent with the leaf shape but smaller, deciduous. Flowers slightly irregular, sepals 4, triangular, the lobes 1.5–2.5 mm long, ascending to spreading, acute to aristate; petals 4, convolute in bud, distinct, bright lavender-rose, the blades asymmetrically ovate to oblong, 12–18 mm long, hirsute on the outer surface, the apices broadly truncate, the midvein excurrent, margins entire, bases short clawed; stamens 8, subequal, in 2 whorls, inserted on the inner hypanthial rim, the filaments 5–7 mm long, curved out or down, reddish, anthers bright yellow, straight to curvate, 1-celled at anthesis, basifixed or with a short spur, surface smooth, dehiscing by an apical pore; carpels 4, the ovary 4-loculed, with a single, slightly curved or straight style 12–14 mm long; hypanthium cylindrical at anthesis, becoming urceolate at maturity, 2-layered, the outer bearing the calyx, the inner the stamens and petals, 0.8–1.3 cm long, usually hirsute, with an upward expanding neck which is about as long as the body, the lower portion enclosing most of the ovary; pedicels 1–3 mm long. Fruit capsular, surrounded by the persisting hypanthial body, with irregular, loculicidal dehiscence. Seeds cochleate, 0.4–0.6 mm long, longitudinally ridged along the crest. (n = 22) Jun–Sep. Moist to wet sandy or sandy clay of roadbanks, swampy forest clearings, or sandy prairie; KS: Cherokee, Crawford; MO: Barton, Jasper, Vernon: (IN to MO, s to LA & TX). *R. interior* Penn.—Fernald.

Reference: Kral, R., & P.E. Bostick. 1969. The genus *Rhexia* (Melastomataceae). Sida 3(6): 387–440.

81. CORNACEAE Dum., the Dogwood Family

by T. Van Bruggen

1. CORNUS L.

Small trees, shrubs or perennial herbs with opposite, exstipulate leaves. Leaf blades firm, entire, the pinnate veins curving upward nearly parallel to each other, the upper pair ending near the apex. Inflorescence of open cymes or capitate heads surrounded by a 4-parted corollalike involucre of bracts. Flowers small, regular, perfect; sepals 4, minutely lobed; petals and stamens 4, inserted on the margin of an epigynous disk; ovary inferior, 2-carpeled, with (1)2 locules; style 1. Fruit drupaceous, (1)2-seeded.

Reference: Wilson, J.S. 1965. Variation of three taxonomic complexes of the genus *Cornus* in Eastern United States. Trans. Kansas Acad. Sci. 67(4): 747–817.

1 Plants herbaceous; leaves falsely whorled in groups of 4–6(7) .... 2. *C. canadensis*
1 Plants small trees or shrubs; leaves opposite.
   2 Inflorescence capitate, subtended by 4 showy bracts; fruit red .... 4. *C. florida*
   2 Inflorescence cymose, lacking showy bracts; fruits white to blue.
      3 Youngest twigs pubescent; sepals usually 1 mm or longer; pith of twigs brown; fruit blue .... 1. *C. amomum*
      3 Youngest twigs glabrous or scabrous; sepals less than 1 mm long; pith of twigs white or tan; fruit white or light blue.

4 First year stems bright red; leaves with 5–7 pairs of lateral veins, glaucous-whitened on the underside; stone of ripe fruit brown with yellow stripes .... 6. *C. stolonifera*
4 First year stems green to tawny brown or light pink; leaves with less than 6 pairs of lateral veins; stone of ripe fruit beige-pink, lacking yellow stripes.
   5 Young twigs and leaves smooth to touch; leaf blades lanceolate to ovate-lanceolate; lateral veins arising evenly from the midrib throughout the blade; inflorescence pyramidal or convex .... 5. *C. foemina*
   5 Young twigs and leaves rough to touch; leaf blades ovate; lateral veins mostly arising from the lower 1/2 of the blade; inflorescence flat or only slightly convex .... 3. *C. drummondii*

1. **Cornus amomum** P. Mill., pale dogwood. Shrub 1–5 m tall, the branches spreading to ascending; branchlets of the current year green to maroon, densely pubescent with gray-silvery to rusty trichomes; older branches becoming maroon or brown and glabrous; pith of 1-year and older wood brown. Leaves 6–9 cm long, 2–3.5 cm wide, broadly lanceolate to ovate, abruptly acuminate with an obtuse tip, base cuneate; upper surface glabrous to sparsely pubescent with appressed trichomes, lower surface papillose, whitish with curling, y-shaped and t-shaped trichomes, lateral veins of 3–5 pairs, evenly distributed along the midrib; petioles 0.5–3 cm long. Inflorescence a convex to flat-topped cyme of 4 main branches 2.5–8 cm across, the pedicels moderately covered with appressed silvery to rusty-brown trichomes, becoming dull maroon at maturity of the fruits. Flowers cream colored, sparsely to densely covered with appressed trichomes, sepals 4, (0.8)1.0–2.1 mm long, lance-linear to oblong, fused at the base to form a shallow cup, petals 4, 3–4.8 mm long, valvate in bud, spreading to revolute at anthesis, lance-linear; stamens 4, 3.5–7.5 mm long, the filament and anthers cream colored; ovary inferior, (1)2-loculed, the floral tube 1–2 mm long; style 1, 2–4 mm long, arising from a yellow-green epigynous ring. Fruit drupaceous, 1-seeded, dark blue, subglobose, 5–9 mm in diam, stone 4–6 mm in diam. May–Jul, fruits Jul–Oct. Moist woods, stream banks, swamps, & marshes; se GP; (ME to SD, s to AR & OK). *Svida amomum* (P. Mill.) Small—Rydberg.

Wilson, op. cit., refers our material to subsp. *obliqua* (Raf.) J. S. Wils., differing from subsp. *amomum* principally in the following: leaves narrower, lower surface papillose, blades cuneate, and distribution in areas north of the line of maximum glaciation at the northern part of the species range, where it reaches e NE, se SD, nw IA, and s MN. He also reports hybrids from crosses between this subspecies and *C. drummondii*, as well as *C. foemina* subsp. *racemosa* in the eastern part of the GP.

2. **Cornus canadensis** L., bunchberry. Herbaceous subshrub 5–20 cm tall, stems about 1 mm in diam, strigillose, leafless or with 1 or 2(3) pair of whitish bracts below, forming colonies from widely creeping horizontal rhizomes. Leaves opposite, borne in clusters of 4–6(7) at the summit of the stem, in false whorls, subsessile, ovate-elliptic to obovate, 2–8 cm long, acute at both ends of the blade, sparsely strigillose to glabrous-green above, pale glaucous beneath, with 2 or 3 pairs of lateral veins arising from below the middle. Inflorescence a solitary, subcapitate cyme of 10–25 flowers on a peduncle 1–3 cm long subtended by 4 petaloid white to pink or purple-tinged, cruciately arranged bracts, 1–2 cm long, shape similar to the leaves. Flowers 2.5–3.5 mm long; sepals 4, fused below, rudimentary, less than 1 mm long, each tipped by a deciduous bristle; petals 4, 1–1.5 mm long, valvate, white or tinged with green or purple; stamens 4, the anther as long as the filament, both less than 2 mm long; style solitary, equaling the petals, purple brown; ovary 2-loculed, ovules 1 in each locule. Fruit a reddish subglobular drupe 5–7 mm long, the stone 2-celled, 1/2 the size of the drupe. May–Jun. Moist woods & shaded canyons; n GP; MN, MT, ND, BH, e CO; (Lab. to AK, s to MN, CO, & NM). *Chamepericlimemum canadense* (L.) Aschers & Graebn.—Rydberg.

3. **Cornus drummondii** C. A. Mey., rough-leaved dogwood. Shrub 2–6(12) m tall, sometimes approaching treelike proportions, often forming clumps by stems growing from

shallow, spreading roots; first year branchlets scabrous, green-olive, becoming light brown, older branches gray-brown; pith of 1 year and older wood light tan to brown. Leaves 6-11 cm long, 2-8.5 cm wide, broadly ovate to lance-ovate, abruptly tapered to prolonged tips, cuneate to truncate at the base; upper surface scabrous, rough, with curling or Y-shaped hairs; lower surface pilose woolly, whitish, papillose; lateral veins 3-5, arising from the basal 1/2 of the blade. Inflorescence a flat-topped to convex cyme usually with 4 main branches, 3-7 cm across, the pedicels 2-7 mm long, liberally covered with appressed and T-shaped trichomes. Flowers white or cream colored, densely covered with appressed hairs; floral tube 1-2 mm long; sepals 4, fused at their bases, the lobes 0.5-1(1.2) mm long; petals 4, valvate in bud, lanceolate, spreading to revolute in anthesis, 2.5-4 mm long; stamens 4, 2.5-6.5 mm long, inserted at the outer margin of the epignyous disk; ovary inferior, (1)2-loculed, style 1, cream colored or light green, 2.5-3.5 mm long, the stigma capitate. Fruit drupaceous, 1-seeded, white or rarely light blue, subglobose, 4-7 mm in diam, stone 3-5 mm in diam. Apr-Jul, fruits Jul-Oct. Swamps, wet to dry woods, marshes, edges of lakes & rivers; se GP, n to se SD; (MI to NE, s to GA & TX). *Svida asperifolia* (Michx.) Small — Rydberg.

Wilson, op. cit., reports that this species hybridizes with *C. foemina* subsp. *racemosa* in the northern part of its range and with *C. amomum* subsp. *obliqua* wherever the parent taxa come into contact, producing plants that are difficult to assign to species in either case.

4. *Cornus florida* L., flowering dogwood. Small tree to 12 m tall, often with several trunks from old root crowns, much branched, the bark red-brown to black on older trees, forming a checkerboard pattern; young branches greenish. Leaves 6-10 cm long, 3-8 cm wide, ovate to widely elliptic, acute to acuminate, tapering or rounded at the base, dark green and pubescent to glabrous above, pale strigillose or glabrous below, appearing after the flowers, lateral veins of 4-6 pairs, evenly distributed along the midrib, petioles 0.3-1.2 cm long. Inflorescence a compact cymose head of 12-25 sessile flowers, 1-1.5 cm across, on terminal branches of current year wood. Flowers subtended by 4 showy, cruciately arranged petallike bracts, 2-5 cm long and 1.5-3.5 cm wide, white to light pink, often unequal, the many paired lateral veins curving inward to a notched apex; flowers greenish-yellow, 4-6 mm long, strigose, sepals 4, the lobes 0.5-1 mm long, erect, persistent; petals 4, 3-5 mm long, green to white, narrowly lanceolate, valvate, spreading at anthesis; stamens 4, inserted on the margin of an epignyous disk, slightly exserted from the corolla; ovary inferior, 2-celled; style 3-4 mm long, from the center of the disk, the stigma only slightly enlarged. Fruit yellow to red, drupaceous, 2-seeded, ellipsoid, 1-1.5 cm long, 4-7 mm in diam. (n = 11) Mar-May. Acidic woods, slopes, bluffs, & ravines on various substrata; se GP; s MO; KS: Cherokee; OK: Craig, Creek, Ottawa, Seminole; (ME to GP, s to FL, TX; Mex.). *Cynoxylon floridum* (L.) Raf.—Rydberg.

5. *Cornus foemina* P. Mill., gray dogwood. Shrubs or small trees up to 5 m tall, often forming dense thickets by proliferation from the roots; branchlets of the current year glabrous or with a few T-shaped trichomes, green, becoming tan on older wood; old branches gray to gray-brown; pith white or tan. Leaves opposite, entire, 3.5-9 cm long, 2.0-5.5 cm wide, lanceolate to lance-ovate or elliptic, acuminate to obtuse tip, cuneate or truncate at base; upper surface sparsely to densely covered with variously shaped curling trichomes, lower surface smooth (shade plants) to slightly scabrous; lateral veins of 3 or 4(5) pairs, evenly spaced and curving towards the apex; petioles smooth, 3-15 mm long. Inflorescence a convex to pyramidal cyme of 4(5) main branches, 2.5-5.5 cm across, the pedicels glabrous to sparsely pubescent. Flowers cream colored to light green, sparsely to densely covered with appressed trichomes; floral tube 1-2 mm long, topped by 4 erect sepal lobes 0.4-1.0 mm long, fused at their bases to form a shallow cup; petals valvate, spreading at anthesis, white to cream colored, 2.2-4.5 mm long, lanceolate, glabrous to sparsely pubescent; stamens 2.3-6.0

mm long, alternate with the petals, arising from the outer rim of the fleshy, variously colored epigynous disk; ovary inferior, (1)2-loculed; style 1.5–3.5 mm long. Fruit white, 4.5–8.5 mm in diam, mostly subglobose but wider than long; stone 3.6 mm in diam, slightly compressed, 1-seeded. Flowers Apr–Jul, fruits Jul–Oct. Thickets, riverbank forests, & swampy to dry, low open areas; e GP; (Que. to Ont., s to NC & MO). *Svida foemina* (P. Mill.) Rydb.—Rydberg.

Wilson, op. cit., refers our material to subsp. *racemosa* (Lam.) J. S. Wils. differing from subsp. *foemina* in having young branches tan instead of maroon, inflorescence convex to pyramidal rather than flat-topped, and fruit white rather than light blue. Additionally, subsp. *foemina* has a distribution s and e of the GP, FL to TX, n to s MO. The same author reports hybrids from crosses between *C. foemina* subsp. *racemosa* and *C. drummondii* as well as *C. amomum* subsp. *obliqua* wherever their ranges overlap s and e of the GP. Both hybrids exhibit characters intermediate between both sets of parents.

6. **Cornus stolonifera** Michx., red osier. Shrub 1–4 m tall, the branches many, erect to often procumbent and rooting at the apex or at nodes near the apex, stoloniferous, the young twigs glabrous to strigillose, red, the older stems deep red, smooth; pith white, 1/3 or more the diam of the branch. Leaves 5–11 cm long, 2.5–5.5 cm wide, oblong-lanceolate to ovate, entire, acute to acuminate with an obtuse tip, base cuneate; upper surface sparsely strigose to glabrous above, glaucous-whitened with glabrous to appressed-pilose beneath, lateral veins of 5–7 pairs evenly spaced along the midrib; petioles 0.5–2.5 cm long. Inflorescence a flattish-topped or slightly convex cyme of 4–6 main branches 2–5 cm across, densely flowered, the pedicels reddish, tomentose early, becoming glabrous. Flowers white to cream colored, ovoid in bud; sepals minute to 0.5 mm long, the floral tube 1.5–2.5 mm long; petals 2–3 mm long, lanceolate, erect to spreading at anthesis with inrolled margins; stamens as long as to longer than the petals, inserted at the outer margin of the doughnut-shaped epigynous disk; ovary inferior, (1)2-loculed; style 1, 2–3 mm long. Fruit white, drupaceous, 6–9 cm in diam, globose; seed 1, stone of ripe fruit brown with yellow stripes, 4–6 mm in diam. May–Jul. River banks, lakeshores, swampy & wet places; GP, except the s part; (Newf., Yukon s to WV, NM). *Svida stolonifera* (Michx.) Rydb., *S. baileyi* (Coult. & Evans) Rydb., *S. instolonea* A Nels., *S. interior* Rydb.—Rydberg.

82. GARRYACEAE Dougl. ex Lindl., the Silk Tassel Family

by Ralph E. Brooks

1. GARRYA Dougl. ex Lindl.

1. ***Garrya ovata*** Benth. subsp. ***goldmannii*** (Woot. & Standl.) Dahling, Mexican silk tassel. Evergeen, dioecious shrubs to 2 m tall; stems tomentose when young and becoming glabrate with age, branching decussate; crown sprouts numerous, well developed, and nearly indistinguishable from the main axis; tap root long with many secondaries. Leaves persistent, simple, petiolate, opposite; blade strongly undulate, narrowly ovate to elliptical or sometimes oblong to obovate, (2.1)3.5–9(10.8) cm long, (1)1.5–4(4.9) cm wide, ca 2 × longer than wide, upper surface glabrous or with sparse, variable appressed pubescence, lustrous, lower surface lanate or sometimes tomentose, white-gray or yellow-green, apex mucronate, acute, or obtuse, base obtuse to subcuneate, margins entire below and more or less muriculate above the middle; petioles connate at the base, adnate to the stem, 0.8–2 cm long. Staminate inflorescences catkinlike, mostly axillary and fasciculate in 3s, rarely solitary, pendulous,

1-4 cm long, internodes inconspicuous; floral bracts green to red-brown, ovate to lanceolate, shorter than the flower, apices truncate to obtuse, tomentose on the abaxial surface; opposite bracts connate; flowers solitary in the bract axils; pedicels minute; perianth segments 4, connate at the bases; stamens free, short, and alternating with the perianth segments, 2-3 mm long; pollen yellow-green, tricolporate, reticulate. Pistillate inflorescence racemose, terminal or sometimes axillary, usually fasciculate, pendulous, 4-8 cm long, internodes conspicuous; floral bracts ovate to lanceolate, the lowermost foliaceous, ca 10 mm long at the inflorescence tip, apices acute, acuminate or obtuse, tomentose; opposite bracts connate; flowers solitary in the bract axils, pedicels minute, apetalous; ovaries bicarpellate (rarely tricarpellate), unilocular, subglobose, subsessile, sometimes with bracts opposite the styles, glabrous; styles 2, rarely 3, persistent. Fruit a berry, dark blue, subglobose, 6-7 mm diam, glabrous; dehiscence irregular; seeds 2, rarely 3, dark blue or black, globose, parietal placentas. (n = 11) Mar-Apr. Steep slopes & canyons in limestone & igneous areas; TX: Briscoe; (sw TX, se NM, & n Mex.).

Reference: Dahling, G. V. 1978. Systematics and evolution of *Garrya*. Contr. Gray Herb. 209: 1-104.

While the relationship of *Garrya* to the Cornaceae is well documented, the bracteate and amentiferous nature of the infloresences and much reduced flowers would seem to be sufficient evidence to warrant the recognition of a distinct family, the Garryaceae.

83. SANTALACEAE R. Br., the Sandalwood Family

by T. Van Bruggen

1. COMANDRA Nutt.

1. *Comandra umbellata* (L.) Nutt. Erect, glabrous, perennial herbs or subshrubs from extensive shallow to deep-seated horizontal rhizomes 1-5 mm thick; nutrition hemiparasitic on subterranean parts of a number of species of many different plant families; stems 0.7-5 dm tall, usually clustered, branched or unbranched, often sterile; not overwintering. Leaves alternate, simple, many distributed along the entire stem, subsessile to short petiolate, 0.7-4(6) cm long, linear or lanceolate to elliptic or ovate, varying from green and very thin with conspicuous veins to gray-green and thick with veins inconspicuous; blade base acute to attenuate, the tip acute to obtuse, exstipulate. Inflorescence loosely paniculate to corymbose, of several terminal or subterminal 3- to 6-flowered cymules, sessile or the pedicels to 4 mm long, the peduncles subtended by foliaceous bracts, each flower subtended by a bracteole. Flowers mostly perfect, floral tube green with (4)5 white, occasionally pink, petaloid sepals, the lobes lanceolate to ovate, 2-4.5 mm long, inner surface with epidermal hairs that adhere to the anthers; petals none; stamens 5, equal to the number of sepals and opposite them, filaments ca 1 mm long, alternate with the lobes of the intrastaminal disk; anthers 0.4-0.7 mm long, longitudinally dehiscent; ovary inferior, unilocular, the ovules 3(2) with free central placentation; style filiform, 2-3 mm long; stigma capitate, at the level of the anthers. Fruit drupaceous, 1-seeded, subglobose to globose, 4-7 mm in diam, smooth to slightly roughened, green early but maturing to chestnut brown or purplish brown; floral tube forming a slightly constricted neck above, the sepals often persisting. (n = 14) Apr-Jul. Dry, sandy or rocky soils, open to partly open wooded areas, occasionally under deciduous or coniferous trees; GP; (temp. N. Amer.; Balkan States).

In the GP two subspecies are recognized by Piehl, M. A. (Mem. Torrey Bot. Club 22(1): 1-97. 1965).

1a. subsp. *pallida* (A.DC.) Piehl. Rhizome cortex blue; plants gray-green, usually glaucous. Lateral leaf veins not conspicuous, leaf margins not revolute upon drying. Sepals 2.7–4.5 mm long. Mature fruits 6 mm or more in diam. GP to e limit. *C. pallida* (A.DC.) Nutt.—Fernald.

1b. subsp. *umbellata*. Rhizome cortex whitish or beige; plants green. Leaves often lighter beneath, lateral veins conspicuous; leaf margins tending to be revolute on herbarium material. Sepals 2–3 mm long. Mature fruits less than 6 mm in diam. e GP; MN, e ND, e SD, IA, MO. *C. richardsiana* Fern.—Fernald.

These two subspecies intergrade over a wide area in Ont., Sask., ND, SD, and NE. The zone of interfertility is narrow in the south, becoming more extensive northward.

Thesium linophyllon L., a member of the Santalaceae and an Old World native, was first collected in Towner County in n ND by Leary, 1943. Subsequent collections in 1944 by Weiland and 1974 by Stevens indicate that it is naturalized in that location. The literature suggests that this species is widely distributed in central Europe, but apparently it is a waif in the GP.

A short description follows: Erect, perennial, hemiparasitic herb from a thick, woody, stoloniferous rootstock with several to many stems in a cluster; stems green with 8–12 longitudinal furrows, the ridges slightly scabrous; 2–6 dm tall, ca 1 mm in diam, leafy to the base, essentially unbranched. Leaves green to yellow-green, linear to lance-linear, alternate, inconspicuously 1- to 3-nerved, 0.5–3.5 cm long, 1–2 mm wide, the margins papillose, sessile. Inflorescence racemose, lax, the flowers on short branches, each subtended by a leaflike bract and 2 smaller bracteoles. Perianth yellow-green, campanulate, (4)5-lobed, ca 1 mm long, with inrolled edges, persisting in fruit; petals none; stamens (4)5, equal and opposite to the sepal lobes, included; ovary inferior, ovule solitary. Fruit a greenish ellipsoidal nut, 2–3 mm long, longitudinally veined.

84. VISCACEAE Miers, the Christmas Mistletoe Family

by T. Van Bruggen and Ralph E. Brooks

1. PHORADENDRON Nutt.

Evergreen, dioecious shrubs, hemiparasites on trees and shrubs producing haustoria which penetrate the host; stems brittle, much branched. Leaves opposite, simple, entire, exstipulate. Inflorescence (in ours) axillary, spikelike with 2–7 segments, each segment with a tight cluster of flowers, 1 apical and the lower 3-ranked; staminate inflorescence usually 15- to 60-flowered; pistillate inflorescence 4- to 11-flowered. Flowers unisexual, small, sessile and sunken along the rachis, calyx usually 3-lobed, sepals distinct and scalelike, incurved or seldom erect, deltoid; apetalous; staminate flower with a sessile 2-chambered anther opposite the base of each sepal; pistillate flower epigynous, stigma capitate, style 1, ovary 1-chambered with the calyx adnate and persistent in fruit. Fruit a drupe, 1-seeded with a mucilaginous mesocarp.

Reference: Wiens, D. 1964. Revision of the acataphyllous species of *Phoradendron*. Brittonia 16: 11–54.

1 Foliage dark green, glabrous or the younger growth stellate-pubescent; inflorescences mostly less than 2.5 mm diam at anthesis; plants of e OK, se KS, or s MO .... 1. *P. serotinum*
1 Foliage yellow-green, stellate-pubescent, infrequently becoming glabrate; inflorescences mostly more than 2.5 mm in diam at anthesis; plants of s-cen or sw OK and southwestward .... 2. *P. tomentosum*

1. **Phoradendron serotinum** (Raf.) M. C. Johnst., eastern mistletoe. Plants with 1-many stems, forming spherical bushes to 1 m in diam, glabrous or infrequently stellate-pubescent on young growth; internodes 1-5 cm long, 1-7 mm diam. Leaves dark green and shiny, succulent; blade obovate to widely elliptic, 2-5 cm long, 1.5-3 cm wide, apex rounded to obtuse, rarely acutish, margins entire, base cuneate or broadly so; petiole 2-7 mm long. Inflorescences occasionally lightly pubescent, bracts ciliate; staminate inflorescence 2-3 cm long, 1.5-2.5 mm diam, segments 3-6, each segment 15- to 40(60)-flowered; pistillate inflorescence 1-3 cm long, 1.5-2.5 mm in diam, segments 3-5, each segment 4-to 11-flowered. Flowers yellowish, about 1 mm across. Fruit white (becoming black on drying), globose, 4-6 mm diam, glabrous; seeds flattened elliptic, 2.5-3 mm long, 1.8-2 mm wide, smooth, mucronate. (n = 14) (Mar)Jul-Sep. Primarily on *Ulmus americana* or infrequently on *Maclura pomifera* or other deciduous hosts especially in wooded flood plains, infrequently on upland trees; extreme se KS & s MO, e 1/3 OK; (NJ sw to se KS, s to FL & e TX). *P. flavescens* (Pursh) Nutt.—Fernald; *P. tomentosum* (DC.) A. Gray, in part—Atlas GP.

Literature accounts quickly reveal that the separation of *P. serotinum* and *P. tomentosum* has been troublesome to many botanists. Field and herbarium studies indicate that *P. serotinum* is a less robust plant than *P. tomentosum* with regard to leaf size (color also differs) and inflorescence diameter. In addition the taxa are geographically separated for the most part and show a high degree of host specificity even where hosts for the other taxa may be present. Delbert Wiens, University of Utah (pers. comm., 1982), feels his 1964 treatment should be used in favor of his treatment in Correll & Johnston (1970), and we concur.

2. **Phoradendron tomentosum** (DC.) Engelm. ex A. Gray, hairy mistletoe. Similar to no. 1 except as follows: Plants, especially the stems, stellate-pubescent, infrequently becoming glabrate. Leaves yellow-green; blade obovate or narrowly so, infrequently elliptic, 2-6 cm long, 1.5-2(2.5) cm wide, apex obtuse to acutish, margins entire, base cuneate to attenuate; petiole 2-7 mm long. Inflorescences pubescent; staminate inflorescence 2-3.5 cm long, 2.5-3 mm in diam; pistillate inflorescence 1.5-3.5 cm long, 2.5-3.5 mm in diam. (n = 14) (Oct)Nov-Dec. Primarily on *Prosopis* & *Celtis* in flood plains or other lowlands; s-cen & sw OK, s TX panhandle; (s-cen & sw OK & s TX panhandle s to Mex.). *P. serotinum* var. *pubescens* (Engelm.) M. C. Johnst.—Johnston, Southwest Nat. 2: 46. 1957.

85. RAFFLESIACEAE Dum., the Rafflesia Family

by T. Van Bruggen

1. PILOSTYLES Guill.

1. **Pilostyles thurberi** A. Gray. Ours fleshy, thalloid plants parasitic on the lower stems and branches of various shrubs, mostly the genus *Dalea;* the vegetative parts entirely within the host plant, with only the small reddish-brown flowers and sometimes a few subtending scalelike leaves exserted from the host tissues. Plants monoecious, or more usually unisexual by abortion, 3-5 mm tall and 2-4 mm across, budlike; leaves, if present, scalelike, fleshy, broadly ovate, 1-1.2 mm long, somewhat imbricate. Flowers solitary, on short, fleshy peduncles 0.5-1.5 mm long; floral bracts 4-7, broadly ovate, fleshy; calyx segments epigynous, 4(5), yellow-brown, 1-1.5 mm long, similar to the 2-5 subtending floral bracts; petals none; staminate flowers with many sessile anthers arranged in 1-3 series around a

thick column that is expanded at its apex into a fleshy disk about 1 mm in diam; pistillate flowers epigynous; ovary 1-celled, about 1 mm in diam and 1–1.5 mm long; stigma sessile on the ovary, ringlike, 1 mm in diam; ovules numerous, on several parietal placentae. Fruit a globose capsule 1–1.5 mm in diam; seeds many, minute, 0.25–0.5 mm long. May–Jul. Parasitic on *Dalea formosa* in our range; s GP; n & w TX, e NM; (TX to CA, s to Mex.).

86. CELASTRACEAE R. Br., the Staff Tree Family

by Ronald L. McGregor

Ours woody vines, shrubs, or small trees. Leaves simple, alternate or opposite; petiolate; stipules absent, or minute and fugaceous. Inflorescence of axillary cymes or terminal racemes or panicles; pedicels jointed; flowers perfect or unisexual, regular, 4- or 5-merous; calyx with sepals united at base; petals distinct and usually imbricated in bud; stamens 4or 5, inserted on margin of a prominent disk; ovary inserted on disk or disk partially surrounding ovary. Fruit a loculicidal capsule; seeds completely enclosed in a red, fleshy aril.

1 Leaves alternate; stems sprawling or twining; flowers in terminal panicles or racemes .... 1. *Celastrus*
1 Leaves opposite; erect shrubs or small trees; flowers in axillary cymes .... 2. *Euonymus*

1. CELASTRUS L., Bittersweet

1. *Celastrus scandens* L., American bittersweet. Climbing or sprawling woody vines, spreading by root suckers; stems to 18 m long, primary stem to 2.5 cm in diam. Leaves alternate, simple; blades elliptic, or ovate-oblong, acuminate, serrulate, (3)5–8(12) cm long, (1.5)3–5(7) cm wide; petioles 1–3 cm long; stipules 0.6 mm long, linear, membranaceous, early deciduous. Flowers unisexual, greenish, in racemes or panicles 3–8 cm long; pedicels jointed; sepals 5; petals 5; stamens 5, inserted on margin of disk; ovary inserted on disk, 3-celled, each cell with 1 or 2 ovules; stigma 3-lobed. Fruit a globose, 3-valved, loculicidal, orange or yellow capsule, 8–12 mm in diam, upon splitting exposing the fleshy scarlet to crimson aril-covered seeds; seeds 1 or 2 per cell, ellipsoid, reddish-brown at maturity, 5–6 mm long. (n = 23) May–Jul. Frequent in woodlands, rocky hillsides, thickets, fence rows, roadsides; GP but absent in sw; (Que. to Man. WY, s to NC, TN, TX).

This species is sometimes grown as an ornamental for the clusters of red fruits that persist until midwinter. While birds readily eat the fruits, it is possible that they may be poisonous to humans. Cases have been reported of horses having been poisoned from eating the leaves. In very shaded places the plant often fails to produce flowers.

Celastrus orbiculatus Thunb., Oriental bittersweet, is sometimes grown in the se GP. It is a native of e Asia and differs in having leaves suborbicular to broadly oblong-obovate and flowers few in small axillary cymes much shorter than subtending leaves. It might be found as an escape.

2. EUONYMUS L., Spindle Tree

1. *Euonymus atropurpureus* Jacq., wahoo. Shrub or small tree with one trunk, erect, 2–4(6) m tall; branches purplish, becoming greenish, usually quadrangular. Leaves opposite; blades elliptic to lance-ovate or oblong-oval, rarely lanceolate, 5–9(12) cm long, acute to acuminate or long attenuate, acute at base, finely serrulate, glabrous above, persistently fine-puberulent beneath; petioles 1-2 cm long; stipules linear, to 1 mm long, promptly deciduous. Peduncles axillary, 2-5 cm long, bearing 7-15 flowers in cymes; sepals 4, united, lobes 1-1.5 mm long, often unequal, purplish or greenish-purple; petals 4, imbricate in bud, distinct, 3.3-3.8 mm long, 3.5 mm wide, red-purple to brownish-purple; disk 4-lobed, purple; stamens 4, sessile or nearly so on ends of lobes; disk cohering with calyx to conceal the ovary; ovary 4-lobed, or by abortion 1- to 3-lobed; 2 ovules per cell; style obsolete. Capsule deeply 4-lobed or 1 or 2 lobes aborted, smooth, red or yellowish-red, 1.5 cm in diam, upon splitting exposing the fleshy red aril-covered seeds; seeds yellowish-brown at maturity, 6-7(8.5) mm long, 4-5 mm in diam, smooth. (n = 16) May-Jul. Wooded areas, bluffs, stream banks, thickets; se ND, e SD, MN, e 1/2 NE & KS, OK; (s Ont., to ND, s to FL, TX).

Euonymus americanus L. with 5-merous flowers, fruits strongly tuberculate and petioles 1-3 mm long has been reported for KS: Coffey, based on an old specimen with incomplete data. Otherwise, the species is known no closer than se MO, AR, and se OK.

87. AQUIFOLIACEAE Bartl., the Holly Family

by Ronald L. McGregor

ILEX L., Holly

1. ***Ilex decidua*** Walt., deciduous holly, possum-haw. Polygamo-dioecious bushy shrub or small tree to 5 m tall. Leaves deciduous, alternate, cuneate-oblong, oblanceolate to narrowly obovate, 3-7 cm long, 1-3 cm wide, finely crenate, glabrous above, pubescent at least on veins below. Flowers in clusters of 5-12 from ends of dwarf branches; sepals 4, entire, 1 mm long; petals 4, united at base, pale yellow to white, 3-4 mm long; stamens 4, those of staminate flowers slightly shorter than petals, staminodes of pistillate flowers 3/4 as long as petals; ovary usually 4-celled; stigma large, sessile, capitate. Drupes bright red to orange, to 7.5 mm in diam; stones usually 4, irregularly ribbed on back, to 5 mm long (n = 20). Apr-May. Low wet woods along streams & rocky upland woods; extreme se GP; (MD, s IN, s IL, MO, se KS, s to FL, TX).

88. EUPHORBIACEAE Juss., the Spurge Family

by Ronald L. McGregor

Ours herbaceous, monoecious or dioecious, annual, biennial, or perennial herbs. Leaves simple or palmately lobed, alternate, opposite or rarely verticillate; usually stipulate, but stipules commonly very small or caducous, or in form of glands. Inflorescence highly variable. Flowers always unisexual; petals present or absent; pistillate flowers usually with a lobate disk; ovary 3(1)-celled, each cell with a separate style; stamens 1-10, 1 in *Euphorbia*. Fruit a 3-celled capsule (1 cell well developed in *Croton monanthogynous*), or a 1-seeded utricle in *Crotonopsis;* seeds 1 or 2 per cell.

1 Flowers without perianth; cyathium present .... 6. *Euphorbia*
1 Flowers with a calyx; cyathium, or involucre absent.
  2 Leaves palmately lobed .... 3. *Cnidoscolus*
  2 Leaves not palmate.
    3 Flowers axillary; seeds 2 per locule.
      4 Fruits 1–3 mm wide; stamens 3 .... 7. *Phyllanthus*
      4 Fruits 6–8 mm wide; stamens 2 .... 8. *Reverchonia*
    3 Flowers in spikes or glomerules, or few and scattered; seeds 1 per locule.
      5 Styles simple.
        6 Plants glabrous .... 9. *Stillingia*
        6 Plants pubescent; with stinging hairs .... 10. *Tragia*
      5 Styles divided.
        7 Plants with stellate or forked hairs or lepidote scales.
          8 Fruit 1-seeded, indehiscent .... 5. *Crotonopsis*
          8 Fruit usually 3-seeded; always dehiscent .... 4. *Croton*
        7 Plants with simple or malpighaceous hairs or glabrous.
          9 Pistillate flowers usually subtended by a folded, dentate, foliaceous bract or fruits papillate-echinate; hairs not malpighiaceous .... 1. *Acalypha*
          9 Pistillate flowers not subtended by a folded bract; usually some hairs malpighiaceous .... 2. *Argythamnia*

1. ACALYPHA L., Three-seeded Mercury

Ours erect, annual, monoecious herbs from taproots. Leaves alternate, with 2 lateral veins from base, petioled, stipules small, lanceolate. Flowers in axillary or terminal spikes or racemes, each flower cluster subtended by a bract; petals none; disk none; staminate flowers with 4 valvate sepals, 8–16 stamens subtended by an inconspicuous lanceolate bract; pistillate flowers with 3–5 sepals, styles irregularly branched, subtended by a lobed foliaceous bract. Capsule usually 3-celled, 1 seed per cell, seeds ovoid, carunculate.

1 Pistillate spikes terminal, staminate spikes axillary; fruit echinate when mature and pubescent .... 2. *A. ostryaefolia*
1 Spikes with pistillate flowers at base and staminate flowers at apex; fruit pubescent.
  2 Fruit 1-seeded .... 1. *A. monococca*
  2 Fruit 3-seeded.
    3 Pistillate bracts with (8)10–14(16) lanceolate lobes; leaves narrowly rhombic to broadly lanceolate .... 4. *A. virginica*
    3 Pistillate bracts with (5)7–9(11) oblong-lanceolate lobes; leaves ovate to broadly rhombic .... 3. *A. rhomboidea*

1. *Acalypha monococca* (Engelm.) P. Mill. Erect, annual herb, 1.5–4 dm tall. Stems simple or branched, with sparse to somewhat dense recurved hairs. Leaves with blades linear or lanceolate, slightly crenate to entire, 2–7 cm long, 3–12 mm wide, with sparse short stiff hairs; petioles 2–10 mm long; stipules about 1 mm long. Spikes axillary with both pistillate and staminate flowers, or rarely a few entirely staminate. Pistillate bracts 1–3 at base of spike, 5–13 mm long, 5–15 mm wide, with (7)9–13(15) deltoid lobes, short stiff hairs sparse to dense, margins with long spreading hairs, often with sparse red short-stalked glands and white long-stalked glands. Capsule 1-seeded, pubescent and occasionally sparsely glandular; seeds shallowly pitted, brown to mottled, 1.5–2 mm long. May–Oct. Rocky, open woodlands, rocky prairies, pastures, roadsides, & thickets; se GP; (s IL, AR, KS, s into TX). *A. gracilens* A. Gray—Rydberg.

    Though *A. gracilens* has been frequently reported in the GP, all specimens are *A. monococca*.

2. *Acalypha ostryaefolia* Ridd. Erect, annual herb, 1-7 dm tall. Stems simple to usually branched, with sparse to dense recurved short hairs and long-stalked white glands. Leaf blades 3-8 cm long, 1.5-5 cm wide, ovate, often somewhat cordate at base, 3-5 main veins, dentate, usually with some short stiff hairs; petioles 1.5-7 cm long; stipules to 1.5 mm long. Pistillate spikes terminal, bracts 3-7 mm long, 5-12 mm wide, lobes (9)13-17(19) and filamentous; staminate spikes axillary, 5-32 mm long. Capsule 3-seeded, echinate near apex; seed 1.5-2.3 mm long, brown, tuberculate, (n = 7) Jun-Oct. Open areas along streams & valleys, fields & waste places; se GP; (PA s to FL, IA, KS, OK, to w TX, NM, & AZ).

3. *Acalypha rhomboidea* Raf., rhombic copperleaf. Erect, annual herb, 1-6 dm tall. Stems simple or frequently branching above, densely pubescent above with recurved hairs, sparsely hairy below. Leaf blades 1.5-10 cm long, 1-5 cm wide, ovate to commonly broadly rhombic, crenate-serrate, glabrate, 3 main veins at base; petioles 0.5-7 cm long; stipules lanceolate, to 1 mm long. Spikes axillary, staminate above, pistillate below; pistillate bracts 1-3, 4.5-5.5 mm long, 6.5-30 mm wide, lobes (5)7-9(11), oblong-lanceolate, long-ciliate, often hirsute and with a few stalked white glands. Capsule 3-seeded, pubescent and often glandular above, glabrous below; seed 1.3-2 mm long, with shallow pits in rows, brown and often mottled. Jul-Oct. Open woodlands, stream valleys, prairies, fields, roadsides, & waste places; ND: Stutsman; se NE; e 1/2 KS, to s & sw; (ME, ND, s to GA, LA, & TX).

4. *Acalypha virginica* L. Erect, annual herb, 1-5 dm tall. Stems rarely branching, recurved hairs sparse to dense, long spreading hairs usually present at least above. Leaf blades 2-10 cm long, 5-35 mm wide, narrowly rhombic to broadly lanceolate, crenate to nearly entire, with sparse, short stiff hairs; petioles 2-60 mm long; stipules to 1 mm long. Spikes axillary; pistillate bracts 1-3, at base of spike, 4.5-7.5 mm long, 6-20 mm wide, with (8)10-14(16) lanceolate lobes, short hairs sparse, long hairs sparse below, ciliate, glands sparse; staminate upper portion of spike to 2 cm long. Capsule 3-seeded, pubescent, usually glandular; seed usually 1.5 mm long, shallowly pitted in rows, brown, rarely mottled. (n = 14) Jul-Oct. Open woodland, stream valleys, thickets, fields, roadsides, & waste places; se SD; to e 1/2 KS & s; (NH, NE, s to SC, GA, LA, TX).

2. ARGYTHAMNIA P. Br., Wild Mercury

Ours monoecious perennial herbs from a woody rootstock. Stems erect, unbranched to branching, ascending, spreading or trailing, with appressed hairs attached at middle, or glabrous. Leaves alternate, entire or serrate. Flowers in axillary bracteate racemes, lower 1-3 flowers pistillate, upper staminate; staminate flowers with 5 sepals and 5 petals, 5 glands, 7-10 stamens; pistillate flowers with 5 sepals, 0-5 petals, 5 glands opposite sepals and inserted on disk. Capsule 3-celled; 1 seed per cell; seed ecarunculate.

1 Inflorescences shorter than leaves; seeds 2.5 mm long, 4 mm wide .... 1. *A. humilis*
1 Inflorescences equalling or longer than leaves; seed about 5 mm long and wide .... 2. *A. mercurialina*

1. *Argythamnia humilis* (Engelm. & Gray) Muell, Arg. Perennial herbs with a woody caudex. Stems to 4 dm long, usually trailing or spreading, freely branching, usually pubescent. Leaves elliptic to oblanceolate, 1-5.5 cm long, 5-15 mm wide, usually pubescent, nearly entire, petiole 2-3 mm long. Pistillate flowers with sepals 3 mm long; petals often absent or 1-5; styles bifid 1/2 of free part. Fruit about 5-7 mm wide; seeds ovoid-spheroidal, 2.5 mm long, 4 mm wide, slightly roughened to reticulately ridged. Jun-Aug. Dry rocky or sandy prairies, ditches, & ravines; sw KS & s; (sw KS, s to TX, NM, & Mex.). *Ditaxis humilis* (Engelm. & Gray) Pax — Rydberg.

Entirely glabrous plants are referred to *A. humilis* var. *laevis* (Torr.) Shinners and have been reported from OK: Cimarron. Otherwise, the variety is south of our area.

2. **Argythamnia mercurialina** (Nutt.) Muell. Arg. Perennial from a woody rootstock. Stems several, ascending, unbranched, 3–7 dm tall, pubescent. Leaves nearly sessile, lanceolate or elliptic to elliptic-ovate or elliptic-obovate, 3–8 cm long, 1–4 cm wide, pubescent. Pistillate flowers with sepals 4.5 mm long; petals usually absent or 1–5; styles bifid only at tip. Fruit about 10 mm wide; seed subspheroidal, 5 mm long and wide, faintly irregularly reticulate. May–Jul. Rocky prairie hillsides; s 1/2 KS & s; (s KS to TX, NM, AZ). *Ditaxis mercurialina* (Nutt.) Coult.—Rydberg.

3. CNIDOSCOLUS J. Pohl, Bull Nettle

1. **Cnidoscolus texanus** (Muell. Arg.) Small, bull nettle. Perennial herb 3–5(10) dm tall and diffuse, to 1 m across, with milky sap and stinging hairs, from roots to 1 m long and 2 dm thick. Stems several from roots, branching below and above ground, stinging hairs with elongate white bases. Leaves alternate; blades deeply palmately 3- to 5-lobed, 6–15 cm wide, stinging hairs sparse; petiole shorter to longer than blade; stipules inconspicuous. Inflorescence terminal, cymose, pedunculate, few-flowered; branches of inflorescence dichotomous toward the end; terminal flower pistillate; staminate flowers on paired ultimate branches. Staminate flowers with a single perianth whorl, white, corollalike, fragrant, trumpet-shaped, tube 15–20 mm long, longer than lobes, stamens 10; pistillate flowers with single perianth whorl, whitish, 10–17 mm long, 5-parted to near base, styles 3, united basally, 3 times dichotomous. Capsules 15–20 mm long, hispid; seeds 3, oblong, 14–18 mm long, smooth, brownish-white, caruncle prominent. Apr–Sep. Sandy areas; cen OK & s; (LA, OK, AR, TX, s into Mex.).

4. CROTON L., Croton

Ours monoecious or dioecious annual or perennial herbs with stellate hairs or peltate scales on at least some parts. Leaves alternate, entire or serrate; petiolate; stipules small. Flowers in axillary or terminal spikes or racemes; staminate flowers: calyx (4)5(6)-parted, petals (4)5(6), glands of disk alternate with petals, stamens 5 or more; pistillate flowers: calyx 5- to 9-cleft or parted, petals none or minute; ovary 3-celled (fewer in *C. monanthogynus*), each cell with one ovule, styles 3 or 2, each 1 or more times dichotomous. Fruit a capsule; seeds carunculate.

References: Ferguson, A. M. 1901. Crotons of the United States. Annual Rep. Missouri Bot. Gard. 12: 33–74; Johnston, M. C. 1959. The Texas species of *Croton* (Euphorbiaceae). Southw. Naturalist 3: 175–203.

1 Plants dioecious; all flowers apetalous.
  2 Plants perennial; leaves scaly silvery-canescent beneath, dark grayish above .... 2. *C. dioicus*
  2 Plants annual; leaves green to yellowish-green, never at all silvery .... 7. *C. texensis*
1 Plants monoecious; at least staminate flowers with petals.
  3 Plants perennial .... 6. *C. pottsii*
  3 Plants annual.
    4 Leaves distinctly serrate .... 3. *C. glandulosus*
    4 Leaves entire or essentially so.
      5 Styles 2; 1 seed per capsule .... 5. *C. monanthogynus*
      5 Styles 3; 3 seeds per capsule.

6 Styles only once-dichotomous .... 4. *C. lindheimerianus*
6 Styles more than once-dichotomous .... 1. *C. capitatus*

1. **Croton capitatus** Michx., woolly croton. Monoecious annual herbs, stellate-tomentose, (2)3–8(10) dm tall, usually umbellately branched above. Leaves alternate, narrowly oblong, lanceolate to ovate or lance-elliptic, blades 3–10 cm long, usually blunt but slightly apiculate, basally usually rounded, rarely subcordate, petioles nearly as long at the top of the plant as at the middle, stipules subulate, early deciduous. Male and female flowers in dense terminal spikelike racemes 1–3 cm long; staminate flowers: sepals 5, nearly distinct, subulate to deltoid, 1 mm long; petals 5, oblanceolate, 1 mm long; stamens 10–14; pistillate flowers: calyx deeply 6- to 9-lobed, lobes linear or oblong, 2–3 mm long at anthesis, later accrescent and nearly as long as fruit; petals absent; styles 3, each 2–3 × dichotomous, 1 mm long, the stigmas thus 12–24. Capsule nearly globose, 6–10 mm long, 3-celled, usually 3-seeded; seeds 3.4–5 mm long. (n = 10) Jul–Oct. Dry sandy or calcareous prairies & open woodlands, fields, pastures, roadsides; se 1/3 GP, less frequent in n & w; (NY to se NE, sw KS, s to GA & TX).

Our plants as here described are the var. *capitatus*. The var. *lindheimeri* (Engelm. & Gray) Muell. Arg., with leaf blades mostly acute, long tapered from near base to apex, petioles long at midstem but shorter above, has been reported for KS: Miami, but otherwise is to the se of our area.

2. **Croton dioicus** Cav. Dioecious perennial herb with several erect stems from the base (but flowering first year), 1.5–5 dm tall, lepidote. Leaves linear-lanceolate, less often ovate-oblong, or elliptical, to rarely broadly ovate, (1)2–4(6) cm long, acute or rounded, base rounded, margin entire, scaly silvery beneath, dark grayish above; petiole 3–6(10) mm long; stipules 1 mm or less long. Staminate flowers in racemes 2–6 mm long: calyx 3 mm wide, hemispheric, with 5 deltoid lobes, petals absent, stamens 10–12; pistillate flowers: in racemes 1 cm long, calyx 3 mm wide, hemispheric, with 5 deltoid lobes; petals absent; ovary globose, lepidote, styles 0.5–1.5 mm long, 2–3 × dichotomous. Capsules globose, 5–6 mm long; seeds 3.5–5 mm long, caruncle 0.3 mm long. Jul–Oct. Dry uplands, roadsides; s TX panhandle; (TX, NM, & s in Mex.).

3. **Croton glandulosus** L., tropic croton. Monoecious annual, stellate pubescent herb 1–5 dm tall, somewhat umbellately branched. Leaves oblong to ovate-oblong or linear, rarely somewhat lanceolate, 1–6.5 cm long, serrate, rounded apically, rounded to cuneate at base, the base with a whitish cartilaginous saucer-shaped gland on each side of the petiolar attachment; stipules minute. Male and female flowers in terminal androgynous racemes about 1 cm long; staminate flowers: sepals 5(4), nearly distinct; petals 5(4); stamens 7–13; pistillate flowers: sepals 5, nearly distinct, 1.5 mm long, becoming much-accrescent; essentially apetalous; styles 3–4 mm long, grayish, often mottled. (n = 8) Jun–Oct. Infrequent to common in sandy open woods, prairies, pastures, fields, roadsides & waste places; se 1/3 GP, less frequent n & w; (PA, IN, IA, NE, s to TX & FL).

Our plants as here described are the var. *septentrionalis* Muell. Arg. and are highly variable. Plants averaging only 1–2 dm tall, with larger leaves about 25 mm long, and central process of trichomes mostly shorter than the radii have been segregated as var. *lindheimeri* Muell. Arg.

4. **Croton lindheimerianus** Scheele. Monoecious annual stellate-tomentose herbs to 5 dm tall. Leaves suborbicular to rhombic-ovate or oblong, usually less than twice as long as broad, apically rounded to broadly acute, basally rounded, marginally entire. Flowers few in terminal androgynous spikelike racemes 10–14 mm long; staminate flowers: calyx 2–2.5 mm wide, with 5(4) nearly distinct sepals; petals with 5 indistinct lobes opposite the sepals; stamens 7–9; pistillate flowers: calyx 3 mm long, with 5 nearly distinct sepals; petals absent; ovary subglobose, styles 3, 2–3 mm long, each bifid to base. Capsules 4–5 mm long;

seeds 3-3.5 mm long, shiny, mottled brown. Jun-Oct. Sandy loam, & rocky clay hillsides, ravines, & roadsides; KS: Barber, Harper, Kiowa; OK, TX; (AR, KS, OK, TX, s to Mex.).

5. *Croton monanthogynus* Michx., one-seeded croton. Monoecious annual somewhat widely dichotomously branched, the lowest branches usually 3-5 together, stellately-tomentose herbs 1-5 dm tall. Leaves ovate, ovate-oblong to nearly round below to frequently elliptic above, 1-4 cm long, marginally entire, apically blunt or apiculate, rounded to truncate at base, often silvery-green. Flowers in terminal androgynous racemes to 1 cm long; staminate flowers: calyx 1.5-2.5 mm wide, sepals 4(5), nearly distinct; petals 4(5); stamens (4)5; pistillate flowers: sepals 5, subequal, 1.5-2 mm long, about 2/3 as long as capsule at maturity; petals absent; ovary ovoid or subglobose, 2-celled with 1 cell large and 1-ovulate, the other small and ovule abortive; styles 2, bifid almost to base. Capsules ovoid, tapering from below middle to apex, about 4 mm long, 1-seeded; seed about 1 mm long. (n = 8) Jun-Oct. Infrequent to abundant in calcareous soils of prairies, open woodland, pastures, roadsides, waste places; s 1/2 GP, becoming infrequent n & w; (MD, OH, IN, NE, s to FL, TX, NM, & n Mex.).

6. *Croton pottsii* (Kl.) Muell. Arg., leather-weed. Monoecious perennial herb 1-5 dm tall, somewhat rhizomatose; main stems several from a stout, often branching caudex, thinly to densely tomentose. Leaves ovate-oblong, obovate-oblong, to nearly orbicular, 2-5 cm long, apically acute to rounded, basally rounded to subtruncate, marginally entire, densely tomentose. Flowers in terminal androgynous spikelike racemes 1-2.5 cm long; staminate flowers; calyx of 5(4) nearly distinct sepals, petals 5(4), somewhat longer than sepals; stamens 11-18; pistillate flowers: calyx 1.5-3 mm long, of 5 nearly distinct lobes, about 1/2 as long as capsule at maturity; petals absent; styles 3, bifid to near base. Capsules oblong to ovoid-oblong, 4-6 mm long; seed 3.5-4.5 mm long. Jun-Oct. Dry, mostly calcareous uplands & hillsides; s TX panhandle; (w TX, cen NM, se AZ, s in Mex.).

7. *Croton texensis* (Kl.) Muell. Arg., Texas croton. Dioecious densely stellate-tomentose annual herb 2-8 dm tall, often branched above. Leaves oblong-linear to narrowly ovate-oblong, (1)2-4(8) cm long, apically rounded or acute, basally rounded to obtuse. Staminate flowers: in racemes 1-2 cm long, calyx 2-4 mm wide, with 5 deltoid acute lobes, petals absent; pistillate flowers: in racemes 1 cm long, calyx 2.5-4 mm wide, with 5 deltoid acute lobes; petals absent; styles 3, each divided nearly to base into 4 or more branches. Capsules globose or globose-ovoid, 4-6 mm long, densely stellate-tomentose, often warty; seeds 3.5-4 mm long, caruncle 1 mm long. Jun-Oct. Infrequent to abundant in sandhills & sandy soils where it is something of a weed; cen SD & s in GP; (SD, e WY, s to TX, NM; w of Rocky Mts. from UT to n Mex.).

5. CROTONOPSIS Michx., Rush-foil

1. *Crotonopsis elliptica* Willd. Annual monoecious herbs with a taproot; stems 1-3 dm tall, solitary at base, branches dichotomous and trichotomous near middle and ascending or erect, covered by stellate trichomes. Leaves alternate, sessile or with petioles 1-2 mm long; blades elliptic to linear-lanceolate, 1-4 cm long, to 12 mm wide, lower surface covered rather densely by stellate trichomes in which radii are more or less coalescent at bases, upper surface with less dense stellate trichomes with distinct radii; stipules minute, soon deciduous. Flowers in androgynous axillary spikes to 1 cm long, with 1 or 2 pistillate flowers below and several staminate ones above; staminate flowers with calyx 1 mm long, deeply 4- to 5-lobed, stellate-pubescent outside, petals shorter than calyx; stamens usually 6; pistillate flowers with calyx 1 mm long, 4- to 5-lobed, disk suppressed; ovary with 1 ovule;

style branches 3. Fruit ovate in outline, radii of sparse stellate trichomes coalescent to near tips; achene 2.5–3 mm long. Jun–Oct. Locally common in sandy or rocky prairies & open wooded areas; s MO, se KS, e OK; (CT, NJ, PA, IN, IL, MO, se KS, s to FL & e TX).

6. EUPHORBIA L., Spurge

Ours annual or perennial herbs with milky acrid juice; very diverse in form, habit, and pubescence. Flowers unisexual, greatly reduced, the staminate consisting of a single stamen; the pistillate of a single stipitate pistil with a 3-celled 3-ovulate ovary; styles 3, each usually bifid. Flowers borne in a cupuliform calyxlike involucre (cyathium) with 4 or 5 lobes, at least 1 bearing a large gland, often with petaloid appendages. Each cyathium contains one pistillate flower and 2–15 or more staminate flowers. Fruit a 3-locular capsule, each locule usually 1-seeded. *Chamaesyce* S. F. Gray, *Galarrhoeus* Haw., *Lepadena* Raf., *Poinsettia* Grah., *Tithymalopsis* Kl. & Garcke, and *Zygophyllidium* Small—Rydberg.

References: Burch, D. 1966. The application of the Linnaean names of some New World species of *Euphorbia* subgenus *Chamaesyce*. Rhodora 68: 155–166; Norton, J. B. S. 1900. A revision of the American species of *Euphorbia* of the section *Tithymalus* occuring north of Mexico. Annual Rep. Missouri Bot. Gard. 11: 85–144; Richardson, J. W. 1968. The genus *Euphorbia* of the high plains and prairie plains of Kansas, Nebraska, South and North Dakota. Univ. Kansas Sci. Bull. 48: 45–112; Wheeler, L. C. 1941. *Euphorbia* subgenus *Chamaesyce* in Canada and the United States exclusive of southern Florida. Rhodora 43: 97–154, 168–205, 223–286.

Euphorbia peplus L., petty spurge, a native of Europe has been collected from a flower garden in North Dakota in 1911. It is not included in this treatment.

1 Glands of cyathia without petaloid appendages, lobed crescent-shaped or entire.
  2 Glands usually 1 or 2, cup-shaped with inflexed margin (subgenus *Poinsettia*).
    3 Main stem leaves all or mostly opposite, pubescent on both sides; upper leaves not fiddle-shaped. .... 7. *E. dentata*
    3 Main stem leaves all or mostly alternate, glabrous above; upper leaves sometimes fiddle-shaped .... 5. *E. cyathophora*
  2 Glands 4 or 5, flat or convex, entire or crescent-shaped (subgenus *Esula*).
    4 Plants annual; stem leaves cuneate-spatulate or obovate-oblong.
      5 Leaf margins entire; seeds with numerous pits .... 15. *E. longicuris*
      5 Leaf margins serrulate at least on upper 1/2; seeds with low reticulate ridges throughout .... 26. *E. spathulata*
    4 Plants perennial; stem leaves not cuneate-spatulate or obovate-oblong.
      6 Stem leaves cordate and often amplexicaul; distinctly pinnate veined .... 1. *E. agraria*
      6 Stem leaves not cordate or amplexicaul; lateral veins usually not visible or very obscure.
        7 Stem leaves about as wide as long .... 23. *E. robusta*
        7 Stem leaves much longer than wide.
          8 Appendages with merely narrow minutely scalloped margins; involucre, ovary, and capsule minutely strigose-puberulent .... 28. *E. strictior*
          8 Appendages crescent-shaped and often with 2 horns; involucre, ovary and capsule not puberulent.
            9 Cauline leaves 1–3 cm long, narrowly linear, those of axillary branches densely crowded .... 6. *E. cyparissias*
            9 Cauline leaves 3–10 cm long, those of axillary branches not densely crowded.
              10 Cauline leaves oblanceolate, widest above the middle, rounded at the apex .... 8. *E. esula*
              10 Cauline leaves widest at or below the middle or of nearly uniform width, tapering to apex or sharply acute.

11 Stems usually branched from base with several sterile branches; cauline leaves narrowly linear, usually less than 1.5 mm wide, acute .... 29. *E. uralensis*
11 Stems usually sparingly branched from middle or above; cauline leaves widest near middle, 3-5 mm wide, tapering to apex .... 21. *E. x pseudovirgata*
1 Glands of cyathia with petaloid appendages.
12 Leaves with symmetrical bases; stipules absent or glandlike (subgenus *Agaloma*).
13 Stem leaves all opposite, petiolate, linear to elliptic lanceolate .... 12. *E. hexagona*
13 Stem leaves alternate, sessile to subpetiolate, ovate, oblong, or elliptical, rarely somewhat linear.
14 Plants perennial; mature capsule usually glabrous; leaves and bracts of inflorescence reduced, glands 5, without white margins; seeds smooth .... 4. *E. corollata*
14 Plants annual; mature capsule usually pubescent; leaves and bracts of inflorescence with prominent white to rarely pinkish petaloid margins; glands (3)4(5); seeds with tubercles in a reticulate pattern .... 17. *E. marginata*
12 Leaves with usually inequilateral bases; stipules well developed (subgenus *Chamaesyce*).
15 The two stipules on each side of the stem at each node united into a whitish or reddish-white membranaceous scale which is entire or lacerate.
16 Plants annual; stems often rooting at nodes; appendages about as long as gland is wide .... 24. *E. serpens*
16 Plants perennial; stems not rooting at nodes; appendages 1–3 × longer than gland is wide .... 2. *E. albomarginata*
15 Stipules otherwise, or if appearing united into a scale then only on one side of stem, deeply lobed or dissected.
17 Herbage, inflorescence, ovaries, and capsules essentially glabrous.
18 Leaves linear, more than 6 × longer than wide, not serrulate.
19 Staminate flowers 24–53 per cyathium; appendages 1.5–4 × longer than as gland is wide; styles bifid 1/2 their length .... 18. *E. missurica*
19 Staminate flowers 3–8 per cyathium; appendages about as long as gland is wide; styles entire or shortly notched .... 22. *E. revoluta*
18 Leaves not linear or if narrow than serrulate or less that 6 × longer than wide.
20 Plants perennial .... 9. *E. fendleri*
20 Plants annual.
21 Capsules (4.5)57 mm long .... 3. *E. carunculata*
21 Capsules less than 2.8 mm long
22 Leaves entire as viewed under lens .... 10. *E. geyeri*
22 Leaves mostly serrate or serrulate at least toward the apex.
23 Plants usually erect, most mature leaves more than 15 mm long .... 19. *E. nutans*
23 Plants usually prostrate or decumbent with mature leaves usually less than 15 mm long.
24 Seeds with definite transverse ridges or corrugated; stems not noticeably flattened near apex .... 11. *E. glyptosperma*
24 Seeds sometimes with shallow transverse wrinkles but these faint; stems often winged toward the tip ... 25. *E. serpyllifolia*
17 Herbage and inflorescence with some pubescence.
25 Ovary and capsule glabrous.
26 Stems usually crisply hairy at least toward tips .... 19. *E. nutans*
26 Stems only with lines of short hairs .... 11. *E. glyptosperma*
25 Ovary and capsule variously pubescent.
27 Plants perennial; stems crisply pubescent to glabräte .... 14. *E. lata*
27 Plants annual; stems shaggy-pubescent, villous or pilose.
28 Seeds punctately pitted and mottled, the base depressed-truncate, apex acute; styles entire .... 27. *E. stictospora*
28 Seeds neither punctately pitted nor mottled, the base obtuse, apex blunt; styles bifid at least briefly.

29 Capsules primarily pubescent on angles; styles 0.1 mm long, bifid nearly to
   base .... 20. *E. prostrata*
29 Capsules rather uniformly pubescent; styles 0.3–0.7 mm long, bifid 1/4–1/3 length.
   30 Plants often rooting at nodes; styles 0.5–0.7 mm long, slender .... 13. *E. humistrata*
   30 Plants never rooting at nodes; styles ca 0.3 mm long, clavate .... 16. *E. maculata*

1. **Euphorbia agraria** Bieb. Herbaceous, glabrous perennial herbs from stout caudex, extensive rhizomes and deep roots; stems erect, 3–9 dm tall, unbranched at base, 0–3 axillary nonflowering branches, and 3–20 axillary flowering branches above. Leaves alternate, (1.5)2.5–4.5(8) cm ong, 1–2 cm wide, triangular-ovate to oblong, cordate, somewhat amplexicaul, entire, distinctly pinnately veined; ray leaves like cauline; raylet leaves triangular-subreniform. Terminal umbel open, rays (6)8–14, up to 2–3 × dichotomous; cyathia 2–3 mm long, glands emarginate or with 2 short horns. Capsules 2.5–3 mm long, deeply sulcate, grandulate-rugose on angles; seeds 2–2.5 mm long, gray or brownish. May–Jun. Roadsides, fields, & waste places; NE: Jefferson; KS: Marshall; WY: Converse, Platte; (reported from PA, NY; GP; Europe).

2. **Euphorbia albomarginata** T. & G. Perennial, usually prostrate herbs, often rooting at nodes; taproot becoming woody; stems 10–40 cm long, mostly annual, but some persistent and perennial. Leaves opposite, orbicular to oblong, 2.5–10 mm long, nearly as wide, rounded at apex, inequilateral and rounded at base, marginally entire, often with narrow elongate median reddish splotches; petioles 0.5–1.3 mm long; stipules on both sides of stem united into a white membranaceous scale 1–2 mm long and entire to lacerate. Cyathia single at nodes and in forks, ca 1 mm long to gland base; glands 4, oblong, shallowly cupped, 0.5–1 mm long; appendages normally conspicuous, white, 1–3 × as long as width of gland; staminate flowers 15–30 per cyathium; pistillate flower with 3 styles 0.5–0.7 mm long, bifid almost entire length, the divisions slender. Capsules 1.3–2 mm long; seed 1.2–1.7 mm long, apically acute, basally truncate, nearly smooth or minutely punctate in transverse lines, pale brown with a white coat. May–Oct. Scattered to common in dry, clay, rocky, or sandgravelly prairies, roadsides, & fields; w OK, se CO, w TX, & NM; (w OK to se CO, w TX, NM to NV, s into Mex.).

3. **Euphorbia carunculata** Waterfall. Prostrate, glabrous, annual herbs; stems 1–10(15) dm long, somewhat succulent, internodes 1–10 cm long, usually distinctly longer than subtending leaves. Leaves opposite; blades ovate to more commonly elliptic-oblong, 10–25 mm long at midstem, ca half as wide, acute and mucronate at apex, slightly asymmetrical basally and truncate to very shallowly cordate, marginally entire; petioles 2–8 mm long; stipules lanceolate, 1–1.8 mm long, usually bifid. Cyathia single in forks and at upper nodes, 2–2.3 mm long to glands; glands 4, short stalked, erect, 0.3–0.9 mm long, flat or cupped; appendages absent, forming a narrow margin, or prominent and erect to 1 mm long; staminate flowes 15–25 per cyathium; pistillate flower with 3 styles 0.7–0.8 mm long, bifid 1/3 length. Capsule ovoid, narrowed above, narrowly truncate basally, 4.5–7 mm long, 4–5 mm wide, deeply 3-lobed; seed laterally compressed, never angulate, acuminately narrowed from base to micropylar end, 4–5 mm long, grayish-white to mottled reddish-brown, apex enlarged or apiculate. Aug–Oct. Locally common on sand dunes; KS: Clark; w OK, w TX; (OK, TX, & Chih.).

4. **Euphorbia corollata** L., flowering spurge. Perennial glabrous or variously pubescent herbs, 2–10 dm tall from a deep root; stems 1 or few from caudex, usually erect, usually simple below, umbellately or paniculately branched above. Leaves alternate, those of midstem oblong, elliptic, or less often somewhat linear, 2–6 cm long, rounded to rarely acute at apex, rounded or narrowed basally, marginally entire, sessile or subpetiolate; stipules minute, glandlike. Cyathia numerous and in corymbiform or paniculiform cymes to 3.5 dm wide;

cyathia with 5-cupped, transversely linear glands, ca 0.6 mm long, appendages 1.5–4.5 mm long, white, conspicuous; staminate flowers ca 10–15 per cyathium; pistillate flower with 3 styles ca 0.7 mm long, bifid ca 1/2 length, divisions clavellate. Capsule 2.5–4.5 mm long; seed ca 2.5 mm long, ovoid, white, smooth. Jun–Oct. Locally common to infrequent in dry rocky prairies, open woodlands, fields, roadsides, & waste places; s MN, se SD, s to w 1/2 OK; (NY, s Ont., NM, NE, s to FL & TX). *Tithymalopsis corollata* (L.) Small — Rydberg; *E. corollata* var. *mollis* Millsp. — Richardson, op. cit.

Plants with stems and leaves variously pubescent have been recognized as var. *mollis* Millsp. The amount and type of pubescence, however, varies from dense and woolly to nearly glabrate. Some plants have glabrous stems and one or both leaf surfaces variously hairy, while others have glabrous herbage but are pubescent in the inflorescence. These forms are rare to common in colonies of otherwise glabrous individuals and are not given taxonomic recognition here.

5. **Euphorbia cyathophora** Murray, fire-on-the-mountain, painted euphorbia. Annual erect herbs to 1 m tall from yellowish taproot; stem usually branched, glabrous to sparsely hairy. Leaves alternate at midstem, highly variable even upon same plant; blades linear to linear-lanceolate, ovate, broadly obovate, upper wider leaves lobed and often panduriform, 5–15 cm long, marginally serrate or entire, thin, usually glossy green, usually with scattered hairs on lower surface, some upper leaves often with reddish or yellowish splotches near base; petioles 3–14 mm long, sparsely hairy; stipules lacking or minute and glandlike. Cyathia at summit of stem and branches; gland usually 1, deeply cupped, sessile, without appendages; staminate flowers 30–50 per cyathium; pistillate flower with styles ca 1.0 mm long, bifid for 1/2 length. Capsule 3–4 mm long, 6–8 mm wide, glabrous; seed ovoid or subglobose, not angular, 2.5–3.0 mm long, caruncle absent or minute, dark brown but covered with low pale tubercles. (n = 28) Jun–Sep. Infrequent to locally common in moist woodlands, brushy ravines, roadsides, & waste places; MN, se SD, NE, e 1/2 KS, e OK, scattered elsewhere; (VA, IN, WI, MN, SD, s to FL & TX). *Poinsetta heterophylla* (L.) Small — Rydberg; *E. heterophylla* L., *E. heterophylla* var. *graminifolia* (Michx.) Engelm — Fernald.

The name *E. heterophylla* L. has long been applied to our plants, but that is a tropical species found only as far north as extreme southern Texas.

6. **Euphorbia cyparissias** L., cypress spurge. Perennial erect densely tufted or colonial herbs from creeping rhizomes and stout roots; stems 2–4 dm tall, usually sterile and fertile stems present. Leaves usually numerous, crowded, linear or linear-spatulate, pale green, 1–3 cm long, 0.5–3 mm wide, 1-nerved. Umbel with usually 10 or more rays, floral leaves reniform or cordate; cyathia ca 3 mm long. Capsule 3 mm long (rarely developed in our area); seeds 1.5–2 mm long, smooth. (n = 10, 20) Apr–Aug. Roadsides, waste places in old cemeteries; infrequent & scattered in n 2/3 of GP; (ME, MN, WI, s to PA, IL, KS, CO; Europe). *Galarrhoeus cyparissias* (L.) Small — Rydberg.

This species has been planted in gardens, cemeteries, etc., from which it has escaped and persisted. It is rare that mature seeds are produced, and thus the plant has not become a noxious weed in our area.

7. **Euphorbia dentata** Michx., toothed spurge. Annual herb 1–6 dm tall; stems erect, usually with ascending decussate branches, with dense short somewhat retrorse strigose hairs and scattered white multicelled hairs. Leaves of midstem usually opposite, ovate, lanceolate, to linear, 15–60 mm long, acute or blunt at apex, acuminate and short attenuate at base, marginally dentate, irregularly serrate, to subentire, sparsely to densely pubescent on both surfaces; petioles 5–25 mm long, pubescent like stems; stipules absent or minute and glandlike. Inflorescences congested at summit of stems, the clusters subtended by opposite leaves, these pale at base. Involucres nearly sessile, campanulate, ca 3 mm long, nearly

glabrous; glands (1)2, with central involucre rarely with 5 glands; appendages absent; staminate flowers 25–40 per cyathium; pistillate flowers with styles 1–1.5 mm long, bifid 1/2 length. Capsule 3–5 mm long, glabrous to sparsely strigose; seeds 2.5–3.0 mm long, sharply angled, finely and densely tuberculate, white to brown or nearly black; caruncle-heart-shaped. May–Oct. Prairie ravines, fields, roadsides, waste places, in a variety of soils; s 2/3 GP; (NY, MN, SD, WY, s to AZ & Mex.). *Poinsetta cuphusperma* (Boiss.) Small, *P. dentata* (Michx.) Small—Rydberg.

Plants with leaves linear to narrowly lanceolate and capsules strigose have been segregated as forma *cuphusperma* (Engelm.) Fern. The species, however, is extremely variable in leaf shape and pubescence so several forms usually can be found in a given colony, making taxonomic distinctions of no value.

8. *Euphorbia esula* L., leafy spurge. Perennial herbs, 3–9 dm tall from stout caudex, forking rhizomes, and deep roots; stems not crowded, simple below, but often freely umbellately branched above, glabrous. Leaves of main stem oblanceote or narrowly oblanceolate-oblong, widest above the middle, usually 3–10 mm wide, 3–10 cm long, rounded at the apex, essentially 1-nerved with lateral veins very obscure; leaves subtending umbel lanceolate to ovate, those of umbel broadly cordate or reniform; rays of primary umbel 5–17, 1–2 × dichotomous; cyathia 1.5–3 mm long; glands 4, emarginate or with 2 horns; appendages absent; staminate flowers 12–25 per cyathium; pistillate flower with 3 styles 1.5–3.0 mm long, bifid 1/3 its length. Capsule 2.5–3.5 mm long, granulate on keels; seeds 2.2–3.0 mm long, ovoid or oblong-cylindrical, smooth, silver-gray, brown, often mottled, caruncle prominent. (n = 10,30,32) May–Sep. Fields, roadsides, stream valleys, open woodlands, waste places, in a variety of soils; n 3/4 GP, less frequent s; N. Engl. to WA, s to MD, IN, NE, ID, CO; Eurasia). *Naturalized. Galarrhoeus esula* (L.) Rydb.—Rydberg.

This species and *E.* x *pseudovirgata* (a presumed hybrid between *E. esula* and *E. virgata*) are aggressive noxious weeds in the northern GP and appear to be increasing in many areas.

9. *Euphorbia fendleri* T. & G., Fendler's euphorbia. Perennial, glabrous, herb; stems many, prostrate to erect, 3–15 cm long, from usually buried caudex, basal parts of annual stems often also buried. Leaves opposite, blades ovate, 1–2x as long as wide, 2–6 mm long, acute at apex, truncate to rounded or slightly cordate at base, marginally entire, rather firm, often somewhat glaucous; petioles 0.5–2 mm long; stipules to 1 mm long, distinct to united basally, segments narrowly lanceolate to linear. Cyathia 1 at nodes or forks; involucre turbinate to campanulate, 1–2 mm long; glands 4, sessile; appendages absent or shorter than gland is wide; staminate flowers (3)8–12(22) per cyathium; pistillate flowers with styles 0.3–0.7 mm long, bifid 1/2 length, divisions slender or slightly clavate. Capsules 2 mm long; seed 1.0–2 mm long, quadrangular, essentially smooth or with faint wrinkles, coat white to pinkish-brown. Apr–Oct. Common on dry sandy or rocky prairies, roadsides, waste places, & alluvial deposits; w GP from sw SD, w 1/3 KS & s; (sw SD, WY, s to TX, UT). *Chamaesyce fendleri* (T. & G.) Small, *C. greenei* (Millsp.) Rydb.—Rydberg.

10. *Euphorbia geyeri* Engelm., Geyer's spurge. Annual, glabrous, prostrate herbs; stems 0.5–4.5 dm long. Leaves opposite; blades oblong to ovate-oblong, 4–12 mm long, ca 2 × as long as wide, obtuse to emarginate at apex, often mucronate, base oblique and obtuse, rounded or rarely slightly cordate, marginally entire; petiole 1–2 mm long; stipules distinct or the ventral united, 1–1.5 mm long, mostly divided into 2–5 filiform segments. Cyathia solitary in the upper forks, somewhat clustered by the shortening of uppermost internodes; involucres turbinate or broadly campanulate, 1–1.5 mm long; glands 4, broadly oval to subrotund, 0.2–0.6 mm long, appendages white to reddish, 1/2–2 × as long as gland is wide; staminate flowers 5–27 per cyathium; pistillate flower with styles 0.2–0.5 mm long, bifid 1/3–1/2 length. Capsule 1.5–2.0 mm long; seeds 1.3–1.6 mm long, smooth, reddish-

brown to nearly white. Jun–Oct. Sand dunes or sandy soils of roadsides & waste places; GP; (WI, MN, IA, ND, MT, s to TX, NM). *Chamaesyce geyeri* (Engelm.) Small — Rydberg.

11. *Euphorbia glyptosperma* Engelm., ridge-seeded spurge. Annual prostrate glabrous herb, with stems 5–40 cm long and much branched, rarely with indistinct lines of hairs on stems. Leaves opposite; blades oblong, rarely obovate-oblong or ovate-oblong, 3–13 mm long, often subfalcate, inequilateral and rounded or truncate at base; marginally serrulate (under lens) at least at apex, rarely entire, with larger side often serrulate to base; petioles to 1 mm long; stipules distinct, dissected into 2 or more filiform segments. Cyathia solitary at upper nodes or in forks; involucre turbinate, 0.5–1.0 mm long; glands 4, minute, 0.2 mm long; appendages white, margin with median notch; staminate flowers (1)4(7) per cyathium; pistillate flower with styles 0.1–0.4 mm long, bifid 1/3–1/2 length. Capsule 1.2–1.8 mm long; seeds oblong-triangular, 1–1.3 mm long, all facets with 3–6 definite transverse ridges, ridges more or less passing through angles, coat white to tan. May–Oct. Frequent in a variety of soils & habitats throughout GP; (widespread in N. Amer. *Chamaesyce glyptosperma* (Engelm.) Small — Rydberg; *C. glyptosperma* var. *integrata* Lunell.—Lunell, Amer. Midl. Naturalist 3: 142. 1913.

12. *Euphorbia hexagona* Nutt., six-angled spurge. Annual erect herbs 2–5(10) dm tall, yellowish-green; stem one from base, decussately and pseudodichotomously branched, branches minutely strigose; leaves opposite; blades linear to oblong or lanceolate to narrowly elliptic, 1–7 cm long, acute at both ends, marginally entire; petioles 1–3 mm long; stipules absent or minute glands. Cyathia solitary in upper axils and forks; involucre 1.5–2 mm long, strigulose; glands 5; appendages white or greenish-white, deltoid; staminate flowers 20–40 per cyathium; pistillate flower with styles 0.5–1 mm long, bifid 1/2 length. Capsule 3–5 mm long; seeds 2.5–3.3 mm long, ovoid to oblong, papillose, tuberculate, or roughened, not angular or pitted, whitish, brown, or nearly black. Jun–Sep. Sand dunes, sandy prairies, fields, roadsides, & waste places; s 3/4 GP; (MN, IA, SD, WY, s to TX, NM). *Zygophyllidium hexagonum* (Nutt.) Small — Rydberg.

13. *Euphorbia humistrata* Engelm., spreading spurge. Annual, prostrate to ascending herbs, often rooting at nodes; stems to 3.5 dm long, crisply and sparsely villous or pilose, sometimes glabrate below; leaf blades oval, to oblong-ovate on main stems, often narrower on branches, 4–15 mm long, mostly glabrous above, sparsely crisply villous to glabrate below, with median reddish splotch above, base inequilateral, apex rounded, marginally entire to remotely serrulate; petiole 1–1.5 mm long; stipules distinct, to usually united to a membranaceous scale that is dissected into filiform segments. Cyathia solitary at nodes or usually in dense clusters on short lateral branches; involucre 0.6–0.8 mm long; glands 0.1–0.3 mm long; appendages narrow, white or pink, crenulate or entire; staminate flowers 3–5 per cyathium; pistillate flowers with styles 0.5–0.7 mm long, bifid 1/2 length. Capsules 1–1.5 mm long, strigose to glabrate; seeds quadrangular, 1 mm long, facets nearly smooth, coat white to reddish-brown. Jun–Oct. Moist alluvial ground along streams, ponds, low fields; se GP; (OH, IL, se KS, s to AL, TX). *Chamaesyce humistrata* (Engelm.) Small — Rydberg.

This species is extremely close to *E. maculata* L., and some specimens are difficult to place definitely in either species. The tendency for *humistrata* to root at the nodes and styles averaging only half the length of those of *maculata* are ultimately the best distinguishing characters. It is doubtful whether *humistrata* should be maintained at the specific level.

14. *Euphorbia lata* Engelm., hoary euphorbia. Perennial herb 5–16 cm tall with stout branching caudex and shallowly buried rhizomes to 3 dm long, these with shoots at intervals; stems solitary or several at each shoot, erect or ascending, reddish-brown, crisp white-pubescent, branching above, with internodes 4–25 mm long and densely fine white-hairy.

Leaves opposite; blades narrowly to broadly deltoid, somewhat acuminate, usually falcate, 4–12 mm long, 3–9 mm wide, truncate at base, minutely crisply white-hairy above, white-strigose beneath; petioles 0.5–1 mm long, white-strigose; stipules united, 1 mm long, subulate. Cyathia solitary in upper forks; involucres 1–1.3 mm long, crisply white-hairy; glands 4 or 5th very minute, 0.5–0.8 mm long; appendages usually well developed, semilunate, ca as long as width of gland; staminate flowers (12)25–35 per cyathium; pistillate flower with styles to 0.8 mm long, bifid nearly to base, divisions clavate. Capsule 2–2.5 mm long, white-strigose; seed oblong, somewhat quadrangular, 1.7–2.0 mm long, ventral facets usually concave and smooth, dorsal facets convex, coat white to brownish. May–Sep. Dry calcareous or sandy prairies, roadsides, & waste places; sw GP; (KS, CO, s to TX & Coah.). *Chamaesyce lata* (Engelm.) Small — Rydberg.

15. *Euphorbia longicuris* Scheele. Annual herb to 20 cm tall; stem simple below with internodes 2–5 mm long, becoming trichotomous above, and above that with 3 branches 1–3 mm long, these repeatedly dichotomous with very short internodes giving a congested appearance. Leaves of stem alternate, cuneate-spatulate below, obovate above, 5–15 mm long, apex mucronate, obtuse or retuse, narrowed at base, marginally entire, stipules absent or very small and glandlike; leaves of inflorescence overlapping, reniform or suborbicular, 4–8 mm long, 7–13 mm wide, slightly connate, inequilateral basally. Cyathia solitary in forks of inflorescence; involucre 1.5–2 mm long; glands 4, crescent-shaped, an erect horn at each end; appendages absent; staminate flowers (5)10–15 per cyathium; pistillate flower with styles 0.5 mm long, bifid. Capsule 2.5 mm long; seeds oblong, 1.5 mm long, with prominent projection from chalaza end, surface with numerous pits, the diam of each less than 1/2 the width of space between them. Apr–Jun. Dry calcareous prairies; w OK & adj. TX; (OK, TX, to N.L.).

16. *Euphorbia maculata* L., spotted spurge. Annual herb; stems usually prostrate or decumbent, 5–45 cm long; usually somewhat shaggy villous, young stems often subtomentose, often glabrate below. Leaves opposite, blades oblong-ovate, elliptic-ovate to linear-oblong, 4–17 mm long, apex obtuse to acute, basally inequilateral and truncate, marginally serrulate to subentire, sometimes with a reddish splotch above, sparsely villous, often glabrate above; petioles 1–1.5 mm long; stipules 2- or 3-parted, 1–1.5 mm long. Cyathia solitary at upper nodes but mostly on short congested lateral branches; involucre ca 1 mm long; glands 4, 2 of them 0.4–0.6 mm long, other ones 0.2–0.3 mm long; appendages white or reddish, 0.2–1.4 mm long; staminate flowers 2–5 per cyathium; styles 0.3–0.4 mm long, bifid 1/4 length, divisions clavate. Capsule 1.4 mm long, strigose; seeds oblong-quadrangular, apically acute, basally truncate, 1–1.2 mm long, facets with irregular low transverse ridges often extending into angles, coat white to pale brownish. Jun–Oct. Frequent in disturbed soils of prairies, fields, roadsides, & waste places; GP but infrequent to absent n & w; (e 1/2 U.S., w to ND, s to TX, introduced CA to OR). *Chamaesyce maculata* (L.) Small — Rydberg; *E. supina* Raf.—Steyermark.

17. *Euphorbia marginata* Pursh, snow-on-the-mountain. Annual herb; main stem to 1 m tall, usually unbranched below rays of terminal inflorescence, villous to glabrate. Leaves alternate, sessile, whorled at base of umbel; oblong to ovate or elliptical, 3–10 cm long, glabrous to pubescent, apex acute, base rounded and sometimes amplexicaul, marginally entire; stipules absent or minute. Inflorescence umbellike, 3- to 4(5)-rayed, each ray again branched several times; branches glabrate to woolly; floral leaves and bracts with conspicuous broad white to pinkish margins. Cyathia solitary in branches of inflorescence; involucre 4 mm long, green, villous; glands (3)4(5), 1 mm long; appendages white, 2–4 mm long; staminate flowers 35–60 per cyathium; styles 1–2.5 mm long, bifid 2/3 length. Capsule 4–6 mm long, pubescent; seed 3–4 mm long, ovoid to globose, light to dark gray, surface with tubercles, these often in a reticulate pattern. Jun–Oct. Infrequent to locally abundant with preference for calcareous soils of prairies, roadsides, pastures, & waste places; GP but infrequent n & se; (MN to MT, s to TX, NM & escaped from cultivation elsewhere). *Lepadena marginata* (Pursh) Nieuw.—Rydberg.

This species is often cultivated for its showy bracts and has frequently escaped. It seems likely that ND records were escapes, and the plant has become more common in e GP since times of settlement. Contact with the plant causes a dermatitis in some persons.

18. *Euphorbia missurica* Raf., Missouri spurge, prairie spurge. Annual glabrous herb; stems to 1 m tall, ascending to erect, often arcuate-ascending, much branched; leaves opposite; blades at midstem linear to oblong, 10–30 mm long, 1–5 mm wide, rounded to truncate or emarginate at apex, sometimes acute and mucronate, base narrowed and often slightly asymmetrical, marginally entire, often somewhat revolute; petioles 1–3 mm long; stipules distinct to partly united, apex dissected into 1–5 linear lobes. Cyathia solitary in upper forks or appearing cymose by shortening of upper nodes; involucre 1.5–2 mm long; glands 4; appendages white or pinkish, 0.5–2.5 mm long; sometimes emarginate; staminate flowers 24–53 per cyathium; styles 0.2–1.5 mm long, bifid 1/2–3/4 length, the divisions widely spreading. Capsule 2–2.5 mm long; seeds ovoid to ovoid-triangular, 1.5–2.2 mm long, white or brownish, smooth to slightly granular roughened to rugose with low wrinkles. Jun–Nov. Infrequent to locally common in a variety of habitats ranging from rocky open woodlands, flood plains, dry rocky & sandy prairies, roadsides, fields, waste places; GP but infrequent in n; (MN to MT, s to TX, NM). *Chamaesyce nuttallii* (Engelm.) Small, *C. petaloidea* (Engelm.) Small—Rydberg; *E. missurica* var. *intermedia* (Engelm.) Wheeler—Fernald; *E. missurica* var. *calcicola* (Shinners) Waterfall—Waterfall, Rhodora 54: 127. 1952.

19. *Euphorbia nutans* Lag., eyebane. Annual herb; stems 3–8(10) dm tall, main stem erect, simple, soon producing long erect to ascending pseudodichotomous branches, upper nodes often crisply white-pubescent on one or two sides. Leaves opposite; blades oblong-lanceolate to oblong, often curved or falcate, 8–40 mm long, often with red splotch above, rounded or acute at apex, inequilateral and rounded or somewhat truncate at base, marginally serrate, usually glabrous above, mostly pilose beneath at least near base; petioles of main stem and larger branches 1–2 mm long, or essentially absent; stipules triangular to subulate, to 1.5 mm long. Cyathia solitary in the forks or in cymose clusters; involucres 0.5–1 mm long; glands, 4, 0.3–0.5 mm in diam; appendages very small to longer than width of gland, entire or irregularly lobed; staminate flowers 5–11(28) per cyathium; styles 0.6–1 mm long, bifid 1/2 length. Capsule 1.6–2.3 mm long; seed ovoid, 1–1.6 mm long, finely and irregularly wrinkled, white to dark brown. May–Oct. Found as a weedy species in a wide variety of habitats with a preference for moist soil; e 1/2 GP from se SD s, scattered elsewhere; (NH to ND, s to FL, TX, NM). *Chamaesyce hyssopifolia* (L.) Small—Rydberg; *E. preslii* Guss.—Gleason & Cronquist; *E. maculata* L.—Steyermark.

The leaves of this species have the habit of folding late in the day. Reports indicate that this species can poison livestock, especially when included in hay, and that it causes more cases of poisoning than other species in the genus.

20. *Euphorbia prostrata* Ait. Annual pubescent herbs; stems prostrate, 4–40 cm long, crisply short-villous to glabrate below. Leaves opposite; blades oblong to ovate-oblong, 3–15 mm long, rounded at apex, inequilateral and rounded at base, serrate especially toward apex, short-villous below, glabrate above; petioles to 1 mm long; stipules 0.3–1 mm long, dorsal ones usually distinct, ventral ones usually united below, narrowly deltoid, subulate. Cyathia solitary at upper nodes but mostly on short lateral branches; involucre 0.4–0.7 mm long, sparsely villous; glands 4, nearly round to 0.3 mm long; appendages white or pinkish, usually shorter than gland is wide; staminate flowers (2)4(5) per cyathium; styles 0.1 mm long, bifid to the base. Capsules 1–1.5 mm long, strigose or crisply villous on angles, less so on sides or these glabrous; seed oblong, quadrangular, 1 mm long, sharply angled, the facets with 5–7 low, sharp transverse ridges, pale pinkish-brown, with a white coat. Jun–Oct. Infrequent to common on disturbed soils in prairies, pastures, fields, alluvial deposits, roadsides, & waste places; s 1/2 GP but infrequent to absent in w; (s U.S., n to MO, SD, introduced elsewhere). *E. chamaesyce* L.—Gleason & Cronquist.

Through error this species has been known as *E. chamaesyce* L., which is a species of the Old World.

21. Euphorbia x pseudovirgata (Schur) Soo, hybrid leafy spurge. Similar to *E. esula* but cauline leaves widest at or below the middle, 3–5 mm wide, tapering towards the apex and narrowing towards the base. May–Jul. Fields, pastures, roadsides, creek valleys, waste places; MN, ND, MT, s to KS, CO; (MA, NJ, MT, s to MO, KS; Europe). *Naturalized*. *E. podperae* Croizat—Richardson—op. cit.

This presumed hybrid between *E. esula* and *E. virgata* Wadlst. & Kit. is the more common leafy spurge in the GP, although *E. virgata* has not been found in North America. It appears to be an unusually stable hybrid.

22. Euphorbia revoluta Engelm. Annual herb; stem erect, 3–20 cm tall, simple below, divergent branching above, ultimate branches dark purple or black. Leaves opposite, blades narrowly linear, 1–2.5 cm long, acute at apex and base, marginally entire, revolute; petioles to 1.5 mm long; stipules distinct, to 0.8 mm long, entire, linear subulate; cyathia solitary in forks or appearing as a cymose-cluster by shortening of upper internodes; involucre 0.6–1 mm long; glands 4; appendages white to dark purple, about as long as gland; staminate flowers 3–8 per cyathium; styles 0.5 mm long, nearly entire. Capsule 1.3–1.8 mm long; seed ovoid, sharply angled 1.2–1.5 mm long, with usually 3 prominent transverse ridges, brownish, covered with a thick white coat. Jul–Sep. Dry rocky prairies & breaks; CO: Baca; (CO s to Chih., w to AZ).

23. Euphorbia robusta (Engelm.) Small. Perennial herb with a deep taproot and branched caudex; stems erect, 1–3 dm tall, glabrous or rarely puberulent, light green. Leaves thick and somewhat fleshy, marginally entire, lower ones alternate, nearly sessile, ovate to oblong, 1–2 cm long, acute or abruptly pointed at apex; floral leaves 1–2 cm long, cordate-ovate, opposite or whorled; stipules absent. Inflorescence umbelliform, 3- to 5-rayed, each ray may be dichotomously branched several times; involucres to 3 mm long, turbinate; glands 4, semilunate, horns short; appendages absent; staminate flowers 10–15 per cyathium; styles 0.5–1 mm long, bifid 1/2 length. Capsule 4–4.5 mm long; seed 2–3 mm long, ovoid, gray or brown, smooth or shallowly reticulate-pitted. Jun–Sep. Dry prairies, rocky hills, & ridges; MN, sw SD, WY, w NE, CO; (MT, SD, MN s to NM, AZ). *Galarrhoeus robustus* (Engelm) Rydb.—Rydberg.

24. Euphorbia serpens H.B.K., round-leaved spurge. Annual glabrous herbs; stems prostrate, 5–40(50) cm long, often rooting at nodes, much branched. Leaves opposite; blades ovate-orbicular, oblong to orbicular, 2–8 mm long, rounded at apex, inequilateral and rounded at base, entire; petioles to 1 mm long; stipules on both sides of stem united into a white or pinkish membranaceous scale to 1.3 mm long, entire or lacerate; cyathia solitary at nodes and in forks; involucres 0.7 mm long; glands 4, 1 mm long; appendages white to pink, ca as long as glands are wide, rarely absent; staminate flowers 3–8(12) per cyathium; styles 0.5 mm long, bifid 1/2 length or only notched. Capsules 1–1.5 mm long; seed narrowly oblong, quadrangular, 0.9–1 mm long, facets essentially smooth, brownish with white coat. Jul–Oct. Common in a wide variety of habitats with a preference for calcareous soils; GP but less frequent w & n; (Ont. to TN & FL, w to MT, AZ, widespread in trop. Amer.). *Chamaesyce serpens* (H.B.K.) Small—Rydberg.

25. Euphorbia serpyllifolia Pers., thyme-leaved spurge. Annual glabrous herb; stems prostrate to ascending, 5–30 cm long, much branched, some of upper nodes winged or flattened. Leaves opposite; blades highly variable, commonly oblong or oblong-ovate or obovate-oblong, sometimes appearing falcate, 3–15 mm long, commonly with reddish splotch on upper surface, rounded at apex, marginally serrate at least at apex and on longer side to near base; petioles to 1 mm long, usually pinkish; stipules distinct, to 2 mm long, entire to dissected into 2 or 3 subulate divisions. Cyathia solitary at nodes and upper forks or appearing cymose by shortening of upper internodes; involucre (0.5)0.8–1.2 mm long; glands 4; appendages as long as glands are wide or shorter; staminate flowers 5–12(18) per cyathium; styles 0.2–0.7 mm long, notched or bifid 1/3 length. Capsules 1.5–2 mm long; seeds 1–1.6 mm long, narrowly oblong-quadrangular, facets smooth or with a few indistinct transverse wrinkles, white, light tan, pinkish-white, or brown. Jun–Sep. Dry prairies, open wooded areas, alluvial deposits, fields, roadsides; with a preference for disturbed soils; n 1/2 GP; (s B.C., Alta., e to NM & w TX s to Mex.). *Chamaesyce serpyllifolia* (Pers.) Small, *C. aeguata* Lunell, and *C. albicaulis* Rydb.—Rydberg; *C. erecta* Lunell—Lunell, Amer. Midl. Naturalist 1: 204. 1910.

26. Euphorbia spathulata Lam. Annual glabrous herbs; stems 5–50 cm tall, erect, main stems 1–3(4–10) from base, branches several above, alternate. Leaves on stems and branches alternate, obovate-oblong, oblong-spatulate, or oblanceolate, 1–4.5 cm long, sessile, obtuse at apex, tapering to base, serrulate at least on upper 1/2; stipules absent; leaves of the umbel shorter and somewhat wider, oblong to ovate or deltoid ovate. Rays of primary umbel usually 3, repeatedly dichotomous in well-developed plants. Cyathia solitary in forks of inflorescence; involucre ca 1 mm long; glands 4(5), minute; appendages and horns absent; staminate flowers 5–8(10) per cyathium; styles 1 mm long, bifid 1/2 length. Capsule 2–3 mm long, smooth except for many warts near apex and along lobes, these rarely absent; seeds roundish, 1.3–2 mm long, with low irregular reticulate ridges, brown. May–Jun. Infrequent to common in rocky, open woods, prairies, roadsides, waste places; GP; (MN to WA, s to AL, TX, & Mex.) *Galarrhoeus arkansanus* (Engelm. & Gray) Small, *G. missouriensis* (Nort.) Rydb., *G. obtusatus* (Pursh) Small—Rydberg; *E. dictyosperma* Fisch. & Mey.—Gleason & Cronquist.

>This is a very variable species in our area, but attempts to recognize subspecific taxa have failed. A report of *E. commutata* Engelm. from Geary Co., KS, by Gates was based on a specimen of *E. spathulata*.

27. Euphorbia stictospora Engelm., mat spurge. Annual pubescent herbs; stems prostrate, 3–30 cm long, sparse to profusely branched, crisply villous. Leaves opposite; blades oblong, somewhat orbicular or oblong-obovate, 3–10 mm long, rounded or obtuse at apex, inequilateral, truncate or rounded at base, serrulate especially toward apex, crisply villous below to glabrate above; petioles to 1.5 mm long; stipules variable, often united on lower side and distinct above, 0.5–1 mm long. Cyathia solitary at nodes but usually on short congested leafy lateral branches; involucre 0.7–1 mm long; glands 5; appendages yellowish-white to reddish, 1–2 × as long as gland width; staminate flowers 3–9 per cyathium; styles 0.2–0.5 mm long, entire or notched. Capsule 1.4–2.3 mm long, strigose or spreading pubescence, more dense on angles; seed oblong-quadrangular, 1–1.5 mm long, sharply angled, with shallow irregular pits or irregular low ridges, pale brown with thin white coat, sometimes appearing mottled. Jun–Oct. Infrequent to common in prairies, open woodlands, fields, roadsides, & waste places, with preference for calcareous soils; GP but becoming infrequent to absent n & e; (ND, SD, WY, s TX, AZ, & Mex.). *Chamaesyce stictospora* (Engelm.) Small—Rydberg.

28. Euphorbia strictior Holz. Perennial herb from deep roots; stems erect, 2–9 dm tall, often branched near base, branches rigid, straight, alternate below, opposite or 2–3 ×

dichotomous above. Lower leaves alternate, 2.5-6 cm long, to 5 mm wide, entire, sessile or subsessile; stipules absent or obsolete and glandlike. Cyathia solitary in upper axils and forks, peduncles 3-10 mm long, erect; involucre ca 3 mm long; glands (4)5, ca 1 mm long, 1.3-1.8 mm wide; appendages yellowish-white, merely a narrow scalloped margin 0.2 mm long and width of gland; staminate flowers 10-15 per cyathium; styles to 1 mm, bifid 1/3 length. Capsule 4-6 mm long, minutely strigose to glabrate, smooth; seed 3-4 mm long, ovoid, very obscurely pitted, gray. Jul-Sep. Infrequent to locally common in dry rocky or sandy prairies & roadsides; TX panhandle & e NM; (perhaps *endemic* in the area).

29. Euphorbia uralensis Fisch. ex Link, narrow-leaf leafy spurge. Similar to *E. x pseudovirgata* but stems usually branched from the base with numerous sterile branches; cauline leaves narrowly linear, 0.5-1.5(3) mm wide, sharply acute. May-Jul. Fields, pastures, roadsides, waste places; rare in NE, KS, WY, CO, (GP & ID; Russia & Asia). *Introduced.*

This plant is known from very few collections in the GP, and careful observations of our leafy spurges are necessary to determine its status.

7. PHYLLANTHUS L., the Leaf-flower

Ours monoecious (rarely dioecious in 1 sp.) annual or perennial herbs; branches persistent or deciduous; leaves entire; petioles ca 1 mm or less long; stipules deciduous or persistent; flowers axillary, solitary or in cymules, apetalous, sepals united; staminate flowers with 2 or 3 stamens; pistillate flowers pedicellate or subsessile, calyx lobes 5 or 6, carpels 3, ovules 2 per locule, styles distinct or united. Fruits capsular, elastically dehiscent; seeds usually 2 per locule.

Reference: Webster, Grady L. 1970. A revision of *Phyllanthus* (Euphorbiaceae) in the continental United States. Brittonia 22(1): 44-76.

```
1  Plants annual; leaves distichous on branches.
   2  Leaves of main stem reduced to scales; seeds longitudinally striate .......... 1. P. Abnormis
   2  Main stem leafy; seeds verruculose ............................................. 2. P. caroliniensis
1  Plants perennial; all leaves spirally arranged ..................... 3. P. polygonoides
```

1. Phyllanthus abnormis Baill. Annual erect monoecious herbs, 1-5 dm tall; main stems often somewhat thickened below and appearing perennial, with spirally arranged scale leaves above, 0.5-1.5 mm long; branches deciduous, 3-6(12) cm long, glabrous or hispidulous, with 15-30 distichous leaves; leaves elliptic to oblong, obtuse or emarginate at apex, cuneate to subcordate at base, usually smooth, 3-10 mm long; stipules ovate to linear-lanceolate, acuminate, 0.6-1.5 mm long. Flowers only on branchlets; first cymules with a pair of staminate flowers, upper cymules with 1 staminate and 1 pistillate flower; staminate flowers with pedicels 0.7-1.5 mm long, calyx lobes 4(5-6), stamens 2(3); pistillate flowers with pedicels 1-3(4) mm long; calyx lobes 5 or 6, often unequal, 0.7-1.2 mm long; styles spreading, bifid, 0.2 mm long. Capsules 2.3-2.7 mm in diam; seeds brownish, 1-1.5 mm long, finely longitudinally ribbed. May-Oct. Sandy prairies, ravines, roadsides; sw OK, TX; (OK, TX, FL, Mex.)

2. Phyllanthus carolinienses Walt. Ours annual monoecious herbs; stems 1-4 dm tall, usually with several lateral branches, glabrous. Leaves distichous; blades 6-25(3) mm long, elliptic or oblong to obovate, obtuse or rounded and somewhat apiculate at apex, acute at base, glabrous or rarely scabridulous; petioles ca 1 mm long; stipules acute to acuminate, 0.7-2 mm long. Cymules axillary, with 1 male and (1)2 or 3(4-5) female flowers; staminate flowers with pedicels 0.5-1 mm long, calyx lobes usually 6, 0.5-0.7 mm long, stamens 3; pistillate flowers with reflexed pedicels ca 1 mm long, calyx lobes (5)6(7), linear-lanceolate or narrowly spatulate, 0.8-1.4 mm long; styles to 0.3 mm long, bifid. Capsule 1.5-2 mm wide, smooth; seeds gray or brown, 0.8–1.1 mm long, verruculose. Jun–Oct. Moist alluvial ground, gravel bars, low fields, moist depressions; extreme se GP; (PA, IL, e KS, s to FL, TX).

3. **Phyllanthus polygonoides** Nutt. Perennial glabrous herb., somewhat suffruticose, 1–4 dm tall, usually monoecious but unisexual plants not uncommon; leaves spirally arranged, 5–10 mm long, narrowly oblong to obovate, acute or mucronulate at apex, obtuse at base, glaucous below; petioles to 0.7 mm; stipules usually 1–2 mm long, acuminate, auriculate at base. Cymules unisexual or bisexual, with usually 1 female flower and/or several male flowers; staminate flowers with pedicels 1.5–3.5 mm long, calyx lobes 6(5), oblong to obovate, 0.7–1.4 mm long, 3 stamens; pistillate flowers with pedicels 2.5–7.5 mm long, calyx lobes 6(5), ovate to obovate, 1.5–2.6 mm long in fruit, styles 3, 0.2–3 mm long, bifid. Capsules 2.7–3.3 mm in diam; seeds dark brown, 1.1–1.5 mm long, irregularly verruculose. Apr–Aug. Infrequent to locally common on rocky calcareous prairies or sandy soils; s GP; (MO, AR, OK, TX, s into Mex.)

8. REVERCHONIA A. Gray

1. **Reverchonia arenaria** A. Gray. Annual monoecious glabrous herbs, 2–5 dm tall; main stem smooth, glaucous-white; lower branches 2–3 dm long, shorter above; leaf blades elliptic to narrowly oblong-elliptic or almost linear, (15)20–35(45) mm long, apiculate at apex, narrowed to base, marginally entire; petiole 1–3 mm long, stipules lanceolate, persistent, 1(2) mm long. Flowers reddish, only on lateral branches, axillary to leaves, each cymule usually with 1 central pistillate flower and 4–6 lateral staminate flowers; pedicels of staminate flower 1.5–2.5 mm long, calyx lobes 4, 1.5–2.5 mm long, stamens 2; pistillate flower with stout pedicel, 1.5–2 mm long becoming 9 mm in fruit; calyx lobes 6(5), styles erect 0.5–0.8 mm long, united ca 1/2 length. Capsule 7–9 mm in diam, smooth, yellowish; seeds trigonous, dark or reddish-brown, smooth on the back, papillate on lateral facets, (4.5)4.7–6.2(6.6) mm long. (2n = 16) May–Sep. Sand dunes, sandy prairies, & sandy stream banks; KS: Clark; w OK, TX; (OK, KS, NM, UT, AR, TX, Chih.).

Reference: Webster, Grady L., & Kim I. Miller. The genus *Reverchonia* (Euphorbiaceae). 1963. Rhodora 65: 193–207.

9. STILLINGIA L.

1. **Stillingia sylvatica** L., queen's delight. Perennial glabrous monoecious herbs with milky sap, from stout woody caudex; stems 3–6(8) dm tall; leaves alternate, ascending, highly variable, blades narrowly elliptic to lanceolate or oblanceolate, 3.5–7(12) cm long, acute at both ends, serrulate or crenulate with a small deciduous gland in each notch or less often on margins; petioles 1–7 mm long; stipules glandlike. Flowers in terminal androgynous spiciform thyrses; staminate flowers in many-flowered cymules; each with a cuplike calyx, stamens 2; pistillate flowers with a deeply 3-lobed calyx, lobes 0.7–2 mm long; petals absent; styles 3, 4–5 mm long, simple. Capsule ca 12 mm long, 3-celled; 3-seeded; seeds ovate-oblong, 8 mm long not including caruncle, ca 6 mm wide, whitish, smooth. May–Aug. Sand dunes, sandy prairies, stream banks, roadsides, & waste places; s 1/4 GP; (VA to FL, w to LA, TX, n to KS, NM). *S. salicifolia* (Torr.) Raf.—Rydberg.

10. TRAGIA L., Noseburn

Ours erect, decumbent, or trailing monoecious herbs from woody crown of taproot; stems usually several, alternately branched; pubescence of stinging hairs and soft spreading hairs. Leaves alternate, petiolate, stipules lanceolate, persistent. Inflorescences appearing as though opposite leaves at upper nodes; racemes androgynous; flowers apetalous, calyx lobes 3–6, disk absent. Capsules explosively dehiscent; seeds 1 per locule.

Reference: Miller, Kim I., & Grady L. Webster. 1967. A preliminary revision of *Tragia* (Euphorbiaceae) in the United States. Rhodora 69: 241–305.

1 Calyx lobes of pistillate flowers longer than pistil at anthesis; staminate flowers (14)20–75 per raceme, usually compactly arranged; well-developed midstem leaves ovate to sublanceolate, usually distinctly cordate, with petioles (5)10–25 mm long .... 1. *T. betonicifolia*
1 Calyx lobes of pistillate flowers shorter than pistil at anthesis; staminate flowers (2)5–20(30) per raceme, usually not compactly arranged; well-developed midstem leaves narrowly oblong to sublinear, usually truncate or obtuse at base, with petioles 1–5(10) mm long .... 2. *T. ramosa*

1. **Tragia betonicifolia** Nutt., noseburn. Perennial herbs with stinging hairs; stems 1–4(6) dm tall, erect or decumbent to slightly twining, solitary or several from woody caudex. Leaves alternate; blades ovate to ovate-lanceolate or triangular-lanceolate, 10–40 mm long, apically acute, basally cordate to truncate, serrate, pubescent, ciliate; petioles (5)10–25 mm long; stipules lanceolate to ovate, 1.2–4.5 mm long. Racemes androgynous, opposite leaves at upper nodes, lowermost 1 or 2 nodes pistillate, the upper 14–75 staminate, bracts of pistillate flowers 1.5–2 mm long, those of staminate flowers 1–2 mm long; staminate flowers with pedicels 0.7–1 mm long, the lower persistent part 0.3–0.6 mm long, calyx lobes 3–4(5), 1–2.4 mm long, stamens 3 or 4(5); pistillate flowers with pedicels 0.7–1 mm long, 3–4 mm long in fruit, calyx lobes 6, 2–3 mm long, 3–5 mm long in fruit, styles united only at base, stigmatic surfaces papillate. Capsules 4–5 mm long, 7–9 mm wide; seeds 3–4 mm long, nearly spherical, brownish-black or yellowish, smooth. May–Sep. Infrequent to common in various soils of rocky prairies, open woodlands, roadsides, & waste places; se GP; (MO, e 1/2 KS, s through OK, AR, TX). *T. nepetaefolia* Cav.—Rydberg; *T. urticifolia* Michx.—Fernald.

This and the following species are often difficult to distinguish from each other, especially in the absence of good flowers and fruit. The two species are sympatric in much of our area and often can be found together on the same prairie hillside. As yet suspected hybrids have not been found, but the two species merit more critical study.

2. **Tragia ramosa** Torr., noseburn. Perennial herbs with stinging hairs; stems 1–3(5) dm tall, erect, decumbent to rarely slightly twining, solitary or several from woody caudex. Leaves alternate; blades narrowly oblong to sublinear, rarely somewhat ovate, usually truncate or obtuse at base, serrate, pubescent, ciliate; petioles 1–5(10) mm long; stipules lanceolate to ovate, 1–4.5 mm long. Racemes androgynous with lower most 1 or 2 nodes pistillate, the remaining nodes staminate; bracts of pistillate flowers lanceolate, 1.5–2 mm long; bracts of staminate flowers lanceolate, 1.2–1.8 mm long; staminate flowers with pedicels 0.7–2 mm long, persistent part 0.4–1.5 mm long, calyx lobes 3 or 4(5), oblanceolate, 1–2 mm long, stamens (2)3–6(7–10); pistillate flowers with pedicels 1–1.5 mm long, to 2.5 mm in fruit, calyx lobes (5)6(7), 1–2.5 mm long at anthesis, 1.5–3 mm long in fruit; styles united to 1/3 length, stigmatic surfaces not papillate. Capsule 3–4 mm long, 6–8 mm wide; seeds nearly spherical, 2.5–3.5 mm in diam, yellowish to brownish-black. (2n = 44) May–Sep. Infrequent to common in various soils of prairies, open woodlands, roadsides, & waste places; NE: Franklin; KS, CO, MO, OK, TX, NM; (MO, AR, w to NE, CO, NM, CA, s into Mex.)

Tragia amblyodonta (Muell. Arg.) Pax & K. Hoffm. is very similar to *T. ramosa*, but the plant is copiously to densely pubescent, especially on young stems. It is reported from ne NM, but otherwise is south of our area.

89. RHAMNACEAE Juss., the Buckthorn Family

by Ronald L. McGregor

Perfect, polygamous, or sometimes dioecious shrubs or small trees, unarmed or end of twigs spinescent. Leaves simple, alternate or opposite, stipules usually present, minute, early deciduous. Inflorescence of axillary or supra-axillary cymes, or dense panicles or corymbs at summit of flowering branches or peduncles, or reduced to solitary flowers. Flowers small, inconspicuous (except in *Ceanothus*); calyx 4- to 5-lobed, usually forming a short basal tube or hypanthium; petals 5(4) or wanting, usually clawed and cucullate; stamens alternate with sepals and equal in number, borne at margin of perigynous disk, often enfolded by petals; ovary 1, on or immersed in disk, 2-4 carpellary. Fruit a capsule or drupe.

1 Fruit capsular, essentially dry at maturity; leaves 3-nerved from near the base .... 1. *Ceanothus*
1 Fruit a 1- to 4-stoned indehiscent drupe; leaves not 3-nerved from bases, if so then branches covered with a dense grayish or whitish waxlike bloom.
   2 Drupes with 2-4 stones, without a bloom on branches .... 2. *Rhamnus*
   2 Drupes with a solitary stone; waxlike bloom present on branches .... 3. *Ziziphus*

1. CEANOTHUS L.

Deciduous or evergreen shrubs. Leaves simple, alternate, 3-nerved. Inflorescence of axillary to terminal umbels or panicles. Calyx 5-lobed, incurved, the lower portion adnate to the thick disk and ovary, lobes deciduous; petals 5, long clawed, spreading, hooded; stamens 5, free, exert, opposite the petals, separated from pistil by the flat, lobed disk; ovary 3-celled, style 3-lobed. Fruit a capsule, subtended by persistent hypanthium, eventually loculicidally dehiscent into 3 1-seeded, dehiscent carpels.

1 Leaves evergreen; glutinous and shining above .... 4. *C. velutinous*
1 Leaves deciduous; usually neither glutinous nor shining.
   2 Mature leaves 1-2.5 cm long; margin entire or remotely glandular-serrate .... 2. *C. fendleri*
   2 Mature leaves (2)3-10 cm long; margin obviously serrate.
      3 Leaves mostly ovate or ovate-oblong; inflorescences terminating axillary peduncles .... 1. *C. americanus* var. *pitcheri*
      3 Leaves elliptic or elliptic-lanceolate; inflorescences terminating leafy shoots of the year .... 3. *C. herbaceous* var. *pubescens*

1. Ceanothus americanus L. var. *pitcheri* T. & G., New Jersey tea. Erect shrub to 1 m tall, often branched above. Leaves usually broadly oblong-ovate, 5-10 cm long, 2.5-6 cm wide, broadly cuneate to rounded or subcordate at base, acute or blunt at tip, irregularly serrate, densely soft-pubescent below. Peduncles axillary on new growth, 4-10 cm long, progressively becoming shorter upward; panicle elongate, with scattered corymbs. Flowers perfect; calyx lobes 5, incurved, about 1.3 mm long, white; petals 5, about 1.6 mm long, claw 1 mm long; stamens 5; central disk 10-lobed, surrounding ovary. Capsules 4-5 mm wide, 3-lobed, each lobe crested, black, hypanthium persistent; seed dark red-brown, surface smooth, glossy. (n = 12) May-Jul. Rocky prairie hillsides & ravines, open woodlands, roadsides; se 1/4 GP; (Que. to MN, s to FL, TX).

The dried leaves of this species have been used as a tea substitute, and the roots were once used for their astringent properties.

2. Ceanothus fendleri A. Gray. Low shrub or dense bush 3-8 dm tall, branches usually ending in spines, canescent, pruinose, gray-green. Leaves alternate, simple, 2-2.5 cm long, elliptic or oblong, usually nearly glabrous above, densely gray-tomentulose or silky-canescent below, entire or with remote glandular serrations. Inflorescence of umbellate clusters terminating stem or branches. Calyx lobes 5, white, about 1.5 mm long, incurved; petals 5, white, 1.5 mm long, claw half the length; stamens 5; central disk about 2 mm wide, often 6-lobed. Fruit 4–5 mm wide, 3-lobed; seed brown, 2.5–3 mm long. Jun–Aug. Rocky open hillsides & ledges; SD: Custer, Lawrence; CO: Las Animas; TX: Childress, Cottle, Foard, Hardeman, Motley; (SD, CO, s to Chih. & Coah.).

3. *Ceanothus herbaceous* Raf. var. *pubescens* (T. & G.) Shinners, New Jersey tea. Bushy shrub to 1 m tall. Leaves oblong to elliptic, or lance-oblong to oblanceolate, 2–6 cm long, blunt to subacute, crenate-serrate with young teeth glandular, veins arising unevenly just above leaf base, upper surface glabrescent, lower surface villous. Panicles several, terminating leafy branches of the season, on peduncles 1–2(5) cm long. Flowers white; calyx lobes 5, 1.6 mm long, incurved; petals 5, hooded, 1.5 mm long, clawed 1/2 length; stamens 5; ovary 3-lobed; surrounded by disk 1.5 mm wide and usually 10-lobed. Capsules 3–4.5 mm wide, 3-lobed, without crests; seeds about 2 mm long, surface smooth, glossy, brownish. Apr–Aug. Rocky, open wooded hillsides, prairies, roadsides; GP but absent in ne 1/3 & w-cen; (s Can & e 1/2 U.S. to GP). *C. ovatus* Desf., *C. pubescens* (T. & G.) Rydb — Rydberg.

4. *Ceanothus velutinous* Dougl., mountain balm. Spreading evergreen shrub 0.5–2.5 m tall, unarmed, usually in dense colonies. Leaves alternate, stipules 1 mm long; blades ovate to ovate-elliptic, 4–8 cm long, glabrous, glutinous-varnished and shining on upper surface, pale and velutinous or canescent below, closely and finely glandular-serrate. Inflorescence paniculate, dense. Flowers white; calyx lobes 5, 1.5–2 mm long, incurved; petals 5, hooded, 2–2.5 mm long, recurved, narrowed abruptly to claw; stamens 5, inserted below disk, curved upward, disk usually 10-lobed. Capsules 5–6 mm wide, 3-lobed, slightly crested above middle; seeds about 2.5 mm long, dark brown, smooth and glossy. Jun–Jul. Dry, open, rocky wooded hillsides; rare to locally common in BH, WY, CO; (SD, CO w to CA, B.C.).

2. RHAMNUS L., Buckthorn

Dioecious or polygamo-dioecious shrubs or small trees; unarmed or with some branchlets ending in thorns. Leaves simple, alternate or opposite, entire or serrate; stipules small and deciduous. Calyx 4- or 5-cleft, tube campanulate, lined or filled with a disk; petals 4, 5, or none, small, short clawed, notched at the end, with the sides folded or wrapped around the stamens; stamens 4 or 5, inserted on the disk, opposite the petals; ovary 2- to 4-locular, partly immersed in the disk. Fruit a berrylike drupe, containing 2–4 1-seeded stones.

1 Leaves opposite; some branches ending in short thorns; introduced.
  2 Leaves usually 1.5× longer than wide; lateral veins commonly 3 on each side .... 2. *R. cathartica*
  2 Leaves usually 2 to 3× longer than wide; lateral veins commonly 4 or more on each side .... 3. *R. davurica*
1 Leaves alternate; branches unarmed; native.
  3 Sepals 4; petals 4; fruits with 2 stones; leaves finely serrulate .... 4. *R. lanceolata* var. *glabratus*
  3 Sepals 5; petals absent; fruits with 3 stones; leaves crenate-serrate .... 1. *R. alnifolia*

1. *Rhamnus alnifolia* L'Her., alder buckthorn. Polygamodioecious low shrub, sometimes trailing, 1.5–8 dm tall, branches few, upright, not spinescent at tip. Leaves oblong-ovate to oblong-elliptic, blades 6–10 cm long, closely crenate-serrate, main lateral veins usually 5–7; stipules oblong, 3–6 mm long, early deciduous. Inflorescence of sessile axillary umbels,

1- to 5-flowered. Flowers functionally imperfect, vestiges of opposite sex present; hypanthium shallowly campanulate; calyx lobes (4)5, 0.5–1 mm long, slightly folded; petals 0; stamens 5; ovary about 1/2 enclosed by the greenish disk. Drupe 6–9 mm in diam, bluish-black, 3-seeded; stone 3.5–4.5 mm long, abruptly pointed at base, inner face with a prominent narrow ridge, outer surface with a broader rounded ridge, brown and roughened. May–Jun. Moist woodlands, edge of swamps, moist brushy ravines; rare in ne ND & BH; (ME, Que to B.C., s to WY, MT, ND, CA, CA).

2. **Rhamnus cathartica** L., common buckthorn. Polygamo-dioecious small tree or much-branched shrub, 1–5 m tall, with spine-tipped branches. Leaves simple, opposite, broadly elliptic, oblong, or elliptic-oblong, 3–6 cm long, 2.5–5.5 cm wide, lateral veins usually 3 per side, upcurved, glabrous, irregularly serrate. Staminate flowers in clusters of 2–8 on spur branch of season, calyx lobes 4, 2.5–3 mm long, petals 4, 1–1.3 mm long, stamens 4, attached below rim of campanulate hypanthium, pistil vestigial; pistillate flowers 2–15 on spur branch of the season, hypanthium as long as wide, calyx lobes soon deciduous, petals 4, 0.6 mm long, stamens abortive. Drupe black 6–8 mm in diam, commonly with 4 stones; stones 4–5 mm long, pointed at base, inner face with a prominent ridge, outer with a conspicuous sulcus, surface greenish to black. May–Jun. Commonly planted & occasionally escaped to various habitats; ND, e SD, se WY, & NE; (throughout e U.S. to GP; Europe).

This species has been planted for hedges, but it is an alternate host for oat rust and after escaping it sometimes becomes somewhat noxious. It was once the source of a dye and also had limited use in medicine.

3. **Rhamnus davurica** Pall. Very similar to *R. cathartica* but leaves are more elliptic, more acute or acuminate at base, and commonly with 4 or more lateral veins on each side; the ovary more deeply embedded in hypanthium and with only the apex exposed. This somewhat recent introduction has been found as an escape in ND: Burleigh, Oliver, Ransom, Richland; SD: Minnehaha; NE: Dawes; (sporadic in Can. & n U.S.; e Asia).

Some of the broader-leaved plants are the var. *davurica,* while the narrow-leaved plants may be the var. *nipponica* Makino.

4. **Rhamnus lanceolata** Pursh var. **glabratus** Gl., lance-leaved buckthorn. Polygamo-dioecous unarmed shrubs to 3(6) m tall, or a small tree. Leaves simple, alternate, lanceolate to elliptic or oblong-lanceolate, 3–8 cm long at maturity, finely serrulate, glabrous beneath or glabrate, 6–9 lateral veins on each side. Flowers of 2 types on different plants; staminate in axillary clusters of (1)3, calyx campanulate, 4-lobed, lobes about 2 mm long, petals 4, 1.3 mm long, stamens 4, style not exserted and ovary abortive; pistillate flowers 1(3) in leaf axils, calyx lobes about 1.6 mm long, petals about 1 mm long, 4 stamens abortive, pistil exserted. Drupes black, with 2 stones; stones about 4.4 mm long, 2.5 mm in diam, deeply grooved on ventral surface. Apr–May. Rocky, open wooded hillsides, prairie ravines, with preference for calcareous soils; se SD, s to e 1/3 KS; (OH, KY, SD, NE, s to AR, OK).

Records of var. *lanceolata* and of *R. carolinianus* Walt. for the GP have been based on young specimens of var. *glabratus* that have pubescent twigs.

3. ZIZIPHUS P. Mill.

1. **Ziziphus obtusifolia** (T. & G.) A. Gray, lotebush. Shrubs 0.5–2 m tall; branches usually covered with a gray or whitish waxlike bloom, terminating in stout thorn-tips. Leaves alternate, grayish-green, to 3 cm long, from deltoid to ovate or oblong to nearly linear, broader leaves usually serrulate above, some narrow leaves slightly emarginate. Flowers in

axillary clusters, 3 mm wide; calyx campanulate, cup filled with a large disk, sepals 5, triangular; petals 5, 1 mm long, hooded, claw nearly obsolete, early deciduous; stamens 5; ovary 2-celled, 2-ovulate, embedded in but not adherent to disk. Drupe 7–9 mm long, black, containing 1 2-celled stone; stone 6–7 mm long. May–Jul. Rocky prairie hillsides; sw OK, TX, NM; (OK, TX, NM, AZ s to V.C.).

90. VITACEAE Juss., the Grape Family

by Ronald L. McGregor

Usually polygamous woody vines or semishrubs. Leaves simple or compound, alternate, with deciduous stipules; tendrils opposite the leaves. Inflorescences opposite the leaves, paniculate or cymose. Flowers numerous, 4- or 5-merous, yellow-green or green, small, regular, perfect or unisexual, more or less perigynous, and usually with an annular or cupulate disk; calyx minute or almost lacking; petals small, valvate in bud; stamens as many as valvate petals and opposite them, abortive in pistillate flowers; carpels 2, style short or none, stigma slightly 2-lobed; ovary superior, 2-locular. Fruit a 1- to 4-seeded berry.

Reference; Bailey, L. H. 1934. Grapes of North America. Gentes Herb. 3: 151–244.

1 All of the leaves simple.
   2 Petals united at apex, corolla deciduous; fruit bluish or blackish; seeds pyriform, bark of old stems exfoliating in shreds .... 4. *Vitis*
   2 Petals separate, falling singly; fruit green, becoming orange, rose, and finally turquoise-blue; bark of old stems tight, not loosening into shreds .... 1. *Ampelopsis*
1 Some or all of the leaves divided into leaflets.
   3 Flowers 4-merous; disk deeply 4-lobed; leaves divided into 3 leaflets or deeply 3-parted, fleshy or succulent .... 2. *Cissus*
   3 Flowers 5-merous; disk with entire or crenulate margin; leaves divided into (3)5(7) leaflets, not fleshy or succulent .... 3. *Parthenocissus*

1. AMPELOPSIS Michx.

1. *Ampelopsis cordata* Michx., raccoon grape. High climbing vine with close bark; covered by lenticels; pith white; tendrils with slender tips, few or none in flowering branches, nearly glabrous. Leaves petiolate, broadly ovate to suborbicular-ovate, cordate or truncate at base, acuminate at apex, coarsely and sharply toothed, rarely some shallowly 3-lobed. Inflorescence of paniculate cymes, long peduncled. Functional staminate flowers with 5 minute calyx lobes; petals 5, ovate, 2.5 mm long; stamens 5, opposite petals, attached at base of irregularly lobed, cup-shaped disk; pistil vestigial. Functional pistillate flowers with 5 calyx lobes to 0.2 mm long; petals 5, greenish, to 2.8 mm long; vestigial stamens 5; ovary about 8 mm long, half imbedded in cup-shaped disk, style to 1.8 mm long. Berries depressed-globose, 7–10 mm in diam, first green, then orange-pink, becoming turquoise-blue, 1- to 3-seeded. (n=20) May–Jul. Rocky wooded hillsides, stream valleys, fence rows; se NE, e 3/4 KS, OK, TX; (OH, IA, NE, s to SC, FL, TX).

The fruits of this species are frequently confused with grapes, but they are not edible.

2. CISSUS L., Possum Grape

1. *Cissus incisa* (Nutt.) Des Moul., possum grape, marine ivy. Perfect or polygamo-monoecious stout scrambling or climbing vine to 10 m long; young stems usually 6-ridged becoming somewhat quadrate; lenticels prominent; orange-red, becoming warty; from

tuberous roots. Leaves deciduous or semievergreen, alternate, somewhat succulent, extremely variable, from simple and broadly ovate to trilobed or usually trifoliate, to 8 cm long, petiolate; leaflets ovate to obovate, cuneate, margins coarsely and irregularly toothed. Inflorescence an umbelliform cyme, axillary. Flowers perfect or unisexual, regular; calyx campanulate, 4-lobed, greenish; petals 4, spreading, somewht cucullate, greenish; stamens 4, opposite petals; disk deeply 4-lobed, free from ovary except at base. Berry obovoid, 1- to 4-seeded, black, 6–9 mm long, on recurved pedicels; seeds trigonous-obovoid, 5–7 mm long, brownish, 2-grooved at apical end. May–Jul. Rocky wooded hillsides, stream banks, prairie ravines, roadsides, & waste places; KS: Chautauqua, Cherokee; OK, TX; (MO, KS, s to FL, TX, n Mex.).

The plant has a disagreeable, pungent, nitrogenous odor similar to that of jimson-weed. The tuberous roots of the related *C. trifoliata* of Arizona and Mexico are reported to be poisonous and to cause dermatitis in some persons who contact the plant. Thus, *C. incisa* should be treated with caution.

3. PARTHENOCISSUS Planch., Virginia Creeper

Perfect or rarely polygamous scrambling or climbing woody vines with branched tendrils. Leaves palmately compound; leaflets coarsely serrate. Inflorescence cymosely compound. Flowers regular; calyx minute, slightly 5-toothed; petals (4)5, free, reflexed or spreading, 3–5 mm long, concave, thick; stamens (4)5, opposite the petals; disk indistinct; ovary 2-celled; cells 2-ovuled. Fruit a 1- to 4-seeded berry.

1 Tendrils with 3 to 8 branches that terminate in adhesive disks; inflorescence with a central axis; leaves dull on the upper surface .... 1. *P. quinquefolia*
1 Tendrils with 3–5 diskless branches; inflorescence dichotomously forked; leaves glossy on upper surface .... 2. *P. vitacea*

1. *Parthenocissus quinquefolia* (L.) Planch., Virginia creeper. Woody vine high climbing or scrambling by means of 3–8 branched tendrils with adhesive disks. Leaves palmately compound, long-petiolate; leaflets (3)5(7), elliptic or obovate, dull above, glabrous or occasionally pubescent beneath, acuminate, coarsely serrate chiefly above the middle, cuneate to the base, petiolulate. Inflorescences terminal and from upper axils, forming a panicle of cymes, with a well-marked central axis. Flowers 25–200 per inflorescence; calyx flat, usually without lobes; petals 5, separate, yellowish-green, 2–3 mm long; disk indistinct or small and adnate to ovary; stamens 5. Drupes black or dark blue, globose, 5–8 mm in diam; seeds 1–3(4), obovoid, 3.5–4 mm long, surface lustrous brown. (n = 20) May–Jul. Rich & open woods, valleys, rocky bluffs, prairie ravines, thickets, fence rows; se SD, e NE, KS, OK, TX, but becoming rare westward in GP; (ME, OH, SD, s to FL, TX). *Psedera hirsuta* (Donn) Greene, *P. quinquefolia* (L.) Greene—Rydberg.

Some plants with lower surface of the leaflets more or less pubescent have been named forma *hirsuta* (Donn) Fern. The plant is frequently cultivated, and a number of minor horticultural forms have been named. It has been reported that the fresh berries are poisonous to children.

2. *Parthenocissus vitacea* (Knerr) Hitchc., woodbine, thicket creeper. Similar to the preceding species but more loosely climbing or scrambling. Tendrils with 3–5 slender-tipped twining branches without disks. Leaves lustrous green above and pale to thinly pubescent below. Inflorescence without a prolonged central axis, with a pair of nearly equally forking spreading branches which are more or less dichotomously branched; flowers 10–60 per inflorescence. Fruit 10–12 mm in diam; seeds about 5 mm long, surface brown and somewhat rugulose. May–Jul. Open woods, rocky hillsides, thickets, ravines; GP but more common westward; (Que. to Man., MT s to WY, TX). *Psedera vitacea* (Knerr) Greene—Rydberg; *Parthenocissus inserta* (Kerner) Fritsch—Stevens.

Plants with lower surface of leaflets more or less pubescent have been named forma *dubia* Rehd. It has been suggested that our two species of *Parthenocissus* intergrade in some characters and may hybridize. While it may be possible, I have not found evidence to support this view.

4. VITIS L., Grape

Polygamo-dioecious viny shrubs, often climbing by the coiling of slender-tipped tendrils; bark usually loosening and exfoliating in ropy shreds; without lenticels; pith brown, interrupted at the nodes by diaphragms. Leaves simple, mostly rounded and cordate, normally lobed. Inflorescence a compound thyrse, opposite a leaf. Flowers fragrant; calyx short and usually with entire margin or none; petals 5, separating only at base, and falling without expanding; disk hypogynous and consisting of 5 nectariferous glands alternate with the stamens; ovary 2-celled; cells 2-ovuled. Berry pulpy, 2- to 4-seeded; seeds usually pyriform with a beaklike base and 2 grooves on the ventral side.

The identification of *Vitis* species is difficult because of a lack of careful field studies, the generally poor quality of herbarium specimens, and the need for monographic studies. In collecting specimens an effort should be made to obtain material with young growing tips, sterile branches with mature leaves, and fruiting branches. Specimens of flowering branches should be split through the lowest nodes to reveal the diaphragms.

Vitis rotundifolia Michx. was reported by Rydberg for Kansas as *Muscadinia rotundifolia* (Michx.) Small, but the nearest localities are in se MO and se OK.

1 Lower surface of fully grown leaves more or less covered with long and cobwebby hairs, continuous or in patches, lying parallel to the surface, or if somewhat glabrous the lower surface either distinctly whitened or silvery bluish-green.
  2 Lower surface of mature leaves either whitened or silvery bluish-green below, with or without cobwebby pubescence; leaves with shallow acute or rounded and abruptly pointed teeth that are about as wide as long.
    3 Leaves of flowering branches shallowly or deeply lobed, the deeply lobed ones with entire sinuses .... 2. *V. aestivalis*
    3 Leaves of flowering branches unlobed or shallowly lobed with the lobes toothed to the base .... 3. *V. cinerea*
  2 Lower surface of mature leaves usually retaining some cobwebby pubescence especially along the veins; leaves with coarse uneven acute or acuminate teeth that are usually longer than wide .... 1. *V. acerifolia*
1 Lower surface of fully grown leaves green and glabrous, except the veins usually hairy, or rarely with short erect hairs on the surface, not cobwebby.
  4 Tendrils none, or produced only opposite the uppermost leaves; leaves usually reniform, wider than long .... 5. *V. rupestris*
  4 Tendrils normally opposite most leaves or lacking opposite each third leaf; leaves cordate, ovate, or triangular.
    5 At least some leaves usually deeply lobed; diaphragm interrupting the pith at lowest nodes of flowering branched 0.5–2 mm thick; mature berries glaucous ... 4. *V. riparia*
    5 Leaves entire or only shallowly lobed; diaphragm 2–6 mm thick; mature berries not glaucous .... 6. *V. vulpina*

1. ***Vitis acerifolia*** Raf., bush grape. Stocky much-branched, bushy, short-jointed vines, often covering shrubs or rocks, rarely climbing; young branches tomentose, becoming glabrous; diaphragms 1–3 mm thick; tendrils usually rare, short, simple or with short branches; stipules 5–7 mm long, dropping early. Leaves thick-textured to subcoriaceous; blades suborbicular to broadly triangular-ovate, acute to long-acuminate, 6–10 cm long, often wider than long, basal sinus from nearly closed to broad, irregularly and sharply and coarsely toothed, sometimes shallowly lobed, upper surface with cobwebby hairs but becoming glabrate, lower surface retaining cobwebby pubescence at least along the veins; petiole 2–5

cm long, cobwebby, becoming glabrous. Thyrse 3-6 cm long, usually simple, peduncle floccose, becoming glabrous. Mature berries 8-12 mm in diam, black with a heavy bloom; seeds 5-6 mm long, 4.5-5 mm wide, abruptly short beaked, pale brown. May, fruits Jul-early Aug. Locally common along streams, dry hillsides, rocky banks, prairie ravines; w 1/2 KS & s in GP; (KS, CO, s to TX). *V. longii* Prince — Rydberg.

In some colonies the mature fruits are sweet, while in others, often nearby, the fruits will be strongly acrid. If often forms dense colonies and is excellent cover for wildlife.

2. Vitis aestivalis Michx., pigeon grape. Vines climbing to a height of 10 m or more or sprawling over bushes and banks or forming bushy clumps; young branches glabrous or with reddish-brown or white flocculent-deciduous tomentum, rarely with velutinous pilosity; diaphragms dividing pith 1-4 mm thick. Leaves of fertile branches suborbicular-ovate, cordate-ovate, or quadrate-rotund; blade 7-20 cm long, about as wide as long, unlobed or merely shouldered to deeply 3- to 5-lobed, margins irregularly and not deeply sinuate-toothed or coarsely and shallowly toothed, upper surface dull green and essentially glabrous, lower surface with subpersistent but loose flocculent tomentum of cobwebby hairs or densely floccose or rarely glaucous; petioles glabrous or tomentose. Thyrse 5-15 cm long, the axis more or less tomentose or woolly or with cobwebby hairs. Berries 5-14 mm in diam, black or dark purple, with a thin bloom; seeds 5-8 mm long, red-brown. May-Jul, fruits Sep-Oct. Dry, rocky & upland woods, along bluffs, fence rows, ravines; NE: Brown & se; e KS & s in GP; (MA, NY, MI, WI, s to KS, TX). *V. bicolor* Le Conte — Rydberg; *V. lincecumii* Buckl. var. *glauca* Munson, *V. aestivalis* Michx. var. *argentifolia* (Munson) Fern. — Fernald.

Named variants of this species freely grade into one another even on the same hillside. The fruits also vary in taste from delicious to acrid.

3. Vitis cinerea Engelm., graybark grape. Somewhat lax high-climbing vine; growing tips and branches angled, permanently close-pubescent with ashy-white or gray hairs; nodal diaphragm 2.5-4 mm thick; stipules 2-4 mm long. Leaves of fertile branches suborbicular to broadly ovate, with a tapering triangular apex, 1-2 dm long, unlobed or with short shoulders, or rarely with 2 or 4 prolonged lobes, upper surface floccose but becoming glabrate, lower surface canescent-pilose or grayish-floccose with cobwebby hairs, the looser hairs often deciduous; petioles canescent or floccose, shorter than blade. Thyrse rather open, canescent or floccose, 6-15 cm long. Berries 4-9 mm in diam, blackish or purplish, with a slight bloom; seeds 4-5 mm long, red-brown. May-Jul, fruits Sep-Oct. Low woods, along streams, open wooded hillsides, thickets; se NE, e KS, OK; (VA, OH, IL, NE, s to FL, TX).

4. Vitis riparia Michx., river-bank grape. Vine climbing to 25 m; young branches green or dull red-brown, glabrous or pubescent; nodal diaphragm 0.5-2 mm thick. Leaves of fertile branches with glabrous petioles, blade 7-15 cm long, about as wide as long, cordate-ovate, apex acuminate, basal sinus broad, margins with coarse acuminate teeth, one or both sides of some teeth concave, usually conspicuously ciliate, glabrous or glabrate, with usually 2 or more erect and prolonged lobes 1-4.5 cm long; leaves of vegetative branches often more palmate-lobed. Thyrse 4-12 cm long, its axis sparsely and loosely long-puberulent to glabrous or nearly so. Berries crowded, 7-11 mm in diam, purple-black, with a heavy bloom, acrid; seeds 4.5-5.5 mm long, rounded to a short beak, red-brown. May-Jun, fruits Jul-Sep. Low rich woodlands, stream banks, ravine, thickets, fence rows, open hillsides; GP but becoming infrequent n & westward; (Que. to Man., MT, s to TN, MO, NM, TX). *V. vulpina* — Rydberg.

The commonly found plant is the variety *riparia* with petioles glabrous or nearly so and lower surface of leaf mainly glabrous except for some hairs on veins. Rare plants with petioles and lower leaf surface more or less densely hairy have been named var. *syrticola* (Fern. & Wieg.) Fern. and are found in e NE and KS. A variety flowering in April and fruiting in June is known in MO: Clay, Jackson; KS: Wyandotte; and is named var. *praecox* Engelm. It has a thyrse 4–6 cm long, berries 6–7 mm in diam and seeds about 4 mm long. The latter variety is in need of more careful study.

5. ***Vitis rupestris*** Scheele, sand grape. Low bushy plant, seldom climbing, trailing over rocks and shrubs; nodal diaphragm 2–3 mm thick; tendrils reddish, absent or only on tips of fertile branches; stipules 4–5 mm long; leaves reniform to reniform-ovate; blade usually conduplicate, 5–10 cm long, to 1.5 dm wide, base with a broad open sinus to truncate, margin with a few coarse teeth and sometimes shallowly lobed, glabrous and shiny on both surfaces; petiole glabrous. Thyrse 2–4.5 cm long. Berries 6–12 mm in diam, black, glaucous; seeds 4–5 mm long. May–Jun, fruits Jul–Aug. Rocky banks, stream banks; OK: Cleveland, Comanche; (s IL, OK, s to TN, AR, TX).

6. ***Vitis vulpina*** L., winter grape. Vine climbing to 25 m; with stout trunk; young branches glabrous or glabrate; nodal diaphragm 2–3 mm thick; stipules 5–8 mm long. Leaves of fertile branches cordate-acuminate, with a deep basal sinus; blade coarsely and sharply irregularly toothed, one or both sides of some teeth convex, unlobed or merely with angled shoulders, upper surface bright green, lower surface light green, usually glabrous except for hairs on veins; petiole usually with short spreading hairs. Thyrse open, to 1.4 dm long; pedicels sometimes with an early deciduous tendril; berries black and shining, 5–10 mm in diam, sometimes with a slight bloom, edible after frost; seed 5–6 mm long, somewhat acuminate, gray-brown. May–Jun, fruits Sep–Oct. Low woods, stream banks, base of bluffs, thickets; scattered in s SD, s in GP but rare westward; (NJ, s WI, s SD, NE, s to FL, TX). *V. cordifolia* Michx.—Rydberg.

In the Atlas GP all records for Montana and North Dakota of *V. vulpina* are incorrect, as the data represent *V. riparia*. Northward in the GP the distinction between the two species often becomes somewhat obscure, and more adequate specimens are necessary for identification and clarification of the problem.

91. LINACEAE S. F. Gray, the Flax Family

by Ronald L. McGregor

LINUM L., Flax

Annual or perennial herbs. Leaves cauline, simple, alternate or opposite, entire, acute; stipular glands present or none. Inflorescence, a panicle of racemes or cymes. Flowers perfect, regular, 5-merous; sepals 5, imbricate; petals 5, distinct or united at base, convolute, early fugaceous, blue or yellow, rarely white; stamens 5, alternate with petals; pistil 1, ovary superior, 5 carpellate; styles 5, united or nearly distinct. Fruit a capsule, dehiscing into 10, or along false septa, into 5 parts.

References: Rogers, C. M. 1963. Yellow-flowered species of *Linum* in eastern North America. Brittonia 15: 97–122; Rogers, C. M. 1968. Yellow-flowered species of *Linum* in central and western North America. Brittonia 20: 107–135.

1 Flowers blue or white.
  2 Inner sepals with ciliate margins; stigmas slender .... 9. *L. usitatissimum*
  2 Inner sepals entire; stigmas capitate.
    3 Perennial; petals 10–15 mm long .... 4. *L. perenne* var. *lewisii*
    3 Annual; petals 5–10 mm long .... 5. *L. pratense*

1 Flowers yellow.
    4 Styles distinct or nearly so; fruit finally dehiscing into 10 1-seeded segments.
        5 Outer sepals entire; styles separate; annual or perennial .... 3. *L. medium* var. *texanum*
        5 All sepals with glandular teeth; styles united at base; annual .... 8. *L. sulcatum*
    4 Styles united to above the middle; fruit finally dehiscing into 5 2-seeded segments.
        6 Sepals not glandular-toothed; flowers terminating branches .... 2. *L. hudsonioides*
        6 Sepals glandular-toothed; inflorescence paniculate.
            7 Plants grayish-pubescent throughout .... 6. *L. puberulum*

            7 Plants glabrous or nearly so throughout.
                8 Plant broomlike; leaves small, the lower tending to be hidden among the branches .... 1. *L. aristatum*
                8 Plant not broomlike; leaves evident .... 7. *L. rigidum*

1. *Linum aristatum* Engelm., broom flax. Annual herb. Stems 1–4.5 dm tall, much branched throughout. Leaves linear, 5–20 mm long, alternate or the lowermost opposite and nearly hidden in branches; stipular glands usually present. Inflorescence a diffuse panicle with flowers on long, stiffly spreading, ascending branches. Sepals 5, narrow, 6–9 mm long, glandular-toothed; petals yellow, 8–12 mm long; styles united nearly to summit, 4.5–7 mm long. Capsule elliptic, 3.5–4 mm long, dehiscing into 5 segments; seeds 2.5–3 mm long. May–Sep. Very sandy open places; TX: Hutchinson; (TX, n Mex. to UT).

2. *Linum hudsonioides* Planch. Annual herb. Stems 11–23 cm tall, hirsutulous on the angles above, branching from the base, branches ascending to erect. Leaves narrow, 6–10 mm long, opposite below, alternate above, imbricated throughout; stipular glands none. Inflorescence few-flowered, commonly conspicuously exserted beyond leaves. Sepals 4.5–7 mm long, persistent; petals 5, yellow, sometimes with a red base, 8–12 mm long; obovate; styles 2.7–6.3 mm long, united nearly to summit. Capsule broadly ovate, 2.7–3.5 mm long, dehiscing into 5 segments; seeds 2.0–2.7 mm long, wedge-shaped. Apr–Jun. Sandy or gravelly prairies; KS: Saline, Sedgwick; sw OK, TX; (KS, OK, NM, TX).

3. *Linum medium* (Planch.) Britt. var. ***texanum*** (Planch.) Fern., sucker flax. Annual or perennial herb. Stems erect, solitary or clustered, simple below the inflorescence, 2–8 dm tall. Leaves narrowly lanceolate to oblanceolate, rounded or acute, entire, alternate or lower opposite; stipular glands none. Inflorescence with more or less elongate, stiffly spreading ascending branches; pedicels articulating 0.7–1.5 mm below the fruit. Sepals 3.5–5 mm long, outer entire, inner glandular-toothed, all sometimes appressed hairy inside; petals lemon-yellow, 5–8 mm long, obovate; styles separate 1–3 mm long. Capsules depressed-globose, about 2 mm high, separating into 10 segments; seeds 1.5 mm long (n = 18) May–Aug. Rocky, open woods & prairies; KS: Cherokee, Crawford; (ME to IA, s KS, FL, TX).

4. *Linum perenne* L. var. ***lewisii*** (Pursh), Eat. & Wright, blue flax. Perennial glabrous herb. Stems branched at base, 2–8 dm tall. Leaves linear to linear-lanceolate, 1–3 cm long, usually crowded below, alternate above; stipular glands absent. Inflorescence paniculate, few-branched, pedicels spreading to recurved. Sepals 3.5–5 mm long, entire; petals blue (white), 10–15 mm long; styles distinct, 4–9 mm long, stigmas capitate. Capsules ovoid, 5–7 mm long, separating into 10 segments; seeds 4–5 mm long. (n = 9) May–Aug. Prairies & open rocky wooded hillsides; ND, w SD, & s along w GP; (w N. Amer.). *L. lewisii* Pursh — Rydberg.

    Collections from e NE and KS consistently have flowers of two kinds on the same plant (one with styles conspicuously longer than the stamens, the other with stamens the longer). These are interpreted as sporadic garden escapes of the heterostylic Eurasian *L. perenne* L. var. *perenne*.

5. **Linum pratense** (Nort.) Small, Norton's flax. Annual herb. Stems usually branched at base, 5–40 cm tall. Leaves linear to linear-lanceolate, 1–2 cm long, crowded below, scattered above; stipular glands none. Inflorescence paniculate, fruit on spreading or recurved pedicels. Sepals 3.5–4.5 mm long, entire; petals blue or whitish, 5–10 mm long; styles separate, 1–3 mm long, stigmas capitate. Capsules ovoid, 4–6 mm long separating into 10 segments; seeds 3–5 mm long (n = 9) Apr–Jun. Sandy prairies & open places, often on calcareous soils; w 1/2 KS, & s; (KS, OK, TX, NM, CO, s to cen Mex.).

6. **Linum puberulum** (Engelm.) Heller, plains flax. Annual herb. Stems 4–20 cm tall, usually branched at base, densely gray-puberulent. Leaves alternate, linear, 7–13 mm long, puberulent, entire or upper with a few glandular teeth; stipular glands conspicuous. Inflorescence paniculate. Sepals lanceolate, 4.5–7 mm long, acute to acuminate, glandular-toothed; petals obcordate or obovate, yellowish-orange to salmon, with reddish base, 9–12 mm long; styles united nearly to summit, 3–7 mm long. Fruit ovate, 3.5–4.0 mm long; seeds ovate, reddish-brown, 2.5–3.0 mm long. May–Sep. short-grass prairies, sandy or calcareous soils; sw WY, NE: Cheyenne, along w edge GP; (w NE to UT, s to n Mex.). *Cathartolinum puberulum* (Engelm.) Small—Rydberg.

7. **Linum rigidum** Pursh. Erect annual, glabrous throughout or puberulent near the base, 5–50 cm tall. Leaves alternate or the lower ones apparently opposite, linear to linear-lanceolate, entire or the upper ones sparingly toothed, 10–30 mm long and up to 4 mm wide, sometimes with stipular glands present. Inflorescence open, paniculate to ± flat-topped; sepals lanceolate, 5–10 mm long, 1- or 3-nerved, conspicuously glandular-toothed, the inner often with scarious margins; petals yellow, or orange to brick-red toward the base, ovate to broadly ovate, mostly 7–16 + mm long, pubescent at the base; stamens (4)6–8(9) mm long, anthers elliptic to narrowly elliptic; styles united to above the middle. (2.5)3.0–9(11) mm long. Fruit elliptic or triangular-ovate, dehiscing into 5 2-seeded segments, mostly 3.5–4.5 mm tall and 2.5–3.5(4) mm across, the wall thin and ± translucent to thickish and opaque; seeds brownish or reddish-brown, narrowly ovate, ± 3 mm long and 1.0–1.5 mm wide. (n = 15)

In the GP this species consists of three reasonably distinct varieties which, however, do intergrade, and some specimens are difficult to assign to any one variety (cf. Rogers 1968, op. cit.). The varieties are generally distinguishable as follows:

1 Styles less than 5 mm long.
  2 Stipular glands present .... 7c. var. *rigidum*
  2 Stipular glands absent .... 7b. var. *compactum*
1 Styles more than 5 mm long.
  3 Fruit thin-walled, ± translucent, the base rounded .... 7c. var. *rigidum*
  3 Fruit thick-walled, opaque, abruptly tapering at the base .... 7a. var. *berlandieri*

7a. var. **berlandieri** (Hook.) T. & G., Berlandier's flax. Compact herb (5)12–24(40) cm tall, mostly with prominent stipular glands on the mid-leaves. Petals orange to brick colored at the base; styles (6.0)7.0–8.5(9.1) mm long. Fruits triangular-ovate, thickish and opaque; seeds reddish-brown. Apr–Jun. Sandy or gravelly prairies; NE: Dundy; e CO, w 1/2 KS, & southward; (NE, CO s to TX). *Cathartolinum berlandieri* (Hook.) Small—Rydberg.

Most of the Kansas plants of this variety are small, 0.5–15 cm tall, and are more compact, leafy, and branched than the more southern colonies. These plants are similar to var. *compactum* from which they are distinguished by the long styles, over 5 mm long, and thick-walled fruits which taper abruptly to the base.

7b. var. **compactum** (A. Nels.) Rogers, compact stiffstem flax. Compact herb (6)12–21(32) cm tall, with stipular glands absent. Petals yellow; styles (2.5)3–3.6(4.2) mm long. Fruits ± ovate; seeds reddish-brown. May–Sep. Sand or gravelly prairies & high plains; GP but rare or absent in the e 1/4; (Alta., Sask. s to TX). *Cathartolinum compactum* (A. Nels.) Small—Rydberg.

Frequent specimens are difficult to separate from var. *rigidum*, particularly those from the n GP, where the style lengths are about equal and where some specimens of var. *compactum* have stipular glands on the lower leaves.

7c. var. *rigidum*, stiffstem flax. Mostly 20–40 + cm tall, with spreading, ascending branches. Stipular glands absent, or present only in some northern populations. Petals yellow; styles (4.7)6.0–8.8(11) mm long. Fruits elliptic, ± translucent; seeds brownish. May–Oct. Sandy or rarely calcareous prairies, open wooded hillsides; GP but rare to absent southeastward; (Man. to Alta. s to Mex.). *Cathartolinum rigidum* (Pursh) Small—Rydberg.

In sw KS and OK var. *rigidum* intergrades with var. *berlandieri* to some extent where the two are found together. From ne CO n to MT and e to MN, plants are found that have the habit of var. *rigidum* but have the styles only 3–6 mm long and stipular glands usually present on the lower leaves. At some sites this is the only type present, while in other sites it is found mixed with var. *rigidum* and var. *compactum*.

8. **Linum sulcatum** Ridd., grooved flax. Erect glabrous annual (22)36–71(80) cm tall, simple below, but with ascending branches above middle. Leaves linear to narrowly lanceolate, entire, (7)13–23(30) mm long, sharp-pointed, lower opposite, decreasing in size upward, stipular glands usually present. Inflorescence open paniculate; sepals lanceolate, acuminate, glandular-toothed, (2.3)3.3–5(7.3) mm long; petals obovate, pale yellow, 5–10 mm long, pubescent at base inside. Fruit globose or ovoid (2.5)2.7–3.1(3.3) mm high, dehiscing into 10 sharp-pointed, 1-seeded segments; seeds elliptic 1.6–2.1 mm long, reddish-brown. May–Sep. Prairies & open woodlands; e 1/2 GP but becoming rare westward; (MI, Man., s to KY, NE, TX). *Cathartolinum sulcatum* (Ridd.) Small—Rydberg.

9. **Linum usitatissimum** L., common flax. Erect glabrous annual, usually branched at base, 3–10 dm tall. Leaves linear to narrowly lanceolate, 12–30 mm long, stipular glands none. Inflorescence paniculate. Sepals 6–8 mm long, outer entire, inner with fringed margin; petals blue (white), 10–15 mm long; styles distinct, 3–5 mm long. Fruit broadly ovoid, 6–8 mm high, dehiscing into 10 segments; seeds 5 mm long. (n = 15) May–Sep. Roadsides & waste places; e 1/2 GP, more common northward; (s Can; throughout U.S.; Europe).

Introduced as a cultivated crop, this species is commonly grown in MN and the Dakotas and to a limited extent southward. Sporadic escapes are found in the n GP, but have become rare southward where the crop is less frequently planted. Escapes do not long persist.

92. POLYGALACEAE R. Br., the Milkwort Family

by Ronald L. McGregor

1. POLYGALA L., Milkwort

Annual herbs or tufted perennials with simple, entire, alternate, opposite, or whorled leaves and no stipules. Inflorescence a raceme or spike. Flowers usually irregular, pink, greenish, or white; sepals 5, the uppermost and lower 2 small and greenish, the 2 lateral (wings) larger and colored like the petals; petals 3, hypogynous, united into a tube, 3-lobed at apex, the middle lobe keel-shaped and often crested on back; stamens (6)8, united to corolla tube in 2 rows. Capsule 2-locular, 2-seeded; seeds 0.5–3 mm long, usually pubescent and usually arillate.

1 Perennials; stems usually several from the base.
  2 Plants glabrous .... 1. *P. alba*
  2 Plants puberulent at least on stems.
    3 Lowest leaves reduced or scalelike .... 4. *P. senega*
    3 Lowest leaves broader than others .... 5. *P. tweedyi*
1 Annuals; stems usually single.
  4 Leaves alternate.
    5 Stems glaucous; corolla 7–10 mm long, about 3× longer than wings; racemes 1–4 cm long, 5–8 mm in diam .... 2. *P. incarnata*
    5 Stems not glaucous; corolla less than 5 mm long; racemes headlike, rounded or short cylindric, 1 cm in diam .... 3. *P. sanguinea*
  4 Leaves whorled at least to middle of stem .... 6. *P. verticillata*

1. *Polygala alba* Nutt., white milkwort. Perennial from a stout vertical root. Stems usually several, simple to rarely branched, 1–4 dm tall, glabrous. Leaves narrowly oblanceolate below, becoming linear upward, 1–2 mm wide. Racemes on slender peduncles, 3–8 cm long, 4–8 mm in diam, tapering to the tip. Flowers white with greenish center; wings elliptic, 2.2–4 mm long, barely exceeding corolla; keel 3 mm long, the crest with 4 lobes on each side. Capsule elliptic to oblong-elliptic, 2.5–3 mm long, 1.5 mm in diam; seeds pilose, 2.4 mm long, aril 0.8–1.5 mm long, the 2 lobes oblong, appressed. May–Aug. Rocky prairie hillsides, ravines, thickets; GP, becoming rare or absent in the e & se; (MN, ND, WA, s to TX, AZ, s Mex.).

2. *Polygala incarnata* L., slender milkwort. Annual. Stems slender, simple or rarely branched, glaucous, 1–6 dm tall. Leaves sparse, alternate, erect, linear, 4–12 mm long, 0.5–1 mm wide, early deciduous. Peduncles 2–4 cm long; racemes dense, 6–38 mm long. Flowers rose-purple (white); wings linear-oblong, 3 mm long, somewhat undulate-convolute; keel 7 mm long, united with staminal tube and upper petals into a trough 5 mm long; crest on each side of 3 lobes, each variously again lobed or cleft. Capsule suborbicular-ovate, 2.5 mm long; seed plump, pilose, 2.2 mm long; aril about 1 mm high, membranaceous, equitant, erect, hardly lobed. Jun–Aug. Dry prairies, open woodlands, & moist meadows; e 1/2 KS & OK; (NY, MI, WI, IA, KS, s to FL, TX).

3. *Polygala sanguinea* L., blood polygala. Annual. Stems erect, simple or branched, 1–4 dm tall. Leaves numerous, linear to elliptic-linear, 0.5–4 cm long, 1–5 mm wide, erect or ascending, papillose-serrulate on margin with subglandular teeth. Peduncles 4–25 mm long; racemes capitate to thick-cylindric, very obtuse and dense, 5–13 mm in diam, the axis 6–35 mm long. Flowers greenish and pinkish (white); wings ovate-oval, 4.8–6 mm long, 2.5–3.5 mm wide, rounded at apex; keel 2.5–2.7 mm long, crest on each side of a lamella and a cuneate lobe, these sometimes connate. Capsule cuneate-suborbicular, base sterile; seed subglobose-pyriform, rounded at apex, pointed at base, short-pilose, 1.5–1.7 mm long; aril 1–1.2 mm long, the 2 lobes linear, scarious, appressed. Jun–Sep. Sandy prairies, open woodlands, moist meadows; rare in BH; e 1/2 NE & KS; (N.S. to s Ont, MN, SD, s to TN, LA, ne TX). *P. viridescens* L.—Rydberg.

4. *Polygala senega* L., Seneca snakeroot. Perennial. Stems usually several from a thick crown, simple, 1–5 dm tall, minutely puberulent. Leaves numerous, alternate, lowest reduced or scalelike, becoming linear-lanceolate to lance-elliptic or lance-ovate above, serrate, 1.5–3.5 cm wide. Racemes terminal, solitary, conic or cylindric-conic. Flowers white; wings suborbicular, 3.3–3.7 mm long, tapered to claws; corolla lacerate. Capsules plump, rounded, 2.5–4.2 mm long; seeds black, 2.5–3.5 mm long, sparsely pubescent; aril with 1 projection, as long or longer than body. May–Jul. Open woods, prairies, with preference for calcareous soils; ne 1/3 ND, BH, NE: Antelope; KS: Cherokee; (N.B. to Alta., s to GA, AR, se KS).

The plants of KS and some in the BH have larger leaves and wings. These have been recognized as *P. senega* var. *latifolia* T. & G., but may be mere variants of moist habitat.

5. **Polygala tweedyi** Britt. Perennial. Stems spreading to erect, woody below, 7-25 cm tall, incurved puberulous. Leaves uniform or of two sorts, linear to lanceolate or elliptic to ovate-lanceolate, lowest broader than upper, coriaceous, reticulate, 4-25 mm long, incurved puberulous. Peduncles 1 cm or less, terminal or sometimes from base of stems; racemes 3- to 20-flowered, usually geniculate, 4-40 mm long. Flowers purple or lavender (whitish); wings obovate, 4-5.3 mm long, obtuse to rounded at apex; keel 3.5-5 mm long, partly yellow, sometimes puberulous, beak 0.5-1.3 mm wide. Seed silky, 2.7-3.4 mm long; aril corneous, 1-2.2 mm long, with 2 linear appressed lateral lobes. May-Aug. Rocky hillsides & ravines; w OK & TX; (OK, TX, to s AZ, n Mex.).

6. **Polygala verticillata** L., whorled milkwort. Annual, erect, stem simple, usually branched above, 5-35 cm tall, glabrous. Leaves in whorls or rarely alternate above, linear to linear-oblong or linear-elliptic, 1-2(3) cm long, 1-3 mm wide. Racemes conic or cylindric-conic, tapering to apex, flowering portion 5-25 mm long. Flowers white or greenish-pink; wings obovate-oval, 1.6-2 mm long, rounded at apex, short clawed; keel 1.2-1.4 mm long with a crest on each side of the lamella, with 1 or 2 lobes. Capsules 1.8-2.4 mm long; seeds black, pilosulous, 1.5-2.2 mm long; aril 0.5-1 mm long, the two lobes linear-oblong, 1/3-1/2 as long as seed. May-Oct. Dry prairies, open woodlands, sand dunes, & open sterile places; GP but absent w 1/3; (MA, s Man., to s FL, TX, UT).

The common GP plant is 5-20 cm tall, with spreading branches, leaves 1-2.5 cm long, and capsules to 1.8 mm long. This is the var. *isocycla* Fern. Toward the se are rare plants 20-30 cm tall, branches spreading, leaves 2-3 cm long, and capsules 1.8-2.4 mm long that have been referred to var. *sphenostachya* Penn. These two varieties occur together and intergrade. A more rare plant in the e GP has the upper leaves alternate and the plant less branched above. These have been named var. *ambigua* (Nutt.) Wood.

93. KRAMERIACEAE Dum., the Ratany Family

by Ronald L. McGregor

1. KRAMERIA L., Ratany

1. ***Krameria lanceolata*** Torr., ratany. Herbaceous perennial from woody, branching caudex; stems decumbent or trailing over vegetation, 2-5(10) dm long, silky-pubescent. Leaves alternate, simple, entire, linear, linear-elliptic, or some elliptic-oblong, 6-20 mm long, acute or usually with a short brownish spine at apex; silky-strigose on both surfaces; stipules absent. Flowers axillary, perfect, bilaterally symmetrical; pedicels 0.5-3 cm long, pubescent; sepals 4 or 5, distinct, unequal, ovate-lanceolate, 8-10 mm long, pubescent and usually purplish; petals 5, unequal, the upper 3 distinct, or nearly so, long clawed, reddish, the other 2 small, thick, sessile, greenish and glandlike; stamens 4; ovary 1-celled, ovules 2, densely pubescent. Fruit globose, 6-9 mm in diam, woolly, 1-seeded, body armed with straight sharp prickles. (n = 6) May-Jul. Infrequent to locally common on rocky, gravelly or sandy prairie hillsides, ravines, roadsides; sw GP; (KS to AZ, s to TX, Chih. & Coah.). *K. secundiflora* DC. — Rydberg.

This genus has been included in the Caesalpiniaceae, (or the subfamily Caesalpinioideae of the Leguminosae), as a genus in the family Polygalaceae, and in the Krameriaceae of the order Polygalales.

94. STAPHYLEACEAE Lindl. the Bladdernut Family

by Ronald L. McGregor

1. STAPHYLEA L., Bladdernut

1. *Staphylea trifolia* L., American bladdernut. Shrub to 3 m tall; forming thickets from root suckers, or rarely a small tree; branches with smooth striped bark. Leaves opposite, trifoliate; petiole 6–10 cm long; stipules early deciduous; leaflets elliptic, ovate to obovate, 4–8 cm long, 1.5–4.5(6) cm wide, cuspidate, serrulate, lateral 2 subsessile, terminal with petiolule 1–3 cm long, glabrous above, finely pubescent below. Inflorescence a dropping racemelike cluster 5–6 cm long, axillary or terminating the branchlets. Flowers perfect, white, pedicels 6–11 mm long; calyx campanulate, deeply 5-parted, lobes erect, 7–8 mm long; petals 5, erect, 8–10 mm long, spatulate; stamens 5, inserted at margin of a disk, alternate with petals; pistil of 3 carpels united in the axis, styles 4–5 mm long, more or less connate. Capsule 3-locular, inflated, 3–5 cm long, 1–4 seeds per locule, dehiscent along inner side near the top; seeds oval, 5.5–6 mm long, surface smooth, semiglossy, light brown. ($2n = 78$) Apr–May. Rich woodlands, bluffs, & river valleys, with a preference for calcareous soils; extreme e NE, e 1/3 KS, e OK; (Que., MN, e NE, s to GA, OK).

95. SAPINDACEAE Juss., the Soapberry Family

by Ronald L. McGregor

Polygamo-dioecious, flowers functionally unisexual, rarely perfect, trees, shrubs or vines. Leaves alternate, ternately or pinnately compound, exstipulate. Inflorescence a terminal panicle, or in axillary racemose-paniculate clusters. Flowers regular or irregular; sepals and petals 4 or 5; petals often with a scale or gland at base; stamens 8–10; disk usually present. Fruit an inflated capsule or berry.

1 Vines; leaves ternate to variously compound .... 1. *Cardiospermum*
1 Trees or large shrubs; leaves even-pinnate .... 2. *Sapindus*

1. CARDIOSPERMUM L., Balloon Vine

1. *Cardiospermum halicacabum* L., common balloon vine. Annual much-branched vine with axillary tendrils. Leaves alternate, usually ternate or twice ternate; leaflets ovate-lanceolate to rhombic-lanceolate or narrowly lanceolate, acuminate at apex, decurrent on petiolules, toothed or incisely lobed, to 8 cm long. Flowers irregular, 4 mm long; sepals 4, 2 large and 2 small; petals 4, whitish, often somewhat unequal, each with a petaloid appendage at base; stamens 8; disk extra-staminal, ovary 3-celled, 1 ovule per cell. Fruit a bladdery-inflated, 3-celled and 3-lobed capsule, 3–4.5 cm in diam; seeds black, about 5 mm in diam. ($n = 11$) Jun–Sep. Moist thickets, riverbanks, waste places; scattered in cen KS, OK, TX; (NJ, PA, OH, MO, KS, s to FL, TX; Mex.; tropical Amer.) *Introduced.*

This is often grown as an ornamental vine and occasionally escapes in our area. Most colonies rarely persist for long, though one colony in se Kansas has perpetuated itself for over 20 years.

2. SAPINDUS L., Soapberry

1. **Sapindus saponaria** L. var. **drummondii** (H. & A.) L. Benson, soapberry, Chinaberry. Polygamous tree to 15 m high, solitary or in thickets. Leaves alternate, even-pinnate; leaflets 6-10 pairs, elliptic-lanceolate to lanceolate, alternate, entire, acuminate, falcate, to 9 cm long and 4 cm wide. Inflorescence a large dense terminal panicle. Flowers white, regular, 4-5 mm wide with a basal disk; calyx deeply 5-lobed, lobes glabrate but ciliate; petals 4 or 5, white, obovate, with a pilose claw attached below the disk; stamens 7-10, with long hairs on filaments, inserted on disk. Fruit a globose berry, about 1.3 cm in diam, yellowish, 1-seeded (developed from 1 of 3 carpels); seeds black, 8.5-9 mm long, surface appearing smooth but minutely pitted. May-Jul. Along woodland margins, rocky hillsides, prairie ravines, edge of fields, & waste places; commonly in groves; s 1/3 GP; (s KS, NM, s to LA, TX, Mex.) *S. drummondii* H. & A. — Rydberg.

The plant contains saponin and the ground fruits have been used as a soap substitute. It is reported that some persons are allergic to the fruits and develop a dermatitis when handling them.

96. HIPPOCASTANACEAE DC., the Buckeye Family

by Ronald L. McGregor

AESCULUS L., Buckeye

1. **Aesculus glabra** Willd. var. **arguta** (Buckl.) Robins., western buckeye. Shrub or small tree, 3-4(9) m tall. Leaves palmately compound, opposite; leaflets (5 or 6)7(8-11), elliptic-lanceolate, lanceolate, or obovate; apex acute or long acuminate; base acute, cuneate, or acuminate; margin entire at base becoming serrate, often doubly and unequally serrate; upper surface glabrous, lower surface pubescent on veins to puberulous or tomentose; petiolules 0-10 mm long. Inflorescence a terminal panicle. Flowers yellow, irregular, andromonoecious, the perfect flowers only on basal branches of inflorescence or scattered throughout; calyx campanulate, 3-8 mm long; sepals 5, lobes subequal; corolla 9-17 mm long, of 4 petals; 2 upper petals with villous claw equaling the spatulate blade in length; 2 lateral petals slightly shorter than upper pair, the claw less than half of petal length; stamens 7, exserted, 13-21 mm long. Fruit ovoid or obovoid, 3-5 cm in diam; pericarp tuberculate-spiny, the spines often deciduous or rarely lacking; seeds chestnut brown, 1-3(4), 2-3 cm in diam. Apr-May. Rocky wooded hillsides, stream banks, lowland woods, thickets in prairie ravines; se NE, e 1/2 KS; (NE, KS, MO, s to TX). *A. arguta* Buckl. — Rydberg; *A. glabra* var. *sargentii* Rehd. — Gates.

In eastern Kansas occasional plants are found with 5-7 leaflets, oval-oblong, or elliptic-obovate with acute or acuminate apices. These have been referred to var. *glabra*, but are always rare plants in stands of var. *arguta*. The western buckeye exhibits considerable variation in number, shape, and size of leaflets. Many specimens from eastern Kansas appear intermediate between the two varieties and were the basis for var. *sargentii* Rehd.

The raw seeds of this plant are poisonous to humans, livestock, and domestic animals. After boiling or roasting, the seeds are harmless and were eaten by the Indians. Livestock also have been poisoned by eating the early leaves.

97. ACERACEAE Juss., the Maple Family

by Ronald L. McGregor

1. ACER L., Maple

Polygamo-dioecious or dioecious trees or rarely shrubs. Leaves opposite, simple and palmately lobed or pinnately compound. Inflorescence terminal on short leafy branches or from lateral buds; corymbose, paniculate, or racemose. Flowers regular, usually completely or functionally unisexual, somewhat perigynous, with a prominent or indistinct to absent disk; calyx colored, usually 5-lobed; petals wanting or as many as calyx lobes; stamens 4–10, often 8; ovary superior, 2-celled, flattened and prolonged into 2 lobes, ovules 2 per cal1; style 1, usually short, stigmas 2. Fruit a schizocarp composed of 2 1-seeded samaras which eventually separate.

```
1  Leaves pinnate, with 3–7(9) leaflets .................................................... 2. A. negundo
1  Leaves simple (rarely ternate in A. glabrum).
   2  Leaf blades less than 6 cm long ...................................................... 1. A. glabrum
   2  Leaf blades 7 cm long or longer.
      3  Sinuses between main leaf lobes rounded; flowers appearing with or after the leaves.
         4  Leaves flat or nearly so; lower surface gray, silvery, or
            blue-green ................................................................... 5. A. saccharum
         4  Leaves with drooping sides; lower surface green or yellowish-green ... 3. A. nigrum
      3  Sinuses between main leaf lobes sharp and forming a definite angle; flowers appearing
         long before the leaves ................................................... 4. A. saccharinum
```

1. *Acer glabrum* Torr., mountain maple. Tree or shrub 3–6 m tall; twigs grayish, rarely reddish. Leaves broadly cordate, 3- to 5-lobed or 3-parted on suckershoots; middle lobe the largest, oblong or broadly cuneate; lateral lobes acute, irregularly toothed; blade often with reddish glands on veins. Flowers glabrous, corymbose in leaf axils; sepals (4)5(6); petals same number as sepals, equal to or smaller than sepals, rarely lacking; stamens usually twice number of sepals or 7–9, inserted at edge of lobed disk. Samaras 2–3 cm long, the pairs narrowly to widely divergent, often reddish. May–Jun. Wooded hillsides, ravines, & along streams; NE: Sioux; sw through WY & CO; (s Alta., nw NE, NM to AK, CA).

2. *Acer negundo* L., box elder. Dioecious tree to 15 m tall; young branches green and glabrous, glaucous, or puberulent. Leaves of fertile branches with 3–5(7) leaflets, pubescent when young, becoming glabrate; leaves of vigorous shoots often with more numerous leaflets; terminal leaflet elliptic to obovate, lateral ones narrower and coarsely few-toothed or entire. Flowers greenish, unisexual (very rarely perfect), produced just before or with the leaves; staminate flowers fascicled and pendulous on slender pedicels, usually 7–15; pistillate flowers racemose, 4–9; petals and disk absent. Samaras 2.5–4 cm long, glabrous or pubescent, greenish, yellowish or reddish, the pairs strongly ascending. (n = 13) Apr–May. Widespread in N. Amer.

The following varieties are in the GP:

```
1  Young branches green or olive-green, glabrous ..................................... 2b. var. negundo
1  Young branches either distinctly glaucous or pubescent.
   2  Young branches glaucous ........................................................ 2d. var. violaceum
   2  Young branches velutinous pubescent.
      3  Mature samaras pubescent ................................................... 2c. var. texanum
      3  Mature samaras glabrous ................................................... 2a. var. interius
```

2a. var. *interius* (Britt.) Sarg. Young branches velutinous pubescent; leaflets usually with prominent tufts of hairs in vein axils below. Mature samaras glabrous. May. Along streams, canyons; w ND, MT, w SD, ne NE, WY & s in mts. of CO & NM. *Negundo interius* (Britt.) Rydb.—Rydberg.

2b. var. *negundo*. Young branches green, olive-green, glabrous, rarely slightly glaucous. Apr. Stream banks, low woodlands, base of bluffs, old farmsteads, & waste places; e GP & locally persisting from plantings westward.

2c. var. *texanum* Pax. Young branches velutinous pubescent. Mature samaras pubescent to nearly glabrous. Apr. Stream banks, low woodlands; w MO, e KS & southward (including reports of var. *interius* of MO & e KS which are based on plants with near or glabrous samaras & may be hybrids with var. *negundo*).

2d. var. *violaceum* (Kirchn.) Jaeg. Young branches greenish or purplish, distinctly glaucous. Apr–May. Stream banks, low woods, canyons, old farmsteads; GP but less common to absent westward. *Negundo nuttallii* (Nieuw.) Rydb.—Rydberg.

In the e GP there is considerable intergradation between this variety and var. *negundo*.

3. Acer nigrum Michx. f., black maple. Tree to 20 m tall; bark dark, shallow or deeply furrowed; young branches orange-brown. Leaves broadly ovate to subreniform, usually 3(5)-lobed; blades yellow-green to fulvous below, the sides usually somewhat drooping; margins with obtuse or rounded teeth. Samaras (2.5)3.5(4.5) cm long. Apr–May. Moist rich woodlands; MO; (Que. NH, WI, MN, s to NC, TN, MO).

Prior reports for SD, NE, and KS were based on specimens inadequate for determination and probably should be referred to *A. saccharum*.

4. Acer saccharinum L., silver maple, soft maple. Tree to 25 m tall with wide spreading branches which are often pendulous; bark light gray, separating into long plates. Leaves 5-lobed to beyond the middle, sinuses narrow, acute; blades usually deltoid-ovate, lobes acuminate and more or less sharply toothed or with smaller lobes along the sides, silvery-white below, glabrous or slightly pubescent when young. Flowers from clusters of lateral buds, each cluster unisexual, opening before the leaves; calyx gamosepalous with 5 shallow lobes; stamens usually 5, long-exert; ovary densely pubescent, style branched about 3.6 mm long. Samaras falling before leaves are mature, 3.5–6 cm long, sparsely pubescent or glabrate, wings divergent and falcate. (n = 26) Feb–Apr. Along streams, low moist woods, & relics around old farmsteads; e 1/2 GP but rare north & westward where introduced; (N.B., Que., MN, SD, s to FL, TN, OK).

5. Acer saccharum Marsh., sugar maple. Tree to 25 m tall, bark gray and becoming deeply furrowed; young branches brown. Mature leaf blades flat, 3- to 5-lobed, 7–15 cm long, 8–18 cm wide, cordate or subcordate with open sinus; usually glabrous below except for tufts of hairs in axils of main veins; petioles usually glabrous. Flowers unisexual in umbels from terminal or uppermost lateral buds, expanding with the leaves; calyx gamosepalous, campanulate, 5-lobed, 3–6 mm long, yellowish, ciliate, glabrous to slightly hirsute; petals absent; stamens 5–8 from center of disk; ovary nearly glabrous; style branches 4–5 mm long. Samaras 2.5–(4) cm long, the seed portions diverging at right angles to the pedicel. Apr–May. Dry rocky hillsides, rich woods, canyons, especially on calcareous soils; extreme e GP & with disjunct colonies in OK: Caddo, Canadian, Comanche; (Que, Man; ND, s to NJ, n GA, TN, MO).

The record for TX: Hemphill in the Atlas GP was in error.

The common plant in the GP with leaf blades glaucous below and hairs only in axils of main veins is f. *glaucum* (F. Schmidt) Pax. A rarely encountered plant with veins or lower surface densely hairy is f. *schneckii* Rehd.) Deam.

98. ANACARDIACEAE Lindl., the Cashew Family

by Ronald L. McGregor

Shrubs or woody vines. Leaves alternate, ternately or once-pinnately compound; essentially exstipulate. Inflorescence a terminal thyrse, or axillary panicle, or of catkinlike terminal and lateral racemes. Flowers polygamous, small, regular, 5-merous; stamens 5, inserted beneath a disk; pistil 1, 3-carpellate, superior, but ovary 1-locular an 1-ovuled. Fruit a drupe.

1 Leaves odd-pinnate and flowers in terminal thyrsoid-panicle; or leaves ternate and flowers from catkins developed late in the previous season; fruits red .... 1. *Rhus*
1 Leaves ternate or palmate and flowers in axillary, lateral thyrsoid-panicles or racemes; fruits whitish or dun colored .... 2. *Toxicodendron*

1. RHUS L., Sumac

Ours polygamo-dioecious shrubs often forming thickets or rarely small trees. Leaves alternate, odd-pinnate or trifoliate (rarely unifoliate); stipules none. Inflorescence a terminal or axillary thyrse, panicle, or a terminal compound spike; flowers appearing before or with the leaves, yellow, each flower subtended by a lanceolate caducous bract (*R. copallina* & *R. glabra*) or each flower subtended by a persistent deltoid bract and 2 similar bracteoles; sepals 5, united below; petals 5, distinct; fertile stamens 5; pistillate flowers with 5-10 staminal vestiges, stamens separated from ovary by a flat, lobed, disk; ovary 1-celled, style short, 3-parted just below stigmas. Fruit a drupe, red or yellowish-red at maturity, surface with glandular and hyaline hairs intermixed.

Reference: Barkley, Fred A. 1937. A monographic study of *Rhus*. Ann. Missouri Bot. Gard. 24: 265-498.

1 Leaf rachis winged.
   2 usually spinescent shrubs; leaflets 6 mm wide or less .... 4. *R. microphylla*
   2 Not spinescent; leaflets 10 mm or more wide .... 2. *R. copallina*
1 Leaf rachis not winged.
   3 Leaves pinnate .... 3. *R. glabra*
   3 Leaves trifoliate .... 1. *R. aromatica*

1. *Rhus aromatica* Ait., fragrant sumac, polecat bush. Straggly to upright shrubs with ascending branches to 2(3) m tall, fragrant when bruised; branches puberulent, glabrate, or densely pilose. Leaves trifoliate, petiolate; leaflets glabrate, puberulent or densely pubescent, variable in shape, lobing, and margins. Inflorescence a terminal compound spike of catkins produced late in previous season; flowers yellow or pale yellow. Fruit red, subglobose, 5-7 mm in diam, pubescent with red glandular and hyaline hairs intermixed; stone globose or compressed.

This is a highly variable species which consists of five geographic varieties in our area, with others found to the west and southwest. The extremes of each of our varieties appear as "good" species, but when their ranges overlap intermediates are common or are the prevailing plants present. Much more field work must be done to obtain specimens in winter condition, in flower and with mature leaves and fruits before more than a provisional treatment can be provided.

1 Terminal leaflet elliptic to rhombic-ovate, tapering about equally to base and apex, not lobed above, apex acute or acutish, base subcuneate .... 1a. var. *aromatica*
1 Terminal leaflet flabelliform-ovate or rhombic-ovate, 3- to 7-lobed above, apex rounded or broadly truncate, base cuneate.
   2 Leaves and young stems densely pilose .... 1c. var. *pilosissima*
   2 Leaves glabrous or pubescent but not densely pilose.
      3 Undersurface of leaflets hairy .... 1d. var. *serotina*
      3 Undersurface of leaflets glabrous or minutely puberulent when young.
         4 Fruit densely long-hairy, stones plumpish, not flattened; plants of cen TX, n to s OK .... 1b. var. *flabelliformis*
         4 Fruit sparsely long-hairy; stone somewhat flattened; plants of ND, SD, w KS to w TX .... 1e. var. *trilobata*

1a. var. *aromatica*. Leaves at first pubescent on both sides, glabrate to rarely pubescent in age; terminal leaflet elliptic to rhombic-ovate, tapering about equally to subcuneate base and acute or acutish at apex. Flowers preceding expansion of leaves; flowering bracts glabrous within ciliate margin at apex. Fruits densely long-hairy; stones 3.8–4.5 mm long. May–Apr, fruits May–Jul. Rocky open woods, thickets, roadsides; extreme se GP; (sw Que., w VT to e KS, s to FL, n LA). *R. aromatica* var. *illinoensis* (Greene) Rehd.—Fernald; *R. crenata* (P. Mill.) Rydb.—Rydberg.

In our area this variety flowers about 2 weeks before var. *serotina* and grows in more mesic habitats. Specimens with densely pubescent leaves have been referred to var. *illinoensis* but in our area these appear to be intermediates with var. *serotina*.

1b. var. *flabelliformis* Shinners. Mature leaves glabrous; terminal leaflet cuneate-obovate, obtuse or with wide, almost truncate tip, variously lobed above. Flowers preceding expansion of leaves or with the young leaves; flowering bracts glabrous to glabrate within ciliate margin at apex. Fruits densely long-hairy, stones 3.5–4.5 mm long. Feb–Apr, fruits May–Jul. Chiefly in calcareous open wooded areas but also in sandy woodlands & bushy ravines; s OK, cen & w TX: (OK, TX).

This variety intergrades with var. *serotina* n and e and with var. *trilobata* in w TX.

1c. var. *pilosissima* (Engl.) Shinners. Young branches and leaves densely pilose; in NM plant pubescence appressed but becomes spreading on plants to the north and eastward. Leaf characters intergrade with var. *serotina*; fruit characters intergrade with var. *trilobata*. Apr–May, fruits in May–Aug; rocky open wooded hillsides, canyons, rocky or gravelly prairie hillsides; sw KS, CO, w OK, TX panhandle, NM; (se KS, se OK, s to trans-Pecos). *R. osterhoutii* Rydb.—Rydberg.

1d. var. *serotina* (Greene) Rehd. Terminal leaflet flabelliform-ovate, variously lobed above, apex rounded or blunt, cuneate at base, usually pubescent at least below. Flowering before the leaves in some plants and with unfolding leaves in others, often in same colony; flowering bracts glabrous, glabrate, or densely hairy within ciliate border at apex. Fruits densely long-hairy; stones 3.8–4.5 mm long. Apr–May, fruits Jun–Aug. Open rocky wooded hillsides, prairie ravines & hillsides, roadsides; se SD, e NE, KS, OK, e TX; (IL, IA, se SD, s to AR, TX). *R. nortonii* (Greene) Rydb.—Rydberg; *R. trilobata* var. *serotina* (Greene) Rehd.—Fernald.

This is the most variable taxon and intergrades with all other varieties where they overlap in distribution. It appears reasonably distinct in much of KS and OK.

1e. var. *trilobata* (Nutt.) A. Gray. Leaves essential glabrous to sparsely puberulent. Flowering before the leaves or rarely with very young leaves. Flowering bracts densley pubescent within ciliate margin at apex. Fruit red, with sparse long hairs, prominently pruinose, stones 4.5–6.2 mm long. Apr–Jun, fruits May–Sep. Rocky wooded hillsides, canyons, prairie ravines, sometimes on sandy gravelly soils; w 1/2 GP; (w 1/2 ND, s to TX panhandle, w to OR, CA). *R. trilobata* Nutt.—Rydberg.

In some of our range this variety is quite distinct, but it intergrades freely with var. *serotina* in w KE, e CO, and south where it also intergrades with var. *pilosissima* and var. *flabelliformis*.

2. **Rhus copallina** L., dwarf sumac. Shrub or small tree 1.5–3(10) m tall; young branches puberulent, older branches glabrate and with prominent lenticels. Leaves odd-pinnately compound, to 35 cm long, the rachis wing-margined between leaflets; petioles 3–6 cm long, puberulent; leaflets 7–11, sessile or terminal one with blade abruptly narrowed and appearing petiolulate; leaflet blades elliptic-lanceolate to ovate-lanceolate, margins entire but slightly revolute, rarely serrate-margined, glabrous or sparsely pilose above, glandular-

hairy below, to 80 cm long and 4 cm wide, obtuse, acute or often acuminate at apex, rounded to inequilaterally cuneate at base. Inflorescence a terminal thyrse to 1.5 dm long and 1 dm wide; flowers greenish-white. Drupes (3.5)4(5) mm in diam, red, with erect short white hairs and scattered longer glandular hairs; seed 3 mm long, 2 mm wide, smooth to slightly roughened. Jun–Jul, fruits Aug–Sep. Rocky open wooded hillsides, prairies, thickets, abandoned fields, roadsides, generally in acid soils; se NE s to e TX; (ME to MI, se NE s to GA, LA, e TX). *R. copallina* var. *latifolia* Engl.—Fernald.

This species is occasionally planted for its lustrous leaves, which in autumn become vermillion-red or orange-red.

3. **Rhus glabra** L., smooth sumac. Shrubs, rarely a small tree, to 3(5) m tall, forming dense thickets from underground rootsuckers; stems and branches glabrous, glaucous. Leaves alternate, odd-pinnate, 3–5 dm long; petioles 4–7 mm long; stipules none; leaflets 11–23(31), lanceolate, oblong-lanceolate, to ovate-lanceolate, 7–9 cm long, acute and often acuminate at apex, base acute or rounded, often inequilateral, margin sharply serrate, upper surface dark green and lustrous, lower surface glaucous, petiolules 0–2 mm long. Inflorescence a dense terminal panicle 10–25 cm long; flowers greenish. Drupes compressed-globose, 3.5–4.5 mm diam, red, viscid pubescent; seed yellowish, 3–3.5 mm long, smooth. May–Jun, fruits Aug–Sep. Open woodlands & borders, thickets, prairies, roadsides, & waste places; w 2/3 GP; (MN, Que., B.C., s to OR, UT, FL, TX, NM, AR).

The somewhat similar staghorn sumac (*R. typhina* L.) of the eastern United States is occasionally cultivated in the e GP. It may be distinguished from *R. glabra* by its densely pilose branches, petioles, and inflorescences.

4. **Rhus microphylla** Engelm., desert sumac. Much-branched shrubs, 0.5–5 m tall; branches becoming spinescent, puberulent when young. Leaves odd-pinnate, alternate, petioles 1–5 mm long, pilose; stipules absent; leaflets 5–9(11), sessile, pilose, rachis winged, pilose, blades less than 2 cm long and 6 mm wide, apex rounded or acute, rounded or narrowed to base. Flowers appearing before leaves; in axillary and terminal inflorescences. Drupes red at maturity, glandular-pubescent, (4.5)5(6) mm in diam, stone somewhat flattened, 4.5–5 mm long. Apr–May, fruits Jul–Aug. Dry rocky hillsides, canyons, uplands, alkali flats, riverbanks; sw OK, TX panhandle, e NM; (sw OK, w 3/4 TX, NM, AZ, & n Mex.).

2. TOXICODENDRON P. Mill, Poison Ivy, Poison Oak

Ours poisonous dioecious woody vines, shrubs, or semishrubs often proliferating from rhizomes; aerial roots present or absent. Leaves alternate, palmately trifoliate, exstipulate; blades of leaflets entire or coarsely toothed, undulate, or variously lobed; lower surface glabrous, or scattered-strigose to densely pilose or velutinous; upper surface glabrous or usually sparsely pubescent. Inflorescence a lateral paniculate thyrse, ultimate clusters of 3 or 4 flowers; sepals 5, united below, imbricate in bud, lower 1/2 greenish, upper 1/2 cream colored; petals 5, free, imbricate in bud, cream colored, usually with prominent purplish veins; stamens 5, from a substaminal disk; ovary 3-carpellate, only 1 carpel fertile, surrounded at base by disk. Fruit a globose drupe with chartaceous exocarp, fibrous-waxy mesocarp, and a bony endocarp; glabrous or hairy; white or dun colored.

Reference: Gillis, W. T. 1971. The systematics and ecology of poison-ivy and the poison-oaks (*Toxicodendron*, Anacardiaceae). Rhodora 73: 72–159; 161–237; 370–443; 465–540.

1 Fruits pubescent or papillose .... 3. *T. toxicarium*
1 Fruits glabrous, rarely with occasional hairs.
   2 Petioles of mature leaves glabrous or essentially so; shrub or subshrub; aerial roots lacking .... 2. *T. rydbergii*

2   Petioles of mature leaves puberulent or pilose, if glabrous leaflets lobed or deeply cut; vines or shrubs; aerial roots usually present .... 1. *T. radicans*

1. ***Toxicodendron radicans*** (L.) O. Ktze., poison ivy. Vine, shrub, often with aerial roots; rhizome creeping. Leaflets 3, ovate to elliptic, entire or irregularly serrate, dentate, lobed, or incised; acute to acuminate or cuspidate at apex; glabrous to puberulent above, glabrous to pilose, scattered-strigose, or velutinous below; lateral leaflets often inequilateral; blade of terminal leaflet 3–15 cm long. Inflorescence a lateral paniculate thyrse. Drupe hard, cream, yellow, or tan; globose.

Three subspecies are present in the GP:

1 Leaflets with lobed or deeply cut margins .... 1c. subsp. *verrucosum*
1 Leaflets with entire undulate, notched or serrate margins.
   2   Leaflets usually appressed to scattered-strigose on lower surface or glabrous .... 1a. subsp. *negundo*
   2   Leaflets densely pilose to velutinous on lower surface; pubescence erect 1b. subsp. *pubens*

1a. subsp. *negundo* (Greene) Gillis, poison ivy. Blades of leaflets ovate to elliptic, usually serrate, notched, or with 1 lobe on each side; glabrate to strigose on lower surface, or pubescence confined to veins and veinlets; glabrous on upper surface or scattered-strigose, usually with small curly hairs on lower midvein and lower lateral veins; lateral leaflets inequilateral. Drupes glabrous, (2.5)4.5(5) mm in diam (n = 15) May. Low or upland dry or wet woods, stream banks, bluffs, thickets, fence rows, roadsides, & waste places; se SD & s in e 1/2 of GP; (s Ont., NY, MI, se SD, s to WV, KY, TN, OK, w to e KS, TX). *T. negundo* Greene — Rydberg.

This is a highly variable taxon in the GP and intergrades freely with the following subspecies. In some habitats the plants grow as shrubs to 3 m tall.

1b. subsp. *pubens* (Engelm.) Gillis, poison ivy. Very similar to the preceding subsp. from which it differs chiefly by having lower surface of the leaflets densely strigose, hirsute, or velutinous, and pubescence erect.

An analysis of Kansas specimens, annotated by Gillis as subsp. *pubens,* reveals that the lower leaflet surfaces have more or less scattered-strigose pubescence with some pilose (erect) hairs. In none of the GP specimens have I found the presence of dense, erect hairs in the interveinal areas on the lower leaf surfaces. The inclusion of subsp. *pubens* in the GP flora is tentative. More careful field studies, particularly in e and se KS, is necessary to help resolve the problem.

1c. subsp. *verrucosum* (Scheele) Gillis, poison ivy. Similar to above subsp. but leaflets usually regularly and deeply incised dentate or acute-angled lobed; lower leaf surface glabrous to substrigose with hairs confined to the primary and secondary veins. May. Open woods, rocky banks, roadsides; MO: Jackson; OK: Garfield, Love, Murray, Woods; s in TX; (MO, se KS, TN, AR, s to Gulf).

2. ***Toxicodendron rydbergii*** (Small) Greene, poison ivy. Shrub 3 dm to 2 m tall; stem simple to upright branched, without aerial roots, often forming thickets from much-branched subterranean stolons. Leaves usually near summit of stem; terminal leaflet blade broadly ovate, rhomboid, or suborbicular, usually acuminate; margins dentate, undulate, or notched; glabrous or glabrate above except for minute pubescence on midrib; glabrous or appressed-strigose below; lateral leaflets inequilateral; petiole of mature leaves glabrous or with scattered hairs. Flowers in small axillary clusters. Drupes globose, yellowish, (4)5–6(7) mm in diam, smooth, often persistent into following season. (n = 15) May–Jun, sometimes again Aug–Sep. Rocky prairie hillsides; stream banks, flood plains, open woods, roadsides, sandy & calcareous soils; GP but absent from much of the se; (across n U.S. & s to n IL, IA, se NE, w KS, w TX, AZ). *T. desertorum* Lunell, *T. fothergilloides* Lunell — Rydberg.

Along a line from OK: Beaver to SD: Union and in some of the stream valleys westward, frequent plants appear to intergrade with *T. radicans* subsp. *negundo.* These are subshrubs or rarely climbing vines; aerial roots rarely present; petioles glabrous to definitely hairy; and fruits 4–5 mm in diam. The fruits of this species which remain on plants through the winter are frequently eaten by quail during period of snow cover.

3. **Toxicodendron toxicarium** (Salisb.) Gillis, eastern poison oak. Shrub to 1 m tall often forming thickets from branching subterranean stolons. Leaflets ovate to oblong or oblong-obovate, lobate-dentate or lyrate, sinuate-pinnatifid with 3–7 rounded, blunt, or rarely subacute lobes, occasionally only undulate; petiole hispid or villous, pilose, strigose, hirsute, or velutinous above, strigose, or velutinous below; lateral leaflets inequilateral. Drupe globose-reniform, tan or yellowish, 3–5 mm in diam. (n = 15) May. Open sandy scrub-oak woodlands, & ravines; KS: Chautauqua; e 1/2 OK & s; (NJ to FL, w to TX, s MO, se KS). *Rhus toxicodendron* L.—Fernald.

99. SIMAROUBACEAE DC., the Quassia Family

by Ronald L. McGregor

1. AILANTHUS Desf., Tree of Heaven

1. *Ailanthus altissima* (P. Mill.) Swingle, tree of heaven, smoke tree. Polygamous tree to 20 m tall, often colonizing by root sprouts; bark smooth with pale stripes; young branches minutely pubescent. Leaves odd-pinnately compound, alternate, 3–6 dm long, leaflets 11–41, lanceolate, falcate, acuminate, entire except for 1–5 rounded basal teeth, each with a prominent gland beneath near the apex, puberulent and glandular above and below. Inflorescence a terminal pyramidal panicle 1–4 dm long. Flowers regular, greenish or yellowish; sepals 5, 0.8–1.2 mm long, lobes imbricated; petals 5, 1.5–3 mm long, infolded, valvate; stamens 10 in staminate flowers, 2 or 3 in perfect flowers, 0 or rudimentary in pistillate flowers; a purplish-brown lobed disk present between the stamens and ovary; carpels 2–5, united at the axis. Fruit a schizocarp with 2–5 narrowly oblong membranaceous samaroid mericarps, each with a single seed in the middle. Mid May–Jun. Usually found in waste areas such as alleyways, roadsides, fence rows, & occasionally in open woodlands; scattered in NE, KS, e & southward; (Ont., MA s in e U.S., w to GP, s Rocky Mts. to CA; Asia). *Naturalized,* but weakly persisting in GP. *A. glandulosa* Desf.—Rydberg.

This is a rapidly growing tree that withstands drought, is nearly free of diseases and insect damage, and tolerates smoke and other gases (origin of name smoke tree). Unfortunately, the plant becomes weedy and often noxious. It is reported that some persons develop a dermatitis after contact with the leaves, and cases of hay fever have been attributed to the plant.

100. RUTACEAE Juss., the Citrus Family

by Ronald L. McGregor

Ours unarmed or armed perennial herbs, subshrubs, shrubs or small trees; leaves simple or compound, alternate, with prominent or obscure oil glands at least on underside; stipules absent; flowers in racemes, racemose cymes, cymes, or in small axillary clusters, perfect or unisexual; sepals 4 or 5(6) or sometimes gamosepalous, or wanting; petals 4 or 5(6); stamens 4 or 5 or 8; disk present between stamens and ovary; ovaries superior; ovaries 1 or 3–5. Fruit a capsule, samara or follicle.

1 Leaves simple; herbs or subshrubs .... 2. *Thamnosa*
1 Leaves trifoliate or compound; shrubs or small trees.
  2 Leaves trifoliate; fruit a samara .... 1. *Ptelea*
  2 Leaves odd-pinnate; fruit of fleshy follicles .... 3. *Zanthoxylum*

1. PTELEA L., Hop-tree

1. *Ptelea trifoliata* L., hop-tree, wafer ash. Unarmed, deciduous, erect or spreading shrubs or small trees, 1–3 m tall, with pallid epidermis and bark. Leaves alternate, trifoliate, glabrous or variable in pubescence, glandular, long petioled; leaflets sessile, 2–6 cm long. Flowers appearing with the leaves; inflorescences in terminal cymes; flowers greenish white, mostly unisexual by abortion, with vestigial parts of opposite sex usually present; sepals 4 or 5(6), 1–2 mm long, soon deciduous; petals 4 or 5(6), distinct, usually 4–6 mm long, broadly elliptic to ovate or linear oblong, hirsute on inside; stamens 4 or 5(6), alternate to petals; disk lobed, acting as a gynophore; ovaries compressed 2(3)-locular; ovules 2 per locule but one aborting. Fruit an indehiscent samara in a single plane, with a thin wing all around; pericarp tough and coriaceous at maturity.

References: Bailey, V. L. 1962. Revision of the genus *Ptelea* (Rutaceae). Brittonia 14: 1-45.

Three completely intergrading subspecies occur in our region as follows:

1 Leaves with glands 0.15–0.25(0.3) mm in diam; fruits often 3-carpellate, seed bearing bodies often 2(3) mm thick and sometimes below middle of wings.
  2 Lateral leaflets nearly equilateral; blades flexible-herbaceous; margins serrulate to irregularly serrate-dentate .... 1a. subsp. *angustifolia*
  2 Lateral leaflets commonly strongly inequilateral; blades firmly herbaceous to subcoriaceous, margins sightly crenate to entire .... 1b. subsp. *polyadenia*
1 Leaves with glands less than 0.1 mm in diam; fruits mostly 2-carpellate, seed bearing bodies usually less than 1 mm thick and usually near center or above middle of wings .... 1c. subsp. *trifoliata*

1a. subsp. *angustifolia* (Benth.) V.L. Bailey var. *persicifolia* (Greene) V.L. Bailey. Lateral leaflets nearly equilateral, the angles between the lower midveins usually not more than 50°; blades flexible-herbaceous, rather thin, with a distinctive luster on both sides, thinly hairy beneath with long slender hairs; veins inconspicuous; glands rather sparse but prominent; leaflet margins serrulate to irregularly serrate-dentate. Mar–May. Rather rare in rocky stream banks, canyons, or wooded ravines; ne & cen OK; (Edwards Plateau of TX, OK, to Ouachita Mts. of AR).

1b. subsp. *polyadenia* (Greene) V. L. Bailey. Lateral leaflets usually strongly inequilateral, angles between lower margins and midvein as much as 90°; blades firmly herbaceous to subcoriaceous, upper surface sparsely short-pubescent and strongly gland-dotted, lower surface more or less densely villous or tomentose, margins crenate to entire. Mar–May. Canyons, & rocky hillsides, creek banks; OK, Texas panhandle, & plains; (AR, OK, TX, s NM, AZ).

1c. subsp. *trifoliata*. Lateral leaflets more or less inequilateral at the base, angle between lower margins and the midvein averaging more than 45–55°, blades herbaceous, usually glabrous or glabrate below, glands inconspicuous, margins irregularly and shallowly crenate, or crenate-serrate to entire. Mar–May. Rocky prairie hillsides, canyons, wooded ravines & valleys, usually on calcareous soils; se NE, e 1/4 KS, MO, OK; (NY, Que., Ont., NE, s to FL, AL, TX).

2. THAMNOSA Torr. & Frem., Dutchman's Breeches

1. *Thamnosa texana* (A. Gray) Torr. Unarmed perennial herbs or subshrubs; stems 0.5–3 dm tall, bluish-green, glandular. Leaves alternate, simple, nearly linear, entire 5–15 mm long, somewhat fleshy, strongly glandular dotted. Flowers perfect, in short racemes or cymose clusters; sepals 4, ovate, 0.6–1 mm long; petals 4, yellowish to purplish, ovate or oblong, 3–6 mm long, erect and incurved at anthesis, appearing somewhat urceolate; disk well developed; stamens 8; ovary 2-celled, 2-lobed, stipitate; ovules several per cell. Fruit a

leathery 2-celled and 2-lobed capsule 3–7 mm long, with shape of inflated dutchman's breeches with legs projecting upward, prominently glandular dotted; seeds pinkish-gray, 1–1.5 mm long, rugose. Mar–May. Rocky or sandy plains, roadsides, stream valleys; TX: Bailey; (TX, NM, AZ, n Mex.).

3. ZANTHOXYLUM L., Prickly Ash

1. *Zanthoxylum americanum* P. Mill., prickly ash, toothache tree. Dioecious thicket-forming shrubs, rarely small trees; stems 0.5–3(6) m tall, armed with stout stipular spines, bark gray-brown with light areas. Leaves alternate, odd-pinnate, (5)10–20(25) cm long; leaflets 5–11(13), oblong, to elliptic or ovate, 3–6(8) cm long, entire or finely crenate, glandular dotted, glabrate in age. Flowers in umbellike axillary clusters on branches of previous year; sepals 0; petals 4 or 5, 1.7–2.4 mm long, greenish and often with reddish fringed tip; ovaries (2)3–5, distinct, stipitate. Follicles stipitate, ellipsoid, 4.5–6 mm long, reddish-brown, the surface pitted; seeds 4–4.5 mm long, ca 2.5 mm diam, finely pitted, glossy-black. (n = 34) Mar–Apr. Open & rocky woodlands, thickets in prairie ravines, fence rows, roadsides, usually in calcareous soils; e 1/2 GP; (Que. to MN, ND, s to GA, AL, OK, MO, KS).

Zanthoxylum hirsutum Buckl., a shrub with twigs and foliage usually more or less hirsutulous or pilosulous, flowers in ample terminal panicles, and 5 sepals, petals, and stamens, is known from OK: Cotton, Jefferson, Love, Marshall, and Tillman, at the edge of our range.

101. ZYGOPHYLLACEAE R. Br., the Caltrop Family

by Ronald L. McGregor

Ours prostrate to decumbent, diffusely radially branching annual herbs with taproots; stems pubescent, sometimes becoming glabrate. Leaves opposite, even pinnate, 1 of each pair alternately smaller or sometimes abortive; leaflets 3–8 pairs, inequilateral; stipules present. Flowers solitary, perfect, regular or nearly so, peduncles from axils of the alternately smaller leaves; sepals 5, distinct, imbricate in bud, persistent to deciduous; petals 5, distinct, apex rounded to slightly notched; stamens 10, in 2 whorls of 5 each, outer whorl opposite petals and basally adnate to them, intrastaminal glands present or absent; ovary 5- or 10-lobed and loculed, ovules 1 or 3–5 per locule; style usually persisting to form a beak on fruit apex. Fruit of 5–10 indehiscent mericarps.

1 Mericarps 10, tubercled, 1 seeded; beak of fruit persisting after mericarps fall; intrastaminal glands absent .... 1. *Kallstroemia*
1 Mericarps 5, spiny, 3- to 5-seeded; beak of fruit falling with mericarps; intrastaminal glands present .... 2. *Tribulus*

1. KALLSTROEMIA Scop.

1. *Kallstroemia parviflora* Norton. Annual herbs with taproots; stems prostrate to decumbent or ascending, to 1 m long, coarsely hirsute and sericeous, becoming glabrate. Leaflets (3)4 or 5(6) pairs, elliptical to oblong or oval, appressed-hirsute, veins and margins sericeous, 8–19 mm long, 3.5–9 mm wide, inequilateral; stipules lanceolate, 5–7 mm long, 1–3 mm wide. Flowers solitary, peduncles usually longer than subtending leaves, 1–4 cm long in flower; sepals 5, lanceolate, 4–7 mm long, 1–2 mm wide, hispid, persistent; petals 5, orange, drying white to yellow, narrowly obovate, 5–11 mm long, 3.5–6 mm wide, marcescent;

stamens as long as style, intrastaminal glands absent; ovary pubescent, ovoid, 1 mm in diam. Fruit ovoid, 3–4 mm high, 4–6 mm wide, strigose, beak 3–9 mm long, persisting after mericarps fall; mericarps 3–4 mm high, 1 mm wide, abaxially rugose to tuberculed, sides pitted. Jul–Sep. Infrequent to common along roadsides, prairies, stream valleys, waste areas; s 1/3 GP; (KS, CO, se CA, s to TX, Mex., introduced elsewhere). *K. intermedia* Rydb. — Rydberg

Reference: Porter, Duncan M. 1969. The genus *Kallstroemia* (Zygophyllaceae). Contr. Gray Herb. 198: 41–153.

Kallstroemia hirsutissima Vail was listed for KS and CO by Rydberg, but the species is well to the sw of our area.

2. TRIBULUS L.

1. *Tribulus terrestris* L., puncture vine, goat head. Annual herbs with taproots; stems prostrate to 1.5 mm long, usually branched, hirsute and appressed-sericeous, becoming glabrate. Leaves opposite, even-pinnate; leaflets (3)4–6(7) pairs, oblong to ovate, 4–11 mm long, 1–4 mm wide, inequilateral, densely to sparsely sericeous below, sparsely so above, becoming glabrate; stipules lanceolate, 3–6 mm long, 0.5–1.3 mm wide. Flowers solitary, peduncles usually shorter than subtending leaves, pubescent; sepals 5, ovate 2–3 mm long, 1.5–2 mm wide, pubescent; petals 5, obovate, apex rounded or lobed, 3–5(6) mm long, 2–3 mm wide, yellow, rarely white, drying whitish; stamens shorter to longer than style, intrastaminal glands free; ovary ovoid, pubescent, 1 mm in diam, 5-lobed and 5-loculed, ovules 3–5 per locule. Fruit 1 cm in diam, separating into 5 indehiscent mericarps, each with 2–4 stout dorsal spines and sometimes with smaller spines and aristate tubercles, beak falling with mericarps; each mericarp internally divided into 3–5 1-seeded compartments. (n = 12, 24) May–Oct. Often abundant in sandy or gravelly disturbed soils; s 2/3 GP, scattered in n; (widely distributed in warm temp. regions of the world, Medit. area).

102. OXALIDACEAE R. Br., the Wood Sorrel Family

by Ronald L. McGregor

1. OXALIS L., Wood Sorrel

Caulescent or acaulescent perennial or annual herbs with rhizomes, stolons, or scaly bulbs. Leaves alternate or basal, ternately compound; leaflets 3, usually obcordate, folding together at night. Inflorescence an umbel or cyme. Flowers regular, perfect; sepals 5, usually imbricate; petals 5, sometimes united at base; stamens 10, of 2 lengths, usually monadelphous at base; pistil with 5 carpels, styles distinct. Capsule longitudinally dehiscent; seeds red or brown, enclosed in a transparent aril.

1 Plants caulescent; flowers yellow.
  2 Stems creeping, rooting at nodes; stipules broad, conspicuous .... 1. *O. corniculata*
  2 Stems ascending, not rooting at nodes; stipules narrow, inconspicuous.
    3 Septate hairs absent on all parts; capsules appressed-pubescent; flowers in umbels .... 2. *O. dillenii*
    3 Septate hairs present on stems, petioles or pedicels; capsules glabrous or with a few nonappressed, septate hairs; flowers in cymosely unequally branched

umbels .... 3. *O. stricta*
1 Plants acaulescent; flowers rose-violet (white) .... 4. *O. violacea*

1. **Oxalis corniculata** L., creeping ladies sorrel. Stems from a slender taproot, prostrate and rooting at the nodes, glabrous to pubescent. Leaves in fascicles at nodes of horizontal stems or on short nodal shoots, often purplish; often with broad brownish or purplish auriculate stipules. Inflorescence 1-flowered or umbellate; pedicels reflexing in fruit. Capsules 0.8-2 cm long, sparsely to densely pubescent; seeds brown, reticulate or transversely ridged. (n = 22, 24, 36, 42) All year in greenhouses; May–Oct. Rarely found in greenhouses & gardens in GP; (scattered in U.S.). *Xanthoxalis corniculata* (L.) Small — Rydberg.

This is a semicosmopolitan weed that has been reported for the GP frequently, but is usually based on incorrect determinations. Though rarely found around greenhouses, the plant does not persist for long outside.

2. **Oxalis dillenii** Jacq., gray-green wood sorrel. Plants usually cespitose; branching at base; rhizomes very short or absent. Stems 1-2.5 dm tall, ascending or decumbent, usually densely pubescent with nonseptate hairs. Inflorescence an umbel with 2-9 flowers; peduncle usually longer than leaves; pedicels strigose with nonseptate hairs. Sepals 3-7 mm long; petals 8-10 mm long, yellow. Capsule 1-2.5 cm long, gray-canescent with nonseptate hairs; seeds with white markings on conspicuous transverse ridges. (n = 18, 24) Mar–Nov. Open woodlands, prairie ravines, gardens, & waste places; GP but uncommon in n & absent sw; (nearly a cosmopolitan weed). *Xanthoxalis stricta* (L.) Small — Rydberg.

3. **Oxalis stricta** L., yellow wood sorrel. Plant to 5 dm tall, erect or becoming decumbent. Stems single from main root or from an underground rhizome; subglabrous to densely pubescent, with septate hairs and short appressed nonseptate hairs. Leaves often purplish; at least some petioles with septate hairs at base and nonseptate ones; stipules none. Inflorescence 1-flowered, umbellate, subcymose or cymose. Flowers orange-yellow or lemon; sepals 3-5 mm long; petals 11 mm long or less. Capsules 5-15 mm long, glabrous or with a few septate hairs; seeds brown, with inconspicuous white markings on transverse ridges. (n = 18, 24) Apr–Oct. Open woods, prairie ravines, stream banks, gardens, waste places; GP becoming uncommon & absent in w 1/4; (nearly a cosmopolitan weed). *Xanthoxalis bushii* Small, *X. cymosa* Small — Rydberg.

A number of infraspecific taxa have been named largely on the basis of variations in pubescence, but none seem worthy of taxonomic recognition. The leaves of this species and *O. dillenii* are used infrequently in green salads, and the leaves are often chewed for their sour taste. These two species are aecial hosts for maize rust, sorghum rust, and *Andropogon* rust. Aecia have been found on specimens in Kansas, Nebraska, and South Dakota.

4. **Oxalis violacea** L., violet wood sorrel. Acaulescent, colonial, finely stoloniferous (seldom noted), glabrous herbs with scaly bulbs. Leaves 6-12 cm long; leaflets 3, 1-2.5 cm wide, lobes rounded. Scapes 1.2-2.5 dm tall, usually 2× length of leaves, umbellately 4- to 15-flowered. Sepals 4-6 mm long, tipped by orange callosities; petals 1-2 cm long, pinkish-purple (white). Capsules 4-6 mm long; seeds reticulate. (n = 14) Apr–Jun, again Sep–Oct. Open woodlands, prairies, roadsides, & waste places; e 1/2 GP & scattered westward; (MA, MN, ND, s to FL, TX). *Inoxalis violacea* (L.) Small — Rydberg.

Occasional plants are found with white flowers, and a rare form with glandular hairs on the petiole has been found in e KS. Colonies flowering in the fall are usually without leaves.

103. GERANIACEAE Juss., the Geranium Family

by Ronald L. McGregor

Annual or perennial herbs with basal, alternate, or opposite stipulate, simple or compound leaves. Inflorescence cymose or a single flower. Flowers perfect, regular or somewhat irregular, 5-merous, hypogynous; sepals imbricated in bud; glands of the disk 5, alternate with petals; stamens, including sterile filaments, as many to 2 × as many as sepals; carpels 5, 2-ovuled, prolonged at maturity into beaks, separating elastically from the axis.

1 Fertile stamens 5; ripe carpels fusiform, the beaks tightly twisted when dry .... 1. *Erodium*
1 Fertile stamens usually 10; ripe carpels plump, the beaks outwardly coiled .... 2. *Geranium*

1. ERODIUM L'Her., ex Ait., Stork's-bill

Winter annual, biennial, or perennial herbs forming rosettes with spreading to ascending branches. Leaves petiolate, entire and palmately lobed or pinnatifid or pinnately compound. Peduncles axillary with flowers in an umbel; sepals 5; petals 5, obovate, upper 2 sometimes smaller than lower three; stamens in two series, outer sterile or reduced, inner fertile and opposite the sepals. Ripe carpels fusiform, hispid, tardily dehiscent, beaks separating elastically and becoming spirally twisted when dry.

1 Leaf blades pinnate-pinnatifid .... 1. *E. cicutarium*
1 Leaf blades simple, basally cordate and shallowly or deeply palmately lobed .. 2. *E. texanum*

1. *Erodium cicutarium* (L.) L'Her., filaria. Winter annual; stems at earliest flowering short, with basal leaves from winter rosette, becoming prostrate or ascending, much branched from the base, with branches to 5 dm long. Mature leaves elongate-oblanceolate, pinnate-pinnatifid. Umbels long-peduncled, 2- to 8-flowered; pedicels 1–2.5 cm long. Flowers 1 cm wide; sepals elliptic, 2–6 mm long; petals pink, elliptic-obovate, slightly longer than sepals; style column to 5 cm long. Seed brown, 2–3 mm long, ellipsoid. Mar–Sep. Lawns, fields, roadsides, & waste places; s 1/2 GP & becoming more frequent northward. (Que., MI, IL, KS, s to VA, TN, AR; Mex; Europe). *Naturalized.*

2. *Erodium texanum* A. Gray. Prostrate or ascending, branched from the base, branches to 5 dm long. Leaves ovate, to 3.5 cm long, lobes rounded, crenate. Inflorescence axillary, pedunculate, in an umbel. Flowers purplish-red; sepals elliptic, apiculate, 6–11 m long; petals broadly obovate, 10 mm or more long, exceeding the sepals; style column to 70 mm long. Seed 4–5 mm long. Mar–Apr. Prairies & open areas; sw OK & TX; (sw OK, TX, to UT, se CA).

2. GERANIUM L., Cranesbill

Annual or perennial herbs with palmately lobed, cleft, or divided stipulate leaves. Flowers usually in pedicillate pairs on axillary peduncles, regular or somewhat irregular. Sepals 5, imbricate; petals 5, pink, white, or bluish, imbricate, obovate to obcordate, alternating at base with 5 glands; stamens 10, usually all fertile, the longer 5 with glands at their base; carpels 5, stylar portions remaining attached at their apex, free portion coiling outwardly.

References: Fernald, M. L. 1935. *Geranium carolinianum* and allies of northeastern North America. Rhodora 37: 295–301. Jones, G. N. & F. F. Jones; 1943. A revision of the perennial species of *Geranium* of the United States and Canada. Rhodora 45: 5–26, 32–53.

1   Plants annual or biennial; petals less than 12 mm long.
    2   Leaves completely divided into separate pinnately cleft segments .... 6. *G. robertianum*
    2   Leaves not completely divided.
        3   Sepals terminating in subulate tips.
            4   Pedicels at maturity more than 2× longer than the calyx; stylar beak of fruit 3–5 mm long .... 1. *G. bicknellii*
            4   Pedicels at maturity 2× longer than calyx or shorter; stylar beak of fruit 1–2.5 mm long .... 2. *G. carolinianum*
        3   Sepals blunt or acute, or with minute callous tips .... 4. *G. pusillum*
1   Plants perennial from a stout rootstock; petals 12 mm or more long.
    5   Petals glabrous except for the ciliate base; fruiting pedicels erect .... 3. *G. maculatum*
    5   Petals pilose on inner surface for 1/4–1/2 their length; fruiting pedicels spreading or reflexed and ultimately bent upward.
        6   Petals white or pale pink; inflorescence pilose with glandular, purplish-tipped hairs; beak of stylar column and free lobes of stigma about equal in length .... 5. *G. richardsonii*
        6   Petals pinkish-lavender to purple; inflorescence glandular with yellowish hairs; stylar column definitely longer than free lobes of stigma .... 7. *G. viscosissimum*

1. ***Geranium bicknellii*** Britt., Bicknell's cranesbill. Annual or biennial erect herb to 5 dm tall with ascending branches. Principal leaves angulate rotund, cleft nearly to base, with normally 5 segments, these deeply incised. Peduncles with 2 flowers; pedicels glandular-villous. Sepals 7–9 mm long at anthesis; petals pink-purple, about as long as sepals. Fruit 15–25 mm long, beak 4–5 mm long, the body 3 mm long, sparsely hirsute; seed reticulate with elongate areolae. Jun–Sep. Rare in upland woods; MN; ND: Bottineau, Emmons; IA; BH of SD; (Newf., Que., B.C., s to NY, PA, IN, IO, SD, WY, MT, UT, WA).

2. ***Geranium carolinianum*** L., Carolina cranesbill. Annual herbs. Stems erect, simple or several from the base, freely branched, 1–5(7) dm tall, pubescence spreading or retrorsely hirsute to pilose and often somewhat glandular at least in the inflorescence. Principal leaves rotund-reniform, 3–7 cm wide, deeply cleft into 5- to 9-toothed or lobed segments. Inflorescence congested, peduncles no longer than pedicels which usually do not exceed calyx. Sepals 4–10 mm long, narrowly to broadly ovate, 3(5)-nerved, bristle-tipped 1–1.5 mm long; petals white, pink, or rose, about equaling the sepals, rounded to obtuse; fertile stamens 10; stylar column 1.5 cm long, the beak, including stigmas, 2–2.5 mm long, body hirsute with antrorse hairs. Seeds oblong to subspherical, 1–1.5(2–2.5) mm in diam, faintly reticulate-alveolate with elongate alveolae. (n = 26) May–Sep. Open woodlands, prairie ravines, fields, roadsides, & waste places; scattered in n 1/2 GP, becoming common in e 3/4 of KS & southward; (a near cosmopolitan weed). *G. sphaerospermum* Fern., *G. carolinianum* var. *confertiflorum* Fern–Fernald.

Plants with congested inflorescences of 5–25 flowers and spreading hairs about 1 mm long on stems and petioles have been called var. *confertiflorum* Fern. Some plants from eastern KS have the hairs but not the congested inflorescence, while others have the inflorescence but lack the longer hairs. In the BH some large plants have broadly ovate sepals which are faintly 5-nerved and seeds to 2.5 mm in diam. These have been named *G. sphaerospermum* Fern., but this phase grades into *G. carolinianum*.

3. ***Geranium maculatum*** L., wild cranesbill. Erect perennial herb 2–5 dm tall, from a stout rhizome. Stem retrorsely pubescent, with 1 pair of leaves 0.5–1.5 dm wide; these deeply 5-parted with cuneate lobed and incised segments; basal leaves similar, long-petioled. Inflorescence a terminal corymb of a few to several flowers; pedicels erect in flower and fruit. Sepals 1 cm long; petals entire, rose-purple (white) 1.5–2.4 cm long, 1 cm wide. Fruit erect; style column 2–3 cm long, minutely pubescent; seeds 2.5–3 mm long, glabrous, finely reticulate. Apr–May. Rich or rocky woods; infrequent in e GP; ND: Cass; SD: Marshall, Roberts; NE: Brown, Douglas, Lancaster, Sarpy; KS: Cherokee, Doniphan, Jefferson, Leavenworth, Wyandotte; (ME, w to Man., SD, s to SC, GA, AR, OK).

4. **Geranium pusillum** L., small cranesbill. Winter annual. Stems slender, weak, diffusely branched, spreading or ascending, 1-5 dm long. Leaves reniform, 1.2-3 cm long, 7- to 9-cleft, divisions lobed, upper becoming progressively smaller. Peduncles from upper axils; pedicels densely but minutely glandular. Sepals at anthesis 2.5-4 mm long, calloustipped; petals pale pink, about as long as sepals; style column 6-9 mm long. Seeds brown to black, 1.5-1.7 mm long, smooth. (n = 13, 17) Apr-Sep. Lawns, fields, waste places; scattered & infrequent in GP; (MA to B.C., s to NC, KS, AR & OR; Europe). *Naturalized.*

5. **Geranium richardsonii** Fisch. & Trautv., Richardson's cranesbill. Perennial herb, 2-8 dm tall, glabrous to rarely sparsely strigulose below, purplish glandular-pilose in inflorescence. Leaf blades 6-12 cm wide, cleft 3/4 the length or more. Sepals 6-11 mm long, setose tips 1.5-2 mm long; petals 12-17 mm long, white to pinkish, pilose 1/2 their length on inner surface; beak of stylar column about equal to free lobes of stigma. Seed brownish, 3 mm long, smooth to faintly roughened. (n = 26) May-Aug. Rich wooded hillsides & ravines; frequent in BH; (w N. Amer.).

Occasional specimens with light lavender flowers and more pubescence may be hybrids with G. *viscosissimum,* which grows with this species.

6. **Geranium robertianum** L., herb-Robert. Annual or winter annual herb; strong scented. Stems weak, spreading, branched, villous, to 6 dm long. Leaves triangular in outline, divided into 3-5 distinct pinnatifid leaflets, the terminal stalked. Flowers pink to red-purple (white); sepals awned; petals long clawed, exceeding the sepals; carpel bodies promptly separating from styles. Seeds about 2 mm long, dark brown, smooth to faintly rugulose or reticulate. May-Sep. Rich woods; cited for NE (Rydberg); (Newf. to Man; s to WV, OH, IN, IL; Eurasia). *Naturalized.*

While specimens are not present in GP herbaria, the species may yet be found.

7. **Geranium viscsissimum** Fisch. & Mey., viscid cranesbill. Perennial herb. Stems 3-9 dm tall, hirsute and glandular-puberulent below, glandular-villous above, rarely somewhat glabrous. Leaf blades 5-10 cm wide, strigulose to hirsutulous and somewhat glandular, parted over 3/4 their length into usually 5 sharply toothed divisions. Sepals 8-12 mm long, setose tips (1)2(3) mm long; petals pinkish-lavender to purplish (white), 14-20 cm long, pilose at base for 1/4 of length; stylar column 3-5 cm long, beak including stigmas 10-14 mm long. Fruits glandular-hirsute; seeds brownish, 3-3.5 mm long, faintly reticulate. May-Aug. Open woods & stream banks; BH where infrequent but abundant in local colonies; (Sask. to B.C., s to w SD, CO, UT, NV).

Some specimens in the BH lack glands on stems and have glabrous lower petioles. These may be referred to var. *nervosum* (Rydb.) C. L. Hitchc. While the extremes are easily separated, the intergradations are complete.

104. BALSAMINACEAE A. Rich., the Touch-me-not Family

by Ronald L. McGregor

1. IMPATIENS L., Touch-me-not

Annual herbs. Stems hollow, glabrous, often succulent; nodes more or less swollen. Leaves simple, alternate, exstipulate, Flowers in pedunculate clusters, or 1-3 in leaf axils, often of two sorts; small cleistogamous ones which regularly produce fruit, and large ones which seldom produce seed; large flowers perfect, irregular; sepals 3, petaloid, 2 lateral ones small, the middle or lower one saccate and spurred at bottom; petals appearing to be 3 with the upper broader than long, each of the lateral 2-lobed and interpreted as 2 united petals; stamens 5, anthers united around stigma; pistil 1, ovary 5-celled, superior. Capsule 5-celled, elastically dehiscent into 5 spirally coiled valves.

1 Flowers orange to reddish, usually heavily reddish-brown spotted; spur 6 mm or more long, bent back and parallel to the body .... 1. *I. capensis*
1 Flowers pale yellow or yellow and white, more or less dotted with red; spur 5 mm or less long, bent at right angle to body .... 2. *I. pallida*

1. *Impatiens capensis* Meerb., spotted touch-me-not. Stems freely branched above, 0.5–1.5 m tall, slightly glaucous. Leaves pale or glaucous beneath, usually green above, ovate, elliptic, or elliptic-ovate, 3–10 cm long, crenately serrate with usually mucronate teeth. Flowers irregular, orange to red, usually with crimson or variously colored spots; spurred sepal longer than width of mouth, spur 6–9 mm long, bent back and parallel to the body. Seeds 4–5 mm long, with 4 corky, longitudinal ridges, surface areolate and often papillose, mottled green to brown. (n = 10) May–Oct. Low moist woodlands, stream banks, marshes, & wet places; e 1/3 ND & SD, BH, NE, e 1/4 KS; (Newf., Que., Sask., s to SC, OK, TX). *I. biflora* Walt.—Rydberg, Atlas GP; *I. nortonii* Rydb.—Rydberg.

About 2/3 of the GP populations have the spurred sepal 2 × longer than wide and tapering gradually to the spur, while the remainder have the spurred sepal 2/3 as long as wide and tapering abruptly to the spur. Both types are to be found in the same colony and intermediate forms are somewhat frequent. A number of color variation forms have been described, but do not appear to be present in the GP.

2. *Impatiens pallida* Nutt., pale touch-me-not, jewel weed. Stems freely branched above, 0.5–2 m tall, glaucous. Leaves conspicuously glaucous above, 5–13 cm long, ovate or elliptic-ovate, crenately serrate, usually with mucronate teeth. Flowers irregular, pale yellow to whitish, with some reddish spots; spurred sepal as wide as long, spur 5 mm long, bent at right angles to the body which tapers abruptly to the spur. Seeds 5–6 mm long with 4 corky longitudinal ridges, surface smooth to rugulose and reticulate, green to brown. (n = 10) Jun–Oct. Low moist woodlands, stream banks, & wet places; e GP with a few scattered stations westward; (Que., Sask., s to NC, TN, e OK).

This species is much less common than *I. capensis* and is found less in the open. Occasional specimens have the spurred sepal pale yellow and the upper and lateral petals white. It has been reported that this and the previous species will prevent poison ivy when the leaves are rubbed on the skin.

105. ARALIACEAE Juss., the Ginseng Family

by Ronald L. McGregor

Aromatic perennial herbs. Leaves usually compound, alternate or solitary. Flowers regular, epigynous, perfect or unisexual, small, umbellate; sepals 5 or obsolete; petals 5, valvate, usually distinct, early fugaceous; stamens usually 5, inserted on a disk; ovary 2- to 5-locular, with 1 ovule per cell, styles fused or distinct. Fruit a berry or drupe.

1 Leaves pinnately decompound, alternate or basal; umbels 2 or more; carpels 5 ...... 1. *Aralia*
1 Leaves digitately once-compound, in a single whorl; umbel solitary; carpels 2 or 3 . 2. *Panax*

1. ARALIA L., Sarsaparilla, Spikenard

Leaves compound or decompound, alternate or solitary, with toothed leaflets. Flowers in racemose or paniculate umbels, greenish-white; sepals 5; petals 5; stamens 5; ovary 5-celled, styles 5, fused basally or separate, stigmas capitate. Drupe 5-lobed, 5-celled, purple or black; seeds 5, flattened.

1 Plant scapose, solitary leaf and peduncle arising from the rhizome .... 1. *A. nudicaulis*
1 Plant not scapose, leaves several .... 2. *A. racemosa*

1. *Aralia nudicaulis* L., wild sarsaparilla. Acaulescent perennial herbs with the leaves and peduncles arising from a long rhizome. Leaves usually solitary, erect, ternately compound with each division 3- to 5-foliate; petiole to 5 dm tall; leaflets lance-elliptic to obovate, 5–13 cm long, acuminate, finely serrate. Peduncles usually solitary, 0.8–2.5 dm long, bearing (2)3(6) umbels; sepals 0.2–0.5 mm long; petals green or white, 1–2 mm long; styles distinct. Drupes purplish-black, 6–10 mm in diam; seeds 5–7 mm long. May–Jul. Moist or dry woodlands; ND, BH & e SD; NE: Cherry, Dixon; (Newf., B.C., s to IN, NE, CO).

The rhizomes and roots of this plant have been used as one of the ingredients of root beer.

2. *Aralia racemosa* L., spikenard. Perennial herb 1–2 m tall, often somewhat woody below; stem purplish or spotted. Leaves spreading, pinnately compound or ternately and pinnately decompound; leaflets 3–7, commonly ovate, to 2 dm long and 1.5 dm wide but variable in size on the same leaf, usually glabrate, acuminate, serrate or doubly serrate, obliquely cordate at base. Inflorescence a panicle with many umbels, 0.4–4 dm long, terminal and axillary. Sepals 0.3 mm long; petals white, 1 mm long; styles fused at base. Drupes blackish, 5–6 mm in diam; seeds 2.5–4 mm long. Jul. Wooded hillsides & ravines; e SD; e NE & Cherry; ne KS; (Que., N.B., MN, SD, s to GA, TX, AZ, n Mex.).

This is a rare plant in the e GP. The roots have been used as one of the ingredients of root beer and as a remedy for respiratory ailments.

2. PANAX L., Ginseng

1. *Panax quinquefolium* L., ginseng. Aromatic perennial herb with fusiform root and solitary stem 1–5 dm tall. Leaves palmately compound, whorled, petiolate, 3 or 4; leaflets 3–5, oblong-obovate, 6–15 cm long, 8 cm wide, acuminate, serrate, petiolulate. Umbel terminal, solitary; peduncle 1–12 cm long. Flowers greenish-white, mostly perfect; sepals 0.2 mm long or obsolete; petals 0.5–1 mm long, white to greenish; styles usually 2, 1–2 mm long. Drupes 10 mm in diam, bright red. Jun–Jul. Rich woods; e SD, e NE; (Que., MN, SD, s to GA, OK). *Endangered* in GP.

This well-known but dubiously medicinal plant has been reported for ne KS, but specimen evidence is lacking. It has rarely been collected in e GP and may be extirpated.

106. APIACEAE Lindl., the Parsley Family

by Ronald L. McGregor

Herbaceous, rarely woody at base, annuals, biennials, or perennials with commonly hollow stems, often aromatic. Leaves alternate or basal, simple or compound, usually much incised or divided, with usually sheathing petioles. Inflorescence of simple or compound umbels or in heads; rays sometimes subtended by bracts forming an involucre; umbellets usually subtended by bractlets forming an involucel. Flowers small perfect or some flowers in the umbellet only staminate, 5-merous; calyx tube completely adnate to ovary, calyx teeth usually obsolete or minute, persistent; corolla usually regular, petals usually with an inflexed tip; stamens inserted on an epigynous disk; ovary inferior, bilocular, styles 2, sometimes swollen at the base, forming a stylopodium. Fruit a schizocarp, consisting of two seedlike mericarps, attached until maturity by their inner face (commissure), compressed or flattened dorsally, laterally, or terete; at maturity they separate and are often suspended from a slender stalk, the carpophore; each mericarp with 5 primary ribs and rarely with secondary ribs; primary ribs winged, or prominently or obscurely rounded; oil tubes obsolete or present in spaces between the ribs, and on the commissural surface.

Reference: Mathias, M. E., & L. Constance. 1944–45. Umbelliferae. North Amer. Fl., 28B: 43–297.

The following species have been reported for our area in various publications, but are based on cultivated specimens, or temporary garden escapes, and are not known to persist in our area. They are excluded in this treatment of the family. *Ammi majus* L., bishops-weed; *Conioselinum chinense* (L.) B.S.P., hemlock parsley; *Coriandrum sativum* L., coriander; *Falcaria sioides* (Wibel) Asch., sickleweed; *Foeniculum vulgare* P. Mill., fennel; *Petroselinum crispum* (P. Mill) Mansf., parsley; and *Scandix pectenveneris* L., venus' comb.

1 Leaves all simple.
  2 At least upper leaves perfoliate .... 4. *Buplerum*
  2 None of the leaves perfoliate.
    3 Leaf blades round-reniform .... 16. *Hydrocotyle*
    3 Leaf blades not round-reniform .... 13. *Eryngium*
1 At least some leaves compound, dissected or deeply lobed.
  4 Inflorescence capitate or not appearing as a true umbel.
    5 Foliage pubescent .... 30. *Torilis*
    5 Foliage glabrous or essentially so.
      6 Plants with a rosette of pinnately dissected leaves on a
        pseudoscape .... 10. *Cymopterus*
      6 Plants leaf-stemmed.
        7 Flowers in dense capitate heads, all perfect and sessile .... 13. *Eryngium*
        7 Perfect flowers sessile or essentially so in glomerules, mixed with pedicellate
          staminate ones .... 25. *Sanicula*
  4 Inflorescence a true umbel or a compound umbel.
    8 Ovaries and fruits distinctly bristly or tuberculate.
      9 Leaves once-palmately divided into 3–7 segments .... 25. *Sanicula*
      9 Leaves once-pinnate or twice-compound or decompound.
        10 Foliage glabrous .... 27. *Spermolepis*
        10 Foliage pubescent.
          11 Bracts of involucre pinnately divided .... 11. *Daucus*
          11 Bracts of involucre not pinnately divided .... 30. *Torilis*
    8 Ovaries and fruits glabrous, pubescent, coarsely scabrous, or inconspicuously
      papilliate.
      12 Leaves much dissected or decompound, the leaflets not immediately obvious,
        usually divided into ultimate segments less than 1 cm wide.
        13 Plants annual.
          14 Ultimate leaf segments filiform.
            15 Primary rays of umbel distinctly unequal .... 27. *Spermolepis*
            15 Primary rays nearly equal.
              16 Flowers yellow; involucre usually wanting .... 2. *Anethum*
              16 Flowers white; involucre present.
                17 Nodes, petioles, and umbel branches
                  scaberulous .... 14. *Eurytaenia*
                17 Nodes, petioles, and umbel branches smooth ..... 24. *Ptilimnium*
          14 Ultimate leaf segments linear or broader.

       18 Fruits without secondary ribs; stems sparsely retrorse hispid
           below .... 6. *Chaerophyllum*
       18 Fruits with secondary ribs; stems glabrous .... 1. *Ammoselinum*
  13 Plants biennial or perennial.
    19 Plants from a globose tuber .... 12. *Erigenia*
    19 Plants without globose tubers.
      20 Plants 1–3 m tall; stems purple-spotted .... 8. *Conium*
      20 Plants less than 1.2 m tall; stems not purple-spotted.
        21 Plants acaulescent or short caulescent.
          22 Fruits without wings .... 19. *Musineon*
          22 Fruits prominently winged .... 18. *Lomatium*
        21 Plants distinctly caulescent.
          23 Flowers yellow; fruit with lateral wings thick and corky .... 23. *Polytaenia*
          23 Flowers white or pinkish; fruit not winged.
            24 Plants with taproots .... 5. *Carum*
            24 Plants with fusiform or tuberously thickened roots .... 22. *Perideridia*
12 Leaves with distinct and separate leaflets nearly uniform in shape, these often, but not
   always, 2 cm or more wide.
  25 Plants acaulescent or subacaulescent.
    26 Fruits without wings .... 19. *Musineon*
    26 Fruits conspicuously winged.
      27 Inflorescence capitate or subcapitate .... 10. *Cymopterus*
      27 Inflorescence of loose compound umbels .... 18. *Lomatium*
  25 Plants distinctly caulescent.
    28 Leaflets entire, without lobes or teeth.
      29 Leaflets ovate to oblong or elliptic .... 28. *Taenidia*
      29 Leaflets linear.
        30 Leaflets cross-septate; upper leaves without axillary
          bulbils .... 17. *Limnosciadium*
        30 Leaflets not cross-septate; upper leaves with axillary bulbils .... 7. *Cicuta*
    28 Leaflets variously toothed or lobed.
      31 Leaves pinnate or bipinnate.
        32 Flowers yellow; fruit dorsally flattened and winged .... 21. *Pastinaca*
        32 Flowers white; fruit laterally flattened, not winged.
          33 Leaflets of upper leaves irregularly incised .... 3. *Berula*
          33 Leaflets regularly toothed.
            34 Primary lateral veins of leaflets directed to the sinuses between teeth;
              primary leaves usually bipinnate .... 7. *Cicuta*
            34 Primary veins not directed to the sinuses; leaves once-
              pinnate .... 26. *Sium*
      31 Leaves palmately or ternately divided or the basal entire.
        35 Terminal leaflets palmately lobed, 1–4 dm long and wide; fruit 8–12 mm long,
          nearly as wide .... 15. *Heracleum*
        35 Terminal leaflets not palmately lobed, much smaller; fruits narrower or much
          smaller.
          36 Flowers white; fruit linear, linear-oblong to narrowly clavate.
            37 Leaves trifoliate, fruit glabrous .... 9. *Cryptotaenia*
            37 Leaves twice ternately compound; fruit sparsely to densely appressed-
              bristly .... 20. *Osmorhiza*
          36 Flowers yellow; fruit ovoid, ovate, or oblong.
            38 Central flower of each umbellet sessile; fruit without wings ..... 31. *Zizia*
            38 Central flower of each umbellet pedicelled; fruit with prominent
              wings .... 29. *Thaspium*

1. AMMOSELINUM T. & G., Sand-parsley

1. ***Ammoselinum popei*** T. & G. Annual herbaceous, slender, erect or diffuse, often branching from the base. Stems 1–3.5 dm tall, from slender taproots. Leaves ternate, biternate,

or ternate-pinnately dissected; ultimate divisions linear, obtuse, slightly mucronulate, glabrous or somewht roughened with small callous teeth. Inflorescence of compound umbels, peduncles axillary and terminal, 0.5-4 cm long; involucre 0 or of a single bract; involucels of several linear, roughened, and basally scarious bracts, longer than pedicel; rays 2-10, unequal, to 2.5 mm long; pedicels 2-11, to 10 mm long. Flowers white, calyx teeth obsolete, stylopodium low-conic. Fruit oblong-ovoid, 3-6 mm long, to 3 mm wide, usually roughened, ribs rounded, the lateral ribs with corky appendages. Mar-Jun. Sandy or calcareous prairies, canyons, stream banks; cen KS, to s & w; (KS, OK, TX, NM, n Mex.).

A. butleri (S. Wats.) Coult. & Rose, of east OK and southward, has been found in a lawn and city park in KS: Cherokee, Crawford. It is only 3-6 cm tall, umbels sessile, fruit glabrous, and lateral ribs without corky appendages.

2. ANETHUM L., Dill

1. ***Anethum graveolens*** L., dill. Annual erect, caulescent, alternately branching above, striate, glabrous and glaucous, herbaceous, 4-17 dm tall; with strong anise odor; from subfusiform roots. Leaves finely pinnately dissected, to 30 cm long and 20 cm wide; upper much reduced. Inflorescence of compound umbels; peduncles 7-15 cm long, exceeding the leaves; involucre and involucel lacking; rays 10-40, 3-10 cm long; pedicels 6-10 mm long; flowers yellow, calyx teeth obsolete, stylopodium short, conic; carpophore divided to base. Fruit ovoid, 4-6 mm long, 2 mm wide, flattened dorsally, ribs narrowly winged. (n = 11) Apr-Aug. Sporadically escaped from cultivation in GP; (throughout U.S.; Europe).

3. BERULA Hoffm., Water-parsnip

1. ***Berula erecta*** (Huds.) Cov. var. ***incisum*** (Torr.) Cronq. Slender, erect or reclining glabrous, branching perennial herb from fascicled roots, 2-10 dm tall. Leaves oblong, pinnate; leaflets oblong, 9-23, subentire to serrate or lobed, upper ones usually lanceolate and deeply incised. Inflorescence of compound umbels; peduncles terminal and axillary, 2-7 cm long, the lateral ones usually overtopping the terminal; involucre of 6-8 unequal foliaceous scarious-margined entire to toothed bracts, 5-14 mm long; involucel of 4-8 linear-lanceolate entire bractlets, 1-5 mm long; rays 5-15, subequal; pedicels 2-5 mm long. Flowers white; calyx teeth minute, stylopodium conic; carpophore divided to base. Fruit oval, 1.5-2 mm long, compressed laterally, with thick corky pericarp. (n = 6) May-Oct. Around springs, cool shallow water; GP but rare northward & absent in se; (NY, s Ont., MN, B.C., s to FL, TX, Mex.). *B. pusilla* (Nutt.) Fern.—Fernald.

4. BUPLERUM L., Thoroughwax

1. ***Buplerum rotundifolium*** L. Glabrous and often glaucous annual, 2-6 dm tall, erect or spreading. Lower and basal leaves oblong to ovate-lanceolate, to 7 cm long and 5 cm wide, rounded at apex, subpetiolate or somewhat perfoliate at base; upper stem leaves, ovate, perfoliate. Inflorescence of compound umbels; involucre lacking; involucel of 5 or 6 broadly ovate acuminate bractlets, 8-12 mm long, 6-10 mm wide, united at base, longer than flowers; rays 4-10, 5-15 mm long; pedicels 10-12, equaling or shorter than fruits. Flowers yellow. Fruit oblong-oval, to 3 mm long, purplish-brown, smooth. (n = 8) Apr-Jun. Waste places; KS: Coffey, Wabaunsee; MO: Clay; (NY, s IL, MO, KS, s to NC, AR, TX; Europe). *Adventive.*

5. CARUM L., Caraway

1. ***Carum carvi*** L., caraway. Erect, glabrous biennials, 3-6 dm tall, from taproot. Leaves 3-4× pinnatifid, with filiform or narrowly linear divisions. Umbels compound, lateral and terminal, peduncled; involucre of 0-3 linear bracts; involucel usually absent; rays few to many. Flowers white (pink). Fruit elliptic to oblong, 3-4 mm long, with conspicuous ribs. (n = 10) May-Jul. Waste places; sporadic & locally weedy in n GP; (Newf. to B.C., s to VA, IN, MO, IA, CA; Eurasia).

6. CHAEROPHYLLUM L., Chervil

Erect or decumbent weedy annuals; stems glabrous or pubescent. Leaves pinnate-ternately dissected; leaflets pinnatifid, with linear or oblong lobes; petiole sheath ciliate. Umbels compound; involucre usually absent; involucel of several bractlets, variable in length, spreading or reflexed in fruit; pedicels 1-5 mm long. Flowers white, 3-10 per umbellet; calyx teeth obsolete; stylopodium conic. Fruit slenderly ovoid to linear-lanceolate, 6-8 mm long, notched at base, with beak about 0.5 mm long.

1 Stems usually spreading to decumbent, glabrous to slightly pubescent at base; pedicels of fruits filiform; ribs on fruits narrower than spaces between ribs .... 1. *C. procumbens*
1 Stems usually erect, solitary, sparsely hispid above, densely hispid below to glabrate; pedicels clavate; ribs on fruits wider than spaces between ribs .... 2. *C. tainturieri*

1. ***Chaerophyllum procumbens*** (L.) Crantz, wild chervil. Stems spreading, weak, usually branched from base, 1-6 dm tall, glabrous or sparsely hairy at base. Leaves ternate-pinnately decompound, glabrous, lobes of leaflets oblong, bluntish. Umbels simple or with 2 or 3 rays; involucel of several elliptic to narrowly obovate bractlets. Fruits on filiform pedicels, widest at or near the middle; ribs on fruit narrower than spaces between the ribs; body of fruit glabrous (rarely hairy), contracted above into a thick neck (rarely not contracted) below the erect stylopodia. Mar-May. Moist woods, creek banks, valleys, thickets, roadsides; se GP; (NY, s MI, IA, KS, s to GA, AR).

Rare plants of MO and e KS with fruits not contracted into a neck and glabrous to minutely hairy have been called var. *shortii* T. & G.

2. ***Chaerophyllum tainturieri*** Hook., chervil. Stems erect, solitary, usually branched near base, 1.5-9 dm tall, densely retrorsely hispid below to glabrate, sparsely hispid above. Leaves ternate-pinnately decompound, lobes linear to ovate, obtuse to acute, glabrous to more or less hispid. Umbels simple or with 2 or 3 rays; involucel of several conspicuous ovate rounded to acute ciliate-margined bractlets, spreading to reflexed in fruit. Fruits 3-10, on clavate pedicels widest near the base; ribs on fruit wider than spaces between ribs, body of fruit glabrous, beaked or narrowed toward apex. (n = 11) Mar-Jun. Rocky prairies, open woodlands, thickets; se GP; (se VA, s IN, MO, KS, s to FL, TX). *C. reflexum* Bush, *C. texanum* Coult. & Rose — Rydberg.

Rare plants with ovaries and fruits conspicuously pubescent have been called var. *dasycarpum* S. Wats. They are found growing with the more characteristic phase.

7. CICUTA L., Water Hemlock

Erect, branching, glabrous perennials or biennials from a somewhat tuberous base bearing fibrous, fleshy-fibrous, or fleshy-tuberous roots. Tuberous-thickened base of stem hollow, with well-developed transverse partitions; stem sometimes mottled with purple below. Leaves 1-3 × pinnate or ternate-pinnate; leaflets usually serrate or incised to nearly entire; petioles sheathing. Inflorescence of loose compound umbels; peduncles terminal and lateral, exceeding the leaves; involucre of inconspicuous, narrow bracts, or wanting. Flowers white; calyx teeth evident; carpophore 2-cleft to base, deciduous. Fruit oval or ovoid to orbicular or ellipsoid, flattened laterally and constricted or not constricted at the commissure, glabrous, ribs usually prominent, obtuse, corky; oil tubes solitary in the intervals, 2 in the commissure. All species POISONOUS.

Reference: Mulligan, Gerald A. 1980. The genus *Cicuta* in North America. Can. J. Bot. 58: 1755-1767.

1 Axils of at least upper leaves bearing bulblets; leaflets with narrowly linear segments to 5 mm wide; flowers usually abort .... 1. *C. bulbifera*
1 Axils of leaves without bulblets; leaflets with segments usually over 5 mm wide; flowers usually form mature fruit .... 2. *C. maculata*

1. *Cicuta bulbifera* L., bulbous water hemlock. Slender, erect, glabrous biennials or perennials with or without thickened roots. Stems usually solitary, not much thickened at base, 3-10 dm tall, at least upper axils bearing bulblets. Leaves 2- or 3-pinnate; leaflets linear to linear-lanceolate, 1-8 cm long, 1-5 mm wide, from nearly entire to sparsely toothed, rarely incised; petioles 1-1.5 dm long. Peduncles 1-5 cm long; involucre of a few small filiform bracts or wanting; involucel of a few linear to lanceolate bractlets, 1-3 mm long, or wanting. Umbels often wanting or present and not bearing fruit as flowers usually abort. Fruit orbicular, ca 2 mm long, constricted at the commissure; ribs low and wider than intervals; oil tubes much narrower than intervals. (n = 11) Jul-Sep. Marshes, wet meadows, edge of streams & ditches; ne ND to n-cen NE; (Newf. to AK, s to NC, KY, n IA, cen NE, MT, OR).

2. *Cicuta maculata* L., common water hemlock. Stout, erect, glabrous, often glaucous perennials or biennials from a fascicle of fleshy tuberous roots. Stems 0.5-2 m tall, often mottled with purple below; base usually hollow and with transverse partitions. Leaves 2- or 3-pinnate; leaflets narrowly lanceolate to lance-oblong, 2-12 cm long, 5-40 mm wide, remotely to sharply serrate; principal petioles 1-3 dm long. Peduncles 2-10 cm long; involucre of a few narrow bracts or wanting; involucel of several linear to lanceolate, scarious-margined bractlets 2.5 mm long. Fruit oval to orbicular 2-4.5 mm long, glabrous, constricted or not constricted at the commissure; dorsal corky ribs equaling to slightly exceeding size of oil tubes or these smaller than oil tubes. (n = 11)

Three intergrading varieties in our area:

1 Style usually less than 1 mm long; fruit usually about as wide as long, constricted at the commissure; principal stem leaflets more than 5 × longer than wide ..... 2a. var. *angustifolia*
1 Style usually more than 1 mm long; fruit longer than wide, constricted or not constricted at the commissure; principal stem leaflets less than 5 × longer than wide.
  2 Fruit constricted at the commissure; dorsal corky ribs much smaller than oil tubes .... 2b. var. *bolanderi*
  2 Fruit not constricted at the commissure; dorsal corky ribs equaling to slightly exceeding size of oil tubes .... 2c. var. *maculata*

2a. var. *angustifolia* Hook. Principal stem leaflets usually more than 5 × longer than wide. Style usually less than 1 mm long. Fruit usually about as long as wide; abruptly constricted at the commissure. Jul-Aug.; fruits Aug-Sep. Locally common in marshes, along stream banks, ditches; cen ND & nw KS, MT, s to NM, rare in IA, NE, KS, OK, TX panhandle; (Ont. to AK, s to WI, IA, AR, OK, n Mex., CA). *C. douglasii* (DC.) Coult. & Rose—Atlas GP and regional reports.

2b. var. *bolanderi* (S. Wats.) G. A. Mulligan. Principal stem leaflets usually less than 5 × longer than wide. Styles over 1 mm long. Fruit a little longer than wide, abruptly and unevenly constricted at the commissure; dorsal corky ribs much smaller than oil tubes; lateral corky ribs larger than dorsals and about 1/2 size of oil tubes. Jun-Aug; fruits Aug-Sep. Infrequent in marshes, edge of ponds, stream banks, ditches; IA, NE, MO, KS, OK, TX; (NC, IN, MN, NE, s to GA, TX, CA, Mex.). *C. maculata*, in part—Atlas GP.

2c. var. *maculata*. Principal stem leaflets usually less than 5 × longer than wide. Styles over 1 mm long. Fruits evidently longer than wide, not constricted at the commissure; dorsal corky ribs equaling or slightly larger than oil tubes; lateral corky ribs larger than dorsals and larger than oil tubes. Jul–Aug; fruits Aug–Oct. Frequent in marshes, edge of ponds, stream banks, ditches, wet prairie depressions; GP, less frequent w & sw; (e 1/2 U.S., scattered elsewhere).

Without mature fruits it is impossible to distinguish between the three varieties of *C. maculata*. Since most herbarium specimens are of flowering material, the real relationships of the taxa are still uncertain. A number of plants have the short styles of var. *angustifolia* but leaflet length-width ratios of var. *maculata* and the reverse is also common.

8. CONIUM L., Poison Hemlock

1. ***Conium maculatum*** L., poison hemlock. Stout to slender, erect, branching biennials to 3 m tall; from stout taproots; stems glabrous, glaucous, usually purplish-spotted. Leaves broadly ovate, 1.5–3 dm long, pinnately decompound, the ultimate divisions pinnately incised; petioles sheathing. Inflorescence of compound umbels; involucre of short ovate-acuminate bracts; involucel of several bractlets, shorter than pedicels. Flowers white; calyx teeth obsolete; stylopodium depressed conic; carpophore entire. Fruit broadly ovoid, 2–2.5 mm long, flattened laterally, glabrous; ribs prominent, undulate, crenate. (n = 11) May–Jul. Roadsides, thickets, stream banks, low waste places; GP but less common northward & absent sw; (throughout U.S., Europe). *Naturalized*. POISONOUS.

This species has been known as a poisonous plant from ancient times, and reputedly an extraction of it was used to put Socrates to death. All parts of the plant are said to be poisonous to humans and all classes of livestock if eaten, especially the flowers, fruits, and leaves. The plant has increased in abundance in the central GP in recent years.

9. CRYPTOTAENIA DC., Honewort

1. ***Cryptotaenia canadensis*** (L.) DC., honewort. Erect, glabrous, leafy-stemmed perennials, 3–10 dm tall, from slender fascicled roots. Leaves petiolate, alternate, ternate; leaflets ovate, obovate, to ovate-lanceolate, pointed, doubly serrate, two outer often deeply lobed; petioles sheathing. Inflorescence of loose compound umbels; peduncles terminal and lateral, normally paniculate; involucre lacking or of one short narrow bract; involucel lacking, or of a few minute bractlets. Flowers white; calyx obsolete or minute; styles erect or reflexed, stylopodium slender-conic; carpophore 2-cleft to base. Fruit linear-oblong, often somewhat falcate, 4–6 mm long, flattened laterally; ribs filiform, obtuse. (n = 10) May–Aug. Rich or rocky woods, ravines, along streams, bluffs; e 1/4 GP; (Que., N.B., Man., s to GA, AL, TX).

10. CYMOPTERUS Raf.

Ours low perennial acaulescent or subcaulescent herbs from thickened rootstocks; subterranean stems often elongated into pseudoscapes bearing leaf clusters and peduncles at summit. Leaves pinnate, bipinnate, or occasionally ternate-pinnate; petioles sheathing. Inflorescence of spreading, compound umbels; peduncles terminal; involucre wanting or present, a low inconspicuous sheath, or of linear-oblong bracts; involucel of conspicuous bracts. Flowers white, pinkish, or purplish; calyx teeth small or obsolete; styles slender, spreading, stylopodium lacking; carpophore divided to base. Fruit ovoid to ovoid-oblong, flattened dorsally; the lateral ribs broadly winged, wings membranaceous to spongy-corky. *Phellopterus* Nutt.—Rydberg.

1 Bractlets of involucel conspicuously foliaceous .... 1. *C. acaulis*
1 Bractlets of involucel conspicuously scarious.
    2 Mature peduncles equaling or longer than the leaves; fruit wings not conspicuously enlarged at base .... 2. *C. macrorhizus*
    2 Mature peduncles shorter than or equaling the leaves; fruit wings conspicuously enlarged at base .... 3. *C. montanus*

1. ***Cymopterus acaulis*** (Pursh) Raf. Plants 3–30 cm tall, mature pseudoscape 1–7 cm long, usually not reaching above ground level, glabrous. Leaves clustered on the pseudoscape, petiolate, oblong to oblong-ovate, the blade 1.5–7 cm long, green, bipinnate or divided 3 × into rather small and narrow ultimate segments. Involucre wanting or vestigial; involucel of prominent, narrow, basally connate, green or greenish, seldom scarious-margined bractlets; peduncle usually shorter than or equaling leaves. Umbels compact, fertile rays 3–6, 2–10 mm long; pedicels to 1 mm long; flowers white; calyx teeth inconspicuous, somewhat scarious, or to 1 mm long. Fruit ovoid to broadly oblong, 5–10 mm long, 3–8 mm wide, wings prominent, constricted at base, equaling or narrower than body, often slightly wavy; oil tubes 5–17 in the intervals, 5–13 on the commisure, sometimes solitary at wing base. Apr–Jun. Dry, often rocky, upland prairies, & hillsides; GP to n OK & absent in e 1/2 KS; (Man. to Alta., s to MN, NE, n OK, CO, ID, OR).

2. ***Cymopterus macrorhizus*** Buckl. Plants developing a psuedoscape to 3(3.5) dm tall. Leaves ovate-oblong, blades 1.5–7.5 cm long, pinnate to bipinnate, pallid, somewhat fleshy, leaflets entire to pinnately lobed, lobes obtuse, commonly mucronate, 1–4 mm long. Peduncles longer than leaves, to 3 dm long, usually minutely roughened at base of umbel; involucre absent or of one or more linear bracts; involucel of prominent scarious bractlets, fringed at the apex, bractlets white with a median dark nerve. Flowers pinkish or nearly white, fertile rays 1–6, inner umbels sterile. Fruit ovoid to ovoid-oblong, 4–10 mm long, 3–8 mm wide, wings about equaling the body; oil tubes 3–8 in the intervals, 4–10 in the commissure. Mar–May. Open prairies & hillsides; sw OK, TX; (no doubt *endemic* in area).

3. ***Cymopterus montanus*** T. & G. Plants developing a pseudoscape to 3 dm tall. Leaves ovate-oblong, blade 1.5–8 cm long, pinnate, bipinnate or sometimes ternate-pinnate, somewhat fleshy, pallid; leaflets entire to pinnately lobed, lobes obtuse, mucronate. Peduncles shorter than or equaling the leaves, to 1 dm long, minutely roughened throughout or only at base of umbel; involucre lacking or of a low sheath or of linear-oblong bracts; involucel of prominent scarious ovate-oblong bractlets, sometimes fringed at apex, white with a green median nerve. Flowers white to purplish, fertile rays 3–6. Fruit ovoid to ovoid-oblong, 5–12 mm long, 4–10 mm wide; wings enlarged at base, twice as wide as body; oil tubes 1–4 in the intervals, 2–6 on the commissure. Mar–May. Open prairies & rocky ravines; w 1/2 GP, absent in ND; (SD, WY, s to TX, NM). *Phellopterus montanus* Nutt.—Rydberg.

11. DAUCUS L., Carrot

Annual or biennial erect, branching plants from taproots. Leaves pinnately decompound, the ultimate segments linear to narrowly lanceolate. Inflorescence of compound umbels, terminal and lateral; involucre of dissected or rarely entire bracts; involucel of toothed or entire bractlets. Flowers white or central flower of each umbellet purple; calyx teeth obsolete; stylopodium conic. Fruit ovoid to ellipsoid, 3–5 mm long, the bristly primary ribs filiform, secondary ribs winged; wings with a single row of prickles; oil tubes solitary under secondary ribs, 2 on the commissure.

1 Bracts pinnately divided into elongate filiform divisions; rays 3–7.5 cm long; carpel widest at the middle; teeth of fruit not prominently barbed apically; central flower of umbellet rose or purplish .... 1. *D. carota*
1 Bracts pinnately divided into short linear or lanceolate divisions; rays 1–4 cm long; carpel commonly widest below the middle; teeth of fruit barbed apically; central flower of umbellet white .... 2. *D. pusillus*

1. ***Daucus carota*** L., wild carrot, Queen Anne's lace. Biennial slender to coarse plants 4–15 dm tall; stems solitary, usually freely branched, glabrous to retrorsely bristly hispid. Leaves ovate-lanceolate, 5–20 cm long, ultimate divisions linear to lanceolate, 2–12 mm long; acute, mucronate, entire, or few-cleft; peduncles 7–50 cm long or longer; involucre of filiform, elongate, pinnately divided (entire) bracts, 4–40 mm long, commonly deflexed; involucel of linear, acuminate, entire (pinnate), ciliate bractlets, about equaling the pedicels; rays many, unequal, 3–7.5 cm long, compact in fruit. Flowers white to yellowish, the central flower of each umbellet normally pinkish or purple. Fruit ovoid, 3–4 mm long, 2 mm wide, widest at the middle; bristles simple or slightly uncinate. (n = 9) Apr–Jun. Roadsides & waste places; e-cen GP & sporadic elsewhere; (throughout U.S., Europe). *Naturalized.*

2. ***Daucus pusillus*** Michx., rattlesnake weed. Annual, slender plants, 1–8 dm tall; stems usually unbranched, retrorsely hispid. Leaves ovate-lanceolate to oblong, 5–10 cm long; ultimate divisions linear, 1 mm wide, acute, hispid. Peduncles 10–40 cm long, retrorsely hispid; involucre of foliaceous bipinnately or pinnately-ternately divided bracts, 12–20 mm long; involucel of linear-acute bractlets; rays few to many, unequal, 4–40 mm long, compact in fruit. Flowers white. Fruit oblong, 3–5 mm long, 2 mm wide, usually widest below the middle, the commissural surface with 2 rows of hispidulous hairs. (n = 11) Apr–Jun. Rocky prairies, open woodlands, pastures, roadsides; se GP; (SC, MO, KS, OK, s to FL, TX, CA, Mex.).

12. ERIGENIA Nutt., Harbinger of Spring

1. ***Erigenia bulbosa*** (Michx.) Nutt., harbinger of spring. Low, slender, erect, nearly acaulescent, glabrous perennial from globose tubers, 5–25 cm tall. Leaves petiolate, ternately decompound, ultimate divisions linear to spatulate, 3–12 mm long, obtuse, distinct; petioles 1.5–2 cm long, sheathing. Inflorescence of loose compound umbels; peduncles terminal and axillary, 3–10 mm long, equaling or longer than leaves; involucre replaced by a single reduced leaf; involucel of oblong or spatulate, foliaceous, entire or toothed bractlets, longer than flowers and fruit; rays 1–4, 2–3 cm long. Flowers white (pinkish); calyx-teeth obsolete; petals without inflexed apex; styles long and slender, recurving, stylopodium lacking. Fruit 2–3 mm long, 3–5 mm wide, glabrous; seed flattened laterally. Feb–May. Rich, rocky wooded hillsides, lowland woods, & along streams; extreme se GP; (NY, s MI, MN, s to AL, OK, KS).

This is the earliest of our herbaceous native plants to bloom. It is frequently mature in early April and is easily overlooked.

13. ERYNGIUM L., Eryngo

Creeping to erect, caulescent or acaulescent, usually glabrous, herbaceous annual, biennial, or perennial from taproots or rootstocks. Leaves entire to pinnately or palmately lobed to divided, often ciliate to spinose; venation parallel or reticulate; petioles sheathing. Inflorescence capitate; heads solitary or in cymes; involucre of 1 or more series of entire or lobed bractlets subtending the flowers; coma bractlets present or absent. Flowers white to purple, sessile; sepals ovate to lanceolate, obtuse to acute, persistent; stylopodium and carpophores lacking. Fruit globose to obovoid, variously covered with scales or tubercles, ribs obsolete; commissure wide; oil tubes 5, inconspicuous.

Eryngium prostratum Nutt., a low prostrate or weakly erect, unarmed, plant with heads solitary at nodes on slender peduncles has been reported from KS: Cherokee based on a single plant found in a marshy area in an open pasture woodland.

Eryngium planum L., a stout, caulescent perennial with basal leaves rosulate, coriaceous, oblong-oval, blades 10–15 cm long, rounded or cordate at base and flowers bright blue, has been reported for SD: Lake, where several plants were found at Lake Madison. It is a native of Eurasia.

1 Leaves parallel-veined .... 3. *E. yuccifolium*
1 Leaves reticulate-veined.
  2 Head 2–3.5 cm long; coma bractlets 4–8, prominent, 1–3.5 cm long .... 2. *E. leavenworthii*
  2 Heads 8–12 mm long; coma bractlets absent .... 1. *E. diffusum*

1. **Eryngium diffusum** Torr., diffuse eryngo. Diffusely branched prostrate to usually erect, 1–4 dm tall, glabrous annual or biennial from a slender taproot. Basal leaves nearly sessile, obovate to cuneate, to 5 cm long and 2 cm wide, deeply palmately parted, divisions spinulose-dentate or lobed, venation palmately reticulate; stem leaves similar to basal. Inflorescence successively trifurcate, or lateral branches with a 1-branched cyme; heads 8–12 mm long, numerous, very short-pedunculate, bluish; bracts 10–12, rigid, spreading, 10–15 mm long, exceeding the heads; bractlets 5 mm long; coma lacking. Fruit globose-ovoid, 2.5–3 mm long, covered with white scales 1–2 mm long. May–Aug. Sandy prairies, pastures, roadsides; w OK & TX; *(endemic)*.

2. **Eryngium leavenworthii** T. & G., Leavenworth eryngo. Erect, 5–10 dm tall, glabrous, purplish (white), annual or winter-annuals from a slender to stout taproot; stems divaricately branching; lower stem leaves short-petiolate, broadly oblanceolate, to 6 cm long and 2 cm wide; upper stem leaves sessile, broadly ovate to orbicular, deeply palmately parted, divisions pinnatifid, with pungent lobes, venation reticulate. Inflorescence sparingly cymose; heads usually few, short-pedunculate, 2–3.5 cm long, purplish to red (white); bracts 4–8, 3–4 cm long, spinose-pinnatifid, about equaling the heads; bractlets 1 cm long; coma of 4–8 prominent spinescent bractlets 1–3.5 cm long. Fruit oblong, 2–4 mm long, covered with white scales 1–2 mm long. Jul–Oct. Rocky prairies, open woodlands, with decided preference for calcareous soils; e KS, e OK; (KS to TX).

3. **Eryngium yuccifolium** Michx., button snakeroot. Stout, glabrous perennials from a fascicle of tuberous roots, with a monocotyledonous appearance. Stems slender, solitary, to 1 m or more tall, branching above, glaucous. Leaves rigid, broadly linear, 1–8 dm long, 1–3 cm wide, parallel-veined, margins weakly spinose with 1(2–5) weak spines; sheaths short, leaves becoming reduced above. Inflorescence cymosely branched; heads globose-ovoid, 1–2.5 cm in diam; bracts 5–10, spreading ascending, to 16 mm long, usually entire, shorter than heads; bractlets exceeding the fruit; coma absent. Fruit oblong 4–8 mm long; angles with lanceolate flattened scales, 1–3 mm long, scales of faces smaller or obsolete. Jun–Sep. Prairies & rocky open woodlands; se GP; (VA, IN, MN, KS, s to GA, FL, TX). *E. yuccifolium* var. *synchaetum* Coult. & Rose—Waterfall.

14. EURYTAENIA T. & G., Texas Spread-wing

1. *Eurytaenia texana* T. & G., Texas spread-wing. Slender, erect, 3–12 dm tall, branching, herbaceous annuals from slender taproots. Basal leaves ovate, to 10 cm long and 5 cm wide, lobed or pinnatifid with obtuse crenate or serrate lobes, petiolate; stem leaves pinnately or ternate-pinnately dissected, divisions oblong-lanceolate to linear-filiform, petioles sheathing. Inflorescence of loose compound umbels; peduncles terminal and lateral; involucre of 5 3-cleft bracts, 5–10 mm long; involucel of several bractlets similar to bracts; rays 8–16, spreading ascending, unequal, to 8 cm long, scaberulous; pedicels 5–8 mm long. Flowers white; calyx teeth prominent; styles reflexed. Fruit orbicular to ovoid, 4–6 mm long, minutely scaberulous, dorsal ribs filiform or very narrowly winged, lateral ribs thick-winged, wings narrower than body, distinctly nerved on the commissural side; oil tubes large, solitary in the intervals, 2 on the commissure. (n = 7) Apr–Jun. Sandy prairies, ravines, roadsides; w OK, TX; (*endemic* to area).

15. HERACLEUM L., Cow Parsnip

1. *Heracleum sphondylium* L. subsp. *montanum* (Schleich.) Briq., cow parsnip, eltrot. Stout, erect perennials to 3 m tall. Stems grooved, tomentose, spreading-pubescent to somewhat scabrous. Leaves ternate, orbicular to reniform, blades 2–5 dm long, 2–5 dm wide, leaflets ovate to orbicular, 1.5–4 dm long, 1–3.5 dm wide, cordate, coarsely serrate and variously lobed, tomentose to variously hairy; petioles 1–4 dm long; upper leaves with conspicuously dilated sheaths. Inflorescence of loose compound umbels; peduncles terminal and lateral; involucre lacking or of fugaceous bracts 5–20 mm long; involucel of numerous, narrow, entire bractlets or lacking. Flowers white or purplish; calyx-teeth minute or obsolete. Fruit obovate to obcordate, 8–12 mm long, 6–9 mm wide, strongly flattened dorsally, somewhat pubescent to nearly glabrous; dorsal ribs filiform, lateral thin-winged; oil tubes large, easily visible from dorsal surface, solitary in the intervals, 2–4 on commisure, extending about 1/2 way from stylopodium to base of mericarp. (n = 11) May–Jul. Moist rich woods, along streams, in thickets; n 1/2 GP; (Lab. to AK, s to GA, KS, AZ). *H. lanatum* Michx.—Rydberg; *H. maximum* Bartr.—Fernald.

Cases of dermatitis have been reported in persons who came into contact with the foliage and cattle have been poisoned by eating the leaves. Indians and others, however, have eaten the root and young leaves.

16. HYDROCOTYLE L., Water-pennywort

1. *Hydrocotyle ranunculoides* L.f., somewhat fleshy, aquatic or subaquatic perennials. Stems slender, glabrous, floating or creeping, rooting at the nodes; leaves roundish-reniform, with a sinus at base; blades 3- to 7-cleft, to 8 cm long, lobes crenate; petioles to 4 dm long. Inflorescence a simple umbel, axillary; peduncle shorter than leaves. Flowers white, greenish, or yellowish. Fruit suborbicular, 1–3 mm long, 2–3 mm wide, dorsal surface rounded; ribs obsolete. Apr–Jul. Shallow water of ponds & marshes; KS: Kingman; OK, & TX; (PA, DE, s to FL, TX, KS, AZ, & WA, s to S. Amer.).

17. LIMNOSCIADIUM Math. & Const.

1. *Limnosciadium pinnatum* (DC.) Math. & Const. Erect, slender, annuals to 8 dm tall, branching above. Basal leaves linear-lanceolate, 5–18 cm long, 3–25 mm wide, entire and

septate, or pinnate; upper leaves pinnate, with 2-8 linear or linear-lanceolate divisions 3-10 cm long, 1-6 mm wide. Inflorescence of compound umbels; peduncles terminal and axillary; involucre of several linear or linear-lanceolate reflexed bracts, 2-6 mm long; involucel of several linear bractlets 1-5 mm long. Flowers white; calyx teeth 0.5 mm wide. Fruit with dorsal ribs filiform, low, the lateral corky-winged; oil tubes solitary in interval, 2 on the commisure. May-Jul. Low wet places, roadsides, & marshes; KS: Cherokee; (IA, KS, s to LA, TX). *Cynosciadium pinnatum* DC.—Rydberg.

18. LOMATIUM Raf., Wild Parsley

Acaulescent or short-caulescent herbaceous perennials from a thickened or swollen taproot. Leaves chiefly basal, petiolate, pinnately, quinately, or ternate-pinnately decompound; petioles sheathing at base. Inflorescence of loose compound umbels; peduncles usually solitary and terminal, equaling or exceeding the leaves; involucre lacking or inconspicuous; involucel of filiform to obovate, foliaceous to subscarious bractlets. Flowers white, yellow, or purplish; calyx-teeth obsolete or minute. Fruit elliptic, oblong to ovate-oblong, flattened dorsally, glabrous or pubescent; dorsal ribs filiform or obsolete, the lateral with wings narrower than or to 1/2 width of the body; oil tubes 1-5 in the intervals, 2 to several on the commissure. *Cogswellia* Spreng., *Cynomaranthum* (Nutt.) Coult. & Rose—Rydberg.

Lomatium montanum Coult. & Rose was reported for ND by Rydberg but is not known in the GP.

1 Flowers yellow.
   2 Leaves with few or large divisions, the divisions remote .... 3. *L. nuttallii*
   2 Leaves decompound, dissected into numerous small divisions .... 1. *L. foeniculaceum*
1 Flowers white or purplish.
   3 Bractlets of involucel more or less tomentose or villous, not markedly scarious-margined .... 2. *L. macrocarpum*
   3 Bractlets of involucel glabrous, with definite scarious margins .... 4. *L. orientale*

1. Lomatium foeniculaceum (Nutt.) Coult. & Rose, wild parsley. Acaulescent perennials, 1-5 dm tall, villous to glabrate, from a long, slender to thickened taproot. Leaves ovate, obovate, to oblong; blades 1-20 cm long, pinnately compound or ternate, ultimate divisions linear, 1-7 mm long, to 1 mm wide, apiculate; petioles 3-15 cm long, sheathing to about the middle, usually purplish, subscarious. Peduncles exceeding the leaves; involucel of linear to linear-lanceolate, acute to acuminate, scarious-margined bracts, connate below to strongly connate above. Flowers yellow; those of central umbellets often sterile and short-rayed or sessile; ovaries glabrous or pubescent. Fruit ovate-oblong, 5-12 mm long, 4-8 mm wide, the body and wings glabrous or pubescent; oil tubes 1-4 in the intervals, 2-4 on the commissure.

Two intergrading varieties are found in the GP.

1 Ovaries moderately to densely pubescent, fruit pubescent to glabrate . 1b. var. *foeniculaceum*
1 Ovaries and mature fruit glabrous, rarely sparsely pubescent .... 1a. var. *daucifolium*

1a. var. *daucifolium* (T. & G.) Cronq. Plants glabrate to villous. Bractlets of involucel usually connate to above the middle. Ovaries and mature fruit glabrous to rarely sparsely pubescent. (n = 11) Mar-May. Open rocky prairie hilltops, slopes, & dry prairies; se 1/4 GP; e 1/2 NE, KS, w MO, ne TX). *Cogswellia daucifolia* (Nutt.) M. E. Jones—Rydberg.

1b. var. *foeniculaceum*. Plants villous to glabrate. Bractlets of involucel usually connate below. Ovaries pubescent; mature fruit pubescent to glabrate. (n = 11) Mar-Jun. Open rocky prairie hilltops, slopes, & dry prairies; n 1/2 GP, less frequent southward; (Ont. to B.C., s to MO, OK, CO, & TX). *Cogswellia foeniculacea* (Nutt.) Coult. & Rose and *C. villosa* (Raf.) Schult.—Rydberg.

The var. *foeniculaceum* is found in the Dakotas and northern Nebraska, but southward in eastern Nebraska and Kansas it intergrades with var. *daucifolium,* particularly in degree of pubescence of ovaries, mature fruit, and foliage. All Kansas colonies contain plants with either glabrous or pubescent ovaries and young fruits. The percentage of plants in a colony with pubescent ovaries ranges from 4% to 55%. Of over 3,500 plants examined in Kansas, 20% had pubescent ovaries, while colonies in the Dakotas are nearly 100% pubescent.

2. **Lomatium macrocarpum** (H. & A.) Coult. & Rose. Short caulescent perennials from an elongate slender or swollen taproot, usually branched near base. Leaves usually ternate-pinnately dissected; ultimate divisions oblong to linear, to 8 mm long and 2 mm wide; sparsely to densely puberulent or villous-puberulent and grayish or purplish. Peduncles 1-several, spreading or ascending, 1-2 dm long at maturity; involucel of usually dimidiate, linear-lanceolate bractlets equaling to exceeding the flowers. Flowers white or purplish-white. Fruit 9-20 mm long, 2-8 mm wide; ovaries and young fruits glabrous or puberulent, mature fruit glabrous or glabrate; wings narrower than the body; oil tubes 1(2 or 3) in the intervals, 2-6 on the commissure. (n=11) Apr-Jun. Rocky hills & prairies, often on clay flats; n & w ND; SD: Harding; (Man. to B.C., s to SD, CO, CA). *Cogswellia macrocarpa* (Nutt.) M. E. Jones—Rydberg.

3. **Lomatium nuttallii** (A. Gray) Macbr., dog parsley. Acaulescent perennials, 1-4 dm tall, with multicipital caudices usually covered by old leaf sheaths. Leaves 1- or 2-pinnate or ternate, the ultimate divisions remote, linear, 10-45 mm long, 1-2 mm wide. Peduncles stout, usually exceeding the leaves; involucel of distinct or short connate, linear bractlets, longer than or equaling the flowers. Flowers yellow. Fruit narrowly oblong, 9-13 mm long, 3-4 mm wide, the wings 1/2 width of body; oil tubes 3-5 in the intervals, 6-10 on the commissure. May-Jun. Rocky hillsides; NE: Scotts Bluff; WY; (w NE, WY, UT). *Cynomarathrum nuttallii* (A. Gray) Coult. & Rose—Rydberg.

4. **Lomatium orientale** Coult. & Rose. Short acaulescent perennials to 4 dm tall from a slender to thick taproot; grayish soft-puberulent, rarely glabrate. Leaves tripinnate, ultimate divisions crowded, linear, 1-12 mm long, 0.5-2 mm wide. Peduncles exceeding the leaves; involucel of linear-lanceolate to obovate, distinct, scarious-margined bractlets, about equaling the flowers. Flowers white or pinkish-white. Fruit ovate-oblong, 5-9 mm long, 3-6 mm wide, glabrous, wings narrower than body; oil tubes 1-4 in the interals, 2-8 on the commissure. Apr-Jun. Dry prairies; GP, but becoming infrequent southward where known in OK: Texas; TX: Sherman; (Man., MN, MT, s to IA, KS, TX, NM, AZ). *Cogswellia orientalis* (Coult. & Rose) M. E. Jones—Rydberg.

19. MUSINEON Raf.

Caulescent or acaulescent erect or spreading perennials from thickened taproots. Leaves petiolate, 1- to 3-pinnate or ternate; petioles sheathing. Inflorescence a loose or subcompact compound umbel; peduncle terminal, shorter or longer than leaves; involucre wanting or of 1 or 2 inconspicuous bracts; involucel of usually distinct, linear, sometimes scarious-margined bractlets, shorter than or equaling the flowers. Flowers yellow or white; calyx teeth ovate, conspicuous. Fruit ovoid to oblong, flattened laterally and somewhat constricted at the commissure, glabrous to scabrous; ribs acute, prominent, oil tubes 3 or 4 in intervals, 2-6 on the commissure. *Daucophyllum* (Nutt.) Rydb—Rydberg.

1 Plants caulescent; ultimate divisions of leaves oblong and lobed .... 1. *M. divaricatum*
1 Plants acaulescent; ultimate divisions of leaves narrowly linear .... 2. *M. tenuifolium*

1. **Musineon divaricatum** (Pursh) Nutt. Caulescent spreading to erect perennials. Stems usually dichotomously branching, glabrous, to scaberulous below the umbel or scabrous

throughout. Leaves 1- or 2-pinnate or ternate-pinnate; leaflets oblong, pinnately lobed, 5-15 mm long, 3-10 mm wide. Bractlets of involucel linear-lanceolate, 2-4 mm long, nearly distinct, sometimes scarious-margined, shorter than the flowers. Flowers yellow. Fruit ovoid-oblong, constricted at apex, 3-6 mm long, 2 mm wide, glabrous to densely scabrous, oil tubes (1)3 or 4 in the intervals, 4-6 on the commissure. May-Jul. Rocky prairie hillsides, flats, open woodlands; essentially w 1/2 Dakotas, nw NE, MT, WY, NM; (Man., Alta., s to SD, NE, CO, ID, NV). *M. trachyspermum* Nutt.—Rydberg.

Some plants of the w GP have peduncles, rays of umbel, fruits, and rachis of leaves prominently scabrous or granular-roughened. These have been segregated as the var. *hookeri* T. & G., but there is every gradation from glabrous to prominently scabrous forms and recognition of the variety has little taxonomic significance.

2. ***Musineon tenuifolium*** Nutt. Acaulescent, erect, subcespitose perennials, 0.5-3 dm tall. Stems not dichotomously branching. Leaves 1- to 3-pinnate, ultimate divisions distinct, linear, 2-25 mm long, to 1 mm wide. Peduncle longer than leaves, hirsutulous just below umbel; involucre usually wanting; involucel of several linear, green bractlets to 2 mm long, usually longer than flowers. Flowers white or yellowish. Fruit narrowly oblong to ovoid, slightly constricted at apex, 2-4 mm long, 1-2 mm wide, granular-scabrous to glabrous; oil tubes (2)3 in the intervals, 2-4 on the commissure. May-Jun. Dry prairies & open wooded hillsides; sw SD, nw NE, WY, & ne CO; (*endemic* in area). *Daucophyllum tenuifolium* (Nutt.) Rydb.—Rydberg.

20. OSMORHIZA Raf., Sweet Cicely

Erect or decumbent at base, branching, pubescent to glabrate perennials, from thick-fascicled roots. Leaves petiolate, membranaceous, ternate or ternate-pinnate, leaflets lanceolate to orbicular, serrate to pinnately lobed, petioles sheathing. Inflorescence of loose compound umbels; peduncles terminal and lateral; involucre wanting or of a single foliaceous bract, or of several narrow bracts, involucel of several narrow foliaceous reflexed bractlets, or wanting. Flowers white or greenish-white; calyx teeth obsolete. Fruit linear-oblong, linear-fusiform, or clavate, tapering, beaked, or constricted at apex, bristly-hispid to glabrous; ribs filiform; oil tubes obscure or wanting.

1 Involucel lacking or of a single small bractlet; mature styles to 1 mm long, curved outward.
  2 Fruit concavely narrowed to summit .... 1. *O. chilensis*
  2 Fruit convexly narrowed to summit .... 3. *O. depauperata*
1 Involucel of several bractlets; mature styles 1-4 mm long, straight or nearly so.
  3 Styles in flower and fruit to 1.5 mm long, shorter than petals; stipules ciliate-hispid; plants not anise-scented .... 2. *O. claytoni*
  3 Styles in flower 2 mm long, in fruit 2-4 mm long, longer than petals; stipules tomentose-felted at least toward margin; plants anise-scented .... 4. *O. longistylis*

1. ***Osmorhiza chilensis*** H. & A. Perennial from a taproot which may have a branched caudex. Stems usually solitary or sometimes 2 or 3, usually branched above, 3-10 dm tall, more or less hispid. Leaves biternate; leaflets thin, narrowly to broadly ovate, coarsely toothed or sometimes incised, 2-7 cm long, 1-5 cm wide, more or less hirsute to glabrate. Peduncles 5-25 cm long; rays 3-8, spreading ascending, 5-30 mm long; flowers greenish-white; styles 0.2-0.5 mm long. Fruit linear-oblong, 12-20 mm long, concavely tapering toward apex, the terminal 1-2 mm more or less distinctly set off or a beaklike apex. (n=11) Jun-Jul. Woodlands; BH, MT, WY, CO; (w SD to AR, s to CO, AZ, s CA, also N.S., s Que. to ME, NH, & Ont., WI; Chile, Argentina). *O. divaricata* Nutt.—Rydberg.

2. ***Osmorhiza claytoni*** (Michx.) Clarke. Perennial from fusiform and fibrous roots. Stems, usually solitary, villous-pubescent, to glabrate, 3-9 dm tall. Leaves ternate-pinnate; leaflets

ovate to lanceolate, 3–7 cm long, 2–3 mm wide, serrate to incised, more or less densely pilose; petioles 5–12 cm long; stipules ciliate-hispid; peduncles 5–14 cm long; involucel of several linear, attenuate, ciliate bractlets, 3–8 mm long, reflexed; rays 3–5, spreading ascending. Flowers white, styles about 0.5 mm long. Fruit oblong, 20–24 mm long, tapering into an attenuate beak at apex, caudate at base, sparsely hispid on ribs, more densely so below. (n = 11) Jul–Sep. Wooded hillsides; e GP, becoming infrequent s; (Que., Sask., s to NC, AL, AR, ND, KS).

3. *Osmorhiza depauperata* Phil. In most respects very similar to *O. chilensis*. Rays more widely and stiffly divaricate. Fruit 10–15 mm long, convexly narrowed to summit, apex not beaklike. (n = 11) Jun–Aug. Rich woodlands; rare in BH; MT, WY, CO, NM; (w SD, Sask., s to NM, se CA, s NV, & Newf., e Que. to VT, also n MI, ne MN, s Ont.; Chile, Argentina). *O. obtusa* (Coult. & Rose) Fern.—Rydberg.

4. *Osmorhiza longistylis* (Torr.) DC., anise root. Perennial from somewhat fleshy roots to 1 cm thick. Stems to 1 m tall, glabrate to densely villous. Leaves biternate or ternate-pinnate; leaflets ovate or oblong-ovate, 3–10 cm long, 1–5 cm wide, coarsely serrate, incised, or pinnately lobed toward base, sparingly short-pilose to densely villous; petioles 5–16 cm long, glabrate to densely villous. Peduncles 5–13 cm long; involucre of 1–several, linear or lanceolate, ciliate bracts, 5–15 mm long; involucel of several bractlets, 5–10 mm long, reflexed; rays 3–6, spreading ascending, 1.5–5 cm long. Flowers white, styles 2–3 mm long. Fruit oblong, 18–20 mm long, acute at apex, caudate at base, hispid on the ribs.

Two intergrading varieties in the GP as follows:

1 Stems glabrous or nearly so .... 4a. var. *longistylis*
1 Stems villous with spreading hairs .... 4b. var. *villicaulis*

4a. var. *longistylis*. Stem glabrous or glabrate, nodes slightly pubescent. Leaflets sparingly hirsutulous to glabrescent on veins beneath; petioles usually glabrous or glabrate. (n = 11) Apr–Jul. Rich woods & thickets, ravines, stream banks; n 1/2 & se GP; (Que. to Sask. s to GA, CO, TX).

This variety is relatively distinct in the n 1/2 of the GP, but from se SD to e 1/4 of KS it is found with var. *villicaulis*. The latter variety is always rare, but many specimens of eastern Kansas cannot readily be referred to either variety. I doubt if recognition of subspecific taxa is meaningful.

4b. var. *villicaulis* Fern. Stems, petioles, and lower part of branches densely villous. (n = 11) Apr–Jun. Rich woods, thickets, stream banks; se SD, extreme e NE & e 1/4 KS. *O. villicaulis* (Fern.) Rydb.—Rydberg.

21. PASTINACA L., Wild Parsnip

1. *Pastinaca sativa* L., wild parsnip. Stout, glabrous, aromatic biennials usually with a well-developed taproot. Stems 3–10 dm tall, grooved. Basal leaves to 5 dm long and about 1/2 as wide, pinnately compound, leaflets to 13 cm long, coarsely serrate; cauline leaves progressively reduced. Inflorescence of compound umbels; peduncles terminal and lateral; involucre and involucel usually wanting; flowers yellow to reddish; calyx teeth obsolete; carpophore bifid to base. Fruit broadly elliptical, 5–6 mm long, 4–5 mm wide; oil tubes 1 in the intervals, 2–4 on the commissure. (n = 11) May–Jul. Ditches, roadsides, along railroads & waste places; GP but infrequent westward & absent in sw; (throughout U.S.; Europe). *Naturalized.*

The wild plant and the cultivated parsnip with larger roots are considered to belong to the same species. It is reported that some persons develop a dermatitis similar to poison ivy when coming in contact with wild parsnip.

22. PERIDERIDIA Reichb.

Slender, erect, glabrous perennials from more or less tuberous-thickened, often fascicled roots. Stems 3–12 dm tall; leaves pinnate, bipinnate, or ternate-pinnately compound, ultimate divisions linear to linear oblong; petioles sheathing. Inflorescence of loose compound umbels; peduncles terminal and lateral, longer than leaves; involucre of 1–several bracts or wanting; involucel of several linear, scarious or green bractlets, shorter than flowers. Flowers white to pinkish; calyx teeth conspicuous; carpophore 2-cleft to base. Fruit ovoid, orbicular, 2–6 mm long, oil tubes 1 or 3 in the intervals, 2 or 4 on the commissure. *Atenia* H. & A., *Eulophus* Nutt.—Rydberg.

1 Mature fruit 3–6 mm long; plants of se KS .... 1. *P. americana*
1 Mature fruit 2–3 mm long; plants of the BH .... 2. *P. gairdneri*

1. *Perideridia americana* (Nutt.) Reichb. Slender erect perennials from a fascicle of tuberous roots, 7–12 dm tall. Lower leaves pinnately or ternate-pinnately compound, ultimate divisions linear to linear-oblong, 0.5–5 cm long, upper leaves ternate; petioles sheathing. Peduncles slender, to 15 cm long; involucre of 1–several bracts or wanting; involucel of several linear acuminate bractlets to 5 mm long. Flowers white; calyx teeth prominent. Fruit ovate or oblong, glabrous, 3–6 mm long; oil tubes 3 in the intervals, 4 on the commissure. (n = 20) May–Jun. Rocky open woodlands with preference for limestone; se KS; (OH, MI, s to TN, AR, KS). *Eulophus americanus* Nutt.—Rydberg.

2. *Perideridia gairdneri* (H. & A.) Math., squaw-root. Slender erect perennials from a usually single or fascicled root. Stem solitary, glabrous. Leaves few, well distributed along the stem, slightly reduced upwards, pinnate or bipinnate, the ultimate divisions linear, 2–15 cm long, entire or rarely lobed. Umbels 1 to several; involucre wanting or of 1–several setaceous bracts; bractlets of involucel several, linear, scarious or green, 1–4 mm long, or nearly obsolete. Flowers white. Fruit suborbicular, 2–3 mm long, nearly as wide; oil tubes solitary in the intervals, 2 on the commissure. (n = 17) Jul–Aug. Rare in moist meadows & woodlands; SD: Lawrence, Pennington; MT, WY; (e Sask. to B.C., s to SD, NM, s CA). *Atenia gairdneri* H. & A.—Rydberg.

23. POLYTAENIA DC., Prairie Parsley

1. *Polytaenia nuttallii* DC., prairie parsley. Stout, erect perennials from subfusiform taproots; stems 4–10 dm tall, puberulent above. Leaves bipinnate or ternate-pinnate, leaflets ovate to oblong, 2–4 cm long, rounded or cuneate at base, sessile, distinct or terminal confluent; petioles 4–15 cm long; upper stem leaves ternate, with dilated sheaths. Inflorescence of loose compound umbels; peduncles terminal and lateral, longer than blades; involucre wanting; bractlets of involucel linear to filiform, puberulent, 2–5 mm long. Flowers yellow; calyx teeth ovate, acute; carpophore 2-cleft to base. Fruit 5–10 mm long, 4–7 mm wide; dorsal ribs filiform, often obscure, the lateral narrowly corky-winged; oil tubes several in the intervals, usually very indistinct. (n = 11) May–Jun. Dry prairies & open woodlands; se NE, e 1/2 KS, e & sw OK; (MI, WI, NE, s to MS, TX, NM). *Pleiotaenia nuttallii* (A. Gray) Coult. & Rose—Rydberg.

24. PTILIMNIUM Raf., Mock Bishop's Weed

1. *Ptilimnium nuttallii* (DC.) Britt., mock bishop's weed. Slender, erect, branching, glabrous annuals, from fibrous roots, 3–6 dm tall. Leaves pinnately decompound, with few divisions, usually 2 at node on the rachis; ultimate divisions filiform, 10–60 mm long; petioles to 1 cm long, Inflorescence of loose compound umbels; peduncles terminal and axillary; involucre of filiform, entire (rarely 1 or more 3-cleft) bracts, much shorter than rays; involucel of filiform entire bractlets, shorter than pedicels. Flowers white; calyx teeth conspicuous, linear-lanceolate, persistent; anthers purplish; styles longer than stylopodium. Fruit orbicular, 1–1.5 mm long, lateral ribs inconspicuous, dorsal ribs wide and rounded. (n = 7) Jun–Aug. Low moist prairies, swampy areas, roadside ditches; extreme se GP; (MO, KS, s to LA, TX).

Ptilimnium capillaceum (Michx.) Raf. with styles shorter than the stylopodia, calyx teeth triangular, and bracts of the involucre mainly 3-cleft, has been reported for KS in several manuals. All specimens studied have 1 or 2 involucre bracts 3-cleft, but otherwise match *P. nuttallii*. It is possible, however, that an occasional specimen of *P. capillaceum* might be found as an introduction in the se GP.

25. SANICULA L., Black Snakeroot

Erect, branching, glabrous biennials or perennials, from slender fibrous roots, or rootstocks. Leaves palmately divided into 3–5(7) segments, the basal long petioled, the stem leaves progressively reduced. Inflorescence of irregular, spreading, compound umbels; peduncles terminal or terminal and lateral; involucre foliaceous, bracts toothed or lobed; involucel of small scarious or foliaceous bractlets. Flowers greenish-white to greenish-yellow, perfect or staminate; sepals prominent, connate to distinct, persistent. Fruit globose, ovoid to obovoid, slightly flattened laterally, ribs obsolete, densely covered with hooked bristles sometimes arranged in rows; oil tubes irregularly arranged.

1 Staminate flowers 2 or 3; styles shorter than bristles of fruit; biennial .... 1. *S. canadensis*
1 Staminate flowers 12–25; styles longer than bristles of fruit; perennial.
   2 Staminate flowers shorter than the fruits; calyx lobes ovate, strongly connate; fruits prickly to base .... 2. *S. gregaria*
   2 Staminate flowers longer than the fruits; calyx lobes lanceolate, nearly distinct; fruits not prickly at base .... 3. *S. marilandica*

1. Sanicula canadensis L. Erect, glabrous, biennial from a short vertical rootstock bearing fibrous roots. Stems 2–10 dm tall, alternately or dichotomously branched above. Basal leaves long-petiolate, suborbicular, blade 4–10 cm long, 2–5 cm wide, palmately 3-parted, leaflet margins sharply serrate; stem leaves becoming sessile above. Umbels 2 or 3 radiate; involucre of usually 2 ovate-lanceolate, subfoliaceous bracts, 3–8 mm long; involucel of a few minute, ovate bractlets. Umbellets 4- to 6-flowered, the staminate flowers 2 or 3, their pedicels about 2 mm long, calyx lobes linear, evidently connate at base; perfect flowers 2 or 3, pedicellate, calyx lobes similar. Fruits globose to ovoid, (2)3(5) mm long, covered with uncinate prickles, dilated below, and in longitudinal rows. (n = 8) May–Jul. Rich or rocky open woodlands, thickets; GP, less frequent & absent in the n & sw; (VT, Ont., MN, SD, s to FL, TX). *S. canadensis* L. var. *grandis* Fern.—Fernald.

2. Sanicula gregaria Bickn. Erect, glabrous, perennial from a short, oblique rootstock bearing fibrous roots. Stems 2–8 dm tall, usually solitary or several from base, subumbellately branched above the middle. Basal leaves long-petiolate, broadly triangular to ovate, blades 3–10 cm long, 5–18 cm wide, palmately 3- to 5-parted, leaflet margins sharply serrate, petiole 8–25 cm long; stem leaves short-petiolate to sessile, 3- or 5-parted. Umbels 1- to 3-radiate; involucral bracts small, foliaceous, 3-lobed, serrate; involucel of inconspicuous, subscarious oval bractlets. Umbellets with staminate and perfect flowers, or staminate flowers in separate umbellets in axil of stem or branches, staminate flowers 12–25, pedicels 3 mm long, calyx lobes triangular-ovate, connate to the middle; perfect flowers 3–5 in each umbellet, calyx lobes similar to those of staminate flowers. Fruit subglobose to obovoid, 3-5 mm long, 2-3 mm wide, distinctly pedicellate, covered with uncinate prickles, lower prickles reduced, prickles not arranged in rows; styles conspicuous, recurved, exceeding bristles of fruit. (n = 8) May-Jul. Rich or rocky open woods & thickets; e 1/4 GP, absent in ND; (N.S., Que., MN, SD, s to FL, TX).

The correct name for this species may be *S. odorata* (Raf.) Phil.

3. Sanicula marilandica L. Erect, glabrous, perennial from a horizontal rootstock with fibrous roots. Stems 2-10 dm tall, usually solitary, scapose below, umbellately branched above. Basal leaves long-petiolate, orbicular, cordate to cuneate, palmately 5-parted, blades 2-10 cm long, 6-15 cm wide, leaflets doubly mucronate-dentate; cauline leaves short-petiolate to subsessile, 3- to 5-parted. Involucral bracts few, foliaceous; involucel of minute, lanceolate bractlets; umbellets with both staminate and perfect flowers or staminate flowers in separate ones; staminate flowers 12-25, their pedicels 3 mm long, calyx lobes lanceolate, attenuate, slightly connate at base; perfect flowers 3-6, sessile, calyx lobes similar to those of staminate. Fruits ovoid, narrowed toward base, 4-6 mm long, sessile, covered with slender, uncinate, bulbous-based prickles, lower prickles obsolete; styles long, reflexed, exceeding bristles of the fruit. Jun-Sep. Open woods, thickets, prairie ravines; n 1/2 GP; (Newf., Que., to B.C., s to IA, NE, CO, NM).

Prior reports of this species for KS were based on incorrect determinations.

26. SIUM L., Water-parsnip

1. Sium suave Walt., water parsnip. Stout, erect, branching, glabrous perennials, 5-12 dm tall, from fusiform, fascicled roots. Leaves pinnate, rarely simple, leaflets lanceolate to linear, 1-6 cm long, 1-2 cm wide, finely serrate or rarely coarsely serrate; petioles stout, often hollow, 1-5 dm long. Inflorescence of loose compound umbels; peduncles terminal and axillary; involucre of 6-10 lanceolate or linear bracts, 3-15 mm long, entire or incised, reflexed; involucel of 4-8 linear-lanceolate bractlets, 1-3 mm long. Flowers white; calyx teeth minute or obsolete. Fruit oval to orbicular, 2-3 mm long, without prominent ribs. (n = 6) Jun-Aug. Swamps, marshes, & other wet places; n 1/2 GP & extreme e KS; (Newf., to B.C., s to FL, TX). *Sium cicutaefolium* Gmel.—Rydberg.

27. SPERMOLEPIS Raf., Scale-seed

Slender, erect, branching, glabrous annuals from slender taproots. Leaves ternately or ternate-pinnately decompound, the ultimate divisions linear to filiform; petioles sheathing. Inflorescence of loose compound umbels; peduncles terminal and axillary; involucre wanting; involucel of a few linear bractlets. Flowers white; calyx teeth obsolete. Fruit ovoid, flattened laterally, slightly constricted at the commissure, smooth, tuberculate, or echinate; ribs filiform; oil tubes 1-3 in the intervals, 2 in the commissure.

1 Fruit covered with hooked short bristles; leaves ovate .... 2. *S. echinata*
1 Fruit smooth or tuberculate; leaves oblong to oblong-ovate.
    2 Rays 3-7, spreading, subequal .... 1. *S. divaricata*
    2 Rays 5-11, erect, unequal .... 3. *S. inermis*

1. Spermolepis divaricata (Walt.) Britt.,forked scale-seed. Erect, slender, with spreading branches, 1-6 dm tall. Leaves ternately or ternate-pinnately decompound, the ultimate divisions linear, 3-15 mm long; petioles to 3 cm long, sheaths with a winged scarious margin.

Peduncles 1–5 cm long; involucel of linear bractlets with margins scarious and often callous-toothed; rays 3–7, divaricate, subequal. Fruit ovoid, 1.5–2 mm long, tuberculate. (n = 8) Apr–Jun. Sandy soils of fields, prairies, & open woodlands; extreme se GP; (VA, MO, KS, s to FL, TX).

2. *Spermolepis echinata* (Nutt.) Heller. Erect, often spreading, 0.5–3(6) dm tall. Leaves ternately decompound, ultimate divisions filiform, 2–18 mm long. Peduncles 8–60 mm long; involucel of a few filiform bractlets, callous-toothed to glabrous; rays 5–14, suberect, unequal. Fruit ovoid, 1.5–2 mm long, covered with short hooked bristles. (n = 8) Apr–Jun. Sandy prairies, pastures, roadsides; extreme se GP; (SC, GA, IL, MO, OK, s to FL, TX).

3. *Spermolepis inermis* (Nutt.) Math. & Const. Erect, 1–5 dm tall, branching above. Leaves ternately decompound, ultimate divisions filiform, 3–30 mm long. Peduncles 2–7 cm long, involucel of a few narrow bractlets; rays 5–11, erect, unequal, 1–12 mm long. Fruit ovoid, 1.5–2 mm long, tuberculate to smooth. (n = 11) May–Jun. Sandy prairies, pastures, open woodlands, roadsides; se 1/4 GP; (IN, NE, e to NC, w to NM, s to MS, TX, Mex.). *Spermolepis patens* (Nutt.) Robins.—Rydberg.

28. TAENIDIA Drude, Yellow Pimpernel

1. *Taenidia integerrima* (L.) Drude, yellow pimpernel. Erect, slender, branching, glabrous and glaucous perennials, from taproots. Stems usually purplish near the base. Leaves 2 or 3 ternately compound, rarely 2- or 3-pinnate; leaflets lanceolate to ovate, entire, 1–3 cm long, rounded to acute, slightly mucronulate at apex, sessile, slightly decurrent to rounded at base; petioles 4–17 cm long, those of stem sheathing. Inflorescence of loose compound umbels; peduncles terminal and lateral; involucre wanting; involucel wanting. Flowers yellow, central ones sterile; calyx teeth obsolete. Fruit oval-oblong, flattened laterally, glabrous, 3–4 mm long; ribs filiform; oil tubes 3 mostly in the intervals, 4 on the commissure. (n = 11) May–Jun. Rocky, dry woodlands, rarely in prairie openings; extreme e-cen & se GP; (Que., MN, IA, s to GA, KS, TX).

29. THASPIUM Nutt., Meadow Parsnip

Erect, slender, branching, glabrous or pubescent perennials from a fascicle of fibrous roots. Leaves ternately divided, the lower often simple; petioles sheathing. Inflorescence of loose compound umbels; peduncles terminal and lateral, longer than leaves; involucre wanting; involucel of inconspicuous linear to lanceolate bractlets. Flowers yellow or purple; calyx teeth inconspicuous. Fruit oval or oblong, glabrous or puberulent; all or several of the ribs winged; oil tubes solitary in the intervals, 2 on the commissure.

1 Basal leaves 2-ternate, or more dissected; at least upper nodes with a beard of short hairs .... 1. *T. barbinode*
1 Basal leaves simple or 1-ternate; nodes glabrous or somewhat scabrous .... 2. *T. trifoliatum*

1. *Thaspium barbinode* (Michx.) Nutt., meadow parsnip. Plants 5–10 dm tall. Stems green or glaucous, hairy at the nodes. Basal leaves 2- or 3-ternate, leaflets ovate or ovate-lanceolate, 5–15 mm long, distinct, coarsely serrate or incised; petioles 5–10 cm long; stem leaves similar, but shorter petioled and upper sessile, the margins short-hairy. Peduncles 4–8 dm long; involucel of inconspicuous linear bractlets, 1–4 mm long; rays 8–15. Flowers yellow. Fruit oval or oblong, glabrous or puberulent, 3–6 mm long, 2–4 mm wide; several of the ribs prominently winged. (n = 11) Apr–Jun. Rocky, wooded hillsides & along streams; extreme e GP; (NY, Ont., MN, s to FL, KS, OK).

2. Thaspium trifoliatum (L.) A. Gray. Plants 2-6 dm tall, glabrous. Basal leaves cordate to reniform, blades 3-6 cm lon, crenate-dentate, simple or less often ternate; petioles 6-10 cm long; stem leaves ternate, divisions lanceolate, serrate or toothed, margins white and glabrous. Peduncles 3-10 cm long; involucel of 4-8 inconspicuous bractlets, 2-4 mm long; rays 6-10. Flowers yellow or purplish. Fruit oval, 3-5 mm long; all of the ribs winged or dorsal suppressed. (n = 11) Apr-Jun. Prairies, rocky, open woodlands, & thickets; extreme se GP; (RI, MN, NE, s to GA, LA, OK).

All plants found in the GP have yellow flowers and are the var. *flavum* Blake. Prior reports of this taxon from KS were based on *Zizia aurea* (L.) Kock.

30. TORILIS Adans., Hedge Parsley

1. *Torilis arvensis* (Huds.) Link, hedge parsley. Erect, divaricately branched, appressed-hispid annuals, 3-10 dm tall. Leaves 2-3 pinnate, or upper simply pinnate; leaflets ovate to linear-lanceolate, 0.5-6 cm long, 2-18 mm wide, acute or acuminate, regularly incised or divided; petioles sheathing. Inflorescence of loose compound umbels; umbels terminal and lateral; peduncles 2-12 cm long; involucre wanting or of a single small linear bract; involucel of several subulate bractlets. Flowers white; calyx teeth evident to obsolete. Fruit ovoid-oblong, 3-5 mm long, 2-3 mm wide, the mericarps covered with uncinate bristles, these spreading at right angles and about as long as width of fruit. (n = 6) Jun-Aug. Open & waste places, roadsides, & similar sites; se GP where becoming an aggressive weed; (NY, OH, KS, s to FL, TX, CA; Europe). Introduced. *Torilis anthriscus* (L.) Bernh. — Rydberg; *T. japonica* (Houtt.) DC. — Fernald.

Torilis nodosa (L.) Gaertn., with umbels sessile or short-pedunculate, capitate, and often hidden among leaves, is known in eastern OK, but has not been found in our area. Introduced from Europe.

31. ZIZIA Koch, Golden Alexanders

Erect, branching or simple, glabrous perennials, from a fascicle of somewhat fleshy roots. Leaves simple or ternately compound; leaflets serrate or dentate; petioles sheathing. Inflorescence of loose compound umbels, terminal, or terminal and lateral; involucre wanting; involucel of a few linear inconspicuous bractlets; pedicels short, spreading, central flower of each umbellet sessile or nearly so. Flowers yellow; calyx teeth prominent. Fruit oblong or oval, glabrous; ribs filiform; oil tubes solitary in the intervals, to 2 in the commissure.

1 Basal leaves simple, rarely ternate, the margins crenate-dentate .... 1. *Z. aptera*
1 Basal leaves all or nearly all ternately compound, the margins serrate or dentate 2. *Z. aurea*

1. *Zizia aptera* (A. Gray) Fern. Plants 3-7 dm tall, glabrous. Basal leaves cordate, blades 4-7 cm long, 3-5 cm wide, simple or rarely ternate, crenate-dentate; petioles 5-10 cm long; stem leaves ternately divided, the divisions lanceolate, coarsely serrate, and often lobed. Peduncles 6-10 cm long; involucel of inconspicuous, linear, bractlets 1-2 mm long; rays 12-18, unequal, 1-3 cm long. Fruit oblong to oval, (2)3(4) mm long. (n = 11) Apr-Jul. Prairies, open wooded hillsides, thickets; ne GP, BH, mts. to the w & se GP; (NY, Man., B.C., s to GA, TN, KS, CO, ne NV, e WA). *Z. cordata* (Walt.) Koch — Rydberg.

2. *Zizia aurea* (L.) Koch. Plants 3-10 dm tall, usually branching, glabrous. Basal leaves ovate to orbicular, blades 6-10 cm long, biternate or middle leaflet pinnatifid, leaflets ovate to lanceolate, 2-5 cm long, 1-3 cm wide, distinct, serrate; petioles 10-15 cm long; stem leaves similar, ternate or irregularly compound. Peduncles 5-15 cm long; involucel of a few linear bractlets 1-3 mm long; rays 10-15, unequal, 1-4 cm long. Fruit oblong-ovoid, 3-4 mm long, 1.5-2 mm wide. (n = 11) May-Jul. Low prairies, margins of ponds, ditches, open woodlands; e 1/4 GP & BH; (ME, Que., Sask., s to FL, TX).

107. LOGANIACEAE Mart., the Logania Family

by Ronald L. McGregor

1. CYNOCTONUM J. F. Gmel., Miterwort

1. *Cynoctonum mitreola* (L.) Britt. Annual herbs; stems 1.5-7 dm tall, erect, freely branched or rarely strict, glabrous. Leaves opposite, entire, glabrous, elliptic to lance-ovate or ovate, blades 2-6(8) cm long, acute to acuminate or obtuse; petioles 3-15 mm long or subsessile above; stipules connate, with a triangular lobe. Inflorescence of terminal, helicoid cymes, cymosely branched, the ultimate branches secund; flowers regular; sepals 5, ovate to elliptic, united at base, 0.8-1.2 mm long; corolla white with bluish tube, 1.5-2.5 mm long, lobes 0.5-1 mm long, erect to ascending; stamens 5, included; styles 2, carpels separate for 1/2 length; capsules 3-4 mm long, exserted, mitriform, smooth or with sparse papillae, many seeded; seed 0.3-0.5 mm in diam, shiny. (n = 10) May-Sep. Rare but locally common on moist areas between sand dunes, along ditches, pond shores, stream banks; OK: Woods; (se VA, TN, AR, s to FL, TX).

108. GENTIANACEAE Juss., the Gentian Family

by Margaret Bolick

Ours annual, biennial or perennial, mostly glabrous herbs. Leaves simple, opposite or whorled, entire, sessile or clasping at the base, exstipulate. Inflorescence 1-flowered, cymose or corymbose; terminal and/or axillary. Flowers perfect, regular; calyx of 4 or 5 more or less united sepals; corolla of 4 or 5 more or less united petals, lobes convolute in bud; stamens as many as corolla lobes and alternate with them; stigma usually 2-lobed, sometimes capitate, style 1, ovary superior, placentation parietal. Fruit a many-seeded, 2-valved, unilocular capsule.

References: Correll, D. S., & M. C. Johnston. 1970. Gentianaceae, pp. 1204-1210 *in* Manual of the vascular plants of Texas, Texas Res. Foundation, Renner, Texas; Gleason, H. A. & A. Cronquist. 1963. Gentianaceae, pp. 548-554 *in* Manual of vascular plants of northeastern United States and adjacent Canada, D. van Nostrand Co., New York; Harrington, H. D. 1954. Gentianaceae, pp. 425-430 *in* Manual of the plants of Colorado, Sage Books, Denver; Radford, A. E., H. E. Ahles, & C. R. Bell. 1968. Gentianaceae, pp. 835-845 *in* Manual of the vascular flora of the Carolinas, Univ. of North Carolina Press, Chapel Hill.

1 Corollas spurred; flowers small (9-12 mm long), purplish or yellowish-green .... 6. *Halenia*
1 Corollas not spurred.
   2 Corolla lobes with large fimbrillate glands on upper surface .... 8. *Swertia*
   2 Corolla lobes without large fimbrillate glands on upper surface.
      3 Anthers circinately or helically coiled; flowers pink or rarely white.

```
4  Anthers circinately coiled; corollas rotate, lobes much longer than tube ......... 7. Sabatia
4  Anthers helically coiled; corollas salverform, tube much longer than
   lobes .................................................................................................. 1. Centaurium
3 Anthers not coiled; flowers purple, blue, white or rarely pink.
   5  Corolla lobes fringed; flowers 4-merous ............................................ 5. Gentianopsis
   5  Corolla lobes, at most, slightly toothed, not fringed; flowers 5-merous.
      6  Corolla with plicate folds between the lobes ...................................... 3. Gentiana
      6  Corolla without folds or pleats between the lobes.
         7  Corolla and calyx deeply cleft, each fused for less than 1/3 of its length; style
            filiform; stamens inserted on corolla throat .................................. 2. Eustoma
         7  Corolla and calyx only slightly cleft, corolla fused for more than 2/3 of its length,
            calyx fused for more than 1/2 of its length; style stout; stamens inserted on corolla
            tube ................................................................................ 4. Gentianella
```

1. CENTAURIUM Hill, Centaury

Glabrous, freely branching annual herbs with opposite leaves. Inflorescence a terminal spike or cyme. Flowers 4- or 5-merous; calyx narrow, deeply cleft, lobes narrow and appressed to corolla tube; corolla pink or rarely white, salverform or broadly funnelform, tube slender, much longer than lobes; stamens inserted on corolla throat, anthers spirally twisting after dehiscence; stigma capitate or bifid, style slender. Capsule fusiform to oblong-ovoid; seeds minute.

Reference: Munz, P. A., & D. D. Keck. 1968. *Centaurium* Hill, pp. 438–440, *in* A California flora. Univ. Calif. Press, Berkeley.

```
1  Cauline leaves linear ..................................................................... 3. C. texense
1  Cauline leaves oblong, lance-ovate or lanceolate.
   2  Pedicels 1–5 cm long ................................................................. 1. C. exaltatum
   2  Pedicels 3–5 mm long ................................................................ 2. C. pulchellum
```

1. ***Centaurium exaltatum*** (Griseb.) W. Wight ex Piper. Simple or branched annual 1–3.5 dm tall. Cauline leaves oblong-elliptic to oblong-lanceolate, 1–3 cm long, 1.5–10 mm wide. Pedicels 1–5 cm long; calyx 7–10 mm long, lobes subulate, 5–8 mm long; corolla tube 8–10 mm long, lobes pale pink to white, 2.5–4 mm long. (n = 40) Aug. Damp alkaline places; adventive in w GP; (e WA to CA, e to UT).

2. ***Centaurium pulchellum*** (Sw.) Druce. Much-branched, erect annual, 1–2 dm tall. Cauline leaves lance-ovate to lanceolate, 1–2 cm long, 1–4 mm wide. Pedicels 3–5 mm long; calyx 5–9 mm long, lobes 4–5 mm long; corolla tube exceeding calyx, 8–13 mm long, lobes pink, 1–4 mm long. Jun–Sep. Disturbed habitats, wet places; NE: Garden; SD: Yankton; (Introduced locally in ne U.S., GP; Europe).

This species is rather weedy and may be spreading in the GP.

3. ***Centaurium texense*** (Griseb.) Fern. Erect, corymbosely branched to simple annual, 1–3 dm tall. Cauline leaves linear, 5–30 mm long, to 4 mm wide; upper leaves reduced. Pedicels 5–12 mm long; calyx 8–10 mm long, lobes 7–9 mm long; corolla tube 9–11 mm long, lobes pink, 3–7 mm long. (n = 21) Jun–Jul. Dry calcareous soils; se KS; (MO s to TX).

2. EUSTOMA Salisb., Prairie Gentian, Catchfly Gentian

1. ***Eustoma grandiflorum*** (Raf.) Shinners. Erect annual or short-lived perennial, 25–60 cm tall, with one or several stems, internodes 1.4–6 cm long; tap root elongate. Leaves

opposite, elliptic-oblong to lance-ovate, glaucous, 3-veined, 1.5–7.5 cm long, 0.3–5 cm wide. Inflorescence cymose-paniculate, in clusters of 2–6 flowers, pedicels to 6 cm long. Flowers 5-merous (rarely 6-merous); calyx deeply cleft, lobes keeled, linear-lanceolate, 1.2–2.3 cm long, 2–3 mm wide; corolla campanulate, deeply cleft, blue-purple, pink or whitish, lobes elliptic to obovate, 3.5–5 cm long, 1.5–2.4 cm wide; stamens 5 or 6, anthers 4–5.5 mm long, filaments 10–15 mm long; stigma 2-lobed, style slender, about as long as the ovary. Capsule ellipsoid, to 2 cm long. (2n = 72) Jul–Sep. Moist meadows & prairies; sw SD, NE, w KS, sw 1/2 OK, TX; (sw SD, NE, & e CO, s to TX, Mex.). *E. russellianum* (L.) Griseb.—Rydberg.

Reference: Shinners, L. H. 1957. Synopsis of the genus *Eustoma* (Gentianaceae). Southw. Naturalist 2: 38–43.

The different color-forms have been named as follows: blue-purple (typical), f. *grandiflorum*; white, f. *fisheri* (Standl.) Shinners; white with purple tinge, f. *bicolor* (Standl.) Shinners; pink, f. *roseum* (Standl.) Shinners; yellow, f. *flaviflorum* (Cockll.) Shinners.

3. GENTIANA L., Gentian

Glabrous to minutely puberulent perennial (ours) or rarely annual herbs. Leaves opposite, subsessile to sessile. Inflorescence a more or less dense head. Flowers 5-merous; calyx tubular with a continuous membrane inside the ovate to linear lobes; corolla tubular to funnelform, lobed, with intermediate plaited, toothed folds between the lobes; stamens inserted on the corolla tube, often with cohering anthers; persistent stigmas 2, style short or none.

References: Mason, C. T., & H. H. Iltis. 1965. Preliminary reports on the flora of Wisconsin. Gentianaceae and Menyanthaceae, Gentian and Buckbean families. Trans. Wisconsin Acad. Sci. 54: 295–329; Pringle, J. S. 1967. Taxonomy of *Gentiana*, Section Pneumonanthae, in eastern North America. Brittonia 19: 1–32; Pringle, J. S. 1971. Hybridization in *Gentiana* (Gentianaceae): Further data from J. T. Curtis's studies. Baileya 18: 41–51.

Identification of gentian species may be rendered difficult by hybridization between three of our taxa. *Gentiana alba*, *G. andrewsii* and *G. puberulenta* cross freely. Their named hybrids are: *G. alba* x *G. andrewsii* = *G.* x *pallidocyanea* J. S. Pringle; *G. alba* x *G. puberulenta* = *G.* x *curtisii* J. S. Pringle; *G. andrewsii* x *G. puberulenta* = *G.* x *billingtonii* Farw.

1 Corolla closed, corolla lobes much reduced to obsolete, not extending above the summit of the plaits .... 3. *G. andrewsii*
1 Corolla open, corolla lobes extending beyond the summit of the plaits.
  2 Corollas yellowish-white; calyx lobes deltoid-ovate to ovate-lanceolate .... 2. *G. alba*
  2 Corollas blue-purple; calyx lobes linear to lanceolate.
    3 Corolla less than 3 cm long .... 1. *G. affinis*
    3 Corolla more than 3 cm long .... 4. *G. puberulenta*

1. Gentiana affinis Griseb., northern gentian. Glabrous perennial, 1–3.5 dm tall, internodes 0.5–4.5 cm long. Leaves lance-ovate to lanceolate, 1–3.5 cm long, 0.3–1.5 cm wide. Inflorescence of several flowers arranged in racemose to capitate clusters in axils of upper leaves. Calyx 7–15 mm long, tube 4–7 mm long, lobes narrowly linear (less than 1 mm wide), obsolete to 7 mm long; corolla blue-purple, narrowly funnelform, open, 2–3 cm long; lobes ovate, acute, extending beyond summit of plaits; lobes of plaits acute. Jul–Aug. Moist meadows of mountainous or hilly areas; ND to w SD; (Sask. to B.C., ND, SD, s to CA, CO, & AZ along mts.). *Dasystephana affinis* (Griseb.) Rydb.—Rydberg.

2. Gentiana alba Muhl. Glabrous perennial, 3–9 dm tall, internodes 3–4.5(11) cm long. Leaves ovate to lance-ovate, 3–10 cm long, 1.7–3.5 cm wide, widest near the clasping base. Flowers many in a terminal cluster, sometimes also 1 or 2 on peduncles of upper axils.

Calyx 11–21 mm long, calyx tube 7–11 mm long, lobes ovate to lance-ovate, 6–10 mm long, 2–4 mm wide; corolla open, greenish- or yellowish-white, funnelform, 3.5–4.6 cm long; lobes broadly ovate, exceeding apex of plaits. (n = 13) Aug–Sep. Moist or swampy areas, mesic open woods or prairies; e KS; (Ont. to MN, IA, s to NC, MO, KS). *Dasystephana flavida* (A. Gray) Britton—Rydberg; *G. flavida* A. Gray—Fernald.

3. Gentiana andrewsii Griseb., closed gentian, bottle gentian. Glabrous to minutely puberulent perennial, (1)3–7 dm tall; internodes 1–9.5 cm long. Leaves lanceolate to lance-ovate, 1.8–8.4 cm long, 0.5–3.5 cm wide. Inflorescence a dense cluster, terminal and sometimes also axillary in upper nodes. Calyx tube 10–12 mm long, lobes lanceolate to lance-ovate, spreading, 4–10 mm long; corolla tubular, closed, blue-purple, 2.8–4 cm long; the thin lobes adnate to the plaits; plaits fimbriate, apex incurved, extending beyond the corolla lobes and forming most of the corolla summit. (n = 13) Aug–Sep. Wet meadows, prairies, or woods; ND s to NE, e to MN, IA, MO; (Que. s to NJ, s to VA along mts, w to ND). *Dasystephana andrewsii* (Griseb.) Small—Rydberg.

4. Gentiana puberulenta Pringle, downy gentian, prairie gentian. Minutely puberulent perennial, 1.5–5 dm tall, internodes (6)10–20(40) mm long. Leaves lanceolate, 1–5.5 cm long, 0.3–2 cm wide, obtuse or rounded at the base. Inflorescence a dense cluster of several (3–10) flowers, terminal and sometimes also axillary in the upper nodes. Calyx tube 7–18 mm long, lobes linear to lanceolate, 4–18 mm long; corolla blue-purple to rose-violet, open funnelform, 3–4 cm long; corolla lobes erect or spreading. (n = 13) Sep–Oct. Drier upland woods & prairies; ND, SD, NE, e KS; (OH to WI, Man., ND, s to KY, MO, & KS). *Dasystephana puberula* (Michx.) Small—Rydberg; *G. puberula* of authors.

4. GENTIANELLA Moench

1. Gentianella amarella (L.) Börner subsp. **acuta** (Michx.) J. Gillett. Glabrous annual herb, stems 1.5–5 dm tall, internodes above rosette 2–5 cm long. Leaves opposite, sessile to subsessile, 3- to 5-veined; basal leaves spatulate, cauline leaves lanceolate to ovate, 1.2–3.5 cm long, 0.2–1.2 cm wide. Flowers 4- or 5-merous, these several (2–10), borne terminally and on peduncles from the upper axils. Calyx tubular, with 4 or 5 reduced lobes, lobes lanceolate, 2–6 mm long, margins green, without a calyx membrane; corolla tubular, narrowly funnelform to salverform, varying from pale blue, purple to white, pale yellow, and greenish, 8–15 mm long, corolla lobes acute, without plaits or folds but with continuous fimbriae across the base of the lobes; stamens inserted in lower 1/2 of corolla tube; stigmas 2; pistil sessile or shortly stipitate. Seeds round to slightly flattened. (2n = 36) Jul–Aug. Moist valleys & ravines, open woods, alpine meadows, alkaline areas up to 3000 m; ND, w SD; (Newf. to AK, ND, s along Rocky Mts. to Mex.). *Amarella acuta* (Michx.) Raf., *A. strictiflora* (Rydb.) Greene—Rydberg; *Gentiana amarella* L.—Fernald.

Reference: Gillett, J. M. 1957. A revision of the North American species of *Gentianella* Moench. Ann. Missouri Bot. Gard. 44: 195–269.

G. amarella frequently has a 5-lobed calyx with a 4-lobed corolla.

G. quinquefolia (L.) Small subsp. *occidentalis* (A. Gray) J. Gillett occurs just east of the GP in WI, IA, & MO. It may be distinguished from *G. amarella* by the lack of a fringed membrane across the base of the corolla lobes. *G. quinquefolia* flowers in Sep–Oct; it grows in prairies, on wooded hillsides, preferring wet areas and limestone soils. (n = 18)

G. tenella (Rottb.) Börner occurs just west of the GP in the Rocky Mts. It may be distinguished from *G. amarella* by its small stature, rarely being over 10 cm tall, and by its solitary flowers on long (2–10 cm) peduncles. (2n = 10)

5. GENTIANOPSIS Ma, Fringed Gentian

Ours annual or biennial glabrous herbs. Leaves opposite, sessile to subsessile. Inflorescence an axillary or terminal cyme or flowers solitary (ours). Flowers 4-merous, long-pedicellate; calyx usually 4-angled, lobes with thin scarious or hyaline margins, in 2 dissimilar pairs, each sinus of calyx with a thin inner membrane; corolla funnelform with distinct fringe or teeth on corolla lobes, without plaits or folds between the lobes; stamens inserted on upper 1/3 of corolla tube; stigma large, 2-lobed, style short. Capsule ellipsoid, stipitate; seeds ellipsoid to angular, minutely or strongly papillose.

References: Andreas, B. K., & T. S. Cooperrider. 1981. The Gentianaceae and Menyanthaceae of Ohio. Castanea 46: 102–108; Gillett, op. cit.; Iltis, H. H. 1965. The genus *Gentianopsis* (Gentianaceae): Transfers and phytogeographical comments. Sida 2: 129–154.

Gillett (op. cit.) considered the following taxa to be subspecies of *crinita* and placed them in *Gentianella*.

1 Leaves ovate to lance-ovate, 2–4(6)× longer than broad; apex of corolla lobes fringed, the teeth 1–2.5 mm long .... 1. *G. crinita*
1 Leaves narrowly linear, (6)10–35× longer than broad; apex of corolla lobes toothed, the teeth 0.1–1.1(1.4) mm long .... 2. *G. procera*

1. ***Gentianopsis crinita*** (Froel.) Ma. Annual or biennial herb, 1–6 dm tall, often branched, internodes 1–7.5 cm long. Leaves sessile, clasping, ovate to lance-ovate, 1.5–6 cm long, 0.5–2.4 cm wide, with 1 prominent vein. Flowers solitary on peduncles 1–10(14) cm long. Calyx of 2 dissimilar pairs, tube 10–20 mm long, outer narrower lobes 10–23 mm long, 2–5 mm wide, inner wider lobes 10–20 mm long, 7–10 mm wide; corolla deep blue, lobes deeply fringed, spreading, 3.7–5.2 cm long. Ovary and fruit stipitate. (n = 39) Aug–Oct. Low wet areas, stream banks; ND, w MN; (Que. to Man., s to ND, IA, MN, along mts. to GA). *Anthopogon crinitus* (Froel.) Raf.—Rydberg; *Gentiana crinita* Froel.—Fernald.

2. ***Gentianopsis procera*** (Holm) Ma. Annual or biennial herb, 2–6 dm tall, sparingly branched, internodes 1.2–7.2 cm long. Leaves linear-lanceolate to narrowly linear, (6)10–35 × long as wide, 1.5–7.2 cm long, 2–4 mm wide; with 1 prominent vein. Flowers solitary on usually long (12–20 cm) peduncles; calyx of 2 dissimilar pairs, tube 12–15 mm long, outer narrower lobes 18–22 mm long, 3–4 mm wide, inner wider lobes 15–20 mm long, 5–8 mm wide, corolla deep blue, lobes broadly obovate, toothed along top and sides, 2.8–3.6 cm long. (n = 39) Aug–Sep. Bogs, meadows, wet areas, especially on calcareous soils; rare in ND, w MN, IA; (w NY & s Ont., s to OH, IL, IN, IA). *Anthopogon procerus* (Holm) Rydb.—Rydberg; *Gentiana procera* Holm—Fernald; *Gentianella crinita* (Froel.) G. Don subsp. *procera* (Holm) J. Gillett—Gillett, op. cit.

6. HALENIA Borkh., Spurred Gentian

1. ***Halenia deflexa*** (Sm.) Griseb. Glabrous annual herb, 2–4 dm tall, often branching above. Stems erect, 4-angled, sometimes slightly winged, internodes 2–11 cm long. Leaves opposite, lower spatulate, upper lance-ovate, 2.3–4.2 cm long, 0.4–1.2 cm wide, glaucous below. Inflorescence of terminal and axillary loose cymes of 2–9 yellowish or purplish-green flowers, pedicels to 4 cm long. Flowers 4-merous; calyx tube short, campanulate, with lanceolate, foliaceous lobes, 5 mm long, 1 mm wide; corolla more or less tubular, each lobe spurred at the base, 6–12 mm long, 3–5 mm wide; stamens inserted on the corolla tube; stigma 2-lobed, ovary conic, tapering to a short style. Capsule lanceolate to oblong. Jul–Aug. Moist or wet areas, often sandy soil; n ND, w SD; (boreal; N.S. to B.C., s to Dakotas).

7. SABATIA Adans.

Erect glabrous annual herbs. Leaves opposite, mostly cauline but also basal in some species. Inflorescence terminal, cymose. Flowers 4- to 12-merous (usually 5-merous in ours), pink (rarely white) with a yellow or greenish-yellow spot at the base of each corolla lobe; calyx tube shorter than the lobes (ours); corolla rotate; anthers circinately coiled after dehiscence, bright yellow; stigma 2-lobed, style slender. Capsule ovoid to cylindric.

References: Wilbur, R. L. 1955. A revision of the North American genus *Sabatia* (Gentianaceae). Rhodora 57: 1–33, 43–71, 78–104; Perry, J. D. 1971. Biosystematic studies in the North American genus *Sabatia* (Gentianaceae). Rhodora 73: 309–369.

1 Stem winged; calyx tube unwinged, covering less than 1/3 of corolla tube; usually 2 branches per node at upper nodes .... 1. *S. angularis*
1 Stem not winged; winged or ridged calyx tube covering 2/3 or more of corolla tube; usually 1 branch per node at upper nodes .... 2. *S. campestris*

1. ***Sabatia angularis*** (L.) Pursh, rose-pink. Robust glabrous annual, (1)3–8 dm tall, branches usually 2 per node on upper nodes, internodes 2.2–8 cm long, stem conspicuously winged on angles. Leaves rounded basally, ovate to lance-ovate, 3-nerved, 1.5–4 cm long, 0.7–2.6 cm wide. Inflorescence of many (20–50+) flowers, these in corymbosely to pyramidally arranged cymules. Calyx 1–2 cm long in flower, to 4 cm in fruit, lobes narrowly linear, tube 1/3 to 1/5 length of corolla tube; corolla lobes spatulate to elliptic, 11–20 mm long, 4–6 mm wide. Capsule cylindrical at maturity, to 8 mm long. (n = 19) Jul–Aug. Moist prairies, woods; se KS, OK; (s NY s to n FL, w to IL, KS, OK, TX).

A white-flowered form, f. *albiflora* (Raf.) House, is reported from Cherokee Co., KS.

2. ***Sabatia campestris*** Nutt., prairie rose gentian. Slender glabrous annual, 1–3.5 dm tall, branches usually 1 per node on upper nodes, internodes 1–4.4 cm long; stem lacking conspicuous wings. Leaves broadly clasping the stem, ovate to lance-ovate, 1- or 3-nerved, 0.8–2.5 cm long, 0.5–1.2 cm wide, lower leaves often smaller. Inflorescence of loosely arranged cymules forming a corymbose cluster. Calyx tube 5-ridged or winged, to 8 mm long, 2/3 or more the length of the corolla tube, lobes 3-nerved, lanceolate to linear, 1–2.8 cm long; corolla lobes broadly obovate or spatulate, 12–22 mm long, 7–11 mm wide. Capsule to 9 mm long. (n = 13) Jun–Jul. Sandy prairies, roadsides, waste places; se KS, OK; (IL s to MS, w to IA, KS, OK, TX).

8. SWERTIA L., Green Gentian

1. ***Swertia radiata*** (Kell.) O. Ktze. Stout erect herbaceous perennial, to 1 m or more in height; internodes 4–8 cm long. Leaves whorled in 3s to 6s or 7s, basally constricted and sheathing the stem, lanceolate to linear, 7–25 cm long, 0.7–7 cm wide, upper leaves decreasing in size, with 5–7 prominent veins, glabrous to sparsely puberulent. Inflorescence cymose, terminal and sometimes also axillary. Flowers 4-merous, greenish-yellow with purple splotches; calyx lobes linear-lanceolate, 12–30 mm long; corolla rotate, lobes ovate to elliptic, acuminate, 15–20 mm long, 5–10 mm wide, with 2 fringed elliptic glands closely spaced on each lobe, each gland 5–8 mm long; stamens inserted on base of corolla, filaments connate basally; stigma capitate, style elongate, 2-lobed; ovary ovoid. Capsule ellipsoid-conic, more or less flattened, about 2 cm long; seeds ca 5 mm long. Jun–Aug. Moist wooded areas; BH; (SD to WA, s to w TX, NM, CA, & ne Mex.).

References: St. John, H. 1941. Revision of *Swertia* (Gentianaceae) of the Americas and the reduction of *Frasera*. Amer. Midl. Naturalist 26: 1–29.

109. APOCYNACEAE Juss., the Dogbane Family

by Ronald L. Hartman

Ours erect or trailing, caulescent, perennial herbs with acrid milky juice from a rhizome or woody rootstock or the base fibrous-rooted. Leaves deciduous or evergreen, opposite, alternate, or rarely whorled, sessile or petiolate, the margins entire, often revolute; stipules inconspicuous or none. Flowers solitary in the upper leaf axils or numerous and in corymbose or thyrsiform cymes; sepals 5, united below, the lobes imbricate in bud; corolla of 5 united petals, white or pinkish to blue or purple, cylindrical to campanulate, urceolate, or salverform, convolute in bud; stamens 5, epipetalous, included, alternating with the corolla lobes and often triangular appendages; anther distinct, 2-celled at anthesis, introrse, sagittate, free or adherent by viscid exudates to the stigma, dehiscing longitudinally; pollen granular; carpels 2, each 1-celled with a pareital placenta, distinct below, fused apically into a short and broadly clavate to elongate and filiform style; stigma depressed-capitate with a transverse, membranous wing, or conical; nectaries subtending the pistil 2 or 5 and fleshy, or none. Fruit of 2 (1 by abortion), narrowly cylindrical to fusiform, glabrous follicles in ours; seeds few to numerous, glabrous or with a tuft of long, silky hairs at one end; embryo large, straight.

The family contains many showy ornamentals as well as poisonous species.

1 Corolla salverform, blue to purple, the tube densely pubescent within; style elongate, filiform; seeds glabrous.
  2 Leaves alternate, deciduous; plants erect; flowers numerous in thyrsiform cymes .... 1. *Amsonia*
  2 Leaves opposite, evergreen; plants with trailing stems; flowers solitary in the axils of the upper leaves .... 3. *Vinca*
1 Corolla tubular to campanulate, white to pinkish, the tube glabrous within; style short, broadly clavate; seeds comose .... 2. *Apocynum*

1. AMSONIA Walt., Amsonia, Blue Star

Erect or ascending herbs from a woody rootstock. Leaves deciduous, alternate, broadly elliptic to linear, petiolate or sessile, the margins often revolute; stipules none. Inflorescence of terminal or subterminal, thyrsiform cymes barely surpassing the foliage, dense to open; bracts inconspicuous, scalelike to linear. Flowers erect to drooping; calyx lobes fused in lower 1/5–1/3; corolla salverform, light blue, glabrous or villous externally, the tube enlarged somewhat in upper 1/2 at point of attachment of stamens, densely pilose at the orifice, becoming retrorsely pubescent within the lower portion, appendages none, the lobes widely spreading to nearly erect, linear to ovate; anthers triangular to ovate, fertile throughout, glabrous, free from and converging above the stigma, the connective not enlarged, the base sagittate for 1/3–1/2 the length of anther; filament straight, not enlarged nor markedly compressed upward, glabrous; stigma depressed-capitate, surrounded by a subterminal, cuplike, membranaceous wing; style elongate, filiform; nectaries none. Follicles straight or curved, linear-cylindrical, erect to pendulous, glabrous; seeds numerous, cylindrical with truncated ends, in 1 row, without a coma.

1 Corolla glabrous externally; leaves 5 cm or less long .... 1. *A. ciliata* var. *texana*
1 Corolla villous externally; leaves mostly 6–13 cm long.
  2 Sepals at least sparsely villous; leaves shiny above, subcoriaceous; follicles widely spreading to pendulous .... 2. *A. illustris*

2 Sepals glabrous; leaves dull, thinly membranaceous; follicles erect ... 3. *A. tabernaemontana*

1. ***Amsonia ciliata*** Walt. var. ***texana*** (A. Gray) Coult., Texas amsonia. Perennial herb; stems 2–3.5 dm tall, glabrous, solitary or more often few to several from a somewhat woody rootstock 0.3–0.8 cm in diam, often with bases of old stems present. Leaves ascending, often dimorphic with the upper much narrower than the lower; blade linear to lanceolate, (1)1.5–5 cm long, 0.1–1.2 cm wide, membranaceous to subcoriaceous, mostly glabrous or with villous margins, dark green and shiny above, pale below, obtuse to acuminate at apex, acute at base, sessile or subsessile, often scalelike near base of stem; petiole, if present, less than 3 mm long. Cymes mostly terminal, dense. Calyx lobes narrowly triangular to ovate, often attenuating apically, 0.5–2.5 mm long, glabrous or villous on the narrowly scarious margin; corolla light blue, the tube often much darker than the lobes or sometimes greenish-tinged, glabrous externally; tube 9–10 mm long; lobes oblong to broadly ovate, 3.5–11 mm long. Follicles erect, 6–10 cm long; seeds 5–11 mm long, brown, with corky ridges. Apr–May. Prairies; OK, TX; (OK to s-cen TX).

2. ***Amsonia illustris*** Woods. Perennial herb; stem 6–9 dm tall, glabrous, solitary or few from a woody rootstock 1–1.5 cm in diam, often with bases of old stems present; branches, if present, arising in upper 1/5–1/3, mostly alternate, ascending. Leaves ascending, not dimorphic; blade lanceolate, (4)5–12 cm long, (0.8)1–2 cm wide, subcoriaceous to firmly membranaceous, glabrous to slightly villous on primary vein, dark green and shiny above, glabrous to slightly villous and pale beneath, acute to acuminate at apex, acute at base, short-petiolate or the lowermost sessile; petiole 2–8 mm long. Cymes terminal or axillary and subterminal, dense. Calyx lobes narrowly to broadly triangular, 0.5–1.7 mm long, sparsely to densely villous, narrowly scarious-margined; corolla light blue, the tube often much darker than the lobes or greenish- or yellowish-tinged, villous externally on the upper 1/2 of the tube and the median portion of the lobes, tube 6–7 mm long; lobes narrowly oblong to elliptic-lanceolate, 6.5–9 mm long. Follicles widely divergent to pendulous at maturity, 8–12 cm long; seeds 7–10 mm long, dark reddish-brown with irregular rows of corky tubercles. Apr–May. Stream beds & flood plains; MO, e KS, e OK; (GP to TX).

3. ***Amsonia tabernaemontana*** Walt., willow amsonia. Perennial herb; stem 5–11 dm tall, glabrous, solitary or few from a woody rootstock 0.7–1 cm in diam, often with bases of old stems present; branches, if present, arising in upper 1/5–1/2, alternate, ascending. Leaves ascending, not dimorphic; blade lanceolate to elliptic-lanceolate or broadly elliptic, (1.5)3–13 cm long, (0.5)1.2–4(5) cm wide, thinly membranaceous, glabrous to slightly pubescent, dark green and dull above, glabrous to moderately pubescent and pale beneath, acute to acuminate at apex, obtuse to acute at base, short-petiolate or sometimes the lowermost sessile; petiole 2–10 mm long. Cymes terminal or axillary and subterminal, dense to loose. Calyx lobes narrowly to broadly triangular, 0.5–1.5 mm long, glabrous, narrowly scarious-margined; corolla light blue, the tube often much darker than the lobes or greenish-tinged, villous externally on the upper 1/2 of the tube and often on the median portion of the lobes, tube 6.5–8 mm long, lobes spatulate to broadly lanceolate, 6–9 mm long. Follicles erect, 8–13 cm long; seeds 6.5–9 mm long, dark brown, with low, irregular, corky ridges. (2n = 22, 32) Mar–May. Wooded areas; MO, e KS, e OK; (MA, IL, KS, s to GA, LA, & TX). *A. salicifolia* Pursh—Rydberg; *A. tabernaemontana* var. *salicifolia* (Pursh) Woods., var. *gattingeri* Woods.—Fernald.

    This species is variable in leaf shape and the presence or absence of leaf pubescence and glaucescence. The two varieties listed above in synonymy are based on extreme forms of this continuum of variability and are of doubtful significance.

2. APOCYNUM L., Dogbane

Erect or ascending, often dichotomously branched, herbs from a rhizome. Leaves deciduous, opposite or rarely whorled, ovate to oblong, lanceolate, or linear-lanceolate, sessile to petiolate, the margins often revolute; stipules inconspicuous, linear to triangular, less than 1 mm long. Inflorescence a terminal or axillary corymbose cyme; bracts subulate to semifoliaceous. Flowers erect to drooping; sepals fused in lower 1/3-2/3; corolla cylindrical to campanulate or urceolate, white or greenish-white to pink, glabrous externally, the tube cylindrical to gradually enlarged upward, glabrous within, with a triangular appendage opposite each lobe, the lobes erect to spreading or reflexed, triangular; stamens attached at the base of the corolla tube; anthers narrowly triangular, fertile in upper 2/3, glabrous, slightly adherent to the stigma, converging to form a cone above it, the base sagittate for about 1/2 the length of anther; filament straight, flattened, and much-enlarged upward, villous on the adaxial surface; stigma terminal, conical; style short, broadly clavate; nectaries subtending the pistil 5, ovate to oblong, alternating with the stamens. Follicles linear-cylindrical, divergent to pendulous, glabrous; seeds numerous, narrowly fusiform, overlapping, each terminated by a coma of long, silky hairs.

Hybridization is frequent in the genus. Plants that are intermediate in the features mentioned in the key are presumably hybrids between *A. androsaemifolia* and *A. cannabinum* and are often called *A. x medium* Greene, although *A. x floribundum* Greene has priority (*A. x medium* var. *floribundum* (Greene) Woods. — Harrington).

1 Corolla (5.2)5.7-8.8 mm long, broadly campanulate, whitish to pink, pink-veined within (at least when fresh), the lobes widely spreading to recurved; leaves horizontal to drooping .... 1. *A. androsaemifolium*
1 Corollas 2.6-4.7 mm long, narrowly campanulate to short-cylindric, white or greenish-white, the lobes erect or nearly so; leaves ascending to erect .... 2. *A. cannabinum*

1. *Apocynum androsaemifolium* L., spreading dogbane. Perennial herb; stems erect or ascending, 1.5-6(9) dm tall, glabrous, often glaucous; branches arising from near the base or in the upper 1/2, alternate, arcuate-spreading. Leaves drooping or sometimes horizontal, short-petiolate; blades broadly ovate to elliptic, 1-7(9) cm long, 0.5-5 cm wide, usually glabrous and dark green above, glabrous to villous or tomentulose and pale beneath, acute to rounded and apiculate at apex, obtuse to subcordate at base. Cymes mostly terminal but also axillary, open; bracts linear to subulate, scarious, subpersistent. Flowers erect to drooping; calyx lobes ovate to ovate-lanceolate, 1-3.1 mm long, glabrous or occasionally sparsely pubescent; corolla whitish to pink, pink-veined within (at least when fresh), broadly campanulate, (5.2)5.7-8.8 mm long, 4-8 mm wide above; lobes 0.3-0.5 × length of tube, spreading to recurved. Follicles pendulose, usually straight, 6-13.5 cm long; seeds 2-3 mm long; coma white to tawny, 1.5-2 cm long. (2n = 16) Jun-Sep. Prairies or along waterways or lakeshores, more commonly on wooded (often pine) slopes or in thickets; n 1/2 GP & s along e & w margins; (Can., ME to WA, s to GA, n AR, w TX, & AZ). *A. ambigens* Greene — Rydberg; *A. androsaemifolium* var. *glabrum* Macoun — Gleason; *A. androsaemifolium* var. *griseum* (Greene) Beg. & Bel., var. *incanum* A. DC. — Scoggan.

2. *Apocynum cannabinum* L., Indian hemp dogbane, prairie dogbane. Perennial herb; stems erect or ascending, 2-10 dm tall, glabrous to villose, often glaucous; branches arising mostly in the upper 1/2, opposite or alternate, ascending to arcuate-spreading. Leaves ascending to erect, short-petiolate to sessile, especially in the lower 1/2 of the main stem; blades ovate to oblong or lanceolate, sometimes narrowly so, 1.5-14 cm long, 0.3-4.5(7) cm wide, glabrous or sometimes villose and pale beneath, acute to rounded and apiculate at apex, acute to cordate at base. Cymes usually terminal, dense; bracts linear to lanceolate, scarious and inconspicuous to semifoliaceous and conspicuous, often caducous. Flowers erect to drooping; calyx lobes linear to lanceolate, 1.2–3(3.5) mm long, glabrous; corolla white to greenish, narrowly campanulate to urceolate or short-cylindric, 2.6–4.7 mm long, 1.5–3 mm wide above; lobes 0.3–0.5 × the length of tube, erect to slightly spreading. Follicles widely divergent to pendulous, straight to curved, 7–19(22) cm long; seeds 3–6 mm long; coma white to tawny, (1.3)1.6–3.7 cm long. (2n = 16, 22) May–Sep. Prairies, open or wooded waterways or lakeshores, disturbed roadsides or fields, & sparsely wooded slopes; GP; (Can., ME to WA, s to FL, TX, & AZ; n Mex.). *A. album* Greene—Rydberg, *A. sibiricum* Jacq., *A. cordigerum* Greene, *A. pubescens* R. Br.—Rydberg; *A. cannabinum* var. *pubescens* (Mitch. ex R. Br.) A. DC., *A. sibiricum* var. *cordigerum* (Greene) Fern.—Fernald; *A. cannabinum* var. *glaberrimum* A. DC.—Gleason; *A. cannabinum* var. *hypericifolium* (Ait.) A. Gray—Boivin; *A. sibiricum* var. *salignum* (Greene) Fern.—Hitchcock, et al.; *A. suksdorfii* Greene var. *angustifolium* (Woot.) Woods.—Woodson, Ann. Missouri Bot. Gard. 17: 119. 1930.

This species is highly variable in leaf shape and pubescence, but the recognition of infraspecific taxa does not appear to be warranted. *Apocynum sibiricum* is often recognized as distinct from *A. cannabinum*. It has been distinguished by the sessile or subsessile (vs. petiolate) leaves with usually cordate-clasping (vs. tapering) bases, at least on the lower half of the main stem, semifoliaceous (vs. scarious) bracts, follicles 4–10 (vs. 12–20) cm long, and seeds with comas 0.8–2 (vs. 2–3) cm long. An analysis of material from throughout the GP shows a complete overlap in coma length (1.2–3.7 vs. 1.6–3.3 cm) and fruit length (7–19 vs. 8–18 cm) between *A. sibiricum* and *A. cannabinum*, respectively, as distinguished by the leaf characters. Furthermore, the bract feature does not appear useful in species delimitation. Because the leaf features are variable and do not appear correlated with other distinguishing features, the two taxa are considered to be conspecific and not worthy of infraspecific recognition.

3. VINCA L., Periwinkle

1. ***Vinca minor*** L., common periwinkle. Trailing herb from a fibrous-rooted base; branches usually simple, erect or ascending, 1–3 dm long, arising at intervals along the trailing stems or stolons. Leaves evergreen, opposite, widely spreading; blades broadly lanceolate to ovate or elliptic, (0.5)2–6 cm long, (0.2)0.8–2.5 cm wide, coriaceous, shiny, glabrous or puberulent on the midvein, acute to rounded at apex, obtuse to rounded at base, short-petiolate or the lowest often sessile and scalelike, the margin revolute; petiole, if present, 1–4(6) mm long; stipules none. Flowers solitary in alternate axils of upper leaves, erect to ascending, long-pedicellate; calyx lobes fused near base, linear to lanceolate, 3–4.5 mm long, glabrous; corolla salverform, blue-violet, rarely white, glabrous externally, the tube becoming funnelform in upper 1/2–2/3, near point of attachment of the stamens, 0.9–1.2 cm long, densely pilose at orifice, becoming retrorsely pubescent lower in the tube, appendages none, the lobes widely spreading, broadly obovate to obtriangular, 0.8–1.4 cm long; anthers obovate, fertile in lower 2/3, sparingly pilose on abaxial surface, separate, converging above the stigma, the broad connective much enlarged at apex, the base sagittate for less than 1/4 the length of anther, filament geniculate in lower 1/4–1/3, compressed, cupulate and much enlarged above, glabrous; stigma depressed-fusiform, surrounded by a subterminal, cuplike, membranaceous wing and densely pilose at apex; style elongate, filiform; nectaries 2, transversely oblong, alternate with the ovaries. Follicles straight, linear-cylindrical or fusiform, spreading, glabrous, 2–3 cm long, rarely produced; seeds few, oblong, subcompressed, in 1 row, black, wrinkled, without a coma. (n = 23) Mar–May. Wooded areas, occasional escape from cultivation; KS, NE; (grown throughout much of N. Amer.; s Europe).

110. ASCLEPIADACEAE R. Br., the Milkweed Family

by Ronald L. Hartman

Caulescent perennial herbs and herbaceous or woody vines usually with milky juice. Leaves usually opposite, sometimes whorled or alternate, the margins entire or sometimes wavy; stipules minute, subulate, often deciduous. Inflorescences of 1 to many, axillary or terminal, usually umbellate cymes or sometimes flowers solitary or paired. Calyx lobes 5, basally connate, with glandular squamellae within near base; corolla rotate to campanulate, 5-lobed, convolute in bud; filaments inserted near base of corolla, connate into a short sheath (column) surrounding the elongate styles, with anthers connate around and adherent to the stigma head (and called the anther head), the combined discoid to columnar structure (stamens, style, and stigma), called the gynostegium, stipitate or sessile (with or without the column free below the base of the corona, respectively) or in *Periploca* filaments distinct but the anthers connivent above the stigma head; anthers basifixed, bilocular, opening apically or longitudinally, each usually with a terminal petaloid outgrowth of the connective (anther appendage) which partially covers the stigma head and lateral, often corneous, winglike margins (anther wings), the pair of wings from adjacent anthers forming a slit peripheral to the stigmatic chamber and often subtended by an entire to bifid fleshy pad; pollen grains firmly coherent in a yellow to golden, spatulate to semicircular mass (pollinium) with a pair of pollinia from adjacent anthers connected by a wishbone-shaped, acellular filament (translator) with a brownish to black, basally cleft, tubular body (corpusculum) in the middle and directly above the stigmatic chamber, this being the unit of pollen dispersal (pollinarium) or in *Periploca* pollen free in the anther chambers and upon dehiscence, adhering to sticky, spoon-shaped stylar appendages alternating with the stamens which are the units of pollen dispersal; flowers with 1 or 2 usually prominent coronas of various sorts, commonly nectariferous, petaloid or fleshy, which surround the base of the gynostegium and/or arise from the summit of the staminal column; gynoecium of 2 ovaries united only by the common stigma head which is usually massive, 5-lobed, and has longitudinal stigmatic surfaces alternating with the anthers, each ovary unilocular, 1-carpellate, wholly superior or slightly sunken into the receptacle; ovules numerous, on a marginal placenta. Fruit a pair of follicles (often only 1 develops), smooth or covered with soft, conical to spinulose processes (turbercles); seeds many, compressed, with a terminal tuft of long, silky hairs (coma).

The pollination mechanism found in the Asclepiadaceae is highly specialized. In *Asclepias speciosa*, the tarsus or another appendage of the pollinator enters the enlarged basal opening between the corneous anther wings. The apices of the anther wings overlap the corpusculum so that the spreading of the wings by the upward movement of the tarsus opens the cliplike corpusculum. Upon contact, the corpusculum snaps onto the tarsus. The pollinarium is then extracted when the pollinator leaves the flower. Once removed, the orientation of both pollinia changes as the translator apparatus dries and a "knee" bend forms in each translator arm. Pollination is completed when the pollinarium is inserted knee first in the anther slit and drawn up into the stigmatic chamber. The pollinia break loose from the translator apparatus with the latter remaining attached to the insect (Bookman, Amer. J. Bot. 68: 675–679. 1981). The pollination mechanism in *Periploca* differs significantly from that described above (see discussion following the description of *Periploca graeca*).

1 Plant an erect to prostrate herb; corona of 5 petaloid or sometimes fleshy blades (hoods) with incurved margins, U-shaped to tubular in cross section, adnate to the staminal column, each often with an acicular to falciform or lingulate appendage (horn) arising from within. .... 1. *Asclepias*
1 Plant a twining vine (except *Matelea biflora* and *M. cynanchoides* with the unique combination of broadly cordate leaves and axillary flowers mostly in pairs); corona a low, entire to deeply lobed rim, a flat or cup-shaped fleshy disk, or 5-bilobed deeply, laminar appendages

at the base of the gynostegium; hoods none (or in *Sarcostemma* inflated, subglobose vesicles 1.3–1.8 mm tall, without appendages).
- 2 Summit of staminal column with 5 inflated, broadly rounded vesicles .... 5. *Sarcostemma*
- 2 Summit of column without vesicles.
    - 3 Plants woody nearly throughout; corona a ring of 5 pairs of broad lobes, each pair alternating with a filiform, apically cleft lobe 5–10 mm long; stamens distinct, the pollen granulose. .... 4. *Periploca*
    - 3 Plants herbaceous or woody only in lower half; corona various, if exceeding 1.5 mm in length then of 5 deeply bilobed appendages; stamens connate around the gynoecium, the pollen firmly coherent.
        - 4 Corona of 5 distinct laminar appendages, each broadly ovate in the lower half, narrowing into a deeply bilobed apex, nearly equaling the corolla lobes in length. .... 2. *Cynanchum laeve*
        - 4 Corona fleshy or if laminar then of 10–15 lobes, much shorter than the corolla in length.
            - 5 Corolla lobes 1.5–2 mm long; corona a fleshy, shallowly lobed cup less than 0.3 mm tall, without appendages .... 2. *Cynanchum nigrum*
            - 5 Corolla lobes 3.5–15 mm long; corona laminar or if fleshy then an irregularly lobed disk or a cup with prominent lobes at least 1.5 mm long or with lingulate appendages within .... 3. *Matelea*

1. ASCLEPIAS L., Milkweed

Caulescent perennial herbs (suffrutescent in *A. macrotis*) with milky juice (except *A. tuberosa*); stems erect to prostrate, 1 to many from an often enlarged and branched rootcrown. Leaves alternate, opposite, or whorled, sessile to petiolate; blade suborbicular to lanceolate, filiform, or acicular; margins entire or wavy; base narrowly acute to rounded or cordate, often with subulate glands on the midrib; stipules minute, subulate, often deciduous. Inflorescence a terminal or axillary, umbellate cyme, or flowers occasionally solitary or in pairs in the axils. Calyx lobes reflexed, with few to many subulate, glandular squamellae within at base; corolla rotate, reflexed or sometimes spreading, white or green to orange, rose, or purple; gynostegium sessile to distinctly stipitate; corona of 5 hoods, adnate at base to the staminal column, each hood consisting of an erect to widely spreading, petaloid or sometimes fleshy blade with incurved margins, U-shaped to circular in cross section, often with an exserted, erect to incurved, acicular to falciform or lingulate appendage (horn) arising from within; fleshy pads entire to bilobed; anther head short-cylindric to truncate-conic or depressed-spheric; anther appendages ovate to deltoid, petaloid; anther wings abruptly angled at base to arcuate, usually prominent, corneous; corpusculum narrowly ovate, reddish-brown to black; pollinia of firmly coherent pollen, more or less asymmetrically spatulate, compressed. Follicle fusiform to ovate, smooth to tuberculate, terete or nearly so.

Several species of *Asclepias* are known to be toxic to livestock, especially sheep, but they are seldom eaten unless no other food is available.

- 1 Horns none or rudimentary and included well within the hood.
    - 2 Hoods appressed to the anther head, obscuring the orifice.
        - 3 Anther wings arcuate, converging apically over top of anther head with the corpuscula horizontal and centrally positioned, the head depressed-spheric; hoods with prominent lateral lobes in lower half, the lobes subtending the anther wings .... 5. *A. engelmanniana*
        - 3 Anther wings distinctly angled, not converging apically, with the corpuscula vertical and laterally positioned; anther head truncate-fusiform; hoods without lateral lobes.
            - 4 Gynostegium distinctly stipitate, the column below hoods 0.5 mm or more tall; hoods overlapping lower 1/3 or less of anther head .... 7. *A. hirtella*

    4 Gynostegium sessile or nearly so, the free portion of column if present 0.2 mm tall or less; hoods overlapping 2/3 or more of anther head.
        5 Anther wings ca 3 mm long, angled above midpoint; inflorescences mostly 2 or more, lateral, in the upper leaf axils, mostly subsessile; flowers 9.5–12.5 mm tall .... 27. *A. viridiflora*
        5 Anther wings ca 1.8 mm long, angled just below midpoint; inflorescences solitary, terminal, usually with an elongated peduncle; flowers 7.5–9.5 mm tall .... 10. *A. lanuginosa*
  2 Hoods abruptly deflexed from the anther head, with broadly rounded or clavate, arcuate-ascending tips, the orifice closed by the incurving of the margins.
    6 Corolla lobes 7–12 mm long; anther head conspicuously wider than tall; anther wings 0.8–1 mm wide, conspicuous .... 3. *A. asperula*
    6 Corolla lobes 13–17 mm long; anther head about as wide as tall; anther wings 0.2–0.4 mm wide, inconspicuous .... 28. *A. viridis*
1 Horns present, exserted from the hoods, subulate to falciform or lingulate (in *A. stenophylla*, represented by the median apical lobe of the 3-lobed hood).
  7 Hoods with 3-lobed apex, the median, narrowly triangular lobe (representing apex of adnate horn) less then 1/2 the length of the lateral pair, the margins with a pair of prominent lateral lobes in lower 1/3, the lobes subtending the anther wings .... 20. *A. stenophylla*
  7 Hoods not as above.
    8 The hoods shorter than to about equal to the anther head in height (slightly longer in *A. amplexicaulis* and *A. arenaria*), the apex not elongated, erect or the very tip slightly spreading to recurved; horns conspicuously exserted and surpassing the hoods in height.
        9 Principal leaves broadly ovate to elliptic, suborbicular, or subquadrate, the apex broadly rounded to truncate or obcordate.
          10 Inflorescences solitary and terminal; leaves sessile; hoods 5–5.5 mm long .... 1. *A. amplexicaulis*
          10 Inflorescences few to several and lateral, from the leaf axils; leaves with petioles 1–18 mm long; hoods 1.3–4 mm long.
            11 Plants markedly and persistently pubescent; petiole 8–18 mm long; hood margins with a pair of broadly rounded subapical lobes and a pair of smaller lobes near the midpoint .... 2. *A. arenaria*
            11 Plants minutely tomentulose when very young, soon glabrate; petiole 1–5 mm long; hood margins entire and truncated apically, sharply angled downward above the midpoint .... 11. *A. latifolia*
        9 Principal leaves filiform to lanceolate, the apex acute, often narrowly so.
          12 Hoods with apex distinctly lower than summit of anther head, the margins with a pair of prominent lobes exceeding the apex in height; horns lingulate.
            13 Inflorescences sessile; anther heads 1–1.5 mm tall; primary leaves 1–3 cm long .... 25. *A. uncialis*
            13 Inflorescences pedunculate; anther heads 2.4–2.7 mm tall; primary leaves 5–15 cm long .... 4. *A. brachystephana*
          12 Hoods with apex nearly equal to the anther head in height, the margins entire; horns acicular.
            14 Leaves mostly lanceolate, 10–45 mm wide, opposite; corolla lobes bright pink .... 8. *A. incarnata*
            14 Leaves linear to filiform, 0.5–4 mm wide, whorled or alternate; corolla lobes white to greenish-white or tinged with rose.
              15 Plants 1–3(4) dm tall; leaves mostly alternate, in a tight spiral .... 16. *A. pumila*
              15 Plants (2)3.5–10 dm tall; leaves mostly whorled.
                16 Secondary clusters of dwarf leaves and branchlets present in at least some leaf axils .... 21. *A. subverticillata*
                16 Secondary clusters of leaves and branchlets none .. 26. *A. verticillata*
    8 The hoods conspicuously surpassing the anther head, the apex often elongated and widely spreading; horns not surpassing the hoods in height.

17 Anther wings arcuate, minutely notched near midpoint; hoods about 2× as long as the anther head, deeply saccate at base, very narrowly oblong and erect in lower 1/2, prominently flaring and somewhat spreading in upper 1/2, the margins usually with a pair of prominent lobes in upper 1/5 .... 14. *A. oenotheroides*
17 Anther wings acutely angled at base, notched near base if at all; hoods various but not with the above combination of characters.
    18 Stems and often leaves densely villous to hirsute with spreading hairs mostly 1–2 mm long; juice not milky; hoods usually bright orange, rarely red or yellow .... 24. *A. tuberosa*
    18 Stems glabrous to tomentose, the hairs less than 1 mm long, appressed to curly; juice milky; hoods white to pink or deep rose, occasionally yellow.
        19 Principal leaves filiform to narrowly lanceolate, 0.1–1 cm wide.
            20 Hoods 3.5–4 mm long, glabrous, the apices narrowly rounded; leaves linear to lanceolate, alternate to subopposite, densely puberulent on the margin .... 9. *A. involucrata*
            20 Hoods 4.5–5.5 mm long, puberulent with spreading hairs, the apex long-attenuate; leaves acicular to filiform, opposite, glabrous .... 12. *A. macrotis*
        19 Principal leaves lanceolate to suborbicular, mostly 1.5–11 cm wide.
            21 Leaves densely pubescent beneath at maturity.
                22 Hoods 9–15 mm long, abruptly narrowed and narrowly lanceolate in upper 1/2; pedicels densely tomentose, 1.1–3 mm wide (including pubescence) 19. *A. speciosa*
                22 Hoods 3.3–7 mm long, ovate to oblong above, not abruptly narrowed; pedicels moderately to densely pubescent to tomentulose, mostly 0.3–0.8 mm wide.
                    23 Margins of hood without prominent lobes .... 17. *A. purpurascens*
                    23 Margins of hood with a pair of sharp, triangular lobes at or near the midpoint.
                        24 Corolla lobes 5–6 mm long, greenish-white, often tinged with purple dorsally; column 0.4–0.6 mm tall; primary leaves 4–8 cm long, 1–4.5 cm wide .... 15. *A. ovalifolia*
                        24 Corolla lobes 6.5–9 mm long, rose to purple; column 1–1.9 mm tall; primary leaves mostly 10–20 cm long, 5–11 cm wide .... 23. *A. syriaca*
            21 Leaves glabrous beneath at maturity (may be sparsely pubescent on midvein and margins, or throughout in *A. hallii*).
                25 Apex of hood truncate to very broadly rounded; leaves sessile or nearly so.
                    26 Corolla lobes pinkish-rose to purple; hoods if fleshy, only slightly so, the margins entire; anther head 3.8–4.2 mm tall, the wings ca 2.5 mm long, notched and prominently spurred at base .... 22. *A. sullivantii*
                    26 Corolla lobes greenish-cream, often tinged with purple dorsally; hoods markedly fleshy, the margins with a pair of teeth in the upper 1/2; anther head 2.6–3 mm tall, the wings ca 1.9 mm long, neither notched nor spurred .... 13. *A. meadii*
                25 Apex of hood narrowly rounded; leaves mostly distinctly petiolate.
                    27 Leaves mostly opposite or whorled; flowers 8–11 mm tall; hoods 3.5–4 mm long .... 18. *A. quadrifolia*
                    27 Leaves mostly alternate or subopposite; flowers 12–15 mm tall; hoods 5–6 mm long .... 6. *A. hallii*

1. **Asclepias amplexicaulis** Sm., bluntleaf milkweed. Perennial herb from a deep rhizome; stems mostly solitary from a simple or branched crown, simple, stout, (2)4–8 dm tall, glabrous, glaucous. Leaves mostly opposite; blade broadly ovate or elliptic to broadly oblong, spreading, 4–12 cm long, 1.8–8 cm wide, thickly membranaceous, glabrous except margins usually puberulent, glaucous; apex obtuse to rounded and usually mucronate, margins wavy, not revolute, base cordate-clasping; petiole none. Inflorescences solitary, terminating

the stem, 18- to 35(60)-flowered; peduncles (3)6–20(30) cm long; pedicels broadly filiform, (2)3.5–4.5 cm long, puberulent. Flowers 15–18 mm tall; calyx lobes green to purple-tinged, lanceolate to ovate-lanceolate, 3–5.5 mm long, glabrous; corolla lobes green and often purple-tinged, lanceolate to elliptic-lanceolate, reflexed, 9–11 mm long, glabrous; gynostegium pale purple or rose, stipitate, glabrous; column cylindric, 1.5–2.1 mm tall, 1.7–2.5 mm wide; hoods oblong, attached near base, erect or somewhat spreading, 5–5.5 mm long, slightly fleshy, freely open above, the apex very broadly rounded to truncate, often irregularly toothed, plane or nearly so, ca 1 mm higher than the anther head, the margins entire or nearly so, the base briefly saccate; horns oblong below, abruptly narrowed into a subulate tip, adnate to lower 2/3 of hood, arching over the anther head, 1.2–1.5 × longer than the hood; fleshy pads bilobed; anther head subcylindric, 3.5–4.5 mm tall, 3–3.2 mm wide; anther appendages ca 1.5 mm long; anther wings acute at base, not notched, prominently spurred, ca 5 mm long; corpusculum ca 0.5 mm long; pollinia ca 1.4 mm long. Follicles fusiform, usually erect on deflexed pedicels, 9–15 cm long, 0.8–1.8 cm thick, without tubercles, puberulent to glabrous, glaucous; seeds broadly ovate, 6.5–9 mm long; coma white to tan, 2.5–6 cm long. May–Jun. Sandy soils of prairies & roadsides; IA, e NE, MO, e 2/3 KS, OK; (NH to MN, NE, s to n FL & TX).

2. *Asclepias arenaria* Torr., sand milkweed. Perennial herb from a deep rhizome; stems solitary from a simple, usually narrow base, usually simple, moderately slender to stout, 2–5 dm tall, moderately to densely pubescent. Leaves opposite; blade subquadrate or occasionally some ovate to ovate-lanceolate, mostly spreading, (2)5–10 cm long, (1.5)3–7.5 cm wide, firmly membranaceous, sparsely to densely puberulent, apex usually broadly rounded to truncate, emarginate, or obcordate, mucronate, margins not revolute, base mostly truncate to subcordate; petiole (0.5)1–1.8 cm long. Inflorescences few to several, scattered in the leaf axils in upper 1/2–2/3 of plant, 25- to 50-flowered; peduncles 0.2–3 cm long or inflorescence sessile in the leaf axil; pedicels slender, 1–3 cm long, moderately to densely woolly-villous. Flowers 11–14 mm tall; calyx lobes green to purplish, lanceolate to ovate-lanceolate, 4.5–7 mm long, moderately to densely villous; corolla lobes pale green, elliptic-lanceolate, reflexed, 7.5–11 mm long, pubescent on dorsal surface; gynostegium white to cream, stipitate, glabrous; column obconic to subcylindric, 1.2–1.6 mm tall, 2–3 mm wide; hoods subquadrate, attached near base, suberect, 3.5–4 mm long, not fleshy, freely open above, the apex truncate to very broadly rounded, slightly recurved, about equal in height to the anther head, the margins with two pairs of lobes, one low, broad, and subapical, the second near the midpoint, the base not saccate; horns falciform but abruptly narrowed into a linear tip, adnate to lower 2/3 of hood, sharply incurved, 1.3–1.5 × longer than the hood; fleshy pads shallowly bilobed; anther head truncate-conic, 3–3.4 mm tall, 3–4 mm wide; anther appendages ca 1 mm long; anther wings acute at base, not notched, minutely spurred, ca 2.4 mm long; corpusculum ca 0.4 mm long; pollinia ca 1.6 mm long. Follicles broadly fusiform, erect on deflexed pedicels, 7–9 cm long, 1.5–2.5 cm thick, without tubercles, puberulent to glabrate; seeds obovate, 10–11 mm long; coma white to tan, 2.5–3 cm long. Jun–Aug. Sandy soils of upland prairies & roadsides; s SD, NE, CO, KS, cen 1/3 OK, TX, e NM (GP s to cen TX & s NM; n Mex.).

3. *Asclepias asperula* (Dcne.) Woods., antelope horns. Perennial herb from a very stout rootstock; stems few to many from a thickened, branched base, simple to sparingly branched, slender to moderately thick, 1–6 dm tall, puberulent to glabrous. Leaves mostly alternate to subopposite; blade narrowly to broadly lanceolate, ascending to spreading, (3)6–15(20) cm long, (0.7)1–3 cm wide, firmly membranaceous, sparsely puberulent to nearly glabrous, apex acute to acuminate, margins not revolute, base obtuse to rounded; petiole 0.2–0.7 cm long. Inflorescences solitary and terminal, 9- to 34-flowered; peduncles to 9 cm long or none and the inflorescence subtended by 1–3 leaves; pedicels moderately thick,

1.5–2.5 cm long, puberulent. Flowers 7–10 mm tall; calyx lobes green to purple-tinged, lanceolate to ovate, 2.5–5.5 mm long, puberulent; corolla lobes pale yellowish-green, sometimes purple-tinged dorsally, elliptic-lanceolate, arcuate-ascending, 7–12 mm long, glabrous; gynostegium greenish-cream to dark purple, sessile, glabrous except minutely puberulent on margins of hoods; column truncate-conic, 1.5–2 mm tall, 1.5–2 mm wide; hoods clavate, attached in lower 1/3, abruptly deflexed and fused along the column with arcuate-ascending tips, 4–6 mm long, somewhat fleshy, closed or nearly so above, the apex broadly rounded, cucullate, ca 1 mm lower than the anther head, with a subterminal, laterally compressed, saccate appendage within, the margins entire, the base not saccate; horns none; fleshy pads entire, broadly rounded; anther head short-cylindric, 2.5–3 mm tall, 4–5 mm wide; anther appendages ca 0.9 mm long; anther wings obtusely and prominently angled in upper 1/3, gradually rounded in lower 1/3, not notched, without spurs, ca 2 mm long; corpusculum ca 0.4 mm long; pollinia ca 1.5 mm long. Follicles fusiform to narrowly so, erect on deflexed pedicels, 4–13 cm long, 1–2.6 cm thick, without tubercles, puberulent; seeds broadly obovate, 6–8 mm long; coma light tan, 2–3.5 cm long.

This species is poisonous to livestock. There are two varieties in our area.

3a. var. *asperula*. Leaves usually narrowly lanceolate. Inflorescences usually long-pedunculate, peduncles mostly 3–10 cm long. Hoods usually dark purple. May–Jun. Sandy or rocky hills; CO: Baca, Las Animas, Prowers; (CO to s ID, s to cen TX & s CA; n Mex.).

3b. var. *decumbens* (Nutt.) Shinners. Leaves usually more broadly lanceolate. Inflorescences sessile or peduncles mostly less than 3 cm long. Hoods usually greenish-cream, at least apically. (2n = 22) Apr–Jul. Sandy or rocky calcareous soils of prairies or badlands; KS, OK, TX; (s GP to s-cen TX). *A. asperula* (Dcne.) Woods. subsp. *capricornu* (Woods.) Woods.—Correll & Johnston; *Asclepiodora decumbens* (Nutt.) A. Gray—Gates.

4. *Asclepias brachystephana* Torr., shortcrown milkweed. Perennial herb from a rhizome; stems 1 to few from a simple to branched base, simple to sparingly branched, slender, 1–4 dm tall, tomentulose to glabrate. Leaves opposite; blade linear to narrowly lanceolate, ascending to spreading, 5–15 cm long, (3)5–10 mm wide, firmly membranaceous, arachnoid to tomentose, glabrate with age, apex acute, margins not revolute, base cuneate to rounded; petiole 0.2–0.7 mm long. Inflorescences few to many, scattered in leaf axils in upper 1/4–3/4 of plant, 5- to 12-flowered; peduncles 0.3–1.5(3) cm long; pedicels slender, 1–1.7 cm tall, white-tomentose. Flowers 4–5.5 mm tall; calyx lobes green to purplish, lanceolate to ovate, 1–2.5 mm long, tomentulose; corolla lobes reddish-purple or violet, elliptic-lanceolate, reflexed, 4–5.5 mm long, sparingly puberulent dorsally; gynostegium pale rose to cream, briefly stipitate to subsessile, glabrous; column subcylindric, 1–1.5 mm tall, 1.2–1.5 mm wide; hoods obovate, attached to lower 2/3, erect to slightly spreading, 1.5–2 mm long, not fleshy, freely open above, the apex broadly rounded, plane, 2–2.5 mm lower than the anther head, the margins with a pair of prominent, lanceolate lobes near the midpoint, the base saccate; horns lingulate, adnate to lower 1/2–2/3 of hood, incurved, 1.7–2 × longer than the hood; fleshy pads bilobed; anther head truncate-conic, 2.4–2.7 mm tall, 2.1–2.3 mm wide; anther appendages ca 1 mm long; anther wings abruptly rounded at base, minutely notched, without spurs, ca 1.8 mm long; corpusculum ca 0.3 mm long; pollinia ca 1 mm long. Follicles fusiform, erect on deflexed pedicels, 5–8 cm long, 1–1.5 cm thick, without tubercles, puberulent to glabrate; seeds obovate, 7–8 mm long; coma light tan, 1.5–3 cm long. Jun–Jul. Sandy or rocky soils of prairies & roadsides; TX: Bailey, Deaf Smith, Randall; (TX to AZ; Mex.).

5. *Asclepias engelmanniana* Woods., Engelmann's milkweed. Perennial herb from a rhizome; stems mostly solitary, from a simple, thickened base, simple or sparingly branched above, slender to moderately stout, 3–12(14) dm tall, sparsely puberulent in lines or glabrous. Leaves mostly alternate to subopposite; blade linear, laxly spreading, (5)10–20 cm long, 1–5 mm wide, firmly membranaceous to subsucculent, sparsely puberulent to glabrate, apex narrowly acute, margins not revolute, base narrowly acute; petiole none. Inflorescences few to many, scattered in leaf axils of upper 1/3–1/2 of plant, 10- to 35-flowered; peduncles 0.1–4 cm long; pedicels moderately slender, 0.8–1.5 cm long, villous. Flowers 7–10 mm tall; calyx lobes green to purple-tinged, lanceolate to narrowly ovate, 2–3.5 mm long, puberulent; corolla lobes pale green, purple-tinged dorsally, elliptic-lanceolate, reflexed, 4.5–6 mm long, glabrous; gynostegium yellowish, briefly stipitate, glabrous; column cylindric, 1.3–2 mm tall, 1.2–1.5 mm wide; hoods oblong, attached in lower 1/2, erect, 2.3–3.2 mm long, somewhat fleshy, the orifice appressed to the anther head and nearly concealed, the apex truncate to broadly retuse, plane, ca 1 mm lower than the anther head, the margins with a pair of prominent lateral lobes near base, the base deeply saccate; horns none; fleshy pads obscure, narrowly bilobed; anther head depressed-spheric, 1.8–2.1 mm tall, 2.7–3 mm wide; anther appendages ca 1 mm long; anther wings arcuate, not notched, without spurs, ca 2 mm long; corpusculum ca 0.4 mm long; pollinia ca 1.3 mm long. Follicles fusiform, erect on deflexed pedicels, 6–12 cm long, 1.3–1.8 cm thick, without tubercles, puberulent to glabrous; seeds broadly obovate, 7–9 mm long; coma white to tan, 3–4 cm long. Jun–Aug. Sandy or rocky calcareous soils of priairies, breaks, or flood plains; NE, KS, CO, OK, TX, NM; (GP w to se UT & s to s TX & s AZ; n Mex.). *Acerates auriculata* Engelm. — Rydberg; *Asclepias auriculata* (Engelm.) Holz. not H.B.K. — Fernald.

6. *Asclepias hallii* A. Gray, Hall's milkweed. Perennial herb apparently from a rhizome; stems usually solitary from a usually simple, thickened base, simple, slender to rather stout, 2–8 dm tall, puberulent to tomentulose, sometimes glabrate. Leaves mostly alternate; blade lanceolate to ovate-lanceolate, ascending to spreading, (4)7–14 cm long, (1)1.5–4.5 cm wide, firmly membranaceous, sparingly pubescent to tomentulose, usually glabrate, sometimes glaucous, apex acute to obtuse, margins not revolute, base obtuse to broadly rounded; petiole 0.2–1.5 cm long. Inflorescences solitary to several, terminal and often subterminal or scattered in leaf axils of upper 1/3 of plant, (8)15- to 35-flowered; peduncles 1–5(8) cm long; pedicels slender, 1.5–2.5 cm long, puberulent to tomentulose. Flowers 12–15 mm tall; calyx lobes green to purple-tinged, ovate-lanceolate to lanceolate, 2.5–3.5 mm long, pilosulous; corolla lobes pale rose to purple, elliptic-lanceolate, reflexed, 6–8 mm long, glabrous; gynostegium pale rose to cream, briefly stipitate, glabrous; column subcylindric, 1.4–1.6 mm tall, 1.8–2.3 mm wide; hoods lanceolate, attached near base, somewhat spreading, 5–6 mm long, not fleshy, freely open above, the apex narrowly rounded, plane, 1.5–2 mm higher than the anther head, the margins entire, the base not saccate; horns falciform, adnate to lower 1/2 of hood, abruptly incurved, 0.7–0.8 × longer than the hood; fleshy pads bilobed; anther head truncate-conic, 2.2–2.5 mm tall, 2.5–3 mm wide; anther appendages ca 0.7 mm long; anther wings abruptly rounded at base, not notched, without spurs, ca 1.6 mm long; corpusculum ca 0.3 mm long; pollinia ca 1.3 mm long. Follicles broadly fusiform, erect on deflexed pedicels, 8–12 cm long, 1–3 cm thick, without tubercles, puberulent to glabrate; seeds broadly obovate, 7–8 mm long; coma white, 2–4 cm long. Jun–Jul. Sandy soils of prairies & roadsides; CO: Denver, El Paso, Pueblo; (CO, s WY to NV, s to s NM & n AZ).

7. *Asclepias hirtella* (Penn.) Woods., prairie milkweed. Perennial herb from a thickened taproot and possibly a rhizome; stems usually solitary from a simple to branched, enlarged base, simple to sparingly branched, slender to moderately stout, 2.5–11 dm tall, puberulent in lines from the leaf bases to evenly so. Leaves mostly alternate; blade linear to narrowly lanceolate, ascending to spreading, 4–14(16) cm long, 0.3–1.5 cm wide, firmly membranaceous, sparsely to moderately puberulent, especially above, apex acute to acuminate, margins not revolute, base acute to obtuse; petiole 0.1–0.7 cm long. Inflorescences few

to many, scattered in leaf axils of upper 1/2 of plant, 25- to 80-flowered; peduncles 0.3-4 cm long; pedicels filiform, 1.5-2 cm long, hispidulous. Flowers 7.5-9 mm tall; calyx lobes green to purple-tinged, lanceolate to ovate, 1.3-2.4 mm long, hispidulous; corolla lobes pale green, purple-tinged dorsally, elliptic-lanceolate, reflexed, 3.8-5.3 mm long, glabrous; gynostegium pale green, stipitate, glabrous; column conic, 1.9-2.5 mm tall, 0.9-1.5 mm wide; hoods oblong-obovate, attached for most of length, erect, 1.6-2.3 mm long, not fleshy, the orifice appressed to the anther head, the apex broadly rounded to truncate, slightly recurved, ca 1.5 mm lower than the anther head, the margins entire, the base deeply saccate; horns none; fleshy pads linear; anther head depressed-fusiform, 1.8-2 mm tall, 2-2.2 mm wide; anther appendages ca 0.5 mm long; anther wings obtusely angled and most prominent above the middle, shallowly and broadly notched, without spurs, ca 1.1 mm long; corpusculum ca 0.3 mm long; pollinia ca 1 mm long. Follicles fusiform, erect on deflexed pedicels, 8-12(15) cm long, 1-1.5 cm thick, without tubercles, puberulent; seeds broadly ovate, 7-9 mm long; coma white to tan, 2.5-4 cm long. May-Sep. Sandy or rocky, calcareous soils of prairies & marshy areas; IA, MO, e KS, e OK; (s Can., MI to MN, KS, s to WV & OK). *Acerates hirtella* Penn.—Rydberg.

8. *Asclepias incarnata* L., swamp milkweed. Perennial herb with a shallow, fibrous root system; stems mostly solitary from the stout base, simple to much branched above, fairly stout, (5)7-20(25) dm tall, glabrous or pubescent in decurrent lines from the leaf bases. Leaves mostly opposite; blade linear-lanceolate to lanceolate or rarely ovate, ascending to spreading, (3)5-15 cm long, (0.5)1-3(4.5) cm wide, membranaceous, sparsely puberulent at least on the veins, apex acute to acuminate, margins inconspicuously revolute, base cuneate to rounded or truncate; petiole 0.3-1.7 cm long. Inflorescences few to many at end of stem and branches, 10- to 40-flowered; peduncles 1-7 cm long; pedicels filiform, 10-17(20) mm long, puberulent. Flowers 9-11 mm tall; calyx lobes green to purple, lanceolate to ovate, 1.3-2.3 mm long, villous; corolla lobes bright pink or rarely white, elliptic to oblanceolate, reflexed, 5-6 mm long, glabrous; gynostegium pale pink or rarely white, stipitate, glabrous; column subcylindric, 1.2-1.8 mm tall, 0.8-1.5 mm wide; hoods oblong, attached near base, slightly spreading, 2-2.7 mm long, not fleshy, freely open above, the apex broadly rounded, plane or nearly so, ca 0.5 mm lower than the anther head, the margins entire, the base not saccate; horns acicular or nearly so, adnate to lower 1/2 of hood, arching over the anther head, 1.3-1.7 × longer than the hood; fleshy pads broadly bilobed; anther head truncate-conic, 2.1-2.5 mm tall, 1.5-1.7 mm wide; anther appendages ca 0.8 mm long; anther wings acute at base, minutely notched, obscurely spurred, ca 1.8 mm long; corpusculum ca 0.3 mm long; pollinia ca 1 mm long. Follicles fusiform, erect on straight to curved pedicels, 5-8 cm long, 0.8-1.1 cm thick, without tubercles, glabrous to sparsely puberulent; seeds broadly ovate, 6.5-9 mm long; coma white, 1.5-2.7 cm long. (2n = 22) Jun-Sep. Banks & flood plains of lakes, ponds, & waterways, marshes, swamps, & other wet areas of prairies; GP except w ND, nw SD, MT; (s Can., ME to ND, s to FL & NM). *A. incarnata* subsp. *incarnata* f. *albiflora* Heller—Fernald.

9. *Asclepias involucrata* Engelm., dwarf milkweed. Perennial herb from a woody, subfusiform rootstock; stems few to many from a branched base, simple to repeatedly branched, slender, ascending to decumbent, 0.3-2.5 dm tall, sparsely to moderately puberulent. Leaves alternate; blade linear to lanceolate, erect to spreading, 1.5-10(12) cm long, 0.3-1 cm wide, firmly membranaceous, glabrate to sparsely puberulent and usually densely so on the margins, apex acute to acuminate, margins not revolute, base acute to obtuse; petiole 0.1-0.3 cm long. Inflorescences 1-3 per stem, terminal or subterminal, 4- to 25-flowered; peduncles 0.2-3 cm long or none and inflorescence sessile in leaf axil; pedicels filiform, 1.5-3 cm long, sparsely puberulent. Flowers 8-9 mm tall; calyx lobes green to purple-tinged, ovate-lanceolate, 2-4 mm long, glabrous to moderately puberulent; corolla lobes pale green to pinkish, elliptic-lanceolate, reflexed, 4.5-7 mm long, glabrous; gynostegium white with pinkish keels, briefly stipitate, glabrous; column obconic, 0.6-1 mm tall, 1.7-2.3 mm wide; hoods ovate, attached near base, widely spreading, 3.5-4 mm long, not fleshy, freely open above, the apex narrowly rounded, slightly recurved, ca 0.5 mm higher than the anther head, the margins abruptly angled downward near the midpoint, the base not saccate; horns falciform, adnate to near midpoint of hood, incurved or ascending, 0.9-1 × longer than the hood; fleshy pads narrowly bilobed; anther head truncate-conic, 2.2-2.3 mm tall, 3.1-3.3 mm wide; anther appendages ca 1.2 mm long; anther wings abruptly rounded at base, not notched, obscurely spurred, ca 1.5 mm long; corpusculum ca 0.3 mm long; pollinia ca 0.9 mm long. Follicles fusiform, erect on deflexed pedicels, 4-7 cm long, 1.5-2 cm thick, without tubercles, sparsely puberulent to glabrate; seeds broadly ovate, 6-10 mm long; coma tan, 2-3 cm long. May-Jun. Sandy or gravelly soils of prairies; CO: Las Animas; KS: Stevens; NM: Curry; OK: Cimarron; TX: Sherman; (GP to se UT & much of NM & AZ; n Mex.).

10. *Asclepias lanuginosa* Nutt., woolly milkweed. Perennial herb from a rhizome; stems solitary or in pairs from a simple to branched, slightly enlarged base, simple, slender, 0.8-2 dm tall, moderately to densely pilose. Leaves alternate to subopposite; blade lanceolate to narrowly ovate, ascending to spreading, 3-8 cm long, 1-2 cm wide, firmly membranaceous, sparsely to moderately pilose, apex acute to narrowly rounded, margins not revolute, base obtuse to rounded; petiole 0.1-0.4 mm long. Inflorescences solitary and terminal, 20- to 60-flowered; peduncles obsolete to 2 cm long; pedicels filiform, 0.8-2 cm long, pilose. Flowers 7.5-9.5 mm tall; calyx lobes green to purple-tinged, lanceolate, 1.5-3 mm long, pilose; corolla lobes pale greenish-yellow, often purple-tinged dorsally, elliptic-lanceolate, reflexed, 4-5.5 mm long, glabrous; gynostegium pale green, sessile, glabrous; column subcylindric, 1.1-1.5 mm tall, 0.8-1 mm wide; hoods oblong-obovate, attached for more than 1/3 of length, erect, 2.5-3.2 mm long, somewhat fleshy, the orifice appressed to the anther head, the apex rounded, slightly recurved, ca 0.5 mm lower than the anther head, the margins with small lateral lobes at base, the base deeply saccate; horns none; fleshy pads scarcely bilobed; anther head truncate-fusiform, 1.9-2.2 mm tall, 2-2.3 mm wide; anther appendages ca 0.9 mm long; anther wings obtusely angled and most prominent just below the midpoint, not notched, without spurs, ca 1.8 mm long; corpusculum ca 0.3 mm long; pollinia ca 1.3 mm long. Follicles narrowly fusiform, the tip long attenuate, erect on deflexed pedicels, 8-11 cm long, 1-1.5 cm thick, lightly sericeous to pilose; seeds obovate, ca 5 mm long; coma white. May-Jul. Sandy or rocky calcareous soils of prairies; MN, ND, SD, IA, NE, KS; IL & WI to GP). *Acerates lanuginosa* (Nutt.) Dcne. — Rydberg.

11. *Asclepias latifolia* (Torr.) Raf., broadleaf milkweed. Perennial herb from a deep rootstock; stems solitary from a simple to branched and thickened base, usually simple, moderately slender to stout, (2)3-6 dm tall, tomentulose, glabrate with age. Leaves opposite; blade broadly oblong to suborbicular, ascending to spreading, 6-15 cm long, 4-12 cm wide, firmly membranaceous to subcoriaceous, tomentulose, glabrate with age, somewhat glaucous, apex broadly rounded to emarginate or obcordate, mucronate, margins not revolute, base obtuse to truncate or cordate-clasping; petiole 0.1-0.5 cm long. Inflorescences few to several, scattered in leaf axils in upper 1/2 of plant, 30- to 55-flowered; peduncles 0.2-1.5 cm long or inflorescence sessile; pedicels slender, 1.5-3.5 cm long, sparsely to densely villous or tomentose. Flowers 12-16 mm tall; calyx lobes greenish, ovate-lanceolate, 3-5 mm long, tomentulose; corolla lobes pale green, sometimes purple-tinged dorsally, elliptic-lanceolate, reflexed, 7.5-10.5(12) mm long, glabrous; gynostegium greenish-white, stipitate, glabrous; column obconic, 1.5-2 mm tall, 2-2.8 mm wide; hoods subquadrate, attached near base, spreading, 2.8-3.8 mm long, somewhat fleshy, freely open above, the apex trun-

cate or broadly retuse, plane, 1–1.5 mm lower than the anther head, the margins abruptly angled downward near midpoint, the base not saccate; horns broadly falciform but abruptly narrowed into a linear tip, adnate nearly the full length of the hood, sharply incurved, 1.1–1.2 × longer than the hood; fleshy pads narrowly bilobed; anther head truncate-conic, 3.5–4 mm tall, 4–4.5 mm wide; anther appendages ca 1.3 mm long; anther wings abruptly rounded at base, not notched, obscurely spurred, ca 3 mm long; corpusculum ca 0.4 mm long; pollinia ca 1.7 mm long. Follicles broadly fusiform, erect on deflexed pedicels, 6–8 cm long, 1.5–3 cm thick, without tubercles, glabrous or nearly so; seeds obovate, 7–8 mm long; coma tan, 2–3 cm long. (2n = 22) Jul–Aug. Sandy, clayey, or rocky calcareous soils of prairies & breaks; s edge NE (also Knox Co.), w 2/3 KS, CO, w 1/2 OK, TX, NM; (GP w to e CA, s to s-cen TX & s AZ).

This species is poisonous to livestock.

12. *Asclepias macrotis* Torr., longhorn milkweed. Suffrutescent perennial from an elongate rhizome; stems usually many, from the slender, branched base, usually repeatedly branched, slender, 1–3 dm tall, glabrous to puberulent in lines from the leaf bases. Leaves opposite; blade filiform to acicular, arcuate-spreading, 2–7 cm long, 0.–2(4) mm wide, coriaceous, glabrous, apex narrowly acute, margins revolute, base narrowly acute; petiole none. Inflorescences 1 to few per stem, in upper leaf axils, 3- to 5-flowered; peduncles 0.3–0.5 cm long; pedicels filiform, 0.7–1 cm long, glabrous to puberulent. Flowers 7.5–9 mm tall; calyx lobes green to purple-tinged, lanceolate to ovate-lanceolate, 1.5–2 mm long, glabrous; corolla lobes pale greenish-yellow, elliptic-lanceolate, reflexed, 4.5–5 mm long, glabrous; gynostegium cream to yellowish, briefly stipitate, sparingly puberulent with spreading, straight hairs; column obconic, 0.6–0.8 mm tall, 1.2–1.6 mm wide; hoods lanceolate, attached near base, widely spreading, 4.5–5.5 mm long, not fleshy, freely open above, the apex long attenuated, plane to incurved, ca 2 mm higher than the anther head, the margins irregularly toothed to undulate, the base not saccate; horns falciform, adnate to lower 1/3 of hood, ascending, 0.5–0.6 × longer than the hood; fleshy pads shallowly bilobed; anther head truncate-conic, 1.6–2 mm tall, 1.9–2.1 mm wide; anther appendages ca 1 mm long; anther wings acute at base, not notched, without spurs, ca 0.9 mm long; corpusculum ca 0.2 mm long; pollinia ca 0.7 mm long. Follicles narrowly fusiform, erect on deflexed pedicels, 4–7 cm long, 0.5–0.7 cm thick, without tubercles, puberulent to glabrate; seeds broadly ovate, ca 6 mm long; coma white to tan, 2.5–3.5 cm long. Jul–Aug. Dry prairies, cliffs, & mesas; CO: Baca, Otero; NM: Curry, Quay; OK: Cimarron; TX: Deaf Smith; (GP w to se AZ, s to trans-Pecos TX; n Mex.).

13. *Asclepias meadii* Torr., Mead's milkweed. Perennial herb from a shallow, slender rhizome; stems usually solitary from an unbranched, slightly thickened base, simple, slender, 2–5 dm tall, glabrous, glaucous. Leaves opposite; blade lanceolate to broadly ovate, ascending to spreading, (3)4–8 cm long, 1–4.5 cm wide, firmly membranaceous or subsucculent, glabrous except margins usually ciliolate, apex acute, margins not revolute, base obtuse to rounded or subcordate; petiole usually none. Inflorescences solitary, terminal, 8- to 18-flowered; peduncles 4–7 cm long; pedicels slender, 1.5–2 cm long, at least sparsely pubescent. Flowers 13–16 mm tall; calyx lobes green to purple-tinged, lanceolate to ovate, 2.5–4 mm long, sparsely to moderately villous; corolla lobes greenish-cream, often purple-tinged dorsally, ovate to elliptic-lanceolate, reflexed, 7–9.5 mm long, glabrous; gynostegium greenish-cream, briefly stipitate, glabrous; column obconic, 1–1.5 mm tall, 3–3.5 mm wide; hoods ovate, attached near base, erect, 4–5.5 mm long, distinctly fleshy, freely open above, the apex very broadly rounded, recurved, ca 1 mm higher than the anther head, the margins with a pair of teeth in upper 1/2, the base saccate; horns falciform, adnate to base of hood, arching over the anther head, 0.8–0.9 × longer than the hood; fleshy pads shallowly bilobed; anther head truncate-conic, 2.6–3 mm tall, 3–4 mm wide; anther appendages ca 1 mm long; anther wings abruptly rounded at base, shallowly notched, without spurs, ca 1.9 mm long; corpusculum ca 0.5 mm long; pollinia ca 1 mm long. Follicles narrowly fusiform, erect on deflexed pedicels, 8–10 cm long, 0.8–1.1 cm thick, without tubercles, glabrous; seeds broadly ovate, 7–8 mm long; coma white, 3–4 cm long. May–Jun. Rare in calcareous soils of prairies; (s IA, MO, e edge KS; s WI & w IL to GP). *Endangered.*

14. ***Asclepias oenotheroides*** Cham. & Schlecht., sidecluster milkweed. Perennial herb from a rhizome; stems few to several from a branching base, usually simple, moderately stout, 1–4 dm tall, moderately to densely pubescent. Leaves opposite or subopposite; blade ovate to oblong or deltoid, ascending to spreading, 2.5–10 cm long, 1.5–6 cm wide, firmly membranaceous, puberulent, apex acute to rounded, often mucronate, margins not revolute, base acute to obtuse or truncate; petiole 0.5–2.5 cm long. Inflorescences few to several, scattered in leaf axils in upper 1/2–2/3 of plant, 6- to 18-flowered; peduncles 0.1–1 cm long or inflorescence sessile; pedicels slender, 1–2 cm long, puberulent; flowers 14–19 mm tall; calyx lobes greenish, ovate, 2–3 mm long, puberulent; corolla lobes greenish-white to yellow, elliptic-lanceolate, reflexed, 9.5–11.5 mm long, glabrous; gynostegium pale greenish-cream, briefly stipitate, glabrous; column subcylindric, 1.5–1.7 mm tall, 2.1–2.3 mm wide; hoods very narrowly oblong in lower 1/2, prominently flaring above, attached in lower 1/2, erect below, spreading somewhat above, 7–9.5 mm long, somewhat fleshy, freely open above, the apex retuse, plane, 3.5–4.5 mm higher than the anther head, the margins usually with a pair of prominent lobes in the upper 1/5, the base deeply saccate; horns subulate, adnate to lower 3/4 of hood, arching over the anther head, ca 1.1 × longer than the hood; fleshy pads obscurely bilobed; anther head truncate-conic, 2.5–3.7 mm tall, 3–4 mm wide; anther appendages ca 1 mm long; anther wings arcuate, minutely notched near midpoint, without spurs, ca 2.7 mm long; corpusculum ca 0.3 mm long; pollinia ca 1.6 mm long. Follicles broadly fusiform, erect on deflexed pedicels, 7–9 cm long, 1.5–2 cm thick, without tubercles, puberulent to glabrate; seeds obovate, 6–8 mm long; coma tan, 2–3 cm long. Jul–Aug. Sandy or rocky calcareous soils of prairies; OK: Blaine, Major; TX: Bailey, Briscoe, Floyd; (OK w to sw NM, s to s TX; Mex. & C. Amer.).

15. ***Asclepias ovalifolia*** Dcne., ovalleaf milkweed. Perennial herb from a shallow, slender rhizome; stems mostly solitary or paired from a simple to branched and somewhat thickened base, simple, slender, (1)2–6 cm tall, sparsely to densely villous. Leaves mostly opposite or subopposite; blade lanceolate to broadly ovate, erect to spreading, (2)4–8 cm long, (1)1.8–4.5 cm wide, firmly membranaceous, sparsely to moderately villous, especially beneath, apex broadly acute to rounded, occasionally mucronate, margins flat to slightly revolute, base obtuse to rounded; petiole 0.2–1 cm long. Inflorescences 1–3, terminal or subterminal, (4)8- to 20-flowered; peduncles 0.5–3 cm long or inflorescence sometimes sessile; pedicels filiform, 15–20 mm long, puberulent. Flowers 8–10 mm tall; calyx lobes green to purple, lanceolate to ovate, 2.3–3.5 mm long, villous; corolla lobes greenish-white, often purple-tinged dorsally, elliptic-lanceolate, reflexed, 5–6 mm long, sparsely to moderately puberulent dorsally; gynostegium greenish-white to cream or yellow, briefly stipitate, glabrous; column obconic, 0.4–0.6 mm tall, 1.2–1.8 mm wide; hoods elliptic-oblong, attached near base, spreading, 3.8–5 mm long, not fleshy, freely open above, the apex rounded, plane, ca 2 mm higher than the anther head, the margins with a pair of triangular lobes below the midpoint, the base not saccate; horns falciform, adnate to lower 1/3 of hood, arching over the anther head, 0.7–0.8 × longer than the hood; fleshy pads obscure, narrowly bilobed; anther head truncate-conic, 1.6–2.5 mm tall, 2.2–3 mm wide; anther appendages ca 1.1 mm long; anther wings abruptly rounded at base, not notched, scarcely spurred, ca 1.8 mm long; corpusculum ca 0.3 mm long; pollinia ca 1.3 mm long. Follicles fusiform, erect on deflexed pedicels, 6–8 cm long, 0.8–1.3 cm thick, without tubercles, densely puberulent; seeds broadly ovate, 5.5–7 mm long; coma tan, 1.8–3.5 cm long. ($2n = 22$) Jun–Jul. Sandy, gravelly, or clayey soils of prairies & woodlands; MN, ND, SD, IA, WY; (s Can., IL & WI to GP).

16. *Asclepias pumila* (A. Gray) Vail, plains milkweed. Perennial herb from a taproot or slender rhizome; stems 1–several arising from a branched base, simple or branched near or below ground, slender, 0.5–3(4) dm tall, puberulent in decurrent lines from the leaf bases. Leaves alternate in a tight spiral, or whorled near base of stem; blade filiform, erect to spreading, 1.5–5(6) cm long, 0.5–1(1.5) mm wide, coriaceous, glabrous to sparsely puberulent, apex narrowly acute, margins tightly revolute, base narrowly acute; petiole none. Inflorescences 1 to many in upper leaf axils, 4- to 20-flowered; peduncles 0.5–2 cm long; pedicels filiform, 4–12 mm long, puberulent. Flowers 5–8 mm tall, calyx lobes green to purple, narrowly triangular to ovate-lanceolate, 1.5–2.6 mm long, sparsely villous; corolla lobes white or tinged with rose or yellow-green, particularly on dorsal surface, oblong to elliptic, reflexed, 2.8–4.2 mm long, glabrous or essentially so; gynostegium greenish-white, stipitate, glabrous; column subcylindric, 0.6–1.1 mm tall, 0.7–0.9 mm wide; hoods oblong, attached near base, erect, 1.6–1.9 mm long, not fleshy, freely open above, the apex very broadly rounded, recurved, about equal in height to the anther head, the margins entire, the base not saccate; horns acicular, adnate to lower 1/2 of hood, arching over the anther head, 1.5–2 × longer than the hood; fleshy pads bilobed; anther head truncate-conic, 1.5–1.7 mm tall, 1.2–1.3 mm wide; anther appendages ca 0.6 mm long; anther wings acute at base, shallowly notched, slightly spurred, ca 1.1 mm long; corpusculum ca 0.2 mm long; pollinia ca 0.8 mm long. Follicles narrowly fusiform, ascending to erect on straight to geniculate pedicels, 4–8 cm long, 0.6–0.8 cm thick, without tubercles, sparsely puberulent; seeds ovate, 4–6 mm long; coma white to tan, 1.2–2.6 cm long. Jul–Sep. Sandy, clayey, or rocky calcareous or gypseous soils of prairies; GP except MN, IA, MO, e 1/3 KS, e 1/2 OK; (GP s to trans-Pecos TX).

This species is poisonous to livestock.

17. *Asclepias purpurascens* L., purple milkweed. Perennial herb from a slender to stout rhizome; stems usually solitary from a simple to branched and thickened base, simple, rather stout, 5–10 dm tall, glabrous to densely villous, especially above. Leaves opposite or rarely in a false whorl of 4; blade broadly lanceolate to ovate, elliptic, or ovate-oblong, ascending to spreading, (4)8–23 cm long, (2)4–10 cm wide, firmly membranaceous, sparsely to densely villous, especially beneath, apex acute to rounded, often mucronate, margins sometimes inconspicuously revolute, base acute to rounded; petiole 0.2–2.5 cm long. Inflorescences 1–several, terminal and subterminal, (12)20- to 50-flowered; peduncles 0.5–9 cm long; pedicels slender, 1.5–3(3.5) cm long, sparsely to densely pubescent. Flowers 12–16 mm tall; calyx lobes green to purple-tinged, lanceolate to ovate-lanceolate, 2.5–3.8 mm long, sparsely to moderately puberulent; corolla lobes deep rose, ovate to elliptic-lanceolate, reflexed, (7)8–9.5 mm long, glabrous; gynostegium deep rose, briefly stipitate, glabrous; column obconic, 1.5–2 mm tall, 1.5–3 mm wide; hoods lanceolate, attached near base, somewhat spreading, 5–7 mm long, somewhat fleshy, freely open above, the apex narrowly rounded, slightly recurved, ca 2 mm higher than the anther head, the margins entire and abruptly angled downward or with a pair of teeth below the midpoint; the base not saccate; horns falciform, adnate to lower 1/2 of hood, abruptly incurved, 0.7–0.8 × longer than the hood; fleshy pads bilobed; anther head truncate-conic, 2.3–2.7 mm tall, 3.7–4 mm wide; anther appendages ca 0.7 mm long; anther wings acute at base, minutely notched, without spurs, ca 2.2 mm long; corpusculum ca 0.3 mm long; pollinia ca 1.3 mm long. Follicles fusiform, erect on deflexed pedicels, 10–15 cm long, 1–2 cm thick, without tubercles, puberulent; seeds broadly ovate, 5–6.5 mm long; coma white, 3.5–4.5 cm long. May–Jul. Sandy or rocky calcareous soils of open deciduous woodlands; IA, MO, e edge KS (also Wallace Co.), e edge OK; (CT & NC w to GP, s to AR).

The inflorescences of this species are reported to be edible when boiled and may be used as a substitute for broccoli.

18. *Asclepias quadrifolia* Jacq., fourleaf milkweed. Perennial herb from a shallow, slender rhizome; stems mostly solitary from the simple to branched and somewhat thickened base, simple, slender, 2.5–6 dm tall, glabrous to pubescent, often in decurrent lines from the leaf bases. Leaves mostly opposite but one internode suppressed to form a false whorl of 4 near midpoint of stem; blade lanceolate to ovate, spreading, (2)5–12 cm long, (1)1.5–4(7) cm wide, thinly membranaceous, sparsely puberulent especially on the veins, apex acute to acuminate, margins not revolute, base cuneate to rounded; petiole 0.2–1(2) cm long. Inflorescences 1–3, terminal or subterminal, (4)15- to 35-flowered; peduncles 0.8–4 cm long; pedicels filiform, 15–22 mm long, puberulent. Flowers 8–11 mm tall; calyx lobes green to purple-tinged, lanceolate to ovate, 1.2–3 mm long, glabrous; corolla lobes pale pink to cream, elliptic-lanceolate, reflexed, 4.5–6 mm long, glabrous; gynostegium white, stipitate, glabrous; column obconic to cylindric, 0.8–1.3 mm tall, 1–1.3 mm wide; hoods narrowly oblong, attached near base, spreading, 3.5–4 mm long, not fleshy, freely open above, the apex rounded, plane, ca 1 mm higher than the anther head, the margins with a pair of prominent teeth below the midpoint, the base not saccate; horns falciform, adnate to lower 1/2 of hood, arching over the anther head, slightly shorter than the hood; fleshy pads bilobed; anther head truncate-conic, 1.7–1.8 mm tall, 2.1–2.3 mm wide; anther appendages ca 0.6 mm long; anther wings abruptly rounded at base, not notched, without spurs, ca 1.3 mm long; corpusculum ca 0.2 mm long; pollinia ca 0.7 mm long. Follicles narrowly fusiform, erect on straight to curved pedicels, 8–14 cm long, 0.6–0.8 cm thick, without tubercles, puberulent to glabrous; seeds broadly oval, 7–8 mm long; coma white to tan, 3.5–4.5 cm long. May–Jun. Cherty soils of woodlands; KS: Cherokee; MO: Barton, Jasper; OK: Ottawa, Tulsa; (s. Can., VT to KS, s to GA, AR, & OK).

19. *Asclepias speciosa* Torr., showy milkweed. Perennial herb from a deep rhizome; stems solitary from a simple to branched, seldom enlarged base, simple or occasionally sparingly branched above, stout, (2.5)5–10 dm tall, densely puberulent above, becoming sparsely so below. Leaves usually opposite; blade broadly lanceolate to ovate, ovate-oblong, or suborbicular, ascending to spreading, (6)8–20 cm long, 2.5–10(15) cm wide, firmly membranaceous, glabrate to sparsely or densely puberulent above, tomentulose beneath, apex acute to obtuse or broadly rounded, usually mucronate, margins not revolute, base obtuse to truncate or cordate; petiole 0.2–1.3 cm long. Inflorescences few–several, in the upper leaf axils, 10- to 40-flowered; peduncles 2–10 cm long; pedicels moderately stout, 1–3.3 cm long, densely woolly-tomentose. Flowers 15–28 mm tall; calyx lobes green to purple-tinged, ovate, 4–6.5 mm long, densely tomentose; corolla lobes purplish-rose, elliptic-lanceolate, reflexed, 9–15 mm long, moderately to densely pubescent dorsally, glabrous to puberulent above near base; gynostegium pale rose or pinkish-cream, briefly stipitate, glabrous; column broadly obconic, 1.1–1.4 mm tall, 2.5–3.3 mm wide; hoods lanceolate, attached near base, widely spreading, 9–15 mm long, fleshy, freely open above, the apex attenuated, plane, 4–6 mm higher than the anther head, the margins with a pair of lobes at or below the midpoint, the base not saccate; horns falciform, adnate to lower 1/2–1/3 of hood, arching over the anther head, 0.4–0.6× longer than the hood; fleshy pads narrowly bilobed; anther head truncate-conic, 4–4.3 mm tall, 4–4.4 mm wide; anther appendages ca 1.5 mm long; anther wings abruptly rounded at base, prominently notched, without spurs, ca 3 mm long; corpusculum 0.7–0.8 mm long; pollinia ca 1.5 mm long. Follicles fusiform to broadly so, erect to ascending on deflexed pedicels, 7–11 cm long, 2–3 cm thick, nearly smooth to densely covered by soft, subulate tubercles, densely tomentose; seeds broadly ovate, 6–9 mm long; coma white to tan, 2–4.5 cm long. ($2n = 22$) May–Aug. Sandy, loamy, or rocky soils on banks & flood plains of lakes, ponds, or waterways or moist areas in prairies; GP except MO, sw IA, e 1/4 KS, e 1/3 OK; (s Can., GP w to WA & w-cen CA, s to s NM).

This species hybridizes with *A. syriaca;* plants intermediate morphologically are presumably hybrids. The species is poisonous to livestock, but the young shoots are reported to be edible when boiled like asparagus, with one or two changes of water.

20. **Asclepias stenophylla** A. Gray, narrow-leaved milkweed. Perennial herb from a stout, vertical rootstock; stems solitary or occasionally paired, from a mostly simple, thickened base, simple or occasionally sparingly branched, slender, 2-10 dm tall, puberulent to glabrate. Leaves mostly alternate to subopposite; blade linear, erect to moderately spreading, (4)8-18 cm long, 1-5(8) mm wide, firmly membranaceous, moderately to sparsely puberulent, apex narrowly acute, margins often revolute, base narrowly acute, petiole, if present, 1-2 mm long. Inflorescences few to several, scattered in leaf axils of upper 1/3-2/3 of plant, 10- to 25-flowered; peduncles 1-4(15) mm long or more commonly none; pedicels slender, 0.5-1.1 cm long, puberulent. Flowers 7.5-9 mm tall; calyx lobes green to purple-tinged, lanceolate, 2-3.3 mm long, puberulent; corolla lobes pale greenish-white to yellow, elliptic-lanceolate, reflexed, 4.5-5.3 mm long, glabrous; gynostegium pale greenish to white, briefly stipitate, glabrous; column conic, 1.2-1.5 mm tall, 1.1-1.2 mm wide; hoods narrowly oblong, attached in lower 1/4, erect, 3.3-3.8 mm long, somewhat fleshy, freely open above, the apex deeply emarginate and appearing 3-toothed or lobed, the shorter median lobe representing the apex of the horn which is adnate the entire length of the hood, plane, ca 0.5 mm lower than the anther head, the margins with a prominent pair of lateral, basal lobes, the base deeply saccate; fleshy pads bilobed; anther head truncate-conic, 2.2-3 mm tall, 2-2.4 mm wide; anther appendages ca 0.6 mm long; anther wings rounded at base, deeply notched, without spurs, ca 1.5 mm long; corpusculum ca 0.5 mm long; pollinia ca 0.8 cm long. Follicles fusiform, erect on deflexed pedicels, 9-12 cm long, 0.7-0.8 cm thick, without tubercles, puberulent to glabrate; seeds broadly obovate, 5-6 mm long; coma tan, 2.5-3.5 cm long. Jun-Aug. Sandy or rocky calcareous soils of prairies; se MT, ne WY, w & s SD, NE, KS, MO, OK, TX (w IL to n-cen CO, s to w AR, TX). *Acerates angustifolia* (Nutt.) Dcne.—Rydberg.

21. **Asclepias subverticillata** (A. Gray) Vail, poison milkweed. Perennial herb from an often deep-seated rhizome; stems 1 to few arising from a branched rootcrown, simple or sparingly branched above with at least a few dwarf axillary branchlets, moderately slender, (3)4-9 dm tall, glabrous to sparsely puberulent in decurrent lines from the leaf bases. Leaves verticillate to subverticillate, mostly 3 or 4 per node, with secondary clusters of dwarf leaves in at least some of the axils; blade linear to filiform, erect to widely spreading, 3-12 cm long, 1-3(4) mm wide, coriaceous to membranaceous, glabrous to sparsely puberulent, apex narrowly acute, margins revolute, base narrowly acute; petiole none. Inflorescences few to many in upper leaf axils, (6)10- to 25-flowered; peduncles 1-3.5 cm long; pedicels filiform, 7-14 mm long, puberulent. Flowers 6.5-8 mm tall; calyx lobes green or purple-tinged, narrowly lanceolate to ovate, 1.2-2.2 mm long, sparsely villous to glabrate; corolla lobes white, rarely tinged with greenish-purple, elliptic, reflexed, 3.4-4.5 mm long, glabrous; gynostegium white to greenish-white, stipitate, glabrous; column subcylindric, 0.7-1.1 mm tall, 0.6-1 mm wide; hoods broadly oblong, attached near base, erect, 1.6-2 mm long, not fleshy, freely open above, the apex very broadly rounded, slightly recurved, ca 0.5 mm lower than the anther head, the margins entire, the base not saccate; horns acicular, adnate to lower 1/3 of hood, arching over the anther head, 1.5-2 × longer than the hood; fleshy pads bilobed; anther head truncate-conic, 1.5-2 mm tall, 1.2-1.5 mm wide; anther appendages ca 0.8 mm long; anther wings acute at base, minutely notched, without spurs, ca 1.5 mm long; corpusculum ca 0.2 mm long; pollinia ca 0.9 mm long. Follicles fusiform, erect on straight to curved pedicels, 6-9(13) cm long, 5-8 mm thick, without tubercles, puberulent; seeds broadly ovate, 5.5-8 mm long; coma white to tan, 2-3.5 cm long. (2n = 22) Jun-Aug. Sandy or rocky soils of prairies, flood plains, & roadside ditches; w KS, se CO, w OK, TX, NM; (GP to s WY, s to n Mex.). *A. galioides* Am. authors not H.B.K.—Gates.

This species is very poisonous to livestock, especially sheep.

22. *Asclepias sullivantii* Engelm., smooth milkweed. Perennial herb from a deep, fleshy rhizome; stems 1 to few from an enlarged crown, simple or sparingly branched above, fairly stout, 4–11 dm tall, glabrous, somewhat glaucous. Leaves opposite; blade broadly ovate to lanceolate or oblong, ascending to spreading, 6–16 cm long, 2–9 cm wide, thickly membranaceous or somewhat succulent, glabrous or nearly so, slightly glaucous, apex obtuse to rounded or truncate and usually mucronate, margins not revolute, base rounded to cordate-clasping; petiole usually none. Inflorescences 1 to several in the upper leaf axils, 15- to 40-flowered; peduncles (0.5)1–7 cm long; pedicels slender, (2)2.5–4 cm long, glabrous or sparsely pubescent. Flowers 14–20 mm tall; calyx lobes green to purple- or pink-tinged, lanceolate to ovate, 3.5–6 mm long, glabrous; corolla lobes pinkish-rose to purple, lanceolate to oblong-lanceolate, reflexed, 9–11 mm long, glabrous; gynostegium pale rose to pinkish, briefly stipitate, glabrous; column broadly obconic, 0.8–1.5 mm tall, 2.2–3.1 mm wide; hoods ovate-oblong, attached near base, spreading, 5–6.5 mm long, somewhat fleshy, freely open above, the apex very broadly rounded to truncated, slightly recurved, 0.5–1 mm higher than the anther head, the margins entire, the base not saccate; horns falciform, adnate to lower 1/2 of hood, abruptly incurved over the anther head, slightly shorter than the hood; fleshy pads narrowly bilobed; anther head truncate-conic, 3.8–4.2 mm tall, 3.5–4.2 mm wide; anther appendages ca 1.2 mm long; anther wings abruptly rounded at base, notched, prominently spurred, ca 2.5 mm long; corpusculum ca 0.5 mm long; pollinia ca 1.5 mm long. Follicles broadly fusiform, erect or ascending on deflexed pedicels, 8–10 cm long, 2–3 cm thick, with soft, spinose tubercles in upper half, puberulent to glabrous; seeds broadly ovate, 7–8 mm long; coma white, 3.5–4.5 cm long. Jun–Aug. Sandy, loamy, or rocky calcareous soils of prairies & roadsides; MN, e ND, se SD, IA, e 1/2 NE & KS, MO, & OK; (s. Can., OH, WI, IN, IL to GP).

23. *Asclepias syriaca* L., common milkweed. Perennial herb from a deep rhizome; stems usually solitary from a simple to branched and thickened base, usually simple, stout, 6–20 dm tall, sparsely to densely puberulent. Leaves mostly opposite; blade broadly ovate to elliptic or oblong, ascending to spreading, (6)10–19(30) cm long, (3)5–11 cm wide, firmly membranaceous, sparsely to moderately pubescent above, tomentulose beneath, apex obtuse to rounded and mucronate, margins not revolute, base obtuse to subcordate; petiole 0.2–1.4 cm long. Inflorescences few to several in upper leaf axils, 20- to 130-flowered; peduncles 1–14 cm long; pedicels slender, 1.5–4.5 cm long, moderately to densely tomentulose. Flowers 11–17 mm tall; calyx lobes green to purple-tinged, lanceolate, 2.5–4 mm long, densely puberulent; corolla lobes rose to purple, rarely white, elliptic-lanceolate, reflexed, 6.5–9 mm long, moderately to densely puberulent dorsally, glabrous to puberulent ventrally near base; gynostegium pale rose, rarely white, briefly stipitate, glabrous; column broadly obconic, 1–1.9 mm tall, 1.5–2.5 mm wide; hoods ovate, attached near base, spreading, 3.3–5.2 mm long, somewhat fleshy, freely open above, the apex rounded, slightly recurved, ca 1 mm higher than the anther head, the margins with a pair of usually prominent teeth near the midpoint, the base not saccate; horns falciform, adnate to lower 1/3 of hood, arching over the anther head; fleshy pads narrowly bilobed; anther head truncate-conic, 2–2.7 mm tall, 3–3.5 mm wide; anther appendages ca 1 mm long; anther wings abruptly rounded at base, minutely notched, without spurs, ca 2 mm long; corpusculum ca 0.5 mm long; pollinia ca 1.3 mm long. Follicles narrowly to broadly fusiform, erect to ascending on deflexed pedicels, 7–11 cm long, 1.8–3.5 cm thick, smooth to densely covered by soft, subulate tubercles, densely tomentose; seeds broadly ovate, 6–8 mm long; coma white, 3–4 cm long. ($2n = 22, 24$) May–Aug. Sandy, clayey, or rocky calcareous soils of banks or flood plains of lakes, ponds, or waterways, or of prairies, forest margins, roadsides, or waste places; GP except MT, WY, CO, NM, & TX; (s. Can., ME to ND, s to NC, nw GA, ne OK, & TX). *A. kansana* Vail—Rydberg, *A. syriaca* L. f. *leucantha* Dore, *A. syriaca* L. var. *kansana* (Vail) Palm. & Steyerm.—Fernald.

This species hybridizes with *A. speciosa;* plants intermediate morphologically are presumably hybrids. The species is poisonous to livestock, but the young shoots and well-developed follicles are reported to be edible if boiled, with one or two changes of water.

24. *Asclepias tuberosa* L., butterfly milkweed. Perennial herb from a deep, woody rootstock; stems 1 to many arising from a branched, thickened rootcrown, simple to sparingly branched above, slender to moderately stout, 3–9 dm tall, densely and evenly villous to hirsute. Leaves alternate or those subtending the inflorescences often opposite or subopposite; blade linear- to ovate-lanceolate, ascending to spreading (2)5–10 cm long, (0.4)0.7–2.3 cm long, firmly membranaceous, sparsely to densely villous to hirsutulous, especially on the veins beneath, apex accuminate to acute or rounded, margins revolute, often inconspicuously so, base obtuse to truncate or cordate; petiole 0.1–0.5 cm long. Inflorescences 1–many at end of stem and helicoid branches and in upper leaf axils, 6- to 25-flowered; peduncles 0.1–3 cm long or inflorescences sessile in the axils; pedicels filiform, 12–19 mm long, pubescent. Flowers 11–15.5 mm tall; calyx lobes green to purple-tinged, linear to lanceolate, 1.9–3.7 mm long, villous; corolla lobes bright orange, rarely red or yellow, elliptic to lanceolate, reflexed, 5.5–8.5 mm long, glabrous; gynostegium orange, rarely yellow, stipitate, glabrous; column obconic, 1–1.5 mm tall, 1.1–1.5 mm wide; hoods lanceolate, attached near base, slightly spreading, 4.5–5.8 mm long, not fleshy, freely open above, the apex narrowly rounded, slightly recurved, 1.5–2 mm higher than the anther head, the margins with a pair of short lobes below the midpoint, the base not saccate; horns acicular, adnate to lower 1/4 of hood, arching over the anther head, 0.7–1.1 × longer than the hood; fleshy pads shallowly bilobed; anther head truncate-conic, 2.2–2.6 mm tall, 2.1–2.4 mm wide; anther appendages ca 0.5 mm long; anther wings acute at base, not notched, without spurs, ca 1.6 mm long; corpusculum ca 0.2 mm long; pollinia ca 1.2 mm long. Follicles fusiform, erect on deflexed pedicels, 8–15 cm long, 1–1.5 cm thick, without tubercles, puberulent; seeds broadly oval, 5–7 mm long; coma white, 3–4 cm long.

This is our only species without milky juice. It is reported to be poisonous to livestock.

Two subspecies recognized by Woodson (Ann. Missouri Bot. Gard. 41:1–211. 1954) occur in the GP. Although they are of questionable validity, they are included below.

24a. subsp. *interior* Woods. Leaves mostly deeply cordate. May–Aug. Sandy, loamy, or rocky calcareous soils of prairies, roadsides, & waste places; MN, e SD, IA, e 1/2 NE, e 2/3 KS, MO, OK, TX; (se Can., NY to MN, SD, s to MS & TX). *A. tuberosa* L. var. *interior* (Woods.) Shinners f. *lutea* (Clute) Steyerm.—Steyermark.

24b. subsp. *terminalis* Woods. Leaves mostly obtuse to truncate at base, varying to slightly cordate. Jun–Jul. Sandy prairies & open woodlands; CO: El Paso, Pueblo, Yuma; MN: Stearns; NM: Union; OK: Cimarron; SD: Fall River; TX: Hutchinson; (MI, WI, CO, s to UT, AZ; n Mex.).

25. *Asclepias uncialis* Greene, dwarf milkweed. Perennial herb from a slender to stout, vertical rootstock; stems several to many from a branched base, usually simple, slender, 1.5–7 cm tall, puberulent. Leaves alternate to subopposite; blade ovate below to linear-lanceolate above, ascending, 0.5–3 cm long, 2–7 mm wide, membranaceous, tomentulose on margins and veins, apex acute to acuminate, margins not revolute, base rounded to acute; petiole if present to 1 mm long. Inflorescences 1–few, terminal and subterminal, 2- to 12-flowered; peduncles none; pedicels filiform, 1–1.5 cm long, puberulent. Flowers 4.5–5 mm tall; calyx lobes green to purple-tinged, lanceolate, 1.5–2.5 mm long, puberulent; corolla lobes purplish-rose, elliptic-lanceolate, reflexed, 3–4 mm long, glabrous; gynostegium pale rose, sessile or nearly so, glabrous; column subcylindric, 0.5–0.7 mm tall, 0.5–0.8 mm wide; hoods ovate, attached in lower 1/3, spreading, 1.3–1.5 mm long, fleshy, freely open above, the apex broadly rounded, plane, 0.5–0.7 mm lower than the anther head, the margin with a pair of prominent, triangular lobes near the midpoint, the base briefly saccate; horns lingulate, adnate to lower 1/2 of hood, nearly erect, 1–1.2 × longer than the hood; fleshy pads narrowly bilobed; anther head truncate-conic, 1–1.5 mm tall, 2–2.3 mm wide; anther appendages ca 0.5 mm long; anther wings acute at base, not notched, without spurs, ca 1.2 mm long; corpusculum 0.2 mm long; pollinia ca 0.5 mm long. Follicles narrowly fusiform, erect on deflexed pedicels, 4.5–5.5 cm long, 0.6–0.8 cm thick, without tubercles, puberulent; seeds unknown. May. Sandy or rocky prairies; CO: Baca, Cheyenne, Denver, Pueblo, Weld; OK: Texas; (GP to NM).

This species is either rare or overlooked.

26. *Asclepias verticillata* L., whorled milkweed. Perennial herbs with a shallow root system; stems 1–few arising from a sparingly branched fibrous rootcrown, simple or occasionally branched above, moderately slender, (2)3.5–9 dm tall, usually puberulent in decurrent lines from the leaf bases. Leaves mostly verticillate to subverticillate, mostly 3–6 per node; blade filiform to linear, erect to spreading, 1.5–8 cm long, 0.5–1.5(3) mm wide, usually coriaceous, glabrous to puberulent, apex narrowly acute, margins revolute, base narrowly acute; petiole none. Inflorescences usually few to many in upper leaf axils, 6- to 20-flowered; peduncles 1–4.5 cm long; pedicels filiform, 5–11 mm long, puberulent. Flowers 5.5–7.5 mm tall; calyx lobes green to purple-tinged, linear-lanceolate to ovate, 1.2–2.5 mm long, sparsely villous to glabrate; corolla lobes white to greenish-white or purple-tinged, elliptic, reflexed, 3.5–4.5 mm long, glabrous; gynostegium greenish-white, stipitate, glabrous; column subcylindric, 0.7–1.1 mm tall, 0.5–0.8 mm wide; hoods broadly oblong, attached near base, erect, 1.4–2 mm long, not fleshy, freely open above, the apex very broadly rounded, plane to slightly recurved, ca 0.5 mm lower than the anther head, the margins entire, abruptly angled downward near midpoint, the base not saccate; horns acicular, adnate to lower 1/3 of hood, arching over the anther head, 1.5–2 × as long as the hood; fleshy pads bilobed; anther head subcylindric, 1.5–1.8 mm tall, 1.3–1.5 mm wide; anther appendages ca 0.5 mm long; anther wings abruptly rounded at base, minutely notched, without spurs, ca 1.4 mm long; corpusculum ca 0.2 mm long; pollinia ca 1 mm long. Follicles narrowly fusiform, ascending to erect on straight to curved pedicels, 8–10.5 cm long, 0.6–0.8 cm thick, without tubercles, sparsely puberulent; seeds broadly ovate, 5–6 mm long; coma white, 2.5–3.5 cm long. (2n = 22) Jun–Sep. Sandy, clayey, or rocky calcareous soils of prairies, badlands, flood plains, & open woods; GP except WY, CO; (s Can., MA to MT, s to FL & AZ).

This species is poisonous to livestock.

27. *Asclepias viridiflora* Raf., green milkweed. Perennial herb from a vertical rootstock; stems mostly solitary or paired from a simple or branched, thickened base, simple or sometimes sparingly branched, slender to rather stout, 1–6(10) dm tall, puberulent to tomentulose, often glabrate. Leaves opposite or subopposite, or occasionally some alternate; blade extremely variable in shape, linear to lanceolate, ovate, or suborbicular, ascending to widely spreading, 4–11(14) cm long, (0.2)0.8–5(6) cm wide, firmly membranaceous, sparsely puberulent to tomentulose, especially below, often glabrate, apex acute to obtuse or emarginate, often mucronate, margins not revolute, base acute to rounded; petiole 0.1–0.5 cm long or none. Inflorescences 1–several, scattered in leaf axils of upper 1/2 of plant, 20- to 80-flowered; peduncles 0.2–2 cm long; pedicels slender, 0.5–1.5(2) cm long, villous to tomentose. Flowers 9.5–12.5 mm tall; calyx lobes green to purple-tinged, lanceolate, 2.1–3 mm long, puberulent; corolla lobes pale green, elliptic-lanceolate, reflexed, 5.7–6.5(7) mm long, sparingly puberulent dorsally; gynostegium pale green, sessile, glabrous; column obconic, 1.2–1.5 mm tall, 1.2–1.4 mm wide; hoods oblong-lanceolate, attached in lower 1/3, erect, 3.9–5 mm long, fleshy, the orifice appressed to the anther head, the apex rounded, plane, ca 1 mm lower than the anther head, the margins with a pair of small lobes near base, the base deeply saccate; horns none; fleshy pads entire, broadly rounded; anther heads truncate-fusiform, 3–4 mm tall, 2.7–3.1 mm wide; anther appendages ca 1 mm long; anther wings obtusely angled and most prominent above the midpoint, not notched, without spurs, ca 3 mm long; corpusculum ca 0.3 mm long; pollinia ca 2.1 mm long. Follicles broadly fusiform, erect on deflexed pedicels, 7–15 cm long, 1.5–2 cm thick, without tubercles, puberulent to glabrate; seeds broadly obovate, 6–7.5 mm long; coma tan, 3–5 cm long. (2n = 24) Jun–Aug. Sandy or rocky calcareous soils in prairies; GP; (s Can., CT to MT, s to GA & AZ; ne Mex.). *Acerates viridiflora* (Raf.) Eaton — Rydberg; *Asclepias viridiflora* var. *lanceolata* (Ives) Torr., var. *linearis* (A. Gray) Fern. — Fernald; *A. viridiflora* var. *ivesii* Britt. — Gates.

28. **Asclepias viridis** Walt., spider milkweed. Perennial herb from a thickened, cylindrical to fusiform, vertical rootstock; stems solitary or paired, from a simple or branched, thickened base, simple or sparingly branched above, slender to moderately stout (1.5)2.5–6.5 dm tall, glabrous or sparsely pubescent above. Leaves mostly alternate to subopposite; blade lanceolate to ovate or broadly oblong, ascending to spreading, 5–12 cm long, 1–5.5 cm wide, firmly membranaceous, sparsely puberulent to glabrous, apex acute to rounded or emarginate, margins not revolute, base acute to rounded, truncate, or subcordate; petiole 0.3–1 cm long. Inflorescences 1–5(7), terminal and subterminal, 3- to 18-flowered; peduncles 0.2–3 cm long; pedicels moderately thick, 1–2.5 cm long, puberulent. Flowers 10–15 mm tall; calyx lobes green to purple-tinged, lanceolate to narrowly ovate, 2.8–5.5 mm long, puberulent; corolla lobes pale green, elliptic-lanceolate to ovate, arcuate-ascending, 13–17 mm long, glabrous; gynostegium pale purplish-rose, sessile, glabrous except minutely puberulent on margins of hoods; column truncate-conic, 2.5–3.2 mm tall, 1–1.6 mm wide; hoods clavate, attached in lower 1/3–1/2, abruptly deflexed and fused along the column with arcuate-ascending tips, 4.3–5.8 mm long, somewhat fleshy, closed or nearly so above, the apex broadly rounded, cucullate, ca 2.5 mm lower than the anther head, with a subterminal, laterally compressed, saccate appendage within; horns none, fleshy pads entire, broadly rounded; anther head truncate-subspheric, 2.7–3.2 mm tall, 2.7–3.4 mm wide; anther appendages ca 0.6 mm long; anther wings obtusely angled in upper 1/3, gradually rounded in lower 1/3, not notched, without spurs, ca 2.4 mm long; corpusculum ca 0.3 mm long; pollinia ca 1.3 mm long. Follicles fusiform to broadly so, erect on deflexed pedicels, 7–13 cm long, 1.3–2 cm thick, without tubercles, sparingly puberulent; seeds broadly obovate, 7–8 mm long; coma white to light tan, 3–4 cm long. Apr–Aug. Sandy or rocky calcareous soils in prairies; se NE, MO, KS, OK; (OH to NE, s to FL & TX). *Asclepiodora viridis* (Walt.) A. Gray — Fernald.

2. **CYNANCHUM** L., Sand Vine

Caulescent perennial vines with milky juice twining over other vegetation; stems few to several, herbaceous, from a thickened rootstock. Leaves opposite, petiolate; blade triangular to ovate; margins entire; base rounded to cordate, with a few subulate glands on the midrib; stipules minute, subulate, often deciduous. Inflorescence a cyme, often corymbose or umbellate, scattered in the leaf axils. Calyx lobes widely spreading, with linear to triangular glandular squamellae within, at or near the sinuses; corolla short-campanulate to rotate, slightly to widely spreading, white or dark purple; gynostegium sessile or stipitate; corona either a fleshy, shallowly lobed, cup-shaped disk or of 5 distinct, petaloid appendages with deeply bilobed apices, arising at or near base of gynostegium; fleshy pads none; anther head depressed-spheric to conic; anther appendages ovate to suborbicular, petaloid; anther wings straight to abruptly angled at base, obscure to prominent, corneous; corpusculum elliptic to linear, reddish-brown; pollinia of firmly coherent pollen, spatulate to oblong, subterete. Follicle fusiform, smooth, terete.

1 Petals whitish to cream; corona petaloid, of 5 distinct, erect appendages 5–6 mm tall, each appendage broadly ovate in lower 1/2, abruptly narrowed apically into a pair of linear lobes; leaf bases deeply cordate .... 1. *C. laeve*
1 Petals dark purple; corona a fleshy, shallowly lobed, cuplike disk 0.4–0.6 mm tall; leaf bases rounded to subcordate. .... 2. *C. nigrum*

1. **Cynanchum laeve** (Michx.) Pers., sand vine, climbing milkweed. Perennial trailing vine; stems simple or branched, mostly slender, villous in lines. Leaves opposite; blade triangular to broadly ovate, widely spreading, (2)4–11 cm long, (1.5)2–10 cm wide, thinly membranaceous, glabrous to sparingly strigose or villous, especially on the veins, apex acuminate to caudate or apiculate, margins flat to inconspicuously revolute, base deeply cordate; petiole 1–9 cm long. Inflorescence an umbellate to corymbose cyme, few to many in the leaf axils, 5- to 40-flowered; peduncles 0.3–5 cm long; pedicels slender, 3–12 mm long, villous. Flowers dimorphic, 5–8 mm in diam; calyx lobes green or purple-tinged, lanceolate to ovate, 1.5–3 mm long, sparingly villous; corolla lobes whitish to cream, narrowly oblong to oblong-lanceolate, spreading, 4–7 mm long, glabrous; corona petaloid, of 5 distinct, erect appendages 5–6 mm tall, each broadly ovate in lower 1/2, abruptly narrowing into a pair of free or partially fused linear lobes, 1.5–2 × longer than the gynostegium; gynostegium stipitate, often obscurely so, glabrous; column either distinct from the conical anther head and obconic or confluent with it and cylindric, ca 0.5 mm tall, in flowers having an obconic column, anther head 1.5–2 mm tall, 1.8–2 mm wide with anther wings 0.5–0.6(0.7) mm long, in flowers with a cylindric column, anther head 2.5–3 mm tall, 2–2.5 mm wide with anther wings 1.5–2 mm long; anther appendages ca 1 mm long; corpusculum 0.2–0.3 mm long; pollinia ca 0.4 mm long. Follicles fusiform, 8–14 cm long, 1.5–2 cm thick, without tubercles, sparsely puberulent to glabrous; seeds obovate, 7–9 mm long; coma white, 3–4 cm long. Jun–Sep. Sandy, clayey, or rocky calcareous soils of forest margins, thickets, flood plains, or disturbed areas; sw IA, MO, e edge NE, e 2/3 KS, e OK; (PA to NE, s to GA & TX). *Ampelamus albidus* (Nutt.) Britt.—Fernald; *Gonolobus laevis* Michx.—Gates.

2. **Cynanchum nigrum** (L.) Pers., black swallow wort. Perennial twining vine; stems simple or branched, mostly slender, villous in lines. Leaves opposite, blade narrowly to broadly ovate, widely spreading, 3–10 cm long, 1.5–7 cm wide, thinly membranaceous, villous on the veins and leaf margins, apex acuminate, margins flat to inconspicuously revolute, base rounded to subcordate; petiole 0.5–1.3 cm long. Inflorescence a cyme, few–many in the leaf axils, 5- to 12-flowered; peduncles 0.5–4 cm long; pedicels slender, 2–6 mm long, villous. Flowers not dimorphic, 6.5–8 mm in diam; calyx lobes green to purplish, narrowly triangular to ovate, 1.5–2 mm long, sparsely villous, especially apically; corolla lobes dark purple, ovate, widely spreading, 3–3.5 mm long, pubescent with erect hairs above, glabrous beneath; corona a fleshy, shallowly lobed, cupulate disk 0.4–0.6 mm tall, equaling or surpassing the anther head; gynostegium sessile, glabrous; column none; anther head depressed-spheric, 0.3–0.4 mm tall, 1.4–1.5 mm wide; anther appendages ca 0.3 mm long; anther wings obscure, ca 0.3 mm long; corpusculum ca 0.25 mm long; pollinia ca 0.25 mm long. Follicle narrowly fusiform, 5–7 cm long, 1–1.3 cm thick, without tubercles, puberulent to glabrous; seeds ovate, 7–8 mm long; coma white, 3–3.5 cm long. (2n = 22) May–Sep. Disturbed areas; KS: Montgomery; (ME to PA, w to OH, KS; sw Europe). *Naturalized*. *Vincetoxicum nigrum* (L.) Moench—Flora Europaea.

3. MATELEA Aubl., Climbing Milkweed, Anglepod

Caulescent perennial herbs with milky juice; stems prostrate to suberect or twining over other vegetation, few to many from a thickened, branched rootcrown. Leaves opposite, petiolate; blade ovate to suborbicular; margins entire; base cordate, usually with subulate to tuberculate glands on the midrib; stipules none. Inflorescence an axillary, corymbose to umbellate cyme or flowers in pairs in the axils. Calyx lobes widely spreading, with few to many subulate to tubular glandular squamellae within, near base or in the sinuses; corolla campanulate to rotate, slightly to widely spreading, white or yellow to purple or brown; gynostegium sessile or briefly stipitate; corona a cupulate to flat, laminar to fleshy disk surrounding the base of the gynostegium, variously lobed, with or without lingulate appendages within; fleshy pads none; anther head discoid; anther appendages ovate to suborbicular, petaloid; anther wings straight to curved, inconspicuous, corneous; corpusculum narrowly elliptic to rhombic, reddish-brown; pollinia of firmly coherent pollen, oblong to obovate, compressed. Follicle fusiform, smooth to tuberculate, terete to distinctly angled.

1 Stems 0.1–0.5 m long, not twining, prostrate to suberect; leaves 1–6.5 cm long, 1–4 cm wide; inflorescences mostly 2-flowered.
  2 Petals glabrous above, 3.5–4 mm long; corona lobes broad and shallow, each with an abruptly reflexed margin and with an ascending, fleshy, lingulate appendage within which is shorter than the lobe and subequal to the anther head. ..... 3. *M. cynanchoides*
  2 Petals pilose above, 5.5–7 mm long; corona lobes triangular below, attenuating above into oblong, truncate, inflexed apices much longer than the anther head. .... 2. *M. biflora*
1 Stems 1–several m long, twining over other vegetation; leaves mostly 8–18 cm long, 6–12 mm wide; inflorescences mostly 8- to 25-flowered.
  3 Petals glabrous, widely spreading; corona disklike, fleshy, irregularly and shallowly lobed, without appendages; follicles sharply angled, without tubercles. .... 5. *M. gonocarpa*
  3 Petals puberulent beneath, ascending; corona cuplike, not markedly fleshy, of 5 deltoid lobes, with usually 2 subulate appendages in each sinus; follicles indistinctly angled if at all, with numerous, short tubercles.
    4 Corolla lobes brown-purple .... 4. *M. decipiens*
    4 Corolla lobes whitish .... 1. *M. baldwyniana*

1. ***Matelea baldwyniana*** (Sweet) Woods., climbing milkweed. Perennial twining vine; stems simple or branched, moderately slender, hirsute with hairs to 2 mm long intermixed with minute, often purple-tinged, glandular hairs ca 0.1 mm long. Leaves opposite; blade broadly ovate to orbicular, spreading, 6–15 cm long, 4–12 cm wide, thinly membranaceous, moderately pubescent with soft, appressed hairs, apex abruptly acuminate, base deeply cordate; petiole 2–7 cm long. Inflorescences few in upper leaf axils, 4- to 20-flowered; peduncles 1.5–8 cm long; pedicels slender, 1–2.5 cm long, pubescent like the stems. Flowers 1–1.4 cm in diam; calyx lobes green, lanceolate, 2.6–3.5 mm long, moderately to densely strigose; corolla lobes whitish, oblong-lanceolate to spatulate, ascending, 8–12 mm long, glabrous above, moderately puberulent beneath; gynostegium sessile, glabrous; column none; corona cupulate, not markedly fleshy, of 5 deltoid lobes, each lobe ca 1/2 as long as the pair of usually subulate appendages in each sinus, much exceeding the anther head; anther head discoid, ca 0.5 mm tall, ca 1.8 mm wide; anther appendages ca 0.2 mm long; anther wings ca 0.2 mm long; corpusculum ca 0.2 mm long; pollinia oblong, ca 0.3 mm long. Follicles fusiform, 8–9 cm long, 1.5–2 cm thick, with numerous short tubercles, not distinctly angled; seeds ovate, 8–9 mm long; coma tan, 3–3.5 cm long. May–Jun. Rocky soils of thickets & open woods; MO: Jasper; (s MO & OK, s to GA & AL). *Gonolobus baldwynianus* Sweet—Fernald.

2. ***Matelea biflora*** (Raf.) Woods., two-flowered milkvine. Perennial herb from a thickened, vertical rootstock; stems several to many, simple or sparingly branched, slender, 1–4 dm long, pilose with hairs to 1 mm long and intermixed with minute, pale glandular hairs ca 0.1 mm long. Leaves mostly opposite; blade ovate to suborbicular, spreading, 1.5–4(5) cm long, 1–3 cm wide, membranaceous, densely pubescent with soft, spreading to appressed hairs, apex obtuse to acuminate, base shallowly to deeply cordate, petiole 0.3–2.5 cm long. Inflorescences many in the upper leaf axils, mostly 2-flowered; peduncles usually none,

if present then to 2 mm long; pedicels slender, 0.5–1 cm long, pubescent like the stems. Flowers 1.3–1.7 cm in diam; calyx lobes green, lanceolate to ovate, 2–3.5 mm long, villous; corolla lobes reddish-purple to dark brown above, green beneath, oblong-lanceolate, spreading perpendicularly to the axis, 5.5–7 mm long, moderately to densely pilose; gynostegium sessile, glabrous; column none; corona a flat, fleshy, broadly 5-lobed disk, each lobe inflexed, triangular in lower 1/2, attenuating above into an oblong, truncate apex, much exceeding the anther head; anther head discoid, ca 0.5 mm tall, ca 1.8 mm wide; anther appendages ca 0.4 mm long; anther wings ca 0.3 mm long; corpusculum ca 0.3 mm long; pollinia obovate, ca 0.3 mm long. Follicles broadly fusiform, 7–10 cm long, 1.5–2.5 cm thick, with few–many short tubercles, moderately to densely pubescent, not distinctly angled; seeds orbicular, 9–10 mm long; coma white to tan, 2.5–3.5 cm long. May–Jun. Clayey, sandy, or rocky soils in prairies or open woodlands; TX: Bailey, Hardemann; (TX & OK).

3. Matelea cynanchoides (Engelm.) Woods., milkvine. Perennial herb from a thickened, vertical rootstock; stems few to several, simple or sparingly branched below, slender, 1.5–5 dm long, pilose with hairs to 1 mm long and intermixed with minute, pale to purplish glandular hairs ca 0.1 mm long. Leaves opposite; blade mostly narrowly to broadly ovate, ascending to spreading, 1–6.5 cm long, 1–4 cm wide, membranaceous, moderately pubescent with soft erect to spreading hairs, apex obtuse to acuminate, base shallowly to deeply cordate; petiole 0.3–2.5 cm long. Inflorescences few in the upper leaf axils, mostly 2-flowered; peduncles none or if present 0.2–1 cm long; pedicels slender, 0.2–1 cm long, pubescent like the stem. Flowers 0.8–1.2 cm in diam; calyx lobes green, lanceolate to elliptic, 2.5–3.5 mm long, villous; corolla lobes brown to dark maroon above, greenish to brownish beneath, lanceolate to ovate, widely spreading, 3.5–4 mm long, glabrous above, pilose beneath; gynostegium briefly stipitate, glabrous; column obconic, 0.2–0.4 mm tall, ca 1 mm wide; corona a shallowly cupulate, fleshy, broadly and shallowly 5-lobed disk, each lobe with an ascending, fleshy, lingulate appendage within which is shorter than the lobe, and free from it in the upper 1/3, and extending little if any beyond the anther head; anther head discoid, ca 0.7 mm tall, ca 1.8 mm wide; anther appendages ca 0.3 mm long; anther wings ca 0.3 mm long; corpusculum ca 0.3 mm long; pollinia semicircular, ca 0.4 mm long. Follicles fusiform, 7–8 mm long, 2–2.5 cm thick, with few to many short tubercles, sparsely to moderately puberulent, not distinctly angled; seeds orbicular, ca 1 cm long; coma tan, 2–2.5 cm long. May–Jul. Sandy soils of open woods, thickets, & disturbed places; OK: Canadian, Kiowa, Payne; (OK to e TX).

4. Matelea decipiens (Alex.) Woods., climbing milkweed. Perennial twining vine; stems simple or branched, slender to moderately stout, hirsute with hairs 1–2 mm long and intermixed with minute purple glandular hairs less than 0.1 mm long. Leaves opposite; blade broadly ovate to suborbicular, widely spreading, (2)6–15 cm long, 6–11 cm wide, thinly membranaceous, sparsely to moderately hirsutulous, apex long-acuminate to abruptly so, base deeply cordate; petiole 3–9 cm long. Inflorescences few–several in the leaf axils, 5- to 25-flowered; peduncles 3–8 cm long; pedicels slender, 1–2 cm long, densely clothed with minute purple hairs less than 0.1 mm long often with scattered hairs to 1 mm long. Flowers 1–1.8 cm in diam; calyx lobes green, ovate to lanceolate, 2–3 mm long, hirsutulous; corolla lobes brownish-purple, narrowly oblong to linear-lanceolate, ascending, 1–1.5 cm long, glabrous above, hirsutulous beneath; gynostegium sessile, glabrous; column none; corona cupulate, not markedly fleshy, of 5 broadly deltoid lobes 1/2–1/3 as long as the pair of subulate appendages in each sinus, much higher than the anther head; anther head discoid, ca 0.5 mm tall, ca 1.8 mm wide; anther appendages ca 0.2 mm long; anther wings ca 0.2 mm long; corpusculum 0.2–0.3 mm long; pollinia oblong, ca 0.3 mm long. Follicles fusiform, 8–11 cm long, 1.5–2 cm thick, with numerous short tubercles, puberulent, somewhat angled; seeds ovate to suborbicular, 8–9 mm long; coma white, 2–4 cm long. May–Jun. Rocky soils of thickets & open woods; KS: Cherokee; MO: Jasper; (NC to KS & OK, s to SC, LA, & e TX). *Gonolobus decipiens* (Alex.) Perry—Fernald.

5. *Matelea gonocarpa* (Walt.) Shinners, anglepod. Perennial twining vines; stems simple or branched, slender, hirsute with hairs ca 1 mm long and intermixed with minute pale to purple nonglandular hairs less than 0.1 mm long. Leaves opposite; blade broadly ovate to suborbicular, widely spreading, (4)6–17 cm long, 4–11 cm wide, thinly membranaceous, sprasely hirsutulous with additional minute hairs 0.1–0.2 mm long beneath, apex long-acuminate to abruptly so, base deeply cordate; petiole 2–12 cm long. Inflorescences few to several in the leaf axils, 2- to 12-flowered; peduncles 1–2.5 cm long; pedicels slender, 1–3 cm long, usually glabrous. Flowers 1.7–2 cm in diam; calyx lobes green, lanceolate, 2.5–5 mm long, glabrous except at apex; corolla lobes yellow to greenish-purple, linear-lanceolate, widely spreading, 0.8–1 cm long, glabrous; gynostegium briefly stipitate, glabrous; column obconic, 0.2–0.3 mm tall, ca 1.3 mm wide; corona flat, fleshy, irregularly and shallowly lobed, without appendages, much exceeded by the anther head; anther head discoid, ca 0.5 mm tall, ca 2.3 mm wide; anther appendages ca 0.2 mm long; anther wings ca 0.2 mm long; corpusculum ca 0.2 mm long; pollinia broadly spatulate, ca 0.5 mm long. Follicles fusiform, (7)9–15 cm long, 2–2.8 cm thick, without tubercles, glabrous, sharply angled; seeds ovate, 8–10 mm long; coma white, 3.5–4.5 cm long. Jun–Aug. Rocky soils of thickets & open woods; KS: Cherokee; MO: Jasper; (VA to KS & OK, s to FL & TX). *Gonolobus gonocarpos* (Walt.) Perry, *G. suberosus* (L.) R. Br.—Fernald.

4. PERIPLOCA L., Silk Vine

1. *Periploca graeca* L., silk vine. Caulescent perennial vine with milky juice; stems trailing or twining over other vegetation, few–many, simple or branched, slender to moderately stout, woody, glabrous, to 7 m long. Leaves opposite; blade narrowly to broadly elliptic or ovate, ascending to widely spreading, 2.5–12 cm long, 1–6.5 cm wide, membranaceous, glabrous, apex obtuse to acuminate, margins entire, base acute to broadly rounded, without glands on the midribs; petiole 0.2–1.5 cm long; stipules minute, subulate. Inflorescence an axillary, corymbose or paniculate cyme, few–many, 6- to 20-flowered; peduncles 0.2–5 cm long; pedicels slender, 0.2–1.5 cm long, glabrous to sparingly pilose. Flowers 2.2–2.7 cm in diam; calyx lobes green, broadly ovate, widely spreading, 1.5–2.5 mm long, glabrous to densely villous, with 1 to few subulate, glandular squamellae within at or near each sinus; corolla brown-purple, rotate, lanceolate to oblong, 9–11 mm long, densely villous in lines along the margins and with an elliptic patch of minute hairs near the base; gynostegium none; column none; corona a ring of 5 pairs of broad lobes 0.5–0.8 mm tall, each pair alternating with a filiform, apically cleft lobe 5–10 mm tall; fleshy pad none; anther head none, the filaments 0.2–0.4 mm long, distinct, the anthers cordate to sagittate, 1–1.2 mm long, dehiscing by longitudinal slits, connivent at apex above the stigma, densely villous abaxially; anther appendages none; anther wings none; corpusculum none; pollinia of loosely coherent pollen grains, spoon-shaped, ca 0.5 mm long. Follicles linear, 9–15 cm long, 0.6–0.8 cm thick, without tubercles, not angled, glabrous; seeds ovate, 7–8 mm long; coma tan, 2.5–3.5 cm long. (2n = 22, 24) Jun–Sep. Open woods or thickets along streams; KS: Greenwood; (NY s to FL, w to KS, OK; s Europe). *Naturalized.*

This genus is a member of the less specialized subfamily Periplocoideae (often treated as a distinct family, Periplocaceae), whereas the other GP genera of the Asclepiadaceae are in the subfamily Asclepiadoideae (with a pollination mechanism as described following the family description). In this species, the "pollinia" differ markedly from those found in the Asclepiadoideae. They consist of 5 spoon-shaped appendages embedded vertically in the 5 angles of the stigma head. The sticky, orbicular, and somewhat funnelform expanded blade of each appendage is positioned directly beneath the two pollen chambers of adjacent anthers. On dehiscence, the pollen tetrads fall from the anthers and adhere to the expanded blade. The stalk of the appendage, which is also sticky, may adhere to a pollinator and be carried to the stigma of another flower. Handbook of Flower Pollination, Oxford (Knuth, 1909).

5. SARCOSTEMMA R. Br., Twine Vine

Caulescent perennial vines with milky juice; stems trailing or twining over other vegetation, few to several, herbaceous above, woody below, from a thickened rootstock. Leaves opposite, petiolate; blade linear to lanceolate or ovate; margins entire to crispate; base cordate to truncate or hastate, usually with a cluster of subulate glands on the midrib, these sometimes obscured by pubescence; stipules minute, subulate, often deciduous. Inflorescence an umbellate cyme scattered in the leaf axils. Calyx lobes widely spreading, often with subulate, glandular squamellae in the sinuses; corolla rotate-subcampanulate, widely spreading, white, pink, or purple; gynostegium at least briefly stipitate; corona of two parts, a low ring or rim arising from the base of the corolla and surrounding the gynostegium and 5 inflated, broadly rounded vesicles arising from the staminal column and peripheral to the anther head; fleshy pads none; anther head truncate-conic; anther appendages ovate to suborbicular, petaloid; anther wings abruptly angled at base, relatively prominent, corneous; corpusculum ovate, reddish-brown; pollinia of firmly coherent pollen, lanceolate to oblong, compressed. Follicles fusiform, smooth, terete.

1 Leaf margins crispate; corona vesicles with a transverse constriction and keel near base; column 1.5–2 mm long .... 1. *S. crispum*
1 Leaf margins flat; corona vesicles neither transversely constricted nor keeled; column 0.6–0.9 mm long .... 2. *S. cynanchoides*

1. Sarcostemma crispum Benth., waxy-leaf twine vine. Perennial, twining or trailing vine; stems simple or branched, slender, moderately to densely puberulent with curved-appressed hairs or glabrous. Leaves opposite; blade lanceolate to linear, ascending to spreading, 2.5–9 cm long, 0.5–3 cm wide, subcoriaceous, puberulent, often more densely so on the midrib, apex acute to acuminate, margins crispate, base cordate to truncate and often hastate, without glands on the midrib; petiole 0.2–1.5 cm long. Inflorescences few to several in the leaf axils, 3- to 12-flowered; peduncles 0.3–2.5 cm long; pedicels slender, 1–2 cm long, puberulent. Flowers 12–20 mm in diam; calyx lobes green to purple, lanceolate, 3–5 mm long, puberulent; corolla lobes purple, oblong-lanceolate, widely spreading, 6–11 mm long, glabrous above, puberulent beneath, the margins densely ciliate; corona ring 0.7–0.9 mm tall; gynostegium purplish, distinctly stipitate, glabrous; column obconic, 1.5–2 mm tall, 1.3–1.5 mm wide; corona vesicles ovate-oblong with a transverse constriction and keel near base, attached at base, erect, 1.3–1.5 mm long, the apex broadly rounded; anther head truncate-conic, 1.2–1.4 mm tall, 1.7–1.8 mm wide; anther appendages 0.6–0.7 mm long; anther wings rounded at base, minutely notched, 0.7–0.9 mm long; corpusculum ca 0.2 mm long; pollinia narrowly oblong, compressed, ca 0.7 mm long. Follicles narrowly fusiform, 8.5–12.5 cm long, 1.2–1.8 cm thick, without tubercles, puberulent to glabrous; seeds narrowly obovate, 6–7 mm long; coma whitish to tan, 2.5–4 mm long. Jun–Aug. Rocky soils of thickets & open wooded slopes; CO: Baca; OK: Cimarron; TX; (w GP to NM & AZ; Mex.). *S. lobata* Waterfall—Waterfall.

2. Sarcostemma cynanchoides Dcne., arroyo twine vine. Perennial twining or trailing vine; stems simple or branched, slender, sparingly puberulent to glabrous. Leaves opposite; blade lanceolate or triangular to broadly ovate, widely spreading, 1–7 cm long, 0.5–6 cm wide, membranaceous, villous to glabrous or glabrate, apex acute to acuminate or apiculate, margins plane, base cordate to subhastate, with 1 to few linear glands, 0.1–0.5 mm long, on the midrib; petiole 0.3–4 cm long. Inflorescences few to many in the leaf axils, 12- to 36-flowered; peduncles 1–5 cm long; pedicels slender, 0.6–1.7 cm long, puberulent. Flowers 11–15 mm in diam; calyx lobes green to purple-tinged, lanceolate to ovate, 2–3 cm long, puberulent; corolla lobes whitish, pink or purplish near base, ovate, widely spreading, 4–5 mm long, puberulent above and beneath, the margins densely ciliate; corona ring ca 0.3 mm tall; gynostegium whitish, briefly stipitate, glabrous; column obconic, 0.6–0.9 mm tall, ca 2 mm wide; corona vesicles obovate-oblong, not constricted below, attached near base, erect, 1.5–1.8 mm long, the apex broadly rounded; anther head truncate-conic, 1.2–1.4 mm tall, 2.4–2.6 mm wide; anther appendages ca 0.4 mm long; anther wings rounded at base, minutely notched, 1.2–1.3 mm long; corpusculum ca 0.2 mm long; pollinia narrowly lanceolate, compressed, ca 0.7 mm long. Follicle fusiform, 6.5–7.5 cm long, 1–1.5 cm thick, without tubercles, puberulent; seeds obovate, 7–8 mm long; coma tan, 3–3.5 mm long. Jun–Sep. Sandy or rocky soils of thickets & forest margins; OK: Comanche, Jackson, Payne, Tillman; TX; (w GP to NM & AZ; n Mex.).

111. SOLANACEAE Juss., the Potato or Nightshade Family

by Ronald L. McGregor, Johnnie L. Gentry, and Ralph E. Brooks

Herbs or less often shrubs or woody vines. Leaves simple to pinnately compound, alternate, sometimes in pairs (subopposite), exstipulate. Inflorescence cymose, paniculate, racemose or umbellate, sometimes a solitary flower, terminal or pseudoterminal, lateral and opposite the leaves, internodal or axillary. Flowers perfect, actinomorphic or sometimes slightly zygomorphic; calyx gamosepalous, 5-lobed or 5-parted, usually persistent and in some accrescent or inflated in fruit; corolla gamopetalous, 5-lobed; stamens epipetalous, alternate with the corolla lobes, anthers dehiscent by longitudinal slits or by terminal pores, sometimes connivent; carpels 2, stigma entire or slightly 2-lobed, style solitary and terminal, ovary superior, 2-locular, sometimes 4-locular by false septae or 3- to 5-locular, ovules several–many. Fruit a berry or capsule; seeds with a subperipheral, often curved embryo, well-developed endosperm; placentation axile.

This is a family of about 85 genera with some 2,300 species, cosmopolitan but best developed in tropical America. The Solanaceae are of great economic importance, producing ornamentals, foods and drugs.

Lycopersicon esculentum P. Mill. (tomato) and *Nicotiana tabacum* L. (tobacco) occasionally come up from garden refuse but never persist. *Nicotiana quadrivalis* Pursh, apparently based on plants cultivated by Indians of the Great Plains, was reported by Rydberg (1932) as "escaped in North Dakota" although specimen evidence cannot be located. *N. trigonophylla* Dun. is native to northern Mexico and the adjacent U.S. and has been collected in OK: Cleveland, Jackson.

1 Fruit a capsule, dehiscent; corollas funnelform to salverform.
  2 Flowers solitary, in the forks of the branching stem; calyx deciduous, circumscissile near the base; fruit armed with spines .... 2. *Datura*
  2 Flowers several, in racemes, spikes or panicles; calyx persistent, not circumscissile; fruit without spines .... 3. *Hyoscyamus*
1 Fruit a fleshy or dry berry, indehiscent; corollas rotate to campanulate, sometimes funnelform.
  3 Anthers dehiscing by terminal pores .... 8. *Solanum*
  3 Anthers dehiscing by longitudinal slits.
    4 Shrubs, often with thorns .... 4. *Lycium*

4 Herbs, without thorns.
  5 Calyx lobes cordate-sagittate at the base; ovary 3- to 5-locular; corolla blue .... 5. *Nicandra*
  5 Calyx lobes not cordate-sagittate at the base; ovary 2-locular; corolla white, yellow or greenish-yellow, or blue.
    6 Fruiting calyx, bladdery-inflated around and completely enclosing the berry, the calyx contracted above.
      7 Flowers erect at anthesis; corolla violet to purple (rarely white); seeds dull, the surface alveolate or rough reticulate .... 7. *Quincula*
      7 Flowers nodding at anthesis; corolla yellow (rarely white); seeds glossy, the surface minutely pitted .... 6. *Physalis*
    6 Fruiting calyx accrescent, tightly investing the berry .... 1. *Chamaesaracha*

1. CHAMAESARACHA A. Gray

Low perennial herb; stems prostrate or ascending, 9–50 cm long; rhizomatous. Leaves simple, entire to pinnatifid, subsessile to distinctly petiolate. Inflorescence axillary; pedicels elongating and becoming curved in fruit. Flowers actinomorphic; calyx campanulate, accrescent in fruit; corolla white, ochroleucous or yellowish-green, campanulate, the limb slightly 5-lobed, with white tomentose appendages in the throat; stamens inserted near the base of the corolla, anthers longtitudinally dehiscent; ovary 2-locular. Fruit a berry, tightly invested but not enclosed by the accrescent calyx; seeds reniform, rugose-reticulate, embryo strongly curved.

Reference: Averett, J. E. 1973. Biosystematic study of *Chamaesaracha* (Solanaceae). Rhodora 75: 325–365.

1 Plants pubescent with simple and glandular hairs .... 1. *C. coniodes*
1 Plants sparsely pubescent with stellate hairs .... 2. *C. coronopus*

1. *Chamaesaracha coniodes* (Moric. ex Dun.) Britt., chamaesaracha. Stems 4–30 cm long. Leaves lanceolate to broadly lanceolate, 2–6 cm long, 5–20 mm wide, gradually narrowing to an attenuate base, pubescent with simple hairs, occasionally mixed with fewer glandular hairs, subsessile. Flowers 1 or 2 in the axils; calyx 3–4 mm long; corolla 10–15 mm wide. Fruit 6–8 mm in diam. (n = 24, 36) May–Sep. Sandy & rocky prairies, roadsides & pastures; CO, KS, OK, TX, NM; (GP s to Mex.).

Chamaesaracha sordida was treated as a synonym of *C. coniodes* by Rydberg, but Averett, op. cit., demonstrated that they are distinctive taxa. *C. sordida* ranges south of our area, in TX and NM.

2. *Chamaesaracha coronopus* (Dun.) A. Gray, green false nightshade. Stems 10–30 cm long. Leaves linear to linear-lanceolate, 1.5–6.5 cm long, 1.5–10 mm wide, gradually narrowing to an attenuate base, sparsely pubescent with stellate hairs, margin subentire to deeply lobed, subsessile. Flowers 1 or 2 in axils; calyx 2.5–4 mm long; corolla 6–10 mm wide. Fruit 5–8 mm in diam. (n = 12, 24, 35) May–Sep. Roadsides & sandy prairies; CO, NM; OK: Cimarron; TX; (CO, OK, w TX to AZ; Mex.).

2. DATURA L., Thorn-apple, Jimsonweed

Ours annual or perennial, rank, narcotic-poisonous herbs; stems erect, ascending, or decumbent. Leaves alternate; blades ovate or elliptic, entire to sinuate-pinnatifid, pubescent or glabrous; petiolate; exstipulate. Flowers large and showy, solitary, axillary, pediceled, nodding or erect; calyx tubular, shallowly 5-lobed, circumscissile near the base with the upper portion breaking away after flowering and leaving an enlarged and subtended disk under the capsule; corolla funnelform, convolute-plicate in bud; stamens 5, anthers longitudinal; stigma capitate, 2-lobed, style filiform, about equaling the anthers in length, ovary 2-celled or sometimes 4-celled below because of the presence of a false septum. Fruit a globose or ovoid capsule, 2- or 4-valved from the top, sometimes splitting irregularly, prickly or spiny.

References: Avery, A. G., S. Satina, & J. Rietsema. 1959. Blakeslee: The Genus *Datura*. Ronald Press Co., N. Y. 289 p.; Ewan, J. 1944. Taxonomic history of perennial southwestern *Datura meteloides*. Rhodora 46: 317–323.

1 Corolla 15–20 cm long; fruit nodding; foliage cinereous-pubescent .... 1. *D. innoxia*
1 Corolla 4–10 cm long; fruit erect; foliage glabrous to lightly pubescent but not cinereous.
   2 Leaves moderately pubescent; capsules armed with heavy prickles, many more than 10 mm long .... 2. *D. quercifolia*
   2 Leaves nearly glabrous; capsules armed with spines mostly 3–5(9) mm long .... 3. *D. stramonium*

1. ***Datura innoxia*** P. Mill., Indian-apple. Perennial, from a thick caudex to 15 dm tall, the herbage cinereous-pubescent; stems several, erect to occasionally decumbent, widely branching. Leaf blade ovate or widely so, 6–20 cm long, 4–15 cm wide, apex acute to short acuminate, margins entire to sinuate-repand, base rounded to truncate, usually somewhat oblique; petiole 3–18 cm long, shorter than or about equaling the blade. Flowers erect or slightly nodding, pedicels 1.5–3 cm long; calyx 7–10 cm long, the persistent disk usually rotate and sometimes reflexed; corolla white, sometimes tinged with violet, 15–20 cm long, the limb 8–15 cm across with 5 subulate teeth 0.5–2 cm long; stamens 10–15 cm long, anthers white, 10–15 mm long, glabrous to sparsely pubescent; style about equaling the anthers in length. Capsules nodding, 2.5–3.5 cm long, densely prickly and puberulent, the prickles 3–8 mm long; seeds buff to light brown, subreniform, compressed, 4–6 mm long, smooth, margin cordlike. (n = 12) (Jun) Jul–Sep. Sandy soil on flood plains & bottomlands; sw OK, s TX panhandle; (sw OK & w TX, w to CA, s to cen Mex.). *D. meteloides* DC., *D. wrightii* Regel—Atlas GP. POISONOUS.

Some botanists have considered our plants to be distinct from the southern ones and have applied the name *D. meteloides*. Barclay (Bot. Mus. Leaflets, Harvard Univ. 8: 245–272. 1959.) shows substantial evidence that the southern and northern plants are the same, in which case the name *innoxia* has long priority.

This species and the similar *D. metel* L. are occasionally cultivated for their large showy flowers and are sometimes reported as escapes in more northern areas of our region.

2. ***Datura quercifolia*** H. B. K., oak-leaf thorn-apple. Annual from a thick tap-root, to 15 dm tall, herbage green and at least moderately pubescent, the younger parts often downy pubescent; stems erect, mostly branched in the upper portion. Leaf blade ovate to elliptic, 6–20 cm long, 4–12 cm wide, sinuate-pinnatifid to pinnately lobed with the lobes sometimes toothed, apex acute to short acuminate, base cuneate or widely so; petiole about equaling the blade or sometimes longer. Flowers erect, pedicels 1–2 cm long; calyx 2–3 cm long; corolla pale violet to purple, 4–7 cm long, the limb 1.5–2.5 cm across with 5 subulate teeth ca 2 mm long; stamens 3–6 cm long, anthers purple, 3–4 mm long, pubescent; style about equaling the stamens in length. Capsules erect, ovoid, 3–4 cm long, armed with large and very unequal, flattened prickles, some to 20 mm long; seeds dark colored, somewhat scrobiculate-rugose. (n = 12) Aug–Sep. Sandy soil on flood plains & bottomlands; KS: Meade, Morton, Seward; OK: Cimarron; TX: Deaf Smith; (sw KS, w OK & w TX to s NM & n Mex.). POISONOUS.

3. ***Datura stramonium*** L., jimson weed. Annual from a thick tap-root, to 1.2 m tall, herbage dark green, lightly pubescent to glabrate; stems erect, mostly branching in the upper

portion. Leaf blade ovate to lance-ovate, 5–25 cm long, 2.5–15(20) cm wide, apex acuminate, margins irregularly and coarsely sinuate-dentate, base cuneate to subtruncate, usually oblique; petiole 2–9 cm long, shorter than the blade. Flowers erect, pedicels 0.5–1.5 cm long; calyx 3.5–5 cm long; the persistent disk rotate and slightly reflexed; corolla white or sometimes violet, 6–10 cm long, the limb 2.5 cm across and bearing 5 subulate teeth 3–8 mm long; stamens 5–9 cm long, anthers white in white flowers, violet in violet flowers, 3.5–5 mm long, sparsely pubescent; style about equaling the anthers in length. Capsules erect, ovoid, 3.5–5 cm long, puberulent and prickly or occasionally nearly smooth, the prickles mostly 3–5(9) mm long; seeds black, subreniform, compressed, 3–4 mm long, rugulose and finely pitted. (n = 12) Jul–Sep. Farm lots, waste ground, flood plains, & bottomlands; e MN, e ND, e SD, IA, e 2/3 NE, MO, KS, OK, TX; (widely distributed in temp. & trop. regions of the world). *D. tatula* L.—Rydberg; *D. stramonium* var. *tatula* (L.) Torr.—Fernald. POISONOUS.

Plants with violet corollas are sometimes recognized as var. *tatula* although this does not appear justified.

3. HYOSCYAMUS L., Henbane

1. **Hyoscyamus niger** L. Coarse, strong-scented, annual or biennial herb, viscid-villous; stem simple or branched, 3–10 dm tall. Leaves simple, oblong to ovate or ovate-oblong, 5–20 cm long, 2–10 cm wide, coarsely dentate or pinnately lobed, the cauline leaves clasping, sessile, the basal leaves petiolate, forming a rosette. Inflorescence a raceme or spike, with leafy bracts. Flowers slightly zygomorphic; calyx tubular-campanulate, 1–1.5 cm long, the lobes triangular, fruiting calyx urceolate, reticulate veined, accrescent; corolla greenish-yellow with purplish veins, funnelform, 2.5–3.5 cm long, 5-lobed; anthers longitudinally dehiscent. Fruit a capsule, enclosed in the accrescent calyx, circumscissile above the middle; seeds reniform, papillate. (2n = 34) May–Jul. Roadsides, pastures, waste places, & dry open woods; MT, ND, w SD, WY, NE: Dawes; (Que. & s Ont. to N. Eng., w & s to much of the U.S.; temp. Eurasia & n Africa). *Naturalized.*

This genus of about 15 spp. is occasionally grown as an ornamental. Henbane leaves are one of the sources of the drug hyoscyamine. All parts of the plant are poisonous.

4. LYCIUM L., Wolfberry

Ours erect, arch-branching, or climbing shrubs, usually spiny. Leaves alternate on young shoots, fascicled on older ones, simple, entire, petiolate, exstipulate. Flowers axillary, solitary or 2–4 in clusters, pedicelled; calyx cup-shaped to campanulate, 3- to 5-lobed; corolla campanulate or tubular-funnelform, the limb 4- or 5-lobed; stamens 4 or 5, barely exserted, anthers longitudinally dehiscing; stigma capitate or 2-lobed, style filiform. Fruit a dry or fleshy berry, globose or ovoid, subtended by a persistent calyx.

1 Leaves 1–2.5 mm wide .... 1. *L. berlandieri*
1 Leaves (3)5–20 mm wide.
    2 Leaves elliptic or oblong, widest at the middle; corolla violet, pale lavender, or pinkish .... 2. *L. halimifolium*
    2 Leaves mostly oblanceolate to spatulate, widest above the middle; corolla greenish-white, sometimes tinged with purple .... 3. *L. pallidum*

1. **Lycium berlandieri** Dun., silvery wolfberry. Erect shrub to 2.5 m tall, armed with needlelike spines on the younger shoots or nearly unarmed; branches somewhat crooked, glabrous. Leaves 1–3 in a fascicle, linear to elliptic-spatulate, 5–25 mm long, 1–2.5 mm

wide, glabrous, apex rounded to acute, margins entire, base attenuate to a short petiole or subsessile. Flowers solitary or in pairs, pedicels 3–20 mm long; calyx cup-shaped, 1–2 mm long, (3)4- or 5-lobed, the lobes usually shorter than the tube, glabrous except for a tuft of hair at the tip of each lobe; corolla blue, pale lavender, or ochroleucous, campanulate-funnelform, 6–7 mm long, the limb 4- or 5-lobed. Berry red, globose to ovoid, 4–8 mm in diam, glabrous. Apr–Sep. Gravelly, usually gypsiferous soil on hillsides, shrubland flats, or in arroyos; OK: Cimarron, Harmon; TX: Randall; (w TX & OK, w to e NM, s to n Mex.).

2. **Lycium halimifolium** P. Mill., matrimony vine. Shrub to 3 m tall, stems arching, recurving, or occasionally climbing, armed with spines at the nodes on older growth, glabrous. Leaves solitary at the nodes or on older growth sometimes in fascicles of 2–6; blade elliptic or oblong, infrequently obovate to oblanceolate, 2–5 cm long, 0.5–2 cm wide, glabrous, apex acute to obtuse, margins entire, base attenuate; petiole 5–15 mm long. Flowers solitary or in clusters of 2–4, pedicel 8–15 mm long; calyx campanulate, 3–5 mm long, 3(4 or 5)-lobed, the lobes about equaling the tube in length, glabrous; corolla violet to pale lavender or pinkish, short funnelform, 8–12 mm long, the limb (4)5-lobed and the lobes shorter than the tube. Berry red (drying purplish or blackish), ovoid, 15–20 mm in diam, glabrous; seeds yellowish, widely ovate or subreniform, compressed, 2–3 mm long, minutely pitted. (n = 12) Jun–Aug. A frequent escape from cultivation, thickets, waste areas, or around old dwellings; scattered in GP; (much of N. Amer.; Eurasia). *Naturalized.*

Lycium chinense P. Mill., a native of China, has been reported as an escape in the s GP. It is similar to *L. halimifolium* except the corolla lobes are longer than the tube.

3. **Lycium pallidum** Miers, pale wolfberry. Upright-spreading, much-branched shrubs to 20 dm tall, branches lightly pubescent to glabrous, sparingly armed with stout spines. Leaves mostly fascicled, except on young growth; blade oblanceolate or spatulate, 1–4 cm long, (3)5–15 mm wide, glabrous, apex acute to obtuse, margins entire, base attenuate; petiole 5–10 mm long. Flowers solitary or occasionally in pairs, pedicel 8–18 mm long; calyx campanulate, 5–9 mm long, 5-lobed, the lobes about equaling or slightly longer than the tube, glabrous; corolla greenish-white, sometimes tinged with purple, funnelform, 15–20 mm long, the limb 5-lobed. Berry red (drying blackish or purplish), glaucous, subglobose to ovoid, 8–12 mm in diam, glabrous; seeds yellowish, widely ovate to subreniform, 2.5–3 mm long, minutely pitted. Apr–Jun. Gravelly soil on dry hills or in arroyos; CO: Las Animas; NM: Colfax, Union; OK: Cimarron; (w OK, to s CO & UT, s to s CA & Mex.).

5. NICANDRA Adans., Apple of Peru

1. **Nicandra physalodes** (L.) Gaertn. Annual herb; glabrous or sometimes sparsely pilose; stem branching above, 2–10 dm tall. Leaves simple, ovate to broadly ovate, 5–25 cm long, 3–12 cm wide, coarsely sinuate-dentate or shallowly lobed, base attenuate; petiolate. Inflorescence axillary, consisting of a solitary, cernuous flower. Calyx angulate, 1–2 cm long, parted to near the base, the lobes accrescent and reticulate veined in fruit, base cordate-sagittate; corolla blue or pale blue, broadly campanulate, 1.5–3 cm long, the limb 2.5–3.5 cm wide, nearly entire to shallowly lobed; anthers longitudinally dehiscent; stigma with 3–5 prominent stigmatic areas. Fruit a dry berry 1–2 cm in diam, 3- to 5-locular, enclosed in the dry accrescent calyx; seeds suborbicular, 1.5–1.8 mm long, reticulate-foveate. (n = 10) Jul–Sep. Waste places & fields; MN, NE, IA, KS, MO, OK; (N.S. & Ont. to MN, s to OK & FL). *Naturalized.*

This monotypic genus, native to Peru, is occasionally cultivated for ornament.

6. PHYSALIS L., Ground Cherry

Ours annual or perennial herbs, some species rhizomatous; stems erect, ascending or decumbent, commonly widely branching. Leaves alternate but sometimes 2 or 3 together, due to internode reduction; blades ovate to lanceolate, entire to coarsely dentate, sometimes sinuate, or rarely pinnatifid or deeply-lobed, pubescence variable to nearly absent; petiolate; without stipules. Flowers usually solitary on pedicels, nodding at anthesis, axillary; calyx small at anthesis, 5-lobed, the tube promptly enlarging, at maturity papery and bluntly conical, often retuse at base, loosely enclosing the fruit; corolla campanulate to subrotate, shallowly 5-lobed to entire, yellow, rarely white, with 5 dark spots at the base in the tube, or these indistinct; stamens 5, inserted near base of corolla tube, erect, separate; anthers dehiscing by lateral slits, yellow to violet, blue or greenish-blue, or so lined or tinged, filaments filiform to nearly as wide as anthers, rarely clavate. Berry 2-locular, globose, many seeded, often dryish; seeds yellow, glossy, minutely pitted.

References: Waterfall, U. T. 1958. A taxonomic study of the genus *Physalis* in North America north of Mexico. Rhodora 60: 107–114, 128–142, 152–173; Hinton, W. Frederick. 1976. The systematics of *Physalis pumila* ssp. *hispida* (Solanceae). Syst. Bot. 1: 188–193.

In collecting specimens of *Physalis* it is essential to note whether plants are annual or perennial, and whether rhizomatous; color of flowers; color of spots near base of corolla limb; anther color; and habit. In our area *P. ixocarpa* Brot. ex Hornem., the strawberry tomato, is sometimes grown, as is *P. alkekengi* L., another strawberry tomato or bladder cherry. Both are excluded from this treatment as they rarely appear spontaneously and never persist.

1 Plants annual; anthers bluish or violet.
  2 Stems and leaves essentially glabrous; fruiting calyx 10-ribbed or angled ... 1. *P. angulata*
  2 Stems, and sometimes the leaves, velutinous, often with glandular hairs admixed; fruiting calyx strongly 5-angled .... 5. *P. pubescens*
1 Plants perennial; anthers yellow, or with bluish tinge or bluish lines on edges, rarely blue.
  3 Pubescence of stem wholly or partly of simple, reflexed or retrorse hairs, often admixed with spreading hairs .... 8. *P. virginiana*
  3 Pubescence of stem without reflexed or retrorse hairs.
    4 Stellate or branched hairs absent throughout or very rare.
      5 Stems without glandular hairs, usually nearly glabrate.
        6 Hairs of calyx appressed or nearly absent .... 4. *P. longifolia*
        6 Hairs of calyx spreading, stiff or sometimes villous 6a *P. pumila* subsp. *hispida*
      5 Stems with glandular hairs, pubescence usually dense.
        7 Flowering pedicels usually 10–15 mm long; stem pubescence of short, glandular hairs admixed with spreading hairs 1–2 mm long .... 3. *P. heterophylla*
        7 Flowering pedicels usually 3–8(10) mm long; stem pubescence usually of dense, short, glandular hairs or mixed with longer spreading hairs.
          8 Stems and leaves with long spreading multicellular hairs admixed with shorter glandular hairs .... 2c. *P. hederifolia* var. *hederifolia*
          8 Stems and leaves densely short glandular hairy, longer spreading hairs few or absent .... 2a. *P. hederifolia* var. *comata*
    4 Stellate or branched hairs evident on stems and leaves.
      9 Hairs of stems, leaves and calyx, or at least some of them 1- to 3-branched, usually admixed with often more numerous unbranched ones .... 6b. *P. pumila* subsp. *pumila*
      9 Hairs of stems, leaves, and calyx predominantly stellate or these mixed with longer unbranched ones.
        10 Flowering pedicels usually 7–35 mm long short; stellate hairs usually dense throughout, long hairs absent .... 7. *P. viscosa* subsp. *mollis* var. *cinerescens*
        10 Flowering pedicels usually 3–8 mm long; stellate hairs sparse .... 2b. *P. hederifolia* var. *cordifolia*

1. Physalis angulata L., cutleaf ground cherry. Annual herbs with tap-roots; stems 1–5(8) dm tall, erect, branched from base or above, sometimes decumbent, glabrous or with a few short antrorsely appressed hairs especially on younger parts. Leaves variable, principal blades (3)4–10 cm long, 3.5–8 mm wide, ovate to lance-ovate, or broadly to narrowly elliptic, sometimes oblongish, margins deeply and irregularly toothed, incised-toothed to undulate-toothed or entire, surfaces glabrous or rarely with sparse appressed hairs; petioles 1–4(8) cm long. Flowering pedicels 5–15 or 15–40 mm long, 20–30 or 20–45 mm long in fruit; calyx at anthesis 3–5 mm long, lobes 1–3 mm long; corolla yellowish, 4–10(12) mm long, immaculate, or with indistinct spots; anthers bluish or violet, 2–2.5 mm long, filaments slender, 3–4 mm long. Fruiting calyx 20–35 mm long, inflated 10-angled or 10-ribbed; berry 10–12 mm in diam; seeds yellowish, flattened, ovate or broadly elliptical, subsmooth. (n = 12, 24) Jun–Oct. Infrequent to locally common in alluvial soils along streams & valleys, roadsides, moist open woodlands, fields, & waste areas; e 1/2 KS, MO, OK, TX; (trop. Amer., n to se VA, MO, KS, w TX).

Most of our specimens have flowering pedicels 15–40 mm long (20–40 mm long in fruit), calyx teeth 1–1.5 mm long at anthesis, and the mature calyx 2–2.5 cm long. These represent the var. *pendula* (Rydb.) Waterfall (*P. pendula* Rydb.). In western MO some plants have flowering pedicels 5–15 mm long (20–30 mm long in fruit), calyx teeth 2–3 mm long at anthesis and the mature calyx to 3.5 cm long. These plants represent the var. *angulata* of more eastern and se U.S. The plants reported for KS in the Atlas GP as var. *angulata* are more or less intermediate in these distinguishing characteristics.

2. ***Physalis hederifolia*** A. Gray, prairie ground cherry. Perennial herbs from a somewhat woody caudex, sometimes producing shoots from lateral roots. Stems erect or ascending to prostrate, simple to usually much branched, from base and above, (1)2–4(7) dm tall or long; vestiture of stems, leaves and inflorescence, various, including long multicellular hairs, mixed with shorter ones, these often glandular, or with short stellate hairs, or some of longer hairs also glandular. Leaves alternate; blades rounded-ovate or rhombic to nearly orbicular, or ovate to ovate lanceolate, (1.5)2–4(5) cm long, 2–4 cm wide, margins entire, sinuate-dentate, or coarsely toothed; bases acute, cordate, or reniform, sometimes truncate; petioles 3–25 mm long. Pedicels rather stout, 3–8(13) mm long at anthesis, to 15–30 mm in fruit; calyx tube at anthesis 3–4 mm long, lobes 4–6 mm long, lanceolate to triangular; corolla yellow or greenish-yellow, often maculate, 10–15 mm long, limb often reflexed. Fruiting calyx 2–3 cm long, 1.5–2 cm wide, 10-angled or 10-ribbed; anthers usually yellow, 1.5–3.5(4) mm long, filaments flattened and sometimes somewhat clavate; berry reddish-brown to yellowish, 8–10(15) mm in diam, seeds yellowish or yellowish-brown, transversely elliptic to depressed ovate, 1.7–2.3 mm long, minutely reticulate or subsmooth (n = 12).

There are 3 vars. in our area:

1 Pubescence of foliage and calyx of stellate hairs, none of elongated multicellular hairs .... 2b. var. *cordifolia*
1 Pubescence not stellate, some elongate and multicellular, others evidently glandular.
   2 Pubescence of lower leaf surface predominantly of short-stipitate glandular hairs .... 2a. var. *comata*
   2 Pubescence of lower leaf surface with many long multicellular hairs, glandular ones sparse .... 2c. var. *hederifolia*

2a. var. *comata* (Rydb.) Waterfall. Leaf blades ovate to rounded; pubescence of short glandular hairs and longer multicellular ones few or absent; calyx at anthesis 8–11 mm wide; anthers 2–3 mm long. Jun–Sep. Common in sandy or rocky prairies, stream valleys, roadsides, & waste places; w 2/3 NE, e WY, s to w 3/4 KS, nw OK, TX panhandle, NM; (NE, WY, s to KS, OK, TX, NM, AZ). *P. comata* Rydb., *P. rotundata* Rydb.—Rydberg.

2b. var. *cordifolia* (A. Gray) Waterfall. Leaf blades ovate to ovate-lanceolate; pubescence stellate. Jul–Sep. Rare in sandy or rocky prairie hillsides & open woodlands; OK: Cimarron; TX panhandle; NM, s CO; (CA, s NV, s UT, e to CO, NM, OK, TX).

2c. var. *hederifolia*. Leaf blades ovate, ovate-lanceolate to rotund; pubescence of long multicellular hairs mixed with shorter glandular ones; calyx at anthesis 4–8 mm wide; anthers 3–4 mm long. Rare in sandy or rocky plains & mts.; NM: Quay; sw TX; (MT?, CO?, sw TX, NM, UT, e AZ; Mex.).

3. Physalis heterophylla Nees, clammy ground cherry. Perennial herbs with usually deeply buried caudex; stems usually erect, simple or much branched, 1.5–5(9) cm tall. Pubescence of stems, foliage, and inflorescence of varying proportions of short usually glandular hairs and long multicelled hairs 1–2(3) mm long. Leaves alternate, principal ones chiefly ovate but varying to rhombic, (3)5–10 cm long, 3.5–6 cm wide, margins irregularly sinuate-dentate or entire, rounded or subcordate at base, pubescent on both sides; petioles 3–6 cm long. Pedicels ca 1 cm long at anthesis, to 3 cm long in fruit; calyx at anthesis 7–12 mm long, 5–12 mm wide, lobes deltoid or ovate; corolla yellow, maculate but not always with a strong contrast, (10)15–18(20) mm long, 12–18(22) mm wide; anthers yellow, sometimes tinged with blue or violet, 3–4.5 mm long, filaments thickened, often as wide as anthers, usually clavate. Fruiting calyx ovoid (2.5)3–4 cm long, 2–4 cm wide, much inflated, evidently retuse at base; berry yellowish, (8)10–12 mm in diam; seeds yellowish, ovate to transversely elliptic, 2–2.5 mm long, minutely pitted. (n = 12) May–Oct. Infrequent to common in prairies, open woodlands, stream valleys, fields, roadsides, waste places; MN, ND, s to CO, e 3/4 KS, MO, OK, e TX; (Que. & N.S. to ND, CO & UT, s to FL & TX). *P. ambigua* (A. Gray) Britt., *P. nyctaginea* Dun.—Rydberg.

Some plants on erect stems with sparse and uniformly distributed hairs 2 mm long, have been recognized as *P. ambigua*, while similar more prostrate plants have been named *P. nyctaginea*. These distinctions appear to be without taxonomic significance.

4. Physalis longifolia Nutt., common ground cherry. Perennial herbs with deep caudex, and producing shoots from lateral roots. Stems usually single, often purplish, usually branching above, with sparse short antrorse hairs, especially in younger portions, or nearly glabrous. Leaves alternate; blades lanceolate to lance-linear, or elliptic lanceolate, rarely ovate, glabrate, margins entire, undulate, or sinuate-dentate, obtuse, cuneate, (2)3–8(15) cm long; petioles slightly winged, 1–6 cm long. Pedicels slender, 5–15 mm long, becoming longer in fruit; calyx tube at anthesis 7–12 mm long, usually with 10 lines of short antrorse hairs, lobes lanceolate to ovate, acute to obtuse, ca 1/2 length of tube; corolla yellow, dark maculate, 10–15 mm long, anthers yellow, 2–3(4) mm long, filaments flattened, nearly as wide as anther. Fruiting calyx 2–3 cm long; berry subglobose, 8–10(15) mm in diam; seeds yellowish, somewhat glossy, 1.7–2.3 mm long, minutely pitted or subsmooth. (n = 12) May–Sep. Common & too often abundant in prairies & plains, open woods, stream valleys, fields, roadsides, & waste places; GP but rare in ne SD, ND, w MN; (VT, Ont. to MT, s to VA, TN, LA, AZ, perhaps introduced in e & w U.S. from plains region). *P. virginiana* P. Mill var. *sonorae* (Torr.) Waterfall—Atlas GP; *P. macrophysa* Rydb., *P. subglabrata* Mack. & Bush—Rydberg; *P. virginiana* P. Mill. var. *subglabrata* (Mack. & Bush) Waterfall—Waterfall (op. cit.).

P. longifolia var. *longifolia*, as described above, is our common plant, but in shaded or more moist places the var. *subglabrata* (Mack. & Bush) Cronq. may be recognized. The latter has leaves wider, more ovate, thinner, and the bases taper more abruptly to the petiole, and the anthers are often bluish or violet. Reportedly (Waterfall, op. cit.) the var. *subglabrata* is more common in the northeastern United States and var. *longifolia* is the plains form. Field experience over many years fails to convince us that the two variants merit taxonomic consideration in our area.

5. Physalis pubescens L., downy ground cherry. Annual herbs with tap-roots; stems erect and widely branched from the base, (1)2–4(7) dm tall, villous or usually viscid-villous, rarely glabrous. Leaves alternate; blades ovate, 3–10 cm long, entire to shallowly and unevenly repand-dentate, or variously toothed, acute to rounded at base, often inequilateral, surfaces greenish, sparsely to somewhat densely pubescent, with or without sessile glands;

petioles 1-7(10) cm long. Pedicels at anthesis (3)5-7(10) mm long, slightly elongating in fruit; calyx at anthesis 4-10 mm long, lobes 2-4 mm long; corolla yellowish, dark maculate, (6)8-10(12) mm long, 10-15 mm wide; anthers bluish or violet, 1-2 mm long. Fruiting calyx strongly 5-angled, (15)18-25(30) mm long, 10-18(22) mm wide; berry 10-18 mm in diam; seeds yellowish, broadly elliptic to ovate, minutely pitted. (n = 12) Jul-Oct. Infrequent to locally common in moist woodlands, stream & pond banks, low fields, roadsides, & waste areas; IA, se NE, e 1/2 KS, MO, e OK; (ne U.S., PA, IA, NE, s to FL, TX). *P. pruinosa* L., *P. missouriensis* Bush—Rydberg; *P. pubescens* var. *grisea* Waterfall, var. *integrifolia* (Dun.) Waterfall, var. *missouriensis* Mack. & Bush—Atlas GP; *P. pubescens* var. *glabra* (Michx.) Waterfall—Steyermark.

This species exhibits much variation in most colonies. Rare plants that are glabrous or nearly so have been called var. *glabra* (Michx.) Waterfall. Those with leaves gray-green, toothed to near base, and with sessile glands are the var. *grisea* Waterfall. Plants with leaves translucent and with entire or only 3 or 4 teeth per side are var. *integrifolia* (Dun.) Waterfall. A few specimens have leaves seldom translucent and with 5-8 teeth on each side and are recognized as var. *pubescens*. In our area, at least, these variants may all be found in any large colony and their recognition does not seem to be justified.

6. ***Physalis pumila*** Nutt., prairie ground cherry. Perennial herbs, with deep caudex, and often propagating by shoots from lateral roots; stems erect, 1.5-4.5 cm tall, branching from base and above; pubescence of stems and leaves of multicellular hairs 1-2 mm long, some 1- to 3-branched, spreading at right angles to the stem, or herbage subglabrous, having varying amounts of stiff appressed or divergent trichomes ca 1 mm long on at least flower buds, calyx, or leaf margins. Leaves alternate, blades ovate to ovate-lanceolate or lanceolate, elliptic, or sometimes rhombic, tapering to a more or less winged petiole, margins entire or slightly and irregularly sinuate-dentate, blades 4-8 cm long. Pedicels 15-30 mm long at anthesis, to 15(25-40) mm long in fruit; calyx at anthesis 10-15 mm long, lobes 3-5 mm long; corolla yellowish, slightly or faintly maculate, 12-17(20) mm long; anthers yellow, 2.5-3 mm long. Fruiting calyx 2(3-4) cm long, 1.5-2 cm wide, usually much inflated, pedicels reflexed; berry (10)12-15 mm in diam; seeds yellowish, broadly elliptic, 2-2.5 mm long, minutely reticulate or pitted.

Two subspecies in our area as follows:

6a. subsp. *hispida* (Waterfall) Hinton. Stem and foliage without branched trichomes or these exceedingly rare, pubescence of varying amounts of stiff, more or less divergent hairs about 1 mm long, at least in flower buds, calyx, or margins of leaves. (n = 12) May-Sep. Rather common on sandy prairies, stream valleys, fields, roadsides, more rarely in nonsandy habitats; SD: Bennett, Todd; NE, se WY, KS, w MO, OK, NM, TX; (GP to sw AZ). *P. virginiana* P. Mill. var. *sonorae* (Torr.) Waterfall—Atlas GP; *P. longifolia* var. *hispida* (Waterfall) Steyerm.—Steyermark; *P. lanceolata* Michx.—Gates.

6b. subsp. *pumila*. Stem and foliage with branched trichomes frequent to abundant among the vestiture of 1-2 mm long multicellular hairs. (n = 12) May-Oct. Infrequent to locally common on dry prairies, open woodlands, thickets, roadsides, & waste places; IA, se NE, e 1/2 KS, MO, OK; (IL to NE, s to KS, e TX).

Collectors frequently have confused subsp. *hispida* with *P. longifolia* but the differences are readily detected in the field with *P. longifolia* having single stems, more erect, and branching above. Subsp. *hispida* usually forms colonies from the rhizomatous habit, is shorter and often branched from the base. *P. longifolia* also frequently has the stems purplish or greenish-purple while those of *P. pumila* subsp. *hispida* are green or grayish-green.

7. ***Physalis viscosa*** L. subsp. ***mollis*** (Nutt.) Waterfall var. ***cinerescens*** (Dun.) Waterfall, ground cherry. Perennial herbs with somewhat woody, branched, subterranean caudex and rhizomatous; stems erect to usually decumbent to nearly prostrate; stems, foliage, and inflorescence more or less densely covered with stellate hairs, not tomentose, without intermingled longer hairs. Leaves alternate, blades ovate to reniform, variable in size but prin-

cipal ones (2)3–6(9) cm long and about as wide, margins entire, undulate, or sinuate-dentate, lower surface densely to rarely sparsely, finely stellate, less so above except along veins, apex obtuse, rounded, or rarely acute to short-acuminate; petioles 5–35(50) mm long, minutely stellate. Calyx at anthesis (3)5–7(9) mm long, lobes 1.5–2 mm long; corolla yellow, usually darkly maculate, rotate-reflexed, 12–15(17) mm wide, sparsely stellate pubescent on both surfaces; anthers yellow, 3–4 mm long. Fruiting calyx (1.5)2–3.5 cm long, retuse at base; berry (10)12–15 mm in diam; seeds yellow, broadly elliptic to ovate, 1.8–2.2 mm long, minutely reticulate or pitted, somewhat lustrous. (n = 12) May–Sep. Infrequent to common in sandy or rocky prairie plains & hillsides, stream valleys, lakeshores, roadsides, waste areas; sw KS, w 1/2 OK, sw CO, TX, NM; (KS, CO, OK, NM, TX; Mex.). *P. mollis* Nutt.—Rydberg.

This is a highly variable species with several subspecific taxa known to the s and se of our area. Small-leaved specimens in our area were segregated as *P. mollis* Nutt. (Rydberg, 1932). Just se of our area some specimens of subsp. *mollis* var. *cinerescens* integrade with var. *mollis*. Specimens of var. *cinerascens* reported in the Atlas GP for NE are here referred to *P. hederifolia*.

8. *Physalis virginiana* P. Mill., Virginia ground cherry. Perennial herbs from deep caudex, rhizomatous; stems usually erect, (1)3–6 dm tall, forked with ascending branches, hairs on stems wholly or in part more or less reflexed or decurved. Leaves alternate, blades ovate to narrowly lanceolate or elliptic, 2–5 cm long, entire to sinuately toothed, narrowed to base and decurrent on petiole, sparsely to densely hirsutulous on both sides; petioles 3–20 mm long. Pedicels at anthesis (5)10–20 mm long, usually with short, retrorse, hairs; calyx tube at anthesis 3–6 mm long, lobes (2.5)4–6 mm long, hirsutulous with spreading hairs; corolla (10)12–18(20) mm long, maculate; anthers yellow, 2–3(4) mm long, filaments narrowed toward apex. Fruiting calyx (2.5)3–4 cm long, longer than wide, sunken at base, 5-angled; berry globose, 10–15(18) mm in diam; seeds yellow, ovate or broadly elliptic, 1.7–2.2 mm long, minutely reticulate or pitted. (n = 12) May–Sep. Infrequent to locally common in open woodlands, thickets, prairie ravines, stream valleys, roadsides, & waste places; MN, ND, NE, e 1/2 KS, IA, MO, OK; (ME to Que., Man., s to SC, AL, OK, AZ).

7. QUINCULA Raf., Chinese Lantern

1. *Quincula lobata* (Torr.) Raf., purple ground cherry. Perennial rhizomatous herbs; stems erect at first, 5–10 cm tall, soon branching and rebranching from base, forming spreading or decumbent plants 1–8 dm in diam; indument throughout of sparse to dense, short-stipitate, spherical, white crystalline vesicles, which collapse on drying and appear as broad, flat, hairs. Leaves alternate, somewhat fleshy; blades oblanceolate to spatulate or oblong, sometimes ovate-lanceolate to linear-lanceolate, principal ones 4–10 cm long, 0.5–3 cm wide, margins entire, sinuate to pinnatifid, base cuneate to winged petiole; petioles 5–20(40) mm long. Peduncles commonly in pairs from axils of leaves, 5–25(30) mm long; flowers regular, erect during anthesis; calyx at anthesis campanulate, tube (2)3–4(5) mm long, lobes (0.5)1–2 mm long; corolla rotate, (1)1.5–2(3) cm wide, violet to purple or rarely white, often marked with red or purplish veins at throat, not marcescent; anthers yellow (1)1.5–2 mm long; style with a sigmoid bend near base. Fruiting calyx 1.5–2 cm long, pentagonal-ovoid, inflated, reticulate, with 5 converging lobes; berry ovoid, greenish-yellow, 5–8(10) mm in diam; seeds few per fruit, yellowish-brown to dark-brown, not glossy, somewhat flattened and irregular in shape, 1.8–2.2 mm long in diploids, 2.7–3 mm long in tetraploids, many often abortive, surface alveolate or rough reticulate, irregularly tuberculate or rugose on thick margin. (n = 11, 22) Mar–Oct. Rather common in open prairies, canyons, fields, roadsides, waste places; w 3/4 KS, CO, w 1/2 OK, NM, TX; (KS, CO, s to TX, NM, s CA; Mex.). *Physalis lobata* Torr.—Atlas GP.

This monotypic genus is often included in *Physalis*, from which it is readily distinguished by the erect flowers at anthesis, flower form and color, sigmoid style, nonglossy seeds with alveolate or rough reticulate surface and tuberculate or rugose backs, habit, and chromosome number. Other than seed size, no characters could be found to distinguish diploids from tetraploids. The diploids are rare in w OK and TX panhandle and become frequent in e NM and s TX, while the polyploid is found throughout the range of the species in the GP.

8. SOLANUM L., Nightshade

Ours annual or perennial herbs or one introduced species a vine; plants unarmed or prickly. Leaves alternate, simple and entire to bipinnatifid; petiolate; stipules wanting. Inflorescences cymose or racemose or appearing paniculate; peduncles axillary. Flowers perfect; calyx 5(6)-lobed, sometimes accrescent and closely investing the fruit, but not inflated; corolla 5-lobed, plaited in the bud, more or less rotate, white or yellow to blue or purple, regular to slightly irregular; stamens 5, connivent around the style, sometimes dimorphic; anthers dehiscing by apical pores or rarely by introrse slits. Fruit a juicy or semidry, usually 2-celled berry, some taxa containing subglobose concretions of stone cells attached to pericarp or among the seeds. *Androcera* Nutt. — Rydberg.

References: Schilling, Edward E. 1981. Systematics of *Solanum* sect. *Solanum* (Solanaceae) in North America. Syst. Bot. 6: 172–185; Whalen, Michael D. 1979. Taxonomy of *Solanum* section *Androceras*. Gentes Herb. 11: 359–426.

Solanum is of economic importance because it includes the potato *(S. tuberosum)*, some species of noxious weeds, and several species that are poisonous or suspected of being poisonous to humans and livestock. Though *S. jamesii* Torr. and *S. triquetrum* Cav. were attributed to our area by Rydberg, they are excluded in this treatment for lack of specimen evidence.

1 Plants spiny or prickly and usually evidently stellate-pubescent (except *S. citrullifolium*).
  2 Plants annual; stamens dimorphic with one longer than others; leaves once or twice pinnatifid.
    3 Corolla yellow; stem hairs stellate .... 8. *S. rostratum*
    3 Corolla violet or blue; stem hairs mostly simple .... 2. *S. citrullifolium*
  2 Plants perennial; leaves undulate but not at all dissected.
    4 Leaves silvery-white canescent with dense stellate hairs .... 5. *S. elaeagnifolium*
    4 Leaves not silvery-white canescent.
      5 Stellate hairs of lower leaf surface sessile; calyx 5–7 mm long; berry 1–2 cm in diam .... 1. *S. carolinense*
      5 At least some of stellate hairs of lower leaf surface stipitate; calyx 8–13 mm long; berry 2.5–3 cm in diam .... 3. *S. dimidiatum*
1 Plants not spiny or prickly; not stellate-pubescent.
  6 Plants tending to climb or scramble; perennial; rhizomatous .... 4. *S. dulcamara*
  6 Plants herbaceous; annual or short-lived perennials.
    7 Leaves deeply and regularly pinnatifid .... 10. *S. triflorum*
    7 Leaves not deeply pinnatifid.
      8 Calyx accrescent, cupping the lower 1/2 of fruit; plant copiously villous-hirsute, the hairs flattened and often gland-tipped .... 9. *S. sarrachoides*
      8 Calyx scarcely accrescent, not cupping fruit; plant glabrous or appressed to patent, hairy, never viscid.
        9 Stems evidently finely puberulent; globose sclerotic granules usually 2(4) per fruit; seeds usually 1.9–2.3 mm long .... 6. *S. interius*
        9 Stems glabrous or glabrescent; sclerotic granules (3)6–10(15) per fruit; seeds 1.5–1.9 mm long .... 7. *S. ptycanthum*

1. ***Solanum carolinense*** L., Carolina horse-nettle. Perennial herbs with erect subterranean branching rootstocks and usually creeping rhizomes; stems erect, 0.3–1 m tall, branching sparingly, armed with flattish, yellow spines (1)2–4(7) mm long, pubescent throughout

with 4- to 8-rayed, sessile stellate hairs. Leaves alternate, blades ovate to elliptic-ovate or ovate-elliptic, (3)5–10(15) cm long, 3–8(10) cm wide, coarsely sinuate to shallowly few-lobed, usually spiny along main veins, both surfaces with sessile, stellate trichomes with 4–8 spreading rays and an elongate central ray; petioles 1–3 cm long, usually with a few spines and sessile stellate pubescence. Inflorescences 5- to 20-flowered, cymose-racemose, elongating at maturity to form a rather tight racemiform cluster. Calyx tube 2–3 mm long, lobes linear-lanceolate or lance-acuminate, 2.5–4 mm long, becoming 8–9 mm long in fruit, surfaces stellate-pubescent, without spines; corolla purple, pale violet or white, 2–3 cm wide, lobes ovate to triangular, spreading or reflexed, 6–9 mm long, ca as long as tube; anthers yellow, tapering to tip, 6–9 mm long, equal, erect, connivent. Berry globose, yellow at maturity, 1–2 cm in diam, smooth at first but becoming wrinkled late in season; seeds 1.5–2.5(3) mm long, yellowish, flattened, semismooth. (n = 12) May–Sep. Fields, open woodlands, pastures, waste places, with preference for sandy or light soils; se SD, e 1/2 NE, IA, MO, e 1/2 KS, OK, TX: Hemphill; (NY, s Ont., MN, e SD, s to FL, TX, & n Mex.). POISONOUS.

This plant has been declared a noxious weed in SD and is of local economic importance where found. Reports of poisoning in cattle, sheep, and deer have been recorded, and there is one report of a child having died from eating the berries.

2. ***Solanum citrullifolium*** A. Br., melon-leaf nightshade. Annual herbs with tap-roots; stems 3–7 dm tall, much branched and spreading, densely pubescent with simple gland-tipped hairs mixed with sparse stellate hairs, and armed with straight prickles 2–9 mm long. Leaves alternate, 5–15 cm long, irregularly or interruptedly bipinnatifid, lobes rounded or obtuse and repand, main veins armed with yellowish, straight prickles, surfaces glandular-pubescent; petioles 1–7 cm long, armed, glandular-pubescent. Inflorescences cymose or cymose-racemose, few-flowered, pedicels spreading. Calyx campanulate, very prickly; corolla 2.5–3.7 mm wide, violet, irregularly 5-cleft lobes ovate, acuminate; anthers linear-lanceolate, 4 yellow and equal, the 5th one 1–1.6 cm long and tinged apically with violet or purple. Berry enclosed by the close-fitting and somewhat enlarged calyx, yellowish, 2-celled, 1–2 cm in diam; seeds yellowish to dark brown, flattish, 2.5–3.2 mm long, minutely rugose. (n = 12) Apr–Oct. Roadsides, fields, waste places; OK: Comanche; (OK, cen & w TX). *Androcera citrullifolium* (A. Br.) Rydb.—Rydberg. POISONOUS.

Though Rydberg listed this species for Iowa and Kansas, we have been unable to locate specimen evidence. The record in the Atlas GP for Harding Co., NM, is here referred to *S. heterodoxum* Dun. var. *setigeroides* Whalen, which has large anther less than 5 mm long, and corollas 1.5 cm or less wide.

3. ***Solanum dimidiatum*** Raf., western horse-nettle. Perennial herbs with deep branching and running rootstocks; stems erect, 3–10 cm tall, often branched above, with rather sparse, yellowish, flattish prickles 2–7 mm long, cinereous with somewhat scurfy stellate pubescence of 9- to 12-rayed, frequently stipitate hairs. Leaves alternate, blades ovate to elliptic-lanceolate, 6–15 cm long, 5–10 cm wide, irregularly sinuate, lobed or parted, base rounded to truncate or somewhat cordate, both surfaces with both sessile and stipitate stellate trichomes with 5–9 spreading rays, the central one not elongate, midribs with subulate prickles or these nearly absent; petioles short, stout. Inflorescences of terminal racemes, panicles, or laxly 2 or 3 divided cymes; pedicels recurved or reflexed in fruit. Calyx 8–13 mm long, often with 6 short-ovate lobes, each with a long, abruptly acuminate apex, 8–13 mm long; corolla bluish-purple to violet or nearly white, (2)3–5 cm in diam, the 5 ovate lobes spreading; anthers erect, connivent, equal, 8–10(12) mm long. Berries pale yellow, globose, (1)2.5–3 cm in diam, seeds yellowish-brown, oval, 3.8–4.3 mm long, minutely rugose. (n = 12) May–Oct. Prairies, fields, roadsides, river valleys, waste places; w MO, s-cen KS, OK, TX; (SC to FL, w to MO, KS, TX). *S. torreyi* A. Gray—Rydberg. POISONOUS.

4. **Solanum dulcamara** L., climbing nightshade, bittersweet. Perennial herbs with rhizomes; stems often somewhat woody below, unarmed, climbing or clambering, greenish, from glabrous to evidently pubescent with mostly simple, rarely stellate hairs. Leaves alternate, blades ovate, 7–10(12) cm long, 4–9 cm wide, from entire to some shallowly to deeply 1 or 2(4)-lobed at base, glabrous or essentially so, acuminate; petioles 2–5 cm long, minutely winged. Peduncles 1–4 cm long, inflorescences cymose or paniculate, freely branched, with 7–14(30) flowers, pedicels 5–10 mm long. Calyx tube 1–1.5 mm long, lobes broadly triangular, 1 mm long; corolla purple or bluish-purple, with 5 pairs of yellowish spots just below base of lobes, lobes lanceolate, reflexed, 6–9 mm long, ca 3× length of tube; anthers connate, 4–6 mm long. Berry red, subovoid, 8–10(12) mm in diam; seeds brownish, ovoid, flattened, 2.5–3 mm long, minutely reticulate. (n = 12) May–Sep. Rarely encountered in low woods, thickets, roadsides, fence rows; MN: Clay; ND: Cass, Mountrail; SD: Clay, Spink; NE: Dodge, Holt, Lancaster, Sarpy; KS: Atchison, MO: Cass, Jackson; (N.S. to Ont., ND, s to GA, MO, KS, also CA, WA, ID; Europe & Asia). *Introduced.*

Berries are very POISONOUS.

5. **Solanum elaeagnifolium** Cav., silver-leaf nightshade, white horse-nettle. Perennial herbs with deep running rootstocks, silvery-canescent all over with dense layer of pubescence composed of many-rayed stellate hairs that hide the surface; stems erect, 1–10 dm tall, often somewhat woody at base, simple or branched, prickles acicular, small, sparse to copious. Leaves alternate, blades narrowly lanceolate to oblong to oblong-lanceolate, 3–10(15) cm long, subentire to undulate or sinuate, usually tapered at base, mostly obtuse at apex; petioles 1–3 cm long. Inflorescences axillary near ends of branches, cymose-racemose, flowers (1)3–7, pedicels recurved or reflexed in fruit. Calyx 5-angled, tube 4–6 mm long, lobes linear, shorter than to slightly longer than tube; corolla pale to deep blue or lavender, rarely white, (15)20–30(35) mm wide, lobes ovate but abruptly acute, equaling or longer than the tube; anthers erect, equal, connivent, 6–9 mm long. Berry yellowish or reddish, eventually nearly black, globose, 1–15 mm in diam; seeds 3–5 mm long, ovoid or oblong, brown, lustrous, semismooth. (n = 12) May–Oct. Fields, prairies, pastures, stream valleys, roadsides, waste places; MO, KS, sw CO, OK, NM, TX; (MO, KS, s to LA, TX, AZ; n Mex.; *adventive* in e & w U.S.).

This is a weedy species which often forms large colonies and is noxious in the s GP. Loss of life in cattle from consuming mature berries has been reported.

6. **Solanum interius** Rydb., plains black nightshade. Short-lived perennial or annual herbs with taproots and subterranean branching caudex; stems erect; soon divergently branching from base and above, (1)3–5(8) dm tall or long, unarmed, upper internodes densely to sparsely strigose and often with divergent hairs usually less than 1 mm long, eglandular. Leaves alternate, highly variable in size; blades rather firm or membranaceous in shaded or mosit areas, ovate, ovate-lanceolate, to triangular-ovate or rhombic, 3–7(10) cm long, densely to sparsely strigose or hirtellous especially below, margins entire, undulate, or sinuate-dentate, bases cuneate, rounded, to subtruncate or rarely subcordate, apex obtuse, acute, or short acuminate; petioles 3–7(10) cm long, usually winged above, strigose or with ascending hairs. Inflorescences axillary, umbellate or rarely corymbiform; peduncles rather stout, (0.5)1–2(3) cm long, strigose; pedicels rather stout, 5–15 mm long. Calyx campanulate, lobes ovate, 1.5–2 mm long; corolla white, sometimes bluish-purple, or with purple stripes, (4)6–9(11) mm long, lobes usually 3–4 mm long, reflexed; anthers 1.5–2 mm long, pollen 22–25μ in diam; style 3–4 mm long. Mature berries purplish-black, dull, (5)7–10 mm in diam, usually detaching at receptacle or sometimes at junction of pedicel with peduncle; sclerotic granules 2(4); seeds yellowish to brown, (30)40–80(90) per fruit, 1.9–2.3 mm long, minutely pitted or reticulate. (n = 12) Jun–Oct. Infrequent to locally common in prairie ravines, pastures, open woodlands, stream valleys, thickets, rarely in fields; SD: Yankton,

Fall River; IA: Harrison, Potawatomie; NE, e WY, w 3/4 KS, e CO, w OK, TX panhandle, NM; (IA, SD, WY, s to TX, NM). *S. americanum*, in part—Atlas GP; *S. nigrum* var. *interius* (Rydb.) F. C. Gates—Gates. POISONOUS?

In our area this species, along with *S. ptycanthum* and *S. sarrachoides* (all n = 12), are the only representatives of a world-wide polyploid complex of about 30 species, often referred to as the taxonomically difficult *S. nigrum* complex. Though our plants have frequently been referred to the Eurasian *S. nigrum* L. (n = 36), this plant, as now recognized, is unknown in the GP. *Solanum sarrachoides* is a rather distinct and easily recognized member of the complex.

The distinction between *S. interius* and *S. ptycanthum* with herbarium specimens is often difficult. There are differences in pubescence, number of sclerotic granules in the fruit, and seed size. *S. interius* is characteristically a perennial with a branching caudex, which is lacking in *S. ptycanthum*. Mature specimens can be assigned rather easily to one or the other taxa, though a few in cen and e NE and cen KS appear to be intermediate. The pubescence character alone is not enough for identification, for all degrees of intergradation can be found. Thus, the many immature specimens can be named only arbitrarily.

Until more adequate field observations are made, mature specimens collected, and experimental work is done, we prefer to recognize the two taxa as distinct, though *interius* could well be considered as a subspecies of *S. ptycanthum*.

7. Solanum ptycanthum Dun. ex DC., black nightshade. Annual herbs with taproots; stems erect or more commonly divergently branching from base and above, (1)3–6(12) dm tall or long, unarmed, glabrous or sparsely strigose, often rather densely strigose in young branchlets, eglandular. Leaves alternate, highly variable in shape and size; blades usually membranaceous, ovate to oval, ovate-lanceolate, triangular-ovate, lanceolate or elliptic-lanceolate 5–10(17) cm long, margins entire, undulate, or sinuate-dentate, bases cuneate, rounded to subtruncate or subcordate, apex acute, obtuse, or short acuminate, surfaces glabrous to sparsely pubescent especially along veins and near the margins (young leaves may be rather densely strigose); petioles (0.1)3–7(10) cm long, usually winged at least above. Inflorescences axillary, umbellate or rarely corymbiform, usually strigose on all parts; peduncles filiform to rather stout, 1–3 cm long; pedicels often unequal, (1)3–7(10) mm long, soon reflexed. Calyx campanulate, lobes often unequal, 1–1.5 mm long, lanceolate to oblong, acute to obtuse; corolla white, sometimes with a yellow star, rarely bluish or streaked with purple, (3)4.5–7(9) mm long, lobes 3–6 mm long, often reflexed; anthers 1.5–2 mm long, pollen 18–22 μ in diam; style 2.5–3.3 mm long. Mature berries purplish-black, shiny or dull, 5–9 mm in diam, usually detaching at junction of pedicel and peduncle but sometimes at the receptacle; sclerotic granules (3)6–10(15), variable in size; seeds (40)60–90(112) per fruit, 1.5–1.9 mm long, flattened, minutely reticulate or pitted. (n = 12) May–Oct. Infrequent to locally abundant in moist open woodlands, prairie ravines, stream banks, & valleys, fields, gardens, roadsides, & waste places; MN, s 1/2 ND, SD, IA, NE, KS, MO, OK; (s Que. to Man., s to FL, TX). *S. americanum* in part—Atlas GP; *S. nigrum* L.—Rydberg.

Though the name *S. americanum* P. Mill. has long been used for this species, the unfortunate choice of a lectotype for *S. americanum* apparently precludes continued use of that name for our plants. For those who would prefer to consider our plant as a variant of *S. nigrum*, the name *S. nigrum* L. var. *virginicum* L. is available.

Numerous cases of black-nightshade poisoning have been reported from various parts of the United States, and thus the immature fruits and foliage are suspect. However, the mature fruits are supposed to be edible.

Solanum scabrum P. Mill. (2n = 72) is sometimes grown in gardens as the garden huckleberry. It differs from other members of the *S. nigrum* complex in our area in having fruits larger than 12 mm in diam, and the anthers brownish-purple rather than yellow. We have no reports of this species as a persisting escape.

8. Solanum rostratum Dun., buffalo bur, Kansas thistle. Annual herbs with taproots; stems (2)3–7 dm tall, widely branching, well armed with straight yellow spines, and copiously pubescent with sessile or stalked stellate pubescence. Leaves alternate, blades 4–15 cm long, 3–7(10) cm wide, elliptic or obovate, once or twice pinnately lobed, cleft, or parted, surfaces with spines and pubescent with sessile stellate trichomes, each with an elongate central ray; petioles 2–6 cm long. Flowers 5–15, racemose, the rachis elongating in fruit and pedicels ascending. Calyx nearly hidden by spinelike prickles, lobes 2–3 mm long at anthesis, about equaling the tube; corolla bright yellow, 1.6–2.5(3) cm wide, lobes triangular or ovate, widely flared, about equaling the tube, stellate-pubescent; stamens and style much declined, anthers adnate to near the throat of the corolla, 4 of them alike, yellow, 7–9 mm long, the 5th one purplish, much longer and with incurved beak. Berry 7–10 mm long, wholly enclosed by the spiny, close fitting, beaked and much enlarged calyx; seeds silvery-dark brown to black, 2.3–2.7 mm long, distinctly pitted, often cross-corrugate on thickened margins. (n = 12) May–Oct. Often abundant in fields, pastures, feed lots, stream valleys, roadsides, & waste areas; GP; (originally native of plains region but now widely distributed in the U.S. & Mex.). *Androcera rostrata* (Dun.) Rydb.—Rydberg; *S. cornutum* Lam.—Kartesz & Kartesz.

There are reports of swine being poisoned by eating berries and roots.

9. *Solanum sarrachoides* Sendtner, viscid nightshade. Annual herbs with taproots; stems unarmed, erect or decumbent, freely branching, 1–8 dm tall or long, densely glandular-pubescent. Leaves alternate, blades ovate, 5–12 cm long, (2)3–6(8) cm long, margins entire, repand, or dentate, surfaces glandular-pubescent, bases cuneate to truncate; petioles 1–5(9) cm long, glandular-pubescent, minutely winged. Inflorescence umbellate or appearing racemose, 1- to 5(7)-flowered, peduncles and pedicels stout, glandular-pubescent. Calyx tube 1–1.5 mm long, lobes at anthesis 2–3 mm long, lanceolate to triangular; corolla white, (4)6–8(11) mm wide, lobes spreading; anthers connivent, (1.5)2 mm long, equal, yellow. Berry green to nearly black, 6–8(10) mm in diam, glabrescent, lower 1/2 covered by enlarged calyx; usually (5)6(7) globose sclerotic concretions; seeds semismooth, ovoid, 1.7–2.4 mm long. (n = 12) May–Oct. Local in farmyards, fields, pastures, open woodlands, roadsides, gardens, & waste places; sporadic throughout GP; (adventive or naturalized over much of U.S.; S. Amer.). *S. villosum* P. Mill—Atlas GP. POISONOUS?

This plant has been confused with *S. villosum*, a tetraploid species of Eurasia, which is very similar but has a small calyx that is not accrescent. Some of our specimens, referred to *S. villosum* in the Atlas GP, are too young for certain determination, and are here referred either to *S. interius* or *S. sarrachoides*. We know of no valid records of *S. villosum* in the G.P.

10. *Solanum triflorum* Nutt., cut-leaved nightshade. Annual malodorous herbs with taproots; stems (1)3–4 dm tall, commonly branching and rebranching from near base, these spreading and often 4 dm or more long, unarmed, sparsely to evidently pubescent. Leaves alternate, blades oblong, deeply pinnatifid with rounded sinuses, 2–5 cm long and 2 cm wide, the lobes lanceolate to linear, usually extending more than halfway to midrib, surfaces nearly glabrous, to evidently pubescent; petioles 3–15 mm long. Peduncles 1–1.5(2) mm long, the (1)2 or 3 flowers umbellate, pedicels stout, about as long as peduncle, soon reflexed. Calyx tube 1–1.5 mm long at anthesis, lobes narrowly lanceolate, (1)1.5–2(2.5) mm long at anthesis, but much accrescent and up to 6 mm long in fruit; corolla white, lobes deltoid, 2–3(4) mm long, variously pubescent without; anthers yellow, equal, 2–2.5 mm long. Berry globose, 9–12 mm in diam, lower 1/3 covered by calyx, greenish but eventually turning black, with several globose concretions among the seeds; seeds yellow, 2–2.5 mm long, flattened, minutely reticulate-pitted. (n = 12) May–Sep. Rocky prairie hillsides, "prairie dog towns," pastures, fields, roadsides, waste places; GP; (B.C. to CA, e to MN, IA, KS, OK, adventive occasionally to e coast). POISONOUS to livestock.

This species is frequently found in prairie dog towns, and some early collectors thought it was restricted to such habitats. We have observed the plant to be a serious weed in some bean fields on the high plains.

112. CONVOLVULACEAE Juss., the Morning Glory Family

by Daniel F. Austin

Ours herbs or vines, the sap milky in some species. Leaves alternate, mostly simple, pinnately lobed or less commonly pectinate, or reduced to scales, exstipulate. Inflorescences axillary, dischasial, solitary, racemose or paniculate to paniculate-thyrsiform. Flowers perfect, regular, small and inconspicuous to large and showy, but mostly evanescent, lasting a single day. Sepals 5, free, or united at least basally, imbricated, equal or unequal, persistent, occasionally accrescent in fruit; corollas sympetalous, tubular, funnelform, campanulate, urceolate, or rotate to salverform, the limb with 5 lobes or teeth or almost entire, mostly induplicate in buds, producing corollas with plicae and interplicae; stamens 5, distinct, the filaments inserted on the corolla tube base alternate with corolla lobes, the anthers mostly linear or oblong, 2-celled, extrorse, disc annular or cupuliform, sometimes 5-lobed, occasionally absent or apparently absent; ovary superior, 2 or 3 carpels, usually 2- or 3-locular, each locule 2-ovulate; styles filiform, simple or bifid or 2 distinct styles; stigmas capitate or 2-lobed or the stigmas 2 and linear, clavate, ellipsoid or globose. Fruits 1- to 4-locular, capsular and dehiscent by valves, transversely or irregularly dehiscent or indehiscent; seeds 1–4, commonly fewer than ovules, glabrous or pubescent, endosperm absent or scanty, cartilaginous, cotyledons mostly foliaceous.

1 Flowers white, with or without tinges of lavender to pink on limb.
  2 Calyx enclosed by 2 foliaceous bracts; corolla funnelform, 4–6 cm long ...... 1. *Calystegia*
  2 Calyx not enclosed, the bracts scalelike; corolla either campanulate, broadly funnelform, funnelform or salverform, 0.5–3 cm long.
    3 Styles 2.
      4 Leaves linear or lanceolate, basally cuneate to attenuate; flowers funnelform, 1.2–3 cm long .... 6. *Stylisma*
      4 Leaves elliptic to lanceolate or ovate-lanceolate; flowers salverform, 0.5–0.65 cm long .... 3. *Cressa*
    3 Style 1.
      5 Stigmas subulate and cylindrical, applanate, apices acute; leaves hastate to sagittate .... 2. *Convolvulus*
      5 Stigmas globose, apices rounded; leaves cordate basally .... 5. *Ipomoea lacunosa*
1 Flowers lavender, blue, red or white with a purple to purple-red throat.
  6 Flowers rotate to broadly campanulate, blue; styles 2, each branch divided into 2 elongate, linear stigmas; suffrutescent, erect herbs .... 4. *Evolvulus*
  6 Flowers funnelform to salverform, lavender, blue, red or white with a purple to purple-red throat; style 1, entire, stigmas globose; herbs, twining, decumbent or rarely erect *(I. leptophylla)* .... 5. *Ipomoea*

1. CALYSTEGIA R. Br., Hedge Bindweed

Herbs; stems twining, perennial, glabrous or pubescent. Leaves ovate, sagittate, hastate or reniform, glabrous or pubescent, petiolate. Flowers axillary, often solitary, with 2 enlarged, foliaceous bracts near the apex of the pedicel, bracts often somewhat inflated basally, covering most or all of the calyx. Sepals subequal, ovate, ovate-lanceolate, or oblong, usually glabrous; corollas funnelform, 30–60 mm long, glabrous or slightly puberulent on the margin of the limb, particularly at the apex of the interplicae, white or pink to pale lavender; stamens included, subequal, the filaments with glandular trichomes basally or these absent, anthers 4–7 mm long, sagittate basally; pollen spheroidal, pantoporate, nonspinulose; ovary ovoid, glabrous, 2-locular basally and 1-locular apically, 4-ovulate, the style 1, entire, glabrous, the stigmas oblong, flattened. Fruits capsular, unilocular, 4-valved, accompanied and surrounded partly by the enlarged sepals and bracts; seeds 1–4, glabrous, black to dark brown, smooth to verrucose. *Convolvulus* sensu auctt., non L., *Convolvulus* sect. *Calystegia* (R. Br.) Benth.

References: Brummitt, R. K. 1965. New contributions in North American *Calystegia*. Ann. Missouri Bot. Gard. 52: 214–216; Lewis, W. H., & R. L. Oliver. 1965. Realignment of *Calystegia* and *Convolvulus* (Convolvulaceae). Ann. Missouri Bot. Gard. 52:217–222; Brummitt, R. K. 1981. Further new names in the genus *Calystegia* (Convolvulaceae). Kew Bull. 35(2): 327–334.

1 Stems erect, not twining .... 3. *C. spithamea*
1 Stems decumbent, either twining throughout or only at tips.
   2 Leaf sinus quadrate, blade tissues not beginning for 2–5(10) mm from petiole attachment .... 4. *C. sylvatica* subsp. *fraterniflora*
   2 Leaf sinus V- or U-shaped, blade tissues beginning at the point of petiole attachment.
      3 Blade of leaf basally 2-angled; plants normally glabrous or with a few trichomes on petioles .... 2. *C. sepium* subsp. *angulata*
      3 Blade of leaf basally rounded; plants normally pubescent on all vegetative parts .... 1. *C. macounii*

1. ***Calystegia macounii*** (Greene) Brummitt. Herbs with erect or sparsely twining stems, these simple or branched, usually from near the base, finely pubescent. Leaves ovate to ovate-lanceolate, finely pubescent, 2–6 cm long, 1–5 cm wide, basally cordate to subsagittate, the lobes usually rounded, less often angled, apically obtuse to rarely acute, the margins entire; petioles 0.5–40 mm long. Flowers solitary, often less than 4 per plant, frequently arising from lower few axils, on peduncles 3–5 cm long, rarely longer; bracts ovate to ovate-oblong, pubescent, sparsely with age, foliaceous, 20–25 mm long, 10–15 mm wide, mucronate, mostly obtuse, sometimes acute, the pedicel absent. Sepals elliptic to ovate, subequal, 10–12 mm long, 5–7 mm wide, thin, transparent at least on margins, acute to acuminate, mucronate, glabrous or ciliate; corollas funnelform, white, 4–5 cm long; stamens subequal, 25–28 mm long, basally glandular pubescent, the anthers 4–5 mm long, basally sagittate; the style 20–23 mm long. Fruits capsular, globose to ovoid, brown. May–Jun (Jul). Disturbed habitats of various types; GP; w MN, IA, nw MO & OK, w to MT, WY, & CO. *Convolvulus interior* House—Rydberg; *Convolvulus sepium* of authors.

2. ***Calystegia sepium*** (L.) R. Br. subsp. ***angulata*** Brummitt, hedge bindweed. Herbs with rhizomatous, twining stems, these cylindrical or angular, glabrous; internodes often 2–7 cm apart. Leaves ovate to ovate-lanceolate, glabrous, 2–15 cm long, 1–9 cm wide, basally cordate-sagittate to hastate, 5-nerved, the auricles obtuse to acute or 2- or 3-dentate, rarely 2-lobed, apically acute to acuminate, the border entire or undulate; petioles 2–7 cm long. Flowers solitary, on peduncles 3–13 cm long; bracts surrounding calyx angular, ovate, convex, glabrous or ciliate, foliaceous, the borders at times pinkish, 14–26 mm long, 10–18 mm wide, mucronate, mostly acute, the pedicels absent. Sepals elliptic to ovate-lanceolate, subequal, 11–15 mm long, 4–6 mm wide, thin, transparent, acute to subobtuse, mucronate, apically ciliate; corollas funnelform, white or tinged at least on limb with rose or pink, 4.5–5.8 cm long; stamens 23–29 mm long, subequal, basally glandular pubescent, the anthers 4.5–6 mm long, basally sagittate; disk annular, 5-lobed; ovary ovoid, glabrous, the style 20–23 mm long. Fruits capsular, 10–13 mm in diam, accompanied and partly surrounded by the enlarged bracts, which reach 30–35 mm long; seeds 4.5–5 mm long, black, glabrous, smooth or granulose. ($2n = 22$) Jun–Aug. Thickets & fence rows; n 3/4 GP; (MA to WA, NC to NM). *Convolvulus sepium* L., *C. americanus* (Sims) Greene—Rydberg.

3. ***Calystegia spithamaea*** (L.) Pursh. Herbs with erect stems, these simple or sometimes branched from near the base, finely pubescent. Leaves oblong-ovate to oblong or pan-

durate, finely pubescent, 3–6(10) cm long, 1–2(5) cm wide, basally cordate to subtruncate, less commonly obtuse, apically obtuse to subacute, the margins entire; petioles 0–12(15) mm long. Flowers solitary, often less than 4 per plant and arising from the lower leaf axils, on peduncles 15–50 mm long, rarely longer; bracts ovate to ovate-oblong, pubescent, at times sparsely, foliaceous, 15–20(25) mm long, 8–10(15) mm wide, mucronate, acute, the pedicels absent. Sepals elliptic to ovate, subequal, 10–12 mm long, 4–6 mm wide, thin, transparent at least on margins, acute to acuminate, mucronate, glabrous or ciliate; corollas funnelform, white, 4.5–5 cm long; stamens subequal, 25–28 mm long, basally glandular pubescent, the anthers 4–5 mm long, basally sagittate; the style 20–23 mm long. Fruits capsular, globose, brown. (2n = 22) (May) Jun–Jul (Aug). Sandy or gravelly roadsides or hillsides; MO: Jackson; (Que. to Ont., SC to MO). *Convolvulus spithamaeus* L.— Rydberg.

> This taxon is included because of a specimen cited by Steyermark, which I have not seen. The habitat where the specimen was collected suits these plants, and there is no reason to believe that it was misdetermined.

4. **Calystegia sylvatica** (Kit.) Griseb. subsp. *fraterniflora* (Mack. & Bush) Brummitt. Herbs with twining stems, these cylindrical to angular, glabrous; internodes 3–6 cm apart. Leaves ovate to ovate-lanceolate, glabrous, 4–8.5 cm long, 2–5 cm wide, basally hastate, 6 or 7-nerved, the auricles obtuse to acute, the sinus quadrate, with the sides of the auricles more or less parallel, apically acute to acuminate, the border entire; petioles 5–7 cm long. Flowers solitary, on peduncles 3.5–7 cm long; bracts surrounding the calyx angular, ovate, obtuse, convex, inflated basally, glabrous or remotely puberulent, foliaceous, 18–22 mm long, 14–16 mm wide, mucronate, mostly obtuse, the pedicel absent. Sepals ovate, subequal, 14–16 mm long, 5–6 mm wide, thin, transparent above, translucent below, obtuse, mucronate, glabrous; corollas funnelform, white, 4–5 cm long; stamens 20–22 mm long, subequal, basally glabrous or glandular pubescent, the anthers 3–4 mm long, basally sagittate; ovary subglobose, glabrous, the style 20–23 mm long. Fruits capsular, ovoid, surrounded by the enlarged bracts; seeds 5–6 mm long, black glabrous, smooth. (2n = 20) May–Sep. Thickets & fence rows; IA, KS, MO; (NH to KS, FL to OK). *Convolvulus sepium* var. *fraterniflorus* Mack. & Bush—Steyermark; *Calystegia sepium* var. *fraterniflorus* (Mack. & Bush) Shinners—Correll & Johnston; *C. fraterniflorus* (Mack. & Bush) Brummitt, 1965, op. cit.

> Another species is present in GP, probably persistent from cultivation and not naturalized as stated by some authors. *Calystegia pellita* (Ledeb.) G. Don is similar to *C. sepium*, but with smaller leaves and bracts. Apparently, all of the plants in the GP are of the double-flowered form, which is sterile. KS: Bourbon, Jefferson, Montgomery, Shawnee; MO: Jackson, Platte; (ME, MI to GP, VA, TN). *Convolvulus japonicus* Rydb., non Thunb., *C. pellitus* f. *anesitus* Fern.—Steyermark.
> Presence of this taxon in the United States has been discussed by Fernald (Rhodora 51: 73–75. 1949). While Fernald pointed out that *Calystegia pubescens* Lindl. (J. Royal Hort. Soc. 1:70. 1846) was the same taxon, he did not list the earlier name *Calystegia pellita*. The earliest name is applied here, even though it strictly conforms with the fertile form of Asia.

2. **CONVOLVULUS** L., Field Bindweed

Herbs or shrubs; stems twining, erect or decumbent, often with tips twining if decumbent, glabrous or pubescent with simple indumentum. Leaves sagittate, hastate or ovate, less commonly ovate-lanceolate to linear, border entire to undulate or irregularly lobed or crenate, rarely laciniate, glabrous or pubescent. Flowers solitary or cymose, axillary; bracts and bracteoles scalelike, linear, elliptic or ovate. Sepals subequal or the inner longer, suborbicular, elliptic or ovate, obtuse to acute, often mucronate, glabrous or pubescent; corollas campanulate, mostly 20–30 mm long, rarely longer, glabrous or puberulent on the margins of the limb, particularly at the apex of the interplicae, white, rose to white on the limb and the throat purplish; stamens included, unequal with 2 shorter than the other 3, glandular pubescent at the filament base, the anthers 1–4 mm long, rarely longer, oblong, basally auriculate; pollen ellipsoid, 3-colpate; ovary ovoid to subglobose, glabrous to pubescent, 2-locular, the style 1, entire, glabrous, the stigmas 2, filiform to cylindric and subulate. Fruits capsular, 2-locular, 4-valved; seeds 1–4, glabrous, black to dark brown, smooth to verrucose or pitted.

References: Lewis, W. H., & R. L. Oliver. 1965. Realignment of *Calystegia* and *Convolvulus* (Convolvulaceae). Ann. Missouri Bot. Gard. 52: 217–222; Sa'ad, F. 1967. The *Convolvulus* species of the Canary Islands, the Mediterranean region and the Near and Middle East. Meded. Bot. Mus. Herb. Rijksuniv. Utrecht 281: 1–288.

1 Leaves almost as broad as long; calyx 3–5 mm long, inconspicuously puberulent or glabrate; perennials from deep creeping root, forming large patches .... 1. *C. arvensis*
1 Leaves usually much longer than broad; calyx 6–12 mm long, densely pubescent; perennials from taproot, sometimes divided at apex but not forming large, creeping patches .... 2. *C. equitans*

1. Convolvulus arvensis L., field bindweed. Herbs with branched, decumbent or twining stems, the roots rhizomatous, widely spreading. Leaves variable, often ovate, ovate-lanceolate to elliptic, 1–10 cm long, 0.3–6 cm wide, entire or with the border somewhat undulate, basally cordate to subtruncate, hastate or sagittate, the lobes obtuse or acute, entire or with 2 or 3 teeth, glabrous or inconspicuously puberulent; petioles 3–40 mm long. Flowers in cymes of 2 or 3 or solitary, on peduncles 1–9 cm long; bracts elliptic, linear or obovate, 2–3(9) mm long, the bracteoles linear, 2–4 mm long; the pedicels 5–18(35) mm long, reflexed in fruit, usually glabrous. Sepals obtuse, or less commonly truncate or emarginate, mucronate, ciliate, the outer elliptic, 3–4.5 mm long, 2–3 mm wide, glabrous or tomentose, the inner suborbicular to obovate, 3.5–5 mm long, 3–5 mm wide; corollas campanulate, white or tinged with pink, 1.2–2.5 cm long; stamens 8–13 mm long, the anthers 2–3.5 mm long; ovary ovoid, glabrous, the style 7–10 mm long. Fruits capsular, subglobose to ovoid, 5–7 mm in diam, glabrous; seeds 1–4, 3–4 mm long, black to dark brown, glabrous, tuberculate. (2n = 48, 50) Jun–Aug. Cult. grounds, fields, roadsides; GP; (throughout temp. U.S.; Europe). *Naturalized.* *C. ambigens* House — Rydberg.

2. Convolvulus equitans Benth. Herbs with branched, prostrate or decumbent stems arising from a taproot, densely pubescent. Leaves variable, ovate-elliptic to triangular-lanceolate or narrowly oblong with projecting basal lobes, most often deeply indented basally, 1–7 cm long, 0.2–4 cm wide, margins toothed or lobed or both, rarely entire, densely pubescent on both surfaces with loosely appressed indumentum; petioles 0.25–5 cm long. Flowers usually solitary, less commonly 2 or 3 and cymose, on peduncles 0.5–10.5 cm long, the pedicels much shorter, 5–24 mm long; bracts and bracteoles scalelike, pubescence like leaves. Sepals oblong to ovate, obtuse to weakly retuse apically, 6–12 mm long, 3–6 mm wide, margins membranaceous, appressed sericeous, subcordate with age; corollas campanulate, white to pink, at times with a reddish center, 1.5–3 cm long, 5-angled, the angles often prolonged as slender points, sericeous on the interplicae; ovary ovoid, glabrous. Fruits capsular, subglobose, 7–8 mm in diam, glabrous; seeds 1–4, 4–4.5 mm long, black, granulate, glabrous. Apr–Sep. Dry plains & hills; NE, CO, KS, OK, TX; (GP s into Mex. & S. Amer.). *C. hermannioides* A. Gray — Rydberg.

This variable species usually has been considered as two distinct taxa by previous authors. There is a great deal of overlap, and Johnston (Texas J. Sci. 11: 191–206. 1969) has suggested that introgression may be involved. Apparently, no one has tested this experimentally. Since the genus is known for variable species, a single wide-ranging, polymorphic species will be recognized here.

3. CRESSA L., Alkali Weed

1. Cressa truxillensis H.B.K. Herbs with erect or decumbent stems, branches appressed pubescent, at times sericeous to hirsute. Leaves elliptic to lanceolate, less commonly ovate-lanceolate, 3-10 mm long, 1.5-3.5 mm wide, entire, acute, subcarnose, with appressed indumentum, sometimes hirsute to sericeous; petioles 0.5-1.5 mm long. Flowers solitary, axillary, clustered toward the apices of the branches; peduncles 2-6 mm long; bracteoles ovate to ovate-lanceolate, acute, 2-3 mm long, 1 mm wide; pedicels absent. Sepals, obtuse to acute, pubescent, the outer elliptic, 3-4 mm long, 2.5-3 mm wide, the inner obovate, 3-4.5 mm long, 2.5 mm wide, the margins scarious; corollas salverform, 5-6.5 mm long, the tube 3-3.5 mm long, the limbs with lobes obtuse to subacute, sparsely pubescent without, 2.5-3 mm long; stamens exserted, equal, 4-6 mm long, with glandular trichomes on the filament bases, the anthers oblong, basally cordate, pollen 3-colpate, ellipsoid; ovary ovoid, apically hirsute, unilocular or subbilocular, the septum incomplete, styles 2, unequal, 3.5-5 mm long, stigmas 2, capitate. Fruits capsular, 4-valved, ovoid, 5-6 mm long, brown, apically pubescent, unilocular; seeds mostly 1, ovoid, 3.5 mm long, brown, glabrous. ($2n = 28$) Apr-Jul. Alkaline ponds, marshes, & lakes; OK: Harmon; TX: Armstrong, Bailey; (OK & TX s into Mex. & S. Amer.). *C. depressa* Goodd.—Correll & Johnston.

Outside the range of this flora there are at least eight other names that have been proposed for these plants. Specimens have been examined from several North American and European herbaria, and it is felt that a single variable taxon is involved. Extreme forms, apparently representing atypical growth patterns, have been chosen by past authors for naming without much previous study.

4. EVOLVULUS L.

1. Evolvulus nuttallianus R. & S., Nuttall's evolvulus. Suffrutescent herb, stems several and arising from a woody base, erect to ascending, 10-15 cm tall, densely villous with an indumentum of ferrugineous, brown, fulvous or gray color. Leaves linear-oblong, narrow-lanceolate to narrow-oblanceolate or rarely oblong, 8-20 mm long, 1.5-5 mm wide, entire, attenuate basally, acute to obtuse apically, densely pubescent like stems; petioles short or absent. Flowers solitary in axils over whole length of stems, peduncles absent; pedicels 3-4 mm long, reflexed in fruit; bracteoles subulate, 1-4 mm long; sepals lanceolate to narrowly lanceolate, long-acuminate, 4-5 mm long, patently villose; corollas rotate to broadly campanulate, 8-12 mm in diam, subentire, purple or blue; anthers 1-2 mm long, oblong, basally auriculate, filaments twice as long as the oblong anthers; ovary subglobose, glabrous. Fruits capsular, ovoid, about as long as the sepals, glabrous; seeds (1)2, brownish, smooth. May-Jul. Sandy prairies & plains; MO to MT, OK to NM; (MT, ND s to MO, KS, TX, AZ). *E. argentus* Pursh (1814), non R. Br. (1810); *E. pilosus* Nutt. (1818). nom. pro syn., Trans. Amer. Philos. Soc. 5: 195. 1837, non Roxb. (1832)

Reference: Van Ooststroom, S. J. 1934. A monograph of the genus *Evolvulus*. Meded. Bot. Mus. Herb. Rijksuniv. Utrecht 14(1934): 1-267.

Perry (Rhodora 37:63. 1939) was apparently the first to point out that the name *E. pilosus* Nutt. was invalid; the next available name is that proposed by Roemer & Schultes.

5. IPOMOEA L.

Herbs, lianas, shrubs or trees; stems twining, prostrate, reclinate to erect, glabrous to pubescent. Leaves scalelike, linear, ovate, lanceolate, to cordate, entire to deeply palmately or pinnately parted, basally often cordate but also acute to obtuse, glabrous to sericeous. Flowers solitary or more commonly dichasial or thyrsiform, rarely racemose, mostly axillary; bracts usually scalelike but foliaceous in a few taxa, bracteoles scalelike, glabrous or pubescent. Sepals equal to unequal, either the inner or outer being longer, variable in shape from suborbicular to lanceolate-linear, glabrous or pubescent; corollas funnelform, salverform or rarely urceolate, very rarely almost campanulate, mostly glabrous but a few taxa puberulent without, lavender to purple, red or orange, less commonly white, rarely yellowish with a purple center; stamens included or excluded, unequal, of 2 or 3 lengths, the filaments glandular pubescent basally, the anthers 1–10 mm long, oblong; pollen pantocolpate, spinulose; ovary ovoid to ellipsoid, glabrous or pubescent, 2- to 4-locular, the style single, the stigmas 2 or 3-globose. Fruits capsular, 4- to 6-valvate, 1- to 4-locular; seeds 1–4 or rarely 6, glabrous to woolly, tan, brown to black, smooth. *Quamoclit* Moench—Rydberg.

References: O'Donell, C. A. 1959. Las especies americanas de *Ipomoea* L. Sect. *Quamoclit* (Moench) Griseb. Lilloa 29: 19–86; Austin, D. F. 1975. Family 164. Convolvulaceae. *In* Flora of Panama. Ann. Missouri Bot. Gard. 62: 157–224; Austin, D. F. 1977. Realignment of the species placed in *Exogonium*. Ann. Missouri Bot. Gard. 64: 330–339; Austin, D. F. 1978. The *Ipomoea batatas* complex. I. Taxonomy. Bull. Torrey Bot. Club 105: 114–129.

The following cultivated species are reported for the GP:

Ipomoea alba L., the moonflower, is cultivated for its nocturnal, fragrant flowers. These plants are cold-sensitive and not likely to become established. KS: Riley.

Ipomoea batatas (L.) Lam., the sweet potato, is cultivated in many parts of the GP. However, most of the area is too cold to permit the plants to become more than adventive.

Ipomoea x *multifida* (Raf.) Shinners, a hybrid between *I. quamoclit* and *I. coccinea*, is cultivated sparingly in the GP. Although plants are self-fertile and seed-set is often high, they have not become established.

Ipomoea nil (L.) Roth, the heavenly blue or Japanese royal morning-glory, is a tropical species that may be planted. This, too, is so cold-sensitive that establishment seems unlikely.

Ipomoea tricolor Cav., the morning-glory, may be planted in gardens and yards. This species is unlikely to become part of the flora.

Gates (1940) suggested that there were hybrids between *I. hederacea* and *I. purpurea* in KS. Of the plants he cited, two are perfectly "normal" *I. hederacea*, and I fail to understand why he thought them hybrids. Another specimen *(Hancin 1766)* might be placed in the tropical taxon *I. purpurea* var. *diversifolia* sensu O'Donell. This taxon I consider unworthy of recognition, but Gates was observant enough to detect the deviation from the norm of plants in the GP. Probably this tropical variant was introduced as an adventive.

1 Leaves deeply pinnately dissected, appearing compound .... 8. *I. quamoclit*
1 Leaves entire to lobed or toothed but appearing simple.
  2 Sepals with a caudate apex subterminal on the outer 2; corollas scarlet; annual.
    3 Leaf blade entire or parted to lobed on the same individual plant (3- to 7-lobed to parted to entire); fruiting pedicels erect to reflexed; capsule 7–8 mm wide, the apiculum 2 mm long .... 2. *I. cristulata*
    3 Leaf blade entire or dentate only; fruiting pedicels always reflexed; capsules 6–7 mm wide, the apiculum 3–4 mm long. .... 1. *I. coccinea*
  2 Sepals acute to obtuse and mucronulate but not caudate; corollas, blue, purple, lavender to white with a purple throat; annual or perennial.
    4 Sepals linear-lanceolate, the tips curved, setose or hispid; corolla blue with a white or pale yellow throat; annual. .... 3. *I. hederacea*
    4 Sepals oblong to ovate, the tips straight, mostly glabrous (hispid in *I. purpurea*, ciliate in *I. lacunosa*); corollas purple, lavender to white with a purple throat; annuals or perennials.
      5 Sepals oblong, hispid; annual; seeds pyriform, short tomentose ...... 7. *I. purpurea*
      5 Sepals ovate to lanceolate, glabrous or ciliate; annual or perennial; seeds rounded to oblong, glabrous or with long woolly indumentum (oblong with short trichomes in *I. leptophylla*).

- 6 Flowers to 2 cm long, white or with pale shades of pink; annuals with lanceolate, ciliate sepals .... 4. *I. lacunosa*
- 6 Flowers 4–7 cm long, if white then with purple or some other color in throat; perennials with ovate, glabrous sepals.
  - 7 Herbs erect, rarely recumbent; flowers lavender; seeds with short, erect brown indumentum .... 5. *I. leptophylla*
  - 7 Herbs with twining stems; flowers white, pink or white with purple throat; seeds with long, woolly brown indumentum.
    - 8 Leaf blades narrowly rhombic to ovate-lanceolate; petioles on flowering branches (except terminal portion) 1–2.5 cm long .... 9. *I. shumardiana*
    - 8 Leaf blades cordate or cordate-ovate; petioles on flowering branches 1–8 cm long .... 6. *I. pandurata*

1. **Ipomoea coccinea** L., red morning-glory. Annual herbs, the stems twining or rarely decumbent, glabrous or rarely slightly puberulent. Leaves ovate, 2–14 cm long, 1–11 cm wide, entire to dentate with 3–5 teeth, basally cordate, the lobes rounded to acute or almost sagittate, often dentate, apically acute to acuminate, glabrous or pilose near the base; petioles 0.6–14 cm long, smooth or muricate above, glabrous or puberulent below. Flowers 2–8 in cymes, rarely solitary; bracts ovate to ovate-lanceolate, 2–4 mm, bracteoles ovate to lanceolate, 1–3 mm, both mucronate and aristate; pedicels 5–15 mm long, smooth or muricate on the angles, erect in flower, reflexed in fruit. Sepals subequal, the outer oblong to elliptic, 3–3.5 mm long, 2–3 mm wide, smooth or slightly muricate, with a subterminal arista 2.5–6 mm long, obtuse and truncate apically, glabrous, the inner oblong, 4.5–6 mm wide, with a subterminal arista 2–5.5 mm long, obtuse to truncate, rarely emarginate, glabrous; corollas salverform, 2–2.5 cm long, the limb 1.7–1.9 cm wide, red or variegated with yellow, glabrous; stamens exserted, 2.7–3 cm long; ovary ovoid, 4-locular, glabrous; style exserted beyond the stamens by 0.5–1 cm. Fruits capsular, subglobose, 6–7 mm in diam, with an apiculum 3–4 mm long, the pedicels reflexed; seeds often 4, 3.5 mm long, black to dark brown, finely tomentose. (2n = 28) Jul–Oct. Thickets & roadsides; KS, MO, OK; (NJ to MO, SC to AR). *I. hederifolia* auctt., non L., *Quamoclit coccinea* (L.) Moench.—Rydberg.

    Similar to, and often confused with, the tropical *I. hederifolia* L., the temperate *I. coccinea* has longer inner sepals, although the caudate aristae may be of equal length in both taxa.

2. **Ipomoea cristulata** Hallier. f. Annual herbs, the stems twining, simple or few-branched, smooth or denticulate on the angles, glabrous or pilose at the nodes. Petioles 1–20 cm long, glabrous or sparsely pilose above. Leaves ovate, typically the lower are entire and the upper 3- to 5-lobed palmately or parted, or all palmately parted or lobed, the margins irregularly dentate, 1.5–10 cm long, 1–7 cm wide, basally cordate to subtruncate, the lobes rounded to acute, apically acute to acuminate, rarely obtuse, mucronate, glabrous or pilose below. Flowers 3–7 in cymes or rarely solitary; bracts linear-lanceolate to ovate, aristate, 1–3.5 mm long, bracteoles ovate to lanceolate, 1–2 mm long; pedicels angular, smooth or muricate on the angles, glabrous, 5–14 mm long, reflexed or erect in fruit. Sepals unequal, the outer oblong, 3–3.5 mm long, 2–2.5 mm wide, obtuse and rounded to subtruncate apically, muricate or smooth, with a subterminal arista 3–5 mm long, glabrous, the inner oblong, 4–5.5 mm long, 3–3.5 mm wide, apically truncate, with a subterminal arista 2.5–3.5 mm long; corollas salverform, 1.8–3.5 cm long, red to red-orange, glabrous, the limb 1.8–2 cm wide; stamens 2.5–3.5 cm long, exserted, anthers 1.5 mm long; ovary ovoid, 4-locular, glabrous; style 2–3 cm long. Fruits capsular, subglobose, 7–8 mm in diam, with an apiculum 2 mm long; seeds 1–4, 3.5–5 mm long, black to dark brown, finely tomentose. Aug–Oct. Thickets & roadsides; IA, KS, TX, NM; (GP, AZ s into Mex.). *I. coccinea* auctt., non L., *I. coccinea* var. *hederifolia* auctt., non (L.) A. Gray.

3. **Ipomea hederacea** Jacq., ivyleaf morning-glory. Annual herbs, the stems twining, with large, setose trichomes, densely to sparsely pubescent throughout. Leaves ovate to suborbicular, 5–12 cm wide and long, entire to 3-lobed, basally cordate, the lobes apically acute to acuminate, pubescent; petioles 5–12 cm long, rarely longer. Flowers 1–3(6) in cymes, on peduncles 5–10 cm long; bracteoles scalelike. Sepals subequal, long-lanceolate, abruptly narrowed from the ovate base, caudate with a narrow linear-lanceolate apex, 12–24 mm long, herbaceous, usually curved at least in fruit, sometimes strongly, densely long-hirsute at least on the basal 1/3; corollas light blue, with the inside of the tube white or pale yellow, 2.5–4.5 cm long, the limb 3–5 cm broad; stamens about 2/3 as long as corolla, white; ovary pubescent; nectary white. Fruits capsular, subglobose, somewhat depressed, 8–12 mm in diam, enclosed within the sepals; seeds often 4, pyriform, black to dark brown, densely pubescent with short trichomes. (2n = 30) Jun–Oct. Cult. ground, roadsides; GP; (ME to SD, FL to Mex.) *I. hederacea* var. *integriuscula* A. Gray—Fernald.

This plant has been spread through the world from its probable homeland in eastern U.S., largely as contaminant in seeds of plants cultivated for food.

4. **Ipomoea lacunosa** L., white morning-glory. Annual herbs, the stems twining to suberect, 1–2(3) m long, glabrous to sparsely pubescent. Leaves broadly ovate, entire, cordate, dentate to deeply 3- to 5-lobed, 3–8 cm long, 2–7 cm wide, basally cordate, the basal lobes rounded to lobed, apically acute to acuminate or obtuse, both surfaces glabrous to sparsely pubescent; petioles 3–8 cm long. Flowers 1–3 in axils, on peduncles mostly shorter than or equaling the petioles, glabrous; bracteoles scalelike. Sepals subequal, (8)11–14 mm long, ovate-lanceolate to lanceolate, acuminate to long-acuminate, mucronate, ciliate, often falcate; corollas funnelform, 1.5–2(2.5) cm long, glabrous, white or with a tinge of pink on the limb, the limb with short obtuse, mucronulate lobes; stamens with purple anthers and white filaments; ovary pubescent; nectary white. Fruits capsular, subglobose, 10–15 mm in diam, bristly pilose with large hirsute trichomes, 2-celled, 4-valved; seeds 4 or rarely fewer, 5–6 mm long, dark brown, ellipsoid, glabrous. (2n = 30) (Jun) Aug–Oct. Fields & low ground; KS, MO, s to TX; (PA to KS, SC to TX).

5. **Ipomoea leptophylla** Torr., bush morning-glory. Perennial herbs from an enlarged root, the stems decumbent to erect, 0.3–1.2 m tall or sometimes slightly taller, glabrous. Leaves linear-lanceolate to linear, 3–15 cm long, 2–8 mm wide, acute, entire, basally acute to subattenuate, glabrous; petioles 1–7 mm long. Flowers 1–3 or rarely more in axillary cymes, on peduncles 7–10 cm long; pedicels 5–10 mm long. Sepals unequal, the inner longer and wider than outer, 5–10 mm long, ovate to elliptic to orbicular-ovate, obtuse; corollas funnelform, 5–9 cm long, purple-red to lavender-pink with a darker throat; stamens unequal, 2–3 cm long, anthers 5–7 mm long, the stamens and styles included; ovary ovoid, glabrous. Fruits capsular, ovoid, 1–1.5 cm long, glabrous; seeds 1–4, oblong-elliptic, covered with dense, short, brown indumentum. (2n = 30) May–Sep. Plains & prairies; SD to CO, OK to NM; (essentially restricted to GP).

6. **Ipomoea pandurata** (L.) G.F. Mey., bigroot morning-glory. Perennial herbs from an enlarged root, the stems trailing, or more commonly twining, glabrate to sparsely pubescent. Leaves cordate-ovate, often pandurate, 3–10 cm long, 2–9 cm wide, glabrous to pubescent beneath; petioles 1–8 cm long. Flowers 1–several, on peduncles 10–20 cm long; pedicels glabrous, 1–2 cm long, somewhat thickened. Sepals coriaceous, oblong-elliptic, 12–15 mm long, glabrous, strongly imbricated, the outer sometimes shorter; corollas funnelform, 5–8 cm long, about as wide, the limb white, the tube lavender to purple-red within; stamens unequal, 2–3 cm long, the anthers 5–7 mm long, both stamens and styles included; ovary ovoid, glabrous. Fruits capsular, ovoid, 1–1.5 cm long, glabrous; seeds 1–4, with long woolly

trichomes on the angles, brown, oblong. (2n = 30) Jun–Sep. Prairies, fields, & cult. ground; MO to KS, NE, TX; (CT to Ont., FL to TX).

7. Ipomoea purpurea (L.) Roth, common morning-glory. Annual herbs, the stems twining, branched or simple, loosely pubescent to tomentose with short, appressed trichomes, retrorse and often large, also with antrorse, oblique or erect trichomes, reaching 4 mm long. Leaves ovate, subtrilobate, trilobate, or rarely 5-lobed, also unlobed, 1–11 cm long, 1–12 cm wide, basally cordate, apically acute to acuminate, rarely obtuse, mucronate, with indumentum similar to other parts; petioles 1–14 cm long, with indumentum similar to branches. Flowers 2–5 in cymes, rarely solitary, on peduncles 0.2–15 cm long, pubescent as other parts; bracts linear to lanceolate, 1.3–9 mm long, pubescent; bracteoles similar to bracts, 4.5 mm long; pedicels 5–16 mm long, erect in flower, but reflexed and enlarged in fruit, reaching 25 mm long, pubescence as elsewhere. Sepals subequal, the outer ovate-lanceolate, narrowly ovate-lanceolate to elliptic, 8–17 mm long, (1.5)2.5–4.5 mm wide, acute to abruptly acuminate apically, pubescence as elsewhere, but more densely concentrated near base, the inner ovate-lanceolate, 8–15 mm long, 2.5–3 mm wide, acute to abruptly acuminate; corolla funnelform, blue, white or purple, white within tube, 2.5–5 cm long, globrous; stamens unequal, the longer 18–25 cm, the shorter 13–22 cm, the anthers 1.5–2 mm long; ovary ovoid, attenuate, glabrous, 3-locular, 6-ovulate, the style 14–22 mm long, stigmas 3, globose. Fruits capsular, 1 cm in diam, 6-valvate; seeds often 6, black to dark brown, 5 mm long, finely tomentose. (2n = 30) Jun–Oct. Disturbed sites & escaped from cultivation, probably naturalized from Mexico; through e 1/2 GP; (widespread in e U.S.; pantrop.).

8. Ipomoea quamoclit L., cypress vine. Annual herbs, the stems twining, glabrous. Leaves ovate to elliptic in outline, 1–9 cm long, 0.8–7 cm wide, deeply pinnatisected, with 9–19 pairs of alternate or opposite segments, these linear, 0.2–1.5 mm wide, acute, mucronulate; petioles 0.2–4.5 cm long. Flowers solitary or 2–5 in cymes, on peduncles 1.5–14 cm long; bracts narrowly elliptic, elliptic to ovate, 0.6–1.5 mm long, mucronate; bracteoles similar to bracts, 0.6–1 mm long; pedicels 8–25 mm long, apically enlarged. Sepals subequal, the outer elliptic to oblong, 3-nerved, 4–6 mm long, 2–3 mm wide, obtuse, mucronate, the subterminal arista 0.25–0.3 mm long, the inner elliptic to oblong, 5–7 mm long, 3–3.8 mm wide, obtuse, the subterminal arista 0.3–0.6 mm long; corolla funnelform, 2–3 cm long, red or white (in cultivated strains), the limb 1.8–2 cm wide; stamens exserted, 2.5–3 cm long, the filaments with inconspicuous glandular trichomes at the base, the anthers 1.5–1.8 mm long; ovary ovoid, glabrous, 4-locular, the style exserted, 2.3–3 cm long, stigmas 2, globose. Fruits capsular, ovoid, 7–9 mm in diam, glabrous, with an apiculum 4.5–6 mm long; seeds black, 4.5–5.5 mm long, with short, scattered tomentum. (2n = 30) Jun–Oct. Disturbed areas; KS, TX; (VA to KS, FL to TX, Mex; cult. throughout tropics). Cult. and *escaped* but perhaps only *adventive*. *Quamoclit vulgaris* Choisy—Rydberg.

9. Ipomoea shumardiana (Torr.) Shinners. Perennial herbs, the stems trailing or twining, glabrous. Leaves deltoid-ovate to narrowly ovate-lanceolate, 3–8 cm long, 1–4 cm wide, truncate to cordate basally, acuminate to acute apically; petioles 5–36 mm long, up to 80 mm on flowering branches. Flowers 1–several, on peduncles 10–20 cm long; glabrous pedicels 1–2 cm long. Sepals coriaceous, oblong-elliptic to oblong-orbicular, 10–15 cm long, glabrous, the outer smaller than inner; corollas funnelform, 5–8 cm long and wide, the limb pink to white with a purple-red throat; stamens unequal, 2–3 cm long, both stamens and styles included; ovary ovoid, glabrous. Fruits capsular, ovoid, 1 cm long or slightly longer; seeds not seen. Jun–Aug. Plains & prairies; KS, OK, TX; *(endemic* GP). *I. carletoni* Hols., *I. longifolia* sensu authors, non Benth., *I. pandurata* sensu authors.

These rare plants occupy a restricted range in the southern GP, where the ranges of *I. pandurata* and *I. leptophylla* overlap. In many ways they resemble *I. longifolia*, but that species is restricted

6. STYLISMA Raf.

1. **Stylisma pickeringii** Torr. var. *pattersonii* (Fern. & Schub.) Myint. Herbs with stems prostrate or trailing, 1–2 m long or more, sparsely to densely villous with appressd hairs. Leaves linear or narrowly linear, 2.5–7 cm long, 1–3 mm wide, margins entire, basally acute to attenuate, apically acute to obtuse, pubescent; petioles absent. Flowers 1–5 in axillary cymes; peduncles 3–7 cm long, pedicels absent or present in central flowers, 4–20 mm long on lateral flowers; bracteoles foliose, 1.5–2.5 cm long. Sepals ovate to ovate-lanceolate, 4–6 mm long, 3–5 mm wide, acute to acutish apically, hoary pubescent; corolla campanulate to funnelform-campanulate, white, hirsute on interplicae, 1.2–1.8 cm long, limb subentire or shallowly lobed; stamens partially exserted; filaments glabrous or rarely with scattered trichomes near base; anthers oblong, basally sagittate; pollen pantocolpate; ovary ovoid, 2-carpellate, 2-locular, densely villous; styles 2, fused nearly to base of stigmas, branches 1–1.5 mm long, unequal; stigmas capitate; stylopodia 1–2 mm long, densely villous. Fruits capsular, ovoid; seeds 1 or 2, brown, smooth. ($2n = 28$) May–Aug. Dry prairies, open sandy woods; KS, OK, TX; (e IA, IL, KS s to TX). *Breweria pickeringii* (Torr.) A. Gray var. *pattersonii* Fern. & Schub.—Fernald.

References: Fernald, M. L., & B. G. Schubert. 1949. Some identities in *Breweria*. Rhodora 51: 35–42; Shinners, L. H. 1962. Synopsis of United States *Bonamia* including *Breweria* and *Stylisma* (Convolvulaceae). Castanea 27: 65–77; Myint, T. 1966. Revision of the genus *Stylisma* (Convolvulaceae). Brittonia 18: 97–116; Lewis, W. H. 1971. Pollen differences between *Stylisma* and *Bonamia* (Convolvulaceae). Brittonia 23: 331–334.

113. CUSCUTACEAE Dum., the Dodder Family

by Daniel F. Austin

1. CUSCUTA L., Dodder, Love Vine

Herbs with twining stems, mostly achlorophyllous and parasitic, but some with small amounts of chlorophyll and autotrophic, attached to hosts by haustoria. Leaves alternate, reduced to scales. Flowers usually in cymose clusters, although these may be branched or compacted so that the cymose nature is obscured, gamosepalous, mostly pentamerous, typically 1.5–5 mm long; stamens epipetalous, with scalelike, often fimbriate appendages attached to filament bases or below filament; ovary 2-celled, each locule 2-ovulate, the styles mostly distinct, rarely united and single, the stigmas capitate to linear-elongate. Fruits capsular, dehiscence circumscissile or irregular; seeds typically 1–3 mm long, brown, glabrous, the embryo acotyledonous, filiform or with one end enlarged.

References: Yuncker, T. G. 1932. The genus *Cuscuta*. Mem. Torrey Bot. Club 18: 111–331; Austin, D. F. 1979. Comments on *Cuscuta*-for collectors and curators. Bull. Torrey Bot. Club 106: 227–228; Austin, D. F. 1980. Studies of the Florida Convolvulaceae—III. Florida Sci. 43(4): 294–302.

The disposition of *Cuscuta planifolia* Ten., as recorded by Rydberg, is uncertain (cf. Yuncker, op. cit.).

1 Flowers subtended by enlarged bracts which resemble the sepals.
   2 Flowers pedicellate, in loose paniculate clusters .... 5. *C. cuspidata*

        2   Flowers sessile, in dense clusters.
            3   Bracts with recurved tips; stems disappearing early in the season; flowers in ropelike clusters .... 8. *C. glomerata*
            3   Bracts with erect tips; stems persisting; flowers clustered, but not ropelike.
                4   Bracts acute .... 13. *C. squamata*
                4   Bracts obtuse .... 2. *C. compacta*
1   Flowers with scalelike bracts not resembling the sepals.
    5   Flowers (3)4(5)-parted.
        6   Capsules with the withered corolla at the top or falling early.
            7   Sepals reaching the sinuses of corolla .... 3. *C. coryli*
            7   Sepals shorter than the corolla tube, not reaching the sinuses .... 1. *C. cephalanthi*
        6   Capsules with the withered corolla surrounding the base .... 12. *C. polygonorum*
    5   Flowers 5-parted.
        8   Sepals triangular, acute.
            9   Corolla tube not enclosed within the short calyx.
                10 Flowers 2–2.5 mm long .... 10. *C. indecora*
                10 Flowers 3–4 mm long .... 14. *C. suaveolens*
            9   Corolla tube about as long as the calyx and more or less enclosed.
                11 Corolla lobes spreading .... 6. *C. epithymum*
                11 Corolla lobes reflexed .... 15. *C. umbellata*
        8   Sepals ovate, obtuse.
            12 Corolla lobes inflexed at least at tips.
                13 Calyx angled; corolla 1.5 mm long .... 11. *C. pentagona*
                13 Calyx rounded and not angled; corolla 2.5–3 mm long .... 7. *C. glabrior*
            12 Corolla lobes spreading to erect.
                14 Corolla lobes erect; seeds 2–2.8 mm .... 4. *C. curta*
                14 Corolla lobes spreading; seeds 1–1.5 mm .... 9. *C. gronovii*

1. **Cuscuta cephalanthi** Engelm., buttonbush dodder. Stems 0.4–0.6 mm in diam. Flowers subsessile or sessile, 2 mm long, in densely paniculate cymes; calyx about 1 mm long, deeply divided, the overlapping lobes oblong-ovate, obtuse, often with somewhat irregular margins; corollas usually 4-parted, less commonly 3- to 5-parted, cylindric-campanulate, somewhat urceolate on the maturing capsules, the lobes ovate, obtuse, erect to spreading, shorter than the tubes; scales fringed with processes, oblong, narrow, reaching the filaments, bridged at 1/4–1/3 of their height; stamens mostly equal to the lobes or slightly shorter, the anthers ovoid to rounded, about equal to the filaments; style equal to or slightly longer than the globose to ovoid ovary. Fruits capsular, globose, about the same thickness throughout, often glandular, capped by the persistent corolla; seeds 1.6 mm long, globose, ovoid to rounded, hilum oblong, linear, oblique. Jul–Oct. Growing on a variety of woody & herbaceous hosts, but often on *Cephalanthus occidentalis*; GP; (ME to WA, VA to CA).

2. **Cuscuta compacta** Juss. Stems 0.6 mm or more in diam. Flowers sessile, in compact clusters or scattered glomerules; calyx of distinct sepals, cupped, orbicular to oval, shorter than the corolla tube, surrounded by 3–5 similar, appressed bracts; corollas 4–5 mm long, cylindrical, excluded from the calyx, becoming urceolate in fruit, the lobes spreading and reflexed, oblong, obtuse; scales with long processes, these about reaching the stamens; stamens shorter than the corolla lobes, exserted, the anthers oval; styles about as long or shorter than the globose ovary, reaching to the corolla lobe sinuses, stigmas capitate. Fruits capsular, globose-conic, slightly pointed, capped by the withered corolla; seeds 2.6 mm long, globose, ovate to angled or flattened, scurfy, hilum oblong, oblique. Aug–Oct. Found on a wide range of woody & herbaceous hosts; se GP; (NH to NE, FL to TX).

3. **Cuscuta coryli** Engelm., hazel dodder. Stems 0.4–0.6 or less in diam. Flowers on pedicels shorter or longer than the flower, in paniculate cymes, usually 4-parted, rarely 5-parted;

calyx equaling or longer than the corolla tube, the lobes triangular, acute; corollas 2 mm long, cylindric-campanulate, the lobes triangular-ovate, crenulate, upright with acute, inflexed tips; scales rudimentary, bifid, toothed, usually forming toothed wings on both sides of the filament attachment, bridged below the middle; stamens about as long as corolla lobes, the anthers ovoid to oblong, on subulate filaments; styles shorter than or equal to the globose-ovoid ovary, the ovary thickened apically. Fruits capsular, globose but becoming depressed-globose, the intrastylar aperture thickened, the withered corolla present at first but falling early; seeds 1.5 mm long, usually 4, globular, to oblique-compressed, scurfy, hilum short, oblong, transverse or oblique. Aug–Sep. Growing on a wide variety of herbaceous & woody hosts, but often on *Corylus*; ND to KS; (N.S. to MT, FL to AZ).

4. *Cuscuta curta* (Engelm.) Rydb. Stems 0.6 mm in diam or slightly more. Flowers on pedicels as long as or usually shorter than the flowers, in cymose paniculate clusters, these becoming globose by the growth and crowding of the capsules; calyx reaching to about the middle of the corolla tube, the lobes ovate, obtuse, overlapping, the margins rarely serrulate; corollas 2–3 mm long, campanulate, the lobes triangular, obtuse, spreading and reflexed in fruit; scale shorter than the tube, variable but often fringed and truncated apically, bridged at the middle; stamens shorter than or as long as the corolla lobes, anthers ovoid, shorter than to equal to longer than the filaments; ovary globose-conic, styles short and about 1/4 the length of the ovary. Fruits capsular, globose-conic, with a short beak, the withered corolla mostly surrounding the apex but at times the base also, the intrastylar aperture large, styles mostly convergent; seeds 2–2.8 mm long, slightly rosulate, hilum transverse to oblique. Jun–Aug. Hosts include *Convolvulus, Scutellaria, Salix,* & *Linum;* MN, ND, SD, NE; (MN to WY, CO, & NM).

5. *Cuscuta cuspidata* Engelm., cusp dodder. Stems 0.4–0.6 mm in diam. Flowers pedicellate or subsessile, in loose or dense, paniculate cymes, the whole inflorescence bracteate; calyx of distinct or slightly united segments, the lobes obtuse, acute and cuspidate, sometimes glandular and thickened medially, the margins thinner and serrulate; bracts subtending flowers, ovate, obtuse to orbicular, acute and sometimes cuspidate; corollas 4 mm long, funnelform, the lobes oblong to ovate, shorter than the tube, obtuse to acute, sometimes mucronulate to cuspidate, usually glandular medially; scales fringed with medium length processes, shorter than the tube or reaching the filament base, bridged near the middle; stamens shorter than the corolla lobes, the anthers ovoid, cordate, shorter than the filaments; styles slender and longer than the globose to conic ovary. Fruits capsular, globose, with a somewhat thickened ridge around the intrastylar aperture, frequently glandular, the withered corolla persistent at apex; seeds 1.4 mm long, obovate, hilum short, oblong to oval, oblique to almost transverse. Aug–Oct. Found on a wide variety of Asteraceae; SD to TX; (IN to CO, AR to UT).

6. *Cuscuta epithymum* Murray. Stems less than 0.4 mm in diam, sometimes red to purple. Flowers sessile and numerous in dense, compact cymose clusters; calyx as long as or shorter than the corolla tube, the lobes triangular, acute, purplish; corollas campanulate, 3 mm long, the lobes triangular, acute, spreading, shorter than the tube; scales spatulate, shorter than tube, fringed on the upper part, bridged at 1/3 their height; stamens shorter than corolla lobes, the filaments longer than the anthers; ovary globose, the apex slightly thickened, the styles with the stigmas about twice as long as the ovary, stigmas filiform, longer than styles; ovary globose. Fruits capsular, globose, capped by the withered corolla; seeds 1 mm long, rough, angled, usually 4, compressed ovoid, hilum short, oblong, transverse. Jul–Sep. Hosts a variety of Fabaceae; ND, SD, NE; (ME to SD, NJ to NE; Europe). *Naturalized* and spread around the world as a contaminant in the seed trade.

7. Cuscuta glabrior (Engelm.) Yunck. Stems 0.4–0.6 mm in diam. Flowers subsessile or on pedicels longer than flowers, in compact or loose, globular cymose clusters; calyx shorter than or about equaling the corolla tube, the lobes oval-ovate, not overlapping, the sinuses often obtuse, apically obtuse to rounded; corollas globular and saccate between the points of stamen attachment, reddish, the lobes triangular, acute or acuminate, about equaling the tube, spreading to reflexed but with the tips inflexed; stamens shortly exserted from sinuses of corolla, the filaments about the same length as the ovoid anthers; scales ovate, fringed, exserted; the styles equal to or exceeding the length of the globose ovary. Fruits capsular, depressed-globose, with a large intrastylar opening, the withered corolla almost enclosing the capsule; seeds 1 mm long, globose, hilum terminal, short, oblong. Jul–Sep. Found on a wide variety of herbaceous hosts; OK, TX, NM; (LA to NM; Mex.).

8. Cuscuta glomerata Choisy. Stems 0.6 mm or more in diam, disappearing from between the dense, straw-colored, ropelike inflorescence clusters. Flowers sessile, mostly forming endogenously, usually in 2 parallel rows on opposite sides of the stems, imbricated with numerous, scarious, lacerated, cupped, oblong, obtuse or acute bracts with mostly recurved tips; calyx of distinct, oblong-oval, obtuse to acute sepals, the tips spreading, but not ordinarily recurved, otherwise similar to the bracts; corollas 4–5 mm long, cylindrical, the lobes spreading or sometimes reflexed, oblong to lanceolate, obtuse to acute, sometimes mucronate, often with a row of glandular cells along the middle, shorter than the tube; scales shorter than the tube, oblong, fringing more abundant near apex, bridged at the middle or above; stamens shorter than the corolla lobes, the elliptical to oblong anthers equal to or shorter than the filaments; styles capillary, much longer than the flask-shaped ovary, stigmas capitate. Fruits capsular, globose-pointed, or flask-shaped, with the withered corolla persisting at the apex; seeds 0–2 per capsule, 1.7 mm long, oval, globose, hilum oblong, oval, transverse. Jul–Sep. Mostly growing on various Asteraceae; SD, NE, TX; (MI to SD, MS to TX). *C. paradoxa* Raf.—Rydberg.

9. Cuscuta gronovii Willd. Gronovius' dodder. Stems 0.4–0.6 mm or more in diam. Flowers 2–4 mm long, on pedicels shorter than or about equaling the flowers, in loose or densely panicled cymes; calyx lobes broad, ovate, orbicular or oblong, obtuse, overlapping, shorter than corolla tube; corollas narrowly campanulate, almost tubular, the lobes mostly shorter than the tube, ovate, obtuse, spreading; scales mostly equaling the tube, usually oblong, fringed; stamens about as long as the lobes, filaments longer than the oval anthers; styles mostly shorter than the globose-conic ovary. Fruits capsular, globose-conic to obpyriform, beaked, occasionally glandular, enveloped at base by withered corolla; seeds 1.5 mm long, 2–4 per fruit, ovate, rosulate, hilum linear, oblique or transverse. Aug–Oct. Found on a number of herbs & shrubs; GP; (N.S. to ND & Man., FL to TX; also W.I.).

10. Cuscuta indecora Choisy, large alfalfa dodder. Stems 0.4–0.6 mm in diam. Flowers 2–2.5 mm long, white, fleshy, papillate, on pedicels longer than the flowers; calyx lobes triangular-ovate, acute to somewhat obtuse, shorter than the corolla lobes; corollas campanulate, the lobes erect to spreading, triangular, acute, with inflexed tips; scales as long as or longer than the corolla tube, ovate to somewhat spatulate, deeply fringed; stamens shorter than the lobes, anthers broad, oval, about equal to the filaments; styles as long as or slightly longer than the globose, pointed ovary, becoming divaricate in fruit. Fruits capsular, circumscissile, thickened at top and enlarged by the withered corolla; seeds 2–4, about 1.7 mm long, rounded or broader than long, scurfy, hilum small, oval, transverse or somewhat oblique. Jul–Oct. On a wide range of woody & herbaceous hosts; GP; (IL to SD, FL to CA; Mex., C. & S. Amer.).

11. Cuscuta pentagona Engelm., field dodder. Stems 0.35–0.4 mm or rarely larger in diam. Flowers 1.5–2 mm long, inconspicuously glandular, on pedicels about equaling the flowers, in loose, cymose clusters; calyx mostly enclosing the corolla tube, loose, lobes mostly broader than long, broadly ovate, obtuse, broadly overlapping at sinuses, giving the calyx an angled appearance; corollas with narrow, lanceolate, spreading lobes, the acute tips inflexed, about equaling the campanulate tube; stamens shorter than the lobes, filaments slender, longer than or equaling the oval anthers; styles about equaling or shorter than the globose ovary, stigmas small, globose. Fruits capsular, globose or depressed-globose, protruding from the withered corolla; seeds 1 mm long, globose, hilum short, oblong, terminal, transverse. Jul–Oct. Hosts variable; often cult. Fabaceae; GP; (MA to MT, FL to CA; Mex. & W.I.).

12. *Cuscuta polygonorum* Engelm., smartweed dodder. Stems 0.4 mm in diam. or less. Flowers mostly 4-parted, subsessile in compact, dense, glomerulate clusters; calyx as long as or longer than the corolla tube, the lobes triangular, often unequal, obtuse; corollas 2–2.5 mm long, glabrous, campanulate, the lobes triangular-acute, upright; scales oblong, reaching the filaments or shorter, mostly bifid, the processes few, short and clustered near the apex, bridged at 1/4 of their height or less; stamens shorter or equaling the corolla lobes, rarely longer, anthers ovoid, shorter than the subulate filaments which are attached in or near the corolla lobe sinuses; styles shorter than the depressed-globose ovary, subulate and divergent. Fruits capsular, globose to obpyriform, often depressed, the intrastylar aperture large and rhombic; seeds 1.3 mm long, rounded, rostrate, and compressed, hilum oblong, linear, transverse to oblique. Jul–Oct. Hosts include *Polygonum* & several other herbs; IA to ND, MO to TX; (PA to MN, DE to AR).

13. *Cuscuta squamata* Engelm. Stems less than 0.4 mm in diam. Flowers sessile, few to several in separate or clustered glomerules, subtended by 2–10 ovate, acute, often cuspidate, serrulate, closely appressed bracts; calyx lobes distinct, ovate, acute, cuspidate, serrulate, equaling the corolla tube; corollas cylindric, the lobes ovate-lanceolate or oblong, acute, sometimes cuspidate, spreading or reflexed; scales reaching the filaments, oblong, the processes of medium length or short and scattered, bridged near the middle; stamens shorter than the corolla lobes, the filaments as long as or shorter than the oblong-ovoid anthers; ovary globose to conic, with two small apical lobes, the styles longer than the ovary, the stigmas capitate. Fruits capsular, globose, conical to ovoid, apically thickened, glandular near the apex, retaining the withered corolla at the top; seeds 1.5 mm long, 1 or 2 per capsule, globose or compressed, hilum short, linear, oblique. Jul–Sep. Hosts are mostly Asteraceae; TX, NM; (TX, NM; n Mex.).

14. *Cuscuta suaveolens* Ser. Stems 0.4–0.6 mm in diam. Flowers on pedicels mostly shorter than the flowers, sometimes longer, in racemose clusters, usually glandular, membranaceous; calyx lobes shorter than the corolla tube, triangular-ovate, acute, not overlapping, sinuses rounded, often revolute; corollas 3–4 mm long, cylindrical-campanulate, becoming globular and surrounding the capsule, lobes ovate-triangular, upright, the acute tips inflexed, about 1/2–3/4 as long as the tube; scales usually not reaching the stamens, oblong-ovate or triangular, fringed with medium processes, bridged below the middle; stamens shorter than the lobes, the subulate filaments about as long as the ovoid to oblong anthers; styles about equal to length of ovary, rarely longer, stigmas capitate. Fruits capsular, globose, not circumscissile, surrounded by corolla; seeds 2–4, rounded, 1.5–2 mm long, hilum oblong, perpendicular. Jul–Oct. Normally growing on Fabaceae, especially *Medicago sativa*; SD; (ME to SD, TX to CA; S. Amer.). *Adventive or naturalized* in many parts of the world. *C. racemosa* var. *chiliana* Engelm.—Rydberg.

15. *Cuscuta umbellata* H.B.K. Stems 0.3–0.4 mm in diam. Flowers glabrous to rarely slightly puberulent, 2–3 mm long, on pedicels longer or shorter than flowers, in dense, compound cymes, the ultimate umbellate division of 3-7 flowers; calyx turbinate, as long as or longer than the campanulate corolla, the lobes triangular-ovate, acute to acuminate; corolla lobes as long as or longer than the tube, reflexed, lanceolate, acute to acuminate; scales somewhat obovate or spatulate, moderately fringed, reaching the filaments or slightly exserted; stamens shorter than the lobes, anthers oblong or oval, shorter than or equaling the filaments; styles longer than the globose ovary. Fruits capsular, depressed-globose, with a ring of thickened knobs around the intrastylar aperture, tardily circumscissle, surrounded by the withered corolla; seeds 1 mm long, angled, oblique, oval, hilum oblong, linear, transverse. Aug-Oct. Typically on halophytic hosts; KS, OK, TX; (GP s to trop. Amer.).

114. MENYANTHACEAE Dum., the Buckbean Family

by G. E. Larson

1. MENYANTHES L. Buckbean

1. *Menyanthes trifoliata* L. Glabrous, rhizomatous perennial with flowering scapes 1.5-3.5 dm tall; rhizome thick, covered with old leaf bases. Leaves all basal, with the sheathing petiole bases arranged alternately on the rhizomes, the blades palmately 3-foliolate, the leaflets elliptic to oblong or oblanceolate to obovate, 3-10 cm long, 1-5 cm wide, entire or sometimes coarsely undulate-dentate. Inflorescence a scapose, bracteate raceme, surpassing the leaves; bracts mostly 3-5 mm long; pedicels 4-20 mm long. Flowers perfect, regular, 5(4-6)-merous, often dimorphic, some flowers with exserted stamens and included style, others with exserted style and included stamens; calyx deeply divided, the oblong-ovate lobes 1.5-3 mm long; corolla whitish or pinkish, salverform, lobed to near or below the middle, 8-12 mm long, the lobes eventually recurved, conspicuously fringed on the inner surface; stamens usually 5, epipetalous; stigma 2-lobed, style elongate, ovary ca 1/3 inferior, 1-celled. Fruit a globose, corky-walled capsule, 6-10 mm in diam, rupturing irregularly to release many shiny, yellowish-brown seeds. (2n = 54, 108) Jun-Aug. Bogs & swamps; MN: Becker, Ottertail; ND: McHenry, Ransom; SD: Brookings, Todd; (circumboreal, in N. Amer., s to DE, VA, OH, MO, SD, CO, & CA).

115. POLEMONIACEAE Juss., the Polemonium Family

by Dieter H. Wilken

Annual, biennial to perennial herbs or suffrutescent subshrubs. Leaves alternate or opposite, simple to commonly palmatifid, pinnatifid or pinnately dissected, the margins entire to serrulate. Flowers axillary or terminal, sessile to pedicellate, solitary, paired or clustered, the basic inflorescence a cyme, these arranged corymbosely or paniculately in thyrses. Calyx (4)5-merous, gamosepalous, cylindrical to ovoid, either herbaceous throughout, accrescent and chartaceous in fruit, or with herbaceous costae separated by hyaline to scarious membranes that are ruptured or distended by the developing capsule, the lobes deltoid to linear, entire to trifidly acerose, acute to spinose; corolla (4)5-merous, gamopetalous, salverform to rotate, companulate or funnelform, regular to slightly zygomorphic; stamens (4)5, exserted or included, equal or unequal in length, equally or unequally adnate to the corolla; pistil 1, the ovary superior, with (2)3 locules, the style exserted to included, with (2)3 stigmatic lobes. Capsule loculicidally dehiscent to rarely indehiscent or circumscissile, with 1–10 ovules per locule, the seeds in some species becoming mucilaginous when wet.

The author acknowledges the aid and criticism of Alva Day and Dale M. Smith.

1 Calyx tube herbaceous throughout, accrescent and chartaceous in fruit.
  2 Annual; leaves entire to remotely lobed; corolla salverform .... 1. *Collomia*
  2 Perennial; leaves pinnately dissected; corolla campanulate .... 9. *Polemonium*
1 Calyx tube with the herbaceous costae separated by hyaline or scarious membranes which are ruptured or distended by the developing capsule.
  3 Leaves entire; stamens very unequally inserted.
    4 Annual; seeds mucilaginous when wet; upper leaves and bracts alternate .... 6. *Microsteris*
    4 Perennial; seeds not mucilaginous when wet; all basal and cauline leaves opposite, alternate only in the inflorescence .... 8. *Phlox*
  3 Leaves palmatifid to pinnatifid; stamens equally to unequally inserted in some species.
    5 Leaves palmatifid, the cauline leaves mostly opposite.
      6 Annual; leaf segments herbaceous .... 5. *Linanthus*
      6 Perennial; leaf segments pungent to spinose .... 4. *Leptodactylon*
    5 Leaves pinnatifid, the cauline leaves alternate.
      7 Calyx lobes unequal; bracts of the inflorescence charataceous, spinose .... 7. *Navarretia*
      7 Calyx lobes equal: bracts of the inflorescence herbaceous, acute to cuspidate.
        8 Corolla rotate to subsalverform, if the latter then the tube length no more than 3× the width of the throat .... 2. *Gilia*
        8 Corolla salverform, the tube length 5× or greater than the throat width .... 3. *Ipomopsis*

1. COLLOMIA Nutt., Collomia

1. ***Collomia linearis*** Nutt. Annuals, 5–60 cm tall, with erect, simple to branched stems. Stems glandular-puberulent above, glabrous to puberulent below. Leaves simple, alternate, the upper lanceolate to ovate, the lower linear to lanceolate, 1–8 cm long, 1–12 mm wide, glabrous to viscid above, puberulent to glandular below, particularly on the midrib, apices acute, margins entire to remotely serrate or lobed, sessile to clasping. Inflorescence terminal and on upper axillary branches, the bracts ovate to linear, glandular-pubescent. Flowers 8–many, cymosely disposed in corymbiform or capitate clusters; calyx 5-merous, campanulate, accrescent, 4–7 mm long at anthesis, 6–9 mm long in fruit, glandular-pubescent, the tube herbaceous throughout, becoming membranaceous or chartaceous with age and not ruptured by the developing capsule, ca equal to the lobes in length, lobes subequal, narrowly triangular, acute in fruit; corolla 5-merous, white to pink, salverform, the tubes 8–15 mm long, lobes 1–4 mm long, sparsely glandular; stamens 5, unequal, the filaments inserted unequally in the tube, anthers included or 1 or 2 partly exserted; ovary ovoid, with a style slightly shorter than the corolla tube and terminated by 3 stigmatic lobes. Ovules and seeds 1 per locule, seeds oblong, becoming mucilaginous when wet. ($2n = 16$) Apr–Aug. Dry to mesic, sandy or gravelly soils of prairie, meadows, or disturbed sites; MT e to MN, s to CO, NE, & MO; (s Can., WA to Que., CA to NM, MO).

    Reference: Wilken, D. H. 1978. Vegetative and floral relationships among Western North American populations of *Collomia linearis* Nuttall (Polemoniaceae). Amer. J. Bot. 65: 896–901.

2. GILIA R. & P.

Erect to diffusely branched, taprooted annuals, biennials or suffrutescent perennials. Leaves basal to alternate, the cauline well developed or much reduced, the blades pinnately toothed to pinnatifid or pinnately dissected, the 3–many segments herbaceous or acerose, weakly cuspidate to pungent, sessile to subsessile. Flowers terminal or axillary in the upper branches, either solitary, unevenly paired or in small clusters at the ends of paniculate branches, flowers within the pairs or clusters ebracteate. Calyx 5-merous, the lobes equal to subequal, acute to weakly cuspidate, the herbaceous costae separated by hyaline or scarious membranes which are ruptured by the developing capsule; corolla subsalverform to rotate, the 5 lobes obtuse, white to blue or violet; stamens exserted in ours, equal to subequal in length, regular to declinate, the filaments inserted equally in the throat; style exserted in ours, terminated by 3 stigmatic lobes. Capsule broadly ovoid to oblong, ours with 2–6 seeds per locule, the seeds mucilaginous when wet.

1 Corollas subsalverform to campanulate; stamens declinate; leaves abruptly reduced above the basal rosette .... 1. *G. pinnatifida*
1 Corollas rotate; stamens regular; the cauline leaves gradually reduced upwards .... 2. *G. rigidula*

1. *Gilia pinnatifida* Nutt. ex A. Gray. Biennial to monocarpic, short-lived perennial 10–60 cm tall, stems glandular-pubescent, simple and erect but often diffusely branching in flower. Basal leaves pinnatifid, 3–6 cm long, the rachis 1–2(3) mm wide, the 8–15 segments linear to narrowly oblong, 3–8 mm long, glaucous to glandular-puberulent, cuspidate; stem leaves entire to unequally bifid or trifid, 8–12 mm long, glandular-puberulent. Flowers terminating the paniculate branches, the bracts linear, cuspidate. Calyx cylindrical to ovoid, tubes 2–3 mm long at anthesis, the herbaceous costae glandular, the lobes 1/3 or less the length of the calyx; corolla white to blue or lavender, subsalverform to campanulate, the tubes exserted, 3–6 mm long, lobes 2–4 mm long, orbicular to oval; stamens exserted, declinate and sternotribal; ovary ovoid to oblong. Mature capsule 3–5 mm long. (2n = 16) May–Jul. Dry sandy or gravelly open sites of prairie, stream beds, & outcrops; WY s to NM, extreme w NE & KS; (UT e to NE, KS, CO, & NM).

This species has been associated with the name *G. calcarea* M. E. Jones — Rydberg. *G. calcarea* is a distinct taxon, primarily distributed in the intermountain region west of the GP, and closely related to *G. mcvickerae* M. E. Jones (A. Day, pers. comm.).

2. *Gilia rigidula* Benth. Suffrutescent perennial, sometimes blooming the first year, 8–25 cm tall, diffusely branched from the base, stems glandular-puberulent. Leaves pinnatifid to pinnate, 6–20(25) mm long, unequally divided, the 2–7 segments acerose to linear-oblong, 2–12 mm long, cuspidate, glandular-puberulent. Flowers solitary to loosely glomerate, the bracts linear to trifid. Calyx cylindrical to ovoid, tubes 2–4(5) mm long at anthesis, the herbaceous costae glandular, the lobes 1/2–2/3 the length of the calyx; corolla blue-violet to purple with yellow throat, rotate, the tubes included in the calyx, 3–4 mm long, lobes 3–5 mm long, orbicular to oval; stamens exserted, regular; ovary globose to ovoid. Mature capsule 3–5 mm long. (2n = 36) Apr–Aug. Dry sandy to rocky prairie; se CO to w KS, NM to OK & TX; (CO, KS s to OK, TX, NM, AZ, Mex.). *Giliastrum acerosum* (A. Gray) Rydb., *Gilia acerosa* A. Gray, *G. rigidula* Benth. subsp. *rigidula*, subsp. *acerosa* (A. Gray) Wherry — Rydberg.

The basionym *G. acerosa* was applied to compact plants less than 10 cm tall and with leaves less than 2 cm long. Infraspecific variation in the *Gilia rigidula* complex is associated with geographical intergradation and polyploidy (A. Day, pers. comm.).

3. IPOMOPSIS Michx.

Erect to diffusely branched, taprooted annuals, biennials and monocarpic, short-lived perennials. Basal leaves in a compact rosette, gradually reduced and alternate on the stem, pinnately to palmately divided, the ultimate segments herbaceous, linear to narrowly oblong, cuspidate to setose. Flowers terminal or axillary, solitary to cymosely arranged in compact to lax, capitate, corymbiform or thyrsoid inflorescences, bracteate, the bracts often pinnatifid. Calyx 5-merous, lobes equal to subequal in length, acute to acuminate, cuspidate to setose, the herbaceous costae separated by hyaline or scarious membranes which are ruptured or distended by the developing capsule; corolla regular to slightly zygomorphic, salverform, the 5 lobes obtuse, shorter than the tube, white to red; stamens exserted to included, equal to unequal in length, the filaments inserted equally and adnate to the tube or in the sinuses of the lobes; style exserted to included, terminated by 3 stigmatic lobes. Ovules 2–many per locule; capsule ovoid to oblong, the mature seeds weakly mucilaginous when wet.

Ipomopsis Michx. represents a distinct phyletic line from *Gilia* sens. str., where it usually has been treated. *Ipomopsis* was circumscribed to include taxa primarily and collectively distinguished by well-developed or gradually reduced cauline leaves, leaf segments with prominently cuspidate to mucronate tips, bracts subtending each flower and salverform corollas (Grant, Aliso 3: 351–362, 1956). The distinction is supported by the chromosome number of $X = 7$, Grant, op. cit., and by flavonoid chemistry (Smith, et al., Biochem. Syst. and Ecol. 5: 107–115, 1977). *Gilia*, in contrast, is delimited by a combination of abruptly reduced cauline leaves, leaf segments acute to cuspidate, single bracts subtending small groups of 1–3 ebracteate flowers, rotate to funnelform corollas, and a basic chromosome number of $X = 8, 9$. Because of apparent morphological convergence among some members of each group, the taxonomic distinction has not been adopted widely. In our area, however, the two genera are separated with relative ease.

1 Flowers pedicellate in open corymbiform panicles or thyrses; corolla tube 10 mm or more long, the lobes 4 mm or more long.
  2 Corolla scarlet-red; all 5 stamens exserted .... 5. *I. rubra*
  2 Corolla white to light bluish-violet; stamens included or at most only 2 exserted.
    3 Corolla tubes usually 10–20 mm long, the lobes 4–6 mm long .... 2. *I. laxiflora*
    3 Corolla tubes usually 27–41 mm long, the lobes 8–11 mm long .... 3. *I. longiflora*
1 Flowers sessile to subsessile in compact, capitate to corymbiform clusters or spicate thyrses; corolla tube usually 10 mm or less long, the lobes 4 mm or less long.
  4 Plants annual; the corolla tube exserted from the calyx tube for over 1/2 its length .... 4. *I. pumila*
  4 Plants biennial to perennial, the corolla tube mostly included in the calyx tube.
    5 Filaments shorter than the anthers; the style included, about 1/2 the length of the tube. .... 6. *I. spicata*
    5 Filaments longer than the anthers; the style equaling the corolla tube or exserted .... 1. *I. congesta*

1. *Ipomopsis congesta* (Hook.) V. Grant, ball-head. Monocarpic, short-lived perennial, 7–20(25) cm tall, stems densely arachnoid-pubescent, erect and usually branched from the base. Basal and stem leaves entire to mostly pinnatifid, 2–6 cm long, distally glaucous, proximally arachnoid-pubescent, the 3–7 segments linear to narrowly oblong, 1–6(9) mm long, cuspidate. Flowers in dense capitate to corymbiform clusters terminating the stems, the outer bracts pinnatifid to nearly palmatifid with 3–5 segments, the inner bracts mostly entire. Calyx cylindrical to ovoid, the tubes 1–2 mm long, the costae densely arachnoid-pubescent, the lobes 1/2 or less the length of the calyx; corolla white, the tubes 2–4 mm long, the lobes oval to orbicular, 1–2 mm long; the stamens slightly exserted, the filaments inserted in the sinuses of the lobes, longer than the rounded anthers; style exserted. Capsule with 1 or 2 seeds per locule. ($2n = 14$) Apr–Jul. On dry sandy to rocky sites of prairie

& slopes; MT to w ND, s to n CO & w NE; (OR & CA e to MT & NE). *Gilia congesta* Hook., *G. iberidifolia* Benth.—Rydberg.

In the GP there exist 2 well-defined subspecies—Day, Madroño 27: 111–112, 1980.

1a. subsp. *congesta*. Basal leaves pinnatifid and mostly senescent at anthesis, 2–3 cm long. Compact inflorescences corymbosely disposed by means of 2–4 short peduncles at the summit of each stem. (2n = 14) Apr–Jul. Dry, sandy to rocky sites of prairie & slopes; range of the species.

1b. subsp. *pseudotypica* (Const. & Roll.) A. Day. Basal leaves densely crowded and persistent, mostly entire and linear, 3–6 cm long. Compact inflorescences solitary and terminating the stems. May–Jun. On sandy or rocky soils of prairie & hillsides; BH, ne WY, & e SD.

This subspecies may be superficially confused with *I. spicata* (Nutt.) V. Grant, and it occurs in the area of sympatry between the latter and *I. congesta* subsp. *congesta*. However, there is no clear evidence of hybridization or primary intergradation among the three taxa.

2. ***Ipomopsis laxiflora*** (Coult.) V. Grant. Annual to biennial, 10–40 cm tall, simple and erect to commonly diffusely branched throughout, the stems glabrous to glandular-puberulent. Basal and stem leaves pinnately divided, 1–3(4) cm long, the 3–5 segments filiform to narrowly linear, 5–15 mm long, sparsely arachnoid-pubescent, cuspidate. Flowers pedicellate, the pedicels 1–2 cm long at anthesis, solitary, paired or in loosely aggregated cymes, the upper 1/2 of the plant appearing corymbiform at anthesis, bracts linear, cuspidate. Calyx ovoid, the tubes (2)3–4 mm long at anthesis, the herbaceous costae sparsely glandular, the lobes 2–3 mm long, cuspidate to setose; corolla white to lightly bluish-violet, the tubes (8)10–20(25) mm long, lobes 4–6(7) mm long; stamens unequal, mostly included but sometimes with 1 or 2 exserted, the filaments unequally inserted in the corolla tube; style with stigmatic lobes in the throat or slightly exserted, ovary ovoid. Mature capsule 7–10 mm long with 4–6 seeds per locule. May–Jul. Dry sandy soil of stream beds & sandhills; CO to KS s to NM, TX, & e OK; (CO & NM e to KS & TX). *Gilia laxiflora* (Coult.) Osterh.—Rydberg.

3. ***Ipomopsis longiflora*** (Torr.) V. Grant. Annual to biennial, (20)30–60 cm tall, diffusely branched from near the base, the stems glabrescent to sparsely glandular. Basal and stem leaves as in *I. laxiflora*, but (1)2–5(6) cm long, the 3–7 segments 10–30 mm long, glabrous. Inflorescence and flowers as in *I. laxiflora*, but with calyx tubes 3–5 mm long, the lobes 1–2 mm long; corolla tubes (25)27–41 mm long, the lobes (6)8–11 mm long. Mature capsule (8)9–13 mm long with 6–8 seeds per locule. (2n = 14) May–Aug. Dry sandy soil of stream beds & sandhills; CO to NE s to NM, TX, & OK; (UT & AZ e to NE, OK, & TX). *Gilia longiflora* (Torr.) G. Don—Rydberg.

4. ***Ipomopsis pumila*** (Nutt.) V. Grant. Diffusely branched annual, 6–20 cm tall, the stems arachnoid-pubescent. Basal and stem leaves pinnatified to pinnately divided, 8–30 mm long, glabrous to sparsely arachnoid-pubescent, the 2–5 segments linear, cuspidate. Flowers axillary or in compact clusters terminating the diffuse, corymbosely disposed branches, the bracts pinnatifid to entire, cuspidate to setose. Calyx ovoid, obscured by the bracts, tubes 3–4 mm long at anthesis, the costae arachnoid-pilose, the lobes setose, 1–2(3) mm long; corolla white to commonly light bluish-violet, the tubes 7–9 mm long, the lobes lanceolate, 2–3 mm long, ca 1 mm wide; stamens exserted, equal to subequal, adnate in the sinuses of the lobes, the inwardly curving filaments longer than the anthers; styles slightly exserted, ovary ovoid to oblong. Capsule with 4–6 seeds per locule. (2n = 14) Apr–Jun. Uncommon in our area, on dry sandy soils of plains; CO: Otero & Pueblo; s to NM & w TX, dubiously reported from w NE & s WY—Correll & Johnston; (UT & w WY to AZ, NM, CO, & TX). *Gilia pumila* Nutt.—Rydberg.

This annual species is often confused with young flowering specimens of *I. congesta*, although the two taxa are easily separated by floral characters. *I. pumila* is more common in the intermountain region west of the continental divide but occurs in Wyoming as far east as the Bighorn Basin.

5. Ipomopsis rubra (L.) Wherry, standing cypress. Erect biennial with a prominent rosette, the single stem 1-2 dm tall, sparsely pubescent. Leaves pinnately divided, 4-8 cm long, sparsely pubescent along the midribs, the 10-15 segments linear to filiform, 5-20 mm long, cuspidate. Flowers in small clusters, these horizontally presented in an elongate thyrse. Calyx tubes 3-4 mm long, the costae glabrescent, viscid, the attenuate lobes 4-6 mm long; corolla scarlet-red, the lobes often flecked with pink or white, the tubes 20-25 mm long, the lobes ovate to elliptical, 9-11 mm long, 4-6 mm wide; stamens unequally exserted, the lower 3 stamens about 1-3 mm longer than the upper 2, the filaments included in the throat; style exserted, the ovary ovoid. Mature capsule oblong, 8-10 mm long, with 10-12 seeds per locule. (2n = 14) Jun-Sep. Open sites & gravelly soils of arroyos; TX to sw OK; (TX e to NC & FL). *Gilia rubra* (L.) Heller, *G. coronopifolia* Pers.—Rydberg.

6. Ipomopsis spicata (Nutt.) V. Grant. Monocarpic, short-lived perennial, 5-35 cm tall, the 1-3 stems arachnoid-pubescent to tomentose, erect. Basal and stem leaves entire to unequally pinnatifid, 1-4(6) cm long, distally glabrous, proximally arachnoid-pubescent, the 3-5 segments linear, the terminal segment often the longest, cuspidate. Flowers in a spicate thyrse, beginning at about the midpoint of the stem, the bracts unequally trifid to entire. Calyx ovoid, the tubes 1-3 mm long, the costae arachnoid-pubescent, the lobes 1-2 mm long, cuspidate; corolla white, persistent, brown when dry, the tubes 5-9 mm long, the lobes narrowly lanceolate, 2-4(5) mm long; stamens slightly exserted, subsessile, the filaments adnate in the sinuses of the lobes and shorter than the elliptical anthers; style included, about 1/2 the length of the tube. Capsule with 1 or 2 seeds per locule. (2n = 14) May-Jul. Sandy to gravelly soils of knolls or open sites; WY to w SD, s to NM & w KS; (UT e to SD, & KS). *Gilia spicata* Nutt.—Rydberg.

This species is closely related to *I. congesta* and sometimes confused with *I. congesta* subsp. *pseudotypica*.

4. LEPTODACTYLON H. & A.

Cespitose to subshrubby, taprooted perennials, the flowering stems suffrutescent, decumbent or erect. Leaves opposite basally, alternate to subalternate in the inflorescence, blades sessile, commonly with axillary fascicles, palmatifid to subpinnatifid, the 3-5 segments linear to subulate, pungent. Flowers terminal, solitary or in congested cymes, sessile. Calyx tubular, the 4 or 5 lobes subulate, pungent, the thickened costae separated by hyaline or scarious membranes which are ruptured or distended by the developing capsule; corolla funnelform, the lobes obtuse, ochroleucous to pink or salmon; stamens equal, included in the throat, the short filaments inserted equally in the upper 1/3 of the tube or in the throat; style included, ovary with few ovules per locule, typically with 1 or 2 seeds per locule at maturity.

1 Calyx and corolla with mostly 4 lobes; stamens usually 4; plants pulvinate .... 1. *L. caespitosum*
1 Calyx and corolla with mostly 5 lobes; stamens usually 5; plants caespitose to subshrubby, with erect, suffrutescent flowering stems .... 2. *L. pungens*

1. Leptodactylon caespitosum Nutt. Matted or pulvinate perennial, the flowering stems erect to decumbent, 1-3 cm long, the numerous, densely crowded internodes obscured by the leaves and axillary fascicles. Blades palmatifid from near the base, usually with 3 segments, these 3-8(11) mm long, about equal in length or the central segment slightly longer, sparsely glandular-pubescent. Flowers solitary and terminal. Calyx tube 5-7 mm long, the lobes 1-2 mm long; corolla ochroleucous to salmon, tubes 6-8 mm long, the 4 lobes 2-3 mm long and 1-2 mm wide; stamens 4, anthers and filaments equal in length, included and inserted in the throat; ovary 2 locular, style included. Jan-Jul. Dry rocky soils or in crevices of outcrops; se WY & ne CO to sw NE; (UT to NE & CO).

This species appears similar in habit to and is often confused with either *Phlox hoodii* or *P. bryoides*.

2. **Leptodactylon pungens** (Torr.) Nutt. Cespitose or subshrubby perennial, the flowering stems suffrutescent and erect; branching from near the base, 1–6 dm tall, the internodes glandular-puberulent. Blades palmatifid to subpinnatified from near the base, with 3–7 segments, the lateral ones 2–6 mm long, the central or terminal one 5–10 mm long, glandular-pubescent in the axils, aromatic. Flowers solitary or 2 or 3 in compact terminal clusters of shortened axillary shoots. Calyx tube 6–10 mm long, the lobes unequal, 2–3 mm long; corolla ochroleucous to pink or light violet, tubes 10–20(25) mm long, the 5 lobes 7–11 mm long; stamens 5, anthers and filaments equal in length, included and inserted in the throat; ovary 3 locular, style included. (2n = 18) May–Jul. On dry sandy or gravelly soils of plains & hillsides; e WY to w NE, s to ne CO; (e WA to e MT s to NV, UT, e to NE).

5. LINANTHUS Benth.

1. **Linanthus septentrionalis** Mason. Annual, 5–10(25) cm tall, with erect, simple to diffusely branched stems, glabrous to puberulent. Leaves palmatifid, opposite, the 3–7 segments 3–15 mm long, glabrous to puberulent. Flowers solitary or paired, pedicellate in the leaf axils, the pedicels 12–30 mm long; calyx 5-merous, 2–4 mm long, the herbaceous costae separated by hyaline membranes which are ruptured or distended by the mature capsule; corolla 5-merous, white to bluish, rotate to funnelform, 2–5 mm long, the tube pilose near the insertion of the filaments; stamens 5, slightly exserted and equally inserted near the summit of the tube; ovary 3-locular, the style equaling the stamens. Mature capsule equaling the calyx, with 2–8 seeds per locule. Apr–Jul. Sandy or gravelly soil of ephemerally moist sites; e MT to w SD s to n WY; (B.C. to Alta., s to UT, SD, & CO).

6. MICROSTERIS Greene

1. **Microsteris gracilis** (Hook.) Greene. Annual, 1–10 cm tall, diffusely branched with erect to mostly decumbent stems. Stems glandular to glandular-puberulent above, glabrous to pilose below. Leaves simple, the upper alternate, linear to lanceolate, the lower obovate to oblanceolate, 1–5 cm long, 3–8 mm wide, glabrous to viscid above, puberulent to glandular below, apices acute, margins entire to remotely serrulate, sessile to clasping. Flowers mostly in pairs terminating the branches or sometimes in cymose clusters of 3–10 flowers at the summit of the main stem, the bracts linear to linear-lanceolate, glandular. Calyx 5-merous, cylindrical, 5–8 mm long, glandular, the tube ca equal to the lobes in length, the herbaceous costae separated by hyaline or scarious membranes which are ruptured by the developing capsule, lobes subequal, lanceolate, acute; corolla 5-merous, white to commonly pink, sometimes with a yellow tube, salverform, the tubes 5–10 mm long, lobes 1–3 mm long; stamens 5, unequal, the short filaments unequally inserted in the tube, included; ovary globose, with a style slightly shorter than the corolla tube and terminated by 3 stigmatic lobes. Ovules and seeds 1 per locule, the seeds lenticular, becoming mucilaginous when wet. (2n = 14) Apr–Jun. Dry sandy or gravelly soils of prairie, stream beds, or disturbed sites; MT, WY, CO, & w NE; (s B.C. to sw Alta., NE, CO s to Mex.). *M. humilus* (Dougl.) Greene, *M. micrantha* (Kell.) Greene — Rydberg.

Recognition of infraspecific taxa var. *humilior* (Hook.) Cronq. and subsp. *humilus* (Greene) V. Grant, based on habit, appears unjustified (D.M. Smith, pers. comm.).

7. NAVARRETIA R. & P.

1. *Navarretia intertexta* (Benth.) Hook. var. *propinqua* (Suksd.) Brand. Erect annuals, 5–10(20) cm tall, with simple to diffusely branched stems, glabrous to glandular-puberulent. Leaves alternate, pinnatifid to bipinnatifid, the segments subulate to linear, pungent to spinose, glabrous to glandular-pubescent, often pilose basally, sessile. Flowers in dense, bracteate, capitate clusters terminating the stems, the bracts similar to the leaves. Calyx 4- or 5-merous, cylindrical, the herbaceous costae narrow, separated by hyaline membranes that are distended by the developing fruit, the lobes unequal, entire to bifurcately or trifidly spinose, glabrous to pilose within the tube; corolla 4- or 5-merous, funnelform to salverform; stamens 4 or 5, unequally exserted, the filaments equally inserted in the throat near the sinuses; ovary oblong, 2- or 3-locular, the style included to slightly exserted. Mature capsule indehiscent to irregularly circumscissile, with 1–4(6) seeds per locule, the seeds mucilaginous when wet. (2n = 18) Jun–Aug. Vernal pools, buffalo wallows, & ephemerally wet sites; MT to ND s to e CO & w NE; (WA & ND s to CA, AZ, & NE).

A closely related species, *N. minima* Nutt., has been reported from our area in WY and CO. Authors differ widely as to how the two taxa are separated, and none of the treatments consistently serve to distinguish the taxa in our area.

8. PHLOX L., Phlox

Cespitose to erect and diffusely branched, taprooted, rhizomatous or stoloniferous perennials, the flowering stems herbaceous to suffrutescent. Leaves primarily opposite but sometimes alternate in the inflorescence, blades entire, ovate or elliptical to linear or subulate and acerose, sessile to subsessile. Flowers terminal, bracteate, solitary to aggregated in corymbiform or paniculate clusters. Calyx 5-merous, the lobes acute to acuminate, cuspidate to aristate, the herbaceous costae separated by hyaline or scarious membranes which are ruptured by the developing capsule; corolla salverform, white to pink, purple or blue; the 5 lobes obtuse to obcordate; stamens unequal, the short filaments inserted unequally in the corolla tube, the anthers included or some partly exserted; ovary either with a short or conspicuously long style, terminated by 3 stigmatic lobes; ovules few per locule, typically with 1 or 2 seeds per locule at maturity. Capsule ovoid to oblong, the mature seeds not mucilaginous when wet.

Reference: Wherry, E. T. 1955. The genus *Phlox*. Morris Arbor, Monogr. 3. Philadelphia.

1 Fertile stems erect to decumbent, the herbaceous shoots (8)10–200 cm tall with developed internodes, 1/3–3/4 as long as the leaves, blades linear to ovate or elliptic, the apices acute to aristate and herbaceous.
  2 Style as long as the corolla tube, the stigmatic lobes 1/10 or less the length of the style .... 8. *P. paniculata*
  2 Style 1/2 or less the length of the corolla tube, the stigmatic lobes ca 1/2 the length of the style.
    3 Corolla lobes emarginate to obcordate, the fertile herbaceous shoots 12 cm or less long .... 7. *P. oklahomensis*
    3 Corolla lobes obtuse, the fertile herbaceous shoots 15 cm or more long.
      4 Stems ascending to decumbent, from long stolons or rhizomes; blades of sterile shoots elliptic, those of the fertile shoots lanceolate to narrowly oblong .... 4. *P. divaricata* subsp. *laphamii*
      4 Stems erect to ascending, from a short rhizome and without decumbent sterile shoots; blades linear to lanceolate.
        5 Uppermost leaves and bracts alternate; hairs of calyx and upper stems mostly 2–4 mm long .... 6. *P. longipilosa*
        5 Uppermost leaves and bracts mostly opposite; hairs of calyx and upper stems less than 1 mm long .... 9. *P. pilosa*

1 Fertile stems compact, forming cespitose or pulvinate plants, the herbaceous shoots usually 10 cm or less long, with obscure internodes commonly 1/3 or less as long as the leaves, blades linear to subulate or elliptic, the apices cuspidate to pungent.
   6 Largest leaves of fertile shoots 2–5 mm wide; corolla lobes 8–12 mm long . 1. *P. alyssifolia*
   6 Largest leaves 2 mm or less wide; corolla lobes 3–8(9) mm long.
      7 Leaves mostly 10–25(30) mm long; corolla lobes 6–8(9) mm long .... 2. *P. andicola*
      7 Leaves mostly 2–10(12) mm long; corolla lobes 3–5 mm long.
         8 Internodes and leaves densely arachnoid-pubescent to woolly; blades broadly subulate to triangular .... 3. *P. bryoides*
         8 Internodes and leaves glabrous to ciliate or sparsely arachnoid-pubescent; blades linear to linear-subulate .... 5. *P. hoodii*

1. Phlox alyssifolia Greene. Taprooted, cespitose perennial, 3–10 cm tall. Fertile shoots with 2–4 nodes, the herbaceous stems with spreading hairs. Blades linear-oblong to elliptical, (7)10–30 mm long, 2–5 mm wide, usually glabrous, sometimes glandular-pubescent, the margins thickened, conspicuously ciliate proximally, the tips cuspidate. Inflorescence a compact cyme with (1)3(5) flowers; pedicels densely glandular-pubescent, subsessile to 10(20) mm long. Calyx glandular-pubescent, 7–10(12) mm long, the tube and lobes about equal in length, the lobes cuspidate; corolla white to commonly pink or bluish-violet, tubes 10–15 mm long, the lobes obovate, obtuse, 8–10(12) mm long, 4–6 mm wide; style 6–12 mm long. May–Aug. Dry, sandy or gravelly soil of open prairie; GP n of se WY & nw NE; (sw Alta. to Sask., s to GP). *P. alyssifolia* Greene subsp. *alyssifolia* (A. Nels.) Wherry, subsp. *abdita* (A. Nels.) Wherry—Wherry, op. cit.

Wherry (op. cit.) recognizes 2 subspecies in our region, based on the lengths of leaves, pedicels, sepals and corolla tubes. Subspecies *abdita* is characterized by shoots 6–10 mm long and flowers measuring in the upper 1/2 of the quantitative ranges. This subspecies is restricted to the BH and neighboring plains but is sympatric with subsp. *alyssifolia*. Intermediate forms are represented in collections from this region. Hybrids between this species and *P. andicola* or *P. hoodii* are reported from SD by Wherry (op. cit.).

2. Phlox andicola Nutt. ex A. Gray, plains phlox. Rhizomatous, cespitose perennial, 4–10(12) cm tall. Fertile shoots solitary or branching near the base, erect to decumbent, with 5–8(10) nodes, the herbaceous stems puberulent to arachnoid-pubescent. Blades linear to subulate, 10–25(30) mm long, 1–2 mm wide, nearly glabrous to pubescent or arachnoid-ciliate proximally, the midrib prominently thickened, the tips pungent to subacerose. Inflorescence compact with 1–3(5) flowers; pedicels glabrous to weakly pilose, subsessile to 2(5) mm long. Calyx 6–11 mm long, arachnoid-pubescent along the margins of the lobes and near the summit of the tube, the tube about 1/2–2/3 as long as the calyx, the lobes subulate and pungent; corolla white, tubes 6–13(17) mm long, lobes obovate, obtuse, 6–8(9) mm long, 4–6(7) mm wide, style 5–9 mm long. (2n = 42) May–Jul. Dry, sandy or gravelly soil of open prairie; GP only, se MT to sw ND s to e CO & nw KS. *P. andicola* Nutt. subsp. *andicola* Wherry, subsp. *parvula* Wherry, subsp. *planitiarum* (A. Nels.) Wherry—Wherry, op. cit.

Subspecies *planitiarum* represents an infrequent form with glandular hairs. The name subsp. *parvula* was applied to plants with fertile shoots less than 4 cm long and leaves less than 15 mm long. Although unlikely, based on the reported chromosome numbers, some specimens referrable to the latter may represent hybrids involving *P. hoodii*. The three subspecies are mostly sympatric throughout their range.

3. Phlox bryoides Nutt. Taprooted, pulvinate perennial, 2–5 cm tall. Shoots numerous, 1–2 cm long, with 4–7 densely crowded nodes, the internodes completely obscured by the imbricate leaves. Blades broadly subulate to triangular, 2–4 mm long and 1 mm or less wide, distally glabrous, proximally and adaxially arachnoid-pubescent to arachnoid-woolly, the hairs often longer than the leaves, the tips pungent. Flowers solitary, sessile and terminal. Calyx sparsely pubescent to arachnoid-woolly, particularly on the adaxial surface, 3–5(7) mm long, the tube and lobes about equal in length, the lobes linear-subulate to triangular and cuspidate; corolla white, commonly with a yellow throat, tubes 4–8 mm long, lobes obovate to elliptical, 3–5 mm long, 1–3 mm wide; style 2–6 mm long. (2n = 14) Apr–Jun. Dry, gravelly to rocky soil of slopes & flats; MT, WY, CO, & w NE; (ID to UT, e to MT & NE).

This species is clearly related to the tetraploid *P. hoodii* subsp. *muscoides,* with which it has been conspecifically treated (Hitchcock et al.). In our region, however, populations of *P. bryoides* are apparently diploid and ecologically restricted to relatively coarse-grained substrates.

4. ***Phlox divaricata*** L. subsp. ***laphamii*** (Wood) Wherry, blue phlox. Stoloniferous to rhizomatous perennial. Stems ascending to decumbent, sparsely glandular-pubescent, the sterile shoots prostrate to decumbent and rooting at the node, the fertile shoots erect with 4–6 remote nodes, 15–50 cm tall. Blades of the sterile shoots elliptic, those of the fertile shoots lanceolate to narrowly oblong; 20–50(55) mm long, 10–20(25) mm wide; glabrescent to sparsely pubescent or villous, the upper also ciliate, the tips obtuse to acuminate. Inflorescence a compact corymbose or paniculate cyme with 9–25(30) flowers; bracts and pedicels glandular-pubescent; the pedicels 1–15 mm long. Calyx glandular-pubescent, 6–11 mm long, the tube slightly shorter than the lobes, lobes linear-subulate; corolla bluish-violet to purple, lighter in the throat, tubes 11–17(20) mm long, the lobes obovate to oblanceolate, obtuse, 10–15(19) mm long the 6–10(13) mm wide; style 1–3 mm long. (2n = 14) Apr–Jun. Open woods & rocky slopes; extreme e SD to MN s to ne OK & MO; (SD s to OK, e to IL & GA).

The typical subspecies, characterized chiefly by obcordate or marginate petals, occurs east of the Mississippi R. valley.

5. ***Phlox hoodii*** Rich., Hood's phlox. Taprooted, compact, pulvinate to cespitose perennial, 2–8 cm tall. Fertile shoots with 2–4(5) crowded nodes, 1–5 cm long, the herbaceous stems glabrous to puberulent. Blades subulate to linear-subulate, (3)4–10(12) mm long, 1 mm or less wide, distally glabrous and proximally ciliate to arachnoid-pubescent, the tips pungent. Flowers solitary, sessile to pedicellate, the pedicels 2–3(4) mm long. Calyx glabrous to glandular or arachnoid-pubescent, 4–7(8) mm long, the tube about 1/2–2/3 as long as the calyx, the lobes linear-subulate to subulate and cuspidate; corolla white to lavender, commonly with a yellow throat, tubes 4–8 mm long, lobes obovate to elliptical, 3–5 mm long, 2–4 mm wide; style 2–7 mm long. (2n = 28) Apr–Jul. Dry, sandy to gravelly soil of prairie & foothills; MT e to ND, s to CO & nw NE; (se B.C. to CA, e to sw Sask., ND, & NE). *P. hoodii* Rich. subsp. *hoodii,* (E. Nels.) Wherry, subsp. *glabrata* (E. Nels.) Wherry, subsp. *viscidula* Wherry, subsp. *muscoides* (Nutt.) Wherry—Wherry, op. cit.

Considerable morphological variation characterizes *P. hoodii* in the GP. Glabrous to sparsely pubescent leaves distinguish subsp. *glabrata,* whereas glandular-pubescent upper leaves and calyces distinguish subsp. *viscidula.* A combination of arachnoid-pubescent leaves and leaf, calyx and corolla tube lengths in the upper 1/2 of the respective quantitative ranges apply to subsp. *hoodii.* Subsp. *muscoides* is chiefly separated by numerous, crowded shoots less than 1 cm long with calyx and corolla lengths in the lower 1/2 of the quantitative ranges. There appears to be little geographical segregation among these taxa in our region, and it is often difficult to apply a name to samples of any particular population.

6. ***Phlox longipilosa*** Waterfall. Rhizomatous perennial with stems branching at the base or sometimes above, 20–45 dm tall. Herbaceous stems sparsely to prominently pilose, the hairs 2–4 mm long and with 10–18 crowded nodes. Blades linear to linear-lanceolate, the upper leaves mostly alternate; glandular-pubescent to sparsely pilose, 30–50(60) mm long, 3–8 mm wide. Inflorescence compact, corymbiform, with 10–25(36) flowers, the leaves and bracts mostly alternate; pedicels 3–12(15) mm long. Calyx 10–12 mm long, glandular and

pilose, the hairs 2–4 mm long, the tube about 1/3–1/2 the length of the calyx, the lobes attenuate to aristate; corolla pink to purple, tubes glandular-pubescent, 8–12(15) mm long, lobes obovate, obtuse, 11–12 mm long, 8–9 mm wide; style 2–3 mm long. Apr–Jun. On humic soils in open woods; known only from OK: Comanche, Greer, Kiowa, & Washita.

7. **Phlox oklahomensis** Wherry. Rhizomatous to stoloniferous, suffrutescent perennial, usually with numerous branches from below the middle, 8–15(20) cm tall. Herbaceous stems pilose, with 3–5 nodes, the fertile shoots 4–8(12) cm long. Blades linear to lanceolate, 10–30(60) cm long, 2–5 mm wide, the lower ones ciliate proximally, the upper pilose throughout. Inflorescence compact, corymbiform, with 3–6(9) flowers, the bracts glandular pubescent; pedicels 3–10(20) mm long. Calyx pubescent to pilose, 7–10 mm long, the tube and lobes about equal in length, lobes linear-subulate, cuspidate; corolla white to more commonly pink or lavender, tubes 8–12 mm long, lobes obcordate to emarginate, 7–9 mm long, 4–7 mm wide, the notch 1–4 mm deep; style 2–3 mm long. (2n = 14) Apr–May. On humic or gravelly soils of limestone outcrops in prairie; known only from KS: Butler, Chautauqua, Cowley, & Elk; OK: Kay, Woods, & Woodward.

8. **Phlox paniculata** L., fall phlox. Perennial from a short thick rhizome, with several stems 75–200 cm tall. Herbaceous stems glabrous to pubescent, with 14–25(40) nodes. Leaves opposite below, subopposite at the upper nodes; blades narrow-elliptic below to oblong-lanceolate or ovate above, 4–12(15) cm long, 1–4(5) cm wide, glabrous to sparsely pilose or pubescent, the margins ciliolate to serrulate. Inflorescence paniculate, with 45 or more flowers; pedicels subsessile to 8 mm long. Calyx 6–10 mm long, the tube and lobes about equal in length, lobes attenuate, aristate; corolla rarely white, usually pink to purple, tubes 16–26 mm long, lobes obovate to orbicular, 7–12 mm long, 5–12 mm wide; styles 15–25 mm long. (2n = 14) Jun–Oct. On humic to well-drained soils of open woods & bottoms; cen & e NE s to e OK; (NE, KS, OK, IA, & AR e to NY & GA, cult. elsewhere).

9. **Phlox pilosa** L., prairie phlox. Perennial from a stout rootstock, with 1–several stems branching at the base or sometimes above, (20)30–60(75) cm tall, pubescent to glandular-pubescent, with 6–12 nodes. Blades linear to lanceolate or ovate, 30–80(100) mm long, 5–20(30) mm wide, pilose, particularly near the margins and the midrib. Inflorescence compact to open, paniculate, with 12–50(100) flowers; pedicels subsessile to 10 mm long, strongly glandular. Calyx 8–15 mm long, glandular-pubescent, tube and lobes ca equal in length, lobes subulate to linear-attenuate, aristate; corolla white, pink or purple, tubes 8–16 mm long, lobes oblanceolate to obovate, 10–12 mm long, 6–8 mm wide; styles 1–3 mm long. (2n = 14) May–Jul. On humic, sandy, or rocky soils of open woods, meadows, or open slopes; e ND to MN s to OK & MO; (ND, MN, WI s to TX, e to NY & FL).

This species has 9 intergrading but well-defined races recognized as subspecies by Levin, Brittonia 18: 142–162. 1966. Of these, 4 are found in our region.

1 Bracts of the inflorescence copiously glandular-pubescent.
  2 Corolla tubes 8–12 mm long, pilose; upper leaves lanceolate to narrowly ovate .... 9c. subsp. *pilosa*
  2 Corolla tubes 12–16 mm long, glandular-pubescent; upper leaves ovate, often with cordate-clasping bases .... 9b. subsp. *ozarkana*
1 Bracts of the inflorescence pubescent to pilose, rarely glandular.
  3 Bracts of the inflorescence densely pilose to white-villous; corolla tubes 10–14 mm long, pilose .... 9a. subsp. *fulgida*
  3 Bracts of the inflorescence glabrous to sparsely pubescent; corolla tubes (11)14–21 mm long; glabrescent .... 9d. subsp. *pulcherrima*

9a. subsp. *fulgida* (Wherry) Wherry. Lower leaf blades linear, graduating to lanceolate above, the surfaces densely pilose. Bracts of the inflorescence densely pilose to white-villous, lustrous. Calyx 8–10 mm long; corolla tubes 10–14 mm long, pilose. in meadows & open woodland; e ND & MN s to e KS & w MO; (s Sask., WI to IL s to MO, KS).

9b. subsp. *ozarkana* (Wherry) Wherry. Lower leaf blades linear to lanceolate, graduating to ovate above, pilose. Bracts of inflorescence copiously glandular. Calyx 10–13 mm long; corolla tubes 12–16 mm long, glandular-pubescent. On humic soil in woods; e OK & s MO; (MO s to OK, Ozark Mts., LA, e to AL).

9c. subsp. *pilosa*. Lower leaf blades linear, glabrescent, graduating to lanceolate to narrowly ovate above, pilose. Bracts of inflorescence glandular-pubescent. Calyx 8–12 mm long; corolla tubes 8–12 mm long, pilose. In open woods, slopes, & moist meadows; se KS & IA s to OK & MO; (KS, OK, MO & e IA e to NY, s to TX & FL).

9d. subsp. *pulcherrima* Lundell. Lower leaf blades linear, graduating to lanceolate above, glabrous to sparsely pubescent. Bracts of inflorescence glabrous to sparsely pilose. Calyx 8–12 mm long; corolla tubes (11)14–21 mm long, glabrous. On humic soil in open woods, in our region reported from w & nw MO; (e TX to AR & LA).

9. POLEMONIUM L.

Cespitose to erect, taprooted or rhizomatous perennials, the flowering stems herbaceous. Leaves basal to alternate, the basal ones long petioled, the cauline sessile to subsessile; blades pinnately divided or compound, glabrous to glandular-pubescent, with a strong mephitic odor. Flowers pedicellate, cymosely disposed in a terminal lax to congested inflorescence. Calyx 5-merous, campanulate, the tube herbaceous throughout, accrescent, chartaceous in fruit and not ruptured by the developing fruit, lobes subequal to the tube; corolla 5-merous, blue, violet or ochroleucous, funnelform to campanulate; stamens subequal to the corolla, inserted equally in the tube; ovary ovoid, the style exceeding the stamens and terminated by 3 stigmatic lobes. Capsule ovoid to oblong, with 5–7 seeds per locule.

1 Corolla funnelform, ochroleucous to yellow; leaflets appearing verticillate, 3- to 5-parted .... 1. *P. brandegei*
1 Corolla campanulate, blue to violet; leaflets entire and in pairs .... 2. *P. reptans*

1. *Polemonium brandegei* Greene. Perennial, 10–50 cm tall, stems 1–several from a rhizome, erect, glandular-pubescent. Leaves pinnately divided, the 17–45 leaflets appearing verticillate on the rachis, each leaflet 2- or 3-parted, the segments confluent, elliptic to spatulate, (1)3–10 mm long, glandular-pubescent. Flowers pedicellate to subsessile, cymosely disposed in a congested, racemose inflorescence. Calyx (6)8–15 mm long at anthesis, glandular-pubescent; corolla funnelform, yellow to ochroleucous, the tube 10–22(25) mm long, lobes 5–10(12) mm long; style exserted, exceeding the stamens. (2n = 18) Jun–Aug. In our area known only from granitic rock outcrops in the BH; (NV to WY s to UT, CO, & NM). *P. viscosum* Nutt. subsp. *mellitum* (A. Gray) Davidson — Davidson, Univ. Calif. Publ. Bot. 23:209. 1950.

This species, treated here as distinct from *P. viscosum*, represents one of many taxa with close congeners found only in the western North American cordillera.

2. *Polemonium reptans* L., creeping polemonium. Perennial, 15–50(75) cm tall, stems 1–several from a woody caudex, erect, glabrous to pubescent. Leaves compound, the (3)7–19 leaflets mostly in pairs along the rachis, entire, ovate to lanceolate, acute, (10)20–50(70) mm long, glabrous to glandular-pubescent. Flowers pedicellate, cymosely disposed in an open, corymbiform inflorescence. Calyx 3–6 mm long at anthesis, puberulent. Corolla campanulate, blue to violet, the tube 5–7 mm long, lobes 5–8 mm long; style exserted, strongly exceeding the stamens. (2n = 18) Apr–Jun. On moist sites in woods; e NE to IA, s to OK & MO; (NE, MN to NY s to AR, GA, & OK).

116. HYDROPHYLLACEAE R. Br., the Waterleaf Family

by Dieter H. Wilken

Annual, biennial or perennial herbs. Leaves alternate to opposite basally, sometimes rosulate, simple to pinnatifid or pinnately compound, margins entire to irregularly toothed. Flowers commonly in simple to compound, scorpioid cymes or solitary, sessile to pedicellate. Calyx 5-merous, deeply divided but connate at the base, herbaceous throughout and often accrescent, the lobes linear to lanceolate, acute; corolla 5-merous, gamopetalous, funnelform to broadly campanulate, regular; stamens 5, exserted or included, equal or unequal in length, equally or unequally adnate to the corolla near the base of the short tube, the filament bases often alternating with either a pair of scalelike appendages or a gland; pistil 1, the ovary superior, with 1 locule, sometimes appearing as 2 because of the intrusive parietal placentae, the style exserted to included, with 2 stigmatic lobes or deeply 2-cleft to near the base. Capsule loculicidally to septicidally dehiscent by 2 or 4 valves, with 1–20 seeds per locule, the seeds smooth, pitted or alveolate to corrugated.

The author acknowledges the helpful criticism of Lincoln Constance.

1 Flowers solitary or 2 or 3 in small cymes, axillary or terminal.
  2 Leaves simple and entire; style bifid to near the base; stamens unequal .... 3. *Nama*
  2 Leaves pinnatifid to pinnately compound; style divided 1/3–1/2 its length into 2 branches; stamens equal.
    3 Sepals alternating with reflexed, auriculate appendages, these 1–2 mm long; seeds bearing an elaiosome .... 4. *Nemophila*
    3 Sepals without auriculate appendages; seeds lacking an elaiosome .... 1. *Ellisia*
1 Flowers typically in open to dense, terminal, scorpioid or congested cymes.
  4 Plants rhizomatous or taprooted; if the latter, then calyx sinuses with conspicuous, reflexed auricles .... 2. *Hydrophyllum*
  4 Plants taprooted annuals to shortlived perennials; calyx sinuses naked .... 5. *Phacelia*

1. ELLISIA L.

1. *Ellisia nyctelea* L., waterpod. Annuals, 5–40 cm tall, the stems simple and erect to diffusely branched and decumbent, angled, retrorsely hispid, 1–2(3) cm wide, pinnately parted to divided, the 7–13 divisions or segments lanceolate to linear-oblong, these sometimes toothed, sparsely hirsute to hispid. Flowers solitary, axillary, the pedicels 4–8(10) mm long. Calyx accrescent, 3–4 mm at anthesis, 5–7 mm long in fruit, the lobes lanceolate, 2–3 × the length of the tube, sparsely hirsute to hispid near the margins; corolla white to bluish, campanulate, equal to or slightly shorter than the calyx; stamens included, equally inserted in the tube; ovary globose, the style included, cleft into 2 lobes ca 1/3 its length. Capsule globose, 5–6 mm in diam, hispid, with 1 locule and ca 4 dark brown to black, globose seeds. (2n = 20) May–Jul. On sandy soils of open sites of prairie & woods; GP, MN to MT, s to NM & MO; (Alta. e to NJ, s to TX & NC).

2. HYDROPHYLLUM L., Waterleaf

Taprooted biennials or herbaceous perennials from horizontal rhizomes with fleshy roots, the stems erect, simple or branched. Leaves petioate, basal to alternate, shallowly lobed to pinnatifid. Flowers pedicellate, several to many in terminal, lax to densely aggregated, subcapitate cymes. Calyx divided nearly to the base, the entire lobes lanceolate to linear, the sinuses naked or with reflexed auricles; corolla white to purple, campanulate, exceeding the calyx in length; stamens equal in length, exserted, the filaments inserted in the base of the tube, each filament sparsely villous near the midpoint and alternating with a pair of ciliate appendages, each appendage adnate to the corolla tube along one edge; ovary globose with 1 locule, the exserted style with 2 stigmatic lobes. Capsule with 1-3 seeds at maturity, hispid to pubescent.

References: Constance, L. 1942. The genus *Hydrophyllum* L. Amer. Midl. Naturalist 27: 710-731; Beckmann, R. 1979. Biosystematics of the genus *Hydrophyllum* L. (Hydrophyllaceae). Amer. J. Bot. 66: 1053-1061.

1 Cauline leaves palmately to subpinnately lobed; calyx sinuses with conspicuous auricles; stamens exserted 1-3 mm beyond the corolla .... 1. *H. appendiculatum*
1 Cauline leaves pinnatifid; calyx sinuses without auricles; stamens exserted 4-8 mm beyond the corolla .... 2. *H. virginianum*

1. Hydrophyllum appendiculatum Michx., notchbract waterleaf. Taprooted biennial, stems 2-6 dm tall, pubescent to hirsute. Basal leaves pinnatified, 5-16 cm long, 4-11 cm wide, the petiole 5-16 cm long, the 5-9 segments ovate-lanceolate to rhombic-ovate, 1-6 cm long; cauline leaves orbicular, 5-15 cm in diam, with a cordate base, palmately to subpinnately lobed; surfaces of all leaves hispid to sparsely strigose. Inflorescence of 1-several lax cymes, the peduncles 3-9 cm long, branched and exceeding the leaves, pedicels 5-20 mm long. Calyx accrescent, strigose to hispid, densely hispid-ciliate along the lobe margins, lobes triangular to lanceolate, 4-8 mm long, the sinuses with lanceolate, reflexed auricles 1-3 mm long; corolla lavender to violet, tubes 4-7 mm long, the lobes 5-7 mm long; stamens exserted 1-3 mm beyond the corolla; style exserted 1-3 mm. Seeds usually 1 per capsule. ($2n = 18$) May-Jul. Mesic sites in woods; IA & e NE, s to MO & e KS; (MN to PA s to AK, TN, MO, & KS).

2. Hydrophyllum virginianum L., waterleaf. Rhizomatous perennial, stems 1-9 dm tall, glabrescent to retrorse-hispid below and sparsely strigose above. Basal and cauline leaves pinnatifid to pinnately divided, 5-30 cm long, 5-15 cm wide, the petiole 5-25 cm long, the 5(7-9) segments ovate-lanceolate to rhombic-ovate, 2-11 cm long, the lowest pair distinct, the upper segments cleft to parted, sparsely strigose above, glabrescent to sparsely strigose below. Inforescence of 1-several lax to subcapitate cymes, the peduncles 2-20 cm long, branched and exceeding the leaves, pedicels 3-10 mm long. Calyx sparsely strigose, sparsely hispid along the lobe margins, the lobes linear, 4-7 mm long, the sinuses lacking appendages; corolla white to purple, tubes 3-5 mm long, lobes 3-5 mm long; stamens exserted 4-8 mm beyond the corolla; style exserted 5-10 mm. Seeds usually 2 per capsule. ($2n = 18$) May-Jul. Mesic sites in woods; MN to e ND, s to MO & e OK; (ND to Que. s to OK, AR, & VA).

3. NAMA L.

Annuals in our area, with diffusely branched stems. Leaves alternate, simple, with entire to sinuate margins, cauline and not much reduced above. Flowers solitary, axillary or in small, terminal cymes, subsessile. Calyx deeply divided to near the base; corolla white to purple, funnelform to campanulate, exceeding the calyx in length; stamens included, unequal in length, filaments unequally inserted in the corolla tube, the base of the filament either dilated with minute marginal appendages or not; ovary ovoid to globose, with 1 locule but often with intruded placentae, the included style with 2 stigmatic lobes. Capsule with many, minute seeds.

References: Hitchcock, C. L. A taxonomic study of the genus *Nama*. 1933. Amer. J. Bot. 20: 415-430, 518-534; Bacon, J. 1974. Chromosome numbers and taxonomic notes in the genus *Nama* (Hydrophyllaceae). Brittonia 26: 101-105.

1 Filament bases not conspicuously dilated; leaf blades linear-oblong to obovate, plane to slightly revolute .... 1. *N. hispidum*
1 Filament bases dilated into free-margined appendages; leaf blades linear to linear-lanceolate, strongly revolute .... 2. *N. stevensii*

1. Nama hispidum A. Gray. Stems diffusely branched from near the base to above the midpoint, (7)10–50 cm tall, strigose to hispid or hirsute. Leaves linear-oblong to obovate, 1–7 cm long, 1–6(8) mm wide, plane to revolute, tapering at the base, strigose to mostly hispid or hirsute. Calyx lobes linear to linear-lanceolate, subequal, 3–7 mm long, mostly hirsute to hispid near the base; corolla funnelform to campanulate, 8–15 mm long, pink to purple or violet; filaments slightly dilated basally, the adnate portion shorter than the free filament. Seeds yellow and alveolate. ($2n = 14$) Mar–Jul. On sandy or gravelly soil of open sites; sw OK to TX; (CA e to TX & OK s to Mex.). *Marilaunidium angustifolium* (A. Gray) O. Ktze.—Rydberg, p.p.

The several varieties recognized by Hitchcock (op. cit.) appear to intergrade and are geographically sympatric (Bacon op. cit., Correll & Johnston).

2. Nama stevensii C. L. Hitchc. Similar to *N. hispidum* with respect to habit. Leaves linear to linear-lanceolate, 1–3 cm long, 1–3 mm wide, strongly revolute, sessile, strigose to hispid. Calyx lobes 5–8 mm long; corolla 6(8–10) mm long, lavender; filaments dilated basally into appendages with free margins, the appendages equaling the free portion of the filaments in length. ($2n = 14$) May–Jul. Apparently restricted to sandy, gypsum soils of open sites; s KS to TX & w OK; (KS to w TX s to Mex.).

4. NEMOPHILA Nutt.

1. Nemophila phacelioides Nutt. ex Bart. Annuals, with erect, simple to branched stems, 2–6 cm tall, hirsute to glabrescent. Leaves alternate, the lower oblong, pinnatifid to pinnately divided, 6–8 cm long, 2–5 cm wide, the 9–11 ovate segments entire or toothed, petioles equal to or shorter than the blades, the upper leaves ovate to orbicular, all hirsute to hispid. Flowers pedicellate, solitary and axillary or in small cymes terminating the stem. Calyx deeply divided to near the base, lobes oblong to ovate-lanceolate, 6–7 mm long, 2–3 mm wide, with spreading to erect auricles in the sinuses, hispid to hirsute; corolla rotate, lobes and throat blue to purple, paler in the tube, exceeding the calyx, 8–12 mm long, the lobes $2 \times$ the tube length; stamens included and equaling the tube in length, equally inserted near the base of the tube, alternating with pairs of broad, partly free and fimbriate appendages; ovary oval to globose, with 1 locule, and 1 style with 2 stigmatic lobes cleft to near the midpoint. Capsule globose, 5–9 mm in diam, hispid, with ca 4 reddish-brown, pitted, globose seeds bearing elaiosomes. ($2n = 14$) Mar–Jun. In shaded woods; OK: Caddo; (e OK & w AR s to e TX).

Reference: Constance, L. 1941. The genus *Nemophila* Nuttall. Univ. Calif. Publ. Bot. 19: 341–398.

5. PHACELIA Juss.

Taprooted annuals, biennials or perennials with erect to decumbent, simple to diffusely branched stems. Leaves petiolate to sessile, basal to alternate, entire to pinnately lobed, cleft or deeply dissected. Flowers pedicellate to subsessile, several to many in terminal or subterminal, single to compound, scorpioid cymes. Calyx divided nearly to the base, the entire lobes linear to lanceolate; corolla white to purple, rotate to campanulate, usually exceeding the calyx in length; stamens equal to subequal in length, included to exserted, the filaments inserted near the base of the short tube, villous to pubescent proximally, alternating either with a pair of appendages or a gland; ovary globose to ovoid with 1 locule, the style included to exserted, cleft into 2 lobes 1/3–3/4 the entire length. Capsule with 1–20 rugulose, corrugated or pitted seeds.

References: Constance, L. 1949. A revision of *Phacelia* subgenus *Cosmanthus* (Hydrophyllaceae). Contr. Gray Herb. 168: 1–48; Atwood, D. 1975. A revision of the *Phacelia crenulatae* group (Hydrophyllaceae) for North America. Great Basin Naturalist 35: 127–190.

1 Perennial plants from a short, semiwoody, underground caudex .... 3. *P. hastata* var. *leucophylla*
1 Taprooted annuals or biennials.
  2 Basal and cauline leaves linear to lanceolate, either simple or pinnatifid with 2–4 linear to linear-lanceolate lobes shorter than the entire, terminal segment .... 6. *P. linearis*
  2 Basal and cauline leaves oblong to ovate, crenate to shallowly lobed or deeply pinnatifid to bipinnately divided.
    3 Filaments alternating basally with pairs of scaly appendages; the capsule with 1–4 seeds; stamens exserted.
      4 Leaves crenate to sinuate, cleft or shallowly pinnatifid.
        5 Corolla salverform, white to pale lavender .... 8. *P. robusta*
        5 Corolla campanulate, purple to lavender .... 5. *P. integrifolia*
      4 Leaves deeply pinnatifid to bipinnate, clearly divided or dissected to the rachis proximally.
        6 Calyx lobes linear; corolla lobes nearly entire; ultimate leaf segments usually over 5 mm wide .... 1. *P. congesta*
        6 Calyx lobes oval to oblanceolate; corolla lobes dentate to crenulate; ultimate leaf segments usually less than 4 mm wide .... 7. *P. popei*
    3 Filaments alternating basally with a gland, this bordered by minute flaps; the capsule with 6–20 seeds; stamens included.
      7 Basal leaves shallowly toothed or lobed, forming a conspicuous rosette; pedicels shorter than the calyx; filaments sparsely villous .... 9. *P. strictiflora*
      7 Basal leaves pinnatifid to pinnately divided, at most weakly rosulate; pedicels equal to or longer than the calyx; filaments densely villous.
        8 Pubescence of the stems and pedicels strigose and appressed; corolla lobes denticulate to fimbriate .... 2. *P. gilioides*
        8 Pubescence of the stems and pedicels hirsute and spreading; corolla lobes entire .... 4. *P. hirsuta*

1. Phacelia congesta Hook. Annuals, with erect, simple or diffusely branched stems, 1–10 dm tall, glandular and puberulent to hispid. Leaves oblong to ovate, (1)3–9(12) cm long, 1–4 cm wide, pinnatifid to pinnately compound, the 3 terminal segments often not completely dissected to the rachis and often longer than the lower petiolate segments, glandular to strigose. Flowers in terminal, compound, scorpioid cymes, pedicels 3 mm or less long, pubescent. Calyx lobes 3–5 mm long, ca 1 mm wide, glandular to setose or hispid; corolla campanulate, blue to purple, 4–7 mm long, lobes pubescent; stamens exserted 2–4 mm beyond the corolla throat; ovary subglobose to oval, puberulent to glandular, the style exserted, 7–8 mm long, deeply cleft ca 3/4 its length. Capsule with 4 brown seeds, these furrowed on both sides of the ventral ridge. ($2n = 22$) Feb–Sep. On sandy, gravelly, or rocky sites; OK: Caddo & Comanche; (OK s to TX & n Mex.).

2. Phacelia gilioides Brand, hairy phacelia. Annuals, with erect and simple stems or diffusely branching from the base, 1–4 dm tall, strigose to canescent. Lower cauline leaves oblong, the upper oblong to orbicular, 1–5 cm long, 1–3 cm wide, pinnatifid to pinnately compound, the 3 terminal segments not completely divided to the rachis and larger than the lower petiolate segments, strigose. Flowers in terminal and axillary scorpioid cymes, pedicels 5–15 mm long, strigose to strigulose. Calyx lobes 5–8 mm long, ca 1–2 mm wide, strigulose to pustulose-hispid; corolla rotate to campanulate, purple, 8–10(12) mm long, lobes fimbriate to denticulate, pilose; stamens included, 4–6 mm long; ovary subglobose, style included, 5–7 mm long, cleft 1/3–1/2 its length. Capsules with 6–8 brown seeds. Apr–Jun. On limestone soils of open sites & in woods; KS: Crawford; (KS e to sw MO, s to e OK to AR).

3. *Phacelia hastata* Dougl. ex Lehm. var. *leucophylla* (Torr.) Cronq., scorpionweed. Taprooted perennials from a woody caudex, with decumbent to ascending, simple to branching stems 1–5(8) dm tall, hispid to tomentose. Basal leaves narrowly lanceolate to elliptic, 4–9(15) cm long, entire to unequally pinnatifid, the latter with 2 short lobes near the base of the blade, petiolate; the cauline leaves reduced in size, entire to basally 1- or 2-lobed, sessile to subsessile; all leaves strigose to hispid. Flowers in terminal, dense, compound, scorpioid cymes, peduncles and pedicels strigose to hispid. Calyx accrescent, lobes 3–5 mm long at anthesis, 5–8 mm long in fruit, strigose to hispid; corolla campanulate, white to lavender, 4–6 mm long; stamens exserted 4–6 mm beyond the corolla throat; ovary subglobose, style exserted, 5–10 mm long, cleft 1/3–1/2 its length. Capsule with 1 or 2 brown, alveolate seeds (2n = 22, 44) May–Jul. On sandy to rocky soils of open, often disturbed sites; w ND w to MT, s to CO; (WA to ND, s to CA, NV, UT, & CO). *P. leucophylla* Torr.— Rydberg.

This taxon represents part of a diploid-tetraploid complex involving several species of *Phacelia* in western North America. In our region, *P. hastata* var. *leucophylla* is apparently tetraploid (2n = 44) and intergrades with tetraploid *P. heterophylla* (2n = 44), at least in Colorado and Montana (Heckard, Univ. Calif. Publ. Bot. 32: 1–126. 1960). Plants with basal leaves having the terminal leaflet usually 1/3 or less the length of the entire leaf may be assigned to *P. heterophylla* Pursh.

4. *Phacelia hirsuta* Nutt. Annuals with erect, simple to diffusely branching stems, 1–5 dm tall, densely hirsute. Basal leaves oblong, 2–5 cm long, 1/3 cm wide, pinnately dissected, the 5–9 oval to orbicular segments entire to toothed, petiolate; cauline leaves oblong to orbicular with 5–9 linear to oval segments, clasping to sessile; all leaves strigose. Flowers in terminal, simple, scorpioid cymes, pedicels 3–15 mm long. Calyx lobes 5–10 mm long, 1–3 mm wide, unequal, strigose to strigulose; corolla broadly campanulate, bluish-lavender with a paler throat, 5–8 mm long; stamens included, 4–5 mm long; ovary subglobose, the summit hirsute, style included in anthesis, 5–6 mm long, cleft 1/3–1/2 its length. Capsule with 6–8 brown seeds. (2n = 18) Mar–May. On sandy soil of open sites & margins of woods; MO to se KS, s to ne TX; (w MO s to LA, w to KS & TX).

5. *Phacelia integrifolia* Torr., gyp phacelia. Annuals to biennials, with erect to branched stems, 1–6 dm tall, glandular-puberulent and hirsute. Basal leaves oblong to narrowly ovate, 2–7 cm long, 7–25 mm wide, crenate to shallowly cleft or pinnatifid, petiolate; the cauline leaves reduced and sessile; all glandular and strigose. Flowers in terminal, compound, scorpioid cymes, pedicels 2 mm or less long. Calyx lobes elliptic to oblanceolate, 2–7 mm long, 1–3 mm wide, glandular-puberulent and hirsute; corolla campanulate, purple to lavender, 4–7 mm long; stamens exserted, 8–12 mm long; ovary subglobose, glandular-puberulent; the style exserted, 5–9 mm long, deeply cleft ca 3/4 its length. Capsule with 4 black to brown seeds, these furrowed on both sides of the ventral ridge. Mar–Jul (Oct). On gypsiferous or calcareous soils; w OK & n TX; (UT & AZ e to OK & TX). *P. texana* Voss—Correll & Johnston.

Populations in our region are referrable to var. *texana* (Voss) Atwood, which is distinguished principally by centrally corrugated, mature seeds 3 mm or less long, whereas seeds of var. *integrifolia* lack ventral corrugations and are longer than 3 mm.

6. *Phacelia linearis* (Pursh) Holz. Annual with erect, single to branched stems, 1–6 dm tall, densely puberulent to hirsute. Leaves linear to lanceolate, 2–8(11) cm long, 1–10(12)

mm wide, entire to pinnately lobed proximally, strigose. Flowers in crowded cymes in the upper axils or terminal, pedicels 3 mm or less long. Calyx lobes 3–5 mm long, ca 1 mm wide, hirsute to hispid; corolla broadly campanulate, bluish-lavender to purple with a paler throat, 6–10 mm long; stamens included, 5–8 mm long; ovary ovoid, sparsely hirsute, the style included, 5–8 mm long, cleft 1/2–1/3 the length. Capsule with 6–15 brown seeds. (2n = 22) Apr–Jun. On dry sandy & gravelly soils; w SD to MT, s to n WY; (B.C. to SD, s to CA, UT, & WY). *P. franklinii* R. Br.—Rydberg.

7. *Phacelia popei* T. & G. Annuals with erect stems usually branched at the base, 5–35 cm tall, glandular-puberulent to hirsute. Leaves oblong, 2–15 cm long, 1–3 cm wide, 1- or 2-pinnate, the ultimate segments linear to lanceolate, glandular and strigose. Flowers in terminal, branched, scorpioid cymes, sessile. Calyx lobes 2–4 mm long, 1–2 mm wide, glandular and hirsute; corolla campanulate, blue to purple, 3–7 mm long; stamens exserted, 5–10 mm long; ovary globose, glandular and pilose, the style exserted, 6–10 mm long, cleft ca 2/3 its length. Capsule with 4 brown seeds, these furrowed on both sides of the ventral ridge. (2n = 22) Feb–May. On sandy or gravelly, sometimes calcareous soils; TX; (NM & TX s to Mex.).

8. *Phacelia robusta* (Macbr.) Johnst. Annuals to biennials, with erect, simple or basally branched stems, 4–12 dm tall, puberulent, pilose and densely glandular. Basal leaves broadly ovate to orbicular, 2–12 cm long, 1–9 cm wide, crenate to sinuate, petiolate; the cauline leaves gradually reduced and subsessile; all puberulent, glandular and hirsute. Flowers in terminal, compound, scorpioid cymes, pedicels 1–2 mm long. Calyx lobes spatulate, 2–5 mm long, 1–3 mm wide, glandular-puberulent; corolla salverform, white to pale lavender, 5–6 mm long; stamens exserted, 8–12 mm long; ovary subglobose, puberulent to strigose; the style exserted, 5–9 mm long, deeply cleft ca 3/4 the length. Capsule with ca 4 reddish brown seeds, these furrowed on both sides of the ventral ridge. Mar–Aug. On gravelly or clay soils; KS: Barber, Comanche, Harper, Kiowa; (KS s to OK, TX, & Mex.).

9. *Phacelia strictiflora* (Engelm. & Gray) A. Gray. Annuals, with erect to decumbent, simple or branched stems, 5–30 cm tall, hirsute. Rosulate, the basal leaves oblong to oval, 1–6 cm long, 1–3 cm wide, toothed or lobed to pinnately divided, glabrescent to strigose, petiolate; cauline leaves orbicular to linear-oblong, dentate to pinnately lobed, reduced above. Flowers in simple, terminal, scorpioid cymes; pedicels 2–10 mm long. Calyx lobes 5–15 mm long, 1–4 mm wide, pubescent; corolla rotate-campanulate, purple to lavender, 7–10 mm long; stamens included, 5–7 mm long; ovary globose, densely hirsute distally, the style included, 5–9 mm at anthesis, cleft ca 1/2 its length. Capsule with 10–20 black seeds. (2n = 18) Mar–May. On sandy soils of open sites & woods; w OK; (OK to TX, MS, & AL).

The four varieties recognized by Constance, vars. *strictiflora, connexa, robbinsii,* and *lundelliana,* are sympatric and intergrade throughout much of their range in TX and OK (Constance, op. cit.).

117. BORAGINACEAE Juss., the Borage Family

by Robert B. Kaul

Ours annual, biennial, and perennial herbs, the stems and leaves often prominently hairy to bristly, occasionally glabrous. Leaves all cauline, or basal and cauline, petiolate or axillary cymes, the cymes helicoid (flowers borne on one side of the rachis) or scorpioid (flowers borne on both sides of the rachis) and often uncoiling and much elongating with age, the cymes solitary, paired, or variously aggregated into thyrselike or paniclelike inflorescences; occasionally the flowers solitary at the branch tips or in the upper axils. Flowers perfect, regular or occasionally irregular, sometimes heterostylous, hypogynous; sepals 5, equal or unequal, essentially free or partially fused at the base, usually hairy or bristly, occasionally glabrous, persistent in fruit; corolla gamopetalous, mostly funnelform to salverform, often distinctly divided into tube and limb, the limb here including the expanded upper part of the tube, lobes 5, equal or occasionally unequal, pointed to rounded; internal appendages of the upper throat (fornices) often present opposite the lobes; small appendages often present on the inside of the tube base; stamens 5, attached variously in the tube, the filaments usually short, sometimes exserted, the anthers usually included, seldom exserted; pistil 1, superior, carpels 2, 2- or 4-chambered, deeply 4-lobed (except *Heliotropium*); style 1, arising between the lobes from the receptacle (or arising from the top of the ovary in *Heliotropium*), the columnar to pyramidal or low flattened receptacle (gynobase) prominent, especially in fruit. Fruits 1–4 nutlets that usually separate from each other, these smooth, variously sculptured, glochidiate-spiny, or keeled, mostly hard and sometimes bony; seeds with little or no endosperm. Heliotropaceae—Rydberg.

Some introduced taxa are grown as culinary or medicinal herbs and ornamentals in GP gardens. They can be expected from time to time persisting in old gardens or briefly escaping.

Symphytum asperum Lepech. and *S. officinale* L., comfrey. Coarse, large-leaved, summer-flowering perennials with lavender-blue tubular-campanulate corollas 1.5–2.5 cm long, the nutlets with a basal protuberance and, in *S. officinale*, the blades decurrent on the petiole and stem, the stems thus alate.

Borago officinalis L. borage, talewort, cool-tankard. Hispid-setose annual with solanaceous-type flowers, the corolla rotate, the filaments with an elongate appendage, the anthers exserted and connivent about the style, the nutlets with a basal protuberance.

Pulmonaria spp., lungwort. Spring-flowering perennials, the corolla tubular-campanulate, pink changing to blue, the leaves sometimes splotched whitish.

Heliotropium spp. and cultivars, heliotrope. Treated as annuals, with massive inflorescences, the flowers purple, rose, or white, very fragrant.

Plants with Flowers

1 Corolla yellow to orange.
  2 Plants perennial, from stout rootstocks; common in undisturbed prairies and open woodlands .... 9. *Lithospermum*
  2 Plants annual; disturbed roadsides, pastures; rare .... 1. *Amsinckia*
1 Corolla white, lavender, blue, or purplish.
  3 Corolla barely to distinctly irregular.
    4 Corolla strongly irregular, 1–2 cm long; stamens exserted .... 5. *Echium*
    4 Corolla slightly irregular, ca 6–8 mm long, stamens included .... 10. *Lycopsis*
  3 Corolla regular.
    5 Plants glabrous throughout; cymes without bracts.
      6 Flowers small, less than 1 cm long, white to bluish; moist saline soils in the open, n & w GP .... 7. *Heliotropium curassavicum*
      6 Flowers larger, ca 4 cm long, blue (pinkish); cool woods, se GP .... 11. *Mertensia virginica*
    5 Plants variously hairy to prickly; cymes with or without bracts.
      7 Corolla white or whitish.
        8 Flowers solitary in upper axils, not in cymes.
          9 Plant perennial; fornices present .... 9. *Lithospermum latifolium*
          9 Plants annual; fornices absent.
            10 Calyx lobes unequal .... 7. *Heliotropium tenellum*
            10 Calyx lobes equal .... 9. *Lithospermum arvense*
        8 Flowers borne in cymes.

BORAGE FAMILY

    11 Cymes without bracts.
        12 Fornices absent .... 7. *Heliotropium*
        12 Fornices present .... 3. *Cryptantha*
    11 Cymes with at least a few bracts in the floriferous region.
        13 Plants perennial from stoutish or stout rootstocks.
            14 Fornices absent .... 13. *Onosmodium*
            14 Fornices present.
                15 Stamens attached above the middle of the corolla
                   tube .... 6. *Hackelia*
                15 Stamens attached below the middle of the corolla
                   tube .... 3. *Cryptantha*
        13 Plants annual, or apparently so.
            16 Calyx lobes unequal .... 12. *Myosotis verna*
            16 Calyx lobes equal
                17 Lower cauline leaves opposite .... 14. *Plagiobothrys*
                17 Lower cauline leaves alternate.
                    18 Stamens inserted below the middle of the corolla tube .... 3. *Cryptantha*
                    18 Stamens inserted above the middle of the corolla tube.
                        19 Cymes bracteate throughout .... 8. *Lappula*
                        19 Cymes bracteate below, ebracteate above .... 6. *Hackelia*
7 Corolla blue, lavender, purplish, or reddish.
    20 Calyx divided to the base or essentially so.
        21 Cymes without bracts.
            22 Corolla without fornices .... 7. *Heliotropium*
            22 Corolla with fornices.
                23 Corolla to 4 cm long .... 11. *Mertensia*
                23 Corolla less than 1 cm long .... 4. *Cynoglossum*
        21 Cymes partially or completely bracteate.
            24 Cymes bracteate below, ebracteate above .... 6. *Hackelia*
            24 Cymes bracteate throughout .... 8. *Lappula*
    20 Calyx divided 1/3–2/3 to the base.
        25 Flowers solitary in axils, not in cymes .... 2. *Asperugo*
        25 Flowers borne in cymes.
            26 Corolla salverform .... 12. *Myosotis*
            26 Corolla tubular-funnelform .... 11. *Mertensia*

Plants with Fruits

1 Fruit not deeply cleft into 4 nutlets, the style absent or borne at the summit of the nutlets, not from a gynobase; cymes ebracteate .... 7. *Heliotropium*
1 Fruit deeply cleft into 1–4 nutlets, the style borne on the gynobase; cymes bracteate, partially bracteate, or ebracteate.
    2 Gynobase low, not at all pyramidal, the nutlets broadly attached at their base, the scar large; calyx divided to the base or essentially so.
        3 Nutlets 3-angled, rugose; stem and leaves often pustulate-bristly; corolla irregular, blue .... 5. *Echium*
        3 Nutlets ovoid or with a single keel, smooth or roughened; stem and leaves variously pubescent but not pustulate; corolla regular, white, yellow, or orange.
            4 Flowers and fruits borne in cymes; plants perennial.
                5 Plants very coarse; leaves 5–12 cm long, the veins prominent on the underside, flowers whitish .... 13. *Onosmodium*
                5 Plants not especially coarse; leaves mostly less than 5 cm long, the veins not unusually prominent; flowers white, yellow or orange .... 9. *Lithospermum*
            4 Flowers and fruits solitary in axils; plant annual; corolla white .... 9. *Lithospermum arvense*
    2 Gynobase raised, mostly pyramidal, the nutlets attached to it apically, medially, or basally (if basally, the scar small).
        6 Nutlets partially or entirely covered with prominent glochidiate spines, or with a sharp keel all around; corolla blue, purplish, or white.

7 Nutlets with a sharp keel all around .... 12. *Myosotis*
7 Nutlets partially or completely distinctly glochidiate.
  8 Nutlets apically attached to the gynobase (but at first sight apparently attached basally) .... 4. *Cynoglossum*
  8 Nutlets partially or completely ventrally attached to the gynobase.
    9 Cymes bracteate throughout; nutlets attached to the gynobase for most of their ventral length; pedicels in fruit erect to spreading .... 8. *Lappula*
    9 Cymes bracteate below, ebracteate above; nutlets attached to the gynobase only in the middle of their ventral side; pedicels reflexed or recurved in fruit .... 6. *Hackelia*
6 Nutlets smooth or variously wrinkled, rugose, or tuberculate but without prominent glochidiate spines or a keel all around (sometimes with a few tiny dorsal spines in *Plagiobothrys*).
  10 Calyx in fruit greatly enlarged and involucroid; flowers and fruits not borne in cymes; leaves often opposite .... 2. *Asperugo*
  10 Calyx in fruit little to somewhat larger than in anthesis but little altered in shape and not at all involucroid; flowers and fruits borne in cymes.
    11 Cymes ebracteate in the floriferous portion; corolla white or blue (pink); leaves alternate (lower leaves opposite in *Plagiobothrys*).
      12 Plant conspicuously pubescent, hirsute, or bristly; sepals free; corolla white, less than 1 cm long; fornices present .... 3. *Cryptantha*
      12 Plants glabrous to strigose but not conspicuously hirsute or bristly; sepals free or basally connate; corolla blue (pink), to 2(4) cm long .... 11. *Mertensia*
    11 Cymes partially to completely bracteate in the floriferous portion; corolla white, yellow, orange, or blue.
      13 Lower cauline leaves opposite; scar of nutlet low on the ventral side; corolla white .... 14. *Plagiobothrys*
      13 Lower cauline leaves alternate, or at least not obviously and regularly opposite; scar of nutlet various; corolla white, yellow, orange, or blue.
        14 Nutlets usually smooth; corolla white .... 3. *Cryptantha*
        14 Nutlets not smooth; corolla white, yellow, orange or blue.
          15 Nutlets with a large ventral groovelike scar that is open or closed at the base; flowers white .... 3. *Cryptantha*
          15 Nutlets and flowers not as above; scar at the base of the ventral keel; corolla blue, yellow, or orange.
            16 Scar surrounded by a raised ring; nutlets minutely tuberculate, reticulately ridged and angular; corolla blue; ne GP .... 10. *Lycopsis*
            16 Scar elevated, carunclelike, not surrounded by a raised ring; nutlets tuberculate, rugose or muriculate, not very angular; corolla yellow to orange .... 1. *Amsinckia*

1. AMSINCKIA Lehm., Fiddleneck

Ours taprooted annuals of disturbed places. Stem simple or branched, hispid. Leaves basal and cauline, hispid and often pustular, the basal sometimes short-lived, the cauline smaller. Cymes terminal on the stems and axillary in upper axils, helicoid when young, elongating and becoming spikelike with age, ebracteate or very sparingly bracteate below, ebracteate above; flowers very short-pedicellate. Sepals free nearly to the base; corolla funnelform, yellow to orange, the throat marked red, the fornices weakly to strongly developed, the throat then open or closed, respectively; stamens included, inserted near the mouth of the tube or far down the tube, the filaments short; style 1; stigma 1, included. Nutlets 4, ovoid, rather trigonous, the small scar near the base of the ventral side; gynobase short, pyramidal.

1 Fornices large, hairy, occluding the throat; stamens attached at or well below the middle .... 2. *A. lycopsoides*
1 Fornices small, not obvious, not hairy, the throat open; stamens attached near the top of the tube, above the middle .... 1. *A. intermedia*

1. *Amsinckia intermedia* Fisch. & Mey. Plant 1–3 dm tall, the stems 1–several, simple or branched. Stems and leaves hispid, the leaves also pustular above and below. Leaves basal and cauline, the basal linear to oblanceolate, tapering to a petiolelike base, to 15 cm long and 1 cm wide; cauline leaves sessile, linear, oblong or oblanceolate, smaller than the basal leaves. Inflorescences solitary at the branch tips and in upper axils, tightly coiled in anthesis, elongate and spikelike in fruit, essentially ebracteate, the flowers sessile or nearly so. Sepals in flower ca 5 mm long, in fruit 6–12 mm long, hispid; corolla yellow-orange or orange, 3–5 mm wide, the throat with red markings, the fornices not obvious and the throat open; stamens attached near the top of the tube, above the middle; style ca 5 mm long; stigma capitate. Nutlets greenish to black, tuberculate, rugose, or muriculate, much shorter than the persistent calyx. (n = 15, 17, 19) Jun–Jul. Weedy roadsides & fields, seldom collected & only a waif in GP; ND: Cass, Pembina, Pierce; NE: Hall; (WA e to ND, NE s to NM & Mex.).

2. *Amsinckia lycopsoides* Lehm. Plant to 3 dm tall, the stem simple or few-branched. Stems and leaves more or less hispid to hirsute. Leaves basal and cauline, linear to oblong or lanceolate, to 10 cm long and 1.3 cm wide. Inflorescences terminating the branches and in upper axils, coiled when young, elongate and racemelike with age, essentially ebracteate. Sepals ciliate and hispid in fruit, 6–10 mm long; corolla yellow to yellow orange, the throat marked red, the fornices well developed, hirsute, occluding the throat; stamens attached below the middle, well down the tube; style ca 2 mm long; stigma capitate. Nutlets greenish to brown, tuberculate, rugose, or muriculate. (n = 15) Jun–Jul. Dry soil in disturbed places; seldom collected & only a waif in GP; NE: Webster; MO: Jackson; OK: Oklahoma; (B.C. to MT, e to MO, OK s to NV, CA). *A. idahoensis* M.E. Jones—Rydberg; *A. barbata* Greene—Fernald.

2. ASPERUGO L., Madwort

1. *Asperugo procumbens* L. Plant annual from a slender taproot, to ca 6 dm tall. Stem simple or branched, the branches weak, erect, sprawling, or scrambling; stems ribbed, the ribs retrorsely barbed but otherwise glabrous. Basal leaves in a rosette, soon withering; cauline leaves often opposite or nearly so, lower ones petiolate; upper cauline leaves usually sessile; leaves ovate to elliptic or oblanceolate, acute to obtuse, mostly 2–4(7) cm long, 4–15(20) mm wide, strigose-hispid above and below. Flowers subsessile, solitary (paired) in the axils, the pedicel elongating in fruit and recurved; calyx ca 3 mm long in anthesis, grossly 5-lobed, divided to below the middle, increasing in fruit to ca 1.5 cm long and then involucroid and net-veined; corolla blue or purplish, funnelform to campanulate, ca 3 mm wide, the limb exceeding the calyx at first, the fornices prominent, white; anthers included, the filaments very short and inserted at the middle of the tube; stigma 1, capitate, included. Nutlets hidden in the enlarged calyx, brown, elongate-ovoid and dorsiventrally flattened, barely tuberculate or verruculose, attached to the raised gynobase below the middle, the scar small. (n = 24) May–Jul. Weedy waste places, cult. fields; ND: Dunn; NE: Cass; KS: Douglas; & to be sought elsewhere in GP; (over much of N. Amer.; Europe). *Introduced*.

3. CRYPTANTHA Lehm.

Ours annual, biennial, or perennial herbs, conspicuously pubescent to setose throughout and often painful to the touch, the coarser hairs often with flattened, light-colored siliceous bases (pustules). Stems 1–several, slender or stout. Leaves basal and/or cauline, alternate, mostly sessile and rather narrow, the basal ones sometimes petiolate, the cauline ones often reduced above and becoming bracts of the inflorescences. Cymes produced at each branch tip, not always obviously paired, condensed or glomerate when young, uncoiling and elongate with age and then spikelike or racemelike, sometimes collectively thrysoid or paniculate; bracts of the cymes present or absent, or present below but absent above, best observed in elongated inflorescences; flowers sessile to short-pedicellate. Sepals essentially free, equal, lanceolate, bristly-hairy on the outside, pubescent to hispid on the margins and inside, appressed to and much exceeding the nutlets in fruit; corolla tiny to rather conspicuous but not showy, regular, salverform to funnelform, white in ours, the fornices sometimes yellow; stamens attached below the middle of the corolla tube, the filaments very short; ovary deeply 4-lobed; style arising from the gynobase among the lobes; stigma capitate. Nutlets 4 or 1–3 by abortion, equal or unequal, often angled, attached for much of their length to the raised gynobase, the ventral scar groovelike and open or closed at the base, the surfaces smooth and shiny or variously roughened. *Oreocarya* Greene—Rydberg.

References: Johnston, I. M. 1923. Studies in the Boraginaceae. IV. The North American Species of *Cryptantha*. Contr. Gray Herb, n.s. 74: 1–114; Payson, E. B. 1927. A monograph of the section *Oreocarya* of *Cryptantha*. Ann. Missouri Bot. Gard. 14: 211–359.

Mature nutlets are needed for identification of many of our species.

Many more species, including some with yellow flowers, occur w of our range and may be expected now and then just within our w border.

1 Plants annual, from a slender taproot, the stems slender, leaves cauline; corolla 2 mm wide or less, inconspicuous.
  2 Nutlets smooth; stems simple or with a few branches below or above.
    3 Nutlets elongate, the scar open at the base; plants of sandy places, common .... 4. *C. fendleri*
    3 Nutlets ovoid, the scar closed at the base; plants of dry plains and pine slopes, rare with us .... 8. *C. torreyana*
  2 Nutlets not smooth; stem branched from the base, the branches often crowded.
    4 Inflorescences ebracteate or essentially so .... 3. *C. crassisepala*
    4 Inflorescences bracteate throughout .... 6. *C. minima*
1 Plants biennial or perennial, from a woody taproot, often with a branching caudex, the stems rather coarse; leaves basal and/or cauline; corolla 4–9 mm wide, conspicuous.
  5 Plants cespitose, silky strigose; nutlets muricate .... 1. *C. cana*
  5 Plants not cespitose or conspicuously silky; nutlets smooth or rugose-tuberculate.
    6 Nutlets not in contact with each other when ripe, smooth, the scar closed .... 5. *C. jamesii*
    6 Nutlets touching each other when ripe, rugose-tuberculate, the scar open or closed below.
      7 Nutlets dull or somewhat glossy, the scar closed at the base; inflorescences collectively rather narrow .... 2. *C. celosioides*
      7 Nutlets glossy, the scar open at the base; inflorescences collectively broad, round-topped .... 7. *C. thyrsiflora*

1. Cryptantha cana (A. Nels.) Pays. Perennial, to 1.5 dm tall, the stems cespitose, often densely so, from a branching woody caudex that is usually invested with old leaf bases below and spreading- or appressed-hirsute above. Leaves mostly clustered on the caudex, linear to narrowly oblanceolate, acute, to 4 cm long, 1.5–4 mm wide, densely silky-strigose, often weakly pustulate, the few cauline leaves similar. Inflorescences narrow, not much expanding with age, occupying the upper 1/2 of the stems; bracts conspicuous below. Calyx in fruit 5–6 mm long, setose and hirsute; corolla white, ca 4–6 mm wide. Nutlet usually 1, asymmetric, the dorsal face densely muricate, the scar open near the base. Early May–mid Jul. Dry sandy & gravelly soils of range lands & slopes; extreme sw SD, se WY, ne CO, NE panhandle; (SD, NE to cen WY, s to CO).

2. **Cryptantha celosioides** (Eastw.) Pays. Biennial (or perennial) from a tap-root, the caudex simple or branched. Stems 1–several, stout, erect, often from basal rosettes, to 3.5(5) dm tall spreading setose, the setae translucent. Leaves basal and/or cauline; basal leaves mostly spatulate, sometimes oblanceolate or obovate, obtuse, 2–5(8) cm long, 4–15 mm wide, setose, tomentose, and often pustulate above and below, often distinctly gray; cauline leaves not much reduced, oblanceolate to elliptic, less often spatulate, acute to obtuse, greenish, the vestiture less dense than the basal leaves, the cauline leaves greenish-gray. Inflorescences narrow and glomerate at first, paniculate with age, occupying 2/3–3/4 of the upper stem, the bracts below conspicuous. Calyx 8–10 mm long in fruit, setose-hirsute; corolla white, the eye yellow, 6–9 mm wide. Nutlets 4, in contact, dull to somewhat glossy, the dorsal face rugose-tuberculate, the scar closed. May–Jul. Dry pastures & canyons; ND & SD w of 100° long., e MT, e WY, ne panhandle & adj. counties of ne CO; (ND to Alta. & B.C. s to NE, CO, & OR). *C. confusa* Rydb., *Oreocarya affinis* Greene, *O. glomerata* (Pursh) Greene, *O. macounii* Eastw., *O. perennis* (A. Nels.) Rydb.—Rydberg; *C. bradburiana* Pays.—Harrington.

3. **Cryptantha crassisepala** (T. & G.) Greene. Plant annual, from a slender tap-root, to 1.5 dm tall, branching from the base, the branches decumbent to ascending, spreading-setose. Leaves linear to lanceolate, acute to obtuse, 1–3 cm long, 1–3 mm wide, strigose-pustulate above and below. Inflorescences elongate and racemelike with age, ebracteate or with a few slender bracts below, these exceeding the calyx in fruit. Calyx in fruit 3–7 mm long, spreading bristly-hispid on the outside, the bristles translucent, to 2 mm long; corolla white, 1–2 mm wide, inconspicuous. Nutlets 4, of 2 sizes, granulate-tuberculate. May–Jul. Dry prairies & slopes, prairie-dog towns; e CO, s-cen & sw KS, OK panhandle, ne NM; (CO to UT, e to KS, s to TX & Mex.).

This plant strongly resembles *C. minima*, with which it is reported to intergrade in CO.

4. **Cryptantha fendleri** (A. Gray) Greene. Plant annual, from a slender tap-root, 1–3(5) dm tall, the stem branched above and often below, strigose or spreading setose, but not densely so. Leaves cauline, linear to narrowly oblanceolate, acute, to ca 3(5) cm long, 1–5 mm wide, setose-hispid and pustulate above and below. Inflorescences elongate and racemelike with age, ebracteate or with a few bracts below. Calyx in fruit 4–6 mm long, setose-hispid, the setae translucent and larger than those of the stem; corolla white, ca 1 mm wide, inconspicuous. Nutlets 4, elongate, smooth, shiny, the scar open at the base. Jun–Aug. Dry sandy prairies & canyons, often common; sw SD, nw NE, se WY; (SD to Sask. & WA, s to OR, AZ, NM, NE).

5. **Cryptantha jamesii** (Torr.) Pays. Erect perennial from a woody tap-root, the caudices slender, branched. Stem branched from the base and above, 1–3 dm tall, appressed-setose or appressed hirsute. Leaves mostly cauline, linear to oblanceolate or spatulate, acute to obtuse, 3–6 cm long, 2–7 mm wide, the ones above often somewhat larger, strigose, and appressed-setulose and pustulate above, appressed-setulose or sericeous below. Inflorescences elongating to ca 10 cm with age, then ascending, the lower bracts leaflike, the upper reduced or absent. Sepals 5–7 mm long in fruit, tomentose and appressed-sericeous, setose but not coarsely bristly; corolla white, ca 5–7 mm wide. Nutlets 2–4, not in contact with each other at maturity, smooth and glossy, the scar closed. May–Aug (Sep). Dry sandy, gravelly, or clay slopes & dunes; e MT, e WY, sw SD, w NE, w KS, OK & TX panhandles, e CO,

ne NM; (SD to MT s to CA & TX). *Oreocarya suffruticosa* (Torr.) Greene—Rydberg.

6. ***Cryptantha minima*** Rydb. Resembling no. 3, q.v. Annual from a slender tap-root, to 1.5 dm tall. Stems numerous from the base, spreading-hirsute and appressed-hirsute. Leaves basal and cauline, linear to oblanceolate, acute to obtuse, to 2.5 cm long, 1–5 mm wide, coarsely setose and pustulate above and below. Inflorescence in fruit occupying about 2/3 of the stem, bracteate throughout, the bracts like the leaves but shorter. Sepals to 7 mm long in fruit, coarsely spreading-setose on the back, the margins velutinous; corolla white, 1–2 mm wide, inconspicuous. Nutlets tuberculate or papillate. May–Jul. Dry, sandy, & gravelly prairies & roadsides; e MT s through w NE, w KS, OK & TX panhandles, w to Rocky Mts.; (Sask. s to TX, e of Rocky Mts.).

7. ***Cryptantha thyrsiflora*** (Greene) Pays. Robust biennial or perennial from a simple or branching woody caudex that is usually invested with old leaf bases; stems 1–several, 1–4 dm tall, erect, stoutish, densely bristly-setose. Leaves basal and/or cauline, oblong to oblanceolate, obtuse, to 10 cm long, the largest ones tapering to a petiole, often strigose-pustulate and appressed-pubescent above and below; petioles ciliate, the lower leaves sometimes silky canescent. Inflorescences occupying upper 1/2–2/3 of the stems, collectively dense, broad, rounded at first, elongating and dense-paniculate with age; bracts large. Sepals 6–10 mm long in fruit, bristly-setose; corolla white, the fornices yellow, 5–7 mm wide. Nutlets ovoid, in contact, the dorsal face rugose-tuberculate, the scar narrowly open at the base. Jun–Jul. Rocky outcrops, calcareous buttes, open ponderosa pine forests; from Rocky Mts. e to BH, w-cen NE, cen KS, OK & TX panhandles, ne NM; (BH to TX, w to Rocky Mts.). *Oreocarya thyrsiflora* Greene—Rydberg.

8. ***Cryptantha torreyana*** (A. Gray) Greene. Annual, to 4 dm tall, the stem simple or branched, rather slender, spreading-hispid and somewhat sericeous. Leaves cauline, linear to oblong to barely oblanceolate, to 4 cm long, 2–4 mm wide, acute to obtuse, strigose and pustulate above, strigose below. Inflorescences usually paired at the branch tips, ebracteate. Calyx in fruit 5–7 mm long, strigose and bristly, the bristles translucent, larger than those of the stem; corolla white, ca 1 mm wide, inconspicuous. Nutlets 4, ovoid, smooth, or nearly so, shiny, the scar closed. Jul. Dry plains, pine slopes; ND: Bowman; WY: Campbell; (ND, MT to B.C. s to WY, UT, & CA). *C. calycosa* (Torr.) Rydb.—Rydberg.

4. CYNOGLOSSUM L., Hound's Tongue

Biennial and perennial herbs. Leaves basal and/or cauline, the basal and lower cauline ones petiolate, the upper ones sessile, the stem leafy to the inflorescences or naked just below them. Cymes solitary or paired at the stem tips, solitary from upper axils, scorpioid when young, elongate and racemelike in fruit, essentially ebracteate, the flowers short-pedicellate. Sepals divided almost to the base, equal, little or somewht enlarged in fruit; corolla blue, purplish-red (white to lilac), broadly funnelform to salverform, less than 1 cm long, the tube about equaling the calyx, the fornices prominent; anthers included or barely exserted. Nutlets 1–4, rather divergent, oblique, attached to the raised gynobase near the tip (but appearing to be basally attached), the surface densely glochidiate except for the wide scar.

1 Stem leafy to the inflorescences; corolla purplish-red; nutlets dorsally flattened, the scar reaching to the middle of the ventral side; disturbed places in the open, GP .... 2. *C. officinale*

1 Stem naked above; corolla blue; nutlets obovoid, the scar reaching only ca 1/4 to the middle of the ventral side; woods & thickets, ND .... 1. *C. boreale*

1. Cynoglossum boreale Fern., northern wild comfrey. Perennial to 8 dm tall, the stem simple, hirsute, leafy below, naked above. Basal and lower cauline leaves petiolate, the blade elliptic to ovate, acute, to 20 cm long and 7 cm wide; upper cauline leaves sessile, clasping, more or less oblong, acute; leaves sparingly strigose above and below. Cymes 1 or 2, terminating the stem, ebracteate, the peduncle naked, conspicuous. Sepals 1–2.5 mm long in flower and fruit, strigose; corolla blue, funnelform, the tube barely if at all exceeding the calyx, the limb 5–8 mm wide, the lobes ovate and the sinuses open, the fornices exserted; anthers included; style very short, 1–2 mm long. Nutlets 3–5 mm long, obovoid, entirely glochidiate, the scar reaching only about 1/4 of the way to the middle. Jun–Jul. Dry woods & thickets; n & w ND; (Newf. to Man. & B.C. s to CT, NY, IN, IA, ND).

A similar but larger species, *C. virginianum* L., is known at the edge of our area in OK and MO and may be expected occasionally just within our se border. It has the sepals 3–5 mm long, the corolla white to lilac and 8–12 mm wide, the lobes rounded and the sinuses closed, the nutlets 5–7 mm long.

2. Cynoglossum officinale L., hound's tongue. Coarse, erect biennial from a thick woody tap-root, pubescent throughout, emitting a musty odor. Stem 1, unbranched below the inflorescence, hirsute, to 6 dm tall, leafy throughout, Basal and lower cauline leaves petiolate, elliptic to oblanceolate, to 15(20) cm long, 2–5 cm wide, the base tapering; upper leaves numerous, not reduced or even larger, acute, sessile, acute to obtuse; all leaves pubescent above and below. Cymes numerous, solitary in upper axils, collectively forming a thyrse, glomerate-coiled in anthesis, much elongating (to 15 cm) and racemelike in fruit; flowers subsessile and crowded in anthesis, pedicellate and less crowded in fruit, not subtended by bracts. Calyx 4–6 mm long in flower, enlarging to nearly 1 cm in fruit, the sepals elliptic to ovate, rounded to blunt, hirsute; corolla dull purplish-red, salverform or nearly so, the tube shorter than the calyx, the limb 5–8 mm wide and barely exserted from the calyx, the lobes rounded, overlapping, the fornices and anthers exserted, the fornices exceeding the anthers, the filaments attached near the top of the tube; stigma 1, included. Nutlets 1–4, brown, 5–7 mm long, ovoid, flattened above, glochidiate all over, the scar extending to about the middle of the ventral surface. (n = 12) May–Jul. Disturbed pastures, roadsides, forest edges, meadows; GP; (over much of N. Amer.; Europe). *Introduced*.

5. ECHIUM L., Blueweed, Viper's Bugloss

1. Echium vulgare L. Erect., coarse biennial to 9 dm tall, the stem usually 1, simple to the inflorescence, often pustulate-bristly. Leaves basal and cauline, the basal ones oblanceolate, to 25 cm long and 3 cm wide, tapering to a petiolelike base; cauline leaves becoming smaller, sessile above; all leaves bristly, pustulate-bristly, and/or hirsute, painful to the touch and producing dermatitis. Cymes short, coiled at first, elongating to ca 7 cm in fruit, aggregated into an elongate, thyrse, bracteate throughout; flowers sessile, crowded. Calyx 5–6 mm long, divided to the base, strigose; corolla bright blue, funnelform, irregular, much exceeding the calyx, ca 1–2 cm long, without fornices; stamens 5, 4 exserted, the fifth barely so; style filiform, pubescent, exserted, stigmas 2, filiform. Nutlets 1–4, much shorter than the persistent calyx, attached at the base to the flat gynobase, ovoid, 3-angled, rugose. (n = 16, 18) Jun–Aug. Roadsides, weedy pastures & fields, waste places; e 1/3 GP, BH, sporadic elsewhere, apparently not abundant for long periods at a site; (e U.S. to GP, also WA; Europe). *Introduced*.

Pink- and white-flowered forms are known beyond our range and may be expected with us. The plant is sometimes a troublesome weed to the e of our range but only seldom so in the GP.

6. HACKELIA Opiz, Stickseed

Biennials and perennials; stems erect, 1–few, usually branched above, Leaves basal and cauline, alternate, hirsute to strigose, the basal ones soon withering and seldom observed; lower cauline leaves tapering to a petiolelike base, or distinctly petiolate; upper cauline leaves sessile and smaller to larger. Cymes paired at the branch tips, circinate at first, racemelike and expanded with age, bracteate below, naked above, the bracts opposed to the flowers or alternate with them; fruiting pedicels recurved or reflexed. Sepals free or nearly so, persistent in fruit, the lobes then more or less spreading, strigose; corolla white to blue, salverform, barely if at all exceeding the calyx, the tube short, the throat appendages (fornices) at the mouth prominent, papillate, the corolla lobes prominent and rounded; stamens 5, included, attached above the middle of the corolla tube; style 1, slender, the stigma capitate. Nutlets 4, exceeding the style, attached only in the middle to the broadly pyramidal gynobase, bearing 2 marginal rows of prominent hooked (glochidiate) flattened bristles, and sometimes also bristles on the dorsal face.

Reference: Gentry, J. L., & R. L. Carr. 1976. A revision of the genus *Hackelia* (Boraginaceae) in North America, north of Mexico. Mem. New York Bot. Gard. 26: 121–227.

In our area, the similar genus *Lappula* differs from *Hackelia* in having cymes bracteate throughout, the pedicels erect in fruit, the style not set sharply apart from the gynobase, the gynobase rather slender, about equaling the nutlets and bearing 4 distinct keels. Also, *Lappula* is annual and is often found in disturbed, weedy places.

1 Dorsal area of nutlets with bristles about equaling the marginal bristles in length; corolla 2–3 mm wide; e & se GP, BH .... 3. *H. virginiana*
1 Dorsal area of nutlets without bristles bweteen the marginal rows of bristles, or with a few bristles shorter than the marginal bristles; corolla 1.5–8 mm wide; n & nw GP.
  2 Nutlets 2–3 mm long, often with a few short bristles on the dorsal face, the marginal bristles only slightly flared below; corolla 1.5–3 mm wide, inconspicuous; GP from Niobrara R. & ne NE northward .... 1. *H. deflexa*
  2 Nutlets 2–4 mm long, the dorsal face without bristles, the marginal bristles broadly flaring below; corolla 4–8 mm wide, rather conspicuous but not showy; nw ND; e MT, extreme w SD & NE, e WY .... 2. *H. floribunda*

1. ***Hackelia deflexa*** (Wahl.) Opiz. Slender biennial-perennial, to 9 dm tall. Stems often solitary, branched above, strigose above, spreading-hirsute below, the branches all terminating in paired cymes. Lower leaves oblanceolate, 6–12(18) cm lon, 1.5–3.5 cm wide, acute; upper leaves lanceolate or narrowly elliptic, acuminate, smaller to larger than the lower leaves. Inflorescences to 10 cm long with age, the fruiting pedicels 3–7(10) mm long. Sepals 1.5–2.5 mm long, ovate, acute, sparingly strigose-hirsute on the outside, essentially glabrous inside; corolla inconspicuous, 1.5–3 mm wide, blue to white. Nutlets 2–3 mm long, the dorsal face naked or with a few bristles shorter than the marginal bristles, the marginal bristles slightly flaring below, distinct to the base. Flowers and fruits Jun–Aug. Woods & thickets on calcareous soils; GP from n NE northward; (N.B. to N.T., s to VT, IA, NE, CO, WA; Eurasia). *Lappula americana* (A. Gray) Rydb., *L. angustata* Rydb., *L. scaberrina* Piper—Rydberg.

2. ***Hackelia floribunda*** (Lehm.) I. M. Johnst., large-flowered stickseed. Plant biennial or perennial, to 7(10) dm tall. Stems erect, 1–few, branched above, often retrorsely strigose below, antrorsely strigose above. Basal leaves early deciduous, seldom observed; lower cauline leaves oblanceolate, ca 3–18(25) cm long, 5–25(30) mm wide, upper cauline leaves

oblanceolate to linear, acute. Cymes borne in axils of upper leaves and bracts, often in pairs, collectively almost paniculate. Sepals oblong to narrowly ovate, hirsute on the outside, glabrous inside, ca 2.3 mm long in fruit; corolla ca 4–8 mm wide, blue (white), the tube shorter than the calyx lobes. Nutlets 2.5–4 mm long, ± ovoid, on pedicels 2–10 mm long, the dorsal face naked, marginal bristles broadly flaring below, free to somewhat fused basally. (n = ca 12) Jun–Aug. Moist creek banks, open woods, sometimes drier soils; n ND, e MT, e WY, e CO, sw SD & BH, NE panhandle; (Sask. to B.C. s to CA, AZ, NM, w NE; Mex.). *Lappula floribunda* (Lehm.) Greene — Rydberg.

3. ***Hackelia virginiana*** (L.) I. M. Johnst. Biennial to 10(15) dm tall. Stems erect, 1–few, branched above, hirsute and sometimes retrorsely strigose below, hirsute and antrorsely strigose above; branches all terminating in paired spreading cymes. Basal leaves smaller than the cauline leaves and withering early; lower cauline leaves to ca 30 cm long, petiolate, 3–10 cm wide, ovate, acute, the lateral veins evident; upper leaves somewhat smaller, sessile, acute to acuminate. Cymes to 15 cm long with age and spreading; fruiting pedicels recurved-reflexed, 2–10 mm long. Sepals in fruit 2–3 mm long, lanceolate, strigose outside, strigulose to puberulent on the inside; corolla white to pale blue, 2–3 mm wide, the tube shorter than the calyx. Nutlets collectively globose, ca 2–3 mm long, the dorsal surface with numerous bristles as long as the marginal bristles; marginal bristles narrowly flared below, distinct. (n = 12) Jun–Sep. Dry to moist woods & thickets; e & se GP from e ND s through cen NE, w-cen KS, & w-cen OK to our e border, also BH; (N.B. to e ND, s to GA, AL, OK, & ne TX). *Lappula virginiana* (L.) Greene — Rydberg.

7. HELIOTROPIUM L., Heliotrope

Ours glabrous or variously pubescent annual herbs, the stems branched, prostrate to erect. Leaves cauline, petiolate to sessile, linear to almost deltoid. Cymes terminal, with or without bracts, elongating and spiciform with age, or the flowers solitary in the upper axils and at the branch tips. Sepals free or nearly so, equal or unequal, glabrous, hirsute or strigose. Corolla funnelform to salverform, white to blue, without fornices. Stamens and style included; style 1; stigma capitate, flattened or bifid. Nutlets 4, free or united in pairs, sometimes appearing as 2. *Tiaridium* Lehm., *Lithococca* Small, *Euploca* Nutt. — Rydberg.

Various cultivars of Peruvian species are grown as annuals in GP gardens. These have massive inflorescences and fragrant blue, purple, or white flowers.

1 Plants glabrous, rather succulent; damp saline soils and marshes .... 2. *H. curassavicum*
1 Plants variously pubescent, hirsute, or strigose, not succulent; dry or damp soils.
   2 Flowers solitary at the branch tips an in upper axils; leaves linear .... 4. *H. tenellum*
   2 Flowers borne in scorpioid cymes; leaves not linear.
      3 Corolla blue, 2–4 mm wide; cymes elongate, spiciform with age; leaves ovate to nearly deltoid, mostly 4–8 cm long; se GP .... 3. *H. indicum*
      3 Corolla white, 8–20 mm wide; leaves ovate to lanceolate, 1–4 cm long; s & sw GP .... 1. *H. convolvulaceum*

1. ***Heliotropium convolvulaceum*** (Nutt.) A. Gray. Annual to 2.5–(3.5) dm tall, from a slender tap-root, branched from the base and above, the stems prostrate or ascending, strigose. Leaves petiolate, 1–4 cm long, the blade ovate to lanceolate, acute to obtuse, strigose above and below, sometimes weakly pustulate. Cymes at the branch tips, not elongate with age, ebracteate or with a few bracts. Sepals ca 5 mm long, subequal, narrowly lanceolate, strigose; corolla salverform to funnelform, showy, white, the tube much exceeding the calyx and again constricted below, the limb 8–20 mm wide, the tube and midribs of the limb strigose; style long, slender, naked; stigma included, capitate, strigose. Nutlets 4 (sometimes appearing as 2), ca 3 mm long, fused in pairs, the adjacent sides flat, smooth, reniform, the exposed sides rounded, appressed-hirsute, the scar with a keel at either side. Jul–Aug.

Sandy hills, plains, & riverbanks; s-cen KS to n CO s to TX panhandle & cen OK; (KS to UT s to TX & Mex.) *Euploca convolvulacea* Nutt.—Rydberg.

2. **Heliotropium curassavicum** L. Succulent, glabrous annual, reportedly sometimes perennial, from a thick rhizomelike tap-root. Stem to 4 dm tall, branched below and above, the branches prostrate to ascending, rather thick. Lower leaves opposite, subopposite, or alternate; upper leaves alternate; leaves succulent, lanceolate to obovate, 2-6 cm long, 4-20 mm wide, tapering to a petiolelike base, acute to rounded. Cymes 1-several on terminal and axillary peduncles, coiled when young, elongate and spikelike with age, ebracteate, the flowers rather crowded, short-pedicellate to subsessile. Sepals 2-3 mm long, not exceeding the mature fruit; corolla funnelform to subsalverform, white or bluish with a yellow eye, the tube about equaling the calyx, the limb 4-7 mm wide; style very short to absent; stigma flat, expanded, as wide as the ovary and often persisting in fruit. Nutlets 4, 2-2.5 mm long, smooth or faintly ribbed. (n = 13, 14) Jun-Oct. Damp saline soils; persistent in dry marshes; ND, SD, MT, NE panhandle s through e CO, cen & w KS to OK & TX; also se NE: Lancaster; (over much of Can. & w U.S., to Cascade Mts.).

Many of our specimens are var. *obovatum* DC. (*H. spathulatum* Rydb—Rydberg), with spatulate leaves and somewhat larger (6-7 mm wide) flowers.

3. **Heliotropium indicum** L., turnsole. Erect annual from a slender tap-root, the stems branched, to 8 dm tall, hirsute to hispid and bristly. Leaves cauline, usually opposite below, alternate above, petiolate, rugose, ovate to subdeltoid, 4-8(12) cm long, 2-6 cm wide, acute to obtuse, tapering to the petiole. Cymes coiled when young, elongate (to ca 15 cm) and spikelike with age, mostly solitary (2) at the stem tips, ebracteate, the flowers crowded, sessile, the rachis hirsute. Sepals 1-2 mm long, much shorter than the mature fruit, hirsute; corolla salverform, blue, the tube exceeding the calyx, the limb 2-4 mm wide, the tube and outer face of the limb pubescent; style ca 1 mm long; stigma capitate. Nutlets 4, at first paired and appearing as 2, mitriform, each with 1 or 2 dorsal ribs, the veins faintly reticulate. Jul-Oct. Damp soil, silty ditches, lakesides; w-cen MO from Kansas City s, se KS: Cherokee, Crawford; ne OK, & occasionally elsewhere in se 1/4 GP; (VA to w MO & se KS s to FL & e TX; tropics). *Introduced. Tiaridium indicum* (L.) Lehm.— Rydberg.

4. **Heliotropium tenellum** (Nutt.) Torr. Erect annual, the stems to 4 dm tall, slender, much branched, strigose. Leaves cauline, sessile, linear, 1-3 cm long, 1-3 mm wide, white-strigose, the margins often revolute, the tip acuminate. Flowers solitary at the branch tips and in upper axils. Calyx lobes unequal, at least 1 of them equaling the corolla, densely strigose; corolla funnelform, white, the tube not exceeding the shortest sepals, the limb 3-5 mm wide, the tube strigose, the outer face of the limb strigulose-pubescent; style ca 1 mm long; stigma bifid. Nutlets 4, smooth, strigulose-pubescent, much shorter than the persistent sepals. Jun-Aug. Dry, often limey soils in the open or in open woodlands; w-cen MO, se KS, cen & ne OK; (KY to se KS s to AL & e TX). *Lithococca tenella* (Nutt.) Small— Rydberg.

8. LAPPULA Gilib., Stickseed

Erect annuals from a taproot, coarsely pubescent to strigose throughout, the hairs of the leaves and sepals often with a broad flat base (pustular). Stems 1-few, branched above, each branch terminating in inflorescences. Leaves alternate, sessile, narrow. Cymes paired at the branch tips and solitary in upper axils, tightly scorpioid at first, becoming elongate and racemelike with age, bracteate throughout, the pedicels more or less erect to spreading in fruit. Sepals free or nearly so, linear to lanceolate or oblanceolate, acute, coarsely hirsute to strigose, especially outside; corolla white to blue, inconspicuous, salverform to funnelform, the fornices prominent; stamens inserted high on the tube, included; style short, included. Nutlets 4, attached to the elongate gynobase for most of their length on a distinct ventral keel, but free at the base, the dorsal face unadorned, verrucose, or tubercled, bordered by 1 or 2 rows of uncinate bristles, these free or basally united.

Reference: Johnston, I. M. 1924. Studies in the Boraginaceae. II. *Lappula*. Contr. Gray Herb. 70: 47–51.

See *Hackelia* for discussion of differences between that genus and *Lappula*.

```
1  Bristles of the nutlets in 2(3) rows.
   2  Inner bristles longer than the outer, mostly shorter than the body of the nutlet; nutlets
      3–4 mm long; corolla 2–3 mm wide; GP n of n NE ............................ 2. L. echinata
   2  Bristles about equal, about as long or longer than the body of the nutlet; nutlets 4–5
      mm long; corolla 1–2 mm wide; nw GP ........................................ 1. L. cenchrusoides
1  Bristles of the nutlets in 1 row.
   3  Bristles free; GP .................................................................... 3. L. redowskii
   3  Bristles united to form a cup around the dorsal face; w GP .............. 4. L. texana
```

1. *Lappula cenchrusoides* A. Nels. Plant 2–4 dm tall, much branched. Leaves strigose-hispid, 1–2 cm long, oblong to ovate, obtuse. Corolla blue, 1–2 mm wide. Nutlets 4–5 mm long, tubercled on the back; bristles in 2(3) rows around the dorsal face, about as long as the body of the nutlet, bristles free to the base or essentially so. Aug–Sep. Dry soils in the open; restricted to ND & SD w of the Missouri R., e MT, e WY, extreme w NE.

2. *Lappula echinata* Gilib., blue stickseed. Stem branched above, to 8 dm tall, pubescent. Leaves linear to linear-oblanceolate, sessile, ascending, 2(8) cm long, 2–5 mm wide, acute to obtuse, rough-hirsute. Cymes paired at the branch tips. Corolla blue, 2–3(4) mm wide. Nutlets 3–4 mm long, the dorsal face with 2(3) rows of bristles around it, the inner bristles longer than the outer, the bristles free to the base. (n = 24) May–Aug. A weed in open waste places; GP n of KS-NE border, & e KS; (N. Amer. e of the Cascade Mts; Eurasia). Introduced. *L. echinata* Gilib., *L. erecta* A. Nels., *L. fremontii* (Torr.) Greene — Rydberg.

3. *Lappula redowskii* (Hornem.) Greene. Strongly resembling no. 2, the leaves merely pubescent. Corolla blue or white, 1.5–2.5 mm wide. Nutlets 2–3 mm long, the dorsal face with a single row of bristles around it, these essentially free to the base. May–Aug. Dry, open waste places, often on sandy soil; GP; (N. Amer. e of the Cascade Mts.; Eurasia). *L. occidentale* Rydb. — Rydberg.

4. *Lappula texana* (Scheele) Britt., cupseed stickseed. Very similar to no. 3, but the bristles of the nutlet distinctly united at the base to form a cup around the dorsal face. May–Aug. Gravelly valleys & waste places; GP w of ca 90° long., & sporadic elsewhere; (w ND to ID, s to TX & Mex.). *L. foliosa* A. Nels., *L. heterosperma* Greene — Rydberg.

This entity is sometimes called *L. redowskii* (Hornem.) Greene var. *texana* (Scheele) Brand.

9. LITHOSPERMUM L.

Annual and perennial spring-flowering herbs with variously pubescent stems and foliage. Leaves cauline, occasionally also basal, sessile, linear to ovate. Flowers solitary in the axils of upper leaves or crowded in leafy scorpioid cymes, sometimes heterostylous. Calyx lobes nearly free, narrow; corolla white to yellow-orange, usually exceeding the calyx, salverform to funnelform, often with conspicuous fornices at the top of the the throat; stamens inserted near the middle or top of the tube, the anthers included or partly exserted; fruiting pedicels erect. Nutlets (1–3)4, ovoid or angular, very hard, porcelainlike, smooth or rough, white, whitish, or brownish-gray, basally attached, the scar large; gynobase depressed or flat.

References: Govoni, D. 1973. The taxonomy of the genus *Lithospermum* L. (Boraginaceae) in the western Great Plains. PhD. dissertation, Univ. of Nebraska; Johnston, I. M. 1952. A survey of the genus *Lithospermum*. J. Arnold Arbor. 33: 299–366.

1 Flowers in leafy cymes at the branch tips, crowded at least at first; corolla yellow to yellow-orange, sometimes greenish-yellow, much exceeding the calyx.
  2 Petals erose to fimbriate-denticulate, seldom entire; corolla distinctly salverform; nutlets smooth to punctate .... 4. *L. incisum*
  2 Petals entire, corolla salverform to funnelform; nutlets smooth; leaves narrowly to broadly elliptical.
    3 Leaves softly canescent, the tip obtuse, the hairs without a papillose base .... 2. *L. canescens*
    3 Leaves more or less strigose, the tip more or less acute, the hairs with a papillose base .... 3. *L. carolinense*
1 Flowers in axils of upper leaves, not crowded in cymes; corolla white, greenish, or pale yellow, barely if at all exceeding the calyx.
  4 Plant annual from a slender taproot; corolla white or whitish, without fornices; leaves without evident lateral veins; nutlets rugose or tuberculate .... 1. *L. arvense*
  4 Plant perennial from a thick taproot; corolla with or without fornices; leaves with evident lateral veins; nutlets smooth or punctate.
    5 Leaves to 4 cm wide; corolla greenish-white to yellowish; nutlets smooth or essentially so .... 5. *L. latifolium*
    5 Leaves to 1 cm wide; corolla greenish-yellow to pale yellow or yellow-orange; nutlets punctate or smooth .... 4. *L. incisum*

1. ***Lithospermum arvense*** L. Strigose annual from a slender taproot, to 6 dm tall but usually much less. Stems simple or branched from the base. Leaves linear to lanceolate, acute to obtuse, the lateral veins not evident, 1–5 cm long, to 10(15) mm wide. Flowers solitary in axils of upper leaves, nearly sessile. Corolla white to bluish-white, barely exceeding the calyx, to 8 mm long and 4 mm wide, without fornices. Nutlets brownish-gray or gray, rugose or tuberculate, with a single keel. (n = 14) Apr–Jul. A weed of waste places, often on sandy soil; scattered through all but sw GP, apparently most abundant in se NE & e KS, but perhaps often overlooked elsewhere; (much of N. Amer.; Eurasia). *Introduced*.

2. ***Lithospermum canescens*** (Michx.) Lehm., hoary puccoon. Perennial from a thick vertical taproot, 12–35 cm tall, the stems several and simple or branched near the top. Leaves cauline, often ascending, lanceolate to elliptic, obtuse, 20–55 mm long, 4–11 mm wide, softly canescent or sericeous, the hairs without a papillose base. Cymes terminal; the bracts much longer than the calyx. Flowers heterostylous; calyx 4–17 mm long; corolla 7–18 mm long, to 1.4 cm wide, yellow to yellow-orange, the lobes entire, the fornices conspicuous, gibbous; stamens borne near the middle of the tube in long-styled flowers or at the middle of the tube in short-styled flowers. Nutlets smooth, shiny, yellowish white, porcelainlike, ovoid, with a ventral keel. Apr–Jun. Dry prairies & open woods, seldom on sandy soils; cen & e ND & w MN s through extreme e parts of SD, NE, & KS to extreme ne OK, & sw MO; also s tier SD & BH; (PA to Sask., s to TX, MO, GA).

L. multiflorum Torr. ex Gray is reported from OK: Cimarron and might key here or to *L. carolinense*. It has yellow, heterostylous flowers but differs in having bracts equal to calyx lobes.

3. ***Lithospermum carolinense*** (Walt.) MacM., puccoon. Perennial from a thick vertical taproot, 20–45 cm tall, the stems simple or branched above. Leaves cauline, crowded, linear to lanceolate, 2–6 cm long, 3–12 mm wide, strigose, the hairs with a papillose base. Cymes terminal; the bracts much longer than the calyx. Flowers heterostylous; calyx strigose, 7–27

mm long; corolla bright yellow-orange, salverform, 11–22 mm long, the lobes entire, the tube in short-styled flowers exceeding the calyx, in long-styled flowers about equaling the calyx, the fornices weakly invaginate; stamens attached near the middle of the tube in long-styled flowers, near the top of the tube in short-styled flowers; style reaching top of the tube in long-styled flowers, reaching middle of the tube in short-styled flowers. Nutlets smooth, shiny, white, porcelainlike, ovoid, ventrally keeled. (n = 8) May–Jun. Sandy prairies & open woods, especially on sandy soil; s-cen SD, NE, ne CO, e & cen KS, OK (except panhandle), e TX panhandle; (NY to SD & NV, s to FL, TX, & Mex.). *L. carolinense*, *L. croceum* Fern.—Fernald; *L. gmelinii* (Michx.) Hitchc.—Rydberg.

4. *Lithospermum incisum* Lehm. Strigose perennial from a woody taproot; stems to 4 dm tall, usually branched above with age. Leaves cauline, basal leaves sometimes present, linear or narrowly lanceolate, acute, 13–45 mm long, 4–10 mm wide, strigose, the hairs sometimes with a papillose base. Cymes terminal. Early, chasmogamous flowers with calyx 3–15 mm long, corolla pale yellow, yellow-orange, or occasionally yellow-green, salverform, 18–48 mm long, ca 15 mm wide, the lobes usually erose or fimbriate-denticulate, occasionally entire, the tube very slender, the throat with weakly to strongly invaginate fornices; stamens borne only near the top of the tube, the style long and the stigmas barely to strongly exserted. Later flowers smaller, cleistogamous, corollas only 2–6 mm long or absent; calyx much exceeding the corolla; the style short. Nutlets white, shiny, porcelainlike, sometimes pitted, constricted near the base. (n = 12) Apr–Jun. Dry prairies & open woods, on sandy, clay, or loamy soils, sometimes in rather disturbed habitats; GP; (Ont. to B.C., s to UT, TX, MO, IN, n Mex.). *L. brevifolium* Engelm. & Gray, *L. linearifolium* Goldie, *L. mandanense* Spreng.—Rydberg.

Plants of the s GP are lower, bushier, and more often have basal leaves, while in the n GP they are taller, less often have basal leaves, and have flowers more obscured by the foliage.

5. *Lithospermum latifolium* Michx., gromwell. Perennial, the stem simple or branched above, to 8 dm tall, the stems and leaves harshly strigose to scabrous. Leaves not crowded, rather thin, lanceolate to ovate, acuminate, to 4 cm wide, the lateral veins conspicuous. Flowers solitary in axils of upper leaves. Corolla barely if at all exceeding the calyx, to ca 6 mm long and about as wide, greenish-white or yellowish, the throat with 5 appendages; stamens attached near middle of tube; style ca 1 mm long. Nutlets white, shiny, smooth or barely pitted. May–Jun. Moist woodlands; w MN, w IA, nw MO, e-cen KS; (NY to w MN, s to TN, AR, e-cen KS).

10. LYCOPSIS L.

1. *Lycopsis arvensis* L. Annual to 4 dm or more tall. Stems erect or ascending, hirsute-hispid, the hairs basally swollen. Leaves cauline, alternate, sessile, oblong to lanceolate or oblanceolate, mostly 3–10 cm long, 5–20 mm wide, hirsute-hispid above and below, obtuse, the margins entire to undulate. Cymes 1 or 2 at the branch tips, bracteate throughout, the flowers crowded and very short-pedicellate, the pedicels elongating in fruit to ca 1 cm. Calyx divided nearly to the base, 4–5 mm long in flower, about twice that in fruit, hirsute; corolla blue, funnelform, tube bent, limb ca 4–6 mm wide, the lobes unequal, fornices hairy, occluding the throat; stamens attached below the middle of the tube; style 1, stigma 1. Nutlets obliquely ovoid, reticulately ridged, tuberculate; scar oval, on the lower ½ of the ventral side, convex, surrounded by an elevated ring; gynobase with concavities where the nutlets attach. (2n = 48) Jul–Sep. A weed in waste places; reported from ND: Stark, Stutsman, & to be sought elsewhere in ne GP; (naturalized in ne U.S., GP; Eurasia). *Introduced.*

11. MERTENSIA Roth

Ours perennials with 1–several erect or ascending, usually simple stems from a woody caudex. Leaves basal and cauline, petiolate to sessile and amplexicaul, glabrous to strigose, the shapes somewhat inconstant on the plants. Cymes produced from axils of upper leaves, simple or branched, ebracteate, compact but usually expanding with age, the flowers often pendulous. Calyx 5-parted, partly to almost completely divided to the base; corolla with distinct tube and limb, tubular blue, (occasionally white or pink in some individuals), pink in bud, shallowly 5-lobed, the lobe rounded, the limb (including the upper part of the throat) somewhat to distinctly expanded; throat with fornices near the mouth, these often conspicuous, glabrous to pubescent; stamens inserted near the appendages, the filaments slender or expanded; style long and the stigmas usually exserted from the throat. Nutlets laterally attached to the convex gynobase, rugose.

References: Macbride, J. F. 1916. The true mertensias of western North America. Contr. Gray Herb. n.s. 48: 1–20; Williams, L. O. 1937. A monograph of the genus *Mertensia* in North America. Ann. Missouri Bot. Gard. 24: 17–159.

1 Corolla 2–4 cm long, the fornices inconspicuous, filaments not expanded; foliage glabrous, the margins unadorned; w MO, se KS, & cultivated in shady gardens .... 4. *M. virginica*
1 Corolla to 2 cm long, the fornices prominent; filaments expanded; foliage glabrous to strigose, the margins mostly strigose; nw GP.
    2 Leaves to 10 cm wide, glabrous (but the margins usually strigose); lateral veins of leaves evident .... 1. *M. ciliata*
    2 Leaves to 4 cm wide, glabrous to pubescent or strigose; lateral veins, at least of cauline leaves, not strongly evident.
        3 Corolla tube longer than the limb; calyx divided nearly to the base; leaves to 2 cm wide, obtuse to rounded .... 3. *M. oblongifolia*
        3 Corolla tube about equal to the limb; calyx divided 1/2–2/3 to the base; leaves to 4 cm wide, acute to obtuse .... 2. *M. lanceolata*

1. *Mertensia ciliata* (Torr.) G. Don. Plant to 1.2 m tall, stems rather thick, glabrous. Leaves thin, glabrous, the lateral veins evident, the margins usually strigose; basal leaves petiolate, the petiole usually narrow-winged, the blade elliptic to nearly ovate, occasionally subcordate, to 10 cm wide; cauline leaves becoming sessile and reduced above, acute to acuminate. Calyx 1–3 mm long, divided nearly to the base, glabrous, the margins papillate to strigose; corolla 1–1.5(2) cm long, the tube shorter than to longer than the limb; limb moderately expanded, the throat appendages prominent; filaments broad. Nutlets rugose or tuberculate. (n = 12, 24). Jun–Jul. Damp thickets, shady stream sides, moist ledges; reported from a few widely scattered localities in e CO, e MT, w SD, & se WY; (SD to OR, s to NM & CA).

2. *Mertensia lanceolata* (Pursh) A. DC. Plant to 4 dm tall; stems glabrous. Basal leaves petiolate, narrowly ovate to elliptic or lanceolate, glabrous or scantily strigose, sometimes pustulate, to 14 cm long and 4 cm wide, usually much less. Cauline leaves sessile to amplexicaul, somewhat reduced above, without prominent lateral veins, strigose on the margins, acute to obtuse. Calyx 2–9 mm long, glabrous, the margins strigose, divided 1/2–2/3 to the base; sepals lanceolate to narrowly ovate, acute; corolla to ca 15 mm long, the tube about equaling the limb; limb rather expanded; the fornices conspicuous; filaments broad. Nutlets black, rugose. (2n = 24) May–Jul. Open or brushy canyon sides & prairies, often in rocky soil; cen & w ND & SD, NE panhandle, e CO; also reported form e SD near the Missouri R.; (ND, MT, & Sask., s to NE & NM). *M. linearis* Greene, *M. papillosa* Greene—Rydberg.

    The var. *brachyloba* (Greene) Pays., with the corolla limb shorter than the tube, occurs in CO: Larimer, at the foot of the mountains at the very edge of our range.

3. ***Mertensia oblongifolia*** (Nutt.) G. Don. Plant to 4 dm tall; stems glabrous. Leaves rather thick, often strigose above and glabrous below. Basal leaves long-petioled, elliptic to lanceolate, oblanceolate, or narrowly ovate, obtuse to rounded, to ca 10 cm long and 2(6) cm wide; cauline leaves becoming sessile above and smaller, without prominent lateral veins. Calyx to 6 mm long, divided nearly to the base; sepals lanceolate, acute to acuminate, glabrous, the margins strigose; corollas to 2 cm long, the tube longer than the limb, the limb often rather expanded; filaments expanded. Nutlets rugose. May–Jun. Dry or moist meadows & ravines; ND w of Missouri R., w SD, e WY; (ND to WA, s to NV & CO). *M. coronata* A. Nels., *M. foliosa* A. Nels.—Rydberg.

4. ***Mertensia virginica*** (L.) Pers., bluebells. Plant rather succulent, to ca 6 dm tall, glabrous, rather glaucous. Leaves thin, pinnately veined, obtuse. Basal and lower cauline leaves petiolate, the petiole sometimes narrowly winged, the blade broadly ovate to elliptic or spatulate, glabrous, without marginal ornamentation; upper cauline leaves smaller, sessile to amplexicaul. Peduncles rather slender, the cyme about equaling the subtending leaf at flowering time. Flowers often pendulous; calyx to 10 mm long, divided barely more than ½ way to the base; corolla to 4(6) cm long, the limb almost campanulate, the fornices inconspicuous; filaments slender. Nutlets rugose. Apr–May. Moist woods & stream sides; w MO, se KS: Miami (where perhaps extirpated) & introduced well beyond its range in shady gardens & woodlands; (NY to se MN, s to VA, TN, & n AL, across e IA to w-cen MO & se KS).

The entire plant dies back to the ground soon after flowering. White- and pink-flowered forms are known and are sometimes seen in GP gardens. The pink-flowered f. *rosea* Steyerm. has been collected in our area in MO: Cass.

12. MYOSOTIS L., Forget-me-not

Annual and perennial herbs to 6 dm tall, glabrous to hirsute or sparsely strigose throughout. Basal leaves, when present, petiolate; cauline leaves sessile, alternate. Cymes paired at the branch tips, solitary in upper axils, coiled at first, expanding, racemelike and ascending with age, ebracteate, or with a few bracts below and ebracteate above; fruiting pedicels ascending or spreading. Calyx divided part way to the base, strigose or hirsute, the lobes equal or unequal; corolla salverform to nearly funnelform, blue or white, sometimes with a yellow eye, the lobes prominent, rounded, the fornices prominent; anthers and style included. Nutlets 4, attached to the low gynobase by a basilateral scar, smooth, shiny, convex at least on one side, with a distinct sharp keel all around.

1 Lower part of inflorescences with bracts; corolla white, funnelform, 1–2 mm wide; calyx lobes unequal; weedy annual; e & se GP .... 3. *M. verna*
1 Inflorescences without bracts in the floriferous portion; corolla blue, salverform, 4–10 mm wide; calyx lobes equal or nearly so; perennials of moist soils; BH and a few other widely scattered sites.
  2 Calyx divided more than ½ way to the base; stems from a woody caudex .... 2. *M. sylvatica*
  2 Calyx divided less than ½ way to the base; fibrous-rooted, the lower branches often rooting .... 1. *M. scorpioides*

1. ***Myosotis scorpioides*** L. Plant perennial, erect, fibrous-rooted, 2–6 dm tall, with few branches, these becoming basally decumbent and rooting at the base, the stems sparingly strigose. Leaves sessile or essentially so, oblong to oblanceolate or spatulate, acute to obtuse, to 6(8) cm long, ca 5–10 mm wide, sparsely strigose. Cymes solitary or paired at the branch tips, ebracteate; pedicels in fruit divergent, longer than the calyx. Calyx strigose

on the outside, appressed-hirsute on the inside, divided less than ½ way to the base, the lobes about equal, acute; corolla in bud pink, in anthesis blue, the eye yellow, 5-10 mm wide, the tube exceeding the calyx. Nutlets angular-ovoid, blackish when ripe, 1-2 mm long, shorter than or equaling the style. (n = 32) Jun. Cool stream sides; BH; (over much of N. Amer.; Eurasia). *Introduced.*

A similar species, *M. laxa* Lehm., has been collected in NE: Brown (sandbars of the Niobrara R.). It has the corolla 2-5 mm wide, the style shorter than the nutlets.

2. *Myosotis sylvatica* Hoffm., garden forget-me-not. Biennial or perennial to 4 dm tall, the roots fibrous. Stems 1-several from a caudex. Leaves and stems hirsute to strigose; basal leaves elliptic to oblanceolate or spatulate, obtuse, 3-10 cm long, to 14 mm wide, sparingly hirsute; cauline leaves usually ascending, gradually reduced above, linear to oblanceolate, obtuse or acute, their base clasping. Cymes without bracts in the floriferous portion; fruiting pedicels more or less ascending, longer than the calyx. Calyx hirsute to strigose, divided more than ½ way to the base, the lobes lanceolate, acute to acuminate; corolla blue, the eye yellow, 4-8 mm wide. Nutlets ovate in outline, blackish, exceeding the style, ca 1.5 mm long. (n = 12, 24) Jul. Moist soil; BH & probably occasional elsewhere; (escaped from cultivation here & there in N. Amer.; Eurasia) *Introduced*. *M. alpestris* F. Schmidt — Rydberg.

The garden forget-me-not is sometimes cultivated in cooler parts of North America but does not fare well in our warm summers. It is thus seldom seen in GP gardens. It is known in the wild only in BH. White, pink, and purple-flowered cultivars exist.

3. *Myosotis verna* Nutt. Annual to 4 dm tall, from a taproot, hirsute to strigose throughout. Stem simple or branched. Leaves often ascending, linear to oblanceolate or spatulate, ca 1-5 cm long, to 1 cm wide, acute to obtuse, those above not much reduced. Cymes with bracts in the lower floriferous part, ebracteate above; fruiting pedicels erect or ascending, equaling to mostly shorter than the calyx. Calyx hirsute and strigose, nearly bilabiate, the upper 3 lobes shorter than the lower 2; corolla white, 1-2 mm wide, more or less funnelform, barely exceeding the calyx. Nutlets exceeding the style, 1-1.5 mm long, brownish. May-Jun. Wet to rather dry woods & fields, rather weedy; mostly se GP from sw MN, w IA, cen KS, & w OK to our e border, also s-cen SD, nw KS, & probably elsewhere; (ME to MN, SD, WY, & B.C., s to GA, TX, CA). *M. macrosperma* Engelm., *M. virginica* (L.) B.S.P. — Rydberg.

13. ONOSMODIUM Michx., False Gromwell

1. *Onosmodium molle* Michx. Coarse, often clumped perennial with mostly cauline leaves, branching above or below, 5-12 dm tall. Basal leaves, if any, early deciduous. Cauline leaves sessile, the lower soon falling, lanceolate, elliptic, or ovate, 5-12 cm long, 1.5-3 cm wide, 5- to 7-nerved, the nerves prominent below, the indumentum duplex. Cymes terminal, simple or dichotomously forked, compact and scorpioid at first, expanded in fruit, leafy bracteate, the bracts enlarging in fruit. Sepals free nearly to the base, persistent in fruit, narrow, 3.5-9 mm long, their outer surface hirsute, the inner surface strigose; corolla almost tubular, much exceeding the calyx, 7-15 mm long, the outside hairy, the tube narrow, whitish, the lobes erect, pointed, narrow, greenish; style long exserted, 10-18 mm long, persistent. Nutlets 1-4, ovoid, basally attached among the sepals, sometimes constricted near the base, smooth or pitted, white to brownish; gynobase low or depressed. Flowers Jun-Jul, fruits Jul-Aug.

References: Das, T. L. 1965. A taxonomic revision of the genus *Onosmodium* (Boraginaceae). M.Sc. thesis, Kansas State Univ.; Mackenzie, K. K. 1906. *Onosmodium*. Bull. Torrey Bot. Club 32: 495-506.

Two varieties occur in our area:

1a. var. *hispidissimum* (Mack.) Cronq. Plant branched below or above. Upper leaf surface hirsute (the hairs falcate, ascending) the hirtellous (the hairs straight, appressed). Calyx 3.5–7 mm long; corolla 7–9 mm long. Nutlets strongly constricted at the base, 3–3.4 mm long, often pitted. Meadows & woods, dry hillsides, often on stony soils; w MN, w MO; (NY to MN s to VA, OH, MO). *O. hispidissimum* Mack.—Rydberg.

1b. var. *occidentale* (Mack.) Johnst. Plant branched above. Upper leaf surface hispid (the hairs often ascending and appearing combed) and hirtellous, the leaves not grayish. Calyx 8–9 mm long; corolla 11–15 mm long. Nutlets barely if at all constricted near the base. Dry, often sandy or gravelly prairies, arroyos, pastures, & open woods, often on alkaline soils; GP; (IL to Man. s to TX). *O. occidentale* Mack.—Rydberg.

Var. *occidentale* shows variation in hairiness of the leaves from north to south in our area; in OK and KS the longer hairs on the upper leaf surface are hispid and appressed, while northward they are less hispid and more ascending. The variety intergrades in our area with var. *hispidissimum* in w MN, and perhaps also in e KS, at least as far as vestiture is concerned. The var. *subsetosum* (Mack. & Bush) Cronq., with the stems glabrous below, approaches our se border in sw MO and ne OK.

14. PLAGIOBOTHRYS Fisch. & Mey., Popcorn Flower

1. *Plagiobothrys scouleri* (H. & A.) I. M. Johnst. Annual from a taproot, rather delicate for this family, the stems 1–several, slender, to 1(1.5) dm tall, erect to prostrate, sparsely strigose. Leaves cauline, sessile, the lower opposite and perfoliate, the upper alternate and little reduced; leaves 1–5 cm long, ca 5 mm wide, linear to oblanceolate, acute to obtuse, sparingly strigose. Cymes racemelike, lax, terminating each branch, usually bracteate, especially below, the bracts often somewhat removed from the flowers; pedicels 1–2 mm long, coarsely strigose. Calyx divided nearly to the base, the sepals erect, lanceolate, acuminate, 2–4 mm long, coarsely strigose on the outside, glabrous below and often hirsute above on the inside; corolla 1–4 mm wide, white, the eye yellow, exceeding the calyx; stamens and style included. Nutlets 4, to 2 mm long, rugose-tuberculate, with a weak dorsal keel, the scar very low on the ventral side. Jul–Aug. Moist, alkaline or saline soils; e MT, e WY, extreme nw NE, ND, SD, sw MN; (Man. to B.C., s to sw MN, NM, CA). *Allocarya nelsonii* Greene, *A. californica* (Fisch. & Mey.) Greene—Rydberg; *P. scopulorum* (Greene) I. M. Johnst.—Fernald.

118. VERBENACEAE St.-Hil., the Vervain Family

by Ronald L. McGregor

Ours herbs, shrubs, or small trees; stems and branches often tetragonal. Leaves opposite, occasionally whorled, rarely alternate, simple or palmately compound; stipules none. Inflorescences terminal or axillary spikes or heads, or small cymose inflorescences aggregated into a terminal mixed panicle. Flowers perfect or rarely some unisexual; sepals (4)5, united to form a toothed or lobed calyx which may be irregular; petals (4)5, united to form a more or less irregular corolla which is sometimes bilabiate; stamens (2)4 or 5, adnate to corolla-tube, and alternate with lobes; disk wanting or weakly developed; ovary 2-carpellate but 4-celled, with a single ovule in each cell, commonly somewhat 4-lobed and separating in fruit into 4 nutlets, or 2-lobed, 2-celled, 2-ovulate, or 1-celled and 1-ovuled. Fruit a drupe with 2 or 4 stones, or 1-seeded, separating nutlets, or achenes *(Phryma)*. Phrymaceae Schauer—Rydberg.

1 Leaves palmately compound .... 5. *Vitex*
1 Leaves simple.
   2 Plants shrubs .... 1. *Callicarpa*
   2 Plants herbaceous.
      3 Flowers opposite in elongate, long peduncled, interrupted spikelike racemes terminating stem and a few upper axils; fruit an achene .... 3. *Phryma*
      3 Flowers in dense or loose spikes ending the stem and branches or in dense, elongating, pedunculate, axillary spikes; fruit of 2 or 4 1-seeded nutlets.
         4 Fruit of 4 1-seeded nutlets; corolla 5-lobed, usually only slightly irregular; inflorescences terminating stem and branches .... 4. *Verbena*
         4 Fruit of 2 1-seeded nutlets; corolla 4-lobed, evidently 2-lipped; inflorescence of elongating, pedunculate, axillary spikes .... 2. *Lippia*

1. CALLICARPA L., Beautyberry

1. *Callicarpa americana* L., American beautyberry. Shrubs 1–2.5 m tall; stems much branched, twigs densely stellate-scurfy or tomentose. Leaves opposite or rarely ternate; blades thin, ovate to elliptic, 8–23 cm long, 3.5–13 cm wide, acute or acuminate, coarsely serrate or crenate-dentate, narrowed to petiole, stellate-scurfy, becoming glabrate above; petiole 0.5–3.5 cm long. Cymes 1–3.5 cm long and wide, compact, many flowered, usually shorter than petiole; peduncles 3–10 mm long, stellate-scurfy or glabrate; calyx obconic or campanulate, 1.6–1.8 mm long, with 4 very short teeth; corolla bluish, pinkish, reddish or white, funnelform, tube 2.6–2.9 mm long, lobes 4, 1.3–1.5 mm long, blunt. Fruits rose-pink to purple or blue, 3–6 mm long and wide, densely clustered; drupe 4-seeded; seeds ca 2.3 mm long. (n = 18) Jun–Oct. Moist wooded hillsides, thickets, bottom lands, fence rows; OK: Cleveland, Tulsa & e; (sw MD to MO, s to NC, TN, AR, OK, TX, n Mex.).

2. LIPPIA L., Fog- or Frog-fruit

Ours perennial herbs; stems obscurely to evidently 4-angled, prostrate or creeping, often rooting at nodes, branches sometimes ascending to erect, sparsely to densely canescent with malpighaceous hairs. Leaves opposite, variously toothed except at base; petioles to 5 mm long, often obscured by decurrent blade tissue. Flowers in dense, elongating, pedunculate, axillary spikes; bracts small, cuneate-obovate to flabelliform; calyx membranaceous, sepals united, usually compressed or 2-alate, its rim with 2 or 4 short teeth, tube shorter than corolla tube and subtending bract; corolla zygomorphic, pinkish, lavender or rarely white, tube slender, straight or incurved, limb spreading, 4-parted and somewhat 2-lipped; stamens 4, didynamous, inserted at about middle of corolla tube; ovary 2-celled, each cell 1-ovulate; stigma thickened, oblique or recurved. Fruit included in the calyx, dry, dividing into 2 nutlets at maturity. *Phyla* Lour.—Rydberg.

1 Leaf blades mostly wider toward the apex and toothed only near the apex.
   2 Bractlets 4–5 mm long; spikes 8–12 mm in diam at maturity; leaves 2–8 mm wide, acute or subacute at apex; infrequently rooting at nodes .... 1. *L. cuneifolia*
   2 Bractlets 2–3 mm long; spikes 6–9 mm in diam at maturity; leaves 6–25 mm wide, usually rounded or obtuse at apex; often rooting at nodes .... 3. *L. nodiflora*
1 Leaf blades mostly widest at or below the middle, toothed from below the middle to apex .... 2. *L. lanceolata*

1. *Lippia cuneifolia* (Torr.) Steud., wedgeleaf fog-fruit. Perennial herbs with a woody caudex; stems branching from the base, prostrate or procumbent, sometimes rooting at nodes, to 1 m long, often with short and erect branches at nodes, sparsely to densely appressed-strigillose with short white malpighaceous hairs. Leaves opposite, sessile, rigid, thick-textured, often with a fascicle of smaller ones in the axils, linear-oblanceolate or cuneiform, often canescent when young, becoming glabrate, (1)2–4(5) cm long, (2)3–6(8) mm wide, acute or subacute at apex, with 1–4 often salient teeth above the middle on each side, or rarely entire, gradually attenuate to cuneate base, secondary venation not evident on both surfaces. Inflorescence spicate, axillary; peduncles 0.5–6 cm long, sparsely appressed-strigillose; spikes at first globose, later cylindric and elongating to 2 cm, 8–12 mm in diam, bractlets obovate, 4–5 mm long, 2.5–3 mm wide, abruptly acuminate, tip finally somewhat reflexed, scarious on upper margins. Calyx small, membranaceous, shorter than corolla tube; corolla whitish or purplish, tube 4–5 mm long, limb 2–4.5 mm wide. Nutlets 1.7–2.1 mm long, oblong, yellowish. Jun–Oct. Locally common in prairies, stream & pond shores, roadside ditches, waste areas; cen SD, NE, KS, CO, OK, NM, TX; (SD, NE, CO, s to TX, AZ, s CA, n Mex.). *Phyla cuneifolia* (Torr.) Greene—Rydberg.

2. *Lippia lanceolata* (Michx.) Greene, northern fog-fruit. Perennial herbs with weak branching caudex; stems prostrate, procumbent or ascending, usually rooting at nodes, (1)2–6(10) dm long, strigillose or glabrate, branches sometimes erect. Leaves opposite; blades bright green on both sides, lanceolate to lance-elliptic, oblong to oblong-lanceolate or ovate, 1.5–7.5 cm long, 5–30 mm wide, acute or subacute at apex, serrate to below middle, widest at or below the middle, narrowed to cuneate base, venation usually conspicuous. Peduncles 4–8 cm long, usually equaling or surpassing the leaves; spikes at first globose, becoming cylindric and elongating to 3.5 cm and 5–7 mm in diam; bractlets obovate, 2.7–3 mm long. Calyx membranaceous, about as long as corolla tube; corolla pale blue, purplish or white, 2–2.5 mm long. Nutlets 0.9–1.2 mm long, obovate, olivaceous or yellowish. (n = 16) May–Sep. Frequent along margins of streams, ponds, lakes, prairie swales, ditches, low woodlands; w MN, e SD, NE, IA, MO s to TX; (Ont. to MN, SD, s to FL, TX, NM, CA, n Mex.). *Phyla lanceolata* (Michx.) Greene—Rydberg; *L. lanceolata* Michx. var. *recognita* Fern. & Grisc.—Fernald.

The more narrow-leaved forms of this species have been recognized as the var. *recognita*, but both narrow lanceolate and broad-leaved forms are found in the same colony.

3. *Lippia nodiflora* (L.) Michx. Our plants similar to no. 1 but caudex less woody; stems more frequently rooting at nodes. Leaf blades more spatulate to broadly oblanceolate and often rhomboid-elliptic, more rounded or obtuse at apex, larger ones usually with few to several teeth per side. Mature spikes only 5–8(10) mm in diam; bractlets only 2–3 mm long. Nutlets 1 mm long, more ovoid. Jun–Oct. Along small streams in prairie canyons, ponds, & lakeshores, ditches, low wet places; KS: Barber, Stafford, Wabaunsee; OK, TX; (FL to TX, n to VA, OK, KS; subtrop. & trop. regions of Old & New Worlds). *Phyla nodiflora* (L.) Greene—Gleason & Cronquist.

This is a polymorphic species with a number of described variants. Var. *nodiflora* has leaf blades sharply serrate above the middle with numerous appressed antrorse teeth and has been reported for TX: Childress. Var. *reptans* (Spreng) O. Ktze. which differs in being more densely strigose throughout, leaves thinner, teeth more spreading and larger, and more or less prominent venation, has been reported from KS: Wabaunsee; TX: Dallas, Wichita. Most of our specimens have either exceedingly variable leaves as to size and shape with the teeth decidedly spreading (*Phyla nodiflora* var. *texensis* Moldenke) or leaves that are more uniformly shaped (*P. nodiflora* var. *incisa* (Small) Moldenke). Recognition of the latter varieties would require new combinations under *Lippia* which should be done only after careful study.

3. PHRYMA L., Lopseed

1. *Phryma leptostachya* L. Perennial herbs; stems erect, 4–10 dm tall, simple or branched above, pubescent or glabrate, swollen above each node for an area of 0.4–1.2 cm long but

often sunken in drying. Leaves opposite; blades ovate to ovate-lanceolate or elliptic-ovate, pubescent or glabrous, (3)6–16 cm long, 3–10 cm wide, irregularly serrate or crenate-serrate, base truncate to cuneate and attenuate, apex acuminate, acute, or obtusish; median petioles 2–5 cm long, shorter upward and downward. Flowers opposite and horizontal, strongly reflexed in fruit, in elongate, long-peduncled, interrupted spikelike racemes terminating the stem and upper axillary branches; sepals 5, united, bilabiate, upper lip 3-lobed, each lobe 2–3 mm long, setaceous, uncinate, lower lip 2-lobed, each 0.3–0.5 mm long, the tube 2–6 mm long; petals 5, united; corolla bilabiate, upper lip purple or tinged purple, 1.5–2 mm long, lobes acute, lower lip 3-lobed, pink, lavender or white, 2.3–3.5(4) mm long, lobes rounded; stamens 4, included, attached at 2 levels to the corolla tube; ovary superior, 1-celled, 1-ovuled. Fruit an achene, ellipsoid, 3–4 mm long, light brown, enclosed in strongly ribbed calyx. (n = 14) Jun–Sep. Moist woodlands, thickets, stream valleys; GP but absent in sw 1/4; (Que. Man., s. s to FL, AL, LA, TX).

4. VERBENA L., Vervain

Ours annual or usually perennial herbs, sometimes flowering first year; stems and branches erect, procumbent or ascending, variously pubescent or glabrate, often tetragonal. Leaves simple, opposite, toothed or variously lobed, incised or pinnatifid. Inflorescence spicate, ending the stem or branches, usually densely many flowered, sometimes flat-topped and appearing umbellate, sometimes much elongated with scattered flowers; each flower subtended by a bractlet. Calyx usually tubular, 5-angled and 5-ribbed, unequally 5-toothed, little changed in fruit; corolla salverform or funnelform, tube straight or slightly curved, limb flat, weakly 2-lipped, lobes 5; stamens 5, didynamous, inserted in upper 1/2 of tube, usually included, anther connective unappendaged or with glandular-appendage; ovary of 2 carpels, 4-lobed, 4-celled, 1 ovule per cell; style 1, terminal, usually short, 2-lobed, 1 lobe smooth, the other papillose and stigmatic. Fruit mostly enclosed by calyx, separating at maturity into 4 1-seeded nutlets.

Reference: Umber, Ray E. 1979. The genus *Glandularia* (Verbenaceae) in North America. Syst. Bot. 4: 72–102.

Several of our species rarely to frequently hybridize. With experience one can usually recognize such plants, as they are generally more or less intermediate between the presumed parents. Our named hybrids are as follows:

Verbena x *blanchardi* Moldenke = *V. hasatata* x *V. simplex*
Verbena x *deamii* Moldenke = *V. bracteata* x *V. stricta*
Verbena x *engelmannii* Moldenke = *V. urticifolia* x *V. hastata*
Verbena x *illicita* Moldenke = *V. urticifolia* x *V. stricta*
Verbena x *moechina* Moldenke = *V. simplex* x *V. stricta*
Verbena x *oklahomensis* Moldenke = *V. canadensis* x *V. bipinnatifida*
Verbena x *perriana* Moldenke = *V. bracteata* x *V. urticifolia*
Verbena x *rydbergii* Moldenke = *V. hastata* x *V. stricta*

1 Styles 6–20 mm long; nutlets with evident cavity at base (section *Glandularia*).
  2 Calyx 6 mm long; corolla tube 8–10 mm long, limb 3–5 mm wide; nutlets 2–2.5 mm long .... 7. *V. pumila*
  2 Calyx 7–13 mm long; corolla tube 16–26 mm long, limb 8–15 mm wide; nutlets (2)3–3.5 mm long.
    3 Corolla 2–3 cm long, tube 2× longer than the calyx, limb 10–15 mm wide .... 3. *V. canadensis*
    3 Corolla 1–1.5 cm long, tube to 1.5× longer than calyx, limb 7–10 mm wide .... 1. *V. bipinnatifida*
1 Styles 2–3 mm long; nutlets without a cavity at base (section *Verbenaca*).
  4 Bractlets subtending flowers and fruits 8–15 mm long, 2–3× as long as calyx .... 2. *V. bracteata*

4　Bractlets shorter than, equaling or slightly longer than calyx, 2-6 mm long.
　　5　Middle stem leaves once or twice pinnatifid .... 4. *V. halei*
　　5　Middle stem leaves not pinnatifid.
　　　　6　Under surface of leaves prominently marked with white veins, especially near
　　　　　　margins .... 6. *V. plicata*
　　　　6　Under surface of leaf without prominent white veins near margin.
　　　　　　7　Principal stem leaves 3-10(15) mm wide, narrowly lanceolate to
　　　　　　　　oblanceolate .... 8. *V. simplex*
　　　　　　7　Principal stem leaves mostly 1.5-6 cm wide, ovate or broadly elliptic, oblong or
　　　　　　　　lanceolate.
　　　　　　　　8　Flowers and fruits usually remotely spaced on spike, not
　　　　　　　　　　overlapping .... 10. *V. urticifolia*
　　　　　　　　8　Flowers and fruits contiguous or overlapping.
　　　　　　　　　　9　Principal leaves definitely petioled; blades usually glabrous or strigulose on
　　　　　　　　　　　　both sides .... 5. *V. hastata*
　　　　　　　　　　9　Principal leaves sessile or nearly so; blades dense hirsute-villous beneath,
　　　　　　　　　　　　hirsute above .... 9. *V. stricta*

1. **Verbena bipinnatifida** Nutt., Dakota vervain. Perennial herbs, sometimes flowering first year; diffusely branched from base, stems prostrate, decumbent, ascending, or rarely erect, 0.5-6 dm long, sometimes rooting at lower nodes, moderately to densely hispid-hirsute, hairs 1-2 mm long. Leaves opposite, petiolate, 1-6 cm long, 1-6 cm wide, bipinnatifid, pinnatifid, 3-parted or deeply incised, with stiff appressed hairs on both surfaces; margins often revolute. Spikes pedunculate, 1-20 cm long, fasciclelike during anthesis, elongating or compact in fruit; bractlets to 1.2 × longer than calyx, glandular or eglandular, hirsute, ciliate. Calyx 7-10 mm long, lobes unequal and 1-4 mm long, veins hirsute, surface densely glandular to eglandular; corolla pink, to lavender or purple, tube ca 1.5 × longer than calyx, villous on outer surface, limb 7-10(15) mm wide; anther connective glandular appendaged, sometimes minute; style 10-20 × longer than ovary. Nutlets 2-3 mm long, grayish-black, commissural face bare or covered with white smooth papillose ridges, outer surface raised reticulate above to raised striate basally. (n = 10, 15) May-Oct. Frequent in dry plains & prairies, pastures, stream valleys, roadsides; s 2/3 GP; (SD, NE, w MO to CO, s to AR, TX, AZ, Mex.). *V. ambrosifolia* Rydb.—Rydberg; *V. ciliata* Benth., *V. wrightii* A. Gray— Atlas GP; *Glandularia bipinnatifida* (Nutt.) Nutt.—Umber, op. cit.

2. **Verbena bracteata** Lag. & Rodr., prostrate vervain. Annual or short-lived perennial herbs; stems usually several from the base, diffusely branched, prostrate or decumbent, rarely erect, hirsute, 1-5(7) dm long. Leaves opposite, 1-7 cm long, 0.6-3 cm wide, lanceolate to ovate-lanceolate, pinnately incised or 3-lobed, lateral lobes smaller than the large, cuneate-obovate, incisely toothed, central one, hirsute on both surfaces; blade narrowed into a short winged petiole. Spikes terminal, sessile, often ascending, 2-20 cm long, 1-2 cm in diam, hirsute; bractlets 8-15 mm long, 2-3 × longer than calyx, lowermost sometimes incised. Calyx 3-4 mm long, hirsute, short lobes connivent; corolla bluish to lavender or purple, tube slightly longer than calyx, finely pubescent outside, limb 2-3 mm wide. Nutlets 2-2.5 mm long, yellowish to reddish-brown, commissural face densely whitish or yellowish short-papillose, outer surface raised reticulate in upper 1/2, raised striate in lower 1/2. (n = 7, 14) Apr-Oct. Frequent along roadsides, pastures, overgrazed prairies, waste places; GP; (almost throughout U.S. & s Can.). *V. bracteosa* Michx.—Rydberg.

3. **Verbena canadensis** (L.) Britt., rose vervain. Perennial herbs, sometimes flowering first year; stems usually several from base, often diffusely branched, decumbent or ascending, often rooting at nodes, 3-6 dm long, glabrate or irregularly spreading-hirsute. Leaves variable, ovate to ovate-oblong or triangular-ovate to lanceolate, 2.5-9 cm long, 1.5-4 cm wide, acute or acuminate, truncate or cuneate at base and narrowed into a winged petiole,

blade incised or incised-pinnatifid to 3-cleft, strigose or glabrate on both surfaces. Spikes pedunculate, fasciclelike or depressed capitate at anthesis, densely many flowered, elongating in fruit; bractlets linear-attenuate, shorter than or equaling calyx, ciliate. Calyx 10–13 mm long, sparsely glandular-hirsute, the lobes unequal, subulate-setaceous; corolla pink, rose, to blue, lavender, purple, or white, tube 2 × as long as calyx, limb 10–15 mm wide, lobed usually emarginate; anther glands usually prominent; styles 15–20 mm long. Nutlets 3–3.5 mm long, grayish-black to black at maturity, commissural face sparsely to densely covered with smooth papillose ridges and papillae, outer face raised reticulate in upper 1/2 to 2/3, to raised striate below. (n = 15) Mar–Oct. Rocky prairie hillsides, bluffs, pastures, open woodlands, roadsides, waste places; se NE, e 1/3 KS, MO, OK; (NC, KY, IA, se NE, KS, s to FL, e TX). *V. drummondii* (Lindl.) Baxter—Rydberg; *Glandularia canadensis* (L.) Nutt.—Umber, op. cit.

4. **Verbena halei** Small, Texas vervain. Perennial herbs; stems 1-several from a woody base, erect, glabrous or strigillose, 3–8 dm tall, usually branching above, branches erect or ascending. Leaves opposite, basal and lower ones oblong to ovate, tapering into a petiole about as long as blade, irregularly dentate or incised, 2–8 cm long, 1–4 cm wide, middle stem leaves, 1 or 2 pinnatifid, petioles shorter, upper leaves dentate or entire. Spikes loosely arranged, paniculate, slender, elongate; bractlets ca 1/2 as long as calyx. Flowers imbricate to somewhat remote; calyx 2.5–3.5 mm long, strigillose, teeth unequal; corolla bluish to lavender, rarely white, tube scarcely longer than calyx, limb (4)6–7 mm wide, lobes retuse. Nutlets 2–2.2 mm long, yellowish or reddish-brown, commissural face obscurely to densely yellow papillose, outer surface obscurely to clearly raised reticulate in upper 1/4 to 1/2, raised striate in lower portion. (n = 7) Mar–Oct. Rocky hillsides, prairies, pastures, roadsides, waste areas, sandy or calcareous soils; sw OK, n TX; (NC to OK, FL, TX).

5. **Verbena hastata** L., blue vervain. Perennial herbs; stems erect 0.5–2.3 m tall, simple to often branched above, usually rough pubescent with short antrorse hairs. Leaves opposite, lanceolate to lance-oblong or lance-ovate, 4–18 cm long, gradually acuminate, coarsely serrate, doubly serrate, or incised, often hastately 3–lobed at base, glabrous or slightly pubescent, to rarely scabrous above, and conspicuously pubescent beneath, petiolate, blade tissue barely decurrent on petioles; spikes stiffly erect, paniculate, compact, densely many flowered; bractlets lanceolate subulate, usually a little shorter than calyx. Calyx 2.5–3 mm long, pubescent, lobes acute, more or less connivent; corolla purplish-blue, rarely white, tube ca 2 × longer than calyx, pubescent outside, limb 3–4.5 mm wide. Nutlets 1.9–2.1 mm long, reddish-brown, commissural face smooth or with sparse, small whitish strigose hairs, outer surface obscurely to irregularly raised reticulate in upper 1/3, slightly raised striate on lower 2/3. (n = 7) Jun–Oct. Moist meadows, low open woodlands, stream banks, around springs & seepage areas, roadsides; GP but rare or absent in nw & sw; (N.S. to B.C., s to FL, TX, NM, CA).

In more dry habitats plants frequently have rather rigid leaves, which are conspicuously scabrous above and usually conspicuously pubescent beneath. These have been named var. *scabra* Moldenke, but the distinction is arbitrary.

6. **Verbena plicata** Greene, fanleaf vervain. Perennial herbs; stems decumbent or ascending, branched, 1–4 dm tall, hirsute and with shorter hairs, sparsely glandular. Leaves opposite, lower blades elliptic-ovate, 1–4 cm long, narrowed at base into a winged petiole to as long as blade, broadly obtuse at apex, plicate, incised-dentate, often 3–lobed, rugose and sparsely hirsute above, hirsute beneath, veins of teeth near the margin, prominent, whitish; upper leaves similar but smaller. Spikes terminal on stems and branches, not compact; bractlets green, 3.5–5 mm long, herbaceous, ovate-lanceolate, equaling or somewhat

surpassing calyx. Calyx 3.5–4 mm long, glandular-hirsute, teeth subulate; corolla blue to lavender or puple, tube usually a little longer than calyx, limb 4–6 mm wide. Nutlets 2–2.5 mm long, olivaceous to yellowish-brown, commissural face with conspicuous white surface on each side of low median ridge, outer surface obscurely to clearly raised reticulate in upper 1/4, somewhat raised striate below. (n = 7) Apr–Aug. Usually on sandy or gravelly prairies, flats, ravines, hillsides, roadsides; OK: Cimarron, Harmon, Jackson; TX: Childress, Cottle, Hall, Motley; (OK, TX, NM, AZ, n Mex.).

7. *Verbena pumila* Rydb., pink vervain. Annual or short-lived perennial herbs; stems usually several from base, branched, decumbent-ascending, rarely erect, 0.5–2(3) dm long, hirsute and often finely glandular. Leaves opposite, 1.5–3 cm long, trifid, sometimes lobed, divisions variously incised, hirsute on both surfaces. Spikes sessile or short-pedunculate, terminating stems or branches, rather compact, hardly elongating in age; bractlets linear-lanceolate, hispid-hirsute,nearly as long as calyx. Calyx 6 mm long, pubescent, often finely glandular, lobes short, subulate; corolla pink, or lavender to blue, rarely white, tube 8–10 mm long, glabrate outside, limb 3–5 mm wide; anther glands minute or absent. Nutlets grayish-black at maturity, 2–2.8 mm long, commissural face with smooth or papillose ridges, outer face raise reticulate in upper 3/4, obscurely raised striate below. (n = 7) Mar–Aug. Prairies, pastures, brushy hillsides, stream valleys, open woodlands, roadsides; OK, TX; (AR, LA, OK, TX, NM, n Mex.). *Glandularia pumila* Rydb.—Umber, op. cit.

8. *Verbena simplex* Lehm., narrow-leaved vervain. Perennial herbs; stems 1–several from base, erect, (2)3–6 dm tall, sometimes much branched above, usually sparsely strigillose. Leaves linear to narrowly oblong-oblanceolate, lanceolate or spatulate, 3–8(10) cm long, to 15 mm wide, serrate, often only toward acute apex, attenuate to subsessile base, rugose above, and veiny beneath, glabrate or sparsely strigillose. Spikes slender usually solitary on stem and branches, flowers and fruits usually rather dense; bractlets equaling or slightly shorter than calyx. Calyx (2)3–4(5) mm long, sparsely hairy, lobes acuminate; corolla blue, lavender or purple, rarely white, tube scarcely longer than calyx, limb (4)5–6 mm wide. Nutlets (2)2.5–3 mm long, olivaceous to reddish-brown, commissural face smooth or with sparse dense white papillae, outer face raised reticulate above, raised striate below. (n = 7) May–Sep. Rocky prairie hillsides, pastures, open waste areas, roadsides; se MN, IA, se NE, e 1/2 KS, MO, e 1/2 OK; (Ont., VT, MA, to MN, NE, s to FL, AL, LA, OK). *V. angustifolia* Michx.—Rydberg.

9. *Verbena stricta* Vent., hoary vervain. Perennial herbs; stems erect, 2–12(18) dm tall, simple or branched above, densely spreading hirsute and often canescent. Leaves opposite, ovate, elliptic, or suborbicular, 3–7 cm long, sessile or nearly so, sharply serrate or biserrate to incised-serrate, hirsute and rugose above, densely hirsute or villous and prominently veined beneath. Spikes 1–several, terminating stem or each branch, stiffly erect, to 3 dm long, flowers and fruit imbricate; bractlets lanceolate-subulate, about as long as the calyx, hirsute, ciliate. Calyx (3)4–5 mm long, densely hirsute, lobes acuminate; corolla mostly blue or purple, or rarely roseate to white, tube slightly longer than calyx, pubescent outside, limb 8–9 mm wide. Nutlets (2)2.5–3 mm long, grayish-brown, commissural face rather densely white papillate, outer face raised-reticulate in upper 1/2, raised striate in lower 1/2. (n = 7) May–Sep. Common in pastures, prairies, thickets, roadsides, waste areas, GP; (Ont. to MN, MT, WY, s to TX, NM, introduced elsewhere).

10. *Verbena urticifolia* L., nettle-leaved vervain. Perennial herbs; stems solitary, erect, 0.5–2.5 m tall, simple or much branched, hispid or hirtellous to glabrate. Leaves opposite, ovate to lance-ovate, or broadly lanceolate to oblong-ovate, 8–15(20) cm long, 2.5–7(12) cm wide, acute or shortly acuminate, coarsely serrate to biserrate, or crenate-serrate, rounded

at base and decurrent into petiole, glabrous on both surfaces or scattered pilose; petioles 1-4 cm long with blade tissue decurrent at least 1/2 the length. Spikes loosely arranged, paniculate, slender, rather sparsely flowered; bractlets ovate-acuminate, ciliate, 0.5-1.5 mm long. Calyx 1.5-2.3 mm long, pubescent, subequal teeth subulate; corolla white, tube 1.6-2.2 mm long, limb 2 mm wide, lobes obtuse. Nutlets 1.5-2 mm long, commissural face smooth, outer face smooth to obscurely or definitely raised reticulate above, smooth or raised striate below. (n = 7) Jun–Oct. Common in woodlands, pastures, along streams, roadsides, waste areas; GP but rare to absent in w 1/2; (Que. & Ont., to ND, s to FL, TX).

Rare plants with lower surface of the leaves densely but short-hairy, mature calyx shorter (1.7-2 vs 2-2.3 mm long) and nutlets 1.5 mm long and smooth, have been referred to var. *leiocarpa* Perry & Fern. This distinction appears arbitrary in our area.

5. VITEX L., Chaste-tree

1. *Vitex agnus-castus* L., common chaste-tree. Shrub or small tree to 4 m tall; branches densely short-puberulent, resinous-punctate. Leaves opposite, palmately compound; leaflets unequal, (3)5-7(9), elliptic-lanceolate to oblanceolate, 7-15 cm long, 1.2-4 cm wide, acute to acuminate, entire, green above, grayish tomentose and glandular beneath; only 3 largest leaflets petiolulate. Flowers in small cymose inflorescences which may be aggregated into a terminal mixed panicle; bractlets 1-4 mm long, linear-setaceous; calyx broadly companulate, 2-2.5 mm long, densely white-puberulent, rim with 5 short teeth; corolla slightly irregular, lavender or lilac, tube 6-7 mm long, lobes 2-3 mm long; stamens 4, often exserted; ovary with 4 cells and 4 ovules; stigma 2-cleft. Fruit drupaceous. (n = 12, 16) Apr–Oct. Pastures, roadsides, fence rows, stream valleys, waste places; OK: Grady, Marshall; n-cen TX; (culti. & escaped, s U.S., n to MD; Eurasia).

119. LAMIACEAE Lindl., the Mint Family

by Ralph E. Brooks

Ours annual, biennial, or perennial herbs, sometimes aromatic; stems typically square. Leaves opposite or rarely whorled, simple or infrequently pinnatifid, exstipulate. Inflorescences mostly small, loose to compact cymules borne in the axils of foliaceous or modified bracts, opposite pairs forming a verticillaster. Verticillasters solitary and terminal or few to many in congested to interrupted spiciform inflorescences; inflorescences essentially racemose where cymules are reduced to single flowers. Flowers perfect or rarely unisexual, bracteoles present or not; calyx persistent, regular or irregular, usually with 5 teeth or lobes (10 in *Marrubium*), sometimes bilabiate, the upper lip 3-toothed and the lower 2-toothed (upper and lower lip entire in *Scutellaria*), corolla sympetalous, mostly bilabiate, the upper lip 2-lobed (rarely the lip unlobed) and lower lip 3-lobed (rarely all 5 lobes forming the lower lip as in *Teucrium*), occasionally corolla regular and 4- or 5-lobed; stamens epipetalous, 4 and didynamous or 2 by abortion of the upper or lower pair with a pair of staminodes present or not; carpels 2 but each divided in half, at maturity each of the 4 partitions ripens into a hard nutlet; stigma 2-lobed, often unequally so; style gynobasic or sometimes terminal; ovary superior, ovules solitary in each lobe of the ovary, erect, basal-axile. Fruit of 4 1-seeded nutlets with a hard pericarp; endosperm lacking or nearly so.

Labiatae Juss. is an acceptable alternative to the use of the Lamiaceae Lindl. according to the International Code of Botanical Nomenclature (Article 18.6).

MINT FAMILY

The following species have been collected in the GP but to our knowledge do not persist, become established, or spread. Each is differentiated in the key and indicated by an asterisk.

Ajuga reptans L. (bugle) is a native of Eurasia and northern Africa. It is a perennial frequently planted as ground cover and occasionally escapes in the se GP.

Melissa officinalis L. (common balm) is a perennial native to Europe. It infrequently escapes from cultivation; KS: Wyandotte; MO: Jackson, Jasper.

Phlomis tuberosa L. (Jerusalem sage) is a perennial native to Eurasia that was collected in ND: Foster, in 1929.

1 Calyx bilabiate, both lips entire and the tube with a distinct cap or protuberance (scutellum) on the upper side .... 21. *Scutellaria*
1 Calyx regular or bilabiate, one or more of the lips toothed.
  2 Calyx regular, 10-toothed, 5 long teeth alternating with 5 shorter teeth .... 11. *Marrubium*
  2 Calyx bilabiate or regular and 4- or 5-toothed.
    3 Upper lip of the corolla scarcely discernible or its lobes appearing laterally on the margins of the lower lip and the corolla then appearing 1-lipped.
      4 Flowers blue (rarely white); median lobe on the lower corolla lip not much longer than the lateral lobes; plants stoloniferous .... *Ajuga*
      4 Flowers white to rose-purple; median lobe on the lower corolla lip 2–3 × longer than the lateral lobes; plants never stoloniferous .... 23. *Teucrium*
    3 Upper lip of the corolla well developed.
      5 Fertile stamens 2.
        6 Calyx regular or nearly so.
          7 Flowers obviously pedicellate .... 3. *Cunila*
          7 Flowers sessile or nearly so.
            8 Corolla regular, less than 6 mm long .... 10. *Lycopus*
            8 Corolla bilabiate, more than 10 mm long .... 13. *Monarda*
        6 Calyx bilabiate.
          9 Flowers sessile .... 2. *Blephilia*
          9 Flowers pedicellate.
            10 Calyx with a well-defined ring of simple hairs within the throat .... 7. *Hedeoma*
            10 Calyx glabrous within .... 19. *Salvia*
      5 Fertile stamens 4.
        11 Inflorescences axillary in general appearance, verticillasters subtended by foliage leaves.
          12 Flowers and fruits sessile.
            13 Lower leaf surface stellate pubescent .... *Phlomis*
            13 Lower leaf surface never stellate pubescent.
              14 Stems hirsute with multicellular hairs .... 5. *Galeopsis*
              14 Stems glabrous to densely pubescent, the hairs simple.
                15 Plants 5–15 dm tall; anther cells parallel .... 9. *Leonurus*
                15 Plants 1–4 dm tall; anther cells divergent .... 8. *Lamium*
          12 Flowers and fruits pedicellate.
            16 Calyx bilabiate.
              17 Stems 1–3 dm tall; calyx 4–5 mm long .... 20. *Satureja*
              17 Stems 4–8 dm tall; calyx 7–10 mm long .... *Melissa*
            16 Calyx nearly regular.
              18 Leaves reniform to suborbicular .... 6. *Glecoma*
              18 Leaves linear-elliptic to lanceolate or ovate.
                19 Annual; calyx stipitate glandular; anther sacs divergent .... 24. *Trichostema*
                19 Perennial; calyx eglandular or sometimes sessile glandular; anther sacs parallel .... 12. *Mentha*
        11 Inflorescences terminal in general appearance, verticillasters solitary (rarely 2) in dense glomerules or loose, irregular corymbs or numerous in spiciform inflorescences.
          20 Verticillasters mostly solitary in dense glomerules or loose, irregular corymbs .... 18. *Pycnanthemum*

20 Verticillasters numerous in spiciform inflorescences.
  21 Calyx regular.
    22 Lower leaf surface stellate pubescent .... *Phlomis
    22 Lower leaf surface never stellate pubescent.
      23 Stamens exserted from the corolla; anther sacs parallel; seeds about 1.5 mm long, apex hispidulous .... 1. *Agastache*
      23 Stamens included; seeds glabrous.
        24 Calyx faintly nerved, inflated at maturity; anther sacs parallel .... 16. *Physostegia*
        24 Calyx prominently nerved, not obviously inflated at maturity; anther sacs divergent .... 22. *Stachys*
  21 Calyx bilabiate.
    25 Verticillasters 2-flowered, usually the racemose inflorescence appearing 1-sided .... 15. *Perilla*
    25 Verticillasters mostly 4- many-flowered.
      26 Corolla typically white, the lower lip spotted purple; upper and lower calyx lips quite similar except in length .... 14. *Nepeta*
      26 Corolla typically rose, purple, or blue; upper calyx lip with 1 or more teeth distinctly different from those of the lower calyx lip.
        27 Perennial; upper calyx lip with the lobes connate for at least 3/4 their length, the lobes mucronate and the median lobe much wider than the lateral .... 17. *Prunella*
        27 Annual or biennial; upper calyx lip distinctly 3-toothed, the lateral teeth narrowly tirangular and the median oblong to obovate and about 2× wider than the lateral ones .... 4. *Dracocephalum*

1. AGASTACHE Clayt. ex Gronov., Hyssop

Ours erect perennial herbs arising from short (to about 10 cm long) slender rhizomes; stems simple or branched, internodes usually elongate. Leaves with blades ovate to broadly lanceolate, variously pubescent, margins serrate; petiolate. Verticillasters borne in continuous, or occasionally interrupted, cylindrical spikes, the subtending bracts conspicuous or not. Calyx tubular, 15-veined or more, 5-toothed, the teeth nearly equal to slightly bilabiate; corolla exceeding the calyx, upper lip shallowly 2-lobed and directed forward, lower lip 3-lobed and recurved; stamens 4, exserted from the corolla, the lower pair ascending under the upper lip and the posterior pair directed downward and exserted between the lower pair; anther sacs parallel; style exserted; stigmas equally 2-lobed. Nutlets elliptic, somewhat 3-sided, hispidulous at the apex.

Reference: Lint, H., & C. Epling. 1945. A revision of *Agastache*. Amer. Midl. Naturalist 33: 207–230.

1 Calyx hirtellous, at least the teeth violet, rarely whitish; leaves minutely tomentulose below, glabrous above .... 1. *A. foeniculum*
1 Calyx glabrous, the teeth usually greenish, occasionally whitish or rose; leaves glabrous or variously pubescent but never white tomentulose.
  2 Calyx teeth ovate, obtuse or broadly acute; corolla whitish to yellowish; stems essentially glabrous .... 2. *A. nepetoides*
  2 Calyx teeth narrowly deltoid, narrowly acute to acuminate; corolla white tinged with rose to deep rose; stems densely pubescent .... 3. *A. scrophulariaefolia*

1. *Agastache foeniculum* (Pursh) O. Ktze., lavender hyssop. Plants 6–12 dm tall; stems simple or sparingly branched, the nodes canescent or sparsely so, internodes glabrous or the lowermost sometimes lightly canescent. Leaf blades ovate or occasionally broadly lanceolate, (3)4–9 cm long, (1.5)2–5.5 cm wide, gradually reduced upwards, glabrous above, white tomentulose below, apex acute to acuminate, margins serrate, base slightly cordate, truncate, rounded or broadly cuneate; petiole 0.5–2(3.5) cm long, gradually reduced upwards, glabrous or hirsute. Spikes 4–8(12) cm long, 1–2 cm diam, the lowermost verticillasters sometimes remote; bracts ovate, about equaling or shorter than the calyx, vestiture as the leaves, acuminate, serrulate; pedicels 0–2 mm long, pubescent. Calyx hirtellous throughout, the tube greenish near the base and violet above; 4–6(7) mm long, the teeth violet, deltoid or slightly narrower, 1–1.5 mm long and acute; corolla blue to violet, hirsutulous without, 7–10 mm long, lobe margins erose. Nutlets yellow-brown, about 1.5 mm long. (2n = 18) Jul–Sep. Moist woodlands, especially along streams or lakeshores, infrequently open, wet ditches or prairies at higher elevations; MN, ND, e 1/4 MT, nw IA, SD, e 1/5 WY; NE: Sheridan, Sioux; (s Ont. w to Sask., Alta., s to WI, n IA, SD, nw NE, & n-cen CO; naturalized in some areas of the ne U.S. & e Can.). *A. anethiodora* (Nutt.) Britt.—Rydberg.

2. *Agastache nepetoides* (L.) O. Ktze., catnip giant hyssop. Plants 1–2.5 m tall; stems usually branching above, glabrous or the nodes sparsely canescent. Leaf blades narrowly to broadly ovate, infrequently lanceolate or elliptic-lanceolate, (3)5–10(17) cm long, 3–7(12) cm wide, reduced slightly upwards and those of the main axis often much larger than the branch leaves, glabrous above, pubescent to glabrate below, apex acute to acuminate, margins serrate, base rounded to subcordate, sometimes cuneate; petiole 2–8 cm long, scarcely winged, sparsely pubescent. Spikes dense and typically continuous, 4–12(20) cm long, 0.8–1.7 cm diam; rachis pubescent; bracts elliptic-ovate, mostly exceeding the calyx, sparsely pubescent, margins entire and ciliate, apex acuminate; bracteoles lanceolate, 0.5–1 mm long; pedicels 0–1 mm long, pubescent. Calyx glabrous, greenish or lighter, the tube 3–4 mm long, the teeth ovate or broadly deltoid, 1–2 mm long and obtuse or broadly acute; corolla whitish or yellowish, hirsutulous without, 5–7 mm long, lobe margins erose. Nutlets yellow-brown or darker, about 1.5 mm long. (n = 9) Jul–Sep. Rocky, wooded hillsides, riparian woodlands, or occasionally moist, open sites; sw MN, IA, se SD, e 1/4 NE, MO, e 1/2 KS, e 1/2 OK; (s Que. & s Ont. s to VA, w to sw MN, se SD, e NE, e KS, e OK, & ne TX).

3. *Agastache scrophulariaefolia* (Willd.) O. Ktze., purple giant hyssop. Foliage similar to no. 2 except as follows: stems and petioles densely pubescent; upper surface of leaf blade usually pubescent. Spikes continuous or interrupted below, 3–15 cm long, 1–2 cm diam, rachis pubescent; bracts greenish to white and rose tinted, elliptic-ovate to suborbicular, slightly longer than the calyx, sparsely pubescent, margins entire to weakly serrulate, apex acuminate; bracteoles lanceolate, less than 1 mm long; pedicels 0–0.5 mm long, pubescent. Calyx greenish to white and rose tinted, the tube 3–4 mm long, the teeth narrowly deltoid, 1–2.5 mm long and narrowly acute to acuminate; corolla white, tinged with rose to deep rose, sparsely pubescent without, 5.5–7.5 mm long, the lobes entire to erose. Nutlets dark brown, about 1.5 mm long. Jul–Sep. Infrequent in riparian woodlands or open, moist sites; sw MN, IA, e edge NE, MO; KS: Leavenworth; SD: Brookings, Roberts; (NY s to NC, w to e SD, e NE, ne KS, & n MO).

2. BLEPHILIA Raf.

Erect perennial herbs arising from slender rhizomes; stems simple or sparingly branched, internodes elongate. Leaves odoriferous, ovate, lanceolate or elliptic, toothed, sessile to petiolate. Flowers numerous and dense in 1–6 remote to congested verticillasters at the uppermost nodes, subtending bracts conspicuous. Calyx tubular, 13-nerved, 5-lobed, bilabiate, the upper lip much longer than the lower, throat glabrous, the teeth narrowly triangular to subulate; corolla bilabiate, exceeding the calyx, villous without; the tube much longer than the lobes, inflated in the throat, and gradually curved downward; upper lip erect and entire; lower lip spreading or deflexed, 3-lobed, the lateral lobes broad and rounded, the median lobe oblong and longer; stamens 2, exserted; anther sacs divergent; stigma unequally 2-lobed; style exserted. Nutlets ellipsoid, slightly flattened, smooth, apex rounded, base acutish.

1 Upper cauline leaves mostly lanceolate or elliptic and cuneate to attenuate at the base, sessile or the petiole 1–8(14) mm long; lobes of the lower calyx lip barely reaching or surpassing the sinuses of the upper calyx lip; flowering May–Jun. .... 1. *B. ciliata*
1 Upper cauline leaves mostly ovate, broadest at or near the base, petioles 10–15(20) mm long; lobes of the lower calyx lip obviously not reaching the sinuses of the upper calyx lip; flowering Jul–Sep .... 2. *B. hirsuta*

1. *Blephilia ciliata* (L.) Benth., Ohio horse mint. Plants 2.5–6 dm tall; stems simple or sparingly branched, variously hirsute to villose. Leaves sparsely hirsute to glabrate above, hirsute below, especially on the veins, margins crenate-serrate; lowermost leaves early deciduous, the blades suborbicular to broadly ovate, 1.5–3.5 cm long, 1–3 cm wide, apex obtuse or rounded, base broadly cuneate to subcordate, petioles 1–2.5 cm long; upper leaves mostly lanceolate to elliptic, 3–6(10) cm long, 1–3(5.5) cm wide, widest at or near the middle, apex acute to obtuse, base cuneate to attenuate, sessile or the petiole 1–8(14) mm long. Verticillasters solitary and terminal or with 1–4 additional axillary groups; foliaceous bracts typically present only on the lower group(s); bracteoles greenish-white to lavender tinged, ovate-elliptic, 6–11 mm long, 2–4(5) mm wide, acute. Calyx 7–11 mm long, hirsute, lobes 1.3–2.2 mm long, the lobes of the lower lip barely reaching or surpassing the sinuses of the upper lip; corolla pale lavender with dark purple dots on the lower lip, 9–13 mm long, hirsute on the outer surface. Nutlets yellow-brown or darker, about 1 mm long. May–Jun. Rocky wooded hillsides, less frequently in open, moist sites; rare; sw MO, se 1/6 KS; (VT w to WI, s to GA, MS, AR, e OK, & se KS).

2. *Blephilia hirsuta* (Pursh) Benth., wood mint. Plants 5–10 dm tall; stems simple or branching above, variously pubescent to hirsute. Leaves all similar, blades mostly ovate, 4–12 cm long, 2–5 cm wide, widest at or near the base, sparsely strigose to glabrate above, puberulent to hirsute below, apex acute to acuminate, margins serrate to remotely so or subentire, base subcordate, rounded or infrequently broadly cuneate; petioles 10–15(20) mm long. Verticillasters (2)3–7, 1 terminal and those additional axillary, foliaceous bracts present only on the lower group(s); bracteoles greenish-white or purplish tinged, ovate to lanceolate, 6–10 mm long, 3–5 mm wide, acuminate. Calyx 4–6(7) mm long, hirsute, upper lobes 1.2–2 mm long, lower lobes 0.5–1.2 mm long, usually wider than the upper and obviously not reaching the sinuses of the upper lip; corolla whitish and lavender tinged to lavender, the lower lip with dark purple spots, 8–12 mm long, outer surface hirsute. Nutlets yellow-brown or darker, about 1 mm long. Jul–Sep. Moist woodlands, especially ravines & along streams; rare; e edge NE, MO, ne 1/6 KS; (Que. & VT w to MN, s to GA, TN, AR, & e OK).

3. CUNILA L., Dittany, Stone Mint

1. *Cunila origanoides* (L.) Britt., dittany. Erect perennial herb, 3–6 dm tall; stems solitary or several in a clump, freely branched, hirsute to glabrate on the angles; rhizomes slender, to about 4 cm long. Leaves ovate to narrowly so, 2–4 cm long, (0.7)1–2 cm wide, punctate, hirsute to glabrate on the veins, usually glabrous otherwise, apex acuminate to acute, margins serrate, the teeth few and widely spaced, rarely entire, base rounded to subcordate, sessile or subsessile. Inflorescences compound cymes, 3- to 30-flowered, terminal and axillary, usually shorter than the subtending leaf; peduncle 0–2 mm long; pedicels 0.5–2 mm long; bracts linear-lanceolate, about 1 mm long, punctate. Calyx regular, 10-nerved, obconic, 1.5–3 mm long, punctate, the lobes triangular, about 1/4 the length of the calyx, villous on the inner side, corolla rose-purple to whitish, 4–7 mm long, outer surface hirsute and punctate, scarcely 2-lipped, the upper obcordate and erect, the lower shallowly 3-lobed and spreading; stamens 2, well exserted; anther sacs parallel; style elongate, included to barely exserted; stigma equally 2-lobed. Nutlets brownish, ellipsoid, about 1.3 mm long, 0.7 mm wide, smooth. (Aug) Sep–Oct. Rocky, moist, wooded hillsides or stream banks, occasionally in exposed sites; MO: Jasper; KS: Cherokee; OK: Ottawa; (FL to e TX, n to NY, PA, WV, OH, IN, IL, MO, se KS).

4. DRACOCEPHALUM L., Dragonhead

Ours annual or biennial herbs; stems simple or branched above, internodes mostly elongate. Leaves oblong to lanceolate, toothed, short-petiolate. Verticillasters dense or few-flowered and in spikes or somewhat remote; bracts foliaceous and reduced; bracteoles inconspicuous or similar to the bracts. Calyx tubular or tubular-campanulate, 15-nerved, 5-lobed, bilabiate, the upper lip 3-lobed with the median lobe much larger than the lateral, the lower lip 2-lobed; corolla bilabiate, exceeding the calyx, outer surface pubescent, the tubes much longer than the lobes and enlarged towards the throat, upper lip erect, emarginate, lower lip spreading, the median lobe larger than the lateral; stamens 4, about equaling the corolla and curved under the upper lip; anther sacs divergent; stigmas equally 2-lobed; style included. Nutlets ovoid, smooth. *Moldavica* (Tourn.) Adans.—Rydberg.

Dracocephalum moldavica L. (*Moldavica punctata* Moench—Rydberg) is reported for NE (Rydberg, 1932) but the specimens were apparently cultivated. It is a native of Europe that can be distinguished from other GP species by its corolla, which is 2–3× longer than the calyx.

1 Verticillasters mostly crowded at the ends of the branches; calyx hirsute .... 1. *D. parviflorum*
1 Verticillasters more or less remote and forming interrupted spikes on the branches; calyx short pubescent .... 2. *D. thymiflorum*

1. *Dracocephalum parviflorum* Nutt., dragonhead. Annual or biennial; stems 3–8 dm tall, variously pubescent. Leaf blade elliptic to lanceolate, rarely ovate, 3–10 cm long, 0.7–3(5) cm wide, glabrous above, glabrate, pubescent, or hirsute below, apex narrowly acute, margins sharply serrate, ciliate or not, base cuneate to attenuate, occasionally rounded; petiole 3–10(20) mm long. Verticillasters mostly crowded at the ends of the branches; bracteoles shorter than to slightly longer than the calyces; pedicels 1–4 mm long. Calyx 8–15 mm long, hirsute, the lobes lanceolate and 3–6 mm long except for the oblong, slightly longer median lobe of the lower lip; corolla lavender or rose, barely exceeding the calyx. Nutlets castaneous to blackish, 2–2.7 mm long. (Jun) Jul–Sep. Gravelly soil along streams or in open woods, moist wooded hillsides or occasionally in disturbed sites; MN, ND, MT, ne & w SD, WY, NE, introduced elsewhere in the GP; (Que. to AK, s to NY, MI, IL, IA, NE, NM, AZ, & OR). *Moldavica parviflora* (Nutt.) Britt.—Rydberg.

2. *Dracocephalum thymiflorum* L. Annual; stems 3–6 dm tall, simple or sparingly branched, variously pubescent. Leaf blades ovate to ovate-lanceolate, 1–3.5 cm long, 0.5–2 cm wide, glabrous above, pubescent below, apex acute to obtuse, margins serrate-crenate, base cordate to cuneate; petioles 2–10 mm long. Verticillasters more or less remote and forming loose, interrupted spikes; bracteoles lanceolate, 1–2 mm long; pedicels 1–3 mm long. Calyx 7–9 mm long, pubescent, the lobes lanceolate to subulate and 1–2 mm long except for the obovate, slightly longer median lobe of the lower lip; corolla lavender to blue, 7–10 mm long. Nutlets dark brown, 1.2–1.6 mm long. Jun–Aug. Open pine-wooded hillsides, fields; ND: Hettinger, Stark, Walsh; WY: Crook; (scattered across n U.S.; Eurasia). Occasionally introduced in the GP and infrequently becoming naturalized. *Moldavica thymiflorum* (L.) Rydb.—Rydberg.

5. GALEOPSIS L., Hemp-nettle

1. *Galeopsis bifida* Boenn., common hemp-nettle. Annual, 4–8 dm tall; stems simple or branching, hirsute with multicellular hairs, these usually mixed with glandular hairs below each node, nodes widely spaced, usually swollen when fresh. Leaves petiolate; blade ovate to lanceolate, (3)5–10 cm long, 1–3.5 cm wide, hirsute above, pubescent below, apex acuminate, margins crenate-serrate, base cuneate; petiole 0.7–2.5 cm long, hirsute. Flowers sessile, 3–6 in dense verticillasters, the uppermost crowded, more distant below, bracteoles subulate, 3–5 mm long. Calyx regular, strongly 10-nerved with weaker intermediate nerves, hirsute and the lobes occasionally stipitate glandular, weakly hirsute on the veins within, 7–12 mm long and enlarging in fruit to as much as 15 mm long, lobes narrowly triangular, aristate, about equaling the tube in length; corolla purplish with at least the lower lip whitish, 13–17 mm long, outer surface hirsute, glabrous within, the tube well exserted from the calyx, 2-lipped, the upper lip obovate to oblong and erect, the lower 3-lobed with 2 conical projections at the base, the middle lobe emarginate and slightly reflexed; stamens 4, didynamous, included, filaments villous, anther sacs divergent; stigma equally 2-lobed; style elongate, included, glabrous. Nutlets brown with tan mottling, ovoid, about 3.5 mm long, 2 mm wide, smooth. (2n = 32) (Jul) Aug–Sep. Woodlands or moist areas along streams; nw MN, e 1/3 ND; SD; BH; NE: Sarpy; (much of Can. & n & e U.S.; Europe) *Naturalized*. *G. tetrahit* L.—Atlas GP; *G. tetrahit* var. *bifida* (Boenn.) Leg. & Court.—Fernald.

6. GLECOMA L.

1. *Glecoma hederacea* L., ground ivy. Perennial herb; stems repent with upright flowering shoots to 4 dm long. Leaf blade reniform to suborbicular, 1–2.5(4) cm long; 1–2.5(4.5) cm wide, glabrate to sparsely hirsute, punctate, apex rounded to subacute, margins coarsely crenate, usually ciliate, base mostly cordate; petiole of leaves on flowering stem 0.5–2(4) cm long, those on the repent stems to 10 cm long, variously pubescent. Flowers 2–6 in cymules, secund; bracteoles setaceous, 1–1.5 mm long; pedicels 1.5–3(4) mm long, pubescent. Calyx tubular-campanulate, 4–7 mm long, 15-nerved, teeth triangular and aristate, upper teeth about 1/4 as long as the tube, the lower slightly shorter; corolla blue or streaked with white, rarely white, 7–17 mm long, outer surface pubescent, the lower lip bearded within, tube straight and widening towards the apex, upper lip flat and emarginate, lower lip larger and 3-lobed; stamens 4, didynamous, ascending under the upper corolla lip and the longer pair about equaling it; anther sacs divergent; stigma equally 2-lobed; style elongate, exserted. Nutlets brownish, ovoid, about 2 mm long, 1 mm wide, smooth. (n = 9, 12, 18) Apr–Jun. Wooded hillsides, low areas along streams, pastures, roadsides, & lawns; MN & ND, s to MO, e OK, & w KS; (scattered across s Can. & n U.S., especially in e; Europe). *Naturalized*. *G. hederacea* var. *micrantha* Moric.—Fernald; *Nepeta hederacea* (L.) Trevisan var. *parviflora* Benth.—Gates.

7. HEDEOMA Pers., Mock Pennyroyal

Ours aromatic annuals or perennials, variously pubescent. Leaves mostly linear, elliptic, or lanceolate, punctate on one or both surfaces, entire or toothed, sessile or petiolate. Flowers solitary or in 2- to 12-flowered verticillasters, usually well spaced along the axis, occasionally congested and spiciform, usually subtended by a foliaceous bract; bracteoles lanceolate to subulate, minute; pedicels much shorter than the calyx. Calyx bilabiate, 5-lobed, tubular and usually gibbous below, 13-nerved, upper and lower lips subequal or the lower decidedly longer than the upper, an included, well defined ring of simple hairs (the annulus) more or less seals the calyx orifice; corolla bilabiate, the tube slightly expanded outward, upper lip ligulate and straight or concave and subgaleate, lower lip equally 3-lobed or the median lobe larger, spreading; stamens 2, ascending upwards under the upper corolla lip or surpassing it, an adaxial pair of staminodia usually present; stigmas unequally 2-lobed; style about equaling the fertile stamens. Nutlets brown or black, orbicular, narrowly ovate or oblong, weakly areolate, ruminate, or smooth.

References: Irving, R. S. 1980. The systematics of *Hedeoma* (Labiatae). Sida 8: 218–295; Epling, C., & W. S. Stewart. 1939. A revision of *Hedeoma* with a review of allied genera. Repert. Spec. Nov. Regni Veg. Beih. 115: 1–49.

1 Leaves ovate to elliptic, margins irregularly toothed; nutlets subspherical, 0.7–0.9 mm diam .... 3. *H. pulgeoides*
1 Leaves linear to linear-oblong, obovate, or spatulate, margins entire; nutlets narrowly ovoid, 1–1.6 mm long.
  2 Leaves obovate or spatulate; stems densely grayish pubescent; calyx long hirsute .... 4. *H. reverchonii*
  2 Leaves linear, linear-elliptic, or linear-oblong; stems variously pubescent but not densely so or grayish.
    3 Upper and lower calyx teeth convergent, closing the orifice at maturity; plants perennial; bracteoles 1–2 mm long .... 1. *H. drummondii*
    3 Upper calyx teeth spreading and the lower calyx teeth only slightly curved upward leaving the orifice open at maturity; plants annual; bracteoles (1.5)2.5–6.5 mm long .... 2. *H. hispidum*

1. **Hedeoma drummondii** Benth., Drummond false pennyroyal. Perennial, (0.5)1–2(3.5) dm tall; stems solitary or few from a weak taproot or numerous from a branching caudex and forming clumps, the shoots simple or sparingly branched, pubescent. Leaves linear to elliptic-oblong, 5–12(20) mm long, 1–2.5(5) mm wide, punctate above and below, glabrous above, pubescent below, apex obtuse, margins entire, base attenuate; subsessile or the petiole less than 2 mm long. Verticillasters somewhat crowded on the upper 3/4 of the stems, cymules (1)2- to 6-flowered; bracteoles lanceolate to subulate, 1–2 mm long; peduncles 1–1.5 mm long; pedicels 1–3 mm long. Calyx 5–7.5 mm long, finely hirsute, the tube gibbous and tapering to a slender neck, upper calyx teeth subulate or narrowly triangular, 1–1.5 mm long, straight, lower calyx teeth subulate, 1.8–2.3 mm long, recurved and arching over the upper teeth, closing the orifice after anthesis; corolla blue or rose-lavender, 7–9 mm long, pubescent without, the tube 5–8 mm long, sparsely pubescent in the throat, upper lip ligulate, median lobe of the lower lip emarginate and slightly larger than the lateral lobes. Nutlets dark brown, occasionally glaucous, narrowly ovoid, 1.3–1.6 mm long, weakly areolate. (2n = 34, 36) May–Sep. Rocky, well-drained prairie sites, especially wash areas or gravelly outcrops; w 1/2 ND, MT, w 1/2 SD, WY, w 1/2 NE, n-cen & w 1/3 KS, CO, w 1/2 OK, TX, NM; (w ND to MT, s to s TX, sw CA; Mex.). *H. camporum* Rydb.—Rydberg; *H. longiflorum* Rydb.—Rydberg, Bull. Torrey Bot. Club 36: 695. 1909.

2. **Hedeoma hispidum** Pursh, rough false pennyroyal. Annual 0.7–3.5 dm tall; stems simple or branching in the lower portion, pubescent to hirsute. Leaves spreading, linear to linear-elliptic, 7–20 mm long, 1–3 mm wide, punctate, sometimes obscurely so, upper surface glabrous, lower surface variously pubescent to glabrous, apex obtuse to acute, margins entire and usually ciliate, base attenuate, sessile or subsessile. Verticillasters at all but the lowermost nodes, well spaced or congested towards the branch tips; cymules (1)3- to 6(12)-flowered; bracteoles subulate, (1.5)2.5–6.5 mm long; peduncles 0.5–1 mm long; pedicels 1.5–4 mm long. Flowers dimorphic; chasmogamous flowers with calyx 4.5–6 mm long, hirsute, the tube gibbous at the base, upper calyx teeth connate, narrowly triangular, 1–1.6 mm long, spreading outward, lower calyx teeth subulate, 1.5–2.3 mm long, curved slightly upward; corolla blue or whitish, 6–7 mm long, well exserted from the calyx, finely pubes-

cent outside, glabrous within, upper lip subgaleate, lobes of the lower lip subequal, median lobe shallowly emarginate; cleistogamous flowers with calyx similar to the chasmogamous flowers, corolla about equaling the calyx in length and the lobes usually not spreading. Nutlets yellow-brown or darker, sometimes glaucous, narrowly ovoid, 1-1.3 mm long, weakly ruminate or areolate to smooth. (2n = 34) May–Jul (Aug, Sep). Exposed sites in prairies & pastures, roadsides, waste ground; GP except TX panhandle & NM; (VT to MI & Alta., s to MS & e TX).

3. **Hedeoma pulgeoides** (L.) Pers., American false pennyroyal. Annual, 1.2-3 dm tall; stems solitary and usually branching in the mid-portion, pubescent. Leaf blades ovate to elliptic, 1-2.5 cm long, 3-9 mm wide, gradually reduced upwards, glabrous above, punctate and sparsely pubescent below, apex acute to obtuse, margins mostly with a few irregular teeth, base cuneate to attenuate; petiole 1-8 mm long, the longest on the lower leaves. Verticillasters at all but the lowermost nodes, usually remote; cymules (1)2- to 5-flowered; bracteoles lanceolate to narrowly elliptic, 1-3.5 mm long; peduncles 0.5-1.5 mm long; pedicels 1-3 mm long. Calyx 4-5 mm long, sparsely hisute and usually punctate, the tube slightly gibbous, upper calyx teeth connate, deltoid, 0.6-0.9 mm long, spreading, lower calyx teeth subulate, 1-2 mm long, slightly curved upwards; corolla bluish to lavender, 5-6 mm long, pubescent without, glabrous within, the tube barely exceeding the calyx, upper lip ligulate, straight, emarginate, lower lip equally 3-lobed, erect. Nulets yellow-brown or darker, suborbicular, 0.7-0.9 mm diam, weakly ruminate to smooth. (2n = 36) Aug–Sep. Usually rocky soil in woodlands, less frequent in moist sites; sw IA, se NE, MO, e 1/3 KS, e OK: (Que. to IA, s to SC, AR, & e OK).

4. **Hedeoma reverchonii** A. Gray, Reverchon false pennyroyal. Robust, suffruticose perennial, 2-4 dm tall; stems solitary to many arising from a taproot or branching caudex and usually forming clumps, the shoots sparingly branched, densely grayish pubescent, sometimes becoming less so lower on the shoots. Leaves mostly spatulate to obovate, 5-18 mm long, 2-7 mm wide, punctate, glabrous above, pubescent or densely so below, apex rounded or obtuse, rarely acute, margins entire, base cuneate or attenuate; sessile or petiole to 1.5 mm long. Verticillasters moderately crowded on the upper 3/4 of the stems, cymules (1)2- or 3(4)-flowered; bracteoles subulate, 2-3.5 mm long; peduncles 1-1.5 mm long; pedicels 2-3.5 mm long. Calyx 6-8.5 mm long, long whitish or grayish hirsute, the tube tapering slightly at the base and apex, not noticeably gibbous, the upper calyx teeth subulate, 0.5-1.2 mm long, slightly divergent, the lower calyx teeth subulate, 1.7-3.5 mm long, only slightly arching upwards and not closing the orifice after anthesis; corolla white or lavender, 6-12 mm long, pubescent without, the tube 5-9 mm long, sparsely pubescent in the throat and denser in the tube, upper lip ligulate to subgaleate, median lobe of the lower lip emarginate and slightly larger than the lateral lobes. Nutlets yellow-brown or darker, narrowly ovoid, 1.2-1.6 mm long, weakly areolate. (2n = 34) May–Jul (Aug). Exposed outcrops or gravelly prairie hillsides; s-cen & sw OK, TX; (s OK & TX panhandle, s to sw NM & Mex.). *H. drummondii* Benth. var. *reverchonii* A. Gray, var. *serpyllifolium* (Small) Irving—Correll & Johnston; *H. reverchonii* var. *serpyllifolium* (Small) Irving—Irving, Sida 8: 259. 1980.

8. LAMIUM L., Dead Nettle

Ours annual or winter annual herbs with short taproots; stems usually branching at the base, weak, decumbent or subprostrate only at the base. Leaves mostly cordate and 3-lobed or deeply toothed. Verticillasters in the axils of foliaceous bracts distinct along the entire stem to congested on the upper stem; cymules 2- to 6-flowered, sessile or nearly so. Flowers sessile; calyx essentially regular, 5-lobed, tubular to subcampanulate; corolla bilabiate, the tube flared at the throat and much longer than the calyx, upper lip erect, galeate, and subentire to emarginate, lower lip spreading, the median lobe constricted at the base, about as broad as long, and emarginate or 2-lobed, the lateral lobes essentially obsolete, represented by the convex margins of the throat; stamens 4, the abaxial pair ascending under the upper lip, the adaxial pair slightly longer; anther sacs divergent, hairy; stigmas equally 2-lobed; style included. Nutlets brownish, obovoid, trigonous, smooth.

1 Bracts sessile and clasping, rarely short petioled, usually wider than long; calyx densely villose .... 1. *L. amplexicaule*
1 Bracts petiolate, usually longer than wide; calyx sparsely hirsute on the veins .... 2. *L. purpureum*

1. *Lamium amplexicaule* L., henbit. Annual or winter annual, 1–3.5 dm tall, lightly appressed hirsute or pubescent, or glabrate; stems decumbent at the base and erect to flexuous above, often purplish streaked, internodes gradually reduced upwards. Leaf blades orbicular to broadly ovate, 3-lobed or not, 5–16 mm long, about as wide, variously pubescent, apex obtuse to rounded, margins crenate-lobulate, base subcordate to rounded; petiole to 3.5 cm long, rapidly reduced upwards. Verticillasters mostly distinct, only the uppermost sometimes congested; cymules 3- to 6-flowered; bracts foliaceous, rarely 3-lobed, usually wider than long, sessile and clasping. Calyx tubular, 5–7 mm long, densely villose, the teeth narrowly deltoid and shorter or longer than the tube; corolla pinkish-purple, the upper lip often darker, 10–20 mm long, pubescent to pilose without, especially the upper lip, glabrous within, the tube 10–14 mm long, straight, upper lip 3–5 mm long, entire to slightly emarginate, lower lip obcordate, 1.5–2.5 mm long; smaller cleistogamous flowers produced especially from fall to spring, white hairy without. Nutlets light brown to olivaceous, 1.5–2 mm long. (2n = 18) Mar–May (Nov). Waste areas, roadsides, lawns, & fields; s 1/2 GP & scattered n; (much of N. Amer., Europe). *Naturalized*. *L. amplexicaule* f. *clandestinum* (Reichb.) G. Beck—Fernald.

2. *Lamium purpureum* L., purple dead nettle. Annual, 1–4 dm tall; stems decumbent at the base, erect above, pilose, pubescent or glabrous, often purplish, the internode between the last leaf pair and first verticillaster usually much longer than the other internodes. Leaf blades ovate to suborbicular, 0.5–3.5 cm long, about as wide, hirsute, apex obtuse or rounded, rarely acute, margins crenate, base cordate, truncate or rounded; petiole 0.5–4 cm long. Verticillasters distinct below but usually congested towards the branch tips; cymules 2 or 3-flowered, bracts foliaceous, mostly ovate, gradually and only slightly reduced upwards, usually longer than wide, apex acute to obtuse, less often rounded, base cordate to cuneate, petiole 3–10(30) mm long. Calyx tubular to subcampanulate, 5–8 mm long, sparsely hirsute on the veins, occasionally glandular, the teeth narrowly deltoid to aristate, usually 1½–2 × longer than the tube; corolla pinkish-purple, 10–20 mm long, sparsely pubescent without, glabrous within, the tube 7–12 mm long, straight, upper lip 3–6 mm long and entire, lower lip obcordate, 1.5–2.5 mm long, spreading, the lateral lobes linear-lanceolate, about 0.5 mm long. Nutlets olivaceous and usually spotted with white, 1.8–2.3 mm long. (2n = 18) Apr–May. Disturbed, usually shaded sites, including ditches, pastures, stream banks, & lawns; MO, se NE, e 1/2 KS, e 1/2 OK; (many areas of N. Amer.; Europe). *Naturalized*.

9. LEONURUS L., Motherwort

Erect, aromatic, biennial or perennial herbs; internodes distinct but shorter towards the branch tips. Leaves unlobed to deeply 3- to 5-lobed, petiolate. Verticillasters dense, many-flowered, cymules subsessile; bracts foliaceous; bracteoles subulate. Flowers subsessle; calyx scarcely bilabiate, tubular-campanulate, 5- or 10-nerved, 5-toothed; corolla bilabiate, the tube shorter than or slightly longer than the calyx, upper lip erect, subgaleate, entire, lower lip spreading, 3-lobed, the median lobe larger than the lateral lobes; stamens 4, included; anther sacs parallel; style included; stigma equally obovoid, 3-sided, apex usually densely pubescent.

1 Corolla distinctly exceeding the calyx; calyx teeth about as long as the tube .. 1. *L. cardiaca*
1 Corolla not exceeding the calyx; calyx teeth 1/3–1/2 as long as the calyx tube .... 2. *L. marrubiastrum*

1. **Leonurus cardiaca** L., motherwort. Perennial with short rhizomes; stems 0.5–2 m tall, branching, glabrous to sparsely pubescent. Leaf blades broadly ovate to suborbicular, 3–12 cm long, 2–11 cm wide, sparsely pubescent; the larger blades digitately 3- to 5-lobed with these sometimes again shallowly lobed, apices acute, margins irregularly and coarsely serrate, base nearly truncate to cuneate; the smaller blades mostly unlobed, apex acute to obtuse, margins coarsely serrate, base subcordate to cuneate; petioles 1–5 cm long. Cymules 3- to 10-flowered; bracts foliaceous and reduced upwards, the uppermost bracts mostly elliptic with 3 large teeth and cuneate to attenuate at the base; bracteoles subulate to linear-lanceolate, 2.5–10 mm long. Calyx 3.5–8 mm long, 5-nerved, prominently veined within, glabrous to sparsely pubescent, teeth spreading at maturity, narrowly deltoid and spinose, about as long as the tube, the lower teeth deflexed; corolla whitish to pink with dark purple spots, 8–12 mm long, exceeding the calyx, the tube with a ring of hairs at the throat within, upper lip villose on the back; upper stamens about 1/2 the length of the lower pair. Nutlets 1.7–2.3 mm long, apex truncate. (2n = 18) May–Sep. Waste places, woodlands, along streams, low pastures, & ditches; MN & ND, s to MO, KS, & ne CO; (established as a casual weed over much of U.S. & adj. Can.; Eurasia). *Naturalized*. *L. cardiaca* subsp. *villosa* (Desf. ex Spreng.) Hyl.—Tutin, et al. Flora Europaea 3: 149. 1972.

2. **Leonurus marrubiastrum** L. Biennial; stems 0.5–1.5 m tall, branching, pubescent to canescent. Leaf blades ovate to suborbicular, 4–9 cm long, 2–5 cm wide, glabrous above, pubescent below, apex acute or narrowly so, margins coarsely and irregularly crenate or serrate, base cuneate or rounded; petiole 1–3 cm long. Cymules 3- to 10-flowered; bracts foliaceous, reduced upwards, the upper bracts becoming narrowly elliptic and less deeply toothed or subentire, apex acute, base attenuate, petiole 0.5–3 cm long; bracteoles acerose, mostly shorter than the calyx. Calyx 5–7 mm long, indistinctly 10-nerved, pubescent to canescent, the teeth straight or slightly spreading, deltoid and spinose, 1/3–1/2 as long as the calyx tube; corolla pale pink or lavender, 5–7 mm long, not exceeding the calyx, the tube glabrous within, upper lip pubescent on the back. Nutlets 2–2.5 mm long, apex rounded. Jun–Aug. Riverbanks & low woodlands; IA: Harrison; KS: Doniphan, Douglas; e edge NE; SD: Hutchinson; (established in scattered areas of e U.S.; Eurasia). *Adventive* or *locally naturalized* in GP.

10. **LYCOPUS** L., Bugleweed

Perennial herbs with rhizomes or slender stolons; stems erect, simple or branched; tubers produced or not. Leaves linear to ovate-lanceolate, entire to deeply pinnatifid, punctate, variously pubescent, sessile or petiolate. Verticillasters dense, distinct and borne at nearly every node, bracteoles present. Calyx tubular, campanulate (sometimes rotate in fruit), 4- or 5-toothed, the teeth equal or slightly irregular and shorter to longer than the nutlets; corolla usually regular, tubular, funnelform, or campanulate, 4- or 5-lobed, the lobes spreading or erect, bearded in the throat; stamens 2 or the adaxial pair present as reduced staminodes, exserted slightly or included; anther sacs divergent; stigmas equally 2-lobed;

style exserted or included; ovary deeply 4-lobed. Nutlets obovoid, slightly trigonous or flattish on one side and the other convex, glabrous, crest smooth to tuberculate-toothed.

Reference: Henderson, N. C. 1962. A taxonomic revision of the genus *Lycopus* (Labiatae). Amer. Midl. Naturalist 68: 95–138.

1 Calyx shorter than to barely equaling the nutlets, the lobes acute or obtuse at the apex.
  2 Crest of nutlets undulate, scarcely tuberculate-toothed; stamens exserted; corolla with 5 spreading lobes; plants bearing tubers .... 4. *L. uniflorus*
  2 Crest of nutlets distinctly tuberculate; stamens mostly included; corolla with 4 erect lobes .... 5. *L. virginicus*
1 Calyx much exceeding the nutlets, lobes acuminate to subulate tipped.
  3 Lower leaves cuneate and sessile .... 2. *L. asper*
  3 Lower leaves attenuate to a narrow, winged petiole or distinctly petiolate.
    4 Lowest leaves and sometimes the upper pinnatifid; corolla 4-lobed, 2.5–3.5 mm long; crest of nutlets smooth .... 1. *L. americanus*
    4 Leaves merely shallowly serrate; corolla 5-lobed, 2.5–5 mm long; crest of nutlets with 4–6 blunt tubercles .... 3. *L. rubellus*

1. *Lycopus americanus* Muhl. ex Bart., American bugleweed. Stems simple or branching, 3–9 dm tall, glabrous to densely pubescent at least in the upper portion, faces of the internodes usually grooved, angles rounded or ridged; roots not tuberiferous. Leaves variable, linear-elliptic to ovate-lanceolate, 1.5–12 cm long, 0.5–6 cm wide, lower leaves pinnatifid or occasionally merely serrate, upper leaves reduced, lobed, dentate, serrate, or subentire, punctate, glabrous above, sparsely pubescent on the veins below, apex narrowly acute or acuminate, base usually attenuate. Verticillasters 10- to 26-flowered; bracteoles lanceolate to subulate, 1–3 mm long, ciliate. Calyx tubular, equally 5-toothed, 2–3 mm long, exceeding the nutlets, the teeth narrowly deltoid to subulate, glandular ciliate; corolla whitish, regular or sometimes appearing slightly bilabiate, 4-lobed, 2.5–3.5 mm long, barely surpassing the calyx; stamens and style slightly exserted. Nutlets brownish with a lighter, narrow winglike band, 1–1.4 mm long, crest smooth. (2n = 22) Jul–Sep. Moist or wet soil, stream banks, lakeshores, sloughs, & ditches, usually exposed sites; common in GP; (s Can. s to NC, TN, LA, TX, NM, AZ, & CA).

2. *Lycopus asper* Greene, rough bugleweed. Stems simple or sparingly branched, 3.5–13 dm tall, sparsely pubescent, faces of the internodes grooved, angles usually rounded; thick, fusiform tubers usually present (although not commonly collected). Leaves elliptic-lanceolate, 5–8 cm long, 1–2 cm wide, punctate, glabrous, apex acute, margins coarsely serrate with the teeth widely spaced, base cuneate, sessile. Verticillasters 6- to 24-flowered; bracteoles lanceolate, 2.5–3.5 mm long, ciliate. Calyx tubular-campanulate, equally 5-toothed, 2.5–4.5 mm long, much exceeding the nutlets, the teeth narrowly triangular, ciliate or glandular-ciliate; corolla whitish, scarcely bilabiate, 4-lobed, 3.5–5.5 mm long, exceeding the calyx by 1–2 mm, sessile glandular, the lobes subequal and rounded; stamens and style included. Nutlets brownish with a lighter, narrow winglike band, 1.7–2.2 mm long, crest undulate. Jul–Sep. Moist or wet soil, stream banks, sloughs, marshes, lakeshores, & around springs, usually exposed sites; common in n GP, s to nw MO, w NE, & se WY; KS: Marshall; (s Sask. & s Ont., s to nw MO, ne KS, w NE, NM, AZ & n CA). *L. lucidus* of reports—Rydberg.

3. *Lycopus rubellus* Moench, water horehound. Stems simple or branched, 4–10 dm tall, pubescent, faces of the internodes grooved, angles rounded; stolons and slender tubers produced. Leaves elliptic-lanceolate to ovate-lanceolate, 5–12 cm long, 1.5–3.5 cm wide, punctate, glabrous or sparsely pubescent below, apex acute to acuminate, margins shallowly serrate, base attenuate to a short winged petiole. Verticillasters 6- to 20-flowered; bracteoles linear, 1–2 mm long, pubescent. Calyx tubular-campanulate, equally 5(4)-toothed, 1.8–2.5

mm long, exceeding the nutlets, pubescent, the teeth narrowly triangular; corolla whitish, slightly bilabiate, 5-lobed, 2.5–5 mm long, sessile-glandular, pubescent, the lobes spreading, rounded; stamens and style slightly exserted. Nutlets brownish with a lighter, narrow winglike band, 1–1.3 mm long, crest with 4–6 blunt tubercles. Aug–Sep. Shaded, wet stream banks & ditches; sw MO; KS: Cherokee, Linn; (ME to MO & se KS, s to FL & TX).

4. *Lycopus uniflorus* Michx., one flower horehound. Stems simple or sparingly branched, 3–6 dm tall, pubescent, faces of the internodes grooved, angles rounded to ridged; producing stolons and tubers. Leaves narrowly elliptic, elliptic-lanceolate, or ovate-lanceolate, 2–10 cm long, 1–4 cm wide, punctate, glabrous or the veins below sparsely pubescent, apex acute, margins shallowly serrate, bases mostly cuneate to a short, narrowly winged petiole or sessile. Verticillasters 2- to 12-flowered; bracteoles subulate, 0.5–1 mm long, ciliate. Calyx campanulate (becoming rotate in fruit), 5(4)-toothed, 1.3–1.6 mm long, about 1/2 as long as the corolla and barely or not as long as the nutlets, pubescent, the teeth triangular and acute or obtuse, the adaxial tooth usually smaller than the other 4; corolla whitish, nearly regular, 5-lobed, 2.5–3.5 mm long, sessile-glandular, puberulent, the lobes subequal, spreading, and rounded; stamens and style exserted. Nutlets brownish, 1–1.2 mm long, crest undulate to scarcely tuberculate toothed. Jul–Sep. Exposed or shaded sites in marshes, bogs, along streams, or around springs; MN, e 1/2 ND, IA, NE; KS: Pottawatomie; (s Que. to se Sask., s to VA, ne KS, & NE; B.C. s to w MT, n ID, & n CA).

5. *Lycopus virginicus* L., Virginia bugleweed. Stems simple or branching, 3–9 dm tall, pubescent, faces of the internodes flat or slightly grooved with rounded angles; tubers rarely produced, stoloniferous. Verticillasters 6- to 20-flowered; bracteoles linear, 0.5–1 mm long, pubescent. Calyx tubular to campanulate, 4(5)-toothed, 1–2 mm long, not exceeding the nutlets, usually glabrous, the teeth broadly triangular, acute to obtuse; corolla whitish, nearly regular, 4-lobed, 1.8–2.2 mm long, sessile-glandular, puberulent, lobes erect; stamens included; style included or exserted. Nutlets brownish, 1.6–1.9 mm long, crest deeply tuberculate toothed. Aug–Sep. Exposed or shaded river banks, marshy areas, or low woodlands; infrequent in IA, e NE, MO, e 1/2 KS; (MA w to e NE, s to FL & e TX).

Lycopus x sherardii Steele is a hybrid of *L. uniflorus* and *L. virginicus* that is morphologically intermediate between the parental species and reportedly occurs in the se GP and eastward (Henderson, op. cit.). Recognizing the parental species in the GP is sometimes difficult because of an apparent overlap of distinguishing characters. The scarcity of both species in our region further adds to the difficulty of adequately assessing this complex in the GP.

11. MARRUBIUM L., Horehound

1. ***Marrubium vulgare*** L., common horehound. Perennial herb with short, stout rhizomes; main stems erect, 3–7 dm tall, white tomentose or lanate in the lower portion, above becoming less hairy and often mixed with stellate hairs, decumbent or ascending, shorter nonflowering stems often present. Leaves with blades broadly ovate to orbicular, 1.5–5 cm long, 1–4 cm wide, green or gray-green and sparsely tomentose above, stellate hairs present or not, whitish and tomentose to lanate below, stellate hairs usually present (often obscured by simple hairs), apex acutish to rounded, margins crenate to serrate, base cuneate to round; petioles of the lower leaves 0.5–4 cm long, reduced upwards. Verticillasters remote, borne mostly in the upper 1/2–3/4 of the stems, dense and many-flowered; bracts foliaceous and gradually reduced upwards or only the lowest foliaceous and those above markedly reduced and barely or not surpassing the verticillasters; cymules sessile; bracteoles linear, about as long as the flowers, apex spinose and sometimes hooked. Calyx tubular, faintly 10-nerved, unequally (8)10-toothed, the tube 3–5 mm long, villose-tomentose mixed with stellate hairs

or sparsely stellate hairy, the teeth alternately short and long with hooked tips; corolla creamy white, bilabiate, 4–6 mm long, the tube barely exceeding the calyx, pubescent, upper lip straight, deeply bifid, densely pubescent without, lower lip 3-lobed, the median lobe obovate, much larger than the lateral lobes, and usually glabrous, lateral lobes rounded, pubescent without; stamens 4, included; anther sacs diverging; stigma unequally 2-lobed; style included. Nutlets dark brown, obovoid to ellipsoid, 1.8–2.2 mm long, smooth. May–Jul. Waste areas, pastures, old home sites, roadsides; scattered in se GP, n to s NE, infrequent w to CO & n to nw NE & BH; (much of N. Amer.; Europe). *Naturalized.*

Marrubium vulgare, according to Flora Europaea (Tutin et al., 1972), typically has 10 equal calyx teeth, each hooked at the tip. Plants in the GP, as well as most other North American plants examined, have 5 long teeth hooked at the tip, alternating with 5 shorter teeth that may or may not be hooked. On this basis, our plants in Flora Europaea key to *M. alternidens* Rech. f., a segregate of the *M. vulgare* complex described in 1952. However, ignoring the calyx teeth, GP plants do not fit the descriptions of either *M. alternidens* or *M. vulgare,* especially with regard to vestiture. Both taxa are described as tomentose or lanate-tomentose but our plants are consistently a dense mixture of simple and stellate hairs, the stellate hairs so dense they have gone unnoticed by most authors. Further study of this complex would appear necessary.

12. MENTHA L., Mint

Ours perennial, aromatic herbs with creeping rhizomes; stems simple or branched, decumbent, ascending, or erect. Leaves ovate, lanceolate, or narrowly elliptic, margins serrate; petiolate or subsessile. Verticillasters dense, distinct, borne at nearly every node or towards the branch tips in congested or interrupted spikes; bracts foliaceous or greatly reduced and similar to the bracteoles. Calyx tubular or campanulate, 10- to 13-nerved, 5-lobed and slightly bilabiate; corolla weakly bilabiate, tubular, 4-lobed, the upper lobe wider than the lower lobes and usually emarginate, the lower lobes obtuse or rounded; stamens 4, subequal, exserted (sometimes included and nonfunctional, especially in cultivars of hybrid origin); anther sacs parallel; stigma equally 2-lobed; style exserted. Nutlets brownish, ovoid to ellipsoid, mostly smooth.

The following species are rarely collected in the GP and are not described in the text. Each taxon is included in the key and marked with an asterisk.

Mentha x *gentilis* L. A sterile hybrid between *M. arvensis* and *M. spicata.* (2n = 54, 60, 84, 96, 108, 120) Jul–Sep. Occasionally escaping from gardens & persisting for a short time; IA: Sioux; NE: Dawes, Dodge.

Mentha longifolia L., horse mint. (2n = 24) Jul–Sep. Adventive from Eurasia. KS: Sedgwick; MO: Clay, Jackson, Jasper.

1 Bracts subtending the verticillasters foliaceous, well surpassing the verticillasters.
  2 Calyx 2.5–3.3 mm long, pubescent; corolla 4.5–6.5 mm long; anthers usually well exserted; common native species of the GP .... 1. *M. arvensis*
  2 Calyx less than 2.5 mm long, glabrous or the teeth sparsely hairy; corolla less than 4 mm long; anthers mostly included; plants introduced and infrequently collected.
    3 Bracts similar to the leaves, only slightly reduced .... *M.* x *gentilis*
    3 Bracts obviously smaller than the leaves .... 2. *M.* x *cardiaca*
1 Bracts subtending the verticillasters reduced and resembling the bracteoles, usually not surpassing the verticillasters.
  4 Lower surface of leaves densely pubescent .... *M. longifolia*
  4 Lower surface of leaves glabrous or sparsely pubescent on the veins.
    5 Leaves mostly cuneate at the base; calyx 1.5–2.3 mm long .... 3. *M.* x *piperita*
    5 Leaves mostly rounded at the base; calyx 1.2–1.5 mm long .... 4. *M. spicata*

1. **Mentha arvensis** L., field mint. Stems simple or branched, ascending to erect, 3–5(9) dm long, internodes variously pubescent or hirsute, infrequently nearly glabrous, nodes

usually with a ring of hairs; rhizomes extensively creeping. Leaves elliptic-lanceolate, lanceolate, or ovate, 2.5–12 cm long, 0.5–4 cm wide, reduced slightly upwards and passing into the foliaceous bracts, mostly glabrous above and sparsely pubescent on the veins below, apex acute, margins serrate and ciliate, base cuneate to a narrowly winged petiole 3–15 mm long. Verticillasters borne in the axils of foliaceous bracts; cymules 8- to 30-flowered; peduncle 1–4(14) mm long; bracteoles linear-lanceolate to elliptic-lanceolate, 3.5–10(15) mm long, 1–5 mm wide, ciliate; pedicels 1–3 mm long. Calyx tubular or campanulate, 5-lobed, 2.5–3.3 mm long, variously pubescent, occasionally sessile-glandular, the teeth narrowly to broadly triangular, about 1/4 as long as the calyx; corolla whitish to lavender, 4.5–6.5 mm long, glabrous; stamens and style exserted 1–2 mm from the corolla (nonfunctional stamens included). Nutlets yellowish-brown, ovoid to ellipsoid, 0.7–1.3 mm long, smooth. (2n = 24, 72, 90) Jul–Sep. Moist, exposed or shaded sites, stream banks, lakeshores, springs, marshes or ditches; common in GP from e KS northward, scattered w & s except OK; (circumboreal, in N. Amer. s to VA, MO, KS, n TX, NM, & CA). *M. arvensis* var. *glabrata* (Benth.) Fern.—Gleason & Cronquist; *M. arvensis* var. *villosa* (Benth.) S. R. Stewart—Fernald; *M. canadensis* L., *M. glabrior* (Hook.) Rydb., *M. pernardi* (Briq.) Rydb.—Rydberg.

Numerous infraspecific taxa have been recognized in this widespread complex, primarily based on vegetative characters. In the GP, supposed diagnostic charcters completely intergrade and show little or no tendency towards geographic distinction. Recognition of infraspecific taxa does not seem warranted.

2. **Mentha x cardiaca** Gerarde ex Baker. Stems branching, decumbent to ascending, 4–10 dm long, glabrous or sparsely pubescent; extensively rhizomatous. Leaves ovate to elliptic, 1.5–7 cm long, 0.7–3.5 cm wide, reduced upwards, punctate below, glabrous except the veins sometimes sparsely pubescent, apex acute, margins serrate, base cuneate to a narrowly winged petiole 2–10 mm long. Verticillasters mostly borne towards the branch tips in interrupted spikes and subtended by foliaceous bracts which well exceed the verticillasters; cymules 10- to 20-flowered; peduncles 0.5–2 mm long; bracteoles linear-lanceolate, 3–5 mm long, 0.5–1.5 mm wide, usually ciliate; pedicels 1–1.5 mm long. Calyx tubular-campanulate, 5-lobed, 1.5–2.3 mm long, sessile-glandular, tube glabrous, the teeth triangular and sparsely hirsute; corolla whitish to lavender, 2.5–3.5 mm long, glabrous; stamens usually included and nonfunctional, rarely exserted; style exserted. Nutlets not seen, apparently rarely produced. Jul–Sep. Shaded or exposed wet sites, mostly along streams; sporadic in se GP; (Newf. w to MI, s to VA, IL & KS; Europe).

This plant is thought to be of hybrid origin from *M. arvensis* and *M. spicata* and may not be sufficiently distinct from *M. x gentilis* to warrant recognition.

3. **Mentha x piperita** L., peppermint. Similar to No. 2 except as noted. Verticillasters crowded toward the branch tips and forming nearly continuous spikes; bracts not foliaceous, small, not exceeding the verticillasters and resembling the bracteoles. (2n = 66, 72) Jul–Sep. Exposed or shaded wet ditches, stream banks, & lakeshores; rare in se GP; (occasionally escaping from gardens, N.S. to MN, s to FL & OK, also CA; Europe).

This plant is thought to be a hybrid originating from *M. arvensis* and *M. aquatica* L.

4. **Mentha spicata** L., spearmint. Stems branched, ascending to erect, 3–7 dm tall, usually glabrous; rhizomatous. Leaves ovate to ovate-lanceolate, 3–9 cm long, 0.7–3 cm wide, punctate below, glabrous or sparsely pubescent on the veins below, apex acute, margins serrate, base rounded, infrequently cuneate; petiole 1–3 mm long, rarely sessile. Verticillasters crowded toward the branch tips, forming spikes; bracts not foliaceous, mostly linear-lanceolate, 3–7 mm long, 0.5–1 mm wide, barely, if at all, surpassing the verticillasters, ciliate; peduncles 0–1 mm long; bracteoles resembling the bracts but smaller; pedicels 0–1 mm long. Calyx

campanulate, 5-lobed, 1.2–1.5 mm long, sessile-glandular, the teeth triangular, about 1/3–1/2 as long as the calyx, ciliate, corolla whitish to lavender, 1.7–3 mm long, glabrous; in ours the stamens mostly included and apparently nonfunctional; style exserted. Nutlets not seen. (2n = 48) Jul–Sep. Exposed or shaded wet sites including ditches, stream banks, & lakeshores; escaping from cultivation in a few areas of the e-cen & se GP; (scattered sites throughout U.S. & Can.). *Naturalized.*

The origin of this plant is unknown, but it probably arose in cultivation. It is widely cultivated for its aromatic oils and as a potherb and is now naturalized in many areas of the world.

13. MONARDA L., Horse Mint, Beebalm

Erect, aromatic, herbaceous annuals with slender taproots or perennials with creeping rhizomes; stems simple or branched, variously pubescent. Leaves ovate to narrowly elliptic, serrate to subentire, sessile or petiolate. Flowers sessile or subsessile and borne in glomerules, these terminal and solitary or several in an interrupted spike, subtended by foliaceous outer bracts and more modified inner bracts, each flower usually subtended by a linear or subulate bracteole. Calyx tubular, 13- to 15-nerved, 5-toothed, regular, the throat hairy or not, the teeth deltoid to aristate, shorter than or about equaling the tube; corolla bilabiate, upper lip erect and linear or falcate and forming a sickle-shaped corolla, lower lip wider, spreading or deflexed, 3-lobed or with a central projecting tooth; stamens 2, ascending under the upper lip, exserted or included; anther sacs divergent; stigmas unequally 2-lobed; style exserted. Nutlets oblong or obovoid, rounded at the apex and somewhat pointed at the base, smooth.

References: McClintock, E., & C. Epling, 1942. A review of the genus *Monarda* (Labiatae). Univ. Calif. Publ. Bot. 20: 147-194; Fernald, M. L. 1944. The confused publication of *Monarda russelliana.* Rhodora 56: 491–493.

1 Flower heads solitary; corolla tube gradually expanded toward the throat, upper lip erect or somewhat arcuate; leaves mostly ovate to ovate-lanceolate.
  2 Calyx teeth 2.5–4 mm long, usually stipitate-glandular .... 1. *M. bradburiana*
  2 Calyx teeth 0.6–1.5 mm long, not stipitate glandular .... 4. *M. fistulosa*
1 Flower heads (1)2–7 in interrupted spikes; corolla tube slender from the base and abruptly funnelform near the throat, upper lip galeate and falcate; leaves mostly lanceolate to narrowly elliptic.
  3 Calyx teeth triangular, about 1 mm long and nearly as wide .... 6. *M. punctata* subsp. *occidentalis*
  3 Calyx teeth aristate, 2–7 mm long.
    4 Bracts densely canescent above .... 2. *M. citriodora*
    4 Bracts essentially glabrous above.
      5 Calyx teeth 2–3(3.5) mm long .... 5. *M. pectinata*
      5 Calyx teeth (3)4–6 mm long .... 3. *M. clinopodioides*

1. ***Monarda bradburiana*** Beck, Bradbury beebalm. Perennial, 2.5–6 dm tall; stems simple or sparingly branched, variously pilose to glabrate; internodes about as long as or longer than the subtending leaves. Leaf blades ovate, deltoid-ovate, or lanceolate, 4–9 cm long, 1–3 cm wide, the lower leaves usually wider and shorter than the median and upper ones, punctate, glabrous or rarely pilose, apex acute to acuminate, margins serrate, ciliate, base rounded to subcordate; petiole 1–5 mm long. Flower heads solitary and terminal, 1.5–3 cm wide (excluding corollas); bracts green or purplish tinged, ovate-lanceolate to linear-lanceolate, 1.5–2 cm long, 3–12 mm wide, apex narrowly acute, margins entire-ciliate; bracteoles mostly linear, 5–10 mm long, ciliate. Calyx tube 6–12 mm long, glabrate without, hirsute in the throat, teeth acicular and the tip sometimes hooked, 2.5–4 mm long, hirsute or not at the base, usually sparsely stipitate-glandular; corolla white or pale lavender,

purple dotted on the lower lip, 2.5-3.5 cm long, lightly pubescent without, the tube hirsute in the throat, upper lip falcate, about as long as the tube and bearded within toward the tip, lower lip shorter than the tube; stamens exserted. Nutlets brownish, 1.5-2 mm long. May-Jun (Jul). Rocky, wooded hillsides; sw MO; KS: Cherokee, Greenwood; (KY w to MO & se KS, s to AL, AR, & e OK).

2. **Monarda citriodora** Cerv. ex Lag., lemon mint, lemon beebalm. Annual or occasionally biennial, 3-8 dm tall; stems mostly branched, retrorsely pubescent; internodes mostly shorter than the subtending leaves. Leaf blades lanceolate to elliptic-lanceolate, 3-6 cm long, 7-15(20) mm wide, punctate, glabrous or sparsely pubescent, apex acute, margins serrate or infrequently subentire, ciliate, base cuneate to attenuate; petiole 2-15(20) mm long, narrowly winged. Flower heads (1)2-6 in an interrupted spike, 1.5-3.5 cm wide (excluding corollas); bracts becoming lanceolate and only the top reflexed, 1.5-3.5 cm long, 3-6 mm wide, densely canescent above, glabrous to sparsely pubescent below, apex abruptly acuminate to a spinose bristle 2-5 mm long, margins entire or the outer bracts sometimes serrulate toward the apex; bracteoles mostly linear, 6-10 mm long, hirsute-ciliate. Calyx tube 6-12 mm long, sparsely puberulent or sometimes hirsute at the base, densely hirsute in the throat, teeth aristate, (2.5)4-7 mm long, antrorsely puberulent and sparsely hirsute; corolla whitish to pale lavender, sometimes purple dotted on the lower lip, 1.5-2.5 cm long, the tube slender toward the base and abruptly expanded and funnelform above, 8-15 mm long, sparsely hirsute at the base of the lower lip, upper lip galeate and falcate, slightly shorter than the tube, emarginate and ciliate at the tip, lower lip about as long as the upper; stamens included under the upper lip. Nutlets yellow-brown, 1.5-2 mm long. (n = 9) May-Jul. Rocky or sandy prairies, pastures, or gravelly hillsides; MO, KS, OK, TX; (MO & KS s to AR, TX, se NM, & ne Mex.). *M. dispersa* Small—Rydberg.

3. **Monarda clinopodioides** A. Gray, basil beebalm. Annual, 2.5-5 dm tall; stems simple or sparingly branched at the base, rarely branched above, retrorsely pubescent; internodes shorter or longer than the subtending leaves. Leaf blades lanceolate to elliptic-lanceolate, the lowermost sometimes elliptic or ovate, 2-5 cm long, 4-10(20) mm wide, punctate, glabrous to pubescent, apex acute, margins serrate to subentire, ciliate, base attenuate to cuneate; petiole 2-15 mm long. Flower heads (1)2-7 in an interrupted spike, 1.5-3 cm wide (excluding corollas); bracts usually purplish, outer bracts ovate-lanceolate to lanceolate, usually reflexed, 1-1.8 cm long, 3-5 mm wide, glabrous above, glabrous to sparsely pubescent below, apex abruptly acuminate to a spinose bristle, margins entire or serrulate toward the apex, inner bracts similar but narrower; bracteoles usually linear, 5-10 mm long, hirsute-ciliate. Calyx tube 6-10 mm long, sparsely hirsute, densely hirsute in the throat, teeth aristate, (3)4-6 mm long, antrorsely puberulent and hirsute-ciliate; corolla white to purplish, 1.5-2(2.5) cm long, the tube slender toward the base, abruptly expanded and funnelform above, 10-15 mm long, sparsely hirsute at the base of the lower lip, upper lip galeate and falcate, shorter than the tube, emarginate and hirsute at the tip without, lower lip about equaling the upper; stamens included under the upper lip or barely exserted. Nutlets brownish, 1.3-1.8 mm long. May-Jul. Sandy or gypsiferous soils in prairies & pastures; s-cen KS, n-cen & s OK; TX: Wichita, Wilbarger; (s-cen KS s to s TX).

4. **Monarda fistulosa** L., wild bergamot. Perennial with slender, creeping rhizomes, 3-12 dm tall, stems simple or branched above, retrorsely pubescent in the upper portion (rarely mixed with longer hairs), usually glabrous on the lower portion. Leaf blades ovate to ovate-lanceolate or lanceolate, 3-10 cm long, 1-3.5(5) cm wide, punctate, sparsely pubescent to glabrous above, pubescent to subtomentose below, apex acute to acuminate, margins coarsely serrate to subentire, base rounded to cuneate, sometimes truncate; petiole 2-20(25) mm long. Flower heads solitary and terminal on the branches, 1.5-3 cm wide (excluding

corollas); bracts green, whitish, or infrequently pinkish, usually reflexed, outer bracts ovate or suborbicular, 1-2.5 cm long, 0.7-1.5 cm wide, apex acute, margins entire, inner bracts mostly narrower; bracteoles present or not, linear, 6-10 mm long, hirsute-ciliate. Calyx tube 5.5-11 mm long, puberulent or pubescent, hirsute in the throat, teeth spinose, 0.6-1.5 mm long; corolla pale to dark lavender, rarely white, 2-3.5 cm long, outer surface puberulent to weakly pilose, glabrous within, tube slender, 15-25 mm long, upper lip usually erect, shorter than the tube, comose without, lower lip spreading to deflexed; stamens exserted from under the upper lip. Nutlets brownish or blackish, 1.5-2 mm long.

This is a variable species in which two geographically distinct varieties may be recognized in our area.

4a. var. *fistulosa*. Stems usually branching in the upper portion. Petioles (8)10-25 mm long. Corolla lavender, rarely white. (n = 16, 18) Jun-Aug. Prairie hillsides, pastures, roadsides, & thickets, occasionally in open woodlands; scattered in n GP, common in the se; MN, ND, sw MT, IA, SD (except BH), NE, MO, e 1/2 KS, e 1/2 OK; (s Ont. w to MT, s to GA, cen TX, & w NE; Mex.). *M. fistulosa* var. *mollis* (L.) Benth.—Fernald; *M. mollis* L.—Rydberg.

4b. var. *menthifolia* (Grah.) Fern. Stems mostly simple. Petioles 2-5(8) mm long. Corolla usually dark lavender to rose-purple, rarely white. (n = 16, 18) Jul-Sep. Moist, mostly exposed sites in meadows, along streams or lakeshores, or in ditches; common in the n & w GP; nw MN, ND, MT, w SD, WY, nw NE; (nw MN & Sask. w to B.C., s to w SD, w TX, AZ, & n ID). *M. comata* Rydb., *M. menthaefolia* Grah.—Rydberg.

Monarda didyma L. is a more robust species with dark red, usually larger flowers that occasionally escapes from cultivation in the se GP.

5. **Monarda pectinata** Nutt., spotted or plains beebalm. Annual, 2-4(5) dm tall; stems mostly branching from the base and again above, often spreading, retrorsely pubescent; internodes longer or shorter than the subtending leaves. Leaf blades lanceolate to elliptic, 2-3.5(4.5) cm long, 5-15 mm wide, punctate, glabrous or sparsely pubescent, apex acute, margins serrate to subentire, ciliate, base cuneate or attenuate; petiole 2-5(15) mm long. Flower heads (1)2-6 in an interrupted spike, 1.2-3 cm wide (excluding corollas); bracts green, infrequently purplish tinged, outer bracts ovate to elliptic-lanceolate, 0.8-1.5(2.5) cm long, 3-5(7) mm wide, glabrous above, variously pubescent below, apex abruptly acuminate to a spinose bristle, margins entire or serrulate at the apex, hirsute-ciliate, inner bracts similar but narrower; bracteoles mostly linear, 4-8 mm long, hirsute-ciliate. Calyx tube 6-8 mm long, puberulent, densely hirsute in the throat, teeth aristate, 2-3(3.5) mm long, antrorsely puberulent and hirsute-ciliate; corolla white to purplish, 12-20 mm long, the tube slender at the base and abruptly funnelform at the throat, 8-14 mm long, sparsely hirsute at the base of the lower lip, upper lip shorter than the tube, galeate and falcate, the tip emarginate and hirsute without, lower lip about equaling the upper; stamens included under the upper lip. Nutlets brownish, 1.2-1.5 mm long. May-Jul (sporadic to Oct). Gravelly or sandy soils in upland prairies, on hillsides, or in pastures; SD: Todd; NE; w 1/6 KS, CO; OK panhandle; TX; (s-cen SD, CO, & s UT, s to cen TX, NM, & AZ).

6. **Monarda punctata** L. subsp. **occidentalis** Epl., dotted beebalm, horse mint. Annual, biennial, or perennial, (1)2-4 dm tall; stems usually branching, retrorsely pubescent; internodes longer or shorter than the subtending leaves. Leaf blades lanceolate to narrowly elliptic, infrequently ovate, 2-4(5) cm long, 3-10(14) mm wide, punctate, usually sparsely pubescent, apex acute, margins serrate to subentire, ciliate, base cuneate to attenuate; subsessile or the petiole to 16 mm long. Flower heads (1)2-3(6) in an interrupted spike, 1.5-2.5 cm wide (excluding corollas); bracts spreading or slightly reflexed, whitish or pinkish, oblong, elliptic, or obovate, infrequently lanceolate, 1.5-3.5(4) mm long, 7-15 mm wide, usually densely puberulent above, variously pubescent to glabrate below, apex acute to short acuminate, not bristle tipped, margins usually serrate near the apex; bracteoles linear,

4–8 mm long, ciliate. Calyx tube 4.5–7 mm long, pubescent, hirsute in the throat, teeth triangular, about 1 mm long and nearly as wide, hirsute-ciliate; corolla white to purplish, often the lower lip with dark purple spots, 1.2–2 cm long, sparsely hirsute at the base of the lower lip, the tube 7–11 mm long, slender at the base and then abruptly funnelform, upper lip galeate and falcate, slightly shorter than the tube, apex scarcely emarginate and hirsute without, lower lip about equaling the upper; stamens included under the upper lip. Nutlets brownish, 1.2–1.5 mm long. (2n = 24) (May) Jun–Jul. Usually sandy soil in prairies, pastures, old fields, & roadsides; sw 1/4 KS; w OK; TX; rare & naturalized in nw CO & w MO; (sw KS s to cen TX & NM).

14. NEPETA L., Catnip

1. *Nepeta cataria* L., catnip. Perennial herb with 1–several stems from a short taproot or sometimes a branched caudex; stems erect or ascending, 3–10 dm tall, branched, white to grayish pubescent to subtomentose especially in the upper portions. Leaf blades ovate to triangular, infrequently ovate-lanceolate, 2–7(10) cm long, 1.5–5(7) cm wide, mostly green and lightly pubescent above, grayish and more densely pubescent to subtomentose below, occasionally lightly pubescent, apex acute to obtuse, margins crenate or serrate, base cordate to subtruncate; petiole 0.5–3(5) cm long. Verticillasters many-flowered, borne in spikelike inflorescences or in lax or dense short-peduncled cymes; bracts ovate-lanceolate to lanceolate, rarely exceeding the verticillasters; bracteoles linear-lanceolate to subulate, 1.5–3 mm long; peduncles 0–5 mm long; pedicels 0–2 mm long. Calyx bilabiate, 5–7 mm long, 15-nerved, the tube straight or slightly inflated at the base, sparsely to densely villose, teeth somewhat spreading, narrowly triangular, 1.5–2.5 mm long, the upper teeth well exceeding the lower; corolla creamy white with purple spots, bilabiate, 7–11 mm long, pubescent to villose without, bearded within near the base of the lower lip, tube slender toward the base and abruptly funnelform at the throat, slightly arcuate, upper lip erect and subgaleate, emarginate or not, lower lip spreading, 3-lobed, the median lobe much larger than the lateral and sometimes erose or repand; stamens 4, didynamous, scarcely exserted; anther sacs diverging; stigma equally 2-lobed. Nutlets brownish, obovoid or ovoid, slightly flattened, 1.3–1.9 mm long, smooth. (n = 16, 18) Jun–Oct. Exposed or shaded, usually disturbed sites including pastures, farm lots, & waste sites, occasionally along stream banks or in thickets; scattered in GP except TX, sw KS, & se CO; (most of N. Amer.; Europe). Naturalized.

15. PERILLA L.

1. *Perilla frutescens* (L.) Britt., common perilla or beefsteak plant. Aromatic annual, often purplish; stems 2–8 dm tall, sparsely pubescent or pilose to glabrate, especially below. Leaf blades broadly ovate to suborbicular, infrequently elliptic, 4–12 cm long, 2–8 cm wide, glabrous or the veins sparsely pubescent, apex acute or abruptly short acuminate, margins coarsely serrate to crenate, base cuneate to attenuate; petiole 1–7 cm long, usually sparsely hairy. Flowers solitary in the axils of small bracts forming an elongate, spikelike, mostly 1-sided raceme 5–10 cm long, lengthening with age to as much as 15 cm long; bracts ovate or elliptic, 3–6 mm long, usually folded, apex acute to acuminate, mostly entire, base rounded or cuneate, sessile or subsessile; peduncle 1–5 mm long. Calyx campanulate, 10-nerved, bilabiate, at anthesis 2–3 mm long, in fruit 8–12 mm long, the tube slightly inflated at the base, villose without, sparsely so within, upper lip erect, with 3 broadly triangular to elliptic teeth united for about 1/3 their length, lower lip with 2 narrowly triangular teeth and about as long as the upper lip; corolla white to lavender, tubular, 2.5–3.5 mm long,

barely exceeding the calyx, nearly regular with 5 broadly rounded lobes, these sometimes hairy without, the lower lobes bearded within; stamens 4, subequal and straight, exserted; anther sacs divergent; stigma unequally 2-lobed. Nutlets reddish-brown or darker, spherical, 1.5–2 mm diam, weakly areolate. (n = 14; 2n = 40) Sep–Oct. Moist soil in pastures, low woodlands, along streams, & occasionally in fields; NE: Richardson; e KS, MO, e OK; (MA w to OH & IA, s to FL & e TX; India). *Naturalized.*

16. PHYSOSTEGIA Benth., Obedient Plant, False Dragonhead, Lionsheart

Ours perennial herbs; stems erect, simple or branching primarily in the inflorescence, glabrous below, densely pubescent in the inflorescence, including the calyces. Leaves mostly narrowly elliptic to lanceolate or oblanceolate, glabrous, sessile or the lower ones short-petiolate. Flowers solitary in the axils of small bracts forming elongate, spiciform inflorescences; subsessile or the lowermost short pedicelled. Calyx nearly regular, tubular-campanulate, 10-nerved (sometimes obscurely so), 5-toothed, the teeth erect and much shorter than the tube; corolla bilabiate, much exceeding the calyx, tubular within the calyx and then gradually widening upward, upper lip erect or slightly arcuate, concave, entire or slightly emarginate, lower lip spreading, 3-lobed, about as long as the upper lip, the median lobe larger than the lateral ones; stamens 4, didynamous, ascending under the upper corolla lip; anther sacs parallel and dorsifixed; stigma equally or subequally 2-lobed, included. Nutlets brown, ovoid and distinctly 3-sided, smooth. *Dracocephalum* L., in part—Rydberg.

All of our species have value as ornamentals.

References: Cantino, P. D. 1981. Change of status for *Physostegia virginiana* var. *ledinghamii* (Labiatae) and evidence for hybrid origin. Rhodora 83: 111–118; Cantino, P. D. 1982. A monograph of the genus *Physostegia* (Labiatae). Contr. Gray Herb. 211:1–105; Godfrey, R. K., & J. W. Wooten. 1981. *Physostegia*, pp. 611–619; *in* Aquatic and wetland plants of the Southeastern United States: Dicotyledons. Univ. Georgia Press, Athens. Philip Cantino at the University of Ohio, Athens, kindly reviewed this treatment and made valuable suggestions.

1 Calyx and rachis of inflorescense bearing stipitate glands.
  2 Corolla 14–23 mm long, eglandular; nutlets 2.8–4 mm long .... 2. *P. ledinghamii*
  2 Corolla 9–12(16) mm long, usually stipitate glandular; nutlets 2.1–3.3 mm long .... 3. *P. parviflora*
1 Calyx and rachis of inflorescence lacking stipitate glands.
  3 Leaves mostly clasping at the base; flowering May to July .... 1. *P. angustifolia*
  3 Leaves sessile or subsessile, not clasping at the base; flowering July to September .... 4. *P. virginiana*

1. *Physostegia angustifolia* Fern., false dragonhead. Stems 3–15 dm tall; arising from an erect caudex. Leaves ascending to nearly appressed, lanceolate, narrowly elliptic, or linear, 5–17 cm long, 5–25 mm wide, apex acute to attenuate, rarely obtuse, margins serrate, base mostly attenuate, sessile and usually clasping. Inflorescences 5–30 cm long, often interrupted towards the base; bracts ovate-lanceolate to lanceolate, about 1/2–2/3 as long as the mature calyx. Calyx 6–8 mm long at anthesis, teeth triangular and acuminate, 1–2.5 mm long, in fruit the calyx to 11(13) mm long; corolla pale lavender to white with purple spots, (1.5)2–3 cm long, puberulent to glabrous without, the tube 1–2.3 cm long, upper and lower lips equal and about 1/3 as long as the tube. Nutlets 2–3 mm long. (2n = 38) May–Jul. Prairies, especially in moist or wet swales & around thickets, ditches, & open stream banks; scattered se GP; sw MO, n-cen & se KS, e OK; (n-cen & se KS & sw MO, s to w GA & cen TX). *P. intermedia* of reports—Atlas GP.

2. Physostegia ledinghamii (Boivin) Cantino. Similar to nos. 3 and 4 but differing as follows: Leaves serrate. Corolla 14–23 mm long, puberulent and occasionally bearing stipitate glands. Nutlets 2.8–4 mm long. (2n = 76) Jul–Aug. Exposed riverbanks, lakeshores, & ditches; rare; ND: Burleigh, McLean; (Man. to Alta. & w ND).

This plant is probably a tetraploid derivative of a hybrid between *P. parviflora* and *P. virginiana* and is not easily distinguished from the parental species (Cantino, 1981, op. cit.).

3. Physostegia parviflora Nutt. ex A. Gray, obedient plant. Stems 2–8 dm tall, arising from a slender, erect caudex or short rhizome. Leaves spreading or ascending, lanceolate to oblong-lanceolate, rarely ovate, 4–12 cm long, 7–20 mm wide, apex acute to obtuse, margins remotely serrate or dentate to subentire, base rounded to subcordate; sessile and clasping. Inflorescences 5–10 cm long, often interrupted at the base; bracts ovate, lanceolate or elliptic-lanceolate, about 1/2 as long as the calyx or shorter. Calyx 3–5(6) mm long at anthesis, to 7 mm in fruit, sparsely stipitate-glandular, teeth triangular, 1–1.5 mm long; corolla dark lavender to rose-purple, 9–12(16) mm long, puberlent and usually bearing stipitate glands, tube 6–9 mm long, upper and lower lips equal, about 1/3 as long as the tube. Nutlets 2.1–3.3 mm long. (2n = 38) Jul–Aug. Meadows & open stream banks; rare; n-cen ND; NE: Scotts Bluff; (sw Man. w to B.C., s to n-cen ND, w MT, n UT, & n OR; WY & w NE). *Dracocephalum nuttallii* Britt.—Rydb.

4. Physostegia virginiana (L.) Benth., Virginia lionsheart. Stems 4–15 dm tall, arising from perennating buds borne along slender rhizomes. Leaves spreading or ascending, elliptic-lanceolate, narrowly elliptic, oblong, or oblanceolate, 4–15 cm long, 1–4 cm wide, apex acute to acuminate, margins serrate, base narrowly cuneate to attenuate, sessile or subsessile, the base not clasping. Inflorescences 5–20 cm long, often interrupted at the base; bracts ovate to ovate-lanceolate, about 1/2 as long as the calyx. Calyx 4–8 mm long at anthesis, to 10 mm in fruit, teeth triangular, 1–2 mm long; corolla deep rose-lavender to white with purple spots, (13)17–25 mm long, puberulent to glabrous, stipitate glands absent, the tube 10–20 mm long, upper and lower lips about equal, about 1/3 as long as the tube. Nutlets 2.5–3.2 mm long. (2n = 39) Jul–Sep. Moist prairies, marshes, bogs, along streams or low woodlands; common ne 1/2 ND, IA, e 1/4 SD, e NE, n MO, & ne 1/4 KS; (Que. to Man., s to VA, TN, IL, & ne KS). *Dracocephalum denticulatum* Ait., *D. formosius* (Lunell) Rydb., *D. speciosum* Sweet, *D. virginianum* L.— Rydberg; *P. formosior* Lunnell—Steyermark; *P. virginiana* var. *speciosa* (Sweet) A. Gray— Fernald.

Our plants as here described are subsp. *virginiana*. Rare plants may be encountered in the se GP (NE: Richardson) which differ by reproducing from basal offshoots (not producing rhizomes) and having larger corollas (25–33 mm long). These large-flowered plants have been called subsp. *praemorsa* (Shinners) Cantino and are distributed primarily s and e of the GP.

17. PRUNELLA L., Self-heal, Heal-all

1. Prunella vulgaris L., self-heal. Perennial herb with a short caudex or short, slender rhizomes; stems branching, ascending to erect and/or prostrate, repent, or decumbent, to 1 m long, 1–3(5) mm diam, often weakly 4-sided, glabrous throughout, glabrous below and villose in lines above, or villose throughout. Leaf blades ovate-lanceolate, lanceolate, or narrowly elliptic, the midcauline to basal blades sometimes ovate or elliptic-oblong, 3–7(10) cm long, (0.5)1–3(4) cm wide, glabrous to sparsely villose, apex acute to obtuse, rarely rounded, margins entire to crenate-serrate, often remotely so, base mostly cuneate to attenuate, occasionally rounded (on the same plant); petiole 0.5–4 cm long, the longest usually on the lower leaves. Verticillasters in dense, terminal spikes 1–7 cm long, the spike usually subtended by a pair of leaves; cymules (2)3-flowered, sessile; bracts depressed ovate to suborbicular, 5-15 mm long, 7-14 mm wide, glabrous or villose-hirsute on the primary veins, apex abruptly short-acuminate or the lowest ones often caudate, margins entire, hirsute-ciliate; bracteoles absent or, if present, subulate and minute. Calyx green or purple, bilabiate, 10-nerved, 6-11 mm long, villose or sparsely so at the base, glabrous or villose in lines above, glabrous within, tube 2.5-3.5 mm long, upper lip with lobes connate for 3/4 or more of their length, lobes mucronate, the median lobe much wider than the lateral ones, lower lip with the lobes connate for about 1/3 their length, the lobes lanceolate or narrowly triangular, about equaling the upper lip in length, ciliate; corolla purple or pale lavender, rarely white, bilabiate, 10-15 mm long, glabrous or sparsely villose without, glabrous within or with a ring of hairs in the throat, tube flaring slightly outwards, 7-12 mm long, exceeding the calyx, upper lip galeate, about 1/3 the length of the tube, usually entire, lower lip deflexed, shallowly 3-lobed, much shorter than the upper lip, erose; stamens 4, included, didynamous, the longer pair usually with 2-pronged filaments, only 1 prong anther bearing, anther sacs divergent; style equally 2-lobed. Nutlets yellow-brown or darker, ovate and slightly flattened, 1.7-2.0 mm long, smooth, carunculate, the outgrowth whitish and triangular. (n = 14, 16) (May) Jun-Oct. Shaded or exposed stream banks, low woodlands, ditches, pastures, lawns, & various waste areas; GP except TX & w OK, common in e GP, especially se, & BH, infrequent elsewhere; (widely distributed in N. Amer.).

This is a nearly cosmopolitan species in which the American and e Asian plants have been recognized as var. *lanceolata* (Bart.) Fern. with mostly narrower, longer leaves with tapering bases. The European plants, var. *vulgaris*, have broader leaves with rounded bases but have not been seen in our region.

Plants growing in the nw GP tend to have villose stems, while those to the e and s are glabrous or villose only in the uppermost portions of the stem. Numerous intermediate individuals occur in the region and it is doubtful that this variant is deserving of taxonomic recognition.

18. PYCNANTHEMUM Michx., Mountain Mint, Basil

Aromatic perennial herbs with slender rhizomes; stems branched mostly in the upper portion, often corymbose. Leaves linear to lanceolate or elliptic, entire or serrate, sessile or short petiolate. Flowers in loose to dense, terminal and solitary glomerules or in loose, irregular corymbs with the branches apparent; bracts and bracteoles present. Calyx regular or bilabiate, tubular, 10- to 13-nerved, the teeth much shorter than the tube; corolla slightly bilabiate, much exceeding the calyx, tube enlarged slightly upwards, the lips about equal in length, upper lip erect, entire or emarginate, lower lip deflexed, 3-lobed, median lobe larger than the lateral ones; stamens 4, didynamous, exserted or not; anther sacs parallel; stigmas unequally 2-lobed. Nutlets obovoid or ellipsoid, 3-sided.

Reference: Grant, E. & C. Epling. 1943. A study of *Pycnanthemum* (Labiatae). Univ. Calif. Publ. Bot. 20: 195-240.

Pycnanthemum torrei Benth. was reported tentatively from KS: Wyandotte by Grant and Epling (1943), who indicated that the plant may have been atypical *P. pilosum* or of hybrid origin. On this basis it would seem justified to omit the taxon from the GP flora, since *P. torrei* is otherwise known only from the middle Atlantic coast of the United States.

1 Leaves ovate, ovate-lanceolate, or elliptic; inflorescences loose and cymose, the branches not concealed .... 1. *P. albescens*
1 Leaves lanceolate or narrower; inflorescences dense glomerules, solitary or several in corymbs.
  2 Stems glabrous; calyx teeth essentially glabrous .... 3. *P. tenuifolium*
  2 Stems hairy at least on the angles; calyx teeth canescent.
    3 Stems pilose on the faces and angles .... 2. *P. pilosum*
    3 Stems pilose only on the angles .... 4. *P. virginianum*

1. *Pycnanthemum albescens* T. & G. ex A. Gray, white mountain mint. Stems 4-8 dm tall, pubescent below, whitish canescent in the upper portions. Leaf blades ovate, ovate-lanceolate, or elliptic, 2.5–5 cm long, 1–2.5 cm wide, green and glabrous to pubescent above, whitish canescent and punctate below, apex acute, margins remotely serrate to subentire, base usually cuneate; petiole 2–6 mm long, narrowly winged. Inflorescences loose and cymose, the branches not concealed, terminal or on short axillary branches, 1.5–4 cm wide, usually subtended by a pair of leaves; cymules and flowers subtended by linear-lanceolate or subulate (rarely wider) bracteoles to 4 mm long. Calyx bilabiate, 3.5–4.5 mm long, canescent, upper lip with teeth connate for about 1/2 their length, deltoid and obtuse, lower lip with teeth free, ovate or oblong and obtuse; corolla whitish or pale lavender, spotted purple on the lower lip, 5–7.5 mm long, sessile glandular and pubescent without, sparsely pubescent within, tube 3–4.5 mm long. Nutlets dark brown, 1.1–1.4 mm long, weakly 3-sided, apex puberulent. (n = 19, 33–36; 2n = 76) Aug–Sep. Woodlands, on dry rocky hillsides or sometimes streams; rare; KS: Cherokee; OK: Ottawa; (FL w to e TX, n to e-cen MO & se KS).

2. **Pycnanthemum pilosum** Nutt., hairy or woods mountain mint. Stems 6–12 dm tall, pilose throughout or less so below. Leaves lanceolate, elliptic-lanceolate, or infrequently elliptic or oblong, 2–7.5 cm long, 4–15 mm wide, glabrate to pilose above, pilose below, apex acute, rarely obtuse, margins entire or with a few remote teeth, ciliate, base narrowly cuneate; sessile or the lower leaves sometimes with petioles to 3 mm long. Inflorescences dense glomerules 7–12 mm wide, terminal and solitary, sometimes disposed in corymbs; bracts subfoliaceous, lanceolate or ovate-lanceolate, 9–16 mm long, apex acute, margins entire; bracteoles linear-lanceolate, 3–5 mm long; peduncles 0.5–1.5 mm long; pedicels 0–1 mm long. Calyx nearly regular, 3.5–4.5 mm long, canescent-tomentose, teeth triangular, about 1 mm long and acute, the upper 3 teeth sometimes connate at their base; corolla whitish to pale lavender, spotted purple on the lower lip, 5–8 mm long, sessile glandular and pubescent without, glabrous or sparsely pubescent within, tube 3–5 mm long. Nutlets brown, 1–1.2 mm long, glabrous or the apex sparsely pubescent. (n = 78) Jul–Sep. Woodlands, on rocky slopes, outcrops, or along stream valleys; NE: Richardson; MO, e 1/6 KS, e OK; (s MI w to cen IA, s to OH, TN, MO, nw AR, & e OK, introduced eastward).

3. **Pycnanthemum tenuifolium** Schrad., slender-leaved mountain mint. Stems 4–11 dm tall, glabrous or the angles sometimes puberulent. Leaves linear or linear-lanceolate, 1.5–4(6) cm long, 1–4.5 mm wide, glabrous, apex narrowly acute, margins entire, ciliolate, sessile. Inflorescences very dense glomerules, 4–9 mm wide, solitary and terminal, disposed in compact corymbs; bracts ovate or ovate-lanceolate, the midvein conspicuously thickened and ending in a spinose tip, 2.5–4 mm long, variously canescent, ciliate; bracteoles similar to the bracts but usually less rigid and shorter; cymules sessile. Flowers sessile, calyx regular, 3.7–5 mm long, tube canescent or sparsely so, teeth narrowly triangular, about 1 mm long, usually glabrous, apex apiculate or subspinose; corolla whitish, occasionally pale lavender, purple spotted on the lower lip, 5–7 mm long, canescent and sessile glandular without, lightly pubescent within, tube 3–4.5 mm long. Nutlets dark brown, 0.7–1 mm long, usually smooth. (n = 18) (Jun) Jul–Sep. Upland prairies, pastures, meadows, open woodlands, & roadsides; sw IA, se NE, MO, e 1/3 KS, e OK; (ME w to e MN & se NE, s to n GA & e TX). *P. flexuosum* B.S.P.—Rydberg.

4. **Pyncanthemum virginianum** Dur. & Jackson ex Robins. & Fern., Virginia mountain mint. Stems 4–9 dm tall, pilose on the angles only. Leaves lanceolate to linear-lanceolate, 2.5–5.5 cm long, 2–10 mm wide, glabrous or sometimes sparsely pubescent below, apex narrowly acute, margins entire or rarely remotely serrate, base cuneate to rounded; sessile. Inflorescences dense glomerules, 6–10 mm wide, terminal and solitary, sometimes disposed in corymbs; bracts ovate-lanceolate, 7–15 mm long, usually glabrous, apex acute, margins entire; bracteoles narrowly ovate to lanceolate, 2.5–4 mm long, canescent; peduncles 0–1.5

mm long; pedicels 0–0.5 mm long. Calyx regular, 3–5 mm long, canescent, teeth triangular, about 0.7 mm long, acute; corolla whitish or pale lavender, spotted purple on the lower lip, 4–6.5 mm long, sessile glandular and pilose without, sparsely pubescent within, tube 2.5–4 mm long. Nutlets brown, 0.8–2.2 mm long, smooth. (2n = 80) Jul–Sep. Prairie sloughs, meadows, & moist, low woodlands; infrequent in e GP; MN; ND: Richland; e 1/4 SD, IA, e 2/3 NE, MO, e 1/3 KS; (s Que. w to se ND, s to VA, TN, n AR, & OK).

19. SALVIA L., Sage

Ours annual or perennial herbs, often aromatic. Flowers borne in few-flowered verticillasters, forming continuous or interrupted, spiciform inflorescences; cymules sessile or subsessile; bracteoles present or not. Calyx campanulate, bilabiate, 10- to 15-nerved, upper lip entire or 3-aristate, straight or slightly arched, lower lip 2-lobed; corolla tubular, bilabiate, pubescent without, glabrous within, upper lip straight and emarginate or galeate and entire, lower lip 3-lobed, about 2 × longer than the upper lip; stamens 2, ascending under the upper lip; connective articulating with the filament, upper arm with a fertile half-anther, lower arm sterile or with an imperfect half-anther; stigmas unequally 2-lobed; style included or barely exserted. Nutlets ovoid, weakly 3-sided, smooth or minutely tuberculate.

Salvia pratensis L. (sage) and *S. sclarea* L. (clary) are European species used in garden plantings that are infrequently collected as escapes in the region.

```
1  Leaves primarily basal, usually lyrate ........................................................ 2. S. lyrata
1  Leaves primarily cauline, unlobed.
   2  Upper calyx lip entire.
      3  Plants annual; corolla less than 10 mm long, the tube scarcely exserted from the
         calyx .................................................................................................. 4. S. reflexa
      3  Plants perennial; corolla 10 mm or more long, the tube well exserted from the
         calyx .................................................................................................. 1. S. azurea
   2  Upper calyx lip 3-aristate ........................................................................ 3. S. nemorosa
```

1. ***Salvia azurea*** Lam., blue sage, Pitcher sage. Perennial, 5–15 dm tall; stems solitary or several from a thick caudex, simple or sparingly branched in the upper portion, cinereous with short recurved pubescence, scattered longer hairs sometimes present, rarely glabrate. Lower leaves usually early deciduous, blades obovate, oblong, or oblanceolate, 3–7 cm long, 1–2.5 cm wide, apex acute to obtuse, margins denticulate to serrate, base usually attenuate to a petiole 0.5–1.5 cm long; middle and upper cauline leaves with blades lanceolate to linear-lanceolate or oblanceolate, 3–7 cm long, 0.5–2 cm wide, variously pubescent, apex narrowly acute, margins serrate to denticulate or subentire, base attenuate, subsessile or the petiole to 10 mm long. Verticillasters 2- to 8-flowered in spiciform inflorescences; bracts linear-lanceolate to subulate, 2–8 mm long; bracteoles similar to the bracts but reduced; pedicels 1–3(5) mm long. Calyx 4.5–10 mm long, pubescent, upper lip 1.5–2.2 mm long, entire, lower lip about as long as the upper, teeth triangular-ovate and acute; corolla deep blue, rarely white, 10–25 mm long, puberulent outside, the upper densely pubescent, tube well exserted from the calyx, 5–12 mm long, upper lip galeate and entire; style densely hirsute. Nutlets brown, 2–2.8 mm long, smooth or resin dotted. (2n = 20) (Jun) Jul–Oct. Rocky or sandy prairies, especially uplands, roadsides, & pastures; MO, s-cen & se NE, e 3/4 KS, OK, TX; introduced e CO & w NE; (SC & FL, w to NE, CO, & TX). *S. pitcheri* Nutt.—Atlas GP.

Our plants may be referred to var. *grandiflora* Benth. which has recurved hairs on the stem, in contrast to the more eastern var. *azurea* with spreading or ascending hairs on the stem. It is doubtful, however, that this distinction is worthy of taxonomic recognition.

2. **Salvia lyrata** L., cancer weed, lyre-leaf sage. Rosulate perennial, 3–8 dm tall; stems scapiform, simple or sparingly branched, lightly pilose; caudex simple and short, sometimes tuberous. Basal leaves numerous, blades oblanceolate or spatulate, vernal leaves sinuate or repand to lyrate-pinnatifid, later leaves usually subentire, 6–15 cm long, 2–7 cm wide, membranaceous, glabrous to lightly pilose, apex rounded or acute, base attenuate to rounded, petioles 2–7 cm long; cauline leaves absent or in 1 or 2 pairs, elliptic-lanceolate or or in 1 or 2 pairs, elliptic-lanceolate or obovate, entire and sessile or lyrate-pinnatifid and petiolate. Verticillasters 3–10 in interrupted spiciform inflorescences; cymules 2- to 5-flowered; bracts linear to lanceolate, 4–7 mm long; bracteoles absent; pedicels 1–4 mm long. Calyx 7–12 mm long, hirsute on the tube, upper lip straight, truncate, 3-aristate, about as long as the tube, lower lip with 2 narrowly triangular teeth, slightly longer than the upper lip; corolla whitish to blue, 18–30 mm long, tube 12–20 mm long, upper lip straight, emarginate, much shorter than the lower lip, both arms of the connective often bearing fertile half-anthers. Nutlets dark brown, about 2 mm long, minutely tuberculate. (2n = 36) May–Jun. Usually sandy soil on open, wooded slopes; rare; KS: Montgomery, Wilson; (NY w to IL, MO, & se KS, s to FL & e TX).

3. **Salvia nemorosa** L., sage. Perennial with a thick creeping rhizome, 5–10 dm tall; stems simple or branched (mostly in the upper portion), densely and usually retrorsely pubescent to hirsute. Basal leaves soon withering, blades ovate to broadly elliptic, 6–12 cm long, 2–6 cm wide, apex acute, margin crenate, base cuneate, petioles to 12 cm long; cauline leaves with blades narrowly triangular to lanceolate, the uppermost sometimes elliptic-lanceolate, 6–12 cm long, 2–4 cm wide, glabrous to sparsely pilose above, pubescent to canescent below, apex acute, margin crenate, base mostly cordate, on uppermost leaves sometimes rounded to cuneate, petioles 4–6 cm long on lower cauline leaves to nearly sessile on the uppermost leaves. Verticillasters (2)4- to 8-flowered in spiciform inflorescences; bracts often dark purple, broadly ovate, 5–10 mm long, 3–10 mm wide, apex abruptly acuminate; cymules sessile; bracteoles, if present, subulate, 1–3 mm long. Calyx 5–8 mm long, pubescent to villose mostly on the nerves, slightly recurved, 1.5–2 mm long, 3-aristate, lower lip straight, teeth triangular to lanceolate, slightly longer than the upper lip; corolla mostly dark purple, blue, or rose, 9–12 mm long, puberulent outside, tube about equaling the calyx, upper lip galeate and entire, lower lip spreading, median lobe broadly elliptic to suborbicular, much larger than the oblong lateral lobes. Nutlets dark brown, 1.5–1.8 mm long, smooth. (2n = 12) Jun–Sep. Pastures, ditches, & other disturbed sites; escaping from gardens, naturalized in n GP, adventive in s GP; ND, se SD, IA, e 1/4 NE; KS: Sumner; (scattered in n-cen & ne U.S. & adj. Can.; Eurasia). *S. sylvestris*, misapplied—Atlas GP.

4. **Salvia reflexa** Hornem., Rocky Mountain or lance-leaved sage. Annual, 1–7 dm tall; stems branched above the base, sparsely pubescent to glabrous. Leaves with blades lanceolate to narrowly oblong or elliptic, infrequently ovate, usually glabrous above, pubescent or glabrous below, apex obtuse or rounded, margins entire to remotely serrate, base cuneate to attenuate; petiole 3–20 mm long. Verticillasters 2(6)-flowered in interrupted, spiciform inflorescences; bracts lanceolate or ovate-lanceolate, 2–6 mm long, much shorter than the calyx; pedicels 1–3 mm long. Calyx at anthesis 4–6 mm long, to 8 mm at maturity, pubescent on the veins, upper and lower lips about 1/2 as long as the calyx, ciliate, upper lip entire, acutish, lower lip teeth triangular, acute; corolla dark to pale blue or white, 6–9 mm long, puberulent outside, glabrous within, tube about equaling the calyx, upper lip galeate, 2.5–3 mm long, densely bearded without, lower lip reflexed, 4.5–5 mm long; style glabrous or sparsely hirsute. Nutlets light tan or buff, mottled with darker brown, 2–2.5 mm long, smooth. (2n = 20) Jun–Oct. Disturbed habitats including pastures, roadsides, fields, & wash areas in prairies; GP; (WI w to MT, s to AR, TX, AZ, & Mex.; adventive eastward). *S. lanceolata* Willd.—Rydberg.

20. SATUREJA L.

1. *Satureja arkansana* (Nutt.) Briq., Arkansas calamint. Perennial, 1-3 dm tall, aromatic; stems usually branching, glabrous; sometimes producing leafy stolons; rhizomes slender, about 2 mm in diam. Leaves linear, linear-oblanceolate, or narrowly elliptic, 7-25 mm long, 1-5 mm wide, punctate, glabrous, apex acute or obtuse, margins entire or rarely serrulate, base attenuate or narrowly cuneate, sessile; stoloniferous. Leaves obovate, spatulate, or elliptic, 5-10 mm long, often purplish below. Flowers solitary or in 4- to 6-flowered verticillasters, usually well-spaced along the axis and subtended by a pair of foliaceous bracts; bracteoles linear or subulate, 1.5-3 mm long; pedicels 3-10 mm long. Calyx tubular, bilabiate, 13-nerved, 5-toothed, 4-5 mm long, glabrous and punctate without, tube 2.5-3 mm long, hirsute in the throat, upper lip straight, 1-1.5 mm long, the 3 teeth connate at the base, triangular and acuminate, lower lip straight and as long as the upper, the 2 teeth not connate, narrowly triangular and acuminate; corolla whitish to lavender, 7-10 mm long, puberulent outside and within, tube widened toward the throat, 6-8 mm long, well exceeding the calyx, upper lip subgaleate, weakly emarginate, 1-1.5 mm long, lower lip 3-lobed, the median lobe slightly larger than the lateral ones, equal to or slightly longer than the upper lip; stamens 4, ascending under the upper lip; anther sacs divergent; stigmas unequally 2-lobed; style included. Nutlets brownish, obovoid or ellipsoid, about 1 mm long, 3-sided, 2 sides flat and 1 convex. May-Jul. Open woods or prairies in shallow, rocky soil; rare; KS: Douglas; (Ont. w to MN, s to NY, OH, IN, IL, AR, & TX). *Calamintha arkansana* Nutt.; *C. nuttallii* Benth.—Rydberg; *S. glabella* (Michx.) Briq. var. *angustifolia* (Torr.) Svens.—Gleason & Cronquist.

21. SCUTELLARIA L., Skullcap

Contributed by Thomas M. Lane

Nonaromatic herbs to subshrubs, all but one of ours perennial; taprooted or rhizomatous with or without underground stolons; stems quadrangular, procumbent to erect, simple to variously branched. Leaves opposite, simple, small and chiefly entire or larger and crenate to serrate; basal leaves usually petiolate, upper ones sessile or petiolate. Inflorescence of axillary or terminal racemes with flowers arising from reduced bracts, or flowers solitary from axils of foliage leaves, borne on pedicels 2-4 mm long. Calyx zygomorphic, accrescent and closing in fruit, 2-parted, upper lobes with an ascending shieldlike protrusion (scutellum); corolla blue to purple, rarely white or pink, falling after anthesis, zygomorphic with a well-developed straight or more often sigmoid tube, upper lip galeate, lower lip flattened and expanded with a variously spotted white patch extending to the tube; stamens didynamous and usually covered by the galea, basally abaxial pair adnate ca 1/2 their length, curving up to become the longer pair, adaxial pair adnate 2/3-3/4 their length, anthers 2-celled, ciliate or villose on margins and longitudinally dehiscent. Stigma inserted between pairs of anthers or, in one species, exserted from galea, style usually free and gynobasic, ovary 4-lobed. Fruit 4 mericarps, yellowish to brown or black, spherical to ovoid, 1-1.8 mm in diam, verrucose to covered with prominent papillae, with or without a median gland.

References: Leonard, E. C. 1927. The North American species of *Scutellaria*. Contr. U.S. Natl. Herb. 22: 703-748; Epling, C. 1942. The American species of *Scutellaria*. Univ. Calif. Publ. Bot. 20: 1-146.

This is a world-wide genus of ca 300 species, found in both temp. and trop. areas.

1 Flowers in axillary or terminal racemes, each arising from the axil of a reduced bract.
  2 Flowers chiefly in axillary racemes; corolla straight, 6-7 mm long; mericarps yellowish .... 5. *S. lateriflora*
  2 Flowers in terminal racemes; corolla sigmoid, 15-23 cm long; mericarps brown to black.

3   Leaves cordate to truncate at the base; stigma inserted between the pairs of anthers, not exserted from tube; corolla without an internal ring of hair. .... 6. *S. ovata*
3   Upper leaves obtuse and somewhat attenuate at the base; stigma/style exceeding stamens in length, stigma exserted 1-2 mm from corolla; corolla with an internal ring of hair .... 4. *S. incana*
1   Flowers 2 per node, each arising from the axil of a foliage leaf.
   4   Leaves thin, shallowly serrate; plants fibrously rooted from lower nodes, rhizomatous; mericarps yellowish or buff, covered with sessile glands; of marshy habitats. .... 3. *S. galericulata*
   4   Leaves slightly thickened, mostly entire; plants taprooted or from thickened rhizomes; mericarps brown to black; of upland habitats.
      5   Plants from thickened rhizomes; mericarps with prominent papillae, usually banded.
         6   Flowers less than 1.2 cm long; plants with a moniliform rhizome; mericarps brown with cylindrical blunt papillae, banded. .... 7. *S. parvula*
         6   Flowers more than 1.8 cm long; plants with a thickened rhizome; mericarps dark brown to black with conical papillae, sometimes banded .... 1. *S. brittonii*
      5   Plants taprooted; mericarps either somewhat lamellate (imbricate) or finely papillate.
         7   Plants perennial; with minute conical hairs; mericarps finely and evenly papillate; flowers more than 1.5 cm long .... 8. *S. resinosa*
         7   Plants annual, with long spreading capitate hairs; mericarps somewhat lamellate; flowers less than 1.3 cm long .... 2. *S. drummondii*

1. *Scutellaria brittonii* Porter, Britton's skullcap. Ascending perennial herbs, (6)8–20(33) cm tall, arising from a thickened horizontal rhizome; retrorsely puberulent to clothed with longer retrorse hairs or long spreading capitate hairs; stems often branched at the base, rarely above. Leaves slightly thickened, ovate to lanceolate to narrowly elliptic, median leaves 1.5–3.3 cm long and 0.5–1 cm wide, reduced and narrower above; petiole most pronounced on basal leaves, 5–10 mm long, subsessile above. Flowers 2 per node in upper 1/2–1/3 of plant, arising from axils of foliage leaves; pedicel 2–4 mm long. Calyx 4–6 mm long in flower, 6–8 mm long in fruit, clothed with long hairs and sessile glands, scutellum 3–4 mm tall in fruit; corolla 19–31 mm long, tube ca 2 mm in diam at the base, lower lip ca 11 mm across. Mericarps dark brown to black, ovoid, ca 1.7 mm in diam, covered with prominent flattened conical papillae, usually banded. May–Jul. Upland prairies & open pine woods, rocky &/or sandy soil; extreme w GP; NE; KS: Cheyenne; WY, CO, NM; (GP, scattered localities just w of area). *S. brittonii* var. *virgulata* (A. Nels.) Rydb.—Rydberg.

2. *Scutellaria drummondii* Benth., Drummond's skullcap. Small upland annual, rarely perennial, 5–20(28) cm tall, densely capitate pilose, especially on younger parts, with or without less conspicuous retrose or spreading eglandular hairs; stems often branched at base, less often above, arising from a slender taproot. Leaves ovate, less often elliptic, larger ones 1.1–1.5(2.4) cm long and 0.6–1(1.6) cm wide, apex rounded-acute, margin entire, lowest leaves sometimes with few shallow teeth, base attenuate; petiole 0–9(20) mm long, longest on lower leaves, ca 1–2 mm long on upper ones. Flowers 2 per node in upper 1/2–2/3 of plant, arising from axils of foliage leaves; pedicel 2–3 mm long. Calyx 2–4 mm long in flower, ca 5 mm long in fruit, scutellum arising at ca 45° angle from upper lobe; corolla 6–10(13) mm long, tube ca 1 mm in diam at the base, lower lip 3–5 mm across. Mericarps dark brown to black, ovoid, ca 1.3 mm in diam, covered with large flattened conical papillae and lamellae (especially on ventral surface). Apr–Jun. Upland prairies or open woodlands, often in sandy or rocky soils; rare in GP; OK; TX; (OK, TX s to Mex.).

3. *Scutellaria galericulata* L., marsh skullcap. Delicate upright or procumbent perennial herbs, (1.3)2–6(10) dm tall, glabrate to sparsely retrorsely strigose; stems simple to freely branching above, fibrously rooted from lower nodes, rhizomatous. Leaves thin, lanceolate-ovate, truncate to sub-cordate at the base, shallowly serrate, (1.8)3–6(9) cm long and

(0.5)1-2(3) cm wide, strigose on underside, reduced upwards; petiole 0-4 mm long. Flowers axillary, 2 per node in upper part of plant; pedicel 2-3 mm long. Calyx strigose and covered with sessile glands, 3-4 mm long in flower and 4-6 mm long in fruit, scutellum ca 3 mm tall in fruit, impressed; corolla (14)16-19(21) mm long, tube ca 2 mm in diam at the base, lower lip ca 7 mm across. Mericarps yellowish to buff, ovoid, ca 1.5 mm in diam, uniformly covered with short, blunt papillae and sessile glands. (2n = 32) Jun-Aug. Aquatic habitats, especially marshes; GP except MO, KS, & OK; TX: Hemphill; (Newf. s to NC, w to AK & AZ). *S. epilobiifolia* Hamilt.—Rydberg.

This is a widespread species of North America and Eurasia. Some taxonomists have called the American specimens *S. epilobiifolia* and the Eurasian *S. galericulata*. Epling (Amer. J. Bot. 26: 17-24, 1939) studied these two assemblages and concluded that populations on each continent show wide variation with respect to degree of pubescence, and leaf and mericarp morphologies. Because of overlapping ranges in morphological variation, all specimens here are referred to *S. galericulata*.

4. *Scutellaria incana* Biehler, hoary skullcap. Tall perennial herb, (5)6-8(12) dm tall, usually unbranched below the inflorescence, minutely ascending-tomentose and dotted with sessile glands throughout. Leaves ovate, 6.5-9.5 cm long and 3.5-5 cm wide, margin crenate-serrate, base subcordate to obtuse on basal leaves, obtuse and somewhat attenuate on the upper ones; petiole 1-2(3) cm long, shortest above. Inflorescence a panicle of 3-many terminal racemes; flowers single from axils of bracts; bracts small, lanceolate, largest ca 6 mm long and 2 mm wide, margin entire. Calyx 2-3 mm long in flower, 5-6 mm long in fruit, scutellum ca 4 mm tall in fruit, impressed; corolla 15-23 mm long with an internal ring of hairs, tube base 2-3 mm in diam; stigma/style longer than stamens and exserted 1-2 mm from corolla. Mericarps dark brown, spherical-ovoid, ca 1.5 mm in diam, covered chiefly with short papillae, giving a verrucose appearance. (2n = 30) Jun-Aug. Upland woods; MO: Jasper; KS: Cherokee; (NJ, NY s to AL, w to KS). *S. canescens* Nutt.—Rydberg.

This species is widespread e of our range but rarely has been collected in the GP.

5. *Scutellaria lateriflora* L., mad-dog or blue skullcap. Delicate upright (procumbent) perennial herb, (1)2-6(10) dm tall; stems glabrate, simple to variously branched above, fibrously rooted from lower nodes, rhizomatous. Leaves thin, ovate, 3-11 cm long and 1.5-5.5 cm wide, progressively reduced upwards, crenate to serrate, obtuse to truncate at the base, with scattered hairs usually on veins; petiole ca 1/4 as long as blade. Flowers chiefly in racemes, arising from axils of reduced leaflike bracts; basal bracts ca 13 mm long and ca 5 mm wide, greatly reduced upward. Calyx ca 2 mm long in flower and ca 3 mm long in fruit, scutellum ca 2 mm tall in fruit, not impressed, clothed with coarse antrorse hairs; corolla 6-7 mm long, tube straight, lower lip less than 2× as wide as tube. Mericarps yellowish, ovoid, ca 1.3 mm in diam, covered with short papillae which may be irregularly coalesced. (2n = 88) Jul-Sep. Riparian habitats, usually stream banks; GP except MT & WY; CO: Yuma; (Newf. s to GA, w to B.C., GP, & CA).

6. *Scutellaria ovata* Hill, eggleaf skullcap. Perennial herb 1.8-6(10) dm tall, rhizomatous with slender underground stolons; stems capitate pilose with eglandular hairs. Leaves thin to slightly thickened or puckered; clothed above with long spreading capitate or eglandular hairs and much shorter eglandular hairs, clothed below with long capitate or eglandular hairs, veins densely covered with slightly shorter eglandular hairs; cordate to ovate, median leaves (2.5)3.5-6(7) cm wide and (4)5-8(12) cm long, dentate-serrate, truncate to cordate at the base; petiole 0.8-4(6) cm long, 0.3-0.8× as long as blade. Flowers in terminal racemes from axils of reduced leaflike bracts, sometimes arising from top 1 or 2(3) nodes; bracts ovate-cordate 4-9 mm long and 3-6 mm wide. Calyx ca 4 mm long in flower, ca 6 mm long in fruit, scutellum ca 4 mm tall in fruit, impressed; corolla 17-23 mm long, base of tube ca 2 mm in diam, lower lip ca 10 mm across. Mericarps brown (black), ovoid,

evenly covered with prominent conical papillae, with or without a median band laterally. Jun–Jul. Woods & open woods, often in sandy soil; extreme se GP; MO, KS, OK; (MD s to TN, w to KS, TX; e Mex.).

This is a widespread, variable species of the e U.S. Leonard, op. cit., listed three varieties, while Epling, op. cit., recognized 12 subspecies. Epling has annotated GP specimens as subsp. *bracteata*, subsp. *versicolor*, and subsp. *mississippiensis*. There is no consistent morphological difference between any of these subspecies in our range. If there is need to recognize infraspecific taxa, the older varietal names (Fernald) circumscribing larger morphological variation are more appropriate. *S. ovata* var. *bracteata* (Benth.) Blake may be recognized by bracts that exceed the calyx in length and slightly thickened (puckered) leaves. Var. *versicolor* (Nutt.) Fern. can be recognized by virtue of its bracts shorter than the calyx and relatively thin leaves.

7. Scutellaria parvula Michx., small skullcap. Small perennial herbs, 8–25(30) cm tall, glabrate to appressed scabrescent or capitate pilose, often branched at base, infrequently above; fibrously rooted from a moniliform rhizome. Leaves slightly thickened, broadly ovate to lanceolate, median leaves 9–20 mm long and 5–15 mm wide, narrowing above; sometimes with sessile glands; veins anastomosing near margins in pilose plants; basal leaves with petioles 2–14 mm long. Flowers 2 per node on upper 1/2 of plant, arising from axils of foliage leaves; pedicel 3–4 mm long. Calyx sometimes with sessile glands, 2–3 mm long in flower, 3–5 mm long in fruit, scutellum impressed, arising at ca 45° angle from upper calyx lobe; corolla 6–11 mm long, tube 1–2 mm in diam, lower lip 4–7 mm across. Mericarps brown, ovoid, ca 1 mm in diam, with a prominent band and long, conical, blunt papillae. Apr–Jun. Upland prairies & open woodlands, often on limestone ledges & rocky-sandy soil; e GP; (Que. s to FL, w to GP, TX).

This is a widespread species of the e U.S. Three varieties can be recognized in the GP.

1 Plants capitate pubescent; leaves ovate to broadly ovate.
  2 Cauline capitate pubescence ± 1/3 as long as stem is wide, usually dense; abaxial leaf surface dotted with sessile glands and with many eglandular hairs along veins; median leaves ovate to narrowly ovate .... 7c. var. *parvula*
  2 Cauline capitate pubescence ± 1/2 as long as stem is wide, usually sparse; abaxial leaf surface with long spreading, usually capitate hairs mostly along veins; median leaves broadly ovate to ovate .... 7a. var. *australis*
1 Plants appearing glabrate, or with few antrorse-appressed eglandular hairs; median leaves narrowly ovate to lanceolate .... 7b. var. *leonardii*

7a. var. *australis* Fassett, southern small skullcap. Plants with long spreading capitate pubescence, ± 1/2 as long as stem is wide. Leaves broadly ovate to ovate, median ones ca 1.25 × longer than wide, veins anastomosing to form a vein along the leaf margin, sparsely clothed with long capitate hairs above and below, especially along veins, margin slightly, if at all revolute. MO, e KS, & OK. *S. parvula* Michx.—Rydberg.

This var. is fairly common within its range in the GP. It can be found growing at the same locality with var. *leonardii*, although few intermediate plants have been collected. In TX many eglandular plants, otherwise the same as GP var. *australis*, have been collected.

7b. var. *leonardii* (Epl.) Fern., Leonard's small skullcap. Plants glabrate to antrorse-appressed scabrescent, with or without scattered capitate hairs near the base. Leaves ovate to lanceolate, median ones ca 2.2 × longer than wide, veins rarely branched or joining at leaf margins, margin quite revolute. MN, IA, MO, ND, SD, NE, KS, OK: Rogers. *S. leonardii* Epl.—Epling, op. cit.; *S. ambigua* Nutt.—Rydberg.

This is the most common var. in the GP, especially in the e and ne. It may occur at the same locality with var. *australis*, although few intermediates have been collected.

7c. var. *parvula*. Plants with dense, spreading capitate pubescence ± 1/3 as long as stem is wide, usually with shorter pubescence of retrorse hairs, calyx and abaxial leaf surface dotted with sessile glands, abaxial leaf veins more densely covered with eglandular hairs. Leaves ovate to narrowly ovate, median ones ca 1.7 × longer than wide, veins subanastomosed near margins, margin revolute. MO, se KS, & OK. *S. parvula* Michx.—Rydberg.

This is the least common of the three vars. in the GP. It is rarely found growing near the others. Infrequently, plants have been collected that lack sessile glands on the abaxial leaf surface.

8. **Scutellaria resinosa** Torr., resinous skullcap. Stiff upland perennial herbs (subshrubs), strongly taprooted, (0.6)1–3(5) dm tall, clothed with minute, conical, spreading to strongly retrorse hairs, and covered with sessile glands, rarely with scattered spreading capitate hairs; many stems arising from a branched woody crown, stems persisting to next season. Leaves slightly thickened, ovate to elliptic, 6–10(12) mm wide and (7)9–13 mm long, reduced above, apex rounded, margin entire, base obuse to attenuate; petiole 1–2(5) mm long, most pronounced on basal leaves. Flowers 2 per node in upper 1/2–3/4 of plant, arising from axils of foliage leaves; pedicel 2–3 mm long. Calyx 2–3 mm long in flower, 4–5 mm long in fruit, scutellum ca 3 mm tall in fruit, impressed; corolla (11)14–16(21) mm long, sigmoid, tube 1–2 mm in diam at the base, lower lip 8–9(12) mm across. Mericarps dark brown to black, ovoid, ca 1.3 mm in diam, uniformly covered with small papillae, giving a pebbled appearance. Apr–Jun. Dry rocky-sandy soil in upland short to midgrass prairies; w-cen KS, OK, TX; (GP to CO & AZ). *S. wrightii* A. Gray—Correll & Johnston.

22. STACHYS L., Hedge-nettle

Ours annual or rhizomatous perennial herbs. Verticillasters 2- to several-flowered in spiciform inflorescences, lower verticillasters remote and subtended by foliaceous bracts, the upper ones more congested and the bracts greatly reduced; cymules sessile or subsessile; bracteoles present or not. Calyx campanulate, nearly regular, equally 5-toothed, 5- to 10-nerved (15 in *S. arvensis*); corolla bilabiate, the tube slender, about equaling the calyx or slightly longer, upper lip straight and plane or subgaleate, entire or shallowly emarginate, lower lip deflexed, longer than the upper lip, 3-lobed, the median lobe much larger than the lateral ones; stamens 4, ascending under the upper lip; anther sacs diverging; stigma equally 2-lobed; style included. Nutlets ovoid or oblong, weakly 3-ribbed, smooth.

Reference: Epling, C. 1934. Preliminary revision of American *Stachys*. Feddes Repert. Spec. Nov. Beih. 80: 1–75.

Stachys annua (L.) L. is an annual plant with yellow flowers and the calyx teeth much shorter than the tube and was collected in KS: Shawnee in 1896. It is a native of Europe.

Stachys arvensis (L.) L. is a European native reported by Rydberg (1932) for NE. The plant is an annual with pink flowers and the calyx teeth about as long as the tube.

1 Middle and upper internodes spreading pilose or retrorsely hirsute on the angles and faces, sometimes mixed with glandular hairs .... 1. *S. palustris* subsp. *pilosa*
1 Middle and upper internodes entirely glabrous or the angles sparsely hirsute or strigose, glandular hairs absent .... 2. *S. tenuifolia*

1. **Stachys palustris** L. subsp. **pilosa** (Nutt.) Epling, hedge-nettle, marsh betony. Perennial; stems simple or sparingly branched, 2.5–10 dm tall, middle and upper internodes spreading pilose and sometimes mixed with shorter glandular hairs or retrorsely hirsute on the angles and faces; rhizomatous and often producing whitish tubers. Leaves lanceolate, ovate, elliptic, or elliptic-oblanceolate, 3–10(15) cm long, 1–3(5) cm wide, densely pilose or sometimes weakly so on both surfaces, ciliate, apex acute, infrequently rounded, margins crenate-serrate to serrate, base rounded to subcordate, sometimes cuneate, sessile or subsessile, infrequently middle and lower leaves on petioles 1–4(10) mm long. Verticillasters 2- to 8-flowered; bracts reduced, foliaceous, slightly exceeding the verticillasters; bracteoles absent; pedicels 0–1.5 mm long. Calyx 6–9(10) mm long, pilose and usually mixed with shorter glandular hairs, teeth narrowly triangular and nearly aristate, 2/3–1 × as long as the tube, hirsute-ciliate; corolla lavender or rose-purple, mottled with lighter and darker

spots, 10–15 mm long, sparsely puberulent outside, glabrous within, tube 7–10 mm long, slightly exceeding the calyx, upper lip subgaleate, bearded without, glandular hairs present or not. Nutlets dark brown, ovoid, 1.7–2.2 mm long. (2n = ±64, 102) Jun–Aug (Sep). Prairie sloughs, ditches, stream banks, & lakeshores, sometimes in upland prairies, usually exposed sites; common n GP, infrequent se MN w to MT, s to MO, ne KS, NE, & n-cen CO; (boreal, in N. Amer. s to N. Eng., ne KS, NE, NM, AZ, NV, & OR). *S. ampla* Rydb., *S. pustulosa* Rydb., *S. schweinitzii* Rydb.—Rydberg; *S. palustris* var. *homotricha* Fern., var. *pilosa* (Nutt.) Fern., var. *phaneropoda* Weath., var. *nipigonensis* Jennings—Fernald.

Plants in the GP typically have stems that are spreading pilose (sometimes mixed with glandular hairs) and sessile or subsessile leaves. Along the e-cen edge of the GP, however, the plants usually have retrorsely hirsute stems and short petiolate leaves. While numerous North American variants have been named, intermediates are numerous and geographical distinction is obscurely evident. On this basis it seems best to recognize the North American plants as a single, highly variable complex species *pilosa*, perhaps distinct from the European species *palustris*.

2. **Stachys tenuifolia** Willd., thinleaf or slenderleaf betony. Perennial; stems usually branched, 3–10 dm tall, middle and upper internodes glabrous or the angles sparsely hirsute or strigose, glandular hairs absent; rhizomatous. Leaves with blades lanceolate or ovate to elliptic, 3–12 cm long, 0.7–4 cm wide, membranaceous, glabrous, apex acute, margins serrate to crenate, base cuneate to rounded; petioles 0.5–3.5 cm long. Verticillasters 2- to 6(8)-flowered; bracts reduced, foliaceous, barely or not exceeding the verticillasters; bracteoles subulate, 0.5–1 mm long; pedicels 0–1 mm long. Calyx 4–7 mm long, glabrous or rarely sparsely pubescent and the teeth ciliate, teeth usually spreading, narrowly triangular, 2/3–1 × as long as the tube, usually aristate; corolla pale lavender or darker, mottled with purple spots, 9–12 mm long, glabrous or the upper lip glandular pubescent without, tube 5–7 mm long, barely exceeding the calyx, upper lip usually straight, entire or minutely erose. Nutlets dark brown or blackish, ovoid, 1.4–1.7 mm long. (n = 16) Jul–Oct. Wooded lakeshores or flood plains, occasionally ditches or prairie swales; sw IA, se NE, MO, e 1/3 KS, e OK; (NY w to MN, s to s-cen LA & e TX). *S. ambigua* of reports—Gates.

In e ND and rarely elsewhere in the extreme e GP, plants occur that superficially resemble *S. tenuifolia* but may better be referred to *S. hispida* Pursh (= *S. aspera* Michx.—Stevens). These plants have stems pubescent to glabrous on the faces; leaves sparsely hirsute to glabrous, petiolate; and calyces long hirsute. The plants are so few in our region with characters less than consistent that certain disposition of them cannot be made without further field studies.

23. TEUCRIUM L., Germander, Wood Sage

Ours perennial herbs. Leaves serrate, crenate, or pinnatifid; subsessile or short-petiolate. Flowers solitary in the axils of foliaceous bracts or verticillasters (2)4- to 6-flowered in spiciform inflorescences; the bracts reduced, foliaceous; bracteoles absent. Calyces campanulate, bilabiate to nearly regular, 5-toothed, 10-nerved; corolla bilabiate, the tube equal to or shorter than the calyx, upper lip cleft into 2 lobes, about equal to or smaller than the lateral lobes of the lower lip, lower lip 3-lobed, median lobe longer than the tube and much larger than the lateral lobes; stamens 4, exserted from between the lobes of the upper lip; anther sacs divergent; filaments pubescent towards the base; stigma equally 2-lobed; style exserted. Nutlets ovoid, glabrous or sparsely pubescent, rugose. *Melosmon* Raf.—Rydberg.

References: McClintock, E., & C. Epling. 1946. A revision of *Teucrium* in the New World, with observations on its variation, geographical distribution and history. Brittonia 5: 491–510; Shinners, L. H. 1963. The varieties of *Teucrium canadense* (Labiatae). Sida 1: 182–183.

1 Leaves merely serrate to crenate .... 1. *T. canadense*
1 Leaves pinnatifid .... 2. *T. laciniatum*

1. *Teucrium canadense* L., American germander, wood sage. Colonial, rhizomatous perennial, sometimes producing whitish tubers; stems simple or sparingly branched, 3–10(14) dm tall, retrorsely pubescent to villose (sometimes sparsely so) or spreading pilose. Leaf blades ovate, ovate-lanceolate, or lanceolate, 3–12(16) cm long, 1–4(6) cm wide, lightly appressed-pilose to glabrate above, sparsely to densely canescent below, apex acute or short acuminate, margins serrate, base usually cuneate; petioles 4–15(25) mm long. Verticillasters (2)4- to 6-flowered in spiciform inflorescences 6–20(30) cm long; bracts lanceolate to subulate, shorter than to slightly exceeding the verticillasters; peduncles 0–1 mm long; pedicels 1–3(4) mm long. Calyx often purplish, bilabiate, 5–9 mm long, variously hairy, tube 4–7 mm long, glabrous within or the throat weakly hirsute, glandular hairs sometimes present, upper 3 teeth broadly ovate to triangular, 1–2 mm long, obtuse to acute, lower 2 teeth narrowly triangular, slightly longer than the upper teeth; corolla light rose, lavender, or purple, infrequently white, 10–18 mm long, puberulent and usually sessile or stipitate glandular outside, tube 4–8 mm long, about equaling the calyx, lobes of the upper lip appearing lateral, erect, triangular or narrowly so, 1.5–3.5 mm long, lateral lobes of the lower lip spreading, similar but usually smaller than the upper lip lobes, median lobe reflexed, broadly obovate and short-clawed, 6–10 mm long. Nutlets light reddish brown, 1.5–2.4 mm long, glabrous, rugose.

This ia a polymorphic complex for which various authors have recognized segregate taxa on both the specific and infraspecfic levels. In the GP two varieties may be distinguished.

1a. var. *boreale* (Bickn.) Shinners. Stems mostly spreading pilose. Calyx spreading pilose, mixed with glandular hairs, teeth usually triangular and acute. (2n = 32) Jul–Sep. Exposed, usually moist sites including stream banks, lakeshores, prairie swales, & ditches; common n GP, infrequent s of NE; MN, ND, MT, SD, WY, nw IA, NE, w KS, ne CO, TX(?); (Que. w to B.C., s to NY, OH, IL, NE, w KS, NM, & CA). *T. boreale* Nutt.—Rydberg; *T. canadense* var. *occidentalis* (A. Gray) McCl. & Epl.—Atlas GP; *T. occidentale* var. *boreale* A. Gray—Fernald.

1b. var. *canadense*. Stems mostly retrorsely pubescent or canescent, scattered longer hairs sometimes present, eglandular or infrequently sessile glandular, teeth usually broadly ovate to triangular, mostly obtuse. (2n = 32) Jun–Aug. Open or shaded usually moist sites including stream banks, lakeshores, marshes, prairie swales, ditches, sometimes in pastures; common se GP, scattered n & w; SD: Union; IA, se 1/2 NE, MO, e 3/4 KS, OK, TX; (N.S. w to MN & NE, s to FL & TX). *T. canadense* var. *virginicum* (L.) Eat.—Atlas GP.

Var. *virginicum* does not seem to be sufficiently distinct from var. *canadense* to warrent taxonomic recognition.

2. *Teucrium laciniatum* Torr., cutleaf germander. Cespitose perennial with deep creeping roots, the vertical shoots branching several cm below and again at ground level; stems rarely branched above the base, 6–15(20) cm tall, retrorsely scabrous on the angles. Leaves deeply pinnatifid into 2 or 3(4) pairs of lobes, 0.8–4(5.5) cm long, 0.6–2(3) cm wide, glabrous to sparsely pubescent, the lobes simple or sometimes branched, 1–3(4) mm wide, margins entire, often ciliate, base attenuate to an obscure petiole or subsessile. Flowers solitary in the axils of foliaceous bracts and crowded towards the tips of the stems; peduncles 1–3(4) mm long. Calyx nearly regular, 7–9(12) mm long, glabrous or scabrous on the veins, teeth narrowly triangular to lanceolate, equal to or longer than the tube, sometimes ciliate; corolla creamy white, sometimes with purplish veins, 10–20 mm long, usually glabrous outside, bearded in the throat or sparsely so on the lower lip, tubes 2–3 mm long, much shorter than the calyx, lobes of the upper lip spreading to reflexed, ovate to elliptic, 2.5–4.5 mm long, lower lip reflexed, lateral lobe similar to but slightly longer than the upper lip lobes, median lobe obovate to spatulate, 8–11(14) mm long. Nutlets dark brown or reddish brown,

2.7–3.5 mm long, glabrous, rugose. (Apr)May–Jul. Sandy soil in prairies, along flood plains, & roadsides, less often in gypsiferous or limestone soils or gravelly hillsides or upland prairies; scattered sw GP; KS: sw 1/6 & Cheyenne; se CO, w 1/2 OK, TX, NM; (w KS & se CO, s to cen TX & NM). *Melosmon laciniatum* (Torr.) Small—Rydberg.

24. TRICHOSTEMA L., Blue Curls

1. *Trichostema brachiatum* L., false pennyroyal. Annual herb; stems solitary and freely branched in the upper 1/2–2/3 with the lower portion usually naked, 1.5–5 dm tall, stipitate glandular, pubescent or in the se GP sometimes mixed with longer hairs. Leaves linear-elliptic to lanceolate, infrequently elliptic, 1–4(5) cm long, 3–10(20) mm wide, sessile glandular, variously pubescent, apex acute, margins entire, base cuneate to attenuate, sessile, subsessile, or petiole 1–5 mm long; bracteoles linear-elliptic, 1–4 mm long; pedicels 1–15 mm long. Calyx nearly regular, campanulate, deeply 5-lobed, 10-nerved, 2.5–4 mm long, increasing to 3.5–8 mm in fruit, glandular and variously pubescent, the tube 1–1.5 mm long, becoming 1.5–3 mm long, the teeth narrowly deltoid, 1.5–2.5 mm long, becoming 2.5 mm long; corolla bluish, bilabiate, 1.5–4.5 mm long, not much exceeding the calyx, sparsely pubescent, the tube straight and slightly flaring, the upper 4 lobes spreading or ascending, the lower ligulate and deflexed, slightly longer than the upper; stamens 4, barely exserted from the corolla tube; anther sacs divergent; stigmas equally 2-lobed; style about equaling the stamens. Nutlets brownish, ovoid, 2.3–3 mm long, lightly pubescent at the apex, prominently areolate-reticulate. (2n = 14) Jul–Sep. Shallow, rocky soil, woodlands, prairie banks, outcrops; s MN, se SD, IA; NE: e & Sioux; MO, e 1/2 KS, e OK; (NY to s MN & se SD, s to FL, AL, AR, & s-cen TX; AZ). *Isanthus brachiatus* (L.) B.S.P.—Atlas GP.

Reference: Lewis, H. 1945. A revision of the genus *Trichostema*. Brittonia 5: 276–303.

For the reasons outlined by Lewis (op. cit.) the recognition of the monotypic genus *Isanthus* does not seem warranted.

120. HIPPURIDACEAE Link, the Mare's Tail Family

by T. Van Bruggen

1. HIPPURIS L., Mare's Tail

1. *Hippuris vulgaris* L. Aquatic or amphibious perennial, with glabrous, simple stems 0.7–2.5 cm thick, erect or ascending, up to 6 dm tall, from slender, creeping rhizomes. Leaves (5)6–13, simple, divergent in whorls 8–12 mm apart on the stem, glabrous, the emersed ones linear-attenuate, sessile, 1–3(4) cm long; submersed leaves linear-filiform or reduced to short, scalelike projections. Flowers perfect or sometimes unisexual, epigynous, clustered in the axils of emersed leaves at the middle and upper part of stem; calyx barrel-shaped, fused to the ovary; petals absent; stamen 1, with a short filament; anther 2-celled, 1–2 mm long; pistil 1, 1-celled with 1 ovule, 1.5–2.0 mm long, with a filiform style stigmatic its entire length. Fruit 1.5–3.0 mm long, ellipsoid, hard, indehiscent, nutlike, 1-seeded; Jun–Aug. Rooted in mud of quiet water of lakes, ponds, sloughs & ditches; ND, MN, SD, NE; (circumboreal, Greenl. to AK, s to IN, NE, & NM; Eurasia).

121. CALLITRICHACEAE Link, the Water Starwort Family

by Ralph E. Brooks

1. CALLITRICHE L., Water Starwort

Small aquatic or terrestrial herbs, monoecious or rarely dioecious, annual or perennial. Leaves simple, opposite, highly polymorphic in some species, mostly membranaceous, lacking stipules. Flowers 1-3 in the axil of a foliage leaf, sessile or subsessile. Staminate flowers minute, consisting of a single stamen; anther 4-locular with longitudinal dehiscence. Pistillate flowers minute, consisting of a single pistil with 2 carpels but 4-celled, each cell with 1 ovule; styles 2, often persistent, elongate. Fruit compressed and 4-lobed, separating into 4 mericarps at maturity; mericarps flattened, usually winged, smooth, papillose or reticulate, each with 1 seed; seeds pendulous, embryo slender, straight or slightly curved.

Reference: Fassett, N. C. 1951. *Callitriche* in the New World. Rhodora 53: 137-155; 161-182; 185-194; 209-222.

1 Plants terrestrial; leaves uniform, obovate-oblanceolate to spatulate, 2-3.5 mm long .... 3. *C. terrestris*
1 Plants aquatic; leaves uniform and linear, or dimorphic with linear and spatulate leaves, usually more than 5 mm long.
   2 Leaves uniform linear-lanceolate, the bases clasping or not, but never connate .... 1. *C. hermaphroditica*
   2 Leaves uniform or more often polymorphic with spatulate floating leaves, and linear-lanceolate submersed leaves, leaf bases connate.
      3 Fruit length equaling or exceeding width; carpels wingless; reticulations on mericarps not in vertical rows .... 2. *C. heterophylla*
      3 Fruit length exceeding width by at least 0.2 mm; carpels winged at the summit; reticulations on the mericarps tending to run in vertical lines .... 4. *C. verna*

1. *Callitriche hermaphroditica* L. Aquatic perennial herbs forming submersed mats; stems leafy for their length, to 35 cm long, rooting at the lower nodes; rhizomes slender. Leaves uniform, linear-lanceolate, (3)5-13 mm long, 0.5-1.3 mm wide, membranaceous, with a single midvein, glabrous, apex obtuse to acute, margins entire and hyaline, base sessile and usually clasping but never connate. Fruit 1-2.5 mm long and about as wide; mericarps strongly compressed on the outer edge, obscurely or irregularly pitted, margins narrowly winged, style sharply reflexed and usually breaking away. (2n = 6) Jun-Aug. Shallow water in streams, ponds, & sloughs; n GP; MN w to MT, s to n NE & WY; (Greenl., s Can., AK, & n U.S., s in w to NE, NM, & CA; Europe). *C. autumnalis* L.—Rydberg.

2. *Callitriche heterophylla* Pursh. Aquatic perennial herbs; stems leafy throughout, slender; rhizomes slender. Leaves polymorphic; the lower submersed leaves linear to linear-lanceolate, 10-20 mm long, 5-12 mm wide, membranceous, 1-nerved, glabrous, apex bidentate, margins entire, bases connate; transition to floating leaves gradual; floating leaves in a compact rosette, obovate to spatulate, 6-15 mm long, 3-7 mm wide, 3-nerved from near the base, glabrous, apex rounded, margins entire, base attenuate and connate. Fruit cordate, 0.6-1.2 mm long, equally as wide; mericarps more broadly rounded at the apex than at the base, convex on the face, surface reticulate, margin wingless or very narrowly winged, styles caducous or persistent. (2n = 20) Apr-Aug. Cool shallow water in marshes, quiet streams, ponds, & ditches; MO w to KS, s to OK & TX, IA, se MN; SD: Harding; (N.S. w to s MN, s to FL & TX; w MT & WA s to CA & C. Amer.).

Previous reports from ne WY are *C. verna*.

3. **Callitriche terrestris** Raf. Tiny annual herbs; stems leafy throughout, to 6 cm long, prostrate or sometimes repent. Leaves uniform, spatulate to oblanceolate or sometimes obovate, 2–3.5 mm long, 0.5–1 mm wide, membranaceous, 3-nerved, glabrous, apex obtuse or rounded, margins entire, base usually attenuate. Fruit 0.4–0.7 mm long, 0.6–0.9 mm wide, pedicel 0.2–0.5 mm long; mericarps slightly cordate, ventral side flat, dorsal side rounded and scarcely winged, style reflexed and usually persistent. (2n = 10) May–Jun. Damp wooded draws & slopes, low moist pastures; e KS & MO; (ME w to e KS, s to VA, GA, LA, & e TX). *C. deflexa* A. Br. var. *austini* (Enlgem.) Hegelm.—Fernald.

4. **Callitriche verna** L. Aquatic perennial herbs; stems leafy throughout; rhizomes slender. Leaves polymorphic; the lower submersed leaves linear to linear-lanceolate, 4–11(16) mm long, 3–8 mm wide, membranaceous, 1-nerved, glabrous, apex obtuse to bidentate, margins entire, base connate; transition to floating leaves gradual; floating leaves in a compact rosette, obovate to spatulate, 7–13 mm long, 4–7 mm wide, 3-nerved from near the base, glabrous, apex rounded, margins entire, bases attenuate and connate. Fruit oval, 0.8–1.7 mm long, 0.6–1.4 mm wide, length always exceeding the width and width greatest just above the middle; mericarps with distinctly reticulate surface, reticulations in vertical rows, margins distinctly winged at least near the summit, styles usually caducous. (2n = 20) May–Aug. Shallow water in small ponds, streams, sloughs, & ditches; MN w to MT, s to cen NE & WY; (Greenl. w to AK, s to WV, OH, IL, IA, NE, NM, AZ, & CA).

Fassett (op. cit.) listed *C. verna* for western KS but cited no specimens and none have been located since.

122. PLANTAGINACEAE Juss., the Plantain Family

by Ronald L. McGregor and Ralph E. Brooks

1. PLANTAGO L., Plantain

Ours annual, biennial, or perennial, acaulescent or caulescent herbs from fibrous roots or taproots, sometimes with short erect or branched caudex; stems obsolete to short or well developed in one species. Leaves usually rosulate or alternate on short stems, or opposite in one species; petioles absent to indistinct or well marked; exstipulate. Flowers small, perfect or unisexual (cleistogamous in one sp.), hypogynous, regular or slightly irregular, in long-peduncled bracted terminal spikes or axillary in one species; sepals 4, slightly connate at base, persistent, the 2 next to the bract often different from 2 next to the axis; petals 4, corolla gamopetalous, salverform, scarious, long-persistent after anthesis, its tube covering summit of capsule, its lobes reflexed or spreading, or erect and connivent; stamens 4 or 2, distinct, inserted on corolla tube, included or exserted; anthers cordate or horned; ovary superior, 2-celled, 1–several ovules per locule, axile or basal; style 1. Fruit a membranaceous circumscissile capsule; seeds often mucilaginous when wet.

1 Plants distinctly caulescent; leaves opposite .... 1. *P. arenaria*
1 Plants acaulescent and leaves rosulate or if with short stems, leaves alternate.
   2 Lower bracts of spike much longer than calyces.
      3 Plants dark green; leaves glabrous on upper surface .... 2. *P. aristata*
      3 Plants yellowish-green; leaves pubescent on upper surface .... 7. *P. patagonica*
   2 Lower bracts not evidently longer than calyces.
      4 Sepals apparently 3, the 2 next to the bract united into 1 with a double midvein .... 5. *P. lanceolata*

4 Sepals 4, lobes nearly distinct.
   5 Bracts or sepals or both pubescent to long-villous.
      6 Pubescence of middle peduncle appressed.
         7 Bracts triangular-ovate, broadly scarious-margined except at apex; leaves glabrous above .... 11. *P. wrightiana*
         7 Bracts narrowly triangular-lanceolate to lance-linear, narrowly scarious-margined near base; leaves pubescent above .... 7. *P. patagonica*
      6 Pubescence of middle peduncle spreading.
         8 Bracts 1–2.5 mm long; corolla lobes 0.8–2.3 mm long; seeds 1.2–2.1 mm long, yellow-brown to black, not hyaline-margined .... 10. *P. virginica*
         8 Bracts 3–4.5 mm long; corolla lobes 2–3 mm long; seeds 2–2.8 mm long, bright red to reddish-black, with hyaline margin .... 8. *P. rhodosperma*
   5 Bracts and sepals glabrous or inconspicuously ciliate, sometimes glabrate.
      9 Leaves narrowly lanceolate or linear, 0.5–2 mm wide .... 3. *P. elongata*
      9 Leaves lanceolate to broadly ovate, usually well over 1 cm wide.
         10 Leaf blades much longer than wide; crown usually brown woolly at summit; bracts and sepals scarcely keeled .... 4. *P. eriopoda*
         10 Leaf blades 1/2–2/3 as wide as long; crown not brown woolly; bracts and sepals prominently raised-keeled.
            11 Capsules circumscissile near the middle; bracts usually broadly ovate; base of petiole usually greenish, rarely reddish .... 6. *P. major*
            11 Capsules circumscissile well below the middle; bracts usually narrowly lance-triangular; petioles usually reddish at base .... 9. *P. rugelii*

1. **Plantago arenaria** Waldst. & Kit., Indian plantain. Annual herb with a taproot; stems erect, 1–5(8) dm tall, usually with ascending straight branches, pubescent, sometimes minutely glandular above. Leaves opposite, linear or linear-oblanceolate, 1–8 cm long, 1–3(5) mm wide, hirsute, margin entire, ciliate. Peduncles axillary, 1–6 cm long, hirsute; spikes 5–15 mm long, dense, 6–10 mm in diam; lowest 2 bracts 6–10 mm long; upper bracts 3.5–4.5 mm long, hirsute, ovate-orbicular, sometimes wider than long, scarious-margined; sepals unequal, the anterior 3.5–4 mm long, obovate, the posterior 3–3.5 mm long, ovate-lanceolate; corolla zygomorphic; lobes 2 mm long, suborbicular to ovate-lanceolate, often crenulate, reflexed at anthesis; style equal to stamens in length; anthers conspicuously horned. Capsule 3–4 mm long, circumscissile near the middle; seeds usually 2 per capsule, red-brown to purplish-black, (2.5)3(4) mm long, oblong-elliptical, smooth. (2n = 12) Jul–Sep. Rare in weedy areas; MN: Clay, & perhaps elsewhere; (e U.S.; native in Europe). *P. indica* L.—Fernald; *P. psyllium* L.—Gleason & Cronquist.

In Flora Europaea 4: 43. 1976, *P. indica* L. is listed as nom. illegit. and *P. psyllium* L. as nom. ambig.

2. **Plantago aristata** Michx., bracted plantain, buckhorn. Annual herbs with taproots; stems obsolete or 2–3(6) cm long in older specimens; leaves rosulate or alternate, usually erect, dark green, (3)6–17(20) cm long, 2–7(10) mm wide linear to narrowly oblanceolate, lower surface hirsute to pilose; margins entire, ciliate. Peduncles 1–several from base, erect, hirsute to pilose or villous, hairs spreading or appressed-ascending, 3–20(25) cm long; spikes, excluding bracts, cylindric (2)7–15 cm long, 6–8(10) mm in diam; bracts aristate to linear and foliaceous, usually hirsute, the lowest 1–3 cm long, progressively reduced upwards; sepals narrowly oblong-obovate, pilose, rounded at apex, 2–2.5 mm long; corolla zygomorphic; petals 1.8–2.2 mm long, limb cordate, involute, tip of posterior limb rounded, spreading after anthesis; stamens 4. Capsules 2.8–3.5 mm long, 1.3 mm in diam, circumscissile just below the middle; seeds 2 per capsule, elliptic, brown or reddish-tan, 2.3–3 mm long, convex on outer side, concave on inner side, the cavity surrounded by a pale stripe. (2n = 20) May–Aug. Locally common in dry prairies, rocky open woodlands, roadsides, pastures, waste areas; e 1/2 NE & KS, IA, MO, OK; (native from IL, NE, s to LA, TX, now naturalized over most of e U.S. & adj. Can.).

3. **Plantago elongata** Pursh, slender plantain. Annual acaulescent herbs with taproots; leaves narrowly linear, 3–7(10) cm long, 0.5–2 mm wide, glabrous or appressed pubescent. Peduncles 1–several from base, commonly longer than leaves, erect or ascending, sparsely appressed pubescent; spikes 2–7(10) cm long, rather loosely flowered and axis exposed; bracts ovate, 2–2.5 mm long, glabrous, margins hyaline; sepals 2–2.5 mm long, ovate; anterior sepals inequilateral, with narrow midvein and wide scarious margins; posterior sepals sharply keeled; corolla lobes 0.5–1 mm long, spreading or reflexed with age; stamens 2. Capsule ovoid, 1.5–3.5 mm long, circumscissile just below the middle; seeds (3)4 or 5(6), elliptic-oblong, dark brown, finely pitted, 1.5–2.5 mm long. (2n = 12) Apr–Jun. Infrequent but locally common in dry sandy prairies, salty or alkaline flats; GP but absent extreme e; (Man. to B.C., s to TX, CO, UT, CA).

In the e 1/2 KS, MO, OK there is a very similar, but often smaller, plant with corolla lobes mostly erect in age and forming a beak; seeds 4 per capsule, 0.7–1.8 mm long. These plants have been referred to *Plantago pusilla* Nutt. We find the distinction between *elongata* and *pusilla* to be rather arbitrary in our area.

P. heterophylla Nutt., of the southeastern U.S. was reported for Kansas by Rydberg but specimen evidence is lacking. It differs from *elongata* and *pusilla* in capsules having 10–25(30) seeds only 0.5–0.8 mm long.

4. **Plantago eriopoda** Torr., alkali plantain. Perennial herbs with taproots and stout erect caudex which is often lobed or short branched, caudex usually conspicuously brown-woolly at summit. Leaves somewhat fleshy, with blades oblanceolate, lance-oblong, or elliptic, sometimes remotely denticulate, 3- to 9-nerved, pubescent, 0.5–2 dm long; petioles 1–6(10) cm long. Peduncles 1–3 dm long, glabrous or glabrate; spikes (2)5–15(20) cm long, flowers dense above, loose below with axis exposed; bracts 2–4 mm long, glabrous, ovate, acute to rounded at apex, somewhat keeled below, with narrow scarious margins; sepals oblong-ovate, subequal to bracts, flat, glabrous, with thin scarious margins; corolla lobes ovate, 1.2–2 mm long, reflexed at maturity. Capsules ovoid, 3–4 mm long, circumscissile below the middle; seeds 2–4 per capsule, 2–2.5 mm long, dark purplish-black, with thin hyaline margin at one or both ends. (2n = 24) Jun–Sep. Locally common in low, often alkaline or salty prairies, stream valleys, salt marshes, roadsides, waste places; w MN, ND, s to s-cen NE, MT, CO; (N.S. to Que., s to ND, MN, IA, NE, CO, NV, WA).

5. **Plantago lanceolata** L., English plantain, buckhorn. Perennial herbs, often flowering first year, fibrous rooted but often with taproots and erect, often branched caudex, 4–6 cm long which is often tan-woolly at summit. Leaves villous to glabrate, 3- to several-nerved, narrowly elliptic or lance-elliptic, 1–4 dm long, 1–4 cm wide, margin entire or remotely denticulate. Peduncles 1–several from base, striate-sulcate, strigose or hirsutulous, 1.5–4(6) dm tall; spike dense, ovate-conic at first, becoming cylindric, 1.5–5(8) cm long, to 1 cm in diam; bracts thin, ovate, 1.5–2.2 mm long, acuminate, scarious, with a green, glabrous midrib; flowers pistillate, staminate or perfect; sepals 2–2.5 mm long, the 2 sepals next to bract fused into a 2-tipped, 2-ribbed unit; sepals scarious, with a villous-ciliate green midnerve; corolla-lobes 2–2.5 mm long, ovate-lanceolate, spreading or reflexed; stamens 4, exserted, anthers cordate at base. Capsule 3–4 mm long, circumscissile near base; seeds (1)2 per capsule, dark brown or blackish, shining, 1.6–2 mm long, concave on adaxial face. (2n = 12, 24, 48) May–Oct. Locally common in lawns, pastures, fields, roadsides, waste places; GP but less common n & w; (a cosmopolitan weed; Eurasia). *P. lanceolata* var. *sphaerostachya* Mert. & Koch — Fernald.

6. **Plantago major** L., common plantain. Perennial herbs, often flowering first year, with fibrous roots and a short, stout, erect, caudex which is not woolly at summit. Leaves rosulate, blades broadly elliptic to broadly ovate or cordate-ovate, usually 4–15(18) cm long, 2.5–11 cm wide, strongly 3- to several-nerved, margin entire or slightly undulate or toothed, glabrate

or hirsutulous; blade abruptly contracted to petiole; petiole winged, shorter to longer than blade, glabrate to hirsutulous, usually green at base. Peduncles 1–30 from base, 5–25 cm long, glabrate to hirsutulous; spikes slender, dense above, axis often exposed below, 5–25 cm long; bracts ovate-lanceolate, 1.5–2.3 mm long, usually shorter than sepals, glabrous, sometimes keeled; sepals 1.5–2.5 mm long, ovate, with scarious margins, glabrous, sometimes keeled; corolla lobes narrowly deltoid, 1 mm long, eventually reflexed; anthers often purplish, distinctly horned, exserted. Capsules 2–3(4) mm long, circumscissile near the middle; seeds (6)8–16(30), brownish-black, 0.7–1 mm long, finely reticulate (use 40 ×) or with irregular ridges. (2n = 12) May–Nov. Locally common in open, rather dry lawns, fields, pastures, roadsides, & waste areas; GP but infrequent s 1/2; (a cosmopolitan weed nearly throughout the world; probably native of Eurasia). *P. major* var. *pilgeri* Domin, var. *scopulorum* Fries & Broberg—Fernald; *P. asiatica* L.—Rydberg.

This taxon is a highly plastic and variable species in which a number of infraspecific taxa have been named but none seem worthy of recognition. In our region it is often difficult to separate from *P. rugelii* unless a careful study of well-grown plants is made.

7. *Plantago patagonica* Jacq., Patagonian plantain. Annual, winter annual, biennial or short-lived perennial (?) with taproots terminated by inconspicuous crown or this rounded and congested to a short 2- to 4-branched caudex, each branch with a rosette; stem obsolete or to 1–4 cm long; plants usually woolly-villous throughout. Leaves of winter rosettes 0.5–3 cm long, narrowly oblanceolate to nearly obovate; principal leaves linear-oblanceolate, pubescent on both sides, 2–15(20) cm long, 0.5–7(15) mm wide, acute to acuminate and usually with a callous apiculate tip, 1- to 3-veined. Peduncles 1–20, 2–26 cm long, appressed pubescent, shorter to longer than leaves; spikes dense, (2)5–10(15) cm long; bracts narrowly triangular-lanceolate to lance-linear, narrowly scarious margined near base, 1–10(14) mm long, progressively shorter upward; sepals 1.4–2.5 mm long, narrowly obovate, with scarious margins; corolla lobes zygomorphic, 1–2 mm long, suborbicular to ovate-lanceolate, spreading after anthesis; stamens 4, slightly exserted to included. Capsule 3–3.5 mm long, circumscissile at or just below middle; seeds 2 per capsule, reddish-tan, 2.5–3 mm long. (2n = 20)

Three intergrading varieties are found in our region.

1 Peduncles 1–6.5 cm long; spikes 2–10 cm long, usually longer than peduncles .... 7a. var. *breviscapa*
1 Peduncles 6–26 cm long; spikes 2–15 cm long, usually shorter than peduncles.
  2 Bracts in lower portion of spike shorter or slightly longer than calyces .... 7b. var. *patagonica*
  2 Bracts evidently much longer than calyces .... 7c. var. *spinulosa*

7a. var. *breviscapa* (Shinners) Shinners. Peduncles 1–6.5 cm long; spikes 2–10 cm long, usually longer than peduncles; bracts slightly to much longer than calyces. Plants appearing as dwarf forms of the following varieties. Mar–Jun, sporadic later. Rather uncommon on dry, sandy, or rocky prairies, plains, hillsides, badlands; w 1/2 GP; (ND, MT, s to TX, NM).

7b. var. *patagonica*. Peduncles 6–26 cm long; spikes 2–15 cm long, often shorter than peduncles; bracts in lower portion of spike shorter to slightly longer than calyces. May–Aug. Locally common on sandy, rocky prairies, plains, pastures, roadsides, waste places; GP but less common eastward; (Sask. to B.C., s to IL, TX, CA; S. Amer.). *P. purshii* R. & S.—Rydberg; *P. patagonica* Jacq. var. *gnaphaloides* (Nutt.) A. Gray—Gleason & Cronquist.

7c. var. *spinulosa* (Dcne.) A. Gray. Similar to var. *patagonica* but bracts of lower spike evidently much longer than calyces. May–Aug. Locally common on sandy or rocky prairies, plains, pastures, roadsides, waste places; w 2/3 GP, rare in e 1/2; (ND, MT, s to TX, NM, AZ). *P. spinulosa* Dcne.—Rydberg.

In their extreme forms the three varieties are easily separated, but all three have been found at the same location and the recognition of varieties is of questionable taxonomic merit. The group merits experimental study, particularly since a few specimens, where contact with *P. aristata* occurs, appear to be hybrids or introgressants of the two taxa.

8. Plantago rhodosperma Dcne., red-seeded plantain. Annual herbs with slender taproots; leaves rosulate, elliptic to elliptic-lanceolate, or usually narrowly to broadly oblanceolate, (1–3)6–20(35) cm long, margins entire, remotely dentate and rarely with evident lateral lobes, surfaces pubescent, acute to obtuse or rounded at apex, with 3–7 veins, long-cuneate at base to evidently petiolate; petioles, when present, 1–7 cm long, pubescent, often reddish at base. Peduncles 1–several from base, (1)3–15(20) cm long, hirsute, shorter to much longer than leaves; spikes dense, (3)6–15(25) cm long; bracts lanceolate, 3–4.5 mm long, with conspicuous keel, keel hirsute, apex blunt to apiculate, margins narrow to broadly scarious; sepals (2)2.5–3.2 mm long, unequal, ovate-lanceolate, slightly keeled, hirsute on keel, with wide scarious margins; corolla lobes zygomorphic, lanceolate to ovate-lanceolate, 2–3 mm long at maturity, strictly erect and connivent after anthesis. Capsules (2)3(4) mm long, circumscissile below the middle; seeds 2, bright-red to reddish-black, 2–2.8 mm long, nearly flat on both sides, with a hyaline margin. Apr–Jun. Locally common in dry sandy, rocky, prairies, plains, pastures, roadsides, & waste places; e 2/3 KS; MO: Jackson; OK, TX; (TN, MO, KS, s to TX, AZ, introduced elsewhere).

9. Plantago rugelii Dcne., Rugel's plantain. Perennial herbs but usually flowering first year, roots usually fibrous or plant with slender to stout taproots, usually with erect, short, caudex. Leaves rosulate; blades broadly elliptic to oval or ovate, glabrous, 5–20 cm long, 1/2–2/3 as wide, narrowed to base, apex rounded or acute, major veins 5–9, margin entire or with remote, small pectinate teeth; petioles 2–15(20) cm long, winged, usually glabrous and reddish at base. Peduncles (3)5–20(30) cm long, 1–several from base, glabrous or glabrate; spikes (3–5)6–20(35) cm long, dense or axis visible below, 5–12 mm in diam; pedicels to 1 mm long; bracts 1.5–2.3 mm long, lanceolate to lance-triangular, narrowed from base to the sometimes attenuate tip, with acute keel, glabrous; sepals 2–2.5 mm long, ovate to oblong, acute, keel wider than the scarious margin, glabrous; corolla lobes deltoid, 0.7–1.2 mm long, reflexed after anthesis. Capsule (4)6–8(10) mm long, circumscissile well below the middle; seeds 4–10 per capsule, 1.8–2.4 mm long, somewhat angular, nearly black. (2n = 24) May–Nov. Locally common in shaded lawns, parks, stream valleys, pastures, roadsides, waste places; GP but more common in e-cen & se, rare in n & w, absent sw; (e 1/2 U.S., w to ND, s to FL, TX).

Immature or sterile material of this species is often impossible to distinguish from *P. major*. Usually *P. rugelii* grows in more moist, and often shaded, habitats.

10. Plantago virginica L., pale-seeded plantain. Winter annual to usually annual herbs with weak taproots; stems very short and compact. Leaves rosulate, oblanceolate, obovate to spatulate, (2)5–10(15) cm long, hirsute or hirsutulous, with 3 major veins, obtuse to acute at apex, long-cuneate to petiolate basally, margins entire or inconspicuously toothed. Peduncles 1–several from base, (3)5–15(25) cm long, hirsute; spikes dense, 3–25 cm long; bracts 1–2.5 mm long, ovate-lanceolate, obtuse to acute, hispid; sepals 1.5–2.7 mm long, oblong-ovate, acute, keel hirsutulous, margin broad and scarious; corolla lobes 0.8–2.3 mm long, sharply acute, erect and connivent after anthesis, ca 2 mm long. Capsules ovoid or rhombic-ovate, 2–2.5 mm long, circumscissile at the middle; seeds 2 per capsule, dull yellow-brown to nearly black, 1.2–2.1 mm long, without hyaline margin. (2n = 20) May–Jun. Locally common in pastures, rocky open woodlands, fields, lawns, roadsides, waste ground; se 1/4 GP; (MA, NY, WI, IA, se SD, s to FL, TX, introduced elsewhere). *P. virginica* var. *viridescens* Fern.—Fernald.

Plants as described above are the usually apomictic form, bearing only cleistogamous flowers with abortive anthers and styles. In some plants the spikes have both cleistogamous and chasmogamous flowers. In chasmogamous flowers the anthers and styles are well developed and the bracts and sepals are more rounded at apex. *P. virginica* is a short-lived plant, usually lasting a little less than 2 months.

11. **Plantago wrightiana** Dcne., Wright's plantain. Perennial herbs, but flowering first year, with taproots and short erect, sometimes branched, caudex; stems obsolete or to 6 cm long. Leaves alternate, crowded, erect, linear to linear-oblanceolate, obtuse or acute at apex, entire, 3–12(20) cm long, 2–7(10) mm wide, glabrous above, pubescent beneath. Peduncles 1–many, rather stout, (5)10–15(25) cm long, appressed pubescent; spikes 1–10 cm long, dense, to 9 mm in diam, villous at base and on axis; bracts rigid, 2–3 mm long, triangular-ovate, herbaceous with broad scarious margins, villous to glabrate; sepals narrowly obovate, 2.7–3.2 mm long, pilose, herbaceous and dark green, becoming brownish; corolla lobes ovate, obtuse, 3 mm long, 2.5 mm wide, spreading and reflexed. Capsules 3.5–4.2 mm long, circumscissile just below middle; seeds 2 per capsule, reddish or reddish-brown, 2.8–3.2 mm long, very slightly constricted just below the middle, finely pitted. Apr–Jul. Sandy, gravelly or rocky prairie hillsides, open woodlands, stream valleys; KS: Coffey, Crawford; OK, TX, NM; (OK, TX, NM, AZ).

This species has been referred to as an annual, but our plants are perennial though flowering the first season.

123. OLEACEAE Hoffmsg. & Link, the Olive Family

by Ronald L. Hartman

Dioecious, polygamo-dioecious, or polygamous shrubs or trees. Leaves opposite or occasionally subopposite, simple or odd-pinnately compound; stipules none. Flowers unisexual or sometimes perfect, in congested clusters or panicles from axils of previous years leaves; sepals 4, distinct or fused into a cup with a laciniate to subentire margin, or none; petals none (or 4 and united in the cultivated species mentioned below); stamens 2–6; anthers dehiscing longitudinally; pistil 1, of 2 united carpels; style 1, capitate or 2-lobed; ovary 2-locular with 2 pendulous ovules in each locule. Fruit 1-locular, 1(2)-seeded, a samara with a prominent, terminal wing or a drupe.

Several introduced species of this family are widely cultivated in the GP. These include *Forsythia viridissima* Lindl. (golden bells), *Ligustrum vulgare* L. (common privet), and *Syringa vulgaris* L. (common lilac). Although they may persist in abandoned homesteads, they usually do not become established.

1 Leaves simple; fruit a wingless drupe .... 1. *Forestiera*
1 Leaves pinnately compound with 5–11 leaflets; fruit a samara with a prominent terminal wing .... 2. *Fraxinus*

1. FORESTIERA Poir.

Dioecious or polygamo-dioecious shrubs (or small trees); branchlets terete; terminal bud with 4–7 pairs of exposed, imbricate scales in 4 rows, the scales never foliar; lateral buds often superposed, resembling the terminal bud; leaf scar elevated, semicircular; vascular bundle scar 1, often obscure. Leaves simple, opposite or subopposite, petiolate; blade usually paler and minutely porulus beneath, the margins entire to serrate; petiole grooved adaxially. Flowers unisexual or sometimes perfect, appearing before the leaves; calyx of 4 minute, caducous sepals, or none; corolla none; staminate flowers with 2–6 functional stamens surrounding a rudimentary pistil; pistillate flowers with 1–4 abortive stamens at base of the functional pistil, or none, the style slender, stigma capitate to clavate. Fruit a 1(2)-seeded drupe, maturing slowly; endocarp thin-walled, longitudinally ribbed; seed narrowly elliptic to oblong, terete.

1 Leaf blades mostly 4-10 cm long, the apex acuminate; petiole 7-14 mm long; drupe 9-18 mm long .... 1. *F. acuminata*
1 Leaf blades 1.3-3.7 cm long, the apex obtuse to rounded; petiole 1-3 mm long; drupe 5-7 mm long .... 2. *F. pubescens*

1. **Forestiera acuminata** (Michx.) Poir., swamp privet. Dioecious or polygamo-dioecious shrub (or small tree) to 6 m tall; bark gray-brown; branchlets gray-brown to blackish, initially often densely pubescent, glabrescent, the lateral branchlets often spinescent; terminal bud gray-brown, ovate, 1.5-3 mm long. Leaf blade ovate to elliptic, (3)4-10 cm long, (1)1.5-3.5(5) cm wide, pubescent when young, usually glabrescent; apex acuminate; margins usually serrate in upper 1/2; base cuneate; petiole 7-14 mm long. Flowers unisexual or sometimes perfect; staminate flowers 4-8 in a congested cluster surrounded by 3 or 4 obovate bracts 4-5 mm long, the inflorescence axes glabrous, the sepals minute or none, the stamens 2-6, with anthers oblong, 0.7-1.1 mm long and filaments 3-4.5 mm long; pistillate flowers 18-32 in a compact panicle, the bracts obovate to spatulate, soon deciduous, the inflorescence axes glabrous, the sepals, if present, 0.1-0.2 mm long, caducous, the ovary 0.8-1.2 mm long with style 1-1.5 mm long. Drupe dark blue to blackish, glaucous, elliptic to oblong, often curved, 9-18 mm long. (n = 23) Apr. Swampy ground or alluvial soils of flood plains; KS: Cherokee, Labette; MO: Jasper (SC, s IN, n MO, s to FL, LA, s TX). *Adelia acuminata* Michx.—Rydberg.

2. **Forestiera pubescens** Nutt., elbow bush. Dioecious shrub to 2 m tall; bark gray-brown, branchlets gray to brown, initially densely pubescent, glabrescent, the lateral branchlets usually not spinescent; terminal bud gray to reddish-brown, ovate, 1.5-2 mm tall. Leaf blade ovate to broadly elliptic or obovate, 1.3-3.7 cm long, 0.7-2 cm wide, sparsely to densely pubescent; apex obtuse to rounded; margins obscurely serrate to crenate; base cuneate to rounded; petiole 1-3 mm long. Flowers unisexual; staminate flowers 3-8 in a congested cluster surrounded by 3 or 4, obovate to orbicular, enlarged, inner bud scales 2-3 mm long, the inflorescence axes glabrous, the sepals none, the stamens 2-6 with anthers broadly elliptic to suborbicular, 0.5-1 mm long and filaments 2-4.5 mm long; pistillate flowers 3-12 in a compact, usually ebracteate corymb or panicle, the inflorescence axes moderately to densely pubescent, the sepals 0.1-0.2 mm long, caducous, the ovary 0.4-1 mm long with a style 0.5-1.5 mm long. Drupe purple to blackish, glaucous, elliptic, symmetrical or nearly so, 5-7 mm long. Mar. Prairies, pastures, & open slopes; OK: Caddo; TX: Armstrong, Bailey, Briscoe, Motley, Randall, Roberts; (TX w to CA; n Mex.).

2. FRAXINUS L., Ash

Dioecious or polygamous trees; branchlets terete; terminal bud with 1 pair of exposed, valvate scales or 2 or 3 pairs of 4-ranked scales, the outer of which are imbricate, the apex of the outermost pair often foliar (obscurely pinnate); lateral buds not superposed, the scales 2, valvate, often indistinct; leaf scars elevated, suborbicular to semicircular or crescent-shaped; vascular bundle scars numerous, in an arc. Leaves odd-pinnately compound, opposite, petiolate; leaflets usually paler or obviously whitened but not porulus beneath, sessile or petiolulate, the margins serrate to undulate or subentire; petiole flat to grooved adaxially. Flowers unisexual or perfect, appearing before or with the leaves; calyx of 4 distinct or connate sepals, sometimes caducous, or none; corolla none; staminate flowers of 2(3) stamens; pistillate flowers with 2 abortive stamens at base of the functional pistil, or none, the style slender to somewhat broadened, compressed, with a narrowly bilobed stigma. Fruit a 1-seeded samara, maturing rapidly; wing terminal, prominent; seed narrowly elliptic to oblong, terete or flat.

This a genus containing attractive ornamental trees and several species which are important sources of commercial lumber. Being wind pollinated, the trees produce tremendous quantities of pollen, which often is a contributing cause of hay fever.

1 Body of samara terete or nearly so, 0.8–3.3 mm wide at widest point; calyx present on fruit, an irregularly toothed to subentire cup 1–1.6 mm long; anthers linear to linear-oblong, apiculate at apex.
  2 Terminal bud usually blunt, wider than tall; body of samara mostly ca 3 mm wide at widest point, the wing decurrent less than 1/3 of length; leaf scars, except on most recent year's growth, mostly crescent-shaped, the upper margin deeply concave .... 1. *F. americana*
  2 Terminal bud acute to acuminate, distinctly taller than wide; body of samara mostly 2 mm wide or less at widest point, the wing decurrent to or below midpoint; leaf scars mostly semicircular, the upper margin truncate or slightly concave .... 3. *F. pennsylvanica*
1 Body of samara flat, 4.5–8 mm wide at widest point; calyx not present on fruit (bases of 2 filaments often present); anthers oblong to elliptic, blunt or split at apex.
  3 Branchlets terete; lateral leaflets sessile .... 2. *F. nigra*
  3 Branchlets 4-angled or narrowly 4-winged; lateral leaflets petiolulate .... 4. *F. quadrangulata*

1. *Fraxinus americana* L., white ash. Dioecious trees to 20(30) m tall; trunk to 0.8(1)m in diam, the bark gray-brown to dark gray, deeply furrowed; branchlets dark green to gray- or orange-brown, terete, usually glabrous; terminal bud rusty to dark brown, blunt, wider than tall, usually contiguous with pair of lateral buds; lateral buds triangular; leaf scars, except on most recent year's growth, crescent-shaped, the upper margin deeply concave. Leaves elliptic to suborbicular in outline, (10)18–40 cm long, 10–15 cm wide; petiole flat to grooved adaxially; leaflets 5–9, mostly 7, usually obviously whitened beneath, lanceolate to ovate, elliptic, or oblanceolate, petiolulate, 5–15 cm long, 2–7(8) cm wide, sparsely to moderately puberulent beneath or often densely so along the veins, or glabrous; apex acute to acuminate; margins serrate, often irregularly so, to undulate or subentire; base cuneate to rounded, often oblique; lateral petiolules 4–10(15) mm long. Flowers unisexual; calyx cupulate, irregularly toothed in upper 1/4–1/3, persistent on the fruit, usually cleft on 1 side; anthers linear to linear-oblong, 2.5–4 mm long, apiculate at apex; filaments 0.1–1.3 mm long. Samaras greenish-tan to brown, narrowly lanceolate, 3–4.5(5) cm long, 0.4–0.7 mm wide; apex rounded to acute, apiculate, or emarginate; body terete or nearly so, narrowly elliptic, (2.2)2.6–3.3 mm but mostly ca 3 mm wide at widest point, 1/4–2/5 as long as the entire fruit, with wing decurrent less than 1/3 of length. (n = 23, 46, 69) Mar–May. Deciduous forests, stream banks, & flood plains, often cultivated; sw IA, e edge NE, e 1/3 KS, MO, e 1/4 OK, with scattered localities w to cen NE, KS, & OK; (se Can., ME to e MN, s to FL, LA, & e TX).

This is a handsome ornamental and a very valuable timber-yielding tree.

2. *Fraxinus nigra* Marsh., black ash. Polygamous or dioecious trees to 25 m tall; trunk to 0.5 m in diam, the bark light gray, with shallow furrows; branchlets tan to gray, terete, glabrous; terminal bud dark brown to nearly black, usually acuminate, about as tall as wide to slightly taller, usually separated by a space from the uppermost pair of lateral buds; lateral buds rounded; leaf scars semicircular to circular, the upper margin convex to truncate or slightly concave. Leaves elliptic to oblong or broadly obovate in outline, 25–40 cm long, 13–26 cm wide; petioles terete to grooved adaxially; leaflets (7)9–11, paler green beneath, elliptic to broadly lanceolate, the lateral ones sessile, (5)12–16(20) cm long, (1.5)3–5(6) cm wide, nearly glabrous to sparsely pubescent beneath, usually densely so along the midvein and at base; apex acute to acuminate; margins serrate; base cuneate to rounded, often oblique; lateral petiolules none. Flowers perfect or appearing so or unisexual; calyx minute, caducous, or none; anthers oblong to narrowly elliptic, 1.5–2 mm long, blunt or split at apex; filament 0.1–0.3 mm long. Samara tan to greenish, oblong to elliptic-oblong or spatulate, 2.5–4.5 cm long, 0.7–1.1 cm wide; apex retuse to apiculate; body flat, narrowly elliptic, 4.5–7 mm wide at widest point, 1/2–3/5 as long as the entire fruit, with wing decurrent 1/2 or less of length. May. Swamps, bogs, low banks of streams & lakes; MN: Clay, Otter Trail; ND: Cass, Pembina; (Can., ME to ND, s to n VA, s IN, & cen IA).

3. **Fraxinus pennsylvanica** Marsh., red or green ash. Dioecious trees to 20 m tall; trunk to 0.5 m in diam, the bark dark gray to brown, with shallow furrows; branchlets greenish to brown or gray, terete, glabrous or sparsely to densely pubescent; terminal bud reddish-brown to nearly black, acute to acuminate, distinctly taller than wide, usually contiguous with a pair of lateral buds; lateral buds rounded; leaf scars semicircular, the upper margin truncate or slightly concave. Leaves elliptic to orbicular in outline, 11–30 cm long, 8–18 cm wide; petiole flat to grooved adaxially; leaflets 5–7(9), paler green beneath, lanceolate to ovate or elliptic, petiolulate, 6–15 cm long, 1.3–5 cm wide, sparsely to densely puberulent, especially along the midvein; apex acute to acuminate; margins serrate, sometimes obscurely so, undulate, or subentire; base cuneate to rounded, often oblique; lateral petiolules 2–7 mm long. Flowers unisexual; calyx cupulate, irregularly toothed to subentire, persistent on the fruit, often cleft on 1 side; anthers linear to narrowly oblong, 2.5–4 mm long, apiculate at apex; filament 0.1–0.6 mm long. Samara tan, narrowly oblanceolate, (2)2.5–4.5(5.5) cm long, 0.4–0.6(0.8) mm wide; apex acute to rounded or emarginate, often apiculate; body terete or nearly so, linear to narrowly oblong, 0.8–2(2.2) mm wide at widest point, 2/5–1/2 as long as the entire fruit, with wing decurrent 1/2 or more of length. (n = 23) Apr–May. Alluvial soils of flood plain forests, along stream or lake margins, & in ravines in prairies, often cult.; GP except TX, w OK, s 1/2 CO (s Can., ME to cen MT, s to FL, LA, e 1/2 TX). *F. campestris* Britt., *F. lanceolata* Borkh.—Rydberg; *F. pennsylvanica* var. *austinii* Fern.—Boivin; *F. pennsylvanica* var. *campestris* (Britt.) F. C. Gates—Gates; *F. pennsylvanica* var. *lanceolata* (Borkh.) Sarg.—Budd & Best; *F. pennsylvanica* var. *subintegerrima*, (Vahl) Fern.—Fernald.

Individuals with glabrous (vs. densely pubescent) twigs and inflorescences are often called var. *subintegerrima*, but this recognition is of dubious taxonomic significance.

4. **Fraxinus quadrangulata** Michx., blue ash. Polygamous trees to 30 m tall; trunk to 0.8 m in diam, the bark light gray, with shallow furrows; branchlets tan to gray-brown, 4-angled or narrowly 4-winged, glabrous or pubescent at nodes when young, usually dull; terminal bud rusty to grayish or dark brown, usually blunt, about as broad as tall, usually contiguous with pair of lateral buds; lateral buds rounded to flattened; leaf scars crescent-shaped to semicircular, the upper margin truncate to concave. Leaves ovate to suborbicular in outline, 15–30 cm long, 15–22 cm wide; petioles mostly flattened adaxially; leaflets (5)7–11, paler green beneath, lanceolate to ovate, petiolulate, (3)5–12 cm long, (1.5)2.2–4.5(6) cm wide, nearly glabrous to pubescent beneath, commonly densely so along the veins, often punctate; apex acute to acuminate; margins serrate; base obtuse to broadly rounded, often oblique; lateral petiolules 2–8(10) mm long. Flowers mostly perfect; calyx minute, caducous, or none; anthers elliptic to oblong, 1–1.2 mm long, blunt or split at apex; filaments 0.1–0.5 mm long, or none. Samara tan to greenish, broadly oblong to spatulate, 2.5–3.5 cm long, 0.8–1.1 cm wide; apex deeply notched to truncate or apiculate; body flat or nearly so, narrowly elliptic, 5–8 mm wide at widest point, 3/5–2/3 as long as the entire fruit, with wing decurrent more than 1/2 of length. (2n = 46) Mar–Apr. Calcareous soils in deciduous forests; se KS, ne OK (extreme s Can., OH, s MI, & s WI s to nw GA, AL, AR, KS, & OK).

This is a large timber-producing tree, the inner bark and twigs of which yield a water-soluble blue dye.

124. SCROPHULARIACEAE Juss., the Figwort Family

by Noel H. Holmgren

Ours annual, biennial or perennial herbs, some genera hemiparasitic. Leaves entire to pinnately or palmately lobed or parted, opposite, alternate, basal or sometimes whorled, exstipulate. Inflorescence a spike, raceme, thyrsoid panicle, or the flowers solitary in the axils of leaves; bractlets present in some. Flowers perfect, typically 5-merous or often 4-merous by reduction, bilabiate to nearly regular in some, ours entomophilous; calyx persistent, of 5, 4, or 2 segments (sepals) parted nearly to the base, or as lobes or teeth of a tube (gamosepalous); corolla sympetalous, 5- or 4-lobed, usually bilabiate; stamens commonly 4 or 2, or 4 fertile and 1 sterile in *Penstemon* and *Scrophularia*, or 5 fertile in *Verbascum*, epipetalous and alternate with the corolla lobes; carpels 2, united into a 2-loculed ovary, the anatropous ovules on axile placentae, the style slender and entire with a single stigma or forked with 2 stigmas. Fruit a capsule, septicidal or loculicidal or both (4-valved), sometimes opening by pores or irregularly bursting; seeds usually many and small, rarely few, the embryo small in a well-developed endosperm.

Reference: Pennell, F. W. 1935. The Scrophulariaceae of eastern temperate North America. Acad. Nat. Sci. Philadelphia Monogr. 1: i–xv, 1–650.

1 Leaves opposite, whorled or all basal.
  2 Fertile stamens 4.
    3 Calyx segments (sepals) parted nearly to the base.
      4 Corolla with a spur at the base ventrally; capsule dehiscing by apical pores or ruptures; leaves opposite only at the base of the stems.
        5 Stems erect or ascending; leaves entire, sessile.
          6 Corolla 7–24 mm long, not including the 2–17 mm spur; flowers in a terminal raceme; capsule symmetrical .... 16. *Linaria*
          6 Corolla 4.5–6 mm long, not including the 1.7–2.8 mm spur; flowers solitary in the axils of leaves, stems often floriferous to near the base; capsule asymmetrical, the anterior cell much larger .... 8. *Chaenorrhinum*
        5 Stems prostrate, twining or trailing; leaves hastate or dentate at the base, petiolate .... 13. *Kickxia*
      4 Corolla not spurred; capsule dehiscing septicidally, sometimes also loculicidally at the apex (4-valved); leaves all opposite.
        7 Staminode (sterile posterior stamen) present and well developed.
          8 Staminode slender, elongate, bent forward against the ventral side of the throat; corolla 10–55 mm long .... 22. *Penstemon*
          8 Staminode flattened, broad, appressed against the dorsal side of the throat; corolla 5–12 mm long .... 23. *Scrophularia*
        7 Staminode not present.
          9 Leaves broadly obovate to suborbicular, entire; plants prostrate, often aquatic and floating .... 4. *Bacopa*
          9 Leaves oblanceolate, serrate or pinnately divided; plants erect, terrestrial.
            10 Leaves serrate apically; plant a glabrous perennial; pedicels with a pair of bractlets at the base .... 18. *Mecardonia*
            10 Leaves pinnately or bipinnately divided; plant a hispid annual; pedicels without bractlets .... 14. *Leucospora*
    3 Calyx segments united into a well-developed tube.
      11 Plants essentially acaulescent (stolons often present), the leaves in a basal tuft; corolla rotate-campanulate, nearly regular; anthers 1-celled .... 15. *Limosella*
      11 Plants caulescent and leafy throughout; corolla not rotate-campanulate, mostly bilabiate; anthers 2-celled, except in *Buchnera*.
        12 Stamens enclosed by a hooded upper corolla lip or a folded lower corolla lip.
          13 Lobes of the upper corolla lip wholly united and forming a hood (galea)

enclosing the anthers, the lower lip not as below .... 21. *Pedicularis*
13 Lobes of the upper corolla lip free, the lower lip with the median lobe folded downward between the lateral lobes forming a pouch that encloses the stamens .... 9. *Collinsia*
12 Stamens not enclosed by either the upper or lower corolla lip, sometimes more or less included in the throat or tube.
14 Calyx pleated; lower corolla lip with 2 elevated ridges forming a palate that sometimes closes off the throat .... 19. *Himulus*
14 Calyx rounded, not pleated; lower corolla lip without a raised palate, the throat rounded and open.
15 Corolla yellow; leaves, at least the lower ones, pinnatifid to bipinnatifid with lanceolate to ovate divisions; plants mostly 10–20 dm tall.
16 Corolla 38–50 mm long; calyx 10–18 mm long; anthers villous, 2.8–5.4 mm long; capsule 15–20 mm long .... 3. *Aureolaria*
16 Corolla 14–16 mm long; calyx 6–8(10) mm long; anthers gabrous, 2.0–2.2 mm long; capsule 6–11 mm long .... 11. *Dasistoma*
15 Corolla purple, violet, magenta or pink; leaves entire (or pinnately divided with linear to filiform divisions in *Tomanthera densiflora*); plants smaller, 1.5–9(12) dm tall.
17 Corolla more or less campanulate; anthers villous, 2-celled; bractlets absent.
18 Calyx 10–16 mm long with lobes 6–18 mm long; leaves lanceolate in outline, entire, auriculate or pinnately divided .... 24. *Tomanthera*
18 Calyx 3–10 mm long with lobes 0.4–5(6) mm long; leaves linear to filiform or sometimes lanceolate, entire, rarely auriculate .... 1. *Agalinis*
17 Corolla salverform; anthers glabrous, 1-celled; bractlets present at base of receptacle .... 6. *Buchnera*
2 Fertile stamens 2.
19 Corolla 5-lobed; stamens included; stigmas 2-lobed and lamellate; capsule septicidal, sometimes also loculicidal (4-valved).
20 Bractlets present at the summit of the pedicel (just below the calyx); anther cells on a membranaceous expansion of the connective .... 12. *Gratiola*
20 Bractlets absent; anther cells not on a membranaceous expansion of the connective .... 17. *Lindernia*
19 Corolla 4-lobed; stamens exserted; stigmas united and capitate; capsule loculicidal, rarely septicidal as well.
21 Leaves 3–7 to a whorl; calyx 5- or sometimes 4-parted; corolla salverform, the tube longer than the lobes; capsule more or less terete .... 27. *Veronicastrum*
21 Leaves opposite; calyx 4-parted; corolla subrotate, the tube shorter than the lobes (except in *V. longifolia*); capsule laterally compressed .... 26. *Veronica*
1 Leaves alternate.
22 Fertile stamens 5; capsule dehiscence septicidal .... 25. *Verbascum*
22 Fertile stamens 4 or 2; capsule dehiscence loculicidal by longitudinal splits or apical pores or ruptures.
23 Corolla tube gibbous or spurred at the base ventrally, the upper lip 2-lobed; capsule dehiscence by terminal pores or ruptures.
24 Corolla 4–24 mm long, spurred at the base ventrally.
25 Stems erect or ascending; leaves entire, sessile.
26 Corolla 7–24 mm long, not including the 2–17 mm spur; flowers in a terminal raceme; capsule symmetrical .... 16. *Linaria*
26 Corolla smaller, 4.3–6 mm long, not including the 1.7–2.8 mm spur; flowers solitary in the axils of leaves and the stems floriferous to near the base; capsule asymmetrical, the anterior cell much larger than the posterior one .... 8. *Chaenorrhinum*
25 Stems prostrate, twining or trailing; leaves hastate, dentate or palmately lobed, petiolate.

27 Leaves hastate to dentate at the base and pinnately veined; spur of the corolla 4–6 mm long .... 13. *Kickxia*
27 Leaves palmately 5(9)-lobed and palmately veined; spur of the corolla 1.3–2.2 mm long .... 10. *Cymbalaria*
24 Corolla longer, (27)30–42 mm, merely gibbous at the base ventrally .... 2. *Antirrhinum*
23 Corolla tube not gibbous or spurred at the base, the upper lip entire; capsule dehiscence by loculicidal splitting.
28 Fertile stamens 2; upper corolla lip flat (corolla absent in *Besseya wyomingensis*).
29 Leaves all cauline; inflorescence usually indistinct, appearing as solitary flowers in the axils of the leaves, often floriferous to near the base .... 26. *Veronica*
29 Leaves basal and cauline, the basal leaves well developed and the cauline reduced and bractlike; inflorescence a dense terminal spike or spikelike raceme .... 5. *Besseya*
28 Fertile stamens 4; upper corolla lip hooded, enclosing the anthers.
30 Galea curved forward; cells of the anther equal; leaves pinnatifid, bipinnatifid or pinnately lobed, the divisions or lobes numerous .... 21. *Pedicularis*
30 Galea erect and straight; cells of the anther unequal, the longer cell attached by its middle and the smaller by its apex; leaves entire or pinnately 3- to 5(7)-divided.
31 Calyx 13–40 mm long; corolla 18–55 mm long, the galea (5)7–24 mm long; plants mostly perennial .... 7. *Castilleja*
31 Calyx 6–8 mm long; corolla 9–12(14) mm long, the galea 2.5–4 mm long; plants slender annuals .... 20. *Orthocarpus*

1. AGALINIS Raf., Gerardia

Ours annual herbs, often hemiparasitic; stems erect, usually 4-angled by decurrent lines from the leaf bases, branched, glabrous or usually with ascending-scabridulous pubescence. Leaves opposite or subopposite, sessile, ours entire, linear to filiform, sometimes lanceolate, ours usually scabridulous or scabrous on the upper surface and glabrous below, often revolute, the leaves of the fascicles and branchlets smaller and narrower. Inflorescence a raceme; bracts leaflike below and progressively reduced upwards, sometimes becoming alternate on the branches; pedicels often expanded at the receptacular attachment. Flowers weakly bilabiate; calyx gamosepalous, nearly regular, somewhat accrescent, campanulate to hemispherical, 5-lobed or -toothed, these usually shorter than the tube, triangular to lanceolate or subulate, acute or acuminate, the posterior one often slightly larger than the others; corolla pink to magenta or pale purple, with 2 yellow lines and purplish or reddish spots within the throat ventrally, the tube campanulate, often somewhat distended ventrally (ampliate), the throat open, the lobes shorter than the tube, slightly irregular, ciliate, the upper 2 usually somewhat smaller, arched, spreading to somewhat reflexed or sometimes projecting, the lower 3 spreading, external in bud; stamens 4, didynamous, usually included, the lower pair longer, the filaments pubescent, at least toward the base, the anther cells 2, parallel, obtuse to caudate at the base, villous; stigma solitary, flattened, more or less elongate. Capsule globose or subglobose, sometimes ellipsoid, loculicidal; seeds numerous, triangular, reticulate. *Gerardia* L.

References: Canne, J. M. 1979 [1980]. A light and scanning electron microscope study of seed morphology in *Agalinis* (Scrophulariaceae) and its taxonomic significance. Syst. Bot. 5: 281–296; Canne, J. M. 1981. Chromosome counts in *Agalinis* and related taxa (Scrophulariaceae) Canad. J. Bot. 59: 1111–1116; Pennell, F. W. 1928. *Agalinis* and allies in North America—I. Proc. Acad. Nat. Sci. Philadelphia 80: 339–449; Pennel, F. W. 1929. *Agalinis* and allies in North America—II. Proc. Acad. Nat. Sci. Philadelphia 81: 111–249.

Agalinis is the largest and most complex genus of Scrophulariaceae in the eastern half of the United States.

1 Plants relatively dark green and often purple-tinged, tending to blacken in drying; calyx tube slightly or not veiny; anthers 1.5–4 mm long (Sect. *Chytra*).
2 Calyx lobes 3–6 mm long, mostly longer than the tube; leaves linear-lanceolate to lanceolate, the principal leaves 2–6 mm wide, the lower sometimes 3-cleft to laciniate at

the base (Subsect. *Heterophyllae*) .... 3. *A. heterophylla*
2 Calyx lobes 0.2–2(3) mm long, decidedly shorter than the tube; leaves linear to filiform, rarely over 2 mm wide, all entire.
3 Calyx lobes acute, obtuse or nearly rounded, the sinuses V- to narrowly U-shaped, seldom broadly rounded, densely puberulent within and on the margins; capsule 7–11 mm long, oblong to ellipsoid, decidedly longer than broad; calyx 5–9 mm long (Subsect. *Asperae*) .... 1. *A. aspera*
3 Calyx lobes acuminate, often with a subulate tip, the sinuses becoming broadly rounded or broadly V-shaped, not or only slightly pubescent within; capsule 4–6(7) mm long, subglobose; calyx 2–5(6) mm long.
4 Pedicels 2–4(5) mm long, as long to shorter than the calyx; corolla (16)20–32 mm long (Subsect. *Purpureae*).
5 Axillary fascicles lacking or weakly developed; stems smooth or sparsely scabridulous, 4-angled .... 4. *A. purpurea*
5 Axillary fascicles well developed; stems scabridulous, the main stem weakly 4-angled to subterete .... 2. *A. fasciculata*
4 Pedicels 7–20 mm long, longer than the calyx; corolla 10–15 mm long (Subsect. *Tenuifolieae*) .... 6. *A. tenuifolia*
1 Plants yellowish-green, not tending to blacken in drying; calyx tube evidently veiny; anthers 0.5–1.3 mm long (Sect. *Chloromone*).
6 Capsules subglobose, 3.5–5 mm long; bracts usually surpassed by the pedicels at anthesis .... 5. *A. skinneriana*
6 Capsules obovoid, 5–6.5 mm long; bracts usually exceeding the pedicels at anthesis .... 7. *A. viridis*

1. ***Agalinis aspera*** (Dougl. ex Benth.) Britt. Scabridulous herb with dark green herbage, tending to blacken in drying; stems 2–6(8) dm tall with ascending branches and well-developed axillary fascicles. Leaves ascending, entire, narrowly linear, sharp-tipped, 2–4 cm long, 0.8–1.5 mm wide, scabrous above, usually revolute, the leaves of the fascicles smaller. Pedicels strongly ascending, slender, as long or longer than the calyx, 5–11 mm long at anthesis and to 18 mm in fruit. Calyx campanulate, 5–7 mm long at anthesis and to 9 mm in fruit, the lobes lanceolate to broadly lancolate, acute to obtuse or nearly rounded, 1.2–2(3) mm long, the sinuses V- to narrowly U-shaped, densely puberulent within and on the margins; corolla 18–25 mm long, its lobes spreading, rounded to emarginate, 3–6 mm long; anthers 2.0–2.6 mm long. Capsule oblong to ellipsoid, decidedly longer than wide, 7–11 mm long; seeds dark brown to blackish, trapezoid or elliptical, 1.0–1.2 mm long, the reticulations deep and irregular. Aug–Sep. Dry upland prairies & open woodlands; e 2/3 GP; (s Man., WS, MN, & ND, s to IL, MO, & OK, possibly n-cen TX & extreme e CO). *A. greenei* Lunell—Rydberg; *Gerardia aspera* Dougl. ex Benth.—Fernald.

2. ***Agalinis fasciculata*** (Ell.) Raf. Stout, scabridulous herb with dark green herbage, tending to blacken in drying; stems 4–7(10) dm tall, the main stem weakly 4-angled to subterete, with well-developed axillary fascicles. Leaves curling, arching or somewhat ascending, entire, linear, 1–3.5 cm long, 1–2 mm wide, the leaves of the fascicles smaller. Pedicels spreading-ascending, shorter than the calyx, 2–4(4.5) mm long. Calyx broadly campanulate to hemispherical, 3.5–5(5.5) mm long, the lobes acuminate, 0.4–1.5(2) mm long, the sinuses broadly rounded; corolla (16)20–31 mm long, its lobes spreading, rounded to truncate, 5.5–8 mm long; anthers 2.3–3.5 mm long. Capsule subglobose, 4.5–6(7) mm long; seeds dark brown to blackish, trapezoid or triangular, 0.7–1.1 mm long, the reticulations irregular and elongate. (2n = 28) Aug–Sep. Moderately moist prairies; se KS & adj. OK & MO; (se VA to se KS, s to the Gulf of Mex. from FL to e TX). *Gerardia fasciculata* Ell.—Fernald; *G. fasciculata* f. *albiflora* E.J. Palm.—Steyermark.

3. ***Agalinis heterophylla*** (Nutt.) Small. Plants dark green, often purple-tinged, tending to blacken in drying; stems 3–10 dm tall with stiff, spreading branches above, scabridulous

to glabrous. Leaves linear-lanceolate to lanceolate, attenuate, the principal leaves 1.5–3(4) cm long, 2–6 mm wide, the lower occasionally 3-cleft or laciniate at the base, scabrous above and on the margins. Pedicels ascending, very short, 1–3 mm long. Calyx campanulate, 6–10 mm long, the lobes lanceolate, acute, stiff, 3–6 mm long, usually longer than the tube, the sinuses narrow; corolla 20–32 mm long, its lobes spreading, broadly rounded, 5–9 mm long; anthers 2.8–3.5 mm long. Capsule subglobose, 5–8 mm long; seeds dark brown to blackish, trapezoid or elliptical, 0.8–1.1 mm long, the reticulations deep and irregular. (2n = 28) Sep–Oct. Moist prairies, plains & open woods, often along streams; e KS & e OK; (s MO & e KS, s to LA & e TX, also in w AL). *Gerardia crustata* Greene— Greene, E. L. Leafl. Bot. Observ. Crit. 2: 108. 1910; *G. heterophylla* Nutt.—Fernald.

4. *Agalinis purpurea* (L.) Penn. Glabrous to sparsely scabridulous herb with dark green herbage, tending to blacken in drying; stems (1.5)4–12 dm tall, much branched above, the branches widely spreading, 4-angled, often with some weakly developed axillary fascicles. Leaves spreading to arcuate-ascending or curled, entire, linear, acute, 1–4(5.5) cm long, 0.8–2(3) mm wide, the leaves of the fascicles smaller. Pedicels spreading, as long as to shorter than the calyx, 2–5 mm long. Calyx hemispherical, 4–5.5(6) mm long, the lobes triangular-lanceolate, acuminate, 1.0–2.2 mm long, the sinuses broadly V-shaped or rounded; corolla (17) 20–32 mm long, its lobes spreading, 5–9 mm long; anthers 2.5–3.2(4) mm long. Capsule subglobose, 4–6 mm long; seeds dark brown to blackish, trapezoid or triangular, 0.7–1.2 mm long, the reticulations irregularly elongate. (2n = 28) Aug–Sep. Wet meadows, margins of ponds, & moist roadsides; e NE, e KS, & MO; (ME to MN & NE, s to the Gulf of Mex., from FL to TX). *Gerardia purpurea* L.—Fernald.

Approaching the region in Minnesota is *A. paupercula* (A. Gray) Britt., a species closely related to *A. purpurea*, differing in its smaller stature, 1.5–4(6) dm height, smaller corolla, 11–18 mm long, and smaller anthers, 1.4–2.0 mm long.

5. *Agalinis skinneriana* (Wood) Britt. Plants yellowish-green, not tending to blacken in drying; stems 1.5–5(7) dm tall, striate-angled, simple to abundantly branched, the branches ascending to spreading. Leaves ascending, linear to filiform, acutish to acuminate, 1–3 cm long, 0.5–1.2 mm wide, scabridulous above. Racemes 1- to 8-flowered, the lateral branches often only 1-flowered, the bracts surpassed by the pedicels at anthesis, the pedicels ascending, 5–32 mm long. Calyx hemispherical, 3–4.5(5) mm long, the tube evidently reticulate-veiny, the lobes triangular-lanceolate, 0.6–1.4 mm long, the sinuses broadly rounded; corolla lavender, 7–13(16) mm long, the lobes truncate, 1.5–4(5) mm long; anthers 0.5–1.3 mm long. Capsule subglobose, 3.5–5 mm long; seeds yellow or yellowish-brown, bluntly trapezoid or triangular, 0.7–0.9 mm long, the reticulations irregularly elongate. Aug–Sep. Dry prairies & open oak-hickory woods; se NE, e KS, MO, & e OK; (s Ont., MI & OH to s MN & NE, s to AL, LA, & e TX). *A. gattingeri* (Small) Small—Atlas GP; *Gerardia gattingeri* Small, *G. skinneriana* Wood—Fernald.

6. *Agalinis tenuifolia* (Vahl) Raf. Glabrous or subglabrous herb with dark green herbage, tending to blacken in drying; stems to about 5 dm tall, much branched, with or without axillary fascicles. Leaves spreading to arched-ascending, entire, linear, acuminate, 3–7 cm long, 1–2 mm wide, finely scabridulous to glabrous. Pedicels widely divaricate, filiform, longer than the calyx, 7–20 mm long. Calyx hemispherical, 3.5–5.5 mm long, the lobes broadly triangular, apiculate, sometimes subulate, 1–2 mm long, the sinuses broadly rounded; corolla 10–15 mm long, glabrate or glabrous except for the ciliate margins of the lobes, its lobes 3–5 mm long, the upper lip concave-arched, projecting forward over the anthers and stigma; anthers 1.5–2.2 mm long, densely to sparingly villous. Capsule globose, 4–6 mm long; seeds dark brown to blackish, bluntly trapezoid or triangular, 0.7–0.9 mm long, the reticulations irregularly elongate. (2n = 28) Aug–Oct. Moist woods or prairies, roadside ditches, lakeshores, & stream banks; throughout GP except w ND & MT; (ME & s Que. to s. Man., e ND, & e WY, s to FL & e TX, OK, & e CO). *Gerardia tenuifolia* Vahl — Fernald.

Agalinis tenuifolia is a complex species in which several weak varieties have been recognized. The typical variety approaches us from the east in Missouri and can be distinguished by its shorter calyx lobes, 0.2–1.0 mm long, and its smaller capsule, 3–4 mm long. It has the densely villous anthers and the near absence of axillary fascicles of var. *macrophylla*. The following two varieties are present in the region.

6a. var. *macrophylla* (Benth. ex Hook.) Blake. Leaves and branches spreading; axillary fascicles not or only slightly developed. Anthers densely villous. IA & e NE to MO & e OK; (PA, OH, s MI, WS, & MN, s to NC & e OK). *A. besseyana* (Britt.) Britt. — Rydberg; *Gerardia tenuifolia* var. *macrophylla* Benth. ex Hook. — Fernald.

6b. var. *parviflora* (Nutt.) Penn. Leaves and branches ascending; axillary fascicles conspicuously developed. Anthers sparingly villous. Throughout GP, except w ND & MT; (VT & Que. to s. Man., e ND, SD, & e WY, across the Great Lakes states & s to MO, OK, & e CO). *Gerardia tenuifolia* var. *parviflora* Nutt. — Fernald.

7. ***Agalinis viridis*** (Small) Penn., green gerardia. Plants yellowish-green, not tending to blacken in drying; stems 1.5–6 dm tall, much branched, the branches widely spreading. Leaves erect-ascending, linear, acuminate, 1.2–2.5 cm long, to 1.3 mm wide, finely scabridulous or smooth on the upper surface. Racemes somewhat elongate, (1)2- to 9-flowered, the bracts usually exceeding the pedicels at anthesis; pedicels slender, 4–15 mm long at anthesis and to 25 mm in fruit, spreading. Calyx campanulate, 3.5–4.5 mm long at anthesis and to 5.5(6) mm in fruit, the tube evidently reticulate-veiny, the lobes triangular-lanceolate to lanceolate, 1.0–1.8 mm long, the sinuses broadly rounded; corolla 8–11(13) mm long, the lobes truncate to erose, 2.5–4 mm long; anthers 0.9–1.3 mm long. Capsule obovoid, 5–6.5 mm long; seeds yellow or yellowish-brown, bluntly trapezoid or triangular, 0.7–0.9 mm long, the reticulations irregularly elongate. Sep–Oct. Moist prairies & open woods; MO: Jasper; (s MO, AR, LA, & e TX). *Gerardia viridis* Small — Fernald.

2. ANTIRRHINUM L.

1. ***Antirrhinum majus*** L., snapdragon. Short-lived perennial herb, usually glandular-pubescent above and glabrous below; stems erect, often much branched, 4–8 dm tall. Leaves alternate, shortly or scarcely petiolate, the blades entire, linear to lanceolate, 3–7 cm long, 4–14 mm broad, pinnately veined. Inflorescence a raceme; bracts alternate, ovate to broadly lanceolate, short; pedicels ascending, becoming stout, 3–9 mm long at anthesis and to 15 mm in fruit. Flowers showy; calyx deeply 5-parted, the segments more or less unequal, broadly ovate, rounded or obtuse, 4.5–8 mm long; corolla blue or purple to white or partly yellow, (27)30–42 mm long, the tube ventrally strongly gibbous near the base, distinctly bilabiate, the upper lip external in bud, the throat closed by a prominent palate; stamens 4, didynamous, the anther cells 2, divaricate, 1.0–1.2 mm long, the filaments often dilated toward the apex; stigmas united and capitate. Capsule ovoid, 10–15 mm long, distinctly asymmetrical, the larger cell wholly in front of the pedicel, dehiscing to resemble a face with 2 terminal pores (eyes) on the large cell and a simple terminal elongate pore (mouth) on the small cell, the base of the style persistent (nose); seeds numerous, deeply reticulate, the corky ridges jagged. (2n = 16, 18, 32). Jun–Sep. Along railroads & waste places near gardens; s GP; (scattered throughout N. Amer.). Escaped from cultivation.

The snapdragon is commonly cultivated in flower gardens and only rarely becomes established outside.

3. AUREOLARIA Raf., False Foxglove

1. Aureolaria grandiflora (Benth.) Penn. var. *serrata* (Torr. ex Benth.) Penn., big-flower gerardia. Large, cinereous perennial herb, hemiparasitic; stems erect, to 15 dm tall, widely branched. Leaves opposite, the lower ovate in outline, more or less pinnatifid, progressively reduced in size and dissection upward, the uppermost lanceolate to ovate, serrate to entire. Inflorescence a raceme of opposite or subopposite flowers; bracts serrate-dentate to entire; pedicels 5–12 mm long, stout, upcurved. Flowers large, weakly bilabiate; calyx gamosepalous, campanulate, 10–15(18) mm long, 5-lobed, the lobes lanceolate, often unequal, the longer 5–11 mm long; corolla yellow, campanulate, 38–50 mm long, the tube somewhat oblique, the throat open, the 5 lobes subequal, shorter than the tube, broadly rounded, spreading, ciliolate, the lower 3 lobes external in bud; stamens 4, didynamous, the lower pair longest, the filaments slender, villous, the anther cells 2, parallel, 4.8–5.4 mm long, short-awned at the base, villous; style slender, the stigma solitary, ovoid-capitate. Capsule ovoid, acute to acuminate, 15–20 mm long, loculicidal; seeds few to several, winged or wingless. Aug–Oct. Open woods, often along streams; se KS & e OK; (nw IN to se MN & e IA, s to w LA & e TX). *A. grandiflora* var. *cinerea* Penn.—Gleason; *A. serrata* (Torr.) Rydb.—Rydberg; *Gerardia grandiflora* var. *cinerea* (Penn.) Cory—Fernald.

Reference: Pennell, F. W. 1928. *Agalinis* and allies in North America—I. Proc. Acad. Nat. Sci. Philadelphia 80: (*Aureolaria*) 372–427.

Our plants belong to var. *serrata* (Torr. ex Benth.) Penn., which includes the often recognized var. *cinerea* Penn. The characters that have been used to separate these varieties appear too weak and unreliable to distinguish taxonomic units.

4. BACOPA Aubl., Water Hyssop

1. Bacopa rotundifolia (Michx.) Wettst. Aquatic or subaquatic perennial herb with prostrate or floating stems, pilose to hispid in young parts, rooting at the lower nodes. Leaves opposite, sessile, entire, obovate-cuneate to suborbicular, 12–27(40) mm long, (8)12–23 mm wide, palmately veined, clasping. Inflorescence of 1–4 flowers in each axil of the upper leaves; bractlets absent; pedicels usually shorter than the subtending leaves, 6–17(23) mm long. Flowers weakly bilabiate; calyx deeply 5-parted, the segments 3–4.5 mm long at anthesis and to 6 mm in fruit, the 3 outer ovate to subrotund and the 2 inner lanceolate; corolla white with a yellow throat, campanulate, 4.5–7 mm long, the 5 lobes subequal, the upper 2 external in bud; stamens 4, didynamous, the anthers of 2 parallel cells, 0.6–0.7 mm long, versatile; style forked with 2 dilated stigmatic lobes. Capsule subglobose, 3.5–5.5 mm long, septicidal and loculicidal (4-valved), membranaceous; seeds numerous and small, ellipsoid to cylindrical, ca. 0.5 mm long, with longitudinally arranged reticulations. ($2n = 56$) Jul–Sep. Floating in shallow ponds, lakes, & slow-moving streams, & on muddy banks; GP; (IN to ND, s to MS, LA & TX, sporadically recurring in ID & CA). *Bramia rotundifolia* (Michx.) Britt.—Winter; *Hydranthelium rotundifolium* (Michx.) Penn.—Stevens; *Macuillamia rotundifolia* (Michx.) Raf.—Rydberg.

Reference: Pennell, F. W. 1946. Reconsideration of the *Bacopa-Herpestis* problem of the Scrophulariaceae. Proc. Acad. Nat. Sci. Philadelphia 98: 83–98.

5. BESSEYA Rydb., Kitten Tails

Perennial herb with well-developed basal leaves and bracteate-leaved scapes, arising from a fibrous root system. Leaves mostly basal, the blades cordate-ovate to oblong, crenate to toothed, petiolate, the cauline leaves bractlike, sessile, alternate. Inflorescence a dense cylindrical spike or spikelike raceme. Flowers bilabiate; calyx deeply (1)2- to 4-parted, sometimes variously united; corolla wanting, vestigial or present, ours yellowish with an entire, concave upper lip and a shorter, entire or more or less 3-lobed lower lip, this external in bud; stamens 2 (postero-lateral pair), in flowers lacking a corolla the filaments arise from a hypogynous disk, conspicuously colored, the anther cells 2, equal, parallel or nearly so, versatile; style slender, the stigma solitary, capitate. Capsule notched to entire, somewhat laterally compressed, loculicidal; seeds numerous, flat.

Reference: Pennell, F. W. 1933. A revision of *Synthyris* and *Besseya*. Proc. Acad. Nat. Sci. Philadelphia 85: 77–106.

1 Calyx 4-lobed; corolla, stamens and styles pale yellow or greenish-yellow; basal leaves broadly ovate to suborbicular, 4–9 cm wide, palmately veined; sw MN, eastward .... 1. *B. bullii*
1 Calyx 2(3)-lobed; corolla absent, the violet or purple coloration from the stamens and styles; basal leaves broadly lanceolate to ovate, 1.5–4 cm wide, pinnately veined; high plains of MT, WY, NE & BH of SD, westward .... 2. *B. wyomingensis*

1. *Besseya bullii* (Eat.) Rydb. Plants (1.5)2.5–4 dm tall, hispid-hirsute. Leaves of 2 types, the basal petiolate with blades broadly ovate to suborbicular, rounded apically, truncate to cordate at base, 5–10 cm long, 4–9 cm wide, dentate to crenate, the primary veins palmate, the petiole (1.5)4–7 cm long, the cauline leaves ovate, serrate, sessile. Calyx segments 4, lanceolate, acuminate, 3.5–6 mm long, ciliolate; corolla pale yellow or greenish-yellow, 3.5–6 mm long, the upper lip broad and entire, the lower lip entire or 3-lobed; stamens exserted, the filaments pale yellow, the anthers 0.8–1.5 mm long; styles pale yellow. Capsule often as wide as long, notched apically, 3.5–6 mm long, puberulent; seeds flat, orbicular, ca 1.5 mm across. (2n = 24) May–Jun. Sandy or gravelly barrens & grassy slopes; sw MN; (s MI to s MN, s to OH, IN, IL, & IA). *Wulfenia bullii* (Eat.) Barnh. — Fernald.

2. *Besseya wyomingensis* (A. Nels.) Rydb. Plants 1–2(3) dm tall, white-hairy. Leaves of 2 types, the basal petiolate with blades broadly lanceolate to ovate, obtuse apically, cuneate to truncate or rarely subcordate basally, 2.5–7 cm long, 1.5–4 cm wide, crenate to serrate, pinnately veined, the petioles 2–8 cm long, the cauline leaves broadly lanceolate to ovate, toothed to entire, sessile. Calyx 2(3)-lobed, united anteriorly and placed on anterior side of the capsule and not attached around the back, 3.5–7 mm long; corolla wanting or vestigial; stamens prominently violet or purple, the filaments 3–8(12) mm long; styles violet or purple, 4–6.5 mm long. Capsule about as long as wide, entire to scarcely notched apically, 3–6.5 mm long, villous; seeds flat, orbicular, 0.7–1.6 mm across. May–Jun. Open slopes; BH, nw NE & high plains of WY; (sw Sask. to se B.C., s to nw NE, n CO, & n UT). *Besseya cinerea* (Raf.) Pennell — Harrington; *Synthyris rubra* Winter, not (Hook.) Benth., *S. wyomingensis* (A. Nels.) Heller — Winter.

6. BUCHNERA L., Bluehearts

1. *Buchnera americana* L. Perennial, hemiparasitic herb, ascending-hispid, the hairs from pustulose bases; stems erect, 3–9 dm tall, commonly unbranched. Leaves opposite, sessile, progressively reduced upwards, the lowest obovate, the mid and upper lanceolate, 3–6 cm long, 1–1.5(2) cm wide, toothed, distinctly and palmately 3-veined. Inflorescence a spike of opposite to alternate flowers, the bracts progressively reduced upwards, with small, linear-lanceolate bractlets. Flowers weakly bilabiate; calyx gamosepalous, tubular, 6–9 mm long, shortly 5-lobed, the lobes triangular-acute, 1–2 mm long; corolla violet to purplish, salverform, 16–23 mm long, the tube narrow, curved, villous within, the 5 lobes subequal, oblong, rounded or emarginate, 5–6.5(9) mm long, widely spreading, the lower 3 external in bud; stamens included, 4, slightly didynamous, the filaments short, the anthers narrowly lanceolate, reduced to 1 cell, glabrous; style included, the stigma entire and clavate. Capsule ovoid-oblong, usually gibbous at the base ventrally, (5.5)7-8.5 mm long, loculicidal; seeds numerous, obconic-truncate, 0.6-0.8 mm long, striate ribbed. (2n = 42) Jun-Aug. Pine woods & prairies; se KS, adj. OK & MO; (NY, s Ont., MI, IL, s MO & se KS, s to GA, AL, MS, & e TX).

7. CASTILLEJA Mutis ex L.f., Indian Paintbrush

Perennial or less commonly annual or biennial herb, hemiparasitic; stems decumbent to erect. Leaves alternate, sessile, mostly cauline, entire or pinnately lobed. Inflorescence a short to elongate spike or spikelike raceme; bracts leaflike below and progressively reduced upwards, becoming more incised and brightly colored, usually more conspicuously colored than the flowers. Flowers strongly bilabiate; calyx gamosepalous, tubular, subequally cleft into 4 segments or by various degrees of lateral fusion becoming unequally cleft into 2 lateral primary lobes, each terminating in 2 segments to entire or emarginate, somewhat accrescent; corolla relatively inconspicuously colored greenish to yellowish in ours, tubular, elongate and narrow, bilabiate, the upper lip galeate (hooded), beaklike, erect and straight, its lobes united to the tip and enclosing the anthers, the lower lip somewhat 3-saccate with 3 more or less rudimentary teeth or sometimes subpetaloid lobes, these external in bud; stamens 4, didynamous, the anther cells 2, unequally placed, the outer one attached by its middle and the other suspended by its apex and smaller; stigma solitary, united and capitate, or sometimes shortly 2-lobed. Capsule more or less asymmetrical, ovate, loculicidal; seeds numerous, the loose testa alveolately reticulate.

Castilleja exilis A. Nels. (*C. minor* sensu authors) has been reported for Nebraska (Rydberg, Winter) without confirmation. The earliest report from which all others stem was by Gray (1878. Synoptical Flora of North America 2(1): 295). This is a slender annual with entire, linear-lanceolate leaves and bracts and red bract tips and calyx lobes. Its documented range is from southern British Columbia, south through the mountain valleys of the mountain states to eastern California, southern Nevada, northern Arizona, and northwestern New Mexico.

1 Corolla tube long, falcate and exserted from the calyx, (24)30-45 mm long; corolla purplish to yellow, sometimes cream white, more conspicuously colored than the green or sometimes pink-tipped calyx and bracts .... 8. *C. sessiliflora*
1 Corolla tube mostly shorter, usually concealed by the calyx, 10-30 mm long; corolla pale greenish, except for the colored ventral margins of the galea and often the teeth of the lower lip, the calyx and bracts bearing most of the attractive coloration.
  2 Calyx segments distinct, (0.5)1-16 mm long; plants perennial, often from a woody crown; leaves all cauline.
    3 Leaves mostly divided.
      4 Calyx segments 1-5 mm long; corolla tube 10-16 mm long; high plains of WY.
        5 Spikes yellow, rarely reddish; calyx more deeply cleft in front, 7-17 mm, than in back, 4-12 mm; herbage retrorsely cinereous .... 3. *C. flava*
        5 Spikes red to orange-red, rarely yellowish; calyx less deeply cleft in front, 4-10 mm, than in back, 6-12 mm, herbage hispid and finely puberulent with spreading hairs .... 1. *C. chromosa*
      4 Calyx segments 7-16 mm long; corolla tube 16-28 mm long; s GP .... 7. *C. purpurea*
    3 Leaves mostly entire, sometimes the uppermost divided.
      6 Calyx subequally cleft in front and in back, not as evidently colored as the bracts.
        7 Spikes pale yellow; galea 6-12 mm long; BH .... 9. *C. sulphurea*
        7 Spikes red to reddish-orange; galea (10)12-20 mm long.
          8 Herbage glabrous to thinly villous; calyx segments 2-8 mm long; galea ca 1/2 the corolla length; ND: Pembina .... 6. *C. miniata*
          8 Herbage densely tomentose to lanate, the leaves glabrous or glabrate on the

upper surface; calyx segments 8–16 mm long; galea ca 1/3 the corolla length; se CO & adj. NM .... 4. *C. integra*
6 Calyx decidedly deeper cleft in front, 10–22 mm, than in back, 2–8(12) mm, and bearing most of the attractive coloration .... 5. *C. linariifolia*
2 Calyx segments wholly united laterally, forming entire or (at most) emarginate primary lobes, never cleft laterally for more than 1 mm; plants annual or biennial from weak roots; basal rosette of leaves present as well as cauline .... 2. *C. coccinea*

1. **Castilleja chromosa** A. Nels. Perennial herb from a woody crown, hispid and finely puberulent, becoming villous above; stems 1.5–3.5(4.5) dm tall, usually several in a cluster, ascending to erect, simple or rarely branched above. Leaves narrowly (ob)lanceolate to linear, 2.5–6(7) cm long, with 1 or 2(3) pairs of lateral lobes. Spikes bright red to orange-red, sometimes yellowish, the bracts lanceolate, with 1 to 3 pairs of rounded lateral lobes. Calyx (17)20–27 mm long, the primary lobes more deeply cleft in back, 6–12 mm, than in front, 4–10 mm, the segments obtuse to rounded, (1.5)2–5 mm long; corolla 21–32 mm long, the galea (9)10–18 mm long, 1/2 the corolla length, the lower lip much reduced with incurved teeth, the tube 10–14 mm long. Capsule 9–12(17) mm long. ($2n = 24, 48$) Apr–Jun. Dry sagebrush-grass desert; high plains of WY: Niobrara & the Powder R. Basin; (WY to e OR, s to nw NM, n AZ, NV, & e CA).

2. **Castilleja coccinea** (L.) Spreng. Annual or biennial from a weak root system, often producing only a basal rosette the first year and sending up a flowering stem the second, villous; stems 1.5–5(7) dm tall, usually solitary, ascending to erect, simple or sometimes branched. Leaves of the basal rosette oblanceolate to obovate, entire, often persisting until anthesis, the cauline leaves lanceolate to linear-lanceolate, 3–7(8) cm long, with 1 or 2 pairs of lateral lobes. Spikes bright red, red-orange, or sometimes yellow or white, congested at first, elongating later, the bracts lanceolate with 1(2) pair(s) of lateral lobes, rounded to obtuse. Calyx 17–23(25) mm long, the primary lobes more deeply cleft in front, 6–10 mm, than in back, 5–8 mm, united laterally to a truncate, entire or emarginate tip; corolla 21–27 mm long, the galea 7–10 mm long, the lower lip relatively well developed, 2–3.5 mm long, the tube 15–18 mm long. Capsule 8–10 mm long. ($2n = 24$) Apr–Jun. Moist prairies, hay meadows, & roadside ditches; e GP from w MN to e KS & e OK, eastward; (s ME to se Sask., s to SC, n GA, MS, LA, & e OK, recurring in FL). *C. coccinea* f. *alba* Farw., *C. coccinea* f. *lutescens* Farw.—Fernald.

Populations of plants seem to disappear from one site after a few years and appear in another. Approaching our region in southeastern Oklahoma (Wagoner Co.) is *C. indivisa* Engelm., a closely related annual lacking the cluster of basal leaves and with broadly obovate, entire bracts and a much reduced lower corolla lip less than 2 mm long.

3. **Castilleja flava** S. Wats. Perennial herb from a woody crown, retrorsely cinereous below, villous to tomentose above; stems 1.5–4(5.5) dm tall, usually several to a cluster, erect or ascending, often branched above. Leaves linear, 2.5–5(6.5) cm long, with 1(2) pair(s) of lateral lobes. Spikes yellow, sometimes reddish, the bracts lanceolate, with 1 or 2 pairs of lateral lobes. Calyx 14–23 mm long, the primary lobes more deeply cleft in front, 7–17 mm, than in back, 4–12 mm, the segments lanceolate, acute, 1–4(5) mm long; corolla 18–28 mm long, the galea (5)6–12 mm long, the lower lip relatively reduced, 1.2–3.0 mm long, the tube ca 12–16 mm long. Capsule 8–15 mm long. ($2n = 48$) Apr–Jun. Dry sagebrush-grass desert; WY: nw Powder R. Basin; (sw MT to s-cen ID, s to nw CO, ne UT, & ne NV).

4. **Castilleja integra** A. Gray. Perennial from a woody crown; stems 1–3(4.5) dm tall, usually several arising from short, woody basal stems, erect or ascending, sometimes branched above,

whitish-tomentose or lanate. Leaves linear to linear-lanceolate, 2–6(8) cm long, entire, often involute, glabrous or subglabrous on the upper surface and tomentose or lanate on the lower. Spikes orange-red, villous and finely glandular-pubescent on the bracts, congested at first, the bracts obovate to oblanceolate, broadly rounded, entire or with a pair of small lateral lobes near the apex. Calyx 23–34(38) mm long, the primary lobes more deeply cleft in back, 8–14(17) mm, than in front, 6–12(16) mm, the segments relatively long, obtuse or rounded, sometimes acute, 8–14(16) mm long; corolla relatively long, (26)29–50 mm long, the galea 10–16(18) mm long, the lower lip 1.3–2.8 mm long, with teeth 0.6–1.2(2.5) mm long, the tube 17–30 mm long. Capsule 12–16 mm long. (2n = 24, 48) Jun–Aug. Dry woodland & high prairies; sw GP in w TX, se CO, & NM; (se CO, s through w TX, NM, & AZ to n Mex.).

5. *Castilleja linariifolia* Benth., Wyoming paintbrush. Perennial from a woody crown, glabrous to puberulent below, hispid to villous or sometimes glabrous in the inflorescence; stems 3–7(10) dm tall, few to several to a cluster, erect or ascending, often branched above. Leaves linear to filiform, rarely narrowly lanceolate, 2–6(8) cm long, entire or sometimes the uppermost with a pair of narrow lateral lobes, often arcuate. Spicate raceme red, orange-red, or rarely yellow, becoming elongate with remote flowers; bracts with a pair of lateral lobes. Calyx 18–28(32) mm long, slightly S-curved with the tube curved forward and the segments flared back, the primary lobes decidedly unequally cleft, only 2–8(12) mm in back, but 10–19(22) mm in front, the segments linear to lanceolate or triangular when very short, acute, 0.5–3(5) mm long; corolla relatively long, 24–40(44) mm, bent forward through the anterior calyx cleft, the galea 10–20(24) mm long, more or less 1/2 the corolla length, the lower lip with reduced incurved teeth, the tube 11–23 mm long. Capsule 9–13 mm long. (2n = 24, 48) Jul–Aug. Dry scrub & open woodlands; WY: Powder R. Basin; (WY to e OR, s to nw NM, n AZ, NV, & CA).

This striking paintbrush is the Wyoming state flower.

6. *Castilleja miniata* Dougl. ex Hook. Perennial herb from a woody crown, glabrous to pilosulose, becoming villous above with some glandular puberulence; stems 2.5–7(10) dm tall, few–several from short, branched, woody basal stems, erect or ascending, often branched above. Leaves narrowly lanceolate or lanceolate, 3–7(8) cm long, entire or sometimes the uppermost 3-lobed. Spikes red to red-orange, rarely yellowish, the bracts lanceoalte with 1 or 2 pairs of lateral lobes, the lower pair often departing from near the middle, acute. Calyx 20–30 mm long, the primary lobes more deeply cleft in front, 9–17 mm, than in back, 8–13(15) mm, the segments linear to lanceolate, acute, (2)3–8 mm long; corolla relatively long, 25–44 mm, the galea (11)14–20 mm long, the lower lip much reduced with incurved teeth, the tube 14–26 mm long. Capsule 10–13 mm long. (2n = 24, 48, 72, 96, 120) Jun–Aug. Moist meadows & stream banks; ND: Pembina; (w Alta. to AK panhandle, s to ND, NM, AZ, & CA).

7. *Castilleja purpurea* (Nutt.) G. Don. Perennial from a woody crown, villous-lanate; stems 2–3(4) dm tall, several, clustered, erect or ascending, simple to few-branched. Leaves linear to lanceolate, with 1 or 2(3) pairs of divergent, narrow lateral lobes. Spikes greenish-yellow, bright yellow, or purple to purplish-red, the bracts with 1 or 2 pairs of lateral lobes, lanceolate to broadly lanceolate. Calyx (20)25–34 mm long, the primary lobes (10)13–22 mm long, the segments linear to lanceolate, 7–16 mm long; corolla 25–40 mm long, the galea 9–13 mm long, the lower, somewhat protruding lower lip 1.5–7 mm long, the tube 16–28 mm long. Apr–May. Dry, often calcareous hills & prairies; s KS & OK; (sw MO to KS, s to TX).

Two varieties are found in the area:

7a. var. *citrina* (Penn.) Shinners. Spikes greenish-yellow to bright yellow. Corolla with a prominent, widely flaring lower lip, 3–7 mm long. S-cen KS & w OK; (s-cen KS through w OK to cen TX). *C. citrina* Penn.—Rydberg.

7b. var. *purpurea*. Spikes purple to purplish-red. Corolla with a shorter lower lip, 1.5–3(4) mm long. KS: Montgomery; OK: Harper, Tulsa; (sw MO, se KS, & e OK to cen TX).

8. ***Castilleja sessiliflora*** Pursh, downy paintbrush. Perennial from a woody crown, villous, the stems and inflorescence sometimes somewhat lanate; stems 1–3 dm tall, usually simple, clustered, ascending to erect, often decumbent at the base. Leaves linear, all entire or the upper with a pair of divergent, narrow lobes. Spikes pale yellowish, sometimes pinkish or purplish, bracts green or sometimes pink-tipped, with a pair of lateral lobes. Calyx 25–40 mm long, the primary lobes 12–20 mm long, the segments long and linear, 8–14 mm long; corolla purplish to yellow, sometimes cream-white, 35–55 mm long, conspicuously exserted beyond the bracts, the galea 9–12 mm long, the prominent lower lip 5–6 mm long with flaring lobes, the falcate tube (24)30–45 mm long. ($2n = 24$) Apr–Jun (Sep). Dry plains & hills; GP; (sw Man. & se Sask., s to n IL, nw MO, e OK, w TX, NM, & se AZ; n Mex.). *C. sessiliflora* f. *purpurina* Penn.—Scoggan.

9. ***Castilleja sulphurea*** Rydb. Perennial herb from a woody crown, scabrid-puberulent, the hairs spreading to retrorsely curved, becoming villous and glandular-puberulent above; stems (2)2.5–5.5(7) dm tall, few to several to a cluster, erect or ascending, often branched above. Leaves linear-lanceolate to lanceolate, 2–5.5(8) cm long, entire. Spikes pale yellow, the bracts broadly lanceolate to ovate, entire or with a pair of small lateral lobes near the apex. Calyx 13–23(28) mm long, the primary lobes more deeply cleft in front, (6)8–13 mm, than in back, (5)6–10(11) mm, the segments obtuse to acute, 1–4 mm long; corolla 18–30 mm long, the galea 6–12 mm long, the lower lip 1–2.5 mm long, the teeth short and blunt, ciliate, the tube 10–20 mm long. Capsule 9–11 mm long. ($2n = 24, 48, 96$) Jun–Jul. Moist mountain meadows; BH; (s Alta. & adj. B.C., s through the Rocky Mts. to n NM & UT & e in the BH). *C. rhexifolia* Rydb.—Rydberg; *C. septentrionalis* Lindl.—Fernald.

8. CHAENORRHINUM (DC.) Reichb.

1. ***Chaenorrhinum minus*** (L.) Lange, Wilk. & Lange, dwarf snapdragon. Annual, glandular-pubescent herb with erect, much-branched stems from a short taproot. Leaves alternate or the lowermost opposite, entire, linear to narrowly oblanceolate, 0.5–3 cm long, 1–3 mm wide. Inflorescence of solitary flowers in the axils of the leaves; floriferous to near the base, pedicels 4–13 mm long. Flowers small; calyx subequally and deeply 5-parted, 2–3.5 mm long at anthesis, to 4.5 mm in fruit, narrowly oblanceolate, obtuse to rounded; corolla pale lavender and white with a yellow throat, tubular or campanulate, 4.5–6 mm long, excluding the 1.7–2.8 mm spur from the base of the tube ventrally, strongly bilabiate, the upper lip nearly straight, 2-lobed, external in bud, the lower lip with 3 spreading lobes and a well-developed palate not closing off the throat; stamens 4, the anther cells 2, divaricate, ca 0.3 mm long; stigmas united and capitate. Capsule subglobose, asymmetrical with the anterior cell much larger and projecting in front of the pedicel, each locule dehiscing by a single large irregular terminal pore, membranaceous, glandular-puberulent; seeds numerous, dark brown, longitudinally ribbed, ca 0.5–0.8 mm long. ($2n = 14, 28$) May–Jun (Aug). Along railroads & waste places; e GP from e ND to e KS, also in w NE; (se Can. & s B.C., ME to e ND, s to VA & e KS; s Europe). *Naturalized*.

9. COLLINSIA Nutt., Blue-eyed Mary

Annual herb with erect stems. Leaves opposite (or partly whorled), entire to crenulate or toothed, the upper usually sessile and clasping, the lower petiolate. Inflorescence a raceme, the flowers borne in pairs or in fascicles from the axils of opposite (sometimes whorled) foliaceous bracts. Flowers bilabiate; calyx gamosepalous, campanulate, somewhat irregularly 5-lobed; corolla gibbous or saccate at the base dorsally, strongly bilabiate (papilionaceous), the upper lip (banner) 2-lobed, erect, external in bud, the lower lip 3-lobed, extended horizontally with the median lobe (keel) folded downward forming a pouch that encloses the style and stamens, shorter than the lateral lobes (wings); stamens 4, didynamous, the anther cells 2, divaricate, confluent at the tip; staminode (vestigial posterior stamen) often present, but much reduced; stigma solitary and capitate or slightly 2-lobed. Capsule ovoid, septicidal and loculicidal (4-valved), thin-walled; seeds 1–many in each cell, dark brown, grooved on one side, smooth to reticulate.

Reference: Newsom, V. M. 1929. A revision of the genus *Collinsia* (Scrophulariaceae). Bot. Gaz. 87: 260–301.

1 Corolla 4–7 mm long; principal leaves short petiolate; pedicels filiform, widely spreading to reflexed; nw part of the GP .... 1. *C. parviflora*
1 Corolla 9–17 mm long; principal leaves sessile, clasping; pedicels slender, ascending to spreading; se corner of the GP.
   2 Corolla lobes retuse or emarginate, the upper lip about as long as the lower; seeds 2.5–3.2 mm long, (2)4 to a capsule; principal leaves widest just above the base .... 2. *C. verna*
   2 Corolla lobes deeply notched, the upper lip much shorter than the lower; seeds 1.2–1.6 mm long, 6–12 to a capsule; principal leaves widest near the middle or well above the base .... 3. *C. violacea*

1. ***Collinsia parviflora*** Dougl. ex Lindl., blue lips. Plants 0.5–2(3) dm tall, puberulent, often glandular-puberulent in the inflorescence and glabrous below; stems simple or branched, often decumbent at the base. Leaves, the principal ones oblanceolate, obtuse to rounded, 1.5–3(5) cm long, 3–5(12) mm wide, tapering to a short petiole, the lowermost orbicular and distinctly petiolate. Pedicels 4–10 mm long at anthesis and to 30 mm in fruit, then often becoming reflexed. Calyx (3.5)4.5–5.5 mm long at anthesis and to 7 mm in fruit, the lobes narrowly lanceolate, acuminate, 1.7–3(5) mm long, the sinuses rounded; corolla blue with a white upper lip, 4–5.5(7) mm long, the lobes of the upper lip erect, those of the lower extended forward; anthers 0.3–0.4 mm long, the filaments all glabrous. Capsule oblong-ovoid to ellipsoid, 4–5.5 mm long; seeds ovoid or elliptic, 1.8–2.2 mm long, smooth, 2 per capsule. ($2n = 14$) Apr–Jun. Moderately moist slopes, open woods, & prairies; w ND, w SD & nw NE, westward; (s Yukon & AK panhandle, s to sw Sask., s Alta. & B.C., & in the U.S. from w ND to WA, s to n NM, n AZ, & e CA, with scattered stations in s Ont., n MI, & w VT).

2. ***Collinsia verna*** Nutt., blue-eyed Mary. Plants 1–3.5(4.5) dm tall, glandular-puberulent above, glabrous below; stems often decumbent at the base. Leaves, the principal ones broadly ovate, widest just above the base, obtuse or sometimes acute, 1.5–3.5(5) cm long, 1–2.5 cm wide, sessile, clasping, entire to remotely toothed, the lower leaves broadly ovate to rotund and petiolate. Pedicels 9–15(25) mm long at anthesis and to 40 mm in fruit. Calyx 3.5–7.5 mm long at anthesis and to 10 mm in fruit, the lobes narrowly lanceolate to lanceolate, ciliate, 2–5(7) mm long; corolla blue with a white or pale blue upper lip, 10–14(17) mm long, the lobes emarginate to retuse, the upper lip nearly as long as the lower; anthers 0.5–0.6 mm long, the posterior pair with pubescent filaments, peltate-explanate. Capsule globose, 4–6.5 mm long; seeds ovoid to broadly elliptic, 2.5–3.2 mm long, (2)4 per cap-

sule. (2n = 14) Apr–Jun. Moist, rich woods & shady places; MO & adj. KS; (cen NY to s WI, se IA, MO & e KS, s to w VA, KY, & n AR).

3. **Collinsia violacea** Nutt., collinsia. Plants 1–3(4.5) dm tall, glabrous below, glandular-puberulent above; stems simple or branched. Leaves, the principal ones lanceolate or elliptic, more uniform in width, broadest near the middle or well above the base, 1.5–3(4.5) cm long, 4–8(16) mm wide, sessile and clasping, the lowermost leaves ovate, petiolate. Pedicels 6–14 mm at anthesis and to 25 mm in fruit. Calyx apparently not accrescent, 5–7 mm long, the lobes lanceolate, 2.3–4 mm long; corolla with a white upper lip and a violet, purple, or sometimes also white lower lip, 9–13 mm long, the lobes deeply notched, the upper lip much shorter than the lower; anthers 0.5–0.6 mm long, the posterior pair with pubescent filaments. Capsule globose, 4–5 mm long; seeds 1.2–1.6 mm long, 6–12 per capsule. (2n = 14) Apr–May. Open woodlands & fields; se KS & OK; (s MO to se KS, s to AR, & ne TX).

10. CYMBALARIA Hill

1. **Cymbalaria muralis** Gaertn., Mey. & Scherb., Kenilworth ivy. Glabrous annual herb with twining or trailing stems. Leaves alternate, long petiolate, suborbicular to reniform in outline, 1–2.5(3) mm long, 1–3.5(4) cm wide, palmately 5(9)-lobed, the lobes broadly rounded with a mucronate tip, palmately veined, the petiole 1.5–5.5 cm long. Inflorescence of solitary flowers in the axils of the leaves; pedicels elongate, flexuous, 1–3.5 cm long at anthesis and to 6 cm in fruit. Flowers small; calyx deeply 5-parted, the segments subequal, lanceolate, 1.5–2 mm long at anthesis and to 3 mm in fruit; corolla blue or blue-violet with a yellow palate, 5–9 mm long, excluding the 1.3–2.2 mm long spur at the base ventrally, strongly bilabiate, the upper lip with 2 erect lobes, external in bud, the lower lip with 3 spreading lobes, the throat closed by the prominent palate; stamens 4, didynamous, the anther cells 2, divaricate, 0.4–0.5 mm long; stigmas solitary and capitate. Capsule globose, 2.5–3.5 mm long, symmetrical, first rupturing distally as 2 pores, then later extending nearly to the base; seeds numerous, ellipsoid to globose, 0.7–1.0 mm long, corky-winged. (2n = 14) May–Sep. Shady rocks & woods, usually disturbed areas; infrequent in SD & NE; (sw & se Can., ne U.S. from MA to SD & NE, s to SC & MO; Europe). *Naturalized.*

11. DASISTOMA Raf., Mullein Foxglove

1. **Dasistoma macrophylla** (Nutt.) Raf. Robust hemiparasitic perennial; stems erect, 10–20 dm tall, branched, retrorsely pubescent above. Leaves opposite, the lower broadly ovate in outline, 15–35 cm long, 8–22 cm wide, deeply pinnatifid or bipinnatifid, with broadly lanceolate, serrate or dentate divisions, the upper progressively reduced in size, becoming lanceolate and sometimes eventually entire. Inflorescence an elongate spicate-raceme, the bracts leaflike, opposite; pedicels 1–4 mm long, clavate. Flowers weakly bilabiate; calyx gamosepalous, campanulate, 6–8 mm long at anthesis and to 10 mm in fruit, 5-lobed, the lobes ovate, 2–4(5) mm long; corolla lasting only 1 day, yellow, narrowly campanulate, 14–16 mm long, the tube ampliate, longer than the lobes, densely villous within, the throat open, the lobes subequal and widely spreading, broadly rounded, auriculate at the base, 5–6 mm long, ciliolate, the lower lobe external in bud; stamens 4, didynamous, scarcely exserted, the filaments villous, the anther cells 2, parallel, oblong, somewhat cuspidate at the base, 2.0–2.2 mm long, completely dehiscent, glabrous; style short, thick, the stigmas united and capitate or bilobed. Capsule dark brown, veiny, globose-ovoid, acuminate, short stipitate, 6–11 mm long, loculicidal, each valve terminating in a short, flat, triangular beak;

seeds dark brown, numerous, angular, reticulate, 2–2.5 mm long. Jul–Sep. Rich woodlands, often along streams; se GP from se NE to e OK, eastward & southward; (WV & OH to s WS, IA & NE, s to n SC, GA, AL, MS, & LA & ne TX). *Afzelia macrophylla* (Nutt.) O. Ktze.—Winter; *Seymeria macrophylla* Nutt.—Fernald.

Reference: Pennell, F. W. 1928. *Agalinis* and allies in North America—I. Proc. Acad. Nat. Sci. Philadelphia 80: 427–431.

12. GRATIOLA L., Hedge Hyssop

Annual, biennial, or perennial herb with erect stems. Leaves opposite, sessile, palmately veined, entire or toothed. Inflorescnece of flowers borne singly or in pairs in the axils of the leaves; pedicels, ours with 2 bractlets at the summit, just below the calyx. Flowers bilabiate; calyx deeply 5-parted, ours more or less equal in shape and length; corolla 5-lobed, the tube quadrangular or narrowly campanulate, the lobes of the upper lip united nearly to the summit, external in bud, pubescent at the bases within the tube, the lobes of the lower lip rounded to emarginate; fertile stamens 2 (the postero-lateral pair), included, the anthers 2-celled, the cells more or less parallel on a membranaceous expansion of the connective; staminodes, when present, 2 and reduced, near the base of the corolla tube on the ventral side; style 2-lobed, the stigmas lamellate. Capsule 2-celled, ovoid, septicidal and generally also loculicidal (4-valved), membranaceous or coriaceous; seeds numerous, small, with longitudinally arranged reticulae.

1 Calyx segments subequal (not to be confused with the 2 bractlets subtending the calyx); flowers pedicellate; capsule globose to broadly ovoid; plant glabrous to glandular-puberulent.
  2 Rhizomatous perennials; corolla limb golden-yellow; leaves glandular-punctate; capsule 3–3.5 mm long; e ND & MN .... 1. *G. aurea*
  2 Annuals; corolla limb white to pale yellow; leaves not or only weakly glandular-punctate; corolla limb white; capsule 3.5–6 mm long.
    3 Pedicels 8–20 mm long, slender; plant glandular-puberulent, at least above; seeds 0.5–0.6 mm long; GP except w KS, w OK, & w TX .... 2. *G. neglecta*
    3 Pedicels 1–4(12) mm long, stout in fruit; plant glabrous; seeds 0.7–0.9 mm long; e KS & e OK, eastward .... 3. *G. virginiana*
1 Calyx segments distinctly unequal; flowers subsessile; capsule narrowly ovoid, acuminate; plant villous-hirsute; just outside the GP region in OK .... *G. pilosa* Michx.

1. *Gratiola aurea* Pursh, golden pert. Rhizomatous perennial, glabrous below and glandular-puberulent above; stems 0.5–2.5(3.5) dm tall. Leaves sessile, lanceolate to narrowly lanceolate, 1–2(2.5) cm long, 2–7 mm wide, entire to sparingly toothed, usually clasping at the base, glandular-punctate. Pedicels (5)10–20 mm long, the bractlets similar to but shorter than the calyx segments 2–3.5 mm long. Calyx segments narrow lanceolate, 3.5–4.5 mm long at anthesis and to 5.5 mm in fruit; corolla golden yellow, 10–15 mm long, unlined; stamens included. Capsule subglobose, 3–3.5 mm long; seeds brown, ca 0.5 mm long, relatively coarsely reticulate. (2n = 28) May–Sep. Wet places around ponds & lakes & along streams; ND: Grand Forks; (s Newf. & N.S. to WI, s to MD, PA, & n IL with scattered occurences in GA, FL, & e ND). *G. lutea* Raf.—Stevens.

2. *Gratiola neglecta* Torr., hedge hyssop. Glandular-puberulent annual; stems 0.5–2(4) dm tall, erect or diffuse, fistulose below. Leaves oblanceolate to obovate, 1–3(5) cm long, serrate apically or sometimes entire, slightly clasping at base. Pedicels filiform, 8–20 mm long, the bractlets similar to the calyx segments, equal or longer. Calyx segments lanceolate, acute to obtuse, rarely rounded, 3.2–5 mm long at anthesis and to 7 mm in fruit, subequal; corolla (7)8–12 mm long, the tube yellow, the limb white to pale yellow or sometimes pale lavender, the lobes of the upper lip nearly completely united, the lobes of the lower lip broad and emarginate; stamens included, the anthers ca 0.7 mm long, the connective usually wider and longer than the pair of anther cells. Capsule ovoid-globose, acuminate, 3.5–5.5 mm long; seeds yellowish, 0.5–0.6 mm long, finely reticulate. Jun–Aug. Moist places, often in mud at edge of ponds & streams & in marshes & swamps; GP except w KS, w OK, & w TX; (s Can., ME to WA, s to n GA, e TX, n NM, n AZ, NV, & ne CA). *G. lutea* of Gates, not Raf.; *G. virginiana* of Winter, not L.

3. Gratiola virginiana L., hedge hyssop. Glabrous annuals; stems 1–4(5) dm tall, erect to decumbent, simple or branched, fistulose below. Leaves (ob)lanceolate, (1.5)2–5 cm long, (5)8–15(20) mm wide, palmately 3- to 5-veined, entire to serrate, tapering to a petiolar base or cordate-clasping. Pedicels 1–4(12) mm long, becoming stout, the bractlets similar to the calyx segments, 1.5–5.5(9) mm long. Calyx segments lanceolate, acuminate, 3.5–4.5 mm at anthesis and to 7.5 mm in fruit; corolla white with purple lines, 8–13 mm long; stamens included, the anthers ca 0.5 mm long, the connective as wide as the pair of anther cells but not as long. Capsule globose, 3.5–6 mm long; seeds yellowish, narrow, 0.7–0.9 mm long, finely reticulate. (2n = 16) Apr–Jun. Rooted in shallow water of streams & marshes; e KS & e OK; (s NJ & MD to IL, se IA, MO, & e KS, s to FL & e TX).

13. KICKXIA Dum.

1. Kickxia elatine (L.) Dum., canker root. Villous annual with prostrate or sprawling, much-branched stems. Leaves simple, alternate or the lowermost opposite or subopposite, short-petioled, the blade broadly ovate, hastate or dentate at the base, pinnately veined. Inflorescence of solitary flowers on long, slender pedicels in the axils of the leaves, the pedicels usually glabrous for most of their length. Flowers small; calyx deeply 5-parted, the segments equal, lanceolate, 2.5–3.5 mm long at anthesis and to 4 mm in fruit; corolla cream or yellow, the upper lip purple, 4–6 mm long, excluding the 3.5–6 mm spur at the base of the tube ventrally, strongly bilabiate, the lips much longer than the tube, the upper more projecting, external in bud, the throat closed by the well-developed palate; stamens 4, the anther cells 2, divergent, ciliate with stiff hairs; stigma solitary and capitate. Capsule subglobose, 3–4 mm long, symmetrical, each cell circumscissile on the side near its summit, nearly 1/2 the valve becoming separated, finely pubescent; seeds numerous, rounded or ovoid-truncate, 0.8–1.2 mm long, with irregular, winglike convolutions. (2n = 18, 36) May–Sep. Moist disturbed stream banks, roadsides, & waste places; se KS & adj. MO; (MA to IL, MO, & se KS, s to FL, LA, & OK, also scattered localities in sw B.C., w OR, & nw CA; Europe). *Introduced.*

Kickxia spuria (L.) Dum., another native of the Old World, may be found as an introduction in the region. It differs from *K. elatine* in having entire leaves that are rounded to cordate at the base and pedicels that are pubescent their entire length.

14. LEUCOSPORA Nutt., Leucospora

1. Leucospora multifida (Michx.) Nutt. Small annual herb with erect stems, 1–1.5(2) dm tall, hispid throughout. Leaves opposite, bipinnatifid to pinnatifid, the divisions oblanceolate to linear-cuneate, 1–2.5 cm long. Inflorescence of flowers borne singly or in pairs in the axils of the leaves; pedicels 3–7 mm long, without bractlets. Flowers bilabiate, relatively small; calyx deeply 5-parted, the segments lanceolate to linear, 2.8–4 mm long at anthesis and to 5.5 mm in fruit; corolla pink or lavender, the throat and tube greenish-yellow inside with a purplish ring at the mouth, tubular, 3–5 mm long, 5-lobed, the lobes slightly spreading, much shorter than the tube, the upper shorter than the lower and overlapping the lower in bud; stamens 4, didynamous, included, the anther cells more or less parallel; style 2-lobed, the stigmas cuneiform. Capsule 2-celled, ovoid, acute, 3.5–4.5 mm long, septicidal, membranaceous; seeds numerous, pale greenish-yellow, 0.2–0.4 mm long, faintly striate-reticulate. Jun–Sep. Wet places along streams & shores; s NE, KS, & OK; (s Ont. & OH to s NE, s to nw GA, n AL, MS, LA, & e TX). *Conobea multifida* (Michx.) Benth.— Fernald.

15. LIMOSELLA L., Mudwort

1. *Limosella aquatica* L. Small, glabrous, tufted, stoloniferous perennial herb with a very short stem. Leaves in a basal tuft, entire, slender, long petiolate, the blade narrowly spatulate to broadly oblong-elliptic, rounded, 1–3 cm long, 3–12 mm wide, palmately 3(5)-veined, the petiole 3–10(20) cm long, sometimes as long as the water is deep, with floating blades, broadening into a hyaline, stipulelike base. Inflorescence of slender pedicels arising from the axils of the leaves in the basal tuft, the pedicels without bractlets, usually shorter than the leaves. Flowers nearly regular; calyx gamosepalous, equally 5-lobed, campanulate, 2–2.5 mm long, the tube as long as or longer than the lobes, the lobes pale green with purplish spots beneath the sinuses, broadly triangular, apiculate, 0.5–0.8 mm long, corolla white or pinkish, rotate-campanulate, 5-lobed, slightly longer than the calyx, the lobes shorter than the tube, oblong, acute, ca 1 mm long, scarcely pilose within, the upper 2 lobes external in bud; stamens 4, didynamous, the filaments flattened, the anthers dark purple, 1-celled (by confluence of the ancestral pair), circular, ca 0.3 mm across; stigma solitary and capitate. Capsule distally 1-celled (the septum not extending to the tip), obovoid, ca 3.2 mm long, septicidal, membranaceous; seeds dark brown, numerous, 0.6–0.7 mm long, with longitudinally arranged reticulations. (2n = 36, 40) Jun–Sep. Muddy shores of ponds or submerged in water up to 15 cm deep; n 2/3 GP, from se MN, ND, & MT to w-cen MO, NE, & CO; (circumboreal, nearly throughout Can., MN to WA, s to NM, AZ, & CA).

16. LINARIA Mill., Toadflax

Glabrous annual or perennial; stems erect, some coarse, usually simple. Leaves alternate or the lowermost sometimes opposite, sessile, pinnately veined, ours entire. Inflorescence a raceme or spike with alternate, reduced bracts. Flowers often showy; calyx deeply 5-parted, the segments subequal; corolla yellow to blue-violet or white, the tube ventrally spurred at the base, strongly bilabiate, the upper lip 2-lobed, external in bud, the lower lip 3-lobed, these usually reflexed, the throat nearly closed by the prominent palate; stamens 4, didynamous, the anther cells 2, divergent, glabrous; stigma solitary and capitate. Capsule ovoid to cylindric or subglobose, symmetrical, thin-walled, dehiscent by distal transverse loculicidal ruptures; seeds mostly numerous, angled, tuberculate, or with thin wings.

Abnormal flowers occasionally occur in species of *Linaria* in which the corolla is actinomorphic with 3–5 spurs or no spurs at all. This teratological feature in flower development is called peloria, and in *L. vulgaris* such monstrosities have been referred to f. *peloria* (L.) Rouleau.

1 Corolla yellow; plants stout perennials, 3–10 dm tall; capsule 5–9 mm long.
  2 Calyx 2.5–3.5 mm long; leaves linear, tapering to a subpetiolar base, 2–6 mm wide .... 3. *L. vulgaris*
  2 Calyx 5–9 mm long; leaves lanceolate to ovate, 10–35 mm wide, broadest near the base .... 2. *L. dalmatica*
1 Corolla bluish; plants slender annuals, 1–4(5) dm tall; capsule 2.8–3.5 mm long .... 1. *L. canadensis*

1. **Linaria canadensis** (L.) Dum. Slender annual (winter annual), glabrous, sometimes glandular-pubescent in the inflorescence; stems 1–4(5) dm tall, erect or ascending, 1–several, simple or branched above, but usually with a rosette of short, prostrate stems, these sometimes developing into upright flowering stems when moist conditions persist long enough. Leaves of 2 types, those of the short basal stems opposite or ternate, or sometimes approximate, oblanceolate to spatulate, 0.5–1 cm long, 2–3.5 mm wide, those of the erect flowering stems alternate, linear, 1–2.5(3) cm long, 1–2 mm wide. Pedicels 1–7 mm long. Calyx segments subequal, broadly lanceolate, 2.8–3.4 mm long; corolla pale blue-violet or bluish-purple with whitish palate, 7–13 mm long, excluding the curved, 2–11 mm spur, the upper lip ca 3.5–5 mm long, with erect lobes, the lower lip 7–11 mm long with round, spreading lobes. Capsule subglobose, 2.8–3.5 mm long; seeds truncate, angular, 0.3–0.5 mm long, tuberculate to smooth. ($2n = 12, 24$) Apr–Jun. Relatively dry places, often in sandy soil; GP; (se Can., e 2/3 of U.S., n Mex. & the Pacific Coast from sw B.C. to n Baja Calif.).

Two varieties can be distinguished in GP.

1a. var. *canadensis*, old-field toadflax. Corolla 8–10 mm long, excluding the 2–6 mm spur. Seeds less evidently tuberculate to sometimes smooth. ($2n = 12$). Sandy soil; infrequent in GP; (se Can. & ME to SD, s to FL & e TX, also Pacific Coast from Vancouver Island to CA).

1b. var. *texana* (Scheele) Penn., Texas toadflax. Corolla (7)10–13 mm long, excluding the 6–11 mm spur. Seeds densely and evidently tuberculate. Sandy soil; GP; (se & midwest states & Pacific Coast from sw B.C. to n Baja Calif.; n Mex.). *L. texana* Scheele — Rydberg.

2. **Linaria dalmatica** (L.) Mill. Glabrous, glaucous perennial herb, spreading by horizontal rootstalks; stems 4–7(10) dm tall, erect, branched above, the branches with smaller leaves. Leaves ovate to lanceolate, sharply acute, 2–4(5) cm long, 10–16(35) mm wide, sessile, clasping, palmately veined, rigid, crowded and overlapping. Raceme elongate, the pedicels 2–4 mm long at anthesis and to 6 mm in fruit. Calyx segments subequal, broadly lanceolate to ovate, sharply acute, 5–7.5(9) mm long, rigid; corolla bright yellow with a densely white- to orange-bearded palate, 14–24 mm long, excluding the 9–17 mm spur, glandular-pubescent on the sides, the upper lip (7)10–15 mm long, the lower lip 5–11 mm long with a well-developed palate closing off the throat; anthers ca 1.2–1.3 mm across after dehiscence. Capsule subglobose, 6–7(8) mm long; seeds tetragonal to discoid-compressed, 1.2–2 mm long, rugulose, the angles winged. ($2n = 12$) Jul–Aug. Roadsides & waste places; GP; (widely scattered throughout much of s Can. & U.S.; Europe). *Naturalized.*

Linaria genistifolia L. has been reported from MN: Pope (Atlas GP). Plants of this species would key more readily to *L. dalmatica* from which they differ only in having a corolla 9–11(13) mm long (excluding the 6–10 mm spur) and lanceolate leaves.

3. **Linaria vulgaris** Hill, butter-and-eggs. Perennial herb, glabrous to sprasely pilose or sometimes glandular-puberulent in the inflorescence; stems 3–6(8) dm tall, erect or ascending, simple to branched, 1 to several, clustered on a taproot; leaves linear, 2.5–4(5) cm long, 2–6 mm wide, crowded, tapering to a petiolar base. Racemes congested, the pedicels 1–4 mm long. Calyx segments subequal, lanceolate to broadly lanceolate, 2.5–3.5 mm long; corolla bright yellow with an orange-hairy palate, 10–14(18) mm long, excluding the nearly straight 8–14 mm spur, the upper lip 8–12 mm long, the lower lip 6–9 mm long with a well-developed palate closing off the throat; anthers ca 1 mm across after anthesis. Capsule subglobose, 5–9 mm long; seeds discoid with broad wings, ca 2 mm long, the central portion tuberculate. ($2n = 12$) Jun–Aug. Disturbed habitats in open places; GP; (widespread in temp. N. Amer., Europe). *Naturalized. L. vulgaris* f. *peloria* (L.) Rouleau — Fernald.

17. LINDERNIA All., False Pimpernel

1. **Lindernia dubia** (L.) Penn. Small, glabrous annual herb; stems 0.9–2(3) dm tall, erect or ascending, simple or branched. Leaves opposite, (ob)lanceolate to (ob)ovate, 1–2(3) cm long, 0.5–1 cm wide, 3- to 5-nerved, crenate, serrate or sometimes entire. Inflorescence of flowers borne singly in the axils of the leaves; bractlets absent; pedicels 0.5–2.5 cm long. Flowers sometimes small and cleistogamous; calyx regular, deeply 5-parted, the segments linear, 2.8–4.5 mm long at anthesis and to 5.5 mm in fruit; corolla blue to lavender, tubular-campanulate, 4–10 mm long, bilabiate, the throat with 2 yellow-hairy ridges ventrally, the upper lip erect and smaller than the lower, shallowly 2-lobed and overlapping the lower in bud, the lower lip projecting to somewhat deflexed, the 3 lobes well developed; fertile stamens 2 (postero-lateral pair), short, included, the anther cells 2, more or less divergent, ca 0.6 mm long; staminodes 2 (sterile anterior stamens), 2-pronged, projecting as forked appendages from the 2 ridges of the corolla palate, in each the erect fork tapered to a point and the forward projecting fork terminating in a knob; style 2-lobed, the stigmas lamellate. Capsule 2-celled, narrowly ovoid to ellipsoid, 3.5–6 mm long, septicidal, the septum persistent as a thin plate, membranaceous; seeds pale yellow, numerous, small, 0.2–0.4 mm long, faintly reticulate to smooth. (2n = 18, 32) Jul–Aug. Moist, often muddy banks of streams & ponds; e GP, from se ND through e parts of SD, NE, & KS to e OK; (se Can., ME to ND, s to the Gulf States & ne Mex., recurring in cen CO, ID, s B.C., WA, OR, & CA; S. Amer.). *Ilysanthes inaequalis* (Walt.) Penn. and *I. dubia* (L.) Barnh.—Rydberg; *Lindernia anagallidea* (Michx.) Penn.—Atlas GP; *L. dubia* var. *riparia* (Raf.) Fern.—Fernald; *L. dubia* subsp. *major* (Pursh) Penn.—Gates.

Lindernia dubia is here regarded in a broad sense to include *L. anagallidea* (Michx.) Penn. As characterized in most floras, the two are sympatric throughout most of their ranges. The distinguishing characters of color, size, and shape of the seeds, length and shape of the leaves, length of the petiole, and the presence or absence of cleistogamous flowers used by Pennell (Acad. Nat. Sci. Philadelphia Monogr. 1: 140. 1935) are inconsistent in the Great Plains material. Cooperrider & McCready (Castanea 40: 191–197. 1975) found this to be true in Ohio material as well, and they reduced *L. anagallidea* to a synonym of *L. dubia*. Reconsidering the problem, Cooperrider (Castanea 41: 224. 1976) revived it as *L. dubia* var. *anagallidea* (Michx.) Cooperrider on the strength of his observations that material from the Atlantic States appears to be distinct. I have not found any consistent way of distinguishing the two in the GP material seen in this study. Such a problem should lend itself well to a dissertation study of reproductive biology using specimens from throughout the range. The false pimpernel is a small annual with a relatively short reproductive cycle.

18. MECARDONIA R. & P.

1. **Mecardonia acuminata** (Walt.) Small. Glabrous perennial herb with much-branched, erect or ascending, 4-angled stems, 2–5(6.5) dm tall. Leaves opposite, sessile, oblanceolate, cuneate, 2–4(5) cm long, pinnately veined, serrate apically. Inflorescence of solitary flowers borne in the axils of the upper leaves; bractlets 2, at base of the pedicel, narrow-lanceolate; pedicels 1–3.5 cm long. Flowers bilabiate, relatively small; calyx deeply 5-parted, the segments 6–8(10) mm long, subequal in length but distinctly unequal in width, the 3 outer broadly lanceolate and the 2 inner linear; corolla white, often with purplish lines and a yellow throat, tubular-campanulate, 5-lobed, the lobes shorter than the tube, the upper lip united beyond the middle, external in bud, the lower lip 3-lobed, the tube villous within posteriorly; stamens 4, didynamous, the anthers of 2 more or less parallel cells; style forked with 2 dilated stigmatic lobes. Capsule narrowly ovoid, 5–8 mm long, septicidal, membranaceous; seeds numerous, small, 0.5–0.6 mm long, reticulate. (2n = 42) Aug–Oct. Moist woods & banks of streams & ponds; KS: Cherokee & MO: Jasper; (MD & DE to se KS,

s to FL, AL, MS, LA, & e TX). *Bacopa acuminata* (Walt.) Robins.—Atlas GP; *Pagesia acuminata* (Walt.) Penn.—Pennell, Acad. Nat. Sci. Philadelphia Monogr. 1: 65. 1935.

19. MIMULUS L., Monkey Flower

Annual or perennial herb, often glandular-pubescent to slightly viscid or pilose, less often glabrous; stems erect or ascending, sometimes prostrate, sometimes rhizomatous or stoloniferous. Leaves opposite, palmately or pinnately veined, entire or toothed, sometimes laciniate, petiolate or sessile. Inflorescence of flowers borne in pairs in the axils of the leaves or in a bracteate raceme; bractlets not present; bracts leaflike below, progressively reduced upwards; pedicels well developed in ours. Flowers slightly to strongly bilabiate; calyx gamosepalous, tubular or campanulate, often inflated, the tube strongly 5-angled (pleated), decidedly longer than the lobes, equally or unequally 5-lobed; corolla yellow through red to purple or bluish, campanulate, the upper lip 2-lobed, erect to reflexed, external in bud, the lower lip 3-lobed, spreading or deflexed, the throat with 2 elevated ridges forming a palate (ours), this sometimes partially or completely closing the orifice, usually bearded; stamens 4, didynamous, the anthers 2-celled, the cells divergent; style 2-lobed, the stigmas lamellate. Capsule 2-celled, usually cylindrical, included in the calyx, loculicidal, sometimes also splitting down the septum dividing the placentae, membranaceous to coriaceous, glabrous; seeds mostly yellowish, numerous, small, oblong to oval or fusiform, reticulate to almost smooth, the reticulae arranged in longitudinal rows.

Reference: Grant, A. L. 1924. A monograph of the genus *Mimulus*. Ann. Missouri Bot. Gard. 11: 99–388.

1 Corolla blue, purple, violet or sometimes lavender; stems 4-angled (sect. *Mimulus*).
  2 Leaves sessile; pedicels 20–45 mm long, longer than the calyx .... 5. *M. ringens*
  2 Leaves petioled; pedicels 2–8(14) mm long, shorter than the calyx .... 1. *M. alatus*
1 Corolla yellow, often with red or brown markings; stems terete.
  3 Fruiting calyx more or less inflated, the posterior lobe decidedly longer than the others (sect. *Simiolus* Greene).
    4 Lateral calyx teeth blunt and relatively short, sometimes nearly obsolete, not folded inward in fruit; corolla throat more or less open; upper leaves usually wider than long; stems usually prostrate .... 3. *M. glabratus* var. *fremontii*
    4 Lateral calyx teeth more or less acute and tending to fold inward in fruit; corolla throat constricted by the well-developed palate; upper leaves usually longer than wide; stems erect or ascending .... 4. *M. guttatus*
  3 Fruiting calyx not inflated, the lobes more or less equal in length; entering the region in the BH (sect. *Paradanthus* Grant) .... 2. *M. floribundus*

1. Mimulus alatus Ait., sharpwing monkey-flower. Stoloniferous, glabrous perennial; stems 3–7 dm tall, ascending to erect, simple or branched above, 4-angled, the angles more or less winged. Leaves petiolate, the blade broadly lanceolate to ovate, acute, 5–8(12) cm long, 2.5–4 cm wide, the margins sometimes scabrid-ciliolate, serrate, tapering to a narrowly winged petiole 1–2 cm long. Pedicels 2–8 mm long at anthesis and to 14 mm in fruit, shorter than the calyx. Calyx 11–18 mm long, the tube not inflated, puberulent within, the lobes more or less equal, broad, mucronate or becoming awned by extension of the nerve, 0.8–2.6 mm long, the margins sometimes scabrid-ciliolate; corolla lavender, violet or purplish-blue, 20–28 mm long, strongly bilabiate, the upper lip erect to reflexed, the lower lip spreading, glandular-pubescent within and without, the palate nearly closing the throat; stamens included, the anthers 0.8–1.3 mm long; style included. Capsule ovoid, 8–11 mm long, distally loculicidal; seeds light brown, oblong, ca 0.3 mm long, finely reticulate. ($2n = 22$) Jul–Sep. Moist banks of streams & lakes, often in wooded areas; n & e NE, e KS, & e OK; (CT, NY, & s Ont. to NE, s to n FL & e TX).

2. **Mimulus floribundus** Dougl. ex Lindl. Delicate annual, glandular-pubescent, sometimes viscid and slimy; stems 0.3–2.2(4) dm tall, erect to decumbent, simple or branched. Leaves petiolate, the blade ovate to lanceolate, rounded to cordate at base, (0.3)1–2(3) cm long, (1)5–13(20) mm wide, thin, sparingly toothed, pinnately to subpalmately veined, the petiole 1–12 mm long. Pedicels 5–14(18) mm long. Calyx cylindric, 4–7 mm long at anthesis and to 9 mm in fruit, the lobes subequal, rounded-triangular, lanceolate, acute to rounded, 0.8–1.6(2) mm long, ciliate; corolla yellow, often with reddish spots, 5–10(14) mm long, weakly bilabiate, the lobes of the upper lip slightly longer, the palate ridges puberulent; stamens included, the anthers ca 0.3 mm long; stigma lips rounded. Capsule obovoid to elliptic, 3.5–5 mm long; seeds yellowish-brown, broadly ovoid, 0.3–0.4 mm long, minutely tuberculate to nearly smooth. (2n = 32, 36) Late May–Jul. Moist habitats, often along streams in woodlands & forests; BH; (sw Alta. & s B.C., s through the mt. states to NM, AZ, & CA, e to BH; n Mex.).

3. **Mimulus glabratus** H.B.K. var. *fremontii* (Benth.) A. L. Grant, roundleaf monkey-flower. Low, creeping to decumbent, aquatic perennial, rooting at the lower nodes, glabrous to sparsely viscid-hispid or glandular-pubescent. Leaves rotund to reniform, rounded to cordate at the base, 1–3 cm long, 1.5–3 cm wide, the upper usually broader than long, palmately veined, dentate, sessile, the lower ones ovate, cuneate, usually broadly and shortly petiolate. Pedicels 10–40 mm long. Calyx campanulate, 5–11 mm long at anthesis and to 16 mm in fruit, becoming inflated, the lobes unequal, the posterior one largest, blunt, rounded, broadly obtuse, sometimes mucronate, 2 × as long as the lateral and anterior lobes, these tending to become obsolete; corolla yellow, often red-brown spotted, 8–12 mm long, strongly bilabiate, the palate heavily bearded, not closing the throat; anthers ca 0.6 mm long. Capsule broadly ovoid, 5–9 mm long; seeds brownish, elliptic-ovoid. (2n = 28, 30, 60, 62, 90, 92) May–Aug. Shallow water of streams or ponds or on muddy or moist banks; GP; (s Can. from Que. to Sask., s through the cen states to TX, NM, & AZ; Mex. & S. Amer.). *M. geyeri* Torr.—Rydberg.

Our plants belong to var. *fremontii*, which is gradually replaced by var. *glabratus* to the south of the GP region. Our variety has somewhat broader leaves and smaller flowers, and the stems are more repent.

4. **Mimulus guttatus** DC., common yellow monkey-flower. Annual or perennial, rooting at the lower nodes, sometimes rhizomatous or stoloniferous, glabrous to puberulent; stems 0.5–5.5(9) dm tall, erect or ascending, sometimes decumbent at the base; simple or branched, often fistulose. Leaves broadly ovate to rotund, 1.5–8.5 cm long and nearly as wide, palmately veined, coarsely and irregularly toothed, sometimes lobed at the base, the upper leaves usually sessile, the lower petiolate, these to 5.5 cm long. Pedicels 10–35(55) mm long. Calyx campanulate, 6–16 mm long at anthesis and to 20 mm in fruit, becoming inflated, the lobes broadly triangular, the posterior one much longer and lanceolate, obtuse, 2–3 × as long as the lateral and anterior lobes, these folding over the orifice of the calyx in fruit; corolla yellow, often red-brown spotted, 9–23(30) mm long, strongly bilabiate, the lobes of the upper lip reflexed, the lower lip longer with spreading lobes, the palate densely bearded, nearly closing the throat; anthers 0.5–1.0 mm long. Capsule oblong-obovoid, rounded, narrowed to a stipitate base, 9–12 mm long; seeds brown, 0.4–0.5 mm long. (2n = 16, 28, 30, 32, 48, 56) Jun–Aug. Wet marshy places, often emergent, along streams & shores; scattered localities in ND & SD; (s Yukon & AK, s to Alta. & B.C. & through the mt. states to NM, AZ, & CA; n Mex. & escaped from cultivation in s Sask., VT, CT, NY, ND, & SD).

5. **Mimulus ringens** L., Alleghany monkey-flower. Rhizomatous or stoloniferous, glabrous perennial; stems 2–13 dm tall, ascending to erect, simple or branched above, 4-angled,

the angles sometimes narrowly winged. Leaves sessile, (ob)lanceolate to narrowly oblong, acute, 2.5–8(10) cm long, 6–20(35) mm wide, often narrowed at the base, clasping or auriculate, pinnately nerved. Pedicels 20–35(45) mm long, divergent. Calyx tubular, 10–16(20) mm long, the tube not inflated, puberulent inside, the lobes more or less equal, triangular, acute, (2)2.5–6 mm long, ciliolate, the sinuses broadly rounded to angled; corolla blue or blue-purple to lavender, 20–27(30) mm long, strongly bilabiate, the upper lip erect to slightly reflexed, the lower spreading, surpassing the upper, the palate nearly closing the throat; stamens included to slightly exserted. Capsule ovoid, 10–12 mm long; seeds light brown, oblong, 0.3–0.4 mm long, papillate. (2n = 24) Jul–Sep. Stream sides & edges of ponds, often emergent; e 1/2 of GP; (s Can. from N.S. to s Sask., ME to ND, s to SC & TX, with widely separated stations in CO & ID). *M. ringens* var. *minthodes* (Greene) Grant—Fernald.

20. ORTHOCARPUS Nutt., Owl Clover

1. *Orthocarpus luteus* Nutt. Annual herb, mostly hemiparasitic, glandular-pubescent with an admixture of longer, nonglandular hairs; stems slender, erect, 0.8–3(5) dm tall, usually simple. Leaves alternate, sessile, linear to linear-lanceolate, 1.5–3.5 cm long, entire or the uppermost 3-lobed. Inflorescence a narrow spike or spicate raceme, the bracts green, usually leaflike below and progressively reduced upwards, 3-lobed. Flowers strongly bilabiate; calyx gamosepalous, tubular-campanulate, 6–8 mm long, 4-lobed, the lateral clefts 1–2.5 mm deep and the median clefts 2–3.5 mm deep in front and 3–5 mm in back; corolla yellow, elongate and narrow, 9–12(14) mm long, puberulent, the upper lip galeate (hooded), erect and straight, beaklike, 2.5–4 mm long, its lobes united to the tip and enclosing the anthers, the lower lip equaling the galea in length with a more or less saccate-inflated pouch, minutely 3-toothed, external in bud; stamens 4, didynamous, the anther cells 2, unequally placed, the outer one attached by its middle, the other suspended by its apex and smaller; stigmas united and capitate. Capsule more or less symmetrical, elliptic-ovoid, loculicidal; seeds several, 1.2–1.5 mm long, the testa reticulate or alveolate, often loose. (2n = 28) Jul–Aug. Upland prairies, dry meadows, & open woodlands; n & w GP; (w-cen Ont. & Man. to B.C., s to w MN, n & w SD, e NE, CO, NM, n AZ, n NV, & ne CA).

21. PEDICULARIS L., Lousewort

Perennial hemiparasitic herb from fibrous or sometimes tuberous roots; stems usually erect. Leaves alternate or sometimes opposite, from equally basal and cauline to nearly all basal or all cauline, serrate-toothed to more often pinnatifid or bipinnatifid, ours never entire. Inflorescence usually a spicate raceme, capitate to elongate, the bracts foliaceous below and progressively reduced upwards. Flowers strongly bilabiate and irregular; calyx gamosepalous, campanulate to tubular, more or less oblique, irregularly cleft into 5, 4 (by reduction of the posterior), or only 2 (by fusion of the lateral) lobes, more deeply cleft in front than in back, accrescent; corolla yellow or white to purple or red, the upper lip galeate (hooded), rounded apically and enclosing the anthers, extending into a beak in some, the lower lip usually shorter, 3-lobed, external in bud; stamens 4, didynamous, the anthers distinct, the anther cells 2, parallel, similar in size and position, glabrous; stigma solitary and capitate. Capsule asymmetrical, somewhat laterally flattened, loculicidal, opening chiefly or wholly on the dorsal side; seeds usually numerous.

1   Galea with 2 lateral teeth just below the apex; calyx lobes entire, not crenate; principal leaves deeply lobed, the sinuses 1/2 to all the way to the midrib; capsule long, 10–17 mm, evidently exserted from the calyx.

2 Calyx 7–9 mm long, with 2 short posterior lobes; corolla 18–25 mm long; plants 1–3(4) dm tall; e GP & high plains, nw NE & cen CO, & adj NM .... 1. *P. canadensis*
 2 Calyx 10–16 mm long, with 5 lobes 3–5 mm long; corolla 25–36 mm long; plants 5–12 dm tall; BH .... 3. *P. procera*
1 Galea truncate at the apex, with no lateral teeth; calyx lobes 2 lateral, crenate; principal leaves shallowly lobed, the sinuses less than 1/2-way to the midrib; capsule 8–11 mm long, hardly exserted from the calyx .... 2. *P. lanceolata*

1. *Pedicularis canadensis* L., common lousewort, wood betony. Perennial, sometimes stoloniferous, villous, becoming woolly in the inflorescence; stems 1–3(4) dm tall, erect, simple, few to several in a cluster. Leaves chiefly basal, alternate, petiolate, the blade (ob)lanceolate to narrowly oblong, 4–15 cm long, 8–25 mm wide, pinnatifid to bipinnatifid, the sinuses 1/2–2/3 the length of the midrib, the segments oblong to ovate, crenate, the upper leaves progressively reduced, short petioled to subsessile. Spikes dense and short at first, elongating in fruit, the bracts oblanceolate, crenate to toothed. Calyx 7–9 mm long, the lobes united laterally and behind, resulting in an oblique tube, deeply cleft in front and 2-toothed in back; corolla pale yellow, sometimes with a purple galea or rarely wholly purple, 18–25 mm long, the galea 11–15 mm long, narrowed to a truncate apex, bearing 2 slender teeth just below the apex, the lower lip 7–10 mm long, with widely spreading lobes, the lateral ones rounded and longer than the midlobe; stamens, the filaments pubescent at the base, the anthers 2.3–2.8 mm long. Capsule lance-oblong, scarcely beaked at the apex, 10–17 mm long, dehiscing along the upper suture. (2n = 16) Apr–Jun. Moist open woods & prairies; e GP; (ME & s Que. to s Man., s to n FL, AL, MS, LA, & e TX, recurring in the high plains (?) & mts. from e-cen CO, through cen NM to n Mex.). *P. canadensis* L. var. *dobsii* Fern.—Waterfall.

The Coloradan, New Mexican, and adjacent Mexican plants are mostly in the mountains from 6000–10,500 ft elevation, but may extend down onto the high plains of our region. They have been considered to be distinct [*P. fluviatilis* Heller or *P. canadensis* var. *fluviatilis* (Heller) Macbr.]. The distinguishing characters are not obvious to me from herbarium specimens. Because of the distributional gap and difference in habitat, these two races should be studied. There is also a collection from northwestern Nebraska (*Williams s.n.*, War Bonnet Canyon, Sioux Co.) that should be examined more closely.

2. *Pedicularis lanceolata* Michx., swamp lousewort. Perennial, glabrous or slightly hairy below, becoming villous in the inflorescence; stems 3–8 dm tall, erect, simple or few-branched. Leaves opposite to subopposite, subsessile, oblong-lanceolate, 5–12 cm long, 14–25(30) mm wide, shallowly pinnately lobed, the sinuses less than 1/2-way to the midrib, and segments crenately toothed. Spicate racemes crowded, terminating the main stem and lateral branches, the bracts auriculate near the base. Calyx 8–12 mm long, with an oblique tube bearing 2 lateral foliose lobes, these 2–4 mm long, crenate; corolla pale yellow, 15–23(27) mm long, the galea 8–13 mm long, narrowed to a short, truncate beak at the apex, the lower lip 7–12 mm long, with appressed-ascending or scarcely spreading lobes; stamens glabrous throughout, the anthers 2.3–3.0 mm long. Capsule obliquely ovoid, slender-beaked at the apex, 8–11 mm long. Aug–Oct. Moist meadows, bogs, & marshes; e 2/3 of ND, SD, & NE, eastward; (MA to s Sask., s to VA, e NC, MO, & NE).

3. *Pedicularis procera* A. Gray, Gray's lousewort. Stout perennial, glabrous below and puberulent in the inflorescence; stems 5–12 dm tall, sometimes branched above in especially robust plants. Leaves alternate, basal and cauline, the basal long-petiolate, the midcauline (10)15–30 cm long, pinnatifid to bipinnatifid, the sinuses divided to the midrib or nearly so, the divisions serrate. Spicate raceme 10–35(65) cm long, the bracts linear, attenuate, caudate, often longer than the flowers, entire or toothed apically. Calyx 10–12(16) mm long, the lobes 5, lance-linear, 3–5 mm long, the posterior one shorter than the others; corolla pale yellow, sometimes streaked with red, 25–36 mm long, the galea 9–11 mm long, narrowed to a truncate apex bearing 2 acute teeth just below the apex, the lower lip 7–12 mm long with broadly rounded, erose, appressed-ascending lobes; anthers 3–3.7 mm long. Capsule subsymmetrical, ovoid, the mucronate tip erect, 10–16 mm long, dehiscing throughout. Jul–Aug. Moderately moist woods & edges of meadows; BH; (BH of w NE, s-cen WY, s through w CO & e UT to w NM & cen AZ). *P. grayi* A. Nels.—Rydberg.

22. PENSTEMON Mitchell, Beardtongue

Contributed by Craig C. Freeman

Ours herbaceous or suffrutescent perennials; stems decumbent or erect, solitary to many from a herbaceous or woody caudex surmounting a taproot. Leaves opposite, entire to toothed; basal leaves absent to well developed or tufted, subpetiolate or more frequently petiolate; cauline leaves filiform to broadly ovate, sessile and frequently clasping. Inflorescence a compact to open thyrse with few to many flowers, with indistinct to distinct verticillasters, seldom secund; bracts much reduced to prominent. Flowers nearly regular to strongly bilabiate; calyx equally to subequally 5-lobed, sepals entire to erose; corolla pink, purplish to blue or white, salverform, funnelform, or tubular-funnelform, slightly to prominently ampliate, occasionally ventricose, the upper lip 2-lobed, the lower lip 3-lobed, unlined or more commonly lined internally with nectar guides, palate rounded to plicate and glabrous to villose; staminode (posterior sterile stamen) slender and elongate, frequently flattened and bearded distally, inserted on the corolla at roughly the height of the ovary, bent forward against the anterior surface of the corolla throat, included to prominently exserted; fertile stamens 4, didynamous, inserted at or near the base of the corolla, anthers with 2 locules, the anther sacs becoming divaricate and dehiscing longitudinally, either completely or incompletely, occasionally becoming explanate, filaments arching; stigmas united and capitate, style slender and elongate. Capsule ovoid, septicidal, the walls firm to flexible; seeds numerous (±50), slightly rounded to angular, finely reticulate.

Reference: *Penstemon* in Pennell, F. W., Acad. Nat. Sci. Philadelphia Monogr. 35: 196–274. 1935.

1 Anthers hirsute; corolla glabrous externally (sect. *Glabri*) .... 10. *P. glaber*
1 Anthers glabrous or if anthers pubescent then the corolla also glandular-pubescent externally.
    2 Corolla salverform; staminode glabrous and well included; leaves filiform; plant suffrutescent (sect. *Ambigui*) .... 2. *P. ambiguus*
    2 Corolla funnelform, tubular-funnelform, tubular-salverform, ampliate or otherwise, but never merely salverform; staminode sparsely to densely bearded and included or exserted; leaves linear or wider; plant herbaceous.
        3 Corolla glandular-pubescent externally or if glabrous then the inflorescence capitate or elongate-cylindrical with dense verticillasters; leaves glabrous to canescent but never obviously glaucous and firm.
            4 Seeds 0.5–1.3 mm long, tan to dark brown; throat of the corolla slightly to distinctly flattened and scarcely to prominently plicate anteriorly within, or if the throat not flattened and plicate then the stems glabrous below (sect. *Penstemon*).
                5 Corolla glabrous externally, 6–11 mm long, deep blue to violet-blue .... 19. *P. procerus*
                5 Corolla glandular-pubescent externally, usually exceeding 10 mm in length, white to pink, purplish, or bluish.
                    6 Stems glabrous toward the base.
                        7 Corolla glandular-pubescent internally, the throat barely inflated, unlined and unridged anteriorly within; anther sacs becoming explanate .... 21. *P. tubaeflorus*
                        7 Corolla sparsely to moderately pubescent internally but the hairs eglandular, the throat abruptly inflated, lined with reddish-purple guidelines

and barely 2-ridged anteriorly within; anther sacs not becoming explanate .... 7. *P. digitalis*
6 Stems puberulent to villose toward the base.
   8 Corolla white or yellowish-white and unlined internally, 24–32 mm long .... 17. *P. oklahomensis*
   8 Corolla white to pink, purplish or bluish, and lined internally, 10–30 mm long.
      9 Stems glandular-villose toward the base, the stems and leaves often appearing velvety .... 18. *P. pallidus*
      9 Stems merely puberulent toward the base, the stems and leaves glabrous to puberulent and not appearing velvety.
         10 Staminode 15–20 mm long from its point of attachment to the corolla and prominently exserted; corolla 20–30 mm long, peduncles ascending or spreading .... 15. *P. laxiflorus*
         10 Staminode 8–12 mm long from its point of attachment to the corolla and included or barely exserted; corolla 10–22 mm long, peduncles appressed or erect.
            11 Corolla pale to dark blue or violet, the throat weakly 2-ridged anteriorly within, stems arising from a much-branched, well-developed suffrutescent caudex .... 22. *P. virens*
            11 Corolla pale lavender to mauve, the throat prominently 2-ridged anteriorly within, stems arising from a usually slender herbaceous caudex .... 11. *P. gracilis*
4 Seeds 2–3.5 mm long, dark brown to black; throat of the corolla rounded and not plicate anteriorly within; stems glabrate to villose-canescent (sect. *Cristati*).
   12 Throat of the corolla glandular internally but not pilose; staminode sparingly bearded and included or slightly exserted.
      13 Corolla 35–55 mm long, white to pink, lilac, or pale violet-purple; the throat abruptly much inflated; cauline leaves 3.5–15 cm long, 1–5.4 cm wide .... 6. *P. cobaea*
      13 Corolla 12–20 mm long, white or occasionally faint pink or violet; the throat funnelform and moderately ampliate; cauline leaves 2.5–6.5 cm long, 0.3–2 cm wide .... 1. *P. albidus*
   12 Throat of the corolla pilose internally or at the orifice; staminode densely bearded and slightly to prominently exserted.
      14 Ovary and style glabrous; stems glabrate to pubescent.
         15 Corolla 24–35 mm long; the throat ventricose-ampliate; anther sacs becoming explanate .... 14. *P. jamesii*
         15 Corolla 16–24 mm long, the throat moderately ampliate; anther sacs not becoming explanate .... 4. *P. auriberbis*
      14 Ovary and occasionally the proximal 1/4–1/2 of the style glandular-puberulent; stems canescent to villose-canescent .... 8. *P. eriantherus*
3 Corolla glabrous externally; the inflorescence thrysoid; leaves glaucous and firm (sect. *Coerulei*).
   16 Sepals 7–13 mm long at anthesis; throat of the corolla distinctly inflated and ventricose posteriorly.
      17 Corolla 35–48 mm long; sepals lanceolate to lance-ovate; capsule 16–20(25) mm long .... 12. *P. grandiflorus*
      17 Corolla 23–25 mm long; sepals linear to linear-lanceolate; capsule 13–16 mm long .... 13. *P. haydenii*
   16 Sepals 3.5–7(8) mm long at anthesis; throat of the corolla scarcely to moderately ampliate and neither distinctly inflated nor ventricose posteriorly.
      18 Throat of the corolla barely ampliate and slightly decurved, distinctly decurved in mature unopened corollas just prior to anthesis; thyrse elongate and distinctly interrupted; upper cauline leaves trullate, widely spaced, and normally much shorter than the internodes .... 9. *P. fendleri*
      18 Throat of the corolla slightly to moderately ampliate, not distinctly decurved in open or unopened corollas; thryse compact to elongate; upper cauline leaves lanceolate to ovate but seldom trullate, compact to moderately spaced and normally longer than the internodes.

19 Inflorescence secund, loose to moderately compact .... 20. *P. secundiflorus*
19 Inflorescence cylindrical and not secund, distinctly to moderately compact.
   20 Bracts lance-ovate to orbiculate; corolla usually lavender but occasionally pale pink to very pale blue, never deep blue; thyrse elongate and frequently distinctly interrupted .... 5. *P. buckleyi*
   20 Bracts lanceolate to lance-ovate; corolla deep blue to blue or less commonly lavender to pink; thyrse elongate to moderately compact and seldom distinctly interrupted.
      21 Cauline leaves linear to lanceolate or lance-ovate, short to long acuminate or acute; anther sacs (0.9)1.1–1.5 mm long .... 3. *P. angustifolius*
      21 Cauline leaves lanceolate to ovate, acuminate or more frequently mucronate; anther sacs 0.7–1.2 mm long .... 16. *P. nitidus*

1. **Penstemon albidus** Nutt., white beardtongue. Perennial herb; stems erect or ascending, (1)1.5–5(5.5) dm tall, retrorsely puberulent below and glandular-pubescent near the inflorescence, 1–5 stems arising from a short-branched caudex surmounting a taproot. Leaves entire to serrate, nearly glabrous to puberulent or scabrous; basal leaves (ob)lanceolate to obovate, 2–8.5(11) cm long overall, (0.4)0.7–1.8(2) cm wide, acute to obtuse, petiolate; cauline leaves lanceolate to lance-ovate, 2.5–6.5 cm long, (0.3)0.7–1.9(2.1) cm wide, acute, sessile and clasping. Thyrse 4–24(30) cm long, with (2)3–9(10) verticillasters, scarcely to distinctly interrupted; individual cymes 2- to 7-flowered, peduncles appressed or erect; bracts lanceolate, acute. Calyx glandular-pubescent, lobes lanceolate to lance-ovate, 4–7 mm long, 1.5–3 mm wide, acute, entire; corolla (12)16–20 mm long, funnelform, weakly bilabiate, white to faintly pink or violet, glandular-pubescent externally and glandular internally, throat (4)6–8 mm broad, moderately ampliate, lined internally on the anterior surface with red or reddish-purple nectar guides, palate densely glandular-pubescent and rounded, lobes of the upper lip spreading, lobes of the lower lip spreading or slightly projecting, the limb often appearing flat; staminode included, flattened only slightly at the tip, the distal 2–6 mm sparsely to moderately bearded with tortuous sordid-yellow to yellow hairs to 1 mm long; anther sacs 0.7–0.9 mm long, black, smooth, dehiscing the entire length and across the connective, becoming explanate; style glabrous. Capsule 8–12 mm long; seeds 2–3 mm long, angular, dark brown to black. (n=8) Apr–Jul. Sandy loam to sand or gravel in open prairies & hills; throughout the GP except the se 1/4; (s Man. & Alta., s to w OK, n TX, & ne NM).

    This is the most widespread of the GP penstemons and is common in prairie plant communities throughout its range. There is relatively little phenotypic variation, although specimens may vary noticeably as to flower color, serration of leaves and amount of puberulence on stems and leaves.

2. **Penstemon ambiguus** Torr., gilia penstemon. Long-lived suffrutescent perennial; stems erect or ascending, (2)3–4(6) dm tall, slender, subglabrous to retrorsely scabrous below and glabrous to puberulent upward, much branched from a woody base surmounting a thick, woody creeping rootstock. Leaves filiform, (3)5–30(40) mm long, 0.5–1(2.5) mm wide, entire, acuminate to mucronate, glabrous to scabrid-puberulent, margins involute with a very fine scarious edge. Thyrse 6–15 cm long, with 6–10 verticillasters, interrupted, loose; cymes with 2 or 3 flowers, peduncles ascending; bracts filiform. Calyx glabrous, lobes ovate, 1.5–3.5 mm long, 1–1.5 mm wide, acuminate to mucronate, green with broad white edges, margins broadly scarious and occasionally erose; corolla (14)16–22(28) mm long, salverform, scarcely bilabiate, glabrous externally, the tube exceptionally slender, throat (3)4–5 mm broad, barely ampliate, pale to deep pink, lined internally with reddish-purple nectar guides and two pubescent reddish or magenta guides on the anterior surface, pubescent laterally at the orifice, the face of the limb very pale milky pink to milky white, lustrous, oblique to the throat and appearing flat, palate rounded and pubescent, lobes of the upper lip rounded and reflexed, lobes of the lower lip rounded and projecting-spreading;

staminode well included, glabrous; anther sacs 0.5–0.6 mm long, glabrous, dehiscing the entire length and across the connective, becoming explanate; style glabrous and very slender. Capsule 7–9 mm long, seeds 1.2–2 mm long, angular, very dark brown. (n = 8) May–Aug. Sandy plains & hills; sw 1/4 GP; (e CO s to w OK & w TX, w to e 1/2 NM). The species has been reported from WY: Albany (Atlas GP), but specimen evidence is lacking.

Our plants are referable to var. *ambiguus*. South and westward beyond our range, it is replaced by var. *laevissimus* (Keck) N. Holmgren. The latter variety differs from the var. *ambiguus* by having glabrous stems and leaf margins that tend to be smooth or only minutely scabrous.

3. **Penstemon angustifolius** Nutt. ex Pursh, narrow beardtongue. Slender to stout herbaceous perennial, stems erect to assurgent, (1)1.5–4.5(6.5) dm tall, glabrous or scabrid-puberulent and usually distinctly glaucous, 1–5(10) stems arising from a woody crown or short-branched woody caudex surmounting a taproot. Leaves entire, glabrous to sparingly scabrid-puberulent and usually glaucous, thick; basal leaves linear to spatulate or oblanceolate, (2.5)4–9 cm long, 0.2–1.8 cm wide, acute to obtuse, subsessile to petiolate, the petioles usually winged; cauline leaves linear to lanceolate or lance-ovate, 3–11 cm long, 0.2–2.4(4) cm wide, acuminate to acute or caudate, sessile to cordate-clasping, upper cauline leaves equaling or commonly much longer than the internodes. Thyrse 4–30(37) cm long, with (3)5–15(26) verticillasters, distinctly interrupted to compact, cylindrical and not secund, cymes (2)4- to 8(10)-flowered; bracts lanceolate to lance-ovate or seldom ovate, gradually reduced upward, acute to long-acuminate, bases scarcely clasping to cordate-clasping and overlapping, lower bracts occasionally concealing the pedicels in wide-bracted plants. Calyx glabrous and glaucous to scarcely scabrid-puberulent, lobes lanceolate to lance-ovate, 4–8 mm long, 1–2.5 mm wide, acute or more frequently acuminate, margins scarious, particularly near the base, entire to sub-erose; corolla 14–20(23) mm long, tubular-salverform, bilabiate, pink to lavender or blue to deep blue, glabrous externally, throat 4–6 mm broad, moderately ampliate and scarcely ventricose anteriorly, pale internally and lined on the anterior and posterior surfaces with violet or reddish-purple guidelines, lobes of the upper and lower lips projecting to spreading, palate glabrous or sparingly pubescent with whitish eglandular hairs; staminode reaching the orifice, broadly flattened and recurved distally, densely bearded at the tip with golden-yellow hairs to 1 mm long and more sparingly bearded for slightly more than 1/2 its length; anther sacs (0.9)1.1–1.5 mm long, papillose along the sutures, divergent, dehiscing nearly to the apices and across the connective, not becoming explanate; style glabrous. Capsule 9–14(15) mm long; seeds 2.5–3.5 mm long, angular, brown to dark brown. (n = 8) May–Jun. Sandy to gravelly soil in open prairies & sandhills; se ND w to e MT & cen WY, s to w KS, w OK, & n NM; (ND w to e MT & cen WY, s to OK, n NM, w CO, e UT, & nw AZ).

This species is morphologically quite variable and has a rich nomenclature, cf. Holmgren (Brittonia 31: 217–242. 1979). Two varieties are recognized in the GP:

3a. var. *angustifolius*. Cauline leaves linear to linear-lanceolate; stem and leaves commonly scabrid-puberulent; corolla 14–18 mm long; bracts mostly gradually tapering from the base to an acuminate or acute tip. se ND w to e MT & cen WY, s to s CO on the e slope of the Rocky Mts.

This variety is commonly encountered in the w 1/2 of the n GP, where the plants are consistently narrow-leaved and the populations have little variation in flower color, the flowers typically being blue to deep blue, or occasionally pink. Plants of this variety tend to be shorter in stature than their more southern relatives.

3b. var. *caudatus* (Heller) Rydb. Cauline leaves lanceolate to lance-ovate; stem and leaves glabrous or very rarely scabrid-puberulent; corolla 16–20(23) mm long; bracts usually broadened above the base and tapering to a short or long-acuminate tip. nw NE w to the foothills of the Rocky Mts., s through extreme w KS & w OK to n NM. *P. caudatus* Heller — Gates; *P. angustifolius* subsp. *caudatus* (Heller) Keck — Barkley.

This variety ± intergrades with var. *angustifolius* where the ranges of the two meet and overlap; however, southward it is quite distinct. Flower color tends to be more variable than in var. *angustifolius*.

4. ***Penstemon auriberbis*** Penn. Compact herbaceous perennial; stems erect to assurgent, 1–3(3.5) dm tall, retrorsely puberulent below and glandular-pubescent near the inflorescence, 1–6(15) stems arising from a simple or branched woody caudex surmounting a taproot. Leaves entire or occasionally denticulate, glabrous to puberulent; basal leaves linear to lanceolate and usually narrower than the cauline leaves, (1.5)3–6(10) cm long overall, (1)2–5(7) mm wide, acute to obtuse, subpetiolate to petiolate, tufted; cauline leaves linear to linear-lanceolate, (2.5)4–8 cm long, 2–7 mm wide, attenuate to acute, sessile to slightly clasping. Thyrse (5)7–23 cm long, with 3–8 verticillasters, compact to interrupted, somewhat secund, leafy-bracted, cymes 2- to 4-flowered, peduncles appressed or erect; bracts linear to linear-lanceolate, acuminate to acute. Calyx densely glandular-pubescent, lobes linear-lanceolate to ovate, (6)7–9 mm long, 1–2 mm wide, long-acuminate, entire, margins narrowly scarious towards the base; corolla (16)18–22(24) mm long, tubular-funnelform, scarcely bilabiate, pale lilac to purplish-blue, the tube and inside paler, glandular-pubescent externally, throat 7–9 mm broad, scarcely inflated and moderately ampliate, lined internally with magenta or bluish-purple guidelines on the anterior and posterior surfaces, palate white or pale lavender and pilose near the base of the lobes of the lower lip with eglandular straw-yellow hairs, lobes of the upper and lower lip spreading, the limb appearing rather flat; staminode barely to conspicuously exserted, distally flattened and recurved, bearded most of its length with stiff or twisted yellow-orange hairs to 2.5 mm long; anther sacs 1.2–1.5 mm long, papillose along the sutures, widely divaricate, dehiscent nearly the entire length and across the connective, not becoming explanate; style glabrous. Capsule (6)8–10 mm long; seeds 2–3 mm long, angular, black. ($n = 8$) May–Aug. Sandy loam to sand in high plains, foothills, & on sagebrush slopes; restricted to se 1/4 CO & extreme ne NM.

Common to locally abundant throughout its range, this species frequently forms a striking lilac or bluish carpet.

5. ***Penstemon buckleyi*** Penn., Buckley's penstemon. Stout herbaceous perennial; stems erect or ascending, 1.5–5.5(8.2) dm tall, glabrous and glaucous, 1–5 stems from a thick crown or short-branched woody caudex surmounting a taproot. Leaves entire, thick, firm, glabrous and glaucous; basal leaves oblanceolate to spatulate, 1.9–11.6(15) cm long overall, 0.3–2.2(3.1) cm wide, acute to obtuse and often mucronate, petiolate, the petioles usually winged; lower cauline leaves lanceolate to lance-ovate, upper cauline leaves lance-ovate to ovate, 2–9.5 cm long, 1–3 cm wide, clasping to cordate-clasping, lower cauline leaves frequently crowded and longer than the internodes. Thyrse (4)10–35(57) cm long, with (2)4–20(35) verticillasters, usually very elongate, scarcely to distinctly interrupted, narrow, cylindrical and not secund, cymes (2)3- to 5(11)-flowered; bracts lance-ovate to ovate or orbiculate, much reduced above, short-acuminate to acute, bases clasping to cordate-clasping and overlapping, the lower bracts concealing the pedicels. Calyx glabrous and glaucous, lobes lance-ovate to ovate, 3.5–6 mm long, 1.5–2.5 mm wide, acuminate to acute, margins broadly scarious; corolla (12)14–20 mm long, tubular-salverform, scarcely bilabiate, pale pink or lavender to very pale blue, glabrous externally and internally, throat 4–6 mm wide, slightly ampliate, scarcely ventricose anteriorly, lined internally on the anterior surface with prominent reddish or reddish-purple guidelines, lobes of the upper lip spreading or only slightly reflexed, lobes of the lower lip spreading, the limb usually appearing relatively flat; staminode included or reaching the orifice, broadly flattened distally, the tip abruptly recurved, densely to moderately bearded with golden-yellow hairs for 1/2 its length, the hairs 1.5 mm long; anther sacs 0.8–1.2 mm long, externally minutely papillose, divergent, dehiscing nearly to the apices and across the connective, not becoming explanate; style glabrous. Capsule 12–18(20) mm long; seeds 2.5–3.5 mm long, angular, brown to dark brown. ($n = 8$) Apr–Jun. Sandy soil, particularly dunes; restricted to n-cen & sw KS w to extreme se CO, s to cen TX & e NM.

This species is quite similar to *P. angustifolius* var. *caudatus*, particularly in the herbarium, but field studies indicate the two taxa are distinct and virtually allopatric in their ranges. Nearly all the specimens cited as *P. angustifolius* var. *caudatus* for Kansas in the Atlas GP are referable to *P. buckleyi*.

Penstemon buckleyi is distinguished morphologically from *P. angustifolius* var. *caudatus* by its slightly smaller flowers, bracts that are more ovate and normally lacking a long-acuminate tip, and a more elongate inflorescence that is frequently interrupted. Probably the most distinguishing character, however, is flower color. Within populations of *P. buckleyi* flower color tends to be quite consistent while interpopulational variation ranges from pale pink to very pale blue, but is most frequently lavender. This variation is far less than that exhibited by populations of *P. angustifolius* var. *caudatus* in which it is common to find populations with flower colors ranging from pink to deep blue.

6. *Penstemon cobaea* Nutt., cobaea penstemon. Robust herbaceous perennial; stems erect or assurgent, (1.5)2.5–6.5(10) dm tall, retrorsely puberulent below and glandular-pubescent near the inflorescence, 1–3(4) stems arising from a frequently massive, woody, creeping rootstock. Leaves subentire to sharply serrate, glabrous to puberulent or occasionally pubescent, thick; basal leaves frequently absent or withering, when present (ob)lanceolate to spatulate, 3.5–18(26.4) cm long overall, 0.8–5.5(7.6) cm wide, acute to rounded, subsessile to petiolate, the petioles frequently winged; cauline leaves lanceolate to ovate, 3.5–12(15) cm long, 1–4.5(5.4) cm wide, acute, the lower ones sessile, becoming cordate-clasping above. Thyrse 10–30(52) cm long, with 3–6(8) distinct to indistinct verticillasters, cymes 2- to 6-flowered, peduncles appressed or erect; bracts mostly ovate, acute, glabrous to glandular-pubescent. Calyx densely glandular-pubescent and viscid, lobes lanceolate to lance-ovate, (8)10–16 mm long, 2.5–4.5 mm wide, acute, entire, herbaceous throughout or with narrowly scarious margins near the base; corolla 35–55 mm long, strongly bilabiate, white or pink to pale violet-purple, glandular-pubescent and viscid externally, throat (15)18–25 mm broad, abruptly much inflated and venticose-ampliate, glandular internally, constricted very slightly at the orifice, lined internally with prominent magenta or violet guidelines, palate glandular-pubescent but not pilose, lobes of the upper lip arched-projecting, lobes of the lower lip spreading to reflexed; staminode included or slightly exserted, tip flattened and recurved, sparingly bearded nearly the entire length with golden-yellow hairs, hairs at the tip tortuous and 3.5 mm long, medial hairs shorter and stiff, retrorse; anther sacs 1.3–1.6 mm long, light brown to brown, glabrous, divaricate, dehiscent throughout and across the connective, becoming explanate; style glabrous. Capsule 13–18 mm long; seeds 2.5–3.5 mm long, angular, black. (n = 8) Apr–Jun. Calcareous or gypsiferous loams or sandy loams in open prairies, eroded pastures, & hills; restricted to se 1/4 GP; (s NE, s to s-cen MO, n-cen & sw AR, OK & e TX; frequently cult. and escaping elsewhere).

7. *Penstemon digitalis* Nutt. ex Sims, smooth penstemon. Robust herbaceous perennial; stems mostly erect, 2.5–9 dm tall, glabrous and shiny or somewhat glaucous, frequently reddish near the base, 1–3 stems arising from a thick caudex surmounting a short taproot. Leaves entire to serrate or denticulate, glabrous; basal leaves lanceolate to (ob)ovate or spatulate, 3–18(22) cm long overall, 0.4–3.8(6) cm wide, acute to obtuse or rounded, subpetiolate to petiolate; cauline leaves linear-lanceolate to ovate, 2.6–17 cm long, 0.4–4.8 cm wide, acuminate to acute, sessile to distinctly clasping, lower cauline leaves commonly crowded and large, upper cauline leaves widely spaced and much reduced, the upper 1/2 of the stem appearing rather naked except for the inflorescence. Thyrse 7–26(34) cm long, with (2)3–6 verticillasters, individual cymes 3- to many-flowered, peduncles much branched and ascending or spreading; bracts normally greatly reduced. Calyx glandular-pubescent, lobes lance-ovate to ovate, 4–8 mm long, 2–3 mm wide, acuminate to acute, entire or occasionally suberose, noticeably scarious along the margins; corolla (18)20–30 mm long, only slightly bilabiate, white, glandular-pubescent externally, throat 8–12 mm broad, abruptlly inflated and slightly ampliate, lined internally with faint reddish-purple guidelines, barely 2-ridged anteriorly within, palate sparsely to moderately bearded with white eglandular hairs, lobes of the upper lip spreading, lobes of the lower lip moderately projecting; staminode reaching the orifice, the tip recurved and flattened, the terminal 6–8 mm sparsely to moderately bearded with stiff yellow or sordid-yellow hairs to 1 mm long; anther sacs 1.4–1.7 mm long, usually pubescent with few to many white hairs, the external surface minutely papillose, lined with white along the sutures, sutures slightly to moderately papillose, widely divergent, dehiscing nearly to the apices, not becoming explanate; style glabrous. Capsule 8–14 mm long; seeds 1–1.3 mm long, slightly rounded to angular, tan to dark brown. (n = 48) Apr–Jul. Loamy to sandy loam soil in prairies, meadows, & open woodlands; extreme se SD, e 1/3 NE, KS, & OK; (ME & Que. w to Ont. & SD, s to AL & LA). *P. digitalis* f. *baueri* Steyermark — Steyermark.

Pennell, op. cit., indicated that *P. digitalis* is probably native to the Mississippi River Basin and that it has been introduced elsewhere, especially to the ne and e, thus it closely parallels the occurrence of *P. tubaeflorus*, another polyploid species of the GP.

8. ***Penstemon eriantherus*** Pursh, crested beardtongue. Herbaceous perennial; stems erect or ascending, 1–4(4.5) dm tall, villose-canescent throughout or merely canescent below, stems 1–5(10) arising from a woody caudex surmounting a taproot. Leaves entire to serrate, moderately to densely pubescent; basal leaves (ob)lanceolate to spatulate, 3.5–8(11) cm long overall, 0.5–3(3.7) cm wide, acute to obtuse, subpetiolate to petiolate; cauline leaves (ob)lanceolate to oblong or spatulate, 2.5–9 cm long, 0.4–1.8(2.5) cm wide, acute, sessile and scarcely clasping. Thyrse 4–13(27) cm long, with (2)3–7(9) verticillasters, compact, leafy-bracted, cymes 3- to 6-flowered; bracts lanceolate to oblong, pubescent. Calyx viscid-pubescent, lobes lanceolate, (6)8–12(13) mm long, 1.5–3 mm wide, acuminate to acute, herbaceous throughout with entire margins; corolla (20)22–35(42) mm long, strongly bilabiate, lavender to pale purple or pinkish, viscid-pubescent externally, throat 8–14 mm broad, abruptly inflated, ventricose-ampliate, lined internally with deep reddish-purple nectar guides on the anterior and posterior surfaces, eglandular to glandular-pubescent internally and on the lateral lobes of the lower lip, palate densely pilose with straw-yellow hairs, lobes of the upper lip arched-projecting, lobes of the lower lip spreading or projecting; staminode prominently exserted, distally expanded and recurved, bearded on the terminal 13–15 mm with golden-yellow hairs, hairs at the tip tortuous and to 4 mm long, medial hairs shorter and stiff, retrorse; anther sacs (1.1)1.2–1.6(1.8) mm long, minutely papillose along the sutures, tan, dehiscing the entire length and across the connective, becoming explanate; ovary and occasionally the proximal 1/4–1/2 of the style sparingly glandular-pubescent. Capsule (6)9–12(13) mm long; seeds 2–2.8 mm long, dark brown to black, angular. (n = 8) May–Jul. Dry sandy or gravelly soil in open prairies, slopes, sides of buttes, & into the lower elevations of the mts. in the nw part of its range; w & s-cen ND w to MT, s to nw NE & extreme n-cen CO; (w & s-cen ND w to B.C., WA, & OR, s to extreme n-cen CO).

This is an exceedingly variable species throughout its range, with plants differing in a number of characters, including corolla size and color, leaf size and shape, pubescence, bearding of the staminode, size of the anther sacs, and degree of anther sac explanation. Such variability has resulted in the naming of a number of subspp. and vars. Our plants are referable to var. *eriantherus*, which is also found in nw MT, se B.C., and WA.

9. ***Penstemon fendleri*** T. & G. Herbaceous perennial; stems mostly erect, (1.5)2–5.5(6) dm tall, glabrous and glaucous, 1 or 2(4) stems arising from a woody subterranean caudex surmounting a taproot. Leaves entire, firm, glabrous and glaucous; basal leaves oblanceolate or spatulate, 2–10 cm long overall, 0.4–2.4 cm wide, acute to obtuse and occasionally mucronate, subsessile or short-petiolate, the petioles often winged; cauline leaves lanceolate to ovate or often trullate midway up the stem, (1.4)2.3–9.5 cm long, (0.4)0.6–3.1 cm wide,

scarcely clasping below to distinctly clasping above, lower cauline leaves normally crowded, upper cauline leaves widely spaced, the upper 1/2 of the stem with leaves normally much shorter than the internodes. Thyrse (5)11–27 cm long, with (3)4–10 remote verticillasters, elongate, distinctly interrupted, cylindrical and not secund, cymes 2- or 3(5)-flowered; bracts much reduced above and resembling the cauline leaves below, trullate to ovate, short-acuminate to acute, bases clasping to distinctly cordate-clasping, the lower bracts usually concealing the pedicels. Calyx glabrous and glaucous, lobes lance-ovate to ovate, 4.5–7 mm long, 1.5–3.5 mm wide, acuminate to acute, margins broadly scarious and frequently suberose to erose, corolla 14–23(28) mm long, tubular-salverform, scarcely bilabiate, purple to violet or pale blue, externally glabrous, tube slender, slightly decurved in mature open flowers and distinctly decurved in mature unopened corollas just prior to anthesis, throat 4–6 mm broad, barely ampliate, lined internally on the anterior surface and to a lesser extent on the posterior surface with prominent violet or reddish-purple guidelines, lobes of the upper lip arched-spreading or reflexed, lobes of the lower lip projecting to spreading, palate glabrous or sparsely bearded with white eglandular hairs; staminode reaching the orifice, broadly flattened distally, the tip abruptly recurved and bearded with golden-yellow hairs to 1.5 mm long; anther sacs 1–1.3 mm long, externally minutely papillose, divergent, dehiscing nearly to the apices and across the connective, not becoming explanate; style glabrous. Capsule 10–15 mm long; seeds 2.5–3.5 mm long, angular, brown to dark brown. (n = 8) Mar–Jun (Jul). Sandy to gravelly, gypsiferous or calcareous soil on prairies, or at low elevations in mts.; sw KS sw to NM & se AZ, s to sw OK & TX; (sw KS sw to NM & se AZ, s to sw OK, w TX, & Chih.)

In n-cen and ne NM, *P. fendleri* is sympatric with *P. secundiflorus*, a closely related species of sect. *Coerulei*. In this area of sympatry, Nisbet and Jackson (Univ. Kansas Sci. Bull. 41: 691–759. 1960) indicate there may be some introgressive hybridization between the two species.

10. *Penstemon glaber* Pursh. Stout herbaceous perennial; stems assurgent, (1)5–6.5(8) dm tall, glabrous to puberulent or pubescent, one to many arising from a well-developed and often much-branched woody caudex surmounting a taproot. Leaves entire, glabrous or pubescent, thick; basal leaves frequently wanting or smaller than the cauline leaves, (ob)lanceolate to obovate, 2–8(15.5) cm long overall, 0.5–2(4.5) cm wide, acute to obtuse and occasionally mucronate, subsessile to petiolate, the petioles usually winged; cauline leaves linear-lanceolate to lanceolate, 3–12(15) cm long, (0.7)1.2–3.5(4.8) cm wide, acute to obtuse, sessile below to broadly clasping above. Thyrse 8–26(30) cm long, with (5)8–12 verticillasters, congested, secund, leafy-bracted at the base, cymes 2- to 4-flowered; bracts lanceolate to lance-ovate. Calyx glabrous to puberulent, lobes lance-ovate to ovate or orbiculate, 2–7(10) mm long, 1.5–3(4) mm wide, broadly rounded at the apex or with an abruptly short-acuminate to long-acuminate tip, margins usually broadly scarious and distinctly erose; corolla 26–35(40) mm long, strongly bilabiate, posteriorly deep blue to bluish-purple or occasionally pink, anteriorly light blue or white, glabrous externally, throat 8–13(18) mm wide, moderately inflated and ventricose-ampliate anteriorly, pale internally and lined on the anterior surface with pale or deep reddish-purple nectar guides, palate glabrous to pubescent or villose with white eglandular hairs, especially near the base of the lobes of the lower lip, lobes of the upper lip arched-projecting, lobes of the lower lip spreading to strongly reflexed; staminode included or slightly exserted, distally expanded and slightly recurved, the tip rounded or distinctly bifurcate, glabrous or the terminal 2 mm sparingly bearded with pale yellow hairs to 1.5 mm long; anther sacs 1.9–2.5(2.8) mm long, sparingly to moderately hirsute or rarely glabrous, papillate along the sutures, divergent, dehiscent nearly to the apices but not across the connective, not becoming explanate, occasionally somewhat twisted; style glabrous. Capsule 10–15 mm long, thin and quite pliable; seeds 2.5–3.2 mm long, angular-elongate, dark brown. (n = 8) Jun–Sep. Sandy to gravelly soil on high plains & mts.

Penstemon glaber varies greatly, and three varieties are recognized in our region.

1. Sepals 2–4 mm long, lance-ovate to orbiculate and broadly rounded or with an abruptly short-acuminate tip; corolla glabrous or pubescent internally on the anterior surface; staminode rounded at the tip .... 10c. var. *glaber*
1. Sepals 4–7(10) mm long, ovate and usually with a long-acuminate tip; corolla glabrous to pubescent or villose internally on the anterior surface; staminode rounded to distinctly bifurcate at the tip.
   2. Staminode rounded or obscurely bifurcate at the tip; corolla 24–35 mm long; stems glabrous to pubescent .... 10a. var. *alpinus*
   2. Staminode usually distinctly bifurcate at the tip; corolla 30–40 mm long; stems puberulent to pubescent .... 10b. var. *brandegei*

10a. var. *alpinus* (Torr.) A. Gray. Herbage glabrous to puberulent. Sepals 4–7 mm long, ovate and long-acuminate; corolla 24–35 mm long, glabrous to pubescent or villose internally on the anterior surface; staminode rounded or obscurely bifurcate at the tip. Occasionally encountered on the high plains along the mts. from sw WY s to cen CO; (mts. & occasionally high plains from se WY s to s-cen CO).

This is a quite variable var. with respect to length of the corolla and pubescence of the stems, leaves, palate of the corolla, and the anther sacs. Plants from the n 1/2 of the range frequently exhibit puberulent to pubescent stems and leaves and have been referred to f. *riparius* (A. Nels.) Penn. Plants from the s 1/2 of the range are normally glabrous to glabrate.

10b. var. *brandegei* (Porter ex Rydb.) C. C. Freeman. Herbage puberulent to pubescent. Sepals 4–10 mm long, ovate and long-acuminate; corolla 30–40 mm long, glabrous to pubescent internally on the anterior surface; staminode usually distinctly bifurcate at the tip. s-cen CO s to ne NM.

This var. barely enters the sw GP. Nisbet & Jackson, op. cit., cite specimens from Colfax and Union Counties in NM, and the taxon may be sporadically encountered on the high plains of s-cen CO although it is most abundant in the mts. Hybrids between var. *alpinus* and var. *brandegei* have been reported from s-cen CO.

10c. var. *glaber*. Herbage glabrous. Sepals 2–4 mm long, lance-ovate to orbiculate and broadly rounded or with an abruptly short-acuminate tip; corolla glabrous externally, occasionally pubescent internally on the anterior surface; staminode rounded distally. sw ND & n-cen SD w to nw WY, s to w NE & se WY.

In se and e-cen WY, var. *glaber* is sympatric with var. *alpinus* and morphological intergradants are observed. Puberulent-stemmed specimens of var. *glaber* may be referred to f. *pubicaulis* Penn.

11. **Penstemon gracilis** Nutt., slender beardtongue. Slender herbaceous perennial; stems erect or assurgent, (1.5)2–5 dm tall, commonly reddish above, retrorsely puberulent below and glandular-pubescent near the inflorescence, stems 1–4(6) arising from a short, typically slender herbaceous caudex surmounting a taproot. Leaves subentire to serrate, glabrous to sparingly puberulent, seldom densely puberulent; basal leaves (ob)lanceolate to ovate, 2.5–7.5 cm long overall, 0.4–1.5 cm wide, acute to obtuse, subsessile or petiolate, the petioles usually winged; cauline leaves linear to lanceolate, 2.5–8(9) cm long, (0.2)0.4–1(1.5) cm wide, acuminate to acute and somewhat clasping; bracts linear-lanceolate and resembling the upper cauline leaves. Thyrse (3)5–17(21) cm long, with (2)3–5(7) verticillasters, interrupted to congested, cymes 2- to 6-flowered, peduncles appressed or erect. Calyx glandular-pubescent, lobes lanceolate to lance-ovate, 4–6 mm long, 1.5–2 mm wide, acuminate to acute, entire, herbaceous throughout or with scarious margins toward the base; corolla (14)15–22 mm long, bilabiate, pale lavender to mauve externally and lighter within, glandular-pubescent externally, tube slender, throat 4–6 mm broad, slightly inflated and flattened, barely ampliate, lined internally with purple or mauve nectar guides, strongly 2-ridged anteriorly within, palate slightly up-arched and bearded with whitish eglandular hairs extending back to 1/2 the length of the throat, lobes of the upper lip spreading to reflexed, lobes of the lower lip projecting and extending beyond the upper lip; staminode reaching the orifice or barely exserted, slightly expanded distally, the terminal 7–9 mm densely bearded with stiff golden-yellow hairs to 1.5 mm long; anther sacs 1–1.3 mm long,

glabrous, papillose along the sutures, divergent, dehiscing nearly to the apices and across the connective, not becoming explanate; style glabrous. Capsule 6–8 mm long; seeds 0.6–0.8 mm long, rounded to slightly angular, dark brown. (n = 8) May–Aug. Sandy or gravelly soil in prairies, valleys, & at lower elevations in mts.; n 1/2 GP; (sw Ont. w to Alta. & ne B.C., s to WI, nw IA, s NE, & along the e slope of the Rocky Mts. from MT s to NM).

Throughout most of its range, *P. gracilis* has leaves that are glabrous or nearly so on the adaxial surface. However, some plants from ND have puberulent leaves, similar to plants from the Driftless Area of WI and have been designated as subsp. *wisconsinensis* (Penn.) Penn. Frequently, these puberulent-leaved plants from ND occur along with glabrous-leaved plants in the same population, and perhaps leaf pubescence is merely a freely varying character state.

12. *Penstemon grandiflorus* Nutt., large beardtongue. Stout herbaceous perennial; stems mostly erect, (4)5–9.5(12) dm tall, glabrous and glaucous, 1 or 2(3) stems arising from a woody and seldom branched subterranean caudex surmounting a taproot. Leaves entire, thick, firm, glabrous and glaucous; basal leaves spatulate or obovate and petiolate, 3–16 cm long overall, 0.6–5 cm wide, acute to obtuse; cauline leaves spatulate to orbicular, 1.8–9(11) cm long, 1.5–5 cm wide, scarcely clasping below to cordate-clasping above. Thyrse 12–30(40) cm long, with 3–7(9) verticillasters, interrupted, leafy-bracted below, cymes 2- or 3-flowered; bracts ovate to orbicular, acute to short-acuminate, bases cordate-clasping and usually overlapping, concealing the peduncles and most pedicels. Calyx glabrous and glaucous, lobes lanceolate to lance-ovate, 7–11 mm long, 2.5–4 mm wide, acute to acuminate, entire, herbaceous throughout or with narrow scarious margins toward the base; corolla 35–48 mm long, distinctly bilabiate, pink to bluish-lavender or pale blue, glabrous internally and externally, throat 15–18 mm broad, abruptly inflated and ventricose posteriorly, moderately ampliate, lined internally on the anterior surface with magenta nectar guides, lobes of the upper lip spreading to reflexed, lobes of the lower lip projecting or spreading; staminode included or reaching the orifice, strongly flattened distally, the tip abruptly recurved and bearded with golden-yellow hairs to 0.5 mm long; anther sacs 1.4–1.7 mm long, widely divergent, dehiscing the full length and across the connective, not becoming explanate, minutely papillose along the sutures; style glabrous. Capsule 16–20(25) mm long; seeds 2.5–4 mm long, angular, brown to dark brown. (n = 8) Apr–Jul. Sandy to loamy soil in prairies; GP although scarce in the sw 1/4 & nw 1/8; (IN w to n-cen ND & ne WY, s to cen TX). *P. bradburii* Pursh — Correll & Johnston.

Penstemon grandiflorus exhibits an irregular distribution throughout its range. It occurs most commonly in the cen GP but is also represented by a number of widely disjunct populations, especially in OK and TX. These southern populations probably represent relicts of a wider, more southern range during the WI glaciation.

Discussions concerning the correct name for the species may be found in Cronquist et al. (Rhodora 58: 23–24. 1956), Cronquist (Rhodora 59: 100. 1957), Reveal (Rhodora 70: 25–54. 1968), and Shinners (Rhodora 57: 290–293. 1955; Rhodora 58: 281–289. 1956).

13. *Penstemon haydenii* S. Wats., Hayden's penstemon. Stout perennial herb; stems decumbent to ascending, (1.5)2–4.5 dm tall, glabrous, 1 to many from a subterranean caudex surmounting a deep taproot. Leaves entire, glabrous and somewhat glaucous, firm; basal and lower cauline leaves linear to linear-lanceolate, (2.5)5.5–11(13) cm long, 0.3–1 cm wide, acuminate to acute, sessile and clasping, upper cauline leaves linear to occasionally lanceolate, 6–11(12) cm long, 0.7–3 cm wide, acuminate with a long narrow tapering tip, sessile and clasping, the vegetative shoots with long linear leaves. Thyrse 6–16 cm long, with (4)6–10(12) verticillasters, very compact, cylindrical and not secund, leafy-bracted, individual cymes with 4–6 fragrant flowers; 2–8 mm long; bracts longer than broad and very distinct from cauline leaves, the lower bracts tapering to a long narrow acuminate or caudate tip, bases cordate and broadly overlapping, concealing the peduncles and most pedicels. Calyx glabrous, lobes subequal, linear to linear-lanceolate, 8–13 mm long, 1–3 mm wide, acuminate, entire, herbaceous throughout or with scarious margins near the base; corolla 23–25 mm long, distinctly bilabiate, milky blue to milky lavender, glabrous internally and externally, the throat 9–11 mm broad, well inflated and ventricose posteriorly, moderately ampliate, lined internally with magenta nectar guides anteriorly in mature unopened or freshly opened flowers, lobes of the upper lip arched-projecting, lobes of the lower lip projecting to spreading; staminode included, distally flattened, minutely bifurcate and slightly recurved, densely pubescent near the tip with golden-yellow hairs to 1 mm long; anther sacs 1.8–2 mm long, widely divergent, dehiscing the full length and across the connective, prominently papillose along the sutures; style glabrous. Capsule 13–16 mm long; seeds 2.5–4 mm long, discoid, light brown to brown. (n = 8) May–Jul. Endemic to dune blowouts in the sandhills of NE; NE: Cherry, Garden, Hooker, Thomas.

Morphologically, *P. haydenii* is one of the most striking members of sect. *Coerulei*, due to its compact cylindrical inflorescence with prominent long-acuminate bracts and its habit of forming large multistemmed clumps. The stems of *P. haydenii* root adventitiously, thus maintaining the plant in shifting sands of dune blowouts. The species is apparently unique in the *Coerulei* in that its flowers possess a distinctive fragrance.

14. ***Penstemon jamesii*** Benth. Herbaceous perennial; stems erect or ascending, 1.4–5(5.2) dm tall, glabrate to retrorsely puberulent below and glandular-pubescent above, 1–7(10) stems arising from a simple or branched woody caudex surmouting a taproot. Leaves entire to serrate, glabrous to puberulent; basal leaves linear to (ob)lanceolate, 2–8(10.5) cm long overall, (0.2)0.5–1(1.3) cm wide, acute to obtuse, subsessile to petiolate, the petioles occasionally winged; cauline leaves linear to lanceolate, 2–10(11) cm long, 0.5–1.5 cm wide, acute or rarey obtuse, sessile and scarcely clasping. Thyrse 5–20(24) cm long, with 2–8 verticillasters, compact to elongate and interrupted, secund, leafy-bracted, cymes 2- to 5-flowered, peduncles appressed or erect; bracts lanceolate, glandular-pubescent. Calyx glandular-pubescent, lobes lanceolate to lance-ovate, 8–12 mm long, 2–3 mm wide, entire, acuminate to acute, herbaceous throughout or with narrow scarious margins towards the base; corolla 24–32(35) mm long, strongly bilabiate, pinkish or pale lavender to violet-blue, viscid glandular-pubescent externally, throat 9–15 mm broad, abruptly much inflated and ventricose-ampliate, usually sparsely glandular-puberulent internally, lined internally on the anterior and posterior surfaces with magenta or violet-blue guidelines, palate moderately to densely pilose with whitish hairs, rounded, lobes of the upper lip arched-projecting, lobes of the lower lip spreading to reflexed; stmainode conspicuously exserted, flattened somewhat distally and slightly recurved, the terminal 10–14 mm bearded, hairs at the tip golden-yellow and tortuous, to 3.5 mm long, medial hairs shorter and retrorse; anther sacs 1–1.2 mm long, glabrous, widely divaricate, dehiscent throughout, becoming explanate; style glabrous. Capsule 14–17 mm long; seeds 2–3 mm long, angular, black. (n = 8) Apr–Jun. Sandy loam to sand or gravel in high plains or gently sloping hills; CO: Las Animas; KS: Morton; NM: Colfax, Curry, Harding, Quay, Union; TX: Bailey, Lamb; (extreme sw KS & adj. CO & NM s to w TX).

Penstemon jamesii is a variable species throughout its range with respect to length and width of the corolla, pubescence of stems and leaves, length of stems, degree of serration of the leaves, and compactness of the thyrse.

15. ***Penstemon laxiflorus*** Penn. Herbaceous perennial; stems erect or assurgent, 2.5–6.5(7) dm tall, spreading or retrorsely puberulent below and glandular-pubescent near the inflorescence, 1–5 stems arising from a simple or branched herbaceous caudex surmounting a taproot. Leaves subentire or more commonly serrate to dentate, nearly glabrous to puberulent, slightly lighter beneath; basal leaves (ob)ovate to spatulate, 2.5–9 cm long overall, 0.8–2.5 cm wide, acute to obtuse, subsessile or petiolate, the petiole sometimes winged; cauline leaves lanceolate to oblanceolate, (2)3–9(11) cm long, (0.2)0.5–2.2 cm wide,

acute to obtuse, sessile and frequently clasping. Thyrse 5-26(32) cm long, with 3-7 verticillasters, interrupted, cymes 2- to 6-flowered, peduncles ascending or spreading, pedicels lax; bracts much reduced and resembling the upper cauline leaves. Calyx glandular-pubescent, lobes lance-ovate to ovate, 2.5-5.5 mm long, 2-3 mm wide, acuminate to acute, entire,, slightly to distinctly scarious margined; corolla 20-28(30) mm long, bilabiate, white or more commonly suffused with pink or mauve, glandular-pubescent externally, throat 4-8 mm broad, slightly to moderately inflated and gradually ampliate, slightly flattened, lined internally with reddish-purple nectar guides, prominently 2-ridged within anteriorly, palate up-arched and partially restricting the orifice, bearded with whitish or pale yellow eglandular hairs, lobes of the upper lip spreading, those of the lower lip projecting well beyond the upper; staminode prominently exserted, flattened and occasionally slightly recurved at the tip, the distal 8-10 mm densely bearded with stiff yellow-orange hairs to 1.5 mm long; anther sacs 1-1.3 mm long, the entire external surface minutely papillose, prominently papillose along the sutures, divaricate, dehiscent nearly to the apices and across the connective, not becoming explanate; style glabrous. Capsule 8-10 mm long; seeds 0.7-1 mm long, slightly angular to rounded, brown to dark brown. (n = 8) Apr-Jun. Sandy or sandy loam, acid soil at the edge of woodlands; OK; (GA & FL w to OK, AR s to LA & TX).

16. *Penstemon nitidus* Dougl. ex Benth. Herbaceous perennial; stems erect or assurgent, (0.5)1-3.5(4) dm tall, glabrous and glaucous, 1-7 stems arising from a thick crown or short-branched woody caudex surmounting a taproot. Leaves entire, thick, firm, glabrous and often heavily glaucous; basal leaves linear-lanceolate to oblanceolate or spatulate, 1.5-10 cm long overall, 0.2-2.7 cm wide, acute or ovate or frequently mucronate, often tufted and reddish, petiolate, the petioles occasionally winged; cauline leaves lanceolate to lance-ovate below, lance-ovate to ovate above, (1.1)1.8-8.5 cm long, (0.3)0.5-2.8(3.2) cm wide, acuminate to acute or frequently mucronate, clasping to cordate-clasping. Thyrse (2)5-17 cm long, with (2)4-10 verticillasters, compact to elongate, scarcely to distinctly interrupted, narrow, cylindrical and not secund, cymes 2- to 5-flowered; bracts resembling the upper cauline leaves below, much reduced above, acuminate to acute, bases clasping to cordate-clasping. Calyx glabrous and somewhat glaucous, lobes lanceolate to lance-ovate, 3-8 mm long, 1-3 mm wide, acuminate, margins narrowly scarious towards the base, entire to slightly erose; corolla (10)13-15(18) mm long, tubular-salverform, bilabiate, deep blue or rarely pink, glabrous externally, throat 4-6 mm broad, moderately ampliate, lined internally on the anterior and posterior surfaces with violet or purple guidlines, lobes of the upper lip spreading, lobes of the lower lip projecting, palate glabrous or sparingly bearded with white eglandular hairs; staminode reaching the orifice or slightly exserted, flattened distally and recurved, densely bearded at the tip with golden-yellow hairs to 1.5 mm long, more sparingly bearded away from the tip for 1/3-1/2 its length; anther sacs 0.7-1.2 mm long, externally minutely papillose, particularly along the sutures, divergent, dehiscing nearly to the apices and across the connective, not becoming explanate; style glabrous. Capsule 9-13 mm long; seeds 2.5-3.5 mm long, angular, brown to dark brown. (n = 8) May-Jun. Rocky to gravelly soil in prairies, hillsides, & at low elevations in mts.; w ND w to MT, s to nw SD & n WY; (s Man. w to se B.C., s to nw SD & n WY). The species has been reported from CO: Washington (Atlas GP) but specimen evidence is lacking.

17. *Penstemon oklahomensis* Penn. Herbaceous perennial; stems mostly erect, (1.5)3.5-5.5 dm tall, spreading or retrorsely puberulent below, glandular-pubescent midway up the stem, 1 or 2(4) stems arising from a short, typically slender herbaceous caudex surmounting a taproot. Leaves subentire to serrate or denticulate, puberulent, paler beneath; basal leaves (ob)lanceolate to obovate or spatulate, 2.5-8(11) cm long overall, 0.5-2.4(3.2) cm wide, acute to obtuse, subpetiolate to petiolate; cauline leaves linear to lanceolate, (2.5)6-12 cm long, 0.4-2 cm wide, acute to acuminate, somewhat clasping. Thyrse (5)8-18(25) cm long,

with (2)3–6 verticillasters, interrupted, cymes 2- to 4(6)-flowered, peduncles appressed or erect; bracts much reduced, linear to linear-lanceolate and resembling the cauline leaves. Calyx glandular-pubescent, lobes lanceolate to ovate, 5–7 mm long, 2–3 mm wide, acute to acuminate, entire, margins scarious particularly near the base; corolla 24–32 mm long, bilabiate, white to yellowish-white and unlined within, glandular-pubescent externally, tube slender, throat 6–8 mm broad, scarcely inflated and barely ampliate, slightly flattened, prominently 2-pleated internally on the anterior surface, palate strongly up-arched and closing the orifice, bearded with dense yellow hairs passing slightly onto the lobes of the lower lip, lobes of the upper lip spreading, lobes of the lower lip projecting and extended well beyond the upper lip; staminode slightly exserted, flattened and densely bearded on the distal 14–16 mm with stiff yellow-orange hairs to 1.5 mm long; anther sacs 1–1.5 mm long, the entire external surface minutely papillose, scarcely to prominently papillose along the sutures, divaricate, dehiscent throughout and across the connective, not becoming explanate; style glabrous. Capsule 8–13 mm long; seeds 0.7–1 mm long, rounded to slightly angular, brown to dark brown. (n = 8) Apr–Jun. Sandy loamy to loam soil in rolling prairies or open woodlands; OK.

This species is restricted almost wholly to the Osage Plains of OK: Osage, se to Pushmataha and Bryan, w to Comanche. It is the only member of sect. *Penstemon* in the GP having its orifice totally obstructed by a strongly up-arched palate.

18. ***Penstemon pallidus*** Small, pale penstemon. Herbaceous perennial; stems erect or ascending, 2.5–5.5(6.5) dm tall, retrorsely puberulent with eglandular hairs and with glandular-villose pubescence below and glandular-pubescent near the inflorescence; 1–6 stems arising from a typically short and slender herbaceous caudex atop a taproot. Leaves subentire to serrate or dentate, pubescent above and below, appearing somewhat velvety, usually distinctly lighter beneath; basal leaves (ob)ovate to spatulate, 2–12(18) cm long overall, 0.5–3.5(4) cm wide, acute to obtuse, subsessile or petiolate, the petioles sometimes winged; cauline leaves lanceolate to lance-ovate or ovate, 2.2–10 cm long, 0.4–2.4 cm wide, narrowly to broadly acute, sessile and often distinctly clasping. Thyrse 5–26(30) cm long, with 3–8 verticillasters, interrupted and triangular in outline when pressed, individual cymes 2- to 8(16)-flowered, peduncles ascending or spreading; bracts much reduced, linear-lanceolate to lanceolate and resembling the upper cauline leaves. Calyx glandular-pubescent, lobes ovate, 2.5–5 mm long, 1.5–3 mm wide, acuminate to acute, entire or occasionally obscurely erose, margins slightly to distinctly scarious, seldom herbaceous throughout; corolla 16–22 mm long, bilabiate, white or tinged mauve, glandular-pubescent externally, tube slender, throat 4–7 mm broad, slightly to moderately inflated and gradually ampliate, flattened somewhat, lined internally on the anterior surface with reddish-purple guidelines, prominently 2-ridged within on the anterior surface, palate somewhat up-arched and bearded with whitish to yellow eglandular hairs, lobes of upper lip spreading, those of the lower lip projecting or spreading slightly, extended beyond the upper lip; staminode distinctly exserted, flattened and slightly recurved distally, pubescent on the terminal 8–9 mm, moderately to densely bearded at the tip with stiff yellow to golden-yellow hairs to 1.5 mm long, more sparingly bearded away from the tip; anther sacs (0.8)1–1.2 mm long, the entire external surface minutely papillose, slightly to prominently papillose along the sutures, widely divaricate, dehiscent nearly to the apices and across the connective, not becoming explanate; style glabrous. Capsule 5–7 mm long; seeds 0.5–0.7 mm long, rounded to slightly angular, brown to dark brown. (n = 8) Apr–Jul. Loamy to sandy loam soil in prairies & rocky, deciduous forests; KS: Douglas, Jefferson, Wilson; MO: Barton, Cass, Vernon; (ME w to ne MN, s to GA & e to KS).

Pennell, op. cit., suggested *P. pallidus* has extended its range in a manner similar to *P. digitalis* and *P. tubaeflorus,* although it is impossible to delimit precisely the native range of the species. It is thought that the species originally occupied the Mississippi R. and Ohio R. valleys from n AR into IA and IL and east up the Ohio R. into New England.

19. ***Penstemon procerus*** Dougl. ex Grah. Slender herbaceous perennial; stems erect or assurgent, (0.5)1–4.5(7) dm tall, glabrate to minutely puberulent, solitary to many and tufted from a somewhat suffrutescent aerial caudex. Leaves entire, glabrous to puberulent; basal leaves often poorly developed or wanting, when present oblanceolate to elliptic or ovate, 2–6(10) cm long overall, 3–15(18) mm wide, acute to obtuse; cauline leaves lanceolate to oblanceolate, 2–5(8) cm long, 5–10(21) mm wide, acute, mostly sessile and clasping. Thyrse (1.5)2–15(23) cm long, with (1)2–5(11) dense vertcillasters, elongate and cylindrical, the lower verticillasters distinctly interrupted, cymes many-flowered, peduncles appressed or erect; bracts linear-lanceolate to lanceolate and resembling the upper cauline leaves. Calyx glabrous to puberulent and occasionally densely so, lobes ovate to obovate, 3–6 mm long, 1–2 mm wide, long-attenuate, nearly entire to erose, margins scarious; corolla 6–11 mm long, slightly bilabiate, deep blue to violet-blue and lighter anteriorly and within, glabrous externally, tube slender, throat 2–3 mm wide, scarcely ampliate, unlined or obscurely to distinctly lined internally with violet-blue guidelines, obscurely 2-ridged anteriorly within, palate sparingly to densely bearded with whitish-yellow or yellow eglandular hairs, lobes of the upper lip arched-projecting to spreading, lobes of the lower lip spreading; staminode included and slightly expanded distally, glabrous or more commonly bearded on the distal 0.5–1 mm with golden-yellow hairs to 0.7 mm long; anther sacs 0.4–0.8 mm long, glabrous, divergent, dehiscing throughout and across the connective, becoming explanate or nearly so; style glabrous. Capsule 3–5 mm long; seeds 0.6–1 mm long, rounded to slightly angular, finely dark brown. (n = 8, 16) Jun–Aug. Loamy to rocky loam soil in meadows, brushy slopes, & open forests in the mts.; MT: Daniels, Phillips, Valley; ND: Burke; (nw ND w to WA & OR, s AK s to s CO).

P. procerus is a variable species with as many as six subspp. recognized by some authors. The species barely enters the GP in the far nw. Our plants are referable to subsp. *procerus*, which is most abundant in the Rocky Mts.

20. ***Penstemon secundiflorus*** Benth. Herbaceous perennial; stems erect or assurgent, (1.5)2–4.5(5) dm tall, glabrous and somewhat glaucous, 1–5(8) stems arising from a thick crown or short-branched woody caudex surmounting taproot. Leaves entire, firm, glabrous and slightly glaucous; basal leaves oblanceolate to spatulate, 2–8(10.2) cm long overall, 0.2–2.5 cm wide, acute to obtuse or occasionally mucronate, often tufted, especially in vegetative shoots, petiolate, the petioles occasionally winged; cauline leaves lanceolate to lance-ovate or ovate, (1.6)2–7.8 cm long, 0.3–2.4 cm wide, clasping to cordate-clasping, gradually reduced upward, the upper 1/2 of the stem with leaves normally equaling or longer than the internodes. Thyrse 6–24(31) cm long, with (2)3–10(12) verticillasters, elongate, loose to moderately compact, secund, cymes few- to many-flowered, peduncles erect or slightly spreading; bracts resembling the upper cauline leaves below, much reduced above, acuminate, bases clasping to cordate-clasping. Calyx glabrous and somewhat glaucous, lobes lance-ovate to ovate, 4–6(7) mm long, 2.5–4 mm wide, short-acuminate to acute, margins broadly scarious and usually erose; corolla 15–25 mm long, tubular-salverform, scarcely bilabiate, pink to lavender or pale to deep blue, glabrous externally, throat 4–7 mm broad, slightly ampliate, scarcely ventricose anteriorly, lined internally on the anterior and posterior surfaces with prominent reddish or reddish-purple nectar guides, lobes of the upper and lower lips projecting or more commonly spreading, the limb usually appearing reltively flat, palate sparsely bearded with white eglandular hairs; staminode included or reaching the orifice, rarely exserted, very broadly flattened distally, the tip abruptly recurved and bifurcate, densely bearded with pale yellow to golden-yellow hairs to 1/3–1/2 its length, the hairs to 2 mm long; anther sacs 1–1.4 mm long, externally minutely papillose, divergent, dehisching nearly to the apices and across the connective, not becoming explanate; style glabrous. Capsule 9–12 mm long; seeds 2.5–3.5 mm long, angular, brown to dark brown. (n = 8) May–Jul. Rocky to gravelly soil on prairies, foothills, mesas, & more frequently at low elevation in mts.; restricted to se WY, e CO, & ne NM.

21. **Penstemon tubaeflorus** Nutt., tube penstemon. Robust herbaceous perennial; stems mostly erect, 2.5–8.5(10) dm tall, glabrous, stems 1 or 2(4) from a short stout herbaceous cuadex surmounting a short taproot. Leaves entire to obscurely serrate, glabrous; basal leaves obovate to spatulate, 2.5–11 cm long overall, 0.6–3.8 cm wide, obtuse to rounded, tapering to a petiolar base; cauline leaves lanceolate to lance-ovate, 1.5–10(13.5) cm long, 0.4–2(3.8) cm wide, acute to obtuse, sessile to distinctly clasping, lower cauline leaves large and separated by short internodes, upper cauline leaves much reduced and separated by longer internodes, the upper 1/2 of the stem appearing somewhat naked except for the inflorescence. Thyrse 8–30(40) cm long, with 4–8(12) distinct verticillasters, cymes 3- to 9-flowered or occasionally many-flowered and extremely congested in robust plants, peduncles appressed or ascending; bracts greatly reduced. Calyx glandular-pubescent, lobes lance-ovate to ovate, 2.5–5 mm long, 1.5–2.5 mm wide, acuminate to acute, entire, herbaceous or with narrowly scarious margins; corolla 15–22 mm long, barely bilabiate, funnelform and slightly decurved, glistening white with viscid glandular hairs externally and internally, throat 4–6 mm broad, barely inflated and moderately ampliate, unlined and unpleated internally, lobes of the upper lip spreading, lobes of the lower lip projecting slightly or spreading; staminode included or reaching the orifice, flattened only slightly distally and distinctly recurved, the terminal 3–4 mm sparsely bearded with yellow or sordid-yellow hairs to 0.8 mm long; anther sacs 0.8–1 mm long, glabrous, widely divaricate, dehiscent throughout and across the connective, becoming explanate; style glabrous. Capsule 7–10 mm long; seeds 1–1.3 mm long, rounded to slightly angular, tan to dark brown. (n = 16) May–Jul. Rich loam or sandy loam soil from open prairie to deciduous woodlands, disturbed areas; e NE, e 2/3 of KS, MO, & e 1/2 of OK; (ME w to ne NE, s to n MS, nw LA, & e TX).

Plants in the Midwest and GP, referable to var. *tubaeflorus*, probably occupy the native range of *P. tubaeflorus*, whereas those found in New England and surrounding areas are undoubtedly adventive (see Pennell, op. cit.). These naturalized plants in New England tend to be taller and more slender and have been designated var. *achoreus* Fern. (Fernald, Rhodora 51: 70–85. 1949). In the GP, the species shows some westward extension of its presumed original range, now found at numerous localities in w-cen and n-cen KS and e NE where it was unknown as recently as the 1930s. Most of these new localities are disturbed habitats such as roadsides, old fields, and pastures where it occasionally forms extensive stands.

22. **Penstemon virens** Penn. Cespitose herbaceous perennial; stems erect or assurgent, 1–4 dm tall, slender, spreading to retrorsely puberulent in lines below and glandular-pubescent near the inflorescence, arising from a branching suffrutescent caudex. Leaves entire to serrulate or denticulate, glabrous and bright green; basal leaves lanceolate to oblanceolate or spatulate, 2–10.2 cm long overall, 0.4–1.5 cm wide, petiolate, tufted; cauline leaves lanceolate to lance-ovate, 1.8–5(7) cm long, 0.3–1.4 cm wide, acuminate to acute, sessile or the upper clasping. Thyrse 5–18 cm long, with 3–6(8) distinct or indistinct verticillasters, cymes 2- to 5-flowered, erect; bracts somewhat prominent below and much-reduced above. Calyx glandular-pubescent, lobes lance-ovate to ovate, 2–4.5 mm long, 1.5–2.5 mm wide, acuminate to acute, with margins narrowly scarious near the base; corolla 10–16(18) mm long, bilabiate, pale to dark blue-tinged violet, typically somewhat lighter anteriorly and near the tube, pale internally, glandular-pubescent externally, throat 3–5 mm broad, slightly inflated and moderately ampliate, flattened slightly, lined internally with reddish-purple or bluish-purple guidelines, weakly 2-ridged within anteriorly, palate moderately bearded with whitish eglandular hairs, lobes of the upper lip spreading to somewhat recurved, lobes of the lower lip spreading and exceeding the upper lip; staminode included or reaching the orifice, barely if at all flattened distally, the tip feebly recurved, the distal 4–5 mm pubescent, densely bearded at the tip with stiff golden-yellow hairs to 1.3 mm long, much more sparingly bearded away from the tip; anther sacs 0.6–0.8 mm long, the entire external surface minutely papillose, divaricate, dehiscent nearly to the apices and not across

the connective, not becoming explanate; style glabrous. Capsule 5–7 mm long; seeds 1–1.3 mm long, rounded to slightly angular, dark brown. (n = 8) May–Aug. Gravelly or rocky, frequently granitic, soil on wooded or brushy slopes in foothills & mts.; CO: Douglas, Elbert, El Paso; WY: Albany, Laramie; (n-cen WY s to s-cen CO).

This species is occasionally found on the high plains near the Front Range. There it frequently forms extensive mats and is often found in habitats with *P. secundiflorus.*

23. SCROPHULARIA L., Figwort

Perennial herb; stems coarse, quadrangular. Leaves simple (ours), opposite, petiolate, the blades irregularly serrate to divided. Inflorescence a relatively large, thyrsoid panicle; bracts much-reduced, alternate or subopposite. Flowers small; calyx deeply and subequally 5-parted, the segments broad, obtuse to rounded; corolla greenish-yellow or greenish-purple to dark reddish-brown, the tube globular to urceolate, strongly bilabiate, the upper lip 2-lobed, flat, erect, external in bud, the lower lip shorter with vertically projecting lateral lobes and a deflexed middle lobe; fertile stamens 4, only slightly didynamous, the anthers 1-celled (by confluence of the ancestral pair); staminode (posterior sterile stamen) flattened, narrowly clavate to broadly flabelliform, attached to the corolla high on the tube and appressed against the dorsal side; stigmas united and capitate. Capsule ovoid or subglobose, septicidal, the walls firm; seeds numerous, oblong-ovoid, turgid, furrowed.

1   Staminode yellow-green, flabelliform, 1.1–2.0 mm wide, usually wider than long; capsule ovoid, acuminate, 5.5–10 mm long, dull brown .... 1. *S. lanceolata*
1   Staminode brownish or purple, clavate, 0.8–1.2 mm wide, usually longer than wide; capsule broadly ovoid to subglobose, 4.5–6 mm long, usually shiny brown .... 2. *S. marilandica*

1. Scrophularia lanceolata Pursh. Plants 8–15 dm tall, glandular-pubescent, the stems clustered. Leaves petiolate, the blade lanceolate to lance-ovate, acuminate, rounded, truncate or cordate at base, 8–13 cm long, 3–5(7) cm wide, simple- or double-serrate, the petiole 1–3(5) cm long. Panicle narrow and elongate, 4–8 cm wide. Calyx 2–3.8 mm long, the segments ovate to broadly rounded, erose-margined; corolla 8–12 mm long, pale brownish or reddish-brown, the median anterior lobe with a greenish cast, the throat slightly constricted; staminode yellow-green, flabellate, 1.1–2.0 mm wide, usually wider than long. Capsule dull brown, ovoid, acuminate, 5.5–10 mm long; seeds 0.8–0.9 mm long. (2n = 92–96) May–Jul (Aug). Moist stream banks, thickets, & open woods; GP, except w parts of KS & OK; (s Can. from N.S. to B.C., ME to WA, s to VA, OH, IN, IL, MO, OK, n NM, UT, & n CA). *S. dakotana* Lunell, Amer. Midl. Naturalist 5: 240. 1919; *S. leporella* Bickn.—Winter; *S. occidentalis* (Rydb.) Bickn.—Rydberg.

2. Scrophularia marilandica L. Plants up to 3 m tall, glandular-pubescent above, glabrous below. Leaves petiolate, the blades lanceolate to ovate, acuminate, cuneate to truncate or sometimes subcordate at base, 8–15(20) cm long, 3–7(9) cm wide, sharply serrate to dentate-serrate, the petioles 1.5–5(8) cm long. Panicles tending to be pyramidal, 5–18 cm wide. Calyx 2–3(3.5) mm long, the segments rounded, about as long as the tube, erose margined; corolla 5–9(10) mm long, reddish-brown, pale green anteriorly; staminode brown or brownish-purple, clavate, 0.8–1.2 mm wide, longer than wide. Capsule usually shiny brown, broadly ovoid to subglobose, acute, 4.5–6 mm long; seeds 0.6–0.7 mm long. (Jul) Aug–Sep. Rich woods & moist shady ravines; se SD, NE, KS, & OK, eastward; (s ME & adj. Que. to se SD & NE, s to SC, GA, AL, MS, LA, & ne TX). *S. marilandica* f. *neglecta* (Rydb. ex Small) Penn.—Steyermark; *S. neglecta* Rydb. ex Small—Rydberg.

24. TOMANTHERA Raf.

Annual, hemiparasitic herb; stems erect, retrorsely hispid. Leaves opposite, lanceolate to ovate in outline, pinnatifid, auriculate or entire. Inflorescence a spike, the bracts leaflike below and progressively reduced upwards. Flowers weakly bilabiate; calyx gamosepalous, regular or nearly so, 5-lobed, the lobes ovate to lanceolate and longer than the tube; corolla purple to magenta with red-purple spots inside the throat ventrally, the tube campanulate, subequally 5-lobed, the lower lip external in the bud; stamens 4, didynamous, included, the lower pair longer and with larger anthers, the filaments villous, the anther cells 2, parallel, blunt basally, villous; style elongate, the stigma solitary, linear. Capsule rounded at the summit with a mucro, loculicidal; seeds numerous, reticulate.

1 Leaves entire or the upper with a pair of spreading or ascending auricles at the base, lanceolate; corolla 20–27 mm long; capsule 10–14 mm long .... 1. *T. auriculata*
1 Leaves pinnatifid with 1–3 pairs of linear to filiform lateral lobes; corolla 25–33 mm long; capsule 8–10 mm long .... 2. *T. densiflora*

1. *Tomanthera auriculata* (Michx.) Raf., earleaf gerardia. Scabrous annual; stems 2–6(8) dm tall, simple or sparingly branched, retrorsely hispid. Leaves lanceolate, acute or acuminate, the upper auriculate at the base, 2.5–6 cm long, 0.5–1.5(2) cm wide, ascending-scabrous on the upper surface. Calyx campanulate, 10–16 mm long, the tube retrorse-pubescent and its lobes ascending-scabrous, the lobes ovate or widely lanceolate, longer than the tube, unequal in width but not in length, 6–9(13) mm long; corolla pink with dark purple spots inside the throat, tubular-campanulate with spreading lobes, 20–27 mm long, the upper lobes longer than the lower, broadly rounded, 6–10 mm long, ciliate; anthers 2.0–2.5 mm long; style pubescent. Capsule broadly ovoid, 10–14 mm long; seeds ovoid, 1.2–1.6 mm long, deeply reticulate. ($2n = 26$) Aug–Sep. Moist to dry prairies & woodlands, often in disturbed places; MO, e KS & e OK; (NJ to s MN, s to WV & cen TX, recurring in AL & MS). *Gerardia auriculata* Michx.—Fernald; *Otophylla auriculata* (Michx.) Small—Rydberg.

2. *Tomanthera densiflora* (Benth.) Penn., fineleaf gerardia. Leafy annual; stems 2–5 dm tall, much branched, finely retrorsely puberulent and coarsely spreading-hirsute. Leaves pinnatifid, 2–3.5 cm long, ascending-spreading, the 3–7 segments linear to filiform. Calyx narrowly funnelform, 12–16 mm long, the tube minutely scabro-puberulent to glabrate, its lobes ascending-scabro-puberulent, ciliate, broadly lanceolate, sharply acute, 7–10 mm long; corolla pink to lavender with dark purple spots inside the throat, tubular funnelform, 25–33 mm long, its lobes broadly rounded, 6–8 mm long, ciliate, the upper lobes arched and the lower spreading; anthers 2.5–3.0(3.5) mm long; style glabrous. Capsule obovoid, 8–10 mm long; seeds ovoid-triangular, 1.5–1.8 mm long, deeply reticulate. ($2n = 26$) Aug–Sep. Dry prairies & bluffs, often on limestone soils; cen KS through cen OK, southward; (a line from cen KS to s-cen TX). *Gerardia densiflora* Benth.—Barkley; *Otophylla densiflora* (Benth.) Small—Rydberg.

25. VERBASCUM L., Mullein

Biennial, taprooted herb (ours); stems usually tall, coarse, simple or virgately branched. Leaves basal and/or cauline, the basal forming a rosette, the cauline alternate, sessile, clasping or somewhat decurrent, pinnately veined. Inflorescence usually a simple raceme or spike, or sometimes a congested panicle; bracts reduced. Flowers nearly regular; calyx deeply 5-parted, the segments equal; corolla yellow or sometimes white, the tube very short, the limb rotate, slightly irregular, the upper 2 lobes external in bud and slightly shorter than the lower 3; stamens 5 and usually all anther-bearing (unusual for Scrophulariaceae), the anthers 1-celled (by confluence of the ancestral pair); the filaments usually villous and more or less alike, or the lower pair glabrous to sparsely hairy; style flattened distally, the stigma solitary and capitate. Capsule ovoid to globose, septicidal, thick-walled; seeds numerous, ours obconic (obovoid, truncate distally), with thick, sinuous longitudinal ribs which sometimes anastomose forming reticulae.

1 Leaves glabrous; filaments of all 5 stamens villous with violet hairs; inflorescence a raceme .... 1. *V. blattaria*
1 Leaves woolly-tomentose; filaments of the upper 3 stamens villous with yellow hairs, the lower 2 glabrous or nearly so; inforescence a congested spike or spikelike panicle .... 2. *V. thapsus*

1. Verbascum blattaria L., moth mullein. Biennial producing a rosette of basal leaves the first year, glandular-pubescent above with simple hairs and glabrous below; stems of second-year's growth 4–15 dm tall, usually simple or branched below. Leaves of the rosette oblanceolate, 5–20 cm long, 1–3 cm wide, short-petiolate, double-crenate, denticulate and often lobed to pinnatifid, the cauline leaves lanceolate, progressively reduced upwards, coarsely dentate to sinuous dentate or crenate, glabrous. Inforescence a simple, elongate raceme; pedicels 1–2(2.5) cm long. Calyx segments lanceolate, 5–8 mm long; corolla yellow or sometimes white, 25–30 mm diam; stamens, the lower 2 with decurrent anthers and violet-hairy filaments, the upper 3 with white and violet hairs on the filaments. Capsule globose, (5)7–8 mm long, glandular-puberulent, the style persistent; seeds 0.6–1.0 mm long. (2n = 18, 30, 32) Jun–Aug. Roadsides & weedy lots; se NE, KS, & e OK; (established as weed throughout most of Can. & U.S.; Europe). *Naturalized.*

A white flowered form has been called f. *albiflora* (G. Don) House.

2. Verbascum thapsus L., common mullein. Stout biennial, producing a rosette of basal leaves the first year, densely woolly-tomentose with dendritic-branched hairs; stems of second-year's growth 3–20 dm tall, simple or sometimes branched above. Leaves of the rosette obovate to oblanceolate, obtuse, 8–50 dm long, 2.5–14 cm wide, tapering to a petiolar base, entire to shallowly crenate, the cauline leaves oblanceolate, progressively reduced upwards, sessile, decurrent. Inflorescence a congested spike or spikelike panicle; bracts ovate to lanceolate, acuminate, 12–18 mm long; pedicels short or wanting. Calyx segments lanceolate, (5)8–12 mm long; corolla yellow, 12–20(35) mm in diam, ciliate with stellate hairs; stamens, the lower 2 glabrous to sparsely villous, the upper 3 filaments yellow-villous. Capsule broadly ovoid, 7–10 mm long, densely covered with stellate or dendritic-branched hairs, the persistent style glabrous above; seeds 0.7–0.8 mm long. (2n = 32, 34, 36) Jun–Jul. Disturbed roadsides, weedy lots, & fields; throughout the GP, but more common in e 1/2; (established as a weed throughout temp. N. Amer.; Europe). *Naturalized.*

26. VERONICA L., Speedwell

Annual, biennial, or perennial herb; stems erect, decumbent or prostrate. Leaves opposite, all cauline (not to be confused with the alternate, leaflike floral bracts in some), entire, crenate, serrate or deeply lobed. Inflorescence a terminal or lateral raceme, the bracts leaflike to reduced, alternate or sometimes subopposite below. Flowers weakly bilabiate; calyx deeply 4-parted, united only at the very base, the segments equal or the anterior pair longer in some; corolla blue, violet to pink or white, subrotate, the tube usually very short, irregularly 4-lobed, the posterior (upper) lobe largest (formed by the fusion of 2 lobes) and the lowermost the smallest, the lower 3 external in bud; stamens 2 (posterolateral pair), each with 2 equal, more or less parallel anther cells, exserted; stigma solitary and capitate. Capsule ovoid, orbicular or often emarginate to deeply notched into a heart shape, strongly laterally compressed to subterete, loculicidal, rarely septicidal at apex as well; seeds few to many, flattened, plano-convex or cup-shaped, smooth to rugose.

References: Pennell, F. W. 1921. "Veronica" in North and South America. Rhodora 23: 1–22, 29–41; Walters, S. M., & D. A. Webb. 1972. *Veronica. In:* Flora Europaea 4: 242–251.

Nine of the following 14 species are introductions from the Old World. The *Flora Europaea* keys (cited above) or other Old World treatments should be consulted for weedy collections that do not appear to be accommodated in the following key.

Veronica latifolia L. is sometimes cultivated in the region. Collections from SD have been reported but no specimens were seen. It is included in the key only.

1 Racemes axillary and opposite (alternate in *V. scutellata*), never terminating the main stem.
  2 Herbage pubescent with nonglandular hairs; plants of dry habitats (sect. *Veronica*).
    3 Leaves sessile or subsessile, cordate to truncate at the base, coarsely dentate; calyx segments unequal, the longer lobes 3.2–4 mm long in fruit; corolla (5.5)6–8.5 mm long; capsule pubescent with nonglandular hairs; stems ascending to erect .... *V. latifolia*
    3 Leaves narrowed to a petiolar base, crenate-serrate; calyx segments mostly equal, 2–3(3.8) mm long in fruit; corolla 3.3–4.6 mm long; capsule glandular-pubescent; stems decumbent, ascending at apex .... 9. *V. officinalis*
  2 Herbage glabrous or sparsely glandular-puberulent; plants of wet habitats (sect. *Beccabunga*).
    4 Leaves lanceolate to ovate, 1.5–5× longer than broad; capsule 2.5–4 mm long and about as wide, entire or notched less than 0.3 mm deep; seeds 0.3–0.7 mm long.
      5 Leaves all petiolate .... 2. *V. americana*
      5 Leaves sessile, rarely the lower subpetiolate.
        6 Leaves broadly lanceolate to ovate, 1.5–3× longer than wide, succulent; racemes usually more than 30-flowered, the pedicels upcurved to ascending; capsule longer than wide, scarcely notched; corolla 5–10 mm across, blue or pale violet .... 3. *V. anagallis-aquatica*
        6 Leaves lanceolate, (2.5)3–5× as long as wide, thin; racemes less than 25-flowered, the pedicels more or less straight and spreading; capsule slightly wider than long with a more evident notch, this 0.1–0.3 mm deep; corolla 3–5 mm across, whitish, sometimes pale bluish .... 6. *V. catenata*
    4 Leaves narrow lanceolate to linear, 4–20× longer than broad; capsule distinctly wider than long, 2.3–3.2(3.5) mm long, 3.6–4.5 mm wide, deeply notched, 0.4–1.0 mm deep; seeds 1.3–1.8 mm long .... 12. *V. scutellata*
1 Racemes terminal, sometimes with additional axillary racemes, then alternate, or rarely in opposing pairs in the axils of the leaves.
  7 Plants tall and stout, 4–12 dm tall; racemes of closely crowded flowers; corolla tube longer than wide; leaves 4–15 cm long; capsule longer than wide, the style 4–10 mm long (sect. *Pseudolysimachium*) .... 8. *V. longifolia*
  7 Plants less than 4 dm tall; racemes less crowded; corolla tube wider than long; leaves less than 3.5 cm long; capsule mostly wider than long, the style 0.1–3.0 mm long.
    8 Perennial from a rhizomatous base; raceme well defined and terminal (sect. *Veronicastrum*) .... 13. *V. serpyllifolia*
    8 Annual from a short tap or fibrous roots; raceme indistinct, the flowers appearing solitary in the axils of leaves, often floriferous from near the base of the stem (sect. *Pocilla*).
      9 Pedicels 0.5–1.5(2) mm long, shorter than the calyx; seeds more or less flat and smooth, numerous, small, 0.5–1.0 mm long; calyx linear to lanceolate.
        10 Notch of the capsule 0.2–0.5 mm deep; style 0.1–0.4 mm long; calyx segments subequal; corolla whitish .... 10. *V. peregrina*
        10 Notch of the capsule 0.5–0.8 mm deep; style 0.4–1.0 mm long; calyx segments unequal, the outer pair longer; corolla blue to blue-violet .... 4. *V. arvensis*
      9 Pedicels longer, 4–30 mm long, usually longer than the calyx; seeds cup-shaped (with one face convex and the other deeply concave) and usually rugose on the outer surface, relatively few per locule, 1.2–3.1 mm long; calyx broadly (ob)lanceolate to ovate.
        11 Pedicels 15–30 mm long; locular lobes of the capsule divergent, the tips pointed outward; style 2–3 mm long .... 11. *V. persica*

11 Pedicels 4–12(15) mm long; locular lobes of the capsule more or less parallel, the tips erect; style 0.6–1.6 mm long.
  12 Leaves crenate or serrate, pinnately or sometimes subpalmately veined; notch of the capsule 1.0–3.5 mm deep.
    13 Notch of the capsule 1.0–1.6 mm deep; stems prostrate or weakly ascending; pedicels becoming recurved at maturity; calyx segments obtuse to rounded .... 1. *V. agrestis*
    13 Notch of the capsule 2.0–3.5 mm deep; stems more or less erect; pedicels remaining straight; calyx segments acuminate .... 5. *V. biloba*
  12 Leaves palmately 3- to 7-lobed and veined; notch of the capsule 0.2–1.1 mm deep.
    14 Capsule 4.2–5.5 mm long, the notch 0.8–1.1 mm deep and the style 1.2–1.6 mm long; seeds 1.5–2 mm long, 6–12 per locule .... 14. *V. triphyllos*
    14 Capsule 3–4 mm long, the notch 0.2–0.3 mm deep and the style 0.6–0.9 mm long; seeds 2.3–3.1 mm long, 2 per locule .... 7. *V. hederaefolia*

1. Veronica agrestis L., field speedwell. Pubescent annual; stems prostrate or weakly ascending, 0.5–2.5 dm long, often branched. Leaves petiolate, the blade ovate to rotund, 4–20 mm long, 2–12 mm wide, rounded to obtuse apically, truncate or subcordate basally, crenate-serrate, the petiole 1–5 mm long. Raceme terminal, indistinct, with leaflike bracts; pedicels 5–15 mm long, usually recurved at maturity. Calyx segments broadly (ob)lanceolate to (ob)ovate, obtuse to rounded, 3.5–6.5 mm long, veiny; corolla blue or white, inconspicuous, ca 2.5–3 mm long. Capsule glandular-puberulent, obcordate, the lobes more or less parallel, 3–4 mm long, 3.5–5.5 mm wide, the notch 1.0–1.6 mm deep, the style 0.8–1.3 mm long; seeds yellowish-brown, 6–12 per locule, cup-shaped, ovate, 1.3–1.7 mm long, rugose on convex surface. (2n = 14, 20, 28) Apr–Aug. Weed in lawns & waste places; se NE, KS, & OK; (ne U.S. & adj. Can. from Newf. to Ont. & MI, s to NC, OH, & cen U.S. from MO & NE to LA & e TX; recurring in n FL & n Mex.; Europe). *Naturalized.* *V. didyma* Ten.—Gates, *V. polita* Fries—Fernald.

    Veronica agrestis is considered here in the broad sense to include the very similar *V. polita* Fries (*V. didyma* Ten.) and *V. opaca* Fries. In *Flora Europaea* (op. cit.), where these three are segregated, *V. agrestis* is distinguished by having a whitish corolla with a blue or pink upper lobe rather than all blue as in *V. polita* and *V. opaca,* and a sparsely glandular-hirsute capsule with none of the eglandular hairs characteristic of the other two. *Veronica polita* is considered to have broadly ovate calyx segments that often overlap at the base rather than the nonoverlapping, oblong-lanceolate to (ob)lanceolate segments of *V. agrestis* and *V. opaca.* These characters seem to be poorly defined in our material.

2. Veronica americana (Raf.) Schwien. ex Benth., brooklime speedwell. Aquatic, glabrous perennial; stems erect, ascending, usually decumbent at the base and rooting at the lower nodes, 0.5–3.5(6) cm long, usually branched. Leaves petiolate, the blade lanceolate to ovate, (0.5)1.5–3(5) cm long, (3)7–20(30) mm wide, acute to obtuse apically, cuneate, rounded or sometimes subcordate basally, crenate-serrate or sometimes subentire below. Racemes axillary, 10- to 25-flowered, the bracts reduced, linear; pedicels 5–10 mm long, divaricate. Calyx 2.5–4.5(5.5) mm long, the segments (ob)lanceolate to (ob)ovate; corolla blue, 5–10 mm across. Capsule usually slightly wider than long, 2.5–3.8 mm long, 3–4 mm wide, entire or scarcely notched, the style 1.7–3(4) mm long; seeds brownish, numerous, planoconvex, 0.5–0.7 mm long. (2n = 16, 36) Jul–Aug. Moist, muddy or inundated soils around springs, in slow-moving streams, lakeshores, or meadows; GP; (Newf. to AK, s to NC, TN, MO, TX, NM, AZ, CA, & n Mex.; ne Asia).

3. Veronica anagallis-aquatica L., water speedwell. Aquatic, rhizomatous perennial; glabrous or sometimes glandular-puberulent in the inflorescence; stems erect or ascending,

1-6(10) dm tall, simple to much branched at the base. Leaves sessile and clasping or the lowermost sometimes subpetiolate, the upper leaves elliptic-lanceolate to ovate, 1.5-3 × longer than wide, 2-6.5(8) cm long, 5-25(40) mm wide, acute to acuminate, crenate-serrate, the lower oblanceolate to obovate, obtuse to acute apically, cordate, rounded or truncate basally. Racemes axillary, 30- to 60-flowered, the bracts reduced, lanceolate to linear; pedicels 4-8 mm long, usually upcurved or ascending. Calyx segments broadly lanceolate, 3-5.5 mm long; corolla blue or pale violet with purplish guidelines, 5-10 mm across. Capsule both septicidal and loculicidal, 2.8-4 mm long, 2.5-3.8 mm wide, slightly longer than wide, and scarcely notched, the style 1.5-3 mm long; seeds numerous, more or less planoconvex, 0.3-0.5 mm long. (2n = 18, 36) Jun-Aug. Wet meadows, stream banks, & sand bars or in water of springs & slow-flowing streams; GP; (throughout most of temp. N. Amer.; Europe). *Naturalized*.

Sterile plants have been found in NE that are thought to be hybrids between *V. anagallis-aquatica* and *V. catenata* Penn. (Brooks, Rhodora 78: 773-775. 1976).

4. *Veronica arvensis* L., corn speedwell. Annual from a short taproot, somewhat villoushirsute or glandular-puberulent; stem erect, often decumbent at the base, sometimes wholly procumbent, 0.5-2(3) dm long, simple to much-branched. Leaves sessile or the lower ones petiolate, the blades ovate to broadly ovate, 0.5-1.5(2) cm long, 3-13 mm wide, rounded apically, truncate to subcordate basally, crenate-serrate. Raceme terminal, becoming elongate, the bracts leaflike at the lower nodes, becoming reduced to narrow elliptic and entire distally; pedicels 0.5-1(2) mm long. Calyx segments lanceolate, unequal, the outer pair longer, 3-6 mm long; corolla blue to blue-violet, inconspicuous, ca 2-3 mm across. Capsule glandular-pubescent on the margins, strongly flattened, obcordate, 2.5-3.5 mm long, 3-4 mm wide, with a notch 0.5-0.8 mm deep, the style 0.4-0.8(1.0) mm long, included to scarcely exceeding the lobes; seeds 8-12 per locule, flat, 0.7-1.0 mm long. (2n = 14, 16) May-Jun. Weed in open, moderately moist meadows, lawns, & gardens; se GP, from w IA & se SD, through e KS to MO & OK; (widespread throughout most of temp. N. Amer.; Eurasia). *Nauralized*.

5. *Veronica biloba* L. Glandular annual from a short taproot; stems erect or loosely ascending, 0.7-2(3) dm tall, simple or branched below. Leaves short-petiolate, the blade (ob)lanceolate to ovate, acute, 0.8-2.5(3.5) cm long, 3-12(20) mm wide, crenate-serrate. Raceme terminal, glandular-pubescent, relatively elongate, with more or less reduced bracts; pedicels 4-8 mm long. Calyx segments prominent, broadly lanceolate, acuminate, 4-8(12) mm long, entire or irregularly toothed, often veiny, bractlike; corolla blue, inconspicuous, 2-4 mm across. Capsule ciliate with long-stalked, gland-tipped hairs, 3-5 mm long, 4-6.5(8) mm wide, the lobes more or less parallel, the notch (2.0)2.5-3.5 mm deep, reaching past the middle, the style 0.8-1.4 mm long; seeds pale yellow, 2-5 per locule, cup-shaped, 1.5-2.6 mm long, rugose on convex side. (2n = 14, 28, 42) May-Jun. Weed in cult. fields & disturbed places; CO, along the Rocky Mts. front; (MT to WA, s to CO & UT; w Asia). *Naturalized*.

6. *Veronica catenata* Penn. Aquatic rhizomatous perennial, glabrous or glandularpuberulent; stems erect or ascending, 1-3(6) dm tall, much branched. Leaves sessile and clasping, lanceolate, acute, (2.5)3-5 × longer than broad, 2.5-5(9) cm long, 5-15(25) mm wide, entire or subcrenate, usually thin in pressed specimens. Racemes axillary, 15- to 25-flowered, the bracts much reduced, narrowly lanceolate; pedicels 3-7 mm long, more or less straight and spreading. Calyx segments broadly lanceolate to ovate, 2.5-3.5 mm long, obtuse to acutish; corolla white to pink or pale bluish, 3-5 mm across. Capsule wider than long, 2.5-3 mm long, 3-5 mm wide, bilobed, the notch 0.1-0.3 mm deep, the style ca 1.3-2.0 mm long; seeds yellow-brown, numerous, plano-convex, 0.3-0.7 mm long. (2n = 16, 36) Jun-Aug. Wet meadows, ditch banks, shores of lakes & ponds & in slow-flowing streams; GP; (NH & VT, across s Can. to s B.C., s to NJ, PA, OH, TN, MO, OK, NM, AZ, & CA; Europe).

6a. var. *catenata*. Upper stems and inflorescence glabrous. Capsule usually broader, 3.5–5 mm wide. n 2/3 GP; (s Man. & s Sask. to WA, s to MO, KS, NM, AZ, & CA). *V. comosa* var. *glaberrima* (Penn.) Boivin—Boivin; *V. connata* subsp. *glaberrima* Penn.—Stevens.

6b. var. *glandulosa* (Farw.) Penn. Upper stems and inflorescence glandular-puberulent. Capsule usually narrower, 3–3.5 mm wide. GP; (NY, VT, s Ont., & s Man., s to NJ & OK). *V. comosa* Richter—Fernald; *V. comosa* Richter var. *glandulosa* (Farw.) Boivin—Boivin; *V. connata* Raf.—Stevens, *V. salina* Schur.—Gleason.

7. **Veronica hederaefolia** L., ivy-leaved speedwell. Annual from a short taproot, sparsely to moderately spreading-hirsute; stems prostrate or weakly ascending, 0.5–2(3.5) dm long, branched at the base. Leaves petiolate, the blades broadly ovate, suborbicular, elliptic to ovate, 0.6–1.5 cm long and about as broad, palmately 3(5)-lobed, the lobes rounded, broadly and shallowly cordate, palmately veined, the petiole mostly 4–6 mm long. Raceme terminal, glabrous, elongate and indistinct, with leaflike bracts; pedicels 8–13(15) mm long. Calyx segments accrescent, broadly triangular-acute, ovate, (3)4–7 mm long, truncate at base with a short petiole, conspicuously ciliate, serrate; corolla pale bluish, ca 3–6 mm across. Capsule glabrous, plump, subglobose, 3–4 mm long, 4–5.2 mm broad, the lobes more or less parallel, the notch 0.2–0.3 mm deep, the style 0.6–0.9 mm long; seeds blackish, 1 or sometimes 2 per locule, cup-shaped, 2.3–3.1 mm long, transversely rugose. (2n = 18, 28, 36, 54, 56) Apr–May. Weed in disturbed habitats; SD: Brookings; (widely introduced in the ne U.S. & sporadic in SD, w & sw Can.; Europe). *Naturalized.*

8. **Veronica longifolia** L. Robust, erect perennial, 4–12 dm tall, puberulent to glabrous below. Leaves opposite or in whorls of 3(4), petiolate, the blades narrowlly lanceolate to lanceolate, 4–12(15) cm long, 1–3 cm wide, acute to acuminate apically, cuneate, truncate or subcordate basally, coarsely and sharply serrate, the petioles 3–15 mm long. Racemes 1 to few, erect, 7–22 cm long, with closely crowded flowers, the axils nonglandular-puberulent; pedicels 0.5–1.5(2) mm long. Calyx segments ovate, 2–3.5 mm long; corolla blue to lilac, 5.8–7(8) mm long, the tube well developed, longer than wide, densely villous within, the lobes 2.5–4.5 mm long; anthers 1.0–1.5 mm long. Capsule glabrous, little flattened, broadly ovoid, longer than wide, 2.3–3.3 mm long, 2–2.9 mm wide, emarginate, the notch 0.1–0.4 mm deep, the 4–10 mm long, dehiscing both loculicidally and septicidally (4-valved) at apex; seeds pale brown, numerous, flat, ca 0.7–0.8 mm long. (2n = 30, 34, 64–70) Jun–Jul. Dry roadsides & borders of fields & thickets; MN: Otter Tail; ND: McHenry, Wells; (scattered localities in se Can., cen & ne U.S. & w of the Cascades in WA & OR; Europe). *Naturalized.* *V. maritima* L.—Rydberg.

9. **Veronica officinalis** L., common speedwell. Pilose perennial with lower stems creeping, flowering branches ascending, rooting at the lower nodes. Leaves petiolate, the blades elliptic-obovate, 2–4(5) mm long, 1.5–2(3) cm wide, rounded apically, cuneate basally, crenate-serrate, the petiole 3–6 mm long. Racemes axillary, the bracts reduced; pedicels 0.5–2 mm long, ascending. Calyx segments lanceolate, subequal or sometimes the anterior pair slightly longer, 2–3 mm long at anthesis and to 3.8 mm in fruit; corolla light blue, sometimes with violet lines, 3.3–4.6 mm long; anthers 0.7–0.9 mm long. Capsule glandular-pubescent, triangular-obcordate, wider than long, 3.5–4.3 mm long, 4–4.8 mm wide, bilobed, the notch 0.1–0.3 mm deep, the style 2.5–4.5 mm long; seeds yellowish, 6–12 per locule, flat, 0.8–1.1 mm long. (2n = 18, 32, 34, 36) Jun–Jul. Relatively dry fields & woods; SD: Brookings; & e ND; (Newf. to Ont., s to NC, TN, IL, WS, SD, ND, & in the w from s B.C. to w OR; Europe). *Naturalized.*

10. **Veronica peregrina** L., purslane speedwell. Annual from a short taproot, glandular-pubescent, sometimes the leaves and bracts glabrous or the plant wholly glabrous; stem erect or ascending, 0.5-2(3) dm tall, simple or branched at the base, sometimes branched in the inflorescence in robust plants. Leaves sessile or the lowermost narrowed to a petiolar base, the blades narrow oblong to oblanceolate, 0.5-2.2 cm long, 0.5-5 mm wide, entire or irregularly crenate-serrate. Racemes terminal, indistinct, with leaflike bracts, these gradually reduced upwards; pedicels 0.5-1.5 mm long. Calyx segments narrowly elliptic to lanceolate, 3-6 mm long; corolla whitish, inconspicuous, ca 2-3 mm across. Capsule more or less obcordate, 3-4(4.5) mm long, 3.5-4.5(5) mm wide, with a broad notch, 0.2-0.5 mm deep, the style very short, 0.1-0.4 mm long; seeds numerous, flat, 0.5-1.0 mm long. (2n = 52) Jun-Aug. Moderately moist meadows, stream banks, & shores of lakes or ponds, often in muddy soil; GP; (temp. N. Amer., Mex., C. Amer., S. Amer., e Asia, & Europe).

Two varieties are recognized in the GP.

10a. var. *peregrina*. Plants glabrous. Wet habitats; e GP from MN & e SD to MO & OK; (P.E.I. & N.B. to s Ont., s to n FL, AL, MS, LA, & e TX & w of the Cascades from s B.C. to nw OR).

10b. var. *xalapensis* (H.B.K.) St. John & Warren. Plants glandular-puberulent. Wet habitats; GP; (s Can. & s AK, s to KY, MO, TX, NM, AZ, & CA; Mex. & Guat.). *V. peregrina* L. subsp. *xalapensis* (H.B.K.) Penn. — Gleason.

11. **Veronica persica** Poir., bird's-eye speedwell. Pilose annual from a short taproot; stems loosely ascending, often decumbent at the base and sometimes rooting at the lower nodes, 1-3 dm tall, simple or branched at the base. Leaves short petiolate, the blades ovate, broadly ovate to suborbicular, 0.7-2.5 cm long, 5-20 mm wide, obtuse to rounded apically, broadly cordate to truncate basally, deeply crenate-dentate, sometimes the lowermost entire. Raceme terminal, indistinct, elongate and lax, with leaflike bracts, these gradually reduced upwards; pedicels 15-27(30) mm long. Calyx segments prominent, broadly lanceolate, 4.5-7(8) mm long, veiny, ciliate; corolla blue, relatively large, 8-12 mm across. Capsule 3.5-4.5 mm long, (5)6-7 mm wide, the lobes divergent with the tips distinctly pointing outward, the notch prominent, 0.7-1.2 mm deep, the style 2-3 mm long; seeds brown, 5-11 per locule, cup-shaped, 1.2-2.0 mm long, transversely rugose on the convex side. (2n = 28) May-Jun. Weed in lawns & cult. fields; IA: Lyon, Sioux; SD: Lawrence; (Newf. to AK, s through much of U.S.; sw Asia). *Naturalized*.

12. **Veronica scutellata** L., marsh speedwell. Aquatic rhizomatous perennial, glabrous throughout; stems decumbent to erect, 1-4 dm long. Leaves sessile, narrowly lanceolate to linear, acute, 4-20 × longer than wide, 2-7(9) cm long, 2-8(15) mm wide, remotely toothed to entire. Racemes axillary, alternate, loosely 5- to 20-flowered, the bracts reduced; pedicels 6-12(15) mm long, filiform, becoming reflexed near the apex. Calyx segments broadly lanceolate, 2-2.7 mm at anthesis, to 3.5 mm in fruit; corolla pale violet or pink to blue violet or sometimes white, 4-5 mm long; anthers 0.6-0.7 mm long. Capsule much wider than long, 2.3-3.2(3.5) mm long, 3.6-4.5 mm wide, strongly 2-lobed, the notch 0.4-0.8(1.0) mm deep; seeds 5-9 per locule, flat, oval, 1.3-1.8 mm long. (2n = 18) May-Sep. Wet habitats in meadows & stream banks & shores; n parts of MN, ND, & MT; (circumboreal, extending s to VA, WV, OH, IN, IL, IA, ND, CO, ID, & CA).

13. **Veronica serpyllifolia** L., thyme-leaved speedwell. Rhizomatous perennial, finely puberulent or glabrous; stems ascending, often with decument bases or sometimes wholly procumbent, 0.8-2(3) dm tall or long, simple or short-branched below. Leaves subsessile to short-petiolate on the lower ones, the blade elliptic to broadly ovate, rounded to obtuse,

1-2(2.5) cm long, 8-15 mm wide, obscurely crenate to entire. Raceme terminal, distinct, 2.5-12 cm long, glandular-pubescent, the bracts lanceolate; pedicels 2.5-5(7.5) mm long. Calyx segments ovate or oblong, slightly unequal, 2.5-4 mm long; corolla white (ours), 4-8 mm across, the tube pubescent within. Capsule sparingly glandular-pubescent, not exceeding the calyx, 2.8-3.7 mm long, 3.5-5 mm wide, the notch evident, 0.3-0.8 mm deep, the style 2.2-3.0 mm long; seeds pale brown, numerous, flat, ca 0.6-0.7 mm long. (2n = 14, 28) Jun-Aug. Moist meadows, stream banks, gravel bars, & shores of lakes & ponds; KS: Saline; & NE: Lincoln; (cosmopolitan, widespread in Can. & U.S.).

Our plants belong to the European introduction, var. *serpyllifolia*, which differs from var. *humifusa* (Dickson) Vahl (a native of cooler climates in Can., adj. ne U.S., and the mts. of w U.S.) in having a pubescence of minute, upcurved (not spreading or glabrous) hairs, a white (not blue) and smaller corolla and smaller capsule.

14. Veronica triphyllos L. Glandular annual, floriferous from near the base; stems erect or often decumbent at the base, 0.5-1.7 dm tall, becoming much branched. Leaves petiolate or sessile, the blades ovate in outline, 8-15 mm long, 6-12 mm wide, palmately 3- to 5(7)-lobed or parted, the lowermost leaves sometimes merely crenate. Racemes terminal, indistinct, with leaflike bracts, these deeply 3- to 5-lobed; pedicels 4-12 mm long. Calyx segments broadly oblanceolate, rounded, the anterior pair longer than the posterior, 5.5-7.5(9) mm long in fruit; corolla deep blue, inconspicuous, 2.5-3.5 mm long; anthers ca 0.6 mm long. Capsule flattened, obcordate, 4.2-5.5 mm long, 4.8-6.4 mm wide, the lobes more or less parallel, the notch 0.8-1.1 mm deep, the style 1.2-1.6 mm long; seeds brown, 6-12 per locule, cup-shaped, 1.5-2 mm long, rugose on convex side. (2n = 14) Jun-Aug. KS: Cloud, Geary, Kingman, & Saline; OK: Oklahoma, Woods, & Cleveland; (e-cen KS to cen OK; NC; Europe). *Introduced.*

27. VERONICASTRUM Fabr., Culver's Root

1. Veronicastrum virginicum (L.) Farw. Stout perennial herb, 8-15 dm tall, glabrous or villous; stems erect, sometimes branched. Leaves 3-7 to a whorl, all cauline, lanceolate, acuminate apically, tapering to a short-petiolar base, (4)7-14 cm long, 1-3 cm wide, serrate, sometimes double-serrate. Inforescence 1 or more slender, densely flowered terminal spikes or spikelike racemes; bracts minute; pedicels 0.3-1.2 mm long. Flowers weakly bilabiate; calyx (4)5-parted, the segments lanceolate, 1.5-3 mm long, the 2 segments of the lower lip longer than the (2)3 of the upper; corolla pink or white, salverform, 4-5.5(6.5) mm long, nearly regular, 4-lobed, the lobes 1-1.8 mm long, much shorter than the tube, the upper lobe wider than the lower 3, the lower external in bud; stamens 2 (postero-lateral pair), long-exserted, the anther cells 2, equal, parallel, contiguous at the apex where often confluent, 1-2 mm long; style persistent, slender, about equaling the stamens, the stigma 1, minute. Capsule ovoid or ellipsoid, acute, more or less terete, 2.5-4.5 mm long, dehiscing by 4 short terminal slits; seeds light brown, numerous, terete, 0.5-0.7 mm long, minutely reticulate. Jun-Aug. Rich woods & moist prairies; e GP, in the easternmost parts of ND, SD, NE, & KS; (VT, s Ont., & se Man., s to nw FL, AL, MS, n LA, & ne TX). *Leptandra virginica* (L.) Nutt.—Winter.

The name f. *villosum* (Raf.) Penn. has been used for plants with densely pubescent lower leaf surfaces.

125. OROBANCHACEAE Vent., the Broomrape Family

by T. Van Bruggen

1. OROBANCHE L., Broomrape

Perennial, herbaceous plants, achlorophyllous, parasitic on roots of various vascular plants; stems fleshy with a caudex partly underground, 3–30 cm tall, yellow to brown or purple, solitary or clustered, scaly at their bases, more or less glandular or pubescent throughout above ground. Leaves alternate, sessile, reduced to white or brown appressed or ascending scales 0.5–1.5 cm long. Flowers solitary on axillary pedicels or in simple to branched terminal spikes; calyx of (4)5 basally united sepals equally or subequally lobed; corolla off-white to purple, withering, persistent, the tube slightly curved, bilabiate, the lips mostly erect, upper one 2-lobed, lower one 3-lobed; stamens 4, didynamous, inserted on the corolla, included or slightly exserted; ovary 1-celled with 2–4 parietal placentae; style elongate, solitary, the enlarged stigma capitate or lobed. Fruit a 2-valved capsule; seeds numerous, minute.

1. Flowers sessile in dense spikes 2–4.5 cm in diam, or on pedicels shorter than the corolla; bracts (1)2 at the base of the calyx; corolla lobes rounded or pointed at the apex.
  2. Corolla lobes obtusely rounded, yellow to purple; style persistent in fruit; calyx (9)10–19 mm long, exceeding the fruit .... 3. *O. multiflora*
  2. Corolla lobes triangularly acute, light to deep purple; style deciduous; calyx 8–12(14) mm long, shorter than to equaling the fruit .... 2. *O. ludoviciana*
1. Flowers solitary on naked, scapose pedicels 4–15 cm long; bracts at the base of the calyx lacking; corolla lobes orbicular.
  3. Caudex subligneous to woody, the 2–several branches above ground with several (5–10) scaly, leaflike bracts; axillary pedicels several, usually not more than twice the length of the caudex; calyx lobes shorter to equal to the tube .... 1. *O. fasciculata*
  3. Caudex herbaceous, short, the 1 or 2(3) branches with few (1–5) scaly leaflike bracts; pedicels 1–3, elongate, slender, several times longer than the caudex; calyx lobes longer than the tube .... 4. *O. uniflora*

1. *Orobanche fasciculata* Nutt. Erect, simple or several-stemmed plants 3–10 cm tall, not including the several erect pedicels, parasitic on the roots of various species, but especially *Artemisia frigida* Willd.; lower portion of caudex usually forked, becoming sublignous to woody, 1–2 cm in diam, with 5–10 scaly, overlapping, simple, erect to spreading bracts, the free portion 0.5–1.0 cm long, brownish, glabrous below soil level. Cauline leaves 3–7, lance-ovate, similar in structure to those below ground except the upper ones acuminate, alternate, not overlapping, glandular pubescent. Flowers solitary on 2–12 erect pedicels from leaf axils, 3–8 cm long, from shorter to not more than 2× the length of stem and caudex; calyx tube 3–4.5 mm long, campanulate, the lobes triangular, acuminate, glandular pubescent, 2–4 mm long, shorter to equaling the tube, persisting in fruit; corolla 14–30 mm long, usually purple, occasionally yellowish, the tube curved, pubescent, slightly bilabiate, the lobes 2–4 mm long, orbicular, spreading; stamens 4, didynamous, the filaments 5–8 mm long, inserted on the lower 1/3 of the corolla tube, mostly included; ovary 2-carpelled, glabrous, the style simple, equaling the ovary, tapered upward to a 2-lobed or crateriform stigma; locule 1, with (2)4 parietal placentae. Fruit a 2-valved capsule 0.8–1.2 mm long, the style persisting; seeds numerous, minute, less than 0.5 mm long, ovoid, rugose. Jun–Jul. Dry prairies, sandy soils; GP; (MI to Yukon, s to IN, NM, & CA). *Anoplanthus fasciculatus* (Nutt.) Walp.—Rydberg.

2. *Orobanche ludoviciana* Nutt., broomrape. Plants erect, of simple or clustered fleshy, stout stems 8–25 cm tall, 5–20 mm in diam, white to pink or yellow, drying dull brown, aboveground portion viscid-pubescent, parasitic on the roots of various vascular plants, especially *Ambrosia, Artemisia,* and *Xanthium* in the Asteraceae. Leaves scalelike, pubescent, 1–1.5 cm long, ascending-spreading, ovate to acute. Inflorescence 2–4.5 cm in diam, consisting of 1/2–1/3 of the stem height, spicate or the lower flowers racemose on pedicels up to 1 cm, many-flowered, dense; each flower subtended by (1)2 linear bracts 4–8 mm long; calyx 8–12 mm long, the basal 2–4 mm fused, the lobes linear-lanceolate, erect, subequal, viscid pubescent; corolla light to deep purple, rarely yellowish, curved, 14–25 mm long, bilabiate, with 2 yellow bearded folds in the open throat, the upper lip of 1 lobe, 4–7 mm long, erect, acute, the lower lip 4-lobed, slightly reflexed; stamens 4, inserted in the lower 1/3 of the tube, the filaments glabrous or pubescent, extended to the throat of the corolla, anthers white; ovary ovoid, glabrous, elongating to 1 cm or more in fruit, 2-carpelled, with 1 locule; style 6–9 mm long, curved, equaling the stamens. Fruit a 2-valved capsule with many small, angular, ovoid seeds. Jun–Aug. Dry, sandy upland prairies; GP; (Man. to B.C., s to IN & TX). *Myzorrhiza ludoviciana* Nutt.—Rydberg.

3. *Orobanche multiflora* Nutt. Plants stout, 0.5–3 dm tall, usually single-stemmed, thick, fleshy, viscid pubescent aboveground, glabrous belowground, the caudex enlarged, parasitic on various species of *Artemisia, Heterotheca, Hymenopappus* and others in the Asteraceae. Cauline leaves scaly, 5–11 cm long, lance-ovate with obtuse tips, appressed-ascending, inrolled, vestiture same as the stems. Inflorescence dense, spicate, rarely branched, 2–4.5 cm in diam, consisting of 1/3–1/2 of the stem height; upper flowers sessile, lower ones on short pedicels; (1)2 lance-linear bracts basally adnate to or immediately subtending the calyx, usually opposite; calyx (9)10–19 mm long, the lobes erect, attenuate, nearly equal, glandular or viscid-pubescent on the outside, violet on the inner surface; corolla 13–30 mm long, the lips 4–10 mm long, yellow to rose or pale-purple; upper lip erect, the lobes 3–5 mm long, obtusely rounded or slightly mucronate, stamens extending to the throat of the corolla, the anthers glabrous to white wooly; ovary ovoid, tapered to a curved style 5–9 mm long, equaling the stamens, elongating to 1 cm or more in fruit. Capsule longitudinally dehiscent, shorter than the calyx lobes, topped by the persistent style; seeds dark, angular, minute. Jun–Aug. Dry, sandy soils of beaches, dunes, or on gypsum ridges; s GP; CO, w KS, OK, TX; (CO to WA, s to TX & CA).

4. *Orobanche uniflora* L., one-flowered cancer-root. Plants simple-stemmed or clustered, slender caudex mostly subterranean, slightly prolonged above ground, 1–3 cm long, with 1–3(4) 1-flowered erect pedicels from the short, scaly stem, parasitic on roots of various trees and shrubs. Leaves 1–5, scaly, white to pale brown, overlapping, oblong-obovate, glabrous. Flowers solitary on slender off-white to yellow pedicels 0.5–2.5 dm tall, viscid or glandular-pubescent; sepals 5, the lobes triangular-lanceolate, 3-nerved, 6–9 mm long, exceeding the lower campanulate tubular portion, glandular-pubescent, bractless; corolla off-white to lilac, the tubular portion 15–18 mm long, with 2 yellow bearded folds in the throat, slightly curved, the lobes 4–6 mm long, nearly equal, wide spreading, orbicular or round-obovate, their margins minutely ciliolate; stamens 8–11 mm long, included, anthers white, glabrous to slightly hairy; ovary lance-ovoid, glabrous, tapering to a simple style that equals or exceeds the stamens, persisting in fruit, stigma two-lobed or crateriform. dehiscent, longer than the persisting calyx lobes; seeds numerous, angular. Apr–Jun. Woods, thickets, & bluffs; se GP; KS, MO, OK, TX; (Que. to MT, s to FL & TX). *Anoplanthus uniflorus* (L.) Endl.—Rydberg.

126. ACANTHACEAE Juss., the Acanthus Family

by Ralph E. Brooks

Ours annual or perennial herbs. Leaves simple, opposite, exstipulate, pinnately netveined. Flowers solitary or in congested few-flowered cymes or spikes, perfect, hypogynous, irregular to nearly regular, calyx persistent, equally 5-lobed; corolla gamepetalous, bilabiate or 5-lobed; stamens 4 and didynamous, or only 2 with 2 staminodes sometimes present; anthers 2-celled, longitudinally dehiscent; ovary 2-celled, 2–10 ovules per locule; style simple, filiform; stigmas 2. Fruit a longitudinally dehiscent capsule, flattened or clavate to fusiform, 2-celled.

1. Inflorescence a crowded, often capitate, spike on a peduncle at least 4 cm long ... 3. *Justicia*
1. Flowers axillary in few-flowered cymes or solitary and peduncled or not.
    2. Flowers nearly regular, 3 cm or more long .... 4. *Ruellia*
    2. Flowers bilabiate, less than 2.5 cm long.
        3. Perennial; stamens 4; anther sacs mucronate at the base. .... 2. *Dyschoriste*
        3. Annual; stamens 2; anther sacs rounded at the base. .... 1. *Dicliptera*

1. DICLIPTERA Juss.

1. ***Dicliptera brachiata*** (Pursh) Spreng. Annual herbs; stems diffusely branched, 3–7 dm tall, lineolate, glabrous to moderately hirsute. Leaves broadly elliptic to ovate, infrequently lanceolate, (2)5–15 cm long, (1.5)2–7 cm wide, sparsely pubescent, the veins below sometimes hirsute, apex acuminate or infrequently acute or obtuse, margins entire to remotely and obscurely toothed, ciliate, base attenuate; petiole (0.5)2–7 cm long, narrowly winged and densely ciliate. Flowers in axillary, sessile or subsessile, few-flowered cymes, near the branch tips the inflorescence appearing spicate where the leaves are absent or reduced, each cyme subtended by a pair of bracts; bracts obovate, 4–7 mm long, 2–5 mm wide, hirsute or sparsely so, acute to acuminate, occasionally apiculate, ciliate. Flowers perfect; calyx equally 5-lobed (rarely 4), 2.5–4 mm long, the tube 1–1.8 mm long, lobes narrowly triangular to subulate, 1.5–2.5 mm long, membranaceous, ciliate or not; corolla pale lavender to whitish with darker spots especially on the lower lip and throat, bilabiate, upper lip erect and entire or shallowly 2-lobed, lower lip usually entire, pubescent externally, with scattered glandular hairs sometimes present, 11–15 mm long, tube 3–6 mm long and only slightly expanded outward, the lips about equal in length; stamens 2, adnate to near the mouth of the corolla tube, slightly exserted or nearly so, filaments sparsely hirsute, anther sacs narrowly ellipsoid, about 1.7 mm long; style filiform, 10–15 mm long, hirsute; stigma bifurcate, the lobes about equal; much reduced cleistogamous flowers often produced. Capsules ovate or suborbicular, flattened, about equaling the bracts, sessile or subsessile, 2- to 4-seeded; seeds dark red-brown to blackish, suborbicular, flattened, about 2.5 mm diam, muricate. ($2n$ = ca 80) Aug–Oct. Moist, usually shaded woodlands, thickets, stream banks or lakeshores; s MO, se KS, e OK; (VA w to IN & se KS, s to FL & e TX).

2. DYSCHORISTE Nees

1. ***Dyschoriste linearis*** (T. & G.) O. Ktze. Perennial herb from a short, slender branching caudex, plant appearing tufted but sometimes producing rhizomes; stems (10)15–25(40) cm tall, frequently branching at the lower nodes, puberulent to hirsute at first and sometimes becoming glabrate. Leaves narrowly obovate to oblanceolate (1)1.5–4.5(6.5) cm long, 2–7(11) mm wide, glabrous or hirsute on the veins below, apex rounded or mucronulate, margins

ciliate, sessile. Flowers solitary or in few-flowered (mostly 3) cymes, axillary and congested, subtended by foliaceous bracts and bracteoles, these narrowly oblanceolate and often equaling the leaves in size. Flowers perfect; calyx deeply 5-parted, fused for about 1/3–1/2 their length, 15–22 mm long, hispid on the veins, the lobes subsetaceous and often flexuous toward the tip, 10–15 mm long, 0.5–1 mm wide, lineolate, long ciliate; corolla lavender, bilabiate, upper lip 2-lobed, lower lip 3-lobed, pubescent externally, 15–25 mm long, the tube 4–7 mm long, slightly shorter than the abruptly ampliated limb, lobes spreading, rounded, 4–7 mm long; stamens 4, didynamous, united at the base and adnate to the base of the corolla tube, filaments pubescent, the shorter ones 10–15 mm long, the longer ones 13–18 mm long, anthers about 2 mm long, anther sacs oblong, mucronate at the base; ovary 2-celled; style filiform, about equaling the stamens in length, pubescent, posterior lobe of stigma rudimentary, anterior lobe oblique and slightly flattened. Capsules (not seen) included in the calyx, oblong-linear, glabrous, 4-seeded, separating into 2 valves; seeds flat. (Apr) May–Jun (Aug). Sandy or gravelly soil, mostly on sparsely vegetated prairie slopes; OK: Cotton, Jackson, Tillman; TX: Floyd; (OK, w 1/2 TX to e NM, s to n Mex.).

3. JUSTICIA Houst. ex L.

1. *Justicia americana* (L.) Vahl, water willow, American dianthera. Perennial herb with widely creeping cordlike rhizomes, usually forming extensive colonies; stems simple or branched, 3–10 cm tall, glabrous, often rooting at the lower nodes, the lowermost internodes usually swollen. Leaves mostly ascending, linear-elliptic or lanceolate to elliptic, occasionally oblanceolate, (4)6–15(20) cm long, (0.5)0.8–2.5(3) cm wide, subcoriaceous or not, glabrous, apex narrowly acute or occasionally broadly so to obtuse, margins entire to obscurely and remotely crenate, base attenuate, sessile or with a short petiole to 1 cm long. Inflorescence a usually crowded, often capitate spike of opposite flowers, 1–3 cm long, axillary, peduncle erect, 4–17 cm long, commonly exceeding the subtending leaf in length. Flowers perfect, sessile or nearly so; each subtended by 3 minute triangular bracts; calyx nearly regular, 5-lobed, 5–7 mm long, glabrous, the tube about 1 mm long, lobes lanceolate, 4–6.5 mm long, entire and ciliolate, apex acute; corolla bilabiate, pale violet to white, the upper lip often darker violet and the lower lip with scattered darker spots, (6)9–12 mm long, pubescent or not within, the upper lip emarginate and reflexed, the lower lip 3-lobed with the lateral lobes spreading and the lower deflexed and its margin wrinkled; stamens 2, about equal to or slightly longer than the corolla in length, anther sacs separate, 1 terminal and traverse and the other lateral, dark violet or white-and-violet mottled, rounded at the base; style sparsely pubescent or not; stigmas unequal. Capsule clavate, 11–18 mm long, glabrous, the base stipelike and about equaling the calyx in length, 2-celled and 4-seeded, the valves reflexed upon splitting; seeds brown or reddish brown, flat and reniform-orbicular, 2–3.5 mm long, densely verrucose. ($2n = 26$) (May) Jun–Sep. Shallow water & mud of rocky stream beds or along rivers, ponds, or lakeshores, mostly exposed sites; MO, e 1/3 KS, e 1/2 & sw OK; (w Que. w to MI & WI, sw to GA, e TX & e KS). *Dianthera americana* L.—Rydberg; *Justicia americana* var. *subcoriacea* Fern.—Fernald.

4. RUELLIA L.

Ours perennial herbs from a short, weak, sometimes branched caudex. Flowers axillary, solitary or in few-flowered cymes, or solitary and long-peduncled, nearly regular, large and showy; corolla funnelform, the narrow tube often longer than the campanulate throat, the limb of 5 nearly equal obtuse spreading lobes; stamens 4, didynamous, anther sacs rounded at the base; stigmas unequally 2-lobed. Capsules clavate to fusiform, constricted below into a stipelike base.

1   Calyx lobes linear-attenuate, less than 2 mm wide, much longer than the capsule .... 1. *R. humilis*
1   Calyx lobes lanceolate to linear-lanceolate, 2-4 mm wide, shorter to slightly longer than the capsule .... 2. *R. strepens*

1. **Ruellia humilis** Nutt., fringeleaf ruellia. Stems erect and simple at first, later becoming diffusely branched, to 6 dm long, occasionally 4-angled, villous-hirsute to glabrate; rhizomes short and thin, if present. Leaves broadly ovate to narrowly so, less often obovate, elliptic or lanceolate, 2-5(7) cm long, 1-2.5(4) cm wide, leaves generally equaling or shorter than the internodes, subcoriaceous, variously hirsute to villous, apex mostly acute but occasionally obtuse or rounded, margins entire, pilose-ciliate, base mostly cuneate to rounded, rarely truncate, sessile or leaves of the main axis with petioles to 3 mm long. Inflorescences (or flower, if solitary) mostly at the middle and upper nodes, subtended by lanceolate bracts 10-18 mm long and 0.7-1.5 mm wide, these hirsute-ciliate; calyx hirsute, especially the lobes, the tube 2-3 mm long, the lobes linear-attenuate, 18-30 mm long, much longer than the capsule, 0.7-1.5 mm wide; corolla lavender, 3-8 cm long, the narrow tube mostly much longer than the throat and limb, variously pilose externally, the limb 2-3.5 cm wide, pilose-ciliate; anthers 2.5-3.5 mm long, filaments 2-5 cm long, glabrous; style 2.5-6 cm long, frequently persisting for a time on the maturing capsule after the corolla has fallen; the longer stigma lobe usually recurved; small closed cleistogamous flowers are occasionally produced in the fall. Capsules tan to dark brown, 10-16 mm long, glabrous; seeds brown or red-brown, orbicular and flattened, about 3 mm diam, occasionally hirsute on the margins. (2n = 34) Jun-Sep. Dry prairies, rocky banks, or dry open woodlands; sw IA, se NE, MO, e 1/2 KS, OK except panhandle; TX: Wheeler; (PA & WV w to MI, se NE, e KS, & s to FL & TX). *R. carolinensis* of reports—Gates; *R. ciliosa* of reports—Rydberg; *R. humilis* var. *expansa* Fern., var. *frondosa* Fern., var. *longifolia* (A. Gray) Fern.—Fernald; *R. humilis* f. *grisea* Fern.—Steyermark.

2. **Ruellia strepens** L., limestone ruellia. Stems erect and simple or sparingly branched, 3-10 dm tall, slightly 4-angled, mostly glabrous to occasionally pilose-hirsute; caudex short, thin and sparingly branched, short rhizomes infrequently produced. Leaves mostly ovate, less often ovate-lanceolate, elliptic or obovate, 5-13(16) cm long, (1.5)2.5-5(6) cm wide, leaves generally 1-1/2 to 2 × longer than the internode above, membranaceous, glabrous to sparsely hirsute, apex acute or short acuminate, occasionally obtuse, margins entire to remotely crenulate, mostly ciliate, base attenuate or occasionally cuneate or rounded; petiole up to 2 cm long, usually narrowly winged. Inflorescence a congested, usually subsessile cyme (these flowers usually cleistogamous) subtended by a pair of lanceolate bracts 10-15 mm long, 1.5-2.5 mm wide and glabrous to sparsely hirsute or flowers solitary, pedunculed (to 6 cm long), and subtended by a pair of foliaceous bracts; flowers of the cymes each subtended by a pair of minute obovate bractlets; calyx glabrous to sparsely hirsute, frequently with glandular hairs, the tube 1-3 mm long, the lobes lanceolate to linear-lanceolate, 9-20 mm long, (1.5)2-4 mm wide, shorter than to slightly longer than the capsule, ciliate with long obviously multicellular hairs; corolla lavender to dark violet, rarely white, (3)4-7 cm long, glabrous or the veins pubescent externally, the narrow tube about 1/2 the length of the corolla, the limb 2-4 cm broad; ciliate or not; anthers 2.5-3 mm long, filaments 2-3.5 cm long; pubescent toward the base; style 2-4.5 cm long, glabrous to sparsely pubescent, rarely remaining attached after the corolla has fallen; cleistogamous flowers tubular and unexpanded, prolific seed producers. Capsules tan to brown, 10-15(20) mm long, glabrous; seeds brown to red-brown, orbicular and flattened, about 2.8 mm diam, occasionally sparsely hirsute. (2n = 34) Jun-Sep (Oct). Low, rich woodlands & shaded stream banks, occasionally drier upland sites; NE: Richardson; MO, e 1/2 KS, e 1/2 OK; (PA w to e IA, se NE, & e KS, s to SC & e TX). *R. strepans* var. *cleistantha* (A. Gray) S. McCoy—Fernald.

127. PEDALIACEAE R. Br., the Unicorn-plant Family

by Ralph E. Brooks and Ronald R. Weedon

1. PROBOSCIDEA Schmid., Unicorn Plant

1. *Proboscidea louisianica* (P. Mill.) Thell., devil's claw, unicorn plant. Fetid annuals with strong horsey odor, densely covered with clammy, obviously articulate glandular hairs, erect or sometimes decumbent, 1.5–6(10) dm tall, diffusely branched to the base; stems thick, spreading up to 1 m; arising from a well-developed, short taproot. Leaves opposite or occasionally alternate in the upper plant; blade reniform or less often suborbicular to broadly ovate, 3–20 cm long, width about equaling or slightly greater than the length, venation palmate, apex rounded, obtuse or rarely broadly acute, margins irregularly sinuate to entire, base cordate; petiole 3–25(30) cm long. Flowers 4–26 in open terminal racemes that mostly exceed the foliage, 8–30 cm long; pedicels 2–4 cm long at anthesis, lengthening slightly and thickening considerably in fruit; pedicel bracts linear-lanceolate, 6–11 mm long, deciduous. Flowers zygomorphic, protandrous; calyx spathaceous and unequally 5-lobed, split below to the base, deciduous, 10–22 mm long, calyx lobes obtuse or sometimes acute and about 1/2 the length of the calyx; 1 or several calyx bracts usually present, linear to oblanceolate, 7–15 mm long; corolla 5-lobed and somewhat bilabiate, tubular-cylindric for 2–5 mm then becoming broadly campanulate, gibbous, dull white to purplish throughout, mottled with yellow or reddish-purple, the inner tube conspicuously reddish-purple spotted and the lower portion yellow striped, tube 2–3.5 cm long, 1.5–2 cm wide, lobes 1.5–2 cm long, 1.5–3 cm wide; fertile stamens 4 and didynamous, accompanied by 1 staminode, or sometimes 2 and 3, respectively, adhering by their edges, epipetalous, about 1/3 as long as the corolla tube; anthers gland-tipped and with 2 divergent cells; filaments arcuate; carpels 2, united to form a 1-celled ovary with 2 parietal placentae, stigma with 2 flat lobes, style 1 and much longer than the anthers. Fruit a drupaceous capsule, bivalved and dehiscent, crested ventrally and sometimes dorsally, fruit body stout, to 1 dm long, 2–3 cm thick, terminated by an incurved beak that exceeds the fruit body in length, at maturity the beak splitting to form a 2-horned "claw"; seeds dull black, narrowly ovoid with 2 sides flattened and 1 convex, 6–11 mm long, corky-tuberculate. (2n = 30) Jun–Oct. Usually in sandy soil in pastures, fields, & waste areas; s IA w to CO, s to MO, OK, TX, & NM; scattered n to w SD; (native of s U.S. & Mex. but occurring n to ME, MN, SD, & CO). *Martynia louisianica* P. Mill.—Gates.

This plant has been cultivated in many parts of the United States.

128. BIGNONIACEAE Juss., the Bignonia Family

by Ronald R. Weedon

Ours trees or shrubs, climbing or twining woody vines. Leaves opposite, petioled, simple or pinnately compound, exstipulate. Inflorescence usually of dichasial cymes compounded into a monochasium, with bracts and bracteoles. Flowers perfect, sympetalous, large and showy; calyx with 5 teeth or lobes, or sometimes bilabiate; corolla tubular or campanulate, usually imbricate, 5-lobed, irregular or 2-lipped, deciduous; stamens epipetalous, didynamous, the 5th or posterior one sterile or vestigial, anthers of 2 diverging locules;

ovary 1- or 2-chambered with many ovules and usually parietal placentation, style long and simple, stigma 2-lobed. Fruit a capsule with large, thin, winged seeds.

This is mostly a tropical to subtropical group distinguished from other families by the absence of endosperm, the structure of the fruit, and the often conspicuously winged seeds.

Prepared with the assistance of Daniel G. Schneider.

1 Leaves compound, divided into 5-13 sharply toothed leaflets; climbing woody vines .... 1. *Campsis*
1 Leaves simple; deciduous shrubs to trees.
   2 Leaves sessile to short-petioled, linear-lanceolate, glabrous, entire .... 3. *Chilopsis*
   2 Leaves petioled, broadly ovate, often pubescent beneath, the larger ones sometimes lobed .... 2. *Catalpa*

1. CAMPSIS Lour.

1. ***Campsis radicans*** (L.) Seem., trumpet-creeper, trumpet vine. Deciduous, woody perennial vine, capable of climbing high by means of aerial rootlets but commonly seen sprawling on low shrubs or fences. Leaves opposite, pinnately compound, to 30 or more cm long, stipules as tufts of hairs; leaflets 5-13, ovate-lanceolate with short petiolules, the blades long-acuminate, rounded or cuneate at the base, serrate, and glabrous but for a fine pubescence along the veins beneath. Inflorescences terminal corymbose clusters of 8-15 flowers. Calyx tubular-campanulate, coarsely 5-toothed, the triangular teeth much shorter than the tube, 2-2.5 cm long, orangish-brown, leathery; corolla 4 × longer than calyx, tubular-funnelform, orange to orange-red outside, with spreading orange-red to scarlet lobes; stamens 5, 1 vestigial, didynamous, glabrous, included, the anthers with divergent pollen sacs dehiscing downward, appressed snugly against the upper side of the corolla tube; ovary with 2 chambers, subtended by a massive annular nectariferous ring; style elongate beyond the anthers at maturity with 2 spreading, flattened, ovate stigma lobes. Fruit a short-stalked capsule, cylindric-oblong, 1-2 dm long, beaked, keeled along the sutures, loculicidally dehiscent, the inside divided by a flat partition; seeds numerous, compressed, with 2 large wings. ($2n = 40$) Jun-Sep. Thickets, fence rows, roadsides, stream banks, flood plains, rocky hillsides, & open woods; n & e TX, AR, MO, sw IA, OK, e 2/3 of KS, se NE; SD: Clay, Union; ND: Cass; (NJ w to GP, s to FL; escaping from cultivation in many areas, becoming an aggressive weed southward; *naturalized* to CT & MI). *Tecoma radicans* (L.) Juss. — Rydberg.

Flowers are cross-pollinated primarily by hummingbirds and long-tongued bees. Trumpet vine is unique among temperate zone plants in that it has 5 distinct nectary systems, 4 of them extra-floral, illustrating the ant-guard symbiosis associated with many tropical and subtropical species. Ours are likely to be well populated with ants. (Elias & Gelband. Science 189: 289-291. 1975).

2. CATALPA L., Indian Cigar Tree

1. ***Catalpa speciosa*** Warder, northern catalpa, hardy catalpa, Catawba-tree. Deciduous tree to 30 m tall, the crown pyramidal; bark red-brown, broken into thick scales; branches and twigs coarse. Leaves opposite, simple, exstipulate, long petioled, ovate, 15-30 cm or more long, 10-20 cm wide, long acuminate, truncate to broadly cordate at the base, bright green and glabrous above, paler, densely pubescent and with well-developed nectaries below, odorless. Inflorescence a relatively few-flowered panicle 15-20 cm long, glabrous, terminal, with irregularly spaced linear pubescent bracts along the floral branch. Calyx lobes 2, distinct, ovate to obovate, apiculate, 9-11 mm long; corolla tubular, to 5(1-7) cm long and 4 cm wide, gibbous on the lower side, the lobes undulate, 2 above, 1 on each side, and

1 extended, notched lobe below, white with purple spots and streaks, and 2 yellow ridges on the lower lobe; stamens 2, epipetalous on the lower corolla lobe, their filaments 2 cm long, bent sharply upward near the yellow anthers, which are 4 mm long, 3 vestigial stamens epipetalous on upper lobes; pistil with cylindric green ovary, white arched style 2 cm long, stigma with 2 flat ovate lobes 2 mm long. Fruit a cylindric loculicidal capsule 25–45(60) cm long and 1–1.5 cm thick, brown, irregularly ridged, often curved, the valves remaining semiterete after dehiscence; seeds flat, 3–4 cm long, 4–5(8) mm wide, winged on 2 sides, the wings terminating in a tuft of hairs, whitish tan. ($2n = 40$) May–Jul. Stream banks, lowland woods, & uplands; planted extensively both for ornamental purposes & for post & pole wood; known to escape cultivation in GP; IA, MO, OK, e 2/3 of KS, e 1/3 of NE; SD: Clay, Lincoln, Union; ND: Cass; (sw IN to IL & MN, ND s to TN & TX, escaped from cultivation or naturalized in much of the se U.S., n to New England).

The similar *Catalpa bignonioides* Walt., southern catalpa or common catalpa, is known only in cultivation in the GP, although it may escape rarely (e 1/2 KS, e 1/3 of NE; ND: Cass). Although its original distribution is uncertain, the species is probably native in sw GA, nw FL, AL, and MS and is widely naturalized northward to s New England and NY w to OH, s MI, GP, and e TX. The species is occasionally difficult to distinguish from *C. speciosa* with herbarium specimens, suggesting that the relationships of these two taxa deserve further study. Nonetheless, horticulturists and others have made much of the differences between the two regarding the quality of the wood and the ornamental value of the trees, favoring *C. speciosa* in our area.

The principal characteristics for identifying *C. bignonioides* are: Leaves 10–25 cm long, apices abruptly short-acuminate, odor of crushed leaves unpleasant or musky. Inflorescence to 30 cm long with many flowers. Smaller flowers, the corolla wrinkled and creased, conspicuously spotted with purple, the lower petals entire, not notched. Capsules shorter and thinner, 15–40 cm long and 1 cm thick; valves of capsules flattening after dehiscence; seeds 2.5–4.5 mm wide, the tufts of hairs at their ends coming together to form a point.

3. CHILOPSIS D. Don

1. ***Chilopsis linearis*** (Cav.) Sweet, desert willow. Spreading deciduous shrub or small tree up to 5 m or more tall. Leaves opposite or alternate, simple, sessile to short petioled, linear to linear-lanceolate, 10–15(30) cm long and up to 1 cm wide, glabrous, often viscid. Inflorescence a terminal raceme, usually densely woolly, each pedicel subtended by a linear bract equal to or exceeding the pedicel in length; flowers fragrant. Calyx glabrous to pubescent, bilabiate, 1–1.5 cm long, broadly ovoid; corolla funnelform-campanulate, the tube scarcely longer than the calyx, 5-lobed, slightly bilabiate, usually whitish tinged with pale purple, or purplish-red with purple stripes; stamens 4, epipetalous, didynamous, included; anthers with 2 oblong-ovate cells, at maturity diverging $180°$ from each other, ventrally dehiscent, opening out wide; staminode 1; pistil ±2.5 cm long, the ovary narrowly cylindrical, the stigma spatulate, slightly divergent into 2 thin plates. Fruit a linear 2-valved capsule, up to 3 dm long and 1.5 cm thick; seeds numerous, oval, flat, winged with a fringe of hair at each end. ($2n = 40$) May–Sep. Riverbanks, floodplains, ravines, & dry stream beds; shelter belts; TX; OK: planted in Woodward, Payne; KS: Comanche, planted in Kingman, Montgomery; (CA e 1/2 of Mohave Desert s into Baja Calif., e to s NV, AZ, KS, s into Mex.). Sometimes *escaped* or *naturalized*.

References: Fosberg, F. Raymond. 1936. Varieties of the Desert Willow, *Chilopsis linearis*. Madrono 3: 362–366; Shinners, Lloyd H. 1961. Nomenclature of Bignoniaceae of the Southern United States. Castanea 26(3): 109–118.

129. LENTIBULARIACEAE Rich., the Bladderwort Family

by G. E. Larson

1. UTRICULARIA L., Bladderwort

Aquatic or semiterrestrial rootless herbs, annual or perennating by winter buds; stems elongate, creeping in mud or floating, often very slender and delicate. Leaves alternate, branched or finely dissected into few to many linear or filiform segments, some or all of the leaves bearing small pear-shaped or ovoid bladders which entrap tiny aquatic invertebrates; stem apices commonly producing winter buds in late summer and autumn. Inflorescence of 1 flower or usually 2-many flowers in a scapose, sparsely bracteate raceme borne above the water surface; bracts scalelike. Flowers perfect, strongly irregular; calyx 2-parted nearly to the base into an upper and lower segment, the upper one somewhat broader; corolla yellow, bilabiate, the upper lip erect, entire or shallowly 2-lobed, the lower lip entire or slightly 3-lobed, usually arched toward the base to form a conspicuous palate, the corolla tube prolonged backward into an elongate or saccate spur; stamens 2, inserted near the base of the tube; stigma unevenly 2-lobed, style short or obsolete, ovary superior, maturing into a 2-valved, many-seeded capsule with parietal placentae.

References: Ceska, A., & M. A. M. Bell. 1973. *Utricularia* (Lentibulariaceae) in the Pacific Northwest. Madroño 22: 74-84; Rossbach, G. R. 1939. Aquatic utricularias. Rhodora 41: 113-128.

1 Ultimate leaf segments flat, nearly to fully as wide as the primary ones.
  2 Bladders borne on leafless branches which are distinct from the leafy branches; leaf segments spinulose-toothed on the margins .... 3. *U. intermedia*
  2 Bladders borne on leafy branches that are only weakly differentiated from bladderless branches; leaf segments entire .... 4. *U. minor*
1 Ultimate leaf segments terete, filiform, the segments progressively narrower in successive branchings.
  3 Stems floating beneath the water surface, ca 1 mm thick; leaves pseudopinnately branched into many segments, usually with many bladders; flowers (3)6-20 per stalk .... 5. *U. vulgaris*
  3 Stems mostly creeping at the bottom of shallow water, less than 1 mm thick; leaves dichotomously branched into 2-several segments, with 1-few bladders; flowers 1-4 per stalk.
    4 Lower corolla lip 3.5-7 mm long, distinctly exceeding the thick, very blunt spur; leaves mostly once dichotomous .... 2. *U. gibba*
    4 Lower corolla lip 8-10 mm long, about equaling or shorter than the conic spur; leaves mostly twice dichotomous .... 1. *U. biflora*

1. *Utricularia biflora* Lam. Stems filiform, creeping on the bottom in shallow water or floating on debris, rather sparsely leafy, seldom over 15 cm long. Leaves very delicate, mostly twice dichotomous, with 3 or more segments and with 1-few bladders, rarely more than 5 mm long; bladders 0.5-1.5 mm long, 0.5-1 mm wide; winter buds absent. Inflorescence of 1-4 yellow flowers; scape 3-16 cm long; bracts 1-3, to 3 mm long; pedicels straight, to 15 mm long. Calyx lobes 2.5-4 mm long; lower lip of corolla 8-10 mm long, about equaling or shorter than the conic spur, the upper lip shorter than to about as long as the lower, the palate prominent. Capsule globose, 3.5-4 mm diam. (n = 14) Jul-Oct. Shallow water of ponds; barely entering our range in OK: Cleveland, Comanche, Ottawa; (e MS to OK, s to FL & TX).

2. *Utricularia gibba* L., conespur bladderwort. Very similar to no. 1 and differing mainly as follows: Leaves mostly only once dichotomous, with only 2 filiform segments and 1-few bladders, 3-10 mm long. Inflorescence of 1-3 flowers; scape to 10 cm long; pedicels 5-10 mm long. Calyx lobes 1.5-2.5 mm long; lower lip of corolla 3.5-7 mm long, distinctly exceeding the thick, very blunt spur. Capsule ca 5 mm in diam. (n = 14) Jul-Sep. Shallow water of marshes, ponds, & lakes; se & s-cen KS; (Que. to WI, s to FL, KS, TX; & along the Pacific Coast from s B.C. to CA; C. Amer. & W.I.).

3. *Utricularia intermedia* Hayne. Stems very slender, usually creeping along the bottom in shallow water, the leafy stems 1-5 dm long, the bladder-bearing branches subterranean, leafless, distinct from the leafy branches. Leaves numerous on aquatic branches, commonly trichotomous at the base and then 1-3 × dichotomous, 0.5-3 cm long, the segments slender, flat, with a central vein, not much reduced with each branching, the ultimate segments blunt-tipped, with 2-10 minute, spinulose teeth on each margin; bladders 1.5-4.5 mm long, 1-3 mm wide; winter buds ovoid or ellipsoid, 2-15 mm long. Inflorescence a lax raceme of 2-5 bright yellow flowers; scape 6-20 cm long; bracts 1.5-4 mm long; pedicels suberect, 3-15 mm long. Calyx lobes 2.5-3.5 mm long; lower lip of corolla prominent, 8-12(18) mm long, nearly 2 × longer than the upper lip, palate well developed, spur straight, nearly as long as the lower lip. Capsule globose, ca 3 mm in diam. Jul-Aug. Shallow water of springs, bogs, & swamps; ND: McHenry, Pembina; (circumboreal, in N. Amer. s to DE, IN, IA, ND, & CA).

4. *Utricularia minor* L., lesser bladderwort. Similar in habit to no. 3. Bladder-bearing branches subterranean, with fewer leaves than the aquatic branches; leafy stems 1.5-7.5 dm long. Leaves numerous on aquatic branches, commonly trichotomous at the base and then dichotomous or irregularly forked 1-3 ×, mostly 0.3-1 cm long, the segments slender, flat, entire, not much reduced with each branching, the ultimate segments strongly acuminate, sometimes toothed at the tip, lacking a central vein; bladders 1.5-2 mm long, 0.7-1.5 mm wide; winter buds obovoid to globose, 2-5 mm diam. Inflorescence of 2-10 pale yellow flowers in a lax raceme; scape 4-15 cm long; bracts 1-4, purple, 1-2 mm long, auriculate; pedicels recurved, 2-10 mm long. Calyx lobes 0.5-2.5 mm long; lower lip of corolla 4-8 mm long, 2 × longer than the upper lip, palate scarcely developed, spur small, saccate, to 1/2 as long as the lower lip. Capsule globose, ca 2-2.5 mm in diam. (2n = 36-40) Jul-Aug. Shallow water of fens & bogs; ND: Kidder, Pembina; (circumboreal, in N. Amer. s to NJ, IN, ND, & CA).

5. *Utricularia vulgaris* L., common bladderwort. Stems free-floating, all leafy, often extensive, 2-15 dm long, ca 1 mm thick, some branches occasionally lacking bladders. Leaves numerous, mostly dichotomous at the base and then repeatedly and unequally dichotomous so that segments appear in a pseudo-pinnate arrangement, 1-5 cm long, the segments terete, progressively reduced with branching, the ultimate segments filiform, attenuate; bladders usually numerous, borne near the bases of leaf branchings, those of primary branches larger than those of secondary branches, 1-4 mm long, 0.5-4 mm wide; winter buds ovoid or ellipsoid, 0.7-3 cm long. Inflorescence a lax raceme of (3)6-20 bright yellow flowers; scape stout, 10-30(40) cm long; bracts 1-5, 2.5-8 mm long; pedicels 6-30 mm long, recurved after anthesis. Calyx lobes 3-6 mm long; lower lip of corolla mostly 10-20 mm long, sometimes much smaller (5 mm) on later reduced flowers, upper lip about equaling the lower, the palate prominent, as large as the upper lip, spur bent forward, ca 2/3 to nearly as long as the lower lip. Capsule globose, ca 6 mm in diam. (n = 21, 2n = 36-40) Jun-Aug. Shallow, standing water of lakes, ponds, marshes, & ditches, often among rushes or cattails; GP; (circumboreal, in N. Amer. s to GP, FL, TX, AZ, & CA). *U. macrorhiza* Le Conte — Rydberg; *Lentibularia vulgaris* var. *americana* Nieuw. & Lunell, Amer. Midl. Naturalist 5:9. 1917.

130. CAMPANULACEAE Juss., the Bellflower Family

by Ralph E. Brooks

Ours herbaceous plants. Leaves simple, alternate and usually spirally arranged, estipulate. Flowers solitary or in racemes or spikes, perfect, epigynous, usually 5-merous (except carpels); calyx 5-lobed, usually persistent; corolla sympetalous, regular or irregular; stamens distinct or united; carpels 2-5, stigma capitate or 2- to 5-lobed, style 1, ovary at least partially inferior, 1- to 5-celled. Fruit a capsule opening by valves or pores, placentation axile or parietal in species with unicellular ovaries; seeds minute and numerous.

1 Corolla regular; anthers and filaments distinct; capsule opening by lateral pores.
  2 Flowers on definite peduncles or pedicels, all similar .... 1. *Campanula*
  2 Flowers sessile, dimorphic, the lowermost being cleistogamous .... 3. *Triodanis*
1 Corolla irregular; anthers and filaments united; capsules opening irregularly or from the apex by valves .... 2. *Lobelia*

1. CAMPANULA L., Bellflower

Our plants small- to medium-sized annual or perennial herbs; stems erect and simple to sparingly branched, or weak, often reclining and dichotomously branched. Inflorescence spiciform, racemose, or paniculate or flowers solitary. Flowers normally all open; corolla campanulate and shallowly 5-lobed or rotate and deeply 5-lobed, valvate in bud; stamens 5, distinct and attached at the base of the corolla, filaments dilated and ciliate at the base; carpels 3-5, stigma 3-5, style elongated, ovules many. Capsules turbinate to obconic or clavate, opening by pores.

Reference: Shelter, S. G. 1963. A checklist and key to the species of *Campanula* native or commonly naturalized in North America. Rhodora 65: 319-337.

1 Plants annual; corolla rotate; capsules erect .... 1. *C. americana*
1 Plants perennial; corolla campanulate; capsules usually nodding.
  2 Calyx lobes less than 3 mm long; stems weak and more or less reclining, 0.5-1 mm diam .... 2. *C. aparinoides*
  2 Calyx lobes 4-8 mm long; stems usually stiff and erect, 1-6 mm diam.
    3 Lower cauline leaves ovate to lanceolate and petioled, 1-5 cm wide .... 3. *C. rapunculoides*
    3 Lower cauline leaves oblanceolate, narrowly elliptic, or lanceolate, mostly sessile, 3-6(9) mm wide .... 4. *C. rotundifolia*

1. *Campanula americana* L., American or tall bellflower. Annual plant to 2 m tall, glabrous to moderately hirsute; stems robust, simple or sparingly branched (usually from the middle or upper portion of the central axis), leafy throughout. Leaves lanceolate to ovate-oblong, 6-15(20) cm long, 1.5-6 cm wide, gradually reduced upward, apex acuminate-attenuate, margins serrate, base attenuate to the narrowly winged petiole. Inflorescence spiciform with flowers solitary and subsessile in the axils or 2-4 (mostly 3s) in axillary cymes, the lower bracts leaflike and upward becoming reduced and linear-lanceolate or often mere rudiments. Flowers with calyx lobes persistent, linear and spreading, 5-10 mm long; corolla rotate, blue with a whitish center, deeply 5-lobed, the lobes broadly lanceolate to triangular-ovate, 6-14 mm long, 3-6 mm wide, usually hirsute on the veins without; style uniquely declined and upcurved. Capsules stiffly erect, slender obconic, 5-11 mm long, 2-5 mm diam, dehiscence by 3-5 roundish pores near the apex; seeds chestnut-brown, lustrous, nearly round to elliptic and biconvex, about 1 mm long, often with a thin, winged edge. (2n = 58) Mid Jun-Sep. Open woods, ditches, stream banks, & wetland sites; sw MN & se SD, s to MO & e 1/3 OK, infrequent in nw KS & w NE; (s Ont. w to MN & w NE, s to nw FL, AL, & OK). *C. americana* var. *illinoensis* (Fresen.) Farw.—Fernald; *Campanulastrum americanum* (L.) Small—Rydberg.

2. ***Campanula aparinoides*** Pursh, marsh bellflower. Perennial; stems weak and more or less reclining, branching mostly in the upper portion, leafy throughout, 1.5–5 dm long, 0.5–1 mm diam, retrosely hispid to glabrate on the angles; rhizomes filiform. Leaves linear, linear-lanceolate, or oblanceolate, 1–4.5 cm long, 1–5 mm wide, often greatly reduced above, 1-veined, glabrous or hispid on the vein below, apex acute to narrowly acuminate, margins entire to remotely serrate and rarely ciliate, base sessile. Flowers solitary on long slender pedicels on the upper portion of the stems; calyx lobes slightly spreading, triangular to lanceolate, 1–2.3 mm long, 0.5–1 mm wide, entire; corolla pale blue, campanulate, 3–8 mm long, lobes about 1/2 the corolla length and slightly spreading, glabrous; stamens included, anthers linear and about 3 × longer than the filaments; style straight and slightly exceeding the anthers, 3-lobed; pedicels lax and often tortuous, 1–4 cm long. Capsules obovoid, 2–3 mm long, 1.5–2.5 mm diam, opening by basal to median pores equal in number to the locules; seeds pale chestnut-brown, shiny, ellipsoid, about 1.2 mm long. Jul–Aug. Moist meadows & marshes, occasionally in shallow streams; wooded hillsides in the BH; MN; SD: Day; sw 1/4 SD, sw to e-cen NE; (N.S. w to Sask., s to GA, KY, MO, & NE). *C. aparinoides* var. *grandiflora* Holz.—Gleason & Cronquist; *C. uliginosa* Rydb.—Rydberg.

3. ***Campanula rapunculoides*** L., creeping or rover bellflower. Perennial plants to 1.5 dm tall; stems simple or sparingly branched, 2–6 mm diam, hispid to glabrate; rhizomes widely creeping, 1–3 mm diam. Leaves ovate to lanceolate, the lowermost tending to be wider than those above, 3–15 cm long, 1–5 cm wide, gradually reduced upward, glabrous above and often hispid below, apex acute to acuminate, margins irregularly double serrate, the lower leaves with a petiole to 6 cm long and the base cordate to rounded, the upper leaves usually sessile or subsessile and cuneate at the base. Inflorescence racemose, the lowermost flowers on pedicels 1–2(6) cm long and subtended by foliaceous bracts, pedicels becoming shorter and the bracts reduced upward. Flowers usually nodding; calyx lobes lanceolate and slightly spreading, 5–8 mm long, scabrous to nearly glabrous, entire; corolla campanulate, blue, 1.5–3 cm long, lobes mostly triangular, about 1/3–1/2 the corolla length, apex acute and usually ciliate. Capsules usually nodding, nearly globose, 5–8 mm diam, opening by basal pores equal in number to the locules; seeds chestnut-brown, lustrous, elliptic and biconvex, about 1.4 mm long. ($2n = 68$, 102) Jun–Aug. Exposed stream banks & disturbed sites; infrequent MN & ND s to e & cen NE; KS: Scott; MO: Clay; WY: Weston; (N.S. to ND, s to DE & NE; Eurasia). *Naturalized* in the northern GP but occurring only as a waif from garden plantings in the southern plains.

4. ***Campanula rotundifolia*** L., harebell. Perennial plant to 1.5–7 dm tall; glabrous or sometimes hirtellous below in lines on the stem and leaf bases; stems somewhat soboliferous, simple or with short branches near the top, 1–3 mm diam; rhizomes much branched, 1–3 mm diam. Leaves dimorphic; basal leaves not often collected, blade nearly round to ovate or narrowly elliptic, 1–3 cm long, 0.5–2 cm wide, apex rounded to acute, margins coarsely serrate to remotely so, base cordate to cuneate, petiole 1–7 cm long; cauline leaves remotely serrate to entire and mostly sessile, those toward the base narrowly elliptic to oblanceolate, 2–9 cm long, 3–6(9) mm wide, and upward soon becoming linear-lanceolate or linear, 1–7 cm long, 0.5–3 mm wide. Flowers pendulous in a loose raceme or panicle or solitary, terminal or axillary on short pedicels; calyx lobes mostly appressed to the corolla, subulate, 4–8 mm long, entire; corolla blue, campanulate, 10–20 mm long, lobes about 1/4–1/3 the length of the corolla, flaring slightly, glabrous or rarely minutely ciliate; stamens included, anthers linear and 3–4 × longer than the filaments; style straight and exceeding the anthers, 3-lobed; pedicels filamentous, 0.5–6 cm long. Capsules turbinate to obconic 4–10 mm long, 3–6 mm diam, opening by basal pores equal in number to the locules;

seeds chestnut-brown and shiny, narrowly elliptic and flattish, about 1 mm long. (n = 17, 28, 34, 68, 102) Jun–Aug (Sep). Dry woods, cliffs, meadows, & stream banks; MN w to MT, s to WY, w 1/2 SD & nw & n-cen NE; CO: Elbert, El Paso; (circumboreal, s in N. Amer. to WV, MO, NE, w TX, NM, AZ & CA; Mex.). *C. intercedens* Witasek; *C. petiolata* A. DC. — Rydberg.

2. LOBELIA L., Lobelia

Small- to medium-sized annual and perennial herbs with acrid, usually milky juice. Inflorescence a crowded to loose raceme. Flowers inverted in anthesis, the pedicel twisted; corolla irregular, the tube usually split to the base on one side, the limb 5-lobed with 3 lobes forming the "upper" lip and 2 lobes (1 each side of the cleft) erect or recurved, the 2 next to the cleft often separating incompletely from the tube from below upward, making the tube fenestrate; anthers united, forming a ring around the style, 2 distinctly smaller and with a tuft of hairs at the tip, the 3 larger anthers pubescent on the backs or at the tip; filaments united above for 1/2–1/3 their length; stigma 2-lobed with a ring of hairs below the apex; ovary 2-celled with axial placenta; ovules numerous, anatropous. Capsules 2-valved; seeds numerous and minute.

References: McVaugh, R. 1936. Studies in the taxonomy and distribution of eastern North American species of *Lobelia*. Rhodora 38: 241–298; McVaugh, R. 1940. Campanulaceae *in* Contributions towards a flora of Panama. Ann. Missouri Bot. Gard. 27: 347–349.

The following species are of doubtful occurrence in our region:

Lobelia halei Small (= *L. flaccidifolia* Small) was reported by McVaugh (1936) from KS: Labette. The collection data may be in error since the taxon is otherwise restricted to the Gulf Coastal Plain. It differs from *L. spicata* by having bracteoles at or above the middle of the pedicels, a character that is infrequently observed on se GP plants.

Lobelia puberula Michx. is reported by Rydberg (1932) for KS but specimen evidence is lacking.

1 Flowers more than 15 mm long.
   2 Corolla crimson; filament tube (15)18–30 mm long .... 1. *L. cardinalis*
   2 Corolla blue (rarely white); filament tube 12–15 mm long .... 4. *L. siphilitica*
1 Flowers less than 15 mm long.
   3 Cauline leaves linear-lanceolate to linear-oblanceolate, 0.5–3 mm wide; pedicels bracteolate above the middle .... 3. *L. kalmii*
   3 Cauline leaves wider, at least some 5 mm wide or more; pedicels bracteolate at or near the base.
      4 Corolla white with bluish lobes, 3–5(6) mm long; capsules 6–10 mm long; plants annual and frequently branched .... 2. *L. inflata*
      4 Corolla blue, 4–10 mm long; capsules 3–5 mm long;; plants perennial and the stem simple .... 5. *L. spicata*

1. **Lobelia cardinalis** L., cardinal flower. Perennial to 1.5 m tall; stems usually simple and leafy throughout, 2–10 mm diam, glabrous. Leaves ovate-lanceolate, lanceolate, or oblanceolate (especially the lower leaves), 2.5–20 cm long, 0.5–5 cm wide, the largest leaves toward the lower midsection on the stem, gradually reduced upward and less so downward, glabrous or rarely remotely hirsute, margins denticulate to irregularly serrate, apex acute to acuminate, or the lowermost leaves sometimes obtuse, base attenuate, mostly sessile above and with a petiole 0.5–1.5 cm long below. Inflorescence a moderately dense raceme, 1–7 dm long, 1 flower per node; bracts foliaceous below and reduced upward, becoming shorter than the flowers. Flowers 3–4.5 cm long; calyx conic to campanulate, 9–15 mm long, 3–5 mm wide, strongly ribbed, glabrous or remotely hispid; calyx lobes persistent, narrowly triangular to subulate, 6–15 mm long, 0.5–2.5 mm wide, entire; soon after anthesis calyx

becoming cup-shaped and enlarged, the lobes up to 25 mm long; corolla crimson, 2-3.5 cm long, glabrous or remotely pubescent within; corolla tube fenestrate; corolla lobes of the lower lip spreading and deflexed, narrowly ovate to narrowly elliptic, about 1/3-1/2 as long as the tube; the 2 upper lobes erect and linear-lanceolate, usually slightly shorter than the lower lobes; anther tube 3-5 mm long; filament tube red, 15-30 mm long and exceeding the corolla tube, straight or slightly curved, where the filaments are separate at the base their margins ciliate; pedicels mostly glabrous to hispid, 3-10 mm long, 2 minute bracteoles at the base. Capsules cup-shaped, 5-9 mm long, 6-10 mm diam; seeds chestnut-brown, ovoid to ellipsoid, about 0.6 mm long, areolate and frequently more or less tuberculate. (n = 7) Aug-early Oct. Wet, open or wooded areas including marshes, stream banks, around springs, & seepage areas; MO, s edge NE, KS, OK, & TX; CO: Las Animas; (FL w to CA, n to N.B., Ont., MN, s NE, & se CO). *L. cardinalis* subsp. *graminea* (Lam.) McVaugh var. *phyllostachya* (Engelm.) McVaugh—Correll & Johnston; *L. splendens* Willd.—Fernald.

Our plants have been referred by some authors to subsp. *graminea* var. *phyllostachya* (Engelm.) McVaugh, which is distinguished by its short filaments (20-24 mm long) and anther tubes (3-4 mm long) and its narrower leaves. The more eastern subsp. *cardinalis* usually has filaments 24-33 mm long, anther tubes 4-5.5 mm long and wider leaves. While many of the GP plants fit var. *phyllostachya*, numerous individuals from both the GP and other regions display any number of character combinations from both taxa, making infraspecific determination impossible. It would seem that recognition of var. *phyllostachya* is not warranted.

2. ***Lobelia inflata*** L., Indian tobacco. Annual, 2-8 dm tall; stems simple or with many racemose axillary branches, 1-5 mm diam, narrowly winged, densely pubescent or hirsute on the lower portion but often becoming glabrous above. Lower cauline leaves elliptic, often broadly so, 4-9 cm long, 1.5-3(3.5) cm wide, glabrous to remotely strigose above, pubescent at least on the veins below, apex acute to obtuse, margins serrate to crenulate or subentire, base cuneate or sometimes rounded, sessile or with a winged petiole to 7 mm long, decurrent; upper cauline leaves gradually reduced in size, mostly narrower, and sessile. Inflorescence a narrow raceme terminating the main axis and branches if present, 1 flower per node; lower bracts foliaceous and upward gradually reduced to rudiments. Flowers inconspicuous; calyx 3-6 mm long, 1.5-2 mm wide, glabrous; calyx lobes persistent, subulate, 2-4 mm long, (nearly equaling the corolla), about 1 mm wide, entire; calyx becoming gibbous rapidly after anthesis; corolla white with the lobes bluish, 3-5(6) mm long; the lower corolla lobes deflexed, oblong, 1/4-1/3 as long as the corolla tube, pubescent at the base within; upper lobes lanceolate and about as long as the lower lobes; anther tube 1.5-2 mm long; filaments 2-3 mm long, pubescent at the base; pedicels 1-7 mm long, puberulent, with 2 minute bracteoles at the base. Capsule ovoid to ellipsoid, 6-10 mm long, 4-6 mm diam; seeds chestnut-brown, ellipsoid to ovoid, about 0.5 mm long, areolate. (n = 7) Jul-Oct. Rich soil in woodlands; IA: Fremont, Ringgold; NE: Burt, Richardson; e 1/5 KS, MO; (N.S. w to MN & e NE, s to GA, AL, MS, & OK).

3. ***Lobelia kalmii*** L., Kalm's lobelia. Slender perennial, 1-5 dm tall, glabrous; stems simple or sparingly branched above the base, 0.5-1.5 mm diam, leafy throughout; rhizome fragile, about 0.3 mm diam, not often collected. Basal rosette present early with spatulate to oblanceolate leaves, 1-2 cm long, 2-3 mm wide, apex rounded and the margins entire; cauline leaves linear-lanceolate to linear-oblanceolate, 0.5-5 cm long, 0.5-3 mm wide, gradually reduced upward, apex acute, margins subentire with remote callose teeth, sessile. Inflorescence a loose raceme of 2-12 flowers terminating the main axis, the branches or flowers, if present, solitary and terminal, flowers 1 per node; lower bracts foliaceous and reduced upward. Flowers often nodding; calyx 3-6 mm long, 1-2 mm wide; calyx lobes subulate, 2.5-4 mm long, about 1 mm wide, entire; corolla pale to dark blue and often with white spots at the base of the lower lip, 7-10 mm long; lower lobes deflexed, about

1-1/2× as long as the corolla tube, broadly ovate to spatulate; upper lobes erect and lanceolate to oblanceolate, about as long as the corolla tube; anther tube 1.3–1.7 mm long; filaments 2.5–3.3 mm long, pubescent at the very base; pedicels flexuous or appressed, filamentous, 3–17 mm long, with 2 bracteoles at or above the middle. Capsules obconic to subglobose, 4–9 mm long, 3–6 mm diam; seeds chestnut-brown, ellipsoid, about 0.7 mm long, weakly areolate. (n = 7) Jul–Aug. Meadows & around springs; MT: Sheridan; ND (except sw 1/6), MN; IA: Palo Alto; (N.S. w to Alta., s to PA, IN, IL, IA, ND, & MT; CO). *L. strictiflora* (Rydb.) Lunnell — Rydberg.

McVaugh (1936) indicated this species occurred in n-cen SD but did not provide specimen citation.

4. ***Lobelia siphilitica*** L., blue cardinal flower, great lobelia. Perennial (1)3–10(15) dm tall; stems mostly simple or occasionally branching above the base, leafy throughout, 2–10 mm diam, glabrous or sparsely hispid. Leaves lanceolate to oblanceolate, 2–15 cm long, 0.6–4.5 cm wide, largest at the lower mid-section of the stem, reduced upward and less so downward, glabrous or rarely scabrous to hispid below on the veins, apex acute to acuminate, on wide leaves sometimes obtuse, margins irregularly serrate, denticulate, crenulate or subentire, base attenuate or sometimes in the upper leaves auricled and clasping, decurrent. Inflorescence a raceme (infrequently paniculate at the lower nodes) terminating the main axis and branches, if present, mostly (3)10- to 40(60)-flowered, (0.5)1–5 dm long, 1 flower per node; bracts foliaceous below and reduced upward, becoming much shorter than the flowers. Calyx mostly campanulate, 8–20 mm long, 4–8 mm diam, glabrous to hispid; calyx lobes persistent, narrowly triangular to oblanceolate, 5–15 mm long, 1–4 mm wide, apex acuminate to attenuate, margins often long-ciliate, entire or irregularly toothed at the base; calyx enlarging soon after anthesis; corolla deep blue-and-white striped in the throat, rarely all white or pale blue, 1.5–3(3.5) cm long, usually hispid on the veins without; corolla tube fenestrate; corolla lobes of the lower lip deflexed, narrow-ovate, about 1/3–1/2 the length of the corolla, apex acute; lobes of the upper lip erect or recurved, narrowly triangular, about 1/2 the length of the corolla, apex acuminate; anther tube 3.5–5 mm long; filament tube blue, 12–15 mm long, about equaling the corolla tube in length, filaments ciliate at the base where separated; pedicels glabrous to hispid, 3–13 mm long, with 2 minute bracteoles at the base or median. Capsules cup-shaped to spherical, 6–9 mm long, 6–10 mm diam; seeds chestnut-brown, ellipsoid, about 0.6 mm long, areolate and often more or less tuberculate. (n = 7) Aug–Oct. Moist soil in open or wooded areas, stream banks, marshes, or meadows; common in cen GP from MN w to sw SD, s to MO & s-cen KS, infrequent in n SD, ND, ne CO; (MA w to Man. & ND, s to NC, TX, & CO). *L. siphilitica* var. *ludoviciana* A. DC., f. *albiflora* Britt., f. *laevicalyx* Fern. — Fernald.

Our plants have been referred to var. *ludoviciana* which differs from typical *L. siphilitica* by being smaller (plants less than 6 dm), glabrous rather than hirsute, and having leaves lanceolate and often auricled instead of wider and attenuate. While plants from the n & w GP are easily distinguished as var. *ludoviciana* many plants from w IA, e NE, e KS, & w MO display intermediate characteristics and are difficult to assign to either variety with certainty.

5. ***Lobelia spicata*** Lam., palespike lobelia. Perennial plants 2–8(10) dm tall; stems simple or occasionally branched above the base, leafy throughout, 1–4 mm diam, near the base hispid or pubescent all around, only on the angles, or rarely glabrous, vesiture decreasing upward on the stem. Basal leaves ovate to elliptic or oblanceolate, 1.5–6 cm long, 0.5–2.7 cm wide, glabrous to strigose or puberulent, apex usually round or obtuse, occasionally acute, margins entire to irregularly denticulate, ciliate, base cuneate to attenuate and decurrent along the angles of the stem, petiole to 1.5 cm long or leaves subsessile; cauline leaves mostly oblanceolate to elliptic, less often lanceolate or oblong, 2–6(9) cm long, 0.4–2(2.7) cm wide, the largest toward the base and sometimes short petioled, upward gradually reduced, and becoming narrower, glabrous to sparingly puberulent, apex acute to obtuse

or rounded, margins subentire to irregularly dentate or serrate, often ciliate, base mostly cuneate to attenuate, rarely roundish and slightly clasping. Inflorescence a loose to moderately dense raceme terminating the main axis and branches, if present, 5-40 cm long, 1(2) flower per node; bracts lanceolate in the lower inflorescence and becoming subulate above, 5-15 mm long, 0.5-3 mm wide, vestiture as on the cauline leaves. Flowers 7-13 mm long; calyx 5-9 mm long, glabrous to scabrous on the tube only or wholly scabrous or puberulent; calyx lobes subulate to linear-lanceolate, slightly spreading, 3-7 mm long, margins entire or with 1 or 2 large teeth per side, often ciliate (especially in n GP), base auricled, auricles inconspicuous, roundish and minute to filiform, to 2 mm long and deflexed, or infrequently broad triangular, flatish and to 1.5 mm long; corolla mostly light blue, occasionally darker, 4-10 mm long, glabrous except for pubescence at the base of the lower lip within; lobes of the lower lip ovate and slightly reflexed, about 1/3-1/2 the length of the corolla; upper lobes lanceolate and curved upward, slightly shorter that the lower lobes; anther tube blue or occasionally white (the pollen abortive in these), 1-2 mm long; filaments 2.3-3.5 mm long, ciliate at the base; pedicels 1-5 mm long, glabrous to scabrous, 2 minute bracteoles at the base or rarely almost median on the same plant. Capsules subglobose, 3.5-5 mm diam; seeds chestnut-brown, ovoid to ellipsoid, about 0.5 mm long, areolate. (n = 7) May–Jul (Aug). Mostly native prairies & meadows, occasionally open woodlands in the se GP; MN w to nw MT, s to MO, e 1/4 KS, & NE; (se Can. w to Sask., s to GA, AL, MS, LA, e TX, & w NE). *L. hirtella* (A. Gray) Greene, *L. leptostachya* A. DC.—Rydberg; *L. spicata* var. *hirtella* A. Gray, *L. spicata* var. *leptostachya* (A. DC.) Mack. & Bush—Fernald.

Lobelia spicata is a highly variable species occurring through much of the GP. While several rather weak segregates have been recognized by some authors, intermediate individuals abound regardless of the criteria used to separate the "races." In extreme se GP (KS: Cherokee; s MO; e OK) the range of *L. spicata* overlaps that of *L. appendiculata* A. DC. The latter species is distinguished from *L. spicata* by having a glabrous stem (or nearly so), wider leaves that are usually clasping at the base, flat nearly foliaceous auricles on the calyx lobes, and large seeds (ca 0.7 mm long).

3. TRIODANIS Raf., Venus' Looking Glass

Small annual plants, simple or branching from the base or rarely to the middle of the central axis, the branches ascending and frequently equaling the central axis; stems leafy below passing to leaflike bracts above, mostly 5-angled, the angles continuous with the decurrent leaf and bract bases. Leaves usually glabrous above and with vestiture below, at least on the principal veins. Inflorescence spiciform, 1-3 flowers per axil, bracts leaflike but sometimes glabrous on both surfaces. Flowers dimorphic, those in the upper axils chasmogamous and open, while the lower flowers are cleistogamous. Chasmogamous flowers with calyx expanded and foliaceous, 5-lobed, the lobes all alike, entire except for 1 callose-glandular tooth on each edge near the base; corolla short funnel-form for about 1/3 its length and with 5 spreading, elliptic and abruptly acuminate lobes, glabrous or with a few bristles near the apex of the lobes; stamens distinct, normally 5, filaments divided into a narrow, linear distal portion and a gradually dilated or abruptly rounded, broad ciliate base, anthers linear, glabrous, longer than the filaments; style pubescent distally for 1/2 its length, stigmas 3 to 5-lobed. Cleistogamous flowers smaller than the open ones and producing smaller capsules, the corolla, stamens, and style reduced to rudiments. Capsules clavate or obovoid to ellipsoid, subulate in some cleistogamous flowers, tipped by the calyx lobes at maturity, pores of the capsule mostly equal in numbers to the locules, distal and opening from the base toward the apex; seeds numerous, the majority being produced by cleistogamous flowers, brownish, lenticular, less than 1 mm long, longer than wide, placenta axile except in *T. leptocarpa* where it is parietal. *Specularia* (Heist.) Fabr.—Rydberg.

Reference: McVaugh, R. 1945. The genus *Triodanis* Rafinesque, and its relationships to *Specularia* and *Campanula*. Wrightia 1: 13-52.

1. Flower bracts linear to lanceolate, at least 6× longer than wide; capsules of cleistogamous flowers usually falcate .... 4. *T. leptocarpa*
1. Flower bracts ovate or broader, less than 3× longer than wide; capsules of cleistogamous flowers straight.
   2. Pore on capsule median or slightly higher, linear, 0.2-0.4 mm wide .... 2. *T. holzingeri*
   2. Pore on capsule basal or distal, if median then elliptic or broader, 0.5-2.0 mm wide.
      3. Seeds 0.8-1 mm long .... 3. *T. lamprosperma*
      3. Seeds 0.4-0.65 mm long.
         4. Pores at the distal extremity of the capsule, oval to nearly round, about 1 mm diam; leaves and bracts elliptic to narrowly ovate, longer than wide .. 1. *T. biflora*
         4. Capsule pores basal to sometimes median, elliptic, 1.3-2 mm long; leaves and bracts cordate to broadly ovate, about as wide or wider than long .... 5. *T. perfoliata*

1. **Triodanis biflora** (R. & P.) Greene. Plants 1-4.5 dm tall; stems 0.5-2 mm diam, the angles hispid to scabrous at least on the lower stem. Basal leaves soon deciduous and not often found on herbarium sheets, elliptic to narrowly obovate, 8-20 cm long, 4-10 mm wide, mostly glabrous above and hispid to scabrous at least on the veins below, apex obtuse to rounded, margins nearly entire to crenate, base sessile or rarely short pedicelled; cauline leaves and bracts mostly elliptic to narrowly ovate, 6-20 mm long, 2-10 mm wide, usually the leaves and bracts becoming slightly smaller on the upper stem, vestiture as the basal leaves, apex acute or rarely obtuse, margins crenate, base sessile or subsessile. Flowers at nearly all the nodes of stems, each stem terminated by 1 open flower, axillary flowers 1(2-3) per node and usually cleistogamous. Open flowers with calyx tube 3-6 mm long and only slightly inflated, lobes narrowly triangular to lanceolate, 5-8 mm long, glabrous, apex usually acuminate, margins entire, base usually slightly narrowed; corolla purple or occasionally with white streaks, glabrous or hirtellous on the veins without, tube 1-2 mm long, lobes 5-10 mm long, 2-3 mm wide, ciliate near the acute apex. Cleistogamous flowers with calyx tube 3-7 mm long, the lobes erect at first and usually becoming slightly reflexed, narrowly triangular, varying in length on the same flower, 0.7-2 mm long, apex mostly acute. Capsules clavate to ovoid, 4.5-8 mm long, 1-1.7 mm diam, glabrous to sparsely hispid; pores at the distal extremity of the capsule, oval to nearly round, about 1 mm diam or less; seeds chestnut-brown, elliptic or broadly so and biconvex, 0.4-0.65 mm long, slightly less wide than long, smooth and shiny. ($2n = 56$) May-Jun. Usually sandy soil in meadows or open woods, occasionally in disturbed sites; s MO, se 1/6 KS, e OK; NE: Johnson, Pawnee; (VA w to se NE, s to FL, TX, s OR s to Mex.; S. Amer.). *Specularia biflora* (R. & P.) A. Gray—Rydberg; *T. perfoliata* var. *biflora* (R. & P.) Bradley—Bradley, Brittonia 27: 110-114. 1975.

2. **Triodanis holzingeri** McVaugh. Similar to *T. perfoliata* except as follows: Flowers bluish-purple. Capsules ellipsoid or oblong to narrowly obovoid, 5-9 mm long, 0.7-2 mm diam, scabrous to glabrous; pores approximately median or slightly higher on the capsule, linear to narrowly oblong, 1.4-2.5 mm long, 0.2-0.4 mm wide; seeds chestnut-brown (often dark), elliptic or ovate and biconvex, 0.3-0.5 mm long, lustrous and minutely low-tuberculate. May-Jun. Usually sandy soil in prairies & pastures, frequently in disturbed sites; MO w to cen & sw KS, s to cen OK; NE: Franklin, Thomas, Webster; WY: Platte; CO: Baca; TX: Wheeler; (w MO nw to se WY, s to OK & s TX). *Specularia holzingeri* (McVaugh) Fern.—Fernald.

3. **Triodanis lamprosperma** McVaugh. Plants similar to *T. perfoliata* except as follows: Cauline leaves and bracts mostly ovate to broadly so, about as long as wide, 6-14 mm long,

3-11 mm wide, glabrous or occasionally the veins on the lower surface scabrous, apex acute, margins crenate to nearly entire, sessile. Capsules usually ellipsoid or narrowly so, 4-10 mm long, 1.5-2 mm diam, glabrous to scabrous on the veins; pores distal to nearly median, broadly elliptic, 1.4-2 mm long, 0.7-1.5 mm wide; seeds chestnut-brown, broadly elliptic and biconvex, 0.7-1 mm long, smooth and lustrous. May-Jul. Sandy soil in prairie meadows, pastures, & occasionally open disturbed sites; extreme se GP; KS: Cherokee, Montgomery; MO; Jasper; (s MO, se KS, w AR, & e OK). *Specularia lamprosperma* (McVaugh) Fern.—Fernald.

4. *Triodanis leptocarpa* (Nutt.) Nieuw. Plants 1-7 dm tall; stems 1-3.5 mm diam, the angles hispid to scabrous. Basal leaves soon withering, narrowly elliptic to oblanceolate, 8-27 mm long, 4-7 mm wide, glabrous above, hispid to scabrous below on the veins, apex acute to obtuse, margin ciliate-crenate, short-petioled. Cauline leaves elliptic to lanceolate, 10-25(30) mm long, 2-8(10) mm wide, pubescence like the basal leaves, apex acute to obtuse, margins entire to crenate, base mostly sessile. Primary bracts mostly lanceolate, 10-25 mm long, 1.5-7 mm wide, pubescence as on leaves, apex acute to acuminate, margins entire, base sessile and often clasping. Flowers borne at most of the nodes, 1-3 per node, chasmogamous flowers usually borne on about upper 1/3-1/2 of the nodes, cleistogamous flowers borne on lower nodes and usually with some of the lower chasmogamous flowers. Open flowers with calyx tube 6-14 mm long, 1-1.5 mm diam, scabrous at least on the veins, calyx lobes usually recurved, narrowly lanceolate to subulate, 5-10 mm long, 0.5-1 mm wide, glabrous or the margins scabrous-ciliate; corolla purple, occasionally white streaked, glabrous or scabrous on the veins without, tube 0.5-1.5 mm long, lobes 3-7 mm long, 1.5-4 mm wide, margins entire or sometimes ciliate at the apex, apex acute to mucronate. Cleistogamous flowers with calyx tubes 10-15 mm long, 1-2 mm diam, usually scabrous, calyx lobes narrowly triangular to subulate, spreading but usually not sharply recurved as in open flowers, 2-6(8) mm long, 0.5-1 mm wide. Capsules dimorphic, hispid to retrorsely scabrous or glabrous; those of open flowers straight and erect, linear, 15-25 mm long, 1.3-1.8 mm diam, unilocular with 1 apical pore (rarely imperfectly bilocular with 2 pores); capsules of cleistogamous flowers subulate and falcate, 8-23 mm long, 0.7-2.5 mm diam, unilocular with parietal placentation, dehiscence in the distal 1/2 by longitudinal slits alternating with the calyx lobes or, less frequently, by a single pore near the apex; seeds light brown and lustrous, ellipsoid and biconvex, 0.7-1 mm long, 0.4-0.6 mm wide. May-Jun. Usually sandy soil in prairies, pastures, & frequently disturbed sites; common in s GP & less frequent n; sw 1/4 ND & se MT, sw to MO & s OK; (IN & MN w to MT, s to AR & TX). *Specularia leptocarpa* (Nutt.) A. Gray—Rydberg.

5. *Triodanis perfoliata* (L.) Nieuw. Plants 1-10 dm tall; stems 0.6-3 mm diam, the angles usually hispid. Basal leaves soon deciduous or withered, broadly ovate and sessile to broadly elliptic and short petioled; cauline leaves and bracts cordate to broadly ovate and usually about as wide or wider than long, 5-20(27) mm long, 8-20(25) mm wide, usually becoming slightly small on the upper stem, upper surface glabrous, lower surface scabrous to densely hispid, especially those on the lower stems, apex broadly acute to rounded, margins crenate to serrate, sessile. Flowers at nearly all the nodes of each stem, 1-3 per node, flowers primarily cleistogamous but usually 1 per node on the upper 1/3-1/2 of the stem open. Chasmogamous flowers with calyx tube 2.5-5 mm long and usually slightly inflated, calyx lobes narrowly triangular to lanceolate, 4-9 mm long, glabrous or ciliate; corolla purple or occasionally streaked with white or wholly white (drying purple), glabrous or hispid on the veins without, tube 1-2.5 mm long, lobes narrowly triangular to elliptic, 4.5-7 mm long, 2-3 mm wide, smooth or ciliate at the apex, apex acute to mucronate. Cleistogamous flowers with calyx tube 3-5 mm long, slightly inflated, calyx lobes narrowly triangular, 2-3 mm long, apex acute to acuminate. Capsules mostly ovoid, 3.5-6 mm long (to 8 mm in open flowers), 1.5–2.5 mm diam, glabrous to hispid on the veins; pores median or lower on the capsule, rarely higher, elliptic, (1)1.3–2 mm long, 0.5–1.5 mm wide; seeds brown, elliptic and biconvex, 0.5–0.6 mm long, finely muricate and dull or occasionally smooth and shiny. (2n = 56) May–Jul. Sandy or gravelly soil, in prairies, pastures, & various disturbed sites, occasionally in woodlands; common in se GP & scattered n; sw MN, w ND & nw WY s to MO, OK; TX: Wheeler; (Ont. w to B.C., s to FL, e TX, w KS, e CO, w MN, AZ, & n CA; Mex.). *Specularia perfoliata* (L.) A. DC.—Rydberg.

Seeds from this taxon in the southern GP are primarily brown and finely muricate with infrequent plants (in the same population) having chestnut-brown shiny seed coats. Northward through KS and NE, plants with smooth, shiny seeds become more frequent until they dominate in the northern plains. From central KS to central NE both seed types can be found in nearly any population examined. At least in our region, the seed character does not vary consistently with bract width or pore location, as has been suggested by several previous authors.

Trent (Trans. Kansas Acad. Sci. 45: 152–164. 1942.) found that in open flowers approximately 47% of the seed produced is viable, while less than 1% of the seeds produced by cleistogamous flowers is viable. Although the seeds of cleistogamous flowers appear normal and have fully developed endosperm, they are usually without embryo.

131. RUBIACEAE Juss., the Madder Family

by Ronald L. McGregor and Ralph E. Brooks

Ours usually annual or perennial herbs, one species a shrub or small tree. Principal leaves opposite or whorled, simple, entire; stipules interpetiolar, or stipules appearing as leaves and the leaves appearing whorled. Flowers perfect, regular, epigynous, gamopetalous, usually in panicles, cymes or heads, or solitary; sepals 4(6) or wanting, more or less united with ovary, usually persistent; corolla of 4(5) petals, funnelform, salverform or rotate; stamens epipetalous, usually as many as petals; ovary inferior or half-inferior; ovary 2(3)-celled, with 1–many ovules per cell. Fruit capsular and dehiscent or variously separating into drupes or nutlets.

1 Shrub or small tree with flowers in spherical, peduncled heads .... 1. *Cephalanthus*
1 Herbaceous plants; flowers not in spherical peduncled heads.
  2 Leaves appearing verticillate, the stipules conspicuous, resembling the leaves ... 3. *Galium*
  2 Leaves opposite; stipules inconspicuous or not resembling leaves.
    3 Flowers on terminal peduncles or in cymes; ovules few to many per cell; stipules lacking bristles or with only 1 or 2 at tip on each side .... 4. *Hedyotis*
    3 Flowers in axillary clusters; ovules 1 per cell; stipules with several to many bristles.
      4 Flowers 2–6 per node; leaves, stems or fruits with hairs .... 2. *Diodia*
      4 Flowers 10–40 per node; plant glabrous .... 5. *Spermacoce*

1. CEPHALANTHUS L., Buttonbush

1. *Cephalanthus occidentalis* L., common buttonbush. Shrub 1–3 m tall, usually with several stems from base, or rarely a small tree 5 m tall, branches slender, brown or grayish, glabrous. Leaves opposite or in whorls of 3(4), simple, entire; blades ovate-oblong, lance-oblong, to broadly elliptic, (6)9–15(19) cm long, 3–5(9) cm wide, base cuneate, subcordate to truncate, apex abruptly short to long acuminate, bright green above, glabrous below or a few hairs in vein axils; petioles 0.5–3(5) cm long, glabrous or puberulent; stipules 2–4(6)

mm long, deltoid, acute to acuminate, glandular-dentate. Peduncles terminating stem or from upper axils, simple or usually branched, stout, to 1 dm long, glabrous; heads 1.5–3 cm in diam; flowers many, intermingled with filiform-clavate bractlets which are usually pilose above; calyx 1 mm long, limb 4-toothed; corolla white, tube 5–9(13) mm long, the 4(5) lobes ovate to oval, sparsely pubescent within, 2–2.4 mm long; style 12–15 mm long, stigma capitate; stamens 4, inserted on corolla throat, included; ovary 2-celled, 1–3 ovules per cell. Fruit obconic, angular, 4–8 mm long, at length splitting from base upward into indehiscent nutlets; seed solitary, brown, with a white aril, 2–4 mm long. (n = 22) Jun–Sep. Locally common along ponds or lakeshores, stream banks, low swamps, woods, prairie marshes; e NE, e 2/3 KS, MO, OK; (N.B., Que. to MN, s to FL, TX).

2. DIODIA L., Buttonweed

1. *Diodia teres* Walt., rough buttonweed. Annual herbs with taproots; stems erect, with prostrate or ascending branches, 2–8 dm long, glabrous or puberulent to hirsute on angles. Leaves opposite, stiff, linear to narrowly lanceolate or elliptic, 2–5 cm long, sessile, rounded to somewhat clasping at base, apex acute to acuminate and aristate, scabrous, ciliate; stipules sheathing, bristles numerous, filiform, usually reddish-brown, often equaling the flowers and longer than fruits. Flowers 1–3 in axils of leaves, 2–6 per node, sessile; sepals 4, lanceolate, 1.8–2.2 mm long, ciliate; corolla funnelform, 4–6 mm long, whitish to pinkish-purple; style undivided; exserted; stamens 4, inserted on corolla tube, usually exserted. Fruit obovate-turbinate, 4–5 mm long, bicarpellate, hairy, crowned by persistent sepals, splitting into 2 nutlets. (n = 14) Jun–Oct. Rather frequent in dry, rocky or sandy soils of prairies, pastures, fields, stream valleys, waste areas; e 1/2 KS, MO, OK; (VT, NY, PA, IL, IA, KS, s to FL, TX).

Most of our specimens have young leaves with slender apical bristles and fall under var. *setifera* Fern. & Grisc., but too many plants lack this character. Recognition of the variety appears unwarranted.

A similar species, *D. virginiana* L., has been reported from MO: Jasper, & KS: Cherokee, along the edge of our area. It differs in having only 2 sepals, corolla salverform, and style cleft into 2 linear segments.

3. GALIUM L., Bedstraw, Cleavers

Annual herbs with slender taproots or usually perennial herbs with simple or branched caudex and often spreading by rhizomes; stems erect, spreading, or ascending, usually 4-angled. Leaves whorled, sessile or short petioled; stipules foliaceous, appearing as another set of leaves. Flowers mostly in axillary or terminal cymes or panicles, or solitary in axils; pedicels present and usually jointed with calyx or obsolete; calyx-tube ovoid or globose, teeth obsolete; corolla rotate, with 3 or 4 short lobes; styles 2, stigmas capitate; stamens 4, usually shorter than corolla; ovary 2-celled, with 1 ovule per cell. Fruit dry, of 2 globose carpels, separating at maturity into 2 seedlike, indehiscent, 1-seeded schizocarps.

Two other genera with verticillate leaves have been sparingly introduced in or near our region. *Sherardia arvensis* L., the field madder, has been reported at the margin of our area in cen OK. It has leaves in whorls of 6, linear to narrowly elliptic, 1 cm long; sepals triangular; corolla pink or blue. *Asperula orientalis* Boiss. & Hohenack., woodruff, was reported by Stevens for ND: Cass as *Sherardia orientalis* Boiss. & Hohenack. and was so included in the Atlas GP. The plant has apparently not persisted in ND. It differs from *Galium* in having a funnelform corolla with the tube and lobes about the same length.

1 Plants annual with short taproots.
  2 Principal leaves in whorls of (6)8 .... 1. *G. aparine*

2 Principal leaves in whorls of 4.
  3 Flowers and fruits nearly sessile in axils of cauline leaves; fruit
    reflexed .... 12. *G. virgatum*
  3 Flowers often terminal on lateral branches; pedicels 5-15(20) mm long; fruits not
    reflexed .... 8. *G. texense*
1 Plants perennial, usually with slender rhizomes.
 4 Flowers bright yellow, rare introduced sp. .... 11. *G. verum*
 4 Flowers white, greenish-white or purplish.
  5 Fruits evidently hairy or bristly.
   6 Principal leaves in whorls of 6; stems prostrate or scrambling ...... 10. *G. triflorum*
   6 Principal leaves in whorls of 4; stems usually erect.
    7 Principal leaves lance-linear .... 2. *G. boreale*
    7 Principal leaves elliptic, oval, or ovate-oblong, widest near the middle.
     8 Most of flowers and fruits sessile along axis of
       inflorescence .... 3. *G. circaezans*
     8 All flowers evidently pedicelled and terminating branches of
       inflorescence .... 7. *G. pilosum*
  5 Fruits smooth or variously roughened but not hairy or bristly.
   9 Corolla with 3 lobes .... 9. *G. trifidum*
   9 Corolla with 4 lobes.
    10 Principal leaves in whorls of 6 .... 4. *G. concinnum*
    10 Principal leaves in whorls of 4.
     11 Plants erect or nearly so .... 2. *G. boreale*
     11 Plants scrambling, reclining or ascending, usually somewhat matted.
      12 Leaves spreading or ascending; inflorescences not overtopped by lateral
         branches; fruits 2.5-3.5 mm in diam .... 6. *G. obtusum*
      12 Leaves soon reflexed; inflorescence soon overtopped by erect lateral
         branches; fruits 1-1.5 mm in diam .... 5. *G. labradoricum*

1. **Galium aparine** L., catchweed bedstraw. Annual herbs with slender taproots; stems weak, reclining and usually scrambling, forming dense tangles, 1-10(20) dm long, retrorsely bristly on the angles, hairy at the nodes. Leaves mostly in whorls of 6(8), linear-oblanceolate, 1-nerved, cuspidate, 1-8 cm long, margins retrorsely scabro-ciliate as is lower midrib. Inflorescences mostly of 3-5 flowers, on axillary peduncles, or in 3s on short branches; corollas white, 1-2 mm long. Fruit uncinate-hispid, (1.5)3-4(5) mm in diam. (n = 10, 11, 22, 32) May-Aug. In woods, thickets, prairies, waste ground, roadsides; GP but less common westward; (circumpolar, found over most of temp. N. Amer.). *G. vaillantii* DC. — Rydberg; *G. aparine* var. *vaillantii* (DC.) Koch — Gates; *G. aparine* var. *echinospermum* (Wallr.) Farw. — Gleason & Cronquist.

In dry habitats, plants often have leaves only 1-3(4) cm long, fruits 1.5-3 mm wide, and flowers somewhat yellowish. These have been referred to var. *vaillantii* or var. *echinospermum*, but the differences appear correlated with the habitat and the varieties do not merit taxonomic recognition.

When the mature fruits are dried and roasted, they yield a coffeelike beverage, often reported to be the best substitute for coffee in North America.

2. **Galium boreale** L., northern bedstraw. Perennial herbs with well-developed, creeping rhizomes; stems erect, numerous, short bearded beneath the nodes, otherwise glabrous or sparsely scaberulous, 2-7(10) dm tall. Leaves in whorls of 4, linear to lanceolate, (1.5)3-4(5) cm long, glabrous or scaberulous, margins sometimes ciliate and slightly revolute, obtuse or acute at apex, often with fascicles of smaller leaves in the axils. Flowers many in terminal, rather showy, cymose panicles, white or rarely yellowish; corollas 4-lobed, 3.5-7 mm wide. Fruit about 2 mm in diam, glabrous, or usually with short, straight or curled, inconspicuous hairs. (n = 33) Jun-Sep. Rocky hillsides, prairies, woodlands, roadsides, often abundant; MN, ND, e SD, w SD, n IA, nw NE, MT, WY, CO, NM; (cirumpolar, s in N. Amer. to WV, MO, NE, TX, AZ, CA). *G. boreale* var. *linearifolium* DC. — Rydberg.

European and w Asian plants reportedly have a chromosome number of n = 22, while American and e Asian plants have n = 33.

3. Galium circaezans Michx., woods bedstraw. Perennial herbs with short branching caudex; stems erect or ascending, 2–4(6) dm tall, simple to sometimes branched, one to several from base, glabrate to hairy on the angles. Principal leaves in whorls of 4(6), ovate-lanceolate, ovate-oblong or elliptic, widest near the middle, (1.5)2–5 cm long, 1–2.5 cm wide, 3- to 5-nerved, obtuse at apex, ciliate, surfaces glabrous to pubescent especially below and on veins. Inflorescences terminal and from upper axils, simple or with 1 or 2 divaricate forks; flowers sessile or short pedicelled, remote; corolla greenish, lobes acute or acuminate and usually hairy on the outside, 1 mm long. Fruits eventually deflexed, uncinate-hispid, 3 mm in diam. (n = 11) May–Jul. Frequent in rich or rocky wooded hillsides, bluffs, thickets; e 1/4 NE, e 1/2 KS, IA, MO, OK; (ME, MI, MN, s to FL, NE, TX). *G. circaezans* Michx. var. *hypomalacum* Fern.—Fernald.

Most of our speciments are the var. *hypomalacum* with stems hairy on the angles and the leaves evidently appressed-pilose beneath. Rare plants of our area are nearly glabrous or glabrate and could be referred to the more southern var. *circaezans*. The merit of recognizing these variants is questionable.

4. Galium concinnum T. & G., shining bedstraw. Perennial herbs with slender rhizomes; stems slender, spreading or ascending, 2–5 dm long, usually much branched, sparsely retrorse-scabrous or glabrate on angles. Leaves in whorls of 6 or rarely those of branches in 4s, linear or linear-elliptic, principal ones 1–2 cm long, acute or cuspidate, 1-veined to veinless, margins upwardly scabrous. Inflorescences terminal and from upper axils; peduncles 2- or 3-forked; the terminal panicles delicate, airy, diffuse; flowers sessile or pedicels evident, slender; corolla white, lobes 0.7–1.2 mm long, acute or short cuspidate, glabrous. Fruits smooth, 1.5–2 mm in diam. (n = 11) Apr–Jul. Locally common in dry or moist woodland, wet places in prairie, rocky ledges of bluffs, thickets, roadsides, waste places; e 1/2 NE & KS, IA, MO, OK; (NJ, MN, s to VA, KY, AR, NE, KS, OK).

A somewhat similar plant, *G. asprellum* Michx., was reported from NE by Rydberg, but specimens have not been located to validate the report. It differs from *G. concinnum* in leaves elliptic to oblanceolate and retrorsely scabrous on the margin.

*
5. Galium labradoricum Wieg., Labrador bedstraw. Perennial herbs with very slender rhizomes; stems slender, erect or ascending, simple or branched above, 0.5–4 dm tall, smooth, to pubescent only at the nodes. Principal leaves in whorls of 4, linear-oblanceolate or narrowly spatulate, 5–15 mm long, usually scabrous on margin, midrib glabrous or scabrous, leaves usually recurved or deflexed. Inflorescences few, mostly terminal, usually 3-flowered, soon overtopped by branches; corolla white, lobes 1 mm long. Fruits smooth, 1–1.5 mm in diam. May–Jul. Moist thickets & woods, usually swampy; ND: Ransom; MN; (Newf., Lab., Man., s to MA, MN, IL, ND).

6. Galium obtusum Bigel., bluntleaf bedstraw. Perennial herbs with very slender rhizomes; stems 2–8 dm long, matted, strict and branched from base, or branched throughout, pubescent at nodes, sometimes hispidulous on the angles. Principal leaves usually in whorls of 4(5–6), linear to lanceolate or oblanceolate or elliptic-oblong, obtuse, 1.5–3 cm long, 1–8 mm wide, slightly scabrous or ciliate on margins. Inflorescence terminating stem and branches, cymes with 2–4 flowers on ascending pedicels; corollas white, lobes 1–1.3 mm long, acute. Fruit smooth, 2.5–3.5 mm in diam. (n = 12) May–Jul. Locally common in swamps, moist woodlands, stream banks, wet places in prairies, roadsides; cen & e GP; (N.S., Que., MN, SD, s to FL, TX). *G. obtusum* Bigel. var. *ramosum* Gl.—Gleason & Cronquist.

This species varies considerably in leaf size even within a colony. Var. *ramosum* has leaves averaging 4 × longer than wide and widest at or below the middle, but such plants grade into var. *obtusum* with leaves averaging 7 × longer than wide. The distinction is arbitrary.

7. Galium pilosum Ait., hairy bedstraw. Perennial herbs with short branching caudex; stems tufted, erect or ascending, 1–7(10) dm tall, usually with several basal branches, otherwise simple to inflorescence, sparsely to densely spreading-pilose, especially on lower angles, rarely glabrous. Principal leaves in whorls of 4, rather firm, oval to elliptic or oblong elliptic, usually obtuse-apiculate at apex, 1.5–2.5 cm long, hairy, usually sparsely glandular-punctate below. Inflorescences terminal and from upper axils, paniculate, elongate, the peduncles 2- or 3-forked; pedicels evident; corollas greenish-white or often purplish, lobes 1.2–2 mm long. Fruits 2.5–3.3 mm in diam, uncinate-bristly. Jun–Aug. Locally common in dry, rocky woodlands, thickets, often in shallow soil over sandstone or limestone; e 1/3 KS, MO, OK; (NH, s Ont., MI, s to NC, TN, KS, TX).

8. Galium texense A. Gray, Texas bedstraw. Annual herbs with slender taproots; stems 1–3 dm tall, weakly erect to procumbent, simple or branched, essentially glabrous above, often hispidulous or spreading-hirsute below, branches usually hirsute. Principal leaves in whorls of 4, ovate to ovate-lanceolate or nearly oval, 1 vein, 3–6(10) mm long, acute or abruptly short cuspidate, pilose on both surfaces, ciliate. Peduncles terminating lateral branches and 1-flowered, pedicels 5–15(20) mm long; from axils a shoot bearing a whorl of 4 small leaves and a short pedicel; corolla white, lobes 1 mm long. Fruits 1.7–2.5 mm wide, 2 mm long, uncinulate-bristly or hairs barely hooked. Mar–Jun. Dry open woodlands, rocky prairie hillsides & bluffs; sw OK, TX; (nw AR, s OK, TX).

9. Galium trifidum L., small bedstraw. Perennial herbs with slender creeping rhizomes; stems numerous, slender, 0.5–6 dm long, scrambling and often forming dense mats, usually retrorse-scabrous on the angles or rarely glabrous. Principal leaves in whorls of 4(5–6), linear to narrowly elliptic, 5–15(20) mm long, blunt, 1-nerved, usually retrorsely scabrous on margins and on midrib below. Peduncles terminal or axillary, often 1–3 on axillary branches, 1- to 3-flowered; corolla white, with 3(4) lobes 0.5 mm long, obtuse. Fruit glabrous, 1–2 mm in diam, mature segments almost distinct at maturity. Jun–Sep. Locally common in moist woodlands, moist prairie, waste places; n 1/2 GP; (circumpolar, s in N. Amer. to NY, OH, MI, n IL, MN, NE, CO, CA). *G. claytoni* Michx., *G. subbiflorum* (Wieg.) Rydb. — Rydberg.

Records in the Atlas GP for AR, OK, MO, KS, and most in e NE were based on incorrect determinations, as most specimens are *G. obtusum*. *G. trifidum* is a highly variable species, and specimens with some leaves in whorls of 5 have been referred to *G. tinctorium* L. The complex merits careful study in our area.

10. Galium triflorum Michx., sweet-scented bedstraw. Perennial herbs with slender creeping rhizomes; stems weak, usually scrambling, 2–8(12) dm long, simple or often much branched, retrorsely scabrous or spreading pubescent on angles below to essentially glabrous. Principal leaves in whorls of 6 or 4 or 5 on branches, with the odor of vanilla, 2–6 cm long, elliptic-lanceolate to oblanceolate, usually cuspidate, often antrorsely scabrous on margins and uncinulate-scabrous on midrib below. Inflorescences axillary, sometimes terminal; peduncles elongate and 3-flowered or sometimes branched and several flowered; corolla greenish-white, with 4 lobes 1–1.5 mm long. Fruit 1.5–2.2 mm in diam, densely uncinate-bristly. (n = 11, 22, 33) May–Sep. Common in moist woodlands, along streams, moist ledges, rarely in moist prairies; GP except sw 1/4; (circumpolar, in N. Amer. s to GA, MS, LA, TX, CA).

11. Galium verum L., yellow bedstraw. Perennial herbs with stoutish rhizomes; stems stiff, erect, numerous, 3–9 dm tall, usually densely puberulent, rarely glabrate, often with suppressed axillary branches. Leaves narrowly linear, (6)8(12) in whorls, ascending or divergent, acute, often finely hairy. Flowers in dense panicles; flowers bright yellow, fading whitish.

Fruit smooth or usually so, 1-2 mm in diam. (n = 11, 12) May-Sep. Rarely found in fields & roadsides, not recorded as persistent in GP; ND: Cass, Stark; SD: Union; NE: Douglas; KS: Leavenworth, Pottawatomie; IA: Selby; (e U.S. & sparingly elsewhere; Eurasia).

This is the species of bedstraw reputedly used to fill the Christ Child's manger at Bethlehem.

12. **Galium virgatum** Nutt., southwest bedstraw. Annual herbs with slender taproots; stems erect, 1-several from base, often wiry, 0.5-2(4) dm tall, hispid or glabrate on angles. Principal leaves in whorls of 4, elliptic, lance-oblong, or narrowly lanceolate, (2)4-7(10) mm long, hispid, ciliate, acute or obtuse. Flowers solitary in leaf axils, sessile or subsessile, shorter than leaves; corolla white, lobes 4, 0.5 mm long. Fruits reflexed, 1.5-2.2 mm in diam, 2 parts nearly separating at maturity, densely uncinate-bristly or glabrous. (n = 11) Apr-Jul. Locally common on dry rocky prairies & open woodlands, often on shallow soils over surfacing limestone or sandstone; se KS, MO, OK, TX; (TN, LA, MO, KS, TX).

With a little search, plants will be found in which the fruits are glabrous and the herbage glabrate. These have been referred to var. *leiocarpum* T. & G., but such plants usually grow with the characteristic pubescent plants.

4. HEDYOTIS L., Bluets

Ours annual or perennial herbs with slender or stout taproots; stems erect or spreading, simple or branched, leaves opposite, sessile or short-petiolate, entire; stipules short, adnate to petioles or leaf bases. Flowers solitary on peduncles or in corymbiform cymes; hypanthium globose or obovoid; sepal lobes 4, distinct; corolla funnelform or salverform, lobes 4, valvate; ovary 2-celled, with several ovules per cell; style 1, slender, stigmas 2, linear; stamens 4, included or exserted, linear or oblong. Capsule protruding beyond hypanthium, loculicidal across the top; seeds globular with a deep pit on one side or crescent-shaped with a ridge across the concave face.

1 Plants winter annuals or annual.
  2 Flowers terminal; peduncle erect after anthesis .... 1. *H. crassifolia*
  2 Flowers axillary; peduncle recurved after anthesis .... 2. *H. humifusa*
1 Plants perennial.
  3 Flowers all pedicelled; capsule 1/2 inferior; middle and upper stipules not bristle-tipped .... 3. *H. longifolia*
  3 Flowers partly sessile; capsule 2/3 or more inferior; middle and upper stipules with bristle tips .... 4. *H. nigricans*

1. **Hedyotis crassifolia** Raf., small bluets. Winter annual or annual herbs with slender taproots, sometimes cespitose; stems (1)3-8(12) cm long, erect, ascending or spreading, often branched, glabrous or somewhat scaberulous on angles. Leaves mainly basal, spatulate, ovate to obovate, 3-10 mm long, (3)4-5(8) mm wide, often ciliate, acute; petioles absent or nearly as long as blade; stipules scarious, 0.5-2 mm long, usually apiculate or cuspidate. Flowers solitary, terminal; hypanthium 0.7-1.2 mm long; calyx lobes oblong-lanceolate or broadly elliptic, sometimes scabrous, 2-3 mm long at anthesis; corolla dark blue, light blue or rarely white, salverform; tube 2.5-4(5) mm long, lobes 2.5-4 mm long, oblong, glabrous; style short. Fruit didymous-flattened, glabrous, 3-4(5) mm wide, shorter than sepals; seeds 1 mm long, globular, with a deep pit on one side, surface minutely papillate, brownish-black. (n = 8) Mar-May. Locally common in prairies, pastures, open woodlands, lawns, roadsides, stream valleys, fallow fields; SD: Union; se NE, e 1/2 KS, MO, OK; (VA, IL, SD, KS, s to FL). *Houstonia minima* Beck—Rydberg; *H. pusilla* Schoepf—Gleason & Cronquist.

The very similar *H. rosea* Raf. approaches our area in se GP. It has corolla distinctly pink, rarely white, flowers with lobes 6 mm long. Lewis (Ann. Missouri Bot. Gard. 55(1): 33. 1968.) reported *H. rosea* for KS: Woodson. A study of this specimen (KANU) reveals it to be *H. crassifolia*.

2. **Hedyotis humifusa** A. Gray, rough small bluet. Winter annual or annual herb with slender taproots; stems erect or often spreading, often much branched, 3-8(10) cm long, evidently scaberulous on angles and less so on surfaces. Leaves opposite, lower linear-oblanceolate, becoming linear, (7)10-15(18) mm long, 0.5-3.5 mm wide, mucronate, usually scaberulous; lower leaves often petiolate, sessile upward; stipules scarious, to 2.5 mm long, setiferous. Flowers axillary, heterostylous, peduncle recurved after anthesis; hypanthium 1 mm long; calyx lobes linear attenuate, 2.5-4 mm long; corolla funnelform, 5-8 mm long, white or pinkish to pale purple, lobes 4-7 mm long, papillose within. Capsule didymous, 2-3 mm in diam, distinctly papillose, shorter than sepals. Seeds 1.5 mm long, black, papillate, with an elongated pit on one side. (n = 11) Mar-Sep. Sandy prairies, open woodlands, pastures, dunes, roadsides; sw OK, TX; (sw OK to s TX, se AZ).

3. **Hedyotis longifolia** (Gaertn.) Hook., slender-leaved bluet. Perennial herbs with short, slender, branching caudex; stems erect or ascending, usually several from base, 1-2.5 dm tall, simple or branched above, glabrous to scaberulous, often with short hairs at nodes. Leaves sessile, opposite, linear or linear-oblong, 1-3 cm long, 2-5 mm wide, sometimes revolute, 1-nerved, glabrous or minutely scaberulous, often ciliate; stipules not bristle tipped. Flowers in loose or crowded cymes, pedicels 3-10 mm long, hypanthium 0.8-1.2 mm long; calyx lobes lance-linear or subulate, 1.6-2.5 mm long, glabrous; corolla purplish, pinkish to white, tube 4-5(6) mm long, lobes 2-2.5(3) mm long, hairy within. Fruit globose, 2.3-3 mm long, 1/2 inferior, as long to shorter than sepals; seeds black, 1-1.2 mm long, minutely papillate, with an obscure ridge on concave face. (n = 6, 12) (Apr) Jun-Jul (Sep). Dry sandy prairies, rocky open woodlands; e ND, MN; KS: Cherokee; (ME, Ont., Sask., s to GA, AR, MO, OK). *Houstonia longifolia* Gaertn.—Rydberg.

Rydberg reported *Hedyotis canadensis* (Willd.) Fosb. as *Houstonia canadensis* Willd. for ND, but specimen evidence has not been located. It is similar to *Hedyotis longifolia*, but has distinctly ciliate basal leaves at anthesis while *longifolia* has basal leaves usually absent at anthesis, or eciliate, if present.

4. **Hedyotis nigricans** (Lam.) Fosb., narrowleaf bluet. Perennial herbs with stout, woody, branching caudex, often suffrutescent; stems few to many from base, 1-4(6) dm tall, often branched, glabrous to minutely scaberulous on angles, rarely short hairy below. Leaves opposite, sessile, numerous, linear to rarely linear-spatulate, 1-3(4) cm long, 1-2(3) mm wide, 1-nerved, usually revolute, glabrous or minutely scaberulous, frequently with axillary fascicles. Flowers numerous, forming a crowded panicle, sessile, or pedicels 1-3(5) mm long; hypanthium somewhat turbinate, 0.7-1.2 mm long; calyx lobes lanceolate to narrowly triangular, 1(2) mm long, sparsely ciliate; corolla funnelform, white to light blue or purplish, tube (2)3-4(6) mm long, lobes (1)2-4(5) mm long, pubescent within. Fruit obovoid, 2/3 or more inferior, 2.5-3.5(4) mm long; seeds black, 1-1.3 mm long, minutely papillate, with inconspicuous ridge on concave face. (n = 9, 10) May-Oct. Dry, rocky, prairie hillsides & bluffs, rocky open woodlands, badlands, roadsides, stream valleys; s 1/2 GP; (GA, OH, s MI, IA, NE, s to FL, TX, Mex.). *Houstonia angustifolia* Michx.—Rydberg; *H. nigricans* (Lam.) Fern.—Fernald.

Several varieties described to the south of our area can be found here, but none seem to merit recognition.

5. SPERMACOCE L., Buttonweed

1. **Spermacoce glabra** Michx., smooth buttonweed. Perennial herbs with subterranean to superficial branching caudex, stems several from base, 1-6 dm tall, erect to often decumbent or procumbent, simple or branched, glabrous, somewhat angled. Leaves opposite,

glabrous, elliptic to lanceolate or oblanceolate, 2–8 cm long, acute to acuminate at apex, tapering to a short petiole; stipular sheaths with several filiform bristles, 2–5 mm long. Flowers glomerate in upper axils, 10–40 per node; hypanthium turbinate, 2–2.5 mm long; calyx lobes lance-oblong, 1–3 mm long; corolla white, 2–3 mm long, conspicuously bearded in the throat; filaments and style short; ovary 2-celled, 1 ovule per cell. Fruit 3–4 mm long, smooth, capped by sepals, eventually septicidal, 1/2 retaining septum and closed, the other open on the inner face; seed reddish-black, 2.5–3.2 mm long, minutely reticulate. (n = 14) Jun–Oct. Locally common in low wet woodlands, edge of ponds, streams, moist waste areas; se KS, MO, OK; (OH, IL, MO, se KS, s to FL, TX).

132. CAPRIFOLIACEAE Juss., the Honeysuckle Family

by Ralph E. Brooks

Ours perennial herbs, woody vines, creeping or erect shrubs, or small trees. Leaves opposite and mostly exstipulate. Flowers variously arranged, perfect, regular or irregular, gamopetalous, epigynous; calyx 3- to 5-lobed; corolla 5-lobed or sometimes bilabiate; stamens ordinarily 5 (4 in *Linnaea*), adnate to the corolla and alternating with the lobes; anthers mostly versatile; ovary 2- to 5-locular with 1 or several pendulous ovules in each, sometimes only 1 locule fertile; stigma capitate or 2- to 3(5)-lobed, style elongate to obsolete. Fruit indehiscent and usually fleshy.

1 Leaves pinnately compound .... 3. *Sambucus*
1 Leaves simple.
  2 Erect coarse herbs with large leaves and sessile flowers .... 5. *Triosteum*
  2 Woody vines, shrubs, or small trees.
    3 Stamens 4, corolla lobes 5; stems trailing or repent; leaves mostly less than 2 cm long .... 1. *Linnaea*
    3 Stamens 5; leaves mostly over 2 cm long.
      4 Style very short or obsolete; flowers small and numerous in cymes; fruit a 1-seeded drupe .... 6. *Viburnum*
      4 Style elongate; flowers few in terminal or axillary clusters.
        5 Corolla irregular, usually more than 1 cm long; fruit a several-seeded, red to blue or black berry. .... 2. *Lonicera*
        5 Corolla essentially regular, less than 1 cm long; fruit a 2-stoned berrylike drupe .... 4. *Symphoricarpos*

1. LINNAEA L., Twinflower

1. ***Linnaea borealis*** L., twinflower. Small evergreen plants; stems woody, repent, 5 dm or more long, about 1 mm diam, pubescent or becoming glabrate in age, occasionally glandular when young, emitting numerous short (to 9 cm long), erect to suberect branches. Leaves opposite but often appearing whorled on young branches with congested nodes, obovate to broadly elliptic or suborbicular, 1–2 cm long, 5–12 mm wide, pinnately veined, sparsely pubescent, especially on the upper surface near the margins and on the veins below, apex broadly acute to rounded, margins with a few shallow teeth toward the apex or rarely entire, base cuneate to attenuate; petiole 1–3(5) mm long, sparingly long ciliate. Flowers regular, borne in pairs at the end of a long naked peduncle terminating the upright branches; calyx 2–4 mm long, hirsute and glandular, deeply 5-lobed, the lobes lanceolate; cor-

olla pink or pinkish, slender-campanulate or funnelform, 7–11 mm long, flaring just above the calyx and with 5 nearly equal, rounded shallow lobes, villose within; stamens 4, didynamous, included and attached near the base of the corolla, anthers about 1.5 mm long, filaments 5–9 mm long; stigma obscurely lobed, style elongate and occasionally exceeding the corolla, ovary 3-locular, 2 of the cells usually containing abortive ovules and the other with a single, pendulous, normal ovule; pedicel 3–18 mm long. Fruit dry, indehiscent, unequally 3-locular, 1-seeded, red-brown, ovoid, about 2.5 mm long and 1.3 mm diam; seed pale yellow-brown, ovoid to ellipsoid, about 1.5 mm long, 0.8–1 mm in diam, apiculate, with a single shallow longitudinal furrow. (n = 16) Jun–Aug. Moist, often rocky woodlands, occasionally in drier exposed sites; ND: Bottineau, Dunn; SD: BH, Harding; sw MT; (circumboreal, s in N. Amer. to WV, IN, n MN, ND, w SD, NM, AZ, & CA). *L. americana* Forbes—Rydberg; *L. borealis* var. *americana* (Forbes) Rehd.—Fernald.

2. LONICERA L., Honeysuckle

Woody vines or shrubs. Leaves simple, opposite, often coriaceous, pinnately veined, entire, exstipulate; petiole short or leaves perfoliate. Flowers borne in pairs on axillary peduncles or in 3-flowered, opposite, and sessile cymules on terminal or axillary rachises; calyx short, 5-toothed; corolla tubular or funnelform, nearly regular and shallowly 5-lobed or bilabiate with a 4-lobed upper lip and 1-lobed lower lip, the tube sometimes gibbous or spurred at the base; stamens 5, about equal in length, mostly exserted, filaments attached about midway on the corolla tube; stigma capitate; style elongate; ovary 2- or 3-locular with several pendulous ovules per cell. Fruit a several-seeded berry, globose; placentation axile. *Xylosteon* Adans.—Rydberg.

```
1  Upright shrubs (the branch tips sometimes twining a few cm).
   2  All leaves subsessile to short petioled, never perfoliate ........................... 7. L. tatarica
   2  At least some of the leaves perfoliate ................................................. 1. L. albiflora
1  Plants twining or trailing.
   3  All leaves subsessile or short-petioled, never perfoliate. ........................ 4. L. japonica
   3  At least some of the leaves perfoliate.
      4  Corolla deep red, scarcely bilabiate, and 3.5–5 cm long .............. 6. L. sempervirens
      4  Corolla white, yellowish, or orange, strongly bilabiate, and usually less than 3.5 cm
         long.
         5  The terminal perfoliate leaves together nearly circular in outline and glaucous
            above; corolla cream colored or yellowish. ............................... 5. L. prolifera
         5  The terminal perfoliate leaves together broadly oblong, elliptic, or rhombic in
            outline and green above.
            6  Corolla yellowish to distinctly rose colored, gibbous near the base; lower leaf
               surface conspicuously glaucous ............................................. 2. L. dioica
            6  Corolla creamy yellow to orange tinged, usually not gibbous near the base;
               lower leaf surface slightly gray-green but not conspicuously glaucous; plants of
               extreme se GP ............................................................................ 3. L. flava
```

1. **Lonicera albiflora** T. & G., white honeysuckle. Plants bushy, to 25 dm tall, the branches occasionally sparsely pubescent and the tips sometimes twining; bark thin and shredding. Leaves yellow- or gray-green above and variously glaucous below, mostly elliptic to suborbicular, rhombic, or obovate, (1.5)2–4(6) cm long, (1)1.5–3.5(4) cm wide, coriaceous, glabrous above and sparsely pubescent to glabrous below, apex broadly acute to rounded, margins entire, base broadly cuneate to rounded or rarely attenuate on juvenile shoots, sessile or subsessile, usually only the terminal pair of leaves perfoliate to form a disk subtending the glomerules. Inflorescence a pair of 3-flowered, sessile cymules, terminal; corolla bilabiate, barely gibbous at the base, white or yellowish, 1.5–2 cm long, inner and outer

surfaces pubescent or not. Fruit reddish-orange, 5-15 mm in diam, glabrous, ripening in late fall; seed pale reddish-brown, ovoid and flattened especially on the ventral surface, about 5 mm long and 2 mm wide, glabrous. Apr-May. Gravelly exposed hillsides & canyons; sw OK, TX; (sw OK, TX, w to NM & AZ, s to n Mex.). *L. albiflora* var. *dumosa* (A. Gray) Rehd.—Correll & Johnston.

Our plants have been referred to var. *dumosa*, which was originally described as having leaves pubescent, at least beneath, and the corolla pilose outside or glabrous. The variability of these characters in our region, however, suggests that the taxon is a dubious one at best.

2. *Lonicera dioica* L., limber or wild honeysuckle. Plants trailing and twining, the stems becoming several m or more long, stems of the season tan or frequently tinged with purple and smooth, older stems grayish and often with shredding bark; adventitious roots occasionally borne at the nodes of trailing stems. Leaves pale to dark green above, conspicuously glaucous below, elliptic to obovate, occasionally narrower or broadly obovate to nearly round, (3)5-12 cm long, 1.5-6 cm wide, subcoriaceous, glabrous or apex acute to obtuse or rounded, margins entire, base cuneate to nearly rounded, short petioled (less than 5 mm) to sessile, the uppermost 1-several pairs perfoliate, the terminal pair together broadly oblong, elliptic, or rhombic in outline. Inflorescence usually a pair of 3-flowered, sessile cymules, terminal; calyx with 5 short, blunt lobes, glandular or not; corolla bilabiate, gibbous at the base, yellowish to distinctly rose-colored, 2-2.5 cm long, glabrous or with mixed glandular and eglandular pubescence on the outer surface, the throat usually pubescent within; stamens well exserted, anthers pubescent or not; style sparsely pubescent to glabrous. Fruit reddish-orange, 5-10 mm in diam, glabrous, ripening midsummer to fall; seed golden yellow or paler, ellipsoid and flattened, about 3 mm long and 2 mm wide, glabrous. ($2n = 18$) Apr-Jul. Wooded hillsides or brushy stream banks; MN, ND, e 1/4 SD & BH, IA, e 1/4 NE, nw MO, ne KS; (Que. w to B.C., s to NC, IN, MO, ne KS, & ND). *L. glaucescens* Rydb.—Rydberg; *L. dioica* var. *glaucescens* (Rydb.) Butt.—Atlas GP.

3. *Lonicera flava* Sims, yellow honeysuckle. Similar to no. 2 except as noted. Leaves slightly gray-green below but not conspicuously glaucous, glabrous. Corolla not gibbous at the base or only slightly so, creamy yellow to orange tinged. ($2n = 18$) Apr-Jun. Rocky or along calcareous outcrops; sw MO, s 1/6 KS; (NC & GA, w to se KS & OK).

4. *Lonicera japonica* Thunb., Japanese honeysuckle. Plants semievergreen, trailing and twining, stems to several m or more long, pubescent, the hairs pale castaneous, older stems reddish-brown and the bark sometimes shredding; adventitious roots frequently borne at the nodes of trailing stems. Leaves never perfoliate, coriaceous, ovate to oblong, 3-8 cm long, 1.5-3 cm wide, upper surface at least pubescent on the midvein, lower surface sparsely pubescent to glabrate, apex acute to obtuse, margins entire and usually ciliate, base broadly cuneate to rounded, occasionally nearly truncate; petiole 2-8(11) mm long, pubescent. Flowers borne in pairs terminating axillary peduncles; bracts foliaceous, ovate; bractlets rotund, minute; calyx glandular and pubescent, lobes triangular to lanceolate, about 1 mm long; corolla bilabiate, not gibbous at the base, white, cream, or pinkish, 2-4.5 cm long, inner and outer surface at least sparsely pubescent, the outer also glandular pubescent; stamens well exserted. Fruit black, 5-8 mm in diam, smooth, ripening in fall; seeds dark brown, ovoid, 2-3.3 mm long, about 2 mm wide, rugose and ridged lengthwise. ($2n = 18$) Apr-Jun (Jul-Sep). Roadsides, woodlands, & various disturbed sites; e NE, MO, e 1/2 KS, e 1/2 OK; (established over much of e & s U.S.; Asia). *Naturalized*.

Japanese honeysuckle is a vigorous weed in some areas of the se GP. In our region the plant primarily reproduces vegetatively; while tremendous numbers of flowers are produced in the spring, fruit set is usually low.

5. **Lonicera prolifera** (Kirchn.) Booth ex Rehd., grape honeysuckle. Similar to no. 2 except as noted. Foliage glaucous above and wholly glabrous, the terminal perfoliate leaves together nearly circular in outline. Corolla cream colored or yellowish. (2n = 18) Apr–Jun. Rocky woods; KS: Bourbon; sw MO(?); (NY w to WI, s to TN, AR, & se KS). *L. prolifera* var. *glabra* Gl.—Gleason & Cronquist.

6. **Lonicera sempervirens** Ait., trumpet honeysuckle. Plants trailing and twining with glabrous foliage, the stems to several m or more long, older stems with reddish-gray shredding bark; occasionally producing adventitious roots at the nodes of trailing stems. Leaves olive-green to dark green above and lighter or sometimes glaucous below, ovate to elliptic or occasionally obovate, 3–6 cm long, 2–4 cm wide, subcoriaceous, apex acute to obtuse, margins entire, base cuneate, the uppermost 1–several pairs perfoliate. Inflorescence a pair of 3-flowered, sessile cymules, terminal; corolla usually scarlet red outside and paler within, shallowly 5-lobed and slightly bilabiate, 3–5 cm long; stamens exserted. Fruit reddish-orange, 6–10 mm in diam, glabrous, ripening by mid-summer; seeds ovoid and flattened, yellow to brownish, 4–5 mm long, 3–3.5 mm wide. (2n = 18) May–Jun (Aug). Woods & thickets; IA: Adair; MO: Barton, Jasper; KS: Cherokee, Neosho; OK: Payne; (native to the e & se U.S., w to e OK, widely escaped from cultivation elsewhere, as in GP).

7. **Lonicera tatarica** L., Tatarian honeysuckle. Plants bushy with glabrous foliage, to 3 m tall; older stems with brownish-gray to gray shredding bark. Leaves ovate to elliptic, rarely obovate, 2.5–4(6) cm long, (1)1.5–2.5(3) cm wide, apex acute to rounded, occasionally short acuminate, margins entire, base cuneate to subcordate, leaves never perfoliate; petiole 2–8 mm long or leaves subsessile. Flowers borne in pairs terminating axillary peduncles; bracts terminating the peduncle linear-lanceolate, 1–6 mm long, glabrous or sparsely pubescent; bractlets subtending each flower minute and broadly ovate; calyx lobes triangular, about 1 mm long; corolla bilabiate, gibbous at the base, white, cream colored, or sometimes pinkish, 11–18 mm long, glabrous outside, glabrous to sparsely pubescent within; stamens exserted. Fruit orange to reddish-orange, 5–8 mm in diam, glabrous, ripening in midsummer; seeds yellowish, ovoid and flattened, 2.5–3 mm long, 2–2.5 mm wide, granular and slightly ridged on each surface, glabrous. (2n = 18) May–Jun. Open woods, stream banks, brushy pastures; MN w to MT, s to IA & ne KS; (common in the GP, infrequent southward; Eurasia). *Naturalized. Xylosteon tataricum* (L.) Medic.—Rydberg.

Lonicera maackii Maxim., Maack honeysuckle, keys here but can be distinguished from *L. tatarica* by its small (2–4 mm diam), reddish fruits. It is a native of Asia and has been collected as an escape in KS: Cowley.

3. SAMBUCUS L., Elderberry

Ours coarse shrubs with pithy stems, spreading by suckers. Leaves imparipinnately compound, opposite, stipules present or not. Flowers regular, small, 5-merous, borne in many-flowered, terminal, compound, basically cymose inflorescences; calyx lobes minute or obsolete; corolla open-urceolate with 5 widely spreading lobes obviously longer than the tube; stigma 3- to 5-lobed; style nearly obsolete; ovary 3- to 5-celled with a single pendulous ovule in each cell. Fruit a juicy, berrylike drupe containing 3 or 4(5) cartilaginous nutlets, each enclosing a seed.

1 Inflorescence umbelliform and flat-topped; fruit dark purple; pith white ..... 1. *S. canadensis*
1 Inflorescence paniculiform and ovoid; fruit bright red; pith brown .... 2. *S. racemosa* subsp. *pubens*

1. **Sambucus canadensis** L., common elderberry. Shrub to 3 m tall; branches glabrous and somewhat glaucous and with a large white pith. Leaves 5- to 9-foliolate, the lowermost

pair sometimes divided again; leaflets lanceolate to ovate or elliptic, 4–13 mm long, 2–5.5(7) cm wide, glabrous above or the midvein puberulent, pubescent to glabrate below, apex mostly caudate to acuminate, margins serrate except at the apex, base nearly rounded and oblique to cuneate; petioles 3–9 cm long, grooved on the upper side, glabrous; petiolules 2–7 mm long; stipules early deciduous. Inflorescence terminal, umbelliform-cymose, 3- to 5-rayed from the base, flat-topped or slightly convex, to 3 dm wide at anthesis; corolla creamy-white, 4–5 mm wide, lobes broadly elliptic. Fruit dark purple, 4–6 mm in diam, glabrous, edible; nutlets yellowish, and slightly flattened obovoid, 2.5–3 mm long, 1.5–2 mm wide, asperous. (n = 18) May–Aug. Moist woods, ditches, fields, or stream banks; MN, e 1/2 ND, e 1/3 SD, IA, NE, MO, KS, OK; TX: Hemphill; (N.S. & Que. w to Man., s to FL, TX, & Mex.; W.I.). *S. canadensis* var. *submollis* Rehd.—Gleason & Cronquist.

2. ***Sambucus racemosa*** L. subsp. ***pubens*** (Michx.) House, stinking elderberry or red-berried elder. Shrubs to 2 m tall; twigs nearly glabrous to pubescent, with a large brownish pith. Leaves 5- to 9-foliolate, the lowest pair infrequently divided again; leaflets narrowly elliptic to lanceolate, the terminal one sometimes oblanceolate to obovate, 4–9(13) cm long, 1.5–4 cm wide, glabrous or puberulent on the midvein above, pubescent or sparsely so below, apex usually acuminate to caudate, sometimes acute, margins serrate except at the apex, base cuneate to nearly rounded, usually oblique; petioles 1.5–5 cm long, pubescent, grooved above; petiolules 1–6 mm long; stipules a pair of glandular appendages or linear and 1–2 mm long, soon deciduous. Inflorescence terminal, paniculiform and ovoid, 3–6 cm wide at anthesis; corolla creamy-white, 3–4 mm wide, lobes broadly elliptic. Fruit bright red or orangish, 4–6(7) mm in diam, glabrous; nutlets yellowish, ellipsoid and slightly flattened, about 2.5 mm long and 1.5 mm wide, asperous. (n = 18) May–Jun. Rich, rocky soil in moist woods or along streams; MN; ND: Richland, Welsh; BH; (Newf., w to Alta., s to GA, TN, IA, e ND, & w SD). *S. microbotrys* Rydb., *S. pubens* Michx.—Rydberg.

4. SYMPHORICARPOS Duham., Snowberry, Coralberry, Wolfberry

Ours low bushy shrubs; bark shredding on older branches. Leaves simple, opposite, subcoriaceous, pinnately veined, exstipulate, and short petioled. Flowers borne in terminal or axillary reduced racemes or spikes; calyx 5(4)-toothed; corolla pink to white, campanulate to salverform, regular, 5(4)-lobed and slightly gibbous at the base; stamens 5(4); stigma capitate or obscurely 2-lobed; style elongate; ovary 4-locular with several abortive ovules in 2 cells and the other cells with 1 pendulous normal ovule in each. Fruit a globose, white or red, berrylike drupe, 4-celled but containing only 2 nutlets, each enclosing a single seed.

Reference: Jones, G. N. 1940. A monograph of the genus *Symphoricarpos*. J. Arnold Arbor. 21: 201–252.

1 Trailing shrub; corolla salverform and 9–12 mm long; leaves averaging less than 2 cm long .... 4. *S. palmeri*
1 Erect shrub; corolla campanulate and less than 9 mm long; leaves averaging more than 2 cm long.
  2 Fruit coral-pink to purplish; corolla 2.5–3(4) mm long; stamens and style included. .... 3. *S. orbiculatus*
  2 Fruit white; corolla 5–8 mm long; stamens and style included or exserted.
    3 Corolla lobes barely spreading; stamens and style included; flowers solitary or borne in few-flowered (2–5) racemes; leaves averaging less than 3 cm long .... 1. *S. albus*
    3 Corolla lobes distinctly spreading at anthesis; stamens and style exserted; flowers borne in spicate clusters (6–12); leaves usually averaging 3 cm or more long. .... 2. *S. occidentalis*

1. ***Symphoricarpos albus*** (L.) Blake, white coralberry. Shrubs to 1 m tall, young branches sparsely pubescent to nearly glabrous; rhizomes occasionally produced. Leaves ovate

to elliptic, occasionally suborbicular, 1–3(5) cm long, 0.7–2(3.5) cm wide, slightly whitened below, glabrous to pubescent, especially below, apex acute to obtuse or occasionally rounded, margins entire or with a few large, irregular, and blunt teeth, base mostly cuneate to rounded; petiole 1–2(4) mm long, usually pubescent. Flowers borne 2 or 3(5) on short racemes; calyx lobes triangular to lanceolate, less than 1 mm long; corolla campanulate with the shallow lobes barely spreading, 5–7 mm long, 3–5 mm wide, outer surface glabrous, densely hirsute within, especially the throat, the lobes rounded; stamens and style included. Fruit white, fleshy, ovoid, 7–9 mm long, 6–8 mm wide, calyx persistent; nutlets white to yellowish ellipsoid and flattened on the ventral side, 4–5 mm long, 2–2.5 mm wide, the ends acute, smooth to minutely striate. (n = ca 27) Jun–Jul. Wooded hillsides, open dry or moist gravelly slopes; MN, ND, MT, s 1/3 SD, n-cen & nw NE, WY; (Que. w to B.C. & AK, s to VA, OH, MN, NE, CO, UT, NV, & CA). *S. pauciflorus* (Robbins) Britt.— Rydberg.

2. ***Symphoricarpos occidentalis*** Hook., western snowberry, wolfberry. Shrubs to 1(1.5) m tall, twigs pubescent to glabrate; rhizomatous, forming large colonies. Leaves ovate, elliptic or suborbicular, infrequently obovate, (1.5)2–6 cm long, (0.7)1–3.5 cm wide, juvenile shoots sometimes with leaves to 10 cm long and 8 cm wide, upper surface glabrate, lower surface whitened and variously pubescent, apex acute to obtuse or rounded, margins entire or with several large, blunt teeth, usually ciliate, base cuneate or broadly so, rounded, or nearly truncate; petiole 2–7 mm long, pubescent to glabrate. Flowers several (up to 12) in spicate clusters; calyx lobes broadly triangular, less than 1 mm long; corolla campanulate with the lobes spreading, 5–8 mm long, often wider than long, outer surface glabrous, the throat densely hirsute within, the lobes rounded; stamens and style exserted. Fruit white (drying blackish or bluish), fleshy, globose, 6–9 mm in diam, calyx persistent; nutlets yellowish, ovoid to ellipsoid and slightly flattened, 2.5–3.5 mm long, 1.5–2 mm wide, essentially smooth. Jun–Jul (Aug). Gravelly hillsides, ravines, open woods, pastures, & prairies; MN w to MT, s to nw MO, w KS, & NM; OK: Cimarron; (Man. w to B.C., s to nw MO, w OK, NM, UT, & WA).

3. ***Symphoricarpos orbiculatus*** Moench, coralberry, buckbrush. Shrubs to 1(1.5) m tall; twigs sparsely pubescent at first and becoming glabrate; rhizomatous and forming large colonies. Leaves elliptic to ovate or suborbicular, (1.5)2–4(5) cm long, 1–3.5 cm wide, upper surface glabrate, lower surface whitened or not and variously pubescent, apex acute to obtuse, sometimes rounded, margins mostly entire, occasionally with several blunt, irregular teeth, base cuneate to rounded; petiole 1–3 mm long, pubescent. Flowers several in spicate clusters; calyx lobes triangular, less than 1 mm long; corolla campanulate with slightly spreading lobes, 2.5–3(4) mm long, 2–3 mm wide, outer surface glabrous, densely hirsute on the throat within, the lobes rounded; stamens and style included. Fruit coral-pink to purplish, pulpy, ellipsoid to subglobose, 4–6 mm in diam, smooth, calyx persistent; nutlets white, ovoid to ellipsoid and flattened on one side, 2.5–3 mm long, about 2 mm wide, smooth. (2n = 18) Jul–Aug. Upland or low-lying areas in woods, pastures, ravines, or along streams; sw IA, se 1/4 NE, MO, e 1/2 KS, e 2/3 OK; (PA w to MN & e NE, s to FL & TX).

4. ***Symphoricarpos palmeri*** G. N. Jones, Palmer's snowberry. Clambering shrubs to about 1 m tall, the sprawling stems to 3 m long with short ascending branches; young stems pubescent. Leaves broadly elliptic to ovate, 7–20 mm long, 5–18 mm wide, upper surface sparsely pilose to glabrate, lower surface pale and pilose at least on the veins, margins entire or with a few large, irregular teeth, petiole 1–3 mm long, pubescent. Flowers short pedicelled, solitary or in pairs; calyx teeth broadly triangular, less than 1 mm long; corolla salverform, 9–12 mm long, 2–3 mm wide, outer and inner surface glabrous, the lobes rounded; stamens and style included. Fruit white, fleshy, ellipsoid, 6–8 mm long; nutlets whitish

ovoid to ellipsoid and slightly flattened, 4–5 mm long, 2–3 mm wide, acutish at the base. May–Jun. Rocky slopes & canyons; OK: Cimarron; (OK, w TX to AZ; CO).

5. TRIOSTEUM L., Horse-gentian, Feverwort

Coarse erect, perennial herbs; stems unbranched, 1–many from a clump, leafy throughout, usually hairy. Leaves simple, opposite, entire, usually hairy, sessile, perfoliate or united by narrow ridge around the stem, exstipulate. Flowers axillary, solitary, or in spicate clusters of 2–4, sessile, bracteate and sometimes bracteolate; calyx lobes linear-lanceolate, foliaceous, persistent; corolla nearly regular, salverform and with 5 shallow, imbricate lobes, the base gibbous, usually hairy within; stamens 5, inserted about halfway up the corolla tube, included or exserted; filaments hairy; stigma capitate or shallowly 3- to 5-lobed; style included or exserted, usually hairy below; ovary 3- to 5-locular, usually only 3 of the cells fertile with a solitary ovule. Fruit a dry drupe containing 3-ribbed stones with a thick bony endocarp.

1 Flowers solitary, subtended by a pair of bracts, the bracts equaling or exceeding the flower (or fruit, including calyx) in length, bracteoles wanting; corolla yellow; style included .... 1. *T. angustifolium*
1 Flowers solitary or in clusters 2–4, the inflorescence subtended by a single bract, the bract about 1/2 as long as the flower (a pair of minute bracteoles subtends each flower); corolla orangish to dark red, rarely yellowish; style usually exserted. .... 2. *T. perfoliatum*

1. **Triosteum angustifolium** L., yellow-flowered horse-gentian. Stems 2.5–5 dm tall, long hirsute, hairs mostly 1–4 mm long and occasionally mixed with shorter glandular hairs. Leaves oblanceolate to narrowly elliptic, occasionally subpanduriform, 8–15 cm long, 3–5(7) cm wide, moderately hirsute, the veins usually canescent and with longer hairs, apex acuminate or sometimes acute, margins ciliate, base usually attenuate and perfoliate with only a narrow ridge of tissue around the stem. Flowers solitary, sessile, each subtended by a pair of bracts; bracts foliaceous, linear-lanceolate to narrowly elliptic, equaling or exceeding the flower (or fruit, including the calyx) in length, sparsely hirsute or pubescent, ciliate; bracteoles wanting. Calyx lobes 7–13 mm long, 1–2 mm wide, sparsely hirsute, apex acute to obtuse, margins entire, ciliate with long and short hairs; corolla yellowish, 12–15 mm long, 5–6 mm wide, outer surface pubescent and glandular; style included. Fruit pale orange, subglobose, 5–6(7) mm in diam, hirsute-canescent; stones ellipsoid and slightly flattened on the ventral side, 4.5–7 mm long, 2.5–4 mm wide. (n = 9) Apr–Jun. Rocky wooded hillsides; KS: Cherokee; (PA w to IL & se KS, s to AL & LA).

2. **Triosteum perfoliatum** L., horse-gentian. Stems 5–13 dm tall, densely to moderately pubescent with glandular and eglandular hairs, longer hairs (to 2.5 mm) present or not. Leaves panduriform and broadly perfoliate at the base to broadly oblanceolate to elliptic, attenuate, and barely perfoliate at the base, (10)12–17(22) cm long, 4–7(10) cm wide, upper surface pubescent to puberulent or nearly glabrate, glandular hairs present or not, lower surface pubescent, soft to the touch, glandular hairs sometimes present, apex acuminate or occasionally acute, margins entire and ciliate. Flowers solitary or 2–4 in spicate clusters, the inflorescence subtended by a single, linear-lanceolate bract about 1/2 as long as the flower and each flower subtended by a pair of bracteoles, these linear-lanceolate and about 2–3 mm long. Calyx lobes 8–15 mm long, 1–2 mm wide, inner surface sparsely pubescent to glabrate, outer surface variously pubescent and glandular, with longer hairs sometimes present, apex acute to obtuse, margins ciliate; corolla dark red to orange, rarely yellow tinged, 8–13 mm long, 2–3 mm wide, outer surface pubescent and glandular; style distinctly exserted. Fruit pale orange to reddish, subglobose, 7–10 mm in diam, pubes-

cent to sparsely hirsute and usually mixed with glandular hairs; stone ellipsoid and flattened on the ventral surface, 6–9.5 mm long, 2.5–4 mm wide. (Apr) May–Jun. Low woods or wooded hillsides, often in rocky soil.

Two intergrading varieties occur in our region and may be distinguished as follows:

2a. var. *aurantiacum* (Bickn.) Wieg., orange-flowered horse-gentian. Stems usually pubescent and at least sparsely glandular hairy, rarely eglandular, these shorter hairs mixed with few to many longer (over 1.5 mm) eglandular hairs. Leaves mostly oblanceolate to elliptic with the base attenuate and barely perfoliate, occasionally broadly perfoliate on 1–3 pairs of leaves. Corolla orangish to dark red. NE: Richardson; e 1/4 KS, MO; (Que. w to MN & NE, s to NC, KY, AR, & e OK). *T. aurantiacum* Bickn., *T. illinoense* (Wieg.) Rydb. — Rydberg; *T. aurantiacum* var. *illinoense* (Wieg.) Palm. & Steyerm. — Atlas GP.

2b. var. *perfoliatum*, horse-gentian. Stems usually densely pubescent and glandular hairy, the hairs mostly less than 1 mm long, longer hairs occasionally present. Leaves mostly panduriform and broadly perfoliate at the base. Corolla mostly dark red. (n = 9). MN: Becker; IA, e 1/5 NE, MO, e 1/3 KS, ne 1/6 OK; (MA w to WI & MN, s to SC, GA, TN, AR, & ne OK).

The *T. perfoliatum* complex has been variously treated as being comprised of several species or, in some cases, fewer species, each with infraspecific taxa. The common plant in the GP is var. *perfoliatum* and it is usually easily recognized. Var. *aurantiacum* is only of scattered occurrence. The characters typically used to distinguish the two varieties, especially stem vestiture and leaf shape, intergrade with some degree of frequency in both GP and extraregional material examined. This suggests that separation of the two entities is dubious. The group warrants further systematic investigation.

6. VIBURNUM L., Viburnum, Arrow-wood

Ours erect shrubs or small trees. Leaves simple, opposite or rarely whorled, petiolate, exstipulate or the stipules partially adnate to the petiole or reduced to glands. Inflorescence terminal, compound, umbellike or paniculate, cymose, bracts and bractlets small, early deciduous. Flowers regular, the marginal flowers sometimes radiate, slightly irregular, and neutral; calyx shallowly 5-lobed, persistent in fruit, corolla open-campanulate to rotate, 5-lobed, white or cream colored, imbricate in bud; stamens 5, inserted at the base of the corolla, included or exserted; stigma obscurely 3-lobed; style short; ovary 3-locular with 2 abortive cells and 1 fertile cell with a single pendulous ovule. Fruit an ellipsoid to globose, 1-seeded drupe with soft pulp.

1 At least some of the leaves 3-lobed.
  2 Flowers of 2 kinds, the marginal ones neutral and much enlarged (to 2.5 cm across), the inner much smaller and perfect with the stamens exserted; stipules linear, 2–6 mm long .... 3. *V. opulus* var. *americanum*
  2 Flowers all perfect and small with included stamens; stipules wanting .... 1. *V. edule*
1 All leaves unlobed.
  3 Leaf margins finely serrate, eciliate.
    4 Leaves mostly obovate, apex rounded or less often acutish .... 6. *V. rufidulum*
    4 Leaves mostly ovate to elliptic, apex acute to acuminate.
      5 Leaf apex acute, rarely acuminate; flowering Apr–May; plants of the se GP .... 4. *V. prunifolium*
      5 Leaf apex usually abruptly acuminate; flowering Jun; plants of the northern GP .... 2. *V. lentago*
  3 Leaf margins dentate and ciliate .... 5. *V. rafinesquianum*

1. **Viburnum edule** (Michx.) Raf., squashberry, mooseberry. Erect or decumbent shrubs to 1.5 m tall; branches grayish, smooth or with sessile glands; winter buds with 2 connate outer scales that often persist well into the summer. Leaf blades typically broadly ovate, 3-lobed, and palmately veined, (3)5–8 cm long, (2.5)4.5–8 cm wide, glabrous above, lower

surface more or less hirsute, at least on the veins, and sessile glandular, lobe apices acute, margins dentate and ciliate, base rounded and with a pair of glandular teeth or projections near junction with the petiole, leaves of the uppermost pair on each branch usually elliptic to obovate, narrower than the lower leaves, unlobed or barely lobed, apex acuminate; petioles 0.5-2.5 cm long, sparsely sessile glandular. Inflorescence a few-flowered (less than 50), umbellike, compound cyme, 2-4 cm across, terminating short axillary branches; peduncle 1.5-2.3 mm long. Flowers all similar; calyx sessile glandular, the lobes triangular, about 0.5 mm long; corolla whitish, campanulate, 2.5-3.5 mm long, lobes rounded and 1-1.5 mm long; stamens included. Fruit reddish-orange, translucent, globose, 7-10 mm in diam, juicy; stone yellow-brown, ellipsoid and flattened, 7-9 mm long, 4-6 mm wide, 1.5-2 mm thick, muricate. Jun. Moist, wooded hillsides & ravines; SD: Lawrence; (Newf., w to AK, s to PA, MN, w SD, CO, ID, & OR). *V. eradiatum* (Oakes) House — Rydberg.

2. **Viburnum lentago** L., nannyberry, sheepberry. Small tree to 5 m tall; branches gray to reddish-brown, usually smooth. Leaf blades ovate to elliptic, rarely suborbicular, (4)5-9 cm long, (2.5)3-6 cm wide, pinnately veined and the veins anastomosing, essentially glabrous, apex acuminate to caudate, margins finely serrate, base cuneate to nearly truncate; petiole 1-2.7 cm long, usually narrowly winged and sparsely scurfy with red-brown stellate hairs, wing undulate and usually revolute. Inflorescence a terminal, sessile, umbellike, compound cyme, 5-12 cm across, rays 3-5(7) and scurfy. Flowers all similar; calyx glabrous, lobes triangular, about 0.5 mm long; corolla whitish, subrotate, 2.5-3.5 mm long, lobes rounded, ciliolate, and 1-1.5 mm long; stamens exserted. Fruit dark blue with a whitish bloom, ellipsoid to globose, 10-14 mm long, pulpy; stone yellowish, ovoid and flattened, 7-11 mm long, 6.5-9 mm wide, 1.5-2.5 mm thick, muricate. May-Jun (Jul). Sandy to rocky soil in open woods, along streams, or occasionally in ditches; MN, ND, e 1/6 & w 1/5 SD, IA, e edge NE plus Dawes; (Que. to Man., s to NJ, GA, MO, NE, & CO).

3. **Viburnum opulus** L. var. **americanum** Ait., highbush cranberry. Shrub or small tree to 4 m tall; branches light gray to brownish, smooth. Leaf blades broadly ovate, suborbicular, or broadly obovate, 3-lobed, palmately veined, (5)6.5-12 cm long, 4-10(12) cm wide, glabrous above, sparsely hirsute on the veins below, lobe apices acuminate, margins nearly entire to dentate, usually ciliate, base cuneate to rounded, occasionally the leaves just below the inflorescence elliptic-lanceolate, unlobed, and pinnately veined; petiole 1-3 cm long, glabrous, with 2 stipitate glands near the summit; stipules adnate to the petiole, linear and slightly enlarged at the apex, 2-6 mm long. Inflorescence terminal, umbellike, compound, and cymose, 7-12 cm across; peduncle 3-9 cm long, sessile glandular. Marginal flowers neutral, whitish, with enlarged, slightly irregular, rotate corollas, 13-20(25) mm wide; the lobes longer than the tube; perfect flowers whitish, subrotate, 2-3 mm long, lobes rounded, margins subentire to crenulate or sparsely glandular-ciliolate; stamens exserted. Fruit reddish-orange, translucent, subglobose, 8-12 mm in diam, juicy; stone yellowish, broadly ovate and flattened, 7-10 mm long, 6-8 mm wide, 1.5-2.5 mm thick, acute, muricate. (2n = 18) Jun. Moist, wooded hillsides or low woodlands; MN, e 1/2 ND; SD: BH & Roberts; NE: Douglas; (Newf., to s B.C., s to PA, IL, e NE, w SD, WY, ID, & WA). *V. trilobum* Marsh. — Rydberg.

4. **Viburnum prunifolium** L., black haw, sweet haw. Similar to no. 2 except as noted: Shrubs to about 2 m tall. Leaf blades ovate to elliptic, 3.5-8 cm long, 2-4.5 cm wide, apex usually acute, infrequently acuminate or rounded; petiole 0.7-2 cm long, barely winged or not, the wing not undulate or revolute. (2n = 18) Apr-Jun. Flood-plain woods or rocky, wooded hillsides; IA: Taylor; MO, e 1/3 KS, ne OK; (NY to MI, IA & e KS, s to FL & TX). *V. prunifolium* var. *bushii* (Ashe) Palm. & Steyerm. — Fernald.

5. Viburnum rafinesquianum Schult., downy arrow-wood. Shrub to 2 m tall; branches reddish-brown to grayish, smooth. Leaf blades ovate to elliptic, 4–9 cm long, 2–7 cm wide, pinnately veined, glabrous to sparsely hirsute above, lower surface hirsute on the veins and otherwise glabrous or entirely hirsute, apex acute, margins coarsely dentate and ciliate, base rounded to subcordate; petiole 2–8 mm long, variously hirsute, stipitate glands sometimes present; stipules adnate to the petiole, linear-lanceolate, 1.5–4 mm long, hirsute. Inflorescence terminal, umbellike, compound, cymose, 3–5 cm across, rays (3)5–7 and glandular; peduncles 0.5–2.5 cm long. Flowers all similar; calyx glabrous or sparsely glandular, lobes triangular, 0.5 mm long, purplish; corolla whitish, subrotate, 2–3 mm long, lobes rounded, about 1 mm long, ciliolate, stamens exserted. Fruit dark blue, elliptic to subglobose, 7–10 mm in diam, pulpy; stones yellowish, ovoid and flattened, 6–7.5 mm long, 4–5.5 mm wide, 1.5–2 mm thick, muricate. ($2n = 20$, 36) Jun. Moist, wooded hillsides or along streams; MN, n-cen & e ND, e edge SD; IA: Auburn, Cherokee; (Que. to Man., s to GA, KY, AR, e SD & n-cen ND; e OK). *V. rafinesquianum* var. *affine* (Bush) House—Atlas GP.

6. Viburnum rufidulum Raf., southern black haw. Similar to no. 2 except as noted: Shrubs or small trees to 3 m tall. Leaf blades obovate or occasionally elliptic, 3.5–6.5 cm long, 2.5–4 cm wide, usually coriaceous, apex obtuse or rounded, infrequently acutish, base cuneate; petiole 0.5–1 cm long, scarcely winged. ($2n = 18$) Apr–May. Wooded uplands, hillsides, or stream banks; s MO, se KS, e 1/2 OK; (FL to e TX, n to VI, OH, IN, IL, s MO, & se KS).

133. ADOXACEAE Trautv., the Moschatel Family

by Ronald R. Weedon

1. ADOXA L.

1. *Adoxa moschatellina* L., moschatel, muskroot. Small delicate perennial herbs, 5–15(20) cm tall with short, musk-scented rhizomes bearing slender roots, 3–5 alternate white scaly leaves, 1–3 long-petioled foliage leaves, and an erect, glabrous, single flowering stem (less frequently 2 or 3). Basal foliage leaves very variable in size but averaging about 8 cm in length, exstipulate, ternately compound, the primary divisions discrete, again 1- to 2-parted, the ultimate segments rather broad and thin, round-toothed and mucronate; cauline leaves an opposite pair, 3-parted, smaller, less divided and shorter petioled, commonly borne a little above the middle of the flowering shoot. Inflorescence an angular capitulumlike cyme composed of 5(3–8) small flowers with no subtending bracts. Flowers perfect, regular or nearly so, sympetalous; calyx 2- to 4-lobed, persistent in fruit; corolla 4- to 6-lobed, united, rotate, the lobes 1.5–3 mm long, yellowish-green; stamens appearing as twice as many as the corolla lobes, each split nearly to the filament base, the anthers thus 1-celled, in epipetalous pairs on the corolla tube and alternate with its lobes; ovary semi-inferior, with 3–5 carpels, style short, distinct, deeply 3- to 5-cleft, with minute stigmas. Fruit a small, dry, green druplet with (1)3–5 lenticular nutlets. ($2n = 36$) May–Aug. Cool, moist wooded valleys, ravines or slopes, most often in shade, on bare or mossy ground; rare & often very local; BH; (circumboreal, s in N. Amer. to n NY, ne IA, SD, WY, & in the Rocky Mts. to NM).

References: Sprague, T. A. 1927. The morphology and taxonomic position of the Adoxaceae. J. Linn. Soc., Bot. 47: 471–487; Cochrane, T. S., & P. J. Salamun. 1974. Preliminary Reports on the

Flora of Wisconsin No. 64. Adoxaceae — Moschatel Family. Trans. Wisconsin Acad. Sci. 62: 247–252. Treatment prepared with the assistance of Daniel G. Schneider.

This monotypic family, worldwide and north temperate, is often now considered to be a specialized offshoot from the Caprifoliaceae and of great biogeographic interest. It is a reasonable hypothesis that *Adoxa* is an old genus whose present-day disjunct range is due to survival in relict isolated microhabitats after a previous cooler interglacial or postglacial period.

134. VALERIANACEAE Batsch, the Valerian Family

by Ronald R. Weedon

Annual or perennial herbs. Leaves in basal rosettes or opposite, exstipulate, their bases often sheathing. Inflorescence a terminal monochasium, thyrse or many-flowered compound dichasial cyme, with bracts and usually bracteoles. Flowers irregular to nearly regular, small, bisexual or unisexual; calyx obsolete or late-developing as fruits mature, adnate to the ovary; corolla tubular and (3 or 4)5-lobed, imbricate, often basally spurred or saccate; stamens (2)3(4), epipetalous, alternating with the corolla lobes, anthers 2-celled, dehiscing longitudinally; pistil with inferior ovary, 3-locular, 1 chamber fertile, the other 2 sterile or suppressed, ovule solitary, pendulous, without endosperm; style 1, stigma simple or with 2 or 3 branches or lobes. Fruit an achene, the calyx often developing into a winged, awned, or plumose pappus.

Reference: Ferguson, I. K. 1965. The genera of the Valerianaceae and Dipsacaceae in the southeastern United States. J. Arnold Arb. 46: 218–231. This treatment was prepared with the assistance of Daniel G. Schneider.

Members of this family often have a very characteristic disagreeable odor, which some have compared with that of a wet dog.

1 Plants annual or short-lived biennials; calyx obsolete or as very short teeth; leaves mostly entire or dentate; fruit 3-locular .... 2. *Valerianella*
1 Plants perennial; calyx lobes 5–15, inrolled in flower, expanded on the mature fruit and appearing as conspicuous plumose pappus; cauline leaves pinnate or deeply parted; fruit 1-locular .... 1. *Valeriana*

1. VALERIANA L., Valerian

Ours perennial herbs from rhizomes or taproots with subscapose or leafy stems that are terete or occasionally quadrangular, glabrous or sparsely hairy with short simple hairs. Leaves decussate, petiolate to nearly sessile, basal and cauline, varying from undivided to pinnate or pinnatifid, frequently decurrent, clasping, entire, serrate, dentate, or repand. Inflorescence determinate, either aggregate-dichasial and thyrsoid or compound cymes, dense, bracteate. Flowers irregular, bisexual or unisexual; calyx inrolled in flower, enlarging and developing in fruit to form a pappus with a short, sessile, membranaceous limb and 5–15 plumose setae; corolla rotate to funnelform or campanulate, the tube usually slightly seccate at the base, more or less hairy on the throat, the 5(3 or 4) lobes equal or subequal; stamens 3(4), epipetalous, inserted toward the top of the corolla tube alternate with the corolla lobes, exserted, anthers 2- to 4-lobed; ovary inferior, 3-carpellate, with two locules supressed and 1 fertile with a solitary pendulous ovule; style 1, the stigma 3-lobed, included or exserted. Fruit a unilocular compressed achene with 3 veins on the outer and 1 vein on the inner surface.

Reference: Meyer, F. G. 1951. *Valeriana* in North America and the West Indies (Valerianaceae). Ann. Missouri Bot. Gard. 38(4): 377–503.

1   Plants with a stout taproot and short, branched caudex; basal leaves tapering gradually to a petiolar base; inflorescence paniculiform .... 3. *V. edulis*
1   Plants with a stout rhizome or caudex and numerous fibrous roots; lower leaves mostly with sharply differentiated blade and petiole; inflorescence corymbiform at anthesis, though it may expand in fruit.
  2   Corolla 4–8 mm long, the lobes about 1/2 as long as the tube, flowers usually all perfect .... 1. *V. acutiloba*
  2   Corolla mostly 2–4 mm long the lobes nearly equal to the tube or just shorter; plants gynodioecious, some with most all flowers pistillate, others with most all flowers perfect .... 2. *V. dioica*

1. *Valeriana acutiloba* Rydb. Perennial with a stout branched rhizome or caudex and fibrous roots, 1–6 dm tall. Leaves bright green, glabrous, the basal ones well developed often on separate short shoots, blades entire, ovate to oblong, up to 8 cm long and 3.5 cm wide, with a short wing-margined petiole; cauline leaves in 1–3 pairs, the lowest often petioled and undivided, the upper mostly pinnatifid with few and reduced lateral lobes. Inflorescence 1.5–5 cm wide at anthesis, often not much enlarged in fruit. Flowers usually all perfect; calyx segments plumose, 10–17; corolla 4–8 cm long, funnelform, white, somewhat hairy outside, the lobes scarcely 1/2 as long as the slightly gibbous tube; stamens well exserted; achenes ovate to oblong-lanceolate, 2.5–6 mm long, 1.5–2 mm wide, glabrous to short-hairy, often purplish. Jun–Jul. Moist areas, including meadows, boggy ground; open, often rocky wooded hillsides & ravines; BH of SD: Lawrence, Pennington, Custer; (sw MT to s OR, s to CA, AZ, WY, SD, CO, & NM). *V. capitata* Pall. subsp. *acutiloba* (Rydb.) F. G. Mey.—Meyer, op. cit.

2. *Valeriana dioica* L. Perennial with a branched rhizome or caudex and fibrous roots, mostly 1–4(6) dm tall, slender, glabrous or nearly so. Basal leaves well developed, petiolate, with blades rarely as long as 8 cm and as wide as 3 cm, mostly entire; cauline leaves in 2–4 pairs, mostly short-petiolate or subsessile, lobed, pinnate to pinnatifid. Plants gynodioecious, some with pistillate, others with perfect flowers. Inflorescence compact and 1.5–3 cm wide at anthesis, elongating and becoming somewhat diffuse, bracts 5–6 mm long glabrous. Calyx segments plumose, mostly 9–15; corolla 2–4 mm long, (or shorter on pistillate flowers), white, rotate to subrotate, the lobes not much, if at all, shorter than the tube, glabrous on the outside; stamens well exserted. Achenes ovate-oblong to lanceolate, 2–5 mm long, less than 2 mm wide, glabrous, tawny. (2n = 26, 32) Jun–Jul. Moist areas, including meadows, rich woods, stream banks, n-facing slopes; BH of SD: Lawrence, Pennington, Custer; to WY: Crook; (across Can. s of 60° lat., s to nw WA, ID, WY, SD). *V. septentrionalis* Rydb.—Rydberg.

This species intergrades west of our range with the more western *V. occidentalis* Heller, cf. Meyer, op. cit.

3. *Valeriana edulis* Nutt. Perennials from a long, stout conical taproot becoming semiwoody, rugose and verrucose with age, and a short-branched caudex, 1–2 dm tall, generally glabrous or nearly so. Leaves predominantly basal, linear to obovate, entire or pinnate to pinnatifid, the blade tapering gradually to the petiole or petiolar base, entire, mostly 7–40 cm long and 7–42(55) mm wide; cauline leaves 2–6 pairs, smaller, almost always pinnatifid with narrow segments, becoming sessile upward. Plants gynodioecious. Inflorescence paniculiform, though often compact at anthesis, expanded and more diffuse in fruit; bracts 3–4 mm long, reduced above. Calyx segments plumose, 8–13; corollas rotate, white to light cream-yellow, those of perfect and staminate flowers mostly 2.5–3.5 mm long, those of pistillate flowers minute, 0.5(1) mm long; stamens and styles exserted but not far.

Achenes oval to ovate-oblong, 2–4.5 mm long, 1.5–3 mm wide, glabrous to densely short-hairy. Jun–Aug. Moist areas, including wooded valleys, stream banks, open meadows; BH of SD: Lawrence, Meade, Pennington, Custer; (s B.C., Ont. s to e WA, MT, SD, ne IA, OH; Mex.). *V. trachycarpa* Rydb.—Rydberg.

2. VALERIANELLA P. Mill., Corn Salad, Lamb's Lettuce

Annual or short-lived biennial herbs, somewhat malodorous when dry, with erect, weakly dichasial branching. Leaves simple, basal leaves in a rosette, petiolate to nearly sessile, entire; cauline leaves connate, dentate. Inflorescence a terminal capitate dichasial cyme, subtended by lanceolate to oblong connate bracts, the latter either entire and glabrous-ciliate or strongly glandularly fimbriate-serrate. Flowers small, nearly regular, bisexual (unisexual), white or bluish-white; calyx obsolete or forming a narrow 5-toothed rim; corolla equally or unequally 5-lobed, funnelform or salverform, the tube basally slightly saccate; stamens 3(2), epipetalous toward the top of the corolla tube, exserted, anthers 4-lobed; style exserted, 3-lobed; ovary and fruit with 3 locules, 2 empty and the remaining 1-seeded. Fruit glabrous or pubescent, 1.5–4 mm long, a groove formed between the sterile locules.

References: Dyal, S. C. 1938. *Valerianella* of North America. Rhodora 40: 185–212; Eggers, D. M. 1969. A Revision of *Valerianella* in North America. Ph.D. dissertation, Vanderbilt Univ., Nashville, Tenn.

1 Plants glabrous, except for tufts of hairs on each side of the leaf bases near the nodes; achenes brownish, white-hirsute, hairs uncinate; inflorescences compact cymose clusters .... 1. *V. amarella*
1 Plants pubescent, in addition to tufts on each side of the leaf bases near the nodes (sometimes only on the leaves and lower parts of the stem); achenes yellowish, glabrous or pubescent; inflorescence loosely cymose .... 2. *V. radiata*

1. ***Valerianella amarella*** (Lindh.) Krok. Stem 1.5–3 dm tall, glabrous. Leaves glabrous except for tufts of uncinate hairs on each side of the base near the nodes; the lower leaves ovate-spatulate, entire; the upper leaves oblong-ovate, sessile. Inflorescence compact, corymbose, cymose; bracts glabrous, ovate-lanceolate, rounded to acute at the apex, usually hyaline-margined. Corolla white, funnelform, 1.5–3 mm long, limb usually as long as the tube and throat combined, a saccate gibbosity at the base of the throat on the ventral side; stamens and style usually long-exserted, sometimes either short or abortive, stigma lobes short. Fruit an achene 1.5–2 mm long, brownish, subglobose, ovoid, hirsute with rather long uncinate white hairs, fertile cell large, sterile cells much smaller, contiguous, the grooves between them very shallow or inconspicuous. Apr–May. Calciphile, on rocky open or wooded hills & prairies, or low grounds & barrens; OK; (cen TX to cen OK).

Dyal, op. cit., cites one collection from KS: Miami, but this specimen has not been seen.

2. ***Valerianella radiata*** (L.) Dufr., corn salad. Stem 1.5–6 dm tall, rather stout and pubescent along the angles. Leaves with hairs on margins and midribs of undersides; the lower oblong-spatulate, entire, connate; the upper oblong-ovate, often coarsely toothed at the base, not connate. Inflorescence loose, corymbose, cymose; bracts lanceolate, the outer slightly ciliate, the inner glabrous. Corolla white, funnelform, 1.5–2 mm long, the tube shorter than the limb, a saccate gibbosity at the base of the throat on the ventral side; stamens and style exserted; stigma lobes short. Fruit an achene 1.5–2 mm long, yellowish, ovoid, glabrous or pubescent, fertile cell as broad or broader than the combined width of the sterile cells; groove between the sterile cells narrow to rather wide and often rather deep, with a shallow groove on each side between the fertile and sterile cells. (n = 45) Apr–May. Low moist ground in woods or in prairie pastures & meadows, old fields, along roadsides, near railroad tracks, often in noncalcareous soils; MO, e KS, s & e OK; (CT & PA s to FL & w to MO, AR, KS, OK, & TX). *V. stenocarpa* (Engelm.) Krok—Rydberg; *V. stenocarpa* var. *parviflora* Dyal—Gates.

Although Eggers (op. cit.) indicates that the distribution of *V. woodsiana* (T. & G.) Walp. extends from e TX and w LA northward through e OK to se KS and sw MO, no specimens have been seen in our regional herbaria. *V. woodsiana* usually can be distinguished by the fertile cell of the fruit that is 1/3–1/2 the combined width of the 2 sterile cells, not fully concealing the sterile cells that extend outward in dorsal view. *V. radiata* has a fertile cell of the fruit that is at least equal to the combined width of the sterile cells, completely concealing the 2 sterile cells in dorsal view. Eggers suggests that future study may show that *V. woodsiana* is merely another fruit form of *V. radiata*.

135. DIPSACACEAE Juss., the Teasel Family

by Ronald R. Weedon

Annual, biennial or perennial herbs. Leaves mostly opposite, exstipulate, usually simple or pinnatifid, prickly. Inflorescences dense, involucrate, cymose, aggregations appearing as heads or spikes. Flowers perfect, more or less irregular, subtended by a calyxlike involucel or epicalyx of fused bracteoles, borne in the axils of imbricate receptacular bracts; calyx small, cupulate, 4- to 5-toothed or divided into 5–25 awns, teeth, or pappuslike segments; corolla tubular, fused through most of its length, 4(5)-lobed, often more or less 2-lipped; stamens (1–3)4, epipetalous near the base of the corolla tube and alternate with its lobes, exserted, anthers 2-celled; ovary inferior, unilocular with a single pendulous ovule, style filiform, stigma simple or 2-lobed. Fruit an achene, with persistent calyx, enclosed by involucel.

References: Ferguson, I. K. 1965. The genera of Valerianaceae and Dipsacaceae in the southeastern United States. J. Arnold Arb. 46: 218–231; Salamun, P. J., & T. S. Cochrane. 1974. Preliminary reports on the flora of Wisconsin No. 65. Dipsacaceae—Teasel Family. Trans. Wisconsin Acad. Sci. 62: 253–260. This treatment prepared with the assistance of Daniel G. Schneider.

The family includes approximately 10 genera with up to 250 species, native to the Old World, mostly eastern Medit. region, with a few naturalized in N. Amer.

1 Stems prickly; heads ovoid to spikelike with the involucral bracts upcurved, spine-tipped, 2–15 cm long; calyx 4-toothed or 4-lobed .... 1. *Dipsacus*
1 Stems not prickly; heads convex, hemispheric or subglobose with the involucral bracts ascending to reflexed, not spine-tipped, 8–13 mm long; calyx of at least 8 elongate teeth or bristles .... 2. *Knautia*

1. DIPSACUS L., Teasel

Tall stout erect biennial (perennial) herbs with rough-prickly stems from taproots. Leaves opposite, sessile or connate to petiolate at stem bases, exstipulate, entire to coarsely pinnatifid. Inflorescence a dense ovoid to cylindric head with 1 or 2 rows of rigid linear to lanceolate, subulate, erect or spreading, spine-tipped involucral bracts. Each flower 8–15 mm long, subtended by a chaffy, rigid, long-tapering receptacular bract and surrounded by a 4-angled, obscurely 4-toothed, tubular calyxlike involucel of fused bracteoles, grooved with a median rib and truncate at the apex; calyx cupulate, 4-angled or 4-lobed with a ciliate margin; corolla funnelform, 8–15 mm long, unequally 4-lobed, the tube whitish and the lobes pale purple; stamens 4, alternate with the corolla lobes, obviously exserted;

ovary unilocular with a solitary ovule, the style filiform, the stigma lateral, entire. Fruit an achene enclosed by the persistent involucel and crowned by the calyx.

The mode of flowering is characteristic for the genus: the flowers open first about 1/2 the length of the head and then develop sequentially both upward and downward.

1 Cauline leaves lanceolate to oblanceolate, their margins toothed and often prickly, becoming entire upward, usually basally connate; involucral bracts linear .... 1. *D. fullonum*
1 Cauline leaves pinnatifid or bipinnatifid, their margins bristly-ciliate, their bases confluent forming a cup; involucral bracts linear-lanceolate to mostly lanceolate .... 2. *D. laciniatus*

1. *Dipsacus fullonum* L., common teasel. Stout biennial 0.5–2(3) m tall, prickly, stems striate-angled. Basal leaves oblong to oblanceolate, arcuate, often dead by the middle of the second season; cauline leaves sessile, lanceolate to oblanceolate, up to 3(4) dm long, 4–10 cm wide, crenate-serrate, often with prickly margins and midribs, becoming entire upward and usually basally connate. Heads ovoid to subcylindric, 3–10 cm high, 3–5 cm wide, on long peduncles, their involucral bracts linear, elongate, subulate, upcurved, often equaling or exceeding the head, 2–15 cm long, 2–5 mm wide. Calyx cup 1 mm high, its base stalklike; corolla (8)13–15 mm long. Achenes 3–4(7) mm long. (2n = 16, 18) Late Jun–Sep. Roadsides, wet ditches, weedy areas of pastures, fields, ravines, cemeteries, & waste areas; scattered s 1/2 GP; (throughout U.S. & s Can.; Europe). *Naturalized*. *D. sylvestris* Huds.—Rydberg.

This species is occasionally cultivated or collected for ornament and is often adventive or locally established due to the activities of man. A white color form is known, but rarely seen.

2. *Dipsacus laciniatus* L., cut-leaved teasel. Resembles *D. fullonum* but basal leaves irregularly cleft, cauline leaves coarsely pinnatifid to bipinnatifid, their margins more or less bristly-ciliate, and the bases of these paired leaves confluent to form prominent moisture-collecting cups. Involucral bracts lanceolate to linear-lanceolate, 2–11 cm long, 3–12 mm wide, mostly shorter than to barely exceeding the mature head. Calyx cup 1 mm high, sunken, its base not stalklike; corolla 13–15 mm long. Achenes 3 mm long. (2n = 16, 18) Jul–Sep. Roadsides, wet ditches, cemeteries, & other habitats similar to those of *D. fullonum*; scattered sites in e KS, IA, apparently less common in GP; (rarely *adventive* in N. Amer.; Europe).

2. KNAUTIA L., Bluebuttons

1. *Knautia arvensis* (L.) Coult., bluebuttons, bluecaps, field scabious. Hirsute perennial herb, 4–10 dm tall, erect, few-branched stems arising from taproot. Leaves opposite, basal ones oblanceolate, simple or lyrate-pinnatifid; lower cauline leaves usually merely coarsely toothed, others variously deeply pinnatifid with narrow lateral and broader terminal segments, reduced upward. Peduncles 4–25 cm long, naked; heads 1.5–4 cm wide, depressed-hemispheric; involucral bracts ovate to narrowly lanceolate, 8–16 mm long, about equaling the heads; receptacle hairy, without bracts. Calyx cupulate, 3–4 mm long, with 8–12 bristlelike teeth, mostly 3–4 mm long; corolla lilac purple, funnelform to narrowly campanulate, 8–18 mm long, 4-lobed, the marginal lobes longer that the central; stamens 4; stigma 4-lobed, extended. Achene 4-ribbed, 5–6 mm long, strongly compressed-ellipsoid, densely hairy, the apex truncate and minutely denticulate, with 8–16 short awns eventually deciduous, enclosed in 4-angled involucel. (2n = 20, 40) Jun–Jul. Escape from gardens, of local occurrence in fields, meadows, roadsides, & waste places; ND: Grand Forks, Ward, Pierce; MN: Clay; SD: Day; to be expected elsewhere; (sporadically established in N. Amer.; Europe). *Scabiosa arvensis* L.—Rydberg.

136. ASTERACEAE Dum., the Sunflower Family

by T. M. Barkley

Ours annual, biennial or perennial herbs or occasionally shrubs. Leaves simple and entire to variously toothed, dissected, parted or sometimes compound, exstipulate, alternate, opposite or rarely whorled, often opposite below and alternate above. Fundamental inflorescence a head of small flowers (here termed "florets") aggregated on a common receptacle and surrounded by few–numerous involucral bracts in 1–several series; florets sessile, sometimes subtended by a separate bract (chaff or chaffy bract), or variously intermixed with bristles, or the receptacle with neither chaff nor bristles (naked); the outermost florets of the head mature first and the innermost last, thus the flowering sequence is indeterminate for the head; the heads variously grouped into determinate inflorescences (with the terminal head maturing first) or indeterminate aggregations of determinate inflorescences, or rarely in strictly indeterminate inflorescences, or sometimes the heads occur singly. Individual florets perfect, unisexual or neutral, epigynous, gamopetalous, regular or zygomorphic, commonly 5-merous, the calyx replaced by a pappus of scales, awns, stiff bristles, long capillary or plumose bristles, or the pappus variously reduced or absent; stamens epipetalous, as many as the corolla lobes and alternating with them; anthers elongate and connate into a tube around the style, dehiscing introrsely with the pollen pushed out of the stamen-tube by the growth of the style, or the anthers rarely free; pistil of 2 carpels, ovary inferior, unilocular with a single basal, erect, anatropous ovule; style 2-cleft (undivided in functionally staminate florets), the stigmatic areas usually restricted to evident lines on the inner faces of the style branches, often the style branches have nonstigmatic appendages toward the tip. Fruit an achene ("cypsela" in some literature), often with the persistent pappus attached. *Ambrosiaceae, Carduaceae, Cichoriaceae*—Rydberg; *Compositae*—Fernald.

Among our species, the florets and heads vary in the following ways: Disk florets have the corolla regular and tubular, and are normally perfect and fertile. Ray florets have the corolla tubular only at the base, and upward it is usually greatly expanded to one side into a prominent, strap-shaped ligule; the stamens are absent and the floret is functionally pistillate. Ligulate florets have the corolla expanded to one side in a ligule, but are perfect and fertile. Ligulate florets occur in heads composed solely of ligulate florets (ligulate head) and are distinctive for the tribe Lactuceae. A head may be composed solely of disk florets (discoid head) or it may have disk florets in the center and whorl of ray florets around the margin (radiate head). Some heads have normal disk florets in the center, and have marginal pistillate florets with the ligule greatly reduced or lacking (disciform head). For convenience, both discoid and disciform heads are termed eradiate heads. The heads are generally obvious although they vary greatly in size. Most species have heads with 5 or more florets, but some extraregional species have the heads reduced to a single floret. *Elephantopus* in our flora has small heads aggregated into secondary heads. Most members of the family are insect pollinated, but members of the Heliantheae-Ambrosiinae are adapted to wind pollination and have many small greenish heads, with separate anthers in the functionally staminate florets.

The Asteraceae, also widely known as the Compositae, are a diverse, world-wide assemblage of more than 15,000 species. The family is broken into a dozen or so tribes (9 with us) which are generally held to be natural groups. The tribes are defined on suites of morphological, cytological, biochemical, and geographical characters, with no single character being of fundamental value. There is great variation and intergradation among the members of the family, and consequently the generic and specific boundaries are often imprecise. This is especially so in our region in the tribes Heliantheae and Astereae. Despite the great diversity, the family remains relatively homogeneous, and there is virtually no confusing a member of the Asteraceae with any other family.

SYNOPSIS OF THE TRIBES

1 Florets all ligulate and perfect; sap milky or colored .... 9. *Lactuceae*
1 At least some of the florets tubular and eligulate; sap watery.
   2 Style typically with a thickened and often hairy ring just below the branches; anthers tailed at the base; receptable bristly or sometimes naked; plants often spiny or prickly .... 8. *Cynareae*
   2 Style without a distinctive thickening or ring of hairs below the branches; anthers not tailed (except in Inuleae); receptacle naked or chaffy, rarely bristly.
      3 Style branches of the fertile florets typically flattish, with evident stigmatic lines extending from the fork to the middle of the branch or beyond; heads often radiate, but sometimes discoid or disciform; corollas often yellow.
         4 Anthers tailed at the base; style branches rounded or truncate, ± papillose but without appendages; pappus of capillary bristles or absent; receptacle naked or chaffy; corollas all tubular; plants white-wooly (except *Pluchea*) .... 5. *Inuleae*
         4 Anthers truncate to sagittate at the base but not tailed; style branches often with a terminal appendage.
            5 Style branches with a terminal appendage that is pubescent on the outer surface but glabrous on the inner (adaxial) surface; receptacle naked; involucral bracts usually in several series; leaves alternate .... 4. *Astereae*
            5 Style branches either with terminal appendages that are pubescent on both the inner and outer surfaces, or unappendaged and the style branches terminating in a ring or tuft of hair; receptacle chaffy or naked.
               6 Pappus of capillary bristles; style branches mostly truncate and unappendaged, with a terminal ring or tuft of hair; receptacle naked; principal involucral bracts in a single series; leaves alternate or opposite .... 3. *Senecioneae*
               6 Pappus of awns, scales, stiff bristles or absent; involucral bracts often in 2–4 series; receptacle chaffy or naked.
                  7 Involucral bracts with scarious or hyaline margins; style branches mostly unappendaged but with hairs at the apex; leaves alternate .... 1. *Anthemideae*
                  7 Involucral bracts herbaceous or thin and membranaceous, but not at all scarious or hyaline; style branches typically appendaged; leaves often opposite, especially the lower ones .... 1. *Heliantheae*
      3 Style branches terete, clavate or filiform, not much flattened, stigmatic toward the base, the stigmatic area not sharply defined morphologically; florets all tubular and perfect, never yellow; receptacle naked.
         8 Style branches ± clavate; obtuse or acutish; anthers rounded at the base; leaves alternate, opposite or whorled .... 6. *Eupatorieae*
         8 Style branches filiform, acute or acuminate; anthers sagittate; leaves alternate .... 7. *Vernonieae*

1. Heliantheae

1 Receptacle chaffy, at least near the margin; the chaffy bracts normally each subtending a floret. (Heliantheae, sens. strict.).
   2 Receptacle chaffy only near the edge; involucral bracts in a single series, prominently keeled on the back and each enclosing a ray achene; rank-scented annuals. (Madiinae) .... 74. *Madia*
   2 Receptacle chaffy throughout or sometimes naked in the center but otherwise not as above.
      3 Corolla of the pistillate florets absent or tubular and reduced; staminate florets with anthers separate or but weakly united and the style undivided; perfect florets none. (Ambrosiinae).

4 Staminate and pistillate florets in the same head .... 63. *Iva*
  4 Staminate and pistillate florets in separate, distinctive heads, the staminate heads usually uppermost.
    5 Involuvral bracts of the staminate heads fused to form a cup; involucre of the pistillate heads forming 1–several series of tubercles or straight spines .... 4. *Ambrosia*
    5 Involucral bracts of the staminate heads separate; involucre of the pistillate heads forming a bur with recurved prickles .... 110. *Xanthium*
3 Corolla of the pistillate florets present and evident, usually ligulate, or the florets all perfect.
  6 Disk florets sterile, with the style undivided. (Melampodiinae).
    7 Achenes short and thick, at most weakly compressed parallel to the radius of the head.
      8 Involucre ± uniseriate, the bracts flat and merely subtending the achenes .... 88. *Polymnia*
      8 Involucral bracts in several series, the inner ones clasping the achenes .... 77. *Melampodium*
    7 Achenes flattened at right angles to the radius of the head.
      9 Ray florets in 2 or 3 series; achenes free from the subtending and nearby bracts .... 97. *Silphium*
      9 Ray florets in 1 series; achene ± adnate to 1 or more adjacent bracts and tending to fall together as a unit.
        10 Rays short and inconspicuous, ligule less than 1.5 mm long, white, or rarely absent .... 81. *Parthenium*
        10 Rays evident, ligule more than 1.5 mm long, yellow.
          11 Ray florets mostly 5; annual herb .... 71. *Lindheimera*
          11 Ray florets mostly 8 or 13; perennials.
            12 Pappus an inconspicuous crown or absent; herbage puberulent, tomentose or scabrous .... 18. *Berlandiera*
            12 Pappus of several scales, with 2 ± longer than the others; herbage densely hispid .... 41. *Engelmannia*
  6 Disk florets fertile, with the style obviously divided at maturity.
    13 Ray corollas persistent on the achenes, becoming papery. (Zinniinae).
      14 Disk achenes ± compressed parallel to the radius of the head; leaves linear .... 112. *Zinnia*
      14 Disk achenes 4-angled; leaves ovate or lanceolate .... 58. *Heliopsis*
    13 Ray corollas deciduous or ray florets absent.
      15 Pappus, at least on some of the disk florets, of 5–many long scales or ± plumose awns. (Galinsoginae).
        16 Leaves opposite; rays present but small .... 49. *Galinsoga*
        16 Leaves alternate and chiefly basal; rays absent .... 75. *Marshallia*
      15 Pappus variously of a few awns or of scales and awns, or reduced to a few teeth or absent, but not of 5–many scales or awns.
        17 Achenes flattened at right angles to the radius of the head, or occasionally 4-angled; chaffy bracts flat or but weakly concave-convex; involucral bracts obviously in 2 series. (Coreopsidinae).
          18 Principal involucral bracts at least partially united .... 103. *Thelesperma*
          18 Involucral bracts free.
            19 Pappus of rigid awns which are usually antrorsely or retrorsely barbed, although sometimes reduced .... 19. *Bidens*
            19 Pappus of 2 awns or teeth, inconspicuously if at all barbed, or sometimes absent .... 34. *Coreopsis*
        17 Achenes flattened parallel to the radius of the head, or weakly angled and subterete; chaffy bracts mostly concave-convex and clasping the achenes, or merely bristlelike; involucral bracts various but not obviously biseriate and dimorphic. (Verbesininae).
          20 Chaff reduced to mere stiff bristles; corollas whitish; rays minute, ligule ±1 mm long .... 39. *Eclipta*

20 Chaffy bracts well developed.
  21 Receptacle conspicuously conic or columnar, often elongating at maturity; ray florets abortive; leaves all alternate.
    22 Disk and ray florets both subtended by chaffy receptacular bracts.
      23 Achenes flattish, 4-angled; leaves deeply dissected; perennials .... 92. *Ratibida*
      23 Achenes subterete; leaves entire or dentate; annual .... 36. *Dracopis*
    22 Disk florets subtended by chaffy bracts; ray florets naked.
      24 Chaffy bracts similar to the involucral bracts, often surpassing the disk corollas, ± spine-tipped; disk corollas expanded below into a bulbous, thickened base .... 38. *Echinacea*
      24 Chaffy bracts distinct from the involucral bracts, at most equaling the disk corollas, acute or but weakly awn-pointed; disk corollas tubular or narrowed toward the base .... 93. *Rudbeckia*
  21 Receptacle flat or convex but not conspicuously elongate; leaves alternate or opposite; ray florets often fertile.
    25 Achenes strongly flattened, often winged with thin margins.
      26 Leaves mostly alternate; receptacle ± flat; disk mostly less than 2 cm across .... 106. *Verbesina*
      26 Principal leaves opposite; receptacle distinctly conic; disk 2–4 cm across. .... 56. *Helianthella*
    25 Achenes somewhat compressed or rhombic in cross section.
      27 Subscapose perennial; ray florets pistillate and fertile .... 16. *Balsamorhiza*
      27 Leafy stemmed herbs or shrubs; ray florets abortive.
        28 Shrubs; pappus absent .... 108. *Viguiera*
        28 Herbs; pappus of 2 early-deciduous awns, sometimes with a few small scales .... 57. *Helianthus*
1 Receptacle naked, or at most with irregular, stiff setae that do not precisely subtend the florets. (Helenieae).
  29 Leaves and/or involucres with conspicuous yellow-brownish oil glands. (Tagetinae).
    30 Style branches of the disk florets short, oblong; involucral bracts all about alike, uniseriate .... 82. *Pectis*
    30 Style branches of the disk florets long and slender; involucral bracts in 2 series .... 37. *Dyssodia*
  29 Herbage not provided with oil glands, although sometimes glandular pubescent or punctate under magnification. (Heleniinae).
    31 Ray corollas persistent on the achenes and falling with them .... 90. *Psilostrophe*
    31 Ray corollas deciduous, or rays absent.
      32 Achenes flattened parallel to the radius of the head, with only marginal ribs .... 83. *Pericome*
      32 Achenes distinctly angled, ribbed or subterete.
        33 Achenes ribbed, columnar; ray florets single per head .... 47. *Flaveria*
        33 Achenes either 3- or 5-angled, ±clavate to obconic; ray florets more than 1 or absent.
          34 Style branches linear, spreading-hispidulous from the apex to the fork; corollas pink, rose to purple, or creamy white.
            35 Pappus of 8–16 scales in 2 series; corollas creamy white; leaves pinnatifid .... 26. *Cheanactis*
            35 Pappus of 7–10 scales; corollas pink, rose to purple; leaves entire or nearly so .... 80. *Palafoxia*
          34 Style branches very short or with a distinct appendage, but not hispidulous from apex to fork; corollas mostly yellow.
            36 Involucral bracts thin and membranaceous on the margins and tip, sometimes ± colored. (tribe Anthemideae) .... 61. *Hymenopappus*

36 Involucral bracts wholly herbaceous or sometimes chartaceous toward the base; margin not scarious.
  37 Involucral bracts reflexing-spreading, especially in age.
    38 Receptacle naked except for a few scattered bristles near the edge .... 55. *Helenium*
    38 Receptacle alveolate, with irregularly placed setae that do not precisely subtend the florets, or the setae reduced to toothlike fimbrillae .... 48. *Gaillardia*
  37 Involucral bracts erect or ascending, not at all reflexed.
    39 Achenes obpyramidal but only 2–3 × longer than broad, pubescent with appressed hairs .... 62. *Hymenoxys*
    39 Achenes narrowly obpyramidal, 3 or more × longer than wide, hispid glandular or glabrous.
      40 Outer involucral bracts flattish or merely arched on the back, stipitate glandular; leaves alternate; annual .... 15. *Bahia*
      40 Outer involucral bracts prominently keeled on the back, canescent; leaves opposite; perennials .... 85. *Picradeniopsis*

2. Anthemideae

1 Pappus of short, separate scales; style branches apically papillose .... 61. *Hymenopappus*
1 Pappus a minute crown or absent; style branches truncate, without appendage.
  2 Receptacle chaffy, at least in the center of the head.
    3 Achenes flattened; heads small and numerous in a corymbiform inflorescence .... 1. *Achillea*
    3 Achenes terete or angled; heads large, solitary on long peduncles .... 7. *Anthemis*
  2 Receptacle naked, or if long-hairy the heads small and eradiate.
    4 Inflorescence racemiform or paniculiform, of numerous small, eradiate heads; receptacle hairy in only 2 species .... 11. *Artemisia*
    4 Inflorescence corymbiform or heads solitary at the ends of erect branches.
      5 Receptacle strongly raised, distinctly hemispheric or conic at maturity .... 76. *Matricaria*
      5 Receptacle flat or merely low-convex.
        6 Heads disciform, the outermost florets pistillate but with a tubular corolla; leaves pinnately dissected .... 101. *Tanacetum*
        6 Heads radiate, or if eradiate, leaves merely toothed .... 28. *Chrysanthemum*

3. Senecioneae

1 Disk florets functionally staminate with style undivided or at most weakly lobed.
  2 Pappus of abundant capillary bristles; imperfectly dioecious .... 84. *Petasites*
  2 Pappus absent; outer florets pistillate, inner staminate .... 2. *Adenocaulon*
1 Disk florets perfect and fertile with style evidently bifid.
  3 Leaves all or chiefly opposite.
    4 Involucral bracts typically 8 or more; principal leaves usually more than 1 cm wide .... 10. *Arnica*
    4 Involucral bracts 4–6; leaves narrowly linear or filiform .... 53. *Haploesthes*
  3 Leaves alternate.
    5 Heads radiate, or discoid in a few species; pistillate florets with a distinct ligule (which may be short); corolla typically yellow or sometimes orange .... 95. *Senecio*
    5 Heads discoid or disciform; corolla white or ochroleucous.
      6 Heads disciform with the outer florets pistillate but with eligulate, tubular-filiform corollas; annual .... 42. *Erechtites*
      6 Heads strictly discoid; all florets tubular and perfect; perennials .... 22. *Cacalia*

4. Astereae

1 Rays yellow and present.
  2 Pappus of a few scales or awns, or reduced to a mere crown on the ray florets.
    3 Pappus of 2–10 ± firm, caducous awns .... 51. *Grindelia*

3 Pappus of prominent scales on the disk florets; that of ray florets reduced, often to a mere crown .... 52. *Gutierrezia*
2 Pappus of capillary bristles.
    4 Pappus of 2 series, the outer distinctly shorter than the inner.
        5 Receptacle with obvious chaffy scales among the achenes .... 109. *Xanthisma*
        5 Receptacle naked or at most weakly alveolate.
           6 Ray florets epappose .... 59. *Heterotheca*
           6 Ray florets with pappus like that of the disk florets .... 29. *Chrysopsis*
    4 Pappus a single series of capillary bristles of nearly uniform size.
        7 Plants fundamentally taprooted; heads mostly large, hemispheric or broadly conic .... 54. *Haplopappus*
        7 Plants fibrous-rooted; heads usually small and numerous; involucre conical or narrowly campanulate.
           8 Inflorescence corymbiform; rays more numerous than the disk florets; at least some of the leaves glandular-punctate .... 45. *Euthamia*
           8 Inflorescence paniculate, thyrselike or of capitate clusters or if corymbiform the ray florets no more than the disk florets; leaves not at all glandular-punctate .... 98. *Solidago*
1 Rays of some color other than yellow, or rays absent.
  9 Dioecious shrubs. .... 14. *Baccharis*
  9 Heads with perfect disk florets; rays present or absent.
    10 Pappus of abundant capillary bristles.
       11 Rays absent and disk florets yellow.
          12 Disk florets 10–20; involucral bracts irregularly imbricated; herbs .... 54. *Haplopappus*
          12 Disk florets mostly 5; involucral bracts imbricated in distinct vertical series; shrubby .... 30. *Chrysothamnus*
       11 Rays white, pink, blue, reddish or purplish; disk florets often the same color as the rays but sometimes yellow.
          13 Pistillate florets numerous and in several series, with the ligule very short and scarcely exceeding the involucre, or sometimes the corolla merely tubular-filiform; disk florets few .... 33. *Conyza*
          13 Pistillate florets few to numerous in 1 or more series, with conspicuous ligules at least on some or most of them, or pistillate florets absent; disk florets normally numerous and prominent.
             14 Pappus bristles, or at least some of them, conspicuously dilated upward. (*S. ptarmicoides*) .... 98. *Solidago*
             14 Pappus bristles cylindrical or tapering upward and not at all clavate.
                15 Involucral bracts subequal or weakly imbricated, often green, not at all leaflike and without an obviously chartaceous base; mostly flowering in spring and summer .... 43. *Erigeron*
                15 Involucral bracts evidently imbricated, often with a green tip and chartaceous base, or if the bracts subequal, the outer ones leafy; mostly flowering in late summer and autumn.
                   16 Plants fundamentally taprooted; leaves frequently sharp-toothed or dissected, and/or spiny but sometimes entire and smooth.
                      17 Heads solitary at the ends of long branches; stems ± clustered from a branching caudex .... 111. *Xylorhiza*
                      17 Heads few to numerous in cymose-corymbiform clusters; stems single or few .... 73. *Machaeranthera*
                   16 Plants fibrous-rooted and sometimes rhizomatous; leaves variously entire or toothed, neither dissected nor spiny.
                   18 Plants depressed, spreading from creeping rootstocks; leaves scalelike .... 69. *Leucelene*
                   18 Plants erect, normally more than 2 dm tall; leaves various but not reduced and scalelike .... 12. *Aster*
    10 Pappus of numerous firm scales or stiff bristles, or of short bristles with or without 2–5 scales or stiff awns, or pappus merely a reduced crown or absent.

19 Pappus essentially absent or merely a minute ring.
  20 Scapose perennial weed; receptacle obviously conic .... 17. *Bellis*
  20 Leafy-stemmed annual; receptacle convex .... 13. *Astranthium*
19 Pappus present, although sometimes rather small.
  21 Pappus of 5 hyaline scales alternating with 5 slender awns .... 27. *Chaetopappa*
  21 Pappus of scales and/or bristles, but not as above.
    22 Achenes 4-angled or subterete; pappus of ± 5 small scales or a minute ciliate crown .... 8. *Aphanostephus*
    22 Achenes flattened, 2-angled or sometimes 3-angled; pappus various.
      23 Plants tyically glabrous, 2–12 + dm tall; achenes ± wing-margined; pappus of several minute bristles and 2(4) prominent awns 20. *Boltonia*
      23 Plants pubescent or glabrate, low, rarely exceeding 2 dm, often acaulescent; pappus of stiff bristles or a low crown of weakly fused squamellae .... 104. *Townsendia*

5. Inuleae

1 Receptacle chaffy .... 46. *Evax*
1 Receptacle naked.
  2 Involucral bracts variously puberulent to glandular-pubescent, but not scarious-hyaline and glabrous toward the tip .... 87. *Pluchea*
  2 Involucral bracts largely scarious-hyaline, especially toward the tip; glabrous above the often woolly base.
    3 Heads all alike with the outer florets pistillate and the inner ones perfect and fertile; heads variously clustered into glomerules .... 50. *Gnaphalium*
    3 Heads of 2 kinds; those on some plants functionally staminate, and those on other plants functionally pistillate; heads typically short-pedunculate or sometimes solitary.
      4 Leaves all cauline; pistillate heads typically with a few central functionally staminate florets; rhizomatous but not stoloniferous .... 5. *Anaphalis*
      4 Largest leaves basal; pistillate heads without any staminate florets; most of our species stoloniferous and mat-forming .... 6. *Antennaria*

6. Eupatorieae

1 Achenes strictly 5-angled .... 44. *Eupatorium*
1 Achenes 10-ribbed.
  2 Basal and lower cauline leaves persistent and evidently the largest; cauline leaves progressively reduced upwards; involucral bracts obscurely or not at all longitudinally ribbed .... 70. *Liatris*
  2 Basal and lower cauline leaves early deciduous and no larger than the middle cauline leaves; at least some of the involucral bracts prominently striate.
    3 Herbs from a stout, strongly taprooted caudex; pappus bristles plumose ... 65. *Kuhnia*
    3 Herbs, subshrubs or shrubs, variously fibrous-rooted or with several thickened, fusiform roots; pappus scabrous to barbellate, plumose in 1 species ...... 21. *Brickellia*

7. Vernonieae

1 Pappus of 5 bristles in a single series; heads aggregated into secondary glomerules .... 40. *Elephantopus*
1 Pappus of numerous bristles in 2 series; heads separate .... 107. *Vernonia*

8. Cynareae

1 Achenes oblique at the base; marginal florets often expanded and falsely raylike .... 25. *Centaurea*
1 Achenes, or at least the central ones, flat at the point of attachment; marginal florets not at all expanded and raylike.
  2 Receptacle alveolate, i.e., fleshy and deeply honeycombed, not bristly or sparingly so .... 79. *Onopordum*
  2 Receptacle flat and densely bristly.
    3 Involucral bracts with recurved-hooked tips; leaves large and not prickly .. 9. *Arctium*
    3 Involucral bracts without hooked tips; leaves variously prickly or spiny on the margin.

4 Pappus of several short unequal scales or often absent; outer involucral bracts expanded and leaflike; corollas yellow .... 24. *Carthamus*
 4 Pappus of elongate bristles; involucral bracts variously narrow or broad but not leaflike; corollas white, pink, blue, red, or purple.
  5 Pappus plumose .... 32. *Cirsium*
  5 Pappus capillary or at most weakly barbellate .... 23. *Carduus*

9. Lactuceae

1 Pappus of simple capillary bristles.
 2 Achenes obviously flattened.
  3 Achenes beaked, or if the beak obscure or absent, the corolla blue to purple; pappus inserted on a prominent, expanded disk; florets normally less than 50 .... 66. *Lactuca*
  3 Achenes beakless and without a prominent pappus-bearing disk; corolla yellow; florets many, more than 80 .... 99. *Sonchus*
 2 Achenes terete or variously angled but not flattened.
  4 Corollas yellow to orange-red, at least when fresh (occasionally fading to pinkish).
   5 Stem leafy, although the upper leaves may be reduced.
    6 Perennials, arising from a rhizome, caudex or crown, with fibrous roots .... 60. *Hieracium*
    6 Annuals, biennials or perennials, arising from a taproot .... 35. *Crepis*
   5 Stem scapose or at most with cauline leaves few and greatly reduced.
    7 Achenes beakless or nearly so; sometimes weakly contracted at the apex.
     8 Head solitary; pappus of broad-based bristles plus a few slender scales .... 78. *Microseris*
     8 Heads usually several in a corymbiform cluster; pappus of bristles only .... 35. *Crepis*
    7 Achenes obviously beaked.
     9 Pappus sordid to dirty-reddish, subtended by a few villous hairs; principal involucral bracts with an apical, keel-shaped appendage . 91. *Pyrrhopappus*
     9 Pappus clearly white; principal involucral bracts unappendaged.
      10 Body of achene spinulose upward; outer involucral bracts shorter than the inner ones and early reflexed .... 102. *Taraxacum*
      10 Body of achene at most striate but not spinulose; involucral bracts subequal .... 3. *Agoseris*
  4 Corollas pink, purple or white, at least when fresh.
   11 Annual with the leaves opposite at the lower 4 or 5 nodes .... 96. *Shinnersoseris*
   11 Perennials with all leaves alternate.
    12 Leaves linear or reduced to bracts; heads terminating the branches .... 72. *Lygodesmia*
    12 Well-developed leaves over 1 cm wide and broader than linear; inflorescence paniculiform or thyrsoid with several–many heads .... 89. *Prenanthes*
1 Pappus of plumose bristle or scales or both, or pappus absent.
 13 Pappus of plumose bristles.
  14 Plume branches of the pappus long and interwebbed; leafy-stemmed weeds with grasslike leaves .... 105. *Tragopogon*
  14 Plume branches of the pappus separate and not interwebbed; plants scapose or leaves not grasslike.
   15 Stem leafy, at least when young, although the uppermost leaves may be reduced.
    16 Perennials with corollas purplish to rose colored .... 100. *Stephanomeria*
    16 Annual hispid-spinescent weeds with corollas yellow .... 86. *Picris*
   15 Leaves chiefly basal, with cauline leaves few and greatly reduced, or none.
    17 Achene with a distinct tubular base; leaves normally pinnatisect; pappus bristles in more than 1 series .... 94. *Scorzonera*
    17 Achene tapering at the base but not distinctly tubular; leaves variously dentate or lobed but rarely clearly pinnatisect; pappus with an inner series of plumose bristles and an outer series of short, stiff setae .... 68. *Leontodon*
 13 Pappus of scales only, of bristles and scales, or none.
  18 Corolla blue to white or sometimes pinkish; papus of small scales only . 31. *Cichorium*

18 Corollas yellow or orange.
    19 Involucre distinctly calyculate; pappus none .... 67. *Lapsana*
    19 Involucre without calyculate bracts; pappus of bristles and scales or of scales, or none .... 64. *Krigia*

ARTIFICIAL KEYS TO THE GENERA

1 Heads with florets all ligulate and perfect; sap milky or colored .... Tribe Lactuceae (see previous key)
1 Heads radiate (with both disk florets and marginal ray florets), or eradiate and discoid (with disk florets only) or eradiate and disciform (with disk florets and marginal pistillate florets with tubular corollas); sap watery.
  2 Shrubs .... Key 1
  2 Herbs, at most with a ligneous caudex.
    3 Heads radiate.
      4 Rays yellow or orange.
        5 Receptacle chaffy or bristly .... Key 2
        5 Receptacle naked .... Key 3
      4 Rays purple, blue, pink, or white.
        6 Receptacle chaffy or bristly .... Key 4
        6 Receptacle naked. .... Key 5
    3 Heads eradiate (some plants with small, inconspicuous rays are keyed here as well as above).
      7 Receptacle chaffy or bristly .... Key 6
      7 Receptacle naked.
        8 Pappus of capillary bristles .... Key 7
        8 Pappus of awns, scales, stiff bristles, reduced to a mere crown or absent .... Key 8

Key 1
Shrubs

1 Heads with yellow ray florets.
  2 Receptacle chaffy .... 108. *Viguiera*
  2 Receptacle naked.
    3 Involucre 3-6 mm tall; involucral bracts in ±3 series; pappus of 8-10 scales .... 52. *Gutierrezia*
    3 Involucre 5+ mm tall; principal involucral bracts in 1 series, often subtended by shorter, irregularly disposed calyculate bracts; pappus of numerous capillary bristles .... 95. *Senecio*
1 Heads discoid or with white rays.
  4 Heads with small white rays; pappus of 2 or 3 awns .... 81. *Parthenium*
  4 Heads discoid; pappus of numerous bristles or absent or reduced to a mere crown.
    5 Involucral bracts imbricated in distinct vertical series .... 30. *Chrysothamnus*
    5 Involucral bracts irregularly imbricated.
      6 Plants dioecious, with the florets of all the heads on any one plant either staminate or pistillate .... 14. *Baccharis*
      6 Plants with at least some of the florets perfect.
        7 Pappus of scabrous bristles .... 21. *Brickellia*
        7 Pappus none or merely an obscure crown .... 11. *Artemisia*

Key 2
Heads radiate; rays yellow or orange;
receptacle chaffy or bristly

1 Pappus of capillary bristles; receptacle merely bristly or with scattered, bractlike chaff near the margin.
  2 Pappus in 2 series, the outer distinctly shorter than the inner .... 109. *Xanthisma*
  2 Pappus a single series of bristles.
    3 Inflorescence basically corymbiform; rays more numerous than the disk florets; at least some of the leaves glandular-punctate .... 45. *Euthamia*

            3   Inflorescence paniculate, thyrselike, or of capitate clusters, or if corymbiform, then the
                rays no more than the disk florets; leaves not at all glandular-punctate .. 98. *Solidago*
    1   Pappus of scales, awns, a reduced crown or absent.
        4   Involucral bracts in a single series, prominently keeled on the back and each enclosing a
            ray achene; chaff present only near the edge of the head; rank-scented
            annuals .... 74. *Madia*
        4   Involucral bracts various but not as above; chaff nearly always throughout the head.
            5   Receptacle strongly conic or elongated, at least at maturity (the "coneflowers").
                6   Chaffy bracts subtending both disk and ray florets.
                    7   Leaves deeply dissected; perennials .... 92. *Ratibida*
                    7   Leaves entire or merely toothed; annual .... 36. *Dracopis*
                6   Chaffy bracts subtending only the disk florets; ray florets naked.
                    8   Chaffy bracts ± spine-tipped, often surpassing the disk corollas 38. *Echinacea*
                    8   Chaffy bracts acute but scarcely spine-tipped, at most equaling but not ex-
                        ceeding the disk corollas .... 93. *Rudbeckia*
            5   Receptacle variously flat, conic or subhemispheric but not prominently elongated.
                9   Disk florets sterile, with ovaries (achenes) obviously smaller than those of the ray florets.
                    10 Ray florets in 2 or 3 series; rays relatively large, ligule often 2–4 +
                        cm long .... 97. *Silphium*
                    10 Ray florets in a single series; rays smaller, ligule rarely exceeding 2 cm.
                        11 Principal leaves opposite; achenes thick and at most weakly compressed
                            parallel to the radius of the head .... 88. *Polymnia*
                        11 Principal leaves alternate; achenes flattened at right angles to the radius of
                            the head.
                            12 Ray florets mostly 5; annual herb .... 71. *Lindheimera*
                            12 Ray florets mostly 8 or 13; perennials.
                                13 Pappus an inconspicuous crown or absent; herbage puberulent,
                                    tomentose or scabrous .... 18. *Berlandiera*
                                13 Pappus of several scales, with 2 of them longer than the others; her-
                                    bage densely hispid .... 41. *Engelmannia*
                9   Disk florets fertile, with the ovaries as large or larger than those of the ray florets.
                    14 Receptacle alveolate, with irregularly placed setae that do not precisely subtend
                        the florets .... 48. *Gaillardia*
                    14 Receptacle with chaffy bracts that ± subtend each floret.
                        15 Leaves all alternate or clustered toward the base.
                            16 Leaves chiefly basal; plant subscapose; pappus absent . 16. *Balsamorhiza*
                            16 Stem leafy.
                                17 Pappus none or a short crown .... 7. *Anthemis*
                                17 Pappus principally of 2 awns.
                                    18 Achenes flattened, with corky wings; pappus of long, thin
                                        awns .... 106. *Verbesina*
                                    18 Achenes stout, compressed or more often rhombic in cross sec-
                                        tion; pappus of scalelike awns, often early-deciduous ....
                                        57. *Helianthus*
                        15 Leaves, or at least the lower ones, opposite.
                            19 Ray corollas persistent on the achenes, becoming papery.
                                20 Leaves linear; disk achenes ± 4-angled but compressed parallel to
                                    the radius of the head .... 112. *Zinnia*
                                20 Leaves ovate or lanceolate; disk achenes 4-angled .... 58. *Heliopsis*
                            19 Ray corollas deciduous.
                                21 Achenes flattened at right angles to the radius of the head, or ±
                                    4-angled; chaffy bracts flat or nearly so; involucral bracts obviously
                                    in 2 series.
                                    22 Principal involucral bracts partially united ..... 103. *Thelesperma*
                                    22 Principal involucral bracts free.
                                        23 Pappus of rigid awns which are usually antrorsely or retrorsely
                                            barbed, but sometimes pappus reduced; achenes
                                            wingless .... 19. *Bidens*

                23 Pappus of 2 awns or teeth that are inconspicuously or not at
                    all barbed, or sometimes absent; achenes winged or
                    wingless .... 34. *Coreopsis*
            21 Achenes flattened parallel to the radius of the head or merely an-
                gled, chaffy bracts mostly concave-convex and clasping the achenes;
                involucral bracts various but not biseriate and dimorphic.
                24 Achenes stout and somewhat compressed, often rhombic in cross
                    section, wingless .... 57. *Helianthus*
                24 Achenes strongly flattened, often obviously winged.
                    25 Ray florets fertile and producing viable achenes; wings of
                        achenes corky; disk mostly less than 2 cm
                        across .... 106. *Verbesina*
                    25 Ray florets sterile; wings of achene thin and ciliate-margined;
                        disk 2–4 cm across .... 56. *Helianthella*

Key 3
Heads radiate; rays yellow or orange;
receptacle naked

1 Pappus of abundant capillary bristles, at least on the disk florets.
    2 Leaves opposite.
        3 Involucral bracts mostly 5; leaves narrowly linear or filiform .... 53. *Haploesthes*
        3 Involucral bracts typically 8; principal leaves more than 1 cm wide .... 10. *Arnica*
    2 Leaves alternate.
        4 Principal involucral bracts of equal length, in a single series; often subtended by
            several small calyculate bracts .... 95. *Senecio*
        4 Principal involucral bracts unequal and imbricated in several series.
            5 Pappus in 2 series; the outer much shorter than the inner.
                6 Ray florets epappose .... 59. *Heterotheca*
                6 Ray florets with pappus like that of the disk florets .... 29. *Chrysopsis*
            5 Pappus a single series of capillary bristles of nearly uniform length.
                7 Plants fundamentally taprooted; heads mostly large and hemispheric or broadly
                    conic .... 54. *Haplopappus*
                7 Plants fundamentally fibrous-rooted; heads usually small and numerous; in-
                    volucre conical or narrowly campanulate.
                    8 Inflorescence corymbiform; rays more numerous than the disk florets; at
                        least some of the leaves glandular-punctate .... 45. *Euthamia*
                    8 Inflorescence paniculate, thyrselike or of capitate clusters, or if corym-
                        biform, then the rays no more than the disk florets; leaves not at all
                        glandular-punctate .... 98. *Solidago*
1 Pappus of scales, awns, stiff and unequal bristles, or pappus none.
    9 Leaves and/or involucres with conspicuous, yellow-brownish oil glands.
        10 Involucral bracts ± alike, in 1 series; style branches of the disk florets short,
            oblong .... 82. *Pectis*
        10 Involucral bracts in 2 series; style branches of the disk florets long and
            slender .... 37. *Dyssodia*
    9 Herbage without conspicuous colored oil glands, at most ± stipitate-glandular or with a
        few translucent oil glands in leaf impressions.
        11 Leaves opposite.
            12 Ray florets 1 per head .... 47. *Flaveria*
            12 Ray florets 3–9 per head.
                13 Pappus a crown of about 8 scales .... 85. *Picradeniopsis*
                13 Pappus of numerous, stramineous, stiff bristles .... 53. *Haploesthes*
        11 Leaves alternate.
            14 Ray corollas persistent on the achenes and falling with them .... 90. *Psilostrophe*
            14 Ray corollas deciduous.
                15 Pappus of 2–several separate, firm, deciduous awns .... 51. *Grindelia*
                15 Pappus of scales, or of basally-united awns.
                    16 Involucral bracts reflexing-spreading, especially in age.

17 Receptacle naked except for a few irregularly scattered hairs near the margin .... 55. *Helenium*
17 Receptacle alveolate, with irregularly-placed setae that do not precisely subtend the florets, or with the setae reduced to toothlike fimbrillae .... 48. *Gaillardia*
16 Involucral bracts erect or ascending, not at all reflexed.
18 Involucral bracts strongly imbricated in ± 3 series; heads very small, 3–5 mm tall, and numerous .... 52. *Gutierrezia*
18 Principal involucral bracts in 1 or 2 series, or if obviously imbricated in several series, then the involucre regularly more than 5 mm tall.
19 Achenes obpyramidal, 2–3× longer than broad, pubescent with appressed hairs; perennials, or if annuals the heads normally solitary at the ends of the peduncles .... 62. *Hymenoxys*
19 Achenes narrowly obpyramidal, 3 or more × longer than broad; annuals with 1–4 heads clustered at the ends of the peduncles .... 15. *Bahia*

Key 4
Heads radiate; rays purple, blue, pink, or white; receptacle chaffy or bristly

1 Leaves, or at least the principal ones, opposite.
  2 Pappus of ray florets fimbriate or awn-tipped scales; that of disk florets reduced or absent .... 49. *Galinsoga*
  2 Pappus of 2 or 3 barbed awns on both ray and disk florets, or pappus reduced to a mere crown or absent.
    3 Pappus of 2 or 3 barbed awns .... 19. *Bidens*
    3 Pappus a mere crown or absent.
      4 Chaff reduced to bristles; ligule of ray florets seldom exceeding 1 mm .... 39. *Eclipta*
      4 Chaffy bracts relatively broad, usually with an obvious midrib; ligule of ray florets 7–13 mm long .... 77. *Melampodium*
1 Leaves alternate.
  5 Ray corollas actually expanded marginal disk florets, with corollas irregularly 5-lobed; receptacle bristly .... 25. *Centaurea*
  5 Ray corollas normal; receptacle chaffy.
    6 Disk florets sterile, style branches undivided; disk achenes reduced .... 81. *Parthenium*
    6 Disk florets fertile, style branches divided; disk achenes as large or larger than those of the ray florets.
      7 Rays mostly 5 or fewer, ligule less than 1 cm long, or if rays ± 10, the ligule less than 2 mm long.
        8 Leaves pinnately dissected .... 1. *Achillea*
        8 Leaves at most merely dentate .... 106. *Verbesina*
      7 Rays more than 5, often more than 1 cm long.
        9 Receptacle alveolate, with irregularly placed setae; pappus of awned scales .... 48. *Gaillardia*
        9 Receptacle chaffy; pappus a crown, toothlike or absent.
          10 Leaves pinnatifid or pinnate-dissected.
            11 Receptacle elongate at maturity; rays purplish-brown .... 92. *Ratibida*
            11 Receptacle conical; rays white .... 7. *Anthemis*
          10 Leaves entire or dentate.
            12 Disk and ray florets both subtended by chaffy bracts; annual .... 36. *Dracopis*
            12 Disk florets subtended by chaffy bracts; ray florets naked; perennials .... 38. *Echinacea*

Key 5
Heads radiate; rays purple, blue, pink, or white; receptacle naked

1 Pappus of scales, stout awns, unequal stramineous bristles, or reduced to a short crown or absent, but not of abundant capillary bristles.

- 2 Pappus reduced to a mere crown or absent.
  - 3 Leaves all basal or nearly so; plant a scapose weed .... 17. *Bellis*
  - 3 Leaves cauline or cauline and basal.
    - 4 Margins or involucral bracts green and not at all scarious.
      - 5 Pappus absent; achenes flattened; annual .... 13. *Astranthium*
      - 5 Pappus a small crown; achenes 4-angled; annuals and perennials .... 8. *Aphanostephus*
    - 4 Upper margins and tips of involucral bracts scarious; achenes angular and striate; annuals and perennials.
      - 6 Receptacle strongly elevated, becoming conic or high-hemispheric at maturity .... 76. *Matricaria*
      - 6 Receptacle flat or merely arched upward .... 28. *Chrysanthemum*
- 2 Pappus of evident scales, stramineous bristles or awns, or if reduced, then of weakly fused squamellae.
  - 7 Receptacle alveolate, with irregularly placed setae or minute fimbrillae among the disk florets .... 48. *Gaillardia*
  - 7 Floor of the receptacle clearly smooth or but obscurely pitted.
    - 8 Pappus of 5 hyaline scales alternating with 5 slender awns .... 27. *Chaetopappa*
    - 8 Pappus of scales, awns, bristles, but not as above.
      - 9 Pappus of about 8 prominent scales, scarious except for a pronounced midrib .... 80. *Palafoxia*
      - 9 Pappus of awns, bristles or small squamellae.
        - 10 Achenes 4-angled or sub-terete; pappus of ± 5 reduced scales or a short, ciliate crown .... 8. *Aphanostephus*
        - 10 Achenes flattened, 2-angled or sometimes 3-angled; pappus otherwise.
          - 11 Plants typically glabrous, 2–12+ dm tall; achenes ± wing-margined; pappus of several minute bristles and 2(4) prominent awns 20. *Boltonia*
          - 11 Plants pubescent to glabrate, low, rarely exceeding 2 dm tall, often acaulescent; pappus of stiff bristles or a low crown of weakly fused squamellae .... 104. *Townsendia*
- 1 Pappus of abundant equal or subequal capillary bristles, sometimes with a few short outer bristles or minute squamellae.
  - 12 Cauline leaves reduced to scales; basal leaves large and deeply incised or toothed .... 84. *Petasites*
  - 12 Stems leafy; basal leaves various.
    - 13 Pappus bristles, or at least some of them, conspicuously dilated upward. (*S. ptarmicoides*). .... 98. *Solidago*
    - 13 Pappus bristles cylindrical or tapering upward and not at all clavate.
      - 14 Ligule of ray florets minute, rarely exceeding the involucre; involucre 3–4(5) mm tall; taprooted annuals .... 33. *Conyza*
      - 14 Ligules conspicuous, or if reduced, involucre regularly more than 4 mm tall and/or plants not taprooted annuals.
        - 15 Involucral bracts subequal or but weakly imbricated, often green throughout and not at all leaflike; mostly flowering in spring and summer .... 43. *Erigeron*
        - 15 Involucral bracts evidently imbricated, often with green tips and chartaceous bases, or if bracts subequal, the outer ones leafy.
          - 16 Plants fundamentally taprooted; leaves frequently sharp-toothed or dissected and/or spiny, but sometimes entire and smooth.
            - 17 Heads solitary at the ends of long branches; stems ± clustered from a branching caudex; perennial .... 111. *Xylorhiza*
            - 17 Heads few to numerous in open clusters; stems single or few.
              - 18 Glabrous annual with leaves entire or nearly so; inflorescence fundamentally paniculate; ligule of ray florets 3–4 mm long; involucral bracts appressed. (*A. subulatus*). .... 12. *Aster*
              - 18 Plants without this combination of characters .. 73. *Machaeranthera*
          - 16 Plants fundamentally fibrous rooted and sometimes rhizomatous; leaves variously entire or toothed but neither divided nor spiny.
            - 19 Plants low, depressed, spreading from creeping rootstocks; leaves scalelike .... 69. *Leucelene*

19 Plants erect, normally more than 2 dm tall; leaves various but not scalelike .... 12. *Aster*

Key 6
Heads eradiate; receptacle chaffy or bristly

1 Pappus of awns, scales, stiff bristles or absent.
  2 Heads unisexual, the staminate ones generally above the pistillate; involucre of the pistillate heads becoming nutlike or burlike at maturity.
    3 Involucre of the pistillate heads forming a bur with numerous, ± hooked prickles; involucral bracts of the staminate heads separate .... 110. *Xanthium*
    3 Involucre of the pistillate heads with 1 or more series of tubercles or straight spines; involucral bracts of the staminate heads fused .... 4. *Ambrosia*
  2 Heads bisexual; involucres neither tuberculate nor prickly.
    4 Pappus absent, or reduced to a mere crown.
      5 Involucral bracts usually only 4, clasping the marginal achenes; strongly scented annual weeds .... 74. *Madia*
      5 Involucral bracts various but more than 4 and not clasping the achenes.
        6 Heads with the central florets staminate only, with the style undivided; marginal florets pistillate and fertile.
          7 Heads in densely woolly glomerules; staminate florets very small, with corollas 4-toothed .... 46. *Evax*
          7 Heads separate or variously crowded but not in woolly glomerules; disk corollas normally 5-toothed.
            8 Inflorescence spikelike, racemiform or paniculate; anthers free or but weakly united at anthesis, filaments ± connate .... 63. *Iva*
            8 Inflorescence corymbiform or cymose; anthers connate.
              9 Principal leaves opposite, but the upper may be alternate; achenes short and thick, little compressed .... 88. *Polymnia*
              9 Leaves alternate; achenes flattened at right angles to the radius of the head .... 81. *Parthenium*
        6 Heads with the central florets fertile, the style normally divided.
          10 Leaves opposite, blade lanceolate to elliptic; low, ± spreading annual . 39. *Eclipta*
          10 Leaves alternate, blade very narrow or pinnatifid with narrow segments; perennials or weak subshrubs .... 11. *Artemisia*
    4 Pappus of awns, scales or stiff bristles.
      11 Leaves opposite, or a few upper ones alternate .... 19. *Bidens*
      11 Leaves alternate or chiefly basal.
        12 Leaves ± spinulose; involucral bracts terminating in a straight spine. .... 24. *Carthamus*
        12 Leaves not spinulose.
          13 Involucral bracts with distinctly recurved-hooked tips .... 9. *Arctium*
          13 Involucral bracts with tips spreading or appressed but not hooked.
            14 Involucral bracts variously fimbriate to erose, or if entire then scarious; achenes oblique at the base .... 25. *Centaurea*
            14 Involucral bracts entire at the apex and not at all scarious.
              15 Leaves chiefly basal and entire .... 75. *Marshallia*
              15 Leaves well distributed along the stem, variously dentate or lobed.
                16 Chaffy bracts prominent, obviously subtending each floret .... 81. *Parthenium*
                16 Receptacle merely alveolate with irregularly placed setae that do not precisely subtend the florets .... 48. *Gaillardia*
1 Pappus of capillary bristles.
  17 Receptacle alveolate, i.e., deeply honeycombed, often with numerous short bristles among the achenes .... 79. *Onopordum*
  17 Receptacle flat and densely bristly.
    18 Achenes oblique at the base; marginal florets often expanded and falsely raylike .... 25. *Centaurea*
    18 Achenes, or at least the central ones, flat at the point of attachment to the receptacle; marginal florets not at all expanded.

19 Involucral bracts with recurved-hooked tips; leaves large and not prickly .... 9. *Arctium*
19 Involucral bracts with straight, usually spiny tips; leaves variously prickly or spiny.
   20 Pappus bristles plumose .... 23. *Carduus*
   20 Pappus bristles merely barbellate .... 32. *Cirsium*

Key 7
Heads eradiate; receptacle naked;
pappus of capillary bristles, at least in part

1 Florets all perfect and alike, heads discoid.
  2 At least the leaf margins spiny.
    3 Receptacle alveolate; heads thistlelike, 2.5–5 cm across; plant spiny throughout .... 79. *Onopordum*
    3 Receptacle smooth; heads 1 cm wide or less; leaf margins spiny. (*M. grindelioides*). .... 73. *Machaeranthera*
  2 Plants not at all spiny.
    4 Corollas yellow or orange-yellow.
      5 Principal involucral bracts of equal length and in a single series, often subtended by a few short calyculate bracts .... 95. *Senecio*
      5 Involucral bracts imbricated in 3 or 4 series .... 54. *Haploppapus*
    4 Corollas purple, pink, white or ochroleucous.
      6 Heads very small and aggregated into secondary heads .... 40. *Elephantopus*
      6 Heads separate.
        7 Pappus in 2 series, an inner whorl of slender capillary bristles, and an outer whorl of short scales or broad bristles .... 107. *Vernonia*
        7 Pappus a single whorl of capillary bristles.
          8 Head with 5 equal involucral bracts and 5 florets .... 22. *Cacalia*
          8 Involucral bracts more numerous, imbricated; florets usually more than 5.
            9 Achenes strictly 5-angled; leaves whorled or opposite, although some of the upper ones may be alternate .... 44. *Eupatorium*
            9 Achenes 10-ribbed.
              10 Basal and lower cauline leaves persistent and evidently the largest, cauline leaves progressively reduced upwards; inflorescence ± spiciform .... 70. *Liatris*
              10 Basal and lower cauline leaves early deciduous or no larger than the cauline leaves; inflorescence corymbiform/paniculate.
                11 Pappus bristles plumose; plants arising from a stout, taprooted caudex; leaves alternate .... 65. *Kuhnia*
                11 Pappus bristles scabrous to barbellate, or plumose in 1 species but then at least some of the leaves opposite; plants fibrous-rooted or arising from thickened, fusiform roots .... 21. *Brickellia*
1 Outer florets, or sometimes all of the florets of some heads, pistillate; heads disciform; pistillate corollas sometimes with minute ligules; or plants dioecious.
  12 Dioecious subshrubs .... 14. *Baccharis*
  12 Herbs.
    13 Well-developed leaves basal or arising in creeping mats on the ground; cauline leaves reduced and sometimes scalelike.
      14 Basal leaves ± reniform or cordate; principal involucral bracts in a single series .... 84. *Petasites*
      14 Basal leaves obovate to oblanceolate; principal involucral bracts imbricated .... 6. *Antennaria*
    13 Leaves primarily cauline, not forming a persistent basal rosette or arising in mats.
      15 Involucral bracts scarious to hyaline and shining, especially toward the tip; herbage tending to be white-woolly.
        16 Heads all alike, with the outer florets pistillate and the inner florets perfect .... 50. *Gnaphalium*
        16 Plants dioecious; heads either wholly staminate, or heads essentially pistillate although with a few central staminate florets .... 5. *Anaphalis*

15 Involucral bracts varying from green to partly dry and subscarious but neither hyaline nor shining; herbage variously pubescent to glabrous, but not white-woolly.
  17 Cauline leaves numerous, usually lobed or toothed, lanceolate to ovate, regularly more than 1.5 cm wide; pistillate corollas tubular-filiform; annuals.
    18 Principal involucral bracts in a single series, subtended by a few short calyculate bracts; herbage glabrate at maturity, or at most with scattered, coarse tomentum .... 42. *Erechtites*
    18 Involucral bracts imbricated in several series; upper herbage glandular-puberulent .... 87. *Pluchea*
  17 Cauline leaves narrowly oblanceolate or narrower, rarely more than 1.5 cm wide; pistillate corollas often terminating in a minute ligule.
    19 Involucre less than 4 mm tall, or if taller the plant low and diffuse-branched; pistillate florets numerous, central perfect florets few; annuals .... 33. *Conyza*
    19 Involucre 4–11 mm tall, plants erect.
      20 Involucral bracts subequal but in 2 or 3 series, the outer ones sometimes longer than the inner ones; taprooted annual. (*A. brachyactis*). .... 12. *Aster*
      20 Involucral bracts ± imbricated, usually purplish toward the tip; fibrous rooted biennial or perennial. (*E. lonchophyllus*). .... 43. *Erigeron*

Key 8
*Heads eradiate; receptacle naked; pappus of awns, scales,
stiff bristles, a mere crown, or pappus absent*

1 Pappus absent or reduced to a mere crown.
  2 Leaves opposite; heads with 3 or 4 disk florets and usually with a single ray floret .... 47. *Flaveria*
  2 Leaves alternate; heads otherwise.
    3 Involucral bracts greenish and thin but not scarious; leaves mostly basal, deltoid to cordate or reniform and white-woolly beneath; achenes stipitate-glandular above .... 2. *Adenocaulon*
    3 Involucral bracts scarious toward the tip; leaves various, often narrow or dissected; achenes smooth, ribbed or wrinkled but not glandular.
      4 Inflorescence racemiform or paniculiform, of numerous small heads . 11. *Artemisia*
      4 Inflorescence corymbiform, or of solitary heads terminating erect branches.
        5 Leaves merely toothed .... 28. *Chrysanthemum*
        5 Leaves pinnately dissected.
          6 Receptacle strongly raised, cone-shaped .... 76. *Matricaria*
          6 Receptacle flat or low-convex .... 101. *Tanacetum*
1 Pappus of awns, scales or stiff bristles.
  7 Involucral bracts and/or upper leaves with conspicuous, translucent oil glands .... 37. *Dyssodia*
  7 Herbage variously glandular and/or pubescent or glabrous, but not provided with oil glands.
    8 Heads very small and aggregated into secondary heads; pappus of 5 bristles .... 40. *Elephantopus*
    8 Heads separate.
      9 Pappus of 2–8 firm, caducous awns .... 51. *Grindelia*
      9 Pappus of scales, sometimes with a few bristles.
        10 Leaves opposite or mostly so; pappus a crown of small scales, sometimes with 1 or 2 thin, caducous bristles .... 83. *Pericome*
        10 Leaves alternate.
          11 Involucral bracts scarious-membranaceous toward the tip .... 61. *Hymenopappus*
          11 Involucral bracts variously thick or thin but not scarious-tipped.
            12 Corollas pale violet to pinkish; annual .... 80. *Palafoxia*
            12 Corollas yellowish or sometimes brownish; rarely violet-pink and if so the plants perennial.

13 Leaves 1-3 × pinnatifid; heads eradiate but the outer disk corollas sometimes enlarged; involucral bracts erect or merely spreading .... 26. *Chaenactis*
13 Leaves variously subentire to pinnatifid; heads often with true ray florets present; involucral bracts reflexing at maturity .... 55. *Helenium*

1. ACHILLEA L., Yarrow

Rhizomatous perennials with alternate leaves and numerous heads in a basically corymbiform inflorescence. Heads radiate or rarely discoid; involucral bracts imbricate in several series, midrib greenish, margins scarious or stramineous; receptacle chaffy; ray florets mostly less than 10, disk florets 10-25(40), perfect and fertile, corolla white or rarely pinkish, style branches flattened and truncate. Achenes flattened tangentially to the axis of the head, glabrous; pappus none. [tribe Anthemidae]

Reference: Mulligan, G. A., & I. J. Bassett. 1959. *Achillea millefolium* complex in Canada and portions of the United States. Canad. J. Bot. 37: 73-79.

1 Leaves bipinnately dissected; involucre slender, cylindrical .... 1. *A. millefolium*
1 Leaves at most pectinate-serrate; involucre ± campanulate .... 2. *A. sibirica*

1. Achillea millefolium L., yarrow. Aromatic perennial 2-6(10) dm tall, woolly-villous, sometimes unevenly so. Stems strict or few-branched, arising singly or loosely clustered from a fibrous-rooted, weakly spreading rhizome. Leaves equally distributed along the stem, but the lower and middle cauline leaves largest, lowermost weakly petiolate; blade lanceolate in outline but bipinnately dissected, with the ultimate segments divergent-flexed above and below the plane of the midrib, overall dimensions 3-15 cm long and 0.5-3 cm wide. Inflorescence with many small heads in a flat-topped or flattened dome-shaped corymbiform cyme; involucre 4-5 mm tall, ray florets ca 5, ligule 2-3 + mm long, white or rarely pink. (x = 9; n = 18, 27) May-Jun (Oct). Widespread in the grasslands & open woods, especially in areas of mild disturbance; throughout the GP, more frequent in e GP; (circumboreal). *A. asplenifolia* Vent., *A. lanulosa* Nutt., *A. occidentalis* Raf.—Rydberg.

This species includes a complex of numerous, ill-defined phases and cytological races, widely distributed throughout the temperate northern hemisphere. The Eurasian materals, which include *A. millefolium* in the narrowest sense, apparently occur in North America only as sporadic but persistent introductions, and they are hexaploid (n = 27). The native GP material is mostly tetraploid (n = 18), and has been treated as subsp. *lanulosa* (Nutt.) Piper. The divergence of the ultimate leaf segments above and below the plane of the midrib in the leaves is ± distinctive for subsp. *lanulosa*.

2. Achillea sibirica Ledeb. Perennial up to 8 dm tall. Leaves narrowly lanceolate, 5-10 cm long and to 1 cm wide, margins incised or pectinate-serrate. Inflorescence a crowded corymbiform cyme; involucre 4-5 mm tall; ray florets ca 10, ligule 1-2 mm long, white. Jul-Aug. Open wooded areas; extreme n GP; MN: Roseau; ND: Bottineau, Cavalier, Pembina, Rolette; (Que. w to Man; ND, ne Asia).

2. ADENOCAULON Hook., Trail Plant

Contributed by Ronald L. McGregor

1. Adenocaulon bicolor Hook. Perennial herbs with short rhizomes. Stems erect, 3-8(10) dm tall, usually floccose below, stipitate-glandular above. Leaves mostly near base, alternate, thin; blades deltoid-ovate to cordate or reniform, 5-15) cm long, 3-10(15) cm wide, entire to usually coarsely toothed or lobed, nearly glabrous above, white-woolly below; petioles shorter to longer than blades, winged. Inflorescence paniculate, its branches stipitate-glandular and subtended by small narrow bracts; involucre small, somewhat turbinate; involucral bracts 1–1.5(2) mm long, ovate, thin, glabrous, reflexed in fruit and eventually deciduous; heads disciform; corollas whitish, tubular, the outer 3–7 pistillate, the inner ones staminate; receptacle naked; anthers sagittate. Achenes clavate, 5–8 mm long, stipitate-glandular above; pappus none. (n = 23) Jun–Sep. Rare in moist woodlands; SD: Lawrence, Pennington; (n MI, s Ont., n MN, BH of SD, nw MT, sw Alta., s B.C., to ID, CA). [tribe Senecioneae]

3. AGOSERIS Raf., False Dandelion

1. *Agoseris glauca* (Pursh) Dietr. Taprooted, scapose perennial, mostly less than 5 dm tall but occasionally taller, glabrous or nearly so and often ± glaucous. Leaves all basal, linear-lanceolate to oblanceolate, 5–30 + cm long and 1–30 mm wide, entire or irregularly sharp-dentate or rarely shallow-lobulate, margins tending to be flat and not at all undulate or crisped. Heads solitary at the ends of long scapes; involucral bracts subequal or imbricate in several poorly defined series, mostly acute, sometimes purplish or purple-spotted; florets all ligulate and fertile, corolla yellow but sometimes drying to pinkish. Achenes normally 10-nerved and hirtellous, the body gradually tapering upward to a stout, striate beak, up to 1/2 the length of the body; pappus of numerous capillary bristles. (n = 9, 18) May–Jun. Open, moist prairies; n 1/4 GP, s along e edge of Rocky Mts. to CO, s in e GP to w IA; (GP to B.C., s to NM & CA). *A. parviflora* (Nutt.) Greene, *A. pumila* (Nutt.) Rydb., *A. scorzoneraefolia* (Schrad.) Greene — Rydberg. [tribe Lactuceae]

This is a complex, widespread species with several more or less discrete varieties in the western cordillera.

Agoseris aurantiaca (Hook.) Greene occurs in the Rocky Mts. and westward, and may be expected along the w edge of our range in ne WY and n. It is distinguished by having orange to brick-colored corollas and achenes which are abruptly contracted to a narrow, obscurely to nonstriated beak.

4. AMBROSIA L., Ragweed

Annual or perennial herbs, or weak subshrubs. Stems arising from a taproot, branching caudex or creeping rootstocks; leafy, with leaves alternate or opposite, entire to more often deeply cleft or lobed to pinnatifid; variously pubescent or glabrous, often resinous with glandular hairs. Inflorescences of unisexual heads with both sexes on the same plant; staminate heads in narrow, elongate, spikelike aggregations at the ends of the branches or in the upper axils; pistillate heads in axillary clusters below the staminate inflorescences. Staminate heads nodding, with few involucral bracts in a single series, more or less fused, receptacle flat and chaffy, florets numerous, with abortive pistil, anthers separate at anthesis; pistillate heads with 1 or 2 florets, involucral bracts fused about the floret(s) to form a hard, indehiscent receptacle ("bur"), with the tips of the bracts projecting as spines or tubercles at maturity. Seeds shed entirely surrounded by involucral body; corolla and pappus absent. [tribe Heliantheae]

References: Payne, W. W. 1964. A re-evaluation of the genus *Ambrosia*. J. Arnold Arbor. 45(4): 401–430, + pl. I-VIII; Payne, W. W. 1970. *Ambrosia*. In Correll, D. S., & M. C. Johnston, Manual of the vascular plants of Texas. Texas Res. Foundation, Renner.

1 Spines of the pistillate inflorescence scattered in several series over the body of the fruiting involucre.
  2 Pistillate heads regularly 2-flowered.
    3 Leaves finely dissected, blades dark green above and white-pubescent beneath .... 8. *A. tomentosa*

  3 Leaves broadly lobed, blades and stems uniformly silvery-gray pubescent or irregularly glabrate in age .... 5. *A. grayi*
 2 Pistillate heads normally 1-flowered, or a few rarely 2-flowered in *A. confertifolia*.
  4 Mature pistillate head 6–10 mm long, spines usually elongate and straight .... 1. *A. acanthicarpa*
  4 Mature pistillate heads 2–5 mm long.
   5 Leaves sessile, green above and white-woolly tomentose beneath; high plains of e CO .... 6. *A. linearis*
   5 Leaves petiolate, both surfaces ± gray-green, strigose, sometimes glabrescent; widespread .... 4. *A. confertifolia*
1 Spines of the pistillate inflorescence in a single series, or reduced and absent.
 6 Leaves 3- to 5-lobed or sometimes entire.
  7 Leaves or the lobes narrow, up to 10 mm wide; stems less than 1 m tall .... 3. *A. bidentata*
  7 Leaves or the lobes more than 10 mm wide; stems often 2–3 m tall .... 9. *A. trifida*
 6 Leaves 1–2× pinnatifid.
  8 Annual, with a taproot .... 2. *A. artemisiifolia*
  8 Perennials with creeping rootstocks.
   9 Leaves petiolate; pistillate inflorescence with 4 short tubercles, or unarmed .... 7. *A. psilostachya*
   9 Leaves sessile; pistillate inflorescence with up to 9 prominent spines . 6. *A. linearis*

1. ***Ambrosia acanthicarpa*** Hook., annual bursage. Branching annual, 2–10(15) dm tall, variously long-pubescent to hispid, or both. Leaves opposite below, alternate above, petiolate or the uppermost becoming sessile; blade variable but broadly ovate to lanceolate in outline, 2–8 cm long, lobed to 1–2× pinnatifid, the lobes sometimes deeply dentate, upper and lower surfaces both green or the upper surface sometimes sparsely white-pubescent. Staminate inflorescence racemiform and often much branched, heads stalked, involucre 3–12 mm across and 3- to 9-lobed; pistillate heads in clusters below the staminate inflorescence, involucre 5–10 mm long and up to 14 mm across, 1-flowered, with several series of flattened, pointed spines or spines occasionally reduced or absent. (n = 18) Aug–Oct. Open, sandy sites; w 1/2 GP, but occasionally adventive eastward; (GP, w to WA, CA, & AZ). *Franseria acanthicarpa* (Hook.) Cov.—Rydberg.

2. ***Ambrosia artemisiifolia*** L., common ragweed, short ragweed. Taprooted annual herb, 3–10 dm tall, branching upward; rough, coarse-pubescent to hispidulous. Leaves opposite below, becoming alternate above, petioles 1–3 cm long, narrowly winged, upper leaves becoming subsessile, blade ovate in outline, 4–10 cm long and up to 7 cm wide, 1–2× deeply pinnatifid, upper leaves usually less divided. Staminate inflorescence of spiciform racemes, heads short-stalked, slightly oblique, ca 3 mm across, margin crenate; pistillate heads in axillary clusters, involucre obovoid, 1-flowered, ca 3 mm long, pubescent or glabrate, spines 5–7, short, surrounding the prominent beak, the beak 1 + mm long. (n = 17, 18) Jul–Oct. Open waste ground or disturbed sites in the prairie; GP, except sw KS, adj. CO, & panhandles of OK & TX; (nearly throughout the U.S. & s Can., except the desert southwest). *A. elatior* L., *A. longistylis* Nutt., *A. media* Rydb.—Rydberg.

 The air-borne pollen of this species is a cause of hay fever.

3. ***Ambrosia bidentata*** Michx., ragweed, southern ragweed. Erect annual herb, 3–10 dm tall, much branched and stems arching upward from a taprooted base; herbage rough-hairy. Leaves mostly alternate, sessile, lanceolate, 3–7 cm long and up to 1 cm wide, usually with a prominent tooth on each side near the base, both upper and lower surfaces green and ± hirsute. Staminate inflorescence of dense, spikelike clusters, heads sessile, involucre glandular, very oblique, turbinate, 2 mm or more across; pistillate heads single or few clustered in the axils of the leaves subtending the staminate spikes, involucre hirsute, elongate,

1-flowered, usually 4-angled with each angle terminating in a sharp, pointed spine 1(+) mm long. (n = 17) Aug–Oct. Weed in open waste places; se 1/4 KS, e OK; (OH to LA, w to our region; adventive elsewhere).

4. *Ambrosia confertifolia* DC. Erect perennial herb 3–6(10$) dm tall, often much branched from the base, spreading by rootstocks and forming large clonal populations; herbage gray-green to whitish pubescent, often somewhat sticky. Leaves alternate, petiolate, the petioles to 15 cm long; blade decurrent down the petiole, narrowly ovate to lanceolate in outline, to 16 cm long and nearly as wide but often much smaller, 1–3 × pinnately lobed, the lobes linear or lanceolate, both upper and lower surfaces green to gray-green, strigose-pubescent to glabrescent. Staminate inflorescence racemose-spicate, heads stalked, involucre to 10 mm across, 5- to 9(+)-lobed; pistillate heads in sessile, often compact clusters in the axils of bracts below the staminate inflorescence, involucre to 5 mm long and 4 mm across or sometimes smaller, 1- or rarely 2-flowered, spines 0–20, mostly on the upper 2/3 of the body, up to 2 mm long and with hooked tips, or sometimes much shorter. (n = 36, 54) Aug–Oct. A weed in disturbed open sites, sw 1/8 GP; sw KS, se CO, s to TX; (GP to CA, to cen & s Mex.; adventive elsewhere). *Franseria tenuifolia* Harv. & Gray — Rydberg.

5. *Ambrosia grayi* (A. Nels.) Shinners, bur ragweed. Erect perennial herb, 3–6 dm tall, spreading by extensive rootstocks and forming large clonal populations; herbage silvery-gray canescent. Leaves alternate, narrowed at the base to a distinct petiole, sometimes with several small lobes; blade ovate to lanceolate in outline, up to 10 cm long and 8 cm wide, irregularly pinnate-lobulate, the lobes serrate. Staminate inflorescence racemose-spicate; heads stalked, involucre up to 5 mm across, 5- to 9-lobed; pistillate heads in clusters or single in the axils of the upper leaves or bracts, involucre up to 7 mm long and 4 mm across, 2-flowered, spines up to 15, scattered, narrowed to a slender, hooked tip. (n = 18) Aug–Oct. Moist or drying, often mildly saline localities, especially in open, fallow fields; sw 1/4 GP, from s NE to n TX; (essentially restricted to the s GP). *Franseria tomentosa* A. Gray — Rydberg.

This species is an aggressive, noxious weed.

6. *Ambrosia linearis* (Rydb.) Payne. Erect perennial subshrub or coarse herb, 2–4 dm tall, branching from the base and perennating by spreading rootstocks. Leaves alternate, sessile with prominent basal lobes; blade ovate in outline, 15–25 mm long and up to 15 mm wide but deeply pinnate-lobed, the lobes nearly linear, margins revolute, upper surface green and with scattered straight hairs, lower surface white woolly-tomentose. Staminate inflorescence racemose-spicate, heads short-stalked, sometimes subtended by bracts, involucre to 5 mm across, 5- to 9-lobed; pistillate heads single or in small clusters in the axils of the upper leaves and in the axils of bracts below the staminate inflorescence, involucre about 3.5 mm long and 2.4 mm across, 1-flowered, spines more or less in a single whorl at the apex, up to 9 in number, ca 1 mm long, with hooked tips. Jun–Aug. Apparently restricted to a few localities in the open high plains of e CO; rarely collected.

7. *Ambrosia psilostachya* DC., western ragweed. Erect perennial herb 3–6(10 +) dm tall, forming extensive clonal colonies from creeping and often deep rootstocks. Stems branching above, variously hirsute to pubescent with short, ascending hairs. Leaves subsessile, lanceolate to ovate in outline, 4–8(10) cm long, usually only once-pinnatifid, with the divisions linear-lanceolate and subentire to toothed; sometimes weakly sticky-pubescent. Staminate inflorescence racemiform, heads oblique, obconic, ca 2.5 mm across, margin crenate; pistillate heads in small axillary clusters below the staminate inflorescence, involucre ca 2.5 mm long, obovoid, 1-flowered, with 4–6 stout tubercles surrounding the single beak, the beak 1 mm long. (n = 18, 36, ca 54, 72) Jul–Oct. Open prairie & waste

places; GP; (throughout cen U.S. & adj. Can., between Appalachian region & Rocky Mts.; adventive elsewhere). *A. coronopifolia* T. & G. — Rydberg.

8. Ambrosia tomentosa Nutt., perennial bursage. Erect or lax perennial herb, 1–4 dm tall, with extensive creeping rootstocks, forming large clonal populations; variously tomentose to glabrate. Leaves alternate, petiolate, the petiole up to 10 cm long and often with several small lobes below the base of the blade; blade ovate-lanceolate in outline, up to 10 cm long and 6 cm wide, irregularly 2–3 × pinnate-lobulate, upper surface dark green and closely scabrous, lower surface white woolly-tomentose, margin revolute. Staminate inflorescence racemose-spicate; heads stalked, involucres up to 7 mm across; pistillate heads in sessile clusters in the axils of bracts subtending the staminate inflorescences; involucres up to 6 mm long and 3.5 mm across, 2-flowered, spines 0–10, ca 1(3) mm long. Jun–Aug. High plains, apparently sporadic in sw SD, extreme w NE, ne CO & w; (w GP to WY & AZ; reputedly a waif in IA, ID, etc.). *Franseria discolor* Nutt. — Rydberg.

9. Ambrosia trifida L., giant ragweed. Taprooted annual, 1–3(4 +) m tall. Stem striate, usually simple, or but sparingly branched upward; herbage scabrous, and somewhat spreading-hairy, often glabrescent below. Leaves mostly opposite, petiolate; blades of the lower and middle leaves ovate to orbicular in outline, 1–2 dm long and 3- to 5-cleft, the major divisions often lobed; margin serrate; upper leaves smaller and only 3-cleft to subentire, or some plants with all leaves simple and subentire throughout. Staminate heads in racemiform, branching panicles, heads saucer-shaped, 3 mm across, with 6–8 lobes; pistillate heads in clusters in the axils of bracts subtending the staminate inflorescences, involucre obovoid, ca 4 mm long, 1-flowered, with 4–8 prominent ridges, each ending in a small, prominent to obsolete tuberculate spine, surrounding the single conic beak, the beak 1(+) mm long. (n = 12) Aug–Oct. Widespread & frequent in moist disturbed sites, especially stream banks, flood plains, & roadside ditches; a weed throughout the GP; (throughout N. Amer. e of the Rocky Mts.; widely adventive). *A. striata* Rydb. and *A. variabilis* Rydb. — Rydberg.

The bulk of the materials from the southern half of our range have less-prominently winged petioles, small fruiting involucres with blunt or obsolete tubercles and strongly scabrous indument; these populations are weakly distinguishable as var. *texana* Scheele.

The air-borne pollen from *A. trifida* is a major cause of hay fever in the autumn months.

5. ANAPHALIS DC., Pearly Everlasting

Contributed by Ronald L. McGregor

1. Anaphalis margaritacea (L.) Benth. & Hook., pearly everlasting. Erect, rhizomatous, polygamodioecious, perennial herbs. Stems solitary or tufted, (2)4–7(9) dm tall, simple or branched below and above, loosely white-wooly, or becoming rusty pubescent in age, sometimes greenish and glabrate below middle. Leaves alternate, equally distributed, simple, entire, sessile or somewhat clasping, linear to elliptic-lanceolate or the lower sometimes oblanceolate, (3)5–9(15) cm long, acuminate to obtuse at apex, 1- to 3-nerved, margins often revolute, tomentose below, often less pubescent to glabrous above, lower ones soon deciduous, those below inflorescence sometimes reduced. Heads usually individually short-pedunculate, 8–10 mm wide, numerous, in cymose panicles; involucre 5–8 mm tall, subglobose, woolly at base; involucral bracts imbricate in several series, the outer series ovate, obtuse, dry, scarious, pearly white, sometimes with a basal dark spot, the inner becoming lanceolate and less scarious; heads of male plants with numerous tubular, functionally staminate flowers, the style short-bifid with astigmatic branches; heads on female plants with numerous pistillate flowers with tubular-filiform corollas, and a few (4) functionally staminate flowers. Achenes 0.7–1 mm long, minutely papillate, olivaceous or brown, pappus of essentially distinct capillary bristles. (n = 14) Jul–Oct. Infrequent, but locally common in open wooded hillsides, valleys, roadsides, waste places; MN, BH, MT, WY, CO, NM; (Newf. to AK, s to NC, MN, SD, NM, AZ, CA; e Asia). *A. subalpina* (A. Gray) Rydb.—Rydberg; *A. margaritacea* (L.) Clarke var. *occidentalis* Greene, var. *angustior* (Mig.) Nakai, var. *intercedens* Hara, var. *subalpina* A. Gray—Fernald. [tribe Inuleae]

This is a highly variable species in which a number of taxa have been segregated, based on number of cauline leaves, leaf shape, pubescence and height of plant. Fernald described four varieties in the BH, but these readily pass into each other and are not worthy of taxonomic recognition. Manuals frequently include KS in the range of the species, but only cult. material is known from that state.

6. ANTENNARIA Gaertn., Pussy-toes, Everlasting

Contributed by Ronald L. McGregor

Ours dioecious, woolly perennial herbs, fibrous-rooted, stoloniferous, matted or colonial, or depressed-cespitose in matted tufts, or with compactly branched caudex and without stolons. Leaves alternate, simple, entire, the largest ones basal and at ends of stolons. Heads unisexual, solitary to fairly numerous in a usually congested inflorescence, disciform, many-flowered; involucral bracts imbricated in several series, scarious at least at the tip, white or colored; receptacle flat or convex, naked; staminate florets with usually undivided style and scanty pappus, the bristles barbellate or clavate; anthers tailed at base; pistillate florets with filiform-tubular corolla, bifid style, and pappus of naked capillary bristles slightly united at base. Achenes small, terete or nearly so, glabrous or papillate. [tribe Inuleae]

Reference: Bayer, R. J., & G. Ledyard Stebbins. 1982. A revised classification of *Antennaria* (Asteraceae: Inuleae) of eastern United States. Syst. Bot. 7: 300–313.

The taxonomy of *Antennaria* is complicated by much apomixis, polyploidy, and hybridization. As a result, specific lines are frequently obscure. However, when a large series of specimens is studied, the basic populations are readily recognized and the frequent intermediates determined. As with other apomictic groups, experience aids in arriving at reasonably satisfactory determinations.

1 Heads solitary, terminal .... 2. *A. dimorpha*
1 Heads several.
  2 Plants without stolons .... 1. *A. anaphaloides*
  2 Plants with short or long stolons.
    3 Basal leaves and those at end of stolons conspicuously 3- to 5-nerved, larger blades 2–6 cm wide .... 6. *A. parlinii*
    3 Basal leaves and those at end of stolons 1-nerved or obscurely 3-nerved, blades less than 2 cm wide.
      4 Upper leaf surface nearly as densely pubescent as the lower, glabrate only in extreme age.
        5 Involucre 7–11 mm high; dry pistillate corollas 5–8 mm long .... 7. *A. parviflora*
        5 Involucre 4–7 mm high; dry pistillate corollas 2.5–4.5 mm long .... 3. *A. microphylla*
      4 Upper leaf surfaces distinctly less pubescent than the lower, becoming glabrate or glabrous from the first.
        6 Mid-cauline leaves tipped by a flat or curled, scarious tip; pistillate and staminate plants equally common .... 4. *A. neglecta*
        6 Cauline leaves with subulate, not scarious tip (may be scarious near inflorescence); staminate plants absent or rare .... 5. *A. neodioica*

1. Antennaria anaphaloides Rydb., anaphalis pussy-toes. Perennial from a stout, compactly branched caudex. Stems 1–several from base, erect or ascending, 2–5 dm tall, densely and persistently tomentose. Basal leaves erect, 5–10(15) cm long (including petioliform base), 5–20 mm wide, elliptic-lanceolate or narrowly oblanceolate, with 3 or more evident to obscure main veins; cauline leaves more linear, becoming sessile, progressively reduced upwards; all leaves loosely tomentose on both sides. Heads numerous in a compact to somewhat open cyme; involucres 6–8 mm tall, tomentose toward base; involucral bracts brown at base with oblong, white, scarious tips, 4–6 mm long in pistillate heads, shorter in staminate ones; dry pistillate corollas 2.5–4(5) mm long; staminate pappus clavate. Achenes 1–1.2 mm long, yellowish brown, striate, glabrous. Jun–Jul. Rare in open wooded hillsides; WY: Albany, Johnson, Natrona, Niobrara; (Alta. to B.C., s to CO, n UT, ne NV).

2. *Antennaria dimorpha* (Nutt.) T. & G., dwarf pussy-toes. Dwarf perennial herbs from a compact, woody, much-branched caudex, depressed cespitose in matted tufts, without stolons. Flowering stems 1–3 cm long, silky tomentose. Leaves 1–3 cm long, 2–3(5) mm wide, linear to oblanceolate or spatulate, obtuse to acutish, tapering to a petiolelike base, rather loosely gray or white tomentose on both sides. Heads solitary, terminating the very short, leafy stems; staminate involucres 5–7 mm tall, involucral bracts blackish green or brownish, margins subhyaline, obtuse or acute at apex; pistillate involucres 10–15 mm tall, involucral bracts linear-lanceolate, attenuate-acuminate at apex, brown or purplish at base, whitish on margins at apex; staminate pappus bristles barbellate upwards, but scarcely clavate. Achenes 3 mm long, brownish, hispidulous. May–Jun. Rare in dry rocky prairie & open woodlands; NE: Dawes, Sioux; WY: Albany, Campbell, Crook, Niobrara, Weston; (MT to B.C., s to NE, CO, UT, s CA).

3. *Antennaria microphylla* Rydb., pink pussy-toes. Perennial, mat-forming, tomentose, stoloniferous herbs; flowering stems erect, (3)5–30(40) cm tall, tomentose, with narrowly lanceolate to oblanceolate leaves; stolons short, procumbent to decumbent. Basal leaves and those at tips of stolons oblanceolate or spatulate, usually 8–25(30) mm long, 2–7 mm wide, the upper surfaces nearly as densely hairy as the lower, glabrate only in extreme age, base often somewhat petiolate. Heads several in a compact or rather loose cyme; pistillate involucres usually 4–7 mm tall; involucral bracts narrowly oblong, rounded, obtuse or acute at apex, the upper scarious portion bright white to dull white or deep pink, sometimes with pink striae; dry pistillate corollas 2.5–4.5 mm long; staminate plants common to infrequent or absent. Achenes 0.8–1.2 mm long, olivaceous to brownish, sometimes lustrous, smooth or rarely minutely papillate. May–Jul. Common in prairies, open woodlands, bluffs, roadsides; MN, ND, SD, w NE, MT, WY, CO; (Ont., to AK, s to SD, w NE, NM, CA). *A. rosea* Greene, *A. oxyphylla* Greene — Rydberg.

This is a highly variable species in which a number of segregates have been named, particularly in the region w of the GP. None appear to warrant taxonomic recognition.

4. *Antennaria neglecta* Greene, field pussy-toes. Perennial fibrous-rooted, stoloniferous herbs, mat-forming or colonial; stolons long, procumbent, tomentose, lateral leaves evidently smaller than those at end of stolons. Flowering stems erect, white tomentose, (3)6–20(30) cm tall. Midleaves tipped by a flat, often curled, scarious appendage; basal leaves and those at end of stolons usually less than 1.5 cm wide, 1-nerved, tapering gradually to a petiolelike base, entire, persistently white-tomentose below, thin and deciduously so above or rarely glabrous and dark green from the first. Heads several in a dense and subcapitate or somewhat open cyme, the pistillate inflorescence often elongating and appearing racemose at maturity; involucral bracts narrow lanceolate, 7–10 mm long, obtuse to acute or acuminate, the outer brownish-red below, the upper scarious portions white; pappus longer than the stigmas; staminate and pistillate plants common. Achenes terete, slightly tapering to the ends, olivaceous to yellow-brown, 1.2–1.5 mm long, minutely papillose. (2n = 28) Mar–Jun. Common in prairies, open woodlands, pastures; GP but largely absent in w 1/2

of s 1/2; (widespread in n N. Amer. s to PA, OH, OK, MT, CO). *A. campestris* Rydb., *A. chelonica* Lunell, *A. longifolia* Greene — Rydberg.

In the BH are found populations with both the basal leaves and those at ends of stolons dark green and glabrate or glabrous from the first. In all other respects they are identical with *A. neglecta*, with both staminate and pistillate plants equally common. The only currently available name for these plants is *A. neglecta* var. *howellii* (Greene) Cronq. of the western United States.

5. **Antennaria neodioica** Greene, northern pussy-toes. Perennial fibrous-rooted, stoloniferous herbs, mat-forming or colonial; stolons short, 5–8 cm long, decumbent. Leaves along the stolons about equal in size to those at end of stolons; basal leaves rather abruptly contracted to a distinct petiole, blades less than 2 cm wide, often evidently 3-veined, entire, persistently dense white-tomentose below, less so above and becoming glabrate; flowering stems erect, white tomentose, (5)10–30(40) cm tall, midleaves with subulate tip or those just under inflorescence with scarious tips. Heads several in a dense or somewhat open cyme, involucral bracts narrow lanceolate, obtuse, (8)10–12 mm long, white or the outer reddish-brown in lower 1/2, white above; pappus longer or rarely shorter than stigmas. Achenes terete or angular, slightly tapering to the ends, minutely papillose, staminate plants have not been seen in GP. (2n = 56, 84) May–Jun. Prairies, open woodlands, pastures; MN, ND, SD, s to cen NE, MT, WY; (Newf. to Alta., s VA, ne IA, MN, NE, WY). *A. neglecta* var. *attenuata* (Fern.) Cronq.

As described above our plants are subsp. *neodioica*. It is possible that the plants with leaves dark green and glabrate to glabrous above, mentioned under *A. neglecta* above, might be *A. neodioica* subsp. *canadensis* (Greene) Bayer & Stebbins.

6. **Antennaria parlinii** Fern., plainleaf pussy-toes. Perennial, mat-forming or loosely colonial, stolonifereous herbs; stolons procumbent or decumbent, tomentose, sparsely leafy or bracteate below. Basal leaves and those at ends of stolons with blades ovate or broadly elliptical, 3–6 cm long, 1.5–4(5) cm wide, prominently 3- to 5-nerved, persistently tomentose below, much less so above and becoming glabrate or glabrous; petioles shorter to longer than blade; pistillate flowering stems 1–4 dm tall, erect, tomentose, sparsely leafy, leaves sessile, lanceolate, acuminate, tip subulate, rarely scarious. Heads usually several, in a compact or loose cyme; mature pistillate involucres 7–10(14) mm tall, involucral bracts narrowly oblong, lanceolate or elliptic, greenish, yellowish or pinkish below, scarious above, often attenuate; staminate involucres smaller, involucral bracts with prominent white, scarious tips; achenes brown, 1.2–1.5 mm long, minutely papillate (2n = 56, 70, 84, 112). Locally common in woodlands, thickets, rarely in prairies; MN, se SD, e NE, e 1/2 KS, OK; (ME to s Ont., s to GA, AL, e TX, se SD, KS). *A. plantaginifolia* Hook., *A. fallax* Greene, *A. occidentalis* Greene — Rydberg.

As described above, our plants are subsp. *fallax* (Greene) Bayer & Stebbins. The subsp. *parlinii*, with upper surfaces of the leaves glabrous, was not found among GP material but may be expected. Specimens mapped in the Atlas GP as *A. plantaginifolia* for ND, all but e SD, cen and w NE, were all *A. neodiocia* in which larger leaves were evidently 3-nerved. *A. plantaginifolia* is not known in the GP.

7. **Antennaria parvifolia** Nutt., pussy-toes. Perennial, mat-forming, stoloniferous, tomentose herbs. Flowering stems erect, (3)5–15(20) cm tall. Basal leaves and those at end of stolons nearly equally hairy above and below, becoming glabrate only in extreme age, blades spatulate or oblanceolate, 1–3.5(4) cm long, 2–10 mm wide; stolons short, decumbent, leafy. Heads closely to somewhat loosely aggregated in a cyme; pistillate involucres 7–11 mm tall, outer involucral bracts oblong, obtuse or rounded, the inner becoming lanceolate, acute or acuminate, the conspicuous scarious portion bright white, rarely dull white or pink; dry pistillate corollas usually 5–8 mm long; staminate plants rare or infrequently collected. Achenes 1–1.3 mm long, minutely papillate, olivaceous to yellowish-brown.

May-Jul. Common in prairies, open woodlands, pastures, roadsides; w MN, ND, SD, NE, MT, WY, CO; (MN, Man., B.C. s to NE, CO, AZ, NV, e WA). *A. aprica* Greene (?) — Rydberg.

The species is similar to *A. microphylla,* but is usually more robust with larger leaves and heads and shorter flowering stems. Some integration is evident but the two species are usually easily recognized.

7. ANTHEMIS L., Chamomile, Dog Fennel

Annual or perennial herbs, branching from the base and often strong-scented. Leaves alternate, pinnately dissected. Heads solitary at the ends of the stem or on conspicuous peduncles, radiate or rarely discoid; involucre hemispherical, involucral bracts numerous, imbricate, margins scarious, receptable conical, chaffy at least toward the center; florets numerous, disk florets perfect and fertile, corolla tube cylindric or somewhat flattened, style branches flattened, truncate, penicillate; ray florets pistillate or sterile. Achenes terete or angled and variously ribbed or striate; pappus none or a very short crown. [tribe Anthemideae]

Anthemis is an Old World genus with a number of highly variable weeds, many of which incorporate several semidistinct phases. Adventive species may occur as casual waifs in our area; for a detailed account, cf. Fernandes, R. 1976. *Anthemis, in* Flora Europaea, vol. 4, T. G. Tutin, V. H. Heywood, et al., eds. Cambridge Univ. Press, Cambridge.

1 Rays yellow, head large, usually with more than 20 rays .... 3. *A. tinctoria*
1 Rays white, heads smaller, with fewer than 20 rays.
  2 Receptacle chaffy throughout, rays fertile; mildly pleasant-scented .... 1. *A. arvensis*
  2 Receptacle chaffy only in the center, rays sterile; plant distinctly
    malodorous .... 2. *A. cotula*

1. ***Anthemis arvensis*** L., corn chamomile. Erect or upward-arching annual weed, 1-3(5) dm tall, bushy-branching from the base; variously light-pubescent to short-villous or subglabrous; mildly pleasant-scented. Leaves 1-2 × pinnatifid, 2-5 cm long, rachis narrowly winged. Involucre lightly pubescent; disk 7-12 mm across; chaff subtending each of the disk florets; ray florets ca 15-20, ligule white, 6-13 mm long. Achene ± 10-nerved; pappus a minute ridge or completely obscure. (n = 9) May-Jul. Sporadic in fields & waste places; known from w MO, NE & ND, but to be expected anywhere in favored sites; (scattered over much of N. Amer.; Eurasia). *Adventive. Naturalized.*

This species has been variously divided by different authors into infraspecific taxa. Our materials have been segregated as var. *agrestis* (Wallr.) DC. by Fernald and others, on the basis of the chaff being distinctly shorter than the disk florets.

2. ***Anthemis cotula*** L., dog fennel. Erect annual weed, 1-5 dm tall, usually glabrous or nearly so, simple or bushy-branched from the base. Leaves 2-6 cm long, 2-3 × pinnatifid with ultimate segments nearly linear. Heads loosely arranged at the ends of the stems; involucre lightly pubescent; the disk 5-10 mm across; chaff subtending only the central florets; ray florets ca 10-20, neutral and infertile, ligule white, 5-11 mm long. Achene ± 10-nerved, minutely tubercalate; pappus none. (n = 9) May-Sep. Fields, waste places & roadsides; e 1/2 GP, but to be expected elsewhere in favored sites; (widespread weed in N. Amer.; Eurasia). *Naturalized. Maruta cotula* (L.) DC. — Rydberg.

3. ***Anthemis tinctoria*** L., yellow chamomile. Short-lived stoloniferous perennial, 3-7 dm tall; short whitish-pubescent, at least above. Leaves 2-5 cm long, pinnatifid, with the segments dentate, pubescent beneath. Heads long-pedunculate, involucre thinly pubescent; disk 12-20 mm across, chaff subtending each of the disk florets, disk corollas ±

compressed; ray florets ca 20–30, pistillate, yellow, ligule 7–15 mm long. Achenes ± quadrangular and striate; pappus a short crown or obscure. (n = 9) Jul–Aug. Weed of fields & waste places in the extreme n GP; ND: Bottineau, Pembina, Rolette; may be expected elsewhere in favored sites; (widespread but scattered in N. Amer.; Eurasia). *Naturalized.* *Cota tinctoria* (L.) Gay — Rydberg.

8. APHANOSTEPHUS DC., Lazy Daisy

Annual or perennial herbs, with stems arising from a taproot or branching woody caudex. Leaves alternate, simple and entire to deeply pinnatifid. Heads solitary and terminal on the upper branches; involucre open-hemispherical to suburceolate; involucral bracts narrowly lanceolate to oblong-lanceolate, acute to acuminate, scarious-margined, imbricate; receptacle hemispherical to low-conical, naked, rough; ray florets pistillate and fertile, ligule white to lavender to rosy-purple; disk florets perfect, fertile, corollas yellow, with the base of the tube sometimes thickened and swollen at maturity, style-branches flattened, with a broadly triangular, acute appendage. Achenes of disk and ray florets similar, subterete to 4-angled, expanded upward, glabrous or sparsely pubescent; pappus an uneven scaly crown or a ring of short, ciliate hairs. [tribe Astereae]

Reference: Shinners, L. H. 1946. Revision of the genus *Aphanostephus.* Wrightia 1: 95–121.

1 Heads apparently crowded, the short peduncles naked for 3–12 mm below the head, or more in age; ray florets 12–18; herbage hispid with long, coarse, jointed, translucent hairs, those of the stem 0.7–2.2 mm long and spreading at right angles, easily broken off ...... 1. *A. pilosus*
1 Heads more loosely arranged, with peduncles naked for 1.5–10 cm below the head, or more in age; ray florets 18–60 + ; herbage soft-pubescent to hispid with hairs widely spreading or apparently deflexed downward, mostly less than 1 mm long.
    2 Pappus an irregular crown of about 5 unequal, acute scales; corolla bases becoming inflated and hardened at maturity .... 4. *A. skirrhobasis*
    2 Pappus a ring of short, subequal cilia or hairs; corolla bases slightly if at all inflated at maturity.
        3 Rays mostly 16–30; plants taprooted annuals .... 2. *A. ramosissimus*
        3 Rays 40–60 + , stems clustered from a branching, woody caudex .... 3. *A. riddellii*

1. ***Aphanostephus pilosus*** Buckl. Erect taprooted annual herb, 6–30 cm tall, branching mostly in the upper 1/2 but sometimes branched from the base; herbage conspicuously hispid-pilose with coarse, jointed, translucent hairs. Lower leaves early deciduous, oblanceolate to linear-lanceolate in outline but parted into narrowly lanceolate segments, or deeply toothed to sometimes subentire, blade 1.5–3.5 cm long and 1–2 cm wide, petiole more or less distinct, 1–2.5 cm long; middle and upper leaves progressively reduced, becoming entire and mere sessile bracts. Heads solitary on the upper branches, somewhat crowded, involucre 4.5–5.5 mm tall and 4–6 mm wide; involucral bracts imbricate in 3 or 4 poorly defined series; ray florets 12–18, ligule 5–7 mm long, white or rosy violet, especially in withering; disk florets with corollas ± 2 mm long, yellow, the lower portion of the tube whitish and somewhat hardened in age, but scarcely inflated. Achenes ca 1.5 mm long, lightly pubescent, 4-angled; pappus a cuplike crown of ± 5 lacerate to ciliate scales, 0.3–0.4 mm long. (n = 4) May–Jul. Open, drying sandy sites, especially shallow ditches; extreme s GP; sw OK & adj. TX panhandle; (s GP & w TX).

2. ***Aphanostephus ramosissimus*** DC. Taprooted annual herb, 5–40 cm tall, stems ascending, branching above; herbage pubescent with soft, spreading or deflexed hairs. Lower leaves early-withering, variously toothed to pinnatifid, blade 1–6 cm long and up to 2 cm wide, petiole up to 3 cm long, sometimes winged and indistinct from the blade; middle and upper cauline leaves progressively reduced, less prominently divided, uppermost reduced

to mere bracts. Heads solitary on long peduncles that are naked for 2-8 cm below the head; involucre broadly campanulate to somewhat urceolate, 3.3-5 mm tall and 5-9 mm wide; involucral bracts imbricate to ca 3 series; ray florets 16-32, ligule 5.5-7.5 mm long, white, or becoming rose to purple, especially on the underside; disk florets with corollas up to 2 mm long, yellow, hardened and thickened toward the base, but not swollen. Achenes ca 1.3 cm long, sparsely pubescent; pappus a minute ciliate crown. (n = 4) May-Sep. Open damp sandy sites; extreme s GP; (GP s through TX, especially Rio Grande Plains, & AZ, n Mex.).

3. *Aphanostephus riddellii* T. & G. Perennial 10-50 cm tall, with stems loosely clustered and branching upward from a subligneous, simple to much-branched caudex; herbage glabrate to somewhat pubescent with widely spreading hispid hairs. Lower leaves crowded, oblanceolate, dentate to deeply pinnatifid, with the lobes entire to toothed, blade 2.5-7.5 cm long and 1-3 cm wide, narrowed basally to a petiole 2-4 + cm long; middle and upper cauline leaves distinctly reduced, the uppermost bract-like. Inflorescence of solitary heads terminating long peduncles that are naked for 2-10 cm below the head when in flower, but elongating to up to 15 cm in fruit; involucre broadly campanulate to suburceolate, 4.5-6.2 mm tall and 9-14 mm wide; involucral bracts imbricate in 3 or 4 series; ray florets 40-60 + , ligule 7-10 mm long, white; disk florets with corolla 2-3 mm long, yellow, slightly hardened and thickened at the base, but not swollen. Achenes 1.5-2.0 mm long, glabrous or nearly so; pappus a minute, ciliate crown, less than 0.2 mm long. (n = 5) Jul-Oct. Open limestone sites; known in the GP from a few collections in the extreme s; TX: Cottle, Donley, Motley; (GP s through TX to n Mex., especially frequent in the Edwards Plateau).

4. *Aphanostephus skirrhobasis* (DC.) Trel. Taprooted annual herb, to 50 cm tall, bushy-branched and up to 50 cm across; herbage gray-pubescent with soft, fine, irregularly spreading or downward-deflexed hairs. Lower leaves with blades oblanceolate to oblong-ovate in outline but variously divided, toothed or subentire, up to 6 cm long and 2 cm wide, petiole more or less distinct, up to 6 cm long; middle and upper leaves progressively reduced, the uppermost sessile and bractlike. Inflorescence of solitary heads terminating long peduncles that are naked for 1-7 cm below the head; involucre 6-8 mm tall and 7-13 mm wide; involucral bracts imbricate in 4 or 5 poorly defined series; ray florets 20-44, ligule 8-15 mm long, white, or reddish to rosy on the lower surface, often the color in streaks; disk florets with corolla 2.0-2.5 mm long, yellow, the base becoming whitened, hardened and bulbous-swollen at maturity. Achenes 1.5-2.2 mm long, grooved; pappus a crown of irregular, lancerate-ciliate or acute scales, 0.3-1 mm long. (n = 3) May-Jul(Aug). Open, sandy sites; s GP, s-cen KS, w OK, & adj. TX; (s GP to FL, TX & n Mex.).

9. ARCTIUM L., Burdock

Contributed by Ronald L. McGregor

1. *Arctium minus* Bernh., common burdock. Biennial, nonspinose herbs, with stout taproots. Stems erect, 0.5-2 dm tall, much branched, usually reddish, ridged, puberulent. Basal leaves ovate, cordate, 3-6 dm long, 3-4 dm wide, with stout hollow petioles 1-4 dm long, green and glabrate above, pale and arachnoid tomentose below, margin undulate or shallowly undulate-lobed, acute or obtuse at apex; upper leaves much smaller, ovate, subcordate, truncate or cuneate at base. Inflorescence racemiform, subthrysoid, or tending to be capitate, clustered at ends of branches; heads subglobose, 1.5-2.5 cm wide, glabrous or slightly glandular, rarely a little arachnoid-tomentose; outer and middle in-

volucral bracts tapering from lanceolate, appressed base to spreading, stiff, uncinate spines, inner involucral bracts somewhat chartaceous, attenuate, tipped by straight or usually uncinate spines; receptable bristly; corollas equaling to usually longer than bracts, pink or purplish, rarely white. Achenes 4–5.5 mm long, oblong, compressed, grayish brown, and with darker mottling, slightly deciduous bristles mostly 1.5–2.5 mm long. (n = 16, 18) Jul–Sep. Local near old dwellings, feed lots, stream valleys, pastures, often in partial shade; GP but rare in sw; (throughout most of U.S. & Can.; Eurasia). [tribe Cynareae]

Arctium tomentosum Miller, the woolly burdock, a native of Eurasia, is uncommon in our region and not weedy. It has the inflorescence regularly corymbiform; involucre evidently tomentose; heads 2–3 cm in diam; corolla abruptly dilated into the limb and limb with glandular hairs. Known from ND: Cass, Grand Forks, Pembina, Ransom; SD: Lawrence; NE: Sherman; MO: Jackson.

Arctium lappa L., great burdock, with heads 3–4.5 cm in diam, is known from ND: Grand Forks, LaMoure; SD: Butte, Lawrence, Pennington, Union; NE: Custer, Sarpy. It is a native of Eurasia and also uncommon in our area.

10. ARNICA L.

Fibrous-rooted, often distinctly rhizomatous perennial herbs. Leaves simple, opposite, the uppermost sometimes reduced to mere bracts or sometimes a few are alternate. Heads solitary to few, large, radiate (ours) or occasional individuals are discoid, ray florets pistillate, yellow, ligule often broad; involucral bracts ± in 2 series. Achenes cylindric or nearly so, 5- or 10-ribbed; pappus of abundant white to stramineous, barbellate, capillary bristles. [tribe Senecioneae]

References: Maguire, B. 1943. A monograph of the genus *Arnica*. Brittonia 4: 386–510; Ediger, Robert I., & T. M. Barkley. 1978. *Arnica*, N. Amer. Flora II. 10: 16–44.

This is a genus of 27 species, best developed in the cordillera of w N. Amer. Apparently it has a base chromosome number of $x = 19$, but there are numerous biological complications relating to polyploidy and apomixis.

```
1   Leaf blades broad, mostly 1–2× longer than wide, cordate at the base; achenes mostly more
    than 6 dm long ............................................................................. 1. A. cordifolia
1   Leaf blades narrow, mostly more than 3× longer than wide, variously rounded to tapering at
    the base; achenes mostly less than 6 mm long.
    2   Heads broadly hemispheric; ray florets 10–23.
        3   Old leaf-bases with dense tufts of long, brown woolly hairs in the axils; disk florets
            with some spreading, glandular hairs as well as being stipitate-
            glandular ............................................................................. 2. A. fulgens
        3   Old leaf-bases without axillary tufts, or the hairs few and white; disk florets stipitate-
            glandular and not otherwise hairy ............................................... 5. A. sororia
    2   Heads turbinate to campanulate but not hemispheric; ray florets mostly 10 or fewer.
        4   Stems solitary or scattered; disk corollas nearly goblet-shaped ...... 3. A. lonchophylla
        4   Stems ± clustered; disk corollas nearly tubular ............................ 4. A. rydbergii
```

1. **Arnica cordifolia** Hook. Herb up to 6 dm tall, glandular puberulent to close-pubescent. Stems arising singly from long, creeping rhizomes. Basal leaves cordate, long-petiolate, dentate and often arising from separate short shoots; cauline leaves mostly 2 or 4 pairs, lowermost obviously the largest, cordate to truncate, blade 4–12 cm long and 2–9 cm wide. Heads 1–3(7), turbinate to campanulate, involucral bracts 10–20 mm long, variously pubescent with long, white hairs, especially toward the base, ± glandular; ray florets ca 13(10–15), ligule 15–30 mm long, or rays occasionally absent. Achenes 6–9 mm long, short-hairy and/or glandular; pappus long, white or whitish, barbellate hairs. Plants apparently polyploid and apomictic. Jun–Aug. Open woodlands; BH, foothills at the nw edge of our range; (AK to AZ, e to SD; disjunct in the Keweenaw Peninsula, MI).

2. **Arnica fulgens** Pursh. Herb 2–6 dm tall, stipitate-glandular and also frequently pubescent, more densely so upward. Stems arising from short, densely scaly and often clustered rhizomes. Leaves with 3–5 prominent veins, entire or nearly so, sparsely to densely pubescent with persistent tufts in the old leaf axils; basal leaves petiolate, oblanceolate to narrowly elliptic, overall dimensions 3–12 cm long and 1–4 cm wide; cauline leaves 2 or 4 pairs, progressively reduced, becoming sessile. Heads 1(3), broadly hemispheric, involucral bracts 10–15 mm long, glandular and pubescent; ray florets ca 13 or 21(10–23), ligule 1–2.5 cm long; disk florets generally glandular and with some spreading, white hairs. Achenes 4–6 mm long, pubescent; pappus whitish or stramineous. Plants mostly sexual diploids. Jun–Jul. Open places, rolling uplands; nw GP; cen ND, w SD, ne CO & nw; (Sask. to ND, CO, w to B.C. & ne CA).

3. **Arnica lonchophylla** Greene. Herb up to 4 dm tall, finely stipitate-glandular and also ± spreading-pubescent. Stems slender, erect to somewhat lax, arising singly or loosely clustered from a scaly, branching rhizome. Lowermost leaves narrowly elliptic to ovate, prominently several-veined, petiolate, overall dimensions 5–15 cm long and 1–3 + cm wide, upper leaves strongly reduced, sessile. Heads 1–7(9), erect, turbinate to campanulate; involucral bracts 1–2 cm long; ray florets ca 8(6–10), ligule 1–2 cm long; disk corollas 6–9 mm long, the slender, tubular lower part 3–4 mm long. Achenes 4.5–6 mm long, variously pubescent. Ours apparently diploid and sexual(?). Jun–Jul. Open places; BH, foothills at the w edge of our range in WY; (unevenly distributed from Newf. to B.C., s to ne MN, SD, & WY).

Our materials are referable to the dubiously distinct var. *arnoglossa* (Greene) Boivin, which differs from the more northern var. *lonchophylla* by having slightly thinner disk corollas, shorter and coarser pubescence, and other poorly defined characters.

4. **Arnica rydbergii** Greene. Herb 1–3 dm tall, variously glandular and pubescent to subglabrous. Stems ± clustered from a scaly, short-branched rhizome. Basal leaves petiolate, oblanceolate to spatulate, with 3–5 prominent veins, overall dimensions up to 7 cm long and 1.5 cm wide, frequently produced on separate short shoots; cauline leaves mostly 3 or 4 pairs, sessile or nearly so, reduced upward, lanceolate to spatulate, entire or nearly so. Heads solitary or few, turbinate-campanulate, involucral bracts 9–13 mm long; ray florets ca 8, ligule 1–2 cm long; disk corollas 7–9 mm long, nearly tubular. Achenes 5–6 + mm long, densely short-villous; pappus of white, barbellate hairs. Apparently all apomictic polyploids. Jun–Jul. Drying open meadows; BH; (Alta. to B.C., s to CO, UT, & CA; disjunct in BH).

5. **Arnica sororia** Green. Similar to *A. fulgens* but generally more slender, with leaves smaller and fewer and the brown woolly hairs absent from the persistent axils of the basal leaves, or white if present. Disk florets stipitate-glandular; pappus white or nearly so. Jun–Jul. Open places in rolling foothills; apparently barely entering our range in the foothills of WY along the w edge of the GP; (WY to B.C., UT, & CA).

11. ARTEMISIA L., Wormwood, Sage, Mugwort

Herbs or shrubs, often aromatic. Stems arising from a taproot, branching caudex or rhizome. Leaves alternate, entire to variously dissected or pinnatifid. Inflorescence spikelike, racemose or paniculate; heads small, discoid or disciform; involucral bracts imbricated, dryish and at least the inner ones with scarious margins; receptacle naked or with long hairs; florets sometimes all perfect and fertile; sometimes the outer florets pistillate while the center florets are perfect, sometimes the outer florets fertile while the inner florets are sterile or functionally staminate; style branches of fertile florets flattened, truncate and penicillate. Achenes mostly glabrous; pappus none or very reduced and obscure. [tribe Anthemideae]

References: Rydberg, P. A. 1916. *Artemisia*, N. Amer. Flora 34(3): 244–285. [a clearly outmoded treatment but significant, for it offers a basic segregation of species-groups and provides nomenclatural references]; Hall, H. M., & F. E. Clements. 1923. The North American species of *Artemisia*, *in* The Phylogenetic Method in Taxonomy, Carnegie Inst. Publ. 326: 31–156. [employs a broader species-concept than do contemporary taxonomists]; Keck, D. D. 1946. A revision of the *Artemisia vulgaris* complex, Proc. Calif. Acad. Sci. 25: 421–468.

1 Heads with florets all perfect; shrubs of the n & w GP.
   2 Leaves linear to linear-elliptic, subentire or weakly lobed but not tridentate; involucre mostly 4–5 mm tall .... 7. *A. cana*
   2 Leaves, or many of them, tridentate or 3- to 6-parted at the apex; involucre mostly 2–4 mm tall .... 15. *A. tridentata*
1 Heads with marginal florets pistillate only; herbs or shrubs.
   3 Disk florets, or at least the central ones, sterile.
      4 Leaves, or the segments, narrowly filiform; erect shrub or subshrub ..... 10. *A. filifolia*
      4 Leaves broader than filiform, or plant not an erect shrub.
         5 Plants low, matted subshrubs; inflorescence an erect, few-flowered spike .... 14. *A. pedatifida*
         5 Plants erect herbs; inflorescence an open panicle.
            6 Leaves mostly entire; rhizomatous perennials .... 9. *A. dracunculus*
            6 Leaves variously lobed-pinnatifid; taprooted biennial or perennial .... 6. *A. campestris* subsp. *caudata*
   3 Disk florets fertile throughout the head.
      7 Receptacle with conspicuous long hairs among the florets.
         8 Stem 4–10 dm tall; leaf segments 1.5–4 mm wide .... 2. *A. absinthium*
         8 Stem 1–4 dm tall; leaf segments ca 1 mm wide or less .... 11. *A. frigida*
      7 Receptacle naked.
         9 Annuals or biennials, taprooted; leaves green and glabrous or nearly so.
            10 Involucre 2–3 mm tall; inflorescence a series of subsessile heads in crowded spikes .... 4. *A. biennis*
            10 Involucre 1–2 mm tall; inflorescence a loose panicle of pedunculate heads .... 3. *A. annua*
         9 Perennials with prominent rhizome or woody caudex.
            11 Plant shrubby; leaves 3-toothed at the apex, sometimes with 1 or 2 additional lateral teeth; marginal florets 1 or 2 and disk florets 3–5 per head; rare in n TX .... 5. *A. bigelovii*
            11 Plants herbaceous, without above combination of characters.
               12 Leaves 2–3 × pinnatifid, thinly tomentose at most, and not silvery-gray in aspect .... 1. *A. abrotanum*
               12 Leaves subentire to 1-pinnatifid or variously toothed, often densely tomentose.
                   13 Stems arising from a woody center, without rhizomes; leaves lance-linear and commonly entire .... 12. *A. longifolia*
                   13 Stems arising from spreading rhizomes; leaves variously toothed, parted or pinnatifid.
                      14 Leaves of the lower stem parted nearly to the midrib into slender lobes, ca 1 mm wide .... 8. *A. carruthii*
                      14 Leaves variously subentire, toothed or incised, the lobes regularly more than 2 mm wide and not extending to near the midrib .... 13. *A. ludoviciana*

1. *Artemisia abrotanum* L., southernwood. Perennial herb, somewhat shrubby below, fragrant, 5–15(+) dm tall, branching upward, glabrous or minutely pubescent. Well-developed leaves 3–6 cm long, green and glabrous above, thinly tomentose below, 2–3 ×

pinnatifid, ultimate segments linear or filiform, 0.5-1.5 mm wide. Inflorescence an open panicle with many heads; involucre 2-3 + mm tall; receptacle naked; outer florets pistillate, center florets perfect. Achenes 4- or 5-angled, glabrous. (n = 9) Aug-Sep. A minor, semipersistent weed in waste places; sporadic in w KS: Gove, Scott, Sheridan, & in w MN: Kittson; (e N. Amer. to GP; Eurasia). Escaped from cultivation and *naturalized* in east, but merely a casual adventive with us.

2. *Artemisia absinthium* L., wormwood. Perennial herb or weak subshrub, fragrant, 4-10 dm tall, finely appressed-pubescent or glabrescent. Leaves silvery-sericeous but often glabrescent above; lower and middle well-developed leaves long-petiolate, 2-3 × pinnatifid, ultimate segments oblong, 1.5-4 mm wide, blade round-ovate in outline, 3-8 cm long; upper leaves reduced, less divided and becoming short-petiolate. Inflorescence an open, diffuse panicle; involucre 2-3 mm tall, sericeous; receptacle with numerous long hairs among the florets; outer florets pistillate, center florets perfect. Achenes glabrous, ± cylindrical. (n = 9) Aug-Sep. Open fields, roadsides, & waste ground; scattered in n GP, especially in e SD & throughout ND; (ne U.S. & adj. Can., to GP; Eurasia). *Naturalized* as a weed, especially in northeast, but adventive elsewhere.

3. *Artemisia annua* L., sweet sagewort. Glabrous, sweet-scented annual, 0.5-1.5 m tall but capable of growing to 3 m in favored sites. Leaves broadly ovate to lanceolate in outline, up to 10 cm long, mostly 1-2 × pinnatifid, the ultimate segments linear and somewhat toothed. Inflorescence an open, broad, loose panicle with the heads pedunculate and often nodding; involucre glabrous, 1-2 mm tall; receptacle naked; outer florets pistillate, center florets perfect. Achene ± cylindrical, obscurely ribbed, glabrous. (n = 9) Sep-Oct. Scattered in waste ground in e KS & w MO; (e U.S. & adj. Can. to GP; Eurasia). *Naturalized* in east; apparently merely adventive with us.

4. *Artemisia biennis* Willd., biennial wormwood. Glabrous, nonaromatic annual or biennial, normally up to 1.5 m tall, but occasionally much taller in favored sites. Leaves ovate in outline, 5-15 cm long, 1-2× pinnatifid, the ultimate segments often toothed. Inflorescence a series of ± dense spikelike clusters of nearly sessile heads; involucre glabrous, 2-3 mm tall; receptacle naked; outer florets pistillate, center florets perfect. Achene ellipsoid and 4- or 5-ribbed, glabrous. Sep-Oct. Mostly in mildly disturbed damp, sandy soil, along stream banks, flood plains, & roadsides; scattered in the GP from KS northward; (scattered throughout the U.S. except the se states, but thought to be native only in the nw states & introduced elsewhere).

5. *Artemisia bigelovii* A. Gray. Small shrub, 2-4 dm tall, with several spreading stems arising from a coarse, woody base. Branches and leaves dull, grayish-white pubescent. Leaves 15-25 mm long and 2-4 mm wide, narrowly cuneate, 3-toothed at the apex. Inflorescence of numerous heads in a dense panicle; involucre 2-2.5 mm tall, involucral bracts 12-15 and imbricated in about 3 series; receptacle naked; florets few, pistillate florets 1 or 2, perfect florets 2-5. Achenes small, ca 1 mm long. (n = 9) Sep-Oct. Open, dry sites; rare in the Texas panhandle, TX: Deaf Smith; (CO to TX, w to AZ).

6. *Artemisia campestris* L. subsp. *caudata* (Michx.) Hall & Clem., western sagewort. An apparent biennial 3-10 dm tall, glabrous or nearly so, inodorous, with stems arising singly from a prominent taproot. Basal leaves numerous, variously persistent to deciduous and absent in the second year, 2-3 × pinnatifid with the ultimate segments linear, up to 2 mm wide, overall dimensions 2-10 dm long, including petiole, and 1-4 cm wide; cauline leaves similar but less divided, reduced upward, the uppermost becoming merely ternate or even simple. Inflorescence of numerous heads in an elongate but narrow panicle, which may

be contracted and subspicate in small individuals; involucre typically glabrous, 2.5–4 mm tall; receptacle naked; outer florets pistillate only but fertile, center florets sterile, with abortive ovary. Mature achenes of the outer florets subcylindric, glabrous. (n = 9) Aug–Sep. Open, especially sandy sites; throughout the GP except se 1/5; (e Can. to FL, w to WY & TX, possibly MT). *A. forwoodii* S. Wats., *A. camporum* Rydb., *A. bourgeauana* Rydb.— Rydberg.

This species forms a complex of numerous semidistinct phases. The materials from the northern GP are somewhat more persistently pubescent than the southern materials, but they are impossible to recognize consistently as a distinctive entity. The species is circumboreal, with subsp. *campestris* native to Eurasia. The New World materials have been treated as a separate species (*A. caudata* Michx.) or as a subsp. of *A. campestris*. Contemporary taxonomy favors the latter. There is a complicated synonymy; cf. Cronquist, A. 1955, *in* C. L. Hitchcock et al., Vascular Plants of the Pacific Northwest, Univ. Wash. Publ. Biol. 17(5): 59–60.

Artemisia campestris subsp. *borealis* (Pall.) Hall & Clem. may barely enter our range along the w edge of the GP in WY and MT. It is similar to subsp. *caudata,* but it is a perennial 1–3 dm tall with several stems arising from a branching, taprooted caudex; the inflorescence is narrow and spikelike, and the involucres are 3–4 mm tall. It is the widespread phase in the western cordillera.

7. Artemisia cana Pursh, dwarf sagebrush. Branching aromatic shrub, mostly less than 1 m tall but occasionally taller; branches and leaves silvery or grayish pubescent. Leaves linear to linear-lanceolate, 2–5(8) cm long, up to 5 mm wide, entire or with 1 or 2 lobes, the lower leaves sometimes apically tridentate. Inflorescence of numerous small spikelike clusters, together forming a contracted, leafy panicle; involucre 4–5 mm tall, receptacle naked, all florets perfect. (n = 9, 18) Aug–Sep. Rocky open sites & flood plains; nw GP, w ND & SD, nw NE, scattered to the e; (GP w to B.C. & CA, s to NM).

8. Artemisia carruthii Wood ex Carruth. Perennial herb 2–6 dm tall, ± white tomentose, fragrant. Stems arising from a creeping rhizome. Leaves up to 5 cm long and 2 cm wide, deeply parted nearly to the midrib, the lobes narrow, 0.5–1.5(+) mm wide; well-developed leaves often with a pair of basal lobes which resemble stipules. Inflorescence of many heads in a broad panicle; involucre 2.5–4+ mm tall; receptacle naked; outer florets pistillate, inner florets perfect. Achenes cylindrical-elliptic. (n = 9) Jul–Aug. Open dry sites; sw GP from w KS s & w; (GP w to UT; reputedly in n Mex.). *A. vulgaris* var. *carruthii* (Wood) F. C. Gates—Gates.

9. Artemisia dracunculus L., silky wormwood. Perennial herb 5–10(15) dm tall, glabrous or scattered villous-short pubescent, variously fragrant to inodorous. Stems often reddish, mostly clustered but sometimes arising singly from a distinct rhizome. Leaves linear to linear-lanceolate, 2–8 cm long and up to 6 mm wide, entire or basally cleft with 1–3 lobes, often at least partially deciduous by autumn. Inflorescence an open panicle, with numerous racemose lateral branches; involucre glabrous or mostly so, 2–3 mm tall; receptacle naked; outer florets pistilate only but fertile, center florets sterile, with aborative ovary. Mature achenes of outer florets ellipsoid, glabrous. (n = 9) Aug–Sep. Dry open places; scattered through the GP; (IL & TX w to B.C. & n Mex.; Eurasia). *A. glauca* Pall., *A. dracunculoides* Pursh, *A. cernua* Nutt.—Rydberg.

This species apparently intergrades to a limited extent with *A. campestris* subsp. *caudata,* especially in the s GP.

10. Artemisia filifolia Torr., sand sagebrush. Weak shrub seldom exceeding 1 m tall, variously close-tomentose to subglabrate. Leaves or their lobes distinctly long-filiform, ± fascicled. Inflorescence of numerous small heads in leafy panicles; involucre pubescent, ca 2 mm tall; receptacle naked; outer florets pistillate only, but fertile and eventually producing mature achenes, center florets sterile but staminate, with abortive ovary. (n = 9)

Jun–Sep. Open sandy areas, often abundant; w GP from SD s through panhandle of n TX; (GP w to NV, AZ, TX, & n Mex.).

11. ***Artemisia frigida*** Willd. Low, spreading shrublet 1–4 dm tall, pleasantly fragrant, whitish or grayish tomentose, arising from a tough, woody crown. Leaves abundant, clustered toward the base and scattered along the stem, with close-appressed tomentum; the lower leaves petiolate, blade up to 12 mm long, 2–3 × divided into linear-filiform segments ca 1 mm wide or less, upper leaves becoming sessile. Inflorescence an open to contracted panicle or narrowed and racemose; involucre 2–3 mm tall, lightly pubescent; receptacle with numerous long hairs among the florets; outer florets pistillate, center florets perfect. Achenes subcylindrical, narrowed toward the base, glabrous. (n = 9) Aug–Sep. Open high plains; nw 1/2 GP from w MN sw to e CO & n NM; (WI to AK & AZ; Eurasia; introduced elsewhere).

12. ***Artemisia longifolia*** Nutt., long-leaved sage. Perennial herb up to 8 dm tall, ± silvery tomentose but sometimes unevenly so. Stems clustered from a woody base, without creeping rhizomes. Leaves linear or lance-linear, strongly tapering and often subcuneate to a long, pointed apex, usually entire but occasionally with a few teeth or short lobes; blade 3–12 cm long and rarely exceeding 1 cm wide, variously tomentose beneath but glabrescent above. Inflorescence a narrow panicle; involucre tomentose, 4–5 + mm tall; receptacle naked; outer florets pistillate, center florets perfect. Jul–Aug. Open alkaline sites in the high plains; nw 1/4 GP; (essentially restricted to the GP, s Sask. & Alta., & w into the intermontane valleys of the Rocky Mts.).

13. ***Artemisia ludoviciana*** Nutt., white sage. Rhizomatous perennial herb 3–7(10) dm tall, aromatic, variously persistent white-tomentose to irregularly glabrescent. Leaves elliptic to narrowly lanceolate, 3–11 cm long and up to 1.5 cm wide, entire to irregularly toothed or sometimes distinctly lobed, but the lobes rather wide, more than 2 mm. Inflorescence broadly to narrowly paniculate; involucre tomentose, 2.5–4 mm tall or sometimes larger; receptacle naked; outer florets pistillate, center florets perfect. Achenes ellipsoid-cylindrical, glabrous. (n = 9, 18). Widespread in N. Amer. from s Ont., IL, & AR to B.C. & n Mex.

This is a complex species with many semidistinct phases and a diversity of taxonomic interpretation. Our materials are separable into two varieties:

13a. var. *ludoviciana*. Leaves entire or subentire to coarsely few-lobed. Inflorescence ± compact but elongate. Aug–Oct. Open upland prairie; throughout the GP but infrequent in se KS, adj. OK, & w MO; (IL w to WT & UT). *A. brittonii* Rydb., *A. diversifolia* Rydb., *A. falcata* Rydb., *A. gnaphalodes* Nutt., *A. herriottii* Rydb., *A. lindheimeriana* Scheele, *A. papularis* (A. Nels.) Rydb., *A. purshiana* Bess., *A. serrata* Nutt. – Rydberg; *A. vulgaris* subsp. *ludoviciana* (Nutt.) Hall & Clem. – Hall & Clements, op. cit.

13b. var. *mexicana* (Willd.) Fern. Leaves, or many of them, deeply parted. Inflorescence open-diffuse, leafy. Aug–Oct. Open prairies & semidisturbed sites; se 1/4 GP, e KS, ne OK, & w MO; (MO to TX & n Mex.). *A. vulgaris* subsp. *mexicana* (Willd.) Hall & Clem. – Hall & Clements, op. cit.

14. ***Artemisia pedatifida*** Nutt. Short, weakly spreading subshrub to 15 cm tall. Stems arising from a branching woody base. Leaves closely appressed-tomentose, the basal ones fascicled, 1–2 cm long and 1–2 × pinnatifid, the ultimate segments short and very narrow; cauline leaves reduced and sometimes entire. Inflorescence few-headed, spicate or racemose; involucre 3–4 mm tall; receptacle naked; outer florets pistillate but fertile; center florets sterile, with abortive ovary. May–Jul. High plains of WY & s-cen MT, extending e barely to enter our range.

15. ***Artemisia tridentata*** Nutt., big sagebrush. Erect, branching shrub 4–20(30 +) dm tall, aromatic and finely appressed-pubescent. Principal leaves narrowly cuneate at the

base and tridentate at the apex, up to 5 cm long. Inflorescence of numerous near-sessile heads in a loose panicle; involucre pubescent, 2–4 mm tall; receptable naked; florets all perfect. (n = 9, 18) Jun–Sep. Open dry plains & hills; w 1/8 GP, w ND, & e MT, extreme nw NE, e CO, & apparently isolated in extreme nw TX; (abundant in w N. Amer. from GP w to B.C. & Baja Calif.).

This is a widespread species, occurring throughout much of the west half of the country. Our materials may be segregated as var. *vaseyana* (Rydb.) Beetle, cf. Beetle, Univ. Wyoming Agric. Exp. Sta. Bull. 368: 54. 1960. However, G. H. Ward, another student of the sagebrushes, includes our materials in var. *tridentata*, cf. Ward, Contr. Dudley Herb. 4: 168–177. 1953.

12. ASTER L., Wild Aster

Perennial herbs arising from a fibrous rooted caudex, crown or creeping rhizomes, or rarely taprooted annuals. Herbage variously pubescent to glabrous, and the hairs, when present, often ± densely disposed in discrete vertical lines on the stem, decurrent from the upper nodes. Leaves alternate, simple. Inflorescence usually of several to numerous heads, in open to variously crowded paniculate or corymbiform clusters at the ends of the stems, or heads rarely solitary; involucre hemispheric or campanulate to subcylindric; involucral bracts in several series, usually distinctly imbricate but subequal, commonly greenish-herbaceous toward the tip and chartaceous below, but sometimes entirely herbaceous or nearly all chartaceous; receptacle flat to weakly convex, naked; ray florets few to numerous, perfect, fertile, corolla with ligule usually prominent and blue to violet, pinkish, or white, sometimes the ligule greatly reduced and inconspicuous; disk florets few to numerous, perfect, fertile, corolla purple or deep pink to yellowish, sometimes white; style branches flat with internal-marginal stigmatic lines and acutish-acuminate, hairy appendages Achenes ± clearly several-nerved, glabrous or glabrate to variously pubescent; pappus either of a single series of capillary hairs or bristles, or double, with an inner series of bristles and an outer whorl of short hairs. [tribe Astereae]

References: Cronquist, A. 1943. Revision of the western North American species of *Aster* centering about *Aster foliaceus* Lindl. Amer. Midl. Naturalist 29: 429–468; Jones, A. 1977. New data on chromosome numbers in *Aster* sect. *Heterophylli* (Asteraceae) and their phylogenetic implications. Syst. Bot. 2: 334–347; Jones, A. 1978. Taxonomy of *Aster* sect. *Multiflori* I. Nomenclatural review and formal presentation of taxa. Rhodora 80: 319–357, and II. Biosystematic investigations. Rhodora 80: 453–490; Jones, A. 1980. A classification of the New World species of *Aster*. Brittonia 32: 230–239; Jones, A., & D. A. Young. 1983. Generic concepts of *Aster:* a comparison of cladistic, phenetic, and cytological approaches. Syst. Bot. 8: 71–84; Semple, J. C. 1979. The cytogeography of *Aster lanceolatus* (synonyms *A. simplex* and *A. paniculatus*) in Ontario with additional counts from populations in the United States. Canad. J. Bot. 57: 397–402; Semple, J. C., & L. Brouillet. 1980a. A synopsis of North American Asters: The subgenera, sections and subsections of *Aster* and *Lasallea*. Amer. J. Bot. 67: 1010–1026; Semple, J. C. & L. Brouillet. 1980b. Chromosome numbers and satellite chromosome morphology in *Aster* and *Lasallea*. Amer. J. Bot. 67: 1027–1039; Shinners, L. C. 1941. The genus *Aster* in Wisconsin. Amer. Midl. Naturalist. 26: 398–420.

Aster is a large genus with the species traditionally grouped into numerous semidistinct subgenera and sections. The plants are abundant and highly variable, thus there is a complicated synonymy. Many species hybridize and intergrade morphologically with their relatives and identification of any one specimen is often difficult. The appearance of intergradation is enhanced in the herbarium, and field workers note that the species are often more discrete and readily recognized in the field. The user of this treatment should have good specimens in hand, complete with underground parts and lower leaves, as well as the flowering materials.

Many botanists have been attracted to *Aster*, giving rise to a rich literature of both revisionary and biosystematic studies. However, there is no current general treatment for the genus. Semple & Brouillet (op. cit., 1980a), provide an overview, and then propose segregating many of the species with a base chromosome number $\times$ = 5 into the resuscitated genus *Lasallea* Greene. This would affect our species: *A. commutatus, A. ericoides* (q.v., for discussion), *A. falcatus, A. novae-angliae,*

A. oblongifolius, A. pansus, A. patens, and *A. sericeus.* Reveal & Keener (Taxon 30: 648-651. 1981) have proposed *Virgulus* Raf. as an earlier name for *Lasallea,* and have made the appropriate nomenclatural adjustments.

1 Basal and lowermost cauline leaves cordate to subcordate and petiolate.
  2 Involucral bracts, or at least some of them, with clearly reflexed tips; ray florets numerous, mostly 20-45. .... 1. *A. anomalus*
  2 Involucral bracts with tips erect and usually appressed; ray florets mostly 10-25.
    3 Leaves entire, subentire or but obscurely toothed, only the lower leaves cordate to subcordate; involucral bracts with a diamond-shaped green tip ..... 21. *A. oolentangiensis*
    3 Leaves, or at least the lower ones, distinctly toothed.
      4 Peduncles and branches of the inflorescence with few or sparsely distributed bracts; peduncles very unequal in length, some 1+ cm long; heads usually fewer than 50 .... 4. *A. ciliolatus*
      4 Peduncles and branches of the inflorescence with abundant bracts, especially when the peduncles approach (or rarely exceed) 1 cm in length; heads often very numerous and appearing crowded.
        5 Plants thinly hairy to glabrescent; green tips of the involucral bracts very narrow; mostly ne 1/3 GP. .... 31. *A. sagittifolius*
        5 Plants rather densely hairy, at least above the middle, with short, spreading hairs; green tips of the involucral bracts narrow but tending to be rhombic; mostly se 1/4 GP .... 7. *A. drummondii*
1 Basal and lower cauline leaves tapering at the base, or sessile to subsessile.
  6 Annuals; pistillate corollas eligulate, or with the ligule at most 3-4 mm long.
    7 Pistillate corollas shorter than the style, ligule vestigial .... 2. *A. brachyactis*
    7 Pistillate corollas exceeding the style, ligule over 1 mm long and rolled outward .... 36. *A. subulatus* var. *ligulatus*
  6 Perennials; pistillate corollas prolonged into an obvious ligule.
    8 Pappus double, i.e., with an inner series of long bristles and an outer series of short setae or bristles (use a hand lens).
      9 Leaves rigid and with but a single prominent midnerve; inner pappus bristles of uniform thickness throughout .... 16. *A. linariifolius*
      9 Leaves flexible, with numerous prominent veins; inner pappus bristles somewhat clavate, i.e., apically thickened .... 29. *A. pubentior*
    8 Pappus simple, not divided into 2 obvious series.
      10 Leaves sessile and distinctly cordate-clasping; involucral bracts well imbricated, hairy and sometimes glandular .... 25. *A. patens* var. *patentissimus*
      10 Leaves sessile or petiolate, sometimes clasping but not cordate, or if apparently auriculate-clasping, the involucral bracts tending to be glabrous and/or subequal or but weakly imbricated.
        11 Leaves few and minute or absent at maturity; upper branchlets reduced to spines; colonial rhizomatous plants of sw GP .... 35. *A. spinosus*
        11 Leaves obviously well developed although sometimes few; stem spineless.
          12 Thin plants with leaves linear to narrowly elliptic; arising from long, very slender rhizomes; cold bogs in n GP .... 13. *A. junciformis*
          12 Plants without the above combination of chracters.
            13 Plants with at least the involucre glandular and sometimes pubescent as well; lower herbage sometimes glandular.
              14 Leaf margins coarsely bristly-ciliate with the bristles ca 1 mm apart; stem glandular-pubescent, especially above .... 11. *A. fendleri*
              14 Leaves entire or dentate, margins at most merely strigose; upper stems glabrous to pubescent and/or glandular.
                15 Leaves dentate; plants of the BH .... 6. *A. conspicuous*
                15 Leaves entire, or nearly so, the margins at most strigose pubescent.
                  16 Stems glabrous; leaves linear to linear-lanceolate, 4-9 cm long; nw GP .... 26. *A. pauciflorus*
                  16 Stems pubescent and/or glandular; leaves usually broader than linear-lanceolate; widespread species.

17 Leaves prominently clasping .... 18. *A. nova-angliae*
17 Leaves sessile but very weakly or not at all clasping .... 19. *A. oblongifolius*
13 Plants not at all glandular.
   18 Leaves silvery in aspect, covered with appressed long silky hairs on both surfaces, leaves entire .... 32. *A. sericeus*
   18 Leaves glabrous or hairy but not silvery silky pubescent; entire or dentate.
      19 Cauline leaves obviously auriculate-clasping.
         20 Involucral bracts, or at least the inner ones, attenuate or long-acuminate.
            21 Stem uniformly hairy, at least below the heads; plant arising from a stout rhizome or caudex, sometimes with a few stolons .... 30. *A. puniceus*
            21 Stem and upper branches pubescent in decurrent lines from the nodes but glabrous or sparingly hispid below the heads; plants arising from long, creeping rhizomes .... 17. *A. lucidulus*
         20 Involucral bracts obtuse, acute or at most acuminate; plants arising from a stout rhizome or branched caudex, rarely with creeping rhizomes .... 14. *A. laevis*
      19 Cauline leaves sessile, sometimes weakly clasping or sheathing but not at all auriculate.
         22 Involucral bracts, or some of them, with a pointed, marginally inrolled subulate tip, somewhat tubular in aspect (use a hand lens); plants arising from a caudex or a short, stout, fibrous-rooted rhizome.
            23 Involucre somewhat urn-shaped to campanulate and about as broad as tall; disk florets usually more than 20 .... 27. *A. pilosus*
            23 Involucre turbinate, taller than broad; disk florets mostly 6–12 .... 24. *A. parviceps*
         22 Involucral bracts flat and without a pointed, tubular tip, although the tip may be spinulose or mucronate.
            24 Heads relatively large, involucre (6)7–12 mm tall; if short, the involucral bracts broad, ca 1 mm or more; ray florets usually bluish, violet or pink, but sometimes white.
               25 Heads 1–few, large, involucre 1–1.5 cm across; rays purple to violet; disk florets with tube longer than the slender limb; plants of higher elevations in BH .... 33. *A. sibiricus* var. *meritus*
               25 Plants without this combination of characters.
                  26 Outer involucral bracts narrower and shorter than the inner ones, grading into the minute peduncular bracts subtending the head . 37. *A. turbinellus*
                  26 Outer involucral bracts little if at all narrower than the inner ones and usually chearly differentiated from the ±leaflike subtending peduncular bracts.
                     27 Pappus very coarse and firm; leaves long-linear .... 22. *A. paludosus* subsp. *hemisphericus*
                     27 Pappus soft and relatively fine; leaves wider than linear.
                        28 Veinlets of the leaf forming a conspicuous reticulum on the underside, with the areolae (areas enclosed by the veinlets) ± isodiametric .... 28. *A. praealtus*
                        28 Veinlets of the leaf variously disposed and not forming a conspicuous reticulum on the underside; the areolae, if evident at all, longer than broad or unevenly formed.
                           29 Pubescence mostly confined to lines on the stem decurrent from the nodes and branch bases .... 12. *A. hesperius*
                           29 Pubescence not clearly in lines, but rather uniformly distributed, or sparse.
                              30 Inflorescence of heads distributed over the upper 1/3–1/2 of the plant; involucral bracts loose, somewhat squarrose, not strongly imbricated; ray florets mostly white or pink .... 8. *A. eatonii*
                              30 Inflorescence often compact at the top of the plant; involucral bracts strongly graduated and imbricated; ray

florets often blue or
pink .... 3. *A. chilensis* subsp. *adscendens*
24 Heads smaller, involucre mostly less than 7 mm tall and the involucral bracts less than 1 mm wide; ray florets frequently white, but sometimes pink or blue.
   31 Involucral bracts, or at least the outer ones, with loose or weakly reflexed bristle tips; leaves entire.
      32 Heads quite small, involucre less than 5 mm tall and ray florets fewer than 20; heads usually secund in arching branches.
         33 Plants colonial from a system of creeping rhizomes; involucre typically cylindrical to campanulate, especially when fresh .... 9. *A. ericoides*
         33 Plants cespitose in clusters, with erect or upward arching stems arising from a caudex with at most very short rhizomes; involucre broadly campanulate .... 23. *A. pansus*
      32 Heads larger, involucre 5–7(8) mm tall, ray florets 20–30+; heads few–numerous and variously arranged but not secund.
         34 Plants colonial from an extensive system of rhizomes and stolons; involucral bracts ± squarrose .... 5. *A. commutatus*
         34 Plants cespitose in clusters; involucral bracts loose to subappressed . 10. *A. falcatus*
   31 Involucral bracts loose-appressed but not at all bristle-tipped; leaves various.
      35 Lobes of the disk corollas ca 1/2–3/4 of the corolla limb; leaves usually pubescent beneath (N.B.: the disk corolla has a slender "tube" at the base and an expanded funnellike "limb" above it; the lobes are the 5 finger- or toothlike petal sections of the limb).
         36 Plants arising from creeping rhizomes; undersides of the leaves usually hairy throughout .... 20. *A. ontarionis*
         36 Plants arising from a caudex or short, stout rhizome; undersides of leaves with hair confined to the midrib .... 15. *A. lateriflorus*
      35 Lobes of the disk corollas less than 1/2 the corolla limb; leaves mostly glabrous on the undersides (except *A. praealtus* var. *nebraskensis*).
         37 Veinlets of the leaf forming a conspicuous reticulum on the underside, with the areolae ± isodiametric; lobes of the disk corollas less than 1/4 the corolla limb .... 28. *A. praealtus*
         37 Veinlets of the leaf variously disposed and not forming a conspicuous reticulum on the underside, the areolae, if evident at all, longer than broad or unevenly formed.
            38 Involucral bracts ± strongly imbricated, the outer mostly less than 2/3 as long as the inner ones.
               39 Involucre mostly less than 5.5 mm tall, ray florets white or sometimes pink to lavender; widespread species .... 34. *A. simplex*
               39 Involucre 5–7 mm tall, ray florets often blue or pink; extreme w GP .... 3. *A. chilensis* subsp. *adscendens*
            38 Involucral bracts subequal and not strongly imbricated, the outer at least 2/3 as long as the inner ones.
               40 Pubescence mostly confined to lines on the stem decurrent from the nodes and branch bases .... 12. *A. hesperius*
               40 Pubescence not clearly in lines but rather uniformly distributed, or sparse .... 8. *A. eatonii*

1. ***Aster anomalus*** Engelm., many-ray aster. Herbaceous perennial 6–10 dm tall, arising from a short, stout rhizome. Stem scabrous to pubescent or occasionally glabrous. Lower leaves ovate to lanceolate, cordate to abruptly contracted, blade 5–9 cm long and (1.5)3–4 cm wide, narrowly acute, upper surface glabrous to scabrous, lower surface scabrous to pubescent, margin entire to subserrate or slightly undulate, petiole winged, 1–8 cm long, sometimes with tufts of smaller leaves in the axil; upper leaves lanceolate, becoming sessile, 10–20 mm long and 1–5 mm wide. Inflorescence open, paniculate, heads (10)30–60(70+); involucre hemispherical, 5–10 mm tall and 4–15 mm wide; involucral bracts loosely imbricated in several series, yellowish-white at the base, with green tips, pubescent, appressed

but the tips ± spreading; ray florets 20–45, ligule bright lavender, (4)6–10(15) mm long; disk florets ca 40, corolla purple, slightly exceeding the pappus. Achenes glabrous, flattish, 1–3 mm long; pappus a single series of numerous, minutely plumose bristles. (n = 9) Aug–Oct. Rocky soils in limestone areas, hillsides, open to dense woodlands; KS: Cherokee, Labette; (IL & MO to AR, se KS, & OK).

2. **Aster brachyactis** Blake, rayless aster. Erect, taprooted annual (1)2–6 + dm tall. Stem glabrous or but remotely pubescent above, often reddish. Leaves linear or very narrowly lanceolate, with a single nerve, (1.5)3–8(15) cm long and 2–4(6 +) mm wide, margin ± ciliolate or obscurely serrulate, acute to acuminate, sessile, the lowermost early deciduous. Heads usually numerous, in an open paniculate cluster or a somewhat contracted racemose inflorescence; involucre 5–11 mm tall, involucral bracts subequal or weakly imbricated in 2 or 3 series, the outer sometimes longer than the inner ones; pistillate florets more numerous that the disk florets, corolla shorter than the style, essentially eligulate; disk florets with corolla violet. Achene brownish, appressed-hairy, ca 1.5 mm long; pappus of abundant soft white capillary bristles. (n = 7) Sep–Oct. Moist, sandy sites along streams & swamps, somewhat tolerant of saline situations; n 1/2 GP, scattered s in e CO, w KS, & w OK; (GP to B.C., WY, & UT; introduced in ne U.S., apparently also in Siberia). *Brachyactis angusta* (Lindl.) Britt.—Rydberg.

3. **Aster chilensis** Nees subsp. **adscendens** (Lindl.) Cronq. Perennial 2–10 dm tall, arising from a rhizome or branching caudex. Stem glabrate or short pubescent, especially toward the top. Lower leaves oblanceolate but early deciduous; cauline leaves linear or narrowly linear or narrowly lanceolate, 2–10 cm long and mostly less than 1 cm wide, usually more than 7 × longer than wide. Inforescence of 1–numerous heads; involucre 5–9 mm tall, 6–12(+) mm wide; involucral bracts imbricated, with green tips, the innermost acute-acuminate; ray florets 15–40, ligule blue to pinkish or rarely white, 6–10 mm long. Achenes ± pubescent; pappus of numerous bristles, white to tawny. (n = 8, 16) Aug–Oct. Open, dry or drying high plains, sometimes in saline sites; barely entering the GP in ne WY; (Sask. to NM, w to WA & CA; reputedly introduced to the e). *A. adscendens* Lindl., *A. nuttallii* T. & G.—Rydberg.

4. **Aster ciliolatus** Lindl. Herbaceous perennial 3–10 dm tall, arising from a rhizome, or sometimes also with a short caudex. Stem mostly glabrous, but puberulent upward, especially in the inflorescence. Lower leaves ovate to lanceolate, at least some of them cordate or abruptly contracted to a distinct petiole, blade 8–15 cm long, 2–6 cm wide, glabrous above but somewhat hairy beneath, acute to acuminate, serrate, petiole 3–12 cm long, ± clasping; leaves and petioles gradually reduced upward, upper cauline leaves becoming lanceolate to linear, entire, sessile and frequently clasping. Inflorescence paniculate with few to numerous heads, usually fewer than 50 and rarely more than 100, peduncles varying greatly in length; involucre hemispherical, 5–7 mm tall and 6–8 mm wide; involucral bracts slender, slightly imbricated, glabrous, yellowish-white at the base, with a diamond-shaped green tip or a green center line; ray florets (12)15–25, ligule bluish-purple, 7–12 mm long; disk florets ca 25, corolla reddish-purple, slightly exceeding the pappus. Achenes yellowish-white, flattened, 1–2 mm long, glabrous or sparsely hairy; pappus of numerous, minutely plumose bristles, 3–6 mm long. (n = 24, 36) Jul–Oct. Rocky, moist sites, especially openings in wooded areas; w MN, ND, s & w to BH, e WY; (Que. & NY to WY, MT, B.C., & N.T.). *A. lindleyanus* T. & G., *A. saundersii* Burgess, *A. wilsonii* Rydb.—Rydberg.

5. **Aster commutatus** (T. & G.) A. Gray. Colonial perennial 3–8(10 +) dm tall. Stems erect or ascending, and usually arising from an extensive, sod-forming system of rhizomes and stolons; herbage ± copiously pubescent. Cauline leaves persistent at flowering time, variable

in size but the prominent ones linear, 1–4 cm long, hispidulous-strigose or glabrate in age. Inflorescence typically diffuse-paniculate with the heads clustered at the ends of the branches; involucre 5–8 mm tall, pubescent; involucral bracts imbricated in several series, ± squarrose, bristle tipped; ray florets 20–35, ligule white or rarely pinkish, 7–10 mm long; disk florets typically more than 20. Achenes brownish to pale purple, 2+ mm long, appressed-puberulent; pappus of numerous capillary bristles, whitish, ± equaling the disk corollas. (n = 10) Jul–Aug. Open, dry prairies & plains; throughout GP except e KS, w MO, & adj. OK; (GP to Alta., s to w TX, AZ, & n Mex.). *A. crassulus* Rydb., *A. polycephalus* Rydb.—Rydberg.

This species is a member of the *Multiflori* complex; cf. the discussion under *A. ericoides*.

6. *Aster conspicuous* Lindl. Rhizomatous perennial 3–7(10) dm tall; herbage obviously glandular, especially toward the top. Leaves sharply dentate or rarely obscurely so; lower ones petiolate and early deciduous; middle cauline leaves well developed, ovate to lanceolate, mostly sessile, 4–14 cm long and 1–5(7) cm wide; uppermost leaves reduced. Inflorescence open, corymbose, heads few to numerous but seldom more than 50; involucre 9–12 mm tall, densely glandular; involucral bracts imbricated in several series, loose or somewhat squarrose; ray florets 15–30, ligule bluish to violet, 8–14 mm long; disk florets with purplish corolla. Achenes brownish, ca 2 mm long; pappus of numerous capillary bristles. Jul–Oct. Rocky roils in open woods; BH; (WY to Sask., Yukon, to B.C. & OR; disjunct in the BH).

7. *Aster drummondii* Lindl., Drummond's aster. Perennial herb 4–12 dm tall, arising from a short, branching caudex or a stout rhizome. Stems erect, sparsely pilose below to densely pilose above the middle. Leaves roughly serrate or shallowly dentate, upper surface scabrous, lower surface ± pilose with short, spreading hairs; basal and lowermost cauline leaves broadly ovate to ovate, cordate, blade up to 14 cm long and 6 cm wide but mostly smaller, petiolate; upper leaves progressively smaller and with shorter, variously winged petioles. Inflorescence an open, paniculate cluster with spreading, bracteate branches; heads numerous, often ± 150; involucre 4.5–7 mm tall; involucral bracts imbricate in several series, whitish at the base but with a prominent elongate-rhombic green area at the tip; ray florets 12–20, ligule light purple to bluish-violet, 5–8 mm long; disk florets with corolla yellowish. Achenes puberulent to glabrate, 2–3 mm long; pappus of abundant white capillary bristles. (n = 16, 18?) Sep–Oct. Rocky or dry openings in wooded areas; se 1/4 GP, scattered n to w IA & adj. MN; (OH to MS w to TX, GP).

8. *Aster eatonii* (A. Gray) Howell. Perennial 4–10 dm tall, arising from a creeping rhizome. Stem uniformly puberulent, especially upward, but sometimes with lines of dense pubescence decurrent from the middle cauline leaf-bases. Lowermost leaves petiolate and early deciduous; cauline leaves ample, scabrous to glabrate, generally entire or subentire, sessile or nearly so, linear to lanceolate, (1)5–15 cm long and 5–10(20) mm wide. Inflorescence a long, narrow leafy-bracted panicle with heads numerous, on erect or ascending branches; involucre 5–10 mm tall and up to 15 mm wide; involucral bracts loose, subequal or but weakly imbricated, at least some of them spreading or subsquarrose; ray florets 20–40, ligule commonly white or pinkish but rarely blue, 5–13 mm long. Achenes pubescent; pappus of abundant capillary bristles, whitish to tawny. Aug–Sep. Moist places in the foothills; barely entering the GP in ne WY; (Alta. to WY, CO, & NM, w to B.C. & CA). *A. oregonus* (Nutt.) T. & G.—Harrington.

The taxonomy of this essentially western montane entity is treated by Cronquist, Amer. Midl. Naturalist 29: 429–468, 1943, under the misapplied name *A. oregonus*. Recent studies suggest that the name *A. eatonii* may be invalid and needs to be replaced.

9. *Aster ericoides* L., white aster. Colonial perennial 5–10 dm tall, arising from an extensive system of rhizomes and stolons. Herbage ± evenly pubescent, or glabrate below. Basal and cauline leaves mostly deciduous before flowering time; persistent leaves mostly near the inflorescence and reduced to subulate or linear, sessile bracts, somewhat grading into the involucral bracts on the peduncles. Inflorescence of numerous small heads, mostly secund on long, recurving branches; involucre 2.5–4.5 mm tall, ± cylindrical to narrowly campanulate; involucral bracts strongly imbricated, ± bristle tipped; ray florets 10–18, ligule white or rarely pinkish, less than 6 mm long; disk florets rarely more than 14, corolla yellowish to purple. Achenes purplish-brown, 1–2 + mm long, appressed-puberulent; pappus of white or whitish capillary bristles, equalling the disk corolla. (n = 5) Sep–Oct. Widespread in open upland prairies & plains; GP except nw 1/5; (ME to MD, w to Man., GP, TX, NM, & n Mex.). *A. exiguus* Rydb., *A. batesii* Rydb.—Rydberg.

This species hybridizes with *A. novae-angliae,* and the hybrid has been called *A. amethystinus* Nutt.; cf. the discussion under *A. novae-angliae.*

Aster ericoides is a part of the *Multiflori* complex which also includes *A. falcatus, A. commutatus,* and *A. pansus.* The group is subject to diverse taxonomic interpretations because the species are abundant and they hybridize or intergrade freely not only with each other, but with several other asters as well. The group has received much attention from Almut Jones and John Semple, whose papers are cited above among the references. Semple has proposed segregating these species, among others, into the genus *Lasallea* (or *Virgulus*), largely on the basis of having chromosome numbers of n = 5, or numbers readily derived from a base of 5.

The *Multiflori* appear to have two sets of characters that vary independently of each other. There are plants with small heads and there are plants with large heads. Also, there are plants with the stems arising in clusters, and plants that are creeping-rhizomatous, with the stems arising singly. Thus there are four basically distinctive groups in the assemblage. It is a matter of taxonomic judgment how these groups should be treated. All four may be regarded as infraspecific taxa within *A. ericoides,* or as 2 species each with 2 infraspecific taxa, or they may be regarded as 4 species. The second choice has often been employed, with a large-headed species (*A. falcatus*) and a small-headed species (*A. ericoides*). However, it would seem just as realistic to segregate species on the basis of habit. Here we elect to recognize the 4 entities as species, with the presumption that this treats the pattern of variation in a logically defensible manner. It is acknowledged that these species are no more distinct from each other than are the species of other similar alliances in *Aster.* The 4 species are separated in the key, and represent the 4 character-states thus: *A. ericoides*—heads small; creeping, colonial habit. *A. pansus*—heads small; clustered habit. *A. commutatus*—heads large; creeping, colonial habit. *A. falcatus*—heads large; clustered habit.

10. **Aster falcatus** Lindl. Cespitose perennial 2–6 + dm tall, forming clusters of few–several stems arising from a caudexlike corm, sometimes interconnected by a few, slender rhizomes. Herbage ± evenly but sparsely pubescent. Basal and lower cauline leaves early pubescent; middle and upper cauline leaves ± persistent, the smaller ones weakly clasping; principal leaves linear or narrowly lanceolate, up to 5 cm long. Inflorescence typically diffuse-paniculate with the heads single or clustered at the ends of the branches; involucre 5–8 mm tall; involucral bracts in several series but not strongly imbricated, loose to subappressed, bristle tipped; ray florets 20–35, ligule white or rarely pinkish, 7–10 mm long; disk florets typically more than 20. Achenes brown or pale purple, 2 + mm long, appressed-puberulent; pappus of numerous capillary bristles, whitish, ± equaling the disk corollas. (n = 10, with a ploidy-series). Aug–Oct. Damp to drying open sites in prairies & plains; nw 1/4 GP, from nw NE to cen ND & w; (GP to AK, s through the cordillera to w NM & AZ).

This species is a member of the *Multiflori* complex; cf. the discussion under *A. ericoides.*

11. **Aster fendleri** A. Gray, Fendler's aster. Suffrutescent perennial 1–4 dm tall, arising from a thick, branching woody caudex. Stems often several, branching from near the base, glandular, hispidulous or sometimes glabrate, especially below. Leaves mostly linear, (1)2–4(5) cm long and (1)2–4 mm wide, sessile but not at all clasping, margin coarsely bristle-ciliate with the bristles ca 1 mm apart, tufts of reduced leaves sometimes present in the axils of the principal leaves; uppermost leaves reduced to mere bracts. Inflorescence paniculate with 3–20 heads per main stem; involucre 5–8 mm tall and 5–12 mm wide, glandular-hispid;

involucral bracts loosely imbricate in several series, the inner ones appressed but the outer ones somewhat spreading to subsquarrose; ray florets numerous, (± 18), ligule bluish-purple, 8–10 mm long; disk florets with corolla reddish-purple. Achenes 1–2 mm long, flattish, pubescent; pappus of abundant capillary bristles, 3–5 mm long, yellow-white, frequently purplish toward the base and sometimes entirely purple. Jul–Oct. Open, rocky sites, especially pastures & breaks; sw 1/5 GP, s-cen NE s through w 1/2 KS to w OK & w; (GP to n TX, NM).

12. *Aster hesperius* A. Gray, panicled aster. Perennial (3)5–10(12 +) dm tall, arising from long rhizomes. Stems often branching above, pubescent in lines decurrent from the leaf bases. Leaves glabrous or somewhat scabrous, entire or sometimes weakly toothed, sessile or the lowermost petiolate and early deciduous, persistent cauline leaves mostly linear or lanceolate, (3.5)5–10(13) cm long and 2.5–20 mm wide. Inflorescence loose, paniculate with ample floral bracts; involucre 5–8 mm tall; involucral bracts subequal or but weakly imbricated; ray florets 20–40(50), ligule blue or sometimes white, 6–10(12) mm long; disk florets with corolla usually violet-blue, corolla lobes constituting less than 1/2 the corolla limb. Achenes brownish, 1 + mm long, pubescent; pappus of numerous white or tawny capillary bristles. (n = 24, 32) Aug–Oct. Open stream banks & ditches; GP from cen KS n to w MN, ND, & WY; absent from both se & sw GP; (WI, MO, & NM w to Alta., ID, & s CA). *A. coerulescens* DC. (in part) — Stevens (misapplied).

Our materials belong to var. *hesperius*. the var. *laetivirens* (Greene) Cronq. occurs in the western montane regions, and has broader, more prominently green-tipped involucral bracts, and usually white ray florets. *Aster hesperius* intergrades with *A. simplex* (q.v.) and morphologically intermediate specimens are to be expected.

13. *Aster junciformis* Rydb. Slender, erect perennial 1.5–7 dm tall, arising from a long thin rhizome, rarely more than 2 mm thick. Stem often reddish and grooved, especially below, glabrous or glabrate below, strigose-pubescent above, mostly in lines decurrent from the leaf bases. Lowermost leaves petiolate but early deciduous; persistent cauline leaves linear or narrowly elliptic to narrowly lanceolate, (1)2–7(10 +) cm long and 1.5–5 + mm wide, entire but margins scabrous, attenuate to acuminate, sessile and often partially clasping, underside with veins forming a distinct reticulum, midrib somewhat puberulent; uppermost leaves reduced to bracts. Inflorescence of few to several heads in an open, corymbiform cluster; involucre 5–7 mm tall and ca 10 mm wide, glabrous; involcural bracts imbricate in 3–5 series, often with purple or roseate tips or margins; ray florets 25–35(50), ligule white to pink or lavender, 7–11 mm long; disk florets with corolla yellowish-brown, corolla lobes constituting ca 1/4 the limb. Achenes yellowish, ca 2 mm long, ribbed, glabrous or lightly pubescent; pappus of numerous white capillary bristles, 2–6 mm long. (n = 16, 27?) Aug–Sep. Cold bogs & swampy sites; n GP, cen NE & n, especially in ND; (Que. & NJ to GP, AK, & CO).

14. *Aster laevis* L., smooth blue aster. Perennial 3–10 dm tall, arising from a stout short rhizome or branching caudex or rarely with a few creeping rhizomes. Herbage glabrous and often glaucous, or sometimes with obscure puberulent lines decurrent from the nodes in the inforescence; lower stem sometimes reddish. Lowermost leaves petiolate and early deciduous; prominent cauline leaves thickish and firm (when fresh), variable in size and shape, broadly ovate to nearly linear, up to 10 cm long and 2 cm wide, but mostly smaller, margin entire to toothed, sessile and at least some prominently auriculate-clasping; uppermost leaves reduced to clasping bracts. Heads several to numerous in an open ± paniculate cluster; involucre 5–7 + mm tall; involucral bracts obtuse, acute or weakly acuminate, imbricate in several series, or obscurely so and apparently subequal, the green tips elongate to short-rhombic or diamond shaped; ray florets 15–25, ligule blue or purple, rarely white, 8–15 mm long. Achene glabrous or glabrate; pappus of capillary bristles,

mostly reddish. (n = 24) Aug–Oct. Open dry or drying sites; n 1/2 GP & s through e 1/3 KS; (Que. to GA, w to Yukon, B.C., & NM).

This species has two varieties that intergrade through the e GP. Var. *laevis* has the involucral bracts clearly imbricate in several series, with the green tip short, broad and rhombic to diamond-shaped. It is the phase of the eastern deciduous woodlands. Much of our material is referable to the essentially nw GP and Cordilleran var. *geyeri* A. Gray, which has subequal and poorly imbricated involucral bracts, with long, rather narrow green tips [*A. geyeri* Howell — Rydberg] (A. Gray). Populations that are not readily referable to either entity are to be expected in the e GP.

15. ***Aster lateriflorus*** (L.) Britt., white woodland aster. Perennial 3–12(15) dm tall arising from a branching caudex or a short, stout rhizome. Stems usually spreading or arching upward, densely pilose or curly villous to glabrate. Basal and lower cauline leaves petiolate and early deciduous; persistent cauline leaves lanceolate or lance-elliptic to linear, the larger ones 5–15 cm long and up to 3 cm wide, entire to serrate, upper surface scabrous or glabrate, lower surface mostly glabrous except for the prominently puberulent midrib; uppermost leaves greatly reduced. Inflorescence of numerous heads, usually secund on lateral branches of a loose panicle; involucre 4–6(7) mm tall, glabrous; involucral bracts imbricate in 3 or 4 series, obtuse or acute, with prominent oblanceolate green or purplish tips; ray florets 9–15, ligule white or occasionally purplish, (3)4–7 mm long; disk florets with corolla 3–5 mm long, corolla lobes prominently recurved, constituting 1/2–3/4 of the limb. Achenes somewhat pubescent, few-nerved; pappus of numerous capillary bristles. (n = 8, 16, 24) Sep–Oct. Infrequent in open wooded sites, especially damp or drying stream banks; e edge of GP from MN to se KS; (Que. to FL, w to GP & e TX). *A. hirsuticaulis* Lindl. — Rydberg.

This is a complicated and frequent species in the chiefly wooded regions of e N. Amer., where several poorly defined varieties have been recognized. Our materials are referable to var. *lateriflorus*.

16. ***Aster linariifolius*** L. Wiry-stemmed perennial (1)2–4(6) dm tall, arising from a short caudex or sometimes from rhizomes. Stem puberulent, becoming tomentose-puberulent upward. Leaves abundant, ± rigid, linear, with a single prominent nerve, scabrous, 1–4 cm long and 0.5–3(4) mm wide, margin entire but scabrous-ciliate, sessile; lowermost leaves similar but early deciduous. Inflorescence an open, loose corymbiform to subracemose cluster of few to numerous heads but rarely more than 30; involucre 6–9 mm tall; involucral bracts strongly imbricated in several series, chartaceous, greenish toward the tip, finely scabrous, the inner ones with ciliate margins and purplish toward the tip; ray florets 6–18, ligule violet or rarely white, 5–11 mm long; disk florets with corolla violet. Achenes long-pubescent, brown; pappus double, the inner series of firm, elongate, uniformly thick bristles, tawny or brownish, ca 5 mm long, the outer series short, 1 mm long or less. (n = 9) Sep–Oct. Drying open sites in wooded areas; extreme se GP; KS: Cherokee; (Que. to FL, w to WI, e GP, & TX). *Ionactis linariifolia* (L.) Greene — Rydberg.

17. ***Aster lucidulus*** (A. Gray) Wieg. Perennial 4–25 dm tall, arising from long, creeping rhizomes. Stem ± ribbed, upper stem pilose, especially in lines decurrent from the leaf bases, but glabrous or nearly so toward the inflorescence. Lowermost leaves early deciduous; prominent cauline leaves lanceolate to ovate-lanceolate, entire or weakly serrate, sessile and distinctly clasping, 4–7 cm long and 1–2.5 cm wide, upper surface glabrous and sometimes shiny, lower leaf surface glabrate but spreading-pubescent on the midrib; uppermost leaves crowded toward the inflorescence and not greatly reduced. Inflorescence an open, paniculate cluster with (few) 60–100 + heads; involucre 6–12 mm tall; involucral bracts loosely imbricate, at least the inner ones attenuate or long-acuminate, whitish toward the base but with elongate-rhombic green tips, margin ciliate; ray florets 30–40, ligule blue to lavender or sometimes white, 10–14 mm long; disk florets with corolla yellow. Achenes

pubescent or glabrate, ca 1.5 mm long; pappus of numerous capillary bristles, 7-8 mm long. Aug-Oct. Wet, open sites; infrequent in n & e GP, IA: Harrison; MN: Lincoln, Otter Tail, Pope, Red Lake; ND: Bottineau, Richland, Rolette; (Que. to WV & NC, w to GP, MO?). *A. puniceus* var. *firmus* (Nees) T. & G.—Fernald.

This species and *A. puniceus* together form a variable, intergrading complex. *Aster lucidulus* differs from *A. puniceus* by having long rhizomes and by its gross aspect of leaves crowded toward the inflorescence.

18. *Aster novae-angliae* L., New England aster. Perennial (3)5-12 dm tall, arising from a branching caudex or stout rhizome, rarely with creeping rhizomes. Stems often clustered, hirsute to hispid, glandular, especially upward. Lowermost leaves early deciduous; persistent cauline leaves lanceolate to oblong, entire, acute, 2-10 cm long and 1-2 cm wide, strigose on both surfaces but softly so beneath, auriculate-clasping at the base; uppermost leaves reduced to mere clasping bracts. Inflorescence a paniculate cluster with numerous to many heads, mostly 30-50; involucre and peduncles densely glandular but with scant pubescence; involucre 6-10 mm tall; involucral bracts loosely imbricated in 2 or 3 series, appressed but the tips sometimes weakly spreading, chartaceous below, the tip green and somewhat purplish-tinged; ray florets ca 50(100), ligule reddish-purple or rarely white, 1-2 cm long; disk florets with corolla reddish-purple and about as long as the pappus. Achenes sericeous-pubescent, ca 2 mm long; pappus of numerous capillary bristles, ca 5 mm long. (n = 5) Sep-Oct. Moist or drying sandy areas, stream banks, & roadsides; e 1/3 GP but scattered w to cen NE, BH, & cen ND; (Que. & VT to SC, w to GP, & Rocky Mts. of WY, CO, & n NM; apparently introduced elsewhere as an escape from cultivation).

The hybrid with *A. erocoides* (q.v.) is widespread and persistent, and has been called *A. amethystinus* Nutt. Both parent species are highly variable, and the hybrid is morphologically intermediate; cf. Wetmore & Delisle, Amer. J. Bot. 26: 1-12. 1939.

19. *Aster oblongifolius* Nutt., aromatic aster. Perennial 1.5-4(7) dm tall, arising from creeping rhizomes or a short caudex. Stems usually much branched, variously hispid-pubescent but rarely densely so, glandular, especially upward. Lowermost leaves early deciduous; persistent cauline leaves oblong to broadly linear, 1-4(8) cm long and 4-6(15) mm wide, sessile and obscurely to evidently auriculate-clasping, margin entire but strigose, sparsely pubescent or glabrous elsewhere, usually minutely glandular; uppermost leaves reduced to mere bracts. Inflorescence paniculate to distinctly corymbiform with few to many heads crowded at the ends of the branchlets; involucre 4-6 mm tall and 5-10 + mm wide; involucral bracts glandular, weakly imbricated in several series, tips long, green, loose or spreading; ray florets ca 30, ligule bluish-purple or rarely white, 7-10 mm long; disk florets with corolla reddish-purple, slightly exceeding the pappus. Achenes sericeous to sparsely so, flattish, 1-2 mm long; pappus of numerous yellow-white capillary bristles, ca 5 mm long. (n = 10) Sep-Oct. Rocky or sandy open sites, often in prairie & pasture hillsides; throughout GP but rare in sw 1/4; (PA to AL, w to GP & NM). *A. kumleinii* Fries—Rydberg.

Our materials belong to var. *oblongifolius*. The taller, less branched, and more pubescent phase of the wooded regions east of the GP is recognized as var. *angustatus* Shinners.

20. *Aster ontarionis* Wieg., Missouri aster. Perennial 5-8(10) dm tall, arising from long, spreading rhizomes. Stem ± evenly spreading-puberulent, especially upward, sometimes glabrate below. Lowermost leaves petiolate and early deciduous; persistent cauline leaves narrowly oblong or lanceolate to broadly linear, (2)4-8 + cm long and 5-25 mm wide, sessile or nearly so, underside variously puberulent to villous, especially along the midrib, rarely subglabrate, upper surface scabrous or glabrate, margin entire to serrate; upper leaves reduced and becoming bractlike. Inflorescence an ample open paniculate cluster, with heads numerous and ± secund on lateral branches; involucre 4-6 mm tall; involucral bracts

glabrous, imbricated in 2 or 3 series, obtuse or acute with an evident green tip; ray florets 10–20, ligule white or light purple to blue, 3–5 mm long; disk florets with corolla violet, corolla lobes prominent, constituting 1/2–3/4 of the limb. Achene lightly pubescent, ca 1 mm long; pappus of numerous capillary bristles, whitish, 3–3.5 mm long. (n = 16) Sep–Oct. Stream banks & flood plains in wooded areas; e 1/5 GP from w MN & ne SD to se KS; (NY & MS w to GP & e OK). *A. missouriensis* Britt.—Rydberg.

This species is similar in aspect to *A. lateriflorus* and *A. simplex*, but it is generally separable from them by the characters in the key.

21. *Aster oolentangiensis* Ridd., azure aster. Suffrutescent perennial, (2)3–10 dm tall, arising from a short caudex or branching rhizome. Stems erect, glabrous or puberulent to hispid. Lowermost leaves cordate or subcordate, lanceolate to ovate, blade 3–10 cm long and up to 7 cm wide, acute or occasionally acuminate, entire or weakly serrate, upper surface scabrous-hispid, lower surface ± pilose, petiole weakly winged, 2–6(13) cm long; lower cauline leaves shallowly cordate or abruptly contracted to tapering at the base, smaller than the basal leaves; middle and upper cauline leaves distinctly smaller, narrower, with shorter petioles and usually tapering at the base; uppermost leaves reduced to sessile bracts. Inflorescence open, paniculate; heads (20)50–100(150), involucre 4.5–8 mm tall; involucral bracts imbricated in 3–5 series, lanceolate, appressed, margin ciliate, white toward the base, with broadly rhombic green tip; ray florets 13–18(20), ligule blue to blue-violet, 6–8(10) mm long; disk florets 15–30, corolla yellow. Achenes glabrous or nearly so, weakly nerved, ca 1 mm long; pappus a single series of capillary bristles, 3–4 mm long, white. (n = 18) Sep–Oct. Prairies & open, dry woods, roadsides; e 1/5 GP from w IA & cen NE s, scattered in e SD & w MN; (NY to AL, w to GP & TX). *A. azureus* Lindl.—Rydberg.

The nomenclature of this species is discussed by Jones, Bull. Torrey Bot. Club 110: 39–42. 1983.

22. *Aster paludosus* Ait. subsp. **hemisphericus** (Alex.) Cronq., single-stemmed bog aster. Perennial 2–6 dm tall, arising from a creeping and sometimes ligneous rhizome. Stem erect, mostly simple, glabrous or nearly so, or at most short-pubescent just beneath the heads. Lowermost leaves early withering and deciduous; persistent cauline leaves firm, glabrous except for scabrous-ciliate margins, entire, linear, 5–11 cm long and mostly 4–6 mm wide, obscurely veined except for the midrib, sessile or nearly so; uppermost leaves reduced. Inflorescence an elongate racemose cluster with heads few, mostly 3–8; involucre 8–12 mm tall, hemispherical, glabrous; involucral bracts imbricated in several series, relatively broad, spreading and at least some loose or subsquarrose; ray florets 20–30+, ligule blue or violet, 1–2 cm long; disk florets with corolla yellowish. Achenes glabrous or strigose; pappus of numerous rigid and ± clavate bristles, tawny or reddish. (n = 9) Jul–Oct. Open, damp or drying sandy sites, often on roadsides; se GP; KS: Cherokee, Crawford, Labette; adj. MO & ne OK; (NC to FL, w to GP & e TX). *A. hemisphericus* Alex.—Fernald.

Our materials are distinguished from subsp. *paludosus* in having a racemose inflorescence and an essentially prairie habitat, while subsp. *paludosus* has a corymbiform inflorescence and occurs in swampy lowlands of se U.S. A case can be made for treating our materials as a distinct species (*A. hemisphericus* Alex.), but it is difficult to separate to two taxa with much consistency except with reference to geography.

23. *Aster pansus* (Blake) Cronq. Cespitose cluster-plant 3–10(20) dm tall, with stems arising from a caudexlike corm, sometimes interconnected by slender horizontal rhizomes. Herbage ± evenly light-pubescent to glabrate. Basal and lower to midcauline leaves mostly early-decidous; persistent cauline leaves linear or but slightly broader, 1–5 cm long and rarely exceeding 3 mm wide, reduced upward to bracts. Inflorescence of numerous small heads; mostly secund on recurving lateral branches; involucre 2.5–4.5 mm tall, ± broadly campanulate; involucral bracts strongly imbricated, bristle tipped; ray florets 10–18, ligule white or rarely pink, less than 6 mm long; disk florets 10–20 but mostly fewer than 15, corolla yellowish to bluish. Achenes pale purple, 1–2 + mm long, appressed-puberulent; pappus of white or whitish capillary bristles. (n = 5) Aug–Sep. Open sandy or loose soils in prairies & plains, sometimes in alkali flats; nw GP from ne MN to e CO, n & w; (GP to n Alta., B.C., UT, & OR). *A. stricticaulis* (T. & G.) Rydb.—Rydberg.

This species is a member of the *Multiflori* complex; cf. the discussion under *A. ericoides*.

24. *Aster parviceps* (Burgess) Mack. & Bush, small white aster. Slender perennial 2–7 dm tall, arising from a short caudex. Stem single, branching in the upper 1/3, somewhat reddish, ± hirsute to glabrous or lightly pubescent in lines decurrent from the nodes. Lowermost leaves oblanceolate and petiolate but mostly early deciduous; persistent cauline leaves sessile or subsessile, linear to lanceolate linear, up to 10 cm long but mostly shorter, 2–6 mm wide, glabrous or hirsute as is the stem, entire to weakly serrate; uppermost leaves reduced and bractlike, numerous, especially on the branches. Inflorescence an ample, paniculate or rarely corymbiform cluster of many heads, often ± secund and crowded on the branches, peduncles long or short, often with numerous bractlike leaves; involucre 3–5 mm tall, ca 3 mm wide, glabrous, slender-turbinate; involucral bracts imbricated in 3 or 4 series, narrow, at least some of them with distinctive, marginally inrolled green tips; ray florets 10–16(18), ligule white, 4–7 mm long; disk florets fewer than the ray florets, mostly 6–12, corolla light violet to yellow. Achenes glabrous or sparsely pubescent; pappus of white capillary bristles, 3–4 mm long. Sep–Oct. Open, dry or drying wooded sites or prairies; se 1/4 KS, adj. MO; (IL & IA to se GP).

25. *Aster patens* Ait. var. ***patentissimus*** (Lindl.) T. & G. Perennial 2–5(10 +) dm tall, arising from a short, fibrous rooted caudex, sometimes also rhizomatous. Stems single, slender, and seemingly brittle, ± loosely pubescent. Basal leaves early deciduous; persistent cauline leaves ovate to oblong, entire, 2–6(7) cm long and up to 2 cm wide, scabrous to pubescent, at least on the underside, sessile and conspicuously cordate-clasping at the base. Inflorescence a widely branching paniculate cluster of single-headed peduncles, often with numerous reduced, bractlike leaves; involucre 5–9 mm tall, pubescent and sometimes glandular; involucral bracts distinctly imbricated in several series, rounded to acute, with spreading green tips; ray florets mostly 20–25, ligule bluish or purplish, or sometimes pink, ca 10 mm long. Achenes short-sericeous; pappus of numerous capillary bristles. (n = 5, 10, 20) Aug–Oct. Rocky soils in open woodlands & fields; se 1/4 KS, adj. MO, e OK; (MA & NH to FL, w to GP & TX). *A. patentissimus* Lindl., *A. tenuicaulis* (Mohr) Burgess—Rydberg; *A. patens* var. *gracilis* Hook.—Fernald.

26. *Aster pauciflorus* Nutt., few-flowered aster. Glabrous perennial 2–4(6) dm tall, arising from a creeping rhizome. Leaves linear to linear-lanceolate, the prominent ones 3–4(9) cm long and 3–4(5) mm wide, glabrous, entire, somewhat firm and fleshy, the uppermost reduced and bractlike. Inflorescence an open, corymbiform cluster of (1)3–8 heads; involucre 4–7 mm tall, glandular; involucral bracts imbricated in 2 or 3 series, lanceolate; ray florets 15–25, ligule blue or purple to light pink, 5–7 mm long; disk florets with corolla yellowish or white. Achenes pubescent, ca 2 mm long; pappus of numerous white bristles, 3–6 mm long. (n = 9) Jul–Sep. Infrequent in dry or drying open sites, especially alkaline areas; w ND & w SD; (s Sask. & e GP, s & w through CO to AZ?).

27. *Aster pilosus* Willd. Perennial (0.5)1–9(15 +) dm tall, arising from a stout, fibrous-rooted caudex. Stem simple below but branching upward, hirsute-hispid to glabrate. Lower leaves often early deciduous but sometimes persistent and forming a basal rosette, oblanceolate to ovate, petiolate, usually less than 10 cm long overall, ± hispid-pubescent; persistent cauline leaves reduced, linear to linear-subulate, 1–2 cm long, stiff, sessile or subsessile, entire or obscurely toothed, apiculate-aristate. Inflorescence an open paniculate cluster, often with numerous pedunculate heads, secund on spreading lateral branches; involucre broadly urn-shaped to campanulate, 3.5-5(6) mm tall and about as broad, glabrous; involucral bracts imbricated in several series, at least some of them with distinctive, subulate, marginally inrolled green tips; ray florets 15-20(35), ligule white or rarely pinkish, 5-10 mm long; disk florets more numerous than the rays, mostly ca 30-35, corolla yellowish. Achenes pubescent, ca 1 mm long; pappus of numerous white capillary bristles. (n = 16, 24) Sep-Oct. Open, dryish sandy sites; se 1/4 GP, scattered n & w to ne SD & cen NE; (Que. & ME to FL, w to GP & LA). *A. glabellus* Nees—Rydberg.

28. *Aster praealtus* Poir., willowleaf aster. Perennial 5-10(12 +) dm tall, arising from long, creeping rhizomes. Stem glabrate to persistently pubescent, branching above. Leaves thickish and firm; prominent cauline leaves linear-lanceolate, mostly 2-8 cm long and 4-8 mm wide but sometimes larger, entire to weakly dentate, sessile and sometimes obscurely clasping, glabrate to scabrous on the upper surface, glabrous to puberulent beneath, veins on the lower surface forming a distinctive reticulum with isodiametirc areolae. Inflorescence an open, ± leafy, paniculate cluster, heads numerous, 60-100 + in well-developed individuals; involucre 5-7(8) mm tall; involucral bracts imbricated in several series, often with a reddish-green tip; ray florets 6-15, ligule blue or purple, sometimes whitish, 7-10 mm long; disk florets with corolla yellowish, corolla lobes constituting less than 1/4 of the limb. Achenes pubescent, 1-2 mm long; pappus of numerous white capillary bristles. (n = 16) Sep-Oct. Low, damp or drying sites.

Two vars. are discernible in the GP:

28a. var. *nebraskensis* (Britt.) Wieg. Stem ± uniformly pubescent, especially upward. Lower surfaces of the leaves persistently spreading-puberulent. Cen GP, mostly n of the range of var. *praealtus*; apparently endemic, but perhaps scattered se to AR. *A. nebraskensis* Britt.—Rydberg.

28b. var. *praealtus*. Stem glabrate to pubescent upward, chiefly in lines of hairs decurrent from the leaf-bases. Leaves glabrate or but scantily scabrous-puberulent. Se GP from se NE to e OK; (MI to GA, w to GP & TX). *A. salicifolius* Lam. (not Ait.), *A. fluviatilis* Osterhout—Rydberg; *A. coerulescens* DC.—Gates.

29. *Aster pubentior* Cronq. Rhizomatous perennial (3)5-15 + dm tall. Herbage variously glabrate or scabrous to puberulent, often ± persistently puberulent on the undersides of the leaves. Prominent leaves flexible, linear-elliptic, 4-8 cm long and 1-2 cm wide, entire, sessile or subsessile. Inflorescence open corymbiform and tending to be flat-topped, with numerous heads; involucre 3-5 mm tall, usually ± puberulent; involucral bracts imbricated in 2 or 3 series; ray florets 5-10(14), ligule white, 5-8 mm long; disk florets with corolla yellowish. Achenes ca 1 mm long, pubescent; pappus in 2 distinct series, the inner of firm whitish bristles, ± thickened toward the apex, the outer of short hairs less than 1 mm long. (n = 9) Aug-Oct. Low damp sites, often in wooded areas; w MN, ne 1/2 ND; IA: Guthrie; NE: Cherry; (MI to e GP & Alta.). *A. umbellatus* var. *pubens* A. Gray—Fernald; *Doellingeria pubens* (A. Gray) Rydb.—Rydberg.

This is a distinctive entity in our region, but eastward in the area of the Great Lakes, it intergrades with and is replaced by *A. umbellatus* Mill., a species with consistently larger heads.

30. *Aster puniceus* L., swamp aster, bog aster. Herbaceous or weakly suffrutescent perennial 4-25 dm tall, arising from a stout, short rhizome or a caudex, sometimes with a few short stolons. Stems simple but much branched upward, glabrate to spreading-hairy, but consistently pubescent toward the inflorescence. Lower leaves petiolate and early deciduous; persistent cauline leaves ovate to narrowly lanceolate or narrow-rhombic, sessile, auriculate-clasping, 2-16 cm long and 1-4 cm wide, ± scabrous on the upper surface, glabrous to weakly pilose beneath. Inflorescence corymbiform, open, leafy, heads 30-50; involucre 6-12

mm tall; involucral bracts subequal or but weakly imbricated in 2 or 3 series, slender, loose, at least the inner ones acuminate-attenuate, often some of the outer ones ± leafy; ray florets 35–50, ligule dark blue or lilac to pinkish, 7–14 mm long; disk florets numerous, corolla yellowish. Achene glabrate or lightly pubescent, ca 1.5 mm long; pappus of abundant white capillary bristles, 5–6 mm long. (n = 8) Aug–Sep. Damp thickets or open swampy sites; e ND, w MN, scattered s to w IA & e NE, also BH; (Que. & Newf. to GA, w to n GP).

31. *Aster sagittifolius* Willd., arrow-leaved aster. Perennial 4–12(20) dm tall, arising from a stout, branching caudex or short rhizome. Stem mostly glabrous or glabrate, or the upper portion pilose, especially in lines decurrent from the nodes. Basal leaves ± persistent, blade ovate to lanceolate-ovate, cordate, 3.5–10 + cm long and 1–7 cm wide, upper surface glabrous or weakly scabrous, lower surface pilose-villous, margin shallow-serrate, petiole long and ± winged; cauline leaves reduced, becoming less prominently cordate and the petiole more prominently winged; uppermost leaves sessile and bractlike. Inflorescence an elongate, paniculate cluster with heads numerous, mostly 70–150, on bracteate branches; peduncles mostly less than 1 cm long and the heads appearing crowded; involucre 4–6 mm tall; involucral bracts imbricated in 3 or 4 series, lanceolate, margin ciliate, with a long, slender narrow greenish tip, or often minutely reddish at the apex; ray florets 10–20, ligule blue or lilac to whitish, 5–10 + mm long; disk florets with corolla whitish to light yellow. Achenes glabrous or glabrate; pappus of abundant white capillary bristles. (n = 9, 18) Jul–Sep. Wooded stream banks, edges of thickets, & sometimes in open, rocky sites; ne 1/3 GP, mostly sw IA, se NE & n; (VT to FL, w to GP).

This species hybridizes with *A. drummondii* wherever the ranges overlap.

32. *Aster sericeus* Vent., silky aster. Perennial 3–6 dm tall, arising from a short, branching, fibrous-rooted caudex. Stems clustered, ± brittle and wiry, glabrate below but thinly and unevenly sericeous above. Leaves entire and persistently appressed silky-pubescent (sericeous) on both surfaces; basal leaves oblanceolate, petiolate and early deciduous; persistent cauline leaves narrowly elliptic to oblong, mostly 1–2 cm long and less than 1 cm wide, sessile, somewhat clustered toward the ends of the branches. Inflorescence a branching cluster of numerous heads; involucre 6–10 mm tall, sericeous; involucral bracts subequal or weakly imbricated but apparently in several series, the tips loose or spreading; ray florets 14–20, ligule purple-violet to pinkish, or rarely white, 12–16 mm long; disk florets with corolla yellowish-brown or rose-tinged. Achenes glabrous, 2–3 mm long; pappus of white capillary bristles 5–7 mm long, becoming golden yellow in age. (n = 5, 10) Sep–Oct. Open, dry or drying upland sites, sometimes in open woods; e 1/3 GP, & BH; (MI to TX & GP; sporadic in relict prairies in TN).

33. *Aster sibiricus* L. var. *meritus* (A. Nels.) Raup, arctic aster. Rhizomatous perennial 1–2.5 dm tall. Stems 1–several, often tufted, simple or little branched, purplish, villous with loose, ascending hairs. Leaves lanceolate to oblong-spatulate (1)3–4(7) cm long, 0.5–2 cm wide, entire to remotely serrate, upper surface glabrate, lower surface glabrate to villous pubescent, sessile to subpetiolate. Heads solitary or few and cymose or rarely more numerous and clustered in cymose panicles; involucre 6–9 mm tall and 1–1.5 cm wide; involucral bracts loosely imbricated in 2 or 3 series, appressed or somewhat squarrose, villous, green, the inner ones purple-margined; ray florets 12–24, ligule purple to violet, 7–12 mm long; disk florets with corolla 5–7 mm long, the limb purplish. Achenes puberulent, ± 2 mm long, brownish; pappus 5–7 mm long, of numerous intermixed capillary bristles of varying lengths. (n = 9) Jun–Aug. Open rocky woods at high elevations; BH; barely outside of our range in ne WY; (SD to OR, s Can.). *A. meritus* A. Nels.—Rydberg.

Var. *sibiricus* has fewer and larger heads with more obviously squarrose involucral bracts, and it is circumboreal. Our plant has been reported as a triploid, with somatic cells having 27 chromosomes, cf. Huziwara, Amer. J. Bot. 49: 116–119. 1962.

34. Aster simplex Willd., panicled aster. Rhizomatous perennial 6–15(20) dm tall. Stems 1–several, erect, stout, branched upward, villous-pubescent in lines decurrent from the upper nodes. Leaves mostly rather thin, linear to lanceolate, sessile or subsessile, variable in size, the chief cauline ones 7–15 cm long and 0.3–3.5 cm wide, entire to serrate, glabrous on both sides or sometimes scabrous on the upper surface, veins on the underside forming an indistinct reticulum, with the areolae obscure or longer than wide and not at all isodiametric; uppermost leaves reduced. Inflorescence a paniculate, elongate leafy cluster with numerous heads; involucre 3–5.5 mm tall; involucral bracts obviously imbricated, appressed, glabrous and frequently with ciliolate margins, whitish but with a green midvein and apex; ray florets 25–35, ligule white or sometimes pink or lavender, 4–7 mm long; disk florets with corolla lobes purplish and constituting less than 1/2 of the limb. Achenes pubescent, ca 1 mm long; pappus of numerous white capillary bristles. (n = 8, with a long polyploid series) Aug–Oct. Damp or drying meadows & other low sites.

This is a highly variable species, and it intergrades with its relatives in the GP, especially *A. hesperius* and *A. praealtus*. A rationale for treating this species is presented by Boivin, Phytologia 23(1): 35–37. 1972. The name *A. lanceolatus* Willd. has been suggested as correct, but the matter is unresolved and here the traditional usage of *A. simplex* is employed. Three vars. are recognized in the GP:

```
1  Heads small, involucre mostly less than 4 mm tall; ligules often lavender; pappus ca 3–3.5
     mm long ............................................................................................. 34a. var. interior
1  Heads larger, involucre 4–5.5 mm tall; ligules white or rarely pink or light blue; pappus ca
     3.5–7 mm long.
   2  Cauline leaves 12–24 × longer than wide, mostly 3–12 mm wide; involucral bracts narrow-
        ly linear ......................................................................................... 34b. var. ramosissimus
   2  Cauline leaves seldom more than 11 × longer than wide, mostly 15–35 mm wide; in-
        volucral bracts linear to linear-oblong ............................................................. 34c. var. simplex
```

34a. var. *interior* (Wieg.) Cronq. Leaves firm. Involucre short, usually 3–4 mm tall. Often on calcareous, rocky sites; e GP from ne NE & adj. IA s to se KS & MO; (NY, s Ont. to OH, w to GP). *A. paniculatus* Lam. (in part)—Rydberg.

34b. var. *ramosissimus* (T. & G.) Cronq. Upper stems often much branched. Leaves ± thin and flexible, linear or linear-lanceolate. Involucre 4 + mm tall. Throughout GP except sw 1/5; (Que. to PA, w to GP). *A. acutidens* Smyth—Rydberg.

34c. var. *simplex*. Leaves ± thin and flexible, oblong or oblong-lanceolate. Involucre 4 + mm tall. E 1/3 GP; (ne Can. to NC, w to Sask. & GP). *A. paniculatus* Lam. (in part), *A. jacobaeus* Lunnell (?)—Rydberg; *A. coerulescens* DC. (in part)—Stevens.

35. Aster spinosus Benth., Mexican devil-weed. Broomlike perennial 5–30 dm tall, arising from deep, woody spreading rhizomes. Stems much branched, glabrous, with some of the lateral branches reduced to spines up to 1.5 cm long. Leaves few, scattered, sessile, linear or linear-spatulate, 1–2.5 cm long and up to 4 mm wide, ± coriaceous; uppermost leaves reduced to scales. Inflorescence an open cluster of few to numerous heads; involucre 4–6 mm tall; involucral bracts imbricated in 3–5 series, lanceolate, appressed, greenish to straw colored with scarious margins; ray florets 15–30, ligule white, 3–4 mm long. Achenes glabrous, 1–1.5 mm long, brownish; pappus of capillary bristles, 4–6 mm long, white. (n = 9) Jul–Sep. Open weedy sites, roadsides; uncommon in extreme s GP; OK: Cleveland, Jackson, Payne; TX: Carson, Hemphill, Oldham; (s GP through TX to Mex., w to CA).

36. Aster subulatus Michx. var. *ligulatus* Shinners, saltmarsh aster. Glabrous, taprooted annual, 1–7(10 +) dm tall, with stems much branched upward. Leaves linear to linear-lanceolate, 1–10(18) cm long and 0.7–1.5(3) cm wide, entire or occasionally remotely serrate, sessile and scarcely clasping; upper leaves reduced. Inflorescence diffuse, paniculate; involucre 5–8 mm tall; involucral bracts imbricated in 3 or 4 series, linear-subulate, acuminate, appressed, greenish toward the apex, the tip reddish or purple; ray florets 15–30 +, ligule purple to pink or sometimes white, 3–4 mm long and often recurved-rolled at maturity; disk florets with corolla yellowish. Achenes appressed-puberulent, light brown,

2-3 mm long; pappus of abundant capillary hairs, 3-5 mm long. (n = 5, 10) Aug-Oct. Low damp or drying sites, especially in saline soils, often weedy; s 1/2 GP from se NE & cen KS to OK & TX; (along e coast from ME to FL, w to s GP & CA, s to C. & S. Amer., mostly in salt marshes). *A. exilis* Ell.—Rydberg.

37. *Aster turbinellus* Lindl., prairie aster. Perennial 3-10 dm tall, arising from a thick, woody branching caudex. Stems 1-several, erect, glabrate or sparingly spreading-pubescent, much branched with the upper branches becoming thin and subfiliform. Lowermost cauline leaves petiolate and early deciduous; persistent cauline leaves linear to linear-lanceolate, (1)4-10 cm long and mostly less than 2 cm wide, entire, glabrous but with the margin and midrib scabrous or scanty-pubescent, sessile or subsessile but not at all clasping; uppermost leaves reduced, subulate and bractlike. Inflorescence paniculate with heads obviously long-pedunculate; involucre glabrous, 7-12 mm tall; involucral bracts imbricated in several series, the outermost narrow, short, and grading into the bracts of the peduncle, the inner comparatively broader and longer, loose, ± squarrose, with greenish tips; ray florets ± 14(20), ligule blue to lavender or purple, 10-18 mm long; disk florets with corolla slightly exceeding the pappus. Achenes 1-1.5 mm long, tawny, appressed-puberulent; pappus of capillary bristles, 3-4 mm long, whitish to reddish-brown. (n = 48-50) Aug-Oct. Rocky open woods; extreme se GP; KS: Cherokee; adj. sw MO & OK; (IL to se GP, AR, & LA).

The colloquial name "prairie aster" seems inappropriate to residents of the GP.

13. ASTRANTHIUM Nutt., Western Daisy

1. *Astranthium integrifolium* (Michx.) Nutt. Slender, diffusely branched, erect or ascending, taprooted annual, 1-4 dm tall; sparsely pubescent, or densely so toward the base. Leaves alternate, simple, entire, the lower ones narrowly oblanceolate to spatulate, 2-7 + cm long, including the short, winged petiole, and up to 2 cm wide, the upper leaves smaller, linear to narrowly elliptic. Heads mostly long-pedunculate; involucre 5-6 mm tall and 8-10 mm wide; involucral bracts subequal, in 2 or 3 series, receptacle convex, naked; ray florets 8-22, pistillate and fertile, ligule 5-10 mm long, white, or blue to purple especially when dry; disk florets perfect and fetile, corolla yellow, style branches with short stigmatic lines and conspicuous, slender, externally papillose appendages. Achenes somewhat flattened, 2-nerved; pappus a minute ring or crown, set well in from the lateral margins of the achene. (n = 8) Apr-Jun. Open sandy sites in wooded areas or along roadsides; extreme se GP, se KS, & cen OK; (GA to s GP & TX; chiefly in the Ozarkian region). *Bellis integrifolia* Michx.—Rydberg. [tribe Astereae]

Reference: DeJong, D. C. D. 1965. A systematic study of the genus *Astranthium*. Publ. Mus. Michigan State Univ., Biol. Ser. 2: 429-528.

Our materials are referable to the weakly defined subsp. *ciliatum* (Raf.) DeJong. However, Correll & Johnston in their Manual of the Vascular Plants of Texas include this subsp. within their conception of var. *triflorum* (Raf.) Shinners. The utility of recognizing these infraspecific taxa is not readily apparent.

14. BACCHARIS L.

Dioecious shrubs or subshrubs, ours glabrous or sometimes the herbage punctate glandular; branches mostly striate-angled or sometimes subterete. Leaves linear to ovate, alternate, entire or weakly to coarsely serrate, 1- or 3-nerved, sessile or short-petiolate, usually reduced upwards. Inflorescence vaguely paniculate or often contracted into a weakly corymbose cluster; heads with either male or female disk florets only; male and female heads on separate plants. Pistillate involucres with involucral bracts in several series, the outer ones ovate to lanceolate, inner ones lanceolate to linear, obtuse to acuminate, usually with margins scarious, midrib prominent or obscure; receptacle pitted to smooth, naked, flat; corolla filiform with 5 distinct lobes or teeth at the apex, ochroleucous or light brown; achene 5- or 10-ribbed, glabrous or hispid; pappus of abundant bristles in 1–several series, equaling or exceeding the exserted style. Staminate involucre with outer involucral bracts ovate to lanceolate, inner ones lanceolate to linear, obtuse to acuminate, margin usually scarious; receptacle pitted to smooth, flat; corolla filiform below but gradually or abruptly enlarged above, funnelform, white to yellowish-brown, with 5 prominent lobes; ovary abortive; pappus of numerous bristles, somewhat plumose-tipped, tending to be barbed, not exceeding the length of the exserted style. [tribe Astereae]

Reference: Mahler, W. F., & U. T. Waterfall, 1964. *Baccharis* (Compositae) in Oklahoma, Texas and New Mexico. Southw. Naturalist 9: 189–202.

This is a group of several hundred species of tropical and subtropical America, and we are at the northern limit of its range. The field worker must collect carefully and note the population structure of the collections, because pistillate plants are necessary for accurate identification.

1 Pappus of pistillate florets in 1 or 2 series; achenes less than 2 mm tall, glabrous.
   2 Achenes 5-ribbed, ± 1 mm tall, pappus uniseriate .... 1. *B. glutinosa*
   2 Achenes 8- to 10-ribbed, 1.2–2 mm tall, pappus biseriate .... 2. *B. salicina*
1 Pappus of pistillate florets many-seriate and tinged reddish-brown; achenes 3–5 mm tall, ± glandular-scabrous.
   3 Pistillate involucre usually 9 mm tall or less, involucral bracts keeled and with the midrib expanded; well-developed leaves mostly more than 1 cm long .... 3. *B. texana*
   3 Pistillate involucre (9)10–12 mm tall, involucral bracts flat or but weakly keeled; most leaves less than 1 cm long .... 4. *B. wrightii*

1. ***Baccharis glutinosa*** (R. & P.) Pers. Shrub 1–3 m tall; branches glabrous and glutinous. Leaves punctate, sessile or weakly petiolate, lanceolate to narrowly elliptic, nearly entire to prominently serrate, 2–8(11) cm long and 1(2) cm wide, distinctly 3-nerved. Inflorescence a terminal corymb or a loose aggregation of several corymbs. Pistillate involucre hemispherical, 4–4.5 mm tall; involucral bracts stramineous, brownish-purple tipped, midrib distinct, margins scarious-erose; receptacle flat, smooth and naked; florets ca 50(+), corolla filiform, 2.0–2.3 mm long with 5 narrow linear lobes, style exserted, bifurcate; achenes ca 1 mm long, 5-ribbed, glabrous; pappus uniseriate, flaccid, 4–5 mm long. Staminate involucre campanulate, ca 4 mm tall; involucral bracts with margins scarious-erose; receptacle flat, weakly pitted, naked; florets 10–20, corolla 3–4 mm long, filiform but the upper 1/2 expanded-funnelform, the 5 lobes lanceolate, ca 1.2 mm long, style exserted, bifurcate; ovary abortive; pappus 3–4 mm long, plumose-tipped, crisped, not exceeding the corolla lobes. (n = 9) Jun–Aug. Along sandy, dry creek beds, barely entering our range in extreme sw GP; TX: Bailey; (sw GP, s & w through TX to CA).

2. ***Baccharis salicina*** T. & G., willow baccharis. Shrub 1–3 m tall; branches glabrous, striate-angled. Leaves sessile or nearly so, oblanceolate-oblong, serrate with forward-projecting distant teeth; blade 3–4(6) cm long and 0.5–1.0(1.5) cm wide, the larger ones tending to be 3-nerved. Pistillate involucres campanulate, ca 6(8) mm tall; involucral bracts spreading at maturity, reddish-brown at the tip, margin scarious, erose; receptacle flat, naked or weakly pitted; florets 25–30, corolla filiform, 3–4 mm long with 5 minute, linear lobes; style exserted, bifurcate; achenes 1.2–2.0 mm long, glabrous, with 8–10 ribs; pappus in 2 series, flaccid, up to 12 mm long, basally united into a ring. Staminate involucres hemispheric, ca 4 mm tall; involucral bracts with scarious, erose margins; receptacle flat, naked, pitted; corolla filiform but gradually funnelform upward, 3.3–4.3 mm long, with 5 linear lobes ca 1 mm long; style exserted, clavate. Jun–Aug. Open sandy flood plains & alluvial soil, especially in mildly saline habitats; sw 1/4 GP from cen KS s & w; (GPs through TX & eNM). *B. neglecta* Britt.—misapplied.

This species is something of a persistent weed in cen KS in the Arkansas R. flood plain and adj. sites of similar habitat.

3. **Baccharis texana** (T. & G.) A. Gray. Low shrub, to 6 dm tall, branching from the base; branches smooth, striate-angled. Leaves alternate, sessile, somewhat punctate, linear to narrowly lanceolate, principal leaves 2–4(5) cm long and 2–4 mm wide, 1-nerved, margin minutely undulate; upper leaves reduced. Inflorescence corymbose; pistillate involucres campanulate, 7–9(10) mm tall; involucral bracts keeled, the midrib expanded, margin narrowly scarious and usually ciliate; receptacle flat, naked, pitted; florets 20–30; corolla filiform, 3.5–4.0 mm long, truncate with 5 obscure teeth ca 0.2 mm long; style bifurcate; achenes 3–4.5 mm long, 5- or 6-ribbed; pappus in several series, tinged somewhat reddish-brown, antrorsely barbellate, 11–13 mm long. Staminate involucres campanulate, 6–7 mm tall; involucral bracts keeled, midrib expanded, margin scarious and usually ciliate; receptacle flat, naked, pitted; corolla tube filiform below but abruptly expanded about halfway up, ca 5 mm long, with 5 lanceolate lobes 1.5–2.0 mm long; style clavellate; ovary abortive; pappus usually plumose-tipped and crisped, about equaling the corolla. Aug–Sep. Apparently highly localized in open mesa country; w OK: Custer, Jackson; (s GP, s through TX & n Mex.).

4. **Baccharis wrightii** A. Gray. Suffruticose perennial 1–7 dm tall, slender and erect to short and branching from the base; branches striate-angled. Leaves alternate, sessile or nearly so, linear to lanceolate, or the lower ones oblanceolate to oblong, entire or nearly so, mostly less than 1 cm long and 1–2 mm wide, but a few larger leaves may be up to 2.5 cm long; upper leaves reduced and the uppermost becoming mere subulate bracts. Pistillate involucre hemispheric, (9)10–12 mm tall; involucral bracts with margins scarious, erose; receptacle flat, naked, smooth; corolla filiform, 3.7–4.7 mm long, with 5 short, linear lobes, 0.3 mm long or less, erose; style exserted beyond corolla, bifurcate; achenes (3)4(5) mm long, glandular, 5- to 10-ribbed; pappus in several series, abundant, antrorsely barbellate, reddish-brown, up to 15 mm long. Staminate involucre hemispheric, 8(9) mm long; involucral bracts with margin weakly serrate to entire, scarious; receptacle flat, naked, smooth; corolla filiform below but gradually to abruptly funnelform above, 4.6–5 mm long, with 5 lanceolate lobes 1–1.6 mm long; style clavellate, bifurcate when exserted; ovary abortive; pappus plumose-tipped, crisped. Aug–Sep. Localized in open, dry plains; sw GP from sw KS to w OK & n TX; (s GP s to n Mex., w to AZ).

15. BAHIA Lag.

1. **Bahia pedata** A. Gray. Annual herb 3–10+ dm tall, variously pubescent and glandular, often with numerous stipitate glands. Stems arising singly from a taproot, erect and branching above. Leaves alternate, often crowded at the base, palmately divided but each segment ± pinnatifid, the ultimate lobes obovate to oblong and 2–5 mm wide; uppermost leaves reduced and entire or parted into linear segments. Heads 1–4 at the ends of the branches; involucre 5–8 mm tall and up to 2 cm across in dried specimens; involucral bracts short stipitate-glandular, 13–17 in 2 series, with 1 or more reduced outer calyculate bracts, margins membranaceous, oblong-lanceolate, obtuse, 4–6 mm long and 1.5–2.5 mm wide, midrib weakly developed or absent; receptacle ± convex and alveolate (honeycombed); ray florets 10–15, pistillate and fertile, ligule 5–6+ mm long, yellow; disk florets 70–110, perfect and fertile, corolla 3–4+ mm long, style branches flattish, with conspicuous stigmatic lines along the upper margins. Achenes narrow, obpyramidal, 2.5–4+ mm long, 4-angled, dark; pappus of 10–15 scarious scales, 1–1.5 mm long. (n = 12) Jul–Oct. Open rocky sites in the grasslands; panhandle of TX & adj. NM; (s GP, s & w to n Mex.). [tribe Heliantheae]

This species has been treated by Ellison in Rhodora 66: 67 et seq. 1964; cf. note under *Picradeniopsis*.

16. BALSAMORHIZA Nutt., Balsam-root

1. **Balsamorhiza sagittata** (Pursh) Nutt. Scapose or subscapose perennial herb to 8 dm tall, silvery tomentose to subvelutinous. Stems arising from a branching crown, surmounting a thick taproot. Basal leaves petiolate, blade cordate-ovate to hastate, mostly entire, acute, 20–40 cm long overall; cauline leaves narrow, few, and reduced. Heads nearly always solitary; involucre 15–30 cm across; involucral bracts in 2–4 series, ovate-lanceolate to lanceolate, acuminate-attenuate, subequal, lanate-tomentose; receptacle broadly convex, chaffy; ray florets ca 13 or ca 21, pistillate and fertile, ligule 2–4 cm long, yellow; disk florets perfect and fertile, style branches slender; chaffy bracts clasping the achene. Achenes 7–8 mm long, ± compressed-quadrangular; pappus absent. (n = 19) May–Jul. Open hillsides & valleys; BH & w edge GP; MT: Rosebud; WY: Crook, Weston; (w through the cordillera to B.C. & CA). [tribe Heliantheae]

Another species, *Balsamorhiza incana* Nutt., may enter the extreme w edge of our region; it differs from *B. sagittata* in having deeply cleft to pinnatifid basal leaves. The species of *Balsamorhiza* are poorly defined, and there is much hybridization and intergradation.

17. BELLIS L. English Dairy

1. ***Bellis perennis*** L. Subcapose perennial 6–15 cm tall, with stems loosely clustered, arising from creeping, fibrous rootstocks; herbage lightly spreading-pilose. Leaves alternate but mostly clustered toward the base, spatulate to obovate, 2–7 cm long, including the poorly defined, winged petiole, and up to 2 cm wide, margin dentate to subentire. Heads solitary at the ends of naked scapes; involucre 4–6 mm tall and 5–8 + mm wide; involucral bracts usually ±13, equal and apparently uniseriate; receptacle naked, conical and somewhat elongating in age; ray florets pistillate and fertile, ligule white to pink, up to 10 mm long; disk florets perfect and fertile, corolla yellow, style branches with inner-marginal stigmatic lines and short, ovate appendages that are externally papillose. Achenes compressed, 2-nerved, somewhat short-pubescent; pappus absent. (n = 9) Apr–Jul (Aug). A casual lawn weed; ND: Cass, but to be expected elsewhere in the GP; (widely introduced as a lawn weed in n U.S.; Europe). *Adventive.* [tribe Astereae]

18. BERLANDIERA DC., Green Eyes

Herbaceous or weakly suffrutescent perennials; herbage pubescent or tomentose to scabrous. Stems erect or subdecumbent, arising from a fleshy taproot. Leaves alternate, petiolate or sessile, variously subentire or dentate to lyrate-pinnatifid. Heads solitary or several in terminal corymbiform clusters; receptacle chaffy throughout; ray florets ca 8(13), pistillate and fertile, ligule yellow, with 9–13 prominent anastomosing veins on the underside; disk florets staminate, with undivided style. Achenes flattened at right angles to the radius of the head, pubescent on the inner surface; pappus an inconspicuous crown of small teeth or absent; mature achene adhering to the 2 adjacent chaffy bracts and associated disk florets plus the subtending involucral bract, and all abscissing together. [tribe Heliantheae]

Reference: Pinkava, Donald J. 1967. Biosystematic study of *Berlandiera* (Compositae). Brittonia 19: 285–298.

1 Veins red or maroon on the underside of the ligule of the ray florets; at least some of the leaves lyrate-pinnatifid; midstem leaves velvety .... 1. *B. lyrata*
1 Veins green on the underside of the ligule of the ray florets; leaves at most serrate or dentate; midstem leaves hirsute to subscabrous .... 2. *B. texana*

1. **Berlandiera lyrata** Benth. Erect or somewhat decumbent herb to 12 dm tall, with stems arising singly or clustered from a ± persistent basal rosette or close-cluster of shortened, leafy branches, surmounting a taproot. Leaves velvety hairy, lower ones long-petiolate, blades or at least those of midstem lyrate-pinnatifid, the terminal lobe crenate or somewhat incised. Heads on long, scabrous peduncles; involucral bracts broadly ovate, herbaceous; disk 13–17 mm across; ray florets with ligule yellow or orange-yellow, 10–14 mm long, with conspicuous red-maroon veins beneath. Achenes 4.5–6 mm long and ca 2.5–3.5 mm wide. (n = 15) May–Aug. Dry, open limestone sites; sw GP; extreme sw KS & CO, southward; (sw GP to n Mex., w to AZ).

2. **Berlandiera texana** DC. Weakly suffrutescent perennial up to 12 dm tall, often much branched. Midstem leaves short-petiolate to sessile, triangular to ovate, 4–15 cm long and 2–6 cm wide, serrate or crenate to dentate, hirsute to scabrous or somewhat loosely hairy, but not velvety. Heads on short, hirsute peduncles; disk 1–2 cm across; involucral bracts relatively broad and leafy; ray florets with ligule 1–2 cm long, yellow or orange-yellow, with green veins beneath. Achenes obovate, 4.5–6 mm long and 3–5 mm wide. (n = 15) May–Aug. Sandy soils in open wooded areas or river banks; sw 1/4 GP; s & cen KS, sw to panhandle of TX, w OK; sw MO just out of our range; (s MO to TX, w to sw KS).

19. BIDENS L., Beggar-ticks

Semiweedy annual or weakly perennial herbs, common in damp sites. Leaves opposite, simple, variously subentire or dissected to pinnatifid. Heads radiate or discoid; involucral bracts in 2 distinct series; outer involucral bracts herbaceous and somewhat leafy, inner involucral bracts ± membranaceous, striate; receptacle flat or weakly convex, chaffy; ray florets neutral or sterile, ligule usually yellow or sometimes whitish, or rays absent; disk florets perfect and fertile, style branches flattish, with an externally hairy, short appendage; chaffy bracts narrow and flat. Achenes flattened at right angles to the radius of the head, or sometimes nearly regularly 4-angled; pappus of 2–4 + awns, which are variously antrorsely or retrorsely barbed, or sometimes smooth, or pappus greatly reduced. [tribe Heliantheae]

References: Hall, Gustav W. 1967. A biosystematic study of the North American complex of the genus *Bidens* (Compositae). Unpublished Ph.D. dissertation, Indiana Univ., Bloomington. Weedon, Ronald R. 1973. Taxonomy and distribution of the genus *Bidens* (Compositae) in the north-central plains states. Unpublished Ph.D. dissertation, Univ. Kansas, Lawrence.

Bidens includes a complex of intergrading species, and the key represents but a general synthesis of the distinguishing features. Several specimens should be studied when identifying *Bidens* spp., for any one specimen may not possess the proper structural features.

Bidens comosa and *B. connata* are here maintained as separate entities, but a good case can be made for treating them as a part of the Old World *B. tripartita* L., with the Old World materials possessing tripartite leaves ± consistently, while our materials are often fully petiolate, subentire, or occasionally tripartite.

1 Achenes strongle dimorphic, the 1–4 outer achenes 5–7 mm long, linear-cuneate, truncate at the apex, papillose-hispidulous and very scabrous; inner achenes 8–12 mm long, narrowly linear and tapering to the apex, glabrous below and hispid above; extreme sw GP .... 1. *B. bigelovii*
1 Achenes essentially alike in shape and surface texture, although the inner achenes may be a little larger; widespread spp.

2   Leaves simple, variously incised to parted but not pinnatifid.
    3   Cauline leaves sessile, or with a very short base and basally connate; heads usually radiate. .... 3. *B. cernua*
    3   Cauline leaves mostly with distinct petioles and weakly if at all connate; heads usually discoid.
        4   Achenes flattened; disk corollas mostly 4-lobed; outer involucral bracts leafy and erect .... 4. *B. comosa*
        4   Achenes 4-angled; disk corollas 5-lobed; outer involucral bracts linear-lanceolate, spreading or loosely ascending .... 5. *B. connata*
2   Leaves pinnatifid, 2–3 × pinnate or trifoliate.
    5   Ray florets absent or very small with the ligule less than 5 mm long.
        6   Achenes linear-oblong; pappus of 3–4 awns; leaves 2- or 3-pinnate .... 2. *B. bipinnata*
        6   Achenes broad and flattened; pappus of 2 awns; leaves 1-pinnate or trifoliate.
            7   Outer involucral bracts (10)13(16); achenes olive colored to brownish-yellow .... 9. *B. vulgata*
            7   Outer involucral bracts (5)8(10); mature achenes black or brown .... 7. *B. frondosa*
    5   Ray florets well developed, the ligule 1 cm or more long.
        8   Achenes oblong-cuneate; outer involucral bracts 6–8, about as long as the disk, margins short-hairy .... 6. *B. coronata*
        8   Achenes mostly ovate or elliptic-ovate; outer involcural bracts 12–20, conspicuously ciliate margined, commonly much exceeding the disk .... 8. *B. polylepis*

1. Bidens bigelovii A. Gray. Annual herb to 8 dm tall, glabrous or nearly so. Leaves petiolate, up to 8 cm long overall, 2- or 3-pinnatifid, the ultimate segments oblong, cuneate. Heads with the disk 6–9 mm across; outer involucral bracts 5–7 mm long, linear; inner involucral bracts somewhat shorter, lanceolate; ray florets very small or absent, whitish. Achenes dimorphic, the 1–4 marginal ones 5–7 mm long, linear cuneate, truncate at the apex, papillose-hispidulous, scabrous, lighter in color than the inner ones; inner achenes 8–12 mm long, narrowly linear, tapering to the apex, glabrous below and at most hispid above; pappus of both inner and outer achenes of 2 or 3 awns, retrorsely barbed. (n = 11) Jul–Sep. Damp soil along streams; extreme s GP; OK: Cimarron; TX: Carson; (s CO to TX, w to AZ, n Mex.).

2. Bidens bipinnata L., Spanish needles. Erect, branching annual, 3–15 + dm tall, glabrous or minutely setose-hispid. Leaves petiolate, 4–20 cm long overall, mostly 2- or 3-pinnatifid, ultimate segments deltoid-lanceolate or oblong, cuneate. Heads with disk 4–6 mm across; outer involucral bracts ca 8, linear, mostly ca 5 mm long; inner involucral bracts ca 8, linear-lanceolate, 5–9 mm long; ray florets with ligule very short, less than 5 mm long, yellow; disk florets with corollas 5-lobed; chaffy bracts slender, deciduous with the achenes. Achene linear, 10–18 mm long or the outermost somewhat shorter, black to brown, tapering to the apex; pappus of (2)3–4 straw-colored, retrorsely barbed awns. (n = 36, but reported as n = 12 from Taiwan; cf. Weedon, op. cit., pp. 27–30). Aug–Oct. A weed in damp, disturbed sites, especially rich, semishaded soils; fence rows, gardens, & waste places; s 1/2 GP, especially in the e portion; (MA to FL, w to GP, e TX; widely adventive elsewhere).

3. Bidens cernua L., nodding beggar-ticks. Erect to somewhat decumbent annual herb, mostly 3–8 dm tall but exceptionally much taller, glabrous or scabrous hispid. Leaves opposite or occasionally whorled, simple, sessile or nearly so and the bases connate; blade linear-lanceolate to oblanceolate, 4–17(20) cm long and up to 4 cm wide, serrate or sometimes subentire. Heads nodding at maturity; disk 1–2 cm across; outer involucral bracts distinctly leafy, 3–4 + cm long and much exceeding the disk; inner involucral bracts shorter, not exceeding the disk; ray florets 6–8, yellow, ligule up to 15 mm long or rarely longer, or rays sometimes absent; disk florets with corollas 5-lobed. Achenes narrowly cuneate, compressed, 4-angled, sometimes tuberculate along the angles, purplish-brown; pappus mostly of 4 retrorsely barbed awns. (n = 12) Aug–Oct. Muddy sites along streams & in swamps, or in drying disturbed areas; weedy; GP, but infrequent in sw 1/4; (across temp. N. Amer.; Eurasia). *B. elliptica* (Wieg.) Gl., *B. filamentosa* Rydb., *B. glaucescens* Rydb., *B. laevis* (L.) B.S.P. [misapplied], *B. prionophylla* Greene – Rydberg.

4. Bidens comosa (A. Gray) Wiegand. Erect annual, 1.5–10+ dm tall, glabrous. Stems ± straw-colored throughout. Leaves opposite, simple, sessile or with a short, winged petiole, blade elliptic-lanceolate, 4–14 cm long and 1–5 cm wide, coarsely dentate to subentire. Heads discoid, the disk 1.5–2 cm across; outer involucral bracts ca 8, leafy, erect, 2–7 cm long and up to 1–5 cm wide; inner involucral bracts ca 8, membranaceous, 1–12 mm long and up to 5 mm wide; ray florets absent; disk florets with corollas pale yellow and 4-lobed, or rarely 5-lobed. Achene flat, cuneate-lanceolate, olive colored or brown but sometimes purplish, margin ciliate and sometimes tuberculate; pappus of (2)3(4) retrorsely barbed awns. (n = 24) Aug–Oct. Sandy damp or drying sites, especially mudflats along streams; e 2/3 GP, but scattered in the w, absent in sw 1/4; (Que. to NC & LA, w to ID & NM). *B. acuta* (Wieg.) Britt. – Rydberg.

This species plus *B. connata* are a part of a complex centering around *B. tripartita* L. and could perhaps be treated as phases of that species. Some populations are not clearly referable to either *B. comosa* or *B. connata*.

5. Bidens connata Muhl. ex Willd. Erect annual, 2–15 dm tall, glabrous, bright green, or the upper stems purplish. Leaves opposite, subsessile or short-petiolate, 5–15+ cm long overall, blade mostly simple or with 2–4 divergent basal lobes, lanceolate to elliptic, margins coarsely serrate or dentate to incised. Heads with disk 1–1.5 cm across; outer involucral bracts 4–5(7), spreading or but loosely ascending, somewhat leafy but narrow, 1–2(4) cm long; inner involucral bracts ca 8, brownish, 7–8 mm long and slightly shorter than the disk; ray florets small, yellow, or rays absent; disk florets with corolla 5-lobed or sometimes 4-lobed. Achene slender, black or sometimes purplish, flattened, 4-angled, marginal achenes sometimes with 3 awns, central achenes typically with 4 retrorsely barbed awns. (n = 24, 36) Sep–Oct. Wet sites, especially stream banks & roadside ditches; e cen GP, but scattered n to e ND, cen NE; (Que. to VA, w to GP).

This species is imperfectly distinct from *B. comosa* (q.v.). Our materials are ± referable to the poorly defined var. *petiolata* (Nutt.) Farw.

6. Bidens coronata (L.) Britt., tickseed sunflower. Annual herb with us, but possibly a biennial elsewhere, 3–15 dm tall, glabrous, the stem often purplish. Leaves opposite, short-petiolate, up to 14 cm long overall, pinnatifid or the lowermost bipinnatifid, (or rarely the leaves undivided), segments 3–7(9), lanceolate, dentate. Heads conspicuously radiate; disk 8–15 mm across; outer involucral bracts 6–11 but usually ca 8, linear, up to 10 mm long, equaling or rarely exceeding the disk, margins at most short-hairy; inner involucral bracts about equaling or slightly shorter than the disk; ray florets (7)8(9), the ligule golden yellow, 1–2.5 cm long; disk florets with corollas 5-lobed; chaffy bracts equaling or slightly exceeding the disk corollas, black-tipped. Achenes cuneate-oblong or the inner ones cuneate-linear, flat or ± 4-angled, up to 9 mm long; pappus of 2 hispid teeth or sometimes of stout erect antrorsely barbed awns, or sometimes reduced to an obsolete crown. (n = 12) Sep–Oct. Damp or drying sand bars, flood plains, especially sandy soils; the sand hills region of cen GP, n-cen NE & adj. SD; (MA & s Ont. to GA, w to s IO & MN; disjunct in GP).

7. Bidens frondosa L., beggar-ticks. Annual herb 2–10+ dm tall, slender, glabrous or lightly scattered-pubescent, stems sometimes purplish. Leaves opposite, petiolate, 5–15(20) cm long overall, pinnatifid with 3–5 lanceolate or lance-ovate segments. Heads numerous, disk ca 1 cm across; outer involucral bracts ca 8, unequal, often very long (3–5 cm) and leafy, linear-spatulate and loosely ascending; inner involucral bracts brownish, ovate to lanceolate, about equaling the disk; ray florets rudimentary, golden yellow, ligule 2–3 mm long, or rays absent; disk florets orange-yellow, corolla usually 5-lobed. Achene flattish, narrowly cuneate, dark brown or black, 1-nerved on each face; pappus of 2 retrorsely barbed awns. (n = 24) Aug–Oct. Moist wooded areas, stream banks, roadside ditches, or sometimes in drying waste areas; GP, except perhaps the extreme sw; (across temp. N. Amer.; Eurasia).

8. *Bidens polylepis* Blake, coreopsis beggar-ticks. Annual herb with us, but perhaps a biennial elsewhere, 3–15 dm tall, glabrous. Leaves opposite, petiolate, 8–15 cm long overall, pinnatifid or bipinnatifid with several (5–7) narrow, linear-lanceolate segments. Heads conspicuously radiate, the disk 1–1.5 cm across; outer involucral bracts numerous, 12–20(30), up to 25 mm long, usually spreading, curling or reflexed, margins conspicuously hispidciliate; inner involucral bracts about equaling or slightly exceeding the disk, lanceolate; ray florets ca 8, ligule up to 2.5 cm long, yellow; disk florets with corolla 5-lobed; chaffy bracts slender, exceeding the disk florets. Achenes flat, brown or nearly black, outer achenes cuneate-obovate, inner achenes cuneate oblanceolate and about 1/4 taller than the outer achenes; pappus of short teeth or reduced to a mere crown, but occasionally of barbed awns. (n = 12) Aug–Oct. Damp lowlands, especially ditches, stream banks, & open marshy sites; se 1/4 GP; se NE, e 1/3 KS & adj. OK; (TN & IL to GP, s to e TX; adventive elsewhere). *B. involucrata* (Nutt.) Britt.—Rydberg.

This species is closely related to *Bidens aristosa* (Michx.) Britt. of the upper Midwest and Great Lakes region. It has been included within that species as var. *retrorsa* (Sherff) Wunderlin f. *involucrata* (Nutt.) Wunderlin (Wunderlin, Ann. Missouri Bot. Gard. 59: 471–473. 1972.).

9. *Bidens vulgata* Greene, beggar-ticks. Coarse annual herb 3–15 dm tall, glabrous to puberulent. Stems often purplish or reddish. Leaves opposite, petiolate, 5–15 cm long overall, pinnatifid with 3–5 lanceolate, serrate segments. Heads discoid or inconspicuously radiate; disk 1.5–2.5 cm across; outer involucral bracts 10–16 but mostly ca 13, leaflike, unequal, loosely ascending, 1–2(4) cm long; inner involucral bracts 8–18, ovate-lanceolate, usually shorter than the disk, olive colored to reddish-brown; ray florets small, ligule up to 3.5 mm long, yellow, or rays commonly absent; disk florets with corollas 5-lobed, yellow; chaffy bracts slender and about as long as the disk florets. Achenes flat, with slender midnerves, ovate-cuneate to oblong-cuneate, brownish to olive colored; pappus of 2 retrorsely barbed awns. (n = 12) Aug–Oct. Damp sites, especially moist woods, & marshes, but also in drying, disturbed places; GP, infrequent in w 1/3 & probably absent in w OK & adj. TX; (Que. to GA, w to WA & CA; adventive elsewhere). *B. puberula* (Wieg.) Rydb.—Rydberg.

This species is closely related to *B. frondosa* and grossly similar but more robust. The ± permanently pubescent individuals have been distinguished as var. *puberula* (Wieg.) Greene, or as f. *puberula* (Wieg.) Fern.

20. BOLTONIA L'Her.

1. *Boltonia asteroides* (L.) L'Her. Glabrous perennial herb 2–12 + dm tall, stems arising from a cluster of stoloniferous fibrous roots. Leaves alternate, entire or nearly so, broadly linear to narrowly lanceolate, sessile, the larger ones 5–15 cm long and 5–20 mm wide, progressively reduced upward. Inflorescence an open, leafy corymbiform cluster; heads campanulate to hemispheric, the disk 6–10 mm wide; involucral bracts imbricate; receptacle hemispheric to conic, naked; ray florets mostly 20–60, pistillate and fertile, ligule 5–10 + mm long, white to pink or blue; disk florets perfect and fertile, yellow, style

branches flattish, with inner-marginal stigmatic lines and short, lanceolate, externally papillose appendages. Achenes flattened, wing-margined; pappus of several minute bristles and 2(4) well-developed awns. (n = 9, 18) Aug–Oct. Open damp or drying sites, stream banks; e 1/2 GP from nw ND to cen OK & e; (NJ to FL, w to GP & TX). [tribe Astereae]

Three vars. are usually recognized; the following 2 are in the GP:

1a. var. *latisquama* (A. Gray) Cronq. Involucral bracts mostly obtuse and broadest above the middle. GP e of a line from the nw corner of ND to cen OK, but disjunct in BH. *B. latisquama* A. Gray—Rydberg.

1b. var. *recognita* (Fern. & Grisc.) Cronq. Involucral bracts linear or nearly so, broadest below the middle, apically acute or acutish. E 1/2 NE & KS, but scattered n to cen ND & e. *B. latisquama* A. Gray var. *recognita* Fern. & Grisc.—Fernald.

21. BRICKELLIA Ell.

Perennial herbs, subshrubs or shrubs with opposite or alternate, petiolate leaves. Inflorescence paniculate or racemose. Heads discoid with all florets tubular and fertile, whitish; involucral bracts imbricated in several series; receptacle naked, corolla tube with 5 inconspicuous teeth at the apex. Achene cylindrical, 10-ribbed, ± pubescent or puberulent; pappus of numerous barbellate or plumose bristles. [tribe Eupatorieae]

References: Robinson, B. L. 1917. A monograph of the genus *Brickellia*. Mem. Gray Herb. 1: 1–151; Flyr, Lowell David. 1970. *Brickellia, in* D. S. Correll & M. C. Johnston. Manual of the Vascular Plants of Texas. Texas Res. Foundation, Renner.

```
1  Petioles less than 1/5 the length of the leaf blade ............................. 1. B. brachyphylla
1  Petioles 1/3–1/2 as long as the leaf blade.
    2  Stems with loose-deciduous papery outer bark; shrubs ..................... 2. B. californica
    2  Stems with ordinary, tightly appressed bark; herbs ......................... 3. B. grandiflora
```

1. Brickellia brachyphylla (A. Gray) A. Gray. Much branched perennial herb or subshrub, up to 8 dm tall, closely stipitate-glandular. Leaves alternate or opposite, lanceolate, the better-developed 2.5–4 cm long and 1–2+ cm wide, margin serrate to subentire, petiole about 1/5 as long as the blade. Inflorescence racemose to somewhat paniculate, rarely a solitary head; involucral bracts 15–20, involucre 7–9 mm tall. Achenes densely pubescent; pappus of numerous plumose bristles. Sep–Oct. Open rocky hilltops & breaks; extreme s GP; panhandle of w OK, n TX, adj. NM, barely entering GP from the s; (TX w to AZ).

2. Brickellia californica (T. & G.) A. Gray. Shrub with widely spreading branches, 6–10 dm tall, glandular short-pubescent, papery exfoliating bark. Leaves alternate, blade broadly ovate to deltoid, truncate to cordate, 1–4 cm long and wide, margin dentate, crenate or shallow-lobed, or sometimes subentire, petiole 1/3 to 1/2 as long as the blade. Inflorescence paniculate or a series of close racemes; involucral bracts ca 20, involucre 5–8 mm tall. Achenes brownish pubescent; pappus of scabrous bristles, Jul–Oct. Open, rocky breaks; extreme s GP; panhandles of w OK & n TX; (widespread in the cordilleran w; WY to TX & n Mex., w to OR, CA, & Baja Calif.).

3. Brickellia grandiflora (Hook.) Nutt. Perennial herb, 3–8 dm tall, short-pubescent with divergent hairs. Stems erect, arising singly or loosely clustered from a prominent crown surmounting a taproot. Leaves opposite or the upper ones alternate, blade triangular to subhastate, 2–7 cm long and 1–4 cm wide, margin crenate-dentate, petiole about 1/2 as long as the blade, or a little longer. Inflorescence open-paniculate, but the heads crowded at the ends of the branches; involucre 7–10 mm tall. Achenes minutely pubescent; pappus bristles scabrous-barbellate, somewhat early-deciduous. (n = 9) Aug–Oct. Open rocky areas, scrub-woodland areas, canyons; sw 1/4 GP from w NE to cen KS to se CO; (AR & TX w to B.C., CA, & n Mex.). *B. umbellata* (Greene) Rydb.—Rydberg.

22. CACALIA L., Indian Plantain

Glabrous perennial herbs with stems arising from a thick, tough caudex. Leaves alternate, strongly and progressively reduced upwards. Inflorescence a simple to variously compound corymbiform cyme with numerous heads; ours with only 5 involucral bracts and 5 florets. Florets tubular (disk) and perfect, corolla whitish to light yellow (ochroleucous) but never truly yellow, or sometime purplish, deeply 5-lobed. Achenes ovoid to obovate, somewhat flattened; pappus of abundant bristles. *Arnoglossum* Raf., *Mesadenia* Raf., *Conophora* Nieuwl. [tribe Senecioneae]

Reference: Pippen, Richard W. 1978. Cacalia. N. Amer. Flora II. 10: 151–159.

1 Basal and lower cauline leaves palmately veined with the veins divergent, involucral bracts without a keel-wing on the back .... 1. *C. atriplicifolia*
1 Basal leaves several-veined but with the veins parallel and converging distally, involucral bracts with a prominent keel-wing on the back .... 2. *C. plantaginea*

1. *Cacalia atriplicifolia* L., pale Indian plantain. Herb 1–2(+) m tall, glabrous, glaucous, especially on the lower stems and the undersides of the leaves. Stems weakly striate or smooth, arising from a thick caudex. Leaves progressively reduced upward, the basal leaves few, petiolate, blade broadly ovate, cordate, up to 30 cm long and wide but frequently smaller, palmately veined with 3–5(7) main veins, margin variously dentate or shallow-lobed, lower cauline leaves similar but the lobing more pronounced, uppermost leaves reduced to bracts. Inflorescence a large, loose, corymbiform cyme with up to 100+ heads; involucre narrowly cylindric, involucral bracts 5, rounded across the back, 7–8(10) mm long; florets 5, white to greenish or light purplish. Achenes 4–5 mm long, 10-ribbed, resinous; pappus 4–6 mm long and early deciduous. (n = 25, 26, 28) Jun–Sep. Open woods; se GP; se NE, e KS, & OK; (NY to FL, w to NE & OK). *Mesadenia atriplicifolia* (L.) Raf.—Rydberg.

2. *Cacalia plantaginea* (Raf.) Shinners, Indian plantain. Stout herb, 5–10 dm tall, glabrous. Stem striate-ribbed, arising from a short, tuberous caudex with many fleshy-fibrous roots. Basal and lower cauline leaves long-petioled, blade elliptic to narrowly ovate, entire, denticulate or slightly crenate with several parallel main veins, 5–15+ cm long and 2–8 cm wide, petiole at least as long as the blade; cauline leaves progressively reduced upward, the uppermost reduced to mere bracts. Inflorescence a broad, corymbiform cyme with ca 50–100 heads; involucre narrow-cylindric, involucral bracts 5, light green, 7–10 mm long, with a distinct keel-wing on the back; florets 5, white or ochroleucous. Achenes 4–5 mm long, 12- to 15-ribbed; pappus 6–8 mm long and early deciduous. (n = 27) Jun–Aug. Damp, rocky prairies; e GP; e NE s to e OK, but locally collected elsewhere; (Ont. to AL, w to MN, NE, & TX). *Cacalia tuberosa* Nutt.—Atlas GP; *Mesadenia tuberosa* (Nutt.) Britt.—Rydberg.

23. CARDUUS L., Plumeless Thistles

Contributed by Ronald L. McGregor

Ours usually winter annual or biennial herbs or rarely flowering first year; with stout, fleshy, taproots. Stems erect, strict or usually much branched, with spiny wings from decurrent leaf bases, glabrous to peduncles or sparsely to densely floccose-tomentose. Rosette leaves large, compact, lobed and spiny; cauline leaves alternate, deeply pinnately divided to nearly entire, spiny-margined, glabrous to variously pubescent, often with silvery margin. Heads discoid, solitary at end of branches or somewhat clustered, 1-7 cm in diam; receptacle flat or convex, densely bristled; involucral bracts with or without a submedian constriction, in several series, imbricated, apical portions spreading or reflexed, mostly spine-tipped; florets perfect, tubular and with long narrow lobes, purplish or reddish, occasionally white; anthers 5, filaments glabrous or pilose, caudate; style with a thickened, often sparsely hairy ring somewhat below the branches. Achenes glabrous, basifixed, oblong but slightly narrowed to base, with longitudinal stripes or lines, with an apical collar; pappus of numerous, minutely barbellate, capillary bristles, deciduous in a ring. [tribe Cynareae]

Reference: Moore, R. J. & C. Frankton. 1974. The thistles of Canada. Canad. Dept. Agric. Monogr. 10: 54-61.

1 Outer and median involucral bracts without a submedian constriction, 1-1.5(2) mm wide at base; heads 1-2 cm in diam .... 1. *C. acanthoides*
1 Outer and median involucral bracts with a submedian constriction, (2)3-10 mm wide at widest point; heads 3-7 cm in diam .... 2. *C. nutans*

1. Carduus acanthoides L., plumeless thistle. Winter annual or biennial herbs, or rarely flowering first year, with stout, fleshy, taproots. Stems 0.3-1.5 m tall, freely branched above, glabrate or with scattered multicellular hairs or villous, with spiny wings 3-15 mm wide extending to heads. Rosette leaves compact, usually 1-2 dm long, oblanceolate to elliptic in outline, with somewhat pinnatifid, spinose lobes, these with silver white or purplish margins in midwinter, upper surface with sparse multicellular hairs, lower surface and especially midvein with sparse to dense 1-celled or multicellular long hairs; cauline leaves alternate, sessile, decurrent, 1-2 dm long, elliptic to lanceolate or oblong, evidently pubescent, deeply and usually irregularly pinnatifid, segments 1-3(4) pointed, marginal spines 1-4(6) mm long. Peduncles usually spiny-winged up to base of heads; heads solitary at ends of branches, or in clusters of 2-5, 1.6-2.7 cm tall, 1-2 cm wide; outer involucral bracts without submedian constriction, spreading-reflexed, narrowly lanceolate, sparsely to densely arachnoid, 1-1.5(2) mm wide at base, tapering gradually to apex, 7-12 mm long; inner involucral bracts to 20 mm long, tips flat, without spines; corollas purplish, rarely white or yellowish, 14-20 mm long, tube 7-10 mm long; anther filaments woolly; stigma lobes 1.8-2.3 mm long. Achenes 2.5-3 mm long, slightly obovate to oblong, somewhat quadrate, a little compressed, stramineous to light brown, usually lustrous, with thin longitudinal lines and a distinct, light, apical collar. (n = 11) May-Aug. Infrequent to locally abundant in pastures, stream valleys, fields, roadsides, waste places; w MN, e ND, e SD, IA, e 1/2 NE; KS: Doniphan, Nemaha; also scattered in n 1/2 of w GP (widely established in U.S.). Introduced.

In se SD and e NE this plant is a serious weed and locally so elsewhere.

A very similar species, *C. crispus* L., the curled or welted thistle, is known in the GP from a single collection in ND: Cass. It differs from *C. acanthoides* principally in having lower leaf surfaces always woolly and a chromosome number of n = 8. It is a native of Europe and Asia and is sparingly found in North America.

2. Carduus nutans L., musk thistle, nodding thistle. Ours winter annual or usually biennial herbs with stout, fleshy, taproots, rarely flowering first year. Stems erect, 0.5-3 m tall, usually much branched, with spiny wings 0.5-2 cm wide from decurrent leaf bases, axis glabrous or sparsely to densely pubescent. Well-developed rosette leaves compact, broadly elliptic to lanceolate, 1.5-5(7) dm long, variously pinnatifid with deltoid, palmate or deeply divided spinose lobes, margins often silvery-white to rose or purplish, glabrous or sparingly to densely pubescent. Cauline leaves alternate, sessile, decurrent, blades lanceolate or oblong-lanceolate or broadly elliptic, lobes triangular or lanceolate to ovate, with numerous marginal spines, glabrous to sparsely or densely pubescent. Heads on well-developed plants terminating branches, globose or ovoid, 3-7 cm in diam, usually nodding; peduncles usually naked for some distance below heads or with 1-few small lanceolate bracts, or peduncles leafy to heads, densely arachnoid; appendages of involucral bracts 2-10 mm wide at widest point, with a shallow median constriction, ovate-lanceolate to lanceolate, apical portion wider or rarely about same width as lower, terminated by a strong spine, spreading or reflexed, inner involucral bracts narrow, with no or weak spines. (n = 8).

There are two subspecies in the GP.

2a. subsp. *leiophyllus* (Petrovic) Stoj. & Stef. Stems glabrous or with very sparse hairs up to the nearly naked, arachnoid peduncles. Leaves glabrous or rarely with scattered hairs on both surfaces. Heads 3.5-5 cm tall, 5-7 cm in diam; appendages of outer and median involucral bracts evidently wider than basal portions, glabrous, appendages (3)4-8(10) mm wide near base, (12)15-20(25) mm long (incl. spine), velvety-puberulent above, ovate-lanceolate to lanceolate, outline slightly curved and not tapering evenly to apical spine or tapering evenly from near base to spine. Achenes 3.5-4 mm long, 1.5 mm in diam, glossy, light brown with longitudinal stripes and a light apical rim. May-Jul, sporadic until frost. Locally abundant to infrequent in pastures, prairie ravines, & hillsides, open wooded stream valleys, fields, roadsides, waste places; GP but less common in n, w, & sw; (widely established in N. Amer.; e Europe, s Russia, Asia Minor, N. Africa). *Introduced.*

This is the common noxious weed in the GP and a member of the polymorphic *C. nutans* complex of Europe. *C. nutans* of Atlas GP; *C. thoermeri* Weinm., Flora Europaea 4: 117. 1976; *C. nutans* subsp. *macrolepis* (Peterm.) Kazmi—Moore & Frankton, op. cit.

The plants in many colonies have the appendages of the involucral bracts more or less contracted to the spine, while plants in other colonies have appendages gradually tapering from the base to the spine. The latter have been referred to subsp. *macrolepis*. I find complete intergradation in this character in populations and even on the same plant, where the later-formed heads have the involucral bracts of *macrolepis*, while the earlier principal heads have the involucral bracts of *leiophyllus*. Plants flowering the first season also normally have involucral bracts of the *macrolepis* type. The recognition of subsp. *macrolepis*, in our area, does not appear to have taxonomic meaning.

2b. subsp. *macrocephalus* (Desf.) Nyman. Stems distinctly pubescent and often arachnoid, leafy to near base of heads or naked for short distance below. Leaves evidently pubescent to nearly arachnoid on both surfaces. Heads 3-4.5 cm tall, 3-5 cm in diam; outer and middle involucral bracts spreading or recurved, with appendages same width as basal portion to usually evidently wider, lightly to densely arachnoid below; appendages usually pubescent, 2-4(6) mm wide near base, the same width as basal portion to somewhat wider, (15)20-40(50) mm long, narrowly ovate-lanceolate, long-subulate or attenuate, tapering evenly to spine from just above the base. Achenes as in subsp. *leiophyllus*. Jun-Sep. Rare but locally common in grasslands & open wooded areas; BH; WY: Crook; MT; (SD, WY, MT, perhaps elsewhere; cen & e Medit. region). *Introduced.*

One colony in WY: Crook appears intermediate between subsp. *leiophyllus* and subsp. *macrocephalus*. The complex merits more careful field work and study.

24. CARTHAMUS L., Safflower

Contributed by Ronald L. McGregor

1. **Carthamus tinctorius** L., safflower. Annual glabrous herbs with taproots. Stems 3-10 dm tall, whitish, striate, simple or branched above. Leaves alternate to subopposite above, sessile and somewhat clasping, simple, broadly elliptic, to ovate or oblong, margins somewhat remotely serrate or dentate spinose, apex with a short spine. Heads discoid, solitary on corymbose branches, 3-4 cm high; outer leaflike involucral bracts about as long as head, spinulose, with a distinct submedian constriction, upper portion wider than basal part; receptacle chaffy; florets all perfect or outer ones pistillate; corollas deeply 5-cleft, bright orange to orange-yellow; filaments subglabrous. Achenes obpyramidal, 4-angled, shiny, glabrous, ivory-white, 6 mm long, 4 mm wide; pappus usually absent or with a few short

rudimentary narrow scales, or sometimes well developed with numerous narrow scales unequal in length and shorter than achene. (n = 17) Jul–Aug. Grown in the w & n GP where fugitives from cultivation have been rarely collected; there is no evidence of plants persisting in the wild; (U.S. cult. & casual escapes; Egypt). *Introduced*. [tribe Cynareae]

This species has been recognized as a dye plant since ancient times. More recently it has been grown for its edible oil, which is not conducive to cholesterol build-up in the blood. Cultivation largely has ceased in the GP.

25. CENTAUREA L., Star-Thistle, Knapweed, Centaury

Contributed by Ronald L. McGregor

Annual, biennial, or perennial herbs. Leaves alternate, usually pinnatifid, sometimes entire, not spinose. Heads discoid, solitary, terminal or axillary, many flowered; florets tubular and perfect or, more often, marginal ones sterile, with enlarged, irregular, falsely radiate corolla; involucre ovoid or globose; involucral bracts imbricated in several series, margins entire, fringed, spinose or tips expanded into a broad appendage; receptacle nearly flat, densely bristly; corollas purple or blue to yellow, pinkish or white, with slender tube and 5 long narrow lobes; anthers usually with caudate appendages; style with a thickened, sometimes hairy ring at base of lobes. Achenes ovoid or oblong, 4-angled or compressed, obliquely or laterally attached; pappus of 1 or 2 or several series of graduated bristles or narrow scales, these often reduced, or absent. [tribe Cynareae]

Except for *C. americana* all of our species of *Centaurea* have been introduced from Eurasia and the Mediterranean region and are either variously naturalized or appear as infrequent nonpersisting escapes or introductions. In addition to the species treated below, *C. nigrescens* Willd. has been reported for NE. It would key to *C. americana*, but has purple flowers, heads only 1.5–2 cm high, involucral bracts with appendages dark brown or black, and the pappus absent or only 0.5 mm long. It is likely that other species of *Centaurea* will be found in the GP.

1 Margins of involucral bracts entire, chartaceous .... 6. *C. repens*
1 Margins of involucral bracts spinose, pectinate, lacerate or fringed.
  2 Some involucral bracts with terminal spines which are longer than lateral ones, if these are present.
    3 Terminal spines of involucral bracts 11–30 mm long.
      4 Florets yellow; pappus ± same length as achene; stems winged .... 8. *C. solstitialis*
      4 Florets pinkish to purplish; pappus about 1/2 as long as achene; stems not winged .... 4. *C. iberica*
    3 Terminal spines of involucral bracts less than 10 mm long; florets usually white or purplish .... 3. *C. diffusa*
  2 Involucral bracts lacking a conspicuous terminal spine.
    5 Body of involucral bracts distinctly articulated into 2 parts, the lower light green and striate, the upper chartaceous and lobed .... 1. *C. americana*
    5 Body of involucral bracts not distinctly articulated.
      6 Principal cauline leaves entire; plants annual .... 2. *C. cyanus*
      6 Principal cauline leaves pinnatifid.
        7 Involucre 10–13 cm high; plants biennial or perennial .... 5. *C. maculosa*
        7 Involucre 16–25 mm high; plants perennial .... 7. *C. scabiosa*

1. ***Centaurea americana*** Nutt., basketflower. Erect annual herbs. Stems (3) 8–15 dm tall, simple below, branched above, grooved, glabrous below to minutely scabrous and glandular above. Leaves alternate, simple, margins entire or remotely denticulate; lower leaves narrowly obovate, scabrous; principal cauline leaves lanceolate to ovate-lanceolate, (2)4–8(10) cm long, glabrous or sparsely scabrous and gland-dotted. Heads solitary at ends of branches, involucre broadly and shallowly campanulate, 3–5 cm tall, involucral bracts in several

series, imbricated, each articulated into 2 portions, the lower part entire, broadly elliptic to linear, light green and striate, the upper part lance-deltoid in outline, yellowish, with a flat subulate tip and margins with 4–7 toothlike lobes or cusps; corollas pink or rarely white, peripheral ones much larger than central ones, marginal ones 2–2.5 cm long. Achenes 4–5 mm long, oblong, grayish-brown or black, obscurely striate, glabrous or with a few long, white hairs, attached at an obliquely leveled area just above the base; pappus in 1 or 2 series, of bristles 8–14 mm long, these minutely barbellate, distinct or weakly united at base. (n = 13) Apr–Aug. Locally common in grasslands, bluffs, stream valleys, roadsides, waste areas; OK, TX, w NM, occasionally found as a nonpersisting waif in e KS & w MO; (se MO to AZ, s to LA, TX, N.L., & Coah.).

This very showy native species is sometimes cultivated.

2. Centaurea cyanus L., bachelor's-button, cornflower. Annual or winter-annual herbs with taproots. Stems 2–10 dm high with ascending branches, slender, green, often loosely white-tomentose or sericeous when young. Cauline leaves linear or lanceolate, entire, white woolly below, to 13 cm long, 2–5(10) mm wide, the lower ones sometimes toothed or lobed. Heads terminating the branches; involucre (11)13–16 mm tall; involucral bracts unarmed, ovate or lanceolate, more or less striate, margins chartaceous and lacerate, body and margins often purple-tinged; corollas mostly blue, purple, pink, or white, the marginal ones enlarged, sterile. Achenes 3.5–4 mm long, compressed, yellowish-brown to black, strigose; pappus 2–4 mm long. (n = 12) May–Aug. Infrequent in fields, roadsides, waste places, sporadic & not long persisting in GP; (widely cult. & escaped over much of N. Amer.; Medit. region). *Introduced.*

3. Centaurea diffusa Lam., diffuse knapweed. Annual or biennial herbs with taproots. Stems 1–6 dm tall, erect, diffusely branched, angled, scabrous-puberulent. Lower leaves obovate in outline, 3–8 cm long, 1–3 cm wide, deeply pinnatifid, scabrous-puberulent under thin arachnoid tomentum, early deciduous; cauline leaves smaller, pinnatifid or upper reduced ones entire. Heads solitary, terminating branches; involucres 8–10 mm tall, 4–5 mm wide, ellipsoid-cylindric; outer involucral bracts stiff, glabrous, ovate-lanceolate, with a terminal spine 1.5–7(8) mm long and with 4–6 pairs of shorter lateral spines; corollas 9–12 mm long, usually white or creamy to purplish, marginal ones not enlarged. Achenes 2.3–2.6 mm long, brown, usually glossy; pappus absent or to 1 mm long. (n = 9) Jul–Sep. Rare in grasslands, roadsides, waste places; CO: El Paso, Kit Carson; (widespread in n U.S. & occasional elsewhere; Medit. region). *Introduced.*

4. Centaurea iberica Trev. ex Spreng., Iberian star-thistle. Biennial herbs with stout taproots. Stems erect, 5–10 dm high, much branched and bushy from below, somewhat angled, glabrous. Rosette leaves oblanceolate, 1–4 dm long, narrowed to a slender petiole, pinnatifid or bipinnatifid, puberulent or thinly arachnoid-tomentose; principal cauline leaves pinnatifid to subentire, the upper sessile, puberulent. Heads borne among the leaves, sessile or short pedunculate, 2–3 cm tall; involucres glabrous; involucral bracts stiff, nerveless, pale green with whitish scarious margins; middle involucral bracts with a stout terminal spine 10–25 mm long and with 1–3 pairs of spines near base; outer and inner involucral bracts spineless or nearly so. Corollas pinkish or purplish, outer ones not enlarged. Achene oblong, 3–4 mm long; pappus bristles in several series, flattened, the longer about 1/2 length of achene. (2n = 16) Jul–Aug. Rare in fields & waste places; KS: Labette; WY: Converse; (sw U.S. to GP; Eurasia). *Introduced.*

Previous reports of *C. calcitrapa* L. for KS and WY were based on specimens of *C. iberica*. *C. calcitrapa* is very similar to *C. iberica* but lacks a pappus.

5. Centaurea maculosa Lam., spotted knapweed. Biennial or short-lived perennials with stout taproots. Stems erect, 1–several from base, branched above, 3–8(12) dm tall, ridged, glabrate, scabrous-sericeous or loosely arachnoid-tomentose. Rosette and basal leaves 5–15 cm long, blades narrowly elliptic to oblanceolate, entire and remotely dentate to usually pinnately 1- or 2-parted, scabrous-puberulent and usually loosely tomentose; principal cauline leaves pinnately divided, segments rather remote, linear, 1–3 mm wide, upper leaves essentially entire, all glabrate to scabrous-puberulent and usually arachnoid-tomentose. Heads solitary, terminating branches, which are cymosely or paniculately arranged; involucre obovate or oblong, 10–15 mm tall, thinly tomentose or glabrate; involucral bracts stiff, striate, the outer and middle ones broadly to narrowly ovate, pale green and brownish below the black-spotted apex, the apex with short dark, pectinate tip; florets pink or purple, rarely white, the outermost enlarged and sterile. Achenes narrowly obovate, 2.5–3.5 mm long, brown or blackish, with longitudinal lines, glabrous or sparsely pilose; pappus usually present, of paleaceous bristles 2–3.5 mm long. (n = 18) Jun–Sep. Pastures, roadsides, fields, waste areas; isolated colonies known in GP; (nearly throughout U.S.; Europe). *Introduced.*

Though widely scattered in the GP, this plant can be a serious pest. At least in the lower GP the species persists for only a few seasons.

6. Centaurea repens L., Russian knapweed. Perennial herbs, forming dense colonies by adventitious shoots from widely spreading black roots. Stems erect, openly branched, 2–10 dm tall, finely arachnoid-tomentose, becoming glabrate and green. Rosette leaves oblanceolate, irregularly pinnately lobed to nearly entire, thinly tomentose or glabrate, 3–8 cm long, 2.5 cm wide, petiole shorter than blade; lower cauline leaves smaller, pinnately lobed; upper leaves 1–3(5) cm long, 2–7 mm wide, narrowed to a sessile base, entire or serrate. Heads numerous, terminating the branches, ovate, 1.5–2 cm tall; involucre pale, 9–15 mm tall, thinly tomentulose or glabrous; the outer and middle involucral bracts broad, striate, with a broad, whitish, chartaceous entire margin, inner ones narrowed to plumose-hairy tip; florets pink or puplish, the marginal ones not enlarged; corolla 12–13 mm long. Achenes 3–3.5 mm long, whitish, slightly ridged, attachment scar sub-basal and lateral at base of fruit; pappus bristles 6–11 mm long, the longer ones subplumose above, all early deciduous. (n = 13) Jun–Sep. Locally common in fields, pastures, roadsides, waste places; widely distributed in GP; (nearly throughout U.S.; Asia). *Introduced.* *C. picris* Pall.—Rydberg.

This noxious weed was first introduced in N. Amer. in 1898. It is persistent in many areas and difficult to eradicate. In recent years it has increased particularly in the n and w GP and merits consistent monitoring. Some botanists place this species in the genus *Acroptilon* where it is known as *A. repens* (L.) DC. *Acroptilon* is distinguished from *Centaurea* by the subbasal, rather than lateral, attachment scar on the achene, and by cytological differences.

7. Centaurea scabiosa L., greater centaurea. Perennial herbs with taproots, summit of caudex covered by persistent leaf-bases of former years. Stems erect, 3–15 dm tall, thinly arachnoid, becoming glabrous. Basal leaves obovate, deeply and irregularly 1- or 2-pinnately divided, segments elliptic to lanceolate, hirsute to glabrous; principal cauline leaves elliptic-obovate in outline, deeply pinnatifid, hirsute or glabrate and thinly arachnoid when young. Heads usually solitary on naked peduncles, hemispheric or ovoid, 20–30 mm tall, 4–6 cm wide; involucre 15–25 mm tall; involucral bracts with a conspicuous blackish, scarious, pectinate-fringed, arachnoid tip. Florets usually purplish, marginal ones usually enlarged, sterile, 40 mm long; inner florets 20 mm long. Achenes 4–5.5 mm long, yellowish-brown, puberulent; pappus 4–5 mm long, grayish or brownish-white. (n = 10) Jul–Sep. Rare in fields & waste places; ND: Barnes, Golden Valley, Ransom, not found in recent years; (sporadic in n U.S. to ND; Europe). *Introduced.*

8. Centaurea solstitialis L., yellow star thistle. Biennial herbs with taproots, sometimes flowering first year. Stems erect, 2–10 dm tall, freely branched, thinly but persistently tomentose, prominently winged by decurrent leaf-bases. Basal and lower leaves lyrate or pinnatifid, oblanceolate, 3-20 cm long, 5-30 mm wide, early deciduous; cauline leaves becoming smaller, linear-oblong to narrowly lanceolate, entire, bases decurrent, forming stem wings; all leaves gray-tomentose. Heads terminating the branches, 18-22 mm tall, constricted above; involucre 10-15 mm tall, 7-9 mm wide; outer and middle involucral bracts ovate, rigid, somewhat arachnoid, spine-tipped, the larger terminal spines 11-30 mm long, bearing 2-4 short lateral spines at base; inner involucral bracts unarmed, with tapering and chartaceous tips; florets yellow, marginal ones not enlarged; corollas 14-17 mm long. Achenes 3.5-4 mm long, yellowish and with brown marbling, sometimes shiny; pappus on marginal flowers usually lacking, that of the others 2-5 mm long. ($n = 8$) Jul-Sep. Infrequently found in fields, roadsides, waste places; widely scattered in GP; (occasional throughout U.S. & weedy on w coast; Eurasia). *Introduced.*

This species is repeatedly introduced in our region but colonies known to me have persisted for only 2-4 years. Usually viable seed is not produced.

26. CHAENACTIS DC.

1. *Chaenactis douglasii* (Hook.) H. & A. var. *achilleaefolia* (H. & A.) A. Nels. Single-stemmed, taprooted perennial herb, mostly 2-5 dm tall, simple or somewhat branched above, densely to sparsely tomentose and sometimes glandular, especially upward. Leaves 2-12 cm long and 1- to 3-pinnatifid, the segments rather thickish and curled or twisted; upper leaves reduced and less dissected. Heads several in a corymbiform cluster; involucre mostly 8-12 mm tall; involucral bracts herbaceous, narrow, subequal or but little imbricated; receptacle naked; ray florets absent but the outer disk florets sometimes with enlarged corollas; disk florets perfect and fertile, corollas creamy white (rarely pinkish?), style branches slightly flattened, elongate, externally pubescent, with submarginal and sometimes obscure stigmatic lines extending nearly to the apex. Achenes somewhat club-shaped; pappus of 8-16 elongate, scarious scales in 2 series. Apr-Aug. Open high plains; scattered in nw 1/5 GP; w ND & MT s to ne WY; (GP & cen CO to B.C. & CA). [tribe Heliantheae]

Reference: Stockwell, Palmer. 1940. A revision of the genus *Chaenactis.* Contr. Dudley Herb. 3: 89-168.

Chaeanactis douglasii is a complex of intergrading, semidistinct varieties, and var. *achilleaefolia* is the widespread, polymorphic phase; cf. Cronquist *in* Hitchcock et al., Vascular Plants of the Pacific Northwest, vol. 5, Univ. Washington Press. 1955. pp. 119-124.

27. CHAETOPAPPA DC.

1. *Chaetopappa asteroides* (Nutt.) DC. Low, slender taprooted annual, 5-15(25) cm tall. Stems freely branching upward, strigose to somewhat hispid-pubescent. Leaves alternate, linear to linear-subulate, mostly 5-10 mm long but the lower ones sometimes longer. Heads solitary at the ends of the upper branches; involucre 3.5-4.5 mm tall and 2-3 mm wide, narrowly cylindrical to weakly conical; involucral bracts imbricate in 3 or 4 series; receptacle flat, naked; ray florets 5-13, pistillate and fertile, ligule white or pinkish, 2-4 mm long; disk florets few, perfect and fertile, corolla yellow, style branches short and flattish, with inner-marginal stigmatic lines and short, blunt, externally short-hairy appendages. Achenes prismatic, 5-nerved, 1.6-2.0 mm long, pubescent; pappus of 5 short, hyaline scales alternating with 5 long, slender awns, 1-3 mm long, or sometimes the pappus much reduced. ($n = 8$) Apr-Jul (Sep). Open, drying sandy or sometimes loose, rocky sites; se KS to sw OK; (s GP to LA, cen TX & n Mex.). [tribe Astereae]

Reference: Shinners, Lloyd H. 1946. Revision of the genus *Chaetopappa* DC. Wrightia 1: 63-81.

28. CHRYSANTHEMUM L., Chrysanthemum

Ours perennial herbs with alternate, toothed to deeply pinnate-lobulate leaves. Inflorescence corymbiform with numerous heads, or heads solitary; heads radiate or discoid; involucral bracts imbricate in several series, tips and margins becoming scarious; receptacle flat to convex, naked; ray florets (when present) in a single whorl, pistillate and fertile; disk florets tubular and perfect, corolla somewhat flattened, style branches flattened, truncate, penicillate. Achenes ± angular, striate; pappus a short crown or none. [tribe Anthemideae]

1 Heads solitary or few, long-pedunculate, disk 10–20 mm across .... 2. *C. leucanthemum*
1 Heads numerous, inflorescence corymbiform, disk 5–10 mm across.
  2 Leaves crenate-dentate, with a few basal lobes; rays absent or very short when present .... 1. *C. balsamita*
  2 Leaves pinnatifid; rays present and well developed .... 3. *C. parthenium*

1. **Chrysanthemum balsamita** L., mint geranium, costmary. Coarse, pleasantly fragrant herb 5–10 dm tall, pubescent above but glabrate below. Leaves silvery pubescent to subsericeous when young, but glabrate in age, margin crenate-dentate, sometimes with a few basal pinnae; basal and lower cauline leaves largest, blade oblanceolate to elliptic, 10–20 + cm long and 2–8 cm wide, petiole about as long as the blade; upper leaves smaller, becoming sessile. Inflorescence corymbiform with numerous heads, disk 4–7 mm across, involucral bracts with conspicuous expanded scarious tips, ray florets nearly always absent but when present very small and white. Achenes subterete, ± 10-ribbed; pappus a minute ridge or crown. Aug–Sep. Roadsides & waste places; scattered in the GP; cen KS, w SD, escaping from cultivation but doubtfully fully established; (e N. Amer., especially in New England & the Great Lakes region, GP; Eurasia). *Balsamita major* Desf.—Rydberg.

2. **Chrysanthemum leucanthemum** L., ox-eye daisy, marguerite. Erect rhizomatous perennial, 2–10 dm tall, strict or few-branched, glabrous or lightly pubescent. Leaves progressively reduced upward; basal and lower cauline leaves oblanceolate to narrowly obovate, 4–12 cm long, including the petiole, margin crenate to lobed or parted; upper leaves becoming sessile and merely toothed to subentire. Inflorescence of a solitary or a few long-pedunculate heads; disk 10–20 mm across, involucral bracts narrow, with brown margin; ray florets 15–30, white, ligule 10–20 mm long, rarely long-tubular or otherwise deformed. Achene ± 10-ribbed; pappus none. (n = 9; + polyploid series of 18, 27, 36) May–Jul, sporadic in autumn. Fields, waste places, roadsides; scattered in GP, especially se 1/4; (ne U.S. & adj. Can.; Eurasia). Escaped from cultivation and now *naturalized* in Northeast, locally established in our region; sometimes distinctly weedy. *Leucanthemum vulgare* Lam.—Rydberg.

Our materials have been segregated as var. *pinnatifidum* Lecoq & Lamotte on the basis of relatively deep lobing of the lower leaves, and the small heads. Neither the utility of the segregation nor the validity of the name are beyond question.

3. **Chrysanthemum parthenium** (L.) Benth., feverfew. Taprooted perennial 3–8 dm tall, pubescent above, but glabrate below. Leaves pinnatifid with the pinnae lobed or incised. Inflorescence corymbiform, with numerous heads, disk 5–10 mm across, involucral bracts narrow, with sharp scarious tips; ray florets 10–20 (more in the "double" forms), white, ligule 4–8 mm long. Achenes subterete, ± 10-ribbed; pappus a low crown or obscure. (n = 9) Aug–Sep. To be expected in damp weedy sites in the GP as an occasional escape from cultivation; (ne U.S. to GP & adj. Can.; Eurasia). *Adventive;* widely cult.

29. CHRYSOPSIS Ell., Golden Aster

Annual or perennial herbs. Stems often clustered, arising from a taprooted caudex and sometimes rhizomatous; herbage variously but usually conspicuously pubescent, the stem-pubescence of 2 classes: a primary pubescence of longer, stiff hairs, and a secondary pubescence of shorter, divergent hairs. Leaves alternate, simple, entire to shallow-dentate; lower leaves petiolate, becoming progressively sessile and sometimes clasping upward. Inflorescence a corymbiform to condensed-paniculate cluster; heads narrow-turbinate or cylindrical to hemispherical; involucral bracts imbricate in 3–5(9) series, the margins scarious; receptacle naked, flat or low-convex; heads radiate (in ours); ray florets pistillate and fertile or sometimes neutral by abortion, ligule spreading, yellow; disk florets perfect and fertile, corolla yellow, tubular with a 5-toothed limb, style branches flattened, with hairy and elongate appendages. Achenes somewhat flattened, pappus in 2 series; the outer pappus of short fimbriate or setose scales, and the inner pappus of numerous elongate capillary bristles. [tribe Asterae]

References: Harms, Vernon L. 1970. *Chrysopsis, in* Correll, D. S., & M. C. Johnston, Manual of the Vascular Plants of Texas. Texas Res. Foundation, Renner; Harms, Vernon L. 1974. A preliminary conspectus of *Heterotheca* section *Chrysopsis*. Castanea 39: 155–165; Semple, John C. 1977. Cytotaxonomy of *Chrysopsis* and *Heterotheca* (Compositae—Astereae): a new interpretation. Canad. J. Bot. 55: 2503–2513; Semple, John C. 1981. A revision of the golden aster genus *Chrysopsis*. Rhodora 83: 323–384.

Chrysopsis has been variously treated in the botanical literature. One extreme has been to merge it with *Heterotheca*, and another has been to regard it as several segregate genera. Semple (op. cit. and in the literature) has presented a case for redefining the boundaries on cytological and other evidence. The treatment presented here is traditional, with *Chrysopsis* conceived broadly, and with *Heterotheca* separated on the technical character of epappose ray florets, plus the less precise features of aspect.

Chrysopsis camporum Greene is similar to *C. villosa* but differs in being more robust (up to 10 dm tall) and in possessing rhizomes and leaves that are generally shallow-dentate. It is native to the prairie areas of WI to e MO and e. In recent years an apparently weedy phase has become prominent in roadside ditches and other disturbed sites, and it may be expected eventually to invade the e GP. *C. camporum* has sometimes been treated as a var. of *C. villosa*, e.g., by Gleason & Cronquist.

1 Annuals; leaves relatively soft and pliant; herbage persistent pilose-pubescent .... 4. *C. pilosa*
1 Perennials; leaves rather firm and not obviously pliant; herbage with at least some rigid hairs and some spreading hairs.
  2 Upper herbage densely hirsute to strigose, nonglandular or with only inconspicuous scattered punctate glands (glandular-resinous in *C. villosa* var. *hispida*).
    3 Upper herbage densely appressed-sericeous; upper primary pubescence fine-hirsute and not at all harsh; upper leaves usually more than 5× longer than wide; heads subtended by peduncular leaves that exceed the involucral bracts; involucre usually less than 7 mm tall; plants rhizomatous and colonial, seldom exceeding 3.5 dm tall .... 1. *C. canescens*
    3 Upper herbage ± coarsely pubescent; upper primary pubescence harshly hirsute to hispid with pustulate-based hairs; upper leaves oblanceolate to lanceolate, often less than 5× longer than wide; peduncular leaves reduced upwards and often grading into the involucral bracts (or prominent in var. *foliosa*); involucre more than 7 mm tall; plants with short rhizomes or none, often exceeding 3.5 dm tall .... 6. *C. villosa*
  2 Upper herbage scattered-pubescent but distinctly green in aspect and with stipitate to sessile glands, or if the herbage is densely pubescent, the upper leaves are coarse and stiff.
    4 Heads apparently pedunculate, with peduncular leaves progressively reduced upwards and grading into the involucral bracts. .... 3. *C. horrida*
    4 Heads apparently sessile, with peduncular leaves quite distinct from the involucral bracts.

5 Principal cauline leaves more than 5× longer than wide, cuneate-sessile; lower stems early-defoliating and becoming glabrous .... 5. *C. stenophylla*
5 Principal cauline leaves less than 5× longer than wide, often subclasping; lower stems defoliating but at least some pubescence remaining .... 2. *C. fulcrata*

1. **Chrysopsis canescens** (DC.) T. & G. Perennial herb, 1.5–3(4) dm tall. Stems simple or much branched upward, arising clustered from a woody caudex or taproot, and often conspicuously rhizomatous; inflorescences and upper leaves covered with a dense, soft, silvery-white, appressed, sericeous pubescence, stems and lower herbage somewhat soft-hirsute and not at all rough-pubescent. Cauline leaves linear to linear-lanceolate, 5× or more longer than wide. Heads closely subtended by narrow peduncular leaves that exceed the involucral bracts; involucre turbinate to cylindrical, 5–7 mm tall and about as wide; involucral bracts imbricate in 3–5 series; ray florets fewer than 20, ligule 6–8 mm long, disk florets with corolla 5–7 mm long. Achenes with outer pappus of small, setaceous scales; inner pappus of numerous bristles. (n = 9, 18) Jul–Sep. Dry, open upland sandy or calcareous sites; s-cen KS, s & w to n TX; (s GP, s & w through TX). *Heterotheca canescens* (DC.) Shinners—Correll & Johnston; *C. berlandieri* Greene—Rydberg; *C. villosa* var. *canescens* (DC.) A. Gray—Atlas GP.

This species is similar to and perhaps intergrades with *C. villosa*, and is sometimes treated as a var. of that species. However, field collectors note that *C. canescens* is as distinct as any species in the group and worthy of recognition.

2. **Chrysopsis fulcrata** Greene. Herbaceous perennial 2–4(8) dm tall. Stems simple or branching upward, erect to somewhat lax-decumbent, arising from a woody caudex surmounting a taproot, stem-pubescence of 2 kinds: primary pubescence dense, often retrorsely hirsute-hispid, deciduous but with at least the hairbases persistent, secondary pubescence dense, hirsute, and usually stipitate-glandular. Lower leaves broadly spatulate, 2–6 cm long and 0.5–2 cm wide, petiole 5–20 mm long; middle and upper cauline leaves becoming sessile to weakly clasping, oblanceolate to oblong, blades variously hirsute to strigose and stipitate-glandular. Inflorescence a loose corymbiform cyme of 1–10+ heads, the heads appearing to be sessile and subtended by several peduncular leaves that are distinctly larger than the involucral bracts; involucre turbinate to subhemispherical, 6–12 mm across and about as tall; involucral bracts lance-linear, imbricate in 3–5 series, glandular to variously strigose-pubescent; ray florets numerous, the ligule 5–13 mm long; disk florets with corolla 5–7 mm long. Pappus with inner bristles 4–9 mm long, outer pappus of minute, fimbriate scales. (n = 9) May–Sep. Open, dry ground, apparently on the w edge of our range in WY; (MT to TX). *Heterotheca fulcrata* (Greene) Shinners—Correll & Johnston.

This is a variable, poorly understood taxon, doubtfully distinct from the next one.

3. **Chrysopsis horrida** Rydb. Herbaceous perennial 2–4+ dm tall; similar in aspect and technical features to *C. fulcrata*, but distinguished by having distinctly pedunculate heads; the peduncles with scattered reduced pedunculate leaves that grade into the involucral bracts; involucre turbinate-cylindrical, less than 8 mm across or tall. (n = 9) Jul–Sep. Scattered across the high plains of the sw 1/4 of the GP; (GP s & w to UT & AZ).

This is a complicated species, perhaps involving more than one entity. At least some of the materials referable here have been called *C. hispida* in some regional floristic works (e.g., Harrington), but that name properly belongs to the *C. villosa* assemblage.

4. **Chrysopsis pilosa** Nutt., soft golden-aster. Taprooted annual 3–5(8) dm tall. Stems arising singly, but much branched in upper 1/2; pubescence of 2 kinds: primary pubescence of long, soft pilose hairs and secondary pubescence spreading, short-puberulent and glandular. Leaves numerous, somewhat soft and pliant-flexible; lower leaves 3–6 cm long, margin

denticulate to weakly incised; upper leaves reduced, 1-2 cm long and becoming entire. Inflorescence a loose cluster of several heads, each distinctly pedunculate; involucral hemispherical, 1-2 cm across and up to 1.5 cm tall; involucral bracts imbricate in 3 or 4 poorly defined series, linear-lanceolate, attenuate. Achenes fusiform, distinctly 10-nerved, silvery pilose; outer pappus scalelike; inner pappus of numerous bristles. (n = 4) Jul-Sep. Open, dry sandy sites; se KS & adj. cen OK; (s MO to LA, se GP & TX). *Heterotheca pilosa* (Nutt.) Shinners—Correll & Johnston.

This is the only member of the genus in our flora with a chromosome number of n = 4, and it is referable to *Chrysopsis* sens. str., in a revised scheme of generic circumscription proposed by J. Semple, op. cit.

5. *Chrysopsis stenophylla* (A. Gray) Shinners. Perennial herb 1-3(4) dm tall. Stems simple or branching upward, arising singly or more often loosely clustered to subcaespitose from a woody, subrhizomatous crown, surmounting a taproot; herbage variously hispid-strigose, with abundant stipitate or sessile resin glands; lower stem defoliating early, becoming glabrous and shiny. Lower leaves petiolate; middle cauline leaves rigid-ascending, linear to narrow-lanceolate, 1-4 cm long and 1-7 mm wide, usually more than 5 × longer than wide. Inflorescence of numerous heads in compact terminal clusters; heads apparently sessile, subtended by several hispid-ciliate peduncular leaves that often exceed the involucre; involucre cylindrical or subhemispherical, 4-8 mm across and nearly as tall; involucral bracts imbricate in 3 or 4 weakly defined series, glandular viscid and sometimes sparsely hairy; ray florets 15-30, ligule about 10 mm long; disk florets with corollas 6-7 mm long. Achenes appressed-villous, outer pappus of conspicuous, narrow scales. (n = 9, 18) Jun-Sep. Open sandy upland sites; sw 1/4 GP, from w-cen KS, s & w; but sporadic n to se SD & IA; (s GP, s through TX). *Heterotheca stenophylla* (A. Gray) Shinners—Correll & Johnston.

6. *Chrysopsis villosa* (Pursh) Nutt., golden aster. Perennial herb (1)3-5(6) dm tall. Stems simple or more often branched upward, arising singly or loosely clustered from a thick woody caudex surmounting a taproot, sometimes short-rhizomatous; herbage with primary pubescence of pustulate-based hairs and somewhat rough, secondary pubescence of variously divergent to subappressed hairs. Lower leaves petiolate, becoming sessile upward; middle cauline leaves oblanceolate, 1-3 cm long and 3-8 mm wide. Inflorescence variously corymbiform to cymose-paniculate, with 3-30 heads surmounting each branch; involucre turbinate to subhemispherical, (6)7-12 mm tall and about as wide; involucral bracts imbricated in 4-9 series; ray florets ± 20-30, ligule 8-12 mm long, disk florets with corolla 5-8 mm long. Achenes with outer pappus of uneven, narrow, fringed scales; inner pappus of numerous bristles. (n = 9, 18) Jul-Sep for all vars. All occur in open, sandy or calcareous upland sites.

This highly variable species has both diploid and tetraploid phases. The morphological extremes in the GP are fererable to the following 4 varieties, but many populations are not clearly referable to any one variety. The varieties have some degree of populational integrity correlated with distribution.

1 Involucre 6-7 mm tall and ± densely strigose; leaves less than 5 mm wide .... 6a. var. *angustifolia*
1 Involucre usually 8-12 mm tall.
  2 Involucre hirsute and glandular-resinous .... 6c. var. *hispida*
  2 Involucre ± strigose-pubescent and sparsely if at all glandular.
    3 Heads sessile or nearly so, subtended by prominent peduncular leaves; stem leaves ± broadly oblong to ovate .... 6b. var. *foliosa*
    3 Heads pedunculate, with peduncular leaves reduced and grading into the involucral bracts; stem leaves oblanceolate to spatulate .... 6d. var. *villosa*

6a. var. *angustifolia* (Rydb.) Cronq. Mostly 2-4 dm tall, herbage tending toward being canescent rather than strongly hirsute. Mostly s 1/2 GP, especially cen NE & KS but sporadic in w OK & adj. TX, apparently in s SD; (s GP, s through TX). *Heterotheca villosa* var. *angustifolia* (Rydb.) Harms—Correll & Johnston; *C. angustifolia* Rydb.—Rydberg.

6b. var. *foliosa* (Nutt.) D.C. Eat. Up to 6 dm, but often 3–4 dm tall, herbage silky-stigose. Mostly in sw 1/4 GP, but n along the w edge of our range; (s GP, w to UT). *Heterotheca villosa* var. *foliosa* (Nutt.) Harms—Correll & Johnston; *C. foliosa* Rydb., *C. imbricata* A. Nels.—Rydberg.

6c. var. *hispida* (Hook.) A. Gray. Mostly 2–3 dm tall, with a short, compact, aspect; herbage hirsute and resinous-glanduliferous. Mostly nw 1/3 GP but occasional s to cen KS, & sporadic to TX; (GP w, apparently to B.C. & CA). *C. hispida* (Hook.) DC.—Rydberg.

6d. var. *villosa*. Mostly 2–4 dm tall, tending toward being grayish-strigose in aspect; freely intergrading with the other vars. Throughout the GP but only sporadic in the s 1/4; (MN & IN, w to B.C. & CA). *Heterotheca villosa* (Pursh) Shinners—Correll & Johnston; *C. hirsutissima* Greene, *C. bakeri* Greene, *C. ballardii* Rydb., *C. asperella* Greene—Rydberg.

30. CHRYSOTHAMNUS Nutt., Rabbit Brush

Much-branched shrubs with erect or upward-arching stems. Herbage glabrous to variously pubescent; twigs pannose-tomentose in 2 species. Leaves alternate, sessile or nearly so, narrow, entire or at most minutely scabrous-serrulate. Inflorescence a series of terminal racemose, cymose or corymbiform clusters. Involucre cylindrical; involucral bracts tending to be imbricated in vertical ranks, chartaceous to coriaceous, or with the tips greenish-herbaceous; receptacle naked; ray florets absent; disk florets mostly 5, perfect and fertile; corolla yellow, tubular-funnelform with the apex parted into 5 prominent lobes; style branches with inner-marginal stigmatic lines and elongate, short-hairy appendages. Achenes slender, somewhat angled, pubescent or nearly glabrate; pappus of copious white to sordid bristles. [tribe Astereae]

References: Hall, H. M., & F. E. Clements. 1923. *Chrysothamnus*, in The phylogenetic method in taxonomy. Publ. Carnegie Inst. Wash. 326: 157–234. [A clearly out-dated work, but a classic summary of this complex genus.] Loran C. Anderson of Florida State University has published a series of very useful anatomical and cytotaxonomic papers on the genus, in several journals.

1 Branches covered with a close, feltlike tomentum of long, entangled hairs, ± infiltrated with resin (note that the tomentum may be mistaken for bark unless scraped with a sharp edge).
  2 Heads in terminal cymose or corymbiform clusters; involucral bracts obtuse to acute but lacking herbaceous or elongate tips; involucre less than 10 mm tall .... 1. *C. nauseosus*
  2 Heads in leafy racemes; outer involucral bracts with elongate, green tips; involucre 10–12 mm tall .... 2. *C. parryi* subsp. *howardii*
1 Branches glabrate or puberulent, but not felted-tomentose.
  3 Involucre 10–12 mm tall, corolla 10–14 mm long, achene normally glabrous .... 3. *C. pulchellus* subsp. *baileyi*
  3 Involucre less than 8 mm tall, corolla less than 7 mm long, achene pubescent .... 4. *C. viscidiflorus*

1. Chrysothamnus nauseosus (Pall.) Britt. Shrub 2–15(20) dm tall; twigs pannose-tomentose, i.e., densely covered with a close gray-green to whitish, felted tomentum that may be infiltrated with resin and may resemble bark, until scraped with a knife-blade. Leaves linear or narrowly linear-lanceolate, 1- to 3(5)-nerved, glabrous to somewhat tomentose, 2–6 cm long and up to 2 mm wide. Heads in terminal, rounded compound-cymose or corymbiform clusters; involucre 6–8 mm tall; involucral bracts 20–25 in vertical ranks; florets 5 per head; corolla 6.5–9 mm long, with lobes 0.5–2 mm long. Achenes ca 5 mm long, 5-angled, pubescent; pappus of numerous capillary bristles. (n = 9) Jun–Sep.

This species is widespread and abundant throughout much of w U.S., with several semidistinct subspecies. Two occur in the GP, and a third, subsp. *albicaulis* (Nutt.) Rydb., may enter our range in ne WY. It is distinguished by possessing abundant, loose, arachnoid tomentum on the branches and leaves.

1a. subsp. *graveolens* (Nutt.) Piper. Shrub, usually 6–15(20) dm tall; involucral bracts ± prominently keeled and glabrous, at least on the outer surface. Open dry hills & plains; scattered in w 1/5 GP, especially in the nw, but also in e CO & n TX; (GP to ID & AZ). *C. nauseosus* var. *glabratus* (A. Gray) Cronq.—Cronquist *in* Hitchcock et al.

1b. subsp. *nauseosus*. Low shrub 2–6 dm tall; involucral bracts weakly if at all keeled, at least the outer ones tomentulose. Open hills & high plains; scattered across the w 1/5 GP from e MT to n TX; (GP to w MT, CO, UT). *C. frigidus* Greene, *C. plattensis* Greene—Rydberg.

2. Chrysothamnus parryi (A. Gray) Greene subsp. **howardii** (Parry) Hall & Clem. Low shrub, 2–6 dm tall; twigs covered with a whitish or greenish pannose tomentum, i.e., a closely felted tomentum resembling bark. Leaves narrowly linear, 2–4 cm long and ca 1 mm wide, with a single prominent midnerve, grayish tomentose, the uppermost usually overtopping the inflorescence. Heads in a terminal, leafy racemose cluster; involucre 10–12 mm tall; involucral bracts 12–20, in vertical ranks, ± pubescent, at least on the margins, midrib prominent, the tip elongate and often spreading; florets 5–7 per head, corolla 8–11 mm long with lobes 0.5–2.5 mm long. Achenes 5–6 mm long, densely appressed-villous; pappus of numerous capillary bristles. ($n = 9$) Aug–Sep. Open, dry plains; scattered in w-cen GP; NE: Kimball, Scotts Bluff; SD: Custer, Fall River; (GP, WY, & CO). *C. howardii* (Parry) Greene—Rydberg.

3. Chrysothamnus pulchellus (A. Gray) Greene subsp. **baileyi** (Woot. & Standl.) Hall & Clem. Low shrub 3–5 dm tall; twigs glabrous, greenish or whitish. Leaves linear or linear oblong, up to 4 cm long and 2 mm wide, with a single prominent nerve, margin scabrous-ciliate, otherwise glabrous. Inflorescence a cymose cluster; involucre 10–12 mm tall; involucral bracts numerous, in 5 vertical ranks, distinctly keeled on the back, greenish and attenuate at the apex; florets usually 5, corolla 10–14 mm long, with a slender tube, expanding upward into an open throat, lobes 1.5–2 mm long. Achenes 6–7 mm long, usually glabrous but reputedly sometimes puberulent; pappus of numerous capillary bristles, exceeding the corolla. ($n = 9$) Aug–Sep. Open high plains, especially in sandy sites; scattered in sw 1/5 GP, sw KS, se CO & s; (GP, w TX, CO, NM, & n Mex.). *C. baileyi* Woot. & Standl.—Rydberg.

4. Chrysothamnus viscidiflorus (Nutt.) Hook. Highly variable shrub 5–10(20) dm tall; twigs pale green or whitish, glabrous or with a little, scattered pubescence. Leaves linear to narrowly lanceolate, often conspicuously twisted, 2–5 cm long and 1–4(5) mm wide, with 1 or 3 prominent nerves, glabrous and usually somewhat viscid. Inflorescence a terminal cymose cluster; involucre 5–7 mm tall; involucral bracts ca 15, rounded on the back and not prominently keeled, arranged in poorly defined vertical ranks; florets usually 5, corolla 4.5–7 mm long, tubular-funnelform and gradually tapering from tube to throat, lobes 1–2 mm long. Achenes 3–4 mm long, densely villous; pappus of numerous ± sordid capillary bristles, equaling or slightly exceeding the corolla. ($n = 9$, 18 + a polyploid series) Jul–Sep. Open dry plains, especially in disturbed sites; extreme w GP; NE: Sioux; e WY, perhaps adj. MT; (GP w to B.C., CA, & AZ).

This is an abundant, highly variable and often somewhat weedy species throughout the western cordillera, and it barely enters the GP. Several subspecies are recognized by Hall & Clements (op. cit.), but they seem to be based largely on morphological variation and are but weakly correlated to populational integrity or distribution. Our materials as described above are referable to subsp. *viscidiflorus*. However, subsp. *lanceolatus* (Nutt.) Hall & Clem. may be expected at the extreme w edge of the GP in WY. It differs in having herbage ± densely puberulent throughout, the larger leaves are 3- to 5-nerved, flat and not twisted.

31. CICHORIUM L., Chicory

1. *Cichorium intybus* L. Erect, branching perennial 5–15 + dm tall, arising from a simple or much-branched taproot, or sometimes subrhizomatous. Leaves alternate, sessile to short-petiolate, the principal ones 8–35 cm long and 2–7 cm wide, lyrate-pinnatifid to subentire, the uppermost reduced to rigid, subentire bracts. Heads numerous, short-pedunculate to sessile, borne singly or in small clusters in the axils of the upper leaves, thus producing a long, interrupted racemiform inflorescence, or some heads terminal on long branches; involucre broadly cylindrical, 1–1.5 cm tall; involucral bracts in 2 series, the outer ca 1/2 as long and fewer than the inner; florets all ligulate and fertile, corolla blue or rarely whitish. Achenes glabrous, obscurely striate, weakly if at all compressed; pappus of 2 or 3 obscure series of reduced scales. (n = 9) Jun–Oct. Occasional as a roadside weed & in other disturbed sites; scattered GP, most frequent in se 1/4; (cosmopolitan weed; Eurasia). *Naturalized.* [tribe Lactuceae]

The species description provided here is for the widespread weedy phase, but there are horticultural phases grown for greens and for the root, which is roasted and used as a coffee substitute or adulterant.

32. CIRSIUM P. Mill., True Thistles

Contributed by Ralph E. Brooks

Ours biennial or perennial herbs. Stems erect, simple, branching above or profusely branched, furrowed, glabrate to arachnose-tomentose. Rosette leaves compact, mostly entire at first and the later ones entire to deeply pinnatifid, margins usually spiny; cauline leaves alternate, entire to pinnatifid, glabrate to white tomentose on one or both surfaces, margins variously spiny, base clasping or not, to strongly decurrent. Heads discoid, solitary or in small clusters at the branch ends, many-flowered; receptacle flat to subconic, densely bristly; involucral bracts imbricated, ovate to lanceolate, mostly spine-tipped. Florets perfect, except in the dioecious *C. arvense;* ray florets absent; disk florets ochroleucous, dark purple to pink, or white, long-tubular, the corolla lobes linear; anthers 5, sagittate at the base, filaments separate, usually papillose and variably hairy; styles mostly filiform and with a thickened, often hairy, ring at the base of the stigmatic portion; pappus of numerous plumose bristles united in a ring at the base, pappus deciduous as a unit. Achenes oblong to obovate and usually flattened. [tribe Cynareae]

References: Moore, R. J., & C. Frankton. 1969. Cytotaxonomy of some *Cirsium* species of the eastern United States with a key to eastern species. Canad. J. Bot. 47: 1257–1275; Moore, R. J., & C. Frankton. 1974. The thistles of Canada. Canad. Dept. Agric. Res. Branch Monog. No. 10. 111 pp.

```
1  Upper leaf surface bearing numerous short, yellowish, appressed prickles ....... 11. C. vulgare
1  Upper leaf surface glabrous to tomentose but lacking prickles.
   2  Plants dioecious (sometimes imperfectly so), perennial with widely creeping roots bearing
      adventitious shoots; involucre 1–2 cm tall, 0.5–1 cm wide ...................... 2. C. arvense
   2  Florets perfect, biennial or perennial, involucres mostly larger.
      3  Outer involucral bracts abruptly acuminate, lacking a spine tip ........ 6. C. muticum
      3  Outer involucral bracts tipped with a spine at least 2 mm long.
         4  Middle and upper cauline leaves strongly decurrent, the wing often extending
            below to the next node.
            5  Involucre 1.5–2.7 cm tall, 1–1.8 cm wide; stems white floccose to glabrate
               mixed with larger multicellular hairs .......................... 8. C. pulcherrimum
            5  Involucre 2.3–4 cm tall, 2–4 cm wide; stems white or grayish tomentose but
               lacking larger multicellular hairs.
               6  Florets ochroleucous (rarely pink); spines of the outer involucral bracts 2–4
                  mm long ................................................................. 3. C. canescens
```

            6   Florets purple to pink (rarely white); spines of the outer involucral bracts 4–10 mm
                long .... 7. *C. ochrocentrum*
    4   Middle and upper cauline leaves sessile, clasping, or if decurrent then for less than 1 cm.
        7   Stems densely white tomentose at least on the middle and upper portions.
            8   Heads in clusters of 2–4 (rarely solitary); restricted to the nw edge of the
                GP .... 8. *C. pulcherrimum*
            8   Heads solitary on the branches; widely distributed in the n GP or throughout.
                9   Cauline leaves ovate to lanceolate, mostly shallowly lobed to subentire; achenes 5–7
                    mm long, 2–3 mm wide; flowering Jun–Jul (Aug); propagating by shoots from
                    deep taproots .... 10. *C. undulatum*
                9   Cauline leaves elliptic to oblanceolate, shallowly lobed to pinnatifid; achenes
                    3–4(5) mm long, 1.5–2 mm wide; flowering Jul–Sep; propagating by buds borne
                    on horizontal roots .... 5. *C. flodmanii*
        7   Stems green and glabrate, sparsely arachnose, or pilose with large multicellular hairs.
            10 Involucres mostly 3–5.5 cm tall and about as wide; restricted to the
                BH .... 4. *C. drummondii*
            10 Involucres 1.5–3.5 cm tall and about as wide.
                11 Plant from a napiform taproot, flowering in May or early
                    Jun .... 9. *C. texanum*
                11 Plant from a slender to stout, fleshy taproot, flowering from Aug–early
                    Oct .... 1. *C. altissimum*

1. ***Cirsium altissimum*** (L.) Spreng., tall or roadside thistle. Biennial from a fleshy taproot, 1–2.5 m tall. Stems freely branching above, green or greenish brown, lightly pilose or floccose and often mixed with larger multicellular hairs. Leaves green and glabrous or nearly so above, densely white tomentose below; first year rosette leaves unlobed to occasionally variously lobed, oblanceolate to elliptic, 10–30 cm long, 4–11 cm wide, margins spinose-serrulate, base gradually tapering and the petiole winged; rosette leaves of the second year and cauline leaves of the main axis similar or sometimes pinnatifid, the petiole shorter or leaves sessile; branch leaves gradually reduced upwards, mostly sinuate to unlobed, narrowly elliptic to oblanceolate, margins irregularly spinose-serrate to -serrulate, base sessile and clasping or not. Heads solitary and terminal on the branches; involucre 2–3.5 cm tall, 2–3.8 cm wide; involucral bracts in 8–10 rows; outer ones (5–7 rows) ovate to ovate-lanceolate, 3–7 mm long, 1.8–3 mm wide, glabrate to weakly arachnose on the margins, acute to acuminate and tipped by a small spreading spine 2–3 mm long; inner ones lanceolate, to 15 mm long, apex distinctly dilated and erose to occasionally attenuate, flexuous, and minutely serrulate. Corolla dark purple or lighter, infrequently white, 22–32 mm long, the lobes 6–9 mm long; anthers 7.5–10 mm long; style 25–36 mm long. Achenes pale brownish or darker with a yellow apical ring, 4.5–6 mm long, 1.5–2 mm wide; pappus white or grayish white, 17–27 mm long. (2n = 18) Aug–Sep (Oct). Open lowland areas, ditches, roadsides, & waste areas; MN s to MO, e 1/3 ND, e 1/3 SD, NE, e 2/3 KS, & e 1/2 OK; (ME w to e ND, s to FL & TX). *C. altissimum* f. *moorei* Steyerm.—Fernald; *C. iowense* (Pammel) Fern.—Rydberg.

Often reported for the GP is *C. discolor* (Muhl. ex Willd.) Spreng., its range approaching the eastern edge of the GP from MN south to AR. While it may occur here, no specimens have been seen that clearly belong to that taxon. *Cirsium discolor* is distinguished from *C. altissimum* by having more deeply pinnatifid cauline leaves with linear-lanceolate lobes and having attenuate, not dilated, apices on the inner involucral bracts. In the GP it is not uncommon to find plants with unlobed leaves and attenuate involucral bracts that are otherwise clearly *C. altissimum*, indicating that characters normally used to separate these closely related species break down in our region.

2. ***Cirsium arvense*** (L.) Scop., Canada or field thistle. Dioecious perennial. Stems branching above, 3–12 dm tall, glabrate to sparsely arachnose, white tomentose on the upper portions when leaves are white tomentose on the lower surface; spreading profusely by deep horizontal lateral roots bearing adventitious shoots. Lower cauline leaves shallowly lobed

to pinnately lobed, occasionally unlobed, oblong to narrowly elliptic or oblanceolate, 5-18 cm long, 1.5-6 cm wide, both surfaces glabrous or the upper lightly arachnose-floccose and the lower densely white tomentose, the lobes rounded to acute and with margins bearing short fine spines or strong spines to 5 mm long and these varying in abundance, leaf bases sessile or occasionally with a petiole to 1 cm long, clasping or not to short decurrent (10-20 mm) upper cauline leaves similar but progressively reduced, often less lobed (especially branch leaves), and sessile. Heads mostly in loose, corymbiform clusters terminating the branches; involucre 1-2 cm tall, 0.5-1 cm wide, male heads sometimes slightly shorter, involucral bracts in 5 or 6 rows; outer ones ovate, 2-6 mm long, 0.7-1.2 mm wide, glabrous to arachnose, especially on the margins and with a narrow glandular dorsal ridge, subulate-tipped, the spine less than 1 mm long; inner ones lanceolate, to 11 mm long, attenuate to slightly dilated and erose at the apex. Corolla pink to purple, rarely white. Staminate corollas 12-14 mm long, the lobes 3-4 mm long; anthers 3.5-4 mm long; apparently normal pistillate parts sometimes present but with only a vestigial ovary. Pistillate corollas 19-24 mm long, the lobes 2-3 mm long; stigma joint about 2 mm from top; vestigial anthers sometimes present. Achenes straw colored or light brown, 2.5-4 mm long, 1-1.5 mm long; pappus white or grayish white, at anthesis about 15 mm long, later to 25 mm long. ($2n = 34$) Jun-Aug. Open areas including pastures, ditches, bottomlands, & waste areas; MN w to MT, s to MO, KS, & CO; (widely established in n U.S. & Can.; Eurasia & N. Africa). *Naturalized.* *C. arvense* var. *horridum* Wimmer & Grab. — Gleason & Cronquist; *C. arvense* var. *mite* Wimmer & Grab., *C. arvense* f. *albiflorum* (Rand & Redf.) R. Hoffm. — Fernald; *C. setosum* (Willd.) Bieb. — Rydberg; *C. arvense* var. *vestitum* Wimmer & Grab. — Scoggan.

This is an extremely variable complex with regard to leaf division and vestiture, and it has been treated as several species, numerous varieties, or as a single highly polymorphic species.

3. ***Cirsium canescens*** Nutt., Platte thistle. Monocarpic perennial from a deep, slender to stout taproot. Stems 4-8 dm tall, simple or branched above, tomentose. Leaves arachnoid and greenish above, densely white tomentose below; first basal leaves narrowly elliptic with entire or undulate margins, later leaves larger, 12-30 cm long, 3-5(7) cm wide, and more deeply lobed, the lobes in 5-12 pairs and mostly at right angles to the midvein, oblong or narrowly so, 3-6 × longer than wide, each lobe tipped with a spine and the margins with weaker spines; cauline leaves gradually reduced upwards, narrowly elliptic to lanceolate, 3-7 cm long, 0.5-2.5 cm wide, the lowest deeply pinnatifid and upward becoming shallowly lobed to subentire, margins weakly spinose, decurrent with the wings usually extending downward to the next node and usually provided with stout spines. Heads at first terminal and large, smaller subsessile heads are sometimes produced later on axillary shoots; first involucres 3-4 cm tall, 2.5-4 cm wide, the later ones as small as 1.5 cm tall and equally wide; involucral bracts in 6-8 rows, lanceolate, glabrate with arachnose margins; the outer ones (3 or 4 rows) 7-17 mm long, 1.5-2.5 mm wide, with a dark, glandular dorsal ridge and yellow spine tip 2-4 mm long; inner ones to 23 mm long, glandular dorsal ridge less distinct, apex brownish, flexuous, slightly dilated, and erose. Corolla ochroleucous, rarely pale lavender, 24-28 mm long, the lobes 5.5-7 mm long; anthers 9-10 mm long; style 25-33 mm long, stigma joint 5-6 mm from tip. Achenes straw colored and usually with brownish streaks, 5-7 mm long, 2.3-2.7 mm wide, smooth; pappus white, 18-30 mm long. ($2n = 34, 36$) Late May-Jun (Jul). Sandy or gravelly soil in upland prairies & especially disturbed sites; NE except se, sw SD, e 1/3 WY, ne 1/4 CO, e 1/4 MT; (NE & sw SD, w to UT & s to cen CO). *C. nelsonii* (Pammel) Rydb., *C. nebraskense* (Britt.) Lunell, *C. plattense* (Rydb.) Cockll. — Rydberg.

This species has been reported for IA: Osceola and ND: McLean, but specimens have not been seen and may be misdetermined.

4. *Cirsium drummondii* T. & G., Drummond's thistle. Biennial herbs, 2-6(7) dm tall. Stems simple, fleshy, robust, 7-15 mm diam, pilose (multicellular hairs) or becoming glabrate; taproot short, slender. Leaves ascending and most numerous at or near ground level, oblanceolate to linear-lanceolate, 8-25 cm long, 1.5-6.5 cm wide, the largest leaves just above the base, reduced upwards, subentire (mostly the basal leaves) to deeply sinuate-pinnatifid with 7-10 pairs of triangular to oblong lobes, sparingly pilose especially on the veins (multicellular hairs), lower surface sometimes appearing lighter green than the upper, margins beset with numerous yellowish spines 1-5 mm long, the longest at lobe apices, bases clasping the stem. Heads terminal, solitary or 2-5 on short peduncles in a close group at the stem apex, usually subtended by 1-3 narrow leaves which may or may not exceed the head; involucre 3-5.5 cm tall and about as wide; involucral bracts in 4-7 rows, glabrous, glandular, the glands sessile and difficult to see on dried specimens, most abundant on the inner involucral bracts, margins ciliate; the outer 2-4 rows narrowly ovate to ovate-lanceolate, 1-2 cm long, 4-6 mm wide, apex usually dark purplish, acuminate, and with a spine 2-3 mm long; inner ones ovate-lanceolate to lanceolate, 2-3.5 cm long, 3-4 mm wide, apex tapering to a broadly dilated, brownish, chartaceous, and erose tip. Corolla rose-purple, 38-44 mm long, lobes 4-7 mm long; anthers 7-9 mm long; style 42-50 mm long, stigma joint about 6-7 mm from the tip. Achenes straw colored with purplish streaks and a yellow apical rim, 4-5.5 mm long by 1.3-2 mm, smooth; pappus white, 30-40 mm long, shorter than the corolla. (n = 17) Jun-Jul (Aug). Moist meadows or pine woodlands; SD: Custer, Lawrence, Pennington; WY: Crook, Weston; (Ont. w to B.C., disjunct in w SD & ne WY). *C. coccinatum* Osterhout—Osterhout, Torreya 34: 45, 1934; *C. foliosum* of GP reports.

5. *Cirsium flodmanii* (Rydb.) Arthur, Flodman's thistle. Perennial; stems usually branching in the upper portion, 3-10 dm tall, white tomentose; readily propagating by buds borne along horizontal roots. Juvenile leaves subentire or occasionally shallowly lobed, narrowly elliptic to oblanceolate, gray tomentose on both surfaces but more so on the lower, apex acute to obtuse, margins irregularly spinose toothed; later rosette leaves of first year and those of second year green and floccose above and gray tomentose below, pinnatifid, remotely lobed or entire, narrowly elliptic to oblanceolate, 12-22(30) cm long, 3-7(10) cm wide, the lobes bifid or not and directed slightly toward the leaf apex, segments lanceolate, tipped by a spine 3-6 mm long, margins with irregular spinose teeth, petiole winged; cauline leaves similar to the rosette leaves or all leaves entire, progressively reduced upwards, sessile and usually clasping at the base. Heads solitary and terminal on the branches; involucre 2-3 cm tall, 1.5-2.5 cm wide; involucral bracts in 6 or 7 rows; outer ones (4 or 5 rows) ovate or ovate-lanceolate, 5-9 mm long, 2.5-3 mm wide, glabrate or the margins slightly arachnose, dorsal ridge purplish and strongly glandular, tipped by a diverging spine 2-4 mm long; inner ones lanceolate, to 18 mm long, apex flexuous and mostly unarmed. Corolla deep purple or infrequently pink, rarely white, 21-36 mm long, corolla lobes 5-9 mm long; anthers 8-11 mm long; style 28-33 mm long, stigma joint 5-6 mm from tip. Achenes brownish, 3-4(5) mm long, 1.5-2 mm wide, with a whitish-yellow apical ring, smooth; pappus white or tawny, 20-30 mm long. (2n = 22, 24) Jul-Sep. Moist, open sites, meadows, pastures, & waste places; MN w to MT, s to IA, n-cen KS, & WY; (s Man. w to s B.C., s to IA, KS, CO, & AZ; introduced in ne U.S.). *C. oblanceolatum* (Rydb.) K. Schum.—Rydberg.

6. *Cirsium muticum* Michx., swamp thistle. Biennial from a fleshy taproot. Stems sparingly branched above, to 18 dm tall; glabrous or with scattered multicellular hairs especially just beneath the heads. Leaves green and sparsely villose above (multicellular hairs), lower surface more or less floccose and with scattered multicellular hairs on the veins; rosette

and lower cauline leaves deeply pinnatifid, with 4-8 pairs of lobes mostly elliptic, 10-25(55) cm long, 3-10(20) cm wide, the lobes lanceolate, ovate, or broadly so, occasionally bifid but mostly with several secondary lobes, each tipped with short weak spines, margins with scattered minute spines, petiole winged; upper cauline leaves progressively smaller and sessile, otherwise similar to the lower leaves. Heads solitary or in clusters of 2 or 3 and terminal on the branches; involucres 1.7-2.5 cm tall and about as wide; involucral bracts in 6-8 rows; outer ones (4-6 rows) mostly ovate, 4-9 mm long, 2-4 mm wide, arachnoid to tomentose especially on the margins, dorsal ridge glandular, apex acute to abruptly acuminate and lacking a spine tip, margins entire or serrulate at the apex; inner ones lanceolate or narrowly so, to 15 mm long, apex flexuous or not and usually serrulate. Corolla dark purple to lavender, 20-27 mm long, the lobes 4-7 mm long; anthers 7-9 mm long; styles 23-30 mm long. Achenes dark brown with a yellow apical rim, 4-6 mm long, 1.5-2 mm wide, smooth; pappus whitish, 13-20 mm long. (2n = 20, 22, 30, 31) Mid-Aug-Sep. Moist low woodland margins, thickets, riverbanks, or meadows; nw MN, ND: e 1/4 & Bottineau; (Newf. w to Sask., s to FL & e TX).

7. Cirsium ochrocentrum A. Gray, yellowspine thistle. Biennial or short-lived perennial (?), sometimes reproducing by tuberous offsets. Stems 4-10(15) dm tall, simple or sparingly branched above, densely white tomentose; taproot slender to stout, rarely branching at its apex. Leaves greenish or grayish and arachnose-floccose above, white tomentose below; juvenile leaves narrowly obovate to elliptic, apex obtuse to acute, margins irregularly spinose; later basal leaves oblanceolate to narrowly elliptic; pinnatifid, 8-22 cm long, 3-6 cm wide, the lobes mostly directed towards the apex, triangular to widely ovate, often bifid, acute and tipped by a yellow spine 4-12 mm long, margins irregularly toothed and these spine-tipped, petiole narrowly winged; cauline leaves arcuate or less often spreading, upwards gradually reduced and becoming less deeply lobed, narrowly oblong to ovate, at least middle and upper leaves decurrent for more than 1 cm along the stems, frequently to the next lower node. Heads terminal and solitary on the branches; involucre 2.3-4 cm tall, 2-3.5 cm wide; involucral bracts in 5-7 rows; arachnose at least on the margins; outer ones (3 or 4 rows) ovate to narrowly so, 6-21 mm long, 2-3 mm wide, with a dark, glandular dorsal ridge and spreading yellow spine tip 4-10 mm long; inner ones lanceolate, to 30 mm long, glandular dorsal ridge less developed, apex brownish, weakly dilated or not, erose. Corolla purple to pink, rarely white, 30-38 mm long, stigma joint 7-9 mm from tip; anthers 11-13 mm long; style 30-45 mm long, the branches 4-7 mm long. Achenes tan to brownish, 6-7 mm long, 2.5-3 mm wide, smooth; pappus white or tawny, 17-26 mm long. (2n = 30, 31, 32, 34, 35) Jul-Aug (Sep). Dry sandy or gravelly soil in prairies, pastures, & open disturbed sites; s-cen & w NE, se WY; SD: Fall River; w 1/2 KS, CO, w OK, TX, NM; (s-cen NE w to se WY, s to nw TX, NM, & AZ).

8. Cirsium pulcherrimum (Rydb.) K. Schum. Perennial herbs, 4-7 dm tall. Stems simple or sparingly branched above, finely white floccose to glabrate, mixed with scattered large multicellular hairs; taproot slender. Leaves spreading to slightly ascending, pale green and glabrous to sparingly floccose above, white tomentose below, narrowly oblanceolate or occasionally narrowly elliptic, 12-22 cm long, 1.5-6.5 cm wide, the largest at the base, reduced upwards, deeply sinuate-pinnatifid or less often shallowly lobed, the lobes in 4-8 pairs, triangular to oblong and sometimes irregularly divided, and with rounded, spine-tipped apices, margins entire and spineless or with numerous, small spines, bases clasping on the lower leaves but frequently short decurrent (1-3 cm) on the upper leaves. Heads in clusters of 2-4 (rarely solitary) terminating the branches, peduncles 2-10(20) mm long; involucre 1.5-2.7 cm tall, 1-1.8 cm wide; involucral bracts in 4-6 rows, appressed, lanceolate and spine-tipped, 7-19 mm long, 1-2 mm wide, with sparse sessile glands, pilose along the entire margin and glabrous otherwise, spine tip 2-8 mm long. Corolla pink to purple,

18–25 mm long, stigma joint 4–6 mm from the tip; anthers 7–9 mm long; style 21–26 mm long, the branches 3–5 mm long. Achenes pale tan with a yellow apical rim, 5–6 mm long, 2–3 mm wide, smooth; pappus grayish white, 15–18 mm long. (2n = 34) Late Jun–Aug. Rocky upland prairies or open pine woodlands; NE: Scotts Bluff; MT: Dawson, Garfield; WY: Campbell; (s MT s to WY, w to ID & UT; NV?).

9. *Cirsium texanum* Buckl., Texas thistle. Biennial or weak perennial from napiform taproot (not often collected). Stems 0.5–2 m tall, simple or branching above, grayish tomentose to glabrate. Rosette leaves (at least later ones) oblanceolate to oblanceolate-elliptic, 13–20 cm long, 4–9 cm wide, pinnatifid with 6–8 pairs of lobes, the lobes ovate to nearly round and mostly with several irregular weakly spine-tipped secondary lobes, green or glabrous or infrequently floccose above, grayish or white floccose-tomentose below, margins with scattered small spinose teeth, the base gradually tapering to a winged petiole; cauline leaves similar but reduced upwards and often absent from the uppermost portion of the flowering branches, becoming less lobed, and often decurrent along the stem. Heads solitary and terminal on the branches; involucre 1.5–2.5 mm tall and about as wide, much wider in age; involucral bracts in 6–8 rows; outer bracts lanceolate, 4–10 mm long, 1–1.5 mm wide, thinly floccose to glabrate, entire, apex with a spreading spine tip 2–4 mm long; inner bracts narrowly lanceolate, to 13 mm long, apex weakly spinose, flexuous. Corolla pink to rose-purple, 18–23 mm long, the lobes 5–6 mm long; anthers 5–6 mm long; style 20–25 mm long. Achenes pale tan, 3–4 mm long, about 2 mm wide, smooth; pappus white, 15–18 mm long. May (Jun). Disturbed, usually sandy soils of pastures, roadsides, & other waste areas; OK: Harmon, Jackson; TX: Wichita; (sw OK s to Tam., N.L., & Coah.).

10. *Cirsium undulatum* (Nutt.) Spreng., wavy-leaf thistle. Perennial with a short, thick deep subterranean taproot, sometimes reproducing by tuberous offsets. Stems 4–10(15) dm tall, simple or sparingly branched above, densely white tomentose; propagating by shoots from usually deep subterranean taproots to form a perennial clone. Leaves usually densely tomentose on both surfaces, the upper surface sometimes less so and greenish or becoming glabrate, in age juvenile leaves elliptic, apex obtuse to acute, margins regular to irregularly spinose toothed; later basal leaves elliptic to broadly oblanceolate, pinnatifid and undulate, 10–30(45) cm long, 2–7(15) cm wide, the lobes mostly directed toward the leaf apex, broadly ovate and sometimes bifid, tipped by a yellow spine to about 5(8) mm long, the margins otherwise with few spines and these usually small, petiole winged and the base clasping; cauline leaves spreading, ascending or sometimes arcuate, gradually becoming reduced upwards, ovate to lanceolate, shallowly lobed to subentire, sessile and clasping or short decurrent (less than 1 cm). Heads terminal and solitary on the branches; involucre 2–4 cm tall, 2–3.5 cm wide; involucral bracts in 6–8 rows; outer ones ovate-lanceolate, 6–17 mm long, 2–4.5 mm wide, margins entire and arachnose, apex usually tipped with a spine 2–5(8) mm long and this usually spreading; inner ones lanceolate, to 33 mm long, apex usually flexuous. Corolla purple to pale purple or white, 30–40 mm long, the lobes 7–10 mm long; anthers 12–14 mm long; style 35–45 mm long, stigma joint 4–6 mm from the tip. Achenes straw colored to brown, 5–7 mm long, 2–3 mm wide, smooth; pappus white, 22–40 mm long. (2n = 26) Jun–Jul (Aug). Dry prairies, pastures, roadsides, & other open disturbed sites; common throughout GP; (s Man. w to B.C., s to MI, MO, TX, & Mex.). *C. megacephalum* (A. Gray) Cockll.—Rydberg; *C. undulatum* var. *megacephalum* (A. Gray) Fern.—Fernald.

11. *Cirsium vulgare* (Savi) Ten., bull thistle. Plants biennial from a short, fleshy taproot, 0.5–2 m tall, bearing many spreading branches. Stems green or brownish, sparsely arachnoid (multicellular hairs), at least in the upper portion irregularly and spiny winged. Leaves green, glabrous, and bearing fine short yelowish appressed prickles above, lower surface green or grayish villose (multicellular hairs), especially on the veins; juvenile leaves elliptic to obovate, unlobed at first but soon with shallowly to pinnately lobed leaves, margins irregularly toothed and these tipped with weak spines 1–5 mm long; petiole winged; later basal leaves oblanceolate to narrowly elliptic, 12–30(50) cm long, 3–15 cm wide, deeply pinnatifid with 4–6 pairs of lobes, the lobes frequently bifid, the segments mostly lanceolate to triangular, apex acute to acuminate and tipped with a spine to about 10 mm long, margins entire or with a few remote spines; cauline leaves similar to the rosette leaves but progressively smaller, the lobes often linear-lanceolate with stouter spines to 17 mm long, and the leaf bases strongly decurrent. Heads numerous, terminal and solitary on the branches, often appearing clustered where the branches are poorly developed; involucres 2–3 cm tall, 2–4.5 cm wide; involucral bracts in 5–8 rows; lanceolate to narrowly so, 7–30 mm long, 1–2 mm wide, progressively longer and narrower from outer to inner ones, at least the margins arachnose and sometimes with scattered glandular hairs at the apex, apex spreading, spine-tipped, margins entire. Corolla dark purple, 27–35 mm long, the lobes 5–6 mm long; anthers 7–8 mm long; styles 28–37 mm long, stigma joint 4–5 mm from the tip. Achenes whitish or pale yellow with dark brown streaks, 3–4 mm long, 1.3–1.6 mm wide; pappus white to tawny, 20–28 mm long. (2n = 68) Jul–Sep. Roadsides, pastures, & other waste areas; MN w to MT, s to MO, e OK, nw KS, & WY; no records from sw GP but to be expected; (widely established as a weed in N. Amer.; Eurasia). *Naturalized.*

33. CONYZA Less.

Annual weedy herbs with stems arising from a taproot. Leaves alternate, narrow, entire or the larger ones toothed. Inflorescence of numerous small heads, inconspicuously radiate (ours) or disciform; involucral bracts imbricate, herbaceous; receptacle flat, naked; pistillate florets numerous, corolla slender and prolonged into a very short, narrow, inconspicuous, white or pinkish ligule, about equaling or barely exceeding the pappus; disk florets few, tubular and perfect, corolla yellowish to light pinkish-purple; style branches flattened, with inner-marginal stigmatic lines and short, externally pubescent appendages. Achenes ± compressed and 2-ribbed; pappus of capillary bristles. [tribe Astereae]

Reference: Cronquist, A. 1943. The separation of *Erigeron* from *Conyza*. Bull. Torrey Bot. Club 70: 629–632.

1 Plants erect and simple, or branching above, with a well-defined central axis; herbage usually spreading-hirsute .... 1. *C. canadensis*
1 Plants low, bushy-diffuse branching from near the base; stems with antrorsely appressed hairs .... 2. *C. ramosissima*

1. *Conyza canadensis* (L.) Cronq., horse-weed. Annual (1)3–15(25) dm tall; stem erect, simple or nearly so to the inflorescence, or sometimes branching above due to injury; herbage mostly coarsely spreading-hirsute but sometimes glabrescent. Leaves abundant, narrowly oblanceolate to linear, sessile, entire or the lower ones coarsely toothed, progressively reduced upward, prominent cauline leaves up to 8 cm long and 1 cm wide, basal leaves larger but early deciduous. Inflorescence an elongate terminal cluster; involucre 3–4 mm tall; involucral bracts greenish, strongly imbricate; pistillate florets in a single series, numerous, 20–40, ligule white or light-pinkish, equaling or slightly exceeding the style and the pappus. Achenes ± hirsute; pappus abundant, dull white. (n = 9) Late Jun–Sep. A weed in open cult. & waste ground; throughout the GP; (throughout temp. N. Amer.; introduced widely elsewhere). *Erigeron canadensis* L.—Fernald; *Leptilon canadense* (L.) Britt.—Rydberg.

Three varieties may be recognized, but the vast bulk of our materials are clearly referable to var. *canadensis*. Just south of the GP in TX & NM, var. *canadensis* is largely replaced by var. *glabrata*

(A. Gray) Cronq., which is glabrous or nearly so, and has stramineous involucral bracts. Var. *pusilla* (Nutt.) Cronq. occurs in the Coastal Plain of se U.S., and it has glabrous stems and pinkish-red tipped involucral bracts.

2. ***Conyza ramosissima*** Cronq., spreading fleabane. Diffuse-branched annual up to 3 dm tall; scattered-pubescent mostly with antrorsely-appressed hairs. Leaves ascending, narrowly linear, 1–2(4) cm long and up to 2 mm wide, the upper ones reduced to mere bracts. Heads numerous, borne at the ends of the branches across the top of the plant; involucre 3–4(5) mm tall; outer involucral bracts pubescent, inner ones glabrous; ray florets in a single series, minute, the ligule purplish-pink. Achenes ± hirsute; pappus dull white. (n = 9) Late Jun–Sep. A weed in lawns & open disturbed sites; GP, especially se 1/4 but scattered elsewhere; (OH to AL, w to GP). *Erigeron divaricatus* Michx.—Fernald; *Leptilon divaricatum* (Michx.) Raf.—Rydberg.

34. COREOPSIS L.

Annual or short-lived perennial herbs. Leaves opposite or occasionally a few alternate; variously pinnatifid or trifoliate to entire. Heads with involucral bracts in 2 series and dimorphic, ± fused at the base; outer involucral bracts shorter, narrower and more herbaceous than the somewhat membranaceous and striate inner involucral bracts; receptacle flat to low-convex, chaffy; ray florets ca 8, neutral, ligule yellow (and with a red basal spot in *C. tinctoria*); disk florets perfect and fertile, style branches flattish, the appendage short and subtruncate or elongate and subcuneate; chaffy bracts slender, thin and flat. Achenes flattened at right angles to the radius of the head, variously winged with prominent, entire wings, to erose or wingless; pappus of 2 awns or teeth, or reduced to a mere crown or absent. [tribe Heliantheae]

Reference: Smith, Edwin B. 1976. A biosystematic survey of *Coreopsis* in eastern United States and Canada. Sida 6: 123–215.

Coreopsis is closely related to *Bidens* and imperfectly separated from that genus. Beyond our range these two generally intergrade into other related genera, forming a sizeable complex of indistinct groups.

1 Disk corollas mostly 4-lobed; style appendages short and blunt; rays yellow but with a reddish spot at the base .... 5. *C. tinctoria*
1 Disk corollas mostly 5-lobed; style branches acute and often with a cusp; rays yellow.
  2 Chaffy bracts flattened near the base and caudate-attenuate above; leaves pinnatifid to entire.
    3 Leaves best developed on the lower 1/2 of the stem; naked peduncles prominent; rare in extreme e GP .... 2. *C. lanceolata*
    3 Leaves well developed along the stem; peduncles less than 1/2 as long as the leafy portion of the stem.
      4 Leaves pinnately parted into narrow, linear or lance-linear segments; se 1/4 GP .... 1. *C. grandiflora*
      4 Leaves simple or with 1 or 2 pairs of broad, lateral lobes; extreme se KS & adj. MO .... 4. *C. pubescens*
  2 Chaffy bracts linear or weakly clavate, blunt to acute; leaves (or many of them) trifoliate or at least 3-parted.
    5 Leaves clearly petiolate and trifoliate .... 6. *C. tripteris*
    5 Leaves apparently sessile or with an indistinct, winged petiole, merely 3-parted .... 3. *C. palmata*

1. ***Coreopsis grandiflora*** Hogg ex Sweet, bigflower coreopsis. Erect perennial 3–10 dm tall, spreading ciliate-pubescent to glabrate. Stems usually clustered, arising from a short caudex. Leaves about evenly scattered along the stem; the naked peduncles less than 1/2 as long as the leafy portion of the stem. Leaves mostly pinnatifid with narrow, linear or

lance-linear segments, the lateral segments mostly less than 5 mm wide; lowermost leaves often entire. Heads few or sometimes solitary on naked peduncles up to 1.5 dm long; outer involucral bracts lanceolate-subulate, up to 10 mm long, margins whitish, ciliate; ray florets with ligule 1.3-2.5 cm long; disk florets with corolla 5-lobed; chaffy bracts 6-7 mm long. Achenes ca 2.5 mm long, with thin, flat, entire wings; pappus usually greatly reduced or absent. (n = 13, with 0-4 B chromosomes) May-Jul. Open, sandy or calcareous hillsides & prairies, roadsides, & semidisturbed sites; se 1/4 GP; e 1/2 KS & OK; (scattered; NC & GA w to se GP, TX; adventive elsewhere as an escape from cultivation as an ornamental).

This species forms a complex of several semidistinct phases. Our materials are referable to var. *harveyana* (A. Gray) Sherff, by E. B. Smith, op. cit., or more conservatively to var. *grandiflora*. *C. grandiflora* is similar and closely related to *C. lanceolata*.

2. Coreopsis lanceolata L. Erect perennial 2-7 dm tall, spreading ciliate-villous to glabrate. Stems clustered, arising from a short caudex; leafy below but elongate and naked above. Leaves spatulate to narrowly lanceolate, simple and entire or with 1 or 2 pairs of small, lateral lobes, well-developed middle and lower leaves long-petiolate, 5-20 cm long overall and up to 2 cm wide, upper leaves reduced and becoming sessile. Heads few or solitary, on prominent naked peduncles up to 4 dm long, the peduncles equaling or exceeding the leafy portion of the stem; outer involucral bracts lanceolate and somewhat scarious-margined, up to 10 mm long; rays yellow, ligule 1.5-3 cm long; disk florets with corollas yellow, 5-lobed; chaffy bracts flat and stramineous below, long-attenuate above. Achene 2-3+ mm long, with thin, flat wings; pappus of 2 short teeth. (n = 13, with 0-2 B chromosomes) Mar-Jul. Open sandy prairies & roadsides; se GP; KS: Chautauqua (rare elsewhere as a waif); MO: Clay, Jasper; OK: Caddo, Stephens; (MA & CT to FL, w to WI, OK, & TX; adventive elsewhere as an escape from cultivation). *C. crassifolia* Ait.—Rydberg.

3. Coreopsis palmata Nutt., finger coreopsis. Rhizomatous perennial 4-9 dm tall, glabrous or with some scattered pubescence, especially at and near the nodes. Leaves all cauline, firm in texture, sessile or subsessile, 3-8 cm long, prominently trifid to about the middle or slightly deeper, lobes oblong, up to 7 mm wide, the central lobe sometimes lobed again. Heads few or solitary, on short peduncles; disk 1-1.5 cm across; outer involucral bracts linear-oblong and about equaling the inner; ray florets with ligule 1.5-3 cm long; disk florets with corolla 5-lobed; chaffy bracts linear-clavate, ± acute. Achene 5-6.5 mm long, with narrow wings; pappus of 2 callous teeth or greatly reduced. (n = 13) Jun-Jul. Open wooded sites, disturbed prairies, roadsides, & rocky ridges; e GP, from w MN & extreme e SD through e 1/5 KS to adj. OK; (MI to LA, w to GP & AR).

4. Coreopsis pubescens Ell. Erect perennial to 10 dm tall; short spreading-hairy to glabrate. Stems clustered, arising from a branching, fibrous-rooted crown or a short caudex. Leaves short-petiolate, blade elliptic to ovate, 4-10 cm long and up to 4 cm wide, entire or often with a pair of lobes near the base. Heads few, on naked peduncles 5-20 cm long; outer involucral bracts narrowly lanceolate-triangular, at least 1/2 as long as the inner involucral bracts; ray florets with ligule yellow, 1-2.5 cm long; disk florets with corollas yellow, 5-lobed; chaffy bracts caudate-attenuate above. Achene 2.5-3 mm long, with thin, spreading wings; pappus of 2 short chaffy teeth. (n = 13, with 0-2 B chromosomes) Jul-Sep. Alluvial soil in wooded sites; extreme se GP; KS: Cherokee; MO: Barton, Jasper; (VA to FL, w to se GP, AR, & OK).

5. Coreopsis tinctoria Nutt., plains coreopsis. Erect annual or perhaps occasionally a short-lived perennial, 6-12 dm tall. Stems arising singly, but much branched. Leaves subsessile or short-petiolate, 5-10 cm long, pinnate or bipinnate, or the uppermost undivided, segments (or blades) linear to linear-lanceolate. Heads numerous and terminating the slender upper branches; outer involucral bracts linear-oblong to triangulate, less than 1/2 as long as the inner involucral bracts; ray florets with ligule 1–1.5 cm long, yellow but usually with a prominent reddish spot at the base; disk florets with corolla 4-toothed, style branches short and blunt; chaffy bracts flattened toward the base and tapering above to a long, thin apex. Achenes often wingless but sometimes with narrow to prominently broad, thin wings; pappus of minute awns or reduced to an obscure crown. (n = 12) Jun–Sep. Seasonally damp, disturbed sites, especially roadside ditches & low, sandy ground; throughout the GP but less frequent in the w & in the ne 1/4; (apparently native to the prairies & plains region, but widely adventive elsewhere). *C. cardaminifolia* (DC.) T. & G. — Rydberg.

This species is widely cultivated as an ornamental, and freely escaping. It is sometimes known in the horticultural trade as "calliopsis," which is an old generic name.

6. ***Coreopsis tripteris*** L., tall coreopsis. Erect perennial 1–2+ m tall, glabrous and sometimes glaucous, occasionally scattered short-pubescent. Stems arising singly from a short, stout rhizome. Leaves numerous, prominently petiolate; blades mostly trifoliate with firm, elliptic to lanceolate segments, 5–10 cm long and up to 2.5 + cm wide, the terminal segment sometimes divided again; uppermost leaves reduced and sometimes entire. Heads numerous; outer involucral bracts linear-oblong, less than 1/2 as long as the inner ones; ray florets with ligule 1–2.5 cm long, yellow; disk florets yellow but becoming purplish to deep red, corolla 5-lobed; chaffy bracts linear to weakly clavate. Achenes 4–7 mm long; pappus of 2 short awns or a few small bristles. (n = 13) Jul–Sep. Open wooded stream banks, roadsides, & wooded damp prairies; w IA to extreme e KS & MO; (PA to FL, w to e GP, & LA).

35. CREPIS L., Hawk's-beard

Taprooted annual, biennial or perennial herbs with milky juice; herbage variously tomentose, glandular or glabrate. Basal leaves well developed, entire to pinnatifid; cauline leaves alternate and mostly reduced, the upper-most bractlike. Heads (1) few to numerous in an open corymbiform or weakly paniculiform inflorescence; involucre with an outer series of reduced involucral bracts and an inner series (or 2 series) of prominent, elongate involucral bracts; receptacle naked or ciliate; florets all ligulate, perfect and fertile, corolla yellow, sometimes deeply so, but sometimes fading to ochroleucous. Achenes terete, fusiform to columnar, ± prominently ribbed, constricted or attenuate upward, sometimes obscurely short-beaked; pappus of numerous white or whitish capillary hairs. [tribe Lactuceae]

Reference: Babcok, E. B. 1947. The genus *Crepis*. Univ. Calif. Publ. Botany, vols. 21 & 22. This is a classic monograph, incorporating experimental data from cytology and cytogenetics into an exceptionally detailed treatment.

Crepis acuminata and *C. occidentalis* are a part of a complex with sexual diploid forms and apomictic polyploid forms. Along the extreme w edge of the GP, in the foothills of WY, one may expect *C. modocensis* Greene, which is also a member of this complex and which is separable from *C. acuminata* and *C. occidentalis* by its conspicuously setose involucres with no glandular hairs. The three species intergrade, and the polyploid, apomictic intergradants form populations that have been called *C. intermedia* A. Gray. The morphological intergradation plus apomictic populations make the specific determination of individual specimens exceedingly difficult.

1 Introduced annual or weakly biennial weeds of lawns and disturbed waste sites.
   2 Inner involucral bracts finely hairy on the inner surface; mature ahcenes purplish-brown, mostly more than 2.5 mm long .... 5. *C. tectorum*
   2 Inner involucral bracts glabrous within; achenes yellowish-brown, mostly 1.5–2.5 mm long .... 2. *C. capillaris*
1 Native perennials, not at all weedy.

3 Herbage glabrous or at most weakly hispid .... 4. *C. runcinata*
3 Herbage grayish tomentose, especially when young.
    4 Heads numerous, up to 100, narrow in aspect, with 8 or fewer inner involucral bracts and mostly 5–10 florets .... 1. *C. acuminata*
    4 Heads seldom more than 20, broader in aspect, with 8–13(+) inner involucral bracts and 10–40 florets .... 3. *C. occidentalis*

1. Crepis acuminata Nutt. Perennial 2–7 dm tall, arising from a taproot; herbage grayish-tomentose, especially when young, sometimes glabrate in age. Basal and lower cauline leaves mostly 1–4 dm long overall, pinnately lobed with the lobes variously entire to dentate, blade tapering to an indistinct petiole; middle and upper leaves few, and the middle similar to the basal ones but smaller, the upper reduced to linear bracts. Inflorescence a branching corymbiform cyme of numerous to many heads, often up to 100; involucre 8–16 mm tall, glabrous to tomentose, outer involucral bracts less than 1/2 as long as the 5–8 prominent inner ones; florets mostly 5–10, all ligulate and fertile, corolla yellow, 10–18 mm long. Achenes yellowish to brown, mostly ca 7–8 mm long; pappus of abundant capillary bristles. (n = 11, + a polyploid series) May–Jul. Dry, open places, especially hillsides & broken slopes; scattered in extreme nw GP from e-cen MT to e WY & nw NE; (nw GP to NM, w to WA & CA).

2. Crepis capillaris (L.) Wallr. Annual or possibly biennial herb, 1–9 dm tall; herbage variously glabrous to hispidulous with short, yellowish hairs. Basal leaves lanceolate to oblanceolate, 3–30 cm long and 0.5–5 cm wide overall, subentire or denticulate to weakly pinnatifid, tapering to an indistinct petiole; cauline leaves progressively reduced upward, the upper ones becoming sessile and auriculate-clasping. Heads several to numerous; involucre 5–8 mm tall, outer involucral bracts about 1/2 as long as the inner ones, inner involucral bracts 8–16, tomentose and sometimes with glandular black hairs on the outer surface but glabrous on the inner surface, receptacle glabrous; florets 20–60, all ligulate and fertile, corolla yellow, 8–12 mm long. Achenes yellowish to pale brown, 1.5–2.5 mm long, smooth to somewhat scabrous; pappus of numerous white capillary bristles. (n = 3) May–Oct. Occasional waif in lawns & gardens; infrequently collected in the GP but to be expected, especially in the ne 1/4; (sparingly established in e U.S. but a frequent weed along the Pacific Coast; Eurasia). *Adventive.*

    A similar weedy species, *C. nicaeensis* Balb. ex Pers., also may be expected in the GP. It is distinguished by having the involucre 8–10 mm tall, ciliate receptacle and heads that become campanulate at maturity rather than turbinate as does *C. capillaris.* Neither species is likely to be abundant or persistent.

3. Crepis occidentalis Nutt. Perennial 1–3(4) dm tall, herbage densely gray tomentose, sometimes glabrate in age, frequently glandular-hairy upward and often with gland-tipped black hairs on the involucre. Lower leaves 1–3 dm long overall, deeply toothed to pinnatifid with toothed lobes, tapering to a winged petiole; upper leaves similar but reduced, the uppermost bractlike. Inflorescence of 2–20 heads in an open, corymbiform cluster; involucre 1–2 cm tall, outer involucral bracts mostly less than 1/2 as long as the inner ones; prominent inner involucral bracts 8–13(+); florets 10–20(40), ligulate and fertile, corolla yellow, up to 2 cm long. Achenes brown, 6–10 mm long; pappus of whitish to yellowish capillary hairs. (n = 11, + a polyploid series) May–Jul. Open, dry sites, especially foothills & broken slopes; nw 1/4 GP, from w ND to nw NE & w; (Sask., nw GP to NM, w to B.C. & CA).

    Our materials are more or less referable on morphological grounds to the poorly defined subsp. *costata* (A. Gray) Babc. & Stebbins. However, subsp. *costata* is known only as an apomictic polyploid, *fide* Babcock, (op. cit.), and the cytological structure and breeding patterns for the GP materials remain to be determined.

4. Crepis runcinata (James) T. & G. Perennial 1–5(7) dm tall, glabrous or weakly hispid, or sometimes the involucres glandular-hairy, but the herbage never tomentose; stems 1–3, scapiform. Basal leaves 3–15(30) cm long overall, elliptic to ovate or oblanceolate, entire to denticulate or sometimes subpinnatifid, tapering to a winged petiole; cauline leaves few and much reduced, mostly linear and bractlike. Inflorescence a simple or branching corymbiform cluster, heads 1–12(30); involucre 1–1.6(+) cm tall; outer involucral bracts unequal, the longest up to 3/4 as long as the inner ones; inner involucral bracts 10–16; florets 20–50, ligulate and fertile, corolla yellow or light yellow, 1–2 cm long. Achenes brown, 4–7 mm long, constricted above and sometimes obscurely short-beaked; pappus of abundant white capillary bristles. (n = 11) May–Jul. Open meadows, often in damp or drying sites, sometimes in saline soils; nw 1/2 GP from w MN to e MT, s to cen NE & nw KS; rare s in the panhandle of TX; (GP & s Sask. to NM, w to WA & CA). *C. glauca* (Nutt.) T. & G., *C. glaucella* Rydb., *C. perplexans* Rydb., *C. petiolata* Rydb., *C. platyphylla* Rydb., *C. riparia* Nels. (?) — Rydberg.

This is a variable and widespread species, with seven subspecies recognized by Babcock (op. cit.), two of which are in the GP, although they are imperfectly separated. Specimens with the involucres glandular-pubescent are referable to subsp. *runcinata*, while those with glabrous involucres may be treated as subsp. *glauca* (Nutt.) Babc. & Stebbins. The latter apparently has some ecological distinctiveness, for it occurs in more alkaline sites.

5. Crepis tectorum L. Annual weed to 10 dm tall, herbage variously glabrous to lightly pubescent. Basal leaves early deciduous, up to 15 cm long overall, lanceolate to oblanceolate, denticulate to variously pinnate-lobulate, tapering to a winged petiole; cauline leaves sessile, narrowly lanceolate to linear, auriculate-clasping. Inflorescence a branching cluster of few to numerous heads; involucre 6–9 mm tall, outer involucral bracts unequal, the longest ca 1/3 as long as the inner ones; prominent inner involucral bracts 12–15, tomentose pubescent and sometimes glandular on the outer surface, strigose-puberulent on the inner surface; receptable finely ciliate; florets 30–70, ligulate and fertile, corolla yellow, up to 13 mm long. Achenes purplish brown at maturity, 2.5–4.5 mm long, attenuate and weakly short-beaked; pappus of white capillary bristles, ± deciduous. (n = 4) Jun–Aug. Sparingly persistent in lawns & gardens in w MN, e ND; sporadic elsewhere in the n GP; (sporadic in much of N. Amer., especially in the cooler regions; Eurasia). *Adventive.*

36. DRACOPIS Cass.

1. Dracopis amplexicaulis (Vahl) Cass., coneflower. Annual 3–7 dm tall, herbage glabrous and glaucous. Leaves alternate, blades elliptic-oblong to ovate, up to 10 cm long and 4 cm wide, entire or somewhat toothed, ± cordate-clasping. Heads 1–few; disk 1–1.5 cm across; involucral bracts in a single series, lance-linear, green; receptacle columnar, elongating to 1.5–3 cm at maturity; ray florets 6–10, neutral, ligule 1–3 cm long, yellow to partly orange or purplish, spreading but eventually reflexed-drooping; disk florets perfect and fertile, corolla narrowed to a distinct tube at the base, style-branches slightly flattened with elongate-acuminate hairy appendages; chaffy bracts subtending both ray and disk florets, ciliate-margined near the apex and weakly clasping the achenes. Achenes of the disk florets glabrous, several-nerved and minutely cross-ridged, obscurely 4-angled or subterete, pappus obsolete; achenes of the ray florets abortive, compressed 4-angled, villous, pappus a short crown. (n = 16) May–Jul. Open places, especially in clay soils; se GP; e 1/2 KS through OK; apparently isolated in ND; (AL to KS & TX). *Rudbeckia amplexicaulis* Vahl — Fernald. [tribe Heliantheae]

37. DYSSODIA Cav.

Strong-scented annual or perennial herbs, glabrous to variously pubescent; herbage dotted with brownish translucent oil glands. Leaves opposite or alternate, simple or lobed to pinnatisect. Heads hemispheric to turbinate, involucral bracts in 2 series, fused or free at the base, and often subtended by a whorl of reduced calyculate bracts; receptacle flat or low-convex, bristly or naked but scarcely chaffy; ray florets pistillate, yellow to orange, sometimes reduced and inconspicuous; disk florets perfect, style branches long and slender, tipped with a short, papillate cone. Achenes stout and obpyramidal to slender; pappus of 5-20 + blunt or awned scales, sometimes each scale has several awns or bristles. [tribe Heliantheae]

Reference: Strother, John L. 1969. Systematics of *Dyssodia* Cavanilles (Compositae: Tageteae). Univ. Calif. Publ. Bot. 48: 1-88.

1 Involucral bracts free or nearly so to the base; ray florets inconspicuous .... 3. *D. papposa*
1 Involucral bracts fused for 1/2 or more of their length; ray florets with conspicuous, showy ligules.
  2 Ephemeral annuals; calyculate bracts absent or greatly reduced; leaves alternate, pinnatisect .... 2. *D. aurea*
  2 Perennials or season-annuals; calyculate bracts 5-8, well developed; leaves opposite or alternate, entire to toothed but not pinnatisect.
    3 Plants less than 3 dm tall, compact to suffruticose; leaves needle-shaped, 1-2 cm long .... 1. *D. acerosa*
    3 Plants usually more than 3 dm tall, corymbosely branched; leaves linear, toothed, 4-9 cm long .... 4. *D. tagetoides*

1. **Dyssodia acerosa** DC. Compact, tightly branched or spreading suffruticose perennial, 1-2.5 dm tall, glabrous or with some minute pubescence. Leaves needle-shaped (acerose), mostly opposite but sometimes the upper alternate, 1-2 cm long and up to 2 mm wide, with numerous small oil glands. Heads sessile or short pedunculate; calyculate bracts ca 5; involucre 5-7 mm tall and 3-4 mm across; principal involucral bracts ca 13, fused nearly to the tip; ray florets ca 8, ligule lemon yellow, 5-6 mm long; disk florets 18-25. Achenes 3-3.5 mm long; pappus of ca 20 scales, each parted into 3-5 bristles. (n = 8, 13) Jul-Sep. Open limestone sites; known in GP from TX: Deaf Smith, but to be expected in adj. areas; (TX to UT & NV, s to n Mex.).

2. **Dyssodia aurea** (A. Gray) A. Nels. Erect or spreading ephemeral annual 1-2 dm tall, diffuse-branched from the base. Leaves alternate, 2-4 cm long, pinnately parted with 5-13 linear lobes that are 5-10 mm long, glabrous but glandular-dotted. Heads on peduncles 1-3 cm long; calyculate bracts absent or reduced to 1 or 2 inconspicuous subulate bracts; involucre 5-6 mm tall; principal involucral bracts ca 13, fused for more than 1/2 their length; ray florets 8-12, ligule 4-6 mm long, yellow; disk florets 30-45. Achenes slender-clavate, 3 mm long; pappus of 8-10 blunt scales. (n = 8) Jun-Sep. Open grasslands; sw GP; infrequent in w KS, e CO, & s; (sw GP to NM, TX, & n Mex.). *Thymophylla aurea* (A. Gray) Greene — Rydberg.

3. **Dyssodia papposa** (Vent.) Hitchc., fetid marigold. Erect to spreading annual 1-5 dm tall, glabrous or sparsely pubescent. Leaves opposite or the uppermost becoming alternate, 15-50 cm long, pinnately parted into numerous lobes which are occasionally parted again, irregularly dotted with oil glands. Heads subsessile or short-pedunculate; calyculate bracts 4-9, linear; involucre cylindrical to campanulate, 6-10 mm tall; principal involucral bracts 6-12, free, with several impressed oil-glands; ray florets 8 or fewer, inconspicuous, the ligule less than 2 mm long; disk florets 12-50, corolla dull yellow. Achenes stout, 3-3.5 mm long, pubescent; pappus of ca 20 scales, each divided at the apex into 5-10 bristles.

(n = 13) Jul–Sep. Open fields, roadsides, & disturbed sites; throughout GP except ne 1/4; (s Ont., w NY to MT & AZ, s to cen Mex.). *Boebera papposa* (Vent.) Rydb.—Rydberg.

This is a weedy species, now adventive beyond its natural range.

4. **Dyssodia tagetoides** T. & G. Annual or short-lived perennial 4–9 dm tall, becoming corymbose-branched above, glabrous. Leaves mostly alternate, linear, 4–9 mm long and up to 6 mm wide, coarsely dentate with subopposite teeth, dotted with orange-brownish oil glands. Heads on bracteose peduncles ca 5 mm long; calyculate bracts conspicuous, 5–8, pinnatisect; involucre cylindrical; 9–12 mm tall and 5–8 mm across; principal involucral bracts 10–12, fused nearly to the apex; ray florets 7–12, ligule 10–15 mm long, yellow; disk florets 20–40, corolla brownish-yellow. Achenes 3–3.5 mm long; pappus of 10–12 scales, each with 1–3 awns at the apex. (n = 13) Jun–Jul. Limestone areas; scattered in sw OK; (s GP to cen TX).

38. ECHINACEA Moench, Purple Coneflower

Perennial herbs with stems arising from upright or horizontal rootstocks. Leaves alternate, petiolate below but reduced and sessile upward, blade ovate to lanceolate to elliptic, entire to coarsely toothed, often prominently 3- to 5-nerved. Heads borne singly on long peduncles; involucral bracts in 3 or 4 series, imbricated, foliaceous and intergrading into the chaffy bracts; receptacle conical to hemispherical; ray florets sterile, ligule 2- or 3-lobed at the apex, pink to purple or whitish; disk florets fertile, corolla expanded below into a bulbous-thickened base, tube cylindrical with a 5-lobed limb, yellowish to purplish; pollen yellow or white, styles flattish with slender, acuminate, hairy appendages; chaffy bracts conduplicate, apex prolonged into a sharp or blunt awnlike spine exceeding the disk corollas. Achenes 4-angled; pappus a short, toothed or smooth crown. [tribe Heliantheae]

Reference: McGregor, R. L. 1968. The taxonomy of the genus *Echinacea* (Compositae). Univ. Kansas Sci. Bull. 48: 113-142.

1 Blades of the basal and lower cauline leaves 2–3(+) × longer than wide; stems arising from a woody rhizome or tough caudex .... 4. *E. purpurea*
1 Blades of the basal and lower cauline leaves mostly more than 5 × longer than wide; stems arising from thick, woody taproots.
  2 Ligule of the ray florets dark purple and sharply reflexed so the tips nearly touch the peduncle .... 2. *E. atrorubens*
  2 Ligule pink to light purple, spreading or drooping but not reflexed.
    3 Ligules of the ray florets 4–9 cm long and clearly drooping, pollen white .... 3. *E. pallida*
    3 Ligules 2–4 cm long and spreading, pollen yellow .... 1. *E. angustifolia*

1. **Echinacea angustifolia** DC. Perennial 1–6 dm tall. Stems strict to branched, variously hirsute or tuberculate-hispid to strigose, especially above, sometimes glabrate below. Leaves alternate, blades of the basal and lower cauline leaves narrowly elliptic-lanceolate to oblong, 5–30 cm long and 1–4(+) cm wide, entire or nearly so, petiolate, the upper leaves progressively smaller and becoming sessile. Heads with involucral bracts 6–11 mm long, disk 1.5 cm across; receptacle 1.5–3 cm tall; ray florets with ligule 2–4 cm long and 5–8 mm wide, mostly light pink to light purplish; disk florets with corolla 6–8 + mm long; body of the chaffy bracts 11–13 mm long, with an apical spine 2–3 mm long. Pollen grains yellow. Achene 4–5 mm long; pappus a toothed crown. (n = 11, 22) Jun–Jul. Open rocky prairies & plains; GP; less frequent on the far-western edge; (GP, barely extending e to cen MN & nw IA). *E. pallida* var. *angustifolia* (DC.) Cronq.—Gleason & Cronquist.

This species passes into *E. pallida* to the east. Some of the materials from s-cen KS, cen OK, & n TX tend to have leaves that are ± consistently strigose or strigose-hirsute, stems that are frequently branched, lower stems that are glabrous and remain green on drying. This phase may be recognized as var. *strigosa* R.L. McGreg.

2. **Echinacea atrorubens** Nutt. Perennial 3–9 dm tall. Stems simple or rarely branched, glabrate below, strigose or strigose-hirsute above. Basal and lower cauline leaves petiolate, blade oblong-lanceolate to lance-elliptic, 4–15 cm long and 1–4 cm wide, upper leaves progressively reduced, becoming sessile. Heads with involucral bracts in 3 or 4 series, linear-lanceolate, attenuate, the inner intergrading with the chaffy bracts; disk 2–3.5 cm across; receptacle up to 4 cm tall; ray florets with ligules 2–3.5 cm long and 4–7 mm wide, dark purple or rarely lighter, strongly recurved-reflexed; disk florets with corolla tube 5.5–6.5 mm long; body of the chaffy bracts 7–9 mm long, with an apical spine 2–3 mm long. Achene 4–5 mm long, glabrous; pappus a crown with unequal teeth. (n = 11) May–Jul. Prairies; se GP; in a narrow band from Topeka, KS s through cen OK; (GP s through OK to the vicinity of Houston, TX).

3. **Echinacea pallida** (Nutt.) Nutt., pale echinacea. Perennial 4–9 dm tall. Stems simple or rarely branched, hirsute below and ± densely so above. Leaves alternate, blades of basal and lower cauline leaves oblong lanceolate to long-elliptic, 1–3(+) dm long and 1–4 cm wide, entire, 3-nerved, petiolate; upper leaves reduced and becoming sessile. Heads with involucral bracts in 3 or 4 series, lanceolate to narrowly oblong, 8–17 mm long, intergrading with the chaffy bracts; disk 1.5–3 cm across; ray florets with ligules 4–9 cm long and 5–8 mm wide, purplish-pink to whitish, reflexed-drooping; disk florets with corolla tube 8–10 mm long; body of the chaffy bracts 8–10 mm long, with an apical spine 2.5–3.5 mm long. Pollen grains white. Achene 4–5 mm long, glabrous; pappus a toothed brown. (n = 22) May–Jul. Dry, open rocky sites, mostly in the prairies; se GP; sw IA, s through e KS to ne OK; (IN & WI, se to AR & n TX).

This species passes into *E. angustifolia* through the e GP.

4. **Echinacea purpurea** (L.) Moench. Erect, branching perennial 6–18 dm tall. Stems arising from a woody rhizome or tough caudex; herbage hirsute to glabrous. Basal and lower cauline leaf blades ovate to ovate-lanceolate, up to 2 dm long and 1.5 dm wide, coarsely to sharply serrate, abruptly contracted to subcordate at the base, petiole up to 2.5 dm long; cauline leaves similar but progressively reduced upward. Heads with involucral bracts linear-lanceolate, attenuate, intergrading with the chaffy bracts; disk up to 3.5 cm across; receptacle 1.5–3 cm tall; ray florets with ligule 3–8 cm long, drooping, reddish-purple or rarely pink; disk florets with corolla 4.5–5.5 mm long; body of chaffy bracts 9–13 mm long overall, with apical spine about 1/2 as long as the body. Achene 4–4.5 mm long; pappus a low crown of equal teeth. Pollen grains yellow. (n = 11) Jun–Aug. Rocky prairie sites in open wooded regions; KS: Cherokee; adj. OK; (OH & MI s to GA, w to OK & ne TX).

39. ECLIPTA L.

1. **Eclipta prostrata** (L.) L., yerba de tajo. Lax or weakly spreading annual weed, branched, pubescent, sometimes rooting at the nodes. Leaves opposite, blade mostly lanceolate to narrowly elliptic, 2–10 cm long and 0.5–2.5 cm wide, finely serrate, sessile or short-petiolate. Heads 1–3 on axillary or terminal peduncles; disk 4–6 mm across; involucral bracts in 1 or 2 series, herbaceous, at least toward the apex, subequal or the inner somewhat shorter; ray florets minute, numerous, pistillate and fertile, ligule ca 1 mm long, white; disk florets numerous, perfect and fertile, corolla mostly 4-toothed, style branches flattish with a short,

hairy, obtuse appendage; chaffy bracts slender and somewhat bristlelike, central chaff sometimes absent. Achene thick and warty, 2–2.5 mm long, marginal achenes regularly 3-angled with 1 side facing an involucral bract, central achenes 4-angled to compressed parallel to a radius of the head; pappus none or a very obscure crownlet. (n = 11) Aug–Oct. Damp sandy or muddy sites; se 1/4 GP but sporadic in sw 1/4; (a pantrop. weed, n to MA, WI, & GP in N. Amer.). *E. alba* (L.) Hassk.—Rydberg. [tribe Heilantheae]

40. ELEPHANTOPUS L., Elephant's Foot

1. *Elephantopus carolinianus* Raeusch. Perennial 4–10 dm tall, variously pubescent to substrigose, glabrescent in age. Stems 1–several, arising from a tough rootstock. Leaves cauline, ovate to lanceolate, base attenuate to a clasping petiole, margin crenate to serrate, overall dimensions 9–18 cm long and 5.5–9 cm wide. Inflorescence a corymbiform-paniculate cluster of glomerules, each of these a cluster of several heads; glomerules terminal, up to 2.5 cm across and subtended by 3 cordate bracts, equaling or exceeding the glomerule; heads with 4 florets, involucre cylindric, involucral bracts 6–10 mm long; corolla white to violet, irregularly 5-lobed, 7–8 mm long. Achenes ribbed, pubescent, 3–4 mm long; pappus a single whorl of 5 bristles, weakly expanded at the base. (n = 11) Aug–Oct. Damp bottomland, wooded areas; se KS, adj. OK, & MO; (PA to FL, w to KS & TX). [tribe Vernonieae]

41. ENGELMANNIA A. Gray ex Nutt.

1. *Engelmannia pinnatifida* A. Gray ex Nutt., Engelmann's daisy. Perennial herb 2–5 dm tall, densely hispid. Stems arising singly or clustered from a crown surmounting a stout, woody taproot. Basal leaves (1)2–3 dm long, once-pinnatifid; cauline leaves 8–30 cm long, deeply pinnatifid. Inflorescence of several heads, each on long peduncles; involucre 6–10 mm tall, involucral bracts in several series, subequal, greenish or the innermost partially white-indurate; receptacle flat and chaffy throughout; ray florets ca 8, pistillate and fertile, each subtended by an inner involucral bract, ligule yellow, ca 1 cm long; disk florets staminate, partly enclosed by a chaffy bract, corolla yellow. Ray achenes flattened at right angles to the radius of the head, obovate, pappus of a few scales with 2 somewhat larger than the others. (n = 9) May–Aug. Open limestone or sandy sites in the high plains; sw 1/4 GP; apparently disjunct or adventive in sw SD; (sw GP to w TX & n Mex.). [tribe Heliantheae]

42. ERECHTITES Raf., Fireweed

1. *Erechtites hieracifolia* (L.) Raf. ex DC. Erect annual (5)7–12(20) dm tall, glabrous to scattered coarse-tomentose. Stem striate-ribbed, sometimes turgid, arising from a simple or weakly branching taproot. Leaves alternate, serrate or weakly lobed, ovate-lanceolate with a weakly-defined petiole, overall dimensions (3)6–20 cm long and (1)2–8 cm wide, well distributed along the stem, lowermost early-withering, uppermost often reduced to mere bracts. Inflorescence a loose to congested, elongate to corymbiform cyme with few to many heads; head turbinate but with an extended disklike receptacle when fresh; involucral bracts in a single series, 10–17 mm long; heads disciform, i.e., with outer florets pistillate and with tubular-filiform white corollas, inner florets perfect but sometimes failing to set seed. Achenes prominently ribbed, 2–3 mm long; pappus of abundant soft hairs, but deciduous. (n = 20) Aug–Oct. Prairie ravines, marshy ground, & woodlands, occasionally

adventive as a weed; se GP; se SD, s in e NE, e KS, & OK; (native Newf. s to FL, w to KS; adventive elsewhere, especially the w coast). [tribe Senecioneae]

43. ERIGERON L., Fleabane

Annual, biennial, or perennial herbs, with underground parts varying from a simple taproot or fibrous roots to a tough, woody perennial caudex. Leaves alternate or rarely all basal. Heads generally few but sometimes either solitary or numerous, commonly borne on subnaked peduncles; involucral bracts narrow and usually herbaceous, but if yellowish or chartaceous, then so throughout or especially toward the tip, subequal or sometimes imbricated; receptacle flat and naked, ray florets pistillate, numerous and usually with evident, narrow ligules, or rarely absent, ligule white to varying shades of pink, blue or purple; some species have eligulate pistillate florets (disciform florets) in 1 or 2 series just inside the normally ligulate florets; disk florets perfect and fertile, numerous, corolla yellowish, those of the outer 2 rows of the head often somewhat reduced; style appendages short, lanceolate and acute to merely triangular, style branches with inner-marginal stigmatic lines. Achene commonly flattened and 2-nerved, sometimes 4-nerved or obscurely many-nerved; pappus simple and of capillary bristles, but more often obviously or obscurely double, with an inner series of numerous capillary bristles and an outer series of a few short setae or narrow scales; rarely the pistillate florets have the inner series absent and bear only short scales. [tribe Astereae]

Reference: Cronquist, Arthur. 1947. Revision of the North American species of *Erigeron* north of Mexico. Brittonia 6: 121–302.

Erigeron is closely related to both *Aster* and *Conyza*, and the three are distinguished more by groups of tendencies than by sharp, qualitative key-characters. *Erigeron* is best developed in the western cordillera of North America, and several species barely enter the western edge of the GP. As with many widespread genera in the Asteraceae, the species boundaries are not always sharp, and morphological intermediates are to be expected.

Some species are quite variable in the amount and nature of their pubescences, and in other characters. The key is designed so that these variable species will fall out in more than one place. Sectional names are those of Cronquist (op. cit.).

1 Pistillate florets very numerous, the corolla filiform with narrow, short, erect ligules, these sometimes not exceeding the involucre, or the inner pistillate corollas tubular and eligulate (Sect. *Trimorphaea*).
  2 Pistillate florets that are tubular, filiform, and essentially eligulate present between the bisexual disk florets and the outer, ligulate pistillate florets; inner involucral bracts long-attenuate; inflorescence ± corymbiform .... 1. *E. acris* var. *asteroides*
  2 Pistillate florets all ligulate; inner involucral bracts acute to merely acuminate; inflorescence ± racemose or heads sometimes solitary .... 12. *E. lonchophyllus*
1 Pistillate florets few to numerous (rarely absent), corolla tube cylindrical with the ligule well-developed and spreading or somewhat reduced, but not short, very narrow and erect.
  3 True perennials, often with prominent, tough rhizomes and/or woody caudices.
    4 Achenes mostly 2-nerved, or if with 4+ nerves, the involucral bracts subequal and not conspicuously imbricated (Sect. *Euerigeron*).
      5 Cauline leaves prominent, usually lanceolate to ovate; plants usually tall and erect.
        6 Stem ± equally leafy, upper leaves gradually reduced, middle leaves about equaling the lowermost leaves.
          7 Upper and middle cauline leaves glabrous or nearly so, except for ciliate margins. .... 19. *E. speciosus* var. *macranthus*
          7 Leaves and stems ± spreading-pubescent with long hairs .... 21. *E. subtrinervis*
        6 Stem with the uppermost leaves conspicuously and abruptly reduced; middle cauline leaves commonly smaller than the lowermost leaves.

- 8 Stem and involucre glandular to viscid; stem ± arching upward from the base .... 10. *E. formosissimus*
- 8 Stem and involucre ± hairy but scarcely glandular; stem usually erect .... 11. *E. glabellus* subsp. *pubescens*
- 5 Cauline leaves usually much reduced, linear to narrowly oblanceolate (often broader in *E. caespitosus*), plants low in aspect.
  - 9 Cauline leaves relatively well developed, although smaller than the basal leaves .... 4. *E. caespitosus*
  - 9 Cauline leaves clearly much reduced and often bractlike and/or linear-lanceolate.
    - 10 Basal leaves deeply lobed or dissected .... 6. *E. compositus*
    - 10 Basal leaves entire or at most toothed.
      - 11 Plant with a simple or weakly branched caudex and abundant fibrous roots; ligules numerous, often 100+, ca 1 mm wide .... 11. *E. glabellus* subsp. *pubescens*
      - 11 Plants with caudex variously branched or simple, but without numerous fibrous roots; ligules mostly fewer than 100 and more than 1 mm wide.
        - 12 Pubescence of the stem appressed or ascending, or absent, or at most spreading only on the upper peduncles and among the heads.
          - 13 Petioles and/or margins of the basal leaves with at least some spreading hairs that are unlike the other hairs; leaves with a single prominent midnerve .... 23. *E. vetensis*
          - 13 Petioles and/or margins of the basal leaves without coarse, spreading hairs; prominent leaves often ± 3-nerved.
            - 14 Caudex well-developed and prominently branching; achenes densely pilose, typically 4-nerved .... 17. *E. pulcherrimus* var. *wyomingia*
            - 14 Caudex short and simple, or nearly so; achenes ± hirsute, typically 2-nerved .... 14. *E. ochroleucus*
        - 12 Pubescence of the stem widely spreading, although sometimes scanty.
          - 15 Basal leaves ± strongly 3-nerved.
            - 16 Involucral bracts prominently thickened on the back; basal leaves obtuse to rounded at the apex; stem greenish or but rarely purplish at the base .... 4. *E. caespitosus*
            - 16 Involucral bracts obscurely if at all thickened on the back; basal leaves acute; stem ordinarily purplish at the base .... 7. *E. corymbosus*
          - 15 Basal leaves with a single prominent midrib, or at most only obscurely 3-nerved.
            - 17 Stem and leaves glandular to viscid, sometimes sparsely hirsute; basal leaves linear-lanceolate; caudex branches very stout; pappus single or double .... 23. *E. vetensis*
            - 17 Stem and leaves not glandular, or if so the basal leaves broader and/or the caudex branches thin and not especially stout; pappus clearly double, i.e., in 2 series.
              - 18 Involucre densely and finely villous-hirsute.
                - 19 Pubescence of leaves mostly appressed .... 14. *E. ochroleucus*
                - 19 Pubescence of leaves dense and spreading ... 4. *E. caespitosus*
              - 18 Involucre coarsely hirsute.
                - 20 Plants with a caudex but no prominent taproot .... 10. *E. formosissimus*
                - 20 Plants with a well-developed taproot, usually surmounted by a branching caudex .... 18. *E. pumilus*
- 4 Achenes mostly 4+ nerved; involucre obviously and regularly imbricated (Sect. *Wyomingia*).
  - 21 Achenes pubescent, mostly 4-nerved; old leaf bases inconspicuous or chaffy at most .... 17. *E. pulcherrimus* var. *wyomingia*
  - 21 Achenes glabrous, 8- to 14-nerved; old leaf bases persistent and becoming fibrous .... 5. *E. canus*

3 Annuals, biennials or short-lived perennials, with neither deep, well-developed rhizomes nor woody caudices; at most with shallow short-lived rhizomes or stolons.
  22 Pappus of ray and disk florets alike (Sect. *Olygotrichium*).
    23 Pappus simple, i.e., of a single whorl.
      24 Annual with stems moderately to diffusely branched; leaves linear to narrowly oblanceolate .... 3. *E. bellidiastrum*
      24 Biennials or short-lived perennials with stems simple or few-branched; leaves relatively broad.
        25 Plants with a few superficial rhizomes or stolons; ligules 1+ mm wide .... 15. *E. pulchellus*
        25 Plants without rhizomes or stolons; ligules ca 1/2 mm wide or less .... 15. *E. philadelphicus*
    23 Pappus double, i.e., in 2 whorls.
      26 Ligules very short and barely, if at all, exceeding the involucre; often bearing stolons .... 8. *E. divergens*
      26 Ligules clearly well-developed and exceeding the involucre.
        27 Heads large, disk usually more than 1 cm across; disk corollas 4–5.5 mm long; plants often robust, simple or few-branched and/or with well-developed cauline leaves .... 11. *E. glabellus* subsp. *pubescens*
        27 Heads small, disk 1 cm across or less; disk corollas 3 mm long or less, (disk to 13 mm and disk corollas to 3.5 mm in *E. flagellaris*); plants usually diffuse branching and/or with narrow cauline leaves.
          28 Stem with some or all of the hairs clearly appressed or closely ascending, or glabrescent.
            29 Essentially all of the hairs on the stem clearly appressed or closely ascending .... 9. *E. flagellaris*
            29 Hairs on the lower stem clearly spreading, while the hairs on the upper stem appressed or ascending.
              30 Basal leaves mostly 5 mm wide or less; middle cauline leaves 1–2(3) mm wide .... 13. *E. modestus*
              30 Basal leaves 5–15 mm wide; middle cauline leaves 3–6 mm wide .... 22. *E. tenuis*
          28 Stem with only spreading hairs.
            31 Ray florets at least 75 and usually more .... 8. *E. divergens*
            31 Ray florets mostly 30–65(+) .... 13. *E. modestus*
  22 Pappus of ray and disk florets unlike; pappus of disk florets of both bristles and short outer setae; that of the ray florets of outer setae only (Sect. *Phalacroloma*).
    32 Leaves numerous, plant mostly 7–10(15) dm tall; stem pubescence long and spreading .... 2. *E. annuus*
    32 Leaves rather sparse; plant mostly (3)4–7+ dm tall; stem pubescence usually appressed but some may be spreading .... 20. *E. strigosus*

1. Erigeron acris L. var. **asteroides** (Andrz. ex Bess.) DC. Erect perennial or biennial herb, 3–8 dm tall. Stems arising from a short, simple or slightly branched caudex; herbage subglabrous to spreading-hirsute and somewhat glandular, especially near the heads. Basal leaves narrowly oblanceolate to spatulate, entire or shallow-serrate, up to 10 cm long and 1.5 cm wide; cauline leaves numerous and well-developed, narrowly lanceolate to lance-oblong. Heads several to numerous, on stiffly ascending to upward arching peduncles; involucre 5–10(12) mm tall, finely glandular and/or hirsute; involucral bracts subequal or the outermost evidently shorter, inner ones long-attenuate; pistillate florets numerous, in several series and of 2 kinds, the outer ones with the corolla a long, filiform tube and a narrow, erect, pinkish ligule, 2.5–4.5 mm long, the inner ones with the corolla essentially tubular and eligulate; disk florets with slender corollas 4.2–6+ mm long. Achenes 2-nerved, sparsely hairy; pappus of numerous slender, remotely barbellate bristles, white to obscurely reddish, sometimes with a few inconspicuous short outer setae. (n = 9) Jun–Aug. Open, rocky wooded areas; BH; (circumpolar, across N. Amer., s to ME, MI, BH, UT, OR). *E.*

droebachensis O. Muell. ex Retz., *E. jucundus* Greene — Rydberg; *E. angulosus* var. *Kamtschaticus* (DC.) Hava — Fernald.

The nomenclature applied to this entity is subject to revision, for it is a part of a widespread species with an exceedingly complex taxonomic history. It is called *E. acer* L. in the more recent European literature.

2. **Erigeron annuus** (L.) Pers., annual fleabane. Erect annual herb, 7–10(15) dm tall. Stem sparingly to rather densely hirsute with long, spreading hairs; upper herbage with hairs usually shorter and somewhat appressed. Leaves well-developed; lowermost leaves ± hirsute, blades coarsely toothed, elliptic to broadly ovate or suborbicular, up to 10 cm long and 7 cm wide, or rarely larger, abruptly to gradually contracted to a petiole about as long as the blade; cauline leaves numerous, broadly lanceolate or wider, all but the uppermost dentate or rarely subentire. Heads several to numerous in a broad, open bracteate cluster; involucre 3–5 mm tall, disk 6–10 mm across; involucral bracts subequal or the outermost shorter, acuminate to attenuate, glandular and sparsely villous-hirsute; ray florets 80–125, ligule white or light blue, up to 10 mm long and ±1 mm wide; disk florets with corolla 2–3 mm long, strongly divided about midlength into tube. Achenes 2-nerved, pubescent; pappus double, the outer of short, slender scales; the inner of 10–15 fragile bristles; ray florets with inner pappus absent. (2n = 27, 54; apomictic). May–Aug. Weedy in open, moist or drying disturbed sites; e 1/4 GP from SD to OK, scattered in w SD; (throughout much of the e 1/2 of U.S. & adj. Can.; sporadic in the se U.S. & in w N. Amer.).

3. **Erigeron bellidiastrum** Nutt., western fleabane. Taprooted annual (biennial?), 1–5 dm tall, branching upward from the base and often intricately branched; herbage finely hirsute with incurved hairs that are usually broadest near the base. Leaves numerous, linear to narrowly oblanceolate, entire or sometimes 3-toothed at the apex, or even pinnatifid; blade up to 4 cm long and 1 cm wide; lowermost leaves tapering to a distinct petiole that is about as long as the blade; lower leaves often deciduous by flowering time. Inflorescence of numerous heads; involucre 3–5 mm tall, hirsute with coarse, incurved hairs, somewhat glandular, disk 5–8 mm across; involucral bracts thickish, with a brown midrib, subequal or the outermost shorter; ray florets 30–70, ligule white or pinkish, 4–6 mm long and ±1 mm wide; disk florets with corolla 2.5–3 mm long. Achenes 2-nerved, short-pubescent; pappus of ca 15 fragile, deciduous bristles. (n = 9) Jun–Aug. Open, damp or drying sandy sites; sw 1/2 GP, from NE: Boyd s & w, but apparently sporadic n & w to e MT; (GP to UT & NM).

The materials from the s GP (i.e., OK, n TX, adj. KS) with thick stems, robust habit, and but moderate branching have been recognized as var. *robustus* Cronq.

4. **Erigeron caespitosus** Nutt. Perennial herb 5–30 cm tall. Stems several, arching upward or somewhat decumbent at the base, arising from a branching caudex surmounting a stout taproot; herbage densely canescent to hirtellous with short, spreading hairs, often grayish in aspect. Basal leaves ±3-nerved, oblanceolate to spatulate, rounded or obtuse at the apex, basally tapering to a distinct petiole, up to 12 cm long overall and 1.5 cm wide, but usually smaller; cauline leaves few to numerous, most often oblong-ovate and only slightly smaller than the basal leaves, but sometimes much reduced and linear-lanceolate. Heads terminal, solitary or up to ca 10; involucre 4–7 mm tall, canescent and sparsely to densely glandular; disk 9–18 mm across; involucral bracts slightly to obviously imbricated, firm and thickened on the back; ray florets 30–100, ligule white or blue to pink, 5–15 mm long and 1–2.5 mm wide; disk florets with corolla 3.2–4.4 mm long. Achenes 2-nerved, hairy; pappus double, the inner of 15–25 bristles, the outer of evident, narrow scales. (n = 18) Jul–Aug. Open, dry rocky places; scattered in nw 1/4 GP, especially w NE to w ND, westward; (GP to Sask., AK, AZ, & NM). *E. subcanescens* Rydb. — Rydberg.

5. **Erigeron canus** A. Gray. Perennial herb 5-35 cm tall. Stems erect or nearly so, arising from a branching caudex surmounting a taproot; herbage grayish-strigose with appressed, whitish hairs. Old leaf-bases persistent and becoming fibrous; basal leaves tufted, oblanceolate to narrowly oblanceolate, up to 10 cm long overall and 7 mm wide, but often much smaller; cauline leaves reduced and mostly narrow, less than 3 cm long, few or occasionally numerous. Heads usually solitary at the ends of the stems, sometimes as many as 4; involucre 5-7 mm tall, canescent-hirtellous with short, spreading white hairs, sometimes lightly glandular, disk 9-16 mm across; involucral bracts conspicuously imbricate, narrow, acute, with a dark midrib; ray florets 30-40, ligule blue or white, 7-12 mm long and ±1 mm wide; disk florets with corolla narrow, 4-5.5 mm long. Achenes 8- to 14-nerved, glabrous; pappus double, the inner of ca 20-25 slender, white bristles, the outer of short setae. Jun-Aug. Open, dry or drying sites, especially high plains; w SD & NE, e WY, & CO; OK: Cimarron; & NM: Union; (GP to UT, NM, & AZ).

6. **Erigeron compositus** Pursh. Low, cespitose or pulvinate perennial, up to 20 cm tall, arising from a stout, branching caudex; herbage glandular and variously pubescent or even subglabrate. Leaves crowded at the base, deeply lobed or dissected; cauline leaves few and reduced to mere bracts. Heads solitary at the ends of the stems; involucre 5-10 mm tall, disk 8-20 mm across; involucral bracts subequal, purplish toward the tip; ray florets 20-60, ligule white or pinkish to blue, up to 12 mm long but sometimes reduced and inconspicuous (or absent?); disk florets with corolla 3-5 mm long. Achenes 2-nerved, hairy; pappus a single series of 12-20 bristles, about equaling the corolla. (n = 9, 27) May-Jul. Open, usually drying, rocky or loose, sandy sites; w ND, w SD, e MT, s along the Rocky Mts. to CO; (Greenl. & Que. to AK, s to GP, CO, AZ, & CA). *E. trifidus* Hook.—Rydberg.

The materials from the n GP mostly have 1-lobed basal leaves and are referable to var. *discoideus* A. Gray. Materials with leaves lobed 2 or 3× are treated as var. *glabratus* Macoun, but this phase is apparently confined to the mountain regions w of the GP. The varietal distinctions are imprecise (Cronquist, op. cit.).

7. **Erigeron corymbosus** Nutt. Perennial herb 1-5 dm tall, arising from a simple or weakly branched caudex surmounting a thin taproot; herbage hirsute; stem purplish, especially below, erect or arching upward. Basal and lowermost cauline leaves 3-nerved, elongate, linear-oblanceolate, up to 15 cm long overall and 14 mm wide, tapering to a petiole, margins ciliate with long hairs; cauline leaves few to numerous, often conspicuously reduced, linear or linear-lanceolate and some of the lower ones 3-nerved. Heads 1-10(+); involucre 5-7 mm tall, canescent, villous-hirsute with short, subappressed hairs, somewhat glandular, disk 7-13 mm across; involucral bracts weakly imbricated, obscurely if at all thickened on the back; ray florets 35-65, ligule deep blue to pink, 7-13 mm long; disk florets with corolla (3)4-5+ mm long. Achenes 2-nerved, hairy; pappus double, the inner series of 20-30 bristles, the outer of evident setae or small scales. (n = 9) Jun-Aug. Open, dry sites, often among sagebrush; extreme w GP; WY: Campbell; (w MT to B.C. & OR).

8. **Erigeron divergens** T. & G. Mostly biennial but possibly annual or short-lived, weak perennial herbs 1-4(7) dm tall, arising from a thin taproot; herbage pubescent with abundant short spreading hairs; stems branching from near the base and arching upward. Basal leaves ovate to spatulate, blade up to 2.5 cm long and 1 cm wide or rarely larger in robust individuals, entire to few-toothed, tapering to a petiole up to 5 cm long; lowermost leaves usually deciduous by flowering time; cauline leaves linear to narrowly oblanceolate, numerous, the lower ones petiolate and resembling the basal leaves, upper cauline leaves bractlike, mostly less than 4 cm long and 5 mm wide. Inflorescence of numerous heads in an open, diffuse cluster; involucre 4-5 mm tall, glandular and variously hirsute with spreading hairs, disk 7-11 mm across; involucral bracts about equal or the outer ones slightly

shorter, acuminate with a brown midvein; ray florets 75–150, ligule blue, pink or white, mostly 5–10 mm long but sometimes much reduced and scarcely exceeding the involucre; disk florets with corolla 2–3 mm long. Achenes 2-nerved or at most obscurely 4-nerved, sparsely hairy; pappus double, the inner series of 5–12 fragile bristles, the outer of conspicuous setose scales. (n = 9, 27?) Jun–Aug. Open, rocky or loose-sandy sites; scattered in nw 1/3 GP, cen NE to w ND & w; (GP to B.C., CO, AZ, CA, n Mex.?).

This is an imperfectly understood entity, apparently intergrading to some extend with *E. flagellaris*, and perhaps with *E. modestus* although good evidence is lacking. The materials from w KS & s which have been referred to *E. divergens* var. *cinereus* A. Gray are distinguished by having stoloniform branches and the early heads borne on long naked peduncles, but they clearly intergrade into *E. modestus* of the s GP, and appear to be disjunct from the typical *E. divergens* of cen NE n & w. The name var. *cinereus* is treated here as a synonym under *E. modestus* (q.v.). The several species of this complex clearly need further study.

9. *Erigeron flagellaris* A. Gray. Biennial or short-lived perennial herb, 5–40 cm tall, arising from a thin taproot which may be surmounted by a simple caudex; herbage lightly to moderately hirsute to strigose with appressed hairs, or the hairs at the base of the stem sometimes weakly spreading; stems erect and few-leaved, or some repent, flagelliform stolons, with scattered leaves, and sometimes rooting or bearing a head at the tip. Basal leaves oblanceolate, petiolate, persistent, up to 5 cm long overall and 8 mm wide, but usually smaller, entire to shallow-pinnate lobed; cauline leaves linear to narrowly linear-oblanceolate, (5)10–30 mm long and up to 4 mm wide. Heads at the ends of the branches; involucre 3.5–5 mm tall, glandular, hirsute with appressed (rarely a few spreading) hairs, disk 7–13 mm across; involucral bracts greenish or purplish; ray florets 50–100(125), ligule white, pinkish or blue, 5–10 mm long; disk florets with corolla 2.5–3.5 mm long. Achenes 2-nerved, sparsely pubescent or glabrate; pappus double, the inner series of 10–15 bristles, the outer of inconspicuous setae. (n = 18, 27?) Jun–Sep. Open, drying sites or meadows; BH & adj. NE, e WY, s along e edge Rocky Mts. in CO; (GP to w TX, s B.C., NV, & AZ).

Some specimens and perhaps entire populations approach *E. modestus* of the s GP in aspect.

10. *Erigeron formosissimus* Greene. Perennial herb 1–4(5) dm tall, arising from a simple or branched caudex, bearing fibrous roots. Stems curved and arching upward from the base, sometimes weakly erect, ± glandular, especially upward, variously pubescent or sometimes glabrate below. Lower leaves glabrate or with scattered long hairs, or sometimes conspicuously pubescent, oblanceolate to spatulate or ovate, entire, narrowed at the base to a winged petiole, 2.5–15 cm long overall and up to 15 mm wide; middle cauline leaves lanceolate to ovate, smaller than the basal leaves, glabrous to somewhat hirsute, margins ciliate; uppermost leaves reduced, bractlike, glandular and/or somewhat pubescent. Heads 1–6, involucre 5–8 mm tall, disk 10–20 mm across; involucral bracts about all equal in length, glandular or viscid and often hirsute; ray florets 75–150, ligule blue or sometimes pinkish or white, 8–15 mm long; disk florets with corolla 3.5–4.5 mm long. Achenes 2-nerved, hairy; pappus double, the inner series of 15–25 bristles, the outer of small setae, often scanty. Jul–Aug. Open, high meadows; BH; (Alta. to BH & CO, UT, AZ, & n Mex.).

Specimens with the involucre glandular but without long, glandular hairs, and with the upper leaves glandular but at most sparsely hairy, have been recognized as var. *viscidus* (Rydb.) Cronq. Both this variety and var. *formosissimus* are reported from the BH, but the species is inadequately understood in our region to make certain whether or not we do have two varieties. Our materials are very similar to *E. glabellus* (cf. Cronquist, op. cit.).

11. *Erigeron glabellus* Nutt. subsp. *pubescens* (Hook.) Cronq. Biennial or weak perennial herb, 1–5(7) dm tall, arising from a simple or little-branched caudex bearing fibrous roots. Stems erect or sometimes slightly arching upward; herbage variously hirsute to strigose with appressed or spreading hairs, sometimes sparingly so. Basal and lowermost cauline leaves oblanceolate, entire to somewhat toothed, tapering at the base to a sometimes poorly defined winged petiole, 5-15 cm long overall and 3-15(+) mm wide; middle and upper cauline leaves reduced, linear to lanceolate, becoming bractlike. Heads 1-15, at the ends of long, nearly naked peduncles; involucre 5-9 mm tall, disk 10-20 mm across; involucral bracts subequal or slightly imbricated; ray florets 125-175, ligule blue to pink or sometimes white, 8-15 mm long; disk florets with corolla 4-5.5 mm long. Achenes 2-nerved, hairy; pappus double, the inner series of bristles, the outer of small setae, sometimes scanty. (n = 9, 18, 27) Jun-Jul. Open meadows & damp or drying prairies; n GP, across ND & e MT, scattered s to BH & nw NE; (WI, n GP to AK, s to CO & UT). *E. tardus* Lunell, *E. asper* Nutt., *E. oligodontus* Lunell, *E. drummondii* Greene, *E. oxydontus* Lunell, *E. abruptorum* Lunell — Rydberg.

Subsp. *pubescens*, with the stem pubescence clearly spreading, includes the materials from the n GP n & w through Can. to AK. Specimens from the cordilleran region of CO & UT typically have the stem pubescence appressed or loosely ascending, and are treated as subsp. *glabellus*. Some of the collections from the BH region may approach subsp. *glabellus*.

12. *Erigeron lonchophyllus* Hook. Biennial or short-lived perennial herb, 1-5(6) dm tall, arising from a weak, fibrous root. Stems variously spreading-hirsute. Leaves sparsely to moderately hirsute or the lower ones sometimes glabrate, margins often hirsute-ciliate; basal leaves oblanceolate to spatulate, up to 15 cm long overall and 12 mm wide; cauline leaves usually narrower, becoming linear, but generally long and conspicuous. Heads borne on erect peduncles and at least the lower ones equaled or surpassed by their subtending leaves; involucre 4-9 mm tall, disk 7-17 mm wide; involucral bracts thin and light green but commonly purplish toward the tip, usually imbricated; pistillate florets numerous, ligules about 2-3 mm long, white or sometimes pinkish, barely surpassing the involucre, eligulate pistillate florets absent; disk florets with corolla 3.5-5 mm long. Achenes 2-nerved, sparsely hirsute; pappus with 20-30 obvious whitish bristles, equaling or surpassing the disk corollas, and sometimes with a few short outer setae. Jul-Aug. Open damp or drying meadows & prairies; scattered in ND, BH; (GP to AK, s to NM & CA; sporadic e across Can. to Que.). *E. minor* (Hook.) Rydb. — Rydberg.

13. *Erigeron modestus* A. Gray. Annual, biennial or weak perennial herb 1-4 dm tall, arising from a short, slightly branched caudex. Stem moderately diffuse-branched, stiffly hirsute with the hairs near the base clearly spreading, those near the top often appressed or ascending. Leaves hirsute, the lowermost oblanceolate, entire or with a few teeth, blade 1-2 cm long and 3-5(7) mm wide, tapering to a petiole up to 2 cm long; middle and upper leaves linear or nearly so, sessile, 1-2.5 cm long and up to 3 mm wide. Heads numerous; involucre 3-5 mm tall, ± hispid-hirsute with curved hairs, disk 5-10 mm across; outer involucral bracts shorter than the inner; ray florets 30-65(100 +), ligule white or pink, mostly 4-6 mm long; disk florets with corolla 2-2.5 mm long. Achenes 2-nerved, sparsely hairy; pappus double, the inner series of ± 12 fragile bristles, the outer of short setae or narrow scales. May-Jul. Open, dry or drying calcareous uplands; sw GP from n-cen KS s through w OK, n TX, ne NM; (s GP through w 1/2 TX, to AZ?, n Mex.). *E. divergens* var. *cinereus* A. Gray — Cronquist, op. cit.

This is the southern phase of a widespread complex centering around *E. divergens* (q.v.) and including *E. flagellaris* and *E. tenuis*. At least some populations in the sw GP are morphologically similar to *E. divergens*, and it is possible that the two entities are in reality conspecific.

14. *Erigeron ochroleucus* Nutt. Low perennial herb, up to 4 dm tall, arising from a crown or short branching caudex surmounting a taproot. Stem hirsute-strigose with appressed hairs, but often with loose-spreading lax hairs upward and occasionally so throughout.

Leaves variously strigose to villous-strigose or subglabrous; basal leaves linear to narrowlanceolate, 1–12 cm long overall and up to 5 mm wide, bases ± enlarged and whitish to purplish, somewhat submembranaceous; cauline leaves linear, smaller than the basal ones, progressively reduced upward, or sometimes few and bractlike. Heads solitary or few; involucre 4.5–8 mm tall, ± hirsute-villous with slender, loose hairs and sometimes rather viscid, disk 6–18 mm across; involucral bracts subequal, often with purplish tips; ray florets 20–80, ligule blue, purple or white, 4–12 mm long; disk florets with corolla 3–4.5 mm long. Achenes 2-nerved, hairy; pappus double, the inner series of 10–20 bristles, sometimes shorter than the disk corollas but often clearly longer, the outer pappus series of scales or thick setae, usually conspicuous. (n = 9) May–Aug. Open, dry plains; BH, adj. nw NE & e WY; (Sask. & Alta. to GP, w through MT & WY). *E. laetivirens* Rydb., *E. montanensis* Rydb.—Rydberg.

The diminutive phase with the cauline leaves few and bractlike is separable as var. *scribneri* (Canby ex Rydb.) Cronq. It is mostly 1–6 cm tall, with clearly expanded leaf bases, and apparently it has some degree of biological integrity, although it is found throughout the range with the taller and more robust var. *ochroleucus*.

15. *Erigeron philadelphicus* L. Biennial or short-lived perennial herb (occasionally annual, or at least flowering in the first year?), 2–7 dm tall, arising from a short, simple or weakly branched caudex; herbage variously pubescent with long, spreading hairs to glabrate. Basal leaves narrowly to broadly oblanceolate, toothed to lobed, tapering to a broad-petioled base, mostly up to 15 dm long overall and 3 cm wide, but varying with robustness and habitat; cauline leaves numerous, ± clasping, crenate or dentate to subentire. Heads usually numerous in an open cluster; involucre 4–6 mm tall, hirsute to glabrous, disk 6–15 mm across; involucral bracts subequal, with broad, hyaline to purplish margins; ray florets numerous, ± 150 or more, ligule pink or rose-purple to white, 5–10 mm long; disk florets with corolla 2.5–3 + mm long. Achenes 2-nerved, sparsely pubescent; pappus simple, of 20–30 bristles, often much shorter than the disk corollas. (n = 9) May–Jul. Open moist sites, especially where there is some disturbance; somewhat weedy; e 1/3 GP, scattered w thoughout ND, BH, Platte R. valley, NE, & to sw OK; (throughout much of temp. N. Amer. but infrequent in the cordilleran region). *E. purpureus* Ait.—Rydberg.

16. *Erigeron pulchellus* Michx., robin's plantain. Biennial or short-lived perennial herb 1.5–6 dm tall, arising from a short, simple or subsimple caudex, producing slender stoloniform rhizomes; herbage variously hirsute with long, spreading hairs, or the upper surfaces of the leaves glabrate. Basal leaves oblanceolate to suborbicular, tapering to an indistinct petiole, 2–13 cm long overall and 6–50 mm wide, shallow-toothed above the middle to subentire; cauline leaves lanceolate or oblong to ovate, entire or nearly so, smaller than the basal ones. Heads mostly 1–4; involucre 5–7 mm tall, ± hirsute and somewhat viscid, disk 12–20 mm across; involucral bracts subequal or the outermost somewhat shorter; ray florets 50–80(+), ligule blue to pinkish, 6–10 mm long; disk florets with corolla 4.5–6 mm long. Achenes 2- or sometimes 4-nerved; pappus simple, of ca 30–35 bristles. (n = 9) Mar–Jun. Wooded areas, especially along stream banks; extreme e edge GP; KS: Cherokee; MN: Mahnomen; w MO; (ME to MN, GA to e TX, also e Can.).

17. *Erigeron pulcherrimus* Heller var. *wyomingia* (Rydb.) Cronq. Perennial herb 5–35 cm tall, arising from a much-branched caudex with thin, peeling bark, surmounting a taproot. Stem erect or nearly so; herbage grayish strigose. Basal leaves tufted, narrowly lanceolate, up to 7 cm long overall and 1.5–5 cm wide; cauline leaves reduced, linear, or the lowermost ones linear-oblanceolate. Heads solitary; involucre 6–9 mm tall, hirsutevillous with crinkled hairs, obscurely viscid, disk 10–20 mm across; involucral bracts imbricated, somewhat yellowish at the base, acuminate to subattenuate; ray florets 25–60,

ligule pink, white or violet-bluish, 8-15 mm long; disk florets with corolla 4-5.7 mm long. Achenes (2)4(5)-nerved, densely pilose; pappus of numerous coarse, sordid bristles and sometimes obscurely distinguishable into an inner and outer series of long but ± subequal bristles. Jun-Jul. Dry, open sites; barely entering GP in e WY; (WY to UT, CO, & n NM).

The nomenclaturally typical var. *pulcherrimus* has narrower basal leaves and occurs in NM, UT. This species has a generally spotty distribution, and it may well be associated with seleniferous soils.

18. *Erigeron pumilus* Nutt. Perennial herb 5-30 cm tall, arising from a short to much-branched caudex surmounting a taproot; herbage variously hirsute with spreading hairs, somewhat glandular or slightly viscid, especially near the heads. Leaves oblanceolate to narrowly oblanceolate, up to 8 cm long overall and 5(8) mm wide but usually smaller; the basal and lower cauline ones tufted and persistent. Heads solitary to numerous; involucre 4-7 mm tall, sparsely to densely hirsute, disk 7-15 mm across; involucral bracts subequal; ray florets 50-100(+), ligule mostly whitish, 6-15 mm long; disk florets with corolla 3-5 mm long. Achenes 2-nerved, mostly somewhat pubescent; pappus double, the inner series of 15-30 slender bristles, the outer ones setae-like bristles and not clearly distinct from the inner pappus. May-Aug. Open, dry prairies & plains; w 1/2 GP from cen KS n & w, scattered e in ND; (GP to B.C., s through CO & n NM to CA).

19. *Erigeron speciosus* (Lindl.) DC. var. *macranthus* (Nutt.) Cronq. Perennial herb 1.5-8 dm tall, arising from a branchng, ligneous caudex; herbage glabrous or glabrate, or with some pubescence just under the heads. Leaves often 3-nerved, entire or sometimes with weakly ciliate margins; basal and lower cauline leaves oblanceolate to spatulate, narrowed basally to a winged petiole, 5-15 cm long overall and up to 2 cm wide, often withered by flowering time; middle cauline leaves commonly as large or a little larger than the basal leaves but sessile and broadly oval; upper cauline leaves not strongly reduced, ovate. Heads 1-12; involucre 6-9 mm tall, disk 10-20 mm across; involucral bracts acuminate or attenuate, subequal; ray florets 75-150, ligule bluish or rarely white, 9-18 mm long; disk florets with corolla 4-5 mm long. Achenes 2- or 4-nerved, hairy; pappus double, inner series of numerous bristles, outer series of setae. (n = 9) Jul-Aug. Mostly in open, wooded areas; BH & adj. WY; (BH to Alta., B.C., MT to NM, AZ).

The var. *speciosus* has the upper leaves clearly lanceolate, and some pubescence on the involucre. Both varieties are widespread in the cordilleran region, but var. *speciosus* is absent from the GP.

20. *Erigeron strigosus* Muhl. ex Willd., daisy fleabane. Annual or rarely biennial herb, mostly 4-7 dm tall; herbage finely strigose to hirsute with shortish, ± appressed hairs, or the hairs somewhat spreading on the lower stem. Leaves few; basal leaves mostly oblanceolate to elliptic, tapering to a petiole, entire to toothed, up to 15 cm long overall and 2.5 cm wide, usually somewhat smaller, often early deciduous; cauline leaves linear to lanceolate, commonly smaller than the basal ones, entire or the lower ones shallow-toothed. Heads several to numerous in an open, wide-branching cluster; involucre (2)3-5 mm tall, disk 5-12 mm across; involucral bracts subequal or the outer ones slightly shorter; ray florets 50-100, or rarely absent, ligule white or pinkish to bluish, up to 6 mm long; disk florets with corolla 1.5-2.5 mm long. Achenes 2-nerved; pappus double, the inner series of 10-15 fragile bristles, the outer of setose scales. (2n = 18, 27, 36, 54; apomictic) May-Aug. Open moist or drying prairies & disturbed sites; often weedy; throughout e 3/5 GP but scattered & localized in w 2/5; (throughout most of temp. N. Amer.). *E. ramosus* (Walt.) B.S.P.—Rydberg.

Our materials are nearly all referable to var. *strigosus*, but infrequent populations have diffuse, subnaked inflorescences and tiny heads (involucre 2-3 mm tall) and may be recognized as var. *beyrichii* (Fisch. & Mey.) A. Gray. This variety may be merely an occasional introduction with us, for it is best developed and most frequent on the eastern Coastal Plain and across the south to TX.

21. Erigeron subtrinervis Rydb. Perennial herb 1.5–9 dm tall, arising from a ± branching, woody caudex; herbage pubescent throughout with rather dense, evenly distributed, spreading hairs. Leaves often 3-nerved, entire, margins ciliate; basal and lowermost cauline leaves oblanceolate, narrowed at the base to a winged petiole, 4–13 cm long overall and 5–23 mm wide, often early deciduous; middle cauline leaves as large as the lower ones, narrowly lanceolate to oblong or broadly ovate, sessile, up to 8 cm long; upper leaves not markedly reduced, narrowly lanceolate to ovate. Heads 1–20; involucre 6–9 mm tall, disk 13–20 mm across; involucral bracts subequal, sometimes surpassing the edge of the disk; ray florets 100–150, ligule blue or rose-purple, 7–18 mm long; disk florets with corolla 4–5 mm long. Achenes 2- or 4-nerved, pubescent; pappus double, the inner series of numerous bristles, the outer of setae. Jul–Aug. Open wooded areas & meadows; scattered in w 1/2 ND, e MT, BH, adj. nw NE; in Rocky Mts. of CO just w of GP; (Alta. & B.C, w GP to n NM & UT).

22. Erigeron tenuis T. & G. Annual or biennial herb 1–4 dm tall, erect or subdecumbent but with an evident main axis. Stem short-hirsute, the hairs near the base spreading, those near the top appressed. Basal leaves oblanceolate to suborbicular, blades 1–3 cm long and up to 1.5 cm wide, or sometimes larger, entire or irregularly toothed or even-lobed, tapering or contracted to a distinct petiole 1–3 cm long, sometimes early-deciduous; middle cauline leaves oblong, sessile, entire or somewhat toothed, up to 4 cm long. Heads 1–many in an open cluster; involucre 2.5–4 cm tall, glabrate to moderately hirsute with curved hairs, sometimes viscid or finely glandular, disk 5–10 mm across; involucral bracts subequal; ray florets 60–120, ligule blue or white, 2.5–4.5 mm long; disk florets with corolla 2–3 mm long. Achenes 2- or sometimes 4-nerved, sparsely pubescent; pappus double, the inner series of ca 12 bristles, the outer of short setae. (n = 9, 18) Apr–May. Open woods & prairies; extreme se GP, se KS s & w to s OK; (MO & se GP to LA & e TX).

23. Erigeron vetensis Rydb. Perennial herb 5–25 cm tall, arising from a stout branching caudex to which the old leaf-bases of past seasons may adhere. Stems somewhat glandular and often sparsely pubescent. Leaves ± glandular but sometimes only sparsely so, margins coarsely ciliate with stiff, spreading hairs, surfaces sometimes scattered-hirsute; basal leaves narrowly oblanceolate, to 15 cm long overall and 7 mm wide but usually less, blade gradually tapering to an indistinct petiole; cauline leaves several, reduced, linear-oblong, usually less than 3 cm long and 3 mm wide. Heads solitary on subnaked peduncles; involucre 4–8 mm tall, densely glandular and sparsely hirsute, disk 7–15 mm across; involucral bracts subequal or but weakly imbricated; ray florets 30–40, ligule blue, pink or purplish-pink or rarely white, 6–16 mm long; disk florets with corolla 3.2–5.3 mm long. Achenes 2-nerved, hairy; pappus double, the inner series of 18–25 sordid bristles, the outer mostly setose, but sometimes obscure. Apr–May. Open, dry wooded places; nw NE & adj. e WY; (w GP through WY s to n NM).

44. EUPATORIUM L.

Ours perennial herbs with stems arising from a fibrous-rooted, creeping rootstock, erect crown or a caudex. Leaves whorled or opposite, or sometimes the uppermost alternate, often variously glandular-punctate and rough; simple, usually toothed. Inflorescence a modified corymbiform or paniculate cyme; heads discoid with all florets tubular and perfect; involucre cylindric or weakly campanulate, involucral bracts green to chartaceous or sometimes weakly coriaceous; receptacle naked, flat, or conic in one species; corolla bluish purple to pinkish or white; style branches with minute projections (papillae), elongate, linear or but weakly clavate, obtuse at the apex, with stigmatic lines toward the base. Achenes

5-angled, glabrous or minutely hairy along the angles; pappus a single whorl of capillary bristles. [tribe Eupatorieae]

Reference: Cronquist, A. 1980. *Eupatorium* in Vascular Flora of the Southeastern U.S., vol. I. Univ. North Carolina Press, Chapel Hill.

Eupatorium is a huge group that has been divided by some investigators into numerous segregate genera. Many species are further complicated by low barriers to hybridization and by polyploidy, which together produce abundant localized and variously persistent races. The GP species are essentially western extensions from the expansive complex, centering in the se U.S. The treatment here follows that of Cronquist, op. cit.

1 Leaves whorled, mostly 2–15 cm wide.
  2 Florets 9–22 per head, inflorescence flat-topped .... 3. *E. maculatum* var. *bruneri*
  2 Florets 5–7 per head, inflorescence ± dome-shaped .... 5. *E. purpureum*
1 Leaves mostly opposite, or a few alternate
  3 Florets mostly 5 per head .... 1. *E. altissimum*
  3 Florets 9 or more per head.
    4 Leaves sessile, broad-based and usually connate-perfoliate .... 4. *E. perfoliatum*
    4 Leaves petiolate.
      5 Florets blue to violet, receptacle conic .... 2. *E. coelestinum*
      5 Florets white, receptacle flat.
        6 Involucral bracts imbricate in 2 or 3 series .... 7. *E. serotinum*
        6 Involucral bracts all of about the same length .... 6. *E. rugosum*

1. Eupatorium altissimum L. Erect perennial up to 2 m tall, pubescent with loose, soft spreading hairs, glabrate below. Leaves numerous, opposite, prominently 3-veined, glandular-punctate, narrowly lanceolate, narrowed at the base to a weakly distinct petiole, margin serrate, especially toward the tip, or subentire, 5–12 cm long and up to 3 cm wide. Inflorescence broad and many-headed; heads 4.5–6.5 mm tall, involucral bracts imbricate, rounded or obtuse; florets 5, corolla white. (various levels of ploidy, x = 10) Aug–Sep. Pastures & disturbed sites; e 1/2 GP from sw MN to OK; (PA to AL, w to GP).

2. Eupatorium coelestinum L., mist flower. Rhizomatous perennial, 3–9 dm tall, short-pubescent. Leaves opposite, deltoid to ovoid, sometimes quite narrowly so, crenate to serrate, petiolate, blade 3–10 cm long and 2–5 cm wide, ± prominently 3-veined. Involucre 3–5 mm tall; involucral bracts narrow, long pointed and weakly imbricate; receptacle distinctly conic; florets 35–70, blue or violet, drying purple. Pappus bristles few. (n = 10) Jul–Oct. Wooded areas along stream; se KS, to cen OK, sporadic in ne KS, w MO; (NJ to FL w to KS & TX).

3. Eupatorium maculatum L. var. **bruneri** (A. Gray) Breitung, joe-pye weed, spotted joe-pye weed. Stems erect, 4–10 + dm tall, variously purple-spotted to evenly purplish. Leaves mostly in whorls of 4s or 5s, narrowly lanceolate to lance-ovate, ± contracted at the base to a short petiole, blade 6–15 cm long and 2–8 cm wide, margin serrate. Inflorescence flat-topped; involucre 6.5–9 mm tall; involucral bracts imbricate, obtuse, purplish; florets 8–20 +, corolla purple. (n = 10) Jul–Sep. Moist places in wooded areas; n GP, e 1/2 ND s to w SD, scattered across NE & ne KS; (Newf. s to NC, w B.C., & NM).

4. Eupatorium perfoliatum L., boneset. Perennial 4–15 dm tall, conspicuously villous-pubescent with long, spreading hairs, or sometimes short-pubescent. Stems arising from a stout, weakly spreading rhizome. Leaves opposite, broad based and usually conspicuously connate-perfoliate, tapering upward to an acuminate tip, margin mostly crenate-serrate, blade 7–20 cm long and up to 4 cm wide, often evident-pubescent beneath. Inflorescence flat-topped, involucre 4–6 mm tall; involucral bracts imbricate in several series, acuminate, pubescent; florets 9–23, corolla white. (n = 10) Aug–Sep. Damp low ground; e 1/2 GP, scattered w to w NE, sw KS, to n TX; (e Can. to FL, w to GP & LA).

Highly variable plants with leaves whorled in 3s, not perfoliate, and with variously purple or pinkish flowers are known in nature.

5. **Eupatorium purpureum** L., sweet joe-pye weed. Stems 6–20 dm tall, often slightly glaucous, greenish but spottily purple, especially at the nodes. Leaves mostly in whorls of 3s and 4s, lanceolate to ovate, tapering or abruptly contracted to a short petiole, margin serrate, blade 8–30 cm long and up to 15 cm wide. Inflorescence flattened dome-shaped; involucre 6.5–9 mm tall; involucral bracts imbricate in several series, obtuse to weakly acute; florets 4–7, corolla pale pink to purplish. (n = 10, 20) Jul–Sep. Open wooded areas, often in drying sites; e NE & KS, adj. w IA & MO, scattered in nw NE, w MN; perhaps elsewhere; (NH to GA, w to GP).

This is a variable species, with solid stems and the pith remaining intact. The related *E. fistulosum* Barr. is similar, but is a little larger and coarser, and has soft, hollow stems that are strongly glaucous and more consistently purple. It occurs from IA to TX and e, and may possibly to be found along the e edge of our range.

6. **Eupatorium rugosum** Houtt., white snakeroot. Stems 3–15 dm tall, glabrous or pubescent with short, spreading hairs. Leaves opposite, petiolate, blade ovate to broadly ovate, contracted to subcordate at the base, variously serrate, acuminate, 6–15 + cm long and 3–12 cm wide, smaller upward. Inflorescence flat-topped or flat dome-shaped; involucre 3–5 mm tall, involucral bracts acuminate to obtuse, all about the same length; florets 12–24, corolla white. (n = 17) Aug–Oct. Open woods, especially in areas of some disturbance; e 1/2 GP; ND s to OK, scattered w in NE; (e Can. to FL, w to GP).

This is a widespread, abundant, and variable species. It is poisonous to livestock, and the cause of "milk sickness" when the poisonous principle is transmitted to humans.

7. **Eupatorium serotinum** Michx. Stems 4–15 dm tall, short-pubescent, especially above. Leaves mostly opposite, but the uppermost sometimes alternate, petiolate, blade lanceolate to ovate, serrate, acuminate, 5–16 + cm long and up to 10 cm wide, with 3–5 or more prominent veins. Inflorescence a many-headed corymb; involucre 4–7 mm tall; involucral bracts distinctly imbricate in 2 or 3 series, rounded or obtuse; florets 9–15, corolla white. (n = 10) Aug–Oct. Mostly in open moist woods in bottomlands; se GP, w IA, e NE, s to cen OK; (NY to FL, w to GP).

45. EUTHAMIA Nutt.

Perennial herb with stems arising singly or at most loosely clustered from well-developed creeping rhizomes. Leaves alternate, narrow, numerous, sessile or subsessile, resinous-punctate, often 3-nerved, the lower ones early-deciduous. Inflorescence an open or contracted corymbiform cluster or series of clusters; heads small and numerous; involucre glutinous; involucral bracts yellowish or greenish-tipped, imbricate; receptacle fimbrillate to subalveolate; ray florets 6–35(+), pistillate and fertile, ligule yellow, 1–3 mm long; disk florets 3–9(15), fewer than the rays, perfect and fertile; style branches flattened, with inner-marginal stigmatic lines and a short, externally pubescent appendage. Achenes several-nerved, short-pubescent; pappus of numerous white capillary bristles. [tribe Astereae]

Reference: Sieren, David J. 1981. The taxonomy of the genus *Euthamia*. Rhodora 83: 551–579.

Euthamia usually has been treated as a section of the genus *Solidago*, but its gross aspect, glandular-punctate leaves, and technical characters of the heads and inflorescence combine to make it distinctive. David J. Sieren kindly supplied much advice on how to treat the genus.

1 Inflorescence elongate or rounded and interrupted, with lateral corymbiform clusters arising from the axils of well-developed leafy bracts; plants often more than 1 m tall; extreme w GP & cen NE .... 3. *E. occidentalis*

1 Inflorescence obviously broad and flattish; plants seldom exceeding 1 m tall; widespread in GP.
  2 Heads relatively small, usually with fewer than 20 florets; herbage ± conspicuously resinous-glutinous; leaves often with but 1 prominent nerve, or if 3-nerved, without any additional lateral nerves .... 2. *E. gymnospermoides*
  2 Heads larger, with 20–40+ florets; herbage smooth and not glutinous although the leaves finely glandular-punctate; leaves evidently 3-nerved, or the larger ones often with 1 or 2 additional pairs of less-prominent lateral nerves. .... 1. *E. graminifolia*

1. **Euthamia graminifolia** (L.) Nutt. Rhizomatous perennial 3–9(12) dm tall, glabrous or but rarely short-hirsute, branching prominently in the upper 1/2. Basal and lower cauline leaves early deciduous; persistent cauline leaves numerous, linear or lance-linear to elliptic-lanceolate, acute to acuminate or attenuate, glandular-punctate (under magnification), 4–12(15) cm long and 2–12 mm wide, evidently 3-nerved and sometimes with 1 or 2 pairs of additional but fainter lateral nerves. Inflorescence broad and corymbiform with the heads sessile or nearly so in small clusters; heads campanulate to obconic; involucre 3–5 mm tall; involucral bracts weakly glutinous-glandular, obtuse or rounded, yellowish; receptacle short-fimbrillate; florets 20–35(40); ray florets small with the ligule scarcely spreading, more numerous than the disk florets. Achenes short-pubescent. (n = 9) Aug–Sep.

This is polymorhic species subject to diverse interpretation, and it is sometimes circumscribed to include *E. gymospermoides*, as one or more additional varieties. Two varieties are here recognized for our region.

1a. var. *garminifolia*. The more robust phase, normally exceeding 7 dm tall. Leaves narrowly lanceolate or narrower, 10–20 × longer than wide, acuminate to attenuate, glabrous or nearly so, at most with scabrous margins. Moist or open drying sites, especially along stream banks; ne GP, mostly in extreme w MN but scattered in e 1/2 ND, possibly in sw SD s of BH; (se Can. to NC & TN, w to B.C. & MN). *Solidago graminifolia* (L.) Salisb. var. *graminifolia* — Fernald.

1b. var. *major* (Michx.) Moldenke. Smaller and less robust, 3–7 dm tall. Leaves often lanceolate, usually less than 10 × longer than broad, blunt or subacute, minutely scabrous-hirsute on the veins beneath. Moist or drying open sites, stream banks, or frequently in roadside ditches; scattered in n 1/2 GP from n NE northward; (Newf. & n Que. to n MI, Alta., & n GP; reputedly in Rocky Mts. to NM). *Solidago graminifolia* var. *major* (Michx.) Fern. — Fernald.

2. **Euthamia gymnospermoides** Greene, viscid euthamia. Rhizomatous perennial to 1 m tall, essentially glabrous but with a few scabrous lines and scabrous leaf margins, often glutinous. Stems single or loosely clustered, mostly branching in the upper 1/2–2/3. Basal and lower cauline leaves early deciduous; persistent cauline leaves numerous, sessile, entire, linear or linear-lanceolate, attenuate, strongly glandular-punctate, 4–9 cm long and 1.5–4(8) mm wide, prominently 1-nerved and often obscurely to evidently 3-nerved, but without additional parallel lateral nerves. Inflorescence broad and corymbiform; heads numerous, sessile or nearly so, in small clusters; involucre (4)4.5–6.5 mm tall, narrow, somewhat glutinous; involucral bracts obtuse, the inner ones acute, straw colored; receptacle alveolate-fimbrillate; florets 14–20; ray florets small and about twice as many as disk florets. Achenes hairy. (n = 18) Aug–Oct. Open, drying sites, especially in sandy or rocky soils & along roadsides; GP except nw 1/4 & extreme sw; (IL & MN to GP; introduced elsewhere). *Solidago gymnospermoides* (Greene) Fern. — Fernald; *S. graminifolia* var. *media* (Greene) S.K. Harris, *S. graminifolia* var. *gymnospermoides* (Greene) T.B. Croat — Atlas GP; *E. camporum* Greene — Correll & Johnston.

3. **Euthamia occidentalis** Nutt. Rhizomatous perennial (0.5)1–1.5(2) m tall, glabrous or nearly so. Leaves linear to linear-lanceolate, entire, sessile, glandular-punctate, (1)3–10 cm long and up to 1 cm wide, margin weakly scabrous. Inflorescence leafy-bracted, ± interrupted, elongate or broadly rounded, with a terminal corymbiform cluster and few to many

corymbiform clusters arising laterally along the upper stem; heads pedicellate; involucre 3–5 mm tall; involucral bracts narrowly lanceolate; receptacle fimbrillate; ray florets 15–30, small, more numerous than the disk florets. Achenes sparingly or ± densely pubescent. (n = 9) Aug–Oct. Open moist ground, especially along stream banks; to be expected along the w edge of the GP; apparently adventive in the Platte R. valley; NE: Hall, Kearney, Scotts Bluff; (GP to B.C., CA, & NM). *Solidago occidentalis* (Nutt.) T. & G.—Harrington.

46. EVAX Gaertn., Rabbit-tobacco

Contributed by Ronald L. McGregor

1. *Evax prolifera* Nutt. ex DC., rabbit-tobacco. Winter annual or annual gray floccose-woolly herbs with slender taproots. Stems erect, 2–10(15) cm tall, simple or branching from base. Rosette leaves 3–6 mm long, usually spatulate, soon disappearing; cauline leaves alternate, narrowly oblanceolate to spatulate, 3–15 mm long, 1-nerved, entire. Heads in very dense woolly glomerules terminating stems and branches; glomerules subtended by leaves 6–12 mm long, and with internal leaves whose tips protrude from between the heads; often 1–4 branches ultimately emerging from axils of leaves subtending glomerules and bearing a terminal glomerule of heads; true involucre absent, the apparent bracts being chaff of the receptacle; receptacle slightly raised to somewhat conical, chaffy nearly throughout with thin paleae that are semipersistent; ray florets absent; outer peripheral florets pistillate and fertile; pistillate florets with a minute tubular corolla; the few infertile central florets with a 4-toothed corolla and 4 very small caudate anthers, the style undivided, ovary not functional. Achenes oblong-elliptic in outline, compressed, 0.6–0.9 mm long, yellowish-brown and somewhat translucent, the edges sharp, surface sometimes minutely papillose; pappus none. Apr–Jul. Infrequent to locally common in prairies, pastures, stream valleys; sw SD, w NE, w 1/2 KS, e CO, OK, TX; (SD, CO, KS, s to TX, AZ). *Diapteria prolifera* Nutt.—Rydberg. [tribe Inuleae]

In southwest OK and s TX panhandle the scarcely distinguishable *E. multicaulis* DC. is infrequent. It differs from *E. prolifera* in glomerules of heads subtended by leaves 3–10 mm long and very small leaves among and about equaling the heads, achenes more blunt-edged, and central florets without abortive ovary. Immature specimens of these species are impossible to distinguish.

47. FLAVERIA Juss.

1. *Flaveria campestris* J. R. Johnst. Coarse annual 3–9 dm tall, glabrous or scattered-villous, especially at and near the nodes. Leaves opposite, sessile, lanceolate, 3–7 cm long and up to 2 cm wide, margins serrulate with shallow, distant teeth. Inflorescence a cluster of numerous small heads, subtended by leafy bracts; heads cylindric, involucre of 3 primary bracts, somewhat membranaceous, 5–7 mm long, subtended by 2 much reduced calyculate bracts 1–3 mm long; ray florets solitary, ligule ca 2 mm long, yellow; disk florets 3 or 4, corolla yellow; receptacle naked. Achenes 2.5–3 mm long, that of the ray floret slightly larger than the others; pappus absent. Jul–Sep. Low waste places & disturbed sites; s 1/5 GP, mostly in sw KS, w OK, & n TX; (GP s & w through CO, TX, & NM; occasionally adventive as a weed elsewhere). [tribe Heliantheae]

48. GAILLARDIA Foug., Blanket Flower

Taprooted annual or perennial herbs with leaves alternate or basal, entire to variously toothed or pinnatifid. Heads terminating the branches; involucral bracts in 2 or 3 series, herbaceous toward the tip, spreading, especially at maturity; receptacle subglobose or convex, alveolate with long setae that do not precisely subtend the florets and that are irregularly fused at the base, or the setae reduced to small toothlike fimbrillae and the receptacle floor merely rugose; ray florets nearly always present, neutral, ligule variously reddish or brownish-purple to yellow, 3-parted at the apex; disk florets perfect and fertile, style branches flattish and with a distinct, elongate, externally hairy appendage, or the appendage short and glabrous in *G. suavis*. Achene broadly obpyramidal, ± covered with long hairs; pappus of several thin, bractlike ovate scales, with a prominent apical awn. [tribe Heliantheae]

Reference: Biddulph, Susann Fry. 1944. [1945]. A revision of the genus *Gaillardia*. Res. Stud. State Coll. Wash. 12(4): 195–256.

1 Receptacle with only small, irregular, soft toothlike fimbrillae representing chaffy bracts, or merely rugose .... 1. *G. aestivalis*
1 Receptacle chaffy with long, stiff setae.
   2 Style branches short and glabrous, stems scapiform. .... 5. *G. suavis*
   2 Style branches elongate and hispidulous, at least the lower stem leafy.
      3 Some or all of the basal leaves pinnate-parted; pappus-awns about 1/2 as long as the basal scale or less; distinctly perennial; sw 1/4 GP .... 3. *G. pinnatifida*
      3 Basal leaves variously toothed, lobed or entire but rarely somewhat pinnate-parted; pappus awns equalling or exceeding the basal scale; annual or weakly perennial.
         4 Rays yellow, reddish-purple at the base; pappus awns about 2× as long as the basal scale; perennial; northern plains .... 2. *G. aristata*
         4 Rays reddish-purple, yellow toward the apex; pappus awns about as long as the basal scale; mostly annual but capable of perennating; chiefly s 1/2 GP .... 4. *G. pulchella*

1. **Gaillardia aestivalis** (Walt.) Rock, prairie gaillardia. Short-lived perennial or annual, 3–6 dm tall, variously scattered to closely pubescent or somewhat glandular. Stem stiff-erect, branching above. Leaves mostly sessile, oblong to lanceolate, entire to deeply toothed, the lowermost sometimes weakly petiolate and pinnatifid. Heads with involucral bracts acute to acuminate, 8–12 mm long, disk 2–2.5 cm across; receptacle with short, soft, toothlike, flattish fimbrillae or merely rugose; ray florets with ligule (8)10–15 mm long, yellow and often reddish toward the base or rays sometimes absent; disk florets with corolla brown or dark purple. Achene ca 2 mm long, covered with long hairs arising from the base; pappus of several scales; body of the pappus-scale lanceolate, 5–6 mm long and tapering to a slightly longer aristate awn. (n = 17, 18) May–Aug. Open wooded areas & brushy grassland, especially in sandy soil; s-cen KS, OK; (s-cen GP to Gulf Coast in TX). *G. fastigiata* Greene, *G. lanceolata* Michx.—Rydberg.

This species includes a complex of several weakly distinct entities that are unsatisfactorily treated by Biddulph, op. cit. Our materials may eventually be referable to the taxon *fastigiata* at an infraspecific rank.

2. **Gaillardia aristata** Pursh, blanket flower. Apparently a perennial but probably able to flower during the first year, 3–6 dm tall. Stems single or sometimes clustered, hirsute with moniliform hairs, arising from a slender taproot. Leaves oblong to lanceolate-ovate, 5–15(20) cm long, including the weakly defined petiole, and up to 2.5 cm wide, entire to variously toothed or even subpinnatifid, scattered along the stem or occasionally mostly basal. Heads solitary or few on long peduncles; disk 1.5–3 cm across; involucral bracts acuminate to attenuate, 1–2 cm long, usually loose-hairy; receptacle covered with irregular but well-developed setae, exceeding the achenes; ray florets with ligule 1–3 cm long, yellow and somewhat purplish toward the base; disk florets with corollas purple or brownish-purple, densely villous toward the apex; style appendages elongate. Achene ca 4 mm long, covered with long hairs; body of pappus 6–7 mm long and abruptly acuminate to an awn 2× as

long as the body. (n = 18) Jun–Aug. Open plains & prairies; n GP, ND, & e MT, s along the edge of the Rocky Mts. to CO; (GP w to Alta., B.C., & OR, s to NM & UT).

3. **Gaillardia pinnatifida** Torr. Short perennial up to 4 dm tall, ± uniformly covered with short, moniliform hairs. Leaves largely confined to the lower 1/2 of the stem, mostly oblanceolate in outline, petiolate to subsessile, some or all of the blades distinctly pinnate-parted. Heads solitary or few, long-pedunculate, disk ca 2 cm across; receptacle with stiffish setae about as long as the achenes; ray florets with ligules yellow, somewhat pinkish at the base and purplish along the veins; disk florets with corollas yellowish-brown to purplish. Achene 3 mm long and covered with silky hairs; body of the pappus scale lanceolate, ca 5 mm long, contracted to an awn that is no more than 1/2 as along as the body. (n = 18) Apr–Aug. Apparently sporadic in open grasslands; extreme sw GP; panhandles of OK & TX, adj. CO & NM; (sw GP s to n Mex., w to UT & AZ).

4. **Gaillardia pulchella** Foug., rose-ring gaillardia, Indian blanket flower. Taprooted annual or short-lived perennial, 1–6 dm tall; herbage variously glandular-villous with moniliform hairs. Stems simple or freely branched, erect or somewhat lax. Leaves lanceolate or oblong in outline, sessile and often clasping or the lowermost petiolate, variously entire or weakly toothed to subpinnatifid, mostly less than 8 cm long and 3 cm wide. Heads solitary or few; disk 1.5–2(+) cm across; involucral bracts lanceolate; receptacle covered with setae somewhat longer than the achenes; ray florets with ligule 1–2 cm long, reddish-purple but yellow at the tip, or infrequently entirely yellow; disk florets with corolla brownish-purple. Achene ca 2 mm long, hairy from the base; body of the pappus scale 5–6 mm long, contracted above into an aristate awn about equaling the body in length. (n = 17, 18) May–Sep. Open, loose soil, especially beaches; s 1/2 GP; s NE to TX & w; (s GP to TX & NM, e to MO & scattered to VA & FL). *G. drummondii* DC.—Rydberg.

5. **Gaillardia suavis** (Gray & Engelm.) Britt. & Rusby. Perennial scapiform herb 3–8 dm tall. Leaves basal, 5–15 cm long, mostly oblanceolate and variously entire or toothed to subpinnatifid, somewhat loosely pubescent. Heads solitary; disk 1.5–2 (+) cm across, spherical; involucral bracts ovate-lanceolate; receptacle covered with stiffish, subulate setae; ray florets with short, brownish inconspicuous ligules, or ray florets sometimes absent; disk florets with the corollas reddish-brown, exceeded by the setae of the receptacle; style branches with foreshortened, glabrous appendages. Achenes ca 2.5 mm long, covered with long, white hairs; pappus scales ca 4–5 mm long, ovate-lanceolate, contracted to an aristate awn of about equal length. (n = 15) Mar–Jun. Open, sandy or gypsiferous prairie sites; s 1/5 GP; s-cen KS, e & w OK, TX; (GP to n Mex.).

49. GALINSOGA R. & P., Quickweed

Freely branching, erect annual herbs with opposite leaves. Heads radiate; involucral bracts few, in 2 or 3 poorly defined series, several-nerved; receptacle ± conic, chaffy; ray florets few, pistillate and fertile, generally subtended by an involucral bract which is often basically fused to 2 adjacent chaffy bracts, ligule short, broad, white or seldom pinkish; disk florets perfect and fertile, style branches flattish, with short, hairy appendages; chaffy bracts membranaceous, narrow and nearly flat. Achenes 4-angled, the outer ones ± flattened parallel to the involucral bracts; pappus of the ray florets of several or many scales, fimbriate or awn-tipped; that of the disk florets tending to be reduced, or absent. [tribe Heliantheae]

Reference: Canne, J. M. 1977. A revision of the genus *Galinsoga* (Compositae: Heliantheae). Rhodora 79: 319–389.

Our 2 species are widespread weeds and are biologically quite distinct, although morphologically similar. Canne, op. cit., pp. 377–378, provides a discussion and list of characters separating the two.

1. Outer involucral bracts 1 or 2, with margins herbaceous, deciduous; inner chaffy involucral bracts entire or but weakly 2- or 3-parted, early deciduous; ligule of ray florets up to 2.5 mm long .... 2. *G. quadriradiata*
1. Outer involucral bracts 2–4, margin scarious, persistent; inner chaffy bracts deeply 3-parted and late-deciduous; ligule of ray florets seldom more than 1.5 mm long ...... 1. *G. parviflora*

1. **Galinsoga parviflora** Cav. Weedy annual herb up to 5 dm tall. Stem erect to lax, sparsely pubescent to glabrate. Leaves petiolate, blade ovate to ovate-lanceolate, 1–7 + cm long and about 1/2 as wide. Heads 3.5–5 mm tall and up to 6 mm across; involucre glabrous, involucral bracts in 2 or 3 weakly defined series; outer involucral bracts 2–4, 1.2–2.2 mm long, with scarious margins; inner involucral bracts 2.5–3.5 mm long; receptacle 0.6–1.7 mm tall; outer chaffy bracts 2–4, elliptic to obovate, often 2 or 3 fused at the base or along the edges and appearing as one, 2–2.5 mm tall, 1- to 4-lobed at the apex, joined at the base to an adjacent involucral bract and these clusters persistent; inner involucral bracts up to 3 mm long and shallowly to deeply 3-parted or rarely 2-parted to entire at the apex; ray florets (3)5(8), ligule 1.5 + mm long or absent, white, ray achene 1.5–2.5 mm long, with a pappus of 5–8 unequal scales up to 1 mm long, or pappus absent; disk florets numerous. Disk achene 1.2–2 mm long with a pappus of 15–20 fimbriate scales up to 2 mm long, or pappus absent. (n = 8) Jun–Oct. Uncommon; known as a sporadic weed in the GP; (scattered across U.S., Mex., & C. Amer.). *Naturalized* as a widespread weed in temperate and tropical regions.

2. **Galinsoga quadriradiata** R. & P., fringed quickweed. Weedy annual herb up to 5 dm tall. Stems erect or somewhat lax, variously pilose-strigose. Leaves petiolate, blade broadly to narrowly ovate, 2–7 + cm long and up to 5 cm wide, both surfaces sparsely to densely pilose, margin ciliate, serrulate to crenate-serrate. Heads 3–8 mm tall and as much as 10 mm across; involucre glabrous or scattered pubescent, the bracts in 2 or 3 weakly defined series; outer involucral bracts 1 or 2(3), 1–3 mm long, margins entire or minutely incised, herbaceous; inner involucral bracts 2.5–4 mm long; receptacle 1–2 + mm tall, chaffy; outer chaffy bracts 2–3 mm long and joined in groups of 2 or 3 at the base to an adjacent involucral bract; inner chaffy bracts up to 3 mm long, entire or shallowly 3-parted at the apex; ray florets (4)5(8), ligule 1–2.5 mm long, white or infrequently pinkish, rarely absent, ray achenes with a pappus of a few bristles or of numerous fimbriate or aristate scales, or pappus absent; disk florets numerous, disk achenes with a pappus of narrow, lanceolate, fimbriate scales, often ca 20, to 1.5 mm long, or pappus absent. Achenes of both ray and disk florets 1–1.5 mm long. (n = 16, 24, 32) Sep–Oct. Uncommon; scattered as a field & lawn weed throughout the GP; (e U.S., adj. Can. to GP, Mex., & C. Amer.). *Naturalized* as a widespread weed in temp. and trop. regions. *G. ciliata* (Raf.) Blake—Rydberg.

50. GNAPHALIUM L., Cudweed, Everlasting

Contributed by Ronald L. McGregor

Annual, biennial, or rarely perennial herbs with taproots, generally more or less white-woolly throughout, sometimes glandular. Stems erect, simple or much branched from base and above. Leaves alternate, entire, sessile, often decurrent. Heads cymosely clustered or in dense glomerules at ends of the branches, sometimes spiciform-thyrsoid or elongate; heads disciform with yellowish, whitish or rarely pinkish florets; involucre ovoid or campanulate, involucral bracts ± imbricated in several series, usually scarious toward the tip, often whitish, receptacle flat to subconic, naked; ray florets absent; outer florets pistillate and fertile, in several series, corolla tubular-filiform and equaling pappus; central florets few, perfect; anthers caudate; style branches slender, minutely papillate on outside. Achenes terete or flattish, pappus of capillary bristles, these sometimes thickened at apex, distinct or sometimes united in a ring at base. [tribe Inuleae]

1 Pappus bristles united basally, deciduous as a unit .... 4. *G. purpureum*
1 Pappus bristles distinct, falling separately.
    2 Glomerules of heads leafy-bracted, overtopped by subtending leaves; involucres 2-4 mm tall.
        3 Leaves linear or narrowly oblanceolate; tomentum appressed; rare introduced species .... 5. *G. uliginosum*
        3 Leaves oblanceolate to oblong or spatulate; tomentum loose; native species .... 3. *G. palustre*
    2 Glomerules of heads not conspicuously leafy-bracted; involucres usually 4-7 mm tall.
        4 Leaves neither decurrent nor adnate-auriculate; plants of southeast GP .... 2. *G. obtusifolium*
        4 Leaves shortly but distinctly decurrent or adnate-auriculate at base; plants of sw or nw GP.
            5 Upper surface of leaves coarsely glandular-hairy, not tomentose ..... 6. *G. viscosum*
            5 Upper surface of leaves tomentose, not glandular.
                6 Leaves usually adnate-auriculate .... 1. *G. chilense*
                6 Leaves obsoletely decurrent .... 7. *G. wrightii*

1. Gnaphalium chilense Spreng., cotton-batting. Annual or biennial, herbage loosely tomentose throughout, not glandular. Stems erect, 2-8 dm tall, often with several simple stems from the base. Leaves mainly cauline, narrow, 2-6 cm long, 2-5 mm wide, gray-tomentose above and below, sessile and conspicuously adnate-auriculate. Heads in dense, terminal glomerules, clusters somewhat flat-topped; involucres 4-6 mm tall, woolly at base; involucral bracts scarious nearly throughout, shiny yellowish-white, the tips rounded, erose; corollas yellowish. Achenes glabrous, yellowish-brown, 0.6-0.8 mm long; pappus bristles not united basally. Jun-Oct. Rare in rocky prairies & fields; OK: Comanche, Kiowa; TX: reported for Wheeler; (MT to WA, s to TX, AZ, CA).

2. Gnaphalium obtusifolium L., fragrant cudweed or fragrant everlasting. Somewhat aromatic annual or winter-annual herbs with taproots. Stems usually erect, 1-few from base, often branched above, (1)3-8(10) cm tall, white-woolly but sometimes becoming subglabrous, rarely a few glands below. Rosette leaves, when present, oblanceolate or spatulate, 2-5 cm long; cauline leaves alternate, sessile but not decurrent, lance-linear to lanceolate, or narrowly oblanceolate, 3-10 cm long, 2-10 mm wide, often undulate-margined, narrowed to base, acute to acuminate at apex, white-woolly below, green and glabrous to slightly glandular or sparsely woolly above. Inflorescence ample, branched and many-headed, in well-developed plants, or round-topped or elongate, the final clusters with heads glomerate; involucre 5-7 mm tall, yellowish-white, soon with a rusty-tinge, woolly near the base; involucral bracts acutish to obtuse or nearly rounded. Achenes glabrous, 0.8-1.2 mm long, olivaceous to yellowish-brown; pappus bristles distinct. (n = 14) Jul-Oct. Common in prairies, open woodlands, roadsides; e 1/2 of s 1/2 GP; (N.S. to Man., s to FL, TX).

    The plant has a rather characteristic balsamic odor.

3. Gnaphalium palustre Nutt., diffuse cudweed. Annual herbs with taproots. Stems (3)5-15(20) cm tall, becoming diffusely branched from the base and above, rarely simple, loosely floccose-tomentose. Leaves alternate, entire, sessile or nearly so, 1-2.5(3.5) cm long, 2-8(10) mm wide, oblanceolate or oblong, loosely floccose-tomentose. Heads in small leafy-bracted glomerules, in the axils and at the ends of stems and branches; involucres 3–4 mm tall, densely woolly below; involucral bracts linear or oblong, not much imbricated, brown or whitish at apex. Achenes 0.6–0.8 mm long, glabrous, often shiny, yellowish-brown; pappus bristles distinct. Jun–Oct. Rare, but locally common in low areas of fields, pastures, prairies, dried beds of vernal pools; nw ND, MT, scattered localities in SD, WY, NE; (ND to B.C., s to NE, CO, NM, CA).

Except for nw ND our specimens may largely have been introduced from the west where the species is common.

4. Gnaphalium purpureum L., purple cudweed. Annual or biennial herbs with taproots, thinly woolly throughout. Stems 1–4 dm tall, erect, simple or with several ascending stems from the base. Lower leaves spatulate or oblanceolate, 2–6(10) cm long, 5–15(20) mm wide, rounded at tip, often mucronate, lower surfaces white pannose, upper surfaces usually green and sparsely pubescent, upper leaves progressively smaller. The glomerules of heads nearly sessile in axils of upper dm of the stem, often appearing as a leafy-bracteate, sometimes interrupted spike; involucre 3–5 mm tall, woolly below; involucral bracts imbricate, acute to acuminate, brown, sometimes tinged with pink. Achenes 0.8–1 mm long, minutely papillate, olivaceous to yellowish-brown, often lustrous; pappus-bristles united at base. (n = 7, 14) Apr–Jul. Locally common in sandy or rocky prairies, open woodlands, pastures, fields, waste places; e 1/2 KS, MO, OK; (ME to B.C., s to Mex. & S. Amer.).

5. Gnaphalium uliginosum L., low cudweed. Annual herbs with taproots. Stems 0.5–2.5 dm tall, diffusely branched or subsimple, densely appressed woolly; leaves numerous, alternate, spatulate-oblanceolate or linear, not decurrent, 1–4 cm long, 1–5 mm wide. Heads glomerate in many small clusters in leaf axils and at ends of branches, over-topped by subtending leaves; involucres 2–3(4) mm tall, woolly below; involucral bracts only slightly imbricate, greenish or brownish, usually paler at tips, acute or obtuse. Achenes 0.7–1 mm long, olivaceous, glabrous or minutely papillate; pappus bristles distinct. (n = 7) Jul–Oct. Rare in cult. fields, pastures, stream banks, waste places; MN; SD: Custer, Lawrence; (Newf., B.C., s to VA, TN, SD, OR; Europe). *Introduced. G. exilifolium* A. Nels., *G. grayi* A. Nels. & Macbr.—Rydberg.

This species is well established in ne U.S. and Can. and may be more frequent in our area than collections indicate.

6. Gnaphalium viscosum H.B.K., clammy cudweed. Annual or biennial herbs with taproots. Stems stiffly erect, more or less glandular-pilose, becoming woolly in inflorescence, 3–10 dm tall, simple below, branched only at top. Leaves alternate, simple, 3–10 cm long, to 1 cm wide, sessile, distinctly decurrent at base, bright green (often drying brown) and usually glandular-pubescent on upper surface, the lower whitish-tomentose. Heads campanulate, subglobose, 5–6 mm tall, with many florets; involucral bracts stramineous to pale brownish, strongly graduated, scarious throughout, tips shiny. Achenes 0.7–1 mm long, glabrous or very minutely papillate, olivaceous; pappus bristles not united basally. Jul–Oct. Infrequent in open places; SD: Lawrence, Pennington; WY: Albany, Crook; (Que., PA, WV, TN, widespread in w U.S.). *G. macounii* Greene—Rydberg.

7. Gnaphalium wrightii A. Gray. Biennial or short-lived perennial(?) herbs with taproots. Stems erect, simple or much branched from base, branched above, 2.5–5 dm tall, white tomentose, not glandular. Leaves alternate, lanceolate or oblanceolate or upper linear, 2–4 cm long, to 10 mm wide, sessile or only obsoletely decurrent, gray-tomentose on both sides. Heads many in an open panicle, often glomerate at end of branches; involucres 4–6 mm tall, woolly at base; involucral bracts imbricated, scarious nearly throughout, tips white or stramineous, obtuse to acute or abruptly pointed. Achenes 0.7–0.8 mm long, olivaceous, glabrous; pappus bristles not united basally. Jul–Oct. Rare in rocky, open wooded hillsides & bluffs; CO: Baca; OK: Cimarron; (CO, w OK, s to TX, CA, n Mex.).

51. GRINDELIA Willd., Gumweed

Contributed by Mark A. Wetter

Ours taprooted, herbaceous biennials or perennials; stems erect, 1–several from base. Leaves alternate, sessile, often clasping, callous-serrulate to more sharply toothed, or entire, the lower ones sometimes pinnatifid, strongly to obscurely resinous punctate. Heads several to many, hemispheric, medium size to rather large (disk 0.7–3 cm wide), terminating the branches, radiate or discoid, yellow; involucral bracts ± resinous, firm, herbaceous-tipped, imbricate or subequal; rays 12–45, pistillate and fertile; disk florets numerous, perfect, inner and sometimes outer sterile; tubular, the lower half of the corolla abruptly constricted into a slender tube, the limb equally 5-toothed; receptacle flat or convex, naked but ± pitted, sometimes alveolate; style branches flattened, with ventro-marginal stigmatic lines and externally hairy, linear-lanceolate or occasionally very short appendages. Achenes compressed to subquadrangular, scarcely nerved; pappus of 2–several firm, caducous, entire to serrulate awns. [tribe Astereae]

1 Heads radiate.
  2 Involucral bracts ascending to loosely spreading, not markedly imbricate; pappus awns the same length as disk florets .... 2. *G. lanceolata*
  2 Involucral bracts strongly recurved, reflexed or squarrose, imbricate in several series; pappus awns shorter than disk florets.
    3 Leaves serrate or serrulate with bristle-tipped teeth, rarely entire, scabridulous, especially on the margins .... 3. *G. revoluta*
    3 Leaves callous-serrulate to sometimes more coarsely toothed or even entire, glabrous .... 4. *G. squarrosa*
1 Heads discoid.
  4 Leaves oval to ovate or broadly oblong, regularly callous-serrulate .... 4. *G. squarrosa*
  4 Leaves oblong to oblong-oblanceolate, serrate to serrulate with bristle-tipped teeth, rarely entire .... 1. *G. inornata*

1. *Grindelia inornata* Greene. Perennial, glabrous; stems 1–several from a crown, 2.5–8 dm tall. Leaves oblong to oblong-oblanceolate, 3–7 cm long, 8–20 mm wide, acute to acuminate, serrate or serrulate with bristle-tipped teeth, or rarely entire, scarcely-numerously punctate. Heads several to numerous; the disk 1.5–2 cm wide; involucral bracts strongly resinous, imbricate in several series, the tips strongly reflexed to revolute; rays none. Achenes 3–6 mm long; pappus awns 2 or 3, finely serrulate or subentire, 1/2–7/8 as long as disk florets. (n = 6) Jul–Oct. Waste places, roadsides, pastures, & prairies; CO: Denver & El Paso; (cen CO).

Although primarily distributed in the foothills of the Rockies in central Colorado, this taxon does appear to be a native in the vicinity of Denver and Colorado Springs. At the latter site members of this species quite often grow intermingled with *G. revoluta* in disturbed habitats and some hybrids are known.

2. *Grindelia lanceolata* Nutt., spinytooth gumweed. Monocarpic perennial, glabrous or sparsely hairy below; stems 1–several from a crown, 1.5–3 dm tall. Leaves linear or lanceolate-oblong, 2.5–11 cm long, 4–28 mm wide, acute to acuminate, sharply serrate or serrulate with bristle-tipped teeth, or sometimes entire, scarcely punctate. Heads several; disk 1.0–2.3 cm wide; involucral bracts slightly resinous, subequal or at least not markedly imbricate, the tips ascending to loosely spreading, not squarrose-reflexed; rays 12–36, 10–16 mm long,

1.5–3 mm broad. Achenes 4–6 mm long; pappus awns mostly 2, entire, 5–6 mm long, equaling to slightly exceeding the disk florets. (2n = 12) Aug–Oct. Rocky prairies, roadsides, pastures, usually on limestone; se KS, MO, OK, TX; (AL to TX, cen MO to Edwards Plateau).

Some of our material from central OK with the involucral bracts less than 1/2 the height of the disk has been separated as *G. texana* Scheele or as *G. lanceolata* var. *texana* (Scheele) Shinners. But this material intergrades with the more typical material over a wide geographical range, and is not recognized in this treatment.

3. **Grindelia revoluta** Steyerm. Perennial; stems 1–several from a crown, 5–8 dm tall. Leaves broadly oblong to oblong-oblanceolate, 3.5–7.5 cm long, 6–30 mm wide, obtuse to acutish, serrate or serrulate with bristle-tipped teeth, or rarely entire, scabridulous especially on the margins, numerously punctate. Heads several to numerous; the disk 1.3–1.7 cm wide; involucral bracts moderately to abundantly resinous, imbricate in several series, the tips strongly reflexed to revolute; rays 21–37, 4–11 mm long, 1.5–3 mm broad. Achenes 3–4 mm long; pappus awns 2–4, scantily to numerously setulose, 1/2–3/4 as long as disk florets. (n = 12) Jul–Oct. Prairies, waste places, roadsides, railroads, & pastures; CO: El Paso, Fremont, Huerfano, & Pueblo; NM: Colfax, Union.

This species is restricted to the plains and lower foothills of the Rockies in southern Colorado and adjacent New Mexico. It usually grows sympatrically with *G. inornata* and is probably the radiate ancestor of that discoid species. Some specimens of *G. revoluta* produce very short rays (4 mm) which may represent a step toward the discoid habit.

4. **Grindelia squarrosa** (Pursh) Dun., curly-top gumweed. Biennial or weak perennial, glabrous; stems 1–several, often solitary, from a herbaceous base, 1–10 dm tall. Leaves ovate or oblong, linear-oblong to oblanceolate, 1.5–7 cm long, 4–20 mm wide, obtuse to acutish, regularly callous-serrulate to sometimes more coarsely toothed, or even entire, abundantly punctate. Heads several to numerous, the disk 0.7–3 cm wide; involucral bracts strongly resinous, imbricate in several series, the tips (especially the outer) squarrose-reflexed; rays 12–37, 7–15 mm long, 0.75–2.5 mm broad. Achenes 2–3 mm long; pappus awns 2–8, finely serrulate or subentire, 1/2–7/8 as long as disk florets. (n = 6, 2n = 24) Jul–Oct. Waste places, roadsides, railroads, & pastures; GP; (cordillera & GP, sporadically introduced in e U.S.).

This is a highly polymorphic species which can be separated into 3 well-marked varieties in our area:

1 Heads discoid .... 4a. var. *nuda*
1 Heads radiate.
    2 Leaves entire or remotely serrulate, or especially the lower, coarsely and irregularly toothed or somewhat pinnatifid; mostly short-lived perennials .... 4b. var. *quasiperennis*
    2 Leaves closely and evenly serrulate or crenate-serrulate; biennials .... 4c. var. *squarrosa*

4a. var. *nuda* (Wood) A. Gray. Leaves oval to broadly oblong, closely and rather evenly crenate. Disk 1.3–2.3 cm broad; rays none. Restricted to sw KS, OK, TX, NM. *G. nuda* Wood—Gates.

4b. var. *quasiperennis* Lunell. Leaves (main cauline) narrowly oblong to oblanceolate, entire to remotely serrulate, the lower coarsely and irregularly toothed or pinnatifid. Disk 0.7–1.6 cm broad; rays 20–32. MN, MT, ND, SD, WY, CO; (Can. to n GP). *G. perennis* A. Nels—Fernald.

4c. var. *squarrosa*. Leaves ovate-oblong to broadly oblong, linear-oblong to oblanceolate, closely and evenly crenate-serrate with obtuse teeth or serrulate with teeth rather sharp. Disk 0.8–1.5 cm broad; rays 12–37. SD, KS, OK, CO, TX; (s GP s to TX, introduced in e U.S.). *G. serrulata* Rydb.—Rydberg.

This variety sometimes has been further subdivided: the typical variety with the leaves 2–4 × longer than wide, and var. *serrulata* (Rydb.) Steyerm. with leaves 5–8 × longer than wide. But in many areas of the western GP the distinctiveness of these two varieties is hard to establish, with plants from any one population showing characteristics of either variety and with many intermediates also present. Possibly this represents a case of previously distinct geographical races now in the process of merging.

Var. *serrulata* is centered in nw CO, se WY, w to UT, and the typical material occurs over the greater part of the southern GP. Most of our material is, therefore, probably best assigned to *G. squarrosa* var. *squarrosa*.

52. GUTIERREZIA Lag.

Semi-weedy to weedy annual herbs or perennial subshrubs. Leaves alternate, narrow, entire, usually resinous. Heads relatively small, in close or loose clusters at the ends of the branches; involucre turbinate or urceolate to campanulate; involucral bracts in about 3 series, imbricated, mostly thick and rather stramineous; receptacle flat or convex, naked but somewhat alveolate; ray florets pistillate and fertile, ligule yellow, prominent, longer than the tube; disk florets perfect and fertile or staminate only, corolla yellow, limb 5-lobed. Achenes obpyramidal, appressed to ascending-pubescent; pappus of ray achenes reduced or a mere crown; pappus of disk achenes of prominent scales or of basally united awns. [tribe Astereae]

Reference: Lane, M. A. 1979. Taxonomy of the genus *Amphiachyris*. Syst. Bot. 4: 178–189. Lane, M. A. 1982. Generic limits of *Xanthocephalum, Gutierrezia, Amphiachyris, Gymnosperma, Greenella* and *Thurovia*. Syst. Bot. 7: 405–416.

The generic boundaries of this complex have been the subject of numerous investigations, and a good case can be made for treating *G. dracunculoides* as *Amphiachyris dracunculoides* (cf. Lane, op. cit.). However, it appears to be a matter of taxonomic outlook, and here we elect to take the traditional position.

1 Suffruticose perennials with stems branching upward from the base .... 2. *G. sarothrae*
1 Annuals with stems branching mostly in the upper 1/2.
  2 Disk florets sterile and with the pappus of 5 or more basally united awns, about equaling the corolla in length .... 1. *G. dracunculoides*
  2 Disk florets fertile, with the pappus of several separate, pointed scales .... 3. *G. texana*

1. Gutierrezia dracunculoides (DC.) Blake, broomweed. Annual, 3–8+ dm tall. Stems arising singly from an erect taproot, bushy-branched in the upper 1/2, glabrous. Leaves alternate, linear or narrowly linear-lanceolate, 1–5(6) cm long and 1–2(3) mm wide, entire, glandular-punctate or somewhat glutinous. Heads numerous, clustered in terminal inflorescences; involucre urceolate-turbinate, 3–5 mm tall; involucral bracts imbricate, shiny, somewhat indurate, herbaceous only toward the apex; ray florets mostly 7–8, the ligule 4–6 mm long, yellow; disk florets 10–25, staminate only and infertile. Ray achenes 1.5–2.2 mm long, purple-black with several narrow, greenish stripes, finely pubescent, pappus absent or a minute crown; pappus of the sterile disk achenes of 5 or more basally united awns. (n=4, 5) Aug–Oct. Widespread in open, upland sites; se NE & adj. MO, s & w to NM, TX, & OK; (SC, AL to s GP, w to AZ). *Xanthocephalum dracunculoides* (DC.) Shinners—Correll & Johnston; *Amphiachyris dracunculoides* (DC.) Nutt.—Rydberg.

This species sometimes forms huge populations of thousands of individuals in heavily grazed, calcareous pastures.

2. Gutierrezia sarothrae (Pursh) Britt. & Rusby, snakeweed. Perennial subshrub 2–10 dm tall. Stems bushy-branched upward from the woody base, herbage glabrous or lightly scabrous-puberulent, often sometimes resinous. Leaves alternate, linear to linear-filiform, 5–60 mm long and 1–3 mm wide. Heads numerous, clustered at the ends of the branches, giving the aspect of a single, large corymb; involucre 3–6 mm tall, ±2 mm across, turbinate; involucral bracts narrow, green at the apex and along the midnerve; ray florets 3–8, fertile, the ligule 1–3 mm long, yellow; disk florets 2–6, fertile. Achenes with pappus of 8–10 acute scales, those of the ray achenes about 1/2 as long as those of the disk achenes.

(n = 4) Aug–Oct. Dry, open plains & upland sites; throughout GP except e 1/5; (Man. to n Mex., w to WA & CA). *G. filifolia* Greene, *G. diversifolia* Greene, *G. juncea* Greene, *G. linearis* Rydb.—Rydberg; *Xanthocephalum sarothrae* (Pursh) Shinners—Correll & Johnston.

This species is particularly abundant in heavily grazed pasture lands in the high plains, where it is toxic to livestock.

3. *Gutierrezia texana* (DC.) T. & G. Annual 1–10 dm tall. Stems arising singly from an erect taproot, bushy-branched in the upper 1/2. Leaves lance-linear, 1–5 cm long and up to 5 mm wide, glabrous but somewhat glutinous. Inflorescence of numerous loosely clustered heads, terminating the branches; involucre turbinate to campanulate, 3–4 mm tall; ray florets 10–15, fertile, the ligule up to 7 mm long; disk florets 10–20, fertile. Achenes minutely pubescent; pappus of ray achenes minute or absent, pappus of disk achenes of several pointed scales. Jul–Oct. Rocky upland prairies; extreme s GP; OK: Comanche; TX: Hardeman; (widespread s in TX, n Mex.). *Xanthocephalum texanum* (DC.) Shinners—Correll & Johnston.

53. HAPLOESTHES A. Gray, False Broomweed

1. *Haploesthes greggii* A. Gray var. *texana* (Coult.) I. M. Johnst. Erect suffrutescent perennial, 1–4 dm tall, sometimes subsucculent, glabrous or nearly so. Stems much branched, arising from a tough, woody taproot. Leaves opposite, numerous, linear-filiform, 2–5 cm long. Heads numerous in a dense, corymbiform terminal cluster; involucre campanulate, involucral bracts 5, 3–4 mm long, broadly overlapping; ray florets 3–5, pistillate, ligule 1–2 mm long, yellow to ochroleucous; disk florets perfect and fertile, numerous. Achenes ca 1.5 mm long, usually 10-ribbed; pappus of persistent, stramineous stiffish bristles. Jun–Sep. Open, gypsiferous plains; s-cen KS, cen OK, n TX; (TX & NM, s to Mex.; reaching its northern limit in GP). [tribe Senecioneae]

References: Turner, B. L. 1975. Taxonomy of *Haploesthes*. Wrightia 5: 108–115; Strother, J. L. 1978. *Haploesthes*. N. Amer. Flora II. 10: 44–46.

54. HAPLOPAPPUS Cass., Goldenweed

Annual or perennial herbs or weak subshrubs; herbage often glandular or somewhat resinous. Leaves alternate, or basal leaves sometimes closely tufted, entire to lobed, or sometimes weakly pinnatifid. Heads variously arranged in terminal inflorescences, occasionally solitary; involucre hemispheric to broadly conical; receptacle naked, but sometimes weakly alveolate with the walls of the alveoli broken into irregular, chafflike scales; ray florets prominent and often numerous, pistillate and fertile or occasionally sterile, ligule yellow, or ray florets sometimes absent; disk florets bisexual and fertile, corolla funnelform to subcylindric, 5-toothed to prominently 5-lobed, style branches flattened, with marginal stigmatic lines, the appendages prominent, acute to somewhat obtuse, minutely papillose. Achenes subcylindrical and weakly striate to prismatic and 4- or 5-angled, glabrous to variously pubescent; pappus of numerous to few capillary bristles of unequal length. [tribe Astereae]

References: Hall, H. M. 1928. The genus *Haplopappus*. Publ. Carnegie Inst. Wash. No. 389. [This monograph is a monumental contribution to systematic botany, for it is the forerunner of those studies where the purpose is to elucidate the natural relationships among the various taxa. Hall's conclusions are now dated, but his work remains the basic treatment for the genus.]; Hartman, R. L. 1976. A conspectus of *Machaeranthera* and a biosystematic study of the section *Blepharodon*. Unpublished Ph.D. dissertation, Univ. of Texas, Austin.

Haplopappus is a diverse group that has been treated as numerous separate, semidistinct genera by some workers, or as a single large genus with several sections. The treatment presented here is patterned after Hall (op. cit.), with his sect. name (the genus names for some authors) supplied in brackets. Hartman (op. cit.) included sect. *Blepharodon* in the genus *Machaeranthera*.

1 Ray florets absent; plants clustered-stemmed, woody-based perennials.
  2 Plants 1-3 dm tall; involucre 8-10 mm tall and 6-8 mm wide .... 4. *H. engelmannii*
  2 Plants 3-6+ dm tall; involucre small, 4-6 mm tall .... 6. *H. heterophyllus*
1 Ray florets normally present; plants variously annual or perennial.
  3 Leaves toothed to pinnatifid, the teeth or lobes spinulose-tipped, only the reduced, uppermost leaves subentire.
    4 Perennials with well-developed, ligneous underground parts.
      5 Involucre hemispheric, 12-18 mm wide; largest leaves basal and subpinnatisect, mostly 8-15 cm long .... 7. *H. lanceolatus*
      5 Involucre rather narrower, 9-12 mm wide; largest leaves midcauline, 1.5-6 cm long; basal leaves often early-deciduous .... 9. *H. spinulosus*
    4 Annuals from a taproot.
      6 Heads hemispheric, 20-25 mm wide; achenes glabrous .... 3. *H. ciliatus*
      6 Heads 10-15 mm wide; achenes sericeous .... 1. *H. annuus*
  3 Leaves entire, or with shallow teeth that are not spinulose-tipped.
    7 Annual with stems arising singly from a taproot .... 10. *H. validus*
    7 Perennials with stems frequently clustered from a woody roostock.
      8 Involucre large, 12-25 mm tall and 20-30 mm wide; cauline leaves linear; extreme w GP .... 5. *H. fremontii*
      8 Involucres smaller, seldom more than 10 mm tall.
        9 Plants with stems single or loosely clustered, 1.5-6+ dm tall .... 7. *H. lanceolatus*
        9 Plants usually short and tufted in aspect, rarely exceeding 1.5 dm tall.
          10 Heads broadly campanulate; ray florets 10-12, ligule 10-12 mm long .... 2. *H. armerioides*
          10 Heads narrowly campanulate; ray florets 6-8, ligule 5-8 mm long .... 8. *H. multicaulis*

1. *Haplopappus annuus* (Rydb.) Cory, [sect. Blepharodon]. Annual or weakly biennial herb, 3-7+ dm tall; herbage sparsely to densely glandular. Stem single, erect, often purple-blotched on main stem and larger branches, leafy throughout, arising from a narrow, woody taproot. Leaves simple; basal leaves early deciduous; cauline leaves oblanceolate to narrowly obovate, 2.5-5+ cm long and 0.5-1.5 cm wide, about 3 × longer than wide, dentate-aristate with 7-10 prominent teeth, each with an apical bristle; uppermost leaves reduced, becoming sessile, subentire and somewhat densely glandular. Inflorescence a terminal, paniculate or subcorymbiform cyme of few to many heads, with lateral reduced inflorescences arising in the axils of the middle and upper leaves; heads hemispheric, 6-8 mm tall and 10-15 mm wide; involucral bracts ca 60-75, densely glandular, recurved, oblanceolate, outermost bracts 4-5 mm long, increasing in length inward, apically tapering to a curled arista; ray florets 25-40, ligule 7-10 mm long, yellow; disk florets 80-100, corolla 3-5 mm long. Achenes turbinate, ca 2 mm long, densely sericeous; pappus of numerous capillary bristles. (n = 6) Aug-Sep. Roadsides, fence rows, sandy prairies, flood plains, & waste areas; scattered in sw 1/4 GP, from w NE to cen OK, s & w; (GP sw to TX & NM). *H. phyllocephalus* subsp. *annuus* Hall — Harrington; *Sideranthus annuus* Rydb. — Rydberg; *Machaeranthera annua* (Rydb.) Shinners — Correll & Johnston.

2. *Haplopappus armerioides* (Nutt.) A. Gray, [sect. Stenotus]. Cespitose subshrub, 5-15+ cm tall, essentially glabrous. Stems numerous, arising from a much-branched, stout woody caudex, surmounting a prominent taproot. Leaves persistent, mostly basal, sessile, narrowly oblanceolate-acuminate, 2-8(10) cm long and 3-10 mm wide, entire; margins sometimes scabrous, or often resinous, cauline leaves few and reduced. Inflorescence usually a single head on a subscapose peduncle, or sometimes with 2–3(5) heads; involucre broadly campanulate, 10–12 mm tall and about as wide; involucral bracts imbricate in 3 or 4 series, obtuse to acuminate, with a conspicuous greenish region on the distal 1/3–1/4; ray florets (8)10–12(15), ligule 10–12 mm long, yellow; disk florets ca 40, corolla ± 5 mm long, yellow. Achenes 4–5 mm long, somewhat flattened, villous; pappus of numerous soft, white bristles, 5 mm long. (n = 9) May–Jul. Upland prairies & open woodland sites; nw 1/4 GP from e & cen MT & w ND s to nw KS, e CO; (GP w to MT & AZ). *Stenotus armerioides* Nutt.—Rydberg.

3. Haplopappus ciliatus (Nutt.) DC., goldenweed, [sect. Prionopsis]. Annual herb, 3–8(10) dm tall, glabrous, with stems arising singly from a taproot. Leaves mostly on upper 2/3 of stem, oblong to narrowly obovate, 3–5(8) cm long and 1–2(4) cm wide, sessile, spinulose-dentate to serrate, the teeth with mucronate to cuspidate tips. Heads few in an open cyme, or sometimes solitary; involucre 20–25 mm wide and not quite so tall, broadly hemispheric; involucral bracts imbricated in 4 or 5 series, the outermost squarrose-spreading, linear-lanceolate, the exposed parts greenish; ray florets ca 25–30, ligule 12–15 + mm long, yellow; disk florets perfect and fertile or some of them sterile, corolla ca 7 mm long. Achenes 2–3(4) mm long, ellipsoid to oblong, glabrous, laterally compressed; pappus of numerous somewhat rigid bristles, yellow to reddish brown, ca 7 mm long, tardily deciduous. (n = 6) Aug–Oct. Dry, open sandy or rocky sites; s 1/3 GP, from NE: Nemaha s & w; (GP s through TX & NM). *Prionopsis ciliata* (Nutt.) Nutt.—Rydberg.

4. Haplopappus engelmannii (A. Gray) Hall, Englemann's goldenweed, [sect. Oonopsis]. Perennial herb 10–30 cm tall, glabrous or somewhat scabrous throughout. Stems numerous, arising from a subligneous, branching woody caudex; striate, woody toward the base, unbranched and leafy throughout. Leaves narrowly linear, erect, 3–6 + cm long and 1–3 mm wide, only the midvein prominent, margins entire but often somewhat involute, sessile or nearly so; upper leaves short scabrous-ciliolate toward the base, often equaling or exceeding the inflorescence. Heads 3–15 in a compact, corymbiform terminal cluster; involucre turbinate-campanulate, 8–10 mm tall and 6–8 mm wide; involucral bracts imbricate in 3 + series, abruptly cuspidate, mostly chartaceous but greenish on the exposed surfaces; ray florets absent; disk florets 15–20, corolla ca 4 mm long, yellow. Achenes nearly prismatic, 5–6 mm long, glabrous or scarcely scabrous; pappus of scanty rigid brownish bristles. (n = 10) Aug–Sep. Infrequent in the high plains of e CO & w KS; known from CO: Lincoln; KS: Hamilton. *Oonopsis engelmannii* Greene—Rydberg.

5. Haplopappus fremontii A. Gray, [sect. Oonopsis]. Perennial herb 10–15(30) cm tall, glabrous to sparsely pubescent, especially so on the stem. Stems several or rarely solitary, yellowish-white, smooth, arising from a woody caudex. Cauline leaves linear, narrowly oblong to lanceolate, 4–8 + cm long and 5–15 mm wide, entire, sessile with a narrow, subclasping base. Heads several in a narrow cyme or sometimes solitary; involucre 12–25 mm tall and 20–30 mm wide, hemispheric; involucral bracts imbricate but all about equal in length, acuminate to cuspidate, the outer ones distinctly greenish; ray florets 15–25, ligule 8–15 mm long, yellow; disk florets numerous. Achenes narrowly turbinate, 5–7 mm long, glabrous; pappus of scanty rigid brownish bristles. (n = 10) Jun–Aug. Dry, sandy or rocky sites; along the w edge of the GP in cen CO & WY; (w GP & foothills of the Rocky Mts.).

This species was attributed to KS by Hall (op. cit.). In e & cen WY this entity has smaller heads, with the disk only 12–16 mm tall, the ray florets absent, and n = 5. This phase has been recognized as subsp. *wardii* (A. Gray) Hall.

6. Haplopappus heterophyllus (A. Gray) Blake, [sect. Isocoma]. Perennial subshrub 3–6(8) dm tall, glabrous or somewhat glandular to scabrous. Stems numerous, erect or arching upward from a taprooted woody caudex. Leaves about equally distributed along the stem, basal leaves early deciduous, cauline leaves narrowly linear-oblanceolate, erect, 4-7 cm long and 2-4 + mm wide, entire, with numerous resin-dots, sessile; uppermost leaves reduced to mere bracts. Inflorescence a corymbiform cluster of few to numerous heads; involucre 4-6 mm tall, hemispheric; involucral bracts imbricate in 3 or 4 series, narrowly ovate-oblanceolate, glabrous but somewhat glutinous, acuminate, margins papery or sometimes sparsely ciliate; ray florets absent; disk florets 10-15, corolla yellow, 4-6 mm long. Achenes tomentose, 2-3 mm long; pappus of abundant, unequal coarse bristles. (n = 24) Jul-Aug. Open disturbed sites, roadsides, & flood plains, especially in sandy soils; extreme s GP in cen TX panhandle & s; (GP to n Mex., w to AZ). *Isocoma pluriflora* (T. & G.) Greene, *I. wrightii* (A. Gray) Rydb. — Rydberg; *Haplopappus pluriflorus* (A. Gray) Hall — Harrington.

7. **Haplopappus lanceolatus** (Hook.) T. & G., [sect. Pyrrocoma]. Perennial herb 1.5-3.5(5) dm tall, tomentose, especially toward the base, in the leaf axils and near the inflorescence; leaves glabrescent. Stems 1-5, erect or arching upward from a branching caudex surmounting a thick, woody taproot. Leaves mostly basal, subpersistent, narrowly lanceolate to lanceolate, the larger ones 8-15 cm long and 0.5-2.5 cm wide, tapering to a short petiole or subsessile; cauline leaves 3-14, oblanceolate to lanceolate, sessile and nearly clapsing, the uppermost reduced to mere bracts. Inflorescence variously corymbiform to paniculate, with up to 20(+) heads, or sometimes reduced to a single head; involucre hemispheric, 7-10 mm tall and 12-18 mm wide (when pressed); involucral bracts imbricate in 3 or 4 series, greenish toward the apex but straw-yellow below, lanceolate-acuminate, margin often ciliate to villous; ray florets (13)21-34 + , ligule yellow, 5-10 mm long; disk florets with corolla yellow, 5 mm long, 5-parted, lobes ± 1 mm long. Achenes 3-4 mm long, subcylindrical, sericeus-tomentose; pappus of abundant sordid capillary bristles. (n = 12, 18) Jun-Aug. Open flats, edges of northern prairie sloughs, & roadside ditches; ne GP, nw ND, e MT, barely entering our range southward in WY, along the e flank of the Rocky Mts.; (nw GP w through the Intermountain Region to NV). *Pyrrocoma lanceolata* (Hook.) Greene — Rydberg.

Hall (op. cit.) divided *H. lanceolatus* into 5 subspecies which "are not sharply set off from one another." The GP materials are best referred to the nomenclaturally typical subsp. *lanceolatus* (termed subsp. *typicus* by Hall) and subspp. *vaseyi* (Parry) Hall. In both subsp. the heads are large, as described here, but subsp. *lanceolatus* is characterized by having numerous heads in an open, corymbiform cluster, while subsp. *vaseyi* is characterized by having 1-5 heads in a narrower, racemose inflorescence. Subsp. *vaseyi* barely enters our range, if at all, in e-cen WY.

8. **Halopappus multicaulis** (Nutt.) A. Gray, [sect. Oonopsis]. Perennial herb 5-10 cm tall, villous-tomentose, but becoming glabrate and smooth. Stems tufted from a much-branched, rhizomatous woody caudex. Leaves alternate, sessile, 3-10 per stem, elongate-oblanceolate, acuminate, 2-8 cm long and 2-4 mm wide, the largest leaves equaling or exceeding the stem length, uppermost leaves reduced to mere bracts. Heads terminal on the branches, sometimes forming a corymb of 3 or 4 heads; involucre campanulate, 7-10 mm tall and about as wide; involucral bracts ca 13, imbricated, ovate-acuminate to oblanceolate-acuminate, 5-7 mm long, margins irregularly ciliate; ray florets 6-8, ligule yellow, 5-8 mm long; disk florets ± 18, corolla yellow, 6-7 mm long. Achenes ±2 mm long, turbinate, tomentose, pappus of scanty capillary bristles, somewhat shorter than the disk corollas. (n = 4) Jun-Jul. Upland prairies & roadsides; nw GP; MT: Carter; SD: Fall River; WY: Converse, Natrona, Weston; (nw GP, w through WY). *Oonopsis argillacea* A. Nels. — Rydberg.

This little-known species barely enters our range.

9. **Haplopappus spinulosus** (Pursh) DC., cutleaf ironplant, [sect. Blepharodon]. Perennial herb, 3-8 dm tall, minutely glandular and tomentose, especially the upper 1/3 of the

stem, or sometimes glabrous. Stems numerous or rarely solitary, simple and erect below but branching upward, arising from a woody, branching caudex. Leaves oblong to linear-subspatulate, 1.5–6 cm long and 2–10 mm wide, sessile or nearly so, dentate with bristle-tipped teeth, or pinnate-parted with bristle-tipped lobes, uppermost leaves reduced. Heads solitary at the ends of the branches; involucre 5–8 mm tall and 9–12 mm wide; involucral bracts imbricate in 4–6 series, acute or attenuate and usually minutely bristle-tipped; receptacle weakly alveolate and with a few irregular chafflike scales derived from the broken walls of the alveoli; ray florets 15–30+, ligule yellow, 8–10 mm long; disk florets numerous, corolla yellow, 4–5 mm long. Achenes narrow-turbinate, 2–2.5 mm long, weakly striate, sparsely appressed-pubescent; pappus of scanty, yellow-brown bristles, 4–5 mm long. (n = 4, 4+ fragments, 6, + a ploidy series). May–Sep. Open prairies & plains; throughout GP but infrequent in e 1/5, absent or rare in se 1/4 NE & e 1/4 KS; (GP; s-cen Can. to TX, n Mex., w to CA). *Machaeranthera pinnatifida* (Hook.) Shinners—Correll & Johnston; *Sideranthus spinulosus* (Pursh) Sweet, *S. glaberrimus* Rydb.—Rydberg.

This remarkably complex species includes numerous semidistinct phases that have been given various recognition in the past. Four of these phases occur in the GP, apparently with some populational integrity. The treatment presented below is purely provisional, and the phases are listed as subspecies as a matter of convenience. The taxonomic position and the choice of names is subject to revision, and no key is provided.

9a. subsp. *australis* (Greene) Hall. Herbage spreading-pubescent and obscurely glandular. ± bushy-branched and pale green in aspect. Extreme sw GP; KS: Morton; CO: Baca; panhandle of TX.

9b. subsp. *cotula* (Small) Hall. Herbage stipitate-glandular. Stems quite leafy, arching upward, producing a short, bushy aspect. (n = 4, 8). Mostly on sandy, gravelly, gypsiferous, or red clay soils; sw KS, w OK, panhandle of TX.

9c. subsp. *glaberrimus* (Rydb.) Hall. Herbage glabrous or sparsely pubescent. Stems few, erect or nearly so. (n = 4). E & cen GP, e of the range of subsp. *spinulosus*.

9d. subsp. *spinulosus* (subsp. *typicus* of Hall, op. cit.). Herbage ± tomentose, at least among the heads. Stems numerous, erect or arching upward. (n = 4). Common in the high plains of the w GP, extending ne through SD & ND.

Hartman and Turner (Wrightia 5: 308–315. 1976.) treat this species as *Machaeranthera pinnatifida*, with 2 subspecies, one with 3 and the other with 4 varieties. Under their concept, our subspp. *spinulosus*, *australis* and *cotula* are included in *M. pinnatifida* subsp. *pinnatifida* var. *pinnatifida*, and our subsp. *glaberrimus* is treated as *M. pinnatifida* subsp. *pinnatifida* var. *glaberrima* (Rydb.) B. L. Turner & R. Hartman.

10. **Haplopappus validus** (Rydb.) Cory, slender goldenweed, [sect. Isopappus]. Annual herb, 4–7 dm tall, stipitate-glandular, especially above. Stems simple below but diffuse-branched above, arising singly from a taproot. Leaves persistent, oblanceolate to spatulate, 4–7 cm long and 4–10 mm wide, entire to irregularly serrate. Inflorescence open-paniculate, with numerous heads; involucre 5–8 mm tall and 6–12 mm wide; involucral bracts strongly graduated, imbricate in several series, narrowly lanceolate and short stipitate-glandular; ray florets mostly 13 or 21, ligule yellow, 7–10 mm long; disk florets numerous, corolla 5–6 mm long, with dark brown or reddish lines separating the tube into 5 lateral sections. Achenes somewhat flattened, obtriangular, 2–2+ mm long, sericeous-canescent; pappus of numerous whitish bristles, ca 4 mm long. (n = 5) Jul–Oct. Dry sandy soils & open prairies, especially the "sandy lands" of s-cen KS; s GP, s-cen KS, cen & w OK, & adj. n TX; (s GP through TX). *Croptilon divaricatum* var. *hookerianum* (T. & G.) Shinners—Correll & Johnston; *Haplopappus divaricatus* (Nutt.) A. Gray—misapplied by various authors, e.g., Fernald, and by Hall (op. cit.); *Isopappus validus* Rydb.—Rydberg.

This entity has been widely known as *H. divaricatus* in the past. The taxonomy has been elucidated by E. B. Smith, Rhodora 67: 217–238. 1965; Canad. J. Genet. Cytol. 8: 14–36. 1966. However, Smith has recently indicated support for recognizing this entity as *Croptilon hookerianum* var. *validus* (Rydb.) E. B. Sm.; Sida 9: 59–63. 1981.

55. HELENIUM L., Sneezeweed

Annual or perennial herbs. Stems erect and often branched upward. Leaves alternate, essentially sessile and sometimes decurrent, lowermost often pinnate-dissected, variously resinous-punctate. Heads on long peduncles; receptacle globose or hemispheric, naked or with a few irregularly scattered bristles around the edge; involucral bracts about 16 in 2 whorls of ca 8, the outer ± reflexed at maturity; ray florets ca 8, pistillate and fertile, or neutral, ligule 3-lobed at the apex, often reflexed, or ray florets sometimes absent; disk florets perfect and fertile, style branches truncate and penicillate, unappendaged. Achenes obpyramidal, short, appressed-hairy, at least below; pappus of ca 5 thin scales which are each ± prolonged into an awnlike apex. [tribe Heliantheae]

References: Rock, H. F. L. 1957. A revision of the vernal species of *Helenium* (Compositae). Rhodora 59: 101, et seq.; Bierner, M. W. 1972. Taxonomy of *Helenium* sect. *Tetrodus* and a conspectus of North American *Helenium* (Compositae). Brittonia 24: 331–355.

1 Leaf bases not decurrent along the stem, the stem thus smooth and unwinged.
  2 Lobes of the disk corollas yellow .... 1. *H. amarum*
  2 Lobes of the disk corollas reddish-brown .... 3. *H. badium*
1 Leaf bases decurrent and the stem thus winged.
  3 Ray florets neutral and sterile, sometimes absent; perennials .... 4. *H. flexuosum*
  3 Ray florets pistillate and fertile.
    4 Fibrous-rooted or subrhizomatous perennial .... 2. *H. autumnale*
    4 Taprooted annual .... 5. *H. microcephalum*

1. *Helenium amarum* (Raf.) Rock, bitter sneezeweed. Taprooted annual 1–4(8) dm tall. Stems striate-ribbed, bushy-branching upward. Leaves numerous, linear to filiform, up to 8 cm long and 2 mm wide, the lowermost sometimes pectinate-pinnatifid, glabrous but densely glandular-punctate. Heads numerous on short peduncles, extending beyond the leaves; receptacle 6–12 mm across, involucral bracts linear, herbaceous, reflexed at maturity; ray florets ca 8, pistillate, ligule 5–10 mm long, yellow; disk florets with yellow corollas. Pappus of translucent scarious scales with the midrib prolonged into an awn about as long as the body of the scale. (n = 15) Jun–Aug (Oct). Open prairies & fields, waste places, & open wooded sites, especially on sandy soil; se GP; e 1/2 KS & OK, but scattered n & w; (VA to FL, w to GP & TX). *H. tenuifolium* Nutt. — Rydberg.

This species is odoriferous and bitter and known to flavor the milk from cows that graze on it.

2. *Helenium autumnale* L., sneezeweed. Perennial 3–10 + dm tall. Stems erect, glabrous or finely pubescent, arising from fibrous rooted, subrhizomatous rootstocks. Leaf blades lance-linear to elliptic-ovate, 4–15 cm long and up to 4 cm wide, narrowed to a subsessile base and decurrent down the stem, not much reduced in size upward. Heads in an open leafy cluster, receptacle hemispheric to globose, disk ca 1–2 cm across; involucral bracts narrow, acuminate-attenuate, deflexed at maturity; ray florets 10–20, pistillate and eventually fertile, ligule yellow; disk florets with yellow corollas. Achenes ca 1.5 mm long; pappus scales lanceolate, tapering to a slender awn-tip usually less than 1 mm long overall. (n = 16, 17, 18) Aug–Oct. Moist open low sites, scattered throughout the e 1/4 GP but sporadically w in NE & n ND, e MT & WY; (Que. to FL, w to B.C. & CA, s to ne Mex.). *H. altissimum* Link, *H. latifolium* P. Mill., *H. montanum* Nutt. — Rydberg.

This is a polymorphic and complex species with numerous weakly definable local races.

3. *Helenium badium* (A. Gray) Greene. Similar in morphology to *H. amarum* except that the lobes of the disk florets are reddish-brown, giving the heads a very different aspect. May–Jul (Oct). Apparently restricted to disturbed, calcareous soils; extreme s GP; sw OK & adj. TX; (GP s & w through TX).

Current taxonomic opinion favors treating this entity as a distinct species, but it has been regarded as a variety of *H. amarum*.

4. **Helenium flexuosum** Raf. Perennial 3–10 dm tall. Stems arising from a fibrous-rooted, branching crown, simple below but much branched above, winged. Basal leaves linear-lanceolate to elliptic, up to 20 cm long, variously subentire to pinnatifid and usually withered by flowering time, cauline leaves persistent, linear-lanceolate to oblong, 3–12 cm long and 5–20 mm wide, entire or nearly so, glandular punctate and short-pubescent, sessile and contracted to a prominently decurrent base, the uppermost much reduced. Heads numerous in an open, corymbiform cluster; involucral bracts lanceolate to lance-linear, acuminate, reflexed at maturity; disk ca 5–15 mm across, globose or broadly ovoid; ray florets ca 8 or ca 13, neutral and sterile, ligule 5–20 mm long, yellow and sometimes purplish at the base or sometimes absent; disk florets with corollas purple or brownish-purple. Achenes 1–1.5 mm long, pappus scales acute, with an apical awn-tip. (n = 14) Jun–Sep. Open moist sites, extreme se KS, adj. MO & OK; (MA to FL, w to se GP & TX; introduced elsewhere). *H. nudiflorum* Nutt.—Fernald; *H. polyphyllum* Small,—Rydberg.

5. **Helenium microcephalum** DC. Taprooted annual 3–6(8) dm tall. Stems simple below but bushy-branched above, winged. Leaves narrowly elliptic to oblong-elliptic, 3–8 cm long and 0.5–2 cm wide, subentire or shallow-dentate to undulate-serrate, subsessile and decurrent. Heads with receptacle usually globose, 8–15 mm across; ray florets 7–13, ligule 3–5 + mm long, yellow; disk florets with corollas reddish-brown. Achenes up to 1 mm long; pappus scales ca 0.5 mm long and frequently less. (n = 13) Jul–Sep. Low, moist areas in waste & disturbed sites; extreme s GP; sw OK & adj. TX; (GP to n Mex., w to s NM).

Our materials are referable to var. *microcephalum*.

56. HELIANTHELLA T. & G.

1. **Helianthella quinquenervis** (Hook.) A. Gray, false sunflower. Stout perennial herb 5–15 dm tall, lightly pubescent to glabrate. Stems arising singly or loosely clustered from a stout taproot. Leaves leathery, the principal ones opposite, of about 4 cauline pairs, ovate-lanceolate to elliptic-lanceolate, up to 5 dm long overall, entire, acuminate, tapering to a distinct petiole, usually with 5 prominent veins. Heads solitary, nodding on a long peduncle, sometimes with a few reduced lateral heads present; involucre hemispheric, ca 2 cm tall and 4–5 cm across, exclusive of rays; involucral bracts broadly ovate-lanceolate, subequal, ciliate-margined, becoming black on drying; receptacle chaffy; ray florets pistillate but infertile, ligule 2.5–3.0 cm long, yellow; disk florets perfect and fertile, yellow; chaffy bracts soft and scarious, subtending and weakly enfolding the developing achenes. Achenes 8–10 mm long, compressed parallel to the radius of the head, thin-edged and ciliate-margined, appressed-pubescent on the sides; pappus of 2 slender awns 4–5 mm long and a crown of short scales. (n = 15) Jun–Aug. Meadows, often in boggy seeps, in coniferous woodlands; BH, ne WY; (cordilleran, MT to NM, w to OR & NV). [tribe Heliantheae]

Reference: Weber, W. A. 1952. The genus *Helianthella* (Compositae). Amer. Midl. Naturalist 48: 1–35.

57. HELIANTHUS L., Sunflower

Annual or perennial herbs. Stems simple or branched upward, variously glabrous or pubescent to scabrous, arising from a taproot or creeping, branching rootstocks. Leaves simple, the lowermost opposite, middle and upper ones sometimes alternate. Inflorescence of heads borne singly on the ends of long branches, or in open, few-flowered panicles to reduced, racemiform-spiciform clusters; heads radiate, involucral bracts in several series and subequal or graduate and strongly imbricated, receptacle flat to convex, chaffy with the chaff closely subtending the disk achenes; ray florets in a single series, pistillate but infertile, ray corolla yellow, with 3 apical teeth, or ray florets rarely absent; disk florets numerous, perfect and fertile, corolla tubular, yellowish or with the teeth and upper tube reddish to brown or purple. Disk achenes moderately compressed parallel to the radius of the head and often nearly rhombic in cross-section; pappus of 2 early-deciduous awns, occasionally with some smaller scales between the awns. [tribe Heliantheae]

   Reference: Heiser, C. B. 1969. The North American sunflower *(Helianthus)*. Mem. Torrey Bot. Club 22(3): 1–218.
   Species boundaries are indistinct for many sunflowers because of hybridization and morphological intergradation. Hybrids occasionally give rise to swarms with some populational integrity and are, therefore, difficult to identify satisfactorily, short of experimental study. Some of the named natural hybrids in our flora and their purported parents are: *H.* x *intermedius* Long *(H. maximilianii* x *grosseserratus)*, *H.* x *kellermanii* Long *(H. grosseserratus* x *salicifolius)*, *H.* x *laetiflorus* Pers. *(H. rigidus* x *tuberosus)* and *H.* x *orygaloides* Cockll. *(H. maximilianii* x *H. salicifolius)*. Many species of sunflower have been experimentally studied by C. Heiser and his students, and the reference noted above cites these investigations.

1 Taprooted annuals.
  2 Involucral bracts ovate to oblong-ovate, regularly more than 4 mm wide; central chaffy bracts inconspicuously hirsute .... 1. *H. annuus*
  2 Involucral bracts lanceolate to subulate, tapering, usually less than 4 mm wide, central chaffy bracts conspicuously white-hairy at the apex .... 8. *H. petiolaris*
1 Perennials with tough creeping rootstocks or erect branching crowns.
  3 Herbage blue-green glaucous, leaf margins distinctly ciliate; rhizomes long, slender and very extensive .... 2. *H. ciliaris*
  3 Herbage green but not glaucous and plant otherwise not as above.
    4 Involucral bracts ± appressed to the disk and clearly imbricate in several series; disk florets usually red or reddish-purple but sometimes yellow (including *H.* x *laetiflorus*) .... 10. *H. rigidus*
    4 Involucral bracts with loose-spreading tips and all of about the same length; disk florets yellow (except *H. salicifolius*).
      5 Principal leaves cordate, truncate or rounded at the base, sessile or with a short, usually wingless petiole.
        6 Leaves sessile and cordate, densely white-villous on both surfaces and the upper stem .... 6. *H. mollis*
        6 Leaves short-petiolate and not at all cordate.
          7 Leaves broadly rounded to subtruncate at the base; stem pubescent .... 4. *H. hirsutus*
          7 Leaves ± abruptly narrowed or occasionally rounded at the base; stem glabrous or nearly so .... 12. *H. strumosus*
      5 Principal leaves tapering or gradually narrowed at the base, usually with a well-defined and evidently winged petiole.
        8 Leaves broadly lanceolate or broader, seldom more than 3× longer than wide.
          9 Middle and lower stems glabrous or nearly so; tubers absent. .... 12. *H. strumosus*
          9 Stem ± pubescent; underground tubers present .... 13. *H. tuberosus*
        8 Leaves lanceolate or narrower, at least 3× longer than wide, usually less than 4 cm wide (except in *H. grosseserratus*).
          10 Leaves very narrow, usually 10(+)× longer than wide; disk florets red to purple .... 11. *H. salicifolius*
          10 Leaves generally less than 10× longer than wide; disk florets yellow.
            11 Stems arising from a tough, erect, taprooted crown, spreading rootstocks absent; extreme w GP .... 9. *H. pumilus*

11 Stems arising from rhizomes, tubers or crowns; widespread.
   12 Middle leaves, or at least some of them, folded lengthwise, leaves scabrous above; inflorescence racemiform-spiciform .... 5. *H. maximilianii*
   12 Middle leaves flat, leaves glabrous to pubescent above but not clearly scabrous; inflorescence paniculate.
      13 Middle and lower stems glabrous; leaves coarsely serrate and often more than 4 cm wide .... 3. *H. grosseserratus*
      13 Middle and lower stems glabrous to hispid; leaves entire to shallow-dentate, less than 4 cm wide .... 7. *H. nuttallii*

1. ***Helianthus annuus*** L., common sunflower. Coarse, taprooted annual, 6–25(+) dm tall, upwardly branched, rough-hairy. Lowermost leaves opposite but most of the leaves alternate, especially on robust plants, long-petiolate, blade ovate to broadly ovate, cordate or truncate, usually 10–40 cm long and about 1/2 as wide, margin toothed to subentire. Heads borne on long peduncles; disk 2 + cm across, involucral bracts ovate to ovate-lanceolate, attenuate, margin usually ciliate, 3–5 + mm wide, ray florets 17 or more, ligule 2.5 cm long or more, disk florets reddish to purple or rarely yellow, chaffy bracts deeply 3-toothed. Achenes 3–5(15) mm long, glabrate; pappus of 2 principal awns but sometimes small, secondary scales present. (n = 17) Jul–Sep. Abundant in open sites; throughout GP; (across N. Amer., s Can. & Mex., widespread as a weed). *H. aridus* Rydb., *H. lenticularis* Dougl. — Rydberg.

    This is a huge, polymorphic complex encompassing numerous wild and weedy races, plus the tall, strict, large-headed cultivated phases. Plant breeders in both N. Amer. and Europe have produced many cultivars for seed, fodder, and ornament.

2. ***Helianthus ciliaris*** DC., Texas blueweed. Perennial 3–7 dm tall, glabrous or nearly so, glaucous and blue-green in aspect. Stems arising from an extensive system of thin, creeping rootstocks. Leaves mostly opposite, blade narrow to broadly lanceolate, margin usually ciliate, 3–7 + cm long and 0.5–2 + cm wide, sometimes rather hispid. Heads few, disk 1.5–2.5 cm across, involucral bracts ovate to oblong, rounded to mucronate at the apex, appressed, somewhat imbricate, shorter than the disk; ray florets 10–18, ligule ca 1 cm long but sometimes smaller or rays occasionally absent; disk florets reddish, chaffy bracts entire or 3-toothed. Achenes ± 3 mm long, pappus of 2 scales. (n = 34, 51) Jun–Sep. Drying or damp sandy disturbed sites; sw GP, scattered in n TX & w OK, sporadic in sw KS & n to s NE; (sw GP w to s CA, n Mex.).

    The extensive creeping rootstocks and aggressive habit make this plant a serious weed in some localities.

3. ***Helianthus grosseserratus*** Martens, sawtooth sunflower. Perennial 1–3(+) m tall. Stems arising from a tough woody, rhizomatous rootstock, glabrous and often glaucous below but with appressed pubescence above, especially in the inflorescence. Leaves mostly alternate, blade lanceolate to lance-ovate, 10–20(30) cm long and 2–5(10) cm wide, serrate or sometimes subentire, scabrous above and puberulent to tomentose beneath, tapering at the base to a prominent petiole. Heads numerous, often in loose paniculate clusters; disk 1.5–2.5 cm across, involucral bracts linear-lanceolate, slightly exceeding the disk, loose and spreading; ray florets 10–20, ligule 3–4 cm long, yellow; disk florets yellow, chaffy bracts entire to shallowly 3-toothed, short-pubescent. Achenes 3–4 mm long. (n = 17) Jul–Oct. Open bottomlands & damp prairies; e 1/4 GP but extending w to cen NE & apparently BH; (NY to TX & Sask., perhaps ME).

    This species intergrades with *H. nuttallii* in the n GP and with *H. giganteus* e of our range. It hybridizes with other perennial diploid sunflowers (n = 17) where opportunity permits.

4. ***Helianthus hirsutus*** Raf., hairy sunflower. Rhizomatous perennial 1–1.5 m tall. Stems simple or few-branched upward, scabrous to hirsute. Leaves mostly opposite or the upper-

most alternate, blades divergent or ascending, lanceolate to ovate, rounded or contracted at the base, 7–18 cm long and 1–6 cm wide, 3-nerved, scabrous or hirsute above and hirsute beneath, serrate to subentire, subsessile to short-petiolate. Heads 1–several on short, stout peduncles; disk 2–3 cm across; involucral bracts linear-lanceolate, attenuate; ray florets 10–15, ligule ca 2 cm long; disk florets yellow; chaffy bracts shallowly 3-toothed, pubescent at apex and along the keel. Achenes 4(+) mm long. (n = 34) Jul–Sep. Dry, open wooded sites; e 1/4 KS & adj. OK, sporadic n to w MN; (PA to FL, w to GP & TX). *H. chartaceous* E. E. Wats.—Rydberg.

This species is closely related to *H. strumosus* and imperfectly distinct from it.

5. **Helianthus maximilianii** Schrad., Maximilian sunflower. Perennial 5–25 dm tall. Stems single or often loosely clustered from short, thick rhizomatous rootstocks, light green or sometimes reddish, scabrous to hispidulous, especially upward. Leaves mostly alternate, lanceolate, 7–20(30) cm long and 1–3(5) cm wide, somewhat folded along the midrib and falcate, both surfaces scabrous to hispidulous, margin entire or sometimes shallow-serrate, gradually tapering at the base to a short but distinct petiole. Heads sometimes few, but when numerous, usually in a spiciform or racemiform cluster; disk 1.5–2.5 cm across, involucral bracts linear-lanceolate and long-attenuate, exceeding the disk, loose-spreading; ray florets 10–25, ligule up to 4 cm long; disk florets yellow, chaffy bracts acuminate and entire to obscurely 3-toothed, pubescent at the apex. Achenes 3–4 mm long. (n = 17) Aug–Oct. Dry or damp open prairies, waste ground, often in sandy sites; GP; (ME to NC, w to Rocky Mts., s Can. to TX; introduced elsewhere).

6. **Helianthus mollis** Lam., ashy sunflower. Perennial 5–12 dm tall. Stems single or often clustered from stout rhizomes, hirsute to villous. Leaves opposite or a few of the uppermost alternate, ovate to ovate-lanceolate, acute to acuminate, cordate or nearly so at the base, sessile and sometimes clasping, blade 6–15 cm long and 2–7 cm wide, margin entire to shallow-serrate, densely ashy-hispid and gray-green in aspect. Heads solitary or in contracted racemiform to spiciform clusters; disk 2–3 cm across, involucral bracts linear-lanceolate to lanceolate, acute to attenuate, equaling or exceeding the disk, appressed below but spreading above; ray florets 15–30, ligule up to 3 cm long, light to deep yellow; disk florets pale to deep yellow; chaffy bracts linear, entire, hairy on the apex. Achene ca 4 mm long, pubescent at apex when young. (n = 17) Jul–Oct. Open drying places, especially sandy prairie sites; se 1/4 GP, se NE s through e 1/3 KS to ne OK; (PA to GA, w to GP & e TX; introduced elsewhere).

7. **Helianthus nuttallii** T. & G., Nuttall's sunflower. Perennial 5–30 dm tall, stems arising from a cluster of enlarged, tuberous roots with short, thick rhizomes, glabrous or lightly hispid-pubescent or hirsute, occasionally glaucous. Leaves opposite or sometimes nearly all alternate, blade ovate to narrowly lanceolate, 4–20 cm long and 1–4 cm wide, entire or serrate, glabrous to hispid or tomentose above and rough hispid to villous-tomentose beneath, tapering at the base to a short, winged petiole. Heads few, at the ends of long peduncles, disk 1–2 + cm across; involucral bracts linear-lanceolate, equaling or slightly exceeding the disk; ray florets 10–17, ligule up to 3 cm long; disk florets yellow. Achene 4–5 mm long. (n = 17) Jul–Sep. (Across N. Amer. from the Great Lakes region to the Pacific Northwest, s in the cordillera to NM, AZ, & CA).

This species is a polymorphic entity with numerous, weakly distinctive phases. It may be regarded as the western element of a near-continuum in variation, passing from the essentially eastern *H. giganteus* through the midwestern *H. grosseserratus* to the western *H. nuttallii*. Two subspecies are recognizable in our flora; a third subspecies is restricted to CA.

7a. subsp. **nuttallii**. Stems mostly 20–30 dm tall. Leaves mostly alternate, lanceolate, acute to acuminate. Largely in the Rocky Mts., uncommon & scattered in the n GP in n-cen NE, SD, ND & w.

7b. **subsp. *rydbergii*** (Britt.) Long. Stems 5–20(+) dm tall. Leaves mostly opposite, lanceolate to ovate, acute to obtuse. N GP from cen NE northward; our common phase, largely replacing the related *H. grosseserratus* in the n & w GP. *H. nitidus* Lunell, *H. rydbergii* Britt.—Rydberg.

8. ***Helianthus petiolaris*** Nutt., plains sunflower. Taprooted annual, 5–12 dm tall. Stem simple or branched, sometimes from near the base, strigose or occasionally glabrate. Leaves nearly all alternate; blade deltoid-ovate to lanceolate, 4–15 cm long and 1–8 cm wide, entire to shallow serrate, strigose on both surfaces, frequently becoming bluish in age, petiolate. Heads often few, long-pedunculate; disk 1–2.5 cm across; involucral bracts lanceolate to ovate lanceolate, 3–4+ mm wide; ray florets 15–30, ligule ca 2 cm long; disk florets reddish purple or very rarely yellow; chaffy bracts 3-forked, the middle cusp exceeding the disk florets and those in the center of the head densely white-hairy at the apex. Achenes 3.5–4.5 mm long, lightly pubescent. (n = 17) (May) Jun–Sep. Open sandy sites; GP; (ME & MA to GA, w to s Alta., UT, & AZ; probably introduced e of WI & IL; introduced in CA & B.C.).

Hybrid swarms with *H. annuus* are known from s-cen KS and elsewhere.

9. ***Helianthus pumilus*** Nutt. Perennial up to 1 m tall with stems arising from a tough, erect, taprooted crown, scabrous-hispid to strigose, simple or branched above. Leaves opposite, blade ovate to lanceolate, tapering or rarely subtruncate at the base, 4–15 cm long and 1–4 cm wide, entire to serrate, ashy-green, distinctly petiolate. Heads few, on short peduncles, disk 1–1.5 cm across, involucral bracts lanceolate to ovate-lanceolate, acute to short-acuminate, equaling or shorter than the disk; rays ± 10, ligule 1–2+ cm long; disk florets yellow or sometimes reddish-tinged; chaffy bracts entire or shallowly 3-toothed. Achene 3–4 mm long. (n = 17) Jul–Aug. Dry, open sites; extreme w GP at the base of the Rocky Mts., s MT, WY, & CO; (a species of the Rocky Mts., barely entering our range).

10. ***Helianthus rigidus*** (Cass.) Desf., stiff sunflower. Perennial 3–20 dm tall. Stems arising from stout rhizomes, variously light to densely scabrous-hispid, green or reddish, strict or few-branched above. Leaves opposite or the uppermost sometimes alternate; blade linear-lanceolate to ovate, 5–20(30) cm long and 2–6 cm wide, thickish, scabrous, gray-green to light green, serrate to subentire, petiolate with winged petiole or becoming subsessile upward, the uppermost leaves often reduced to mere bracts. Heads single or several on long peduncles; disk 1.5–2(3) cm across; involucral bracts elliptic to oblong-ovate, apex acute or obtuse, tightly appressed and obviously imbricate or graduated in length, shorter than the disk; ray florets 10–20, ligule up to 3.5 cm long; disk florets reddish-purple or yellow with reddish lobes, or rarely entirely yellow; chaffy bracts entire or 3-forked, often lightly pubescent at the apex and on the keel. Achenes 5–6 mm long, glabrous or nearly so; pappus often with secondary scales between the 2 prominent awns. (n = 51) Jul–Sep (Oct). (Across e N. Amer. from N.B. to GA, w to Alta. & TX; the e coast occurrences probably represent introductions from a basically midwestern distribution).

This is a complex species with numerous variable and semidistinct races. Two subspecies and one named hybrid are recognized as follows:

10a. **subsp. *rigidus*.** Plants 1–2 m tall. Leaves in 7 or more pairs, blade oblong-lanceolate to narrowly ovate, 8–20+ cm long, apex long-acuminate. Disk florets deep red to almost entirely yellow. Dry to drying prairies; se 1/4 GP; e 1/2 NE & KS, sporadic nw to w ND. *H. laetiflorus* var. *rigidus* (Cass.) Fern.—Fernald.

10b. **subsp. *subrhomboideus*** (Rydb.) Heiser. Plants usually less than 1.2 m tall. Leaves in 5–7 pairs, blade rhombic to lanceolate, 5–12 cm long, acute to obtuse. Disk florets deep red to purplish. Dry to drying open prairies & plains; n 1/2 GP, from cen NE northward, largely replacing subsp. *rigidus* in SD, ND & w. *H. subrhomboideus* Rydb.—Rydberg.

This subsp. is similar to *H. nuttallii* subsp. *rydbergii*, which has a similar geographic distribution.

10c. **Helianthus x laetiflorus** Pers. Apparently a series of persistent hybrids between *H. rigidus* and *H. tuberosus*, and ± intermediate between the 2 parents. *Helianthus* x *laetiflorus* is distinguishable from *H. rigidus* in that it has consistently yellow disk florets and more acute involucral bracts, and from *H. tuberosus* by the relatively shorter involucral bracts and consistently more scabrous leaves. Dry to drying prairies; se 1/4 GP (widely scattered in e N. Amer.). Cf. Heiser, op. cit., for discussion.

11. **Helianthus salicifolius** A. Dietr., willow-leaved sunflower. Perennial 15–25(+) dm tall. Stems arising from coarse, spreading rhizomatous rootstocks, erect, branching above, glabrous and sometimes glaucous, pale green to reddish-purple. Leaves mostly alternate, very numerous; blade narrowly linear-lanceolate to linear, flat, 8–20 cm long and usually less than 1 cm wide, apex and base narrowly attenuate, entire or shallow-serrate, subsessile, lowermost usually deciduous. Heads numerous in an open, paniculate cluster, peduncles slender and sometimes short-pubescent; disk 1–1.5 cm across; involucral bracts linear-lanceolate, long-attenuate and at least some of them exceeding the disk, loose-spreading; ray florets 10–20, ligule up to 3 cm long, disk florets reddish purple; chaffy bracts linear, entire to weakly 3-toothed, apex and keel short-pubescent. Achenes 4–6 mm long, glabrous and slender. (n = 17) Aug–Oct. Limestone prairies; w MO, e 1/4 KS & adj. OK; (se GP s to TX).

12. **Helianthus strumosus** L. Perennial 1–2 m tall. Stems arising from well-developed creeping rhizomes, glabrous and sometimes glaucous, or occasionally with a few long hairs. Leaves opposite or the uppermost alternate, blade linear-lanceolate to ovate, acuminate at the apex, subcordate to rounded or abruptly tapering at the base to a slender, distinct petiole, 7–20 cm long and 2–10 cm wide, entire to serrate, ± scabrous above and variously glabrous to tomentose beneath. Heads few to numerous on short, divergent or ascending peduncles; disk 1.5–2(+) cm across; involucral bracts lanceolate, acuminate to attenuate, erect or loose, squarrose, equalling or exceeding the disk; ray florets 10–15, ligule up to 4 cm long; disk florets yellow; chaffy bracts subentire to distinctly 3-toothed, pubescent at apex. Achene ca 5 mm long. (n = 34, 51) Aug–Oct. Disturbed, open wooded areas; extreme e GP; w MN, w IA, w MO, & extreme e KS; (ME & Que. s to FL, w to GP & e TX).

This species includes a complex of poorly defined, semidistinct phases, barely entering our range. It intergrades with *H. hirsutus* and *H. tuberosus*, and the phases are complicated by polyploidy.

13. **Helianthus tuberosus** L., Jerusalem artichoke. Perennial 1–3 m tall. Stems arising from well-developed, tuber-bearing rhizomes, scabrous to hirsute or sometimes glabrate, usually branched above. Leaves opposite but the uppermost alternate, or sometimes alternate on the upper 1/2 of the stem; blade ovate to lanceolate, 10–25 cm long and 6–15 cm wide, abruptly or gradually contracted to a distinct, winged or partially winged petiole, scabrous above and puberulent to tomentose beneath, usually serrate. Heads several to numerous; disk 1–2(+) cm across; involucral bracts linear-lanceolate to lanceolate, acute to acuminate, equaling or slightly exceeding the disk, loose and erect or reflexed at maturity; ray florets 10–20, ligule up to 4 cm long; disk florets yellow, chaffy bracts 3-toothed, pubescent at apex. Achenes 5–7 mm long, glabrous. (n = 51) Aug–Oct. Open or shaded, moist to drying sites; e 2/3 GP; w ND s & e to cen OK; (ME & Que. s to GA & n FL, w to Man. & e TX). *H. besseyi* Bates — Rydberg.

This is a variable species, distinctive in the production of edible tubers. *H. tuberosus* intergrades and/or hybridizes with several other sunflowers (including *H. strumosus*) and it is presumed to be a parent for the hybrid *H.* x *laetiflorus* (cf. *H. hirsutus*).

Our materials tend to be a little shorter, have more of the leaves opposite and with more densely pubescent undersides than the materials from the eastern woodland regions, and have been recognized as the weak var. *subcanescens* A. Gray.

58. HELIOPSIS Pers.

1. *Heliopsis helianthoides* (L.) Sweet var. *scabra* (Dun.) Fern., false sunflower, ox-eye. Perennial herb 8–15 dm tall. Stems arising from a fibrous-rooted caudex or stout rootstocks, scabrous or mostly so. Leaves opposite, blades ovate or broadly lanceolate, 5–15 cm long and 3–8 cm wide, serrate, subtruncate to contracted at the base, petiolate, mostly scabrous on both surfaces, somewhat firm. Heads 1–several on naked peduncles; disk 1–2.5 cm across; involucral bracts in 1 or 2 series, subequal; receptacle conic and elongating at maturity; ray florets 10–16, usually fertile, ligule up to 4 cm long, yellow; disk florets pistillate and fertile, style branches with flattish, short, hairy appendages; chaffy bracts concave and clasping, subtending both ray and disk florets. Ray achenes often 3-angled, disk achenes 4-angled, glabrous; pappus absent or a short, obscure crown. (n = 14) Jul–Oct. Dry open woods, edges of fields, & in waste places; e 1/3 GP but scattered westward in s SD & NE; (Que. to GA, w to B.C. & NM). *H. scabra* Dun.—Rydberg; *H. helianthoides* var. *occidentalis* (Fisher) Steyerm.—Steyermark. [tribe Heliantheae]

Var. *helianthoides* occurs in the woodland regions from MO eastward, and it is characterized by near-glabrous herbage and evidently thinner leaves.

59. HETEROTHECA Cass., Camphor Weed

1. *Heterotheca latifolia* Buckl. Erect to bushy annual, taprooted herb, (2)8–15 + dm tall; glandular, spreading-hairy to long-hirsute, especially below, somewhat aromatic. Leaves alternate, the lower ones petiolate, ovate to elliptic ovate, mostly less than 10 cm long and 1–3(5) cm wide, toothed or subentire; middle and upper leaves progressively reduced upward, becoming sessile and clasping; scabrous or hairy on both surfaces but especially below. Inflorescence a loose, corymbiform cluster; heads turbinate to campanulate, 5–10 mm tall and 5–15 mm wide; involucral bracts imbricate in 4–6 series, short hirsute-pubescent, especially upward along the midrib, usually distinctly glandular; receptacle flat or nearly so, naked; ray florets mostly 20–40, pistillate and fertile, ligule ± 5 mm long, yellow; disk florets perfect and fertile, style branches flat with marginal stigmatic lines and slender, ventrally-pubescent appendages. Ray achenes somewhat triangular, mostly glabrous, pappus absent; disk achenes flattened or weakly 4-angled, villous-hirsute; pappus double, outer pappus of short, flattened scales and the inner pappus of elongate, capillary bristles. (n = 9) Jul–Oct. Open, usually sandy disturbed sites; s 1/3 GP; (GP s & w through TX & NM). *H. subaxillaris* (Lam.) Britt. & Rusby—Fernald. [tribe Astereae]

References: Wagenknecht, B. L. 1960. Revision of *Heterotheca* section *Heterotheca*. Rhodora 62: 61–76, 97–107; Harms, V L. 1965. Biosystematic studies in the *Heterotheca subaxillaris* complex. Trans. Kansas Acad. Sci. 68: 244–257.

Heterotheca latifolia is a poorly defined species within the *H. subaxillaris* assemblage. Our materials are largely referable to var. *macgregoris* Wagenkn. (cf. Wagenknecht, op. cit.), but in the far sw GP passing into var. *latifolia*, which has ± coarsely toothed leaves and a taller stature.

60. HIERACIUM L., Hawkweed

Fibrous-rooted perennial herbs, often with a short, branching caudex or rhizome; juice milky; herbage variously pubescent to glabrate, often with at least some stellate hairs. Leaves alternate or basal, simple, entire or toothed. Heads in a corymbiform or subumbellate cluster, or in an elongate, weakly racemiform-cylindric cluster, or occasionally solitary; involucre cylindric to hemispheric; florets all ligulate, perfect and fertile, corolla yellow to orange, (white in *H. albiflorum*). Achenes ± prismatic, narrowed toward the base and either narrowed upward or truncate; pappus of few to numerous capillary bristles. [tribe Lactuceae]

The native GP species are diploids, which at most may produce an occasional hybrid between species. The Old World representatives form a huge diploid-polyploid-apomictic assemblage with confused species boundaries and a complex taxonomic history. The fact of apomixis has inspired the publication of hundreds of species names. It is possible that some weedy species of hawkweed from the Old World might occasionally be found in the GP, but thus far none are known to persist.

1 Florets white; stellate hairs lacking; rare in the high meadows of BH .... 1. *H. albiflorum*
1 Florets yellow; at least a few stellate hairs present, and often abundant.
   2 Involucral bracts ± imbricated in several series; pappus bristles numerous, unequal; achenes apically truncate.
      3 Leaf margins lacking obconic hairs; lower stem and leaves with long, spreading hairs; leaves 2–5 × longer than wide .... 2. *H. canadense*
      3 Leaf margins with stout, obconic hairs; lower stems and leaves lacking long, spreading hairs; leaves mostly 4–12 × longer than wide .... 5. *H. umbellatum*
   2 Involucre a single series of nearly equal principal bracts, subtended by small calyculate bracts; pappus bristles relatively few and equal; achenes distinctly narrowed toward the summit.
      4 Lower stem shaggy-pubescent with hairs mostly 1 cm or more long; heads large, with 40–60(90) florets .... 4. *H. longipilum*
      4 Lower stem variously pubescent but with hairs nearly always less than 1 cm long; heads smaller, mostly with 20–40 florets .... 3. *H. gronovii*

1. **Hieracium albiflorum** Hook. Perennial 3–8 dm tall, loosely scattered-pubescent, especially toward the base but glabrate upward. Basal and lower cauline leaves persistent and often tufted, the prominent ones 4–12 cm long overall and up to 4 cm wide, entire to shallow-dentate; cauline leaves progressively reduced upward, becoming mere sessile bracts, some of which may be clasping. Inflorescence of few-numerous heads on slender peduncles in an open cluster; involucre 5–10 mm tall; involucral bracts variously glandular-pubescent to glabrous; florets 12–36, all ligulate, corolla white or ochroleucous. Achenes ca 3 mm long; pappus whitish, longer than the achene. (n = 9) Jul–Aug. Scarce in open, high meadows; disjunct in BH; (widespread in the cordillera, BH, AK, & Yukon to CO & CA).

2. **Hieracium canadense** Michx. Perennial (2)5–10(15) dm tall, arising from a short caudex; spreading-pubescent, at least toward the base, often stellate-hairy, especially on the leaves. Basal and lowermost cauline leaves early deciduous; cauline leaves numerous, about equal in size, sessile and often weakly clasping, ovate or lanceolate to oblong, 3–12 cm long and up to 4 cm wide, often irregularly dentate. Inflorescence a loose, corymbiform to subumbellate cluster, or occasionally the heads solitary; involucre 6–12 mm tall; involucral bracts imbricated in several series; florets all ligulate and fertile, corolla yellow. Achenes truncate, ca 3 mm long; pappus of numerous unequal bristles. (n = 9) Jul–Sep. Damp, open or wooded sites, especially in sandy or loose soils; scattered in ne 1/5 GP, plus BH; (Lab., Newf. to NJ, w to B.C. & WA).

Two weakly separated vars. are sometimes recognized. Our materials are referable to var. *canadense*, but some materials may approach var. *fasciculatum*, which is largely restricted to the Great Lakes region and e, and is generally more rubust, 10–15 dm tall, with the inflorescence distinctly subumbellate.

3. **Hieracium gronovii** L. Perennial up to 15 dm tall, arising from a short, erect or weakly rhizomatous caudex; conspicuously spreading-pubescent, especially toward the base, the hairs strigose-puberulent and a few stellate, less than 1 cm long, or herbage glabrate upward. Leaves pubescent with both small stellate hairs and with some scattered, long hairs; basal and lowermost cauline leaves often early deciduous but when persistent, broadly

lanceolate to ovate, up to 20 cm long and 5 cm wide, cauline leaves ± progressively reduced upward, the uppermost sessile and becoming clasping. Inflorescence elongate and racemiform-cylindric, peduncles short-pubescent to stipitate-glandular; involucre 6-9 mm tall; principal involucral bracts in a single series; florets 20-40, all ligulate and fertile, corolla yellow. Achenes 2.5-4 mm long, narrowed toward the summit; pappus bristles rather few and equal in length. (n = 9) Jul-Oct. Dry or drying open woods or disturbed sites, especially on sandy soils; se KS, w MO, & adj. e OK; (MA to FL, w to se GP & TX).

4. *Hieracium longipilum* Torr. Perennial 5-15(20) dm tall, arising from a stout, sometimes branching caudex; densely long-pubescent, especially toward the base, the hairs often approaching 2 cm long, or herbage sometimes glabrate upward. Leaves mostly densely pubescent but the hairs sometimes shorter than those on the lower stem; basal and lower cauline leaves oblanceolate to elliptic, 10-20(30) cm long and up to 4 cm wide, often crowded, the lowermost sometimes early deciduous; middle and upper cauline leaves progressively reduced upward, those of the upper 1/2 of the stem usually reduced to minute bracts. Inflorescence usually elongate and racemiform-cylindric, peduncles stellate-pubescent and usually stipitate-glandular; involucre 7-10 mm tall; principal involucral bracts in a single series, stellate-pubescent and with prominent, black glandular hairs; florets 40-90, all ligulate and fertile, corolla yellow. Achenes 3-4.5 mm long, narrowed toward the summit; pappus bristles rather few and equal in length. (n = 9) Jul-Sep. Drying upland prairies & open woods, often in sandy or loose soils; se NE, w IA, e 1/2 KS, & cen OK; (MI & IN to se GP).

5. *Hieracium umbellatum* L. Perennial 4-8(12) dm tall, arising from a short caudex; stem glabrate toward the base, or but lightly pubescent, often stellate-pubescent upward and frequently with some stout, obconic hairs among the heads. Basal and lower cauline leaves small and early-deciduous; prominent cauline leaves numerous and about equal in size, sessile, lanceolate to narrowly lanceolate, up to 10 cm long and 2 cm wide, 4-12 × longer than wide, margins with at most a few coarse teeth and usually with abundant short, stiff obconic hairs; uppermost leaves reduced to mere bracts. Inflorescence a loose, corymbiform cluster; involucre 6-12 mm tall; involucral bracts imbricated in several series; florets all ligulate and fertile, numerous, corolla yellow. Achenes truncate apically, 3 + mm long; pappus of numerous, unequal bristles. (n = 9) Jul-Sep. Damp sites, especially openings in wooded areas, often on sandy soils; n 1/4 GP, BH, s to nw NE & w; (circumboreal, s in N. Amer. to MI, n GP, CO, & OR).

This entity is closely related to and intergrading with *H. canadense*, and is a part of a widespread species-complex centered in Eurasia, with both sexual and apomictic phases. It is perhaps possible to distinguish infraspecific taxa, and should that prove useful, our plant could be recognized as var. *scabriusculum* Farw.

61. HYMENOPAPPUS L'Her.

Erect biennial or perennial herb. Stems subscapose or leafy from an unbranched erect taproot or from a perennating, branching crown surmounting a woody taproot. Leaves alternate, forming a basal rosette, 1 or 2 × pinnately dissected, reduced or few upward, margins of ultimate lobes often inrolled. Inflorescence a few- to many-headed paniculate cyme; heads discoid; involucral bracts 6-14 in 2 or 3 series, membranaceous toward the apex; receptacle naked; disk florets yellowish or white, corolla expanded above into a campanulate or funnelform throat, corolla lobes reflexed at maturity; style branches obtuse, flattened papillose at the apex. Achenes 4-angled and sometimes with secondary nerves between the angles; pappus of 12-20 obtuse, hyaline scales. [tribe Anthemideae]

Reference: Turner, B. L. 1956. A cytotaxonomic study of the genus *Hymenopappus*. Rhodora 58: 163, et seq.

Hymenopappus is here aligned with the tribe Anthemideae in accord with recent practice by many American synantherologists. Its past disposition has been with the untenable tribe Helenieae.

1 Plants perennial, mature individuals with several leafy crowns at the apex of a woody taproot; cauline leaves usually 8 or fewer .... 1. *H. filifolius*
1 Plants biennial with a single rosette surmounting the taproot; cauline leaves regularly more than 8.
  2 Corolla tube 1.5–2(2.2) mm long, throat campanulate; ultimate segments of basal leaves linear, 0.5–6 mm wide.
    3 Flowers yellow; ultimate leaf segments 2–6 mm wide .... 2. *H. flavescens*
    3 Flowers white; ultimate leaf segments 0.5–1.5 mm wide .... 4. *H. tenuifolius*
  2 Corolla tube 2–3 mm long, throat funnelform; ultimate leaf segments 2–8 mm wide .... 3. *H. scabiosaeus* var. *corymbosus*

1. Hymenopappus filifolius Hook. Subscapose perennial up to 6 dm tall, tomentose to glabrate in age. Basal rosette leaves 5–15 cm long, bipinnately dissected with flattened, linear to filiform, punctate ultimate segments; cauline leaves few, usually fewer than 8, strongly reduced upward. Inflorescence of few to many (40 +) heads; involucral bracts 5–9 mm long and 1–4 mm wide; florets 20–50, yellow or whitish, corolla tube variously glandular. Achenes 4–6 mm long with conspicuous long hairs; pappus of 12–18 linear-oblong scales 1–2 mm long.

This is a complicated species occurring from the GP westward, with 13 varieties recognized by the monographer (Turner, op. cit.). Two varieties occur in our area:

1a. var. *cinereus* (Rydb.) I. M. Johnst. Plants 1.5–4 dm tall; cauline leaves 2–4. Heads 1–6; corolla throat 1.5–2.5 mm long. (n = 17) May–Aug. Rocky limestone canyons; panhandle of n TX & OK; (w from GP to CO, NM, UT, AZ).

Our materials are at the se end of the natural range for this entity, and appear to be a little shorter and with a more clustered aspect than do the great body of specimens from the mountain regions.

1b. var. *polycephalus* (Osterh.) B. L. Turner. Plants 3–6 dm tall; cauline leaves 3–8. Heads 5–50; corolla throat 1.3–1.8 mm long. (n = 34) Jun–Aug. Rolling open prairies & plains; w 1/2 GP w to the Rocky Mts., e to ne NE; (w from GP into the montane grasslands of the Rocky Mts.).

2. Hymenopappus flavescens A. Gray. Biennial 2–9 dm tall, with stems arising singly from a taproot. Basal rosette leaves bipinnately parted, 6–14 cm long, ultimate segments 2–6 mm wide, variously glabrate above but tomentose below; cauline leaves numerous, progressively reduced upward. Inflorescence with 30–100 heads in a congested cyme; involucral bracts 4–5 + mm long; florets 30–70, bright yellow, corolla tube densely to sparsely glandular, 1.5–2 mm long, throat campanulate, 0.8–1.5 mm long. Achenes 3.4–4 mm long, pubescent mostly on the angles, with hairs up to 1 mm long; pappus of 18–20 scales, ca 1 mm long. (n = 17) May–Aug. Sandy soils; sw GP; sw KS & the panhandles of w OK & TX, adj. CO & NM; (from GP sw to w TX, e NM).

Our materials are referable to var. *flavescens*. The semidistinct var. *cano-tomentosus* A. Gray has a more southwestern distribution, plus certain morphological distinctions (cf. Turner, op. cit.).

3. Hymenopappus scabiosaeus L' Her. var. *corymbosus* (T. & G.) B. L. Turner, old plainsman. Biennial 3–6(10) dm tall with stems arising singly from a taproot, or rarely with stems clustered on injured plants, tomentose to glabrate. Basal rosette leaves pinnately dissected, 5–15 cm long, mostly glabrate or sparsely hairy above, but tomentose below, ultimate leaf segments linear, acute, 2–8 mm wide; cauline leaves numerous, progressively reduced upward. Heads numerous, 40–100, discoid; involucral bracts 5–9 mm long, 2–4 mm wide, the upper 1/3 yellowish-white or white membranaceous; florets 20–60, white or creamy, corolla tube sparsely glandular, 2–3 mm long, throat funnelform, 1–1.5 mm long. Achenes 3–4 mm long, short-pubescent; pappus of 14–18 scales, ca 0.2–1 mm long. (n = 17) Apr–Jul. Open prairies, especially on & near limestone outcrops; se 1/4 GP; s NE, e 1/2 KS & cen OK; (GP s through cen TX).

In se KS and e OK our var. *corymbosus* tends toward var. *scabiosaeus*, which is of more eastern distribution. The two differ in that var. *scabiosaeus* regularly has bracted peduncles, longer involucral bracts, plus a series of less precise differences relating to aspect. However, our materials are all referable to var. *corymbosus* (Turner, op. cit.).

4. Hymenopappus tenuifolius Pursh. Biennial 4–10(15) dm tall with stems arising singly from a taproot, much branched above, sparsely white tomentose to glabrescent. Basal rosette leaves 8–15 cm long, bipinnately dissected, ultimate segments linear-filiform, 0.5–1.5 mm wide, punctate; cauline leaves numerous, reduced upward. Inflorescence a paniculate cyme with 20–200 heads; involucral bracts mostly glabrate or glandular, 5–8 mm long, the upper 1/5 yellowish-membranaceous; florets 25–50, white, corolla tube glandular, 1.5–2.2 mm long, throat campunulate, 0.8–1.5 mm long. Achenes 3.5–4.5 mm long, pubescent mostly along the angles with hairs up to 1 mm long; pappus of 16–18 scales, 1–1.5(2) mm long. (n = 17) May–Aug. Open sandy prairies & plains, rocky calcareous hillsides; throughout GP except e 1/4; (GP w through cen NM in the intermontane plains).

62. HYMENOXYS Cass., Bitterweed

Annual or perennial herbs with a taproot, often expanded into a perennating crown or caudex, sometimes becoming coarsely rhizomatous; herbage variously covered with fine, minute resinous bodies, and glabrate to long-villous or pilose. Leaves alternate and/or chiefly basal. Heads solitary on distinct peduncles or cymosely clustered; involucre hemispheric to nearly cylindrical; outer involucral bracts either united at the base, rigid and strongly thickened below, or distinct at the base and not at all strongly thickened; inner involucral bracts distinct to the base and either leathery in texture and clearly in 1 whorl, or ± herbaceous with membranaceous margins and in 2 whorls; receptacle conical, naked; ray florets pistillate, fertile, ligule yellow and reflexed at maturity; disk florets perfect and fertile, corolla yellow, style branches dilated-truncate, pencillate. Achene weakly 4-angled, coarsely pubescent with appressed hairs, 1.5–3.5 mm long; pappus of 5–8 translucent scales. [tribe Heliantheae]

Reference: Parker, K. F. 1970. *Hymenoxys, in* D. S. Correll & M. C. Johnston, Manual of the Vascular Plants of Texas. Texas Res. Foundation, Renner.

1 Involucral bracts in 2 dissimilar series; outer bracts rigid, keeled, thickened and united below, inner bracts separate and leathery, slightly exceeding the outer bracts; stem leafy.
  2 Perennial with a branching, woody caudex; nw GP .... 4. *H. richardsonii*
  2 Annual with a slender taproot; s GP .... 3. *H. odorata*
1 Involucral bracts ± similar, separate, erect and neither rigid not strongly thickened.
  3 Annual with both a basal rosette and well-developed cauline leaves; glabrate to pilose .... 2. *H. linearifolia*
  3 Perennial with a well-developed woody taproot and/or crown.
    4 Plant scapiform with leaves clearly basal, herbage mostly sericeous-villous .... 1. *H. acaulis*
    4 Plant with leaves both basal and on the lowermost portions of the stem, variously pilose to glabrate .... 5. *H. scaposa*

1. Hymenoxys acaulis (Pursh) Parker, stemless hymenoxys. Subcespitose, scapiform perennial, 1–3 dm tall, with a short, weakly spreading, branching caudex, covered by persistent old leaf bases with woolly axils, and eventually derived from a tough, woody taproot. Leaves all basal, linear oblanceolate, up to 6 cm long and 8 mm wide, ± densely sericeous. Heads solitary on naked scapes; involucral bracts 4.5–6 mm long, ray florets ca 8 or ca 13, ligule 6–14 mm long. Pappus scales 5–6(7), 2–3 mm long. (n = 14, 15, 30) May–Aug. Open calcareous soil & rocky breaks; w 1/2 GP; (s Sask. to TX, w to CA). *Tetraneuris acaulis* (Pursh) Greene, *T. simplex* A. Nels.—Rydberg.

This species forms a complex of numerous regionally distinctive phases throughout w N. Amer., differing largely on vestiture and placement of leaves. Our materials appear to be all referable to var. *acaulis*.

2. ***Hymenoxys linearifolia*** Hook. Annual herb up to 4 dm tall, or rarely taller. Stems arising from a basal rosette, early stem little-branched, but becoming bushy-branched later in the season, variously heavily to lightly pilose or glabrate. Rosette leaves deciduous, oblanceolate to linear-lanceolate, subentire or few-lobed; cauline leaves similar but becoming narrowly linear, 3–6(8) cm long and 3–10 mm wide, persistent. Heads on naked peduncles above the leaves; disk 10 + mm across, or less in later heads; involucral bracts 3–4 mm long; ray florets 6–21, variable in number depending upon the size of the head, ligule 5–6 mm long. Achene ca 2 mm long; pappus scales ca 2 mm long, including the narrowed awn-tip. (n = 14, 15) Apr–Jul. Open plains, especially flood plains & disturbed sites with calcareous soil; s GP; w 1/2 OK, adj. TX, sporadic to ne KS; (s GP, TX, NM, & n Mex.). *Tetraneuris linearifolia* (Hook.) Greene—Rydberg.

3. ***Hymenoxys odorata*** DC., bitterweed. Taprooted, bushy-branched annual up to 5 dm tall, aromatic, lightly pubescent to glabrate. Rosette leaves withering and early deciduous; cauline leaves numerous, 2–10 cm long, pinnatisect into 3–15 narrow, filiform divisions. Heads numerous, solitary on terminal or axillary peduncles; involucral bracts 4–6 mm long, outer bracts united at the base and distinctly thickened below, inner bracts exceeding the outer ones, convergent at maturity; ray florets 6–13, ligule up to 10 mm long. Achene 1.5–2 mm long; pappus scales 5(6), 1.5–2 + mm long, including the awn-tip. (n = 11, 15) May–Aug. Open disturbed sites, especially heavily grazed pastures on calcareous soil; sw GP; sw 1/4 KS, adj. CO & NM to TX & w OK; (sw GP to TX, w to CA; n Mex.).

This species is poisonous to livestock, especially sheep.

4. ***Hymenoxys richardsonii*** (Hook.) Cockll., Colorado rubber plant. Bushy, subcespitose perennial, mostly 1–2 dm tall, scattered pubescent but persistent woolly-villous in the lower leaf axils. Stems branched upward, arising from a clustered branching woody caudex, covered with persistent old leaf bases. Leaves basal and densely cauline, forming a tuft around the stem base, 2–10 + cm long, entire or commonly pinnatisect into 3–7 + linear segments. Heads 1–5 on separate peduncles; disk 1–2 cm across; outer involucral bracts 8–16, 5–8 mm long and united at the base to 1/2 their lengths, distinctly thickened below; inner bracts separate, equaling or exceeding the outer bracts; ray florets ca 13, ligule up to 20 mm long. Achenes 3–4 mm long; pappus scales 5–6(7), 3–4 mm long including the awn-tip. (n = 14, 15) Jun–Aug. Open plains; nw GP; w 1/2 ND, adj. MT & WY, s along the edge of the Rocky Mts. in CO; (s Sask. to TX, w to UT & AZ).

Our materials are referable to var. *richardsonii*, but s & w they pass into var. *floribunda* (A. Gray) Parker, with smaller, more numerous heads, and taller, more branched stems. Var. *floribunda* is poisonous to livestock.

5. ***Hymenoxys scaposa*** (DC.) Parker. Subscapose perennial 1–3 + dm tall, densely to lightly tomentose. Stems arising from a thin and herbaceous to a woody and much branched caudex, ultimately derived from a taproot. Leaves basal and densely imbricate-clustered on the lower part of the stem, 2–10 cm long and mostly less than 6 mm wide, linear to linear-lanceolate, expanded to a clasping base, axil densely villous-tomentose. Heads solitary on naked scapes; disk 1–2 cm across; involucral bracts in 2 or 3 series, 4–6 mm long, distinct

and subequal; ray florets ca 13 or ca 21, ligule up to 20 mm long. Achene 2-3.5 mm long; pappus scales 5-7(8), 1-2 + mm long, including the awn-tip. (n = 15) May-Aug. Open, rocky limestone sites in the high plains; (cen GP s to TX & n Mex., NM).

Two varieties are weakly distinguishable in our region:

5a. var. *glabra* (Nutt.) Parker. Plant 0.5-2 dm tall, with a tough, woody, branching caudex. Leaves rarely more than 4 mm wide. Sw NE, w 1/2 KS, OK, & adj. CO, TX, & NM. *Tetraneuris fastigiata* Greene — Rydberg.

5b. var. *scaposa*. Plant 1-3 + dm tall, with a slender taproot and a poorly developed, weak caudex; never bushy or woody. Leaves linear or broader, 2-10 + mm wide. w 1/2 KS, w OK, adj. TX, CO, & NM.

63. IVA L., Marsh Elder

Annual or perennial herbs, glabrous or variously pubescent. Leaves alternate or opposite and in our species entire to toothed but not deeply lobed. Inflorescence spikelike, racemiform or paniculate clusters; heads 2-8 mm across, containing both pistillate and staminate florets, the pistillate florets peripheral and the staminate florets central; involucral bracts 3-9, free or united, receptacle chaffy, the chaffy bracts often large and subtending the florets; pistillate florets 1-9, corolla tubular, up to 6 mm long but rudimentary or absent in *I. xanthifolia*; staminate florets 3-20, corolla funnelform, 5-lobed, filaments connate but anthers weakly united or free at anthesis, ovary abortive or absent. Achenes obovate and ± compressed at right angles to the radius of the head, glabrous or short-hairy; pappus none. [tribe Heliantheae]

Reference: Jackson, R. C. 1960. A revision of the genus *Iva* L. Univ. Kansas Sci. Bull. 41: 793-875.

1 Involucral bracts united, at least at the base, to form a lobed or toothed cuplike involucre.
  2 Taprooted annual; pistillate florets 1 or 2 per head .... 1. *I. angustifolia*
  2 Perennial with creeping rootstocks; pistillate florets 5-8 .... 3. *I. axillaris*
1 Involucral bracts distinct.
  3 Corolla of the pistillate florets rudimentary or absent; heads arising on naked peduncles and not subtended by leafy bracts .... 4. *I. xanthifolia*
  3 Corolla of the pistillate florets prominent; heads arising in the axils of herbaceous bracts .... 2. *I. annua*

1. ***Iva angustifolia*** Nutt. Taprooted annual (sometimes perennating?), 4-10(12) dm tall, erect or arching upward, scattered to densely strigose-pubescent. Leaves opposite below and alternate above, petiolate, the blade linear-lanceolate to lanceolate, 2-5 cm long and up to 1 cm wide, conspicuously 3-nerved, margin entire to serrate. Heads subsessile in the axils of linear bracts; involucre 2-3 mm across and about as tall, involucral bracts united into a cup with toothed margin; chaffy bracts subtending the staminate florets only, staminate florets 1-4, corolla 2-4 mm long; pistillate florets 1 or rarely 2, corolla ca 1.5 mm long. Achenes black, 2-3 mm long. (n = 16) Aug-Oct. Moist low disturbed sites; extreme se KS & adj. OK; (AR, LA, OK, & TX; adventive elsewhere).

2. ***Iva annua*** L., marsh elder. Taprooted annual 4-20 dm tall, glabrous below but strigose to hirsute pubescent above. Leaves nearly all opposite but the uppermost sometimes alternate, petiolate, the blade lanceolate to ovate, 5-15 cm long and 2-7 cm wide, 3-nerved, margin somewhat serrate. Inflorescence of spikelike branches, heads sessile and solitary in the axils of linear to ovate bracts; involucre 4-5 mm across and nearly as tall, involucral bracts 3-5, subtending the mature achenes, chaff subtending the staminate florets linear, chaff subtending the pistillate florets filiform or with clavate tips; staminate florets mostly 9-17, pistillate florets 3-5 and with corolla 1.5 mm long, tending to persist on the maturing achenes. Achene ovate, 2-4 mm long, brownish. (n = 17) Aug-Oct. Weed in moist

disturbed sites; frequent in se 1/4 GP, from cen NE s to OK, adventive elsewhere in GP; (IN to MS, w to GP & NM; now widely scattered in U.S. as a weed). *I. ciliata* Willd.— Rydberg.

3. *Iva axillaris* Pursh, poverty weed. Perennial herb with extensive, creeping rootstocks, 2-6 dm tall, stems tough-herbaceous or weakly ligneous at the base, subglabrous or sparsely strigose-villous. Leaves opposite or the uppermost alternate, subsessile, ovate or elliptic to nearly spatulate, 1-5 cm long, entire, weakly 3-nerved. Heads short-pedicillate, solitary in the axils of reduced upper leaves; involucre hemispherical, 4-5 mm across and 2.5-3 mm tall; 4 or 5 involucral bracts united to form a cup with a toothed or lobed margin, or rarely one involucral bract free; chaff subtending staminate florets oblanceolate, chaff subtending pistillate florets oblanceolate or absent; staminate florets 8-20, pistillate florets 5-8. Achenes 2.5-3 mm long, brownish. (n = 17, 18, 27). Jun-Oct. Dryish, disturbed, especially mildly alkaline soils in open prairies & high plains; nw 1/2 GP, more frequent in w ND, w SD, nw NE, & adj. MT & WY, sporadic elsewhere; (GP w to s B.C. & CA).

4. *Iva xanthifolia* Nutt., marsh elder. Coarse, robust annual, 5-20 dm tall, branching upward, glabrous or somewht villous-pubescent upward. Leaves mostly opposite but the uppermost alternate, long-petiolate, blade ovate to broadly ovate, 5-20(+) cm long and 3-15 cm wide, 3-nerved, coarsely serrate and sometimes lobed, scabrous above and ± sericeous beneath. Inflorescence a large paniculate cluster with numerous subsessile heads, not subtended by bracts; involucre 4-5 mm across and 1.5-3 mm tall, with 5 prominent flattish involucral bracts and 5 or more marginal receptacular bracts (chaff) which partly enclose the achenes; staminate florets 8-20, pistillate florets usually 5, with corollas greatly reduced, often to a mere crown at the base of the style. Achene ovate, ca 3 mm long, dark brown. (n = 18) Jul-Sep. Occasional but widespread weed in sandy, damp or drying sites, especially stream beds and flood plains, GP; (native to middle N. Amer.; now widely introduced from s Can. to MD, MO, TX, & AZ). *Cyclachaena xanthifolia* (Nutt.) Fresen.— Rydberg.

64. KRIGIA Schreb.

Annual or perennial herbs with milky juice, glabrate or variously pubescent, often with gland-tipped hairs, especially near the heads. Leaves alternate or in basal rosettes, or sometimes the cauline leaves appearing opposite because of very short internodes, entire or toothed to pinnatifid. Heads numerous to few or solitary, often on long peduncles and sometimes on naked scapes; involucral bracts about all equal and the head not at all calyculate; florets all ligulate, perfect and fertile, corolla yellow to orange. Achenes oblong or turbinate, ± prominently nerved or ribbed, somewhat transversely wrinkled; pappus of about 5 or more prominent or reduced scales, plus 5-40 long, unequal bristles, or the much reduced to a mere scaly crown, or absent. [tribe Lactuceae]

Reference: Shinners, L. H. 1947. Revision of the genus *Krigia* Schreber. Wrightia 1: 198-206.

1 Involucral bracts 4-8× longer than wide; involucre 7-15 mm tall; perennials.
  2 Plants caulescent, branching upward, with several heads; fibrous-rooted ..... 1. *K. biflora*
  2 Plants scapose, heads solitary; producing small tubers along the
    rhizome .... 3. *K. dandelion*
1 Involucral bracts 3× longer than wide or wider; involucre mostly less than 6 mm tall; annuals.
  3 Pappus well developed, of both scales and bristles; plants scapose with leaves all
    basal .... 4. *K. occidentalis*
  3 Pappus absent or reduced to a series of minute, rudimentary scales; plants caulescent
    with both basal and cauline leaves .... 2. *K. caespitosa*

1. **Krigia biflora** (Walt.) Blake. Perennial 2-6(8) dm tall, arising from a tuft of fibrous roots; glabrous and somewhat glaucous, or with some yellowish hairs near the inflorescence. Basal leaves oblanceolate to elliptic-oblong, the larger ones up to 25 cm long overall and 6 cm wide, entire or toothed, or sometimes weakly lobed; cauline leaves few, reduced, sessile and clasping. Inflorescence of several heads on long, naked peduncles, arising in the axils of the upper cauline leaves; involucre 7-14 mm tall; involucral bracts in a single series; florets all ligulate and fertile, corolla orange and much exceeding the involucre. Achenes 2-3 mm long, obscurely ribbed; pappus of 20-40 unequal bristles, plus an outer whorl of ca 10 short, inconspicuous hyaline scales. (n = 5, 10) May-Sep. Open wooded areas, roadsides; extreme e GP; w MN s to w MO & e OK; (MA to GA, w to GP; CO, AZ).

2. **Krigia caespitosa** (Raf.) Chambers. Slender, branching annual 5-30(45) cm tall, erect or somewhat lax, glabrous or lightly pubescent with yellowish, glandular hairs. Leaves basal and cauline, linear-oblong to narrowly obovate, up to 15 cm long and 2 cm wide, variously entire to weakly pinnatifid, the upper cauline leaves sometimes appearing subopposite. Inflorescence of several to numerous heads, on naked peduncles arising singly or in 2s or 3s in the axils of the upper, bractlike leaves; involucre 3-15 mm tall; involucral bracts in a single series or in 2 weakly defined series, unequal, persistently erect and becoming keeled at maturity; florets 14-24, all ligulate and fertile, corolla yellow. Achenes ca 1.5 mm long, conspicuously ribbed; pappus none or a minute, scaly crown. (n = 4) Mar-Jun (Oct). Open or shaded, moist or drying sites; se GP; se NE, e 1/2 KS, e OK, s & e; (VA to FL, w to se GP, TX). *K. oppositifolia* Raf.—Gleason; *Serinia oppositifolia* (Raf.) O. Ktze.—Rydberg.

3. **Krigia dandelion** (L.) Nutt. Scapose, colonial perennial, 1-5 dm tall, arising from a creeping, tuber-bearing rhizome. Herbage glabrous or nearly so, sometimes pubescent near the head. Leaves all basal or with 1 or 2 lower cauline leaves, narrow-lanceolate to lanceolate, 3-20 cm long overall and 2-25 mm wide, entire or toothed to weakly lobed. Head solitary; involucre 9-15 mm tall; involucral bracts in a single series, reflexing in age; florets 25-35, all ligulate and fertile, corolla light orange to yellowish-orange, much exceeding the involucre. Achenes ca 2.5 mm long, ribbed, appressed-pubescent; pappus of 25-40 unequal capillary bristles, plus an outer whorl of ca 10 hyaline scales, usually less than 1 mm long, with the margins split or lacerate. (n = 58, 60) Apr-Jun. Low, damp sites, meadows & wet prairies, often on loose or sandy soils; extreme se GP; KS: Reno, s & e; (NJ to FL, w to GP & TX). *Cynthia dandelion* (L.) DC.—Rydberg.

4. **Krigia occidentalis** Nutt. Thin, scapose annual, 2-10(16) dm tall, erect or somewhat lax, glabrous or sometimes with scattered, spreading glandular hairs. Leaves in a basal rosette, oblanceolate to ovate, 2-8 cm long overall and up to 1 cm wide, entire to weakly dentate-lobulate. Heads solitary; involucre 4-6 mm tall; involucral bracts in a single series, united at the base, keeled or ribbed on the back (outside), persistently erect and not reflexing at maturity; florets 6-25, all ligulate and fertile, corolla orange to yellow-orange, the head usually closing in midday sun. Achenes less than 2 mm long, ± expanded at the summit, weakly ribbed and angular; pappus of 5 + unequal bristles, plus an outer whorl (or alternating with) 5 + thin, hyaline scales, ca 0.5 mm long. (n = 4, 6) Apr-May. Open, drying prairies, often on sandy or loose soils; extreme se GP; KS: Sedgwick, s & e; (s MO, se GP to AR, LA, & TX). *Cymbia occidentalis* (Nutt.) Standl.—Rydberg.

65. KUHNIA L., False Boneset

Pubescent perennials from a stout, taprooted caudex. Leaves altternate, chiefly cauline, the uppermost reduced. Heads discoid, florets all perfect; involucral bracts imbricate, the outer ones shorter or differently shaped than the inner ones; receptacle naked, corolla dull yellowish-white or creamy, 5-lobed, anthers weakly separating at maturity, style branches clavate, flattened and long. Achenes columnar and prominently 10-ribbed; pappus a single whorl of plumose bristles. [tribe Eupatorieae]

    Reference: Shinners, L. H. 1946. Revision of the genus *Kuhnia* L. Wrightia 1: 122–144.
    This genus is sometimes included within the closely related genus *Brickellia*.

1 Leaves very narrow, seldom exceeding 5 mm wide, entire or nearly so, or at most with a pair of basal lobes; heads single or in loose clusters of 2–5; uncommon in s GP .... 1. *K. chlorolepis*
1 Leaves variously lanceolate, often coarsely toothed, 5–40 mm wide; heads several (3–8) in clusters; widespread .... 2. *K. eupatorioides*

1. *Kuhnia chlorolepis* Woot. & Standl. Perennial 3–7 dm tall, often bushy-branched, especially above; minutely pubescent. Cauline leaves narrowly lanceolate to sublinear, sessile or nearly so, somewhat punctate beneath, well-developed leaves 3-veined, overall dimensions 3–7 cm long, 1.5–5(8) mm wide. Heads few or solitary at the ends of the branches, discoid; involucre 8–12 mm tall; florets 15–34, corolla ca 6–7+ mm long. Achenes 5+ mm long, blackish-brown; pappus bristles numerous, plumose. Aug–Oct. Open, dry sites; infrequent in the open southern high plains, barely entering our region from the s; OK: Cimarron, Texas; TX: Briscoe; (apparently uncommon; w TX to CO & AZ, s to cen Mex.).

2. *Kuhnia eupatorioides* L. Single or several-stemmed perennial, 3–10 dm tall, densely short-pubescent to subglabrous. Stems rising from a tough, subligneous, branching caudex. Leaves cauline, numerous, narrowly to broadly lanceolate, gland-dotted beneath, variously entire to irregularly toothed, blade 2–10 cm long and 0.5–4 cm wide, tapering to a short petiole. Inflorescence a series of small corymbiform clusters terminating the branches; heads discoid, involucre ca 7–14 mm tall, involucral bracts in ca 4 series; florets 7–33, creamy or light yellowish-white, corolla 4.5–6 mm long. Achenes 10-ribbed, 4.5–5+ mm long; pappus bristles 20, dull white to light brown, plumose. ($n = 9$) Aug–Oct. Open prairies & plains, especially in loose, sandy soils; (NJ & OH to FL, w to MT & AZ).

    Several varieties are recognized, three of which occur in the GP:

1 Middle and outer involucral bracts acute or acuminate, mostly less than 3/4 as long as the inner ones, the tips usually erect and appressed .... 2a. var. *corymbulosa*
1 Middle and outer involucral bracts with conspicuous, elongate, ± twisted or filiform tips, at least some of them more than 3/4 as long as the inner ones.
  2 Heads relatively small, with 10–14 florets .... 2b. var. *ozarkana*
  2 Heads large, with 18–33 florets .... 2c. var. *texana*

2a. var. *corymbulosa* T. & G. The widespread phase, occurring throughout the GP. East of our range it passes into var. *eupatorioides*, which has smaller heads, thinner leaves, and less pubescence.

2b. var. *ozarkana* Shinners. Scattered in se KS & adj. MO; primarily restricted to the Ozarkian uplift region of MO & AR.

2c. var. *texana* Shinners. Barely entering our region in cen OK; (OK & n-cen TX).

66. **LACTUCA** L., Lettuce

    Herbaceous annuals, biennials, or perennials; taprooted but may become fibrous at maturity; stems erect, arising singly or rarely loosely clustered, branching, only occasionally branched from base, glabrous, hirsute or with stiff bristles; latex white or brown. Basal leaves early deciduous; lowermost cauline leaves sessile, linear-ovate to obovate or lanceolate, variously dentate or pinnatifid to deeply pinnate-lobed or runcinate; middle and upper cauline leaves about equaling the lower cauline leaves in size but frequently differing in shape; uppermost leaves reduced to bracts. Inflorescence a diffuse conical panicle, sometimes corymbose or virgate. Heads few to many; cylindrical at anthesis, becoming urceolate at maturity; involucral bracts imbricate in 3 or 4 series, progressively longer inward; outer bracts ovate, inner bracts lanceolate, apex dark, tomentose, margin scarious; receptacle flat, naked; florets 9–45; ligule yellow, blue, violet, reddish or rarely white; corolla tube more than 1/2 as long as ligule, with or without ring of hairs at summit; anther bases sagittate. Achene body dorsiventrally compressed with lateral ridges or wings; beak filiform and prominent to stubby or absent; pappus of capillary bristles, white or brown. [tribe Lactuceae]

Reference: Dille, D. P. 1976. A revision of the genus *Lactuca* in the Great Plains. Unpublished master's thesis, Kansas State Univ., Manhattan.

Lactuca is a cosmopolitan genus inhabiting north temp. regions world-wide and trop. Africa. Seven of the 15 North American species occur in the GP.

1 Achene beak distinctly filiform, at least 2 mm long.
  2 Achene at least 2 mm wide; beak no longer than achene body; latex brown.
    3 Achene plus beak 5–6 mm long; involucral bracts 8–12 mm long; pappus bristles 5–6 mm long .... 2. *L. canadensis*
    3 Achene plus beak 7–9 mm long, involucral bracts 13–20 mm long; pappus bristles 8–9 mm long .... 4. *L. ludoviciana*
  2 Achene about 1 mm wide; beak conspicuously longer than achene body; latex white.
    4 Achene body with conspicuous bristles at base of beak; stiff bristles present on lower stem and abaxial midrib of most leaves .... 7. *L. serriola*
    4 Achene body without conspicuous bristles at base of beak; stem glabrous and abaxial midrib of leaves glabrous or merely hirsute .... 6. *L. saligna*
1 Achene beak stubby, less than 2 mm long or absent.
  5 Beak whitish; involucral bracts 12–17 mm long; plants 1 m tall or less at maturity .... 5. *L. oblongifolia*
  5 Beak brown or absent; involucral bracts 8–11 mm tall; plants 1.5–2(+) m tall at maturity.
    6 Pappus white; involucre of 14–17 bracts .... 3. *L. floridana*
    6 Pappus brown; involucre of 23–25 bracts .... 1. *L. biennis*

1. **Lactuca biennis** (Moench) Fern., blue wood lettuce. Herbaceous biennial 1.5–2 m tall; stem glabrous; latex white. Basal and lower cauline leaves sessile, variously pinnatifid to deeply pinnate-lobed with a broadly triangular apex, dentate, base attenuate, sagittate, often with trichomes on abaxial and adaxial surfaces; middle and upper cauline leaves lanceolate, pinnate-lobed to entire, base attenuate, sagittate, glabrous throughout. Inflorescence a diffuse conical panicle of 50–100+ heads. Heads cylindric; involucral bracts ca 24, 8 mm long at anthesis, 11 mm long in fruit; florets 25–40; ligule light blue with purple apex, 4 mm long; corolla tube 3 mm long without a ring of hairs at the summit; anthers blue, exserted ca 1 mm beyond anther tube. Achene body light brown, 4–5 mm long, 1.5–2 mm wide, with 4 or 5 conspicuous ridges on the abaxial and adaxial surfaces; beak brown, stout and 1 mm long, or absent; pappus brown, in 3 or 4 series, each bristle 5–6 mm long, 3 or 4 cells thick, short outer pappus present. (n = 17) Aug–Sep. Found in forest clearings, & along streams & lakes in rich, moist soil; rare in GP; w MN, e ND, e SD, & BH; (Newf. & NC, w to IA, SD, WY, cen B.C., s to CA). *L. spicata* (Lam.) Hitchc.—Rydberg.

2. **Lactuca canadensis** L., wild lettuce. Herbaceous biennial (0.5)1–2(3) m tall; stem glabrous or rarely hirsute; latex brown. Basal and lower cauline leaves highly variable, sessile, often falcate, linear-ovate, obovate to ovate, variously weakly dentate or pinnatifid to deeply pinnate-lobed, margins ciliolate, base sagittate to cuneate, 20–30 cm long, 5–15 cm wide, abaxial midrib often hirsute; middle and upper cauline leaves linear-ovate, obovate to ovate or lanceolate, variously subentire to pinnate-lobed, margin ciliolate or rarely

glabrous, base sagittate to cuneate, abaxial midrib glabrous. Inflorescence a diffuse, conical panicle of 50–100(+) heads. Heads cylindric, involucral bracts 17, 8–9 mm long at anthesis, 11–12 mm long in fruit; florets 17–22; ligule reddish at the apex and becoming yellow at the base or entirely yellow, 3.5–4 mm long; corolla tube 4–4.5 mm long with a ring of hairs at the summit; anthers yellow, exserted ca 2 mm; style branches reddish, exserted ca 1 mm beyond anther tube. Achene body dark brown, 3(4) mm long, ca 2 mm wide, with 1 conspicuous ridge on the abaxial and adaxial surfaces; beak filiform, 2–2.5 mm long; pappus white, in 3 or 4 series, each bristle 5–6 mm long, 3 or 4 cells thick. (n = 17) Jul–Sep. Common in forest clearings & margins; MN, MO, IA, ND; SD: Lawrence, Roberts, Union; NE, KS, OK; (N.S. & n FL w to Man. & e TX). *L. canadensis* var. *latifolia* O. Ktze., var. *obovata* Wieg.—Fernald; *L. sagittifolia* Ell.—Gleason & Cronquist.

Two geographically distinct phases based on flower color exist in this species. In the GP and as far east as St. Louis the flower color is always reddish or orange, but farther east the flower color is brilliant yellow. The areas of sympatry of these two phases have not yet been determined.

3. *Lactuca floridana* (L.) Gaertn., Florida lettuce. Herbaceous biennial 1.5–2(3) m tall; stem glabrous; latex white. Basal and lower cauline leaves sessile, ovate, variously dentate to deeply pinnate-lobed, apex broadly triangular, base attenuate and sagittate, 17(+) cm long, 9(+) cm wide, abaxial veins hirsute; middle and upper cauline leaves ovate to lanceolate, variously dentate to pinnately lobed, apex broadly triangular to acute, base attenuate, sagittate, glabrous throughout. Inflorescence a diffuse panicle of 50–100+ heads. Heads cylindric; involucral bracts 14–17, 8–9 mm long at anthesis, 8–11 mm long in fruit; florets 11–16, ligule light blue or rarely white, 5.5–6 mm long; corolla tube 5 mm long with a ring of hairs at the summit; anther dark blue, exserted ca 3 mm; style branches light blue, exserted ca 1 mm beyond anther tube. Achene body dark brown, 4–5 mm long, 1.5–2 mm wide, with 4 or 5 conspicuous ridges on the abaxial and adaxial surfaces, beak brown, stout and 1 mm long, or absent; pappus white, in 3 or 4 series, each bristle 5–6 mm long, 3 or 4 cells thick, short outer pappus present. (n = 17) Jul–Oct. Common in forest clearings & along streams & lakes in rich, moist soil; e GP; IA, MO, KS; SD: Clay; OK; (ME s to FL, w to e ND & TX). *L. villosa* Jacq.—Rydberg.

This species has several phases based on variations in leaf morphology and flower color. The color is usually light blue, but occasional white flowers are found. This phase has been referred to as f. *leucantha* Fern. (Fernald, Rhodora 42: 498. 1940). The lower leaves always have a broadly triangular apex, but may either be pinnate-lobed or unlobed below the apex. The phase with unlobed leaves is the basis for var. *villosa* (Jacq.) Cronq. (Cronquist, Rhodora 50: 31. 1948).

4. *Lactuca ludoviciana* (Nutt.) Ridd., western wild lettuce. Herbaceous biennial 1–2 m tall; stem glabrous; latex brown. Basal and lowermost cauline leaves sessile, ovate to obovate, variously dentate or pinnatifid to deeply pinnate-lobed or lyrate, margins glabarous or rarely ciliate, base sagittate to clasping, 20–30 cm long, 5–10 cm wide, abaxial midrib often hirsute; middle and upper cauline leaves ovate to obovate, variously dentate or pinnate-lobed to sinuate, margin glabrous, base sagittate to clasping, abaxial midrib glabrous. Inflorescence a diffuse, conical panicle of 50–100(+) heads. Heads cylindric, involucral bracts 17, 13–15(17) mm long at anthesis, 18–20 mm long in fruit; florets 20–30; ligule blue, blue with yellow apex or yellow, 5–5.5 mm long; corolla tube 7–7.5 mm long with a ring of hairs at the summit; anthers white, exserted ca 3 mm; style branches white, exserted ca 1 mm beyond anther tube. Achene body dark brown, 4–5 mm long, 2–3 mm wide, with 1 conspicuous ridge on the abaxial and adaxial surfaces; beak filiform, 3–4 mm long; pappus white, in 3 or 4 series, each bristle 8–9 mm long, 3 or 4 cells thick. (n = 17) Jun–Sep. Common in prairies & open places; MN, IA; MO: Atchison; ND, SD, NE, KS, OK; MT: Richland; CO: Kit Carson, Yuma; TX; (cen Ont. & MS w to cen Sask. & w-cen TX).

This taxon has several phases defined by variations in flower color or leaf morphology. The blue-flowered phase has been called f. *campestris* (Greene) Fern.—Fernald. Most plants have the lower cauline leaves pinnate-lobed to sinuate and the upper cauline leaves ovate and unlobed to sinuate. This is probably an intermediate between the two more "distinct" but less common phases with all leaves obovate, unlobed, or all leaves pinnatifid.

5. ***Lactuca oblongifolia*** Nutt., blue lettuce. Herbaceous perennial, 0.3–1 m tall; stem glabrous, arising singly or rarely loosely clustered from a deeply taprooted rhizomatous caudex; latex white. Lower cauline leaves sessile, linear-ovate to oblong, variously pinnate-lobed to runcinate or entire, base sagittate to cuneate, 9–13 cm long, 1–3 cm wide, abaxial midrib glabrous; middle and upper cauline leaves linear or lanceolate to oblong, pinnate-lobed or entire. Inflorescence a corymbose panicle of 20–50 heads; involucral bracts 18, 12–13(15) mm long at anthesis, 15–17 mm long in fruit; florets 19–21; ligule blue to violet (white), 9–10 mm long; corolla tube 6 mm long with a ring of hairs at the summit; anthers white, exserted ca 5 mm; style branches blue, exserted ca 2 mm beyond anther tube. Achene body reddish, 4 mm long, 1–1.5 mm wide, with 4–6 conspicuous ridges on the abaxial and adaxial surfaces; beak whitish, stout, 1.5 mm long; pappus white, in 3 or 4 series, each bristle 8–10 mm long, 5–9 cells thick. (n = 9) Jun–Sep. Locally common in low, moist, open meadows; throughout GP; OK: Washita; MO: Atchison; (w Ont., MO w to s-cen AK & s-cen CA). *L. pulchella* (Pursh) DC.—Rydberg.

The populations of *L. oblongifolia* are usually very uniform, with blue or lavender flowers, but specimens have been collected from a population that had pure white flowers.

Three phases based on the leaf morphology have been identified, but few data on their distribution are available. The most common phase has oblong leaves with the lower cauline leaves possessing 1–3 pairs of linear lobes close to the leaf base. Of the two less common phases, one has totally linear leaves and the other has ovate-runcinate leaves.

6. ***Lactuca saligna*** L., willow-leaved lettuce. Herbaceous winter annual 0.5–1 m tall; stem whitish, glabrous; latex white. Basal and lower cauline leaves sessile, linear, entire to pinnate-lobed with 1 or 2 pairs of linear lobes near base, or pinnatifid, base sagittate, 15–25 cm long, 0.5–4 cm wide, abaxial midrib often hirsute; middle and upper cauline leaves linear, entire or pinnate-lobed with 1 or 2 pairs of linear lobes. Inflorescence a virgate panicle of 50–100 heads. Heads cylindric; involucral bracts 17, 7 mm long at anthesis, 12 mm long in fruit; florets 10–15; ligule yellow with a dark blue stripe on the abaxial side, 4.5–5 mm long; corolla tube yellow, 3–3.5 mm long with a ring of hairs at the summit; anthers yellow, exserted 2–2.5 mm; style branches yellow, exserted ca 1 mm beyond anther tube. Achene body light brown, 3 mm long, 1 mm wide, with 5–7 conspicuous ridges on the abaxial and adaxial surfaces; beak filiform, 4–4.5 mm long; pappus white, in 2 or 3 series, each bristle 4 mm long, 2 cells thick. (n = 9) Jun–Sep. Becoming frequent in disturbed habitats & waste areas; MO: Callaway, Stone; NE: Lancaster, Richardson; e to cen KS, OK: Oklahoma; (OH to e-cen KS; cen CA; Europe). *Introduced.*

Three phases of *L. saligna* based on leaf morphology occur throughout its range. The phase with all leaves linear and entire, referred to by Fernald as f. *ruppiana* (Wallr.) G. Beck, will often occur in the same population with the phase that has all leaves pinnatifid. The phase with lower cauline leaves pinnatifid and upper cauline leaves linear and entire may be found with one or both of the other phases.

7. ***Lactuca serriola*** L., prickly lettuce. Herbaceous winter annual 0.5–1.5 m tall; stem with stiff bristles on lower 1/3, latex white. Basal and lower cauline leaves sessile, obovate to ovate, variously dentate to deeply pinnate-lobed, base sagittate, 15(20) cm wide, abaxial midrib with stiff bristles; middle and upper cauline leaves lanceolate to lance-ovate, variously dentate to pinnate-lobed; abaxial midrib glabrous or rarely with stiff bristles. Inflorescence a diffuse conical panicle of 50–100(+) heads. Heads cylindric; involucral bracts 17, 7–8

mm long at anthesis, 11–12 mm long in fruit; florets 18–25; ligule yellow with dark blue stripe on abaxial side, 4.5–5 mm long; corolla tube 3–3.5 mm long with a ring of hairs at the summit; anthers yellow, exserted ca 2–2.5 mm; style branches yellow, exserted ca 1 mm beyond anther tube. Achene body light brown, 3 mm long, 1 mm wide, with 5–7 conspicuous ridges on the abaxial and adaxial surfaces and conspicuous bristles at the base of the beak; beak filiform, 4 mm long, 2 cells thick. (n = 9) Jun–Sep. Common in disturbed habitats & wastes areas; widely distributed in GP; especially se; (throughout U.S.; Europe). *Introduced. L. intergrata* (Gren. & Godr.) A. Nels.—Nelson, New Manual Rocky Mt. Bot. 596, 1909; *L. virosa*, sensu Rydb., *L. scariola* L.—Rydberg.

Three phases of *L. serriola* based solely on variations in leaf morphology occur throughout the range and in mixed populations. The phase with unlobed obovate leaves does not appear to be as common as the phase with deeply pinnate-lobed leaves. The third very common phase has the lower cauline leaves variously sinuate to deeply pinnate-lobed and the upper cauline leaves variously unlobed to sinuate.

67. LAPSANA L., Nipplewort

1. ***Lapsana communis*** L. Erect, branching to single-stemmed annual herb, 1.5–10(15) dm tall, hirsute to glabrescent; juice milky. Leaves alternate, petiolate, thin, ovate or rounded-obtuse, varously dentate, or the lowermost lyrate, 2.5–10 cm long overall and up to 7 cm wide, progressively shorter-petiolate upward, the uppermost blades distinctly narrowed. Heads few–numerous in a corymbiform to somewhat elongate paniculiform inflorescence; involucre 5–8 mm tall, cylindric to campanulate-spreading; principal involucral bracts ca 8, subequal, uniseriate, subtended by minute calyculate bracts; receptable naked; florets 8–15, all ligulate and fertile, corolla yellow. Achenes 3–5 mm long, ± terete to weakly flattened, curved, glabrous, with numerous nerves; pappus none. (n = 6, 7, 8) Jun –Sep. Open or lightly wooded waste sites; extreme ne GP (ND: Cass), but to be expected as a weed along the e edge GP; (widely naturalized in ne 1/4 U.S. & adj. Can.; Eurasia). *Adventive.* [tribe Lactuceae]

68. LEONTODON L., Hawkbit

1. ***Loentodon hispidus*** L. Scapose, fibrous-rooted perennial up to 5 dm tall, variously bristly-hispid to glabrous; juice milky. Leaves all basal, oblong-lanceolate, up to 2(+) dm long overall and 3 cm wide, subentire or dentate to pinnatifid. Head solitary, involucre 1–1.5 cm tall, or more at maturity; involucral bracts weakly imbricate in several series; florets all ligulate and perfect, corolla yellow; receptacle ± pitted. Achenes fusiform, 6–12 mm long, sometimes weakly beaked, scabrous; pappus of plumose, broad-based bristles with an outer series of short, stiff setae. (n = 7) May–Sep. Disturbed, fertile soil; to be expected in lawns and gardens along the e edge GP; (e U.S. & adj. Can.; Eurasia). *Adventive. L. hastilis* L. (incl. var. *vulgaris* W.J. Koch)—Fernald. [tribe Lactuceae]

The data are insufficient for assessing the present distribution of this weed in the GP, but it is infrequent and doubtfully persisting.

69. LEUCELENE Greene, White Aster

1. ***Leucelene ericoides*** (Torr.) Greene. Slender, low, branching, perennial herb 5–15 + cm tall, with numerous loosely clustered stems arising from a much-branched, creeping to erect, subligneous rootstock; herbage mostly hirsute to strigose, often granular. Leaves

of the lower stems reduced to mere scales; middle leaves thickish, linear-spatulate or subulate, 2-12(15) mm long and up to 2 mm wide, ascending or appressed, margins often ciliate, uppermost leaves becoming short, subulate bracts. Heads solitary at the ends of the upper branches, and the inflorescence appearing as a loose, corymbiform cluster because of the nearly uniform height of the branches; involucre 5-7 mm tall and 5-10 mm wide, hemispheric to subturbinate; involucral bracts imbricate in 4-7 series, spreading flat at maturity; receptacle flat, naked; ray florets 12-24, pistillate and fertile, ligule white, or rosy on withering, 2-3 mm long; disk florets perfect and fertile, corolla yellow, style branches flattish, with triangular, externally granular-scabrous appendages. Achenes 2-3 mm long, 5-nerved, prismatic or somewhat flattened at maturity, those of the ray florets slightly smaller; pappus of numerous scabrous capillary bristles, 4.5-5.5 mm long. (n = 8) May-Aug. Dry to drying, open sandy sites & rocky, eroding hillsides; sw 1/4 GP, ne to s-cen NE; (GP to TX & n Mex., w to UT & CA). Including *L. alsinoides* Greene — Rydberg; *Aster arenosus* (Heller) Blake — Barkley. [tribe Astereae]

Reference: Shinners, L. H. 1946. Revision of the genus *Leucelene* Greene. Wrightia 1: 82-89. This innocuous little plant enjoys a most complicated synonymy, cf. Shinners, op. cit.

70. LIATRIS Schreb., Gay-feather, Blazing Star

Perennial herbs with stems arising singly or clustered from a thickened, weakly spreading, erect or cormlike rootstock. Stems erect, strict or few-branched upward. Leaves alternate, simple, entire, variously punctate and/or rough, narrow and sessile or tapering to an ill-defined petiole; basal and lower cauline leaves the largest. Inflorescence an elongate spike or raceme; heads variously short pedunculate to sessile, always discoid; involucral bracts imbricate, greenish to pinkish-purple, margins often scarious; receptacle flat, naked; florets all tubular and perfect, pink to pink-purple or occasionally white, style branches elongate-clavate, with an obtuse apex, stigmatic lines inconspicuous. Achenes 10-ribbed; pappus of numerous stout barbellate or distinctly plumose bristles. [tribe Eupatorieae]

References: Gaiser, L. O. 1946. The genus *Liatris*. Rhodora 48: 165 et seq.; Menhusen, B. R. 1963. Variation in the Punctatae series of the genus *Liatris*. Unpublished Ph.D. dissertation, Univ. Kansas, Lawrence.

Liatris forms a conspicuous group with often ill-defined species. Apparently, there is much hybridization, and the hybrid populations persist and back-cross with the parental types. Field workers should make every effort to collect the rootstocks, for herbarium studies suggest that the nature of the underground structures may be highly significant in understanding the genus, cf. the note under *L. mucronata*.

1 Pappus plumose to the naked eye, the lateral cilia many times longer than the diam of the bristle.
  2 Florets 10-60 per head, corolla lobes pubescent within .... 7. *L. squarrosa*
  2 Florets 3-8 per head, corolla lobes glabrous.
    3 Rootstock expanded into a globose corm .... 4. *L. mucronata*
    3 Rootstock elongated into a thickened taproot .... 5. *L. punctata*
1 Pappus merely barbellate, the lateral cilia only 3-4× longer than the diam of the bristles.
  4 Heads large, broadly cylindric to hemispheric, with (16)20-70+ florets.
    5 Heads all about the same size, with 16-35 florets, corolla hairy within .... 1. *L. aspera*
    5 Heads variable in size, the terminal head distinctly larger; heads with 30-70+ florets, corolla glabrous within .... 3. *L. ligulistylis*
  4 Heads small, cylindric to turbinate or weakly campanulate, with 4-18 florets.
    6 Involucral bracts with acuminate, spreading tips .... 6. *L. pycnostachya*
    6 Involucral bracts rounded and appressed at the apex, becoming erose .... 2. *L. lancifolia*

1. Liatris aspera Michx. Perennial 4-10(12) dm tall, short-pubescent, glabrate or glabrous throughout. Leaves reduced upward; lowermost petiolate but becoming sessile, 5-40 cm long and 0.6–4 + cm wide. Inflorescence elongate and spikelike, heads sessile or on short peduncles; heads campanulate to subhemispheric, 8–15 mm tall, involucral bracts loosely spreading, often purplish, with distinct scarious margins which may be lacerate to erose or wavy; florets 16–35, corolla hairy within. Pappus strongly barbellate. (n = 10) Aug–Oct. Dry open places, especially open woods in sandy sites; e 1/2 GP, cen ND s to cen OK; (s Can. to SC, w to GP). *L. scariosa* Willd.—Gates.

2. **Liatris lancifolia** (Greene) Kittell. Perennial (2)4–8(+) dm tall, glabrous or nearly so. Leaves broadly linear, reduced upward, the best-developed 10–30 cm long and up to 1.5 cm wide, tapering equally toward the base and toward the ± obtuse tip. Inflorescence spikelike, with numerous sessile heads; heads cylindric to narrowly campanulate, 7–11 mm tall; involucral bracts erect and not spreading, the tip rounded or obtuse, scarious-margined, often purplish; florets ca 12, corolla glabrous or nearly so within. Pappus barbellate. (n = 10) Jul–Sep. Meadow lands & open slopes, apparently tolerant of saline conditions; sw 1/4 GP, e to KS: Stafford; n to WY & se SD; (GP s & w, TX & NM).

This species is similar to *L. spicata* (L.) Willd. of the eastern third of the U.S., and the name has sometimes been listed in synonymies under that species.

3. **Liatris ligulistylis** (A. nels.) K. Schum. Perennial 2–10 dm tall, glabrous, especially above, but sometimes irregularly and lightly pubescent. Leaves very numerous, the larger ones 8–25 cm long and 0.5–4 cm wide. Inflorescence spikelike, heads short-pedunculate to nearly sessile, terminal head conspicuously the largest; heads hemispheric, 13–20 mm tall, involucral bracts weakly if at all spreading-squarrose, margins scarious and sometimes lacerate; florets 30–70 + , corolla glabrous within. Pappus barbellate. Jul–Sep. Open, often moist sites; ND, e 1/2 SD & BH; (scattered; s Can. to MO, w to CO & NM).

4. **Liatris mucronata** DC. Perennial 1.5–6 dm tall, glabrous or nearly so. Stems arsing singly or sometimes several clustered from a prominent, globose woody corm, 2–4 cm across, either just at or beneath the surface of the ground. Leaves numerous, somewhat overlapping and ascending, 5–8 cm long and 0.2–0.5 cm wide. Inflorescence spikelike, heads cylindric or weakly campanulate, 1.2–1.8 cm tall, involucral bracts mucronate or cuspidate; florets 3–6, corolla lightly pilose. Pappus distinctly plumose. (n = 20, 30) Aug–Oct. Dry, open uplands, especially in loose soil; se NE, e 1/2 KS, cen OK; isolated in sw KS & adj. OK; (TX & Coah. n to GP, e in MO).

This species is similar to *L. punctata* and sometimes not separated from it as a distinct entity. However, the presence of the globose corm is clearly distinctive. J. A. Steyermark (Fl. Missouri, p. 1475, 1963) notes that the corm is a persistent structure, apparently not attributable to environmental factors, and it is, therefore, of genetic consequence. Further study is needed to elucidate the relationship between *L. mucronata* and *L. punctata*.

5. **Liatris punctata** Hook. Perennial 1–8 dm tall, glabrous or nearly so. Stems arising singly or clustered from an erect or weakly spreading, taprootlike rootstock. Leaves numerous, overlapping and arching upward, punctate, linear or a little wider, the largest up to 15 cm long and 0.5 cm wide. Inflorescence spikelike; heads cylindric to weakly campanulate, 1.5–2 cm tall, involucral bracts narrow, long-acute to acuminate; florets 4–8, corolla pilose inside. Pappus distinctly plumose. (n = 10, 20, 30) Jul–Oct. Dry, open upland sites, especially in sandy soils; abundant throughout the GP; (MI to AR, w to Alta. & NM).

Our materials generally have slightly broader, stiffer, and more ciliate leaves (3–5 mm across), and are shorter in stature than the more eastern materials.

6. **Liatris pycnostachya** Michx. Stiff, tall perennial 5–15 dm tall, hirsute at maturity, at least above. Leaves numerous, linear, reduced upward, the best developed 10–40 + cm long and up to 1 cm wide. Inflorescence a long spike of sessile heads; heads cylindric or nearly

so, 8–11 mm tall; involucral bracts acuminate with a spreading-recurved tip; florets 5–7(12), corolla glabrous or nearly so within. Pappus strongly barbellate. (n = 10, 20) Jul–Sep. Open damp prairies; e 1/3 GP, w to cen KS & NE; (KY sw to LA & TX, w to GP).

7. *Liatris squarrosa* (L.) Michx. Perennial 3–6 + dm tall, glabrous or pubescent. Stems often several, arising from a prominent corm. Leaves linear or broadly linear, reduced upward, the best developed 10–25 cm long and 4–12 mm wide. Inflorescence variously compact-racemose to spikelike. Heads often few or even solitary, pedunculate or sessile, 12–25(+) mm tall; involucral bracts with squarrose or loose, spreading acuminate tip; florets (10)25–40 or up to 60 in the large terminal head, corolla pubescent within. Pappus plumose. (n = 10) Jul–Sep. Dryish upland prairies & open woods; (MD to FL, w to SD & TX).

This is a complicated species with several recognizable varieties, some of which have been treated as separate species. Two varieties occur in our area. Var. *squarrosa* occurs only from MO eastward.

7a. var. *glabrata* (Rydb.) Gaiser. The glabrous phase, scattered in sandy sites through the cen GP from s SD, s to cen OK & TX, mostly w of var. *hirsuta. L. glabrata* Rydb.—Rydberg; *L. compacta* (T. & G.) Rydb.—Rydberg.

7b. var. *hirsuta* (Rydb.) Gaiser. Permanently pubescent, at least among the heads in the inflorescence; mostly on compact, calcareous soils. GP from nw IA s to se KS. *L. hirsuta* Rydb.—Atlas GP.

71. LINDHEIMERA Gray & Engelm.

1. *Lindheimera texana* Gray & Engelm. Taprooted annual 1–2(3) dm tall, usually with a single, simple or few-branched stem with coarse, spreading pubescence. Leaves alternate, crowded over most of the stem, obovate to ovate or oblanceolate, well-developed ones 10–20 cm long, coarsely toothed on the upper margin, sessile or nearly so. Heads solitary on peduncles 1–3(4) cm long; involucre 10–15 mm tall, ± hemispheric; involucral bracts in 2 series of 4 or 5 bracts each, outer involucral bracts narrower and more acute than the foliaceous inner bracts; receptacle chaffy throughout and divisable into an outer ring where the ray florets are attached and a central raised area, 1–2 mm tall, where the disk florets are attached; ray florets ca 5, pistillate and fertile, ligule 1 + cm long, yellow; disk florets staminate only. Ray achenes large, flattened at right angles to the radius of the head, and each appressed to an inner involucral bract, wing-margined, with the shoulders of the wings extending upward into a pappus of 2 awns. (n = 8) Mar–May. Open prairies of sw OK; (sw GP through TX & n Mex.). [tribe Heliantheae]

72. LYGODESMIA D. Don, Skeletonweed

Herbaceous perennials arising from a rhizomatous rootstock; stems green, usually striate, little to much branched. Leaves alternate, mostly linear, entire or sparingly laciniate, forming a basal rosette in *L. texana;* cauline leaves reduced upward. Heads solitary at the ends of the branches; involucre cylindrical or nearly so; principal involucral bracts subequal, weakly imbricated, linear, grayish-green with narrow, hyaline margins; calyculate bracts much shorter, in 1–3 weakly defined series; receptacle flattish, scabrous, pitted with achenial scars; florets 5–12, all ligulate and fertile; ligule pink to lavender or white. Achenes columnar to cylindric, not beaked; pappus of numerous, fine capillary and nonplumose bristles. [tribe Lactuceae]

Reference: Tomb, A. S. 1980. Taxonomy of *Lygodesmia*. Syst. Bot. Monographs 1: 1–51.
The species long called *Lygodesmia rostrata* A. Gray is here referred to *Shinnersoseris,* q.v.

1 Lowermost leaves forming a persistent basal rosette; n TX & sw OK .... 3. *L. texana*
1 Lowermost leaves cauline and not forming a basal rosette.

2   Heads large; florets usually 9 per head, involucre 1.8–2.1 cm tall, ligule 16–19 mm long; e WY .... 1. *L. grandiflora*
2   Heads smaller; florets usually 5 per head, involucre 1.3–1.6 cm tall, ligule 10–12 mm long; widespread .... 2. *L. juncea*

1. ***Lygodesmia grandiflora*** (Nutt.) T. & G. Herbaceous perennial 1.1–4.5 dm tall, with white milky juice, arising from prominent vertical rhizome-bearing roots; rhizomes covered with a thin, dark brown periderm. Stems paniculate-branched, glabrous and green. Lower leaves not rosette-forming, linear, 5–16 cm long and 2–4 mm wide, thickish; cauline leaves similar but progressively reduced upward, the uppermost reduced to mere bracts. Heads terminal; involucre 1.8–2.1 cm tall, cylindrical; principal involucral bracts 8 or 9, linear, outer surface scabrate, apex ciliate and with a small keellike appendage; calyculate bracts subulate and apparently in 1 series; florets (6)9(12), ligulate and fertile, corolla pink to lavender; ligule 16–19 mm long. Achenes subcylindric, 10–13 mm long, obscurely angled, striate, somewhat expanded on the outer (abaxial) surface; pappus of numerous capillary bristles, often weakly connate at the base, equaling the body of the achene in length. (n = 9) Jun–Jul. Loose, gravelly or sandy, alkaline sites; barely entering the GP in e WY; (WY, cen & w CO, UT).

2. ***Lygodesmia juncea*** (Pursh) Hook. Herbaceous perennial 1.3–7 dm tall, with yellow milky juice, arising from a woody, erect or horizontal rhizome-bearing rootstock. Stems erect or semidecumbent and arching upward, glabrous or glaucous, intricately branched from the base, often bearing numerous spherical galls, ca 1 cm across, produced by solitary wasps. Leaves few, the lower ones linear to linear-lanceolate, usually shorter than 4 cm; cauline leaves reduced to mere subulate scales. Heads numerous, terminal; involucre 1.3–1.6 cm tall, cylindrical; principal involucral bracts 5–7, linear, subtended by ca 3 series of short, linear calyculate bracts; florets mostly 5, ligulate and fertile, corolla pink to lavender or sometimes whitish, ligule 10–12 mm long. Achenes cylindrical, 6–10 mm long, obscurely striate; pappus of numerous capillary bristles, 6–9 mm long. (n = 9) Jun–Sep. Open high plains & prairies, often in barren alkaline sites; throughout GP except se 1/4 KS & adj. MO & OK; (GP to B.C., NV, & cen NM).

    This species rarely produces seeds, and it maintains itself largely by asexual means, cf. Tomb, op. cit.

3. ***Lygodesmia texana*** (T. & G.) Greene ex Small. Herbaceous perennial 2.5–6.5 dm tall, with white milky juice, arising from a thick, fleshy, branching rhizomatous root. Stems erect, glabrous, few-branched. Basal leaves linear, entire or weakly laciniate, 10–20 cm long and up to 5 mm wide, forming a persistent basal rosette; cauline leaves similar but few and reduced, the uppermost bractlike. Heads terminal, large; involucre 1.8–2.5 cm tall, cylindrical; principal involucral bracts 8–10, linear, with a keellike appendage near the apex; calyculate bracts subulate, ciliate, in ca 2 series; florets 8–12, ligulate and fertile, corolla lavender or rarely white. Achenes subcylindric, 1.3–1.7 cm long, smooth, somewhat expanded below on the outer (abaxial) surface; pappus of sordid capillary bristles, 10–15 mm long. (n = 9) May–Sep. Open or lightly wooded, rocky, calcareous sites; panhandle region of n TX & adj. sw OK; (s GP, e NM s through TX to n Mex.).

73. MACHAERANTHERA Nees

Taprooted annuals, biennials or perennials, the latter often with a woody caudex. Leaves alternate, entire or spinulose-dentate to pinnatifid. Heads few to numerous in open cymose-corymbiform clusters; involucre of several series of imbricated involucral bracts, these often chartaceous or stramineous at the base, but greenish toward the tip, squarrose or spreading;

receptacle naked but sometimes alveolate; ray florets pistillate and fertile, ligule prominent, white or pinkish to purple-blue, or ray florets sometimes absent (or greatly reduced?), disk florets bisexual and fertile, style branches with submarginal stigmatic lines and prominent, externally pubescent appendages. Achenes variously pubescent to glabrate, mostly with several nerves; pappus a series of unequal capillary bristles, often brownish or sordid. [tribe Astereae]

Reference: Cronquist, A., & D. D. Keck. 1957. A reconstitution of the genus *Machaeranthera*. Brittonia 9: 231–239.

Machaeranthera includes some 30 species in temperate North America and nearly all have at one time or another been included in *Aster* and/or *Haplopappus*, or some segregate genus. There is no shortage of published opinion on the generic disposition of the various species-alliances of the complex, but no comprehensive monographic treatment is yet available. The treatment offered here for our few species is therefore provisional and traditional.

Xylorhiza glabriuscula is keyed here for convenience.

1 Leaves, or at least the well-developed ones, pinnatifid or bipinnatifid ..... 4. *M. tanacetifolia*
1 Leaves subentire, dentate or spinulose, but not pinnatifid.
   2 Perennials with a stout, woody taproot, often surmounted by a branching caudex.
      3 Heads numerous in branching inflorescences, discoid or ray florets rarely present ..... 2. *M. grindelioides*
      3 Heads solitary at the ends of the stems, radiate; caudex broad and distinctly woody ..... *Xylorhiza glabriuscula* (q.v.)
   2 Annuals or biennials (or short-lived perennials?) from an elongate, slender taproot.
      4 Leaves cinereous-puberulent, at least beneath, herbage often glandular ..... 1. *M. canescens*
      4 Leaves ± glabrate at maturity, or but lightly spreading-pubescent ..... 3. *M. linearis*

1. **Machaeranthera canescens** (Pursh) A. Gray, hoary aster. Annual or biennial, (perhaps perennial) from a taproot, 1–4(6) dm tall. Stems often several, arching upward from the base, herbage often glandular-canescent. Leaves subentire to toothed, cinereous-puberulent, at least beneath, the prominent leaves linear-oblanceolate to subspatulate, 1–7(10) cm long, mostly less than 5 mm wide, the upper leaves reduced to mere linear bracts. Heads numerous; involucre 6–10 mm tall, canescent or glandular (or both); involucral bracts conspicuously imbricate in several series, with short, green, ± squarrose tips, occasionally somewhat reddish; ray florets 8–25, ligule bright bluish-purple to pink, 5–13 mm long, or rarely the ray florets absent; disk florets with corolla 5–6 mm long. Achenes 3–4 mm long; pappus of capillary bristles. (n=4) Jul–Oct. Open, dry sites; nw 1/4 GP; (GP w to B.C., CO, NM, & CA). *M. ramosa* A. Nels.—Rydberg; *Aster canescens* Pursh—Stevens.

2. **Machaeranthera grindelioides** (Nutt.) Shinners, goldenweed. Herbaceous perennial or weak subshrub, 1–3 dm tall, from a taprooted, woody caudex; herbage tomentose to nearly glabrous. Leaves oblong to oblong-spatulate, dentate to serrate with spinulose teeth, thickish, 2–3.5 cm long and 4–10 mm wide. Heads discoid and numerous, usually in a loose cyme at the ends of the branches; involucre 7–10 mm tall and 8–10 mm across; involucral bracts imbricate in ca 3 series, mostly green; ray florets absent (or sometimes rudimentary?), or reputedly rarely present and then white; disk florets with corollas 5–8 mm long. Achenes 2.5–3 mm long, sericeous canescent; pappus reddish to tan. (n=8) Jun–Sep. Open, dry sites; nw 1/4 GP from nw NE northwestward; (GP to Alta., MT, NM, & AZ). *Haplopappus nuttallii* T. & G.—Harrington; *Sideranthus grindelioides* (Nutt.) Britt.—Rydberg.

3. **Machaeranthera linearis** Greene. Annual or biennial herb, (1.5)3–6 dm tall. Stems arising singly or often few-branched from a taproot, the lower stems often reddish-tinged; herbage glabrate to strigose or puberulent, slightly if at all glandular. Leaves oblanceolate

or linear-oblanceolate to spatulate, dentate to subentire, 3–6(8) cm long and up to 10 mm wide, glabrous to puberulent but never cinereous, lower leaves early deciduous. Heads in open corymbiform clusters; involucre 7–12 mm tall, hemispheric; involucral bracts imbricated in several series, narrow, chartaceous-stramineous below, green at the tip, ± squarrose; ray florets numerous, ligule 7–14 mm long, blue to purple or white; disk florets with corolla ca 5 mm long. Achenes 3–4 mm long, sparingly short-pubescent; pappus of stiffish, ± sordid capillary bristles. (n = 4) Aug–Oct. Open, sandy dry sites, especially in disturbed areas; w edge GP, but e into sandhills of NE & adj. SD, also in extreme sw KS; (GP to UT & AZ, elsewhere?). *M. sessiliflora* (Nutt.) Greene — Rydberg; *Aster rubrotinctus* Blake — Harrington (as to circumscription).

This species is in need of further study. Most populations have blue or purplish ray florets, but those in sw KS have exclusively white rays. The relationships between this species and the allied *M. canescens* (Pursh) A. Gray and *M. commixta* Greene (a cordilleran species) are yet to be clarified.

4. *Machaeranthera tanacetifolia* (H.B.K.) Nees, tansy aster, tahoka daisy. Annual 1–5 dm tall, with stems arising singly or branched from a taproot, well-developed plants bushy branched; herbage glandular, puberulent and villous. Leaves, or at least the larger ones, pinnatifid to bipinnatifid, up to 7 cm long, the segments linear to oblong-lanceolate and terminating in a callous bristle. Heads solitary at the ends of the upper branches, but collectively forming a loose corymbiform cluster; involucre 8–12 mm tall; involucral bracts chartaceous below, green at the tip, spreading or reflexed; ray florets numerous, ligule blue to purple, 1–2 cm long; disk florets with corolla 4–5 mm long. Achenes appressed-pubescent; pappus of stiffish capillary bristles. (n = 4) Jul–Sep. Open dry sites; w 1/5 GP from e MT to n TX, scattered e to cen KS; (GP to Alta., MT, s to AZ, TX, & n Mex.).

This is a highly distinctive species, sometimes cult. as an ornamental.

74. MADIA Mol., Tarweed

1. *Madia glomerata* Hook. Strong-scented annual 1–5 + dm tall, ± appressed-pubescent and unevenly stipitate-glandular. Leaves opposite below but otherwise alternate, linear to narrowly lanceolate, entire, 2–7 cm long and 1–4 mm wide. Heads mostly in small clusters, involucre fusiform, 6–9 mm tall and 2–5 mm wide; involucral bracts mostly 4, herbaceous, about equal in length, clasping the achenes, and the heads thus appearing deeply furrowed; receptacle convex, chaffy near the edge; ray florets 1–3 or sometimes absent, fertile, ligule small, ca 2 mm long, yellow; disk florets perfect but sometimes sterile, style branches flat with acute, pubescent appendages. Achenes angled, or ray achenes flattened; pappus none. (n = 14) Jun–Aug. Open disturbed sites; infrequent in the nw 1/4 GP; (native to the w cordilleran regions & adventive eastward as a weed). [tribe Heliantheae]

75. MARSHALLIA Schreb.

1. *Marshallia caespitosa* Nutt. ex DC. Subscapose, fibrous-rooted annual or short-lived perennial, mostly 2–5 dm tall. Stems arising singly or clustered from a short crown, somewhat hairy, especially upward. Leaves firm, clustered near the base, linear-lanceolate to linear-elliptic, 5–15 cm long and up to 1 cm wide or the lowermost to 1.5 cm wide, tapering to an indistinct petiole. Heads typically solitary, discoid, 2.5–3.5 cm across at maturity; involucral bracts narrow, subequal, in 1 or 2 series, ± herbaceous; receptacle convex, not alveolate, chaffy; florets all tubular and fertile, style branches flattened, elongate with a short, hairy appendage; corolla white or sometimes pinkish, slender, hairy and conspicuously

lobed; chaffy bracts acute to weakly mucronate. Achene 5-angled and 10-ribbed; pappus of 5 scarious, translucent scales 2–2.5 mm long, equaling or exceeding the achene. May–Jun. Open, drying meadows & plains; extreme se KS, adj. MO & ne OK; (se GP to LA & TX). [tribe Heliantheae]

Our materials are referable to var. *caespitosa*. Var. *signata* Beadle & Boynt. occurs farther south in TX and is distinguished by possessing abundant cauline leaves and often several heads.

76. MATRICARIA L.

Ours mostly low, weedy annuals with leaves alternate, pinnatifid to finely divided. Inflorescence of single heads or with heads in a corymbiform cluster; heads radiate or discoid; involucral bracts dry, in 2 or 3 series, margins weakly scarious; receptacle naked, hemispheric to conic; ray florets pistillate and usually fertile, white; disk florets yellowish, corolla 4- or 5-toothed; style branches flattened, truncate, penicillate. Achenes generally ribbed on the lateral margins and ventrally (i.e., on the inner face) but smooth dorsally (i.e., the outer face), variously glabrous or roughened; pappus a short crown or none. [tribe Anthemideae]

1 Ray florets present; disk corollas 5-toothed.
  2 Pleasant-scented; achenes ± 5-ribbed .... 1. *M. chamomilla*
  2 Odorless; achenes strongly 3-ribbed .... 2. *M. maritima*
1 Ray florets absent; disk corollas 4-toothed .... 3. *M. matricarioides*

1. ***Matricaria chamomilla*** L., false chamomile. Pleasant-scented glabrous weedy herb, 2–5(8) dm tall. Leaves bipinnatifid, segments linear, becoming filiform; well-developed leaves 3–6 cm long. Inflorescence of numerous heads; disk 6–10 mm across, receptacle conic; ray florets 10–20, white, the ligule up to 10 mm long; disk corolla 5-toothed. Achenes with 2 marginal and 3 ventral ribs, prominent but not strongly winglike; pappus a short crown or absent. (n = 9) May–Aug. Damp weedy sites, roadsides, & old gardens; scattered in ND, apparently in KS: Riley; potentially present in the GP as an adventive weed; (across N. Amer.; Eurasia). *Naturalized.*

2. ***Matricaria maritima*** L., wild chamomile. Odorless weedy herb, 1–5+ dm tall, glabrous or nearly so, annual but possibly perennating under favorable conditions. Leaves bipinnatifid, segments becoming linear-filiform or sometimes broader; well-developed leaves 4–8 cm long. Inflorescence of numerous heads, disk 8–15 mm across, receptacle hemispheric; ray florets 12–25, ligule up to 12 mm long; disk corolla 5-toothed. Achenes with 2 marginal and 1 prominent ventral rib, strongly developed and nearly winglike; pappus a low crown. (n = 9, 18) Jun–Sep. Roadsides & other weedy sites; scattered in n GP; n & e ND, s NE, & cen KS; (across N. Amer.; Eurasia). *Naturalized. Chamomilla inodora* (L.) Gilib.—Rydberg.

3. ***Matricaria matricarioides*** (Less.) Porter, pineapple weed. Pleasantly pineapple-scented, glabrous, low branching annual weed, 5–30 cm tall. Leaves well developed and numerous, 1–5 cm long including the short petiole, 1–3 times pinnatifid, segments usually short-linear. Inflorescence of numerous heads, or sometimes with a single, prominent terminal head, eriadiate; disk 5–10 mm across, receptacle strongly conic and pointed; involucral bracts with broad scarious margin; disk corolla 4-toothed. Achenes with 2 well-developed marginal ribs and 1 or more less well-defined ventral ribs; pappus a short crown or obscure. (n = 9) May–Aug. Roadsides, gardens, & poorly managed lawns; scattered in GP, especially in e 1/4 & across ND; (across N. Amer., especially Pacific Coast states; Eurasia). Widely *naturalized. Chamomilla suaveolens* (Pursh) Rydb.—Rydberg.

77. MELAMPODIUM L.

1. *Melampodium leucanthum* T. & G., black-foot daisy. Weakly suffrutescent perennial herb 15–50 cm tall, strigose-strigillose or rarely hispid with abudant short hairs. Leaves opposite, sessile, linear-oblong, 2–4.5 cm long and up to 1 cm wide, entire to pinnately lobed. Heads solitary on peduncles 3–7 cm long; involucre 6–8 mm tall; involucral bracts in 2 series, outer involucral bracts ovate, entire-margined and fused for 1/2–3/5 their lengths; inner involucral bracts each enclosing a single ray-achene and expanded upward into a blunt hood; receptacle chaffy; ray florets 8–13, pistillate and fertile, ligule 7–13 mm long, creamy white; disk florets 25–50, functionally staminate, corollas yellow; chaffy bracts oblong, ca 2.5 mm long, scarious, with prominent midrib. Ray achenes ± compressed parallel to the radius of the head, 1.5–2.5 mm long, lateral surfaces somewhat spiny or warty; pappus none. (n = 10, 20) May–Sep. Open limestone soils in the high plains; sw 1/4 GP, from sw KS & se CO southward; (sw GP to cen TX, w to AZ & n Mex.). [tribe Heliantheae]

Reference: Stuessy, T. F. 1972. Revision of the genus *Melampodium* (Compositae: Heliantheae). Rhodora 74: 1–70, 161–219.

78. MICROSERIS D. Don.

1. *Microseris cuspidata* (Pursh) Sch.-Bip. Taprooted scapose perennial, 1–3 dm tall, villous-tomentose upward or glabrate. Leaves all basal, narrow, 1–3 dm long overall and up to 2 cm wide, entire, long-accuminate, margins often weakly curled-crisped and ciliolate. Head solitary; involucre 2–2.5 cm tall; involucral bracts subequal or but weakly imbricated, linear-lanceolate to lanceolate, inner ones and sometimes the outer ones conspicuously acuminate to attenuate, sometimes speckled; receptacle naked; florets all ligulate and fertile, corolla yellow. Achenes 8–10 mm long, slightly tapering upward but not at all beaked, prominently 8- to 10-nerved; pappus of abundant capillary bristles intermixed with numerous slender, attenuate scales. Apr–Jun. Dry or drying open prairies & plains, often on loose, gravelly soil; infrequent in the s but locally common in the nw. GP; (WI to IL w to e edge of Rocky Mts.). *Nothocalais cuspidata* (Pursh) Greene—Rydberg; *Agoseris cuspidata* (Pursh) Raf.—Fernald. [tribe Lactuceae]

Microseris nutans (Hook.) Sch.-Bip. is an essentially cordilleran species to be expected along the nw edge of our range in WY and MT. It is caulescent with several prominent cauline leaves, and it has only 15–20 pappus members, each bearing a long, white plumose bristle.

79. ONOPORDUM L., Scotch Thistle

Contributed by Ronald L. McGregor

1. *Onopordum acanthium* L., Scotch thistle. Biennial herbs from stout, fleshy taproots, or rarely flowering first year. Stems 0.5–3 m tall, simple or usually much branched above, sparsely to very densely tomentose, winged by decurrent leaf bases; wings 5–15(20) mm wide, sinuate-dentate lobed, lobes terminating in sharp spines to 10 mm long, usually with smaller spines between longer ones, wings sparsely to densely tomentose. Rosette leaves oblong to oblanceolate in outline, sparsely to densely tomentose, 1–5 dm long, 0.5–1.5(2) dm wide, with irregular large triangular lobes, these with smaller lobes, tips of lobes acute or with weak spines 1–5 mm long; cauline leaves alternate, sessile, conspicuously decurrent; blades oblong to oblanceolate in outline, 0.5–3 dm long, 0.5–2 dm wide, triangular lobed or toothed, with spines 5–15(20) mm long, tomentose and with sessile glands, often becoming

glabrate. Heads solitary and terminating branches or 2-5 in ± loose clusters; heads subglobose, 2.5-5 cm across; outer involucral bracts linear-subulate, floccose and glandular to glabrate, tipped by semirigid spines 2-5 mm long; receptacle flat, fleshy, alveolate with deep pits and with short bristles on the partitions; ray florets absent; disk florets all fertile; corolla purplish or pinkish-white, 20-25 mm long, tube slender, throat dilated, limb 5-cleft; filaments of stamens pilose, anthers sagittate at base. Achenes 4-5 mm long, marbled grayish-black, transversely rugulose, 4-angled; pappus of slender barbellate bristles 6-10 mm long. (n = 17) Jun-Aug. Infrequent but locally common to abundant in feed yards, pasture ravines, around ponds, roadsides, & waste places; scattered in e WY; e CO: Kit Carson; cen NE, cen KS, w OK, n TX; (found sparingly over much of U.S.; Eurasia). [tribe Cynareae]

Locally, this is a serious weed. *O. tauricum* Willd. is a similar weed which has been found in CO: Pueblo. It differs in having herbage glandular-puberulent, but not tomentose.

80. PALAFOXIA Lag.

Taprooted annuals with stems ascending, few to much branched above. Leaves alternate, firm-membranaceous, petiolate, or the upper ones subsessile. Heads terminal in open, corymbiform clusters; involucre ± hemispheric or turbinate to subcylindric; receptacle flat, naked; involucral bracts apparently in 2 (or 3) series, subequal, thickish, and green or sometimes pinkish toward the apex; ray florets pistillate and fertile, ligule pink or roseate, conspicuously 3-toothed, or ray florets absent; disk florets perfect and fertile, corollas all alike (in our species), style branches linear, spreading hispidulous. Achenes 4-angled, obpyramidal; pappus of 7-10 scarious scales. [tribe Heliantheae]

Reference: Turner, B. L. & M. I. Morris. 1976. Systematics of *Palafoxia* (Asteraceae: Helenieae). Rhodora 78: 567-628.

1 Heads with disk florets only .... 1. *P. rosea*
1 Heads with conspicuous ray florets present .... 2. *P. sphacelata*

1. ***Palafoxia rosea*** (Bush) Cory. Annual herb 1-5 dm tall, somewhat scabrous. Leaves alternate, but the earliest sometimes opposite, mostly narrowly lanceolate, 3-6 cm long and 2-6 mm wide, inconspicuously 3-nerved; petiole 3-15 mm long. Heads terminal on long peduncles; involucre broadly to narrowly turbinate; involucral bracts linear to obovate, 5-10 mm long and 1.2-3.0 mm wide; ray florets absent; disk florets pale violet or pinkish. Achenes 5-8 mm long, covered with short, appressed hairs; pappus of ca 8 scales, scarious except for a prominent midrib. (n = 10).

Two varieties are present in GP:

1a. var. *macrolepis* (Rydb.) Turner & Morris. Pappus long to long-acuminate, 3-8 mm long; involucral bracts 7-10 mm long. Jun-Oct. Open high plains; sw GP; w OK & n TX, apparently isolated in KS: Clark, Comanche, Sumner; (GP s through e NM, cen & s TX).

1b. var. *rosea*. Pappus short, obtuse to acute, 1-3 mm long; involucral bracts 5-7 mm long. Jun-Oct. Mostly in sandy soils; sw 1/4 OK; (s GP to s & cen TX).

Turner & Morris (op. cit.) cite a collection from OK: Beckham (w of Elk City, *Eskew 1502*) as *P. callosa*. However, the specimen at NY is best regarded as *P. rosea* var. *rosea*. *Palafoxia callosa* differs in having narrowly linear involucral bracts, 0.6-1.3 mm wide, pappus less than 2 mm long, and in being generally restricted to calcareous sites. It does occur in the Ozarkian region, so it may be expected in the se GP. The Atlas GP includes several localities for *P. texana* (map 1572) in sw OK and the adj. Panhandle region of TX. However, these materials are referable to *P. rosea* var. *rosea* under the taxonomic concepts employed here. The name *P. texana* is best restricted to those materials with lanceolate leaves 6-15 mm wide, with distinct petioles, with n = 11 and a distribution in s TX & n Mex., cf. Turner & Morris, op. cit.

2. *Palafoxia sphacelata* (Nutt.) Cory. Annual herb 2-5(8) dm tall, variously glandular-pubescent to hispid. Leaves alternate, or a few of the earliest opposite, broadly to narrowly lanceolate, 3-9 cm long and up to 2 cm wide, 3-nerved. Heads 3-20 in a corymbiform cluster, involucre broadly turbinate; principal involucral bracts 10-15, linear-lanceolate, 9-12 mm long, often purplish-margined; ray florets with ligules up to 10 mm long, pale to dark violet, 3-lobed at apex; disk florets with regular, violet corollas. Achenes 6-9 mm long, evenly short-pubescent; pappus of ray florets of ca 8 scarious scales less than 1 mm long, of disk florets similar but 7-9 mm long, attenuate and with a darkened midregion. (n = 12) Jul-Sep. Open, mostly sandy places; sw 1/4 GP from nw NE southward; (GP to TX, NM & n Mex.). *Othake sphacelata* (Nutt.) Rydb.—Rydberg.

81. PARTHENIUM L.

Bitter, aromatic herbs or shrubs. Leaves alternate, entire, toothed or deeply lobed. Heads ± hemispheric, in open to contracted corymbiform clusters; involucral bracts ± imbricate in 2-4 series, the innermost subtending the ray florets; receptacle conic-convex, chaffy; ray florets 5, ligule white, pistillate and fertile, ligule short or sometimes suppressed and the head thus disciform; disk florets staminate, the style undivided and the pistil sterile; chaffy bracts subtending all florets. Ray achenes flattened at right angles to the radius of the head and their subtending chaffy bracts fused at the base to the contiguous pair of sterile disk florets, thus the achene plus the attached sterile florets and chaffy bract fall together at maturity; pappus of 2 or 3 awns or 2 scales, or sometimes greatly reduced. [tribe Heliantheae]

Reference: Rollins, R. C. 1950. The guayule rubber plant and its relatives. Contr. Gray Herb. 1972: 1-73.

1 Shrubs; pappus of 2 or 3 distinct awns .... 3. *P. incanum*
1 Herbs; pappus scalelike or of weak, aristate awns.
  2 Annual weeds; leaves lobed to pinnatisect; pappus of 2 distinct scales .... 2. *P. hysterophorous*
  2 Perennials; leaves merely dentate; pappus of 2 or 3 thin, aristate awns.
    3 Herbage hispid with conspicuous whitish hairs; upper leaves auriculate; rhizomatous .... 1. *P. hispidum*
    3 Herbage merely pubescent with short hairs, glabrate below; upper leaves sessile but hardly clasping; root short, thickened-tuberous .... 4. *P. integrifolium*

1. *Parthenium hispidum* Raf. Perennial herb to 8 dm tall, pilose-hispid with whitish hairs. Stems arising from laterally spreading rhizomes. Lower leaves long-petiolate, blade lanceolate to ovate, 6-20 cm long and 5-10 cm wide, margin crenate to dentate, both surfaces pilose-hispid, petiole up to 15 cm long; cauline leaves progressively reduced upward, becoming sessile and auriculate-clasping. Heads in a corymbiform inflorescence; involucre 4-7 mm tall, disk 6-10 mm across; outer involucral bracts broadly oblong, obtuse to acute and usually apiculate; inner involucral bracts obovate and rounded at the apex; ray florets with ligule white, 1.5-3 mm long. Achenes obovate, blackish, pubescent above and on the inner face, 3-5 mm long; pappus of 2 weak awns. (n = 36) May-Sep. Open wooded areas & prairies; s Flint Hills region of KS & adj. OK; extreme sw MO; (Ozarkian area of s MO & n AR to nw LA).

This species is closely related to *P. integrifolium* and the Appalachian *P. auriculatum* Britt., and these three are variously treated in regional floristic manuals. All three are incorporated into *P. integrifolim* by Mears, Phytologia 31: 463-482. 1975, where *P. hispidum* becomes *P. integrifolium* var. *hispidum* (Raf.) Mears, and our *P. integrifolium* is referable to var. *integrifolium*.

2. *Parthenium hysterophorus* L., Santa Maria. Annual weed 3-10 dm tall, pubescent and somewhat glandular, especially above. Stems arising singly, sparingly to much branched

above. Lowermost leaves forming a rosette, pinnate to bipinnately lobed, petiolate, up to 2 dm long and 1 dm wide overall; upper leaves becoming shallow-lobed to subentire and sessile or nearly so. Heads numerous in an open inflorescence; outer involucral bracts oblong, pubescent; inner involucral bracts broadly ovate and scarious margined; disk 3–4 mm across; ray florets with white ligule, minute. Achenes black, broadly oblanceolate to obovate, glabrous below but papillose toward the apex, 2–2.5 + mm long; pappus of 2 petaloid scales. (n = 17) Jul–Oct. Open, waste ground; sporadic in se 1/4 GP; (widespread as a sporadic weed in the warmer parts of the U.S., trop. Amer.). *Naturalized.*

3. *Parthenium incanum* H.B.K., mariola. Intricately branched shrub to 1 m tall, cottony-pubescent with long, simple hairs, but glabrescent in age. Leaves broadly oblong to obovate in outline but variously lyrate-lobed, 1–6 cm long and up to 2 cm wide. Heads numerous in terminal spreading corymbiform clusters; outer involucral bracts densely puberulent, acute; inner involucral bracts membranous; disk 3–5 mm across; ray florets with ligule white, 1–2 mm long. Achenes black, oblanceolate, 1.5–2 mm long, pubescent especially on the inner surface; pappus of 2 prominent lateral awns and 1 weaker erect inner (ventral) awn, which is sometimes absent. (n = 27, + polyploid series) Jun–Oct. Open, waste ground; barely entering the s GP in TX: Deaf Smith; (w TX, NM, AZ; n & cen Mex.).

4. *Parthenium integrifolium* L. Perennial herb 3–10 dm tall, ± densely pubescent, especially above, sparsely so below. Stems single or sometimes several clustered from a short, thickish-tuberous root. Lower leaves long-petiolate; blade ovate to lanceolate, 10–20 cm long and 4–12 cm wide, margin dentate, both surfaces scabrous; cauline leaves progressively reduced upward, becoming sessile but not auriculate-clasping. Heads numerous in an open corymbiform inflorescence; involucre 3–5 mm tall, outer involucral bracts ovate-obovate, usually apiculate; inner involucral bracts broadly ovate, rounded at the apex; disk 4–6 mm across; ray florets with ligules white, ca 2 mm long. Achenes obovate, blackish, pubescent toward the apex, ca 3 mm long; pappus of 2 or 3 weak awns. (n = 36) Jun–Sep. Open wooded sites; extreme se GP; KS: Cherokee; & w MO; (VA to GA, w to se MN & e GP).

This species is imperfectly separated from *P. hispidum* (q.v.).

82. PECTIS L.

1. *Pectis angustifolia* Torr. Low bushy-branched annual herb 1–2 dm tall; densely leafy toward the ends of the branches, glabrous or nearly so, glandular with yellowish oil glands, aromatic with a lemonlike scent. Leaves opposite, nearly linear, 1–4 cm long and 1–2 + mm wide, expanded and somewhat scarious margined at the base. Heads clustered at the ends of the branches; involucre turbinate, 4–5 mm tall; involucral bracts 8–10, uniseriate and subequal, narrowly linear, strongly keeled to the tip, abruptly truncate, with 1 or 2 prominent subterminal glands and several smaller submarginal glands; receptacle flat, naked; ray florets 8, pistillate and fertile, ligule yellow, 3–5 mm long; disk florets (7)10–20, perfect and fertile, corolla 2.5–3.5 mm long and slightly bilabiate; style hispidulous and with short, obtuse branches. Achenes 2.5–4 mm long, cylindrical, pubescent, black; pappus a low crown, 0.1–0.3 mm tall. (n = 12) Jun–Sep. Low areas in sandy ravines & on sand bars; scattered through sw 1/4 GP from n-cen NE: Cherry southward; (GP to TX, AZ, & n Mex.). [tribe Heliantheae]

83. PERICOME A. Gray

1. *Pericome caudata* A. Gray. Coarse, much-branched suffrutescent perennial, up to 1.5 m tall; finely puberulent to glandular dotted, the leaves ± subscabrous. Leaves numerous, opposite or mostly so, petiolate, blade cordate-hastate to deltoid, 3–8 + cm long overall and 2–4 + cm wide, progressively smaller upward, acuminate-attenuate at the apex, margin variously cleft, toothed or entire, petiole 1–3 cm long. Heads in an open, corymbiform cluster; involucre hemispherical to turbinate, 5–7 mm tall and 5–7 mm across; involucral bracts 16–20, in a single series, linear with scarious margins and a prominent, rounded midrib, ± united for about 1/2 their length; receptacle flat or low-convex, naked; ray florets absent; disk florets perfect and fertile, corolla yellow, style branches linear-filiform, slightly flattened. Achenes black, 3.5–5 mm long, oblanceolate, flattened parallel to the radius of the head; pappus a crown of small scales, sometimes with 1 or 2 delicate, caudcous bristles, 1–4 mm long. (n = 18) Aug–Oct. Occasionally found in the canyons & breaks of se CO, ne NM, & OK: Cimarron; (sw GP to w TX & CA, n Mex.). *P. caudata* var. *glandulosa* (Goodm.) Harrington — Harrington. [tribe Heliantheae]

Reference: Powell, A. M. 1973. Taxonomy of *Pericome* (Compositae — Peritylinae). Southw. Naturalist 18: 335–339.

84. PETASITES P. Mill., Sweet Coltsfoot

Perennial herb from stout, creeping rhizomes. Foliage leaves all basal and appearing with or just after the flowers; cauline leaves few and reduced to mere bracts. Inflorescence corymbiform to racemose with several heads, whitish to pinkish-purple; involucral bracts in a single series, but often subtended by a few reduced calyculate bracts. Plants imperfectly dioecious, with one plant largely male and the other largely or exclusively female. Male plants with florets all functionally staminate, or sometimes with a few outer florets pistillate like those in the female heads. Female plants with florets pistillate and fertile, or sometimes with a few central florets functionally staminate; central pistillate florets tubular, but the marginal ones sometimes ligulate. Achenes linear, 5- to 10-ribbed; pappus of capillary bristles. [tribe Senecioneae]

Reference: Cronquist, A. 1987. Petasites. N. Amer. Flora II. 10: 174–179.

1 Leaves lobed, the sinuses extending ca 2/3 to the petiole .... 1. *P. frigidus*
1 Leaves merely toothed, with 20–30 + teeth on each side .... 2. *P. sagittatus*

1. *Petasites frigidus* (L.) Fries. Stems robust, up to 6 dm tall. Foliage leaves 1–2(4) dm across, deeply lobed, the sinuses extending ca 2/3 to the petiole, sparsely hirsute to glabrous above, but commonly pubescent beneath. Flowers white or pinkish; in pistillate plants the outer with ligules up to 7 mm long. Achenes with a prominent pappus when young; pappus reduced in staminate florets. (2n = 60, 61, 62) Apr–May. Moist places in wooded areas; ND: Bottineau (Turtle Mts.); (circumboreal, across N. Amer., s to MA & mts. of CA).

Our materials are referable to var. *palmatus* (Ait.) Cronq., which is the robust southern phase, reaching its southern limit for our region at the northern edge of the GP.

2. *Petasites sagittatus* (Banks ex Pursh) A. Gray. Similar to *P. frigidus*, but the foliage leaves merely dentate and not at all deeply lobed, cordate-sagittate at the base, variously pubescent, especially on the underside. Ligule of female florets 8–9 mm long. (n = 29, + B chromosome) Apr–May. Wet places in wooded regions; scattered in n & ne ND, plus BH; (Lab. to AK, s to GP).

85. PICRADENIOPSIS Rydb.

Weakly suffrutescent perennials with creeping rootstocks; herbage grayish-strigose and impressed punctate glandular. Leaves opposite, palmately divided with the lobes often divided again; uppermost leaves reduced. Heads in terminal corymbs; involucre distinctly campanulate; involucral bracts in 2 semidistinct series, the outer ones rather prominently keeled and canescent; receptacle naked; ray florets few, pistillate and fertile, ligule yellow; disk florets with corolla tube glandular, style branches flat, with prominent stigmatic lines. Achenes narrowly obpyramidal, either glandular or hispid; pappus a crown of about 8 scales. [tribe Heliantheae]

Reference: Ellison, W. L. 1964. A systematic study of the genus *Bahia* (Compositae). Rhodora 66: 67–86, 177–215, 281–311.

Picradeniopsis has been included in *Bahia* (q.v.) by many taxonomists but current opinion favors regarding it as distinct; cf. Ellison, Southw. Naturalist 11: 456–466. 1966, and Stuessy, Irving, & Ellison, Brittonia 25: 40–56. 1973.

1 Achenes distinctly glandular; pappus scales ± ovate, with the midrib becoming indistinct about halfway up the scale .... 1. *P. oppositifolia*
1 Achenes hairy but not glandular; pappus scales lanceolate, with the prominent midrib often projecting beyond the apex into a short, excurrent bristle .... 2. *P. woodhousei*

1. **Picradeniopsis oppositifolia** (Nutt.) Rydb. Perennial herb or subshrub up to 2 dm tall; variously canescent-puberulent and impressed-punctate glandular. Stems arising singly or loosely clustered from laterally creeping rootstocks, much branched. Leaves opposite, ternately parted into linear segments 1–3 mm wide, the segments entire or again parted or toothed. Inflorescence of 1–6 terminal heads; involucre 6–8 mm tall and 1–1.5 cm across, campanulate; involucral bracts 7–9, apparently in 2 series, 4–8 mm long and 1–4 mm wide, oblong to ovate, ± prominently keeled on the outside; ray florets 4–7, ligule up to 4 mm long, yellow; disk florets 30–40, perfect and fertile. Achenes 3–5 mm long, narrow, obpyramidal; pappus of 8(9) scarious, ovate or obovate to oblong scales usually less than 1.5 mm long, with the midrib short and obscure upward, rarely projecting into an excurrent bristle. (n = 24) Jun–Sep. Open, often gravelly sites in the high plains; w 1/5 GP, e to w ND & SD; (GP, sw to NM, AZ, & n Mex.). *Bahia oppositifolia* (Nutt.) DC.—Stevens.

2. **Picradeniopsis woodhousei** (A. Gray) Rydb. Perennial herb or weak subshrub 5–15 cm tall, herbage grayish appressed-hairy and impressed-punctate glandular. Stems arising singly or several clustered from laterally creeping rootstocks, much branched. Leaves opposite, ternately parted into linear segments 1–2 mm wide, which may be again parted or toothed. Inflorescence of 1–6 terminal heads; involucre 4–8 mm tall and 8–12 mm across, campanulate; involucral bracts 8–10, in 1 or 2 series, 2.5–7 mm long and 1.5–3 mm wide, oblong to obovate and ± keeled with a prominent midrib; ray florets 5–9, pistillate and fertile, ligule up to 4 mm long, yellow; disk florets 25–40, perfect and fertile, 3–4 mm long, narrow and obpyramidal, 4-angled, black or dark brown, hispid-hispidulous; pappus of 8 or 9 scarious lanceolate scales 0.6–2 mm long, with a well-developed midrib often extending into a prominent bristle at the apex. (n = 12) May–Sep. Open high plains; sw 1/4 GP but scattered in w NE & SD; (GP s & w to TX & NM). *Bahia woodhousei* (A. Gray) A. Gray—Harrington.

86. PICRIS L.

1. **Picris echioides** L., ox-tongue. Coarse, hispid-spinescent annual 3–7 dm tall. Leaves alternate, toothed or entire, the lower ones oblanceolate and petiolate, the blade up to 25 cm long and 7 cm wide; middle and upper leaves lanceolate to oblong, progressively more sessile and clasping. Heads several, at the ends of the upper branches; involucre 1–2 cm tall; involucral bracts biseriate, the outer ones 3–5, somewhat loose or spreading, foliaceous, ovate to lance-ovate, 4–8 mm wide; inner involucral bracts narrow, usually spine

tipped; florets all ligulate and fertile, corolla yellow. Achenes distinctly slender-beaked, the body 2–4 mm long and the beak nearly equal or sometimes longer; the outer achenes weakly falcate, pubescent and partly enclosed by the bracts; central achenes brown, rugulose and obscurely nerved; pappus of plumose bristles. (n = 5) Jul–Sep. Rare in our region as a garden weed; ND: Cass, McLean, but to be expected elsewhere in ne GP; (established in ne 1/4 U.S. & adj. Can.; Eurasia). *Adventive*. [tribe Lactuceae]

87. PLUCHEA Cass., Marsh-fleabane, Stinkweed

Contributed by Ronald L. McGregor

1. *Pluchea odorata* (L.) Cass. Annual herbs with more or less fibrous roots. Stems erect, 4–10 dm tall, glabrate below, glandular and puberulent above. Leaves alternate, somewhat succulent (drying thin), blades ovate to lanceolate or elliptic, 4–10(15) cm long, 1–5(7) cm wide, margins evenly to unevenly serrate, or serrate-dentate to nearly entire, acute or acuminate at apex, base rather abruptly to gradually narrowed, surfaces from glabrous to sparsely or densely puberulent, often also glandular; petioles absent or to 3 cm long. Overall aggregation of heads cymiform, the younger branches elongating and exceeding the central ones, producing a flat-topped or layered inflorescence, the cyme bearing branches puberulent or cinereous, usually glandular; involucre 4–7 mm tall; outer bracts puberulent, ciliate, and glandular-viscid, usually pink or purple above, the inner sparsely puberulent at summits; ray florets wanting; outer florets pistillate and fertile, the central ones perfect but sterile, corollas rose-purplish. Achenes 0.9–1.2 mm long, brownish-black, 4- to 6-angled, sparsely atomiferous or setose; pappus a single series of barbellate capillary bristles united at base. (n = 10) Jul–Oct. Locally common in saline marshes & salty stream banks; s-cen KS, OK, TX; (s 1/2 U.S., s to S. Amer.). *P. purpurascens* (Sw.) DC.—Rydberg. [tribe Inuleae]

Our plants as here described are the var. *odorata*. Rare plants from nonsaline marshes and lakeshores in extreme se GP are referred to *P. camphorata* (L.) DC. which is a little larger, sometimes perennial, and with the involucre only atomiferous-glandular.

88. POLYMNIA L., Leaf-cup

Perennial herbs (ours). Principal leaves opposite, or the uppermost alternate. Heads with involucral bracts green, in a single series; receptacle flat, chaffy; outer florets pistillate and fertile, the corolla minute or expanded into a ligulate ray, yellow or whitish; disk florets staminate only, with the style undivided; chaffy bracts subtending the marginal pistillate florets larger than those of the disk florets. Marginal achenes thick; pappus none. [tribe Heliantheae]

Reference: Wells, J. R. 1965. A taxonomic study of *Polymnia* (Compositae). Brittonia 17: 144–159.

1 Achenes 3-angled and 3-ribbed; principal leaves pinnately lobed and veined; ligule of marginal florets whitish, rarely longer than 1 cm, or ligule absent .... 1. *P. canadensis*
1 Achenes impressed-striate with numerous nerves; principal leaves palmately or pinnipalmately lobed and nerved; ligule of the ray florets 1–2(3) cm long, yellow .... 2. *P. uvedalia*

1. *Polymnia canadensis* L. Perennial herb 5–15 dm tall, mostly glabrate below and viscid glandular above. Leaves large, broadly oblong to ovate in outline, but pinnately lobed and toothed, petiolate, up to 3 dm long overall. Heads clustered in terminal cymes; involucral bracts lanceolate to lance-linear, often shorter than the chaffy bracts subtending the ray

achenes; disk 5-12 mm across; fertile marginal florets with corolla minute and tubular or expanded into a short ligule usually less than 10 mm long, whitish or ochroleucous. Functional achenes 3-4 mm long, 3-ribbed and 3-angled, not at all striate. (n = 15) May-Oct. Deep moist woods, especially in limestone areas; infrequent in se GP; KS: Anderson, Cherokee, Miami; MO: Jackson; (VT, s Ont. to GA, w to MN, GP, & AR).

2. **Polymnia uvedalia** (L.) L., bear's foot. Perennial herb 1-3 m tall, glabrous below but glandular or somewhat spreading-hairy among the heads in the inflorescence. Leaves large, regularly up to 30 cm long and sometimes as much as 60 cm, deltoid ovate, ± palmately to pinnipalmately lobed and veined, weakly hispid to glabrate above, finely pubescent to glandular beneath, petiolate to subsessile. Heads clustered in loose, leafy cymes; involucral bracts lance-ovate to elliptic, 1-2 cm long, broader and longer than the outer chaffy bracts; disk ca 1.5 cm across; fertile marginal florets with ligule 1-2(3) cm long, yellow, or rarely reduced and inconspicuous. Functional achenes ca 6 mm long, impressed-striate with numerous nerves. (n = 16) Jul-Oct. Open damp meadows & woods; scattered in se GP; KS: Cherokee; MO: Jasper; OK: Canadian, Cleveland, Oklahoma, Tulsa; (NY to FL, w to GP).

Several varieties have been proposed in the literature, but the most recent student of the genus has concluded that they are not worth recognizing; cf. Wells, Rhodora 71: 204-211. 1969.

89. PRENANTHES L., Rattlesnake-root, White Lettuce

Perennial herbs with milky juice and rather strongly tuberous-thickened roots. Leaves alternate, numerous, and prominent along the stem. Inflorescence paniculiform or thyrsoid, with several to many heads; involucre mostly with ca 8 or ca 13 principal involucral bracts, subtended by several outer bracts, some of which approach the principal bracts in size; florets all ligulate, perfect and fertile, corolla pink to lavender or creamy. Achenes cylindric or weakly angled, usually rather indistinctly ribbed; pappus of numerous capillary bristles. [tribe Lactuceae]

Reference: Milstead, W. L. 1964. A revision of the North American species of *Prenanthes*. Unpublished Ph.D. dissertation, Purdue Univ., Lafayette, Indiana.

Prenanthes in North America occurs chiefly in the eastern deciduous woodlands, where its taxonomy is complicated by hybridization and great morphological variability.

1 Involucre glabrous or nearly so, often with a waxy bloom; inflorescence
    paniculiform .... 1. *P. alba*
1 Involucre pubescent to long-hirsute; inflorescence thyrsoid.
    2 Florets creamy; herbage variously rough-hairy .... 2. *P. aspera*
    2 Florets pink to purple; herbage glabrous except in the
        inflorescence .... 3. *P. racemosa* subsp. *multiflora*

1. **Prenanthes alba** L. Erect coarse herb 5-15(18) dm tall, often glaucous. Leaves glabrous above but sometimes pubescent beneath, highly variable in size and shape, the lower ones distinctly petiolate, variously lobed, to sagittate or hastate and merely coarsely toothed; cauline leaves progressively smaller, less lobed, and less petiolate upward. Inflorescence profuse, elongate, paniculiform, occupying the upper 1/4-1/3 of the plant; heads nodding; involucre 11-14 mm tall, principal involucral bracts ca 8, purplish, glabrous or nearly so but often covered with a waxy bloom; florets mostly 9, ligulate and fertile, corolla pink to lavender. Achenes indistinctly ribbed, 4-6 mm long; pappus of numerous brownish to brownish-red capillary bristles. (n = 16) Aug-Sep. Wooded areas; ne 1/3 ND & w MN, s to w IA; (Que. to NC, w to s Man. & MO, nw AR). *Nabalus albus* (L.) Hook.— Rydberg.

2. ***Prenanthes aspera*** Michx. Erect herb 4–15 + dm tall, herbage rough-hairy or scabrate, especially upward; stem green but sometimes with purple areas. Leaves scabrous-hirsute beneath and sometimes above, entire to variously dentate; lowermost leaves well developed, petiolate, early deciduous; middle and upper cauline leaves progressively reduced upwards, becoming sessile and clasping, the larger ones oblong to elliptic lanceolate, up to 10 cm long and 4 cm wide. Inflorescence elongate, narrow and thyrsoid; heads crowded, mostly weakly ascending; involucre 12–17 mm tall, coarsely and usually densely hirsute; principal involucral bracts mostly 8, the reduced outer bracts up to 6 mm long; florets ca 13, ligulate and fertile, corolla creamy. Achenes 4–6.5 mm long, with 10–12 prominent but irregular ribs; pappus of pale yellow capillary bristles. Aug–Sep. Open prairies & edges of wooded areas; e 1/4 GP from ne SD w to n-cen NE & s; (OH to GP & AR). *Nabalus asper* (Michx.) T. & G. — Rydberg.

3. ***Prenanthes racemosa*** Michx. subsp. ***multiflora*** Cronq. Coarse herb 3–15 dm tall, arising from well-developed fibrous roots; glabrous, except long-hairy among the heads in the inflorescence. Lower leaves petiolate, obovate or elliptic to oblanceolate, 7–40 cm long overall and up to 10 cm wide; cauline leaves progressively reduced upward, becoming sessile and weakly clasping. Inflorescence elongate, narrow and thyrsoid, the heads ascending or occasionally nodding; involucre 10–15 mm tall, purplish or blackish, variously hirsute; principal involucral bracts ca 13; florets all perfect and fertile, ca 21(17–26), corolla pink to purple. Achenes 4.5–6.5 mm long, with 8–12 ribs; pappus of numerous deciduous, pale yellow bristles. ($n = 8$) Aug–Sep. Damp open prairies, meadows, & stream banks; ne 1/4 GP; BH; NE: Sioux; (Que. to NJ, w to Alta. & CO for the entire species). *Nabalus racemosus* (Michx.) DC. — Rydberg.

The var. *racemosa* has smaller heads and occurs in the essentially wooded regions to the e of the GP.

90. PSILOSTROPHE DC., Paper Flower

Taprooted annual or short-lived perennial herbs with stems often clustered, branching upward, densely felted-pubescent to glabrate. Leaves alternate, lower ones petiolate but upper ones sessile, becoming linear or spatulate. Heads in subcorymbiform clusters; involucre campanulate to subcylindric; involucral bracts in 2 series, the outer 4–12, linear-lanceolate, the inner 1–7, smaller and scarious; sometimes a single calyculate bract present; receptacle flat or but weakly convex, naked; ray florets 3–7, pistillate and fertile, ligule yellow, papery, 3-lobed, persistent on the maturing achenes; disk florets perfect, fertile; style branches truncate. Achenes linear, striate; pappus of 4–6 hyaline scales. [tribe Heliantheae]

Reference: Heiser, C. B. 1944. Monograph of *Psilostrophe*. Ann. Missouri Bot. Gard. 31: 279–301.

1 Heads loosely clustered, peduncles mostly more than 5 mm long; ligules 5–9 + mm long, shallow-lobed .... 1. *P. tagetina*
1 Heads densely clustered, peduncles mostly less than 5 mm long; ligules 3–5 mm long, deeply lobed .... 2. *P. villosa*

1. ***Psilostrophe tagetina*** (Nutt.) Greene. Perennial herb 1–5 dm tall, densely to lightly villous or glabrate, somewhat woody at the base. Basal leaves ovate to oblanceolate-spatulate, entire to pinnately lobed, 2–10 + cm long, upper leaves narrower, smaller, 1–7 cm long. Heads in corymbiform clusters, peduncles 5–20 + mm long; involucre usually densely woolly, 5–7 mm tall; ray florets 3–5, ligule 5–9 + mm long, shallowly 3-lobed at the apex; disk florets 6–12. Achenes glabrous or with sparse hairs; pappus scales hyaline, 1/2–2/3 the length of the disk corolla. ($n = 16$) Jun–Aug. Rare in open, sandy sites in extreme sw GP; NM: Curry; TX: Childress; (w TX to AZ, n Mex.).

The species includes var. *lanata* A. Nels., a robust, densely pubescent phase which may be distinctive, but occurs to the south and west of our range.

2. **Psilostrophe villosa** Rydb. Densely to sparsely villous herb 1-6 dm tall. Basal leaves oblanceolate to spatulate, entire or lobed, short-petiolate, 5-10 cm long overall; upper leaves smaller and becoming sessile, infrequently lobed. Heads in small, congested clusters; involucre ca 5 mm tall, densely woolly; ray florets 3 or 4, ligule 3-5 mm long and apically 3-lobed to about 1/2 the length, or rarely 4-lobed; disk florets (5)6-8(12). Achenes glabrous or nearly so; pappus scales linear-lanceolate and about 1/2 as long as the disk corollas, or slightly longer. (n = 17) May-Sep. Open sandy sites, roadsides, & semidisturbed places; sw 1/4 GP, from w-cen KS s & w & to be expected elsewhere as a casual weed; (sw GP to w TX & NM).

91. PYRRHOPAPPUS DC., False Dandelion

Taprooted annual, biennial or perennial herbs with milky juice; herbage variously glabrous to loosely or lanately pubescent. Leaves alternate or basal, dentate or pinnately undulate-parted to subentire, pubescent at least along the midvein on the lower surface and often elsewhere but glabrescent in age; upper leaves reduced to sessile bracts. Inflorescence of 1-several long-pedunculate heads; principal involucral bracts linear-lanceolate, green but usually dark and thickened toward the apex, marginally subscarious; outer calyculate bracts subulate to linear, in 2 or 3 weakly defined series, the outermost 2-6 often prominent and long; all involucral bracts becoming dry and deflexed at maturity; receptacle naked, convex; florets numerous, 45-100, all ligulate and fertile, corolla lemon yellow or infrequently light yellow. Achene with a prominent, ribbed body, contracted below and abruptly tapering above to a narrow, filiform beak about as long or longer than the body; pappus of numerous capillary bristle, sordid or off-white to dirty reddish, subtended by a few villous hairs, pappus bristles persistent, deciduous or falling as a unit with the beak. [tribe Lactuceae]

Reference: Northington, D. K. 1971. Taxonomy of *Pyrrhopappus*, a cytotaxonomic and chemosystematic study. Unpublished Ph.D. dissertation, Univ. of Texas, Austin; Northington, D. K. 1974. Systematic studies of the genus *Pyrrhopappus*. Special publication 6, The Museum, Texas Tech Univ.

Pyrrhopappus rothrockii A. Gray has been reported from n TX, but its occurrence there seems doubtful. It is a perennial arising from a thick, woody taproot, without an enlarged subterranean tuber. Its natural range is w TX to e AZ, s into Mex.

1 Perennial with a slender, vertical taproot terminating 4-10 cm below the ground in a tuber, 1-2 cm across .... 2. *P. grandiflorus*
1 Annual or biennial with tapering taproots.
   2 Prominent cauline leaves usually merely toothed, rarely distinctly lobed; lower stem glabrous or rarely loose-puberulent; basal leaves early deciduous .... 1. *P. carolinianus*
   2 Prominent cauline leaves with 2-3(5) prominent lobes on each side; lower stem loosely puberulent to lanate-pubescent; basal leaves tending to be persistent in a rosette .... 3. *P. multicaulis* var. *geiseri*

1. **Pyrrhopappus carolinianus** (Walt.) DC. Annual or biennial 3-8(10) dm tall, arising from a thick, tapering taproot. Stem single or occasionally 2-5, erect, loosely puberulent, especially above, often glabrate below. Basal leaves in a rosette but mostly early deciduous; cauline leaves oblanceolate, variously lobed, toothed or subentire, progressively reduced upward, the lower, prominent ones 4-15(25) cm long and 1-3 + cm wide overall, toothed or rarely lobed; uppermost ones reduced to weakly clasping bracts. Inflorescence 1-several heads; involucre cylindrical; principal involucral bracts usually ca 21, 15-25 mm long;

calyculate bracts in 2 or 3 series, 8–14 mm long but quite variable, usually with 1 or 2 nearly as long as the principal involucral bracts; florets all ligulate and fertile, corolla yellow. Achenes with a prominent cylindrical body, abruptly tapering to a narrow beak, 14–17 mm long overall; pappus of abundant capillary bristles, off-white to dirty reddish, 7–12 mm long. (n = 6) Mostly May–Jun, infrequently Apr, Jul–Oct. Open, sandy sites, especially in disturbed areas, sometimes weedy; se 1/4 GP, mostly e 1/3 KS, w MO & s; (PA & MD to n FL, w to GP & s TX).

2. **Pyrrhopappus grandiflorus** (Nutt.) Nutt., tuber false dandelion. Perennial 2–3 dm tall, arising from a long, thin cylindrical taproot, terminating 4–10 cm below the ground in a spherical tuber 1–2 cm across. Stems 1 or 2(4), scapose or occasionally with 1–few scattered, bractlike leaves; herbage loosely puberulent, especially upward, or becoming glabrate. Basal leaves in a rosette, blade oblanceolate in outline but variously undulate-parted to deeply dentate, 7–12 + cm long and 2–4 cm wide overall; cauline leaves, when present, few and reduced to entire or dentate bracts. Inflorescence a terminal head; involucre cylindrical; principal involucral bracts ca 13 or 21 in number, 15–20 mm long; outer calyculate bracts in 2 or 3 poorly defined series, reduced and varying in length; florets all ligulate and fertile, corolla yellow. Achenes with a 5-ribbed, cylindrical body, abruptly tapering upward to a long, thin beak, 10–13 mm long overall; pappus of abundant sordid to beige capillary bristles, ca 10 mm long. (n = 12) May–Jun. Open clay or sandy soils, especially in disturbed areas; s 1/4 GP, mostly cen KS, cen OK & n TX; (s GP, s through TX).

3. **Pyrrhopappus multicaulis** DC. var. *geiseri* (Shinners) Northington. Annual 2–6 dm tall, arising from a thick, tapering vertical taproot. Stems 1 or 2(5), erect or ascending, branching upward, loosely pubescent to somewhat lanate, or rarely glabrate, especially below. Basal leaves 4–12 in a rosette, oblanceolate in outline but dentate or pinnate to undulate-parted, 10–15 + cm long and 3–5 cm wide overall; principal cauline leaves 1–3 on the lower stem, similar to the basal leaves, usually with 2 or 3(5) linear lobes on each side; upper cauline leaves reduced but ± pinnatisect. Heads terminal; involucre cylindrical; principal involucral bracts ca 13 or 21 (apparently either one or the other in any given population), 16–20 mm long; outer calyculate bracts shorter, numerous; florets all ligulate and fertile, corolla yellow. Achenes with a 5-ribbed body, contracted upward into a prominent filiform beak, 12–14 mm long overall; pappus of abundant sordid or off-white capillary bristles, 8–10 mm long. (n = 6) Apr–May. Open sandy or clay soils, especially in disturbed sites; cen & sw OK; (s GP, sw AR, w LA, TX).

Var. *multicaulis* is shorter, bushy-branched with numerous stems, and is restricted to cen TX.

92. **RATIBIDA** Raf., Prairie Coneflower

Herbaceous perennials with stems simple or branched, mostly strigose-hirsute, striate with dark green grooves, containing unicellular globular glands. Leaves alternate, mostly variously pinnate or lyrate to pinnatifid, but some may be subentire; strigose-hirsute and punctate-glandular on both surfaces. Heads on naked peduncles, receptacle globular or cylindric-columnar; involucral bracts reflexed, in 2 series, the inner ones short-ovate and the outer ones linear-lanceolate; ray florets neutral, ligule yellow to purplish, spreading or reflexed, 2- or 3-toothed at the apex; disk florets fertile, with corolla yellowish-green to purplish-green; styles flattened, with appendages elongate or short and blunt; chaffy bracts conduplicate, apex densely velutinous and incurved, subtending both ray and disk florets, falling at maturity with the achenes. Achenes compressed parallel to the radius of the head, ± 4-angled, glabrous or with ciliate hairs on the angles; pappus a crown, or of a few teeth, or absent. [tribe Heliantheae]

Reference: Richards, E. L. 1968. A monograph of the genus *Ratibida*. Rhodora 70: 348–393.

1 Plants fibrous rooted, with stout rhizomes or a caudex; pappus none .... 2. *R. pinnata*
1 Plants taprooted or with a taprooted crown; pappus a crown or toothlike.
  2 Receptacle distinctly long-columnar; pappus of toothlike projections .... 1. *R. columnifera*
  2 Receptacle globular; pappus a crown .... 3. *R. tagetes*

1. **Ratibida columnifera** (Nutt.) Woot. & Standl., prairie coneflower. Perennial 3–10 dm tall. Stems arising singly or often clustered from a prominent taproot. Well-developed leaves up to 15 cm long and 6 cm wide, pinnatifid to partly bipinnatifid, with ultimate segments linear to oblong, often very unequal. Heads solitary or several at the ends of long peduncles; receptacle columnar, up to 4.5 cm tall and ± 1 cm across; involucral bracts 5–14, in 2 series; ray florets 4–11, ligule yellow or purplish-yellow to purple, 1–3 + cm long, spreading or reflexed; disk florets with corolla 1.5–2.5 mm long, style branches short and blunt; chaffy bracts with margins ± ciliate, somewhat glandular toward the base. Achenes 1.5–3 mm long, oblong, ciliate on the inner edge; pappus of 1 or 2 teeth, or perhaps rarely absent. (n = 14; also reported as 13, 17–19) Jun–Sep. Prairies & open waste ground, roadsides; throughout GP; (MI to AR & TX, w to Alta. & AZ; casually introduced elsewhere). *Lepachys columnifera* (Nutt.) Rydb.—Rydberg.

    The phase with purple or purplish-yellow ray florets has been distinguished as f. *pulcherrima* Fern.

2. **Ratibida pinnata** (Vent.) Barnh., grayhead prairie coneflower. Perennial 3–12 dm tall. Stems arising singly or clustered from a stout, woody rhizome or a short caudex, branched, ± hirsute. Leaves up to 4 dm long, including the petiole, deeply pinnatifid, the larger segments lance-ovate, acute, coarsely dentate to entire; upper leaves smaller, the uppermost reduced to subentire bracts. Heads several, at the ends of long, naked peduncles; receptacle globular or oblong, 1–2.5 cm tall and 1–2 cm across; involucral bracts 10–14, in 2 series, ray florets 6–13, ligule (2)3–6 cm long, yellow, spreading or reflexed; disk florets with corolla 1.3–3 mm long, ciliate on the inner edge; pappus absent. (n = 14) Jun–Sep. Disturbed prairies & openings in wooded areas, mostly on calcareous soils; e edge GP, ne SD s to ne OK; (s Ont. to GA, w to GP; mostly in cen U.S.). *Lepachys pinnata* (Vent.) T. & G.—Rydberg.

3. **Ratibida tagetes** (James) Barnh., short-ray prairie coneflower. Bushy-branched leafy perennial 2–4(+) dm tall, arising from a taproot. Lower leaves lanceolate, entire or pinnatifid to bipinnatifid, up to 13 cm long, including the indistinct petiole, upper leaves 3- to 5-cleft into narrow segments 5–30 mm long. Heads on short peduncles and often closely clustered; receptacle globular-oblong, 0.8–1.5 cm tall and about 1 cm across; involucral bracts 10–12, in 2 series; ray florets 5–10, ligule 4–8 mm long, yellow to reddish-brown; disk florets with corollas 1.5–2 mm long; chaffy bracts 2–4 mm long, punctate-glandular at the apex. Achene 2–3 mm long, oblong-oblique, glabrous or lightly ciliate at the apex, winged on the outer edge; pappus a thick crown. (n = 16) Jun–Sep. Open high plains & rocky hillsides; sw 1/4 GP, from cen KS s & w; sporadic in ne CO n to extreme sw SD; (sw GP to n Mex., w to AZ). *Lepachys tagetes* (James) A. Gray—Rydberg.

93. RUDBECKIA L., Coneflower

Annual or perennial herbs with alternate, entire to toothed or deeply lobed leaves. Heads with involucral bracts subequal or of irregular lengths, green and herbaceous, mostly spreading or reflexed; receptacle enlarged and hemispheric or columnar, with chaffy bracts subtending the disk florets and partly enclosing the achenes; ray florets 5–21, neutral, ligule prominent, yellow to orange; disk florets perfect and fertile, the corolla narrowed toward the base to a distinct tube; style branches flattened and with a short, blunt appendage or (in *R. hirta*) an elongate-subulate appendage, without distinct stigmatic lines. Achenes 4-angled, glabrous; pappus a short toothed or irregular crown, or absent. [tribe Heliantheae]

Reference: Perdue, R. E. 1957. Synopsis of *Rudbeckia* subgenus *Rudbeckia*. Rhodora 59: 293–299.

1 Style appendages distinctly elongate-subulate; pappus absent; herbage coarsely hirsute .... 2. *R. hirta*
1 Style appendages short and blunt; pappus present but sometimes minute; herbage pubescent to glabrous.
   2 Chaffy bracts glabrous .... 5. *R. triloba*
   2 Chaffy bracts pubescent at the apex with short, viscidulous hairs.
      3 Leaves subentire or merely toothed .... 1. *R. grandiflora*
      3 Well-developed leaves deeply trilobed or pinnatifid.
         4 Stem glabrous or nearly so; disk yellow or grayish, elongating and columnar at maturity .... 3. *R. laciniata*
         4 Stem densely short-pubescent, at least above; disk purple or brown, hemispheric and not elongating .... 4. *R. subtomentosa*

1. ***Rudbeckia grandiflora*** (Sweet) DC., rough coneflower. Perennial herb 5–10 dm tall. Stems arising from a woody caudex or a stout rhizome. Leaves large, blade ovate to ovate-lanceolate or elliptic, up to 25 cm long and 10 cm wide, toothed or subentire, lower ones long-petiolate, the uppermost becoming subsessile, pubescent on both surfaces with long, loose hairs. Heads solitary or few, long-pedunculate; disk 1.5–2.5 cm across, ovoid or hemispheric; involucral bracts spreading or reflexed, subequal; ray florets 10–20, ligule yellow, drooping, up to 5 cm long; disk florets purplish to brownish; chaffy bracts blunt and pubescent at the apex; style appendages blunt. Achenes 4-angled but slightly flattened; pappus a conspicuous but irregular crown. (n = 19) Jun–Aug. Open, dry woods; extreme se KS & adj. OK; (MO & adj. se KS, s to LA & TX).

Our materials are referable var. *grandiflora*. The short-hair, scabrous-puberulent phase of LA & TX is separable as var. *alismaefolia* (T. & G.) Cronq.

2. ***Rudbeckia hirta*** L., black-eyed susan. Biennial or short-lived perennial, but apparently able to flower the first year, 3–10 dm tall; herbage variously hispid-canescent throughout. Stems arising from a taproot or a cluster of fibrous roots. Leaves variable in size and shape, but the well-developed lower leaf blades mostly oblanceolate to elliptic, up to 15 cm long and petiolate, middle leaves lance-linear to oblong and becoming sessile upwards. Heads long-pedunculate; disk up to 2 cm across, hemispheric or ovoid; involucral bracts distinctly hirsute-hispid, sometimes prominently elongate; ray florets 8–21, ligule 2–4 cm long, orange or orange-yellow, sometimes purple-tinged at the base; disk florets dark purple, or the disk sometimes becoming brownish; chaffy bracts acute and hispid-hispidulous at the apex; style branches distinctly elongate-subulate. Achenes quadrangular, the faces flat or weakly bulging; pappus none. (n = 19) May–Sep. Disturbed prairies, roadsides, & waste places; somewhat weedy; mostly e 1/2 GP but scattered w in NE & SD to e WY, & along the edge of the Rocky Mts. in CO; (Newf. to FL, w to B.C. & Mex.). *R. serotina* Nutt.—Fernald.

Our materials are referable to the poorly defined var. *pulcherrima* Farw., which is characterized by entire or finely toothed and rather narrow leaves, and a weedy habitat preference. Var. *hirta* is the eastern phase with more coarsely toothed leaves and an occurrence in undisturbed sites; var. *angustifolia* (T. V. Moore) Perdue is a southern phase with leaves large and mostly basal.

3. ***Rudbeckia laciniata*** L., golden glow. Perennial herb 0.5–2(3) m tall. Stems arising from a coarse, woody base, glabrous and often somewhat glaucous. Leaves large, some or most of them deeply pinnatifid or at least trilobed, margins coarsely toothed to laciniate, both surfaces mostly glabrous, or hirsute-strigose beneath. Heads 1–many; disk 1–2 cm

across, hemispheric but becoming cylindric-columnar at maturity; involucral bracts spreading or reflexed, green and leafy; ray florets mostly ca 8 or ca 13, ligule 3–6 cm long, drooping; disk florets yellow but the disk may become grayish in aspect; chaffy bracts blunt, viscid-pubescent at the apex; style appendages blunt. Achenes equally 4-angled; pappus a short, toothed crown. (n = 18, + a polyploid series; plants often apomictic) Jul–Sep. Moist places; e 1/2 GP but scattered w in BH, e MT, & WY & along edge of the Rocky Mts., CO; (Que. to FL, w to MT & AZ).

Our materials are referable to var. *laciniata*.

4. ***Rudbeckia subtomentosa*** Pursh, sweet coneflower. Perennial 6–20 dm tall, glabrous below but densely short-pubescent above. Stems arising from stout rhizomes. Leaves thickish and firm, blade ovate to elliptic or the larger ones deeply 3-parted, short-hairy on both surfaces and especially so beneath, margin serrate. Heads several, disk 1–1.5 cm across, hemispheric or ovoid and not elongating; involucral bracts narrow, spreading or reflexed, subequal, green; ray florets 12–20, ligule yellow, 2–4 cm long; disk florets purplish or becoming brownish; chaffy bracts obtuse to acute, pubescent toward the apex with short, whitish hairs; style appendages short and blunt. Achenes about equally 4-angled; pappus a minute crown. (n = 19) Jul–Sep. Prairies & low, damp sites; se edge GP; w IA & e NE, s to ne OK; (MI to LA w to WI, GP, & OK).

5. ***Rudbeckia triloba*** L., brown-eyed susan. Short-lived perennial, to 15 dm tall, variously hirsute or strigose to subglabrous. Leaves thin, lower blades broadly ovate, seldom more than 5 cm long, the larger ones deeply 3-lobed or -parted, or rarely pinnatifid, long-petiolate; middle and upper leaves narrower and short-petiolate to subsessile. Heads at the ends of the branches; disk 1–1.5 cm across, hemispheric or ovoid; involucral bracts narrow, subequal, green, spreading or reflexed; ray florets 6–12, ligule 1–2(+) cm long, yellow or orange; disk florets dark purple; chaffy bracts glabrous and abruptly contracted into a short, distinct awn tip that shortly exceeds the disk corollas; style appendages short and broad, ± acute. Achenes about equally 4-angled; pappus a minute crown. (n = 19) Jul–Oct. Open, moist woods; se edge GP; w IA & e NE, s to ne OK, but scattered in cen & w NE; (CT & NY to FL, w to MI, e GP, & OK).

Our materials are referable to var. *triloba*; two other vars. are known farther east.

94. SCORZONERA L., False Salsify

1. ***Scorzonera laciniata*** L. Annual, winter annual or biennial herb with milky juice, arising from a taproot. Stems 1–several, usually branching upward from about the middle, erect or ascending, somewhat striate above; herbage glabrous to scattered floccose. Basal and lower cauline leaves pinnatisect, up to 20 cm long and 4 cm wide overall, the lateral leaf segments ovate to narrowly ovate, sometimes dentate; cauline leaves similar but reduced, sometimes entire. Inflorescence of 1–several pedunculate heads; involucre 7–20 mm tall in flower, becoming up 2 × as tall in fruit; involucral bracts in several series; receptacle naked; florets all ligulate, perfect and fertile, corolla yellowish, equaling or exceeding the involucral bracts. Achenes glabrous, brownish to gray, up to 17 mm long, columnar but with a weakly defined tubular base ca 1/3 as long as the body of the achene; pappus of plumose bristles in several series, about as long as the achene. (n = 7) May–Jul. Occasional to locally abundant as a weed in open disturbed sites; along the edge of the Rocky Mts. in CO; from the vicinity of Ft. Collins s to Colorado Springs, e to nw KS; (CO, KS; Eurasia). *Naturalized. Podospermum laciniatum* (L.) DC. of reports. [tribe Lactuceae]

Reference: Chater, A. O. 1976. *Scorzonera in* Flora Europaea, vol. 4, T. G. Tutin, V. H. Heywood, et al., eds. Cambridge Univ. Press, Cambridge.

This plant was first noted in North America in the middle 1950s around CO: Boulder, and since then it has become a locally abundant weed along the foothills of the Front Range. It has migrated east, especially along U.S. Hwy I-70 and is now known in KS: Cheyenne, Norton, Rawlins, Sherman, Thomas. It is to be expected elsewhere.

95. SENECIO L., Groundsel

Ours herbs or shrubs. Leaves cauline and alternate or basal, blades pinnately veined, variously entire to toothed to deeply divided. Inflorescence of few to many heads, often in a loose to compact corymbiform cyme. Heads radiate or sometimes discoid, yellow or less frequently ochroleucous or orange; principal involucral bracts equal in size and forming a single series, variable in number but approximating the Fibonacci series (5-8-13-21-34), with 13 or 21 bracts the most frequent, often subtended by a few small, calyculate bracts, principal involucral bracts and calyculate bracts sometimes prominently black-tipped; ray florets fertile, in a single series, variable in number but approximating the Fibonacci series; ligules conspicuous or rarely greatly reduced; disk florets perfect and fertile, corolla expanded upward into a campanulate limb with 5 shallow teeth; style branches flattened, truncate, penicillate. Achenes cylindrical, 5- to 10-ribbed, glabrous or variously hispidulous-pubescent, especially along the angles; pappus of abundant long, white, weakly barbellulate hairs. [tribe Seneconeae]

Reference: Barkley, T. M., 1978. *Senecio*. N. Amer. Flora II. 10: 50-139.

This is a world-wide genus with perhaps three thousand species, varying in the extreme from tropical trees to desert succulents to arctic herbs.

1 Leaves progressively reduced in size upward along the stem; perennial herbs.
  2 Plants with an erect caudex, rhizomes or a taproot, plus branching lateral roots; basal leaves entire to toothed but without marginal callous denticles.
    3 Herbage glabrous or at least glabrescent at flowering time.
      4 Basal leaves subcordate to truncate .... 13. *S. pseudaureus* var. *semicordatus*
      4 Basal leaves tapering or rounded at the base.
        5 Basal leaves obovate, obtuse or rounded at the base; plants often with long, slender stolons .... 10. *S. obovatus*
        5 Basal leaves ovate to oblanceolate or narrower; without stolons.
          6 Plants with a taproot; basal leaves narrowly lanceolate, ca 3× longer than wide .... 17. *S. tridenticulatus*
          6 Plants with a fibrous-rooted caudex; basal leaves ovate to lanceolate, 1-3× longer than wide .... 11. *S. pauperculus*
    3 Herbage variously pubescent at flowering time.
      7 Herbage floccose-tomentose, glabrescent at maturity but with at least some conspicuous pubescence in the leaf axils and among the heads .... 12. *S. plattensis*
      7 Herbage densely lanate-pubescent .... 1. *S. canus*
  2 Plants with a thick rhizome or an abruptly shortened caudex with numerous unbranched fleshy-fibrous roots; leaves often with marginal callous denticles.
    8 Plants with an extended rhizome or long caudex.
      9 Ray florets absent; leaves prominently dentate .... 14. *S. rapifolius*
      9 Ray florets present; lower leaves shallow-dentate to subentire, but often with minute callous denticles .... 3. *S. crassulus*
    8 Plants with a short, buttonlike caudex.
      10 Herbage glabrous and glaucous; swamp plants ± 10 dm tall ..... 7. *S. hydrophilus*
      10 Herbage variously pubescent to glabrate; plants of dry sites, seldom exceeding 5 dm tall .... 9. *S. integerrimus*
1 Leaves about equally distributed along the stem, or concentrated upward.

11 Plants ± shrubby, or at least with the upward-branching aspect of a shrub; leaves or their segments linear-filiform.
   12 Herbage irregularly but prominently lanate-tomentose ...... 4. *S. douglasii* var. *longilobus*
   12 Herbage glabrous.
      13 Heads small, disk 3–6 mm across; involucral bracts ca 8; leaves simple .... 16. *S. spartioides*
      13 Heads large, disk 7–12 mm across; involucral bracts ca 13; leaves pinnatilobate .... 15. *S. riddellii*
11 Plants herbaceous and not at all shrubby.
   14 Perennials with a well-developed rhizomatous caudex; leaves pinnate-lacerate .... 5. *S. eremophilus*
   14 Annuals.
      15 Soft-stemed semiaquatics with abundant soft-villous pubescence .... 2. *S. congestus*
      15 Without this combination of characters.
         16 Ray florets absent .... 18. *S. vulgaris*
         16 Ray florets present and prominent.
            17 Stem arising from a cluster of many fibrous roots .... 6. *S. glabellus*
            17 Stem arising from a distinct taproot .... 8. *S. imparipinnatus*

1. **Senecio canus** Hook., gray ragwort. Perennial 1–3(4.5) dm tall, densely lanate-pubescent with subappressed felted hairs, gray or silvery-gray in aspect, sometimes glabrescent above. Stem single or rarely clustered from a rosette of basal leaves; rosettes arising from a creeping or suberect caudex or from a rhizome. Basal leaves petiolate, blade ovate to lanceolate, tapering at the base, entire to weakly dentate, 2.5–5 cm long and (0.5)1–3 cm wide, ca 1.5 × as long as wide or 3–4 × longer than wide in very narrow-leaved individuals, petiole about as long as the blade; cauline leaves strongly reduced upward, lower ones similar to basal leaves, middle and upper ones often clasping and reduced to mere bracts, subentire or weakly dentate to nearly lyrate. Inflorescence a loose to congested corymbiform cyme of 6–12+ heads; principal involucral bracts ca 13 or 21 and (3)5–6(10) mm long; ray florets ca 8 or 13, ligule 7–10 mm long, or rays sometimes absent. Achenes glabrous. (n = 23, 46, ± 48, 69) May–Jul(Aug). Open plains, often in dry, rocky cold sites; n GP; MN to MT, s to extreme nw KS: Cheyenne; (MN to WA, B.C. to CA). *S. purshianus* Nutt.—Rydberg.

    This species is abundant in the montane west but sporadic and localized in the northern plains.

2. **Senecio congestus** (R. Br.) DC., swamp ragwort. Annual (biennial?) with loose arachnoid tomentum, especially among the heads in the inflorescence, often yellowish to brown, sometimes glabrescent, especially below. Stem single, hollow, sometimes pinkish, arising from a shortened, densely fibrous-rooted caudex. Leaves about equally distributed, somewhat larger on the lower 1/2 of the stem, oblanceolate to nearly spatulate, coarsely dentate to weakly pinnatifid, or sometimes subentire, petiole poorly defined, overall dimensions 5–15 cm long and 1–3(5) cm wide, ca 4–5 × longer than wide. Inflorescence 6–20(40) heads in an open corymbiform cyme; involucral bracts ca 21 and 5–10 mm long, sometimes pink-tipped; ray florets ca 21, or fewer, ligule seldom exceeding 8 mm. Achene glabrous, but with abundant long pappus-hairs. (n = 24, 25?) May–Jul. Marshes & stream banks; ND, e SD, w MN, nw IA; (Newf. to AK, s to GP). *S. palustris* (L.) Hook.—Rydberg.

3. **Senecio crassulus** A. Gray. Perennial 2–5 dm tall, glabrous. Stems single or loosely clustered from an erect and frequently branching, weakly lignified caudex. Basal and lower cauline leaves broadly ovate to spatulate, sharply dentate to subentire, but frequently with callous denticles on the margins, blade 3–12+ cm long and 1.5–5 cm wide, ca 1.5 × longer than wide, tapering to a semidistinct petiole, lowermost cauline leaves often larger than the basal leaves; upper cauline leaves reduced, sessile to clasping. Inflorescence a corymbiform cyme of 4–12 heads; involucral bracts 5–9 mm long and often with a conspicuous, black, villous tip, ca (8)13 or 21; ray florets ca 8 or 13, ligule 5–12 mm long. Achenes glabrous.

(n = 20) Jun–Jul. Open woodlands & rocky edges of meadows; BH; (MT to e OR, s to NM).

4. **Senecio douglasii** DC. var. **longilobus** (Benth.) Benson. Shrub 0.3–1 + m tall, tomentose, often with loose, lanate hairs, sometimes irregularly glabrescent. Stems numerous, few-branched, arching upward from a taprooted, woody caudex. Leaves narrow-linear to filiform, simple or deeply pinnatifid into long, narrow segments; fasicles of small leaves sometimes borne in the axils of larger leaves. Inflorescence a compound series of corymbiform or subcorymbiform cymes, with the inflorescence of each branch terminating in 3–10(20) heads, and the clusters of branches aggregated to form a large and showy inflorescence; involucral bracts ca 13(21) and 5–8 mm long, ray florets ca 8, ligule 10–15 mm long, light yellow to nearly ochroleucus. Achenes canescent-hirtellous. (n = 20) May–Aug. Open, sandy or rocky sites; sw KS & se CO, s through adj. NM & OK into TX; (s GP to AZ; Mex.).

Var. *longilobus* is the easternmost extreme of a continuum stretching from California to Kansas. It is distinguished by having smaller heads, minute or obsolete calyculate bracts and long, lanate tomentum. Cf. Barkley, op. cit., pp. 116–118.

5. **Senecio eremophilus** Rich. Perennial 4–8 dm tall, glabrous or nearly so, sometimes pinkish below. Stems single or several loosely clustered from a coarse, woody fibrous-rooted caudex. Leaves equally distributed along the stem, the lowermost sometimes early deciduous; blades ovate to narrowly lanceolate in outline but variously lacerate to pinnate-laciniate, with the lobes dentate, variously petiolate to sessile, overall dimensions (3)6–12(20) cm long and 1.5–5 + cm wide, ca 2–3 × longer than wide. Inflorescence a compound corymbiform cyme with several cymules of 2–6 + heads each, heads totaling 10–40; principal involucral bracts ca 13 and 6–8 mm long, calyculate bracts prominent and at least some ca 3/4 as long as the principal bracts; ray florets ca 8, the ligule 5–10 mm long. Achenes glabrous or minutely hirtellous. (n = 20) Jun–Jul (Sep). Grassy or gravelly sites, especially roadside ditches & stream beds; BH; ND: Bottineau, Rolette; (ND, N. T. & B.C., s to AZ & Mex.?).

This species is separable into three varieties. Our materials belong to var. *eremophilus*, characterized by the relatively large heads with prominent calyculate bracts, and a northern distribution.

6. **Senecio glabellus** Poir., butterweed. Turgid annual 2–6(10) dm tall, glabrous or very lightly tomentose in the leaf axils and among the heads, sometimes pinkish or purplish-tinged. Stems normally single, arising from a cluster of thin fibrous roots; leafy throughout, but the upper leaves smaller; middle and lower leaves obovate-oblanceolate in outline but deeply pinnate-lyrate with the margins of the lobes crenate to dentate, lobes tapering at the base but broadly attached to the midrib; overall leaf dimensions 5–15 cm long and 1–3 + cm wide, without a well-defined petiole, upper leaves sometimes clasping. Inflorescence a subumbelliform cyme of 8–20(+) heads or of several cymules in robust individuals; involucral bracts ca 21 or 13 and 4–7 mm long; ray florets ca 13 or 8, the ligule 6–9 mm long. Achenes lightly pubescent, especially along the angles, or glabrous. (n = 23) Apr–Jun. Damp, open woods or swampy grassland; extreme se NE, e KS, & ne OK; (OH to NE, s to FL & TX).

This is a variable species, barely entering our range from the east, where hybridizes with the eastern *S. aureus*, and possibly perennates in favorable sites.

7. **Senecio hydrophilus** Nutt. Tall perennial (biennial?) 4–10(+) dm tall, distinctly glaucous and sea-green in color. Stem hollow, arising singly from a buttonlike, foreshortened caudex with a cluster of unbranched, fleshy fibrous roots. Leaves thickish-turgid, progressively reduced upward; basal and lower cauline leaves petiolate, blades elliptic to oblanceolate, denticulate to entire, 5–20 cm long and 2–8(10) cm wide, about 2 × longer than wide,

petiole about as long as the blade; middle and upper cauline leaves strongly reduced, becoming mere sessile bracts. Inflorescence a compound corymbiform cyme of 20–40 heads; involucral bracts ca 13 or 8 and 5–8 mm long, frequently black-tipped; ray florets few, ca 5, the ligule 3–8 mm long. Achenes glabrous. (n = 20) Jul–Aug. Swamps, tolerant of standing water & alkaline situations; BH; (SD w through the cordillera to B.C. & CA).

8. **Senecio imparipinnatus** Klatt. Thin annual or winter annual, 1–3(8) dm tall, glabrous or with scattered light tomentum, especially among the heads. Stems arising singly or several clustered from a pronounced taproot. Leaves well developed on the lower 1/3 of the stem, the uppermost gradually reduced; lower and middle leaves oblanceolate to spatulate in outline, but lyrate-pinnate, with the lobes crenate-dentate, lobes distinctly cuneate-contracted at the point of attachment to the midrib, overall leaf dimensions 3–7 cm long and 1–3 cm wide, about 2× longer than wide, petiole indistinct. Inflorescence a corymbiform or subumbelliform cyme of 6–16 heads, or of several cymes in robust individuals; involucral bracts ca 13 or 21 and 4–7 mm long; ray florets ca 8, the ligule 8–12 mm long. Achenes lightly hirtellous, especially along the angles. (n = 23) Apr–May. Low sandy or drying sites, mildly weedy; extreme se KS, e & cen OK; (MS to TX, n to KS; n Mex.).

This is a weed in the Rio Grande R. valley, barely entering our range. It appears as a small *S. glabellus*, but with a distinct taproot and basally constricted leaf lobes.

Recent studies show that the correct name for this species is likely to be *S. tampicanus* DC.

9. **Senecio intergerrimus** Nutt. Biennial or perennial (1)2–7 dm tall, arachnoid villous with crisp, jointed hairs when young, but generally glabrate in age, although with hairs remaining in and near the axils of the leaves and among the heads. Stems single, arising from a short, buttonlike caudex with numerous fleshy-fibrous roots. Basal and lowermost cauline leaves petiolate or with the blade tapering to a broadly winged petiole, entire to irregularly dentate, lanceolate to subspatulate, entire leaf 6–25 cm long and the petiole, when distinct, about equaling the blade; upper cauline leaves strongly and progressively reduced, becoming mere bracts. Inflorescence a corymbiform cyme of (3)6–20 heads, central terminal head often on a distinctly shortened peduncle; involucral bracts 7–15 mm long, often with small, faint black tips, ca 21 or 13; ray florets ca 13 or 8, the ligule 6–15 mm long or sometimes ray florets absent. Achenes glabrous. (n = 20, 40) May–Jun. Open, undisturbed prairie, especially in damp grassy sites; (MN & IA to B.C. & CA).

Five weakly defined varieties are recognized for this widespread species, of which two occur in our region (Barkley, op. cit., pp. 107–110).

9a. var. *integerrimus*. Essentially glabrous at flowering time. Involucral bracts distinctly subulate and minutely, if at all blacktipped. N GP, extending e to w MN, IA, & n KS. Passing into the more western var. *exaltatus* (Nutt.) Cronq., at the foot of the Rocky Mts. uplift. Var. *exaltatus* is merely glabrescent and has distinctly black-tipped involucral bracts.

9b. var. *scribneri* (Rydb.) T. Barkley. Leaves narrow. Heads usually 6 or fewer, involucral bracts 10–15 mm long and without black tips, ray florets ochroleucous. Restricted to open prairies and hillsides in e-cen MT.

10. **Senecio obovatus** Muhl. ex Willd., roundleaf groundsel. Perennial 2–5 dm tall, glabrous or very lightly tomentose when young, with some tomentum remaining in and near the axils of the lower leaves. Stem single or sometimes clustered from a short, erect to horizontal caudex which is usually stoloniferous, stolons slender, 1–2 + dm long or rarely stout, usually unbranched and distally giving rise to a rosette of basal leaves which eventually develops into a new plant. Basal leaves distinctly petiolate, blades obovate or oblong ovate to orbicular, abruptly contracted or tapering at the base, crenate-dentate to serrate, 2–7(15) cm long and 1–4(6+) cm wide, 1–2× longer than wide, petiole 1–2× longer than the blade; cauline leaves reduced upward to mere bracts. Inflorescence a corymbiform cyme

of 3–10(20) heads; peduncles usually with 1–several minute bracteoles irregularly spaced between the stem and the receptacle; principal involucral bracts ca 13 or 21 and 3–6(7) mm long; ray florets ca 8 or 13, the ligule 5–10 mm long or ray florets sometimes absent. Achenes glabrous or lightly hirtellous, especially along the angles. (n = 22) Mar–May. Rocky, wooded hillsides & stream beds; e KS & OK; (VT & NH s to FL, w to KS & TX; n Mex.). *S. rotundus* (Britt.) Small — Rydberg.

11. ***Senecio pauperculus*** Michx., balsam groundsel. Perennial 1–4 dm tall, glabrous or lightly tomentose, especially toward the base and among the heads, but glabrate in age. Stem arising from a short, simple to weakly spreading caudex. Basal leaves petiolate, blades lanceolate or oblanceolate to subelliptic, obtuse to cuneate at the base, weakly crenate to subentire, 3–7 cm long and 1–2(3) cm wide, ca 1.5–3 × longer than wide, petiole about as long as the blade; cauline leaves progressively reduced upward, margins variously dissected, incised to lacerate, or the lower cauline leaves even sublyrate; uppermost cauline leaves frequently reduced to mere bracts. Inflorescence a corymbiform cyme with 3–10 + heads; principal involucral bracts ca 13 or 21 and 4–8 mm long; ray florets ca 8 or 13, the ligule 5–10 mm long, or ray florets occasionally absent. Achenes glabrous or sometimes hispidulous, especially along the angles. (n = 22, 23) May–Jun. Damp, grassy meadows & rich prairies; n GP; w MN & IA, scattered w through SD, ND, & MT; (widespread across N. Amer. from Lab. to GA, w to AK & OR).

This is a highly complex species, intergrading with *S. plattensis* in the ne GP and with other senecios out of our range.

12. ***Senecio plattensis*** Nutt., prairie ragwort. Biennial or short-lived perennial (1)2–5(7) dm tall; floccose-tomentose or irregularly glabrescent. Stems single or rarely 2 or 3 loosely clustered from a short, ascending caudex which is sometimes stoloniferous. Basal leaves petiolate, blade elliptic-ovate to oblanceolate, crenate or serrate-dentate to subentire, (1)2–6(10) cm long and (0.5)1–3(5) cm wide, petiole ca 1–1.5 × as long as the blade; cauline leaves progressively reduced upward, lower and middle ones sublyrate to pinnatisect, uppermost ones irregularly dissected to subentire. Inflorescence a corymbiform cyme of 6–20 heads; involucral bracts ca 13(21), 5–6(7) mm long; ray florets ca 8, ligule ca 10 mm long. Achenes normally hirtellous, especially along the angles, or sometimes glabrous. (n = 46) Apr–Jun. Open prairies & roadsides; GP, more common in e; (w NC, TN to CO; Man. & N.T. to TX).

S. plattensis intergrades with *S. pauperculus* in the ne GP and through the upper Mississippi valley.

13. ***Senecio pseudaureus*** Rydb. var. ***semicordatus*** (Mack. & Bush) T. Barkley. Perennial 2–5 dm tall, thin and slender, glabrous or lightly floccose-tomentose when young, sometimes lightly pinkish tinged. Stem arising from a simple or weakly branched caudex. Basal leaves petiolate, blade ovate to broadly lanceolate, truncate to subcordate or at least abruptly contracted at the base, serrate to crenate-dentate, 2–4 cm long and 1–2 + cm wide but occasionally larger, ca 1.5 × longer than wide, petiole ca 1–2 × longer than the blade; cauline leaves sublyrate or laciniate to subentire, strongly reduced up the stem, the uppermost sessile and sometimes clasping. Inflorescence a subembellate to corymbiform cyme; involucral bracts ca 21 or 13 and 4–6 mm long; ray florets ca 13 or 8, the ligule 6–10 mm long, or ray florets sometimes absent. Achenes glabrous (n = 23 in GP ?, n = 20, 40 in w) May–Jul. Damp, open prairies; e KS n to e ND, BH; (GP, adj. Can., to WI).

Var. *semicordatus* passes morphologically into the typical var. *pseudaureus* in s Can. and in the Rocky Mts. uplift of central MT and WY. It intergrades with the eastern *S. aureus* along the e edge of the GP. Cf. Barkley, op. cit., pp. 62–64.

14. ***Senecio rapifolius*** Nutt. Perennial 3–6 dm tall, glabrous, often glaucous, sometimes purplish-tinged. Stem arising from a suberect to weakly spreading, woody caudex. Basal

and lower cauline leaves ovate to broadly lanceolate, tapering or abruptly contracted to a weakly winged petiole, sharply dentate to incised, blade 4–8(9) cm long and 3–5 cm wide, about 2 × longer than wide, petiole about as long as the blade; middle cauline leaves similar but clasping and subauriculate; uppermost leaves reduced to bracts. Inflorescence an open to congested corymbiform cyme of several (3–10) cymules, for a total of ca 30–60 heads; principal involucral bracts ca 8 or sometimes ca 5 and (2)3–5 mm long; ray florets absent. Achenes glabrous. (n = 20) Aug–Oct. Rocky hillsides & cliffs in coniferous areas; BH; (WY to CO).

15. Senecio riddellii T. & G., Riddell ragwort. Subshrubby perennial 3–10 dm tall, glabrous. Stems numerous, arching and branching upward from a taprooted woody crown. Leaves irregularly pinnate-divided into linear-filiform segments 4–9 cm long and 1–5 mm wide; foliage often dense, lowermost leaves often drying and pendulous on mature plants. Inflorescence a series of corymbiform cymes with ca 5–20 heads each; heads spreading-campanulate, involucral bracts ca 13 and 7–10 + mm long, calyculate bracts, when present, less than 1/4 as long as the principal bracts; ray florets ca 8, the ligule 8–10 + mm long, often early deciduous. Achenes gray-hirtellous. Aug–Oct. Open sites, especially sandy, drying flood plains; sw GP; sw SD s to w OK, TX; (GP sw to AZ).

This plant is poisonous to livestock and, therefore, locally scarce due to efforts at eradication.

16. Senecio spartioides T. & G. Subshrubby perennial 2–8 dm tall, glabrous. Stems numerous, branching upward from an erect, branched, taprooted crown. Leaves narrowly linear and entire, or rarely with 1 or 2 pairs of linear filiform lobes, blade ca 2 mm wide or more, rarely as narrow as 1 mm, normally 5–10 cm long. Inflorescence a corymbiform cyme of several cymules, totaling 10–20 + heads, surmounting the principal branches; heads subcylindrical or very narrow-campanulate; principal involucral bracts ca 8 or infrequently ca 13, 6–9(+) mm long; ray florets ca 5 or 8, the ligule 10 + mm long. Achenes gray-hirtellous. (n = 20) Jul–Sep. Dry, open sites; infrequent in w SD & w NE, occasionally encountered westward in GP; (s WY to NM, w to CA).

17. Senecio tridenticulatus Rydb. Perennial 1–3 dm tall, glabrous or very lightly floccose-tomentose when very young. Stems single or sometimes several loosely clustered from an erect, terminally branched taproot or taprootlike caudex. Leaves thickish and turgid when fresh; basal leaves petiolate, blades lanceolate to narrowly oblanceolate, tapering at the base, subentire to dentate, especially toward the apex, sometimes irregularly pinnatisect, 2–4(+) cm long and 0.5–1 + cm wide, ca 3 × longer than wide, petiole 1–2 × as long as the blade; cauline leaves similar in aspect to the basal leaves but strongly reduced upward. Inflorescence a corymbiform cyme of 4–12(20) heads; involucral bracts ca 13 or sometimes 21 and 6–10 mm long; ray florets ca 8 or 13, ligule 5–8 mm long. Achenes glabrous or hirtellous along the angles. (n = 23) May–Jun. Open plains & foothills; ND to TX, w to the Rocky Mts. uplift; (Man. to TX, w to the geological parks of the Rocky Mts.) *S. densus* Greene; *S. oblanceolatus* Rydb.—Rydberg.

This species is apparently rather localized but to be expected throughout the high plains.

18. Senecio vulgaris L., groundsel. Annual, 1–3(4) dm tall, glabrous or sparsely tomentose when young. Stems erect or sublax, single or branched near the base, arising from a taproot. Leaves obovate to oblanceolate in outline but deeply and irregularly dentate to lobulate, the lobes often dentate-denticulate, overall dimensions 2–10 cm long and 0.5–2 cm wide, lower and middle leaves tapering to a short and poorly defined petiole, upper ones subsessile to weakly clasping. Inflorescence a loose cluster of 8–20 heads, terminating each main branch; involucral bracts ca 21 and 4–6 mm long, frequently with prominent black tips, subtended by 2–6 conspicuously black-tipped calyculate bracts; ray florets absent. Achenes lightly pubescent or sometimes nearly glabrous. (n = 20, 24) Mar–May. A weed of well-watered, rich soils; rare & sporadic in the GP, but to be expected as a waif anywhere; (across N. Amer.; Eurasia). *Naturalized* in e U.S. & on the w coast.

96. SHINNERSOSERIS Tomb

1. *Shinnersoseris rostrata* (A. Gray) Tomb. Taprooted annual 1–4(8) dm tall, with cream-colored milky juice. Stem glabrous, green, paniculately branched in the upper 2/3 but sparingly branched below. Leaves opposite in the lowest 5–8 nodes, not forming a basal rosette, linear, 6–10 cm long and 2–4 mm wide, entire; upper leaves similar but alternate and smaller, the uppermost reduced to mere bracts. Heads terminal, involucre 12–15 mm tall, narrowly cylindrical; principal involucral bracts 7 or 9, subtended by ca 2 series of reduced calyculate bracts; florets 8–11, all ligulate and fertile, corolla lavender with yellow apices, ca 6 mm long, erect (i.e., not reflexed) at anthesis. Achenes subcylindrical, 8–10 mm long, abruptly contracted and scabrate below the summit; pappus of numerous capillary bristles 6–8 mm long, white, somewhat connate at the base. (n = 6) Jul–Sep. Open sandy prairies & plains, stream banks; scattered throughout GP but absent in e 1/3 KS & s; (GP, cen CO). *Lygodesmia rostrata* A. Gray — Rydberg. [tribe Lactuceae]

Reference: Tomb, A. S. 1973. *Shinnersoseris*, gen. nov. Sida 5: 183–189.

This species has long been known in the genus *Lygodesmia*, but Tomb makes a good case for treating it as a distinct genus.

97. SILPHIUM L., Rosin-weed

Coarse, erect perennial herbs, often scabrous hispid; sparingly branched upward. Leaves alternate or opposite, entire or dentate to deeply pinnate-lobed. Heads subhemispheric, radiate; involucral bracts subequal or imbricate, herbaceous to somewhat membranaceous; receptacle flat, chaffy; ray florets pistillate and fertile, ligule yellow; disk florets staminate only, with undivided style; chaffy bracts subtending the florets. Ray achenes glabrous, strongly flattened at right angles to the radius of the head, margin winged; pappus absent or of 2 awns arising from the wings of the achene. [tribe Heliantheae]

1 Leaves connate-perfoliate; stems square in cross section .... 3. *S. perfoliatum*
1 Leaves basally free from each other and not connate-perfoliate.
  2 Leaves deeply pinnate-lobed (laciniate); plants taprooted .... 2. *S. laciniatum*
  2 Leaves subentire to toothed; plants fibrous rooted with a short rhizome or caudex.
    3 Stem glabrate to velvety or scabrous, with nearly all of the hairs less than 0.5 mm long; abundant in se GP .... 1. *S. integrifolium*
    3 Stem coarsely and densely hispid to hispid-scabrous, with the hairs mostly 1 mm long or longer; scattered in OK .... 4. *S. radula*

1. *Silphium integrifolium* Michx. Coarse perennial 5–15 + dm tall, glabrate to variously velvety or scabrous, with most hairs less than 0.5 mm long. Stems few-branched, arising from a fibrous-rooted caudex or short rhizome. Leaves about evenly distributed along the stems, mostly opposite but sometimes alternate or in whorls of 3, sessile and often clasping, ovate or narrowly ovate to elliptic, 7–15 cm long and 2–6 cm wide, margin entire or toothed, scabrous above and scabrous to velvety or glabrate beneath. Heads in a short, open inflorescence; involucre 1–2 cm tall, involucral bracts ovate to elliptic, leafy and somewhat imbricate, margin ciliate; disk 1.5–2.5 cm across; ray florets commonly ca 21 or 34, ligule 2–5 cm long, yellow. Achenes obovate to suborbicular, with wings commonly terminating upward in a sharp tooth on each side of the achene body. (n = 7) Jun–Oct.

Open roadsides & casually disturbed places in the prairie; se 1/4 GP; (MI to MS, w to GP, AR, & possibly TX).

Two semidistinct varieties occur in our region, and they have been treated as separate species in the past.

1a. var. *integrifolium*. A ± persistently hairy phase, with the outside of the involucre, the upper stems and the undersides of the leaves velvety to scabrous. Typically found in the richer soils in the tall grass prairie along the e edge of the GP, but sporadic to cen KS and se SD.

1b. var. *laeve* T. & G. A less-pubescent phase, with the upper herbage glabrate, and sometimes subglaucous. More frequent in the mixed prairie region and less well-watered sites, but sporadic eastward. *S. speciosum* Nutt.—Rydberg.

2. ***Silphium laciniatum*** L., compass plant. Erect, coarse perennial 1-3 m tall, hispid or hirsute with spreading hairs, sometimes ± glandular. Stems arising from a prominent woody taproot. Leaves alternate, deeply pinnatifid to laciniate, the larger lobes sometimes again lobed; lowermost leaves up to 4 dm long, but progressively reduced upward. Heads several in an elongate, racemiform inflorescence; involucre hirsute to hispid, 2-4 cm tall, involucral bracts ovate, acuminate, subequal; disk 2-3 cm across; ray florets 15-24(34), ligule 2-4 + cm long, yellow. Achenes with wings apically expanded into rounded projections, forming a notch 1-3 mm deep. (n = 7) Jun-Sep. Open prairies, especially in areas of mild disturbance; e 1/3 GP from se SD to cen OK; (OH to AL, w to GP).

The basal leaves are often generally aligned in a north-south direction, hence the common name.

3. ***Silphium perfoliatum*** L., cup plant. Coarse perennial 1-2 + m tall, glabrous or lightly spreading-hairy. Stems square in cross section. Leaves mostly opposite, or the uppermost sometimes alternate, the petiolar bases connate-perfoliate; blade ovate to deltoid ovate, up to 3 dm long, margin coarsely toothed, mostly glabrate or short-hairy beneath. Heads numerous in an open inflorescence; involucre 1.5-2.5 cm tall, involucral bracts elliptic to ovate, subequal, margin ciliate; disk 1.5-2.5 cm across; ray florets ca 21 or ca 34, ligule 1.5-4 cm long, yellow. Achenes obovate, with prominent wings and a deep, narrow apical notch. (n = 7) Jul-Sep. Moist low ground, especially roadside ditches & open wooded areas; e 1/5 GP from ne ND to e KS & adj. OK; (s Ont. to NC, w to GP & AR).

4. ***Silphium radula*** Nutt. Coarse perennial to 2 m tall, distinctly hispid-scabrous to hispid with abundant hairs that are mostly 1 mm long or more, but sometimes glabrate in age. Stems sparingly branched above, arising from a short rhizome or caudex. Leaves mostly opposite but the upper often alternate, sessile and entire or toothed, ± evenly distributed over the stem and nearly all equal in size and only the uppermost reduced and bractlike; blades ovate to oblanceolate, 6-12(15) cm long and 3-9 cm wide. Heads in an open subcorymbiform to contracted and nearly racemiform inflorescence; involucre 1-2 + cm tall; involucral bracts ovate to elliptic, leafy and usually somewhat imbricate; disk to 2.5 cm across; ray florets numerous, often ca 34 but sometimes ca 21, ligule 3-4 + cm long, yellow. Achenes prominently winged, with a deep apical notch. (n = 7) Jul-Sep. Prairies or open wooded sites; cen OK; (OK & TX; disjunct in AL & GA). *S. asperrimum* Hook.—Fernald.

Silphium asteriscus L. is similar, but less robust, mostly less than 10 dm tall, and with smaller heads, mostly with ca 13 ray florets. Its var. *scabrum* Nutt. is of Ozarkian distribution and may encroach into the se GP. *S. asteriscus, S. radula,* and *S. integrifolium* collectively form a complex of semidistinct entities whose taxonomy is yet to be fully elucidated.

98. **SOLIDAGO** L., Goldenrod

Perennial herbs with stems arising singly or variously clustered from fibrous rooted rhizomes or a branching caudex. Leaves alternate, variously glabrous to hairy but not resinous-punctate (as in *Euthamia*). Inflorescence of numerous relatively small heads, arranged in axillary or capitate clusters, often thyrselike, or paniculate with recurved-secund branches, or sometimes in terminal, corymbose clusters; involucral bracts imbricate in several series or occasionally subequal, chartaceous to stramineous at the base but commonly greenish toward the tip; receptacle flat or low-convex, normally naked but occasionally with a few chaffy bracts near the margin; ray florets yellow (white in *S. ptarmicoides*), few, rarely more than 13, pistillate and fertile; disk florets seldom more than 20 and mostly fewer, perfect and fertile; style branches flattish, with internal-marginal stigmatic lines and a lanceolate, pubescent appendage. Achenes subterete or angled, sometimes prominently ribbed, glabrous or hairy; pappus of numerous white bristles. [tribe Astereae]

    References: Croat, T. 1967. The genus *Solidago* of the North Central Plains. Unpub. Ph.D. dissertation, Univ. of Kansas, Lawrence; Croat, T. 1972. *Solidago canadensis* complex in the Great Plains. Brittonia 24: 317–326; Cronquist, A. 1980. *Solidago in* Vascular Flora of the Southeastern U.S., vol. 1. Univ. of North Carolina Press, Chapel Hill.

    Goldenrods are a diverse group with many intergrading species, especially in the woodland regions east of the GP. Some of our species appear to hybridize freely and often between species that seem not to be closely related.

    Inflorescence patterns and disposition of the leaves are significant features in the taxonomy of the goldenrods. There are three general patterns for the inflorescence: (1) the heads are in axillary clusters and/or in a somewhat elongate, terminal thyrse that is straight and more-or-less cylindrical, or the inflorescence may have several thyrselike branches; (2) the inflorescence is paniculate (sometimes termed "paniculiform") with at least the lower branches recurved-secund, i.e., the branches drooping and the heads disposed chiefly along one side of the branch, usually the upper side; (3) the inflorescence is corymbose (or "corymbiform"), i.e., a somewhat open, broad, and flat to dome-shaped terminal cluster, but with the branches not at all recurved.

    There are two general patterns for the disposition of the leaves. In some goldenrods the principal leaves are basally disposed. The radical and lowermost cauline leaves are persistent, well developed and petiolate, while the cauline leaves tend to be smaller, progressively reduced upward and becoming sessile. In other species the leaves are chiefly cauline, with the radical and lowermost cauline leaves reduced and usually early-deciduous, so the lower stem is nearly naked by flowering time; the largest persistent leaves are then the middle cauline leaves, and the upper cauline leaves are often reduced. The distinction between these two leaf-habits is not absolute, and a species may range from one habit-type to another, e.g., *S. ulmifolia*.

1 Inflorescence corymbiform or capitate, i.e., the heads disposed in a terminal, flat-topped cluster or in a conspicuous, tight group at the end of the central, upright axis.
    2 Rays white, pappus-bristles, or at least some of them, obviously subclavate, i.e., thickened toward the apex .... 8. *S. ptarmicoides*
    2 Rays yellow, pappus-bristles not at all thickened upward.
        3 Herbage rough-puberulent; principal cauline leaf blades 2–6 × longer than wide, usually single-nerved; widespread in open, dry prairies .... 11. *S. rigida*
        3 Herbage essentially glabrous, or weakly pubescent upward on the stem and on the leaf margins; principal cauline leaf blades 6–10 + × longer than wide and tending to be 3-nerved; wet prairies in extreme ne GP .... 10. *S. riddellii*
1 Inflorescence variously axillary or paniculate to thyrselike, but not a corymbiform or capitate terminal cluster.
    4 Inflorescence of small, axillary clusters, or of a terminal, simple or branched thyrse, i.e., a straight, cylindrical cluster, and neither nodding nor with recurved, secund branches.
        5 Achenes persistently hairy.
            6 Rays 3 or 4; leaves single-nerved or obscurely 3-nerved; upper stem glabrous .... 2. *S. flexicaulis*
            6 Rays 6–18; leaves 3-nerved, herbage variously pubescent to subglabrous.
                7 Rays 9 or fewer; herbage spreading-hirtellous, or glabrescent below .... 5. *S. mollis*
                7 Rays typically 10–18; herbage various, often somewhat scabrous or puberulent .... 1. *S. canadensis*
        5 Achenes glabrous, at least at maturity.
            8 Cauline leaves distinctly smaller than the basal leaves and progressively reduced upward; involucral bracts obtuse to rounded .... 13. *S. speciosa*

        8   Cauline leaves about equaling the basal leaves and scarcely reduced upward; involucral bracts acute to acuminate .... 7. *S. petiolaris* var. *angusta*
 4 Inflorescence terminal, paniculate to unilaterally racemose, nodding at the summit and/or with the lower branches recurved-secund, i.e., with the heads disposed chiefly along one side of the peduncle.
     9 Leaves chiefly basal, with the cauline leaves smaller and progressively reduced upward (note that the lowermost cauline leaves may be withered in specimens collected late in the season).
         10 Herbage densely and persistently fine-puberulent; stems often clustered from a branching caudex .... 6. *S. nemoralis*
         10 Herbage variously and unevenly pubescent, but the upper stem glabrate, especially below the inflorescence; stems often arising singly or a few from a loose caudex or rhizome.
             11 Basal and lower cauline leaves tapering to a weakly defined petiole; plants wholly glabrous, arising from a creeping rhizome or spreading caudex; widespread in GP .... 4. *S. missouriensis*
             11 Basal and lower cauline leaves abruptly contracted to the petiole; plants pubescent at least on the main veins on the underside of the leaves; arising from a short caudex or rhizome; se GP .... 14. *S. ulmifolia*
     9 Leaves chiefly cauline, and the basal leaves, when present, no larger than the cauline leaves.
         12 Leaves with a single prominent nerve or midrib (in addition to the network of smaller veins) .... 14. *S. ulmifolia*
         12 Leaves, or at least the larger ones, prominently 3-nerved.
             13 Herbage essentially glabrous, except for light puberulence among the heads in the inflorescence and sometimes on the main nerves on the undersides of the leaves; upper stem glaucous .... 3. *S. gigantea*
             13 Herbage variously puberulent to scabrous, sometimes unevenly so.
                 14 Rays numerous, (8)10–18 .... 1. *S. canadensis*
                 14 Rays few, 4–8.
                     15 Leaves greenish; pubescence coarse and scabrous; extreme se GP .... 9. *S. radula*
                     15 Leaves pale-gray, canescent.
                         16 Leaves oblanceolate; involucral bracts with distinctly acute tips; scattered along w GP and BH .... 12. *S. sparsiflora*
                         16 Leaves elliptic or broadly lanceolate; involucral bracts rounded to weakly acutish; widespread in GP .... 5. *S. mollis*

1. **Solidago canadensis** L., Canada goldenrod. Perennial herb 2.5–20 dm tall; herbage usually puberulent to spreading pubescent, at least above the middle. Stems arising singly or loosely clustered from an often extensive system of creeping rhizomes, or from a weakly creeping caudex. Leaves numerous, chiefly cauline and about equally distributed along the stem, 3-nerved, narrowly lanceolate to elliptic, 3–15 cm long and 5–20+ mm wide, sessile or very nearly so, usually densely spreading-puberulent on the underside but frequently scabrous or even subglabrous above. Inflorescence paniculate, with prominent recurved, secund lateral branches, or thyrselike and not secund (var. *salebrosa*); involucre 2–4.5+ mm tall; involucral bracts thinnish, narrow, acute to acuminate, yellowish but somewhat greenish toward the tip; ray florets (5)10–18, ligule 1–3 mm long; disk florets 2–8. Achenes short-pubescent. (n = 9, 18, 27) Jul–Sep. Damp or drying open places, often in loose soils, & in clearings in wooded regions.

This is a widespread, highly diverse species throughout much of North America, with several semidistinctive regional and ecological varieties. The treatment here generally follows that of Croat, 1972, op. cit. Five varieties may be distinguished, at least to some extent, in the GP.

 1 Involucre typically less than 3 mm tall.
     2 Leaves thin in texture and usually serrate; rays short, with ligules rarely more than 1.5 mm long; ne GP and BH .... 1a. var. *canadensis*

2   Leaves thickish and firm, entire to serrate; rays long with the ligules often up to 4.5 mm; widespread in e 1/2 GP .... lc. var. *hargeri*
1 Involucre 3 mm tall or more.
3   Leaves thin in texture; inflorescence typically a contracted panicle, becoming thyrselike; scattered in extreme n and w GP, and BH .... 1d. var. *salebrosa*
3   Leaves thickish and firm; inflorescence paniculate with recurved, secund branches; widespread.
4   Leaves scabrous to subglabrous on the upper surface, with hairs fewer, shorter and stiffer than on the lower surface; well-developed plants normally 1–1.5 + m tall .... le. var. *scabra*
4   Leaves spreading-puberulent on the upper surface, with the hairs slightly if at all fewer and stiffer than those on the lower surface; well-developed plants rarely exceeding 1 m tall .... 1b. var. *gilvocanescens*

1a. var. *canadensis*. Stem sparsely pubescent but sometimes densely so. Leaves thinnish, puberulent on the midrib and prominent lateral veins on the underside, serrate or sometimes subentire. Scattered in ne GP & apparently present in BH; (Newf. & NY, w to GP).

1b. var. *gilvocanescens* Rydb. Stems rarely more than 1 m tall, arising singly or widely scattered from a short, weakly creeping rhizome or laterally branching caudex. Leaves thick and firm, the well-developed ones typically less than 7 cm long. Widespread in n 2/3 GP; (MN to MO, w through GP). *S. pruinosa* Greene—Gleason; *S. lunellii* Rydb.—Rydberg.

1c. var. *hargeri* Fern. Stems regularly arising in clumps; herbage densely spreading-pubescent. Leaves thickish and firm, entire to serrate. Heads small; disk florets (1)3 or 4(6). Scattered in e 1/2 GP (CT to VA w to GP & scattered sw to NM). *S. satanica* Lunell—Rydberg.

This variety is similar to var. *scabra*, and sometimes not distinguished from it, but var. *hargeri* is typically diploid (n = 9), and flowers ± 2 weeks prior to var. *scabra*. Furthermore, it has a distinctive aspect in the field; cf. Croat, 1972, op. cit.

1d. var. *salebrosa* (Piper) M. E. Jones. Stem loosely clustered from creeping rhizomes. Leaves thinnish, flexible, and with scattered hairs or subglabrate. Inflorescence typically unlike that of the other vars. of *S. canadensis* in that it is thyrselike and resembles a contracted to spreading cone or cylinder, without recurved, secund branches; heads large; involucre (2)3–5 mm tall. Barely entering the nw GP, but apparently scattered in cen ND, & in the BH; (nw GP w to AK & OR, apparently e through cen Can. to Lab.). *S. dumetorum* Lunell—Rydberg.

1e. var. *scabra* T. & G. Stems regularly 1–1.5(2) m tall, arising from an extensive system of branching rhizomes. Leaves thickish, the prominent ones nearly always 7 cm long or more. Heads large, involucre normally more than 3.5 mm tall. Throughout the GP, but less frequent on the w edge; (Que. to n FL, w to GP). *S. altissima* L.—Rydberg.

This variety is similar to but much more abundant than var. *hargeri*, from which it differs in the characters given in the key and descriptions, plus being typically hexaploid, (n = 27).

2. **Solidago flexicaulis** L., broad-leaved goldenrod. Perennial 3–10 + dm tall. Stems somewhat grooved, glabrous above, arising singly from an elongate rhizome. Leaves chiefly cauline, with a single prominent midrib or obscurely 3-nerved; sharply dentate, acuminate, usually hirsute on the underside, at least on the main veins, glabrous to thinly pubescent above; the prominent midcauline leaves ovate to elliptic, 7–15 cm long and 3–10 cm wide, mostly 1–2 + × longer than wide, abruptly contracted to a winged petiole. Inflorescence a series of short clusters of heads, the lower ones in the axils of the upper leaves, which in turn are progressively reduced upward to mere bracts; the terminal inflorescence appearing as a naked thyrse; involucre 4–6 mm tall, outer involucral bracts obtuse but the inner ones ± rounded; ray florets 3 or 4; disk florets 5–9. Achenes short-hairy. (n = 9, 18) Aug–Oct. Wooded areas; extreme e edge GP; (N.B. & N.S. to VA, w to GP & AR).

3. **Solidago gigantea** Ait., late goldenrod. Large perennial (5)10–15 + dm tall. Stems glabrous and glaucous below, or finely pubescent in the inflorescence, arising singly or loosely clustered from elongate, creeping rhizomes. Leaves chiefly cauline, numerous, prom-

inently 3-nerved; the middle ones acuminate, subentire to serrate-dentate, narrowly elliptic to lanceolate, 6–17 cm long and 1–4.5 cm wide, sessile or weakly petiolate, the underside glabrous or somewhat pubescent on the 3 main veins. Inflorescence paniculate, with conspicuous recurved-secund branches; involucre 2.5–4 mm tall, involucral bracts somewhat blunt and greenish toward the apex; ray florets (8)10–18; disk florets 6–10. Achenes short-hairy. (n = 9, 18, 27) Aug–Oct. Moist, open sites; throughout GP; (Que. to AL, w to B.C. & NM). *S. serotina* Ait.—Rydberg.

Plants with the undersides of the leaves glabrous have been segregated as var. *serotina* (O. Ktze.) Cronq., but these plants appear to occur together with the nomenclaturally typical var. *gigantea*, which has the veins short-hairy on the undersides of the leaves. The relationship between the two presumed entities is yet unclear. *Solidago gigantea* is similar to *S. canadensis* but is basically glabrous, and the involucral bracts are a little more coarse, firm, and blunt.

4. ***Solidago missouriensis*** Nutt., prairie goldenrod. Perennial 4–9+ dm tall, glabrous. Stems arising singly or clustered from a creeping rhizome or a spreading caudex. Leaves thickish and somewhat rigid, prominently 3-nerved; lowermost leaves the largest, but usually early-deciduous; cauline leaves progressively reduced upward, sometimes strongly so; well-developed lower and middle cauline leaves lance-elliptic to broadly linear, entire or serrate, (5)6–12+ cm long and (0.5)1–3 cm wide, sessile or nearly so, often with axillary fascicles of reduced leaves. Inflorescence paniculate, usually broader than tall, with conspicuous, recurved-secund branches; involucre 3–5 mm tall, the involucral bracts broadly rounded to subacute, firm; receptacle often with at least some slender, chaffy bracts among the florets; ray florets 7–13; disk florets 8–13. Achenes glabrous or somewhat sparsely pubescent. (n = 9) Jul–Oct. Open prairies & sparsely wooded areas; throughout GP except extreme sw; (GP s to AR, w to B.C. & WA; elsewhere as a relict (TN?) or a weed. *S. hapemaniana* Rydb., *S. glaberrima* Martens, *S. moritura* Steele—Rydberg.

Our plants are referable to the weakly defined var. *fasciculata* Holz., which is taller and more robust than var. *missouriensis*, which occurs with two other varieties in the Pacific Northwest, cf. Cronquist, *in* Hitchcock et al. Vascular Plants of the Pacific Northwest 5: 307. 1955. [Univ. Wash. Publ. Biology, no. 17].

5. ***Solidago mollis*** Bartl., soft goldenrod. Perennial 1–5(7) dm tall, herbage spreading hirtellous-pubescent throughout or the stem becoming glabrate below, somewhat grayish-green in aspect. Stem single to several loosely clustered, arising from a creeping rhizome. Leaves thickish and firm, chiefly cauline, prominently 3-nerved, the lower cauline leaves early-deciduous; larger midcauline leaves elliptic to lanceolate, 3–8 cm long and 1–3 cm wide, 2.5–5 × longer than wide, irregularly dentate to subentire, sessile or short-petiolate. Inflorescence a dense, rather elongate panicle or sometimes a compact thyrse, the lower branches often weakly recurved-secund; involucre 3.5–6 mm tall; involucral bracts rounded to somewhat acute, ray florets 6–10, often 8; disk florets 3–8. Achenes short or appressed hairy. Jul–Oct. Dry or drying prairies & open woods, especially in fence rows; throughout GP except e 1/3 KS & adj. OK; (GP, s-cen Can.).

This is one of our most distinctive goldenrods, rarely to be confused with any other. However, it is also one of the most variable in aspect. The inflorescence varies from simple and thyrselike to definitely pyramidal with spreading recurved branches. The relative numbers of leaves and the upward diminution of leaves are also highly variable. Robust plants growing in undisturbed, grassy sites often become top-heavy and lax or subprostrate in age.

6. ***Solidago nemoralis*** Ait., gray goldenrod. Perennial 1–10(13) dm tall, herbage densely and finely puberulent with close, spreading hairs. Stems arising singly or often clustered from a branching caudex, commonly forming obvious clumps. Basal leaves well developed, with a single prominent midrib or but obscurely 3-nerved, larger basal leaves oblanceolate

to narrowly ovate, tapering to a winged petiole, 5–20 cm long (including the petiole) and 0.5–4 cm wide, crenate-dentate to subentire; cauline leaves smaller and progressively reduced upward. Inflorescence an elongate to distinctly pyramidal panicle, sometimes merely nodding at the tip, but often with long, spreading recurved-secund branches; involucre (3.5)4.5–6 mm tall; involucral bracts with ciliolate margins; ray florets 4–9; disk florets 3–6(9). Achenes whitish and subsericeus, but sometimes brownish and weakly hirtellous-viscidulous. (n = 9, 18) Aug–Oct. Dry, open places, especially in sandy soils, or sometimes in open scrubby woods; throughout GP except sw 1/4; (N.S. to n FL, w to cen Can. & TX). *S. hispida* Muhl., *S. longipetiolata* A. Nels., *S. pulcherrima* A. Nels.—Rydberg; *S. nemoralis* var. *decemflora* (DC.) Fern.—Fernald.

Our materials as described here are referable to var. *longipetiolata* (Mack. & Bush) Palmer & Steyermark. Var. *nemoralis* and var. *haleana* Fern. both have smaller heads and regularly viscidulous achenes, and are generally restricted to the region of the eastern deciduous woodlands. However, our materials are but weakly separated from these eastern phases, and var. *longipetiolata* is poorly defined.

7. Solidago petiolaris Ait. var. *angusta* (T. & G.) A. Gray, downy goldenrod. Perennial (3)4–15 dm tall. Stem short-hispid above, but glabrescent below, arising singly or clustered from a stout caudex, which sometimes has long, slender rhizomes. Leaves chiefly cauline, numerous, thick and firm, entire or shallow-crenate to serrate, glabrous, somewhat lustrous and velutinous on the upper surface and glabrous to short pubescent on the veins beneath, margin usually ciliolate; well-developed midcauline leaves lance-linear to narrowly ovate, 3–10(15) cm long and 0.5–3 cm wide, short-petiolate to sessile. Inflorescence a narrow, somewhat elongate open thyrse, usually with leafy bracts, the lower clusters elongate but ascending and not secund or recurved; involucre 5–8 mm tall; involucral bracts lightly pubescent to glandular-hairy or sometimes glabrous, tending to be squarrose; ray florets (5)7–11; disk florets 10–16. Achenes glabrous, at least at maturity. (n = 9) Aug–Oct. Open rocky woodlands, especially bluff escarpments or sandy, limestone sites; scattered in se 1/4 GP but sporadic w in sw KS, adj. OK, & n TX; (NC to n FL, w to GP & NM). *S. angusta* T. & G.—Gleason & Cronquist; *S. lindheimeriana* Scheele, *S. wardii* Britt.—Rydberg; *S. petiolaris* var. *wardii* (Britt.) Fern.—Fernald.

Var. *petiolaris* differs in having the leaves hardly glutinous at all, the lower leaf surfaces softly pubescent, and in occurring throughout se U.S. It intergrades into our var. *angusta* in se KS, s MO, & s to LA.

8. Solidago ptarmicoides (Nees) Boivin, sneezewort aster. Perennial 1–7 dm tall, with stems clustered from a branching, subligneous caudex, scabrous to glabrate. Leaves firm, glabrous or somewhat scabrous, entire or nearly so, obscurely to prominently 3-nerved, the lower ones linear-lanceolate and petiolate, persistent, 3–20 cm long (including petiole) and (1.5)3–10 mm wide; cauline leaves progressively reduced upward, becoming mere sessile bracts. Inflorescence an open, flat-topped corymb with 3–60 heads; involucre turbinate-campanulate, 5–7 mm tall, involucral bracts imbricate in 3 or 4 series, greenish toward the apex, with a prominent midrib; ray florets 10–25, the ligule 5–9 mm long and white; disk florets numerous and white. Achenes glabrous; pappus of abundant bristles, at least some of which are clavate-thickened toward the apex. (n = 9) Aug–Sep. Open, drying prairies, often on limestone bluffs or sandy sites; n 1/3 GP; (Que. & NY to MO, w to GP; locally in NC, GA, AR). *Aster ptarmicoides* (Nees) T. & G.—Fernald; *Unamia alba* (Nutt.) Rydb., *U. lutescens* (Lindl.) Rydb.—Rydberg.

This species is unusual in *Solidago* because of its white flowers and superficial aspect of *Aster*, but it hybridizes with other species of *Solidago*.

9. Solidago radula Nutt. Perennial 4–12 dm tall. Stems scabrous to short hirsute, arising singly or loosely clustered from a caudex, sometimes with creeping rhizomes. Leaves chiefly cauline, firm, hirsute or scabrous to subglabrous, distinctly greenish and not at all pale gray, the well-developed leaves obscurely 3-nerved, elliptic to lanceolate, dentate to subentire, 3–8(15) cm long and 1–3 cm wide, mostly 2–5 × longer than wide. Inflorescence paniculate with densely flowered, recurved-secund branches, or sometimes merely nodding; involucre 3.5–6 mm tall; involucral bracts usually obtuse to rounded but sometimes acutish; ray florets 4–7; disk florets 2–5. Achenes short-hairy. (n = 9) Aug–Oct. Open, rocky sites, especially calcareous ledges; KS: Cherokee; (s IL, MO, & adj. se GP, to LA & e TX; disjunct in NC, GA).

10. *Solidago riddellii* Frank. Perennial 4–7(10) dm tall, glabrous or lightly pubescent above. Stems arising from a thick caudex, sometimes with creeping rhizomes. Leaves glabrous, but scabrous on the margin, firm, entire, basal leaves long-petiolate, with the blade tending to be 3-nerved, lanceolate to spatulate, 1–2 cm long and 5–15 mm wide, conduplicate, much exceeded by the petiole, often early-deciduous; cauline leaves numerous, somewhat smaller, conduplicate and falcate, tending to be sheathing at the base, the uppermost sessile but not strongly reduced. Inflorescence a terminal corymb with numerous, crowded heads and the branches not at all secund; involucre 5–6 mm tall; involucral bracts obtuse and striate; ray florets 7–9; disk florets 10–20. Achenes glabrous or with but few inconspicuous hairs. (n = 9) Jul–Oct. Open, wet prairies & roadside ditches; extreme ne GP; (Ont. & OH to ne GP & MO). *Oligoneuron riddellii* (Frank) Rydb.—Rydberg.

A rare hybrid between *S. riddellii* and *S. ptarmicoides* has been named *S.* x *bernardii* Boivin, cf. Boivin, Phytologia 23(1): 21. 1972.

11. *Solidago rigida* L., rigid goldenrod. Variable perennial, 2–16 dm tall, densely pubescent with short, spreading hairs, or rarely subglabrate. Stems arising from a stout, branching caudex. Leaves firm, entire or obscurely dentate; basal and lower cauline leaves prominent and persistent; blade elliptic-oblong or lanceolate to broadly ovate, 5–15(25) cm long and 2–10 cm wide, often much exceeded by the long petiole; cauline leaves progressively reduced and with shorter petioles upward, the middle and upper ones becoming sessile. Inflorescence corymbiform and congested, heads relatively large; involucre 5–9 mm tall; involucral bracts firm, rounded and conspicuously striate; ray florets 7–14; disk florets 19–31. Achenes prominently ribbed, glabrous or with a few short hairs near the summit. (n = 9, 18) Aug–Oct. Widespread in dry or drying prairies, rocky open sites, & sandy soils; throughout GP but infrequent in sw 1/4; (CT & NY to GA, w to GP & NM).

Two varieties occur with us; a third, essentially glabrous phase (var. *glabrata* E. L. Braun) is restricted to se U.S.

11a. var. *humilis* Porter. Plants short, 2–6(8) dm tall. Achenes mostly with a few short hairs near the summit. GP, but apparently more abundant in the cen & w areas. *Oligoneuron canescens* Rydb., *O. bombycinum* Lunell—Rydberg.

11b. var. *rigida*. Plants relatively robust, 6–16 dm tall. Achenes usually entirely glabrous. GP, but more frequent in e 1/2. *Oligoneuron rigidum* (L.) Small—Rydberg.

The morphological distinction between these two vars. is imprecise, but they appear to have some populational integrity of their own. Var. *rigida* extends eastward into the region of continuous woodlands, while var. *humilis* is better developed in the grasslands. Evidence presently suggests that var. *rigida* is a tetraploid (n = 18) and that var. *humilis* is a diploid (n = 9).

12. *Solidago sparsiflora* A. Gray, three-nerved goldenrod. Perennial 3–6(8) dm tall; herbage grayish-canescent. Stems somewhat decumbent or arching upward from a coarse, branching caudex. Leaves firm, oblanceolate to lance-linear, entire to ± crenate-dentate, especially distally; well-developed midcauline leaves prominently to obscurely 3-nerved, 4–10 cm long and 0.5–2 cm wide; uppermost leaves reduced to mere bracts. Inflorescence a variable, often elongate panicle with the tip nodding, the lateral branches recurved-secund and often scarcely exceeding the subtending leafy bracts; involucre 4-6 mm tall; involucral bracts acute to acuminate; ray florets 8 or less; disk florets of about the same number. Achenes hispidulous. (n = 9) Jul-Nov. Open, sandy, coniferous woodlands, rocky slopes, or canyons; BH & scattered along the extreme w edge of GP in WY & s, also OK: Cimarron; (GP s & w to UT, TX, & AZ). *S. trinervata* Greene — Rydberg.

13. **Solidago speciosa** Nutt., showy-wand goldenrod. Perennial (2)3-15(20) dm tall; often puberulent above but otherwise glabrous to scabrous throughout. Stems arising singly or loosely clustered from a stout woody caudex, with numerous fibrous roots. Leaves thick and firm, numerous, entire or the lower ones sometimes shallow-toothed; basal and lower cauline leaves short-petiolate, sometimes the lowermost early-deciduous, blade ovate to lanceolate or oblong, 5-15(30) cm long and 1-4(10) cm wide; middle and upper cauline leaves becoming sessile and smaller upward, or sometimes the cauline leaves all nearly uniform. Inflorescence simple or of numerous crowded, ascending branches, forming a long, thyrselike cluster, or sometimes somewhat open and paniculate but the branches not at all secund; heads conspicuously pedicellate; involucre 3-5(6) mm tall; involucral bracts imbricate and prominently keeled; ray florets 4-8(11); disk florets 4 or 5(11), usually fewer than the ray florets. Achenes glabrous. (n = 9, 18) Aug-Oct. Prairies & especially dry, open wooded sites, often in rocky or clay soils; scattered in e 1/3 GP, but occasionally w in n NE, adj. SD & BH, to e WY; (NJ to GA, w to GP, AR, TX; perhaps introduced elsewhere).

Two varieties occur with us:

13a. var. *rigidiuscula* T. & G. Smaller, usually prominently scabrous, seldom exceeding 8 dm tall, with the leaves smaller and the lowermost often deciduous. Involucre mostly 3-4.5 mm tall. The common phase of the GP; scattered in the e 1/3 GP & sporadic w to e WY; (found in prairie habitats as far e as the w edge of the Appalachian Mts. uplift). *S. speciosa* var. *angustata* T. & G. — Fernald (misapplied); *S. pallida* (Porter) Rydb., *S. rigidiuscula* (T. & G.) Porter — Rydberg.

13b. var. *speciosa*. A robust phase, somewhat smooth and glabrate, often 1 + m tall, with basal leaves up to 30 cm long and 10 cm wide. Involucre 4.5-6 mm tall. Primarily in the region of the eastern deciduous woodland, and known from the extreme e edge of the GP.

14. **Solidago ulmifolia** Muhl., elm leaved goldenrod. Perennial 4-15 dm tall. Stem glabrous or nearly so, at least below the inflorescence, arising singly or at most paired from a stoloniferous caudex or short rhizome. Leaves chiefly cauline, numerous, thin, sharply serrate, glabrous to hirsute on the upper surface, loosely hirsute beneath, especially on the veins; lower leaves early deciduous; prominent midcauline leaves ovate to rhombic-ovate or elliptic, 6-12(20) cm long and 1.2-6 cm wide, short-petiolate or subsessile, upper leaves reduced and bractlike below the inflorescence. Inflorescence paniculate, with few, long, divergent recurved-secund branches; heads numerous and crowded; involucre 2.5-4.5 mm tall; involucral bracts imbricate in 3 or 4 series; ray florets 3-5; disk florets 4-6. Achenes short-hairy, ribbed. (n = 9) Aug-Sep. Dry or drying, open rocky woodlands, especially along stream banks & bluffs; se GP, from sw IA & se NE, through e 1/5 KS to e OK; (e Can. to GA & n FL, w to GP & TX). *S. microphylla* Engelm. — Rydberg.

Nearly all of our materials are referable to var. *ulmifolia*. However, the phase with very narrow, glabrous leaves has been recognized as the poorly defined var. *microphylla* A. Gray. It has been collected in e KS and adj. areas, but not to the exclusion of var. *ulmifolia*.

99. SONCHUS L., Sow Thistle

Ours annual or perennial herbs with milky juice. Leaves alternate, variously dissected to subentire and mostly auriculate-clasping, prickly margined. Heads several to numerous in a terminal corymbiform cluster; involucre ovoid-subglobose to campanulate; involucral bracts imbricated in several series, often thickish toward the base; florets all ligulate, perfect and fertile, numerous, corolla yellow. Achenes flattened, prominently ribbed on each face and mostly transversely rugose, narrowed toward the tip but without a distinct beak; pappus of numerous white capillary bristles. [tribe Lactuceae]

References: Boulos, L. 1972-1974. Révision systématique du genre *Sonchus* s.l., I-VI Bot. Notis. 125: 287-319; 126: 155-196; 127: 7-37, 402-451; Boulos, L. 1976. *Sonchus* in Flora Europaea, vol. 4. T. G. Tutin & V. H. Heywood, et al., eds. Cambridge Univ. Press, Cambridge.

1 Perennial with deep, creeping roots; heads relatively large, 2.5-3.5 cm across ... 1. *S. arvensis*
1 Annuals arising from a taproot; heads smaller, mostly less than 2.5 cm across.
  2 Leaf auricles rounded and not at all acute; mature achenes ribbed but not wrinkled-rugose .... 2. *S. asper*
  2 Leaf auricles with a distinctly acute base; mature achenes several-nerved and transversely wrinkled-rugose .... 3. *S. oleraceus*

1. Sonchus arvensis L., field sow thistle. Perennial 4-15(20) dm tall, arising from an extensive creeping root system; glabrous, at least below the inflorescence, and often somewhat glaucous. Lower and middle cauline leaves pinnate-lobulate to weakly pinnatifid to merely dentate or even subentire, 6-40 cm long and up to 15 cm wide overall, margin prickly; upper leaves progressively less lobed and becoming auriculate-clasping; uppermost leaves bractlike. Heads few-several in a terminal corymbiform cluster, relatively large, 2.5-3.5 cm across in flower; mature involucres 10-22 mm tall, variously glandular-pubescent to glabrous; florets all ligulate and fertile, corolla yellow. Achenes flattened, ribbed and rugose, beakless, 2.5-3.5 mm long; pappus of abundant capillary bristles. (n = polyploid series based on x = 9) Jul-Sep. *Introduced*.

This is a variable Eurasian weed, now widely established throughout the cool temperate regions. Two subspecies are recognized:

1a. subsp. *arvensis*. Involucre variously provided with spreading gland-tipped hairs and sometimes also irregularly tomentose; involucre mostly 15-22 mm tall. A weed in cultivated sites; n 1/4 GP, especially ND, e SD, adj. MN & IA; (scattered across N. Amer.; Eurasia).

1b. subsp. *uliginosus* (Bieb.) Nyman. Involucre glabrous or nearly so and mostly less than 15 mm tall. A weed in cult. sites; n 1/2 GP but mostly in e 1/3 ND, adj. SD & e; (scattered across N. Amer.; Eurasia). *S. arvensis* var. *glabrescens* (Guenth.) Grab. & Wimm.—Cronquist, 1980. Vascular Flora Southeastern U.S., Univ. NC Press. Chapel Hill.

2. Sonchus asper (L.) Hill, prickly sow thistle. Annual 1-10(15) dm tall, arising from a taproot; herbage mostly glabrous but sometimes with a few glandular hairs on and among the heads in the inflorescence. Larger leaves obovate to ovate in outline but variously pinnatifid to lobed or subentire, distinctly prickly, auriculate-clasping with rounded (not acute) auricles; middle and upper leaves progressively reduced and less divided. Heads several in a corymbiform cluster, mostly less than 2.5 cm across; involucre up to 15 mm tall at maturity; florets all ligulate and fertile, ligule shorter than the corolla tube, corolla yellow. Achenes flattened, 2-3 mm long, with 3(5) ribs on each face and not at all wrinkled-rugose, but the margins may have minute, recurved spinules; pappus of white capillary bristles, 6-9 mm long, tending to be deciduous. (n = 9) May-Sep (Nov). A weed in cult. & disturbed ground, often in gardens; GP; (across N. Amer.; Eurasia). *Introduced*.

3. Sonchus oleraceus L., common sow thistle. Annual 1-20 dm tall, arising from a taproot; herbage glabrous except for some spreading, glandular hairs on the involucres and peduncles. Prominent leaves pinnatifid to toothed or rarely subentire, up to 30 cm long and 15 cm wide overall, margins with few to numerous prickles, all but the lowermost leaves conspicuously auriculate-clasping, the auricles rather prominently acute; upper leaves progressively reduced and less divided. Heads several in a terminal corymbiform cluster, mostly

1.5–2.5 cm across; involucre 9–15 mm tall at maturity; florets all ligulate and fertile, the ligule about as long as the corolla tube, corolla yellow. Achenes flattened, 2.5–3 mm long, 3- to 5-ribbed on each face and evidently transverse rugose; pappus of abundant white capillary bristles, 5–8 mm long, tending to be persistent. (n = 8, 16) (Apr) May–Sep (Nov). A cosmopolitan weed in cult. or disturbed ground; e 1/5 GP from e ND & adj. MN to e KS, scattered westward, perhaps more frequent & widespread than the records indicate; (scattered across N. Amer.; Eurasia). *Introduced.*

100. STEPHANOMERIA Nutt., Wire lettuce, Skeletonweed

Ours perennial herbs with milky juice; branching upward, somewhat rigid and rushlike in aspect. Leaves alternate, narrow, entire to runcinate-pinnatifid, the upper ones reduced and bractlike. Inflorescence of few to numerous terminal heads; involucre cylindrical or narrowly campanulate, principal involucral bracts mostly 5, in a single series, subtended by shorter, calyculate bracts; receptacle naked, florets all ligulate, perfect and fertile, corolla pinkish or rose colored. Achenes columnar, beakless, 5-angled or 5-ribbed; pappus of plumose bristles. [tribe Lactuceae]

Reference: Tomb, A. S. 1974. Chromosome numbers and generic relationships in subtribe Stephanomeriinae. Brittonia 26: 203–216.

1 Pappus sordid to tawny, the bristle plumose above but naked or merely weakly scaly for ca 1 mm above the base .... 1. *S. pauciflora*
1 Pappus white or nearly so, the bristles plumose essentially throughout.
   2 Plants seldom more than 2 dm tall; principal leaves clearly runcinate-pinnatifid; achenes evidently rugose-tuberculate .... 2. *S. runcinata*
   2 Plants normally 2–5 dm tall; leaves linear-filiform, entire or merely toothed; achenes ribbed but otherwise smooth .... 3. *S. tenuifolia*

1. ***Stephanomeria pauciflora*** (Torr.) A. Nels. Perennial up to 6 dm tall, arising from a short woody taproot or caudex. Stems light green, rigid, bushy-branched. Leaves linear to narrowly lanceolate, the larger ones up to 7 cm long, toothed or often runcinate-pinnatifid; uppermost reduced and bractlike. Heads few or solitary at the ends of the branches; involucre 8–10(12) mm tall, glabrous; florets all ligulate and fertile, corolla pinkish. Achenes columnar, ribbed, and somewhat wrinkled-rugose; pappus sordid tawny, plumose to ca 1 mm from the base. (n = 8) Aug–Sep. Open dry plains; sw 1/4 GP & n along the edge of the Rocky Mts. to WY: Platte; (GP to TX, w to CA). *Ptiloria pauciflora* (Torr.) Raf.— Rydberg; *Lygodesmia pauciflora* (Torr.) Shinners—Correll & Johnston (changed to *Stephanomeria* in appendix).

2. ***Stephanomeria runcinata*** Nutt. Perennial, mostly 1–2 dm tall, glabrous to weakly scabrous-puberulent, sparingly branched from the base. Larger leaves narrowly lanceolate, up to 7 cm long and 1.5 cm wide overall, runcinate-pinnatifid; upper leaves reduced and bractlike. Heads solitary at the ends of the branches; involucre 9–12 mm tall; florets mostly 5, all ligulate and fertile, corolla pink. Achenes ca 5 mm tall, evidently rugose-tuberculate; pappus white, plumose to the base. (n = 8) Jul–Aug (Sep). Dry, open sites in the high plains; nw NE, e WY, & MT; (n GP, w in the montane valleys of MT & WY). *Ptiloria ramosa* Rydb.—Rydberg.

3. ***Stephanomeria tenuifolia*** (Torr.) Hall. Perennial 2–5+ dm tall, arising from an apparently deep, creeping root system, herbage glabrous to puberulent. Leaves linear to filiform, up to 8 cm long and rarely more than 3 mm wide, entire or sparingly toothed; uppermost leaves reduced and bractlike. Heads terminating the branches; involucre 7–11

mm tall; florets mostly 5, all ligulate and fertile, corolla pink. Achenes 4–6 mm tall, longitudinally ribbed but otherwise smooth or nearly so; pappus white, plumose to very near the base. (n = 8) Jul–Sep. Open rocky sites, barely entering our range in the nw GP; ND: Billings, Slope, Stark; (nw GP to TX(?), NM, w to B.C., CA).

101. TANACETUM L., Tansy

1. **Tanacetum vulgare** L., common tansy. Strong-scented perennial herb 4–10(+) cm tall, glabrous or nearly so. Stems arising singly or clustered from a stout, branching rhizome. Leaves alternate, numerous, 8–18 cm long and about 1/2 as wide, pinnately parted, with the pinnae again dissected or lobed, glandular punctate. Inflorescence a dense, corymbiform cyme with up to 200 heads in well-developed plants; heads disciform, receptacle naked, hemispheric-conic at maturity, disk 5–10 mm across; involucral bracts imbricate, margins ± scarious; outermost florets pistillate, with short, tubular 3-toothed corollas; disk florets perfect and fertile, yellow, corolla 5-toothed, style branches flattened, truncate, and minutely penicillate. Achenes 5-angled or ribbed; pappus a very short crown. (n = 9) Jul–Aug. Mostly roadsides & waste places; widely distributed but infrequent in the GP, from KS n to ND; (across N. Amer.; Eurasia). *Naturalized* in northeast and the Pacific Coast states; scattered elsewhere as an adventive. [tribe Anthemideae]

102. TARAXACUM Wiggers, Dandelion

Scapose, taprooted herbaceous perennials with milky juice. Leaves confined to a basal rosette, variously pinnatifid, dissected to toothed. Heads single at the ends of the 1–few scapes; involucral bracts biseriate, the outer series shorter and early-reflexed; receptacle naked; florets all ligulate and perfect, numerous. Achenes with a prominent obconic body, 4- or 5-angled or ribbed, weakly spiny above and tapering to a long, distinct beak; pappus of numerous capillary bristles, giving the achene its "parachute" aspect. [tribe Lactuceae]

Taraxacum is only introduced to the GP, but in Eurasia it is complicated by apomixis and polyploidy, resulting in imperfect morphological distinctions among the species. This has given rise to the publication of hundreds of specific names, based on the assumption that each morphotype of apomictic plants is entitled specific recognition. The two species recognized here are imperfectly separated, and one could create a defensible case for treating all of our dandelions as a part of a single entity with numerous weakly defined races.

1 Achenes reddish, purplish, or reddish-brown at maturity; terminal lobe of the leaves no larger than the lateral lobes; leaves rather deeply lobed .... 1. *T. laevigatum*
1 Achenes brown or brownish; terminal lobe of the leaves obviously larger than the lateral lobes; leaves variously lobed but often not deeply so .... 2. *T. officinale*

1. **Taraxacum laevigatum** (Willd.) DC., red-seeded dandelion. Similar to and rather poorly differentiated from *T. officinale*. Leaves deeply incised for their entire length, the terminal lobe no larger than the lateral lobes. Heads rarely more than 2 cm tall; principal involucral bracts mostly 11–13, subtending bracts usually less than 1/2 as long. Achene body reddish, purplish, or reddish-brown. (n = 8, 16) Apr–Oct. A weed in disturbed sites, especially lawns, apparently in somewhat drier sites & much less frequent than *T. officinale*; scattered throughout GP; (scattered across N. Amer.; Eurasia). *Introduced*. *T. erythrospermum* Andrz.—Rydberg.

2. **Taraxacum officinale** Weber, common dandelion. Scapose taprooted perennial 4–30(50) cm tall. Leaves lightly pubescent, especially beneath and on the midvein but sometimes

completely glabrous, oblanceolate in outline, up to 25 cm long and 15 cm wide overall, variously lobed, runcinate-pinnatifid or toothed, the terminal lobe usually larger than the others, tapering at the base to a winged and indistinct petiole. Head single on a villous-pubescent to glabrous hollow scape; involucre 1.5–2.5 cm tall; principal involucral bracts 13–20, subtended by a series of shorter and often wider calyculate bracts; the involucral bracts all erect when young but reflexing in age so that the head with the mature fruits form a characteristic, easily-disassembling ball; florets all ligulate, corolla yellow. Achenes with a brown or brownish body 3–4 mm long and somewhat short-spiny above, sharply tapering to a thin beak 2.5–4 × longer than the body; pappus of abundant white capillary bristles. (n = polyploid series based on x = 8) Mostly Apr–Oct, but earlier or later, depending upon local conditions and the year. A common weed in waste & disturbed sites, notably in lawns; GP; (nearly ubiquitous in the Temperate Zone; Eurasia). *Introduced.*

103. THELESPERMA Less.

Erect perennial herbs, often weakly suffrutescent at the base, glabrous or lightly pubescent at the leaf-bases. Stems arising singly or several clustered from a branching crown surmounting a taproot or creeping rootstocks. Leaves opposite, evenly distributed or crowded at the base, variously 1–3 × deeply pinnatisect into narrowly lanceolate to filiform segments, or entire and linear-filliform. Heads on naked peduncles; involucral bracts of 2 sizes, outer involucral bracts herbaceous, relatively short, often with narrow, scarious margins; inner involucral bracts longer, somewhat membranaceous and with broad, transparent scarious margins, basally connate for 1/4–1/2 their length, the upper, free portion lanceolate to ovate, sometimes pubescent at the tip; receptacle flat, chaffy; ray florets, when present, neutral and sterile, ligule yellow; disk florets perfect and fertile, corolla equally or unequally lobed, sometimes deeply so, when unequally lobed one lobe is much longer than the other 4; style branches with papillose appendages; chaffy bracts scarious, with 2 reddish-brown medial veins, oblong to ovate-lanceolate, usually enfolding and falling with the mature achenes. Achenes linear to linear-oblong, somewhat curved, 2–7 mm long, terete or somewhat flattened; pappus of 2 or sometimes 3 awns, or of small teeth or absent. [tribe Heliantheae]

References: Correll, D. S., & M. C. Johnston. 1970. Manual of the Vascular Plants of Texas, Texas Res. Foundation, Renner. *Thelesperma*, pp. 1665–1669; Alexander, E. J. 1955. *Thelesperma, in* N. Amer. Flora II. 2: 65–69.

1 Outer involucral bracts 1/2 as long as the inner; rays present; involucre suburceolate .... 2. *T. filifolium*
1 Outer involucral bracts small, 1–2 mm long and less than 1/4 as long as the inner; rays present or absent; involucre hemispheric or campanulate.
  2 Rays present and conspicuous, rarely absent; disk reddish-brown; n TX .... 1. *T. ambiguum*
  2 Rays absent; or rarely present and then small; disk yellow.
    3 Leaves mostly basal; plant seldom more than 2 dm tall; pappus mostly less than 1.5 mm long; MT & ND .... 3. *T. marginatum*
    3 Leaves well developed along the stem; plants up to 7 dm tall; pappus regularly more than 1.5 mm long; s 1/2 GP .... 4. *T. megapotamicum*

1. ***Thelesperma ambiguum*** A. Gray. Perennial herb 2.5–5 dm tall. Stems 1–several, arising from slender, creeping rootstocks. Leaves variable as to shape and arrangement, but usually in a basal rosette and with cauline leaves mostly restricted to the lower 1/3 of the stem; larger leaves usually 2–3 × pinnatisect, 6–16 cm long overall, cauline leaves smaller and 1–2(3) × pinnatisect. Heads on nearly naked peduncles; involucre 7–15 mm across; outer involucral bracts 4–6, oblong, 1–2+ mm long; inner involucral bracts 8, ovate-

lanceolate, 7–10 mm long, basally connate for ca 1/2 their length; ray florets normally present but occasionally absent, ligule yellow; disk florets with corolla irregularly and deeply lobed, reddish-brown; chaffy bracts 6–9 mm long and partially enfolding the achene. Achenes erect or weakly incurved, 4–6.5 + mm long; pappus of deltoid, retrorsely barbed awns, 1–2 + mm long. (n = 22) Jul–Sep. Open limestone areas; extreme s GP, TX: Deaf Smith; (TX & n Mex.).

2. *Thelesperma filifolium* (Hook.) A. Gray, greenthread. Annual, winter annual or short-lived perennial herb 2–7 dm tall. Stems single or several, simple or often branching from the base, arising from a taproot or weakly branching suberect rootstock. Leaves evenly distributed throughout, basal and lower cauline leaves mostly 5–12 cm long, 1–3 × pinnatisect, ultimate segments narrowly oblanceolate to linear, ca 1 mm wide, upper cauline leaves reduced and less dissected. Heads on mostly naked peduncles, reflexed in bud but erect at anthesis; involucre ± urceolate, 8–14 mm across at the widest point; outer involucral bracts 7–12, narrowly linear, spreading, 3–12 mm long and variable in length; inner involucral bracts 8, lanceolate to ovate-lanceolate, 7.5–10 mm long, basally connate for 1/4–2/5 their length, reddish toward the tip; ray florets 8, ligule 9–22 mm long, yellow or golden yellow, sometimes reddish-tinged toward the base, 3-lobed at the apex; disk florets with corollas yellow or reddish-brown; chaffy bracts 5–7.5 mm long, clasping the achene. Achenes 3.5–6.5 mm long, the outer ones mostly curved, short and thick, the inner ones relatively elongate and slender; pappus of 2 erect, retrorsely barbed awns, up to 2 mm long. (n = 8, 9)

Two weakly defined varieties occur with us; apparently the distinction between the two is more pronounced farther south, in Texas.

2a. var. *filifolium*. Plants relatively robust, 2.5–7 dm tall. Outer involucral bracts usually more than 1/2 as long as the inner; ligule of the ray florets golden yellow. May–Sep. Open, weedy sites; se GP; se 1/4 KS & cen OK; (s through TX). *T. intermedium* var. *rubrodiscum* Shinners—Field & Lab. 18:17–24. 1950 (cf. for nomenclature); *T. trifidum* (Poir.) Britt.—misapplied.

2b. var. *intermedium* (Rydb.) Shinners. Plants short, 2–3(4) dm tall. Outer involucral bracts ca 1/2 as long as the inner; ligule of the ray florets distinctly yellow and not at all golden. Jun–Aug. Disturbed sites in sandy-gravelly soils; sw 1/2 GP; s SD, e WY to w OK & the panhandle region, westward; scattered in e NE, apparently absent from nw KS & cen NE; (sw GP to TX & NM). *T. trifidum* (Poir.) Britt.—Harrington, misapplied; cf. Shinners, op. cit., pp. 18–19.

3. *Thelesperma marginatum* Rydb. Perennial herb to 2 dm tall. Stems arising singly or rarely clustered from a basal rosette surmounting a series of creeping rootstocks. Leaves mostly basal, 1–2 × pinnatisect into 3–9 linear segments, up to 7 cm long overall. Inflorescence subscapose, simple or 1- to 3-branched; heads terminal; involucre 7–10 mm across; outer involucral bracts 3–5 mm long and lanceolate; inner involucral bracts 7–10 mm long, basally connate for 1/3–1/2 their length; ray florets absent; chaffy bracts 4–7 mm long. Achenes ca 5 mm long, somewhat curved, pappus of 2 lanceolate, nearly glabrous teeth. Jun–Aug. Open, high plains; infrequent in nw GP; ND: Divide, Williams; MT: Rosebud; (n GP, s Sask., Alta.).

This species apparently is distinguished from the more southwestern *T. subnudum* A. Gray by its eradiate heads and distribution.

4. *Thelesperma megapotamicum* (Spreng.) O. Ktze. Perennial herb 3–7.5 dm tall. Stems arising singly or clustered from creeping rootstocks. Leaves either evenly distributed along the stem, with rather long internodes, or crowded toward the lower 1/2; basal and lower cauline leaves once or infrequently twice pinnatisect, sometimes merely trifid or simple, 4–9 cm long overall, ultimate segments linear-lanceolate to linear. Involucre campanulate to subhemispheric, 7–14 mm across; outer involucral bracts 4–6, oblong to ovate, 1–2 +

mm long, usually appressed; inner involucral bracts 8, ovate-lanceolate, basally connate for barely 2/3 their length; ray florets rarely present; disk florets with corolla yellow, with reddish-brown veins, deeply and irregularly lobed; chaffy bracts 7-11 mm long, often erose. Achenes 4-8 mm long, erect or somewhat incurved, subterete to weakly flattened; pappus of 2 triangular, retrorsely barbed awns, 1.5-3 mm long. (n = 11) May-Sep. Widespread in open sites; s 1/2 GP from s SD to TX, infrequent in se GP; (GP to TX, UT, & AZ, Mex.; s to S. Amer.). *T. gracile* (Torr.) A. Gray—Rydberg.

104. TOWNSENDIA Hook., Easter Daisy

Taprooted caulescent or acaulescent herbs, pubescent or glabrate, with leaves alternate or all basal. Heads terminal on branches or sessile among the tufts of basal leaves; involucral bracts imbricate in 4-7 series, narrowly ovate to linear-subulate, usually scarious margined; receptacle low-convex, ± shallow-pitted and minutely pubescent; ray florets pistillate and fertile, in a single series, ligule whitish or bluish to somewhat pink, the lower surface often with a pink or lavender stripe; disk florets bisexual and fertile, corolla yellow and frequently pink or purple tinged; style branches ca 1-2.5 mm long, with inner-lateral stigmatic lines. Achenes flattish and 2-angled, or sometimes the ray achenes 3-angled, variously pubescent to subglabrate; pappus of the ray florets of numerous, barbellate, rigid bristles or reduced to a mere crown of weakly connate squamellae; pappus of disk florets a prominent uniseriate whorl of rigid, elongate, barbellate bristles. [tribe Astereae]

Reference: Beaman, J. H. 1957. The systematics and evolution of *Townsendia* (Compositae). Contr. Gray Herb. 183: 1-151.

Townsendia is complicated by much apomixis, where the plants produce viable seeds without benefit of cross fertilization, thus the seeds are genetically identical with the parent plant. Populations of plants can occur that are apparently intermediate between the expected normal species, thereby confusing the taxonomy of the group. The greatest diversity (and taxonomic difficult) occurs west of the GP in the Rocky Mts. and westward.

1 Plants depressed, acaulescent or nearly so, with heads sessile among crowded basal leaves.
   2 Involucral bracts mostly narrow-lanceolate and without a terminal ciliate tuft; disk corollas ± 9-10 mm long; disk pappus conspicuously longer than the disk corollas .... 1. *T. exscapa*
   2 Involucral bracts linear-subulate, with a terminal tuft of tangled cilia; disk corollas ± 5 mm long; disk pappus barely exceeding the disk corollas .... 3. *T. hookeri*
1 Plants distinctly caulescent, with heads terminal on leafy stems.
   3 Involucre 1-2 cm tall; involucral bracts with long, rigid bristly acuminate tips and broad, hyaline margins; wc GP .... 2. *T. grandiflora*
   3 Involucre 0.7-1+ cm tall; involucral bracts acute or slightly acuminate, scarious margined; extreme s GP .... 4. *T. texensis*

1. ***Townsendia exscapa*** (Richards.) Porter. Dwarf acaulescent subcespitose perennial, arising from a coarse, branching caudex surmounting a taproot. Leaves all basal, lanceolate or linear-lanceolate, entire, strigose or subsericeus, 1-5(8) cm long and 2-6 mm wide. Heads sessile or nearly so among the leaves; involucre 1-2+ cm tall; involucral bracts in 4-7 series, narrow-lanceolate and without a terminal tuft of ciliate hairs; ray florets 20-40, ligule white or pinkish, 12-22 mm long and 1-3 mm wide; disk florets with corolla yellow, sometimes pink or purple-tinged at the apex, (6)9-10(12) mm long. Achenes flattened and 2-ribbed, or ray achenes sometimes 3-ribbed, 3.5-6 mm long, pubescent; pappus of disk and ray florets similar, of slender barbellate bristles, disk pappus exceeding the corolla. (n = 9, but 2n = 27-66? in apomictic plants) Mar-May. Open, dry plains; throughout w & cen GP; (GP to e B.C., NV & AZ, n Mex.). *T. sericea* Hook.—Harrington (an illegitimate name).

2. **Townsendia grandiflora** Nutt. Caulescent, taprooted biennial or short-lived perennial, 5-20 + cm tall; stems several, branching at the base, suberect, strigose-pubescent. Leaves variously canescent to subglabrate; basal leaves linear to lanceolate, up to 5 cm long and early deciduous; cauline leaves spatulate to oblanceolate, entire, (2)3-6(9) cm long and up to 1 cm wide, but mostly ± 4 mm wide. Heads terminal, often subtended by 1 or several conspicuous leafy bracts; involucre 1-2 cm tall and 1.5-3 cm wide; involucral bracts in 4-7 series, with bristly, stiff-acuminate tips; ray florets 20-40, ligule white, 1.5-2 + cm long; disk florets with yellow corolla. Achenes compressed, pubescent, 3-4.5 mm long; pappus of ray florets of short squamellae or bristles, not more than 2 mm long; disk pappus of 15-30 stiff bristles, 4-6 mm long. (n = 9, but 2n = 27-36? in apomictic plants) May-Jul. Open dry plains & hillsides; sw SD, w NE, & adj. WY & CO; (sw SD & s along the e edge of the Rocky Mts. to n NM, mostly just w of the GP).

3. **Townsendia hookeri** Beaman. Dwarfish acaulescent subcespitose perennial, arising from thickened branches of a taprooted caudex. Leaves tufted, linear or narrowly lanceolate, often involute, sericeous-canescent, 1-4.5 cm long and ca 2 mm wide. Heads sessile or rarely short-pedunculate among the leaves; involucre 1-1.5 cm tall; involucral bracts linear-subulate with a terminal tuft of ciliate hairs; ray florets 15-30, ligule white above but creamy or pinkish beneath, 9-15 mm long; disk florets yellow or sometimes pinkish-tipped, 4-6 + mm long. Achenes compressed, pubescent; pappus of barbellate bristles; disk pappus equaling or barely exceeding the disk corolla. (n = 9, but 2n = 27-36? in apomictic plants) Apr-Jun. Open, dry plains & hillsides; nw GP, e MT, BH, ne WY, & s along the e edge of the Rocky Mts. in CO; (GP to ID & UT).

This species is similar to *T. exscapa* and was included within that species by many authors prior to Beaman (op. cit.).

4. **Townsendia texensis** Larsen. Caulescent, taprooted biennial 5-20 + cm tall; stems several, few-branched, arching upward. Leaves strigose to glabrate, basal leaves early deciduous; cauline leaves narrowly oblanceolate, 1-6 cm long and ca 5 mm wide. Heads terminal; involucre 7-10 + mm tall and 15-20 mm wide; involucral bracts in 4-6 series, apically acute or slightly acuminate; ray florets 25-40, ligule bluish or less frequently white, 10-18 mm long; disk florets with corolla yellow. Achenes compressed, pubescent; pappus of ray floets a low corona of fused squamellae or of short bristles, ±1(2.5) mm long; disk pappus of 16-25 long bristles. (n = 9) Apr-Jun. Open, dry, calcareous plains; infrequent; s GP, panhandle of TX, adj. OK; (s GP s to cen TX).

105. TRAGOPOGON L., Goat's beard, Salsify

Taprooted biennials with milky juice. Leaves alternate, entire, long-linear and somewhat grasslike. Heads large, terminating the branches; involucre cylindric or obconic when young, becoming narrowly campanulate; involucral bracts essentially uniseriate and subequal; receptacle naked; florets all ligulate, perfect and fertile, corolla yellow or purple. Achenes terete or angled, 5- to 10-nerved, narrowed at the base and constricted upward to a slender beak; outer achenes sometimes only obscurely beaked; pappus a single series of plumose bristles, the plume branches long and interwebbed; involucre and receptacle reflexing at maturity so the achenes form a characteristic "ball" of fruits, similar to that of *Taraxacum* (dandelion). [tribe Lactuceae]

Reference: Ownbey, M. 1950. Natural hybridization and amphiploidy in the genus *Tragopogon*. Amer. J. Bot. 37: 487-499.

The three species recognized here are widely naturalized in North America and are distinctive in our region. They hybridize and form amphiploid populations in the Pacific Northwest.

1 Corolla purple .... 2. *T. porrifolius*
1 Corolla yellow.
   2 Peduncle expanded upward and fistulous at flowering and fruiting time; involucral bracts exceeding the corollas .... 1. *T. dubius*
   2 Peduncles not tapering-enlarged upward; involucral bracts equaling or shorter than the corollas .... 3. *T. pratensis*

1. **Tragopogon dubius** Scop., goat's beard, western salsify. Biennial or short-lived perennial 3–8(10) dm tall, arising from a long taproot; herbage unevenly floccose-tomentose when young but glabrate in age. Leaves long and narrow, up to 30 cm long, gradually tapering from the base to the apex. Heads solitary at the ends of long, fistulous peduncles; involucral bracts typically 13 but sometimes ca 8 in depauperate plants, 2.5–4 cm tall in flower and clearly exceeding the florets, elongating and reflexing at maturity; florets all ligulate and fertile, corolla yellow. Achenes with a slender body, tapering upward to a stout beak, 2.5–3.5 cm long overall; pappus a single series of white to sordid plumose bristles. (n = 6) Mostly May–Jul. A weed on roadsides & disturbed waste sites, especially along the edges of grain fields; GP; (established over much of temp. N. Amer.; Eurasia). *Introduced.* *T. major* Jacq.—Fernald.

2. **Tragopogon porrifolius** L., salsify, vegetable oyster. Biennial up to 10 dm tall, arising from a prominent taproot. Leaves long and narrow, up to 30 cm long, gradually tapering to the apex. Heads terminal on expanded, hollow peduncles; involucral bracts ca 8, 2.5–4 cm long, equaling or exceeding the florets, elongating and reflexing at maturity; florets all ligulate and fertile, corolla purple. Achenes with a prominent body, abruptly contracted to a slender beak, 2.5–4 cm long overall; pappus of brownish plumose bristles. (n = 6) May–Aug. An infrequent weed on roadsides & waste ground; scattered in s 2/3 GP; (sporadic over temp. N. Amer.; Eurasia). *Introduced.*

    This species is cultivated for its edible taproot and sometimes as an ornamental. Our weedy phase may be an escape from cultivation or it may have been introduced as a weed.

3. **Tragopogon pratensis** L., meadow salsify. Biennial or short-lived perennial up to 8 dm tall, arising from a taproot; herbage unevenly floccose-tomentose when young, but glabrate in age. Leaves elongate, narrow, up to 30 cm long, somewhat abruptly contracted a short distance above the base, then gradually tapering to the apex, margin ± crisped-waxy. Heads solitary, peduncles long but not tapering-enlarged upward; involucral bracts ca 8, 12–24 mm long, equaling or shorter than the florets, elongating and reflexing at maturity; florets all ligulate and fertile, corolla deep yellow. Achenes with a prominent body, abruptly contracted to a slender, rather short beak, 15–25 mm long overall; pappus of whitish plumose bristles. (n = 6) May–Jul. Infrequent weed in disturbed sites in the GP; (widely scattered in the cool temp. parts of N. Amer.; Eurasia). *Adventive.*

    This plant is superficially similar to *T. dubius,* but it is infrequently collected in the GP. It may be expected as an occasional weed, especially in the n GP.

106. VERBESINA L., Crownbeard, Wingstem

    Herbs with opposite or alternate simple leaves and some with winged stems. Heads radiate; involucral bracts subequal or weakly imbricate, somewhat herbaceous; receptacle low-conic to flat; ray florets pistillate and fertile or neutral; disk florets perfect and fertile, style branches flattish, with acute papillate or hairy appendages; chaffy bracts partly enclosing the achenes. Achenes flattened parallel to the radius of the head, sometimes distinctly winged; pappus of 2 awns and sometimes with a few short scales. [tribe Heliantheae]

Reference: Olsen, J. 1979. Taxonomy of *Verbesina virginiana* complex. Sida 8: 128–134.

1 Plants taprooted annuals .... 2. *V. enceliodes* subsp. *exauriculata*
1 Plants perennial, with fibrous roots.
    2 Involucral bracts reflexed at maturity, with the achenes spreading to form a loose, globose head; leaves alternate .... 1. *V. alternifolia*
    2 Involucral bracts appressed to loose but not deflexed; achenes remaining erect in the head.
        3 Heads mostly more than 20, leaves alternate, rays white .... 4. *V. virginica*
        3 Heads mostly 1–10, leaves opposite, rays yellow .... 3. *V. helianthoides*

1. Verbesina alternifolia (L.) Britt., wingstem. Erect leafy-stemmed perennial 1–2 m tall, variously hirsute to glabrate. Leaves alternate, blade lanceolate to narrowly ovate, 10–25 cm long and 2–8 cm wide, serrate or subentire, scabrous-hirsute, especially above, narrowed at the base to a decurrent petiole-base, producing wings along the stem. Heads often many (up to 100) in an open inflorescence; disk 1–1.5 cm across, chaffy; involucral bracts few, small, narrow, and soon deflexed; ray florets 2–10, neutral, ligule yellow, 1–3 cm long; disk florets spreading even before anthesis. Achenes widely spreading, forming a globose head, broadly winged or sometimes wingless. (n = 34) Aug–Oct. Open moist wooded areas, stream banks, & bottomlands; se 1/4 GP, from sw IA & extreme ne NE s through e 1/3 KS to e 1/2 OK, rarely westward; (NY & s Ont. to FL, w to GP). *Actinomeris alternifolia* (L.) DC. — Rydberg.

2. Verbesina encelioides (Cav.) Benth. & Hook. subsp. *exauriculata* (Robins. & Greenm.) Coleman, golden crownbeard. Taprooted annual to 7 dm tall, strigose-canescent. Leaves mostly alternate, blade ovate to deltoid ovate, 4–9(12) cm long and 2–5(8) cm wide, coarsely dentate, especially toward the base, petiole a little shorter than the blade. Heads in an open inflorescence; disk 1–2 cm across, chaffy; involucral bracts loose or weakly spreading; ray florets 10–15, pistillate, ligule 1–2 cm long, yellow. Achenes winged, weakly spreading but not reflexed. (n = 17) Jul–Sep. Open, disturbed, & often waste places; scattered in s 1/2 GP; (mostly s GP & sw to AZ, CA, & n Mex.; scattered e to FL & NC; casually introduced elsewhere). *Ximenesia exauriculata* (Robins. & Greenm.) Rydb. — Rydberg.

    Subsp. *encelioides* differs in having distinctly dilated-auriculate petiole bases and somewhat larger heads, and it occurs e of our region on the Gulf Coastal Plains.

3. Verbesina helianthoides Michx. Rhizomatous perennial up to 1 m tall, hirsute. Leaves alternate, blade lanceolate to narrowly ovate, 6–12 + cm long and 2–6 cm wide, short-petiolate to subsessile, decurrent and producing wings on the stem. Heads mostly 10 or fewer; disk 1–1.5 cm across, chaffy; involucral bracts erect or loose but not reflexed; ray florets 8–15, pistillate or neutral, ligule 1–2 cm long, yellow. Achene winged or wingless, weakly spreading. (n = 17) May–Jul. Open prairies in wooded regions; extreme se GP; w MO, se KS, & adj. OK; (OH & NC to GA, w to GP & TX).

4. Verbesina virginica L., frostweed. Coarse perennial up to 2 m tall. Stems arising from fleshy-fibrous roots, finely pubescent. Leaves alternate, blade ovate, 8–18 cm long and 3–9 cm wide, toothed or entire, scabrous to glabrate above and finely appressed-pubescent or velutinous beneath, petiole somewhat decurrent down the stem. Heads numerous, up to 200, in well-developed plants, in an open, corymbiform cluster; disk 3–7 mm across, chaffy; involucral bracts appressed, imbricate; ray florets 1–5, pistillate, ligule up to 1 cm long, white. Achene flat, winged or wingless. (n = 16, 17) Aug–Sep. Damp openings in wooded areas; se GP; se KS & adj. OK; (VA to FL, w to GP & TX).

107. VERNONIA Schreb., Ironweed

Perennial herbs (ours) with stems arising singly or several clustered from a tough, fibrous-rooted base. Leaves alternate, cauline and evenly distributed along the stem. Inflorescence compact to variously loose and irregular corymbiform cymes. Heads discoid, involucres cylindric to hemispheric or campanulate; involucral bracts imbricate with the inner bracts progressively longer and variously tipped; florets numerous, commonly more than 12, corolla purple (or rarely white), style branches elongate, filiform-subulate, short-hairy with stigmatic lines present only toward the base, anthers sagittate. Achenes ribbed or smooth, sometimes resin-dotted between the ribs; pappus in 2 series, an inner series of slender bristles and an outer series of short scales or broad bristles. [tribe Vernonieae]

Reference: Jones, S. B., & W. Z. Faust. 1978. *Vernonia*. N. Amer. Flora II. 10: 180–195.

Vernonias are often weedy in pastures and disturbed habitats. Hybrids are frequent, and local hybrid swarms may obscure the species boundaries.

1 Tips of the inner involucral bracts long-acuminate to filiform .... 1. *V. arkansana*
1 Tips of the middle and inner involucral bracts variously rounded to obtuse or acute to short acuminate, but not linear-filiform.
   2 Leaves conspicuously pitted beneath, the pits appearing as dark spots under low magnification.
      3 Inner involucral bracts rounded to weakly acute-tipped; widespread species .... 3. *V. fasciculata*
      3 Inner involucral bracts acute/acuminate-tipped; sw GP .... 5. *V. marginata*
   2 Leaves variously hairy or glabrous but not pitted beneath.
      4 Heads usually with 30–60 florets .... 6. *V. missurica*
      4 Heads usually with fewer than 30 florets.
         5 Tips of middle and inner involucral bracts recurved or spreading, acute to acuminate; stem pubescent .... 2. *V. baldwinii*
         5 Tips of middle and inner involucral bracts ± erect and not recurved or spreading; stems glabrate or merely lightly puberulent .... 4. *V. gigantea*

1. **Vernonia arkansana** DC. Perennial ca 1(2) m tall, glabrous to minutely puberulent. Stems leafy. Well-developed middle cauline leaves short-petiolate, blade narrowly lanceolate, 9–20 cm long and 1–2.5 cm wide, margin slightly revolute with callous teeth. Inflorescence a variable, compact cluster; heads with ca 50–100 florets; involucre broadly hemispheric, 10–15 mm across; involucral bracts imbricate with long, spreading tips, inner bracts with conspicuous long-flexuous spreading tips 2–4 mm long; corollas 10–11 mm long. Achenes 4–5 mm long, prominently ribbed; pappus brownish-purple, inner bristles 6–7 mm long, scales ± 1 mm long. (n = 17) Jul–Sep. Rocky stream beds & low, open woods; se KS & ne OK; (Ozarkian region, ne to the Great Lakes).

2. **Vernonia baldwinii** Torr., western ironweed. Perennial 8–15 dm tall, thinly pubescent to distinctly tomentose. Leaves numerous, middle cauline leaves lanceolate to narrowly ovate, 3.7–17 cm long and 2–6 cm wide, serrate, very short petiolate; inflorescence a loose, irregular cluster; heads with 17–34 florets, 4–6 mm across; involucral bracts imbricate, greenish-brown to purple, tips recurved or spreading. Achenes puberulent to glabrous, ribbed; pappus light brown to brownish-purple, inner bristles 5–6 mm long, outer scales ca 0.7 mm long. (n = 17) Jul–Sep. (IL to AR & TX, w to CO).

Two subspecies are weakly distinguishable in our area:

2a. subsp. *baldwinii*. Upper stems often distinctly tomentose. Heads with 23–34 florets, tips of involucral bracts recurved. Upland pastures, open woods, & roadsides; se KS, w MO, ne OK.

2b. subsp. *interior* (Small) Faust. Upper stems merely puberulent or lightly pubescent. Heads with 17–27 florets, tips of the involucral bracts appressed to loose and spreading, not recurved. Dry, disturbed, or heavily grazed pastures & open upland sites, roadsides; often weedy; s 1/2 GP, more frequent in tall grass & mixed prairie regions.

3. **Vernonia fasciculata** Michx. Perennial 3–12 dm tall, glabrous or nearly so. Stems often reddish to purplish, especially below, leafy. Middle cauline leaves lanceolate, 3.5–15 cm long and 0.4–4.5 cm wide, deeply pitted and punctate beneath (use hand lens), serrate, sessile or very short petiolate. Inflorescence a flat-topped cluster; heads with 10–26 florets, involucre 3.3–6.8 mm across; involucral bracts imbricate, the inner ones 4.5–7.5 mm long with subacute to rounded tips; corollas 9–11 mm long. Achenes glabrous to puberulent, 3–3.5 mm long; pappus light brownish to purplish, bristles 5–6 mm long, scales 0.5–1 mm long. (n = 17) Jul–Oct. (OH to Sask. & MT, s to TX).

Two subspecies are recognized:

3a. subsp. *corymbosa* (Schwein. ex Keating) S. B. Jones. Mature plants 3–6 dm tall. Middle cauline leaves less than 10 cm long, scabrous above. Inner involucral bracts 2–3 mm wide. Damp open prairies; scattered throughout e 1/2 GP, sporadic in w 1/2; (MO to CO, s Can. to TX).

3b. subsp. *fasciculata*. Mature plants 5–12 dm tall. Middle cauline leaves 8–15 cm long, glabrous or nearly so above. Inner involucral bracts up to 2 mm wide. Damp prairies & disturbed stream banks; e 1/2 KS & NE, w IA & MO, scattered westward; (OH to MN & OK).

4. **Vernonia gigantea** (Walt.) Trel. ex Branner & Cov. Perennial ca 1–2 m tall, glabrescent or lightly puberulent below. Leaves numerous, petiolate, petioles up to 2 cm long; middle cauline leaf blades lanceolate to linear-lanceolate, variously scabrous or puberulent to nearly glabrous, often pubescent on the veins beneath, 6–30 cm long and 1–7 cm wide. Inflorescence open and irregular, heads with 12–30 florets, 2–6 mm across; involucral bracts imbricate, appressed, purplish, inner bracts acute to obtuse or mucronate with tips ± 0.2 mm long; corollas 9–11 mm long. Achenes pubescent on the ribs, ca 3.5 mm long; pappus brownish or sometimes purplish-tinged, inner bristles ca 6 mm long, outer scales less than 1 mm long. (n = 17) Aug–Oct. Damp or drying open woods, becoming weedy in pastures, roadsides, & disturbed sites; extreme e NE & KS; (MI to FL, w to NE & TX).

5. **Vernonia marginata** (Torr.) Raf., plains ironweed. Herb 3–7 dm tall, puberulent above but glabrescent below. Leaves numerous, very short-petiolate; middle cauline leaves broadly linear, glabrous or nearly so, scabrous along the veins above; deeply pitted and punctate with glandular trichomes beneath. Inflorescence variously compact; heads with 8–26 florets, involucre 3.2–7 mm across; involucral bracts imbricate, greenish purple, inner bracts 5.5–7 mm long, with acute to acuminate tips 0.5–1 mm long, corollas 11–12 mm long. Achenes glabrous, sometimes resinous, 3.5–4.2 mm long; pappus purple to brownish-purple, inner bristles 8–10 mm long, outer scales 1.2–1.8 mm long. (n = 17) Jul–Aug. Along streams, in low pastures; extreme sw KS & se CO, s through OK panhandle to TX; (GP s through TX & NM to Coah.).

6. **Vernonia missurica** Raf. Perennial 1–2 m tall, pubescent to tomentose. Leaves numerous, cauline, short petiolate; middle cauline leaves lanceolate to narrow-ovate, 8–20 cm long and 2–6 cm wide. Inflorescence loose; heads with 30–60 florets, involucres ca 7–10 mm tall; involucral bracts imbricate, greenish to brownish, inner bracts 6–10 mm long, with obtuse to subacute tips, 0.1–0.2 mm long. Achenes puberulent to nearly glabrous on the angles, 4–6 mm long; pappus brownish, inner bristles 6–8 mm long, outer scales 0.6–1 mm long. (n = 17) Jul–Oct. Damp prairies & bottom lands, roadsides; w IA & w MO, se NE, extreme e KS; (TN & MS, w to IA & TX).

108. VIGUIERA H.B.K., Golden Eye

1. **Viguiera stenoloba** Blake, resin bush. Branching shrub ca 1 m tall, glabrous to strigillose or even subcancescent. Leaves alternate or a few opposite, 2.5–6 cm long overall, deeply

dissected into 3-7 linear or linear-lanceolate, entire or toothed lobes, 1-5 mm wide; the uppermost leaves merely linear and entire or subentire. Heads single at the ends of the branches; involucre 7-8 mm tall and 8-9(12) mm across; involucral bracts in ca 3 series; receptacle flattish, chaffy; ray florets pistillate but infertile, ca 12-13, ligule 7-14 mm long, yellow; disk florets numerous, perfect and fertile, style branches hispid above, with acute appendages; chaffy bracts firm but scarious, ca 5 mm long, embracing the achenes but persistent after the achenes fall. Achenes somewhat compressed parallel to the radius of the head, nearly rhombic in cross section, ca 3.5 mm long; pappus absent. (n = 17, 34) Aug-Oct. Infrequent in open, dry sites; barely entering our range, NM: Union; (NM & w TX, s in Mex.). [tribe Heliantheae]

Viguiera ovalis Blake, a branched herbaceous perennial with oval or elliptic leaves, may barely enter our range in ne NM.

109. XANTHISMA DC., Sleepy Daisy

1. ***Xanthisma texanum*** DC. subsp. ***drummondii*** (T. & G.) Semple. Erect, taprooted annual herb, 2-7 + dm tall, with stems branching upward from the middle, or sometimes from the base; herbage glabrous or nearly so. Leaves alternate, the lowermost 5-8 cm long, ovate to obovate in outline but variously lobed to pinnatifid; middle leaves becoming sessile, lance-linear to linear, entire or nearly so, 1-3 cm long and 3-8(12) mm wide; the uppermost reduced to mere appressed bracts. Heads solitary at the ends of the upper branches; involucre hemispheric or campanulate to turbinate, 5-10 mm tall and at least as wide; receptacle convex, with subulate chaffy scales among the achenes; involucral bracts firm, imbricate in several series, oblong-lanceolate to ovate, stramineous-margined, contracted apically to a sharp, stiff point; ray florets pistillate and fertile, ligule yellow, up to 10 mm long; disk florets perfect and fertile, corolla yellow, tapering-tubular, and scarcely expanded upward. Achenes of the ray florets curved, triangular in cross section, achenes of the disk florets ± columnar, obovate; both with fine, ascending whitish hairs that darken with age; pappus of both ray and disk achenes double, with an outer series of linear-subulate scales less than 3 mm long, more-or-less alternating with an inner series of long (ca 6 mm) lance-acuminate scales. (n = 4, 4 + 1, 8 + 2) May-Aug. Infrequent in open, sandy places; extreme s GP, cen & w OK, n TX; (GP, e NM, s through TX). *X. texanum* DC. var. *drummondii* (T. & G.) A. Gray—Atlas GP. [tribe Astereae]

Reference: Semple, J. C. 1974. The phytogeography and systematics of *Xanthisma texanum* DC. (Asteraceae); proper usage of infraspecific categories. Rhodora 76: 1-19.

The unusual cytology of this species has attracted much attention and has caused the creation of an extensive bibliography.

110. XANTHIUM L., Cocklebur

Coarse, taprooted annuals. Leaves alternate. Inflorescence of small unisexual axillary heads; staminate heads uppermost with florets surrounded by 1-3 series of separate bracts, receptacle chaffy, conic, florets with minute tubular corolla and 5 free anthers, ovary and style vestigial; pistillate heads in the middle and upper axils, involucre of fused bracts completely enclosing the 2 pistillate florets in a 2-chambered, prickly bur, corolla absent, achenes solitary in the chambers of the bur; pappus absent. [tribe Heliantheae]

1 Nodes usually with a prominent 3-forked axillary spine; leaves tapering at the base .... 1. *X. spinosum*
1 Nodes spineless; leaves cordate or truncate at the base .... 2. *X. strumarium*

1. **Xanthium spinosum** L., spiny cocklebur. Plants 3–10(+) dm tall, strigose or short-pubescent. Leaves alternate, tapering to a short petiole, blades lanceolate, entire or with a few coarse teeth or lobes, 2–6 cm long and 0.5–2.5 cm wide, lightly pubescent or glabrate above and densely silvery-pubescent beneath, with a prominent 3-forked yellow spine 1–2 cm long arising in the axil. Burs solitary or few in the middle and upper leaf axils, ca 1 cm long, finely pubescent and covered with short, hooked prickles. (n = 18) Jul–Oct. Occasional in damp, fertile, disturbed soils & waste places; scattered in cen & w KS, sporadic in NE & TX; (occasional Ont., New England, w to GP, widespread weed in the warmer parts of the world, probably of American origin).

This species apparently has become more frequent in the s GP in recent years, especially where irrigation systems have been introduced.

2. **Xanthium strumarium** L., cocklebur. Plants (2)4–20 dm tall, glabrous to short-strigose. Leaves alternate, long-petiolate, blades broadly ovate to suborbicular, cordate to truncate at the base, up to 15 cm long, margin toothed and sometimes shallow-lobed. Heads in axillary clusters, bur cylindric or ovoid to subglobose (1)2–3 + cm long, covered with stiff, hooked prickles and terminated by 2 prominent incurved beaks. (n = 18) Jul–Oct. Open fields & waste ground, especially drying, sandy stream beds & flood plains, throughout GP; (e U.S. to GP, widespread weed in the warmer & drier parts of the world). *X. chinense* P. Mill., *X. globosum* Shull, *X. pensylvanicum* Wallr., *X. echinatum* Murr., *X. glanduliferum* Greene, *X. commune* Britt., *X. acerosum* Greene, *X. speciosum* Kearney—Rydberg.

This highly variable assemblage has been unsuccessfully segregated into numerous species in the past. Two weak varieties may be recognized among the materials from our area:

2a. var. *canadense* (P. Mill.) T. & G. Plants with the lower part of the bur ± spreading-hairy and the bur brownish, 2–3.5 cm long.

2b. var. *glabratum* (DC.) Cronq. Plants with the burs at most weakly glandular to subglabrous, paler in color, and usually less than 2 cm long.

The var. *strumarium* has yellow-green burs with straight beaks and scant pubescence; it is of the American tropics and does not enter our region.

The young herbage is poisonous to pigs. The burs sometimes cause mechanical injury when ingested by livestock. The burs are known fancifully as "porcupine eggs."

111. XYLORHIZA Nutt.

1. **Xylorhiza glabriuscula** Nutt. Erect perennial herbs 1–3 dm tall, arising from a branching, woody caudex surmounting a tough taproot. Stems several and usually simple, arching upward from the base; herbage variously villous to glabrate. Leaves linear to narrowly oblanceolate (2)3–7 cm long overall and 3–9 mm wide, attenuate at the base, margin entire. Heads solitary at the ends of the branches; involucre campanulate, 8–12 + mm tall; involucral bracts imbricate in 2 or 3 series, scarious-chartaceous below but greenish and attenuate-acuminate at the apex (never squarrose-reflexed); ray florets pistillate and fertile, mostly 12–22(30), ligule white to light blue, 1–2 cm long; disk florets bisexual and fertile, corolla 4.5–6 mm long; style branches linear, with acute, distinct appendages. Achenes linear-clavate, 4–5.5 mm long, weakly compressed, velutinous-sericeus with silky, subappressed hairs, and with subobscure nerves; pappus a series of unequal bristles. (n = 6) May–Jun. Infrequent in open, alkaline sites; nw GP, known from w SD: Butte, Harding, Fall River; WY: Niobrara, Weston; apparently more frequent just w of our range; (GP, s MT, WY, & nw CO; isolated in e UT). *Machaeranthera glabriuscula* (Nutt.) Cronq. & Keck—Atlas GP; *Aster xylorrhiza* T. & G.—Harrington. [tribe Astereae]

Our materials are referable to var. *glabriuscula*. Another taxon, var. *linearfolia* T. J. Wats., is restricted to e UT.

Xylorhiza(also spelled *Xylorrhiza*) has been treated as a part of *Machaeranthera*, or as a part of *Aster.* T. J. Watson (Brittonia 29: 199-216. 1977.) has made a good case for treating it as a distinct genus. Acceptance of *Xylorhiza* does not in itself imply eventual acceptance of the numerous segregate genera that have been proposed for the *Haplopappus-Machaeranthera* (et al.) group.

112. ZINNIA L.

1. *Zinnia grandiflora* Nutt. Rocky Mt. zinnia. Low, branching perennial 1-2 dm tall, with several stems arising from a taprooted, woody caudex. Leaves opposite, linear, ± 3-nerved, 1-3 cm long, sessile, the bases connate to some extent, strigose and somewhat impressed-punctate. Heads terminal on the upper stems, not much elevated above the leaves; involucre narrowly campanulate to subcylindrical, 5-8 mm tall; involucral bracts oblong or obtuse, graduated and imbricate, erose-ciliate and subsquarrose at the apex, often reddish-tipped; receptacle chaffy; ray florets 3-6, pistillate, fertile, ligule up to 18 mm long, yellow; disk florets 18-24, corollas red or green, perfect and fertile, corolla tube 5-toothed but one tooth often larger than the others, style branches velutinous, acute; chaffy bracts membranaceous, scarious, enclosing the achenes. Ray achenes oblanceolate-linear, 3-angled; lateral pappus-awns ± adnate to the ligule, awn of the inner angle minute or absent; disk achenes oblanceolate, angular or compressed, 4-5 mm long; pappus awns (1)2-4, unequal, or awns absent. (n = 21) May-Oct. Open, dry limestone areas; sw 1/4 GP; (sw GP to TX & AZ; n Mex.). [tribe Heliantheae]

Reference: Torres, A. M. 1963. Taxonomy of *Zinnia*. Brittonia 15: 1-25.

Class Liliopsida (Monocots)

137. BUTOMACEAE Rich., the Flowering Rush Family

by Robert B. Kaul

1. BUTOMUS L., Flowering Rush

1. *Butomus umbellatus* L. Emergent paludal perennial with thick, prostrate rhizomes. Leaves erect, ensiform, basally triangular and spongy, to 1 m or more tall, 1 m wide. Inflorescence umbelliform, on a more or less triangular scape to 1.5 m tall; umbel subtended by 3 persistent purplish acute-attenuate bracts; pedicels unequal, to 8 cm long. Flowers numerous, regular, hypogynous, perfect, 2-3 cm wide; sepals 3, pink or rose, tinged greenish, ovate, persistent; petals 3, similar to sepals but slightly larger, persistent; stamens 9, anthers red; pistils (5)6(7), pinkish in anthesis, basally connate, the stigmas dark red; placentation laminar. Follicles basally connate; seeds straight, without endosperm. (2n = 16-42) Jun-Aug. Marshes & muddy shores; ND: Barnes, Bottineau; SD: Faulk; MN: Rice (just beyond our range); (St. Lawrence R. valley & Great Lakes, w to ND, & ID). *Introduced*.

Reference: Anderson, L. C., C. D. Zeis, & S. F. Alam. 1974. Phytogeography and possible origins of *Butomus* in North America. Bull. Torrey Bot. Club 101: 292-296.

This species was first found in the Great Lakes in 1897 and 1918 and only recently discovered near and in our range. It probably will become more common in our northern waters. The submersed f. *vallisnerifolius* (Sagorski) Gluck, nonflowering and with elongate lax leaves, has not been collected in the GP.

138. ALISMATACEAE Vent., the Water Plantain Family

by Robert B. Kaul

Emersed and submersed annual and perennial aquatic herbs with milky juice, ours glabrous, the plants monoecious or dioecious or the flowers perfect. Roots coarse, white, gray, or tan. Rhizomes, stolons, and corms cometimes present. Leaves basal, long petioled, with sheathing bases, the blades simple and entire, or absent and the leaves phyllodial. Inflorescences scapose, simple or branched, umbellate, racemose, or paniculate, erect, inclined, or repent, the flowers often in whorls of 3 and subtended by bracts. Flowers regular, hypogynous, perfect or imperfect; sepals 3, green, persistent; petals 3, white or occasionally cream or pinkish, soon withering; stamens 6–many; pistils 12–numerous, free, simple, in a ring or densely crowded, ours all uniovulate and maturing into achenes, the placentation basal. Seeds horseshoe-shaped, without endosperm.

1. Pistils in one whorl on the low receptacle, mostly fewer than 20; achenes wingless, dorsally grooved; flowers perfect; blades never sagittate, sometimes absent in submersed specimens .... 1. *Alisma*
1. Pistils not in a whorl, usually very numerous and crowded over the globose receptacle; achenes winged or ribbed; flowers perfect or imperfect; blades sagittate, cordate, unlobed, or (in submersed specimens) absent.
   2. Achenes bilaterally compressed, winged; flowers imperfect or perfect; blades sagittate, caudate, ovate, linear, or absent .... 3. *Sagittaria*
   2. Achenes turgid, wingless, ribbed; flowers perfect; blades never sagittate, often cordate or truncate at the base, or absent .... 2. *Echinodorus*

1. ALISMA L., Water Plantain

Perennials, the leaves basal, erect or floating or submersed, ours mostly with blades well developed and the petioles triangular. Inflorescences erect, rarely lax in submersed forms, sparsely or abundantly branched and paniculate. Flowers perfect; petals white or pinkish, involute in the bud; stamens 6–9; pistils 6–28, the styles lateral, maturing into a vaguely 3-sided ring of bilaterally compressed achenes. Achenes with smooth sides, 1 or 2 dorsal grooves, persistent styles. Deepwater forms are sometimes sterile or have cleistogamous flowers.

References: Fernald, M L. 1946. The North American representatives of *Alisma plantago-aquatica*. Rhodora 48: 86–88.; Voss, E. G. 1958. Confusion in *Alisma*. Taxon 7: 130–133.; Pogan, E. 1963. Taxonomical value of *Alisma triviale* Pursh and *Alisma subcordatum* Raf. Canad. J. Bot. 41: 1011–1013.

1. Achenes with 2 dorsal grooves; inflorescence erect to somewhat decumbent, lax, simple or a sparingly branched panicle, to 3 dm long in submersed specimens but much less in emergent specimens .... 1. *A. gramineum*
1. Achenes with a single dorsal groove; panicle erect and often conical in outline, much branched, to 1 m tall.
   2. Petals 3–5 mm long, exceeding the sepals; stamens distinctly exceeding the pistils; pedicels slender, the inflorescence stiff .... 3. *A. triviale*
   2. Petals 1–3 mm long, about equaling the sepals; stamens barely exceeding the pistils; pedicels fine, the inflorescence rather delicate .... 2. *A. subcordatum*

1. *Alisma gramineum* J.G. Gmel. Erect, emergent, with slender-petioled lanceolate to elliptic leaves to 2 dm long, or submersed with floating linear leaves to 9 dm long and 2.5 cm wide, with or without blades. Inflorescence scapose, simple or usually paniculate, to 5 dm long in submerged specimens but shorter in emergent plants, sometimes exceeding the leaves; pedicels slender, to 2(3) cm long, subtended by papery lanceolate bracts. Flowers 5–7 mm wide; sepals 1–2.5 mm long, but equaling the achenes; petals rhombic, pinkish, 2–4 mm long; stamens about equaling the pistils, anthers rounded; style 1/2 the length of the ovary, becoming a beak on the inner edge of the achene. Fruiting head 4–6 mm wide; achenes with 2 dorsal grooves, falling singly from the receptacle. (2n = 14, 16) Jul–Sep. On mud flats, or submersed in brackish calcareous waters, local; GP n of 40° lat.; (MN to WY s to NM & CA). *A. geyeri* Torr.—Rydberg.

2. *Alisma subcordatum* Raf., water plantain. Resembling no. 3, the flowers smaller and the inflorescence more delicate. Leaves to 75 cm long, the blades to 15 cm long and 10 cm wide, ovate to elliptic, cordate or cuneate at the base. Inflorescence, to 1 m tall, abundantly branched, conical or pyramidal, often much exceeding the leaves, the pedicels fine and the panicle delicate. Flowers small; sepals 1.5–2.5 mm long, cucullate, their scarious margins rather narrow; petals 1–3 mm long, suborbicular, white, the base yellow; stamens barely exceeding the pistils. Achenes 1–2 mm long, dorsally rounded, with a single dorsal groove. (2n = 14) Jun–Sep. Muddy shores, stream sides, marshes; e ND, e SD, MN, IA, e NE, e KS, OK; (ND to VT s to FL & TX). *A. plantago-aquatica* var. *parviflorum* (Pursh) Torr.—Waterfall.

3. *Alisma triviale* Pursh, water plantain. Erect, often robust perennial. Leaves to 90 cm long; blades to 35 cm long, 3–12 cm wide, elliptic or ovate, the base rounded to subcordate. Inflorescence usually 1, scapose, paniculate, conical or pyramidal, to 1 m tall and usually much exceeding the leaves, the pedicels slender and the inflorescence rather stiff. Sepals to 4 mm long, cucullate, obtuse, their margins scarious and rather broad; petals 3–5 mm long, exceeding the sepals, white, the base yellow; stamens much exceeding the pistils. Achenes 2–3 mm long, dorsally rounded, with a single dorsal groove. (2n = 28) Jun–Sep. Mud flats, marshes, watersides, rarely submersed; GP n of OK-KS border; (Que. to B.C. s to MD, KS, NM & CA). *A. brevipes* Greene—Rydberg.

This plant is often treated as *A. plantago-aquatica* L. var. *americanum* R. & S.

2. ECHINODORUS Rich., Burhead

Emergent annuals and perennials, sometimes submersed and then nonflowering. Emersed leaf blades lanceolate to ovate, obtuse to rounded, basally cordate to tapering; submersed leaves thin, elongate to subcordate. Inflorescences 1–several, simple or few-branched, equal to or usually much exceeding the leaves, erect or becoming repent; pedicels slender, the flowers in whorls of 3–6(15). Flowers perfect, the receptacle globose or convex; sepals reflexed in fruit; petals white, imbricated in the bud; stamens 6–20(30), filaments elongate; pistils many and crowded over the receptacle, the style terminal. Achenes turgid, ribbed, beakless, or beaked and the head then scabroid or echinate. *Helianthium* Engelm.—Rydberg.

Reference: Fassett, N. C. 1955. *Echinodorus* in the American tropics. Rhodora 57: 133–156; 174–188; 202–212.

1 Achenes 9–15, beakless or nearly so, reddish; blades lanceolate-elliptic; inflorescences umbelliform, the pedicels usually recurving; plant diminutive .... 2. *E. parvulus*
1 Achenes numerous, beaked, brown; blades mostly ovate, cordate-truncate; inflorescences paniculate or verticillate, erect or repent.
  2 Inflorescences erect, branched, the peduncle triangular; fruiting heads echinate; stamens ca 12 .... 3. *E. rostratus*

2   Inflorescences erect at first, becoming prostrate and rooting at the flowering nodes later, verticillate; fruiting heads merely scabroid; stamens usually more than 12 .... 1. *E. cordifolius*

1. Echinodorus cordifolius (L.) Griseb. Coarse emergent perennial with prominently arching scapes, or sometimes submersed and then usually nonflowering. Roots sometimes with abundant cylindric outgrowths. Leaves long petioled, the blade to 18 cm long and 15 cm wide, broadly ovate, obtuse, the base cordate, the arcuate veins prominent on the lower side. Inflorescence simple, the flowers in remote whorls, verticillate, eventually much exceeding the leaves and then arching, becoming prostrate and rooting at the nodes; pedicels to 6 cm long, reflexed in fruit. Flowers 1–2.5 cm wide, sepals obtuse, petals broadly obovate to 10 mm long; stamens 12–20, filaments slender, anthers versatile; pistils very numerous, crowded on the globose receptacle. Achenes forming a spherical scabroid head, brown or tan, few-ribbed, the beak erect or slightly curved; seeds brown, with regular rows of indentations. ($2n = 22$) May–Sep. In shallow water & on mud; se KS, w-cen–se MO, ne OK; (VA to KS, s to FL & TX). *E. radicans* (Nutt.) Engelm.—Rydberg.

2. Echinodorus parvulus Engelm. Plants small, erect, sometimes creeping. Blades lanceolate-elliptic, tapering basally, to 5 cm long, 6 mm wide, the petioles shorter than the blades. Inflorescences 1–6, umbellate, few-flowered, equaling or exceeding the leaves, the scape triangular, 2- to 8-flowered, the pedicels 9–15 mm long, recurved in fruit. Flowers to 6 mm wide; stamens 6–9, pistils ca 14, the style shorter than the ovary. Achenes in a globose head, 8-ribbed, beakless or with a tiny beak, reddish. Jul. Sandy shores; KS: Harvey; (MA to KS, s to FL & TX). *Helianthium parvulum* (Engelm.) Small—Rydberg; *E. tenellus* (Mart.) Buch.—Fernald.

3. Echinodorus rostratus (Nutt.) Engelm., burhead. Rather coarse annual (perennial), emersed or submersed, sometimes with floating leaves. Petiole of emersed and floating leaves triangular, the blade 3–6 cm long, broadly ovate, basally cordate or truncate, the main veins arcuate and prominent below; submersed leaves linear to cordate, thin, undulate to crisped. Panicles 1–7, stiffly erect, to 3 dm or more tall and usually much exceeding the leaves, mostly 3-branched at the first 1–3 nodes, the branches subtended by 3 lanceolate-acuminate bracts 1–3 cm long; peduncle and branches triangular, often ribbed between the angles. Flowers in whorls of 3–9, the pedicels to 1.5 cm long, each whorl subtended by 3 acuminate bracts. Flowers 6–14 mm wide, the petals white, broadly ovate to suborbicular, clawed, 4–7 mm long and exceeding the sepals; stamens ca 12, shorter than the pistils; pistils numerous, crowded, long-styled. Fruiting head spherical, echinate; achenes brown, 2–3 mm long, longitudinally ribbed and fluted, the stylar beak erect or nearly so, stout; seeds brown, irregularly alveolate. Jul–Oct. On mud near lakes & ponds & in clear shallow water, occasionally in deeper water, erratically abundant from year to year; se SD, e 1/3 NE, w IA, e 2/3 KS, OK, TX, NM (Ont. to CA s to FL & Mex.). *E. cordifolius* (L.) Griseb. misapplied—Rydberg, Waterfall; *E. berteroi* (Spreng.) Fassett as to descr.—Waterfall.

Smaller individuals with lanceolate leaves have been called var. *lanceolatus* Engelm. but these are merely forms on less saturated soils. Submersed specimens without floating or emersed leaves are sterile.

3. SAGITTARIA L., Arrowhead

Rhizomatous, often cormous perennials or sometimes annuals, with fibrous roots, ours usually emergent when in flower. Monoecious, dioecious, or the flowers perfect. Roots coarse, white or gray, chambered. Petioles long, spongy, variously convex and angular in section, the blade unlobed or sagittate, acute to acuminate, or absent, the plants often heterophyllous and heteroblastic; petioles bearing unlobed blades held smartly erect, or geniculate near the blade and the blade erect; petioles bearing sagittate blades erect to lax. Inflorescences 1-several, mostly erect but sometimes procumbent, simple or branched, the peduncles long, basally enclosed by the leaf sheaths, the flowers pedicellate in whorls of 3, each subtended by a bract. Flowers mostly imperfect (mostly perfect in no. 3), the males above in monoecious individuals and soon falling; sepals green, appressed or usually reflexed on fruiting heads, reflexed on male flowers; petals white, imbricate in the bud; stamens numerous, the filaments glabrous or pubescent; pistils very numerous, crowded on the globose receptacle. Achenes beaked, often winged on the margins and often winged or ribbed on the faces, bilaterally compressed, often with more or less conspicuous "resin ducts" on the faces.

References: Beal, E. O., J. W. Wooten, & R. B. Kaul. 1982. A review of the *Sagittaria engelmanniana* complex (Alismataceae) with environmental correlations. Syst. Bot. 7: 417–432; Bogin, C. 1955. Revision of the genus *Sagittaria* (Alismataceae). Mem. New York Bot. Gard. 9: 179–233; Wooten, J. W. 1973. Taxonomy of seven species of *Sagittaria* from eastern North America. Brittonia 25: 64–74.

Leaves of most species are variable according to environment and season. In at least 3 of our sagittate-leaved species (nos. 2, 3, 6), emersed leaves with extremely narrow lobes are commonly found on plants growing in 5–20 cm of water, and these narrow-lobed forms often intergrade to progressively broader-lobed forms on nearby shores. The narrow-lobed forms flower less frequently. Bladeless forms are sometimes found submersed in even deeper water and in flowing water, especially in nos. 3, 4, and 5, while no. 4 occasionally produces cordate floating blades.

1 Blade of emersed leaves not sagittate (but sometimes with small caudal lobes).
  2 Female flowers and fruiting heads apparently sessile; peduncle sometimes bent near the lowest flowering node .... 8. *S. rigida*
  2 Flowers and fruiting heads obviously pedicellate; peduncle not bent near the lowest flowering node (but sometimes bent near the base).
    3 Filaments pubescent and basally dilated; bracts ovate and basally united; dorsal wing of achene high-rounded, the beak lateral or absent; e 1/2 GP & w across NE .... 5. *S. graminea*
    3 Filaments glabrous, slender; bracts linear, acuminate, essentially free, papillose; dorsal wing of achene not especially rounded, the lateral beak then near the top; e KS, e OK, sw MO .... 1. *S. ambigua*
1 All or most blades of emersed leaves sagittate, the basal lobes large.
  4 Petioles terete; inflorescences becoming procumbent with age; sepals appressed to the fruiting head; most or all flowers perfect, their pedicels thick .... 3. *S. calycina*
  4 Petioles angular; inflorescences erect or inclined in fruit but not procumbent; few or no flowers perfect, their pedicels not thick.
    5 Achene beak oblique to erect, the dorsal wing not confluent with it; bracts 1–4 cm long.
      6 Dorsal wing of achene entire or unevenly crenulate and often truncate at its distal end, leaving a saddlelike depression between the wing and beak; ventral wing straight; beak recurved, prominent; fruiting heads depressed ...... 2. *S. brevirostra*
      6 Dorsal wing of achene entire, apically rounded; ventral wing usually convex below the beak; beak straight, obscure; fruiting heads globose .... 4. *S. cuneata*
    5 Achene beak horizontal, the dorsal wing confluent with it; bracts less than 1.5 cm long.
      7 Ventral wing of achene somewhat narrower than the dorsal wing; achene faces without keels or ridges; bracts acute to obtuse, essentially free; basal lobes of blade narrow to wide, shorter to longer than the terminal lobe; GP .... 6. *S. latifolia*
      7 Ventral wing of achene much narrower than the dorsal wing or rudimentary; beak sometimes rudimentary; achene faces often with a ridge; bracts lanceolate-attenuate, basally connate; basal lobes of blade narrow and usually much longer than the terminal lobe; s NE to TX .... 7. *S. longiloba*

1. *Sagittaria ambigua* J. G. Sm. Rhizomatous, cormous erect perennial. Leaves to 5 dm long, the petiole to twice the length of the blade; blade ovate to mostly lanceolate, not sagittate, 5–22 cm long, 1–7 cm wide. Inflorescences ordinarily simple and only somewhat exceeding the leaves, bearing up to 10 whorls of flowers, the lowest 1–5 whorls female, the upper whorls male; bracts 7–25 mm long, linear to lanceolate, acuminate, sparsely papillose on the ribs, essentially free. Sepals reflexed, 5–7 mm long, sparsely papillose; petals ovate, somewhat longer than the sepals; filaments slender, glabrous, equal to or slightly longer than the anthers. Fruiting heads mostly less than 1 cm wide; achenes obovate, 1.5–2 mm long, with thin narrow dorsal and ventral wings, the wings confluent with the beak, the sides sometimes with a wing, the tiny lateral beak near the top and horizontal or incurved. Jul–Sep. Shores & other damp places, not common; se KS, e OK, sw MO; (MO, KS, OK).

2. *Sagittaria brevirostra* Mack. & Bush, arrowhead. Erect, often robust perennial from rhizomes, bearing corms in autumn. Monoecious or sometimes dioecious. Leaves to 6 dm tall, petiole angular, the blade sagittate, variable, to 30 cm long and 20 cm wide but mostly smaller, sometimes the basal lobes divergent and often about 1/2 the width of the terminal lobe but equaling its length; specimens from deeper water (5–20 cm) often have very narrow lobes. Inflorescences solitary or few, simple to branched at the lowest node, partly exceeding the leaves; lower whorls female, upper whorls male, or sometimes all flowers male or female; bracts lanceolate, gibbous, 1–4 cm long and sometimes exceeding the flowers in anthesis, firm, stramineous; pedicels of female flowers 1–2 cm long, ascending, those of the male flowers longer. Sepals ovate, reflexed in fruit; petals showy, 1–2 cm long. Fruiting heads to 2 cm wide, often apically depressed; achenes cuneate-obovate, 2–3 mm long, with a distinct entire to crenulate dorsal wing that is often truncate at its distal edge, the facial wings 3, the beak oblique and recurved upward, the achene in profile then showing a distinct apical saddle. Jun–Sep. Muddy lakeshores & stream sides; SD to TX; (OH to SD s to TX, & occasionally far beyond this range?. *S. engelmanniana* J. G. Sm. subsp. *brevirostra* Bogin — Gleason & Cronquist.

The apical saddle and uneven dorsal wing of the achene help to distinguish this species, as does the often apically depressed fruiting head.

3. *Sagittaria calycina* Engelm., arrowhead. Small to large emergent annual without corms or rhizomes. Leaves to 1 m tall, the petiole thick, terete, erect or variously inclined or procumbent; blade sagittate, sagittate-hastate, or sagittate-deltoid, 3–40 cm long, 2–25 cm wide, the lobes linear in smaller leaves and on plants in deeper water (5–20 cm), deltoid and widespreading in large leaves, the basal lobes equaling or longer than the terminal lobe; leaves linear in submersed individuals. Inflorescenes 1–25, stout, shorter than the leaves and variously inclined or procumbent among them, always procumbent in fruit; inflorescence simple or branched from the lowest node; peduncle terete; bracts delicate, colorless, ca 3 mm long, soon withering, their free end acute to obtuse. Flowers perfect or occasionally a few distal flowers male; pedicels rather thick, to 5 cm long, thickening and recurving in fruit; sepals broad, obtuse, 5–12 mm long, reflexed in flower but appressed to the fruiting head and covering more than 1/2 of it; petals white, the base pale yellow, about 1/3 longer than the sepals; filaments linear, puberulent. Fruiting heads to 2 cm wide, oblate or convex; achenes narrow, barely cuneate, mostly 2–3 mm long, 1 mm wide, the dorsal wing narrow and diminished near the beak, the ventral wing narrower and confluent with the beak; beak horizontal or barely oblique, about 0.5 mm long. ($2n = 22$) May–Oct. Locally abundant on muddy shores, abundant some years & absent others; GP except ND, MT, WY; (OH to SD s to TX & beyond). *Lophotocarpus calycinus* (Engelm.) J. G. Sm. — Rydberg; *S. montevidensis* Cham. & Schlect. subsp. *calycina* (Engelm.) Bogin — Gleason & Cronquist.

This species is readily distinguished by its terete petioles, perfect flowers, appressed sepals in fruit, and procumbent inflorescences. Linear-leaved submersed juveniles sometimes produce flowers at the water's surface.

4. Sagittaria cuneata Sheld., arrowhead. Rhizomatous, cormous perennial, emersed or submersed, sometimes with floating leaves. Monoecious or occasionally dioecious. Leaves to 6 dm long but usually much less, erect or inclined, the petioles sometimes curved upward, angular; emersed blades sagittate, floating blades sagittate, cordate, or elliptic. Sagittate blade 5–15(20) cm long, the terminal lobe often deltoid or ovate, the basal lobes somewhat to much shorter than the terminal lobe and about half its width. Inflorescence about equaling the leaves or sometimes exceeding them, simple or branched, sparsely flowered, the lower whorls usually female, the upper male; bracts lanceolate-acuminate, whitish, to 4 cm long; pedicels of female flowers to 1 cm long, those of male flowers longer. Sepals ovate, 4–9 mm long; petals to twice the length of the sepals. Filament glabrous, the anther longer than the filament. Fruiting heads globose. Achenes 1.5–3 mm long, with a prominent, pale, rounded dorsal wing and a somewhat narrower, pale ventral wing, this extending to the beak, the dorsal wing stopping short of the beak, a single wing on each face, the beak erect, tiny. ($2n = 22$) Jun–Sep. Shores & stream sides, especially on sandy soil; GP (N.S. to B.C. s to CA, NM, OK, IL, & NY).

   This species is distinguished readily by the rounded dorsal wing and tiny erect beak of the achene. Submersed rosettes of linear leaves are usually sterile but occasionally produce elliptical to sagittate floating leaves and floating or emersed inflorescences.

5. Sagittaria graminea Michx. Mostly monoecious, rhizomatous, cormous perennial or annual. Emergent leaves very long-petiolate, to 5 dm long, with linear to narrowly ovate mostly 3-nerved blades 2–18 cm long and to 2 cm wide, never sagittate; leaves of submersed specimens lax, oblong-ovate, cuspidate, to 1 cm wide. Inflorescence solitary or few, mostly shorter than the longest emergent leaves, simple, with 2–6(9) whorls of flowers, the lower 1–5 whorls female, the upper whorls male, or rarely the plants dioecious; bracts ovate, slightly connate, about 75 mm long; pedicels slender to filiform, 1–1.5(2) cm long, horizontal or ascending. Sepals ovate, to 5 mm long, reflexed; petals broad, to 1.5 cm long, white to pale pinkish; filaments pubescent, basally dilated. Fruiting heads 4–15 mm wide; achenes obovate, the dorsal wing high-rounded at the top, the sides unadorned or with a few ridges, the beak tiny and laterally inserted or absent, the seed within evident. ($2n = 22$) Jun–Sep. Occasional in or near water; e 1/2 SD, NE, KS, OK; (Newf. to SD, s to FL & TX).

   Our specimens are mostly var. *graminea*, except in se KS and ne OK where var. *platyphylla* (Engelm.) J. G. Sm. occurs. It has thicker and downturned pedicels on the fruiting heads, the leaves broader and sometimes with short basal lobes. *S. platyphylla* (Engelm.) J. G. Sm.—Rydberg.

6. Sagittaria latifolia Willd., arrowhead. Erect, emergent, mostly monoecious perennial bearing rhizomes and, in autumn, corms. Leaves variable, to 1.5 m long with blades to 50 cm long, these mostly sagittate but phyllodial in the rare submersed individuals, the apical and basal lobes very narrow (especially in deeper water forms) to deltoid, the basal lobes parallel or widely divergent, somewhat shorter than the terminal lobe in broader blades but longer than the terminal lobe in narrower blades. Inflorescence solitary or several, simple or branched at the lower nodes, often exceeding the leaves, erect, the peduncle and branches triangular, the bracts free, papery, boat-shaped, obtuse or acute, to 1.5 cm long and not exceeding the flowers in anthesis; lower whorls of flowers female, the upper male; pedicels slender. Sepals reflexed in fruit, to 1 cm long; petals about $2 \times$ as long as the sepals, showy; filaments glabrous, linear, longer than the oblong anthers. Fruiting heads to 3 cm broad, subspherical and not apically depressed; achenes obovate, to 4 mm long, the dorsal wing wider than the ventral wing and often diminished near the beak, the faces without wings or sharp ridges, the laterally placed tapering beak horizontal or barely oblique. ($2n = 22$) Jun–Sep. Muddy shores & ditches; GP; (N.S. to B.C., s to SC, CA; S. Amer.). *S. esculenta* Howell—Rydberg.

The shorter, boat-shaped bracts and spherical fruiting heads distinguish this species from *S. brevirostra*, and the shape of the achenes distinguishes it from that species and *S. cuneata*.

7. ***Sagittaria longiloba*** Engelm., longbarb arrowhead. Rhizomatous erect perennial bearing autumnal corms. Leaves erect or spreading, to 8 dm long; blades consistently sagittate, the linear to lanceolate basal lobes to 22 cm long and 0.5–2 cm wide, usually distinctly longer than the terminal lobe and about its width. Inflorescences to 1.5 m tall, simple or with a single branch at the lowest node; bracts basally connate, ovate-lanceolate, attenuate, mostly less than 1.5 cm long, smooth; pedicels 0.7–3 cm long; lower whorls female, upper whorls male. Sepals 4–7 mm long, reflexed; petals about twice the length of the sepals; filaments linear, glabrous, longer than the linear anthers. Fruiting heads about 1 cm wide; achenes obovate, with a prominent dorsal wing, the ventral wing much narrower; beak triangular, short, laterally inserted, or obsolete. Jun–Sep. Shallow water, swamps, ditches; s-cen NE to TX; (NE s to TX & AZ).

This is our only sagittate-leaved species that has consistent leaf shape. The narrow lobes and the longer length of the basal lobes distinguish it, as does its rather restricted range.

8. ***Sagittaria rigida*** Pursh. Rhizomatous perennial, emergent and erect or submersed and lax. Leaves highly variable, to 9 dm long, bladeless and submersed or with linear to ovate blades to 9 cm long, mostly 1–6(10) cm wide, sometimes 1 or more blades with short, narrow lobes but never truly sagittate; petiole sometimes geniculate near the blade. Inflorescence erect or flexuous, or lax and elongate in submersed forms, often geniculate near the lowest node, not exceeding the largest leaves; lower whorls female, upper whorls male; female flowers sessile or nearly so, with pedicels shorter than the flowers; male flowers with slender pedicels 1–1.5 cm long. Sepals 4–7 mm long, ovate, reflexed; petals about $2 \times$ as long as the sepals; stamens 12 or more, the pubescent filaments widening basally; anthers ovate. Fruiting heads to 1 cm wide, echinate, apparently sessile; achenes obovate or oblong, 2–2.5 mm long, the wings narrow, the dorsal wing diminished near the beak, the beak elongate, recurved upward, laterally inserted. (2n = 22) Jun–Sep. Brackish & alkaline waters; MN, IA, MO, n & e NE, KS: Reno, (Que. to MN s to KS, MO, VA).

This species is readily distinguished from *S. ambigua* and *S. graminea* by its apparently sessile female flowers and fruiting heads and by the often geniculate petioles and peduncles.

139. HYDROCHARITACEAE Juss., the Frog's-bit Family

by Robert B. Kaul

Ours all submersed, dioecious herbs of quiet or gently flowing fresh waters, the flowers appearing at the water surface. Leaves basal or cauline, sessile, linear or oblong, obtuse to acute, entire, minutely serrulate, or minutely denticulate. Inflorescences with a spathe, 1- to many-flowered, pedunculate and floating, or sessile and submersed. Flowers imperfect, regular or nearly so. Sepals 3, green or pale; petals 3, white, or absent. Stamens 2 or 9, or present as staminodia in pistillate flowers; ovary inferior, obscurely 3-carpellate, 1-locular; placentation parietal, ovules several; styles and stigmas 3. Fruits indehiscent, ripening underwater; seeds straight, without endosperm. Elodeaceae – Rydberg.

Rydberg cites *Limnobium spongia* (Bosc) Rich. for KS, but no specimens have been seen.

1 Leaves cauline, opposite or whorled, less than 3 cm long .... 1. *Elodea*
1 Leaves basal, to 1 m long .... 2. *Vallisneria*

1. ELODEA Michx., Waterweed

Submersed herbaceous perennials of clear waters, rooting along the stems and usually much branched. Leaves cauline, sessile, linear, sometimes lime-encrusted, mostly opposite or whorled, 1-nerved, serrulate or minutely denticulate. Spathes sessile or appearing slightly pedunculate, axillary, born singly in the upper axils, 2-lobed. Male flowers 1-several in the spathe and long-pedicellate with the flower opening at the surface, or single in the spathe and breaking from it to float free at the surface and there to open, filaments sometimes connate. Female flowers single in the spathe, sessile but appearing pedicellate, the very long hypanthium elevating the perianth to the surface. *Anacharis* Rich.—Rydberg.

References: St. John, H. 1962. Monograph of the genus *Elodea* (Hydrocharitaceae) Part I. The species found in the Great Plains, Rocky Mountains, and the Pacific states and provinces of North America. Wash. State Univ. Res. Stud. 30: 19–44; St. John, H. 1965. op. cit., Part 4. The species of eastern and central North America, and summary. Rhodora 67: 1–35; 155–180.

1 Leaves all opposite; spathes very elongate; WY, CO, ND, & MT .... 3. *E. longivaginata*
1 Lower leaves opposite or whorled; middle and upper leaves whorled; spathes not especially elongate; ND to KS.
  2 Middle and upper leaves mostly 2–3 cm long, 4 mm wide; only male flowers produced .... 2. *E. densa*
  2 Middle and upper leaves 5–15 mm long, 0.5–3 mm wide; male and female flowers produced on separate plants.
    3 Middle and upper leaves 5–15 mm long, 1.5–3 mm wide; male flowers pedicellate and attached in anthesis; leaves crowded toward branch tips .... 1. *E. canadensis*
    3 Middle and upper leaves mostly 5–10 mm long, 0.5–1.5 mm wide; male flowers breaking from the pedicels in the spathe and floating free to the surface; leaves not crowded .... 4. *E. nuttallii*

1. ***Elodea canadensis*** Michx. Slender-stemmed, creeping or erect, often abundantly branched, sometimes lime-encrusted. Lower leaves opposite or whorled, small; upper leaves in whorls of 3, larger, linear to oblong, minutely serrulate, crowded towards the branch tips, 5–15 mm long, 1.5–3(5) mm wide, sometimes recurved. Staminate spathes appearing pedunculate, inflated, bifid, gaping. Male flowers several in the spathes, long-pedicellate, the perianth floating; sepals 3, dark striate, to 5 mm long; petals 3, white, to 5 mm long, stamens 9, basally connate, the inner 3 elevated. Pistillate spathes cylindric, bifid, 1-flowered, the flower sessile but appearing pedicellate by the greatly elongate filiform hypanthium, the perianth floating. Sepals 3, 2–3 mm long, linear; petals 3, white, 2.5 mm long, spatulate; the 3 tiny staminodia acicular; stigmas 3, to 4 mm long, apically 2-cleft, equaling or exceeding the perianth; ovary ovoid, with 2–5 ovules. ($2n = 24, 48$) Jun–Sep. Locally abundant in still or running calcareous waters; MT to MN s to IA, NE, WY; (Que. to Sask. s to CA, NM, NE, IA & VA). *Anacharis canadensis* (Michx.) Rich.—Rydberg; *A. planchonii* (Casp.) Rydb.—Rydberg, for the male plants, which have slightly narrower leaves (St. John, 1962, op. cit.).

2. ***Elodea densa*** (Planch.) Casp. Coarser than nos. 1, 3, and 4; sparingly branched, the stems about 3 mm thick. Upper leaves in whorls of 4–6, the lower whorls with fewer leaves; leaves sessile, often crowded, especially near the stem tip, linear to linear-lanceolate, acuminate to nearly obtuse, serrulate, to 3 cm long, 4 mm wide. Staminate spathes several-flowered, unilaterally cleft. Male flowers long-pedicellate, the perianth floating; sepals 3, green, elliptic, to 3.5 mm long; petals 3, white, showy, nearly orbicular, to 9 mm long and 7 mm broad; stamens 9, free; filaments papillose; a 3-lobed nectary at the center. Pistillate plants unknown in this country. ($2n = 48$) Jun–Sep. Quiet waters, rare; KS: Douglas; NE: Hall; (N. Eng. to NE, s to FL, LA; S. Amer.). *Introduced*. *Anacharis densa* (Planch.) Vict.—Gleason.

This tropical species is commonly grown in ornamental pools and aquaria and can be expected occasionally in GP waters, but perhaps only as a temporary member of our flora.

3. **Elodea longivaginata** St. John. Stems slender, elongate, sparingly branched. Leaves opposite, the pairs separated by internodes about their length, the terminal leaves not especially crowded, linear to narrowly lanceolate, acute to obtuse, mostly 6–15(20) mm long, 0.5–1.5(2) mm wide, flaccid, the leaves all similar. Staminate spathe 2–15 cm long, inflated near the tip, with 2 prominent apical teeth. Male flowers on filamentous pedicels to 30 cm long; sepals elliptic, 3.5–5 mm long, 1.9–2.5 mm wide; petals white, linear, 5 mm long, 0.6 mm wide; central stamens not connate. Pistillate spathe 3–7 cm long, cylindric, with 2 apical teeth; hypanthium very elongate to 2 dm long, pedicellike, elevating the perianth and styles to the surface beyond the spathe; sepals elliptic, 2.8 mm long, 1.3 mm wide; petals white, spatulate, 4 mm long, 1.3 mm wide; staminodia 3, strap-shaped; style slender, stigmas 3, oblong; ovary becoming greatly enlarged in fruit. Fruit 10 mm long, 3 mm wide; seeds cylindric. Jul–Aug. Quiet waters, lakes, on the plains; ne WY; MT: Philips; CO: Larimer; ND: Burke, Burleigh; (ND to Alta. s to NM, mostly in mt. lakes).

4. **Elodea nuttallii** (Planch.) St. John. Similar to no. 1 but more delicate. Stems slender, often much branched. Leaves of upper whorls in 3s and 4s, the whorls seldom crowded; leaves pale, sessile, flaccid, linear-lanceolate, 5–10(12) mm long and 0.5–1.5(2) mm wide. Staminate spathes sessile, about 2 mm wide, subspherical, opening apically by a single suture flanked by 2 acuminate teeth. Male flower 1, sessile in the spathe and breaking from it, rising to the surface and floating unattached before opening; sepals 3, about 2 mm long, boldly reflexed, elevating the stamens; petals wanting or occasionally present and tiny; stamens 9, the 3 central ones basally connate and appearing elevated. Pistillate spathes to 1.5 cm long, cylindric, narrow, basally inflated, apically 2-toothed, 1-flowered; female flower sessile but appearing long-pedicellate by extension of the hypanthium which elevates the perianth to the surface; sepals 2, about 1 mm long; petals 3 and slightly longer than the sepals. Fruit sessile, axillary, enclosed by the spathe, ovoid to elongate; seeds pilose. (2n = 48) Jun–Aug. Locally abundant in quiet water; SD, MN, IA, MO, NE, KS, OK; (Que. to MN, s to NM, OK, MO, & NC). *Anacharis occidentalis* (Pursh) Vict.—Rydberg; *A. nuttallii* Planch.—Gleason & Cronquist.

2. VALLISNERIA L., Eelgrass, Tapegrass

1. **Vallisneria americana** Michx. Dioecious stoloniferous perennial with basal leaves. Leaves thin, linear, to almost 1 m long, 8 mm wide, irregularly septate, striate except for marginal bands, minutely denticulate. Staminate inflorescences short-pedunculate, remaining submersed among the leaf bases, the spathe ovoid, opening from the top and the freed ends becoming revolute. Male flowers numerous, crowded, tiny, short-pedicellate, breaking from their pedicels in bud and rising to the surface and there floating and opening; sepals 3, ovate, acute, petal 1(2), much smaller than the sepals; stamens 2, their filaments connate, a third stamen represented by a staminodium resembling the petal. Pistillate spathe becoming elevated to the surface by the long peduncle, the spathe tubular, opening apically and exposing the perianth and stigmas of the single sessile flower. Sepals of female flower 3, ovate, thick; petals 3, tiny, lanceolate, scarious; 1 or more staminodia sometimes present; styles 3, short, thick, the stigmas 3, large, bilobed; ovary cylindric; placentation parietal; ovules numerous, straight. Fruit partially retained in the spathe, slightly curved, maturing under water when the peduncle coils after pollination and pulls the flower beneath the surface. Seeds elaborately transversely fluted. Jul–Sep. Sandy soil in shallow water, sometimes in streams; ne SD, MN, nw IA, & reported for NE (sandhills); (N.S. to MN, SD, s to TX & FL).

This plant is sometimes confused with species of *Sagittaria* but can be distinguished from them by the distinct marginal leaf band and the stoloniferous habit.

140. SCHEUCHZERIACEAE Rudolphi, the Scheuchzeria Family

by Robert B. Kaul

1. SCHEUCHZERIA L.

1. *Scheuchzeria palustris* L. var. ***americana*** Fern., the only species in the family. Perennial paludal herb of sedge mats and bogs. Rhizome creeping, stem to 3 dm tall, erect, unbranched, zigzag. Leaves cauline, alternate, with prominent sheaths completely enveloping the lower internodes, but only partially enveloping the upper internodes, a ligule present at the mouth of the sheath; blades tubular, grasslike, a tiny pore at the tips, the upper ones exceeding the terminal inflorescence. Inflorescence racemose, few-flowered, bracteate, the bracts reduced above. Flowers perfect, hypogynous, regular; perianth parts 6 in 2 series, greenish white, ca 3 mm long, ovate-lanceolate; stamens 6, exceeding the perianth and pistils in anthesis. Pistils 3, ovoid, barely united basally, the styles stout, the stigmas blunt. Fruits inflated follicles, 6–10 mm long, 1- or 2-seeded, the seeds ellipsoid, black. (2n = 22) Jun–Jul. Acid bogs; MN: Otter Tail, Pope; (circumboreal, s in N. Amer. to NJ, IA, CA).

141. JUNCAGINACEAE Rich., the Arrowgrass Family

by Robert B. Kaul

1. TRIGLOCHIN L., Arrowgrass

Grasslike tufted perennials of brackish and alkaline marshes. Leaves basal, distichous, sheathing at the base and becoming flat or subterete above, thick, fleshy, and not grasslike to the touch. Inflorescences erect, straight or sinuous, ebracteate spiciform racemes usually exceeding the leaves, the pedicels slender and sometimes decurrent. Flowers numerous, inconspicuous, unevenly spaced on the axis, not subtended by bracts, the floriferous axis about equaling the naked scape in length. Flowers perfect, regular; perianth parts 6, free, greenish or purplish; stamens 6, the anthers nearly sessile and attached to the perianth parts and usually falling with them; pistils 3–6, superior, joined to a central carpophore, unilocular, uniovulate, the styles obscure. Fruits free or united, indehiscent; seeds without endosperm. Scheuchzeriaceae, in part—Rydberg.

The plants contain varying amounts of cyanogenic glycosides that produce hydrocyanic acid, a poison that acts by blocking cellular respiration. Arrowgrasses under drought stress contain more of the poision than those in wetter conditions. The leaves are sometimes eaten by grazing stock in sufficient quantity to cause poisoning.

1 Pistils 3; fruit linear-clavate, each carpel sharp-pointed at its base; carpophore 3-angular .... 3. *T. palustris*
1 Pistils mostly 6; fruit oblong-elliptic or ovate-angular, the carpels not pointed at their bases; carpophore terete.

2   White leaf bases persisting on rhizome; ligules entire, 1-5 mm long; fruit 3.5-4.5 mm long, 2-3.5 mm thick .... 2. *T. maritima* var. *elata*
2   Rhizome with brown fibers; ligules deeply bilobed, 0.5-1 mm long; fruits 3-4 mm long, 1-2 mm thick .... 1. *T. concinna* var. *debilis*

1. ***Triglochin concinna*** Davy var. ***debilis*** (M.E. Jones) Howell. Plant slender, from an elongate, slender rhizome, the base invested with brownish fibers of the leaf bases. Leaves to 18 cm long, the sheaths with membranaceous margins, the bilobed ligule at the mouth of the sheath to 1 mm long. Inflorescences erect, straight or irregularly twisted, the flowers rather remote and the axis visible in the floriferous portions. Perianth parts 6, greenish, each with an attached stamen; pistils 6. Fruit of united carpels, 3-4 mm long, 1-2 mm thick, the carpels separating from the carpophore individually. ($2n = 48, 96$) Jun-Aug. Brackish marshes & stream side meadows; WY, MT, ND, SD; (ND to OR s to AZ, CO, Baja Calif., S. Amer.).

2. ***Triglochin martima*** L. var. ***elata*** (Nutt.) A. Gray. Plant medium- to coarse-textured, from a short, stout rhizome. Leaves linear, thick, to 8 dm long but mostly less, 1.5-3.5 mm wide; sheath prominent, membranaceous-margined, the rounded to subacuminate, entire or shallowly lobed ligule about 5 mm long, the leaf-bases persistent. Inflorescences to 1 m tall, but mostly much less, barely to greatly exceeding the leaves, the scape thickish, straight, the axis straight or gently sinous; flowers numerous, unevenly crowded, mostly hiding the axis; pedicels slender, erect, their bases decurrent. Perianth parts 6, greenish, each with an attached stamen; pistil of (3)6 carpels united to a central carpophore, rounded at the base. Fruit ovoid, 3.5-4.5 mm long, 2-3.5 mm wide, with prominent reflexed longitudinal ridges, the beak revolute; mature carpels falling separately from the carpophore. ($2n = 144$) Apr-Aug. Saline & alkaline marshes; MN to MT s to CO & n KS; (Lab. to AK s to CA, NM, n KS, PA; Mex.).

3. ***Triglochin palustris*** L. More delicate than nos. 1 and 2, from a short ascending rhizome. Leaves slender to acicular, 1-2 mm wide, to 30 cm long but usually much less, the ligule 0.5-1 mm long, cleft to its base into 2 lobes. Inflorescences 1-3, somewhat to mostly much exceeding the leaves, straight or nearly so, the flowers unevenly distributed and the axis visible; pedicels filiform, 2-5 mm long, erect, their bases not decurent. Perianth parts 6, greenish, each with an attached subsessile anther; pistils 3. Fruits linear-clavate, mostly 4-7 mm long, separating below but persisting at the upper end to the 3-angled carpophore; styles blunt, the bases pointed. ($n = 12, 18, 24$; $2n = 24$) Jun-Sep. Brackish or alkaline marshes, on mud & gravel; MT, ND, SD, WY, NE, CO; (Lab. to AK, s to NY, NE, NM, & CA).

142. POTAMOGETONACEAE Dum., the Pondweed Family

by G. E. Larson and William T. Barker

1. POTAMOGETON L., Pondweed

Submersed or floating-leaved, rhizomatous aquatics, perennial from the rhizomes, tubers or by budding from the lower nodes, sometimes reproducing and overwintering by free-floating winter buds. Stems elongate, flexuous, anchored by roots and rhizomes. Leaves alternate, becoming opposite upward in some spp., simple; stipules present, fused to each other along one or both margins or adnate to the leaf blade in some linear-leaved spp., forming an open or closed sheath around the stem, fibrous or membranaceous, often rapidly deteriorating. Submersed leaves filiform to lanceolate, thin and flexuous, usually sessile. Floating leaves produced in some spp., basically elliptic in outline, petioled, rather leathery with a waxy upper surface. Winter buds produced in the leaf axils of some spp., these consisting of tightly compressed apices bound by reduced leaves. Inflorescence an axillary or terminal spike bearing few to many minute whorled flowers; peduncles stout to filamentous, usually lifting the spike above the water surface at anthesis, often recurved with age. Flowers perfect, regular; perianth of 4 sepaloid bracts; stamens 4, each inserted on the claw of a perianth bract; carpels 4, separate, each maturing into a strongly to weakly beaked, drupelike fruit.

References: Fernald, M. L. 1932. The linear-leaved North American species of *Potamogeton* section *Axillares*. Mem. Amer. Acad. Arts 17: 1–183; Haynes, R. R. 1974. A revision of North American *Potamogeton* subsection *Pusilli* (Potamogetonaceae). Rhodora 76: 564–649; Larson, G. E. 1976. The Potamogetonaceae in North Dakota. Prairie Naturalist 8: 1–18; Ogden, E. C. 1943. The broad-leaved species of *Potamogeton* of North America north of Mexico. Rhodora 45: 57–105, 119–163, 171–214; Reznicek, A. A., & R. S. W. Bobbette. 1976. The taxonomy of *Potamogeton* subsection *Hybridi* in North America. Rhodora 78: 650–673; St. John, H. 1916. A revision of the North American species of *Potamogeton* of the section *Coleophylli*. Rhodora 18: 121–138.

1 Stipules adnate to the leaf blade for a distance of 10 mm or more, the free portion projecting as a ligule; floating leaves absent.
  2 Stipular sheaths of the main stem inflated 2–5 × the thickness of the stem; floral whorls 5–12 per spike .... 18. *P. vaginatus*
  2 Stipular sheaths of the main stem about as wide as the stem; floral whorls 2–6 per spike.
    3 Stems dichotomously branched from the base, mostly unbranched above; fruits olivaceous, 2–3 mm long, the beak flat, inconspicuous .... 6. *P. filiformis*
    3 Stems dichotomously branched above; fruits yellowish to brown, 3–4 mm long, apiculate, with a beak usually 0.3–0.5 mm long .... 13. *P. pectinatus*
1 Stipules free of the leaf blade or adnate for a distance of less than 10 mm, often early deteriorating; floating leaves present or absent.
  4 Submersed leaves linear, 6 mm or less wide (sometimes wider in *P. epihydrus*), mostly 20 × or more longer than wide.
    5 Stipules adnate to the submersed leaf blades for mostly 1–4 mm; embryo coil plainly visible through the papery thin walls of the fruit; small elliptic floating leaves 5–40 mm long usually present .... 4. *P. diversifolius*
    5 Stipules free of the leaf blades; fruit walls firm, the embryo coil obscured by the walls of the fruit; floating leaves, if present, mostly larger.
      6 Leaves dimorphic, both floating and submersed leaves produced.
        7 Submersed leaves reduced to phyllodes 1–2 mm wide, these often absent with age; floating leaves rounded to cordate at the base .... 11. *P. natans*
        7 Submersed leaves ribbonlike, mostly 3–6(10) mm wide, with a cellular-reticulate strip on each side of the midvein forming a conspicuous median band 1–2 mm wide; floating leaves tapered to the petiole .... 5. *P. epihydrus*
      6 Leaves all alike, all submersed.
        8 Leaves with many (15–35) fine nerves; mature fruits 4–4.5 mm long .... 19. *P. zosteriformis*
        8 Leaves 3- to 7(9)-nerved; mature fruits 1.5–3.6 mm long.
          9 Fruits with an undulate to dentate dorsal ridge or keel; glands rarely present at the base of the stipules .... 7. *P. foliosus*
          9 Fruits dorsally smooth and rounded; glands usually present at the base of the stipules.
            10 Stipules whitish, fibrous, the oldest often shredding into fibers; peduncles mostly clavate; indurate winter buds often present.
              11 Leaf tips acute, rarely obtuse; leaves 3- to 5(7)-nerved; 0.6–2 mm wide; peduncles mostly terete; stems mostly terete .... 17. *P. strictifolius*

11 Leaf tips rounded to apiculate; leaves 5- to 7(9)-nerved, 1.2–3.2 mm wide; peduncles compressed; stems compressed .... 8. *P. friesii*
10 Stipules tan to brownish-green, delicate, usually decomposing with age; peduncles filiform to cylindric; winter buds seldom present .... 15. *P. pusillus*
4 Submersed leaves linear-lanceolate, lanceolate, oblong or ovate, broader in proportion to length.
12 Leaves all submersed, sessile, weakly to strongly clasping at the base, often undulate-crisped.
13 Leaf margins finely serrate; fruit beak 2–3 mm long; indurate winter buds commonly produced in upper axils .... 3. *P. crispus*
13 Leaf margins entire; fruit beak 1.5 mm or less long; winter buds lacking.
14 Stems whitish; leaves 10–25 cm long; peduncles over 10 cm long; fruits 4–5 mm long .... 14. *P. praelongus*
14 Stems brownish to yellowish-green; leaves less than 10 cm long; peduncles 2–10 cm long; fruits 2.5–3.5 mm long .... 16. *P. richardsonii*
12 Floating leaves commonly present by flowering time, occasionally lacking; submersed leaves sessile or petiolate, not clasping the stem, flat or falcate, not undulate-crisped.
15 Upper submersed leaves falcately folded, 25- to 50-nerved; mature fruits 4–5 mm long .... 2. *P. amplifolius*
15 Upper submersed leaves more or less symmetrical and not folded, 3- to 17(19)-nerved; mature fruits 1.7–4 mm long.
16 Fruits tawny-olive; floating leaves often lacking, thin and delicate, the blade tapering indistinctly into a short petiole; submersed foliage reddish-tinged .... 1. *P. alpinus*
16 Fruits brown, reddish-brown or greenish; floating leaves leathery, the blades distinct from the petioles; submersed foliage dark green to brownish-green.
17 Submersed leaves commonly disintegrating by fruiting time; tapering to petioles (2)4 cm long or often much longer, acute to blunt-tipped; mature fruits brownish to reddish-brown, 3–4 mm long .... 12. *P. nodosus*
17 Submersed leaves usually persistent, sessile or tapering to petioles up to 4 cm long, acute to abruptly acuminate or apiculate; mature fruits green or olive, 1.7–3.5 mm long.
18 Stems usually freely branched, 0.5–1 mm thick; submersed leaves 3–10(15) mm wide, 3- to 7(9)-nerved; floating leaf blades 2–9 cm long, 1–3.5 cm wide; fruiting spikes 1.5–3.5 cm long; fruits 1.7–2.8 mm long, the lateral keels obscure .... 9. *P. gramineus*
18 Stems simple or once-branched, 1–5 mm thick; submersed leaves (1)1.5–4 cm wide, 9- to 17(19)-nerved; floating leaf blades 4–14(19) cm long, 2–7 cm wide; fruiting spikes 2–6 cm long; fruits 2.7–3.5 mm long, the lateral keels strong .... 10. *P. illinoensis*

1. **Potamogeton alpinus** Balbis. Stems terete, 1–2 mm thick, simple or rarely branched above, to 1 m long; foliage reddish-tinged, especially the upper leaves and peduncles. Submersed leaves linear-lanceolate to linear-oblong, 4–18 cm long, 5–15(20) mm wide, usually 7-(to 11)-nerved, blunt and obtuse to rarely acutish at the tip, narrowed to a sessile base. Floating leaves often lacking, thin and delicate, obovate or oblanceolate to elliptic-oblanceolate, mostly 4–6 cm long, 1–2 cm wide, 7- to 15-nerved, obtuse, tapering indistinctly into a short petiole. Stipules free, membranaceous, 1–2.5(4) cm long; spikes cylindric, with 5–9 crowded whorls of flowers, 1.5–3.5 cm long; peduncles about as thick as the stem, 3–15 cm long. Fruits tawny-olive, obliquely obovoid, 2.5–3.5 mm long; dorsal keel usually narrow and prominent; lateral keels absent or low and rounded; beak short, curved backward. (2n = 26, 52) Jul–Sep. Cold streams & lakes in the BH; SD: Custer, Lawrence, Pennington; (circumboreal, in Amer. from Newf. to AK, s to MA, NY, WI, MN, SD, CO, UT, & CA).

Our plants belong to var. *tenuifolius* (Raf.) Ogden, which differs from Eurasian *P. alpinus* in having narrower leaves and a more curved fruit beak.

2. *Potamogeton amplifolius* Tuckerm., largeleaf pondweed. Stems terete, 2–4 mm thick, simple or occasionally branched above, to 1 m long. Upper submersed leaves broadly lanceolate to ovate, falcately folded and often arcuate, 8–20 cm long, 2–7 cm wide, 25- to 50-nerved; lower submersed leaves often decayed by fruiting time, lanceolate, often not folded, 19- to 25-nerved; both submersed leaf types obtuse to broadly acute, tapering to petioles 1–6 cm long. Floating leaves seldom lacking at flowering time, ovate to elliptic, 5–10 cm long, 3–7 cm wide, 25- to 45-nerved, obtuse or abruptly acute, cuneate or rounded at the base; petioles 5–15 cm long. Stipules open and free of the petioles, 5–12 cm long, fibrous and persistent. Spikes cylindric, dense, 4–8 cm long in fruit; peduncles broadening upward, 5–20(30) cm long. Fruits greenish-brown to brown, obliquely obovoid, 4–5 mm long; dorsal keel prominent, the lateral keels less distinct; beak to 1 mm long. (n = 26) Jun–Aug. Uncommon & local in quiet water of streams & lakes; MN, ND, SD, IA, NE, KS, & OK; (Que. to B.C., s to AL, OK, & CA).

3. *Potamogeton crispus* L., curly muckweed. Stems slightly compressed, mostly 1–2 mm thick, usually branching, mostly 4–8 dm long. Leaves all submersed, sessile, slightly clasping, linear-oblong to linear-oblanceolate or oblong to oblanceolate, 3–8 cm long, 3–15 mm wide, 3- to 5-nerved, rounded at the tip, narrowed at the base, the margin usually undulate-crisped, finely serrate; stipules slightly adnate at the base, 4–10 mm long, early shredding. Winter buds commonly produced in some leaf axils, indurate, ca 1–2 cm long. Spikes dense, short-cylindric, 1–2 cm long; peduncles terete, about as thick as the stem, 2–5(7) cm long. Fruits brown, ovoid, the body 2–3 mm long, the beak very prominent, 2–3 mm long; keels low, rounded. (2n = 36, 42, 50, 52, 72, 78) Apr–Jun. Shallow water of lakes, ponds, & slow-moving streams; spreading throughout the GP; (N. Amer. along both coasts & inland; Europe). *Naturalized.*

4. *Potamogeton diversifolius* Raf., waterthread pondweed. Stems slender, terete, 0.5–1 mm thick, to 8 dm long, usually with short lateral branches. Submersed leaves linear, flat, 1–8 cm long, 0.3–1.5 mm wide, 1- to (3)-nerved, obtuse to long-acuminate; stipules of submersed leaves 2–18 mm long, membranaceous, adnate to the leaf blade for mostly 1–4 mm, usually adnate for less than 1/2 the total length. Floating leaves sometimes lacking, the blades elliptic to elliptic-lanceolate or elliptic-oblanceolate, 5–40 mm long, 5–20 mm wide, 3- to 17-nerved, acute to rounded at the tip, cuneate to rounded at the base; petioles mostly 5–40 mm long; stipules of floating leaves free of the leaf bases, 2–25 mm long, membranaceous to weakly fibrous. Spikes dimorphic, the lower submersed spikes capitate to ellipsoid, 1.5–6 mm long, few to several-fruited, the upper spikes ellipsoid to cylindric, 5–30 mm long, usually many-fruited; peduncles slightly clavate, 3–32 mm long. Fruits olive to stramineous, round and flattened, with a prominent winged dorsal keel and slightly ridged to winged lateral keels, the keels entire to toothed, the embryo coil plainly visible through the papery thin walls of the fruit, the beak minute. Jun–Sep. Shallow water of ponds & marshes; scattered in the GP, mostly in the s, n to s MN, s ND & ne WY; (CT to MT & OR, s to FL, TX, AZ, CA, & into Mex.).

Reports of the similar *P. spirillus* Tuckerm. in our range were based upon misdetermined specimens of *P. diversifolius.*

5. *Potamogeton epihydrus* Raf., ribbonleaf pondweed. Stems somewhat flattened, 1–2 mm thick, simple or sparingly branched, to 2 m long. Submersed leaves linear, ribbonlike, 5–20 cm long, 3–6(10) mm wide, 5- to 7(13)-nerved, with a cellular-reticulate strip on each side of the midvein forming a conspicuous median band 1–2 mm wide, acute to rather blunt, slightly tapered to the sessile base. Floating leaves usually present, elliptic or oblong-elliptic, (2)3–8 cm long, (5)10–20 mm wide, mostly 11- to 25-nerved, obtuse to bluntly mucronate at the tip, tapering to flattened petioles which are usually shorter than the blades;

stipules free, rather membranaceous and delicate, 1-3 cm long. Spikes dense, cylindric, usually 2-4 cm long; peduncles about as thick as the stem, 3-8 cm long. Fruits olivaceous to brown, broadly and obliquely obovate, concave on the sides, 2-3 mm long; dorsal keel prominent, thickly winged, lateral keels low, mostly rounded; beak minute. (n = 13, 2n = 26) Jul-Sep. Stream pools & lakes in the BH; SD: Custer, Pennington; (Newf. & Que. to AK, s to GA, MO, SD, CO, ID, & CA). *P. epihydrus* var. *nuttallii* (Cham. & Schlect.) Fern.—Fernald.

6. *Potamogeton filiformis* Pers., slender pondweed. Stems subterete, to 1 mm wide, branching dichotomously from the base, mostly unbranched above, 1-5 dm long. Leaves all submersed, filiform to narrowly linear, 5-12 cm long, 0.2-2 mm wide, 1(3)-nerved, acute to obtuse; stipules adnate to the base of the leaf blade, 1-4 cm long, forming a tight sheath around the stem, the free portion projecting as a ligule 2-10 mm long. Spikes elongate, 1-5 cm long, with 2-5 remote to adjacent whorls of flowers; peduncles slender, 2-15 cm long. Fruits olivaceous, obovoid, 2-3 mm long, rounded on the back, the beak flat, inconspicuous. (2n = 78) Jul-Aug. Shallow, standing or flowing water; MN: Douglas, Marshall; ND: Divide; SD: Fall River, Pennington; NE: Lincoln; (circumboreal, in Amer. s to ME, PA, MI, MN, NE, CO, AZ & CA; also Africa & Australia). *P. interior* Rydb.—Rydberg; *P. filiformis* var. *borealis* (Raf.) St. John, var. *macounii* Morong—Fernald.

7. *Potamogeton foliosus* Raf., leafy pondweed. Stems compressed, mostly 0.5-1 mm wide, freely branched, to 8 dm long. Leaves all submersed, linear, 1.3-8.2 cm long, 0.3-2.3 mm wide, 1- to 3(5)-nerved, acute to apiculate at the tip, tapered to the sessile base; stipules free, greenish to brown, mostly 0.5-2 cm long, membranaceous to slightly fibrous, eventually deteriorating; glands rarely present at the base of the stipules. Winter buds uncommon, lateral, 1-2 cm long. Spikes capitate to short-cylindric, 1.5-7 mm long; floral whorls 1 or 2, the whorls 0.6-1.2 mm apart when 2; peduncles usually clavate, recurved, 0.3-1 cm long. Fruits olive to greenish-brown, obliquely obovoid, 1.5-2.7 mm long; dorsal keel ridged or winged with an undulate to dentate margin 0.1-0.4 mm high; sides rounded to centrally depressed. (2n = 28) Jun-Aug. Shallow water of rivers, streams, lakes, & ponds; GP, especially common s & e; (Newf. to B.C., s throughout most of the U.S. & into Mex., C. Amer.; also HI & W. I.). *P. foliosus* var. *macellus* Fern.—Fernald.

8. *Potamogeton friesii* Rupr. Stems compressed, mostly 0.5-1 mm wide, simple or branched, to 1-1.5 m long. Leaves all submersed, linear, 2.3-6.5 cm long, 1.2-3.2 mm wide, 5- to 7(9)-nerved, rounded to apiculate at the tip, the margin flat to eventually revolute, tapered to the sessile base; stipules free, white and fibrous, often shredding above, 5-20 mm long; glands present at the base of the stipules. Winter buds common, terminal or lateral, 1.5-5 cm long; inner leaves reduced, arranged into a fan-shaped structure; outer leaves 2 or 3 per side, apiculate to acute, indurate and corrugated at the base. Spikes cylindric, 7-16 mm long; floral whorls 2-5, 1.5-5 mm apart; peduncles slightly clavate, mostly 1.5-4(7) cm long. Fruits olive-green to brown, ovoid to obovoid, 1.8-2.5 mm long, rounded on the back. (2n = 26) Jun-Aug. Shallow, fresh to brackish water of lakes & ponds; MN, IA, ND, SD, NE, more frequent in the n part; (Newf. to AK, s to PA, n IN, n IA, NE, & UT).

9. *Potamogeton gramineus* L., variable pondweed. Stems subterete, 0.5-1 mm thick, usually freely branched, to 8 dm long. Submersed leaves linear to linear-lanceolate or broadly lanceolate, sometimes oblanceolate, 3-9 cm long, 3-15 mm wide, 3- to 7(9)-nerved, acute to acuminate, tapered to the sessile base. Floating leaves rarely absent, elliptic or oblong-elliptic, 2-9 cm long, 1-3.5 cm wide, 11- to 19-nerved, obtuse or abruptly acute at the tip, rounded to cuneate at the base; petioles 2-10(15) cm long, shorter than to exceeding

the length of the blades; stipules free, persistent, mostly 0.5–4 cm long. Spikes dense, cylindric, 1.5–3.5 cm long; peduncles stout, usually broadening upward, 2–10(20) cm long. Fruits dull green, obliquely obovoid, 1.7–2.8 mm long, dorsal keel sharp, lateral keels obscure. (n = 26, 2n = 52) Jun–Aug. Shallow, usually standing water of ponds, lakes, marshes, & ditches; MN, n IA, ND, SD, NE; KS: Harvey, most frequent in the n part; (circumboreal, in Amer. s to NY, IA, KS, CO, AZ, & CA). *P. heterophyllus* Schreb.—Rydberg, misapplied; *P. gramineus* var. *maximus* Morong, var. *myriophyllus* Robbins—Fernald.

Apparent hybrids between this species and *P. illinoensis* are frequently observed among collections from the Nebraska Sandhills, where their ranges overlap and where they often grow together. The hybrids are typically intermediate between the parents in morphology, but may be mistaken for extremes of either parent. Flowers should be checked for abortive pollen grains to confirm a suspected hybrid.

10. *Potamogeton illinoensis* Morong, Illinois pondweed. Stems subterete, (1)1.5–5 mm thick, simple or branched, to 2 m long. Submersed leaves elliptic or oblong-elliptic to lanceolate or linear-lanceolate, sometimes somewhat arcuate, 5–20 cm long, (1)1.5–4 cm wide, mostly 9- to 17(19)-nerved, acute to mucronate, tapered to the subsessile or petioled base, the petioles up to 2–4 cm long. Floating leaves often lacking, oblong-elliptic or ovate-elliptic to broadly elliptic, 4–14(19) cm long, 2–7 cm wide, 13- to 29-nerved, obtuse to bluntly mucronate at the tip, rounded to cuneate at the base; petioles 2–9 cm long, shorter than the blades; stipules free, persistent, mostly 2–8 cm long. Spikes dense, cylindric, 2–6 cm long; peduncles usually thicker than the stem, 4–20(30) cm long. Fruits olive-green or gray-green, obliquely obovoid, 2.7–3.5 mm long, dorsal and lateral keels prominent. (n = 52) Jun–Sep. Shallow to fairly deep water of sandy lakes & ponds; MN, s SD, IA, NE, MO, & KS; (N.S. to Ont. & B.C. s to FL, KS, CA, & into Mex.). *P. angustifolius* Berch. & Presl, *P. lucens* L.—Rydberg.

See discussion under no. 9.

11. *Potamogeton natans* L., floatingleaf pondweed. Stems subterete, 0.8–2 mm thick, simple or rarely branched, to 2 m long. Submersed leaves reduced to linear, bladeless phyllodes, these often disintegrating with age, 10–20 cm long, 1–2 mm wide, tapering to an obtuse tip. Floating leaves ovate-lanceolate to ovate-elliptic, 3–10 cm long, 1–5 cm wide, mostly 19- to 35-nerved, acute to obtuse at the tip, rounded to cordate at the base; petioles usually much exceeding the blade in length, usually forming an angle with the blade at their juncture; stipules free, fibrous, persistent or shredding with age, 4–10 cm long. Spikes dense, cylindric, 2–5 cm long; peduncles thicker than the stem, 3–10 cm long. Fruits greenish-brown to brown, obliquely elliptic-obovoid, 3–5 mm long, often pitted on the sides, the dorsal keel sharp or rounded with age, the lateral keels obscure. (2n = 52) Jul–Aug. Shallow to rather deep water of lakes & ponds; MN; ND: Bottineau, Rolette; n IA, ne SD, NE; KS: Rice; (circumboreal, in Amer. s to PA, IN, IA, KS, NM, AZ & CA).

12. *Potamogeton nodosus* Poir., longleaf pondweed. Stems subterete, 1–2 mm thick, simple or seldom branched, to 1.5 m long. Submersed leaves commonly deteriorating by fruiting time, linear-lanceolate to elliptic-lanceolate, 10–20(30) cm long, 1–2(3) cm wide, 7- to 15-nerved, acute to blunt-tipped, gradually tapering to petioles mostly (2)4–10 cm long. Floating leaves elliptic to oblong-elliptic, 5–13 cm long, (1.5)2–4.5 cm wide, 15- to 25-nerved, acute to nearly rounded at the tip, sometimes obtusely mucronate, acute to somewhat rounded at the base; petioles winged, mostly 2–3 mm wide, 5–20 cm long, usually longer than the blades; stipules free, those of the submersed leaves often decaying early, those of the floating leaves persistent, 3–10 cm long. Spikes dense, cylindric, 2–6 cm long; peduncles

thicker than the stem, 3-15 cm long. Fruits reddish-brown to brown, obovoid, 2.7-4.3 mm long, the dorsal keel sharp, the lateral keels low. (n = 26) Jun–Oct. Usually in shallow water of rivers, streams, lakes, ponds, & marshes; GP; (cosmopolitan). *P. americanus* Cham. & Schlect. — Rydberg.

13. ***Potamogeton pectinatus*** L., sago pondweed. Stems terete, ca 1 mm thick, sparingly branched at the base, becoming freely dichotomously branched above, 3-10 dm long. Leaves all submersed, filiform to narrowly linear, 3-12 cm long, usually 0.2-1 mm wide, 1- to 3-nerved, acute, sometimes wider with obtuse tips early in the growing season or on plants from running water; stipules adnate to the base of the leaf blade for 1-3 cm, forming a sheath about as wide as the stem, occasionally wider on the main stem with age. Spikes elongate, 1-5 cm long, with 2-5(7) unevenly spaced floral whorls; peduncles lax, filiform, to 15 cm long. Fruits yellowish to tawny, drying brown, obliquely obovoid, 2.7-4 mm long, rounded on the back, apiculate due to the style beak which is usually 0.3-0.5 mm long. (2n = 42, 78) Jun–Sep. Common in shallow to moderately deep, fresh to subsaline water; GP; (nearly cosmopolitan). Including *P. interruptus* Kit. — Rydberg.

14. ***Potamogeton praelongus*** Wulf., whitestem pondweed. Stems whitish, slightly compressed, 1.5-4 mm thick, sparingly branched, to 2-3 m long, the shorter internodes often zigzag. Leaves all submersed, oblong-lanceolate, 10-25(35) cm long, 1-3 cm wide, with 3-5 primary nerves, rounded and cucullate at the tip, the margin entire and somewhat undulate, sessile and weakly to strongly cordate-clasping at the base; stipules free, whitish, 1-3 cm long, fibrous, early shredding. Spikes dense, cylindric, 2.5-5 cm long; peduncles elongate and thickening upward, 10-40 cm long. Fruits greenish-brown, obovoid, 4-5 mm long, the dorsal keel sharp, the lateral ones obscure. (2n = 52) Jun–Jul. Deep water of cold, clear lakes; MN; ND: Bottineau, Rolette; ne & BH in SD, n IA, n NE; (circumboreal, in Amer. s to CT, NY, IN, NE, CO, UT, & CA).

15. ***Potamogeton pusillus*** L., baby pondweed. Stems terete to subterete, 0.1-0.7 mm thick, simple to freely branching, 2-15 dm long. Leaves all submersed, linear, 0.9-6.5 cm long, 0.2-2.5 mm wide, 1- to 3(5)-nerved, acute to obtuse or apiculate at the tip, tapered to the sessile base; stipules free, brownish-green, 3-9 mm long, delicate and nonfibrous, soon decomposing; glands usually present at the base of the stipules. Winter buds sometimes produced, lateral or terminal, 0.9-3.2 cm long; inner leaves rolled into an indurate fusiform structure; outer leaves 1-3 per side, acute to obtuse, without corrugations at the base. Spikes short-cylindric to cylindric, 1.5-10 mm long; floral whorls 1-3(4), 1.2-4.7 mm apart; peduncles filiform to cylindric, 0.5-6 cm long. Fruits green to brown, obliquely obovoid, 1.5-2.2 mm long, rounded on the back, often concave on the sides. (2n = 26) Jun–Oct. Common in shallow, fresh to brackish water of lakes, ponds, marshes, ditches, & sluggish streams; GP; (Newf. to N. T., s throughout most of the U.S. & into Mex.; also Eurasia).

Two varieties occur in our region:

15a. var. *pusillus*. Leaves with up to 2 rows of lacunae along the midrib, apex acute, rarely apiculate; stipules mostly connate. Spikes usually of 2-4 verticels; peduncles filiform to cylindric, usually 1-3 per plant. Mature fruit widest above the midline, the sides concave and the beak forward. GP, the prevalent form. *P. pusillus* var. *minor* (Biv.) Fern. & Schub. — Fernald; *P. panormitanus* Biv., including var. *major* G. Fischer and var. *minor* Biv. — Fernald, op. cit.

15b. var. *tenuissimus* Mert. & Koch. Leaves with 1-5 rows of lacunae along the midrib, apex acute to obtuse; stipules mostly convolute. Spikes mostly of 1 or 2 adjacent verticels; peduncles cylindric, usually more than 3 per plant. Mature fruit mostly widest at or below the middle, the sides rounded and the beak central. Of limited occurrence; MN; ND: Cass, Steele, Stutsman; n IA; NE: Cherry; WY: Albany, Crook. *P. berchtoldii* Fieb., incl. var. *polyphyllus* (Morong) Fern., var. *colpophilus* Fern., var. *acuminatus* Fieb., var. *tenuissimus* (Mert. & Koch) Fern., var. *lacunatus* (Hagstrom) Fern. — Fernald;

P. pusillus var. *lacunatus* (Hagstrom) Fern., var. *cuspidatus* G. Fischer, var. *polyphyllus* Morong, var. *colpophilus* Fern.—Fernald, op. cit.

16. Potamogeton richardsonii (Benn.) Rydb., claspingleaf pondweed. Stems brownish to yellowish-green, terete, 1-2.5 mm thick, sparingly to freely branched, mostly 3-10 dm long, the shorter internodes rarely zigzag. Leaves all submersed, ovate-lanceolate to lanceolate, 2-10 cm long, 1-2.5 cm wide, with 13-25 prominent nerves, rounded to acute and not cucullate at the tip, the margin entire and undulate-crisped, sessile and strongly cordate-clasping at the base; stipules free, 1-2 cm long, early shredding into whitish fibers. Spikes dense, cylindric, 1.5-4 cm long, peduncles strongly recurved in fruit, often thickening upward, 2-10 cm long. Fruits green to brown, obliquely obovoid, 2.5-3.5 mm long, rounded to faintly keeled dorsally. (n = 26) Jun-Aug. Shallow to moderately deep, fresh to brackish water of lakes, ponds, marshes, ditches, & sluggish streams; MN, n IA, ND, SD, NE, MT, WY, CO, most common in the n part; (Lab. to AK, s to PA, IN, NE, CO, UT, & CA). *P. perfoliatus* L.—Stevens, misapplied.

17. Potamogeton strictifolius Benn. Stems mostly terete, 0.4-0.8 mm thick, simple or branched, to 1 m long. Leaves all submersed, linear, 1.2-6.3 cm long, 0.6-2 mm wide, 3- to 5(7)-nerved, acute to attenuate at the tip, the margin often revolute, tapered to the sessile base; stipules free, white, fibrous, shredding at the tip, 6-16 mm long; glands present at the base of the stipules. Winter buds common, terminal or lateral, 2.5-4.8 cm long; inner leaves undifferentiated from the outer ones; outer leaves 3-4 per side, acute, mostly without or rarely with corrugations at the base. Spikes cylindric, 6-13 mm long; floral whorls 3-4, 1.5-4.2 mm apart; peduncles cylindric, rarely slightly clavate, 1-4.5 cm long. Fruits greenish-brown, obovoid, 1.9-2.1 mm long, rounded on the back. (2n = 52) Jul-Aug. Shallow water of lakes, ponds, & slow streams; MN; ND: McHenry; NE: Cherry; WY: Albany; (Newf. & Que. to Alta. & N. T., s to NY, OH, n IN, MN, NE, WY & n UT). *P. rutilus* Wolfgang—Rydberg, misapplied; *P. strictifolius* var. *rutiloides* Fern.—Fernald.

18. Potamogeton vaginatus Turcz., sheathed pondweed. Stems terete, mostly 1-2 mm thick, freely branched above, to 1.5 m long. Leaves all submersed, filiform to narrowly linear, 2-8(30) cm long, 0.5-2 mm wide, 1(3)-nerved, the tip acute to obtuse or sometimes retuse; stipules adnate to the base of the leaf blade for 1-5 cm and sheathing the stem, the sheaths along the main stem inflated 2-5 × the thickness of the stem. Spikes elongate, 3-6 cm long, with 5-12 evenly spaced floral whorls; peduncles slender and lax, up to 10 cm long, often much surpassed by the upper leaves. Fruits dark green, obliquely obovoid, ca 3 mm long, rounded on the back; stigma sessile, forming a low beak. (2n = 78, ca 88) Jul-Aug. Deep water of cold, clear lakes; ND: Bottineau, Rolette; SD: Day; WY: Crook; (Newf. to AK, s to NY, WI, MN, SD, WY, & OR; also Eurasia & Africa). *P. moniliformis* St. John—St. John, Rhodora 18: 130. 1916.

19. Potamogeton zosteriformis Fern., flatstem pondweed. Stems strongly compressed and winged, mostly 2-3 mm wide, freely branched, to 1 m long. Leaves all submersed, linear, 5-15(20) cm long, 2-5 mm wide, many (15-35)-nerved, acute to cuspidate at the tip, sessile and slightly narrowed at the base; stipules free, usually whitish, fibrous and shredding with age, 1-4 cm long. Spikes densely flowered, usually uncrowded in fruit, cylindric, 1-2.5 cm long; peduncles compressed, 1.5-10 cm long. Fruits dark green to brown, obliquely elliptic-ovoid, 4-4.5 mm long, the dorsal keel sharp and somewhat undulate or dentate, the lateral ones obscure. (2n = 28) Jul-Aug. Shallow to deep water of lakes, ponds, & marshes; MN, n & w IA, ND, SD, n NE; KS: Harvey; MT & WY; (Que. & N.B. to B.C., s to VA, IN, KS, MT, ID, & n CA). *P. compressus* L.—Rydberg.

143. RUPPIACEAE Hutchins., the Ditchgrass Family

by Robert B. Kaul

1. RUPPIA L., Ditchgrass, Widgeon Grass

1. *Ruppia maritima* L. Submersed aquatic herb of brackish and alkaline waters, profusely branched and often forming extensive beds. Stems very slender to filiform, the internodes long or short. Leaves alternate, but often appearing tufted on the branches, the blades filiform to capillary, 2–10(15) cm long, less than 1 mm wide, the sheaths prominent and tapering, sometimes eligulate, the tips rounded or obtuse. Inflorescences axillary, with 2 sessile flowers subopposite to distichously disposed on the slender rachis, emerging from the sheath of the spathe-leaf and elevated to the surface by the peduncle, the peduncle usually coiling after anthesis and retracting the flowers to ripen underwater. Flowers perfect, regular, hypogynous, without perianth; stamens 2, sessile, the anther sacs large and separated by a broad connective, the flower thus appearing 4-staminate; pollen floating on the water surface; pistils (3)4(7), free, sessile in anthesis, uniovulate, styleless, the stigma blunt and peltate, the pistil becoming long-stipitate as the nutlets mature. Fruits pyriform to ovoid, sometimes asymmetric, mostly 2–3 mm long. (2n = 16, 20, 40) Jun–Sep. Locally abundant in quiet water; GP; (cosmopolitan). *R. occidentalis* S. Wats.—Rydberg.

At least 2 varieties may be recognized in our area:

1a. var. *occidentalis* (S. Wats.) Graebn. Sheaths 1.5–3 cm long and red-lineate; fruits symmetrically ovoid. Less common, perhaps more often found in hyperalkaline waters.

1b. var. *rostrata* J. G. Agardh. Sheaths to 1 cm long; fruits obliquely ovoid. Common

The fruits and foliage of *Ruppia* are among the most important of our waterfowl foods.

144. NAJADACEAE Juss., the Naiad Family

by Robert B. Kaul

1. NAJAS L., Naiad, Water-nymph

Submersed monoecious and dioecious branching annual herbs of fresh and brackish waters. Stems slender, often rooting at the nodes, appearing dichotomously branched. Leaves subopposite or appearing whorled but inserted at barely different levels, linear, dentate, denticulate, serrulate, or entire, their sheathing bases conspicuous and eligulate. Flowers axillary; staminate flower consisting only of a single stamen enclosed by two clear membranaceous perianthlike envelopes, the anther protruding from the envelopes in anthesis, 1- or 4-celled; pistillate flower sessile, naked or with a membranaceous sheath, of a single unilocular pistil, the 2–4 stigmas linear, the single ovule basal, anatropous. Fruits fusiform, 1-seeded, indehiscent, the wall decomposing and the seed remaining in the axil; seeds hard, areolate or nearly smooth, without endosperm.

References: Posluszny, U. & R. Sattler, 1976. Floral development of *Najas flexilis*. Canad. J. Bot. 54: 1140-1151; Haynes, R. R. 1977. The Najadaceae in the southeastern United States. J. Arnold Arbor. 58: 161–170; Haynes, R. R. 1979. Revision of North American and Central American *Najas*. Sida 8: 34–56.

Leaves, stems, and seeds are among the most important waterfowl foods, and the plants provide good habitats for fish. Naiads are considered by some to be weeds and are controlled with herbicides.

1. Leaves and sometimes the stems coarsely toothed; seeds distinctly ovoid; plants dioecious .... 3. *N. marina*
1. Leaves entire or only minutely toothed, the stems smooth; seeds not distinctly ovoid, plants monoecious.
   2. Seeds ellipsoid or narrowly ovoid, distinctly shiny; leaves tapering to a point, becoming crowded near the branch tips .... 1. *N. flexilis*
   2. Seeds fusiform, not shiny; leaves linear, not strongly tapered, their tips rounded or acute; leaves about evenly distributed along the stems, not especially crowded toward the branch tips .... 2. *N. guadalupensis*

1. ***Najas flexilis*** (Willd.) Rostk. & Schmidt, naiad. Monoecious, slender-stemmed, bushy or elongate, and often forming extensive beds. Leaves numerous, becoming more crowded toward the branch tips; narrow, 1–3 cm long, mostly less than 1 mm wide, gradually tapering to a sharp point, minutely toothed, each tooth of a single cell; sheath shoulders oblique; toothed. Anthers 1-celled; pistils mostly with 3 stigmas. Fruit and seed narrowly ovoid or elliptic, 2–3 mm long; seed entirely filling the fruit, shiny, reddish-brown, very finely areolate but appearing smooth under low magnification. (2n = 12, 24) Jun–Sep. Still or gently flowing water, locally common; MN, ND, SD, IA, NE; (Newf. to Man., s to MD & NE, possibly also Alta., B.C. to OR).

2. ***Najas guadalupensis*** (Spreng.) Magnus, naiad. Similar to no. 1. Monoecious and highly variable, often elongate and mat-forming, sometimes lime-encrusted. Stems slender, branching, rooting at the lower nodes. Leaves about evenly distributed along the stems, not crowded toward the branch tips, to 2 cm long but mostly 1 cm long and 1 mm wide, linear, acute or sometimes rounded at the tips, the margins finely and remotely toothed or sometimes entire; shoulders of the sheaths sloping, spinulose. Anthers 4-celled. Seeds 2–3 mm long, fusiform, dark brown, dull or not conspicuously shiny, areolate, the areolae in longitudinal rows and easily seen under low magnification. (2n = 12–60), Jun–Sep. Locally abundant in lakes, streams, & ponds; GP except in the far n; (MA to OR s to S. Amer.).

3. ***Najas marina*** L. Dioecious, brittle, stout plants of alkaline water, coarser than nos. 1 and 2. Stems rooting and branching at the base, sometimes prickly. Leaves linear, to 4 cm long, to 3 mm wide, irregularly coarse-toothed, spine-tipped, the midrib sometimes with spines on the lower side; sheaths broad, with rounded shoulders, entire or with a few small teeth. Anthers 4-celled. Seed ovoid, 4–5 mm long, dull, smooth but becoming minutely areolate. (n = 6, 2n = 12, 12 + 8) Jun–Sep. Calcareous lakes, ponds, & quiet streams, rare in our range; MN: Big Stone, Pope, Polk; ND: Richland; SD: Roberts; (sporadic NY, MI to ND; Eurasia, Africa). *Introduced.*

145. ZANNICHELLIACEAE Dum., the Horned Pondweed Family

by Robert B. Kaul

1. ZANNICHELLIA L., Horned Pondweed

1. ***Zannichellia palustris*** L., horned pondweed. Slender, usually much branched, submerged, monoecious aquatic herb, the stems creeping on the bottom or erect, 1–5(8) dm

long. Leaves opposite but sometimes appearing whorled where branches depart, light green, filiform, to 10 cm long and ca 0.5 mm wide, entire, acute; stipules sheathing, membranaceous, to 4 mm long. Inflorescences usually 2 flowered, 1 staminate and 1 pistillate, the whole appearing as a single flower enveloped in bud by a hyaline sheath. Flowers without a perianth; staminate flower consisting of only a single stamen, the filament elongate, slender, exceeding the pistillate flower; pistillate flower of (2)4(6) carpels, slender-ovoid, uniovulate, the style cylindric, the prominent stigmas unevenly funnelform and peltate. Nutlets pedicellate, brown, oblong, curved, flattened, beaked, ridged, and dentate on the back. Seed pendulous, without endosperm. (n = 12; 2n = 12-36) Jun-Nov. Locally abundant on sandy soils in quiet or running waters; GP; (cosmopolitan).

References: Reese, G. 1967. Cytologische and taxonomische Untersuchungen an *Zannichellia palustris* L. Biol. Zentralbl. 86(Suppl.): 277-306.; Tomlinson, P. B., & U. Posluszny. 1976. Generic limits in the Zannichelliaceae (sensu Dumortier). Taxon 25: 273-279.

The foliage and fruits of this species are important waterfowl foods.

146. ARACEAE Juss., the Arum Family

by Steven P. Churchill

Our plants monoecious (rarely dioecious) perennial herbs, from a creeping rhizome or a corm; stems mostly erect, from a basal sheath. Leaves simple or compound, venation parallel or net. Inflorescence consisting of a spadix subtended by a leaflike spathe or spathe appearing as a continuation of the stem. Flowers small, aggregated over the lower portion or all of the spadix; perianth present and inconspicuous, or absent; ovary superior, occasionally embedded in the spadix. Fruit a berry, sometimes leathery, seeds 1-several.

This is a large family of some 110 genera and about 2,000 species distributed mainly in the tropics, with only a minor element extending into the temperate region. Some members of this family, including our native *Arisaema* and the common house plant *Philodendron*, cause poisoning, primarily due to the presence of calcium oxalate crystals.

1 Leaves compound, with 3-5 or more leaflets; spadix concealed by spathe .... 2. *Arisaema*
1 Leaves simple; spadix exposed.
    2 Spathe ovate to cordate, white .... 3. *Calla*
    2 Spathe linear, grasslike, green .... 1. *Acorus*

1. ACORUS L., Sweet Flag, Calamus

1. *Acorus calamus* L. Grasslike aromatic rhizomatous herb, 9-15 dm or more tall. Leaves ensiform, erect and linear, 9-12 dm long, (4)7(10) mm wide, venation parallel. Spathe appearing leaflike and a continuation of the stem, 10-70 cm long; spadix exposed and appearing lateral, 4-9 cm long, from a continuing spathe. Flowers light brown, perfect; perianth of 6 short segments, ca 2 mm long; stamens 6, 2-2.5 mm long, anthers reniform, filaments slender; ovary 2- or 3-locular. Seeds 1-few. (2n = 36, 48, 72) May-Aug. Swamps & marshes; scattered in e GP; (s Can., ME to OR s to FL, TN, GP; Europe). *Naturalized.*

This is a small genus of 2 species distributed in the Northern Hemisphere. The plant was held in high esteem by the eastern plains Indian tribes both for its medicinal and religious uses.

2. ARISAEMA Mart. Jack-in-the-Pulpit, Indian Turnip

Plants monoecious or, by abortion, dioecious; stem erect, from a corm. Leaves 1 or 2(3), pinnately divided with sheathing petiolules, venation net. Spathe various, but concealing the spadix; spadix with pistillate flowers basal, separated or continuous with staminate flowers just above. Staminate flowers with 2-5 nearly sessile stamens; pistillate flowers with 1-locular ovary, stigma short and broad. Fruit a red fleshy berry, in clusters; exposed after spathe and often the leaf have died back.

A large genus of about 150 species distributed throughout much of the Equatorial Old World region and North America and confined to moist subtropical and deciduous forests.

1 Leaves palmately divided into 3 leaflets .... 2. *A. triphyllum*
1 Leaves pedately divided into 5 or more leaflets .... 1. *A. dracontium*

1. *Arisaema dracontium* (L.) Schott, dragonroot. Herb up to 5.5 dm tall, from a corm 1-2.2 cm wide. Leaf single, pedately divided into 5-14 leaflets, each lanceolate to elliptic, 7-28 cm long, 2-5(6.5) cm wide. Peduncle from basal sheath, shorter than petiole; spathe 4-8 cm long, 1.5-2 cm wide; spadix 8-15(20) cm long; sterile portion 6-12(18) cm long, fertile portion 1.2-2.5 mm long. Fruit reddish orange, 5-10 mm long. (2n = 56) May. Moist deciduous woods; MO & se NE s to KS & e OK; (se Can., NE, s to FL & TX). *Muricauda dracontium* (L.) Small — Rydberg.

2. *Arisaema triphyllum* (L.) Schott, jack-in-the-pulpit. Herb up to 6 dm tall, from a corm 1.2-3.5 cm wide. Leaf 1(2 or 3), palmately divided into 3 leaflets, each ovate to elliptic, 6-22 cm long, (3)5-14 cm wide, acuminate, base oblique, underside lighter than upper surface, slightly glaucous. Spathe 9-15.5 cm long, 3-7 cm wide at the hood; inner surface of hood often purple to brownish-red with yellow linear veins running parallel and diverging to apex; spadix (5)6(8) cm long; sterile portion 4-6 cm long, fertile portion 1-2 cm long. Fruit red, 5-7 mm long. (2n = 28, 56) May-Jun. Moist to intermittently dry deciduous woods; e GP; (se Can., ME to MN, e GP, s to FL & LA).

This is a variable species that has been divided into at least 2 forms in our region (Huttleston, Bull. Torrey Bot. Club 76: 407-413. 1949). Populations in our region often exhibit both forms with various intermediates that hardly warrant recognition as separate taxa.

3. CALLA L., Wild Calla

1. *Calla palustris* L. Plants up to 5 dm tall, from a long rhizomatous root. Leaf ovate to cordate, (6)10(12.5) cm long, (4.5)8(10) cm wide, venation net, petiole thick. Spathe white, ovate, 3.5-5.5 cm long, 3-4 cm wide; spadix exposed, 2-3 cm long. Flowers perfect to unisexual; stamens 1-1.5 mm long, anthers short and broad, filament narrow; ovary 1-locular. Fruit a red berry, 6-10 mm long; seeds enclosed by a gelatinous substance. May-Jun. Marshes & swamps, & along margin of ponds & streams; MN: Becker; ND: Pembina, Rolette; (s Can. & n U.S.; Eurasia).

147. LEMNACEAE S.F. Gray, the Duckweed Family

By Robert B. Kaul

Small to minute perennials floating at or below the surface, rootless, or bearing long simple roots, but not often rooting in soil. Plant body thalloid, without clear distinction between stem and leaf. Thallus spherical, ovoid or flattened, rounded to oblong, green, mostly less than 8 mm long, reproducing by budding from lateral or basal pouches and the thalli then at least temporarily colonial. Flowering seldom observed and probably rare in some species. Flowers produced in a pouch, usually 1 pistillate flower, consisting of a single uniovulate pistil, with 1 or 2 staminate flowers, each consisting of 1 stamen, the whole appearing as a tiny perfect flower without a perianth. Fruit a 1- to several-seeded utricle. Growing in quiet, shallow waters, most abundant in clear, warm, nutrient-rich waters and often forming extensive mats, in winter forming turions that sink to bottom.

References: Clark, H. L., & J. W. Thieret. 1968. The duckweeds of Minnesota. Michigan Bot. 7: 67–76; Hartog, C. den, & F. van der Plas. 1970. A synopsis of the Lemnaceae. Blumea 18: 355–368; Landolt, E. 1980. Biosystematic investigation of the family of duckweeds (Lemnaceae), Vol. 1. Veröff. Geobot. Inst. ETH, Stiftung Rübel, Zürich 70: 1–247; McClure, J. W., & R. E. Alston. 1966. A chemotaxonomic study of the Lemnaceae. Amer. J. Bot. 53: 849–860.

The duckweeds and the many invertebrates living in them are important waterfowl foods.

1 Thalli with roots.
  2 Each thallus with a single root .... 1. *Lemna*
  2 Each thallus with more than 1 root .... 2. *Spirodela*
1 Thalli rootless.
  3 Thalli minute, spherical to ovoid, sometimes flattened on top, usually floating at the surface .... 3. *Wolffia*
  3 Thalli to 1 cm long, flattened, oblong, stipitate, often strongly colonial, floating under the surface or near the bottom .... 1. *Lemna trisulca*

1. LEMNA L., Duckweed

Thalli floating on the surface (*L. trisulca* submersed except at flowering time), solitary or colonial, flattened to slightly convex on both sides, or inflated below, each thallus bearing a single basally sheathed root (very young or old thalli sometimes rootless), the root cap often prominent. Flower pouches 2, near the thallus base.

1 Thalli submersed except at flowering time, long-stipitate and strongly colonial, elliptic, the apex erose to dentate .... 4. *L. trisulca*
1 Thalli floating, not stipitate or merely short-stipitate, solitary or in small colonies, orbicular, oblong, or elliptic, entire, sometimes papillate above.
  2 Upper surface smooth or with a few obscure median papillae; thallus obscurely 1-nerved or nerveless .... 5. *L. valdiviana*
  2 Upper surface variously papillate; thallus 3- to 5(7)-nerved, sometimes obscurely so.
    3 Root-sheath winged; thallus never red beneath .... 3. *L. perpusilla*
    3 Root-sheath not winged; thallus sometimes red beneath.
      4 Thallus flat or convex but not inflated below, the apex usually symmetrical; upper surface dark green, not mottled, lower surface sometimes reddish ..... 2. *L. minor*
      4 Thallus convex to obviously inflated below, the apex symmetrical or not; upper surface often yellowish and mottled; both surfaces sometimes reddish .. 1. *L. gibba*

1. *Lemna gibba* L. Thalli solitary or colonial, orbicular to obovate, 3- to 5-nerved, to 6 mm long and 5 mm wide, the apex symmetrical or not, green to yellowish and sometimes mottled red above, the lower surface sometimes inflated (gibbous) and reddish with age. Fruits winged, with 1–3 seeds. ($2n = 64$) Jul–Sep. Locally abundant, GP; (cosmopolitan, except S. Amer. & Australia).

Landolt (op. cit.) distinguishes those specimens with the papillae of nearly equal size as *L. turionifera* Landolt (GP) and those with a prominent apical papilla and gibbous below as *L. gibba* (w NE only).

This species and the next often are difficult to separate on morphological grounds but they are chemotaxonomically distinct (McClure & Alston, op. cit.).

2. **Lemna minor** L. Thalli solitary or colonial, suborbicular to ovate or obovate, 2–4(6) mm long, symmetric to barely asymmetric, obscurely 3-nerved, flat to barely convex on both surfaces, apically papillate, green above and often reddish below. Fruits wingless, 1-seeded. (2n = 40) GP from s SD to TX; (cosmopolitan, except S. Amer.).

Landolt (op. cit.) distinguishes specimens red beneath and with a prominent apical papilla as *L. obscura* (Austin) Daubs.

3. **Lemna perpusilla** Torr. Thalli solitary or colonial, orbicular, obovate, or elliptic, asymmetric, to ca 3 mm long and 2.5 mm wide, rather thick, flat to barely convex, pale green, the upper surface with 1–3 prominent papillae 1- to 3-nerved, the nerves rather obscure; root-sheath winged. Fruit 1-seeded. Aug–Oct. s 1/2 GP & probably elsewhere; (e N. Amer. to GP).

L. perpusilla is found in flower and fruit more often than our other species. Landolt (op. cit.) distinguishes specimens with 2 or 3 papillae above the node and seeds with 35–60 faint ribs as *L. perpusilla*, while specimens with 1 prominent papilla above the node and seeds with 8–22 prominent ribs are *L. aequinoctialis* Welw.

A similar entity with a thinner and more distinctly 3-nerved thallus is sometimes regarded as *L. perpusilla* var. *trinervis* Aust. or *L. trinervis* (Aust.) Sm. It is chemotaxonomically distinct from *L. perpusilla*, according to McClure & Alston (op. cit.).

4. **Lemna trisulca** L., star duckweed. Plants not strongly resembling other duckweeds. Thalli colonial, submersed, the colonies often large and caught in submersed vegetation or floating at the surface at flowering time. Vegetative thalli (not including stipe) elliptic, symmetrical, prominently stipitate, green, becoming pale with age, usually to 10 mm long and 4 mm wide, sometimes larger, 3-nerved, the apex erose to dentate, root falling early and seldom observed; flowering thalli floating, smaller, thicker, reddish. (2n = 44) Jun–Oct. Locally abundant, especially in waters with sandy substrate; ND & SD e of Missouri R., MN, IA, MO, NE mostly n of Platte R., ne WY, e CO; (N. Amer.; Old World).

5. **Lemna valdiviana** Phil. Thalli solitary or colonial, to 5 mm long and 2 mm wide, elliptic to obovate, symmetrical or asymmetrical at the base, sometimes falcate, nerveless or barely 1-nerved, the upper surface smooth or obscurely papillate on the median, pale green. Fruit wingless, 1-seeded. Jun–Sep. Seldom collected & perhaps rare in GP; w sandhills of NE, occasional in KS & OK; (widespread in the New World). *L. cyclostasa* (Ell.) Chev.— Rydberg.

Landolt (op. cit.) refers to those specimens with the thallus 1-1/3–3 × longer than wide and the base asymmetrical as *L. valdiviana* and to those with the thallus only 1–1-2/3 × longer than wide and the base symmetrical as *L. minuscula* Herter. The latter is more common with us, but the two can be found together.

2. SPIRODELA Schleid.

1. **Spirodela polyrrhiza** (L.) Schleid., duckmeat, greater duckweed. Thalli floating on the surface, colonial in groups of 2–5, these marginally or eccentrically attached by short stipes visible from below, orbicular to obovate or oblong, often somewhat asymmetrical, flat to slightly convex below, 3–10 or more nerved, the nerves radiating from an obvious eccentric point, to 10 mm wide, smooth above and often dark reddish below, each bearing 2 or more roots. Flowers seldom observed and perhaps rarely produced. Fruit small-winged; seeds 2. (2n = 40) Infrequent to common, with other duckweeds or by itself, not flourishing in moving waters; GP except the w; (cosmopolitan).

3. WOLFFIA Horkel ex Schleid., Watermeal

Thalli minute, less than 1.5 mm long, floating, rootless, spherical to ovoid, the upper surface sometimes flattened and papillate, not strongly colonial but often occurring in vast numbers as a green, algalike film on the water. Flower pouch 1, basal. Fruit spherical, 1-seeded.

These are the world's smallest flowering plants and are often found with other Lemnaceae in quiet waters.

1 Upper surface of thallus rounded, smooth or minutely papillate; thallus not brown-punctate .... 3. *W. columbiana*
1 Upper surface of thallus flattened, with or without a single large raised central papilla; thallus often brown-punctate.
   2 Upper surface with a prominent central papilla .... 2. *W. brasiliensis*
   2 Upper surface flat or slightly raised at one end .... 1. *W. borealis*

1. **Wolffia borealis** (Engelm.) Landolt. Thallus elliptic as seen from above, often acute to obtuse at one end, the upper surface flattened and emersed, the pointed apical end raised, brown-punctate on both sides (when dry) and dark green above. (2n = ca 40). MN, IA, NE, KS, e CO; (CT to MN & CO s to TX & FL; W.I.). *W. punctata* of reports, non Griseb.—Atlas GP.

2. **Wolffia brasiliensis** Wedd. Thalli ovate to elliptic as seen from above, apically rounded, the upper surface marginally flattened and emersed, the center elevated to a large round papilla (papilla lacking on flowering thalli), brown-punctate on both sides (when dry). se KS & probably elsewhere in se GP; (VA to KS s to TX & FL). *W. papulifera* Thomps.—Atlas GP; *W. punctata* Griseb., nec auct. amer.

3. **Wolffia columbiana** Karst. Thalli symmetrical, spherical to ellipsoid, floating just beneath the surface or partially emersed, the upper surface convex, sometimes minutely papillate on the median, the apex rounded, uniformly green throughout, not punctate. (2n = ca 42) Often very abundant in late summer; mostly e GP but w to CO & probably often overlooked elsewhere; (New England to CO, s to CA, GP; n S. Amer.).

148. COMMELINACEAE R. Br., the Spiderwort Family

by Margaret Bolick

Annual or perennial herbs, often somewhat succulent. Leaves alternate, veins parallel, sheaths tubular. Inflorescence terminal and/or axillary, basically cymose from the axils of foliaceous bracts or spathes. Flowers 3-merous, perfect; usually actinomorphic, sometimes zygomorphic; sepals herbaceous, separate; petals colored or white, ephemeral; stamens 6, rarely 5, all fertile or fertile and sterile mixed; anther locules separate, filaments sometimes bearded with moniliform hairs; stigmas undivided, style solitary, ovary superior, sessile or stipitate, 2- or 3-locular, placentation axile. Capsules loculicidal or indehiscent, 3–several seeded.

References: Correll, D. S., & M. C. Johnston. 1970. Commelinaceae, pp. 356–366, *in* Manual of the vascular plants of Texas, Texas Res. Foundation, Renner, Texas; Radford, A. E., H. E. Ahles, & C. R. Bell. 1968. Commelinaceae, pp. 268–271, *in* Manual of the vascular flora of the Carolinas, Univ. of North Carolina Press, Chapel Hill.

1 Inflorescence subtended by a conspicuous spathe; fertile stamens 3 .... 1. *Commelina*
1 Inflorescence subtended by a normal or reduced foliage leaf, fertile stamens 6 .... 2. *Tradescantia*

1. COMMELINA (Plum.) L., Dayflower

Herbaceous annuals or perennials. Leaves ovate to linear with conspicuous basal sheaths. Flower buds enclosed by a spathe; peduncles 0.2–5 cm long. Flowers zygomorphic; sepals 3, one subequal to the other 2; petals 3, unequal, the upper blue, the lowermost smaller, sometimes paler; fertile stamens 3, sterile stamens 3 or 2, smaller than the fertile, filaments glabrous. Fruits 1 or 2 per spathe. Capsules 2- or 3-celled, lower two cells fertile with 1 or 2 ovules, upper median cell smaller with 1 ovule or abortive.

Reference: Brashier, C. K. 1966. A revision of *Commelina* (Plum.) L. in the U.S.A. Bull. Torrey Bot. Club 93: 1–19.

1 Spathes open across the top and down the adaxial side to spathe stalk.
  2 Upper petals blue, lowermost white; sterile anthers 3; fertile anther locules 1.5–2.0 mm long .... 1. *C. communis*
  2 Upper and lowermost petals blue; sterile anthers 2; fertile anther locules 0.7–1.0 mm long .... 2. *C. diffusa*
1 Spathes open across the top but closed down the adaxial side.
  3 Upper and lowermost petals blue; sheaths with reddish marginal hairs ..... 4. *C. virginica*
  3 Upper petals blue, lowermost white; sheaths with whitish marginal hairs ...... 3. *C. erecta*

1. ***Commelina communis*** L., dayflower. Erect or repent annual; stems 1.5–10 dm long, glabrous to somewhat pubescent at nodes, often rooting at nodes; internodes to 13 cm long; roots fibrous. Leaves lance-ovate, 1.6–9 cm long, 1.1–3.2 cm wide, acute to acuminate, glaucous beneath, scabrous white hairy above; sheaths 4–20 mm long, throat of sheath more or less hairy. Spathes solitary to congested, 12–29 mm long, 6–14 mm wide; margins open across top and down side to spathe stalk. Corolla blue with 1 smaller lanceolate white petal; sterile anthers 3, fertile anthers 3; fertile anther locule (1.2)1.5–2.0 mm long. Capsule 6–7 mm long, about 1/2 as wide, usually 2-celled with 1 or 2 seeds per cell, third cell abortive; seeds foveolate to roughened, 2.5–3.0 mm long. Jul–Sep. Riverbanks, moist places; e GP, s to TX; (e US to GP, TX; e Asia). *Naturalized.*

2. ***Commelina diffusa*** Burm. f., creeping dayflower. Diffusely branched decumbent annual; stems to 6 m long, glabrous, glabrate or with a few hairs at nodes; internodes to 6 cm long; roots fibrous. Leaves lanceolate, 2–6 cm long, 0.8–2.1 cm wide, acute to acuminate, glabrous beneath, glabrous to slightly scabrous above, especially along margin; sheaths to 10 mm, edges hairy. Spathes usually solitary, 12–24 mm long, 5–11 mm wide, glabrous, open across top and down adaxial side to point of attachment to stalk. All petals blue; sterile anthers 2, fertile anthers 3, fertile anther locules 0.7–1.0(1.2) mm. Capsules usually 2-celled, elongate, 6–7 mm long and 1/2 as wide; seeds 2 in each of 2 cells, third cell usually aborted, 2.2–2.5 mm long, reticulate. Aug–Oct. Moist, waste places; e KS, OK; (se U.S. to GP). Perhaps naturalized from Old World. *C. longicaulis* Jacq.—Rydberg.

3. ***Commelina erecta*** L., erect dayflower. Erect or decumbent perennial; stems to 1 m, usually pubescent at least at nodes; internodes to 11 cm; root system thickened, fibrous. Leaves linear-lanceolate to ovate-elliptic, 3–11 cm long, 0.9–3.0 cm wide, acuminate, glaucescent, glabrous to lightly pubescent below, lightly pubescent, rarely scabrous above; sheaths to 22 mm long, margins pubescent with whitish hairs. Spathes more or less hirsute, solitary to few, rarely congested, 1.2–2.5 cm long, 0.8–1.2 cm wide, acute to acuminate, adaxial

side of spathe closed, top open. Corolla blue with one smaller white petal. Capsules 3-celled, 2 cells dehiscent, 1 indehiscent, 3–5 mm long and as wide, 1 seed in each of the dehiscent cells, with or without a single seed in the indehiscent cell; seeds 2.5–3.8 mm long, smooth and covered with whitish dots. May–Oct. Sandy or rocky soils; sw SD, NE, KS, OK, TX; (s NY to e WY, sw SD s to FL & AZ). *C. crispa* Woot.—Rydberg.

This variable species has been divided into two varieties and several forms.

3a. var. *angustifolia* (Michx.) Fern. Widest leaves less than 1.6 cm wide and narrowest leaves less than 0.8 cm wide. Seeds 2.5–3.0 mm long. Most common in sandy soils; w 1/2 GP.

3b. var. *erecta*. Widest leaves more than 1.7 cm wide and narrowest leaves more than 0.8 cm wide. Seeds 3.5–3.8 mm long. Moist rocky soils; from cen GP e.

White-flowered specimens of this var. from Kansas have been named f. *alba* Magrath.

To the e of the GP var. *erecta* f. *intercursa* Fern. has narrower leaves than f. *erecta* and overlaps in leaf and seed size with var. *angustifolia*.

4. **Commelina virginica** L. Erect or decumbent, glabrate to pubescent, rhizomatous perennial; stems to 9 dm long, internodes to 21 cm long. Leaves broadly lanceolate, slightly scabrous above, (7)11–16(18) cm long, 1.5–4(4.5) cm wide; sheaths to 28 mm long, fringed with reddish hairs 2–5 mm long at throat and down open edge. Spathes terminal, congested, rarely solitary and/or axillary, usually glabrous, 20–28 mm long, 14–20 mm wide, acute, closed down the adaxial side, open across top. All petals blue. Capsules 7–8 mm long, 1/2 as wide, 2- or 3-celled, 1 or 2 seeds in each of 2 cells, third cell usually aborted; seeds 3–5 mm long, smooth. Aug–Oct. Lowlands, moist woods; e KS, OK; (NJ to ne KS, s to FL & TX).

2. TRADESCANTIA L., Spiderwort

Subsucculent, tufted perennial herbs with alternate linear or lanceolate leaves. Cymes umbellate, several to many flowered, subtended by 2 leaflike bracts; pedicels unequal. Flowers actinomorphic; sepals 3, nearly equal in size, green or purplish; corolla blue, rose or white, showy, ephemeral; fertile stamens 6, filaments bearded; stigma capitate, style filiform. Capsules subglobose to oblong-ellipsoid; seeds 1 or 2 per locule, gray, somewhat flat and ellipsoid, foveolate.

Reference: Anderson, E., & R. E. Woodson, Jr. 1935. The species of *Tradescantia* indigenous to the United States. Contr. Arnold Arbor. 9: 1–132.

1 Sepals glabrous or with only a tuft or hairs at the tip .... 3. *T. ohiensis*
1 Sepals pubescent.
    2 Sepals with only uniseriate nonglandular hairs .... 4. *T. tharpii*
    2 Sepals with uniseriate nonglandular hairs and uniseriate glandular hairs or with only glandular hairs.
        3 Sepals with mixed glandular and eglandular hairs; sheath width 0.8–2 × blade width; sparingly branching .... 1. *T. bracteata*
        3 Sepals with only glandular hairs; sheath width 2–4 × blade width; freely branching .... 2. *T. occidentalis*

1. ***Tradescantia bracteata*** Small. Stems erect, rarely branching, glabrous to sparsely pilose, 2–4 dm in length; 2–4 nodes, internodes to 20 cm long. Leaves glabrous to sparsely pilose, linear-lanceolate, 8–30 cm long, 7–16 mm wide, rarely folded; sheath usually not inflated, to 35 mm long, 4–12 mm wide, 0.8–2 × the width of the blade. Cymes umbellate, few- to many-flowered (ca 20 flowers), terminal, solitary on main stem; bracts leafy, glabrous to pilose, 6–30 cm long; pedicels to 25 mm long, pubescent with uniseriate glandular and eglandular hairs, 0.5–1.5 mm long. Sepals elliptic, acute to acuminate, 8–14 mm long,

margins sometimes lightly suffused with purple, pubescent with uniseriate glandular and eglandular hairs; petals broadly ovate; filaments pilose; ovary ovoid, pubescent. Seeds 2–6 per fruit, compressed oblongoid, 2–4 mm long. (n = 6) May–Aug. Moist areas, prairies, disturbed sites; e 1/2 GP; ND s to TX, e to Mississippi R.; (MI to MT, s to IA, IL, TX).

T. bracteata and *T. occidentalis* intergrade completely in the central GP; many specimens cannot be easily assigned to one or the other. They are maintained as separate species here only because there is some degree of geographic separation with those plants traditionally called *T. occidentalis* occurring in the western part of the plains or where sandy soil is found in the eastern plains while those plants called *T. bracteata* are most commonly found farther east.

2. *Tradescantia occidentalis* (Britt.) Smyth. Stems erect, often branching, glabrous, glaucous, to 5 dm tall; subsucculent; 2–6 nodes, internodes to 25 cm long. Leaves glaucous, glabrous, linear-lanceolate, often folded, 9–33 cm long, 4–12(15) mm wide, sheath inflated, to 30 mm long, 4–12 mm wide, 2–4 × the width of the blade. Cymes umbellate, few to many-flowered (ca 25 flowers), terminal, solitary at the ends of main stem or branches; bracts leafy, glabrous, 6–20 cm long, pedicels to 20 mm long, glandular with uniseriate glandular hairs 0.5–1.5 mm long. Sepals elliptic, acute to acuminate, 8–13 mm long, margin sometimes suffused with purple, glandular; petals broadly ovate; filaments pilose; ovary ovoid, glandular to glabrous. Seeds 2–6 per fruit, compressed oblongoid, 2–4 mm long. (n = 6, 12) May–Aug. Sandy, often dry soils, prairies, disturbed sites; w 1/2 GP; ND to TX, e to Missouri R., w to Rocky Mts.; (GP to AZ). *T. universitatis* Cockll.—Rydberg. See note under *T. bracteata*.

3. *Tradescantia ohiensis* Raf. Stems erect, sometimes branching, glabrous, glaucous, to 1 m tall, subsucculent; 3–8 nodes; internodes to 23 cm long. Leaves glaucous, glabrous, linear-lanceolate, 11 to 35 cm long, 6–19 mm wide; sheath more or less pilose and inflated, 12–35 mm long, 6–18 mm wide. Cymes umbellate, few to many-flowered (ca 20 flowers), solitary, terminal on main stem or branches; bracts leafy, glaucous, glabrous to pilose at base, 3–20 cm long. Pedicels to 28 mm long, glabrous; sepals elliptic, acute to acuminate, glabrous or with an apical tuft of eglandular hairs, margins sometimes suffused with purple, 8–12 mm long; petals broadly ovate; filaments pilose; ovary ovoid, glabrous. Seeds 2–6 per fruit, oblongoid, 2–4 mm long. (n = 12) Apr–Jul. Disturbed sites, sandy rocky soils, moist or dry woods, prairies; e NE, KS, OK, TX; (throughout e U.S., w to GP). *T. reflexa* Raf.—Rydberg. *T. canaliculata* Raf.—Anderson & Woodson, op. cit.

4. *Tradescantia tharpii* Anders. & Woods. Stems erect, rarely branching, pilose; 1 or 2 nodes, internodes to 9 cm long. Leaves not glaucous, pilose, margins often hyaline or suffused with rose, linear-lanceolate, 14–30 cm long, to 25 mm wide; sheath to 15 mm long, 3 cm wide. Cymes umbellate, many-flowered, terminal, solitary; bracts leafy, pilose, to 26 cm long and 25 mm wide; pedicels 4–6 cm long, pilose, more or less rose colored. Sepals broadly elliptic, acute to acuminate, 12–16 mm long, somewhat petalaceous, usually rose colored, eglandular pilose; petals broadly ovate; filaments pilose; ovary ovoid, eglandular pubescent. Seeds compressed oblongoid, 2–3 mm long. (n = 12) Mar–Apr. Clay or rocky soils, prairies & open woods; cen KS to sw MO, s to n TX; (essentially restricted to GP).

149. JUNCACEAE Juss., the Rush Family

by Steven P. Churchill

Mostly perennial grasslike herbs, often tufted or strongly rhizomatous; stems cylindrical to slightly compressed, usually hollow, from a sheathing base. Leaves basal with a few alternating cauline leaves, cylindrical, ensiform, channeled or flat; mostly sheathing at base, often with prominent auricle at blade and sheath juncture, lightly pubescent or glabrous. Inflorescence an open panicle to compact head, occasionally solitary. Perianth of 2 whorls, each of 3 similar segments, chaffy or somewhat leathery texture; stamens 3 or 6, anthers basifixed; ovary superior, 3-locular, fused 1 or 3 locules, ovules usually many, placentation axile or parietal, or with 3 basal ovules. Fruit a dry capsule, dehiscent; seeds small, usually numerous, starch endosperm with a straight embryo embedded.

Reference: Buchenau, F. 1906. Juncaceae, *in* A. Engler (ed.), Das Pflanzenreich 4: 1–284.

This is a family of some 9 genera with nearly 400 species distributed world-wide, mostly confined to moist or aquatic habitats, in montane regions in the tropics.

1 Leaf blades glabrous; capsule with numerous seeds; sheath generally open .... 1. *Juncus*
1 Leaf blades with scattered hairs; capsule with 3 seeds; sheath closed .... 2. *Luzula*

1. JUNCUS L., Rush

Perennial or rarely annual herbs, cespitose or rhizomatous, with stems erect. Leaves cauline or chiefly basal, glabrous; blade flat, ensiform, channeled or cylindrical, often septate with lateral ridges. Inflorescence terminal or appearing lateral when the lowest bract (involucral bract) is leaflike and projects upward as a continuation of the stem; flowers few to numerous in cymose clusters (heads), each cluster subtended by 1–several ± leafy or scarious bracts. Flowers small, subtended by 1 or 2 bracteoles; perianth parts greenish but often turning straw colored or dark brown after anthesis; stamens 3 or 6; ovary 3-locular or falsely 1-locular. Fruit a capsule, often 3-angled; seeds numerous.

References: Engelmann, G. 1868. A revision of the North American species of the genus *Juncus*, with a description of new or imperfectly known species. Trans. Acad. Sci. St. Louis 2: 424–498; Hermann, F. J. 1975. Manual of the rushes (*Juncus* spp.) of the Rocky Mountains and Colorado Basin. U.S.D.A. Forest Serv. Gen. Tech. Rep. RM-18; Wiegand, K. M. 1900. *Juncus tenuis* Willd. and some of its North American allies. Bull. Torrey Bot. Club 27: 511–527.

There are some 300 species of *Juncus* world-wide, and they have local value as forage, although not of high quality. Mature plants are required for accurate identification.

1 Each flower subtended by a pair of opposite bracteoles; flowers single or few-clustered, not forming a dense head; leaves not septate.
  2 Plants annual; inflorescence large, more than 1/2 the height of the plant; auricles absent .... 8. *J. bufonius*
  2 Plants perennial, cespitose or rhizomatous; inflorescence usually only 1/3 of the height of the plant or less; auricles present or absent.
    3 Inflorescence appearing lateral from a seemingly continuous stem; leaves all basal, reduced to sheath only, auricles absent.
      4 Plants arising from extensive creeping rhizomes; anthers 1.2–2.2 cm long, 2× as long as the filaments; stems smooth and not longitudinally furrowed .... 4. *J. balticus*
      4 Plants tufted, not forming long rhizomes; anthers 1 mm long or less, filaments equaling or shorter than the anthers; stems longitudinally furrowed .... 12. *J. effusus*
    3 Inflorescence appearing terminal, subtended by 1–several leafy bracts; 2 or more cauline leaves present.
      5 Outer perianth segments obtuse-rounded, apex incurved; rhizome horizontal, becoming slender and elongate; leaf sheath extending to the middle of the culm .... 14. *J. gerardii*
      5 Outer perianth segments acute to acuminate, apex not incurved; rhizome short and erect, inconspicuous; leaf sheath confined to the base.
        6 Capsule distinctly 3-locular.

       7 Seeds 1–1.5 mm long, with a white, terminal appendage equaling the body of the seed in length; capsule exceeding the perianth .... 25. *J. vaseyi*
       7 Seeds ca 0.5 mm long, lacking an appendage; capsule shorter than the perianth.
          8 Involucral bracts generally shorter than or rarely exceeding the inflorescence; inforescence distinctly secund; leaf auricles up to 0.5 mm long .... 21. *J. secundus*
          8 Involucral bracts generally exceeding the inflorescence; inflorescence a compact cyme; auricles usually more than 0.5 mm long .... 7. *J. brachyphyllus*
     6 Capsule apparently 1-locular, or incompletely 3-lobular with partial partitions ± halfway to the central axis.
       9 Auricles prolonged into a membranaceous or scarious projection 3–5 mm long .... 22. *J. tenuis*
       9 Auricles shorter, prolonged up to 2 mm beyond the sheath, submembranaceous or cartilaginous.
          10 Auricles cartilaginous, dull to shiny yellow, very rigid; bracteoles blunt to broadly acute .... 11. *J. dudleyi*
          10 Auricles submembranaceous, not rigid; bracteoles acute, acuminate or aristate .... 15. *J. interior*
1 Each flower subtended by a single bracteole; flowers often forming a dense head; leaves septate or not.
   11 Leaves not septate; blades flat.
     12 Perianth segments 5–6 mm long; stamens 6, anthers yellow; capsule long-beaked .... 16. *J. longistylis*
     12 Perianth segments shorter, up to 3.5 mm long; stamens 3, anthers reddish-brown; capsule beakless .... 17. *J. marginatus*
   11 Leaves septate; blades cylindrical or ensiform, often hollow.
     13 Leaf blades ensiform, 3–6 mm wide; septa incomplete . 13. *J. ensifolius* var. *montanus*
     13 Leaf blades cylindrical, less than 3 mm wide; septa complete.
       14 Flowers many in dense spherical heads; perianth segments narrowly linear; seeds without an appendage.
         15 Stamens 3, opposite the outer perianth segments.
           16 Capsule ovoid, shorter than the outer perianth segments .... 5. *J. brachycarpus*
           16 Capsule slender, exceeding the outer perianth segments.
             17 Outer perianth series longer than the inner; leaf blades laterally compressed .... 24. *J. validus*
             17 Outer and inner perianth series equal or nearly so; leaf blades cylindrical .... 20. *J. scirpoides*
         15 Stamens 6, opposite the outer and the inner perianth segments.
           18 Inner perianth segments shorter than the outer segments; outer perianth segments 4–5 mm long .... 23. *J. torreyi*
           18 Inner perianth segments equaling or exceeding the outer segments; inner perianth segments 2.5–3.5(4) mm long .... 19. *J. nodosus*
       14 Flowers few, in hemispherical heads (*J. acuminatus* and *J. canadensis* sometimes have subspherical heads); perianth segments lanceolate, sometimes broadly so; seeds with or without an appendage.
         19 Seeds with a distinct, whitish appendage.
           20 Perianth segments acute, sharply pointed; most heads with more than 5 flowers .... 9. *J. canadensis*
           20 Perianth segments blunt; most heads with fewer than 5 flowers .... 6. *J. brachycephalus*
         19 Seeds lacking an appendage, the ends blunt or at most apiculate.
           21 Stamens 3, opposite the outer perianth segments.
             22 Capsule nearly 2× longer than the perianth; mature perianth segments 2.5–3 mm long .... 10. *J. diffusissimus*

    22 Capsule nearly equal to or barely exceeding the perianth; mature perianth segments 2–4 mm long.
        23 Capsule lance-ovoid to ellipsoid; flowering heads fewer than 50; perianth segments 3–4 mm long .... 1. *J. acuminatus*
        23 Capsule ovoid or broadly ellipsoid; flowering heads numerous, more than 50; perianth segments 2–2.5 mm long .... 18. *J. nodatus*
    21 Stamens 6, opposite the outer and the inner perianth segments.
        24 Outer perianth segments acute, longer than the inner ones; capsule rounded; flowering branches erect .... 2. *J. alpinus*
        24 Outer perianth segments acute to acuminate, equal to or shorter than the inner ones; capsule apex acute; flowering branches spreading .... 3. *J. articulatus*

1. **Juncus acuminatus** Michx. Cespitose plants 2.5–8 dm tall; culm erect. Leaves 2 or 3, septate, auricles obtuse-rounded, 0.5–1.5 mm long. Inflorescence an open to compact cyme, several to many hemispherical to occasionally spherical heads, each 2- to 30(or more)-flowered; involucral bract short, not exceeding inflorescence. Perianth 3–4 mm long, segments lanceolate, outer and inner perianth equal or outer exceeding inner; stamens 3, opposite outer perianth segments, shorter than perianth, anther shorter than filament. Capsule ellipsoid to narrowly ovoid, apex mucronate, equal or exceeding perianth by 0.5–1 mm, 1-locular; seeds ellipsoid, 0.3–0.5 mm long, ends apiculate. (2n = 40) May–Aug. Lake & stream marshes, swamps, wet prairies & ditches, soils often sandy; e GP; IA s to KS, MO, & OK; (ME to B.C. s to GA, MO, AZ, & CA).

2. **Juncus alpinus** Vill. Cespitose plants 1.5–4 dm tall, from a short creeping rhizome; culm erect to decumbent, slender. Leaves 2 or 3, 1/3–1/2 culm length, terete, septate, auricles rounded and short, 0.3–0.5(1) mm long. Inflorescence a cyme, 1- to 6(10)-flowered, branches ascending; involucral bract much shorter than inflorescence. Perianth 2–2.5 mm long, outer series just exceeding inner, outer lanceolate, acute and often apiculate, inner rounded to obtuse and acute apiculate, stamens 6, 1/2–2/3 perianth length, anther shorter than filament. Capsule oblong to obtuse, equal or just exceeding perianth, 1-locular; seeds fusiform, 0.3–0.5 mm long, ends apiculate. (2n = 40, 80) Jun–Aug. Stream & lake marshes, sand bars, wet prairies & ditches; cen & n GP; MN to MT s to NE & MO; (Can. & AK s to PA, MO, CO, & WA; Eurasia). *J. richardsonianus* J.A. Schultes — Rydberg.

   Hamet-Ahti (Ann. Bot. Fennici 17: 341–342. 1980) has pointed out that *J. alpinoarticulatus* Chaix in Vill. is the legitimate name for *J. alpinus*.

3. **Juncus articulatus** L. Plants 1.5–6 dm tall, from a creeping rhizome; culm erect, slender. Leaves 1–3, septate, auricles 1–1.5 mm long, acute. Inflorescence a short diverging cyme, many hemispherical heads, 2- to 10(or more)-flowered; involucral bract very short, not exceeding inflorescence. Perianth segments 2–3 mm long, nearly equal or inner slightly shorter, lanceolate, acute; stamens 6, shorter than perianth, anther equal to filament length. Capsule shiny dark brown, ovate, apiculate, exceeding perianth by 1 mm, 1-locular; seeds obovoid, 0.3–0.5 mm long, ends apiculate. Jul–Sep. Stream marshes; ne GP; SD: Roberts; (s Can. s to WV, IN, MN, SD, & AZ; Eurasia). *J. amblyocarpus* Rydb.—Rydberg.

4. **Juncus balticus** Willd., Baltic rush. Plants 2–9 dm tall, from an extensively creeping tough rhizome; culm stout and rigid, not striate. Leaves lacking a blade and auricles, only sheath present. Inflorescence a loose to clustered cyme; involucral bract appearing as a continuation of culm. Perianth bracteolate, 3.5–6 mm long, with equal to subequal segments or with outer series longer than inner; stamens 6, shorter than perianth, anther much longer than filament. Capsule oblong to broadly lanceolate, acute to mucronate, equal to or more often shorter than inner perianth, 1-locular; seeds ovoid-ellipsoid, 0.5–0.8 mm long, ends

apiculate. (2n = 40, 80) Jun–Aug. Lake & stream marshes, wet prairies & ditches, often found in alkaline sites; GP; MN to MT s to w OK & TX; (widespread throughout the Northern Hemisphere).

Our plants can mostly be assigned to var. *montanus* Engelm. (*J. ater* Rydb.—Rydberg), but var. *littoralis* Engelm. may be found in e GP. Var. *montanus* Engelm. has perianth segments nearly equal, the outer perianth series short acuminate, inner series acute to obtuse, and capsule equal to or shorter than inner perianth series. Var. *littoralis* Engelm. has outer perianth segments longer than inner, outer perianth acuminate, inner acute and capsule longer than inner series.

5. *Juncus brachycarpus* Engelm. Plants 3–7 dm tall, from a thickened white rhizome; culm erect. Leaves 2–4(5), erect to slightly spreading, terete, septate, not exceeding culm, auricles acute, 2–3 mm long. Inflorescence a cyme, several to many spherical heads, 7–10 mm in diam, few to more often many-flowered; involucral bract short, not exceeding inflorescence. Perianth 3–4 mm long, lance-subulate, outer perianth 1/3 longer than inner; stamens 3, opposite outer perianth segments, much shorter than inner perianth series, anther shorter than filament. Capsule ovoid to obovate, much shorter than perianth, acute, 1-locular; seeds oblong, 0.3–0.4 mm long, ends apiculate. (2 = 44) Jun–Sep. Stream marshes, wet prairies & meadows, open woods & ditches, soil often sandy; KS: Bourbon, Cherokee, Montgomery, Wilson; MO: Barton, Jasper, Vernon; (se Can. s to GA & TX).

6. *Juncus brachycephalus* (Engelm.) Buch. Cespitose plants 2.5–6 dm tall; culm erect, slender. Leaves 3–5, terete, septate, auricles acute, to 2 mm long, membranaceous. Inflorescence a diffuse cyme, several to many hemispherical heads, 2- to 5-flowered; involucral bract not exceeding inflorescence. Perianth 2–2.5(3) mm long, outer perianth series slightly shorter than inner series, outer segments acute, inner obtuse or blunt; stamens 6, slightly shorter than perianth, anther much shorter than filament. Capsule ellipsoid, exceeding perianth by 1 mm or more, beak short, 1-locular; seeds ellipsoid, 0.6–0.8 mm long, each end with a white-tailed appendage. (2n = 80) Jul–Aug. Lake & stream marshes; MN: Marshall; (ME & se Can. s to MD, OH, & MN).

7. *Juncus brachyphyllus* Wieg., small-headed rush. Cespitose plants 2.5–6.5 dm tall; culm stout and rigid. Leaves flat to channeled, up to 1/2 culm length, 1–2 mm wide, auricles 0.5–2 mm long, firm and membranaceous. Inflorescence many-flowered compact cyme; involucral bracts equal or mostly exceeding inflorescence. Flowers bracteolate; perianth segments 3.5–6 mm long, outer series longer than inner series, lanceolate-subulate, margins broadly scarious; stamens 6, 1/2 as long as the inner perianth segments, anther equal to filament length. Capsule oblong, equal to slightly less than inner perianth length, completely 3-locular, meeting walls straight, not concave; seeds oblong, 0.2–0.4 mm long, ends shortly apiculate. May–Aug. Moist to wet prairies, salt flats, & slightly drier upland open woods, soils often sandy; NE s to OK; (NE to ID & WA s to TX, NM, & CA). *J. kansanus* Herm.—Fernald.

In the original description of *J. kansanus*, Hermann, op. cit., made note of the similarity between *J. kansanus* and *J. brachyphyllus*, but regarded them as distinct species. Recently Harms (Sida 3: 525–528. 1970) discussed this subject and concluded that both are the same taxon. Harms also provided a distribution map of *J. brachyphyllus*.

8. *Juncus bufonius* L., toad rush. Tufted annual plants 0.5–3 dm tall. Leaves inconspicous, blades involute, 0.5–1.5 mm wide, sheath tapering into blade, margin hyaline, auricles absent. Inflorescence an open cyme, often more than 1/2 plant height; involucral bract exceeding or shorter than inflorescence. Flowers bracteolate, 1(2 or 3) per pedicel, outer perianth series lanceolate-acuminate, 2–5.5 mm long, inner series lanceolate to oblong, shorter than outer series; stamens 6 (rarely 3), much shorter than perianth, anther shorter

than filament. Capsule oblong to nearly globose, obtuse, shorter than perianth, 1-locular; seeds ovoid to ellipsoid, 0.2–0.3 mm long, ends apiculate or truncate. (2n = 60, ca 108, 120) May–Aug. Lake & stream marshes, sand bars, often found in brackish habitats; GP; (almost cosmopolitan, absent in the tropics & arctic).

9. *Juncus canadensis* Gay ex Laharpe. Cespitose plants 3–9 dm tall; culm stiffly erect. Leaves 3 or 4, terete, septate, generally not exceeding culm, auricles acute to rounded, 1 mm long. Inflorescence an open or congested cyme, heads hemispherical to spherical up to 40- (or more)-flowered; involucral bract shorter than or exceeding inflorescence. Perianth segments lanceolate, 2.5–3 mm long, outer perianth series shorter than inner; stamens 3, shorter than perianth, anther shorter than filament. Capsule ovoid to oblong tapered, just exceeding perianth, 1-locular; seeds fusiform, 1.2–1.8 mm long, each end with a white-tailed appendage, 1/2 or more longer than seed proper. (2n = 80) Jul–Sep. Marshy sites along sandy stream banks & ponds or lakeshores; NE: Blaine, Brown, Cherry; (se Can. to MN s to FL, IL, & NE).

10. *Juncus diffusissimus* Buckl., slimpod rush. Tufted plants 2.5–6.5 dm tall; culm erect, slender to thick. Leaves 2 or 3, compressed, septate, auricles obtuse, 1.5–2 mm long. Inflorescence a diffuse cyme, often 1/3–1/2 plant height; heads hemispherical, numerous, 2- to 5-flowered; involucral bract shorter than inflorescence. Perianth 2.5–3 mm long, lance-subulate, segments equal or outer slightly longer than inner; stamens 3, opposite outer perianth segments, much shorter than perianth, anther shorter than filament. Capsule linear-lanceolate, exceeding perianth by 1–3 mm, 1-locular; seeds ovate, 0.3 mm long, ends apiculate. May–Oct. Stream & pond marshes, wet prairies, meadows or ditches; KS & MO s to OK; (VA to IN & MO s to GA & TX).

11. *Juncus dudleyi* Wieg., Dudley rush. Cespitose plants 1.5–9.5 dm tall; culm stout. Leaves 1/2 or less than culm height, blade 0.5–1 mm wide, flat to channeled, rarely involute, auricles dull to glossy yellow or amber, thick and rounded, 0.2–0.5 mm long, cartilaginous. Inflorescence a compact cyme, few-flowered; involucral bract exceeding inflorescence. Flowers braceolate, bracteoles ovate, acute to obtuse; perianth segments 3.5–5.5 mm long, spreading, equal to subequal, outer series lance-subulate, inner lanceolate; stamens 6, 1/2 perianth length, anther slightly shorter than filament. Capsule ovate to oval, shorter than perianth length, 1-locular; seeds oblong, 0.3 mm long, ends apiculate. (2n = 80) May–Sep. Lake & stream marshes, meadows & wet prairies; GP; (s Can. s to VA, MO, NM, & AZ; Mex.).

Some recent floristic treatments have recognized *J. tenuis* Willd. as a highly polymorphic species that encompasses both *J. dudleyi* Wieg. and *J. interor* Wieg. among its forms. This has been justified by the observation that these "forms" intergrade in a small portion of their range (sw US), but no serious research has yet supported such conclusions. Our taxa are phenotypically distinct and do not intergrade, thus they are recognized as distinct species. Further work is needed for the entire *J. tenuis* species-group.

12. *Juncus effusus* L., bog rush. Plants 2–12 dm tall, forming tussocks from a stout, thick rhizome; culm thick, many-striate. Leaves lacking a leaf blade and auricles, only a mucronate-tipped sheath present. Inflorescence a many-flowered diffuse cyme; involucral bract appearing as a continuation of culm. Perianth bracteolate, segments 2–3 mm long, subequal or outer somewhat longer than inner segment series; stamens 3 (rarely 6), opposite outer perianth segments, anther as long as or shorter than filament. Capsule obovoid, equal or subequal to perianth, 1-locular; seeds ellipsoid, 0.2–0.3 mm long, ends short apiculate. (2n = 40, 42) May–Aug. Pond & stream marshes, wet prairies & ditches; NE s to e KS & OK; (se Can. s to FL, KS, & TX).

13. ***Juncus ensifolius*** Wikst. var. ***montanus*** (Engelm.) C.L. Hitchc. Plants 3–7 dm tall, from a creeping rhizome; culm compressed. Leaves equitant, septa incompletely formed, shorter than culm, 2–6 mm wide; auricles inconspicuous, broadly ovate, 1 mm long. Inflorescence paniculate, few-flowered heads, spherical to hemispherical; involucral bracts not exceeding inflorescence. Perianth segments lanceolate, acuminate, 3–4 mm long, subequal or outer longer than inner series; stamens 6 (rarely 3), shorter than perianth, anther equal to shorter than filament; style up to 0.5 mm long. Capsule oblong, equal to just exceeding perianth, 1-locular; seeds fusiform, 1 mm long, ends apiculate, one end or rarely both occasionally slightly prolonged. (2n = 40) Jul–Aug. Along streams at somewhat high elevations; MT; Custer; SD: Custer, Lawrence, Pennington; WY: Crook; (AK & sw Can. s to CO & CA; Mex.). *J. saximontanus* A. Nels.—Rydberg.

14. ***Juncus gerardii*** Lois., blackgrass. Plants 2–5 dm tall, from an elongated rhizome; culm slender, rigid. Leaves few, occasionally surpassing or shorter than culm, blade channeled, 1–2 mm wide; auricles short and rounded, 0.5 mm long, membranaceous. Inflorescence a loose to clustered cyme; involucral bract mostly equal to shorter than inflorescence. Perianth braceolate, segments 2–3.5 mm long, outer and inner series equal, obtuse, outer tips incurved; stamens 6, nearly equal to perianth, anther 3 × longer than filament. Capsule ellipsoid-ovoid, ca equal to slightly longer than perianth, 1-locular; seeds obovoid, 0.3–0.5 mm long, ends apiculate. (2n = 80, 84) May–Jun. Lake & stream marshes; MN: Clay, Kittson; ND: Cass, Richland; (s Can. s to FL, IN, ND, & CO; Eurasia & Africa).

15. ***Juncus interior*** Wieg., inland rush. Cespitose plants 2–8.5 dm tall. Leaves 1/3 to 1/2 culm length, 1–2 mm wide, channeled or involute, auricles 0.3–0.5 mm long, rounded, submembranaceous and hyaline. Inflorescence an open to compact cyme; involucral bracts shorter than or more often longer than inflorescence. Flowers bracteolate, each pair of bracteoles acuminate to abruptly aristate; perianth segments erect to spreading, broadly subulate, inner and outer series 3–6 mm long, equal or outer just exceeding inner; stamens 6, 1/2 perianth length, anther much shorter than filament. Capsule oblong to ovoid, obtuse, shorter than or equal to perianth, 1-locular; seeds oblong to ovoid, 0.3–0.5 mm long, ends apiculate. (2n = 80) May–Aug. Lake & stream marshes, wet prairies & ditches, at border of or in open woods; GP; (MI to WA s to MO & AZ; cen Can.).

16. ***Juncus longistylis*** Torr. Plants 2.5–7 dm tall, from a rhizome; culm slender and compressed. Leaves mostly basal, few reduced upward, dorsiventrally flattened, basal blade shorter than culm, 1.5–3 mm wide, auricles obtuse, 0.4–1 mm long. Inflorescence a cyme, 2–8 or more heads, each 2- to 12-flowered; involucral bract shorter or rarely exceeding inflorescence. Perianth 5–6 mm long, outer series equal to or just exceeding inner series, segments broadly lanceolate, apex broadly acute, midrib of outer segment dark green, turning brown, margins white-hyaline; stamens 6, shorter than perianth, anthers yellow mostly longer than filament. Capsule oblong, rounded to truncate, long-beaked, equal to perianth, 1-locular; seeds oblong, 0.2–0.4 mm long, ends apiculate. Jun–Aug. Stream & lake marshes, swamps, wet prairies & ditches; MN & ND s to NE; (s Can. s to NE, CO, & NM).

17. ***Juncus marginatus*** Rostk., grassleaf rush. Plants 2–7 dm tall, from a knotty rhizome; culm slender, ± compressed. Leaves short, up to 1 dm long, 1–5 mm wide, flat, soft, auricles scariously rounded, 0.5–1 mm long. Inflorescence a cyme, 5–40 heads, each 2-to 10-flowered; involucral bract shorter to occasionally longer than inflorescence. Perianth segments 2–3.5 mm long, inner series slightly longer than outer, outer segments acute, inner blunt to attenuate; stamens 3, shorter than outer segments, anthers rusty red, much shorter than filament. Capsule obovoid, apex rounded to truncate, nearly equal to perianth length, in-

completely 3-locular; seeds ovoid, 0.3–0.5 mm long, ends apiculate. (2n = 38, 40) Jun–Oct. Stream & lake marshes, wet prairies, ditches, & at margin or in open woods; s SD to OK; (se Can., ME to WA s to FL, TX, & CA). *J. setosus* (Cov.) Small — Rydberg.

Juncus biflorus Ell. may be found in the se GP; the distinction between it and *J. marginatus* is not great and further taxonomic investigation is needed in this group.

18. *Juncus nodatus* Cov., stout rush. Cespitose plants up to 1 m tall, from a thick knotty base; culm thick and stout. Leaves 3–5, strongly septate, auricles acute, 2–3 mm long, membranaceous. Inflorescence a diffuse cyme, branches ascending to divergent, many hemispherical heads, each 2- to 8-flowered; involucral bract short, not exceeding inflorescence. Perianth segments 2–2.5 mm long, lance-subulate, outer perianth series longer than or subequal to inner; stamens 3, opposite outer perianth segments, shorter than perianth, anther shorter than filament. Capsule ovoid, short beaked, 1-locular; seeds ellipsoid, 0.5 mm long, ends apiculate. Jun–Aug. Pond marshes, wet prairies & ditches, often in standing water; MO & KS s to OK; (IN to KS s to TX & AL).

19. *Juncus nodosus* L., knotted rush. Plants 1–4 dm tall, from a creeping tuberiferous rhizome; culm slender to moderately stout, erect. Leaves 2 or 3, septate, slender, 1–2 mm wide, often exceeding culm; auricles rounded, 0.5–1 mm long. Inflorescence of few to many spherical heads, each 5–12 mm in diam, up to 20- to 30-flowered; involucral bract generally exceeding inflorescence. Perianth 2.5–3.5(4) mm long, segments lanceolate, long acuminate, outer and inner series equal or inner just exceeding outer perianth; stamens 6, 1/2 perianth length, anther shorter than filament. Capsule lance-subulate, beak tapering, just exceeding perianth, 1-locular; seeds obovoid to oblong, 0.5 mm long, ends abruptly mucronate. (2n = 40) Jun–Sep. Lake & stream marshes, wet prairies, ditches, & open woods; MN to MT s to MO & NE; (Can. & AK s to VA, MO, NM, & CA).

20. *Juncus scirpoides* Lam. Plants 1–4 dm tall, from a thickish white rhizome; culm slender to moderately stout, erect. Leaves 1–3, septate, not exceeding culm, auricles short and acute, 1–2 mm long. Inflorescence a cyme, many spherical heads, each 6–12 mm in diam, 30(or more)-flowered; involucral bract short, not exceeding inflorescence. Perianth 3–3.5 mm long, outer and inner series equal to subequal, erect, subulate; stamens 3, opposite outer perianth segments, equaling perianth length, anther much smaller than filament. Capsule obovate triangular, tapered, 1-locular, just exceeding perianth; seeds ovoid, 0.3–0.5 mm long, ends apiculate. Jun–Oct. Lake & pond marshes, wet prairies, often in wet sandy soils; NE s to OK; (NY to MI & NE s to FL & TX).

21. *Juncus secundus* Beauv. Tufted plants 2.5–4 dm tall; culm wiry and slender. Leaves densely tufted, flat to channeled, 1/3–1/2 culm length, 1–2 mm wide, auricles rounded, 0.5 mm long, membranaceous. Inflorescence an open cyme, branches appearing incurved-ascending, flowers noticeably secund and along inner branches; involucral bract shorter to longer than inflorescence. Perianth bracteolate, bracteoles ovate-oblong, obtuse to cuspidate, perianth segments 3–4 mm long, equal, ascending to slightly spreading, lance-subulate, acute, margins scarious; stamens 6, 2/3 perianth length, anthers slightly longer than filament. Capsule oblong-oval, equal to slightly less than perianth length, distinctly 3-locular; seed oblong, curved, 0.3 mm long, ends apiculate. Jul–Sep. Stream marshes, upland or well-drained prairies, & open dry rocky ground; s MO; (se Can., ME to IN & MO s to GA & AR).

22. *Juncus tenuis* Willd., path rush. Cespitose plants 1–5 dm tall, dull to bright green; culm slightly spreading to erect. Leaves 1/2 to as long as culm, 1–1.5 mm wide, flat or channeled, rarely involute, soft, auricles lance-acuminate and prolonged, (2)3–5 mm long,

thin membranaceous. Inflorescence an open to compact cyme; involucral bracts mostly longer than inflorescence. Perianth bracteolate, bracteoles ovate to obtuse, acute, perianth segments spreading to erect, 3–4 mm long, nearly equal or inner slightly shorter; stamens 6, 1/2 perianth length, anther much shorter than filament. Capsule oblong-ovoid, obtuse, shorter than perianth, 1-locular; seeds oblong, 0.3–0.5 mm, ends bluntly apiculate. (2n = 30, 32, 40, 84) May–Aug. Lake & stream marshes, dry to moist prairies, pastures, fields, ditches, often of disturbed sites; MN & SD s to OK; (distributed over N. Amer.; also C. & S. Amer., Europe & Africa).

23. *Juncus torreyi* Cov., Torrey's rush. Plants (2)3–8 dm tall, from a creeping tuberiferous rhizome; culm stout, stiffly erect. Leaves 2–4, septate, thick, 2–5 mm wide, often exceeding culm, auricles rounded, 1–3.5 mm long. Inflorescence of many spherical heads, each 10–15 mm in diam, 30(or more)-flowered; involucral bracts exceeding inflorescence. Perianth segments 4–5 mm long, long-lanceolate, long-acuminate, the outer exceeding inner series by 0.5–1.0 mm; stamens 6, 1/2 perianth length, anther shorter than filament. Capsule narrowly lance-subulate, 1-locular, the long tapering apex equaling or just exceeding the perianth; seeds obovoid to oblong, 0.3–0.5 mm long, ends apiculate. (2n = 40) Jun–Oct. Lake & stream marshes, sand bars, wet prairies & ditches; GP; (s Can., NY to WA s to VA, AL, TX, & CA; n Mex.).

Inflorescences in the septate species of *Juncus* are often replaced by hornlike galls and this is particularly true of *J. torreyi*. Depauprate forms of *J. torreyi* often resemble *J. nodosus*, but these can be distinguished by the different inner and outer perianth lengths.

24. *Juncus validus* Cov. Tufted plants 0.3–1.2 m tall; culm erect to ascending, stout. Leaves 1–3, septate, laterally compressed, not exceeding culm, aruicles not distinct. Inflorescence an open cyme, few to many spherical heads, each 1–1.5 cm in diam, 30- to 60(or more)-flowered; involucral bract short, not exceeding inflorescence. Perianth 3.5–4.5 mm long, lance-subulate, outer perianth series exceeding inner; stamens 3, not exceeding perianth, anther shorter than filament. Capsule subulate, exceeding perianth, tip often remaining attached at dehiscence; seeds ovoid, 0.5 mm long, ends shortly apiculate. Jun–Sep. Lake marshes & wet prairies; se GP; se KS & OK; (NC to MO & KS s to FL & TX). *J. crassifolius* Buch.—Atlas GP.

25. *Juncus vaseyi* Engelm. Tufted plants 2–8 dm tall, culm slender. Leaves 1 or 2(3), not exceeding culm, 1 mm wide, terete, slightly channeled, auricles rounded, 0.5–2 mm long, membranaceous. Inflorescence a cyme of many-clustered flowers, branches erect; involucral bract just surpassing inflorescence to long exserted. Perianth segments 3.5–4.5 mm long, subequal or inner series slightly shorter, outer perianth lanceolate, acute, inner perianth broadly lanceolate and ± blunt; stamens 6, 1/2 perianth length, anther equal to filament length. Capsule oblong-cylindrical, obtuse, surpassing perianth 1 mm or more, 3-locular; seeds slender, 1–1.5 mm long, ends tailed, each white tail 1/2 seed length. Jun–Sep. Lakeshores, marshes, & woods; ne GP; MN; Polk; (s Can. s to NY, IL, & MN; in w, s to CO & UT).

2. LUZULA DC., Woodrush

Contributed by Janice Coffey

Perennial herbs, usually cespitose, with rhizomes or stolons, or both; stems 0.5–8 dm tall. Basal and stem leaves present, usually pilose along margins, blade mostly plane; ligule a sparse or thick tuft of fine hairs. Inflorescence terminal, bracteate with 1–3 leaflike bracts at every branch and 1 or 2 bracteoles subtending each flower. Sepals 3, petals 3, usually similar in color, texture and shape, not always equal, persistent in fruit; stamens 6; pistil 1, stigmas 3, filiform, style 1, ovary 1-locular, ovules 3, anatropous, attached to basal placentae. Fruit a loculicidal capsule; seeds of glomerulate inflorescences with a white, micropylar appendage (elaiosome).

The report of *L. multiflora* from WY: Johnson (Atlas GP) was not verified.

The distributions noted below are based on those few specimens that could be positively identified. Most collections from the GP are too immature for identification. All species of *Luzula* are probably more common in our area than the existing collections indicate.

1 Flowers borne singly on compound, diffuse inflorescence branches .... 4. *L. parviflora*
1 Flowers borne in few-flowered glomerules.
  2 Glomerules cylindric or subcylindric; perianth shorter than to barely exceeding capsule; secondary peduncles rare; "bulbs" absent .... 3. *L. multiflora*
  2 Glomerules not cylindric, or if cylindric, plant with "bulbs" at base; perianth barely shorter than to obviously exceeding the capsule; secondary peduncles often present.
    3 Glomerules cylindric; usually many white, bulbous bases of undeveloped leaves on the slender rhizomes .... 1. *L. bulbosa*
    3 Glomerules subglobose or ovoid; sometimes with base of flowering stem swollen .... 2. *L. echinata*

1. Luzula bulbosa (Wood) Rydb. Plant weakly cespitose or stem solitary; rhizome short, slender, bearing few to several white, swollen, reduced leaves, often referred to as "bulbs"; stem 0.8–4 dm tall. Basal leaves few, stem leaves 2–4, to 17 cm long and 7 mm wide, with sparsely to densely ciliate margins. Glomerules 3–20, cylindric, 6- to 20-flowered; bracts leaflike, small. Sepals and petals with shining chestnut centers and usually wide, hyaline margins and tips; sepals 2–3 mm long, ca 0.75 mm wide, equaling or usually slightly exceeding the petals; stigmas conspicuously exceeding style. Capsule stramineous to castaneous, lustrous, obovoid with truncate apex, usually exceeding the perianth; seeds dark brown, elliptic, 0.9–1.3 mm long; elaiosome 1/2–2/3 length of seed. (n = 6) Jun–Jul. Dry, open woods & fields, occasional; MO: Jasper, Barton, Vernon; KS: Cherokee, Neosho, Chautauqua, Woodson; (Can., N. Eng., se U.S. w to GP).

2. Luzula echinata (Small) Hermann. Loosely cespitose with shallow knotty rhizomes to 1.5 cm long; stem 1.5–4.5 dm tall, sometimes with swollen base, which along with the knotty rhizomes can resemble the "bulbs" of no. 1. Leaves 4–15 cm long, 2–7 mm wide, margins pilose to sparsely hairy. Glomerules 4–15, broadly conical to nearly globose, often loosely flowered; peduncles to 9 cm long, diverging at angles to 90°; bracts leaflike, not exceeding inflorescence. Sepals and petals greenish to stramineous or brown, 2.6–4 mm long, 0.56–0.75 mm wide, usually conspicuously exceeding the capsule; stigmas 3–4 mm long, style 1–2 mm long. Capsule stramineous to brown, obovoid to subglobose; seeds dark brown, globose; elaiosome 0.5–0.6 mm long, about 1/3 length of seeds. (n = 6) Jun–Jul. Moist woods & bluffs, sometimes in clearings, occasional; IA: Emmet; (Can., N. Eng., se U.S. w to GP).
L. campestris var. *echinata* (Small) Fernald & Wiegand—Gleason.

L. campestris (L.) DC. does not occur in North America.

3. Luzula multiflora (Retz.) Lej. Plant cespitose, often with short rhizomes; stem 2–4 dm tall. Leaves mostly basal, 7–13 cm long, 2–9 mm wide, pilose along margins, more heavily so toward base. Glomerules 4–12, cylindric or subcylindric, 5- to 10-flowered; unbranched peduncles 0.5–7 cm long, ascending; lowermost bract leaflike, overtopping the inflorescence; bracteoles shining, conspicuous in fruit. Sepals and petals stramineous, more or less equal; style less than 1 mm long, shorter than stigmas. Capsule stramineous to brown, subglobose to ovoid or obovoid, shorter than to hardly exceeding the perianth; seeds rufous, 1–1.3 mm long, 0.6–0.7 mm broad; micropylar elaiosome ca 1/2 length of seeds. (n = 12, 18)

Jun–Jul. In open woods, infrequent; MN: Otter Tail, Kittson; SD: Custer, Pennington, Lawrence; (scattered throughout N. Amer., except se U.S.). *L. campestris* var. *multiflora* (Ehrh.) Celak.—Gleason.

4. *Luzula parviflora* (Ehrh.) Desv. Plant with stolons to 5 cm long, covered with diminutive scale leaves, stem 2–8 dm tall. Leaf blades nearly glabrous with few hairs at sheath opening, 3–29 cm long, 3–14 mm wide, abruptly ending or gradually tapering to involute callous tips. Inflorescence with very regular branches, up to 2.3 dm long, and nearly as wide, the main branches diverging at 45°–90° angles (sometimes termed an "anthela"), flowers pedicellate; lowermost bract leaflike, usually less than 5 cm long (to 8 cm). Sepals and petals similar, equal, 1.7–2.4 mm long, light stramineous or with rusty centers and apices; stigmas not conspicuously exceeding perianth. Capsule rufous to blackish, ovate with persistent style base conspicuous, usually exceeding the perianth; seeds shining translucent, varying from rufous to cinnamon, elliptic, 1.1–1.35 mm long, 0.56–0.75 mm wide, with fibrous hairs at base representing the funiculus. (n = 12) Jun–Aug. In montane forests; SD: Lawrence; (throughout N. Amer. except se U.S.).

50. CYPERACEAE Juss., the Sedge Family

by Ole A. Kolstad

Herbs, mostly perennials, a few species annuals. Culms usually triangular, solid or seldom hollow, usually with 3-ranked leaves, blade typically elongate, or some or all of the blades reduced or lacking; sheaths closed. Inflorescence of 1–several spikelets; spikelets 1- to several-flowered. Flowers perfect or often unisexual, each floret subtended by a scale; perianth absent or reduced to bristles or modified scales; stamens 1–3; stigmas 2 or 3; styles 2- or 3-cleft. Fruit a flattened or trigonous achene.

```
1  Achene enclosed in a saclike structure (the perigynium) .................................... 2. Carex
1  Achene not enclosed in a saclike structure.
   2  Spikelets all bisexual and essentially alike.
      3  Spikelets flattened; scales of the spikelet 2-ranked.
         4  Inflorescence terminal; achene not subtended by bristles ................... 3. Cyperus
         4  Inflorescence axillary; achene subtended by bristles ...................... 5. Dulichium
      3  Spikelets not flattened; scales of the spikelet spirally arranged.
         5  Spikelets several-flowered or with several achenes.
            6  Achene crowned with a persistant tubercle.
               7  Culms with bladeless sheaths; spikelet single and terminal ..... 6. Eleocharis
               7  Culms leafy, spikelets 2–many in terminal inflorescences ....... 1. Bulbostylis
            6  Achene not crowned with a persistent tubercle.
               8  Achenes subtended by scales or bristles or both.
                  9  Achenes subtended by bristles.
                     10  Bristles 0–8, usually not exceeding the outer scale ......... 12. Scirpus
                     10  Bristles numerous, elongate ................................. 7. Eriophorum
                  9  Achenes subtended by one or more scales.
                     11  Achene subtended by an inconspicuous hyaline inner
                         scale ......................................................... 10. Hemicarpha
                     11  Achene subtended by 3 stalked ovate-oblong petal scales . 9. Fuirena
               8  Achenes not subtended by scales or bristles.
                  12  Inflorescence bract 1, usually erect, appearing to be a continuation of
                      the culm ............................................................. 12. Scirpus
                  12  Inflorescence bracts 2 or more.
                     13  Style base swollen, not persistent .......................... 8. Fimbristylis
```

    13 Style base not swollen, uniform throughout .... 12. *Scirpus*
 5 Spikelets usually 1- or 2-flowered.
    14 Inflorescence bracts usually white, spikelets flattened .... 4. *Dichromena*
    14 Inflorescence bracts usually green, spikelets not flattened .... 11. *Rhynchospora*
2 Spikelets unisexual, spikelets dimorphic .... 13. *Scleria*

1. BULBOSTYLIS Kunth

1. *Bulbostylis capillaris* (L.) Clarke. Cespitose annual with fibrous roots. Culms filiform, up to 4 dm tall. Leaves setaceous, up to 0.5 mm wide and about 1/3 as long as the culm. Inflorescence umbellate, 5–30 mm long, often reduced; involucral bracts capillary, usually longer than the inflorescence; spikelets ovoid-oblong, 3–5 mm long. Scales spirally imbricate, ovate, dark brown to purple with a keeled green midrib; stamens 2, with anthers up to 0.5 mm long. Achene trigonous, obovate up to 1 mm long, stramineous to pale brown, rugulose, tubercle minute. Moist sandy soils along rivers, ponds, & lake margins; scattered, e 1/2 NE, KS, OK, & w MO; (New England s to FL, w to MN, KS, TX, also NM, AR, & CA). *Stenophyllus capillaris* (L.) Britt.—Rydberg; *B. capillaris* var. *crebra* Fern.—Fernald.

2. CAREX L., Sedge

  Grasslike perennial herbs. Culms mostly triangular, rough or smooth on angles, usually solid. Leaves 3-ranked, long and usually narrow, sheaths closed. Plants monoecious or rarely dioecious with flowers arranged in spikes. Spikes 1–many, pistillate, staminate, androgynous or gynecandrous, sessile or peduncled. Flowers unisexual, solitary in axils of scales; staminate flowers of 3 stamens, rarely only 2; pistillate flowers of 1 pistil with a 2- or 3-branched style and 2 or 3 stigmas, surrounded by a saclike structure or perigynium; at maturity, stigmas, and at times part of the style, protrude through the terminal orifice. Perigynia triangular, inflated or flattened, glabrous or pubescent, beakless or strongly beaked. Achenes triangular, lenticular or plano-convex.

  References: Hermann, F. J. 1954. Addenda to North American carices. Amer. Midl. naturalist 51(1): 265–285; Mackenzie, K. K. 1931–35. *Carex, in* N. Amer. Flora 18: 9–478.

  Carex flaccosperma Dewey has been reported for KS: Crawford by Gibson (Trans. Kansas Acad. Sci. 66: 685–726. 1964).

CAREX SPECIES GROUPS

1 Spikes bisexual; perigynia flattened; stigmas 2.
 2 All or only terminal spike gynecandrous .... Group I
 2 All spikes androgynous.
  3 Beaks of perigynia entire or obliquely cut at apex .... Group II
  3 Beaks of perigynia prominently bidentate at apex .... Group III
1 Upper spikes staminate or gynecandrous; if gynecandrous, stigmas 3; if stigmas 2, then spikes unisexual or lateral spikes androgynous.
 4 Terminal spike bisexual and gynecandrous .... Group IV
 4 Terminal spikes unisexual, androgynous or solitary.
  5 Spikes solitary, androgynous or dioecious .... Group VII
  5 Spikes unisexual, terminal spikes staminate, lateral ones occasionally androgynous.
   6 Beaks of perigynia bidentate; stigmas 3 .... Group VIII
   6 Beaks of perigynia entire or obliquely cut (bidentulate) stigmas 2 or 3.
    7 Stigmas 3 .... Group V
    7 Stigmas 2 .... Group VI

Group I

1 Perigynia spongy-thickened at the base; scarcely wing-margined; pistillate scales often tipped with awn 1/2 the length of the scales .... 39. *C. deweyana*
1 Perigynia not spongy-thickened at the base; wing margin evident; pistillate scales obtuse or acute at apex, never awned.
  2 Lateral spikes wholly pistillate or staminate, terminal spikes usually gynecandrous.
    3 Beak of perigynium only slightly bidentate, about 1/3 the length of body .... 66. *C. interior*
    3 Beak of perigynium sharply bidentate, 1/2 as long to nearly as long as the body .... 115. *C. sterilis*
  2 Spikes all alike, all gynecandrous.
    4 Culms arising solitary or few together from long creeping rhizomes ...... 49. *C. foenea*
    4 Culms loosely to densely cespitose, rhizome (if present) very short.
      5 Lower bracts of inflorescence many times longer than the heads.
        6 Bracts leaflike .... 119. *C. sychnocephala*
        6 Bracts not leaflike .... 13. *C. athrostachya*
      5 Lower bracts of inflorescence slightly (if at all) longer than the heads.
        7 Sheaths green-striate ventrally, except for V-shaped hyaline area at mouth.
          8 Perigynia 7–10 mm long; spikes 1.5–2.5 cm long; heads 5–8 cm long .... 86. *C. muskingumensis*
          8 Perigynia 3–5 mm long; spikes less than 1.5 cm long; heads 2–5 cm long.
            9 Perigynia thin and scalelike, their tips appressed or ascending; spikes ovoid-oblong.
              10 Perigynia lanceolate to ovate-lanceolate, 3–4× as long as wide .... 125. *C. tribuloides*
              10 Perigynia ovate or obovate-orbicular, nearly as broad as long .... 3. *C. albolutescens*
            9 Perigynia thickened, plano-convex, their tips widely spreading and recurved; spikes almost globose .... 36. *C. cristatella*
        7 Sheaths white-hyaline ventrally.
          11 Scales about the same length as perigynia and nearly the same width above, perigynia concealed or nearly so above.
            12 Inflorescences stiff, spikes usually aggregate .... 133. *C. xerantica*
            12 Inflorescences flexuous and moniliform.
              13 Sheaths green-and-white mottled dorsally; beaks flat ..... 1. *C. aenea*
              13 Sheaths not green-and-white mottled dorsally; beaks terete .... 98. *C. praticola*
          11 Scales shorter than perigynia and narrower above, perigynia exposed above.
            14 Beaks slender, terete, slightly serrulate toward tips, obliquely cut dorsally .... 82. *C. microptera*
            14 Beaks flattened and serrulate to tips, bidentate.
              15 Perigynia subulate to narrowly ovate-lanceolate, 3–4× as long as broad; marginal wings narrow their entire length.
                16 Leaf blades 2.5–7 mm wide; sterile leafy culms abundant with spreading leaf blades scattered on culm; perigynia strongly wing-margined to below middle .... 99. *C. projecta*
                16 Leaf blades 1–3 mm wide; sterile leafy culms few with ascending leaf blades clustered at tips of culms; perigynia wing-margined to base .... 110. *C. scoparia*
              15 Perigynia ovate-lanceolate, ovate or orbicular, never more than 2× as long as broad; marginal wings broad for their entire length.
                17 Perigynia ovate-lanceolate or narrowly ovate, 3–4 mm (rarely 5 mm) long.
                  18 Dorsal surface of leaf sheaths green-and-white mottled or white hyaline between the green nerves .... 88. *C. normalis*
                  18 Dorsal surface of leaf sheaths or entire sheath green.
                    19 Spikes aggregated into compact heads; perigynia brown at maturity, the ventral surfaces nerveless .... 16. *C. bebbii*

                    19 Spikes loosely arranged in moniliform heads; perigynia
                       straw colored at maturity, the ventral surfaces
                       nerved .... 120. *C. tenera*
                17 Perigynia suborbicular or orbicular; 3–8.5 mm long.
                    20 Spikes usually separated in a lax, moniliform inflorescence;
                       perigynia to 4 mm long, beaks about as long as the bodies;
                       achenes 1.5 mm long .... 46. *C. festucacea*
                    20 Spikes closely aggregated; perigynia 3.7–8.5 mm long; beaks
                       up to 1/2 as long as the bodies; achenes 1.7–2.5 mm long.
                        21 Perigynia 5.5–8.5 mm long, membranaceous, thin, except
                           where distended over achenes.
                            22 Perigynia 5.5–7 mm long, strongly to few-nerved
                               ventrally .... 18. *C. bicknellii*
                            22 Perigynia 7.5–8.5 mm long, nerveless or nearly so
                               ventrally .... 21. *C. brittoniana*
                        21 Perigynia 1.7–5 mm long, firm textured, thickened, plano-
                           convex in cross section.
                            23 Ventral surfaces of perigynia sometimes nerved,
                               perigynia tapering into beaks; dorsal surface of leaf
                               sheaths green-and-white mottled or white hyaline be-
                               tween the green nerves .... 84. *C. molesta*
                            23 Ventral surface of perigynia nerveless, perigynia
                               abruptly contracted into beaks; dorsal surface of leaf
                               sheaths entirely green .... 20. *C. brevior*

                                          Group II
1 Plants forming colonies; spreading by long creeping rhizomes, the culms arising singly or in
  small clumps.
    2 Ventral leaf sheath green-striate, sheath prolonged into a conspicuous hyaline, tubular
      ligule .... 107. *C. sartwellii*
    2 Ventral leaf sheath hyaline, truncate at base, ligule inconspicuous.
        3 Perigynia lance-ovate; beaks of perigynia as long as or longer than the bodies; heads
          usually dioecious .... 42. *C. douglasii*
        3 Perigynia ovate; beaks of perigynia 1/2 as long as the bodies; heads monoecious,
          androgynous.
            4 Culms obtusely angled, smooth; rootstock 1–2 mm thick, slender; leaves involute,
              1–2 mm broad .... 44. *C. eleocharis*
            4 Culms acutely triangular, roughened above; rootstock 2–6 mm thick, stout; leaves
              flattened, 2–3 mm broad .... 96. *C. praegracilis*
1 Plants cespitose, forming large dense clumps, roots fibrous, rhizome (if present) very short.
    5 Heads simple; spikes single at each node; perigynia densely white-puncticulate.
        6 Spikes androgynous; perigynia usually biconvex .... 41. *C. disperma*
        6 Spikes (at least uppermost) gynecandrous; perigynia plano-
          convex .... 22. *C. brunnescens*
    5 Heads compound; lower spikes paired at each node; perigynia not white-puncticulate.
        7 Ventral surface of leaf sheath white-hyaline; spikes closely aggregated; perigynia shin-
          ing .... 40. *C. diandra*
        7 Ventral surface of leaf sheath copper colored at least at the mouth; lower spikes more
          or less separated; perigynia dull .... 97. *C. prairea*

                                          Group III
1 Spikes single at each node, usually less than 10.
    2 Perigynia spongy-thickened at base, achenes filling only the upper portion of perigynia;
      perigynia radiating or reflexed in all directions.
        3 Beaks of perigynia smooth.
            4 Perigynia biconvex, base striate ventrally .... 101. *C. retroflexa*
            4 Perigynia plano-convex, base nerveless ventrally .... 122. *C. texensis*
        3 Beaks of perigynia serrulate.

5 Perigynia gradually tapering into the beaks; not white-hyaline at orifice; leaves 1–2 mm wide .... 104. *C. rosea*
5 Perigynia abruptly contracted into the beaks; white-hyaline at orifice; leaves 1.5–3 mm wide .... 33. *C. convoluta*
2 Perigynia not spongy-thickened at base; achenes almost filling the bodies of the perigynia.
6 Beaks of the perigynia obliquely cut dorsally, only slightly bidentate . 129. *C. vallicola*
6 Beaks of the perigynia strongly bidentate.
7 Sheaths tight ventrally, slightly (if at all) septate-nodulose or green-and-white mottled dorsally.
8 Heads usually less than 2 cm long, densely capitate.
9 Pistillate scales chestnut-brown; perigynia brown .... 62. *C. hoodii*
9 Pistillate scales green to brown; perigynia green to tan.
10 Perigynia round tapering at base, broadest at middle; sheaths thickened at mouth .... 28. *C. cephalophora*
10 Perigynia truncate, broadest at base; sheaths not thickened at mouth .... 74. *C. leavenworthii*
8 Heads greater than 2 cm long, not capitate.
11 Perigynia hidden in head by scales, scales reddish-brown . 63. *C. hookerana*
11 Perigynia conspicuous in head; scales green to brown.
12 Lower bracts prolonged, 2–3 × length of head, leaflike .... 9. *C. arkansana*
12 Lower bracts not prolonged, seldom exceeding the head, not leaflike .... 85. *C. muhlenbergii*
7 Sheaths loose and easily breaking ventrally; prominently septate-nodulose or green-and-white mottled dorsally.
13 Perigynia light green to brown; scales long acuminate to awn-tipped; spikes crowded .... 54. *C. gravida*
13 Perigynia dark green; scales obtuse to acute; lower spikes usually separate.
14 Sheaths not cross-rugulose .... 2. *C. aggregata*
14 Sheaths cross-rugulose.
15 Spikes crowded into a head; scales obtuse to acutish .... 27. *C. cephaloidea*
15 Spikes remote, in a moniliform head; scales short-cuspidate .... 112. *C. sparganioides*
1 Spikes 2 or more on a branch at the lower node.
16 Body of perigynia tapering into a beak, if abruptly contracted, then culm winged and flattened under pressure.
17 Culms flattened, winged; perigynia ovate, rounded at base, contracted into beaks.
18 Beaks of perigynia about the length of bodies; leaf sheaths smooth ventrally .... 5. *C. alopecoidea*
18 Beaks of perigynia about 1/2 the length of the bodies; leaf sheaths cross-rugulose ventrally .... 32. *C. conjuncta*
17 Culms triangular, slightly (if at all) winged, perigynia truncate-rounded at base, tapering into beaks.
19 Perigynia 6 mm or more long, beaks 2–3 × as long as the bodies .... 37. *C. crus-corvi*
19 Perigynia less than 6 mm long; beaks less than 2 × as long as the bodies.
20 Sheaths green-and-white mottled dorsally, white hyaline ventrally .... 90. *C. oklahomensis*
20 Sheaths sepate-nodulose dorsally, either smooth or cross-rugulose ventrally.
21 Leaf sheath conspicuously cross-rugulose ventrally and prolonged at mouth; pistillate scales acute or mucronulate .... 116. *C. stipata*
21 Leaf sheath usually smooth ventrally, concave at mouth and not prolonged; pistillate scales obviously awned .... 71. *C. laevivaginata*
16 Body of perigynia abruptly contracted into a beak, culms not winged or flattened.
22 Perigynia 3.0–4.5 mm long; pistillate scales acute or cuspidate; beaks of perigynia much shorter than the bodies .... 48. *C. fissa*

22 Perygynia 2.2–3.5 mm long; pistillate scales strongly awned, beaks of perigynia shorter than to equaling the bodies.
   23 Beaks of perigynia about equaling the bodies; sheaths cross-rugulose ventrally and rarely red dotted .... 132. *C. vulpinoidea*
   23 Beaks of perigynia much shorter than the bodies; leaves normally shorter than the culms.
     24 Body of perigynia reniform or orbicular, strongly resinous dotted .... 124. *C. triangularis*
     24 Body of perigynia narrower, slightly resinous dotted .... 7. *C. annectens*

Group IV

1 Stigmas 2.
  2 Terminal spike staminate at base; scales brown to purple, blunt at tip ...... 51. *C. garberi*
  2 Terminal spike staminate throughout; scales whitish, short pointed at tip .... 14. *C. aurea*
1 Stigmas 3.
  3 Style continuous with achene, persistent.
    4 Terminal spike staminate .... 50. *C. frankii*
    4 Terminal spike gynecandrous.
      5 Perigynia with beaks widely radiating; style abruptly bent near base; pistillate scales acute to short awned .... 114. *C. squarrosa*
      5 Perigynia with beaks mostly ascending; style straight at base; pistillate scales obtuse .... 127. *C. typhina*
  3 Style jointed with achene, deciduous.
    6 Perigynia compressed-flattened, 2-keeled, otherwise nerveless; lower spikes long peduncled .... 111. *C. shortiana*
    6 Perigynia obtusely triangular, or rounded in cross section, several- to many-nerved or ribbed; spikes sessile or short peduncled.
      7 Spikes linear, elongate, loosely flowered, lower spikes drooping .... 38. *C. davisii*
      7 Spikes short-cylindric, closely flowered, sessile or erect.
        8 Sheaths and blades pubescent; pistillate scales hyaline with a green midrib.
          9 Terminal spikes staminate .... 123. *C. torreyi*
          9 Terminal spikes pistillate at apex and staminate at base (gynecandrous).
            10 Perigynia dorsiventrally compressed; slightly nerved ventrally.
              11 Leaf blades, leaf sheaths and culms strongly short-pubescent .... 59. *C. hirsutella*
              11 Leaf blades, leaf sheaths and culms glabrate or nearly so .... 30. *C. complanata*
            10 Perigynia rounded to obscurely triangular, strongly nerved.
              12 Sheaths glabrous on ventral side; blades pubescent only at base .... 26. *C. caroliniana*
              12 Sheath pubescent on ventral side; blades usually pubescent .... 23. *C. bushii*
        8 Sheaths and blades glabrous; pistillate scales purple-brown with light-colored midrib.
          13 Terminal spike pistillate, androgynous or all staminate .... 56. *C. hallii*
          13 Terminal spike gynecandrous.
            14 Scales of perigynia awned; sheaths filamentose .... 24. *C. buxbaumii*
            14 Scales of perigynia obtuse, acute or mucronate; sheaths not filamentose.
              15 Perigynia 3 mm or longer; achenes 2 mm or longer; leaves 3–6 mm wide .... 17. *C. bella*
              15 Perigynia less than 3 mm long; achenes less than 2 mm long; leaves 2–3 mm wide .... 92. *C. parryana*

Group V

1 Spikes solitary on each culm.
  2 Spikes dioecious .... 109. *C. scirpiformis*
  2 Spikes monoecious, androgynous.
    3 Staminate scales tightly connate to base.

```
    4  Perigynia subtended by a short, not leaflike scale, shorter than
       perigynia .................................................................................... 75. C. leptalea
    4  Perigynia subtended by a long, foliaceous scale, longer than perigynia.
       5  Scales of pistillate flowers with hyaline margins; staminate flowers 6-20; lowest
          bracts 1-2 mm wide ...................................................... 68. C. jamesii
       5  Scales of pistillate flowers green throughout; staminate flowers about 3; lowest
          bract 3-6 mm wide.
          6  Beaks of perigynia 2-3 mm long; perigynia 4.5-6 mm long, empty at
             summit ............................................................................ 15. C. backii
          6  Beaks of perigynia 0.5-1 mm long, perigynia 4 mm long, tightly filled by
             achene .................................................................... 108. C. saximontana
 3  Staminate scales free to base.
    7  Perigynia puberulent above, beak less than 0.5 mm long; scales obtuse and
       broad ................................................................................... 47. C. filifolia
    7  Perigynia glabrous, beak 0.5 mm long or longer; scales acuminate or cuspidate,
       narrow .................................................................................. 89. C. obtusata
1  Spikes 2 or more on each culm.
 8  Bracts of pistillate spikes bladeless, sheathing only.
    9  Perigynia shining puncticulate .................................................. 43. C. eburnea
    9  Perigynia pubescent or puberulent.
       10  Scales of pistillate flowers cuspidate; bracts of inflorescence with short and
           rudimentary blades ...................................................... 94. C. pedunculata
       10  Scales of pistillate flowers obtuse or acute; bracts of inflorescence bladeless.
           11  Pistillate scales shorter than perigynia, staminate spikes 3-6 mm
               long ........................................................................ 31. C. concinna
           11  Pistillate scales exceeding perigynia; staminate spikes 10-25 cm
               long ....................................................................... 103. C. richardsonii
 8  Bracts of pistillate spikes with well-developed blades.
    12  Pistillate spikes drooping on flexuous peduncles.
        13  Pistillate spike more than 9 mm long, beak of perigynia entire and less than 0.5
            mm long, smooth .................................................................. 76. C. limosa
        13  Pistillate spikes less than 9 mm long, beak of perigynia obliquely cut and more
            than 0.5 mm long, serrulate ............................ 25. C. capillaris var. elongata
    12  Pistillate spikes not drooping.
        14  Beaks of perigynia abruptly contracted or tapering, obliquely cut at apex;
            perigynia pubescent.
            15  Perigynia densely hispidulous, abruptly contracted or
                tapering ............................................................... 11. C. assiniboinensis
            15  Perigynia loosely pubescent, abruptly contracted .................. 60. C. hirtifolia
        14  Beaks of perigynia lacking, minute or very short, tubular and entire; perigynia
            glabrous.
            16  Perigynia tapering into a short tubular beak; pistillate scales with prolonged
                awns.
                17  Sheaths glabrous, prolonged 1-2 mm; perigynia up to 4 mm long, beaks
                    gradually tapering and straight, achenes apiculate and
                    straight .......................................................... 91. C. oligocarpa
                17  Sheaths hispid, usually truncate; perigynia greater than 4 mm long, beaks
                    abruptly contracted and bent; achenes bent-
                    apiculate ...................................................... 61. C. hitchcockiana
            16  Perigynia beakless or beak minute (if tubular, scales not awned); pistillate
                scales obtuse, acute, or mucronate.
                18  Bract of lowest pistillate spikes sheathless or nearly so.
                    19  Terminal spikes staminate ......................................... 123. C. torreyi
                    19  Terminal spikes pistillate at apex and staminate at base
                        (gynecandrous).
                        20  Perigynia dorsiventrally compressed; slightly nerved ventrally.
                            21  Leaf blades, leaf sheaths, and culms strongly short-
                                pubescent .................................................. 59. C. hirsutella
```

21 Leaf blades, leaf sheaths, and culms glabrous or nearly
so .... 30. *C. complanata*
20 Perigynia rounded to obscurely triangular, strongly nerved.
22 Sheaths glabrous on ventral side; blades pubescent only at
base .... 26. *C. caroliniana*
22 Sheaths pubescent on ventral side; blades usually
pubescent .... 23. *C. bushii*
18 Bract of lowest spike with a well-developed sheath.
23 Perigynia 2-ribbed, nerveless or with less than 10 faint nerves.
24 Perigynia beakless, or nearly so, 4.3–6.7 mm long.
25 Pistillate spikes loosely flowered; culms slender; perigynia dark
green .... 121. *C. tetanica*
25 Pistillate spikes closely flowered; culms stout; perigynia yellow-
green .... 79. *C. meadii*
24 Perigynia beaked, 2.4–4.1 mm long.
26 Leaves of the sterile culms 12–40 mm wide; pistillate scales
broadly obovate-orbicular, truncate, culms narrowly wing-
angled .... 4. *C. albursina*
26 Leaves of the sterile culms 2–15 mm wide; pistillate scales
mucronate to long awned; culms slightly or not at all wing-
angled.
27 Staminate spike usually long peduncled, its tip always and its
base usually well above the uppermost pistillate spike; culm
purplish-tinged at base .... 52. *C. gracilescens*
27 Staminate spike sessile or subsessile, its tip frequently exceeded
by the uppermost pistillate spike; culm brownish at
base .... 19. *C. blanda*
23 Perigynia 2-ribbed, and strongly nerved, usually more than 10 nerves.
28 Pistillate spikes sessile or nearly so; perigynia spreading or
squarrose .... 131. *C. viridula*
28 Pistillate spike peduncled (at least lower ones); perigynia ascending.
29 Perigynia with spaces between the nerves little more than 2× the
thickness of the nerves; beakless .... 6. *C. amphibola* var. *turgida*
29 Perigynia with spaces between the nerves several times wider than
thickness of the nerve; with short but distinct beaks.
30 Plants cespitose; staminate spikes short-peduncled or sessile;
upper pistillate spikes aggregate .... 53. *C. granularis*
30 Plants with prolonged root stocks; staminate spikes long
peduncled; pistillate spikes separate.
31 Perigynia many-nerved, beaks short and entire; leaves 1–4
mm wide .... 34. *C. crawei*
31 Perigynia many-ribbed, beaks short and bidentate; leaves
3–6 mm wide .... 81. *C. microdonta*

Group VI

1 Achene blackish at maturity.
2 Terminal spike staminate at base; scales brown to purple, blunt at tip ...... 51. *C. garberi*
2 Terminal spike staminate throughout; scales whitish, short pointed at tip .... 14. *C. aurea*
1 Achene tan to brown at maturity.
3 Achene constricted in the middle .... 35. *C. crinita*
3 Achene not constricted in the middle.
4 Beaks of perigynia bidentate; perigynia 2-ribbed and strongly nerved between
ribs .... 87. *C. nebraskensis*
4 eaks of perigynia entire; perigynia 2-ribbed and faintly nerved or nerveless between ribs.
5 Perigynia obovate, broadest at apex; leaves glaucous; lower leaves
phyllopodic .... 8. *C. aquatilis* var. *altior*
5 Perigynia ovate or elliptical, broadest at the middle or base; leaves not glaucous
but darker green; lower leaves aphyllopodic.

            6 Perigynia ovate, broadest at base, slightly constricted at the middle, straw colored at maturity .... 45. *C. emoryi*
            6 Perigynia ellipitical, broadest at middle, tapered toward both ends, brown or green at maturity.
               7 Perigynia usually flattened, not inflated at apex, green to dark green at maturity; dorsal sheaths occasionally pubescent .... 117. *C. stricta*
               7 Perigynia inflated at apex, brown at maturity; dorsal sheaths glabrous .... 57. *C. haydenii*

Group VII

1 Spikes unisexual; dioecious .... 109. *C. scirpiformis*
1 Spikes bisexual; androgynous.
   2 Staminate scales free and not overlapping below.
      3 Perigynia puberulent at apex .... 47. *C. filifolia*
      3 Perigynia glabrous .... 89. *C. obtusata*
   2 Staminate scales connate or overlapping below.
      4 Lowest pistillate scales shorter than to slightly longer than perigynia, not leaflike; perigynia beakless .... 75. *C. leptalea*
      4 Lowest pistillate scales much longer than perigynia and leaflike; beak of perigynia long, 0.5–3 mm.
         5 Scales of pistillate flowers with hyaline margins; staminate flowers 6–20; lowest bracts 1–2 mm wide .... 68. *C. jamesii*
         5 Scales of pistillate flowers green throughout; staminate flowers about 3; lowest bract 3–6 mm wide.
            6 Beaks of perigynia 2–3 mm long; perigynia 4.5–6 mm long, empty at summit .... 15. *C. backii*
            6 Beaks of perigynia 0.5–1 mm long, perigynia 4 mm long, tightly filled by achene .... 108. *C. saximontana*

Group VIII

1 Spikes on flexuous peduncles, drooping or widely spreading.
   2 Perigynia obscurely nerved or 2-keeled; style jointed with achene .... 113. *C. sprengellii*
   2 Perigynia conspicuously many-nerved, not 2-keeled; style continuous with achene.
      3 Perigynia suborbicular in cross section, ascending or spreading, not reflexed; beaks of perigynia with rigid, erect teeth to 1 mm long .... 65. *C. hystericina*
      3 Perigynia flattened-triangular, reflexed when mature; beaks of perigynia with curved or straight teeth 0.6–2 mm long.
         4 Teeth of perigynia recurved-spreading, 1.3–2 mm long .... 29. *C. comosa*
         4 Teeth of perigynia erect or little spreading, 0.5–1 mm long .... 100. *C. pseudo-cyperus*
1 Spikes on erect short peduncles or sessile.
   5 Style jointed with achene; perigynia slightly nerved or 2-keeled.
      6 Perigynia 2-keeled, otherwise nerveless, puberulent or short pubescent.
         7 Peduncles 2–4 on each culm, the terminal with a staminate spike, the shorter and lateral ones pistillate.
            8 Bract of the lowest nonbasal pistillate spike foliaceous, exceeding the staminate spike .... 105. *C. rossii*
            8 Bract of the lower nonbasal pistillate spike scalelike, not surpassing the staminate spike.
               9 Beaks of perigynia up to 1 mm long, 1/2 the length of the body of perigynia; pistillate scales usually with acute tips; sheaths little if at all filamentose .... 83. *C. microrhyncha*
               9 Beaks of perigynia 1–1.8 mm long, 3/4 the length of the body of perigynia; pistillate scales with acuminate tips; sheaths filamentose .... 128. *C. umbellata*
         7 Peduncle single with 2–4 spikes produced on each culm.
            10 Bodies of the perigynia oblong-ovoid; somewhat longer than wide.
               11 Perigynia much exceeding the scales .... 93. *C. peckii*

    11 Perigynia equaling or barely exceeding the scales .... 10. *C. artitecta*
   10 Bodies of the perigynia globose, length slightly more than or equal to width.
    12 Perigynia bodies globular and about 1.5 mm in diam ...... 95. *C. pensylvanica*
    12 Perigynia bodies broadly ellipsoid, 1.5–2.2 mm in diam .... 58. *C. heliophila*
 6 Perigynia not 2-keeled, nerveless or impressed-nerved, pubescent or glabrous.
   13 Foliage and culms pubescent; scales ciliate at margins .... 60. *C. hirtifolia*
   13 Foliage glabrous, culms glabrous or rough and sheaths pubescent; scales entire-margined.
    14 Perigynia glabrous .... 80. *C. melanostachya*
    14 Perigynia pubescent.
     15 Culms usually smooth; leaves up to 2 mm wide; margin involute toward apex; achenes bent-apiculate .... 73. *C. lasiocarpa* var. *americana*
     15 Culms usually rough; leaves 2–5 mm wide; revolute toward apex; achenes straight-apiculate .... 72. *C. lanuginosa*
5 Style continuous with achene; perigynia several-nerved, not 2-keeled.
  16 Perigynia strongly 7- to 9-nerved, glabrous, inflated.
   17 Scales of pistillate spike with a slender rough-margined awn much longer than body .... 78. *C. lurida*
   17 Scales of pistillate spike lanceolate to ovate, acute, acuminate or cuspidate.
    18 Bracts of pistillate spikes several–many × longer than the inflorescence; perigynia wide spreading, lower ones reflexed .... 102. *C. retrorsa*
    18 Bracts of pistillate spikes shorter than to slightly exceeding the inflorescence; perigynia ascending to slightly spreading.
     19 Base of culms little spongy-inflated; pistillate flowers in 6–8 rows, ascending .... 130. *C. vesicaria* var. *monile*
     19 Base of culms spongy-inflated, pistillate flowers in 8–10 rows, spreading at maturity .... 106. *C. rostrata*
  16 Perigynia 10(or more)-nerved or only slightly nerved and pubescent, slightly to strongly inflated.
   20 Pistillate spikes 15–35 mm wide, not over 3× as long; perigynia 10–20 mm long, inflated, 3.5–8 mm wide.
    21 Style bent or twisted; pistillate spikes cylindrical .... 77. *C. lupulina*
    21 Style straight; pistillate spikes globose or subglobose.
     22 Perigynia dull, cuneate at base, often hispid .... 55. *C. grayi*
     22 Perigynia shining, rounded at base, smooth .... 67. *C. intumescens*
   20 Pistillate spikes 10–15 mm thick, cylindric; perigynia less than 11 mm long, slightly inflated, 3.0 mm or less wide.
    23 Teeth of perigynia less than 1 mm long, erect or slightly curved.
     24 Perigynia pubescent .... 118. *C. subimpressa*
     24 Perigynia glabrous.
      25 Base of plant purple tinged; bladeless; sheaths pinnately filamentose ventrally; leaves grass-green; mature perigynia strongly nerved .... 69. *C. lacustris*
      25 Base of plant white or brown; blades present; sheaths less frequently filamentose; leaves glaucous to bluish-green, mature perigynia nerveless or with impressed nerves .... 64. *C. hyalinolepis*
    23 Teeth of perigynia 1–3 mm long, straight or recurved.
     26 Perigynia pubescent .... 126. *C. trichocarpa*
     26 Perigynia glabrous.
      27 Sheaths pubescent; teeth of perigynia recurved .... 12. *C. atherodes*
      27 Sheaths glabrous; teeth of perigynia straight .... 70. *C. laeviconica*

1. Carex aenea Fern. (Group I) Cespitose perennial, rootstock short, black. Culms 3–12 dm tall, nodding, obtusely triangular. Leaves 2–4 mm wide; sheaths tight, thin-hyaline ventrally, green-and-white mottled dorsally. Heads 3.5–7 cm long; spikes bisexual, gynecandrous, moniliform; pistillate scales acute to acuminate. Perigynia concavo-convex, green to brown, several-nerved dorsally, wing-margined to base, 4–5.9 mm long, 2 mm wide, base round tapering, apex tapering to a serrulate beak. Achenes 2 mm long, 1.5 mm wide, lenticular; stigmas 2. Wooded areas & ravines; ND, SD; (Lab. to B.C., s to PA, SD, & ID).

2. **Carex aggregata** Mack. (Group III) Rhizomatous perennial, rootstocks short creeping, brown to black. Culms 4–10 dm tall, triangular with sides usually sulcate. Leaves 3.5–6 mm wide, sheaths not cross-rugulose and rarely red dotted ventrally, green-and-white mottled dorsally. Heads 2.5–5 cm long, aggregate except for lower spikes; scales ovate, acuminate to cuspidate. Perigynia 3.8–4.4 mm long, 1.8–2.4 mm wide, plano-convex, round-truncate at base, apex tapering to a serrulate beak; nerveless ventrally, few-nerved dorsally. Achenes 2 mm long, lenticular; stigmas 2. Dense to open wooded areas; SD, NE, KS, MO; (NY to SD, s to VA, TN, OK). *C. sparganioides* Muhl. var. *aggregata* (Mack.) Gl.—Gleason & Cronquist.

3. **Carex albolutescens** Schwein. (Group I) Densely cespitose perennial, rootstock short, black. Culms slender, 0.3–1.2 m tall, sharply triangular. Leaves 2–4 mm wide; sheaths loose, strongly green striate ventrally. Heads slightly open and moniliform, 2.5–6 cm long; spikes bisexual, gynecandrous; pistillate scales ovate, acute or short acuminate. Perigynia firm but thin, 3.2–4.5 mm long, 1.5–2.5 mm wide, the body obovate, broadest above the middle, abruptly narrowed into the beak, wing-margined to base, finely nerved on both sides. Achenes lenticular, 1.5 mm long, 0.7 mm wide; stigmas 2. Wet wooded areas; MO; (N.S. to MO, s to FL & TX).

4. **Carex albursina** Sheld. (Group V) Loosely cespitose perennial, rootstocks short. Culms 2–6 dm tall, soft, conspicuously wing-angled, sharply triangular. Leaves up to 30 mm wide; sheaths loose, thin ventrally. Staminate spike usually sessile, 5–17 mm long; pistillate spikes usually 3 or 4, erect, 1–2.5 cm long; pistillate scales broadly obovate-orbicular. Perigynia obovoid, 3–4 mm long, 2 mm wide, obtusely triangular, not inflated, glabrous, many-nerved, abruptly contracted into a minute, bent beak, 0.5 mm long. Achenes obovoid, sharply triangular with concave sides, 2.5 mm long, 1.7 mm wide; stigmas 3. Moist woodlands; IA, MO; (Que. to MN, s to VA & MO). *C. laxiflora* Lam. var. *latifolia* F. Boott—Gleason & Cronquist.

5. **Carex alopecoidea** Tuckerm. (Group III) Cespitose perennial, rootstock short, black. Culms 4–10 dm tall, sharply winged-triangular, concave sides, flattened in drying. Leaves 3–6 mm wide; sheaths tight, purple dotted, and not cross-rugulose ventrally, septate-nodulose dorsally. Heads 1.5–5 cm long; spikes bisexual, androgynous, aggregate; pistillate scales acuminate to cuspidate. Perigynia plano-convex, brownish-yellow, many-nerved dorsally, nerveless ventrally, 3.4–4 mm long, 1.5–2 mm wide, abruptly contracted to serrulate beak equal in length to the body. Achenes 1.5 mm long, lenticular; stigmas 2. Wooded areas & swamps; ND, MN; (Que. to ND, s to NJ, IL, & IA).

6. **Carex amphibola** Steud. var. **turgida** Fern. (Group V) Cespitose perennial, rootstock short. Culms 2–8 dm tall, smooth triangular. Leaves 1.5–10 mm wide; sheaths thin, prolonged. Spikes unisexual, terminal spike staminate, 0.7–3 cm long, sessile or peduncled, lateral spikes pistillate, lower peduncled; pistillate scales awned. Perigynia suborbicular to triangular, light or yellow-green, many impressed nerves, 4–5.5 mm long, 2–2.5 mm wide, tapering to both ends, beakless. Achenes 2.5 mm long, 2 mm wide, triangular with concave sides; stigmas 3. Mostly wooded areas, occasionally ditches & prairies; SD, NE, IA, KS, MO, OK; (NY, Ont. to NE, s to FL & TX). *C. grisea* Wahl.—Rydberg.

7. **Carex annectens** (Bickn.) Bickn. (Group III) Cespitose to somewhat rhizomatous perennial, rootstock short to prolonged, blackish. Culms 4–9 dm tall, acutely triangular; usual-

ly longer than the leaves. Leaves 2–6 mm wide; sheaths tight, cross-rugulose and red-dotted ventrally, Heads 2–7 cm long; spikes bisexual, androgynous; slightly separate; pistillate scales ovate, awned. Perigynia plano-convex, brownish to golden-yellow, few-nerved dorsally, nerveless ventrally, 2.2–3.2 mm long, 1.7–2.7 mm wide; base round or truncate, apex contracted to a serrulate beak, beak much shorter than body. Achenes 1.5 mm long, 1 mm wide, lenticular; stigmas 2.

Two varieties may be recognized in our area:

7a. var. *annectens*. Plants tending toward cespitose. Perigynia yellow-brown, 1.7–2.2 mm wide. Roadside ditches, ungrazed prairies, & swales; KS, MO, OK, TX; (N.B. to Ont., s to FL, AR, & OK).

7b. var. *xanthocarpa* (Bickn.) Wieg. Plants rhizomatous. Perigynia golden-brown, 1.4–1.7 mm wide. Ravines & ungrazed prairies; IA, NE, KS, MO; (N.B. to WI & KS, s to VA, IL, MO). *C. brachyglossa* Mack.—Rydberg.

8. ***Carex aquatilis*** Wahl. var. ***altior*** (Rydb.) Fern. (Group VI) Tufted perennial, rootstock prolonged, brown. Culms 5–10.5 dm tall, acutely triangular, leaves 2.5–8 mm wide; sheaths fargile, white or purple dotted ventrally, septate-nodulose, hispidulose dorsally. Spikes unisexual, terminal spike staminate, 3–5 cm long, lateral spikes pistillate; pistillate scales acuminate. Perigynia flattened-biconvex, body obovate, red-striate and glandular, slightly nerved both sides, 2-ribbed, 2.3–3.3 mm long, 1.5–2.3 mm wide, beak 0.1–0.3 mm long. Achenes 1.5 mm long, 0.5 mm wide, lenticular; stigmas 2. Shores, swamps, & marshes; MT, ND, WY, SD, CO, NE, IA, KS; (Newf. to AK, s to NJ, MO, NM, & CA). *C. substricta* (Kükenth.) Mack.—Rydberg.

9. ***Carex arkansana*** (Bailey) Bailey. (Group III) Cespitose perennial, rootstock short. Culms 2–6 dm tall; obtusely triangular. Leaves 1–2 mm wide, sheaths tight, not septate-nodulose or cross-rugulose. Leafy bracted inflorescence 1.5–3 cm long, bracts much prolonged, 2–3 × exceeding the head, up to 6.2 cm long; spikes aggregate, androgynous; pistillate scales ovate-triangular, acuminate or cuspidate. Perigynia plano-convex, 3.5–4 mm long, 2.5 mm wide, margined to base, nerveless ventrally, few-nerved dorsally, abruptly contracted into a serrulate beak. Achenes 2 mm long, 1.7 mm wide, lenticular; stigmas 2. Damp prairies, roadside ditches, & woodlands; KS, MO, OK; (KS & MO, s to TX & AR).

10. ***Carex artitecta*** Mack. (Group VIII) Cespitose perennial, rootstock short, scaly, brown. Culms 0.5–5 dm tall, triangular. Leaves 0.5–2.5 mm wide; sheaths loose to filamentose ventrally. Terminal spike staminate, unisexual, lateral spikes pistillate, separate; pistillate scales acute-cuspidate. Perigynia body oblong-ovoid and obtusely triangular, yellow-green puberulant, 2-ribbed, 2–3 mm long, 1–1.5 mm wide, beak 1 mm long. Achenes 1.5 mm long, 1.0 mm wide, triangular with convex sides; stigmas 3. Open moist woods; NE, IA, KS, MO, OK; (VT to NE, s to GA & OK; also Que.). *C. varia* Muhl.—Rydberg; *C. nigromargina* Schwein. var. *muhlenbergia* (A. Gray) Gl.—Gleason & Cronquist.

11. ***Carex assiniboinensis*** W. Boott. (Group V) Cespitose perennial, rootstock short. Culms 3.5–7.5 dm tall, weak, compressed triangular. Leaves 1–3 mm wide; sheaths long, tight, yellow-brown ventrally, concave and ciliate at mouth. Spikes unisexual, terminal spike staminate, 2–3 cm long, lateral spikes pistillate, loosely flowered; pistillate scales awned, cuspidate or acuminate, about equaling perigynia. Perigynia suborbicular, green to straw colored, 2-ribbed, densely hispidulous, 5.5–8 mm long, 1.7 mm wide, beak slender and as long as the body. Achenes 2.5 mm long, 1.2 mm wide, triangular with concave sides; stigmas 3. Wooded areas; ND, MN, SD; (Man. to Sask., s to SD, IA, & WI).

12. ***Carex atherodes*** Spreng. (Group VIII) Rhizomatous perennial, rootstock prolonged, scaly. Culms 5–12 dm tall, triangular. Leaves 4–12 mm wide, sheaths soft-hairy to puberulent

dorsally, soft-hairy, brown to purple-tinged at mouth ventrally. Spikes unisexual, terminal spikes staminate; lateral spikes pistillate, pistillate scales rough-aristate. Perigynia lanceolate, suborbicular, inflated, yellow-green to brown, strongly ribbed, 6–11 mm long, 2 mm wide, tapering to a smooth beak, teeth 1.2–3 mm long, recurved. Achenes 2.5 mm long, 1.2 mm wide, triangular; stigmas 3. Wet meadows, shores, marshes, & ditches; MT, ND, MN, SD, NE, IA, MO; (Que. to B.C., s to NY, MO, & CA).

13. *Carex athrostachya* Olney. (Group I) Cespitose perennial, rootstock short, black. Culms 0.5–6 dm tall, triangular. Leaves 1.5–4 mm wide; sheaths tight, hyaline and thin ventrally. Heads 1–2 cm long; spikes bisexual, gynecandrous; pistillate scales acute to cuspidate. Perigynia plano-convex, biconvex or triangular, green to straw colored, lightly nerved dorsally, wing-margined to base, 3–4.5 mm long, 1.2–1.5 mm wide, base round tapering, apex tapering to a serrulate beak. Achenes 1.25–1.5 mm long, 0.75–1 mm wide, lenticular; stigmas 2. Margins of sloughs; ND; (Sask. to B.C., s to ND, AZ, & CA).

14. *Carex aurea* Nutt. (Groups IV & VI) Loosely cespitose and rhizomatous perennial, rootstock prolonged, yellow-brown. Culms 0.5–5.5 dm tall, roughened triangular. Leaves 1–4 mm wide; sheaths concave at mouth. Spikes usually unisexual, terminal spikes staminate, 3–10 mm long, rarely with perigynia at apex, lateral spikes pistillate and lower peduncled, separate; bracts of inflorescence much prolonged; pistillate scales acute and cuspidate. Perigynia flattened, yellow-brown, puncticulate, several-ribbed, 2–3 mm long, 1.5 mm wide, base tapered, beak less than the length of body. Achenes 1.5 mm long, 1.2 mm wide, lenticular, blackish; stigmas 2. (n = 26) Meadows, shores of streams, sand dunes, & woodlands; MT, ND, MN, WY, SD, NE, CO, KS; (Newf. to AK, s to PA, TX, & CA).

15. *Carex backii* F. Boott. (Groups V & VII) Cespitose perennial, rootstock short, dark brown. Culms 0.1–3 dm tall, weak, narrowly winged. Leaves 3–6 mm wide; sheaths thin-hyaline ventrally. Spikes solitary, unisexual, androgynous, aggregate; staminate scales obtuse to cuspidate; pistillate scales bractlike, 3–4 cm long. Perigynia suborbicular, green, obscurely many-nerved, 2-edged, 4.5–6 mm long, 2.5 mm wide, base spongy, tapering to a smooth beak 2–3 mm long. Achenes 2.5 mm long, 1 mm wide, triangular to globose, convex sides; stigmas 3. Wooded areas; BH, ND; (Que. to B.C., s to VT, NY, ND, & OR). *C. durifolia* Bailey—Rydberg.

16. *Carex bebbii* (Bailey) Fern. (Group I) Cespitose perennial, rootstock short, black. Culms 2–8 dm tall, sharply triangular. Leaves 2–4.5 mm wide; sheaths tight, white-hyaline ventrally. Heads 1.5–2.5 cm long; spikes bisexual, gynecandrous, aggregate; pistillate scales acute or acuminate. Perigynia plano-convex, green to brown, finely many-nerved dorsally, usually nerveless ventrally, narrow wing-margined to base, 3–3.5 mm long, 1.5–2 mm wide, base rounded, tapering to a serrulate beak 1/3–1/4 the length of the body. Achenes 1 mm long, 0.6 mm wide, lenticular; stigmas 2. (n = 34) Shores, moist woods, marshy areas, & swamps; ND, MN, WY, SD, NE, IA; (Newf. to AK, s to NJ, CO, & OR).

17. *Carex bella* Bailey. (Group IV) Cespitose perennial, rootstock short. Culms 5–9 dm tall, drooping, roughened triangular. Leaves 3–6 mm wide; sheaths thin, reddish-brown tinged ventrally. Upper spike gynecandrous, lower spikes pistillate, aggregate; pistillate scales obtuse to acute, purplish-brown. Pergynia flattened, white to green, 2-ribbed, 3–4 mm long, 1.7–2 mm wide, beak 0.3 mm long, bidentate. Achenes 2–2.5 mm long, 1.7 mm wide, triangular with concave sides; stigmas 3. (n = 20) Wet soil in mountain areas; SD; (SD & WY, s to NM & AZ; Mex.).

18. *Carex bicknellii* Britt. (Group I) Cespitose perennial, rootstock short, black. Culms 3–12 dm tall, sharply triangular. Leaves 2–4.5 mm wide; sheaths white-hyaline ventrally,

papillate dorsally. Heads 2–6 cm long, spikes bisexual, globose to ovoid, gynecandrous; pistillate scales lance-ovate, obtuse to acute. Perigynia flat and thin, stramineous, strongly 12-nerved dorsally, strongly to few-nerved ventrally, broad, thin, wing-margined to base, 5.5–7 mm long, 2.7–4.8 mm wide, base round-truncate, abruptly contracted to a serrulate beak 1/3–1/4 the length of the body. Achenes 1.7 mm long, 1.5 mm wide, lenticular; stigmas 2. Priairies, woods, prairie ditches, & occasional shores; ND, MN, SD, NE, IA, KS, MO, OK; (ME to Sask., s to NJ, AR, TX).

19. Carex blanda Dew. (Group V) Cespitose perennial, rootstock short, black. Culms 1–6 dm tall, acutely triangular, slightly winged. Leaves 3–15 mm wide; sheaths thin and white ventrally, septate-nodulose dorsally. Spikes usually unisexual, terminal spike staminate or gynecandrous, 5–20 mm long, lateral spikes pistillate and lower peduncled; pistillate scales awned, green-white. Perigynia obscurely triangular, yellow-green, strongly many-nerved, 3–4.5 mm long, 1.7 mm wide, base spongy, beak 0.5 mm long, recurved or bent. Achenes 2.5 mm long, 1.5 mm wide, triangular, style bent apiculate; stigmas 3. Banks of streams & rivers, mostly in woods, occasionally ditches, & rarely in meadows; ND, MN, SD, NE, IA, KS, MO, OK; (Que. to ND, s to GA & TX). *C. laxiflora* Lam. var. *blanda* (Dew.) F. Boott — Gleason & Cronquist.

20. Carex brevior (Dew.) Mack. ex Lunell. (Group I) Cespitose perennial, rootstock short, black. Culms 3–10 dm tall, sharply triangular. Leaves 1–4 mm wide, sheaths tight, white-hyaline ventrally. Heads 2–4.5 cm long; spikes bisexual, ovoid-oblong, gynecandrous, aggregate to moniliform; pistillate scales ovate, obtuse or acute. Perigynia plano-convex, green to greenish-white, strongly several to many-nerved dorsally, usually nerveless ventrally, wing-margined to base, 3–4.5 mm long, 2–3 mm wide, base truncate-rounded, abruptly contracted to a serrulate beak, 1/3 the length of body. Achenes 1.7–2 mm long, 1.5–1.7 mm wide, lenticular; stigmas 2. Meadows, marshes, ditches, & woodlands; GP; (N.B. to B.C., s to FL, TX, AZ, OR).

21. Carex brittoniana Bailey. (Group I) Cespitose perennial, rootstock strongly elongate, thick, black. Culms 3–8 dm tall, sharply triangular. Leaves 3 mm wide; sheaths white-hyaline ventrally. Heads 3–6 cm long, spikes gynecandrous, subglobose; scales narrowly ovate, acute or obtuse. Perigynia very flat, membranaceous, 7.5–8.5 mm long, 5–6 mm wide, the body orbicular, nerveless or nearly so on both faces, base broadly truncate, abruptly narrowed into a serrulate beak 2.5 mm long, 1/3 the length of the body. Achenes lenticular, broadly oblong, 2.5 mm long, 2 mm wide; stigmas 2. Prairies; OK; (OK, TX, & FL).

22. Carex brunnescens (Pers.) Poir. (Group II) Densely cespitose perennial, rootstock stout, blackish or brownish. Culms slender, 0.7–7 dm tall, sharply triangular. Leaves 1.2–5 mm wide; sheaths tight, hyaline and thin ventrally. Heads straight and stiff, 1.5–5 cm long; spikes gynecandrous; pistillate scales ovate, obtuse or acute. Perigynia plano-convex, ovate-elliptic, 2–2.5 mm long, 1–1.5 mm wide, lightly nerved both sides, tapering at apex into a short serrulate beak, 0.5 mm long. Achenes lenticular, ovate-orbicular, 1.5 mm long, 1 mm wide; stigmas 2. (n = 28) Wet wooded areas; MN, SD, OK; (Lab. to AK, s to GA, OK, & CA).

23. Carex bushii Mack. (Groups IV & V) Cespitose perennial, rootstock short. Culms 3–9 dm tall, sparsely pubescent, triangular with concave sides. Leaves 2–4 mm wide, hairy to glabrous; sheaths long, tight, hairy ventrally. Spikes usually 2 to 3, terminal spike gynecandrous, lateral spikes pistillate; pistillate scales lanceolate to cuspidate or awned, often slightly pilose, exceeding perigynia. Perigynia obscurely triangular, turgid, olive-green, strongly ribbed, 3–4 mm long, 1.5–2 mm wide, beak short. Achenes 2.5 mm long, 1.8 mm wide, triangular with concave sides, style bent-apiculate; stigmas 3. Most common in ungrazed prairies, occasionally in ditches & margins of wooded areas; NE, KS, MO, OK; (RI to NE, s to VA, MS, & TX). *C. caroliniana* Schwein. var. *cuspidata* (Dew.) Shinners—Waterfall.

24. Carex buxbaumii Wahl. (Group IV) Rhizomatous perennial, rootstock prolonged. Culms 4–10 dm tall, rough triangular. Leaves 2–3 mm wide; lower sheaths filamentose, upper thin, purple dotted ventrally. Terminal spike gynecandrous; lateral spikes pistillate, aggregate; pistillate scales acute and acuminate, purplish-black. Perigynia flattened to distended, green, finely many-nerved, 2.7–3.7 mm long, 1.5–2 mm wide, beak 0.2 mm long. Achenes 1.7 mm long, 1.5 mm wide, triangular; stigmas 3. (2n = ca 100) Wet meadows; ND, MN, NE, IA, KS, MO; (Newf. to AK, s to GA, OK, & CA).

25. Carex capillaris L. var. *elongata* Olney. (Group V) Cespitose perennial, rootstock short. Culms 1.5–4 dm tall, triangular. Leaves 0.7–2.5 mm wide, basal; sheaths tight, truncate. Spikes unisexual, terminal spikes staminate, 4–8 mm long, pistillate spikes lateral, loosely flowered or drooping-peduncled; pistillate scales obtuse or acute. Perigynia obtusely triangular, green-brown, 2-ribbed, strongly ciliate; 2.4–3.5 mm long, 1.7 mm wide, tapering to a serrulate beak 1 mm long. Achenes 1.5 mm long, 0.7 mm wide, triangular with concave sides; stigmas 3. (2n = 54) Wet meadows; ND, MN, SD; (Newf. to Que., NY w to MN, ND, SD).

26. Carex caroliniana Schwein. (Group IV & V) Cespitose perennial, rootstock short. Culms 3–8 dm tall, triangular with concave sides. Leaves 2–4 mm wide, pubescent at base; sheaths long, tight, glabrous or nearly so, red tinged or spotted ventrally, glabrous or hairy dorsally. Spikes usually 3, terminal spike gynecandrous, lateral spikes pistillate; pistillate scales obtuse to acuminate or cuspidate. Perigynia obscurely triangular, turgid, brown to green, strongly ribbed, 2.2–2.8 mm long, 1.5 mm wide, short beaked, entire. Achenes 1.5–2 mm long, 1.2 mm wide, triangular with concave sides, style bent-apiculate; stigmas 3. Ditches, shores of streams & lakes; KS, MO; (NJ to KS, s to NC & TX).

27. Carex cephaloidea (Dew.) Dew. (Group III) Cespitose perennial, rootstocks short creeping, brownish-black. Culms 0.3–8 dm tall, sharply triangular; sheaths loose, friable ventrally. Heads densely aggregated, 1.5–4 cm long; spikes androgynous; pistillate scales ovate, obtuse to acute. Perigynia dark green, 3–4.5 mm long, 1.5–2.5 mm wide, gradually tapering to a serrulate beak, obscurely few-nerved dorsally. Achenes 1.8 mm long, lenticular, suborbicular stigmas 2. Dry open woodlands; IA; (Ont. s to MD, IL, OH, & NJ). *C. sparganioides* Muhl. var. *cephaloidea* (Dew.) Cary—Gleason & Cronquist.

28. Carex cephalophora Willd. (Group III) Cespitose perennial, rootstock short, black. Culms 3–6 dm tall, triangular. Leaves 2–5 mm wide, sheaths tight; heads dense, 1–2 cm long. Spikes androgynous; pistillate scales ovate. Perigynia plano-convex, 2.5–3.5 mm long, beak 1/3 as long as body, round tapering at base with serrulate margins. Achenes 1–1.5 mm long; stigmas 2. Dry to moist woodland, occasionally open prairies.

Two varieties may be distinguished:

28a. var. *cephalophora*. Body of the scales exceeded by the perigynium. NE, IA, KS, MO; (Que. to Man., s to FL & TX).

28b. var. *mesochorea* (Mack.) Gl. Body of the scales about as long as the broadly ovate perigynium. MO; (MA & NJ, w to TN & MO). *C. mesochorea* Mack.—Rydberg.

29. Carex comosa F. Boott. (Group VIII) Cespitose perennial, rootstock short, stout. Culms 5–15 dm tall, winged triangular. Leaves 6–16 mm wide; sheaths thin and hyaline ventrally. Spikes unisexual, terminal spike staminate, 3–7 cm long, lateral spikes pistillate, short peduncled, lower spreading to drooping, pistillate scales rough awned. Perigynia planoconvex or triangular, yellow-green, shining, strongly 12- to 17-nerved, reflexed, 5.7–7.7 mm long, 1.5 mm wide, tapering to a long smooth beak, teeth divergent, 1.3–2 mm long. Achenes 1.7 mm long, 0.7 mm wide, triangular; stigmas 3. Wet meadows & swamps; MN, SD, NE, IA, MO; (Que. to Ont., s to FL & TX; also ID & west coast).

30. *Carex complanata* Torr. & Hook. (Groups IV & V) Cespitose perennial in small clumps, rootstock short. Culms 2–6 dm tall, triangular with slightly concave sides, glabrous. Leaves 1.5–4 mm wide, sheaths tight, slightly hairy ventrally, concave and short-pilose at the mouth. Spikes usually 3, the terminal spike gynecandrous, lateral spikes pistillate; pistillate scales ovate-triangular. Perigynia flattened-obovoid, 2–3.5 mm long, 1.5 mm wide, obscurely nerved to nerveless ventrally, strongly nerved dorsally, apex beakless. Achenes obovoid, sharply triangular with concave sides, 1.7 mm long, 1.2 mm wide; short thick style, stigmas 3. Dry upland, open wooded slopes; MO; (ME to Alta., MI, & IA, s to AR & TX, also NJ & FL).

31. *Carex concinna* R. Br. (Group V) Cespitose perennial, rootstocks occasionally prolonged, brown-black. Culms 0.3–2 dm tall, triangular. Leaves 2–3 mm wide; sheaths tight, hyaline ventrally. Terminal spike staminate and sessile, 3–6 mm long; lateral spikes pistillate, aggregate; pistillate scales obtuse. Perigynia obovoid-oblong and triangular, red-brown, hirsute, 2-ribbed, several-nerved; 2.5–3 mm long, 1.2 mm wide, beak short. Achenes 1.5 mm long, 1 mm wide, triangular with concave sides; stigmas 3. Dry soil; SD; (Newf. to AK, s to MI, CO, & OR).

32. *Carex conjuncta* F. Boott. (Group III) Cespitose perennial, rootstock short, black. Culms 4–12 dm tall, sharply and narrowly winged-triangular, spongiose. Leaves 3–10 mm wide; sheaths fragile, cross-rugulose and red dotted ventrally, slightly septate-nodulose dorsally. Heads 1.5–7.5 cm long; spikes bisexual, androgynous, lower separate; pistillate scales acuminate to cuspidate. Perigynia plano-convex, greenish to yellow green, 4- to 5-nerved dorsally, several-nerved at base ventrally, 3.5–5 mm long, 1.6–2 mm wide; base roundtruncate, tapering to a serrulate beak about 1/2 the length of body. Achenes 1.5 mm long, lenticular; stigmas 2. Mostly open wooded areas, occasionally wet ditches, & marshy areas; NE, IA, KS, MO; (NY to SD, s to VA, AL, & KS; also Ont.).

33. *Carex convoluta* Mack. (Group III) Cespitose perennial, rootstock short. Culm 3–6 dm tall, sharply triangular. Leaves 1.5–3 mm wide; sheaths tight, septate-nodulose, not cross-rugulose. Heads 3–5 cm long, spikes androgynous, separate below; pistillate scales obtuse, rounded. Perigynia plano-convex, margins slightly raised ventrally, spreading, 3–4 mm long, 1.2–1.7 mm wide, nerveless both sides, abruptly contracted into a serrulate beak, white hyaline at orifice. Achenes 2 mm long, 1.6 mm wide, lenticular; stigmas 2 and twisted. Dry to moist woodlands; MN, SD, NE, KS, MO; (N.S. to Man., s to SC, AL, & KS).

34. *Carex crawei* Dew. (Group V) Rhizomatous perennial, rootstock prolonged. Culms 0.2–4 dm tall, obscurely triangular. Leaves 1–4 mm wide; sheaths tight, hyaline ventrally. Spikes unisexual, terminal spike staminate, 1–3 cm long, lateral spikes pistillate, separated, lower spikes basal; pistillate scales acuminate-cuspidate, red-brown. Perigynia ascending, suborbicular, green to brown, many-nerved, 3–3.5 mm long, 1.2–2 mm wide, beak short, entire. Achenes 1.7–2 mm long, 1.2 mm wide, triangular with concave sides; stigmas 3. Wet ditches, meadows, & prairie swales; ND, MN, SD, NE, IA, KS, OK; (N.S. to B.C., s to NJ, AL, OK, UT, & WA).

35. *Carex crinita* Lam. (Group VI) Cespitose perennial, rootstock short, stout. Culms 3–16 dm tall, acutely triangular. Leaves 4–12 mm wide; sheaths smooth and glabrous ventrally.

Upper 2 spikes staminate at base, 3.5–10 cm long, separate or aggregate, loosely spreading and drooping, subtended by long bracts; pistillate scales long-serrulate, awned. Perigynia suborbicular, inflated and crumpled, green or straw colored, nerveless, 2–4 mm long, 1–3 mm wide, beak 0.2 mm long, entire. Achenes lenticular or triangular and constricted in middle; stigmas 2.

Two varieties are present in GP area:

35a. var. *brevicrinis* Fern. Pistillate spikes often staminate at apex; awns of lower scales about equaling to 2× the length of perigynia; upper scales about equaling or slightly longer than perigynia. Perigynia strongly inflated, 3–4 mm long and 2–3 mm thick. Swampy meadows, river bottoms, & sloughs; MO, NE; (New England to NC, w to KY, MO, & TX).

35b. var. *crinita*. Pistillate spikes rarely, if at all, staminate at apex; awns of lower scales 2–4× as long as perigynia; upper scales longer than perigynia. Perigynia 2–3(3.5) mm long and 1–2 mm thick. Swampy meadows, river bottoms, & sloughs; MO; (Newf. to Man., s to FL & TX).

36. ***Carex cristatella*** Britt. (Group I) Cespitose perennial, rootstock short, black. Culms 3–10 dm tall, sharply triangular with concave sides. Leaves 3–7 mm wide; sheaths loose, green-striate, hyaline ventrally. Heads 2–3 cm long, spikes bisexual, globose, gynecandrous, aggregate; pistillate scales lanceolate, acute-acuminate. Perigynia plano-convex, slightly nerved both sides, strongly winged to below the middle and narrower to base, 3–4 mm long, 1.2–1.5 mm wide, base round tapering to a serrulate beak 1/3–equal the length of the body, with notched or wrinkled margin at base. Achenes 1.5 mm long, 1.5 mm wide, lenticular; stigmas 2. Meadows, woods, swamps, & sand dunes; ND, MN, SD, NE, IA, KS, MO; (Que. to Alta., s to VA & KS).

37. ***Carex crus-corvi*** Shuttlew. ex O. Ktze. (Group III) Cespitose perennial, rootstock short, black. Culms 5–10 dm tall; sharply triangular with concave sides. Leaves 5–12 mm wide; sheaths thin, purplish dotted, not cross-rugulose ventrally, septate-nodulose dorsally. Heads 7–25 cm long; spikes bisexual, androgynous, lower often separate; pistillate scales acuminate to aristate. Perigynia plano-convex, yellowish to brown, strongly many-nerved dorsally, obscurely nerved ventrally, 6–9 mm long, 1.5–2.5 mm wide, base spongy, truncate-cordate, tapering to a slender serrulate beak 2–3× the length of body. Achenes 1.7 mm long, 1 mm wide, lenticular; stigmas 2. Wet areas, ponds, swamps, ditches, & moist open woods; NE, IA, KS, MO, OK; (MD to MN, s to FL & TX).

38. ***Carex davisii*** Schwein. & Torr. (Group IV) Cespitose perennial, rootstock short, dark. Culms 3–9 dm tall, glabrous or pubescent, triangular. Leaves 3–8 mm wide; soft-pubescent; sheaths long, hairy, yellow-brown ventrally. Terminal spike gynecandrous, lateral spikes pistillate and peduncled; pistillate scales awned and cuspidate. Perigynia obscurely triangular, green to yellow-brown, red-brown dotted, strongly ribbed, 4.5–6 mm long, 2–2.5 mm wide, beak bidentate. Achenes 2.5 mm long, 2 mm wide, triangular with concave sides; stigmas 3. Chiefly wooded areas, occasionally ditches & low meadows; ND, NE, IA, KS, MO; (Que. to MN, s to DE, GA, & TX).

39. ***Carex deweyana*** Schwein. (Group I) Cespitose perennial, rootstock usually short. Culms 3–12 dm tall, sharply triangular with flat sides. Leaves 2–5 mm wide; sheaths tight, thin, hyaline ventrally. Heads 2–5 cm long, 1.5–2 mm wide, lateral spikes pistillate, terminal gynecandrous; pistillate scales obtuse to awned. Perigynia plano-convex, light green, obscurely many-nerved at base dorsally, nerveless ventrally, 4.5–5.5 mm long, 1.5–2 mm wide, base rounded-spongy, tapering to a serrulate beak 1/2–1/3 the length of the body. Achenes 2–2.5 mm long, 1.3–1.7 mm wide, lenticular; stigmas 2. (n = 27) Wooded areas; MT, ND, MN, SD; (Newf. to AK, s to PA, IA, CO, AZ, & CA).

40. Carex diandra Schrank. (Group II) Cespitose perennial, rootstock short, black. Culms 4–8 dm tall, sharply triangular. Leaves 1–3 mm wide, sheaths pale-striate, strongly red dotted, not copper tinged dorsally. Heads 2–4 cm long; spikes bisexual, androgynous, aggregate; pistillate scales acute, brown tinged. Perigynia unequally plano-convex, shiny brown, few-nerved dorsally, slightly nerved ventrally, 2–2.8 mm long, 1–1.3 mm wide; base round-truncate tapering or contracted to a serrulate beak shorter than body. Achenes 1 mm long, lenticular; stigmas 2. (n = 30). Swamps, meadows, & shores; ND, NE; (Newf. to AK, s to NJ, MO, & CA).

41. Carex disperma Dew. (Group II) Loosely cespitose perennial, rootstock prolonged, light brown. Culms 0.5–6 dm tall, slender, weak, triangular. Leaves 0.7–1.5 mm wide; sheaths tight, hyaline ventrally. Heads 1.5–2.5 cm long; spikes bisexual, androgynous; lower separate; pistillate scales acuminate to mucronate. Perigynia unequally biconvex, nearly terete, yellow-green, finely many-nerved both sides, white puncticulate, 2–2.8 mm long, base spongy and rounded, abruptly contracted to a minute (0.25 mm or less), smooth beak. Achenes 1.7 mm long; 1 mm wide, lenticular, deep brown to black; stigmas 2. (n = 35) Marshes, swamps, & ravines; ND, MN, SD; (Newf. to AK, s to NJ, IN, SD, NM, & CA).

42. Carex douglasii F. Boott. (Group II) Rhizomatous perennial, rootstock long and slender, brown. Culms 1–3 dm tall, obtusely triangular. Leaves 1–2 mm wide, involute; sheaths tight. Heads 1.5–3.2 cm long, usually dioecious, spikes unisexual, androgynous, separated; pistillate scales acute or acuminate. Perigynia plano-convex, lance-ovate, straw colored to light brown, 3–4 mm long, 1.4–2 mm wide, base round tapering, contracted to a serrulate beak about 1.7 mm long. Achenes 1.4–1.6 mm long, 1–1.5 mm wide, lenticular; stigmas 2. Dry to wet prairies, ditches, & sandy areas; MT, ND, WY, SD, NE, CO; (Man. to B.C., s to MO, NM, & CA).

43. Carex eburnea F. Boott. (Group V) Cespitose perennial, rootstock prolonged, brown. Culms 1–3.5 dm tall, obtusely triangular. Leaves 0.5 mm wide; sheaths tight, white-hyaline and yellow-brown tinged ventrally. Spikes unisexual, terminal spike staminate, 4–8 mm long, pistillate spikes lateral and peduncled; pistillate scales obtuse or acute. Perigynia triangular, green to brown, finely nerved, 2-ribbed, puncticulate, 2 mm long, 1 mm wide, beak short. Achenes 1.8 mm long, 0.7 mm wide, triangular with concave sides; stigmas 3. Primarily wooded areas, but occasionally in adjacent prairies; ND, MN, SD, NE, IA, MO; (Newf. to AK, s to VA, AL, TX, & MT).

44. Carex eleocharis Bailey. (Group II) Rhizomatous perennial, rootstock long, slender, brown. Culms 0.2–2 dm tall, obtusely triangular. Leaves 1–2 mm wide, involute; sheaths tight, thin ventrally. Heads 1–2 cm long; spikes bisexual, androgynous, aggregated, pistillate scales acute to obtuse and cuspidate. Perigynia plano-convex, straw colored to black, striate dorsally, 2.5–3 mm long, 1.5–1.7 mm wide, base round tapering, contracted to a serrulate beak shorter than the body. Achenes 1.7 mm long, 1.5 mm wide, lenticular; stigmas 2. Dry prairies, rocky hilltops, & sandy areas; MT, ND, MN, WY, SD, NE, IA, CO, KS; (Man. to AK, s to IA, KS, NM, & AZ). *C. stenophylla* Wahl. var. *enervis* (C.A. Mey.) Kükenth.—Fernald.

Immature or depauperate plants of this species may be confused with *C. praegracilis*.

45. Carex emoryi Dew. (Group VI) Rhizomatous perennial, rootstock prolonged, brown scaly. Culms 4–10 dm tall, smooth to rough triangular. Leaves 2–5 mm wide; sheaths white-hyaline ventrally, white to yellow-tinged dorsally. Spikes unisexual, terminal spike staminate 2–4.5 cm long, lower pistillate or androgynous, pistillate scales blunt to acute. Perigynia biconvex, ovate, broadest at base, green to straw colored, red dotted, few-nerved both sides,

1.6-3 mm long, 1.5-1.7 mm wide, beak 0.2 mm long. Achenes 1.5 mm long, 1 mm wide, lenticular; stigmas 2. Moist meadows, ditches, & shores; ND, MN, SD, NE, IA, CO, KS, MO, OK; (NY to ND, s to FL, TX & NM; also Man.). *C. stricta* Lam. var. *elongata* (Boeck.) Gl.—Gleason & Cronquist.

46. Carex festucacea Schkuhr. (Group I) Cespitose perennial, rootstock short, black. Culms 5-10 dm tall, sharply triangular. Leaves 2-5 mm wide; sheaths tight, hyaline ventrally, slightly septate-nodulose dorsally. Heads 2.5-6 cm long; spikes bisexual, gynecandrous, moniliform; pistillate scales ovate, acute. Perigynia plano-convex, orbicular green to straw colored, strongly 5-nerved dorsally, nerved on both margins, similar but less nerved ventrally, wing-margined to base, 2.7-4 mm long, 2 mm wide, base rounded and contracted to a serrulate beak 1/2-equal the length of the body. Achenes 1.5 mm long, 1 mm wide, lenticular; stigmas 2. Wooded bluffs; ND, NE, IA, MO; (Newf. to Ont., s to FL & TX).

47. Carex filifolia Nutt. (Groups V & VII) Cespitose perennial in dense tussocks, rootstock short, black. Culms 0.5-3 dm tall, slightly triangular. Leaves 0.25 mm wide; sheaths truncate ventrally. Spikes solitary, 3-5 mm long, bisexual, androgynous, aggregate; pistillate scales obtuse. Perigynia obtusely triangular, white-striate, puberulent above, slightly 2-ribbed, 3-3.5 mm long, 2 mm wide, base rounded, tapering to a short beak 0.2-0.4 mm long. Achenes 2.2-3 mm long, 1.4 mm wide, triangular; stigmas 3. Upland prairies; MT, ND, MN, WY, SD, NE, CO, KS; (Man. to Yukon, s to MN, TX, & CA).

48. Carex fissa Mack. (Group III) Cespitose perennial with a thick, black, short prolonged rootstock. Culms 2.5-7.5 dm tall, bluntly triangular. Leaves 3-5 mm wide, sheaths not septate-nodulose, strongly cross-rugulose ventrally, red dotted and strongly prolonged at mouth. Heads 2.5-4 cm long, spikes androgynous, pistillate scales ovate, acute or cuspidate. Perigynia plano-convex, 3.0-4.5 mm long, 2 mm wide; round-truncate at base, abruptly contracted to a serrulate beak, usually nerveless dorsally, few-nerved ventrally. Achenes 2 mm long, 1.7 mm wide, lenticular; stigmas 2. Sandy roadsides & disturbed areas; KS, OK; (se KS & e OK).

49. Carex foenea Willd. (Group I) Perennial with rootstocks long creeping, slender, covered with brown fibrillose scales. Culms 1-9 dm tall, sharply triangular above. Leaves stiff, 1-3 mm wide; ventral sheaths hyaline-truncate. Heads 2-3.5 cm long, gynecandrous, pistillate scales obtuse to acute. Perigynia plano-convex, 4.5-6 mm long, 1.7-2 mm wide, many-nerved on both sides, winged above, abruptly contracted into a beak nearly or as long as the body. Achenes 2 mm long, lenticular; stigmas 2. (n=35) Dry, open soil of wooded areas; ND, MN, SD, NE, CO; (N.S. & Que., s to VA, w to ND & CO). *C. siccata* Dew.—Rydberg.

50. Carex frankii Kunth. (Group IV) Cespitose perennial, rootstock short. Culms 2-8 dm tall, obtusely triangular. Leaves 3-10 mm wide; sheaths tight, yellow-brown tinged ventrally. Spikes unisexual, terminal spike staminate, 2-3 mm long, lateral spikes pistillate; pistillate scales aristiform. Perigynia squarrose, olive-green, strongly-ribbed, 4-5 mm long, 2-2.5 mm wide, beak 1.3-2.3 mm long, teeth 0.5 mm long. Achenes 1.5-2 mm long, 1 mm wide, triangular, style persistent & straight; stigmas 3. Swamps, low prairies, & ditches; NE, KS, MO, OK; (NY to NE, s to GA & TX).

51. Carex garberi Fern. (Group IV & VI) Rhizomatous perennial, rootstock prolonged. Culms 2-3 dm tall, erect and firm, triangular. Leaves 2-3 mm wide. Terminal spike gynecandrous, 0.8-3 cm long, lateral spikes pistillate, lower separate to aggregate; bract of inflorescence longer than inflorescence; pistillate scales round, ovate or blunt, brown to purple. Perigynia elliptical or triangular, ribbed and granular, puberulent, 2-3 mm long, 1.2 mm wide. Achenes 1.5 mm long, 1.2 mm wide, lenticular; stigmas 2. (n = 25) Swamps & margins of ponds; ND; (Que. to B.C., s to NY, OH, ND, & NV).

52. *Carex gracilescens* Steud. (Group V) Densely cespitose perennial, rootstocks very short. Culms 1.5-6 dm tall, sharply triangular. Leaves 3-8 mm wide; sheaths greenish-white dorsally, truncate at mouth. Staminate spike normally strongly peduncled, lateral spikes pistillate, the lower on long exserted, slender peduncles; pistillate scales ovate or obovate, cuspidate or awned. Perigynia obovoid, obtusely triangular, 2.5-3.5 mm long, 1.5-1.8 mm wide, many-nerved, beak strongly curved. Achenes obovoid, 2-2.5 mm long, 1.5 mm wide, triangular, abruptly bent apiculate, stigmas 3. Dry woods, base of slopes, valleys, ravines; MO; (Que. to Ont., s to VA & TX).

53. *Carex granularis* Muhl. ex Willd. (Group V) Cespitose perennial, rootstock short. Culms 1.5-9 dm tall, obtusely triangular. Leaves 1-12 mm wide; sheaths long, red dotted ventrally, septate-nodulose dorsally. Long pistillate spikes lateral, scattered, all but upper peduncled; pistillate scales acuminate to awned, brown. Perigynia suborbicular, olive-green to brown; slightly inflated, many-nerved, 2-4 mm long, 1.5-2.5 mm wide, beak minute, straight or bent. Achenes 1.7 mm long, 1.2 mm wide, triangular; stigmas 3.

Two varieties may be recognized:

53a. var. *granularis*. Perigynia 2.5-4 mm long, 1.5-2.5 mm wide, broadly ovoid to subglobose; spreading at maturity. Ditches, swamps, & river-bottom woods; ND, MN, SD, NE, IA, KS, MO, OK; (N.B. to Sask., s to FL & TX).

53b. var. *haleana* (Olney) Porter. Perigynia 2-2.8 mm long, 1-1.5 mm wide, narrowly ovoid or oblong, not spreading at maturity. Bogs, swamps, & ditches; ND, NE, KS; (Que. to Sask., s to VA, IA, & KS). *C. shriveri* Boott—Rydberg.

54. *Carex gravida* Bailey. (Group III) Cespitose perennial, rootstock short, black. Culms 3-6 dm tall, sharply triangular. Leaves 4-8 mm wide; sheaths green-and-white mottled and septate-nodulose, usually not cross-rugulose. Heads densely aggregated, 1-3 mm long; spikes androgynous; pistillate scales ovate, acuminate to cuspidate. Perigynia 3-5 mm long, 2-3.5 mm wide, gradually tapering or abruptly narrowed into a serrulate beak; obscurely to few-nerved dorsally. Achenes 2 mm long; style base much enlarged, stigmas 2.

Two varieties are present in GP area:

54a. var. *gravida*. Perigynia gradually tapering into the beak, nerveless or slightly so on dorsal surface, 1/2 as wide as long. Swales, shores, prairies & plains, open woods & roadside ditches; MT, ND, MN, WY, SD, NE, IA, KS, MO, OK; (PA to Sask., s to VA, MO, TX, & NM).

54b. var. *lunelliana* (Mack.) Herm. Perigynia abruptly narrowed into the beak, strongly few-nerved on dorsal surface, 2/3-4/5 as wide as long. Roadside ditches, shores, prairies, & swales; SD, NE, IA, CO, KS, MO, OK; (IN to NE & CO, s to MO, TX, & NM). *C. lunelliana* Mack.—Rydberg.

55. *Carex grayi* Carey. (Group VIII) Cespitose perennial, rootstock short. Culms 3-8 dm tall, rough triangular. Leaves 6-12 mm wide; sheaths loose, white-hyaline ventrally. Spikes unisexual, terminal spike staminate, 2 mm long, lateral 1 or 2 spikes pistillate, globose; pistillate scales acuminate. Perigynia suborbicular, spreading, strongly ribbed, glabrous to hispid, 12-18 mm long, 6-7 mm wide, tapering to a smooth beak 1.5-2.5 mm long. Achenes 4-6 mm long, 3-4 wide, triangular, concave sides below, style persistent and usually straight; stigmas 3. Rich wooded areas; IA, KS, MO; (Newf. to Ont., s to FL & TX). *C. asa-grayi* Bailey—Rydberg; *C. grayii* Carey var. *hispidula* A. Gray—Fernald.

56. ***Carex hallii*** Olney. (Group IV) Rhizomatous perennial, rootstock prolonged, scaly. Culms 1–6 dm tall, obtusely triangular. Leaves 2–4 mm wide; sheaths thin and white-hyaline ventrally, concave at mouth. Spikes unisexual, terminal spikes usually staminate, 1.5–2.5 cm long, lateral spikes pistillate, pistillate scales obtuse to mucronate. Perigynia planoconvex to obtusely triangular, green or white, 2-ribbed, 2–3 mm long, 2 mm wide, beak 0.3 mm long, bidentulate. Achenes 1.7 mm long, 2 mm wide, triangular; stigmas 3. Low prairies, sandy sloughs, & low montane areas; ND, MN, SD, WY, NE; (Man. to Alta., s to MN, NE, CO, & ID).

57. ***Carex haydenii*** Dew. (Group VI) Loosely cespitose perennial, rootstock short. Culms 5–10 dm tall, roughened triangular. Leaves 2–5 mm wide; sheaths white-hyaline to yellow tinged ventrally, purple-red dotted dorsally. Spikes unisexual, terminal, 1–3 spikes staminate, lateral spikes pistillate; pistillate scales long-acuminate. Perigynia biconvex, inflated at apex, brown, resinous dotted, 1 or 2 faintly nerved both sides, 2-edged, 2-ribbed, 2–2.5 mm long, 1.2–1.7 mm wide, beak minute. Achenes 1 mm long, 1.6 mm wide, lenticular; stigmas 2. Ditches, marshes, & wet meadows; ND, SD, NE, IA, MO; (N.B. to Alta., s to VA, MO, CO; also OR, CA, & AZ).

58. ***Carex heliophila*** Mack. (Group VIII) Strongly rhizomatous perennial, rootstock short or prolonged. Culms 1–3 dm tall. Leaves 1–2.5 mm wide; sheaths breaking and filamentose. Terminal spike staminate, occasionally gynecandrous, lateral spikes pistillate, aggregate; pistillate scales obtuse to acute-cuspidate. Perigynia ellipsoid, green, puberulent, 2-keeled, 3–4 mm long, 1.5–2.2 mm wide, base spongy, contracted to a serrulate beak 0.5–1 mm long, 1.75–2 mm wide, triangular with convex sides; stigmas 3. Prairies, pastures, thickets, & occasionally open woods; MT, ND, MN, WY, SD, NE, IA, CO, KS, MO; (Ont. to Alta., s to OH, MO, KS, & NM; also NY). *C. pensylvanica* Lam. var. *digyna* Boeck.—Fernald.

59. ***Carex hirsutella*** Mack. (Groups IV & V) Cespitose perennial, rootstock short. Culms 2–9 dm tall, slender, stiff, acutely triangular with concave sides, pubescent. Leaves 1.5–4 mm wide, pubescent; sheaths long, tight, hairy ventrally. Terminal spike gynecandrous, lateral spikes pistillate; pistillate scales obtuse to acuminate or cuspidate. Perigynia obtusely triangular, strongly nerved dorsally, lightly nerved ventrally, 2.2–3 mm long, 1.5 mm wide, serrulate beak, apex of beak obtuse or minute. Achenes 1.7 mm long, 1.2 mm wide, triangular with concave sides, style bent-apiculate; stigmas 3. Wooded areas; KS, MO, OK; (ME to Ont., s to GA & TX). *C. complanata* Torr. & Hook. var. *hirsuta* (Bailey) Gl.—Gleason & Cronquist.

60. ***Carex hirtifolia*** Mack. (Groups V & VIII) Loosely cespitose perennial, rootstock short. Culms 3–6 dm tall, pubescent, triangular. Leaves 3–7 mm wide, soft hairy; sheaths tight, thin, brown tinged ventrally. Spikes unisexual, terminal spike staminate, 8–20 mm long; lateral spikes pistillate, aggregate; pistillate scales ciliate. Perigynia triangular, loosely pubescent, nerveless, 2.6–3.5 mm long, 1.2 mm wide, beak 0.8–1.3 mm long. Achenes 2 mm long, 1.1 mm wide, triangular; stigmas 3. Dry woods; KS, MO; (N.B. to Ont., s to MD, KY, MO, & KS). *C. hirtiflora* Mack.—Rydberg.

61. ***Carex hitchcockiana*** Dew. (Group V) Cespitose perennial, rootstock short. Culms 1.5–7 dm tall, rough triangular. Leaves 3–7 mm wide; sheaths tight, long, brown tinged, hispidulose ventrally. Spikes unisexual, terminal spike staminate, 1–2.5 mm long, lateral spikes pistillate with lowest separate; pistillate scales rough-awned, keeled. Perigynia obtusely triangular, yellow to gray-green, finely impressed, many undulate nerves, 4.3–5.9 mm long, 2.5 mm wide, base spongy, tapering to base and contracted to apex, beak 1 mm long and entire. Achenes 3 mm long, 2 mm wide, triangular, style bent-

apiculate; stigmas 3. Wooded bluffs; NE, IA, KS, MO; (VT to Que., s to VA, AR, & KS).

62. **Carex hoodii** F. Boott. (Group III) Cespitose perennial, rootstocks short, black. Culms 2.5–8 dm tall, sharply triangular. Leaves 1.5–3.5 mm wide; sheaths tight, not cross-rugulose or septate-nodulose. Heads 1–2 cm long, spikes androgynous, closely aggregated; pistillate scales ovate-triangular, chestnut-brown. Perigynia flat or concave ventrally, rounded and usually nerved dorsally, 3.5–5 mm long, 1.7–2 mm wide, tapering to a serrulate, bidentate beak. Achenes 2 mm long, 1.7 mm wide, lenticular; stigmas 2. (n = 29, 30) Meadows & slopes; MT, WY, SD; (Sask. to B.C., s to w SD, CO, & CA).

63. **Carex hookerana** Dew. (Group III) Cespitose perennial, short black rootstocks. Culms 1.5–4.5 dm tall. Leaves 1–2.5 mm wide, sheaths tight, not cross-rugulose or septate-nodulose. Heads slender, 2–5 cm long; lower spikes separate, androgynous; pistillate scales ovate-triangular, long awned, concealing perigynia. Perigynia plano-convex, 2.7–3.5 mm long, about 1.2 mm wide, nerveless ventrally, little nerved dorsally, abruptly narrowed to a serrulate beak. Achenes 1.7 mm long and slightly narrower, lenticular; stigmas 2. Swamps, moist open woods, & prairies; ND; (Ont. to Alta., s to ND).

64. **Carex hyalinolepis** Steud. (Group VIII) Rhizomatous perennial, rootstock short, scaly. Culms 5–10 dm tall, smooth to rough triangular. Leaves 4–15 mm wide; basal sheaths sometimes filamentose, light brown. Spikes unisexual, upper 2–4 spikes staminate, 1–4 cm long, lower spikes pistillate; pistillate scales acute to aristate. Perigynia flattened to suborbicular, green, impressed-nerved, 6 mm long, 2.5 mm wide, beak 1 mm long, teeth 0.5 mm long. Achenes 2–2.5 mm long, 1.2–1.5 mm wide, triangular with concave sides and persistent style; stigmas 3. Swamps, wet ditches, & margins of wooded areas; NE, KS, MO, OK; (NJ to MN & NE, s to FL & TX; also Ont.). *C. lacustris* Willd. var. *laxiflora* Dew.—Gleason & Cronquist; *C. impressa* (S.H. Wright) Mack.—Rydberg.

65. **Carex hystericina** Muhl. ex Willd. (Group VIII) Cespitose perennial, rootstock short, stout. Culms 2–10 dm tall, rough triangular. Leaves 3–8 mm wide; sheaths white-hyaline, yellow tinged ventrally, lower filamentose. Spikes unisexual, terminal spikes staminate, 1–5 cm long, usually peduncled, lateral spikes pistillate, lower nodding and peduncled, separate to aggregate; pistillate scales not conspicuous. Perigynia spreading, inflated suborbicular, green to straw colored, shining, strongly 12- to 17-ribbed, 5–7.5 mm long, teeth 0.4–1 mm long. Achenes 1.7 mm long, 1 mm wide, triangular with concave sides; stigmas 3. (n = 29) Swamps, shorelines, & ditches; ND, MN, SD, WY, NE, IA, CO, KS, MO, TX; (Newf. to Alta., s to GA, TX, & CA). *C. hystricina* Muhl.—Rydberg.

66. **Carex interior** Bailey. (Group I) Densely cespitose perennial, rootstock not prolonged, dark. Culms 2–5 dm tall, slender, sharply triangular. Leaves 1–2 mm wide; sheaths tight, hyaline ventrally. Heads 1–2 cm long; spikes bisexual, terminal spike gynecandrous; lower spikes separate; pistillate scales ovate-obtuse. Perigynia concavo-convex, serveral-nerved dorsally, nerveless to a few at base ventrally, 2.3–3 mm long, 1.5–2 mm wide; base spongy, truncate-rounded, contracted to a serrulate, bidentate beak 1/3–1/4 the length of a body. Achenes 1.2 mm long, 1.5 mm wide, lenticular; stigmas 2. Dry wooded areas, prairies, & marshes; ND, MN, SD, NE, KS; (Newf. to B.C., s to NJ, MO, CA; also FL, Mex.).

67. **Carex intumescens** Rudge. (Group VIII) Cespitose perennial, rootstock short. Culms 3–8 dm tall, rough triangular. Leaves 3–9 mm wide; sheaths loose, white-hyaline ventrally. Spikes unisexual, terminal spike staminate, 2 mm long, peduncled, lateral 1–3 spikes pistillate, ovoid-globose; pistillate scales obtuse to serrulate-awned. Perigynia suborbicular,

inflated, green, shining, strongly ribbed, 10–16 mm long, 3.5–8 mm wide, beak 2–3.5 mm long, teeth 1 mm long. Achenes 5 mm long, 3 mm wide, triangular with concave sides, persistent style; stigmas 3. Wooded areas; MN, SD; (Newf. to Man., s to FL, MN, & TX).

68. *Carex jamesii* Schwein. (Groups V & VII) Cespitose perennial, rootstock short, dark. Culms 0.5–4 dm tall, weak. Leaves 2–3 mm wide. Spikes solitary, unisexual, androgynous, aggregate; staminate scales truncate, margins overlapping or meeting at base; pistillate scales acuminate, hyaline margins. Perigynia subglobose, green, nerveless, 2-keeled, 4.4–6.4 mm long, 2.5 mm wide, base spongy, contracted to a serrulate beak 3 mm long, flat. Achenes 2.5 mm long, 2.3 mm wide, triangular with convex sides, stigmas 3. Rich, moist wooded areas; NE, KS, MO; (Ont., NY s to MD, TX, & OK).

69. *Carex lacustris* Willd. (Group VIII) Rhizomatous perennial, rootstock prolonged, scaly. Culms 6–13 dm tall, stiff, rough triangular. Leaves 8–15 mm wide; lower sheaths purple tinged, pinnately filamentose ventrally. Spikes unisexual, upper 2 or 3 spikes staminate, 4–7 cm long, lower spikes pistillate, peduncled; pistillate scales awned or acuminate. Perigynia flattened-suborbicular, olive-green, strongly nerved dorsally, 5.5–7 mm long, 2 mm wide, tapering to a smooth beak 1 mm long, teeth 0.4–0.8 mm. Achenes 2.5 mm long, 1.5 mm wide, triangular, with persistent style, stigmas 3. Swamps, shores, & wet meadows; ND, MN, SD, NE, IA, KS, MO; (Newf. to Alta.; s to VA, MO, & KS).

70. *Carex laeviconica* Dew. (Group VIII) Rhizomatous perennial, rootstock short, scaly. Culms 3–12 dm tall, triangular. Leaves 2–8 mm wide; sheaths purple tinged at base and filamentose. Spikes unisexual, upper 2–6 spikes staminate, lower spikes pistillate; pistillate scales acute to aristate. Perigynia suborbicular, inflated, green-yellow, strongly ribbed, 6 mm long, 3.5 mm wide, beak 2 mm long, teeth 1–2 mm long and straight. Achenes 2.5 mm long, 1.5 mm wide, triangular with concave sides; stigmas 3. Swamps, sloughs, wooded ravines, ditches, low meadows, & shores; MT, ND, MN, SD, NE, IA, KS, MO; (Man. to Alta., s to IL, MO, KS, & MT). *C. atherodes* Spreng. var. *longo-lanceolata* (Dew.) Gilly— Gilly, Iowa State Coll. J. Sci. 21: 55. 1946.

71. *Carex laevivaginata* (Kükenth.) Mack. (Group III) Densely cespitose perennial, rootstocks short prolonged, dark. Culms 3–8 dm tall, sharply triangular, slightly wing-angled, weak. Leaves 3–6 mm wide, sheaths septate-nodulose dorsally, usually not cross-rugulose ventrally. Heads linear-oblong or oblong, 1.5–6 cm long; spikes bisexual, androgynous; pistillate scales triangular-lanceolate. Perigynia plano-convex, lanceolate, strongly nerved on both sides, 4.5–7 mm long, 1.5–2 mm wide, tapering into a serrulate beak 1– 2 × length of body. Achenes lenticular, body ovate-orbicular, 1.5–1.8 mm long, 1.3 mm wide, style slender, straight, stigmas 2. Swamps, marshes, wet meadows; IA; (Ont. s to New England, FL, IA, & OK). *C. stipata* Muhl. var. *laevivaginata* Kükenth.— Gilly, op. cit.

72. *Carex lanuginosa* Michx. (Group VIII) Rhizomatous perennial, rootstock prolonged, scaly. Culms 3–10 dm tall, acutely triangular. Leaves 2–5 mm wide; lower sheaths filamentose, purplish tinged dorsally. Spikes unisexual, usually 2 upper spikes staminate, 2–6 mm long, lower 2 or 3 spikes pistillate, aggregate; staminate scales acute to cuspidate, pistillate scales acuminate to awned. Perigynia suborbicular, inflated, brownish-green, densely hairy, many-ribbed, 2.5–5 mm long, 1.2–3 mm wide, beak 1 mm long; teeth 0.3–0.8 mm long. Achenes 1.7–2 mm long, 1.5 mm wide, triangular-obovoid with concave sides; stigmas 3. (n = 39) Swamps, low moist prairies, marshy areas, & shores; MT, ND, MN, WY, SD, NE, CO, KS, MO, OK, TX; (N.S. to AK, s to VA, TX, & CA). *C. lasiocarpa* Ehrh. var. *latifolia* (Boeck.) Gilly—Gleason & Cronquist.

73. Carex lasiocarpa Ehrh. var. *americana* Fern. (Group VIII) Tufted perennial from long creeping rootstocks and stolons. Culms 0.3-1.2 m tall, obtusely triangular. Leaves 2 mm or less wide; sheaths yellowish-brown tinged ventrally, concave at mouth. Staminate spikes usually 2, long peduncled, 2-6 cm long; pistillate spikes 0.8-6 cm long; pistillate scales lanceolate or ovate lanceolate, frequently awn-tipped or cuspidate. Perigynia ovoid-ellipsoid, 3.4-5 mm long, 1.8 mm wide, suborbicular in cross section, densely soft-hairy, the ribs obscure, contracted at apex into a stout bidentate beak up to 1 mm long. Achenes obovoid, triangular with concave sides, 1.7-2 mm long, 1.5 mm wide, style straight or flexuous; stigmas 3. Swamps, sloughs, borders of lakes & ponds; IA; (Newf. to B.C., s to VA, IN, IA, & WA).

74. Carex leavenworthii Dew. (Group III) Cespitose perennial, rootstock short, dark. Culms 1-6 dm tall, sharply triangular. Leaves 1-3 mm wide; sheaths tight, not septate-nodulose or cross-rugulose. Heads 7-20 mm long; pistillate scales ovate, acute to cuspidate. Perigynia plano-convex, broadest at base and truncate-cordate, 2-3.5 mm long, 1.2-1.7 mm wide, flat and nerveless ventrally, slightly nerved dorsally, tapering to a short serrulate beak. Achenes 1-1.5 mm long, lenticular; stigmas 2. Typically open woods, occasionally prairies; NE, KS, MO, OK; (MO to MI & NE, s to FL & TX, also Ont.).

75. Carex leptalea Wahl. (Groups V & VII) Cespitose perennial, rootstock short, slender, dark. Culms 1-7 dm tall, obtusely triangular. Leaves 0.5-1.25 mm wide; sheaths tight, white-hyaline ventrally. Spikes solitary 4-16 mm long, unisexual, androgynous, aggregate; staminate scales obtuse to acute, connate below; pistillate scales obtuse to cuspidate. Perigynia orbicular or slightly flattened, 2-edged, yellow-green, 2.5-5 mm long, 1.5 mm wide, base spongy emarginate, beakless. Achenes 1.5-2 mm long, triangular with concave sides; stigmas 3. (n = 26) Dry to wet woods; ND, SD; (Newf. to AK, s to FL, TX, & CA).

76. Carex limosa L. (Group V) Rhizomatous perennial, rootstock prolonged, brown. Culms 3-5 dm tall, exceeding leaves, sharply triangular. Leaves 1-3 mm wide; sheaths thin and hyaline ventrally. Spikes unisexual, terminal spikes staminate, 1-3 cm long, peduncled, pistillate spikes lateral, 1-3, peduncled and drooping; pistillate scales obtuse to cuspidate. Perigynia flattened-triangular, green, few-nerved, 2.5-4.2 mm long, 2 mm wide, beak minute, entire. Achenes 2.2 mm long, 1.8 mm wide, triangular; stigmas 3. (n = 28, 31, 32) Marshy areas; ND, NE; (Newf. to AK, s to DE, OH, NE, & CA).

77. Carex lupulina Willd. (Group VIII) Cespitose perennial, rootstock short. Culms 3-12 dm tall. Leaves 4-15 mm wide; sheaths white-hyaline ventrally. Spikes unisexual, terminal spike staminate, 3-5 mm long, peduncled, lateral spikes pistillate; pistillate scales acuminate to rough-awned. Perigynia suborbicular, inflated, ascending, green-brown, strongly ribbed, 10-20 mm long, 4-7 mm wide, beak 5-10 mm long, rough, teeth 0.7-2 mm long. Achenes 4 mm long, 2.5-3 mm wide, triangular; style persistent, bent or twisted; stigmas 3. Swamps, ditches, wooded ravines, & shores; NE, IA, KS, MO, OK; (N.S. to Man., s to FL & TX).

78. Carex lurida Wahl. (Group VIII) Cespitose perennial in dense clumps; rootstock short, stout. Culms 1.5-10 dm tall, obtusely triangular. Leaves 2-7 mm wide; sheaths loose, hyaline ventrally, usually yellowish-brown tinged, concave or truncate at mouth. Terminal spike staminate, 1-7 cm long; pistillate spikes subglobose to cynlindric, 1-8 cm long; pistillate scales lanceolate or oblanceolate with long rough awns. Perigynia 6-9 mm long, 2.5-4 mm wide, strongly inflated, the body ovoid or obovoid-globose, suborbicular in cross section, strongly ± 10-nerved, tapering into a slender bidentate beak from 1/2 to nearly as long as the body, teeth slender, erect or spreading, 0.5-1 mm long. Achenes oval-obovoid,

2-2.5 mm long, 1.5 mm wide, triangular with concave sides; style twisted or bent; stigmas 3. Swales, swamps, wet meadows; MO; (N.S. to Ont., s to FL & TX; also CA, Mex.).

79. **Carex meadii** Dew. (Group V) Rhizomatous perennial, rootstock prolonged, whitish. Culms 2-5 dm tall, stiff, triangular. Leaves 2-7 mm wide; sheaths tight, concave at mouth. Spikes unisexual, terminal spike staminate, 1.5-3 cm long, lateral spikes pistillate and peduncled; pistillate scales awned, purple-red or brown, as wide as but shorter than perigynia. Perigynia triangular, in 6 rows, yellow-green or brown; many-nerved, 2-ribbed, 3-5 mm long, 1.7-2.5 mm wide, beak minute and bent. Achenes 2.7-3.5 mm long, 1.5-2.5 mm wide, triangular with concave sides; stigmas 3. Mostly prairies, ungrazed pastures, & occasionally in ditches & woodlands; ND, MN, SD, NE, IA, KS, MO, OK, TX; (Ont. to Sask., s to GA, AR, & TX).

80. **Carex melanostachya** Willd. (Group VIII) Loosely cespitose perennial, forming large colonies by long scaly horizontal rhizomes. Culms 2-5 dm tall, acutely triangular and slightly roughened on angles, purplish-red at base, lower sheaths filamentose. Leaves 3-6 on each fertile culm, blades 1.5-4 dm long, 2-3 mm wide, light green, roughened toward apex; sheaths septate-nodulose dorsally, slightly pubescent, ciliate and concave at mouth ventrally, ligule wider than long. Upper spike staminate, long peduncled, 2-3 cm long, 2-3 mm wide, lower 2 or 3 spikes pistillate, sessile, or lower ones sometimes peduncled, 1.5-3 cm long, 3-6 mm wide, each with a leaflike bract much longer than the spikes; staminate scales red-brown to brown, with ciliate margins, acuminate to short awned; pistillate scales with deep-red to purple margins and 3-nerved green center, short awned and ciliate at upper margins. Perigynia 10-30 to a spike, 3.5-4 mm long, 1.2-1.5 mm wide, ascending, suborbicular to triangular, little inflated, glabrous, many impressed nerves, short-stipitate beak 0.5 mm long, bidentate, orifice purple and ciliate on inner margins of teeth. Achenes 2.4 mm long, 1.3 mm wide, triangular, sides concave below, stipitate at apex, jointed with style; stigmas 3, black. Dry to wet roadsides, ditches, and disturbed prairies; KS: Douglas. *Introduced.*

81. **Carex microdonta** Torr. & Hook. (Group V) Rhizomatous perennial, rootstock prolonged. Culms 1.5-6 dm tall, smooth or rough triangular. Leaves 3-6 mm wide; sheaths tight and hyaline ventrally. Spikes usually unisexual, terminal spike staminate, 2-4 cm long, lateral spikes pistillate to androgynous, lower spikes separate and peduncled; pistillate scales acuminate to cuspidate, red-brown. Perigynia squarrose-ascending, suborbicular, olive-green, many-ribbed, 3-4.5 mm long, 1.5-2 mm wide, beaks short and slightly bidentate. Achenes 2.5 mm long, 1.5-1.7 mm wide, triangular with concave sides, style bent-apiculate; stigmas 3. Heavily grazed prairies, rocky to wet prairies & ditches; KS, MO, OK; (MO to KS, s to LA & TX).

82. **Carex microptera** Mack. (Group I) Cespitose perennial, rootstocks short, black. Culms 3-10 dm tall, obtusely triangular. Leaves 2-6 mm wide; sheaths tight, hyaline ventrally, sterile shoots with leaves mostly clustered at top; 1.2-2.5 cm long. Spikes bisexual, gynecandrous, aggregate; pistillate scales obtuse to acute. Perigynia thin, plano-convex, light green to stramineous, several-nerved dorsally, 3- to 4-nerved at base ventrally, strongly wing-margined to the base, 3.7-5 mm long, 1.5-2 mm wide, base rounded, tapering to a serrulate beak 1/2 the length of body. Achenes 1.5 mm long, 1 mm wide, lenticular; stigmas 2. (n=45) Meadows; SD, CO; (Man. to B.C., s to SD, NM, & CA; Mex.). *C. festivella* Mack.—Rydberg.

83. **Carex microrhyncha** Mack. (Group VIII) Cespitose perennial, rootstocks short, stout, and branching. Culms up to 15 cm tall, sharply triangular. Leaves 1.5-3 mm wide; sheaths

little (if at all) filamentose ventrally. Spikes unisexual, terminal spike staminate, lateral spikes pistillate; pistillate scales obovate, acute or obtuse. Perigynia 2-keeled, triangular-orbicular in cross section, 2.2–3.5 mm long, 1.3 mm wide, short pubescent above, beak up to 1 mm long. Achenes 1.5 mm long, triangular with convex sides. Usually calcareous soils of open woods, prairies, rocky open ground; e 1/6 KS, MO, e OK; (KS, MO, OK, & TX). *C. abdita* Bickn.—Steyermark.

84. Carex molesta Mack. (Group I) Cespitose perennial, rootstock short, black. Culms 3–10 dm tall, sharply triangular. Leaves 2–3 mm wide, sheaths tight, white-hyaline ventrally, green-and-white mottled dorsally. Heads 2–3 cm long; spikes bisexual, rounded at apex, truncate at base, gynecandrous, aggregate; pistillate scales ovate, obtuse or acute. Perigynia plano-convex, green, few-nerved dorsally, nerveless to strongly few-nerved ventrally, strongly wing-margined to base, 4.5 mm long, 2–3 mm wide, base rounded, tapering to a serrulate beak 1/2 the length of body. Achenes 1.7 mm long, 1.2 mm wide, lenticular; stigmas 2. Woodlands, ditches, shores, & prairies; ND, MN, SD, NE, IA, CO, KS, MO; (VT to Sask., s to VA, KS, & CO, also CA).

85. Carex muhlenbergii Willd. (Group III) Rhizomatous to cespitose perennial. Culms 2–10 dm tall; sharply triangular. Leaves 2–5 mm wide; sheaths tight, not septate-nodulose, occasionally cross-rugulose. Heads 2–4 cm long, aggregate; pistillate scales ovate, acute to acuminate or cuspidate. Perigynia plano-convex, 3–4 mm long, 2–3 mm wide, several-nerved dorsally, usually nerveless ventrally. Achenes 2–2.5 mm long, 1.7–2 mm wide, lenticular; stigmas 2.

Two varieties are known in the GP:

85a. var. *australis* Olney. Plants cespitose. Perigynia ascending, 3.5–4 mm long; scales equal to or exceeding the entire perigynia. Prairies, woodlands, & roadside ditches; NE, IA, KS, MO, OK; (IA to NE, s to AR & TX). *C. austrina* Mack.—Rydberg.

85b. var. *enervis* F. Boott. Plants rhizomatous. Leaf sheaths cross-rugulose. Perigynia spreading, 3–3.5 mm long; scales about as long as the body of the perigynium. Open moist areas & sandy shores. NE, KS, MO, OK; (NE, s to OK & MO). *C. plana* Mack.—Rydberg.

86. Carex muskingumensis Schwein. (Group I) Cespitose perennial, rootstock short, black. Fertile culms 5–10 dm tall, sharply triangular concave sides, numerous leafy sterile culms. Leaves 3–5 mm wide; sheaths loose, green-striate ventrally. Heads 5–8 cm long; spikes bisexual, fusiform, pointed at apex, gynecandrous, approximate; pistillate scales oblong-ovate, acute to obtuse. Perigynia flat, scalelike, lanceolate, membranaceous, straw colored, few-nerved both sides, strongly wing-margined below middle and narrower, 7–10 mm long, 3.5 mm wide, base round tapering, apex tapering to a serrulate beak 1/3 the length of body. Achenes 2.5 mm long, 0.7 mm wide, lenticular; stigmas 2. Moist wooded areas; KS, MO; (Ont. to Man., s to KY, AR, & OK).

87. Carex nebraskensis Dew. (Group VI) Rhizomatous perennial, rootstock prolonged, stout, brown or straw colored, scaly. Culms 2.5–12 dm high, stout, rough triangular. Leaves 3–8 mm wide; sheaths yellow-brown tinged ventrally. Spikes unisexual, terminal spike staminate, lateral spikes pistillate; pistillate scales obtuse to acuminate, purple to brown-black. Perigynia flattened plano-convex or biconvex, straw colored, red dotted, several-nerved both sides, 3–3.5 mm long, 2 mm wide, beak 0.5–1 mm long, bidentate. Achenes 1.5 mm long, 1.2 mm wide, lenticular; stigmas 2. Swamps, wet meadows, & margins of lakes or ponds; MT, ND, WY, SD, NE, CO, KS; (Alta. to B.C., s to KS, e MO, NM, & CA).

88. Carex normalis Mack. (Group I) Cespitose perennial, rootstock short to prolonged, black. Culms 3–15 dm tall. Leaves 3–6 mm wide; sheaths loose, prolonged, hyaline ventrally, green-and-white mottled dorsally, sterile shoots with leaves at summit. Heads 1.5-5 cm long; spikes bisexual, gynecandrous, moniliform to aggregate; pistillate scales ovate, acute to obtuse. Perigynia plano-convex, green to straw colored, many-nerved dorsally, 7-nerved ventrally, broad wing-margined to base, 3-4.4 mm long, 1.5-2 mm wide, base rounded, apex tapering to a serrulate beak 1/2 the length of body. Achenes 1.7 mm long, 1 mm wide, lenticular; stigmas 2. Primarily woodlands, occasionally prairies & ditches; MT, SD, NE, IA, KS, MO; (N.B. to Man. s to NC & OK).

89. **Carex obtusata** Lilj. (Groups V & VII) Rhizomatous perennial, rootstock prolonged, purplish-black. Culms 0.5-2 dm tall. Leaves 1-1.5 mm wide. Spikes solitary, bisexual, androgynous, aggregate; staminate and pistillate scales acuminate or cuspidate. Perigynia suborbicular to triangular, dark chestnut or blackish-brown, finely many sulcate, 3-3.5 mm long, 1.7-2 mm wide, base round tapering, beak short, 0.5-1 mm. Achenes 1.7 mm long, 1 mm wide, triangular with prominent angles; stigmas 3. Prairies, sloughs, & uplands; ND, MN, SD; (Man. to AK, s to MN, SD, NM, & UT).

90. **Carex oklahomensis** Mack. (Group III) Cespitose perennial, rootstock short, dark. Culms 3.5-8 dm tall, sharply triangular. Leaves 2.5-5 mm wide; sheaths tight, not red dotted nor cross-rugulose ventrally, green-and-white mottled and slightly septate-nodulose dorsally. Heads 4-7 cm long; spikes bisexual, androgynous, aggregate; pistillate scales ovate or lance-ovate. Perigynia plano-convex, greenish to straw colored, 7- to 10-nerved dorsally, few-nerved ventrally, sharp margined to base; 4-5 mm long, 1.7 mm wide, base truncate-subcordate, tapering to a serrulate beak, shorter than body. Achenes 1.7 mm long, 1.5 mm wide, lenticular; stigmas 2. Sedge swales & wooded areas; KS, MO; (MO to KS, s to AR & TX).

91. **Carex oligocarpa** Willd. (Group V) Cespitose perennial, rootstock short. Culms 2-5 dm tall, rough triangular. Leaves 2-4 mm wide; sheaths tight, prolonged, white hyaline ventrally. Spikes unisexual, terminal spike staminate, 1-3 cm long, lateral spikes pistillate and separate; pistillate scales rough-awned and keeled. Perigynia obtusely triangular, grayish-green; impressed undulate many-nerved, 3.5-4.3 mm long, 2 mm wide, tapering to both ends, base spongy, beak 0.7 mm long and entire. Achenes 2.2 mm long, 1.7 mm wide, triangular with a minute straight beak; stigmas 3. Rich, moist woods; NE, IA, KS, MO, OK; (DE to Ont., s to GA, AR, & TX).

92. **Carex parryana** Dew. (Group IV) Rhizomatous perennial, rootstock prolonged, scaly. Culms 1.5-4 dm tall, stiff, obtusely triangular. Leaves 2-3 mm wide; sheaths thin, white-hyaline ventrally, concave at mouth, terminal spike gynecandrous. Lateral spikes pistillate; pistillate scales obtuse to mucronate. Perigynia obtusely triangular, straw colored, purplish tinged, rough granular at apex; 2-ribbed, 2-2.5 mm long, 1-1.5 mm wide, beak 0.1-0.2 mm long, bidentulate. Achenes 1.5-1.7 mm long, 1-1.2 mm wide, triangular; stigmas 3. Swampy areas; WY; (Man. to B.C., s to CO & VT).

93. **Carex peckii** Howe. (Group VIII) Cespitose to rhizomatous perennial, rootstock short, scaly. Culms 1-6.5 dm tall, triangular. Leaves 1.5-3 mm wide; sheaths tight ventrally. Inflorescence 8-20 mm long, spikes unisexual, terminal spikes staminate, lateral spikes pistillate or androgynous; pistillate scales mucronate to obtuse. Perigynia body oblong-ovoid and obtusely triangular, yellow-green, hirsute-pubescent, 2-ridged, 3-4 mm long, 1 mm wide, base spongy, beak 0.5 mm long. Achenes 2 mm long, 1 mm wide, triangular with convex sides; stigmas 3. Rich moist woods; ND, MN, SD, NE; (Que. to AK, s to NJ, MI, MN, & NE). *C. nigromarginata* Schwein. var. *elliptica* (F. Boott.) Gl. — Gleason & Cronquist.

94. ***Carex pendunculata*** Willd. (Group V) Somewhat rhizomatous perennial, rootstock prolonged, woody. Culms 0.5–3 dm tall, triangular. Leaves 2–5 mm wide; sheaths loose. Upper spike staminate or with a few basal perigynia, 6–15 mm long, lower spikes pistillate and long peduncled; pistillate scales rough cuspidate. Perigynia triangular, green, puberulent, 2-ridged, 3.5–4.5 mm long, 1.5 mm wide, base spongy, beak minute and bent. Achenes 2.5–3.5 mm long, 1 mm wide, triangular with concave sides; stigmas 3. Woodlands; MN, SD; (Newf. to B.C., s to VA, IA, & SD; also GA).

95. ***Carex pensylvanica*** Lam. (Group VIII) Cespitose or rhizomatous perennial, rootstock short to prolonged, reddish. Culms 2–4 dm tall, triangular. Leaves 1.5–3 mm wide; sheaths puberulent to filamentose ventrally. Spikes unisexual, terminal spikes staminate, lateral spikes pistillate, aggregate; pistillate scales obtuse, acuminate to cuspidate. Perigynia subglobose to obtusely triangular, yellow-green, puberulent, 2-keeled, 2.2–3.3 mm long, 1–1.5 mm wide, beak 0.5–0.7 mm long. Achenes 1.5–2 mm long, 1.2 mm wide, triangular with concave sides; stigmas 3. Open woodlands, moist thickets; ND, MN, SD, IA, MO; (Que. to B.C., s to GA, MO, KS, & WY).

96. ***Carex praegracilis*** W. Boott. (Group II) Rhizomatous perennial, rootstock long, stout black. Culms 2–7 dm tall, acutely triangular. Leaves 2–3 mm wide, flattened; sheaths truncate. Heads 1–3 cm long; spikes bisexual, androgynous, separated; pistillate scales aristate to acuminate. Perigynia plano-convex, straw colored to brown-black, few-nerved dorsally 3–4 mm long, 1.2–1.5 mm wide, base spongy stipitate, tapering into a serrulate beak 1/2 the length of body. Achenes 1.4–1.8 mm long, 1–1.2 mm wide, lenticular; stigmas 2. Moist spots & swales in prairies & plains, also saline & sandy soils; MT, ND, MN, SD, WY, NE IA, CO, KS, MO, OK; (Man. to AK, s to MI, MS, TX, & CA; also New England, Mex.)

97. ***Carex prairea*** Dew. (Group II) Cespitose perennial, rootstock short, black. Culms 5–10 dm tall, sharply triangular. Leaves 2–3 mm wide; sheaths tight, red dotted and copper tinged at prolonged mouth. Heads 3–8 cm long; spikes bisexual, androgynous, upper aggregate; pistillate scales reddish-brown, ovate-acuminate. Perigynia plano-convex, brown few-nerved dorsally, few-nerved at base ventrally, 2.5–3 mm long, 1.2–1.5 mm wide, base truncate, tapering to a serrulate beak almost as long as body. Achenes 1 mm long, lenticular; stigmas 2. Marshes, wet meadows, & swamps; ND, MN, SD, NE, IA; (Que. to Alta., s to NJ, IL, & NE).

98. ***Carex praticola*** Rydb. (Group I) Cespitose perennial, rootstock short, black. Culms 2–6 dm tall, sharply triangular, nodding. Leaves 2–3 mm wide; sheaths tight, white hyaline ventrally. Heads 1.5–5 cm long; spikes bisexual or gynecandrous, separate, moniliform; pistillate scales acute. Perigynia flattened plano-convex, light-green or whitish, lightly many-nerved dorsally, nerveless ventrally, wing margined to base, 4.5–6 mm long, 1.5–2 mm wide, base rounded, tapering to a serrulate beak 1/3–1/2 the length of body. Achenes 1.5–2 mm long, 1.5 mm wide, lenticular; stigmas 2. High open meadows; SD; (Lab. to AK, s to ME, MI, SD, CO, & CA).

99. ***Carex projecta*** Mack. (Group I) Cespitose perennial, rootstocks short to prolonged, black. Culms 5–9 dm tall, sharply triangular, roughened on angles. Leaves 3–7 mm wide; sheaths loose, septate-nodulose and green-and-white mottled dorsally, narrowly white-hyaline ventrally and easily broken. Inflorescence moniliform, 3–5 cm long, spike gynecandrous, subglobose, 4–8 mm long, pistillate scales lanceolate or ovate-lanceolate, obtuse to acute. Perigynia flat, lanceolate, 3–5 mm long, 1.5 mm wide, strongly wing-margined and serrulate to below middle, membranaceous, several-nerved both sides, tapering into a beak 1/3–1/2 the length of body. Achenes lenticular, oblong-oval, 1.5 mm long, 0.5 mm wide;

style straight, slender; stigmas 2. Wet prairies & open moist woodlands; IA, MO; (Newf. to Sask. s to WV & AR).

100. Carex pseudo-cyperus L. (Group VIII) Cespitose perennial, rootstock short, stout. Culms 3-10 dm tall, rough triangular. Leaves 5-15 mm wide; sheaths thin-hyaline ventrally, yellow tinged dorsally. Spikes unisexual, terminal spike staminate, 3-7 cm long, lateral spikes pistillate, lower drooping, peduncled; pistillate scales rough-awned. Perigynia flattened-triangular, strongly 12- to 17-nerved, spreading to reflexed, 4.2-6.2 mm long, 1.5 mm wide, tapering to a smooth beak 1 mm long, teeth straight, 0.5-1 mm long. Achenes 1.7 mm long, 0.7 mm wide, triangular; stigmas 3. Swamps, shores, & marshes; ND, MN; (Newf. to Alta., s to PA, IN, MN, ND, ID, & CA).

101. Carex retroflexa Willd. (Group III) Perennial with rootstock short prolonged, dark colored. Culms 2-4 dm tall, sharply triangular. Leaves 1-3 mm wide; sheaths tight, not septate-nodulose, not cross-rugulose. Heads 1-4 cm long, spikes androgynous, lower spikes separate, subtended by a setaceous bract; pistillate scales ovate, acuminate, or cuspidate. Perigynia biconvex, 2.5-3 mm long, 1.7 mm wide, rounded and spongy at base, many striate ventrally at base, apex contracted into a smooth margined beak. Achenes 1.3 mm long, rounded at base and tapering to apex, lenticular; stigmas 2. Dry wooded areas; KS, MO, OK; (New England to KS, s to FL & TX).

102. Carex retrorsa Schwein. (Group VIII) Cespitose perennial, rootstock short. Culms 2-10 dm tall, obtusely triangular. Leaves 3-10 mm wide; sheaths long, loose, yellow-brown ventrally. Spikes unisexual, upper 1-4 spikes staminate, lower spikes pistillate; pistillate scales acute to obtuse or cuspidate. Perigynia suborbicular, inflated, yellow-green to green, shining, coarsely nerved, 7-10 mm long, 3 mm wide, spreading or reflexed, tapering or contracted to a smooth or serrulate beak, beak 2-3.5 mm long, teeth 0.5 mm long. Achenes 2.5 mm long, 1.2 mm wide, triangular, style twisted or bent, persistent; stigmas 3. Sloughs, swamps, & shores; ND, MN, SD, IA; (Newf. to B.C., s to NJ, IL, SD, CO, & NV).

103. Carex richardsonii R. Br. (Group V) Loosely cespitose perennial, rootstock prolonged, brown-black. Culms 1.5-3.5 dm tall, triangular. Leaves 2-2.5 mm wide; sheaths tight. Terminal spike staminate, 10-25 mm long, lateral spikes pistillate, aggregate; pistillate scales acute. Perigynia triangular, straw colored to light brown, appressed pubescent, 2-keeled, 2.5-3 mm long, 1.5 mm wide, beak 0.5 mm long. Achenes 1.5-2 mm long, 1-1.2 mm wide, triangular with convex sides; stigmas 3. Low prairies, ditches, & hillsides; ND, SD; (Ont. to Alta., s to NY, OH, IA, SD). *C. richardsonii* R. Br. f. *exserta* Fern.—Fernald.

104. Carex rosea Willd. (Group III) Cespitose perennial, rootstock short, dark. Culms 2-5 dm tall; slender, sharply triangular. Leaves 1-2 mm wide; sheaths tight, not septate-nodulose, not cross-rugulose. Heads 3-5 cm long, spikes androgynous, lower ones separate; pistillate scales triangular-ovate. Perigynia plano-convex, with raised margins ventrally, spreading, 3-3.5 mm long, 1.5 mm wide, nerveless to slightly nerved both sides, margins serrulate above, spongy and round at base, apex tapering to a contracted beak, not hyaline at orifice. Achenes 1.7 mm long, 1.2 mm wide, lenticular; stigmas 2 and usually not twisted. Mosit to dry, open woodlands; ND, MN, SD, NE, IA, KS, MO; (Newf. to Man., s to GA, LA, OK).

105. Carex rossii F. Boott ex Hook. (Group VIII) Cespitose perennial, rootstock short, lignescent. Culms 1-3 dm tall, triangular. Leaves 1-2.5 mm wide; sheaths minutely hispidulose dorsally. Spikes unisexual, terminal spikes staminate, 3-15 mm long, lateral spikes pistillate; pistillate scales obtuse, cuspidate or awned. Perigynia triangular, green, short pubescent, 2-keeled, 3–4.5 mm long, 1 mm wide, base 0.5–1.5 mm long, tapering to a serrulate beak 1.3–1.7 mm long. Achenes 2.3 mm long, 1 mm wide, obtusely triangular with concave sides; stigmas 3. Open woodlands; ND, SD, WY, NE; (Man. to AK, s to MI, MN, NE, CO, AZ, & CA).

106. *Carex rostrata* Stokes ex Willd. (Group VIII) Cespitose perennial, rootstock short. Culms 3–12 dm tall, bluntly triangular, bases spongy-inflated. Leaves 2–12 mm wide; sheaths thin, white-hyaline ventrally. Spikes unisexual, terminal 2–4 spikes staminate, 3–4 mm long, lower spikes pistillate; pistillate scales acute to awned. Perigynia suborbicular, inflated, yellow-green to brown, strongly nerved, 3.5–8 mm long, 2.5–3.5 mm wide, ascending to reflexed, beak 1–2 mm long, teeth 0.5–0.7 mm long. Achenes 2 mm long, 1.2 mm wide, triangular, style persistent, twisted or bent; stigmas 3. (n = 36, 37, 38, 41) Sloughs, bogs, & shores; ND, MN, SD, IA; (Newf. to AK, s to DE, KS, NM, & CA).

107. *Carex sartwellii* Dew. (Group II) Rhizomatous perennial, rootstock long creeping, black, fibrillose. Culms 4–8 dm tall, sharply triangular. Leaves 2.5–4 mm wide; sheaths green-striate ventrally with connate margins. Heads 3–6 cm long; spikes bisexual, androgynous (upper entirely staminate in some) aggregate; pistillate scales obtuse to cuspidate. Perigynia plano-convex, tan to brown, nerved both sides, 2.5–3 mm long, 1.5–1.7 mm wide, body contracted into a serrulate beak 1/4 the length of body. Achenes 1.5 mm long, less than 1 mm wide, lenticular; stigmas 2. Wet meadows, sloughs, & shallow marshes; ND, MN, SD, NE, IA, KS, MO; (Ont. to B.C., s to PA, MO, NE, & CO).

108. *Carex saximontana* Mack. (Groups V & VII) Cespitose perennial, rootstock short. Culms 3.5 dm tall, narrowly winged. Leaves 3–5 mm wide; sheaths thin, hyaline ventrally. Spikes solitary, unisexual, androgynous, separate or aggregate; staminate scales obtuse to cuspidate, margins connate; lower pistillate scales leaflike, 7–35 mm long. Perigynia suborbicular, green, faintly many-nerved, 2-keeled, 4 mm long, 2 mm wide, base spongy, contracted to a serrulate beak 0.5–1 mm long. Achenes 3 mm long, 1.5 mm wide, triangular to globose with concave sides; stigmas 3. Dry to wet woods, ND, MN, SD, NE; (Man. to B.C., s to MN, NC, CO, & NV).

109. *Carex scirpiformis* Mack. (Groups V & VII) Rhizomatous perennial, rootstock prolonged, purplish-black. Culms 2–4.5 dm tall, triangular. Leaves 1.5–4 mm wide; sheaths hyaline, brown tinged, pubescent dorsally, ciliate at mouth ventrally. Spikes single, dioecious, 2–4 cm long; pistillate scales obtuse or acute, pubescent. Perigynia compressed-triangular, green to yellow-brown, hirsute, 2-ribbed, 2–5 mm long, 1.2 mm wide, beak 0.5 mm long. Achenes 1.5 mm long, 0.7 mm wide, triangular; stigmas 3. Mt. slopes & dunes; ND, MN; (Man. to Alta., s to ND & MT, also UT). *C. scirpoidea* Michx. var. *scirpiformis* (Mack.) O'Neill & Duman — Fernald.

110. *Carex scoparia* Schkuhr ex Willd. (Group I) Cespitose perennial, rootstock short, dark brown. Culms 2–10 dm tall, sharply triangular. Leaves 1–3 mm wide; sheaths tight, white-hyaline ventrally, sterile shoots with leaves at apex. Heads 1–5 cm long; spikes bisexual or gynecandrous; aggregate to moniliform; pistillate scales lance-ovate. Perigynia flat, greenish-white, strongly to slightly many-nerved dorsally, several-nerved ventrally, wing-margined to base, 4–6.5 mm long, 1.2–2 mm wide, beak 1/3 the length of body. Achenes 1–1.5 mm long, 0.5–0.7 mm wide, lenticular; stigmas 2. Wooded areas, sandy shores of lakes, low meadows, prairies, open swamps, or wet places; ND, MN, SD, NE, IA, KS, MO; (Newf. to B.C., s to FL, TX, & CA).

111. ***Carex shortiana*** Dew. (Group IV) Cespitose perennial, rootstock short, dark. Culms 4–8 dm tall, stiff, triangular. Leaves 4–8 mm wide; sheaths red tinged ventrally, greenish-white and septate-nodulose dorsally. Spikes bisexual and unisexual, terminal spike gynecandrous, lateral spikes pistillate, lower on peduncles; pistillate scales acute or mucronate. Perigynia compressed-triangular, green to brown, 2-ribbed on angles, transversely corrugated, 1.8–2.6 mm long, 2 mm wide, beak 0.2 mm long. Achenes 1.7 mm long, 1 mm wide, triangular and granular; stigmas 3. Wooded areas, ditches, & wet waste places; KS, MO; (PA to IA, s to VA, IN, & OK).

112. ***Carex sparganioides*** Willd. (Group III) Cespitose perennial, short to elongate rootstock. Culm 0.4–1 m tall, triangular, not sulcate. Leaves 5–8.5 mm wide, sheaths cross-rugulose and rarely red dotted dorsally, green-and-white mottled and septate-nodulose ventrally, loose. Heads 3–15 cm long, lower spikes separate, 0.5–6 cm long; pistillate scales short-cuspidate. Perigynia plano-convex, 3.4–4.2 mm long, 1.8–2.2 mm wide, nerves faint on both sides, round tapering at base, apex abruptly contracted into a serrulate beak. Achenes 1.5 mm long, lenticular; stigmas 2. Dry to moist woods & thickets; SD, NE, IA, KS, MO; (N.B. to Ont., s to NC, TN, & OK).

113. ***Carex sprengelii*** Dew. ex Spreng. (Group VIII) Rhizomatous perennial, rootstock prolonged, stout, matted, brown-fibrillose. Culms 3–8 dm tall, weakly triangular. Leaves 2–4 mm wide; sheaths long, thin, white-hyaline ventrally. Spikes usually unisexual, upper 1 to 3 spikes staminate, occasionally androgynous, pistillate spikes lateral, pendulous, separate and loosely flowered; pistillate scales acute, cuspidate or awned. Perigynia globose, green to straw colored, shining, 2-ribbed, 5–6 mm long, 2 mm wide, beak equal or longer than the body, bidentate. Achenes 2.2–4.2 mm long, 1.7 mm wide, triangular; stigmas 3. Prairies & sandy areas, but primarily wooded areas; ND, MN, WY, SD, NE, IA; (Que. to B.C., s to DE, AR, & CO).

114. ***Carex squarrosa*** L. (Group IV) Cespitose perennial, rootstock short, black. Culms 3–8 dm tall, rough triangular. Leaves 2–6 mm wide; sheaths brown-hyaline ventrally. Terminal spikes gynecandrous, pistillate spikes often lacking; pistillate scales acute to awned. Perigynia squarrose, inflated, suborbicular, green-brown, strongly ribbed above, 4–5 mm long, 3 mm wide, beak 2.5–3.5 mm long, teeth 0.25 mm long. Achenes 3 mm long, 1.5 mm wide, triangular, style persistent and bent; stigmas 3. Wooded & waste areas; NE, KS, MO; (Que. to Ont., s to GA, AR, & OK).

115. ***Carex sterilis*** Willd. (Group I) Densely cespitose perennial, rootstocks short, black. Culms 3–6 dm tall, sharply triangular, roughened on angles above. Leaves 1.2–5 mm wide, sheaths tight, hyaline ventrally, thickened, concave and yellowish-brown tinged at mouth. Heads 2–3 cm long, spikes approximate or scattered, variable, pistillate, staminate, androgynous or gynecandrous; pistillate scales ovate, obtuse or somewhat cuspidate. Perigynia plano-convex, ovate or ovate-lanceolate, 2.5–3 mm long, 1.5–1.7 mm wide, thick, spongy, and truncate at base, several-nerved both sides, abruptly contracted into a beak about 1/2 the length of body, margins at the body and beak serrulate. Achenes lenticular, broadly ovate, 1.5 mm long, 1 mm wide, style slender, straight; stigmas 2. Swampy meadows; ND; (Newf. to Ont., s to NJ, OH, IA, ND, & MT). *C. muricata* L. var. *sterilis* (Mack.) Gl.— Gleason & Cronquist.

116. ***Carex stipata*** Muhl. (Group III) Cespitose perennial, rootstock short, dark. Culms 3–12 dm tall, slightly winged, concave, spongy, triangular. Leaves 4–8 mm wide; sheaths cross-rugulose ventrally, septate-nodulose dorsally. Heads 3–10 cm long; spikes bisexual, androgynous, lower often separate; pistillate scales ovate-triangular, acuminate to cuspidate.

Perigynia plano-convex, yellow, strongly several-nerved dorsally, strongly few-nerved ventrally, 4–5 mm long, 1.6–2 mm wide, base spongy, truncate-rounded, tapering to a serrulate beak, beak about 2 × the length of body. Achenes 1.5–2 mm long, 1.2–1.7 mm wide, lenticular; stigmas 2. Low grounds, sedge swales, swamps, & moist woods; ND, MN, SD, NE, IA, KS, MO, OK, TX; (Newf. to AK, s to FL, TX, & CA).

117. *Carex stricta* Lam. (Group VI) Cespitose perennial, rootstock short to prolonged, stout, scaly, brown. Culms 3–8 dm tall, rough triangular. Leaves 2–5 mm wide; sheaths white to red-brown tinged ventrally, filamentose. Spikes unisexual, terminal spike staminate, 2–4 cm long, lateral spikes pistillate or androgynous, pistillate scales obtuse to acuminate. Perigynia usually flattened, dark green; few-nerved both sides, 2-edged, 2-ribbed, 2.2–2.7 mm long, 1.5 mm wide, beak less than length of body. Achenes 1.7 mm long, 1.2 mm wide, lenticular; stigmas 2. Shoreline of swamps & marshes; ND, MN, WY, SD, NE, IA, KS, OK; (Newf. to Man., s to NC, TN, & TX). *C. strictior* Dew.—Rydberg.

118. *Carex subimpressa* Clokey. (Group VIII) Loosely cespitose to rhizomatous perennial, rootstock prolonged, scaly. Culms 3–6 dm tall, rough triangular. Leaves 3–6 mm wide; sheaths tight, concave, dull brown. Spikes unisexual, upper 3 spikes staminate, 2.5–5 cm long; lower 2 spikes pistillate; pistillate scales rough-awned. Perigynia suborbicular, usually inflated, densely pubescent, many-ribbed, 5 mm long, 2.5 mmm wide, beak 1.5 mm long, flattened, bidentate, teeth of beak 0.5 mm long, hispid within. Achenes 1.7 mm long, 1.5 mm wide, triangular with concave sides, style persistent; stigmas 3. Wet ditches of flood plains; KS; (MI & IL, w to MO & KS).

This species is an apparent hybrid between *C. lanuginosa* and *C. hyalinolepis*.

119. *Carex sychnocephala* Carey. (Group I) Cespitose perennial, rootstock short, black. Culms 0.5–6 dm tall, sterile culms with long leaves at top, obtusely triangular. Leaves 1.5–4 mm wide; sheaths tight, white-hyaline ventrally. Heads 1.5–3 cm long; spikes bisexual, gynecandrous, aggregate; pistillate scales acuminate or cuspidate. Perigynia flat, scalelike, green to straw colored, slightly nerved dorsally, narrow wing-margined to base, 5–6.5 mm long, 1 mm wide, base spongy, apex tapering or contracted to a serrulate beak 3–4.5 mm long. Achenes 1 mm long, 0.5 mm wide, lenticular; stigmas 2. Sandy shores & river banks; ND, MN, SD, IA; (Ont. to AK, s to PA, IA, SD, CO, & WA).

120. *Carex tenera* Dew. (Group I) Cespitose perennial, rootstock short, black. Culms 3–7.5 dm tall, sharply triangular. Leaves 0.5–2.5 mm wide, sheaths septate-nodulose, hyaline ventrally. Heads 2.5–5 cm long; spikes bisexual, gynecandrous, moniliform, flexuous or nodding; pistillate scales ovate. Perigynia plano-convex, straw colored, strongly 3- to 7-nerved dorsally, less-nerved ventrally, wing-margined to base, 3–4.5 mm long, 1.5–1.8 mm wide, base rounded and contracted to a serrulate beak 1/2 the length of the body. Achenes 1.2 mm long, 1 mm wide, lenticular; stigmas 2. Woodlands, uplands, & occasional meadows; ND, MN, WY, SD, NE, MO; (Que. to Alta., s to NC, AR, OK, NM, & OR).

121. *Carex tetanica* Schkuhr. (Group V) Cespitose perennial, rootstock short and slender, whitish. Culms 3–6 dm tall, rough triangular. Leaves 2–5 mm wide; sheaths long, tight, overlapping, white or yellow hyaline ventrally. Spikes unisexual, terminal spike staminate, 1.5–3 cm long; lateral spikes 2 or 3, widely separate; pistillate scales acuminate to awned, purple-brown tinged. Perigynia obtusely triangular, 3-ranked, deep green, 2-keeled, faintly nerved, 2–4 mm long, 1.5–2 mm wide, broadest above middle and tapering to base and tip, beak minute. Achenes 2–2.5 mm long, 1.2 mm wide, triangular with concave sides; stigmas 3. Swamps, wet meadows, & ditches; ND, MN, NE, IA, MO; (Ont. to Alta., s to VA, AR, & KS).

122. **Carex texensis** (Torr.) Bailey. (Group III) Perennial with rootstocks short to prolonged, dark colored. Culms 1–3 dm tall, very slender, sharply triangular. Leaves 0.7–1.5 mm wide; sheaths tight, not septate-nodulose, not cross-rugulose. Heads 0.6–3 cm long, spikes androgynous, lower spikes separated; pistillate scales ovoid, acuminate to cuspidate. Perigynia plano-convex, 3 mm long, 1 mm wide, rounded and strongly spongy at base, tapering into a smooth-margined beak. Achenes 1.3 mm long, rounded at base and tapering to apex, lenticular; stigmas 2. Moist to dry, open woodlands; NE, KS; (NJ to NE, s to GA & TX). *C. retroflexa* Muhl. var. *texensis* (Torr.) Fern.—Gleason & Cronquist.

123. **Carex torreyi** Tuckerm. (Groups IV & V) Cespitose perennial, rootstock short. Culms 1.5–5 dm tall, triangular with convex sides. Leaves 1.5–3 mm wide, soft-pilose; sheaths tight pubescent, cinnamon-brown ventrally. Spikes unisexual, terminal spike staminate 8–16 mm long, lateral spikes pistillate; pistillate scales short cuspidate. Perigynia obscurely triangular, yellow-green, several-nerved, 2.3–3 mm long, 2 mm wide, base narrowed, beak 0.2–0.5 mm long. Achenes 2.5 mm long, 1.7 mm wide, triangular with concave sides; stigmas 3. Moist, open meadows & wooded areas; ND, MN, SD; (Man. to Alta., s to MN, SD, & CO). *C. abbreviata* Prescott—Rydberg.

124. **Carex triangularis** Boeck. (Group III) Cespitose perennial, rootstock short, black. Culms 4–9 dm tall, acutely triangular. Leaves 2–4.5 mm wide; sheaths tight, cross-rugulose, slightly red dotted ventrally, prolonged, reddish-brown tinged dorsally. Heads 3–5 cm long; spikes bisexual, androgynous, densely aggregate; pistillate scales ovate-triangular. Perigynia plano-convex, brownish-yellow, red dotted and punctate, few-nerved dorsally, usually nerveless ventrally, 3–3.5 mm long, 2.5–3 mm wide, base truncate-cordate, abruptly contracted to a serrulate beak 1/3 length of body. Achenes 1.7 mm long, 1.2 mm wide, lenticular; stigmas 2. Prairie ditches; KS, TX; (MO & KS, s to MS, LA, & TX).

125. **Carex tribuloides** Wahl. (Group I) Cespitose perennial, rootstock short, black. Culms 6–9 dm tall, sharply triangular with concave sides. Leaves 2.5–7 mm wide, sheaths loose, green-striate, white-hyaline ventrally. Heads 2.5–5 cm long; spikes bisexual with a blunt apex, gynecandrous, aggregate; pistillate scales acute-acuminate. Perigynia flattened, plano-convex, reniform or orbicular, green to straw colored, several-nerved both sides, narrow wing-margined to base, 3.5–5 mm long, 1–1.5 mm wide, base rounded, apex tapering to a serrulate beak 1/3 the length of body. Achenes 1.5 mm long, 0.5–0.7 mm wide, lenticular; stigmas 2. Swamps, shores, ditches, & prairies; NE, IA, KS, MO, OK; (Que. to B.C., s to FL & TX).

126. **Carex trichocarpa** Schkuhr. (Group VIII) Loosely cespitose and stoloniferous perennial, stolon long, slender, scaly. Culms 5–12 dm tall, sharply triangular and rough above. Leaves 3–8 mm wide, sheaths tight, deeply concave, purple tinged at mouth, not breaking or filamentose. Staminate spikes usually 3, narrowly linear, 2–5 cm long; pistillate spikes 2–4, sessile or peduncled, 2–8 cm long; scales lanceolate or ovate-lanceolate, awned, acuminate or acute. Perigynia oblong-ovoid, 6–10 mm long, 2–3 mm wide, suborbicular in cross section, somewhat inflated, loosely white-short-hirsute, strongly many-ribbed, tapering into a bidentate beak 1/2 as long as the body, teeth 2 mm long, erect or spreading. Achenes elliptic, 2.5 mm long, 2 mm wide, triangular with concave sides, style slender, straight; stigmas 3. Marshes & wet meadows; IA, NE; (Que. to Ont., s to MD, IN, MO, NE).

127. **Carex typhina** Michx. (Group IV) Cespitose perennial, rootstock short, thick, black. Culms 3–9 dm tall, sharply triangular, very rough on the angles above. Leaves 3–10 mm wide, sheath usually loose, brownish-hyaline ventrally, concave at mouth. Spikes 1–6, usually 3, the lower portion of terminal one staminate, the remainder pistillate, pistillate portion

of terminal spike oblong-cylindric, 2–4 cm long, the lateral spikes shorter, usually peduncled; pistillate scales lanceolate, obtuse at maturity. Perigynia conic-obovoid or obconic, crowded, the body 3–5 mm long, about 3 mm wide, inflated, suborbicular in cross section, several-ribbed, glabrous, abruptly contracted into a conic beak, 2–3.5 mm long, sparingly serrulate. Achenes obovoid, 2.5 mm long, 1.5 mm wide, triangular with concave sides, style straight below or bent at summit; stigmas 3. Marshes & wet wooded areas; MO; (Que., s to GA, LA, & TX).

128. *Carex umbellata* Schkuhr ex Willd. (Group VIII) Densely cespitose perennial, rootstock short, stout. Culms very short to 15 cm tall, sharply triangular, rough or angled. Leaves 1.5–3 mm wide, sheaths smooth dorsally, little breaking or filamentose ventrally. Terminal spike staminate, sessile or short peduncled, 5–10 mm long, usually a short pistillate spike or staminate spike, peduncled, basal spikes 2 or 3, 4–10 mm long; pistillate scales ovate, abruptly acute, acuminate or cuspidate. Perigynia 2.2–3.5 mm long, the body subglobose, triangular-orbicular in cross section, 1–1.3 mm wide, short pubescent, 2-keeled, otherwise nerveless or obscurely nerved, abruptly contracted into a short beak, 3/4 the length of body. Achenes triangular with convex sides, 1.5 mm long, style slender; stigmas 3. Usually sandy soils in dry upland woods, or around sandstone ridges; e 1/2 KS, MO; (Newf. to B.C., s to VA & MO). *C. umbellata* Schkuhr f. *vicina* (Dew.) Wieg.—Steyermark.

129. *Carex vallicola* Dew. (Group III) Cespitose perennial, short prolonged rootstock. Culms 2.5–6 dm tall, sharply triangular. Leaves 1 mm wide, sheaths tight, not septate-nodulose or cross-rugulose. Heads 1.5–3 cm long, aggregated; spikes androgynous; pistillate scales acute to cuspidate. Perigynia plano-convex, 3.5 mm long, 2 mm wide, slightly nerved dorsally, abruptly narrowed to a serrulate beak. Achenes 2 mm long, 2 mm wide, lenticular; stigmas 2. Dry hillsides & thickets; SD; (MT to OR, s to CA, also w NE; Mex.).

130. *Carex vesicaria* L. var. ***monile*** (Tuckerm.) Fern. (Group VIII) Cespitose perennial rootstock short. Culms 2–10 dm tall, sharply triangular. Leaves 2–5 mm wide; sheaths fibrillose ventrally. Spikes unisexual, terminal spike staminate, 2.5–4 mm long, 2 or 3 lateral spikes pistillate, 2–7.5 mm long; pistillate scales long-acuminate. Perigynia globose-ovoid, inflated, yellow-green to brown, strongly nerved, 5–8 mm long, 3 mm wide, beak 2 mm long, teeth 0.5–1 mm long. Achenes 2.5 mm long, 1.7 mm wide, triangular, persistent bent style; stigmas 3. Roadside marshes; MN, SD, IA, MO; (Newf. to B.C., s to DE, IN, MO, NM, & CA).

131. *Carex viridula* Michx. (Group V) Cespitose perennial, rootstock short, brown. Culms 1–4 dm tall, stiff, obtusely triangular. Leaves 1–3 mm wide; sheaths thin, white ventrally. Spikes unisexual, terminal spike staminate, 3–15 mm long, lateral spikes pistillate; pistillate scales obtuse to cuspidate. Perigynia obtusely triangular, yellow-green, ribbed; 2–3.6 mm long, 1.2 mm wide, spreading, beak 0.5–0.7 mm long, slightly bidentate. Achenes 1.2 mm long, 1 mm wide, triangular with concave sides; stigmas 3. (2n = ca 70) Shorelines, marshes, & wet meadows; MT, ND, MN, SD; (Newf. to AK, s to PA, IL, MN, SD, NM, & CA).

132. *Carex vulpinoidea* Michx. (Group III) Cespitose perennial, rootstock short, blackish. Culms 3–9 dm tall, sharply triangular, usually shorter than leaves, at least in young plants. Leaves 2–4 mm wide; sheaths tight, cross-rugulose, greenish-white ventrally. Heads 4–8 cm long; spikes bisexual, androgynous, aggregate to separate in lower spikes; pistillate scales ovate, aristate. Perigynia plano-convex, yellow-green, several-nerved dorsally, usually nerveless ventrally, 2–2.7 mm long, 1–1.5 mm wide, base round, tapering to a serrulate beak, beak about equal to body. Achenes 1 mm long, lenticular; stigmas 2. (n = 26, 27) Hillsides, ravines, swampy areas, shores of ponds & lakes, & ditches; MT, ND, MN, WY, SD, NE, IA, KS, MO, OK; (Newf. to B.C., s to FL, TX, AZ, & OR; Mex.).

133. Carex xerantica Bailey. (Group I) Cespitose perennial, rootstock short, black. Culms 3–6 dm tall, sharply triangular. Leaves 2–3 mm wide; sheaths tight, white-hyaline ventrally. Heads 3–5 cm long; spikes bisexual, gynecandrous, approximate; pistillate scales ovate-acute. Perigynia flat plano-convex, green to golden-yellow, many-nerved dorsally, nerveless ventrally, strongly wing-margined to base, 4–5.5 mm long, 2–2.5 mm wide, base round tapering, apex tapering or contracted to a serrulate beak, beak 1/4 to equal the length of the body. Achenes 2.5 mm long, 1.5–1.7 mm wide, lenticular; stigmas 2. Prairies & woody slopes; ND, SD, NE; (Man. to Alta., s to MN, w NE, NM, & AZ; Mex.).

3. CYPERUS L., Umbrella Sedge

Cespitose annual or loosely rhizomatous to tufted perennial. Culms triangular, solid, with closed sheath. Leaves involute or flat, mostly at the base. Inflorescence of terminal umbellike spikes or heads, subtended by 1–several bladelike involucral bracts; spikelets several to many, few- to many-flowered, usually laterally compressed. Scales 2-ranked, imbricate, keeled, stamens 1–3, styles 2- or 3-cleft. Achenes lenticular or trigonous.

References: Corcoran, Sister Mary Lucy. 1941. A revision of the subgenus *Pycreus* in North and South America. Catholic Univ. Amer. Biol. Ser. 37: 1–68; Marcks, B. G. 1974. Preliminary reports on the flora of Wisconsin. No. 66, Cyperceae II.—Sedge Family II. The genus *Cyperus*—The umbrella sedges. Wisconsin Acad. Sci., Arts and Letters 62: 261–284; McGivney, Sister M. Vincent De Paul. 1938. A revision of the subgenus *Eucyperus* found in the United States. Catholic Univ. Amer. Biol. Ser. 26: 1–74.

1 Achene lenticular, compressed to biconvex, style 2-cleft.
  2 Spikelets with only 1 flower, spikelets slightly longer (2–3×) than broad .... 19. *C. tenuifolius*
  2 Spikelets with several flowers (6–50), spikelets much longer (4–12×) than broad.
    3 Style slightly or not exserted, cleft nearly to the middle .... 13. *C. rivularis*
    3 Style conspicuously exserted, cleft nearly to the base .... 3. *C. diandrus*
1 Achene 3-sided (trigonous), style 3-cleft.
  4 Rachilla of spikelets continuous, scales gradually deciduous, from base of rachilla to apex.
    5 Stamens 1 or 2, rachilla not winged or narrowly winged.
      6 Achenes 0.6–0.8 mm long, culms with retrorse projections ..... 18. *C. surinamensis*
      6 Achenes 0.9–1.5 mm long, culms smooth or with horizontal or antrorse projections.
        7 Stamens 3, styles 0.1–0.5 mm long .... 7. *C. fuscus*
        7 Stamens 1, styles 0.5–1 mm long.
          8 Achene linear, 0.2–0.3 mm wide; scales linear; perennial .... 12. *C. pseudovegetus*
          8 Achene oblong, 0.4–0.5 mm wide; scales ovate; annual ... 1. *C. acuminatus*
    5 Stamens 3; rachilla winged.
      9 Scales 1.2–1.6 mm long; achenes 0.6–1.2 mm long .... 5. *C. erythrorhizos*
      9 Scales 1.4–4 mm long; achenes 1–2 mm long.
        10 Scales 1.4–2.9 mm long; rhizomes scaly .... 6. *C. esculentus*
        10 Scales 3–4 mm long; rhizomes not scaly.
          11 Leaflike bracts of involucre 3 or 4, about equaling the inflorescence; scales 3–3.5 mm long .... 14. *C. rotundus*
          11 Leaflike bracts of involucre 5–13, usually longer than the inflorescence; scales 3.8–4 mm long .... 16. *C. setigerus*
  4 Rachilla of spikelets articulated; scales persistent and then falling all at once from the rachilla.
    12 Rachilla articulating at the base of each scale.

13 Scales approximate, tip of scale overlapping the base of the one above it .... 10. *C. odoratus*
13 Scales remote, tip of scale not overlapping the base of the one above it .... 4. *C. engelmannii*
12 Rachilla not separating into joints (short segments).
   14 Scales with strongly recurved acuminate tips .... 2. *C. aristatus*
   14 Scales with incurved or straight blunt tips.
      15 Rachilla of spikelet essentially wingless.
         16 Mucro of the scales 0.3–1.5 mm long, culms scabrous, rarely smooth, achene 2.2–2.6 mm long .... 15. *C. schweinitzii*
         16 Mucro of the scale up to 0.3 mm long, culm smooth; achene 1.4–2.2 mm long .... 9. *C. lupulinus*
      15 Rachilla of spikelet with wings up to 1.2 mm broad.
         17 Spikelets in dense, symmetrical, globose to oblong spikes; spikelets 1- to 6-flowered.
            18 Each wing of the spikelet axis thickened and clasping the achene; rachis of spike obvious between the spikelets .... 20. *C. uniflorus*
            18 Each wing not thickened and usually not clasping the achene; rachis of spike not obvious between the spikelets .... 11. *C. ovularis*
         17 Spikelets loose, or if dense, the spikes irregular in shape; spikelets with more than 6 florets.
            19 Spikelets, at least the lowermost, reflexed or pointing downward .... 8. *C. lancastriensis*
            19 Spikelets ascending and horizontally spreading .... 17. *C. strigosus*

1. Cyperus acuminatus Torr. & Hook. Cespitose annual. Culms slender, 2–36 cm tall, usually stramineous. Leaves few, borne at the base, 0.5–2.6 mm wide, nearly equaling or slightly longer than the culm. Involucral bracts 2–4, unequal, longer than the inflorescence. Inflorescence of numerous dense globose spikes; spikelets ovate to oblong, 1.5–3 mm wide, 2.4–7 mm long, strongly flattened, closely imbricate, 10- to 24-flowered. Scales 1.4–2.6 mm long, outwardly curved at the tip, strongly 3-nerved; stamen 1; style 3-cleft. Achene trigonous, 0.5–1.2 mm long, 0.4–0.5 mm broad. Shorelines, stream banks, & other low, wet areas; GP; (VA to ND, s to FL & TX, also WA & OR).

2. Cyperus aristatus Rottb. Cespitose annual, sweet-scented. Culms slender, smooth, 2–13 cm tall, purplish tinged at base. Leaves few, near base, 1.6–3.4 mm wide, about as long as culm. Inflorescence a cluster of sessile spikelets or an umbel of 1–3 rays; involucre very long, exceeding the inflorescence; spikelets in dense cluster, flattened, ovate, oblong, 3.6–5.6 mm long, 1–2 mm wide, 8- to 16-flowered. Scales oblong, 7- to 9-nerved, outwardly curved tip, 0.8–1.6 mm long; stamen 1; style 3-cleft. Achenes oblong, triangular, light brown, 0.6–1 mm long, 0.2–0.4 mm wide. Low wet areas of streams & shores; GP; (N.B. to B.C., s to FL, MO, & TX). *C. inflexus* Muhl.—Rydberg.

3. Cyperus diandrus Torr. Annual with fibrous root. Culms 9–30 cm tall, smooth. Leaves 1.4–2.4 mm wide, nearly equaling the culm. Inflorescence usually with 1–5 unequal peduncles bearing loose spikes, up to 10 spikelets; spikelet 5–10.2 mm long, 10- to 18-flowered. Scales ovate, loosely imbricate, 1.8–2.6 mm long; stamens 2; style 2-cleft nearly to base, persistent, exserted up to 4 mm beyond the scale. Achene lenticular, 0.8–1.2 mm long, 0.4–0.7 mm wide. Wet areas along shores & banks of lakes & streams; MN, NE, IA, MO; (N.B. to MN, s to SC, IA, MO, & NM).

4. Cyperus engelmannii Steud. Annual with fibrous roots. Culms 2–60 cm tall, smooth, trigonous. Leaves up to 6 mm wide, usually exceeding the culm. Inflorescence of several unequal peduncles, either simple or compound. Spikes with numerous spikelets, spreading horizontally or ascending; spikelets 6- to 18-flowered, up to 8–24 mm long, breaking easily into sections, each composed of a scale, the next lower internode, the attached wings and achene. Scales ovate-lanceolate, 2.1–2.8 mm long, the tip of each barely reaching the base of the one above it (giving the spikelet a zigzag appearance); stamens 3; style 3-cleft. Achene trigonous, 1.5–2.5 mm long, 0.3–0.6 mm wide, linear, oblong, brownish. Sandy, muddy shorelines of ponds, lakes, marshes; ND, MN, SD, IA, NE; (MA to ND, s to VA & NE).

5. *Cyperus erythrorhizos* Muhl. Cespitose annual from fibrous blood-red roots. Culms erect, 1–7 dm tall, smooth, bluntly triangular. Leaves several, 2–10 mm wide, about as long as the culm, with scabrous margins. Inflorescence a compound umbel with 1–several sessile spikes and numerous divergent rays; several bladelike involucral bracts, most longer than the inflorescence; each spike 5–40 mm long, with numerous crowded spikelets, 8- to 30-flowered, 4–12 mm long, 0.8–1.5 mm wide; rachilla narrowly hyaline-winged, wings deciduous. Scales reddish-brown with a green keel, 1.2–1.6 mm long, densely aggregate; stamens 3; style 3-cleft. Achene trigonous, ovoid, 0.6–1.2 mm long, 0.3–0.6 mm wide. (n = 48) Flood banks, sandy river shores, river sand bars, boggy depressions; ND, MN, SD, IA, NE, KS, MO, OK; less frequent n & w; (MA to WA, s to FL, OK, CA).

Although considerable variation occurs within this species, it may be distinguished by its small reddish-brown scales and ivory-to-white trigonous achene.

6. *Cyperus esculentus* L. Perennial from many slender, scaly rhizomes that terminate in small hard tubers. Culms 1–9 dm high, acutely 3-angled, smooth. Leaves crowded, about as long as the culm, mostly 3–10 mm wide. Inflorescence a compound umbel of several subsessile spikes and 1–10 rays; involucral bracts broad and longer than the inflorescence; spikelets slender, 5–20 mm long, 0.8–1.8 mm wide, 9–25 on a spike, 8- to 20-flowered. Scales overlapping, 1.4–2.9 mm long, several-nerved, golden-brown; rachilla winged, stamens 3, styles 3-cleft. Achenes trigonous, oblong-obovoid, 0.9–1.5 mm long, 0.5–0.8 mm wide. (n = 9, ca 54) Margins of lakes, streams, & ditches, frequently in cult. soil & wet prairies; ND, MN, SD, IA, NE, CO, KS, MO, TX, OK; (N.S. to WA, s to FL, TX, AZ; Mex.).

7. *Cyperus fuscus* L. Annual from fibrous roots. Culms 2–30 cm tall, flattened, trigonous. Leaves few per culm, 1–25 cm long, 1–4 mm wide. Inflorescence a hemispheric head of 6–24 spikelets; involucral bracts 3, 1–12 cm long, up to 0.5 mm wide; spikelets 3–6 mm long, 1–1.5 mm wide, linear, compressed, 8- to 18-flowered. Scales 1–1.2 mm long, oval, 3-nerved, dark reddish-brown; stamens 3; style 3-cleft. Achene 0.9–1.1 mm long, 0.5 mm broad, trigonous, ovoid. Sandy soil; NE: Lincoln; (MS to NY, VA, also NE). *Introduced.*

8. *Cyperus lancastriensis* Porter. Perennial with short rhizomes less than 1 cm long. Culms 3–90 cm tall, trigonous, smooth. Leaves shorter than the culm, glabrous, up to 10 mm wide. Inflorescence of 4–10 peduncled short spikes; spikelets 6–26 mm long, 1–1.5 mm wide, 3- to 14-flowered; bracts 6–10, longer than inflorescence. Scales 3–5 mm long, 9- to 13-nerved, yellowish; stamens 3; style 3-cleft. Achene 2–3 mm long, 0.5–0.8 mm wide, trigonous, oblong, brown. Rocky, open slopes, moist woods; MO, OK; (NJ to OK, s to GA).

9. *Cyperus lupulinus* (Spreng.) Marcks. Cespitose perennial from tuberous-thickened rhizomes. Culms very slender and wiry, trigonous, glabrous, up to 40 cm tall. Leaves conduplicate or flat, margins scabrous, up to 3.5 mm wide. Inflorescence a single, sessile, dense hemispherical to nearly spherical head, up to 25 mm in diam, rarely with 1 or 2 additional rays; involucral bracts several, mostly deflexed; spikelets radiate and densely congested, up to 25-flowered, up to 20 mm long. Scales oblong, up to 3.5 mm long, short mucronate, obviously nerved; stamens 3; style 3-cleft. Achene trigonous up to 2.2 mm long, up to 1.1 mm wide, dark brown or black. Dry prairie, moist sandy or limestone soil.

Two subspecies are recognized by Marcks, op. cit.

9a. subsp. *lupulinus*. Spikelets 6- to 22-flowered. Scales 2.5–3.5 mm long, fitting loosely over the achene, achene 1.8–2.2 mm long. NE, IA, CO, KS, MO, OK, TX; (New England to SD & CO, s to FL & TX, also nw U.S.). *C. bushii* Britt., *C. filliculmis* Vahl—Rydberg.

9b. subsp. *macilentus* (Fern.) Marcks. Spikelets 3- to 7-flowered. Scales 1.8–2.5 mm long, fitting firmly over the achene, achene 1.2–1.8 mm long. SD, NE, IA; (New England to SD & NE, s to VA). *C. filiculmis* Vahl var. *macilentus* Fern.—Fernald.

10. Cyperus odoratus L. Annual with fibrous roots, or behaving as a tufted perennial in southern part of range. Culms up to 1 m tall, trigonous, smooth. Leaves shorter than culm, up to 10 mm wide. Inflorescence of numerous unequal peduncles, either simple or compound; spikes with numerous spreading spikelets, 4- to 30-flowered, up to 25 mm long; spikelet breaking easily into sections, each comprised of a scale, the next lower internode, the attached wings and achene. Scales ovate, 1.5–2 mm long, the tip of each overlapping the base of the one above it; stamens 3; style 3-cleft. Achenes trigonous, 1–1.7 mm long, red-brown to brown. Wet sand & mud, riverbanks, ponds, sloughs, marshes; ND, MN, SD, IA, NE, CO, KS, MO, OK, TX; (MA to OR, s to AL, TX, & CA). *C. speciosus* Vahl—Rydberg; *C. ferruginescens* Boeck.—Fernald.

This species is probably the most abundant and variable of our Cyperus.

11. Cyperus ovularis (Michx.) Torr. Perennial from short rhizomes. Culms 2.5 to 9 cm tall, smooth, trigonous, with tubers (tuberlike enlargements) at the base. Leaves 2–10 mm wide, shorter than to about equaling the culm. Inflorescence with 3–12 unequal peduncles, each with a globose dense spike 5–20 mm in diam; spikelets 3–9 mm long, numerous, radiating from a central point, 2- to 4-flowered; bracts 4–7, the longer much exceeding the inflorescence. Scales with blunt tips, up to 4 mm long; stamens 3; style 3-cleft. Achene trigonous, 1.8–2.2 mm long, up to 0.7 mm wide, narrowly oblong, brown. Low prairies, roadside ditches, wooded areas; MO, KS, OK; (NY to KS, s to FL & TX). *C. ovularis* (Michx.) Torr. var. *sphaericus* Boeck.—Steyermark; *C. ovularis* (Michx.) Torr. var. *cylindricus* (Ell.) Torr.—Waterfall.

12. Cyperus pseudovegetus Steud. Tufted perennial from rhizomes. Culms 40–80 cm tall, smooth. Leaves few to several, 2–5 mm wide, some equaling or longer than culms. Inflorescence of 3–10 unequal rays, heads several, with 15–50 spikelets; bracts 3–6, most surpassing the inflorescences; spikelets 2.5–6 mm long, 3 mm broad, narrowly ovate with 6–18 flowers, flattened. Scales 1.8–2.5 mm long, tips excurrent; stamen 1; style 3-cleft. Achene linear, 0.8–1.4 mm long, 0.3 mm wide, bluntly trigonous. Shorelines, ditches, wet low prairies; KS, OK, MO; (NJ to KS, s to FL & TX). *C. virens* Michx.—Fernald.

13. Cyperus rivularis Kunth. Annual with fibrous roots. Culms 7–33 cm tall, smooth. Leaves 0.8–2.0 mm wide, shorter than culms. Inflorescence usually with 1–5 unequal peduncles with loose spikes, up to 10 spikelets; bracts 1–3, divergent, spikelets 10- to 30-flowered, 6–20 mm long. Scales ovate, closely imbricate, 1.8–2.4 mm long, dark purplish-brown from midvein to margins (rarely unpigmented); stamens 2; style 2-cleft to about the middle, deciduous. Achene 0.8–1.2 mm long, 0.6–1 mm wide, lenticular, reddish-brown. Wet ground or margins of streams, lakes, marshes, moist meadows; ND, MN, SD, IA, NE, CO, KS, MO, OK; (ME to ND, s to GA & TX, CA; Mex.). *C. niger* R. & P. var. *castaneous* (Pursh) Kükenth.—Correll & Johnston.

Specimens from Kansas identified as *C. flavescens* L. var. *poaeformis* (Pursh) Fern. belong to *C. rivularis*.

14. Cyperus rotundus L. Perennial with long rhizomes with fibrous-coated tuberlike thickenings. Culms 8–50 cm tall, triangular, sometimes compressed. Leaves 3–6 mm wide, about as long as the inflorescence. Inflorescence of 3–8 peduncles, each bearing a cluster of 2–12 spikelets; bracts 2–4, longer or shorter than the inflorescence; spikelets compressed, 4–40 mm long, 1–2 mm wide, 12- to 36-flowered. Scales 3.0–3.5 mm long, 7-nerved, reddish-brown; stamens 3; style 3-cleft. Achene 1.5–2 mm long, 1 mm wide, trigonous, black, shining. Flood plains, wet meadows, loamy soils; OK; (OK & MO s to TX).

15. *Cyperus schweinitzii* Torr. Perennial from short, hard, cormlike rhizomes. Culms up to 8 dm tall, trigonous, antrorsely scabrous on the angles near the tip. Leaves up to 8 mm wide, seldom as long as the inflorescence. Inflorescence with a cluster of oblong sessile spikes and up to 8 unequal, ascending pedunculate rays, up to 12 cm long; several involucral bracts as long as or exceeding the inflorescence and strongly ascending, spikelets ascending, ovate to oblong, compressed, up to 18-flowered, up to 25 mm long. Scales ovate, firm, conspicuously nerved, mucro up to 1.0 mm long; stamens 3; style 3-cleft. Achene trigonous, up to 2.6 mm long, up to 1.3 mm wide, brown or black. Sandy soils of lowland prairie, or sandhills; throughout GP; (Que. to Sask., s to NJ, TX, NM).

Specimens from the GP previously identified as *C. houghtonii* Torr. belong to this taxon or *C. lupulinus*.

16. *Cyperus setigerus* Torr. & Hook. Perennial with creeping rhizomes. Culms 70–110 cm tall, sharply triangular. Leaves 4–7 mm wide, much shorter than the culm. Inflorescence of 9–13 peduncles, compound clusters or short spikes, 10–30 spikelets per cluster; spikelets 6–40 mm long, 1.5–2 mm wide, 6- to 40-flowered, much compressed; bracts 5–13, some 2 to 3 × length of inflorescence. Scales reddish-brown to yellowish-brown, 3–4 mm long, 5- to 7-nerved; stamens 3; style 3-cleft. Achene 1.5 mm long, 0.5 mm wide, trigonous, dark brown. Scattered moist prairies, ditches, shoreline of ponds & lakes; KS, OK, MO, TX; (KS, MO s to OK & TX). *C. hallii* Britt.—Rydberg.

17. *Cyperus strigosus* L. Tufted perennial. Culms up to 1 m tall, smooth, trigonous, from a tuberlike base. Leaves 2–8 mm wide, longer ones about equaling the culm. Inflorescence with 4–12 unequal peduncles of simple or compound spikes 10–50 mm long, each spike with 20–90 spreading spikelets; bracts 3–10, surpassing the inflorescence; spikelets 5–25 mm long, 1–2 mm wide, 5- to 20-flowered. Scales 3.7–4.5 mm long with 7–9 nerves, straw colored to golden-brown, stamens 3; style 3-cleft. Achene 1.5–2.2 mm long, 0.5–0.7 mm wide, trigonous, linear-oblong, brown. Wet areas of lakes, marshes, ponds; MN, SD, NE, KS, MO, OK, TX; (Que. to SD, s to FL & TX, also WA & CA).

18. *Cyperus surinamensis* Rottb. Perennial from fibrous roots. Culms 10–60 cm tall, trigonous, downwardly scabrous. Leaves about as long as the culm, 1–3 mm wide. Inflorescence of several hemispheric heads with 8–10 spikelets; spikelets 3–14 mm long, 2 mm wide, flattened. Scales 1–1.5 mm long, 3-nerved; stamen 1; styles 3. Achenes 0.6–0.8 mm long, 0.2–0.3 mm wide, bluntly trigonous, narrowly ovoid-oblong, reddish-brown. Wet places; KS, OK; (SC to FL, w to KS & TX).

19. *Cyperus tenuifolius* (Steud.) Dandy. Cespitose annual with fibrous roots. Culms 1–30 cm tall, smooth. Leaves 1–3 mm wide, usually shorter than culms. Inflorescence of a 3-lobed sessile spike with numerous spikelets; bracts 3, spreading or reflexed; spikelets 2-ranked, 2–2.5 mm long, with 2 scales, the upper scale sterile. Scales 1.5–2.4 mm long in fertile flowers; stamens 2; style 2-cleft. Achene lenticular, 0.8–1.1 mm long, 0.5 mm wide, dark brown. Marshy banks of streams, margins of ponds, wet places in prairies, wet open woods; MO, KS; (NY to KS, s to FL & TX; Mex.).

20. ***Cyperus uniflorus*** Boeck. Tufted perennial from a fibrous root system. Culms 8–75 cm tall, trigonous, smooth with a slightly tuberous thickened base. Leaves 1–3 mm wide, usually shorter than culm. Inflorescence of 3–6 unequal peduncles, each bearing a spike of 5–50 spikelets; bracts 3–6, surpassing the inflorescence; spikelets 2–20 mm long, up to 1 mm wide, 1- to 5-flowered. Scales 2.5–4.5 mm long, with 9–11 nerves; stamens 3; style 3-cleft. Achene oblong, 1.8–2.5 mm long, 0.6–1 mm wide, trigonous, dark brown. Sandy areas, ravines, moist soil along streams; OK, TX; (AR to AZ, s to TX & Mex.).

4. DICHROMENA Michx., White-top Sedge

1. ***Dichromena nivea*** (Boeck.) Britt. Tufted perennial, rhizomes not present. Culms erect, 1–3 dm tall, up to 1 mm thick. Leaves on lower part of culms, blades up to 1 mm wide basally and becoming arcuate-filiform at tip, up to 7 cm long. Bracts 2 (occasionally 3), filiform with a white spot at the base, up to 37 mm long. Inflorescence a dense terminal head with several spikelets; spikelets usually whitish, 4–7 mm long. Scales spirally imbricate, the terminal ones enclosing a fertile floret, the lower ones staminate or empty. Achenes lenticular, transversely rugulose, crowned with a tubercle. Prairie ravines & creek banks; rare in s-cen OK; (s OK, AR, & TX).

5. DULCHIUM Rich. ex Pers.

1. ***Dulchium arundinaceum*** (L.) Britt. Perennial with rhizomes. Culms arising singly from the rhizomes, up to 8 dm tall, hollow, terete or obtusely triangular with short internodes. Leaves cauline, sheathed and 3-ranked, blades 5–8 cm long, 3–6 mm wide. Inflorescence axillary from the leaf sheaths; spikelets 2-ranked, 13–18 mm long, flattened, 5–7 perfect flowers, scales brownish-green, 6–9 mm long, acute or acuminate, several-nerved. Achenes yellowish, 2.5–3.5 mm long, lenticular, bifid style, persistent bristles usually 8. (n = 16) Marshes, swamps, & wetlands; NE, MN, MO; (Newf. to B.C., s to FL, IL, MO, NE, CA).

6. ELEOCHARIS R. Br., Spikerush

Annual or perennial herb ranging up to 1.5 m tall. Culms just below spikelet terete, flattened, 4-sided or -angled, with leaves at base consisting of bladeless sheaths. Spikelets solitary and terminal with few–many perfect flowers; scales spirally imbricate or distichous, variously shaped and colored, the lower 1–3 scales sterile in most species. Flowers usually consist of 3 stamens, 6–9 bristles (reduced in number or lacking in some species) and a 2- or 3-cleft style enlarged into a persistent base (tubercle), either distinct or confluent with the achene. Achene body trigonous to biconvex, at times almost terete, with various surface texture and color, ranging up to 3.0 mm long.

References: Drapalik, D. J., & R. H. Mohlenbrock, 1960. A study of *Eleocharis*, Series Ovatae. Amer. Midl. Naturalist 64 (2): 339–341; Harms, L. J. 1968. Cytotaxonomic Studies in *Eleocharis* Subser. Palustres: Central United States Taxa. Amer. J. Bot. 55: 966–974; Svenson, H. K. 1957. *Elocharis*. N. Amer. Flora 18(9): 509–540.

1 Culms 4-sided, culm and spikelet about the same thickness .... 13. *E. quadrangulata*
1 Culms triangular, terete or flattened, culm just below spikelet smaller than the spikelet.
  2 Style 2-branched; achenes biconvex or lenticular.
    3 Plants cespitose annuals, with fibrous roots.
      4 Achenes yellow to glistening brown when mature, tubercle compressed.

          5 Spikelets broadly ovoid to cylindric; scales rounded or obtuse at
             apex .... 10. *E. obtusa*
          5 Spikelets lanceolate, acuminate; scales acute at apex .... 7. *E. lanceolata*
       4 Achenes olivaceous to purplish-black when mature.
          6 Achene body 0.7–1 mm long .... 3. *E. caribaea*
          6 Achene body 0.5–0.7 mm long .... 2. *E. atropurpurea*
    3 Plants stoloniferous perennials.
       7 Scale (1) at the base of the spikelet sterile, completely encircling the base; culms up to
          1.5 mm wide .... 5. *E. erythropoda*
       7 Scales (2 or 3) at the base of the spikelet sterile and not encircling the base; culms up
          to 5 mm wide.
          8 Culms flattened, rigid; tubercles deltoid .... 18. *E. xyridiformis*
          8 Culms terete or subterete; tubercles conic.
             9 Basal sheaths with prominent V-shaped (oblique) sinuses; culms soft and in-
                flated .... 15. *E. smallii*
             9 Basal sheaths truncate to slightly oblique at apex; culms
                rigid .... 8. *E. macrostachya*
 2 Styles 3-branched; achenes trigonous or terete.
    10 Tubercle not well differentiated from the body of the achene, but appearing confluent.
       11 Achenes 1.0–1.3 mm long; plants small, usually less than 7 cm
          tall .... 11. *E. parvula*
       11 Achenes 2–3 mm long; plants larger, rarely less than 12 cm tall.
          12 Culms compressed, up to 2 mm wide .... 14. *E. rostellata*
          12 Culms not compressed, up to 1.0 mm wide .... 12. *E. pauciflora*
    10 Tubercle obviously differentiated from the body of the achene, forming an apical cap.
       13 Achenes with many longitudinal ribs and numerous cross ridges.
          14 Culms usually compressed, 2-edged, 0.5–1.5 mm wide .... 17. *E. wolfii*
          14 Culms not compressed or 2-edged, up to 0.5 mm wide .... 1. *E. acicularis*
       13 Achenes without longitudinal ribs and cross ridges.
          15 Culms flattened, scales acuminate .... 4. *E. compressa*
          15 Culms several-angled to terete; scales obtuse to acute.
             16 Surface of the achene smooth to minutely punctate.
                17 Achenes olive-green .... 6. *E. intermedia*
                17 Achenes yellowish or brown .... 9. *E. montevidensis*
             16 Surface of the achene distinctly warty or pitted .... 16. *E. verrucosa*

1. *Eleocharis acicularis* (L.) R. & S. Tufted perennial from slender rhizomes 0.2–0.5 mm thick. Culms 2–21 cm tall, 0.1–0.4 mm thick, with sheaths reddish or straw colored and often oblique at the apex. Spiklets 2–9 mm long, 3- to 15-flowered; scales 1.5–2.2 mm long with a green midrib and white to red margins. Bristles usually 3, but frequently fewer or absent; styles 3-branched on obovoid, usually trigonous, achenes, 0.5–0.8 mm long, pearly-white to yellowish with numerous longitudinal ridges and a number of trabeculae; tubercle conical, up to 0.2 mm long. (n = 10–29). Wet, sandy or muddy shorelines, flood plains, low prairie areas, roadside ditches, edges of marshes & sloughs; throughout GP; (Greenl. to AK, s to FL, GP; Mex.). *E. acicularis* (L.) R. & S. var. *gracillescens* Svens.—Waterfall.

2. *Eleocharis atropurpurea* (Retz.) Kunth. Cespitose annual from fibrous roots. Culms erect or arcuate, terete, up to 15 cm tall; sheath oblique at apex and firm. Spikelet ovoid, up to 8 mm long, many-flowered; scales ovate, membranaceous with a green midrib and deep brown sides. Bristles several, translucent, slender, shorter than the achene or reduced; stamens 1–3; styles 2-branched, on a lenticular, obovoid achene, about 0.5 mm long, smooth, shining, lustrous black; tubercle conic and minute. Sandy, moist areas, low prairies, margins of ponds; NE, KS, TX; (FL to TX, n to IA, NE, CO, & WA).

3. *Eleocharis caribaea* (Rottb.) Blake. Cespitose annuals without rhizomes or stolons. Culms thick, terete, furrowed, up to 40 cm tall, sheaths oblique and firm at the apex. Spikelets subglobose, ovoid, obtuse, up to 6 mm long, many-flowered; scales ovate-orbicular, up to 2 mm long, membranaceous with a pale midrib, yellow to pale brown with hyaline margins. Bristles usually equaling the tubercle; stamens 2 or 3; styles 2-branched on biconvex, obovoid achenes 0.7–1 mm long, shiny black to purplish, tubercle short, conic. Damp sandy areas; KS, OK; (Ont. to MI, OK, s to FL, CA). *E. geniculata* (L.) R. & S.—Fernald.

4. *Eleocharis compressa* Sulliv. Cespitose perennial from short, thick, forking rhizomes. Culms slender, erect, strongly compressed up to 30 cm tall and up to 2 mm wide; sheaths firm and truncate at apex. Spikelets oblong-ovoid, up to 12 mm long, many-flowered; scales ovate-lanceolate, chestnut-brown with a light margin, apex commonly bifid. Bristles 1–5, usually shorter than the achene; styles 3-branched on obovoid, trigonous achenes 0.8–1.5 mm long, golden to brown, reticulate; tubercle depressed to globose-conic. ($2n = 24, 36$). Low, wet prairies, sandy flood plains, marshy areas; ND, MN, SD, IA, NE, CO, KS, MO, OK; (Que. to Sask., s to GA & TX). *E. acuminata* (Muhl.) Nees—Rydberg.

5. *Eleocharis erythropoda* Steud. Cespitose perennial with reddish rhizomes and stolons. Culms slender, terete, up to 7 dm tall, sheaths red or castaneous, oblique at summit. Spikelet lanceolate or narrowly ovoid, up to 17 mm long, many-flowered; fertile scales oblong to ovate, obtuse, closely appressed; sterile basal scale orbicular or rounded, completely encircling the stem. Bristles 0–4, up to the length of the achene. Achene lenticular, narrowly obovoid, up to 1.7 mm long, tubercle conical. ($n = 18, 19, 20$). Wet areas, usually standing in water of roadside ditches, shores, marshy meadows; throughout GP; (Man. to Que., s to VA, TX, GP, NM). *E. calva* Torr., *E. uniglumis* (Link) J. A. Schult.—Rydberg.

This is one of four species belonging to the *Palustres* complex, the others being *E. macrostachya*, *E. smallii* and *E. xyridiformis*. The species of the complex are recognized and treated according to Harms, op. cit.

6. *Eleocharis intermedia* (Muhl.) J. A. Schult. Plants cespitose from fibrous roots. Culms numerous, ascending or reclining, up to 4 dm tall. Sheath apex thin, usually with an acute projection present. Spikelets ovoid, acute, up to 7 mm long, up to 20-flowered; scales obtuse or acute with a green keel and brown sides, the lowest scale sterile, obtuse and encircling the culm. Bristles usually 6 and slightly larger than the achene body; styles 3-branched on obovoid trigonous achenes, up to 1.2 mm long, olive-green and puncticulate; tubercle conic-subulate, up to 1/2 as long as the achene body. Muddy shorelines of rivers & creeks; MN; (Que. to MN, s to TN, WV, IA).

7. *Eleocharis lanceolata* Fern. Densely cespitose annual. Culms slender, erect, up to 2 dm tall, 1.0 mm thick, sheath firm and oblique at apex. Spikelets ovate-lanceolate, acuminate, up to 8 mm long, many-flowered; scales ovate, acute at apex, with a green midrib and brownish margins. Bristles 4 or 6, usually exceeding the achene body; styles 2-branched on obovoid, biconvex achenes, up to 1.2 mm long and brownish; tubercle forming a crown confluent with the achene body. Wet, sandy areas; KS, MO; (s GP to AR).

8. *Eleocharis macrostachya* Britt. Perennial from reddish rhizomes. Culms filiform, soft to rigid, frequently flattened, up to 1.2 m tall; sheaths truncate to slightly oblique at apex, pale brown. Spikelets subcylindric to lanceolate, up to 3 cm long, few- to many-flowered; basal scales 2 or 3 and sterile; fertile scales narrowly ovate, membranaceous, pale brown or purplish. Achenes obovate, yellowish or pale brown, glistening, up to 1.8 mm long; tubercle depressed deltoid; bristles up to 8 or wanting. Shorelines of lakes & ponds, marshy meadows, roadside ditches; throughout GP; (MN to Sask. & B.C., s to LA, TX, CA; Mex.). *E. mamillata* Lindb.—Rydberg.

9. *Eleocharis montevidensis* Kunth. Perennial from extensively creeping, thick rhizomes. Culms erect, soft, up to 50 cm tall, sheath firm and truncate at the apex, dark brown. Spikelets ovoid to oblong, blunt at the apex, up to 14 mm long, many-flowered; scales ovate to oblong, obtuse, brownish or yellowish with scarious margins. Bristles 4–6, equaling or shorter than the achene; styles 3-branched on triangular obovoid achenes, up to 1.2 mm long, with a glossy, puncticulate surface; tubercle conic and short. (n = 5) Sandy meadows, margins of lakes & streams; KS, OK, TX; (s GP w to OR, ID, e to GA & SC). *E. arenicola* Torr.— Rydberg.

10. *Eleocharis obtusa* (Willd.) J. A. Schult. Cespitose annuals. Culms erect up to 50 cm tall, sheaths firm and oblique at apex. Spikelets ovoid, up to 16 mm long, many-flowered; scales oblong or ovate, obtuse, purplish-brown with a green midrib and scarious margins. Bristles usually 6 or 7, equaling or exceeding the achene, or lacking; styles 2- or 3-branched, on lenticular, obovoid achenes, 0.7–1.2 mm long, yellow to brown; tubercle flat, deltoid.

Two varieties are recognized by Drapalik and Mohlenbrock, op. cit.

10a. var. *obtusa*. Tubercle more than 2/3 the width of the achene; bristles longer than the achene, or rarely lacking. Muddy shorelines of ponds, marshes, & streams; MO.

10b. var. *ovata* (Roth) Drapalik & Mohlenbrock. Tubercle 1/2–2/3 the width of the achene; bristles usually longer than the tubercle. Wet ground of marshes, ditches, ponds, lakes; ND, MN, SE, NE, IA, KS, OK, TX; (Newf. to MN s to FL & TX, also WA coast). *E. ovata* (Roth) R. & S., *E. engelmannii* Steud.— Rydberg.

11. *Eleocharis parvula* Link ex Boff. & Fingerbr. Cespitose perennial from short rhizomes, forming dense mats. Culms filiform, up to 7 cm tall, terete or occasionally flattened, sheaths short and membranaceous. Spikelets ovoid, up to 4 mm long, 2- to 9-flowered; scales ovate, up to 2 mm long, green to brown, the lowest one empty. Bristles equaling or larger than the achene, or reduced or absent; styles 3-branched on ovoid to obovoid trigonous achenes up to 1.3 mm long, smooth or roughened, straw colored to black, narrowing into a confluent tubercle.

Two varieties may be distinguished:

11a. var. *anachaeta* (Torr.) Svens. Bristles lacking. (n = 4, 5) Sandy or muddy shorelines of lakes & ponds; ND, MN, SD, IA, WY, KS, OK, TX; (GP to OR, LA; Mex.). *E. pygmaea* Torr.— Rydberg.

11b. var. *parvula*. Bristles present. Sandy margins of coal strip mines; MO; (Newf. to MN, s to NY, MI, MO, CA, also B.C.).

12. *Eleocharis pauciflora* (Lightf.) Link. Perennial from stoloniferous stout rhizomes, usually with conspicuous terminal buds. Culms erect up to 3 dm tall, slender, up to 1 mm thick, usually angled, with sheaths brownish and truncate. Spikelets ovoid up to 8 mm long, 2- to 9-flowered; scales ovate, acute, up to 5.5 mm long, deep brown with hyaline margins. Bristles about as long as the achene, rarely absent; styles 3-branched on obovoid, trigonous achenes, 2–3 mm long, including style base, gray-brown to olive color; tubercle not well differentiated from the body of the achene. (n = 10) Sandy, moist meadows; NE, ND, MN; (Newf. to B.C., s to n New England, PA, IN, IA, NE, & CA). *E. pauciflora* (Lightf.) Link var. *fernaldii* Svens.— Fernald.

13. *Eleocharis quadrangulata* (Michx.) R. & S. Cespitose perennial from coarse fibrous roots. Culms stout, up to 15 dm tall, 2–4 mm thick, sharply 4-angled, sheaths brown or dark red with an oblique apex. Spikelets 2–5.5 cm long, cylindric, usually as thick as the culm, with 50–100 flowers; scales elliptic to obtuse, up to 6 mm long, straw colored with scarious margins. Bristles usually 6, equal to or longer than the achene; stamens 3; styles

2- or 3-branched, on biconvex, obovoid achenes, 1.6–3 mm long, brown; tubercle conical, up to 1.5 mm long. Shallow water & muddy shorelines; KS, MO, OK; (MA to WI, KS, s to FL & TX; Mex.). *E. quadrangulata* (Michx.) R. & S. var. *crassior* Fern.—Fernald.

14. *Eleocharis rostellata* (Torr.) Torr. Cespitose perennial on short, erect rhizomes. Culms erect, up to 10 dm tall and 2 mm wide, distally flattened, occasionally arching over and rooting from the apex. Spikelets lanceolate, acute, up to 20 mm long, many-flowered; scales ovate to obtuse, up to 5 mm long, brownish. Bristles firm, about as long as the achene and tubercle; styles 3-branched on obovoid, trigonous achenes up to 3 mm long, olive-brown, narrowing to a confluent tubercle, about 1/3 the length of the achene. Wet sedge meadows, seeping areas, marshy ground; SD, ND, NE, KS, OK, TX; (N.S. to B.C. s to FL & TX; Mex.).

15. *Eleocharis smallii* Britt. Perennial with reddish to blackish rootstock. Culms firm, slender and stout, to soft and inflated, up to 5 mm wide, up to 1 m tall; sheath firm with V-shaped sinuses, usually black bordered at apex. Spikelets slender to ovoid, up to 2 cm long, loosely flowered; fertile scales acute or acuminate with two purple bands and a green midrib; basal scales 2 or 3, sterile and not encircling the culm. Achene obovoid, the body up to 20 mm long, yellowish to dark brown; tubercle depressed conic, as long as wide; bristles as long as the achene or wanting. (2n = 16, 36) Shallow water of lakes, ponds, creeks, ditches, & marshes; ND, MN, SD, NE, CO, KS, MO; (Newf. to Ont., s to WV, OK, CO, & WY).

16. *Eleocharis verrucosa* (Svens.) L. Perennial from thick, long rhizomes. Culms slender, 4- or 5-angled, up to 50 cm tall and 0.6 mm thick, sheaths tight, truncate at the apex. Spikelets ovate, lance-ovoid, up to 10 mm long, with 20–40 flowers; scales ovate, to obovate, obtuse or acute, brown to black with a green midrib and scarious margins. Bristles short, rarely persisting; styles 3-branched on trigonous, obovoid achenes, 0.6–1 mm long, olivaceous, surface warty or pitted, with depressed pyramidal tubercle. (2n = 20) Low wet ground, open oak wooded areas, moist upland prairies, roadsides; NE, KS, MO, OK; (PA to cen NE, s to VA & TX).

17. *Eleocharis wolfii* (A. Gray) Patt. Sparsely tufted from slender, creeping rhizomes. Culms erect, up to 4 dm tall, 0.5–1.5 mm wide, strongly flattened and often inrolled, sheaths oblique at the apex. Spikelets ovoid-lanceolate, 4–10 mm long, 15- to 35-flowered; scales ovate-lanceolate, acute, up to 3 mm long, red-purple with a green midvein and stramineous margins. Bristles absent; styles 3-branched on obovoid, slightly trigonous to terete achenes, 0.8–1.0 mm long, white to light brown with numerous longitudinal ridges and a number of trabeculae; tubercle conical, up to 0.1 mm long. Wet river & lake margins; MN, ND, NE, KS; (Alta. to IA, s to LA, MO, KS, w to CO).

18. *Eleocharis xyridiformis* Fern. & Brackett. Perennial from blackish rhizomes. Culms rigid, compressed and often twisted, up to 5.5 dm tall, sheath reddish at the base. Spikelet narrowly lanceolate, up to 2 cm long, densely flowered; basal scale orbicular, straw colored, up to 2.5 mm broad, fertile scales elliptic or ovate, obtuse to rounded at the apex, light colored with white margins. Achenes obovoid, straw colored to dark brown, up to 1.8 mm long and 1.4 mm broad; tubercle deltoid, up to 0.5 mm long. Shorelines of lakes, ponds, marshes, low swamps, wet ditches; MT, SD, WY, ND, NE, KS, TX; (ND, WY s to TX; Mex.).

7. ERIOPHORUM L., Cottongrass

Tufted or stoloniferous perennial. Culms triangular, terete or flattened, up to 10 dm tall. Leaves with linear-elongate, grasslike blades, some reduced to bladeless sheaths. Spikelets single and terminal or several in headlike clusters or umbelliform cymes; inflorescence subtended by foliaceous-elongate bracts or small scalelike involucral bracts; scales of the spikelets membranaceous, spirally arranged, not awned. Flowers perfect, perianth of numerous bristles, becoming greatly elongate at maturity; stamens 3; style 3-cleft, slender. Achene trigonous, oblong to obovate.

Reference: Fernald, M. L. 1905. The North American species of *Eriophorum*. Rhodora 7: 81–92; 129–136.

The presence of *E. virginicum* L. in the GP, noted by Rydberg, cannot be verified.

1 Spikelets solitary; leafy bracts absent .... 1. *E. chamissonis*
1 Spikelets 2 or more; leafy bracts present.
  2 Leaflike bracts 2 or more, the longest usually longer than the inflorescence.
    3 Midrib of the scale prominent to the tip; sheaths not dark girdled at apex; leaves triangular-channeled only at the apex .... 4. *E. viridicarinatum*
    3 Midrib of the scale prominent below the membranaceous tip; sheaths dark girdled at apex, leaves triangular-channeled above the middle .... 3. *E. polystachion*
  2 Leaflike bract solitary, usually shorter than the inflorescence .... 2. *E. gracile*

1. *Eriophorum chamissonis* C. A. Mey. Stoloniferous colonial perennials. Culms somewhat triangular, solitary or a few together, up to 8 dm tall. Leaves triangular-channeled, up to 6.5 cm long and 2 mm wide, upper 1 or 2 sheaths bladeless. Spikelet solitary and terminal, in fruit up to 4 cm long. Sterile scales few, brownish to blackish; bristles reddish-brown. Achenes oblong-obovate, trigonous, up to 2.7 mm long, spiculate. (n = 29) Marshes and other wet places; ND; (Lab. to AK, s to N.S., ND, & WA).

2. *Eriophorum gracile* Koch. Perennial with weak, slender culms, terete to triangular above, up to 6 dm tall. Leaves on lower culm well developed, up to 2 mm wide, triangular-channeled, upper leaves round-tipped and shorter than sheath. Inflorescence of 2–5 pedunculate spikelets in an umbelliform cyme, subtended by a solitary, erect involucral bract up to 2 cm long. Scales ovate, obtuse, blackish, with a prominent midrib, bristles bright white. Achenes narrowly obovate, up to 3.5 mm long. (n = 30, 38) Swamps & wet meadows; NE, CO; (Lab. to AK, s to PA, NE, CO, & CA).

3. *Eriophorum polystachion* L. Colonial perennial with culms obtusely angled, up to 8 dm tall. Leaves triangular-channeled above the middle and flat toward the base, up to 15 cm long and 6 mm wide, the upper sheaths dark girdled at apex. Inflorescence of 2–8 pedunculate spikelets in umbelliform cymes, subtended by 2 or more involucral bracts. Spikelets up to 10, peduncles mostly glabrous, often drooping, up to 4.5 cm long in fruit, scales up to 10 mm long, thin, ovate-lanceolate, with the distinct midrib not extending into the membranaceous tip. Achenes compressed-trigonous, blackish, 2–3 mm long, bristles numerous and white. (n = 29, 30) Marshes, wet meadows, bogs; MT, ND, MN, SD, NE, CO; (Lab. to AK, s to MA, NE, CO, & WA). *E. angustifolium* Honck., *E. angustifolium* Honck. var. *majus* Schult.—Fernald.

4. *Eriophorum viridicarinatum* (Engelm.) Fern. Tufted perennial with slender, triangular culms, up to 9 dm tall. Leaves up to 15 cm long, 6 mm wide, triangular-channeled only at the very tip, otherwise flat. Inflorescence of 3–many, peduncled, drooping spikelets in an umbelliform cyme, subtended by 2 or more involucral bracts. Spikelets up to 3 cm long in fruit, scales ovate-lanceolate, blackish-green, up to 6 mm long with a prominent midrib extending to the tip and at times beyond. Achenes trigonous, dark brown, narrowly obovate, up to 3.5 mm long, bristles numerous, white or buff. Marshes, bogs, & wet meadows; ND, MN; (Lab. to AK, B.C., s to CT, MN, SD).

8. FIMBRISTYLIS Vahl

Annual or perennial. Culms solitary or in tufts with a few leaves below. Leaves filiform to narrowly or broadly linear, flat or involute, with or without a ligule, sheaths closed. Inflorescence of sessile or peduncled spikelets, subtended by a leafy involucre; spikelets lanceolate, oblong or ovoid, several-flowered; scales spirally imbricate, glabrous or pubescent. Florets perfect; perianth none; stamens 1–3; style 2- or 3-cleft, the base enlarged but not persisting as a tubercle. Achene lenticular or trigonous, surfaces smooth, reticulate, warty or ribbed.

References: Kral, R. 1971. A treatment of *Abildgaardia, Bulbostylis* and *Fimbristylis* (Cyperaceae) for North America. Sida 4(2): 57–227.; Svenson, H. K. 1957. Cyperaceae. Tribe 2, Scirpeae. N. Amer. Flora 18(9): 505–556.

1 Achene trigonous; style 3-branched .... 2. *F. autumnalis*
1 Achene lenticular or biconvex; style 2-branched.
  2 Spikelets (at least 1 or more) peduncled; achenes 1.0 mm or more long.
    3 Plants annual, cespitose; achene horizontally ribbed, usually warty .... 1. *F. annua*
    3 Plants perennial with culms solitary or a few together; achenes not horizontally ribbed, not warty .... 3. *F. puberula*
  2 Spikelets sessile, all close together in a capitate cluster, achenes about 0.5 mm long .... 4. *F. vahlii*

1. ***Fimbristylis annua*** (All.) R. & S. Cespitose annual. Culms decumbent, ascending or erect, up to 5 dm tall. Leaves narrow-linear, flat, up to 2 mm wide; shorter than or nearly as long as the culms, glabrous to tomentose; sheaths broad, smooth or pubescent; ligule a line of short hairs. Inflorescence a simple or compound umbellate cyme, of few to many spikelets; spikelets ovoid to obovoid, up to 10 mm long. Scales obtuse to ovate, the apex obtuse to acute, glabrous, the midrib conspicuous. Stamens 1 or 2; style 2-branched. Achene obovoid, lenticular, up to 1.3 mm long, pale gray to stramineous, with about 10 horizontal and about 11 longitudinal ribs, usually warty. (n = 10, 15) Moist, sandy or clay soils in ravines, sloughs, & depressions in prairies & pastures; KS, MO, OK; (Mex. n to KS & MO, e to PA & FL). *F. baldwiniana* (Schult.) Torr.—Fernald; *F. dichotoma* (L.) Vahl—Steyermark.

2. ***Fimbristylis autumnalis*** (L.) R. & S. Cespitose annual. Culms erect or diffuse, up to 2 dm tall, slender, glabrous. Leaves linear, glabrous, flat, up to 4 mm wide, shorter than to nearly as long as the culms; sheaths broad, keeled; ligule a line of short hairs. Inflorescence a simple or compound umbellate cyme, subtended by 2 or 3 involucral bracts; spikelets ovoid to linear oblong, up to 10 mm long, sessile or pedunculate. Scales ovate-lanceolate, keeled, mucronate to acuminate. Stamens usually 2; style 3-branched. Achene obovoid, trigonous, pale brown, shining, about 1 mm long, smooth to faintly reticulate. (n = 5) Moist to dry sandy prairies, shores of streams & ponds; SD, NE, IA, KS, MO, OK; (Mex. n to NE & IA, e to N. Eng., NC, FL). *F. autumnalis* (L.) R. & S. var. *mucronulata* (Michx.) Fern.—Fernald; *F. mucronulata* (L.) R. & S.—Gates.

3. ***Fimbristylis puberula*** (Michx.) Vahl. Perennial from thick to slender, short rhizomes. Culms solitary or in small tufts, stiff, up to 1 m tall. Leaves shorter than to nearly equaling the culms, narrowly linear, involute to flat, smooth or pubescent; sheaths hard, thick, fibrous; ligules incomplete or absent. Inflorescence an umbellate cyme, compact to open, subtended by several involucral bracts, the longest being shorter or longer than the inflorescence; spikelets lance-ovoid, ovoid, cylindrical or ellipsoid, up to 1 cm long. Scales ovate, smooth to ciliate, nerves inconspicuous. Stamens 3; style 2-branched. Achenes lenticular, obovoid, up to 1.8 mm long, reticulate, with several longitudinal ribs and finer horizontal lines.

Two varieties are recognized (Kral, op. cit.):

3a. var. *interior* (Britt.) Kral. Base of the culms rarely bulbous; numerous, very slender, twisted, orangish rhizomes, arising from a base that is usually not thickened. Longest bract of the inflorescence usually surpassing the inflorescence; scales of the spikelet smooth; ligule inconspicuous or present as a narrow line of short ascending hairs. (n = 10) Wet, sandy clay prairies, subirrigated meadows, shorelines of lakes & ponds; NE, KS, OK; (TX n to NM & NE). *F. interior* Britt. — Rydberg.

3b. var. *puberula*. Base of the culm bulbous; rhizomes thick with old leaf bases often persisting as sheath remnants. Longest bract of the inflorescence usually much shorter than inflorescence; scales of the spikelet ciliate; ligule inconspicuous, incomplete or absent. (n = 10, 20) Low, wet prairies, sandy shores of lakes & ponds; NE, KS, MO, OK; (TX n to NE, MI, e to NJ, NC, FL). *F. castanea* (Michx.) Vahl var. *puberula* (Michx.) Britt. — Gates; *F. castanea* (Michx.) Vahl, *F. caroliniana* (Lam.) Fern.; *F. drummondii* Boeck. — Fernald; *F. spadicea* (L.) Vahl — Waterfall.

4. **Fimbristylis vahlii** (Lam.) Link. Cespitose low-growing annual. Culms slender, up to 15 cm tall. Leaves filiform, as long as or exceeding the culms, up to 1 mm wide, somewhat channeled; sheaths broad, smooth or slightly pubescent; ligule absent. Inflorescence a dense, terminal, capitate cluster of 3–8 sessile spikelets subtended by several involucral bracts; spikelets oblong-cylindric, obtuse, up to 10 mm long. Scales lanceolate to oblong-lanceolate, glabrous with a prominent midrib. Stamen 1; style 2-branched. Achene obovoid, lenticular, up to 0.7 mm long, reticulate with 5–7 horizontal rows of cells on each side. (n = 10) Very moist, sandy areas along margins of ponds & lakes; KS, MO, OK; (Mex. n to KS & MO, w to CA, e to SC & FL).

9. FUIRENA Rottb., Umbrella-Grass

1. **Fuirena simplex** Vahl. Perennial from rhizomes or annual without rhizomes. Culms angled, single, tufted or from rhizomes; ascending or spreading; slender to stout, mostly up to 6 dm tall. Leaf sheaths variously hispid or hirsute. Leaf blades linear to lance-linear, erect or spreading, up to 25 cm long, up to 9 mm wide, leaf surface varying from glabrous to hispid or pilose. Spikelets in clusters of 1–3(5), the lower on slender peduncles, the cluster subtended by 1–3 leaflike bracts, the larger usually longer than the cluster of spikelets, 0.5–0.8 cm long; fertile scales 2.0–5.0 mm long, usually hairy, with scabrous or hispid awns, nerves 3–7. Floret of 3 retrorsely barbellate bristles alternating with 3 stalked, ovate-oblong, awn-tipped, petal scales; stamens usually 3, anthers up to 1.2 mm long. Achene trigonous, faces concave to flat, shiny yellowish-brown to white, tapering to a beak.

Reference: Kral, R. 1978. A synopsis of *Fuirena* (Cyperaceae) for the Americas north of South America. Sida 7(4): 309–354.

Recognition is given to two varieties by Kral:

1a. var. *aristulata* (Torr.) Kral. Annual without rhizomes, usually tufted; plants shorter, up to 3 dm tall. Nerves of fertile scale usually 3; anthers up to 0.8 mm long. Low marshy prairies, usually river & stream bottoms; NE, KS, OK, MO, TX; (Mex. n to NE & nw MO).

1b. var. *simplex*. Perennial from slender rhizomes, culms solitary or following rhizomes; plants up to 6 dm tall. Nerves of fertile scale usually 5–7; anthers 0.9–1.2 mm long. Sandy soils, damp ground, banks of creeks, ditches, swales, & other wet places; KS, OK, & TX; (Mex. n to KS & MO).

10. HEMICARPHA Nees & Arn.

Small tufted annual. Culms slender, terete or compressed, up to 20 cm tall. Leaves hairlike and few in number. Inflorescence a cluster of 1–several small spikes subtended by 2 or 3 involucral bracts, the longest appearing to be a continuation of the culm, causing the inflorescence to appear lateral; spikes composed of numerous spirally arranged, 2-scaled, sessile spikelets; spikelets with a single, perfect floret, subtended by a well-developed outer scale and a thin, smaller or absent inner scale (interpreted to be a modified perianth). Flowers with a single stamen and a 2-cleft style on a cylindric to compressed achene with many low papillae.

> Reference: Friedland, S. 1941. The American Species of *Hemicarpha*. Amer. J. Bot. 28: 855–861. This genus is included in *Scirpus* by some authors.

1 Inner scale ("perianth") as long as or longer than and partly enclosing the achene .... 1. *H. drummondii*
1 Inner scale ("perianth") much shorter than the achene, or lacking .... 2. *H. micrantha*

1. Hemicarpha drummondii Nees. Tufted annuals with slender culms up to 10 cm tall. Leaves narrow. Inflorescence with 1–3 sessile spikes; each spike composed of numerous spikelets of a single floret with outer scales oblong to broadly obovate, with the mucro never exceeding the length of the body and usually shorter; the inner scale as long or longer than and enclosing the achene. Achene obovoid, black when mature. Wet to moist sandy soil, shorelines; ND, NE, KS, MO, OK; (NE s to KS, MO, OK, AR, & TX). *H. micrantha* var. *drummondii* (Nees) Friedl.—Friedland, op. cit.

2. Hemicarpha micrantha (Vahl) Britt. Dwarf tufted annual with slender culms up to 20 cm tall. Leaves narrow. Inflorescence with 1–3 sessile spikes; each spike composed of numerous spikelets of a single perfect floret with outer scales oblong or narrowly obovate, with a mucro shorter than the body; the inner scale shorter than the achene, usually bifid and often lacking. Achene cylindrical, iridescent brown when mature. Wet, sandy areas, sand bars, sandstone outcrops; ND, NE, KS, MO, OK; (Mex. n to WA, NE, MN, WI, & New England, s to FL). *H. micrantha* var. *aristulata* Cov.—Friedland, op. cit.; *H. aristulata* (Cov.) Smyth—Rydberg.

11. RHYNCHOSPORA Vahl, Beaked Rush

Perennial with triangular, leafy solid culms. Leaves narrow with closed sheaths. Inflorescence of terminal and axillary clusters either sessile or pedicellate; spikelets subglobose to narrowly ovoid; with several spirally imbricate scales, the lowest scale striate, the upper with perfect or sometimes imperfect (staminate) flowers. Stamens usually 3; bristles usually 6 or 8(20) or absent, very short to longer than the achene. Achene lenticular to flattened, with a conspicuous tubercle, style usually 2-cleft.

> Reference: Gale, Shirley. 1944. *Rhynchospora*, Section *Eurhynchospora*, in Canada, the United States and the West Indies. Rhodora 46(544): 88–134, 159–197, 207–245, 255–278.
>
> *Rhynchospora corniculata* listed for Kansas by Gates cannot be verified.

1 Achenes 4–5 mm long; tubercle 14–22 mm long .... 5. *R. macrostachya*
1 Achenes less than 3 mm long; tubercles less than 3 mm long.
  2 Bristles antrorsely barbellate, rarely smooth or lacking.
    3 Achene surface with transverse wavy ridges, with cells larger than broad; tubercle deltoid, compressed .... 3. *R. globularis* var. *recognita*
    3 Achene surface without transverse wavy ridges, honeycombed with cells of about the same diam; tubercle conical, apiculate with a subterete base .... 4. *R. harveyi*
  2 Bristles retrorsely barbellate, rarely smooth.
    4 Tubercle more than 1/2 as long as the achene body; bristles about equaling the length of achene and tubercle; leaves up to 3.0 mm wide .... 2. *R. capitellata*
    4 Tubercle 1/2 as long as the achene body or shorter; bristles equaling or shorter than the achene only; leaves up to 1.4 mm wide .... 1. *R. capillacea*

1. Rhynchospora capillacea Torr. Cespitose perennials. Culms capillary, up to 40 cm tall. Leaves involute, narrowly linear, up to 0.4 mm wide, the upper usually exceeding the inflorescence. Inflorescence a terminal fascicle of usually 2 glomerules with 2-10 erect spikelets, the lateral glomerules subsessile or short-peduncles with 1-4 spikelets; spikelets sessile or nearly so, up to 7 mm long, 1- to 5-fruited. Bristles 6, equaling or shorter than the achene, retrosely barbellate or smooth. Achene oblong-obovoid, up to 2.6 mm long, less than 1/2 as wide; tubercle lanceolate, up to 1.6 mm long. Moist, sandy soil; SD, IA; (Newf. to Sask. s to VA, MO, & SD).

2. **Rhynchospora capitellata** (Michx.) Vahl. Cespitose perennial. Culms slender, bluntly triangular, up to 80 cm tall. Leaves up to 3 mm wide. Inflorescence a terminal fascicle of 2-several glomerules, the lateral glomerules subsessile or short-peduncled; spikelets up to 5 mm long, 2- to 5-flowered. Bristles 6, retrorsely barbellate, about equaling the tubercle. Achene pyriform or obovate up to 1.6 mm long, narrowly winged; tubercle widened at the base, up to 1.4 mm long. Sandy soils; MO; (N.S. to Ont., s to FL, MO, TX, also GA & OR).

3. **Rhynchospora globularis** (Chapm.) Small var. **recognita** Gale. Small cespitose perennial. Culms stiff, erect, obtusely triangular to almost terete, glabrous, up to 90 cm tall. Leaves flat, up to 4 mm wide, sheaths glabrous. Inflorescence a terminal fascicle with 7 peduncled glomerules; spikelets crowded, ovoid, up to 4 mm long, with 1-4 flowers and 1- to 3-fruited. Bristles up to 6, antrorsely barbellate, shorter than the body of the achene. Achene obovoid to subglobose, transversely ridged, up to 1.6 mm long; tubercle deltoid, up to 0.6 mm long. Moist, rocky soils; KS, MO, OK; (NJ to KS, s to FL & TX, also CA).

4. **Rhynchospora harveyi** F. Boott. Cespitose perennials. Culms erect, up to 60 cm tall, obtusely triangular. Leaves flat, up to 3 mm wide. Inflorescence a terminal fascicle of 1-4 glomerules and 1 or 2 lateral glomerules; spikelets 2-flowered, 1-fruited, up to 3 mm long. Bristles 6, antrorsely barbellate, about 1/2 to nearly as long as the achene. Achenes ovoid to suborbicular, up to 1.8 mm long and 1.6 mm wide, surface honeycombed with cells of the same diam; tubercle conical, up to 0.5 mm long, not flattened. Moist, sandy or clay soils & shallow water; KS: (VA to KS, s to FL & TX).

5. **Rhynchospora macrostachya** Torr. Cespitose perennial with base bulbous, thickened. Culms erect, glabrous, triangular, up to 1 m tall. Leaves glabrous, up to 1.2 cm wide, sheaths glabrous and becoming fibrous. Inflorescence of a terminal fascicle of sessile to subsessile glomerules with numerous spikelets; spikelets up to 3 cm long, reddish-brown. Bristles 6, antrorsely barbellate, some up to 1 cm long. Achene obovoid, flattened on faces, up to 5 mm long and 3 mm wide; tubercle subulate-serrulate, up to 2 cm long and 2 mm broad at base. Sand bars, muddy silt; KS, OK, MO; (ME to KS, s to FL, MO, & OK).

12. SCIRPUS L., Bulrush

Annual or perennial herb. Culms usually solid triangular or terete. Leaves with closed sheaths, blades well developed or reduced or absent. Inflorescences variable, spikelets 1-many, each few- to many-flowered; inflorescence subtended by 1-several leafy or scalelike bracts. Flowers perfect, each with a single subtending scale, perianth of 1-6 variable bristles; stamens 2 or 3; style 2- or 3-cleft. Achene usually apiculate, plano-convex, lenticular or trigonous.

References: Beetle, A. A. 1947. *Scirpus*. N.Amer. Flora 18(8): 481-504; Koyama, T. 1958. Taxonomic study of the genus *Scirpus* Linne. J. Fac. Sci. Univ. Tokyo, Sec. 3, Bot. 7: 271-366; Schuyler, A. E. 1967. A taxonomic revision of North American leafy species of *Scirpus*. Proc. Acad. Nat. Sci., Philadelphia 119: 295-323.; Schuyler, A. E. 1974. Typification and application of the names *Scirpus americanus* Pers., *S. olneyi* Gray, and *S. pungens* Vahl. Rhodora 76: 51-52.; Ward, R. L. 1973. A cytotaxonomic study of the *Scirpus lacustris* L. complex in the northern plains states. North Dakota Univ., Fargo, ND. Unpublished Ph.D. dissertation, 169 pp.

Scirpus deltarum Schuyler has been reported infrequently for se KS and nw MO. Its presence has been attributed to dispersal by waterfowl (Schuyler, Not. Nat. 427: 1-4, 1970). It may key to *S. pungens*.

1 Involucral bracts leaflike; inflorescence appearing terminal; culms with well-developed leaves.
  2 Spikelets small, 3-10 mm long, 2-4 mm wide, very numerous; culms mostly obtusely triangular.
    3 Bristles without teeth and equaling or exceeding the length of the scales.
      4 Scales with prominent green midribs; bristles not exceeding the scales in length .... 15. *S. pendulus*
      4 Scales with inconspicuous midribs; bristles exceeding the length of the scales.
        5 Scales reddish-brown to dark brown (sometimes blackish); spikelets sessile or subsessile in glomerules .... 5. *S. cyperinus*
        5 Scales greenish-black, spikelets solitary and pediceled .... 3. *S. atrocinctus*
    3 Bristles with retrorse or antrorse teeth, shorter than to exceeding the scales, occasionally lacking.
      6 Styles mostly 2-cleft; achenes lenticular to plano-convex .... 12. *S. microcarpus*
      6 Styles mostly 3-cleft; achenes trigonous.
        7 Bristles 0-3, shorter than the achenes; sheaths and blades usually not cross-septate .... 7. *S. georgianus*
        7 Bristles usually 5 or 6, shorter than or little longer than the achenes; sheaths and blades cross-septate.
          8 Scales dark green, becoming blackish or brownish, usually 1.4-2.1 mm long with a mucronate tip less than 0.4 mm long .... 4. *S. atrovirens*
          8 Scales usually blackish, 1.8-2.8 mm long with an awnlike tip usually more than 0.4 mm long .... 14. *S. pallidus*
  2 Spikelets larger, 10-25 mm long, 6-12 mm wide; few in number; culms sharply triangular.
    9 Achene lenticular; style 2-cleft; leaf sheaths truncate or concave at the apex .... 11. *S. maritimus* var. *paludosus*
    9 Achene trigonous; style 3-cleft; leaf sheaths convex at the apex .... 6. *S. fluviatilis*
1 Involucral bract culmlike; inflorescence appearing lateral; culms with leaves usually reduced.
  10 Tufted annuals with slender, mostly terete culms.
    11 Achenes transversely ridged.
      12 Styles 2-cleft; achene plano-convex .... 8. *S. hallii*
      12 Styles 3-cleft; achene strongly trigonous .... 17. *S. saximontanus*
    11 Achenes smooth or papillate.
      13 Achenes trigonous, papillate; scales boat-shaped and strongly keeled .... 10. *S. koilolepis*
      13 Achenes plano-convex, smooth; scales not boat-shaped and strongly keeled .... 18. *S. smithii*
  10 Perennials with rhizomes; culms triangular or terete and stout.
    14 Spikelets few in number (15 or less) or solitary and sessile or subsessile; culms triangular to subterete.
      15 Achenes apiculate; culms triangular.
        16 Achenes plano-convex or unequally biconvex.
          17 Achenes 2.5-3.4 mm long; lower scales of spikelets enlarged and sterile; spikelets solitary or up to 8 .... 16. *S. pungens*
          17 Achenes 1.8-2.5 mm long; lower scales of spikelet not enlarged and fertile; spikelets up to 15 .... 2. *S. americanus*
        16 Achenes distinctly triangular .... 19. *S. torreyi*
      15 Achenes not apiculate; culms subterete to obscurely triangular ... 13. *S. nevadensis*
    14 Spikelets several to many (more than 15), usually in clusters on elongate branches.
      18 Achenes trigonous, 3.2 mm long; styles 3-cleft; spikelets usually one per pedicel .... 9. *S. heterochaetus*
      18 Achenes lenticular, 2.4 mm long; style 2-cleft; rarely 3-cleft; spikelets usually in glomerules of 2 or more per pedicel.

19 Spikelets usually in glomerules of 3-8, sessile or on short stiff pedicels; scales shorter than or equal to achenes; culms firm and dark green .... 1. *S. acutus*
19 Spikelets usually in glomerules of 2, seldom 3 or single, on long, lax pedicels; scales exceeding the achenes; culms soft and light green .... 20. *S. validus*

1. ***Scirpus acutus*** Muhl. Perennial with light brown to dark brownish-black stout rhizomes. Culms erect, slender, to 3.5 mm tall, terete, green to very dark green, firm to hard. Leaves from upper sheaths long tapering. Involucral bract solitary, rarely 2 or 3, appearing as a continuation of culm. Inflorescence paniculate, generally stiff, spikelets sessile, forming dense glomerules, or pendulous with short to long pedicels, spikelets in glomerules of 3-8(2-15), rarely solitary; spikelets numerous, ± 10-15 mm long, 4 mm wide, ovoid to linear-cylindric, acute; scales to 8 mm long, to 2.6 mm wide, base rounded, tip mucronate, shorter than or equal to the achene. Bristles 6, shorter than or about equaling culm. Achene 2.4 mm long, 1.75 mm wide, beak to 0.3 mm long, plano-convex, apiculate, style 2-cleft. Sloughs, deep-water marshes, ponds, lakes, & deep roadside ditches; throughout GP; (Newf. to B.C., s to NE, TX, CA).

2. ***Scirpus americanus*** Pers. Perennial from long rhizomes. Culms erect, up to 15 dm tall, sharply triangular, the sides often concave. Leaves few, up to 2 dm long if present, most sheaths leafless. Bract solitary, appearing as a continuation of the stem, up to 5 cm long. Spikelets up to 15, sessile in a dense cluster, up to 15 mm long, ovoid, obtuse, many-flowered; scales broadly rounded with a short mucro. Bristles usually 4, about equaling the achene, retrorsely barbellate. Achenes obovate, plano-convex, up to 2.5 mm long, 2.0 mm wide, apiculate, style 2-cleft. (n = 39) Marshes, low wet places; KS: Meade; OK; (N.S. to WA, s to FL & CA). *S. olneyi* A. Gray—Rydberg.

3. ***Scirpus atrocinctus*** Fern. Densely tufted slender perennial from fibrous rootstock. Culms up to 1.5 m tall, triangular. Leaves mostly basal, 3-5 mm broad. Involucral bracts 3-5, leaflike, exceeding the inflorescence, black at base. Inflorescence a compound spreading umbel with numerous spikelets, mostly pedicelled, spikelet up to 6 mm long; scales ovate-lanceolate, greenish-black. Bristles 6, much exceeding the achene. Achenes 1 mm long, white or yellowish, sharply beaked. (n = 34) Swamps & boggy areas; MN; (Newf. to MN & WA, s to WV).

4. ***Scirpus atrovirens*** Willd. Cespitose perennial with short tough firbrous rhizomes. Culms slender, somewhat trigonous, up to 1.5 m tall. Leaves up to 20 mm wide, leaves and sheaths usually nodulose-septate. Bracts several, leaflike. Inflorescence once or usually twice branched, spikelets ovate or narrowly ovate, 2-5 mm long, 1-2.5 mm wide, usually in glomerules; scales elliptic or broadly elliptic, mucronate. Bristles 6, shorter than to slightly exceeding the achene. Achenes 1-1.3 mm long, elliptic or obovate, flattened to trigonous. (n = 28) Marshes, shores of streams, wet meadows; ND, SD, NE, IA, KS, MO, OK; (Que. to Sask., s to GA & TX).

5. ***Scirpus cyperinus*** (L.) Kunth. Cespitose perennial from short rhizomes. Culms up to 20 dm tall, terete. Leaves mostly basal, numerous, 3-10 mm wide. Involucral bracts leaflike, unequal, spreading, little if at all surpassing the inflorescence. Inflorescence of numerous spikelets, sessile to pedicellate on several ascending rays. Spikelets ovoid to cylindric, many-flowered, 3-6 mm long; scales ovate-lanceolate, obtuse and subacute, reddish-brown to dark brown. Bristles several, smooth, much exceeding the scales. Achenes up to 1 mm long, oblong-apiculate, flattened to trigonous. Wet or boggy places; MN, SD, IA, MO; (N.S. to Sask., s to FL, OK; Mex.).

6. ***Scirpus fluviatilis*** (Torr.) A. Gray. Perennial from a thick rhizome. Culms stout, sharply triangular, usually with concave sides, up to 15 dm tall. Leaves several, distributed along

the culm, up to 20 mm wide. Involucral bracts several, surpassing the inflorescence. Spikelets 10–25 mm long, ovoid to fusiform, sometimes all sessile, or in small capitate clusters on single flexuous peduncles; scales acute or acuminate with a short awn. Bristles 6, retrorsely barbellate, usually about as long as the achene. Achenes obovoid, trigonous, apiculate, up to 5 mm long, style 3-cleft. Marshes, shorelines of rivers, ponds, & lakes; ND, MN, SD, NE, IA, KS, MO; (N.B. to WA, s to VA, KS, & CA).

7. *Scirpus georgianus* Harper. Cespitose perennial from short, tough rhizomes. Culms erect, trigonous, up to 15 dm tall. Leaves numerous, long-linear, up to 20 mm wide, deep green. Involucral bracts leaflike. Inflorescence paniculate, spikelets up to 200, ovoid to cylindric, up to 4 mm long, 1–2 mm wide, in glomerules, many-flowered; scales ovate, acute, mucronate. Bristles lacking or 1, 2, or 3, shorter than achene. Achenes oblong, 0.8–1.2 mm long, trigonous. (n = 25, 26, 27) Swamps & wet meadows; NE, KS, MO; (Newf. to MN, s to GA & OK). *S. atrovirens* Willd. var. *georgianus* (Harper) Fern.—Fernald.

8. *Scirpus hallii* A. Gray. Cespitose slender annual. Culm up to 4 dm tall, terete. Leaves few, usually basal and short. Primary bract appearing as a continuation of the culm, other bracts below the inflorescence and inconspicuous. Spikelets 2–several, sessile or short-peduncled, lance-ovate, up to 11 mm long, many-flowered; scales ovate, acuminate-acute. Bristles variable. Achenes up to 1.5 mm long, transversely ridged, plano-convex; style 2-cleft. Sandy shores or prairies; NE, KS; (MA, GA, w to IL, TX). *S. supinus* L. var. *hallii* (A. Gray) A. Gray—Beetle, op. cit.

9. *Scirpus heterochaetus* Chase. Perennial with brown to reddish-brown stout rhizomes. Culms erect, 1–2.3 m tall, terete, green to dark green, firm to hard. Leaves from upper sheaths long tapering. Involucral bract single, appearing to be a continuation of the culm. Inflorescence paniculate, loosely spreading, spikelets solitary on slender pedicels. Spikelets numerous, to 17 mm long, 3–4.6 mm wide, cylindric to ovoid; scales to 5.2 mm long, 2.3 mm wide, rounded toward base, tip mucronate. Bristles 2(4), irregular in length and about equaling the length of the achene, retrorsely barbed. Achene 3.2 mm long, 2 mm wide, trigonous, beak to 1 mm long, style 3-cleft. Slough, marshes, ponds, roadside ditches; ND, MN, SD, NE, IA, KS, MO, OK, TX; (MA to WA, s to NY, KY, TX, & OR).

10. *Scirpus koilolepis* (Steud.) Gl. Small cespitose annual. Culms slender, compressed, up to 20 cm tall. Leaves basal, setaceous, up to 5 cm long. Bract solitary, appearing as a continuation of the culm. Spikelets 1–3, sessile, ovoid, 3–7 mm long; scales boat-shaped, ovate, acuminate, sharply keeled. Bristles absent. Achenes obovoid, trigonous, to 1.5 mm long, beakless, papillate, style 3-cleft. Shores of streams, sandy soils; KS, MO, OK; (TN to CA, s to AL, OK; Mex.).

11. *Scirpus maritimus* L. var. *paludosus* (A. Nels.) Kükenth. Rhizomatous perennial from stout, extensive, tuber-bearing rhizomes. Culms sharply triangular, up to 15 dm tall. Leaves several, scattered on culm or on lower half, up to 1 cm wide; mouth of the leaf-sheath truncate or concave, the veins diverging well below the summit, leaving a thin, easily torn triangle. Involucral bracts several, leaflike, ascending or spreading. Spikelets 1–many, up to 2 cm long, ovoid to ovoid-cylindric, clustered, sessile, subsessile or peduncled; scales thin margined, puberulent, short awned from a notched tip. Bristles 2–6. Achenes lenticular, obovate or obovoid, 3–4 mm long, 2 mm wide, short beaked, style 2-cleft. Shores & marshes, alkaline or saline places; MT, ND, MN, WY, SD, NE, KS, OK, TX; (Que. to B.C., s to NY, TX, NM, CA; Mex.). *S. paludosus* A. Nels.—Rydberg.

12. *Scirpus microcarpus* Presl. Perennial from thick, reddish, long creeping rhizomes. Culms solitary, stout, triangular, up to 15 dm tall. Leaves several, to 15 mm wide, sheaths

reddish-tinged. Involucral bracts usually 3, leaflike, often exceeding the inflorescence. Spikelets numerous, narrowly to broadly ovate, acute, 3–8 mm long, 1–3.5 mm wide, in glomerules; scales ovate, marked with green and black. Bristles 4, retrorsely barbellate, length variable. Achenes lenticular to plano-convex (sometimes 3-angled), obovate, 0.7–1.6 mm long, apiculate, style usually 2-cleft. (n = 33) Marshy places, along streams, wet low areas, meadows; MT, ND, MN, WY, SD, NE, CO; (Newf. to Sask., s to WV, NE, & CO). *S. rubrotinctus* Fern.—Fernald; *S. microcarpus* Presl var. *rubrotinctus* (Fern.) M.E. Jones—Van Bruggen.

13. *Scirpus nevadensis* S. Wats. Rhizomatous or slightly cespitose perennial from tough elongated rhizomes. Culms slender, subterete, smooth, up to 4 dm tall. Leaves basal, narrow, to 2 mm wide. Involucral bract erect, appearing to be a continuation of the culm, with additional scalelike bracts subtending inflorescence. Spikelets several, ovoid-oblong, to 18 mm long, sessile in a compact cluster; scales brown, smooth and shining. Bristles 1–3, retrorsely barbellate, length variable. Achenes plano-convex, scarcely beaked, minutely reticulate. Wet places, especially alkaline soils; ND, NE; (Sask. to B.C., s to NE, UT, & CA).

14. *Scirpus pallidus* (Britt.) Fern. Cespitose perennial from short stout rhizomes. Culms trigonous, up to 15 dm tall. Leaves several, up to 18 mm wide. Bracts several and leaflike. Inflorescence an umbel of up to 8 primary rays, spikelets 3–4 mm long, numerous, often in large compound clusters. Scales 1.8–2.8 mm long, blackish with awnlike tips. Bristles usually 6, equaling or shorter than the achene, retrorsely barbellate. Achenes to 1 mm long, white, trigonous, apiculate, style 3-cleft. Marshes, streams, & ditches; MT, ND, MN, WY, SD, NE, IA, CO, KS, MO, OK, TX; (Man. to WA, s to AR, MO, & TX). *S. atrovirens* Willd. var. *pallidus* Britt.—Gleason & Cronquist.

15. *Scirpus pendulus* Muhl. Plants perennial, cespitose. Culms slender, firm, up to 1.5 cm tall. Leaves several, scattered on stem, up to 8 mm wide. Involucral bracts several, leaflike. Inflorescence of 1 terminal, sessile cluster of spikelets and sometimes several peduncled spikelets. Spikelets oblong-cylindric, 6–10 mm long, 2–3 mm wide; scales ovate, acute, acuminate, with prominent green midrib. Bristles glabrous about 2 × the length of the achene. Achenes biconvex to slightly trigonous, beak short, 0.8–1.3 mm long. (n = 20) Swampy soil, marshes, moist meadows; SD, NE, IA, KS, MO, OK, TX; (ME to MT, s to FL, TX, & NM, also OR; Mex.). *S. lineatus* Michx.—Rydberg.

16. *Scirpus pungens* Vahl. Perennial from long creeping, reddish-brown, rhizomes. Culms numerous, up to 12 dm tall, sharply triangular, flat, concave or convex sides. Leaves 2–4, all near the base of culm, 2–4 mm wide. Main bract up to 15 cm long, appearing as a continuation of the culm, up to 2 smaller bracts usually present and resembling enlarged scales. Spikelets 1–8, sessile, ovoid to lance-ovoid, up to 20 mm long, many-flowered; scales ovate to obovoid with a mucro or short awn. Bristles 4–6, little if at all exceeding the achene, retrorsely barbellate. Achenes lenticular or trigonous up to 3.4 mm long, 2.3 mm wide, apiculate, style 2- or 3-cleft. (n = 38, 39, 60) Marshes, sloughs, wet areas; throughout GP; (Newf. to B.C., s to FL, TX, & CA). *S. americanus* Pers.—Rydberg; *S. americanus* Pers. var. *longispicatus* Britt.—Gates.

17. *Scirpus saximontanus* A. Gray. Cespitose annual. Culms slender, up to 4 dm tall, terete. Leaves few, usually basal with or without blade. Involucral bract appearing as a continuation of the culm, other small bracts below the inflorescences. Spikelets 2–several, usually sessile, slender-ovoid, many-flowered; scales ovate, cuspidate-acuminate. Bristles variable. Achenes up to 1.5 mm long, ridged transversely, strongly trigonous; style 3-cleft. Shorelines, ditches, clay soil; SD, NE, KS, OK; (GP w to CO, s to Mex.). *S. supinus* L. var. *saximontanus* (Fern.) Fern.—Koyama, op. cit.

18. Scirpus smithii A. Gray. Cespitose annual. Culms slender, terete or somewhat angled, up to 4 dm tall. Leaves short, basal, with sheaths frequently bladeless. Involucral bract erect or nearly so, appearing as a continuation of the culm. Spikelets 1–10, ovoid, acute, to 10 mm long, 3–5 mm wide in a capitate cluster; scales oblong-ovate, blunt or mucronulate. Bristles variable in number and length. Achenes obovoid, plano-convex, smooth, 1.4–2 mm long, minutely beaked, style 2-cleft. Sandy or muddy areas; NE; (Que. to Ont., s to VA & NE). *S. debilis* Pursh—Rydberg.

19. Scirpus torreyi Olney. Perennial from short brownish rhizomes. Culms usually solitary, erect, sharply 3-angled with concave sides, up to 1.5 dm tall. Leaves narrowly linear, several, firm. Involucral bract erect, appearing as a continuation of the culm. Spikelets 1–4, ovoid or fusiform, up to 11 mm long in a capitate cluster; scales ovate, acute or mucronate. Bristles longer than the achene. Achenes trigonous, obovoid, 3–4 mm long, beaked, style 3-cleft. Swamps, shores; NE; (N.B. to Man., s to PA & NE).

20. Scirpus validus Vahl. Perennial with light reddish-brown stout rhizomes. Culm erect, stout, to 3.9 m tall, terete, pale to light green, soft. Leaves of upper sheath with long tapering blades. Involucral bract single, very rarely 2, appearing as a continuation of culm. Inflorescence paniculate, spreading to loosely spreading, lax, spikelets often solitary or in glomerules of 2, rarely 3 or more. Spikelets numerous, up to 10.2 mm long, 4 mm wide, ovoid to cylindric; scales to 5.8 mm long, 2 mm wide, mucronate tip, exceeding the length and width of the achene. Bristles 6, equaling or often exceeding the length of the achene. Achene 2.4 mm long, 1.5 mm wide, beak to 0.3 mm long, apiculate, lenticular, style 2-cleft. Sloughs, marshes, shallow lake edges, roadside ditches; throughout GP; (Newf. to AK, s to FL & CA). *S. validus* Vahl var. *creber* Fern.—Fernald.

13. SCLERIA Berg., Nut-Rush

Perennial from stout rhizomes. Culms erect, triangular, usually less than 1 m tall. Leaves linear and flat with closed sheaths. Inflorescence of terminal and axillary fascicles, the axillary fascicles peduncled; involucral bracts usually exceeding the inflorescence; spikelets unisexual, the 1-flowered pistillate spikelets usually intermixed with clusters of staminate spikelets. Stamens 1–3, style 3-branched. Achene globular to ovoid, white, bony or crustaceous, subtended by a hypogynium bearing 3–6 tubercles, sometimes on a rough white crust.

References: Fairey, John E., III. 1967. The genus *Scleria* in the Southeastern United States. Castanea 32: 37–71; Fairey, John E., III. 1969. *Scleria pauciflora* Muhl. and its varieties in North America. Castanea 34: 87–90.

1 Hypogynium covered with a rough white crust .... 3. *S. triglomerata*
1 Hypogynium not covered with a rough white crust.
  2 Hypogynium bearing 6 tubercles; achene 1–2.5 mm long .... 2. *S. pauciflora*
  2 Hypogynium bearing 3 tubercles; achene 2–3 mm long .... 1. *S. ciliata*

1. Scleria ciliata Michx. Perennial from stout, elongate rhizomes. Culms erect, sharply triangular, glabrous to hairy, up to 7 dm tall. Leaves up to 7 mm wide, up to 40 cm long, glabrous or pubescent. Sheaths glabrous or pubescent. Inflorescence of 1–3 fascicles, terminal and axillary. Bracts lanceolate, ciliate to glabrous; scales lanceolate, pubescent or glabrous. Achene globose, up to 3 mm long, verrucose-tuberculate, white; hypogynium an obtusely 3-angled border, bearing 3 entire, 2-lobed tubercules.

Two varieties are recognized by Fairey, 1967:

1a. var. *ciliata*. Culms slender, hairy. Leaves rather narrow, 1–3.5 mm wide, ciliate, bracts ciliate. Tubercles entire or lobed. Prairies & open woods; KS, OK; (VA to KS, OK s to FL & TX).

1b. var. *elliottii* (Britt.) Fern. Culms stout, pubescent or ciliate. Bracts densely ciliate; scales pubescent; pistillate scales usually with a conspicuous midrib projecting as a keel toward the base. Tubercles usually strongly 2-lobed. Prairies & open woods; MO; (VA to MO, OK s to FL & TX).

2. **Scleria pauciflora** Muhl. ex Willd. Perennial from thick forking rhizomes. Culms erect, trigonous, glabrous or pilose up to 5 dm tall. Leaves narrowly linear up to 4 mm wide, up to 20 cm long, glabrous or pubescent. Sheaths short pubescent or villous. Inflorescence of 1–3 small terminal fascicles and sometimes peduncled axillary ones, usually few-flowered; bracts foliaceous, erect, much longer than the inflorescence; scales ovate, acuminate to aristate. Achenes subglobose, up to 2 mm long, papillate, white; hypogynium a narrow 3-angular border, with 6 rounded tubercles in pairs.

Two varieties are recognized by Fairey, 1969:

2a. var. *caroliniana* (Willd.) Wood. Plants up to 40 cm tall, copiously villous-ciliate with spreading hairs 0.5–1 mm long on the culms, leaves, and bracts. Leaves slightly shorter than or equaling the culms. Upland prairies & open woods; KS, MO; (NH to MI, s to FL, MO, & KS).

2b. var. *pauciflora*. Plants up to 50 cm tall, glabrous or hairy but not copiously villous-ciliate. Leaves shorter than the culms. Sandy prairies or open woods; KS, MO; (NJ to KS, s to FL & TX).

3. **Scleria triglomerata** Michx. Perennial from clustered, stout, knotty rhizomes. Culms usually cespitose, up to 10 dm tall, sharply trigonous, up to 6 mm thick at base. Leaves up to 9 mm wide, up to 30 cm long, slightly pubescent, scabrous on margins and keel. Sheaths glabrous or nearly so, not winged, ligule rigid, hairy or glabrous. Inflorescence in about 3 fascicles, terminal or axillary, few- to many-flowered; bracts foliaceous, lanceolate, ciliate or glabrous; staminate scales lanceolate, acuminate; pistillate scales ovate with midrib prolonged into an awn. Achene subglobose to ovoid, up to 3 mm long, obtuse, smooth, shining, white to buff; hypogynium low, shallowly 3-angled, surface with a white papillose crust. Sandy native prairies, open wooded areas; KS, OK, TX, IA, MO; (NY to MN, s to FL & TX).

151. POACEAE Barnh., the Grass Family

by David Sutherland

Herbaceous annuals or perennials. Stems (culms) with swollen nodes and intercalary meristems at the base of each internode, fistulose to solid, branched to unbranched, sometimes modified as stolons, rhizomes, or, rarely, the base swollen and cormlike. Leaves alternate, each one consisting of a basal, enclosing sheath with a meristem at its base, open or fused by its margins and a terminal blade which is folded, C-shaped in cross section, or rolled in the bud and flat, involute, or folded at maturity; the junction of the blade and sheath usually bearing an erect fringe of hairs or a membrane (ligule) and sometimes also earlike projections of the blade margins (auricles). Inflorescence made up of very reduced spikes (spikelets), borne in clusters which are named as if the spikelets represented individual flowers (i.e., panicles, racemes, or spikes), rarely, the spikelet clusters enclosed in burs or involucres; spikelets bearing 1–numerous perfect to imperfect or sterile flowers and associated protective bracts, the flowers and bracts clearly distichous on the spikelet axis; each spikelet normally subtended by 2 protective awned or awnless bracts (glumes), one or both of these sometimes reduced or absent; individual flowers normally each subtended by 2 bracts, a lower, enclosing, awned or awnless bract (lemma), and an upper, normally awnless bract (palea), the palea sometimes inconspicuous or absent; each flower and its (1)2 subtending bracts collectively called the floret; florets of the spikelet borne on a central axis (the rachilla) the tip of which may or may not be prolonged beyond the uppermost floret as a bristle; spikelets and florets, at maturity, disarticulating in a variety of patterns, the glumes falling with the spikelet or remaining behind, the rachilla remaining intact or breaking into segments, sometimes a portion of each rachilla segment modified as a hardened sharp point, the callus, which remains attached to the floret and is here measured as a part of the lemma, or, in some cases, the axis of the whole inflorescence disarticulating and spikelets falling singly or in groups attached to rachis segments; spikelet variously constituted, all of its 1–numerous florets perfect, all of them imperfect, or with (a) sterile or staminate floret(s) borne either above or below (a) perfect one(s). Flowers lacking a well-developed perianth, this represented only by 2 minute lodicules which function in floret opening and closure. Anthers 3, or, rarely, only 1 or 2, borne on slender filaments, usually dangling outside the floret at maturity, but sometimes retained. Pistil with 1 locule and 1 ovule; styles 1 or 2; stigmas 2, plumose, usually exserted at maturity, but sometimes retained within the floret. Fruit a 1-seeded grain, the pericarp usually attached to the seed coat. *Gramineae* Juss.

Some workers have recently adopted a controversial "genome based" generic concept for grasses belonging to the tribe Triticeae. If adopted, this plan would change the generic names for many of the taxa here assigned to x *Agrohordeum, Agropyron, Elymus, Hordeum,* and *Sitanion.* These proposals are summarized in Barkworth and Dewey, Amer. J. Bot. 72:767–776. 1985.

The following species, which are rare in our range, belong to genera that are not described in the text. However, all of these genera are differentiated in the key and indicated with an asterisk.

Arrhenatherum elatius (L.) Presl (tall oatgrass) is known from scattered introductions in the GP (CO: El Paso; IA: Dickinson, Page; MO: Jasper; NE: Box Butte, Douglas, Lancaster) but is doubtfully persistent.

Brachiaria ciliatissima (Buckl.) Chase enters our range in TX: Cottle. It is a perennial of open, sandy soil.

Enneapogon desvauxii Beauv. is a low perennial of dry plains and slopes. It is known from the s part of our range in OK: Cimarron, and TX: Deaf Smith.

Erianthus ravennae (L.) Beauv. is a tall ornamental grass which has escaped from cultivation at several locations in KS: Harvey & Wyandotte (Brooks, McGregor, & Hauser, Techn. Publ. State Biol. Surv. Kansas 1: 11. 1976).

Holcus lanatus L. is a velvety-pubescent perennial of low stature. It has been widely cultivated as a pasture grass and appears to be established in MO & e KS.

Leucopoa kingii (S. Wats.) W. A. Weber (listed as *Hesperochloa kingii* (S. Wats.) Rydb. in the Atlas of the Great Plains Flora, 1977) is a dioecious, rhizomatous perennial of open ridges and slopes. It enters our range in WY, w NE, & w KS.

Limnodea arkansana (Nutt.) L. H. Dewey is a low perennial whose distribution is mostly in the se coastal plain. It enters our area in sw OK.

Miscanthus sacchariflorus (Maxim.) Hack. is a tall, rhizomatous perennial that is cultivated as an ornamental and has escaped from cultivation in scattered locations in IA & NE.

Sclerochloa dura (L.) Beauv. is a diminutive annual introduced to this continent from s Europe. So far, it has been collected at scattered locations in KS, MO, & OK.

Scleropogon brevifolius Phil. is a dioecious, stoloniferous perennial with strikingly dimorphic staminate and pistillate plants. It is found in dry, open ground in the s part of our range (TX: Bailey; OK: Cimarron).

Secale cereale L. (rye) is cultivated in our range, and, although it evidently does not persist, it may be collected as a temporary waif at the edges of fields, along railroads, etc.

Trichachne californica (Benth.) Chase is a distinctive perennial of the sw region. It enters our range in w OK and n TX. The genus is closely related to *Digitaria* and has been combined with it by some authors.

Trisetum includes several species reported from in or near our range. *T. interruptum* Buckl., an annual species with its spikelets disarticulating below the glumes, is known from OK and TX. *T. flavescens* (L.) Beauv., an introduced perennial species with a relatively loose, yellowish panicle and spikelets disarticulating above the glumes, is known from KS: Riley, and MO: Jackson. *T. spicatum* (L.) Richt., a perennial with a very dense panicle and spikelets disarticulating above the glumes, is known from the Black Hills of SD.

Triticum aestivum L. (wheat) does not normally persist after cultivation in our range, but it is rather commonly collected as a waif at the edges of fields, roads, and railroad tracks.

1 Spikelets borne in a burlike structure, this bur falling intact at maturity.
  2 Plants perennial, stoloniferous, mat-forming; leaf blades up to about 2 mm wide .... 13. *Buchloë*
  2 Plants annual, often with the culms geniculate at the base and rooting at the lower nodes, but not really stoloniferous or mat-forming; leaf blades averaging well over 2 mm wide .... 17. *Cenchrus*
1 Spikelets not borne in a bur which falls intact at maturity.
  3 Inflorescence with highly dimorphic spikelets (staminate and pistillate) borne in clearly marked regions, one above the other.
    4 Pistillate spikelets beadlike, unawned, borne toward the bases of most inflorescence branches; plants rhizomatous perennials .... 75. *Tripsacum*
    4 Pistillate spikelets not beadlike, awned, borne in the upper part of the inflorescence; plants robust, nonrhizomatous annuals .... 76. *Zizania*
  3 Inflorescence not as above, the spikelets either not clearly of 2 kinds or, if so, the 2 kinds mixed in the inflorescence or borne on separate plants.
    5 Plants clearly dioecious (both staminate and pistillate plants available for examination.)
      6 Pistillate spikelets bearing lemma awns 5–10 cm long .... *Scleropogon*
      6 Pistillate spikelets awnless or bearing lemma awns less than 1 cm long.
        7 Staminate and pistillate inflorescences distinctly different, the staminate ones consisting of 2 or 3(5) unilateral spicate branches, the pistillate spikelets borne in burlike clusters .... 13. *Buchloë*
        7 Staminate and pistillate inflorescences not markedly different, both consisting of open to condensed and headlike panicles.
          8 Plants creeping annuals; lemmas plainly 3-nerved .... 32. *Eragrostis*
          8 Plants perennials; lemmas with more than 3 nerves, the nerves distinct to faint.
            9 Lemmas 7- to 11-nerved; ligule a short fringed membrane 0.1–0.4 mm long .... 28. *Distichlis*
            9 Lemmas 5-nerved; ligule longer, erose-ciliolate to entire or lacerate.
              10 Leaf blades folded when young, the tips prow-shaped at maturity; lemmas often keeled or villous or both .... 58. *Poa*
              10 Leaf blades rolled when young, the tips not prow-shaped; lemmas scabrous, scarcely keeled .... *Leucopoa*
    5 Plants not clearly dioecious (or dioecious, but only one sex available for examination).
      11 Plants stoloniferous annuals with scabrous or pubescent culms; leaves in fascicles; spikelets in headlike clusters or in small groups among the leaves of a fascicle.
        12 Blades usually folded, sharp-pointed, the margins thick and whitish; spikelets in small groups among the leaves; plants bisexual .... 51. *Munroa*
        12 Blades not as above; spikelets elevated above the leaves in capitate or subcapitate clusters; plants unisexual .... 32. *Eragrostis*
      11 Plants nonstoloniferous, or if stoloniferous, then culms not pubescent, leaves not in fascicles, or spikelets not in clusters.
        13 Inflorescence consisting of 1 to numerous spikes or spikelike racemes (examine the inflorescence carefully to make sure it is not a condensed panicle).
          14 Inflorescence a solitary bilateral spike or raceme.

15 Inflorescence slender, streamlined, and terete, with spikelets fitted into concavities of the broad rachis.
  16 Spikelets borne singly; glumes awned, especially in the upper part of the inflorescence .... 1. *Aegilops*
  16 Spikelets in pairs of 1 fertile and 1 rudimentary, the rudimentary one-pediceled, the fertile one sessile; glumes not awned .... 48. *Manisurus*
15 Inflorescence thick, more open, the spikelets not fitting tightly into concavities of the rachis.
  17 Spikelets in 3s at each node, the 2 lateral ones of each group normally staminate or sterile, the central one perfect and 1-flowered.
    18 Central spikelet next to the axis, mostly concealed by the lateral spikelets; spikelet group pilose at the base .... 39. *Hilaria*
    18 Central spikelet plainly visible, not concealed by the lateral spikelets; spikelet group not pilose at the base .... 40. *Hordeum*
  17 Spikelets solitary, paired, or 3 or more per node, but if with 3, the central one with more than 1 floret.
    19 Rachis disarticulating at maturity and breaking into sections attached to spikelets or spikelet groups.
      20 Glumes, together with their awns, mostly 3-9 cm long, the awns becoming divergent at maturity; spikelets paired at most nodes .... 66. *Sitanion*
      20 Glumes, together with their awns, mostly less than 3 cm long, the awns not divergent; spikelets 1-3 at each node.
        21 Spikelets usually paired at the lower nodes, single, but with 3 glumes at the middle nodes, and single with 2 glumes at the upper nodes .... 2. x *Agrohordeum*
        21 Spikelets (2)3 at most nodes .... 31. *Elymus*
    19 Rachis continuous at maturity, not breaking into sections.
      22 Spikelets solitary at each rachis node, rarely paired at some nodes.
        23 First glume wanting in all but the terminal spikelet; back of the first lemma facing the rachis .... 46. *Lolium*
        23 First glume present in all spikelets, the lemmas and the glumes placed at right angles to the rachis.
          24 Plants perennial .... 3. *Agropyron*
          24 Plants annual.
            25 Glumes broad, 3-nerved .... *Triticum*
            25 Glumes subulate, 1-nerved .... *Secale*
      22 Spikelets 2 or more at nearly all nodes.
        26 Glumes absent or irregularly developed in some spikelets as 1 or 2 short scales or bristles; spikelets widely spreading when mature .... 41. *Hystrix*
        26 Glumes relatively conspicuous, usually as long as the spikelets; spikelets ascending when mature .... 31. *Elymus*
14 Inflorescence consisting of 2—numerous spikes or spikelike racemes, or solitary, but then clearly unilateral.
  27 Each spikelet containing only 2 staminate florets, the pistillate spikelets on separate plants; plants stoloniferous perennials; inflorescences consisting of 2 or 3(5) short, unilateral spicate branches .... 13. *Buchloë*
  27 Many or all of the spikelets containing 1 or more perfect florets; plants otherwise not as above.
    28 Spikelets each containing more than 1 perfect floret or with 1 perfect floret and 1 or more sterile florets above it.
      29 Perfect florets 2-several in each spikelet.
        30 Inflorescence of 1-7 flattened branches, these mostly digitately arranged, the largest ones 3.5-5.5 mm wide .... 30. *Eleusine*
        30 Inflorescence not as above in all respects.
          31 Lemmas 5-nerved; dwarf annuals with the leaf blades folded in the bud and prow-shaped at the tip .... *Sclerochloa*
          31 Lemmas 3-nerved; annuals or perennials with leaf blades rolled in the bud and not obviously prow-shaped at the tip.

32 Plants annual; spikelets unilateral on the branches of the
   inflorescence .... 44. *Leptochloa*
 32 Plants perennial; spikelets not unilateral on the branches of the
   inflorescence .... 32. *Eragrostis*
29 Perfect floret 1 in each spikelet, 1 or more sterile or staminate florets above it.
 33 Inflorescence branches digitate or approximately so.
  34 Lemma of perfect floret awnless .... 21. *Cynodon*
  34 Lemma of perfect floret awned .... 19. *Chloris*
 33 Inflorescence branches distributed along a relatively long axis, or rarely, solitary.
  35 Spikelets widely spaced, appressed along the slender
    branches .... 36. *Gymnopogon*
  35 Spikelets closely crowded, not appressed .... 10. *Bouteloua*
28 Spikelets each containing 1 perfect floret, with or without a sterile or staminate floret below
 it.
 36 Ligule absent .... 29. *Echinochloa*
 36 Ligule present.
  37 Lemma of perfect floret firmer than the glume or glumes and sterile lemma, clasping
    the palea; spikelets flattened dorsally.
   38 Each spikelet subtended by a cuplike thickening; first glume apparently
    absent .... 33. *Eriochloa*
   38 Each spikelet not subtended by a cuplike thickening; first glume present or
    absent.
    39 Fertile lemma with its back turned toward the branch axis.
     40 Fertile lemma membranaceous-margined, clasping the palea but not further
       inrolled.
      41 Inflorescence branches digitate or nearly so .... 27. *Digitaria*
      41 Inflorescence branches disposed along an elongate
        axis .... *Trichachne*
     40 Fertile lemma firm at the margin, clasping the palea and also further
       inrolled .... 54. *Paspalum*
    39 Fertile lemma with its back turned away from the branch
      axis .... *Brachiaria*
  37 Lemma of perfect floret similar to or more delicate than the glumes, not tightly clasp-
    ing the palea; sterile lemma absent or similar to the fertile one in texture; spikelets
    terete or flattened laterally.
   42 Inflorescence branches not bearing spikelets all on one side; spikelets in pairs of 1
    sessile and 1 pediceled or 1 short- and 1 long-pediceled, the lower spikelet of each
    pair usually bearing a sterile lemma below the fertile one.
    43 Inflorescence branches terete, streamlined, the spikelets fitting into concavities
     in the rachis .... 48. *Manisurus*
    43 Inflorescence branches more open, the spikelets not fitted into concavities in
     the rachis.
     44 Spikelets all alike, unequally pedicellate, both members of a pair
       perfect .... *Miscanthus*
     44 Spikelets unlike, the lower one of the pair sessile and perfect, the upper
       one usually sterile or staminate (rarely perfect) and
       pediceled .... 6. *Andropogon*
   42 Inflorescence branches bearing spikelets all on one side; spikelets otherwise not as
    above.
    45 Spikelets disarticulating below the glumes; grasses of moist or wet places.
     46 Glumes unequal, narrow .... 69. *Spartina*
     46 Glumes about equal, broad, inflated .... 9. *Beckmannia*
    45 Spikelets disarticulating above the glumes; grasses of dry ground.
     47 Inflorescence branches digitate; rachilla prolonged beside the
       floret .... 21. *Cynodon*
     47 Inflorescence branches disposed along an elongate axis; rachilla not
       prolonged .... 62. *Schedonnardus*

13 Inflorescence a condensed to open panicle, this sometimes reduced to a nonspiciform raceme.
  48 All or most spikelets consisting solely of a solitary floret or of a solitary floret subtended by 1 or 2 glumes.
    49 Both glumes absent.
      50 Lemma 5-nerved, awnless .... 43. *Leersia*
      50 Lemma 3-nerved, tapering into a scabrous awn .... 50. *Muhlenbergia*
    49 One or both glumes present.
      51 Spikelet subtended by a cup-shaped thickening .... 33. *Eriochloa*
      51 Spikelet not subtended by a cup-shaped thickening.
        52 Lemma awned.
          53 Spikelets disarticulating above the glumes (the glumes sometimes falling later in some species).
            54 Lemma bearing 3 awns or a single 3-branched awn .... 7. *Aristida*
            54 Lemma bearing 1 unbranched awn.
              55 Callus bearing a prominent tuft of long, straight hairs; lemma awned from the back .... 14. *Calamagrostis*
              55 Callus glabrous to pilose; awns borne from the tip or near the tip of the lemma.
                56 Glumes small, sometimes only 1 developed; rachilla prolonged and present beside the floret; floret obviously stipitate within the glumes .... 11. *Brachyelytrum*
                56 Glumes larger or the rachilla not prolonged; floret not obviously stipitate.
                  57 Lemmas indurate at maturity.
                    58 Lemma awn very conspicuous, twisted, persistent .... 72. *Stipa*
                    58 Lemma awn less conspicuous, usually not much twisted, deciduous .... 52. *Oryzopsis*
                  57 Lemmas not indurate at maturity.
                    59 Plants delicate annuals .... 4. *Agrostis*
                    59 Plants perennials .... 50. *Muhlenbergia*
          53 Spikelets disarticulating below the glumes.
            60 Each spikelet flanked by 1 or 2 feathery structures (the pedicels of obsolete spikelets) .... 67. *Sorghastrum*
            60 Spikelets not flanked by feathery pedicels.
              61 Many of the spikelets paired on unequal pedicels, falling mostly in pairs with the pedicels attached .... 47. *Lycurus*
              61 Spikelets falling individually, not with the pedicels attached.
                62 Glumes awned .... 59. *Polypogon*
                62 Glumes awnless.
                  63 Panicle dense, cylindrical; glumes united below .... 5. *Alopecurus*
                  63 Panicle open to somewhat contracted, but not cylindrical; glumes not united.
                    64 Plants low annuals; awns of lemmas more than 5 mm long .... *Limnodea*
                    64 Plants tall perennials; awns less than 5 mm long .... 20. *Cinna*
        52 Lemmas awnless.
          65 Fertile floret subtended by 2 reduced sterile florets, these usually represented by 2 scales appressed to the surface of the fertile floret and not immediately apparent .... 55. *Phalaris*
          65 Fertile floret solitary.
            66 Panicles cylindrical, dense, not lobed.
              67 Glumes awned .... 56. *Phleum*
              67 Glumes awnless.
                68 Glumes united below .... 5. *Alopecurus*
                68 Glumes not united below .... 71. *Sporobolus*

66 Panicles condensed to open, but not cylindrical and unlobed.
  69 Spikelets disarticulating above the glumes.
    70 Lemma bearing a prominent tuft of hairs from the callus, 1-nerved .... 15. *Calamovilfa*
    70 Lemma sometimes pubescent at the base, but not bearing a prominent tuft of hairs from the callus, 1- to 5-nerved.
      71 Ligule predominantly a fringe of hairs .... 71. *Sporobolus*
      71 Ligule membranaceous.
        72 Palea minute or absent .... 4. *Agrostis*
        72 Palea present and conspicuous.
          73 Plants with open panicles and with the glumes surpassing the lemmas .... 4. *Agrostis*
          73 Plants either with slender contracted panicles or with the glumes shorter than the lemmas or both .... 50. *Muhlenbergia*
  69 Spikelets disarticulating below the glumes.
    74 Spikelets flattened dorsally (2-flowered, but the first glume sometimes lacking and then the lower, sterile lemma simulating a glume, the sterile palea present as a membrane to 0.5 mm long, but easily overlooked) .... 45. *Leptoloma*
    74 Spikelets flattened laterally, 1-flowered.
      75 Glumes 1-nerved; lemmas 1-nerved .... 71. *Sporobolus*
      75 Glumes 1- to 3-nerved; lemmas 3- to 5-nerved.
        76 Spikelets nearly orbicular, the glumes inflated .... 9. *Beckmannia*
        76 Spikelets not nearly orbicular; glumes not inflated.
          77 Lemmas less than 2 mm long; low plants of stream banks and irrigation ditches .... 4. *Agrostis*
          77 Lemmas more than 2 mm long; tall plants of moist places, usually found in woods .... 20. *Cinna*
48 Spikelets with 1 perfect floret and 1 or more staminate or sterile florets (often represented only by bracts) or spikelets with 2-numerous similar florets.
  78 Main spikelets with 1 perfect floret and 1 or more staminate or sterile florets, these differing markedly in appearance from the perfect floret (sometimes sterile or staminate spikelets are present in addition to the perfect ones).
    79 Staminate or sterile florets borne above the perfect ones.
      80 Lowest lemma bearing about 9 plumose awns .... *Enneapogon*
      80 Lowest lemma awnless or bearing 1 awn.
        81 Spikelets short pediceled and appressed along the panicle branches; lowest lemma awned from the tip; upper (sterile) spikelet consisting of a pediceled awn .... 36. *Gymnopogon*
        81 Spikelets long pediceled and not appressed along the panicle branches; lower lemma awnless .... *Holcus*
    79 Staminate or sterile florets borne below the fertile ones.
      82 Staminate floret single, awned from near its base, its awn longer than the short awn of the pistillate floret above; both lemmas firm .... *Arrhenatherum*
      82 Staminate or sterile floret(s) 1 or 2, not awned from near the base; spikelets otherwise not as above.
        83 The single perfect floret flanked by 2 well-developed staminate florets or by 2 reduced sterile florets, the 3 florets falling from the glumes as a unit.
          84 Perfect floret flanked by 2 well-developed staminate florets .... 38. *Hierochloë*
          84 Perfect floret flanked by 2 scales (sterile florets), these sometimes appressed to it and not immediately apparent .... 55. *Phalaris*
        83 The single perfect floret with a single staminate or sterile floret below it; disarticulation below the glumes in most species.
          85 Mature lemma of perfect floret hyaline, more delicate than the glumes, long-awned.

86 Spikelets paired, 1 spikelet of each pair sessile on the panicle branch, the other pedicellate.
   87 Spikelets of a pair alike, both perfect .... *Erianthus*
   87 Spikelets unlike, the pediceled one staminate .... 68. *Sorghum*
86 Spikelets not paired, but each flanked by 1 or 2 featherlike appendages (representing the pedicels of obsolete spikelets) .... 67. *Sorghastrum*
85 Mature lemma of perfect floret not hyaline, becoming indurate, blunt to acuminate or short awned.
  88 Spikelets subtended by prominent bristles; inflorescence cylindrical .... 65. *Setaria*
  88 Spikelets not subtended by prominent bristles, although bristlelike hairs sometimes present; inflorescence open to condensed, but not cylindrical.
    89 Spikelets more or less covered with long, silky hairs .... *Trichachne*
    89 Spikelets not covered with long, silky hairs.
      90 Each spikelet subtended by a prominent cup-shaped thickening .... 33. *Eriochloa*
      90 Each spikelet not subtended by a cup-shaped thickening.
        91 Plants annual; ligules absent .... 29. *Echinochloa*
        91 Plants annual or perennial; ligule present (absent in a few perennial species).
          92 Perfect floret with lemma margins hyaline, clasping the palea but not further inrolled; ligule membranaceous .... 45. *Leptoloma*
          92 Perfect floret with lemma margins firm, usually clasping the palea and also further inrolled; ligule principally a fringe of hairs in most species but sometimes principally or entirely membranaceous or even absent.
            93 Plants annual .... 53. *Panicum*
            93 Plants perennial.
              94 Plants either strongly stoloniferous or strongly rhizomatous .... 53. *Panicum*
              94 Plants neither strongly stoloniferous nor rhizomatous, the culms sometimes arising from a somewhat knotty base.
                95 Sheaths compressed .... 53. *Panicum*
                95 Sheaths not compressed.
                  96 Spikelets acute to acuminate-tipped; plants neither producing a winter rosette of short, broad leaves nor producing long, exserted anthers early in the season and shorter, retained anthers later in the season .... 53. *Panicum*
                  96 Spikelets blunt or obtuse to somewhat acute; plants producing either a winter rosette of short, broad leaves or producing spikelets with relatively long, exserted anthers early in the season and shorter, retained anthers later in the season (or both) .... 26. *Dichanthelium*
78 Spikelets containing 2–many similar florets, these usually perfect, but sometimes all or part of them imperfect or sterile.
  97 Plants both with the lemmas plainly 3-nerved *and* with the ligule consisting of a fringe of hairs or a prominently fringed membrane.
    98 Plants tall reeds with plumose panicles; rachilla internodes with long, silky hairs .... 57. *Phragmites*
    98 Plants not tall reeds or not otherwise as above.
      99 Nerves of the lemma pubescent or puberulent, at least at the base, or the callus pubescent.
        100 Callus of each lemma bearing a tuft of hairs; nerves of the lemma glabrous.
          101 Lemmas unawned .... 61. *Redfieldia*

          101 Lemmas each bearing 3 awns .... *Scleropogon
     100 Callus not bearing a tuft of hairs; nerves of lemma pubescent, at least toward the base.
          102 Plants slender annuals; palea ciliate on its upper margins ...... 74. *Triplasis*
          102 Plants perennials; palea not ciliate on its upper margins .... 73. *Tridens*
  99 Nerves of the lemma not pubescent or puberulent but sometimes scabrous or glandular; callus not pubescent.
     103 Spikelets with florets widely spaced on the rachilla, each rachilla internode more than 1 mm long .... *Scleropogon
     103 Spikelets closely crowded and overlapping, the rachilla internodes shorter.
          104 Spikelets papery, whitish, and often also purplish-tinged; plants perennials, glaucous; panicles contracted; most spikelets 4–7 mm long ...... 73. *Tridens*
          104 Spikelets not papery and whitish or otherwise not as above .... 32. *Eragrostis*
97 Plants either with the lemmas more than 3-nerved *or* with the ligules membranaceous and not fringed (at most ciliolate)
  105 Tip of the first glume surpassing the tip of the lowest lemma (exclusive of awns).
     106 Lemmas awnless or awned from a nonbifid tip; spikelets under 10 mm long .... 50. *Muhlenbergia*
     106 Lemmas awned from the back or from between 2 teeth at the tips (awnless in cultivated forms of *Avena*, which have spikelets more than 10 mm long).
          107 Ligule a fringe of hairs .... 23. *Danthonia*
          107 Ligule a membrane.
               108 Glumes more than 15 mm long; lowest lemma more than 13 mm long, usually 7-nerved .... 8. *Avena*
               108 Glumes less than 15 mm long; lowest lemma less than 13 mm long, usually 5-nerved, but the nerves sometimes obscure.
                    109 Spikelets disarticulating below the glumes .... *Trisetum
                    109 Spikelets disarticulating above the glumes.
                        110 Spikelets more than 1 cm long, mostly 3- to 5-flowered .... 37. *Helictotrichon*
                        110 Spikelets smaller, usually 2-flowered.
                             111 Rachilla prolonged beyond the upper floret .... 24. *Deschampsia*
                             111 Rachilla not prolonged beyond the upper floret .... 14. *Calamagrostis*
  105 Tip of the first glume not surpassing the tip of the lowest lemma.
     112 Lemma awned from the back at about midlength with a strong, geniculate awn .... 37. *Helictotrichon*
     112 Lemma awned from at or near the tip, often from a bifid apex, or awnless.
          113 Disarticulation of the spikelet below the glumes (disarticulation of the uppermost floret may sometimes precede the fall of the rest of the spikelet).
               114 Second glume clearly broader than the first .... 70. *Sphenopholis*
               114 Second glume somewhat narrower than the first .... 49. *Melica*
          113 Disarticulation of the spikelet above the glumes and between the florets.
               115 Lemmas 3-nerved.
                    116 Plants stoloniferous, growing in or near water; leaf blades prow-shaped at the tip, folded in the bud .... 16. *Catabrosa*
                    116 Plants seldom stoloniferous, but if so, then not otherwise as above.
                        117 Lemmas membranaceous at the margins, the 3 nerves clustered in the leathery center region, terminating in a cusp; grain pyriform, large, its beak protruding at maturity ... 25. *Diarrhena*
                        117 Lemmas membranaceous throughout; grain not pyriform, not with a protruding beak .... 50. *Muhlenbergia*
               115 Lemmas 5- to many-nerved, the nerves sometimes obscure.
                    118 Spikelets large, 15–25 (34) mm long, with the lowest 1 or 2 lemmas normally empty; plants tall, woodland perennials .. 18. *Chasmanthium*
                    118 Spikelets smaller or not otherwise as above.

119 Callus of each floret bearded or hirsute.
    120 Sheaths closed to near the summit, even the ligule sometimes tubular ... 63. *Schizachne*
    120 Sheaths open, the ligules not tubular.
        121 Lemmas awnless, the tips usually erose .... 64. *Scolochloa*
        121 Lemmas awned from between or near the points of the bifid apex .... *Trisetum
119 Callus not bearded, but lemma base may sometimes bear long, cobwebby hairs.
        122 Spikelets crowded at the ends of the panicle branches and turned to one side .... 22. *Dactylis*
        122 Spikelets not crowded at the ends of the panicle branches and not turned to one side.
            123 Leaf blades plainly prow-shaped at the tips.
                124 Blades heavily ridged above; sheaths open to the base; second glume about as long as the first lemma; most spikelets 2-flowered .... 42. *Koeleria*
                124 Blades or other features not as above .... 58. *Poa*
            123 Leaf blades not prow-shaped at the tips.
                125 Spikelet terminating in a club-shaped rudiment .... 49. *Melica*
                125 Spikelet with uppermost floret usually reduced, but not modified into a club-shaped rudiment.
                    126 Plants tough-stemmed, mat-forming perennials of saline areas; blades stiff, pungent; spikelets all pistillate or all staminate on a given plant .... 28. *Distichlis*
                    126 Plants not as above in all features.
                        127 Sheaths open to the base.
                            128 Lemmas awned from a strongly bifid apex .. *Trisetum
                            128 Lemmas awnless or awned from a scarcely bifid or nonbifid apex.
                                129 Lemmas with the nearly parallel nerves ending before the scarious margins, awnless .... 60. *Puccinellia*
                                129 Lemmas with the 5 nerves converging at the apex, awned or awnless .... 34. *Festuca*
                        127 Sheaths closed, at least toward the base.
                            130 Nerves of the lemma running to near the margins or ending before the margins and not strongly convergent; lemmas unawned.
                                131 Nerves faint; plants of saline soils .... 60. *Puccinellia*
                                131 Nerves prominent; plants of moist ground .... 35. *Glyceria*
                            130 Nerves of the lemma convergent; lemmas awned or awnless.
                              132 Lemmas awned .... 12. *Bromus*
                              132 Lemmas awnless.
                                133 Spikelets large, mostly more than 15 mm long .... 12. *Bromus*
                                133 Spikelets smaller .... 58. *Poa*

1. AEGILOPS L., Goatgrass

1. **Aegilops cylindrica** Host, jointed goatgrass. Weedy annual 2–8 dm tall. Culms erect or geniculate at base, hollow. Blades flat, rolled in the bud, glabrous to pilose, 2–12 cm long, 1–3.5 mm wide; auricles inconspicuous or absent; sheaths open, glabrous and ciliate to sparsely pilose; ligules membranaceous, 0.2–0.6 mm long. Inflorescence a compact, terminal, terete spike 7–16 cm long which disarticulates at the nodes; spikelets borne singly, appressed flatwise and fitting closely into the curvature of the rachis, 2- to 3(5)-flowered;

glumes indurate, asymmetrical, many-nerved, 6–10 mm long, the first one usually slightly shorter than the second, both awned, the awns 5–18 mm long in the lower spikelets, much longer in the uppermost spikelet; lemmas of lowest florets 8–10 mm long, unawned or with awns to 3.5 mm long, much longer in the uppermost spikelet. Anthers 2–4 mm long. ($2n = 28$) May–Jun. Waste places, railroads, roadsides, & fields; GP, infrequent in n; (adventive throughout much of the U.S.; Europe). *Naturalized. Triticum cylindricum* (Host) Ces., Pass. & Gib.—Gould, Grasses of Texas, p. 175. Texas A. and M. Univ. Press, College Station. 1975.

2. X AGROHORDEUM G. Camus ex. A. Camus

1. **X *Agrohordeum macounii*** (Vasey) Lepage. Cespitose sterile hybrid perennial, 3–10 dm tall. Culms erect, hollow. Blades flat at maturity, rolled in the bud, glabrous to scabrous, 4–18 cm long, 1.5–5 mm wide; auricles small or absent; sheaths open, smooth; ligule membranaceous, 0.2–0.9 mm long. Inflorescence a terminal spike, 6–15 cm long, which readily disarticulates at the nodes when mature; spikelets usually borne in pairs at the lower nodes, singly, but with 3 glumes, at the middle nodes, and singly, with 2 glumes, at the upper nodes, (1) 2- or 3(4)-flowered; glumes 1- to 3-nerved, slender, the upper ones 5–10 mm long, tapering into awns (1)4–10(16) mm long; lemmas smooth, indistinctly several-nerved, the lower lemmas of the upper spikelets 7–10 mm long, tapering into awns (1.5)6–11(22) mm long. Anthers 0.8–1.6 mm long, pollen abortive. ($2n = 28$) Jun–Aug. Disturbed to moderately disturbed habitats, often near water; MT, ND, MN, WY, SD, NE, IA, KS; (of widespread but scattered occurrence where putative parents occur together). *Hybrid of native and naturalized parents. Elymus macounii* Vasey—Rydberg.

Reference: Boyle, W. S., & A. H. Holmgren. 1955. A cytogenetic study of natural and controlled hybrids between *Agropyron trachycaulum* and *Hordeum jubatum*. Genetics 40: 539–545.

This taxon represents a collection of F_1 hybrids between *Agropyron* and *Hordeum*. The presumed parents are *A. caninum* and *H. jubatum*, but the presence of occasional plants with rhizomes suggests that other *Agropyron* parents may sometimes be involved. The plants are usually found with one or both parents, often in habitats intermediate between the parental habitats.

3. AGROPYRON Gaertn., Wheatgrass

Cespitose to rhizomatous perennials. Culms hollow to solid. Blades rolled in the bud, flat to involute at maturity; sheaths open; ligules membranaceous, short (less than 1.2 mm long); auricles usually present. Inflorescence a solitary, bilateral, terminal spike which does not disarticulate at the nodes when mature; spikelets borne singly or rarely some of them paired, flatwise to the rachis, disarticulating above the glumes, 2- to 11(15)-flowered; glumes subequal or often the first slightly shorter than the second, 1- to 9-nerved, awned or awnless; lemmas rounded, indistinctly to distinctly several-nerved, awnless, or bearing straight or geniculate terminal awns.

Reference: Hitchcock, C. L. 1969. *Agropyron*, pp. 447–461, *in* C. L. Hitchcock, A. Cronquist, M. Ownbey, & J. W. Thompson. Vascular plants of the Pacific Northwest, Part 1. Univ. of Washington Press, Seattle.

Species of *Agropyron* may hybridize naturally with several other genera, notably *Elymus, Sitanion,* and *Hordeum*, to produce sterile offspring. One such hybrid, x *Agrohordeum macounii*, is of relatively common occurrence and is here afforded a separate treatment. x *Agroelymus* and x *Agrositanion* are also known from within our range but are evidently very uncommon. See Bowden, Canad. J. Bot. 45: 711–724. 1965, and Holmgren & Holmgren, *in* Intermountain Flora 6: 295–296. 1977.

1 Spikelets crowded, usually some of them more than 4× the length of the internode, more or less strongly divergent .... 2. *A. cristatum*

1 Spikelets less crowded, erect, appressed to the rachis.
  2 Anthers short, under 2.3 mm long, often retained at maturity; plants cespitose .... 1. *A. caninum* subsp. *majus*
  2 Anthers longer, more than 2.3 mm long; plants cespitose to rhizomatous.
    3 Glumes and lemmas blunt-tipped (sometimes mucronate); culms usually solid.
      4 Plants nonrhizomatous; leaf-blades involute, the upper surface heavily 6- to 8-ridged .... 4. *A. elongatum*
      4 Plants strongly rhizomatous; leaf blades often flat, or sometimes involute, the upper surface with more numerous, fine ridges .... 5. *A. intermedium*
    3 Glumes and lemmas acute, acuminate or awned at the tip; culms hollow or very rarely pith-filled.
      5 Rhizomes absent or very short; spikelets overlapping each other very little, most of them equal to or shorter than the internodes of the rachis (disregarding awns); lemmas usually awned, the awns strongly divergent .... 8. *A. spicatum*
      5 Rhizomes strong and conspicuous; spikelets more crowded; lemmas sometimes awned, but the awns not strongly divergent.
        6 Leaf blades flat, with the upper and lower surfaces unridged or about equally ridged, the widest blades 5–9 mm wide; herbage seldom very glaucous .... 6. *A. repens*
        6 Leaf blades involute to (seldom) flat, the upper surface more obviously ridged than the lower, the widest blades normally less than 5 mm wide, even when unrolled.
          7 Glumes tapering from below the middle into an awn-tip; lemmas often glabrous; lower sheaths often glabrous .... 7. *A. smithii*
          7 Glumes normally widest above the middle, the tip acute to acuminate; lemmas strongly pubescent; lower sheaths usually pubescent, although often very finely so .... 3. *A. dasystachyum*

1. **Agropyron caninum** (L.) Beauv. subsp. **majus** (Vasey) C. L. Hitchc., slender wheatgrass. Tufted perennials, 3–11 dm tall. Culms erect, hollow. Blades flat to involute, scabrous to pilose, 3–27(39) cm long, 1–6(8) mm wide; sheaths glabrous to pilose; ligules 0.3–1.1 mm long; auricles absent to present and conspicuous. Spikes compact, 5–22 cm long; spikelets usually more than 2× as long as the internodes, closely appressed, 3- to 7(9)-flowered; glumes 7–14(16) mm long, bearing awns up to 12.5 mm long or, more frequently, awnless or short-awned; lowest lemma (7)8–14 mm long, the lemmas awnless or with awns up to 30 mm long. Anthers 0.8–2.1 mm long, often retained within the spikelet. (2n = 28) Jun–Aug. In a variety of moist to relatively dry habitats.

There are 2 relatively weakly defined varieties in our region:

1a. var. *majus*. Lemmas awnless or with short awns up to 4 mm long. In the n GP; (Newf. to AK, s to NY, NM, & CA). *A. tenerum* Vasey, *A. biflorum* (Brign.) R.& S.—Rydberg; *A. trachycaulum* (Link) Malte—Fernald.

1b. var. *unilaterale* Vasey. Lemmas with awns more than 4 mm long, some of them usually as long as 12 mm. Sporadic in SD, WY, IA, NE, CO; (ME to Yukon, s to NY, NE, CO, OR). *A. subsecundum* (Link) Hitchc.—Stevens, misapplied; *A. richardsonii* (Trin.) Schrad.—Rydberg, misapplied.

This is a widespread and variable, largely cleistogamous taxon whose elements have been recognized under numerous names. The name *A. trachycaulum* (Link) Malte has been used by authors who consider American plants to belong to a different species from European plants, but Hultén (Arkiv Bot. 7: 1–474. 1968) has pointed out that *A. pauciflorum* (Schwein.) Hitchc. is an older available name. *A. pauciflorum* Schur, 1859, does not interfere with its use, since that name is invalid.

Agropyron caninum ssp. *majus* evidently hybridizes with other species of *Agropyron* and with species of other genera as well, causing considerable confusion in the herbarium. Such specimens may have highly abortive pollen. The name *A. pseudorepens* applies to specimens of a cross between *A. caninum* and a rhizomatous species, probably *A. smithii* (Pohl, Rhodora 64: 143–147. 1962) or *A. dasystachyum* (Bowden, Canad. J. Bot. 43: 1421–1448. 1965).

2. *Agropyron cristatum* (L.) Gaertn., crested wheatgrass. Cespitose perennials (2)3.5–7(11) dm tall. Culms erect, hollow to pith-filled. Blades flat, pubescent to glabrous on the upper surface, 2–17 cm long, 1.5–7 mm wide, ligules 0.1–1 mm long; auricles present. Spikes dense, 2–8 cm long, the spikelets closely overlapping, several times longer than the internodes, strongly divergent, not appressed to the rachis, 3- to 6(9)-flowered; glumes 2–6 mm long, their awns lacking or up to 3.5 mm long; lowest lemma 4–8 mm long, the lemmas awnless or bearing awns up to 5 mm long. Anthers 2.4–4(4.7) mm long. (2n = 14, 28) Jun–Aug. Dry roadsides, pastures, & revegetated rangeland; GP, less common in s part; (cult. & widely escaped in w, & parts of e U.S.; Russia). *Naturalized*. *A. pectiniforme* R. & S.—Atlas GP.

Crested wheatgrass has been widely introduced to revegetate pastures and rangeland. Although several forms have been introduced, the separation of our plants into more than one taxon proves impractical. In the Nevski treatment of this group in the Flora of the U.S.S.R. (V. L. Komarov, ed., N. Landau, tr., Vol. 2, pp. 653–661, Smithsonian Inst., Washington D.C. 1963) most of our plants key to *A. pectiniforme* with a few keying to several other species (*A. cristatum*, *A. desertorum*, and *A. sibiricum*). Most of the material does not match the description Nevski provided for *A. pectiniforme*, however, since it varies toward *A. cristatum* in pubescence features and awn length.

3. *Agropyron dasystachyum* (Hook.) Scribn. Widely rhizomatous, glaucous perennials 1.5–9 dm tall. Culms erect, hollow. Blades usually involute, finely scabrous to pilose or glabrous, mostly 2–25 cm long, 0.8–4.0 mm wide; at least the lower sheaths usually finely strigillose; ligules 0.1–0.6 mm long; auricles often present and conspicuous. Spikes compact, 3–12(20) cm long; spikelets (2)4- to 7(9)-flowered; glumes broadest above the middle, narrowed to an acute or acuminate tip, pubescent to scabrous or glabrous, (3)4.5–8.5(11) mm long, the first usually slightly shorter than the second; lemmas normally strongly pubescent, 6.5–10 mm long, sometimes bearing an awn tip to 2(10) mm long. Anthers 3–5.5 mm long. (2n = 28) Jun–Aug. Dry to moist prairie, roadsides, railroads; ND, MT, SD, WY, NE, CO; (Ont. to AK, s to IL, NE, NM, & CA). *A. riparium* Scribn. & Sm. *A. subvillosum* (Hook.) E. Nels.—Rydberg.

This taxon needs careful study. Its variability is great, many of the specimens approaching *A. smithii* Rydb., a species with which it is said to be intersterile (Gillett & Senn, Canad. J. Bot. 38: 747–760. 1960). A fairly large number of plants examined from our region had rather high percentages of abortive pollen.

Long-awned plants which otherwise resemble *A. dasystachyum* have been assigned by Rydberg and others to *A. albicans* Scribn. & Sm. and *A. griffithsii* Scribn. & Sm. These taxa were said by C. L. Hitchcock (op. cit.) to represent probable hybrids of *A. dasystachyum* and *A. spicatum*. Such an origin was demonstrated convincingly by Dewey (Amer. J. Bot. 57: 12–18. 1970).

Forms of *A. dasystachyum* with glabrous to scabrous lemmas have been recognized as var. *riparium* (Scribn. & Sm.) Bowden. Such plants are occasional in our range.

4. *Agropyron elongatum* (Host) Beauv., tall wheatgrass. Robust, cespitose, glaucous perennials 8–16 dm tall. Culms solid, erect. Blades involute, the upper surface strongly 6- to 8-ridged, glabrous to sparsely pilose, mostly 8–50 cm long, (1.5)2.5–7 mm wide; sheaths ciliate-margined, otherwise glabrous; ligules 0.1–0.7(1.2) mm long; auricles conspicuous. Spikes relatively open, especially below, (5)20–43 cm long; spikelets (2)5- to 11(13)-flowered; glumes blunt-tipped, the first (5.5)6–9 mm long, the second (6)7–10 mm long; lemmas blunt-tipped, the lowest one 7–12 mm long; glumes and lemmas both awnless. Anthers 3–6.5 mm long. (2n = 14, 42, 56, 70, plus a number of aneuploid counts) Jun–Sep. Waste ground, roadsides, & ditches, especially in alkaline soils; GP, somewhat more common in cen part; sporadically introduced in the U.S.; Medit.). *Naturalized*.

5. *Agropyron intermedium* (Host) Beauv., intermediate wheatgrass. Rhizomatous perennials (4)6–12 dm tall. Culms erect, solid. Blades stiff, involute to flat, bearing rather

numerous, narrow ridges on the upper surface, glabrous to scabrous or, rarely, pilose, 5–28 cm long, 2–8(10) mm wide; sheaths glabrous to scabrous, often ciliate-margined; ligules 0.2–1 mm long; auricles usually pronounced. Spikes relatively open, (4)13–22(27) cm long; spikelets (2)4- to 7(10)-flowered; glumes blunt-tipped, glabrous to pubescent, the first one 5–8(9) mm long, the second one 6–9(10) mm long; lemmas blunt, or, especially the upper ones, acute or mucronate, the lowest one (5)8–11 mm long; glumes and lemmas awnless. Anthers 2.5–5.5(8) mm long. (2n = 42, 56) Jun–Sep. Of sporadic occurrence in rangeland, waste ground, railroads, roadsides, & sandy flats.

Two varieties of this European grass have been introduced into the GP:

5a. var. *intermedium*. Lemmas glabrous. ND, MT, SD, WY, NE, CO; sporadically introduced to w & cen U.S.; Eurasia). *Naturalized*.

5b. var. *trichophorum* (Link) Halac. Lemmas hirsute. Presently known from NE: Phelps, Red Willow, Scottsbluff, Box Butte, but to be expected elsewhere; (occasionally introduced in w & cen U.S.; Eurasia). *Naturalized*.

6. *Agropyron repens* (L.) Beauv., quackgrass. Strongly rhizomatous perennials 5–11 dm tall. Culms erect to decumbent, hollow. Blades flat, glabrous to pilose, 8–30 cm long, (2)5–9(18) mm wide; sheaths glabrous to pilose; ligules 0.1–0.5(0.8) mm long; auricles usually conspicuous. Spikes 4–19(26) cm long; spikelets 3- to 5(8)-flowered; glumes acute-tipped, 5–13 mm long, the first one usually slightly shorter than the second, unawned, or bearing awns up to 3(6) mm long; lemmas unawned or awn-tipped, the awns up to 5(7) mm long, the lowest lemma body 6–12 mm long. Anthers 3–5.5 mm long. (2n = 42) May–Aug. Common as a weed in gardens, along ditch banks, & in other relatively moist places; GP, most common in n; (introduced from Newf. to AK, s to NC, OK, CO, & CA; Europe & Asia). *Naturalized*. *A. leersianum* (Wulfen) Rydb.—Rydberg, misapplied.

7. *Agropyron smithii* Rydb., western wheatgrass. Strongly rhizomatous, glaucous, often glabrous, perennials (2.5)5–9(11) dm tall. Culms erect, hollow. Blades usually stiff, involute, strongly ridged above, 4–22 cm long, 1–5.5 mm wide; sheaths usually glabrous, but rarely strongly pubescent; ligules 0.2–0.8 mm long, auricles conspicuous. Spikes 3–16(24) cm long, some of the nodes sometimes bearing 2 spikelets; spikelets 3- to 8(15)-flowered; glumes broadest below the middle, tapering to an acute, acuminate, or awned tip, the overall length (5)7–13(18) mm, the first one slightly shorter than the second; lemmas 7–15 mm long, often with a short awn up to 2 mm in length. Anthers 3–5 mm long. (2n = 56) May–Sep. Common in dry prairie, where sometimes dominant, also along roads, railroads, & in waste ground; GP; (Ont. to B.C., s to KY, TX, CA). *A. palmeri* (Scribn. & Sm.) Rydb.—Rydberg.

According to Dewey (Amer. J. Bot. 62: 524–530. 1975) *A. smithii* probably arose through chromosome doubling following hybridization between *A. dasystachyum* and *Elymus triticoides*, both of which it resembles. This may explain why it is sometimes very difficult to separate from the former.

8. *Agropyron spicatum* (Pursh) Scribn. & Sm., bluebunch wheatgrass. Strongly cespitose perennials sometimes producing short rhizomes. Culms erect, hollow to partially pith-filled, 1.5–7.5(10) dm tall. Blades involute to flat, glabrous or sometimes pubescent, particularly on the upper surface, 2–19(25) cm long, 1–4 mm wide; sheaths glabrous to pubescent; ligules 0.1–0.8 mm long; auricles prominent. Spikes open, (3)6–13(17) cm long; spikelets, exclusive of the awns, often shorter than or very little longer than the internodes of the rachis, (2)3- to 7(9)-flowered; glumes acute-tipped or tapering to a short awn tip on some spikelets, the first one 4–9.5 mm long and the second (5)6–10.5 mm (measurements including tips); lemmas awned with a strongly divergent awn (2.5)6–15(20) mm long or, rarely, unawned, the body of the lowest lemma 7–11 mm long. Anthers 3.5–6 mm long. (2n = 14, 28) Jun–Aug. Dry prairies; w ND, MT, w NE, WY; (MI to AK, s to w TX & n CA). *A. inerme* (Scribn. & Sm.) Rydb.—Rydberg (the awnless form).

This is an important grass of the bunchgrass prairies of the Pacific Northwest, but it barely enters our range. See the discussion under *A. dasystachyum*.

4. AGROSTIS L., Bentgrass

Delicate annuals to tufted, stoloniferous, or rhizomatous perennials. Culms erect or decumbent at the base, hollow. Blades folded or rolled in the bud, flat to involute at maturity; sheaths open; ligules membranaceous; auricles none. Inflorescence an open to rather condensed panicle; spikelets 1-flowered, usually disarticulating above the glumes; glumes subequal, usually the first just slightly longer than the second, 1(3)-nerved, acute or acuminate; lemmas shorter than the glumes, delicate, 3- to 5-nerved, awnless, or with a delicate awn arising from below the tip; paleas well developed to rudimentary or absent.

1 Plants delicate annuals; lemmas with threadlike awns arising from just below the tips .... 1. *A. elliottiana*
1 Plants tufted or rhizomatous to stoloniferous perennials; lemmas normally unawned.
  2 Palea present, conspicuous, usually at least 1/2 as long as the lemma.
    3 Spikelets disarticulating below the glumes; anthers usually under 0.5 mm long; inflorescence short, normally less than 8 cm long; panicle branches in crowded verticels .... 6. *A. semiverticillata*
    3 Spikelets disarticulating above the glumes; anthers normally more than 0.5 mm long; inflorescence usually more than 8 cm long; panicle branches not in crowded verticels .... 7. *A. stolonifera*
  2 Palea absent or rudimentary.
    4 Panicle elongate, condensed, the branches ascending and often bearing spikelets to near their bases .... 2. *A. exarata* subsp. *minor*
    4 Panicles open, at least when mature, some of the branches spreading, not bearing spikelets to near their bases.
      5 Panicle branches forking near the middle or below the middle .... 4. *A. perennans*
      5 Panicle branches forking toward the tips.
        6 First glume less than 2 mm long .... 3. *A. hyemalis*
        6 First glume more than 2 mm long .... 5. *A. scabra*

1. ***Agrostis elliottiana*** Schult. Delicate annual 0.5–4 dm tall. Culms slender, erect to decumbent. Blades folded or with the 2 edges approximate in the bud, flat to involute or folded at maturity, the tip prow-shaped, scabrous on both surfaces, 0.8–8 cm long, (0.3)0.5–1.0(1.2) mm wide; sheaths smooth to slightly scabrous; ligules 0.8–5 mm long. Panicles open, 3–19 cm long, the branches lacking spikelets near their bases but branching near or well below the middle; spikelets usually clustered in groups near the branch tips glumes about equal, the first one usually just slightly longer, 1.1–2.4 mm long; lemma 0.8–2 mm long, awned from about 0.2 mm below the tip with a delicate, twisted, threadlike awn 3–10 mm long. Anthers 0.1–0.8 mm long. Apr–Jun. Pastures, waste ground, grassy areas, along ditches, & in open woods; MO, e KS, e OK; (MA to KS, s to GA & TX, introduced farther n; e Mex.).

2. ***Agrostis exarata*** Trin. subsp. ***minor*** (Hook.) C. L. Hitchc., spikebent. Tufted perennials usually without rhizomes, 4–10 dm tall. Culms erect to decumbent at the base. Blades rolled in the bud, flat at maturity, scabrous, 8–15 cm long, (1)2–5(7) mm wide; sheaths smooth to finely scabrous; ligules 2–6 mm long. Panicles condensed, (8)13–19(28) cm long, the branches erect, bearing spikelets to near their bases; glumes 1.9–3 mm long, subequal, scabrous on the nerve; lemma acute, 1.5–2.5 mm long, usually unawned; palea absent. Anthers 0.3–0.5 mm long. (2n = 42, n = 28 also reported) Jul–Aug. Moist areas; scattered in w SD, w NE, WY, CO, w OK; (Alta. to AK, s to much of w U.S.). *A. grandis* Trin., *A. asperifolia* Trin.—Rydberg.

3. ***Agrostis hyemalis*** (Walt.) B.S.P., ticklegrass. Delicate tufted perennials 2.5-6 dm tall. Culms slender, erect. Blades rolled in the bud, usually involute at maturity, finely scabrous, often about 7-ribbed above, 3-10 cm long, 0.5-2.5 mm wide; sheaths glabrous; ligules 0.5-3 mm long. Panicles open and diffuse at maturity, 11-25 cm long, the branches long, flexuous, rebranching usually only near the tips; glumes acute or acuminate, the first one 1-2 mm long, the second 0.8-1.6 mm long; lemma delicate, unawned, 0.5-1.5 mm long; palea absent. Anthers 0.2-0.4 mm long. (2n = 14, 28) Apr-Jul. Moist soil in pastures, prairies, & ditches; SD, WY, IA, NE, CO, MO, KS, OK; (MA to SD & WY, s to FL, CO, & TX).

This species is very closely related to *A. scabra*, from which it is sometimes difficult to distinguish. When the two species grow together, *A. hyemalis* blooms earlier.

4. ***Agrostis perennans*** (Walt.) Tuckerm., autumn bent. Clumped perennials 1.5-9 dm tall. Culms weak, erect to somewhat decumbent. Blades rolled in the bud, flat at maturity, glabrous to slightly scabrous, 3-18(21) cm long, 0.5-3(4) mm wide; sheaths smooth; ligules erose, 0.7-4.1 mm long. Panicles open, (4)8-20(30) cm long, the branches rebranched at or below the middle; glumes slightly scabrous on the nerves, the first 1.6-2.4(3) mm long, the second 1.4-2.0(2.4) mm long; lemma 1.2-2.3 mm long. Anthers 0.3-0.8 mm long. (2n = 42) Jul-Oct. Woods, bottomland, stream banks, & waste ground; MN, ND, e SD, IA, e KS, e OK; (Que. to ND, s to FL, TX, & Mex.). *A. oreophila* Trin.—Rydberg.

5. ***Agrostis scabra*** Willd., ticklegrass. Delicate tufted perennials 3-7 dm tall. Culms slender, erect. Blades rolled in the bud, flat to folded or somewhat involute at maturity, scabrous, often about 10-ridged above, 4-12 cm long, 1-3 mm wide; sheaths smooth; ligules 0.5-4 mm long. Panicles open and diffuse at maturity, 11-30 cm long, the branches slender, flexuous, rebranching only near the tips; glumes acute to acuminate, the first glume 2-3.1 mm long, the second 1.5-2.5 mm long; lemma delicate, 1-2 mm long; palea absent. Anthers 0.3-0.5 mm long. (2n = 42) (May)Jun-Sep. Roadsides, ditches, pastures, & open woods, often where moist; MN, ND, MT, SD, IA, NE, WY, CO, NM; (Newf. to AK, s to FL, TX, & CA). *A. hyemalis* (Walt.) B.S.P. var. *tenuis* (Tuckerm.) Gl.—Gleason.

See the discussion under *A. hyemalis*.

6. ***Agrostis semiverticillata*** (Forsk.) C. Chr., water bentgrass. Loosely cespitose perennials 1-5 dm tall. Culms often decumbent at the base, sometimes rooting at the nodes. Blades rolled in the bud, flat at maturity, smooth to scabrous, 1-12 cm long, 1-5(7) mm wide; sheaths smooth; ligules 1.9-3.2 mm long, truncate, pubescent. Panicles 3-8 cm long, densely flowered, the branches mostly verticillate; spikelets disarticulating below the glumes; glumes scabrous, subequal, 1-2.1 mm long; lemmas and paleas both about 0.8-1.8 mm long. Anthers 0.2-0.5 mm long. (2n = 28) Mar-Jul. Moist areas, often on stream banks; OK: Cimarron; TX: Armstrong, Hartley, & Hutchinson; NM: Quay; to be expected elsewhere in the s part of our range; (introduced in w & sw U.S.; Old World). *Naturalized.*

7. ***Agrostis stolonifera*** L., redtop. Strongly to weakly rhizomatous perennials, sometimes also stoloniferous, 3-10(12) dm tall. Culms erect to decumbent, stout to relatively weak. Blades rolled in the bud, flat at maturity, smooth to scabrous, 4-24 cm long, 1.5-6.5 mm wide; sheaths smooth; ligules 1.5-7 mm long. Panicle often purplish when mature, open and loose to rather contracted, (5)8-22(30) cm long; glumes acute, scabrous on the keels, subequal, 1.5-3 mm long; lemma 1.2-2.5 mm long; palea 0.5-1.5 mm long. Anthers 0.5-1.3 mm long. (2n = 28, 42, and assorted other counts) Jun-Aug. Abundant in low moist ground; ground; GP; (introduced to most of the cooler parts of N. Amer.; Eurasia). *Naturalized*. *A. alba* L.—Fernald, misapplied; *A. gigantea* Roth—Fernald.

This taxon is variously separated by authors into several varieties or even species. Although these are difficult to recognize from pressed specimens, fresh material may be separated as follows:

1 Panicles condensed, the branches appressed, even at maturity; plants strongly stoloniferous .... var. *palustris* (Huds.) Farw.
1 Panicles open, the branches spreading; plants stoloniferous or rhizomatous.
  2 Plants strongly rhizomatous, sometimes also with stolons, robust .... var. *major* (Gaud.) Farw.
  2 Plants weakly rhizomatous, strongly stoloniferous, slender .... var. *stolonifera*

The majority of our material appears to belong to var. *major* if sorted in this way, although the identification of these taxa is considered too difficult to afford them formal recognition in this flora.

5. ALOPECURUS L., Foxtail

Slender to robust annuals or rhizomatous to nonrhizomatous perennials of wet habitats. Culms hollow, erect to decumbent, sometimes rooting at the lower nodes. Blades rolled in the bud, flat at maturity; sheaths open, air chambered and clearly cross-septate in transmitted light; ligules membranaceous; auricles none. Inflorescence a cylindrical spikelike panicle; spikelets 1-flowered, laterally compressed, disarticulating below the glumes; glumes about equal, usually keeled, normally connate at the base, with acute to obtuse tips, 3-nerved, awnless, villous; lemmas approximately as long as the glumes, faintly 3- to 5-nerved, awned from the back, the awn sometimes inconspicuous or, rarely, absent; paleas absent.

1 Anthers more than 1.5 mm long; glumes more than 2.5 mm long; lemmas often more than 2.5 mm long.
  2 Awns elongate, geniculate, 5–10 mm long, the exserted portion 2–6 mm long; plants nonrhizomatous, although some stems tending to root at the lower nodes .... 5. *A. pratensis*
  2 Awns shorter, 0.3–2(6) mm long, seldom exserted as much as 2 mm; plants strongly rhizomatous .... 2. *A. arundinaceus*
1 Anthers less than 1.5 mm long; glumes usually less than 2.5 mm long; lemmas often less than 2.5 mm long.
  3 Awn less than 2.5 mm long, usually straight, arising from near the middle of the back of the lemma .... 1. *A. aequalis*
  3 Awn of the lemma more than 2.5 mm long, usually geniculate, arising from near the base.
    4 Plants perennial; anthers more than 0.6 mm long .... 4. *A. geniculatus*
    4 Plants annual; anthers less than 0.6 mm long .... 3. *A. carolinianus*

1. **Alopecurus aequalis** Sobol., short-awn foxtail. Tufted perennials, 2–6.5 dm tall. Culms erect to decumbent and rooting at the nodes. Blades flat, sharply ridged and finely scabrous on the upper surface, 5–15 cm long, 0.9–5 mm wide; sheaths smooth; ligules 2.5–7 mm long. Panicle 2.5–7 cm long, 2–6(8) mm wide; glumes obtuse at the tip, 1.5–2.5 mm long, united below for less than 0.5 mm, the keel and lateral nerves villous; lemma subequal to the glumes, the usually straight awn arising from near the middle of the back and 0.5–2 mm long (or, rarely, absent), the exserted portion not over 1.8 mm long. Anthers 0.5–0.8 mm long. (2n = 14) Jun–Aug. Wet places, sometimes growing partly submerged in water; MN, ND, MT, SD, WY, IA, NE, CO, MO, KS, NM; (Greenl. to AK, s to PA, KS, NM, & CA; Eurasia). *A. aristulatus* Michx.—Rydberg.

2. **Alopecurus arundinaceus** Poir., creeping foxtail. Strongly rhizomatous perennials, 7–11 dm tall. Culms erect, stout. Blades flat, smooth to scabrous, at least on the upper surface, 5–30 cm long, 3–10 mm wide; sheaths smooth to finely scabrous; ligules 1–3 mm long. Panicles 3–9.5 cm long, 6–10 mm wide; glumes acute to acuminate-tipped, 2.5–5 mm long, united below for more than 1 mm, rather uniformly villous or the hairs virtually confined

to the midnerve; lemmas 2–4 mm long, awned from below the middle to rather near the base, the awn 0.3–2(6) mm long, straight to slightly geniculate, unexserted or exserted for as much as 2 mm. Anthers 1.5–2.5 mm long. (2n = 28) Jun–Jul. Hay meadows, road banks, & cult. fields; ND: Burleigh, Cass, Emmons, Logan, McLean, & probably spreading; (introduced in Can. & ND; Eurasia). *Naturalized. A. ventricosus* Pers.—Fernald.

3. *Alopecurus carolinianus* Walt., Carolina foxtail. Tufted annual, 0.5–4.5 dm tall. Culms stiffly erect to geniculate at base. Blades sharply ridged and finely scabrous on the upper surface, 1–10 cm long, 0.9–3.5 mm wide; ligules 1.5–5 mm long. Panicles 1–5 cm long, 2–5 mm wide; glumes obtuse-tipped, scarcely united below, 1.5–2.5 mm long, the keel and lateral nerves villous; lemma about equal to the glumes, its usually geniculate awn arising from well below the middle, 2.5–5 mm long, the exserted portion 1–3.5 mm long. Anthers 0.2–0.5 mm long. May–Jul. Moist places; GP, more common in the s part; (B.C. to WA, s to FL, TX, & CA).

See the discussion under *A. geniculatus.*

4. *Alopecurus geniculatus* L., marsh foxtail. Tufted to rather widely spaced perennials, 3–7 dm tall. Culms erect, geniculate at the base, or decumbent and rooting at the nodes. Blades flat, sharply ridged and scabrous above, 5–15 cm long, 1.5–4.5 mm wide; sheaths smooth; ligules 2.5–5 mm long. Panicles 2.5–7 cm long, 2–7 mm wide; glumes obtuse-tipped, 2–3 mm long, slightly united below, the keel and the lateral nerves villous; lemma subequal to the glumes, the usually geniculate awn arising from well below the middle, 3–4.5 mm long, the exserted portion 1–3 mm long. Anthers 0.7–1.5 mm long. (2n = 28) Jun–Aug. Moist places; MN, ND, SD; (Newf. to AK, s to SD, AZ, & CA).

This species is often confused with *A. carolinianus* in the herbarium, probably because the perennial or annual nature of these aquatic or subaquatic grasses is not always easy to judge from a pressed specimen. The anthers provide a useful supplementary character, and a few trapped ones can often be found on these dense panicles, even when the plants are rather mature.

5. *Alopecurus pratensis* L., meadow foxtail. Nonrhizomatous perennials, 4–9(12) dm tall. Culms stout, erect, sometimes rooting at the lower nodes. Blades flat, smooth or finely scabrous above, 4–20(33) cm long, 2–7 mm wide; ligules 1–3.5 mm long. Panicles 3–10 cm long, 5–11 mm wide; glumes acute-tipped, villous, especially on the keel and lateral nerves, 3.5–6 mm long, united below for about 1 mm; lemma 3.5–5.5 mm long, its usually geniculate awn 5–10 mm long, the exserted portion 2–6 mm long. Anthers 1.6–3.5 mm long. (2n = 28) Apr–Jun. Moist meadows & marshy areas; scattered locations in MN: Douglas; ND: Bottineau & McKenzie; NE: Cuming, Douglas, Gage, Lancaster, & Sarpy; probably more common than the collections suggest; (introduced from Newf. to AK & sporadically to many of the n states; Europe). *Naturalized.*

Alopecurus myosuroides Huds. is known from KS: Riley and is to be expected in other parts of our range. It will key here to *A. pratensis* but differs in being an annual with its panicle strongly tapered at both ends and its glumes short-ciliate rather than villous.

6. ANDROPOGON L., Bluestem

Strongly cespitose to widely rhizomatous perennials. Culms erect, solid, often grooved on one side. Blades flat to folded or revolute at maturity, rolled or folded in the bud; sheaths rounded to strongly keeled, open; ligules membranaceous, ciliate or fringed on the margin. Flowering culms branched or unbranched and terminating in spicate racemes or in panicles with racemose branches which are rarely rebranched; spikelets in pairs, one sessile and perfect, the other pediceled and usually staminate or neuter, or sometimes rudimentary (rarely, perfect); rachis of raceme or racemose branch disarticulating at maturity, each segment remaining attached to a spikelet pair; sessile spikelets with 2 florets, the upper one perfect, the lower neuter or vestigial; glumes of sessile spikelet unawned, firm, the first one dorsally flattened, often strongly 4-nerved, and partially enclosing the second, which is somewhat compressed or keeled; lemmas and paleas delicate, hyaline, enclosed by the glumes, the lemma of the fertile floret usually bearing a twisted, geniculate awn, this sometimes rudimentary or lacking. *Schizachyrium* Nees, *Bothriochloa* O. Ktze.

Reference: Gould, F. W. 1975. *Andropogon*, pp. 579–86, *Bothriochloa*, pp. 591–605, *Schizachyrium*, pp. 605–611, *in* The grasses of Texas. Texas A. and M. Univ. Press, College Station.

1 Each of numerous branches of the flowering culm terminating in a single raceme .... 7. *A. scoparius*
1 Culms or branches of the culms terminating in panicles, each with 2 to many racemose branches.
  2 Anthers more than 2.5 mm long; pediceled spikelets usually over 5.5 mm long.
    3 Plants nonrhizomatous to weakly rhizomatous; leaf blades about equally ridged on both surfaces, with about every fourth ridge larger than the intervening ridges; ligules usually less than 3 mm long; awns of fertile lemmas over 8 mm long; anthers usually less than 3.8 mm long .... 3. *A. gerardii*
    3 Plants strongly rhizomatous; leaf blades more prominently ridged on the upper surface than on the lower, with most ridges about equal in prominence; ligules usually more than 3 mm long; awns of fertile lemmas absent, or present but less than 8 mm long; anthers usually more than 3.8 mm long .... 4. *A. hallii*
  2 Anthers less than 2.5 mm long; pediceled spikelets less than 5.5 mm long.
    4 Pedicels of reduced spikelets neither strongly flattened nor grooved on both sides, although sometimes slightly flattened and often grooved on one side; sheaths strongly keeled.
      5 Pediceled spikelet vestigial or lacking; anthers usually less than 1.1 mm long; awns of fertile spikelets (7.5)9–16(20) mm long .... 10. *A. virginicus*
      5 Pediceled spikelet 1.5–3.6 mm long; anthers usually more than 1.1 mm long; awns of fertile spikelets (11)16–25 mm long .... 9. *A. ternarius*
    4 Pedicels of reduced spikelets strongly flattened and grooved on both sides, hence the central portion thin, even membranaceous, and easily ruptured with a probe; sheaths rounded or only slightly keeled.
      6 Pediceled spikelets equal to or nearly equal to the sessile spikelets.
        7 Panicles elongated, the central axis normally longer than the branches .... 2. *A. bladhii*
        7 Panicles with the central axis shorter than the branches .... 5. *A. ischaemum* var. *songaricus*
      6 Pediceled spikelets markedly shorter and narrower than the sessile spikelets.
        8 Main panicle branches usually fewer than 9, borne on a realtively short axis less than 5 cm long; nodal hairs conspicuous, spreading, often 3–7 mm long; glumes 4.5–7 mm long .... 8. *A springfieldii*
        8 Main panicle branches more than 9, borne on a relatively long axis usually over 5 cm long in well-developed plants; nodal hairs inconspicuous, appressed to ascending, often less than 3 mm long (rarely lacking); glumes usually less than 4.5 mm long in our material.
          9 Inflorescence somewhat flabellate, broadest toward the top; glumes 3.5–5 mm long; mean pollen diam greater than 45 μm .... 1. *A. barbinodis*
          9 Inflorescence broadest near the center or base; glumes 2.5–4.5 mm long; mean pollen diam less than 45 m .... 6. *A. saccharoides* var. *torreyanus*

1. ***Andropogon barbinodis*** Lag., cane bluestem. Cespitose perennial 6–9(11) dm tall. Culms solid, grooved on one side, bearded at the nodes, the hairs usually less than 3 mm long and appressed to ascending. Blades flat to revolute, not strongly keeled but the midribs prominent, usually smooth, often pilose in the collar region, 4–26 cm long and (1)3–5.5

mm wide; sheaths smooth, only slightly keeled; ligule a fringed membrane 1–2.5 mm long. Inflorescence broadest near the top, 7–13 cm long, the numerous branches concealing the relatively elongate axis, the rachis internodes and margins of the pedicels densely villous; glumes of sessile spikelets 3.5–5 mm long in ours, borne on flattened pedicels which are grooved on both sides; pediceled spikelets 2.5–4 mm long; awns of fertile lemmas 11–25 mm long. Anthers 0.9–1.2 mm long. (2n = 180) Jun–Sep. Roadsides, waste ground, & gravelly places; s OK, NM, & TX; (OK to CA, s in Mex.) *Bothriochloa barbinodis* (Lag.) Herter var. *barbinodis*, var. *perforata* (Trin. ex Fourn.) Gould—Gould, op. cit.

Some authors recognize two varieties, based on the presence or absence of a glandular pit in the first glume of the sessile spikelet. Such variation is not here given formal recognition, although both phases occur in the s GP. We have this species only at the northern edge of its range, and perhaps for this reason the measurements for our materials are smaller on the average (except for pollen) than those reported by Gould, op. cit.

2. ***Andropogon bladhii*** Retz., Caucasian bluestem, Australian bluestem. Cespitose perennial (3.9)5–9(10.5) dm tall. Culms solid, grooved on one side, bearded at the nodes with hairs mostly appressed and less than 2 mm long, or nodes glabrous. Blades flat to revolute the midrib prominent, not strongly keeled, rolled in the bud, 3–25(38) cm long and 1–5.5 mm wide; sheaths only slightly keeled, smooth; ligule a fringed membrane 0.5–1.5 mm long. Panicle with the axis normally longer than the branches, the branches sometimes rebranching; rachis internodes and margins of the pedicels villous; glumes of sessile spikelets (2)2.5–4 mm long, the second glume sometimes slightly longer than the first, the first glume scabrous to thinly strigose, villous toward the base; pediceled spikelets only slightly smaller, 1.6–3.8 mm long, borne on narrow pedicels which are grooved on both sides; awns of fertile lemmas 11–17 mm long. Anthers 1–1.5(1.8) mm long. (2n = 40, 60, 80) Jul–Oct. Roadsides, waste ground, pastures, sometimes planted for forage; extreme s NE; CO: Kit Carson; KS, OK; (GP to TX; s Asia, Australia, Pacific islands). *Naturalized* here and elsewhere. *Bothriochloa bladhii* (Retz.) S. T. Blake—Gould op. cit.; *B. intermedia* (R. Br.) A. Camus— Correll & Johnston; *A. intermedius* R. Br.—Gould, Madroño 14: 18–29. 1957.

The name *A. intermedius* R. Br. has commonly been applied to this species but is antedated by *A. bladhii* Retz., cf. Blake, Proc. Roy. Soc. Queensland 80: 55–84. 1968.

3. ***Andropogon gerardii*** Vitman, big bluestem. Stout nonrhizomatous or short-rhizomatous perennials (5)6–14(20) dm tall. Culms solid, grooved on one side. Blades with prominent midribs but not strongly keeled, flat to revolute or involute, rolled in the bud, about equally ridged on both surfaces with about every fourth vein clearly larger than the intervening veins, usually pilose, at least near the collar, 5–50 cm long, 2–9 mm wide; sheaths only slightly if at all keeled, smooth to pilose; ligule a fringed membrane 0.4–2.5 mm long. Inflorescence with 2–7 floriferous branches on a short axis at the end of each main branch, the floriferous branches not rebranched, the longest of them 4–11 cm long; rachis internodes and margins of the pedicels thinly to rather densely villous, the hairs relatively short, usually less than 3 mm long, seldom at all yellowish; glumes of sessile spikelets scabrous (seldom ciliate), subequal, (5)6–11 mm long; pediceled spikelets almost as long, (3.5)6–10(12) mm, borne on pedicels which are not grooved on both sides; awns of fertile lemmas 8–14(19) mm long. Anthers 2.5–3.8(4.4) mm long. (2n = 60, but 2n = 20, 70 and n = 40 are also reported). Jul–Oct. Prairies & roadsides, especially abundant in lowland prairie; GP, more common in e part; (s Can., ME to MT, s to FL, NM, & Mex.) *A. provincialis* Lam.— Rydberg; *A. furcatus* Muhl.—Stevens.

This species intergrades rather freely with *A. hallii* (see comments).

4. ***Andropogon hallii*** Hack., sand bluestem. Strongly rhizomatous, glaucous perennial (4)6–15(20) dm tall. Culms solid, grooved on one side. Blades with prominent midribs,

but not strongly keeled, flat to revolute or involute, rolled in the bud, strongly ridged on the upper surface but not below, the ridges about equal in prominence, often pilose, at least near the collar, 3–40(51) cm long and (1.5)2–10 mm wide; sheaths slightly if at all keeled; ligule a fringed membrane (0.9)3–4.5 mm long. Inflorescence with 2–7 floriferous branches on a short axis at the end of each main branch, the floriferous branches not rebranched, the longest of them 4–7(9) cm long; rachis internodes and margins of the pedicels thinly to rather densely villous, the hairs relatively long, those of the pedicels usually 3–5.5 mm long, whitish to strongly yellowish; first glume of sessile spikelet often ciliate, glumes subequal, (5)6.5–12 mm long, those of the pediceled spikelets almost as long, (3.5)6–10(12) mm, borne on pedicels which are not grooved on both sides; awns of fertile lemmas absent or present and less than 8 mm long. Anthers (2.3)4–6 mm long. (2n = 60, but counts of 2n = 70 and 100 also reported). Jul–Oct. An important component of sandhills prairie, growing elsewhere in sandy soils; GP, more common in w part; (ND to MT, s to IA, TX, & AZ). *A. paucipilus* Nash, *A. chrysocomus* Nash–Rydberg; *A. gerardii* var. *paucipilus* (Nash) Fern., var. *chrysocomus* (Nash) Fern.–Gould, op. cit.

Although *A. hallii* is distinguishable from *A. gerardii* by a large number of characteristics, the 2 species intergrade at some localities and many specimens show some degree of intermediacy. Pollen stainability varies widely in the 2 taxa, and unstainability does not appear correlated with any particular morphology, a pattern suggestive of introgression. The suggestion of Gould, op. cit., that the taxa are best recognized under the single species name *A. gerardii* appears sound biologically. However, the names he adopts (from Fernald, Rhodora 45: 255–258. 1943) are not the oldest varietal epithets available (see Hackel, Sitzungsber. Kaiserl. Akad. Wiss., Math–Naturwiss. cl., Abt 1 89: 127–128. 1884; and Hackel, *in* De Candolle, Monogr. Phanerog. 6: 444. 1889). Also, it seems unnecessary to recognize as Gould does the form with yellowish hairs on the pedicels as a separate entity, since such plants occur sporadically throughout much of the range of *A. hallii.*

5. *Andropogon ischaemum* L. var. *songaricus* Rupr. ex Fisch. & Mey., King Ranch bluestem, Turkestan bluestem. Slender perennial, clumped to widely rhizomatous, (3.2)4–9.5 dm tall. Culms solid, grooved on one side, glabrous to bearded at the nodes with short appressed hairs, sometimes decumbent at the base and rooting at the lower nodes. Blades flat to folded or revolute, slightly keeled, the midrib very prominent, rolled in the bud, scabrous to thinly pilose, especially near the ligule, 1–22 cm long and 1–3(4.5) mm wide; sheaths slightly keeled, smooth; ligule a fringed membrane 0.5–1.5 mm long. Inflorescence with a short axis and 2–8 main branches 2–7 cm long; rachis internodes and margins of the pedicels long-villous; glumes of sessile spikelets about 3–4.5 mm long, the outer scabrous to strigose, villous toward base; pediceled spikelets as large as or slightly larger than the sessile spikelets, borne on pedicels which are flattened and grooved on both sides; awns of fertile lemmas 9–17 mm long. Anthers 1–2 mm long (2n = 40, 50, 60) Jul–Oct. Pastures, roadsides, waste ground; KS, e CO, OK, TX; (s GP; cen Europe & Asia). *Naturalized. Bothriochloa ischaemum* (L.) Keng var. *songarica* (Rupr.) Celerier & Harlan–Gould, op. cit.

6. *Andropogon saccharoides* Sw. var. *torreyanus* (Steud.) Hack., silver bluestem. Cespitose perennial (3.7)6–10(10.7) dm tall. Culm solid, grooved on one side, often with short, appressed hairs at the nodes. Blades with a prominent midrib, glabrous to thinly pilose at the collar region, glaucous but often brownish toward the margin, especially on the upper surface, flat or folded to revolute, rolled in the bud, 2–25 cm long and (1.4)2.5–7 mm wide; sheaths smooth, only slightly keeled. Inflorescence silky, with an elongate central axis, broadest near the center or the base, main branches often rebranched; rachis internodes and margins of the pedicels villous; glumes of sessile spikelets 2.5–4.5 mm long, equal or the second glume a little shorter, the first glume villous toward the base; pediceled spikelets 1.5–3(3.5) mm long, borne on pedicels which are flattened and grooved on both sides; awns of fertile lemmas 11–16.6(19) mm long. Anthers 0.7–1.2 mm long.

(2n = 60) Jul–Oct. Dry, often sandy soil of roadsides, prairies, & waste ground; GP s of s NE; (AL & MO to s CO, s to LA & AZ; n Mex.). Probably introduced in n part of range. *Bothriochloa saccharoides* (Sw.) Rydb. var. *torreyana* (Steud.) Gould—Gould, op. cit.

7. ***Andropogon scoparius*** Michx., little bluestem. Cespitose perennial, sometimes with short rhizomes, (2.8)4–9.5 dm tall. Culms solid, slightly flattened, not grooved. Blades folded, sometimes revolute, glabrous to pilose, especially near the collar, folded in the bud, 4–30 cm long and 1.5–4 mm wide; sheaths keeled, usually glabrous; ligule a fringed membrane 0.6–2.3 mm long. Each of numerous branches of the flowering culm terminating in a single spicate raceme 2–6 cm long; racemes straight to undulate at maturity, the rachis joints and margins of the pedicels hairy to villous; glumes of sessile spikelets (5.5)6–8(9.5) mm long; pediceled spikelets 3–6 mm long, the pedicels flattened but not grooved; awns of fertile lemmas 9–16 mm long. Anthers 2.4–3.8 mm long. (2n = 40) Jul–Oct. Prairies, often a dominant or co-dominant species; GP; (s Can., ME to ID, s to FL & AZ; Mex.). *Schizachyrium scoparium* (Michx.) Nash—Gould, op. cit.; *A. scoparius* var. *frequens* Hubb., var. *neomexicanus* (Nash) Hitchc.—Fernald.

In the GP, variation in this wide-ranging species is clinal (McMillan, Amer. J. Bot. 51: 1119-1128. 1964) and varieties are not readily recognized. Most of our material, however, could be assigned to var. *frequens* Hubb., with some specimens assignable to (or approaching) var. *neomexicanus* (Nash) Hitchc.

8. ***Andropogon springfieldii*** Gould. Cespitose perennial 4–9.5 dm tall. Culms solid, grooved on one side, bearded at the nodes with spreading hairs which are often more than 3 mm long. Blades flat to revolute, 3–20(30) cm long, and (0.8)2–3.5(4.6) mm wide, glabrous to densely pilose on the upper surface, the lower surface glaucous, the margins often brownish, the midrib prominent but the blades not strongly keeled; sheaths scarcely keeled, smooth; ligules 1–2.5 mm long. Inflorescence densely silky, with a short central axis, broadest near the top, the panicle with 3–9 main branches, the rachis and pedicel margins copiously villous; glumes of sessile spikelets 4.5–7 mm long, the first glume hairy on the lower part; pediceled spikelets 3.5–5.5 mm long, borne on strongly flattened pedicels which are grooved on both sides; awns of fertile lemmas (15)19–23 mm long. Anthers 1–1.5 mm long. (2n = 120) Jul–Sep. Prairies, roadsides, & waste ground; TX, NM; (w TX to AZ). *Bothriochloa springfieldii* (Gould) Parodi—Gould, op. cit.

This species appears amply distinct and is characterized by poor pollen production. The anthers are so readily available in the spikelets that one suspects cleistogamy.

9. ***Andropogon ternarius*** Michx., splitbeard bluestem. Cespitose perennial, 4–11 dm tall. Culms solid, grooved on one side. Blades glabrous, or scabrous to pilose near the collar region, folded or flat, sometimes revolute, keeled, the midrib prominent, 13–30(39) cm long and 1–3 mm wide, folded in the bud, sheaths smooth to scabrous or sometimes long-pilose, strongly keeled and compressed; ligule a ciliate membrane 0.4–1.5 mm long. Flowering culms branched above, each branch terminating in (1)2 villous spicate branches (2)3–5.5 cm long; rachis joints and pedicels long-villous, especially on the margins; glumes of sessile spikelets 4.5–7(8.4) mm long, the first glume scabrous or ciliate on marginal veins near the tip; pediceled spikelets rudimentary, 1.5–3.6 mm long, borne on pedicels which are flattened but not grooved; awns of fertile lemmas (11)16–25 mm long. Anthers 1.2–2.3 mm long. (n = 20, 2n = 60 also reported but doubtful) Aug–Oct. Prairies, woodland borders, pastures, & waste ground; MO, se KS, OK; (MD to KS, s to FL & e TX).

10. ***Andropogon virginicus*** L. broomsedge bluestem. Tufted perennial 5.5–12(14) dm tall. Culms solid, grooved on one side. Blades folded to flat or revolute, keeled, midrib very prominent, glabrous to scabrous or sparsely pilose, especially near the collar, folded in the

bud, 8–45 cm long and 1.5–5 mm wide; sheaths glabrous to sparsely long-pilose, very strongly keeled, compressed; ligule a ciliate membrane 0.3–0.9 mm long. Flowering culms branched above, the ultimate branches bearing (usually) 2 flowering branches which are partly included in the sheath of the subtending leaf; rachis joints and pedicels slender, long-villous; glumes of sessile spikelets 2.5–4.2 mm long; pediceled spikelets absent or when present less than 0.5 mm long; awns of fertile lemmas (7.5)9–16(20) mm long. Anthers 0.6–1 mm long. (2n = 20) Sep–Nov. Prairies, pastures, roadsides, & waste ground, often where sandy or rocky; MO, se KS, e OK; (MA to MI, KS, s to FL & e TX).

A similar species, *A. glomeratus* (Walt.) B.S.P., has been collected in OK: Caddo, Cleveland, & McLain. It will key here but can be distinguished from *A. virginicus* by the abundantly branched and rebranched broomlike flowering culm.

7. ARISTIDA L., Three-awn

Low annuals or nonrhizomatous perennials. Culms hollow or sometimes solid above, erect. Blades often involute, slender, seldom more than 3 mm wide, prominently ridged on the upper surface, C-shaped in cross section or slightly rolled in the bud stage; sheaths open; ligule a short fringe of hairs from a basal membrane, seldom more than 0.8 mm long. Inflorescence a panicle or sometimes a raceme of large spikelets; spikelets disarticulating above the glumes; glumes 1- to 3-nerved, acute, acuminate or awn-tipped (if present, awns are included in glume measurements) nearly equal or the second one larger, or rarely the first larger; lemmas 3-nerved, firm, terete, the callus hardened, sharp-pointed, and often bearded, the upper part terminating in a short to elongate column which bears 3 awns, these about equal in length or the central one clearly longer than the 2 lateral ones; paleas small, 2-nerved.

References: Gould, F. W. 1975. *Aristida*, pp. 382–406, *in* The Grasses of Texas. Texas A. and M. Univ. Press, College Station; Holmgren, A. H., & N. H. Holmgren. 1977. *Aristida*, pp. 452–456, *in* Intermountain Flora, vol. 6. Columbia Univ. Press, N. Y.; Pohl, R. W. 1966. *Aristida*, pp. 457–465, *in* Grasses of Iowa. Iowa State J. Sci 40: 341–566.

In addition to the species described below, *A. desmantha* Trin. & Rupr. was collected several times long ago in NE: Kearney. It can easily be recognized by its annual habit and by having the lemma articulate with the awn column. Rydberg's reports of *A. desmantha* and *A. tuberculosa* Nutt. are both believed to be based on these specimens of *A. desmantha*.

1 Plants annual.
  2 Central awn more than 2.5 cm long; first glume more than 16 mm long .... 7. *A. oligantha*
  2 Central awn less than 2.5 cm long; first glume usually less than 16 mm long.
    3 Glumes nearly equal in most spikelets, sometimes the second slightly larger.
      4 Central awn, measured from the tip of the lemma, shorter in length than the glumes, with a spiral coil at the base .... 4. *A. dichotoma*
      4 Central awn longer than the glumes, sometimes deflexed, but lacking a spiral coil at the base .... 6. *A. longespica*
    3 Glumes markedly unequal, the second decidedly longer than the first.
      5 Central awn with a spiral coil at the base; awns very unequal, the lateral ones less than 3/4 the length of the central one .... 3. *A. basiramea*
      5 Central awn lacking a spiral coil at the base; awns more nearly equal, the lateral ones more than 3/4 of the length of the central one .... 1. *A. adscensionis*
1 Plants perennial.
  6 Panicle open, its branches rather stiffly spreading; awns usually less than 2 cm long.
    7 Pedicels and branchlets spreading; calluses visible at branchlet bases ... 2. *A. barbata*
    7 Pedicels and branchlets appressed; calluses not present at branchlet bases .... 5. *A. divaricata*
  6 Panicle contracted to open but, if open, the branches not stiffly spreading.

8 Glumes almost equal, or sometimes the first slightly longer than the
  second .... 8. *A. purpurascens*
8 Glumes markedly unequal, the first one clearly shorter than the second. ..... 9. *A. purpurea*

1. **Aristida adscensionis** L., sixweeks three-awn. Tufted annuals 1.5–7.5 dm tall. Culms wiry, hollow, erect to slightly geniculate at the lower nodes. Blades C-shaped in cross section in the bud, involute at maturity, glabrous beneath, scabrous to short-pilose above, usually with 2 of the lateral veins as prominent as the midrib beneath and more prominent above, 2–15 cm long, 0.5–2 mm wide; ligules 0.4–0.7 mm long. Inflorescence 5–18 cm long, contracted, the branches ascending to appressed; glumes acute to acuminate or awn-tipped, unequal, the first 4.5–8.5 mm long, the second 5.5–11 mm long; lemma scabrous in lines above, the callus bearded, tapering to the awns, 5.5–11 mm long to the base of the awns; awns about equally divergent and nearly equal in length, the central one 11–15(22) mm long and the lateral ones 9–13(20) mm long. Anthers 0.8–1.1 mm long. (2n = 22) Jul–Sep. Waste ground, often in dry, sandy soil; NE (uncommon), CO, MO, KS, OK, TX, NM; (MO, NE to CA, s to TX; trop. Amer.). *A. fasciculata* Torr.—Rydberg.

2. **Aristida barbata** Fourn. Cespitose perennials 1.5–3.5 dm tall. Culms wiry, scabrous, hollow, erect. Blades involute, smooth or finely scabrous beneath, scabrous and ridged above, the ridges more or less equal in size or the marginal 1 or 2 clearly larger, 3–12 cm long, 0.4–2 mm wide; sheaths glabrous, but often villous marginally near the ligule; ligule 0.2–0.4 mm long. Inflorescence 8–17 cm long, open, the branches and branchlets spreading and with obvious calluses at their bases; glumes nearly equal, the first 8–12 mm long, the second 9–12.5 mm long; lemmas gradually narrowed into a twisted or slightly twisted column, bearded on the callus, 7–12 mm long to the base of the awns; central awn (11)14–20 mm long, the lateral awns almost as long, (11)13–18 mm long. Anthers 0.5–1 mm long. (2n = 22, 44) Jul–Oct. Waste ground, roadsides, prairies, & pastures; se CO, extreme w OK, TX, NM; (se CO to TX & AZ; cen Mex.).

3. **Aristida basiramea** Engelm. ex Vasey, forktip three-awn. Tufted annuals 2–4.5(5.5) dm tall. Culms wiry, often scabrous, hollow, erect. Blades rolled or C-shaped in cross section in the bud, involute at maturity, with thickened ridges at the margins, the midribs not prominent except toward the base, sometimes pilose, especially near the collar, scabrous, especially on the upper surface, 4–16(18)cm long, 0.5–2 mm wide; sheaths glabrous to scabrous; ligules 0.2–0.5 mm long. Inflorescence 4–10(16) cm long, rather narrow, stiff to flexuous, branches and pedicels ascending; glumes unequal, the first 6–10(12) mm long, the second (8.5)10–16(19) mm long; lemmas pilose on the callus, 6–11 mm long to the base of the awns; the central awn with a spiral coil at the base, (5.5)9–21 mm long, the lateral awns shorter, 1–12 mm long. Anthers usually very reduced and difficult to observe, but 2.6–3.1 mm long when, as rarely, they are fully developed. Jul–Sep. Roadsides, pastures, & waste ground.

I agree with Shinners (Amer. Midl. Naturalist 23: 633. 1940) that the so-called *A. dichotoma* var. *curtissii* should be considered a variety of *A. basiramea*, since it matches perfectly with that species in all characters except the length of the lateral awns and differs from *A. dichotoma* in many respects. All 3 taxa are largely cleistogamous (Pohl, op. cit.):

3a. var. *basiramea*. Lateral awns more than 1/3 the length of the central awns, 6–12 mm long. IA, NE, MO, KS, OK; (ME to NE, s to KY, OK, CO).

3b. var. *curtissii* (A. Gray) Shinners. Lateral awns short, less than 1/3 the length of the central awns, 1–6 mm long. MN, SD, IA, NE, MO, KS, OK; (MD & VA to SD, s to MO, CO, also in FL). *A. dichotoma* Michx. var. *curtissii* A. Gray—Fernald; *A. curtissii* (A. Gray) Nash—Rydberg.

4. **Aristida dichotoma** Michx., churchmouse three-awn. Slender annuals 2.5–5 dm tall. Culms wiry, usually branched, hollow only at the base, glabrous to scabrous. Blades C-

shaped in cross section in the bud, involute at maturity, with thickened ridges at the margins, scabrous, especially on the upper surface, sometimes thinly pilose, especially in the collar region, 2–11 cm long, 0.5–1.5 mm wide; sheaths glabrous to scabrous; ligules 0.1–0.3 mm long. Inflorescence 4–9 cm long, spikelike, the (branchlets and) pedicels appressed; glumes about equal, 5–10 mm long; lemmas often strigose, 4.5–7 mm long to the base of the awns; central awn with a spiral coil at the base, deflexed, 3.5–7 mm long, the lateral awns erect, 0.5–2 mm long. Anthers usually reduced and not easily observed but (according to Pohl, op. cit.) about 3 mm long when developed. Aug–Oct. Waste ground, often in sandy soil; MO, e KS, e OK; (ME to WI, KS, s to FL & TX).

5. **Aristida divaricata** Humb. & Bonpl. ex Willd., poverty three-awn. Cespitose perennials 2–7 dm tall. Culms erect, smooth to scabrous. Blades involute, rather evenly ridged and scabrous above, 7–15(23) cm long, 0.9–2 mm wide; sheaths glabrous to scabrous, but often long-villous at the collar; ligules 0.2–0.7 mm long. Inflorescence (14)17–30(34) cm long, open, the main branches stiffly spreading, the pedicels and minor branchlets appressed, the branchlets without caluses, glumes approximately equal, 8.5–15 mm long; lemmas gradually narrowed into a twisted awn column, 10–13.5(15.5) mm long to the base of the awns; central awn (11)12–16 mm long, the lateral awns almost as long, (9)11–14 mm. Anthers 0.7–1.3 mm long. (n = 11) Jul–Sep. Waste ground, roadsides, prairies, & pastures; sw KS, CO, TX, NM; (KS to CA, s to C. Amer.)

6. **Aristida longespica** Poir., slimspike three-awn. Slender tufted annuals 2.5–7 dm tall. Culms wiry, hollow, erect to geniculate, smooth to scabrous. Blades C-shaped in cross section in the bud, involute at maturity, 1 or 2 marginal veins on each side usually thicker than the others, scabrous above, scabrous to glabrous beneath, 3–13(16) cm long, 0.5–2 mm wide; sheaths smooth or finely scabrous; ligules 0.1–0.5 mm long, the membranaceous basal portion conspicuous. Inflorescence 7–20 cm long, stiff to flexuous, the pedicels and branchlets appressed to ascending; glumes approximately equal in most spikelets, 3–9.5 mm long; lemmas bearded on the callus, scabrous above, about equal to the glumes or slightly longer, 2.9–8.5 mm long to the base of the awns; central awn 5–24 mm in length, the lateral ones varying from much shorter to only slightly shorter, 0.9–17 mm long. Anthers usually reduced and not easily observed, but 2–2.5 mm long when fully developed. Aug–Nov. Scattered, perhaps overlooked by collectors, in dry, sandy, or rocky prairie & in waste ground.

This taxon needs intensive study. It appears to be principally cleistogamous (Pohl, op. cit.). Although the 2 varieties are said to be broadly intergradient elsewhere (Gould, op. cit.), our material can be separated rather readily.

6a. var. *geniculata* (Raf.) Fern. Glumes 5.5–9.5 mm long; lemmas 5.5–8 mm long; central awn not usually reflexed, (13)15–24 mm long, the lateral awns more than 1/2 as long, 9–17 mm. IA, NE, KS, OK, TX; (MI to NE, s to FL & TX). *A. intermedia* Scribn. & Ball — Rydberg; *A. necopina* Shinners — Shinners, Rhodora 56: 30. 1954.

6b. var. *longespica*. Glumes 3–6 mm long; lemmas 2.9–6(8.5) mm long; central awn usually reflexed, 5–15 mm long, the lateral awns seldom much more than 1/2 as long, 0.9–8 mm long. KS, OK, TX; (NH to MI, KS, s to FL & TX). *A. gracilis* Ell. — Rydberg.

Rydberg's report of *A. ramosissima* Engelm. ex Gray for KS may have been based on this taxon. *A. ramosissima* is similar but has much longer lemmas (ca 2 cm long).

7. **Aristida oligantha** Michx., oldfield three-awn, prairie three-awn, slender to robust tufted annual 3–7 dm tall. Culms much branched, hollow, wiry, glabrous to scabrous. Blades rolled to C-shaped in cross section in the bud, involute at maturity, the marginal ridges thicker than the central ones, scabrous above, sometimes thinly pilose, especially near the ligule, 3–20(27) cm long, 0.9–2 mm wide; sheaths glabrous to slightly scabrous; ligules

0.1–0.5 mm long. Inflorescence 13–22(33) cm long, rather open, the short pedicels spreading; spikelets very large; first glume 17–25(33) mm long, the second glume usually longer, 22–30(37) mm long; lemma pilose on the callus, 13–20 mm long to the base of the awn; central awns 2.5–7 cm long, the lateral ones only slightly shorter. Anthers usually reduced and difficult to observe but 3.5–4.5 mm long when, as rarely, developed. (n = 11) Aug–Oct. Abundant on various soils but often where dry & sandy in waste ground; most of GP, but apparently not in ND & MT, & only in extreme se SD; (MA to SD, s to FL & TX, also scattered in w states).

8. *Aristida purpurascens* Poir., arrowfeather three-awn. Tufted perennials 3.5–8 dm tall. Culms solid to hollow, stiffly erect. Blades rolled in the bud, involute to flat at maturity, sometimes plainly thickened at the margins, scabrous above or on both surfaces, sometimes sparsely pilose near the ligule, 8–45 cm long, 1–3 mm wide; sheaths smooth to sparsely hirsute; ligules 0.1–0.4 mm long. Inflorescence spikelike, 16–30(40) cm long; glumes nearly equal, the first 7.5–11.5 mm long, the second 6–11.5 mm long; lemmas bearded on the callus and scabrous above, 3–10 mm long to the base of the awns; central awn 16–25(30) mm long, the lateral ones not much shorter, 13–23 mm. Anthers (0.7)1.2–2.7 mm long. Jul–Oct. In open woodland & sandy prairie; NE: Jefferson; MO, e KS, e OK; (MA to WI, NE, s to FL & TX).

Aristida arizonica Vasey has been collected in OK: Cimarron & Roger Mills. It will key here to *A. purpurascens* but may be distinguished by its obvious twisted awn column and by its longer glumes (more than 12 mm) and longer lemmas (also more than 12 mm).

9. *Aristida purpurea* Nutt. Cespitose perennials 1.2–8 dm tall. Culms erect, hollow, glabrous. Blades rolled to C-shaped in cross section in the bud, involute to flat at maturity, straight to flexuous, usually scabrous above, sometimes hirsute, 2–30 cm long, 0.6–2.5 mm wide; sheaths glabrous to scabrous but often villous in tufts at the collar; ligules 0.1–0.4 mm long. Inflorescence 4–30(33) cm long, strict to rather loose and open, the branches stiff to flexuous; glumes markedly unequal, the first 4–14 mm long, the second 7.5–25 mm long; lemmas 7.5–15 mm long to the base of the awns, scabrous in lines above, short-bearded on the callus, the column not obvious to well developed and twisted; the 3 awns nearly equal, 10–95 mm long. Anthers 0.8–3.1 mm long. Dry prairies & waste ground.

The approach to this complex used here follows that used by Holmgren & Holmgren (op. cit.), and the reader is referred to that reference for a detailed rationale. Briefly, the 4 taxa below have traditionally been placed in 5 separate species. They are united here because many of the characters commonly used vary with the growing conditions and the age of the plant and because the intergradation between the forms is extensive. Most plants can be separated as follows:

1 Panicle branches nodding, flexuous, the pedicels slender; awns less than 5.5 cm long .... 9c. var. *purpurea*
1 Panicle branches straight and ascending or appressed, or if, as rarely, somewhat flexuous, then the awns more than 5.5 cm long.
  2 Awns 5–9.5 cm long; leaves not principally in a basal tuft but distributed up the culm .... 9d. var. *robusta*
  2 Awns shorter or the leaves mostly all in a basal tuft (or both).
    3 Inflorescence elongate, normally more than 10 cm long; lemmas usually with a clearly marked, often twisted, awn-column .... 9a. var. *glauca*
    3 Inflorescence usually shorter, normally less than 10 cm long; lemmas not with a clearly marked awn column .... 9b. var. *longiseta*

9a. var. *glauca* (Nees) A. Holmgren & N. Holmgren, blue three-awn. Plants (1.8)2.5–7.7 dm tall. Blades mostly 2–17(24) cm long, basally clustered or distributed up the stem. Inflorescence 10–23(33) cm long, stiffly erect; first glume (4)6–11 mm long, second glume 8–15 mm long; lemma 9–15 mm to the base of the awns, usually beaked, the beak often twisted; awns 1–3.5 cm long. (2n = 22,44) May–Aug. In s KS, s CO, OK, TX, NM; (s KS to CA, s to TX, Mex.). *A. glauca* (Nees) Walp.—Gould—op. cit.; *A. wrightii* Nash—Rydberg.

9b. var. *longiseta* (Steud.) Vasey, Fendler three-awn. Plants 1–4 dm tall. Leaves nearly all borne in a basal tuft, often crisped, the blades short, 2–11 cm long. Inflorescence usually narrow, short, normally 4–10(14) cm long; first glume 5.5–10(13) mm long, second glume (9)11–16(19) mm long; lemma 9.5–13 mm to the base of the awns, unbeaked or the beak not clearly defined; awns usually 2–5 cm long, sometimes longer. (n = 11, 22) Jun–Sep. In w SD, w NE, WY, w KS, CO, TX, NM; (Alta. to B. C. SD, s to TX & CA, most common southward). *A. fendleriana* Steud.—Rydberg.

9c. var. *purpurea*, purple three-awn. Plants 2–8 dm tall. blades 3–25 cm long. Inflorescence (8)11–30 cm long, open, the axis flexuous, the branchlets and pedicels often slender and curving, at least at anthesis; first glume 5–11 mm; second glume 11–16 mm; lemma 7.5–13 mm long to the base of the awns, usually lacking an obvious column; awns (2.5)3–5.5 cm long. (n = 11, 22, n = 33, 44 also reported) May–Aug. In sw NE, KS, OK, TX, NM; (sw NE to CA, s to TX, Mex.).

9d. var. *robusta* (Merrill) A. Holmgren & N. Holmgren, red three-awn. Plants 2–7 dm tall. Leaves distributed well up the stem; blades 5–22(30) cm long. Inflorescence strict to somewhat open, sometimes the axis even curved or the branchlets curved or flexuous; first glume 7–14 mm long; second glume 12–25 mm long; lemma 10–15 mm to the base of the awns, lacking a clearly defined column; awns (4.5)5–10 cm long. (n = 11, 2n = 44, 66, 88) Jul–Sep. GP, more common in the w part; (Alta. to B. C., ND, s to TX & CA, most common northward). *A. longiseta* Steud.—Rydberg, misapplied.

Aristida roemeriana Scheele has been collected in TX: Hutchinson. It will key to *A. purpurea* var. *purpurea*, but it may be distinguished by the smaller lemmas (usually less than 8 mm long to the base of the awns) and the shorter awns (usually less than 3 cm long).

8. AVENA L., Oats

1. *Avena fatua* L. wild oats. Annuals 6–8(10.2) dm tall. Culms erect, hollow. Blades rolled in the bud, flat at maturity, 10–25 cm long and 4.5–8(18) mm wide; sheaths open; ligules membranaceous, 1.5–5 mm long; auricles none. Inflorescence an open panicle 13–25(40) cm long, the spikelets pendulous; spikelets 2- to 3-flowered, the disarticulation above the glumes; glumes longer than the florets, subequal, the first 7-nerved, 17–23 mm long, the second 9-nerved, 19–24 mm long; lemmas pubescent, the lowest lemma usually 7-nerved, 14–21 mm long, awned from about 5–7 mm above the base with a geniculate awn 2.5–4 cm long, its twisted basal portion about 1 cm long; paleas 2-nerved, ciliate on the nerves, shorter than the lemmas. Anthers 1.5–3 mm long. (2n = 42) Jun–Jul. Waste ground, cornfields, railroads; MN, ND, SD, IA, CO, NE, KS, more common in n; (ME to WA, s to NY, KS, CA; Europe) *Naturalized*.

Cultivated oats, *Avena fatua* var. *sativa* (L.) Hausskn., is often seen as a waif in waste ground near cultivated areas. It can be distinguished by the glabrous lemmas and by the absent or short awns (under 3 cm long).

9. BECKMANNIA Host, Sloughgrass

1. *Beckmannia syzigachne* (Steud.) Fern., American sloughgrass. Robust, glabrous, sometimes stoloniferous annual (2)5–9(11) dm tall. Culms erect, hollow. Blades rolled in the bud, flat at maturity, 8–17(25) cm long, 2.5–8.5 mm wide; sheaths open, cross-septate; ligules membranaceous, acuminate, usually 5–10 mm long (unless the tip is bent back or broken); auricles none. Inflorescence a rather narrow condensed panicle (7)12–25(28) cm long, its branches spicate, with spikelets closely overlapping in 2 rows. Spikelets laterally compressed, nearly orbicular, disarticulating below the glumes, 1-flowered; glumes subequal, 3-nerved, laterally compressed, inflated, apiculate-tipped, mostly 2–3 mm long; lemmas faintly 5-nerved, lanceolate, not filling the glumes, 2.4–3.5 mm long, including the acuminate tip. Anthers 0.5–1.1 mm long. (2n = 14) Jun–Aug. Ditches, marshes, edges of ponds & lakes, sometimes in standing water; n GP, in w MN, ND, MT, WY, CO, nw IA, nw NE; (Man. to AK, s to NY, IA, NM, & CA). *B. erucaeformis* (L.) Host—Winter, misapplied.

10. BOUTELOUA Lag., Grama Grass

Tufted annuals to tufted, strongly cespitose, or sometimes rhizomatous or stoloniferous perennials. Culms solid or sometimes hollow. Blades rolled or C-shaped in cross section in the bud, flat to involute at maturity; sheaths open; ligules predominantly a fringe of hairs or sometimes principally membranaceous. Inflorescence consisting of 1–many spicate branches; spikelets sessile in 2 rows on one side of each branch rachis; each spikelet consisting of 1 perfect floret below 1 or 2 rudimentary florets, disarticulating above the glumes; glumes 1-nerved, unequal, the first much narrower and shorter than the second; fertile lemma 3-nerved, its nerves ending in awns, the central awn longest; sterile lemmas usually also 3-awned, the awns often longer than those of the fertile lemma; fertile palea 2-nerved.

Reference: Gould, F. W. 1979. The genus *Bouteloua* (Poaceae). Ann. Missouri Bot. Gard. 66: 348–416.

1 Spicate branches deciduous as a unit, each branch bearing fewer than 10 spikelets.
   2 Central awns of fertile lemmas less than 1.0 mm long or lacking; awns of sterile lemmas usually not more than 4 mm long; spicate branches more than 10 in number .... 2. *B. curtipendula*
   2 Central awns of fertile lemmas longer than 1.0 mm; awns of sterile lemmas more than 4 mm long; spicate branches usually fewer than 10 in number .... 6. *B. rigidiseta*
1 Spicate branches not deciduous as a unit, the individual spikelets disarticulating above the glumes; each spicate branch bearing more than 10 spikelets.
   3 Plants annual; second glume less than 3 mm long; anthers less than 1 mm long .... 1. *B. barbata*
   3 Plants perennial; second glume more than 3 mm long; anthers usually more than 1 mm long.
      4 Lower culm internodes woolly; second glume lacking pustular-based hairs .... 3. *B. eriopoda*
      4 Lower culm internodes not woolly; second glume often bearing pustular-based hairs.
         5 Rachis of spicate branch projecting beyond the terminal spikelet ...... 5. *B. hirsuta*
         5 Rachis of spicate branch not projecting beyond the terminal spikelet .... 4. *B. gracilis*

1. *Bouteloua barbata* Lag., sixweeks grama. Slender tufted annuals, 0.5–3.5 dm tall. Culms hollow, often geniculate at the base. Blades rolled in the bud, flat at maturity, smooth to slightly scabrous below and scabrous above, sometimes also pilose with pustular-based hairs, 1–6 cm long, 0.7–2(3) mm wide; sheaths smooth to slightly scabrous; ligule 0.4–1 mm long, consisting of a fringe of hairs from a membranaceous base. Inflorescence with (2)4–8 slender curved branches, each branch persistent, (10)15–20 mm long, bearing numerous spikelets, glumes unequal, the first about 1–1.5 mm long, the second about 2–2.5 mm long; fertile lemmas 1–2.5 mm long, the central awn less than 2 mm long; sterile rudiments often 2, the lower rudiment with awns 1–2.5 mm long, the upper rudiment awnless, minute. Anthers 0.4–0.7 mm long. (2n=20) Aug–Oct. Sandy soil, waste ground, railroads; KS: Harvey & Reno; OK: Cimarron, Harmon; TX: Deaf Smith; (KS, CO to NV, s to TX, CA, & Mex.; perhaps introduced in our range).

Bouteloua simplex Lag. has been collected several times at the edge of our range (NM: Harding; WY: Laramie; CO: Kiowa; TX: Deaf Smith). It is an annual species; but it has only a solitary spike, and the spikelets are larger than those of *B. barbata*.

2. **Bouteloua curtipendula** (Michx.) Torr., sideoats grama. Rhizomatous perennials (0.8)4–7(8) dm tall. Culms solid, erect, arising in clumps or relatively few together. Blades rolled in the bud, flat at maturity, usually smooth beneath and scabrous above, often thinly to more or less densely pilose, the hairs on the margin near the ligule pustular-based, blades mostly 2–30 cm long, (1.4)2.5–5(6.5) mm wide; sheaths glabrous or pilose near the summit; ligule a fringed or erose membrane 0.3–0.7 mm long. Inflorescence elongate, consisting of (12)20–45(60) reflexed branches; each branch deciduous as a unit, 5–14(19) mm long, bearing 3–8 spikelets; glumes unequal, the first 2.5–5(6) mm long, the second 4–7(8.5) mm long; fertile lemma 3–6.5 mm long, its central awn less than 0.5 mm long; sterile rudiment usually consisting of a lemma 0.4–3.5 mm long, its central awn 1.5–4 mm long or sometimes the rudiment made up only of awns. Anthers (1.5)2–3(3.5) mm long. (2n = 40, or aneuploid from 41–64, 2n = 20 also reported) Jun–Aug. Open grassland & woodland openings, often dominant or co-dominant, increasingly used in native-grass lawn mixtures; GP; (Ont. to Man., throughout cen & e U.S. to n-cen Mex.).

Our plants are apparently *B. curtipendula* var. *curtipendula* since they are almost always rhizomatous, although this is difficult to verify on some of the herbarium sheets. According to the map given by Gould & Kapadia (Brittonia 16: 182–207. 1964) the nonrhizomatous var. *caespitosa* Gould & Kapadia may enter our range in the extreme sw part.

3. **Bouteloua eriopoda** (Torr.) Torr., black grama. Tufted nonrhizomatous perennials (2.4)3.4–6.5(7.3) dm tall. Culms solid, slender, often decumbent at the base and rooting at the nodes, at least the lower internodes long-woolly pubescent. Blades C-shaped in cross section to rolled in the bud, flat to involute at maturity, often 4 or 5 of the veins slightly larger than the others, finely scabrous above, often with pustules or pustular-based hairs along the margins, 2.5–6 cm long, 0.5–2 mm wide; sheaths smooth; ligule a fringe of hairs 0.2–0.6 mm long. Inflorescence with 1–6 slender branches, each branch white-woolly at the base, persistent, 14–30(40) mm long, bearing numerous spikelets; glumes unequal, the first 2–4.5 mm long, the second 4.5–8 mm long; fertile lemma 4–7 mm long, its central awn 1.5–4 mm long; rudiment consisting of 3 awns 4–7.5(9) mm long arising from a short, pilose base. Anthers 1.5–3 mm long. (2n = 20, 2n = 21, 28 also reported) Jul–Oct. Rocky or sandy soil; s CO, sw KS, w OK, TX, NM; (KS to WY, UT, s to TX, AZ, n Mex.).

4. **Bouteloua gracilis** (H.B.K.) Lag. ex Griffiths, blue grama. Mat-forming perennial 1.6–5 dm tall, forming short rhizomes. Culms solid, slender, often geniculate below. Blades C-shaped in cross section in the bud, flat at maturity to involute, especially toward the tip, short-pubescent to scabrous above, smooth to lightly scabrous below, sometimes also sparsely pilose on both surfaces, the margins not particularly thickened, 2–12(19) cm long, 0.5–2(2.5) mm wide; sheaths smooth to sparsely pilose, especially on the margins near the ligule; ligule a short fringe of hairs 0.1–0.4 mm long. Inflorescence with 1–3 curved branches, each branch persistent, 14–30(40) mm long, bearing numerous spikelets; glumes unequal, the first 1.5–3.5 mm long, the second 3.5–5(6) mm long, glabrous to scabrous but with pustular-based hairs on the midnerve; fertile lemma 3.5–5 mm long, its central awn 0.4–1.5 mm long; lemma of sterile rudiment 0.9–3 mm long, its central awn 1–3 mm long. Anthers 1.7–2.9 mm long. (2n = 20, 40, 60 plus aneuploid counts of 2n = 28, 35, 42, 61, 77, 84) Jun–Aug. Dry prairie, often dominant, also in waste ground, becoming increasingly common in native lawn mixtures; GP; (Man. to Alta., & N.T., s through cen & w U.S. to Mex.).

5. **Bouteloua hirsuta** Lag., hairy grama. Cespitose perennial 1–4(7) dm tall. Culms solid, erect to somewhat geniculate below. Blades C-shaped in cross section in the bud, flat at maturity, with obviously thickened margins, pustular-based hairs scattered on the lower margins and often on both surfaces as well, otherwise smooth to scabrous, 1–13(19) cm

long, 1-2.5 mm wide; sheaths glabrous to finely scabrous or puberulent, pilose near the ligule; ligule a short fringe of hairs 0.2-0.5 mm long. Inflorescence with 1 or 2(4) relatively straight branches, each branch persistent, (10)15-25(33) mm long, bearing numerous spikelets, the rachis prolonged beyond the terminal spikelet as an obvious point; glumes awn-tipped, unequal, the first 1.4-3.5 mm long, the second 3-5(6) mm long, beset with pustular-based hairs on the midnerve; fertile lemma 2-4.5 mm long, its central awn 0.4-1.7 mm long; lemma of rudimentary floret 0.5-2 mm long, its central awn 2-4.5 mm long. Anthers 1-2.8 mm long. ($2n = 20$, 40, 60 plus aneuploid counts of $2n = 21$, 22, 24, 28, 29, 36, 42, 46, 50) Jul–Oct. Prairies & pastures, often where sandy or rocky; GP, becoming uncommon in the n; (Sask. to Alta., s throughout most of U.S. e of Rocky Mts., Mex.).

Bouteloua pectinata Featherly is reported from within our range in OK: Comanche. It will key to *B. hirsuta* but differs in having a tuft of hairs at the base of the rudimentary floret, and its leaves are mostly in a basal cluster.

6. *Bouteloua rigidiseta* (Steud.) Hitchc. Tufted nonrhizomatous to short-rhizomatous perennials 1.5-3 dm tall. Culms solid, erect. Blades flat at maturity, scabrous above, often sparingly pilose, about 5-20 cm long, 1-2.5 mm wide; sheaths smooth, pilose on the margin near the ligule; ligule a short fringe of hairs 0.2-0.4 mm long. Inflorescence with (2)5- to 8(10) wedge-shaped branches, each branch deciduous, 8-15 mm long, bearing 4 or 5 spikelets; glumes unequal, the first 3-4.5(6.5) mm long, the second 6-8 mm long; fertile lemma about 4-6 mm long, its central awn 2-4 mm long; rudimentary floret(s) 1 or 2, the lower with a lemma about 3.5-5 mm long, its central awn 7-8 mm long, the upper, if present, somewhat smaller. Anthers (0.6)1.5-3 mm long. ($2n = 40$; $2n = 28$, 35 also reported) Apr–May. Prairies, pastures, woodland openings, & waste ground; OK, TX; (OK, TX, ne Mex.).

11. BRACHYELYTRUM Beauv.

1. *Brachyelytrum erectum* (Schreb.) Beauv. Perennials 5.5-9 dm tall, arising from short, knotty rhizomes. Culms slender, erect, solid, glabrous or retrorsely pubescent, especially near nodes. Blades glabrous or scabrous or, more commonly, thinly to strongly pilose, especially below, rolled in the bud, flat at maturity, 3-21 cm long, (4)7-16 mm wide; sheaths open to the base, thinly to densely retrorse-pilose, especially at the collar region; ligules membranaceous, scabrous to pubescent, truncate to acute, 1-4 mm long; auricles none. Inflorescence a slender panicle 7-17 cm long; spikelets 1-flowered, disarticulating above the glumes, floret clearly pediceled within the glumes; glumes extremely variable in size, even on the same plant, the first one usually very small (or lacking), less than 1.1 mm long, the second longer, 1.5-8 mm long, sometimes aristate; lemmas rather strongly 5-nerved, pubescent with appressed to ascending hairs, lanceolate, 9-13 mm long, tapering into a straight awn 13-26 mm long; palea almost as long as the lemma body; rachilla prolonged beyond the palea as a slender, naked bristle about 6-8 mm long. Anthers 3.6-6 mm long. ($2n = 22$) Late Jun–Aug. Moist woods; MN, IA, e NE, e KS, e OK; (Newf. to MN, s to GA & OK).

12. BROMUS L., Brome Grass

Annuals to rhizomatous or nonrhizomatous perennials. Culms erect to ascending, hollow. Blades rolled in the bud, flat at maturity; sheaths closed to near the top; ligules membranaceous, sometimes lacerate; auricles present or absent. Inflorescence a panicle, or sometimes racemose in depauperate plants; spikelets large, bearing 3-many florets, disarticulating above the glumes and between the florets; glumes unequal to, rarely, subequal, the first one generally the shorter, both of them shorter than the lowest lemma; first glume 1- to 7-nerved; second glume 3- to 9-nerved; lemmas 5- to 13-nerved, acute, often awned, the awn, when present, arising from just below the tip, the apex often bifid; paleas equal to or shorter than the lemmas, 2-nerved.

References: Soderstrom, T. R., & J. H. Beaman. 1968. The genus *Bromus* (Gramineae) in Mexico and Central America. Publ. Mus. Michigan State Univ., Biol. Ser. 3: 465–520; Wagnon, H. K. 1952. A revision of the genus *Bromus*, section *Bromopsis*, of North America. Brittonia 7: 415–480.

1 Spikelets flattened, the glumes and lemmas strongly compressed-keeled ...... 13. *B. unioloides*
1 Spikelets terete to somewhat flattened, the glumes and lemmas only slightly keeled to, more commonly, rounded on the back.
  2 Plants annual.
    3 First glume 1-nerved; awns of the lowest lemmas usually more than 7 mm long. .... 12. *B. tectorum*
    3 First glume 3- to 5-nerved; awns of the lowest lemmas rarely more than 7 mm long.
      4 Fruit thick, lunate in cross section, the lemma margins strongly inrolled at maturity, exposing the rachilla of the spikelet; palea subequal to or even slightly exceeding the lemma; anthers usually more than 1 mm long; lemmas usually less than 8 mm long .... 10. *B. secalinus*
      4 Fruit thinner, flat or curved, the lemma margins not inrolled, or not so strongly as to expose the rachilla; if plants immature, then usually differing from the above in some other respect.
        5 Nerves of the lemmas elevated and prominent; inflorescence contracted, usually with most of the branches shorter than the spikelets; lemmas 5.5–8(10) mm long; anthers 0.4–1(1.6) mm long .... 7. *B. mollis*
        5 Nerves of the lemmas neither particularly elevated nor especially prominent; inflorescence usually more open, with many branches longer than the spikelets; lemmas usually 7–10 mm long; anthers variable, but often 1 mm long or longer.
          6 Spikelets large, often nodding, usually more than 7 mm wide; upper lemmas of the spikelet strongly rhombic; awns divaricate ...... 11. *B. squarrosus*
          6 Spikelets variable but often either less than 7 mm wide or not nodding; upper lemmas obovate-lanceolate to somewhat rhombic; awns divaricate to erect.
            7 Panicles loose, the branches flexuous, the spikelets often nodding at maturity, the awns divaricate .... 4. *B. japonicus*
            7 Panicles rather stiff, the branches mostly not flexuous, the spikelets not nodding, the awns not divaricate .... 2. *B. commutatus*
  2 Plants perennial.
    8 First glume normally 1-nerved on all spikelets.
      9 Plants with strong rhizomes; awns of upper lemmas absent or most of them under 2 mm long .... 3. *B. inermis* subsp. *inermis*
      9 Plants lacking rhizomes; awns of the upper lemmas often more than 2 mm long.
        10 Lemmas with hairs mainly restricted to the marginal regions, sometimes also on the back at the very base .... 1. *B. ciliatus*
        10 Lemmas with pubescence general on the back or, rarely, glabrous.
          11 Plants with numerous nodes (often more than 10) which are mostly concealed by the overlapping sheaths; auricles present and conspicuous .... 6. *B. latiglumis*
          11 Plants with fewer nodes, many of them visible; auricles absent .... 9. *B. pubescens*
    8 First glume 3-nerved on many or all of the spikelets.
      12 Second glume 3-nerved .... 8. *B. porteri*
      12 Second glume 5-nerved .... 5. *B. kalmii*

1. *Bromus ciliatus* L., fringed brome, Nonrhizomatous perennials 5–11(16) dm tall. Culms often pubescent at the nodes. Blades flat, glabrous to pilose, especially on the upper surface, 10–28 cm long, 3.5–10(12) mm wide; sheaths glabrous to pilose; ligules 0.3–1.5 mm long; auricles absent. Inflorescence a loose panicle 7–18(33) cm long, the branches usually somewhat drooping. Spikelets 4- to 10-flowered, 14–25 mm long, 3.5–10 mm wide; first glume 1-nerved, 4.5–9.5 mm long; second glume 3-nerved, 5.5–11(14) mm long; lowest lemma 8–15 mm long, its awn 1.5–5(13) mm long; longest awns of spikelet 2–6(13) mm long. Anthers extremely variable in length, 0.7–2(4.6) mm long. (2n = 14, 28) Jul–Aug. Meadows to thickets & woodlands, often near water; MN, ND, MT, SD, IA, NE, WY, CO, NM; (Newf. to WA, s to NJ, NE, TX, s CA). *B. ciliatus* L. var. *intonsus* Fern.—Fernald; *B. richardsonii* Link—Wagnon, op. cit.

Although the large-anthered forms of this species are reportedly tetraploid and might logically be referred to a separate species *(B. richardsonii)*, other characteristics separating the two are weak, and many specimens would be problematical. Characteristics used by Mitchell & Wilton (Brittonia 17: 278–284. 1965) are not bimodal in our material.

2. ***Bromus commutatus*** Schrad., hairy chess. Annuals 3–10.5 dm tall. Culms glabrous or pubescent. Blades flat, usually pilose, 6–25 cm long, 1–6(9) mm wide; sheaths long-pilose; ligules 0.5–2.5 mm long; auricles absent. Inflorescence a relatively compact, but open, panicle 6–17(24) cm long, the branches mostly longer than the spikelets, stiffly spreading, seldom flexuous, the spikelets not drooping; spikelets 6- to 11-flowered, 6–22(25) mm long, 4–9 mm wide; first glume 3- to 5-nerved, 4.5–6.5(7) mm long; second glume 5- to 9-nerved, 6–7.5(8) mm long; lowest lemma 7–10 mm long, its awn (0)1.5–7 mm long, the longest awns of the spikelet up to 10.5 mm long; awns not divaricate at maturity; palea shorter than or equal to the lemma. Anthers 0.7–2 mm long. (2n = 14, 28, 56) May–Jun. Not common, in dry to moist waste ground; MT, NE, CO, MO, KS, OK, TX; (sporadic throughout much of the n U.S.; Europe). *Naturalized*.

This species is distinguished from the much more common *B. japonicus* by rather vague characteristics, and it also appears to intergrade with *B. secalinus*. Depauperate specimens are often identified as *B. racemosus* (here treated as a glabrous phase of *B. mollis*), but that entity is rather readily distinguished by its raised lemma nerves.

3. ***Bromus inermis*** Leyss. subsp. ***inermis***, smooth brome. Strongly rhizomatous perennials 4–10(12) dm tall. Culms, including the nodes, glabrous. Blades flat, glabrous, 9–21(33) cm long, 4.5–10(13) mm wide; sheaths glabrous; ligules 0.5–2 mm long; auricles absent or present but relatively small and inconspicuous. Panicles 12–19(23) cm long, only moderately open, the branches ascending, or sometimes very open in shade forms, with some branches even reflexed: spikelets 4- to 10-flowered, 18–40 mm long, 2.5–5(7) mm wide; first glume usually 1-nerved, 5–8 mm long; second glume usually 3-nerved, 6.5–10 mm long; lowest lemma 9–12(14) mm long; lemmas all unawned, or a few with awns up to about 2 mm long. Anthers 3–6 mm long. (2n = 28, 42, 56, 70, plus a variety of aneuploid counts) Mostly May–Jul, sometimes isolated plants blooming again in the fall. Widely planted for cover, pasture, & hay, & escaped into a variety of habitats; GP, less common in the extreme s part; (introduced from Newf. to B. C., s to NJ, GP, AZ, & CA; Europe, Siberia, China). *Naturalized*.

Bromus inermis Leyss. subsp. *pumpellianus* (Scribn.) Wagnon, the form native to N. Amer., is common in the Rocky Mts. and has been collected from the edge of our area (SD: Custer, Day, Lawrence, and Pennington; NE: Cherry and Sioux; WY: Crook). It may be recognized by its pubescent nodes, its usually awned lemmas (awns up to 6 mm long), and its often pubescent leaf blades.

Bromus biebersteinii R. & S. is now being seeded for cover and pasture in NE and perhaps elsewhere in the GP. A native of the U.S.S.R., it is similar to *B. inermis* subsp. *inermis*, but the lemmas have awns 3–5 mm long. It does well in NE and may be expected to escape.

4. ***Bromus japonicus*** Thunb. ex Murr., Japanese brome. Annuals 2–6(9) dm tall. Blades flat, usually pilose, 2–20 cm long, 1–4.5(10) mm wide; sheaths densely long-pilose; ligules

0.5–2 mm long; auricles absent. Inflorescence an open panicle (3)6–17(24) cm long with spreading, often flexuous branches, these often longer than the spikelets: spikelets 6- to 10(13)-flowered, 15–22(28) mm long, 4–7 mm wide; first glume usually 3- to 5-nerved, 3.5–6 mm long; second glume 5- to 9-nerved, 5–8.5 mm long; lemmas oblanceolate to slightly rhombic, their margins not strongly inrolled in fruit, the lowest lemma 7–9 mm long, its awn 1.5–7 mm long, the longest awns of the spikelet up to 13 mm long; awns divaricate at maturity; paleas 1–3 mm shorter than the lemmas. Anthers 0.7–1.5 mm long. (2n = 14, 28, n = 21 also reported) May–Jul. Dry to moist waste ground, abundant; GP; (widely introduced in n temp. regions & common throughout much of the U.S.; Old World). *Naturalized. B. patulus* Mert. & Koch — Rydberg.

This species appears to intergrade with the much less common *B. squarrosus* and with *B. commutatus*. It takes considerable intuition to identify members of this group of annual bromes, since depauperate or immature specimens may key to the wrong species. These taxa, as they are found in our area, would merit close study.

Bromus arvensis L., field brome, has been reported from ND: Pembina, and from OK: Comanche. It should key here to *B. japonicus*, but it may be distinguished by its long anthers (3–4 mm).

5. ***Bromus kalmii*** A. Gray. Nonrhizomatous perennials 5–9(10.5) dm tall. Culms glabrous to short-pilose on the internodes, usually retrorsely pilose on the nodes. Blades flat, long-pilose to merely scabrous, 7–25 cm long, 1.5–5(8) mm wide; sheaths usually retrorsely long pilose, rarely glabrous; ligules usually under 0.6 mm long, ciliolate; auricles none. Panicles 5–10 cm long, open, the branches often very slender and flexuous; spikelets 5- to 9-flowered, 9–20(24) mm long, 4–7 mm wide; glumes and lemmas uniformly pubescent; first glume 3-nerved, 5–6.5 mm long, second glume 5-nerved, 6–7.5 mm long; lowest lemma 7.5–9 mm long, its awn 1–2.5 mm long, the longest awns of the spikelet up to 3.0 mm long. Anthers 1.2–2.4 mm long. (2n = 14) Jun–Aug. Meadows, moist swales, & open woods; MN, IA, sw SD (BH); (NH to MN, s to MD & SD). *B. purgans* L. — nom. confus.

J. McNeill (Taxon 25: 611-616. 1976) has proposed that the name *B. purgans* L., which properly belongs to this species (see Baum, B. R., Canad. J. Bot. 45: 1845–1852. 1967), be rejected as a persistent source of confusion. I heartily concur and have tentatively retained the name *B. kalmii*, long applied to this taxon.

6. ***Bromus latiglumis*** (Scribn. ex Shear) Hitchc. Nonrhizomatous perennials 5.5–13(16) dm tall. Culms mostly glabrous, the internodes rather numerous, usually more than 10, the nodes usually concealed by the overlapping sheaths. Blades flat, glabrous to scabrous or, rarely, thinly appressed-pubescent, 10–30 cm long, 3.5–11(13) mm wide; sheaths glabrous to densely retrorse-pilose; auricles present and conspicuous (unless broken off); ligules mostly less than 1.5 mm long, often pubescent. Inflorescence 10–21(24) cm long, open, the branches spreading to ascending, somewhat flexuous; spikelets 5- to 10(12)-flowered, 15–33 mm long, 4–9 mm wide; glumes and lemmas usually pubescent; first glume 1-nerved, (3.8)5–7(8.5) mm long; second glume 3-nerved, (5.4)6.5–9(11.5) mm long; lowest lemma 8.5–11.5 mm long, its awn 1–5 mm long; longest awns of spikelet 3–5 mm long. Anthers 1.5–2.8 mm long. (2n = 14) Jul–Oct. Moist woods; MN, ND, SD (evidently rare), IA, NE, nw MO, extreme ne KS; (ME & s Que. to Alta., s to DC & KS). *B. purgans* L. — Wagnon, op. cit., misapplied.

7. ***Bromus mollis*** L., soft chess, bald brome. Annuals 2–5(7.5) dm tall. Culms glabrous to retrorsely pubescent, especially at the nodes. Blades flat, usually pilose, at least on the upper surface, 2–11(17) cm long, 1–3.5(4.5) mm wide; sheaths usually retrorse-pilose; ligules 0.5–2 mm long; auricles absent. Panicles rather short, 3–7(11) cm long, usually contracted, with most branches equal to or shorter than the spikelets; spikelets 4- to 9-flowered, 11–19 mm long, 4–7 mm wide; glumes and lemmas glabrous to pubescent; first glume 3- to

5-nerved, 4–6.5(7.5) mm long; second glume 7- to 9-nerved, 5–7.5(8.5) mm long; lemmas with prominently raised veins, the margins not strongly inrolled at maturity, the lowest one 5–8(10) mm long, its awn 2.5–5(6) mm long; awns straight, the longest awns of the spikelet up to about 7.5 mm long; palea often subequal to the lemma at maturity. Anthers 0.5–1(1.6) mm long. (2n = 14, 28) Jun–Jul. Waste places; sparingly introduced in GP, known with certainty from KS, NE, CO; (scattered in N. Amer. from N.S. to AK, s to NC, NE, & CA; Europe). *Naturalized. B. racemosus* L.—Rydberg.

Because of the similarity in measurements, and because plants with glabrous spikelets occur mixed with plants with pubescent spikelets, I follow several recent authors in considering *B. racemosus* a glabrous phase of *B. mollis*. Specimens identified in our herbaria as *B. racemosus* are often either immature *B. japonicus* or depauperate *B. commutatus*.

Bromus briziformis Fisch. & Mey. has been collected in our range and may be established (ND: Kidder; NE: Scotts Bluff, Thomas; SD: Lawrence). It may key here, but it can be at once recognized by its inflated, awnless, mucronate lemmas.

8. *Bromus porteri* Rydb. nodding brome. Nonrhizomatous perennials (4)6–8.5(9.5) dm tall. Culms subglabrous, but usually appressed pubescent at and just below the nodes. Blades flat, glabrous to rather thinly long-pilose, 8–20(30) cm long, 1–5 mm wide; sheaths long-pilose to glabrous; ligules 0.5–2.5 mm long, erose-ciliolate; auricles none. Inflorescence 5–17 cm long, open, the branches often flexuous; spikelets 5- to 10-flowered, 17–27(31) mm long, 4–8.5 mm wide; glumes and lemmas mostly rather uniformly pubescent, the glumes sometimes more sparsely so; first glume 3-nerved or sometimes 2-nerved on some spikelets, (5)5.5–7.5 mm long; second glume 3-nerved, 6.8–8.5 mm long, rarely mucronulate; lowest lemma 8.4–11 mm long, its awn 0.3–2.5 mm long, the longest awns of the spikelet up to 4 mm long. Anthers (1.7)2–2.8(3.1) mm long. (2n = 14) Late Jun–Aug. Mostly in open woods; nw ND, MT, sw SD, w NE, WY, & CO; (Sask. & ND, s to NE, through the Rocky Mts. to Mex.). *B. anomalus* Rupr. ex Fourn.—Hitchcock, Manual of the grasses of the United States, U.S.D.A. Miscl. Publ. No. 200, p. 46, 1950, in part.

The recognition of this species as distinct from *B. anomalus* seems to rest upon the importance attributed to the 3-nerved glumes and to rather minor differences in measurements of spikelet parts. Wagnon (op. cit.) recognized both species but failed to distinguish them clearly, according to Soderstrom & Beaman, (op. cit.), who consider them amply distinct in Mexico and Central America. *B. anomalus* in the sense of Soderstrom and Beaman apparently does not occur in our range.

9. *Bromus pubescens* Muhl. ex Willd., Canada brome. Tall, nonrhizomatous perennials 8–13 dm tall. Culms with the nodes usually fewer than 10, many often visible, thinly or densely pilose, especially at the nodes. Blades flat, pilose, often densely so, especially above, 15–28 cm long, (3.4)5–14 mm wide; sheaths pilose; ligules erose, 0.5–2.2 mm long; auricles absent. Panicles 10–24 cm long, very open, the branches erect to widely spreading, often flexuous; spikelets 4- to 8-flowered, 15–30 mm long, 3–8(10) mm wide; glumes and lemmas often rather uniformly pubescent; first glume 1-nerved, 5–8 mm long; second glume 3(5)-nerved, 6.9–9(10.5) mm long; lowest lemma 8–10.5(12) mm long, its awn 3–6.5 mm long, the longest awns of the spikelet up to 7.5(9) mm long. Anthers (2.2)2.5–4.1 mm long. (2n = 14) May–Jun. Woodlands; e GP with scattered locations westward; (VT to ND & WY, s to GA & TX). *B. purgans* L.—misapplied to this species by numerous authors.

Bromus lanatipes (Shear) Rydb., which has been collected in the se part of our range (OK: Cimarron; TX: Deaf Smith) and is evidently abundant w of our range in CO & NM, will key to this species, but it may be distinguished by its glabrous glumes in combination with pubescent lemmas and by its narrower blades (usually less than 5 mm wide).

10. *Bromus secalinus* L., cheat. Annuals (2)5–11(12.5) dm tall. Culms glabrous or retrorsely pubescent at the nodes. Blades flat, thinly pubescent to sometimes glabrous, (4)9–22(36) cm long, 2.4–7(9) mm wide; sheaths glabrous or sometimes pilose with retrorse hairs; ligules

0.9-2.2(3) mm long, erose; auricles absent. Panicles open, the branches spreading to ascending, (6)10-20 cm long; spikelets 5- to 12-flowered, the rachilla visible in mature material, 14-21(24) mm long, 3.6-9 mm wide; first glume 3- to 5-nerved, 4.5-5.5(6.5) mm long; second glume 5- to 7-nerved, 5-7(8.5) mm long; lemmas involute at maturity, closely investing the thickened fruit, the lowest one (5.5) 6.5-8(9) mm long, awnless, or with an awn up to 6.5 mm long, the longest awns of the spikelet 0-8 mm long; awns straight or flexuous, but not especially divaricate; palea subequal to the lemma. Anthers 0.9-2 mm long. ($2n = 14, 28$) May–Jul, rarely later. Waste ground; GP, most common in e NE, e KS, e OK; (widely introduced in n temp. regions & throughout most of the U.S.; Europe). *Naturalized.*

This species, although very distinctive in the fruiting condition, is readily confused with other annual bromes, especially *B. commutatus,* when immature.

11. **Bromus squarrosus** L. Annuals 2.5-5 dm tall. Culms glabrous or pubescent at the nodes. Blades flat, pilose to glabrous, 3-15 cm long, 1-4.5 mm wide; sheaths glabrous to pilose; ligules 1-2.6 mm long; auricles absent. Panicle open but not much branched, (3)6.5-14(17) cm long; spikelets 11- to 16(23)-flowered, slightly inflated, 22-34 mm long, 6.5-10 mm wide; glumes markedly unequal, the first 5- to 7-nerved, 5-6.5 mm long, the second most often 9-nerved, 6.8-9 mm long; lemmas, especially the upper ones, broadly rhombic, the angular exposed margin not inrolled; lowest lemma usually 8-10 mm long, its awn 1.5-4.5 mm long; awns of the spikelet divaricate at maturity, the longest ones up to 10.5 mm long. Anthers 0.9-1.3 mm long. ($2n = 14$) Jun–Jul. Waste ground; uncommon in GP, known from ND, MT, SD, NE; (rather sporadically introduced in U.S., principally in the n-cen region; Europe). *Naturalized.*

This species appears to intergrade with *B. japonicus,* which see.

12. **Bromus tectorum** L., downy brome. Annuals 2-6 dm tall, usually retrorse-pubescent to soft-pilose throughout, except less so on culms and upper sheaths, the spikelets rarely glabrous. Blades flat, 1-10(19) cm long, 1-5 mm wide; ligules 0.9-3.5 mm long; auricles lacking. Panicle 4-18(21) cm long, often nodding; spikelets 4- to 7-flowered, 11-20(24) mm long, 3-6 mm wide; first glume 1-nerved, 4.5-7.5 mm long; second glume 3-nerved, (7)8-10.5 mm long; lowest lemma 8.5-12 mm long, its awn 7-13 mm long, the longest awns of the spikelet 7-17 mm long. Anthers 0.4-0.8 mm long, usually retained. ($2n = 14$) May–Jun. Very common in waste ground; GP; (common in most of U.S.; Europe). *Naturalized.*

13. **Bromus unioloides** H.B.K., rescue grass. Annuals 1-4.5(6) dm tall. Culms glabrous to retrorse pubescent just below the nodes. Blades flat, usually pubescent, 3-15(18) cm long, 1.5-4(6) mm wide; lower sheaths pilose, the upper ones often glabrous; ligules whitish, rather long, 1-5 mm; auricles absent. Inflorescence usually contracted, the branches often erect, 3-13(26) cm long; spikelets 4- to 10(12)-flowered, compressed-keeled, glabrous, 10-35 mm long, 3.5-11 mm wide; first glume 5- to 7(9)-nerved, 7.5-10(12) mm long; second glume 7- to 9(11)-nerved, 8-12.5 mm long; lemmas keeled, the lowest one 11-14.5 mm long, awnless or with an awn under 1(2) mm long; longest awns of the spikelet mostly less than 2(3) mm long. Anthers 0.5-1 mm long, usually retained. ($2n = 28, 42$) Apr–Jul. Pastures, roadsides, & waste ground; w MO, s KS, OK, TX, NM; introduced in many s states as a forage grass; s. Amer.). *Naturalized.*

Soderstrom & Beaman (op. cit.), noting wide variation in South American material, retain the name *B. unioloides* for this grass, in spite of the contention of Raven (Brittonia 12: 219-221. 1960) that our material should be called *B. willdenowii* Kunth.

Bromus carinatus H. & A. is known from several collections at the edge of our range (ND: Slope; NE: Dawes, Sheridan, & Sioux; SD: Pennington & Lawrence), and was reported by Rydberg as *B.*

breviaristatus Buckl. It is similar to *B. unioloides*, but it may be distinguished by its perennial habit (in ours) and by its longer awns (more then 3 mm long).

13. BUCHLOË Engelm., Buffalo Grass

1. ***Buchloë dactyloides*** (Nutt.) Engelm., buffalo grass. Strongly stoloniferous, dioecious (rarely, monoecious), mat-forming perennials 0.3–2.0 dm tall, the staminate plants averaging slightly taller than the pistillate plants. Culms solid. Blades flat, rolled in the bud, glabrous to sparsely hispid, 1–10(14) cm long, 0.8–2 mm wide; sheaths open, often glabrous except in the collar region; ligule a short fringe of hairs less than 0.9 mm long, often flanked by long hairs; auricles lacking. Staminate inflorescences on slender culms, each consisting of 2 or 3(5) unilateral spicate branches, each branch 7–11(13) mm long, bearing spikelets in 2 rows; staminate spikelets 2-flowered, retained or tardily disarticulating above the glumes; glumes unequal, 1(2)-nerved, the first 1.4–3 mm long, the second 2–4.5 mm long, sometimes mucronate; lower lemma 3.5–5 mm long, 3-nerved; palea subequal to the lemma, 2-nerved; anthers 2.2–3.3 mm long. Pistillate spikelets borne in 2 or 3 burlike clusters closely subtended by more or less modified foliage leaves, usually with 2–4 spikelets per bur, the bur falling as a unit; pistillate spikelets usually 1-flowered; first glume small, membranaceous (often obsolete in all but 1 spikelet of each bur), 0.5–4.5 mm long; second glume indurate at base, united with the indurate axis of the bur, enveloping the lemma and terminating in (usually) 3 awnlike points; lemma also 3-nerved and with 3 short awns, 3.5–6 mm long. (2n = 20, 40, 60) Mostly Apr–Jun, but staminate plants often collected (with retained spikelets) until late summer or fall. Prairies, often dominant in short-grass areas of w GP, mostly confined to exposed, well-drained locations farther e; GP; (MN to MT, s to LA & AZ).

14. CALAMAGROSTIS Adans., Reedgrass

Rhozomatous perennials. Culms erect, hollow. Blades rolled in the bud, flat to involute at maturity; sheaths open; ligules prominent, membranaceous; auricles none. Inflorescence an open or contracted panicle; spikelets 1-flowered, disarticulating above the glumes; rachilla prolonged beyond the glumes, bearded; glumes about equal, acute to acuminate, often scabrous, particularly on the keel, the first 1-nerved, the second 3-nerved; lemma shorter than the glumes, rather delicate, 5-nerved, awned from the back with a delicate or stout straight or geniculate awn, the callus bearing a conspicuous tuft of straight hairs; palea prominent, 2-nerved.

Reference: Stebbins, G. L. 1930. A revision of some North American species of *Calamagrostis*. Rhodora 32: 35–57.

1 Plants low, usually less than 4.5 dm tall; awns strongly geniculate, projecting sideways from between the lemmas .... 2. *C. montanensis*
1 Plants taller, normally more than 4.5 dm tall; awns slender, straight to slightly geniculate.
    2 Panicle open, the branches readily distinguished; mature plants mostly more than 9 dm tall; blades usually flat, the furrows on the upper surface not closely spaced and not normally more than 1/2 the section in depth .... 1. *C. canadensis*
    2 Panicle condensed, most branches obscured by the crowded spikelets; mature plants normally less than 9 dm tall; blades flat or involute, the closely-spaced furrows of the upper surface often more than 1/2 the section in depth, especially in robust plants .... 3. *C. stricta*

1. ***Calamagrostis canadensis*** (Michx.) Beauv., bluejoint. Rhizomatous perennials (7)10–13.5 dm tall. Culms stout, glabrous, erect. Blades flat to slightly involute, the ridges rather widely

spaced, the furrows not usually pronounced, glabrous to scabrous, 8–30(38) cm long, 1.5–6(8) mm wide; sheaths glabrous or scabrous; ligules 3–7 mm long on the culm leaves, usually lacerate, inflorescence an open to somewhat contracted panicle, its branches mostly plainly visible and not obscured by spikelets; panicle 7–18(23) cm long, 1–6(16) cm wide when pressed; glumes acute or acuminate, nearly equal or the second slightly shorter, scabrous on the keel, 1.5–4 mm long; lemma 1.5–3 mm long, the callus hairs 1.4–2.6 mm long, the awn slender and usually straight, 0.2–2.2 mm long, arising from 0.2–2 mm behind the tip. Anthers 1–1.7 mm long. ($2n = 42, 56, 2n = 28, 49, 51, 52$ also reported) Jun–Aug. Wet places, marshes, sloughs, & ravine bottoms; GP, except TX, OK; (Greenl. to AK, s to NC, KS, & CA). *C. canadensis* var. *macouniana* (Vasey) Steb.—Fernald; *C. langsdorfii* (Link) Trin.—Rydberg, misapplied; *C. macouniana* Vasey—Rydberg.

This is an exceedingly variable species in which numerous varieties have been proposed. Our plants key mostly to var. *canadensis* and var. *macouniana* (Vasey) Steb., but since they are predominantly intermediate between those two taxa, recognition of varieties becomes, for us, an inconvenience.

2. *Calamagrostis montanesis* (Scribn.) Scribn., plains reedgrass. Rhizomatous perennials 1–4.5(6.5) dm tall. Culms erect, usually scabrous below the panicle. Blades involute, rather stiff and erect, scabrous, ridges and furrows pronounced on the upper surface and closely spaced, 3–18(24) cm long, 1–3.5 mm wide; sheaths smooth; ligules mostly 2–4.5 mm long on the culm leaves, usually lacerate. Inflorescence a spikelike panicle 3–10 cm long, 0.6–2.2 cm wide when pressed; glumes scabrous, (2.5)4–5.5(7) mm long, the second often slightly shorter than the first; lemma 2.5–4.2(5) mm long, the callus hairs 1.4–3.2 mm long, the awn stout and geniculate, 1.3–3.0 mm long, arising from 1.9–3(3.6) mm behind the tip, protruding from between the glumes when mature. Anthers 1.4–2.5 mm long. ($2n = 28$) Jun–Jul. Low areas to slopes, mostly in native prairie; MN, ND, MT, SD, CO; (Man. to se B. C., s to GP).

Calamagrostis purpurascens R. Br. is known from SD: Lawrence and Pennington and may key here. It may be distinguished from *C. montanensis* by its long awn, which is exserted 1–3 mm beyond the glume tips.

3. *Calamagrostis stricta* (Timm.) Koel. Rhizomatous perennials 5–10 dm tall. Culms smooth to scabrous, erect. Blades involute or, rarely, flat, scabrous to pilose above and smooth to scabrous below, the furrows and ridges strong and closely spaced on the upper surface, 4–36 cm long, 1–4.5(6) mm wide; sheaths smooth or, rarely, somewhat scabrous; ligules 0.8–6 mm long on the culm leaves, sometimes lacerate. Panicle contracted, spikelike or lobed, 5.5–16(22) cm long, 0.6–3(5) cm wide when pressed, the individual branches mostly obscured by the crowded spikelets; glumes scabrous to glabrous, 1.8–4(5.5) mm long, the second often slightly shorter than the first; lemmas glabrous to scabrous, 1.5–3.5(4.5) mm long, the callus hairs 1.2–3.5 mm long, the awn usually straight, rather weak, 1–3 mm long, arising from 0.8–2.5(3.2) mm behind the tip. Anthers 1.2–2.6 mm long. ($2n = 28, 42, 56, 70, 84$, plus some aneuploid counts) Jun–Aug. (Sep). Wet places, stream banks, marshes, etc.; GP, except OK, TX, rare in KS: (Newf. to AK, s across n U.S.). *C. inexpansa* A. Gray; *C. americana* (Vasey) Scribn; *C. neglecta* (Ehrh.) Gaertn., Mey. & Scherb.—Rydberg.

This taxon has traditionally been divided into 2 species—*C. neglecta* and *C. inexpansa*. *C. inexpansa* A. Gray has been held by Holmgren & Holmgren (Intermountain Flora 6: 270. 1977) to be inseparable from *C. neglecta* by any of the characteristics regularly used. *C. inexpansa* reportedly has longer ligules, more pubescent lemmas, and broader, more scabrous blades, but these and other characteristics occur in all combinations and the taxa are as fully intergradient in our region as in the Intermountain area.

Calamagrostis neglecta (Ehrh.) Gaertn., Mey. & Scherb., although evidently a taxonomic synonym of *C. stricta* (Timm.) Koel., is based on an illegitimate name and must, therefore, be treated as new by Gaertn., Mey. & Scherb.; hence it is not the oldest name available for this species. For complete discussions of this see Love, Taxon 19: 299-300. 1970, and Voss, Michigan Bot. 11: 28-29. 1972.

15. CALAMOVILFA (A. Gray) Hack. ex Scribn. & Southworth, Sandreed

Stout perennials with strong creeping rhizomes. Culms erect, glabrous, solid, or hollow near the base. Blades rolled in the bud, flat near the base at maturity, but tapering into involute, filiform tips; sheaths open, not keeled; ligule a short fringe of hairs; auricles lacking. Inflorescence an open to rather narrow panicle; spikelets 1-flowered, disarticulating above the glumes; glumes 1-nerved, unequal, the second somewhat longer than the first; lemma 1-nerved, usually somewhat shorter than the second glume, bearing a conspicuous tuft of long hairs at the base; palea 2-nerved, about equal to the lemma in size and similar in texture.

1 Lemmas villous on the back (in addition to the basal tuft of hairs); anthers usually over 4.6 mm long .... 1. *C. gigantea*
1 Lemmas glabrous on the back above the basal tuft of hairs; anthers usually under 4.6 mm long .... 2. *C. longifolia*

1. *Calamovilfa gigantea* (Nutt.) Scribn. & Merr., big sandreed. Robust, often glaucous plants (9)13-29 dm tall or taller. Blades glabrous, mostly 20-80 cm long, (3)7-11 mm wide near the base; sheaths usually glabrous or sometimes pubescent in the vicinity of the ligule; ligule 0.8-2.3 mm long. Panicles mostly very open, 22-65 cm long; glumes acute, the first one (4)5-8 mm long, the second 6-9.5 mm long; lemma acute, subequal to or slightly exceeding the second glume in length; palea firm, slightly longer than the lemma. Anthers 4.7-5.6 mm long. (2n = 40) Jul-Sep. Sandy hills, dunes, & stream margins; s KS, OK, TX; (KS to UT, s to TX & AZ).

2. *Calamovilfa longifolia* (Hook.) Scribn., prairie sandreed. Plants often glaucous, (3.5)8-14(17) dm tall. Blades glabrous, mostly 10-60 cm long, (2)4-8 mm wide near the base; sheaths mostly glabrous but often villous marginally at the collar; ligule 0.5-2.2 mm long. Panicles open to mostly narrow, 10-40 cm long; glumes acute, the first 4-6(7) mm long, the second 5-7(8) mm long; lemma acute, usually slightly shorter than the second glume; palea firm, often equal to or slightly shorter than the lemma. Anthers 2.9-4.5 mm long. (2n = 40) Jul-Sep. In a variety of sandy habitats, often dominant in sandhills prairie; MN, ND, MT, SD, WY, IA, NE, CO, n KS; (MI to Alta., s to IN, CO, & ID).

16. CATABROSA Beauv.

1. *Catabrosa aquatica* (L.) Beauv., brookgrass. Glabrous stoloniferous perennials. Culms hollow, erect to decumbent, the decumbent lower culms rooting at the nodes, mostly 2-8 dm long. Blades folded in the bud, flat to folded at maturity, the tips prow-shaped, mostly 3-19 cm long and (2.5)6-11(12.5) mm wide; sheaths rather loose, closed for 1/2 or more of their length, often plainly cross-septate; ligules membranaceous, mostly 2.5-5.5 mm long; auricles lacking. Inflorescence an open panicle (7)10-25(27) cm long, often partly included in the sheath of the uppermost leaf; spikelets disarticulating above the glumes, (1)2-flowered; glumes scarious, irregularly toothed, the first nerveless to 1-nerved near the base, 0.4-2 mm long, the second nerveless or 1- to 3-nerved near the base, 1-2.2 mm long; lemmas truncate, strongly 3-nerved, the nerves approximately parallel, the lowest lemma 1.7-3 mm long; palea approximately equal to the lemma and similar in texture. Anthers 1-2 mm long. (2n = 20) Late May-Aug. In & near creeks, springs, lakes, & ponds; ND, MT, w SD, WY, w NE, CO; (Newf. to Alta.; s to WI, CO, AZ, & OR).

17. CENCHRUS L., Sandbur

Annuals. Culms solid or sometimes hollow below, geniculate at base. Blades folded or rolled in the bud, folded to flat at maturity, somewhat keeled, especially near the base; sheaths flattened, keeled, open; ligule predominantly a short fringe of hairs; auricles lacking. Inflorescence spikelike, consisting of several to many spikelet-containing burs sessile on a stout axis; each bur pubescent and beset with retrorsely barbed spines, disarticulating from the axis at maturity, closely enclosing 1-3 spikelets; spikelets 2-flowered, the upper floret perfect, the lower staminate or sterile; glumes acute or acuminate, membranaceous, plainly nerved, the first one short, 1-nerved, the second one longer, 3- to 5-nerved; lemma of the lower floret 3- to 5-nerved, resembling the glumes in texture; palea of the lower floret elongate in the largest of the spikelets, often equal to or longer than its lemma, 2-nerved, membranaceous, and containing 3 enlarged stamens, but often reduced or absent and not containing stamens in the smaller spikelets; lemma of the upper floret firm, indistinctly 5-nerved, with a U-shaped ridge near the base, long-beaked, its edges overlapping the 2-nerved palea; the upper floret perfect, but the anthers only about 1/2 the length of those in the staminate lower florets, the stamens of the largest spikelets thus clearly dimorphic.

1 Spines usually fewer than 30 on each bur, the broadest ones over 0.9 mm wide at the base .... 1. *C. incertus*
1 Spines usually more than 30 on each bur, the broadest ones under 0.9 mm wide at the base .... 2. *C. longispinus*

1. Cenchrus incertus M. A. Curtis. Plants 2-5 dm tall. Blades glabrous, folded in the bud, folded to flat at maturity, 3-14(17) cm long, 2-4.5 mm wide; sheaths usually ciliate on the margins, otherwise glabrous or only sparsely pilose; ligules 0.6-1.4 mm long. Burs with 15-30 mostly rather stout spines, the largest ones 3-5 mm long and 1.0-1.5 mm wide at the base; spikelets (1)2 or 3 per bur; largest spikelet with its first glume 1.8-2.8 mm long, its second glume 3.8-5.0 mm long, 3- to 5-nerved, the lemma of its lower floret 4-5.5 mm long, the palea of that floret usually as long as the lemma, enclosing a staminate flower, the lemma of its upper floret 4.3-5.5 mm long; the 1 or 2 additional spikelets with glumes and lemmas averaging smaller, their lower florets often sterile and with greatly reduced or absent paleas. Anthers of upper florets 0.5-0.8 mm long, those of the lower florets, when present, 1.3-1.6 mm long. (2n = 34, n = 16 also reported) May-Sep. Waste ground; s-cen KS, OK, NM, TX; (VA to KS, s to FL, TX, & NM).

2. Cenchrus longispinus (Hack.) Fern. Plants 2-5.5(8.5) dm tall. Blades rolled, rarely folded, in the bud, flat at maturity, 4-14(27) cm long, 2.8-6.5(7.5) mm wide; sheaths not usually ciliate, but sometimes sparsely so, normally glabrous; ligules 0.6-1.3(1.8) mm long. Burs with 30-60 (or more) spines of various sizes, the largest ones mostly 3-5 mm long and 0.5-0.8 mm wide at the base; spikelets (1)2 or 3 per bur; largest spikelet with its first glume 0.8-3 mm long, its second glume 4-6 mm long, 5-nerved, the lemma of its lower floret 4-6.1 mm long, the palea of that floret usually as long as the lemma, enclosing a staminate flower, the lemma of its upper floret 5-7 mm long; the 1 or 2 additional spikelets with glumes and lemmas averaging smaller, their lower florets often sterile and with reduced or absent paleas. Anthers of the upper florets 0.7-1 mm long, those of the lower florets, when present, 1.5-2 mm long. (n = 17, 2n = 36 also reported) Jul-Sep. Waste ground; GP, less common in n part; (Ont.) to OR, s throughout most of U.S. to Mex. & trop. Amer.; adventive elsewhere). *C. pauciflorus* Benth.—Rydberg, misapplied.

18. CHASMANTHIUM Link

1. **Chasmanthium latifolium** (Michx.) Yates, wild oats. Rhizomatous perennials (6)8–13 dm tall. Culms simple to sparingly branched, glabrous, striate, hollow. Blades rolled in the bud, flat at maturity, with a moderately pronounced midrib, glabrous to thinly pilose near the ligule, (3)9–21 cm long, (6.5)10–21 mm wide; sheaths glabrous, open, rounded to very slightly keeled; ligule a short ciliate membrane 0.3–1.0 mm long; auricles absent. Inflorescence a nodding panicle (12)16–21(25) cm long; spikelets (5)7- to 17-flowered, laterally compressed, large, 15–25(34) mm long and (4.5)10–19 mm wide, disarticulating above the glumes and between the florets; glumes subequal, keeled, the keels scabrous, 5- to 7-nerved, 4.5–7.5 mm long; lemmas strongly keeled, the keels usually scabrous above and ciliate below, 9- to 15-nerved, the lowest 1 or 2 empty, 6–8.5 mm long, the upper ones fertile and longer, the longest central ones 9–13 mm long; paleas 2-keeled, shorter than the lemmas. Anther 1, short (0.4–0.8 mm long) and the floret cleistogamous or longer (1.2–3 mm long) and then the floret often chasmogamous. (2n = 48) (Jun) Jul–Oct. In moist soil, usually in the woods, often along streams; e & s KS, e OK (PA to KS, s to nw FL & e TX). *Uniola latifolia* Michx.—Rydberg.

This species is usually placed in *Uniola* and was so assigned in the Atlas GP. However, it clearly must be transferred to *Chasmanthium* for reasons discussed by Yates in two detailed papers (Southw. Naturalist 11: 145–189 and 415–455. 1966). Briefly, the genus *Uniola* as traditionally constituted consists of disparate elements which properly belong in separate subfamilies.

19. CHLORIS Sw., Windmill Grass

Annuals or nonrhizomatous perennials. Culms solid, at least above, sometimes flattened. Blades folded to rolled in the bud, folded or flat at maturity, often keeled, especially near the base; sheaths open, usually keeled; ligule a short, ciliate membrane, auricles none. Inflorescence a panicle of often digitately arranged spicate branches, the spikelets closely crowded to rather distantly spaced in two rows on one side of each branch; each spikelet consisting of 1 perfect floret below 1 rudimentary floret, disarticulating above the glumes; glumes 1-nerved, usually unequal, the second longer, lanceolate, sometimes short-awned; lemmas of lower florets 3-nerved, awned; rudimentary floret in ours consisting of a cylindrical to inflated, 3- to 5-nerved, awned lemma.

1 Rudimentary floret strongly inflated, awns of both fertile and rudimentary florets less than 2 mm long .... 1. *C. cucullata*
1 Rudimentary floret cylindrical, awns of both fertile and rudimentary florets more than 2 mm long.
   2 Plants perennial; axis of inflorescence 1–4 cm long, the branches disposed along it and not digitate; blades folded in the bud .... 2. *C. verticillata*
   2 Plants annual; axis of the inflorescence very short, the branches essentially digitate; blades rolled in the bud .... 3. *C. virgata*

1. **Chloris cucullata** Bisch., hooded windmill grass. Nonrhizomatous perennials 1.4–6.5(9) dm tall. Culms solid, glabrous, somewhat flattened. Blades folded at maturity and in the bud, glabrous to scabrous, keeled near the base. 1–20(27) cm long, 1.8–3.5(4.6) mm wide; sheaths 2-ranked, keeled, glabrous; ligule 0.2–0.9 mm long, truncate, ciliate. Inflorescence consisting of 6–20 branches disposed along a short axis 2–6(10) mm long, the longest branch 2–5.5 cm long; spikelets crowded; first glume 0.5–1.5 mm long, the second 1.2–2.2 mm long; fertile lemma 1.5–2.1 mm long, awned from below the tip with a short awn 0.3–1.8 mm long; rudiment strongly inflated, 5-nerved, its upper margin inrolled, 1–1.5 mm long, its awn 0.3–1.2 mm long. Anthers 0.3–0.6 mm long. (2n = 40) May–Sep. Pastures, road

sides, & waste places, often in sandy or rocky soil; KS: Reno; cen & s OK; TX, NM; (KS to TX & NM; Mex., perhaps naturalized at its more northern locations). *C. brevispica* Nash—Rydberg.

D. Anderson (*in* Gould, Grasses of Texas, Texas A. and M. Univ. Press, College Station, p. 321. 1975) reports that *C. verticillata* may hybridize with *C. cucullata* where the two grow together. Such contact is possible throughout our part of the range of *C. cucullata*.

2. **Chloris verticillata** Nutt., windmill grass. Perennials 1–3.5(4.1) mm tall. Culms solid, erect above, geniculate and decumbent below, not uncommonly rooting at the lower nodes, flattened. Blades glabrous to scabrous, folded in the bud and at maturity, keeled, especially toward the base, 1–12 cm long, 1.8–3.8 mm wide; sheaths keeled, glabrous, 2-ranked, the uppermost one often dilated; ligules 0.4–1 mm long, truncate, with long hairs behind and at the margins. Inflorescence consisting of 6–20 branches well distributed, often at least partly in verticels, along an axis 1–5 cm long; spikelets rather widely spaced; first glume 1.7–2.5(3) mm long, the second 2.4–4.4 mm long; fertile lemma 2–3 mm long, awned from below the tip, the awn 4–9 mm long; rudiment 3-nerved, truncate at the apex, cylindrical, 1–2.1 mm long, its awn 3–5(6.2) mm long. Anthers 0.3–0.6 mm long. ($2n = 40$, $2n = 63$ also reported) May–Sep. Lawns, roadsides, & waste areas; se SD, NE, MO, KS, CO, OK, TX, NM; (MO to SD, CO, s to LA & AZ, introduced elsewhere).

See comments under *C. cucullata*.

3. **Chloris virgata** Sw., showy chloris. Robust, usually tufted, annuals, 2–9 dm tall. Culms solid above, often hollow below, somewhat flattened, glabrous, mostly decumbent at the base. Blades flat, glabrous to pilose, especially near the ligule, flattened in the bud, but clearly rolled and not folded, weakly keeled near the base, rather clearly bordered, up to 23 cm long, 3–8 mm wide; sheaths glabrous, only slightly keeled, the uppermost sometimes dilated; ligule in ours a short ciliate membrane 0.4–1.4 mm long. Panicles with 6–20 digitate, crowded branches, the longest of these 4.5–9 cm long; spikelets crowded; first glume 1.5–2.8 mm long; second glume including its short awn 2.7–4(4.5) mm long; lemma of fertile floret gibbous, pubescent at the base and on the keel and long-ciliate on the upper margin, 2.4–3.4 mm long, its awn (3)4–8(9.5) mm long; rudiment conical, 3-nerved, 1.5–2.7 mm long, its awn 2.5–5.5(8.5) mm long. Anthers 0.3–0.7 mm long. ($2n = 20$, $2n = 26$, 40 also reported) Jul–Aug. Pastures, roadsides, & waste ground; s KS, CO, OK, TX, NM, also known from a few locations in ND & NE; (KS to TX & s CA, rather commonly introduced farther n). *C. elegans* H.B.K.—Rydberg.

20. CINNA L., Woodreed

Rhizomatous or nonrhizomatous perennials. Culms rather slender, often geniculate at the base, hollow. Blades flat, lax, rolled in the bud; sheaths open to the base; ligules membranaceous, brownish; auricles lacking. Inflorescence a condensed to relatively open panicle; spikelets 1-flowered, disarticulating below the glumes, the floret stipitate, the rachilla prolonged as a minute bristle; glumes keeled, 1- or 3-nerved, the second somewhat longer than the first; lemmas keeled, shorter than the second glume, 3-nerved, usually bearing a short, straight awn just below the apex, palea 1-nerved, shorter than the lemma.

1 Anthers more than 1 mm long at maturity; second glume usually 3-nerved and more than 3.7 mm long; panicle rather dense, the branches ascending .... 1. *C. arundinacea*
1 Anthers less than 1 mm long at maturity; second glume 1-nerved, usually less than 3.7 mm long; panicle often open (rarely condensed), its branches spreading .... 2. *C. latifolia*

1. **Cinna arundinacea** L., woodreed. Nonrhizomatous to weakly rhizomatous perennials 5.8–13(17) dm tall. Culms erect or somewhat decumbent below, often bulbous-based. Blades

flat, rather lax, glabrous to scabrous, 9-34 cm long, 5.9-12(14) mm wide; sheaths glabrous; ligule prominent, thin and lacerate, 2.5-11 mm long. Panicle dense, 10-25(30) cm long, the base often partly included in the sheath of the uppermost leaf, the branches ascending; glumes lanceolate, sometimes slightly aristate, the first 1-nerved and 3.2-4.2 mm long, the second 3-nerved and 3.8-5.4 mm long; lemma 3.5-4.5 mm long, often bearing a straight awn 0.2-1.5 mm long 0.2-0.9 mm back of the tip. Anthers 1.1-1.7 mm long. (2n = 28) Jul-Sep. Moist woods, ditches, stream edges, & riverbanks; nw MT, n & e ND, e SD, MN, NE, IA, KS, MO, & OK; (ME to MT, s to GA & e TX).

2. *Cinna latifolia* (Trev. ex Goepp.) Griseb., drooping woodreed. Rather weakly rhizomatous perennials 6-13 dm tall. Culms slender, sometimes decumbent below, not bulbous-based. Blades flat, lax, scabrous or glabrous, mostly 10-23 cm long and (4)6-13(16) mm wide; ligules prominent, thin and lacerate, 2-7 mm long. Panicles 12-30(37) cm long, open, with branches spreading, or sometimes rather condensed, the branches ascending, the base sometimes included in the sheath of the uppermost leaf; glumes lanceolate, sometimes slightly aristate, both 1-nerved, the first 1.9-3(3.4) mm long, the second 2.3-3.5(3.8) mm long; lemma 2-3.5 mm long, often with a straight awn 0.3-1.5 mm long less than 0.5 mm back of the tip. Anthers 0.6-0.8 mm long. (2n = 28) Jul-Aug. Wet woods; MN, ND, sw SD; (Newf. to AK, s to CT, NC, SD, AZ, & CA).

21. CYNODON Rich.

1. *Cynodon dactylon* (L.) Pers., Bermuda grass. Rhizomatous, stoloniferous perennials 1.4-4.5(6.5) dm tall. Culms decumbent below, usually hollow. Blades rolled in the bud, flat to loosely involute or folded at maturity, glabrous except near the ligule to, infrequently, long-pilose, 1-10(15) cm long, 1-3.5 mm wide; sheaths open, slightly keeled, glabrous except for long-pilose regions near the ligules; ligule a short, strongly ciliate membrane 0.1-0.4 mm long, usually with numerous long hairs behind and on either side of it; auricles lacking. Inflorescence of (3)4 or 5 digitately arranged spicate branches 2-6 cm long, these bearing spikelets closely imbricated in 2 rows on one side of the rachis; spikelet containing 1 perfect floret, the rachilla prolonged behind it and often bearing a tiny rudimentary second floret, disarticulating above the glumes; glumes 1-nerved, subequal, lanceolate, the first curved, 1-2 mm long, the second nearly straight, 1-2.1 mm long; lemmas strongly flattened laterally, boat-shaped, 3-nerved, the lateral nerves near the margins, the nerves pubescent or scabrous, 1.7-2.8 mm long; prolonged rachilla 0.8-1.2 mm long; rudimentary floret, when present, up to 0.4 mm long. Anthers 0.9-1.3 mm long. (2n = 18, 36, with counts of 2n = 27, 54, 30, 40, and n = 14 also reported) Jun-Oct. Pastures, lawns, waste ground, railroads, & stream banks; MO, KS, OK, TX, also reported from NE: Dawson; (MD to KS, s to FL & TX, w to CA, originally from trop. & subtrop. Africa). *Naturalized.*

This grass has become a common lawn and pasture grass in the s GP. *C. transvaalensis* Davy, a more delicate species with fewer inflorescence branches, shorter glumes, and narrower leaves has been introduced as a lawn grass at NE: Lancaster and has persisted there as a weed for at least a decade.

22. DACTYLIS L.

1. *Dactylis glomerata* L., orchard grass. Clumped, nonrhizomatous perennials (3.5)6-10(11) dm tall. Culms erect, glabrous, hollow, often slightly compressed. Blades soft, folded in the bud, usually flat (to folded) at maturity, scabrous to glabrous, the midvein prominent beneath, mostly (3.5)10-43 cm long, 3-9 mm wide; sheaths closed toward the base with the margins overlapping and fused, glabrous, keeled, laterally compressed; ligules mem-

branaceous, lacerate at maturity, 2.5–13 mm long; auricles none. Inflorescence a panicle with relatively few major branches, the spikelets clustered in 1-sided groups on short pedicels; spikelets compressed, 2- to 4(6)-flowered, disarticulating above the glumes and between the florets; glumes varying in texture and size depending partly upon the position in the spikelet cluster, 2.5–7 mm long, approximately equal or markedly unequal (either one may be reduced), 1- or 2(3)-nerved, often markedly asymmetrical, acute to acuminate or short awned, glabrous to scabrous or hispid, especially on the main nerve; lemmas rather faintly 5-nerved, acuminate to short-awned, scabrous to pubescent, the lowest one 4.5–7(8) mm long, with an awn up to 1.3(1.8) mm long; paleas shorter than the lemmas. Anthers 1.8–3.7 mm long. (most counts 2n = 14, 28, but 2n = 42 and various aneuploid counts also reported) May–Oct. In a variety of habitats such as pastures, disturbed woodland, & waste places; GP, apparently more common in e; (widely introduced in U.S. as a forage grass; Europe & Asia). *Naturalized.*

23. DANTHONIA Lam. & DC., Oatgrass

1. *Danthonia spicata* (L.) Beauv. ex R. & S., poverty oatgrass. Cespitose perennials 2–6 dm tall. Culms smooth, erect, hollow. Blades rolled in the bud, involute at maturity, or flat, but then often with involute margins, pilose beneath or on both surfaces or, rarely, glabrous, 1–10(17) cm long, 0.5–2.2(3) mm wide; sheaths open, glabrous to pilose, usually densely long-pilose near the ligule; ligule a short fringe of hairs 0.1–0.4 mm long; auricles lacking. Inflorescence a sparingly branched panicle or a raceme 2–6.5 cm long; spikelets 4- to 9-flowered, disarticulating above the glumes and between the florets; glumes glabrous, lanceolate, 3(5)-nerved, longer than the florets, the first 8.5–12 mm long, the second 7.5–11 mm long; lemmas pilose over the back, broadly ovate, the lowest 3–5.5 mm long, membranaceous toward the tip, obscurely 5-nerved, the nerves anastomosing above, strongly 2-toothed, the teeth up to 2 mm long and acute to acuminate, a strong awn about 5–8 mm long arising from between the teeth, this awn with a flattened brownish base which is coiled or twisted when dry and a pale, straight terminal portion (2.5)4–6 mm long. Anthers often abortive and less than 0.2 mm long, but sometimes developed, 1.4–2.5 mm long. Small ovate to pyriform cleistogamous florets sometimes present in the axils of basal leaves. (2n = 36) May–Jul. Rocky slopes & open or dry woods; entering our range but uncommon in n MN, nw ND, MT, IA, sw SD, WY, MO, se KS, & OK; (Newf. to B.C., s to most of U.S.). *D. thermale* Scribn. — Rydberg.

The presence of relatively few well-developed stamens has led many authors to suggest the probability of widespread apomixis in *Danthonia*. If this is true, then the combination of some taxa presently recognized on apparently minor characteristics might well be in order. Findlay and Baum (Canad. J. Bot. 52: 1573–1582. 1974) have provided a key to Canadian taxa based entirely upon nontraditional characters of the lodicules and lemma setae and have named a new species, *D. canadensis*, which may well extend to within or near our range. I have chosen to retain the traditional circumscription of *D. spicata*, since it seems evident that study of the genus south of Canada is badly needed and the characters emphasized by Findlay and Baum appear to correlate poorly with other features in our material.

Danthonia intermedia Vasey enters our range in the BH of SD. It may be distinguished by its lemmas, which are longer than those of *D. spicata* and pilose only on the margins and the callus, and by its longer glumes. *D. unispicata* (Thurb.) Macoun has been reported from WY: Crook & Weston. It is a low-growing species (under 3 dm tall) with usually a single spikelet only, this much larger than the spikelets of *D. spicata*

24. DESCHAMPSIA Beauv., Hairgrass

1. *Deschampsia cespitosa* (L.) Beauv., tufted hairgrass. Strongly cespitose perennials 2.5–10.5 dm tall. Culms erect, hollow, smooth. Leaves mostly basal; blades folded in the bud, folded to involute or, rarely, flat at maturity, with a relatively small number of very

prominent narrow ridges on the scabrous upper surface, scabrous to glabrous beneath, 2–33 cm long, 1–3 mm wide; sheaths glabrous or scabrous, slightly keeled, open; ligules membranaceous, acute, 2–8 mm long, often rather clearly nerved; auricles none. Inflorescence an open panicle 8–22(28) cm long, its branches often spreading or drooping; spikelets 2(3)-flowered, disarticulating above the glumes; glumes greenish to tawny or purplish, both acute, the tips sometimes jagged, the first 1-nerved, 2–4 mm long, the second 3-nerved, 2.3–4.5 mm long; lemmas membranaceous, the lower one 2–3.5 mm long, very faintly 5-nerved or apparently nerveless, villous on the base, truncate, erose at the tip, bearing an awn 0.7–3.5 mm long attached about 0.2–0.8 mm from the base; rachilla prolonged beyond the attachment of the upper floret 0.7–1.8 mm. Anthers 1–1.6 mm long. (2n = 26, 2n = 28, 52, 27, 28, 24 also reported) Jun–Aug. Wet meadows & prairies, ditches, marshes, & boggy areas; MN, ND, MT, BH, WY, CO, NM; (Greenl. to AK, s to GA, SD, AZ, CA; Eurasia, also Mex.).

Our plants belong to var. *cespitosa*. The specific epithet is often spelled "caespitosa" but "cespitosa" is the original spelling and may not be changed.

25. DIARRHENA Beauv., Beakgrain

1. ***Diarrhena americana*** Beauv. var. ***obovata*** Gl., American beakgrain. Rhizomatous perennials 4–12 dm tall. Culms glabrous, hollow, erect, covered by overlapping sheaths. Leaves distributed mostly toward the base of the plant; blades rolled in the bud, flat at maturity, glabrous to scabrous or short-hirsute, mostly (10)25–60 cm long, 7–17 mm wide; sheaths glabrous to short-hirsute, open; ligule a truncate, ciliolate membrane 0.2–0.6 mm long; auricles lacking. Inflorescence a slender erect or arching panicle (3.5)7–16(22) cm long, its branches mostly erect; spikelets 2- to 5-flowered, the uppermost floret usually rudimentary, disarticulating above the glumes and between the florets; glumes acute, leathery, unequal, the lower 1.5–3 mm long and 1- to 3-nerved, the upper 2–4.5 mm long and (3)5-nerved; lemmas 4.5–7.5 mm long, leathery at the center, membranaceous at the nerveless margins, with 3 convergent nerves terminating in a cusp; paleas shorter, similar in texture to the lemmas, strongly 2-nerved. Stamens 2 or 3; anthers 1.4–2.5 mm long; grain 4–7 mm long, pyriform, the beak protruding strongly at maturity. Jun–Oct. Rich woods, often near water; se SD, IA, e NE, e KS, MO, e OK; (VA to SD, s to TN & e TX). *D. arundinacea* (Zea ex Lag.) Rydb. — Rydberg.

26. DICHANTHELIUM (Hitchc. & Chase) Gould

Cespitose, nonstoloniferous, usually nonrhizomatous, perennials. Early culms unbranched or sparingly branched, usually becoming more branched later in the season, erect or spreading to somewhat decumbent, hollow. Blades rolled in the bud, flat at maturity, those of the main culm often the largest and also relatively broad, those of the later branches often markedly shorter and narrower, usually a winter rosette of leaves with short, broad blades forming late in the season; sheaths open; ligule a fringe of hairs, a fringed membrane, or absent; auricles none. Early inflorescences usually pyramidal panicles terminating the main culms, later inflorescences usually considerably reduced and often partly included in the leaf sheaths; spikelets ovate to obovate, seldom beaked, disarticulating below the glumes, bearing 2 florets; glumes unequal, the first one greatly reduced in most species, nerveless or 1- to 7-nerved, the second one normally nearly as long as the spikelet, mostly 5- to 9-nerved; lower floret usually neuter, rarely staminate, the lower lemma similar to the second glume in size, texture, and nervation, the lower palea membranaceous (rarely absent); upper floret perfect, near the second glume and the lower lemma in size, but smooth, shiny, and indurate, the lemma closely clasping the palea. Anthers variable in size, the largest ones often forming in the earliest flowers and exserted, the smallest ones forming in the later flowers and retained.

References: Gould, F. W., & C. A. Clark. 1978. *Dichanthelium* (Poaceae) in the United States and Canada. Ann. Missouri Bot. Gard. 65: 1088–1132; Hitchcock, A. S., & A. Chase. 1910. The North American species of *Panicum*. Contr. U. S. Natl. Herb. 15: 1–396.

The contention that *Dichanthelium* merits recognition as a genus distinct from *Panicum* appears to have a sound morphological and physiological basis. See the rationale in Gould and Clark (op. cit.). The treatment below follows closely that of Gould and Clark.

1 Main culm leaves rather prominently basally distributed, long and narrow, mostly 5–22 cm long and not more than 4 mm wide; plants not forming a winter rosette of short, broad leaves .... 5. *D. linearifolium*
1 Main culm leaves distributed up the culm, variable, but either some of them clearly more than 4 mm wide or the plant possessing remnants of a winter rosette of short, broad leaves.
  2 Spikelets 2.5 mm or more in length.
    3 Blades broad, those of the main culm mostly 1.3–3.5 cm wide.
      4 Lower floret normally staminate; first glume usually more than 1/2 the length of the spikelet .... 4. *D. leibergii*
      4 Lower floret neuter; first glume usually less than 1/2 the length of the spikelet.
        5 Sheaths hispid with pustular-based hairs .... 2. *D. clandestinum*
        5 Sheaths glabrous to softly pubescent .... 3. *D. latifolium*
    3 Blades narrower, under 1.3 cm wide.
      6 Ligules very short, primarily membranaceous, 0.1–0.4 mm long; first glume more than 1.5 mm long; lower spikelet normally staminate .... 4. *D. leibergii*
      6 Ligules longer, principally made up of a fringe of hairs; first glume usually shorter; lower spikelet neuter.
        7 Culm nodes prominently retrorsely bearded .... 6. *D. malacophyllum*
        7 Culm nodes glabrous to pubescent but not prominently retrorsely bearded.
          8 Blades long-pubescent on both surfaces, less than 5 mm wide; spikelets 2.4–3.3 mm long .... 10. *D. wilcoxianum*
          8 Blades less strongly pubescent, many of them more than 5 mm wide (at least the main culm leaves); spikelets 2.8–3.8 mm long .... 7. *D. oligosanthes* var. *scribnerianum*
  2 Spikelets less than 2.5 mm in length.
    9 Culm nodes prominently retrorsely bearded.
      10 Culms with a broad glabrous or glandular band below each node; spikelets 2–2.5 mm long .... 8. *D. scoparium*
      10 Culms not with a glabrous or glandular band below each node; spikelets 2.3–3.3 mm long .... 6. *D. malacophyllum*
    9 Culm nodes glabrous or pubescent but not retrorsely bearded.
      11 Ligule absent or consisting of a few weak hairs .... 9. *D. sphaerocarpon*
      11 Ligule present and conspicuous.
        12 Ligules 0.3–1.6 mm long; spikelets 2.4–3.2 mm long .... 10. *D. wilcoxianum*
        12 Ligules very conspicuous, more than 1.6 mm long on at least some of the leaves, or if, as rarely, as short as 1 mm, then spikelets well under 2.4 mm long .... 1. *D. acuminatum*

1. **Dichanthelium acuminatum** (Sw.) Gould & Clark. Tufted perennials 0.5–7.5 dm tall. Culms at first simple, often becoming much branched in age, glabrous to pubescent. Blades variously pubescent, those of the main culm mostly 4–10 cm long, 3–9 mm wide, but those of later branches shorter, narrower, and more crowded, and those of the well-developed winter rosette shorter but relatively broad; sheaths glabrous to variously pubescent; ligule a prominent fringe of hairs (1.5)2–6 mm long. Early panicles pyramidal, 3–9 cm long, later ones reduced, often partly enclosed in the leaf sheaths; spikelets ovate to elliptic, usually pubescent, 1.2–2.4 mm long, the first glume 0.3–1.1 mm long, the second glume and sterile

lemma just slightly shorter than the spikelet; lower palea inconspicuous, 0.3–1.2 mm long; upper floret 1–2 mm long, 0.6–1.2 mm wide. Anthers 0.2–0.8 mm long, the shorter ones usually retained within the spikelet, the longer ones associated with chasmogamous florets. (2n = 18) mostly May–Jun, continuing to produce secondary panicles until fall. Gould and Clark (op. cit.) recognize the following 4 varieties for this region:

1 Culms and sheaths glabrous or nearly so; spikelets mostly 1.3–1.6 mm long in our plants .... 1c. var. *lindheimeri*
1 Clums and sheaths pubescent to puberulent; spikelets variable in size.
  2 Hairs of the sheath soft and spreading, 2–5 mm long; spikelets mostly 1.7–2.2 mm long .... 1d. var. *villosum*
  2 Hairs of the sheath rather stiff, not over 2 mm long; spikelets variable in size.
    3 Spikelets 1.2–1.6 mm long .... 1b. var. *implicatum*
    3 Spikelets (1.4)1.6–2.4 mm long .... 1a. var. *acuminatum*

1a. var. *acuminatum*. Culms usually pubescent. Blades and sheaths variously pubescent with ascending to spreading hairs or, rarely, almost glabrous; blades of main culms mostly 2–13 cm long, 3.5–8.5 mm wide; ligules 2–5.5 mm long. Primary inflorescence 3–9 cm long; spikelets 1.4–2.4 mm long, pubescent; first glume about 0.4–0.6 mm long; upper lemma about 1.3–2 mm long, 0.7–1.1 mm wide. In a wide variety of open & woodland habitats; GP, more common in e part; (Que. to B. C., throughout U.S. to e Mex., C. Amer., n S. Amer., the Antilles). *Panicum lanuginosum* Ell.—Hitchc. & Chase (op. cit.); *P. lanuginosum* var. *fasciculatum* (Torr.) Fern.—Fernald; *P. subvillosum* Ashe, *P. huachucae* Ashe *P. tennesseense* Ashe, *P. scoparioides* Ashe—Rydberg.

1b. var. *implicatum* (Scribn.) Gould & Clark. Culms usually pubescent. Blades and sheaths variously pubescent to nearly glabrous; blades of main culms mostly 2–8 cm long, 2.5–8 mm wide; ligules 1.5–4.5 mm long. Primary panicles 3–8 cm long; spikelets 1.2–1.5(1.6) mm long, pubescent, elliptic; first glume about 0.3–0.6 mm long; upper lemma about 1–1.5 mm long, 0.6–0.9 mm wide. In a variety of habitats in woods & in the open; MN, ND, IA, e NE, MO, e KS, e OK; (N. S. to ND, s to FL, TX, Cuba). *Panicum albemarlense* Ashe—Rydberg; *P. lanuginosum* Ell. var. *implicatum* (Scribn.) Fern.—Fernald.

This variety is separated rather arbitrarily from var. *acuminatum* on the basis of spikelet size (see Gould & Clark, op. cit.). Most GP specimens of this were mapped as *Panicum lanuginosum* var. *fasciculatum* in the Atlas GP.

1c. var. *lindheimeri* (Nash) Gould & Clark. Culms glabrous or very sparingly pubescent. Sheaths and blades essentially glabrous except sometimes on the margins; blades of the main culm leaves mostly (2)3–8 cm long, 2–7 mm wide, those of the branches markedly smaller, on the average; ligules 2.5–4.5 mm long. Primary panicles 3–6 cm long; spikelets mostly 1.3–1.6 mm long, usually puberulent; first glume about 0.3–0.6 mm long; upper lemma about 1–1.5 mm long, usually puberulent; first glume about 0.3–0.6 mm long; upper lemma about 1–1.5 mm long, 0.5–0.9 mm wide. In a variety of habitats; MO, e KS, e OK, TX; (s N.B. to Man. s to FL & TX, ne Mex. scattered elsewhere). *Panicum lindheimeri* Nash—Rydberg; *P. lanuginosum* Ell. var. *lindheimeri* (Nash) Fern.—Fernald.

1d. var. *villosum* (A. Gray) Gould & Clark. Culms pubescent. Blades pilose, the principal culm blades 2–8 cm long, 3–8 mm wide; sheaths pubescent with soft hairs 2–5 mm long; ligules 2–5 mm long. Primary panicles 2–6 cm long; spikelets in ours mostly 1.7–2.2 mm long, reportedly sometimes longer, pubescent; first glume about 0.4–1.1 mm long; fertile lemma about 1.4–1.7 mm long, 0.8–1.2 mm wide. In various habitats, in our area principally in prairie; e GP; (CT & NY to ND, s to FL & TX, also CA). *Panicum praecocius* Hitchc. & Chase, *P. villosissimum* Nash—Rydberg.

2. Dichanthelium clandestinum (L.) Gould, deertongue dichanthelium. Stout perennials 3.5–10(14) dm tall, cespitose to slightly rhizomatous, or the bases of the culms merely decumbent and rooting at the nodes. Culm internodes glabrous to scabrous or papillose-puberulent. Blades glabrous to scabrous, very broad, basally cordate, those of the main culms 12–20 cm long, (1.3)1.8–3.2 cm wide, those of the later side branches generally somewhat smaller; sheaths, at least many of them, papillose-hispid; ligule a short ridge or membrane 0.2–0.6 mm long. Terminal panicles pyramidal, mostly 5–10 cm long; spikelets 2.6–3.6 mm long, elliptic, finely puberulent; the first glume 0.7–1.8 mm long, the second glume and lower lemma nearly as long as the spikelet; lower palea prominent, 1.3–2 mm long; lower floret neuter; upper floret about 2.3-3 mm long, 1.2-1.7 mm wide. Anthers 0.3-1.4 mm long, the longer ones associated with chasmogamous spikelets, the shorter ones usually retained within the spikelets. (n = 18) Mostly Jun, the secondary panicles continuing to bloom until fall. Woodlands, brushy areas, & prairies, often in moist, sandy or rocky soil; MO, se KS, e OK (se Can. & ME to IA, s to FL & TX). *Panicum clandestinum* L. — Rydberg.

This species is closely related to *D. latifolium*, which see.

3. **Dichanthelium latifolium** (L.) Gould. Stout perennials 3.5-9(13) dm tall from a knotty base. Culms glabrous to puberulent. Blades basally cordate, glabrous to scabrous or more or less thinly pilose, those of the main culm 2-15 cm long, (0.5)1.8-3.5 cm wide; sheaths nearly glabrous to variably soft-pubescent but never papillose-hispid; ligule a short fringed membrane 0.2-0.8 mm long. Terminal panicles pyramidal, mostly 6-11 cm long; spikelets 2.8-3.7(3.9) mm long, elliptic, finely pubescent, the first glume 0.9-1.8 mm long, the second glume and lower lemma nearly as long as the spikelet; lower palea prominent, 1.5-2.2 mm long; lower floret neuter; upper floret about 2.3-3.1 mm long, 1.2-1.6 mm wide. Anthers 0.4-1.6 mm long, the longer ones associated with chasmogamous spikelets, the shorter ones retained within the spikelets. Mostly May-Jun, continuing to bloom sparingly from lateral panicles until fall. Wooded areas, often in rocky or sandy soil; IA, se NE, MO, e KS; (e Can. & ME to MI, NE, s to GA & AR). *Panicum latifolium* L. — Rydberg.

This species is closely related to *D. clandestinum*, from which it differs principally in the absence of hispid pubescence on the sheaths. It also tends to have somewhat shorter leaves and slightly larger spikelets.

Dichanthelium boscii (Poir.) Gould & Clark has been reported from the edge of our range (KS: Montgomery, Cherokee; MO: Jackson, Jasper; OK: Ottawa). It may key here to *D. latifolium*, but it differs in having most spikelets more than 4 mm long and in its bearded culm nodes.

4. **Dichanthelium leibergii** (Vasey) Freckmann, Leiberg dichanthelium. Tufted perennials 2-8 dm tall. Culms erect to geniculate below, pilose to scabrous, branching mostly from the base in the autumnal phase and hence not becoming bushy-branched. Blades usually papillose-puberulent, mostly 3-13 cm long, 4-12(21) mm wide on the principal culms, not much reduced on the later branches; sheaths papillose-pubescent; ligule a short ciliated membrane 0.2-0.4 mm long or sometimes nearly lacking. Primary panicles 4-9(13) cm long, pyramidal, those of the later branches reduced and slender; spikelets 3.1-4.1(4.3) mm long, papillose-pubescent; first glume often relatively long, 1.5-2.5 mm; second glume and lower lemma about equal, 2.5-3.6 mm long; lower palea prominent, 2.2-2.9 mm long; the lower floret usually staminate; upper floret 2.4-3.4 mm long, 1.3-2 mm wide. Anthers of lower floret 0.9-1.8 mm long, those of the upper floret 0.6-1.3 mm long, the spikelets evidently chasmogamous. May-Jun, producing secondary inflorescences until early fall. Prairies & open woods; MN, e ND, IA, SD, e NE, MO, e KS; (Ont. to Sask., s to PA & KS). *Panicum leibergii* (Vasey) Scribn. — Fernald.

This species is more clearly marked than most. Judging from the material I have seen, its winter rosette is only weakly developed. The usual presence of a staminate lower floret and the evident chasmogamy of even the later inflorescences are both unique features in the dichantheliums of our region.

Specimens identified as *Panicum xanthophysum* A. Gray (*Dichanthelium xanthophysum* (A. Gray) Freckmann) from the BH of SD are perhaps best considered to be atypical material of *D. leibergii*. They match it in all particulars except that the leaf surfaces are nearly glabrous, and they do not closely resemble specimens of *D. xanthophysum* from farther east.

5. **Dichanthelium linearifolium** (Scribn.) Gould, slimleaf dichanthelium. Tufted perennials 1-6 dm tall. Culms bearing most of the leaves and, later, the branches near the base, glabrous to puberulent. Blades varying from more or less pilose to nearly glabrous, those

of the main culm leaves mostly 5–21 cm long and 1–4 mm wide, the plants not forming a winter rosette with short, broad blades; sheaths usually pilose; ligule a fringe of hairs 0.4–1.4 mm long. Inflorescences slender to pyramidal, the primary ones exserted, 3–8 cm long, the secondary ones reduced and forming near the base of the plant; spikelets elliptic, puberulent to glabrous, 2.3–3.5(3.8) mm long; first glume 0.6–1.9 mm long; second glume and lower lemma 2–3.1 mm long; lower palea 0.9–1.7 mm long; fertile lemma about equal to the sterile lemma in length, 1.1–1.7 mm wide. Anthers 0.3–1.5 mm long, the longer ones associated with chasmogamous spikelets, the shorter ones usually retained within the spikelets. Mostly May–Jun, with secondary panicles produced until fall. Prairies & open woods; MN, ND, SD, IA, e & cen NE, MO, e KS, e OK; (Que. to Ont. & ND, s to FL, TX, & NM). *Panicum linearifolium* Scribn., *P. perlongum* Nash, *P. werneri* Scribn. — Rydberg.

 The related *D. depauperatum* (Muhl.) Gould has the spikelet somewhat beaked and the upper floret clearly shorter than the second glume and lower lemma. It also has spikelets which are, on the average, larger. Although many GP specimens approach *D. depauperatum*, either in spikelet size or shape, few of them may be assigned there with certainty.

6. Dichanthelium malacophyllum (Nash) Gould. Perennials mostly (2)3–6 dm tall. Culms pilose, strongly bearded at the nodes, becoming branched above in late-season plants. blades of primary culms pubescent, mostly 3–8 cm long, 4.5–11.5 mm wide, averaging smaller on the later branches; sheaths pilose; ligule a rather prominent fringe of hairs 0.6–1.5 mm long. Primary panicles pyramidal, 3–9 cm long, the later ones axillary and reduced; spikelets elliptic, pilose, 2.3–3.0(3.3) mm long; second glume and lower lemma subequal, only slightly shorter than the spikelet; lower palea 1–1.8 mm long, 1.1–1.6 mm wide. Anthers 0.2–0.9 mm long, the smaller ones usually retained within the spikelets, the larger ones associated with the early chasmogamous spikelets. Mostly May–Jun, the lateral panicles blooming until fall. Open woods; MO, e KS, e OK; (MO & KS, s to TN, AR, & TX). *Panicum malacophyllum* Nash — Fernald.

7. Dichanthelium oligosanthes (Schult.) Gould var. *scribnerianum* (Nash) Gould, Scribner dichanthelium. Tufted perennials 1–7 dm tall. Culms glabrous to scabrous or pilose with ascending to spreading hairs, simple at first but normally becoming branched above later in the season. Blades of primary culms glabrous or with very limited pubescence, mostly 3–10 cm long, 3–12(13) mm wide, those of the later branches shorter, narrower, and crowded, those of the winter rosette short and relatively broad; sheaths glabrous to long-pilose; ligule a short fringe of hairs 0.6–1.5 mm long. Primary panicles pyramidal, 3.5–8 cm long, the later ones greatly reduced; spikelets glabrous to variously pubescent, 2.8–3.8 mm long; first glume 0.8–1.6 mm long; second glume and lower lemma subequal, 2.2–3.3 mm long; lower palea well developed, 1.2–2 mm long; lower floret neuter; upper floret 2–3 mm long, 1.3–1.9 mm wide. Anthers 0.2–1.1 mm long, the longer ones tending to be characteristic of the early chasmogamous spikelets, the smaller ones usually retained within the spikelets. (2n = 18) Apr–Jun, the secondary panicles blooming until fall. Open prairie, disturbed ground, sometimes in woods; GP; (Ont. to B. C., throughout U.S., n Mex.). *Panicum scribnerianum* Nash — Rydberg; *P. oligosanthes* Schult. var. *scribnerianum* (Nash) Fern. — Fernald.

 D. oligosanthes var. *oligosanthes* has also been reported from within the s part of our range, but I have seen no material (see Gould & Clark, op. cit.) It may readily be distinguished from var. *scribnerianum* by its longer ligule (more than 1.6 mm long) and its tomentose lower leaf surface.

8. Dichanthelium scoparium (Lam.) Gould, velvet dichanthelium. Stout perennials 4–13 dm tall. Culms simple at first but becoming thick and much branched in late-season plants, internodes mostly velvety-pubescent except for a glabrous or glandular band just below each node, nodes retrorsely bearded. Blades usually densely soft-pubescent or velvety, those of the primary culm mostly 5–20 cm long and 6–15(19) mm wide, those of the later branches mostly less than 5 cm long and only 2–6 mm wide, crowded, those of the winter rosette mostly less than 5 cm long and relatively broad; sheaths usually soft-pubescent; ligule a tuft of hairs 0.6–1.8 mm long. Primary panicles pyramidal, 6–12 cm long, the secondary panicles greatly reduced; spikelets 2–2.5 mm long, ovate, usually puberulent; first glume variable, 0.3–1.5 mm long; second glume and lower lemma subequal, about equal to the spikelet in length; lower palea about 0.5–1 mm long; lower floret neuter; fertile lemma about 1.8–2.2 mm long, 1.1–1.5 mm wide. Anthers 0.2–0.9 mm long, the longer ones normally associated with the early chasmogamous spikelets, the shorter ones often retained within the spikelets. (n = 9) Mostly Jun–Jul, but continuing to bloom on secondary panicles until autumn. Moist areas, usually in woodlands, relatively uncommon; MO, extreme e KS, e OK; (MA to KS, s to FL, TX, & Mex; Antilles). *Panicum scoparium* Lam.— Rydberg.

9. ***Dichanthelium sphaerocarpon*** (Ell.) Gould. Perennials with culms 2–5 (reportedly to 8) dm tall, simple at first, but branching in late-season plants. Blades nearly glabrous except on the margins near the base, somewhat cordate below, those of the primary culms mostly 3–10 cm long, 5–12 mm wide, those of the later branches smaller, on the average, and more crowded, those of the winter rosette short and relatively broad; sheath nearly glabrous to, often, ciliate marginally; ligule absent or consisting of a few very short hairs. Primary panicles pyramidal, 3–11 cm long, the later ones much reduced; spikelets subcircular to obovoid, 1.4–2.2 mm long, pubescent to glabrous, usually purplish; first glume 0.3–0.7 mm long, second glume and lower lemma just slightly shorter than the spikelet; lower palea 0.4–1.0 mm long; lower floret neuter; upper floret mostly 1.4–1.8 mm long, 0.8–1 mm wide. Anthers 0.2–0.9 mm long, the longer ones associated with the early, chasmogamous spikelets, the smaller ones often retained. (2n = 18) May–Jun, continuing to bloom from secondary panicles until fall. Woodlands, fields, & disturbed ground; MO, e KS, e OK; (ME to MI, KS, s to FL & TX). *Panicum sphaerocarpon* Ell.—Rydberg.

10. ***Dichanthelium wilcoxianum*** (Vasey) Freckmann, Wilcox dichanthelium. Tufted perennials 1.5–2.4(3) dm tall. Culms pubescent, branching mainly near the base, not becoming much branched above. Blades hirsute on both surfaces, narrow, those of the main culm mostly 3–10 cm long, 2–4(5) mm wide, those of the winter rosette much shorter but similar in width; sheaths hirsute; ligule a short fringe of hairs 0.4–1.6 mm long. Panicles, even the primary ones, often not much exserted beyond the leaf clusters, the early ones mostly 2–5 cm long; spikelets pubescent, elliptic, 2.4–3.3 mm long; first glume 0.6–1.4 mm long; second glume and lowest lemma subequal, 2–2.8 mm long; lower palea 0.8–1.4 mm long; lower floret neuter; fertile floret about 2.1–2.6 mm long, 1–1.4 mm wide. Anthers 0.2–0.9 mm long, the longest ones usually associated with the early, chasmogamous spikelets, the smaller ones often retained. Mostly May–Jun, continuing to bloom from the secondary panicles until fall. Well-drained prairies & pastures; MN, ND, SD, IA, NE, e & cen KS; (Man. to Alta., s to IL, KS, CO, & NM). *Panicum wilcoxianum* Vasey—Rydberg; *Dichanthelium oligosanthes* (Schult.) Gould var. *wilcoxianum* (Vasey) Gould & Clark—Gould & Clark, op. cit.

Although this species overlaps *D. oligosanthes* var. *scribnerianum* in most measurements, it can be separated on a combination of features. The habit and blade pubescence are different enough so that there is almost no confusion of the taxa in the field or herbarium. Indeed, *D. wilcoxianum* is more likely to be confused with *D. linearifolium,* if its rosette leaves are not discerned, than it is with *D. oligosanthes.*

27. DIGITARIA Heist. ex Fabr., Crabgrass

Tufted annuals. Culms hollow, relatively slender, stiffly erect to, more commonly, decumbent and rooting at the lower nodes. Blades rolled in the bud, flat at maturity; sheaths open; ligules membranaceous; auricles lacking. Inflorescence a panicle of spikelike branches, these digitate or disposed along a short axis, each branch with spikelets arranged in groups of 2 or 3, their pedicels of different lengths and appressed, the groups borne in 2 rows on one side of a flattened or triangular straight to sinuate rachis; spikelets dorsally compressed, 2-flowered, but with only the upper floret functional, disarticulating below the glume or glumes; first glume obsolete or very short; second glume short to as long as the spikelet, herbaceous, 3-nerved, variously pubescent; lower (sterile) lemma usually as long as the spikelet, like the second glume in texture, 5-nerved; upper (fertile) lemma cartilaginous, smooth, often darkening with age, about equal to and enclosing the palea. *Syntherisma* Walt.—Rydberg.

References: Ebinger, J. E. 1962. Validity of the grass species *Digitaria adscendens*. Brittonia 14: 248–253; Gould, F. W. 1963. Cytotaxonomy of *Digitaria sanguinalis* and *D. adscendens*. Brittonia 15: 241–244; Gould, F. W. 1975. *Digitaria*, pp. 406–417, *in* The grasses of Texas. Texas A. and M. Univ. Press, College Station.

1 Plants stiffly erect, the culms not decumbent and rooting at the lower nodes; rachis of the inflorescence branch not flattened, usually under 0.4 mm wide .... 2. *D. filiformis*
1 Plants decumbent and rooting at the lower nodes; rachis of the inflorescence branch flattened, usually over 0.4 mm wide.
  2 Second glume as long as or nearly as long as the spikelet .... 3. *D. ischaemum*
  2 Second glume not over 80% the length of the spikelet.
    3 Lateral nerves of the sterile lemma not scabrous; rachises of the inflorescence branches less than 0.8 mm wide; upper surfaces of the culm leaves not usually pustular-pilose; second glume more than 1/2 the length of the spikelet .... 1. *D. ciliaris*
    3 Lateral nerves of the sterile lemma scabrous; rachises of the inflorescence branches often 0.8 mm wide or wider; upper surfaces of the culm leaves usually pustular-pilose; second glume 30–60% as long as the spikelet .... 4. *D. sanguinalis*

1. Digitaria ciliaris (Retz.) Koel., southern crabgrass. Annuals 1.5–8.5 dm tall. Culms glabrous, ascending, usually geniculate to decumbent and rooting at the lower nodes; blades, flat, the midrib obvious, usually glabrous or scabrous, sometimes pustular-pilose near the throat, very rarely the blades more generally pubescent, especially those near the base of the plant, 2–12 cm long, 3–6(8.5) mm wide; sheaths keeled, pustular-pilose; ligules thin, membranaceous, irregularly toothed, 1–1.5 mm long. Panicle branches digitate or disposed along a short axis, the longest ones 5–11(14) cm long, the flattened rachis about 0.5–0.7 mm wide; first glume less than 0.6 mm long, second glume 1.5–2.5 mm long, lanceolate, 1/2–3/4 the length of the spikelet, pubescent; sterile lemma acute, about equal to the spikelet, 2.5–3.5 mm long, variously pubescent near the margin with appressed to spreading hairs and, rarely, also with sturdy bristles, but the nerves smooth; fertile lemma 2.5–3.5 mm long, cartilaginous. Anthers 0.5–1.1 mm long. (2n = 54 in ours, n = 9, 18; 2n = 60, 70, 72 reported from elsewhere) Jul–Oct. Weed in waste ground, lawns, gardens, & fields; se NE, KS, OK, TX; (s VA to se NE, s to TX, throughout Mex. to S. Amer., warmer parts of Old World). *Naturalized. D. adscendens* (H.B.K.) Henr.—Ebinger, op. cit.; *Syntherisma marginatum* (Link) Nash—Rydberg, misapplied; *D. sanguinalis* var. *ciliaris* (Retz.) Parl.—Fernald.

Although this species is amply distinct as shown by Ebinger (op. cit.) and Gould (1963), the correct name has been somewhat in doubt (see Howell, Wasmann J. Biol. 29: 101. 1971.). Pollen from our plants has a mean diameter of 35–42 µm, suggesting that they are hexaploid (Gould, 1963).

2. Digitaria filiformis (L.) Koel., slender crabgrass. Slender annuals 2.5–7.5 dm tall. Culms erect, glabrous. Blades flat, glabrous or scabrous to sparsely pustular-pilose, 2–18 cm long and 1–4 mm wide; sheaths keeled, pustular-pilose to, rarely, glabrous; ligules irregularly toothed, 0.3–1.1 mm long. Panicle branches 2–6, filiform, unequal, digitate or disposed along a short axis, the longest one 3–10 cm long, the triangular rachis about 0.2–0.4 mm wide; first glume absent, the second 1–2 mm long, 75–95% the length of the spikelet, subglabrous or pubescent with gland-tipped hairs; sterile lemma about equal to the spikelet, 1.3–2.0 mm long, subglabrous to glandular-puberulent; fertile lemma 1.5–2 mm long, cartilaginous, dark at maturity. Anthers 0.3–0.6 mm long. (2n = 36, 54) Aug–Oct. Sandy soil, often in woods; MO, e KS, e OK; (NH to IA s to FL, TX, & Mex.). *Syntherisma filiforme* (L.) Nash — Rydberg; *D. filiformis* var. *villosa* (Walt.) Fern. — Fernald.

3. **Digitaria ischaemum** (Schreb. ex Schweigg.) Schreb. ex Muhl., smooth crabgrass. Annuals 2–5.5(7) dm tall. Culms slender, ascending but usually geniculate to decumbent and rooting at the lower nodes, glabrous. Blades flat, the midrib absent, glabrous to very sparsely pustular-pilose near the ligule, 2.5–11(15) cm long and 2.5–9 mm wide; sheaths keeled, glabrous; ligules often smooth-margined, 0.6–1.8 mm long. Panicle branches digitate or disposed along a short axis, the longest one 2.5–8.5 cm long, the flattened rachis 0.7–1.1 mm wide; first glume absent or vestigial, second glume 1.5–2.5 mm long, nearly equal the length of the spikelet, subglabrous to rather densely pubescent; sterile lemma similar to the second glume in length and pubescence; fertile lemma smooth, dark at maturity. Anthers mostly 0.4–0.6 mm long. (2n = 36) Aug–Oct. Weed in waste ground or lawns, gardens, & fields; MN, ND, e SD, IA, e & cen NE, MO, KS, e OK; (Que. to GA, w to WA & CA; widely introduced in temp. & subtrop. areas; Europe). *Naturalized. Syntherisma ischaemum* (Schreb.) Nash — Rydberg; *D. ischaemum* var. *mississippiensis* (Gatt.) Fern. — Fernald.

4. **Digitaria sanguinalis** (L.) Scop., hairy crabgrass. Annuals 2–7(11.2) dm tall. Culms ascending, usually geniculate to decumbent and rooting at the lower nodes, glabrous. Blades flat, the midrib obvious, usually pilose with pustular-based hairs on both surfaces, 2–11(14) cm long and 3–8(12) mm wide; sheaths pustular-pilose, keeled; ligules thin, membranaceous, irregularly toothed, 0.5–1.2 mm long. Panicle branches digitate or disposed along a short axis, the longest one 4–14 cm long, the flattened rachis about 0.8–1.2 mm wide; first glume less than 0.5 mm long, the second 1–1.7 mm long, lanceolate, 30–60% the length of the spikelet, pubescent; sterile lemma acute, about equal to the spikelet, 2.2–3.2 mm long, variously pubescent near the margins with appressed to spreading hairs or merely scabrous, but the nerves clearly scabrous in any case; fertile lemmas about 2.1–3 mm long, cartilaginous. Anthers 0.5–0.9 mm long. (2n = 36 in ours, but 2n = 28, 34, 54 also reported) Aug–Oct. Weed in waste ground of lawns, gardens, & fields; most of GP, except extreme nw part; (introduced from Can., throughout U.S. to Mex. & elsewhere in temp. to subtrop. regions; Europe). *Naturalized. Syntherisma sanguinale* (L.) Dulac. — Rydberg.

It may be seen from the above descriptions that *D. sanguinalis* differs from *D. ciliaris* in a large number of characteristics, many of which are, unfortunately, overlapping. Our plants of *D. sanguinalis* appear to be tetraploid, since the pollen grains have a mean diameter of 30–35 μm (Gould, 1963).

28. DISTICHLIS Raf., Saltgrass

1. **Distichlis spicata** (L.) Greene var. **stricta** (Torr.) Beetle, inland saltgrass. Mat-forming, dioecious, strongly rhizomatous perennials 0.8–3.5 dm tall, the pistillate plants on the average shorter than the staminate. Culms stiffly erect, glabrous, solid to hollow. Blades folded to C-shaped in cross section in the bud, involute to flat at maturity, tapered, rather stiffly erect, evenly ridged above, distichous, smooth below and scabrous to pilose above, 2–12 cm long, 1–3.5 mm wide; sheaths not keeled, glabrous to sometimes pilose, especial-

ly above, open; ligule a short fringed membrane 0.1–0.4 mm long, often flanked by long hairs, sometimes with long hairs directly behind it; auricles lacking. Inflorescence a short, contracted panicle 2–7 cm long; spikelets 8–17(22) mm long, disarticulating above the glumes and between the florets, 5- to 13(17)-flowered; glumes and lemmas at first membranaceous with green markings, later becoming stramineous; glumes firm, slightly keeled, acute, the first 1- to 7-nerved, but the lateral nerves often indistinct, (1.5)2.5–4.5(6) mm long, the second 5- to 11-nerved, the lateral nerves sometimes indistinct, (2.3)3–5.5(6.8) mm long; lemmas acute, slightly keeled, 7- to 11-nerved, (3)4–7(8) mm long; paleas about equal to the lemmas, strongly 2-nerved, the nerves serrate. Anthers of staminate plants 2–3(4) mm long. (2n = 40, but 2n = 38, 72 also reported) May–Aug. Moist ground in low prairie, along roads, near creeks & lakes, often in sandy, alkaline, or saline ground; GP; (GP w to Pacific Coast; S. Amer., also Australia, Tasmania, Tahiti). *D. stricta* (Torr.) Rydb.—Rydberg.

Reference: Beetle, A. A. 1955. The grass genus *Distichlis*. Revista Argent. Agron. 22: 86–94.

29. ECHINOCHLOA Beauv.

Tufted or solitary-stemmed annuals. Culms prostrate or spreading to erect, often slightly flattened, hollow to pith-filled, sometimes rooting at the lower nodes. Blades lax, rolled in the bud, flat at maturity, the midribs prominent; sheaths open, compressed, somewhat keeled; ligules absent. Inflorescence a panicle, the primary branches either simple or, usually, with rather crowded short branchlets, the spikelets often crowded and turned to one side; spikelets somewhat dorsally compressed, disarticulating below the glumes, 2-flowered, the upper floret perfect, the lower one neuter; glumes usually pubescent or setose, unequal, the first one short, obtuse to acute or acuminate, bearing 3 converging nerves, the second as long as the spikelet, acuminate or short-awned, usually with 5 converging nerves; lower lemma similar to the glumes, with 3 principal converging nerves and usually with additional shorter nerves near the acuminate or awned apex; lower palea vestigial to well developed and conspicuous, membranaceous; upper (fertile) lemma acuminate, indistinctly nerved, the body firm, smooth, shiny, the tip firm or membranaceous and sometimes finely pubescent; palea of upper floret about equal to the lemma, enclosed by the lemma margins, but becoming free at the tip.

Reference: Gould, F. W., M. A. Ali, & D. E. Fairbrothers. 1972. A revision of *Echinochloa* in the United States. Amer. Midl. Naturalist 87: 36–59.

1 Palea of lower floret not well developed, either absent or vestigial and less than 1 mm long .... 2. *E. crus-pavonis* var. *macera*
1 Palea of lower floret well developed, more than 1.5 m long.
    2 Lemma of fertile floret with a withering, membranaceous tip set off from the body of the lemma by a line of minute hairs clearly visible under 25 × magnification .. 1. *E. crusgalli*
    2 Lemma of the fertile floret with a gradual transition from the body to a firm, stiff tip, the tip not set off from the body by a zone of minute hairs .... 3. *E. muricata*

1. *Echinochloa crusgalli* (L.) Beauv., barnyard grass. Erect to prostrate or ascending annuals usually 3–10(12) dm long. Culms glabrous to, rarely, strigose at the nodes. Blades and sheaths glabrous to, rarely, thinly pilose, often bearing a few pustular-based hairs on the margin in the collar region; blades mostly 8–22(31) cm long, (1.4)6–12 mm wide; sheaths somewhat compressed, keeled. Inflorescence 5–27 cm long, the primary branches appressed or spreading, bearing long setae which may be longer than the spikelets; glumes and the lemma of the sterile floret variably pubescent to nearly glabrous; first glume acute to acuminate, about 1–2 mm long, second glume 2.5–4.5 mm long, acuminate to short-awned, the awn usually not over 1 mm long; lemma of the sterile floret 1–4 mm long, acuminate or with an awn up to 40 mm long; palea of sterile floret about 2–3.5 mm long; fertile lemma 2–4 mm long, acuminate, but the tip membranaceous and withering, sharply differentiated from the body of the lemma and set off by a line of tiny hairs visible under 25 × magnification. Anthers 0.5–0.9 mm long. (2n = 54) Jun–Sep. In a variety of disturbed, usually moist habitats; GP, but apparently less common in nw part; (throughout much of N. Amer., widespread in Europe & Asia). *Naturalized*. *E. occidentalis* (Wieg.) Rydb.—Rydberg.

Echinochloa colonum (L.) Link enters our range in KS: Lyon; in OK: Cleveland & Osage; & in several n TX counties. It will key either to *E. crusgalli* or *E. muricata* here but is readily recognized by its simple inflorescence branches, its absence of long setae in the inflorescence, and by its small spikelets with the fertile lemma averaging less than 3 mm long.

The distinction between *E. muricata* and *E. crusgalli* is a fine one which some workers have been reluctant to accept. In our material, there appears to be a rather clear difference between the taxa, with each character having its own range of variability in each species. Interestingly, our material of *E. crusgalli* approaches *E. muricata* var. *muricata* rather closely in anther size but is more like *E. muricata* var. *microstachya* in other spikelet measurements.

2. **Echinochloa crus-pavonis** (H.B.K.) Schult. var. **macera** (Wieg.) Gould. Annuals 3–10(13) dm long. Culms erect to decumbent at the base, glabrous, hollow. Blades and sheaths nearly glabrous, but sometimes with pustular-based cilia in the collar region; blades usually 5–25(30) cm long, 4–10 mm wide; sheaths slightly keeled, compressed. Inflorescence narrow and strict, the branches ascending, lacking long setae among the spikelets; glumes and sterile lemmas rather evenly pubescent with stiff, ascending, nonpustular hairs; first glume 1–2.5 mm long, acuminate, second glume about 3–4 mm long, acuminate, or bearing an awn up to 1.2 mm long; sterile lemma 2.3–4.2 mm long, acuminate or with an awn up to 2(11) mm long; sterile palea absent or vestigial, under 0.5 mm long when visible; fertile lemma 2.5–4 mm long, with a clearly differentiated, withering, membranaceous tip. Anthers 0.7–1.2 mm long. (2n = 36) Jul–Sep. Moist soil, often near water; s KS, OK, TX, NM; (MS to KS, s to LA & AZ, w to CA, also in n Mex.). *E. zelayensis* (H.B.K.) Schult.—Rydberg, misapplied.

3. **Echinochloa muricata** (Beauv.) Fern. Tufted to single-stemmed annuals 1.5–12(16) dm long. Culms glabrous, pith-filled to hollow, slightly compressed, decumbent to ascending or erect, sometimes rooting at the lowest nodes. Blades mostly 4–35(50) cm long, 2–20 mm wide; sheaths compressed and keeled. Inflorescence 7–22(34) cm long, the primary branches appressed or spreading, the branches glabrous to variously pubescent, but setae, if present, seldom exceeding the spikelets in length; glumes and sterile lemma variable, but usually the second glume and the sterile lemma both bearing pustular-based setae; first glume 1–2.8 mm long, obtuse to acute or acuminate; second glume 2.8–5 mm long, acuminate or bearing an awn up to 2.5 mm long; sterile lemma 2.3–4.5 mm long, acuminate or with an awn up to about 16 mm long; sterile palea about 1.5–3.5 mm long; fertile lemma 2.3–5 mm long, acuminate, with a firm, nonwithering tip which is neither membranaceous, not separated from the lemma body by a line of hairs. Anthers 0.4–1 mm long. (2n = 36)

This species is closely similar to *E. crusgalli*. See the discussion under that species. There are 2 well-marked but somewhat intergradient varieties.

3a. var. *microstachya* Wieg. Spikelets less than 3.5 mm to the base of the awn or to the acuminate tip of the sterile lemma; first glume 0.9–1.6 mm long; second glume 2.8–3.8 mm long, acuminate, or, rarely, with an awn under 1 mm long; sterile lemma about 2.3–3.6 mm long, acuminate or bearing an awn, but the awn not usually more than 3 mm long; sterile palea 1.5–2.5 mm long; fertile lemma 2.4–3.5 mm long. Anthers mostly 0.4–0.7 mm long. Jun–Oct. Waste ground, often where moist; GP; (nearly throughout Can. & U.S. to Mex.). *E. microstachya* (Wieg.) Rydb.—Rydberg.

3b. var. *muricata*. Spikelets 3.5 mm or more to the base of the awn or to the acuminate tip of the sterile lemma; first glume 1–2.6 mm long, second glume 3–5 mm long, acuminate or with an awn up to about 2.5 mm long; sterile lemma about 3–4.5 mm long, rarely acuminate, usually with an awn 1–16 mm long; sterile palea 2–3.5 mm long; fertile lemma 2.5–5 mm long. Anthers 0.5–1 mm long. Jul–Oct. Waste ground, often where moist; IA, e NE, e KS, MO, e OK; (e Can. & ME to MN, NE, s to FL & TX). *E. pungens* Rydb.—Rydberg.

30. ELEUSINE Gaertn.

1. *Eleusine indica* (L.) Gaertn., goosegrass. Tufted annual 1–7.5 dm long. Culms prostrate to ascending or erect, nearly solid, slightly flattened, glabrous. Blades folded to slightly rolled in the bud, flat to folded at maturity, the tips obtuse to rounded, glabrous except sometimes for long hairs on the upper surface and margins near the ligule, mostly 5–15(21) cm long, 2–6(8) mm wide; sheaths keeled, glabrous or with marginal hairs near the ligule, open; ligules membranaceous, lacerate, 0.4–1.4(2.5) mm long; auricles none. Inflorescence consisting of 1–7 spicate branches, these mostly digitately arranged, but sometimes 1 or 2 branches borne beneath the apical whorl, the rachis of each branch flattened, white margined, bearing crowded spikelets in two rows, the largest branches (2)4–7(10) cm long, 3.5–5.5 mm wide; spikelets laterally compressed, 2- to 7-flowered, disarticulating above the glumes and between the florets; glumes and lemmas glabrous except for the usually scabrous, whitish, raised keel, membranaceous marginally and greenish centrally, obtuse to acute-tipped; first glume 1.5–3(3.5) mm long, 1-nerved, second glume 2.5–4.5 mm long, several-nerved, the lateral nerves crowded in the green area near the midnerve; lowest lemma 2.5–4.5 mm long, 3- to 9-nerved, the lateral nerves mostly crowded in the greenish area near the midnerve, but 2 faint nerves sometimes visible in the membranaceous potion; paleas 2-nerved, shorter than the lemmas. Anthers 0.3–0.7 mm long. (2n = 18) Jul–Oct. Lawns, waste ground, pastures, roadsides; se ND, IA, NE, CO, MO, KS, OK, & TX; (MA to ND & OR, s to FL, TX, & CA, more common in southern part of range; world-wide in warmer regions). *Naturalized.*

31. ELYMUS L., Wild Rye

Contributed by Ralph E. Brooks

Ours cespitose or rhizomatous perennials. Culms hollow or solid. Blades rolled in the bud, flat to involute at maturity; ligules membranaceous, minute to prominent; auricles present or not. Inflorescence a spike with spikelets mostly 2 or 3 per node, sometimes solitary or some with up to 6; rachis continuous except in *E. junceus*; spikelets sessile, with 2–6 florets more or less dorsiventral to the rachis; glumes subequal or sometimes one of a pair much shorter or absent, setaceous to lanceolate, acute to awned; lemmas rounded on the back, faintly nerved, tapering to a short or long awn, the awn straight or divergent, sometimes awnless; palea about as long as the lemma body; lodicules 2 and usually pubescent at the tip; stamens 3 with long anthers; caryopsis linear-oblong, furrowed in the back and pubescent at the apex, adhering to the palea.

References: Bowden, W. M. 1964. Cytotaxonomy of the species and interspecific hybrids of the genus *Elymus* in Canada and neighboring areas. Canad. J. Bot. 42: 547-601.; Church, G. L. 1967. Taxonomic and genetic relationships of eastern North American species of *Elymus* with setaceous glumes. Rhodora 69: 121-162.

Species of *Elymus* will hybridize naturally with several other genera, including *Agropyron, Hordeum,* and *Sitanion,* and any of these might be found in our range. The name *Elymus vulpinus* Rydb. is based on sterile plants from NE: Grant, which may represent sterile hybrids between some species of *Elymus* and a rhizomatous *Agropyron.*

1   Glumes linear to linear-lanceolate, mostly 1-2 mm wide, the margins hyaline .. 4. *E. glaucus*
1   Glumes setaceous and less than 1 mm wide or if wider the margins never hyaline.
    2   Lemma awns mostly longer than the lemma bodies.
        3   Glumes of each spikelet usually very unequal in length, one or both of a pair sometimes absent .... 3. *E. diversiglumis*
        3   Glumes of each spikelet subequal or equal, both members present.
            4   Glumes widest at the base, 0.2-0.6 (0.8) mm wide .... 8. *E. villosus*
                5   Plants with strong rhizomes more than 10 cm long; rare, w edge GP .... 7. *E. simplex*
                5   Plants lacking rhizomes or, if present, rhizomes weak and less than 5 cm long.
                    6   Glume bodies longer than the first lemma body; lemma awns at maturity straight; auricles minute or lacking .... 9. *E. virginicus*
                    6   Glume bodies shorter than the first lemma body; lemma awns at maturity usually divergent to recurved; auricles prominent, 1.5-4 mm long .... 1. *E. canadensis*
    2   Lemma awns mostly shorter than the lemma bodies.
        7   Glumes (0.8)1-3 mm wide .... 9. *E. virginicus*
        7   Glumes 0.5 mm wide or less.
            8   Ligules 3-7 mm long; spikes 10-17 cm long .... 2. *E. cinereus*
            8   Ligules 0.2-2 mm long; spikes 6-11 cm long.
                9   Plants lacking rhizomes; nodes glabrous; rachis disarticulating at maturity .... 6. *E. junceus*
                9   Plants strongly rhizomatous; nodes villose; rachis continuous at maturity .... 5. *E. innovatus*

1. **Elymus canadensis** L., Canada wild rye. Tufted perennial normally lacking rhizomes, rare individuals with short (to 4 cm long), weak rhizomes 1-2 mm in diam. Culms erect, hollow, 8-15(18) dm tall. Blades ascending to spreading, flat to involute (especially in drier habitats) green or glaucous, 10-25(40) cm long, 4-10(15) mm wide, mostly shorter and narrower in the n GP, glabrous or occasionally scabrous in the s GP; sheaths usually glabrous but occasionally sparsely hirsute or hirsute-ciliate in the se GP; ligules 0.5-1(1.5) mm long, entire to erose or rarely ciliate in se GP; auricles prominent, 1.5-3(4) mm long, usually clasping the culm. Spikes arching or erect (especially in the sw GP), exserted from the leaf sheath, (5)8-18(31) cm long; rachis minutely scabrid-ciliate to distinctly so, internodes about 1/2 the length of the lemma bodies making a rather loose spike, or internodes much shorter and the spike dense (the latter most common from NE s); spikelets (1)2(4) per node, appressed or only slightly diverging from the rachis (awns may be recurved), each with 3-5(7) florets; glumes linear to subsetaceous, (11)15-25(30) mm long, 0.8-1.5 mm long wide, glume bodies shorter than the lemma bodies, glabrous to scabrous on the veins, margins smooth to ciliate, awns at maturity nearly straight to divergent or recurved; lemma bodies (7)9-12(15) mm long, usually hirsute, frequently scabrous to glabrous and the margins ciliate from s SD southward, lemma awns (10)15-30(40) mm long, at maturity divergent to sharply recurved; paleas about equaling or slightly shorter than the lemma bodies in length, nerves scabrous to scabrid-ciliate toward the apex, apex acute and usually bifid; anthers 2-2.5 mm long. (2n = 28) Jun-Aug. Dry or moist sandy, gravelly, rocky, or alluvial soil of prairies, stream banks, lakeshores, ditches, & various disturbed sites, most often in open areas; GP; (Que. w to AK, s to NC, TX, NM, AZ, & CA). *E. brachystachys* Scribn. & Ball, *E. philadelphicus* L., *E. robustus* Scribn. & Sm.—Rydberg; *E. canadensis* var. *brachystachys* (Scribn. & Ball) Farw., var. *robustus* (Scribn. & Sm.) Mack & Bush—Gates; *E. canadensis* f. *glaucifolius* (Muhl.) Fern.—Fernald.

As interpreted here, *E. canadensis* includes forms that have been given varietal and even specific status. Extremes of these forms can certainly be recognized and some of these occur with greater frequency in some areas than in others. However, intermediates of all degrees are so bountiful in the GP that practicable recognition of infraspecific taxa seems superfluous.

Church (1967) attributed *E. weigandii* Fern. [*E. canadensis* var. *weigandii* (Fern.) Bowden] to the northern GP, listing it from w MN, ND; SD: Roberts; & WY: Crook. He distinguishes it from *E. canadensis* by its abruptly pendent spikes and wide (12-22 mm) leaf blades that are commonly pilose above. A few unusual specimens from ND and SD were examined, but these had glabrous leaves and spikes that were no more pendent than often seen in *E. canadensis*. Without further study of our plants in the northern plains, it seems best to include these plants as variants of Canada wild rye.

Hybrids between *E. canadensis* and *E. virginicus* (*E.* x *maltei* Bowden) are infrequently encountered in the se GP. They usually have erect spikes that are much longer than those of either parent and the glume bodies and lemma bodies are of about equal length. Their pollen is less than 5% stainable and seed set is normally 0% (weak F2's were produced is garden studies). *Elymus canadensis* also hybridizes with *E. diversiglumis* in the GP (see no. 3).

2. Elymus cinereus Scribn. & Merr., basin wild rye. Robust, usually tufted perennial lacking rhizomes or occasionally with short, thick rhizomes. Culms solid, 1-2 m tall, glabrous or scabrous below, except at the nodes which are usually pubescent, the culm becoming pubescent just below the spike. Blades flat (rarely involute), 20-50 cm long, 5-12 mm wide, glabrous or remotely scabrous, strongly nerved; sheaths mostly pubescent; ligules 3-7 mm long; auricles lacking or poorly developed. Spikes erect, stiff, mostly exserted, 10-17 cm long, often interrupted below, appressed pubescent to sparsely so throughout; spikelets 2-4(6) per node, 9-16 mm long, each containing (2)3-5 florets; glumes subulate and tapering from the base, subequal, usually shorter than the spikelet, 6-12 mm long, 0.2-0.4 mm wide, nerves mostly obscure, base weak; lemma body 8-11 mm long, awnless or with a short awn to 3 mm long; anthers 3.5-5.5 mm long. (2n = 28, 56) Jun-Jul. Sandy or clay soils, prairie flats or hillsides, open woodlands, or ravines, roadsides; MT; SD: Pennington, Walworth; WY; NE: Antelope, Dawes, Sioux; (Sask. w to B.C., s to CO, n AZ, & CA, e to SD, NE). *E. condensatus* misapplied—Rydberg; *E. piperi* Bowden—Scoggen; *Leymus cinereus* (Scribn. & Merr.) Löve—Löve, Taxon 29: 168. 1980.

3. Elymus diversiglumis Scribn. & Ball. Tufted perennial lacking rhizomes, green or glaucous; culm hollow, 8-15 dm tall, glabrous. Blades flat, 8-40 cm long, 8-13(17) mm wide, pilose on the nerves above, glabrous below; sheaths glabrous; ligules 1-2.3 mm long; auricles inconspicuous to 2 mm long. Spikes flexous and arching at maturity, 8-24 cm long; rachis flattened, margins scabrid- to hirsute-ciliate, internodes mostly 6-9 mm long, giving the spike a rather loose appearance; spikelets 1-3 per node, each containing 2-4 florets; glumes setaceous or sometimes lacking, subequal to one glume much reduced, 2-15 mm long, 0.2-0.6 mm wide, scabrous at least toward the tip, base terete and indurated; lemma body 7-12 mm long, pubescent to hirsute, awn 2-3.5 cm long, divergent at maturity; palea 8-9 mm long, slightly shorter than the lemma body, apex obtuse and with a median tuft of short dense hairs; anthers 2-3.2 mm long. (2n = 28) Late May-July. Streambanks, wooded moist hillsides or lowlands; n MN, ne ND & ND: Montrail; BH, WY: Crook; (Ont. w to Sask., s to WI, IA, w SD & ne WY). *E. interruptus* of American authors, in part, not *E. interruptus* Buckl.—Atlas GP

A few specimens from the n GP that are actually *E. canadensis* will key here. They can be separated from *E. diversiglumis* by their glabrous, wide (to 2.5 cm) leaves and bidentate palea apices. See discussion under *E. canadensis*.

Church (1969) reported a probably hybrid between *E. diversiglumis* and *E. canadensis* from Spearfish Canyon in SD: Lawrence (*Martin 477*, US). A similar specimen from the same area (*Stephens 7729*, KANU) was examined and found to have abortive pollen with 0% stainability.

4. Elymus glaucus Buckl., blue wild rye. Tufted perennial lacking rhizomes or with short rhizomes 1-3 mm in diam. Culms hollow, 5-10(15) dm tall. Blades mostly flat, usually glaucous, 6-20(35) cm long, 3-7(12) mm wide, pilose on the veins above, scabrid or glabrous below; sheaths glabrous or sometimes those at the base hirsute; ligules 0.5-1 mm long, entire or ciliate; auricles well developed, 1-2.5 mm long, usually clasping the culm. Spikes erect to arched-pendulous, exserted, 6-12(16) cm long; rachis scabrid to glabrous; spikelets 1 or 2(3) per node, each with 3-6 florets; glume bodies linear-lanceolate to linear, 7-14 mm long, 1-2 mm wide, shallowly keeled or flat, margins hyaline, apex sharp pointed or with an awn to 5 mm long; lemma bodies 8.5-13 mm long, glabrous or scabrous, awn usually straight, 10-25(32) mm long; anthers 1.5-3 mm long. (2n = 28) Jul-Aug. Open woodlands & occasionally meadows; ND: McHenry; MT; SD: BH, Day, Harding, Roberts; MO: Barton, Jasper; (Ont. w to AK, s to NY, MI, AR, SD, NM, AZ, & CA).

5. **Elymus innovatus** Beal, fuzzyspike wild rye. Rhizomatous perennial usually forming clumps, the rhizomes 2-3 mm in diam. Culms hollow, 5-12 dm tall, enveloped at the base by fibrous remains of old leaf sheaths, nodes usually villous, internodes glabrous or sparsely pubescent. Blades flat or rarely involute, glaucous or not, 6-20(30) cm long, 2-8 mm wide, glabrous, margins scabrous-ciliate; ligules 1-2 mm long, ciliate; auricles pronounced on younger leaves. Spike stiff and erect, well exserted, 6-10 cm long, sometimes interrupted below; rachis villous-ciliate; spikelets (1)2 or 3 per node, 1-1.8 cm long, each containing 3 or 4(5) florets; glumes rigid, setaceous or linear-subulate, one glume usually much reduced, rudimentary or absent, (0)0.2-7 mm long, to 0.5 mm wide, appressed pubescent; lemmas longer than the glumes, 7-9 mm long, awnless or with an awn to 3 mm long, villous to appressed pubescent, the hairs towards the apex often purplish; anthers yellow or purplish, 3.5-6 mm long. (2n = 28) Jun-Jul. Open pine woodlands & associated prairies; SD: Custer, Lawrence, Pennington, Shannon; (Ont. w to AK, s to B.C., MT, & w SD). *E. flavescens*, misapplied—Hitchcock, U.S.D.A. Miscl. Publ. 200. 1935.

6. **Elymus junceus** Fisch., Russian wild rye. Densely tufted perennial lacking rhizomes. Culms hollow, 4-10 dm tall, enveloped at the base by fibrous remains of many old leaf sheaths, glabrous or scabrous just below the spike. Blades primarily basal, mostly flat, 7-30 cm long, 1.5-4 mm wide, glabrous; ligules 0.2-1 mm long, truncate, ciliate; auricles poorly developed. Spikes erect and stiff, dense, 6-11 cm long, 5-9 mm wide; rachis hirsute-ciliate, disarticulating between the spikelets at maturity; spikelets (2)3 per node, 6-11 mm long, each with 2 or 3 florets; glumes subulate, usually shorter than the lemma bodies, subequal, 3-9 mm long 0.2-0.4 mm wide, hirsute, ciliate, bases indurate; lemmas 6.5-9 mm long with an awn 0.5-2 mm long, hirsute; anthers 3.5-4.5 mm long. (2n = 14) May-Jun. Roadsides, prairie pastures, badlands areas; ND: Sargent; MT: McCone; SD: Shannon; WY: Goshen, Laramie; CO: Lincoln; (becoming naturalized in N. Amer.; Asia). *Introduced.*

7. **Elymus simplex** Scribn. & Williams, alkali wild rye. Perennial from extensively creeping rhizomes, the rhizomes up to 5 cm long; culms 2-5(7) dm tall. Leaves mostly basal with only a few on the culm; blades rigid, usually involute, 5-20 cm long, 3-5 mm wide, upper surface scabrous, lower surface glabrous; sheaths glabrous; ligules about 0.5 mm long, truncate, ciliolate; auricles poorly developed or to 1 mm long. Spikes stiff and erect, (5)8-12(16) cm long, sometimes interrupted toward the base; rachis flattened, scabrous; spikelets 1 or 2 per node; if 2, then 1 sessile and 1 short pedicellate, 11-17 mm long and each contraining 5-7 florets; glumes subulate, asymmetrical, subequal, 5-8 mm long, glabrous to scabrous, obscurely nerved; lemma body 5-8 mm long with an awn 3-14 mm long, body glabrous with ciliate margins; paleas 5-7 mm long, apex bifid; anthers 3.5-5 mm long. (2n = 28) Jun-Aug. Sandy soils along rivers & alkaline flats; WY: Crook, Laramie; CO: El Paso; (WY, cen CO, & ne UT). *Leymus triticoides* (Buckl.) Pilger subsp. *simplex* (Scribn. & Williams) Löve—Löve, Taxon 29: 168. 1980.

8. **Elymus villosus** Muhl. ex Willd., hairy wild rye. Tufted perennial lacking rhizomes. Culms hollow, 4-15 dm tall. Blades flat, 5-17(30) cm long, 3-9(13) mm wide, pilose on

the veins above, glabrous below; sheaths glabrous or occasionally sparsely pilose; ligules 0.3–1 mm long, entire or erose; auricles pronounced, 1.5–2.5 mm long, clasping the stem. Spikes straight to arching, exserted, 4–10(13) cm long; rachis hirsute- or rarely scabrous-ciliate; spikelets diverging from the rachis at about a 45° angle, (1)2(3) per node, each with 1 or 2(3)) florets, glumes straight, setaceous or subsetaceous, 15–30 mm long, 0.2–0.6(0.8) mm wide, the body wider than the awn, hispid or hirsute, rarely scabrous; lemma bodies 6–8(9) mm long, hispid to pilose or rarely glabrous or scabrous, lemma awns straight, (10)13–30(36) mm long; palea 6.1–7.2 mm long; anthers 2–3.5 mm long. (2n = 28) Late May–Jul. Sandy, shaley, or loamy soils in woods or at the edges, or on protected stream banks; MN w to ND, s to MO & e 2/3 OK; WY: Crook; (s Ont. w to ND & ne WY, s to SC & e TX). *E. arkansanus* Scribn. & Ball, *E. striatus* Willd.—Rydberg; *E. villosus* var. *arkansanus* (Scribn. & Ball) Hitchc.—Gates; *E. villosus* f. *arkansana* (Scribn. & Ball) Fern.—Fernald.

9. ***Elymus virginicus*** L., Virginia wild rye. Tufted perennial lacking rhizomes. Culms lax to stiffly erect, hollow, 3–15 dm tall. Blades ascending to spreading, flat or sometimes becoming involute, green or occasionally glaucous, 5–25(30) cm long, 4–18 mm wide, glabrous, scabrous or pilose; ligules 0.3–1 mm long; auricles lacking or minute. Spikes stiff and erect, partially included in the leaf sheath to well exserted (as in the type), 4–16 cm long; rachis scabrid- to villous-ciliate; spikelets (1)2(3) per node, appressed to diverging at about 45° from the rachis, each with 2–4(6) florets; glumes linear-lanceolate to subulate, 1–4 cm long, (0.8)1–3 mm wide, awnless to long awned, glume bodies longer than the lemma bodies, glabrous to villous-hirsute, bases moderately to strongly indurate, curved and terete; lemma bodies 7–12 mm long, sharp pointed or with an awn to 4 cm long; palea about as long as the lemma body; anthers 2–3 mm long. (2n = 28) May–Jul. Bottomlands, streambanks, low prairies, & edges of woods; common in se GP, less so to the n & w; no records seen from e CO or ne NM but surely occurs in both areas; (Ont. w to B.C. s to FL, TX, NM, ID, & WA).

1 Lemmas awnless or with awns less than 4 mm long; flowering from early July to mid August .... 9b. var. *submuticus*
1 Lemma awns (5)8–40 mm long; flowering before July.
   2 Glumes 27–40 mm long; lemma awns 20–40 mm long.
      3 Plants (9)10–14 dm tall; spikes 9–16 cm long; flowering from mid to late June .... 9a. var. *glabriflorus*
      3 Plants 6–9 dm tall; spikes 6–9 cm long; flowering from mid-May to first week of June .... 9c. var. *virginicus*
   2 Glumes 16–26 mm long; lemma awns 8–20 mm long .... 9c. var. *virginicus*

9a. var. *glabriflorus* (Vasey) Bush. Plants (9)10–14 dm tall with robust culms. Spikes well exserted from the leaf sheaths, spikes 9–16 cm long; glumes 28–40 mm long; lemma glabrous to hirsute, awns 25–40 mm long. Mid Jun–late Jun. Common in s MO, se 1/6 KS & e OK, infrequent n to NE: Lancaster; (se U.S., n to ME, PA, s IL, s IA & se NE, w to e TX). *E. glabriflorus* (Vasey) Scribn. & Ball, *E. australis* Scribn. & Ball—Rydberg; *E. virginicus* var. *australis* (Scribn. & Ball) Hitchc.—Gates; *E. virginicus* f. *australis* (Scribn. & Ball) Fern.—Fernald.

9b. var. *submuticus* Hook. Spikes partially inserted in the leaf sheaths, infrequently exserted, 7–12 cm long; glumes 10–15 mm long; lemmas glabrous to scabrous, awnless or with awns to 2(4) mm long. Early Jul–mid Aug. GP, more frequent in n & w; (Ont. w to B.C., s to PA, KY, MO, OK, CO, ID & WA). *E. curvatus* Piper—Rydberg.

9c. var. *virginicus*. Spikes partially inserted to well exserted from the leaf sheaths, 4–13.5 cm long; glumes 16–33 mm long; lemmas glabrous to hirsute, awns 8–32 mm long. Mid May–late Jun. GP, common in se, infrequent n & w; (Ont. w to Sask., s to GA & through GP to TX, NM, & AZ). *E. jejunus* (Ramaley) Rydb., *E. hirsutiglumis* Scribn. & Sm.—Rydberg; *E. virginicus* var. *intermedius* (Vasey) Bush—Gates; *E. virginicus* f. *hirsutiglumis* (Scribn. & Fern.) Fern. and var. *jejunus* (Ramaley) Bush—Fernald.

Within this complex is a race of plants that in garden and field studies consistently flower from mid May to very early June rather than in mid and late June as does typical var. *virginicus*. The race occurs from extreme e KS and w MO south to ne TX and is characterized by weak culms, well-exserted spikes, glumes 27–33 mm long, and lemma awns 20–32 mm long. Typical var. *virginicus* has shorter glumes and lemma awns with variable culm habit and spike exsertion. While it appears the race warrants taxonomic recognition, I feel this may be premature without further study, in view of past taxonomic unrest in the *E. virginicus* complex.

32. ERAGROSTIS v. Wolf, Lovegrass

Tufted or creeping annuals to cespitose or short-rhizomatous perennials. Culms erect to more or less decumbent, solid to hollow, often solid above and hollow below. Blades rolled in the bud, flat to often involute, especially marginally, at maturity; sheaths open, somewhat keeled; ligule a short fringe of hairs usually less than 1 mm long, rarely to 2 mm long, sometimes flanked or backed by longer hairs; auricles lacking. Inflorescence an open panicle of 2- to many-flowered spikelets, these disarticulating in a variety of ways, often the glumes falling first, followed by the lemmas, grains, and, finally, by the paleas, or the paleas persistent on the rachilla, sometimes the glumes persistent or the rachilla disarticulating; glumes subequal or unequal, the second longer than the first, unawned, (nerveless) 1- to 3(7)-nerved; lemmas 3-nerved, the nerves inconspicuous to very prominent, unawned; paleas 2-nerved, conspicuous, the nerves sometimes scabrous or ciliate. Anthers usually very short (under 0.5 mm long) but long in a few species. *Neeragrostis* Bush.

References: Harvey, L. 1975. *Eragrostis*, pp. 177–201, in Gould, F. W. The grasses of Texas. Texas A. and M. Univ. Press, College Station; Koch, S D. 1969. The *Eragrostis pectinacea-pilosa* complex in North and Central America. Illinois Biol. Monogr. 48: 1–75.

This genus is notoriously difficult, especially the annual species, and identification of immature material is risky. The tiny anthers of many species are often found entangled in the stigmas of the mature fruits, suggesting at least some degree of autogamy.

1 Culms widely creeping, rooting at the nodes.
  2 Plants dioecious; anthers more than 1.3 mm long in the staminate flowers .... 12. *E. reptans*
  2 Plants perfect-flowered; anthers less than 0.4 mm long .... 7. *E. hypnoides*
1 Culms not widely creeping, although sometimes geniculate and decumbent at the base, not normally rooting at the nodes.
  3 Plants annual.
    4 Mature spikelets small, the largest ones 2- to 4(6)-flowered and less than 3.5 mm long.
      5 Grain strongly grooved on the side opposite the embryo; panicle very open, the pedicels mostly more than 2× the spikelet length, the panicle usually considerably more than 1/2 the height of the whole plant .... 2. *E. capillaris*
      5 Grain often slightly flattened or even very shallowly indented on the side opposite the embryo, but not strongly grooved; panicle more dense, many of the pedicels less than 2× the length of the spikelet, the panicle usually not more than 1/2 the height of the whole plant .... 6. *E. frankii*
    4 Mature spikelets larger, the largest ones 4- to many-flowered and more than 3.5 mm long.
      6 Plants sparingly, if at all glandular, lacking conspicuous indented glands near the summit of the internodes or wartlike glands on the keels of the lemmas or the margins of the leaves.
        7 Lowest glume very short, less than 1/2 the length of the first lemma and hence its tip touching that lemma well below midlength; mean pollen diam less than 28 μm .... 11. *E. pilosa*
        7 Lowest glume reaching or surpassing the midpoint of the first lemma on some or all spikelets; mean pollen diam 28 μm or more.
          8 Spikelets lying more or less parallel to the main panicle branches; lemmas mostly 1.5–2.3 mm long; anthers 0.2–0.4 mm long; grains 0.7–1.1 (averaging about 0.8) mm long .... 10. *E. pectinacea*

        8   Spikelets on spreading pedicels; lemmas mostly 1.2–1.7 mm long; anthers 0.2–0.3 mm long; grains 0.5–0.9 (averaging about 0.7) mm long .... 6. *E. frankii*
   6   Plants with glandular regions near the summit of the internodes or with wart-shaped glands on the leaf margins or the keels of the lemmas.
        9   Dish-shaped glands conspicuous at the summit of at least some internodes, these sometimes fused into a glandular ring; wart-shaped glands not present on the lemma keels or on the margins of the leaves.
            10   Longer spikelets with more than 10 florets; glands near the summit of the internodes fused, often forming a ring; first glume more than 0.7 mm long .... 1. *E. barrelieri*
            10   Longer spikelets usually with fewer than 10 florets; glands near the summit of the internodes very scattered; first glume less than 0.7 mm long .... 11. *E. pilosa*
        9   Dish-shaped glands present or absent at the summit of the internodes; wart-shaped glands present on the leaf margins and usually also on the lemma keels.
            11   Spikelets mostly 2.1–3.6 mm wide at maturity .... 3. *E. cilianensis*
            11   Spikelets mostly less than 2 mm wide at maturity .... 9. *E. minor*
3   Plants perennial.
    12   Spikelets sessile and borne on simple panicle branches .... 14. *E. sessilispica*
    12   Spikelets pedicellate.
        13   Spikelets and their pedicels conspicuously appressed on stiff, spreading branches .... 4. *E. curtipedicellata*
        13   Spikelets on spreading pedicels on stiff to flexuous branches, or the panicles densely flowered and lacking spreading branches.
            14   Panicle branches densely flowered; spikelets mostly more than 2.5 mm wide; grains relatively long, mostly 1–1.5 mm long, but anthers short, 0.3–0.6 mm .... 13. *E. secundiflora* subsp. *oxylepis*
            14   Panicle branches loosely flowered; spikelets mostly less than 2.5 mm wide; anthers and grains various, but plants not with a combination of long grains and short anthers.
                15   Anthers and grains both short, the anthers less than 0.5 mm long, the grains mostly 0.6–0.8 mm long.
                    16   Nerves of the lemma obscure; grains more or less clearly grooved on the side opposite the embryo .... 8. *E. intermedia*
                    16   Nerves of the lemma conspicuous; grains not grooved on the side opposite the embryo .... 15. *E. spectabilis*
                15   Anthers and grains both relatively long, the anthers more than 0.8 mm long, the grains mostly 0.8–1.6 mm long.
                    17   Glumes unequal, the first about 1–2 mm long, the second 2–3 mm long; grains somewhat flattened, but not grooved on the side opposite the embryo, pale in color .... 5. *E. curvula*
                    17   Glumes subequal, both 1.8–4 mm long; grain clearly grooved on the side opposite the embryo, dark brown at maturity .... 16. *E. trichodes*

1. *Eragrostis barrelieri* Daveau, Mediterranean lovegrass. Tufted or solitary-stemmed annuals (0.8)1.5–4.5(6) dm tall. Culms hollow to pith-filled, erect but often geniculate below, glabrous, usually with a row of dish-shaped glands near the top of at least each of the upper internodes, these often partly or fully fused into a ring. Blades flat to involute, especially at the margins, glabrous to thinly pilose above, 1–8 cm long, 1.5–4.5 mm wide; sheaths usually long-pilose at the collar but otherwise glabrous, slightly keeled; ligules mostly 0.3–0.6 mm long, rarely longer. Inflorescence open to somewhat condensed, often with glandular patches or rings below some branches, 4–12 cm long; spikelets mostly (4)6–16(18) mm long, about 1–2 mm wide, (4)10- to 17(23)-flowered, the glumes dropping first, followed by the lemmas and the grains, the rachilla and paleas persistent; glumes unequal, acute and 1-nerved with the nerve scabrous near the tip, or the first glume sometimes nerveless; first glume 0.7–1.3 mm long; second glume 1.1–1.9 mm long; lemmas scabrous on the midnerve,

1.4–2.5 mm long; paleas ciliolate on the nerves. Anthers about 0.2–0.3 mm long, often retained within the spikelet. Grain 0.7–1 mm long, oblong, rusty-brown to tan at maturity. (2n = 60) Jul–Oct. Waste ground, often in sandy soil; cen & sw KS, NE, OK, TX, NM; (from cen GP throughout most of TX, also FL, Mex., W.I.; Medit. region). *Naturalized.*

2. **Eragrostis capillaris** (L.) Nees, lacegrass. Delicate tufted or single-stemmed annuals, 1–6 dm tall. Culms slender, hollow to pith-filled, glabrous, erect, not geniculate. Blades glabrous to scabrous or long-pilose above, flat to involute, especially at the margins, 5–20(30) cm long, 0.7–4 mm wide; sheaths keeled, usually long-pilose at the collar and on the margins; ligules 0.1–0.6 mm long. Inflorescence very open and diffuse, the spikelets all borne on spreading pedicels more than 2× the length of the spikelets, (12)19–35(43) cm long, usually well over 1/2, averaging about 3/4, of the total height of the plant; spikelets small, mostly about 1.5–3.3 mm long, 0.9–2 mm wide, (1)2- or 3(4)-flowered, the glumes falling first, followed by the lemmas and grains, the rachilla and paleas persistent; glumes acute to acuminate, scabrous on the single nerve, subequal, the first 0.7–1.6 mm long, the second 0.9–1.7 mm long; lemmas acute, the lateral nerves somewhat obscure, the midnerve scabrous, the lowest one 1.2–1.8 mm long; paleas scabrous on the nerves. Anthers mostly 0.2–0.3 mm long, exserted or retained within the spikelet. Grains strongly grooved on the side opposite the embryo, 0.5–0.8 mm long. (2n = 50, 100) Jul–Oct. Usually on dry ground in open woods; IA, e NE, MO, e KS, OK; (ME to WI, NE, s to GA & TX).

This species can be distinguished from members of the *E. pectinacea-pilosa* complex by the strongly grooved grains, but it is closely similar to the perennial species *E. intermedia*, and the relationship to that species should be investigated. In our region *E. capillaris* normally has smaller pollen (mean diameter 23–26 μm for *E. capillaris* and 27–31 μm for *E. intermedia*), and its anthers and spikelets are also usually smaller (see descriptions).

3. **Eragrostis cilianensis** (All.) E. Mosher, stinkgrass. Tufted annuals 1–5 dm tall. Culms hollow to pith-filled, often geniculate below, glabrous or with a ring of glandular pits below the nodes. Blades flat to somewhat involute, glabrous, but with wartlike glands on the margins and often also on the midribs, 3–16(28) cm long, 1.5–5(7) mm wide; sheaths glabrous, but frequently with a line or tuft of long hairs at the collar, often bearing scattered dish-shaped glands; ligules 0.4–0.9 mm long. Panicle 2–12 cm long, relatively dense, the spikelets considerably exceeding their pedicels in length; spikelets large, (3)5–12(15) mm long, 2–3.6 mm wide, (5)9- to 24(33)-flowered, the glumes falling early, but the lemmas and paleas usually retained beyond fruit maturation; glumes 1- to 3-nerved, scabrous and often glandular on the midnerve with wartlike glands, the first one 1.3–2.1 mm long, the second one 1.4–2.3 mm long; lemmas 3-nerved, the lateral nerves prominent, the midnerves usually both scabrous and glandular, the lowest lemma 1.7–2.4 mm long; paleas ciliolate on the nerves. Anthers 0.2–0.4 mm long, retained or exserted. Grains dark, ellipsoidal, 0.4–0.7 mm long. (2n = 20) Jul–Oct. Waste ground; GP; (introduced in s Can., most of U.S., Mex., W. I., much of S. Amer.; Europe). *Naturalized. E. megastachya* (Koel.) Link—Fernald.

4. **Eragrostis curtipedicellata** Buckl., gummy lovegrass. Tufted perennials 2–7.5 dm tall. Culms pith-filled or hollow, often viscid, especially below the nodes, erect to somewhat geniculate below. Blades flat to involute, glabrous to long-pilose near the ligule, mostly 6–16 cm long, 2.5–5.5 mm wide; sheaths keeled, often viscid, usually pilose in a line or tuft near the collar, sometimes also with additional pubescence; ligules short, usually only 0.1–0.5 mm long, but sometimes flanked and backed by longer hairs and thus apparently longer. Panicle (13)19–33(41) cm long, open, with the spikelets and their pedicels appressed to the stiff, spreading branches, the rachis and the larger axes viscid; spikelets 3–6 mm long, 1–2 mm wide, 4- to 11-flowered, disarticulating irregularly, usually with a few lem-

mas or the glumes falling first, but the rachilla also breaking up and, ultimately, the entire spikelet deciduous; glumes acute to acuminate, scabrous on the single nerve, the first 0.9–1.8 mm long, the second 1.1–2 mm long; lemmas acute, scabrous on the midnerve, the lateral nerves conspicuous, the lowest lemma 1.4–2 mm long; paleas ciliolate on the nerves. Anthers mostly 0.2–0.3 mm long, often retained within the spikelet. Grains oblong, 0.6–0.8 mm long. (2n = 40) Jun–Sep. Sandy & clay soil, in pastures, along roadsides, & in waste ground; sw KS, OK, TX, NM; (se AR & s KS to TX, e NM, & ne Mex.).

5. *Eragrostis curvula* (Schrad.) Nees, weeping lovegrass. Cespitose perennials (4.5)6–12(14) dm tall. Culms pith-filled to hollow, erect, often somewhat geniculate below, glabrous. Blades involute, flexuous, glabrous to scabrous except for long hairs behind the ligule, (5)20–35(50) cm long, mostly 1.5–3 mm wide; sheaths keeled, the lowest ones usually pubescent with ascending hairs, the upper ones usually glabrous except in the collar region; ligules 0.8–1.3(2) mm long, often backed by longer hairs. Panicles open, 16–36(40) cm long, nodding, the branches somewhat flexuous; spikelets mostly 4–7.5 mm long, 1–2 mm wide, 3- to 10-flowered, disarticulating in various patterns, but often one or both glumes and 1 or 2 lemmas falling before the ultimate break-up of the rachilla, or sometimes the glumes persisting and falling last; glumes 1-nerved, acute to acuminate, unequal, the first one 1–2 mm long, the second 2–3 mm long; lemmas obtuse, rather plainly 3-nerved, the lowest one 2–3 mm long; paleas scabridulous on the keels. Anthers long, 0.9–1.5 mm, not retained in the mature spikelets. Grains ovate to elliptic, rather long, 1.2–1.6 mm, pale, with a darker region surrounding the embryo, slightly flattened. (2n = 40, 50) May–Aug. Roadsides & waste ground, often in sandy soil; cen & s KS, OK, TX, NM; (cult. & escaped in s U.S.; S. Africa). *Naturalized.*

6. *Eragrostis frankii* C. A. Mey. ex Steud., sandbar lovegrass. Delicate annuals 1–3(5) dm tall. Culms slender, pith-filled to hollow, sometimes with viscid regions or with a few depressed glands below the nodes. Blades flat to involute, sometimes pilose on the upper surface near the base, mostly 2–14(17) cm long, 0.9–4 mm wide; sheaths glabrous except for tufts of long hairs in the collar region, keeled, rarely with a few glandular pits; ligules about 0.2–0.8(1) mm long. Panicles open to rather diffuse, 7–26 cm long, the pedicels spreading; spikelets very small, mostly 1.5–3.5 mm long, 1–2.5 mm wide, and 2- to 4(6)-flowered or larger in some plants, up to 6(8) mm long and 9(17)-flowered, the glumes usually falling first, followed by the lemmas and grains, the paleas and rachillas persisting; glumes slightly unequal to subequal, scabrous on the single nerve, the first one 0.7–1.5 mm long, the second one 1–1.7 mm long; lemmas acute, scabrous on the midnerve, the lateral nerves obscure or only moderately conspicuous, the lowest one 1–1.7 mm long. Anthers 0.2–0.3 mm long, often retained within the spikelet at maturity. Grains ellipsoidal to almost spherical, but flattened or even very shallowly indented on the side opposite the embryo, brown, 0.4–0.7(0.9) mm long. (2n = 40, 80) Aug–Sep. Stream banks & moist ground; MO, e KS, e OK; (NH to MN, s to FL & OK).

The form of *E. frankii* with more diffuse panicles, long spikelets, and more numerous florets normally has larger pollen. It is probably the octoploid form discussed by Koch (op. cit.). This form is common in e KS and is perhaps distinctive enough to warrant taxonomic recognition.

A few plants which differ from *E. pectinacea* in no way except that they have spreading pedicels may key to *E. frankii*. Such plants will not match the above description particularly well because they have, on the average, wider spikelets and larger spikelet parts and the lateral nerves of the lemmas are relatively conspicuous. They are here tentatively assigned to *E. pectinacea*, although they have been called *E. tephrosanthos* Schult. or *E. arida* Hitchc. (see Koch, op. cit.).

A rather robust annual, *E. mexicana* (Lag.) Link, has been reported from the edge of our range in the TX panhandle (as *E. neomexicana* Vasey). Although the reports are believed to be based on erroneous identifications, it seems possible that this grass may yet be found in the s GP. It will key here to *E. frankii* (large form) but should be easily recognizable because of its strongly grooved grains.

7. ***Eragrostis hypnoides*** (Lam.) B.S.P., teal lovegrass. Mat-forming annual with creeping, branched, geniculate culms bearing erect flowering branches, 0.2–2.5 dm tall. Culms hollow, or pith-filled in the upper parts, glabrous. Blades flat to involute, pilose with ascending hairs, especially above, or glabrous beneath and merely scabrous above, 0.5–5 cm long, 0.8–2.8 mm wide; sheaths slightly keeled, pilose with ascending hairs, these most prominent at, or sometimes confined to, the node and collar, or, rarely, the sheaths glabrous; ligules 0.3–0.9 mm long. Inflorescence fairly compact to moderately open, 1–6 cm long; spikelets 3–8(16) mm long, 1.4–3 mm wide, and (5)13- to 20(44)-flowered; glumes and lemmas long-persistent but eventually falling, the glumes usually falling first, followed by the lemmas and grains, the rachillas and paleas persistent; glumes unequal, acute to acuminate, the single nerve often scabrous; first glume 0.4–1.1 mm long; second glume 0.7–1.5 mm long; lemmas acute to acuminate, plainly 3-nerved, the midnerve scabrous, the lowest one 1.5–2.0 mm long; paleas ciliolate on the nerves. Anthers about 0.2–0.3 mm long, exserted or retained within the spikelet. Grains 0.4–0.6 mm long, tan, somewhat flattened laterally. (n = 10) Jun–Oct. Sandy to muddy ground near streams, rivers, ponds, & lakes; MN, e ND, e SD, IA, NE, nw CO, MO, e KS, e OK; (e Can., most of U.S., but not common in the Rocky Mts. & w GP region; Mex., W. I., S. Amer.).

8. ***Eragrostis intermedia*** Hitchc., plains lovegrass. Tufted perennials 2–7(11) dm tall. Culms erect or geniculate below, pith-filled to hollow, glabrous. Blades flat to involute, usually nearly glabrous, but sometimes sparsely long-pilose on both sides or on the upper surface only, especially near the ligule, mostly 5–15(35) cm long, 1.5–5 mm wide; sheaths keeled, nearly glabrous or long-pilose at the margin near the top or more or less pilose throughout; ligules 0.2–0.6 mm long. Panicles very open and diffuse but seldom more than half the height of the plant, (9)17–25(50) cm long; spikelets small, 2.5–5(7) mm long, 1–2 mm wide, 2- to 5(9)-flowered, the glumes usually falling first, followed by the lemmas and grains, the paleas and rachillas long persistent but eventually deciduous; glumes unequal, acute, scabrous on the solitary nerve, the first one 0.7–1.5 mm long, the second one 1.2–1.8 mm long; lemmas with the lateral nerves obscure, scabrous on the midnerve, acute, the lowest one 1.2–2 mm long; paleas scabrous on the nerves. Anthers mostly 0.3–0.4 mm long, exserted or retained within the spikelet. Grains brownish to tan, 0.6–0.8 mm long, shallowly to deeply grooved on the side opposite the embryo. (2n = 60, 80, 100, 120, 2n = 72 also reported) Jul–Oct, rarely also in spring. Open woods, rocky hillsides, & waste ground; MO, e KS, OK; (MO to AZ, s to GA, LA, & Mex.).

The group of species to which this variable taxon belongs needs serious study. *E. intermedia* is very similar to *E. capillaris*, which see.

Eragrostis hirsuta (Michx.) Nees, another close relative of *E. intermedia*, enters our range in the s part (MO: Barton & Jasper, OK: Cimarron, Cleveland, Creek, & Kay). Plants of *E. hirsuta* from near our range have, on the average, longer stamens, slightly longer glumes, and fewer-flowered spikelets than *E. intermedia*. Reportedly, this species is also taller, frequently exceeding 1 m in height. Some plants from e OK appear to be intermediate between *E. intermedia* and *E. hirsuta*.

9. ***Eragrostis minor*** Host, little lovegrass. Tufted annual 0.5–3.1 dm tall. Culms glabrous, often with glandular depressions forming a partial or complete ring below some or all of the nodes, sometimes with additional scattered dish-shaped glands, hollow or pith-filled, often geniculate below. Blades flat to involute, especially at the margins, bearing wart-shaped glands on the margins and sometimes also on the midribs, glabrous to long-pilose, especially near the ligule, 1–8 cm long, 1–3.5 mm wide; sheaths long-pilose at the collar, often also with additional hairs, especially near the margins, with or without scattered glands; ligules 0.2–0.5 mm long. Inflorescence rather open, 5–12(19) cm long, ovate, usually with dish-shaped and irregular glandular patches on the rachis and branches; spikelets mostly 4–8(10) mm long, 1.4–2.4 mm wide, 5- to 12(19)-flowered, the glumes usually falling first, followed by the lemmas and grains, the paleas and rachillas normally persistent; glumes

slightly scabrous on the solitary nerve, that nerve often also bearing wart-shaped glands, acute to obtuse, the first one 1–1.6 mm long, the second 1.3–1.8 mm long; lemmas strongly 3-nerved, not very keeled, but the midnerve often glandular and sometimes slightly scabrous, obtuse, the lowest one 1.4–2 mm long; paleas ciliolate on the nerves. Anthers mostly 0.2–0.3 mm long, often retained. Grains elliptic, light brown, 0.4–0.7 mm long. (2n = 40) Jun–Aug. Waste ground; known from scattered locations in IA, NE, MO, KS, & OK, but apparently not common; (sparingly introduced in various parts of e, cen, & s U.S.; Europe). *Naturalized. E. poaeoides* Beauv.—Rydberg.

Since the author of this genus has been determined to be N. M. von Wolf and not P. de Beauvois, the earlier name *E. minor* Host must replace *E. poaeoides* Beauv. See S. D. Koch (Rhodora 80: 397. 1978) for a full explanation.

10. *Eragrostis pectinacea* (Michx.) Nees, Carolina lovegrass. Tufted annuals 1–6 dm tall. Culms hollow to pith-filled, glabrous, spreading to erect, often geniculate below. Blades flat to involute, glabrous to scabrous, 2–15(17) cm long, 0.5–3.5(4.5) mm wide; sheaths keeled, glabrous, except for marginal tufts of white hairs at the collar; ligules 0.2–0.7 mm long. Panicles open, ovate, (3)5–19(22) cm long, the primary branches usually spreading, the secondary and tertiary ones normally appressed, the spikelets tending to lie parallel to the branches; spikelets mostly 3.5–7.5 mm long, 1–2 mm wide, and (3)6- to 13-flowered, the glumes often falling first, followed by lemmas and grains, the rachillas and paleas persistent or eventually deciduous; glumes acute to acuminate, scabrous on the solitary nerve, the first one 0.7–1.5 mm long, the second 1–1.6 mm long; lemmas with the lateral nerves relatively conspicuous, acute, scabrous on the midnerve, the lowest one 1.4–2.2 mm long. Anthers mostly 0.2–0.4 mm long, exserted or included in the mature spikelet. Grains oblong, brown, slightly flattened laterally, mostly 0.7–1.1 mm long. (2n = 60) Late Jun–Oct. Waste ground; very common in GP, but less so in nw part; (throughout s Can. & the U.S. but uncommon in the w; Mex. & C. Amer.). *E. diffusa* Buckl.—Harvey, op. cit.; *E. purshii* Schrad.—Rydberg; *E. tephrosanthos* Schult.—Koch, op. cit.; *E. arida* Hitchc.—Hitchcock, Manual of the grasses of the United States, U. S. D.A. Misc. Publ. No. 200, p. 158. 1950.

Forms of this species having spreading pedicels may key to *E. frankii*. See the discussion following that species.

11. *Eragrostis pilosa* (L.) Beauv., India lovegrass. Tufted annuals mostly 1.5–5 dm tall. Culms glabrous, sometimes geniculate below, hollow to pith-filled. Blades flat to involute, glabrous, 2–12(18) cm long, 1–3.5 mm wide; sheaths keeled, glabrous except for marginal tufts of hairs in the collar region, or these sometimes absent; ligules 0.2–0.5 mm long. Panicles 6–20 cm long, open and diffuse at maturity, ovate, the primary branches spreading, the secondary ones appressed to spreading; spikelets 2–7 mm long, 0.7–1.7 mm wide, 3- to 11-flowered, glumes falling first, followed by the lemmas, the grains, and, eventually, by the paleas, the sinuate rachillas long-persistent; glumes markedly unequal, the first 1-nerved or nerveless, usually not reaching the midpoint of the first lemma, mostly 0.3–0.8 mm long, the second scabrous on the single nerve, 0.7–1.4 mm long; lemmas with rather inconspicuous lateral nerves, the midnerve scabrous, the lowest lemma 1.2–2.2 mm long; paleas scabrous on the nerves. Anthers 0.1–0.4 mm long (averaging around 0.2 mm), often included at maturity. Grains smooth, ovoid, slightly flattened, 0.5–0.9 mm long. (2n = 40) Jul–Aug.

There are 2 varieties:

11a. var. *perplexa* (Harvey) S. D. Koch. Culms bearing several to many scattered glandular depressions on the upper portions of many of the internodes; glands also present in the inflorescence, at least on the main rachis, and often on some sheaths and blade midribs as well. Moist ground or waste ground, apparently rare or overlooked; known only from widely scattered locations in s-cen ND, s-cen SD, w NE, se WY, w KS, e & cen CO, n TX; (GP, also known from a single collection in Mex.). *E. perplexa* Harvey—Harvey, Bull. Torrey Bot. Club 81: 405–410. 1954.

11b. var. *pilosa*. Plants eglandular or sparingly glandular, the glands, when present, confined to limited areas on the panicle rachis and to 1 or 2 pits on the culm internodes. Waste ground; MO, OK, TX; (ME to MI, s to FL & TX, sparingly introduced elsewhere).

12. ***Eragrostis reptans*** (Michx.) Nees. Dioecious, mat-forming annuals with creeping, branched, often geniculate culms, the erect flowering branches 0.3–2.6 dm tall. Culms hollow, usually densely pilose with ascending, sometimes gland-tipped, hairs, rarely glabrous. Blades and sheaths similarly pubescent, or rarely glabrous; blades flat, 1–4 cm long, 1–3 mm wide; sheaths keeled; ligules 0.3–0.6 mm long. Inflorescence normally dense and headlike in both staminate and pistillate plants, usually not much over 3 cm long; spikelets 3–12(14) mm long, 1.8–4 mm wide, and 4- to 19(35)-flowered, sometimes 1 or both glumes deciduous early, especially in the staminate plants, lemmas of the staminate plants retained, those of the pistillate plants eventually deciduous; staminate spikelets with the first glume 1-nerved, 1.5–3.5 mm long, the second glume 1- to 3-nerved, 2–4 mm long, the lemmas plainly 3-nerved, acute, the lowest one 2.1–4 mm long; pistillate spikelets with the first glume nerveless or 1-nerved, 0.2–1.5 mm long, the second glume 1- to 3-nerved, 1–2 mm long, the lemmas plainly 3-nerved, acute to acuminate, the lowest one 1.5–2.5 mm long. Anthers 1.4–2.2 mm long. Grains brownish, 0.4–0.6 mm long. (2n = 60) Jul–Sep. Along streams & at the edges of ponds & lakes; e SD, IA, NE, MO, KS, OK, relatively uncommon; (KY to SD, s to LA & TX, also FL, Mex.). *Neeragrostis reptans* (Michx.) Nicora — Gould, Grasses of Texas, p. 202. Texas A. and M. Univ. Press, College Station. 1975.

S. D. Koch (Rhodora 80: 390–403. 1978) presents an argument for retaining this grass within the genus *Eragrostis*.

13. ***Eragrostis secundiflora*** Presl subsp. ***oxylepis*** (Torr.) S. D. Koch, red lovegrass. Tufted perennials 1.5–6(7.5) dm tall. Culms glabrous, hollow to pith-filled, erect, sometimes geniculate below. Blades involute, nearly glabrous, but often long-pilose on the upper surface near the ligule, mostly 5–25 cm long, 1.2–3.2 mm wide; sheaths not prominently keeled, nearly glabrous except for long hairs flanking the ligule at the summit; ligules short, only 0.3–0.6 mm long but often flanked and backed by longer hairs. Panicles often rather dense, the main branches ascending, the secondary branches and pedicels appressed; spikelets reddish-tinged, large, about 5–13 mm long, 2–4.5 mm wide, and (5)9- to 19(29)-flowered; rachilla disarticulating above the glumes and between the florets; glumes 1-nerved, acuminate, subequal, rather long, about 2–3.5(3.8) mm; lemmas acute, the lateral nerves very prominent, the lowest lemma shorter than the shortest glume, about 2.2–3.5 mm long, but the lemmas becoming considerably longer above the spikelet base; paleas ciliolate on the nerves. Anthers 0.3–0.6 mm long, sometimes retained at maturity. Grains tan, narrowly ovate, 1.0–1.4 mm long, slightly flattened laterally. (2n = 40) Mostly Jul–Sep, sometimes also in the spring. Dry, often sandy pastures, prairies, & waste ground; cen & s KS, OK, TX, NM; (TN to CO, s to w FL, TX, n Mex.). *E. oxylepis* (Torr.) Torr. — Correll & Johnston; *E. beyrichii* J. G. Sm. — Hitchcock, op. cit., p. 146.

The typical subspecies is found in s Mexico and is disjunct in S. Amer. (Koch, Rhodora 80: 390–403. 1978).

14. ***Eragrostis sessilispica*** Buckl., tumble lovegrass. Sturdy perennials 2.5–9.5 dm tall. Culms glabrous, hollow, erect or arching. Blades involute to flat, usually nearly glabrous except for silky pubescence on the upper surface near the ligule, sometimes sparsely long-pilose throughout, mostly 5–30 cm long, 1.5–3 mm wide; sheaths not keeled or only slightly keeled, glabrous except, usually, for long tufts of hairs at the margin in the collar region; ligules short, only 0.2–0.4 mm long, but flanked and backed by long silky hairs and hence often difficult to measure. Panicle 18–56(72) cm long, often breaking free of the plant at maturity, the branches widely spaced, stiffly spreading, mostly simple, bearing widely spaced

sessile spikelets; spikelets 8–12 mm long, 1.4–3 mm wide, and 3- to 9-flowered, disarticulating above the glumes and between the florets; glumes indurate, acuminate, the first one usually 1-nerved, 2.9–6 mm long, the second 3- to 5(7)-nerved, 3–7 mm long; lemmas indurate, lanceolate-acuminate, the lateral nerves conspicuous, the lowest lemma 3–6 mm long; paleas gibbous at the base, ciliolate on the keels. Anthers 0.3–0.5 mm long, often retained at maturity. Mature grains brown, tapering toward the apex, slightly flattened laterally, 1.2–1.5 mm long. (2n = 40) Mostly May–Jul. Sandy prairies & roadsides; sw KS, w OK, TX, NM; (s GP to n Tam.). *Acamptocladus sessilispicus* (Buckl.) Nash — Rydberg.

15. ***Eragrostis spectabilis*** (Pursh) Steud., purple lovegrass. Tufted perennials with knotty short-rhizomatous bases, 2.5–6.5(8.5) dm tall. Culms erect or arching, glabrous, hollow, or sometimes solid above. Blades flat to involute, usually nearly glabrous, but often sparsely to densely pilose above near the ligule and rarely pilose on both surfaces, mostly 10–30 cm long, 2.5–8 mm wide; sheaths not keeled or scarcely so, pilose with spreading to ascending hairs to nearly glabrous, but almost always with lines of long hairs at the collar region; ligules short, 0.1–0.4 mm long, but often backed and flanked by longer hairs. Panicles 15–30(60) cm long, very open and diffuse, the branches rather stiff and spreading at maturity when the whole panicle may sometimes break free of the plant; spikelets usually purplish, (3)4–7 mm long, 1–2.5 mm wide, and (5)7- to 10(12)-flowered, the rachilla disarticulating above the glumes and between the florets, the glumes ultimately deciduous as well; glumes acute, the first 1-nerved, 1–2 mm long, the second 1- to 3-nerved, 1.4–2.2 mm long; lemmas acute, the lateral nerves prominent, the lowest one 1.5–2.5 mm long; paleas prominently ciliolate. Anthers 0.3–0.4 mm long, often retained at maturity. Grains brown, ellipsoid, mostly 0.6–0.7 mm long. (2n = 42) Mostly Aug–Oct. Dry to moist pastures, railroads, waste ground; MN, se ND, e SD, IA, NE, MO, KS, se CO, OK, TX; (ME to MN & ND, s to FL & AZ, also W. I., e Mex.). *E. pectinacea* (Michx.) Steud. — Rydberg, misapplied; *E. spectabilis* (Pursh) Steud. var. *sparsihirsuta* Farw. — Fernald.

16. ***Eragrostis trichodes*** (Nutt.) Wood, sand lovergrass. Strongly cespitose perennials 3–12 dm tall. Culms solid or hollow below, erect. Blades flat to somewhat involute, the midrib usually prominent, glabrous or pilose on the upper surface near the ligule, mostly 20–46(54) cm long, 1.5–6(8) mm wide; sheaths not keeled or scarcely so, usually glabrous except for tufts of hairs marginally at the collar; ligules mostly 0.2–0.4 mm long. Panicles very open and diffuse, many of the branches somewhat flexuous; spikelets 3.8–10 mm long, 1.4–2.5(4) mm wide, and (3)5- to 10(13)-flowered, the rachilla disarticulating above the glumes and between the florets or sometimes the rachilla rather long-persistent and the glumes and lemmas dropping first, not infrequently the grains dropping before the break-up of the spikelet; glumes subequal, lance-acuminate, 1.8–4 mm long, 1-nerved; lemmas with the lateral nerves moderately conspicuous, acute, the lowest one 2.2–3.5 mm long, sometimes surpassed by the first glume; paleas scabrous on the nerves. Anthers 0.9–1.6 mm long, not normally retained. Grains dark brown, strongly grooved on the side opposite the embryo, 0.8–1.1 mm long. (2n = 40) Jul–Oct. Sandy prairies & open sandy woods; GP from s SD s; (IL to s SD, s to sw AR & TX). *E. trichodes* (Nutt.) Wood var. *pilifera* (Scheele) Fern. — Fernald; *E. pilifera* Scheele — Hitchcock, op. cit., p. 163.

33. ERIOCHLOA H.B.K., Cupgrass

1. ***Eriochloa contracta*** Hitchc., prairie cupgrass. Tufted annuals 1.5–8 dm tall. Culms glabrous to short-pubescent, especially near the nodes, hollow, ascending, but usually geniculate and decumbent below. Blades rolled in the bud, flat at maturity, usually short-pubescent, the midrib not prominent, mostly 4–17(34) cm long, 3–8(10) mm wide; sheaths

usually short-pubescent, not keeled, open; ligule a dense fringe of hairs from a short membranous base, 0.5-1 mm long. Inflorescence a panicle 5-11(15) cm long, the spikelets in 2 rows and short-pediceled on one side of each principal branch; spikelets 3.9-5 mm long, including the acuminate tip, somewhat dorsally compressed, disarticulating below the glumes, with 2 florets but the lower floret consisting solely of a sterile lemma; glumes strongly unequal, the first not recognizable as a glume but much reduced and united with the upper part of the pedicel to form a beadlike cup or disk, the second 5-nerved, pubescent, awned or acuminate, 3.2-4.5 mm long including the tip; sterile lemma similar to the second glume but with 3 principal converging nerves, sometimes with additional nerves near the acuminate tip, 2.5-4.2 mm long; fertile lemma oblong, firm, pale, finely rugose, the margins inrolled, 2-2.8 mm long, the tip bearing a firm, straight or twisted awn 0.3-0.9 mm long, this rather readily breaking off. Anthers exserted or sometimes included at maturity, 0.6-0.9 mm long. (2n = 36) Jul-Oct. Ditches, roadsides, & waste ground, usually where moist; se NE, KS, MO, TX, OK; (MO to NE, s to LA, TX, AZ, & Mex.). *E. punctata* (L.) Desv. ex Hamilt. — Rydberg, misapplied.

Two other species of *Eriochloa* are known from the edge of our range. *E. sericea* (Scheele) Munro, a perennial species with stiffly erect culms, is known from KS: Montgomery; & OK: Canadian, Cleveland, Greer, & Tillman. *E. lemmonii* Vasey & Scrib. is known from OK: Tillman, & from TX: Hale. It is an annual species, but it has long hairs on the pedicels and an awnless fertile floret. These specimens were referred to *E. gracile* [err. for *gracilis*] (Fourn.) Hitchc. in the Atlas GP, but that species cannot satisfactorily be separated from *E. lemmonii* (see Gould, Leafl. W. Bot. 6: 50-51. 1950).

34. FESTUCA L., Fescue

Tufted or solitary stemmed annuals or rhizomatous to nonrhizomatous perennials. Culms hollow, erect to geniculate below. Blades rolled in the bud, flat to involute at maturity; sheaths open to the base; ligules short, membranaceous, often highest at the sides, sometimes ciliate; auricles absent or present. Inflorescence a panicle, sometimes racemose in depauperate plants; spikelets bearing 2-numerous florets, disarticulating above the glumes and between the florets; glumes unequal, the first one shorter, both of them normally shorter than the first lemma, the first one shorter, both of them normally shorter than the first lemma, the first glume 1-nerved, the second 3-nerved; lemmas mostly acute, 5-nerved, the nerves plainly visible to very obscure, awnless or with an awn arising at or near the tip; paleas equal to or exceeding the lemmas. *Vulpia* C. C. Gmel.

```
1  Plants annual ............................................................................. 3. F. octoflora
1  Plants perennial.
   2  Auricles prominent; blades mostly 2-8 mm broad and flat; coarse introduced grasses.
      3  Auricles ciliate-margined ................................................. 1. F. arundinacea
      3  Auricles not ciliate-margined ............................................ 6. F. pratensis
   2  Auricles absent; blades variable, but often involute.
      4  Anthers more than 2 mm long; second glume more than 5.5 mm long, the first lem-
         ma often shorter; ligules ciliate-margined, the fringe as long as the
         membrane ...................................................................... 7. F. scabrella
      4  Anthers variable but usually 2 mm or less in length; second glume often less than 5.5
         mm long, but usually longer than the lowest lemma; ligules sometimes ciliolate, but
         lacking cilia as long as the membranaceous portion.
         5  Spikelets unawned; most blades more than 2 mm wide, usually flat.
            6  Inflorescence very open, the lower panicle branches with 2-7 spikelets; plants
               of woodland .......................................................... 2. F. obtusa
            6  Inflorescence less open, the lower branches bearing more numerous spikelets;
               plants of prairie or prairie edge ........................... 5. F. paradoxa
         5  Spikelets usually awned; blades more or less filiform,
            involute .......................................................... 4. F. ovina var. rydbergii
```

1. **Festuca arundinacea** Schreb., tall fescue. Stout cespitose to rhizomatous perennials (4)7–14 dm tall. Culms erect, glabrous. Blades flat, 5–45 cm long, mostly 3–8 mm wide; sheaths smooth to scabrous; ligules collarlike, 0.2–0.6 mm long; auricles prominent, ciliate on the margins. Inflorescence open to relatively narrow, (9)15–25(36) cm long; spikelets 3- to 6(9)-flowered, 7–13 mm long, 2.5–5.5 mm wide; first glume 2.5–5(6) mm long; upper glume 3.5–5.5(6.5) mm long; lowest lemma 5.5–8 mm long, awnless or with an awn up to 1.5 mm long borne just back of the tip. Anthers 2.3–4 mm long. (2n = 42, 2n = 28, 56, 63, and 70 also reported) May–Oct. Along roads & ditches & in pastures, often where moist; GP, especially the e part; (sporadic throughout much of U.S.; Europe, widely planted & escaped elsewhere). *Naturalized.*

This grass has been rather commonly confused with *F. pratensis*, to which it is very similar (see Terrell, Rhodora 70: 564–567. 1968). It overlaps that species in most features, except for the ciliated auricles, but is generally larger in height, spikelet size, glume length, and lemma length. It is normally hexaploid, and the mean pollen diameter is 35–40 μm in our material. This is in contrast to our material of *F. pratensis* which has the mean pollen diameter 30–34μm. Many of the dots under *F. pratensis* in the Atlas GP, especially those representing recent collections, have proven to be specimens of *F. arundinacea* rather than *F. pratensis*, but this does not change the overall range given for the 2 species.

Festuca arundinacea and *F. pratensis* are closely related to *Lolium perenne*, q.v.

2. **Festuca obtusa** Biehler, nodding fescue. Nonrhizomatous perennials 4–10(11) dm tall. Blades flat, glabrous to scabrous or thinly pubescent above, the larger ones 10–20(25) cm long, 3–7 mm wide; sheaths smooth or thinly pubescent; ligules collarlike, 0.3–0.6(1) mm long, brownish; auricles lacking. Panicle very open, often nodding at maturity, the spikelets overlapping and mostly near the branch tips, only 2–7 of them on each of the lower branches; spikelets 2- to 5-flowered, 4–7 mm long, 1–3(4.5) mm wide; first glume 2–4 mm long; second glume 2.4–4.5 mm long; lemmas smooth, acute-tipped, the nerves very obscure, the lowest one 3–5 mm long. Anthers 0.7–1.4 mm long. (2n = 42) Jun–Aug. Moist woodlands; MN, e ND, e SD, IA, NE, e KS, MO, e OK; (N.S. to s Ont. & Man., s to FL, MS, & TX). *F. nutans* Willd.—Winter.

3. **Festuca octoflora** Walt., sixweeks fescue. Solitary-stemmed or tufted annuals, 0.5–4 dm tall. Culms erect, smooth. Blades involute, glabrous to finely pubescent above, 1–5 cm long, up to about 1.2 mm wide; sheaths smooth or finely pubescent; ligules 0.1–0.7 mm long, highest on the sides; auricles absent. Inflorescence a slender panicle with ascending branches or sometimes a raceme, 1.5–9 cm long; spikelets 7 to 11(13)-flowered, about 6–9 mm long; first glume 1.8–3.6 mm long; second glume 2.7–4.2 mm long; lemmas obscurely nerved, the lowest one 3–5 mm long, bearing an awn tip up to about 2.5 mm long, the upper spikelets often with slightly longer awns. Anther 1, 0.1–0.6 mm long, usually retained within the spikelet. (2n = 14) Apr–Jun. Prairies, pastures, & waste ground; GP; (Que. to B. C., throughout U.S., Mex., adventive elsewhere). *Vulpia octoflora* (Walt.) Rydb. var. *octoflora*, var. *tenella* (Willd.) Fern., var. *glauca* (Nutt.) Fern.—Fernald.

Although some authors have recognized several varieties in this species based on spikelet size and awn length (Lonard and Gould, Madroño 22: 217–230. 1974), the recognition of these is not entirely satisfactory. Pubescence features may be quite variable in the same population and even on the same plant.

Two other annual fescues have been collected in our region. Both may be readily distinguished from *F. octoflora* because their spikelets are mostly fewer than 6-flowered. *F. sciurea* Nutt. is known from MO: Jasper; & from OK: Blaine, Kingfisher, & Payne. It has small spikelets, the first glume less than 3.5 mm long, and the lemmas are appressed-pubescent over the back. *F. myuros* L. is known from KS: Douglas; OK: Cleveland; & TX: Carson. It may be recognized by its very short first glume, which is less than half the length of the second glume.

4. **Festuca ovina** L. var. **rydbergii** St. Yves, sheep's fescue. Strongly cespitose perennials 3–7(11) dm tall. Culms glabrous, most of the leaves distributed toward the base. Blades

glabrous to scabrous, involute, 1–10(20) cm long, 0.5–1.2 mm wide, ligules 0.1–0.3 mm long, highest on the sides; auricles absent. Panicles usually rather condensed, 1–9(11) cm long; spikelets 3- to 6-flowered, 4–5.5(8) mm long, 1–3 mm wide; first glume 1.5–3 mm long; second glume 2.5–4.5 mm long; lowest lemma 2.5–5 mm long, obscurely nerved, bearing a terminal awn 0.4–1.5 mm long, the upper lemmas often with somewhat longer awns. Anthers 0.8–1.7(2.3) mm long. (Chromosome number for this variety perhaps not reported, but 2n = 14, 28, 42, and various aneuploid counts known for the species) Jun–Aug. In a variety of habitats, but mostly on open prairie slopes; MN, ND, MT, WY, CO, rare in NE; (Newf. to B. C., s to MN, NE, CO, UT, & OR). *F. saximontana* Rydb. — Rydberg.

Festuca ovina var. *rydbergii* is relatively common in the n GP, but many of the more southern locations dotted in the Atlas GP probably represent other varieties of *F. ovina* (especially *F. ovina* var. *duriuscula* (L.) Koch and varieties of the similar *F. rubra* L. which have been introduced in lawn mixtures. *F. rubra* is rather readily distinguished from *F. ovina* by its decumbent, somewhat rhizomatous base with the lowest sheaths becoming reddish or brownish and tearing into fibers. It also has, on the average, larger spikelets and longer anthers.

Two other perennial species of *Festuca* are known from the w edge of our region; both will probably key here. *F. subulata* Trin. has been collected in SD: Lawrence, and *F. idahoensis* Elmer has been collected in CO: El Paso. *F. subulata* may be recognized by its broad, flat blades (mostly over 3 mm wide), its drooping panicles, and its long awns (mostly longer than the body of the lemma). *F. idahoensis* is a strongly cespitose species which normally has longer awns than *F. ovina* (2–5 mm long). It also has longer glumes, longer lemmas, and longer anthers.

5. ***Festuca paradoxa*** Desv., cluster fescue. Nonrhizomatous perennials 6–13 dm tall. Culms glabrous. Blades flat, scabrous, 10–35 cm long, 4–10 mm wide; sheaths glabrous; ligules 0.3–0.9 mm long; auricles absent. Panicles open, the branches somewhat ascending, the lowest ones each normally bearing more than 7 spikelets, the spikelets not as prominently clustered toward the end of the branch as in *F. obtusa*; spikelets ovate, 2- to 6-flowered, about 4–7 mm long, 1.5–5 mm wide; first glume 2–3.5 mm long; second glume 3–4.5 mm long; lemmas obtuse, the lowest 3–5 mm long, awnless. Anthers mostly 0.8–1.1 mm long. May–Jul. Prairies, woodland margins, & open woods; IA, se NE, e KS, MO; (PA to WI & NE, s to SC & e TX). *F. shortii* Kunth — Rydb.

6. ***Festuca pratensis*** Huds., meadow fescue. Stout, cespitose to short-rhizomatous perennials (3)5–9(11) dm tall. Culms erect, glabrous. Blades flat, (3)5–20 cm long, mostly 2–5(7.5) mm wide; sheaths smooth to scabrous; ligules collarlike, 0.3–0.5(0.7) mm long; auricles prominent, not ciliate. Panicles relatively narrow, (7)10–18(20) cm long; spikelets 4- to 10-flowered, 10–15 mm long, 2.5–5.5 mm wide; first glume 1.5–3.5(4) mm long; second glume 2.9–5.0 mm long; lemmas awnless or with short awns up to 2 mm long, the lowest lemma 5–7 mm long. Anthers 2–3.5 mm long. (2n = 14, 2n = 28, 42 also reported) Jun–Oct. In pastures, along roads, & in ditches; GP, especially in the e part; (sporadic throughout much of U.S.; Europe, widely naturalized elsewhere). *Naturalized*. *F. elatior* L. — Rydberg.

This grass has often been called *F. elatior*, but that name should be regarded as a *nomen confusum*. Please see the discussion following *F. arundinacea*.

7. ***Festuca scabrella*** Torr. in Hook; rough fescue. Strongly cespitose perennials (1.5)4–7.5 dm tall. Culms smooth to dull and scabrous, especially above, erect, with the leaves distributed mostly toward the base. Blades stiff, scabrous on both surfaces, strongly ridged above, involute, (5)15–37 cm long, 0.5–2(2.5) mm wide; ligule collarlike, highest on the sides, 0.1–0.6 mm long, long-ciliate; auricles lacking. Panicle condensed, 4–14(16) cm long; spikelets mostly 2- or 3-flowered, 6–9 mm long, 1.5–3.5 mm wide; first glume 5–8.5 mm long; second glume 6.5–8(9) mm long; lemmas acuminate to awn-tipped, the lowest one often shorter than the second glume, normally only 5.5–7 mm long, its awn, when present, up to 1 mm long. Anthers 2.2–3.6 mm long. (2n = 28, 56) Jun–Jul. Open prairie

slopes; n ND, MT; (Newf. to B. C., s to ND, MT, CO, ID, & OR). *F. campestris* Rydb.— Rydberg.

Festuca versuta Beal, a broad-leaved species of s distribution, approaches our range from the s and has been collected in KS: Labette. Because of its large spikelets, it will key here to *F. scabrella*, but it is a tall, broad-bladed species with a very open panicle.

35. GLYCERIA R. Br., Mannagrass

Rhizomatous perennials. Culms erect to partially decumbent and rooting at the nodes, hollow. Blades rolled or somewhat folded in the bud, flat to folded at maturity, soft, ridged between the veins, with both air chambers and cross-septae, the septae usually visible in transmitted light, midrib usually conspicuous; sheaths closed, at least in the lower part, also with air chambers and cross-septae, rounded to slightly keeled; ligules membranaceous, over 1 mm long; auricles lacking. Inflorescence an open panicle with stiffly spreading to nodding branches; spikelets 3- to 13-flowered, disarticulating readily between the florets and above the glumes; glumes shorter than the lowest lemma, 1-nerved; lemmas awnless, rounded, 5- to 9-nerved, the nerves running to near the margins and not strongly convergent; paleas 2-keeled, about equal to the lemmas. Stamens 2 or 3.

The report of *Glyceria canadensis* (Michx.) Trin. for KS by Rydberg is believed to be erroneous.

1 Spikelets mostly over 9 mm long, 7- to 13-flowered; lowest lemma over 3 mm long .... 1. *G. borealis*
1 Spikelets under 9 mm long, mostly 3- to 7-flowered; lowest lemma under 2 mm long.
   2 Lowest lemma more than 2 mm long; second glume more than 1.5 mm long; anthers usually 3 .... 2. *G. grandis*
   2 Lowest lemma less than 2 mm long; second glume less than 1.5 mm long; anthers usually 2 .... 3. *G. striata*

1. Glyceria borealis (Nash) Batch., northern mannagrass. Rhizomatous perennials 6–11(13.5) dm tall. Culms erect, but often decumbent and rooting at the lower nodes, smooth. Blades flat to folded, soft, smooth below and finely papillate above, 10–21(28) cm long, (1.2)3.5–6(8.5) mm wide; sheaths glabrous, slightly keeled, closed to near the throat or open above; ligules acute, 5–12 mm long, the upper portion usually withered at maturity. Inflorescence 23–40 cm long, at first narrow but later opening up, the spikelets ascending and appressed along the major branches; spikelets mostly 7- to 13-flowered, subterete, 7–17 mm long and 1.7–2.9 mm wide. Glumes oblong to lanceolate, thin, the lower one 1–3 mm long, the upper 2.4–3.7 mm long; lemmas 7-nerved, the nerves scabridulous, the margin scarious, the lowest one 3.3–4.3 mm long; paleas slightly shorter than the lemmas. Anthers 3, 0.5–0.9 mm long. (2n = 20) Jun–Aug. Marshes, creeks, ditches, & wet meadows, often partly submerged; MN, ND, IA, w SD, & nw NE; (Newf. to AK s to PA, IA, AZ, CA).

Glyceria fluitans (L.) R. Br., a similar species with more scabrous, longer lemmas 4–7 mm long, has been collected in SD: Custer.

2. Glyceria grandis S. Wats. ex A. Gray, tall mannagrass. Rhizomatous perennials 9–13(16) dm tall. Culms erect, stout. Blades relatively flat, smooth to scabrous, soft, 15–45 cm long, 7–14 mm wide; sheaths smooth, relatively loose, closed to near the summit or open above, not or only slightly keeled; ligules mostly 2–5 mm long, often acute or even acuminate. Inflorescence 24–35 cm long, open, the branches often nodding; spikelets mostly 4- to 7-flowered, about 4–6 mm long and 1.5–2.5 mm wide. Glumes lanceolate-ovate, papery, the lower one 1–2 mm long, the upper 1.7–2.4 mm long; lemmas obtuse, 7-nerved to near the margin, scabridulous, often reddish-tinged at maturity, the lowest mostly 2.1–2.9 mm long; palea about equal to the lemma. Anthers usually 3, 0.7–1.1 mm long. (2n = 20) Jun–Aug. Damp ground, often at the edge of water; MN,ND, MT, SD, CO, IA, NE, WY (e Can. to AK, s to VA, NE, AZ, & OR).

3. *Glyceria striata* (Lam.) Hitchc., fowl mannagrass. Rhizomatous perennials 5–10(14.5) dm tall. Culms stout to slender, erect to decumbent at the base and rooting at the lower nodes, smooth; blades smooth to scabrous, flat to folded, 10–40 cm long, (2)3–6 mm wide; sheaths scabrous, slightly keeled, closed to near the throat; ligules 1–4.5 mm long, sometimes closed in front. Inflorescence mostly 8–25 cm long, open, nodding; spikelets 3- to 6-flowered, 2.5–3.8 mm long and 1.3–2.4 mm wide; glumes ovate, papery, the first 0.6–1 mm long, the second 0.7–1.4 mm long; lemmas obtuse, 7-nerved to the summit, scabridulous, the lowest 1.3–2.0 mm long; paleas equal to or exceeding the lemmas. Anthers 2, 0.3–0.5 mm long. (2n = 20) May–Jul. Wet ground along streams, ponds, & lakes & in wet woods; GP; (Newf. to AK, s throughout most of the U.S., not common in the s). *G. nervata* (Willd.) Trin., *G. rigida* (Nash) Rydb. — Rydberg

36. GYMNOPOGON Beauv., Skeleton Grass

1. *Gymnopogon ambiguus* (Michx.) B.S.P., bearded skeleton grass. Perennial 2–6 dm tall from a knotty, short-rhizomatous base. Blades spreading, rolled in the bud, flat to somewhat involute at maturity, glabrous except for the pubescent collar, lacking a midrib, 3–9 cm long, 4–11 mm wide, abruptly narrowed at the base; sheaths smooth, long-pubescent at the collar, open to the base, the upper ones often somewhat inflated, not keeled; ligule a minute membrane 0.2–0.3 mm long; auricles none. Inflorescence an open panicle of spikelike racemes, its overall length 12–18(25) cm, its central axis about 5–10 cm long, its individual branches stiff, ascending at first, then becoming reflexed, the longest 8–15(18) cm long, each branch often with a hirsute callus at its base; spikelets short pediceled, appressed, borne in 2 rows on one side of each branch; spikelets 2-flowered, the lower floret perfect, the upper one rudimentary, disarticulation above the glumes, the 2 florets falling together; glumes narrow, stiff, 1-nerved, the first one 3–5 mm long, the second one often slightly longer, 3.5–5.5 mm; lemmas more delicate than the glumes, 3-nerved, 3–4 mm long, bearing a tuft of hairs at the base and often additional pubescence as well, awned from the tip, the awn 3–8.5 mm long; paleas slightly longer than the lemmas; rudimentary upper floret consisting solely of an awn (0.4)1.5–4.5(6.5) mm long on a stipe 1–2.5 mm long. Anthers 0.6–0.9 mm long. (2n = 40) Aug–Nov. Usually in openings in dry woods in rocky or sandy soil; se KS, MO, e OK; (NJ to KS, s to FL & TX).

37. HELICTOTRICHON Bess. ex J. A. Schult. & J. H. Schult., Spike Oat

1. *Helictotrichon hookeri* (Scribn.) Henr., spike oat. Cespitose perennials 1–7.5 dm tall. Culms erect, smooth, hollow. Blades glabrous to scabridulous, thickened and whitish on the margins, with the midrib prominent and whitish below, folded in the bud and usually also at maturity, the tip boat-shaped, mostly 2–20 cm long, 1–4.5 mm wide when unfolded; sheaths glabrous, keeled, open to the base; ligules membranaceous, whitish, usually lacerate, 3–7 mm long on the culm leaves; auricles lacking. Inflorescence a congested panicle or raceme (4)6–11(13) cm long, most of its branches bearing only 1 or 2 large spikelets; spikelets 3- to 5-flowered, disarticulating above the glumes and between the florets, the rachilla bearded; glumes about as long as the spikelet, acute-tipped, rather thin and membranaceous, the first 3-nerved, 9–13 mm long, the second 3- to 5-nerved, (9)11–14 mm long; lemmas firm, 5-nerved, the lowest one 10–12 mm long, awned from about midlength

with a strong, geniculate, twisted awn 12–16 mm long. Anthers 2.7–4.5 mm long. (2n = 14) Jun–Aug. Open prairie slopes; MN, ND, MT; (Man. to Alta., s to MN, MT, & NM). *Avena hookeri* Scribn.—Rydberg.

38. HIEROCHLOË R. Br., Sweetgrass

1. *Hierochloë odorata* (L.) Beauv., sweetgrass. Rhizomatous perennials 1–7(8.5) dm tall. Culms erect, hollow, glabrous. Blades rolled in the bud, flat at maturity, usually nearly glabrous, mostly 10–30 cm long, but those of the culm leaves much shorter, (1)3–5(6) mm wide; sheaths glabrous or puberulent, especially at the collar, open to the base; ligule membranaceous, obtuse to acute, mostly 1–3 mm long on the culm leaves; auricles absent. Inflorescence a panicle (2)4–9 cm long; spikelets 3-flowered, the 2 lowest florets staminate, the uppermost one perfect, disarticulating above the glumes, the 3 florets falling as a unit. Glumes glabrous, broad, membranaceous, the lower one 3.5–5.6 mm long, 1- to 3-nerved, the upper one 3.8–6 mm long, 3-nerved; lemmas of the 2 staminate florets 5-nerved, strongly pubescent, especially on the margin, the lower one 3.4–5 mm long, sometimes bearing a short awn less than 1 mm long, the upper one only slightly shorter; lemma of the perfect floret 3- to 7-nerved, pubescent, especially above, 2.5–3.5 mm long, very rarely with a short awn tip. Anthers of the lowest staminate floret about 1.6–2.2 mm long, those of the perfect floret about 1–1.6 mm long. (2n = 28, 42, 56) May–Jul. Wet meadows, low prairies, & edges of sloughs & marshes; MN, ND, MT, SD, & nw IA; (Lab. to AK, s to NJ, IN, IA, NM, & AZ; Eurasia).

Our material was assigned by Weimarck (Bot. Notiser 124: 129–175. 1971) to *H. hirta* (Schrank) Borbas ssp. *arctica* (Presl) G. Weim., but that taxon is distinguished from *H. odorata* on rather nebulous, nonquantitative characteristics.

39. HILARIA H.B.K.

1. *Hilaria jamesii* (Torr.) Benth., galleta. Rhizomatous perennials 2.5–5.5(7.5) dm tall. Culms solid, ridged, glabrous to scabrous or minutely pubescent, often long-pilose at the nodes. Blades rolled in the bud, involute to flat at maturity, scabrous to minutely pubescent or long-pilose near the ligule, strongly ridged on both surfaces, but the ridges below very closely crowded, mostly 1–15(21) cm long, 1.5–3.3(4.2) mm wide; sheaths grooved, smooth to scabrous, open to the base, often long-pilose on the margins in the region of the ligule; ligules membranaceous, ciliate, usually lacerate, 0.8–3.6 mm long; auricles lacking. Inflorescence a bilateral spike (2)3.5–7 cm long with spikelets borne in groups of 3, disarticulating at the base of each group, each group pilose at its base with hairs (1.5)3–5 mm long; lateral spikelets (1)2(3)-flowered, staminate, borne in front of the central 1-flowered perfect spikelet; lateral spikelets with firm, lanceolate glumes, the lowest (medial) glume 3- to 5-nerved, 4.5–7 mm long, bearing a stout awn 2–3.5 mm above its base, the awn 3–6.2 mm long, the upper (lateral) glume 4- to 5-nerved, unawned or merely mucronate, 5.1–7.2(9) mm long; the lemmas of the lateral spikelets 3-nerved, the lower one 4.4–7 mm long; paleas of the lateral spikelets 2-nerved, subequal to the lemmas; glumes of the central spikelet subequal, short, usually 5.5–8(9) mm long including the 5–10 awns, slightly keeled, one or more of the awns arising from the keel; lemma of the central spikelet 3-nerved, 5.8–7.6 mm long, bearing an awn 1.2–2.8(3.9) mm long just back of the apex. Anthers of lateral and central spikelets subequal, about 2.5–4.5 mm long. (2n = 18, 36, 38) May–Oct, mostly Jun–Jul. Dry prairies & slopes; CO, w KS, extreme w OK, TX, NM; (WY & UT to TX & CA).

Hilaria mutica (Buckl.) Benth. (tobosa) is a similar species known from the edge of our range (OK: Harmon, Jackson; TX: Armstrong). It may be distinguished by the broadened, flabellate glumes of the staminate spikelets.

40. HORDEUM L., Barley

Tufted or solitary-stemmed annuals or cespitose perennials. Culms hollow, erect. Blades rolled in the bud, flat at maturity; sheaths open to the base; ligule a short, truncate, collarlike membrane; auricles present or absent. Inflorescence bilateral and spikelike, disarticulating at each node at maturity; spikelets usually 1-flowered, normally borne in groups of 3 at each node, the central one of each group perfect, sessile, the 2 lateral ones short-pedicellate, rudimentary and neuter or, rarely, staminate; each spikelet group falling attached to the internode below; glumes narrow, awned, or the whole glume awnlike, borne in front of the floret; fertile spikelet with a rachilla behind its floret, this terminating in a bristle or, rarely, in a small rudimentary floret; fertile lemma obscurely 5-nerved, awned; palea about as long as the body of the lemma.

Hordeum jubatum forms natural sterile hybrids with several other genera. Hybrids with *Agropyron* are common enough in our range to afford full treatment to the genus x *Agrohordeum* in this flora. Hybrids with *Elymus*, although less commonly formed, have been reported from within our range (Bowden, Canad. J. Bot. 45: 711–724. 1967). Hybrids with *Elymus canadensis* [x *Elyhordeum dakotense* (Bowden) Bowden = *Hordeum pammeli* Scribn. & Ball — Rydberg] and *Elymus virginicus* [x *Elyhordeum montanense* (Scribn.) Bowden] have been reported from the Dakotas, and the hybrid with *Elymus villosus* (x *Elyhordeum iowense* R. Pohl) also seems likely to occur. Such hybrids might key to this genus, because they usually have 3 spikelets per node and a disarticulating rachis, but the spikelets have 1–4 florets and are completely sterile.

1 Plants perennial; glumes, including the awns, more than 2 cm long .... 1. *H. jubatum*
1 Plants annual; glumes, including the awns, not more than 2 cm long .... 2. *H. pusillum*

1. ***Hordeum jubatum*** L., foxtail barley. Tufted perennials (1.5)3–8 dm tall. Culms hollow, erect to geniculate below, glabrous. Blades flat, rather evenly ridged above, scabrous to finely pubescent, 2–13 cm long, 1.4–4.5(5.5) mm wide; sheaths smooth; ligules ciliolate, 0.2–0.6(1.2) mm long; auricles small or absent, sometimes present on some leaves but not on others. Inflorescence nodding, sometimes partly included in the uppermost leaf sheath, 7.5–12.5(14) cm long, including the awns; glumes setaceous, scabrous, those of the fertile central spikelet 2.5–6.8 cm long, those of the lateral spikelets not much shorter; lemma of central spikelet narrow, tapering into a scabrous awn, the body (4)5–8 mm long, the awn 2.5–6.2 cm long; lemmas of lateral spikelets acuminate or awned, 3.6–8(12) mm long, including the tip. Anthers 0.8–1.6 mm long, exserted, or sometimes included, at maturity. (2n = 28) Mostly Jun–Aug. Roadsides, pastures, & waste ground; GP; (most of N. Amer., except the se states).

Hordeum brachyantherum Nevski, a smaller perennial species with shorter glumes, grows at the nw edge of our range (MT: Carter, Daniels, McCone, Powder River, Prairie, & Rosebud). *H. caespitosum* Scribn. of Rydberg's manual probably represents a hybrid between this species and *H. jubatum* (Mitchell & Wilton, Madroño 17: 269–280. 1964.)

Hordeum jubatum also hybridizes with certain species of other genera (see the discussion above).

2. ***Hordeum pusillum*** Nutt., little barley. Tufted or solitary-stemmed annuals 1–4 dm tall. Culms erect to geniculate below, glabrous. Blades scabrous to densely pubescent, rather evenly ridged above, 1–10 cm long, 1.5–4.5 mm wide; sheaths smooth to densely pubescent, especially the lower ones; ligules finely ciliolate, 0.1–0.6 mm long; auricules none. Inflorescence erect, rather stiff, sometimes partly included in the uppermost leaf sheath, 2.5–7 cm long; glumes of the fertile central spikelet alike, scabrous to pubescent, indistinctly

3-nerved, bowed out at the base, with a lanceolate body 3.4–5.5 mm long, tapering into a scabrous awn 2.5–7.0 mm long; glumes of the 2 lateral pedicellate spikelets dimorphic, the lower ones like the glumes of the central spikelet, the upper ones setaceous, 5–11 mm long; lemma of central floret scabrous to pubescent, with a body 4.5–6.5 mm long, tapering into an awn (1.5)3.5–8 mm long; lemma of rudimentary floret acuminate to awn-tipped, 3.3–5.8 mm long, including the tip. Anthers 0.6–1.2 mm long, often retained within the spikelets. (2n = 14) May–Jun. Waste ground; GP, less common in the n part; (most of the U.S., n Mex.).

Hordeum vulgare L., cultivated barley, does not become established after cultivation in our region, but may persist for a time. It differs from *H. pusillum* in its much larger size, its continuous, nondisarticulating rachis, its fertile lateral spikelets, and in many other features.

41. HYSTRIX Moench, Bottlebrush Grass

1. ***Hystrix patula*** Moench, bottlebrush grass. Tufted or solitary-stemmed perennials (5)7.5–12 dm tall. Culms erect or geniculate below, hollow, smooth. Blades rolled in the bud, flat at maturity, usually thinly pilose above and also finely scabrous, mostly 12–15(30) cm long, 5.5–10(12.4) mm wide; sheaths open, not keeled, usually smooth but rarely pilose; ligule a short, truncate, ciliolate membrane 0.3–0.7 mm long; auricles present and conspicuous to, rarely, inconspicuous or absent. Inflorescence a loose spike with distant, deciduous, divergent spikelets borne mostly in pairs (1–3) on both sides of a persistent rachis, 9–16(18) cm long, including awns; spikelets 2- to 4-flowered, usually glumeless, but some of the spikelets sometimes bearing 1 or 2 short, scalelike glumes or bristlelike glumes up to 1 cm long, these seldom associated with all spikelets and then not normally more than 1 per spikelet; lemmas lanceolate, pubescent to glabrous, obscurely 5-nerved, 8–11 mm long, tapering into a scabrous awn 1.8–3.8 cm long. Anthers 2.4–4.2 mm long. (2n = 28) Jun–Oct, mostly Jul. In rich soil in woods; MN, e ND, se SD, IA, e NE, MO, e KS; (N.S. to ND, s throughout much of e U.S. to GA & AR). *H. patula* var. *bigeloviana* (Fern.) Deam — Fernald; *Elymus hystrix* L., *E. hystrix* L. var. *bigelovianus* (Fern.) Bowden — Bowden, Canad. J. Bot. 42: 586–587. 1964.

42. KOELERIA Pers., Junegrass

1. ***Koeleria pyramidata*** (Lam.) Beauv., Junegrass. Strongly cespitose perennial 2–6 dm tall. Culms hollow, erect glabrous to, usually, minutely puberulent in the inflorescence and near the nodes. Leaves mostly basally distributed; blades folded in the bud, flat, somewhat involute or folded at maturity, glabrous to weakly or densely short-hispid, thickish, finely striate beneath, with relatively few, broad ribs above, these separated by rather deep, narrow furrows, the tip boat-shaped, 2–18 cm long, 1–3(3.7) mm wide; sheaths glabrous to short-hispid, open to the base; ligules whitish, membranaceous, 0.3–1.5 mm long, sometimes highest on the sides, ciliolate, often pubescent. Inflorescence a congested panicle 3–9(15) cm long of (1)2(4) flowered spikelets; spikelets flattened laterally, obovate, disarticulating above the glumes and between the florets, variable in size upon each panicle branch; glumes usually scabrous, at least on the keel, the first one lanceolate, 1-nerved, 1.9–4.3(4.8) mm long, the second oblanceolate, 1- to 3-nerved, 2.5–5.5 mm long; lemmas 5-nerved, the lateral nerves faint, scabrous, at least on the midnerve, awnless, 2.5–4.8 mm long; lowest lemma usually reaching the tip of the longest glume. Anthers 1.1–2 mm long. (2n = 14, 28, 2n = 56, 70, 84 also reported) May–Aug, mostly Jun. Common in dry prairie & open woods; GP; (throughout temp. Northern Hemisphere). *K. gracilis* Pers., *K. nitida* Nutt., *K. latifrons* (Domin) Rydb.—Rydberg; *K. cristata* Pers.—Fernald; *K. macrantha* (Ledeb.) Schult.—Voss, Michigan Flora 1: 166. 1972.

43. LEERSIA Sw.

Rhizomatous, weak-stemmed perennials. Culms hollow, pubescent at the nodes, often geniculate. Blades rolled in the bud, flat at maturity; sheaths open; ligules membranaceous, firm, flanked by and fused with sheath extensions at each margin; auricles otherwise none. Inflorescence an open panicle with spikelets appressed along the branches. Spikelets 1-flowered, laterally compressed, disarticulating at the base; glumes absent; lemma broad, folded, 5-nerved, enclosing the margins of the narrow, 3-nerved palea. Anthers 2, 3, or sometimes only 1.

In addition to the species described below, *Leersia lenticularis* Michx. is known from several collections in our range (KS: Linn, Miami; MO: Bates, Clinton, Jasper, Vernon). The described species have the spikelets under 2 mm broad; *L. lenticularis* has them 3 mm broad or broader.

1 Sheaths with air-chambers, the chambers abundantly cross-septate, the septae visible in transmitted light; most lemmas over 4 mm long .... 1. *L. oryzoides*
1 Sheaths lacking air-chambers, not cross-septate; most lemmas under 4 mm long .... 2. *L. virginica*

1. ***Leersia oryzoides*** (L.) Sw., rice cutgrass. Perennial, tufted, but with long slender rhizomes. Culms 4–11 dm long, terete, retrorse-pubescent at the nodes, often decumbent below. Blades flat, harsh, strongly scabrous, especially on the margins, (4)10–31 cm long, (5)7–10(12) mm wide; sheaths slightly keeled, cross-septate, glabrous to usually scabrous; ligules ciliolate, 0.4–2.0 mm long. Panicle open, 14–22(30) cm long, sometimes partly included in the sheath of the uppermost leaf, the lowest branches, when visible, usually whorled. Spikelets 3.5–5.2 mm long, the lemma and palea hispid on the nerves, especially the keels. Anthers 3(1 or 2), included or exserted, extremely variable in size, 0.2–2.4 mm long, the short anthers normally associated with cleistogamous florets. (2n = 48) Jul–Oct, mostly Aug–Sep. Along ditches, streams, ponds, lakes, & marshes; GP; (Que. to WA, s throughout most of U.S.; Europe).

2. ***Leersia virginica*** Willd., whitegrass. Rhizomatous perennial, the rhizomes short, dark, with closely placed, overlapping scales. Culms 4.5–12 dm long, mostly scrambling or decumbent, sometimes rooting at the lower nodes, hollow, somewhat flattened, retrorse-pubescent at the nodes; blades flat, usually scabrous but not particularly harsh, 4–20 cm long, 3–10(12.5) mm wide; sheaths keeled, not cross-septate, glabrous to scabrous; ligule 0.4–1.5 mm long, often somewhat jagged. Panicle open, 7–18(22) cm long, sometimes partly included in the sheath of the uppermost leaf, lowest branches not whorled; spikelets 2.7–4.1 mm long, the nerves, especially the keels, of the lemma and palea often hispid. Anthers (1)2(3), 1.1–1.8 mm long. (2n = 48) Jun–Oct, mostly Sep–Oct. Usually in moist ground in the woods; MN, extreme se ND, se SD, IA, NE, KS, MO, & OK; (Que. to ND, s to FL & TX). *L. virginica* var. *ovata* (Poir.) Fern.—Fernald.

44. LEPTOCHLOA Beauv., Sprangletop

Tufted annuals. Culms hollow, smooth. Blades rolled in the bud, flat at maturity, the midrib obscure to very prominent and whitish on the upper surface of the blade; sheaths open, slightly keeled; ligules membranaceous; auricles none. Inflorescence an elongate panicle with unbranched, spikelike, racemose branches, the spikelets moderately spaced in 2 rows on one side of each branch; spikelets 2- to 9-flowered, disarticulating above the glumes and between the florets; glumes lanceolate, membranaceous, 1-nerved, varying from shorter than the lowest lemma to longer than the entire floret group; lemmas 3-nerved, unawned and obtuse or retuse to awned from a bifid tip; paleas 2-nerved, slightly shorter than the lemmas. Including *Diplachne* Beauv.

In addition to the 2 annual species described below, a perennial species, *Leptochloa dubia* (H.B.K.) Nees, is known from the s edge of our range (TX: Armstrong, Briscoe, & Randall).

1 Lemmas of lowest florets more than 2.5 mm long, often short awned, not surpassed by the glumes .... 1. *L. fascicularis*
1 Lemmas of lowest florets less than 2.5 mm long, obtuse to retuse, unawned, often surpassed by the glumes .... 2. *L. filiformis*

1. ***Leptochloa fascicularis*** (Lam.) A. Gray, bearded sprangletop. Tufted annual 2–12 dm tall. Culms relatively stout, erect to partially or almost entirely decumbent. Blades flat, scabrous, slightly keeled, especially near the base, the midrib prominent and whitish, especially on the upper surface near the base, often several of the lateral ridges also clearly enlarged and whitish, mostly 6–51 cm long, 2–6 mm wide; sheaths slightly keeled, glabrous to scabrous; ligules lacerate, 2–6(8.5) mm long. Inflorescence 10–46 cm long, often partly included in the upper sheaths; spikelets 7–11 mm long, 5- to 9-flowered; glumes unequal, the first one 1.4–3.5 mm long, the second 2.5–4.6(6) mm long; lemmas bifid at the tip, often bearing a short awn up to 1.5 mm long, pubescent on the lower portions of all 3 nerves, the lowest lemma 3.2–4.5(6) mm long. Anthers 0.1–0.4 mm long, included or exserted. (2n = 20, 40) Jul–Oct, mostly Aug–Sep. In muddy or wet sandy soils in marshes, along ditches, & as a weed in gardens & parking lots; GP; (most of U.S. & trop. Amer. to Argentina). *Diplachne acuminata* Nash — Rydberg; *D. fascicularis* (Lam.) Beauv. — Fernald.

Although there are fundamental floral and vegetative similarities between our species of *Leptochloa*, the argument has been presented by McNeill (Brittonia 31: 399–404. 1979) for assigning them to 2 different genera based principally on the characters of the lemma and caryopsis. If this approach is followed (the usual treatment outside N. Amer.), then this species should be called *Diplachne fascicularis* (Lam.) Beauv.

McNeill (op. cit.) considers *Leptochloa acuminata* (Nash) Mohlenbrock (as *D. acuminata* Nash) to be a good species distinct from *L. fascicularis*, distinguished from that species by its larger glumes and lemmas. Keys provided by McNeill and by Mohlenbrock (Illustrated Flora of Illinois, Grasses: *Panicum* to *Danthonia*, S. Illinois Univ. Press, p. 288. 1973) to separate these species do not work well with our GP specimens, since the characters used are not bimodal. Because material from the type collection of *D. acuminata* Nash, seen at NY, does not even key unequivocally to that species, this matter appears to need further study.

2. ***Leptochloa filiformis*** (Lam.) Beauv., red sprangletop. Tufted annual 0.3–6.6(9.6) dm long. Culms slender, erect to decumbent or partially decumbent. Blades flat, usually thinly pustular-pilose, a whitish midrib obvious in the larger ones, especially on the upper surface near the base, several lateral ridges sometimes also clearly enlarged and whitish, mostly 2–20(24) cm long, (1.3)3–9.3 mm wide; sheaths slightly keeled, glabrous to thinly pustular-pilose; ligules 0.2–1.4(2.3) mm long. Inflorescence (2.4)13–31(40) cm long, often partly included in the uppermost leaf sheath; spikelets 1.2–3(3.8) mm long, 2- to 4-flowered; glumes subequal to unequal, 1.2–3(3.8) mm long, one or both of them normally as long as the spikelet; lemmas obtuse to retuse, glabrous to obscurely pubescent on the nerves, short, the lowest one only 0.7–1.8 mm long. Anthers 0.05–0.25 mm long, included or exserted. (2n = 20) Mostly Jul–Oct. Sandy or muddy ground near rivers, streams, & ponds & in disturbed places; e KS, MO, e OK, reported also from se SD, where doubtfully persistent; (VA to e KS, s to FL & sw to s CA). *L. attenuata* (Nutt.) Steud. — Fernald.

45. LEPTOLOMA Chase

1. ***Leptoloma cognatum*** (Schult.) Chase, fall witchgrass. Perennials (1.5)2.5–4.5(6) dm tall from a short-rhizomatous base. Culms erect to ascending, often decumbent and rooting at the lower nodes, hollow, glabrous. Blades glabrous to thinly long-pilose, rolled in the

bud, flat at maturity, one or both margins often somewhat crisped, 2–8(11) cm long, 1.2–5 mm wide; sheaths open to the base, the lowest ones usually densely long-pilose, the upper ones usually thinly pilose; ligule a firm obliquely truncate membrane 0.5–2 mm long, flanked by thickened edges which appear to represent an extension of the sheath; true auricles none. Inflorescence an open panicle with spreading branches and pedicels, (6.5)9–18(21) cm long; spikelets 2-flowered, the lower floret sterile, the upper perfect, the spikelet disarticulating below the glume or glumes; first glume absent or, more commonly, rudimentary, up to 0.4 mm long; second glume 2.1–2.9 mm long, strongly 3- to (5)-nerved, densely villous to nearly glabrous between the nerves; sterile lemma like the second glume in texture and pubescence, 5- to (7)-nerved, 2.3–3.1 mm long; sterile palea usually apparent as a short membrane up to 0.5 mm long; fertile lemma brownish, thin-margined, cartilaginous, smooth, obscurely 5-nerved, the margins somewhat inrolled and partly enclosing the palea. Anthers 0.3–0.7 mm long, exserted or retained within the floret. (2n = 72, 2n = 36, 70, 20 also reported) May–Oct, with 2 blooming peaks, one in early summer and the other in fall. Sandy soil, often in disturbed ground; se NE, MO, e KS, OK, TX, NM; (NH to MN, s to FL, TX, & AZ).

46. LOLIUM L., Ryegrass

Cespitose perennials or tufted to solitary-stemmed annuals. Culms hollow. Blades rolled to folded in the bud, usually flat at maturity; sheaths open; ligules membranaceous, short (mostly less than 1.5 mm long); auricles usually present. Inflorescence a solitary, terminal, bilateral spike which does not disarticulate at the nodes when mature; spikelets borne singly, rarely some of them paired, placed edgewise to the rachis, 4- to 15-flowered, disarticulating above the glume(s); first glume absent except in the terminal spikelet, the second glume prominent, stiff, 5- to 9-nerved, equal to or longer than the first lemma; first lemma with its back to the rachis, 5-nerved, awned or awnless.

1 Plants perennial or, rarely, annual; glumes usually less than 10 mm long, but if longer, not equaling or exceeding the spikelet; the body of the first lemma usually less than 8 mm long .... 1. *L. perenne*
1 Plants annual; glumes more than 10 mm long, equaling or exceeding the spikelet; the body of the first lemma usually more than 8 mm long .... 2. *L. persicum*

1. *Lolium perenne* L., ryegrass. Short-lived perennials, biennials, or, rarely, annuals 3–10 dm tall. Culms erect to geniculate below. Blades glabrous to scabrous above; rolled to folded in the bud, flat to slightly involute at maturity, mostly 3–26 cm long, 2–7.3 mm wide; sheaths rounded to slightly keeled; ligules membranaceous, 0.3–1.5 mm long; auricles usually present and conspicuous, sometimes lacking. Inflorescence 6–24(39) cm long; spikelets 4- to 11(15)-flowered; glume (4.5)6–10(11) mm long, normally not exceeding the spikelet; lemmas crowded, 5-nerved, the lowest one 5–7.5 mm long, awnless to short awned, the upper ones sometimes bearing awns up to 8 mm long. Anthers (2.3)3–4 mm long.

There are 2 varieties in our region:

1a. var. *aristatum* Willd., Italian ryegrass. Plants lacking numerous basal innovations. Blades usually rolled in the bud, relatively broad at maturity, (2)3–7 mm wide. Glumes 4.5–9 mm long; lemmas, at least the upper ones, awned. (2n = 14) May–Jul (Sep). Pastures, roadsides, & waste areas; of scattered occurrence in MN, ND, NE, MO, KS, OK; (ME to B.C., introduced throughout the U.S.; Europe). *Naturalized*. *L. multiflorum* Lam.—Rydb.

1b. var. *perenne*, perennial ryegrass. Plants with numerous basal innovations. Blades often folded in the bud, relatively narrow, (1.8)2–3.5(6) mm wide. Glumes 4.5–11 mm long; lemmas unawned. (2n = 14, 28) May–Jul (Nov). Pastures, lawns, roadsides, & disturbed areas; GP, most common in the s-cen. part; (Newf. to AK, s throughout most of U.S.; Europe). *Naturalized*.

Lolium perenne is closely related to *Festuca arundinacea* and *Festuca pratensis* and may form hybrids with either species (Stebbins, Evolution 10: 235–245. 1956; Terrell, Bot. Rev. (Lancaster) 32: 138–164. 1966). Such hybrids could occur in our region.

2. ***Lolium persicum*** Boiss. & Hohenack. Slender tufted or solitary-stemmed annuals, 2–6.5 dm tall. Culms erect to somewhat geniculate, smooth. Blades folded or rolled in the bud, flat to somewhat involute at maturity, nearly glabrous to scabrous above, mostly 6–16 cm long, 2–4(6) mm wide; sheaths smooth; ligules membranaceous, 0.4–0.9 mm long; auricles prominent. Inflorescence (4)6–15 cm long; spikelets 5- to 8-flowered; glumes lanceolate, stiff, 5- to 7-nerved, long, almost equal to, equal to, or exceeding the length of the entire spikelet, 10–16(18) mm long; lemmas 5-nerved, awned from near the tips, the awns 2–16 mm long, straight to somewhat scabrous, sinuate, the body of the lowest lemma 8.4–11 mm long. Anthers 1.8–3(3.5) mm long. (2n = 14) Jun–Jul (Aug). Weed in cult. fields, ditches, & roadsides; ND; (Ont. to Alta., ND; Russia). *Naturalized.*

Lolium temulentum L. has been collected from scattered locations in our range (KS: Cherokee, Riley, Wilson; MO: Jackson and Platte; ND: Richland, Stark, and Walsh; OK: Payne) but is not believed to be long-persistent. It will key here to *L. persicum,* but may be recognized by its smaller, more turgid spikelets.

47. LYCURUS H.B.K.

1. ***Lycurus phleoides*** H.B.K., wolftail. Tufted perennials 2–6 dm tall. Culms wiry, solid, ascending to decumbent below, somewhat geniculate, dull-puberulent, compressed. Leaves distichous; blades folded in bud and at maturity, the tips prow-shaped and acuminate, the margins and midnerve of the lower surface whitish and thickened, 1–13 cm long, 0.9–2.6 mm wide when unfolded; sheaths compressed, sharply keeled, scarious-margined, dull-puberulent toward the base, closed; ligule a whitish, lacerate to 3-lobed acuminate membrane (1)2.5–6.5 mm long; auricles none. Inflorescence a spikelike panicle (1.5)3–10.5 cm long, 0.4–1.3 cm thick, normally most of the ultimate panicle branches dividing into 2 unequal pedicels near the base, each pedicel bearing only 1 spikelet; spikelets all 1-flowered, but those on the longer pedicels usually perfect and those on the shorter pedicels either sterile or staminate (rarely perfect), most spikelets falling in pairs, the panicle branch disarticulating at its base, some of the uppermost and lowermost spikelets, however, often borne singly and falling singly, attached to their pedicels; glumes subequal, the body about 0.8–2.0 mm long, the first one 2(3)-nerved, bearing 2(3) equal or unequal awns (0.7)1.5–4 mm long, the second 1(2)-nerved, bearing 1(2) awn(s) 1.2–5 mm long; lemmas longer than the glumes, pubescent toward the margins, 3-nerved, the body about 2.5–4 mm long, tapering to a single awn 1.2–4 mm long; palea slightly shorter than the body of the lemma, pubescent. Anthers 1.4–2.2 mm long, normally longest in the perfect spikelets. (2n = 40) Mostly Jul–Sep. Dry hillsides & hilltops, often in rocky or sandy soil; extreme sw KS to CO, OK, NM, TX; (CO to UT, s to TX, AZ, & Mex.).

48. MANISURUS L., Jointtail

1. ***Manisurus cylindrica*** (Michx.) O. Ktze., Carolina jointtail. Perennials with short, knotty rhizomes. Culms mostly 5–11 dm tall, glabrous, solid. Blades folded in the bud, flat to V-shaped in cross section or folded at maturity, sometimes revolute marginally, nearly glabrous, mostly 4–30 cm long and 1.4–3.5 mm broad; sheaths glabrous, open, rounded to somewhat keeled; ligules membranaceous, ciliate, truncate, mostly 0.3–0.7 mm long; auricles none. Inflorescence consisting of 1–several spikelike, cylindrical racemes 6–15 cm

long, each raceme disarticulating at the nodes when mature; spikelets in pairs, each pair fitting into a concavity on the thickened internode above and falling attached to that internode, each spikelet pair consisting of a rudimentary spikelet 1–2.8 mm long borne on an arcuate ribbonlike pedicel 3–5.2 mm long and a sessile, fertile, 2-flowered spikelet, the fertile spikelet pressed sidewise against the rachis so that only the lower glume is visible when the spikelet is unopened; lower glume of fertile spikelet 3.5–5.5 mm long, thickened, about 7- to 9-nerved, usually with depressions more or less obvious in rows on the exposed surface; upper glume thinner, about 3- or 4-nerved, acuminate; lower floret of fertile spikelet sterile, consisting of a delicate, membranaceous sterile lemma slightly shorter than the second glume; lemma and palea of the fertile upper floret similar in texture to and only slightly shorter than the sterile lemma. Anthers 2–3 mm long, shorter in the occasional cleistogamous florets. (2n = 18) Mostly Jun–Jul. Prairies & woods; e OK; (NC to MO, s to FL & TX). *Coelorachis cylindrica* (Michx.) Nash — Gould, The Grasses of Texas, p. 620. Texas A. and M. Univ. Press, College Station. 1975.

49. MELICA L., Melic

1. **Melica nitens** (Scribn.) Nutt. ex Piper, threeflower melic. Rhizomatous perennials 5.5–9.5(13) dm tall. Culms glabrous, hollow, erect to decumbent below. Blades rolled in the bud, flat at maturity, glabrous, scabrous, or often pubescent above with somewhat crisped hairs, mostly 10–25 cm long, (3)5–10 mm wide; sheaths thin, closed to near the summit, slightly keeled above, glabrous to scabrous; ligule a whitish, lacerate membrane, 1.2–5.3 mm long; auricles none. Inflorescence a panicle 11–26 cm long, the branches ultimately spreading, the spikelets tending to be secund, stramineous at maturity, 8–12 mm long, 2- or 3(4)-flowered, with a sterile, club-shaped, erect, rudimentary floret at the summit, disarticulating below the glumes; glumes membranaceous-margined, irregularly 5- to 7-nerved, the first broadly ovate, 5–9 mm long, the second narrower, 6–10 mm long; lemmas rather plainly striate with 7–11 scabrous nerves, sometimes with additional faint nerves between the major ones, ovate-lanceolate, the lowest one 6.5–11.5 mm long; paleas 2-keeled, shorter than the lemmas, ciliate on the keels. Anthers 2–3.2 mm long. (2n = 18) Mostly May–Jun. Usually in dry to moist woodlands, sometimes growing in the open, not common; IA, se NE, MO, e KS, e OK; (PA to NE, s to VA & w TX).

Reference: Boyle, W. S. 1945. A cytotaxonomic study of the North American species of *Melica*. Madroño 8: 1–32.

Several additional species of *Melica* enter or approach our range. *M. mutica* Walt. is very similar to *M. nitens*, but its rudimentary floret is bent at an angle to the axis of the spikelet. It approaches the GP in se OK and has been reported from Kiowa, Comanche, and Canadian counties. *M. porteri* Scribn. was reported in Atlas GP from two Kansas counties, but those specimens have been reassigned to *M. nitens*. *M. smithii* (Porter) Vasey has been reported for SD: Pennington, but I have not seen specimens. Two bulbous-based species, *M. bulbosa* Geyer and *M. subulata* (Griseb.) Scribn., are also known from the edge of our range (Joe Thomasson, pers. communication). *M. subulata*, which has acuminate lemmas, is known from SD: Lawrence, and *M. bulbosa*, with acute to obtuse lemmas, is known from WY: Crook.

50. MUHLENBERGIA Schreb., Muhly

Rhizomatous to nonrhizomatous perennials, rarely annuals. Culms usually erect, simple to branched, smooth to scabrous or puberulent, solid to hollow. Blades rolled, C-shaped in cross section, or folded in the bud, flat, folded or involute at maturity; sheaths open, strongly to slightly keeled or rounded, smooth to scabrous, sometimes incised with narrow grooves between the veins; ligule a short to elongate membrane, rarely with a pronounced fringe of hairs; auricles none. Inflorescence an open to condensed panicle; spikelets 1(2)-flowered, disarticulating, at least initially, above the glumes; glumes membranaceous, 1(3)-nerved to nerveless, blunt to acute or sometimes awned; lemmas 3-nerved, the lateral nerves faint to prominent, awned or awnless, usually longer than the glumes but sometimes shorter or about equal to them.

References: Pohl, R. W. 1969. *Muhlenbergia*, subgenus *Muhlenbergia* (Gramineae) in North America. Amer. Midl. Naturalist 82: 512–542, Reeder, C. G. 1975. *Muhlenbergia*, pp. 246–285, *in* F. W. Gould, The Grasses of Texas. Texas A. and M. Univ. Press, College Station. This was written with the assistance of James Kurtz.

In addition to the species described and discussed below, *M. minutissima* (Steud.) Swall., a very delicate annual with a diffuse panicle, is of rare occurence at the w edge of our range (NE: Scottsbluff; SD: Pennington; WY: Weston). This species was erroneously referred to *Sporobolus confusus* (Fourn.) Vasey by Rydberg.

1 Panicles open and diffuse, the branches spreading.
  2 Ligules prominently fringed with hairs, the hairs usually as long as or longer than the membranaceous basal portion .... 10. *M. pungens*
  2 Ligules membranaceous, at most erose-ciliolate, but not fringed.
    3 Lemma less than 2 mm long, usually awnless; ligules short, less than 1 mm long, truncate; outer layers of culm readily separating and peeling off ..... 2. *M. asperifolia*
    3 Lemma more than 2 mm long, awned; ligules usually more than 2 mm long; outer layers of culm not readily separating and peeling off.
      4 Blades long, the shortest usually more than 10 cm long, filiform-tipped, but more than 1.4 mm wide at the base, rolled to C-shaped in cross section in the bud; sheaths not keeled .... 4. *M. capillaris*
      4 Blades short, the longest less than 10 cm long, usually less than 1.4 mm wide, folded in the bud; sheaths somewhat keeled.
        5 Blades short, the longest ones less than 5 cm long, arcuate, mostly crowded at the base of the culm, few or none of the internodes visible .... 16. *M. torreyi*
        5 Blades, at least some of them, more than 5 cm long, not strongly arcuate, distributed up the culm, some internodes visible .... 1. *M. arenicola*
1 Panicles condensed to very slender, the branches appressed.
  6 Glumes reduced to minute nerveless scales less than 0.4 mm long .... 13. *M. schreberi*
  6 Glumes conspicuous, one or both of them more than 0.5 mm long, midnerve present.
    7 Blades slender, most of them less than 2 mm wide, folded, rolled, or C-shaped in cross section in the bud, flat to involute or folded at maturity.
      8 Ligules less than 1 mm long; culms and sheaths not nodulose-roughened; blades folded in the bud, flat or folded at maturity .... 5. *M. cuspidata*
      8 Ligules more than 1 mm long; culms and sheaths smooth to roughened; blades rolled or C-shaped in section (rarely folded) in the bud, flat or involute at maturity.
        9 Plants tufted to stoloniferous annuals; culms and sheaths not nodulose-roughened; glumes obtuse; anthers under 1 mm long .... 6. *M. filiformis*
        9 Plants rhizomatous perennials; culms and sheaths nodulose-roughened; glumes acute; anthers over 1 mm long .... 12. *M. richardsonis*
    7 Blades broader, most of them well over 2 mm wide, rolled in the bud, flat at maturity.
      10 Glumes awn-pointed, both normally more than 3 mm long, including the awn, 1/3 or more longer than the awnless (or short awned) lemma.
        11 Ligules mostly more than 0.6 mm long; grains more than 1.4 mm long; anthers 0.4–0.8 mm long; plants usually much branched above .... 11. *M. racemosa*
        11 Ligules mostly less than 0.6 mm long; grains mostly less than 1.4 mm long; anthers more than 0.8 mm long; plants seldom branched above .... 8. *M. glomerata*
      10 Glumes awnless or awn-pointed but not usually over 3 (rarely up to 4) mm in overall length, shorter than, equal to, or slightly longer than the awned to awnless lemma.

12 Internodes dull and puberulent, especially near the summit.
   13 Ligules conspicuous, (1.2)1.5–2.6 mm long, projecting above the sheath summit; leaf sheaths rounded to slightly keeled; anthers mostly 0.6 mm or more in length .... 15. *M. sylvatica*
   13 Ligules inconspicuous, barely visible from the side, 0.4–1.2(1.5) mm long; leaf sheaths abruptly keeled; anthers often shorter than 0.6 mm .... 9. *M. mexicana*
12 Internodes smooth and shining.
   14 Plants with many axillary inflorescences, most of these partly included in leaf sheaths.
      15 Most ligules less than 0.7 mm long; leaves of side branches often much smaller than those of the main branches; glumes usually conspicuously shorter than the lemmas .... 3. *M. bushii*
      15 Most ligules more than 0.7 mm long; leaves of the side branches often similar to those of the main branches; glumes usually subequal to or even slightly longer than the lemmas .... 7. *M. frondosa*
   14 Plants without many axillary inflorescences or these abundant and exserted on long peduncles.
      16 Ligules more than 1 mm long; lemmas mostly more than 2.4 mm long, normally awned .... 15. *M. sylvatica*
      16 Ligules less than 1 mm long; lemmas mostly less than 2.4 mm long, normally awnless .... 14. *M. sobolifera*

1. **Muhlenbergia arenicola** Buckl., sand muhly. Nonrhizomatous or short-rhizomatous perennials 2.5–5(6.5) dm tall. Culms hollow to pith-filled, erect, often decumbent at the base, puberulent to strongly scabrous below the nodes, at least some of the internodes exposed. Leaves mostly basal but some, at least, distributed up the culm; blades folded in the bud, flat or folded at maturity, scabrous, about equally ridged on both surfaces, mostly 4–10 cm long, 0.6–1.2 mm wide; sheaths slightly keeled, smooth to scabrous, often puberulent near the base; ligules membranaceous, often splitting, 2.5–8 mm long. Panicles diffuse, (12)16–25(32) cm long; spikelets 1(2)-flowered; glumes lanceolate, acute to jagged, acuminate, or awn-tipped, 1-nerved, scabrous, 1.5–3.3 mm long; lemma 3-nerved (1.2)2.5–4.4 mm long, bearing a slender scabrous awn 1.5–4.2 mm long; palea about as long as the body of the lemma. Anthers 1.3–2 mm long. Grains mostly 2–2.3 mm long. May–Oct, mostly midsummer. Open slopes, often in dry, sandy soil; se KS to s & e OK, TX, & NM; (KS to CO, s to TX, AZ, & n Mex.).

Muhlenbergia porteri Scribn. occurs at the sw edge of our range (OK: Cimarron and Texas; TX: Deaf Smith). It will key here to *M. arenicola*, but it is very different in habit, having much-branched, spreading culms, the branches strongly divergent.

2. **Muhlenbergia asperifolia** (Nees & Mey.) Parodi, scratchgrass. Strongly rhizomatous perennials 1.5–6.5 dm tall. Culm solid, decumbent at the base, slightly flattened, smooth, its outer layer readily separating and peeling away. Leaves well distributed along the culms; blades folded in the bud, flat to folded at maturity, scabrous above, 2–10 cm long, 1–2(2.3) mm wide; sheaths smooth, slightly keeled; ligules truncate, membranaceous, erose-ciliolate, 0.3–0.9 mm long. Panicles open and diffuse, 10–25(30) cm long, often partly included in leaf sheaths below; spikelets 1(2)-flowered; glumes lanceolate, acute, subequal, 1-nerved, scabrous on the nerve, 0.7–1.9 mm long; lemma faintly 3-nerved, rounded or mucronate-tipped, 1.1–1.7 mm long. Anthers 0.7–1 mm long. Grain 0.8–1.2 mm long, falling with or before the lemma and palea. (2n = 20, n = 11, 14 also reported) Jun–Oct. Usually in low, moist soil; GP; (MN to B.C., s to TX, CA, & Mex.; s S. Amer.). *Sporobolus asperifolius* Nees & Mey.—Rydberg.

3. **Muhlenbergia bushii** R. Pohl. Strongly rhizomatous perennials 3.5–9.5 dm tall. Culms hollow, much branched above, internodes smooth and shining. Blades rolled in the bud,

flat at maturity, smooth to scabrous, 1–15 cm long and (1.5)2–6(7) mm wide, those of the side branches usually clearly shorter and narrower than those of the main culm; sheaths slightly keeled to rounded, glabrous; ligules membranaceous, truncate, ciliolate, 0.3–0.7(1.1) mm long. Panicles numerous, slender, about 3–10 cm long, the lateral ones mostly included at the base in leaf sheaths; glumes strongly 1(2)-nerved, acute to acuminate, subequal, (0.5)1–2.5 mm long; lemma plainly 3(4)-nerved, pilose at the base, 2.2–3.8 mm long, awnless or with an awn tip 0.1–1.4(2.1) mm long. Anthers 0.3–0.5 mm long, sometimes included at maturity. Grain 1.4–2 mm long. (2n = 40) Aug–Oct. Wooded areas & woodland edges, often where somewhat disturbed; se NE, sw IA, MO, e KS, e OK; (MD to WI, NE, s to NC & TX). *M. brachyphylla* Bush — Fernald.

4. *Muhlenbergia capillaris* (Lam.) Trin., hairgrass. Cespitose perennials (3.5)4–9 dm tall. Culms hollow to pith-filled, erect, glabrous to puberulent below the nodes, the internodes not usually exposed. Blades rolled or C-shaped in cross section in the bud, flat to involute at maturity, scabrous, the ridges closely spaced on both surfaces, 15–35(50) cm long, 1.3–3.5(4) mm wide at the base but tapering to filiform tips; sheaths not keeled, scabrous, particularly in the finely incised, narrow grooves; ligules membranaceous, firm, often obtuse, 1.8–6.4(8) mm long, the longest ones normally on the culm leaves. Panicles 19–41 cm long, open and diffuse, the pedicels very slender; glumes indistinctly 1-nerved, erose, subequal, 1–2.3 mm long, the first usually awnless, the second often bearing an awn up to 1.8 mm long; lemma scabrous toward the tip, slightly pubescent at the base with appressed hairs, 2.9–5 mm long, bearing a slender, flexuous, scabrous awn 4.8–11(22) mm long. Anthers 1–1.8 mm long. Grain 2.2–3 mm long. Jul–Nov, mostly Sep–Oct. Open woods, prairies, & pastures, e KS, sw MO, e OK; (MA to IN & KS, s to FL, TX, W.I., & eMex.).

5. *Muhlenbergia cuspidata* (Torr.) Rydb., plains muhly. Cespitose perennials from knotty, thickened culm bases, 3–6 dm tall. Culms solid, rather stiffly erect, flattened, crisped-puberulent below the nodes, often branched above. Blades folded in the bud, flat to folded at maturity, scabrous above, with closely crowded ridges, slightly keeled, 2–22 cm long, 0.6–2(3.5) mm wide; sheaths keeled, glabrous to scabrous; ligules membranaceous, truncate, short, only 0.2–0.5(0.8) mm long. Panicles 4–14 cm long, slender, the branches erect; spikelets 1(2)-flowered; glumes 1-nerved, subequal, 1.2–3 mm long, acute, scabrous on the nerves; lemma 3-nerved, acute, 2.5–3.6 mm long. Anthers 1.2–1.8 mm long. Grain 1.6–2.3 mm long. Jun–Oct, but only the later specimens with anthers or grains. Dry prairies; MN, ND, SD, WY, IA, NE, CO, MO, KS; (MI to Alta., s to KY, OK, & NM).

6. *Muhlenbergia filiformis* (Thurb.) Rydb. Tufted or somewhat stoloniferous annuals 1–3(5) dm tall. Culms erect, somewhat flattened, slender, solid, branching below. Blades C-shaped to folded in the bud, flat to involute at maturity, often puberulent above, 1–6 cm long and 0.6–1.4 mm wide; sheaths glabrous, not nodulose roughened, not keeled; ligules membranaceous, acute, 2–3.5 mm long. Glumes obtuse, erose, the first 0.6–1.1 mm long, the second 0.7–1.2 mm long; lemma acuminate or short awned, 1.9–3.2 mm long including the tip, 3(5)-nerved. Anthers 0.4–0.9 mm long. Grain not seen. (2n = 18) Aug–Oct. On riverbanks, lakeshores, & on sandbars, probably often overlooked by collectors; KS: Finney; NE: Cherry, Keya Paha, & Thomas; SD: Pennington; (Alta. to B.C., s to KS, AZ, & CA). *M. simplex* (Scribn.) Rydb. — Rydberg.

7. *Muhlenbergia frondosa* (Poir.) Fern., wirestem muhly. Strongly rhizomatous perennials (2.5)4–10 dm tall. Culms hollow, glabrous, erect to sometimes decumbent, usually branched above. Blades rolled in the bud, flat at maturity, glabrous to scabrous, 3–20 cm long, 2–7 mm wide; sheaths glabrous, rounded to slightly keeled; ligule membranaceous, erose-ciliolate, (0.7)1–1.2(1.5) mm long. Panicles numerous, about 2–13 cm long, the lateral ones mostly partly included in leaf sheaths, relatively slender to congested and often lobed, glumes strongly 1-nerved, acute, acuminate or awn tipped, variable in length even on the same plant, subequal, or the first one shorter, (0.5)1–4.1 mm long; lemma pubescent toward the base, 3-nerved, exceeding, equaling, or slightly shorter than the glumes, the body 2.3–4 mm long, awnless or with an awn 0.1–13 mm long. Anthers 0.2–0.6 mm long, sometimes included at maturity. Grains mostly 1.6–1.9 mm long. (n = 20) Jul–Nov, mostly Sep–Oct. Wooded areas & woodland edge, often where disturbed; e GP; extending relatively far w in NE & KS; (se Can., ME to ND, s to GA & TX). *M. mexicana* (L.) Trin.—Rydberg, in part (misapplied), *M. commutata* (Scribn.) Bush—Rydberg.

This species is very close to *M. bushii* in both habit and habitat. The two species are sometimes difficult to separate when using herbarium material.

8. *Muhlenbergia glomerata* (Willd.) Trin. Strongly rhizomatous perennials mostly (3.5)4.5–6.5 dm tall. Culms slender, hollow, erect, unbranched to sparingly branched, puberulent, especially near the summit of the internode. Blades rolled in the bud, flat at maturity, glabrous to minutely scabrous, 2–12 cm long, (1.5)2–4 mm wide; sheaths slightly keeled, glabrous to scabrous, sometimes puberulent near the base; ligules membranaceous, ciliolate, 0.3–0.6 mm long. Panicle condensed, lobed, 2–7 cm long; glumes subequal, 1-nerved, narrow, tapering gradually to a scabrous awn, the length, including the awn, 3–6.5 mm; lemma 3-nerved, pubescent toward the base, 1.9–3.1 mm long, sometimes with a short awn tip to about 1 mm long. Anthers 0.8–1.2 mm long. Grain 1–1.4 mm long. (n = 10) Jul–Oct, mostly Aug. Relatively uncommon in wet meadows, bogs, & spring areas; MN, ND, NE; (Que. to N.T. & B.C., s to WV, IN, UT, & OR).

Some authors have merged *M. glomerata* with *M. racemosa*. Although the 2 species are closely similar, the distinctions seem sharp and have been well elucidated by Pohl (op. cit.). Pohl reports that hybridization of *M. glomerata* with *M. mexicana* is common in some regions.

9. *Muhlenbergia mexicana* (L.) Trin., wirestem muhly. Strongly rhizomatous perennials 3.9–9.3(14) dm tall. Culms hollow, erect, branched above, the upper ones, especially, somewhat flattened, the internodes dull and puberulent, especially in the upper part. Blades rolled in the bud, flat at maturity, minutely scabrous to glabrous, (2)7–30(38) cm long, (1)2–6(7.5) mm wide; sheaths strongly keeled, flattened, glabrous to minutely scabrous; ligules membranaceous, ciliolate, (0.4)0.6–1.1(1.5) mm long. Panicles 4.5–17 cm long, dense and lobulate to very slender and elongated, the lateral and terminal ones usually both well exserted; glumes acuminate to awn tipped, strongly 1-nerved, subequal, 1.5–3.1 mm long; lemma about equal to the glumes, 3-nerved, pilose toward the base, awnless, or with an awn tip 0.1–8(10.0) mm long. Anthers 0.3–0.7 mm long, sometimes included at maturity. Grain 1.1–1.6 mm long. (n = 20) Aug–Oct. In disturbed to undisturbed woodlands, sometimes also in the open; GP, but rare or seldom collected in much of the w region; (Que. & ME to B.C. & WA, s to NC, NM, & CA). *M. foliosa* Trin., *M. polystachya* Mack. & Bush—Rydberg.

The name *mexicana* is a misnomer, since the species does not grow there. In herbaria it is most often confused with *M. sylvatica* when it has slender panicles and with *M. racemosa* when it has dense lobulate panicles. From the former is easily distinguished by its shorter ligules and its keeled sheaths, from the latter by its shorter glumes. The glumes of *M. mexicana* may be strikingly longer than the lemmas on some of the spikelets, but not on all of the spikelets, as in *M. racemosa*.

Muhlenbergia tenuiflora (Willd.) B.S.P. occurs rarely on the e edge of our range (MO: Jackson and Jasper; NE: Douglas and Sarpy; OK: Comanche). Although it will probably key here to *M. mexicana*, it differs in several features. Its sheaths are not keeled and its anthers are much longer (1.1–2.2 mm long).

10. *Muhlenbergia pungens* Thurb., blowout grass, sand muhly. Rhizomatous perennials 2–5.3 dm tall, forming extensive mats. Culms solid, internodes woolly pubescent near the summit. Blades C-shaped in cross section in the bud, flat to involute at maturity, scabrous below, scabrous and pubescent above, with closely crowded ribs on both surfaces, pungent-tipped, 2–8 cm long, 1.2–2.2 mm wide; sheaths not strongly keeled, incised with narrow grooves, scabrous and often also woolly-pubescent toward the base; ligule 0.5–0.8(1.2) mm long, consisting of a fringed membrane with the fringed portion usually exceeding the membranaceous portion in length. Panicles open and diffuse, 8–19 cm long; spikelets long pediceled; glumes 1-nerved, scabrous, ovate-acuminate, subequal, 1.2–2.6 mm long; lemmas obscurely 3-nerved, often purplish, 2.6–3.6 mm long, with a short, firm awn 0.5–1.5 mm long; paleas 2-awned at the tip. Anthers 1.7–2.1 mm long. Grain 1.8–2.5 mm long, often falling before the lemma and palea. (2n = 42, 60) Jul–Oct, mostly Sep. Sandy prairie, usually in blowouts & loose sand; sw SD, NE, WY, CO; (SD to WY & UT, s to TX, NM, & AZ).

11. *Muhlenbergia racemosa* (Michx.) B.S.P., marsh muhly. Strongly rhizomatous perennials 3–8(10.5) dm tall. Culms erect, hollow, branched above, the internodes glabrous and shining to, not uncommonly, puberulent-roughened, especially near the summit. Blades rolled in the bud, flat at maturity, glabrous or minutely scabrous, 3–17(21) cm long, (1)2.2–5(5.8) mm wide; sheaths keeled, glabrous to minutely scabrous, ligules membranaceous, (0.6)1–1.5(1.7) mm long. Panicles dense, often lobed, mostly 3–13 cm long, mostly exserted on peduncles, sometimes included partly in leaf sheaths; glumes 1-nerved, slender, tapering gradually to a scabrous awn, subequal, the overall length 3.5–6 mm long; lemma 3-nerved, pilose at the base, 2.5–3.8 mm long. Anthers 0.4–0.8 mm long. Grain 1.4–2.1 mm long. (n = 20) Jul–Oct, mostly Aug–Sep. In disturbed ground along roads & railroads, in pastures, & in woods; GP; (Man. to Alta. & MI to WA, s to MO, NM, & AZ).

See the discussions after *M. mexicana* and *M. glomerata*.

12. *Muhlenbergia richardsonis* (Trin.) Rydberg. Clumped but widely rhizomatous delicate perennials 1–7 dm tall. Culms erect, slender, solid, slightly flattened, minutely nodulose-roughened. Blades rolled to C-shaped in cross section in the bud, flat to involute at maturity, usually strongly scabrous above, 1–5 cm long, 0.6–1.5(2.1) mm wide; sheaths nodulose-roughened, not keeled; ligules membranaceous, acute, 1.1–3.1 mm long. Spikelets 1(2)-flowered; glumes acute, 1-nerved, about equal, 0.6–1.8 mm long; lemma acute to acuminate, 2–3 mm long, 3(5)-nerved, not pubescent. Anthers 1–1.6 mm long. Grain about 1.2–1.5 mm long. (2n = 40) Jul–Sep. Moist to dry prairies; MI, ND, MT, scattered in SD, NE, WY, CO; (N.B. & ME to Alta., s to MI, NE, NM, AZ, & Baja Calif.).

Muhlenbergia filiculmis Vasey is common w of our range and may be expected to occur within it. It will key here to *M. richardsonis*, but it is a strongly cespitose, nonrhizomatous species, and it lacks the nodulose-roughened culms and sheaths.

13. *Muhlenbergia schreberi* J. F. Gmel., nimblewill. Nonrhizomatous, erect to sprawling perennials 2–7.5 dm tall. Culms in late season decumbent, geniculate, and rooting at many of the lower nodes, solid, flattened, glabrous, slender. Blades rolled in the bud, flat at maturity, glabrous or minutely scabrous, usually pubescent near the collar, 1–10 cm long, 1.4–4.3 mm wide; sheaths glabrous to minutely scabrous, keeled, flattened, pubescent on the margins at the collar, sometimes slightly puberulent near the base; ligules membranaceous, 0.2–0.4 mm long. Panicles condensed and lobulate to very slender, (3)6–22 cm long, exserted or included in leaf sheaths at the base; glumes nerveless, very short (under 0.5 mm) and easily overlooked, the first usually shorter than the second; lemma 3-nerved, somewhat pilose at the base, 1.8–2.8 mm long, tapering into a scabrous awn 1.7–4.6 mm long. Anthers 0.2–0.5 mm long, sometimes included at maturity. Grain 1–1.4 mm long. (n = 20) Aug–Oct. Moist woods & other shady habitats, often becoming a serious weed in lawns & gardens; IA, e NE, MO, e KS, e OK; (NH to WI & e NE, s to FL & e TX).

14. ***Muhlenbergia sobolifera*** (Muhl.) Trin., rock muhly. Strongly rhizomatous perennials 3.5–9.5 dm tall. Culms erect, branched above, hollow, internodes glabrous, smooth and shining. Blades rolled in the bud, flat at maturity, glabrous to minutely scabrous, 4–16(19) cm long, (1)3.5–7 mm wide; sheaths glabrous, only slightly keeled; ligules membranaceous, 0.4–0.7(1) mm long. Panicles mostly 4–14(19) cm long, very slender, both the terminal and lateral ones exserted on very long peduncles; glumes subequal, acuminate, 0.9–2.3 mm long, both 1-nerved or the second rarely 3-nerved; lemma awnless or with a very short awn point, 3-nerved, pubescent toward the base, 1.6–2.4 mm long. Anthers 0.4–0.9 mm long. Grain 1–1.4 mm long. (n = 20) Aug–Sep. Upland deciduous woods, often in rocky soil; s IA, se NE, MO, e KS, e OK, e TX; (NH to WI, s to VA, se NE, TX).

15. ***Muhlenbergia sylvatica*** (Torr.) Torr. in A. Gray, forest muhly. Strongly rhizomatous perennials 4–10 dm tall. Culms erect, hollow, becoming branched above, internodes usually puberulent, especially on the upper parts, but not infrequently glabrous in our range. Blades rolled in the bud, flat at maturity, 1–20 cm long, 1.4–5.6(7.5) mm wide; blades and sheaths glabrous to minutely scabrous; sheaths rounded to slightly keeled; ligules membranaceous, relatively long, (1.1)1.4–2.6 mm long. Panicles slender, 4–13 cm long, often becoming long-exserted; glumes strongly 1-nerved, acuminate, variable in length, even on the same plant, subequal, 1.2–1.8 mm long; lemma plainly 3-nerved, pilose toward the base, 2.3–3 mm long, tapering into a scabrous awn (1.5)5–9 mm long. Anthers 0.4–0.9 mm long. Grain 1.4–1.8 mm long. (n = 20) Aug–Oct. Upland woods, not common; MN, e SD, IA, e NE, MO, e KS, e OK; (ME & Que. to MN, SD, s to NC & TX). *M. torreyi* (Kunth) A.S. Hitchc.—Rydberg, misapplied; *M. umbrosa* Scribn.—Winter.

Muhlenbergia sylvatica may easily be mistaken for the much more common *M. mexicana*, which see. The plants of *M. sylvatica*, with glabrous or nearly glabrous internodes, can be distinguished readily from *M. sobolifera* by the longer, awned lemmas and the longer ligules.

Muhlenbergia tenuiflora, a species of rare occurrence on the edge of our range, might possibly key here. See the discussion under *M. mexicana*.

16. ***Muhlenbergia torreyi*** (Kunth) A.S. Hitchc. ex Bush, ring muhly, ring grass. Nonrhizomatous or short-rhizomatous, mat-forming perennials 1.5–4.5 dm tall. Culms hollow, often decumbent below, puberulent below the internodes, but none or very few of the internodes exposed. Leaves mostly basal, crowded; blades folded in the bud, tightly folded or involute at maturity, arcuate, 0.5–4 cm long, 0.3–1.1 mm wide when unfolded; sheaths slightly keeled, smooth to minutely scabrous, often puberulent toward the base; ligules membranaceous, 2–4.5 mm long. Panicles diffuse, (8)10–21 cm long; glumes lanceolate, acuminate to awn-tipped, 1-nerved, 1.3–2.5(3) mm long; lemma 3-nerved, 2–3.1 mm long, bearing a slender awn 1.9–4 mm long. Anthers 1.2–1.6 mm long. Jul–Sep. In dry, often sandy, soil; s WY, CO, w KS, w OK, TX, NM; (w KS to WY, s to TX, AZ, & n Mex.). *M. gracillima* Torr.—Rydberg.

51. MUNROA Torr.

1. ***Munroa squarrosa*** (Nutt.) Torr., false buffalo grass. Stoloniferous, mat-forming annuals, the ascending culms not over 1 dm tall, but the mats 4–40 cm wide. Culms scabrous, hollow, with a few, long, exposed internodes, many of which arch over or rest on the ground, and numerous very short internodes, concealed by the crowded fascicles of leaves. Blades scabrous, folded in the bud, folded to flat at maturity, with conspicuous white, thickened margins and pungent tips, 1–3 cm long and 1.2–2.6 mm wide; sheaths glabrous to scabrous, open, ciliate on the margins and with prominent tufts of hair at the collar; ligule a fringe of hairs 0.4–0.8 mm long; auricles none. Inflorescence consisting of several small groups

of spikelets borne among the leaves of a fascicle, the subtending leaves mostly similar to normal vegetative leaves, but usually some of them reduced and bearing double, divaricate, awnlike blades; each spikelet group consisting of (2)3 more or less dimorphic spikelets, the lower ones sessile and bearing subequal glumes and rigid lemmas, the upper one pediceled and bearing unequal glumes (the lower glume sometimes obsolete) and less rigid lemmas; all spikelets disarticulating above the glumes; lower spikelets 1- to 4-flowered, the glumes about 3–5.2 mm long, 1-nerved, the lemmas coriaceous, 3-nerved, the body often with tufts of hair on the margins at midlength, the lowest about 4.9–6.2 mm long, the midvein often extended as a stout awn up to 2 mm long, the lateral veins also sometimes protruding; upper spikelets 3- to 5-flowered, the glumes 1-nerved, the lower glume obsolete or up to 2.1 mm long, the upper glume 2.1–3.9 mm long; lemmas similar to those of the lower spikelets but membranaceous, the body of the lowest 4–4.8 mm long and its central awn tip 1–2 mm long. Anthers 0.8–1.4 mm long. (2n = 16) May–Sep, mostly Jul and Aug. Open ground, often in disturbed soil; throughout w GP; (Alta. & ND to MT s to TX, AZ, NV, & Mex.).

52. ORYZOPSIS Michx., Ricegrass

Cespitose or tufted nonrhizomatous or weakly rhizomatous perennials. Culms erect to decumbent at the base, hollow or sometimes solid. Blades rolled to C-shaped in cross section in the bud, flat to involute at maturity; sheaths open; ligule membranaceous, collarlike and truncate or acute or obtuse and elongate; auricles none. Inflorescence a panicle or raceme of 1-flowered spikelets; spikelets disarticulating above the glumes; glumes equal to slightly unequal, membranaceous or firm, 3- to 7(11)-nerved; lemmas nearly equaling or considerably shorter than the glumes, indurate, glabrous, strigose, or hirsute, terminally awned, the awn deciduous, straight or flexuous; palea subequal to the lemma, firm. *Eriocoma* Nutt.—Rydberg.

This is a heterogeneous group in our range, with the four species falling into all the three sections of the genus (Johnson, Bot. Gaz. 107: 1–32. 1945). The genus has much in common with *Stipa*, and *Oryzopsis* species, especially *O. hymenoides*, may hybrydize with various species of *Stipa* to produce sterile hybrids.

In addition to the species described below, two others have been found on the periphery of our range. Both have strigose lemmas and small spikelets and hence will not readily key below. *O. pungens* (Torr.) Hitchc. (known from SD: Lawrence) may be distinguished by its short awn (less than 2 mm long) and its short ligules (less than 2 mm long). *O. exigua* (known from MT: Custer) has longer awns and ligules.

1 Lemma glabrous and shining, less than 3 mm long at maturity; glumes less than 4 mm long .... 3. *O. micrantha*
1 Lemma strongly strigose to hirsute, usually more than 3 mm long; glumes more than 4 mm long.
    2 Lemma hirsute, the hairs long and silky, often concealing the body of the lemma; anthers under 2 mm long; most ligules more than 2 mm long .... 2. *O. hymenoides*
    2 Lemma strigose; anthers more than 2 mm long; ligules very short, less than 1 mm long.
        3 Culm leaves with greatly reduced blades or bladeless; mature lemma pale, its awn under 13 mm long .... 1. *O. asperifolia*
        3 Culm leaves large; mature lemma brown, its awn more than 13 mm long .... 4. *O. racemosa*

1. ***Oryzopsis asperifolia*** Michx., rough-leaved ricegrass. Cespitose perennials 3–10 dm tall. Culms smooth to scabrous, hollow, erect, often overtopped by the largest leaves. Blades of the culm leaves reduced or obsolete, those of the innovations stiff, rolled in the bud, flat to slightly involute at maturity, rather strongly many-ridged above, scabrous, mostly

20–40(56) cm long, 2–7(9) mm wide, even the broadest tapering to a narrow, involute base; sheaths smooth to scabrous; ligule a short, uneven, ciliate membrane 0.1–0.4 mm long. Inflorescence a narrow raceme or sometimes a panicle with the main branches spreading, 4–8 cm long; glumes broad, nearly equal, 7(11)-nerved, 5.4–7 mm long, lemma pale, appressed-pubescent, with longer, denser, more spreading hairs on the callus, the lemma normally completely enclosing and concealing the palea. Anthers 2.5–4 mm long. (2n = 46,48) May–Jun. In dry to moist, deciduous or coniferous woodlands; MN, ND, w SD, CO; (Newf. to B. C. s to WV, IN, SD, NM, & ID).

2. *Oryzopsis hymenoides* (R. & S.) Ricker, Indian ricegrass. Strongly cespitose perennials 3.5–8 dm tall. Culms erect, hollow to pith-filled, glabrous. Blades C-shaped in cross section to rolled in the bud, involute at maturity, strongly ribbed and scabrous to minutely hirsute above, smooth below, mostly 8–40 cm long, 0.8–2.9 mm wide; sheaths usually glabrous but ciliate on the overlapping margins and sometimes bearing more or less inconspicuous tufts of long hairs at the collar; ligules acute to acuminate, 3–9 mm long. Panicle 8–20 cm long, very open, the slender branches and flexuous pedicels spreading divaricately, the branching at least in part dichotomous; glumes ovate and acuminate, 3(5)-nerved, usually somewhat unequal, the first 4.9–8.2 mm long, the second 4.7–7.5 mm long; lemma 2.9–4.5 mm long, usually more or less covered with numerous hairs which approach the lemma in length, the awn of the lemma usually firm and straight, 4–8 mm long. Anthers 0.6–1.2 mm long. (2n = 48) May–Jul. Open grassland, especially in sandy or rocky soil; w MN, w ND, MT, w SD, w NE, WY, w KS, CO, OK, TX, NM; (Man. to B.C., s throughout w U.S. to n Mex.). *Eriocoma hymenoides* (R. & S.) Rydb.—Rydberg.

   Natural sterile hybrids between this species and various species of *Stipa* are known throughout most of the w U.S. Plants from the badlands of ND, assigned by A. Chase to *O. bloomeri* (Bol.) Ricker, have been shown to represent sterile hybrids of *O. hymenoides* and *S. viridula* (Johnson and Rogler, Amer. J. Bot. 30: 49–56. 1943).

3. *Oryzopsis micrantha* (Trin. & Rupr.) Thurb., little-seed ricegrass. Strongly cespitose perennials 2–8 dm tall. Culms hollow, glabrous, slender, sometimes partly decumbent. Blades rolled in the bud, flat to involute at maturity, scabrous, mostly 6–30 cm long, 0.9–2 mm wide; sheaths smooth to scabrous or puberulent; ligule a short, truncate membrane, sometimes higher on the sides than at the back, about 0.2–1.1 mm long. Panicle delicate, open (to condensed) but the spikelets rather crowded along the main branches. Glumes membranaceous, about equal, 3(5)-nerved, 2.4–3.6 mm long; lemmas firm, smooth and shining, dark at maturity, 1.5–2.5 mm long, the awn 4.7–8.4 mm long. Anthers 0.4–1 mm long. (2n = 22, 24) Jun–Aug. Open woods; w ND, MT, w SD, w & n NE, CO, w OK, TX; (Sask. to ND & MT, s to NM, NV, & CA).

4. *Oryzopsis racemosa* (J. E. Sm.) Ricker, black-seed ricegrass. Tufted perennials from short, knotted rhizomes, mostly 5–9 dm tall. Culms erect, hollow, usually scabrous to appressed-pubescent, especially at the nodes. Blades rolled in the bud, flat at maturity, the midvein usually obvious, mostly thinly pilose above and nearly glabrous or scabrous beneath, the lower blades greatly reduced, but the upper ones 15–30 cm long, 6–15 mm wide; sheaths glabrous to scabrous, but often appressed-pubescent at the base and near the collar, or the pubescence more general; ligules short, uneven, truncate, sometimes highest at the sides, 0.2–0.6 mm long. Panicle reduced to a raceme or, more commonly, with spreading branches, 7–25 cm long; glumes (5)7(11)-nerved, 6.5–8.5 mm long; lemma appressed-pubescent, dark brown or almost black at maturity, 6.5–8 mm long, the palea enclosed by it but partially visible at maturity, the somewhat flexuous awn 16–21 mm long. Anthers 2.5–4 mm long. (2n = 46, 48) Jul–Sep. Deciduous woods, MN, e ND, e SD, IA, e NE; (Que. to ND, s to VA, KY, NE).

53. PANICUM L., Panic grass

Rhizomatous to nonrhizomatous perennials or tufted annuals. Culms hollow or, rarely, solid. Blades rolled in the bud, normally flat at maturity; sheaths open, keeled to rounded; ligule predominantly a fringe of hairs from a short basal membrane, or sometimes the membrane predominating, the overall ligule short, seldom over 4 mm long; auricles none. Inflorescence a panicle with appressed to spreading branches. Spikelets terete to somewhat dorsally compressed, disarticulating below the glumes, bearing 2 florets, the lower floret neuter or staminate, the upper one perfect; glumes markedly unequal to subequal, the first one usually shorter, 3- to 5(9)-nerved, the second normally almost as long as the spikelet and 5- to 9(11)-nerved; lemma of the lower floret very similar to the second glume in length and texture, 5- to 9(15)-nerved; lemma of the upper floret indistinctly nerved, glabrous, smooth or, rarely, rugose, indurate, clasping the palea; palea of the lower floret membranaceous and conspicuous to sometimes absent; palea of the upper floret indurate, strongly 2-nerved.

Reference: Gould, F. W. 1975. *Panicum*, pp. 433-477, *in* The Grasses of Texas. Texas A. and M. Univ. Press, College Station.

In addition to the species described below *P. fasciculatum* Sw. has been collected from the edge of our range (OK: Comanche, Grady; TX: Deaf Smith, Hardeman. It will not key readily below, but it is an annual species with the spikelets mostly less than 3 mm long, and the upper lemma rugose. *Panicum* is here treated in the restricted sense of Gould. See *Dichanthelium*.

1 Plants annual.
  2 Spikelets mostly more than 4 mm long.
    3 Lemma of upper floret rugose .... 11. *P. texanum*
    3 Lemma of upper floret smooth .... 7. *P. miliaceum*
  2 Spikelets less than 4 mm long.
    4 First glume short, blunt; sheaths glabrous .... 3. *P. dichotomiflorum*
    4 First glume acute or acuminate; sheaths normally pubescent.
      5 Spikelets very short, 1.6–2.2(2.4) mm long and acute or short-acuminate at the tips .... 9. *P. philadelphicum*
      5 Spikelets usually longer, 2–3.6 mm long, acuminate tipped.
        6 Spikelets averaging more than 3 mm long; pulvini of lower panicle branch axils not normally pubescent; anthers mostly more than 1.1 mm long ... 4. *P. flexile*
        6 Spikelets averaging less than 3 mm long; pulvini of lower panicle branch axils pubescent; anthers usually less than 1.1 mm long.)
          7 Upper floret with a crescent-shaped marking at the base; palea of lower floret present .... 6. *P. hillmanii*
          7 Upper floret lacking the crescent-shaped marking; palea of the lower floret absent .... 2. *P. capillare*
1 Plants perennial.
  9 Plants either with conspicuous rhizomes or strongly stoloniferous.
    10 Glumes subequal .... 8. *P. obtusum*
    10 Glumes unequal, the first one much shorter than the second.
      11 Spikelets mostly less than 3.7 mm long, appressed along the major panicle branches; anthers usually less than 1 mm long .... 1. *P. anceps*
      11 Spikelets mostly more than 3.7 mm long, not appressed along the major panicle branches; anthers usually more than 1 mm long .... 12. *P. virgatum*
  9 Plants neither rhizomatous nor stoloniferous.
    12 Lower sheaths compressed and keeled; spikelets less than 2.5 mm long .... 10. *P. rigidulum*
    12 Lower sheaths not compressed and keeled; spikelets more than 2.5 mm long .... 5. *P. hallii*

1. ***Panicum anceps*** Michx., beaked panicum. Stout perennials (3.5)5–10(13) dm tall from thick, sharp-pointed rhizomes. Culms hollow, compressed, glabrous to scabrous. Blades

flat, with a prominent midvein, nearly glabrous to hirsute, especially near the collar, mostly (10)20–50(84) cm long, (4.3)6–9(11) mm wide; sheaths keeled, compressed, glabrous to hirsute or hispid on the margins or more generally; ligule a brownish membrane 0.1–0.5 mm long. Panicles usually open but with the spikelets borne on short pedicels and appressed along the primary and secondary branches; spikelets acuminate, mostly 2.7–3.7 mm long; first glume 3- to 5-nerved, 1–2 mm long; second glume 5- to 7-nerved, 2.4–3.6 mm long; lower lemma 5-nerved, almost equaling the second glume; fertile lemma smooth, about 1.5–2.3 mm long; lower palea present, about 1.3–2 mm long. Anthers normally absent from the lower floret, 0.7–1 mm long in the upper floret. (2n = 18, 36) Aug–Oct. In low, moist areas in the open or in woodland; s MO, se KS, OK; (NJ to KS, s to FL & TX). *P. rhizomatum* Hitchc. & Chase—Hitchcock, Manual of the grasses of the United States, U.S.D.A. Miscl. Publ. No. 200, p. 702. 1950.

2. ***Panicum capillare*** L., common witchgrass. Tufted annuals mostly 2–7 dm tall. Culms hollow, often somewhat spreading and decumbent below, nearly terete to slightly flattened, thinly pubescent near the nodes. Blades flat, hispid on the margins near the base or more or less hispid throughout, mostly (4)6–20(27) cm long, 6–16 mm wide; sheaths papillose-hispid, not keeled; ligule 0.5–2.2 mm long, consisting of a fringe of hairs from a short basal membrane. Panicles open, very diffuse, usually purplish at maturity, 10–30 cm long, often almost as wide as long at full maturity, usually partly included in the uppermost leaf sheath; spikelets acuminate, 2–3 mm long; first glume acute, 3- to 7-nerved, 0.8–1.5 mm long, second glume 7- to 9-nerved, 2–2.8 mm long; lemma of lower floret similar to the second glume and nearly equal to it; lemma of upper floret smooth, 1.3–2 mm long, not bearing a lunate mark at the base; palea of lower floret absent. Anthers absent from the lower floret, 0.7–1.1 mm long in the upper floret. (2n = 18) Jul–Oct. A weed in disturbed ground: GP; (s Can. throughout the U.S., Bermuda). *P. barbipulvinatum* Nash—Rydberg.

3. ***Panicum dichotomiflorum*** Michx., fall panicum. Tufted annuals 3–5 dm tall. Culms thick, often decumbent below, glabrous, terete to slightly flattened. Blades flat at maturity, usually glabrous or nearly so, mostly 5–35(45) cm long, (4)6–16 mm wide; sheaths smooth, not keeled or slightly keeled; ligule 1.1–2.5 mm long, predominantly a fringe of hairs from a membranaceous base. Terminal panicles open, 8–35 cm long, the pedicels of the spikelets more or less appressed along the secondary branches; spikelets 2.2–3.6 mm long, acuminate or acute-tipped; first glume obtuse-tipped to truncate, obscurely several-nerved to nerveless, 0.5–1.2 mm long; second glume 5- to 7(9)-nerved, 2–3.5 mm long; lower lemma similar to the second glume; upper lemma smooth, 1.8–2.3 mm long; lower palea absent or vestigial to well developed, 1–2.8 mm long. Anthers absent from the lower floret, 1–1.4 mm long in the upper floret. (2n = 54) Jul–Oct. In moist, disturbed ground; s MN, se SD, IA, e NE, MO, KS, OK, TX (N. S. to MN, s to FL & TX, also introduced farther w).

4. ***Panicum flexile*** (Gatt.) Scribn., wiry witchgrass. Rather delicate tufted or solitary-stemmed annuals 2–7.5 dm tall. Culms hollow, slender, smooth to pilose, erect. Blades flat, smooth to pilose, mostly 3–32 cm long, 1–7 mm wide; sheaths pilose, not keeled; ligule a rather weak fringe of hairs from a membranaceous base, 0.3–1 mm long. Panicle very open, relatively few-flowered, 5–27(45) cm long; spikelets long-acuminate, 2.5–3.5 mm long; first glume acute to acuminate, 3- to 5-nerved, 1–1.7 mm long; second glume acuminate, 7- to 9-nerved, 2.4–3.4 mm long; lower lemma similar to the second glume but slightly shorter; upper lemma smooth, 1.7–2.2 mm long; lower palea absent. Anthers absent from lower floret, 1.1–1.4 mm long in the upper floret. (2n = 18) Aug–Oct. Open or wooded areas, often in rocky, moist, limestone soil, also in disturbed ground; MO, e KS, OK; (e Can., NY reportedly to ND, s to FL, OK & reportedly to TX).

Hitchcock reports (op. cit., p. 687) this grass from both Dakotas and from IA, but specimens were not seen from those localities. F. W. Gould (op. cit., p. 448) notes that it has evidently not been collected in TX since the last century.

5. **Panicum hallii** Vasey. Tufted perennials 3–10 dm tall. Culms terete, hollow, glabrous or appressed-pubescent at the nodes. Blades flat, often slightly glaucous, nearly glabrous to sparsely papillose-hirsute, especially on the margins near the base, mostly 5–20 cm long, 2–5(6.5) mm wide; sheaths glabrous to sparsely papillose-pubescent; ligule 0.7–1.3 mm long, consisting of a fringe of hairs from a short basal membrane. Inflorescence open, 9–30 cm long, the spikelets borne on pedicels which are appressed along the primary or secondary branches; spikelets acute to acuminate-tipped, 3–3.7 mm long; first glume rather large, 1.7–2.6 mm long, 3- to 5-nerved; second glume and lower lemma each about 2.8–3.6 mm long, 7- to 9-nerved; upper lemma smooth, 1.9–2.4 mm long; lower palea 1.2–1.8 mm long. Anthers absent from the lower floret, 1–1.3 mm long in the upper floret. (2n = 18) May–Sep. In the open on dry, sandy to rocky soil; extreme sw KS, w OK, TX; (OK & CO s to TX, AZ, & Mex.).

Plants reported in the Atlas GP as *P. hirticaule* Presl are depauperate specimens of *P. hallii*. Our plants of *P. hallii* belong to the typical variety but *P. hallii* var. *filipes* (Scribn.) Waller occurs just s of our range and may enter it. This variety has smaller spikelets and, often, glabrous culm nodes.

6. **Panicum hillmanii** Chase. Tufted annual, 1.5–4.5 dm tall. Culms hollow, often papillose-hirsute, erect to more or less spreading and decumbent below. Blades flat at maturity, hispid on the margins to more or less hispid throughout, mostly 5–25 cm long, 3–10 mm wide; sheaths hispid, not keeled; ligule 1–2.5 mm long, consisting of a fringe of hairs from a basal membrane. Panicles open, diffuse, 9–21 cm long, often nearly as broad as long at maturity; spikelets acuminate, 2–3 mm long; first glume acute, 3- to 5-nerved, 0.8–1.4 mm long; second glume 7- to 9-nerved, 2–2.8 mm long; lemma of the lower floret similar to the second glume and nearly equal to it; lemma of upper floret smooth, 1.5–2.1 mm long, bearing a crescent-shaped marking on the base; palea of the lower floret present, 1–1.8 mm long. Anthers absent from the lower floret, 0.7–1.1 mm long in the upper floret. (2n = 18) Jul–Aug. Roadsides & ditches, often in disturbed ground; scattered in KS, OK, TX; (KS to TX, NM, & CA).

7. **Panicum miliaceum** L., broom-corn millet, proso millet. Coarse annuals 3–10 dm tall. Culms stout, glabrous to hirsute, hollow. Blades flat, glabrous to variously pubescent, mostly 5–40 cm long, (4)6–15(18) mm broad; sheaths nearly glabrous to papillose-hispid, rounded to slightly keeled; ligule 1–3 mm long, consisting of a fringe of hairs from a rather prominent basal membrane. Panicle 11–25(40) cm long, contracted to open, somewhat nodding; spikelets 4–5.4 mm long, ovoid, acuminate-tipped; first glume acute to acuminate, about 7-nerved, 2.5–3.6 mm long; second glume 9- to 11-nerved, 3.8–4.8 mm long; lower lemma similar to the second glume in size and texture but 9- to 15-nerved; upper lemma smooth, plump, 2.7–3.7 mm long; lower palea present, 0.9–1.9 mm long, its margins inrolled. Anthers absent in the lower floret, 1.6–2 mm long in the upper floret. (2n = 36) Jul–Sep. Cult., sometimes escaping in disturbed habitats & relatively persistent; MN, ND, SD, IA, NE, CO, MO, KS; (introduced in many parts of the U.S., especially in n states; Old World). *Naturalized*.

8. **Panicum obtusum** H.B.K., vine-mesquite. Perennials 2.5–7(8) dm tall, developing sturdy stolons from a knotty or short-rhizomatous base. Culms hollow to nearly solid, compressed to terete, glabrous or nearly so. Blades flat, usually nearly glabrous, often somewhat glaucous, mostly 4–26 cm long, (2)3–6.3(7) mm wide; sheaths mostly glabrous or nearly so, except for a few long hairs near the ligule, but the basal ones strongly villous, particularly at the

point of attachment to the stolon; ligule a truncate membrane 0.9–2.1 mm long. Panicle narrow, 6–12 cm long, the main branches appressed and sparingly rebranched; spikelets blunt-tipped, 3.3–4.4 mm long; glumes subequal, both nearly as long as the spikelet, 2.5–4.1 mm long, 5- to 9-nerved, lemma of the lower floret about equal to the glumes, usually 5-nerved; lemma of the upper floret smooth, 2.6–3.6 mm long; palea of the lower floret prominent, membranaceous, 2.9–3.8 mm long. Lower floret usually staminate, its anthers 1.8–2.6 mm long; anthers of upper floret 1.5–2.2 mm long. ($2n = 20, 40$) May–Aug. Roadside ditches, prairies, & pastures, usually in wet or moist ground; s KS, s CO, OK, TX, NM; (MO, where introduced, & s KS to UT, s to AR, NM, & n Mex.).

9. *Panicum philadelphicum* Bernh. ex Trin., Philadelphia witchgrass. Tufted annuals 1.5–7 dm tall. Culms slender, hollow, erect to decumbent below, nearly glabrous to hirsute, especially near the nodes. Blades almost glabrous to hirsute more or less throughout, flat, mostly 4–14 cm long, (0.6)2–6(8) mm wide; sheaths pustular-hirsute, not keeled; ligule a fringe of hairs 0.7–1.3 mm long. Panicles open, 5–21(26) cm long, relatively few-flowered; spikelets obtuse to acute-tipped or sometimes slightly cuspidate, very small, 1.6–2.2(2.4) mm long; glumes unequal, the first one 3- to 5-nerved, 0.5–1.2 mm long, the second about 7-nerved, 1.4–2.2 mm long; lower lemma about equal to the second glume, 7- to 9-nerved; upper lemma smooth, 1.3–2 mm long; lower palea absent or vestigial, or rarely up to 0.8 mm long. Anthers absent from the lower floret, those of the upper floret 0.7–1 mm long. ($2n = 18$) Aug–Oct. In sandy or rocky ground, often in woods, uncommon; MN, IA, MO, se KS, OK, TX; (CT to MN, s to GA & TX).

10. *Panicum rigidulum* Bosc. ex Nees, redtop panicum. Tufted perennials 4–10 dm tall. Culms flattened, glabrous, hollow, rather stout. Blades flat or folded, usually glabrous or nearly so, 8–50 cm long, 3–9 mm wide; sheaths keeled, usually glabrous, sometimes with a line of hairs at the collar; ligule a truncate, erose-ciliolate membrane 0.6–1 mm long. Inflorescences normally both lateral and terminal, the terminal one 8–30 cm long, often partly included in the uppermost leaf sheath, its main branches ascending, to, more frequently, spreading, the spikelets crowded, appressed on short pedicels along the primary and secondary branches, the pedicels often bearing a few spreading hairs; spikelets small, mostly 1.5–2 mm long, often purplish; first glume 3- to 5-nerved, 0.8–1.3 mm long; second glume and lower lemma subequal, 1.2–1.7 mm long, both about 5-nerved; upper lemma smooth, 0.8–1.3 mm long; palea of lower floret present, 0.8–1.3 mm long, often purplish. Anthers absent in the lower floret, very short, only 0.2–0.4 mm long, in the upper one, but evidently not included at maturity. ($2n = 18$) Sep–Oct. In low, moist areas, often near water; MO, e KS, OK; (ME to KS, s to FL & TX, introduced elsewhere). *P. agrostoides* Spreng.—Rydberg.

11. *Panicum texanum* Buckl., Texas panicum. Stout annuals 3–10 dm tall. Culms erect to geniculate at the base, hollow. Blades finely pilose, 5–20 cm long, 5–15(18) mm wide; sheaths rounded, more or less finely pilose; ligule 0.8–1.5 mm long, consisting of a fringe of hairs from a short basal membrane. Panicles slender, the branches appressed, finely pilose, with longer stiffer hairs on the pedicels just below the spikelets; spikelets large, 4.5–6 mm long, acute or acuminate-tipped; first glume 3.2–4.2 mm long, 5- to 7-nerved; second glume 4.2–5.2 mm long, about 7-nerved; lower lemma nearly equal to the second glume, about 5-nerved; upper lemma obviously rugose, 3.4–4.1 mm long; palea of lower floret prominent, 2.8–3.6 mm long. Anthers present or absent in the lower floret, when present 1.8–2.4 mm long; anthers of the upper floret 1.3–1.9 mm long. ($2n = 54$, $2n = 36$ also reported) Jul–Oct. A weed in disturbed ground; OK, TX; (NC to OK, s to FL & Mex., probably introduced in the n part of its range).

12. *Panicum virgatum* L., switchgrass. Strongly rhizomatous perennials mostly (3)5–14 dm tall, often forming large clumps. Culms firm, hollow, erect, glabrous or appressed-

pubescent at the nodes. Blades flat to somewhat involute, nearly glabrous to pilose, particularly on the upper surface near the ligule, mostly 15–55 cm long, (2)5–9(11) mm wide; sheaths not keeled, glabrous; ligules (1)2–4 mm long, consisting of a fringe of hairs from a short, membranaceous base. Panicles (14)20–30(45) cm long, very open and diffuse at maturity; spikelets mostly 3.5–6 mm long; glumes acute or acuminate, the first one 2.3–4.2(5) mm long, 3- to 5-nerved, the second one 3.3–5.5(6) mm long, 7- to 9-nerved; lower lemma usually slightly shorter than the second glume, 5- to 7-nerved; upper lemma smooth, 2.4–3.5 mm long; lower palea prominent, 2.4–4.2 mm long. Anthers often present in the lower floret, 1.3–2.3 mm long; anthers of the upper floret 1.6–2.1 mm long. (2n = 18, 36, 54, 72, 90, 108, aneuploid counts of 2n = 21, 25, 30, 32, 55–65 also reported) Jul–Sep. In moist lowland prairies & other moist areas; GP; (N.S. & Ont., ME to ND, s to FL, NV, & AZ, also Mex. & C. Amer.).

54. PASPALUM L.

Tufted or solitary-stemmed annuals to cespitose or rhizomatous perennials. Culms solid or hollow, terete to flattened, erect to often at least in part decumbent and rooting at the nodes. Blades rolled in the bud, flat at maturity; sheaths rounded to, usually, somewhat keeled, open, often air-chambered and cross-septate; ligules membranaceous, fragile, often brownish; auricles none. Inflorescence of 1–many spiciform branches, its spikelets short pediceled, borne singly or in pairs in 2 rows on one side of the flattened branch rachis; spikelet flattened dorsally, consisting of an upper perfect floret and a lower sterile floret subtended by 1 or 2 glumes, the disarticulation below the entire spikelet; first glume absent or very short; second glume usually convex, turned toward the rachis; second glume and lower lemma both about equal in size to the entire spikelet, 2- to 5(7)-nerved; lower lemma turned away from the rachis; lower palea absent or vestigial; lemma of upper floret firm, smooth, its margins inrolled around the broad palea.

References: Banks, D. J. 1966. Taxonomy of *Paspalum setaceum* (Gramineae). Sida 2: 269–284; Chase, A. 1929. The North American species of *Paspalum*. Contr. U.S. Natl. Herb. 28: 1–310; Gould, F. W. 1975. *Paspalum*, pp. 500–527, *in* The Grasses of Texas, Texas A. and M. Univ. Press, College Station.

In addition to the species described and discussed here, *Paspalum dilatatum* Poir. has been reported from OK: Cleveland, Custer, and McLain. It will not key readily in the key below but may be recognized by its broadly ovate, acute-tipped spikelets which are pubescent on the margins with long, silky hairs.

1 Inflorescence branches more than 10; plants annual; rachis of inflorescence branch wider than the 2 rows of spikelets, leaflike .... 2. *P. fluitans*
1 Inflorescence branches fewer than 10; plants perennial; rachis of inflorescence branch not wider than the 2 rows of spikelets.
   2 Spikelets large, averaging 3.5–4.5 mm long and 2.5–3.4 mm wide .... 1. *P. floridanum*
   2 Spikelets smaller.
      3 Spikelets suborbicular and very small, mostly 1.8–2.4 mm long .... 6. *P. setaceum*
      3 Spikelets either larger, averaging more than 2.4 mm long, or not suborbicular.
         4 Inflorescence branches usually 2; anthers more than 1.3 mm long 4. *P. paspalodes*
         4 Inflorescence branches (2)3–6(8); anthers less than 1.3 mm long.
            5 Spikelets borne singly along the rachis; rachises averaging less than 1.5 mm wide .... 3. *P. laeve*
            5 Spikelets borne in pairs along most of the rachis (the second spikelet of the pair sometimes rudimentary); rachises averaging more than 1.5 mm wide .... 5. *P. pubiflorum* var. *glabrum*

1. Paspalum floridanum Michx., Florida paspalum. Stout perennials (4)9–16 dm tall, from short, thick rhizomes. Culms erect or ascending, firm and tough, but fistulose, glabrous.

Blades firm, flat, glabrous to more or less densely hirsute, mostly (6)18–52 cm long, 5–14 mm wide; sheaths glabrous to hirsute, air-chambered, cross-septate, keeled; ligules delicate, 1.0–2.3 mm long. Inflorescence of (2)3–5(8) branches, the longest ones 8–17 cm long; branch rachises mostly 1–1.8 mm wide; spikelets normally borne in pairs; glabrous, widely elliptic to nearly circular, 3.5–4.6 mm long, 2.6–3.4 mm wide; first glume absent; second glume and sterile lemma normally 5-nerved, equaling the spikelet. Anthers 1.5–2 mm long. (2n = 120, 160, ca 160–170) Jul–Nov. Open, low, moist to relatively dry ground; sw MO, e KS, e OK; (NJ to KS, s to FL & TX).

There are 2 weakly defined and intergradient varieties of this species. Both occur in the same general part of our range, but the second variety is more common.

1a. var. *floridanum*. Blades densely to moderately pilose throughout their length, especially on the upper surface.

1b. var. *glabratum* Engelm. ex Vasey. Blades glabrous throughout or hirsute only near the ligule. *P. glabratum* (Engelm.) Mohr—Rydberg.

2. Paspalum fluitans (Ell.) Kunth, horsetail paspalum. Annuals 1.5–10 dm long. Culms erect to decumbent and rooting at the lower nodes, often floating, soft, fistulose, glabrous in the internodes, long-pilose at the nodes. Blades glabrous to scabrous but often with pustular-based hairs on the margin near the ligule, thin, flat, 3–25 cm long and 3–10(18) mm wide; sheaths loose, pustular-pilose, rather thin but chambered and delicately cross-partitioned, often apparently extending upward in 2 points beside the ligule; ligule delicate, 1–2.2 mm long. Inflorescence branches very numerous, the longest ones 2.5–6 cm long, the branch rachises foliaceous, 0.8–1.8 mm wide, wider than the double row of spikelets and prolonged beyond the terminal spikelets; spikelets borne singly, ovate to ovate-lanceolate, acute-tipped, 1.2–1.6 mm long; first glume absent; second glume and sterile lemma as long as the spikelet, delicate, adhering to the fertile floret, 2-nerved (their midnerves lacking), pubescent; sterile palea often present as a V-shaped rudiment which may become dark and show as a pinkish V-shaped region through the translucent lower lemma. Anthers 0.2–0.4 mm long. (n = 10) Aug–Sep. Mud flats along streams, shaded creek banks, & moist woods, often aquatic; sw MO, se KS, e OK; (VA to IL & KS, s to FL & TX, also S. Amer.). *P. mucronatum* Muhl.—Rydberg.

3. Paspalum laeve Michx., field paspalum. Tufted perennials (3)4–10 dm tall. Culms ascending to erect, often somewhat geniculate, slightly compressed, solid above, fistulose below. Blades flat, variously pubescent to nearly glabrous, mostly 12–32 cm long, (3.5)5–9 mm wide; sheaths variously pubescent to nearly glabrous, keeled, usually air-chambered and cross-septate; ligules 1.2–3 mm long, often brownish. Inflorescence of (2)3–6 branches, the largest one 5.5–10 cm long, the branch rachises mostly 0.7–1.5 mm wide; spikelets borne singly, widely elliptic to nearly circular, normally 2.6–3.3 mm long and 2.1–3.1 mm wide; first glume absent; second glume and sterile lemma 5-nerved, equaling the spikelet. Anthers 0.8–1.2 mm long. (2n = 40) Jul–Sep. Pastures, fields, roadsides, & in waste ground.

There are 2 varieties of this species in our range:

3a. var. *circulare* (Nash) Fern. Lower leaf sheaths glabrous to pilose. Spikelets nearly circular, mostly more than 2.7 mm broad. In sw MO, se KS, e OK; (MA to KS, s to GA & TX). *P. circulare* Nash—Hitchcock, Manual of the Grasses of the United States, p. 618. U.S.D.A. Misc. Publ. No. 200. 1950.

3b. var. *pilosum* Scribn. Lower leaf sheaths strongly pilose. Spikelets broadly elliptic, less than 2.7 mm broad KS: Bourbon, Cherokee; sw MO; (NY to KS, s to FL & TX). *P. longipilum* Nash—Hitchcock, op. cit. p. 615.

Although the typical variety of this species has not been reported in our range, it is to be expected at the edge of it in OK or KS. It is very similar to var. *pilosum* except that the lower sheaths are nearly glabrous.

4. **Paspalum paspalodes** (Michx.) Scribn., knotgrass. Stoloniferous, rhizomatous perennials with flowering culms 4–8 dm tall. Culms solid, often decumbent below, somewhat compressed, glabrous or pubescent at the nodes. Blades flat, nearly glabrous to pustular-pilose near the collar, mostly 3–22 cm long, (1.5)3–7 mm wide; sheaths slightly keeled, air-chambered and cross-septate, glabrous to pilose, especially at the throat; ligules membranaceous, delicate, 0.6–2 mm long. Inflorescence branches 2, each about 3–6(13) cm long, the branch rachises mostly 1.1–1.6 mm wide; spikelets borne singly or paired, ovate or ovate-lanceolate, acute-tipped, 2.4–3.1 mm long; first glume absent or present and up to 1.3 mm long; second glume thinly pubescent; second glume and sterile lemma about equaling the spikelet, 3-nerved. Anthers 1.4–2.0 mm long. (2n = 40, 60, 2n = 48 also reported) Jun–Oct. Moist soil, especially near ponds & ditches; se KS to OK, TX; (NJ to FL, w through s U.S. to CA & up the coast to WA, widespread in trop., subtrop., & coastal temp. regions). *P. distichum* L.—Gould, op. cit.

This grass is referred to as *Paspalum distichum* in most manuals. According to Renvoize and Clayton (Taxon 29: 339–340. 1980), the name *P. distichum* L. properly applies to the grass usually called *P. vaginatum* Sw. They propose to reject the name *P. distichum*.

5. **Paspalum pubiflorum** Rupr. ex Fourn. var. **glabrum** Vasey ex Scribn. Perennials 4–11 dm tall. Culms ascending to erect, but often decumbent below and rooting at the lower nodes, firm, somewhat compressed, solid to fistulose, glabrous, but often at least the lower nodes pilose. Blades flat, usually somewhat pilose toward the base, but often otherwise glabrous, mostly 5–27 cm long, 5–14(19) mm wide; sheaths keeled, the lower usually with at least some pustular-based hairs; ligules 0.7–2.6 mm long. Inflorescence of (2)3–5(8) branches, the longest one 4–11 cm long, the branch rachises 1.3–2.4 mm wide; spikelets paired, sometimes one member of each pair rudimentary (or lacking), glabrous or nearly so, obovate, 2.4–3.4 mm long, 1.4–2.3 mm wide; first glume absent; second glume and sterile lemma about equaling the spikelet, 3- to 5-nerved. Anthers 0.8–1.1 mm long. (2n = 60, ca 64) Jul–Oct. Moist ground along streams & ditches, roadsides, often in limestone soil; MO, KS, OK; (NC & KY to KS, s to FL & TX).

6. **Paspalum setaceum** Michx. Tufted perennials 1.5–7(10) dm tall, arising from a knotty base. Culms erect to spreading, glabrous, solid in the upper part and fistulose, but firm; below, the nodes sometimes slightly pubescent. Blades flat, variably pilose and/or puberulent to nearly glabrous, 2–25 cm long, 4–10(15) mm wide; sheaths keeled, pilose on the margins only or throughout, with or without air chambers; ligules short, mostly less than 1 mm long, but often backed by long hairs. Inflorescence branches 1–3 on terminal peduncles, usually solitary on axillary ones, the longest branches (3.5)5–9(12) cm long; the branch rachises mostly 0.6–1.4 mm wide; spikelets usually borne in pairs, glabrous or pubescent, sometimes spotted or somewhat viscid, nearly circular, 1.8–2.4 mm long, 1.6–2.2 mm wide; first glume absent; second glume and sterile lemma equaling the spikelet, both 3-nerved or the glume 4- to 5-nerved or the sterile lemma sometimes 2-nerved (its midnerve lacking). Anthers 0.6–0.9 mm long.

This is a widespread and variable species with 2 rather poorly marked varieties in our range.

6a. var. *muhlenbergii* (Nash) D. Banks. Midnerve of the sterile lemma usually present; spikelets usually glabrous. Leaf blades normally pilose. (2n = 20) Jun–Oct. Open woods, pastures, along roads, often in dry soil; w MO, e KS, OK; (VT & MA to MI, KS, s to FL & TX). *P. muhlenbergii* Nash—Rydberg; *P. ciliatifolium* Michx.—Rydberg, misapplied; *P. pubescens* Muhl.—Hitchcock, Manual of the Grasses of the United States, p. 608. U.S.D.A. Miscl. Publ. No. 200. 1950.; *P. ciliatifolium* Michx. var. *muhlenbergii* (Nash) Fern.—Fernald.

6b. var. *stramineum* (Nash) D. Banks. Midnerve of sterile lemma absent on most spikelets; spikelets pubescent to glabrous. Leaf blades nearly glabrous to variously puberulent or pilose. (2n = 20) May–Sep. Open ground, often in sandy soil; MN, IA, NE, CO, MO, KS, OK, TX; (MI to MN, s to TX & AZ, also in se coastal region & Mex. to Panama, Bermuda, & W. I.). *P. stramineum* Nash, *P. bushii* Nash — Rydberg; *P. ciliatifolium* Michx. var. *stramineum* (Nash) Fern. — Fernald.

55. PHALARIS L.

Tufted annuals or rhizomatous perennials. Culms erect to geniculate below, smooth, hollow. Blades rolled in the bud, flat at maturity; sheaths open, not keeled, sometimes air-chambered and cross-septate; ligules membranaceous, 1-8 mm long, truncate; auricles present or absent. Inflorescence a condensed lobulate or cylindrical panicle. Spikelets laterally compressed, articulating above the glumes, 1-flowered, but the single perfect floret flanked by 2 reduced sterile florets, these usually appressed to the surface of the perfect floret and falling attached to it, often not immediately apparent; glumes usually 3-nerved, subequal, glabrous, often keeled, the keel sometimes winged; sterile florets normally scalelike, pubescent, often subulate; fertile lemma obscurely 5-nerved, indurate at maturity, pubescent; palea subequal to the lemma.

Reference: Anderson, D. E. 1961. Taxonomy and distribution of the genus *Phalaris*. Iowa State J. Sci. 36: 1-96.

1 Plants perennial from scaly rhizomes; inflorescence usually lobed, more than 8 cm long .... 1. *P. arundinacea*
1 Plants annual; inflorescence cylindrical, usually less than 8 cm long.
   2 Sterile florets not subulate; anthers more than 2 mm long; wings on glumes often approaching 1 mm wide at broadest part .... 2. *P. canariensis*
   2 Sterile florets subulate; anthers less than 2 mm long; wings on glumes much narrower .... 3. *P. caroliniana*

1. *Phalaris arundinacea* L., reed canary grass. Strongly rhizomatous perennials (5)9-16(21) dm tall. Culms erect to geniculate at the base, glabrous. Blades flat, glabrous or scabrous, mostly 7-41 cm long, (4)7-16(20) mm wide; sheaths conspicuously air-chambered and cross-septate; ligules (1.5)3-7 mm long. Panicle dense, lobed, often somewhat reddish-tinged during anthesis and becoming stramineous in fruit, 6-18 cm long. Glumes subequal, not winged to very slightly winged, 3.4-5.5 mm long; sterile florets 0.6-2 mm long, subulate, pubescent; fertile floret glabrous to somewhat appressed-pubescent, becoming shiny, 2.7-4.2 mm long. Anthers 1.8-3.4 mm long. ($2n = 14$, 28, 42; counts of $2n = 27$, 29, 30, 31, 35, 48 also reported) Mostly May-Jul. Wet ground in swales, marshes, & ditches, often abundant, even dominant, in many moist locations; common in n GP, becoming rare in OK; (throughout temp. Northern Hemisphere, also introduced in Southern Hemisphere).

2. *Phalaris canariensis* L., canary grass. Tufted or solitary-stemmed annuals 2-7.5(11) dm tall. Culms erect or geniculate below, glabrous. Blades flat, glabrous to scabrous, mostly 4-24 cm long, 4-9(11) mm wide; sheaths glabrous to scabrous, the upper ones conspicuously dilated; ligules 2-7.5 mm long; auricles none or present, but very small. Panicle dense, ovate-cylindrical, 1.8-3.5 cm long. Glumes strigose to glabrous, papery, bearing a wing on the midnerve which tapers from the narrow base to a broad portion approaching 1 mm wide on the larger spikelets, pale, with a dark green band flanking the wing on each side and with dark green flanking the lateral nerves as well; glumes at the center of the panicle 6-9 mm long, those on the lower part of the panicle often becoming much smaller; sterile florets sparsely pubescent, acute, but not subulate, (1.7)2.3-3.6 mm long; fertile lemma strigose, acute, becoming shiny, 3.6-5.4 mm long. Anthers 2.7-3.5 mm long. ($2n = 12$) Mostly Jun-Jul. Uncommon in waste ground; GP; (introduced throughout much of the world as a component of bird seed; s Europe). Naturalized.

3. ***Phalaris caroliniana*** Walt., May grass. Tufted or solitary-stemmed annuals 2–9(12) dm tall. Culms smooth, erect to geniculate below. Blades glabrous to scabrous, mostly 1–18 cm long, (1.3)5–8(12) mm wide; sheaths glabrous, air-chambered and cross-septate, the upper ones dilated; ligules truncate, (1.5)2.5–5 mm long; auricles either absent or 1 or 2 minute ones present. Panicle condensed, cylindrical, 1–6(8) cm long; glumes winged on the midnerve, but the wing not more than about 0.5 mm wide at the widest part, the margin of the wing slightly scabrous; glumes subequal, pale with darker green on the wing and flanking the lateral nerves, 4–6(6.5) mm long; sterile florets pilose, subulate, 0.8–2.5 mm long; fertile lemma pilose, becoming shiny, abruptly acute, 2.9–4.1(4.8) mm long. Anthers 0.5–1 mm long. (2n = 14) May–Jun. Moist swales, sloughs, ditches, & prairies; MO, e KS, OK, & TX; (VA to CO & OR, s to FL, TX, CA, & n Mex.).

56. PHLEUM L.

1. ***Phleum pratense*** L., timothy. Tufted or solitary-stemmed perennials 5.5–10(14) dm tall. Culms erect, often geniculate below, bulbous-based, hollow, glabrous. Blades rolled in the bud, flattened at maturity, glabrous or scabrous, 3–27 cm long, 2.4–8 mm wide; sheaths glabrous, not keeled, open; ligules membranaceous, 1.2–5 mm long; auricles none. Inflorescence a cylindrical, condensed, spikelike panicle 2–15 cm long and 5.5–9 mm wide when pressed; spikelets 1-flowered, disarticulating above the glumes and eventually below the glumes as well, the fruit sometimes retained until the latter event; glumes about equal, laterally compressed, the 3 nerves crowded in the green center, the midrib long-ciliate, the margins membranaceous, the body 1.8–3.2 mm long, the nerves extended into a thickish awn 0.5–1.5 mm long; lemma delicate, membranaceous, 1.3–2.4 mm long, 5-nerved, unawned or the central nerve sometimes extended into a delicate awn point; palea a little shorter than the lemma, membranaceous. Anthers 1.3–1.8 mm long. (2n = 42) May–Aug, mostly Jun–Jul. Pastures, roadsides, lawns, & ditches; GP except TX, increasingly common northward; (throughout U.S.; Eurasia, widely planted & escaped elsewhere). *Naturalized.*

Phleum alpinum L., a smaller species with broader panicles (usually over 1 cm wide when pressed) has been collected from the margins of our range (CO: Arapaho; SD: Pennington; WY: Laramie).

57. PHRAGMITES Trin., Reed

1. ***Phragmites australis*** (Cav.) Trin. ex Steud., common reed. Rhizomatous perennials 13–30(43) dm tall. Culms stout, erect, terete, smooth, hollow. Blades rolled in the bud, flat at maturity, glabrous, smooth, mostly (10)25–55 cm long, 1–4 cm wide; sheaths open, clearly cross-septate in transmitted light, not keeled; ligule a membrane 0.3–1.2 mm long, variously backed and fringed with short to long hairs; auricles none. Inflorescence a dense plumose panicle 20–35(42) cm long, the branches and apex nodding at maturity; spikelets 3- to 6-flowered, disarticulating above the glumes and at the base of each rachilla internode; the internodes between the flowers pilose with long silky hairs; glumes lanceolate, with 3 principal nerves, sometimes with additional secondary nerves, the first glume 2.9–5.5 mm long, the second one 5.5–8 mm long; lemmas 3-nerved, the lowest lemma acuminate but not awned, staminate, 8–12 mm long, the upper ones successively smaller, awned, pistilate to perfect. Anthers 0.9–2.1 mm long. (2n = 36, 48, 72, 96 and 54, 84 also reported) Jun–Oct, mostly Jul–Sep. Streams, lake borders, & marshes; GP; (nearly cosmopolitan). *P. communis* Trin.—Rydberg; *P. communis* Trin. var. *berlandieri* (Fourn.) Fern.—Fernald.

58. POA L., Bluegrass

Annuals or cespitose to stoloniferous or rhizomatous perennials. Herbage nearly glabrous (to scabrous). Culms hollow. Blades folded in the bud, flat to folded at maturity, usually prow-shaped at the tip; sheaths partially closed, rounded to keeled; ligules membranaceous; auricles none. Inflorescence an open to condensed panicle of small (1)2- to 8-flowered spikelets; spikelets disarticulating above the glumes and between the florets; glumes usually keeled, acute, slightly unequal, normally surpassed by the tip of the lowest lemma, the first 1(3)-nerved, the second usually 3-nerved and often longer than the first; lemmas faintly to strongly 5-nerved, the nerves converging, keeled to rounded, acute to obtuse, nearly glabrous to variously scabrous or pubescent, sometimes with a tuft of long cobwebby hairs (cobweb) at the base; palea usually nearly as long as the lemma. Flowers perfect or imperfect in some species.

Reference: Marsh, V. L. 1952. A taxonomic revision of the genus *Poa* of the United States and southern Canada. Amer. Midl. Naturalist 47: 202–250.

Poa is a genus famous for its difficulty, much of which has been ascribed to apomixis. Unfortunately, many of the taxa are separated by rather intergradient qualitative features of lemma shape and pubescence and by few quantitative features. Most of our species are relatively clear-cut, except for the group of species related to *P. sandbergii* (*P. canbyi, P. glaucifolia, P. juncifolia*).

An attempt has been made in the following key to minimize the use of the character "lemma keeled vs. rounded" since this feature appears to account for the majority of the misidentifications of poas in our herbaria.

In addition to the species described, *Poa bulbosa* L. has been collected at the edge of our range and seems likely to spread. It has most of its florets modified into bulblets, and its culms are basally swollen. It has been collected in cen & e KS; ND: Stark; NE: Dawes and Sioux; OK: Cleveland; SD: Lawrence; and WY: Crook.

Rydberg's reports of *Poa languida* Hitchc., *P. alsodes* A. Gray, and *P. glauca* Vahl for, respectively, KS, NE, and ND are believed to be erroneous.

1 Plants annual.
  2 Lemmas with a prominent cobweb at the base; florets retaining their anthers, these under 0.3 mm long .... 5. *P. chapmaniana*
  2 Lemmas pubescent on the mid and lateral nerves, but lacking a cobweb at the base; anthers longer and often exserted .... 1. *P. annua*
1 Plants perennial.
  3 Plants strongly rhizomatous or stoloniferous.
    4 Culms strongly compressed, sharply 2-edged; plants perfect-flowered .. 6. *P. compressa*
    4 Culms slightly compressed or terete, not sharply 2-edged; plants perfect flowered or dioecious.
      5 Cobwebs obvious on the lemmas of at least some florets or plants dioecious.
        6 Plants usually dioecious, perfect-flowered plants only occasional in the populations; lowest lemmas large, mostly 3.5–6.5 mm long; anthers of staminate flowers (1.6)1.8–2.7 mm long .... 2. *P. arachnifera*
        6 Plants perfect-flowered; lowest lemmas shorter, mostly 2–3.8 mm long; anthers usually 0.8–1.7 mm long.
          7 Plants stoloniferous, not rhizomatous; inflorescence diffuse, (9)13–28 cm long .... 12. *P. palustris*
          7 Plants rhizomatous, not stoloniferous; inflorescence more compact, usually 3–13 cm long .... 13. *P. pratensis*
      5 Cobwebs absent, although lemmas sometimes pubescent; plants perfect-flowered.
        8 Lemmas glabrous to scabrous, not sharply keeled .... 11. *P. juncifolia*
        8 Lemmas pubescent, at least on the mid and marginal nerves, sharply to faintly keeled.
          9 Lemmas rather short and broad, mostly 2.8–3.7(4) mm long, usually more than 1/5 as broad as long when viewed from the side .... 3. *P. arida*
          9 Lemmas elongate, slender, mostly (3)3.5–5 mm long, usually less than 1/5 as broad as long when viewed from the side .... 9. *P. glaucifolia*

3 Plants neither rhizomatous nor stoloniferous.
  10 Lemmas webbed at the base, the webbing copious to scant.
    11 Leaves broad, (1.5)2.6–5(6) mm wide; closed spikelets broad, often more than 4/10 as broad as long; lemmas pubescent between as well as on the mid and lateral nerves .... 15. *P. sylvestris*
    11 Leaves narrower, (0.5)1–3(3.8) mm wide; closed spikelets narrower, usually less than 4/10 as broad as long; pubescence of lemmas usually confined to mid and lateral nerves.
      12 Erect-stemmed plants; the inflorescence 5–15 cm long; culm ligules mostly less than 1.6 mm long .... 10. *P. interior*
      12 Weak-stemmed, decumbent-based plants; the inflorescence (9)13–28 cm long; culm ligules more than 1.6 mm long .... 12. *P. palustris*
  10 Lemmas not webbed at the base, although sometimes very pubescent.
    13 Lemmas glabrous to scabrous.
      14 Flowers imperfect, ours normally all pistillate with tiny abortive stamens; inflorescence usually less than 8 cm long; lemmas sharply keeled .... 7. *P. cusickii*
      14 Flowers perfect; inflorescence 7–20 cm long; lemmas rounded to weakly keeled .... 11. *P. juncifolia*
    13 Lemmas pubescent, the pubescence either general, especially toward the base or confined to the mid and marginal nerves.
      15 Lemmas sharply keeled; plants mostly imperfect-flowered (usually pistillate) .... 8. *P. fendleriana*
      15 Lemmas rounded, not more than slightly keeled; plants perfect-flowered.
        16 Culms mostly 1.8–3.5(5.7) dm tall; blades narrow, mostly less than 2 mm wide; leaves principally basal; anthers averaging about 1.6 mm long on most plants .... 14. *P. sandbergii*
        16 Culms mostly (3.5)5–8(9) dm tall; some blades usually more than 2 mm wide; some leaves distributed up the culm; anthers averaging about 2.1 mm long on most plants .... 4. *P. canbyi*

1. **Poa annua** L., annual bluegrass. Tufted, often somewhat spreading, annuals mostly 0.5–2.5(3.2) dm tall. Culms slender, flattened. Blades bright green, usually flat, mostly 1–13 cm long and 0.8–3(4) mm wide; sheaths closed only near the base or up to about 2/3 of their length; ligules obtuse to truncate, 0.8–2.5 mm long. Panicles pyramidal, open, 1–6(15) cm long, the lowest branch(es) often 1 or 2(4); spikelets 2- to 4(6)-flowered, 2.8–5.5 mm long, 0.8–2 mm wide when fully developed but unopened; glumes scarious-margined, unequal, the first narrow, 1.4–2(2.4) mm long, the second broader, usually 3(4)-nerved, 1.7–2.6(2.8) mm long; lemmas scarious-margined, usually rather coarsely pubescent on the nerves, lacking a cobweb at the base, the lowest ones 2.3–3.4 mm long. Anthers 0.7–1.1 mm long. (2n = 14, 28, 2n = 24–26 also reported) Apr–Oct. A weed in lawns, along roads, & in ditches, usually where moist; GP, evidently more common in the e part; (Newf. to AK s to FL & CA; Europe.) *Naturalized*.

2. **Poa arachnifera** Torr., Texas bluegrass. Dioecious, strongly rhizomatous perennials 2.5–8.5 dm tall. Culms erect, flattened but not strongly 2-edged. Blades firm, folded to flat at maturity, mostly 3–30 cm long; 1.4–4.5 mm wide; sheaths lightly keeled, compressed, closed at the base; culm ligules pointed, 1.3–4 mm long. Inflorescence contracted, 3–14 cm long, the lowest branches usually 2(5); spikelets slightly dimorphic, the first glume of both types 1(3)-nerved, the second usually 3-nerved; staminate spikelets (2)4- to 7-flowered, mostly 4–7 mm long, 1.5–2.6 mm wide when mature but unopened, the first glume 2–3.5 mm long, the second 2.6–3.8 mm long, the lowest lemma 3.5–5 mm long, nearly glabrous except for the scant web at or below the base, anthers 1.6–2.7 mm long; pistillate spikelets (1)3- to 5-flowered, mostly 4.5–9 mm long, 1.7–3 mm wide when mature but unopened, the first glume 2.5–4.5(5.6) mm long, the second 3.3–4.8(5.7) mm long, the lowest lemma 4.2–6.4 mm long, pubescent on the mid and lateral nerves and copiously webbed at the

base; spikelets of the rare hermaphrodite plants similar to those of pistillate plants, but bearing stamens. (n = 42, 2n = 42, ca 54, 56, ca 63 also reported) Apr–May (Jun). Roadsides, pastures, & prairies; s KS, OK, TX; (KS to AR & TX, introduced eastward).

3. **Poa arida** Vasey, plains bluegrass. Long-rhizomatous to short-rhizomatous perennials (2)3–8(9.5) dm tall. Culms terete to flattened but not sharply 2-edged. Blades folded to somewhat involute or flat, the tips often not particularly boat-shaped, 1–30(40) cm long, mostly 1.8–3.5(4.5) mm wide; sheaths usually closed only near the base, not keeled; ligules (1.8)3–5 mm long, acute. Panicles rather contracted, 6.5–12.5(18) cm long; spikelets 3- to 7(9)-flowered, 4.5–8 mm long and 1.5–2.5 mm wide when fully mature but unopened; glumes somewhat unequal, 1(3)-nerved, the first 2.4–3.6 mm long, the second 3–4.2 mm long; lemmas mostly 2.9–4 mm long, rather broad, normally more than 1/5 as wide as long in side view, strongly to rather weakly keeled, villous on the mid and lateral nerves and often pubescent between the nerves toward the base, cobweb none. Anthers 1.2–2.1 mm long. (2n = 63, 64, 84, ca 90, ca 103) Apr–Jun (Aug). Dry or moist areas in prairies, pastures, & along roads.& railroads, often where sandy or alkaline; GP; (Man. to Alta., s to IA, TX, & NM). *P. pratensiformis* Rydb., *P. overi* Rydb.—Rydberg.

This species is variable, and some specimens approach *P. glaucifolia* rather closely. The best distinction between the two appears to be the elongate, narrow lemma of *P. glaucifolia*, which is usually less than one fifth as broad as long in side view. In addition, the pubescence of the lemma is less prominent and somewhat finer in *P. glaucifolia*, and the mature but unopened spikelets tend to be narrower. They are often less than one third as wide as long, while those of *P. arida* are usually broader.

4. **Poa canbyi** (Scribn.) Piper, Canby's bluegrass. Tufted perennials (3.5)4.5–8(9.5) dm tall. Culms more or less terete, the leaves not all in a basal tuft, at least a few of them distributed up the culm; blades mostly flat to folded or involute, 4–10(25) cm long, 1–2.5(3.5) mm wide; sheaths closed near the base, not keeled; ligules usually scabrous, those of the culm leaves (0.8)1.3–4 mm long, often acute. Inflorescence normally condensed, 6–13.5(15) cm long, the lowest branches (1)2–3(6) in number; spikelets 2- to 6-flowered, rather elongate, 5–10 mm long and 1–2 mm wide when mature but unopened; first glume 1(3)-nerved, 2.8–5 mm long, second glume 3-nerved, 3.3–5.5 mm long; lemmas rounded on the back, scarcely at all keeled, crisp-puberulent or strigose, especially near the base, 3.6–5.1 mm long. Anthers 1.7–2.4(2.6) mm long. (2n = 72, 82–85, 90, 93, 94, 105, 106) Mid Jun–Aug. Open prairie areas & pastures, often where moist; MN, ND, MT, SD, CO; (Que. to AK, s to SD, CO, & CA). *P. laevigata* Scribn., *P. lucida* Vasey—Rydberg.

This entity is not well marked in our range, many of the specimens being transitional to *P. sandbergii* and a few to *P. juncifolia*. On the average, *P. canbyi* is taller and later blooming than *P. sandbergii*. In addition, it has, on the average, longer, broader leaves, a longer inflorescence, more numerous florets in the spikelet, longer, narrower spikelets, and larger lemmas. It is closely similar to *P. juncifolia* in most quantitative characters, differing principally in the pubescent lemma. A few specimens are intermediate in this feature.

5. **Poa chapmaniana** Scribn., Chapman bluegrass. Tufted annuals 0.6–2.4 dm tall. Culms slender, erect to decumbent at the base, terete to slightly flattened. Blades flat to folded or involute, 1–6 cm long, 0.6–1.7(2.5) mm wide; sheaths closed near the base, not keeled; ligules truncate, those of the culm leaves (0.5)1–2(2.6) mm long. Inflorescence 2–7 cm long, condensed to pyramidal, the lowest branches appressed to spreading or reflexed, 1–4 in number; spikelets 2- to 6-flowered, 2.3–4.1 mm long, 1–2 mm wide, normally always closed; glumes scarious-margined, the first 1-nerved, 1.5–2.5 mm long, the second 3-nerved, 1.8–2.8 mm long; lemmas keeled, the lowest ones 1.9–2.6 mm long, villous on the mid and marginal nerves and with a prominent cobweb at or below the base. Anthers less than 0.3 mm long, retained within the spikelets. Apr–May. In a variety of habitats, often in disturbed ground; extreme s IA, se NE, MO, e KS; (MA to NE, s to FL & TX).

This species appears to be amply distinct from the related *P. annua* in our region. In addition to the obvious difference of lemma pubescence, it has shorter anthers and is evidently cleistogamous. It also has, on the average, narrower leaf blades and shorter lemmas.

6. Poa compressa L., Canada bluegrass. Strongly rhizomatous, glaucous perennials (2)2.5–6(8) dm tall. Culms flattened, sharply 2-edged in the upper parts, wiry. Blades flat to folded, 2–11(16) cm long, 1.2–3.2(4.5) mm wide; sheaths compressed, keeled, closed only near the base; ligules truncate, those of the culm leaves 0.7–2 mm long. Inflorescence compact to open, 1.5–8.5(11) cm long, the lowest branches (1)2(4) in number; spikelets 2- to 6-flowered, 2.8–5.5 mm long, 0.8–2 mm wide when mature but unopened; glumes slightly unequal, the first 1- to 3-nerved, 1.5–3 mm long, the second 3-nerved, 1.8–3.3 mm long; lemmas strongly keeled, pubescent on the mid and marginal nerves but seldom webbed, 2–3 mm long. Anthers 1–1.7 mm long. (2n = 14, 35, 39, 45, 49, 50, 56) Mostly Jun–Aug. Often in rocky soil or waste ground in a variety of habitats; GP, becoming uncommon in s part; (introduced from Newf. to AK, s throughout most of the U.S.; Europe). *Naturalized.*

7. Poa cusickii Vasey. Plants in dense tufts, ours pistillate, 1–5.5 dm tall. Culms erect, slender, terete. Leaves mostly distributed toward the base; blades folded or involute to flat, 1–16 cm long, 0.8–2.2 mm wide; sheaths slightly keeled, often closed about 1/2 the length; ligules 0.6–2.6 mm long, blunt and somewhat erose to acute. Panicles dense, 2.5–8 cm long, the lowest branches normally 1–3 in number. Spikelets (2)3- to 5-flowered, 5–8 mm long, 1.4–2.4 mm wide when mature but unopened; glumes unequal, the first usually 1-nerved, 3–4.6 mm long, the second 3-nerved, 3.4–5.5 mm long; lemmas strongly keeled, nearly glabrous to scabrous, the lowest one 4.2–5.6 mm long. Flowers in ours pistillate, usually bearing abortive anthers 0.3–0.8 mm long. (2n = 28, 42, 56, 59) May–Jul, mostly Jun. Open hillsides & prairies; n & w ND, MT, WY; (Alta. to B.C., s to ND, CO, & CA).

This is a rather uniform and distinctive entity in our region. It is sometimes confused with *P. sandbergii*, with which it may grow. It shares a similarity in blooming time and habit, but differs markedly in spikelet features.

8. Poa fendleriana (Steud.) Vasey, muttongrass. Dioecious (ours usually pistillate) perennials. Plants cespitose, 2.5–6(7.5) dm tall. Blades folded or involute to flat, mostly 2–21 cm long, 1.5–3.5 mm wide; sheaths slightly keeled, closed only at the base; ligules of culm leaves extremely variable, truncate to acute or acuminate, 2–6(9) mm long. Inflorescence usually 4–11 cm long, rather condensed, the lowest branches normally 2 or 3 in number; spikelets 2- to 7-flowered, 5.5–10 mm long, 1.8–3 mm wide when mature but unopened; first glume usually 1-nerved, 2.9–5.1 mm long, second glume usually 3-nerved, 3.3–5.3 mm long; lemmas strongly keeled, silky pubescent on the mid and marginal nerves, 4.2–6 mm long. Flowers in ours almost always pistillate, often with abortive anthers 0.2–0.9 mm long. (2n = 56) May–Aug. Open slopes or dry woods; SD, WY, nw NE, & CO; (Man. to B.C., s to NE, w TX, CA, & n Mex.). *P. longiligula* Scribn. & Williams, *P. brevipaniculata* Scribn. & Williams — Rydberg.

This is a distinctive species, but other taxa are often mistaken for it. At least some of the more eastern locations dotted for it in the Atlas GP are doubtful. The two locations shown in ND, for example, have proven to be misidentified specimens of *P. cusickii*.

Poa rupicola Nash ex Rydb., a plant of meadows and open slopes at high elevations, has recently been collected in SD: Pennington. It will key here to *P. fendleriana* but can be distinguished readily by its much smaller, perfect florets and spikelets.

9. Poa glaucifolia Scribn. & Williams. Short-rhizomatous, often glaucous, perennials, 4–10 dm tall. Culms erect, terete or slightly flattened. Blades folded or involute to flat, mostly 5–35 cm long and 1.3–4 mm wide; sheaths closed only near the base; ligules of culm leaves

1.3–4 mm long, often acute. Inflorescence rather slender, condensed, 8–18 cm long, the lowest branches 2–6 in number; spikelets 2- to 5-flowered, mostly 4–7 mm long and 0.9–2 mm wide when mature but unopened; glumes unequal, the first mostly 1- to 3-nerved, 2.6–4.4 mm long, the second 3(5)-nerved, 3.2–4.8 mm long; lemmas variously pubescent, normally the pubescence most prominent on the mid and marginal nerves, rather weakly keeled, elongate, the lowest one 3–5 mm long and usually less than 1/5 as wide as long when viewed from the side. Anthers 1.5–2.1 mm long. (2n = 50, 56, ca 70, ca 100) Jun–Aug. Meadows, prairies, & open woods, often in wet ground; ND, w SD, nw NE, WY, CO; (Alta. to B.C., s to MN, NE, NM, & AZ). *P. plattensis* Rydb. — Rydberg.

This entity is rather poorly defined in our region. It is variable, and it appears to merge somewhat with *P. arida*, with *P. canbyi*, and, especially, with *P. juncifolia*. It is slightly less common in ND than indicated by the map in the Atlas GP, since many of the ND specimens are more readily assignable to one of the other species mentioned. The specimen dotted for central NE is *P. arida*.

10. Poa interior Rydb., inland bluegrass. Tufted perennials 2–7.5(9.5) dm tall. Culms terete, erect, rarely slightly decumbent at the base, but not stoloniferous. Blades flat to, less frequently, folded or involute, mostly 3–16 cm long, 1.6–2.2 mm wide; sheaths rounded to slightly keeled, closed only near the base or up to about 1/3 of their length; ligules normally truncate, those of the culm leaves 0.6–1.5 (rarely to 2.5) mm long. Inflorescence open and diffuse, pyramidal, its lowest branches 2–5 in number; spikelets (1)2- to 3(4)-flowered, 2.2–4.8 mm long, 0.8–1.5 mm wide when mature but unopened; glumes slightly unequal, the first 1- to 3-nerved, 1.5–3 mm long, the second 3-nerved, 1.8–3.2 mm long; lemmas sharply keeled, 2–3.2 mm long, pubescent on the mid and marginal nerves, webbed on or below the base, the webbing sometimes scant. Anthers 0.8–1.6 mm long. (2n = 28, 34, 42, 56) Jun–Aug (Sep). Prairies, hillsides, & woodlands; MN, ND, MT, w SD, nw NE, WY, CO; (Que. to B.C., s to VT, MN, CO, & AZ). *P. nemoralis* L. var. *interior* (Rydb.) Butt.s & Abbe — Scoggan.

In herbarium material, shade forms of this species are sometimes difficult to distinguish from *P. palustris*, which it closely approaches. It differs so markedly in habit and habitat that this difficulty is not encountered in the field.

The European species *P. nemoralis* L. is very similar to *P. interior*, and some authors have referred our material to that species. *P. nemoralis* has been planted in our area (for example in NE: Douglas) but evidently has not become naturalized. It has shorter culm ligules and first glumes which are markedly narrower than the first lemmas.

11. Poa juncifolia Scribn. Tufted, nonrhizomatous or, rarely, short-rhizomatous perennials 4–10 dm tall. Culms terete, erect. Blades folded or involute to flat, 5–35 cm long, 1.2–3(3.4) mm wide; sheaths closed near the base, usually not for more than 1/4 the distance to the top, rounded to slightly keeled; ligules usually short, truncate to acute, those of the culm leaves 0.6–2.2(2.5) mm long, those of the innovations shorter. Inflorescence normally condensed, slender, 7–19 cm long, the lowest branches 1–3 in number; spikelets 2- to 6-flowered, rather elongate, 4.5–9 mm long, 1–2.2 mm wide when mature but unopened; first glume 1- to 3-nerved, 2.8–4.4(4.8) mm long; second glume 3-nerved, 3.2–5.2 mm long; lemmas glabrous to scabrous, not webbed, rounded to slightly keeled, the lowest one 3.5–5.2 mm long. Anthers 1.6–2.1 mm long. (2n = 28, ca 60,64,62–84) Jun–Jul. w ND, MT, w NE, WY; (ND to AK, s to NE, NM, NV, & CA). *P. confusa* Rydb. *P. truncata* Rydb.— Rydberg; *P. ampla* Merrill — Stevens.

Poa ampla is not separable from *P. juncifolia* in our range, since perhaps the majority of our specimens are intermediate in characteristics. Some of our specimens of *P. juncifolia* appear transitional to *P. canbyi* or to *P. glaucifolia*. See the descriptions and discussions of those species.

12. Poa palustris L., fowl bluegrass. Perennials (3.5)6–12 dm tall. Culms terete, rather weak, usually leaning on other vegetation or partly decumbent, then often rooting at the nodes and the plants stoloniferous. Blades flat to folded, 5–15(25) cm long, 1–3.1(3.8) mm wide; sheaths rounded to slightly keeled, closed near the base; ligules rather long, acute to acuminate, those of the culm leaves mostly (1.5)5–5.5 mm long. Panicle open and diffuse, (9)13–25(28) cm long, the lowest branches (2)3–6(8) in number; spikelets (1)2- to 4(5)-flowered, 2.6–4.4 mm long and 0.7–1.7 mm wide when mature but unopened; glumes slightly unequal, the first 1(3)-nerved, 1.5–3 mm long, the second 3-nerved, 1.9–3.1 mm long; lemmas sharply keeled, 2–3 mm long, pubescent on the mid and marginal nerves, webbed on or below the base, the webbing sometimes sparse. Anthers 0.8–1.2 mm long. (2n = 28, 42, 2n = 21, 30, 32, and 29 also reported) Jun–Aug. In damp ground of marshes & at the edges of streams, ponds, & lakes, also in damp pastures & ditches; MN, ND, MT, SD, WY, IA, NE, CO, MO; (introduced from Newf. to AK, s to VA, MO, NM, NV, & CA; Europe). *Naturalized. P. crocata* Michx.—Rydberg

This species is sometimes confused with *P. interior*, which see. *P. trivialis* L., which is widely planted as a turf grass, is occasionally collected and may be naturalized at some places in our range. It resembles *P. palustris*, but has more conspicuous intermediate nerves and glabrous lateral nerves on its lemmas.

13. Poa pratensis L., Kentucky bluegrass. Strongly rhizomatous mat-forming perennials 1–10 dm tall. Culms erect, nearly terete to slightly flattened. Blades flat to folded, 1–15(37) cm long, 0.9–3.6(5.3) mm wide; sheaths rounded to slightly keeled, closed for about the lower 1/2; ligules truncate, those of the culm leaves 0.7–2.1 mm long. Panicles moderately open to somewhat contracted, 3.5–12.5 cm long, 1.2–2 mm wide when mature but unopened; glumes unequal, scabrous on the keels, the first 1-nerved, 1.7–3.3 mm long, the second 3-nerved, 2–3.8 mm long; lemmas strongly keeled, the lowest one 2.6–3.8 mm long, villous on the keel and marginal nerves, copiously webbed at the base. Anthers 1.1–1.7 mm long. (2n = 25–124) Mostly May–Aug. Very common in a great variety of habitats; GP; (widespread throughout Can. & most of the U.S.; Eurasia). *Probably both native and naturalized. P. agassizensis* Boivin & Löve—Boivin & Löve, Naturaliste Canad. 87: 173–180. 1960.

This is the most popular lawn grass in much of our region and very likely exists here in both native and naturalized forms (see Boiv. & Löve, op. cit.). In spikelet features, however, it does not appear to be much more variable than most of our other taxa, and no attempt is made here to recognize varieties.

14. Poa sandbergii Vasey, Sandberg's bluegrass. Strongly cespitose perennial forming large clumps, 1.5–3.5(5.7) dm tall. Culms terete, erect. Leaves distributed mostly toward the base; blades folded to involute, mostly 1–10 cm long, 0.5–1.8 mm wide; sheaths rounded or slightly keeled, closed at the base; ligules usually acute, those of the culm leaves 1–3.5 mm long. Inflorescence slender, condensed, 2.5–7.5(11) cm long, the lowest branches usually (1)2–3 in number. Spikelets 2- to 3(5)-flowered, elongate, slender, (3.3)4.8–7 mm long, 0.9–2 mm wide when mature but unopened; glumes unequal, the first 1(3)-nerved, 2.2–4(4.8) mm long, the second 3-nerved, 2.8–5 mm long; lemmas rounded on the back, crisp-puberulent, not webbed, 3–4.7 mm long. Anthers 1.2–2 mm long. (2n = 49–90) May–Jul, mostly Jun. Hillsides & prairies; w ND, MT, w SD, WY, nw NE, CO; (ND to Yukon s to NE, NM, & s CA). *P. buckleyana* Nash—Rydberg; *P. secunda* Presl—misapplied by various authors.

This species is similar to and somewhat intergradient with *P. canbyi*, which see.

15. Poa sylvestris A. Gray, woodland bluegrass. Tufted perennials 4.5–9 dm tall. Culms delicate, terete to slightly flattened. Blades flat, 5–15(20) cm long, 1.5–5.3(6) mm wide; sheaths closed about 1/2–3/4 of their length, rounded to slightly compressed and keeled; ligules of the culm leaves 0.5–2.2 mm long, erose, truncate. Inflorescence open, cylindrical, the lowest branches (2)5 or 6(7) in number, usually reflexed at maturity; spikelets 1- to 4-flowered, 3–4.5 mm long and 1.4–2 mm wide when mature but unopened; glumes unequal, the first 1-nerved, 1.2–2.8 mm long, the second 3-nerved, 1.7–3.6 mm long; lemmas strongly keeled, villous on the mid and marginal nerves, pubescent between the nerves basally, prominently webbed at the base, the lowest one 2.2–3.4 mm long. Anthers 0.6–1.2 mm long. (2n = 28) May–Jun. rich woodlands; e SD, IA, NE, MO, e KS, e OK; (NY to WI & SD, s to FL & TX).

Specimens of this species are evidently responsible for erroneous reports of *P. wolfii* Scribn. in NE, including the report for Brown Co. in the Atlas GP.

59. POLYPOGON Desf.

1. ***Polypogon monspeliensis*** (L.) Desf., rabbitfoot grass. Tufted or solitary-stemmed annuals 0.9–6(7) dm tall. Culms hollow, erect or geniculate at the base. Blades scabrous, rolled in the bud, flat at maturity, ridged more deeply on the upper surface than on the lower, but the fine striations of the lower surface much more numerous than the ridges above, mostly 2–18 cm long, (2)3–7(8) mm wide; sheaths open, glabrous to scabrous; ligules membranaceous, acute to acuminate, veined, scabrous to pubescent, 3.5–8 mm long; auricles none. Inflorescence a dense, crowded, uninterrupted panicle 2–9 cm long; spikelets 1-flowered, disarticulating below the glumes; glumes narrow, pubescent, 1-nerved, subequal, 1.5–2.5 mm long, awned from the notched apex, the awn 5–7.5(9.5) mm long; lemma and palea delicate, membranaceous, 0.7–1.3 mm long, the apex of both usually toothed, the lemma awnless or, more commonly, with a delicate deciduous awn 0.4–2.5 mm long. Anthers 0.3–0.6 mm long. (2n = 28) May–Sep. In ditches & at the edges of rivers, lakes, marshes, & streams; GP, less common in the n part; (widespread in N. Amer.; Europe). Naturalized.

Polypogon interruptus H.B.K., a perennial species with lobed panicles, is known to be established in NE: Keith and should be expected elsewhere.

60. PUCCINELLIA Parl., Alkali-grass

Slender tufted perennials. Culms hollow, erect to decumbent, sometimes rooting at the lower nodes. Blades scabrous, C-shaped in cross section, slightly rolled, or folded in the bud, involute to flat at maturity; sheaths closed at the very base, rather tight-fitting, not keeled to slightly keeled, the lowest usually air-chambered; ligule membranaceous, truncate to pointed; auricles none. Inflorescence an open panicle, its branches ascending when young, spreading or reflexed at maturity; spikelets (1)3- to 6(8)-flowered, disarticulating above the glumes and between the florets; glumes unequal, the first shorter than the second, both normally shorter than the lowest lemma, the first 1-nerved, the second usually 3-nerved; lemmas not keeled, bearing 5 nearly parallel nerves which end before the scarious margins, pubescent toward the base; palea subequal to the lemma.

Reference Fernald, M. L., & C. A. Weatherby. 1916. The genus *Puccinellia* in eastern North America. Rhodora 18: 1–23.

1 Anthers 1–2 mm long; lowest lemmas normally 2.5–3.5 mm long .... 1. *P. cusickii*
1 Anthers less than 1 mm long; lowest lemmas usually less than 2.5 mm long.
  2 Lemmas truncate or blunt-tipped, mostly 1.5–2.2 mm long; ligules of culm leaves truncate, often less than 1.5 mm long .... 2. *P. distans*
  2 Lemmas narrowed and usually pointed at the apex, mostly 2–2.5 mm long; ligules of culm leaves usually pointed, often more than 1.5 mm long .... 3. *P. nuttalliana*

1. ***Puccinellia cusickii*** Weath. Tufted perennials 3–9 dm tall. Culms erect, glabrous to finely scabrous. Blades normally C-shaped in cross section in the bud to sometimes slightly

rolled, usually involute at maturity, strongly ribbed and scabrous on the upper surface, mostly 2–14 cm long, 1–2.4 mm wide; sheaths basally closed, tight-fitting, not keeled, at least the lower ones eventually becoming irregularly air-chambered; ligules usually pointed, those of the culm leaves 1.5–3 mm long. Panicles 5–16 cm long, the branches erect to spreading, rarely the lowermost reflexed; spikelets mostly (2)4- to 6(8)-flowered, 4–6(8.5) mm long, 1–2(3.2) mm wide; glumes unequal, the first (0.7)1–1.9(2.2) mm long, the second (1.4)2–2.5(2.8) mm long; lemmas narrowed toward the tip, mostly somewhat pointed, rarely at all truncate, the lowest (2.3)2.7–3.4 mm long. Anthers 1.1–2 mm long; grain 1.4–1.7 mm long. (2n = 28) Jun–Jul. Wet, sandy areas, moist prairies, often where alkaine; w ND to MT & WY; (Alta. to WA, s to ND, WY, & OR).

This species is so similar to *P. nuttalliana* that it might well be considered a form of that species. The diagnostic feature, anther length, is variable in both species, but it correlates well with differences in glume, lemma, and grain length, and is clearly a bimodal character. *P. nuttalliana* has many individuals with small anthers which are retained within the spikelet; if such individuals exist in *P. cusickii*, they would probably be difficult to distinguish from *P. nuttalliana*. This may account for a few individuals seen which have small anthers, as in typical *P. nuttalliana*, but relatively large glumes and lemmas.

2. **Puccinellia distans** (L.) Parl. Tufted perennials 2–6 dm tall. Culms erect or geniculate below, glabrous. Blades loosely folded to somewhat rolled in the bud, flat to somewhat involute or folded at maturity, ribbed, but not as strongly ribbed as our other species, mostly 1–10 cm long, 1.5–4 mm wide; sheaths basally closed, loose to tight-fitting, not keeled, at least the lower ones becoming rather irregularly air-chambered; ligules usually blunt, those of the culm leaves mostly 1–1.5 mm long. Panicles 3–10(19) cm long, the lower branches usually reflexed at maturity; spikelets mostly (1)3- to 5(6)-flowered, 3–5 mm long, 0.9–2 mm wide; glumes unequal, the first 0.5–1.3 mm long, the second 0.7–2 mm long; lemmas greenish or purplish below, ovate, the apex scarious, erose-ciliolate, broad, blunt, the base somewhat pubescent, the lowest one 1.5–2.2 mm long. Anthers 0.5–0.8 mm long; grains mostly 0.9–1.3 mm long. (2n = 14, 28, 42) Jun–Jul. Moist, often alkaline soil; known from the edge of our range in w NE, MN, CO, WY, & to be expected elsewhere; (introduced throughout much of w & ne U.S.; Eurasia). *Naturalized.*

Most of the NE specimens assigned to this species in the Atlas GP have proven to a small-anthered phase of *P. nuttalliana*, which see.

3. **Puccinellia nuttalliana** (Schult.) A. Hitchc. Tufted perennials 2.5–6.5(10) dm tall. Culms erect, glabrous to finely scabrous. Blades C-shaped in cross section in the bud, usually involute at maturity, strongly ribbed and scabrous on the upper surface, mostly 1–15 cm long and 0.8 –2.6 mm wide; sheaths basally closed, tight-fitting, not keeled, the lower ones becoming air-chambered at maturity; ligules usually pointed, those of the culm leaves 1.5–3.0 mm long. Panicles 7–26 cm long, the branches erect to spreading, rarely the lowermost reflexed; spikelets (2)4- to 6(7)-flowered, (3.5)4.5–6(8.3) mm long, 0.9–2 mm wide; glumes unequal, the first 0.8–1.7 mm long, the second 1.3–2.1 mm long; lemmas narrowed toward the tip, usually pointed, rarely at all truncate, the lowest 1.9–2.5(2.9) mm long. Anthers exserted or retained within the spikelet, extremely variable in size, the smaller sizes tending to be retained within the spikelets which contain them, 0.2–0.9 mm long; grains 0.9–1.5 mm long. (2n = 42, 56) Jun–Aug. Moist, sandy ground, often where alkaline; MN, ND, MT, n & w SD, w NE, WY, w KS, CO; (WI to B. C., s to KS & CA, introduced in ne U.S.). *P. airoides* (Nutt.) Wats. & Coult.—Hitchcock, A. S. Manual of the Grasses of the United States, p. 80. U.S.D.A. Miscl. Publ. No. 200. 1950.

Separation of this species from the related *P. distans* based on quantitative features is difficult or impossible in our region; the shape of the lemma and several vegetative features seem to provide more reliable characters.

Vegetatively, this species is virtually indistinguishable from *P. cusickii*, which see.

61. REDFIELDIA Vasey

1. *Redfieldia flexuosa* (Thurb.) Vasey, blowout grass. Perennials 5.5–9(13) dm tall from scaly rhizomes, these often buried deep in the sand and seldom collected. Culms ascending, the basal portion usually buried in the sand and rooting at the nodes, solid. Blades rolled in the bud, involute at maturity, firm, flexuous, smooth, but with evenly spaced narrow furrows on both surfaces and scabrosity usually apparent in the furrows, mostly (16)40–75 cm long, (1.5)2–4 mm wide near the base but tapering into a long filiform tip; sheaths firm, open, smooth, but with evenly spaced narrow furrows and scabrosity apparent in the furrows, not keeled, the lower ones sometimes appressed-pubescent near the base, the ones below ground often splitting into brownish fibers; ligule a dense fringe of hairs from a very short basal membrane, 0.9–1.6 mm long; auricles none. Inflorescence an open panicle with delicate, flexuous branches, 23–46(73) cm long; spikelets (1)2- to 4(5)flowered, disarticulating above the glumes and between the florets; glumes 1-nerved or the second 2- or 3-nerved, acute to acuminate, the first 1.8–4.1 mm long, the second 2.2–4.5 mm long; lemmas acute to acuminate, 3-nerved, each one with a tuft of hairs at the base, the lowest one 4.2–6 mm long. Anthers 2.2–3.6 mm long. ($2n = 25$). Jun–Oct, mostly Jul–Sep. In loose, sandy soil, often in blowouts; ND, SD, NE, KS, CO, OK, TX; (ND to OK, w to UT & AZ).

J. Reeder has studied this grass extensively (Madroño 23: 434–438, 1976) and his work is the source of the chromosome count. The number $2n = 25$ is based on two collections, both from the NE sandhills, and Reeder suggests that this material may be cytologically atypical. This grass has several unusual vegetative features, the most notable one being the narrow grooves in the leaf sheath.

62. SCHEDONNARDUS Steud.

1. *Schedonnardus paniculatus* (Nutt.) Trel., tumblegrass. Tufted perennial, about 1.5–5(7.5) dm tall. Culms solid, scabrous, somewhat flattened, curving or arching, especially above. Blades white-margined, keeled, folded in the bud, Y- or V-shaped in cross section at maturity, scabrous, especially on the margins, 2–11 cm long, 1–2.8 mm wide; sheaths flattened, keeled, glabrous to scabrous, open to the base, the broad membranaceous margin continuous with the ligule; ligule membranaceous, acuminate, whitish, 1–4 mm long. Inflorescence a panicle of widely spaced spikes, the axis and the branches curving at maturity; spikelets appressed in 2 rows on one side of each branch, disarticulating above the glumes, 1-flowered; glumes acuminate, stiff, 1-nerved, the first 1.5–3.1(4.1) mm long, the second (2.1)2.8–4(4.8) mm long; lemmas acuminate, 2.9–5 mm long, 3-nerved. Anthers 0.7–1.4 mm long. ($2n = 20$) Apr–Aug, mostly Jun–Aug. Open moist to dry roadsides, pastures, disturbed ground; GP; (Sask., MN to MT s to LA & AZ; Argentina).

63. SCHIZACHNE Hack.

1. *Schizachne purpurascens* (Torr.) Swall., false melic. Loosely tufted perennials (2)5–10(12) dm tall, sometimes rhizomatous. Culms glabrous, hollow, slender, often rather weak, decumbent at the base. Blades rolled in the bud, flat at maturity, rather clearly many-ribbed on both surfaces, 5–35 cm long, (1.5)2–4(4.8) mm wide; sheaths closed to the summit, ribbed, not keeled; ligules membranaceous, often tubular, highest in front, usually less than 1(1.2) mm long in back; auricles none. Inflorescence a slender panicle or raceme of relatively few spikelets, 5–13(16) cm long; spikelets often purplish, 3–6 flowered, 9–17 mm long, 1.7–3.5 mm wide, disarticulating above the glumes and between the florets; glumes membranaceous, unequal, the first 3- to 5-nerved, (3.8)4.5–6.5 mm long, the second

5-nerved, 5.5–8.5 mm long; lemmas bearded on the callus, hyaline-margined, striate with 7 or more strong converging nerves, the tip prolonged into 2 hyaline teeth 1.2–2.4 mm long from between which arises a stout awn 8–14 mm long; lemmas, including teeth 8–10 mm long; paleas 2-nerved, shorter than the lemmas. Anthers 1.1–1.7 mm long. (2n = 20) May–Aug, mostly Jun. Moist to relatively dry woods; scattered in MN, ND, MT, sw SD, w NE, CO; (Newf. to AK, s to MD, KY, NE, MT, & in the mts. to NM; also Asia).

64. SCOLOCHLOA Link

1. *Scolochloa festucacea* (Willd.) Link, sprangletop. Perennials 8–16 dm tall from thick, soft rhizomes. Culms stout, hollow, glabrous. Blades narrow-tipped, rolled in the bud, flat at maturity, fibrous, strongly many-ribbed on both surfaces, but especially so above, nearly glabrous except for the somewhat scabrous margins or more generally scabrous above, 20–50 cm long, (4)6–10(13) mm wide; sheaths open, glabrous, not keeled, air chambers and cross-septae prominent in transmitted light; ligules membranaceous, lacerate, usually thickened at the margins, 4–10 mm long; auricles none. Inflorescence an open, diffuse panicle, erect to slightly nodding, 13–37 cm long; spikelets (5)7–9(10) mm long, mostly 2- to 4-flowered, disarticulating above the glumes and between the florets; glumes membranaceous-scarious, acute or lacerate, unequal, the lower one (2)3(4)-nerved, (4.4)5–7.8 mm long, the upper one (4)5(6)-nerved, 5.6–8.5(9.4) mm long; lemmas rounded, the tips usually erose, the calluses bearded, usually the (5)7(9) nerves most prominent near the tip and fainter below; the lowest lemma (4.9)6–7.5 mm long; paleas about as long as the lemmas. Anthers 1.9–3.3 mm long. (2n = 28, 42) Jun–Jul (Aug). In potholes, ditches, sloughs, marshes, ponds, & wet meadows, often in standing water; MN, ND, e SD, n IA, n-cen NE; KS: Pottawatomie; (Man. to B.C., s to IA, NE, WY, & e OR; Eurasia). *Fluminea festucacea* (Willd.) Hitchc. — Rydberg.

Reference: Smith, A. L. 1973. Life cycle of the marsh grass *Scolochloa festucacea*. Canad. J. Bot. 51: 1661–1668.

This grass is abundant in ND, Sask. & Man. but is very scattered throughout much of its North American range. This scattered distribution has led some authors (for example Hitchcock, Manual of vascular plants of the Pacific Northwest 1: 695. 1969) to suppose that it is introduced in N. Amer. The relatively abundant collections from the n GP collected over a long period of time make that supposition seem unlikely.

65. SETARIA Beauv.

Tufted annuals or cespitose to short-rhizomatous perennials. Culms solid to hollow, erect, often geniculate below. Blades rolled in the bud, usually flat at maturity; sheaths open, somewhat keeled; ligule a fringe of hairs from a short membranaceous base. Inflorescence a cylindrical contracted panicle with some or all spikelets subtended by 1–several scabrous or barbed bristles; spikelets usually disarticulating below the glumes, bearing 2 florets; lower floret sterile or staminate; upper floret perfect; first glume membranaceous, short, bearing 3 converging nerves; second glume longer, with 3–6 converging nerves; lower lemma like the glumes in texture, 5(7)-nerved; lower palea absent or vestigial to well developed and 2-nerved; upper lemma and palea indurate, smooth to papillate or cross-corrugate, the lemma very obscurely 3- to 5-nerved, partly enclosing the 2-nerved palea.

Reference: Pohl, R. W. 1951. The genus *Setaria* in Iowa. Iowa State J. Sci. 25: 501–508.

In addition to the species described below, *S. reverchonii* (Vasey) Pilg. (*Panicum reverchonii* Vasey) has been reported from the s edge of our range (OK: Harmon and Jackson). It differs from all of our other species in the very slender, much-interrupted inflorescence and in the paucity of its

1   Bristles of the panicle downwardly barbed .... 6. *S. verticillata*
1   Bristles of the panicle upwardly barbed.
   2   Sheaths villous to short-ciliate on the margins; lower spikelet usually sterile, the lower palea somewhat reduced to absent.
      3   Plants perennial; bristles mostly 1(2) per spikelet .... 5. *S. leucopila*
      3   Plants annual; bristles, on the average, more numerous.
         4   Spikelets disarticulating above the glumes .... 4. *S. italica*
         4   Spikelets disarticulating below the glumes.
            5   Spikelets mostly less than 2.4 mm long; blades glabrous above; fertile (upper) lemma minutely papillate .... 7. *S. viridis*
            5   Spikelets mostly more than 2.4 mm long; blades usually pubescent above; fertile lemma papillate and also cross-ridged .... 1. *S. faberi*
   2   Sheaths not villous to short-ciliate on the margins; lower spikelets mostly staminate, the lower palea well developed.
      6   Plants perennial from short rhizomes; spikelets 2.1–3.0 mm long ...... 2. *S. geniculata*
      6   Plants annual; spikelets 2.5–3.5 mm long .... 3. *S. glauca*

1. Setaria faberi Herrm., Chinese foxtail. Stout annuals mostly 5–14 dm tall (shorter where mowed). Culms ascending, often geniculate below, glabrous, hollow. Blades glabrous to scabrous below, normally pilose above, mostly 13–50 cm long, 6–22 mm wide; sheaths somewhat keeled, glabrous to scabrous, ciliate on the margins; ligule 1–3 mm long, consisting of a fringe of hairs from a membranaceous base. Panicles often nodding, cylindrical, 3–20 cm long, 1.4–3.1 cm wide, the bristles upwardly scabrous, usually 3–6 per spikelet; spikelets 2.4–2.8 mm long; first glume 0.8–1.5 mm long; second glume 1.8–2.4 mm long; lower floret usually sterile, rarely staminate, its lemma 2.3–2.7 mm long, its palea 1.4–2.3 mm long; upper floret perfect, about 2.3–2.7 mm long, its lemma clearly papillate and cross-ridged. Anthers 0.5–1.1 mm long. ($2n = 36$) Jul–Oct. Waste ground, cult. fields; MN, se SD, IA, se NE, MO, KS, e OK; (NY to SD, s to NC & AR, introduced here & elsewhere from China). *Naturalized.*

This species has often been confused with *S. viridis*. The differences are made especially clear by Pohl (Brittonia 14: 210–213. 1962).

2. Setaria geniculata (Lam.) Beauv., knotroot bristlegrass. Short-rhizomatous perennials mostly 2–11 dm tall. Culms erect to geniculate below, glabrous, hollow. Blades glabrous to scabrous, often glaucous, mostly 4–29 cm long, 3.5–11 mm wide; sheaths keeled, glabrous or scabrous, not ciliate on the margins; ligule a fringed membrane 0.3–1.4 mm long, the fringed portion often shorter than the membranaceous base. Inflorescence erect, cylindrical, 2–6.5 cm long, 1.2–2.5 cm wide, the bristles upwardly scabrous and usually yellowish, normally more than 5 per spikelet; spikelets 2.2–3.1 mm long; first glume 1–1.8 mm long; second glume 1.4–2.3 mm long; lower floret usually staminate, rarely sterile, its lemma 2.3–2.8 mm long, its palea 2–2.6 mm long; upper floret perfect, about 2.2–2.6 mm long, its lemma plainly cross-corrugate. Anthers 1–1.6 mm long. ($2n = 36, 72$) Jun–Oct. Moist habitats along streams & ditches; se MO, se & s-cen KS, OK, TX; (MA to KS & CA, throughout the s states to trop. S. Amer.).

3. Setaria glauca (L.) Beauv., yellow foxtail. Tufted annuals mostly 4–10 dm tall (shorter where mowed). Culms ascending, usually geniculate below, often somewhat flattened, hollow to pith-filled; blades usually glabrous to scabrous, sometimes thinly long-pilose above or even densely so near the ligule, often glaucous, mostly 9–31 cm long, 3–12.5 mm wide; sheaths keeled, glabrous to scabrous, not ciliate on the margins; ligule 0.5–1.8 mm long, consisting of a fringe of hairs from a short membranaceous base. Inflorescence erect to somewhat nodding, cylindrical, 3–12 cm long, 1–2.2 cm wide, the bristles upwardly scabrous and often yellowish, normally more than 5 per spikelet; spikelets 2.5–3.4 mm long; first glume 0.9–1.8 mm long; second glume 1.4–2.4 mm long; lower floret usually staminate, rarely sterile, its lemma 2.2–3.1 mm long, its palea 2–2.9 mm long; upper floret perfect, about 2.1–3.1 mm long, its lemma plainly cross-corrugate. Anthers 0.9–1.4 mm long. (2n = 36, 72) Jul–Sep. Abundant in disturbed ground; GP; (widely introduced in temp. regions, found throughout the U.S. & Can.; Europe). *Naturalized.* *S. lutescens* (Weig.) Stuntz (err. for Hubb.) — Rydberg.

This grass has appeared in many American floras as *S. lutescens* (Weig.) Hubb. The rationale for restoring the epithet *glauca* is given by Reeder (Rhodora 53: 27–30. 1951) and by Terrell (Taxon 25: 297–304. 1976).

4. ***Setaria italica*** (L.) Beauv., foxtail millet, Stout annuals mostly (3.5)6–11 dm tall. Culms ascending, erect, nearly glabrous, but often appressed-pubescent at the nodes, hollow. Blades glabrous to scabrous, mostly 9–34 cm long, (5)8–16(21) mm wide; sheaths usually short-ciliate on the upper margins and often appressed-pubescent at the collar and the base, otherwise glabrous to scabrous; ligule 0.9–2.5 mm long, consisting of a fringe of hairs from a short membranaceous base. Inflorescence often nodding, cylindrical to lobed, 3–20 cm long, 1.5–4 cm wide, the bristles upwardly scabrous, usually 1–3 per spikelet; spikelets disarticulating above the glumes and sterile floret, the upper floret falling free with its enclosed grain; spikelets 2.6–3.4 mm long; first glume 0.8–1.4 mm long; second glume 2–2.6 mm long; lower floret sterile, its lemma 2.2–2.8 mm long, its palea vestigial or up to 1.5 mm long; upper floret perfect, about 2.2–2.8 mm long, almost orbicular at full maturity, the lemma not cross-corrugate. Anthers 0.5–0.9 mm long. (2n = 18) Jul–Oct. In disturbed ground; scattered throughout much of GP, but not nearly as common as *S. viridis* or *S. glauca;* (escaped from cultivation at many locations in U.S.; Old World). *Naturalized.*

5. ***Setaria leucopila*** (Scribn. & Merrill) K. Schum., plains bristlegrass. Tufted perennials 2–11 dm tall. Culms erect to geniculate below, solid, somewhat flattened and grooved on one side, scabrous and often puberulent near the nodes. Blades usually gabrous to scabrous, flat to folded at maturity, mostly 5–22 cm long, 2–6(9) mm wide; sheaths somewhat keeled, villous to ciliate on the upper margins; ligule 1.3–3 mm long, consisting of a dense fringe of hairs from a basal membrane. Inflorescence erect, spikelike, but often interrupted, 8–13.5(15) cm long, 0.7–1.7 cm wide, the bristles upwardly scabrous and usually 1(2) per spikelet; spikelets 2.2–2.8 mm long; first glume 1.1–1.8 mm long; second glume 1.7–2.4 mm long; lower floret sterile, its lemma 2.2–2.8 mm long, its palea 0.9–1.8 mm long; upper floret perfect, about 2.2–2.8 mm long, its lemma finely papillate and minutely cross-ridged. (2n = 54, 68, 72, ca 108) May–Jul. In sandy prairies & pastures & along roads; s CO, NM, TX, s OK; (s OK to s CO, s to cen Mex.). *S. macrostachya* H.B.K. — Hitchcock, Manual of the Grasses of the United States. U.S.D.A. Miscl. Publ. No. 200. 1950, in part.

The distinctness of this taxon from *S. macrostachya* is discussed by Emery (Bull. Torrey Bot. Club 84: 95–121. 1957).

6. ***Setaria verticillata*** (L.) Beauv., bristly foxtail. Tufted annuals, mostly 3–10(12) dm tall. Culms ascending, often geniculate below, hollow, glabrous. Blades glabrous to scabrous, mostly 10–40 cm long, 5–13(20) mm wide; sheaths keeled, glabrous to scabrous, sometimes slightly ciliate on the upper margins; ligule a fringed membrane 1–2.2 mm long. Inflorescence erect, spiciform but interrupted below, lobulate in larger plants, 6–12(18) cm long, 0.6–1.5 cm wide, the bristles downwardly barbed and adhering to objects, 1 or 2 per spikelet; spikelets 1.8–2.3 mm long; first glume 0.8–1.2 mm long; second glume 1.7–2.2 mm long; lower floret sterile, its lemma 1.8–2.2 mm long, its palea 0.9–1.3 mm long; upper floret perfect, about 1.7–2.1 mm long, its lemma minutely papillate. Anthers about

0.6–0.9 mm long. (2n = 18, 36, 54) Jul–Sep. Cult. ground, waste ground; MN, e ND, SD, IA, NE, ne KS; (introduced nearly throughout the U.S., but most common in the ne; Europe). *Naturalized.*

7. **Setaria viridis** (L.) Beauv., green foxtail. Annuals, 3–10(4) dm tall (shorter where mowed). Culms ascending, usually geniculate below, glabrous, hollow. Blades glabrous to scabrous, 6–23 cm long, 5–10(15) mm wide; sheaths somewhat keeled, glabrous to scabrous, ciliate on the margins; ligule 0.9–3 mm long, consisting of a fringe of hairs from a short membranaceous base. Inflorescence erect or nodding, cylindrical to slightly lobulate in large plants, 2.5–14(22) cm long, 1–2.3(3.5) cm wide, the bristles upwardly scabrous, usually 1–3 per spikelet; spikelets 1.7–2.3 mm long; first glume 0.5–1 mm long; second glume 1.6–2 mm long; lower floret sterile, its lemma 1.6–2.3 mm long, its palea vestigial or absent (up to about 0.8 mm long); upper floret perfect, about 1.6–2.1 mm long, its lemma minutely papillose, but not transversely ridged. Anthers 0.4–0.8 mm long. (2n = 18, 36). Jul–Sep. Abundant in disturbed ground; (widely introduced & abundant throughout much of s Can., U.S., & Mex.; Europe). *Naturalized.*

66. SITANION Raf., Squirreltail

1. **Sitanion hystrix** (Nutt.) J. G. Sm. var. **brevifolium** (J. G. Sm.) C. L. Hitchc., squirreltail. Tufted perennials 1.8–5(6) dm tall. Culms hollow, erect, smooth. Blades rolled in the bud, flat to folded or involute at maturity, strongly ridged above, scabrous to puberulent above, glabrous, scabrous, or puberulent below, 2–16 cm long, (1.1)1.8–3.8 mm wide; sheaths open, glabrous, scabrous, or puberulent; ligule a short, truncate, erose-ciliolate membrane 0.1–0.5 mm long; auricles present, usually prominent. Inflorescence a bilateral spike 7–17 cm long, including the awns, with (1)2(3) spikelets on alternating sides at each node, the axis breaking at maturity into sections and the spikelets of each node falling attached to the internode below; spikelets (1)2- to 3-flowered; glumes subulate, extending into a long, scabrous awn, the glume and awn together 3.1–8.6 cm long, the awn becoming divergent at maturity; lemmas scabrous to puberulent, faintly 5-nerved, awned from the tip with a scabrous awn, the lowest lemma of the spikelet with a body 6.5–10 mm long and an awn 2.7–6.5 cm long; palea about equal to the body of the lemma. Anthers 1–2.1 mm long. (2n = 28) May–Aug, mostly May–Jun. Dry soil of pastures & roadsides, often in waste ground; throughout w GP; (SD to OR, s to TX, AZ, CA, & n Mex.). *S. elymoides* Raf.— Rydberg; *S. longifolium* J. G. Sm.—Correll & Johnston; *Elymus longifolius* (J. G. Sm.) Gould—Gould, Brittonia 26: 59–60. 1974.

Although an excellent case can be made for including this taxon in the genus *Elymus,* an almost equally good case can be made for uniting other genera of the tribe Triticeae. The genus *Sitanion* is retained here for consistency of treatment.

Natural hybrids of this species and species of *Agropyron, Elymus,* and *Hordeum* are known (Bowden, Canad. J. Bot. 45: 711–724. 1966). It seems likely that several of these might be found within our range.

Sitanion hystrix var. *hystrix,* which has the lowest lemmas glumelike and sterile, has been reported at the edge of our range (MT: Garfield and Prairie; ND: Sioux; SD: Pennington and Washabough).

67. SORGHASTRUM Nash

1. **Sorghastrum nutans** (L.) Nash, Indian grass. Short-rhizomatous perennials 6–20 dm tall. Culms erect, hollow, the nodes pubescent with ascending hairs. Blades scabrous, rolled in the bud, flat at maturity, the midrib prominent near the base, 5–40(60) cm long, 3–8(12) mm wide; sheaths glabrous to weakly or strongly pilose, open, not keeled, extended

upward at the collar to form firm pointed projections flanking and joined to the ligule; ligule firm, membranaceous, 1.5-7 mm long; auricles none. Inflorescence a somewhat condensed panicle 11-27 cm long, bearing perfect spikelets each flanked by 1 pedicel of an absent sterile spikelet (by 2 at the end of the branch), the perfect spikelet and its associated sterile pedicel (or pedicels) and rachis joint disarticulating as a unit; both the inflorescence branchlets and the sterile pedicels rather densely pilose with whitish hairs; glumes of fertile spikelets densely pilose with whitish hairs, subequal, 5.4-7(7.8) mm long, brownish or tawny; floret of fertile spikelet only one, the lower floret reduced or rudimentary; lemma of the remaining floret membranaceous, bearing a stout, geniculate awn 13-20 mm long, its basal portion tightly twisted, 4-8 mm long. Anthers 3.2-4.5 mm long. (2n = 20, 40, n = 40) (Jul) Aug-Oct. Open prairies, where often a dominant or co-dominant species, to open woods; GP, but less common in nw part; (Que. & ME to Man. & ND, s to FL, AZ, &Mex.). *S. avenaceum* (Michx.) Nash—Atlas GP.

A complicated nomenclatural shift advocated by Baum (Canad. J. Bot. 45: 1843-1852. 1967) would cause this grass to be known as *S. avenaceum* (Michx.) Nash and would transfer the epithet *nutans* to a species of *Trichachne*. Since this transfer agrees neither with Linnaeus' species description nor with his generic descriptions in *Genera Plantarum,* I follow Gould (Grasses of Texas, p. 577. Texas A. and M. Univ. Press, College Station. 1975) and tentatively reinstate the name *S. nutans* for this species. I agree with Gould that the matter needs further study.

68. SORGHUM Moench

1. **Sorghum halepense** (L.) Pers., Johnson-grass. Rhizomatous perennials (5)8-20 dm tall. Culms stout, terete, solid, more or less appressed-pubescent at the nodes but otherwise glabrous. Blades rolled in the bud, flat at maturity, the midrib prominent, usually glabrous, mostly 10-90 cm long, 8-31 mm wide; sheaths open, rounded to slightly keeled, air-chambered, somewhat cross-septate; ligule membranaceous but with a prominent fringe, usually also backed by long hairs, about 2-3(4) mm long; auricles lacking. Inflorescence an open to somewhat contracted panicle 15-49 cm long bearing perfect sessile spikelets, each adjacent to a staminate pediceled spikelet (or to 2 at the end of the branch); disarticulation variable, but usually the ultimate panicle branches breaking apart and the spikelet pair and a rachis joint (or the 3 terminal spikelets) normally falling together as a unit, or the pediceled spiklets disarticulating separately at the summit of the pedicels; sessile spikelet dorsally compressed, its glumes hardened, shiny, appressed-puberulent, obscurely nerved, the first one somewhat broader then the second and partially enclosing it, the 2 subequal in length, 3.8-5.5 mm long; florets of the sessile spikelet 2, the lower consisting only of a delicate transparent fringed membrane, the upper perfect and with delicate, transparent, fringed lemma and palea, the lemma awnless or with a geniculate, twisted awn up to 13 mm long; pediceled spikelet borne on a pedicel 1.8-3.3 mm long, slightly dorsally compressed, narrower than the sessile spikelet, its glumes not hardened and more conspicuously nerved with about 5-9 nerves, subequal, 3.6-5.6 mm long, its florets 2, similar to those of the sessile spikelet but the upper floret staminate, not perfect, and its lemma unawned. Anthers of both sessile and pediceled spikelets 1.9-2.7 mm long. (2n = 40, 2n = 20 and several aneuploid counts also reported) Jun-Oct. Cult., but now common as a weed of roadsides, ditches, & field margins, often in moist ground; GP, but a serious weed only in the s part; (introduced and widespread in U.S. but only becoming a weed in the warmer regions; s Europe). *Naturalized.*

The cultivated sorghum, *S. bicolor* (L.) Moench, is an annual grown in various forms in the GP and is sometimes collected as a waif in weedy areas. It apparently does not persist in our range.

Weedy hybrids of cultivated *S. bicolor* and wild *S. bicolor* are often found beyond the range of the wild plants. In addition, *S. bicolor* may introgress with *S. halepense* to form both annual and perennial derivatives. Annual wild cane or shattercane, which often becomes a nuisance in cultivated fields, probably developed from cult. *S. bicolor*. It has been recognized, along with other weedy sorghums, as *S. bicolor* subsp. *drummondii* (Steud.) de Wet (de Wet, Amer. J. Bot. 65: 477–484. 1978).

The grass that is here described as *S. halepense* may actually represent a stabilized perennial introgressant of *S. halepense* and *S. bicolor*, according to de Wet (op. cit.). *Sorghum almum* Parodi, described from Argentina, may represent another such introgressant that is somewhat taller (usually over 20 dm) and less strongly rhizomatous. This grass has been introduced to the southwest part of KS and southward and may be found as an escape in our range.

69. SPARTINA Schreb., Cordgrass

Rhizomatous perennials. Clums sturdy, hollow, erect. Blades rolled in the bud, flat to involute at maturity, usually tapered to a nearly filiform tip, flexuous; sheaths firm, open; ligule a short fringe of hairs; auricles none. Inflorescence consisting of 2–many 1-sided spikes, these short peduncled to nearly sessile along an unbranched axis; spikelets closely crowded in 2 rows on 1 side of each spike, laterally compressed, 1-flowered, disarticulating below the glumes; glumes unequal, awned or awnless, keeled, the first glume 1-nerved, the second with 3 nerves crowded together near the center; lemma keeled, 1-nerved, awnless; palea equal to or exceeding the lemma in length, bearing 2 nerves near the center.

Reference: Mobberly, D. G. 1956. Taxonomy and distribution of the genus *Spartina*. Iowa State Coll. J. Sci. 30: 471–574.

1 Second glume awnless or bearing an awn less than 0.8 mm long, its 3 closely spaced nerves about equally prominent, the crease occurring at one of the lateral nerves .... 1. *S. gracilis*
1 Second glume bearing an awn 2–6 mm long, its central nerve very strong, and 2 smaller lateral nerves not immediately obvious, the crease occurring at the central nerve .... 2. *S. pectinata*

1. Spartina gracilis Trin., alkali cordgrass. Perennials 3.5–8(9.5) dm tall, from slender rhizomes. Culms terete to slightly flattened. Blades flat to, usually, involute, strongly ridged and scabrous above, the ones on the main culm 5–40 cm long, (1.2)2.5–4.5(5.5) mm wide near the base, tapering to a filiform tip; sheaths smooth; ligules 0.5–1.5(1.8) mm long. Inflorescence 7–16(18.5) cm long, consisting of 2–7 branches, each branch a spike 1.5–4(7) cm long, the lower ones pedunculate; first glume 1-nerved, narrow, 3.5–6(6.8) mm long, awnless or, rarely, bearing an awn less than 0.5 mm long, second glume with 3 nerves, the 3 all about equally prominent, the crease occurring at the nerve nearest the rachis, this nerve usually conspicuously ciliate, the other nerves scabrous to ciliate, the glume 6–9.5 mm long, awnless or, rarely, with an awn less than 0.5 mm long; lemma 5.5–8 mm long, ciliate on the keel. Anthers 3.2–4.6 mm long. (2n = 40, 2n = 42 also reported) Jun–Sep, mostly Jul–Aug. Alkaline meadows, marshes, & ditches to dry, sandy soil; ND, MT, w SD, WY, w NE, CO, w KS; (Sask. to B.C., s to KS, AZ, CA, & Mex.).

2. Spartina pectinata Link, prairie cordgrass. Perennials from stout, sharp-pointed rhizomes, (7.5)10–20 dm tall. Culms sturdy, terete. Blades flat to involute, scabrous on the margins, strongly ridged above, the ridges of varying sizes, 30–120 cm long, (3.5)5–13 mm wide near the base, but tapering to a nearly filiform tip; sheaths smooth; ligules (1.5)2–4 mm long. Inflorescence 20–45 cm long, consisting of 5–32 branches, each branch a spike (3)4.5–10(13) cm long, the lower ones pedunculate; first glume 1-nerved, narrow, 3.2–7 mm long, bearing a terminal awn 0.9–3(4) mm long, second glume with 3 nerves, the central one very strong, pectinate-ciliate, the smaller lateral nerves very close to it and not immediately obvious, the crease occuring at the central nerve, the glume (7)9–12 mm long, bearing a terminal awn 2-6 mm long; lemma 6-9 mm long. Anthers 4.5-6.2 mm long. (2n = 40, 80, 2n = 28, 42, 84 also reported) Jun-Oct, mostly Aug-Sep. Swales, ditches, & wet prairies; GP; (s Can. & n U.S. as far s as NC, n TX, & n OR; introduced in Europe).

70. SPHENOPHOLIS Scribn., Wedgegrass

1. *Sphenopholis obtusata* (Michx.) Scribn., wedgegrass. Tufted to solitary-stemmed annuals or perennials 1.8-9.2(13) dm tall. Culms glabrous, hollow, erect to geniculate below. Blades rolled in the bud, flat at maturity, scabridulous to pubescent, mostly (2.5)5-20(26) cm long, 1.5-5.7 mm wide; sheaths open, glabrous to scabrous or pubescent; ligules membranaceous, usually lacerate, (0.5)1-3 mm long; auricles lacking. Inflorescence a moderately open to strongly condensed erect to nodding panicle 4-21 cm long; spikelets usually with 2 florets, the rachilla prolonged beyond the upper floret, the disarticulation ultimately below the glumes, but disarticulation of the upper floret often preceding the fall of the entire spikelet; glumes usually scabrous, unlike, the first very narrow, 1-nerved, 1-2.4 mm long, the second 3(5)-nerved, obovate, truncate to obtuse or acute-tipped, sometimes cucullate, 1.5-2.9 mm long; lemmas obscurely nerved, smooth to scabrous, the lower one 1.5-3.1 mm long; palea about equal to the lemma. Anthers 0.2-0.7 mm long. (2n = 14) May-Aug.

This taxon consists of 2 very well-marked varieties. These have often been considered species, but Erdman (Iowa State J. Sci. 39: 259-336. 1965) assembled ample evidence that they are strongly intergradient in many regions. In our range this appears to be particularly true in NE and SD.

1a. var. *major* (Torr.) Erdm., slender wedgegrass. Inflorescence usually nodding and somewhat open; second glume more than 3 × as long as wide (width taken from midvein to margin), not cucullate; lower lemma 1.9-3.1 mm long. Wet ground, often in woods; GP, but less common than the following variety, especially in the sw part; (across s Can., s throughout most of U.S. except extreme w). *S. intermedia* Rydb., *S. pallens* (Spreng.) Scribn.—Rydberg (erroneous for *S. pallens* (Bieh.) Scribn., a name properly applied to an eastern hybrid).

1b. var. *obtusata*, prairie wedgegrass. Inflorescence usually erect, not nodding, moderately to very condensed; second glume less than 3 × as long as wide (width taken from the midvein to the margin), cucullate; lower lemma 1.5-2.5 mm long. Wet or marshy ground, often near lakes, ponds, or streams, most often in the open; GP; (sw Can. most of U.S., Mex.). *S. robusta* (Vasey) Heller—Rydberg.

71. SPOROBOLUS R. Br., Dropseed

Annuals or perennials. Culms solid, erect to decumbent at the base. Blades rolled, C-shaped in cross section, or folded in the bud, flat, involute, or folded at maturity; sheaths open, rounded to slightly keeled, with or without air chambers; ligule a short, prominently fringed membrane, the fringe often exceeding the membranaceous basal portion in length; auricles none. Inflorescence a panicle, its branches spreading to appressed at maturity, often partly to wholly included in the upper leaf sheaths. Spikelets 1-flowered, disarticulating above the glumes and often below the glumes as well; glumes 1-nerved (nerveless), unequal to subequal, the first one usually shorter when unequal, the second one not surpassing the lemma in most species, slightly surpassing it in one species; lemma 1-nerved, glabrous to strigose, acute, unawned; paleas subequal to or longer than the lemmas, 2-nerved. Anthers retained or exserted.

Reference; Reeder, C. G. 1975. *Sporobolus*, pp. 286-311, *in* Gould, F. W. The Grasses of Texas. Texas A. and M. Univ. Press, College Station.

Rydberg's report of *Sporobolus ejuncidus* Nash (a synonym of *S. junceus* (Michx.) Kunth) for KS is believed to be erroneous.

1 Plants annual.
  2 Lemmas strigose; most glumes more than 2.5 mm long; grains mostly 1.8–2.1 mm long .... 9. *S. vaginiflorus*
  2 Lemmas glabrous; most glumes under 2.5 mm long; grains mostly 0.8–1.3 mm long .... 6. *S. neglectus*
1 Plants perennial.
  3 Panicle, at least its exserted portion, open, the branches spreading at maturity.
    4 Spikelets large, the first glume normally more than 2 mm long, the second normally more than 3 mm long, the lemmas more than 3 mm long; blades folded in the bud .... 5. *S. heterolepis*
    4 Spikelets smaller, the first glume normally less than 2 mm long, the second usually less than 3 mm long, the lemmas less than 3 mm long; blades rolled or C-shaped in cross section in the bud.
      5 Pedicels over 5 mm long .... 8. *S. texanus*
      5 Pedicels under 5 mm long.
        6 Panicle short, mostly 4–15 cm long, the branches viscid, the lowest ones whorled .... 7. *S. pyramidatus*
        6 Panicle often longer, its branches not viscid and the lowest ones not whorled (most often the lowest ones included in the uppermost leaf sheath).
          7 Conspicuous tufts of long hairs present at the margin of the sheath at the collar region; blades with the number of ridges about equal on upper and lower surfaces .... 3. *S. cryptandrus*
          7 Conspicuous tufts of long hairs lacking at the margin of the sheath at the collar region, although a few long hairs sometimes present; blades with fewer and coarser ribs above and more numerous, narrow ribs below .... 1. *S. airoides*
  3 Panicle condensed, the branches appressed, often wholly or partly included in the sheath.
    8 Sheaths bearing conspicuous tufts of long hairs at the margins in the collar region; most lemmas less than 3.3 mm long.
      9 Panicle thick, with an exserted portion more than 8 mm wide; most lemmas more than 2.5 mm long; culms often over 1 m tall .... 4. *S. giganteus*
      9 Panicle narrower, not exserted (exserted forms of this species are open-panicled and will key above); most lemmas less than 2.5 mm long; culms often less than 1m tall .... 3. *S. cryptandrus*
    8 Sheaths bearing, at most, a few long hairs at the margins in the collar region; most lemmas more than 3.3 mm long .... 2. *S. asper*

1. Sporobolus airoides (Torr.) Torr., alkali sacaton. Strongly cespitose perennials 3.5–12.5 dm tall. Culms solid, erect, glabrous. Blades firm, tapering to filiform tips, rolled to C-shaped in cross section in the bud, flat to involute at maturity, prominently and heavily ridged above, the ridges scabrous and varying in size, finely ridged with more numerous, less scabrous ridges beneath, usually thinly to densely long-pilose just above the ligule, 3–60 cm long, mostly 1.9–4.4 mm wide; sheaths not keeled, air-chambered, the chambers slightly cross-septate, glabrous except, sometimes, for a few long hairs in the vicinity of the collar; ligule a very short fringed membrane 0.1–0.3 mm long, often flanked and/or backed by long hairs. Panicle open and diffuse, usually partly included in the uppermost sheath, the exserted part 11–47 cm long; glumes acute, membranaceous, 1-nerved or the first one sometimes essentially nerveless, unequal, the first 0.7–2 mm long, the second 1.1–2.4 mm long; lemmas similar to the glumes, 1.5–2.5 mm long; paleas subequal to the lemmas. Anthers 1–1.8 mm long; grains 1–1.4 mm long. (2n = 108, 126, n = ca 45) (Jun) Jul–Oct. Dry to moist sandy or gravelly soil, tolerant of saline conditions; GP, except MN, e ND, SD; (w ND to WA, s to MO, TX, CA, & Mex.).

2. Sporobolus asper (Michx.) Kunth, rough dropseed. Cespitose to solitary-stemmed perennials 2–13 dm tall. Culms smooth, solid, erect. Blades tapered to a filiform tip, C-shaped to rolled in the bud, flat to involute at maturity, scabrous to pilose, mostly 5–70 cm long,

1.5–4.5 mm wide; sheaths glabrous to pilose near the ligule, solid to air-chambered; ligule a short, fringed membrane 0.1–0.4 mm long. Panicles narrow, condensed, the branches appressed, most of the panicles partly to entirely included in the uppermost and upper sheaths, but the terminal one sometimes entirely exserted, up to about 18 cm long; glumes membranaceous, acute, folded, the midnerve usually scabrous, unequal, the first (1.2)2–3.1(4.4) mm long, the second (1.8)2.5–4.5(5.2) mm long; lemmas similar to the glumes in texture, glabrous to strigose, (2.2)3.3–5.5(6.4) mm long; palea subequal to or longer than the lemmas. Anthers extremely variable in length, even on a single panicle, 0.2–3.2 mm long, the shortest ones usually included and borne on sheath-enclosed lateral panicles, the longest ones often exserted and borne on the upper part of the terminal panicle; grains 1.3–2.8 mm long, the surface either papery or somewhat gummy when moistened.

This species can be divided rather clearly into 3 varieties in our range as follows:

1 Lemma strigose; grains mostly 2.4–2.8 mm long .... 2b. var. *clandestinus*
1 Lemma glabrous; grains mostly 1.3–2 mm long.
   2 Culms slender, mostly less than 2 mm wide near the base of the plant; upper sheaths mostly under 3 mm wide when pressed .... 2c. var. *hookeri*
   2 Culms stout, mostly more than 2 mm wide near the base of the plant; upper sheaths mostly more than 3 mm wide when pressed .... 2a. var. *asper*

2a. var. *asper*. Culms stout, mostly more than 2 mm broad near the base of the plant. Blades mostly 1.9–4.5 mm wide, scabrous to pilose; upper sheaths mostly 3–6 mm wide when pressed. Lower glume 2.4–3.7 mm long, upper glume 3.2–4.5(5.2) mm long; lemma 4–5.5 mm long, glabrous. Grains plump, 1.3–1.9 mm long, slightly gummy when moistened. (2n = 54, 108, ca 88) Aug–Oct. Prairies, roadsides, & a variety of other habitats; GP; (VT to Wa, s to MI & AZ). *S. pilosus* Vasey—Rydberg; *S. asper* (Michx.) Kunth var. *pilosus* (Vasey) Hitchc.—Atlas GP.

2b. var. *clandestinus* (Biehler) Shinners. Culms slender, mostly less than 2 mm wide at the base of the plant. Blades mostly scabrous, 1.7–3 mm wide; sheaths glabrous or the lower ones pilose. Lower glume 2.5–4.3 mm long, upper glume 3.2–4.8 mm long; lemma strigose, 4.3–6.3 mm long; fruits elongate, 2.5–2.8 mm long, not gummy when moistened. Aug–Oct. Prairies, oak woodlands, & disturbed ground; MO, e KS, e OK; (CT to WI, s to FL, e KS, & TX). *S. clandestinus* (Biehler) Hitchc., *S. canovirens* Nash—Rydberg.

2c. var. *hookeri* (Trin.) Vasey. Culms slender, mostly less than 2 mm broad at the base of the plant. Blades mostly 1.5–3.0 mm wide, scabrous to pilose, upper sheaths mostly 1–2.5 mm wide when pressed. Lower glume 1.3–3(3.6) mm long, upper glume (1.9)2.4–4.2 mm long; lemma (2.2)3.3–4.4(4.7) mm long, glabrous. Grains plump, 1.5–1.9 mm long, slightly gummy when moistened. Sep–Oct. Prairies, roadsides, & other disturbed sites; MO, e KS, e OK; (AL to KS, s to TX). *S. drummondii* (Trin.) Vasey—Fernald; *S. asper* var. *drummondii* (Trin.) Vasey—C. Reeder, op. cit.

Pilose-leaved material of *S. asper* may be separated out as a fourth variety, var. *pilosus* (Vasey) Hitchc. Since the pilosity is variable in extent and occurs on plants otherwise similar to varieties *asper* and *hookeri*, such an approach is not adopted here. Material from the type collection of *S. pilosus* seen at NY is rather similar to typical *S. asper* except for the pilose leaves and sheaths.

S. asper, var. *clandestinus* is clearly marked and, with further study, may prove to be best recognized as a separate species (see C. Reeder, op. cit.).

Although the epithet *drummondii* is as old as *hookeri*, the latter clearly has priority at the varietal level.

3. **Sporobolus cryptandrus** (Torr.) A. Gray, sand dropseed. Cespitose perennials 3.5–10 dm tall. Culms flattened or grooved on one side, erect to somewhat geniculate below, glabrous. Blades rolled in the bud, flat to involute at maturity, white-margined, glabrous to scabrous, mostly 2–26 cm long, 2–6 mm wide; sheaths not keeled, glabrous to scabrous, often ciliate-margined, bearing a dense tuft of long hairs at the collar, air-chambered; ligule a short fringed membrane 0.3–0.8 mm long. Panicles with the main branches spreading at maturity when exserted, but often partly to entirely sheath-enclosed, the exserted portion, when present, up to 40 cm long, the secondary branches mostly appressed;

glumes membranous, acute, 1-nerved, or the first one sometimes nerveless, usually somewhat scabrous, the first 0.6–1.9 mm long, the second 1.3–2.7 mm long; lemmas similar to the glumes, mostly 1.4–2.4 mm long; paleas slightly shorter than the lemmas. Anthers included or exserted, 0.1–1.1 mm long; grains 0.5–1.2 mm long. (2n = 18, 36, 38, 72) Jun–Sep, mostly Jul–Aug. Along roads, in pastures, etc., often on sandy soil; GP; (ME & Ont. to Alta. & WA, s to NC LA, CA, & Mex.).

This species is similar to *S. giganteus,* which see. *S. flexuosus* (Thurb.) Rydb., which is similar to *S. cryptandrus* but with spreading secondary branches and branchlets, has been reported from the s edge of our range (Atlas GP), but the specimens I have seen were assignable to other taxa.

4. Sporobolus giganteus Nash, giant dropseed. Perennials 8–18 dm tall. Culms stout, glabrous, solid, grooved or flattened on one side, over 3 mm wide at the base. Blades filiform-tipped, white-margined, rolled in the bud, flat at maturity, glabrous to scabrous, mostly 10–40 cm long, (3)5–10 mm wide; sheaths rounded, nearly glabrous but usually bearing prominent tufts of long hairs in the collar region and often also ciliate on the upper margins; ligule a fringed membrane 0.5–1.5 mm long. Inflorescence a thick condensed whitish panicle, usually included below and exserted above, the exserted portion up to about 60 cm long; glumes membranaceous, acute, 1-nerved, the nerves somewhat scabrous, unequal, the first 1–1.6 mm long, the second 2.2–3.8 mm long; lemmas similar, 2.1–3.2 mm long; paleas somewhat shorter than the lemmas. Anthers 0.1–0.6 mm long, usually retained even in the fruiting stage; grains 0.8–1.6 mm long. (n = 18) Jul–Sep. In open ground along roads, stream banks, & pastures, usually in sandy soil; w KS, s CO, w OK, TX, & NM; (w KS to CO, s to w TX & AZ).

At its most typical this grass is rather easily separated from *S. cryptandrus* by its dense panicle, its stouter stems, and by its larger second glumes, lemmas, and fruits. To the s of our range, however, it is separable with considerable difficulty from *S. contractus* Hitchc., which is much closer to *S. cryptandrus* in spikelet size. *S. contractus* has the dense panicle of *S. giganteus* but differs in having a narrower culm base and slightly smaller spikelets (see Hatch, Bot. Soc. Amer. Misc. Ser. Publ. 157: 57–58 *Abstr.* 1979). Future work may show that *S. contractus* and *S. giganteus* should be united or that both should be regarded as races of *S. cryptandrus*. Typical *S. contractus* has been collected from several localities at the s edge of our range (OK: Cleveland; TX: Deaf Smith).

5. Sporobolus heterolepis (A. Gray) A. Gray, prairie dropseed. Cespitose perennials 4–9.5 dm tall. Culms slender, erect, solid, glabrous. Blades slender, folded in the bud, folded, flat, or slightly involute at maturity, glabrous to scabrous, somewhat keeled, mostly 7–31 cm long, 1.4–2.4 mm wide; sheaths firm, slightly keeled, glabrous or the lower ones pubescent; ligule a fringed membrane 0.1–0.3 mm long. Panicle 11–22 cm long, usually well exserted, open, the primary branches spreading, the secondary branches often more or less appressed; spikelets 1(2)-flowered, greyish; glumes membranaceous, the first slender, 1-nerved or nerveless, (1.2)1.8–4.5 mm long, the second broader, 1- or 3-nerved, (2.4)3.2–6 mm long; lemmas blunt to acute, 1-nerved, 3.2–4.2 mm long; paleas about equaling or longer than the lemmas. Anthers dark, (1.1)1.7–3 mm long; grains globose, opaque, 1.4–2.0 mm long. (2n = 72) Jul–Oct. Open woods & upland or lowland prairies; MN, ND, SD, IA, NE, MO, e KS, & e OK; (Que. to Sask., s to CT, e TX, & CO).

6. Sporobolus neglectus Nash, poverty grass. Slender tufted annuals 1.5–4.5 dm tall. Culms wiry, grooved on one side, solid, erect to ascending, often decumbent and geniculate below, glabrous to scabrous. Blades folded to C-shaped in cross section in the bud, flat to involute at maturity, scabrous, sometimes with scattered pustular-based hairs toward the base, mostly 1–10 cm long and 1–2 mm wide; sheaths slightly keeled, glabrous to scabrous, sometimes with scattered long hairs at the collar; ligules 0.1–0.3 mm long. Both terminal and lateral panicles present, these somewhat exserted to mostly sheath-enclosed, slender and with appressed branches when exserted, glumes subequal, membranaceous, 1-nerved, the first

1.2–2.3(2.7) mm long, the second (1.3)1.7–2.6 mm long; lemma glabrous, ovate, 1.7–2.8 mm long; palea subequal to or surpassing the lemma. Anthers variable, 0.4–1.8 mm long, usually the longer ones associated with terminal panicles and exserted, the shorter ones associated with sheath-enclosed panicles and tangled on the stigmas, or some anthers on sheath-enclosed florets less than 0.1 mm long and then apparently nonfunctional on short filaments; grains 0.8–1.3 mm long. (2n = 36) Aug–Oct. Waste ground, often where sandy or rocky, often along streets & in parking lots; GP, more common in e part; (Que. & ME to MT, s to VA, LA, & AZ). *S. vaginiflorus* var. *neglectus* (Nash) Shinners—Atlas GP.

This taxon is close to *S. vaginiflorus* and has been united with that taxon by Shinners (Rhodora 56: 29. 1954). Throughout our range, however, the taxa remain clearly distinct. The reported difference in chromosome number is correlated with a significant difference in pollen size. See *S. vaginiflorus*.

7. *Sporobolus pyramidatus* (Lam.) Hitchc., whorled dropseed. Cespitose perennials 1–6 dm tall. Culms hollow to pith-filled above, erect to geniculate below, glabrous. Blades rolled in the bud, flat at maturity, white-margined, scabrous to sparingly long-pilose, especially near the ligule, mostly 2–18 cm long, 2.4–5.0(5.8) mm wide; sheaths air-chambered below, rounded, glabrous to long-ciliate near the upper margins; ligule predominantly a fringe of hairs 0.5–1.1 mm long. Panicles open, pyramidal, 4–15 cm long, mostly exserted, the main branches spreading, the lowest ones normally verticillate, the secondary branches appressed; panicle branches viscid; spikelets greyish; glumes acute, membranaceous, 1-nerved or the first one nerveless, unequal, the first 0.3–0.9 mm long, the second 1.2–2 mm long; lemmas similar, 1.2–2 mm long; paleas mostly a little shorter than the lemmas. Anthers 0.2–0.5 mm long, often retained within the spikelet; grains 0.6–1.0 mm long. (2n = 24, 36, 54, 2n = 30 also reported) Jun–Sep. Dry, open ground; e KS, e OK, TX, NM (KS to CO, s throughout much of trop. Amer.). *S. argutus* (Nees) Kunth.—Rydberg.

8. *Sporobolus texanus* Vasey, Texas dropseed. Cespitose perennials 2–7.5 dm tall. Culms spreading to erect, smooth, terete to flattened or grooved on one side, solid. Blades rolled in the bud, flat to involute at maturity, prominently ridged above with rather numerous scabrous ridges, glabrous to slightly scabrous beneath, mostly 4–18 cm long, 1.8–4.2 mm wide; sheaths rounded, glabrous to long-pilose; ligule a fringe of hairs from a short basal membrane, 0.2–0.6 mm long. Panicles often partly included, the exserted portion 10–35 cm long, very open and diffuse, the main and secondary branches and pedicels spreading, the pedicels over 5 mm long; glumes acute, unequal, membranaceous, 1-nerved or the lower one nerveless, the first 0.5–1.7 mm long, the second 1.7–2.2 mm long; lemmas similar, 2–2.9 mm long. Anthers 0.2–1 mm long, retained within the floret, or the larger ones exserted; grains 1.1–1.6 mm long. Jul–Oct. Usually in low, moist alkaline habitats; NE: Lancaster; KS, OK, TX; (NE to CO, s to AZ & TX).

9. *Sporobolus vaginiflorus* (Torr. ex Gray) Wood, poverty grass. Slender tufted annuals 2–6 dm tall. Culms slender, wiry, solid, erect to ascending, often decumbent below, grooved or flattened on one side, glabrous to scabrous; blades folded to C-shaped in cross section in the bud, flat to involute at maturity, scabrous, sometimes with scattered pustular-based hairs toward the base, mostly 2–12 cm long, 0.6–2.0 mm wide; sheaths rounded, glabrous but often with scattered long hairs at the collar. Both terminal and lateral panicles present, these somewhat exserted to mostly sheath-enclosed, slender and with appressed branches when exserted; glumes subequal, membranaceous, 1-nerved, the first (2.2)3.0–4.7 mm long, the second (2.3)2.8–4.6 mm long; lemmas strigose, elongate, (2.3)2.8–4.8 mm long; palea subequal to or clearly surpassing the lemma. Anthers variable, 1.4–3.2 mm long, the longer ones usually associated with terminal panicles and exserted, the shorter ones usually associated with included panicles, or some anthers on sheath-enclosed florets less than 0.2 mm long and then evidently nonfunctional on short filaments; grains 1.8–2.1

mm long. (2n = 54) Sep–Oct. In waste ground, often along roads & in rocky or sandy soil; MN, se SD, IA, e NE, MO, KS, OK; (ME & Ont. to MN & SD, s to GA, TX, AZ, & Mex.). *S. vaginiflorus* var. *inequalis* Fern.—Fernald.

A specimen assignable to *S. ozarkanus* Fern. has been recently collected in KS: Montgomery. Although the collection is pollen-sterile, it matches published descriptions of *S. ozarkanus* rather well. This species appears to be somewhat intermediate between *S. vaginiflorus* and *S. neglectus* but has some distinctive features of its own, for example, 3-nerved lemmas and pilose lower sheaths. Cytological and field study of this complex is needed.

72. STIPA L., Needlegrass

Nonrhizomatous, cespitose perennials. Culms hollow, erect, glabrous to pubescent at the nodes. Blades rolled to C-shaped in cross section in the bud, involute to flat at maturity, usually strongly ridged above; sheaths rounded, open to the base; ligules membranaceous, acute to lacerate, truncate, or even higher on the sides than in the center, the culm ligules longer than the basal ones; auricles none. Inflorescence a somewhat condensed panicle; spikelets 1-flowered, disarticulating above the glumes; glumes acuminate to awn-tipped, 3- to 5(9)-nerved; lemma 5-nerved, firm at maturity, attached to a long, sharp-pointed, pubescent callus, enclosing the palea, prominently awned, the awn terminal, firmly attached, stout, twisted, geniculate; the palea slightly to much shorter than the lemma. Anthers greatly reduced and retained within the spikelet to exserted and sometimes very long.

Reference: Barkworth, M. E. 1977. A taxonomic study of the large-glumed species of *Stipa* (Gramineae) occurring in Canada. Canad. J. Bot. 56: 606–625.

1. Glumes under 14 mm long; body of the lemma, including the callus, under 7 mm long; awn under 4 cm long .... 4. *S. viridula*
1. Glumes usually more than 14 mm long; body of the lemma, including the callus, over 7 mm long; awns over 4 cm long.
    2. Body of the lemma, including the callus, over 16 mm long, usually pubescent only at the base and on the margins, most of the glumes, including their tips, over 3 cm long .... 3. *S. spartea*
    2. Body of the lemma, including the callus, under 16 mm long, pubescent throughout or principally at the base and on the margins; most of the glumes, including their tips, under 3 cm long.
        3. Most awns with a relatively straight terminal segment, the overall length of the awn seldom over 10 cm long; ligules of culm leaves usually 2.5 mm or less in length and those of the basal leaves truncate or highest at the sides and not usually over 0.8 mm long .... 2. *S. curtiseta*
        3. Most awns with a flexuous or coiled terminal segment, the overall length of the awn 10–21 cm long; ligules of culm leaves variable but often more than 2.5 mm in length, those of the basal leaves seldom truncate, usually over 0.8 mm long ...... 1. *S. comata*

1. **Stipa comata** Trin. & Rupr., needle-and-thread. Strongly cespitose perennial (3.5)4.5–8(12) dm tall. Culms glabrous to variously pubescent at the nodes. Blades C-shaped in cross section in the bud, usually involute at maturity, strongly ridged and more or less pubescent above, relatively smooth to pubescent below, mostly 7–35 cm long and 1–3(4) mm wide when unrolled; sheaths glabrous to pubescent; ligules often scabrous to puberulent, those of the upper culm leaves usually pointed, 2–6(11) mm long, those of the basal leaves acute to truncate, often splitting, 0.7–2.8 mm long. Panicle contracted, 10–28(56) cm long; glumes acuminate or awn-pointed, 3- to 5-nerved, usually subequal, (1.3)1.9–2.8(3.4) cm long; lemma indistinctly 5-nerved, strongly pubescent on the callus and more or less pubescent throughout, bearing a small crown of hairs at the base of the awn, the body 8–12(14) mm long, including the callus, the awn 1 or more times geniculate, the proximal segment

tightly coiled, the distal segment usually flexuous or loosely coiled, the overall awn length about (10.5)11.5–19(21) cm; palea prominent, enclosed by the lemma, 2-nerved. Anthers either very short, 0.1–0.6 mm, the plants clearly cleistogamous, or less commonly, longer, 3.4–5.4(6) mm long, the plants then normally chasmogamous. (2n = 44, 2n = 38 and 46 also reported) Mostly May–Jul. Prairies & pastures often in sandy or rocky soil; MN, ND, MT, SD, CO, n & w NE, WY, w KS, w OK, TX, NM; (s Man. to s Yukon, s to IN, TX, AZ, & CA).

2. *Stipa curtiseta* (A. S. Hitchc.) Barkworth, needlegrass. Cespitose perennials 3–6.5 dm tall. Culms usually glabrous, but sometimes pubescent at the nodes. Blades C-shaped in cross section in the bud, usually involute at maturity, strongly ridged and glabrous to pubescent above, smooth beneath, mostly 8–26 cm long and 0.6–2.5 mm wide when unrolled; sheaths glabrous; ligules glabrous to puberulent, those of the upper culm leaves usually pointed, 1–2.5 mm long, those of the basal leaves truncate or highest at the sides, 0.2–0.8 mm long. Panicle contracted, 6–20(24) cm long; glumes acuminate or awn-pointed, mostly 5- to 7-nerved, subequal, 1.4–2.6 cm long; lemma indistinctly 5-nerved, strongly pubescent on the callus and lower part and the upper margins, rarely the pubescence more general, a small crown of hairs present at the base of the awn, the body 10.5–14 mm long including the callus, the awn usually twice geniculate, the proximal portion tightly coiled, the distal portion relatively straight, the overall awn length about 5–9(10.5) cm long; palea prominent, enclosed by the lemma, 2-nerved. Anthers very short, 0.1–0.5 mm long, the plants cleistogamous. Jun–Jul. Dry open hillsides & prairies; ND & SD (BH); (se Man. to N.T., s to SD, MT, & B.C.). *S. spartea* var. *curtiseta* A. S. Hitchc.—Hitchcock. Manual of the Grasses of the United States, U.S.D.A. Miscl. Publ. No. 200, p. 450. 1950.

This taxon has been considered a variety of *S. spartea*, but Barkworth (op. cit.) presents ample arguments for considering it a distinct species. It does not intergrade with *S. spartea* in our region and is as frequently identified as *S. comata* as *S. spartea*. It was not dot-mapped in the Atlas GP, but I have seen specimens from the following ND counties: Benson, Bottineau, Burke, Divide, McIntosh, Mountrail, Pierce, Ramsey, Stark, and Ward.

3. *Stipa spartea* Trin., porcupine-grass. Strongly cespitose perennials 4.5–9(14) dm tall. Culms usually nearly glabrous, but somewhat pubescent in lines on the lower nodes. Blades C-shaped in cross section in the bud, involute to flat at maturity, strongly ridged and usually pubescent above, not prominently ridged and glabrous or scabrous below, mostly 15–45(85) cm long and 1.7–3.5(4.8) mm wide when unrolled; sheaths usually glabrous, normally ciliate on the margins; ligules glabrous to puberulent, those of the upper culm leaves usually pointed, 3.5–8 mm long, those of the basal leaves truncate or rounded, 0.7–1.5(2.5) mm long. Panicles contracted, mostly 10–17(25) cm long; glumes acuminate or awn-pointed, (3)5- to 7(9)-nerved, subequal, (2.2)3.1–4.5 cm long; lemma indistinctly 5-nerved, strongly pubescent on the callus and lower part and on the upper margins, a small crown of hairs present at the base of the awn, the body, including the callus, 17–24 mm long, the awn usually twice geniculate, the proximal portion tightly coiled, the distal segment relatively straight, the overall awn length 10–18 cm long; palea prominent, 2-nerved, enclosed by the lemma. Anthers very short to moderately short, 0.3–1.1(1.5) mm long, the plants clearly cleistogamous, or, less commonly, very long, 6–11 mm long, the plants then chasmogamous. (2n = 46) Mostly Jun–Jul. Open hillsides & prairies; MN, ND, MT, SD, IA, n & e NE, MO, e KS, & ne OK; (s Ont. to sw N.T., s to PA, MO, & NM).

Stipa neomexicana (Thunb.) Scribn. may key here but will not match the description. It can be recognized by its plumose terminal awn segment. It has been collected several times at the edge of our range (CO: Otero; OK: Cimarron; TX: Armstrong, Bailey, Deaf Smith).

4. *Stipa viridula* Trin., green needlegrass. Tufted perennials 5.5–10(14) dm tall. Culms glabrous to puberulent just below the nodes. Blades rolled in the bud, flat to somewhat

involute at maturity, rather prominently many-ridged and often scabrous above, smooth below, often pubescent at the margin near the ligule, mostly 20–50 cm long and 2–5.5 mm wide; sheaths often with a zone of pubescence at the collar and ciliate on one margin; ligules pubescent to glabrous, rounded to truncate, the upper ones mostly 0.9–3.5 mm long the lower ones under 1 mm long. Panicles contracted, dense, 13–21(28) cm long; glumes membranaceous, 3-nerved, subequal, 6–10(14) mm long including the slender tips; lemma sericeous, indistinctly 5-nerved, its margins strongly overlapping, the body 4.1–7 mm long including the callus, the awn somewhat twisted at the base and often twice geniculate, the overall awn length about 2.2–3.8 cm; palea reduced, less than 1/2 as long as the lemma and completely concealed by it. Anthers extremely variable in size, from 0.3–3.8 mm long, those of cleistogamous spikelets mostly 0.3–1.2 mm long, those of chasmogamous ones mostly (1.4)2–3.8 mm long; panicles entirely cleistogamous (the commonest form), entirely chasmogamous, or, infrequently, mixed, with chasmogamous spikelets above and cleistogamous below, some central spikelets in such panicles sometimes with mixed anther lengths. (2n = 82) Jun–Jul. Dry soil in prairies & open woodland; MN, ND, MT, SD, WY, IA, NE, nw KS, CO; (Alta. & Sask. to IL, KS, NM, & AZ, also introduced farther e).

Stipa virdula is known to hybridize with *Oryzopsis hymenoides* in our range. See discussion under that species.

Several additional species of *Stipa* have been collected at the edge of our range. All should key here to *S. viridula*. *S. robusta* (Vasey) Scribn. (*S. vaseyi* Scribn.— Rydberg), known from SD: Fall River, Custer, Pennington, & from KS: Greeley, is very similar to *S. viridula* in vegetative features and in the overall appearance of the inflorescence, but may be distinguished by several microscopic features, particularly by the long, pubescent palea (3–6 mm long) (Mondrius, Proc. Iowa Acad. Sci. 85: 84–87. 1978; Thomasson, Southw. Naturalist 26: 211–214. 1981). *S. occidentalis* Thurb. (SD: Custer, Lawrence, Pennington) has a spikelet much like that of *S. robusta* and is vegetatively unlike both *S. robusta* and *S. viridula*, lacking the zone of pubescence on the collar and the ciliate sheath margin characteristic of those two species. It also has a much more delicate, slender panicle. *S. scribneri* Vasey, which has been collected in OK: Cimarron, may be distinguished from *S. viridula* by its obviously unequal glumes and its shorter awns. *S. richardsonnii* Link is unlike *S. viridula* in having an open, diffuse panicle whose branches bear spikelets only toward their tips. It is known from SD: Lawrence, Pennington.

73. TRIDENS R. & S.

Nonrhizomatous to somewhat rhizomatous perennials. Culms erect to ascending, solid to firm but fistulose, terete to compressed. Blades flat to involute or folded, the midrib often prominent, rolled to folded in the bud; sheaths rounded to keeled, open; ligule a short ciliate membrane, the membranaceous portion often shorter than the fringe of hairs; auricles none. Inflorescence an open or contracted panicle or raceme of 3- to 15-flowered spikelets; spikelets disarticulating above the glumes and between the florets. Glumes subequal or the first one shorter, the first 1(3)-nerved, the second 1(9)-nerved; glumes unawned, shorter than, equal to, or exceeding the lowest lemma; lemmas prominently 3-nerved, rounded on the back, emarginate to bluntly bilobed, the midnerve (and sometimes the lateral nerves) often excurrent, the nerves prominently hairy on the lower part (the hairs obscure or absent in *T. albescens*); palea subequal to or equal to the lemmas. Anthers usually short, less than 1.2 mm long.

1 Inflorescence open and diffuse; anthers mostly more than 0.7 mm long .... 2. *T. flavus*
1 Inflorescence contracted; anthers mostly less than 0.7 mm long.
  2 Lemma nerves glabrous or only slightly pubescent at the very base; spikelets whitish .... 1. *T. albescens*
  2 Lemma nerves prominently pubescent below; spikelets of various colors.
    3 Inflorescence short, less than 5 cm long; lemma margins densely long-ciliate .... 4. *T. pilosus*
    3 Inflorescence longer; lemma margins sometimes pubescent but not long-ciliate.

4　Spikelets normally much longer than broad; glumes not usually exceeding the spikelet in length; second glume 3- to 5(0)-nerved .... 3. *T. muticus* var. *elongatus*
4　Spikelets often nearly as broad as long or broader than long; glumes usually equaling or exceeding the spikelet in length; second glume 1-nerved .... 5. *T. strictus*

1\. **Tridens albescens** (Vasey) Woot. & Standl., white tridens. Tufted perennials (2)4–9(12) dm tall. Culms solid to firm but fistulose, somewhat compressed, erect to ascending, often geniculate below, usually glabrous. Blades glabrous, glaucous, rolled in the bud, flat to involute at maturity, the white midrib usually obvious toward the base on the upper surface, the tip usually narrow, mostly 5–40(56) cm long, 2–5.5 mm wide; sheaths only slightly keeled, glabrous, air-chambered between the veins; ligule a fringe of hairs from a short basal membrane, 0.5–1.2 mm long. Panicle contracted, 7–22 cm long, its branches appressed; spikelets papery, whitish, but often slightly purple-tinged, 6- to 13-flowered, mostly 3.8–7.1 mm long and 2–3.6 mm wide; glumes 1-nerved, ovate, acute-tipped, 2.6–5.7 mm long; lemmas glabrous or minutely pubescent at the very base, blunt and notched at the tip, the midvein slightly excurrent or not, the lowest lemma 2.6–4.2 mm long. Anthers exserted or included, 0.2–0.6 mm long. (2n = 60, n = 36 also reported) Mostly May–Aug. Low prairies & roadside ditches, often in sandy or rocky soil; sw KS, OK, TX, NM; (KS to CO, s to TX & NM). *Rhombolytrum albescens* (Vasey) Nash — Rydberg.

2\. **Tridens flavus** (L.) Hitchc., redtop. Tufted perennials, sometimes forming short rhizomes, 6–15 dm tall. Culms stout, terete to slightly compressed, hollow, at least below, glabrous. Blades rolled in the bud, flat at maturity, the midrib prominent toward the base, tapered to a narrow tip, glabrous to sparsely pilose, usually somewhat pilose at least near the ligule, mostly 20–50(70) cm long, 5–13 mm wide; sheaths keeled, pubescent in a triangular zone at the collar, otherwise glabrous; ligule a fringe of hairs from a short basal membrane, 0.3–0.7 mm long. Inflorescence open, the branches flexuous and drooping at maturity, 16–40 cm long; spikelets purplish or reddish when fully mature, 3- to 8-flowered, 5–9 mm long, 1.5–3 mm wide; glumes 1-nerved, the nerve sometimes minutely excurrent, the first one 2.2–4.4 mm long, the second 2.9–4.6 mm long; lemmas with the nerves pubescent to above the middle, notched, the midnerve and often also the lateral nerves minutely excurrent, the lowest one 3.4–4.8 mm long. Anthers 0.7–1.2 mm long, exserted or sometimes retained within the spikelet. (2n = 40) Jul–Oct. Open woods, pastures, & roadsides, in a variety of soils; IA, se NE, MO, e KS, OK, TX; (NH to NE, s to FL & TX). *Triodia flava* (L.) Smyth, *T. flava* f. *cuprea* (Jacq.) Fosb. — Fernald.

3\. **Tridens muticus** (Torr.) Nash var. **elongatus** (Buckl.) Shinners, slim tridens. Cespitose perennials 2–8 dm tall. Culms erect, solid, terete to somewhat flattened, glabrous or scabrous, often pubescent at the nodes. Blades rolled in the bud, involute to flat at maturity, tapering to a narrow tip, midrib obvious to not especially prominent, glabrous to scabrous or sparsely pilose, 2–23 cm long, 1.4–3.5 mm wide; sheaths scabrous, the lower ones pubescent, not keeled; ligule a lacerate, fringed membrane 0.4–1 mm long. Panicle narrow, elongate, 6–21 cm long, the short branches appressed; spikelets often purplish, 3- to 11-flowered, 7–12 mm long, 2–5 mm wide; glumes acute, the first glume 1(3)-nerved, 3.2–8.4 mm long, the second (3)5(9)-nerved, 3.8–8.4 mm long; lemmas pubescent on the nerves to near or above the middle, the tips blunt to emarginate, the midnerve minutely excurrent in at least the lowest florets, the lowest lemma 4–5.8 mm long. Anthers usually retained within the floret, 0.2–0.5 mm long. (2n = 40) Jun–Oct. Prairies, pastures, & open woods, in sandy, rocky, or clay soils; MO, e KS, OK, TX; (MO w to CO, s to TX & AZ). *T. elongatus* (Buckl.) Nash — Rydberg; *Triodia elongata* (Buckl.) Scribn. — Fernald.

4\. **Tridens pilosus** (Buckl.) Hitchc., hairy tridens. Tufted perennials to 2.5 dm tall. Culms slightly compressed, solid, glabrous to scabrous or puberulent, the leaves distributed principally toward the base. Blade pilose, folded in the bud, folded to flat at maturity, abruptly acute to blunt-tipped, the margins conspicuously thickened, whitish, the midrib also thickened and whitish below; mostly 1-8 cm long, 0.5-2 mm wide; sheaths keeled, often pilose at the collar, with scarious, finely ciliate margins; ligule a fringe of hairs 0.2-0.5 mm long. Inflorescence a short contracted panicle or raceme 1-4 cm long; spikelets 6- to 14-flowered, 10-15 mm long, 3.5-7.5 mm wide; glumes acuminate, 1-nerved, the nerves sometimes excurrent as short awns, especially on the second glume; first glume 3.4-6.0 mm long, second glume 3.8-7.0 mm long; lemmas acute to slightly notched, the midrib excurrent as a short awn up to about 1 mm long, densely pilose with long white hairs toward the base and on the margins above, the lowest one 4.8-6.5 mm long. Anthers 0.2-0.5 mm long. (2n = 16, 32) May–Sep, mostly May–Jul. Pastures, prairies, & roadsides in dry, sandy or rocky soil; sw KS, s CO, OK, TX, NM; (sw KS to NV, s to TX, AZ & cen Mex.). *Erioneuron pilosum* (Buckl.) Nash—Rydberg.

The removal of this species and several closely similar ones to the genus *Erioneuron* has been advocated by Tateoka (Amer. J. Bot. 48: 565-573. 1961) on morphological grounds.

5. *Tridens strictus* (Nutt.) Nash, longspike tridens. Tufted perennials 5-15 dm tall. Culms erect, glabrous, slightly compressed, firm-walled but fistulose. Blades firm, rolled in the bud, flat to involute at maturity, tapering to a narrow tip, the whitish midrib prominent on the upper surface toward the base, usually glabrous except for prominent hairs near the ligule, mostly 4-65 cm long, 2-6(8) mm wide; sheaths not keeled or only slightly keeled, air-chambered between the veins, glabrous except, sometimes, for hairs at the collar; ligule 0.3-0.8 mm long, consisting of a fringe of hairs from a short basal membrane. Panicles dense and contracted, 10-31 cm tall; spikelets 6- to 9-flowered, 4.5-7 mm long, 3.5-6.5 mm wide; glumes 1-nerved, often acuminate, subequal, often as long as or exceeding the floret cluster in length, 3.8-7 mm long; lemmas with the nerves pubescent to above the middle, notched, the midnerve and sometimes also the lateral nerves minutely excurrent, the lowest one 2.9-4.1 mm long. Anthers 0.4-0.8(1) mm long, normally not retained. (2n = 40) Aug–Oct. Open woods, prairies, pastures, & roadsides, mostly in low, moist ground; se KS, OK; (NC to IL & KS, s to GA & TX). *Triodia stricta* (Nutt.) Benth.—Fernald.

74. TRIPLASIS Beauv.

1. *Triplasis purpurea* (Walt.) Chapm., sandgrass. Delicate annual 1.5-5.5 dm tall. Culms erect to ascending, wiry, solid, usually plainly flattened, concave, and bent outward opposite each sheath, scabrous, appressed-pubescent at the nodes. Blades scabrous, pustular-pilose on the margin, rolled in the bud, flattened to involute at maturity, 1-11 cm long, 0.9-2.8 mm wide; sheaths usually both pilose and scabrous, most of them clearly swollen near the base to accommodate cleistogamous florets; ligule a fringe of hairs 0.4-1.4 mm long; auricles none. Inflorescence a few-branched panicle 3-10(16) cm long, consisting mostly of chasmogamous spikelets but supplemented by branches of mostly cleistogamous spikelets included in the sheaths. Chasmogamous spikelets (1)2- or 3(4)-flowered; glumes membranaceous, 1-nerved, the apex often somewhat jagged, the first glume 1-3.5 mm long, the second 2-4 mm long; lemmas 3-nerved, villous on the nerves, the apex bifid to a depth of 0.3-1 mm, bearing an awn from the notch 0.7-1.1 mm long, the overall length about 3-4.3 mm; paleas 2-keeled, long-silky on the keels; anthers 1.5-2.7 mm long. Cleistogamous spikelets borne in the sheaths at all nodes, but increasingly modified and reduced in number toward the base of the plant, those in the upper sheaths numerous and similar to the exserted spikelets but with fewer florets on the average and with greatly reduced stamens, only 0.1-0.4 mm long, those in the lower sheaths 1-several on short branches bearing 1-several spathelike leaves, single-flowered, and with glumes very reduced or absent, the first (or only) spathelike

protective leaf usually bony in texture, closely fitting into the concavity of the culm node above. (2n = 40) Jul–Oct, mostly Aug–Sep. In sandy ground in prairies & pastures & along roadsides; se SD, NE, CO, MO, KS, OK, TX, also known from ND: Ransom; (Ont., ME to MN, s to FL, CO, & TX, also reported from C. Amer.).

75. TRIPSACUM L., Gammagrass

1. *Tripsacum dactyloides* (L.) L., eastern gammagrass. Stout perennials from short, thick rhizomes, mostly 11–20 (reportedly to 30) dm tall. Culms glabrous, solid, somewhat flattened, erect to geniculate below. Blades rolled in the bud, flat at maturity, the midrib prominent, glabrous, but scabrous on the margins and sometimes pilose above near the ligule, mostly 10–70 cm long and 7–20 mm broad; sheaths open, prominently keeled, glabrous; ligule a short fringe of hairs from a basal membrane, 0.4–2 mm long; auricles none. Inflorescences usually both terminal and axillary, consisting of 1–4 spikelike branches, these usually bearing staminate spikelets above and few to numerous pistillate spikelets toward the base; the staminate spikelets paired, both spikelets of the pair sessile or one slightly pedicellate, the pairs borne in two rows on one side of the flattened rachis; pistillate spikelets solitary, sunken in depressions on the same side of the rachis as the staminate spikelets; the staminate portion of the branch often breaking free as a unit at maturity, the pistillate portion breaking into beadlike 1-spikelet segments; staminate spikelets 2-flowered, the outer glumes firm, obscurely to distinctly 9- to 16-nerved, usually bearing 2 keels near the margin, 5.5–10.0 mm long, inner glumes about equal to the outer in size but more delicate and more convex, faintly 5- to 12-nerved, the lemmas and paleas delicate, membranaceous, anthers normally present in both flowers, 3–7 mm long. Pistillate spikelets pyriform, also 2-flowered, but the lower floret sterile, the glumes both pyriform, shining, the outer one enclosing the inner, the outer glume strongly indurate, obscurely 11- to 22-nerved, 5–8.5 mm long, the inner glume thinner, membrane-lined and margined, faintly 9- to 13-nerved; lemmas and paleas of both florets delicate, membranaceous. (2n = 36, 72, 2n = 54 also reported) May–Oct. Moist grassland sites; s IA, se NE, MO, KS, OK, & TX; (throughout e U.S. to GP; Mex., C. Amer., S. Amer. to Bolivia & Paraguay).

The expanded range given for this species is from Randolph (Brittonia 22: 305–337. 1970).

76. ZIZANIA L., Wild Rice

1. *Zizania aquatica* L., wild rice. Stout, solitary-stemmed, monoecious annuals (4)7–22 dm tall. Culms terete, hollow, erect, nearly glabrous, but usually appressed-pubescent at the nodes. Blades rolled in the bud, flat at maturity, the lowest one sometimes floating, 5–65(100) cm long, 5–35 mm wide; sheaths open, rather loose, abundantly and conspicuously cross-septate, not keeled, often bearing a zone of appressed hairs at the base and also at the collar; ligules membranaceous, 7–15(23) mm long; auricles lacking. Inflorescence a much-branched panicle with staminate spikelets below and pistillate spikelets above, the main branches at first all ascending, but the staminate branches spreading or drooping at maturity and the pistillate ones remaining erect, the overall inflorescence 15–60 cm long, the pistillate portion 12–33 cm long; spikelets 1-flowered, lacking glumes, the disarticulation below the spikelet; staminate spikelets usually reddish, the lemma 5-nerved, thin, 5–11 mm long, acute to acuminate, awnless or with an awn up to 3 mm long, the palea 3-nerved, similar to the lemma in size and texture, the stamens 6, the anthers 4.1–6.3 mm long; pistillate spikelets usually paler, seldom reddish, the rigid lemma strongly 5-ribbed and rather indistinctly 3(5)-nerved, its margins thickened and strongly involute, its body 12–15.5 mm long, its terminal sinuate awn 2.5–6.5 cm long, the palea 2-nerved, its margins sharply

reflexed at the nerves, the nerves fitting tightly into the involute margin of the lemma. (2n = 30) Mostly Jul–Aug. Margins of streams, lakes, & ponds, often partly in standing water; MN, e ND, e SD, IA, e & cen NE, MO; (most of s Can., U.S. e of the Rocky Mts.). *Z. aquatica* var. *interior* Fassett—Fernald; *Z. interior* (Fassett) Rydb., *Z. palustris* L.—Rydberg.

152. SPARGANIACEAE Rudolphi, the Bur-reed Family

by Robert B. Kaul

1. SPARGANIUM L., Bur-Reed

Perennial, colonial, monoecious, emergent or floating and usually rhizomatous aquatic herbs. Leaves basal and cauline, alternate, 2-ranked, stiff and erect or limp and floating, linear, internally septate; erect leaves often triquetrous or V-shaped in cross section; floating leaves flat, sometimes rounded below. Flowering stems erect or floating, simple or branched, the many-flowered globular heads scattered along the zig-zag rachises, the branches and some heads subtended by leafy bracts; lower heads pistillate, sessile or pedunculate, often both in the same inflorescence; upper heads staminate, sessile, somewhat crowded, opening before the pistillate, soon falling and the naked axes often persisting. Flowers imperfect, individually inconspicuous but the heads obvious; each staminate flower consisting only of several irregular scales subtending a group of about 5 stamens, the filaments long and slender, the anthers ellipsoid or clavate; each pistillate flower with 3–6 linear to spatulate perianth parts surrounding the single pistil; pistil 1- or 2-locular; stigmas 1 or 2 and sometimes persistent in fruit. Fruit an achene with a persistent style-beak, 1- or 2-seeded, collectively forming burred heads before falling separately.

The species are rather variable; species with normally emergent leaves are sometimes found with floating leaves, and deepwater and swiftwater sterile forms with attenuate leaves occur. Emergent specimens can be distinguished from cattails (*Typha*) by their stiffer, shorter, brighter green leaves that are often V-shaped or triquetrous in section.

1 Stigmas 2; achenes stoutly obpyramidal, lustrous; pistillate heads sessile at the lowest nodes of the branches and usually ebracteate, or the lowest one pedunculate and subtended by a bract; common and widespread in GP .... 6. *S. eurycarpum*
1 Stigma 1; achenes fusiform, lustrous or dull; pistillate heads sessile or pedunculate at or above the nodes, the pistillate nodes usually bracteate; nw, ne, & se GP.
  2 Leaves and stem usually floating; leaves mostly 5 mm or less wide, sometimes inflated at the base; inflorescence simple; e SD, w MN .... 3. *S. angustifolium*
  2 Leaves and stem erect; leaves mostly 5–15 mm wide; inflorescence simple or branched.
    3 Inflorescence simple; at least the lowest pistillate head supra-axillary or pedunculate; nw & ne GP.
      4 Leaves little or not at all inflated at the base; achenes lustrous, greenish; plant slender; ne GP .... 4. *S. chlorocarpum*
      4 At least the cauline leaves inflated at the base, achenes dull brown; plant rather stout; nw GP .... 5. *S. emersum* var. *multipedunculatum*
    3 Inflorescence usually branched; pistillate heads sessile at the nodes; se GP.
      5 Leaves firm and often keeled; branches of inflorescence without pistillate heads; achenes lustrous, pale brown .... 2. *S. androcladum*
      5 Leaves flaccid and usually flat; inflorescence branches mostly with at least 1 pistillate head; achenes dull, dark brown .... 1. *S. americanum*

1. ***Sparganium americanum*** Nutt. Plant somewhat stout, erect, to 1 m tall, rhizomatous. Leaves flaccid, thin, soft, to 1 m long and 5–12(18) mm wide, flat or barely keeled, those

of the flowering stem somewhat dilated and scarious-margined near the base. Inflorescence usually branched, the branches strict and usually bearing both pistillate and staminate heads; pistillate heads sessile, axillary, often subtended by a leafy bract. Stigma 1. Fruiting head 1.5–2.5 cm in diam; achene fusiform, stipitate, dull, dark brown, barely constricted about the middle, the body 3–5 mm long, the beak 4–6 mm long, prominent. May–Jul. Shallow, often running waters; KS: Douglas, Franklin; MO: Jackson, Jasper; (Newf. to ne MN, s to TX & FL).

2. *Sparganium androcladum* (Engelm.) Morong. Plant stout, erect, to 1 m tall, not conspicuously rhizomatous. Leaves firm and usually carinate to keeled on the back, 6–12(15) mm wide. Inflorescence usually branched, the branches without pistillate heads; pistillate heads sessile on the main axis, their bract ascending. Stigma 1. Head in fruit ca 3 cm in diam.; achene fusiform, lustrous, pale brown, the body 5–7 mm long, barely constricted about the middle, the beak 2–5 mm long. May–Jul. Quiet waters; w-cen MO, ne OK; (ME to e MN, s to ne OK, KY, & VA).

3. *Sparganium angustifolium* Michx. Plant submersed, the stems and leaves floating at the surface, or occasionally emersed and the leaves erect. Leaves often very long, 2–5(8) mm wide, flat above and rounded below, the upper ones basally inflated and their basal margins narrowly scarious. Inflorescence simple, the foliaceous bracts inflated at the base, the lowest pistillate head(s) usually long peduncled and sometimes supra-axillary, the upper sessile and axillary; staminate heads somewhat crowded. Stigma 1. Fruiting heads 1–2 cm wide; achene dull, fusiform, ca 5 mm long, constricted about the middle. (2n = 30) Jun–Aug. Shallow waters or occasionally in deeper water or stranded on mud; e SD, w MN; (Lab. to AK, s to CA, NM, MN, & PA).

4. *Sparganium chlorocarpum* Rydb. Plant slender, erect, to ca 7 dm tall, the leaves much exceeding the inflorescence. Leaves strongly ascending, stiff, to 1 cm wide, flat or barely keeled, scarious-margined near the base but barely inflated there. Inflorescence usually simple, the bracts ascending; lowest pistillate head often peduncled and supra-axillary, the others sessile. Stigma 1. Fruiting heads 1.5–2.7 cm wide; achene fusiform, stipitate, lustrous, ribbed near the top, greenish at maturity, the body 4–6 mm long, the beak 2–4 mm long. Jun–Aug. Shallow water & on mud; ND, nw MN, ne SD; (Newf. to ND, s to IA, IN, NC, & reported w of the Rocky Mts.). *S. acaule* (Beeby) Rydb.—Rydberg.

Smaller, shorter-stemmed northern specimens have been distinguished as var. *acaule* (Beeby) Fern.—Fernald.

5. *Sparganium emersum* Rehmann var. *multipedunculatum* (Morong) Reveal. Plant rather stout, erect, to ca 7 dm tall. Leaves 4–8(12) mm wide, often much exceeding the inflorescence, flat or nearly so, prominently scarious-margined near the base. Inflorescence simple, the leafy bracts basally inflated and distinctly scarious-margined; lowest pistillate head pedunculate and/or supra-axillary, the upper ones sessile and supra-axillary. Stigma 1. Fruiting heads 2–3 cm wide; achene stipitate, fusiform, usually constricted about the middle, dull brownish, the body 4–6 mm long, the beak shorter than to about as long as the body. (2n = 30) Jun–Aug. Uncommon in shallow waters; MT: Daniels; BH; (Lab. to AK, s to CA, NM, SD, & PA). *S. multipedunculatum* (Morong) Rydb.—Rydberg; *S. simplex* of authors, nom. illegit.

This entity is doubtfully distinct from no. 3 and no. 4.

6. *Sparganium eurycarpum* Engelm. Plant stout, erect, often robust, to 1 m tall. Leaves to 1 m long, 8–15 mm wide, distinctly V-shaped in section, especially near the base. Inflorescence usually branched and equaling or exceeding the basal leaves, the main axis bear-

ing only staminate heads, the branches each bearing 1 pistillate and several staminate heads; pistillate heads ebracteate, sessile at the nodes or sometimes appearing pedunculate by abortion or dehiscence of the rachis above. Perianth parts spatulate; style 1 and persisting as the achene beak; stigmas 2, sometimes persisting on the beak. Fruiting heads 2–5 cm wide; achenes sessile, 2-seeded, thick, hard, obpyramidal, lustrous, pale, the beak stout and finally dehiscent. (2n = 30) Jun–Aug. Locally abundant in quiet waters; GP except TX, sw OK, se CO; (Que. to B.C., s to CA, UT, MO, KY, & VA).

153. TYPHACEAE Juss., the Cat-tail Family

by S. G. Smith and Robert B. Kaul

1. Typha L., Cat-tail

Perennial monoecious wetland herbs. Rhizomes stout, creeping just below the soil surface to form dense colonies, the erect shoots unbranched, to about 3 m tall. Leaves erect, distichous, basal on vegetative and cauline on flowering shoots, the blades nearly linear, plano-convex or concave-convex near the base, becoming nearly flat near the apex; basal sheaths bearing numerous dotlike mucilage glands on adaxial surfaces. Inflorescence ("spike") terminal, erect, protogynous, sheathed in the bud by several bractlike deciduous leaves; staminate portion terminal and contiguous with or separated from the pistillate portion below, usually deciduous after flowering. Flowers numerous, minute, densely packed, lacking a perianth; staminate flowers of 2–7 sessile, linear anthers on a single slender stalk, intermixed with numerous scalelike papery bracteoles, usually soon falling after anthesis; pistillate flowers of a single 1-celled ovary on a stipe to which are attached numerous hairs; many flowers with abortive ovaries lacking ovules; style and stigma 1, white to green in anthesis but drying brown, persistent but often broken off in late season. Fruit a fusiform thin-walled achene falling with the attached stipe, hairs, and style. The only genus of the family, with about 16 species world-wide, 3 (and their hybrids) in our area.

References: Hotchkiss, N., & H. L. Dozier. 1949. Taxonomy and distribution of North American cat-tails. Amer. Midl. Naturalist 41: 237–254; Lee, D. W. 1975. Population variation and introgression in North American *Typha*. Taxon 24: 633–641; Lee, D. W., & D. E. Fairbrothers. 1969. A serological and disk electrophoretic study of North American *Typha*. Brittonia 21: 227–243; McNaughton, S. J. 1966. Ecotype function in the *Typha* community-type. Ecol. Monogr. 36: 297–325; Smith, S. G. 1967. Experimental and natural hybrids in North American *Typha* (Typhaceae). Amer. Midl. Naturalist 78: 257–287.

Cat-tails often form extensive, nearly pure stands in marshes and other wetlands. Although they are excellent cover for many kinds of wildlife, their aggressive nature and robust habit cause some people to regard them as weeds. Numerous morphological and physiological ecotypes of *Typha* have been described in our area and elsewhere. These ecotypes, as well as the hybrids, make specific determination difficult or impossible for some specimens.

1 Staminate and pistillate parts of spike usually contiguous, not separated by a naked segment of the axis; pistillate bracteoles lacking; stigmas lanceolate to ovate-lanceolate; pollen grains in tetrads .... 3. *T. latifolia*
1 Staminate and pistillate parts of spike usually separated by a naked portion of the rachis; pistillate bracteoles present (but in hybrids visible only with 20–30× magnification); stigmas linear to linear-lanceolate; pollen entirely or partly single-grained.
    2 Living pistillate spikes brown at anthesis; pistillate bracteoles visible with 10× magnification, dark to light brown at tips and broader than the linear stigmas; pollen single-grained.

3 All leaf sheaths of flowering shoots auricled at summit; mucilage glands absent from leaf blade; pistillate bracteoles darker brown than stigmas; spike dark brown, slender; GP .... 1. *T. angustifolia*
3 Uppermost or all leaf sheaths tapered to the blades; brown-punctate mucilage glands (visible when dry or in late season) extending above the sheath onto the inner surface of the blade; pistillate bracteoles lighter brown than stigmas; spike lighter brown, thick; cen and s GP .... 2. *T. domingensis*
2 Living pistillate spikes green at anthesis; pistillate bracteoles visible only with 20–30 × magnification, pale brown to nearly colorless at tip, narrower than the linear-lanceolate stigmas; pollen mostly abortive, in groups of 1, 2, 3, and 4 .... 4. *Hybrids: T. angustifolia* x *latifolia* and *T. domingensis* x *latifolia*

1. *Typha angustifolia* L., narrow-leaved cat-tail. Leaves much exceeding the spike, the largest when fresh about 6–10 mm wide, the sheath summits membranaceous and auricled, mucilage glands brown and evident by midsummer, restricted to the inner surface of the sheath. Pistillate and staminate parts of the spike separated by about 1–8 cm of naked rachis, the pistillate about 1–2 cm thick in fruit, dark brown at all stages, the compound pedicels peglike, about 0.5–0.7 mm long. Pistillate flowers subtended by slender bracteoles with dark brown, obtuse to rounded, enlarged apices (visible with a 10 × hand lens) that are broader than the stigmas and lie among the brown, slightly enlarged tips of the hairs of the pistil; stigmas linear, exceeding the pistil-hairs, white at first, drying uniformly dark brown. Staminate bracteoles narrowly to broadly linear, simple to bifid, brownish, equaling or exceeding anthers at anthesis; pollen grains single. (n = 15) Late May–Jul. Locally abundant, especially in unstable habitats, where fugitive; tolerant of deeper water than our other species; apparently more salt tolerant than *T. latifolia* but dwarfed in extremely saline soils; GP; (N.S. to Man. & CO, s to MO, TX, & cen CA; Eurasia).

This species forms fertile hybrids and intergrades with *T. domingensis* in regions of sympatry. The hybrids are intermediate between the parents and can be identified by their acuminate, brown pistillate bracteoles that slightly exceed the tips of the pistil hairs as well as by the lighter brown color of the pistillate inflorescence. *Typha angustifolia* x *T. latifolia* hybrids (*T. glauca* Godr. *pro sp.* q.v.), closely resemble *T. angustifolia* in macroscopic features but are somewhat larger and often have persistent staminate flowers that do not shed pollen.

2. *Typha domingensis* Pers. Resembling no. 1 but often somewhat taller. Leaves about equaling the spike, to 15 mm wide, the sheath summits on all or at least the uppermost cauline leaves not auricled, the mucilage glands extending 1–10 cm onto the adaxial surface of the blade, brown and evident late in season or when dry. Pistillate spike yellowish to cinnamon-brown at all stages, separated from the staminate spike, 2–3 cm wide in fruit. Pistillate bracteole apices light brown, acuminate to apiculate, paler than dry stigmas, exceeding the slightly swollen straw-colored tips of the pistil hairs; stigmas drying uniformly bright brown. Staminate bracteoles linear to cuneate, mostly irregularly branched toward the apex, stramineous to bright cinnamon-brown; pollen grains single. Late May–Jul. (n = 15) Locally abundant, especially in brackish soils; GP s of the Platte R. (NE); (Gulf Coast, Atlantic Coast n to MD & DE, inland to s NE, UT, & CA).

This species apparently forms hybrids with both *T. angustifolia* and *T. latifolia*. It is a worldwide, variable species of the tropics and warm temperate regions with hot, dry summers.

3. *Typha latifolia* L., broad-leaved cat-tail. Leaves moderately exceeding the spike, the largest when fresh about 10–23 mm wide, the sheath summits firm and obtusely tapered to the blade or slightly auricled, mucilage glands not normally visible without staining, restricted to inner surface of sheath. Pistillate and staminate parts of the spike contiguous or in occasional clones separated by as much as 4 cm; pistillate spike about 1.7–3 cm thick in fruit, green when young and fresh, becoming bright to blackish-brown as the stigmas dry, the whitish hairs of the pistils often prominent in fruit, the compound pedicels in fruit about 1.5–3.5 mm long, including the hairlike tip. Pistillate flowers lacking bracteoles the stigmas broadly ovate to lanceolate, much exceeding the linear and colorless hairs, at first fleshy and greenish and forming the entire surface of the young spike, drying bright to dark brown or with blackish tips. Staminate bracteoles filiform, unbranched, nearly colorless, exceeded by the anthers at anthesis; pollen shed in tetrads. Late May–Jul. (n = 15) Abundant & widespread in nonsaline habitats throughout GP; (Newf. to AK, s to FL, TX, CA, & Mex.; Eurasia).

This species apparently hybridizes with both *T. angustifolia* and *T. domingensis* in our area.

4. HYBRIDS: *Typha angustifolia* x *latifolia* (T. x *glauca* Godr., *pro sp.*) and *T. domingensis* x *latifolia*. These hybrids are apparently intermediate between the parents in all characters. The most diagnostic structures are the pistillate bracteoles, but they are visible only under ca 20–30× magnification and are easily confused with the hairs of the pistil. In the hybrids, the bracteoles are more slender than the linear-lanceolate stigmas, whereas in both *T. angustifolia* and *T. domingensis* they are broader at their apices than the linear stigmas. The mostly abortive pollen is shed in various mixtures of monads, dyads, triads, and tetrads. No or very few fruits are formed. Although the two hybrid combinations are very similar, *T. domingensis* x *latifolia* has somewhat lighter and brighter-colored spikes as well as usually tapered leaf sheath apices.

These hybrids are locally abundant in the GP wherever the parent species occur. They most commonly grow in disturbed places such as roadside ditches and borrow-pits but also grow in natural prairie marshes where water levels vary greatly. Although mostly sterile, they may develop extensive pure stands by rhizomatous growth. They also may form some F2 offspring as well as backcrosses to *T. angustifolia* and *T. domingensis*. Plants intermediate between the F1 hybrids and *T. angustifolia* are fairly frequent in the GP, especially in very disturbed habitats. Plants intermediate between the hybrids and *T. latifolia*, which are occasional in California, have yet to be observed in the GP. Introgression of *T. latifolia* genes into *T. angustifolia* and *T. domingensis* probably occurs very locally. In regions where all three species occur, e.g., in NE, KS, OK, trihybrids may be expected.

154. PONTEDERIACEAE Kunth, the Pickerel-weed Family

by Robert B. Kaul and Charles N. Horn

Herbaceous annuals and perennials of shallow water and muddy shores. Stems creeping or erect. Leaves alternate, basally sheathing, linear and bladeless or petiolate with blades, the veins not prominent. Inflorescences axillary but usually appearing terminal, a bladeless sheath forming the spathe; flowers single or numerous in a raceme or spiciform panicle; peduncle becoming bent downward after flowering. Flowers perfect, hypogynous, regular or irregular, trimerous, the perianth of 6 rather similar lobes in 2 series, the parts connate below into a tube; stamens 3 and equal, or 3 or 6 and unequal, the filaments adnate to the perianth tube, the anthers introrse; ovary 1- or 3-locular; placentae parietal; ovules 1–many, anatropous; style 1, stigmas 3- or 6-parted. Fruit a capsule or 1-seeded utricle.

1 Plant robust, erect, emersed; leaves with blades; flowers blue; stamens 6, the anthers versatile .... 2. *Pontederia*
1 Plant small to moderate-sized, erect or creeping, emersed or submersed; leaves with or without blades; flowers yellow, white, or blue; stamens 3, the anthers basifixed.
   2 Flowers yellow, solitary; stamens alike; plant often submersed; leaves linear, without blades .... 3. *Zosterella*

2   Flowers white to blue, solitary or grouped; stamens dissimilar; plant emersed or partially submersed; leaves with or without blades .... 1. *Heteranthera*

1. HETERANTHERA R. & P., Mud Plantain

Erect or creeping glabrous or glandular-pubescent annual and perennial herbs of muddy places or submersed in shallow water with the blades floating. Leaves sheathing, bladeless or with the blades linear, ovate, or reniform to cordate, thin or firm. Flowers 1–many in an axillary spathe. Perianth white or blue, salverform, divided into 6 linear lobes, regular or irregular; stamens 3, dissimilar; anthers basifixed, sagittate or ovate, sometimes both in the same flower. Fruit a 3-locular, many-seeded capsule.

1   Leaves linear, without distinction of blade and petiole; plant glandular-pubescent .... 2. *H. mexicana*
1   Leaves with blade and petiole; blades lanceolate, ovate, reniform, or cordate; plant glabrous
   2   Flower solitary, blue or white, 10–30 mm wide; blades ovate to nearly ellipsoid .... 1. *H. limosa*
   2   Flowers few to many in the spathe, blue, 5–10 mm wide; blades reniform to cordate .... 3. *H. peduncularis*

1. ***Heteranthera limosa*** (Sw.) Willd. Plant submersed with stems elongate, or emersed with stems contracted. Leaves erect or the blades floating; petioles 5–20 cm long; blades lanceolate to mostly ovate, the base subcordate or truncate but tapering on narrower blades, the tip rounded. Inflorescence pedunculate, the sheathing spathe 2–4 cm long, acuminate, enclosing a single flower. Flower 1–3 cm wide, salverform, 1–4 cm long; perianth lobes linear-lanceolate, white or blue with the base of the upper 1–3 yellow; stamens 3, unequal, 2 with short yellow anthers, the third with an elongate, light blue or yellow anther; ovary incompletely 3-celled, the 3 parietal placentae intruding; ovules numerous. Fruit a many-seeded cylindrical capsule retained in the spathe. May–Oct. In shallow water or on mud, erratically abundant from year to year; MN & s SD to CO, s to TX; (MN to CO, s to Mex. & MS).

2. ***Heteranthera mexicana*** S. Wats. Erect, grasslike glandular-pubescent herb, to 40 cm tall. Leaves basal and cauline, sheathing, linear, 5–14 cm long and 3–10 mm wide, bladeless. Inflorescence an open spike exserted from a spreading spathe. Flowers light to dark blue, salverform, irregular, 1–4 cm long, with a yellow spot at the base of the upper 1–3 lobes; stamens 3, unequal, the 2 shorter with inflated filaments and yellow, ovate anthers, the longer with a slender filament and blue, sagittate anther. Fruit a many-seeded capsule. Jun–Oct. Wet places, temporary pools; TX: Carson, Swisher; (TX & n Mex.). *Eurystemon mexicanum* (S. Wats.) Alex.—Atlas GP.

3. ***Heteranthera peduncularis*** Benth. Plant typically creeping on mud, sometimes floating, the stem elongate. Leaves petiolate; blade reniform to cordate, 2–5 cm wide. Peduncle less than 2 cm long; inflorescence spiciform, 5- to 16-flowered, the lower 1–5 flowers included in the spathe. Flowers pale blue to violet, 5–10 mm wide; perianth tube 5–12 mm long; perianth lobes narrow, equal, regular; stamens 3, unequal, the 2 shorter with pilose filaments and ovoid anthers, the longer glabrous with an elongate, sagittate anther; ovary incompletely 3-locular, the 3 parietal placentae intruding; ovules numerous. Fruit a capsule. Jun–Oct. Muddy shores & shallow waters; s NE & n MO, s to TX; (NE, s to TX & Mex., e to MS & FL). *H. reniformis*—Atlas GP, non R. & P.

    The collections from the GP appear to be intermediate between *H. reniformis*, typical of the eastern United States, and *H. peduncularis* of the Mexican highlands, but they clearly have closest affinity with the Mexican material. The type collection of *H. peduncularis,* from Aguas Calientes, Mexico, has distinctly cordate leaves and the inflorescence has more than 10 flowers, of which all but one or two are exserted from the spathe.

2. PONTEDERIA L.

1. ***Pontederia cordata*** L., pickerel weed. Robust, erect, glabrous perennial herb to 1 m tall, from creeping rhizomes. Basal leaves long-petiolate, the cauline leaves on flowering stems mostly shorter-petiolate; sheaths conspicuous; blades variable, firm, 8–18 cm long, 2–14 cm wide, lanceolate to ovate to deltoid, the bases cordate to truncate, the tips rounded. Inflorescence spiciform, exserted from a bladed sheath (the spathe) terminating the stem above the single cauline leaf, whose blade it barely exceeds. Flowers blue; perianth funnelform, bilabiate, 10–18 mm long; stamens 6, the upper 3 included and usually unequal, the lower 3 exserted; anthers blue, versatile; ovary 3-locular but only 1 locule fertile and bearing a single ovule; styles short, medium, or long (tristylous). Fruit a 1-seeded achenelike, beaked and ridged utricle enveloped in the perianth tube. (n = 8, 2n = 16) Jun–Oct. Shallow water & wet places; KS: Cherokee, Lyon; sw IA, MO, OK; (N.S. to s Ont. & e MN, s to TX & SC).

3. ZOSTERELLA Small

1. ***Zosterella dubia*** (Jacq.) Small, water stargrass. Grasslike submersed and trailing glabrous plant with the internodes conspicuous, or stranded and the stems contracted. Stems slender, branching, rooting at the nodes. Leaves alternate or (at the flowering nodes) opposite; sheaths with 2 apical linear-acute lobes; leaves linear, thin, pellucid, 3–12 cm long and 2–7 mm wide. Spathes sessile, axillary, 1–6 cm long, 1-flowered. Flowers yellow, the filiform tube 1–20 cm long and elevating the limb to the surface; perianth lobes linear, spreading, 6–15 mm long; stamens 3, equal, exserted; filaments dilated, anthers elongate, coiled after anthesis, ovary 1-celled, placentae 3, parietal; ovules several. Fruit a relatively few-seeded capsule; seeds with 10–16 longitudinal membranous wings. (2n = 30) Jun–Oct. Submersed in quiet, often calcareous waters & often forming beds, or stranded on muddy or sandy shores; MN & ND s to OK; (Que. to OR, s to OK, Mex., & FL). *Heteranthera dubia* (Jacq.) MacM.—Atlas GP.

Where abundant, this species is an important wildlife food. It often grows with *Potamogeton,* which is resembles.

155. LILIACEAE Juss., the Lily Family

by Steven P. Churchill

Mostly perennial herbs, forming various underground storage organs including bulb, corm, or rhizome; stems if present, erect, solitary or branched. Leaves alternate or whorled, rarely opposite, cauline or basal, linear, lanceolate or ovate-lanceolate with parallel veins or broadly elliptical to ovate with net veins, leaves rarely reduced to scales. Inflorescence a raceme, cyme, or umbel. Flowers mostly of 6(4) similar perianth segments in 2 whorls of 3 segments each, free or partially adnate into a tube; stamens 6(4), opposite perianth segments, attached to side or base of perianth, anthers extrorse or introrse; stigma 1–3, style 1–3, ovary superior or inferior, (1)3-locular, placentation axil, ovules numerous. Fruit a dry capsule or fleshy berry, 1–many seeded.

The classification of the Liliaceae employed here (Cronquist, 1981. An integrated system of classification of flowering plants.) is in the broad sense to include one additional family, the Amaryllidaceae (*Cooperia, Hypoxis,* and *Zephyranthes* from our area). As treated in this broad concept, the Liliaceae includes some 310 genera and about 3800 species that are widely distributed throughout the world except in the arctic region. This family is both horticulturally and economically important.

1 Leaves whorled on stem, occasionally upper or lowermost leaves alternate or opposite.
  2 Leaves in a single terminal whorl of 3; flowers white, yellow or dark purple brown .... 23. *Trillium*
  2 Leaves in 1–several whorls of more than 3, upper or lower leaves often alternate or opposite; flowers reddish-orange .... 14. *Lilium*
1 Leaves alternate on stem or leaves basal.
  3 Flower or fruit single on a scape.
    4 Ovary superior; stamens 6; leaves 2 .... 9. *Erythronium*
    4 Ovary inferior; stamens 3; leaves more than 2.
      5 Perianth yellow, throat and limb longer than tube; spathe 2–3 cm long .... 25. *Zephranthes*
      5 Perianth white and ± reddish tinged, throat and limb shorter than tube; spathe 4–5 cm long .... 7. *Cooperia*
  3 Flower or fruit 2 or more on a scape or leafy stem.
    6 Leaves reduced to small scales, branchlike leaves filiform; flowers greenish-white, 4–5 mm long .... 3. *Asparagus*
    6 Leaves linear, lanceolate to ovate, or oval, not reduced to scales.
      7 Stems leafy and lacking basal leaves or, if basal leaves present, few and often dying back.
        8 Flowers on short or long pedicels from leaf axil.
          9 Perianth forming a tube, segments separating at tip; flowers usually 2 or more on a peduncle .... 20. *Polygonatum*
          9 Perianth of separate segments; flowers 1 or 2.
            10 Flowers whitish-green, segments 1.5 cm or less long; fruit a berry .... 24. *Streptopus*
            10 Flowers pale yellow, segments 2 cm or more long; fruit a capsule .... 24. *Uvularia*
        8 Flowers terminal on stems or branches.
          11 Perianth of 4 segments; leaves 2 or 3 .... 15. *Maianthemum*
          11 Perianth of 6 segments; leaves usually more than 4.
            12 Flowers many, usually more than 5; paniculate or racemose .... 21. *Smilacina*
            12 Flowers few, usually 1–3; solitary or few in an umbellike cluster.
              13 Leaves ovate to oblong.
                14 Flowers greenish-white; filament longer than anther; stem and pedicels pubescent .... 8. *Disporum*
                14 Flowers pale yellow; filament much shorter than anther; stem and pedicels glabrous .... 24. *Uvularia*
              13 Leaves narrowly linear.
                15 Perianth segments similar, oblong to rhombic .... 10. *Fritillaria*
                15 Perianth segments dissimilar, outer series lanceolate, inner series as broad as long .... 4. *Calochortus*
      7 Stem leaves basal with few or none reduced upward on stem.
        16 Leaves and flowers all appearing basal, produced at ground level; flowers white with a long tube .... 13. *Leucocrinum*
        16 Leaves mostly basal but flowers produced on stem.
          17 Inflorescence an umbel or irregular corymblike umbel.
            18 Perianth segments longer than 5 cm; inflorescence an irregular corymblike umbel .... 11. *Hemerocallis*

LILY FAMILY

18 Perianth segments 3 cm long or less; inflorescence an umbel.
  19 Perianth with a fused tube, separated in upper
    1/3–1/2, blue or violet .... 2. *Androstephium*
  19 Perianth with separated segments, sometimes fused at very base, white, cream, or pink.
    20 Plants with an onion odor; seeds 1 or 2 per locule; perianth pink to white, rarely yellow, erect to patent or withering away at fruiting time .... 1. *Allium*
    20 Plants lacking an onion odor; seeds 3 or more per locule; perianth yellow, reflexed at fruiting time .... 18. *Nothoscordum*
17 Inflorescence a raceme or a panicle.
  21 Style distinctly 3-cleft.
    22 Inflorescence and upper stem pubescent .... 16. *Melanthium*
    22 Inflorescence glabrous .... 26. *Zigadenus*
  21 Style simple or only slightly 3-cleft.
    23 Perianth segments united for 1/2 or more their length.
      24 Perianth pubescent on outer surface, yellow .... 12. *Hypoxis*
      24 Perianth glabrous, blue or white.
        25 Flowers blue; leaves linear; bulb present .... 17. *Muscari*
        25 Flowers white, drying to yellow; leaves broadly oblong to elliptical; rhizome present .... 6. *Convallaria*
    23 Perianth segments separate or united only at the base.
      26 Flowers greenish-white to white, abaxial perianth segment with a green stripe .... 19. *Ornithogalum*
      26 Flowers blue, lacking a green adaxial stripe .... 5. *Camassia*

1. ALLIUM L., Onion

Scapose perennial herbs from a fibrous or membranaceous coated, truncated bulb, often with odor of onion or garlic. Leaves basal or a few confined to lower 1/2 of stem, narrowly linear to long lanceolate, flat or terete and then often hollow. Inflorescence a terminal false umbel, from a spathe that divides into 1–3 bracts, each bract 1- to 7-nerved. Flowers white, pink, or purple, in some cases flowers replaced by bulblets, bulblets sessile, rarely pedicellate; perianth of 6 similar segments, erect to spreading, generally free, presisting and investing capsule or withering way from capsule; stamens 6, adnate to perianth, exserted or included, anthers introrse, filaments often flat and wide, uniting at base; stigma single or shortly trifid, style 1, ovary superior, lobed or crested, 3-locular, ovules 1–few per locule. Fruit a short globose-ovoid capsule; seeds black.

References: Ownbey, M. 1950. The genus *Allium* in Texas. Res. Stud. State Coll. Wash. 18: 181–222; Ownbey, M., & H. C. Aase. 1955. Cytotaxonomic studies in *Allium*. 1. The *Allium canadense* alliance. Res. Stud. State Coll. Wash., Monogr. Suppl. 1: 1–106.

This is a large genus of nearly 500 species distributed throughout the Northern Hemisphere. While the bulbs are edible, one should take care to note the difference between the onion (*Allium*) which has an umbel inflorescence and death camass (*Zigadenus*) which has a raceme inflorescence; see comments under the latter genus. Livestock poisoning has been attributed to several species of *Allium*.

1 Leaves terete, hollow below middle, not confined to base of stem .... 10. *A. vineale*
1 Leaves flat, folded or channeled, most confined to stem base.
  2 Leaves lanceolate-elliptic, 2–5 cm wide, absent when flowering scape is present; ovary deeply 3-lobed, locule 1-seeded .... 9. *A. tricoccum*
  2 Leaves linear, mostly less than 1 cm wide, present with flowering scape; ovary moderately lobed or crested, locule 2 or more seeded.
    3 Bulb coat longitudinally nerved, cellular between nerves, membranaceous, nonfibrous.
      4 Inflorescence nodding, not erect .... 2. *A. cernuum*
      4 Inflorescence erect.
        5 Flowers mostly replaced by bulblets; leaves few, on lower 1/2 of stem .... 6. *A. sativum*

        5 Flowers present, bulblets lacking; leaves confined to base to stem, at ground
           level .... 7. *A. stellatum*
3 Bulb coat reticulated, coarsely fibrous.
    6 Ovary and capsule crested.
        7 Leaves 3 or more; flowers usually pink .... 4. *A. geyeri*
        7 Leaves 2; flowers usually white .... 8. *A. textile*
    6 Ovary and capsule crestless.
        8 Spathe bracts each 1-nerved; perianth segments spreading and becoming rigid in
           fruit .... 3. *A. drummondii*
        8 Spathe bracts each 3- to 7-nerved; perianth segments various.
            9 Perianth segments presisting, investing capsule; flowers produced, no bulblets;
               plants 1-2 dm tall .... 5. *A. perdulce*
            9 Perianth segments withering away from capsule; flowers or bulblets
               produced; plants often over 2 dm tall .... 1. *A. canadense*

1. *Allium canadense* L., wild onion. Slender to stout plants (1.5)2-9 dm tall, from a fine to coarsely fibrous-reticulate bulb. Leaves 2 or more, shorter than scape, 1-3(7) mm wide, channeled. Inflorescence an erect umbel with flowers or bulblets; spathe divided into 3(4) bracts, each bract 3- to 7-nerved. Flowers white to pink, campanulate, withering in fruit; perianth segments elliptic-lanceolate, apex obtuse to acute; stamens just shorter than perianth; ovary crestless.

Five varieties can be distinguished in our area:

1 Bulblets produced, pedicels very short or absent, occasionally a few long pediceled flowers
   present .... 1a. var. *canadense*
1 Bulblets not produced, flowers on pedicels.
    2 Perianth usually white .... 1b. var. *fraseri*
    2 Perianth usually pink or lilac.
        3 Pedicels filiform; plants slender .... 1e. var. *mobilense*
        3 Pedicels stout; plants usually robust.
            4 Flowers fragrant .... 1c. var. *hyacinthoides*
            4 Flowers odorless .... 1d. var. *lavendulare*

1a. var. *canadense*. Basal bulblets lacking; leaves 3 or more, 1-5 mm wide. Spathe becoming 3 bracts, each bract 3- to 7-nerved, pedicels, if present, stout, 2-3 × perianth length. Flowers mostly replaced by bulblets, bulblets sessile, each often long beaked, when flowers present, pedicels 4-6 × the length of the perianth, perianth white or pink, 4-7 mm long (2n = 21, 28) Apr-Jul. Prairies, meadows, open woods, & disturbed sites; e GP; MN & SD s to MO & OK; (se Can. to ND, s to FL & TX).

1b. var. *fraseri* M. Ownbey. Basal bulblets lacking; leaves 3 or more, 1-4 mm wide. Spathe becoming 3 bracts, each bract 3- to 7-nerved; pedicels filiform, 3-4 × perianth length. Flowers white (rarely pink), 4-7 mm long (2n = 14, 21) Apr-Jul. Plains & prairies; w & cen GP; SD & MT, s to OK & TX. Endemic.

1c. var. *hyacinthoides* (Bush) M. Ownbey. Basal bulblets lacking; leaves several, 2-7 mm wide. Spathe becoming 3 or 4 bracts, each bract 3- to 7-nerved; pedicels stout, 2 or 4 × perianth length. Flowers pink, fragrant, 5-7 mm long. (2n = 14) Mar-Apr. Plains & prairies; s GP; OK; (OK & TX).

1d. var. *lavendulare* (Bates) M. Ownbey. Basal bulblets lacking; leaves usually 3, 1-7 mm wide. Spathe usually becoming 3 bracts, each bract 3- to 7-nerved; pedicels stout, 3-5 × perianth length. Flowers lavender-pink (rarely white), odorless, 5-8 mm long (2n = 14, 28) May-Jul. Prairies, & occasionally meadows & open woods; e GP; e SD s to OK; (IA & SD, s to OK & AR).

1e. var. *mobilense* (Regel) M. Ownbey. Basal bulblets 1 or 2; leaves 2 or more, 1-2 mm wide. Spathe becoming 3 bracts, each bract 3- to 5-nerved; pedicels filiform, 2-4 × perianth length. Flowers pink, 4-6 mm long. (2n = 14, 28) Apr-Jun. Se GP; KS: Cherokee; MO: Jasper; OK: Craig, Ottawa, Rogers; (MO & e KS, s to FL & TX). *A. mutabile* Michx.—Rydberg.

Steyermark (Flora of Missouri. 1963. pp. 428-429) has recognized var. *canadense* at the specific level, and placed both var. *lavendulare* and var. *mobilense* as synonyms under *A. mutabile* Michx. Further studies appear to be warranted, particularly for the *A. mutabile* group.

2. **Allium cernuum** Roth, wild onion. Plants 1–7 dm tall, from a membranaceous, longitudinally nerved, coated bulb. Leaves glaucous, 1–4 mm wide, flat or ± channeled. Inflorescence nodding; spathe usually dividing into 2 bracts; pedicels slender, 2–3 × perianth length, in fruit stouter and arching upward. Flowers white to pink, campanulate, perianth 4–6 mm long, segments elliptic-ovate, apex obtuse; stamens exserted; ovary distinctly 6-crested (2n = 16) Jul–Aug. Open deciduous or conifer woods, occasionally in prairies; ne & sw SD, nw NE s to CO & NM; (NY w to B.C., s to GA & NM; Mex.). *A. recurvatum* Rydb.—Rydberg.

3. **Allium drummondii** Regel, wild onion. Plants 1–3 dm tall, from a fibrous coated bulb. Leaves 2 or 3, nearly equal to scape length, 1–3 mm wide, channeled. Inflorescence an erect umbel; pedicels 1–3 × perianth length; spathe 2 or 3 divided, 1-nerved. Flowers white, pink or red (rarely yellow), rotate-campanulate, perianth segments 5–8 mm long, ovate to lanceolate, apex obtuse to acute; stamens shorter than perianth; ovary lobed, not crested. (2n = 14, 18) Apr–Jun. Plains & prairies; s SD & WY s to OK & TX; (SD & WY s to TX & NM; Mex.). *A. nuttallii* S. Wats.—Rydberg.

4. **Allium geyeri** S. Wats. Plants 1–3(5.5) dm tall, from a fibrous-coated bulb. Leaves 3 or more, shorter than scape, 1–5 mm wide, channeled. Inflorescence an erect umbel, pedicels 2 × perianth length or less, becoming rigid and stiffly spreading in fruit, not flexuous; spathe dividing into 2 or 3 bracts, each bract 1-nerved. Flowers pink or white, campanulate, perianth segments (4)6–8(10) mm long, ovate to lanceolate, apex obtuse to acuminate, becoming callous-keeled and investing capsule; stamens shorter than perianth; ovary inconspicuously 6-crested by round knobs, each 0.5 mm high. (2n = 14, 28) Jun–Jul. Moist meadows & along streams in open woods, occasionally in prairies; w GP; SD & WY s to NM; (sw Can., MT & WA, s to SD & AZ).

5. **Allium perdulce** S. V. Fraser, Plants 1–2(2.3) dm tall, from a fibrous-coated bulb, onion odor lacking. Leaves 3 or more, nearly equal or exceeding scape, 1–2 mm wide, channeled. Inflorescence an erect umbel; spathe dividing into 2 or 3 bracts, each 5-nerved; pedicels about equal to perianth length, becoming longer in fruit and ± flexuous. Flowers deep rose, fading purple, urceolate, perianth segments 7–10 mm long, lanceolate, apex obtuse to acute, becoming callous-keeled and investing capsule; stamens shorter than perianth length; ovary crestless. (2n = 14, 28). Apr–Jun. Prairies & plains; cen & s GP; IA & SD s to OK & TX; (*endemic* to the s & cen U.S.).

6. **Allium sativum** L., garlic. Stout plants 5–10 dm tall or more, from a membranaceous-coated thick bulb. Leaves sheathing and extending up to 1/2 of scape length, 0.5–1.5 cm wide, flat. Inflorescence an erect umbel, flowers replaced by bulblets, bulblet often long beaked, if flowers present then perianth a whitish-green; spathe single and long beaked, persisting; stamens longer than perianth segments, filaments 3-cleft, 2 segments awnlike, 1 segment bearing anther; ovary apex emarginated. (2n = 16) May–Jul. Escaping from cultivation & found in disturbed areas, fields, & meadows; MO & e KS: (widespread in e U.S., Eurasia;). *Naturalized.*

7. **Allium stellatum** Ker., pink wild onion. Slender plants 2–6 dm tall, from a membranaceous coated bulb. Leaves several, shorter than scape, 1–4 mm wide, channeled, mostly withering at anthesis. Inflorescence an umbel, cernuous but becoming erect; spathe becoming 2 bracts or remaining united, each bract 7-nerved, becoming reflexed; pedicels 2–3 × perianth length. Flowers deep pink, stellate, withering in fruit, perianth segments 5–8 mm long, elliptic-lanceolate, apex acute; stamens exserted; ovary distinctly 6-crested, crest flattened, entire or with toothed processes. (2n = 32) Jul–Sep. In prairies; e GP; MN & ND s to KS & OK: (IL to s-cen Can., s to TX).

8. ***Allium textile*** A. Nels. & Macbr. Plants (0.5)1–3 dm tall, from a fibrous-coated bulb. Leaves usually 2, equal to or exceeding scape, 1–3 mm wide, channeled. Inflorescence an erect umbel, becoming flexuous and rigid in fruit; spathe dividing into (2)3 bracts, each bract 1-nerved. Flowers white, rarely pink, campanulate; perianth segments 5–7 mm long becoming callous-keeled and permanently investing capsule, lanceolate, inner series obtuse to acuminate; stamens shorter than perianth; ovary ± 6-crested, knobs separated or united. (2n = 14, 28) May–Jul. Plains & open, dry coniferous woods; n & w-cen GP; MN w to MT, s to w NE & KS; (cen Can. & MN to WA, s to NM, KS, & IA).

9. ***Allium tricoccum*** Soland., wild leek. Plants 1.5–4 dm tall, from a fibrous-coated bulb. Leaves 2, lanceolate-elliptic, 10–30 cm long, 2–6 cm wide, ± fleshy, tapering at base, withering before anthesis. Inflorescence an erect to spreading umbel; spathe becoming 2 bracts; pedicels 2–3 × perianth length. Flowers white, perianth segments 5–7 mm long, oblong to ovate, apex obtuse; stamens equal to perianth length; ovary deeply 3-lobed, locule 1-seeded. (2n = 16) Jun–Jul. Rich deciduous wooded slopes; uncommon in e GP; MN & e ND s to IA, e NE, & MO; (se Can. to MN, s to GA & NE). *Validallium tricoccum* (Ait.) Small — Rydberg.

Two species have been recognized recently by Jones (Syst. Bot. 4: 29–43. 1979.): *Allium tricoccum* and *A. burdickii*. Both taxa exhibit a notable amount of phenotypic overlap, and it might be best to regard *A. burdickii* as a subspecies of *A. tricoccum,* or at least, when positive identification can not be made, then to use *A. tricoccum* sensu lato, while acknowledging *A. burdickii* as a distinct entity. According to Jones, both taxa are fully reproductively isolated.

10. ***Allium vineale*** L., wild garlic. Plants up to 10 dm tall, from a membranaceous bulb, bulb often clustered. Leaves sheathing up to scape middle, terete and becoming channeled, hollow. Inflorescence an erect umbel; spathe split. Flowers pink to purple, occasionally white, flowers often replaced by bulblets, each bulblet tipped by a small slender leaf; pedicels 1–2 cm long when present; stamens differentiated, inner series a 3-parted filament, 2 segments threadlike and longer than segment bearing anther. (2n = 32) May–Jun. Disturbed areas, lawns, fields, & meadows; MO & e KS s to OK; (widespread in e U.S.; Europe). Introduced.

Both of the following species either are known to escape from or persist after cultivation: *Allium porrum* L., leek, similar to *A. vineale* but the leaves are flat, spathe membranaceous, and flowers white to pink (IA: Mills; KS: Doniphan, Rush, Saline, Washington); and *Allium cepa* L., cultivated onion, characterized by a broadly ovoid bulb, scape and leaves inflated and hollow, and pedicels many times longer than flowers.

2. ANDROSTEPHIUM Torr.

1. ***Androstephium caeruleum*** (Scheele) Torr., blue funnel lily. Scapose perennial plants (1)1.5–1.7(3.4) dm tall, glabrous and glaucous green; from a fibrous coated corm, up to 2.5 cm wide. Leaves 5 or 6, sheathing base, exceeding flowering scape, linear, 12–19(33) cm long, 2–3 mm wide, somewhat folded. Inflorescence an umbel; pedicels erect to spreading, 10–20 mm long, usually shorter than flowers. Flowers 2–9, pale blue to violet, (18)25(32) mm long, perianth tube less than length of oblong lobes; stamens 6, inserted on throat, anthers introrse, basifixed, stigma trifid, ovary superior, 3-locular. Capsule 3-angled, 12–16 mm long; seeds black. Apr–May. Prairies, frequently in rocky areas, plants often hidden among previous season's grasses; cen KS s to OK & TX; (endemic to s-cen U.S.).

This is a genus of 3 species distributed in western United States and Mexico. The name *Bessera* Schult. f. has been used at various times for *Androstephium* Torr.

3. ASPARAGUS L, Asparagus

1. *Asparagus officinalis* L. Perennial plants 1.5–2 m tall, from a rhizome; stems with many finely dissected branches. Leaves reduced to scales, on stem appearing alternate and on branches appearing as if in fascicles. Flowers dioecious, greenish-white, occurring on jointed pedicels; perianth campanulate, each segment 4–6 mm long, oblong; stamens 6, attached to base of perianth, anthers introrse, shorter than the filaments in length; ovary superior, 3-locular. Fruit a red berry, ca 1 cm wide; seeds 3–6, black. ($2n = 20$) May–Sep. Escaping from cultivation & found in moist, often shaded thickets & margin of woods, disturbed areas along ditches, streams, & fields; GP, less frequent in w; (widespread in Can. & U.S.). *Naturalized.*

This is a large genus of some 300 species distributed mainly throughout the arid regions of the Old World. Several species are used as ornamentals.

4. CALOCHORTUS Pursh, Sago Lily, Mariposa

Glabrous erect perennial herbs, from a tunicate, membranaceous-coated bulb. Leaves linear, becoming reduced upward. Inflorescence solitary to pseudoumbellate, bracts shorter to longer than pedicels. Flowers perfect, campanulate; perianth of 6 segments in 2 series, 3 outer segments lanceolate, 3 inner segments broadly obovate, cuneate with a glandular spot on the inner face near base; stamens 6, in 2 series, filaments subulate, basally dilated, ± adherent to perianth base, anthers basifixed; stigma trifid, ovary superior, 3-locular, ovules many per locule. Fruit 3-angled, erect, dehiscent capsule; seeds flattened, hexagonally reticulate.

Reference: Ownbey, M. 1940. A monograph of the genus *Calochortus*. Ann. Missouri Bot. Gard. 27: 371–560.

This is a large genus of some 60 species mostly from western North America. The bulbs were used as a food source by the Plains Indians.

1 Anther obtuse, nearly equal to filament length; petal gland circular .... 2. *C. nuttallii*
1 Anther acute to acuminate, longer than filament length; petal gland transversely oblong .... 1. *C. gunnisonii*

1. *Calochortus gunnisonii* S. Wats. Plants 2.4–5.5 dm tall. Leaves 2–4, 18–35 cm long, 2–8 mm wide. Inflorescence 1(2 or 3)-flowered. Flowers white to slightly purple, a purple band on petal above gland; outer 3 segments 20–33 mm long, 4–8 mm wide; inner 3 segments obovate, cuneate, 30–45 mm long, 27–38 mm wide; gland depressed transversely oblong, slightly arched, covered with short distally bilobed or branched processes around gland, outermost ± united and forming a discontinuous membrane; anthers lanceolate, 7–11 mm long, acute to acuminate, apiculate, filaments 6–8 mm long, usually shorter than anthers. Fruit linear-oblong, 3–5 cm long. ($2n = 18$) Jul–Aug. Dry prairies to open, coniferous or deciduous woods, often on rocky slopes; w SD & MT s to ne NE, WY, CO, & NM; (SD to MT, s to NM & AZ).

2. *Calochortus nuttallii* T. & G. Plants 1.4–4.4 dm tall. Leaves 2 or 3(4), 8–16 cm long, 1–2 mm wide, ± involute. Inflorescence 1(2 or 4)-flowered. Flowers white but yellow at base, outer 3 segments lanceolate, 20–30(35) mm long, 6–12 mm wide, acuminate, inner 3 segments broadly obovate, cuneate, often short-acuminate, 26–45 mm long, 21–34 mm wide, gland circular, depressed, surrounded by a conspicuous fringed membrane, densely covered with short simple or distally branched processes, more slender hairs near gland; anthers oblong, (5)7(10) mm long, obtuse, filaments (5)7(8) mm long, usually equal or subequal to anther length. Fruit linear-lanceolate, acuminate, 3.5–5.5 cm long. ($2n = 16$)

Jun–Jul. Dry prairies, occasionally in open coniferous woods; w ND & MT s to nw NE & WY; (ND to ID, s to NM, AZ, & CA).

5. CAMASSIA Lindb.

1. **Camassia scilloides** (Raf.) Cory, eastern camass. Perennial scapose erect herbs up to 7 dm tall, from a black to brown scaly coated bulb. Leaves basal, linear, basal 1/3 of blade keeled, up to 6 dm long, 4–20 mm wide. Inflorescence a bracted raceme from a scape; bracts equal to slightly longer than pedicels. Flowers pale blue, blue-violet or white, 6 perianth segments spreading, 7–15 mm long, short clawed, 3- to 5-nerved; stamens 6, anthers introrse, versatile, filament filiform, much longer than anthers; style as long as filaments, stigma distinctly 3-parted, ovary superior, 3-locular, ovules numerous. Fruit a subglobose capsule, seeds black, few. (2n = 30) Apr–Jun. Slightly dry to wet prairies & open woods; MO & e KS s to OK; (PA to WI & KS, s to GA & TX). *C. angusta* (Engelm. & Gray) Blank.—Atlas GP; *C. esculenta* (Ker) Robins.—Rydberg.

References: Gould, F. W. 1942. A systematic treatment of the genus *Camassia* Lind. Amer. Midl. Naturalist 28: 712–742; Steyermark, J. A. 1963. *Camassia*, pp. 434–436, *in* Flora of Missouri, Iowa State Univ. Press.

A small New World genus of 6 or 7 species distributed mostly in western North America.

Steyermark (op. cit.) and others have suggested that there are in fact 2 distinct taxa: *C. scilloides* and *C. angusta*. The latter taxon is distinguished by a somewhat later blooming time, a more robust nature, long raceme and more flowers, but in nearly every feature there appears to be abutting or overlap of both qualitative and quantitative characters. I would agree with Gould's (op. cit.) suggestion that *C. angusta* should be recognized at the subspecies level under *C. scilloides* or as a distinct species. Further populational and biosystematic studies are warranted for this species complex.

6. CONVALLARIA L.

1. **Convallaria majalis** L., lily-of-the-valley. Perennial scapose herbs up to 2.5 dm tall, from a slender rhizome. Leaves 2(3), broadly oblong or elliptic, 10–20 cm long, 3–6 cm wide. Inflorescence a raceme, up to 20 cm long. Flowers white, campanulate, perianth of 6 similar segments, lower 1/2 fused, lobes recurved; stamens 6, inserted at perianth base, anthers introrse, much longer than filaments; stigmas 3-lobed, style stout, ovary superior, 3-locular, ovules few per locule. Fruit a red berry, seeds few. (2n = 19, 38) Apr–May. Occasionally escaping from or persisting after cultivation, in disturbed sites; MO & e KS; (possibly native VA to TN, escaped in various areas U.S., Europe). Introduced.

Convallaria majalis is known to cause poisoning, the primary agent being cardiac glycosides convallarin and convallamarin. This genus consists of but two species distributed in the Northern Hemisphere.

7. COOPERIA Herb.

1. **Cooperia drummondii** Herb., rain-lily. Scapose perennial plants up to 3.5 dm tall, from a large subglobose bulb. Leaves 2–6, linear, ± glaucous, up to 30 cm long. Flower single, lasting but a few days, from a spathe 3.5–5 cm long, perianth white and often red-tinged on midveins and tips, tube long, 10–12.5 cm, limb salverform, lobe ovate, 1–2.5 cm long; stamens inserted on throat, anthers 7–10 mm long, filament equal to 1/2 anther length; stigma 3-parted, ovary inferior, 3-locular. Fruit a capsule, ca 12 mm long, seeds compressed, black, 6–8 mm long. (2n = 48) Jul–Sep. Shallow soil over limestone in prairies or pastures; se KS to OK; (KS to NM, s to LA & TX; Mex.).

This is a genus of 9 species distributed in the New World, eight are found in the United States and northern Mexico, and one species is recorded from Brazil. *Cooperia* is often included in *Zephyranthes*.

8. DISPORUM Salisb.

1. Disporum trachycarpum (S. Wats.) Benth. & Hook., fairybells. Perennial plants 1–6.5 dm tall, from a rhizome; stem pubescent. Leaves ovate to ovate-lanceolate, 3–10 cm long, 1.5–5 cm wide, slightly pubescent beneath, glabrous above, base ± cordate, sessile. Flowers 1 or 2, greenish-white, narrowly campanulate, perianth spreading, oblong-linear segments 9–15 mm long; stamens just exceeding perianth or equal, anthers extrorse, filaments filiform, ca 3 × anther length; stigma 3-lobed, rarely entire, style filiform, ovary superior, 3-locular, smooth when young, becoming papillose with age. Fruit a bright red or reddish-orange berry; seeds usually 4–12. (2n = 22) May–Jul. Rich deciduous or coniferous-deciduous woods; ne GP; ND & MT s to w SD & nw NE; (w Can. s to NE, NM, & AZ).

Reference: Jones, Q. 1951. A cytotaxonomic study of the genus *Disporum* in North America. Contr. Gray Herb. 173: 1–39.

9. ERYTHRONIUM L., Dog's-tooth-violet

Low perennial stemless herbs from a deep, scaled bulb. Leaves 2 in flowering form, single in vegetative form, linear-lanceolate to broadly lanceolate or elliptical, tapered at both ends, sheathing flowering scape. Flower single on terminal scape, white or yellow; perianth of 6 similar lanceolate segments, spreading to recurved; stamens 6, ca 1/2–2/3 perianth length, anthers introrse, basifixed, filaments long; style elongate, ovary superior, 3-locular, obovoid. Fruit a capsule, few to many seeded.

References: Ireland, R. R. 1957. Biosystematics of *Erythronium albidum* and *E. mesochorum*. M.A. thesis, Univ. Kansas. 49 pp; Robertson, K. R. 1966. The genus *Erythronium* (Liliaceae) in Kansas. Ann. Missouri Bot. Gard. 53: 197–204.

This is a Northern Hemisphere genus of some 25 species with their greatest diversity in western North America.

1 Perianth segments yellow, auricles present at base; leaves not glaucous, similar on both surfaces .... 3. *E. rostratum*
1 Perianth segments white, auricles absent; leaves glaucous beneath.
  2 Perianth segments reflexed; leaves nearly always mottled; mature capsule erect to nodding, held off ground .... 1. *E. albidum*
  2 Perianth segments spreading, not reflexed; leaves not mottled; mature capsule bent over onto ground .... 2. *E. mesochoreum*

1. Erythronium albidum Nutt., white dog's-tooth-violet. Plants producing stolons and usually these hidden among leaf litter. Leaves brown-purple mottled, occasionally not so with age, elliptical-lanceolate to ovate-lanceolate, (.6)1–2.5(4) cm wide, flat to slightly folded, often glaucous on both surfaces. Perianth white, abaxial side tinged with pink or lavender, segments completely reflexed in full bloom, lacking auricles and yellow spot at base; stigma trifid, lobes long, style not persisting, ovary sides convex. Capsule generally erect or nodding; seeds 1–3, seldom more. (2n = 44) Mar–May. Moist woods, rarely in transitional prairie-woods; e GP; MN & NE s to OK & MO; (Ont. to MN & NE, s to GA & TX).

2. Erythronium mesochoreum Knerr, white dog's-tooth-violet. Plants not producing stolons. Leaves lanceolate to linear-lanceolate, (0.6)1–2(3) cm wide, folded, occasionally only partially folded, not mottled, glaucous on both surfaces. Flowers white, abaxial side often lavender tinged; perianth segments lacking auricles, yellow spot present at base, spreading to quarter reflexed in full bloom; stigma trifid, lobes long, style not persistent, ovary sides convex. Capsule usually resting on ground; seeds most often more than 3. (2n = 22) Mar–Apr. Prairies & open woods, occasionally found in cut-over woods; IA & NE s to MO & OK; (IA & NE s to TX). *E. albidum* var. *mesochoreum* (Knerr) Rickett — Fernald.

Nearly every aspect of the life history and morphology of *E. mesochoreum* differs from *E. albidum*. The distinctive features of both species that are evident in the field are often not apparent in mounted herbarium specimens. Ants have been observed eating the distinctive white oil body attached to the seed, and are at least one of the agents responsible in the dispersal of seeds (Churchill & Bloom, unpubl. data).

3. Erythronium rostratum C. B. Wolf, yellow dog's-tooth-violet. Plants producing stolons. Leaves broadly lanceolate to elliptic, (2.5)3–4(5) cm wide, purplish-brown mottled on abaxial surface, not glaucous. Flowers yellow, abaxial outer perianth series with small purplish-brown specks, inner segment series with well-developed auricles at base, clasping opposite filament; stigma short and swollen, style persistent, forming a beak on capsule. Fruit an ellipsoid capsule, held erect to nodding at maturity. (2n = 24) Mar–Apr. Rich deciduous woods; se GP; KS: Cherokee; MO: Barton, Jasper; (TN to MO & KS, s to AL, AR, & OK).

10. FRITILLARIA L., Fritillary

Perennial erect herbs, narrow leafy stem from a scaly bulb, often attached to bulblets. Leaves mostly cauline, whorled, opposite or alternate, linear-lanceolate, sessile. Flowers solitary, terminal, perianth segments oblong or rhomboidal, bearing a shallow nectar gland just above base; stamens 6, inserted at base of perianth, anthers extrorse, filaments slender; style entire to trifid, ovary superior, 3-locular, ovules numerous, in 2 rows in each locule.

Reference: Beetle, D. E. 1944. A monograph of the North American species of *Fritillaria*. Madroño 7: 133–159.

This is a widely distributed Northern Hemisphere genus having some 85 species.

1 Perianth segments rhomboid, purplish mottled; styles 3-cleft .... 1. *F. atropurpurea*
1 Perianth segments oblong, yellowish-orange, turning dark red; style 1 .... 2. *F. pudica*

1. Fritillaria atropurpurea Nutt., leopard lily. Plants up to 4 dm tall, from a bulb, bulblets few or absent. Leaves 5 or 6(13), linear, 4–9 cm long, 3–4 mm wide. Flowers open campanulate, purplish mottled, perianth segments rhomboid, 15–20 mm long, 5–8 mm wide, yellow short tufted hairs at apex, gland an indistinct yellowish-brown spot at base; style 3-cleft; anthers 1/2 as long or less than slender filaments. Capsule obovoid, acutely angled. (2n = 24) May–Jun. Grassy slopes in coniferous woods; w GP; w ND & MT s to NE; (ND to OR, s to NM & CA).

2. Fritillaria pudica (Pursh) Spreng., yellow bell. Plants 7–16(30) cm tall, from a thick scaly bulb, bulblets few to several. Leaves 2–4(6), linear to lanceolate, alternate or congested and appearing whorled, 4–10 cm long, 3–7 mm wide. Flowers solitary, pendent, narrowly campanulate, yellow to orange and turning red with age, perianth segments oblong, 10–18 mm long, 3–7 mm wide, gland small, at base; style single; anthers to 1/2 as long as filaments. Capsule elongate-obovoid, not acutely angled. (2n = 24, 26, 36, 78). Apr–Jun. Open, grassy areas in coniferous woods; nw GP; ND: Billings, Morton; (ND to sw Can., s to UT & CA).

11. HEMEROCALLIS L., Daylily

1. *Hemerocallis fulva* L. Scapose perennial herb up to 1-1.5 m tall from a fleshy rootstock, stoloniferous. Leaves basal, linear, 5-10 dm long, 1-3 cm wide. Inflorescence a several flowered raceme or false umbel. Flowers broadly campanulate to funnelform, orange, perianth segments 6, inserted on the perianth, lobes 6-10 cm long, spreading or slightly recurved, tube up to 3.5 cm long; stamens 6, anthers introrse; stigma small, style long and slender, exceeding anthers, ovary superior, 3-locular, ovules numerous. Fruit a capsule; seeds generally not maturing. (2n = 22, 33) May-Aug. Escaping from cultivation, found along roadsides & streams, in pastures & fields, or persisting on abandoned farm sites; e GP; e SD s to MO & e KS; (widespread in e U.S.; Eurasia). *Introduced*.

This is a genus of 20 species distributed in Eurasia, particularly in Japan.

12. HYPOXIS L., Stargrass

1. *Hypoxis hirsuta* (L.) Cov., yellow stargrass. Hirsute perennial herbs up to 3 dm tall, from a subglobose to ellipsoid, membranaceous coated corm, 8-20 mm wide. Leaves 3-6, linear, 5-26 cm long, 1-6 mm wide. Inflorescence ascending to spreading, 2- to 14-flowered, pedicellate, scapes filiform. Flowers yellow, perianth segments 6-14 mm long, 2-6.5 mm wide, pubescent on outer surface; stamens 6, anthers 2-4 mm long, versatile, introrse, filament (1)2-3 mm long; ovary inferior, 3-locular, pilose. Capsule ellipsoid, somewhat pilose; seeds black, 1-1.5 mm in diam. (2n = 11) Apr-Jul. Moist to dry prairies & occasionally in open deciduous woods; e GP; (se Can., ME to MN & ND, s to GA & TX).

This is a large genus of 100 species widely distributed in the New World except for the northern regions.

13. LEUCOCRINUM Nutt., Mountain Lily

1. *Leucocrinum montanum* Nutt. Low growing, acaulescent perennial herbs to 15(20) cm tall, from a deeply buried rootstock. Leaves tufted in a sheathing base, linear, 10-20 cm long, 2-8 mm wide, slightly folded, margins membranaceous. Inflorescence an umbellike cluster, pedicels near ground level. Flowers white, perianth tube 2-4 × longer than limbs, each limb linear-lanceolate, 2-2.5 cm long; stamens 6, inserted near tube top, anthers appearing basifixed, introrse, 4-6 mm long; stigma 3-lobed, ovary superior, 3-locular. Capsule ± 3-angled, 5-7 mm long; seeds black, few. (2n = 22, 26, 28). Apr-Jun. In open coniferous woods & shortgrass prairies; w GP; w ND & e MT s to w NE, e WY & ne CO; (NE to OR, s to CO & CA).

Reference: Ornduff, R., & M. S. Cave. 1975. Geography of pollen and chromosomal heteromorphism in *Leucocrinum montanum* (Liliaceae). Madroño 23: 65-104.

There are two forms (not formally recognized) of this western North American endemic based on chromosome number and pollen type. Our plains material has pollen that is shed as monads and a chromosome number of 2n = 14, whereas to the west of our range plants have pollen that is shed as tetragonal tetrads and a chromosome of 2n = 22 and 26.

14. LILIUM L., Lily

Perennial simple stemmed plants from a scaly bulb. Leaves whorled, often uppermost or lowermost alternate. Flowers few, terminal, nodding or erect; perianth of 6 similar segments, spreading or recurved; stamens 6, anthers extrorse, versatile, filaments long; stigma 3-lobed, style elongate, ovary superior, 3-locular, ovules numerous. Fruit a capsule, subcylindric, 3-angled; seeds in 2 rows per locule, flat.

Reference: Boivin, B., & W. J. Cody. 1956. The variation of *Lilium canadense* Linnaeus. Rhodora 58: 14–20.

This large genus, prized for its horticultural value, contains some 80 species widely distributed in the Northern Hemisphere.

1 Flowers nodding; perianth recurved, spotted to middle or beyond .... 1. *L. canadense*
1 Flowers erect; perianth spreading open-campanulate, spotted at base (lower 1/3) .... 2. *L. philadelphicum*

1. ***Lilium canadense*** L., Turk's cap lily. Stout erect plants 0.7–2 m tall. Leaves whorled, uppermost alternate, lanceolate to elliptic, 5–15 cm long, 5–20 mm wide. Flowers 1–several, nodding from a long ascending or erect pedicels, orange to reddish-orange, purple spotted to middle or beyond; perianth segments strongly recurved, each lanceolate, 3–10 cm long; filament curving outward. Capsule 3–3.5 cm long. (2n = 12) Jun–Jul. Moist prairies & woods; e GP; n MN & SD s to MO & e KS; (e Can. to MN & SD, s to FL & KS). *L. michiganense* Farw.—Rydberg; *L. superbum* L.—Gleason & Cronquist.

Our plants have been called subsp. *michiganense* (Farw.) Boivin & Cody.

2. ***Lilium philadelphicum*** L. wild lily. Plants 3–9 dm tall. Leaves 3–6 in a whorl above, below alternate or opposite, linear to lanceolate, 4–7.5 cm long, 3–10 mm wide. Flowers 1–3(5), erect, spreading open-campanulate, deep red to reddish-orange, purple spotted at base (lower 1/3); perianth segments lanceolate, up to 5 cm long, base tapering to a long claw; filaments spreading, 3–5 cm long, anthers extrorse, 8–10 mm long. Capsule 3–3.5(4) cm long. (2n = 24) Jun–Aug. Moist to dry open woods & prairies; MN & ND s to IA & NE; (Que. to B.C., s to AR, TX, NM, & AZ). *L. umbellatum* Pursh—Rydberg.

Our plants can be assigned to var. *andinum* (Nutt.) Ker.

15. MAIANTHEMUM Weber

1. ***Maianthemum canadense*** Desf., wild lily-of-the-valley. Small perennial plants 8–20 cm tall, from a filiform rhizome; stems sparingly pubescent. Leaves 2(3), ovate to elliptical, (3.5)6(8) cm long, (1.2)2.5–4.5(5.5) cm wide, base cordate, shortly petioled or sessile. Inflorescence a raceme, 1.5–4 cm long. Flowers creamy white, perianth segments 4, ca 2 mm long, 1 mm wide; stamens 4, anthers introrse, filaments ca 2 × anther length; ovary superior, 2-locular. Fruit a red berry, ca 3 mm wide, 1- or 2-seeded. (2n = 36) Jun–Jul. Moist deciduous or coniferous woods; ne GP; BH of SD & WY; (se Can. to ND, s to NC, SD, & WY).

This is a small genus of 3 species distributed in the Northern Hemisphere. Our plants can be referred to var. *interius* Fern., distinguished by leaves that are pubescent beneath and having inconspicuous transverse veins.

16. MELANTHIUM L.

1. ***Melanthium virginicum*** L., bunchflower. Stout perennial plants up to 1–2 m tall, from a thick rootstock. Leaves linear to lanceolate, sheathing lower leaves 2–5 dm or more long, 1–2.5 cm wide. Inflorescence a large panicule, 2–4 dm long with lateral branches, villous-pubescent. Flowers cream color, turning green to light purple, lower flowers perfect, up-

per staminate; perianth ovate, oblong or obovate, 5–9 mm long, clawed, scurfy abaxially, base rounded to hastate, lobe 2–3 × claw length, 2 ovate, dark brown glands at base; stamens 6, attached to middle or at lobe-claw junction, slightly less than lobe length, anthers extrorse; style 3-cleft, ovary superior, 3-locular. Capsule ovoid to ellipsoid, furrowing between convex back, 1–1.8 cm long, 3-beaked; seeds whitish, obovate, 6–10 per locule, 5–8 mm long. Jun–Jul. Lowland prairies & moist open woods; MO & KS; (NY to IA & KS, s to FL & TX).

This is a small genus of two species, distributed in eastern North America. It has been placed within the genus *Zigadenus,* and whether there is any true relationship between them or not, it should be suspected of causing poisoning.

17. MUSCARI P. Mill.

1. *Muscari botryoides* (L.) P. Mill., grape-hyacinth. Scapose perennial plants, from a scaly bulb. Leaves linear-lanceolate to oblanceolate, up to 3 dm long and 2–10 mm wide, channeled. Inflorescence a cylindrical raceme on a scape. Flowers blue to violet, subtended by diverging bracts; perianth subglobose, fused, 3–6 mm long; stamens 6, anthers introrse; ovary superior, 3-locular. Capsule subglobose, angled; seeds 2 per locule, obovoid, rugulose, black. (2n = 18, 36, ca 40) Apr–May. Occasionally escaping from cultivation in our area; e MO; (N. Eng. to MN, s to VA & MO; Europe). *Introduced.*

This is a genus of some 60 species distributed in Europe, the Mediterranean, and western Asia.

18. NOTHOSCORDUM Kunth.

1. *Nothoscordum bivalve* (L.) Britt., false garlic. Scapose perennial herb 1–3.5 dm tall, from a truncate bulb ca 1.5 cm thick. Leaves 2–6, basal, linear, 10–25 cm long, 2–5 mm wide. Inflorescence a terminal umbel, from a many-nerved spathe. Flowers (3)–8(11), pale yellow, perianth of 6 similar segments, each with a single distinct nerve, oblong, 8–10 mm long, 2–4 mm wide; stamens 6, anthers introrse, versatile, filament slender, 2 × anther length; stigma slightly 3-parted or capitate, style slender, ovary superior, 3-locular. Fruit a capsule, 3–5 mm long; seeds black. (2n = 18) Apr–May (Oct). Prairies & open woods, occasionally in pastures, meadows, & ditches; MO & e KS, s to OK; (VA to IL & KS, s to FL & TX).

This is an American genus of 35 species, whose members often have been placed in the genus *Allium.*

19. ORNITHOGALUM L.

1. *Ornithogalum umbellatum* L., star of Bethlehem. Scapose perennial herbs 1.5–3 dm tall, from a membranaceous coated bulb. Leaves basal, linear, 2–5 mm wide when folded. Inflorescence a several-flowered raceme. Flowers erect, white, green nerve on back, perianth of 6 separate segments; stamens 6, anthers versatile, introrse, filament broad, flat; stigma clavate, style single, ovary superior, 3-locular, several ovules per locule. Fruit a capsule, 3-angled, seeds few per locule. (2n = 27, 28, 35, 36, 42, 44, 45, 46, 54) Apr–Jun. Escaping from cultivation & found along roadsides, fields, & occasionally grassy or wooded sites; (Newf. to Ont., s to N. Eng., NC, MS; Europe). *Introduced.*

This is a large Old World temperate genus of 150 species.

20. POLYGONATUM (Tourn.) P. Mill., Solomon's Seal

1. *Polygonatum biflorum* (Walt.) Ell. Perennial herbs up to 12 dm tall; from a rhizome; stems erect to slightly arching. Leaves simple, alternate, ovate-lanceolate to broadly oval, 7–16(20) cm long, 3–9 cm wide, green above, glaucous below, sessile to somewhat amplexicaulis. Inflorescence axillary, peduncle 1–4 cm long, (1)2–3(6)-flowered, pedicels subequal (3)5–6(11) mm long, shorter than peduncle. Flowers greenish white to cream, outer and inner perianth segments united into a cylindric tube with 6 lobes, perianth 12–17 mm long, lobes (2)3(4.5) mm long; stamens 6, inserted on perianth tube, anthers introrse, filaments smooth to papillose; stigma slightly 3-lobed, ovary superior, oblong to globose, 3-locular, ovules several per locule. Fruit a dark blue berry, 7–15 mm long; seeds several to many, 3–4 mm in diam. (2n = 20, 40) Apr–Jul. Moist deciduous woods; GP, less frequent in w; (e N. Amer.). *P. canaliculatum* (Muhl.) Pursh—Fernald; *P. commutatum* (Shultes) Dietr., *P. giganteum* Dietr.—Rydberg.

Reference: Ownbey, R. P. 1944. The liliaceous genus *Polygonatum* in North America. Ann. Missouri Bot. Gard. 31: 373–413.

This is a genus of some 50 species distributed in the north temp. regions.

21. SMILACINA Desf., False Solomon's Seal

Perennial plants from a whitish rhizome, stem erect, simple, from a basal membranaceous sheath; stem and leaves lightly pubescent. Leaves alternate, lanceolate to elliptical, entire, sessile and slightly clasping or short petioled, tapering at both ends, often finely puberulent. Inflorescence terminal, racemose or paniculate. Flowers white to cream, 6 similar perianth segments; stamens 6, each attached to base of perianth, anthers introrse; style 3, ovary superior, 3-locular, 2 ovules per locule. Fruit a berry, 1- to 6-seeded.

Reference: Galway, D. H. 1945. The North American species of *Smilacina*. Amer. Midl. Naturalist 33: 644–666.

This is a genus of 25 species distributed in North and Central America, East Asia, and the Himalayas.

1 Inflorescence paniculate; perianth segments 1–3(4) mm long .... 1. *S. racemosa*
1 Inflorescence racemose; perianth segments (3)4–7 mm long .... 2. *S. stellata*

1. *Smilacina racemosa* (L.) Desf., false spikenard. Plants 3.5–8 dm tall. Leaves 2-ranked, 6–15 cm long, 2–7 cm wide, apex acuminate, base obtuse to rounded, short petioled to sessile. Inflorescence a many-flowered panicle. Flowers white to whitish-green, perianth segments 1–3(4) mm long; stamens longer than perianth, 1.5–3 mm long, filament thick. Fruit a berry, 4–6 mm in diam, early stages red and green, later red (purple). (2n = 36) Apr–Jul. Rich moist thicks & coniferous or deciduous woods; e GP, BH, scattered in extreme w GP; (s Can., ME to AK, s to GA, GP, & CA).

2. *Smilacina stellata* (L.) Desf., spikenard. Plants 1.5–6.5 dm tall; glabrous to lightly pubescent. Leaves often 2-ranked, slightly folded, 4–12(16) cm long, 1–4 cm wide, lanceolate to oblong, apex obtuse to acuminate, sessile. Inflorescence a raceme, sessile or short peduncled. Flowers creamy white to greenish-white, perianth segments (3)4–7 mm long; stamens shorter than perianth, 2–5 mm long, filaments slender. Fruit a berry, 7–9 mm in diam, early stage light green with 6 blue longitudinal stripes, later turning dark blue. (2n = 36) May–Jun. Moist to dry coniferous or deciduous woods, meadows, frequent along streams & rivers; GP; MN to MT, s to n-cen OK; (s Can., ME to WA, s to NC, KS, & CA).

22. STREPTOPUS Michx. Twisted-stalk

1. *Streptopus amplexifolius* (L.) DC. Moderately large rhizomatous perennial herb, glabrous stems 4–10 mm tall. Leaves amplexicaul, ovate to ovate-lanceolate, 5–10 cm long,

2–6 cm wide, acuminate, entire to minutely toothed along margins. Inflorescence axillary, peduncle joined to pedicel and appearing bent, up to 3 cm long. Flowers 1(2), perianth greenish-white, (7)10(15) mm long, spreading to slightly recurved from middle; stamens unequal, outer series shorter than inner, anthers lanceolate, ca 3 mm long, basifixed, extrorse, filament shorter than anther; stigma entire or somewhat 3-lobed, ovary superior, 3-locular, ovules many. Fruit a red berry, ca 10 mm in diam; seeds many, 3 mm long. (2n = 16, 32) Jun–Jul. Moist coniferous-deciduous woods; BH; (s Can., MA to AK s to NC, SD, & AZ).

    Reference: Fassett, N. C. 1935. A study of *Streptopus*. Rhodora 37: 88–113.
    This is a genus of 10 species distributed in the Himalayas, Eurasia, and North America.

23. TRILLIUM L.

    Low perennial herbs from a short stout rhizome. Leaves 3 in a single terminal whorl; ovate to obovate, net venation. Flower single, peduncled or sessile, terminal; perianth segments differentiated, outer series of 3 green segments, persistent, inner series broader, of 3 white, yellow or purplish-brown segments, not persistent with age; stamens 6, anthers linear, basifixed, introrse, filament short; stigma 3, style slender, recurved, ovary superior, 3-lobed or angled, 3-locular. Fruit a berry, seeds several per locule.

    Reference: Freeman, J. D. 1975. Revision of *Trillium* subgenus *Phyllantherum* (Liliaceae). Brittonia 27: 1–62.
    This is a genus of some 30 species distributed in the west Himalayas, East Asia, and North America.

1 Flowers sessile.
  2 Stamens 1/2 as long as inner perianth; leaves often abruptly acute .... 4. *T. sessile*
  2 Stamens 1/4–1/3 as long as inner perianth; leaves often acute to acuminate .... 5. *T. viridescens*
1 Flowers peduncled.
  3 Leaves petioled; ovary 3-lobed, not winged .... 3. *T. nivale*
  3 Leaves sessile or leaf base petiolelike; ovary 6-angled or winged.
    4 Filaments more than 1/2 as long as anthers .... 1. *T. cernuum*
    4 Filaments 1/3–1/2 as long as anthers .... 2. *T. gleasoni*

1. *Trillium cernuum* L., nodding trillium. Plants 3–4(5) dm tall. Leaves broadly rhombic-obovate, 6–11 cm or more long, narrowing to a petiolelike leaf base. Flowers peduncled, 3–4 cm long, reflexed, below whorled leaves; outer perianth segments broadly lanceolate, 1.5–2 cm long, recurved, inner series oval to obovate, white, 1–3 cm long; anthers 4–5 mm long, equal to or somewhat longer than filaments; ovary 6-angled, pink to white. Fruit an ovoid berry. (2n = 10) Jun–Jul. Moist, rich deciduous woods; ne GP; MN & e ND, s to IA & ne SD; (VT w to Sask. & ND, s to GA & AL).

2. *Trillium gleasoni* Fern. Plants up to 4 dm tall. Leaves broadly rhombic, 7–15 cm long, somewhat wider than long, base tapered, not petioled. Flowers peduncled, horizontal to ± declined, 3–12 cm long; outer perianth segments lanceolate, 2.5–3.5 cm long, inner series lance-ovate to obtuse, white, spreading, 2–5 cm long; anthers 6–15 mm long, 2 × or more filament length, ovary 6-angled. (2n = 10) Apr–May. Moist, rich woods; e GP; MN: Clay; SD: Marshall, Roberts; (NY to MN & SD, s to MD, TN, & MO). *T. flexipes* Raf.—Fernald.

3. *Trillium nivale* Ridd., dwarf white trillium. Small plants up to 1.5 dm tall. Leaves ovate to elliptic, 2–4.5 cm long, base tapered to a short petiole, 3–5 mm long. Flowers erect to horizontal, 1–2.5(3) cm long, recurved in fruit; outer perianth segments oblong to lanceolate, shorter than inner series, 1.5-2 cm long, inner segments white, elliptic to oval, 2-4 cm long; anthers 4-6 mm long, just longer than filaments; ovary nearly globose, weakly 3-lobed. Mar-May. Moist, rich deciduous woods; e GP; MN & e SD, s to MO & e NE; (PA to MN, s to MO & NE).

4. **Trillium sessile** L., toadshade. Plants 1.5-2.5(3) dm tall, from a stout, thick rhizome. Leaves broadly ovate, 4-9 cm long, not mottled, apex cuspidate or inrolled and pointed, base rounded, sessile. Flowers sessile; outer perianth segments green, spreading or occasionally erect, broadly lanceolate, 1.5-3 cm long, inner segment series brownish-purple, broadly elliptical, 2-4 cm long, ascending, slightly longer than outer series; stamens 1/2 or less than inner perianth series, anthers 6-15 mm long, much longer than filament; ovary ovoid to globose, 6-angled. Apr-May. Rich deciduous wooded slopes & bottomlands; se GP; MO & e KS, s to OK; (PA to IL & KS, s to VA & AR).

5. **Trillium viridescens** Nutt. Stout plants 2-5 dm tall, from a thick ± erect rhizome. Leaves broadly lanceolate or more commonly ovate, 9-14 cm long, faintly mottled or green, occasionally pubescent beneath, apex acuminate, base sessile. Flower sessile, outer perianth segments spreading, green or tinged with purple, oblong-linear to narrowly lanceolate, 3.5-5.5 cm long, inner perianth segments yellowish-green above, purple below, oblanceolate or elliptic to nearly linear, 4.5-8 cm long, erect; stamens 1/3 inner perianth length, anthers 15-20 mm long; ovary ovoid, 6-angled. (2n = 10) Apr-May. Moist rich deciduous woods; se GP; KS: Cherokee; MO: Newton; OK: Ottawa; (sw MO & se KS to nw AR, OK, & ne TX). *T. viride* Beck.—Atlas GP.

Freeman (op. cit.) recognized a segregate species from *Trillium viride*, *T. viridescens* Nutt., which is the taxon found in our area. It differs in having broad ovate leaves that lack stomata on the adaxial surface, except at the tip, and stamens averaging 2-2.5 × carpel length. It might be best to treat our plants as a subspecies under *T. viride* Beck. based on the data given by Freeman.

24. UVULARIA L., Bellwort

Perennial herbs, from a rhizome with thickened fibrous roots. Stem simple or once branched above. Leaves alternate, perfoliate or simply sessile, narrowly oblong to oblong-ovate. Flowers solitary, terminal but appearing axillary at fruiting with stem continuing to grow; perianth yellow, slenderly campanulate, 6 similar segments, smooth, ± gibbous at base, base narrow with slightly fleshy nectariferous depressions; stamens 6, attached to receptacle shorter than perianth, extrorse, filaments ± flattened, shorter than anthers; style united at base, divided above, ovary superior, 3-locular, ovules 2-6 per locule. Fruit a 3-angled capsule; seeds 1-3 per locule, reddish-brown. *Oaksiella*—Rydberg.

Reference: Wilbur, R. L. 1963. A revision of the North American genus *Uvularia* (Liliaceae). Rhodora 65: 158-188.

This is an eastern North American genus of 5 species associated with mesic deciduous woods.

1 Leaves perfoliate, pubescent beneath; ovary and capsule sessile; rhizomes elongate .... 1. *U. grandiflora*
1 Leaves sessile, glabrous and glaucous beneath; ovary and capsule distinctly stalked; rhizome short and fleshy .... 2. *U. sessilifolia*

1. **Uvularia grandiflora** Sm., large bellwort. Plants glabrous except lower leaf surface, 3.5-6(7) dm tall, from an elongated rhizome. Leaves ovate-oblong to elliptic, 6-12 cm long, 2-6 cm wide, perfoliate, lowest subtending leaf up to 10 cm or more long, base broadly rounded, leaf margins smooth, scariously rimmed. Perianth segments 2.5-4.5 cm long, acute to acuminate, shallow nectariferous depression ca 2 mm long; anthers 10-15 mm

long, filament 3–5 mm long; style united 1/5 to 1/3 from base, style and stigma ca equal to stamen length or less; ovary oblong-cylindric, ± 3-lobed, each segment longitudinally grooved medially, ca 2.5 mm long, sessile. Capsule apex rounded to truncate, 3-lobed, each segment slightly grooved; seeds nearly globose or ± compressed, 3–5 mm in diam. (2n = 7) Apr–May. Moist deciduous woods; e GP; MN & ND, s to SD, MO, & KS; (Que. to ND, s to GA & OK).

2. **Uvularia sessilifolia** L., small bellwort. Plants glabrous, stems and lower leaf surface glaucous, 3.5–5 dm tall, from a fleshy short rhizome. Leaves narrow to broadly elliptic, 4–6.5 cm long, 1.5–3.0 cm wide, sessile, lower branch subtending leaf usually longest, up to 7 cm, apex acute to shortly acuminate, leaf margin scarious or denticulate, base tapered or broadly rounded. Perianth segments 1.5–2.4 cm long; nectariferous depression about 1 mm long; anthers 8–12 mm long, filament 2–3 mm long; style united 4/5 of distance to stigmatic tip, style and stigma ca 7 mm long, ovary 3-lobed, tapering from middle to both apex and base, each side ± concave, 3–5 mm long, stipe ca 1 mm long. Capsule 3-winged, concave; seeds globose, 3–4 mm in diam. (2n = 7) Apr–May. Moist deciduous woods; e GP; MN & e ND, s to se SD; (N.B. to ND, s to GA & MO). *Oakesiella sessilifolia* (L.) Sm.—Rydberg.

25. ZEPHYRANTHES W. Herbert

1. **Zephyranthes longifolia** Hemsl., zephyr-lily. Scapose perennial plants up to 3.5 dm tall, from an ovoid, scaly bulb. Leaves basal, grasslike, linear, 15–20 cm long, 1–2 mm wide. Flower single, bright yellow, from a spathe 2–3 cm long, pedicel shorter than spathe; perianth tube up to 10 cm long, lobes 1–2 cm long; stamens contained within lobes, anthers extrorse, ca 2.5 mm long, filaments ca 10 mm long, stigma distinctly 3-parted, ovary inferior, 3-locular. Capsule globose; seeds few to many per locule, compressed, black. Apr–Jul. Dry plains; n TX; (TX to AZ, s to Mex.).

This is an American and West Indies genus of 35–40 species.

26. ZIGADENUS Michx., Death Camass

Erect perennial herbs, from a deep truncated bulb. Leaves narrowly linear, sheathing stem, flat to folded, basal longest, cauline becoming reduced upward. Inflorescence a raceme, rarely a panicle, flowered pedicels erect to spreading, subtended by bracts. Flowers perfect to unisexual, rotate to widely campanulate, greenish-white, cream or yellow; perianth segments similar, attaching to ovary base or middle, ovate to broadly lanceolate, generally clawed, 1 or 2 dark yellow glands at base; style distinctly 3-parted, ovary superior or partially inferior, 3-locular, ovules numerous; stamens 6, hypogynous or perigynous, anthers basifixed, extrorse. Fruit a capsule, 3-lobed; seeds several, light to dark brown. *Toxicoscordion*—Rydberg.

Several species of *Zigadenus* are known to be poisonous. One should take care to learn the features of this genus, and to differentiate it from other liliaceous plants, particularly the onion genus *Allium* that is edible. *Zigadenus* contains some 15–20 species distributed in both North America and Asia. An orthographic variant of the generic name is *Zygadenus*.

1 Ovary partially inferior, stamens perigynous; gland at perianth base 2-lobed; petals not clawed .... 1. *Z. elegans*
1 Ovary superior, stamens hypogynous; gland at perianth base 1-lobed; petals clawed.
    2 Perianth segment 4–5(6) mm long, with a short claw (especially inner perianth series), 0.5–1 mm long; gland globose .... 3. *Z. venenosus*

2   Perianth segment 6-8 mm long, abruptly narrowed; clawlike, 1-1.5 mm long; gland obovate .... 2. *Z. nuttallii*

1. Zigadenus elegans Pursh, white camass. Stout plants 1-7 dm tall, from a fibrous coated bulb. Leaves 1-3.5 dm long, 0.2-1 cm wide, flat above, folded below, margins scabridulous. Inflorescence a raceme, rarely a panicle; pedicels 10-30 mm long. Flowers greenish-white, rotate; perianth segments oval to obovate, claw lacking, 6-8 mm long, gland 2-lobed, dark green; stamens perigynous; ovary partially inferior. Capsule 15-20 mm long, seeds 4-5 mm long. (2n = 32) Jun-Aug. Plains, prairies, & open coniferous woods; MN to MT, s to IA, NE, & CO; (Man. to AK, s to MN, IA, NE, NM, & AZ). *Anticlea elegans* (Pursh) Rydb., *A. chlorantha* (Richardson) Rydb.—Rydberg.

2. Zigadenus nuttallii A. Gray, death camass. Stout plants 3.5-7 dm tall, from a papery thin coated bulb, not fibrous. Leaves mostly basal, up to 5 dm long, up to 1 cm wide when folded, ± falcate. Inflorescence a raceme, rarely a panicle; pedicels 5-25 mm long. Flowers creamy white to yellow, segments ovate, abruptly narrowed, clawlike, (5)6-9 mm long, gland single, obovate; stamens hypogynous; ovary superior. Capsule 8-12 mm long; seeds few, 1.5-2 mm long. (2n = 32) Apr-Jun. Prairies, often rocky sites; e KS to OK; (TN to KS, s to TX). *Toxicoscordion acutum* Rydb., *T. nuttallii* (A. Gray) Rydb.—Rydberg.

3. Zigadenus venenosus S. Wats., death camass. Plants 1-3.5(4) dm tall, slender to moderately stout, from a fibrous bulb. Leaves linear, up to 3 dm long, 2-6 mm wide when folded, margins smooth or scabridulous. Inflorescence a ± dense raceme; pedicels 5-20 mm long. Flowers cream to white, campanulate; perianth segments ovate, 4-5(6) mm long, inner segments distinctly clawed 0.5-1.0 mm long, outer less so, gland single, globose; stamens hypogynous; ovary superior. Capsule 7-15 mm long, seeds 3-6 mm long. (2n = 22) May-Jul. Dry plains, occasionally in prairies, & open coniferous woods; ND & MT, s to NE & CO; (Sask. to B.C., s to NE, CO, & CA). *Toxicoscordion gramineum* Rydb.—Rydberg.

Our plants can be assigned to var. *gramineus* (Rydb.) Walsh ex Peck.

156. IRIDACEAE Juss., the Iris Family

by Steven P. Churchill

Perennial herbs with erect or slightly spreading stems, from a bulb, rhizome or fibrous roots. Leaves conduplicate, ensiform to linear, mostly 2-ranked, basal, cauline leaves few, reduced upward. Inflorescence terminal, a raceme or panicle. Flowers perfect, arising from spathelike bracts, 6 petaloid or modified segments in 2 series; stamens 3, opposite outer perianth segments, anthers extrorse, filament adnate to perianth segment; stigma 3, style single or 3-cleft, modified or not, ovary inferior, 3-locular, axile placentation. Fruit a capsule; seeds numerous.

This is a family of some 1800 species in nearly 70 widely distributed genera. Many are plants of horticultural value, including *Crocus, Gladiolus,* and *Iris.*

1   Plants rhizomatous; leaves broad, often more than 10 mm wide; flower over 3 cm across.
  2   Outer and inner perianth segments similar, orange with dark red-purple spots; seeds spherical .... 1. *Belamcanda*
  2   Outer and inner perianth segments dissimilar, blue, violet, white etc. but not orange with spots .... 2. *Iris*
1   Plants with either fibrous roots or producing bulbs, not rhizomatous; leaves narrow, mostly less than 10 mm wide; flowers 2 cm or less across.

3 Plants from a bulb; leaves much longer than flowers .... 3. *Nemastylis*
3 Plants from fibrous roots; leaves shorter than flowers .... 4. *Sisyrinchium*

1. BELAMCANDA Adans., Blackberry Lily

1. ***Belamcanda chinensis*** (L.) DC. Rhizomatous herb, 4–10 dm tall. Leaves 25–40 cm long, 1.7–3(3.5) cm wide. Inflorescence cymose-paniculate, branched. Flowers 4–12, orange with red-purple spots; perianth segments equal to subequal, 2.5–3 cm long, broadly elliptic; anthers 5–8 mm long, filaments 8–11 mm long, attached to base of perianth segments; stigma clavate, style 3-cleft. Capsule oblong to pyriform, 3-lobed, each valve recurving to expose seeds; seeds black and fleshy, 4–6 mm in diam. (2n = 32) Jul–Aug. Escaping from or persisting after cultivation, along roadsides, fields, & wooded margins; w IA & se SD, s to MO & e KS; (e U.S. to GP; e Asia). *Naturalized.*

2. IRIS L., Iris, Flag

Rhizomatous perennial herbs with erect flowering stalks. Flowers showy, purple to blue or yellow; perianth of 6 clawed segments, outer segment series ("the fall") spreading and often recurved, bearded, inner series ("the standard") erect or arching; stamens attached at base of outer segment series, anthers linear to oblong, filaments longer than anthers; style united with outer and inner perianth series into a tube, divided distally into 3 petaloid appendages that over-top stamens, tip 2-lobed, stigma beneath the 2-lobed appendage, ovary 3- to 6-angled. Capsule coriaceous or chartaceous, indehiscent or loculicidal; seeds in 1 or 2 rows per locule.

References: Anderson, E. 1936. The species problem in *Iris.* Ann. Missouri Bot. Gard. 23: 457–509; Dykes, W. R. 1913. The Genus *Iris.* Cambridge Univ. Press, 245 pp.; Foster, R. C. 1937. A cyto-taxonomic survey of the North American species of *Iris.* Contr. Gray Herb. 119: 3–82.

This is a genus of some 200 species distributed mainly in the Northern Hemisphere. Several species are planted in our area as garden ornamentals and occasionally escape from cultivation and become established along roadside ditches, or persist after cultivation. Iris is known to cause poisoning in livestock and humans.

1 Perianth yellow, tube not constricted above ovary .... 1. *I. pseudoacorus*
1 Perianth pale to dark blue or violet, tube constricted above ovary.
  2 Leaves narrow, 4–8 mm wide; perianth segments bilobed .... 2. *I. missouriensis*
  2 Leaves broad, 10–30 mm wide; perianth segments entire.
    3 Ovary and capsule 6-angled .... 1. *I. brevicaulis*
    3 Ovary and capsule 3-angled.
      4 Base of tufted leaves purplish; spathe papery or scarious, 3–6 cm long; capsule inner surface lustrous, seeds also lustrous with finely and regularly pitted coat .... 4. *I. versicolor*
      4 Base of tufted leaves buff or pale brown; spathe firm, to 14 cm long; capsule often asymmetrical, inner surface dull, seeds irregularly deep-pitted with a brittle corky coat .... 5. *I. virginica*

1. ***Iris brevicaulis*** Raf., blue flag, lamance iris. Stem distinctly zigzag with 3–6 cauline leaves. Leaves deep green to lightly glaucous, broadly ensiform, 3–6 dm long, 1.5–3 cm wide. Flower single from cauline leaf axil, ca 2 from terminal cluster; spathe lanceolate, acute, outer bract green, inner partly scarious; pedicel short; outer perianth segments ca 7.5–9.5 cm long, 2.5–3 cm wide, inner blade ovate, deep blue to purple, claw cuneate, light yellowish-green with a distinct yellow midrib and yellow to white blotch at union of claw and blade, ± pubescent; inner segment series oblanceolate, light to dark blue; stigma

bilobed, style branches green, up to 4 cm long, style crests subquadrate, up to 1.5 cm long; anthers ca 1.5 cm long, longer than filaments. Capsule ovoid to ellipsoid, 6-angled, 3–5 cm long; seeds large, irregularly circular, coat thick, light brown. (2n = 44) May–Jun. Moist sites at wooded margins, marshes along ponds & ditches; KS: Leavenworth, Wyandotte; MO: Jackson; (OH to KS, s to KY & LA).

2. *Iris missouriensis* Nutt., western blue flag. Plants 2.5–6 dm tall. Leaves light green, glaucous, linear, (18)25–30(45) cm long, 4–8 mm wide, previous year's leaves retained and clothing rhizome. Spathe membranaceous to scarious, 5–7.5 cm long, outer bract often shorter than inner. Flowers (1)2, pale blue; perianth tube constricted above ovary, outer segments 4–6 cm long, blade obovate, lilac to purple with yellowish-white blotch at base, claw yellow to white; anthers 1.2–2 cm long, equal to filament length; stigma bilobed, style crest 8 mm long, ovary 6-angled. Capsule oblong, 6-angled, 3–5 cm long; seeds dark brown, subglobular to pyriform, 4–4.5 mm long. (2n = 38, 48) Jun–Jul. Marshes, seepy meadows, & long streams in open woods or margins of woods, rarely in prairies; s-cen ND, BH, & WY, w NE & CO; (w Can., NE, ND to WA, s to NM & CA).

3. *Iris pseudacorus* L., yellow iris. Stem up to 10 dm tall, from a stout rhizome. Leaves stiff, erect to arching. Spathe equal to subequal in length, outer bract sharply keeled, inner rounded. Flowers yellow; perianth tube not constricted above ovary, 3 outer segments 4.5–8 cm long, arching, yellow with brown markings toward base, blade ovate, claw broad with involute edges, 3 inner segments shorter than outer, yellow, linear, obtuse, 2.5–3 cm long, anthers shorter than filaments; stigma rounded, style keeled, ovary 3-angled. Capsule cylindric-prismatic to ellipsoid, beak 5 mm long, 3-angled, 5–8 cm long; seeds suborbicular, 6–7 mm long. (2n = 34) May–Jul. Escaping from cultivation, found in marshes & margins of ponds & streams; e NE s to e KS; (widely planted in U.S.; native to the Old World temp. region). *Naturalized.*

4. *Iris versicolor* L., blue flag. Plants up to 9.5 dm tall, from a thick rhizome; stems long and simple or branching, up to 6 dm tall. Cauline leaves, 1 or 2, 2–6 dm long, 1–3 cm wide. Pedicel 2–8 cm long, usually exceeding spathes, spathe unequal, outer bract shorter than inner. Flowers blue to purplish-blue; perianth tube constricted above ovary, ca 1–1.2 cm long, outer segment series ovate, blade blue-purple, claw yellowish-green, inner segments similar to outer in color, lanceolate-oblong, 2–4.5 cm long, 1 cm wide; anthers 0.8–1.5 cm long, filaments exceeding anthers; stigma entire, style crest up to 1.5 cm long, ovary narrow, 3-angled, 3–5 cm long. Capsule ovoid to oblong-ellipsoid, 3–5 cm long, 3-angled; seeds shiny dark brown. (2n = 108) Jun–Jul. Marsh or swampy sites; MN: Becker, Cottonwood, Kittson, Pennington, Polk; NE: Sarpy; (se Can., ME to MN, s to VA & NE).

5. *Iris virginica* L., blue flag. Plants up to 7.5 dm tall; stems simple to occasionally branched. Leaves linear-ensiform, grayish-green to bright green, prominent ribs several, 40–52 cm long, 1–3 cm wide. Spathe unequal, to 14 cm long, outer bract shorter than inner, herbaceous, not inflated. Flowers pale blue to purple; perianth tube constricted above ovary, 1–2 cm long, outer segment series 6–7 cm long, 2.5–3 cm wide, blade obovate-ovate, lavender blue, claw broad, 1.5 cm long, yellowish-green, yellow midrib pubescent at base of blade, inner segment series obovate to spatulate, often emarginate, 5–6 cm long, color similar to outer series; anthers 1–1.5 cm long, filament ca 1.5 cm long; stigma 3-angled, style branches up to 3.5 cm long, style crest reflexed and toothed, up to 1.5 cm long, ovary 3-angled. Capsule ovoid to ellipsoid, 3-angled; seeds round or irregular, pitting irregular on coat. (2n = 72) May–Jun. Flood plains along streams, ditches, & pond margins; MN & e NE, s to MO & OK; (se Can., MI, & OH to MN, s to VA, FL, & TX). *I. shrevei* Small—Atlas GP.

Our plants can be assigned to *Iris virginica* var. *shrevei* (Small) Anderson.

3. NEMASTYLIS Nutt., Celestial Lily

1. ***Nemastylis geminiflora*** Nutt., prairie iris. Perennial herb 3–5 dm tall, from a globose bulb, 1.5–3 cm wide. Leaves 2 or 3, linear to ensiform, (13)22–45 cm long, 3–10 mm wide. Spathe subequal, outer bract 30–40(48) mm long, inner (36)40(47) mm long. Flowers 2 or 3, blue, fugacious; perianth of 6 similar segments, 20–25 mm long, 15–20 mm wide; stamens opposite inner perianth segments, anthers linear, 6–13 mm wide, filaments 2–3 mm long, united at base; style 3, each divided for ca 6 mm. Capsule ovoid, truncate, 2–2.5 cm long, 1–1.5 cm wide, apically dehiscent; seeds rusty red, 3–4 mm long. Apr–Jun. Moist to dry prairies; MO & e KS, s to e OK; (TN & MO to KS, s to GA & TX). *N. acuta* (Bart.) W. Herbert — Rydberg.

Reference: Goldblatt, P. 1975. Revision of the bulbous Iridaceae of North America. Brittonia 27: 373–385.

This is a genus of 5 species distributed in south central United States and Mexico.

4. SISYRINCHIUM L., Blue-eyed Grass

Cespitose perennial grasslike herbs, from fibrous roots. Stem narrow, spreading to erect, winged or not. Leaves linear, erect, equal to or shorter than flowering stems. Inflorescence a subumbellate cluster, from a spathe of equal or unequal bracts, spathe sessile or pedunculate. Flowers pale to dark blue or white, perianth of 6 similar segments, rotate, segments oblong, apex slightly cleft to 2-lobed with a cuspidate or aristate tip; stamens 3, filament partially adnate at tip; style branches filiform and alternating with stamens, ovary inferior, 3-locular, ovules several per locule. Capsule globose, dehiscent, somewhat 3-angled; seeds black, globose to subglobose, smooth or slightly pitted.

References: Bicknell, E. P. 1899. Studies in *Sisyrinchium* III. *S. angustifolium* and some related species, new and old. Bull. Torrey Bot. Club 26: 335–349; Bicknell, E. P. 1901. Studies in *Sisyrinchium* — IX. The species of Texas and the Southwest. Bull. Torrey Bot. Club 28: 570–592; Henderson, D. M. 1976. A biosystematic study of Pacific Northwestern blue-eyed grasses (*Sisyrinchium*, Iridaceae). Brittonia 28: 149–176; Mosquin, T. 1970. Chromosome numbers and a proposal for classification in *Sisyrinchium* (Iridaceae). Madroño 20: 269–275; Oliver, R. L. 1970. *Sisyrinchium*, pp. 425–428, *in* D. S. Correll & M. C. Johnston, Manual of the Vascular Plants of Texas. Texas Research Foundation, Renner.

This is a genus of possibly 100 species distributed primarily throughout the n. temp. New World. It is a very problematic genus, each regional or state floristic treatment presenting a different concept of individual species and each author interpreting the work of Bicknell in a slightly different manner. Studies by E. P. Bicknell and T. Mosquinn offer very opposite and somewhat extreme views concerning the species of *Sisyrinchium*, but Mosquin has raised some important issues concerning the various recognized taxa. This present treatment separates phenotypes only, mainly those that have been generally accepted, and therefore will simply perpetuate the existing problems. The only solution to our taxa in the GP, as elsewhere, is an extensive populational and biosystematic study with a reassessment of morphological features throughout the *entire* range of the genus.

1 Spathes sessile and solitary, if pedunculate then some flowering plant spathes sessile.
  2 Outer bract margin free at base; flowers pale blue to white, seldom dark blue; spathes always sessile .... 2. *S. campestre*
  2 Outer bract margin fused; flowers dark blue; spathes sessile or occasionally pedunculate.
    3 Stem wings absent or nearly so; outer bract united for 1–2 mm; spathes always sessile .... 6. *S. mucronatum*
    3 Stem wings present and distinct; outer bract united for 2–4 mm; spathes sessile or pedunculate, populations often mixed .... 5. *S. montanum*

1   Spathes pedunculate, 1–several, from axil of leaflike bract.
   4   Leaves and stem scabrellous on margins; pedicels slenderly flexuous; ovary
glabrous .... 7. *S. pruinosum*
   4   Leaves and stem smooth, not scabrellous on margins; pedicels usually erect spreading, not flexuous, if flexuous then ovary puberulent.
      5   Stem wings broad, usually (1)3–5 mm or more wide, each wing as wide or wider than stem proper .... 1. *S. augustifolium*
      5   Stem wings narrow, usually less than 2.5 mm wide, each wing not as wide as stem proper.
         6   Plants erect; pedicels upright, just exceeding the spathe; ovary puberulent .... 3. *S. demissum*
         6   Plants spreading; pedicels spreading to pendent, ca 5 mm or more longer than spathe; ovary glandular-puberulent .... 4. *S. ensigerum*

1. *Sisyrinchium angustifolium* P. Mill., blue-eyed grass. Plants bright green, drying dull dark green to black, 3–4.5 dm tall. Stems spreading to generally erect, broadly winged, 3–5 mm wide. Leaves not exceeding stem, 3–6 mm wide. Peduncules 1 or 2(3); spathe long pedunculate from a leafy bract 4–10 cm long, (2.5)3–5 mm wide, outer spathe bract just exceeding or 1/3 longer than inner. Flowers pale blue to violet; pedicels erect to more commonly spreading, equaling spathe length; ovary and young fruit pubescent. Capsule subglobose, 5–7 mm long, light dull to dark brown or nearly black. (2n = 96) May–Jun. Open or marginal deciduous woods, & in moist prairies; MO & e KS s to OK & TX; (se Can. w to ID s to FL & TX). *S. graminoides* Bickn.—Rydberg, Gleason.

Various authors have questioned the distinction between *S. angustifolium* P. Mill. and *S. montanum* Greene. Both species produce pedunculate spathes in our area, although this is not consistent with *S. montanum*. Past reports (Atlas GP) of *S. angustifolium* from NE & WY northward are based on pedunculate *S. montanum*.

2. *Sisyrinchium campestre* Bickn., white-eyed grass. Plants glaucous, (1)2–3(4) dm tall, stems erect, 1–2 mm wide, winged. Leaves ca 1/2–2/3 stem length, 1.2–2 mm wide. Spathe sessile, outer bract 2.5–4.5 cm long, margin free at base, inner bract 1/2 outer bract length, base gibbous. Flowers pale blue to white, seldom dark blue; ovary glandular-pubescent; pedicels exserted just beyond spathe. Capsule 3–5 mm long. (2n = 32) Apr–Jun. Prairies & open woods; e GP; MN & e SD, s to MO & OK; (WI to Man., s to LA & TX). *S. kansanum* (Bickn.) Alex.—Gleason; *S. campestre* var. *kansanum* Bickn.—Atlas GP.

Several morphotypes have been recognized at the level of species, variety, and form based on whether spathes are glabrous or scabridulous, and whether flower color is blue or white. These various features can be found in nearly all possible combinations within a single population, at least in Nebraska and Kansas. Therefore, no formal categorical rank is recognized.

3. *Sisyrinchium demissum* Greene. Plants pale green to glaucous, 2–7 dm tall; stems slenderly erect to flexuously curved. Leaves 1/2 or more stem length, 1–2.5 mm wide, edge smooth except upper 1/4 denticulate. Spathe pedunculate, narrow, bracts usually subequal, 1.2–2.5 cm long, outer bract united at base for 3–8 mm. Flowers pale blue-violet; pedicels stiffly erect, barely exceeding bracts; ovary puberulent. Capsule 4–5 mm long. Jun–Aug. In prairies & meadows; s GP; TX; (CO w to OR, s to TX & CA; Mex.).

4. *Sisyrinchium ensigerum* Bickn. Plants glaucous green, 1.5–3(4) dm tall; stem erect to divergently curved, 1.5–3.5 mm wide, winged, margin rough-serrulate. Leaves ca 1/2 stem length, 1.5–4 mm wide. Spathe pedunculate, margin rough-serrulate, spathe 2–2.5 cm long, bracts subequal or outer longer, slender tips often recurved, outer bract fused for 2–4 mm at base, keeled, serrate. Flowers blue to violet; ovary glandular-puberulent; pedicels 5 mm long or more than bracts, spreading to pendent, occasionally erect. Capsule

4–7 mm long. (2n = 32) Apr–May. Prairies & plains; s GP; OK s to TX; (OK s to TX & Mex.).

5. Sisyrinchium montanum Greene. Plants green to somewhat glaucescent, (1)2–3(4.5) dm tall; stem (1.2)2(3) mm wide, winged, wing width less than stem proper. Leaves 1/2 stem length, (1.5)2–2.5(3) mm wide. Spathe sessile or pedunculate, outer spathe bract 2–7 cm long, 1– 2 × inner bract length, united 2–4 mm above base. Flowers violet or dark blue; pedicels erect, shorter than, equal to or just exceeding inner bract; ovary glandular-pubescent. Capsule pale papery brown, 4–6 mm long. (2n = 96) May–Jul. Prairies & plains, meadows & open coniferous woods; cen & n GP; MN to MT, s to CO, NE, & w KS; (s Can. s to NY & IN, w to KS, CO, & AZ).

Our plants exhibit both sessile and pedunculate spathes within populations and frequently even in the same tuft of plants. Various authors have also noted this phenomenon, including Henderson (op. cit.) and Voss (Michigan Flora, Part 1, Gymnosperms and Monocots, Bull. Cranbrook Inst. Sci. No. 55. 1972.) but not in as great a frequency as found in the GP. Further thorough sampling of populations should aid in determining the extent of these morphotypes. See comments under *S. angustifolium* P. Mill. These two species, *S. montanum* and *S. angustifolium*, are two distinct taxa and should not be placed in synonymy.

6. Sisyrinchium mucronatum Michx. Plants dull green to ± glaucous, 1.5–2(3) dm tall; stems slenderly erect, 0.5–1(1.5) mm wide, wings absent or, if present, then not as wide as stem proper. Leaves not exceeding stem, 0.5–1 mm wide. Spathe sessile, erect, slightly united at base for 1–2 mm, rarely free, inner bract 1/2 or less than the outer bract length. Flowers deep dark blue; pedicels exceeding inner but not outer spathe bract. Capsule 2–4 mm long. (2n = 30, 32) May–Jun. Prairie meadows & open mesic woods; ne GP; MN & ND; (ME to ND, s to NC & WI).

The presence of this species in the GP area is somewhat questionable, except for w MN.

7. Sisyrinchium pruinosum Bickn. Plants light green, glaucescent, 1–2(3) dm tall; stems slender, ascending to loosely erect, 1–2.5 mm wide, wings generally distinct. Leaves just over 1/2 stem length, 1–3.5 mm wide. Spathe peduncules 2 or 3, slender, ascending, rough-scabrous, 1/2 to as long as stem length, spathe narrow, bracts subequal, outer spathe bract up to 30 mm long, base united for 3–5 mm or more from base. Flowers violet-blue to purple; pedicels slenderly flexuous, exserted; ovary glabrous. Capsule 3–5 mm long. (2n = 32) Apr–May. Prairies, rarely open woods; s GP; s KS to n TX; (AR & KS s to TX).

The southern GP pedunculate species that include *S. demissum*, *S. ensigerum*, and *S. pruinosum* form a poorly understood complex and are in need of further detailed studies.

157. AGAVACEAE Endl., the Agave Family

by Steven P. Churchill

Our plants subsucculent or xerophytic, evergreen perennials, acaulescent or caulescent, from a semiwoody erect caudex, forming a dense clustered rosette of numerous leaves. Leaves alternate, long linear and grasslike to elongate-lanceolate linear and thickly fibrous or occasionally fleshy; apex tapering and either short pointed or spine-tipped, margin fibrous to spiny. Inflorescence an erect raceme or panicle. Flowers polygamo-dioecious to perfect, cream to white and often somewhat greenish; perianth of 6 similar segments, free, fleshy to papery; stamens 6, included, anthers dorsifixed, introrse, filament base narrow to broad;

stigma 3-lobed, style short, ovary superior, 3-locular or appearing 6-locular. Fruit a dry or fleshy capsule, seeds globose to flat.

This is a family of 18 genera and about 600 species found in dry vegetational regions of the world.

1 Flowers 0.5 cm long; leaves grasslike, not fleshy, lacking a spine tip .... 1. *Nolina*
1 Flowers 2 cm or more long; leaves thick and ± fleshy, spine-tipped .... 2. *Yucca*

1. NOLINA Michx.

1. *Nolina texana* S. Wats., sachuista. Polygamo-dioecious plants 3–6 dm tall, from a woody caudex, trunk forming at surface. Leaves many, basally clustered, spreading to procumbent, linear, 6–12 dm long, 2–5 mm wide, ± triangular toward apex, margin sporadically toothed or smooth. Inflorescence a dense compound racemose panicle, main branches subtended by leaflike bract, base broadly membranaceous; pedicels jointed near base, 4–6 mm long. Flowers white to cream or ± greenish; perianth of 6 similar elliptical, concave segments, spreading to slightly reflexed, 3–3.5 mm long, each 1-nerved; stamens 6, mostly abortive in fertile flowers; style short, recurved, ovary superior, deeply 3-lobed, 2 ovules per locule. Fruit a capsule, ± inflated, 3 mm long, 4–5 mm wide; seed 1 per locule, globose, ± smooth, 2–4 mm in diam. May–Jun. Dry, rocky soil in grasslands; NM: Harding, Quay; OK: Cimarron; (OK & NM s to TX & Mex.).

Reference: Trelease, W. 1911. The desert group Nolineae. Proc. Amer. Philos. Soc. 50: 404–443.

This is a genus of some 25 species distributed throughout the arid region of southwest North America.

2. YUCCA L., Soapweed, Yucca

Acaulescent semiwoody plants from a thick underground caudex. Leaves radiating from a basal rosette, linear to narrowly lanceolate, thick and ± fleshy or thin, fibrous, tip often spiny, margins filiferous, base broad and clasping. Inflorescence a raceme or panicle. Flowers perfect, large, campanulate to globose, perianth of 6 similar segments, oval to lanceolate, outer 3 segments narrower than inner series, greenish-white to cream; anthers sagittate, filament glandular-pubescent, base broad; stigma 3-parted, style short and broad, ovary 3-locular, ovules numerous. Fruit a ± 6-sided dry or fleshy capsule; seeds flat and black, D-shaped.

References: McKelvey, S. D. 1938 & 1947. Yuccas of the southwestern United States. Part I, pp. 1–150, & Part II, pp. 1–192. Arnold Arboretum, Harvard Univ.; Reveal, J. 1977. *Yucca*, pp. 527–536, *in* Cronquist, A. et al. Intermountain Flora, Vol. 6. Columbia Univ. Press.; Trelease, W. 1902. The Yucceae. Report Missouri Bot. Gard. 13: 27–133; Webber, J. M. 1953. Yuccas of the southwest. Agric. Monogr. U.S.D.A. 17: 1–97.

This is a North American genus of 30–40 species, having their greatest diversity in the southwest. *Yucca* had many uses among the Plains Indians. The roots were used in making soap, the leaves were used for fiber, and the spiny tip was used as a needle — often with the fibers still attached serving as the thread.

1 Fruit indehiscent; perianth segments often 7 cm or more long .... 1. *Y. baccata*
1 Fruit dehiscent; perianth segments usually 6 cm or less long.
  2 Leaf blade short, lanceolate, 1.5–4.5 dm long, in cross section concavo-convex; fruit constricted distally, valve apex recurved to spreading .... 3. *Y. harrimanae*
  2 Leaf blade longer, linear to linear-lanceolate, over 4 dm long; fruit not constricted distally, valve apex erect, not spreading.
    3 Inflorescence a panicle, exserted above the leaf rosette; blades 2–4 cm wide .... 4. *Y. smalliana*
    3 Inflorescence a raceme (rarely branched), subequal to slightly exserted above leaf rosette; blade up to 1.5 cm wide .... 2. *Y. glauca*

1. **Yucca baccata** Torr., datil. Plants simple or clumped, rarely caulescent. Leaves flat or slightly incurved, deeply concavo-convex, 3–7 dm long, 3–5.5 cm wide, margin with short coarse recurved fibers or long curly fibers. Inflorescence a panicle, contained mostly among the leaves. Flowers pendent, white to cream, tinged purple; perianth segments lanceolate to oblanceolate, outer series (4)7(10) cm long,1.5–2(2.5) cm wide, inner series (4)6.5(9.5) cm long, (1.8)2.3(3) cm wide; filament 1/2 or more length of perianth; style 4–10 mm long, ovary ca 5 cm long. Fruit (7)12–18(23) cm long; seeds 7–11 mm in diam, 1–2 mm thick. Apr–Jun. Dry plains & hilly grasslands; sw GP; CO: Baca; NM: Union; OK: Cimarron; (CO to s CA, s to TX, NM, & AZ; Mex.).

2. **Yucca glauca** Nutt. Plants simple or often clumped. Leaves glaucous green, linear, plano-convex, flat to ± inrolled, 4–7 dm long, 0.5–1.5 cm wide; margin greenish-white and becoming finely filiferous. Inflorescence a raceme, seldom with an abortive branch, somewhat exserted above leaves except lower most part; subtending bract membranaceous. Flowers globose to campanulate, greenish-white, tinged with purple; perianth segments thick, outer series 3.5–5.5 cm long, 2.5–3 cm wide, acute; filament 10–20 mm long; style 8–12 mm long and nearly as broad, ovary 3–3.5 cm long. Capsule 4.5–6 cm long, oblong-cylindrical, seeds 7–10 mm in diam with a marginal wing. (2n = 52) May–Jul. Plains & prairies, occasionally in open coniferous woods, often in well-drained sites; GP except ne & se; (ND & MT, s to MO, TX, & NM). *Y. arkansana* Trel.—Atlas GP.

A southern form of *Yucca glauca* in the Great Plains (OK to AR, s to TX) is var. *mollis* Engelm. (= *Y. arkansana*) characterized by its broader leaves (up to 1.5 cm wide) that are flattened and flexible, whereas the typical var. *glauca* has stiff leaves ca 1 cm wide. Reveal (op. cit.) has noted the distribution of *Yucca baileyi* Wooton & Standley for the Front Range of CO and this may occur in our area. The distinctions are not great: *Y. glauca* having a dark green style that is tumid and *Y. baileyi* having a white to fairly dark green style that is narrow to slightly swollen.

3. **Yucca harrimaniae** Trel., sensu Reveal. Plants forming open clumps, often cespitose. Leaves dark green, ± glaucous, linear to lanceolate, concavo-convex, ± thin, flexible, (1.2)2(4.5) dm long, (0.7)1.4(2.0) cm wide, margin white, occasionally with fine fibers. Inflorescence a raceme 4.5–7 dm long, scape 1.2–4 dm long. Flowers broadly campanulate, pendent, white to greenish-white, tinged with purple; perianth segments obtuse, 4–6 cm long, inner segments usually wider than outer; filament ca 20 mm long, hirsute; style ± swollen toward middle, pale green to white. Capsule 3–4 cm long, cylindrical, beak short attenuate; seeds thin, very narrow marginal wing. May–Jun. High plains grassland to open conifer woods; se CO to nw OK & ne NM; (CO to UT, s to OK & AZ). *Y. neomexicana* Woot. & Standl.—Atlas GP.

The treatment follows the concept presented by Reveal (op. cit.) and describes the var. *neomexicana* (Woot. & Standl.) Reveal, the more eastern phase of *Y. harrimanae*.

4. **Yucca smalliana** Fern. Leaves green and slightly glaucous, linear-lanceolate, ± thick, flat, 5–10 dm long, 2–4 cm wide, margin smooth or scabrous, often with coarse curly fibers, tip ending in a sharp spine. Inflorescence a panicle, pubescent, 1.5–2(3) m long. Flowers creamy white, somewhat greenish; perianth segments acuminate, (3)5(7.5) cm long; style white, 2–5 mm long, slightly swollen; filament pruinose-pilose. Capsule 4–6 cm long, 2–4 cm wide; seeds 5–7 mm in diam. May–Jul. Escaping from cultivation in our area & found along roadsides, thickets, etc.; se NE s to e KS; (NC to TN, s to FL & LA). *Y. filamentosa* L. (in part)—Gleason & Cronquist.

This is an attractive evergreen perennial commonly found in cultivation in the southern half of the GP. There is also a variegated form of this species.

158. SMILACACEAE Vent., the Catbrier Family

by Steven P. Churchill

1. SMILAX L., Greenbrier, Catbrier

Perennial dioecious herbs or shrubby vines generally climbing by tendrils. Leaves alternate, simple, broad and net veined, leaves often 1/4–1/3 normal size at flowering. Inflorescence an umbel, on axillary peduncle; peduncles longer than subtending petioles. Flowers unisexual, small, yellow-green, the staminate often larger, perianth segments similar, spreading; stamens 6, anthers introrse, basifixed, oblong, filament linear, flat, equal to shorter than anther length, in pistillate flowers reduced to 1–6 filiform staminodes; stigma nearly sessile, thick and ± spreading; ovary superior, 3-locular, 1 or 2 ovules per locule. Fruit a small berry with up to 6 seeds.

References: Mangaly, J. K. 1968. A cytotaxonomic study of the herbaceous species of *Smilax*: section *Coprosmanthus*. Rhodora 70: 55–82, 247–273; Stephens, H. A. 1973. *Smilax*, pp. 20–23, *in* Woody Plants of the North Central Plains. Univ. Press, Kansas.

This is a genus of some 350 species distributed in the tropics and temperate regions of the world. *Smilax*, the largest genus, is one of four genera assigned to the Smilacaceae.

1 Plants herbaceous; stems annual, lacking prickles; ovules generally 2 in each locule.
   2 Peduncles borne below stem leaves, from an axil of leafless bract; plants lacking tendrils or with poorly developed upper tendrils; plants up to 1 m tall .... 2. *S. ecirrhata*
   2 Peduncles borne from axil of stem leaves; plant tendrils present; plants 1–3 m tall .... 3. *S. herbacea*
1 Plants woody; stems perennial, prickles present; ovules single in each locule.
   3 Prickles green, stout and notably flattened; branches angular; leaf margins distinctly thickened .... 1. *S. bona-nox*
   3 Prickles black, slender and slightly flattened; branches terete or slightly angular; leaf margins thin .... 4. *S. hispida*

1. **Smilax bona-nox** L., greenbrier. Vines up to 8 m long, tendrils attaching to shrubs and trees; stems generally 4-angled, angles raised, prickles stout, flattened, and scattered, green. Leaves hastate, deltoid or ovate, blade 3–10(12) cm long, 3–8(10) cm wide, 3–5 main veins arising from petiole, apex cuspidate, margins distinctly thickened, base truncate or broadly rounded. Flowering peduncles slender, flattened. Flowers 15–45, sepals and petals green, those of staminate flowers 4–5 mm long, those of pistillate flowers smaller. Fruit a black berry; seed single, ellipsoid, 4–5 mm in diam. (2n = 16) Apr–May. Deciduous woods, thickets, disturbed wooded sites; se KS to e OK; (MD to IL & KS, s to FL & TX; Mex.).

2. **Smilax ecirrhata** (Engelm.) S. Wats., greenbrier. Herbaceous plant up to 8 dm tall, lacking tendrils or, if present, then poorly developed at upper end. Leaves narrow to broadly ovate, blade 4.5–21 cm long, 3–9.5 cm wide, apex cuspidate, base cordate to truncate, pubescent beneath. Flowering peduncles borne on stem, below leaves, occasionally borne from axil of lowest leaf. Flowers few, perianth segments 4–6 mm long. Fruit a blue berry; seeds 3, 4–5 mm in diam. Apr–Jun. Deciduous woods & thickets; scattered from MN & ND s to more frequent in IA, MO, & e KS; (MI to ND s to SC & KS). *Nemexia ecirrhata* (Engelm.) Small—Rydberg.

3. **Smilax herbacea** L., carrion-flower. Herbs up to 2 dm tall; stems lacking prickles. Leaves ovate to oval, blade 6–12 cm long, 3.5–9 cm wide, apex short acuminate to cuspidate, base cordate to rounded, glabrous or lightly pubescent beneath on and between veins. Peduncles axillary to leaves, ± flattened, bearing umbels with numerous flowers. Flowers carrion-

scented, perianth segments 3–6 mm long; filaments somewhat slender, nearly 2 × anther length. Fruit a blue or black berry, ca 10 mm in diam; seeds 3–6, brown, ca 4 mm long. Moist deciduous to coniferous-deciduous woods and thickets.

Three varieties can be recognized for our area. Both variety *herbacea* and variety *lasioneuron* (Small) Rydb. are common to our area, while variety *puverulenta* (Michx.) A. Gray is known for only a few scattered sites in the e GP.

1 Leaves glabrous and glaucous beneath .... 3a. var. *herbacea*
1 Leaves pubescent beneath.
    2 Leaf blade light green, not shiny beneath; petiole short; berry blue .... 3b. var. *lasioneuron*
    2 Leaf blade dark green, shiny beneath; petiole long; berry black ..... 3c. var. *pulverulenta*

3a. var. *herbacea* Leaves glabrous and glaucous beneath, blade 6–11 cm long, 4.5–8 cm wide, apex cuspidate. Peduncles mostly longer than leaf. Berry blue. (2n = 26) May–Jun. IA s to KS & OK; (se & cen Can., s to GA & OK). *Nemexia herbacea* (L.) Small — Rydberg.

3b. var. *lasioneuron* (Small) Rydb. Leaves light green, pubescent beneath, blade 7–12 cm long, 4–9 cm wide, apex cuspidate. Peduncles mostly shorter or equal to leaf. Berry blue. (2n = 26) Apr–Jul. MN & ND s to NE, WY, MO, & e KS; (Ont. w to Sask. & MT s to GA, AL, OK, & CO). *Nemexia lasioneuron* (Hook.) Rydb. — Rydberg; *S. lasioneura* Hook. — Fernald.

3c. var. *pulverulenta* (Michx.) A. Gray Leaves dark green, blade 6.5–8 cm long, 3.5–8 cm wide, apex acuminate to cuspidate. Peduncles as long as or longer than leaf. Berry black. (2n = 26) May–Jun. Scattered in extreme e GP; (NJ to MN, s to GA & MO). *Nemexia pulverulenta* (Michx.) Small — Rydberg; *S. pulverulenta* Michx. — Fernald.

4. **Smilax hispida** Muhl., bristly greenbrier. Woody vine up to 13 m long, attaching by tendrils to shrubs and young or small trees; stem branches terete or nearly so, prickles slender and ± flattened, black or on new stems yellow. Leaves ovate to oval, blade 5–15 cm long, 3.5–13 cm wide, 5 main veins arising from petiole, apex cuspidate, margin thin, base rounded to truncate. Peduncles slender, 5–8 mm long. Flowers several, the staminate 3.5–5 mm long. Fruit a black berry, 4.5–7 mm long; seeds 1 or 2. Apr–Jun. Deciduous thickets & woods, especially common along wooded riparian sites; se SD & e NE, s to MO & OK; (NY to MN, s to FL & TX). *S. tamnoides* L. var. *hispida* (Muhl.) Fern. — Fernald.

Smilax hispida as well as several other GP *Smilax* species exhibit a tremendous amount of phenotypic variation both in the shape and size of leaves. Occasionally vines can be found that are nearly devoid of prickles, especially younger plants or younger branches of older plants. The report of *Smilax glauca* Walt. for Nebraska given by Rydberg can probably be attributed to *Smilax hispida*.

159. DIOSCOREACEAE R. Br., the Yam Family

by Robert B. Kaul

1. DIOSCOREA L.

1. ***Dioscorea villosa*** L., wild yam. Perennial, dioecious, herbaceous twining vine to 5 m long, from a rhizome. Leaves alternate, glabrous or villosulous; petiole slender, 2–5 cm long; blade cordate, 5–10 cm long, the tip acuminate; veins arcuate, depressed above and prominent below, a midrib lacking. Staminate panicles axillary, branched, glomerulate, 3–8 cm long, with 1–4 flowers in each glomerule; pistillate spikes axillary, 4–8 cm long, the flowers solitary and separated, the bracts tiny. Flowers regular, perianth of 6 similar

greenish-white segments; stamens 6(3?); ovary inferior, ca 6 mm long, styles 3. Fruit a 3-winged capsule 14–24 mm long, with usually 2 seeds in each locule; seeds flat, broadly and somewhat irregularly membranous-winged. (2n = 60) Jun–Jul. Moist woods, frequently nonflowering; sw IA, se NE, e KS, MO, e OK; (CT to e MN & se NE, s to FL & TX). *D. paniculata* Michx.—Rydberg.

Dioscorea batatas Dcne., a native of Asia, occurs in KS: Cherokee. It can be recognized by its panduriform leaves and its much longer than wide fruits.

160. ORCHIDACEAE Juss., the Orchid Family

by Robert B. Kaul

Ours all perennial herbs, erect, terrestrial; autotrophic and green-leaved, or saprophytic and without chlorophyll or normal leaves, mycorrhizal. Plants rhizomatous, cormous or tuberous, the rhizomes sometimes coralloid; roots fibrous, fleshy, or absent. Leaves 1–several, alternate or opposite in ours, basal and/or cauline, simple, entire, often sheathing, sometimes petiolate, smooth or plicate, glabrous or pubescent. Flowering stems mostly simple and erect in ours, the inflorescences 1-flowered scapes or few- to many-flowered racemes or spiciform racemes, bracteate. Flowers mostly perfect, large and colorful to small and inconspicuous, occasionally with pale variants, entomophilous by various elaborate structural and behavioral mechanisms, distinctly zygomorphic and sometimes bizarre; pedicels short; sepals 3, like or unlike the petals, sometimes the lower 2 fused below the lip and then appearing as 1; petals 3, the lowest (the lip or labellum) largest and often elaborated into lobes, grooves, sacs, and spurs, the lateral petals often forming a hood with the dorsal sepal; stamens fused with the style and stigmas to form a prominent column subtended by the lip; fertile stamens 2 *(Cypripedium)* or 1 (all others of ours), a staminodium sometimes present; pollen sacs 2, sometimes with a connective, pollen in 1–4 masses (pollinia) or granular; stigmas 3, sticky, near the tip of the column, one of them sometimes nonfunctional and forming a beaklike rostellum near the tip of the column. Ovary inferior, usually resembling a pedicel, twisting 180° (in ours) as it develops to bear the lip lowermost; ovary 1- or 3-celled, placentae 3, parietal. Capsules in ours dry, dehiscent, sometimes infrequently produced, long-persistent; seeds very numerous, tiny, without endosperm, the embryo undifferentiated.

References: Correll, D. S. 1950. Native orchids of North America North of Mexico, Chronica Botanica, Waltham, MA.; Henderson, J. L. 1977. A taxonomic treatment of the genus *Habenaria* in eleven midwestern states. M.S. thesis, Univ. Missouri-Kansas City. 168 pp.; Luer, C. A. 1975. The native orchids of the United States and Canada, excluding Florida. New York Bot. Gard., Bronx, N.Y. 363 pp.; Magrath, L. K. 1973. The native orchids of the prairies and plains region of North America. Ph.D. dissertation, Univ. Kansas, Lawrence. 284 pp.

The Orchidaceae is one of the world's largest plant families, and its greatest diversity is found in the tropics. None of our species is confined to our area, although *Spiranthes magnicamporum* is limited to the central grasslands of the continent. Many have continental or even circumboreal distribution and are at the edges of their ranges with us, and they are often somewhat smaller here than in the main parts of their distributions.

The disappearance of so much native prairie has made many of our species rare, although some were never common. All orchids should be considered endangered because of their low tolerance for environmental disturbance. Most are difficult to keep in cultivation, and few transplant well. Some of our showier species are available in the trade through commercial propagation and can be sought there. Some states have laws against picking or digging orchids, and good judgment requires that they be left alone.

ORCHID FAMILY   1269

Two orchids have been recently discovered in the se GP: *Calopogon pulchellus* (Salisb.) R. Br. (KS: Cherokee; MO: Jasper) and *Malaxis unifolia* Michx. (KS: Cherokee.)

1 Plants lacking chlorophyll, yellowish, reddish, or brown; leaves reduced to sheaths; rhizomes coralloid.
  2 Lip 3-lobed, free; pollinia 8 .... 9. *Hexalectris*
  2 Lip 1- to 3-lobed, lateral 2 sepals fused basally; pollinia 4 .... 3. *Corallorhiza*
1 Plants with chlorophyll, green; leaves normally developed but not always present at flowering time.
  3 Plants leafless at flowering time.
    4 Lip spurred; BH, e WY .... 12. *Piperia*
    4 Lip not spurred.
      5 Flowers yellow or greenish, tinged with purple or brown; corms present; spring-flowering woodland species; se GP .... 1. *Aplectrum*
      5 Flowers white or whitish; plant without corms; autumn-flowering prairie species .... 14. *Spiranthes (S. magnicamporum)*
  3 Plants with leaves at flowering time.
    6 Lip inflated, saclike, not spurred.
      7 Leaves several, cauline, pubescent; anthers 2 .... 4. *Cypripedium*
      7 Leaf solitary, basal, glabrous; anther 1 .... 2. *Calypso*
    6 Lip not flared or saclike, with or without a spur.
      8 Lip with a distinct basal spur.
        9 Flowers lavender with a white lip .... 6. *Galearis*
        9 Flowers white, green, or yellowish .... 8. *Habenaria*
      8 Lip without a basal spur.
        10 Leaves plicate, ribbed; BH only .... 5. *Epipactis*
        10 Leaves smooth, not plicate or ribbed except for the midrib.
          11 Leaves 2, opposite, basal, erect; a small cormlike pseudobulb present .... 10. *Liparis*
          11 Leaves single or several, basal or cauline but not in a basal pair.
            12 Leaves basal.
              13 Plant rooting from a creeping rhizome; leaves reticulately marked with white or with a white midrib; woodlands .... 7. *Goodyera*
              13 Plant not rooting as above; leaves unmarked; prairies and marshes .... 14. *Spiranthes*
            12 At least some leaves cauline.
              14 Leaves opposite, a single pair near the middle of the stem .... 11. *Listera*
              14 Leaves alternate.
                15 Flowers numerous, in a twisted spike .... 14. *Spiranthes*
                15 Flowers 1–10, not in a twisted spike.
                  16 Flowers solitary (2); leaves elliptic, not clasping; wet meadows .... 13. *Pogonia*
                  16 Flowers 3(1–6); leaves ovate, clasping; woodlands 15. *Triphora*

1. APLECTRUM Nutt.

1. *Aplectrum hyemale* (Muhl. ex Willd.) Torr., putty-root, Adam-and-Eve. Plant 25–60 cm tall, scapose, from a more or less globose corm, the current season's corm often attached by a slender short rhizome to the previous season's corm. Leaf 1(2), basal, appearing in late summer and persisting through the winter, then usually dying by anthesis the following spring; elliptic, plicate, gray-green, the nerves lighter, the lower sides often purplish, 10–20 cm long, 3.5–8 cm wide. Raceme loose, with 6–10 flowers; floral bracts small, subulate, cauline bracts tubular, sheathing about 3/4 of the scape, pale. Flowers greenish or yellowish, tinged or spotted purple or brown; ovary short-pedicellate, 7–10 mm long; sepals and petals similar, 1–1.5 cm long, 2–4 mm wide, oblanceolate; lateral sepals somewhat falcate, obtuse to acute; lip 3-lobed, obovate, 1–1.3 cm long, ca 8 mm wide, white with purple mark-

ings, lateral lobes ovate, obtuse to acute, 2 mm long, central lobe much larger than the laterals, suborbicular, the margin undulate, the disk with 3 ridges near the base; column compressed, 7 mm long, purple-spotted. May-Jun. Moist oak woodlands; e KS, w MO; (se MN to NH, s to GA, AL, MO, & e KS) *A. spicatum* (Walt.) B.S.P.—Rydberg.

Occasional white-flowered forms are known from beyond our range and are to be expected with us.

2. CALYPSO Salisb.

1. *Calypso bulbosa* (L.) Oakes, fairy slipper orchid. Plant glabrous, scapose, 6-22 cm tall, from an ellipsoid, sheathed corm 1-2.5 cm long; roots slender, fleshy to coralloid. Leaf 1, basal, long-petiolate, bluish-green, the blade ovate to subcordate, rounded to acute, the margin undulate, plicate, 2-6.5 cm long, 1.5-5 cm wide, the petiole sulcate, to 6 cm long. Scape purplish, sheathed below by tubular bracts; flower 1, subtended by a lanceolate, acuminate, purplish bract 1-2.5 cm long. Flower showy, pendent, long-lasting, purplish, pink, and white; sepals and petals similar, spreading, rose-pink, linear-lanceolate, acuminate, undulate, the lateral sepals oblique, 1.2-2.3 cm long, 2-3 mm wide; lip pendent, whitish with purple markings, saccate, ovate, 1.5-2.4 cm long, 7-13 mm wide, a 2-horned white apron in front of the sac, the apron yellow-bearded with clavate hairs; column petaloid, suborbicular, inverted in the orifice of the lip, 7-12 mm long and almost as wide; anther 1, on the under side of the column; pollinia 2, bilobed. Capsule erect, ellipsoid, 2-3 cm long. Late May-Jun. Cool coniferous woodlands in litter; BH & ne WY; (AK to Newf., s to NY, MI, WI, & ne MN, and in the mts. to NM & AZ; BH). *Cytheria bulbosa* (L.) House—Rydberg.

Ours have been treated as typical var. *bulbosa* by most authors and as var. *americana* (R. Br.) Luer by Luer. This is the only species in the genus.

3. CORALLORHIZA (Hall.) Chat., Coral-root

Perennial, terrestrial, saprophytic erect herbs, rootless, leafless, largely without chlorophyll, glabrous, rhizomes much branched, corallike, mycorrhizal. Stems simple, succulent, yellow, red, purple, or brown, sheathed by membranaceous bracts. Flowers purplish, greenish, yellowish, or whitish, sometimes spotted or striped purple; perianth often connivent and forward-directed; sepals equal, lateral sepals usually united to form a mentum (spur) adnate to the ovary; lip simple or 3-lobed; column compressed; anther terminal, pollinia 4. Capsule pendent, often long-persistent on the dry stem.

Plants appear sporadically in woodland litter, seeming to disappear for years, and then reappear in suitable seasons. They often occur in groups or large colonies. Twelve species occur in North and Central America.

1 Perianth distinctly striped purple to brownish; lip purplish.
  2 Lip 7.5-16 mm long; perianth 8-18 mm long .... 3. *C. striata* var. *striata*
  2 Lip 3-7.5 mm long; perianth 5-8 mm long; ND only .... 3. *C. striata* var. *vreelandii*
1 Perianth plain or spotted, but not striped.
  3 Lip entire, without a pair of lateral lobes near the base.
    4 Lip yellow or white, unspotted or only lightly spotted; perianth yellow or yellowish.
      5 Lip yellow .... 3. *C. striata* var. *ochroleuca*
      5 Lip white, sometimes lightly spotted red .... 4. *C. trifida*
    4 Lip distinctly spotted pink or purple; perianth greenish-purple or purple.
      6 Stem bulbous-based .... 2. *C. odontorhiza*
      6 Stem normal, not bulbous-based .... 5. *C. wisteriana*
  3 Lip 3-lobed or with a pair of small lateral teeth.

7 Lip distinctly spotted purpled, perianth purple to greenish .... 1. *C. maculata*
7 Lip unspotted or occasionally lightly spotted red, perianth yellow or yellowish .. 4. *C. trifida*

1. **Corallorhiza maculata** (Raf.) Raf. Spotted coral-root. Stem variable in stature, slender to stout, purplish, reddish, yellowish, or brown, 15–60 cm tall, often in groups; bracts 2–10 cm long, sheathing the stem for much of their length. Raceme 5–20 cm long, loosely or densely flowered; flowers up to 40, spreading; floral bracts tiny, subulate, 1.5–3 mm long. Pedicels 2–5 mm long; perianth the color of the stem, purple or greenish, occasionally yellow, usually spotted purple; sepals linear to lanceolate, 7–9 mm long, 1.5–2.5 mm wide, obtuse to acute; lateral sepals united at the base and forming a small mentum at the base of the lip; petals oblanceolate, oblique, 5.5–8 mm long, 1.5–3 mm wide; lip ovate, unequally 3-lobed, erose, 5–8.5 mm long, 3.5–5 mm wide, white or white spotted-purple, the small lateral lobes curved forward, 1–1.5 mm long, the middle lobe much larger; column yellow, spotted, curved, compressed, 4–5.5 mm long. Capsule 1–2.4 cm long, 5–8 mm wide. (n = 21; 2n = 42) Jun–Aug. Dry coniferous woods, often on slopes, in leaf litter; nw NE, w SD, CO: Elbert, ne WY, w & n ND, w MN; (B.C. to Newf., s to NC, NE, NM, CA, & Mex.; Guat.).

Various color forms may be recognized. In our area two such forms occur: f. *maculata* (lip white with purple spots; all other plant parts purplish to brownish; the common form), and f. *flavida* (lip white, unspotted, all other plant parts lemon or pale yellow; only BH in our area; rare).

2. **Corallorhiza odontorhiza** (Willd.) Nutt., late coral-root. Plant inconspicuous, slender, 6–32(40) cm tall, greenish-yellow to light brown to purple, rhizome much branched. Stem base distinctly bulbous-thickened, cauline bracts almost entirely enclosing the stem. Raceme 2–6 cm long; floral bracts minute, 1 mm long. Flowers small, greenish purple to purple, pedicels 1–3 mm long; perianth parts more or less convergent over the lip; sepals linear to lanceolate, 2.5–4.5 mm long, 1–1.5 mm wide; petals elliptic, subacute, margins more or less erose, 2.5–4.3 mm long, 1.1–2 mm wide; lip slender-clawed, arcuate, obovate or ovate, obtuse to retuse, unevenly crenulate, white, spotted purple, 2.5–4.5 mm long, 2–3.5(5) mm wide, the disk with two short basal lamellae; column slender, compressed, 2 mm long. Capsule ovoid to ellipsoid, 5–8 mm long, 2.5–4 mm wide. Aug–Oct. Dry to moist slopes in oak or pine woods; BH; se NE: Richardson, Sarpy; e 1/4 KS, nw MO, e & cen OK; (BH, NE to NH, s to FL & e TX; C. Amer.).

3. **Corallorhiza striata** Lindl., striped coral-root. Stem 15–45 cm tall, yellowish, often tinged purple or brown; cauline bracts 1–14 cm long. Raceme 5–16 cm long, usually loosely flowered, the flowers to 35; floral bracts tiny, 2–3.4(4) mm long. Flowers nodding, yellow, tinged and striped purple, or only yellow; pedicels 1–3 mm long, perianth parts spreading or connivent; sepals with 3 purple veins or unmarked, linear-oblong, rounded to subacute, 5.5–18 mm long, 2.3–5 mm wide; lateral sepals not forming a mentum at the base; petals elliptic-oblanceolate, about equaling the sepals, veined purple or unmarked; lip purplish, whitish-yellow, or yellow, veined purple or unmarked, entire, basally reflexed, fleshy, involute-margined, elliptic-obovate, 3–16 mm long, 3–7.5(8.5) mm wide; disk with a fleshy, bilobed lamella near the middle of the base; column slender, curved, 2–5 mm long. Capsule ellipsoid, 1.2–2 cm long. (2n = 42) Magrath, op. cit., recognizes 3 varieties in our area.

1 Lip yellow, 4–7.5(8) mm long; perianth parts more or less connivent, yellow, without purplish veins or nearly so; dorsal sepal 6.5–9.5 mm long .... 3a. var. *ochroleuca*
1 Lip purplish or purplish-maroon, 3–16 mm long; perianth parts spreading (seldom connivent), yellowish with prominent purplish striations; dorsal sepal 5.5–18 mm long.
  2 Lip 3–7(7.5) mm long; perianth 5–8 mm long .... 3c. var. *vreelandii*
  2 Lip 7.5–16 mm long; perianth 8–18 mm long .... 3b. var. *striata*

3a. var. *ochroleuca* (Rydb.) Magrath. Perianth parts, including lip, yellow, unmarked or only barely marked purplish; dorsal sepal 6.5–9.5 mm long; lip 4–7.5(8) mm long; perianth parts more or less connivent. Jun–Aug. Pine, pine-spruce, or pine-spruce-aspen woods on limestone, sandy, or clay soils in deep decaying leaf litter; NE: Sioux; SD: BH; (GP, s-cen CO, UT, Ont.) *C. ochroleuca* Rydb.—Rydberg.

3b. var. *striata*. Perianth parts spreading (seldom connivent), yellowish, prominently purplish-striated, 8–18 mm long; lip purplish to maroonish, 7.5–16 mm long. Jun–Aug. Pine or pine-spruce woods on limestone, sandy, or clay soils in deep decaying leaf litter; w-cen & n-cen ND, BH, ne WY, & reported from OK: Jackson; (N.B. to B.C., s to CA, NM, TX, SD, Great Lakes, & NY). Intermediates with 3c are known.

3c. var. *vreelandii* (Rydb.) L. O. Wms. Perianth parts spreading (seldom connivent), yellowish, prominently purplish-striated, 5–8 mm long; lip purplish to maroonish, 3–7(7.5) mm long. Jun. Aspen woods in thick decaying leaf litter; ND: Stark; (ND, WY, CO, NM, & AZ). Intermediates with 3b are known.

4. ***Corallorhiza trifida*** Chat., pale coral-root. Stem greenish to yellow, 4–34 cm tall, cauline bracts 1.5–7 cm long, mostly sheathing the stem. Raceme loosely flowered with up to 20 flowers; floral bracts less than 2 mm long. Flowers small, erect or spreading, greenish to yellow; pedicels 1–2 mm long; sepals linear-oblanceolate, 4–6.5 mm long, 1–1.8 mm wide; lateral sepals forming a small mentum; petals linear-oblanceolate, 4–4.5 mm long, 1.2–2 mm wide; lip white, sometimes with light red spots, obovate, a pair of small lateral lobes beside the large middle lobe, 3.5–5 mm long, 2–3 mm wide; disk with a pair of small lamellae; column barely clavate, curved, 3.5–5 mm long. Capsule 8–12 mm long, 5–6 mm wide. ($2n = 42$) Jun–Aug. Dry to moist coniferous woods & woodland marshes, in sandy soil & litter; BH, n-cen & ne ND; (AK to Newf., s to PA, Great Lakes, ND, CO, NM, & OR; Eurasia).

5. ***Corallorhiza wisteriana*** Conrad, Wister's coral-root. Stem 10–40 cm tall, yellowish to purple, the base barely thickened; cauline bracts 2–11 cm long, mostly sheathing the stem. Raceme with up to 25 flowers, 2.5–14 cm long; floral bracts 1–3 mm long, the lowest one sometimes much longer. Flowers spreading; perianth greenish to purplish brown; sepals linear to lanceolate, 6–9 mm long, 1.2–2.2 mm wide; petals linear to elliptic, 5.5–7.5 mm long, 2–2.5 mm wide; lip white, spotted pink, or purple, short clawed, recurved, 5.5–8 mm long, 4–5 mm wide, broadly ovate, its apex truncate or retuse, the margin undulate, disk with 2 linear lamellae near the base; column compressed, 3–4 mm long. Capsules ovoid, 6–12 mm long, 3–6 mm wide. Apr–Jul. Oak-hickory & pine-spruce woods, in stony soil or decaying litter; ne WY, BH, nw NE, e KS s to cen OK; (WY, NE, KS to PA, s to FL & e TX; Rocky Mts. from s MT to Mex.).

4. CYPRIPEDIUM L., Lady's-slipper

Perennial, terrestrial herbs of forests and wet prairies, roots fibrous, hairy. Stems rhizomatous and erect, ours all leafy. Leaves alternate in ours, plicate, more or less sheathing the stem. Inflorescence 1- to few-flowered, the floral bracts green, conspicuous. Flowers showy, sometimes fragrant, short-pedicellate; sepals 3 but appearing as 2 by fusion of the lower 2 almost to their apices; dorsal sepal erect or inclined over the lip; ventral sepals fused behind the lip and together resembling the dorsal sepal in size and color, the bifid apex, when present, the only evidence of their union; petals spreading, flat or twisting, the color of the sepals but somewhat longer and narrower; lip saclike, inflated into a distinct and prominent horizontal or somewhat declining pouch, in ours white, pink, white striped with pink or purple, or yellow, its aperture dorsal and partly occluded by a colorful petallike staminodium that is partly fused to the column below; column declining in the lip, obscured by the staminodium; stamens 2, laterally fused to the column; pollen granular and exposed; rostellum absent, the stigma convex and not viscid. Capsules erect, ovoid, obovoid, or ellipsoid, infrequently produced.

Our species, all placed in Section *Cypripedium*, have always been uncommon or rare, and they are far less abundant now than in the past. Although long-lived, they withstand almost no environmental disturbances by man. Transplantation is usually unsuccessful and should not be attempted; it is illegal in some states.

1 Lip yellow .... 2. *C. calceolus*
1 Lip white, ivory, or pink.
  2 Lip pink (occasionally white with pink veins or all white); petals flat, not narrow, obtuse; woodlands .... 4. *C. reginae*
  2 Lip white or ivory; petals twisted, narrow, acuminate; wet prairies.
    3 Sepals and petals greenish-yellow; lip 1.6-2.3 cm long; dorsal sepal 2.3-2.7 cm long .... 3. *C. candidum*
    3 Sepals and petals wine-red; lip ca 3 cm long; dorsal sepal 3.5 cm long .... 1. *C.* x *andrewsii*

1. *Cypripedium* x *andrewsii* A. M. Fuller [*C. candidum* Muhl. ex Willd. x *C. calceolus* L. var. *parviflorum* (Salisb.) Fern.]. Plants with characteristics more or less intermediate between their putative parents. Plants small, with the short stature and rather conduplicate leaves of *C. candidum*. Sepals and petals wine-red; dorsal sepal 3.5 cm long; petals 4.5 cm long; lip white or whitish, purple spotted about the opening, 3 cm long. May-Jun. Moist prairies; e ND; (ND e to IL & MI).

This hybrid has been collected only once in our area, but is to be expected wherever the ranges of the parents overlap.

2. *Cypripedium calceolus* L., yellow lady's-slipper, yellow moccasin flower. Plant glandular-pubescent throughout, 20-30 cm tall. Rhizome stout, creeping, often forming large colonies. Leaves 3-6, ovate-lanceolate, plicate, 5-21 cm long, 2-11 cm wide. Flowers 1 or 2, each subtended by a green bract, the bracts ovate-lanceolate, acuminate, 4-11 cm long, to 5 cm wide, the upper one much smaller than the lower, when 2 are present. Pedicels to 10 mm or more long; sepals and petals yellowish, greenish, greenish-brown, or reddish-brown; dorsal sepal ovate-lanceolate, acuminate, usually undulate, not twisted or only weakly so, 2.5-8 cm long, 1-2 cm wide; lateral sepals united below the lip and appearing as a single sepal, 2-7.5 cm long, 0.6-1.8 cm wide; petals colored like the sepals, linear-lanceolate, acuminate, several times twisted or occasionally flat, 3-9 cm long, to 1 cm wide, conspicuously long-hairy near the base; lip yellow, often with spots or veins of purple on the inside, the margin of the opening turned in, 1.8-5 cm long; staminodium yellow, often purple spotted, bluntly triangular, on a thick fleshy stalk. Capsules ellipsoid, 2-3.5 cm long and about 1 cm wide. ($2n = 20$)

Several varieties are recognized throughout the range of the species. Variety *calceolus* is European. All 3 varieties usually recognized in North America are present in our range, as well as intermediates between them, according to Magrath, op. cit.

1 Petals flat, not twisted; wet prairies in ne GP .... 2b. var. *planipetalum*
1 Petals twisted; plants of woodlands and forest openings and margins.
  2 Lip large, 3-5 cm long; perianth greenish, greenish-brown, or yellowish .... 2c. var. *pubescens*
  2 Lip smaller, 1.8-2.5 cm long; perianth reddish-brown .... 2a. var. *parviflorum*

2a. var. *parviflorum* (Salisb.) Fern., small yellow lady's-slipper. Dorsal sepal 2.5-4(4.5) cm long, undulate, abruptly narrowed to subacuminate at the base; petals 3.5-4.5 cm long, spirally twisted or undulate, acute to acuminate; lip 2-2.7(3) cm long. Apr-Jul. Forests, forest edges, brook sides, alder swamps; sometimes locally abundant; nw MN, e & n-cen ND, BH; nw NE: Dawes; ne WY; (Newf. to B.C., s to WA, UT, WY, nw NE, MN, IA, TX, GA, & NJ).

This var. hybridizes with *C. candidum* in ND, the hybrid called *C.* x *andrewsii,* q.v.

2b. var. *planipetalum* (Fern.) Vict. & Rouss. Dorsal sepal 3-3.8 cm long, flat, not undulate, somewhat rounded at the base; petals 3.5-4 cm long, 0.6-0.9 cm wide, flat, linear-lanceolate, obtuse, with a small acute tip; lip ca 3 cm long. Jun-Jul. Bogs, wet roadside ditches in the prairie; ND: Ransom, Rolette; (Newf., Que., MI, ND).

2c. var. *pubescens* (Willd.) Correll, large yellow lady's-slipper. Dorsal sepal 4-7 cm long, spirally twisted or undulate, petals 4.5-9 cm long, spirally twisted or undulate, acute to acuminate; lip 2.7-7 cm long, usually more than 3 cm long. Apr-Jul. Hardwood forests in moist soil & thick litter; se NE near the Missouri R., ne & se KS, nw MN; IA: Dickinson, Guthrie; (Newf. to Man., s to OK, MO, & GA; also in B.C., CO, AZ, NM, & TX).

Most of our specimens are intermediate between this and var. *parviflorum,* but with closer affinities to var. *pubescens,* according to Magrath, op. cit. This var hybridizes with no. 3, *C. candidum,* beyond our range. The hybrid is named *C.* x *favillianum* Curtis and is to be sought in our area where the parental ranges overlap in se NE.

3. **Cypripedium candidum** Muhl. ex Willd., small white lady's-slipper. Plant 15-30 cm tall, sparsely pubescent, rhizomes short, often forming dense clumps. Leaves 3 or 4, appearing crowded about the middle of the stem, inclined to erect, elliptic-lanceolate, acute to acuminate, plicate, 7-15 cm long, 2-4 cm wide. Flowers 1 or 2, appearing with the leaves; bracts large, foliaceous, erect, the same shape as the leaves, 3.5-6 cm long, to 2 cm wide. Flowers somewhat fragrant, nearly sessile; sepals and petals greenish-yellow, usually streaked purple; dorsal sepal ovate-lanceolate, acuminate, sometimes undulate, 2.3-2.7 cm long, 1-1.3 cm wide; lateral sepals similar to the dorsal, 2-2.4 cm long, ca 1 cm wide; petals linear-lanceolate, helically twisted or undulate, 2.5-4 cm long, 3-6 mm wide; lip obovoid, white, the interior with purple veins, 1.6-2.3 cm long, 1-1.5 cm deep and wide, the margins of the orifice turned inward; staminodium on a slender stalk, ovate, yellow, spotted with crimson, 1-1.2 cm long including the stalk, about 2.5 mm wide. Capsule ellipsoid, 2-3 cm long. ($2n = 20$) May-Jun. Moist prairies & marshes, in the open; e ND, w MN, w IA, e & cen NE; (Man. to CT, s to OH, IN, MO, & NE).

This species apparently hybridizes with *C. calceolus* var. *parviflorum* to produce *C.* x *andrewsii,* q.v.

Probably quite abundant in some areas in the past, this species is now becoming rare due to the disappearance or disturbance of its habitat.

4. **Cypripedium reginae** Walt., showy lady's-slipper. Plant usually large, robust, 35-80 cm tall, densely pubescent, the hairs notably irritating to some people, rhizome large. Leaves 3-8, ovate-lanceolate, acute to acuminate, plicate, ribbed, 10-24 cm long, 6-15 cm wide. Flowers 1 or 2(4), showy; floral bracts elliptic-lanceolate, acute or acuminate, 6-12 cm long, 2-5 cm wide. Flowers short-pedicellate; sepals and petals white; dorsal sepal ovate-orbicular, obtuse, 2.8-4 cm long, 1.7-3 cm wide; lateral sepals united, the product ovate-orbicular, obtuse, slightly smaller than the dorsal sepal; petals oblong-elliptic, slightly oblique, flat, not twisted, 2.5-4 cm long, 0.5-1.5 cm wide; lip usually rose-pink streaked with white but sometimes pale pink, or entirely white in f. *album,* inflated, 2.5-4.5 cm long, 2.3-3.2 cm wide, the margins of the orifice infolded; staminodium ovate, on a stout stalk, white with rose, pink, or yellow spots, 1-1.7 cm long. Capsule ellipsoid, 2.5-4.5 cm long. ($2n = 20$) Jun-Aug. Shady swamps & bogs in acid, neutral, or slightly alkaline soil; e ND, nw MN; (se Sask. to Newf., s in the mts. to n AL, OH, IL, MO, cen IA, MN, & e ND). *C. hirsutum* P. Mill.—Rydberg.

This is one of the showiest N. Amer. orchids and is the state flower of MN. The white-flowered form is sometimes called *C. album* Ait.; it often occurs with the typical form elsewhere, but it has not been collected in our area.

5. EPIPACTIS Sw.

1. *Epipactis gigantea* Dougl. ex Hook., helleborine, chatterbox. Plant large, coarse, erect, 30–80 cm tall, mostly glabrous, rhizome short, creeping. Stem stout, leafy, green, sometimes purplish tinged. Leaves 4–12, sheathing, 5.5–22 cm long, 2–6 cm wide, the lower ones reduced to sheaths, the upper ones reduced to nonsheathing floral bracts in the inflorescence. Leaves plicate, erect or inclined, ovate-lanceolate, obtuse to acuminate. Raceme loose, 2- to 8-flowered; bracts lanceolate, acuminate to attenuate, 1–15 cm long, to 2 cm wide, much exceeding the flowers at the base of the raceme, but becoming smaller upward. Flowers showy; ovary arching; sepals ovate-lanceolate, acuminate, greenish to pinkish with purplish nerves, concave or falcate, the lateral sepals oblique, the dorsal sepal erect, 1.5–2.4 cm long, 7–9 mm wide; lip yellowish with purple or brown nerves on the lateral lobes, constricted in the center, sessile at the base of the column, distinctly and unevenly 3-lobed, 1.5–1.8 cm long; lateral lobes concave, bluntly triangular to suborbicular, obtuse to rounded, ca 9 mm long and 8 mm wide; central lobe smaller than the laterals, yellow tinged with purplish, arching, oblanceolate, somewhat involute, 8–11 mm long, 3–4 mm wide, the calli winglike, corrugate, fleshy, column short, erect, arching, 2-horned below the anther, 7–9 mm long. Capsule pendent, ellipsoid, 2–2.5 cm long. (n = 30) Jun–Jul. Stream margins, gravel bars, seepage areas, & springs, often in calcareous soil; s BH; (s B.C. to w MT, s in the mts. to BH; s-cen OK: Arbuckle Mts; w TX, Baja Calif; s Mex.).

6. GALEARIS Raf.

1. *Galearis spectabilis* (L.) Raf., showy orchis. Plant 7–23 cm tall, rather succulent, glabrous, rhizome short, roots fleshy, slender. Leaves 2(3), basal, subopposite, the petioles sheathing, elliptic, oblong-obovate to suborbicular, rounded to obtuse, 2–8 cm wide, subtended by 2 scarious sheathing bracts 1.5–6 cm long. Raceme few-flowered, 2–8 flowers, ca 9 cm long; floral bracts leafy, lanceolate or elliptic, 2–7 cm long, 0.5–2 cm wide, the lower ones much exceeding the flowers, the upper ones equal to or shorter than the flowers. Flowers showy, erect; ovary stout, 1–2 cm long; sepals and petals connivent over the column, purplish, lavender, or pink, rarely white; sepals elliptic to lanceolate, 1.3–2 cm long, 5.5–6.5 mm wide; the lateral sepals oblique; petals linear, 12–15 mm long, 2–4 mm wide; lip white, occasionally lavender-tinged, ovate to suborbicular, obtuse to rounded, the margin wavy, 1–1.6 cm long, 7–15 mm wide, with a tubular basal clavellate spur, the spur 1.3–2 cm long; column stout, 6–8 mm long. Capsule erect, ellipsoid, 1.8–2.5 cm long, 5–7 mm wide. (2n = 42) Mid Apr–early Jun. Moist upland woods; w MN, Missouri R. & tributaries in w IA, se NE, ne KS, nw MO; (MN to N.B., s to n GA, n AL, n AR, & w along the Missouri R. & tributaries to KS & NE). *Galeorchis spectabilis* (L.) Rydb.—Rydberg; *Orchis spectabilis* L.—Gleason & Cronquist.

7. GOODYERA R. Br.

Plants scapose, stoloniferous, roots thick. Stem green, slender, pubescent, scape with distant sheathing green or whitish bracts. Leaves several in a basal rosette, or at least crowded at the base of the scape, alternate, ovate-lanceolate, bluish-green and conspicuously marked with a white midrib or white reticulations, evergreen, but disappearing after flowering of the scape. Raceme spicate, cylindrical or secund, loosely to densely flowered, the floral bracts green and much shorter than the flowers. Flowers small, white or greenish, sometimes tinged pink; dorsal sepal and petals connivent over the column and lip; lip with a small or large saccate nectary at the base; column stout, rostellum pointed, the anther dorsal; pollinia 2. Capsules erect.

1 Leaves plain green or white reticulate, but always prominently white along the midrib, 4–11 cm long, 1.5–3.5 cm wide; lip 5–8 mm long .... 1. *G. oblongifolia*
1 Leaves white-reticulate (green), the midrib area green, 1–4.5 cm long, 6–14 mm wide; lip 3.5–5 mm long .... 2. *G. repens*

1. **Goodyera oblongifolia** Raf., rattlesnake plantain. Plant pubescent or glandular-pubescent above, 10–40 cm tall. Leaves 3–7, oblong-elliptic to ovate or lanceolate, obtuse to acute, mostly tapering at the base, slightly oblique, dark bluish-green, the midrib prominently white, lamina usually distinctly white-reticulate, 4–11 cm long, 1.5–3.5 cm wide. Raceme densely or loosely flowered, the flowers spiral or secund, flowers to 30; floral bracts ovate-lanceolate, acute, 6–9 mm long. Flowers white or greenish-white, ovary stout; dorsal sepal greenish, lanceolate, obtuse, recurved, concave, 6–8 mm long, 3.5–4 mm wide; lateral sepals oblique, ovate, acuminate, 5–8 mm long, 2.5–4 mm wide, greenish-white; petals oblique, spatulate, tapering to the base and tip, white, the midrib green, 6.5–8 mm long, 3–4 mm wide; lip saccate, beaked, the margins involute, 5–8 mm long, beak sulcate; column 4–5.5 mm long, the rostellum long pointed. Capsule obovoid or ellipsoid, 6–10 mm long. (2n = 22) Jul–Aug. Moist or mesophytic coniferous forests on limestone or granite, in leaf litter; BH; ne WY: Crook; (s AK to Que., s to Great Lakes & in the mts. to BH, NM, AZ, cen CA, & Mex.). *G. decipiens* (Hook.) F. T. Hubb.—Rydberg.

2. **Goodyera repens** (L.) R. Br., dwarf rattlesnake plantain. Plant glandular-pubescent above, 7–20 cm tall. Leaves 3–7, in a basal rosette or part way up the stem, more or less ovate, obtuse to acute, dark green, frequently white-reticulate, 1–4.5 cm long, 6–14 mm wide. Raceme 2–7 cm long, flowers spiral or secund; floral bracts lanceolate, 5–12 mm long. Sepals white or greenish-white, 3–3.5 mm long, the dorsal 1–1.5 mm wide, the laterals 2–2.5 mm wide; petals white, sometimes tinged with pink, oblong-spatulate, more or less acute, 3–3.5 mm long, 1–1.5 mm wide; lip distinctly saccate, with an elongate beak, the margins flared and recurved, 3–3.5 mm long, 1.5–2 mm wide; column 1–1.5 mm long, the rostellum shorter. Capsule ellipsoid, 4–7 mm long. (2n = 30) Jul–Aug. Leaf litter of coniferous (deciduous) woods & bogs; BH, sw MN: Becker, Nobles; (AK to Newf., s to NC, Great Lakes, & in the mts. to SD, NM, & AZ; Europe, Asia). *G. ophioides* (Fern.) Rydb.—Rydberg.

The white-reticulate leaf form is much more common in BH; the plain green leaf form occurs with it. This is a somewhat variable species throughout its range.

8. HABENARIA Willd.

Erect, glabrous, terrestrial or semiaquatic herbs; roots fibrous or mostly fleshy. Stems simple, green or yellow, leafy or bracteate. Leaves 1–several, sometimes fugacious, cauline or basal, alternate or subopposite, the upper ones often much reduced, sessile, sheathing the stem below, linear to orbicular. Inflorescence a spiciform raceme, bracteate. Flowers small to medium-sized, ours green, white, yellowish or faintly purplish-tinged; sepals free, the dorsal sepal often connivent with the petals to form a hood over the column; petals free; lip simple or lobed, entire or toothed or fringed, with a distinct basal spur; column short; anther cells 2, separate, the pollen not contained in a pouch. Capsules erect, narrow and cylindrical to ellipsoid. Acid to neutral, usually damp, soils. Including *Blephariglottis, Coeloglossum, Denslovia, Habenaria, Limnorchis, Platanthera*—Rydberg.

This large and polymorphic group is sometimes split into several genera.

1 Lip deeply tripartite and the parts distinctly fringed.
   2 Flowers white or whitish; spur 2–5.5 cm long .... 5. *H. leucophaea*
   2 Flowers green; spur 1–2.3 cm long .... 4. *H. lacera*

3   Leaves 2 and basal, or 1(2) and about halfway up the stem
    4   Leaves 2, basal, spreading on the ground; flowers whitish; BH .... 6. *H. orbiculata*
    4   Leaf 1(2) about halfway up the stem; flowers greenish .... 1. *H. clavellata*
3   Leaves several and cauline.
    5   Spur small, less than one-half the length of the lip, scrotiform or saccate.
        6   Lip shallowly lobed, oblong-spatulate; spur scrotiform or saccate .... 8. *H. viridis*
        6   Lip entire, linear to elliptic; spur scrotiform to clavate-cylindric .... 7. *H. saccata*
    5   Spur cylindric to clavate, half or more the length of the lip.
        7   Lip abruptly dilated at the base; flowers white to pale greenish; spur shorter than the lip, barely clavate .... 2. *H. dilatata*
        7   Lip gradually widened but not abruptly dilated at the base; flowers greenish; spur not usually longer than the lip but if so, then clavate .... 3. *H. hyperborea*

1. Habenaria clavellata (Michx.) Spreng., green woodland orchis. Slender terrestrial herb 8–3.5 cm tall; roots few, fleshy, slender. Stem slightly angular and winged, green. Leaf 1(2), near the middle of the stem, obovate to oblanceolate, keeled, 5–18 cm long, 1–3.5 cm wide. Raceme 3- to 15-flowered, 2–6 cm long; floral bracts linear-lanceolate, acuminate, 3–10 mm long. Flowers small, green or yellow green, oblique; ovary stout, spreading, 1 cm long; sepals ovate, rounded to obtuse, 4–5 mm long, 2.5 mm wide; petals ovate, obtuse, 3–5 mm long, 2 mm wide; lip narrowly oblong, truncate, shallowly tridentate, 3–7 mm long, 3–4 mm wide; spur longer than the ovary, slender, clavate, curved, 8–12 mm long. Capsule horizontal to erect, ellipsoid, 10 mm long, 4 mm wide. (2n = 42) Jul. Swampy forests, wooded bogs; e ND; (ND to Lab., s to GA & e TX). *Denslovia clavellata* (Michx.) Rydb.—Rydberg.

This species is represented in our area by a single specimen taken in 1905; locally abundant e of GP.

2. Habenaria dilatata (Pursh) Hook., white orchis, bog candles. Slender to stout herb, 15–120 cm tall; roots slender, fleshy. Leaves ca 12, cauline, linear to lanceolate, obtuse to acuminate to 30 cm long and 5.5 cm wide. Raceme loosely to densely flowered, 5–15(45) cm long; floral bracts lanceolate, acuminate, exceeding the flowers, to 2.5 cm long, 1.5–4 mm wide. Flowers white, yellowish, or greenish, fragrant; dorsal sepal ovate to elliptic, obtuse, connivent with the petals to form a hood over the column, 3–7 mm long, 2.5–4 mm wide at base; lateral sepals elliptic to lanceolate, obtuse to acuminate, spreading or reflexed, 4–9 mm long, 1–3.5 mm wide; petals ovate-lanceolate to linear-lanceolate, obliquely dilated at base, 4–8 mm long, 2–4 mm wide; lip variably lanceolate, basally dilated, projecting, 5–10 mm long, 2–5 mm wide at base; spur cylindrical, equal to or shorter than the lip, 3–7 mm long. Capsule ellipsoid, erect, 7–12 mm long, 3–4 mm wide. (2n = 42). Jun–Aug. BH; (AK to Greenl., s to PA, SD, NM, & CA). *Limnorchis dilatata* (Pursh) Rydb.—Rydberg.

Two varieties can be recognized in our area: var. *dilatata* (flowers greenish, spur about equaling the lip; moist but not marshy soils) and var. *albiflora* (Cham.) Correll (flowers white or creamy, spur shorter than the lip; marshy soils in cold bogs and streams). Plants intermediate between this species and *H. saccata* and *H. hyperborea* can be found.

3. Habenaria hyperborea (L.) R. Br., northern green orchis. Plant 15–60 cm tall; roots fleshy, elongate, slender. Leaves up to 6, cauline, partially sheathing, linear to elliptic or lanceolate, obtuse to acuminate, 4.5–25 cm long, 0.8–4 cm wide. Raceme loosely to densely flowered, 3–25 cm long; floral bracts lanceolate, acuminate, the lowermost to 4 cm long and greatly exceeding the flowers. Flowers small, green, yellowish-green, or greenish-white, occasionally marked with brownish-purple; dorsal sepal ovate to ovate-elliptic, its apex rounded to obtuse, connivent with the petals and forming a hood over the column, 3–7 mm long, 1.3–4 mm wide below the middle; lateral sepals ovate to elliptic-lanceolate, obtuse or subacute, oblique, spreading or reflexed, 3–9 mm long, 1–3.5 mm wide; petals rather fleshy, lanceolate to ovate-lanceolate, falcate, acute to acuminate, dilated obliquely at the base, 3–7 mm long, 1–3 mm wide; lip fleshy, usually lanceolate, occasionally linear, obtuse, not obviously dilated at base, 3–9 mm long, 1.5–2.5 mm wide; spur cylindrical or clavate, more or less equal to the lip; column broad, thick, 1.5–2.5(3.5) mm long. Capsule obliquely ellipsoid, 1–1.5 cm long, 4–7 mm wide. (2n = 42, 84) Jun–Aug. Cool marshes & roadside ditches; n-cen & nw NE, w SD, ND, nw MN; (AK to Greenl., s to PA, NE, NM, & CA; also Asia, Iceland, & Spain). *Limnorchis huronensis* (Nutt.) Rydb.—Rydberg.

A highly polymorphic species; intermediate forms with *H. saccata* and *H dilatata* can be found.

4. Habenaria lacera (Michx.) Lodd., ragged orchis. Plant slender or stout, 25–60 cm tall; roots fleshy, slender, numerous; stem barely ribbed, green. Leaves 2–5, cauline, erect, linear-oblong to obovate or linear-lanceolate, basally sheathing, 7–21 cm long, 1.5–5 cm wide. Raceme loosely to densely flowered, 3–15 cm long; floral bracts equaling the ovaries, the lower ones barely exceeding the flowers, lanceolate to linear, acuminate, 1–4 cm long. Flowers greenish or greenish-white, ovaries arcuate, 1.5–2 cm long; dorsal sepal ovate to elliptic, obtuse, 4–5 mm long, 3–4 mm wide; lateral sepals obliquely ovate, obtuse, 4–6 mm long, 3 mm wide; petals linear to oblong-spatulate, usually entire, 5–7 mm long, less than 2 mm wide; lip deeply 3-lobed, 1–1.6 cm long, 1.3–1.7 cm wide, lateral divisions 3-parted to near the base, the parts also cut, middle lobe cut less deeply; spur slender, clavellate, equaling or exceeding the ovary, 1–2.3 cm long; column 2–2.5 mm long, 2 mm wide. Capsule 1.5 cm long, 3–4 mm wide. (2n = 42) Jun. Damp meadows; se KS, ne OK; (Man. to Newf., s to GA & e TX). *Blephariglottis lacera* (Michx.) Farw.—Rydberg.

5. Habenaria leucophaea (Nutt.) A. Gray, prairie fringed orchis. Plant stout, 30–80 cm tall; roots fleshy, coarse, numerous; stem barely angular. Leaves 2–5, cauline, oblong-elliptic to lanceolate, sheathing the stem below, 7–16 cm long, 1–4 cm wide, obtuse to acute. Raceme loosely flowered, 5–22 cm long; floral bracts lanceolate, acuminate, 1.5–4 cm long. Flowers white or whitish, occasionally slightly greenish; ovaries 2–3 cm long, recurved; perianth directed forward and forming a hood over the column, dorsal sepal ovate, concave, 9–13 mm long, 5–8 mm wide; lateral sepals obliquely ovate, acute to obtuse, 7–14 mm long, 5–10 mm wide; petals cuneate to flabelliform, broadly rounded to truncate at the apex; lip deeply 3-parted, 2–3 cm long, 2–3.5 cm wide, the parts cut about 1/2 way to the base; spur curved, clavate, usually much longer than the ovary, (2)4–5.5 cm long; column auricled. Capsules ellipsoid, 2–2.5 cm long, 4–6 mm wide. Jun–Jul. Damp meadows; ne OK, e KS, cen & e NE, se SD, se ND, w MN, w IA; (se ND to NY, s to OH & OK, also ME & N.S.). *Blephariglottis leucophaea* (Nutt.) Farw.—Rydberg.

6. Habenaria orbiculata (Pursh) Torr., round-leaved orchis. Plant scapose, 40–50 cm tall; roots fleshy, fusiform; stem stout. Leaves 2, basal and spreading on the ground, subopposite, suborbicular to broadly elliptic, nearly obtuse, 12–16 cm long, 8–10 cm wide. Raceme loosely flowered, 6–10 cm long; floral bracts linear to lanceolate, acute to obtuse, not equaling the flowers, 1–1.5 cm long, 2–3 mm wide. Flowers greenish-white, ovaries slender and pedicellate, 1.5(2.5) cm long; dorsal sepal suborbicular, erect 8–9 mm long, 7 mm wide; lateral sepals ovate, obtuse, oblique, ca 10 mm long, 5–6 mm wide; petals ovate to ovate-lanceolate, obtuse to subacuminate, oblique, reflexed, ca 10 mm long, 2–4 mm wide; lip ligulate or linear-oblong, pendent, 13–15 mm long, 2.5–3 mm wide; spur slender, cylindrical to clavellate; 1.5–2.2 cm long, equaling or longer than the ovary; column 4 mm long. Capsule ellipsoid, curved, 1–1.5 cm long, 3–5 mm wide. (2n = 42) Jul. Damp, shady coniferous woods; BH; (s AK to Newf., s to NC, MN, BH, & WA). *Platanthera orbiculata* (Pursh) Lindl.—Rydberg.

7. Habenaria saccata Greene, slender bog orchis. Plant slender to stout, 27–54 cm tall; roots fleshy, fusiform. Leaves to 10, cauline, the lowermost nearly ovate, the others elliptic to linear lanceolate to oblanceolate, obtuse or acute, barely if at all sheathing the stem, 4–8(14) cm long, 1–2.6(4) cm wide, the upper ones reduced. Raceme loosely to densely flowered, slender, 4–16 cm long; floral bracts linear, acuminate, the lower ones 3 or more cm long and much exceeding the flowers. Flowers small, green, sometimes tinged purplish; sepals thin, petals and lip fleshy; dorsal sepal suborbicular, rounded to obtuse, erect, connivent with the petals to form a hood over the column, 3–4 mm long, 2.5–3 mm wide; lateral sepals spreading or reflexed, triangular-ovate to elliptic-lanceolate, oblique, obtuse, 4–6 mm long, 2–2.5 mm wide; petals triangular-lanceolate to elliptic-lanceolate, falcate, obliquely dilated and auriculate, 3–4 mm long, 1–1.5(2) mm wide; spur scrotiform to cylindric-clavate, shorter than the lip, 2–3 mm long; lip entire, ±linear, column short, thick. Capsule obliquely ellipsoid, ca 1 cm long, 3–4 mm wide. (2n = 42) Aug. Marshy places & along streams near pine woods; BH; (AK s to MT, WY, & CA).

This species resembles slender forms of *H. hyperborea* and intermediate plants can be found.

8. **Habenaria viridis** (L.) R. Br. var. **bracteata** (Muhl.) A. Gray, long-bracted orchis, frog orchid. Plant stout or slender, 23–41 cm tall; roots fleshy, palmate. Leaves cauline, variable, the lower ones obovate to oblanceolate, the upper ones oblanceolate to lanceolate, obtuse to acute, 4–15 mm long, 4 cm wide. Raceme spiciform, usually loosely flowered with up to 20 flowers, 5–15 cm long; floral bracts linear-lanceolate, acuminate, 1.5–4.5 cm long, the lower ones much exceeding the flowers, the upper ones about equaling the flowers. Flowers green, sometimes tinged reddish; ovary pedicellate, stout; dorsal sepal ovate-orbicular to oblong-elliptic, (3)4–7 mm long, 2–3 mm wide; lateral sepals obliquely ovate, obtuse, 4–7 mm long, 2–3(4) mm wide; petals linear-lanceolate, acute to obtuse, 3–5 mm long, lip oblong-spatulate to narrowly cuneate, apex lobed, the middle lobe small, 5–8(10) mm long, 2–3(4) mm wide; sometimes tinged reddish; spur scrotiform, 2 mm long. Capsule ellipsoid, 7–10 mm long, 2.5–4 mm wide. (2n = 42) Jun–Aug. Marshes, bogs, stream edges, often in shade; se, n-cen & nw NE, w SD, e WY, ND, nw MN; (AK to Newf., s to NC, MN, & s in the mts. to NM & WA; Asia, Europe). *Coeloglossum bracteatum* (Willd.) Parl.—Rydberg.

The pink-flowered soldier's plume, *Habenaria psychodes* (L.) Spreng., has supposedly been collected once in our range (NE: Thomas) in the artificial forest planted there near the turn of the century. The specimen is fragmentary and the habitat is unlikely.

9. HEXALECTRIS Raf.

1. **Hexalectris spicata** (Walt.) Barnh., crested coral-root. Plant saprophytic, without chlorophyll, leafless, glabrous, 16–60 cm tall; rhizome annular, coralloid, stout, branching. Stem usually simple, yellow to madder-purple or brownish; bracts wide-spaced, sheathing, scalelike, 5–15 mm long. Raceme scapose, loosely flowered, with up to 25 flowers, 5–35 cm long; floral bracts purplish, ovate, acute, 5–13 mm long. Flowers showy, drooping, yellowish to reddish, ovary ca 9 mm long; sepals and petals spreading, yellow or yellowish or pinkish, with purple or brown striations; dorsal sepal oblong-elliptic, obtuse, 2–2.6 cm long, 5–7 mm wide; lateral sepals ovate-oblong, obtuse, oblique, 1.4–2 cm long, 5.5–8 mm wide; petals oblong-elliptic, somewhat spatulate, subacute, 1.7–2 cm long, 5–7 mm wide; lip shallowly 3-lobed, concave, ovate or obovate, short clawed, recurved, white or yellow, tinged purplish, with purple striations, 1.4–2 cm long, 1–1.4 cm wide, lateral lobes entire, rounded, incurved around the column; central lobe broad, almost truncate, undulate or barely notched; disc with 5–7 crested fleshy nerves; column white, arched, 1.3–1.6 cm long, with apical wings. Pollinia 8. Capsule pendent, ellipsoid, 3-ribbed, 2–2.5 cm long, long-persistent. Late Jun–Jul. Well-drained slopes in oak woods; se KS: Montgomery; s-cen OK: Grady, Caddo, Rogers; (VA s to FL, w to se AZ, n to s OK, se KS, & s IN).

The flowering scapes appear sporadically, abundant in one year and absent another. They sometimes occur in groups.

10. LIPARIS Rich.

1. *Liparis loeselii* (L.) Rich., Loesel's twayblade. Plant inconspicuous, glabrous, scapose, 6–30 cm tall; roots fibrous. Stem with a basal pseudobulb that often persists into the next year, slender, somewhat angular above. Leaves (1)2, erect, basal, glossy, basally sheathing the stem, oblong-elliptic, obtuse to acute, keeled beneath, to 26 cm long, 1–4.5 cm wide; a few sheathing bracts below the leaves. Raceme loosely flowered, 2–11 cm long, flowers 1–12; floral bracts minute. Flowers green or greenish-white or yellowish; ovary slender; dorsal sepal erect, oblong-lanceolate, the base hastate, 5–6 mm long, ca 2 mm wide; lateral sepals oblong-lanceolate, ca 5 mm long, ca 2 mm wide; petals filiform, tubular, curved, ca 5 mm long; lip recurved, unlobed, obovate to oblong, cuneate, apiculate, the margins wavy, 4–5 mm long, 3–3.5 mm wide, thickened down the center; column short, ca 2 mm long. Capsule ellipsoid, erect, 6–17 mm long, 4 mm wide. May–Aug. Open or shady boggy areas, sometimes in drier places; ne ND, w MN, n-cen NE (sandhills), ne KS, nw IA; (Man. to N.S., s in the mts. to NC, w to KS, & n to ND; also AL, WA, & Sask.).

11. LISTERA R. Br.

1. *Listera convallarioides* (Sw.) Nutt., broad-leaved twayblade. Plant often slender, inconspicuous, 5–20 cm tall; rhizome small, roots fibrous. Stem glandular-pubescent above the leaves, its base with a few sheaths. Leaves 2, opposite or subopposite at about the middle of the stem, ovate to elliptic or suborbicular, rounded to obtuse, glabrous, 2–7 cm long, 1.5–6 cm wide. Raceme loosely flowered with up to 20 greenish or yellowish flowers, 2–9 cm long; floral bracts lanceolate to rhombic, 2.5–3.5 mm long. Ovary stout, 4–6 mm long; sepals and petals reflexed, dorsal sepal elliptic, 4.5–5 mm long, ca 2 mm wide; lateral sepals lanceolate, falcate, 4.5–5.5 mm wide; petals linear, falcate, 4–5 mm long, ca 1 mm wide; lip short clawed, cuneate, apically dilated and notched, the lobes rounded, the base above the claw blunt-toothed, 8–13 mm long, 5–7 mm wide; column curving, slender, 3.2–5.5 mm long. Capsule ellipsoid, 5–7 mm long, 2–3 mm wide, pubescent along the veins. Jul. Moist coniferous woods, bogs; ne WY, BH; (B.C. to Newf., s to NY, n MI, ne MN, BH, & in the mts. to UT, CO, & CA).

12. PIPERIA Rydb.

1. *Piperia unalascensis* (Spreng.) Rydb., Alaska orchis. Plant erect, 30–50 cm tall; roots fleshy, short, with a pair of tubers 1–2(4) cm long and 1 cm thick; stem yellowish, sometimes tinged purplish. Leaves 2–4, basal, erect-spreading, often drying before the flowers appear, oblanceolate to lanceolate, rounded to subacuminate, sheathing the stem below, 7.5–17 cm long, 1–3.4 cm wide; cauline bracts translucent. Raceme spiciform, loosely flowered, elongate, to 35 cm long; floral bracts ovate to linear-lanceolate, obtuse to long acuminate, concave, 3–8 mm long, 2–3 mm wide, shorter than the flowers. Flowers small, whitish, greenish, or yellowish, often marked with purple; ovaries arcuate; sepals thin, ovate-elliptic, obtuse to rounded, 2–3(4) mm long, ca 1 mm wide; lateral sepals spreading or reflexed, adhering to the base of the lip; petals fleshy, ovate to elliptic-lanceolate, obtuse to subacute, 2–3 mm long, ca 1 mm wide; lip fleshy, ovate-elliptic, 2–3 mm long, ca 1 mm wide; spur cylindric to clavellate, curved, 2–3 mm long, about equaling the lip; column ca 1 mm

long. Capsule obliquely ellipsoid, 5–7.5(10) mm long, 2–3 mm wide. Jul. Moist or dry pine woodlands & meadows on sandy soil, often on dry slopes; BH & e WY; (AK s to Baja Calif., e to MT, CO, & SD; also in MI, Ont., Que.). *Habenaria unalascensis* (Spreng.) S. Wats.—Van Bruggen.

13. POGONIA Juss.

1. *Pogonia ophioglossoides* (L.) Ker, rose pogonia. Plant 20–30 cm tall, glabrous; roots fibrous. Stem slender, sometimes purplish, a bract near its base. Leaf 1(2), below the middle of the stem, not basal, elliptic to ovate, 4–5 cm long, 1–2 cm wide. Flower 1(2), terminal, subtended by a foliaceous bract that is oblong to ovate, acute, 1.5–2 cm long, 5–7 mm wide. Flower pink, occasionally white, fragrant; sepals oblong-elliptic, subobtuse, 1.8–2.2 cm long, 4–5 mm wide; petals elliptic, obtuse, 1.7–2 cm long, 5–7 mm wide; lip with a dark pink-red fringe, oblong-spatulate, the 3 veins of the disk distinctly yellow-bearded in the middle, becoming reddish-fringed toward the apex, 1.5–2.5 cm long, ca 1 cm wide; column pink, curved, ca 1 cm long, the anther terminal, pollinia 2. Capsule erect, ellipsoid, 2–3 cm long. (2n = 18) Jul–Aug. Bogs & marshy meadows; e ND: Grand Forks; (e ND to Newf., s to FL & e TX; MO).

This species has been collected in our area only once; it is to be sought in suitable habitats in e ND and nw MN.

14. SPIRANTHES Rich., Lady's-tresses

Perennial erect herbs of prairies and woodlands; roots fibrous or fleshy. Stem simple, erect, often solitary but sometimes several from a basal rosette, terminating in an apparently twisted spike, variously pubescent. Leaves mostly basal, sometimes cauline on the lower stem, reduced to sheathing bracts on the upper stem, persistent or disappearing before flowering time; linear, elliptic, oblanceolate, or ovate. Flowers small to medium sized, often white or whitish, sometimes marked on the lip with yellow or green; lateral sepals decurrent, dorsal sepal and petals adherent over the column; lip sessile or short clawed, usually flanked with small tuberosities at the base; lip apex variously embellished; column usually short, terete or clavate, elevating as it ages; anther dorsal on the column, pollinia 2. Capsule erect, ellipsoid to ovoid or obovoid.

This genus contains about 300 species, mostly in the New World. In the GP it is the most common grassland orchid.

```
1  Flowers in a single spiral on the spike.
   2  Leaves linear, lanceolate, or oblanceolate, cauline and/or basal, present at flowering time
      3  Lip white with a yellow-green center ................................................. 1. S. cernua
      3  Lip white ........................................................................... 8. S. vernalis
   2  Leaves elliptic to ovate, basal, often absent at flowering time.
      4  Lip white; root 1 ................................................................. 7. S. tuberosa
      4  Lip white with a yellow-green center; roots fasciculate ..................... 2. S. lacera
1  Flowers in more than 1 spiral on the spike.
   5  Lip yellow or orange; leaves glossy ................................................ 3. S. lucida
   5  Lip white or whitish or with a greenish or yellowish center; leaves shiny but not glossy.
      6  Flowers small, about 5 mm broad; leaves oblanceolate; shaded
         woodlands ...................................................................... 6. S. ovalis
      6  Flowers larger, 7–12 mm broad; leaves linear; prairies and swales, open woodlands
         7  Leaves absent at flowering time; lip not constricted near the
            middle ................................................................. 4. S. magnicamporum
         7  Leaves present at flowering time; lip slightly to distinctly constricted near the middle.
```

8   Lip thin, pandurate, noticeably constricted near the center .... 5. *S. romanzoffiana*
8   Lip thick, not strongly constricted near the center .... 1. *S. cernua*

1. **Spiranthes cernua** (L.) Rich., lady's-tresses. Plant to 60 cm tall, often much less; roots numerous, slender, fleshy, fasciculate. Stem green, glabrous below, pubescent above, the hairs blunt. Leaves mostly basal, linear to lanceolate, acute or acuminate, soft, to 25 cm long and 2.5 cm wide. Spike with usually several(1) twisting tight ranks of flowers; floral bracts green, ovate to long-acuminate, 8-25 mm long. Flowers up to 60, mostly perpendicular to the rachis, fragrant at times, appearing inflated; sepals white, lanceolate, ca 1 cm long and 3 mm wide; petals white or whitish, linear, somewhat falcate, to 1.5 cm long; lip thick, white with yellow-green center, ovate, somewhat constricted at the center, its apex rounded, recurved, undulate or crenulate, basal tuberosities prominent; column green, stout, 3-7 mm long; seeds polyembryonic. (2n = 30, 50). Aug-Nov. Wet or dry prairies, occasionally in open woodlands; cen OK, cen & e KS, NE, e SD, se ND, w MN (U.S. e of 100° long., except FL; s Ont., Que.). *S. cernua* (L.) Rich, *S. ochroleuca* Rydb.—Rydberg.

This is our commonest orchid. It is highly variable and probably hybridizes with other species. In our area it is most similar to *S. magnicamporum*.

2. **Spiranthes lacera** (Raf.) Raf., slender lady's tresses. Plant 18-60 cm tall, sometimes with several stems from a rosette of basal leaves; roots fleshy, fasciculate. Leaves 3-5, all basal, not always persisting at flowering time, ovate to elliptic, acute, lamina 1.5-6.5 cm long, 1-2.5 cm wide, short petioled; cauline bracts sheathing the stem. Spike with up to 40 flowers in a single tight or loose spiral, or secund; floral bracts ovate to ovate-lanceolate, acute to acuminate, 5-10 mm long, 1-3 mm wide. Flowers small, white with a green center; ovary stout, 3 mm long; sepals and petals equal, 4-5.5 mm long; dorsal sepal elliptic-oblong to oblong-lanceolate, obtuse to acute; lateral sepals lanceolate, acute; petals adherent to the dorsal sepal, linear; lip white with green center, oblong, 4-6 mm long, 2-2.5 mm wide, its apex finely lacerate; column 2 mm long. Capsule 3-7 mm long, 2-4 mm wide. (2n = 30) Aug-Oct. Moist meadows, open deciduous woodlands; se NE, e KS, e OK, e TX; (N. Amer. e of 95° long., except peninsular FL, n to s Can.).

Ecotypically variable, the woodland plants of this species have more loosely spiraled inflorescences than the prairie plants. Two varieties may be recognized: var. *lacera* (loosely spiraled flowers; thin, persistent leaves) and var. *gracilis* (Bigel.) Luer (tightly spiraled flowers; thicker, fugacious leaves). Magrath, op. cit., has experimentally shown these to be woodland and prairie forms, respectively.

3. **Spiranthes lucida** (H. H. Eaton) Ames, shining lady's-tresses. Small, slender, glabrous below, sparsely pubescent above, to 35 cm tall but usually much less; roots fleshy, from a thickened base. Leaves in a basal cluster, and a few sometimes on the lower part of the stem; glossy, fleshy, elliptic to lanceolate, acute, sheathing the stem below, 3-12 cm long, 0.5-1.8 cm wide. Flowers in 2 or 3 spiral ranks; floral bracts lanceolate, acute, 1-1.5 cm long. Flowers tubular, nodding, white with a yellow center; dorsal sepal linear, oblong, 4.5-5.5 mm long, 1-1.8 mm wide, forming a hood with the petals; lateral sepals free, not much spreading, linear, obtuse, 5-6 mm long, 1-1.5 mm wide; petals linear to oblanceolate, 5-6 mm long, ca 1-1.5 mm wide; lip yellow (orange), edged with white, subquadrate when spread out, conduplicate, apex crenulate, about 5 mm long, 2-3 mm wide; column white, 2-3 mm long. May-Jul. Wet meadows, gravel bars, disturbed wet areas; e NE: Cass; n-cen KS: Cloud; (NE, KS, & IL to N.S., s to KY & NJ). *S. plantaginea* (Raf.) Torr.—Rydberg.

This species has always been rare and is now possibly extirpated in GP. It has not been seen here since 1930, but is locally abundant in its main range east of our area.

4. ***Spiranthes magnicamporum*** Sheviak, Great Plains lady's-tresses. Plant to 60 cm tall, pubescent, the hairs capitate; roots stout, fleshy, fasciculate. Leaves linear-lanceolate, to 14 cm long and about 1 cm wide, usually dying before the flowers appear. Spike with several twisting tight ranks of flowers; floral bracts ovate, attenuate, to 3 cm long. Flowers numerous, fragrant, slender and not inflated; sepals white, linear-lanceolate, 7–11 mm long, 1.5–3 mm wide; petals linear, 7–11 mm long, 1–2 mm wide, adherent to the dorsal sepal; lip white with the center yellowish and fleshy, oblong-ovate, its margin crisped, not dilated basally, basal tuberosities small; column 3 mm long. Seeds monoembryonic. Sep–Nov. Moist ditches & dry grassy prairies on calcareous chert & gypsum soils; ND to TX e of 100° long.; (eastward in the prairie peninsula to IN; also in UT, CO, NM, OH, & AL). *S. cernua* (L.) Rich., in part—Rydberg.

This species resembles *S. cernua* but begins to flower later, after its leaves have dried, and is more abundant in dry grasslands than *S. cernua*.

5. ***Spiranthes ovalis*** Lindl., oval lady's-tresses. Plant slender, glabrous below, more or less pubescent above, to 45 cm tall; roots few, elongate, fleshy. Leaves 2 or 3, basal or on the lower part of the stem, oblanceolate, obtuse or acute, narrowed below the middle, 5–17 cm long, 6–13 cm wide. Spike with up to 50 flowers in 2 or 3 tight spirals; floral bracts ovate-lanceolate, long acuminate, 4–10 mm long, 1–3 mm wide. Flowers very small, 3–5 mm long; sepals similar, white, lanceolate, acute, 3.5–5 mm long, 1–2 mm wide; petals white, 3.5–5 mm long, 1 mm wide; lip ovate, somewhat constricted above the middle, 4–5.3 mm long, 2.4–4 mm wide, recurved, column about 2 mm long. Capsule 4–7.5 mm long, 3–4 mm wide. Sep–Oct. Uncommon & local in rich soil of woodlands of oak, hickory, & sugar maples; cen OK, e KS; (n MO to VA, s to FL & e TX).

6. ***Spiranthes romanzoffiana*** Cham., hooded lady's-tresses. Stem slender or stout, glabrous below, glandular-pubescent above, 14–30 cm tall; roots long, fleshy. Leaves basal and cauline, the upper ones reduced and sheathing, linear to oblanceolate or oblong-lanceolate, 5–14 cm long, 6–11 mm wide, subacute to acuminate. Flowers to 60, crowded in 2 or 3 spirals, tilted upward; floral bracts lanceolate to ovate, acuminate, 2–2.5 cm long, 6–8 mm wide. Flowers white or cream, tubular, dilated above the middle; dorsal sepal oblong-elliptic to lanceolate, 6–8 mm long, 1–2 mm wide; lateral sepals lanceolate, falcate; petals linear, obtuse, 7–11 mm long, 1–2.5 mm wide; lip thin, pandurate, constricted above the middle, 7–8(11) mm long, 5 mm wide, basal tuberosities small, apex finely lacerate; column 2–3 mm long; capsule 6–7 mm long, 3–4 mm wide. (2n = 60) Jul–Sep. Swamps, wet meadows, open woodlands; w SD, n ND, n NE, nw IA; (AK to Newf., s to New England, Great Lakes, GP, & s in the mts. to NM & CA; Ireland). *S. stricta* Rydb.—Rydberg.

This species resembles *S. cernua* but is more western in distribution in GP, the flowers are tilted upward, the lip is constricted, and the perianth forms a hood.

7. ***Spiranthes tuberosa*** Raf., little lady's-tresses. Slender, glabrous herb, 8–40 cm tall, sometimes with several stems from the same basal rosette of leaves; root 1, fleshy, fusiform, the previous year's root occasionally persisting. Leaves all basal, in a rosette, fugacious, ovate, 2.5–6.5 cm long, 6–15 mm wide; cauline bracts small, sheathing. Inflorescence rachis very slender, flowers up to 30, loosely to tightly spiraled or secund; floral bracts ovate, 3–5 mm long, 0.7–1.5 mm wide. Flowers small, white, in a single spiral; sepals and petals equal, 2–3.5 mm long; dorsal sepal oblong-elliptic to oblanceolate, obtuse; lateral sepals lanceolate, acute, recurved; petals adherent to the dorsal sepal; lip white, ovate to oblong-quadrate, the apex barely erose, 2.3–4 mm long, 1.5–2 mm wide; column 1.5 mm long, green. Capsule obliquely ovoid, 3–4 mm long, 2.5–3 mm wide. Sep–Oct. Dry woods of oak, hickory & pine (mostly blackjack & post oaks) in sandy or cherty soils; e & se KS, e OK; (KS to MA s to FL & e TX).

8. ***Spiranthes vernalis*** Engelm. & Gray, twisted lady's tresses, spring lady's-tresses. Plant 18–117 cm tall, densely pubescent above with pointed articulate hairs; roots coarse, fusiform, fasciculate. Leaves basal and sometimes cauline, more or less erect, linear to narrowly lanceolate, acute to acuminate, to 30 cm long and 1.6 cm wide. Spike with up to 50 whitish flowers, loosely to tightly spiraled in one rank or secund, usually perpendicular to the rachis; rachis with reddish or whitish pointed hairs; floral bracts ovate to oblong-lanceolate, acuminate, 7–23 mm long, 4–7.5 mm wide, the margins scarious. Perianth pubescent on the outside; petals and sepals about equal, 5–10 mm long; dorsal sepal adherent with the petals, oblong-lanceolate, obtuse or acute; lateral sepals lanceolate, acute; petals linear, obtuse, 1–2 mm wide; lip ovate, fleshy, recurved, apically crenulate, 4.5–8 mm long, basal tuberosities stout, pubescent; column greenish, 2 mm long. Capsule obliquely ovoid to elliptic, 6–9 mm long, 3.5–6 mm wide. Jun–Aug. Moist meadows, damp ditches, loamy soils; se SD, NE: Hall, e KS, e OK; (se SD to MA, s to FL & e TX; Mex., Guat.).

15. TRIPHORA Nutt.

1. ***Triphora trianthophora*** (Sw.) Rydb., nodding pogonia. Plant 8–25 cm tall, inconspicuous, gregarious, glabrous; roots white, slender, producing terminal tubers up to 3 cm long. Stem slender, succulent, suffused purple. Leaves 2–8, distant, small, often suffused purple, ovate or cordate, obtuse to acute, clasping the stem, 6–14(20) mm long, 4–10 mm wide, the lower ones reduced to bracts. Inflorescence somewhat racemose, flowers 1–6, borne in the axils of the upper leaves. Flowers nodding or erect, lasting a day, white and/or pink, green-tinged; ovary slender; sepals and petals similar, oblanceolate, 1.3–1.5 cm long, 4 mm wide; lip white, 3-lobed, obovate, narrow-clawed, 1–1.5 cm long, 1 cm wide, the disk with 3 green tuberculate crests, lateral lobes ovate, obtuse, central lobe orbicular, sinuate; column white, 1 cm long; anther terminal, pollinia purple. Capsule ellipsoid, 1–2 cm long, pendent or erect. (2n = 44) Mid Aug–Sep. Moist upland woods; se NE, e-cen KS, cen OK; (s WI to ME, s to FL & e TX, w in Missouri R. valley to NE & KS; OK).

These plants are sporadic in appearance, being present in an area one year and absent another.

Abbreviations for Nomenclatural Authorities

Compiled by Ralph E. Brooks

Entries are arranged alphabetically according to the abbreviation and not to the author's full name. This compilation is based in part on those included in Munz and Keck, *A Flora of California* (1959) and Correll and Johnston, *Manual of the Vascular Plants of Texas* (1970). Other references too numerous to cite, as well as personal communications with countless individuals, contributed to this compilation. Michael T. Stieber, Hunt Institute for Botanical Documentation, was especially helpful in the final stages of the preparation.

Adans.   Michel Adanson, 1727–1806, French botanist and explorer; author of *Families des Plantes* (1763); originated some 1600 generic names.

Aellen   Paul Aellen, 1896-1973, professor in Basel, Switzerland; student of Chenopodiaceae.

Agardh   Carl Adolf Agardh, 1785-1859, noted Swedish algologist, professor at Lund (1807–1835); later bishop of Karlstad.

Agardh, J. G.   Jakob Georg Agardh, 1813–1901, son of C. A. Agardh, professor of botany, Lund; noted algologist.

Ahles   Harry E. Ahles, 1924–1981, University of North Carolina, Chapel Hill; later University of Massachusetts, Amherst; author (with A. E. Radford and C. Ritchie Bell) of *Guide to the Vascular Flora of the Carolinas* (1964) and *Manual of the Vascular Flora of the Carolinas* (1968).

Aiken   Susan G. Aiken, 1937–, Biosystematic Research Institute, Ottawa, Canada; *Myriophyllum*.

Ait.   William Aiton, 1731–1793, English botanist, Royal Gardener at Kew; issued *Hortus Kewensis* (1789).

Alex.   Edward Johnston Alexander, 1901–, New York Botanical Garden; authority on botany of the southeastern United States; Iridaceae (*Iris*).

All.   Carlo Allioni, 1725(8?)–1804, Italian physician and botanist, professor of botany, Turin; author of *Flora Pedemontana* (1785).

Al-Shehbaz   Ihsan A. Al-Shehbaz, fl. 1973, University of Bagdad, Bagdad, Iraq; *Thelypodium*.

Anders. & Woods.   Edgar Shannon Anderson, 1897–1969, Missouri Botanical Garden, St. Louis; experimental taxonomy, author of *Introgressive Hybridization* (1949), and R. E. Woodson.

Anderss.   Nils Johan Andersson, 1821–1880, Swedish botanist, director of the Botanical Museum, Stockholm; collected in California in 1852; *Salix, Andropogon*.

Andr.   Henry C. Andrews, fl. 1794-1830, English botanical artist and engraver; *Geranium, Erica,* and *Rosa*.

Andrz.   Antoni Lukianowicz Andrzejowski, 1785–1868, professor of botany, Vilna, Lithuania.

Angstr.   Johan Angstrom, 1813–1879, Swedish bryologist.

Antoine   Franz Antoine, 1815–1886, Austria

1285

Arcang. Giovanni Arcangeli, 1840–1921, Italian cryptogamic botanist.
Ard. Pietro Arduino, 1728–1805, Italian botanist and agriculturist, Padua.
Arn. George Arnott Walker Arnott, 1799–1868, Scottish botanist.
Arthur Joseph Charles Arthur, 1850–1942, authority on plant rusts, Purdue University; editor of *Botanical Gazette.*
Asch. Paul Friedrich August Ascherson, 1834–1913, professor of botany, Berlin.
Asch. & Graebn. P. F. A. Ascherson and K. O. Graebner.
Ashe William Willard Ashe, 1872–1932, North Carolina botanist and forester, U.S. Forest Service; *Panicum, Asarum, Crataegus, Polycodium.*
Atwood Nephi Duane Atwood, 1938–, Uinta National Forest, Provo, Utah; *Phacelia,* flora of the Intermountain Region.
Aubl. Jean Baptiste Christophore Fusee Aublet, 1720–1778; French botanical collector and author of a French Guiana flora (1775).
Aust. Coe Finch Austin, 1831–1880, American botanist; Lemnaceae, author of *Musci Applachiani* (1870).
Ave-Lall. Julius Leopold Eduard Ave-Lallemante, 1803–1867, German associate of the botanical garden in St. Petersburg.
Averett John Earl Averett, 1943–, Dept. of Botany, University of Texas, Austin, later University of Missouri, St. Louis; *Chamaesaracha* (Solanaceae).
Bacig. Rimo Carlo Felice Bacigalupi, 1901–, Jepson Herbarium, Berkeley; Saxifragaceae, flora of California.
Backeb. Curt Backeburg, 1894–1966, Germany; Cactaceae.
Bailey Liberty Hyde Bailey, 1858–1954, Cornell University; eminent horticultuist and author, *Standard Cyclopedia of Horticulture* (1914–1917), *Manual of Cultivated Plants* (1925), *Hortus* (1930), etc., founder of *Gentes Herbarum;* student of *Carex, Rubus,* palms, Cucurbitaceae.
Bailey, V. L. Virginia Long Bailey, 1908–, Detroit Institute of Science; *Ptelea.*
Baill. Henri Ernest Baillon, 1827–1895, Paris botanist and physician; author of *Histoire des Plantes* (1866–1895) and many other works.
Baker John Gilbert Baker, 1834–1920, Keeper of the Herbarium, Kew (1890–1899); student of ferns, Amaryllidaceae, Bromeliaceae, Iridaceae, Liliaceae, Asteraceae, and the flora of tropical Africa.
Balbis Giovanni Battista Balbis, 1765–1831, professor of botany, Turin.
Baldw. William Baldwin, 1779–1819, Pennsylvania physician and botanist who collected and studied plants of Delaware, Georgia, and South America, died while on expedition to Rocky Mts.
Balf. John Hutton Balfour, 1808–1884, physician and botany professor at Glasgow, Scotland, then at Edinburgh; author of *Class Book of Botany* (1855–1871).
Ball, C. R. Carleton Roy Ball, 1873–1958, Senior agronomist, U.S. Dept. of Agriculture, Washington; *Salix.*
Ball, P. W. Peter William Ball, 1932–, Hartley Botanical Laboratories, University of Liverpool, England.
Ball & Heywood P. W. Ball and Vernon Hitton Heywood, 1927–, director of Plant Sciences Laboratories, University of Reading, England; Mediterranean flora, *Digitalis,* Asteraceae, Apiaceae.
Banks Sir Joseph Banks, 1743–1820, British naturalist, traveler, philanthropist, and scientist, accompanied Capt. Cook in his first voyage of circumnavigation in 1768; early director of Kew; president of the Royal Society.
Banks, D. Donald Jack Banks, 1930–, Oklahoma State University, Stillwater; Poaceae (*Paspalum*).
Barker William Thomas Barker, 1941–, North Dakota State University; student of the northern Great Plains flora.
Barkley, F. Fred Alexander Barkley, 1908–, Northeastern University, Boston (1965–), formerly University of Texas, Austin (1943–1947), U.S. State Dept., Medellin, Colombia (1947–1949), Inst. Miguel Lillo, Tucuman, Argentina (1949–1951), industrial chemicals (1951–1961), University of Baghdad, Iraq (1963–1965); Anacardiaceae, *Begonia.*
Barkley, T. Theodore Mitchell Barkley, 1934–, Kansas State University; New World *Senecio, Manual of the Flowering Plants of Kansas* (1968).
Barkworth Mary Elizabeth Barkworth, 1941–, Utah State University, Logan; Stipeae and Triticeae in North America.
Barneby Rupert Charles Barneby, 1911–, New York Botanical Garden; student of the North American Intermountain flora; *Astragalus, Cassia, Dalea, Oxytropis.*
Barnh. John Hendley Barnhart, 1871–1949, American bibliographer, New York Botanical Garden; Lentibulariaceae.
Barr. Joseph Barratt, 1796–1882, Connecticut, geologist and physician; *Salix.*
Bart William Paul Crillon Barton, 1786–1856, U.S. Navy surgeon and professor of botany, Pennsylvania; author of *A Flora of North America* (1820–1823).

Bartal.   Biagio Bartalini, 1746–1822, professor of botany, Siena, Italy.
Bartl.   Friedrich Gottlieb Bartling, 1798–1875, professor of botany, Gottingen.
Bartr.   William Bartram, 1739–1823, American botanist in Philadelphia; author of *Travels through North and South Carolina, etc.* (1791).
Batch.   Fredrick William Batchelder, 1838–1911, New Hampshire; ornithologist, floristics of Manchester, NH.
Bates   Rev. John Mallery Bates, 1846–1930, Nebraska clergyman, botanist, and ornithologist who studied plants of the state; Cyperaceae.
Batsch   August Johann Georg Karl Batsch, 1761–1802, German botanist and horticultural writer, Jena.
Bauh.   Caspar Bauhin, 1560–1624, professor medicine, Basel; author of the *Pinax*; maker of the first distinct diagnoses of species.
Baum   Bernard Rene Baum, 1937–, Paris-born Canadian botanist, Biosystematics Research Institute, Ottawa, Canada; *Avena, Tamarix.*
Baumg.   Johann Christian Gottlob Baumgarten, 1765–1843.
Bayer & Stebbins   Randall J. Bayer, graduate student at Ohio State University in early 1980's; *Antennaria*; and G. L. Stebbins.
Beadle   Chauncey Delos Beadle, 1866–1950, Canadian-born botanist with Biltmore Herbarium, North Carolina (1890–1916); botany of southern United States; *Crataegus.*
Beadle & Boynt.   C. D. Beadle and F. E. Boynton.
Beal, W.   William James Beal, 1833–1924, Michigan State University, East Lansing, agrostologist, author of *Michigan Flora* (1904).
Beal, E. O.   Ernest Oscar Beal, 1928–, North Carolina State University, Raleigh; later Western Kentucky University, Bowling Green; Nymphaeaceae (*Nuphar*), Lemnaceae, experimental taxonomy of aquatic plants.
Beaman   John Homer Beaman, 1929–, Michigan State University, East Lansing; Asteraceae, high altitude floras of Mexico and Central America.
Beauv.   Baron Ambroise Marie François Joseph Palisot de Beauvois, 1752–1820, French naturalist; Poaceae.
Bebb   Michael Schuck Bebb, 1833–1895, Illinois; *Salix*, author of *Botany of the Northern and Middle States* (1853).
Beck   Lewis Caleb Beck, 1798–1853, Rutgers College, New Jersey, then Albany Medical College, New York.
Beck, G.   Gunther Beck von Mannagetta und Lerchenau, 1856–1931, Czechoslovakian paleobotanist.

Beeby   William Haddon Beeby, 1849–1910, English botanist.
Beetle   Alan Ackerman Beetle, 1913–, professor of agronomy, University of Wyoming, Laramie; Poaceae, Cyperaceae.
Beg. & Bel.   Augusto Beguinot, 1875–1940, director of the Botanical Garden in Padua (1915–1921), then professor in Genoa, Italy; *Apocynum*, floristics of Padua, and Nicola Belosersky (Belozersky), fl. 1913, Russian.
Benn.   Arthur Bennett, 1843–1939, English builder and amateur botanist.
Benson, L.   Lyman David Benson, 1909–, professor of botany, Pomona College, California; author (with Robert A. Darrow) of *A Manual of Southwestern Desert Trees and Shrubs* (1945) and *Plant Classification* (1957); *The Cacti of the United States and Canada* (1982); *Ranunculus*, Cactaceae.
Benth.   George Bentham, 1800–1884, long-time president of the Linnaean Society; outstanding English taxonomist; author of *Plantae Hartwegianae . . . enumerat* (1839), *Handbook of the British Flora* (1858, with later editions), *Flora Hongkongensis* (1861), *Flora Australiensis* (1863–1878), and monographic works on Leguminosae, Lamiaceae, Scrophulariaceae, Asteraceae, etc.
Benth. & Hook.   G. Bentham and Sir J. D. Hooker; authors of *Genera Plantarum* (1862–1883).
Bercht. & Presl   Friedrich Graf von Berchtold, 1781–1876, Czechoslovakian physician and botanist, and K. B. Presl.
Berger   Alwin Berger, 1871–1931, German, curator of the garden at La Mortola, Italy; student of Cactaceae and other succulents.
Berl.   Jean Louis Berlandier, 1805–1851. French-Swiss pharmacist who collected plants in northern Mexico and Texas; Grossulariaceae.
Bernh.   Johann Jacob Bernhardi, 1774–1850, professor of botany, Erfurt, Germany.
Berth.   Sabin Berthelot, 1794–1880, French consul on Teneriffe.
Bess.   Wilibald Swibert Joseph Gottlieb von Besser, 1784–1842, Austrian botanist, professor in the Wolynien Lyceum (Poland); student of the flora of Galicia and southwest Russia.
Best   George Newton Best, 1846–1926, physician at Rosemont, New Jersey; bryology and Canadian flora.
Beurl.   Pehr Johan Beurling, 1800–1866, Swedish botanist.
Bickn.   Eugene Pintard Bicknell, 1859–1925, New York banker and amateur botanist.
Bieb.   Baron Friedrich August Marschall von Bieberstein, 1768–1826, Stuttgart; German explorer and author of works on the flora of southern Russia and the Caucasus.

Biehler   Johann Friedrich Theodor Biehler, fl. 1807, Halle, Germany.
Bigel.   Jacob Bigelow, 1787–1879, professor of botany, Boston; author of *Florula Bostoniensis* (1814), and *American Medical Botany* (1817).
Bisch.   Gottlieb Wilhelm Bischoff, 1797–1854, professor of botany, Heidelberg, Germany.
Biv.   Baron Antonio Bivona-Bernardi, 1778–1837, Messina, Sicily.
Bl.   Carl Ludwig von Blume, 1796–1862, Dutch botanist, director of the botanic garden at Batavia and writer (with J. B. Fischer) on the flora of Java (1828–1851).
Blake   Sidney Fay Blake, 1892–1959, U.S. Dept. of Agriculture, Beltsville, Maryland; Asteraceae, *Polygala*, bibliographer; author of *Geographical Guide to the Floras of the World* (with Alice C. Atwood), Pt. I (1942), Pt. II (1961).
Blanch.   William Henry Blanchard, 1850–1922, New England teacher.
Blank.   Joseph William Blankinship, 1862–1938, botanist and plant pathologist, Montana State University, Bozeman, later with smelting companies.
Blasd.   Robert Ferris Blasdell, 1929–, Canisius College, Buffalo, New York; *Cystopteris*.
Boeck.   Johann Otto Boeckeler, 1803–1899, apothecary-botanist of Oldenburg, Germany; Cyperaceae.
Boehm.   George Rudolf Boehmer, 1723–1803, German botanist, Leipzig.
Boenn.   Clemens Maria Franz von Boenninghausen, 1785–1864, Dutch-born, practiced homeopathic medicine at Munster, Germany.
Bogin   Clifford Bogin, 1920–, New York Botanical Garden; later Woodmere Academy, Long Island, New York; *Sagittaria*.
Boiss.   Pierre Edmond Boissier, 1810–1885, Swiss botanist and traveler, Geneva; one of the outstanding systematists of the 19th century; author of the monumental *Flora Orientalis* (1867–1884); monographer of the Plumbaginaceae and *Euphorbia*.
Boiss. & Hohenack.   P. E. Boissier and R. F. Hohenacker
Boivin   Joseph Robert Bernard Boivin, 1916–1985, Biosystematics Research Institute, Ottawa, Canada; *Thalictrum*, author of *Flora of the Prairie Provinces*.
Boivin & Löve   J. R. B. Boivin and A. Love.
Bol.   Henry Nicholas Bolander, 1831–1897, California State Superintendent of Schools (1871–1875), then to Portland, Oregon in 1883; made extensive collections in California.
Bong.   August Heinrich Gustav Bongard, 1786–1839, professor of botany, St. Petersburg; monographer of Brazilian plants; describer of Merten's Alaskan collection.
Booth, B.   William Beattie Booth, 1804–1874, English gardener and author.
Boott, F.   Francis Boott, 1792–1863, American physician who settled in London; *Carex*.
Boott, W.   William Boott, 1805–1887, Boston; *Polypodiaceae*.
Bor.   Alexandre Boreau, 1793–1875, French botanist who wrote a flora of central France (1857).
Borbas   Vincze von Borbas, 1844–1905, Hungarian botanist.
Borkh.   Moritz Balthasar Borkhausen, 1760–1806, Germany, forester and naturalist.
Borner   Carl Borner, 1880–1953, Naumburg, Germany; *Gentianella*.
Bosc   Louis Augustin Guillaume Bosc, 1759–1828, French naturalist.
Bowden   Wray Merrill Bowden, 1914–, Dept. of Agriculture, Canada; cytogenetics of Poaceae.
Boyle   William Sidney Boyle, 1915–, professor of botany, Utah State University, Logan; Poaceae (*Melica*).
Boynt.   Frank Ellis Boynton, 1859–?, botanical collector for the Biltmore Herbarium, Biltmore, North Carolina; superintendent of Biltmore Estates to about 1935.
Br., A.   Alexander Carl Heinrich Braun, 1805–1877, professor of botany and director of the botanical garden, Berlin; Characeae, *Selaginella, Marsilea*.
Br., P.   Patrick Browne, 1720–1790, Irish physician and Naturalist, explorer of Jamaica, and an author of its natural history (1756).
Br., R.   Robert Brown, 1773–1858, British botanist, librarian, and first keeper of botany, British Museum; a chief exponent of the Natural System; noted morphologist and cytologist.
Brackett   Amelia Ellen Brackett, 1896–1926, Gray Herbarium, Harvard University.
Bradley   Ted Ray Bradley, 1940–, George Mason University, Fairfax, Virginia; *Triodanis*.
Brainerd   Ezra Brainerd, 1844–1924, President, of Middlebury College, Vermont; *Viola*.
Brand   August Brand, 1863–1930, German student of Polemoniaceae, Hydrophyllaceae, and Boraginaceae.
Branner   John Casper Branner, 1850–1922, American geologist, President of Stanford University (1913–1915).
Braun, E. L.   Emma Lucy Braun, 1889–1971, American ecologist.
Breitung   August Johann Breitung, 1913–, Canadian taxonomist; *Salix*.
Brew.   William Henry Brewer, 1828–1910,

geologist and botanist, leader of field parties in the California State Geological Survey.

Brign. Giovanni de Brignoli di Brunnhoff, 1774–1857, professor of botany, Modena; Poaceae.

Briot Pierre Louis Briot, 1804–1888, chief gardener at Petit-Trianon and Grand-Trianon, France.

Briq. John Isaac Briquet, 1870–1931, Swiss botanist, director of the Conservatoire Botanique, Geneva; Lamiaceae, Apiaceae, Asteraceae; noted for his work to advance modern nomenclature through the Botanical Congresses.

Britt. Nathaniel Lord Britton, 1859–1934, director-in-chief, New York Botanical Garden (1896–1930); flora of North America, West Indies, Bolivia; a prolific writer.

Britt. & Rendle N. L. Britton and Alfred Barton Rendle, 1865–1938, keeper of botany, British Museum.

Britt. & Rose N. L. Britton and J. N. Rose; authors of *The Cactaceae*, 4 vols. (1919–1923).

Britt. & Rusby N. L. Britton and Henry Hurd Rusby.

Brongn. Adolphe Theodore Brongniart, 1801–1876, Paris, France, physician, paleobotanist, taxonomist, and anatomist.

Brooks, R. E. Ralph Edward Brooks, 1950–, University of Kansas, Lawrence; student of the Great Plains flora, pteridophytes, *Juncus*.

Brot. Felix d'Avellar Brotero, 1744–1828, professor of botany, Coimbra, Portugal; author of *Flora Lusitanica* (1804); *Physalis*.

Brouillet Luc Brouillet, 1954–, McGill University Herbarium, MacDonald College, Ste-Anne-de-Bellevue, Quebec; *Aster*.

Broun Maurice Broun, 1906–, Massachusetts, ornithologist; author of *Index to North American Ferns* (1938).

Brummitt Richard Kenneth Brummitt, 1937–, botanist, Royal Botanic Garden, Kew, England; Convolvulaceae.

B.S.P. N. L. Britton; Emerson Ellick Sterns, 1846–1926, United States; Justus Ferdinand Poggenburg, 1840–1893, United States.

Buch. Franz George Phillip Buchenau, 1831–1906, professor in Bremen; Alismataceae, Juncaceae.

Buckl. Samuel Botsford Buckley, 1809–1884, state geologist of Texas; collected plants, shells, and insects from Georgia to Texas.

Bullock Arthur A. Bullock, 1906–, English botanist, Kew; Asclepiadaceae and Rubiaceae of Africa, nomenclature.

Bunge Alexander Andrejewitsch von Bunge, 1803–1890, Russian botanist and explorer, professor of botany, Dorpat; author of the flora of Russia and central Asia (1851); *Astragalus*.

Burgess Edward Sanford Burgess, 1855–1928, professor of science, Hunter College, New York City, a student of the flora of Chautauqua County, New York; Asteraceae.

Burm. f. Nicholaas Laurens Burman, the son, 1734–1793, Dutch physician and botanist.

Burnett Gilbert Thomas Burnett, 1800–1835; Brassicaceae.

Bush Benjamin Franklin Bush, 1858–1937, storekeeper, Courtney, Missouri; amateur botanist.

Butler Bertram Theodore Butler, 1872–fl. 1908, Columbia University; *Betula*.

Butt. Fredrick King Butters, 1878–1945, professor of botany, University of Minnesota.

Butt. & Abbe F. K. Butters and Ernst Cleveland Abbe, 1905–, American morphologist and taxonomist; Betulaceae, *Rorippa*.

Butt. & St. John F. K. Butters and H. St. John.

Camb. Jacques Cambessedes, 1799–1863, French botanist.

Camp Wendell Holmes Camp, 1904–1963, curator, New York Botanical Garden (1936–1949), then professor of botany, University of Connecticut, Storrs, a leader in the founding of the American Society of Plant Taxonomists; Ericaceae (*Vaccinium*).

Camus A. Aimee Antoinette Camus, 1879–1965, French taxonomist; Orchidaceae, Poaceae, flora of Indochina.

Camus, G. Edmond Gustar Camus, 1852–1915, France; Poaceae, Cyperaceae.

Canby William Marriott Canby, 1831–1904, Delaware businessman, accumulator of a large herbarium.

Cantino Philip Cantino, 1948–, University of Ohio; *Physostegia*.

Carey John Carey, 1797–1880, English engraver and taxonomist, came to U.S.A. in 1830; *Carex*, *Salix*.

Carr. Elie Abel Carriere, 1818–1896, French horticulturist, Museum d'Histoire Naturelle, Paris; editor of *Revue Horticole*; Coniferae.

Carruth James H. Carruth, 1807–1896; clergyman and state botanist in Kansas.

Casp. Johann Xaver Robert Caspary, 1818–1887, professor of botany, Königsberg; flora of Russa, especially aquatic plants.

Cass. Alexandre Henri Gabriel (Comte de) Cassini, 1781–1832, French botanist; Asteraceae.

Cav. Antonio Jose Cavanilles, 1745–1804, Spanish botanist, professor of botany and director of the botanic gardens, Madrid; author of *Icones et descriptiones plantarum* (1791–1801).

Cavara & Grande Fridiano Cavara, 1857–1929, and Loreto Grande, 1878–1965; Brassicaceae.

Celak. Ladislav Josef Celakovsky, 1834–1902, professor of botany, Prague, Czechoslovakia; author of *Prodromus der Flora von Bohmen* (1867).

Celarier Robert P. Celarier, 1921–1959, Oklahoma State University, Stillwater; cytogenetics, experimental taxonomy; Andropogoneae.

Celarier & Harlan R. P. Celarier and J. R. Harlan.

Cerv. Vincente de Cervantes, 1759?–1829, professor of botany and director of the botanic garden, Mexico City.

Cas., Pass. & Gib. Vincenzo Cesati, 1806–1883, professor of botany, Naples, Italy; Giovanni Passerini, 1816–1893, professor of botany at Parma; Giuseppe Gibelli, 1831–1898, physician and professor of botany at Bologna, then at Turin.

Chaix Dominique Abbe Chaix, 1730–1799, student of the flora of the French Alps.

Cham. Ludolf Adalbert von (formerly Louis Charles Adalaide Chamisso de Boncourt) Chamisso, 1781–1838, poet-naturalist, Berlin; botanist on the ship *Rurik*, which visited California in 1816.

Cham. & Schlect. A. L. Chamisso and D. F. L. von Schlechtendal.

Chambers, K. Kenton L. Chambers, 1929–, Oregon State University, Corvallis; biosystematics of Asteraceae, floristics of Pacific Northwest.

Chapm. Alvan Wentworth Chapman, 1809–1899, American botanist, Florida; author of *Flora of the Southern United States* (1860; 2nd ed. 1883; 3rd ed. 1897).

Chase Mary Agnes (Merrill) Chase, 1869–1963, Custodian of Grasses, U. S. National Herbarium, Washington; eminent agrostologist; author (with Cornelia D. Niles) of *Index to Grass Species* (1962).

Chat. Jean Jacques Chatelain, 1736–1822, published work in 1760, Switzerland.

Cheney, R. H. Ralph Holt Cheney, 1896–?, paleobotanist, Brooklyn College and Long Island University, New York; *Delphinium, Coffea*, medicinal and drug plants.

Chev. Francois Fulgis Chevallier, 1796–1840, French mycologist.

Ching Ren-Chang Ching, 1899–, Chinese pteridologist.

Chodat Robert Hippolyte Chodat, 1865–1934, Switzerland; *Cleome*.

Choisy Jacques Denys (Denis) Choisy, 1799–1859, Swiss philosopher, clergyman and botanist, professor in Geneva; collaborator in de Candolle's *Prodromus*; Clusiaceae, Convolvulaceae, Nyctaginaceae.

Chr., C. Carl Frederick Albert Christensen, 1872–1942, Danish student of ferns; author of *Index Filicum* (1905–1912).

Christ Konrad Hermann Heinrich Christ, 1833–1933, Swiss authority on ferns; Coniferae, *Carex, Rosa*; author of the classic *Pflanzenleben der Schweiz* (1879).

Claph. Arthur Roy Clapham, 1904–, English ecologist and taxonomist.

Clark Robert Brown Clark, 1914–, Missouri Botanical Garden, St. Louis, Missouri; *Bumelia*.

Clark, C. James Curtis Clark, 1951–, California State Polytechnic University, Pomona; *Lesquerella, Encelia,* and *Eschscholtzia*.

Clarke Charles Baron Clarke, 1832–1906, superintendent of the Royal Botanic Gardens, Calcutta; student of the flora of India; Cyperaceae.

Clausen Robert Theodore Clausen, 1911–1981, professor of botany, Cornell University, Ithaca, New York; Ophioglossaceae, *Sedum*.

Clayt. John Clayton, 1685–1773, physician in Virginia; collector for Gronovius.

Clem. & Clem. Frederic Edward Clements, 1874–1945, American plant ecologist and climatologist, Carnegie Institute; and Edith Gertrude (Schwartz) Clements, 1877–1971, ecologist; authors of *Rocky Mountain Flowers* (3rd ed. 1928).

Clem., I. Ian Duncan Clement, 1917–, American taxonomist, director, Atkins Garden and Research Laboratory of Harvard University, Soledad, Cuba (1948–1962), later National Science Foundation, Washington; Malvaceae, tropical economic botany.

Clewell Andre F. Clewell, 1934–, botanist;, Florida State University, Tallahassee; Fabaceae (*Lespedeza*).

Clokey Ira Waddell Clokey, 1878–1950, Colorado, Nevada, and California.

Clute Willard Helson Clute, 1869–1950, professor of botany, Butler University, Indianapolis, Indiana, a founder of the American Fern Society; editor of the *Fern Bulletin* and the *American Botanist*; author of *The Fern Allies of North America north of Mexico* (1905; 2nd ed. 1928), and *Our Ferns: Their Haunts, Habits and Folklore* (1938).

Cockll. Theodore Dru Alison Cockerell, 1866–1948, professor of biological sciences in Colorado and New Mexico, intrepid student of natural history.

Coleman Nathan Coleman, 1825–1887, Michigan.

Collins Lawrence Turner Collins, 1937–,

botanist, University of Wisconsin, Milwaukee, then Evangel College, Springfield, Missouri; Orobanchaceae.

Conrad   Soloman White Conrad, 1779–1831, Philadelphia bookseller, publisher, and botanist.

Const.   Lincoln constance, 1909–, professor of botany, University of California, Berkeley; Apiaceae, Hydrophyllaceae, etc.

Const. & Roll.   L. Constance and R. Rollins.

Cooperrider   Tom Smith Cooperrider, 1927–, Kent State University, Ohio.

Corb.   François Mariel Louis Corbiere, 1850–1941, French bryologist and taxonomist; *Vicia*.

Correll   Donovan Stewart Correll 1908–1983, Botanical Museum, Harvard University (1938–1946), U.S. Dept. Agriculture, Beltsville, Maryland (1946–1956), Botanical Laboratory, Texas Research Foundation, Renner, Texas (1956–1970's), later Fairchild Tropical Garden, Florida; ferns, orchids, *Solanum*, flora of southwestern United States; author (with M. C. Johnston) of *Manual of the Vascular Plants of Texas* (1970).

Cory   Victor Louis Cory, 1880–1964, Southern Methodist University, Dallas, Texas, botanist and collector of Texas plants.

Coss.   Ernest Saint-Charles Cosson, 1819–1889, French botanist.

Coss. & Germ.   E. S. Cosson and Jacques Nicholaus Ernest Germain de Saint-Pierre, 1815–1882, Paris physician; authors of a flora of Paris (1845).

Coult.   John Merle Coulter, 1851–1928, professor of botany, University of Chicago, founder and longtime editor of the *Botanical Gazette*; author of *Botany of Western Texas* (1891–1894) and *Manual of the Botany of the Rocky Mountain Region* (1885, rev. by A. Nelson, 1909).

Coult. & Evans   J. M. Coulter and Walter Harrison Evans, 1863–1941, U.S. Dept. of Agriculture; Cornaceae, Alaskan vegetation.

Coult. & Rose   J. M. Coulter and J. N. Rose.

Court.   Richard Joseph Courtois, 1806–1835, Belgian physician and botanist at Liege.

Cov.   Frederick Vernon Coville, 1867–1937, curator, U.S. Natural Herbarium (1893–1937); botanist of the Death Valley Expedition in 1891.

Crantz   Heinrich Johann Nepomuk von Crantz, 1722–1797, professor of medicine, Vienna; author of a flora of Austria; Brassicaceae, Apiaceae.

Crep.   François Crepin, 1830–1903, director of the botanic garden, Brussels.

Croat, T. B.   Thomas Bernard Croat, 1938–, Missouri Botanical Garden; *Solidago* in the Great Plains, flora of Panama; author of *Flora of Barro Colorado Island* (1979).

Croizat   Leon Camille Marius Croizat, 1894–, Botanical Institute, Caracas, Venezuela; Euphorbiaceae, Cactaceae.

Cronq.   Arthur (John) Cronquist, 1919–, curator, New York Botanical Garden; Compositae; author (with H. A. Gleason) of *Manual of Vascular Plants of Northeastern United States and Adjacent Canada* (1963), (with other authors) *Vascular Plants of the Pacific Northwest* (1955–1964), *An Integrated System of Classification of Flowering Plants* (1981), and numerous other works.

Cronq. & Keck   A. J. Cronquist and D. D. Keck.

Crosswhite   Frank Samuel Crosswhite, 1940–, University of Wisconsin, Madison; later Boyce Thompson Southwestern Arboretum, Superior, Arizona; Scrophulariaceae (*Penstemon*).

Curt.   William Curtis, 1746–1799, English botanist, entomologist and editor, founder of *Botanical Magazine*; author of *Flora Londinensis* (1777–1798).

Curtis, M. A.   Moses Ashley Curtis, 1808–1872; author of *The Shrubs and Woody Vines of North Carolina* (1860, reprinted 1946).

Cutak   Ladislaus Cutak, 1908–1973, horticulturalist, Missouri Botanical Garden, St. Louis; Cactaceae, Bromeliaceae, Araceae.

Cutler   Hugh Carson Cutler, 1912–, Missouri Botanical Garden, St. Louis; *Ephedra*, economic botany.

Czern.   V. M. Czernajew, 1796–1871.

Cyr.   Domenico Maria Leone Cirillo (Cyrillo), 1739–1794, Naples, Italy; *Stellaria*.

Dahling   Gerald V. Dahling, fl. 1978, graduate student at Harvard University; *Garrya*.

Danby.   Henry Danvers, Earl of Danby, 1573–1644, English statesman and botanical patron.

Dandy   James Edgar Dandy, 1903–1976, botanist, British Museum of Natural History, London; Magnoliaceae, Hydrocharitaceae, Potamogetonaceae.

Daniels   Francis Potter Daniels, 1869–1947, botanist, ordained minister and linguist, Georgia State College for Women (1923–1935); collected and published on Colorado plants (1911).

Danser   Benedictus Hubertus Danser, 1891–1943, Dutch botanist, professor of botany in Groningen; Polygonaceae.

Darby   John Darby, 1804–1877, New York botanist; author of *Botany of the Southern States* (1855).

Darl.   Josephine Darlington, 1905–, special fellow in botany at the Missouri Botanical Garden about 1930; *Mentzelia*.

Daubs Edwin Horace Daubs, fl. 1965, graduate student at the University of Illinois, Urbana; Lemnaceae.

Daveau Jules Alexandre Daveau, 1852–1929, director, Botanic Garden, Lisbon.

Davis, L. I. Louie Irby Davis, 1897–, collected plants in Mexico and Texas in 1941–1965; published on *Verbena* in 1941.

Davis, R. J. Ray Joseph Davis, 1895–?, professor of botany, Idaho State College, Pocatello; author of *Flora of Idaho* (1952).

Davy Joseph Burtt Davy, 1870–1940, English taxonomist and forester.

Day & V. Grant Alva Day (Grant), and V. E. Grant.

DC. Augustin Pyramus de Candolle (also Decandolle), 1778–1841, Swiss botanist, professor of botany, Geneva; first in an illustrious line of systematists; founder of the *Prodromus*, a fundamental work in the development of modern taxonomy.

DC., A. Alphonse Louis Pierre Pyramus de Candolle, the son, 1806–1893, Geneva; author of the last 10 vols. of the *Prodromus*; founder of *Monographiae Phanerogamarum*, a continuation and revision of the *Prodromus*.

Dcne. Joseph Decaisne, 1807–1882, Belgian botanist, director, Jardin des Plantes, Paris; Asclepiadaceae, Plantaginaceae.

Deam Charles Clemon Deam, 1865–1953; floristics of Indiana.

DeBruyn Ary Johannes de Bruyn (Bruijn), 1811–1895.

DeJong Diederik Cornelius Dignus DeJong, 1931–, University of Cincinnati; Asteraceae.

Desf. Rene Louiche Desfontaines, 1750–1833, French botanist, professor in the Jardin des Plantes, Paris; author of *Flora Atlantica* (1798–1799).

Des Moul. Charles Robert Alexandre Des Moulins, 1798–1875, Bordeaux, France.

Desr. Louis Auguste Joseph Desrousseaux, 1753–1838, French cloth manufacturer; contributor to Lamarck's *Encyclopedia*.

Desv. Nicaise Auguste Desvaux, 1784–1856, French botanist, professor of botany, Angers; editor of *Journal de Botanique*.

Detl. LeRoy Ellsworth Detling, 1909–1967, professor of botany, University of Oregon, Eugene; Brassicaceae *(Dentaria, Descurainia)*.

Dew. Rev. Chester Dewey, 1784–1867, professor, University of Rochester, New York; specialist in *Carex*.

Dewey, L. H. Lyster Hoxie Dewey, 1865–1944, Michigan botanist, published on flora of islands in The French River, Ontario (1939).

Dickson James Dickson, 1738–1822, Scotch nurseryman; writer on cryptogams.

Dietr. Friedrich Gottlieb Dietrich, 1768–1850, garden director at Eisenach, Germany.

Dietr., A. Albert Gottfried Dietrich, 1795–1856, German gardener, curator of the Berlin botanical garden.

Dipp. Leopold Dippel, 1827–1914, Germany, dendrologist.

Dode Louis Albert Dode, 1875–1943, France, botany and forestry; *Populus*.

Dole Eleazer Johnson Dole, 1888–?, professor of botany, University of Vermont.

Döll Johann Christoff Döll (Also Doell), 1808–1885.

Domin Karel Domin, 1882–1954, director of the Botanical Institute and Gardens, Prague.

Don, D. David Don, brother of George, 1799–1841, British botanist, professor in King's College, London; librarian to the Linnaean Society.

Don, G. George Don, 1798–1856, British plant collector for the Royal Horticultural Society in Brazil, West Indies and Africa.

Donn James Donn, 1758–1813, British gardener under Aiton at Kew, later curator of Cambridge Garden; author of *Hortus Cantabrigiensis* (1796), which went through 13 editions.

Dougl. David Douglas, 1798–1834, ardent Scottish collector in northwestern America, taking nearly 500 species in California alone for the Royal Horticultural Society; collected extensively along the Columbia River.

Drapalik & Mohlenbrock Donald Joseph Drapalik, 1934–, Georgia Southern College, Statesboro; Asclepiadaceae *(Matelea)*, and Robert H. Mohlenbrock, 1931–, Southern Illinois University, Carbondale; Fabaceae, Cyperaceae, flora of Illinois.

Drew William Brooks Drew, 1908–, Michigan State University, East Lansing; *Ranunculus*.

Druce George Claridge Druce, 1850–1932, professor at Oxford.

Drude Carl Georg Oscar Drude, 1852–1933, Germany.

Duchn. Antoine Nicholas Duchesne, 1747–1827, France; author of works on useful plants, especially strawberries *(Fragaria)*.

Dudley, T. R. Theodore R. Dudley, 1936–, U.S. National Arboretum, Washington, D. C.; *Alyssum, Ilex, Viburnum*.

Dudley, W. R. William Russell Dudley, 1849–1911, Stanford University.

Dufr. Pierre Dufresne, 1786–1836, Geneva, Switzerland.

Duham. Henri Louis Duhamel de Monceau, 1700–1781, French agronomist, forester, and botanist.

Dulac Joseph Dulac, 1827–1897.

ABBREVIATIONS FOR NOMENCLATURAL AUTHORITIES

Dum. Count Barthelemy Charles Joseph Dumortier, 1797–1878, Belgian botanist, president of the Belgian Chamber of Deputies.

Dumont Georges Louis Marie Dumont de Courset, 1746–1824, French agronomist and horticultural writer.

Dun. Michel Felix Dunal, 1789–1856, French botanist, professor of botany, Montpellier; Vacciniaceae, Solanaceae, contributor to de Candolle's *Prodromus*.

Duncan Wilbur Howard Duncan, 1910–, University of Georgia, Athens.

Dunn David Baxter Dunn, 1917–, University of Missouri; *Lupinus*.

Dur & Jackson (Theodore) Theophile Alexis Dur, 1855–1912, director of the Brussels Botanical Garden (1901–1912); and Benjamin Daydon Jackson, 1846–1927, British botanist and eminent bibliographer.

Durand Elias Magloire Durand, 1794–1873, Philadelphia pharmacist.

Durazz. Ippolito Durazzo, 1750–1818, Italy.

Durieu Michel Charles Durieu de Maisonneuve, 1796–1878, French botanist, director of the botanic gardens, Bordeaux; student of the flora of southern France, Spain, and Algeria.

Du Roi Johann Philipp Du Roi, 1741–1785, physician at Braunschweig, Germany; North American trees.

Dusen Per Karl Hjalmar Dusen, 1855–1926, Swedish civil engineer who made extensive botanical trips to various countries including Brazil, where he became secretary of the Museum of Rio de Janeiro.

Dyal Sarah Creecie (later Nielsen) Dyal, 1907–, Cornell University, Ithaca, New York; *Plectritis, Valerianella*.

Dykes William Rickatson Dykes, 1877–1925, secretary of the Royal Horticultural Society; author of *The Genus Iris* (1925).

Eames Edwin Hubert Eames, 1865–19?; *Dicentra*.

Eames & Boivin E. H. Eames and J. R. B. Boivin.

Eastw. Alice Eastwood, 1859–1953, Canadian-born California botanist, curator of botany, California Academy of Sciences (1892–1950); student of the West American flora and cultivated plants.

Eat. Amos Eaton, 1776–1842, American botanist, lecturer and writer, New York; produced first botanical manual in America which had descriptions in English.

Eat. & Wright A. Eaton and John Wright, 1811–1846, professor of physiology, Rensselaer Institute.

Eat., A. A. Alvah Augustus Eaton, 1865–1908, New England and Florida; student of ferns and early collector in Florida.

Eat., D. C. Daniel Cady Eaton, grandson of Amos, 1834–1895, professor of botany, Yale University, New Haven, Connecticut; authority on ferns.

Eat., H. H. Hezekiah Hulbert Eaton, 1809–1892, professor of chemistry, Transylvania University, son of Amos Eaton.

Eckenw. James E. Eckenwalder, 1949–, University of Toronto; *Populus*.

Ediger Robert Ike Ediger, 1937–, California State University at Chico; *Senecio*.

Eggers Donna Marie Eggers (Ware), 1942–, William & Mary College, Williamsburg, Virginia; *Valerianella*.

Eggert Heinrich (Karl Daniel) Eggert, 1841–1904, American botanical collector.

Eggl. Willard Webster Eggleston, 1863–1935, botanist, U.S. Dept. of Agriculture, Washington; student of poisonous and drug plants, prodigious collector of American plants.

Ehrh. Friedrich Ehrhart, 1742–1795, German botanist of Swiss origin, pupil of Linnaeus; advocate of monomial nomenclature.

Eifert Imre Janos Eifert, 1934–, formerly Department of Botany, University of Texas; Leguminosae (*Hoffmanseggia, Caesalpinia*).

Ell. Stephen Elliott, 1771–1830, American botanist, professor in Charleston; author of *A Sketch of the Botany of South Carolina and Georgia* (1816–1829).

Elmer Adolph Daniel Edward Elmer, 1870–1942, collector in California, Washington, and the Philippine Islands.

Endl. Stephen Friedrich Ladislaus Endlicher, 1804–1849, Austrian botanist, professor of botany, and director of the botanic garden, Vienna; author of *Genera Plantarum* (1836–1840), and many other large works; student of Coniferae.

Engelm. George Engelmann, 1809–1884, German-born physician and eminent botanist in St. Louis; painstaking student of North American *Cuscuta, Juncus*, Euphorbiaceae, *Isoetes, Yucca*, Cactaceae, *Pinus, Abies, Juniperus, Agave*, and *Vitis*.

Engelm. & Bigel. G. Engelmann and J. Bigelow.

Engelm. & Gray G. Engelmann and A. Gray.

Engl. (Heinrich Gustaf) Adolf Engler, 1844–1930, director of the botanic garden and museum, Berlin; founder and editor of *Botanische Jahrbücher, Die Vegetation Der Erde*, and *Das Pflanzenreich*; Araceae, monographer of *Saxifraga*.

Ensign Margaret Ruth Ensign (later Lewis), 1919–, Pomona College, Claremont, California; *Forsellesia*.

Epl. Carl Clawson Epling, 1894–1968, University of California, Los Angeles; North American Lamiaceae.

Erdm. Kimball S. Erdman, 1939–, Slippery Rock College, Pennsylvania; dendrologist.

Erickson Ralph Orlando Erickson, 1914–, formerly Missouri Botanical Garden, St. Louis; later University of Pennsylvania; *Clematis*.

Ewan Joseph Andorfer Ewan, 1909–, professor of botany, Tulane University, New Orleans, Louisiana; *Delphinium*, botanical historian, author of *Rocky Mountain Naturalists* (1950), etc.

Fabr. Philipp Conrad Fabricius, 1714–1774, German physician and botanist, professor at University of Helmstedt.

Farw. Oliver Atkins Farwell, 1867–1944, consulting botanist for Parke, Davis & Co., Detroit, Michigan.

Fassett Norman Carter Fassett, 1900–1954, professor of botany, University of Wisconsin, Madison; author of *A Manual of Aquatic Plants* (1940, revised by E. C. Ogden in 1957), *Spring Flora of Wisconsin* (1947), etc.

Faust Zack Faust, 1939–, Colombus College, Columbus, Georgia; *Vernonia*.

Featherly Henry Ira Featherly, 1893–?, Oklahoma State University, Stillwater; author (with Clara E. Russell) of *The Ferns of Oklahoma* (1939), Poaceae.

Fedde Friedrich Karl Georg Fedde, 1873–1942, professor and editor in Berlin; founder and editor of *Feddes Repertorium Specierum Novarum Regni Vegetabilis* (1905–), *Mahonia*, Papaveraceae.

Fee Antoine Laurent Apollinaire Fee, 1789–1874, French botanist, professor at Strasbourg; noted student of cryptogams.

Fenzl Eduard Fenzl, 1808–1879, director, Botanical Garden, Vienna; Caryophyllaceae.

Fern. Merritt Lyndon Fernald, 1873–1950, director of Gray Herbarium (1937–1947); a founder of *Rhodora*; noted plant geographer and systematist; *Potamogeton*; author of Gray's *Manual of the Botany . . .* , 8th ed. (1950).

Fern. & Brack. M. L. Fernald and A. E. Brackett.

Fern. & Grisc. M. L. Fernald and Ludlow Griscom, 1890–1959, Harvard University research ornithologist.

Fern. & Schub. M. L. Fernald and B. G. Schubert.

Fern. & Weath. M. L. Fernald and C. A. Weatherby.

Fern. & Wieg. M. L. Fernald and K. M. Wiegand.

Fieb. Franz Xaver Fieber, 1807–1872, Czechoslovakian entomologist and botanist.

Fiori Adriano Fiori, 1865–1950, physician and professor of botany at Florence, Italy; author (with G. Paoletti) of *Flora of Italy*.

Fisch. Friedrich Ernst Ludwig von Fischer, 1782–1854, director of the botanic garden, St. Petersburg (1823–1850).

Fisch. & Ave-Lall. F. E. L. von Fischer and J. L. E. Ave-Lallemant.

Fisch. & Mey. F. E. L. von Fischer and C. A. Meyer.

Fisch. & Trautv. F. E. L. von Fischer and Ernst Rudolph von Trautvetter, 1809–1889, Russia.

Fischer, G. Gustav Adolph Fischer, 1886–1946, German physician at Berlin; collected in East Africa.

Fisher, T. T. Richard Fisher, 1921–, Bowling Green State University, Ohio; Asteraceae.

Fluegge Johann Fluegge, 1775–1816, German physician and botanist, botanical garden in Hamburg; Poaceae, *Salix*.

Focke Wilhelm Olbers Focke, 1834–1922, physician in Berne, Switzerland; *Rubus*.

Forsk. Petrus Forsskal (also Pehr Forskal), 1732–1763, Danish student of Linnaeus who traveled to Arabia and wrote a flora of Egypt and Arabia; died on the desert of starvation and exposure after repeated encounters with bandits.

Forst., T. F. Thomas Furly Forster, 1761–1825, English student of floristics; author of *Flora Tonbrigansis* (1816).

Fosb. Francis Raymond Fosberg, 1908–, U.S. Geological Survey and Smithsonian Institution, Washington; student of the California, South American, and Polynesian floras; Rubiaceae.

Foug. Auguste Denis Fougeroux de Bondaroy, 1732–1789, France.

Fourn. Eugene Pierre Nicolas Fournier, 1834–1884, physician in Paris; Asclepiadaceae.

Fourr. Pierre Jules Fourreau, 1844–1871, Lyon, France; floristics.

Franch. & Sav. A. R. Franchet, 1834–1900, French botanist; and Paul Amedee Ludovic Savatier, 1830–1891, French marine medical officer and botanist.

Frank Joseph C. Frank, 1782–1835, American botanist; author of *Rastadt's Flora*.

Fraser, S. V. Samuel Victorian Fraser, 1890–1972, clergyman; student of the flora of central Kansas; *Allium*.

Freckmann Robert W. Freckmann, 1939–, University of Wisconsin, Stevens Point; *Dichanthelium*, *Panicum*, central Wisconsin flora.

Freeman, C. C. Craig Carl Freeman, 1955–, University of Connecticut, Storrs; *Penstemon*.

Freeman, F. L. Florence L. Freeman, 1912–,

assistant at Arnold Arboretum (1934–1939); *Psoralea.*
Fresen. Johann Baptist Georg Wolfgang Fresenius, 1808–1866, German physician and botanist.
Friedl. Solomon Friedland, 1912–, New York Botanical Garden; *Hemicarpha.*
Fries Elias Magnus Fries, 1794–1878, professor of botany, Uppsala; noted mycologist and student of *Hieracium.*
Fries & Broberg E. M. Fries and Hans Broberg, fl. 1764, d. 1795, master gardener at the Uppsala (Sweden) Botanical Garden.
Fritsch Karl F. Fritsch, 1864–1934, Graz, Austria; *Parthenocissus.*
Froel. Joseph Aloys von Froelich, 1766–1841; *Gentiana, Hieracium.*
Fryxell Paul A. Fryxell, 1927–, Texas A & M University, College Station; Malvaceae.
Fuchs, H. P. Hans Peter Fuchs, 1928–, Swiss taxonomist and historian of science at Tignuppa, Switzerland; alpine flora, *Isoetes, Dryopteris.*
Fuller, A. M. Albert M. Fuller, 1900–1981, Milwaukee Public Museum, Orchidaceae of Wisconsin, *Rubus.*
Gaertn. Joseph Gaertner, 1732–1791, physician near Stuttgart; structure of fruit and seeds.
Gaertn., Mey. & Scherb. J. Gaertner, G. F. W. Meyer, and J. Scherbius.
Gagnebin Abraham Gagnebin, 1707–1800, Swiss botanist.
Gaiser Lulu Odel Gaiser, 1896–1965, Canadian botanist, Barnard College, New York, and Gray Herbarium, Harvard University, and MacMaster University, Hamilton, Ontario; *Liatris,* flora of Lambton County, Ontario.
Gale Shirley Gale (Cross), 1915–, Gray Herbarium, Harvard University; Cyperaceae (*Rhynchospora*).
Gand. Michel Gandoger, 1850–1926, French botanist; author of *Flora Europae* (27 vols.); voluminous writer; amasser of a huge herbarium now at Lyon; a "splitter" who named thousands of unacceptable microspecies.
Garcke Christian August Friedrich Garcke, 1819–1904, curator of the herbarium, Berlin; author of *Flora von Nord- und Mittell-Deutschland* (1849) that went through 20 editions.
Garden Alexander Garden, 1730–1791, physician and botanist at Charleston, South Carolina, for whom Linnaeus named the genus *Gardenia.*
Gardner, R. Robert Carl Gardner, 1946–1982, Baylor University, Waco, Texas; *Cirsium,* Hawaiian *Lipochaeta.*
Garrett Albert Osmun Garrett, 1870–1948, high-school teacher in Salt Lake City.
Gates, F. C. Frank Caleb Gates, 1887–1955, Kansas State University, Manhattan; flora of Kansas.
Gatt. Augustin Gattinger, 1825–1903, author of *The Flora of Tennessee and a Philosophy of Botany* (1901), etc. His herbarium, housed at the University of Tennessee, was regrettably destroyed by fire.
Gaud. Jean Francois Aimee (Gottlieb) Philippee Gaudin, 1766–1833, Swiss clergyman and agrostologist, Nyon.
Gaudich. Charles Gaudichaud-Beaupre, 1789–1854, French naval pharmacist, made many voyages and collected widely.
Gay Jacques Etienne Gay, 1786–1864, French student of the flora of Switzerland and the Pyrenees, systematist, and morphologist.
Gay & Durieu J. E. Gay and M. C. Durieu de Maisonneuve.
Gerard John Gerard, 1545–1612, English physician, herbalist, and horticulturalist; author of *Herbal* (1597).
Gilbert Benjamin Davis Gilbert, 1835–1907, American fern student.
Gilg Ernst Friedrick Gilg, 1867–1933, Botanical Museum, Berlin.
Gilib. Jean Emmanuel Gilibert, 1741–1814, French botanist, professor in Vilna, Lithuania, later in Lyon, France; author of *Flora Lithuanica* (1785).
Gill. & Arn. John Gillies, 1747–1836, Scottish physician who resided some years in Argentina and collected in Chile, and G. A. W. Arnott.
Gillett, J. John Montague Gillett, 1918–, Dept. of Agriculture, Canada; Gentianaceae.
Gillis William Thomas Gillis, 1933–1979, Michigan State University, then Fairchild Tropical Garden, Miami, Florida; *Toxicodendron.*
Gillman Henry Gillman, 1833–1915, Irish scientist who immigrated to Detroit, Michigan in 1850; *Geodesic Survey of the Great Lakes* (1851–1869).
Gilly Charles Louis Gilly, 1911–1970, Michigan State University, East Lansing.
Ging. Frederic Charles Jean de Gingins, dit de Gingins-La Sarraz, 1790–1863, Swiss botanist; Violaceae, Lamiaceae.
Gl. Henry Allan Gleason, 1882–1975, assistant-director and head curator, the New York Botanical Garden; author of *New Britton & Brown Illustrated Flora* (1952), etc.
Gl. & Cronq. H. A. Gleason and A. Cronquist.
Glox. Benjamin Peter Gloxin, 1765–1794, French physician at Colmar; author of *Observations Botanical* (1785).

Gluck Christian Maximilian Hugo Gluck (Glueck), 1868–1940, Germany; hydrophytes.

Gmel., C. C. Carl Christian Gmelin, 1762–1837, physician in Karlsruhe, Germany; author of *Flora Badensis Alsatica* (1805–1826).

Gmel., J. F. Johann Friedrich Gmelin, 1748–1804, professor in Tübingen, then in Göttingen, Germany; editor of the 13th ed. of Linne's *Systema Naturae* (1788–1793). Nephew of J. G. Gmelin.

Gmel., J. G. Johann Georg Gmelin, 1709–1755, German botanist who traveled in Siberia and Kamchatka (1733–1743), and gave first account of their floras, author of the classic *Flora Sibirica* (1747–1769); later professor in Tübingen, Germany

Gmel., S. G. Samuel Gottlieb Gmelin, 1743–1774, traveled with Pallas in southeastern Russia; collaborated with J. G. in producing *Flora Siberica*, Nephew of J. G., cousin of F. J.

Godfrey Robert Kenneth Godfrey, 1911–, professor of botany, Florida State University, Tallahassee; floristics, southeastern United States, Compositae; author (with Herman Kurz) of *Trees of Northern Florida* (1962), etc.; collector par excellence.

Godr. Dominique Alexandre Godron, 1807–1880, French botanist of Nancy.

Goepp. Johann Heinrich Robert Goppert, 1800–1884; *Cinna*.

Goldie John Goldie, 1793–1886, Scotsman who traveled through eastern Canada and New England in 1819.

Goodd. Leslie Newton Goodding, 1880–1967, student of Aven Nelson, botanist, U.S. Soil Conservation Service, Arizona; forest pathology, southwestern flora, conservation.

Goodd., C. Charlotte Olive Goodding (later Reeder), the daughter, 1916–, Yale University, later University of Wyoming, now at University of Arizona; Poaceae *(Bouteloua, Muhlenbergia)*.

Goodmn. George Jones Goodman, 1904–, University of Oklahoma, Norman; West American genera of Polygonaceae, flora of the Redlands.

Gord. George Gordon, 1806–1879, superintendent of the Horticultural Gardens, Chiswick, near London.

Gord. & Glend. G. Gordon and Robert Pince Glendinning, 1840–1906, of the Chiswick Nursery, authors of *The Pinetum* (1858; 2nd ed. 1875).

Gould Frank Walton Gould, 1913–1981, Texas A & M University, College Station, agrostologist; author of *Texas Plants—A Checklist and Ecological Summary* (1962), *Grass Systematics* (1968), etc.

Gould & Clark F. W. Gould and Carolyn A. Clark, fl. 1979, graduate student at Texas A & M University, College Station.

Gould & Kapadia F. W. Gould and Zarir Jamasji Kapadia, 1935–, Bangalore, India; Poaceae *(Bouteloua)*.

Graebn. Karl Otto Robert Peter Paul Graebner, 1871–1933, professor of botany, Berlin.

Grah. Robert Graham, 1786–1845, professor botany, Edinburgh, Scotland.

Grant Adele Lewis Grant, 1881–1969, Missouri Botanical Garden, later professor of botany, University of Southern California, Los Angeles; *Mimulus*.

A. & V. Grant Alva (Day) Grant, 1920–, Rancho Santa Ana Botanic Garden, California; later California Academy of Sciences; Polemoniaceae; and V. Grant.

V. Grant Verne Edwin Grant, 1917–, cytogeneticist, Rancho Santa Ana Botanic Garden, Claremont, California, later University of Texas at Austin; Polemoniaceae.

Gray, A. Asa Gray, 1810–1888, professor of botany, Harvard University, preeminent American systematist; author of *Manual of the Botany of the Northern States* (1848, now through 8 editions); (Gamopetalae of California) in Brewer and Watson's *Botany of California* (1876); *Synoptical Flora of North America* (1878–1897); botanical textbooks, numerous reports of collections and revisions, etc.

Gray & Engelm. A. Gray and G. Engelmann.

Gray, S. F. Samuel Frederick Gray, 1766–1828, British pharmacist and naturalist; author of *Natural Arrangement of British Plants* (1821), a work much in advance of its time.

Greene Edward Lee Greene, 1843–1915, first professor of botany, University of California (1885–1895), Berkeley, then at Smithsonian Institution in Washington; later at Catholic University of America; editor of *Pittonia* and *Leaflets of Botanical Observation and Criticism*; believer in absolute priority in nomenclature, the names of his microspecies are called "chloronyms."

Greene & Godr. E. L. Greene and D. A. Godron.

Greenm. Jesse More Greenman, 1867–1951, curator of the herbarium, Missouri Botanical Garden (1913–1948); *Senecio*.

Gremli August(e) Gremli, 1833–1899.

Gren. & Godr. Jean Charles Marie Grenier, 1808–1875, and D. A. Godron, authors of a flora of France (1848–1856).

Grev. & Hook. Cf. Hook. & Grev.

Griseb. August Heinrich Rudolph Grisebach, 1814–1879, professor of botany, Göttingen, Germany; noted for his *Vegetation der Erde*

(1872; 2nd ed. 1884) and studies on the West Indian flora (1857–1866); Gentianaceae.

Gronov. Johannes Fridericus Gronovius, 1690–1762, senator in Leiden, friend of Linnaeus, author of *Flora Virginica* (1743; 2nd ed. 1762).

Guill. Jean Baptiste Antoine Guillemin, 1796–1842, French botanist who traveled in Brazil.

Gurke Robert Louis August Max Gurke, 1854–1911, Germany; Caryophyllaceae.

Guss. Giovanni Gussone, 1787–1866, professor of botany, Naples, author of valuable works on the flora of southern Italy and Sicily.

H. & A. W. J. Hooker and G. A. W. Arnott; authors of *The Botany of Captain Beechey's Voyage* (1830–1841).

H. & B. Baron Friedrich Wilhelm Heinrich Alexander von Humboldt, 1769–1859, German geographer, naturalist, and explorer, and Aimee Jacques Alexandre Bonpland, 1773–1858, French botanist; authors of the classic *Voyage aux Regions Equinoctiales du Nouveau Continent, etc.* (1805–1837).

Haage Ferdinand Haage, Jr., 1859–1930, Enfurt Germany, owned horticultural firm; Verbenaceae, author of *Haage's Cacteenkultur* (2nd ed. 1900).

Haage & Schmidt F. Haage and Friedrich Schmidt, 1832–1908, Germany; Verbenaceae.

Hack. Eduard Hackel, 1850–1926, noted Austrian agrostologist.

Hagstrom Johan Oskar Hagstrom, 1860–1922; Sweden; *Potamogeton, Ruppia*.

Halac. Eugen von Halacsy, 1842–1913, Hungarian-Austrian botanist at Vienna; Poaceae.

Hall Harvey Monroe Hall, 1874–1932, professor of botany, University of California, Berkeley, then Carnegie Institution of Washington, Stanford; pioneer in experimental taxonomy; student of the Asteraceae (*Haplopappus*, Madiinae); author (with Carlotta Hall) of *A Yosemite Flora* (1912).

Hall & Clem. H. M. Hall and F. E. Clements; authors of *The Phylogenetic Method of Taxonomy* (1923), in which North American *Artemisia, Chrysothamnus*, and *Atriplex* are monographed.

Haller Albrecht von Haller, 1708–1777, Swiss botanist, physician, poet, and statesman; professor in Göttingen, Germany, later in Berne, Switzerland.

Hallier f. Johann Gottfried Hallier ("Hans"), 1868–1932, Dutch botanist of Buitenzorg and Leiden, son of E. Hallier; *Ipomea*, phylogeny.

Hamilt., A. Arthur Hamilton, Swiss botanist, Geneva, author of a monograph on *Scutellaria* in 1832.

Handel-Mazzetti Heinrich von Handel-Mazzetti, 1882–1940, Austrian botanist; *Cardaria*.

Hara Hiroshi Hara, 1911–, taxonomist, cytologist, synantherologist, Botanic Gardens, Koishikawa, Japan.

Harlan Jack Rodney Harlan, 1917–, University of Illinois, Urbana; crop plants.

Harms, L. Lawrence Harms, fl. 1965, north Dakota State University, Fargo (until 1968); Cyperaceae.

Harms, V. Vernon L. Harms, 1930–, professor of botany, University of Alaska, then University of Saskatchewan, Saskatoon, Canada; Asteraceae (*Heterotheca, Petasites*), northwestern U.S. and Canada, *Sparganium*.

Harper Roland McMillan Harper, 1878–1966, Maine-born Alabama naturalist; field botanist and author of works on botany, geology, ecology, etc., of southeastern U.S.

Harrington Harold David Harrington, 1903–1981, professor of botany, Colorado State University, Ft. Collins; author of *Manual of the Plants of Colorado* (1954).

Harris, S. K. Stuart Kimball Harris, 1906–, Boston University; *Solidago*.

Hartm. Carl Johan Hartman, 1790–1849, Swedish botanist.

Hartman, E. Emily Lou Hartman, fl. 1960, University of Colorado (Denver); *Equisetum*.

Hartman, R. Ronald Hartman, 1945–, University of Wyoming; floristics of Wyoming; Asteraceae, Caryophyllaceae, Apiaceae.

Harv. William Henry Harvey, 1811–1866, professor of botany and keeper of the herbarium, Trinity College, Dublin; made known the California collections of Thomas Coulter; author (with O. W. Sonder) of *Flora Capensis* (1859–1865).

Harv. & Gray W. H. Harvey and A. Gray.

Hassk. Justus Carl Hasskarl, 1811–1894, German botanist at Buitenzorg, Java.

Hausskn. Heinrich Karl Haussknecht, 1838–1903, professor in Weimar; monographer of *Epilobium*.

Haw. Adrian Hardy Haworth, 1767–1833, English gardener and entomologist; student of succulents, *Mesembryanthemum*.

Hayne Friedrich Gottlob Hayne, 1763–1832, professor of botany, Berlin.

H.B.K. Baron F. W. H. A. von Humboldt, A. J. A. Bonpland, the two forming the famous scientific expedition to tropical America, and C. S. Kunth, who wrote the text of their descriptive work *Nova genera et species Plantarum* (1815–1825).

Heer & Regel   Oswald Heer, 1809–1883, and Edward August von Regel, 1815–1892.

Henrickson   James Henrickson, fl. 1970s; California State University, Los Angeles.

Heimerl   Anton Heimerl, 1857–1942, professor in Vienna, Austria; Nyctaginaceae, *Achillea*.

Heiser   Charles Bixler Heiser, Jr., 1920–, professor botany, Indiana University, Bloomington; Asteraceae, Solanaceae.

Heist.   Lorenz Heister, 1683–1758, professor in Helmstedt, Germany.

Heller   Amos Arthur Heller, 1867–1944, Pennsylvania botanist, noted collector of western American plants; founder and editor of *Muhlenbergia*.

Hemsl.   William Botting Hemsley, 1843–1924, keeper of the herbarium, Kew (1899–1908); author of 5 vol. work on botany in *Biologia Centrali-Americana* (1879–1888).

Henr.   Jan Theodoor Henrard, 1881–1974, Dutch pharmacist and agrostologist, Rijks Herbarium, Leiden; *Aristida*.

Henry, A.   Augustine Henry, 1857–1930, Irish physician and dendrologist, at Cambridge (1907–1913), then Dublin (1913–1926); collected plants in China and Formosa.

Herbert, W.   William Herbert, 1778–1847, English politician and botanist later clergyman; Amaryllidaceae.

Herm.   Frederick Joseph Hermann, 1906–, U.S. Dept. of Agriculture, Beltsville, Maryland, later U.S. Forest Service, Washington; *Carex*, bryophytes.

Herrm.   Rudolf Albert Wolfgang Herrmann, b. 1885, fl. 1911, German botanist; *Setaria*.

Herter   Wilhelm Gustav Herter, 1884–1958, German botanist who resided in Uruguay after 1924; author of an illustrated flora of Uruguay (1939–1957).

Heynh.   Gustav Heynhold, 1800–1850, Saxony; *Hieracium, Stenosiphon*.

Hiern   William Philip Hiern, 1839–1925, British Museum; flora of tropical Africa and India.

Hieron.   Georg Hans Emmo Wolfgang Hieronymus, 1846–1921, professor in Berlin; student of the Argentine and Andean flora.

Hiitonen   Henrik Ilmari Augustus Hiitonen, 1898–, University of Helsinki, Finland; flora of Finland, *Salix*.

Hildebr.   Friedrich Hermann Gustav Hildebrand, 1835–1915, professor at Freiburg in Breisgan; *Abies*.

Hill   John Hill, 1716–1775, London apothecary and naturalist, author of herbals and nature books; produced the first flora of England on the Linnaean system.

Hill, A. F.   Albert Frederick Hill, 1889–1977, American economic botanist, Botanical Museum, Harvard University (1939–1957); author of *Economic Botany* (1937).

Hill, E. J.   Ellsworth Jerome Hill, 1833–1917, American clergyman, bryologist, and highschool teacher in Chicago; *Crataegus, Quercus, Salix, Prunus*.

Hinton   W. Frederick Hinton, fl. 1980, George Mason University, Fairfax, Virginia; *Physalis*.

Hitchc., A.   Albert Spear Hitchcock, 1865–1935, botanist, Kansas State University, later U.S. Dept. of Agriculture, Washington, leading American agrostologist; author of *Manual of Grasses of the United States* (1935) and other works.

Hitchc. & Chase   A. S. Hitchcock and M. A. Chase.

Hitchc., C. L.   Charles Leo Hitchcock, 1902–, University of Washington, Seattle; *Lycium, Draba, Lepidium, Lathyrus, Sidalcea*, etc.; senior author of *Vascular Plants of the Pacific Northwest* (1955–1964).

Hitchc., E.   Edward Hitchcock, 1793–1864, professor of chemistry and natural history, and president, Amherst College, Massachusetts.

Hoch   Peter C. Hoch, fl. 1980, Missouri Botanical Garden; *Epilobium*.

Hoffm., G.   Georg Franz Hoffmann, 1760–1826, professor of botany in Moscow (1804–1826); student of lichens, *Salix*, and Apiaceae.

Hoffm., R.   Reinhold Hoffmann, b. 1885, fl. to 1931, professor at Konigsburg (Prussia); plant growth and cultivation.

Hoffmsg. & Link.   Johann Centurius Graf von Hoffmannsegg, 1766–1849, Dresden, and J. H. F. Link; authors of *Flore Portugaise* (1809–1840).

Hogg   Thomas Hogg, 1778–1855, English-born horticulturalist and nurseryman who established a plant business in New York in 1821.

Hohenack.   Rodolph Friedrich Hohenacker, 1798–1874, Switzerland.

Hollick & Britt.   Charles Arthur Hollick, 1857–1933, paleobotanist, the New York Botanical Garden, and N. L. Britton.

Holm   Herman Theodore Holm, 1854–1932, botanist of Denmark, Greenland, and United States.

Holmgren, A.   Arthur H. Holmgren, 1912–, Utah State University, Logan; flora of the Intermountain Region, Poaceae.

Holmgren, N.   Noel Herman Holmgren, 1937–, botanist, New York Botanical Garden; Scrophulariaceae, *Castilleja*

Holz.   John Michael Holzinger, 1853–1929, German-born American bryologist, teacher in State Teachers College, Winona, Minnesota.

Honck. Gerhard August Honckeny, 1724–1805, Germany; author of *Synopsis Plantarum Germania* (1792–1793).

Hook. Sir William Jackson Hooker, 1785–1865, director of Kew (1841–1865): author of *Flora Boreali-Americana* (1833–1840) and many other illustrious works; founder and editor of *Journal of Botany* and *Icones Plantarum*; editor of *Botanical Magazine*.

Hook. f. Sir Joseph Dalton Hooker, the son, 1817–1911, British Botanist and explorer, director of Kew (1865–1885); talented editor and student of New Zealand, Himalayan, and Indian floras; visited America in 1877.

Hook. & Baker W. J. Hooker and J. G. Baker.

Hook. & Grev. W. J. Hooker and Robert Kaye Greville, 1794–1866, professor in Edinburgh; authors of *Icones Filicum* (1827–1832).

Hopffer Carl Hopffer, 1810–?, interested in Cactaceae.

Hopk., M. Milton Hopkins, 1906–, Oklahoma (1936–1945), then editor for publishing houses; *Arabis*, *Cercis*.

Horkel Johann Horkel, 1769–1846, professor of physiology at Berlin, Germany; Lemnaceae.

Hornem. Jens Wilken Hornemann, 1770–1841, professor of botany, Copenhagen.

Hort. Hortorum, of gardens, used with plants of presumed origin in cultivation, or for a name used by gardeners but which has no formal botanical standing.

Host Nicolaus Thomas Host, 1761–1834, imperial physician in Vienna; author of *Flora Austriaca* (1827–1831).

House Homer Doliver House, 1878–1949, New York State botanist (1914–1948); author of *Wild Flowers of New York* (1918 and 1934).

Houtt. Maarten Houttuyn, 1720–1798, Dutch physician and naturalist.

Howe Elliot Calvin Howe, 1828–1899, physician at Yonkers, New York; *Carex*.

Howell Thomas Jefferson Howell, 1842–1912, Portland, Oregon; author of *A Flora of Northwest America*.

Hubb., F. T. Frederick Tracy Hubbard, 1875–1962, librarian in Economic Botany and editor, botanical Museum, Harvard University; agrostologist.

Huds. William Hudson, 1730–1793, London apothecary and botanist; author of *Flora Anglica* (1762, 1st of 3 ed.).

Hutchins. John Hutchinson, 1884–1972, Kew; author of *Families of Flowering Plants* (1959); a new system of phylogeny proposed in 1926, 1934.

Huth Ernst Huth, 1848–1897, professor of botany at Frankfort, Germany.

Hylander Nils Hylander, 1904–1970, Institution of Systematic Botany, Uppsala, Sweden.

Iljin Modese Mikhailovich Iljin, 1889–1967, U.S.S.R.; Asteraceae, *Corispermum*.

Iltis Hugh Hellmut Iltis, 1925–, Czechoslovakian-born American botanist, University of Wisconsin, Madison; Capparaceae, flora of Wisconsin.

Irving Robert Stewart Irving, 1942–, University of Montana, Missoula; Lamiaceae.

Irwin Howard Samuel Irwin, 1928–, formerly president of the New York Botanical Garden, later professor of botany at Long Island University, New York; Leguminosae; flora of central Brazil.

Isely Duane Isely, 1918–, Iowa State University, Ames; Leguminosae.

Ives Eli Ives, 1779–1861, professor at Yale.

Jackson, R. C. Raymond C. Jackson, 1928–, Texas Tech University, Lubbock, Texas; Asteraceae.

Jacq. Nicolaus Joseph Baron von Jacquin, 1727–1817, professor of botany and director of the botanic garden, Vienna; noted systematist.

Jaeg. Herman Jaeger, 1815–1890, court gardener in Weimar and horticultural writer.

Jalas Jaakko (Arvo Jaakko Juhanni) Jalas, 1920–, director, Botanical Museum, University of Helsinki, Finland.

James Edwin James, 1797–1861, surgeon-naturalist, first botanical collector in Colorado and probably first botanical collector in Texas, with Major S. H. Long's expedition to the Rocky Mts. (1819–1820).

Jarmalenko A. V. Jarmalenko, 1905–1944, Russian; *Cardaria*.

Jaub. Hippolyte Francois Jaubert, 1798–1874, French agriculturalist and botanist.

Jennings Otto Emery Jennings, 1877–1964, curator at Carnegie Museum of Natural History (Pittsburgh), paleobotanist, and taxonomist; *Stachys*.

Jeps. Willis Linn Jepson, 1867–1946, professor of botany, University of California; author of *A Flora of California* (1909–1939), *A Manual of the Flowering Plants of California* (1923, 1925, 1950), and other works; a founder of the California Botanical Society.

Johnst., I. M. Ivan Murray Johnston, 1898–1960, botanist, Harvard University; authority on world floras, Boraginaceae, plant explorer, prolific author.

Johnst., J. R. John Robert Johnston, 1880–1953, American plant pathologist; student of the plants of Guatemala and Venezuela.

Johnst., M. C. Marshall Conring Johnston, 1930–, University of Texas, Austin; author with (D. S. Correll) of *Manual of the Vascular Plants* of *Texas* (1970); Rhamnaceae, Euphorbiaceae, flora of Chihuahua, Mexico.

Jones, G. N.   George Neville Jones, 1904–1970, University of Illinois, Urbana; author of floras on the Olympic Peninsula (1936), Mt. Rainer (1938), and Illinois (1945).

Jones, M. E.   Marcus Eugene Jones, 1852–1934, Utah mining consultant; assembled extensive herbarium of Great Basin plants now at Pomona College; published his botanical observations in a private journal, *Contributions to Western Botany*, that is marked by his cutting colorful criticism of almost all his contemporaries.

Jones, S. B.   Samuel B. Jones, 1933–, University of Georgia; Vernonieae.

Jonsell   Bengt Edvard Jonsell, 1936–, University of Stockholm, Stockholm, Sweden; Brassicaceae.

Jord.   Alexis Jordan, 1814–1897, Lyon, France; demonstrated the existence of many genetically distinct races (which he called species) in such complexes as *Erophila verna;* such microspecies are now often called "jordanons."

Juss.   Antoine Laurent de Jussieu, 1748–1836, professor in the Jardin des Plantes, Paris; first characterizer of natural families; expounder of the taxonomic system prepared by his uncle, Bernard de Jussieu.

Juss., A.   Adrien Henri Laurent de Jussieu, the son of Antoine, 1797–1853, professor in the Jardin des Plantes, Paris.

Kalm   Pehr (Peter) Kalm, 1715–1779, professor of natural science, Abo, Finland, pupil of Linnaeus who traveled in eastern North America (1747–1749), and published in 1765 the first part of *Flora Fennica*.

Karst.   Gustav Karl Wilhelm Hermann Karsten, 1817–1908, German botanist, professor in Vienna; author of *Flora von Deutschland, Oesterreich und der Schweiz* (1985).

Kaul   Robert B. Kaul, 1935–, University of Nebraska-Lincoln; monocots, angiosperm evolution, morphology, and biogeography.

Kaulf.   Georg Friedrich Kaulfuss, 1786–1830, professor in Halle an der Saale, Germany; student of ferns.

Kazmi   Syed M. A. Kazmi, 1926–, National Herbarium, Mogadishu, Somalia; *Carduus*.

Kearn.   Thomas Henry Kearney, 1874–1956, U.S. Dept. of Agriculture and California Academy of Sciences; taxonomist and cotton breeder, Malvaceae.

Keating   Richard Clark Keating, 1937–, Southern Illinois University, Edwardsville.

Keck   David Daniels Keck, 1903–, New York Botanical Garden; later National Science Foundation, Washington; Asteraceae (Madiinae), California flora, *Penstemon*, experimental taxonomy.

Kell.   Albert Kellogg, 1813–1887, San Francisco physician and botanist; a founder of the California Academy of Sciences.

Kelso   Leon Hugh Kelso, 1907–, U.S. Biological Survey in Colorado and Wyoming (1930s); *Salix*, Poaceae.

Keng   Yi Li Keng, 1898–1975, professor of botany, Nanking University, China; Poaceae.

Ker   John Bellenden Ker, or John Ker Bellenden, or (before 1804) John Gawler, 1764–1842, British botanist, first editor of *Edwards' Botanical Register*. (Sometimes cited as Ker-Gawl.)

Kindb.   Nils Conrad Kindberg, 1832–1910, professor in Linkoping, Sweden; *Spergularia*.

King, R. M.   Robert Merrill King, 1930–, botanist, U.S. National Herbarium, Washington; Asteraceae (Eupatorieae).

King & Rob.   R. M. King and Harold Ernest Robinson, 1932–, botanist, U.S. National Herbarium, Washington; Asteraceae, bryophytes.

Kirchn.   Ernst Otto Oskar von Kirchner, 1851–1925, Germany.

Kit.   Paul Kitaibel, 1757–1817, professor of botany and chemistry, and director of the botanical garden, Budapest.

Kittell   Sister Teresita Kittell, 1892–, Catholic University of America, Washington, later Holy Family College, Manitowoc, Wisconsin; author of *Critical Revision of the Compositae of Arizona and New Mexico* (1941) (cf. Tidestrom).

Kl.   Johann Friedrich Klotzsch, 1805–1860, curator of the herbarium, Berlin; monographer of Begoniaceae.

Kl. & Gke.   J. F. Klotzsch and F. A. Garcke.

Klatt   Friedrich Wilhelm Klatt, 1825–1897, Hamburg; Iridaceae.

Klinge   Johannes Christoph Klinge, 1851–1902, librarian at the Royal Botanical Gardens at St. Petersburg.

Knerr   Ellsworth Brownell Knerr, 1861–1942, Atchison, Kansas, physician; *Erythronium*.

Koch   Wilhelm Daniel Joseph Koch, 1771–1849, German botanist, professor of botany, Erlangen; author or editor of three floras of Germany and Switzerland; Apiaceae.

Koch, K.   Karl Heinrich Emil Koch, 1809–1879, German traveler in the Orient, dendrologist, professor in Berlin; author of *Hortus Dendrologicus* (1853).

Koch, S.   Douglas Stephen Douglas Koch, 1940–, Colegio de Postgraduados Chapingon, Mexico; *Eragrostis*, Poaceae of Mexico.

Koehne   Bernhard Adalbert Emil Koehne, 1848–1918, professor in Berlin; Lythraceae.

Koel.   Georg Ludwig Koeler, 1765–1807, professor in Mainz; author of a treatise on the grasses of Germany and France.

Kohli & Packer   B. Kohli, 1940–, University of Alberta, Edmonton, Canada; Poaceae (India), flora of western Canada; and J. G. Packer.

Kolstad   Ole A. Kolstad, 1930–, Kearney State College, Nebraska; Cyperaceae, flora of Nebraska.

Koyama   Tetsuo Koyama, 1933–, Japanese-born American botanist, New York Botanical Garden; earlier Botanical Institute, University of Tokyo, Japan; Cyperaceae.

Kral   Robert Kral, 1926–, Vanderbilt University, Nashville, Tennessee; Cyperaceae, Xyridaceae, Eriocaulaceae, flora of Alabama and middle Tennessee.

Kral & Bostick   Robert Kral and P. E. Bostick, 1939–, Emory University, Atlanta, Georgia (1966–1971), then Kennesaw College, Marietta, Georgia; *Rhexia*.

Krapov.   Antonio Kaprovickas, 1921–, director of Instituto de Botanica del Noroeste, Corrientes, Argentina; cytotaxonomy of Malvaceae and *Arachis*.

Krause, E. H. L.   Ernst Hans Ludwig Krause, 1859–1942, marine staff doctor, Kiel; *Rubus, Silene*.

Krok   Thorgny Ossian Bolivar Napoleon Krok, 1834–1921, Swedish botanist and botanical bibliographer, Stockholm; *Valerianella;* author of *Bibliotheca Botanica Suecana* (1925).

Kruschke   Emil Paul Kruschke, 1908–1973, Milwaukee Public Museum; *Crataegus*.

Ktze., O.   Carl Ernst Otto Kuntze, 1843–1907, German traveler and botanist, advocate of strict priority in nomenclature; author of *Revisio Generum Plantarum* (1891), in which the names of over 30,000 species were changed.

Kuhn   Friedrich Adalbert Maximilian Kuhn, 1842–1894, German student of ferns.

Kükenth.   Georg Kükenthal, 1864–1956, clergyman of Coburg, Germany; Cyperaceae.

Kunth   Carl Sigismund Kunth, 1788–1850, professor botany, Berlin; voluminous writer.

Kunze   Gustav Kunze, 1793–1851, German botanist and physician, director of the Leipzig botanical garden; student of ferns.

L.   Carl Linnaeus (later von Linne), 1707–1778, Uppsala, Sweden; founder of binomial nomenclature and the sexual system on classification in *Species Plantarum* (1753); the "Father of Taxonomy," as we know and practice it today; a prodigious author.

L. f.   Carl Linnaeus, the son, 1741–1783, successor to his father in the professorship of botany in Uppsala.

Lag.   Mariano Lagasca y Segura, 1776–1839, professor and director of the botanic garden, Madrid; his collections were destroyed by a mob, and he was exiled to England.

Lag. & Rodr.   M. Lagasca y Segura and Jose Demetrio Rodriguez, 1780–1846, director of the botanic garden, Madrid.

Laharpe   Jean Jacques Charles de Laharpe, 1802–1863.

Lahman   Bertha Marion Lahman (later Sherwood), 1872–?, United States; Cactaceae.

Lam.   Jean Baptiste Antoine Pierre Monet, Chevalier de Lamarck, 1744–1829, famous French student of natural history who propounded an early theory of evolution; first to use dichotomous keys in natural history in his *Flore Françoise* (1778), also author of *Encyclopedie Methodique, Botanique* (1783–1817) and *Histoire naturelle des Vegetaux Classes par Familles* (1803).

Lamb.   Aylmer Bourke Lambert, 1761–1842, an original member and vice-president of the Linnean Society in London, patron of botany; author of *A Description of the Genus Cinchona* (1797), *A Description of the Genus* Pinus (1803–1842, in 5 ed.), etc.

Landolt   Elias Landolt, 1926–, director of Geobotanisches Institut, Zurich, Switzerland; Lemnaceae.

Lange   Johan Martin Christian Lange, 1818–1898, professor of botany, Copenhagen.

Larisey   Mary Maxine Larisey, 1909–, School of Pharmacy, Medical College of South Carolina; Leguminosae.

Larsen   Esther Louise Larsen (later Doak), 1901–, Missouri Botanical Garden (1927–1928); Asteraceae (*Townsendia*).

Larson, G.   Gary E. Larson, 1950–, South Dakota State University; aquatic vascular plants in the northern Great Plains.

Laws.   Peter Lawson, d. 1820, and Sir Charles Lawson, the son, 1794–1873; Edinburgh nurserymen.

Lawson   George Lawson, 1827–1895, Scottish-born Canadian, professor of chemistry and natural history at Queens College, Kingston, Ontario, then at Dalhousie University, Halifax.

Laxmann   Erik G. Laxmann (also Eric), 1737–1796, Finnish-born clergyman, professor of economics and chemistry at St. Petersburg; *Koelreuteria*.

Leavenw.   Melines Conklin Leavenworth, 1796–1862, student of southern U.S. botany.

Le Conte   John Eatton Le Conte, 1784–1860, Philadelphia, Pennsylvania, topographical engineer who studied botany and entomology.

Lecoq & Lamotte   Henri Lecoq, 1802–1871, and Martial Lamotte, 1820–1883, authors of a catalog of plants of central France (1848).

Ledeb.   Carl Friedrich von Ledebour, 1785–1851, professor in Dorpat; author of *Flora*

Altaica (1829-1833) and *Flora Rossica* (1842-1853).

Leggett   William Henry Leggett, 1816-1882; *Lechea*.

Lehm.   Johann Georg Christian Lehmann, 1792-1860, director of the botanic garden, Hamburg; authority on *Potentilla* and other genera.

Leib.   John Bernhard Leiberg, 1853-1910, founder of the famous garden at Baden-Baden; introducer of numerous ornamental species.

Lej.   Alexander Louis Simon Lejeune, 1779-1858, Belgian physician at Verviers.

Lej. & Court.   A. L. S. Lejeune and R. S. Courtois, 1806-1835, Belgium.

Lemm.   John Gill Lemmon, 1832-1908, pioneer California botanist and schoolteacher in Sierra Valley, early correspondent of Asa Gray; Coniferae.

Lepage   Abbe Ernest Lepage, 1905-1981, Rimouski, Quebec, Canada, bryologist.

Lepech.   Ivan Ivanovic Lepechin, 1740-1802, St. Petersburg.

Less.   Christian Friedrich Lessing, 1809-1862, German physician; Asteraceae.

Levl.   Augustin Abel Hector Leveille, 1863-1918, French clergyman, professor at Pondicherry, India; flora of China.

Levl. & Van.   A. B. H. Leveille and E. Vaniot.

Lewis & Szweykowski   (Frank) Harlan Lewis, 1919-, University of California, Los Angeles, and Jerzy Szweykowski; Onagraceae.

Lewis, W. H.   Walter Hepworth Lewis, 1930-, Canadian-born U.S. botanist, Washington University and Missouri Botanical Garden, St. Louis; Rosaceae (*Rosa*), Rubiaceae (*Hedyotis*), Portulacaceae.

Ley   Frances Arline Ley (Later Fitch), 1919-, Claremont Graduate School, California; *Holodiscus*.

Leyss.   Friedrich Wilhelm von Leysser, 1731-1815.

L'Her.   Charles-Louis L'Heritier de Brutelle, 1746-1800, celebrated French magistrate and botanist; author of *Sertum Anglicum* (1788).

Liebm.   Frederick Michael Liebman, 1813-1856, Danish professor of botany and director of the Botanic Gardens, Copenhagen, Denmark; collected 40,000 plant specimens in Mexico.

Lightf.   John Lightfoot, 1735-1788, English clergyman; author of *Flora Scotica* (1777).

Lilj.   Samuel Liljeblad, 1761-1815, Swedish physician and professor of applied botany at Uppsala.

Lindb.   Sextus Otto Lindberg, 1835-1889, Finland; noted student of mosses.

Lindberg   Harold Lindberg, 1871-1963, curator of botany at the Helsinki Museum, Finland, son of Sextus Otto Lindberg; *Cystopteris*, paleobotany, and taxonomy.

Lindh.   Ferdinand Jakob Lindheimer, 1801-1879, German-born collector of Texas plants (1836-1879) and correspondent of Asa Gray and George Engelmann; resided in New Braunfels, Texas.

Lindl.   John Lindley, 1799-1865, professor of botany, London; editor of *Edward's Botanical Register* (1829-1847), horticulturist, eminent orchidologist, textbook writer.

Link   Johann Heinrich Friedrich Link, 1767-1851, professor of natural science and director of the botanic garden, Berlin.

Link & Otto   J. H. Link and C. F. Otto.

Lloyd   James Lloyd, 1810-1896, London and Nantes; student of the flora of western France.

Lodd.   Conrad Loddiges, 1738-1826, English nurseryman.

Loefl.   Pehr (Peter) Loefling, 1729-1756, Swedish student of Linnaeus, who traveled to Venezuela for him and died there.

Lois.   Jean Louis Augusta Loiseleur-Deslongchamps, 1774-1849, French physician; author of *Flora Gallica* (1806-1807; 2nd ed. 1828), *Juncus*.

Lojacono   Michele Lojacono-Pojero di Palermo, 1853-1919, professor at Instituto Tecnico, Messina, Sicily; bryologist.

Long   Robert W. Long, 1928-1976, University of South Florida; co-author (with Olga Lakela) of *A Manual of Tropical Florida* (1971).

Loud.   John Claudius Loudon, 1783-1843, English horticulturist, prolific author and editor of garden books.

Lour.   João de Loureiro, 1710-1791, Portuguese missionary and naturalist in Asia.

Löve & Löve   Askell Löve, 1916-, Icelandic botanist; formerly at University of Lund, Sweden (1940-1945), University of Iceland, Reykjavik (1945-1951), University of Manitoba, Winnipeg (1951-1956), University of Montreal, Canada (1956-1964), University of Colorado, Boulder, later in San Jose, California; Polygonaceae (*Rumex*), arctic-alpine-plants, and Doris (Mrs. Askell) Löve, 1918-; Caryophyllaceae, arctic-alpine plants.

Lucanus   Anonymous citizen of Lucca, fl. 1753; accidentally provided technical validation of *Dalea* "Linn."

Luer, C.   Carlyle A. Luer, 1922-, The Marie Selby Botanical Gardens, Sarasota, Florida; author of *The Native Orchids of Florida* (1972) and author of *The Native Orchids of the United States and Canada* (1975).

Lundell   Cyrus Longworth Lundell, 1907–, founder and director of Texas Research Foundation, Renner; author (with collaborators) of *Flora of Texas* (1942–1969, vols. 1, 2, and 3 publ.), *The Vegetation of Peten–* (1936), *The genus Parathesis of the Myrsinaceae* (1966), student of floras of Texas, Mexico, and Central America; Celastraceae, Myrsinaceae; author of numerous papers in agriculture and botany.

Lunnell   Joel Lunnell, 1851–1920, North Dakota botanist.

Ma   Yu-Chuan Ma, 1916–, University of Inner Mongolia, Huhehot, People's Republic of China; Gentianaceae, Brassicaceae, Apiaceae.

Macbr.   James Francis Macbride, 1892–1976, Gray Herbarium, Harvard University, later Chicago Natural History (Field) Museum; student of the West American flora; author (with collaborators) of *Flora of Peru* (1936–).

Mack.   Kenneth Kent Mackenzie, 1877–1934, New York City corporation attorney and noted botanist; *Carex*.

Mack. & Bush   K. K. Mackenzie and B. F. Bush.

MacM.   Conway MacMillan, 1867–1929, state botanist, Minnesota; author of *Minnesota Plant Life* (1899).

Macoun   John Macoun, 1831(?2)–1920, eminent Irish-born Canadian botanist, professor of botany, Albert College, Belleville, Ontario; author of *Catalogue of Canadian Plants* (1883-1902), etc.

Magnus   Paul Wilhelm Magnus, 1844–1941, German taxonomist; *Najas*.

Magrath   Lawrence K. Magrath, 1943–, University of Science & Arts of Oklahoma, Chickasha; Orchidaceae, flora of Oklahoma.

Maguire   Bassett Maguire, 1904–, New York Botanical Garden; flora of the Great Basin and northeastern South America.

Makino   Tomitaro Makino, 1862–1957; Japanese botanist; Rhamnaceae.

Malte   Oscar Malte, 1880–1933, Swedish-born, chief botanist, National Herbarium of Canada (1921–1933); Poaceae.

Marcks   Brian G. Marcks, fl. 1974, University of Wisconsin; *Cyperus*.

Marsh.   Humphrey Marshall, 1722–1801, Pennsylvania, the father of American dendrology; author of *Arbustum Americanum* (1785).

Mart.   Karl Friedrick Philipp von Martius, 1794–1868, German botanist and traveler, professor in Munich; founder of the classic *Flora Brasiliensis* (1840–1906) and prolific writer on systematic botany and zoology.

Martens   Martin Martens, 1797–1863, Belgian physician and taxonomist, University of Louvain, collected in Mexico.

Martin   Robert Franklin Martin, 1910–, Division of Plant Exploration and Introduction, U.S. Dept. of Agriculture, Beltsville, Maryland; later lawyer in West Virginia; Papaveraceae.

Martin, F. L.   Floyd Leonard Martin, 1909–, Claremont Graduate School, and teacher, Mark Keppel High School, Alhambra, California; Rosaceae (*Cercocarpus*).

Mason   Herbert Louis Mason, 1896–, professor of botany, University of California, Berkeley; Polemoniaceae, fossil floras of California, author of *Flora of the Marshes of California* (1957).

Mast.   Maxwell Tylden Masters, 1833–1907, England; editor of *The Gardener's Chronicle*, contributor to Martius' *Flora Brasiliensis* and Oliver's *Flora of Tropical Africa*.

Math. & Const.   Mildred Esther Mathias (later Hassler), 1906–, University of California, Los Angeles; Apiaceae; and L. Constance.

Mattf.   Johannes Mattfeld, 1895–1951, Berlin, Germany; Asteraceae, Caryophyllaceae, Cyperaceae.

Maxim.   Carl Johann (Karl Ivanovich Maksimovich) Maximowicz, 1827–1891, Russian botanist, director of the botanic garden, St. Petersburg.

Maxon   William Ralph Maxon, 1877–1948, curator, U.S. National Herbarium, Washington; eminent American pteridologist, prolific author and longtime editor (1934–1947) of the *American Fern Journal*.

McCl.   Elizabeth May McClintock, 1912–, botanist, California Academy of Sciences; Lamiaceae.

McCl. & Epl.   E. M. McClintock, and C. C. Epling.

McCoy, S.   Scott McCoy, 1897–, collected in Indiana from 1929–1942; *Ruellia*

McGreg., R. L.   Ronald Leighton McGregor, 1919–, professor of botany, University of Kansas, Lawrence; *Echinacea*, student of the Great Plains flora.

McKelvey   Susan Adams McKelvey (nee Delano), 1883–1964.

McNeill   John McNeill, 1933–, Biosystematics Research Institute, Agriculture Canada, Ottawa, later University of Ottawa; weeds, Caryophyllaceae.

McVaugh   Rogers McVaugh, 1909–, professor of botany (Emeritus), University of Michigan, Ann Arbor; Campanulaceae, student of Mexican flora, botanical history, etc.

Mears   James Austin Mears, 1944–, botanist, Academy of Natural Sciences, Philadelphia; Amaranthaceae, *Parthenium* (Sect. *Bolophyta*).

Medic. Friedrich Kasimir Medicus, 1736–1808, German botanist.

Meerb. Nicolaas Meerburgh, 1734–1814, Dutch gardener.

Meisn. Carl Friedrich Meisner (also Meissner), 1800–1874, Swiss botanist in Basel; Polygonaceae, Lauraceae, Ericaceae, Convolvulaceae.

Merrill Elmer Drew Merrill, 1876–1956, director of New York Botanical Garden, later Arnold Arboretum, Harvard University; bibliographer, prolific contributor on the flora of the Philippine Islands, China, and Indo-Malaysia.

Mert. Franz Carl Mertens, 1764–1831, professor in Bremen.

Mert. & Koch F. C. Mertens and W. D. J. Koch.

Mertens Thomas Robert Mertens, 1930–, botanist, Ball State University, Muncie, Indiana; *Polygonum*.

Mett. George Heinrich Mettenius, 1823–1866, German pteridologist, professor of botany, Leipzig; *Salvinia*, Selaginellaceae.

Mey., B. Bernhard Meyer, 1767–1836; *Armoracia*.

Mey., C. A. Carl Anton Andreevitch von Meyer, 1795–1855, director of the botanic garden, St. Petersburg.

Mey., E. Ernst Heinrich Friedrich Meyer, 1791–1858, German botanist, professor in Konigsberg; Juncaceae.

Mey., F. G. Frederick Gustav Meyer, 1917–, American taxonomist, Missouri Botanical Garden, later U.S. National Arboretum, Washington, D. C.; taxonomy of cultivated woody plants; *Valeriana*.

Mey., G. F. Georg Friedrich Wilhelm Meyer, 1782–1856, German botanist, professor in Gottingen; author of *Flora of Hannover* (1849).

Michx. Andre Michaux, 1746–1802, French botanist and explorer of North America; author of *Flora Boreali-Americana* (1803), *Quercus*.

Michx. f. Francois Andre Michaux, the son, 1770–1855, French botanist; author of *The North American Sylva* (1817–1819, first English ed.).

Mickel John Thomas Mickel, 1934–, curator of ferns, New York Botanical Garden; student of ferns.

Miers John Miers, 1789–1879, London; student of the South American flora.

Milde Carl August Julius Milde, 1824–1871(?2), German student of Pteridophyta.

Mill., P. Philip Miller, 1691–1771, British gardener; author of *The Gardener's Dictionary* (1731), which went through eight editions.

Mill., L. Lillian Wood Miller, 1937–, Jacksonville University, Jacksonville, Florida; Euphorbiaceae *(Acalypha)*.

Miller, E. S. Elihu Sanford Miller, 1848–1940, superintendent for John Lewis Childs, seedsman and nurseryman at Floral Park, New York, 1888–1902.

Millsp. Charles Frederick Millspaugh, 1854–1923, curator, department of botany, Field Museum, Chicago (1894–1923); student of Yucatan flora (1903–1904), author (with N. L. Britton) of *The Bahama Flora* (1920), prolific writer on floras of West Virginia, California, Florida, West Indies, etc.

Miq. Frederik Anton Wilhelm Miquel, 1811–1871, Dutch botanist, professor of botany, Utrecht; Urticaceae, Primulaceae, prolific writer.

Mirb. Charles François Brisseau de Mirbel, 1776–1854, professor at Paris; physiological plant anatomy.

Mitch. John Mitchell, 1676–1768; English-born Virginia physician.

Moench Conrad Moench, 1744–1805, botany professor in Marburg, Germany.

Mohlenbrock Robert H. Mohlenbrock, 1931–, Southern Illinois University, Carbondale; flora of Illinois.

Mohr Charles Theodore Mohr, 1824–1901, German chemist who eventually settled in Alabama; author of *Plant Life of Alabama* (1901).

Mol. Juan Ignacio (Giovanni Ignazio) Molina, 1740–1829, Chilean Jesuit; author of a natural history of Chile (1782).

Moldenke Harold Norman Moldenke, 1909–, the New York Botanical Garden, later Trailside Museum, New Jersey; Verbenaceae, Avicenniaceae, Eriocaulaceae, author of *American Wild Flowers* (1949), *Plants of the Bible* (1952); editor and publisher of *Phytologia*.

Moore Thomas Moore, 1821–1887, curator, Chelsea Botanic Garden, England; student of ferns and orchids.

Moore, T. V. Thomas Verner Moore, 1877–1969, American priest, educator and physiologist, Catholic University of America, Washington; published on *Rudbeckia*.

Moq. Christian Horace Benedict Alfred Moquin-Tandon, 1804–1863, French botanist; pupil of A. P. de Candolle; Chenopodiaceae, Amaranthaceae.

Moric. Moise Etienne Moricand, 1779–1854, Swiss botanist, Geneva.

Morong Rev. Thomas Morong, 1827–1894, Massachusetts minister and amateur botanist; Najadaceae, Potamogetonaceae.

Morton Conrad Vernon Morton, 1902–1972, curator of ferns, U.S. National Herbarium,

Washington; eminent pteridologist, prolific author and longtime editor (1945–1972) of the *American Fern Journal*.
Mosher, E. Edna Mosher, American entomologist; Poaceae.
Mosquin Theodore Mosquin, 1932–, Canadian botanist.
Muell. Arg. Jean (Argoviensis, i.e., of Aargau) Mueller, 1828–1896, Swiss botanist; Euphorbiaceae, Buxaceae, Resedaceae.
Muench. Otto Freiherr von Muenchhausen, 1716–1744, German botanist and baliff at Kalemburg.
Muhl. Henry Muhlenberg (Gotthilf Heinrich Ernst Muehlenberg, also Heinrich Ludwig Muehlenberg), 1753–1815, German-educated Lutheran minister and pioneer botanist of Pennsylvania.
Muhlenpfordt F. Muhlenpfordt, German physician at Hannover, horticulturist (Cactaceae) of middle 19th century.
Mull. Otto Fridrich Muller, 1730–1784, Danish privy-councilor at Copenhagen; one author of *Flora Danica* (1775–1782).
Mulligan Gerald A. Mulligan, 1928–, Biosystematic Research Institute, Ottawa, Canada; *Cicuta*.
Munro William Munro, 1818–1880, English general and agrostologist; Bambusoideae.
Munson Thomas Volney Munson, 1843–1913, grape breeder of Denison, Texas.
Munz Philip Alexander Munz, 1892–1974, Rancho Santa Ana Botanic Garden, Claremont, California; author of *Manual of Southern California Botany* (1935), *A California Flora* (1959); Onagraceae, *Aquilegia*.
Murr Josef Murr, 1864–1932, professor at Innsbruck, Austria; *Hieracium*, *Viola*.
Murray Johann Andreas Murray, 1740–1791, Swedish professor of medicine and botany, Göttingen, Germany (1760–1791); student of Linnaeus and editor of some of his later editions.
Mutis Jose Celestino Mutis, 1732–1808, Spanish botanical explorer in Colombia; correspondent of Linnaeus.
Myint Tin Myint, 1936–, Burmese botanist, Mandalay; published on *Stylisma* (Convolvulaceae) in 1966.
Nakai Takenoshin Nakai, 1882–1952, professor of botany and director of the National Science Museum at the University of Tokyo; author of *Flora Koreana* (1909–1911) and *Flora Sylvatica Koreana* (1915–1939).
Nash George Valentine Nash, 1864–1921, head gardener, the New York Botanical Garden; agrostologist.
Necker Noel Joseph de Necker, 1730–1793, Lille France, then elector at Mannheim; bryology, author of a flora of French Flanders (1768).
Nees Christian Daniel Nees von Esenbeck, 1776–1858, German botanist, professor in Breslau; prolific botanical writer; Acanthaceae, Cyperaceae, Poaceae.
Nees & Mey. C. G. D. Nees von Esenbeck and Franz Julius Ferdinand Meyen, 1804–1840, German physician and botanical artist who collected in Brazil and Peru.
Neilr. August Neilreich, 1803–1871, judge of the Tribunal at Vienna University; phytogeography, Austrian flora.
Nels., A. Aven Nelson, 1859–1952, University of Wyoming, Laramie; student of the Rocky Mountain flora, reviser of Coulter's *New Manual of Botany of the Central Rocky Mountains* (1909).
Nels., A. & Cockll. A. Nelson and T. Cockerell.
Nels., A. & Macbr. A. Nelson and J. F. Macbride.
Nels., E. Elias Emanuel Nelson, 1876–1949, Swedish-born American agriculturist, U.S. Dept. of Agriculture Experimental Farm, Bend, Oregon; horticulture, forage plants for arid regions, editor, author of *The Shrubs of Wyoming* (1902).
Nevski Sergei Arsenjevic Nevski, 1908–1938, senior agrostologist at Botanical Institute of Academy of Science in U.S.S.R., Leningrad.
Newm. Edward Newman, 1801–1876, English naturalist, entomologist, ornithologist, student of ferns.
Nicora Elisa G. Nicora, 1912–, Instituto de Botanica Darwinion, San Isidro, Argentina; Poaceae.
Nielsen Etlar Lester Nielsen, 1905–1969, professor of agronomy, University of Wisconsin; *Amelanchier*, grassland ecology.
Nieuw. Julius Aloysius Arthur Nieuwland, 1878–1936, professor of botany (1904–1918) and professor of organic chemistry (1918–1936), Notre Dame University, Indiana; founder and first editor of *The American Midland Naturalist*.
Nort. John Bitting Smith Norton, 1872–1966, taxonomist and plant pathologist, Kansas State University, later University of Maryland, College Park.
Northington David Knight Northington, 1944–, Texas Tech University, Lubbock; systematics of phanerogams, *Pyrrhopappus*.
Nutt. Thomas Nuttall, 1786–1859, English-American naturalist, botanist, and ornithologist who resided in the U.S. (1808–1841), collector of western American plants; author of *The Genera of North American Plants* (1818) and *The North American Sylva* (1842).

Nyman   Carl Fredrik Nyman, 1820–1893, conservator at the Museum in Stockholm (1855–1889), explored Sicily, Malta, and Austria; *Carduus.*
Oakes   William Oakes, 1799–1848, Massachusetts; student of Vermont flora.
Oeder   George Christian von Oeder, 1728–1791, Danish botanist; professor in Copenhagen; first editor of *Flora Danica* (1761–1771).
Ogden   Eugene Cecil Ogden, 1905–, state botanist, New York; palynology, Potamogetonaceae.
Ohwi   Jisaburo Ohwi, 1905–1976, National Science Museum, Tokyo; student of flora of Japan, Poaceae and Cyperaceae of eastern Asia.
Olney   Col. Stephen Thayer Olney, 1812–1878, Providence, Rhode Island; *Carex.*
O'Neill & Duman   Hugh Thomas O'Neill, 1894–1969, curator of the Langlois Herbarium, The Catholic University of America, Washington; Cyperaceae, and Maximilian George Duman, 1906–, St. Vincent College, Latrobe, Pennsylvania; North American *Carex.*
Opiz   Philipp Maximilian Opiz, 1787–1858, zealous Bohemian botanist who named great numbers of "species" which have not been accepted.
Ort.   Casimiro Gomez Ortega, 1740–1818, Spanish botanist, director of the botanical garden, Madrid.
Osterh.   George Everett Osterhout, 1858–1937, Colorado lumberman and amateur botanist.
Otth   Karl Adolph Otth, 1803–1839.
Otto   Christoph Friedrich Otto, 1783–1856.
Ownbey, G.   Gerald Bruce Ownbey, brother of Marion, 1916–, University of Minnesota; *Argemone, Corydalis, Cirsium.*
Ownbey, M.   (Francis) Marion Ownbey, 1910–1975, Washington State University, Pullman; *Calochortus, Allium, Tragopogon;* co-author of *Vascular Plants of the Pacific Northwest* (1955–1964).
Packer   John G. Packer, 1929–, University of Alberta, Edmonton, Canada; arctic and alpine plants, flora of Alberta.
Paine   John Alsop Paine, 1840–1912, professor at various institutes in the U.S. and abroad, curator at the Metropolitan Museum of Art, New York (1889–1912), studied a wide variety of subjects including botany, archeology, and geology.
Pall.   Peter Simon Pallas, 1741–1811, German botanist, student of the Russian and Siberian floras; early monograph of *Astragalus* (1800), also eminent as zoologist.
Palm., E. J.   Ernest Jesse Palmer, 1875–1962, field collector for Missouri Botanical Garden and Arnold Arboretum (1913–1948); specialist on *Crataegus.*
Palm. & Steyerm.   E. J. Palmer and J. A. Steyermark.
Pammel   Louis Hermann Pammel, 1862–1931; professor of botany, Iowa State University, Ames; student of poisonous plants, flora of Iowa, etc.
Parker   Kittie Lucille (Fenley) Parker, 1910–, U.S. National Herbarium, later George Washington University, Washington; Asteraceae (*Hymenoxys*), weeds of Arizona
Parks   Harris Braley Parks, 1879–1958, apiculturist, Texas A & M University, College Station; collector of Texas plants and author.
Parl.   Filippo Parlatore, 1816–1877, Italian botanist, professor in Florence and founder of the herbarium; Gnetaceae, Coniferae.
Parodi   Lorenzo Raimundo Parodi, 1895–1966, agrostologist and professor of botany at Buenos Aires, Argentina.
Parry   Charles Christopher Parry, 1823–1890, Colorado and Iowa, botanist with the Mexican Boundary Survey.
Patt.   Harry Norton Patterson, 1853–1919, Oquawka, Illinois, printer and botanist; published a catalogue of Illinois plants in 1876.
Pax   Ferdinand Albin Pax, 1858–1942, professor of botany, Breslau; Aceraceae, Primulaceae.
Pax & K. Hoffm.   F. A. Pax and Kaethe Hoffmann, 1883–?, professor in Breslau; made the huge contribution to Das Pflanzenreich on the Euphorbiaceae.
Payne   Willard William Payne, 1934–, botanist, University of Illinois, Urbana, later director of Cary Arboretum, New York; Asteraceae (*Ambrosia*).
Pays.   Edwin Blake Payson, 1893–1927, University of Wyoming, Laramie; Brassicaeae, *Cryptantha.*
Peck   Morton Eaton Peck, 1871–1959, Williamette University, Salem, Oregon; author of *A Manual of the Higher Plants of Oregon* (1941).
Penn.   Francis Whittier Pennell, 1886–1952, curator of botany, Academy of Natural Sciences of Philadelphia; authority on Scrophulariaceae; author of *The Scrophulariaceae of Eastern Temperate North America* (1935).
Perdue   Robert Edward Perdue, Jr., 1924–, botanist, U.S. Dept. of Agriculture, Beltsville, Maryland; *Rudbeckia.*
Perry   Lily May Perry, 1895–?, Canadian-born taxonomist, Arnold Arboretum, Harvard University; *Verbena.*
Pers.   Christian Hendrik Persoon, 1761(?2)–1836, bizarre individual and brilliant my-

cologist, born in South Africa, studied in Germany, lived in Paris; author of *Synopsis Plantarum* (1805–1807) and other valuable botanical works.

Peterm.   Wilhelm Ludwig Petermann, 1806–1855, professor at Leipzig, student of the flora of Germany.

Petrovic   Sava Petrovic, 1839–1889, Yugoslavian botanist and physician in Belgrade.

Pfeiffer   Norma Etta Pfeiffer, 1889–?, Boyce Thompson Institute for Plant Research, Yonkers, New York; *Isoetes*.

Phil.   Rudolf Amandus Philippi, 1808–1904, Chilean botanist, director of Museo Nacional, and professor of botany, Santiago; student of botany, zoology, and paleontology of Chile.

Phillips   Lyle L. Phillips, 1923–, University of Washington, Seattle; *Lupinus*.

Pick.   Charles Pickering, 1805(?6)–1878, American physician and botanist on the Wilkes Expedition to explore the western slope of the Rocky Mts. in the Columbia River area; plant geographer, ethnologist, historian.

Pilg.   Robert Knud Friedrich Pilger, 1876–1953, director of the botanic garden and museum, Berlin; Plantaginaceae, Coniferae, Poaceae.

Piper   Charles Vancouver Piper, 1867–1926, professor of botany and zoology, Washington State University, specialist in agricultural plants.

Planch.   Jules Emile Planchon, 1823–1888, French botanist, assistant to W. J. Hooker (1844–1848), professor of botany, Ghent, Nancy, and Montpellier; Ulmaceae, Vitaceae.

Plum.   Charles Plumier, 1646–1704, French Franciscan monk who explored and wrote on the flowering plants and ferns of tropical America.

Pohl   Johann Emanuel Pohl, 1782–1834, Austrian botanist and traveler, collected in Brazil.

Pohl, R.   Richard Walter Pohl, 1916–, Iowa State University, Ames; Poaceae.

Poir.   Jean Louis Marie Poiret, 1755–1834, French botanist and traveler in North Africa; completed Lamarck's *Encyclopedie Methodique Botanique* (1783–1817).

Pollard   Charles Louis Pollard, 1872–1945, Vermont librarian and plant collector.

Pollard & Ball   C. L. Pollard and C. R. Ball.

Porter   Thomas Conrad Porter, 1822–1901, professor of botany, Lafayette College, Pennsylvania; author of *Flora of Pennsylvania* (1903).

Porter, D. M.   Duncan MacNair Porter, 1937–, Virginia Polytechnic Institute and State University, Blacksburg; Zygophyllaceae, Galapagos Islands flora.

Prain   David Prain, 1857–1944, British botanist at Kew.

Prantl   Karl Anton Eugen Prantl, 1849–1893, German botanist, professor of botany, Breslau; prepared (with H. G. A. Engler) the classic work *Die Naturlichen Pflanzenfamilien* (1887–1915).

Prescott   John D. Prescott, 17?–1837, a merchant in St. Petersburg; his herbarium, rich in collections of Douglas, Nuttall, and Scouler, is now at Oxford.

Presl   Karel Boriwag Presl, 1794–1852, professor of natural history, Prague; in *Reliquiae Haenkeanae* (1825–1835) he described the collections made along the western side of the American continent by Thaddeaus Haenke, the first botanist to visit California.

Presl, J.   Jan Swatopluk Presl, the brother, 1791–1849, professor in Prague.

Prince   William Robert Prince, 1795–1869, American nurseryman.

Pringle   Cyrus Guernsey Pringle, 1838–1911; noted for his extensive collections of plants in Mexico.

Pritz.   George August Pritzel, 1815–1874, bibliographer, Academy of Sciences, Berlin; Lycopodiaceae, *Anemone;* author of the invaluable *Thesaurus Literaturae Botanicae* (1851; 2nd ed. 1872–[1877]).

Pursh   Frederick Traugott Pursh, 1774–1820, born in Saxony, settled in Philadelphia; author of *Flora Americae Septentrionalis* (1814).

R. & P.   Hipolito Ruiz Lopez, 1750–1815, and Jose Antonio Pavon, 1750–1844, Spanish explorers and authors of a flora of Peru and Chile (1794 and 1798–1802).

R. & S.   J. J. Roemer and J. A. Schultes, produced an edition of Linnaeus' *Systema Vegetabilium* (1817–1830).

Rabenh.   Gottlob Ludwig Rabenhorst, 1806–1881, German author of cryptogamic floras.

Raf.   Constantine Samuel Rafinesque (or Rafinesque-Schmaltz), 1783–1840, born in Constantinople, lived in Sicily and Kentucky; brilliant, eccentric pioneer naturalist; profligate author of binomials, with many "species" quite untraceable.

Raim.   Rudolph Raimann, 1863–1896, Vienna.

Ramaley   Francis Ramaley, 1870–1942, Minnesota, Colorado, and Wyoming.

Rand & Redf.   Edward Lathrop Rand, 1859–1924, Massachusetts, and John Howard Redfield, 1815–1895, New England.

Raup   Hugh Miller Raup, 1901–, research associate, Arnold Arboretum; botanical explorer in northwestern Canada.

Rausch.   Ernst Adolf Rauschel, fl. 1772–1797, Germany; author of *Nomenclator Botanicus* (1797).

Raven Peter Hamilton Raven, 1936–, director of the Missouri Botanical Garden, St. Louis; Onagraceae of western North America; flora of Chiapas.

Raven & Gregory P. Raven and David P. Gregory, fl. 1972, Keene State College, New Hampshire; *Gaura*.

Ray James Davis Ray, Jr., 1918–, professor of botany, University of South Florida, Tampa; *Lysimachia*.

Rech. f. Karl Heinz Rechinger, 1906–, Natural History Museum, Vienna; flora of the Mediterranean and the Near East; *Rumex*, Lamiaceae.

Reed, E. L. Edward Looman Reed, 1878–1946, Texas A & M University, then John Tarleton College, then Texas Technological College, Lubbock; collector and student of Texas botany.

Regel Eduard August von Regel, 1815–1892, director of the botanic garden, St. Petersburg; editor of *Gartenflora*, Betulaceae.

Rehd. Alfred Rehder, 1863–1949, German-born American botanist, curator of the herbarium, Arnold Arboretum, Harvard University; author of *Manual of Cultivated Trees and Shrubs Hardy in North America* (1927; 2nd ed. 1940).

Rehmann Anton Rehmann, 1840–1917, professor of geography at Leinberg.

Reich. Johann Jacob Reichard, 1743–1782, German physician and botanist, Frankfurt am Main.

Reichb. Heinrich Gottlieb Ludwig Reichenbach, 1793–1879, German naturalist, professor in Dresden; first editor of *Icones Florae Germanicae et Helveticae* and author of many other extensive works on plants and animals.

Retz. Anders Jahan Retzius, 1742–1821, Swedish botanist, professor in Lund; prolific writer on botany and zoology.

Reveal James Lauritz Reveal, 1941–, University of Maryland, College Park; intermountain flora, *Eriogonum*.

Rich. Louis Claude Marie Richard, 1754–1821, French botanist and collector in South America and the West Indies.

Rich., A. Achille Richard, 1794–1852, the son; physician and botanical demonstrator to the faculty of medicine, Paris.

Richard. Sir John Richardson, 1787–1865, Scottish biologist attached to Capt. Sir John Franklin's expedition to arctic America.

Richt. Karl (Carl) Richter, 1855–1891, Vienna.

Ricker Percy Leroy Ricker, 1878–1973, agronomist, U.S. Dept. of Agriculture, Washington; Poaceae, Leguminosae.

Ridd. John Leonard Riddell, 1807–1865, author of *Catalogus Florae Ludovicianae* (1852).

Robbins James Watson Robbins, 1801–1879, physician at Uxbridge, Massachusetts; studied the New England flora; *Potamageton*.

Robins. Benjamin Lincoln Robinson, 1864–1935, curator of Gray Herbarium, Harvard University (1892–1935); co-author of *Gray's Manual of Botany*, 7th ed.; Asteraceae.

Robins. & Fern. B. L. Robinson and M. L. Fernald.

Robins. & Greenm. B. L. Robinson and J. M. Greenman.

Rock Howard Francis Leonard Rock, 1925–1964, professor of biology, Vanderbilt University, Nashville, Tennessee; Asteraceae (*Helenium*).

Roehl. Johann Christoph Roehling, 1757–1813; German clergyman; author of *Deutschlands Flora* (1796), with later editions.

Roem. Johann Jacob Roemer, 1763–1819, physician and professor of botany at Zurich, Switzerland; active editor, with J. A. Schultes published 16th ed. of Linnaeus' *Systema Vegetabilium*.

Rogers Claude Marvin Rogers, 1919–, Wayne State University, Detroit, Michigan; Linaceae.

Roll. Reed Clark Rollins, 1911–, Gray Herbarium, Harvard University, director 1948–1980; Brassicaceae, *Parthenium*.

Rose Joseph Nelson Rose, 1862–1928, botanist, U.S. Dept. of Agriculture, later U.S. National Herbarium, Washington; Cactaceae, Umbelliferae, Crassulaceae.

Rosend., Butt. & Lak. Carl Otto Rosendahl, 1875–1956, professor of botany, University of Minnesota; Saxifragaceae; and F. K. Butters and Olga Korhoven Lakela, 1890–?, University of Minnesota-Duluth, later University of South Florida, Tampa; author (with Robert Long) of *A Flora of Tropical Florida* (1970).

Rostk. Friedrich Wilhelm Gottlieb Rostkovius, 1770–1848, physician in Stettin, (now Poland); author (with W. L. E. Schmidt) of *Flora Sedinensis* (1824).

Rostk. & Schmidt F. W. G. Rostkovius and F. W. Schmidt.

Roth Albrecht Wilhelm Roth, 1757–1834, German physician and botanist.

Rothmaler Werner Hugo Paul Rothmaler, 1908–1962, contemporary botany of Griefswald, Germany; Brassicaceae.

Rottb. Christen Friis Rottboll (Rottboell), 1727–1797, professor of botany and director of the botanical garden, Copenhagen, Denmark.

Rouleau Joseph Albert Ernest Rouleau, 1916–, University of Montreal, Quebec, Canada; Canadian flora, bibliography.

ABBREVIATIONS FOR NOMENCLATURAL AUTHORITIES

Rowlee   Willard Winfield Rowlee, 1861–1923, American student of *Salix*.

Roxb.   William Roxburgh, 1751–1815, Scottish physician and botanist, director of the Royal Botanic Gardens, Calcutta; author of *Flora Indica* (1820–1824 and later editions).

Rudge   Edward Rudge, 1763–1846, British botanist and antiquarian.

Rudolph, J. H.   Johann Heinrich Rudolph, 1744–1809, German botanist at St. Petersburg.

Rudolphi   Carl Asmund Rudolphi, 1771–1832, Swedish-born physician and professor at Berlin, Germany, "father of Helminthology."

Rugel   Ferdinand Rugel, 1806–1878, German-born plant explorer in southeastern United States.

Rupr.   Franz Joseph Ruprecht, 1814–1870, Austrian-born Russian physician and botanist, curator of the herbarium of the Academy of Science, St. Petersburg; Poaceae, Apiaceae, *Botrychium*, algae.

Rusby   Henry Hurd Rusby, 1855–1940, dean of the Columbia College of Pharmacy, New York; an active collector in South America.

Russell   Norman Hudson Russell, 1921–, professor of botany, Central State University, Edmond, Oklahoma; Violaceae.

Rydb.   Per Axel Rydberg, 1860–1931, Swedish-born American botanist, curator, New York Botanical Garden; author of *Flora of Colorado* (1906), *Flora of the Rocky Mountains and Adjacent Plains* (1917), *Flora of the Prairies and Plains of Central North America* (1932), etc.

Sagorski   Ernst Sagorski, 1847–1929, German teacher in Naumberg.

Salisb.   Richard Anthony (born Markham) Salisbury, 1761–1829, British botanist; early proponent of the natural system of classification.

Salm-Dyck   Joseph Franz Maria Anton Hubert Ignatz, Furst zu Salm-Reifferscheid-Dyck, 1773–1861, German amateur botanist and horticulturalist; succulents; especially Cactaceae.

Sarg.   Charles Sprague Sargent, 1841–1927, botanist and dendrologist, curator and director of Arnold Arboretum, Harvard University; author of *The Silva of North America* (1891–1902), *Manual of the Trees of North America* (1905), etc.

Sauer   Jonathan Deininger Sauer, 1918–, University of Wisconsin, Madison, later University of California, Los Angeles; *Amaranthus, Canavalia*.

Savi   Gaetano Savi, 1769–1844, Italian botanist; author of a flora of Pisa (1798).

Sch. Bip.   Carl Heinrich Schultz, Bipontinus (i.e., of Zweibrucken), 1805–1867, Germany; Asteraceae.

Schaffn., J. H.   John Henry Schaffner, 1866–1939, professor of botany, Ohio State University; *Equisetum*, phylogeny, author of *Field Manual of the Flora of Ohio and Adjacent Territory*, etc. (1928).

Scheele   Georg Heinrich Adolf Scheele, 1808–1864, German botanist who described plants from Texas (1849).

Scherb.   Johannes Scherbius, 1769–1813, German physician, professor of botany and medicine.

Scheutz   Nils Johan Wilhelm Scheutz, 1836–1889, Swedish botanist; studied the flora of Smaland.

Schinz   Hans Schinz, 1858–1941, director, botanical garden and museum, Zurich, Switzerland; *Kochia*.

Schleich.   Johann Christoph Schleicher, 1768–1834, German-Swiss botanist.

Schinz & Thell.   H. Schinz and A. Thellung.

Schkuhr   Christian Schkuhr, 1741–1811, German botanist and "university mechanic" in Wittenberg; student of the German flora.

Schlecht.   Diederich Franz Leonard von Schlechtendal, 1794–1866, German botanist and botanical editor, professor in Halle an der Saale; Elaeagnaceae.

Schleich.   Johann Christoph Schleicher, 1768–1834, German-Swiss botanist.

Schleid.   Matthias Jakob Schleiden, 1804–1881, Germany; author of botanical handbooks, co-proponent of the cell theory.

Schmid., C.   Casimir Christoph Schmidel, 1718–1792, German physician and botanist at Erlangen.

Schmidt, F.   Friedrick W. Schmidt, 1832–1908, St. Petersburg paleontologist who studied the eastern Siberian flora.

Schneid.   Camillo Karl (formerly Carl Camillo) Schneider, 1876–1951, Austria and Germany; dendrologist who explored for plants in China.

Schott   Heinrich Wilhelm Schott, 1794–1865, Austrian botanist, director of the royal garden in Schonbrunn, Vienna; Araceae.

Schrad.   Heinrich Adolph Schrader, 1767–1836, German botanist, professor in Göttingen; *Verbascum*.

Schrank   Franz von Paula von Schrank, 1747–1835, German botanist, professor in Munich; author of floras of Bavaria, Salisburg, and Monaco.

Schreb.   Johann Christian Daniel von Schreber, 1739–1810, German botanist, professor in Erlangen; editor of the 8th ed. of Linnaeus' *Genera Plantarum* (1789–1791).

Schub.   Bernice Giduz Schubert, 1913–, Gray Herbarium, Harvard University (1941–1950), Brussels (1950–1952), U.S. Dept. of Agriculture, Beltsville (1952–1961), Arnold Arboretum, Harvard University (1962–), editor,

Journal of the Arnold Arboretum (1963–); *Desmodium, Begonia, Dioscorea.*

Schult., J. A.   Joseph August Schultes, 1773–1831, Austrian botanist, professor of botany in Vienna, Cracow, and Landeshut.

Schult., J. H.   Julius Hermann Schultes, 1804–1840, son of J. A.; *Helictotrichon.*

Schult. & Schult.   J. A. Schultes and J. H. Schultes.

Schultz, F.   Friedrick Wilhelm Schultz, 1804–1876, German physician, brother of Karl Heinrich (Schulz-Bipontinus); student of the flora of the Upper Rhine.

Schulz   Otto Eugen Schulz, 1874–1936, German taxonomist; Brassicaceae.

Schum., K.   Karl Moritz Schumann, 1851–1904, German botanist, curator of the herbarium, Berlin; Cactaceae.

Schur   Phillip Johann Ferdinand Schur, 1799–1878, Germany, flora of Transylvania, studied pharmacy at Konigsberg, director of a chemical firm in Hermannstadt.

Schuyler   Alfred Ernest Schuyler, 1935–, Academy of Natural Sciences of Philadelphia, Pennsylvania; Cyperaceae.

Schwarz, O.   Otto Schwarz, 1900–.

Schweigg.   August Friedrich Schweigger, 1783–1821, German professor of botany at Königsberg; flora of Erlangen.

Schwein.   Lewis David von Schweinitz, 1780–1834, German-born Pennsylvania clergyman; noted student of fungi.

Schwein. & Torr.   L. D. von Schweinitz and J. Torrey.

Scop.   Johann Anton (Giovanni Antonio) Scopoli, 1723–1788, Austrian botanist, physician, and professor of natural history, Pavia.

Scribn.   Frank Lamson Scribner, 1851–1938, agrostologist, U.S. Dept. of Agriculture, Washington.

Scribn. & Ball   F. L. Scribner and C. R. Ball.

Scribn. & Merr.   F. L. Scribner and E. D. Merrill.

Scribn. & Sm.   F. L. Scribner and J. G. Smith.

Scribn. & Williams.   F. L. Scribner and T. A. Williams.

Seem.   Berthold Carl Seemann, 1825–1871, German naturalist and world traveler living in England, editor of *Bonplandia,* founder of the *Journal of Botany;* author of *Botany of the Voyage of the H. M. S. Herald* (1852–1857).

Seland.   Nils Sten Edvard Selander, 1891–, Swedish ecologist; *Urtica.*

Semple   John Cameron Semple, 1947–, University of Waterloo, Ontario, Canada; North American Astereae.

Sendtner   Otto Sendtner, 1813–1859, professor of botany, Monaco.

Ser.   Nicolas Charles Seringe, 1776–1858, French botanist, at first curator of the de Candolle herbarium, later professor of botany in Lyon; important collaborator in de Candolle's *Prodomus;* Caryophyllaceae, Rosaceae, Cucurbitaceae, *Salix, Aconitum.*

Ses.   Martin de Sessé y Lacasta, 1788–1809, Spanish botanist, director of the botanic garden in Mexico City.

Ses. & Moc.   M. Sessé y Lacasta and J. M. Mociño, 1757–1820, Mexican physician.

Shafer   John Adolf Shafer, 1863–1918, New York Botanical Museum.

Sharp   Ward McClintic Sharp, 1904–, U.S. Fish and Wildlife Service, State College, Pennsylvania.

Shaver   Jesse Milton Shaver, 1888–1966; *Ferns of Tennessee.*

Shaw   George Russell Shaw, 1848–1937; *Pinus;* author of *Pines of Mexico* (1909).

Shaw, E.   Elizabeth Anne Shaw, 1938–, Gray Herbarium, Harvard University; Brassicaceae.

Shear   Cornelius Lott Shear, 1865–1965, plant pathologist and agrostologist, U.S. Dept. of Agriculture, Washington; collected widely in the Rocky Mts., published with P. A. Rydberg on grasses (1897).

Sheld.   Edmund Perry Sheldon, 1869–?, resident first Minnesota, later in Portland, Oregon, forestry; author of *The Forest Wealth of Oregon* (1904), *Astragalus.*

Sherff   Earl Edward Sherff, 1886–1966, botanist, Chicago Teachers College; student of *Bidens* and the Hawaiian flora.

Sheviak   Charles John Sheviak, 1947–, New York State Museum, Albany; Orchidaceae.

Shinners   Lloyd Herbert Shinners, 1918–1971, Canadian-born botanist, professor of botany, Southern Methodist University; student of floras of north-central Texas, southeastern U.S., Asteraceae, Convolvulaceae, Caryophyllaceae, author of *Spring Flora of the Dallas-Fort Worth Area Texas* (1958), founder and editor of journal *Sida,* etc.

Short & Peter   Charles Wilkins Short, 1794–1863, medical botanist, professor in Louisville, Kentucky, and Robert Peter, 1805–1894; *Ludwigia.*

Shuttlew.   Robert James Shuttleworth, 1810–1874, English botanist and conchologist who resided most of his life in Bern; amassed a herbarium of 170,000 specimens now in the British Museum.

Sibth.   John Sibthorp, 1758–1796, professor of botany, Oxford, England.

Sibth. & Sm.   J. Sibthorp and J. E. Smith.

Sieb. & Zucc.   Philipp Franz von Siebold, 1796–1866, German botanist, made several

trips to study the botany, agriculture, and ethnography of Japan, and J. G. Zuccarini; authors of a flora of Japan (1835–1870).
Sim   Robert Sim, 1791–1878, Scottish-born English nurseryman; fern horticulture, bryology.
Sims   John Sims, 1749–1831, England; for 25 years editor of *Curtis' Botanical Magazine.*
Sm., J. E.   Sir James Edward Smith, 1759–1828, British botanist, founder and for 40 years president of the Linnaean Society; purchaser of the Linnaean herbarium and library (1784), now at the Linnaean Society of London.
Sm. & Rydb.   J. G. Smith and P. A. Rydberg.
Sm., E. B.   Edwin Burnell Smith, 1936–, University of Arkansas, Fayetteville; Asteraceae.
Sm., J. G.   Jared Gage Smith, 1866–1925, agrostologist, U.S. Dept. of Agriculture, Washington, later Hawaii; *Sagittaria.*
Small, E.   Ernest Small, 1940–, Biosystematic Research Institute, Ottawa, Canada; Cannabaceae, *Daucus, Medicago.*
Small, J.   John Kunkel Small, 1869–1938, American botanist, head curator, New York Botanical Garden; author of *Flora of the Southeastern United States* (1903), *Manual of the Southeastern Flora* (1933), etc.
Smyth   Bernard Bryan Smyth, 1843–1913, collector of Kansas plants and curator of State Museum of Kansas.
Smyth & Smyth   B. B. Smyth and Lumina C. Riddle Smyth, wife of B. B. Smyth.
Sobol.   Gregory Fedorovitch Sobolewski, 1741–1807; author of *Flora Petropolitana* (1799).
Soland.   Daniel Carl Solander, 1733–1782, England; gifted Swedish student of Linnaeus; accompanied Sir Joseph Banks on Capt. Cook's first voyage of circumnavigation.
Solbrig   Otto Thomas Solbrig, 1930–, Argentine-born U.S. botanist, University of Michigan, Ann Arbor, later Gray Herbarium, Harvard University; cytotaxonomy and genetics of plants.
Sojak   Jiri Sojak, 1936–, National Museum of Prague, Czechoslovakia; *Potentilla.*
Somes   Melvin Philip Somes, 18?–1925, Iowa; *Claytonia.*
Soo   Karoly Rezso Soo von Bere, 1903–1980, Budapest; Hungarian flora.
Spach   Edouard Spach, 1801–1879, French botanist; author of *Histoire Naturell des Vegetaux Phanerogames* (1834–1848).
Spreng.   Curt Polykarp Joachim Sprengel, 1766–1833, professor of medicine and botany, Halle; author of works on the flora of Halle, Apiaceae, and history of botany, editor of the 18th edition of Linnaeus' *Systema Vegetabilium* (1825–1828).
Spring   Antoine Frederich Spring, 1814–1872, professor in Luttich, Belgium; Lycopodiaceae.
St.-Hil.   August François César Prouvençal, de Saint-Hilaire, 1779–1853, French botanist and explorer; collected 7,000 species of plants during extensive travels in Brazil and Paraguay, published many works.
St.-Hil., J.   Jean Henri Jaume Saint-Hilaire, 1772–1845, French botanist.
St. John   Harold St. John, 1892–, professor of botany and curator, Bishop Museum, University of Hawaii, formerly at Washington State University, Pullman; student of the flora of eastern Washington and the Pacific Islands.
St. John & Warren   H. St. John and Fred Adelbert Warren, 1902–, American; flora of the northwestern United States.
St. Yves   Alfred Marie Augustine St. Yves, 1855–1933, France, agrostology.
Standl.   Paul Carpenter Standley, 1884–1963, curator, U.S. National Herbarium (1909–1928), Chicago (Field) Natural History Museum (1928–1950), then Zamorano, Honduras; student of the Mexican and Central American floras, author of *Plants of Glacier National Park* (1926), *Trees and Shrubs of Mexico* (1920–1924), etc.
Stanford   Ernest Elwood Stanford, 1888–?, professor of botany, University of the Pacific; *Polygonum.*
Stapf   Otto Stapf, 1857–1933, Austrian botanist at Kew from 1890, keeper from 1909–1922; contributor to Harvey and Sonder's *Flora Capensis* and Oliver's *Flora of Tropical Africa.*
Steb.   George Ledyard Stebbins, Jr., 1906–, professor of genetics, University of California, Davis; cytogeneticist and cytotaxonomist; *Crepis, Antennaria,* Poaceae.
Steele   Edward Strieby Steele, 1850–1942, American; *Lycopus.*
Steud.   Ernst Gottlieb von Steudel, 1783–1856, German physician and botanical bibliographer; agrostologist.
Stevens, O.   Orvin Alva Stevens, 1885–1979, North Dakota State University, Fargo; *Handbook of North Dakota Plants* (1963).
Stevens, W.   William Chase Stevens, 1861–1955, University of Kansas, Lawrence; plant anatomy, author of *Kansas Wild Flowers* (1948).
Stewart, S. R.   Sara R. Stewart (later Hinckley), 1913–, Hanover, Pennsylvania; *Mentha.*
Steyerm.   Julian Alfred Steyermark, 1909–, curator, Chicago (Field) Natural History Museum (1937–1958), botanist, Instituto Botanico of the Ministry of Agriculture,

Caracas, Venezuela (1958–); botanical explorer and student of Central and South American floras, *Grindelia*, author of *Flora of Missouri* (1963), floristic works on Central and South American countries, etc.

Stoj. & Stef.   Nikolai Andreev Stojanov, 1883–1968, University of Sofia; *Flora of Bulgaria* (1924–1925) and Carlo de Stefani, 1851–1924.

Stokes   Susan Gabriella Stokes, 1868–1954, high-school teacher in San Diego; *Eriogonum*.

Stuckey   Ronald Lewis Stuckey, 1938–, Ohio State University, Columbus; taxonomy and distribution of angiosperms; *Rorippa*.

Stuessy   Tod Falor Stuessy, 1943–, Ohio State University, Columbus; Asteraceae (Heliantheae: *Melampodium, Lecocarpus*, etc.).

Sudw.   George Bishop Sudworth, 1864–1927, chief dendrologist, U.S. Forest Service, Washington; author of *Checklist of the Forest Trees of the United States, Their Names and Ranges* (1898, 1927), *Forest Trees of the Pacific Slope* (1908), etc.

Suksd.   Wilhelm Nikolaus Suksdorf, 1850–1932, German-born American botanist, Bingen, Washington, whose valuable herbarium was bequeathed to Washington State University, Pullman.

Sulliv.   William Starling Sullivant, 1803–1873, eminent Ohio bryologist for whom the Sullivant Moss Society (now American Bryological Society) was named.

Sutherland   David Michael Sutherland, 1940–, University of Nebraska-Omaha; Poaceae.

Svens.   Henry Knute Svenson, 1897–?, Swedish-born American, Brooklyn Botanic Garden, American Museum of Natural History, U.S. Geological Survey; *Eleocharis*.

Sw.   Olof Peter Swartz, 1760–1818, Swedish botanist, professor in Stockholm, student of Linneaus; organizer of the botany of the West Indies in his *Flora Indiae Occidentalis*, 3 vols. (1797–1806).

Swall.   Jason Richard Swallen, 1903–, curator, U.S. National Herbarium, Washington; Poaceae.

Sweet   Robert Sweet, 1783–1835, England, horticulturist and ornithologist; monographer of the Geraniaceae (1820–1830).

Swezey.   Goodwin Deloss Swezey, 1851–1934, professor of astronomy, Nebraska.

Szweyk.   Jerzey Szweykowski, 1925–; *Gayophytum*.

T. & G.   J. Torrey and A. Gray; authors of *A Flora of North America* (1838–1840, 1843).

Tausch   Ignaz Friedrich Tausch, 1793–1848, Bohemian botanist; *Hieracium, Salsola*.

Taylor, J. & C.   Raymond John Taylor, Jr., 1930– and Constance Elaine Southern Taylor, 1937–, Southeastern Oklahoma State University, Durant; flora of Oklahoma.

Ten.   Michele Tenore, 1780–1861, Italian botanist, professor in Naples.

Terscheck   Adolph Terscheck, fl. 1843, gardener at the Royal Palace in Pillnitz, Germany; Cactaceae.

Thell.   Albert Thellung, 1881–1928, keeper in the Botanical Institute of the University of Zurich; Brassicaceae.

Thieret   John William Thieret, 1926–, Northern Kentucky University, Highland Heights; *Baptisia*, Scrophulariaceae, Poaceae.

Thomps., C.   Charles Henry Thompson, 1870–1931, Missouri Botanical Garden, St. Louis, then University of Massachusetts, Amherst; Lemnaceae.

Thomps. & Zavortink   H. J. Thompson and J. R. Zavortink.

Thomps., H. J.   Henry Joseph Thompson, 1921–, University of California, Los Angeles; Loasaceae, *Dodacatheon*.

Thornb.   John James Thornber, 1872–1962, Northern Arizona University, Flagstaff; student of flora of Arizona and Nebraska.

Thou.   Louis Marie Aubert du Petit-Thouars, 1758–1831, French taxonomist and botanical writer.

Thuill.   Jean Louis Thuillier, 1757–1822, French botanist in Paris; author of a flora of Paris (1790).

Thunb.   Carl Peter (Pehr) Thunberg, 1743–1828, professor of botany, Uppsala, student of Linneaus; collected in South Africa, Japan, Ceylon, prolific author of notable works including *Flora Japonica* (1784) and *Flora Capensis* (1820).

Thurb.   George Thurber, 1821–1890, New York; botanist and quartermaster with the Mexican Boundary Survey (1850–1853); editor of the *American Agriculturist*.

Tidest.   Ivar T. Tidestrom, 1864–1956, Swedish-born American botanist, professor of botany, Catholic University of America, Washington; author of *Flora of Utah and Nevada* (1925) and (with Sister T. Kittell) *A Flora of Arizona and New Mexico* (1941).

Timm   Joachim Christian Timm, 1754–1805, German pharmacist, mayor of Malchin, Mecklenberg; flora of Mecklenberg.

Todaro   Agostino Todaro, 1818–1892, director of the botanical garden, Palermo, Sicily.

Tomb   Andrew Spencer Tomb, 1943–, Kansas State University, Manhattan; Asteraceae (Lactuceae).

Torr.   John Torrey, 1796–1873, physician, professor of chemistry and botany, College of

Physicians and Surgeons, New York; collector and describer of many western American plants; an outstanding systematist and botanical author.

Torr. & Hook.　J. Torrey and J. D. Hooker.

Tourn.　Joseph Pitton de Tournefort, 1656–1708, French botanist, professor in the royal garden, Paris; first clear characterizer of genera, his botanical system was the highest pre-Linnaean development.

Towner & Raven　Howard Frost Towner, 1943–, Stanford University, Stanford, California; Onagraceae (*Calylophus*); and P. H. Raven.

Trautv.　Ernst Rudolph von Trautvetter, 1809–1889, Russian taxonomist and agronomist.

Tratt.　Leopold Trattinnick, 1764–1849, Austria; Rosaceae.

Trel.　William Trelease, 1857–1945, director, Missouri Botanical Garden, St. Louis (1885–1912), then professor of botany, University of Illinois, Urbana (1913–1926), first president of the Botanical Society of America; *Yucca, Agave, Epilobium, Quercus, Piper.*

Trev.　Ludolph Christian Treviranus, 1779–1864, professor of botany, Bonn, Germany.

Trevisan　Vittore Benedetto Antonio Trevisan de Saint-Leon, 1818–1897, Italy.

Trin.　Carl Bernhard von Trinius, 1778–1844, German-born Russian court physician, also poet and noted agrostologist.

Trin. & Rupr.　C. B. von Trinius and F. J. Ruprecht.

Tryon, A.　Alice Faber Tryon, 1920–, Gray Herbarium, Harvard University; Polypodiaceae (Tribe Cheilantheae, *Pellaea, Notholaena, Cheilanthes*).

Tryon, R.　Rolla Milton Tryon, 1916–, curator, Missouri Botanical Garden, St. Louis, later curator of ferns, Gray Herbarium, Harvard University; author (with others) of *The Ferns and Fern Allies of Wisconsin* (1940, 1953), *The Ferns and Fern Allies of Minnesota* (1954), etc..

Tuckerm.　Edward Tuckerman, 1817–1886, professor at Amherst, Massachusetts; lichenologist.

Turner, B. L.　Billie Lee Turner, 1925–, professor of botany, University of Texas, Austin; plant taxonomy, chromosomal studies of higher plants; author of *Legumes of Texas* (1959), *Biochemical Systematics* (1963, with Ralph E. Alston), and numeous papers, poems, and essays.

Turner & Morris　B. L. Turner and Michael I. Morris, 1939–, College of Southern Idaho, Twin Falls; *Palafoxia.*

Ugent　Donald Ugent, 1933–, Southern Illinois University, Carbondale; Solanaceae, *Epilobium.*

Uline & Bray　Edwin Burton Uline, 1867–1933, New York high school principal, and William L. Bray, 1865–1953, adjunct and later professor of botany, University of Texas at Austin (1898–1907) where he was first teacher of botany, then forest pathologist, U.S. Dept of Agriculture, later botanist and dean of the graduate school, Syracuse University, New York; student and collector of Texas and southwestern U.S. plants.

Umber　Ray Ernest Umber, 1948–, formerly University of Wyoming; Verbenaceae, Scrophulariaceae of the Rocky Mts.

Underw.　Lucien Marcus Underwood, 1853–1907, professor of botany, Columbia University, New York; student of ferns; author of *Our Native Ferns and Their Allies* (1880 and subsequent editions).

Urban　Ignatz Urban, 1848–1931, professor at Berlin; generic authority on the flora of tropical America; monographer of *Medicago*, Loasaceae, author of *Symbolae Antillanae* (1898–1908).

Urban & Gilg　I. Urban and Ernst Friedrich Gilg.

Vahl　Martin Hendriksen Vahl, 1749–1804, Danish botanist, professor in Copenhagen; student of Linnaeus and one of the editors of *Flora Danica.*

Vail　Anna Murray Vail, 1863–1955, librarian, New York Botanical Garden; Asclepiadaceae, Leguminosae.

Van.　Eugene Vaniot, fl. late 19th century, d. 1913, professor at LeMans, France; flora of China.

Van Bruggen　Theodore Van Bruggen, 1926–, University of South Dakota, Vermillion; author of *Vascular Plants of South Dakota* (1976).

Vasey　George Vasey, 1822–1893, English-born American botanist, eminent agrostologist, curator, U.S. National Herbarium, Washington.

Vavilov　Nikolai Ivanovic Vavilov, 1887–1942, Leningrad; *Cannabis*, economic plants.

Vent.　Etienne Pierre Ventenat, 1757–1808, professor of botany, Paris; horticulturist.

Vict.　Frere Marie-Victorin (born Conrad Kirouac), 1885–1944, director of the Botanical Institute, Montreal; author of *Flore Laurentienne* (1935), the first flora in which chromosome numbers appeared.

Vict. & Rouss.　F. Marie-Victorin and Joseph Jules Jean Jacques Rousseau, 1905–.

Vill.　Dominique Villars, 1745–1814, physician and professor in Grenoble, then Strasbourg; author of a basic work on the flora of the West Alps.

Vilm.　Several generations of the family of Vilmorin in Paris; seedsmen and authors of many books and memoirs on botany and horticulture; Pierre Philippe Andre Leveque de Vilmorin, 1776–1862; Pierre Louis François Leveque de Vilmorin, 1816–1862; Charles Phillippe Henry Leveque de Vilmorin, 1843–1899; August Louis Maurice Leveque de Vilmorin, 1849–1918; Joseph Marie Leveque de Vilmorin, 1872–1917.

Vilm., E.   Elisa (Bailly) de Vilmorin, 18?–1868.

Vitman   Fulgenzio Vitman, 1728–1806, Italian botanist and clergyman in Pavia and Milan, founded a botanical garden in Milan.

Voss, A.   Andreas Voss, 1857–1924, German student of conifers.

Voss, E.   Edward Groesbeck Voss, 1929–, University of Michigan, Ann Arbor; botanical nomenclature.

Wagenkn.   Burdette Lewis Wagenknecht, 1925–, William Jewell College, Liberty, Missouri; Asteraceae.

Wagner, W. L.   Warren Lambert Wagner, 1950–, Bernice P. Bishop Museum, Honolulu, Hawaii; *Oenothera*.

Wagner, Stockhouse & Klein   W. L. Wagner; Robert Erland Stockhouse, fl. 1980, Pacific University, Oregon; and William McKinley Klein, Jr., 1933–, Morris Arboretum, Philadelphia; *Oenothera*.

Wagnon   Harvey Keith Wagnon, 1916–, plant pathologist with the California Dept. of Agriculture, Sacramento; *Bromus*.

Wahl.   Goran (Georg) Wahlenberg, 1780–1851, Swedish botanist, professor in Uppsala, plant geographer, author of *Flora Lapponica* (1812), *De Vegetatione et Climate Helvetiae* (1813), *Flora Carpatorum Principalium* (1814), *Flora Upsaliensis* (1820), *Flora Suecica* (1824–1826).

Wahl, H. A.   Herbert Alexander Wahl, 1900–1975, professor of botany, Pennsylvania State University; *Chenopodium*.

Waldst. & Kit.   Count Franz de Paula Adam von Waldstein-Wartemberg, 1759–1823, Austria, and P. Kitaibel, authors of *Descriptions et Icones Plantarum Rariorum Hungariae* (1802–1812).

Waller   Floyd R. Waller, fl. 1974, then at Texas A & M University, College Station; *Panicum*.

Wallr.   Carl Friedrich Wilhelm Wallroth, 1792–1857, Danish physician and botanist.

Walp.   Wilhelm Gerhard Walpers, 1816–1853, German botanist; author of *Repertorium Botanices Systematicae* (1842–1853), *Annales Botanices Systematicae* (1848–1868), etc.

Walsh   Full name unknown, in *Manual of the Higher Plants of Oregon* (1941) M. E. Peck lists *Zigadenus venenosus* var. *gramineus* (Rydb.) Walsh, an apparent new combination.

Walt.   Thomas Walter, 1740–1789, British-American botanist, Charleston planter; author of *Flora Caroliniana* (1788).

Wang.   Friedrich Adam Julius von Wangenheim, 1749–1800, German forester.

Warnock, M. J.   Michael James Warnock, 1956–, Sam Houston State University, Huntsville, Texas; *Delphinium*.

Waterfall   Umaldy Theodore Waterfall, 1910–1975, professor of botany, Oklahoma State University, Stillwater; Solanceae *(Physalis)*, Malvaceae *(Callirhoe)*; author of *Keys to the Flora of Oklahoma* (1953–1960, 4th ed. 1969).

Wats., S.   Sereno Watson, 1826–1892, assistant to Asa Gray, curator of Gray Herbarium, Harvard University (1888–1892), critical student of western American plants.

Wats. & Coult.   S. Watson and J. M. Coulter.

Wats., E. E.   Elba Emanuel Watson, 1871–1936, high-school biology teacher in Michigan and New York; *Helianthus*.

Wats., T. J.   Thomas J. Watson, Jr., fl. 1977, graduate student at the University of Texas, Austin, later University of Montana, Missoula; *Xylorhiza*.

Waugh   Frank Albert Waugh, 1869–1947, professor of horticulture, Amherst.

Watt   David Allan Poe Watt, 1830–1917, Montreal, Canada; student of ferns.

Weath.   Charles Alfred Weatherby, 1875–1949, Gray Herbarium, Harvard University; student of ferns, bibliographer.

Webb. & Moq.   Phillip Barker Webb, 1793–1854, English traveler and classical scholar, author of *Histoire Naturelle des Iles Canaries* (1835–1850); and C. H. B. A. Moquin-Tandon.

Wedd.   Hugh Algernon Weddell, 1819–1877, British botanist and traveler, collector in South America.

Weedon   Ronald R. Weedon, 1939–, Chadron State College, Nebraska; *Bidens*.

Wehmer   C. F. W. Wehmer, fl. 1911, Germany; *Cannabis*.

Weim., G.   Gunner Weimarck, 1936–, University of Lund, Sweden; *Hierochloa*, *Hieracium*.

Weinm.   Johann Anton Weinmann, 1782–1858, director, botanic garden, St. Petersburg, Russia.

Welw.   Friedrich Martin Josef Welwitsch, 1806–1872, Austrian-born, director of the botanical gardens in Lisbon and Coimbra.

Wemple   Don Kimberly Wemple, 1929–, Iowa State University (mid 1960s); Leguminosae.

Wendl.   Hermann Wendland, 1825–1903, director of the royal garden at Herrenhausen, Hannover, Germany; Palmae.

Wettst.   Richard Ritter Wettstein von Wester-

sheim, 1863–1931, director of the botanic garden, Vienna; outstanding systematist, proposer of a phylogenetic system.
Whalen   Michael D. Whalen, 1950–, Liberty Hyde Bailey Hortorium, Cornell University, Ithaca, New York; Solanaceae.
Wheeler   Louis Cutter Wheeler, 1910–?, professor of botany, University of Southern California, Los Angeles; *Euphorbia* and other genera of Euphorbiaceae.
Wheelock   William Efner Wheelock, 1852–1926, New York physician.
Wherry   Edgar Theodore Wherry, 1885–1982, professor of botany, University of Pennsylvania (1930–1955); student of ferns, Polemoniaceae, author of *Guide to Eastern Ferns* (1942), *The Genus Phlox* (1955), *The Southern Fern Guide* (1964), etc.
White   Theodore Greeley White, 1872–1901, student at New York Botanical Garden; Fabaceae (*Lathyrus*).
Wibel   August Wilhelm Eberhard Christoph Wibel, 1775–1814, physician in Wertheim.
Wieg.   Karl McKay Wiegand, 1873–1942, professor of botany, Cornell University, Ithaca, New York; author (with A. J. Eames) of *The Flora of the Cayuga Lake Basin*, New York (1926).
Wiens   Delbert Wiens, 1932–, University of Utah, Salt Lake City; Viscaceae, Loranthaceae.
Wiggers   Friedrich Heinrich (Wichers) Wiggers, 1746–1811; author of *Flora of Holstein* (1780).
Wight & Hedr.   W. F. Wight and Ulysses Prentiss Hedrick, 1870–1951, pomologist, director of Agricultural Experiment Station, Geneva, New York.
Wight, W.   William Franklin Wight, 1874–1954, U.S. Dept. of Agriculture, Palo Alto, California; botanical and agricultural collector in South America, author of *Origin, Introduction and Primitive Culture of the Potato* (1916).
Wikst.   Johan Emanuel Wikstrom, 1789–1856, director of the Botanical Museum, Stockholm, Sweden; *Juncus.*
Willd.   Carl Ludwig von Willdenow, 1765–1812, German botanist, director of the Berlin Botanical Garden (1801–1812); produced the 4th ed. of Linnaeus' *Species Plantarum* (1797–1810).
Wilken   Dieter H. Wilken, 1944–, Colorado State University, Fort Collins; floristics of Colorado, Polemoniaceae, Hydrophyllaceae.
Williams   Thomas Albert Williams, 1865–1900, U.S. Dept. of Agriculture, agrostologist.
Wils., J. S.   James Stewart Wilson, 1932–1985, Emporia State University, Emporia, Kansas; *Cornus, Polygonum.*
Wimm.   Christian Friedrich Heinrich Wimmer, 1803–1868, German philologist and botanist, school official in Breslau.
Wimmer & Grab.   C. F. H. Wimmer and Heinrich Emanuel Grabowski, 1792–1842, Breslau, Germany.
Witasek   Johanna A. Witasek, 1865–1910, Austrian, taxonomist, teacher in secondary schools in Vienna.
Wms., L. O.   Louis Otho Williams, 1908–, botanist, Botanical Museum, Harvard University, then Escuela Agricola Panamericana, Zamorano, Honduras, then U.S. Dept. of Agriculture, Beltsville, later Chicago (Field) Natural History Museum; orchidologist, student, and collector of Latin American flora, author of *Orchids of Mexico*, *Ceiba*, Vol. 2 (1951), etc.
Wolf, C. B.   Carl Brandt Wolf, 1905–, Rancho Santa Ana Botanic Garden, California (1930–1945); *Erythronium, Rhamnus.*
Wolf, T.   Franz Theodor Wolf, 1841–1924, German geologist in Ecuador until 1891, then a teacher in Dresden; *Potentilla.*
Wolf, v.   Nathanael Matthaeus von Wolf, 1724–1784, German physician and botanist.
Wolf, W.   Wolfgang Wolf, 1872–?, St. Bernard College, Alabama; student of Liliaceae (*Erythronium,* 1941).
Wolfgang   Johann Friedrich Wolfgang, 1776–1859.
Wood   Alphonso Wood, 1810–1881, principal, Brooklyn Female Academy, New York, author of *Class-Book of Botany* (1845; 2nd ed. 1861), in which dichotomous keys were first employed in America, collected in California in 1866.
Woods.   Robert Everard Woodson, 1904–1963, Missouri Botanical Garden, St. Louis; Asclepiadaceae, Apocynaceae.
Woot.   Elmer Otis Wooton, 1865–1945, professor of biology, New Mexico State University (1890–1911), then U.S. Dept. of Agriculture, Washington (1911–1935); student of the New Mexican flora.
Woot. & Standl.   E. O. Wooten and P. C. Standley; authors of *Flora of New Mexico* (1915).
Woynar   Heinrich Karl Woynar, 1865–1917, Austrian pharmacist; pteridology, flora of Tyrol and Styria.
Wright, S. H.   Samuel Hart Wright, 1825–1905, New York; published on *Dichromena* (Cyperaceae).
Wulf.   Franz Xaver (Freiherr) von Wulfen, 1728–1805, professor at Klagenfurt, Hungary.
Yates   Harris Oliver Yates, 1934–, David Lipscomb College, Nashville, Tennessee; Poaceae.

Yunck. Truman George Yuncker, 1891–1962, De Pauw University, Indiana; *Cuscuta, Piper, Peperomia.*

Zabel Karl Hermann Zabel, 1832–1912, German dendrologist.

Zavortink Joyce Roberts Zavortink, 1930–, University of California, Los Angeles; Loasaceae.

Zea Francisco Antonio Zea, 1770–1822, Colombian statesman, director of the botanical gardens in Madrid and king's botanist; in 1803, when Colombia became a republic, he was made vice-president.

Zenkert Charles Anthony Zenkert, 1886–?, American; *Hybanthus.*

Zucc. Joseph Gerhard Zuccarini, 1797–1848, Munich, Germany; Oxalidaceae, flora of Japan.

Glossary

Compiled by Eileen K. Schofield

A-   A prefix meaning without.
Abaxial   The side of an organ away from the axis.
Abortive   Imperfectly developed; barren.
Acaulescent   Stemless, or appearing so; stem sometimes subterranean.
Accessory   A fruit with fleshy parts not derived from the pistil (e.g., receptacle in Rosaceae).
Accrescent   Enlarging with age.
Accumbent   Cotyledons lying with their edges against the radicle.
Acerose   Needle-shaped.
Achene   A dry, indehiscent, 1-seeded fruit.
Acicular   Slender and needle-shaped.
Acropetal   Developing in succession from the base toward the apex.
Actinomorphic   Symmetrical, regular; divisible into equal halves in two or more planes.
Acuminate   Tapering gradually to a terminal point.
Acute   Terminating in a sharp point.
Adaxial   The side of an organ toward the axis.
Adnate   Fused together, usually unlike organs.
Adventitious   Developing irregularly or accidentally.
Adventive   Of occasional occurrence, not thoroughly naturalized.
Aecium (pl. aecia)   A structure producing aeciospores; one stage in the life cycle of a rust fungus.
Aestival   Appearing in the summer.
Aggregate   A fruit formed by coherence of several pistils that were separate in the flowers.
Alate   Winged.
Alternate   Placed singly one above the other on the axis.
Alveolate   Honeycombed surface.
Ament   A catkin, usually unisexual and pendulous.
Amentiferous   Bearing aments.
Amplexicaul   Clasping the stem.
Ampliate   Enlarged.
Anastomosing   Connecting by cross-veins to form a network.
Anatropous   An inverted ovule with micropyle near point of attachment of funiculus.
Androecium   Collective term for stamens.
Androgynous   An inflorescence with pistillate flowers at the base and staminate at the apex.
Anthocarp   A structure in which a fruit is united with perianth or receptacle (Nyctaginaceae).
Annual   Living one year.
Annular   Arranged in a ring.
Annulus   A ring; in Lamiaceae, a ring of hairs inside calyx.
Anther   Pollen-bearing part of a stamen.
Anthesis   Flowering, or time when pollination takes place.
Anthocyanin   A red or purplish pigment.
Antrorse   Directed forward or upward.
Apetalous   Without petals.
Apex   The tip.
Aphyllopodic   Lacking leaves at the base.
Apical   Relating to the apex.
Apiculate   Ending in a short, abrupt, flexible point (apiculum).
Apomixis   Nonsexual reproduction.
Apophysis   Swelling or enlargement of the surface of an organ.
Appressed   Flatly and closely pressed against.
Arachnoid   Cobweblike, with entangled, slender, loose hairs.
Arcuate   Curved or arching.
Arenaceous   Sandlike or growing in sand.
Areolate   Divided into small angular spaces.
Areole   Small space marked out on a surface; the spine-bearing area on a cactus.
Aril   An appendage growing on a seed, in the area of the hilum.

Arillate Having an aril.
Arista A stiff awn or bristle.
Aristate Bearing stiff awns or bristles.
Armed Provided with sharp thorns, spines, etc.
Aromatic Having a fragrant odor.
Articulate Having nodes or joints.
Asperous Rough to the touch.
Assurgent Ascending.
Atomiferous Bearing very fine glandular hairs.
Atropurpureous Dark purple.
Attenuate Gradually tapering to very slender tip.
Auricle An ear-shaped appendage.
Auriculate Having auricles.
Autotrophic Able to make its own food by photosynthesis.
Awn A terminal, bristlelike appendage.
Axil An angle formed between two organs (as petiole and stem).
Axile In the axil, especially placentation.
Axillary On or related to the axil.
Axis (pl. axes) The main longitudinal support on which parts are arranged.
Banner The upper petal or standard of a corolla in Fabaceae.
Barbed Armed with bristles having sharply reflexed hooks.
Barbellate Finely barbed.
Basifixed Attached by the base.
Basipetal Developing in succession from the apex toward the base.
Basiscopic Facing basally.
Beak A long, firm, slender point.
Bearded Bearing long or stiff hairs in a line or tuft.
Berry A pulpy, indehiscent fruit with few to many seeds.
Bi- A Latin prefix meaning two.
Biennial Living for two years.
Bifid Two-cleft.
Bifurcate(d) Forked or Y-shaped.
Bilabiate Two-lipped, especially corolla.
Bilocular Having two cavities.
Bipinnate Twice compound, with leaflets arranged on both sides of the axis.
Bladder A hollow, membranaceous appendage on the root that traps insects *(Utricularia)*.
Blade The expanded portion of a leaf.
Bract A modified, reduced leaf.
Bracteate Having bracts.
Bracteole A bract borne on a secondary axis (e.g., pedicel).
Bracteose Having numerous or conspicuous bracts.
Branchlets The ultimate divisions of a branch.
Bristle A stiff hair.
Bulb An underground bud with thick, fleshy scales.

Bulbil, bulblet A small bulb.
Caducous Falling off early.
Calcareous Containing limestone or chalk.
Callosity A thickened, raised area.
Callous Having the texture of a callus.
Callus A hard protuberance.
Calyculate Calyxlike; usually bracts resembling an outer calyx.
Calyx The outer whorl of the perianth, composed of sepals.
Campanulate Bell-shaped.
Canescent Having gray or whitish pubescence.
Capillary Hairlike.
Capitate Headlike, aggregated into a dense cluster.
Capitellate Aggregated into small, dense cluster.
Capsule A dry, dehiscent fruit with more than one carpel.
Carinate Having a central, longitudinal projection on the lower surface.
Carnose Fleshy, succulent.
Carpel A foliar, ovule-bearing unit of a compound pistil, or a simple pistil.
Carpophore A slender stalk supporting pendulous ripe carpels (Apiaceae).
Cartilaginous Tough and firm, but flexible.
Caruncle An appendage near the hilum of a seed.
Castaneous Chestnut colored (dark brown).
Catkin A spikelike inflorescence of unisexual, apetalous, bracteate flowers.
Caudate Bearing a slender tail-like appendage.
Caudex (pl. caudices) The tough, persistent base of an otherwise herbaceous stem; the main axis of a plant consisting of root and stem.
Caulescent Having an obvious stem.
Cauline Belonging to the stem.
Cernuous Nodding or drooping.
Cespitose, caespitose Growing in tufts or mats.
Chaff Small, dry, membranaceous bracts or scales.
Chalaza The basal part of the ovule where it attaches to the funiculus.
Chartaceous Having a papery texture.
Chasmogamous A flower that is open at the time of pollination.
Cicatrix (pl. cicatrices) A permanent leaf scar of the stem.
Cilia Marginal hairs.
Ciliate Fringed with marginal hairs.
Ciliolate Minutely ciliate.
Cinereous Light gray or ash colored.
Circinate Coiled from the top downward, with apex at center.
Circumscissile Dehiscing by a circular line around the organ.

GLOSSARY

Clavate Club-shaped, or thickened toward the top.
Claw The long, narrow base of a petal (or sepal).
Cleistogamous Small flowers that remain closed and are self-pollinated.
Clone A group of individuals of the same genotype, usually propagated vegetatively.
Cochleate Coiled like a snail shell.
Collar The area between belowground and aboveground portions of the axis; the area separating the sheath and the blade of a grass leaf.
Colliculose Covered with small, rounded elevations.
-colpate A suffix referring to pollen grains having grooves (colpi).
-colporate A suffix referring to pollen grains having grooves and pores.
Column A group of united filaments (Malvaceae) or filaments and style (Orchidaceae).
Columnar Column-shaped.
Coma A tuft of hairs (usually at tip of seed).
Comose Bearing a tuft of hairs.
Commisure The surface where organs are joined (as in carpels).
Compound Made up of two or more parts.
Conduplicate Folded together lengthwise.
Cone A dense, usually elongated collection of sporophylls and bracts on a central axis.
Connate Joined or united, usually similar structures.
Connective Tissue connecting the locules of an anther.
Connivent Coming together or converging, but not securely fused; usually like parts.
Contiguous Touching, but not fused; like or unlike parts.
Contrary In an opposite direction or at right angles to (as compression of silicles in Brassicaceae).
Convolute Rolled up longitudinally.
Cordate Heart-shaped, with rounded lobes and a sinus at the base.
Coriaceous With a leathery texture.
Corm The fleshy, bulblike base of a stem, usually underground.
Cormous Having a corm.
Corneous Horny in texture.
Corniculate Bearing a small horn or horns.
Corolla The inner whorl of the perianth, composed of the petals.
Corona A crownlike structure; appendage between corolla and stamens (Asclepiadaceae, Passifloraceae).
Corpusculum A dark, basally cleft, tubular body above the stigmatic chamber, connected to pollinia (Asclepiadaceae).
Corrugate Having wrinkles or folds.
Cortex The tissue between the stele and epidermis of a stem; bark.
Corymb A flat-topped inflorescence, progressively flowering from the margin inward.
Corymbiform Shaped lke a corymb.
Corymbose In corymbs or corymblike.
Costa A rib; the midvein of a leaf.
Cotyledon The primary leaf of the embryo; seed leaf.
Crateriform In the shape of a saucer; shallow and hemispherical.
Crenate Toothed with shallow, rounded teeth; scalloped.
Crenulate Finely crenate.
Crown The portion of a stem at the surface of the ground.
Crispate, crisped Curled.
Cruciate, cruciform Cross-shaped.
Crustaceous With a brittle, hard texture.
Cryptogams Plants without flowers or seeds, often reproducing by free spores.
Cucullate Hood-shaped.
Culm The stem of grass or sedge.
Cuneate, cuneiform Triangular or wedge-shaped, with narrow end at point of attachment.
Cup, cupule The cuplike structure at the base of a fruit (acorn).
Cupuliform Cupule-shaped.
Cusp An elongated, sharp tip.
Cuspidate Bearing an elongated, sharp, and firm point at the tip.
Cyathium An inflorescence with a cuplike involucre bearing unisexual flowers *(Euphorbia)*.
Cyme A broad, flat inflorescence with central flower blooming first.
Cymose Bearing cymes.
Cymule A small, few-flowered cyme.
Cypsela Achene derived from an inferior ovary and covered by fused calyx.
Cystolith An intercellular, stonelike concretion, usually calcium carbonate.
Deciduous Not persistent; falling at end of growing season.
Declinate, declined Bent forward or downward.
Decompound More than once compound.
Decumbent Reclining, but with the end ascending.
Decurrent Extending downward from point of insertion and adnate (leaf base).
Decurved Curved downward *(Penstemon)*.
Decussate Pairs of opposite organs alternating at right angles at successive levels.
Deflexed Bent downward.
Dehiscence Method of opening.
Dehiscent Opening regularly by slits, valves, etc.

Deliquescent With the primary axis much branched above; softening or wasting away.
Deltate, deltoid Triangular.
Dendritic Treelike, as in branching.
Dentate With sharp, spreading teeth.
Denticulate Minutely dentate.
Depauperate Stunted.
Determinate An inflorescence with the terminal flower opening before those below.
Diadelphous Stamens arranged in two sets, often unequal in number.
Diaphragm Dividing membrane.
Dichasium A cyme with two lateral axes.
Dichotomous Forked regularly in pairs.
Dicotyledons Flowering plants having two cotyledons, net venation, and flower parts usually in 5s.
Didymous Occurring in pairs.
Didynamous Stamens arranged in two sets of different lengths.
Diffuse Widely spreading.
Digitate Compound with members arising from one point; palmately compound.
Dilated Expanded into a blade, as though flattened.
Dimidiate Halved, as if one half is missing.
Dimorphic Occurring in two forms.
Dioecious Having staminate and pistillate flowers on different plants.
Disc, disk A fleshy or elevated development of the receptacle; in Asteraceae, the central portion of the flowering head.
Disciform Shaped like a disc; in Asteraceae, a head with disk florets in center and marginal florets with ligule reduced or lacking.
Discoid Resembling a disc; in Asteraceae, a head composed entirely of disk florets.
Discrete Separate.
Disk floret Small flower with tubular corolla, perfect, fertile, in disk portion of head (Asteraceae).
Dissected Divided into narrow segments.
Distal Remote from the place of attachment.
Distichous Arranged in two vertical ranks.
Divaricate Very widely spreading.
Diverging Spreading broadly, less so than divaricate.
Dolabriform Ax-shaped, as a hair attached in middle; malpighaceous.
Dorsal Referring to the back or outer surface of an organ; the lower surface of a leaf.
Drupaceous Resembling a drupe.
Drupe A fleshy, indehiscent fruit with 1 seed enclosed in a stony endocarp.
Drupelet Small drupe; one part of an aggregate drupe (raspberry).
Duplex Double, as in pubescence composed of two kinds of hairs.

E-, ex- Latin prefixes denoting that parts are missing.
Early leaves In *Populus*, leaves that overwinter in the bud and quickly expand in the spring.
Echinate Having stout, straight, pricklelike hairs.
Elaiosome A micropylar appendage on some seeds *(Luzula)*, morphologically a caruncle.
Elliptic, ellipsoid Oval in shape, widest at middle and tapering equally to both rounded ends.
Embryo The rudimentary plant in the seed.
Endocarp The inner layer of the pericarp (fruit wall).
Endogenous Produced deep within another body.
Endosperm The reserve food (often starch) stored around the embryo in the seed.
Ensiform Sword-shaped.
Entire Whole; with a continuous margin.
Entomophilous Pollinated by insects.
Ephemeral Lasting only one day.
Epi- A Greek prefix meaning upon.
Equitant Overlapping in two ranks (*Iris* leaves).
Eradiate Lacking ray florets (discoid or disciform heads, Asteraceae).
Erose With the margin appearing eroded or gnawed.
Estipulate, exstipulate Lacking stipules.
Evanescent Of short duration.
Excurrent Extending beyond the tip or margin.
Exocarp The outer layer of the pericarp (fruit wall).
Explanate Spread out flat.
Exserted Projecting beyond (e.g., stamens from a corolla).
Extrorse Facing outward.
Falcate Sickle-shaped.
Fall Outer, spreading and often recurved, bearded perianth segment in *Iris*.
Farinose Covered with meal-like powder.
Fascicle A condensed bundle or cluster.
Fasciculate Congested in bundles or clusters.
Fastigiate With branches erect and close together.
Fenestrate Perforated with holes (windows).
Ferrugineous Rust colored.
Fibrillose Having small fibers or appearing finely lined.
Fibrous Resembling or composed of fibers.
Filament The stalk of the stamens that supports the anther; thread.
Filamentous, filamentose Composed of filaments or threads.
Filiferous With coarse marginal threads.
Filiform Long and very slender, threadlike.
Fimbriate Fringed; margined with long hairlike appendages.
Fimbrillate With a minute fringe (fimbrillae).

GLOSSARY

Fistulose Hollow and cylindrical.
Flabellate, flabelliform Fan-shaped.
Flaccid Limp, lax.
Flagelliform Lashlike; resembling a runner.
Flange A stipulelike wing at the base of a petiole *(Ranunculus)*.
Flexuous Having a zigzag form; curving alternately in opposite directions.
Floccose Covered with tufts of soft, woolly hairs.
Flocculent Minutely floccose.
Floret Small flower of dense inflorescence.
Floricane The flowering stem *(Rubus)*.
Floriferous Bearing flowers.
Foliaceous Leaflike.
Foliolate Having leaflets.
Foliose Bearing numerous leaves.
Follicle A dry dehiscent fruit, opening by one suture.
Fornices Internal appendages in upper throat of corolla (Boraginaceae).
Foveolate Minutely pitted.
Frond The leaf of a fern.
Fugacious Withering very early.
Fulvous Tawny, brownish-yellow.
Funicle, funiculus The stalk attaching ovule to ovary wall or placenta.
Funnelform Shaped like a funnel.
Fuscous Grayish-brown.
Fusiform Spindle-shaped, swollen in middle and tapering to both ends.
Galea A helmet-shaped portion of a perianth *(Aconitum)*.
Galeate Having a galea.
Gametophyte The haploid, gamete-bearing generation of a plant.
Gamopetalous Having the petals at least partially united; sympetalous.
Gamosepalous Having the sepals at least partially united.
Geniculate Bent sharply, like a knee.
Gibbous Swollen on one side.
Glabrate, glabrescent Nearly glabrous, or becoming glabrous.
Glabrous Smooth and not hairy.
Glandular, glanduliferous Having glands or secretory organs.
Glaucescent Slightly glaucous.
Glaucous Covered with a whitish waxy bloom that rubs off easily.
Glochid An apically barbed bristle or hair.
Glochidiate Having glochids.
Glomerate In dense or compact cluster.
Glomerulate Arranged in small, dense clusters.
Glomerule A small, compact cluster.
Glume A small, chaffy bract; a sterile bract at the base of grass spikelet.
Grain General term for the fruit of grass, a dry, indehiscent, 1-seeded fruit in which seed and pericarp are fused.
Granulate Covered with minute, grainlike particles.
Gynecandrous Having staminate and pistillate flowers in same inflorescence, with pistillate at the top.
Gynobase An enlargment of the receptacle bearing the ovary.
Gynobasal, gynobasic Referring to or having a gynobase.
Gynodioecious Dioecious, but with some flowers perfect and others pistillate.
Gynostegium Combined connate stamens, style, and stigma in Ascelpiadaceae.
Gypsiferous, gypseous Containing gypsum (calcium sulfate).
Habit The general appearance of a plant.
Hastate Shaped like an arrowhead, with narrow basal lobes standing out at wide angles.
Haustoria Suckerlike attachment organs of parasitic plants.
Head A short, dense cluster of sessile or nearly sessile flowers.
Helicoid Spiraling like a snail shell, with lateral branches on one side.
Hemi- A Greek prefix meaning half.
Herb A plant lacking persistent woody parts aboveground.
Herbage The green vegetative parts of a plant.
Herbaceous Having the character of a herb.
Hetero- A Greek prefix meaning other, various, or having more than one kind.
Heterophyllous Having more than one form of leaf.
Heterostylous Having styles of two different lengths.
Hilum The scar on a seed, indicating the point of attachment.
Hirsute Having coarse or stiff, long hairs.
Hirsutulous, hirsutulose Minutely hirsute.
Hirtellous Minutely hirsute.
Hispid Having bristly or rigid hairs.
Hispidulous, hispidulose Minutely hispid.
Hoary Covered with a fine white or grayish pubescence.
Homo- A Greek prefix meaning alike or very similar.
Hood An erect to spreading petaloid blade with incurved margins (Asclepiadaceae).
Horn An exserted appendage on the hood (Asclepiadaceae).
Humic Consisting of or derived from humus (organic portion of soil).
Humistrate Laid flat on the soil.
Husk The outer covering of some fruits (e.g., *Physalis*).
Hyaline Translucent or colorless.
Hygroscopic Changing shape or position because of changes in moisture.

Hypanthium An enlarged receptacle below the calyx, often including fused floral envelope and androecium parts (Rosaceae).
Hypogeous Below the ground.
Hypogynium A perianthlike structure below the ovary (Cyperaceae).
Hypogynous Borne on the receptacle, under the ovary.
Imbricate Overlapping.
Immaculate Not spotted.
Imparipinnate Unequally or odd-pinnate, with single terminal leaflet.
Imperfect Having either functional stamens or functional pistils; unisexual.
Incanous Gray or hoary.
Incised Cut irregularly, sharply and more or less deeply.
Included Not protruding or exserted, as stamens within the corolla.
Incumbent Cotyledons lying with the back of one against the radicle.
Indehiscent Not opening by sutures, pores, etc.
Indument, indumentum A hairy covering, particularly if dense.
Induplicate Folded or rolled inward.
Indurate Hard.
Indusium The covering of the sporangia on fern fronds.
Inferior An ovary that appears to be below the point of insertion of the perianth, often adnate to the calyx.
Inflorescence The mode of arrangement of flowers, or the flowering portion of the plant.
Infra- A Latin prefix meaning below.
Infructescence The inflorescence in the fruiting stage.
Inserted Attached to or placed on.
Integument The covering of an organ (e.g., ovule).
Inter- A Latin prefix meaning between.
Internode The portion of a stem between two nodes.
Introrse Turned inward, toward the axis.
Involucel A secondary, often reduced involucre.
Involucre A whorl of bracts under a flower or flower-cluster.
Involute Rolled inward, toward the upper side.
Irregular Showing inequality in similar parts; asymmetrical.
Keel A central dorsal ridge; the two united front petals of a flower (Fabaceae).
Labellum Lip; the enlarged upper petal that appears to be the lower petal because of twisting of pedicel (Orchidaceae).
Lacerate Irregularly cut, as if torn.
Laciniate Slashed or cut into narrow, pointed lobes.
Lacuna A hole or cavity; an air space in tissue.

Lamella A flat, thin plate.
Lamellate Made up of flat, thin plates.
Lamina An expanded part of blade.
Lanate Woolly, with intertwined, curly hairs.
Lanceolate Lance-shaped; much longer than broad, widest near the base and tapering to the apex.
Late leaves In *Populus,* leaves that expand during the growing season in which they are found.
Latisept With broad partitions in the fruits (Brassicaceae).
Latrorse Dehiscing laterally and longitudinally (anther).
Lax Loose.
Leaflet One part (blade) of a compound leaf.
Legume A bilaterally symmetrical fruit produced from a unilocular ovary, dehiscing into two valves, with seeds attached along with ventral suture (Fabaceae).
Lemma The lower of two bracts enclosing the flower of grass.
Lenticel A corky spot on the epidermis or bark.
Lenticular Lens-shaped.
Lepidote Covered witih small scales.
Liana A vinelike plant.
Lignescent Somewhat woody or becoming woody.
Ligneous Woody.
Ligulate floret Small flower with corolla expanded into ligule, but perfect and fertile (Lactuceae, Asteraceae).
Ligule A strap-shaped limb or body.
Liguliform Strap-shaped.
Linear Long and narrow with sides more or less parallel.
Lingulate Tongue-shaped.
Lobate Having lobes.
Lobe A partial division of an organ, especially if rounded.
Lobulate Having small lobes or lobules.
Locular Having locules.
Locule Compartment or cavity of an organ, especially of an ovary.
Loculicidal Dehiscent into the locule, on the back, more or less halfway between the partitions.
Loment A leguminous fruit, constricted between the seeds, each 1-seeded segment separating at maturity.
Lunate Crescent-shaped.
Lurid Dirty, dingy.
Lyrate Pinnatifid, with enlarged terminal lobe and smaller lower lobes.
Maculate Mottled or blotched.
Malpighaceous Resembling the Malpighiaceae, especially in possessing straight hairs attached by the middle; dolabriform.

GLOSSARY

Marcescent Withering, but persistent on the plant.
Massula Mucilaginous large cells enclosing or attached to fern spores *(Azolla)*.
Median Pertaining to the middle.
Mega- A Greek prefix meaning large.
Megaspore A large spore giving rise to the female gametophyte *(Azolla)*.
Membranaceous, membranous Thin, pliable, more or less translucent.
Mentum An extension of the base of the column, projecting in front of the flower (Orchidaceae).
Mephitic Having an offensive odor.
Mericarp A portion of a dry dehiscent fruit that splits away as if separate (Apiaceae).
Meristem Undifferentiated tissue, capable of developing into various organs.
-merous A Greek suffix referring to the number of parts.
Mesocarp The middle layer of the fruit wall.
Micro- A Greek prefix meaning small.
Micropyle The opening between integuments into an ovule.
Microspore A spore giving rise to the male gametophyte.
Mitriform Shaped like a miter, or bishop's hat.
Monadelphous Stamens with their filaments united into one group.
Moniliform Constricted at regular intervals, resembling a string of beads.
Mono- A Greek prefix meaning one.
Monocotyledons Flowering plants having one cotyledon, parallel venation and flower parts usually in 3s.
Monoecious Having staminate and pistillate flowers on the same plant.
Monopodial With an evident single and continuous axis.
Mucilaginous Slimy.
Mucro A sharp, short and abrupt tip.
Mucronate Terminated by a mucro.
Mucronulate Terminating with a small mucro.
Multicipital With many heads.
Multiple A single fruit formed by the coalescence of several fruits from separate flowers.
Muricate Rough, with hard, short points.
Muriculate Minutely muricate.
Muticous Blunt, lacking a point.
Mycorrhiza A symbiotic relationship between fungi and roots of plants.
Napiform Turnip-shaped (roots).
Natant Floating underwater; immersed.
Nectariferous Having or producing nectar.
Nectary The organ from which nectar (sweet fluid) is secreted.
Nerve A simple vein or rib.
Node The place on the stem where a leaf is borne.
Nodule A small, knoblike enlargement.
Nodulose Having small, knobby nodes or knots.
Nut An indehiscent, hard, 1-seeded fruit.
Nutlet A small nut.
Ob- A Latin prefix meaning inverted.
Obconic Cone-shaped with attachment at apex.
Oblanceolate Lanceolate with broadest part above the middle and tapering toward the base.
Oblate Flattened at the poles.
Oblique Slanting.
Oblong Longer than broad, with sides nearly parallel.
Obovate, obovoid Egg-shaped, with broader part toward the top.
Obsolescent Nearly obsolete.
Obsolete Not apparent, rudimentary, extinct.
Obtuse Rounded, blunt.
Ochroleucous Yellowish-white.
Ocrea A papery sheath formed by the fusion of stipules (e.g., around nodes of *Polygonum*).
Ocreola Small and secondary stipular sheath, especially at the nodes in an inflorescence.
Olivaceous Olive-green in color.
Opposite Arranged two at each node, on opposite sides of the axis.
Orbicular Circular.
Orifice An opening.
Orthotropous Erect with micropyle at apex (ovule).
Oval Broadly elliptical.
Ovary The part of the pistil containing the ovules.
Ovate, ovoid Egg-shaped with the broader part near the base.
Ovulate Bearing ovules.
Ovule The body that when fertilized becomes the seed.
Palate A rounded projection on the lower lip of 2-lipped corolla, closing the throat.
Palea The small, upper bract enclosing the flower of grass.
Paleaceous Chaffy.
Pallid Pale in color.
Palmate Divided in a palmlike or handlike manner.
Paludal Growing in marshes.
Pandurate, panduriform Fiddle-shaped.
Panicle An irregularly compound inflorescence with pedicillate flowers.
Paniculate Borne in a panicle.
Paniculiform Resembling a panicle.
Pannose Having the texture or appearance of woolen cloth.
Pantocolpate A pollen grain with grooves scattered on the surface.

Pantoporate   A pollen grain with pores scattered on the surface.
Papilionaceous   Butterflylike corolla with standard, wings, and keel (Fabaceae).
Papilla   Small, pimplelike projection.
Papillate, papillose   Bearing papillae.
Pappus   Modified perianth forming a crown on an achene (Asteraceae).
Parasite   A plant that lives on and derives its food from another plant.
Parietal   Borne on the wall (as ovule on wall of ovary).
Patelliform   Disk-shaped.
Patent   Spreading.
Pectinate   Pinnatifid with very narrow, close division.
Pedate   Palmately divided with the lateral segments again divided.
Pedicel   The stalk of a single flower.
Pedicellate   Borne on a pedicel.
Peduncle   The stalk of a flower cluster or of one flower when it is the only member of an inflorescence.
Pellucid   Clear, almost transparent.
Peloria   An irregular flower that becomes regular by the abnormal development of some irregular parts.
Peltate   Shield-shaped; attached to the stalk by the lower surface, often near the center.
Pendent, pendulous   Hanging down.
Penicillate   Brushlike.
Pentagonal   Shaped like a pentagon (5-sided).
Pentamerous   Having the parts in 5s.
Perennial   Lasting several years.
Perfect   Having both functional stamens and pistils.
Perfoliate   Referring to a sessile leaf whose base completely surrounds the stem.
Perforate   Having translucent dots that look like small holes, or pierced through.
Peri-   A Greek prefix meaning around.
Perianth   The floral envelope, corolla and calyx.
Pericarp   The wall of a fruit.
Perigynium   The papery sheath surrounding the pistil in *Carex*.
Perigynous   Arising from or borne around the ovary.
Perisperm   The nutritional material of a seed formed outside the embryo sac.
Perisporium, perispore   A thin envelope enclosing a spore.
Petal   One division of the corolla.
Petaloid   Resembling a petal.
Petiolate   Having a petiole.
Petiole   The stalk of a leaf.
Petiolule   The stalk of a leaflet in a compound leaf.
Phyllodium, phyllode   Leaflike petiole serving as the blade.

Pilose   Covered with soft, distinct, thin hairs.
Pilosulous   Minutely pilose.
Pinna   One primary division of a pinnate leaf.
Pinnate   Compound leaf with leaflets arranged on both sides of the axis; odd-pinnate if terminal leaflet is present, even-pinnate if terminal leaflet is absent.
Pinnatifid   Cleft in a pinnate manner.
Pinnatisect   Pinnately divided to the midrib.
Pistil   The seed-bearing organ of a flower; ovary, style, and stigma.
Pistillate   Having pistils but no functional stamens.
Pith   The soft, spongy center of a stem.
Placenta   A part of the ovary where ovules are attached.
Placentation   Arrangement of ovules within the ovary.
Plait   A lengthwise fold.
Plicate   Folded into plaits (plicae).
Plumose   Having fine, elongate hairs; featherlike.
Pod   A dry, dehiscent fruit.
Pollen   Grains borne in the anther, containing the male gametophyte.
Pollinarium   The unit of pollen dispersed in the Asclepiadaceae, consisting of pollinia, translator, and corpusculum.
Pollinium (pl. pollinia)   A mass of coherent pollen grains (Orchidaceae, Asclepiadaceae).
Poly-   A Greek prefix meaning many.
Polygamous   Having bisexual and unisexual flowers on the same plant.
Polygamodioecious   Dioecious plants having some perfect flowers.
Polygamomonoecious   Monoecious plants having some perfect flowers.
Pome   A fleshy fruit formed from an inferior ovary with several locules (apple).
Porulus   Somewhat porous (pierced with small, round holes).
Precocious   Developing very early.
Prehensile   Clasping.
Prickle   A small, sharp outgrowth of epidermis.
Primocane   The first year's shoot of woody biennials *(Rubus)*.
Prismatic   Shaped like a prism; angular with flat sides.
Procumbent   Lying or trailing on the ground.
Prophyllate   Having prophylla.
Prophyllum (pl. prophylla)   Bracteole at base of individual flower *(Juncus)*.
Prostrate   Lying flat on the ground.
Protandrous, proterandrous   Having the anthers mature before the pistils in the same flower.
Pruinose   Having a bloom on the surface, a waxy, powdery secretion.
Pseudo-   A Greek prefix meaning false.

Puberulent Minutely pubescent with hairs hardly visible.
Pubescent Covered with short, soft, downy hairs; a general term for any kind of hairiness.
Pulvinate Cushion-shaped.
Pulvinus The swollen base of a petiole.
Punctate Having colored or translucent dots, or pits.
Puncticulate Minutely punctate.
Pungent Ending in a sharp, stiff point.
Pustule A blisterlike elevation.
Pyriform Pear-shaped.
Quadrate Nearly square.
Raceme A simple, elongated inflorescence with pedicellate flowers.
Racemiform In the form of a raceme.
Racemose Having racemes or racemelike inflorescences.
Rachilla A secondary axis; in grass and sedge, the flower-bearing axis.
Rachis The axis of a compound leaf or an inflorescence.
Radiate Spreading from a common center; in Asteraceae, a head with disk florets in center and a whorl of ray florets around the edge.
Radicle That portion of the embryo below the cotyledons.
Raphe A portion of the funiculus that is adnate to the integument, appearing as a ridge.
Raphide Needle-shaped crystal in a plant cell.
Raphidulous Resembling or having raphides.
Ray Outer floret of Asteraceae, with straplike corolla, no stamens, functionally pistillate; the branch of an umbel.
Receptacle The expanded end of the axis bearing flower parts.
Reclinate Bent or turned downward.
Recumbent Leaning or reclining.
Recurved Curved backward or downward.
Reflexed Abruptly bent or turned downward.
Regular Uniform or symmetrical in shape; actinomorphic.
Remote Distantly spaced.
Reniform Kidney-shaped.
Repand Slightly uneven or sinuate.
Repent Creeping, prostrate and often rooting.
Replum The partition between two halves of a fruit (Brassicaceae).
Resinous Producing or containing resin.
Reticulate Like a network.
Retrorse Directed backward or downward.
Retuse With a slight notch at a rounded apex.
Revolute Rolled backward, toward the lower side.
Rhizomatose Resembling a rhizome.
Rhizomatous Possessing a rhizome.
Rhizome An underground stem, usually lateral and rooting at the nodes.

Rhombic, rhomboid Shaped like an equilateral, oblique-angled figure.
Rib A primary or prominent nerve or vein.
Rootstock Underground stem; rhizome.
Roseate Rosy or pinkish in color.
Rosette A cluster of organs arranged in a compact circle.
Rostellum A small beak; in Orchidaceae, an extension from the upper edge of the stigma.
Rostrate Having a beak.
Rosulate In rosettes.
Rotate Wheel-shaped.
Rotund Nearly circular.
Rufescent Becoming reddish-brown.
Rufous Reddish-brown.
Rugose Wrinkled.
Rugulose Somewhat wrinkled.
Ruminate Mottled, appearing as though chewed.
Runcinate Sharply incised or serrate, with teeth pointing backward.
Saccate Bag-shaped.
Sagittate Shaped like an arrowhead, with basal lobes pointing downward.
Salient Projecting forward.
Salverform Corolla with a slender tube abruptly expanded into a flat limb.
Samara An indehiscent, winged fruit.
Saprophyte A plant that derives its food from dead organic matter.
Scaberulous Minutely roughened.
Scabridulous Slightly rough.
Scabrous Rough; feeling rough to the touch.
Scale Any thin, dry, appressed organ (usually leaf or bract).
Scandent Climbing.
Scape A leafless flowering stem arising from the ground.
Scapiform Resembling a scape.
Scapose Bearing flowers on a scape.
Scarious Thin, dry, membranaceous, not green (leaf or bract).
Schizocarp A dry, dehiscent fruit that splits into two halves.
Sclerenchyma Tissue composed of cells with thick, hard walls.
Sclerenchymatous Having sclerenchyma.
Sclerotic Stony in texture.
Scorpioid Inflorescence coiled circinately with flowers in two rows along the outer side.
Scrobiculate Marked by minute or shallow depressions.
Scrotiform Pouchlike.
Scurfy With scalelike particles.
Scutellum A shieldlike protrusion on the calyx (Lamiaceae).
Secund Directed to one side.
Seleniferous Containing selenium.

Semi- A Latin prefix meaning half.
Senescent Ageing or aged.
Sepal One division of the calyx.
Sepaloid Resembling a sepal.
Septate Divided by partitions.
Septicidal Dehiscing along partitions.
Septifragal Breaking away at the partitions, as the valves of a capsule.
Septum A partition.
Seriate Arranged in a series of rows.
Sericeous Silky, with appressed, soft hairs.
Serrate With sharp teeth pointing forward.
Serrulate Finely serrate.
Sessile Without a stalk.
Seta (pl. setae) A bristle.
Setaceous Bristlelike or having bristles.
Setiform Bristle-shaped.
Setose Covered with bristles; similar to hispid.
Setulose Having minute bristles.
Sheath A tubular structure surrounding part or all of an organ; the portion of a grass leaf that surrounds the stem.
Sigmoid Curved like the letter S.
Siliceous Containing or composed or silica.
Silicle A short silique, about as long as wide.
Silique An elongate, dry, dehiscent fruit with a septum separating the two valves (Brassicaceae).
Simple Not compound or not branched.
Sinuate, sinuous With margin strongly wavy.
Sinus The space between two lobes.
Sobole A shoot, especially one from the ground.
Soboliferous Producing basal shoots, clump-forming.
Sordid Dirty white in color.
Sorus (pl. sori) A cluster of sporangia on a fern frond.
Spadix A fleshy or thick spike of minute flowers (Araceae).
Spathaceous Resembling a spathe.
Spathe A leaflike bract surrounding an inflorescence (Araceae) or smaller bract below flower *(Iris)*.
Spatulate Spoon-shaped.
Spicate Spikelike.
Spiciform Shaped like a spike.
Spicule A secondary spike; a fine, erect point.
Spike A simple, elongated inflorescence with sessile flowers.
Spikelet A small or secondary spike.
Spine A sharp, rigid, outgrowth, usually from the wood of a stem.
Spinescent Ending in a spine.
Spiniferous, spinose Having spines.
Spongiose Soft, spongy.
Sporangium (pl. sporangia) A body bearing spores.

Sporangiophore An appendage holding a sporangium.
Spore A reproductive body capable of developing into a new individual (ferns).
Sporocarp An organ containing sporangia.
Sporophyll A leaf that bears sporangia.
Spur A hollow, tubular projection from a petal or sepal, usually containing nectar.
Squamella A tiny or secondary scale.
Squarrose With parts spreading or recurved at the end.
Stamen The pollen-bearing organ of a flowering plant.
Staminate Having stamens but no functional pistil.
Staminode, staminodium A sterile stamen or a structure resembling one.
Standard Upright large petal of Fabaceae flower; erect or arching perianth segment in *Iris* flower.
Stellate Star-shaped.
Sternotribal Flowers in which anthers are positioned to dust pollen on underside of thorax of insects.
Stigma The receptive part of the pistil that receives the pollen.
Stigmatic Belonging to the stigma.
Stipe The stalk of a pistil or gland; the petiole of a fern frond.
Stipel The stipule of a leaflet.
Stipitate Borne on a stipe.
Stipule An appendage at the base of a petiole, usually in pairs.
Stolon A horizontal stem that roots at the tip or at the nodes; runner.
Stoma (pl. stomata), stomate A minute opening between two guard cells in the epidermis of a leaf, through which gases are exchanged.
Stomatiferous Bearing stomata.
Stramineous Straw colored.
Striate Marked with fine longitudinal lines or ridges.
Strict Very straight.
Strigose Having sharp, stiff, straight and appressed hairs that are often swollen at base.
Strigillose, strigulose Minutely strigose.
Strophiole An appendage at the hilum of a seed (Portulacaceae).
Strophiolate Having a strophiole.
Style The usually elongated part of the pistil between the ovary and the stigma.
Stylopodium A disklike expansion at the base of the style (Apiaceae).
Sub- A Latin prefix meaning "beneath" but sometimes signifying "slightly" or "somewhat."
Subulate Awl-shaped, tapering toward the apex.
Succulent Fleshy, juicy.

GLOSSARY

Suffrutescent   Slightly shrubby.
Suffruticose   Plants woody at the base and herbaceous above.
Sulcate   Furrowed or grooved lengthwise.
Sulcus   A groove or furrow.
Superior   An ovary free from the perianth, which is inserted below it.
Supra-   A Latin prefix meaning above.
Suppressed   Failing to develop.
Surculose   Producing suckers, shoots arising from underground parts.
Suture   A line of dehiscence; a groove marking a junction.
Symmetrical   Regular in number and size of parts.
Sympetalous   With the petals at least partially united; gamopetalous.
Taproot   The primary descending root.
Tawny   Dull brownish-yellow; fulvous.
Tendril   A slender, twisting organ by which a plant clings to a support.
Tepal   A segment of a perianth not clearly differentiated into sepals and petals.
Teratological   Distinctly abnormal.
Terete   Circular in transverse section.
Ternate   In 3s.
Tetradynamous   Having 4 long and 2 shorter stamens.
Tetragonal   4-angled.
Tetrahedal   4-sided.
Thallus (pl. thalli)   A flat, leaflike organ; in cryptogams, the vegetative body, which is not differentiated into stem and leaves.
Theca   The pollen sac of an anther.
Throat   The opening of a gemopetalous corolla.
Thyrse   A compact, usually compound panicle.
Thyrsoid   Resembling a thyrse.
Tomentose   Densely pubescent with short, woolly hairs.
Tomentulose   Slightly or finely tomentose.
Tomentum   A covering of dense, woolly hairs.
Tortuous   Twisted.
Torulose   Irregularly swollen at close intervals; knobby.
Torus   Receptacle; ring subtending another organ.
Trabecula   A transverse partition.
Translator   Wishbone-shaped acellular filament connecting pairs of pollinia from adjacent anthers to the corpusculum (Asclepiadaceae).
Trapezoid   A body with 4 unequal sides.
Tri-   A Latin prefix meaning "three."
Trichome   A hair or bristle growing from epidermis.
Trifid   3-cleft.
Trifurcate   Forked into 3 parts.
Trigonous   3-angled, with plane faces between.
Triquetrous   3-angled with concave faces between, angles projecting forward.
Trullate   Trowel-shaped, with straight margins and widest below the middle.
Truncate   Ending abruptly, as if cut off nearly straight across.
Tuber   A thick, short branch, usually subterranean, with numerous buds.
Tubercle   A small tuber or tuberlike body.
Tuberculate   Covered with knobby projections.
Tuberiferous   Bearing tubers.
Tumid   Swollen.
Tunicate   Having concentric layers.
Turbinate   Inversely conical or top-shaped.
Turgid   Swollen by pressure from within.
Turion   A scaly, swollen offshoot of a rhizome.
Umbel   A flat-topped or rounded inflorescence in which pedicels or peduncles arise from a common point (Apiaceae).
Umbellet   A secondary umbel.
Umbelliform   In the shape of an umbel.
Umbo   A conical projection on the surface.
Umbonate   Bearing an umbo in the center.
Uncinate   Hooked obtusely at the tip.
Uncinulate   Minutely uncinate.
Undulate   Unevenly wavy on the surface or margin.
Uni-   A prefix meaning one.
Urceolate   Urn-shaped.
Utricle   A bladderlike, 1-seeded, indehiscent fruit.
Valvate   Opening by valves; meeting at the edge, but not overlapping.
Valve   A separable part; one of the units into which a capsule splits.
Vein   A thread of fibrovascular tissue in a leaf.
Velutinous   Velvety.
Venation   The pattern of veins.
Ventral   Referring to the front or inner surface of an organ; the upper surface of a leaf.
Ventricose   Swollen or inflated on one side.
Vernal   Appearing in the spring.
Vernation   The arrangement of leaves in a bud.
Versatile   Anther attached in center and able to move freely on filament.
Verrucose   Covered with wartlike projections.
Verruculose   Finely verrucose.
Verticel   A whorl.
Verticillaster   A false whorl composed of pairs of opposite cymes (Lamiaceae).
Vesicle   A small cavity or bladder.
Vesture, vestiture   Any covering on a surface making it other than glabrous.
Villose, villous   Having long, soft hairs, not matted.
Villosulous   Minutely villose.
Virgate   Long, slender, and straight; wand-shaped.
Viscid   Sticky.

Viscidulous   Somewhat sticky.

Whorl   Arranged in a circle, as leaves around the stem at a single node.

Wing   A thin, membranaceous extension of an organ; the lateral petal of a Fabaceae flower.

Woolly   Having curly, soft hairs, usually matted; lanate.

Zygomorphic   Irregular; divisible into equal halves in only one plane.

Index to Latin and Common Names

Compiled by Eileen K. Schofield

Names of plants, families, and higher groups mentioned in keys and descriptions are in roman type. Names mentioned as synonyms or in discussions are in italics.

Abies concolor	74	*lanulosa*	854	*modestum*	51
Abronia	145	millefolium	854	pedatum	51
fragrans	145	ssp. *lanulosa*	854	*Adicea fontana*	129
micrantha	152	occidentalis	854	*opaca*	129
Abutilon	241	sibirica	854	Adoxa moschatellina	832
incanum	241	*Acmispon*	461	Adoxaceae	832
parvulum	241	*americanus*	462	*Adlumia fungosa*	114
theophrasti	241	*Acnida*	180	Aegilops cylindrica	1122
Acacia	408	*altissima*	184	Aesculus	568
angustissima v. hirta	408	v. *prostrata*	184	*arguta*	568
prairie	408	v. *subnuda*	184	glabra v. arguta	568
Acaciella hirta	408	*subnuda*	184	v. *glabra*	568
Acalypha	536	*tamariscina*	184	v. *sargentii*	568
gracilens	536	Aconitum	85	*Afzelia macrophylla*	765
monococca	536	columbianum	85	Agalinis	753
ostryaefolia	537	*porrectum*	85	aspera	754
rhomboidea	537	*ramosum*	85	besseyana	756
virginica	537	tenue	85	fasciculata	754
Acamptocladus sessilispicus	1178	Acorus calamus	1042	*greenei*	754
Acanthaceae	800	*Acrolasia*	270	heterophylla	754
Acanthus Family	800	*compacta*	272	*paupercula*	755
Acer	569	*Acroptilon repens*	900	purpurea	755
glabrum	569	Actaea	86	skinneriana	755
negundo v. interius	569	*alba*	86	tenuifolia	755
v. negundo	570	arguta	86	v. *macrophylla*	756
v. texanum	570	brachypoda	86	v. *parviflora*	756
v. violaceum	570	pachypoda	86	viridis	756
nigrum	570	rubra	86	Agastache	710
saccharinum	570	f. *neglecta*	86	*anethiodora*	711
saccharum	570	*Actinomeris alternifolia*	1016	foeniculum	710
f. *glaucum*	570	Adam-and-Eve	1269	nepetoides	711
f. *schneckii*	570	Adder's-tongue	48	scrophulariaefolia	711
Aceraceae	569	limestone	48	Agavaceae	1263
Acerates angustifolia	627	northern	48	Agave Family	1263
auriculata	620	southern	48	Agoseris	855
hirtella	621	Adder's-tongue Family	45	*aurantiaca*	855
lanuginosa	622	*Adelia acuminata*	748	*cuspidata*	979
viridiflora	631	Adenocaulon bicolor	854	glauca	855
Achillea	854	Adiantum	50	*parviflora*	855
asplenifolia	854	capillus-veneris	50	*pumila*	855

1329

scorzoneraefolia	855	v. *stolonifera*	1129	sativum	1245		
Agrimonia	366	Ailanthus altissima	575	stellatum	1245		
grypospala	366	*glandulosa*	575	textile	1246		
microcarpa	367	Aizoaceae	152	tricoccum	1246		
parviflora	367	*Ajuga reptans*	709	vineale	1246		
pubescens	367	Alabama lip fern	54	*Allocarya californica*	701		
rostellata	367	Alaska orchis	1280	*nelsonii*	701		
striata	367	*Albizia julibrissin*	407	Alnus	142		
Agrimony	366	Alder	142	incana ssp. rugosa	142		
downy	367	buckthorn	555	*rugosa*	142		
hooked	366	smooth	142	serrulata	142		
many-flowered	367	speckled	142	*tenuifolia*	142		
striate	367	Alfalfa	465	Alopecurus	1129		
woodland	367	dodder, large	664	aequalis	1129		
× *Agrohordeum macounii*	1123	wild	475	*aristulatus*	1129		
× *Agroelymus*	1123	Alisma	1022	arundinaceus	1129		
Agropyron	1123	*brevipes*	1023	carolinianus	1130		
albicans	1125	*geyeri*	1023	geniculatus	1131		
biflorum	1124	*gramineum*	1023	*myosuroides*	1130		
caninum ssp. majus	1124	*plantago-aquatica*	1023	pratensis	1130		
v. majus	1124	v. *americanum*	1023	*ventricosus*	1130		
v. unilaterale	1124	v. *parviflorum*	1023	Alpine milk-vetch	424		
cristatum	1125	subcordatum	1023	Alsike clover	484		
dasystachyum	1124	triviale	1023	*Alsinaceae*	192		
v. *riparium*	1125	Alismataceae	1022	*Alsine*	211		
desertorum	1125	Alkali blite	172	*longifolia*	213		
elongatum	1125	Alkali cordgrass	1223	*media*	213		
griffithsii	1125	Alkali-grass	1215	*Alsinopsis*	193		
inerme	1126	Alkali milk-vetch	434	*nuttallii*	194		
intermedium	1125	Alkali plantain	744	Alternanthera caracasana	179		
v. intermedium	1126	Alkali sacaton	1225	Althaea	242		
v. trichophorum	1126	Alkali weed	656	officinalis	242		
leersianum	1126	Alkali wild rye	1161	rosea	242		
palmeri	1126	Alleghany monkey-flower	771	Alumroot	358		
pauciflorum	1124	Alliaria	296	Alyssum	297		
pectiniforme	1125	*officinalis*	297	alyssoides	297		
pseudorepens	1124	petiolata	296	desertorum	297		
repens	1126	Allionia	146, 147	hoary false	303		
richardsonii	1124	*bracteata*	148	*minus* v. *micranthus*	297		
riparium	1125	*carletonii*	148	pale	297		
sibiricum	1125	*decumbens*	148	Amaranthaceae	179		
smithii	1124, 1126	*diffusa*	150	Amaranthus	180		
spicatum	1126	*glabra*	149	albus	181		
subsecundum	1124	*hirsuta*	150	v. *pubescens*	181		
subvillosum	1125	incarnata	146	arenicola	181		
tenerum	1124	*linearis*	150	*blitoides*	182		
trachycaulon	1124	*nyctaginea*	151	caudatus	179		
× *Agrositanion*	1123	*ovata*	151	graecizans	181		
Agrostemma githago	193	Allium	1243	hybridus	182		
Agrostis	1127	*burdickii*	1246	palmeri	182		
alba	1128	canadense	1244	powellii	183		
asperifolia	1127	v. canadense	1244	retroflexus	183		
elliottiana	1127	v. fraseri	1244	rudis	183		
exarata ssp. minor	1127	v. hyacinthoides	1244	spinosus	184		
gigantea	1128	v. lavendulare	1244	tamariscinus	184		
grandis	1127	v. mobilense	1244	*torreyi*	181, 182		
hyemalis	1128	cepa	1246	tuberculatus	184		
v. *tenuis*	1128	cernuum	1245	*Amarella acuta*	607		
oreophila	1128	drummondii	1245	*strictiflora*	607		
perennans	1128	geyeri	1245	Ambrosia	799, 855		
scabra	1128	*mutabile*	1244	acanthicarpa	856		
semiverticillata	1128	*nuttallii*	1245	artemisiifolia	856		
stolonifera	1128	perdulce	1245	bidentata	856		
v. major	1129	*porrum*	1246	confertifolia	857		
v. palustris	1129	*recurvatum*	1245	coronopifolia	858		

elatior	856	*pitcheri*	420	septentrionalis	344
grayi	857	Amphicarpaea bracteata	420	v. *puberulenta*	345
linearis	857	Amsinckia	686	v. *subulifera*	345
longistylis	856	barbata	687	*simplex*	344
media	856	*idahoensis*	687	Androstephium coeruleum	1246
psilostachya	857	intermedia	687	Anemone	86
striata	858	lycopsoides	687	berlandieri	87
tomentosa	858	Amsonia	610	canadensis	87
trifida	858	ciliata v. texana	611	candle	87
v. *texana*	858	illustris	611	Carolina	87
variabilis	858	*salicifolia*	611	caroliniana	87
Ambrosiaceae	838	tabernaemontana	611	cylindrica	87
Amelanchier	368	v. *gattingeri*	611	*decapetala*	87
alnifolia	368	v. *salicifolia*	611	false rue	94
v. dakotensis	368	Texas	611	*globosa*	88
arborea	369	willow	611	*heterophylla*	87
canadensis	369	*Amygdalaceae*	364	*hudsoniana*	88
carrii	368	Amygdaloideae	265	meadow	87
humilis	369	Anacardiaceae	571	multifida	88
v. *campestris*	369	*Anacharis*	1029	patens	88
v. *compacta*	369	*canadensis*	1029	quinquefolia	88
v. *exserrata*	369	*densa*	1029	v. *interior*	88
laevis	369	*nuttallii*	1030	*riparia*	89
leptodendron	368	*occidentalis*	1030	rue	89
macrocarpa	368	*planchonii*	1029	tall	88
sanguinea	369	*Anagalis gattingeri*	755	tenpetal	87
American beakgrain	1156	Anagallis arvensis	343	virginiana	88
American beautyberry	702	f. *arvensis*	344	wood	88
American bellflower	808	f. *coerulea*	344	Anemonella thalictroides	89
American bittersweet	534	Anaphalis margaritacea	858	Anemopsis californica	79
American bladdernut	567	v. *angustia*	859	Anethum graveolens	587
American bugleweed	719	v. *intercedens*	859	Anglepod	632, 635
American chestnut	134	v. *occidentalis*	859	Anise root	598
American dianthera	801	v. *subalpina*	859	Annonaceae	77
American elm	122	pussy-toes	859	Annual bluegrass	1210
American false pennyroyal	716	*subalpina*	859	Annual bursage	856
American germander	739	*Androcera*	647	Annual eriogonum	216
American milfoil	493	*citrullifolium*	648	Annual fleabane	927
American milk-vetch	424	*rostrata*	650	Annual knawel	204
American pillwort	70	Andropogon	1130	Anoda cristata	242
American potato bean	420	barbinodis	1131	*Anogra*	516
American sloughgrass	1139	bladhii	1132	*albicaulis*	517
American vetch	487	chrysocomus	1133	*bradburiana*	517
Ammania	494	furcatus	1132	*cinerea*	522
auriculata	495	gerardii	1132	*coronopifolia*	519
v. *arenaria*		v. *chrysocomus*	1133	*latifolia*	522
f. *brasiliensis*	495	v. *paucipilus*	1133	*perplexa*	517
coccinea	495	glomeratus	1134	Anoplanthus fasciculatus	798
robusta	495	hallii	1132	*uniflorus*	799
Ammi majus	585	intermedius	1132	Antelope horns	618
Ammoselinum	586	ischaemum v.		*Antelophragma alpinum*	424
butleri	587	songaricus	1133	Antennaria	859
popei	586	*paucipilus*	1133	anaphaloides	859
Amorpha	418	*provincialis*	1132	*aprica*	862
canescens	418	saccharoides v.		*campestris*	861
f. *glabrata*	419	torreyanus	1133	*chelonica*	861
fragrans	419	scoparius	1134	dimorpha	860
fruticosa	419	v. *frequens*	1134	*fallax*	861
v. *angustifolia*	419	v. *neomexicanus*	1134	*longifolia*	861
v. *tennesseensis*	419	springfieldii	1134	microphylla	860
nana	419	ternarius	1134	neglecta	860
Ampelamus albidus	632	virginicus	1134	v. *attenuata*	861
Ampelopsis cordata	557	Androsace	344	v. *howellii*	861
Amphiachyris dracunculoides	945	occidentalis	344	neodioica	861
Amphicarpa comosa	420	*puberulenta*	345	ssp. *canadensis*	861

ssp. *neodioica*	861	Aquifoliaceae	535	*platyceras* v. *hispida*	111
occidentalis	861	Aquilegia	89	polyanthemos	111
oxyphylla	860	*brevistylus*	89	squarrosa	112
parlinii	861	canadensis	89	*Argentina*	382
ssp. *fallax*	861	v. *hybrida*	89	*anserina*	383
ssp. *parlinii*	861	v. *latiuscula*	89	Argythamnia	537
parvifolia	861	*latiuscula*	89	humilis	537
plantaginifolia	861	Arabidopsis thaliana	297	v. *laevis*	538
rosea	860	Arabis	298	mercurialina	538
Anthemideae	842	*brachycarpa*	299	Arisaema	1043
Anthemis	862	canadensis	298	dracontium	1043
arvensis	862	*confinis*	299	triphyllum	1043
v. *agrestis*	862	*connexa*	299	Aristida	1135
cotula	862	*dentata*	302	adscensionis	1136
tinctoria	862	divaricarpa	299	*arizonica*	1138
Anthopogon crinitus	608	drummondii	299	barbata	1136
procerus	608	fendleri	299	basiramea	1136
Anthyllis vulneria	416	v. fendleri	300	v. basiramea	1136
Anticlea chlorantha	1258	v. spatifolia	300	v. curtissii	1136
elegans	1258	glabra	300	*curtisii*	1136
Antirrhinum majus	756	hirsuta v. pycnocarpa	300	*desmantha*	1135
Anychia	202	holboellii	301	dichotoma	1136
canadensis	202	v. collinsii	301	v. *curtissii*	1136
polygonoides	203	v. pinetorum	301	divaricata	1137
Apache-plume	375	laevigata	301	*fendleriana*	1139
Aphanostephus	863	shortii	301	*glauca*	1138
pilosus	863	virginica	302	*gracilis*	1137
ramosissimus	863	Araceae	1042	*intermedia*	1137
riddellii	864	*Arachis hypogaea*	416	longespica	1137
skirrhobasis	864	Aralia	584	v. geniculata	1137
Apiaceae	584	nudicaulis	584	v. longespica	1137
Apinus	75	racemosa	584	*longiseta*	1139
flexilis	76	Araliaceae	583	*necopina*	1137
Apios americana	420	Arctic aster	884	oligantha	1137
f. *cleistogama*	421	Arctium	864	purpurascens	1138
f. *pilosa*	421	*lappa*	865	purpurea	1138
v. *turrigera*	421	minus	864	v. glauca	1138
tuberosa	421	*tomentosum*	865	v. longiseta	1139
Aplectrum hyemale	1269	Arctostaphylos uva-ursi	334	v. purpurea	1139
spicatum	1270	v. adenotricha	334	v. robusta	1139
Apocynaceae	610	v. coactilis	334	*ramosissima*	1137
Apocynum	612	Arenaria	193	*roemeriana*	1139
album	613	drummondii	194	*tuberculosa*	1135
ambigens	612	fendleri	194	*wrightii*	1138
androsaemifolium	612	hookeri	194	Arkansas calamint	733
v. *glabrum*	612	v. hookeri	194	Aristolochia	80
v. *griseum*	612	v. pinetorum	194	*durior*	80
v. *incanum*	612	lateriflora	195	*macrophylla*	80
cannabinum	612	patula	195	serpentaria	80
v. *glaberrimum*	613	f. *meadia*	195	tomentosa	80
v. *hypericifolium*	613	v. *patula*	195	Aristolochiaceae	80
v. *pubescens*	613	f. *pitcheri*	195	Armoracia	302
cordigerum	613	v. *robusta*	195	*lapathifolia*	302
× *floribundum*	612	rubella	195	rusticana	302
× *medium*	612	serpyllifolia	196	Arnica	865
v. *floribundum*	612	v. *tenuior*	196	cordifolia	865
pubescens	613	stricta	196	fulgens	866
sibiricum	613	ssp. *dawsoniensis*	196	lonchophylla	866
v. *cordigerum*	613	ssp. texana	196	v. *arnoglossa*	866
v. *salignum*	613	v. *texana*	196	v. *lonchophylla*	866
suksdorfii v.		Argemone	111	rydbergii	866
angustifolium	613	*albiflora*	112	sororia	866
Apple	396	hispida	111, 112	*Arnoglossum*	895
Apple of Peru	641	*intermedia*	112	Aromatic aster	880
Apricot	390	*mexicana*	111	*Arrhenatherum elatius*	1114

INDEX

Arrowfeather three-awn		1138
Arrowgrass		1031
Arrowgrass Family		1031
Arrowhead	1024, 1026,	1027
longbarb		1028
violet		264
Arrow-leaved aster		884
Arrow-wood		830
downy		832
Arroyo twine vine		636
Artemisia	799,	866
abrotanum		867
absinthium		868
annua		868
biennis		868
bigelovii		868
bourgeauana		869
brittonii		870
campestris		868
ssp. *borealis*		869
ssp. *campestris*		869
ssp. *caudata*		868
camporum		869
cana		869
carruthii		869
cernua		869
diversifolia		870
dracunculoides		869
dracunculus		869
falcata		870
filifolia		869
forwoodii		869
frigida	798,	870
glauca		869
gnaphalodes		870
herriottii		870
lindheimeriana		870
longifolia		870
ludoviciana		870
v. ludoviciana		870
v. mexicana		870
papularis		870
pedatifida		870
purshiana		870
serrata		870
tridentata		870
v. *tridentata*		871
v. *vaseyana*		871
vulgaris v. *carruthii*		869
ssp. *ludoviciana*		870
ssp. *mexicana*		870
Artichoke, Jerusalem		957
Arum Family		1042
Asarum		81
acuminatum		81
canadense		81
reflexum		81
Asclepias		615
amplexicaulis		617
arenaria		618
asperula		618
v. asperula		619
ssp. *capricornu*		619
v. *decumbens*		619
auriculata		620

brachystephana	619
engelmanniana	619
galioides	627
hallii	620
hirtella	620
incarnata	621
ssp. *incarnata*	
f. *albiflora*	621
involucrata	621
kansana	629
lanuginosa	622
latifolia	622
macrotis	623
meadii	623
oenotheroides	624
ovalifolia	624
pumila	625
purpurascens	625
quadrifolia	626
speciosa	626
stenophylla	627
subverticillata	627
sullivantii	628
syriaca	628
v. *kansana*	629
f. *leucantha*	629
tuberosa	629
ssp. *interior*	629
v. *interior* f. *lutea*	629
ssp. terminalis	629
uncialis	629
verticillata	630
viridiflora	630
v. *ivesii*	631
v. *lanceolata*	631
v. *linearis*	631
viridis	631
Asclepiadaceae	614
Asclepiodora decumbens	619
viridis	631
Ascyrum hypericoides v. multicaule	237
Ash	748
black	749
blue	750
green	750
mountain	405
prickly	577
red	750
wafer	576
white	749
Ashy sunflower	955
Asimina triloba	77
Asparagus officinalis	1247
Aspen	273
bigtooth	279
quaking	280
Asperugo procumbens	687
Asperula orientalis	817
Asplenium	51
platyneuron	51
f. *serratum*	52
pycnocarpon	54
resiliens	52
rhizophyllum	52

septentrionale		52
trichomanes		52
viride		52
Aster		871
acutidens		885
adscendens		875
amethystinus		880
anomalus		874
Arctic		884
arenosus		972
aromatic		880
arrow-leaved		884
azure		881
azureus		881
batesii		877
bog		883
brachyactis		875
canescens		976
chilensis ssp. *adscendens*		875
ciliolatus		875
coerulescens	878, 883,	885
commutatus		875
conspicuous		876
crassulus		876
drummondii		876
Drummond's		876
eatonii		876
ericoides		876
exiguus		877
exilis		886
falcatus		877
fendleri		877
Fendler's		877
few-flowered		882
fluviatilis		883
geyeri		879
glabellus		883
golden	903,	905
hemisphericus		881
hesperius		878
v. *hesperius*		878
v. *laetivirens*		878
hirsuticaulis		879
hoary		876
jacobaeus		885
junciformis		878
kumleinii		880
laevis		878
v. *geyeri*		879
v. *laevis*		879
lanceolatus		885
lateriflorus		879
v. *lateriflorus*		879
linariifolius		879
lindleyanus		875
lucidulus		879
many-ray		874
meritus		884
Missouri		880
missouriensis		881
nebraskensis		883
New England		880
novae-angliae		880
nuttallii		875
oblongifolius		880

v. *angustatus*		880	adsurgens v. robustior	423	Athyrium		53
v. *oblongifolius*		880	agrestis	424	filix-femina		53
ontarionis		880	alpinus	424	v. angustum		53
oolentangiensis		881	americanus	424	v. asplenioides		53
oregonus		876	barrii	425	v. cyclosorum		53
paludosus ssp.			bisulcatus	425	v. michauxii		
hemisphericus		881	bodini	425	f. rubellum		53
ssp. *paludosus*		881	canadensis	426	pycnocarpon		53
paniculatus		885	caryocarpus	427	thelypteroides		53
panicled	878,	885	v. *trichocalyx*	427	Atragene		90
pansus		881	ceramicus v. filifolius	426	*tenuiloba*		91
parviceps		882	*chandonetti*	424	Atriplex		161
patens		882	crassicarpus	426	argentea		162
v. *gracilis*		882	v. crassicarpus	426	ssp. *argentea*		162
v. patentissimus		882	v. paysoni	427	ssp. *expansa*		162
patentissimus		882	v. trichocalyx	427	canescens		162
pauciflorus		882	distortus	427	confertiflora		163
pilosus		882	drummondii	427	dioica		163
polycephalus		876	flexuosus	427	*gardneri*		164
prairie		886	gilviflorus	428	*hastata*		165
praealtus		883	*goniatus*	424	heterosperma		163
v. nebraskensis		883	gracilis	428	hortensis		163
v. praealtus		883	*parviflorus*	428	*lapathifolia*		165
ptarmicoides		1005	hyalinus	428	*nitens*		164
pubentior		883	kentrophyta	429	nuttallii		164
puniceus		883	lindheimeri	429	*oblanceolata*		164
v. *firmus*		880	*longifolius*	426	oblongifolia		164
rayless		875	lotiflorus	430	*patula*	161,	165
rubrotinctus		977	*cretaceous*	430	v. *hastata*		161
sagittifolius		884	*nebraskensis*	430	powellii		164
salicifolius		883	miser v. hylophilus	430	rosea		164
saltmarsh		885	missouriensis	430	subspicata		165
saundersii		875	mollissimus	431	*tridentata*		164
sericeus		884	v. *earlei*	431	Aureolaria		757
sibiricus v. meritus		884	neglectus	431	grandiflora		757
v. *sibiricus*		884	nuttallianus	431	v. *cinerea*		757
silky		884	v. austrinus	432	v. serrata		757
simplex		885	v. nuttallianus	432	*serrata*		757
v. interior		885	pectinatus	432	Australian bluestem		1132
v. ramosissimus		885	plattensis	432	Austrian field cress		326
v. simplex		885	praelongus v. ellisiae	433	Autumn bent		1128
single-stemmed bog		881	puniceus	433	Autumn willow		290
small white		882	v. *gertrudis*	433	Avena fatua		1139
smooth blue		878	v. *puniceus*	433	v. *sativa*		1139
sneezewort		1005	purshii	433	*hookeri*		1184
soft golden		904	racemosus	434	Avens		376
spinosus		885	v. *longisectus*	434	heartleaf		379
stricticaulis		882	v. *racemosus*	434	large-leaved		378
subulatus v. ligulatus		885	sericoleucus	434	purple		379
swamp		883	shortianus	435	rough		378
tansy		977	spatulatus	435	water		379
tenuicaulis		882	*striatus*	424	white		378
turbinellus		886	tenellus	435	yellow		377
umbellatus		883	*triphyllus*	428	Axyris amaranthoides		165
v. *pubens*		883	vexilliflexus	436	Azolla		71
white	876,	971	*viridis*	429	*caroliniana*		71
white woodland		879	Astranthium integrifolium	886	mexicana		71
wild		871	ssp. *ciliatum*	886	Azure aster		881
willowleaf		883	ssp. *triflorum*	886			
wilsonii		875	Atelophragma	421	Baby pondweed		1038
xylorrhiza		1020	*aboriginum*	423	Baby's breath		200
Asteraceae	663, 664, 665,	838	*forwoodii*	423	Bacopa		757
Astereae		842	*glabriusculum*	423	*acuminata*		770
Astragalus		421	*herriottii*	423	rotundifolia		757
aboriginum		423	Atenia gairdneri	599	Baccharis		886

INDEX

glutinosa	887	*Batidophaca cretacea*	430	Beefsteak plant	726
neglecta	887	lotiflora	430	Beggar's-ticks	890, 892, 893
salicina	887	nebraskensis	430	coreopsis	893
texana	888	*Batodendron arboreum*	335	nodding	891
willow	887	*Batrachium*	96	Begonia, wild	235
wrightii	888	divaricatum	106	Belamcanda chinensis	1259
Bachelor's-button	899	longirostre	103	Bellflower	808
Bahia	888	Beach heather	254	American	808
oppositifolia	984	Beaded lip fern	56	creeping	809
pedata	888	Beaked hazelnut	144	marsh	809
woodhousei	984	Beaked panicum	1200	rover	809
Bald brome	1145	Beaked rush	1106	tall	808
Ball-head	669	Beaked willow	285	Bellflower Family	808
Ball mustard	324	Beakgrain	1156	*Bellis*	889
Balloon vine	567	American	1156	*integrifolia*	886
common	567	Bean, American potato	420	perennis	889
Balm, common	709	buck	481	Bellwort	1256
mountain	555	*common*	416	small	1256
Balm-of-Gilead	279	slick-seed	479	large	1256
Balsam groundsel	997	wild	478	Belvedere	175
Balsaminaceae	582	Bean Family	416	Bent, autumn	1128
Balsamita major	902	Bearberry	334	Bent-flowered milk-vetch	436
Balsamorhiza	889	Bearded skeletongrass	1183	Bentgrass	1127
incana	889	Bearded sprangletop	1188	water	1128
sagittata	889	Beardtongue	774	*Benzoin aestivale*	78
Balsam poplar	277	crested	780	Bequilla	477
hybrid	277	large	783	Berberidaceae	107
Balsam-root	889	narrow	777	*Berberis*	108
Baltic rush	1052	slender	782	repens	108
Bamboo, Japanese	228	white	776	swaseyi	108
Mexican	228	Bear's foot	986	trifoliolata	108
Baneberry	86	Beautyberry	702	Bergamot, wild	724
Baptisia	436	American	702	*Bergia texana*	236
australis v. minor	436	Bechtel crab	396	Texas	236
× *bicolor*	437	Beckmannia	1139	Berlandiera	889
bracteata	437	*erucaeformis*	1140	lyrata	890
v. *bracteata*	437	syzigachne	1139	texana	890
v. glabrescens	437	Beckwith's clover	483	Berlandier evening primrose	500
lactea	437	Bedstraw	817	Berlandier's flax	563
leucantha	437	bluntleaf	819	Bermuda grass	1154
leucophaea	437	catchweed	818	Berry, globe-	268
minor	437	hairy	820	service-	368
vespertina	437	Labrador	819	shad-	369
Barbarea	303	northern	818	Berteroa incana	303
americana	303	shining	819	*mutabilis*	303
orthoceras	303	small	820	Berula erecta v. incisum	587
vulgaris	303	southwest	821	*pusilla*	587
Barberry	108	sweet-scented	820	*Bessera*	1246
Texas	108	.Texas	820	Besseya	757
Barberry Family	107	woods	819	bullii	758
Barley	1185	yellow	820	*cinerea*	758
foxtail	1185	Bee plant	291	wyomingensis	758
little	1185	Rocky Mountain	292	Bessey's locoweed	467
Barnyard grass	1164	Beebalm	723	Betony, marsh	737
Barrel cactus	155	basil	724	slenderleaf	738
Barr orophaca	425	Bradbury	723	thinleaf	738
Basketflower	898	dotted	725	wood	773
Basil	729	lemon	724	Betula	142
beebalm	724	plains	725	*alba*	143
Basin wild rye	1168	spotted	725	*fontinalis*	143
Bassia	165	Beech	135	*glandulifera*	143
five-hook	165	blue	141	glandulosa v.	
hirsuta	166	Beech fern	67	glandulifera	142
hyssopifolia	165	broad	67	*lutea*	143
Basswood	239				

INDEX

nigra	143	Bird's-foot trefoil	461	lesser	807
occidentalis	143	Bird's-foot violet	261	Bladderwort Family	806
papyrifera	143	Birthwort Family	80	Blanket flower	937, 938
v. *cordifolia*	143	Bishop's cap	360	Indian	939
pendula	143	*Bishop's-weed*	585	Blazing star	270, 972
platyphylla	143	mock	599	Bleeding hearts	117
Betulaceae	141	Bistort	230	*Blephariglottis*	1276
Bicknell's cranesbill	581	*Bistorta*	220	*lacera*	1278
Bidens	890	*vivipara*	230	*leucophaea*	1278
acuta	892	Bitter cress	306, 307	Blephila	711
aristosa	893	smallflower	306	*ciliata*	712
v. *retrorsa* f.		Bitter dock	233	*hirsuta*	712
involucrata	893	Bitter sneezeweed	951	Blind eyes	113
bigelovii	891	Bitternut hickory	132	Blood polygala	565
bipinnata	891	Bittersweet	534, 649	Bloodleaf	187
cernua	891	American	534	Bloodroot	114
comosa	892	*Oriental*	534	Blowout grass	1195, 1217
connata	892	Bitterweed	962, 963	Blue ash	750
v. *petiolata*	892	Black ash	749	Blue aster, smooth	878
coronata	892	Black cherry, wild	395	*Blue beech*	141
elliptica	892	Black currant, wild	353	Blue cardinal flower	812
filamentosa	892	Black dalea	442	Blue cohosh	109
frondosa	892	Black-eyed Susan	991	Blue curls	740
glaucescens	892	Black-foot daisy	979	Blue-eyed grass	1261, 1262
laevis	892	Black grama	1141	Blue-eyed Mary	763
latisquama	894	Black haw	831	Blue false indigo	436
v. *recognita*	894	southern	832	Blue flag	1259, 1260
polylepis	893	Black hickory	133	western	1260
prionophylla	892	Black Hills spruce	74	Blue flax	562
tripartita	892	Black locust	477	Blue funnel lily	1246
vulgata	893	Black maple	570	Blue grama	1141
v. *puberula*	893	Black medic	465	Blue larkspur	93
f. *puberula*	893	Black mustard	305	Blue lettuce	970
Biennial wormwood	868	Black nightshade	650	Blue lips	763
Big bluestem	1132	plains	649	Blue mustard	308
Big-flower gerardia	757	Black oak	141	Blue phlox	675
Big sagebrush	870	Black poplar	280	Blue prairie violet	261
Big sandreed	1150	Black raspberry	403	Blue sage	731
Big shellbark hickory	132	Black-seed ricegrass	1199	Blue skullcap	735
Big-tree plum	393	Black snakeroot	600	Blue star	610
Bigflower coreopsis	915	Black-stem spleenwort	52	Blue stickseed	695
Bignonia Family	803	Black swallow wort	632	Blue three-awn	1138
Bignoniaceae	803	Black walnut	134	Blue vervain	706
Bigroot morning-glory	659	Black willow	288	Blue violet, downy	264
Bigtooth aspen	279	Blackberries	400	Blue waxweed	496
Bilderdykia	220	Blackberry, common	401	Blue wild rye	1168
convolvulus	228	creeping	404	Blue wood lettuce	968
dumetorum	229	dwarf	404	Bluebells	699
scandens	229	high-bush	403, 404	Blueberry	334
Bindweed, field	654, 655	lily	1259	hillside	335
hedge	652, 653	Blackgrass	1055	Bluebunch wheatgrass	1126
Birch	142	Blackjack oak	139	Bluebuttons	837
bog	142	Bladder campion	209	Bluecaps	837
canoe	143	*Bladder cherry*	642	Bluegrass	1209
Japanese	143	Bladder fern, bulblet	56	annual	1210
mountain	143	Tennessee	58	Canada	1212
paper	143	Bladdernut	567	Canby's	1211
river	143	American	567	Chapman	1211
water	143	Bladdernut Family	567	fowl	1213
weeping	143	Bladderpod	318, 322	inland	1213
yellow	143	double	324	Kentucky	1214
Birch Family	141	oval-leaf	323	plains	1211
Bird cherry	394	Bladderwort	806	Sandberg's	1214
Bird vetch	488	common	807	Texas	1210
Bird's-eye speedwell	796	conespur	806	woodland	1214

INDEX

Bluehearts	758	lunaria	46	Bristly greenbrier	1267
Bluejoint	1148	v. *minganense*	47	Britton's skull cap	734
Bluestem	1130	f. *minganense*	47	Broad beech fern	67
Australian	1132	matricariifolium	47	Broadleaf milkweed	622
big	1132	minganense	47	Broad-leaved cat-tail	1238
broomsedge	1134	multifidum	47	Broad-leaved goldenrod	1003
cane	1131	*neglectum*	47	Broad-leaved twayblade	1280
Caucasian	1132	*obliquum*	46	Brome, bald	1145
King Ranch	1133	*silaifolium*	47	Canada	1146
little	1134	simplex	47	downy	1147
sand	1132	virginianum	48	*field*	1145
silver	1133	Bottle gentian	607	fringed	1143
splitbeard	1134	Bottlebrush grass	1186	grass	1142
Turkestan	1133	Boulder raspberry	402	Japanese	1144
Bluet, narrowleaf	822	Bouncing Bet	204	nodding	1146
rough small	821	Bouteloua	1140	smooth	1144
slender-leaved	822	barbata	1140	Bromus	1142
small	821	curtipendula	1141	*anomalus*	1146
Bluets	821	v. *caespitosa*	1141	*arvensis*	1145
Blueweed	691	v. *curtipendula*	1141	*biebersteinii*	1144
Texas	954	eriopoda	1141	*breviaristatus*	1148
Bluntleaf bedstraw	819	gracilis	1141	*briziformis*	1146
Bluntleaf milkweed	617	hirsuta	1141	*carinatus*	1147
Blunt-lobed woodsia	69	*pectinata*	1142	ciliatus	1143
Bodin milk-vetch	425	rigidiseta	1142	v. *intonsus*	1144
Boebera papposa	921	*simplex*	1140	commutatus	1144
Boehmeria cylindrica	128	Box elder	569	inermis ssp. inermis	1144
Boerhaavia erecta	146	*Brachiaria ciliatissima*	1114	ssp. *pumpellianus*	1144
spicata	147	*Brachyactis angusta*	875	japonicus	1144
Bog aster	883	Brachyelytrum erectum	1142	kalmii	1145
single-stemmed	881	Bracken	66	*lanatipes*	1146
Bog birch	142	Bracted plantain	743	latiglumis	1145
Bog candles	1277	Bradbury beebalm	723	mollis	1145
Bog orchis, slender	1278	Brake	66	*patulus*	1145
Bog rush	1054	*Bramia rotundifolia*	757	porteri	1146
Bog violet, northern	259	Brasenia schreberi	83	pubescens	1146
Bog willow	289	Brassica	304	*purgans*	1145, 1146
Bog yellow cress	327	*arvensis*	305	*racemosus*	1146
Boisduvalia glabella	499	campestris	304	*richardsonii*	1144
Boltonia asteroides	893	hirta	304	secalinus	1146
v. latisquama	894	juncea	304	squarrosus	1147
v. recognita	894	kaber	305	tectorum	1147
Boneset	934	*napus*	304	unioloides	1147
false	966	nigra	305	*wildenowii*	1147
Borage	684	*rapa*	304	Brook cinquefoil	389
Borage Family	683	Brassicaceae	293	Brookgrass	1150
Boraginaceae	683	*Brayulinea densa*	186	Brooklime speedwell	793
Borago officinalis	684	Breadroot scurf-pea	473	Brookweed	350
Bothriochloa	1131	little	474	Broom, sweet-	452, 453
barbinodis v.		*Breweria pickeringii* v.		Broom-corn millet	1202
barbinodis	1132	*pattersonii*	661	Broom flax	562
v. *perforata*	1132	Brickellia	894	Broomrape	798, 799
bladhii	1132	brachyphylla	894	Broomrape Family	798
intermedia	1132	californica	894	Broomsedge bluestem	1134
ischaemum v. *songarica*	1133	grandiflora	894	Broomweed	945
saccharoides v.		*umbellata*	895	false	946
torreyana	1134	Brier, catclaw sensitive	411	Broussonetia papyrifera	126
springfieldii	1134	sensitive	410, 411	Brown-eyed Susan	992
Botidophaca	421	western sensitive	411	Buchloë dactyloides	1148
Botrychium	46	Bristlegrass, knotroot	1219	Buchnera americana	758
dissectum	46	plains	1220	Buck bean (buckbean)	481, 666
v. dissectum	46	Bristly buttercup	101	prairie	481
v. *obliquum*	46	Bristly crowfoot	104	Buckbean Family	666
f. *obliquum*	46	Bristly foxtail	1220	Buckbrush	828
lanceolatum	46	Bristly gooseberry	356	Buckeye	568

western	568	Bush clover	457	*neglecta*	1149
Buckeye Family	568	Chinese	458	*purpurascens*	1149
Buckhorn	743, 744	Bush grape	559	*stricta*	1149
Buckley's penstemon	778	Bush morning-glory	659	Calamint, Arkansas	733
Buckthorn	555	Bushy cinquefoil	387	*Calamintha arkansasa*	733
alder	555	Bushy peppergrass	318	*nuttalli*	733
common	556	Bushy seedbox	514	Calamovilfa	1150
lance-leaved	556	Bushy vetchling	456	*gigantea*	1150
woolly	341	Bushy wallflower	315	*longifolia*	1150
Buckthorn Family	554	Butler's quillwort	41	Calamus	1042
Buckwheat	219	Butomaceae	1021	*Calceolaria*	256
climbing	228	*Butomus umbellatus*	1021	*verticillata*	257
false	229	f. *vallisnerifolius*	1022	Calla palustris	1043
James' wild	218	Butter-and-eggs	768	wild	1043
matted wild	219	Buttercup	95	Callicarpa americana	702
nodding wild	216	bulbous	98	Callirhoe	243
spreading wild	217	bristly	101	*alcaeoides*	243
wild	215, 228	early	99	*digitata*	244
yellow wild	217	early wood	97	v. *stipulata*	244
Buckwheat Family	214	hooked	104	*involucrata*	244
Buffalo bur	650	Macoun's	103	v. *lineariloba*	244
Buffalo clover	485	marsh	101	*leiocarpa*	244
running	486	prairie	104	*papaver* v. *bushii*	245
Buffalo currant	355	shore	99	Callitrichaceae	741
Buffalo-gourd	268	small yellow	101	Callitriche	741
Buffalo grass	1148	tall	98	*autumnalis*	741
false	1197	threadleaf	100	*deflexa* v. *austini*	742
Buffaloberry	491	Buttercup Family	84	*hermaphroditica*	741
Bugle	709	Butterfly milkweed	629	*heterophylla*	741
Bugleweed	718	Butterfly pea	438	*terrestris*	742
American	719	Butterfly weed	509	*verna*	742
rough	719	Butternut	134	Calochortus	1247
Virginia	720	Butterweed	995	*gunnisonii*	1247
Bugseed	174	Button snakeroot	593	*nuttallii*	1247
Bulblet bladder fern	56	Buttonbush	816	Caltha palustris	89
Bulbostylis capillaris	1060	common	816	Caltrop Family	577
v. *crebra*	1060	dodder	662	Calycocarpum lyonii	109
Bulbous buttercup	98	Buttonweed	817, 822	Calylophus	500
Bulbous water hemlock	589	rough	817	*berlandieri*	500
Bull nettle	538	smooth	822	ssp. *berlandieri*	501
Bull thistle	913			ssp. *pinifolius*	501
Bulrush	1107	Cabomba caroliniana	83	*drummondianus*	
Bumelia lanuginosa	341	Cabombaceae	83	ssp. *berlandieri*	501
v. *albicans*	342	Cacalia	895	ssp. *drummondianus*	501
v. *oblongifolia*	341	*atriplicifolia*	895	*hartwegii*	501
Bunchberry	528	*plantaginea*	895	ssp. *fendleri*	501
Bunchflower	1252	*tuberosa*	895	ssp. *filifolius*	501
Bundleflower, Illinois	408	Cactaceae	153	ssp. *lavandulifolius*	502
slender-lobed	409	Cactus, barrel	155	ssp. *pubescens*	502
Buplerum rotundifolium	587	hedgehog	155, 156	*lavandulifolius*	502
Bur-clover, small	465	lace	155	*serrulatus*	502
Bur cucumber	269	nipple	156, 160	Calypso bulbosa	1270
Bur oak	138	pincushion	154	v. *americana*	1270
Bur ragweed	857	Cactus Family	153	v. *bulbosa*	1270
Bur-reed	1235	Caesalpinia jamesii	412	Calystegia	652
Bur-reed Family	1235	Caesalpinia Family	411	*fraterniflorus*	654
Burdock	864	Caesalpiniaceae	411	*macounii*	653
common	864	Calamagrostis	1148	*pellita*	654
great	865	*americana*	1149	*pubescens*	654
woolly	865	*canadensis*	1148	*sepium* ssp. *angulata*	653
Burhead	1023, 1024	v. *macouniana*	1149	v. *fraterniflorus*	654
Burnet	405	*inexpansa*	1149	*spithamaea*	653
prairie	405	*langsdorfii*	1149	*sylvatica* ssp.	
Bursage, annual	856	*macouniana*	1149	*fraterniflora*	654
perennial	858	*montanensis*	1149	Camass, death	1257, 1258

INDEX

eastern	1248	pensylvanica	307	crawei	1074
white	1258	Cardaria	307	crinita	1074
Camassia	1248	chalepensis	307	v. brevicrinis	1075
angusta	1248	draba	307	v. crinita	1075
esculenta	1248	v. *repens*	307	cristatella	1075
scilloides	1248	pubescens	308	crus-corvi	1075
Camelina	305	v. *elongata*	308	davisii	1075
microcarpa	305	Cardinal flower	810	deweyana	1075
sativa	305	blue	812	diandra	1076
Campanula	808	Cardiospermum halicacabum	567	disperma	1076
americana	808	*Carduaceae*	838	douglassii	1076
v. *illinoensis*	809	Carduus	895	*durifolia*	1071
aparinoides	809	acanthoides	896	eburnea	1076
v. *grandiflora*	809	*crispus*	896	eleocharis	1076
intercedens	810	nutans	896	emoryi	1076
petiolata	810	ssp. leiophyllus	897	*festivella*	1083
rapunculoides	809	ssp. macrocephalus	897	festucacea	1077
rotundifolia	809	ssp. *macrolepis*	897	filifolia	1077
uliginosa	809	*thoermeri*	897	fissa	1077
Campanulaceae	808	Carex	1060	*flaccosperma*	1060
Campanulastrum		*abbreviata*	1091	foenea	1077
americanum	809	*abdita*	1084	frankii	1077
Camphor weed	958	aenea	1068	garberi	1077
Campion	205	aggregata	1069	gracilescens	1078
bladder	209	albolutescens	1069	granularis	1078
snowy	208	albursina	1069	v. granularis	1078
starry	209	alopecoidea	1069	v. haleana	1078
white	208	amphibola v. turgida	1069	gravida	1078
Campsis radicans	804	annectens	1069	v. gravida	1078
Camptosorus	51	v. annectens	1070	v. lunelliana	1078
rhizophyllus	52	v. xanthocarpa	1070	grayi	1078
Canada bluegrass	1212	aquatilis v. altior	1070	v. *hispidula*	1078
Canada brome	1146	arkansana	1070	*grisea*	1069
Canada goldenrod	1002	artitecta	1070	hallii	1079
Canada milk-vetch	426	*asa-grayi*	1078	haydenii	1079
Canada-plum	390	assiniboinensis	1070	heliophila	1079
Canada thistle	909	atherodes	1070	hirsutella	1079
Canada tickclover	446	v. *longo-lanceolata*	1081	*hirtiflora*	1079
Canada wild rye	1167	athrostachya	1071	hirtifolia	1079
Canary grass	1207	aurea	1071	hitchcockiana	1079
reed	1207	*austrina*	1084	hoodii	1080
Canby's bluegrass	1211	backii	1071	hookerana	1080
Cancer-root, one-flowered	799	bebbii	1071	hyalinolepis	1080
Cancer weed	732	bella	1071	hystericina	1080
Candle anemone	87	bicknellii	1071	*hystricina*	1080
Cane bluestem	1131	blanda	1072	*impressa*	1080
Canker root	766	*brachyglossa*	1070	interior	1080
Cannabaceae	123	brevior	1072	intumescens	1080
Cannabis sativa	124	brittoniana	1072	jamesii	1081
ssp. *indica*	124	brunnescens	1072	lacustris	1081
ssp. *sativa*	124	bushii	1072	v. *laxiflora*	1081
v. *sativa*	124	buxbaumii	1073	laeviconica	1081
v. *spontanea*	124	capillaris v. elongata	1073	laevivaginata	1081
Canoe birch	143	caroliniana	1073	lanuginosa	1081
Caper Family	291	v. *cuspidata*	1073	lasiocarpa v. americana	1082
Capparaceae	291	cepahloidea	1073	v. *latifolia*	1081
Capparidaceae	291	cephalophora	1073	*laxiflora* v. *blanda*	1072
Caprifoliaceae	823	v. cephalophora	1073	v. *latifolia*	1069
Capsella bursa-pastoris	305	v. mesochorea	1073	leavenworthii	1082
Caragana arborescens	437	comosa	1073	leptalea	1082
Caraway	588	complanata	1074	limosa	1082
Cardamine	306	v. *hirsuta*	1079	*lunelliana*	1078
bulbosa	306	concinna	1074	lupulina	1082
concatenata	306	conjuncta	1074	lurida	1082
parviflora v. arenicola	306	convoluta	1074	meadii	1083

melanostachya	1083	texensis	1091	chromosa	760
mesochorea	1073	torreyi	1091	citrina	761
microdonta	1083	triangularis	1091	coccinea	760
microptera	1083	tribuloides	1091	f. alba	760
microrhyncha	1083	trichocarpa	1091	f. lutescens	760
molesta	1084	typhina	1091	exilis	759
muhlenbergii	1084	umbellata	1092	flava	760
v. australis	1084	f. vicina	1092	indivisa	760
v. enervis	1084	vallicola	1092	integra	760
muricata v. sterilis	1089	varia	1070	linariifolia	761
muskingumensis	1084	versicaria v. monile	1092	miniata	761
nebraskensis	1084	viridula	1092	minor	759
nigromargina v.		vulpinoidea	1092	purpurea	761
muhlenbergia	1070	xerantica	1093	v. citrina	761
v. elliptica	1085	Carleton four-o'clock	148	v. purpurea	762
normalis	1084	Carolina anemone	87	rhexifolia	762
obtusata	1085	Carolina clover	483	septentrionalis	762
oklahomensis	1085	Carolina cranesbill	581	sessiliflora	762
oligocarpa	1085	Carolina foxtail	1130	f. purpurina	762
ovularis	1096	Carolina horse-nettle	647	sulphurea	762
parryana	1085	Carolina jointtail	1190	Catabrosa aquatica	1150
peckii	1085	Carolina lovegrass	1176	Catalpa	804
pedunculata	1085	Carolina moonseed	110	bignonioides	805
pensylvanica	1086	Carolina poplar	277	common	805
v. digyna	1079	Carolina willow	286	hardy	804
plana	1084	Carpetweed	191	northern	804
praegracilis	1085	Carpetweed Family	191	southern	805
prairea	1086	Carpinus betulus	141	speciosa	804
praticola	1086	caroliniana	141	Catawba-tree	804
projecta	1086	Carpogymnia dryopteris	60	Catbrier	1266
pseudo-cyperus	1087	Carrion-flower	1266	Catbrier Family	1266
retroflexa	1087	Carrot	591	Catchfly	205
v. texensis	1091	wild	592	forked	206
retrorsa	1087	Carthamus tinctorius	897	gentian	605
richardsonii	1087	Carum carvi	588	night-flowering	208
f. exserta	1087	Carya	131	royal	209
rosea	1087	× brownii	132	sleepy	206
rossii	1087	buckleyi	133	smooth	206
rostrata	1088	cordiformis	132	Catchweed bedstraw	818
sartwellii	1088	glabra	131	Catclaw	409
saximontana	1088	illinoensis	132	sensitive brier	411
scirpiformis	1088	laciniosa	132	Catgut	480
scirpoidea v.		× laveyi	132	Cathartolinum berlandieri	563
scirpiformis	1088	ovata	132	compactum	564
scoparia	1088	pecan	132	puberulum	563
shortiana	1089	texana	133	rigidum	564
shriveri	1078	tomentosa	133	sulcatum	564
siccata	1076	Caryophyllaceae	192	Catnip	726
sparganioides	1089	Cashew Family	571	giant hyssop	711
v. aggregata	1069	Cassia	412	Cat's-claw mimosa	409
v. cephaloidea	1073	chamaecrista	412	Cat-tail	1237
sprengelii	1089	v. robusta	413	broad-leaved	1238
squarrosa	1089	v. rostrata	413	narrow-leaved	1238
stenophylla v. enervis	1076	fasciculata	413	Cat-tail Family	1237
sterilis	1089	marilandica	413	Caucasian bluestem	1132
stipata	1089	medsgeri	413	Caulophyllum thalictroides	109
v. laevivaginata	1081	nictitans	413	Ceanothus	554
stricta	1090	obtusifolia	413	americanus v. pitcheri	554
v. elongata	1076	occidentalis	413	fendleri	554
strictior	1090	roemeriana	414	herbaceous v. pubescens	555
subimpressa	1090	tora	413	ovatus	555
substricta	1070	Castanea dentata	134	pubescens	555
sychnocephala	1090	mollissima	134	velutinous	555
tenera	1090	ozarkensis	134	Cedar, red	73
tetanica	1090	Castilleja	759	salt	265, 266

Celandine	111	*holosteoides*	199	*petaloidea*	548
Celastraceae	534	*longipedunculatum*	199	*serpens*	549
Celastrus	534	nutans	199	*serpyllifolia*	550
orbiculatus	534	v. *brachypodum*	198	*stictospora*	550
scandens	534	*oreophilum*	197	Chameperilimemum	
Celestial lily	1261	*strictum*	197	*canadense*	528
Celosia argentea	179	*velutinum*	197	Chamomile	862
cristata	179	*viscosum*	198	corn	862
Celtis	120, 533	*vulgatum*	198,199	false	978
canina	121	f. *glandulosum*	199	wild	978
crassifolia	121	v. *hirsuta*	199	yellow	862
georgiana	121	Ceratoides lanata	166	*Chamomilla inodora*	978
laevigata	120	Ceratocephalus	106	*suaveolens*	978
v. *laevigata*	120	*testiculatus*	106	Chapman bluegrass	1211
v. *reticulata*	121	Ceratophyllaceae	84	Charlock	305
v. *smalli*	120	Ceratophyllum demersum	84	Chasmanthium latifolium	1152
v. *texana*	120	*echinatum*	84	Chaste-tree	708
mississippiensis	120	Cercis canadensis	414	common	708
occidentalis	121	f. *glabrifolia*	414	Chatterbox	1275
v. *canina*	121	Cercocarpus montanus	369	Cheat	1146
v. *crassifolia*	121	v. *argenteus*	369	Cheilanthes	54
v. *occidentalis*	121	v. *glaber*	369	alabamensis	54
v. *pumila*	121	Cevallia sinuata	270	*cancellata*	62
pumila	121	Chaenactis douglassii		*cochisensis*	62
v. *georgiana*	121	v. *achillaefolia*	901	eatonii	54
reticulata	121	Chaenorrhinum minus	762	f. *castanea*	55
rugosa	121	Chaerophyllum	588	feei	55
rugulosa	121	procumbens	588	fendleri	55
smallii	120	v. *shortii*	588	*integerrima*	62
tenuifolia	121	*reflexum*	588	lanosa	55
v. *georgiana*	121	tainturieri	588	lindheimeri	55
Cenchrus	1151	v. *dasycarpum*	588	*standleyi*	62
incertus	1151	*texanum*	588	tomentosa	56
longispinus	1151	Chaetopappa asteroides	901	*vestita*	55
pauciflorus	1151	Chaff-flower	179	wootonii	56
Centaurea	898	mat	179	*Cheirinia*	313
americana	898	Chaffweed	345	*argillosa*	314
calcitrapa	899	Chamaecrista	412	*aspera*	314
cyanus	899	*fasciculata*	413	*asperrima*	314
diffusa	899	nictitans	414	*cheiranthoides*	314
greater	900	Chamaenerion	504	*elata*	314
iberica	899	latifolium	504	*inconspicua*	315
maculosa	899	Chamaerhodos	370	*repanda*	315
nigrescens	898	erecta	370	*syrticola*	315
picris	900	v. *erecta*	370	Chelidonium majus	111
repens	900	v. *parviflora*	370	Chenopodiaceae	160
scabiosa	900	*nuttallii*	370	Chenopodium	166
solstitialis	900	Chamaesaracha	638	*albescens*	172
Centaurium	605	coniodes	638	album	168, 169, 171
exaltatum	605	coronopus	638	ambrosioides	168
pulchellum	605	sordida	638	atrovirens	168
texense	605	Chamaesyce	541	berlandieri	169
Centaury	605, 898	aequata	550	v. *zschackei*	169
Centunculus minimus	345	albicaulis	550	*boscianum*	172
Cephalanthus occidentalis		erecta	550	botrys	169
	662, 816	fendleri	545	bushianum	169
Cerastium	197	geyeri	546	captitatum	169
arvense	197	glyptosperma	546	cycloides	170
v. *villosum*	197	v. *integrata*	546	*dacoticum*	169
v. *viscidulum*	197	greenei	545	desiccatum	170, 172
brachypetalum	198	humistrata	546	*ferulatum*	169
brachypodum	198	hyssopifolia	548	fremontii	170
campestre	197	lata	547	gigantospermum	170
fontanum ssp. *triviale*	199	maculata	547	glaucum	171
glomeratum	198	nuttallii	548	*humile*	172

hybridum	170	dwarf	140	bulbifera	589
v. *gigantospermum*	170	Chittimwood	341	*douglasii*	589
incanum	171	Chloris	1152	maculata	589, 590
v. incanum	171	*brevispica*	1153	v. angustifolia	589
leptophyllum	171	cucullata	1152	v. bolanderi	589
missouriense	171	*elegans*	1153	v. maculata	590
overi	170	showy	1153	Cinna	1153
paganum	169	verticillata	1153	arundinacea	1153
pallescens	172	virgata	1153	latifolia	1154
petiolare	169	Choke cherry	395	Cinquefoil	381, 386, 387, 388
pratericola	172	Cholla	156	brook	389
rubrum	172	pencil	158	bushy	387
salinum	171	tree	158	Norwegian	387
standleyanum	172	Chorispora tenella	308	old-field	389
strictum	172	Christmas fern	65	shrubby	385
ssp. glaucophyllum	173	Christmas Mistletoe Family	532	silvery	383
subglabrum	173	Chrysanthemum	902	sulphur	388
watsonii	173	balsamita	902	tall	383
Cherry	390	leucanthemum	902	three-toothed	389
bird	394	v. *pinnatifidum*	902	Circaea	503
bladder	642	*Chrysobotrya odorata*	355	alpina	503
choke	395	Chrysopsis	903	ssp. *alpina*	503
clammy ground	644	*angustifolia*	906	canadensis v. *virginiana*	504
common ground	644	*asperella*	906	× *intermedia*	503
cutleaf ground	642	*bakeri*	906	lutetiana	503, 504
downy ground	644	*ballardii*	906	ssp. canadensis	503
dwarf	394	berlandieri	904	quadrisulcata	
ground	642, 645	*camporum*	903	v. *canadensis*	504
mahaleb	393	canescens	904	Cirsium	908
perfumed	393	*foliosa*	906	altissimum	909
pin-	394	fulcrata	904	f. *moorei*	909
prairie ground	643, 645	*hirsutissima*	906	arvense	909
purple ground	646	hispida	904, 906	f. *albiflorum*	910
sand	394	horrida	904	v. *horridum*	910
sour	390	*imbricata*	906	v. *mite*	910
sweet	390	parthenium	902	v. *vestitum*	910
Virginia ground	646	pilosa	904	canescens	910
wild black	395	stenophylla	905	coccinatum	911
Chervil	588	villosa	905	*discolor*	909
wild	588	v. angustifolia	905	drummondii	911
Chess, hairy	1144	v. *canescens*	904	flodmanii	911
soft	1145	v. foliosa	906	foliosum	911
Chestnut, American	134	v. hispida	906	iowense	909
Chinese	134	v. villosa	906	megacephalum	913
oak, yellow	139	Chrysothamnus	906	muticum	911
Chick-pea	416	*baileyi*	907	nebraskense	910
Chickasaw plum	392	*frigidus*	907	nelsonii	910
Chickweed	211	*howardii*	907	oblanceolatum	911
common	213	nauseosus	906	ochrocentrum	912
common mouse-ear	199	ssp. *albicaulis*	906	*plattense*	910
forked	202, 203	v. *glabratus*	907	pulcherrimum	912
giant	211	ssp. *graveolens*	907	setosum	910
jagged	200	ssp. nauseosus	907	texanum	913
mouse-ear	197	parryi ssp. howardii	907	undulatum	913
nodding	199	*plattensis*	907	v. *megacephalum*	913
prairie	197	pulchellus ssp. baileyi	907	vulgare	913
Chicory	908	viscidiflorus	907	Cissus incisa	557
Chilopsis linearis	805	ssp. *lanceolatus*	907	*trifoliata*	558
Chimaphila umbellata	337	ssp. *viscidiflorus*	907	Cistaceae	253
occidentalis	337	Churchmouse three-awn	1136	Citrus Family	575
Chinese bush clover	458	Cicely, sweet	597	Clammy cudweed	942
Chinese chestnut	134	*Cicer arientinum*	416	ground cherry	644
Chinese foxtail	1219	*Cichoriaceae*	838	-weed	292
Chinese lantern	646	Cichorium intybus	908	Clapweed	77
Chinkapin oak	139	Cicuta	588	*Clary*	731

INDEX

Clasping peppergrass	317	red	485	diffusa	1047
Claspingleaf pondweed	1039	round-headed prairie-	443	erecta	1047
Claytonia	188	running buffalo	486	v. angustifolia	1048
robusta	188	silky prairie-	445	v. erecta	1048
virginica	188	slimleaf prairie-	444	f. erecta	1048
Clearweed	129	small bur-	465	f. intercursa	1048
Cleavers	817	small hop-	483	longicaulis	1047
Clematis	90	strawberry	484	virginica	1048
dioscoreifolia	92	sweet	466	Commelinaceae	1046
drummondii	91	water	70	Common balloon vine	567
fremontii	90	western water	70	Common balm	709
v. riehlii	90	white	485	Common bean	416
Fremont's	90	white prairie-	440	Common blackberry	401
hirsutissima	90, 91	white sweet	466	Common bladderwort	807
ligusticifolia	90	yellow sweet	466	Common buckthorn	556
maximowicziana	92	Clubmoss	39	Common burdock	864
missouriensis	92	Clubmoss Family	39	Common buttonbush	816
pitcheri	91	Clusiaceae	236	Common chickweed	213
Pitcher's	91	Cluster fescue	1181	Common catalpa	805
pseudoalpina	91	Clustered mallow	248	Common chaste-tree	708
tenuiloba	91	Cnemidophacos	421	Common dandelion	1010
terniflora	91	pectinatus	432	Common elderberry	826
virginiana	92	Cnidoscolus texanus	538	Common evening primrose	
western	90	Cobaea penstemon	779		517, 525
Cleome	291	Cocculus carolinus	110	Common flax	564
hassleriana	291	Cockle, corn	193	Common ground cherry	644
lutea	291	cow-	213	Common hemp-nettle	714
serrulata	292	sticky	208	Common hops	125
spinosa	291	white	208	Common horehound	720
yellow	291	Cocklebur	1019, 1020	Common juniper	72
Cleomella angustifolia	292	spiny	1020	Common lespedeza	460
eastern	292	Cockscombs	179	Common lilac	747
Cliff-brake	63	Cockspur hawthorn	372	Common lousewort	773
dwarf	64	Coeloglossum	1276	Common mallow	247, 248
purple-stemmed	63	bracteatum	1279	Common milkweed	628
smooth	63	Coelorachis cylindrica	1191	Common morning-glory	660
Wright's	64	Coffee-tree, Kentucky	415	Common mouse-ear	
Climbing buckwheat	228	Cogswellia	595	chickweed	199
Climbing milkweed	632, 633, 634	daucifolia	595	Common mullein	791
Climbing nightshade	649	foeniculacea	595	Common perilla	726
Climbing rose	400	macrocarpa	596	Common periwinkle	613
Clitoria mariana	438	orientalis	596	Common plantain	744
Cloak-fern	61	villosa	595	Common plum	390
false	61	Collinsia	763, 764	Common polypody	65
Fendler's	61	parviflora	763	Common privet	747
long	62	verna	763	Common purslane	189
star	62	violacea	764	Common ragweed	856
Closed gentian	607	Collomia linearis	667	Common reed	1208
Clover	481	Colorado rubber plant	963	Common scouring rush	44
alsike	484	Coltsfoot, sweet	983	Common sow thistle	1008
Beckwith's	483	Columbine	89	Common speedwell	795
buffalo	485	wild	89	Common stitchwort	212
bush	457	Comandra	531	Common St. John's-wort	238
Carolina	483	pallida	532	Common sunflower	954
Chinese bush	458	richardsiana	532	Common tansy	1010
crimson	484	umbellata	531	Common teasel	837
golden prairie-	440	ssp. pallida	532	Common vetch	489
ladino	485	ssp. umbellata	532	Common water hemlock	589
low hop	483	Comarum	382	Common witchgrass	1201
massive spike prairie-	441	Combleaf evening primrose	519	Common yellow monkey-	
nine-anther prairie-	441	Comfrey	684	flower	771
owl	772	northern wild	691	Compact stiffstem flax	563
Persian	486	Commelina	1047	Compass plant	1000
purple prairie-	444	communis	1047	Compositae	838
rabbit-foot	482	crispa	1048	Coneflower	919, 990

INDEX

grayhead prairie	990	bigflower	915	americana	144
prairie	989	*cardaminifolia*	917	v. *indehiscens*	144
purple	921	*crassifolia*	916	v. *missouriensis*	144
rough	991	finger	916	*avellana*	144
short-ray prairie	990	grandiflora	915	cornuta	144
sweet	992	v. *grandiflora*	916	*rostrata*	144
Conespur bladderwort	806	v. *harveyana*	916	Coryphantha	154
Conioselinum chinense	585	lanceolata	916	missouriensis	154
Conium maculatum	590	palmata	916	v. caespitosa	154
Conobea multifida	767	plains	916	v. missouriensis	154
Conophora	895	pubescens	916	vivipara	154
Consolida ambigua	93	tall	917	v. *radiosa*	155
Convallaria majalis	1248	tinctoria	916	v. *vivipara*	155
Conringia orientalis	308	tripteris	917	Costmary	902
Convolvulaceae	652	*Coriander*	585	*Cota tinctoria*	863
Convolvulus	654, 663	*Coriandrum sativum*	585	Cotton-batting	941
ambigens	655	Corispermum	173	Cottonflower	186
americanus	653	*emarginatum*	174	dense	186
arvensis	655	*hyssopifolium*	173	Cottongrass	1102
equitans	655	*marginata*	174	Cottonwood	273, 278
hermannioides	655	*nitidum*	174	lanceleaf	275
interior	653	*orientale*	174	narrowleaf	276
japonicus	654	*sibericum*	174	Cow-cockle	213
pellitus f. *anestitus*	654	*simplicissimum*	174	Cow parsnip	594
sepium	653	*villosum*	174	Cowherb	213
v. *fraterniflorus*	654	Corn chamomile	862	Cowlily	82
spithamaeus	654	Corn cockle	193	Coyote willow	287
Conyza	914	Corn-flower	899	Crab, Bechtel	396
canadensis	914	Corn salad	835	Iowa	396
v. *canadensis*	914	Corn speedwell	794	Crabapple	396
v. *glabrata*	914	Corn spurry	210	Crabgrass	1162
v. *pusilla*	915	Cornaceae	527	hairy	1163
ramosissima	915	Cornus	527	slender	1162
Cool-tankard	684	amomum	528	smooth	1163
Coontail	84	ssp. *amomum*	528	southern	1162
Cooper milk-vetch	431	ssp. *obliqua*	528	*Cracca leucosericea*	481
Cooperia drummondii	1248	canadensis	528	*virginiana*	481
Copperleaf, rhombic	537	drummondii	528	Crack willow	287
Coptidium	96	florida	529	Cranberry, highbush	831
Coralberry	827, 828	foemina	529	Cranesbill	580
white	827	ssp. *foemina*	530	Bicknell's	581
Corallorhiza	1270	ssp. *racemosa*	530	Carolina	581
maculata	1271	stolonifera	530	Richardson's	582
f. *flavida*	1271	Coronilla varia	438	small	582
f. *maculata*	1271	Correll's eriogonum	217	wild	581
ochroleuca	1271	*Corrigiolaceae*	192	viscid	582
odontorhiza	1271	Corydalis	114	Crassulaceae	356
striata	1271	aurea	115	Crataegus	370
v. *ochroleuca*	1272	ssp. *aurea*	115	*berberifolia*	372
v. *striata*	1272	v. *aurea*	115	calpodendron	371
v. *vreelandii*	1272	ssp. *occidentalis*	115	v. *globosa*	372
trifida	1272	v. *occidentalis*	115	v. *hispidula*	372
wisteriana	1272	campestris	117	v. *obesa*	372
Coral-root	1270	crystallina	116	chrysocarpa	374
crested	1279	curvisiliqua ssp.		coccinioides	372
late	1271	grandibracteata	116	*collina* v. *callicola*	374
pale	1272	flavula	116	v. *secta*	374
spotted	1271	golden	115	columbiana	374
striped	1271	mealy	116	crus-galli	372
Wister's	1272	micrantha	117	v. *barrettiana*	372
Cordgrass	1223	v. *australis*	117	v. *pyracanthifolia*	372
alkali	1223	v. *micrantha*	117	*dasyphylla*	372
prairie	1223	*montanum*	115	discolor	372
Coreopsis	915	pale	116	disjuncta	373
beggar-ticks	893	Corylus	143, 663	*douglasii*	370

INDEX

engelmannii	372	hoary	307	thyrsiflora	690
globosa	371	lens-padded hoary	307	torreyana	690
hannibalensis	372	mouse-ear	297	Cryptogams, vascular	39
lanuginosa	372	northern winter	303	*Cryptogramma acrostichoides*	49
lasiantha	373	rock	298, 300, 301, 302	*crispa* ssp. *acrostichoides*	49
macrantha v. *colorado*	374	smallflower bitter	306	*Cubelium concolor*	256
v. *occidentalis*	374	smooth rock	301	Cucumber, bur	269
v. *pertomentosa*	374	spring	306	creeping	269
mackenzii v. *bracteata*	373	winter	303	wild	268
mollis	373	yellow	325, 327	Cucumber Family	267
monogyna	370	yellow creeping	328	*Curcurbita foetidissima*	268
munita	372	yellow spreading	328	*perennis*	268
munsoniana	370	Cressa	656	Cucurbitaceae	267
palmeri	373	*depressa*	656	Cudweed	940
pruinosa	373	*truxillensis*	656	clammy	942
punctata	373	Crested beardtongue	780	diffuse	941
regalis v. *paradoxa*	372	Crested coral-root	1279	fragrant	941
reverchoni v. *discolor*	372	Crested shield fern	58	low	942
rotundifolia	374	Crested wheatgrass	1125	purple	942
stevensiana	372	Crimson clover	484	Culver's root	797
succulenta	374	Cristatella	293	*Cunila origanoides*	712
v. *occidentalis*	374	*jamesii*	293	Cup plant	1000
v. *pertomentosa*	374	*Crocanthemum bicknellii*	254	Cupgrass	1178
tantula	372	*Croptilon divaricatum*		prairie	1178
vallicola	372	v. *hookerianum*	950	Cuphea	496
viridis	375	v. *validus*	950	*petiolata*	496
v. *lanceolata*	375	Crossosoma Family	406	*viscosissima*	496
v. *lutensis*	375	Crossosomataceae	406	Cupressaceae	71
v. *ovata*	375	*Crotalaria sagittalis*	438	Cupseed	109
Creamy poison-vetch	434	Croton	538	stickseed	695
Creek plum	395	capitatus	539	Curly dock	232
Creeper, thicket	558	v. *capitatus*	539	Curly muckweed	1035
Virginia	558	v. *lindheimeri*	539	Curly-top gumweeed	944
Creeping bellflower	809	dioicus	539	Currant	352
Creeping blackberry	404	glandulosus	539	buffalo	355
Creeping cucumber	269	v. *lindheimeri*	539	swamp	354, 356
Creeping dayflower	1047	v. *septentrionalis*	539	western red	353
Creeping foxtail	1129	lindheimerianus	539	wild black	353
Creeping juniper	72	monanthogynus	540	Currant Family	352
Creeping ladies sorrel	579	one-seeded	540	Cursed crowfoot	105
Creeping lespedeza	459	pottsii	540	Cuscuta	661
Creeping polemonium	677	Texas	540	*cephalanthi*	662
Creeping yellow cress	328	texensis	540	*compacta*	662
Crepis	917	tropic	539	*coryli*	662
acuminata	918	woolly	539	*curta*	663
capillaris	918	*Crotonopsis elliptica*	540	*cuspidata*	663
glauca	919	Crowfoot	95	*epithymum*	663
glaucella	919	bristly	104	*glabrior*	664
intermedia	917	cursed	105	*glomerata*	664
modocensis	917	white water	103, 106	*gronovii*	664
nicaeensis	918	Crown vetch	438	*indecora*	664
occidentalis	918	Crownbeard	1015	*planifolia*	661
ssp. *costata*	918	golden	1016	*paradoxa*	664
perplexans	919	*Cryptotaenia canadensis*	590	*pentagona*	664
petiolata	919	Cruciferae	293	*polygonorum*	665
platyphylla	919	Cryptantha	688	*racemosa* v. *chiliana*	665
riparia	919	*bradburiana*	689	*squamata*	665
runcinata	919	*calycosa*	690	*suaveolens*	665
ssp. *glauca*	919	cana	688	*umbellata*	665
ssp. *runcinata*	919	celosioides	689	Cuscutaceae	661
tectorum	919	*confusa*	689	Cusp dodder	663
Cress, Austrian field	326	crassisepala	689	Custard-apple Family	77
bitter	306, 307	fendleri	689	Cutgrass, rice	1187
bog yellow	327	jamesii	689	Cutleaf germander	739
garden	318	minima	690	Cutleaf ground cherry	642

Cutleaf ironplant	949	tenuifolius	1097	James'	442
Cut-leaved evening primrose	521	uniflorus	1098	lanata	442
Cut-leaved grape-fern	46	*virens*	1096	*laxiflora*	441
Cut-leaved nightshade	651	Cypress, mock	175	leporina	443
Cut-leaved teasel	837	spurge	544	multiflora	443
Cyanococcus vacillans	336	standing	671	nana	443
Cyclachaena xanthifolia	965	summer	175	purpurea	444
Cyclanthera dissecta	268	vine	660	v. arenicola	444
Cycloloma atriplicifolium	174	Cypress Family	71	v. purpurea	444
Cylindric-fruited seedbox	514	Cypripedium	1272	silktop	440
Cymbalaria muralis	764	*album*	1274	tenuifolia	444
Cymbia occidentalis	966	× andrewsii	1273	*tenuis*	439
Cymopterus	590	calceolus	1273	villosa	445
acaulis	591	v. *calceolus*	1274	woolly	442
macrorhizus	591	v. *parviflorum*	1273	Dame's rocket	315
montanus	591	v. *planipetalum*	1274	*Damson plum*	390
Cynanchum	631	v. *pubescens*	1274	Dandelion	1010
laeve	632	candidum	1274	common	1010
nigrum	632	× calceolus v.		false	855, 988
Cynareae	844	parviflorum	1273	krigia	966
Cynoctonum mitreola	604	× *favillianum*	1274	red-seeded	1010
Cynodon dactylon	1154	*hirsutum*	1274	tuber false	989
transvaalensis	1154	reginae	1274	Danthonia	1155
Cynoglossum	690	Cystopteris	56	*candensis*	1155
boreale	691	bulbifera	56	*intermedia*	1155
officinale	691	fragilis	57	spicata	1155
virginianum	691	v. *dickieana*	57	*thermale*	1155
Cynomaranthum	595	v. *fragilis*	57	*unispicata*	1155
nuttallii	596	v. *protrusa*	58	*Dasiphora*	382
Cynosciadium pinnatum	595	v. *simulans*	58	*fruticosa*	385
Cynoxylon floridum	529	f. *simulans*	58	Dasistoma macrophylla	764
Cynthia dandelion	966	v. *tennesseensis*	58	*Dasystephana affinis*	606
Cyperaceae	1059	protrusa	57	*andrewsii*	607
Cyperus	1093	tennesseensis	58	*flavida*	607
acuminatus	1094	*Cytheria bulbosa*	1270	*puberula*	607
aristatus	1094			Datil	1265
bushii	1096	Dactylis glomerata	1154	Datura	638
diandrus	1094	Daisy, black-foot	979	innoxia	639
engelmannii	1094	Easter	1013	*metel*	639
erythrorhizos	1095	Engelmann's	923	*meteloides*	639
esculentus	1095	English	889	quercifolia	639
ferruginescens	1096	fleabane	932	stramonium	639
filiculmis	1096	lazy	863	v. *tatula*	640
v. *macilentus*	1096	ox-eye	902	*tatula*	640
flavescens v. *poaeformis*	1096	sleepy	1019	*wrightii*	639
fuscus	1095	Tahoka	977	*Daucophyllum*	596
hallii	1097	western	886	*tenuifolium*	597
houghtonii	1097	Dakota vervain	705	Daucus	591
inflexus	1094	Dalea	439	carota	592
lancastriensis	1095	*alopecuroides*	443	pusillus	592
lupulinus	1095	aurea	440	Dayflower	1047
ssp. *lupulinus*	1096	black	442	creeping	1047
ssp. *macilentus*	1096	candida	440	erect	1047
niger v. *castaneous*	1096	v. *candida*	440	Daylily	1251
odoratus	1096	v. *oligophylla*	441	Dead nettle	716
ovularis v. *cylindricus*	1096	*compacta* v. *pubescens*	439	purple	717
v. *sphaericus*	1096	cylindriceps	441	Death camass	1257, 1258
pseudovegetus	1096	dwarf	443	Deciduous holly	535
rivularis	1096	enneandra	441	Deer pea vetch	488
rotundus	1096	v. *pumila*	441	Deer vetch	461
schweinitzii	1097	formosa	442, 534	Deerberry	336
setigerus	1097	foxtail	443	Deertongue dichanthelium	1158
speciosus	1096	frutescens	442	Delphinium	92
strigosus	1097	hare's-foot	443	ajacis	93
surinamensis	1097	jamesii	442	bicolor	93

INDEX

carolinianum	94	Devil's head	155	*virginiana*	817
v. *crispum*	94	Devil-weed, Mexican	885	Dioscorea	1267
geyeri	94	Dewberries	400	*batatas*	1268
menziessii	93	Dewberry, northern	402	*paniculata*	1268
nelsonii	93	southern	404	*villosa*	1267
nuttallianum	93	Diamond willow	286	Dioscoreaceae	1267
penardii	94	*Dianthera americana*	801	Diospyros virginiana	342
tricorne	94	American	801	v. *platycarpa*	342
virescens	94	Dianthus armeria	199	v. *pubescens*	342
ssp. *penardii*	94	*Diapteria prolifera*	937	*Diplachne*	1188
ssp. *wootonii*	94	Diarrhena americana		*acuminata*	1188
wootonii	94	v. *obovata*	1156	*fascicularis*	1188
Dense cottonflower	186	*arundinacea*	1156	Diplotaxis muralis	310
Denslovia	1276	Dichanthelium	1156	Dipsacaceae	836
clavellata	1277	acuminatum	1157	Dipsacus	836
Dentaria laciniata	306	v. acuminatum	1158	*fullonum*	837
Deptford pink	199	v. implicatum	1158	*laciniatus*	837
Deschampsia cespitosa	1155	v. lindheimeri	1158	*sylvestris*	837
v. *cespitosa*	1156	v. villosum	1158	*Disella*	249
Descurainia	308	*boscii*	1159	*hederacea*	249
intermedia	309	clandestinum	1158	Disporum trachycarpum	1249
magna	309	deertongue	1158	Distichlis spicata v. stricta	1163
pinnata	309	*depauperatum*	1160	*stricta*	1164
ssp. brachycarpa	309	latifolium	1159	*Ditaxis humilis*	537
v. *brachycarpa*	309	Leiberg	1159	*mercurialina*	538
ssp. halictorum	309	leibergii	1159	Ditch stonecrop	357
ssp. intermedia	309	linearifolium	1159	Ditchgrass	1040
v. *intermedia*	309	malacophyllum	1160	Ditchgrass Family	1040
v. *osmiarum*	309	oligosanthes	1160	*Dithyrea wislizenii* v. *palmeri*	310
richardsonii	309	v. *oligosanthes*	1160	Dittany	712
sophia	310	v. *scribnerianum*	1160	Dock	230
Desert seepweed	178	v. *wilcoxianum*	1161	bitter	233
Desert sumac	573	scoparium	1160	curly	232
Desert willow	805	slimleaf	1159	golden	233
Desmanthus	408	sphaerocarpon	1160	great water	234
cooleyi	408	velvet	1160	pale	232
illinoensis	408	Wilcox	1161	patience	234
leptolobus	409	wilcoxianum	1161	water	235
Desmodium	445	*xanthophysum*	1159	western	234
acuminatum	448	Dicentra	117	willow-leaved	233
bracteosum	448	*canadensis*	118	yard	232
longifolium	448	cucullaria	117	Dodder	661
canadense	446	f. *purpuritineta*	118	buttonbush	662
canescens	447	Dichromena nivea	1098	cusp	663
hirsutum	447	Dicliptera brachiata	800	field	664
ciliare	447	Dicots	77	Gronovius'	664
cuspidatum	447	Didiplis diandra	496	hazel	662
v. *longifolium*	448	Diffuse cudweed	941	large alfalfa	664
dillenii	450	Diffuse eryngo	593	smartweed	665
glabellum	450	Diffuse knapweed	899	Dodder Family	661
glutinosum	448	Digitaria	1162	Dodecatheon	345
illinoense	448	*adscendens*	1162	*amethystinum*	346
marilandicum	449	ciliaris	1162	meadia	346
nudiflorum	449	filiformis	1162	v. *brachycarpum*	346
obtusum	450	v. *villosa*	1163	v. *meadia*	346
paniculatum	450	ischaemum	1163	*pauciflorum*	346
v. dillenii	450	v. *mississippiensis*	1163	pulchellum	346
v. paniculatum	450	sanguinalis	1163	*radicatum*	346
pauciflorum	451	v. *ciliaris*	1162	*salinum*	346
perplexum	450	*Diholcos*	421	*thornense*	346
rigidum	450	*bisulcatus*	425	*Doellingeria pubens*	883
rotundifolium	451	Dill	587	Dog fennel	862
sessilifolium	451	Dimorphocarpa palmeri	310	Dog mustard	313
tweedyi	449	Diodia teres	817	Dog parsley	596
Devil's claw	803	v. *setifera*	817	*Dog rose*	397

Dogbane	612	rough	1225	Earleaf gerardia	790
Indian hemp	612	sand	1226	Early buttercup	99
prairie	612	Texas	1228	Early meadow rue	107
spreading	612	whorled	1228	Early wood buttercup	97
Dogbane Family	610	Drosera	252	Easter daisy	1013
Dogberry	354	*annua*	253	Eastern camass	1248
Dog's-tooth-violet	1249	*brevifolia*	253	Eastern cleomella	292
white	1249	*rotundifolia*	253	Eastern mistletoe	533
yellow	1250	Droseraceae	252	Eastern gammagrass	1234
Dogwood, flowering	529	Drummond false pennyroyal	715	Eastern poison oak	575
gray	529	Drummond milk-vetch	427	Eastern prickly pear	157
pale	528	Drummond's aster	876	Eaton's lip fern	54
rough-leaved	528	Drummond's skullcap	734	Ebenaceae	342
Dogwood Family	527	Drummond thistle	911	Ebony Family	342
Dollarleaf	451	*Drymocallis*	382	Ebony spleenwort	51
Dotted beebalm	725	*agrimonioides*	384	Echinacea	921
Double bladderpod	324	*fissa*	385	*angustifolia*	921
Downy agrimony	367	*glandulosa*	386	v. *strigosa*	922
Downy arrow-wood	832	*pseudorupestris*	386	*atrorubens*	922
Downy blue violet	264	Dryopteris	58	*pale*	922
Downy brome	1147	*austriaca* v. *spinulosa*	60	*pallida*	922
Downy gentian	607	*carthusiana*	60	v. *angustifolia*	921
Downy goldenrod	1005	*cristata*	58	*purpurea*	922
Downy ground cherry	644	× *spinulosa*	59	Echinocactus	155
Downy hawthorn	373	*disjuncta*	60	*simpsonii*	160
Downy milk-pea	452	*filix-mas*	59	*texensis*	155
Downy paintbrush	762	*goldiana*	59	Echinocereus	155
Downy yellow violet	262	*hexagonoptera*	67	*baileyi*	156
Draba	311	*marginalis*	59	*caespitosus*	155
aurea	311	*spinulosa*	60	*reichenbachii*	155
aureiformis	311	*thelypteris* v. *pubescens*	67	v. *albispinus*	156
brachycarpa	311	*Duchesnea indica*	365	v. *perbellus*	156
cana	312	Duckmeat	1045	v. *reichenbachii*	156
caroliniana	312	Duckweed	1044	*viridiflorus*	156
coloradensis	312	greater	1045	Echinochloa	1164
cuneifolia	312	star	1045	*colonum*	1165
lanceolata	312	Duckweed Family	1043	*crusgalli*	1164
lutea	312	Dudley rush	1054	*crus-pavonis* v. *macera*	1165
micrantha	312	*Dulchium arundinaceum*	1098	*microstachya*	1165
milk-vetch	435	Dutchman's breeches	117, 576	*muricata*	1165
nemorosa	312	Dwarf blackberry	404	v. *microstachya*	1165
reptans	312	Dwarf cherry	394	v. *muricata*	1165
v. *micrantha*	312	Dwarf chinkapin oak	140	*occidentalis*	1165
v. *reptans*	312	Dwarf cliff-brake	64	*pungens*	1166
surculifera	311	Dwarf dalea	443	*zelayensis*	1165
shortpod	311	Dwarf hackberry	121	Echinocystis lobata	268
stenoloba	312	Dwarf juniper	72	Echinodorus	1023
wedgeleaf	312	Dwarf larkspur	94	*berteroi*	1024
Dracocephalum	713, 727	Dwarf locoweed	469, 470	*cordifolius*	1024
denticulatum	728	Dwarf milkwort	621, 629	*parvulus*	1024
formosius	728	Dwarf pussy-toes	860	*radicans*	1024
moldavica	713	Dwarf rattlesnake plantain	1276	*rostratus*	1024
nuttallii	728	Dwarf sagebrush	869	v. *lanceolatus*	1024
parviflorum	713	Dwarf scouring rush	45	*tenellus*	1024
speciosum	728	Dwarf snapdragon	762	*Echium vulgare*	691
thymiflorum	713	Dwarf St. John's-wort	238	Eclipta	922
virginianum	728	Dwarf sumac	572	*alba*	923
Dracopis amplexicaulis	919	Dwarf white trillium	1255	*prostrata*	922
Dragonhead	713	Dwarf wild indigo	419	Eelgrass	1030
false	727	*Dyschoriste linearis*	800	Eggleaf skullcap	735
Dragonroot	1043	Dyssodia	920	Elaeagnaceae	490
Drooping woodreed	1154	*acerosa*	920	Eleagnus	490
Dropseed	1224	*aurea*	920	*angustifolia*	490
giant	1227	*papposa*	920	*commutata*	491
prairie	1227	*tagetoides*	921	Elatinaceae	236

Elatine		236	× *Elyhordeum dakotense*		1185	Torrey		77
americana		236	*iowense*		1185	torreyana		77
brachysperma		236	*montanense*		1185	Ephedra Family		76
triandra		236	Elymus		1166	Ephedraceae		76
Elbow bush		748	*arkansanus*		1170	Epilobium		504
Elderberry		826	*australis*		1170	*adenocladon*		508
common		826	*brachystachys*		1167	*alpinum*		504
stinking		827	*canadensis*		1167	*americanum*		505
Elder, box		569	v. *brachystachys*		1167	*anagallidifolium*		504
marsh	964,	965	f. *glaucifolius*		1167	angustifolium ssp.		
red-berried		827	v. *robustus*		1167	circumvagum		505
Eleocharis		1098	v. *wiegandii*		1168	ciliatum		505
acicularis		1099	*cinereus*		1168	ssp. ciliatum		505
v. *gracillescens*		1099	*condensatus*		1168	ssp. glandulosum		506
acuminata		1100	*curvatus*		1170	coloratum		506
arenicola		1101	diversiglumis		1168	*drummondii*		509
atropurpurea		1099	*flavescens*		1169	halleanum		506
calva		1100	*glabriflorus*		1170	hornemannii		507
caribaea		1099	glaucus		1168	*latifolium*		504
compressa		1100	*hirsutiglumis*		1170	leptophyllum		507
engelmannii		1101	*hystrix*		1186	*occidentale*		506
erythropoda		1100	v. *bigelovianus*		1186	palustre		507
geniculata		1100	innovatus		1169	paniculatum		508
intermedia		1100	*interruptus*		1168	saximontanum		508
lanceolata		1100	*jejunus*		1170	strictum		504
macrostachya		1100	junceus		1169	× *wisconsinensis*		506
mamillata		1100	*longifolius*		1221	*wyomingense*		508
montevidensis		1101	*macounii*		1123	Epipactis gigantea		1275
obtusa		1101	*philadelphicus*		1167	Equisetaceae		42
v. obtusa		1101	*piperi*		1168	Equisetum		42
v. ovata		1101	*robustus*		1167	*affine*		44
ovata		1101	simplex		1169	arvense		43
parvula		1101	*striatus*		1170	f. *boreale*		43
v. anachaeta		1101	*triticoides*		1126	f. *diffusum*		43
v. parvula		1101	villosus		1169	f. *ramulosum*		43
pauciflora		1101	f. *arkansana*		1170	f. *varium*		43
v. *fernaldii*		1101	v. *arkansanus*		1170	× ferrissii		43
pygmaea		1101	virginicus		1170	fluviatile		43
quadrangulata		1101	v. australis		1170	f. *linnaeanum*		43
v. *crassior*		1102	f. *australis*		1170	hyemale		44
rostellata		1102	v. glabriflorus		1170	v. *affine*		44
smallii		1102	v. *hirsutiglumis*		1170	v. *elatum*		44
uniglumis		1100	v. *intermedius*		1170	v. *intermedium*		43
verrucosa		1102	v. *jejunus*		1170	v. *pseudohyemale*		44
wolfii		1102	v. submuticus		1170	*intermedium*		43
xyridiformis		1102	v. virginicus		1170	kansanum		44
Elephantopus carolinianus		923	*vulpinus*		1166	f. *ramosum*		44
Elephant's foot		923	*wiegandii*		1168	laevigatum		44
Eleusine indica		1166	Echanter's nightshade		503	ssp. *funstonii*		44
Ellisia nyctelea		678	*Endolepis dioica*		163	palustre		44
Elm		121	*suckleyi*		163	*praealtum*		44
American		122	*Enemion biternatum*		95	pratense		44
-leaved goldenrod		1007	Engelmannia pinnatifida		923	*robustum*		44
Siberian		122	Engelmann's daisy		923	scirpoides		45
rock		123	Engelmann's evening			sylvaticum		45
slippery		123	primrose		519	variegatum		45
winged		122	Engelmann's goldenweed		948	Eragrostis		1171
Elm Family		119	Engelmann's milkweed		619	*arida*	1174,	1176
Elodea		1029	English daisy		889	barrelieri		1172
canadensis		1029	English oak		137	*beyrichii*		1177
densa		1029	English plantain		744	capillaris		1173
longivaginata		1030	*Enneapogon desvauxii*		1114	cilianensis		1173
nuttallii		1030	Ephedra		76	curtipedicellata		1173
Elodeaceae		1028	antisyphilitica		77	curvula		1174
Eltrot		594	*coryi*		76	*diffusa*		1176

frankii		1174	minor		930	v. pauciflorum	219
hirsuta		1175	modestus		930	shortstem	216
hypnoides		1175	montanensis		931	tenellum	219
intermedia		1175	ochroleucus		930	trichopes	219
megastachya		1173	v. ochroleucus		931	visheri	219
mexicana		1174	v. scribneri		931	Visher's	219
minor		1175	oligodontus		930	winged	216
neomexicana		1174	oxydontus		930	*Erioneuron pilosum*	1233
oxylepis		1177	philadelphicus		931	Eriophorum	1102
pectinacea	1176,	1178	pulchellus		931	*angustifolium*	1103
perplexa		1176	pulcherrimus		931	v. *majus*	1103
pilifera		1178	v. *pulcherrimus*		932	chamissonis	1103
pilosa		1176	v. wyomingia		931	gracile	1103
v. perplexa		1176	pumilis		932	polystachion	1103
v. pilosa		1177	*purpureus*		931	*virginicum*	1103
poaeoides		1176	ramosus		932	viridicarinatum	1103
purshii		1176	speciosus v. macranthus		932	Erodium	580
reptans		1177	v. *speciosus*		932	cicutarium	580
secundiflora ssp.			strigosus		932	texanum	580
oxylepis		1177	v. *beyrichii*		932	Eruca sativa	313
sessilispica		1177	v. *strigosus*		932	*versicaria* ssp. *sativa*	313
spectabilis		1178	subcanescens		927	Erucastrum gallicum	313
v. *sparsihirsuta*		1178	subtrinervis		933	*pollichii*	313
tephrosanthos	1174,	1176	*tardus*		930	Eryngium	592
trichodes		1178	tenuis		933	diffusum	593
v. *pilifera*		1178	trifidus		928	leavenworthii	593
Erechtites hieracifolia		923	vetensis		933	*planum*	593
Erect dayflower		1047	Eriochloa		1178	*prostratum*	593
Erect knotweed		223	contracta		1178	yuccifolium	593
Eremogone		193	*gracile*		1179	v. *synchaetum*	593
Erianthus ravennae		1114	lemmonii		1179	Eryngo	592
Ericaceae		334	punctata		1179	diffuse	593
Erigenia bulbosa		592	sericea		1179	Leavenworth	593
Erigeron		924	*Ericoma*		1198	Erysimum	313
abruptorum		930	hymenoides		1199	asperum	314
acer		927	Eriogonum		215	capitatum	314
acris v. asteroides		926	alatum		216	cheiranthoides	314
angulosus v.			v. *glabriusculum*		216	inconspicuum	314
kamtschaticus		927	annual		216	repandum	315
annuus		927	annuum		216	Erythronium	1249
asper		930	brevicaule		216	albidum	1249
bellidiastrum		927	*campanulatum*		216	v. *mesochoreum*	1250
v. *robustus*		927	cernuum		216	mesochoreum	1249
caespitosus		927	correllii		217	rostratum	1250
canadensis		914	Correll's		217	*Eulophus americanus*	599
canus		928	effusum		217	Euonymus	534
compositus		928	v. effusum		217	*americanus*	535
v. *discoideus*		928	ssp. *fendlerianum*		217	atropurpureus	534
v. *glabratus*		928	v. *rosmarinoides*		217	Eupatorieae	844
corymbosus		928	*fendlerianum*		217	Eupatorium	933
divaricatus		915	flavum		217	altissimum	934
divergens v. *cinereus*			v. *crassifolium*		218	coelestinum	934
928,	929,	930	gordonii		218	maculatum v. bruneri	934
droebachensis		927	Gordon's		218	perfoliatum	934
drummondii		930	helichrysoides		217	purpureum	935
flagellaris		929	jamesii	217,	218	rugosum	935
formosissimus		929	lachnogynum		218	serotinum	935
v. *formosissimus*		929	Lindheimer's longleaf		218	Euphorbia	541
v. *viscidus*		929	longifolium v.			agraria	543
glabellus		929	lindheimeri		218	albomarginata	543
ssp. *glabellus*		930	*microthecum*		217	carunculata	543
ssp. pubescens		929	pauciflorum		218	*chamaesyce*	549
jucundus		927	v. canum		219	commutata	550
laetivirens		931	v. gnaphalodes		219	corollata	543
lonchophyllus		930	v. nebraskense		219	v. *mollis*	544

INDEX

cyathophora	544	Evening primrose	526	False nettle	128
cyparissias	544	Berlandier	500	False nightshade, green	638
dentata	544	combleaf	519	False pennyroyal	740
f. *cuphusperma*	545	common	517, 525	American	716
dictyosperma	550	cut-leaved	521	Drummond	715
esula	545	large-flowered cut-leaved	520	Reverchon	716
fendleri	545	Engelmann's	519	rough	715
Fendler's	545	floating	515	False pimpernel	769
geyeri	545	fourpoint	524	False rue anemone	94
glyptosperma	546	Fremont's	523	False salsify	992
heterophylla	544	gumbo	518	False Solomon's seal	1254
v. *graminifolia*	544	Hartweg	501	False spikenard	1254
hexagona	546	hoary	523	False sunflower	952, 958
hoary	546	Missouri	523	Fameflower	189, 190
humistrata	546	narrow-leaved	522	prairie	190
lata	546	Oklahoma	523	Fanleaf vervain	706
longicuris	547	pale	517, 522	Fanwort	83
maculata	547, 548	showy white	524	Feather plume	442
marginata	547	spotted	518	Fendler three-awn	1139
missurica	548	stemless	525	Fendler's aster	877
v. *calcicola*	548	white-stemmed	524	Fendler's cloak-fern	61
v. *intermedia*	548	Evening Primrose Family	498	Fendler's euphorbia	545
nutans	548	Everlasting	859, 940	Fendler's lip fern	55
painted	544	fragrant	941	*Fennel*	585
peplus	541	pea	455	Fennel, dog	862
podperae	549	pearly	858	Fern, Alabama lip	54
preslii	548	Evolvulus	656	beaded lip	56
prostrata	548	*argentus*	656	beech	67
× pseudovirgata	549	nuttallianus	656	broad beech	67
revoluta	549	Nuttall's	656	bulblet bladder	56
robusta	549	*pilosus*	656	Christmas	65
serpens	549	Eyebane	548	crested shield	58
serpyllifolia	550			Eaton's lip	54
spathulata	550	Fabaceae	416, 663, 665	Fendler's lip	55
stictospora	550	Fagaceae	134	fragile	56, 57
strictior	550	Fagopyrum esculentum	219	glade	53
supina	547	*Fagus*	135	Goldie's	59
uralensis	551	Fairy slipper orchid	1270	hairy lip	55
virgata	545, 549	Fairybells	1249	holly	65
Euphorbiaceae	535	*Falcaria sioides*	585	Lindheimer's lip	55
Euploca	693	Fall panicum	1201	lip	54
convolulacea	694	Fall phlox	676	lowland fragile	57
European hornbeam	141	Fall witchgrass	1188	maidenhair	50, 51
Eurotia lanata	166	Fallugia paradoxa	375	male	59
Eurystemon mexicanum	1240	False alyssum, hoary	303	marginal shield	59
Eurytaenia texana	594	False boneset	966	marsh	67
Eustoma	605	False broomweed	946	mosquito	71
grandiflorum	605	False buckwheat	229	oak	60
f. *bicolor*	606	False buffalo grass	1197	ostrich	60
f. *fisheri*	606	False chamomile	978	rattlesnake	48
f. *flaviflorum*	606	False cloak-fern	61	resurrection	64
f. *grandiflorum*	606	False dandelion	855, 988	royal	48
f. *roseum*	606	tuber	989	sensitive	62
russellianum	606	False dragonhead	727	slender lip	55
Euthamia	935	False flax	305	spinulose wood	60
camporum	936	small-seeded	305	sword	66
graminifolia	936	False foxglove	757	Tennessee bladder	58
v. *graminifolia*	936	False garlic	1253	Venus'-hair	50
v. *major*	936	False gromwell	700	walking	52
gymnospermoides	936	False indigo	419, 436	water	71
occidentalis	936	blue	436	wood	58
viscid	936	False loosestrife	513	woolly lip	56
Evax	937	False mallow	251	Fescue	1179
multicaulis	937	red	251	cluster	1181
prolifera	937	False melic	1217	meadow	1181

nodding		1180	*drummondii*		1105	Forked catchfly	206
rough		1181	*interior*		1105	Forked chickweed	202, 203
sheep's		1180	*mucronulata*		1104	Forked scale-seed	601
sixweeks		1180	puberula		1104	Forked spleenwort	52
tall		1180	v. interior		1105	Forktip three-awn	1136
Festuca		1179	v. puberula		1105	*Forsellesia planitierum*	407
arundinacea		1180	*spadicea*		1105	*Forsythia viridissima*	747
campestris		1182	vahlii		1105	Four-o'clock	147
elatior		1181	Fineleaf gerardia		790	Carleton	148
idahoensis		1181	Finger coreopsis		916	hairy	149
myuros		1180	Finger poppy mallow		244	narrowleaf	150
nutans		1180	Fire-bush, Mexican		175	smooth	149
obtusa		1180	Fire-on-the-mountain		544	spreading	151
octoflora		1180	Fireweed	175, 504,	923	tall	148
ovina		1180	Five-hook bassia		165	trailing	146
v. *duriuscula*		1181	Flag		1259	white	147
v. *rydbergii*		1180	blue	1259,	1260	wild	150
paradoxa		1181	western blue		1260	Four-O'Clock Family	145
pratensis	1181,	1190	Flatstem pondweed		1039	Four-wing saltbush	162
rubra		1181	Flaveria campestris		937	Fourleaf milkweed	626
saximontana		1181	Flax		561	Fourpoint evening primrose	524
scabrella		1181	Berlandier's		563	Fowl bluegrass	1213
sciurea		1180	blue		562	Fowl mannagrass	1183
shortii		1181	broom		562	Foxglove, false	757
subulata		1181	common		564	mullein	764
versuta		1182	compact stiffstem		563	Foxtail	1129
Fetid marigold		920	false		305	barley	1185
Feverfew		902	grooved		564	bristly	1220
Feverwort		829	Norton's		563	Carolina	1130
Few-flowered aster		882	plains		563	Chinese	1219
Few-flowered tickclover		451	small-seeded false		305	creeping	1129
Fiddleneck		686	spurge-		498	dalea	443
Field bindweed	654,	655	stiffstem		564	green	1221
Field brome		1145	sucker		562	marsh	1130
Field cress, Austrian		326	Flax Family		561	meadow	1130
Field dodder		664	Fleabane		924	millet	1220
Field horsetail		43	annual		927	short-awn	1129
Field madder		817	daisy		932	yellow	1219
Field milk-vetch		424	marsh-		985	Fragaria	375
Field mint		721	spreading		915	americana	376
Field paspalum		1205	western		927	× *ananassa*	376
Field pennycress		332	Fleshy stitchwort		212	*chiloensis*	376
Field peppergrass		317	Flixweed		310	*glauca*	376
Field poppy		113	Floating evening primrose		515	*grayana*	376
Field pussy-toes		860	Floatingleaf pondweed		1037	*pauciflora*	376
Field scabious		837	Flodman's thistle		911	*pumila*	376
Field snake-cotton		185	Florida lettuce		969	*vesca*	376
Field sow thistle		1008	Florida paspalum		1204	v. *americana*	376
Field speedwell		793	Flower-of-an-hour		246	v. *vesca*	376
Field thistle		909	Flowering dogwood		529	*virginiana*	376
Fig-marigold Family		152	Flowering plants		77	v. *glauca*	376
Figwort		789	Flowering rush		1021	v. *illinoensis*	376
Figwort Family		751	Flowering Rush Family		1021	Fragile fern	56, 57
Filaria		580	Flowering spurge		543	lowland	57
Filix bulbifera		57	*Fluminea festucacea*		1218	Fragrant cudweed	941
fragilis		57	*Foeniculum vulgare*		585	Fragrant everlasting	941
Fimbristylis		1104	Fog-fruit		702	Fragrant sumac	571
annua		1104	northern		703	Fragrant white waterlily	83
autumnalis		1104	wedgeleaf		702	*Franseria acanthicarpa*	856
v. *mucronulata*		1104	Forest muhly		1197	*discolor*	858
baldwiniana		1104	Forestiera		747	*tenuifolia*	857
caroliniana		1105	acuminata		748	*tomentosa*	857
castanea		1105	pubescens		748	Fraxinus	748
v. *puberula*		1105	Forget-me-not		699	americana	749
dichotoma		1104	garden		700	*campestris*	750

INDEX

lanceolata	750	*esula*	545	hairy		512
nigra	749	*missouriensis*	550	large-flowered		510
pennsylvanica	750	*obtusatus*	550	longiflora		510
v. *austinii*	750	*robustus*	549	*michauxii*		510
v. *campestris*	750	Galearis spectabilis	1275	neomexicana ssp.		
v. *lanceolata*	750	Galeopsis bifida	714	coloradensis		511
v. *subintegerrima*	750	*tetrahit*	714	*parviflora*		510
quadrangulata	750	v. *bifida*	714	f. *glabra*		511
Fremont goosefoot	170	Galeorchis spectabilis	1275	v. *lachnocarpa*		511
Fremont's clematis	90	Galinsoga	939	*pitcheri*		511
Fremont's evening primrose	523	*ciliata*	940	scarlet		510
Fringed brome	1143	*parviflora*	940	sinuata		511
Fringed gentian	608	*quadriradiata*	940	sinuate-leaved		511
Fringed loosestrife	348	Galium	817	suffulta		512
Fringed orchis, prairie	1278	*aparine*	818	v. *suffulta*		512
Fringed quickweed	940	v. *echinospermum*	818	triangulata		512
Fringeleaf ruellia	802	v. *vaillantii*	818	velvety		511
Fritillaria	1250	*asprellum*	819	villosa		512
atropurpurea	1250	*boreale*	818	v. *arenicola*		513
pudica	1250	v. *linearifolium*	818	ssp. *villosa*		513
Fritillary	1250	*circaezans*	819	Gaurella		516
Froelichia	184	v. *circaezans*	819	*canescens*		519
campestris	185	v. *hypomalacum*	819	Gay-feather		972
floridana	185	*claytoni*	820	Gayophytum		513
v. *campestris*	185	*concinnum*	819	diffusum ssp.		
v. *floridana*	185	*labradoricum*	819	parviflorum		513
gracilis	185	*obtusum*	819	*humile*		513
Frog-fruit	702	v. *obtusum*	819	*nuttallii*		513
Frog orchid	1279	v. *ramosum*	819	*racemosum*		513
Frog's-bit Family	1028	*pilosum*	820	Gentian		606
Frosty hawthorn	373	*subbiflorum*	820	bottle		607
Frostweed	253, 1016	*texense*	820	catchfly		605
Fuirena simplex	1105	*tinctorium*	820	closed		607
v. *aristulata*	1105	*trifidum*	820	downy		607
v. *simplex*	1105	*triflorum*	820	fringed		608
Fumaria	118	*vaillantii*	818	green		609
officinalis	118	*verum*	820	horse-		829
vaillentii	118	*virgatum*	821	northern		606
Fumariaceae	114	v. *leiocarpum*	821	prairie	605,	607
Fumewort	114	Galleta	1184	prairie rose		609
slender	117	Galpinsia	500	spurred		608
Fumitory	118	*fendleri*	501	yellow-flowered horse-		829
Fumitory Family	114	*interior*	502	Gentian Family		604
Fuzzyspike wild rye	1169	*lavandulaefolia*	502	Gentiana		606
		Gambel's oak	138	*affinis*		606
Gaillardia	937	Gammagrass	1234	*alba*		606
aestivalis	938	eastern	1234	*amarella*		607
aristata	938	Garden cress	318	*andrewsii*		607
drummondii	939	Garden forget-me-not	700	× *billingtonii*		606
fastigiata	938	Garden pea	416	*crinita*		608
lanceolata	938	Garlic	1245	× *curtisii*		606
pinnatifida	939	false	1253	*flavida*		607
prairie	938	mustard	296	× *pallidocyanea*		606
pulchella	939	wild	1246	*procera*		608
rose-ring	939	Garrya ovata ssp.		*puberula*		607
suavis	939	goldmannii	530	puberulenta		607
Galactia	452	Garryaceae	530	Gentianaceae		604
mississippiensis	452	Gaura	509, 512	Gentianella		607
regularis	452	*biennis*	509	amerella ssp. acuta		607
volubilis	452	*coccinea*	510	crinita ssp. *procera*		608
v. *mississippiensis*	452	v. *coccinea*	510	quinquefolia ssp.		
v. *volubilis*	452	v. *glabra*	510	occidentalis		607
Galarrhoeus	541	v. *parviflora*	510	*tenella*		607
arkansanus	550	*filipes*	510	Gentianopsis		608
cyparissias	544	*glabra*	510	crinita		608

INDEX

procera	608	Giant dropseed	1227	Goat's rue	480
Geoprumnon	421	Giant hyssop, purple	711	Goatgrass	1122
crassicarpum	427	catnip	711	jointed	1122
plattense	433	Giant ragweed	858	Golden Alexanders	603
succulentum	427	Gilia	668	Golden aster	903, 905
trichocalyx	427	*acerosa*	668	soft	904
Geraniaceae	580	*acerosum*	668	*Golden bells*	747
Geranium	580	*calcarea*	668	Golden corydalis	115
bicknellii	581	*congesta*	670	Golden crownbeard	1016
carolinianum	581	*coronopifolia*	671	Golden dock	233
v. *confertiflorum*	581	*iberidifolia*	670	Golden eye	1018
maculatum	581	*laxiflora*	670	Golden glow	991
mint	902	*longiflora*	670	Golden pert	765
pusillum	582	*mcvickerae*	668	Golden prairie-clover	440
richardsonii	582	penstemon	776	Golden selenia	329
robertianum	582	pinnatifida	668	Goldenrod	1000
sphaerospermum	581	*pumila*	670	broad-leaved	1003
viscosissimum	582	rigidula	668	Canada	1002
v. *nervosum*	582	ssp. *acerosa*	668	downy	1005
Geranium Family	580	ssp. *rigidula*	668	elm-leaved	1007
Gerardia	753	*rubra*	671	gray	1004
aspera	754	*spicata*	671	late	1003
auriculata	790	*Gillenia stipulata*	381	prairie	1004
big-flower	757	Ginger, wild	81	rigid	1006
crustata	755	Ginseng	584	showy-wand	1007
densiflora	790	Ginseng Family	583	soft	1004
earleaf	790	Glade fern	53	three-nerved	1006
fasciculata	754	*Glandularia bipinnatifida*	705	Goldenweed	946, 948, 976
f. *albiflora*	754	*canadensis*	706	Englemann's	948
fineleaf	790	*pumila*	707	slender	950
gattingeri	755	Glaucium corniculatum	112	Goldie's fern	59
grandiflora v. *cinerea*	757	Glaux maritima	347	Gold-of-pleasure	305
green	756	Glecoma hederacea	714	*Gonolobus baldwynianus*	633
heterophylla	755	v. *micrantha*	714	*decipiens*	635
purpurea	755	Gleditsia triacanthos	414	*gonocarpos*	635
shinneriana	755	f. *inermis*	415	*laevis*	632
tenuifolia	756	Glinus lotoides	191	*suberosus*	635
v. *macrophylla*	756	Globe-berry	268	Goodyera	1275
v. *parviflora*	756	Globe mallow	251	*decipiens*	1276
viridis	756	narrowleaf	251	*oblongifolia*	1276
Germander	738	Glossopetalon planitierum	406	*ophioides*	1276
American	739	Glyceria	1182	*repens*	1276
cutleaf	739	borealis	1182	Gooseberry	352
Geum	376	*canadensis*	1182	bristly	356
aleppicum	377	*fluitans*	1182	Missouri	355
v. *strictum*	377	grandis	1182	Goosefoot	166
camporum	378	*nervata*	1183	Fremont	170
canadense	378	*rigida*	1183	maple-leaved	170
v. *camporum*	378	striata	1183	oak-leaved	171
laciniatum	378	*Glycine max*	416	pitseed	169
v. *trichocarpum*	378	Glycyrrhiza lepidota	452	sandhill	170
macrophyllum	378	v. *glutinosa*	452	Goosefoot Family	160
v. *macrophyllum*	378	Gnaphalium	940	Goosegrass	1166
v. *perincisum*	378	chilense	941	Gordon's eriogonum	218
oregonense	378	*exilifolium*	942	*Gossypianthus*	186
perincisum	378	*grayi*	942	*lanuginosus*	186
rivale	379	*macounii*	942	*tenuiflorus*	186
strictum	377	obtusifolium	941	Gourd	268
triflorum	379	palustre	941	buffalo-	268
v. *ciliatum*	379	purpureum	942	Grama, black	1141
v. *triflorum*	379	uliginosum	942	blue	1141
vernum	379	viscosum	942	grass	1140
virginianum	378	wrightii	942	hairy	1141
Geyer's spurge	545	Goat head	578	sideoats	1141
Giant chickweed	211	Goat's beard	1014		

INDEX

sixweeks	1140	Gray-green wood sorrel	579	purple	646		
Gramineae	1114	Gray poplar	278	Virginia	646		
Grape	559	Gray ragwort	994	Ground ivy	714		
bush	559	Gray's lousewort	773	Ground pine	39		
-fern	46	Grayhead prairie coneflower	990	Ground-plum	426		
cut-leaved	46	Grease-bush	406	Ground rose, little	370		
leathery	47	Greasewood	177	Groundnut	420		
little	47	*Great burdock*	865	Groundsel	993, 998		
matricary	47	Great lobelia	812	balsam	997		
graybark	560	Great Plains ladies'-tresses	1283	roundleaf	996		
honeysuckle	826	Great spurred violet	264	Grouseberry	336		
-hyacinth	1253	Great St. John's-wort	239	Grove sandwort	195		
Oregon	108	Great water dock	234	Guilleminea	186		
pigeon	560	Greater centaurea	900	densa	186		
possum	557	Greater duckweed	1045	v. *aggregata*	186		
raccoon	557	Greater St. John's-wort	238	v. *densa*	186		
river-bank	560	Green ash	750	lanuginosa	186		
sand	561	Green eyes	889	v. *sheldonii*	186		
winter	561	Green false nightshade	638	v. *tenuiflora*	186		
Grape Family	557	Green foxtail	1221	Gumbo evening primrose	518		
Grass, alkali-	1215	Green gentian	609	Gumbo lily	518		
barnyard	1164	Green gerardia	756	Gum-elastic	341		
bearded skeleton	1183	Green haw	375	Gummy lovegrass	1173		
Bermuda	1154	Green milkweed	630	Gumweed	943		
blowout	1195	Green needlegrass	1230	curly-top	944		
blue-eyed	1261, 1262	Green orchis, northern	1277	spinytooth	943		
bottlebrush	1186	Green parrot's feather	493	Gutierrezia	945		
brome	1142	Green pigweed	182	*diversifolia*	946		
buffalo	1148	Green spleenwort	52	*dracunculoides*	945		
canary	1207	Green violet	256	*filifolia*	946		
false buffalo	1197	nodding	256	*juncea*	946		
grama	1140	Green woodland orchis	1277	*linearis*	946		
hooded windmill	1152	Greenbrier	1266	*sarothrae*	945		
Indian	1221	bristly	1267	*texana*	946		
Johnson-	1222	Greenthread	1012	Gymnocarpium dryopteris	60		
lace	1173	Grindelia	943	Gynmocladus dioica	415		
May	1208	inornata	943	Gymnopogon ambiguus	1183		
-of-Parnassus	361	lanceolata	943	Gymnosperms	71		
northern	362	v. *texana*	944	Gyp phacelia	682		
small-flowered	362	*nuda*	944	Gypsophila	200		
orchard	1154	*perennis*	944	*elegans*	200		
panic	1200	*revoluta*	944	*muralis*	200		
porcupine-	1230	*serrulata*	944	*paniculata*	200		
poverty	254, 1227, 1228	squarrosa	944				
rabbit foot	1215	v. *nuda*	944	Habenaria	1276		
reed canary	1207	v. quasiperennis	944	clavellata	1277		
rescue	1147	v. *serrulata*	944	dilatata	1277		
ring	1197	v. squarrosa	944	v. *albiflora*	1277		
scratch	1193	*texana*	944	v. *dilatata*	1277		
skeleton	1183	Gromwell	697	hyperborea	1277		
umbrella	1105	false	700	lacera	1278		
white-eyed	1262	Gronovius' dodder	664	leucophaea	1278		
whitlow	311	Grooved flax	564	orbiculata	1278		
widgeon	1040	*Grossularia cynosbati*	354	*psychodes*	1279		
windmill	1152, 1153	*hirtella*	354	saccata	1278		
Grass Family	1113	*missouriensis*	355	unalascensis	1281		
Grassleaf rush	1055	*oxyacanthoides*	356	viridis v. bracteata	1279		
Gratiola	765	*setosa*	356	Hackberry	120, 121		
aurea	765	Grossulariaceae	352	dwarf	121		
lutea	765, 766	Ground cherry	642, 645	netleaf	121		
neglecta	765	clammy	644	Hackelia	692		
virginiana	766	common	644	deflexa	692		
Graybark grape	560	cutleaf	642	floribunda	692		
Gray dogwood	529	downy	644	virginiana	693		
Gray goldenrod	1004	prairie	643, 645	Hairgrass	1155, 1194		

tufted	1155	summer	373	badium	951		
Hairy bedstraw	820	sweet	831	flexuosum	952		
Hairy chess	1144	Hawkbit	971	*latifolium*	951		
Hairy crabgrass	1163	Hawk's-beard	917	microcephalum	952		
Hairy four-o'clock	149	Hawkweed	958	v. *microcephalum*	952		
Hairy gaura	512	Hawthorn	370, 372	*montanum*	951		
Hairy grama	1141	cockspur	372	*nudiflorum*	952		
Hairy lespedeza	458	downy	373	*polyphyllum*	952		
Hairy lip fern	55	frosty	373	*tenuifolium*	951		
Hairy mountain mint	730	hillside	373	Heliantheae	839		
Hairy phacelia	681	northern	374	Helianthemum bicknellii	253		
Hairy pinweed	254	Palmer's	373	*Helianthium*	1023		
Hairy prickly poppy	111	succulent	374	*parvulum*	1024		
Hairy sunflower	954	urn-tree	371	Helianthella quinquenervis	952		
Hairy tridens	1232	woolly	372	Helianthus	952		
Hairy vetch	489	Hayden's penstemon	783	annuus	954		
Hairy wild rye	1169	Hazel dodder	662	aridus	954		
Halberd-leaved rose mallow	245	Hazelnut	143, 144	besseyi	957		
Halenia deflexa	608	beaked	144	*chartaceous*	955		
Halerpestes	96	Heal-all	728	ciliaris	954		
cymbalaria	99	Heartleaf avens	379	giganteus	954		
Hall's milkweed	620	Heartwing sorrel	232	grosseserratus	954		
Haloragaceae	492	Heath Family	334	hirsutus	954		
Hamosa	421	Heather, beach	254	× *intermedius*	953		
leptocarpa	432	Heavenly blue morning-glory	657	× *kellermanii*	953		
Haploesthes greggii v.		Hedeoma	714	× *laetiflorus*	953, 957		
texana	946	*camporum*	715	v. *rigidus*	956		
Haplopappus	946	drummondii	715	*lenticularis*	954		
annuus	947	v. *reverchonii*	716	maximilianii	955		
armerioides	947	v. *serpyllifolium*	716	mollis	955		
ciliatus	948	hispidum	715	*nitidus*	956		
divaricatus	950	*longiflorum*	715	nuttallii	955		
engelmannii	948	pulgeoides	716	ssp. nuttallii	955		
fremontii	948	reverchonii	716	ssp. rydbergii	956		
ssp. *wardii*	948	v. *serpyllifolium*	716	× *orygaloides*	953		
heterophyllus	948	Hedge bindweed	652, 653	petoliaris	956		
lanceolatus	949	Hedge hyssop	765, 766	pumilus	956		
ssp. *lanceolatus*	949	Hedge mustard	329, 330	rigidus	956		
ssp. *typicus*	949	tall	330	ssp. rigidus	956		
ssp. *vaseyi*	949	Hedge-nettle	737	ssp. subrhomboideus	956		
multicaulis	949	Hedge parsley	603	*rydbergii*	956		
nuttallii	976	Hedgehog cactus	155, 156	salicifolius	957		
phyllocephalus ssp.		Hedgehog prickly poppy	112	strumosus	957		
annuus	947	Hedyotis	821	*subrhomboideus*	956		
pluriflorus	949	canadensis	822	tuberosus	957		
spinulosus	949	crassifolia	821	v. *subcanescens*	957		
ssp. australis	950	humifusa	821	Helictotrichon hookeri	1183		
ssp. cotula	950	longifolia	822	Heliopsis helianthoides	958		
ssp. glaberrimus	950	nigricans	822	v. *helianthoides*	958		
ssp. spinulosus	950	*rosea*	822	v. *occidentalis*	958		
ssp. *typicus*	950	Hedysarum	452	v. scabra	958		
validus	950	alpinum	453	*scabra*	958		
Harbinger of spring	592	v. *americanum*	453	Heliotropaceae	684		
Hardy catalpa	804	v. *philoscia*	453	Heliotrope	684, 693		
Harebell	809	*americanum*	453	Heliotropium	684, 693		
Hare's-ear mustard	308	boreale	453	convolvulaceum	693		
Hare's-foot dalea	443	v. *boreale*	454	curassavicum	694		
Hare's locoweed	468	v. *cinerascens*	454	v. *obovatum*	694		
Harry Lauder's walking stick	144	*cinerascens*	454	indicum	694		
Hartmannia	516	*mackenzii*	454	*spathulum*	694		
speciosa	525	*occidentale*	453	tenellum	694		
Hartweg evening primrose	501	Helenium	951	Helleborine	1275		
Haw, black	831	altissimum	951	Hemerocallis fulva	1251		
green	375	amarum	951	Hemicarpha	1105		
southern black	832	autumnale	951	aristulata	1106		

INDEX

drummondii	1106	canadense	958	Tartarian	826
micrantha	1106	v. *canadense*	958	trumpet	826
v. *aristulata*	1106	v. *fasciculatum*	958	white	824
v. *drummondii*	1106	gronovii	958	wild	825
Hemlock, bulbous water	589	longipilum	960	yellow	825
common water	589	umbellatum	960	Honeysuckle Family	823
parsley	585	v. *scabriusculum*	960	Hood's phlox	675
poison	590	Hierochloë	1184	Hooded ladies'-tresses	1283
water	588	*hirta* ssp. *arctica*	1184	Hooded windmill grass	1152
Hemp	124	odorata	1184	Hook-spurred violet	258
-nettle	714	High-bush blackberry	403, 404	Hooked agrimony	366
common	714	Highbush cranberry	831	Hooked buttercup	104
tall water	184	High mallow	248	Hop-clover, low	483
Hemp Family	123	Hilaria jamesii	1184	small	483
Henbane	640	*mutica*	1185	Hop-hornbeam	144
Henbit	717	Hillside blueberry	335	Hop-tree	576
Heracleum	594	Hillside hawthorn	373	Hops	124
lanatum	594	Hill's oak	137	common	125
maximum	594	Hippocastanaceae	568	Japanese	125
sphondylium ssp.		Hippuridaceae	740	Hordeum	1185
montanum	594	Hippuris vulgaris	740	*brachyantherum*	1185
Herb-Robert	582	Hoary aster	976	*caespitosum*	1185
Hesperis matronalis	315	Hoary cress	307	*jubatum*	1185
Hesperochloa kingii	1114	lens-padded	307	*pammelii*	1185
Heteranthera	1240	Hoary euphorbia	546	*pusillum*	1185
limosa	1240	Hoary evening primrose	523	*vulgare*	1186
mexicana	1240	Hoary false alyssum	303	Horehound	720
peduncularis	1240	Hoary pea	480	common	720
reniformis	1240	Hoary puccoon	696	water	719
Heterotheca	799, 958	Hoary skullcap	735	Hornbeam, European	141
canescens	904	Hoary tickclover	447	Horned pondweed	1041
fulcrata	904	Hoary vervain	707	Horned Pondweed Family	1041
latifolia	958	Hoary vetchling	455	Horned poppy	112
v. *latifolia*	958	Hoary willow	286	red	112
v. *macgregoris*	958	Hoffmannseggia	415	Hornwort	84
pilosa	905	*densiflora*	416	Hornwort Family	84
stenophylla	905	drepanocarpa	415	Horse-gentian	829
subaxillaris	958	glauca	416	yellow-flowered	829
villosa	906	*jamesii*	412	Horsemint	721, 723, 725
v. *angustifolia*	906	Hog peanut	420	Ohio	712
v. *foliosa*	906	Hog plum	395	Horse-nettle, Carolina	647
Heuchera	358	*Holcophacos*	421	western	648
americana	359	*distortus*	427	white	649
hispida	359	*Holcus lantatus*	1114	Horse purslane	153
hirsuticaulis	359	Holly	535	Horse-weed	914
richardsonii	359	deciduous	535	Horseradish	302
v. *grayana*	360	fern	65	Horsetail	42
v. *hispidior*	360	Holly Family	535	field	43
v. *richardsonii*	360	Hollyhock	242	marsh	44
Hexalectris spicata	1279	Holosteum umbellatum	200	meadow	44
Hibiscus	245	*Homalobus*	421	paspalum	1205
laevis	245	*caespitosus*	435	water	43
lasiocarpos	246	*dispar*	436	wood	45
militaris	246	*hylophilus*	430	Horsetail Family	42
moscheutos	246	*stipitatus*	436	Hound's tongue	690, 691
trionum	246	*tenellus*	436	*Houstonia angustifolia*	822
Hickory	131	*vexilliflexus*	436	*canadensis*	822
big shellbark	132	Honewort	590	*longifolia*	822
bitternut	132	Honey locust	414	*minima*	821
black	133	Honey mesquite	410	*nigricans*	822
mockernut	133	Honeysuckle	824	*pusilla*	821
shagbark	132	grape	826	Huckleberry, mountain	335
shellbark	132	Japanese	825	Hudsonia tomentosa	254
Hieracium	958	limber	825	Humulus	124
albiflorum	958	Maack	826	*americanus*	125

japonicus	125	*virginicum* v. *fraseri*	239	*Ionactis linariifolia*		879
lupulus	125	*Hypopithys lanuginosa*	341	Iowa crab		396
ssp. *americanus*	125	Hypoxis hirsuta	1251	*Ipomoea*		656
v. *lupuloides*	125	Hyssop	710	*alba*		657
v. *neomexicanus*	125	catnip giant	711	*batatas*		657
v. *pubescens*	125	hedge	765, 766	*carletoni*		660
Hyacinth, grape-	1253	lavender	710	*coccinea*		658
Hybanthus	256	purple giant	711	v. *hederifolia*		658
concolor	256	water	757	*cristulata*		658
f. *subglaber*	256	Hyssopleaf tickseed	173	*hederacea*		659
linearis	257	*Hystrix patula*	1186	v. *integriuscula*		659
verticillatus	256	v. *bigeloviana*	1186	*hederifolia*		658
Hybrid balsam poplar	277			*lacunosa*		659
Hybrid leafy spurge	549	Iberian star-thistle	899	*leptophylla*		659
Hydrangea arborescens	351	*Ibervillea lindheimeri*	268	*longifolia*		660
wild	351	*Ilex*	535	× *multifida*		657
Hydrangea Family	351	*decidua*	535	*nil*		657
Hydrangeaceae	351	Illinois bundleflower	408	*pandurata*	659,	660
Hydranthelium rotundi-		Illinois pondweed	1037	*purpurea*		660
folium	757	Illinois tickclover	448	v. *diversifolia*		657
Hydrocharitaceae	1028	*Ilysanthes dubia*	769	*quamoclit*		660
Hydrocotyle ranunculoides	594	*inaequalis*	769	*shumardiana*		660
Hydrophyllaceae	678	*Impatiens*	582	*tricolor*		657
Hydrophyllum	678	*biflora*	583	*Ipomopsis*		669
appendiculatum	679	*capensis*	583	*congesta*		669
virginianum	679	*nortonii*	583	ssp. *congesta*		670
Hymenoppapus	799, 960	*pallida*	583	ssp. *pseudotypica*		670
filifolius	961	India lovegrass	1176	*laxiflora*		670
v. *cinereus*	961	Indian-apple	639	*longiflora*		670
v. *polycephalus*	961	Indian blanketflower	939	*pumila*		670
flavescens	961	Indian cigar tree	804	*rubra*		671
v. *cano-tomentosus*	961	Indian grass	1221	*spicata*		671
v. *flavescens*	961	Indian hemp dogbane	612	*Iresine*		187
scabiosaeus v.		Indian mallow	241	*celosia*		187
corymbosus	961	Indian milk-vetch	423	*rhizomatosa*		187
v. *scabiosaeus*	962	Indian mustard	304	Iridaceae		1258
tenuifolius	962	Indian paintbrush	759	*Iris*		1259
Hymenoxys	962	Indian-physic	381	*brevicaulis*		1259
acaulis	962	Indian pipe	340, 341	*missouriensis*		1260
v. *acaulis*	963	Indian Pipe Family	340	*lamance*		1259
linearifolia	963	Indian plantain	743, 895	prairie		1261
odorata	963	pale	895	*pseudacorus*		1260
richardsonii	963	Indian ricegrass	1199	*shrevei*		1260
v. *floribunda*	963	Indian rush-pea	416	*versicolor*		1260
v. *richardsonii*	963	*Indian strawberry*	265	*virginica*		1260
scaposa	963	Indian tobacco	811	v. *shrevei*		1261
v. *glabra*	964	Indian turnip	1043	yellow		1260
v. *scaposa*	964	Indigo	454	Iris Family		1258
stemless	962	blue false	436	Ironplant, cutleaf		949
Hyoscamus niger	640	dwarf wild	419	Ironweed		1017
Hypericaceae	237	false	419, 436	plains		1018
Hypericum	237	long-bracted wild	437	western		1017
ascyron	239	plains wild	437	Ironwood		144
canadense	238	white wild	437	*Isanthus brachiatus*		740
cistifolium	239	*Indigofera*	454	*Isnardia*		513
drummondii	238	*leptosepala*	454	*palustris*		515
kalmianum	237	*miniata* v. *leptosepala*	454	*Isocoma pluriflora*		949
majus	238	Inland bluegrass	1213	*wrightii*		949
mutilum	238	Inland rush	1055	Isoetaceae		41
perforatum	238	Inland saltgrass	1163	*Isoetes*		41
pseudomaculatum	239	*Inoxalis violacea*	579	*butleri*		41
punctatum	238, 239	Intermediate scouring rush	43	*melanopoda*		41
pyramidatum	239	Intermediate wheatgrass	1125	f. *pallida*		41
sphaerocarpum	239	Inuleae	844	*Isopappus validus*		950
subpetiolatum	239	*Iodanthus pinnatifidus*	315	*Isopyrum biternatum*		94

INDEX 1359

Italian ryegrass	1189	brachycarpus	1053	Kidney-leaved violet	263
Iva	964	brachycephalus	1053	King Ranch bluestem	1133
annua	964	brachyphyllus	1053	Kingnut	132
angustifolia	964	bufonis	1053	Kiss-me-over-the-garden-gate	227
axillaris	965	canadensis	1054	Kitten tails	757
ciliata	965	*crassifolius*	1057	Knautia arvensis	837
xanthifolia	965	diffusissimus	1054	Knawel	204
Ivy, ground	714	dudleyi	1054	annual	204
Kenilworth	764	effusus	1054	Knapweed	898
marine	557	ensifolius v. montanus	1055	diffuse	899
poison	573, 574	gerardii	1055	Russian	900
Ivyleaf morning-glory	659	interior	1055	spotted	899
Ivy-leaved speedwell	795	*kansanus*	1053	*Kneiffia*	516
		longistylis	1055	*perennis*	516
Jack-in-the-pulpit	1043	marginatus	1055	*pratensis*	516
Jagged chickweed	200	nodatus	1056	*spachiana*	516
James' dalea	442	nodosus	1056	Knotgrass	1206
James' nailwort	203	*richardsonianus*	1052	Knotroot bristlegrass	1219
James rush-pea	412	*saximontanus*	1055	Knotted rush	1056
James' saxifrage	363	scirpoides	1056	Knotweed	220, 222, 223
James' wild buckwheat	218	secundus	1056	erect	223
Japanese bamboo	228	setosus	1056	Kochia scoparia	175
Japanese birch	143	tenuis	1056	v. *culta*	175
Japanese brome	1144	torreyi	1057	f. *trichophylla*	175
Japanese honeysuckle	825	validus	1057	*siversiana*	175
Japanese hops	125	vaseyi	1057	Koeleria	1186
Japanese lespedeza	460	June-berry	368, 369	*cristata*	1186
Japanese rose	399	Junegrass	1186	*gracilis*	1186
Japanese royal morning-glory	657	Juniper	71	*latifrons*	1186
Jasmine, northern rock	344	common	72	*macrantha*	1186
rock	344	creeping	72	*nitida*	1186
western rock	344	dwarf	72	pyramidata	1186
Jerusalem artichoke	957	one-seeded	72	Korean lespedeza	459
Jerusalem oak	169	Pinchot	73	Krameria lanceolata	566
Jerusalem sage	709	Rocky Mountain	73	*secundiflora*	566
Jewel weed	583	Juniperus	71	Krameriaceae	566
Jimsonweed	638, 639	communis	72	Krigia	965
Joe-pye weed	934	v. *depressa*	72	biflora	966
spotted	934	horizontalis	72	caespitosa	966
sweet	935	monosperma	72	dandelion	966
Johnny-jump-up	263	pinchotii	73	occidentalis	966
Johnson-grass	1222	scopulorum	73	*oppositifolia*	966
Jointed goatgrass	1122	v. *columnaris*	73	Kudzu-vine	476
Jointtail	1190	virginiana	73	Kuhnia	966
Carolina	1190	*Jussiaea*	513	chlorolepis	967
Jointweed	220	*diffusa*	515	eupatorioides	967
Juglandaceae	131	*repens* v. *glabrescens*	515	v. *corymbulosa*	967
Juglans	133	Justicia americana	801	v. *eupatorioides*	967
cinerea	134	v. *subcoriacea*	801	v. *ozarkana*	967
microcarpa	134			v. *texana*	967
nigra	134	Kallstroemia	577		
regia	133	*hirsutissima*	578	Labrador bedstraw	819
rupestris	134	*intermedia*	578	Lace cactus	155
sieboldiana	133	parviflora	577	Lace grass	1173
Juncaceae	1049	Kalm's lobelia	811	Lactuca	967
Juncaginaceae	1031	Kansas thistle	650	biennis	968
Juncus	1050	Kenilworth ivy	764	canadensis	968
acuminatus	1052	*Kentrophyta*	421	v. *latifolia*	969
alpinus	1052	*montana*	429	v. *obovata*	969
amblyocarpus	1052	*viridus*	429	floridana	969
articulatus	1052	Kentrophyta, Nuttall's	429	f. *leucantha*	969
ater	1053	Kentucky bluegrass	1214	v. *villosa*	969
balticus	1052	Kentucky coffee-tree	415	*hastilis* v. *vulgaris*	971
v. *littoralis*	1053	Kickxia elatine	766	*intergrata*	971
v. *montanus*	1053	*spuria*	766	ludoviciana	969

f. *campestris*	970	Large-flowered cut-leaved		*Leek*	1246
oblongifolia	970	evening primrose	520	Leek, wild	1246
pulchella	970	Large-flowered gaura	510	*Leersia*	1187
sagittifolia	969	Large-flowered stickseed	692	*lenticularis*	1187
saligna	970	Large-flowered tickclover	448	oryzoides	1187
f. *ruppiana*	970	Large-leaved avens	378	virginica	1187
scariola	971	Large pondweed	1035	v. *ovata*	1187
serriola	970	Large yellow lady's slipper	1274	Leiberg dichanthelium	1159
spicata	968	Larkspur	92	*Lemna*	1044
villosa	969	blue	93	*aequinoctialis*	1045
virosa	971	dwarf	94	*cyclostasa*	1045
Lactuceae	845	little	93	*gibba*	1044
Ladies'-tresses	1281, 1282	prairie	94	minor	1045
Great Plains	1283	rocket	93	*minuscula*	1045
hooded	1283	-violet	261	*obscura*	1045
little	1283	*Larrea densiflora*	416	*perpusilla*	1045
oval	1283	*jamesii*	412	v. *trinervis*	1045
shining	1282	*Lasallea*	871	*trinervis*	1045
slender	1282	Late coral-root	1271	*trisulca*	1045
twisted	1284	Late goldenrod	1003	*turionifer*	1044
Ladino clover	485	*Lathyrus*	454	*valdiviana*	1045
Lady-fern	53	*decaphyllus*	456	Lemnaceae	1043
northern	53	*hapemanii*	456	Lemon beebalm	724
southern	53	*incanus*	456	Lemon mint	724
western	53	*latifolius*	455	Lemon scurf-pea	474
Lady's-fingers	416	*macranthus*	455	*Lens esculenta*	416
Lady's-slipper	1272	*ochroleucus*	455	Lens-padded hoary cress	307
large yellow	1274	*odoratus*	416	*Lentibularia vulgaris* v.	
showy	1274	*palustris*	455	*americana*	807
small white	1274	v. *pilosus*	455	Lentibulariaceae	806
small yellow	1273	*polymorphus*	455	*Lentil*	416
yellow	1273	ssp. *incanus*	456	Leonard's small skullcap	736
Lady's thumb	227	ssp. *polymorphus*	456	Leontodon hispidus	971
Lamance iris	1259	*pusillus*	456	*Leonurus*	717
Lamb's quarters	166, 168	*stipulaceous*	456	*cardiaca*	718
Lamb's lettuce	835	venosus v. intonsus	456	ssp. *villosa*	718
Lamiaceae	708	v. *venosus*	456	*marrubiastrum*	718
Lamium	716	Lauraceae	78	Leopard lily	1250
amplexicaule	717	Laurel Family	78	*Lepachys columnifera*	990
f. *clandestinum*	717	Laurel-leaved willow	289	*pinnata*	990
purpureum	717	*Lavauxia*	516	*tagetes*	990
Lanceleaf cottonwood	275	*flava*	520	*Lepadena*	541
Lance-leaved buckthorn	556	*triloba*	525	*marginata*	548
Lance-leaved sage	732	*watsonii*	525	Lepidium	316
Laportea canadensis	128	Lavender hyssop	710	austrinum	316
Lappula	694	Lavender leaf primrose	502	*bourgeauanum*	317
americana	692	Lazy daisy	863	campestre	317
angustifolia	692	Lead plant	418	densiflorum	317
cenchrusoides	695	Leadwort Family	235	*fletcheri*	317
echinata	695	Leaf-cup	985	latifolium	317
erecta	695	Leaf-flower	551	neglectum	317
floribunda	693	Leafy pondweed	1036	oblongum	317
foliosa	695	Leafy rose	399	perfoliatum	317
fremontii	695	Leafy spurge	545	ramosissimum	318
heterosperma	695	Leather-weed	540	sativum	318
occidentale	695	Leathery grape-fern	47	virginicum	318
redowskii	695	Leavenworth eryngo	593	*Leptandra virginica*	797
v. *texana*	695	Leavenworth's vetch	488	*Leptilon canadense*	914
scaberrina	692	Lechea	254	*divaricatum*	915
texana	695	intermedia	254	Leptochloa	1187
virginiana	693	mucronata	254	*acuminata*	1188
Lapsana communis	971	stricta	255	*attenuata*	1188
Large alfalfa dodder	664	tenuifolia	255	*dubia*	1188
Large beardtongue	783	*villosa*	255	fascicularis	1188
Large bellwort	1256	*Ledum groenlandicum*	334	filiformis	1188

INDEX

Leptodactylon	671	white	986	*Limnodea arkansana*	1114
caespitosum	671	wild	968	*Limnorchis*	1276
pungens	672	willow-leaved	970	*dilatata*	1277
Leptoglottis	410	wire	1009	*huronensis*	1278
nuttallii	411	*Leucanthemum vulgare*	902	Limnosciadium pinnatum	594
Leptoloma cognatum	1188	Leucelene	971	Limonium limbatum	235
Lespedeza	457	*alsinoides*	972	Lomosella aquatica	767
capitata	457	ericoides	971	Linaceae	561
× *violacea*	457	Leucocrinum montanum	1251	Linanthus septentrionalis	672
v. *vulgaris*	458	*Leucopoa kingii*	1114	Linaria	767
common	460	Leucospora multifida	766	canadensis	768
creeping	459	*Leymus cinereus*	1168	v. canadensis	768
cuneata	458	*triticoides* ssp. *simplex*	1169	v. texana	768
frutescens	460	Liatris	972	dalmatica	768
hairy	458	aspera	972	*genistifolia*	768
hirta	458	*compacta*	974	vulgaris	768
intermedia	460	*glabrata*	974	f. *peloria*	767, 768
Japanese	460	*hirsuta*	974	Linden	239
Korean	459	lancifolia	973	Linden Family	239
leptostachya	458	ligulistylis	973	Lindera benzoin	78
× *longifolia*	458	mucronata	973	Lindernia	769
× *manniana*	458, 460	punctata	973	*anagallidea*	769
× *nuttallii*	458	pycnostachya	973	dubia	769
prairea	460	*scarioso*	973	v. *anagallidea*	769
prairie	460	*spicata*	973	ssp. *major*	769
procumbens	459	squarrosa	974	v. *riparia*	769
× *virginica*	457	v. *glabrata*	974	Lindheimer milk-vetch	429
repens	459	v. *hirsuta*	974	Lindheimera texana	974
round-head	457	v. *squarrosa*	974	Lindheimer's lip fern	55
sericea	458	Licorice	452	Lindheimer's longleaf	
× *simulata*	458	wild	452	eriogonum	218
slender bush	461	*Ligustrum vulgare*	747	Linnaea	823
slender spike	458	*Lilac, common*	747	americana	824
stipulacea	459	Liliaceae	1240	borealis	823
striata	460	Liliopsida	1021	v. *americana*	824
stuevei	460	Lilium	1251	Linum	561, 663
tall bush	460	canadense	1252	aristatum	562
trailing	459	ssp. *michiganense*	1252	hudsonioides	562
violacea	460	*michiganense*	1252	medium v. texanum	562
× *virginica*	457	philadelphicum	1252	perenne v. lewisii	562
virginica	461	v. *andinum*	1252	v. *perenne*	563
Lesquerella	318	*superbum*	1252	pratense	563
alpina	319	*umbellatum*	1252	puberulum	563
arenosa	319	Lily	1251	rigidum	563
v. arenosa	320	blackberry	1259	v. berlandieri	563
v. argillosa	320	blue funnel	1246	v. compactum	563
argentea	322	celestial	1261	v. rigidum	564
auriculata	320	gumbo	518	sulcatum	564
fendleri	320	leopard	1250	usitatissimum	564
gordonii	321	-of-the-valley	1248	Lionsheart	727
gracilis ssp. nuttallii	321	wild	338, 1252	Virginia	728
ludoviciana	322	mountain	1251	Liparis loeselii	1280
montana	322	rain-	1248	Lip fern	54
ovalifolia	323	sago	1247	Alabama	54
ssp. alba	323	sand	270	beaded	56
ssp. ovalifolia	323	Turk's cap	1252	Eaton's	54
repanda	321	wild	1252	Fendler's	55
Lesser bladderwort	807	zephyr-	1257	hairy	55
Lettuce	967	Lily Family	1240	Lindheimer's	55
blue	970	Limber honeysuckle	825	slender	55
blue wood	968	Limber pine	76	woolly	56
Florida	969	Limestone adder's-tongue	48	Lippia	702
lamb's	835	Limestone ruellia	802	cuneifolia	702
prickly	970	*Limnobium spongia*	1028	lanceolata	703
western wild	969	*Limnobotrya lacustris*	355	v. *recognita*	703

nodiflora	703	Locoweed	467	purple	497
v. *nodiflora*	703	Bessey's	467	tufted	349
v. *reptans*	703	dwarf	469	whorled	349
Listera convallarioides	1280	Hare's	468	winged	497
Lithococca	693	purple	469	Loosestrife Family	494
tenella	694	showy	471	*Lophotocarpus calycinus*	1026
Lithophragma	360	slender	468	Lopseed	703
bulbifera	360	white	470	Lotebush	556
parviflora	360	woolly	431	Lotus	81, 461
Lithospermum	695	Locust	477	*americanus*	462
arvense	696	black	477	*corniculatus*	461
brevifolium	697	honey	414	milk-vetch	430
canescens	696	Lodgepole pine	75	*purshianus*	461
carolinense	696, 697	Loeflingia squarrosa	201	*tenuis*	462
croceum	697	*texana*	201	Lotus Family	81
gmelini	697	Loesel's twayblade	1280	Lousewort	772
incisum	697	Logania Family	604	common	773
latifolium	697	Loganiaceae	604	Gray's	773
linearifolium	697	Lolium	1189	swamp	773
mandanense	697	*multiflorum*	1189	Lovegrass	1171
multiflorum	696	*perenne*	1189	Carolina	1176
Little barley	1185	v. *aristatum*	1189	gummy	1173
Little bluestem	1134	v. *perenne*	1189	India	1176
Little breadroot scurf-pea	474	*persicum*	1190	little	1175
Little grape-fern	47	*temulentum*	1190	Mediterranean	1172
Little ground rose	370	Lomatium	595	plains	1175
Little ladies'-tresses	1283	*foeniculaceum*	595	purple	1178
Little larkspur	93	v. *daucifolium*	595	red	1177
Little lovegrass	1175	v. *foeniculaceum*	595	sand	1178
Little prickly pear	156	*macrocarpum*	596	sandbur	1174
Little-seed ricegrass	1199	*montanum*	595	teal	1175
Little vetch	488	*nuttallii*	596	tumble	1177
Little walnut	134	*orientale*	596	weeping	1174
Live oak	141	Lombardy poplar	280	Love vine	661
Lizard's tail	79	London rocket	330	Low cudweed	942
Lizard's Tail Family	79	Long-bracted orchis	1279	Low hop-clover	483
Loasaceae	269	Long-bracted wild indigo	437	Low service-berry	369
Lobelia	810	Long cloak-fern	62	Lowland fragile fern	57
appendiculata	813	Long-leaf tickclover	447	Ludwigia	513
cardinalis	810	Long-leaved sage	870	*alternifolia*	514
ssp. *cardinalis*	811	Long-leaved stitchwort	213	v. *pubescens*	514
ssp. *graminea* v.		Long-stalked stitchwort	213	*decurrens*	513
phyllostachya	811	Longbarb arrowhead	1028	*glandulosa*	514
flaccidifolia	810	Longhead poppy	113	*natans stipitata*	515
great	812	Longhorn milkweed	623	*palustris*	514
halei	810	Longleaf eriogonum, Lind-		v. *americana*	515
hirtella	813	heimer's	218	*peploides*	515
inflata	811	Longleaf pondweed	1037	ssp. *glabrescens*	515
kalmii	811	Longspike tridens	1233	*polycarpa*	515
Kalm's	811	Lonicera	824	*repens*	515
leptostachya	813	*albiflora*	824	Lungwort	684
palespike	812	v. *dumosa*	825	Lupine	462
puberula	810	*dioica*	825	Platte	463
siphilitica	812	v. *glaucescens*	825	rusty	464
f. *albiflora*	812	*flava*	825	silky	464
f. *laevicalyx*	812	*glaucescens*	825	silvery	462
v. *ludoviciana*	812	*japonica*	825	small	464
spicata	812	*maackii*	826	tailcup	463
v. *hirtella*	813	*prolifera*	826	Lupinus	462
v. *leptostachya*	813	v. *glabra*	826	*aduncus*	463
splendens	811	*sempervirens*	826	*argenteus*	462
strictifolia	812	*tartarica*	826	v. *argenteus*	463
Loco, pendulous-pod	468	Loosestrife	347, 348, 496	v. *parviflorus*	463
red	467	false	513	v. *stenophyllus*	463
white	477	fringed	348	v. *tenellus*	463

caudatus		463	*dacotanum*		497	rotundifolia	247,	248
floribundus		463	*lanceolatum*		497	sylvestris		248
leucopsis		464	salicornia		497	v. *mauritiana*		248
parviflorus		463				verticillata		248
perennis ssp. *plattensis*		464	Maack honeysuckle		826	v. *crispa*		248
plattensis		463	Machaeranthera		975	Malvaceae		240
pusillus		464	*annua*		947	Malvastrum		248
sericeus		464	canescens		976	*angustum*		249
Luzula		1057	*conmixta*		977	*coccineum*		252
bulbosa		1058	*glabriuscula*		1020	hispidum		248
campestris v. *echinata*		1058	grindelioides		976	Malvella		249
v. *multiflora*		1059	linearis		976	leprosa		249
echinata		1058	*pinnatifida*		950	sagittifolia		250
multiflora		1058	ssp. *pinnatifida* v.			Mammillaria	154,	156
parviflora		1059	*glaberrima*		950	heyderi		156
Lychnis	201,	205	v. *pinnatifida*		950	*missouriensis*		154
alba		209	*ramosa*		976	*vivipara*		154
chalcedonica		201	*sessiliflora*		977	Manisurus cylindrica		1190
drummondii		207	tanacetifolia		977	Mannagrass		1182
scarlet		201	Maclura pomifera	126,	533	fowl		1183
Lycium		640	Macoun's buttercup		103	northern		1182
berlandieri		640	*Macuillamia rotundifolia*		757	tall		1182
halimifolium		641	Mad-dog skullcap		735	Many-flowered agrimony		367
chinense		641	*Madder, field*		817	Many-ray aster		874
pallidum		641	Madder Family		816	Manyseed seedbox		515
Lycopersicon esculentum		637	Madia glomerata		977	Maple		569
Lycopodiaceae		39	Madwort		687	black		570
Lycopodium		39	Magnoliophyta		77	-leaved goosefoot		170
annotinum		39	Magnoliosida		77	mountain		569
complanatum		39	*Mahonia aquifolium*		108	silver		570
dendroideum		39	Mahogany, mountain		369	soft		570
obscurum		39	Maianthemum canadense		1252	sugar		570
Lycopsis arvensis		697	v. *interius*		1252	Maple Family		569
Lycopus		718	Mahaleb cherry		393	Mare's tail		740
americanus		719	Maidenhair		379	Mare's Tail Family		740
asper		719	fern	50,	51	Marginal shield fern		59
lucidus		719	spleenwort		52	Marguerite		902
rubellus		719	Malaceae		364	Marigold, fetid		920
sherardii		720	*Malcolmia africana*		308	marsh		89
uniflorus		720	Male fern		59	Marijuana		124
virginicus		720	Mallow		246	*Marilaunidium angustifolium*		680
Lycurus phleoides		1190	clustered		248	Marine ivy		557
Lygodesmia		974	common	247,	248	Mariola		982
grandiflora		975	false		251	Mariposa		1247
juncea		975	finger poppy		244	Marrubium		720
pauciflora		1009	globe		251	*alternidens*		721
rostrata		999	halberd-leaved rose		245	*vulgare*		720
texana		975	high		248	Marshallia caespitosa		977
Lyre-leaf sage		732	Indian		241	v. *caespitosa*		978
Lysimachia		347	narrowleaf globe		251	v. *signata*		978
ciliata		348	pink poppy		243	Marsh bellflower		809
hybrida		348	poppy		243	Marsh betony		737
lanceolata		347	purple poppy		244	Marsh buttercup		101
longifolia		349	red false		251	Marsh elder	964,	965
nummularia		349	rose	245,	246	Marsh fern		67
punctata		347	small-fruited		247	Marsh-fleabane		985
quadrifolia		349	Venice		246	Marsh foxtail		1130
radicans		347	Mallow Family		240	Marsh horsetail		44
thrysiflora		349	Maloideae		265	Marsh marigold		89
Lythraceae		494	*Malus ioensis*		396	Marsh muhly		1196
Lythrum		496	Malva		246	Marsh rosemary		235
alatum		497	*crispa*		248	Marsh seedbox		514
v. alatum		497	*neglecta*		247	Marsh skullcap		734
v. lanceolatum		497	parviflora		247	Marsh speedwell		796
californicum		497	pusilla		248	Marsh St. John's-wort		239

Marsh vetchling	455	
Marsh-violet, northern	260	
Marshmallow	242	
Marsilea	70	
mucronata	70	
quadrifolia	70	
tenuifolia	70	
uncinata	70	
vestita	70	
Marsileaceae	70	
Martynia louisianica	803	
Maruta cotula	862	
Maryland senna	413	
Maryland tickclover	449	
Massive spike prairie-clover	441	
Mat chaff-flower	179	
Mat spurge	550	
Matelea	632	
baldwyniana	633	
biflora	633	
cynanchoides	634	
decipiens	634	
gonocarpa	635	
Matricaria	978	
chamomilla	978	
maritima	978	
matricarioides	978	
Matricary grape-fern	47	
Matrimony vine	641	
Matted wild buckwheat	219	
Matteuccia struthiopteris	60	
Maximilian sunflower	955	
May-apple	109	
May grass	1208	
May-pop	266	
Maximowiczia lindheimeri	269	
Meadow anemone	87	
Meadow beauty	526	
Meadow fescue	1181	
Meadow foxtail	1130	
Meadow horsetail	44	
Meadow parsnip	602	
Meadow rue	106	
early	107	
purple	107	
Meadow salsify	1015	
Meadow-sweet	405, 406	
Meadow willow	289	
Mead's milkweed	623	
Mealy corydalis	116	
Mecardonia acuminata	769	
Medicago	465	
lupulina	465	
v. *glandulosa*	465	
minima	465	
sativa	465, 665	
ssp. falcata	466	
ssp. sativa	466	
Medic, black	465	
prickly	465	
Medick	465	
Mediterranean lovegrass	1172	
Megapterium	516	
argophyllum	523	
brachycarpum	521	

fremontii	523	
missouriense	523	
oklahomense	524	
Meibomia	445	
acuminata	448	
bracteosa	448	
canadensis	447	
canescens	447	
dillenii	450	
illinoense	449	
longifolia	448	
marilandica	449	
michauxii	451	
nudiflora	449	
paniculata	450	
pauciflora	451	
pubens	450	
rigida	450	
sessilifolia	452	
Melampodium leucanthum	979	
Melandrium	205	
album	209	
drummondii	207	
Melanthium virginicum	1252	
Melastomataceae	526	
Melastome Family	526	
Melic	1191	
false	1217	
threeflower	1191	
Melica	1191	
bulbosa	1191	
mutica	1191	
nitens	1191	
porteri	1191	
smithii	1191	
subulata	1191	
Melilotus	466	
alba	466	
indica	467	
officinalis	466, 709	
Melon-leaf nightshade	648	
Melosmon	738	
laciniatum	740	
Melothria	269	
chlorocarpa	269	
pendula v. *chlorocarpa*	269	
pendula	269	
Menispermaceae	109	
Menispermum canadense	110	
Mentha	721	
aquatica	722	
× arvensis	721	
arvensis v. *glabrata*	722	
v. *villosa*	722	
canadensis	722	
× cardiaca	722	
gentilis	721	
glabrior	722	
longifolia	721	
pernardii	722	
× piperita	722	
spicata	722	
Mentzelia	270	
albescens	271	
albicaulis	271	

decapetala	271	
dispersa	272	
laevicaulis	272	
multiflora	272	
nuda	272	
oligosperma	273	
pumila	272	
reverchonii	273	
ten-petal	271	
Menyanthaceae	666	
Menyanthes trifoliata	666	
Mercury, three-seeded	536	
wild	537	
Meriolix	500	
intermedia	502	
melanoglottis	501	
oblanceolata	502	
serrulata	502	
Mertensia	698	
ciliata	698	
coronata	699	
foliosa	699	
lanceolata	698	
v. *brachyloba*	698	
linearis	698	
oblongifolia	699	
papillosa	698	
virginica	699	
f. *rosea*	699	
Mesadenia	895	
atriplicifolia	895	
tuberosa	895	
Mesquite	410	
honey	410	
vine-	1202	
Mexican bamboo	228	
Mexican devil-weed	885	
Mexican fire-bush	175	
Mexican plum	393	
Mexican silk tassel	530	
Mexican tea	76, 168	
Mezereum Family	498	
Micrampelis lobata	268	
Micranthes	362	
texana	363	
Microphacos	421	
gracilis	428	
parviflorus	428	
Microseris cuspidata	979	
nutans	979	
Microsteris	672	
gracilis	672	
v. *humilior*	672	
ssp. *humilis*	672	
humilis	672	
micrantha	672	
Midland quillwort	41	
Mignonette, wild	333	
Mignonette Family	333	
Mild water pepper	226	
Milfoil, American	493	
water	492, 493	
Milk-pea	452	
downy	452	
Milk vetch	421	

INDEX

alkali	434	Millet, broom-corn	1202	Moccasin flower, yellow	1273
alpine	424	foxtail	1220	Mock bishop's weed	599
American	424	proso	1202	Mock cypress	175
bent-flowered	436	Mimosa	407, 409	Mock pennyroyal	714
Bodin	425	biuncifera	409	Mockernut hickory	133
Canada	426	borealis	409	*Moehringia*	193
Cooper	431	cat's-claw	409	*lateriflora*	195
draba	435	pink	409	*Moldavica*	713
Drummond	427	Mimosa Family	407	*parviflora*	713
field	424	Mimosaceae	407	*punctata*	713
Indian	423	*Mimulus*	770	*thymiflorum*	713
Lindheimer	429	*alatus*	770	Molluginaceae	191
lotus	430	*floribundus*	771	*Mollugo verticillata*	191
Missouri	430	*geyeri*	771	*Monarda*	723
Ozark	427	*glabratus* v. *fremontii*	771	*bradburiana*	723
painted	426	v. *glabratus*	771	*citriodora*	724
Platte River	432	*guttatus*	771	*clinopodioides*	724
pliant	427	*ringens*	771	*comata*	725
pulse	435	v. *minthodes*	772	*didyma*	725
Pursh	433	Mint	721	*dispersa*	724
Short's	435	field	721	*fistulosa*	724
slender	428	geranium	902	v. *fistulosa*	725
small-flowered	431	hairy mountain	730	v. *menthifolia*	725
standing	423	horse	721, 725	v. *mollis*	725
stinking	433	lemon	724	*menthaefolia*	725
tine-leaved	432	Ohio horse	712	*mollis*	725
Trinidad	433	mountain	729	*pectinata*	725
woodland weedy	430	slender-leaved mountain	730	*punctata* ssp.	
Milkvine	634	stone	712	*occidentalis*	725
two-flowered	633	Virginia mountain	730	*Moneses uniflora*	339
Milkweed	615	white mountain	729	Moneywort	349
bluntleaf	617	wood	712	Monkey-flower	770
broadleaf	622	woods mountain	730	Allegheny	771
butterfly	629	Mint Family	708	common yellow	771
climbing	632, 633, 634	*Minuartia*	193	roundleaf	771
common	628	*drummondii*	194	sharpwing	770
dwarf	621, 629	*michauxii* v. *texana*	196	Monkshood	85
Engelmann's	619	*rubella*	196	Monocots	1021
fourleaf	626	*stricta*	196	*Monolepis nuttalliana*	175
green	630	*Mirabilis*	147	*Monotropa*	340
Hall's	620	*albida*	147	*hypopithys*	340
longhorn	623	v. *uniflora*	148	*uniflora*	341
Mead's	623	*carletonii*	148	Monotropaceae	340
narrow-leaved	627	*decumbens*	148	Moonflower	657
ovalleaf	624	*diffusa*	150	Moonpod	151
Plains	625	*exaltata*	148	spreading	151
poison	627	*gausapoides*	149	Moonseed	110
prairie	620	*glabra*	149	Carolina	110
purple	625	*hirsuta*	149	Moonseed Family	109
sand	618	*jalapa*	147	Moonwort	46
shortcrown	619	*linearis*	150	Mooseberry	830
showy	626	v. *subhispida*	149	Moraceae	126
sidecluster	624	*nyctaginea*	150	Mormon tea	76
smooth	628	*oxybaphoides*	151	Morning-glory	657
spider	631	*pauciflora*	147	bigroot	659
swamp	621	*Miscanthus sacchariflorus*	1114	bush	659
whorled	630	Missouri aster	880	common	660
woolly	622	Missouri evening primrose	523	heavenly blue	657
Milkweed Family	614	Missouri gooseberry	355	ivyleaf	659
Milkwort	564	Missouri milk-vetch	430	Japanese royal	657
sea	347	Missouri spurge	548	red	658
slender	565	Missouri willow	286	white	659
white	565	Mist flower	934	Morning Glory Family	652
whorled	566	Mistletoe, eastern	533	*Morus*	127
Milkwort Family	564	*Mitella nuda*	360	*alba*	127

nigra		127	sand	1193,	1195	Naiad	1040,	1041
rubra		127	wirestem	1194,	1195	Naiad Family		1040
Moschatel		832	Mulberry		127	Nailwort		202
Moschatel Family		832	red		127	James'		203
Mosquito fern		71	paper		126	Najadaceae		1040
Moss rose		188	white		127	Najas		1040
Motherwort	717,	718	Mulberry Family		126	flexilis		1041
Moth mullein		791	Mullein		790	guadalupensis		1041
Moundscale		164	common		791	marina		1041
Mountain ash		405	foxglove		764	Nama		679
Mountain balm		555	moth		791	hispidum		680
Mountain birch		143	Multiflora rose		399	stevensii		680
Mountain huckleberry		335	Munroa squarrosa		1197	Nannyberry		831
Mountain lily		1251	*Muricauda dracontium*		1043	Narrow beardtongue		777
Montain mahogany		369	*Muscadinia rotundifolia*		559	Narrow-leaved cat-tail		1238
Mountain maple		569	Muscari botryoides		1253	Narrow-leaved		
Mountain mint		729	Musineon		596	evening primrose		522
hairy		730	divaricatum		596	Narrow-leaved milkweed		627
slender-leaved		730	v. *hookeri*		597	Narrow-leaved poison-vetch		432
Virginia		730	tenuifolium		597	Narrow-leaved trefoil		462
white		729	*trachyspermum*		597	Narrow-leaved vervain		707
woods		730	Musk thistle		896	Narrow-leaved willow-herb		507
Mountain ninebark		380	Muskroot		832	Narrowleaf bluet		822
Mouse-ear chickweed		197	Mustard		304	Narrowleaf cottonwood		276
common		199	ball		324	Narrowleaf four-o'clock		150
Mouse-ear cress		297	black		305	Narrowleaf globe mallow		251
Mouse-tail		95	blue		308	Nasturtium officinale		323
Muckweed, curly		1035	dog		313	microphyllum		324
Mud plantain		1240	garlic		296	*Naumburgia thyrsiflora*		350
Mudwort		767	hare's-ear		308	Navarretia intertexta v.		
Mugwort		866	hedge	329,	330	propinqua		673
Muhlenbergia		1191	Indian		304	minima		673
arenicola		1193	tall hedge		330	Needle-and-thread		1229
asperifolia		1193	tansy	308,	309	Needlegrass	1229,	1230
brachyphylla		1194	tower		300	green		1230
bushii		1193	tumbling		330	*Neeragrostis*		1171
capillaris		1194	white		304	*reptans*		1177
commutata		1195	Mustard Family		293	*Negundo interius*		569
cuspidata		1194	Muttongrass		1212	*nuttallii*		570
filiculmis		1196	Myosotis		699	*Neltuma glandulosa*		410
filiformis		1194	alpestris		700	Nelumbo lutea		81
foliosa		1195	laxa		700	nucifera		81
frondosa		1194	macrosperma		700	Nelumbonaceae		81
glomerata		1195	scorpioides		699	Nemastylis		1261
gracillima		1197	sylvatica		700	*acuta*		1261
mexicana		1195	verna		700	geminiflora		1261
minutissima		1192	*virginica*		700	*Nemexia ecirrhata*		1266
polystachya		1195	Myosoton		211	herbacea		1267
porteri		1193	aquaticum		212	lasioneuron		1267
pungens		1195	Myosurus		95	pulverulenta		1267
racemosa		1196	aristatus		95	Nemophila phacelioides		680
richardsonis		1196	minimus		95	*Neobesseya missouriensis*		154
schreberi		1196	Myriophyllum		492	*Neomamillaria*		154
simplex		1194	brasiliense		494	missouriensis		154
sobolifera		1197	exalbescens		493	radiosa		155
sylvatica		1197	heterophyllum		493	similis		154
tenuiflora	1195,	1197	pinnatum		493	Nepeta cataria		726
torreyi		1197	spicatum		493	hederacea v. *parviflora*		714
umbrosa		1197	v. *exalbescens*		493	Neptunia lutea		410
Muhly		1191	verticillatum		493	Neslia paniculata		324
forest		1197	*Myzorrhiza ludoviciana*		799	Netleaf hackberry		121
marsh		1196				Nettle		130
plains		1194	*Nabalus albus*		986	bull		538
ring		1197	asper		987	dead		716
rock		1197	racemosus		987	false		128

hedge-		737	Notchbract waterleaf	679	Oats	1139
-leaved vervain		707	*Nothocalais cuspidata*	979	wild	1139, 1152
purple dead		717	Notholaena	61	Obedient plant	727, 728
stinging		130	dealbata	61	Oenothera	516
weak		130	fendleri	61	albicaulis	517
wood		128	sinuata	62	biennis	517
Nettle Family		127	v. cochisensis	62	v. *canescens*	526
New England aster		880	v. integerrima	62	ssp. *centralis*	518
New Jersey tea	554,	555	standleyi	62	caespitosa	518
Nicandra physalodes		641	Nothoscordum bivalve	1253	ssp. *caespitosa*	518
Nicotiana tabacum		637	Nuphar	82	ssp. *eximia*	518
quadrivalis		637	advenum	82	ssp. *macroglottis*	518
trigonophylla		637	luteum	82	ssp. *purpurea*	518
Night-flowering catchfly		208	ssp. macrophyllum	82	canescens	518
Nightshade		647	ssp. variegatum	82	coronopifolia	519
black		650	*variegatum*	82	elata ssp. hirsutissima	519
climbing		649	Nut-rush	1112	engelmannii	519
cut-leaevd		651	*Nuttallia*	270	flava	520
enchanter's		503	decapetala	272	fremontii	523
green false		638	*nuda*	273	grandis	520
melon-leaf		648	stricta	273	*greggii*	502
plains black		649	Nuttall's evolvulus	656	*harringtonii*	518
silver-leaf		649	Nuttall's kentrophyta	429	*hartwegii fendleri*	501
viscid		651	Nuttall's sunflower	955	*hookeri*	519
Nightshade Family		637	Nuttall's violet	260	ssp. *hirsutissima*	519
Nimblewill		1196	Nyctaginaceae	145	howardii	521
Nine-anther prairie clover		441	Nymphaea	82	jamesii	521
Ninebark		380	odorata	83	laciniata	521
mountain		380	tuberosa	83	v. *grandiflora*	521
Nipple cactus	156,	160	Nymphaeaceae	81	latifolia	522
Nipplewort		971			*lavandulaefolia*	502
Nits-and-lice		238	Oak	134	linifolia	522
Nodding beggar-ticks		891	black	141	macrocarpa	523
Nodding brome		1146	blackjack	139	ssp. *fremontii*	523
Nodding chickweed		199	bur	138	ssp. *incana*	523
Nodding fescue		1180	chinkapin	139	ssp. *macrocarpa*	523
Nodding green violet		256	dwarf chinkapin	140	ssp. *oklahomensis*	523
Nodding pogonia		1284	eastern poison	575	v. *oklahomensis*	524
Nodding saxifrage		363	English	137	*missouriensis*	523
Nodding thistle		896	fern	60	v. *oklahomensis*	524
Nodding trillium		1255	Gambel's	138	*muricata*	518
Nodding wild buckwheat		216	Hill's	137	nuttallii	524
Nolina texana		1264	Jerusalem	169	*pallida* ssp. *latifolia*	522
Northern adder's-tongue		48	-leaf thorn-apple	639	*perennis*	516
Northern bedstraw		818	-leaved goosefoot	171	*pilosella*	516
Northern bog violet		259	live	141	rhombipetala	524
Northern catalpa		804	northern pin	137	*serrulata*	502
Northern dewberry		402	pin	139	*oblanceolata*	502
Northern fog-fruit		703	poison	573	spachiana	516
Northern gentian		606	red	137	speciosa	524
Northern grass-of-Parnassus		362	Schneck	140	*strigosa* ssp. *canovirens*	526
Northern green orchis		1277	shin	139	ssp. *strigosa*	526
Northern hawthorn		374	shingle	138	triloba	525
Northern lady-fern		53	shinnery	138	*watsonii*	525
Northern mannagrass		1182	Shumard's	140	villosa	525
Northern marsh-violet		260	swamp white	137	ssp. *strigosa*	526
Northern pin oak		137	wavy-leaf	140	ssp. *villosa*	526
Northern pussy-toes		861	white	137	Ohio horse mint	712
Northern rock jasmine		344	yellow chestnut	139	Oklahoma evening primrose	523
Northern stitchwort		212	Oak Family	134	Oklahoma plum	392
Northern wild comfrey		691	*Oaksiella sessilifolia*	1256	Old plainsman	961
Northern winter cress		303	Oat, spike	1183	Old-field cinquefoil	389
Norton's flax		563	Oatgrass	1155	Old-field toadflax	768
Norwegian cinquefoil		387	poverty	1155	Oldfield three-awn	1137
Noseburn	552,	553	tall	1114	Oleaceae	747

Oleaster	490	*pedunculatum*	475	Osmundaceae	48
Oleaster Family	490	Orchard grass	1154	Ostrich fern	60
Oligoneuron bombycinum	1006	Orchid, fairy slipper	1270	*Ostrya virginiana*	144
canescens	1006	frog	1279	v. *lasia*	144
riddellii	1006	Orchid Family	1268	*Othake sphacelata*	981
rigidum	1006	Orchidaceae	1268	*Otophylla auriculata*	790
Olive, Russian	490	Orchis, Alaska	1280	*densiflora*	790
Olive Family	747	green woodland	1277	Oval ladies'-tresses	1283
Onagra nuttallii	524	long-bracted	1279	Oval-leaf bladderpod	323
Onagraceae	498	northern green	1277	Ovalleaf milkweed	624
One-flowered cancer-root	799	prairie fringed	1278	Owl clover	772
One-flowered wintergreen	339	ragged	1278	Oxalidaceae	578
One-seeded croton	540	round-leaved	1278	*Oxalis*	578
One-seeded juniper	72	showy	1275	*corniculata*	579
One-sided wintergreen	339	slender bog	1278	*dillenii*	579
Onion	1243	*spectabilis*	1275	*stricta*	579
pink wild	1245	white	1277	*violacea*	579
wild	1244, 1245	Oregon grape	108	Ox-eye	958
Onobrychis viciaefolia	416	Oregon woodsia	69	Ox-eye daisy	902
Onoclea sensibilis	62	*Oreocarya*	688	Ox-tongue	984
v. *obtusilobata*	63	*affinis*	689	*Oxytropis*	467
Onopordum acanthium	979	*glomerata*	689	*besseyi*	467
tauricum	980	*macounii*	689	*campestris* v. *gracilis*	468
Onosmodium	700	*perennis*	689	v. *dispar*	468
hispidissimum	701	*suffruticosa*	690	*deflexa* v. *sericea*	468
molle	700	*thrysiflora*	690	*dispar*	468
v. *occidentale*	701	Oriental bittersweet	534	*gracilis*	468
v. *hispidissimum*	701	Oriental poppy	113	*hookeriana*	469
v. *subsetosum*	701	*Ornithogalum umbellatum*	1253	*involuta*	469
occidentale	701	Orobanchaceae	798	*lagopus* v. *atropurpurea*	468
Oonopsis argillacea	949	*Orobanche*	798	*lambertii*	469
engelmannii	948	*fasciculata*	798	v. *articulata*	469
Ophioglossaceae	45	*ludoviciana*	799	v. *bigelovii*	469
Ophioglossum	48	*multiflora*	799	v. *lambertii*	469
engelmannii	48	*uniflora*	799	*macounii*	468
vulgatum	48	*Orophaca*	421	*multiceps*	469
v. *pseudopodum*	48	*argophylla*	429	*nana*	470
v. *pycnostichum*	48	*caespitosa*	428	*plattensis*	469
Opium poppy	113	*sericea*	435	*pinetorum*	470
Opuntia	156	Orophaca, Barr	425	*richardsonii*	471
compressa	157	plains	428	*sericea*	470
davisii	160	silky	434	v. *sericea*	470
erinacea v. *utahensis*	159	summer	428	*splendens*	471
fragilis	156	*Orthilia secunda*	339	*villosa*	468
humifusa	157	*Orthocarpus luteus*	772	Ozark milk-vetch	427
imbricata	158	*Oryzopsis*	1198		
leptocaulis	158	*asperifolia*	1198	*Pachylophus*	516
lindheimeri	158	*bloomeri*	1199	*caespitosus*	518
macrorhiza	158	*exigua*	1198	*canescens*	518
v. *macrorhiza*	158	*hymenoides*	1199	*eximus*	518
v. *pottsii*	158	*micrantha*	1199	*montanus*	518
phaeacantha	158	*pungens*	1198	*Pagesia acuminata*	770
v. *camanchica*	159	*racemosa*	1199	Paintbrush, downy	762
v. *major*	159	Osage orange	126	Paintbrush, Indian	759
v. *phaeacantha*	159	*Osmorhiza*	597	Paintbrush, Wyoming	761
polyacantha	159	*chilensis*	597	Painted euphorbia	544
v. *polyacantha*	159	*claytoni*	597	Painted milk-vetch	426
v. *trichophora*	159	*depauperata*	598	*Palafoxia*	980
rutila	159	*divaricata*	597	*callosa*	980
tortispina	158	*longistylis*	598	*sphacelata*	981
tunicata v. *davisii*	159	v. *longistylis*	598	*rosea*	980
Orache	161, 163	v. *villicaulis*	698	v. *macrolepis*	980
red	164	*obtusa*	598	v. *rosea*	980
Orange, osage	126	*villicaulis*	598	Pale alyssum	297
Orbexilum	471	*Osmunda regalis* v. *spectabilis*	48	Pale coral-root	1272

INDEX

Pale corydalis	116	*villosissimum*	1158	*bushii*	1207		
Pale dogwood	528	virgatum	1203	*ciliatifolium*	1206		
Pale dock	232	*werneri*	1160	v. *muhlenbergii*	1206		
Pale echinacea	922	*wilcoxianum*	1161	v. *stramineum*	1207		
Pale evening primrose	517, 522	*xanthophysum*	1159	*circulare*	1205		
Pale Indian plantain	895	Pansy, wild	263	*dilatatum*	1204		
Pale penstemon	786	Papaver	113	*distichum*	1206		
Pale-seeded plantain	746	dubium	113	field	1205		
Pale smartweed	226	orientale	113	Florida	1204		
Pale touch-me-not	583	rhoeas	113	floridanum	1204		
Pale wolfberry	641	somniferum	113	v. *floridanum*	1205		
Palespike lobelia	812	Papaveraceae	110	v. *glabratum*	1205		
Palm-leaved scurf-pea	473	Paper birch	143	fluitans	1205		
Palmer's hawthorn	373	Paper flower	987	*glabratum*	1205		
Palmer's pigweed	182	Paper mulberry	126	horsetail	1205		
Palmer's snowberry	828	Parietaria pensylvanica	129	laeve	1205		
Panax quinquefolium	584	Parnassia	361	v. *circulare*	1205		
Panic grass	1200	*americana*	362	v. *pilosum*	1205		
Panicled aster	878, 885	glauca	361	*longipilum*	1205		
Panicled tickclover	450	palustris	362	*mucronatum*	1205		
Panicum	1200	v. *neogaea*	362	*muhlenbergii*	1206		
albemarlense	1158	parviflora	362	paspalodes	1206		
anceps	1200	Paronychia	202	*pubescens*	1206		
agrostoides	1203	canadensis	202	pubiflorum v. *glabrum*	1206		
barbipulvinatum	1201	depressa	202	setaceum	1206		
beaked	1200	*diffusa*	203	v. *muhlenbergii*	1206		
capillare	1201	fastigiata	203	v. *stramineum*	1206		
clandestinum	1159	v. *paleacea*	203	*stramineum*	1207		
dichotomiflorum	1201	jamesii	203	*vaginatum*	1206		
fall	1201	sessiliflora	203	Pasque flower	88		
fasciculatum	1200	wardii	203	*Passerina annua*	498		
flexile	1201	Parosela alopecurioides	443	Passiflora	266		
hallii	1202	aurea	440	incarnata	266		
v. *filipes*	1202	enneandra	441	lutea v. *glabriflora*	267		
hillmanii	1202	*jamesii*	442	Passifloraceae	266		
hirticaule	1202	lanata	443	Passion-flower	266, 267		
huachucae	1158	nana	444	Passion-flower Family	266		
lanuginosum	1158	Parrot's feather	494	Pastinaca sativa	598		
v. *fasciculatum*	1158	Parrot's feather, green	493	Pasture rose	399		
v. *implicatum*	1158	*Parsley*	585	Patagonian plantain	745		
v. *lindheimeri*	1158	Parsley, dog	596	Path rush	1056		
latifolium	1159	hedge	603	Patience dock	234		
leibergii	1159	hemlock	585	Pawpaw	77		
lindheimeri	1158	prairie	599	Pea, breadroot scurf-	473		
linearifolium	1160	sand-	586	butterfly	438		
malacophyllum	1160	Parsley Family	584	downy milk-	452		
miliaceum	1202	Parsnip, cow	594	everlasting	455		
obtusum	1202	meadow	602	*garden*	416		
oligosanthes v.		water-	587, 601	hoary	480		
scribnerianum	1160	wild	595, 598	lemon scurf-	474		
perlongum	1160	Parthenium	981	little breadroot scurf-	474		
philadelphicum	1203	auriculatum	981	milk-	452		
praecocius	1158	hispidum	981	palm-leaved scurf-	473		
redtop	1203	hysterophorus	981	scarlet	454		
reverchonii	1218	incanum	982	scurf-	471		
rhizomatum	1201	integrifolium	982	scurfy	475		
rigidulum	1203	Parthenocissus	558	sensitive partridge	413		
scoparioides	1158	*inserta*	558	showy partridge	412		
scoparium	1161	quinquefolia	558	-shrub	437		
scribnerianum	1160	f. *hirsuta*	558	Siberian	437		
sphaerocarpon	1161	vitacea	558	silver-leaf scurf-	472		
subvillosum	1158	f. *dubia*	559	slimleaf scurf-	475		
tennesseense	1158	Partridge pea, sensitive	413	tall-bread scurf-	472		
texanum	1203	showy	412	vetch, deer	488		
Texas	1203	Paspalum	1204	yellow	481		

INDEX

Peach	390, 394	buckleyi	778	*coccinea*	225
Peachleaf willow	285	Buckley's	778	*hydropiper*	226
Peanut	416	*caudatus*	777	*hydropiperoides*	226
Peanut, hog	420	cobaea	779	*iowensis*	225
Pear	396	digitalis	779	*lapathifolia*	227
Pearlwort	203	f. *baueri*	780	*longistyla*	225
trailing	203	eriantherus	780	*maculosa*	227
Pearly everlasting	858	v. *eriantherus*	780	*mesochora*	225
Pecan	131, 132	fendleri	780	*nebraskensis*	225
Pectis angustifolia	982	gilia	776	*omissa*	227
Pedaliaceae	803	glaber	781	*opelousana*	226
Pedicularis	772	v. alpinus	782	*orientalis*	227
canadensis	773	f. *riparius*	782	*pensylvanica*	227
v. *dobsii*	773	v. brandegei	782	*pratincola*	225
v. *fluviatilis*	773	v. glaber	782	*psycrophila*	225
fluviatilis	773	f. *pubicaulis*	782	*punctata*	228
grayi	774	gracilis	782	*rigidula*	225
lanceolata	773	grandiflorus	783	*setacea*	226
procera	773	haydenii	783	*tomentosa*	227
Pediocactus simpsonii	160	Hayden's	783	*vestita*	225
Pediomelum	471	jamesii	784	Persimmon	342
cuspidatum	473	laxiflorus	784	Peruvian spikemoss	40
esculentum	473	nitidus	785	*Petalostemon*	439
hypogaeum	474	oklahomensis	785	*arenicola*	444
Pellaea	63	pale	786	*candidus*	441
atropurpurea	63	pallidus	786	*compactus*	441
v. *bushii*	64	procerus	787	*mollis*	444
fendleri	62	ssp. *procerus*	787	*multiflorus*	443
glabella	63	secundiflorus	787	*occidentale*	441
v. glabella	63	smooth	779	*oligophyllus*	441
v. occidentalis	64	tubaeflorus	788	*porterianus*	445
mucronata	64	v. *achoreus*	788	*pulcherrimum*	439
pumila	64	v. *tubaeflorus*	788	*purpureus*	444
ternifolia v. *wrightiana*	64	tube	788	*standsfieldii*	439
wrightiana	64	virens	788	*tenuifolius*	445
Pellitory	129	Penthorum sedoides	357	*tenuis*	439
Pennsylvania	129	*Peplis diandra*	496	*villosus*	445
Pencil cholla	158	*Pepo foetidissimus*	268	Petasites	983
Pencil-flower	480	Peppergrass	316, 317, 318	frigidus	983
Pendulous-pod loco	468	bushy	318	v. *palmatus*	983
Peniophyllum	516	clasping	317	sagittatus	983
linifolium	523	field	317	Petrophytum caespitosum	380
Pennsylvania pellitory	129	Peppermint	722	*Petroselinum crispum*	585
Pennsylvania smartweed	227	Pepperwort	70	Petty spurge	541
Pennycress	332	Pepperwort Family	70	*Phaca*	421
field	332	Perennial bursage	858	*americana*	425
perfoliate	333	Perennial ryegrass	1189	*bodinii*	425
Pennyroyal, American false	716	Perennial sweetpea	455	*longifolia*	426
Drummond false	715	Perfoliate pennycress	333	*neglecta*	431
false	740	Perfumed cherry	393	Phacelia	680
mock	714	Pericome caudata	982	congesta	681
Reverchon false	716	v. *glandulosa*	983	*franklinii*	683
rough false	715	Periderida	599	gilioides	681
Pennywort, water-	594	americana	599	gyp	682
Penstemon	774	gairdneri	599	hairy	681
albidus	776	Perilla	726	hastata v. leucophylla	682
ambiguus	776	common	726	*heterophylla*	682
v. *ambiguus*	777	frutescens	726	hirsuta	682
v. *laevissimus*	777	Periploca graeca	635	integrifolia	682
angustifolius	777	*Peritoma luteum*	292	v. *integrifolia*	682
v. angustifolius	777	serrulatum	292	v. *texana*	682
v. caudatus	777	Periwinkle	613	leucophylla	682
ssp. *caudatus*	777	common	613	linearis	682
auriberbis	778	Persian clover	486	popei	683
bradburii	783	*Persicaria*	220	robusta	683

INDEX

strictiflora	683
v. *connexa*	683
v. *lundelliana*	683
v. *robbinsii*	683
v. *strictiflora*	683
Phalaris	1207
arundinacea	1207
canariensis	1207
caroliniana	1208
Phaseolus coccinea	416
polystachios	416
vulgaris	416
Phegopteris	67
dryopteris	60
hexagonoptera	67
Phellopterus	591
montanus	591
Philadelphia witchgrass	1203
Phleum	1208
alpinum	1208
pratense	1208
Phlomis tuberosa	709
Phlox	673
alyssifolia	674
ssp. *abdita*	674
ssp. *alyssifolia*	674
andicola	674
ssp. *andicola*	674
ssp. *parvula*	674
ssp. *planitiarum*	674
blue	675
bryoides	674
divaricata ssp. laphamii	675
fall	676
hoodii	675
ssp. *hoodii*	675
ssp. *glabrata*	675
ssp. *muscoides*	675
ssp. *viscidula*	675
Hood's	675
longipilosa	675
paniculata	676
pilosa	676
ssp. fulgida	676
ssp. ozarkana	677
ssp. pilosa	677
ssp. pulcherrima	677
plains	674
prairie	676
Phoradendron	532
flavescens	533
serotinum	533
v. *pubescens*	533
tomentosum	533
Phragmites australis	1208
communis	1208
v. *berlandieri*	1208
Phyrma leptostachya	703
Phyrmaceae	702
Phyla	702
cuneifolia	703
lanceolata	703
nodiflora	703
v. *incisa*	703
v. *texensis*	703

Phyllanthus	551
abnormis	551
caroliniensis	551
polygonoides	552
Physalis	642
alkekengi	642
ambigua	644
angulata	642
v. *angulata*	643
v. *pendula*	643
comata	643
hederifolia	643
v. comata	643
v. cordifolia	643
v. hederifolia	644
heterophylla	644
ixocarpa	642
lanceolata	645
lobata	646
longifolia	644
v. *hispida*	645
v. *longifolia*	644
v. *subglabrata*	644
macrophysa	644
missouriensis	645
mollis	646
nyctaginea	644
oklahomensis	676
pendula	642
pruinosa	645
pubescens	644
v. *glabra*	645
v. *grisea*	645
v. *integrifolia*	645
v. *missouriensis*	645
v. *pubescens*	645
pumila	645
ssp. hispida	645
ssp. pumila	645
rotundata	643
subglabrata	644
virginiana	646
v. *sonorae*	644, 645
v. *subglabrata*	644
viscosa ssp. mollis v. cinerascens	645
v. *mollis*	646
Physaria brassicoides	324
didymocarpa	324
Physocarpa	380
intermedia	381
monogyna	380
Physocarpus	380
monogynus	380
opulifolius	380
v. *intermedius*	381
Physostegia	727
angustifolia	727
formosior	728
intermedia	727
ledinghamii	728
parviflora	728
virginiana	728
ssp. *praemorsa*	728
v. *speciosa*	728

ssp. *virginiana*	728
Phytolacca americana	144
decandra	144
Phytolaccaceae	144
Pickerel weed	1240
Pickerel-weed Family	1239
Picradeniopsis	983
oppositifolia	984
woodhousei	984
Picea glauca	74
pungens	75
Picris echioides	984
Pigeon grape	560
Pigeon wings	438
Pignut	416
Pigweed	180
green	182
Palmer's	182
Powell's	183
prostrate	181
rough	183
Russian	165
sandhills	181
slender	182
spiny	184
winged	174
Pigweed Family	179
Pilea	129
fontana	129
opaca	129
pumila	129
v. *deamii*	129
Pillwort	70
American	70
Pilostyles thurberi	442, 533
Pilularia americana	70
Pimpernel	343
false	769
water	350, 351
yellow	602
Pinaceae	74
Pin-cherry	394
Pincushion cactus	154
Pinchot juniper	73
Pine	75
drops	341
ground	39
limber	76
lodgepole	75
pinyon	75
ponderosa	76
prince's	337
Pine Family	74
Pineapple weed	978
Pinesap	340
Pink	199
deptford	199
mimosa	409
poppy mallow	243
pussy-toes	860
smartweed	225
vervain	707
wild onion	1245
Pink Family	192
Pin oak	139

northern	137	*pusilla*	744	annua	1210
Pinophyta	71	rhodosperma	746	arachnifera	1210
Pinus	75	rugelii	746	arida	1211
banksiana	75	spinulosa	745	*brevipaniculata*	1212
contorta v. latifolia	75	virginica	746	*buckleyana*	1214
echinata	75	v. *viridescens*	746	*bulbosa*	1209
edulis	75	wrightiana	747	canbyi	1211
flexilis	76	Plantain	742	chapmaniana	1211
murrayana	75	alkali	744	compressa	1212
ponderosa	76	bracted	743	*confusa*	1213
scopulorum	76	common	744	*crocata*	1214
Pinweed	254, 255	dwarf rattlesnake	1276	cusickii	1212
Pinyon pine	75	English	744	fendleriana	1212
Piperia unalascensis	1280	Indian	743, 895	*glauca*	1209
Pipevine, woolly	80	mud	1240	glaucifolia	1212
Pisophaca	421	pale Indian	895	interior	1213
elongata	428	pale-seeded	746	juncifolia	1213
flexuosa	428	Patagonian	745	*laevigata*	1211
Pisum sativum	416	rattlesnake	1276	*languida*	1209
Pitcher sage	731	red-seeded	746	*longiligula*	1212
Pitcher's clematis	91	robin's	931	*lucida*	1211
Pitseed goosefoot	169	Rugel's	746	nemoralis v. *interior*	1213
Plagiobothrys	701	slender	744	*overi*	1211
scopulorum	701	water	1022, 1023	palustris	1213
scouleri	701	Wright's	747	*plattensis*	1213
Plainleaf pussy-toes	861	Plantain Family	742	pratensis	1214
Plains beebalm	725	Platanaceae	119	*pratensiformis*	1211
Plains black nightshade	649	*Platanthera*	1276	*rupicola*	1212
Plains bluegrass	1211	orbiculata	1278	sandbergii	1214
Plains bristlegrass	1220	Platanus occidentalis	119	*secunda*	1214
Plains coreopsis	916	f. *attenuata*	119	sylvestris	1214
Plains flax	563	v. *glabrata*	119	*truncata*	1213
Plains ironweed	1018	Platte lupine	463	*wolfii*	1215
Plains lovegrass	1175	Platte River milk-vetch	432	Poaceae	1113
Plains milkweed	625	Platte thistle	910	Podophyllum peltatum	109
Plains muhly	1194	*Pleiotaenia nuttallii*	599	*Podospermum laciniatum*	992
Plains orophaca	428	*Pleurophragma*	332	Pogonia	1281
Plains phlox	674	*lilacinum*	332	nodding	1284
Plains prickly pear	158, 159	Pliant milk-vetch	427	ophioglossoides	1281
Plains reedgrass	1149	Pluchea	985	rose	1281
Plains sunflower	956	*camphorata*	985	*Poinsettia*	541
Plains wild indigo	437	odorata	985	*cuphusperma*	545
Plains yellow primrose	502	v. *odorata*	985	*dentata*	545
Plane-tree	119	*purpurascens*	985	*heterophylla*	544
Planeleaf willow	290	Plum	390	Poison hemlock	590
Plantaginaceae	742	big-tree	393	Poison ivy	573, 574
Plantago	742	*Canada-*	390	Poison milkweed	627
arenaria	743	chickasaw	392	Poison oak	573
aristata	743	*common*	390	eastern	575
asiatica	745	creek	395	Poison suckleya	179
elongata	744	*damson*	390	Poison-vetch, creamy	434
eriopoda	744	ground-	426	narrow-leaved	432
heterophylla	744	hog	395	Pokeberry	144
indica	743	Mexican	393	Pokeweed	144
lanceolata	744	Oklahoma	392	Pokeweed Family	144
v. *sphaerostachya*	744	sandhill	392	Polanisia	292
major	744	sugar-	368	dodecandra	292
v. *pilgeri*	745	wild	391	ssp. *dodecandra*	293
v. *scopulorum*	745	wild goose	392, 394	ssp. trachysperma	292
patagonica	745	Plumbaginaceae	235	*graveolens*	293
v. *breviscapa*	745	Plumeless thistle	895, 896	jamesii	293
v. *gnaphaloides*	745	Poa	1209	*trachysperma*	293
v. *patagonica*	745	*agassizensis*	1214	Polecat bush	571
v. *spinulosa*	745	*alsodes*	1209	Polemoniaceae	666
psyllium	743	*ampla*	1213	Polemonium	677

brandegei	677	
creeping	677	
reptans	677	
viscosum ssp. *mellitum*	677	
Polemonium Family	666	
Polycodium neglectum	336	
stamineum	336	
Polygala	564	
alba	565	
blood	565	
incarnata	565	
sanguinea	565	
senega	565	
v. *latifolia*	566	
tweedyi	566	
verticillata	566	
v. *ambigua*	566	
v. *isocycla*	566	
v. *sphenostachya*	566	
viridescens	565	
Polygalaceae	564	
Polgonaceae	214	
Polygonatum biflorum	1254	
canaliculatum	1254	
commutatum	1254	
giganteum	1254	
Polygonella americana	220	
Polygonum	220, 665	
achoreum	222	
amphibium	225	
v. *emersum*	225	
v. *stipulaceum*	225	
arenastrum	222	
aviculare	222	
v. *angustissimum*	222	
v. *erectum*	223	
bicorne	225	
buxiforme	222	
cespitosum v. longisetum	226	
coccineum	225	
compactum	229	
convolvulus	228	
cuspidatum	228	
v. *compactum*	229	
douglasii	222	
erectum	223	
exsertum	223	
fluitans	225	
hartwrightii	225	
hydropiper	226	
hydropiperoides	226	
v. *hydropiperoides*	226	
v. *persicarioides*	226	
lapathifolium	226	
latum	223	
leptocarpum	223	
longistylum	225	
natans	225	
neglectum	222, 223	
orientale	227	
pensylvanicum	227	
persicaria	227	
persicarioides	226	
prolificum	223	
punctatum	227	
ramosissimum	223
v. *prolificum*	223
v. *ramosissimum*	223
reynoutria	229
sachalinense	229
sagittatum	229
sawatchense	223
scandens	229
v. *dumetorum*	229
v. *scandens*	229
tenue	223
virginianum	229
viviparum	230
Polymnia	985
canadensis	985
uvedalia	986
Polypodiaceae	49
Polypodium	64
amorphum	64
hesperium	64
polypodioides v. michauxianum	64
virginianum	65
vulgare v. *virginianum*	65
Polypody	64
common	65
western	64
Polypogon monspeliensis	1215
Polystichum	65
acrostichoides	65
f. *incisum*	65
lonchites	65
munitum	66
Polytaenia nuttalii	599
Ponderosa pine	76
Pondweed	1032
baby	1038
claspingleaf	1039
flatstem	1039
floatingleaf	1037
horned	1041
Illinois	1037
large	1035
leafy	1036
longleaf	1037
ribbonleaf	1035
sago	1038
sheathed	1039
slender	1036
variable	1036
waterthread	1035
whitestem	1038
Pondweed Family	1032
Pontederia cordata	1240
Pontederiaceae	1239
Poorman's weatherglass	343
Popcorn flower	701
Poplar	273
balsam	277
black	280
Carolina	277
gray	278
hybrid balsam	277
lombardy	280
silver	276
Poppy	113
field	113
hairy prickly	111
hedgehog prickly	112
horned	112
longhead	113
mallow	243
finger	244
pink	243
purple	244
opium	113
Oriental	113
prickly	111
red horned	112
Poppy Family	110
Populus	273
× acuminata	275
alba	276
× *andrewsii*	280
angustifolia	276
balsamifera	277
v. *subcordata*	277
× *berolinensis*	274
besseyana	279
bolleana	276
× brayshawii	277
× canadensis	277
v. *eugenei*	278
candicans	280
× canescens	278
deltoides	278
ssp. *deltoides*	279
ssp. *monilifera*	278
ssp. *wislizenii*	279
× *eugenei*	278
× *gileadensis*	280
grandidentata	279
× *jackii*	279
maximowiczii	274
nigra	280
'Italica'	280
v. *italica*	280
sargentii	279
v. *texana*	279
simonii	274
tacamahaca	277
tremula	276, 278
tremuloides	280
v. *aurea*	281
trichocarpa	274, 280
virginiana	279
Porcupine-grass	1230
Porteranthus stipulatus	381
Portulaca	188
grandiflora	188
mundula	189
neglecta	189
oleracea	189
parvula	189
retusa	189
umbraticola	189
Portulacaceae	187
Possum grape	557
Possum-haw	535
Post-oak	140

Potamogeton	1032	*argyrea*	387	nine-anther	441
alpinus	1034	*atrovirens*	388	purple	444
v. *tenuifolius*	1034	biennis	384	round-headed	443
americanus	1038	*bipinnatifida*	388	silky	445
amplifolius	1035	*camporum*	386	slimleaf	444
angustifolius	1037	concinna	384	white	440
berchtoldii v.		v. *concinna*	384	Prairie coneflower	989
acuminatus	1038	v. *divisa*	384	grayhead	990
v. *calophilus*	1038	v. *macounii*	384	short-ray	990
v. *lacunatus*	1038	diversifolia	384	Prairie cordgrass	1223
v. *polyphyllus*	1038	v. *perdissecta*	385	Prairie cupgrass	1178
v. *tenuissimus*	1038	*effusa*	387	Prairie dogbane	612
compressus	1039	*finitima*	388	Prairie dropseed	1227
crispus	1035	fissa	385	Prairie fameflower	190
diversifolius	1035	*flabelliformis*	386	Prairie fringed orchis	1278
epihydrus	1035	fruticosa	385	Prairie gaillardia	938
v. *nuttallii*	1036	*glabella*	388	Prairie gentian	605, 607
filiformis	1036	glandulosa	385	Prairie goldenrod	1004
v. *borealis*	1036	v. *glandulosa*	386	Prairie ground cherry	643, 645
v. *macounii*	1036	v. *intermedia*	386	Prairie iris	1261
foliosus	1036	v. *pseudorupestris*	386	Prairie larkspur	94
v. *macellus*	1036	*glaucophylla*	385	Prairie lespedeza	460
friesii	1036	gracilis	386	Prairie milkweed	620
gramineus	1036	v. *glabrata*	386	Prairie parsley	599
v. *maximus*	1037	hippiana	386	Prairie phlox	676
v. *myriophyllus*	1037	*lasiodonta*	388	Prairie primrose	517
heterophyllus	1037	*millegrana*	389	Prairie ragwort	997
illinoensis	1037	*monspeliensis*	387	Prairie rose	400
interior	1036	*nicolletii*	387	gentian	609
interruptus	1038	norvegica	387	white	399
lucens	1037	*nuttallii*	386	Prairie sandreed	1150
natans	1037	paradoxa	387	Prairie spurge	548
moniliformis	1039	pensylvanica	387	Prairie star	360
nodosus	1037	v. *bipinnatifida*	388	Prairie trefoil	461
panormitanus v. *major*	1038	v. *glabrata*	388	Prairie three-awn	1137
v. *minor*	1038	*pentandra*	389	Prairie-turnip	473
pectinatus	1038	plattensis	388	Prairie violet	261
perfoliatus	1039	*propinqua*	387	blue	261
praelongus	1038	recta	388	yellow	260
pusillus	1038	rivalis	389	Prairie wild rose	398
v. *calpophilus*	1039	v. *millegrana*	389	Prairie willow	287
v. *cuspidatus*	1039	v. *pentandra*	389	Prenanthes	986
v. *lacunatus*	1039	v. *rivalis*	389	alba	986
v. *minor*	1038	simplex	389	aspera	987
v. *polyphyllus*	1039	v. *argyrisma*	389	racemosa ssp. multiflora	987
v. *pusillus*	1038	v. *calvescens*	389	v. *racemosa*	987
v. *tenuissimus*	1038	strigosa	388	Prickly ash	577
richardsonii	1039	*sulphurea*	388	Prickly lettuce	970
rutilus	1039	tridentata	389	Prickly medic	465
spirillus	1035	*viridescens*	386	Prickly pear	156, 158
strictifolius	1039	*Poteridium annuum*	405	eastern	157
v. *rutiloides*	1039	Poverty grass 254, 1227,	1228	little	156
vaginatus	1039	Poverty oatgrass	1155	plains	158, 159
zosteriformis	1039	Poverty three-awn	1137	Texas	158
Potamogetonaceae	1032	Poverty weed 175,	965	Prickly poppy	111
Potato	647	Powell's pigweed	183	hairy	111
Potato bean, American	420	Powell's saltbush	164	hedgehog	112
sweet	657	Prairie acacia	408	Prickly sida	250
Potato Family	637	Prarie aster	886	Prickly sow thistle	1008
Potentilla	381	Prairie buck bean	481	Prickly wild rose	397
anserina	383	Prairie burnet	405	Primrose	350
argentea	383	Prairie buttercup	104	Berlandier evening	500
arguta	383	Prairie chickweed	197	combleaf evening	519
v. *arguta*	384	Prairie-clover, golden	440	common evening	517, 525
v. *convallaria*	384	massive spike	441	cut-leaved evening	521

Engelmann's evening	519	
evening	516	
floating evening	515	
fourpoint evening	524	
Fremont's evening	523	
gumbo evening	518	
Hartweg evening	501	
hoary evening	523	
large-flowered, cut-leaved evening	520	
lavendar leaf	502	
Missouri evening	523	
narrow-leaved evening	522	
Oklahoma evening	523	
pale evening	517, 522	
plains yellow	502	
prairie	517	
showy white evening	524	
spotted evening	518	
stemless evening	525	
water	515	
white-stemmed evening	524	
Primrose Family	342	
Primula incana	350	
Primulaceae	342	
Prince's pine	337	
Prince's plume	331	
Prionopsis ciliata	948	
Privet, common	747	
swamp	748	
Proboscidea louisianica	803	
Proso millet	1202	
Prosopis	410, 533	
chilensis glandulosa	410	
glandulosa	410	
Prostrate pigweed	181	
Prostrate vervain	705	
Prunella vulgaris	728	
v. *lanceolata*	729	
v. *vulgaris*	729	
Prunus	390	
americana	391	
v. *lanata*	393	
angustifolia	392	
v. *varians*	392	
v. *watsonii*	392	
armeniaca	390	
avium	390	
besseyi	396	
cerasus	390	
demissa	396	
domesticus	390	
gracilis	392	
hortulana	392	
insititia	390	
lanata	393	
mahaleb	393	
melanocarpa	396	
mexicana	393	
munsoniana	394	
nana	396	
nigra	390	
× *orthosepala*	392	
× *palmeri*	393	
pensylvanica	394	
persica	394	
pumila v. *besseyi*	394	
rivularis	395	
rugosa	392	
serotina	395	
× *slavinii*	392	
virginiana	395	
v. *deamii*	396	
v. *melanocarpa*	396	
v. *virginiana*	396	
watsonii	392	
Psedera hirsuta	558	
quinquefolia	558	
vitacea	558	
Psilostrophe	987	
tagetina	987	
v. *lanata*	988	
villosa	988	
Psoralea	471	
argophylla	472	
robustior	472	
collina	472	
cuspidata	472	
digitata	473	
v. *parvifolia*	473	
esculenta	473	
floribunda	476	
hypogaea	474	
v. *hypogaea*	474	
v. *scaposa*	474	
lanceolata	474	
linearifolia	475	
micrantha	474	
psoralioides v. *eglandulosa*	475	
tenuiflora	475	
v. *floribunda*	476	
v. *tenuiflora*	476	
Psoralidium	471	
argophyllum	472	
batesii	476	
collinum	472	
digitatum	473	
floribundum	476	
lanceolatum	474	
linearifolium	475	
micranthum	474	
tenuiflorum	476	
Ptelea trifoliata	576	
ssp. *angustifolia* v. *persicifolia*	576	
ssp. *polyadenia*	576	
ssp. *trifoliata*	576	
Pteretis nodulosa	61	
pennsylvanica	61	
Pteridium aquilinum	66	
v. *latiusculum*	66	
v. *pseudocaudatum*	66	
v. *pubescens*	67	
Pteridophytes	39	
Pterospora andromedea	341	
Ptilimnium	599	
capillaceum	600	
nuttallii	599	
Ptiloria pauciflora	1009	
ramosa	1009	
Puccinellia	1215	
airoides	1216	
cusickii	1215	
distans	1216	
nuttalliana	1216	
Puccoon	696	
hoary	696	
Pueraria lobata	476	
Pulmonaria	684	
Pulsatilla	86	
ludoviciana	88	
Pulse milk-vetch	435	
Puncture vine	578	
Purple avens	379	
Purple coneflower	921	
Purple cudweed	942	
Purple dead nettle	717	
Purple giant hyssop	711	
Purple ground cherry	646	
Purple-leaved willow herb	506	
Purple locoweed	469	
Purple loosestrife	497	
Purple lovegrass	1178	
Purple meadow rue	107	
Purple milkweed	625	
Purple poppy mallow	244	
Purple prairie clover	444	
Purple rocket	315	
Purple-stemmed cliff-brake	63	
Purple three-awn	1139	
Pursh milk-vetch	433	
Purslane	188	
common	189	
horse	153	
sea	152	
slenderleaf	189	
speedwell	796	
water	496	
Purslane Family	187	
Pussy-toes	859, 861	
anaphalis	859	
dwarf	860	
field	860	
northern	861	
pink	860	
plainleaf	861	
Pussy willow	286	
western	290	
Putty-root	1269	
Pycnanthemum	729	
albescens	729	
flexuosum	730	
pilosum	729, 730	
tenuifolium	730	
torrei	729	
virginianum	730	
Pyrola	337	
americana	339	
asarifolia	338	
chlorantha	340	
elliptica	338	
picta	338	
rotundifolia	339	
secunda	339	

uliginosa	338	midland	41	v. *falsus*	102
uniflora	339	Quillwort Family	41	v. hispidus	102
virens	339	Quincula lobata	646	v. *marylandicus*	102
Pyrolaceae	336			v. nitidus	102
Pyrrhopappus	988	Rabbitberry	492	inamoenus	102
carolinianus	988	Rabbit brush	906	laxicaulis	103
grandiflorus	989	Rabbit-foot clover	482	*limosus*	101
multicaulis v. geiseri	989	Rabbit foot grass	1215	longirostris	103
v. *multicaulis*	989	Rabbit-tobacco	937	macounii	103
rothrockii	988	Raccoon grape	557	micranthus	104
Pyrrocoma lanceolata	949	*Radicula*	325	obtusiusculus	103
Pyrus	396	*columbiae*	326	*ovalis*	105
angustifolia	396	*sessiliflora*	328	pensylvanicus	104
communis	396	*sinuata*	328	*purshii*	101
coronaria	396	Radish	324	v. *dissectus*	101
ioensis	396	Rafflesia Family	533	v. *geranioides*	101
malus	396	Rafflesiaceae	533	v. *polymorphus*	100
		Ragged orchis	1278	v. *radicans*	101
Quackgrass	1126	Ragweed	855, 856	v. *schizanthus*	100
Quaking aspen	280	bur	856	recurvatus	104
Quamoclit	657	common	856	*reptans*	100
coccinea	658	giant	858	v. *ovalis*	100
vulgaris	660	short	856	rhomboideus	104
Quassia Family	575	southern	856	*rivularis*	104
Queen Anne's lace	592	western	857	sardous	105
Queen's delight	552	Ragwort, gray	994	sceleratus	105
Quercus	134	prairie	997	v. *multifidus*	105
alba	137	Riddell	998	v. *sceleratus*	105
bicolor	137	swamp	994	*septentrionalis*	102
borealis	137	Rain-lily	1248	v. *caricetorum*	102
v. borealis	137	Ranunculaceae	84	v. *pterocarpus*	102
v. maxima	137	Ranunculus	95	subridigus	106
coccinea	137	abortivus	97	testiculatus	106
ellipsoidalis	137	v. *acrolasius*	98, 104	*texensis*	103
gambelii	138	f. *giganteus*	98	trichophyllus	103
havardii	138	acris	98	*waldronii*	101
imbricaria	138	*aquatilis*	103	Rape	304
macrocarpa	138	arvensis	98	Raspberries	400
v. *depressa*	139	bulbosus	98	Raspberry, boulder	402
v. *olivaeformis*	139	cardiophyllus	99	black	403
mandanensis	139	*caricetorum*	102	red	402
margaretta	140	*carolinianus*	102	Raphanus sativus	324
marilandica	139	*circinatus*	103	Ratany	566
maxima	137	v. *subrigidus*	106	Ratany Family	566
mohriana	139	cymbalaria	99	Ratibida	989
muehlenbergii	139	f. *hebecaulis*	99	columnifera	990
palustris	139	v. *saximontanus*	99	f. *pulcherrima*	990
prinoides	140	*delphinifolius*	100	pinnata	990
v. *acuminata*	139	*falcatus*	106	tagetes	990
rubra	137	fascicularis	99	Rattlebox	438
robur	137	v. *apricus*	100	Rattlepod	438
shumardii	140	flabellaris	100	Rattlesnake fern	48
v. schneckii	140	f. *riparius*	100	Rattlesnake plantain	1276
v. shumardii	140	flammula	100	dwarf	1276
stellata	140	v. *ovalis*	100	Rattlesnake-root	986
v. *margaretta*	140	glaberrimus	100	Rattlesnake weed	592
texana	140	v. *ellipticus*	101	Rayless aster	875
undulata	140	v. *glaberrimus*	101	Red ash	750
velutina	141	gmelinii	101	Red-berried elder	827
virginiana v. fusiformis	141	v. *hookeri*	101	Red cedar	73
Quicksilver-weed	107	v. *limosus*	101	Red clover	485
Quickweed	939	v. *prolificus*	101	Red currant, western	353
fringed	940	v. *terrestris*	101	Red false mallow	251
Quillwort	41	hispidus	101	Red horned poppy	112
Butler's	41	v. *caricetorum*	102	Red loco	467

Red lovegrass		1177	v. *serotina*	572	Rorippa		325
Red morning-glory		658	*typhina*	573	*armoracia*		302
Red mulberry		127	Rhynchospora	1106	*austriaca*		326
Red oak		137	*capillacea*	1106	*calycina*		326
Red orache		164	*capitellata*	1107	*curvipes*		326
Red osier		530	*corniculata*	1106	*hispida*		327
Red raspberry		402	*globularis* v. *recognita*	1107	*islandica*		327
Red scale		164	*harveyi*	1107	v. *fernaldiana*		327
Red-seeded dandelion		1010	*macrostachya*	1107	v. *hispida*		327
Red-seeded plantain		746	Ribbonleaf pondweed	1035	*lyrata*		329
Red sprangletop		1188	Ribes	352	*nasturium-aquaticum*		323
Red three-awn		1139	*americanum*	353	*obtusa*	327,	329
Redbud		414	*cereum*	353	*palustris*		327
Redfieldia flexuosa		1217	v. *inebrians*	353	ssp. glabra v.		
Redtop	1128,	1231	*cynosbati*	354	*fernaldiana*		327
panicum		1203	*hirtellum*	354	v. *glabrata*		327
Reed, bur-		1235	*inebrians*	353	ssp. hispida v.		
canary grass		1207	*lacustre*	354	*elongata*		327
common		1208	*missouriense*	355	v. *hispida*		327
Reedgrass		1148	*odoratum*	355	*sessiliflora*		327
plains		1149	*oxyacanthoides*	355	*sinuata*		328
Rescue grass		1147	*setosum*	356	*sylvestris*		328
Reseda lutea		333	*triste*	356	*tenerrima*		328
Resedaceae		333	Rice cutgrass	1187	*truncata*		329
Resin bush		1018	wild	1234	Rosa		396
Resinous skullcap		737	Ricegrass	1198	*acicularis*		397
Resurrection fern		64	black-seeded	1199	ssp. *acicularis*		398
Reverchon false pennyroyal		716	Indian	1199	ssp. *sayi*		398
Reverchonia arenaria		552	little-seed	1199	*alcea*		398
Rhamnaceae		554	rough-leaved	1198	*arkansana*		398
Rhamnus		555	Richardson's cranesbill	582	v. *arkansana*		398
alnifolia		555	Riddell ragwort	998	v. *suffulta*		398
carolinianus		556	Ridge-seeded spurge	546	*blanda*		398
cathartica		556	Rigid goldenrod	1006	*bourgeauiana*		398
davurica		556	Ring grass	1197	*canina*		397
v. *davurica*		556	Ring muhly	1197	*carolina*		399
v. *nipponica*		556	River-bank grape	560	v. *carolina* f.		
lanceolata var. glabratus		556	River birch	143	*glandulosa*		399
v. *lanceolata*		556	Roadside thistle	909	v. *grandiflora*		399
Rheum rhabarbarum		230	Robinia	477	v. *villosa*		399
rhaponticum		230	*hispida*	477	*conjuncta*		398
Rhexia		526	*pseudo-acacia*	477	*eglanteria*		397
interior		527	Robin's plantain	931	*engelmannii*		398
mariana var. interior		526	Rock cress	298, 300, 301, 302	*fendleri*		400
Rhombic copperleaf		537	smooth	301	*foliolosa*		399
Rhombolytrum albescens		1232	Rock elm	123	*lunellii*		398
Rhubarb		230	Rock jasmine	344	*lyoni*		399
wild		233	northern	344	*macounii*		400
Rhus		571	western	344	*multiflora*		399
aromatica		571	Rock muhly	1197	*pimpinellifolia*		397
v. *aromatica*		572	Rock sandwort	196	*polyantha*		398
v. *flabelliformis*		572	Rock spikemoss	40	*pyrifera*		400
v. *illinoensis*		572	Rock-spiraea	379	*rubifolia*		400
v. *pilosissima*		572	Rocket, dame's	315	*rubiginosa*		397
v. *serotina*		572	larkspur	93	× *rudiuscula*		398
v. *trilobata*		572	London	330	*serrulata*		399
copallina		572	purple	315	*setigera*		400
v. *latifolia*		573	-salad	313	v. *setigera*		400
crenata		572	sand	310	v. *tomentosa*		400
glabra		573	Rockpink	190	*spinosissima*		397
microphylla		573	Rockrose Family	253	*subblanda*		398
nortonii		572	Rocky Mountain bee plant	292	*subglauca*		398
osterhoutii		572	Rocky Mountain juniper	73	*suffulta*		398
toxicodendron		575	Rocky Mountain sage	732	*terrens*		400
trilobata		572	Rocky Mountain woodsia	69	*woodsii*		400

Rosaceae	364	deliciosus	402	hymenosepalus	233
Rose	396	*enslenii*	402	maritimus	233
climbing	400	flagellaris	402	v. *fueginus*	233
dog	397	*frondosus*	404	mexicanus	233
gentian, prairie	609	*hancinianus*	402	obtusifolius	233
Japanese	399	*hispidus*	405	occidentalis	234
leafy	399	idaeus ssp. sachalinensis	402	orbiculatus	234
little ground	370	v. *idaeus*	403	patientia	234
mallow	245, 246	v. *sachalinensis*	402	*persicarioides*	233
halberd-leaved	245	*laudatus*	403	stenophyllus	234
moss	188	*melanolasius*	403	venosus	235
multiflora	399	*mollior*	404	verticillatus	235
pasture	399	*nigrobaccus*	402	Running buffalo clover	486
-pink	609	occidentalis	403	Ruppia maritima	1040
pogonia	1281	*occidualis*	402	v. *occidentalis*	1040
prairie	400	*oppositus*	402	v. *rostrata*	1040
wild	398	*orarius*	404	*occidentalis*	1040
prickly wild	397	ostryifolius	403	Ruppiaceae	1040
-ring gaillardia	939	parviflorus	403	Rush	1050
Scotch	397	pensilvanicus	404	Baltic	1052
smooth wild	398	*plicatifolium*	402	beaked	1106
vervain	705	pubescens	404	bog	1054
western wild	400	v. *pilosifolius*	404	common scouring	44
white prairie	399	*strigosus*	403	Dudley	1054
Rose Family	364	trivialis	404	dwarf scouring	45
Rosemary, marsh	235	Rudbeckia	990	flowering	1021
Rosoideae	265	*amplexicaulis*	919	-foil	540
Rosin-weed	999	grandiflora	991	grassleaf	1055
Rotala ramosior	498	v. *alismaefolia*	991	inland	1055
v. *interior*	498	v. *grandiflora*	991	intermediate scouring	43
Rough avens	378	hirta	991	knotted	1056
Rough bugleweed	719	v. *angustifolia*	991	nut-	1112
Rough buttonweed	817	v. *hirta*	991	path	1056
Rough coneflower	991	v. *pulcherrima*	991	-pea	415
Rough dropseed	1225	laciniata	991	Indian	416
Rough false pennyroyal	715	v. *laciniata*	992	James	412
Rough fescue	1181	serotina	991	sicklepod	415
Rough-leaved dogwood	528	subtomentosa	992	scouring	42
Rough-leaved ricegrass	1198	triloba	992	slimpod	1054
Rough pigweed	183	v. *triloba*	992	small-headed	1053
Rough small bluet	821	Rue anemone	89	smooth scouring	44
Round-head lespedeza	457	false	94	stout	1056
Round-headed prairie clover	443	goat's	480	toad	1053
Round-leaved orchis	1278	Ruellia	801	Torrey's	1057
Round-leaved spurge	549	*carolinensis*	802	variegated scouring	45
Round-leaved sundew	253	*ciliosa*	802	Rush Family	1049
Round-leaved wintergreen	338, 339	fringeleaf	802	Russian knapweed	900
Roundfruit St. John's-wort	239	*humilis*	802	Russian olive	490
Roundleaf groundsel	996	v. *expansa*	802	Russian pigweed	165
Roundleaf monkey-flower	771	v. *frondosa*	802	Russian-thistle	177
Rover bellflower	809	v. *grisea*	802	Russian wild rye	1169
Royal catchfly	209	v. *longifolia*	802	Rusty lupine	464
Royal fern	48	limestone	802	Rusty woodsia	68
Royal Fern Family	48	strepens	802	Rutaceae	575
Rubacer	401	v. *cleistantha*	802	*Rye*	1114
parviflorum	404	Rugel's plantain	746	Rye, alkali wild	1169
Rubber plant, Colorado	963	Rumex	230	basin wild	1168
Rubiaceae	816	acetosella	231	blue wild	1168
Rubus	400	*alluvius*	235	Canada wild	1167
aboriginum	402	altissimus	232	fuzzyspike wild	1169
allegheniensis	401	*brittanica*	234	hairy wild	1169
alumus	404	crispus	232	Russian wild	1169
argutus	403	*cristatus*	234	Virginia wild	1170
baileyanus	402	domesticus	232	wild	1166
		hastatulus	232	Ryegrass	1189

INDEX

Italian	1189	v. *perrostrata*	285	western	1015
perennial	1189	candida	286	Salsola	176
		caroliniana	286	collina	176
Sabatia	609	*cordata*	288	iberica	177
angularis	609	discolor	286	*kali* v. *tenuifolia*	177
f. *albiflora*	609	v. *latifolia*	286	*pestifer*	177
campestris	609	v. *overi*	286	Salt cedar	265, 266
Sabina	71	drummondiana	281	Salt-marsh sand spurry	210
horizontalis	72	eriocephala	286	Saltbush	161
scopulorum	73	exigua	287	four-wing	162
virginiana	74	ssp. exigua	287	Powell's	164
Sabulina	193	ssp. interior	287	silver-scale	162
dawsonensis	196	fragilis	287	spiny	163
patula	195	*geyeriana*	281	Saltgrass	1163
propinqua	196	glauca	281	inland	1163
texana	196	*gooddingii*	289	Saltmarsh aster	885
Sacaton, alkali	1225	*gracilis*	289	Saltwort	176
Sachuista	1264	v. *textoris*	289	Salvia	731
Safflower	897	humilis	287	azurea	731
Sage	731, 732, 866	v. *hyporhysa*	288	v. *azurea*	731
blue	731	v. *microphylla*	288	v. *grandiflora*	731
Jerusalem	709	v. *rigidiuscula*	288	*lanceolata*	732
lance-leaved	732	*interior*	287	lyrata	732
long-leaved	870	v. *pedicellata*	287	nemorosa	732
lyre-leaf	732	f. *wheeleri*	287	*pitcheri*	731
pitcher	731	v. *wheeleri*	287	pratensis	731
Rocky Mountain	732	*lasiandra* v. *caudata*	281	reflexa	732
white	166, 870	*linearifolia*	287	sclarea	731
wood	738, 739	*longipes* v. *wardii*	286	*sylvestris*	732
Sagebrush, big	870	lucida	288	Salviniaceae	71
dwarf	869	lutea	288	Sambucus	826
sand	869	v. *famelica*	288	canadensis	826
Sagewort, sweet	868	v. *platyphylla*	288	v. *submollis*	827
western	868	maccalliana	288	*microbotrys*	827
Sagina decumbens	203	*missouriensis*	287	*pubens*	827
ssp. *occidentalis*	204	*monticola*	290	racemosa ssp. pubens	827
v. *smithii*	204	*nelsonii*	290	Samolus	350
Sagittaria	1024	nigra	288	*cuneatus*	351
ambigua	1025	v. *lindheimeri*	289	ebracteatus v. cuneatus	350
brevirostra	1026	*padophylla*	290	*floribundus*	351
calycina	1026	pedicellaris	289	parviflorus	351
cuneata	1027	pentandra	289	Samson's snakeroot	475
engelmanniana ssp.		*perrostrata*	285	Sand bluestem	1132
brevirostra	1026	petiolaris	289	Sand cherry	394
esculenta	1027	v. *angustifolia*	289	Sand dropseed	1226
graminea	1027	*phylicifolia* ssp.		Sand grape	561
v. *graminea*	1027	*planifolia*	290	Sand lily	270
v. *platyphylla*	1027	planifolia	290	Sand lovegrass	1178
latifolia	1027	v. *monica*	290	Sand milkweed	618
longiloba	1028	v. *nelsonii*	290	Sand muhly	1193, 1195
montevidensis ssp.		*prinoides*	286	Sand-parsley	586
calcycina	1026	pseudomonticola	290	Sand puffs	151
platyphylla	1027	v. *padophylla*	290	Sand rocket	310
rigida	1028	*rigida* v. *angustata*	288	Sand sagebrush	869
Sago lily	1247	v. *rigida*	287	Sand spurry	210
Sago pondweed	1038	v. *vestita*	287	salt-marsh	210
Sainfoin	416	v. *watsonii*	288	Sand verbena	145
Salicaceae	273	scouleriana	290	sweet	145
Salicornia rubra	176	serissima	290	Sand vine	631
Salix	281, 663	*tristis*	288	Sandalwood Family	531
alba	285	wardii	286	Sandbar lovegrass	1174
alba v. vitellina	285	*wheeleri*	287	Sandbar willow	287
amygdaloides	285	Salsify	1014, 1015	Sandberg's bluegrass	1214
babylonica	281	false	992	Sandbur	1151
bebbiana	285	meadow	1015	Sandgrass	1233

Sandhill goosefoot	170	Schedonnadrus paniculatus	1217	common	44
Sandhill plum	392	Scheuchzeria palustris v.		dwarf	45
Sandhills pigweed	181	americana	1031	intermediate	43
Sandreed	1150	Scheuchzeria Family	1031	smooth	44
big	1150	Scheuchzeriaceae	1031	variegated	45
prairie	1150	Schizachne purpurascens	1217	Scratch grass	1193
Sandwort	193	*Schizachyrium*	1131	Scrophularia	789
grove	195	*scoparium*	1134	dakotana	789
rock	196	Schneck oak	140	lanceolata	789
thyme-leaved	196	Schrankia	410	*leporella*	789
Sanguinaria canadensis	114	nuttallii	411	marilandica	789
Sanguisorba annua	405	occidentalis	411	f. *neglecta*	789
Sanicula	600	*uncinata*	411	*neglecta*	789
canadensis	600	Scirpus	1107	*occidentalis*	789
v. *grandis*	600	acutus	1109	Scrophulariaceae	751
gregaria	600	americanus	1109	Scurf-pea	471
marilandica	601	v. *longispicatus*	1111	breadroot	473
odorata	601	atrocinctus	1109	lemon	474
Santalaceae	531	atrovirens	1109	little breadroot	474
Santa Maria	981	v. *georgianus*	1110	palm-leaved	473
Sapindaceae	567	v. *pallidus*	1111	silver-leaf	472
Sapindus	568	cyperinus	1109	slimleaf	475
drummondii	568	*debilis*	1112	tall-bread	472
saponaria v.		*deltarum*	1108	Scurfy pea	475
drummondii	568	fluviatilis	1109	Scutellaria	663, 733
Sapodilla Family	341	georgianus	1110	*ambigua*	736
Saponaria officinalis	204	hallii	1110	brittonii	734
vaccaria	214	heterochaetus	1110	v. *virgulata*	734
Sapotaceae	341	koilolepis	1110	canescens	735
Sarcobatus vermiculatus	177	*lineatus*	1111	drummondii	734
Sarcostemma	636	maritimus v. paludosus	1110	*epilobiifolia*	735
crispum	636	microcarpus	1110	galericulata	734
cynanchoides	636	v. *rubrotinctus*	1111	incana	735
lobata	636	nevadensis	1111	lateriflora	735
Sarothra	237	olneyi	1109	*leonardii*	736
drummondii	238	pallidus	1111	ovata	735
Sarsaparilla	584	*paludosus*	1110	ssp. *bracteata*	736
wild	584	pendulus	1111	v. *bracteata*	736
Saskatoon service-berry	368	pungens	1111	ssp. *mississippiensis*	736
Sassafras albidum	78	*rubrotinctus*	1111	ssp. *versicolor*	736
variifolium	78	saximontanus	1111	v. *versicolor*	736
Satureja arkansasa	733	smithii	1112	parvula	736
glabella v. *angustifolia*	733	*supinus* v. *hallii*	1110	v. *australis*	736
Saururaceae	79	v. *saximontanus*	1111	v. *leonardii*	736
Saururus cernuus	79	torreyi	1112	v. *parvula*	736
Sawtooth sunflower	954	validus	1112	resinosa	737
Saxifraga	362	v. *creber*	1112	*wrightii*	737
cernua	363	Scleranthus annuus	204	Sea blite	178
occidentalis	363	Scleria	1112	Sea milkwort	347
simulata	363	ciliata	1112	Sea purslane	152
texana	363	v. *ciliata*	1113	*Secale cereale*	1114
Saxifragaceae	358	v. *elliottii*	1113	Sedge	1060
Saxifrage, James'		pauciflora	1113	umbrella	1093
nodding	363	v. *caroliniana*	1113	white-top	1098
Saxifrage Family	358	v. *pauciflora*	1113	Sedge Family	1059
Scabiosa arvensis	837	triglomerata	1113	Sedum	357
Scabious, field	837	Sclerochloa	1218	lanceolatum	357
Scale-seed	601	*dura*	1114	nuttallianum	357
forked	601	festucacea	1218	pulchellum	358
Scandix pectenveneris	585	*Scleropogon brevifolius*	1114	stenopetalum	357
Scapose tickclover	449	Scorpionweed	682	Seedbox	513
Scarlet gaura	510	Scorzonera laciniata	992	bushy	514
Scarlet lychnis	201	*Scotch rose*	397	cylindric-fruited	514
Scarlet pea	454	Scotch thistle	979	manyseed	515
Scarlet runner	416	Scouring rush	42	marsh	514

INDEX

Seepweed	178	Setaria	1218	Sida	249,	250
desert	178	faberi	1219	*hederacea*		250
Selaginella	39	geniculata	1219	*lepidota* v. *sagittaefolia*		250
densa	40	glauca	1219	*leprosa* v. *hederacea*		250
engelmannii	40	italica	1220	v. *sagittaefolia*		250
peruviana	40	leucopila	1220	physocalyx		250
rupestris	40	*lutescens*	1220	prickly		250
sheldonii	40	*macrostachya*	1220	spinosa		250
underwoodii	40	*reverchonii*	1218	Sidecluster milkweed		624
Selaginellaceae	39	verticillata	1220	Sideoats grama		1141
Selenia aurea	329	viridis	1221	*Sideranthus annuus*		947
golden	329	*Seymeria macrophylla*	765	*glaberrimus*		950
Self-heal	728	Shad-berry	369	*grindelioides*		976
Selinocarpus diffusus	151	Shad-bush	368	*spinulosus*		950
Seneca snakeroot	565	Shagbark hickory	132	*Sidopsis hispida*		249
Senecio	993	Sharpwing monkey-flower	770	*Sieversia*		376
canus	994	Sheathed pondweed	1039	*ciliata*		379
congestus	994	Sheep sorrel	231	*triflora*		379
crassulus	994	Sheep's fescue	1180	Silene		205
densus	998	Sheepberry	831	*alba*	208,	209
douglassii v. longilobus	995	Shellbark hickory	132	antirrhina		206
eremophilus	995	big	132	f. *apetala*		206
v. *eremophilus*	995	Shepherdia	491	f. *bicolor*		206
glabellus	995	argentea	491	f. *deaneana*		206
hydrophilus	995	canadensis	492	cserei		206
imparipinnatus	996	Shepherd's-purse	305	*cucubalus*		210
intergerrimus	996	*Sherardia arvensis*	817	dichotoma		206
v. *exaltatus*	996	*orientalis*	817	drummondii		207
v. integerrimus	996	Shield fern, crested	58	*fabaria*		206
v. scribneri	996	marginal	59	*latifolia*		210
oblanceolatus	998	Shingle oak	138	menziesii		207
obovatus	996	Shining bedstraw	819	v. *viscosa*		208
palustris	994	Shining ladies'-tresses	1282	nivea		208
pauperculus	997	Shining willow	288	noctiflora		208
plattensis	997	Shinnersoseris rostrata	999	pratensis		208
pseudaureus	997	Shinnery oak	138	stellata		209
v. *pseudaureus*	997	Shin oak	139	v. *scabrella*		209
v. *semicordatus*	997	Shooting star	345, 346	regia		209
purshianus	994	Shore buttercup	99	vulgaris		209
rapifolius	997	Short-awn foxtail	1129	Silk tassel, Mexican		530
riddellii	998	Short ragweed	856	Silk Tassel Family		530
rotundus	997	Short-ray prairie coneflower	990	*Silk tree*		407
spartioides	998	Short's milk-vetch	435	Silktop dalea		440
tridenticulatus	998	Shortcrown milkweed	619	Silkvine		635
vulgaris	998	Shortpod draba	311	Silky aster		884
Senecioneae	842	Shortstem eriogonum	216	Silky lupine		464
Senna	412	Showy chloris	1153	Silky orophaca		434
Maryland	413	Showy ladies'-slipper	1274	Silky prairie clover		445
two-leaved	414	Showy locoweed	471	Silky wormwood		869
Sensitive brier	410, 411	Showy milkweed	626	Sillscale		163
catclaw	411	Showy orchis	1275	Silphium		999
western	411	Showy partridge pea	412	*asperrimum*		1000
Sensitive fern	62	Showy-wand goldenrod	1007	asteriscus		1000
Sensitive partridge pea	413	Showy white evening primrose	524	v. *scabrum*		1000
Sericea lespedeza	458	Shrubby cinquefoil	385	integrifolium		999
Serinia oppositifolia	966	Shumard's oak	140	v. integrifolium		1000
Service-berry	368, 369	Siberian elm	122	v. laeve		1000
low	369	Siberian pea-shrub	437	laciniatum		1000
Saskatoon	368	*Sibbaldiopsis*	382	perfoliatum		1000
Serviceberry willow	290	*tridentata*	390	radula		1000
Sesbania	477	Sicklepod	298	*speciosum*		1000
exaltata	477	rush-pea	415	Silver bluestem		1133
macrocarpa	477	*Sickleweed*	585	Silver-leaf nightshade		648
Sessile-leaved tickclover	451	*Sicydium lindheimeri*	269	Silver-leaf scurf-pea		472
Sesuvium verrucosum	152	Sicyos angulatus	269	Silver maple		570

Silver poplar	276	Slender goldenweed	950	*glauca*	1267
Silver-scale saltbush	162	Slender ladies'-tresses	1282	herbacea	1267
Silverberry	491	Slender-leaved bluet	822	v. herbacea	1267
Silverweed	383	Slender-leaved mountain mint	730	v. lasioneuron	1267
Silvery cinquefoil	383	Slender lip fern	55	v. pulverulenta	1267
Silvery lupine	462	Slender-lobed bundleflower	409	*lasioneura*	1267
Silvery wolfberry	640	Slender locoweed	468	hispida	1267
Simaroubaceae	575	Slender plantain	744	*pulverulenta*	1267
Sinapsis	304	Slender milk-vetch	428	tamnoides v. *hispida*	1267
alba	304	Slender milkwort	565	Smoke tree	575
arvensis	305	Slender pigweed	182	Smooth alder	142
Single-stemmed bog aster	881	Slender pondweed	1036	Smooth blue aster	878
Singletary vetchling	456	Slender snake-cotton	185	Smooth brome	1144
Sinuate-leaved guara	511	Slender spike lespedeza	458	Smooth buttonweed	822
Sisymbrium	329	Slender tickclover	447	Smooth catchfly	206
altissimum	330	Slender wheatgrass	1124	Smooth cliff-brake	63
irio	330	Slenderleaf betony	738	Smooth crabgrass	1163
loeselii	330	Slenderleaf purslane	189	Smooth four-o'clock	149
officinale	330	Slick-seed bean	479	Smooth milkweed	628
v. leiocarpum	331	Slim tridens	1232	Smooth penstemon	779
v. officinale	331	Slimleaf dicanthelium	1159	Smooth rock cress	301
Sisyrinchium	1261	Slimleaf prairie clover	444	Smooth scouring rush	44
angustifolium	1262	Slimleaf scurf-pea	475	Smooth sumac	573
campestre	1262	Slimpod rush	1054	Smooth wild rose	398
v. *kansanum*	1262	Slimspike three-awn	1137	Smooth yellow-violet	262
demissum	1262	Slippery elm	123	Snailseed	110
ensigerum	1262	Sloughgrass	1139	Snake-cotton	184
kansanum	1262	American	1139	field	185
montanum	1263	Small bedstraw	820	slender	185
mucronatum	1263	Small bellwort	1256	Snakeroot, black	600
pruinosum	1263	Small bluet	821	button	593
Sitanion	1221	rough	821	Samson's	475
elymoides	1221	Small bur-clover	465	seneca	565
hystrix v. brevifolium	1221	Small cranesbill	582	Virginia	80
v. *hystrix*	1221	Small-flowered grass-of-		white	935
longifolium	1221	Parnassus	362	Snakeweed	945
Sium	601	Small-flowered milk-vetch	431	Snapdragon	756
cicutaefolium	601	Small-fruited mallow	247	dwarf	762
suave	601	Small-headed rush	1053	Sneezeweed	951
Six-angled spurge	546	Small hop-clover	483	bitter	951
Sixweeks fescue	1180	Small lupine	464	Sneezewort aster	1005
Sixweeks grama	1140	Small-seeded false flax	305	Snow-on-the-mountain	547
Sixweeks three-awn	1136	Small skullcap	736	Snowberry	827
Skeleton grass	1183	Leonard's	736	Palmer's	828
bearded	1183	southern	736	western	828
Skeletonweed	974, 1009	Small white aster	882	Snowy campion	208
Skullcap	733	Small white lady's-slipper	1274	Soapberry	568
blue	735	Small yellow buttercup	100	Soapberry Family	567
Britton's	734	Small yellow lady's-slipper	1273	Soapweed	1264
Drummond's	734	Smallflower bitter cress	306	Soapwort	204
eggleaf	735	Smallflower wallflower	314	Soft chess	1145
hoary	735	Smartweed	220, 226	Soft golden-aster	904
Leonard's small	736	dodder	665	Soft goldenrod	1004
mad-dog	735	pale	226	Soft maple	570
marsh	734	Pennsylvania	227	Solanaceae	637
resinous	737	pink	225	Solanum	647
small	736	swamp	225	*americanum*	650
southern small	736	water	225, 227	carolinense	647
Sleepy catchfly	206	Smilacaceae	1266	citrullifolium	648
Sleepy daisy	1019	Smilacina	1254	*cornutum*	650
Slender beardtongue	782	racemosa	1254	dimidiatum	648
Slender bog orchis	1278	stellata	1254	dulcamara	649
Slender bush lespedeza	461	Smilax	1266	elaeagnifolium	649
Slender crabgrass	1162	bona-nox	1266	heterodoxum v.	
Slender fumewort	117	ecirrhata	1266	setigeroides	648

interius	649	v. *glabrata*	1006	prickly	1008
jamesii	647	v. *humilis*	1006	Soybean	416
nigrum	650	v. *rigida*	1006	Spanish needles	891
v. *interius*	650	*rigidiuscula*	1007	Sparganiaceae	1235
v. *virginicum*	650	*satanica*	1003	Sparganium	1235
ptycanthum	650	*serotina*	1004	americanum	1235
rostratum	650	sparsiflora	1006	androcladum	1236
sarrachoides	651	speciosa	1007	angustifolium	1236
scabrum	650	v. *angustata*	1007	chlorocarpum	1236
torreyi	648	v. *rigidiuscula*	1007	v. *acaule*	1236
triflorum	651	v. *speciosa*	1007	emersum v.	
triquetrum	647	*trinervata*	1007	multipedunculatum	1236
tuberosum	647	ulmifolia	1007	eurycarpum	1236
villosum	651	v. *microphylla*	1007	*multipedunculatum*	1236
Solomon's seal	1254	v. *ulmifolia*	1007	*simplex*	1236
false	1254	*wardii*	1005	Sparkleberry	335
Solidago	1000	Sonchus	1007	Spartina	1223
altissima	1003	asper	1008	gracilis	1223
angusta	1005	arvensis	1008	pectinata	1223
× *bernardii*	1006	ssp. arvensis	1008	Spatterdock	82
canadensis	1002	v. *glabrescens*	1008	Spearmint	722
v. canadensis	1003	ssp. uliginosus	1008	Spearscale	165
v. gilvocanescens	1003	oleraceus	1008	Spearwort	100
v. hargeri	1003	*Sophia*	308	water plantain	103
v. salebrosa	1003	*brachycarpa*	309	Speckled alder	142
v. scabra	1003	*intermedia*	309	Spectacle pod	310
dumetorum	1003	*magna*	309	*Specularia*	813
flexicaulis	1003	*multifida*	310	*biflora*	814
gigantea	1003	*pinnata*	309	*holzingeri*	814
v. *gigantea*	1004	*richardsoniana*	310	*lamprosperma*	814
v. *serotina*	1004	Sophora nuttalliana	477	*leptocarpa*	814
glaberrima	1004	sericea	478	*perfoliata*	816
graminifolia v.		Sorbus	405	Speedwell	791
graminifolia	936	aucuparia	405	bird's-eye	796
v. *gymnospermoides*	936	scopulina	405	brooklime	793
v. *major*	936	Sorghastrum	1221	common	795
v. *media*	936	avenaceum	1222	corn	794
gymnospermoides	936	nutans	1221	field	793
hapemaniana	1004	Sorghum	1222	ivy-leaved	795
hispida	1005	*almum*	1223	marsh	796
lindheimeriana	1005	*bicolor*	1222	purslane	796
longipetiolata	1005	ssp. *drummondii*	1223	thyme-leaved	797
lunellii	1003	halepense	1222	water	793
microphylla	1007	Sorrel	230	Spergula arvensis	210
missouriensis	1004	creeping ladies	579	v. *sativa*	210
v. *fasciculata*	1004	gray-green wood	579	Spergularia marina	210
v. *missouriensis*	1004	heartwing	232	v. *leiosperma*	211
mollis	1004	sheep	231	*rubra*	211
moritura	1004	violet wood	579	*salina*	211
nemoralis	1004	wood	579	Spermacoce glabra	822
v. *decemflora*	1005	yellow wood	579	Spermolepis	601
v. *haleana*	1005	Sour cherry	390	divaricata	601
v. *longipetiolata*	1005	Southern adder's-tongue	48	echinata	602
v. *nemoralis*	1005	Southern black haw	832	inermis	602
occidentalis	937	Southern catalpa	805	*patens*	602
pallida	1007	Southern crabgrass	1162	Sphaeralcea	249, 251
petiolaris v. angusta	1005	Southern dewberry	404	angustifolia v. cuspidata	251
v. *petiolaris*	1005	Southern lady-fern	53	*angusta*	249
v. *wardii*	1005	Southern ragweed	856	coccinea	251
pruinosa	1003	Southern small skullcap	736	*cuspidata*	251
ptarmicoides	1005	Southernwood	867	fendleri	252
pulcherrima	1005	Southwest bedstraw	821	Sphaerophysa salsula	478
radula	1005	Sow thistle	1007	Sphenopholis	1224
riddellii	1006	common	1008	*intermedia*	1224
rigida	1006	field	1008	obtusata	1224

v. major	1224	airoides	1225	spotted	547
v. obtusata	1224	*argutus*	1228	spreading	546
pallens	1224	asper	1225	thyme-leaved	550
robusta	1224	v. asper	1226	toothed	544
Spice bush	78	v. clandestinus	1226	Spurge Family	535
Spider flower	291	v. *drummondii*	1226	Spurred gentian	608
Spider milkweed	631	v. hookeri	1226	Spurred violet, great	264
Spiderling	146	v. *pilosa*	1226	Spurry	210
Spiderwort	1048	*asperifolius*	1193	corn	210
Spiderwort Family	1046	*canovirens*	1226	salt-marsh sand	210
Spike oat	1183	*clandestinus*	1226	sand	210
Spikemoss	39	*confusus*	1192	Squashberry	830
Peruvian	40	cryptandrus	1226	Squaw-root	599
rock	40	*drummondii*	1226	Squirreltail	1221
Underwood's	40	*ejuncidus*	1224	Stachys	737
Spikemoss Family	39	*flexuosus*	1227	*ambigua*	738
Spikenard	584, 1254	giganteus	1227	*ampla*	738
false	1254	heterolepis	1227	*annua*	737
Spikerush	1098	*junceus*	1224	*aspera*	738
Spindle tree	534	neglectus	1227	*arvensis*	737
Spinulose wood fern	60	ozarkanus	1229	*hispida*	738
Spiny cocklebur	1020	*pilosus*	1226	palustris ssp. pilosa	737
Spiny pigweed	184	pyramidatus	1228	v. *homotricha*	738
Spiny saltbush	163	texanus	1228	v. *nipigonensis*	738
Spinytooth gumweed	943	vaginiflorus	1228	v. *phaneropoda*	738
Spiraea	405	v. *inequalis*	1229	v. *pilosa*	738
alba	405	v. *neglectus*	1228	*pustulosa*	738
betulifolia	406	Spotted beebalm	725	*schweinitzii*	738
v. *betulifolia*	406	Spotted coral-root	1271	tenuifolia	738
v. *corymbosa*	406	Spotted evening primrose	518	Staff Tree Family	534
v. *lucida*	406	Spotted joe-pye weed	934	Standing cypress	671
densiflora	405	Spotted knapweed	899	Standing milk-vetch	423
latifolia	405	Spotted spurge	547	St. Andrew's cross	237
lucida	406	Spotted St. John's-wort	238	Stanleya	331
prunifolia	405	Spotted touch-me-not	583	*bipinnata*	331
rock-	379	Sprangletop	1187, 1218	*glauca*	331
salicifolia	405	bearded	1188	integrifolia	331
thunbergii	405	red	1188	pinnata	331
tomentosa	405	Spreading dogbane	612	v. *integrifolia*	331
vanhouttei	405	Spreading fleabane	915	v. *pinnata*	331
wild	406	Spreading four-o'clock	151	Staphylea trifolia	567
Spiraeoideae	265	Spreading moonpod	151	Staphyleaceae	567
Spiranthes	1281	Spreading spurge	546	Star cloak-fern	62
cernua	1282	Spreading wild buckwheat	217	Star duckweed	1045
lacera	1282	Spreading yellow cress	328	Star of Bethlehem	1253
v. *gracilis*	1282	Spring beauty	188	Star-thistle	898
v. *lacera*	1282	Virginia	188	Iberian	899
lucida	1282	Spring cress	306	yellow	900
magnicamporum	1283	Spruce	74	Stargrass	1251
ochroleuca	1282	Black Hills	74	water	1240
ovalis	1283	white	74	yellow	1251
plantaginea	1282	Spurge	541	Starwort, water	741
romanzoffiana	1283	cypress	544	Starry campion	209
stricta	1283	-flax	498	*Steironema ciliatum*	348
tuberosa	1283	flowering	543	*hybridum*	349
vernalis	1284	Geyer's	545	*pumilum*	348
Spirodela polyrrhiza	1045	hybrid leafy	549	*quadriflorum*	349
Spleenwort	51	leafy	545	*verticillatum*	349
black-stem	52	mat	550	Stellaria	211
ebony	51	Missouri	548	aquatica	211
forked	52	*petty*	541	*borealis*	212
green	52	prairie	548	calycantha	212
maidenhair	52	ridge-seeded	546	v. *floribunda*	212
Splitbeard bluestem	1134	round-leaved	549	v. *isophylla*	212
Sporobolus	1224	six-angled	546	crassifolia	212

INDEX

graminea	212	cordatus	331	sawtooth	954
laeta	213	hyacinthoides	331	stiff	956
longifolia	213	Streptopus amplexifolius	1254	tickseed	892
longipes	213	Striate agrimony	367	willow-leaved	957
media	213	Striped coral-root	1271	Sunflower Family	838
Stemless evening primrose	525	Strophostyles	478	*Svida amomum*	528
Stemless hymenoxys	962	helvola	479	*asperifolia*	529
Stenophyllus capillaris	1060	v. *missouriensis*	479	*baileyi*	530
Stenosiphon linifolius	526	leiosperma	479	*foemina*	530
virgatus	526	*missouriensis*	479	*instolonea*	530
Stenotus armerioides	948	*umbellata*	480	*interior*	530
Stephanomeria	1009	St. John's-wort	237	*stolonifera*	530
pauciflora	1009	common	238	Swallow wort, black	632
runcinata	1009	dwarf	238	Swamp aster	883
tenuifolia	1009	great	239	Swamp currant	354, 356
Stickleaf	270, 273	greater	238	Swamp lousewort	773
Stickleaf Family	269	marsh	239	Swamp milkweed	621
Stickseed	692, 694	roundfruit	239	Swamp privet	748
blue	695	spotted	238	Swamp ragwort	994
cupseed	695	St. John's-wort Family	236	Swamp smartweed	225
large-flowered	692	Stylisma pickeringii v.		Swamp thistle	911
Sticky cockle	208	pattersonii	661	Swamp white oak	137
Stiff sunflower	956	Stylosanthes	480	Sweet-broom	452, 453
Stiffstem flax	564	biflora	480	*Sweet cherry*	390
compact	565	v. *hispidissima*	481	Sweet cicely	597
Stillingia	552	*riparia*	480	Sweet clover	466
salicifolia	552	Suaeda	178	white	466
sylvatica	552	*calceoliformis*	178	yellow	466
Stinging nettle	130	depressa	178	Sweet coltsfoot	983
Stinkgrass	1173	erecta	178	Sweet coneflower	992
Stinking elderberry	827	intermedia	178	Sweet flag	1042
Stinking milk-vetch	433	moquinii	178	Sweet haw	831
Stinkweed	985	nigrescens v. glabra	178	Sweet joe-pye weed	935
Stipa	1229	suffrutescens	178	*Sweet-pea*	416
comata	1229	v. *detonsa*	178	Sweat-pea, perennial	455
curtiseta	1230	v. *suffrutescens*	178	Sweet potato	657
neomexicana	1230	*torreyana*	178	Sweet sagewort	868
occidentalis	1231	Succulent hawthorn	374	Sweet sand verbena	145
richardsonii	1231	Sucker flax	562	Sweet-scented bedstraw	820
robusta	1231	Suckleya	179	*Sweetbriar*	397
scribneri	1231	poison	179	Sweetgrass	1184
spartea	1230	suckleyana	179	Swertia radiata	609
v. *curtiseta*	1230	Sugar maple	570	Switchgrass	1203
vaseyi	1231	Sugar-plum	368	Sword fern	66
viridula	1230	Sugarberry	120	Sycamore	119
Stitchwort	211	Sulphur cinquefoil	388	Sycamore Family	119
common	212	Sumac	571	Symphoricarpos	827
fleshy	212	desert	573	albus	827
long-leaved	213	dwarf	572	occidentalis	828
long-stalked	213	fragrant	571	orbiculatus	828
northern	212	smooth	573	palmeri	828
Stone mint	712	Summer cypress	175	*pauciflorus*	828
Stonecrop	357	Summer haw	373	*Symphytum asperum*	684
ditch	357	Summer orophaca	428	*officinale*	684
Stonecrop Family	356	Sundew	252	*Syndesmon thalictroides*	89
Stork's-bill	580	round-leaved	253	Syntherisma	1162
Stout rush	1056	Sundew Family	252	*filiforme*	1163
Strawberry	375	Sunflower	952	*ischaemum*	1163
blite	169	ashy	955	*marginatum*	1162
clover	484	common	954	*sanguinale*	1163
Indian	265	false	952, 958	Synthyris rubra	758
tomato	642	hairy	954	*wyomingensis*	758
wild	376	maximilian	955	*Syringa vulgaris*	747
woodland	376	Nuttall's	955		
Streptanthus	331	plains	956	Taenidia integerrima	602

1386　　　　　　　　　　　　　　　　　　　　　　　　　　　　　　　　　　　　　　　INDEX

Tahoka daisy	977	virginiana	480	Thermopsis	481
Tailcup lupine	463	Tetragoniaceae	152, 191	arenosa	481
Talewort	684	Tetraneuris acaulis	963	rhombifolia	481
Talinum	189	*fastigiata*	964	v. *divaricata*	481
auranticum	190	*linearifolia*	963	v. *rhombifolia*	481
calycinum	190	*simplex*	963	Thesium linophyllon	532
parviflorum	190	Teucrium	738	Thicket creeper	558
rugospermum	190	*boreale*	739	Thimbleberry	402, 403
teretifolium	190	canadense	739	Thinleaf betony	738
Tall anemone	88	v. boreale	739	Thistle, bull	913
Tall-bread scurf-pea	472	v. canadense	739	Canada	909
Tall bellflower	808	v. *occidentalis*	739	common sow	1008
Tall bush lespedeza	460	v. *virginicum*	739	Drummond	911
Tall buttercup	98	laciniatum	739	field	909
Tall cinquefoil	383	*occidentale* v. *boreale*	739	field sow	1008
Tall coreopsis	917	Texas amsonia	611	Flodman's	911
Tall fescue	1180	Texas barberry	108	Iberian star-	899
Tall four-o'clock	148	Texas bedstraw	820	Kansas	650
Tall hedge mustard	330	Texas bergia	236	musk	896
Tall mannagrass	1182	Texas bluegrass	1210	nodding	896
Tall oatgrass	1114	Texas blueweed	954	Platte	910
Tall thistle	909	Texas croton	540	plumeless	896
Tall water hemp	184	Texas dropseed	1228	prickly sow	1008
Tall wheatgrass	1125	Texas panicum	1203	roadside	909
Tall white violet	259	Texas prickly pear	158	Russian-	177
Tamaricaeae	265	Texas spread-wing	594	Scotch	979
Tamarix	265	Texas thistle	913	sow	1007
chinensis	265	Texas toadflax	768	star-	898
gallica	265	Texas vervain	706	swamp	911
parviflora	265	Texas walnut	134	tall	909
pentandra	265	Thalictrum	106	Texas	913
ramosissima	266	dasycarpum	107	wavy-leaf	913
Tamarix Family	265	v. *hypoglaucum*	107	yellow-spine	912
Tanacetum vulgare	1010	dioicum	107	yellow star-	900
Tansy	1010	*hypoglaucum*	107	true	908
aster	977	*lunellii*	107	Thlaspi	332
common	1010	*megacarpum*	107	*arvense*	332
mustard	308, 309	*nigromontanum*	107	*fendleri*	333
Tapegrass	1030	*occidentale*	107	perfoliatum	333
Taraxacum	1010	*thalictroides*	89	Thorn-apple	638
erythrospermum	1010	venulosum	107	oak-leaf	639
laevigatum	1010	Thamnosa texana	576	Thoroughwax	587
officinale	1010	Thaspium	602	Threadleaf buttercup	100
Tarweed	977	barbinode	602	Three-awn	1135
Tassel flower	179	trifoliatum	603	arrowfeather	1138
Tatarian honeysuckle	826	v. *flavum*	603	blue	1138
Tea, Mexican	76, 168	Thelesperma	1011	churchmouse	1136
Mormon	76	ambiguum	1011	Fendler	1139
New Jersey	554, 555	filifolium	1012	forktip	1136
Teal lovegrass	1175	v. filifolium	1012	oldfield	1137
Tear-thumb	229	v. intermedium	1012	poverty	1137
Teasel	836	*gracile*	1013	prairie	1137
common	837	*intermedium* v.		purple	1139
cut-leaved	837	*rubrodiscum*	1012	red	1139
Teasel Family	836	marginatum	1012	sixweeks	1136
Tecoma radicans	804	megapotamicum	1012	slimspike	1137
Telesonix	363	*subnudum*	1012	Three-nerved goldenrod	1006
heucheraeformis	364	*trifidum*	1012	Three-seeded mercury	536
jamesii	363	Thelypodium	332	Three-toothed cinquefoil	389
v. *heucheriformis*	364	integrifolium	332	Threeflower melic	1191
Tennessee bladder fern	58	*lilacinum*	332	Thyme-leaved speedwell	797
Ten-petal mentzelia	271	wrightii	332	Thyme-leaved sandwort	196
Tenpetal anemone	87	Thelypteris	67	Thyme-leaved spurge	550
Tephrosia	480	hexagonoptera	67	Thymelaea passerina	498
leucosericea	481	palustris	67	Thymelaeaceae	498

INDEX

Thymophylla aurea	920	grandiflora	1014	slim	1232
Tiardium	693	hookeri	1014	strictus	1233
indicum	694	sericea	1013	white	1232
Tickclover	445	texensis	1014	*Trientalis borealis*	343
Canada	446	Toxicodendron	573	Trifolium	481
few-flowered	451	*desertorum*	574	arvense	482
hoary	447	*fothergilloides*	574	beckwithii	483
Illinois	448	*negundo*	574	campestre	483
large-flowered	448	radicans	574	carolinianum	483
long-leaf	447	ssp. negundo	574	dubium	483
Maryland	449	ssp. pubens	574	fragiferum	484
panicled	450	ssp. verrucosum	574	ssp. *bonannii*	484
scapose	449	rydbergii	574	ssp. *fragiferum*	484
sessile-leaved	451	toxicarium	575	hybridum	484
slender	447	*Toxiscordion*	1257	ssp. *elegans*	484
Ticklegrass	1128	acutum	1258	ssp. *hybridum*	484
Tickseed, hyssopleaf	173	*gramineum*	1258	incarnatum	484
sunflower	892	*nuttallii*	1258	*medium*	485
Tick-trefoil	445	*Toxylon pomiferum*	127	pratense	485
Tidestromia lanuginosa	187	*Tracaulon*	220	*procumbens*	483
Tilia americana	239	*sagittatum*	229	reflexum	485
Tiliaceae	239	Tradescantia	1048	v. *glabrum*	485
Timothy	1208	bracteata	1048	repens	485
Tine-leaved milk-vetch	432	canaliculata	1049	resupinatum	486
Tiniaria	220	occidentalis	1049	stoloniferum	486
convolvulus	228	ohiensis	1049	Triglochin	1031
scandens	229	*reflexa*	1049	concinna v. debilis	1032
Tithymalopsis	541	tharpii	1049	maritima v. elata	1032
corollata	544	*universitatis*	1049	palustris	1032
Tium	421	Tragia	552	Trillium	1255
drummondii	427	*amblyodonta*	553	cernuum	1255
racemosum	434	betonicifolia	553	dwarf white	1255
Toad rush	1053	*nepetaefolia*	553	*flexipes*	1255
Toadflax	767	ramosa	553	gleasonii	1255
old-field	768	*urticifolia*	553	nivale	1255
Texas	768	Tragopogon	1014	nodding	1255
Toadshade	1256	dubius	1015	sessile	1256
Tobacco	637	*major*	1015	*viride*	1256
Tobacco, Indian	811	porrifolius	1015	viridescens	1256
rabbit-	937	pratensis	1015	Trinidad milk-vetch	433
Tomanthera	790	Trail plant	854	Triodanis	813
auriculata	790	Trailing-four-o'clock	146	biflora	814
densiflora	790	Trailing lespedeza	459	holzingeri	814
Tomato	637	Trailing pearlwort	203	lamprosperma	814
strawberry	642	Tree cholla	158	leptocarpa	814
Toothache tree	577	Tree of heaven	575	perfoliata	815
Toothcup	306, 494, 498	Trefoil	461	v. *biflora*	814
Toothed spurge	544	bird's-foot	461	*Triodia elongata*	1232
Torch flower	379	narrow-leaved	462	*flava*	1232
Torilis	603	prairie	461	f. *cuprea*	1232
anthriscus	603	tick-	445	*stricta*	1233
arvensis	603	Triadenum fraseri	239	Triosteum	829
japonica	603	*virginicum*	239	angustifolium	829
nodosa	603	Trianthema portulacastrum	153	aurantiacum	830
Torrey ephedra	77	Tribulus terrestris	578	v. *illinoense*	830
Torrey's rush	1057	*Trichachne californica*	1114	*illinoense*	830
Touch-me-not	582	Trichostema brachiatum	740	perfoliatum	829
pale	583	Tridens	1231	v. aurantiacum	830
spotted	583	albescens	1232	v. *perfoliatum*	830
Touch-Me-Not Family	582	*elongatus*	1232	Triphora trianthophora	1284
Tovara	220, 229	flavus	1232	Triplasis purpurea	1233
virginiana	229	hairy	1232	Tripsacum dactyloides	1234
Tower mustard	300	longspike	1233	Tripterocalyx micranthus	151
Townsendia	1013	muticus v. elongatus	1232	*Trisetum flavescens*	1114
exscapa	1013	pilosus	1232	*interruptum*	1114

Entry	Page
spicatum	1114
Triticum aestivum	1114
cylindricum	1123
Tropic croton	539
True Fern Family	49
True thistles	908
Trumpet-creeper	804
Trumpet honeysuckle	826
Trumpet vine	804
Tube penstemon	788
Tuber false dandelion	989
Tufted hairgrass	1155
Tufted loosestrife	349
Tumble lovegrass	1177
Tumblegrass	1217
Tumbleweed	176, 177, 181
Tumbler ringwing	174
Tumbling mustard	330
Turkestan bluestem	1133
Turk's cap lily	1252
Turnip	304
Turnip, Indian	1043
prairie-	473
wild	304
Turnsole	694
Twayblade, broad-leaved	1280
Loesel's	1280
Twine vine	636
arroyo	636
waxy-leaf	636
Twinflower	823
Twist-flower	331
Twisted ladies'-tresses	1284
Twisted-stalk	1254
Two-grooved vetch	425
Two-flowered milkvine	633
Two-leaved senna	414
Typha	1237
angustifolia	1238
× latifolia	1239
domingensis	1238
× latifolia	1239
× glauca	1239
latifolia	1238
Typhaceae	1237
Ulmaceae	119
Ulmus	121
alata	122
americana	122, 533
fulva	123
pumila	122
racemosa	123
rubra	123
thomasi	123
Umbrella-grass	1105
Umbrella sedge	1093
Unamia alba	1005
lutescens	1005
Underwood's spikemoss	40
Unicorn plant	803
Unicorn-plant Family	803
Uniola latifolia	1152
Urn-tree hawthorn	371
Urtica	130
chamaedryoides	130
dioica	130
ssp. *dioica*	130
ssp. *gracilis*	130
v. *gracilis*	130
v. *procera*	130
v. *procera*	130
gracilenta	130
gracilis	130
procera	130
viridis	130
Urticaceae	127
Urticastrum divaricatum	128
Utricularia	806
biflora	806
gibba	806
intermedia	806
macrorhiza	807
minor	807
vulgaris	807
Uvularia	1256
grandiflora	1256
sessilifolia	1256
Vaccaria pyramidata	213
segetalis	214
vaccaria	214
vulgaris	214
Vaccinium	334
arboreum	335
membranaceum	335
neglectum	336
pallidum	335
scoparium	336
stamineum	336
vacillans	336
Valerian	833
Valerian Family	833
Valeriana	833
acutiloba	834
capitata ssp. *acutiloba*	834
dioica	834
edulis	834
occidentalis	834
septentrionalis	834
trachycarpa	835
Valerianaceae	833
Valerianella	835
amarella	835
radiata	835
stenocarpa	836
v. *parviflora*	836
woodsiana	836
Validallium tricoccum	1246
Vallisneria americana	1030
Variable pondweed	1036
Variegated scouring rush	45
Vascular cryptogams	39
Vegetable oyster	1015
Velvet dichanthelium	1160
Velvet-leaf	241
Velvety gaura	511
Venice mallow	246
Venus' comb	585
Venus'-hair fern	50
Venus' looking glass	813
Verbascum	790
blattaria	791
f. *albiflora*	791
thapsus	791
Verbena	704
ambrosifolia	705
angustifolia	707
bipinnatifida	705
× *blanchardii*	704
bracteata	705
bracteosa	705
canadensis	705
ciliata	705
× *deamii*	704
drummondii	706
× *engelmannii*	704
halei	706
hastata	706
v. *scabra*	706
× *illicita*	704
× *moechina*	704
× *oklahomensis*	704
× *perriana*	704
plicata	706
pumila	707
× *rydbergii*	704
sand	145
simplex	707
stricta	707
sweet sand	145
urticifolia	707
v. *leiocarpa*	708
wrightii	705
Verbenaceae	701
Verbesina	1015
alternifolia	1016
encelioides ssp. exauriculata	1016
ssp. *encelioides*	1016
helianthoides	1016
virginica	1016
Vernonia	1017
arkansana	1017
baldwinii	1017
ssp. baldwinii	1017
ssp. interior	1017
fasciculata	1018
ssp. corymbosa	1018
ssp. fasciculata	1018
gigantea	1018
marginata	1018
missurica	1018
Vernonieae	844
Veronica	791
agrestis	793
americana	793
anagallis-aquatica	793
arvensis	794
biloba	794
catenata	794
v. catenata	795
v. glandulosa	795
comosa	795
v. *glaberrima*	795

INDEX

v. *glandulosa*	795	pliant milk-	427	*arvensis*	263
connata	795	pulse milk-	435	canadensis	259
ssp. *glaberrima*	795	Pursh milk-	433	v. *canadensis*	259
didyma	793	Short's milk-	435	v. *rugulosa*	259
hederaefolia	795	slender milk-	428	*conspersa*	259
latifolia	792	small-flowered milk-	431	*cucullata*	257
longifolia	795	standing milk-	423	*emarginata*	264
maritima	795	stinking milk-	433	*eriocarpa*	262
officinalis	795	tine-leaved milk-	432	v. *leiocarpa*	262
opaca	793	Trinidad milk-	433	*incognita*	259
peregrina	796	two-grooved	425	*kitaibeliana* v. *rafinesquii*	263
v. peregrina	796	woollypod	489	*lanceolata*	257
v. *xalapensis*	796	woodland weedy milk-	430	*lovelliana*	260
ssp. *xalapensis*	796	Vetchling	454	macloskeyi	259
persica	796	bushy	456	v. *macloskeyi*	259
polita	793	hoary	455	*missouriensis*	262
salina	795	marsh	455	nephrophylla	259
scutellata	796	singletary	456	nuttallii	260
serpyllifolia	796	yellow	455	v. *vallicola*	260
v. *humifusa*	797	Viburnum	830	*pallens*	259
v. *serpyllifolia*	797	edule	830	palmata	260
triphyllos	797	*eradiatum*	831	v. *palmata*	260
Veronicastrum virginicum	797	lentago	831	v. *triloba*	260
f. *villosum*	797	opulus v. americanum	831	palustris	260
Vervain	704	prunifolium	831	*papilionacea*	262
blue	706	v. *bushii*	831	pedata	261
fanleaf	706	rafinesquianum	832	v. *lineariloba*	261
hoary	707	v. *affine*	832	pedatifida	261
narrow-leaved	707	rufidulum	832	*pensylvanica*	262
nettle-leaved	707	*trilobum*	831	v. *leiocarpa*	262
pink	707	Vicia	486	pratincola	261
prostrate	705	americana	487	pubescens	262
rose	705	v. americana	487	v. *eriocarpa*	263
Texas	706	v. minor	487	v. *peckii*	262
Vervain Family	701	*angustifolia*	489	v. *pubescens*	262
Vetch	486	*caroliniana*	487	rafinesquii	263
alkali milk-	434	cracca	488	renifolia	263
American	487	*dasycarpa*	489	v. *brainerdii*	263
bent-flowered milk-	436	*dissitifolia*	487	*retusa*	260
bird	488	exigua	488	*rugulosa*	259
Bodin milk-	425	leavenworthii	488	sagittata	264
Canada milk-	426	ludoviciana	488	*sarmentosa*	257
common	489	v. laxiflora	489	selkirkii	264
Cooper milk-	431	v. ludoviciana	489	*septentrionalis*	257
creamy poison-	434	*oregana*	487	sororia	264
crown	438	sativa	489	× *pratincola*	264
deer	461	v. angustifolia	489	*striata*	257
deer pea	488	v. sativa	489	*subvestita*	259
draba milk-	435	*sparsifolia*	487	triloba v. *dilatata*	260
Drummond milk-	427	*trifida*	487	*vallicola*	260
alpine milk-	424	villosa	489	*viarum*	261
American milk-	424	v. glabrescens	489	*villosa*	257
field milk-	424	ssp. *varia*	489	Violaceae	255
hairy	489	v. villosa	490	Violet	257
Indian milk-	423	Viguiera	1018	arrowhead	264
Leavenworth's	488	*ovalis*	1019	bird's-foot	261
Lindheimer milk-	429	stenoloba	1018	blue prairie	261
little	488	Vinca minor	613	dog's-tooth-	1249
lotus milk-	430	*Vincetoxicum nigrum*	632	downy blue	264
milk	421	Vine-mesquite	1202	downy yellow	262
Missouri milk-	430	Viola	257	great spurred	264
narrow-leaved poison-	432	adunca	258	green	256
Ozark milk-	427	v. *adunca*	259	hook-spurred	258
painted milk-	426	v. *minor*	259	kidney-leaved	263
Platte River milk-	432	*affinis*	262	larkspur-	261

nodding green	256	smallflower	314	Waxy-leaf twine vine	636
northern bog	289	western	314	Weak nettle	130
northern marsh-	260	wormseed	314	Weatherglass, poorman's	343
Nuttall's	260	Walnut	133	Wedgegrass	1224
prairie	261	black	134	Wedgeleaf draba	312
smooth yellow	262	little	134	Wedgeleaf fog-fruit	702
tall white	259	Texas	134	Weed, poverty	172
white dog's-tooth-	1249	Walnut Family	131	*Weeping birch*	143
wild white	259	Water avens	379	Weeping lovegrass	1174
wood-	260	Water bentgrass	1128	Western blueflag	1260
wood sorrel	579	Water birch	143	Western buckeye	568
yellow dog's-tooth-	1250	Water chinkapin	81	Western clematis	90
yellow prairie	260	Water clover	70	Western daisy	886
Violet Family	255	western	70	Western dock	234
Viorna	90	Water crowfoot, white	103, 106	Western fleabane	927
fremontii	90	Water dock	235	Western horse-nettle	648
pitcheri	91	great	234	Western ironweed	1017
scottii	90	Water fern	71	Western lady-fern	53
Viper's bugloss	691	Water Fern Family	71	Western polypody	64
Virginia bugleweed	720	Water hemlock	588	Western pussy willow	290
Virginia creeper	558	bulbous	589	Western ragweed	857
Virginia ground cherry	646	common	589	Western red currant	353
Virginia lionsheart	728	Water-hemp	183	Western rock jasmine	344
Virginia mountain mint	730	tall	184	Western sagewort	868
Virginia snakeroot	80	Water horehound	719	Western salsify	1015
Virginia spring beauty	188	Water horsetail	43	Western sensitive brier	411
Virginia wild rye	1170	Water hyssop	757	Western snowberry	828
Virgin's bower	90, 92	Water milfoil	492, 493	Western wheatgrass	1126
Virgulus	872	Water Milfoil Family	492	Western wallflower	314
Viscaceae	532	Water-nymph	1040	Western water clover	70
Viscid cranesbill	582	Water-parsnip	587, 601	Western wild lettuce	969
Viscid euthamia	936	Water-pennywort	594	Western wild rose	400
Viscid nightshade	651	Water pepper	226	*Wheat*	1114
Visher's eriogonum	219	mild	226	Wheatgrass	1123
Vitaceae	557	Water pimpernel	350, 351	bluebunch	1126
Vitex agnus-castus	708	Water plantain	1022, 1023	crested	1125
Vitis	559	spearwort	103	intermediate	1125
acerifolia	559	Water Plantain Family	1022	slender	1124
aestivalis	560	Water primrose	515	tall	1125
v. *argentifolia*	560	Water purslane	496	western	1126
bicolor	560	Water shield	83	White ash	749
cinerea	560	Water Shield Family	83	White aster	876, 971
cordifolia	561	Water smartweed	225, 227	small	882
lincecumi v. *glauca*	560	Water speedwell	793	White avens	378
longii	560	Water stargrass	1240	White beardtongue	776
riparia	560	Water starwort	741	White camass	1258
v. *praecox*	561	Water Starwort Family	741	White campion	208
v. *riparia*	560	Water willow	801	White clover	485
v. *syrticola*	560	Watercress	323	White cockle	208
rotundifolia	559	Waterleaf	678, 679	White coralberry	827
rupestris	561	notchbract	679	White dog's-tooth-violet	1249
vulpina	560, 561	Waterleaf Family	678	White evening primrose,	
Vulpia	1179	Waterlily	82	showy	524
octoflora	1180	fragrant white	83	White-eyed grass	1262
v. *glauca*	1180	white	83	White four-o'clock	147
v. *octoflora*	1180	Waterlily Family	81	White honeysuckle	824
v. *tenella*	1180	Watermeal	1046	White horse-nettle	649
		Waterpod	678	White lady's-slipper, small	1274
Wafer ash	576	Waterthread pondweed	1035	White lettuce	986
Wahlenbergella	205	Waterweed	1029	White loco	477
drummondii	207	Waterwort	236	White locoweed	470
Wahoo	534	Waterwort Family	236	White milkwort	565
Walking fern	52	Wavy-leaf oak	140	White morning-glory	659
Wallflower	313	Wavy-leaf thistle	913	White mountain mint	729
bushy	315	Waxweed, blue	496	White mulberry	127

INDEX

White oak	137	long-bracted	437	serviceberry	290
White mustard	304	plains	437	shining	288
White oak, swamp	137	white	437	water	801
White orchis	1277	Wild leek	1246	western pussy	290
White prairie-clover	440	Wild lettuce	968	yellow	288
White prairie rose	399	western	969	yellowstem white	285
White sage	166, 870	Wild licorice	452	Willow Family	273
White snakeroot	935	Wild lily	1252	Willowleaf aster	883
White spruce	74	Wild lily-of-the-valley	338, 1252	Wind flower	86
White-stemmed evening		Wild mercury	537	Windmill grass	1152, 1153
primrose	524	Wild mignonette	333	hooded	1152
White sweet clover	466	Wild oats	1139, 1152	Winged elm	122
White-top sedge	1098	Wild onion	1244, 1245	Winged eriogonum	216
White tridens	1232	pink	1245	Winged loosestrife	497
White trillium, dwarf	1255	Wild pansy	263	Winged pigweed	174
White violet, tall	259	Wild parsley	595	Wingstem	1015, 1016
wild	259	Wild parsnip	598	Winter cress	303
White water crowfoot	103, 106	Wild plum	391	northern	303
White waterlily	83	Wild rice	1234	Winter fat	166
fragrant	83	Wild rose, prairie	398	Winter grape	561
White whitlowort	312	prickly	397	Wintergreen	337
White wild indigo	437	smooth	398	one-flowered	339
White willow, yellowstem	285	western	400	one-sided	339
White woodland aster	879	Wild rhubarb	233	round-leaved	338, 339
Whitegrass	1187	Wild rye	1166	Wintergreen Family	336
Whitestem pondweed	1038	alkali	1169	Wire lettuce	1009
Whitetop	308	basin	1168	Wirestem muhly	1194, 1195
Whitlow grass	311	blue	1168	Wiry witchgrass	1201
Whitlow-wort	202	Canada	1167	Wister's coral-root	1272
Whitlowort, white	312	fuzzyspike	1169	Witchgrass, common	1201
yellow	312	hairy	1169	fall	1188
Whorled dropseed	1228	Russian	1169	Philadelphia	1203
Whorled loosestrife	349	Wild sarsaparilla	584	wiry	1201
Whorled milkweed	630	Wild spiraea	406	Wolfberry	640, 827, 828
Whorled milkwort	566	Wild strawberry	376	pale	641
Widgeon grass	1040	Wild turnip	304	silvery	640
Wilcox dichanthelium	1161	Wild white violet	259	Wolffia	1046
Wild alfalfa	475	Wild yam	1267	borealis	1046
Wild aster	871	Willow	281	columbiana	1046
Wild bean	478	amsonia	611	brasiliensis	1046
Wild begonia	235	autumn	290	*papulifera*	1046
Wild bergamot	724	baccharis	887	*punctata*	1046
Wild black cherry	395	beaked	285	Wolftail	1190
Wild black currant	353	black	288	Wood anemone	88
Wild buckwheat	215, 228	bog	289	Wood betony	773
James'	218	Carolina	286	Wood buttercup, early	97
matted	219	coyote	287	Wood fern	58
nodding	216	crack	287	spinulose	60
spreading	217	desert	805	Wood horsetail	45
yellow	217	diamond	286	Wood lettuce, blue	968
Wild calla	1043	-herb	504, 505	Wood mint	712
Wild carrot	592	narrow-leaved	507	Wood nettle	128
Wild chamomile	978	purple-leaved	506	Wood sage	738, 739
Wild chervil	588	hoary	286	Wood sorrel	578
Wild columbine	89	laurel-leaved	289	gray-green	579
Wild comfrey, northern	691	-leaved dock	233	violet	579
Wild cranesbill	581	-leaved lettuce	970	yellow	579
Wild cucumber	268	-leaved sunflower	957	Wood Sorrel Family	578
Wild four-o'clock	150	meadow	289	Wood-violet	260
Wild garlic	1246	Missouri	286	Woodbine	558
Wild ginger	81	peachleaf	285	Woodland agrimony	367
Wild goose plum	392, 394	planeleaf	290	Woodland aster, white	879
Wild honeysuckle	825	prairie	287	Woodland bluegrass	1214
Wild hydrangea	351	pussy	286	Woodland orchis, green	1277
Wild indigo, dwarf	419	sandbar	287	Woodland strawberry	376

INDEX

Woodland weedy milk-vetch	430	
Woodreed	1153	
drooping	1154	
Woodruff	817	
Woodrush	1057	
Woods bedstraw	819	
Woods mountain mint	730	
Woodsia	68	
blunt-lobed	69	
ilvensis	68	
× *kansana*	69	
mexicana	68	
obtusa	69	
× *oregana*	69	
Oregon	69	
oregana	69	
Rocky Mountain	69	
rusty	68	
scopulina	69	
Woolly buckthorn	341	
Woolly burdock	865	
Woolly croton	539	
Woolly dalea	442	
Woolly hawthorn	372	
Woolly lip fern	56	
Woolly locoweed	431	
Woolly milkweed	622	
Woolly pipevine	80	
Woollypod vetch	489	
Wormseed wallflower	314	
Wormwood	866,	868
biennial	868	
silky	869	
Wright's cliff-brake	64	
Wright's plantain	747	
Wulfenia bullii	758	
Wyoming paintbrush	761	
Xanthisma	1019	
texanum ssp.		
drummondii	1019	
v. *drummondii*	1019	
Xanthium	799,	1019
acerosum	1020	
chinense	1020	
commune	1020	
echinatum	1020	
glanduliferum	1020	
globosum	1020	
pensylvanicum	1020	
speciosum	1020	
spinosum	1020	
strumarium	1020	
v. canadense	1020	
v. glabratum	1020	
v. *strumarium*	1020	
Xanthocephalum		
dracunculoides	945	
sarothrae	946	
texanum	946	
Xanthoxalis bushii	579	
corniculatus	579	
cymosa	579	
stricta	579	
Ximenesia exauriculata	1016	

Xylophacos	421	
missouriensis	431	
purshii	434	
shortianus	435	
Xylorhiza glabriuscula	1020	
v. *glabriuscula*	1021	
v. *linearifolia*	1021	
Xylosteon	824	
tataricum	826	
Yam, wild	1267	
Yam Family	1267	
Yard dock	232	
Yarrow	854	
Yellow avens	377	
Yellow bedstraw	820	
Yellow bell	1250	
Yellow birch	143	
Yellow buttercup, small	101	
Yellow chamomile	862	
Yellow chestnut oak	139	
Yellow cleome	291	
Yellow cress	325,	327
bog	327	
creeping	328	
spreading	328	
Yellow dog's-tooth-violet	1250	
Yellow-flowered horse-		
gentian	829	
Yellow foxtail	1219	
Yellow harlequin	116	
Yellow honeysuckle	825	
Yellow iris	1260	
Yellow lady's-slipper	1273	
large	1274	
small	1273	
Yellow moccasin flower	1273	
Yellow monkey-flower,		
common	771	
Yellow pea	481	
Yellow pimpernel	602	
Yellow prairie violet	260	
Yellow primrose, plains	502	
Yellow-puff	410	
Yellow-spine thistle	912	
Yellow stargrass	1251	
Yellow star thistle	900	
Yellow sweet clover	466	
Yellow vetchling	455	
Yellow violet, downy	262	
smooth	262	
Yellow whitlowort	312	
Yellow wild buckwheat	217	
Yellow willow	288	
Yellow wood sorrel	579	
Yellowstem white willow	285	
Yerba de tajo	922	
Yerba mansa	79	
Yucca	1264	
arkansana	1265	
baccata	1265	
baileya	1265	
glauca	1265	
v. *glauca*	1265	
v. *mollis*	1265	

harrimaniae	1265
v. *neomexicana*	1265
filamentosa	1265
neomexicana	1265
smalliana	1265
Zannichellia palustris	1041
Zannichelliaceae	1041
Zanthoxylum americanum	577
hirsutum	577
Zephyr-lily	1257
Zephyranthes longifolia	1257
Zigadenus	1257
elegans	1258
nuttallii	1258
venenosus	1258
v. *gramineus*	1258
Zinnia grandiflora	1021
Zizania aquatica	1234
v. *interior*	1235
interior	1235
palustris	1235
Zizia	603
aptera	603
aurea	603
cordata	603
Ziziphus obtusifolia	556
Zosterella dubia	1240
Zygophyllaceae	577
Zygophyllidium	541
hexagonum	546

89012841862

б8901284186 2а